P9-BTN-855

PERIODIC TABLE OF THE ELEMENTS

Key: 1 — Atomic number; H — Symbol; Hydrogen — Name; 1.0079 — Average atomic mass

Metals / Metalloids / Nonmetals

	1 1A	2 2A	3 3B	4 4B	5 5B	6 6B	7 7B	8 8B	9 8B	10 8B	11 1B	12 2B	13 3A	14 4A	15 5A	16 6A	17 7A	18 8A
1	1 **H** Hydrogen 1.0079																	2 **He** Helium 4.0026
2	3 **Li** Lithium 6.941	4 **Be** Beryllium 9.0122											5 **B** Boron 10.811	6 **C** Carbon 12.011	7 **N** Nitrogen 14.007	8 **O** Oxygen 15.999	9 **F** Fluorine 18.998	10 **Ne** Neon 20.180
3	11 **Na** Sodium 22.990	12 **Mg** Magnesium 24.305											13 **Al** Aluminum 26.982	14 **Si** Silicon 28.086	15 **P** Phosphorus 30.974	16 **S** Sulfur 32.065	17 **Cl** Chlorine 35.453	18 **Ar** Argon 39.948
4	19 **K** Potassium 39.098	20 **Ca** Calcium 40.078	21 **Sc** Scandium 44.956	22 **Ti** Titanium 47.867	23 **V** Vanadium 50.942	24 **Cr** Chromium 51.996	25 **Mn** Manganese 54.938	26 **Fe** Iron 55.845	27 **Co** Cobalt 58.933	28 **Ni** Nickel 58.693	29 **Cu** Copper 63.546	30 **Zn** Zinc 65.38	31 **Ga** Gallium 69.723	32 **Ge** Germanium 72.63	33 **As** Arsenic 74.922	34 **Se** Selenium 78.96	35 **Br** Bromine 79.904	36 **Kr** Krypton 83.798
5	37 **Rb** Rubidium 85.468	38 **Sr** Strontium 87.62	39 **Y** Yttrium 88.906	40 **Zr** Zirconium 91.224	41 **Nb** Niobium 92.906	42 **Mo** Molybdenum 95.96	43 **Tc** Technetium [98]	44 **Ru** Ruthenium 101.07	45 **Rh** Rhodium 102.91	46 **Pd** Palladium 106.42	47 **Ag** Silver 107.87	48 **Cd** Cadmium 112.41	49 **In** Indium 114.82	50 **Sn** Tin 118.71	51 **Sb** Antimony 121.76	52 **Te** Tellurium 127.60	53 **I** Iodine 126.90	54 **Xe** Xenon 131.29
6	55 **Cs** Cesium 132.91	56 **Ba** Barium 137.33	57 **La** Lanthanum 138.91	72 **Hf** Hafnium 178.49	73 **Ta** Tantalum 180.95	74 **W** Tungsten 183.84	75 **Re** Rhenium 186.21	76 **Os** Osmium 190.23	77 **Ir** Iridium 192.22	78 **Pt** Platinum 195.08	79 **Au** Gold 196.97	80 **Hg** Mercury 200.59	81 **Tl** Thallium 204.38	82 **Pb** Lead 207.2	83 **Bi** Bismuth 208.98	84 **Po** Polonium [209]	85 **At** Astatine [210]	86 **Rn** Radon [222]
7	87 **Fr** Francium [223]	88 **Ra** Radium [226]	89 **Ac** Actinium [227]	104 **Rf** Rutherfordium [265]	105 **Db** Dubnium [268]	106 **Sg** Seaborgium [271]	107 **Bh** Bohrium [270]	108 **Hs** Hassium [269]	109 **Mt** Meitnerium [276]	110 **Ds** Darmstadtium [281]	111 **Rg** Roentgenium [280]	112 **Cn** Copernicium [285]	113 **Nh** Nihonium [284]	114 **Fl** Flerovium [289]	115 **Mc** Moscovium [290]	116 **Lv** Livermorium [293]	117 **Ts** Tennessine [294]	118 **Og** Oganesson [294]

6 Lanthanides	58 **Ce** Cerium 140.12	59 **Pr** Praseodymium 140.91	60 **Nd** Neodymium 144.24	61 **Pm** Promethium [145]	62 **Sm** Samarium 150.36	63 **Eu** Europium 151.96	64 **Gd** Gadolinium 157.25	65 **Tb** Terbium 158.93	66 **Dy** Dysprosium 162.50	67 **Ho** Holmium 164.93	68 **Er** Erbium 167.26	69 **Tm** Thulium 168.93	70 **Yb** Ytterbium 173.05	71 **Lu** Lutetium 174.97
7 Actinides	90 **Th** Thorium 232.04	91 **Pa** Protactinium 231.04	92 **U** Uranium 238.03	93 **Np** Neptunium [237]	94 **Pu** Plutonium [244]	95 **Am** Americium [243]	96 **Cm** Curium [247]	97 **Bk** Berkelium [247]	98 **Cf** Californium [251]	99 **Es** Einsteinium [252]	100 **Fm** Fermium [257]	101 **Md** Mendelevium [258]	102 **No** Nobelium [259]	103 **Lr** Lawrencium [262]

We have used the U.S. system as well as the system recommended by the International Union of Pure and Applied Chemistry (IUPAC) to label the groups in this periodic table. The system used in the United States includes a letter and a number (1A, 2A, 3B, 4B, etc.), which is close to the system developed by Mendeleev. The IUPAC system uses numbers 1–18 and has been recommended by the American Chemical Society (ACS). Although we show both numbering systems here, we use the IUPAC system exclusively in the book.

Element	Symbol	Atomic Number	Average Atomic Mass[a]
Actinium	Ac	89	[227]
Aluminum	Al	13	26.982
Americium	Am	95	[243]
Antimony	Sb	51	121.76
Argon	Ar	18	39.948
Arsenic	As	33	74.922
Astatine	At	85	[210]
Barium	Ba	56	137.33
Berkelium	Bk	97	[247]
Beryllium	Be	4	9.0122
Bismuth	Bi	83	208.98
Bohrium	Bh	107	[270]
Boron	B	5	10.811
Bromine	Br	35	79.904
Cadmium	Cd	48	112.41
Calcium	Ca	20	40.078
Californium	Cf	98	[251]
Carbon	C	6	12.011
Cerium	Ce	58	140.12
Cesium	Cs	55	132.91
Chlorine	Cl	17	35.453
Chromium	Cr	24	51.996
Cobalt	Co	27	58.933
Copernicium	Cn	112	[285]
Copper	Cu	29	63.546
Curium	Cm	96	[247]
Darmstadtium	Ds	110	[281]
Dubnium	Db	105	[268]
Dysprosium	Dy	66	162.50
Einsteinium	Es	99	[252]
Erbium	Er	68	167.26
Europium	Eu	63	151.96
Fermium	Fm	100	[257]
Flerovium	Fl	114	[289]
Fluorine	F	9	18.998
Francium	Fr	87	[223]
Gadolinium	Gd	64	157.25
Gallium	Ga	31	69.723
Germanium	Ge	32	72.63
Gold	Au	79	196.97

Element	Symbol	Atomic Number	Average Atomic Mass[a]
Hafnium	Hf	72	178.49
Hassium	Hs	108	[269]
Helium	He	2	4.0026
Holmium	Ho	67	164.93
Hydrogen	H	1	1.0079
Indium	In	49	114.82
Iodine	I	53	126.90
Iridium	Ir	77	192.22
Iron	Fe	26	55.845
Krypton	Kr	36	83.798
Lanthanum	La	57	138.91
Lawrencium	Lr	103	[262]
Lead	Pb	82	207.2
Lithium	Li	3	6.941
Livermorium	Lv	116	[293]
Lutetium	Lu	71	174.97
Magnesium	Mg	12	24.305
Manganese	Mn	25	54.938
Meitnerium	Mt	109	[276]
Mendelevium	Md	101	[258]
Mercury	Hg	80	200.59
Molybdenum	Mo	42	95.96
Moscovium	Mc	115	[290]
Neodymium	Nd	60	144.24
Neon	Ne	10	20.180
Neptunium	Np	93	[237]
Nickel	Ni	28	58.693
Nihonium	Nh	113	[284]
Niobium	Nb	41	92.906
Nitrogen	N	7	14.007
Nobelium	No	102	[259]
Oganesson	Og	118	[294]
Osmium	Os	76	190.23
Oxygen	O	8	15.999
Palladium	Pd	46	106.42
Phosphorus	P	15	30.974
Platinum	Pt	78	195.08
Plutonium	Pu	94	[244]
Polonium	Po	84	[209]

Element	Symbol	Atomic Number	Average Atomic Mass[a]
Potassium	K	19	39.098
Praseodymium	Pr	59	140.91
Promethium	Pm	61	[145]
Protactinium	Pa	91	231.04
Radium	Ra	88	[226]
Radon	Rn	86	[222]
Rhenium	Re	75	186.21
Rhodium	Rh	45	102.91
Roentgenium	Rg	111	[280]
Rubidium	Rb	37	85.468
Ruthenium	Ru	44	101.07
Rutherfordium	Rf	104	[265]
Samarium	Sm	62	150.36
Scandium	Sc	21	44.956
Seaborgium	Sg	106	[271]
Selenium	Se	34	78.96
Silicon	Si	14	28.086
Silver	Ag	47	107.87
Sodium	Na	11	22.990
Strontium	Sr	38	87.62
Sulfur	S	16	32.065
Tantalum	Ta	73	180.95
Technetium	Tc	43	[98]
Tellurium	Te	52	127.60
Tennessine	Ts	117	[294]
Terbium	Tb	65	158.93
Thallium	Tl	81	204.38
Thorium	Th	90	232.04
Thulium	Tm	69	168.93
Tin	Sn	50	118.71
Titanium	Ti	22	47.867
Tungsten	W	74	183.84
Uranium	U	92	238.03
Vanadium	V	23	50.942
Xenon	Xe	54	131.29
Ytterbium	Yb	70	173.05
Yttrium	Y	39	88.906
Zinc	Zn	30	65.38
Zirconium	Zr	40	91.224

[a]Average atomic mass values for most elements are from *Pure Appl. Chem.* (2011) 83, 359. Those for B, C, Cl, H, Li, N, O, Si, S, and Tl are from *Pure Appl. Chem.* (2009) 81, 2131, and are within the ranges cited in the first reference. Atomic masses in brackets are the mass numbers of the longest-lived isotopes of elements with no stable isotopes.

SIXTH EDITION

Chemistry

The Science in Context

Thomas R. Gilbert
NORTHEASTERN UNIVERSITY

Rein V. Kirss
NORTHEASTERN UNIVERSITY

Stacey Lowery Bretz
MIAMI UNIVERSITY

Natalie Foster
LEHIGH UNIVERSITY (EMERITA)

W. W. NORTON & COMPANY
Independent Publishers Since 1923

W. W. Norton & Company has been independent since its founding in 1923, when William Warder Norton and Mary D. Herter Norton first published lectures delivered at the People's Institute, the adult education division of New York City's Cooper Union. The firm soon expanded its program beyond the Institute, publishing books by celebrated academics from America and abroad. By midcentury, the two major pillars of Norton's publishing program—trade books and college texts—were firmly established. In the 1950s, the Norton family transferred control of the company to its employees, and today—with a staff of five hundred and hundreds of trade, college, and professional titles published each year—W. W. Norton & Company stands as the largest and oldest publishing house owned wholly by its employees.

Printed in Canada

Editor: Erik Fahlgren
Senior Associate Managing Editor, College: Carla Talmadge
Developmental Editor: John Murdzek
Associate Director of Production: Benjamin Reynolds
Managing Editor, College: Marian Johnson
Media Editor: Cailin Barrett-Bressack
Associate Media Editor: Grace Tuttle
Media Project Editor: Jesse Newkirk
Media Editorial Assistant: Sarah McGinnis
Managing Editor, College Digital Media: Kim Yi
Ebook Production Manager: Danielle Vargas
Marketing Manager, Chemistry: Stacy Loyal
Design Director: Jillian Burr
Designer: Lisa Buckley
Director of College Permissions: Megan Schindel
College Permissions Assistant: Patricia Wong
Photo Editor: Stephanie Romeo
Composition: GraphicWorld/Gary Clark, Project Manager
Illustration Studio: Imagineeringart.com/Alicia Elliott, Project Manager
Manufacturing: Transcontinental—Beauceville, QC

Permission to use copyrighted material is included at the back of the book.

ISBN: 978-0-393-67403-3

W. W. Norton & Company, Inc., 500 Fifth Avenue, New York, NY 10110

wwnorton.com

W. W. Norton & Company Ltd., 15 Carlisle Street, London W1D 3BS

1 2 3 4 5 6 7 8 9 0

Brief Contents

1 Particles of Matter: Measurement and the Tools of Science 2

2 Atoms, Ions, and Molecules: Matter Starts Here 46

3 Stoichiometry: Mass, Formulas, and Reactions 86

4 Reactions in Solution: Aqueous Chemistry in Nature 148

5 Properties of Gases: The Air We Breathe 212

6 Thermochemistry: Energy Changes in Chemical Reactions 270

7 A Quantum Model of Atoms: Waves, Particles, and Periodic Properties 334

8 Chemical Bonds: What Makes a Gas a Greenhouse Gas? 392

9 Molecular Geometry: Shape Determines Function 444

10 Intermolecular Forces: The Uniqueness of Water 506

11 Solutions: Properties and Behavior 548

12 Solids: Crystals, Alloys, and Polymers 600

13 Chemical Kinetics: Reactions in the Atmosphere 648

14 Chemical Equilibrium: How Much Product Does a Reaction Really Make? 708

15 Acid–Base Equilibria: Proton Transfer in Biological Systems 754

16 Additional Aqueous Equilibria: Chemistry and the Oceans 802

17 Thermodynamics: Spontaneous and Nonspontaneous Reactions and Processes 850

18 Electrochemistry: The Quest for Clean Energy 898

19 Nuclear Chemistry: Applications in Science and Medicine 946

20 Organic and Biological Molecules: The Compounds of Life 986

21 The Main Group Elements: Life and the Periodic Table 1044

22 Transition Metals: Biological and Medical Applications 1082

Contents

List of Applications xv

List of Animations xvii

About the Authors xix

Preface xx

1 Particles of Matter: Measurement and the Tools of Science 2

1.1 How and Why 4

1.2 Macroscopic and Particulate Views of Matter 5
Classes of Matter 5 • A Particulate View 7

1.3 Mixtures and How to Separate Them 9

1.4 A Framework for Solving Problems 11

1.5 Properties of Matter 12

1.6 States of Matter 14

1.7 The Scientific Method: Starting Off with a Bang 16

1.8 SI Units 18

1.9 Unit Conversions and Dimensional Analysis 20

1.10 Evaluating and Expressing Experimental Results 22
Significant Figures 22 • Significant Figures in Calculations 23 • Precision and Accuracy 27

1.11 Testing a Theory: The Big Bang Revisited 32
Temperature Scales 32 • An Echo of the Big Bang 34

Summary 37 • Particulate Preview Wrap-Up 37 • Problem-Solving Summary 38 • Visual Problems 38 • Questions and Problems 40

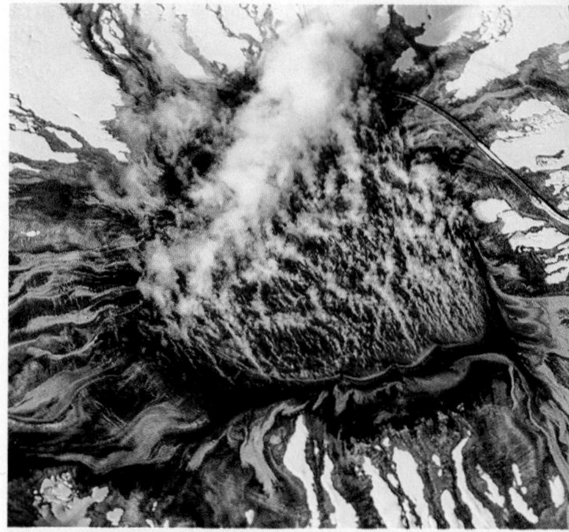

2 Atoms, Ions, and Molecules: Matter Starts Here 46

2.1 Atoms in Baby Teeth 48

2.2 Discovering the Structure of Atoms 49
Electrons 49 • Radioactivity 51 • Protons and Neutrons 52

2.3 Isotopes 54

2.4 Average Atomic Mass 56

2.5 The Periodic Table of the Elements 58
Navigating the Modern Periodic Table 59

2.6 Trends in Compound Formation 61
Molecular Compounds 62 • Ionic Compounds 62

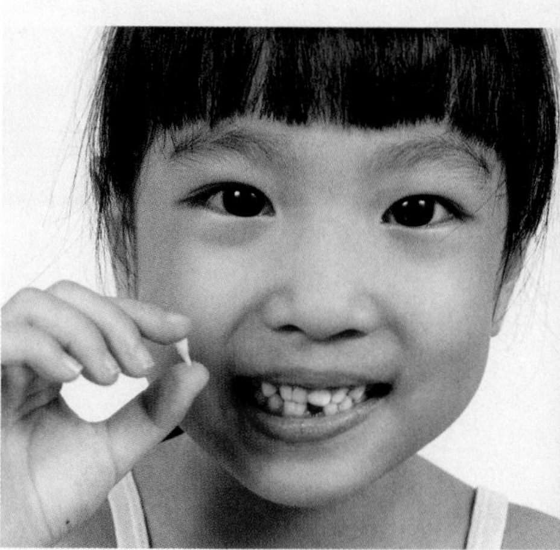

2.7 Naming Inorganic Compounds and Writing Their Formulas 64
Binary Molecular Compounds 65 • Binary Ionic Compounds 66 • Compounds of Metals That Form More than One Cation 67 • Polyatomic Ions 68 • Acids 69

2.8 Organic Compounds: A First Look 70
Hydrocarbons 71 • Heteroatoms and Functional Groups 72

2.9 Nucleosynthesis: The Origin of the Elements 74
Primordial Nucleosynthesis 74 • Stellar Nucleosynthesis 75

Summary 77 • Particulate Preview Wrap-Up 78 • Problem-Solving Summary 78 • Visual Problems 79 • Questions and Problems 80

3 Stoichiometry: Mass, Formulas, and Reactions 86

3.1 Air, Life, and Molecules 88
Chemical Reactions and Earth's Early Atmosphere 89

3.2 The Mole 91
Molar Mass 93 • Molecular Masses and Formula Masses 95 • Moles and Chemical Equations 98

3.3 Writing Balanced Chemical Equations 100

3.4 Combustion Reactions 105

3.5 Stoichiometric Calculations and the Carbon Cycle 107

3.6 Limiting Reactants and Percent Yield 112
Calculations Involving Limiting Reactants 113 • Actual Yields versus Theoretical Yields 116

3.7 Determining Empirical Formulas from Percent Composition 119

3.8 Comparing Empirical and Molecular Formulas 124
Molecular Mass and Mass Spectrometry 127

3.9 Combustion Analysis 128

Summary 134 • Particulate Preview Wrap-Up 134 • Problem-Solving Summary 134 • Visual Problems 136 • Questions and Problems 139

4 Reactions in Solution: Aqueous Chemistry in Nature 148

4.1 Ions and Molecules in Oceans and Cells 150

4.2 Expressing Concentrations 153
Concentration Units 153 • Molarity 154

4.3 Dilutions 159
Determining Concentration 162

4.4 Electrolytes and Nonelectrolytes 163
Ions in Solution 163

4.5 Acid–Base Reactions: Proton Transfer 165

4.6 Titrations 171

4.7 Precipitation Reactions 175
Precipitation Formation 176 • Using Precipitation in Analysis 180 • Saturated Solutions and Supersaturation 182

4.8 Oxidation–Reduction Reactions Electron Transfer 184
Oxidation Numbers 184 • Changes in Oxidation Numbers in Redox Reactions 186 • Electron Transfer in Redox Reactions 187 • Balancing Redox Reactions by Using Half-Reactions 188 • The Activity Series of Metals 191 • Redox in Nature 193

Summary 198 • Particulate Preview Wrap-Up 199 • Problem-Solving Summary 199 • Visual Problems 200 • Questions and Problems 202

5 Properties of Gases: The Air We Breathe 212

5.1 Air: An Invisible Necessity 214
5.2 Atmospheric Pressure and Collisions 215
5.3 The Gas Laws 220
Boyle's Law: Relating Pressure and Volume 220 • Charles's Law: Relating Volume and Temperature 223 • Avogadro's Law: Relating Volume and Quantity of Gas 225 • Amontons's Law: Relating Pressure and Temperature 226
5.4 The Ideal Gas Law 228
5.5 Gases in Chemical Reactions 232
5.6 Gas Density 235
5.7 Dalton's Law and Mixtures of Gases 238
5.8 The Kinetic Molecular Theory of Gases 243
Explaining Boyle's, Dalton's, and Avogadro's Laws 244 • Explaining Amontons's and Charles's Laws 245 • Molecular Speeds and Kinetic Energy 246 • Graham's Law: Effusion and Diffusion 249
5.9 Real Gases 250
Deviations from Ideality 251 • The van der Waals Equation for Real Gases 252

Summary 255 • Particulate Preview Wrap-Up 256 • Problem-Solving Summary 256 • Visual Problems 257 • Questions and Problems 261

6 Thermochemistry: Energy Changes in Chemical Reactions 270

6.1 Sunlight Unwinding 272
6.2 Forms of Energy 273
Work, Potential Energy, and Kinetic Energy 273 • Kinetic Energy and Potential Energy at the Molecular Level 276
6.3 Systems, Surroundings, and Energy Transfer 279
Isolated, Closed, and Open Systems 279 • Exothermic and Endothermic Processes 281 • *P–V* Work and Energy Units 283
6.4 Enthalpy and Enthalpy Changes 286
6.5 Heating Curves, Molar Heat Capacity, and Specific Heat 289
Hot Soup on a Cold Day 289 • Cold Drinks on a Hot Day 294
6.6 Calorimetry: Measuring Heat Capacity and Enthalpies of Reaction 297
Determining Molar Heat Capacity and Specific Heat 297 • Enthalpies of Reaction 299
6.7 Hess's Law 304
6.8 Standard Enthalpies of Formation and Reaction 308
6.9 Fuels, Fuel Values, and Food Values 313
Alkanes 313 • Fuel Value 316 • Food Value 317

Summary 321 • Particulate Preview Wrap-Up 321 • Problem-Solving Summary 322 • Visual Problems 323 • Questions and Problems 325

7 A Quantum Model of Atoms: Waves, Particles, and Periodic Properties 334

7.1 Rainbows of Light 336
7.2 Waves of Energy 339
7.3 Particles of Energy and Quantum Theory 341
Quantum Theory 341 • The Photoelectric Effect 343 • Wave–Particle Duality 344

7.4 The Hydrogen Spectrum and the Bohr Model 345
The Hydrogen Emission Spectrum 345 • The Bohr Model of Hydrogen 347

7.5 Electron Waves 350
de Broglie Wavelengths 350 • The Heisenberg Uncertainty Principle 352

7.6 Quantum Numbers and Electron Spin 354

7.7 The Sizes and Shapes of Atomic Orbitals 359
s Orbitals 359 • *p* and *d* Orbitals 361

7.8 The Periodic Table and Filling the Orbitals of Multielectron Atoms 362

7.9 Electron Configurations of Ions 370
Ions of the Main Group Elements 370 • Transition Metal Cations 372

7.10 The Sizes of Atoms and Ions 373
Trends in Atom and Ion Sizes 374

7.11 Ionization Energies 376

7.12 Electron Affinities 379

Summary 381 • Particulate Preview Wrap-Up 382 • Problem-Solving Summary 382 • Visual Problems 382 • Questions and Problems 384

8 Chemical Bonds: What Makes a Gas a Greenhouse Gas? 392

8.1 Types of Chemical Bonds and the Greenhouse Effect 394
Forming Bonds from Atoms 395

8.2 Lewis Structures 398
Lewis Symbols 398 • Drawing Lewis Structures 399 • Lewis Structures of Molecules with Double and Triple Bonds 401 • Lewis Structures of Ionic Compounds 404

8.3 Polar Covalent Bonds 405
Polarity and Type of Bond 407 • Vibrating Bonds and Greenhouse Gases 408

8.4 Resonance 409

8.5 Formal Charge: Choosing among Lewis Structures 414
Calculating Formal Charge of an Atom in a Resonance Structure 415

8.6 Exceptions to the Octet Rule 418
Odd-Electron Molecules 418 • Molecules in Which Atoms Form More than Four Bonds 420 • Lewis Structures: Atoms with More than an Octet 421 • Lewis Structures: Atoms with Less than an Octet 423 • The Limits of Bonding Models 425

8.7 The Lengths and Strengths of Covalent Bonds 426
Bond Length 426 • Bond Energies 427

Summary 432 • Particulate Preview Wrap-Up 432 • Problem-Solving Summary 432 • Visual Problems 433 • Questions and Problems 435

9 Molecular Geometry: Shape Determines Function 444

9.1 Biological Activity and Molecular Shape 446

9.2 Valence-Shell Electron-Pair Repulsion (VSEPR) Theory 447
Central Atoms with No Lone Pairs 448 • Central Atoms with Lone Pairs 452

9.3 Polar Bonds and Polar Molecules 457

9.4 Valence Bond Theory 461
Bonds from Orbital Overlap 462 • Hybridization 463 • Tetrahedral Geometry: sp^3 Hybrid Orbitals 463 • Trigonal Planar Geometry: sp^2 Hybrid Orbitals 465 • Linear Geometry: *sp* Hybrid Orbitals 467

9.5 Shape, Large Molecules, and Molecular Recognition 470

Drawing Larger Molecules 470 • Molecules with More than One Functional Group 472 • Chirality and Molecular Recognition 473

9.6 Molecular Orbital Theory 477
Molecular Orbitals of Hydrogen and Helium 478 • Molecular Orbitals of Homonuclear Diatomic Molecules 480 • Molecular Orbitals of Heteronuclear Diatomic Molecules 484 • Molecular Orbitals of N_2^+ and Spectra of Auroras 486 • Using MO Theory to Explain Fractional Bond Orders and Resonance 486 • A Bonding Theory for SN > 4 487 • Metallic Bonds and Conduction Bands 490 • Semiconductors 491

Summary 495 • Particulate Preview Wrap-Up 496 • Problem-Solving Summary 496 • Visual Problems 496 • Questions and Problems 498

10 Intermolecular Forces: The Uniqueness of Water 506

10.1 Intramolecular Forces versus Intermolecular Forces 508
10.2 Dispersion Forces 509
The Importance of Shape 510
10.3 Interactions Involving Polar Molecules 511
Ion–Dipole Interactions 511 • Dipole–Dipole Interactions 512 • Hydrogen Bonds 513
10.4 Vapor Pressure of Pure Liquids 519
Vapor Pressure and Temperature 520 • Volatility and the Clausius–Clapeyron Equation 521
10.5 Phase Diagrams: Intermolecular Forces at Work 523
Phases and Phase Transitions 523
10.6 Some More Remarkable Properties of Water 526
Surface Tension, Capillary Action, and Viscosity 526 • The Densities of Cold Water and Ice: Their Impact on Aquatic Life 528
10.7 Polarity and Solubility 529
Combinations of Intermolecular Forces 532
10.8 Solubility of Gases in Water 534

Summary 538 • Particulate Preview Wrap-Up 539 • Problem-Solving Summary 539 • Visual Problems 539 • Questions and Problems 541

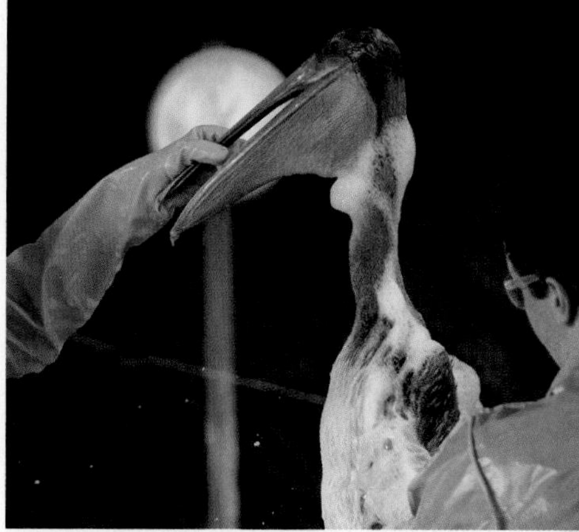

11 Solutions: Properties and Behavior 548

11.1 Interactions between Ions 550
11.2 Energy Changes during Formation and Dissolution of Ionic Compounds 553
Calculating Lattice Energies by Using the Born–Haber Cycle 556 • Enthalpies of Hydration 559
11.3 Vapor Pressure of Solutions 561
Raoult's Law 562
11.4 Mixtures of Volatile Solutes 564
Vapor Pressures of Mixtures of Volatile Solutes 564
11.5 Colligative Properties of Solutions 569
Molality 570 • Boiling Point Elevation 572 • Freezing Point Depression 573 • The van 't Hoff Factor 575 • Osmosis and Osmotic Pressure 579 • Reverse Osmosis 583 • Using Osmotic Pressure to Determine Molar Mass • 585
11.6 Ion Exchange • 587

Summary 590 • Particulate Preview Wrap-Up 590 • Problem-Solving Summary 590 • Visual Problems 592 • Questions and Problems 595

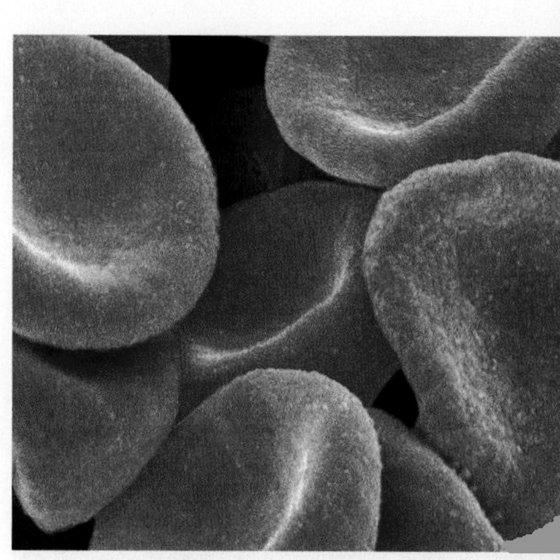

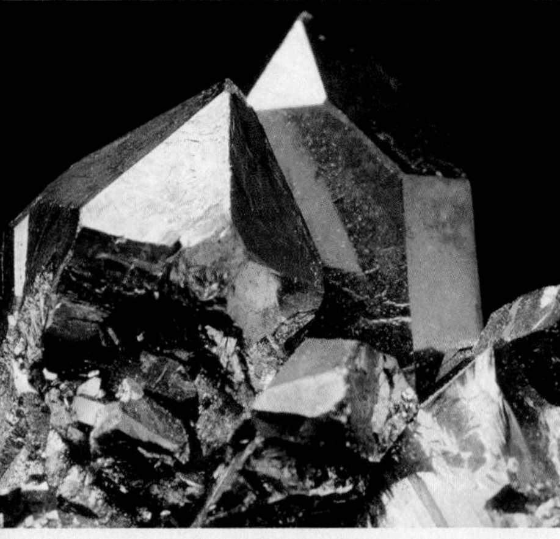

12 Solids: Crystals, Alloys, and Polymers 600

12.1 The Solid State 602

12.2 Structures of Metals 604
Stacking Patterns and Unit Cells 605 • Unit Cell Dimensions 607

12.3 Alloys and Medicine 612
Substitutional Alloys 613 • Interstitial Alloys 614

12.4 Ionic Solids and Salt Crystals 616

12.5 Allotropes of Carbon 620

12.6 Polymers 622
Polymers of Alkenes 622 • Polymers Containing Aromatic Rings 625 • Polymers of Alcohols and Ethers 626 • Polyesters and Polyamides 628

Summary 635 • Particulate Preview Wrap-Up 636 • Problem-Solving Summary 636 • Visual Problems 637 • Questions and Problems 640

13 Chemical Kinetics: Reactions in the Atmosphere 648

13.1 Cars, Trucks, and Air Quality 650

13.2 Reaction Rates 652
Experimentally Determined Reaction Rates 654 • Average Reaction Rates 656 • Instantaneous Reaction Rates 656

13.3 Effect of Concentration on Reaction Rate 659
Reaction Order and Rate Constants 659 • Integrated Rate Laws: First-Order Reactions 664 • Reaction Half-Lives 668 • Integrated Rate Laws: Second-Order Reactions 670 • Zero-Order Reactions 672

13.4 Reaction Rates, Temperature, and the Arrhenius Equation 674

13.5 Reaction Mechanisms 680
Elementary Steps 680 • Rate Laws and Reaction Mechanisms 682 • Mechanisms and Zero-Order Reactions 686

13.6 Catalysts 687
Catalysts and the Ozone Layer 687 • Catalysts and Catalytic Converters 690 • Enzymes: Biological Catalysts 691

Summary 694 • Particulate Preview Wrap-Up 695 • Problem-Solving Summary 695 • Visual Problems 696 • Questions and Problems 698

14 Chemical Equilibrium: How Much Product Does a Reaction Really Make? 708

14.1 The Dynamics of Chemical Equilibrium 710

14.2 The Equilibrium Constant 712

14.3 Relationships between K_c and K_p Values 717

14.4 Manipulating Equilibrium Constant Expressions 719
K for Reverse Reactions 719 • *K* for an Equation Multiplied or Divided by a Number 721 • Combining *K* Values 722

14.5 Equilibrium Constants and Reaction Quotients 724

14.6 Heterogeneous Equilibria 726

14.7 Le Châtelier's Principle 728
Effects of Adding or Removing Reactants or Products 729 • Effects of Pressure and Volume Changes 731 • Effect of Temperature Changes 734 • Catalysts and Equilibrium 736

14.8 Calculations Based on *K* 736

Summary 744 • Particulate Preview Wrap-Up 745 • Problem-Solving Summary 745 • Visual Problems 746 • Questions and Problems 748

15 Acid–Base Equilibria: Proton Transfer in Biological Systems 754

15.1 Acids and Bases: A Balancing Act 756

15.2 The Molecular Structures and Strengths of Acids and Bases 757
Strong and Weak Acids 758 • Strong and Weak Bases 763

15.3 Conjugate Pairs and Their Complementary Strengths as Acids and Bases 764
Recognizing Conjugate Pairs 765 • Relative Strengths of Conjugate Acids and Bases 766

15.4 pH and the Autoionization of Water 767
The pH Scale 768 • pOH, pK_a, and pK_b Values 771

15.5 K_a, K_b, and the Ionization of Weak Acids and Bases 772
Weak Acids 772 • Weak Bases 776

15.6 Calculating the pH of Acidic and Basic Solutions 778
Strong Acids and Strong Bases 778 • Weak Acids and Weak Bases 779 • pH of Very Dilute Solutions of Strong Acids 781

15.7 Polyprotic Acids 783
Acid Rain 783 • Normal Rain 785

15.8 Acidic and Basic Salts 788

Summary 793 • Particulate Preview Wrap-Up 794 • Problem-Solving Summary 794 • Visual Problems 796 • Questions and Problems 798

16 Additional Aqueous Equilibria: Chemistry and the Oceans 802

16.1 Ocean Acidification: Equilibrium under Stress 804

16.2 The Common-Ion Effect 806

16.3 pH Buffers 809
Buffer Capacity 812

16.4 Indicators and Acid–Base Titrations 816
Acid–Base Titrations 818 • Titrations with Multiple Equivalence Points 824

16.5 Lewis Acids and Bases 827

16.6 Formation of Complex Ions 830

16.7 Hydrated Metal Ions as Acids 832

16.8 Solubility Equilibria 834
K_{sp} and *Q* 839

Summary 842 • Particulate Preview Wrap-Up 843 • Problem-Solving Summary 843 • Visual Problems 844 • Questions and Problems 846

17 Thermodynamics: Spontaneous and Nonspontaneous Reactions and Processes 850

17.1 Spontaneous Processes 852
17.2 Entropy and the Second Law of Thermodynamics 855
17.3 Absolute Entropy and the Third Law of Thermodynamics 859
Entropy and Structure 862
17.4 Calculating Entropy Changes 863
17.5 Free Energy 864
17.6 Temperature and Spontaneity 869
17.7 Free Energy and Chemical Equilibrium 872
17.8 Influence of Temperature on Equilibrium Constants 877
17.9 Driving the Human Engine: Coupled Reactions 880
17.10 Microstates: A Quantized View of Entropy 884
Summary 888 • Particulate Preview Wrap-Up 888 • Problem-Solving Summary 889 • Visual Problems 889 • Questions and Problems 892

18 Electrochemistry: The Quest for Clean Energy 898

18.1 Running on Electrons: Redox Chemistry Revisited 900
18.2 Voltaic and Electrolytic Cells 903
Cell Diagrams 903
18.3 Standard Potentials 907
18.4 Chemical Energy and Electrical Work 911
18.5 A Reference Point: The Standard Hydrogen Electrode 914
18.6 The Effect of Concentration on E_{cell} 916
The Nernst Equation 916 • $E°$ and K 919
18.7 Relating Battery Capacity to Quantities of Reactants 920
Nickel–Metal Hydride Batteries 921 • Lithium-Ion Batteries 923
18.8 Corrosion: Unwanted Electrochemical Reactions 925
18.9 Electrolytic Cells and Rechargeable Batteries 928
18.10 Fuel Cells and Flow Batteries 931
Summary 936 • Particulate Preview Wrap-Up 936 • Problem-Solving Summary 937 • Visual Problems 937 • Questions and Problems 939

19 Nuclear Chemistry: Applications in Science and Medicine 946

19.1 Energy and Nuclear Stability 948
19.2 Unstable Nuclei and Radioactive Decay 950
19.3 Measuring and Expressing Radioactivity 956
19.4 Calculations Involving Half-Lives of Radionuclides 957
19.5 Radiometric Dating 959
19.6 Biological Effects of Radioactivity 962
Radiation Dosage 963 • Evaluating the Risks of Radiation 965

19.7 Medical Applications of Radionuclides 967
Therapeutic Radiology 967 • Diagnostic Radiology 968

19.8 Nuclear Fission 968

19.9 Nuclear Fusion and the Quest for Clean Energy 971

Summary 977 • Particulate Preview Wrap-Up 977 • Problem-Solving Summary 978 • Visual Problems 978 • Questions and Problems 980

20 Organic and Biological Molecules: The Compounds of Life 986

20.1 Molecular Structure and Functional Groups 988
Families Based on Functional Groups 989

20.2 Organic Molecules, Isomers, and Chirality 991
Chirality and Optical Activity 995 • Chiral Mixtures 1000

20.3 The Composition of Proteins 1001
Amino Acids 1001 • Zwitterions 1003 • Peptides 1006

20.4 Protein Structure and Function 1009
Primary Structure 1010 • Secondary Structure 1011 • Tertiary and Quaternary Structure 1012 • Enzymes: Proteins as Catalysts 1013

20.5 Carbohydrates 1016
Molecular Structures of Glucose and Fructose 1016 • Disaccharides and Polysaccharides 1017 • Energy from Glucose 1019

20.6 Lipids 1020
Function and Metabolism of Lipids 1022 • Other Types of Lipids 1024

20.7 Nucleotides and Nucleic Acids 1025
From DNA to New Proteins 1027

20.8 From Biomolecules to Living Cells 1029

Summary 1032 • Particulate Preview Wrap-Up 1032 • Problem-Solving Summary 1032 • Visual Problems 1033 • Questions and Problems 1035

21 The Main Group Elements: Life and the Periodic Table 1044

21.1 Main Group Elements and Human Health 1046

21.2 Periodic Properties of Main Group Elements 1049

21.3 Major Essential Elements 1050
Sodium and Potassium 1051 • Magnesium and Calcium 1054 • Chlorine 1055 • Nitrogen 1057 • Phosphorus and Sulfur 1060

21.4 Trace and Ultratrace Essential Elements 1065
Selenium 1066 • Fluorine and Iodine 1066 • Silicon 1067

21.5 Nonessential Elements 1067
Rubidium and Cesium 1067 • Strontium and Barium 1067 • Germanium 1068 • Antimony 1068 • Bromine 1068

21.6 Elements for Diagnosis and Therapy 1068
Diagnostic Applications 1069 • Therapeutic Applications 1071

Summary 1074 • Particulate Preview Wrap-Up 1075 • Problem-Solving Summary 1075 • Visual Problems 1076 • Questions and Problems 1078

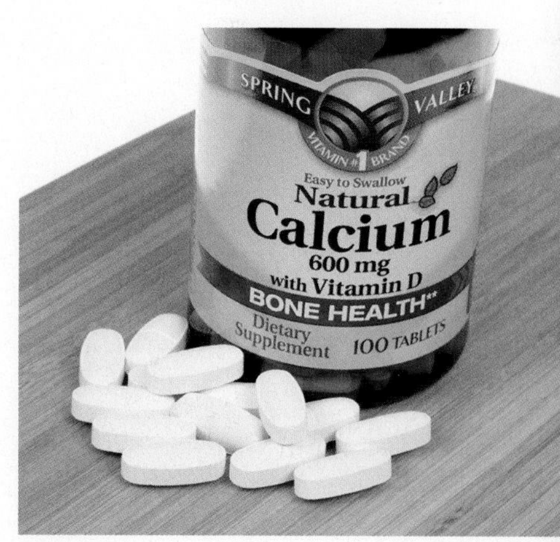

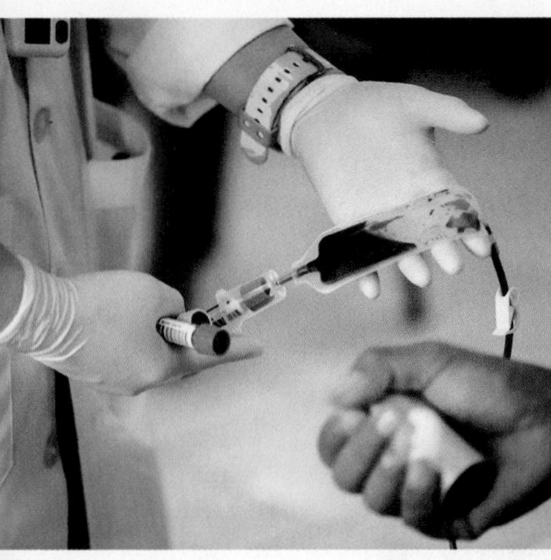

22 Transition Metals: Biological and Medical Applications 1082

22.1 Transition Metals in Biology: Complex Ions 1084

22.2 Naming Complex Ions and Coordination Compounds 1088
Complex Ions with a Positive Charge 1088 • Complex Ions with a Negative Charge 1090 • Coordination Compounds 1092

22.3 Polydentate Ligands and Chelation 1093

22.4 Crystal Field Theory 1097

22.5 Magnetism and Spin States 1102

22.6 Isomerism in Coordination Compounds 1104
Enantiomers and Linkage Isomers 1106

22.7 Coordination Compounds in Biochemistry 1108
Manganese and Photosynthesis 1108 • Transition Metals in Enzymes 1109

22.8 Coordination Compounds in Medicine 1113
Transition Metals in Diagnosis 1113 • Transition Metals in Therapy 1116

Summary 1121 • Particulate Preview Wrap-Up 1121 • Problem-Solving Summary 1122 • Visual Problems 1122 • Questions and Problems 1125

Appendices APP-1

Glossary G-1

Answers to Particulate Review, Concept Tests, and Practice Exercises ANS-1

Answers to Selected End-of-Chapter Questions and Problems ANS-15

Credits C-1

Index I-1

Applications

Blood centrifugation 9
Electrophoresis of blood proteins 10
Air filtration 10
Seawater distillation 10
Cosmic microwave background 35
Driving the Mars rover *Curiosity* 36
The Baby Tooth Survey 48
Big Bang 74
Star formation 75
Medical imaging 77
Miller–Urey experiment 88
Earth's interior layers 89
Volcanic eruptions 90
Dental fillings 93
Natural gas stoves 105
Photosynthesis, respiration, and the carbon cycle 108
Fossil fuels and atmospheric carbon dioxide 109
Power plant emissions 110
Asthma inhalers 120
Anticancer drugs (Taxol) 132
Water on Mars 150
Poly(vinyl chloride) (PVC) pipes 156
Great Salt Lake 157
Saline intravenous infusion 161
Stalactites and stalagmites 168
Chemical weathering 168
Drainage from abandoned coal mines 171
Antacids 174
Rock candy 182
Iron oxides in rocks and soils 193
Shelf stability of drugs 197
Barometers 215
Weather maps 217
Manometers 218
Tire pressure 227
Aerosol cans 227
Breathing 230
Weather balloons 231
Compressed oxygen for mountaineering 231
Oxygen masks in airplanes 233
Air bag inflation 234
Dieng Plateau gas poisoning disaster 235
Nitrogen narcosis 242
Gas mixtures for scuba diving 254
Hydrogen-powered vehicles 279
Diesel engines and hot-air balloons 284
Resurfacing an ice rink 287
Car radiators 293
Chilled beverages 294
Comparing fuels 316
Calories in food 318
Recycling aluminum 319
Rainbows 336
Remote control devices 344
Lasers 359
Signal flares 369
Fireworks 380
Greenhouse gases 394
Oxyacetylene torches 403
Atmospheric greenhouse effect 408
Atmospheric ozone 409
Moth balls 431
Polycyclic aromatic hydrocarbon (PAH) intercalation in DNA 473
Cilantro 473
Ripening tomatoes 473
Spearmint and caraway aromas 475
Auroras 477
Bar-code readers and DVD players 492
Pheromones 494
Hydrogen bonds in DNA 516
Supercritical carbon dioxide and dry ice 525
Water striders 526
Aquatic life in frozen lakes 528
Petroleum-based cleaning products 532
Surfactants 532
High-altitude endurance training 536
Drug efficacy 537
Antifreeze 563
Maple syrup 572
Radiator fluid 574
Osmosis in red blood cells 579
Saline and dextrose intravenous solutions 583
Desalination of seawater by reverse osmosis 584
Water softeners and zeolites 587
Eggs 589
Nanoparticles 602
Brass and bronze 612
Shape-memory alloys in stents 613
Stainless steel and surgical steel 614
Diamond and graphite 620
Graphene, fullerenes, and carbon nanotubes 621
Polyethylene: LDPE, HDPE, and UHMWPE materials 622
Teflon in cookware and surgical tubing 624
Polypropylene products 624
Polystyrene and Styrofoam 625
Plastic soda bottles 626
Artificial skin and dissolving sutures 629
Synthetic fabrics: Dacron, nylon, and Kevlar 630
Camping lanterns 634
Photochemical smog 650
Chlorofluorocarbons (CFCs) and the ozone layer 687
Catalytic converters 690
Biocatalysis 692
Smog simulations 693
Fertilizers 710
Hindenburg airship disaster 717
Limestone kilns 727
Colors of hydrangea blossoms 756
Lung disease and respiratory acidosis 757
Food digestion 770
Liquid drain cleaners 778
Carabid beetles 779

Acid rain and normal rain 783
Bleach 789
pH of human blood 792
Atmospheric carbon dioxide and ocean acidification 804
Swimming pool pH test kits 816
Sapphire Pool in Yellowstone National Park 826
Milk of magnesia 834
Impact of ocean acidification 842
Instant cold packs 854
Engine efficiency 869
Energy from glucose; glycolysis 881
Prehistoric axes and copper refining 887
Alkaline, nicad, and zinc–air batteries 909
Lead–acid car batteries 917
Hybrid vehicles and nickel–metal hydride batteries 921
Electric vehicles and lithium–ion batteries 923
Statue of Liberty 925
Rechargeable batteries 928
Electroplating 930
Hydrogen-fueled vehicles and fuel cells 931
Redox flow batteries 934
Corrosion at sea 935
Scintillation counters and Geiger counters 956
Radiometric dating 959
Chernobyl and Fukushima 964
Radon gas exposure 965
Therapeutic and diagnostic radiology 967
Nuclear weapons and nuclear power 969
Solar fusion 971
Tokamak reactors and ITER 973
Radium paint and the Radium Girls 975
Rice and beans 1003
Aspartame 1007
Sickle-cell anemia and malaria 1010
Silk 1012
Alzheimer's disease 1012
Hemoglobin and keratin 1013
Enzymes 1013
Lactose intolerance 1013
Thalidomide 1014
Blood type and glycoproteins 1016
Ethanol production from cellulose 1019
Saturated fats, unsaturated fats, and trans fats 1020
Olestra 1023
Cholesterol and cardiovascular disease 1025
DNA and RNA 1025
Origin of life on Earth 1029
Hydrogenated oils 1031
Dietary reference intake (DRI) 1048
Ion transport across cell membranes 1051
Osteoporosis and kidney stones 1054
Chlorophyll 1054
Teeth, bones, and shells 1054
Acid reflux and antacid drugs 1056
Bad breath, skunk odor, and smelly shoes 1063
Toothpaste and fluoridated water 1066
Goiter and Graves' disease 1067
Colonoscopy 1073
Prussian blue pigment 1088
Food preservatives 1096
Anticancer drugs (cisplatin) 1104
Cytochrome proteins 1110
Thalassemia and chelation therapy 1116
Water quality in swimming pools 1120

Animations

Dimensional Analysis 21
Significant Figures 23
Scientific Notation 23
Precision and Accuracy 27
Temperature Scale 32
Temperature Conversion 33
Cathode-Ray Tube 49
Parts of the Atom 49
Millikan Oil-Drop Experiment 50
Rutherford Experiment 52
Simplified Mass Spectrometer 54
Lanthanides and Actinides in the Periodic Table 60
NaCl Reaction 63
The Synthesis of Elements 75
Avogadro Constant 91
Balancing Equations 101
Carbon Cycle 109
Limiting Reactant 113
Percent Composition 120
Combustion Analysis 129
Molarity 154
Dilutions 160
Ions in Solution 164
Strong vs. Weak Acids 170
Balancing Redox Reactions 189
Measuring Gas Pressure 216
Manometer 218
Ideal Gas Law 230
Dalton's Law 238
Molecular Speed 247
State Functions and Path Functions 274
Internal Energy 283
Pressure–Volume Work 284
Heating Curves 290
Calorimetry 298
Hess's Law 305
Light Diffraction 337
Absorption of Light 339
Electromagnetic Radiation 340
The Photoelectric Effect 343
Emission Spectra and the Bohr Model of the Atom 346
de Broglie Wavelength 350
Quantum Numbers 356
Electron Configuration 365
Periodic Trends 375
Bonding 395
Lewis Structures 400
Bond Polarity and Polar Molecules 407
Vibrational Modes 408
Greenhouse Effect 409
Resonance 411
Estimating Enthalpy Changes 428
Hybridization 468
Structure of Benzene 472
Optical Activity 476
Molecular Orbitals 478
Intermolecular Forces 511
Interactions Involving Polar Molecules 516
Phase Diagrams 523
Surface Tension 526
Capillary Action 527
Henry's Law 536
Dissolution of Ammonium Nitrate 554
Born–Haber Cycle 556
Vapor Pressure 561
Fractional Distillation 564
Raoult's Law 567
Boiling and Freezing Points of Solutions 572
Osmotic Pressure 579
Unit Cell 606
Alloys 612
Allotropes of Carbon 620
Polymers 622
Reaction Rate 652
Reaction Order 659
Collision Theory 660
Arrhenius Equation 675
Reaction Mechanisms 680
Equilibrium 712
Equilibrium in the Gas Phase 714
Le Châtelier's Principle 729
Solving Equilibrium Problems 737
Acid–Base Ionization 759
Acid Strength and Molecular Structure 762
Conjugate Acids and Bases 765
Autoionization of Water 768
pH Scale 768
Acid Rain 783
Common-Ion Effect 807
Buffers 810
The Buffer System 811
Indicators 817
Acid–Base Titrations 818
Titrations of Weak Acids 820
Hydrated Metal Ions 832
Selective Precipitation 840
Spontaneous Processes 852
Entropy 855
Reversible Processes 857
Gibbs Free Energy 865
Equilibrium and Thermodynamics 873
Microstates 886
Zinc–Copper Cell 902
Electricity and Water Analogy 902
Voltaic vs. Electrolytic Cells 904
Alkaline Battery 909
Standard Hydrogen Electrode (SHE) 914
Cell Potential 916
Lead–Acid Battery 917
Cell Potential, Equilibrium, and Free Energy 919
Concentration Cell 919
Fuel Cell 931
Modes of Radioactive Decay 950
Belt of Stability 951
Balancing Nuclear Equations 952
Geiger Counter 956
Activity Example 956

Half-Life 957
Radiation Penetration 963
Transmutation 968
Induced Fission and Chain Reactions 969
Fusion of Hydrogen 972
Naming Branched Alkanes 991
Chiral Centers 996
Condensation of Biological Polymers 1007
Fiber Strength and Elasticity 1009
Formation of Sucrose 1017
Naming Coordination Compounds 1088
Crystal Field Splitting 1098

About the **Authors**

Thomas R. Gilbert has a BS in chemistry from Clarkson and a PhD in analytical chemistry from MIT. After 10 years with the Research Department of the New England Aquarium in Boston, he joined the faculty of Northeastern University, where he is currently associate professor of chemistry and chemical biology. His research interests are in chemical and science education. He teaches general chemistry and science education courses and conducts professional development workshops for K–12 teachers. He has won Northeastern's Excellence in Teaching Award and Outstanding Teacher of First-Year Engineering Students Award. He is a Fellow of the American Chemical Society and in 2012 was elected to the ACS Board of Directors.

Rein V. Kirss received both a BS in chemistry and a BA in history as well as an MA in chemistry from SUNY Buffalo. He received his PhD in inorganic chemistry from the University of Wisconsin, Madison, where the seeds for this textbook were undoubtedly planted. After two years of postdoctoral study at the University of Rochester, he spent a year at Advanced Technology Materials, Inc., before returning to academics at Northeastern University in 1989. He is an associate professor of chemistry with an active research interest in organometallic chemistry. He has been awarded Northeastern's Excellence in Teaching Award and received the John A. Timm Award from the New England Association of Chemistry Teachers in 2019.

Stacey Lowery Bretz is a University Distinguished Professor in the Department of Chemistry and Biochemistry at Miami University in Oxford, Ohio. She earned her BA in chemistry from Cornell University, MS from Pennsylvania State University, and a PhD in chemistry education research from Cornell University. She spent one year at the University of California, Berkeley, as a postdoc in the Department of Chemistry. Her research expertise includes the development of assessments to measure chemistry students' thinking with multiple representations (particulate, symbolic, and macroscopic) and to promote meaningful and inquiry learning in the chemistry laboratory. She is a Fellow of the American Chemical Society and a Fellow of the American Association for the Advancement of Science. She has been honored with both of Miami University's highest teaching awards: the E. Phillips Knox Award for Undergraduate Teaching and the Distinguished Teaching Award for Excellence in Graduate Instruction and Mentoring. Stacey won the prestigious, international award from the American Chemical Society for Achievement in Research for the Teaching and Learning of Chemistry in 2020.

Natalie Foster is emerita professor of chemistry at Lehigh University in Bethlehem, Pennsylvania. She received a BS in chemistry from Muhlenberg College and MS, DA, and PhD degrees from Lehigh University. Her research interests included studying poly(vinyl alcohol) gels by NMR as part of a larger interest in porphyrins and phthalocyanines as candidate contrast enhancement agents for MRI. She taught both semesters of the introductory chemistry class to engineering, biology, and other nonchemistry majors and a spectral analysis course at the graduate level. She is the recipient of the Christian R. and Mary F. Lindback Foundation Award for Distinguished Teaching and a Fellow of the American Chemical Society.

Preface

Dear Student,

We wrote this book with three overarching goals in mind: to make chemistry interesting, relevant, and memorable; to enable you to see the world from a molecular point of view; and to help you become an expert problem-solver. You have a number of resources available to assist you to succeed in your general chemistry course. This textbook will be a valuable resource, and we have written it with you, and the different ways you may use the book, in mind.

If you are someone who reads a chapter from the first page to the last, you will see that the Sixth Edition introduces the chemical principles within a chapter by using contexts drawn from daily life as well as from other disciplines, including biology, environmental science, materials science, astronomy, geology, and medicine. We believe that these contexts make chemistry more interesting, relevant, and memorable.

Chemists' unique perspective of natural processes and insights into the properties of substances, from high-performance alloys to the products of biotechnology, are based on understanding these processes and substances at the particulate level (the atomic and molecular level). A major goal of this book is to help you develop this microscale perspective and link it to macroscopic properties.

With that in mind, we begin each chapter with a **Particulate Review** and **Particulate Preview**. The goal of these tools is to prepare you for the material in the chapter. The **Particulate Review** assesses important prior knowledge that you need to interpret particulate images in the chapter. The **Particulate Preview** asks you to speculate about new concepts you will see in the chapter and is meant to focus your reading.

If you want a quick summary of what is most important in a chapter to direct your studying on selected topics, check the **Learning Outcomes** at the beginning of each chapter. Whether you are reading the chapter from first page to last, moving from

PARTICULATE **REVIEW**

Particles in the Gas Phase

In Chapter 5 we focus on the properties of gases, including those that serve as fuels in combustion reactions, as described in Chapter 3, and in other forms of energy production. One such fuel is hydrogen gas, which can be produced by passing an electric current through water, causing molecules of liquid H_2O to decompose into molecules of H_2 and O_2 gas.

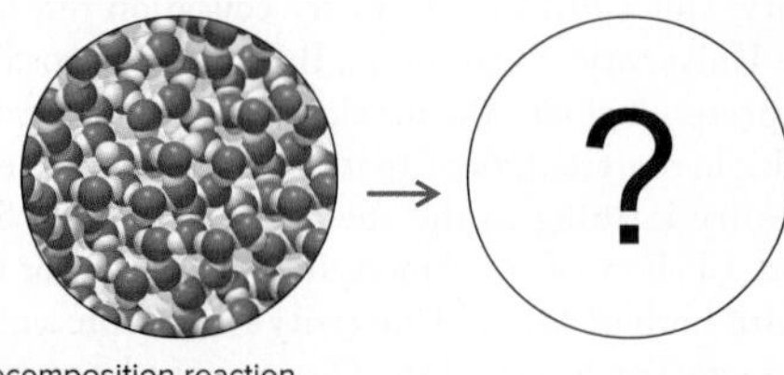

- Write a balanced chemical equation describing this decomposition reaction.
- In the circle on the right, draw the products that would be produced by decomposition of six of the water molecules on the left.
- Classify the products as elements, compounds, or a mixture. Choose all that apply.

(Review Sections 1.1, 1.2, and 3.3 if you need help.)

(Answers to Particulate Review questions are in the back of the book.)

PARTICULATE **PREVIEW**

Pressure, Volume, and Temperature

As you read Chapter 5, look for ideas that will help you answer these questions.

- Draw particulate images of the helium in the tank and in one of the balloons. How do these drawings differ?
- Suppose the tank and several helium-filled balloons are placed in the trunk of a car on a hot summer day.
  - How would your particulate image for the helium in the balloon change?
  - How would your particulate image for the helium in the tank change?

topic to topic in an order you select, or reviewing material for an exam, the Learning Outcomes can help you focus on the key information you need to know and the skills you should develop.

In every section, you will find **key terms** in boldface in the text and in a **running glossary** in the margin. We have inserted the definitions throughout the text, so you can continue reading without interruption but quickly find key terms when doing homework or reviewing for a test. All key terms are also defined in the Glossary in the back of the book.

Approximately once per section, you will find a **Concept Test**. These short, conceptual questions provide a self-check opportunity by asking you to stop and answer a question relating to what you just read. We designed them to help you self-assess, and you will find answers to Concept Tests in the back of the book.

CONCEPT **TEST**

Which graph in **Figure 5.18** correctly describes the relationship between the value of *V*/*n* as *n* is increased at constant *P* and *T*?

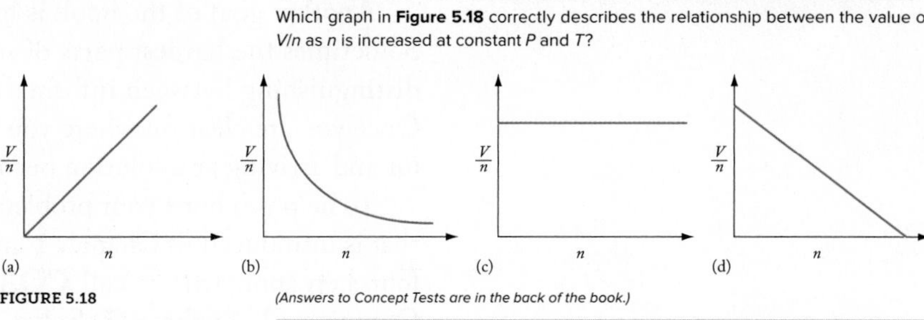

FIGURE 5.18

(Answers to Concept Tests are in the back of the book.)

New concepts naturally build on previous information, and you will find that many concepts are related to others described earlier in the book. We point out these relationships with **Connection** icons in the margins. These reminders will help you see the big picture and draw your own connections between concepts in the book.

CONNECTION In Chapter 3 the number of particles in a mole was defined as the Avogadro constant, in honor of Amedeo Avogadro's early work with gases that led to determining atomic masses.

At the end of each chapter are **Visual Problems** that ask you to interpret atomic and molecular views of elements and compounds, along with graphs of experimental data. The last Visual Problem in each chapter contains a **visual problem matrix**. This grid consists of nine images followed by a series of questions that will test your ability to identify the similarities and differences among the macroscopic and particulate images.

5.22. Use representations [A] through [I] in Figure P5.22 to answer questions (a)–(f). The pink balloons contain hydrogen, the yellow balloons contain nitrogen, and the gray balloons contain oxygen.

a. Identify three different changes that could be responsible for the change in size of the pink balloon from [A] to [C].
b. If the smaller pink balloon in [A] corresponds to the particulate view in [B], which of the changes identified in part (a) would result in the larger pink balloon in [C] also corresponding to the particulate view in [B]?
c. If the smaller yellow balloon in [D] corresponds to the particulate view in [E], which particulate view corresponds to the larger yellow balloon in [F] if no additional nitrogen has been added?
d. If the larger gray balloon in [I] corresponds to the particulate view in [E], which particulate view represents the gas at a lower temperature?
e. If each gray balloon contains 1 mol of gas at 25°C, in which balloon are the collisions between the oxygen molecules and the inside of the balloon more frequent?
f. Which balloon contains the gas with the shortest mean free path?

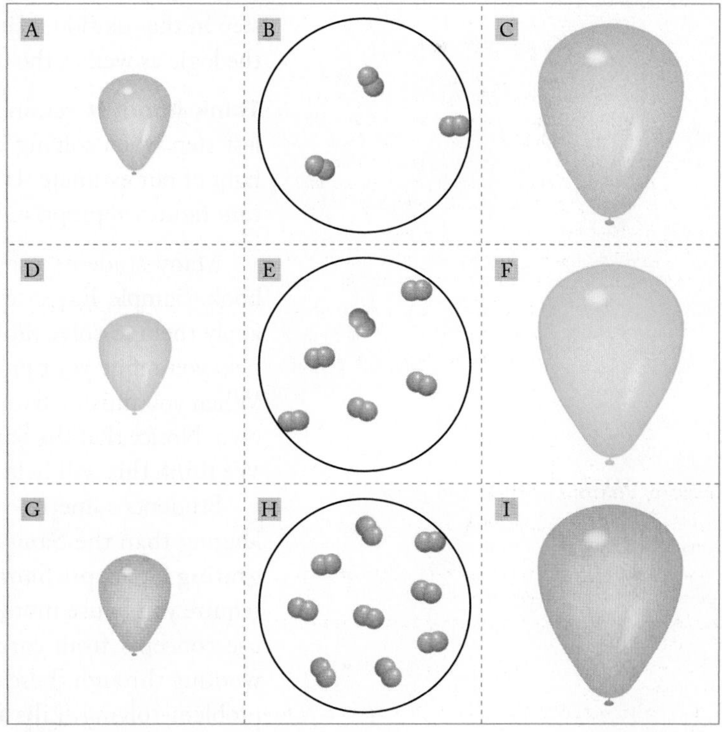

FIGURE P5.22

CHEMTOUR

STEPWISE ANIMATION

If you're looking for additional help visualizing a concept, we have nearly 140 animations and simulations, denoted by the **ChemTour** and **Stepwise Animation** icons. These animations, available at digital.wwnorton.com/chem6, show chemical concepts and processes to help you visualize events at the macro, micro, and symbolic level. Many of the ChemTours are interactive, allowing you to manipulate variables and observe changes in a graph or a process. Questions at the end of ChemTours offer step-by-step assistance in solving problems and provide useful feedback.

Another goal of the book is to help you improve your problem-solving skills. Sometimes the hardest parts of solving a problem is knowing where to start and distinguishing between information that is relevant and information that is not. Once you are clear on where you are starting and where you are going, planning for and arriving at a solution become much easier.

To help you hone your problem-solving skills, we have developed a framework that is introduced in Chapter 1 and used consistently throughout the book. It is a four-step approach we call **COAST**, which is our acronym for (1) **C**ollect and **O**rganize, (2) **A**nalyze, (3) **S**olve, and (4) **T**hink About It. We use these four steps in *every* Sample Exercise and in the solutions to *odd-numbered* problems in the Student Solutions Manual. They are also used in the hints and feedback embedded in the Smartwork5 online homework program. To summarize the four steps:

Collect and Organize helps you understand where to begin. In this step we often point out what you must find and what is given, including the relevant information that is provided in the problem statement or available elsewhere in the book.

Analyze is where we map out a strategy for solving the problem. As part of that strategy we often estimate what a reasonable answer might be.

Solve applies our strategy from the second step to the information and relationships identified in the first step as we solve the problem. We walk you through each step in the solution, using dimensional analysis consistently, so that you can follow the logic as well as the math.

Think About It reminds us that calculating or determining an answer is not the last step when solving a problem. We check whether the answer is reasonable in light of our estimate. Is it realistic? Are the units correct? Is the number of significant figures appropriate?

Many students use the **Sample Exercises** more than any other part of the book. Sample Exercises take the concepts being discussed and illustrate how to apply them to solve problems. We hope that repeated application of COAST will help you refine your problem-solving skills and become an expert problem-solver. When you finish a Sample Exercise, you'll find a **Practice Exercise** to try on your own. Notice that the Sample Exercises and the Learning Objectives are connected. We think this will help you focus efficiently on the main ideas in the chapter.

Students sometimes comment that the questions on an exam are more challenging than the Sample Exercises in a book. To address this, we have an **Integrating Concepts Sample Exercise** near the end of each chapter. These exercises require you to use more than one concept from the chapter and may expect you to use concepts from earlier chapters to solve a problem. Please invest your time working through these problems because we think they will further enhance your problem-solving skills and give you an increased appreciation of how chemistry is used in the world.

If you use the book mostly as a reference and problem-solving guide, we have a learning path for you as well. It starts with a **Summary** and a **Problem-Solving**

Summary at the end of each chapter. The first is a brief synopsis of the chapter, organized by Learning Outcomes and referencing sections from the chapter. Key figures have been added to this Summary to provide visual cues as you review. The Problem-Solving Summary organizes the chapter by problem type and summarizes relevant concepts and equations you need to solve each type of problem. The Problem-Solving Summary also points you back to the relevant Sample Exercises that model how to solve each problem and cross-references the Learning Outcomes at the beginning of the chapter.

PROBLEM-SOLVING SUMMARY

Type of Problem	Concepts and Equations	Sample Exercises
Calculating pressure of any gas; calculating atmospheric pressure	Divide the force by the area over which the force is applied, using the equation $P = \frac{F}{A}$ (5.1)	**5.1, 5.2**
Calculating changes in *P*, *V*, or *T* in response to changing conditions	Rearrange $\frac{P_1V_1}{T_1} = \frac{P_2V_2}{T_2}$ (5.18) for whichever variable is sought and then substitute given values. (*T* must be in kelvins, and *n* must be constant.)	**5.3, 5.4, 5.5, 5.6**
Determining *n* from *P*, *V*, and *T*	Rearrange $PV = nRT$ (5.15) for *n* and then substitute given values of *P*, *T*, and *V*. (*T* must be in kelvins.)	**5.7, 5.8, 5.9**
Calculating the density of a gas and calculating molar mass from density	Substitute values for pressure, absolute temperature, and molar mass into the equation $d = \frac{P\mathcal{M}}{RT}$ (5.21) Substitute values for pressure, absolute temperature, and density into the equation $\mathcal{M} = \frac{dRT}{P}$ (5.22)	**5.10, 5.11**
Calculating mole fraction for one component gas in a mixture	Divide the number of moles of the component gas by the total number of moles in the mixture: $X_x = \frac{n_x}{n_{total}}$ (5.24)	**5.12**
Calculating partial pressure of one component gas in a mixture and total pressure in the mixture	Substitute the mole fraction of the component gas and the total pressure in the equation $P_x = X_xP_{total}$ (5.25) Solve the equation $P_{total} = P_1 + P_2 + P_3 + P_4 + \ldots$ (5.23) for the partial pressure of the component gas and then substitute given values for other partial pressures and total pressure.	**5.13, 5.14**

Following the summaries are groups of questions and problems. The first group is the **Visual Problems. Concept Review Questions and Problems** come next, arranged by topic in the same order as they appear in the chapter. Concept Reviews are qualitative and often ask you to explain why or how something happens. Problems are paired and can be quantitative, conceptual, or a combination of both. **Contextual problems** have a title that describes the context in which the problem is placed. **Additional Problems** can come from any section or combination of sections in the chapter. Some of them incorporate concepts from previous chapters. Problems marked with an asterisk (*) are more challenging and often require multiple steps to solve.

5.87. Biological Effects of Radon Exposure Radon is a naturally occurring radioactive gas found in the ground and in building materials. It is easily inhaled and emits α particles when it decays. Cumulative radon exposure is a significant risk factor for lung cancer.
a. Calculate the density of radon at 298 K and 1.00 atm of pressure.
b. Are radon concentrations likely to be greater in the basement or on the top floor of a building?

*5.88. Four empty balloons, each with a mass of 10.0 g, are inflated to a volume of 20.0 L. The first balloon contains He; the second, Ne; the third, CO_2; and the fourth, CO. If the density of air at 25°C and 1.00 atm is 0.00117 g/mL, which of the balloons float in this air?

5.155. Anesthesia A common anesthesia gas is halothane, with the structure shown in Figure P5.155. Liquid halothane boils at 50.2°C and 1.00 atm. If halothane behaved as an ideal gas, what volume would 10.0 mL of liquid halothane (d = 1.87 g/mL) occupy at 60°C and 1.00 atm of pressure? What is the density of halothane vapor at 55°C and 1.00 atm of pressure?

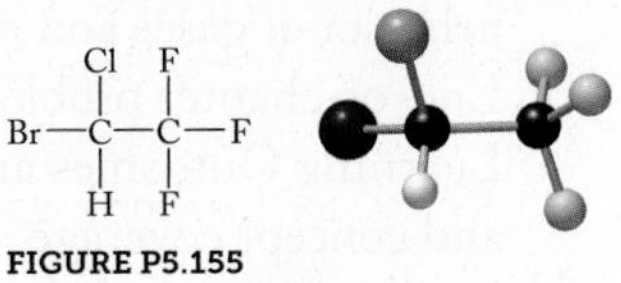

FIGURE P5.155

We want you to have confidence in using the answers in the back of the book as well as the Student Solutions Manual, so we continue to use a rigorous triple-check accuracy program for the Sixth Edition. Each end-of-chapter question and problem has been solved independently by at least three PhD chemists. For the Sixth Edition the team included solutions manual author Don Carpenetti and two additional chemistry educators. Don compared his solutions to those from the reviewers and resolved any discrepancies. This process is designed to ensure clearly written problems and accurate answers in the appendices and solutions manual.

No matter how you use this book, we hope it becomes a valuable tool for you and helps you not only understand the principles of chemistry but also apply them to solving global problems, such as diagnosing and treating disease or making more efficient use of Earth's natural resources.

Changes to the Sixth Edition

Dear Instructor,

As authors of a textbook we are very often asked: "Why is a new edition necessary? Has the science changed that much since the previous edition?" Although chemistry is a vigorous and dynamic field, most basic concepts presented in an introductory course have not changed dramatically. However, two areas tightly intertwined in this text—pedagogy and context—have changed significantly, and those areas drive this new edition. Here are some of the most noteworthy changes we made throughout this edition:

- Adopters enthusiastically embraced the visualization pedagogy that was introduced in the Fifth Edition and then asked for additional tools to help them use this research-based pedagogy to assess student learning. The teaching tools accompanying the Sixth Edition include:
  - In-class activities, such as pre-made handouts, help you facilitate active learning, with a focus on visualization, in any size classroom.
  - Projected visualization questions help you assess students on their understanding and support the Particulate Review, Particulate Preview, and Visual Matrix problems in the book by providing color images that you can project in your class accompanied by questions that you can use on your exam or for in-class quizzes.
  - The new Interactive Instructor's Guide (IIG), an easily searchable online resource, contains all the teaching resources for the Sixth Edition. Instructors can search by chapter, phrase, topic, or learning objective to find activities, animations, simulations, and visualization questions to use in class.
- Based on her chemistry education research, Stacey Lowery Bretz evaluated and revised the art in the book and the media to be more pedagogically effective and address student misconceptions.
- The gases chapter, now Chapter 5, comes before thermochemistry (Chapter 6), which allows instructors to introduce pressure–volume work and its impact of the internal energy of gaseous systems after students have explored the behavior of gases and the inverse relationship between pressure and volume.
- End-of-chapter problems have been refined to better align with the text's Learning Outcomes and provide a more balanced selection in terms of level and concept coverage. We estimate that between 10%–15% of the end-of-chapter problems have been replaced or revised.

- Smartwork5 now offers more than 5000 problems, including 600 new to the Sixth Edition. New Smartwork5 problems include more end-of-chapter problems, algorithmic problems, and pooled problems.
- Adaptive functionality in Smartwork5, powered by Knewton, has been continuously improved, resulting in more intuitive experiences for students as they work through personalized learning paths. It also offers the availability of better data for instructors to review their students' mastery of chosen learning objectives.
- Revised ChemTour animations, also found in the ebook and Smartwork5, are now easier to navigate and individual sections of each animation can now be easily viewed and assigned. ChemTours are available streaming in the ebook, in Smartwork5 questions, as remediation in adaptive assignments, and are in the IIG, which also includes discussion questions and activity ideas for each animation.
- Our new partnership with Squarecap classroom response system provides you with book-specific clicker questions in an easy-to-use system at a low cost to students.

Teaching and Learning Resources

Smartwork5

digital.wwnorton.com/chem6

Smartwork5 is the most intuitive online tutorial and homework management system available for general chemistry. The many question types, including graded molecular drawings, math and chemical equations, ranking tasks, and interactive figures, help students develop and apply their understanding of fundamental concepts in chemistry. Adaptive functionality, powered by Knewton, gives students personalized coaching, allowing them to reach mastery of assigned concepts at their own pace.

Every problem in Smartwork5 includes response-specific feedback and general hints often organized by the steps in COAST. Links to the ebook version of *Chemistry*, Sixth Edition, take students to the specific place in the text where the concept is explained. All problems in Smartwork5 use the same language and notation as the textbook.

Smartwork5 also features Tutorial Problems. If students ask for help in a Tutorial Problem, the system breaks the problem down into smaller steps, coaching them with hints, answer-specific feedback, and probing questions within each part. At any point in a Tutorial, a student can return to and answer the original problem.

Assigning, editing, and administering homework within Smartwork5 is easy. It's tablet compatible and integrates with the most common campus learning management systems. Smartwork5 allows the instructor to search for problems by text sections, learning objectives, question type, difficulty, and Bloom's taxonomy. Instructors can use premade assignment sets provided by Norton authors, modify those assignments, or create their own. Instructors can also make changes in the problems at the question level. All instructors have access to our WYSIWYG (What You See Is What You Get) authoring tools—the same ones Norton authors use. Those intuitive tools make it easy to modify existing problems or to develop new content that meets the specific needs of a course.

Wherever possible, Smartwork5 makes use of algorithmic variables so that students see slightly different versions of the same problem. Assignments are graded automatically, and Smartwork5 includes sophisticated yet flexible tools for managing class data. Instructors can use the class activity report to assess students' performance on specific problems within an assignment. Instructors can also review individual students' work on problems.

Smartwork5 for *Chemistry*, Sixth Edition, features the following problem types:

- End-of-Chapter Problems. These problems, which use algorithmic variables when appropriate, all have hints and answer-specific feedback to coach students through mastering single- and multiple-concept problems based on chapter content. They make use of all of Smartwork5's answer-entry tools.
- ChemTour Problems. Many of the ChemTour animations have been redesigned to make them easier than ever to navigate, making it simple to assign Smartwork5 questions on specific interactives.
- Visual and Graphing Problems. These problems challenge students to identify chemical phenomena and to interpret graphs. They use Smartwork5's drag-and-drop functionality.
- Ranking Task Problems. These problems ask students to make comparative judgments between items in a set.
- Nomenclature Problems. New matching and multiple-choice problems help students master course vocabulary.
- Multistep Tutorials. These problems offer students who demonstrate a need for help a series of linked, step-by-step subproblems to work. They are based on the Concept Review problems at the end of each chapter.
- Math Review Problems. These problems can be used by students for practice or by instructors to diagnose the mathematical ability of their students.

Ebook

digital.wwnorton.com/chem6

An affordable and convenient alternative to the print text, Norton Ebooks let students access the entire book and much more: they can search, highlight, and take notes with ease. The Norton Ebook allows instructors to share their notes with students. The ebook can be viewed on most devices—laptop, tablet, even a public computer—and will stay synced between devices.

The online version of *Chemistry*, Sixth Edition, also provides students with one-click access to our approximately 130 simulations and animations.

The online ebook is available bundled with the print text and Smartwork5 at no extra cost, or it may be purchased bundled with Smartwork5 access.

Norton also offers a downloadable PDF version of the ebook.

Student Solutions Manual

by Don Carpenetti, Craven Community College

The Student Solutions Manual provides students with fully worked solutions to select end-of-chapter problems using the **COAST** four-step method (**C**ollect and **O**rganize, **A**nalyze, **S**olve, and **T**hink About It). The Student Solutions Manual contains several pieces of art for each chapter, designed to help students visualize ways to approach problems. This artwork is also used in the hints and feedback within Smartwork5.

Test Bank

by Chris Harmon, Humboldt State University, and Jamie Schneider, University of Wisconsin River Falls

Norton uses an innovative, evidence-based model to deliver high-quality and pedagogically effective quizzes and testing materials. Each chapter of the Test Bank is structured around an expanded list of student learning objectives and evaluates student knowledge on six distinct levels based on Bloom's Taxonomy: Remembering, Understanding, Applying, Analyzing, Evaluating, and Creating.

Questions are further classified by text section and difficulty, making it easy to construct tests and quizzes that are meaningful and diagnostic, according to each instructor's needs. New questions are marked and sortable, making it easy to find brand new problems to add to exams. More than 2500 questions are divided into multiple choice and short answer.

The Test Bank is available with ExamView Test Generator software, allowing instructors to effortlessly create, administer, and manage assessments. The convenient and intuitive test-making wizard makes it easy to create customized exams with no software learning curve. Other key features include the ability to create paper exams with algorithmically generated variables and export files directly to Blackboard, Canvas, Desire2Learn, and Moodle.

Projected Visualization Questions

by Chris Harmon, Humboldt State, and Jamie Schneider, University of Wisconsin River Falls

Stacey Lowery Bretz believes that you must include visualization questions on exams if you want students to take learning this skill seriously. To overcome the challenge of not being able to distribute four-color exams to her class, she includes several problems on her exams that require students to look at a particulate, macro, or symbolic image which is projected at the front of the room. Our new projected visualization problems allow instructors to use this approach to ask questions about full-color images on exams or in-class activities.

Instructor's Solutions Manual

by Don Carpenetti, Craven Community College

The Instructor's Solutions Manual provides instructors with fully worked solutions to every end-of-chapter Concept Review and Problem. Each solution uses the **COAST** four-step method (**C**ollect and **O**rganize, **A**nalyze, **S**olve, and **T**hink About It).

Interactive Instructor's Guide

iig.wwnorton.com/chem6/full
by Kevin Braun, Virginia Military Institute, Nicole Heldt, SUNY Canton, and Mohammed Shahin, Fullerton College

The Interactive Instructor's Guide will help instructors use our unique visualization pedagogy in class and for assessment by compiling the many valuable teaching resources available with *Chemistry*, Sixth Edition, in an easily searchable online format. In-class activities, backed by chemical education research and written by active learning experts, emphasize visualization and are designed to promote collaboration. New projected visualization problems, based on how

Stacey Lowery Bretz gives her own exams, allow instructors to ask exam questions using full-color images. The ChemTours are also accompanied by new activity ideas, discussion questions, and clicker questions, making them easier than ever to use. All resources are searchable by chapter, keyword, or learning objective.

Clickers in Action: Increasing Student Participation in General Chemistry

by Margaret Asirvatham, University of Colorado, Boulder

An instructor-oriented resource providing information on implementing clickers in general chemistry courses, *Clickers in Action* contains more than 250 class-tested, lecture-ready questions, with histograms showing student responses, as well as insights and suggestions for implementation. Question types include macroscopic observation, symbolic representation, and atomic/molecular views of processes.

Downloadable Instructor's Resources

digital.wwnorton.com/chem6

This password-protected site for instructors includes:

- Stepwise Animations and classroom response questions. Developed by Jeffrey Macedone of Brigham Young University and his team, these animations, which use native PowerPoint functionality and textbook art, help instructors to "walk" students through nearly 100 chemical concepts and processes. Where appropriate, the slides contain two types of questions for students to answer in class: questions that ask them to predict what will happen next and why, and questions that ask them to apply knowledge gained from watching the animation. Self-contained notes help instructors adapt these materials to their own classrooms.
- Lecture PowerPoints.
- All ChemTours.
- Test Bank in PDF, Word, and ExamView Assessment Suite formats.
- Solutions manual in PDF and Word, so that instructors may edit solutions.
- All of the end-of-chapter Questions and Problems, available in Word along with the key equations.
- Photographs, drawn figures, and tables from the text, available in PowerPoint and JPEG format.
- Clicker questions, including those from *Clickers in Action*.
- Course cartridges. Available for the most common learning management systems, course cartridges include access to the ChemTours and Stepwise Animations as well as links to the ebook and Smartwork5.

Acknowledgments

We begin by thanking the users and reviewers who provided the feedback necessary to make the Sixth Edition possible. In addition to written reviews, the comments we receive at meetings, in focus groups, in emails, and from office visits with the Norton sales, editorial, and marketing staff help us identify what works well and what needs to be improved. Those suggestions and critical feedback encouraged us to make the content, context, and pedagogy work better and maximize learning for students.

Our colleagues at W. W. Norton remain a constant source of inspiration and guidance. Their passion for providing accurate and reliable content sets a high standard that motivates us to create an exceptional and user-friendly set of resources for instructors and students. The people at W. W. Norton with whom we work most closely deserve much more praise than we can possibly express here. Our editor, Erik Fahlgren, continues to offer his wisdom, guidance, energy, creativity, and most impressively, an endless amount of patience, as he simultaneously leads and pushes us to meet deadlines. Erik's leadership is the single greatest reason for this book's completion, and our greatest thanks are far too humble an offering for his unwavering vision and commitment. He is the consummate professional and a valued friend.

We have enjoyed and benefited from the humor and wisdom of our developmental editor, John Murdzek. Carla Talmadge has been the project editor on many editions and we trust that she will help synchronize our words and images on every page. Thanks to our production manager, Ben Reynolds, who handled his first chemistry rodeo like a pro; copyeditor Marjorie Anderson who was punctual, accurate, and pleasant to work with; proofreader Chris Curioli whose work is always thorough and steady; Stephanie Romeo and Julie Tesser who helped us find just the right photos; Grace Tuttle who did a masterful job at creating and revising the teaching and learning resources for this edition; Cailin Barrett-Bressack for her work on Smartwork5 and the other media that help students learn and practice and instructors assess that learning; and to Stacy Loyal who continues to creatively market the book and work with the Norton team in the field.

This book has benefited greatly from the care and thought that many reviewers, listed here, gave to their readings of earlier drafts. We owe an extra-special thanks to Don Carpenetti for his dedicated and precise work on the solutions manual. He, along with Andy Ho and Sri Kamesh Narasimhan are the triple-check accuracy team who solved each problem and reviewed each solution for accuracy. We also greatly appreciate Thomas Berke, Kevin Boyd, Carmin Burrell, Mark Campbell, Gabriel Cook, Jim Davis, John Elliff, Matthew Hurst, Rebecca Laird, Edgar Lee, Brian Leskiw, Carol Martinez, Drew Meyer, James Nyachwaya, Julie Pigza, Wendy Schatzberg, Daniel Swart, and Jordan Vincent for checking the accuracy of the myriad facts that form the framework of our science.

Thomas R. Gilbert
Rein V. Kirss
Stacey Lowery Bretz

Sixth Edition Reviewers

Joseph Awino, *Iowa State University*
Mohammad Bahrami, *California State University—Los Angeles*
Suzanne Bart, *Purdue University*
Thomas Berke, *Brookdale Community College*
Glenn Bobo, *Bevill State Community College*
Kevin Boyd, *University of Minnesota*
Carmin Burrell, *Ivy Tech Community College*
Derek Bussan, *McNeese State University*
Mark Campbell, *United States Naval Academy*
Don Carpenetti, *Craven Community College*
Gabriel Cook, *Oklahoma State University*
Brandon Cruickshank, *Northern Arizona University*
Jim Davis, *South Alabama University*
Bryan Davis, *Shenandoah University*
Logan Dewayne, *Baton Rouge Community College*
John Elliff, *Kirkwood Community College*
John Farrar, *North Kentucky University*
Karen Fortune, *Houston Community College*
Eric Goll, *Brookdale Community College*
Thomas Greenbowe, *University of Oregon*
Chris Hamaker, *Illinois State University*
Hill (William) Harman, *University of California, Riverside*

Christopher Harmon, *Humboldt State University*
Antony Hascall, *Northern Arizona University*
Andy Ho, *Lehigh University*
Matthew Hurst, *Humboldt State University*
Scott Kirkby, *East Tennessee State University*
Larry Kolopajlo, *Eastern Michigan University*
Rebecca Laird, *University of Iowa*
Dan Lawson, *University of Michigan*
Edgar Lee, *Northern Arizona University*
Brian Leskiw, *Youngstown State University*
Joanne Lin, *Houston Community College*
Tamika Madison, *University of Pittsburgh*
Carol Martinez, *Central New Mexico Community College*
Dan Matthew, *Northern Arizona University*
Drew Meyer, *Case Western Reserve University*
Smita Mohanty, *Oklahoma State University*
Kamesh Narasimhan, *Corning Community College*
Ross Nord, *Eastern Michigan University*
James Nyachwaya, *North Dakota State University*
Julie Pigza, *University of Southern Mississippi*
Wendy Schatzberg, *Dixie State University*
David Schiraldi, *Case Western Reserve University*
Erik Sneddon, *McNeese State University*
Lothar Stahl, *University of North Dakota*
Uma Swamy, *Florida International University*
Daniel Swart, *Minnesota State University—Mankato*
Paul Szalay, *Muskingum University*
Amy Toole, *University of Toledo*
Steve Trail, *Elgin Community College*
Jordan Vincent, *California State University—Los Angeles*
Ralph Zehnder, *Angelo State University*

Previous Edition Reviewers

William Acree, Jr., *University of North Texas*
R. Allendoefer, *University at Buffalo*
Thomas J. Anderson, *Francis Marion University*
Sharon Anthony, *The Evergreen State College*
Jeffrey Appling, *Clemson University*
Marsi Archer, *Missouri Southern State University*
Margaret Asirvatham, *University of Colorado, Boulder*
Robert Balahura, *University of Guelph*
Ian Balcom, *Lyndon State College of Vermont*
Anil Banerjee, *Texas A&M University, Commerce*
Sandra Banks, *Mills College*
Mikhail V. Barybin, *University of Kansas*
Mufeed Basti, *North Carolina Agricultural & Technical State University*
Shuhsien Batamo, *Houston Community College, Central Campus*
Robert Bateman, *University of Southern Mississippi*
Kevin Bennett, *Hood College*
H. Laine Berghout, *Weber State University*
Lawrence Berliner, *University of Denver*
Mark Berry, *Brandon University*
Narayan Bhat, *University of Texas, Pan American*
Eric Bittner, *University of Houston*
David Blauch, *Davidson College*
Robert Boggess, *Radford University*
Ivana Bozidarevic, *DeAnza College*
Chris Bradley, *Mount St. Mary's University*
Michael Bradley, *Valparaiso University*
Richard Bretz, *Miami University*
Timothy Brewer, *Eastern Michigan University*
Karen Brewer, *Hamilton College*
Ted Bryan, *Briar Cliff University*
Donna Budzynski, *San Diego Community College*
Julia Burdge, *Florida Atlantic University*
Robert Burk, *Carleton University*
Andrew Burns, *Kent State University*
Diep Ca, *Shenandoah University*
Sharmaine Cady, *East Stroudsburg University*
Chris Cahill, *George Washington University*
Dean Campbell, *Bradley University*
Kevin Cantrell, *University of Portland*
Nancy Carpenter, *University of Minnesota, Morris*
Patrick Caruana, *SUNY Cortland*
David Cedeno, *Illinois State University*
Tim Champion, *Johnson C. Smith University*
Stephen Cheng, *University of Regina*
Tabitha Chigwada, *West Virginia University*
Allen Clabo, *Francis Marion University*
William Cleaver, *University of Vermont*
Penelope Codding, *University of Victoria*
Jeffrey Coffer, *Texas Christian University*
Claire Cohen-Schmidt, *University of Toledo*
Renee Cole, *Central Missouri State University*
Patricia Coleman, *Eastern Michigan University*
Shara Compton, *Widener University*
Andrew L. Cooksy, *San Diego State University*
Elisa Cooper, *Austin Community College*
Brian Coppola, *University of Michigan*
Richard Cordell, *Heidelberg College*
Chuck Cornett, *University of Wisconsin, Platteville*
Mitchel Cottenoir, *South Plains College*
Robert Cozzens, *George Mason University*
Mark Cybulski, *Miami University*
Margaret Czrew, *Raritan Valley Community College*
Shadi Dalili, *University of Toronto*
William Davis, *Texas Lutheran University*
John Davison, *Irvine Valley College*
Jalison de Lima, *Vanier College*
Laura Deakin, *University of Alberta*
Mauro Di Renzo, *Vanier College*
Anthony Diaz, *Central Washington University*
Klaus Dichmann, *Vanier College*
Kelley J. Donaghy, *State University of New York, College of Environmental Science and Forestry*
Michelle Driessen, *University of Minnesota*
Theodore Duello, *Tennessee State University*
Dan Durfey, *Naval Academy Preparatory School*
Bill Durham, *University of Arkansas*
Stefka Eddins, *Gardner-Webb University*

Gavin Edwards, *Eastern Michigan University*
Kimberly Hill Edwards, *Oakland University*
Alegra Eroy-Reveles, *San Francisco State University*
Dwaine Eubanks, *Clemson University*
Lucy Eubanks, *Clemson University*
Jordan L. Fantini, *Denison University*
Nancy Faulk, *Blinn College, Bryan*
Tricia Ferrett, *Carleton College*
Matt Fisher, *St. Vincent College*
Amy Flanagan-Johnson, *Eastern Michigan University*
George Flowers, *Darton College*
Richard Foust, *Northern Arizona University*
David Frank, *California State University, Fresno*
Andrew Frazer, *University of Central Florida*
Cynthia Friend, *Harvard University*
Karen Frindell, *Santa Rosa Junior College*
Barbara Gage, *Prince George's Community College*
Rachel Garcia, *San Jacinto College*
Simon Garrett, *California State University, Northridge*
Nancy Gerber, *San Francisco State University*
Brian Gilbert, *Linfield College*
Jack Gill, *Texas Woman's University*
Arthur Glasfeld, *Reed College*
Samantha Glazier, *St. Lawrence University*
Stephen Z. Goldberg, *Adelphi University*
Pete Golden, *Sandhills Community College*
Frank Gomez, *California State University, Los Angeles*
John Goodwin, *Coastal Carolina University*
Steve Gravelle, *Saint Vincent College*
Stan Grenda, *University of Nevada*
Nathaniel Grove, *University of North Carolina, Wilmington*
Tammy Gummersheimer, *Schenectady County Community College*
Kim Gunnerson, *University of Washington*
Margaret Haak, *Oregon State University*
Todd Hamilton, *Adrian College*
Robert Hanson, *St. Olaf College*
David Hanson, *Stony Brook University*
Holly Ann Harris, *Creighton University*
David Harris, *University of California, Santa Barbara*
Donald Harriss, *University of Minnesota, Duluth*
Dale Hawley, *Kansas State University*
Sara Hein, *Winona College*
Brad Herrick, *Colorado School of Mines*
Vicki Hess, *Indiana Wesleyan University*
Paul Higgs, *University of Tennessee, Martin*
K. Joseph Ho, *University of New Mexico*
Donna Hobbs, *Augusta State University*
Angela Hoffman, *University of Portland*
Matthew Horn, *Utah Valley University*
Zhaoyang Huang, *Jacksonville University*
Donna Ianotti, *Brevard Community College*
Tim Jackson, *University of Kansas*
Tamera Jahnke, *Southern Missouri State University*
Shahid Jalil, *John Abbot College*
Eugenio Jaramillo, *Texas A&M International University*
Kevin Johnson, *Pacific University*
David Johnson, *University of Dayton*
Lori Jones, *University of Guelph*
Martha Joseph, *Westminster College*
David Katz, *Pima Community College*
Jason Kautz, *University of Nebraska, Lincoln*
Phillip Keller, *University of Arizona*
Resa Kelly, *San Jose State University*
Vance Kennedy, *Eastern Michigan University*
Michael Kenney, *Case Western Reserve University*
Mark Keranen, *University of Tennessee, Martin*
Robert Kerber, *Stony Brook University*
Elizabeth Kershisnik, *Oakton Community College*
Angela King, *Wake Forest University*
Edith Kippenhan, *University of Toledo*
Sushilla Knottenbelt, *University of New Mexico*
Tracy Knowles, *Bluegrass Community and Technical College*
John Krenos, *Rutgers University*
C. Krishnan, *Stony Brook University*
Stephen Kuebler, *University of Central Florida*
Liina Ladon, *Towson University*
Richard Langley, *Stephen F. Austin State University*
Timothy Lash, *Illinois State University*
Sandra Laursen, *University of Colorado, Boulder*
Richard Lavrich, *College of Charleston*
David Laws, *The Lawrenceville School*
Edward Lee, *Texas Tech University*
Willem Leenstra, *University of Vermont*
Alistair Lees, *Binghamton University*
Scott Lewis, *Kennesaw State University*
Joanne Lin, *Houston Community College, Eastside Campus*
Matthew Linford, *Brigham Young University*
George Lisensky, *Beloit College*
Jerry Lokensgard, *Lawrence University*
Boon Loo, *Towson University*
Karen Lou, *Union College*
Leslie Lyons, *Grinnell College*
Laura MacManus-Spencer, *Union College*
Roderick M. Macrae, *Marian College*
John Maguire, *Southern Methodist University*
Susan Marine, *Miami University, Ohio*
Albert Martin, *Moravian College*
Brian Martinelli, *Nevada State College*
Diana Mason, *University of North Texas*
Laura McCunn, *Marshall University*
Ryan McDonnell, *Cape Fear Community College*
Garrett McGowan, *Alfred University*
Thomas McGrath, *Baylor University*
Craig McLauchlan, *Illinois State University*
Lauren McMills, *Ohio University*
Heather Mernitz, *Tufts University*
Stephen Mezyk, *California State University, Long Beach*
Rebecca Miller, *Lehigh University*
John Milligan, *Los Angeles Valley College*
Timothy Minger, *Mesa Community College*
Ellen Mitchell, *Bridgewater College*
Robbie Montgomery, *University of Tennessee, Martin*

Joshua Moore, *Tennessee State University*
Stephanie Myers, *Augusta State College*
Richard Nafshun, *Oregon State University*
Steven Neshyba, *University of Puget Sound*
Melanie Nilsson, *McDaniel College*
Mya Norman, *University of Arkansas*
Sue Nurrenbern, *Purdue University*
Gerard Nyssen, *University of Tennessee, Knoxville*
Ken O'Connor, *Marshall University*
Jodi O'Donnell, *Siena College*
Jung Oh, *Kansas State University, Salinas*
Joshua Ojwang, *Brevard Community College*
MaryKay Orgill, *University of Nevada, Las Vegas*
Robert Orwoll, *College of William and Mary*
Gregory Oswald, *North Dakota State University*
Jason Overby, *College of Charleston*
Greg Owens, *University of Utah*
Maria Pacheco, *Buffalo State College*
Stephen R. Parker, *Montana Tech*
Jessica Parr, *University of Southern California*
Pedro Patino, *University of Central Florida*
Vicki Paulissen, *Eastern Michigan University*
Trilisa Perrine, *Ohio Northern University*
Giuseppe Petrucci, *University of Vermont*
Julie Peyton, *Portland State University*
David E. Phippen, *Shoreline Community College*
Alexander Pines, *University of California, Berkeley*
Prasad Polavarapu, *Vanderbilt University*
John Pollard, *University of Arizona*
Lisa Ponton, *Elon University*
Pete Poston, *Western Oregon University*
Gretchen Potts, *University of Tennessee, Chattanooga*
Robert Pribush, *Butler University*
Gordon Purser, *University of Tulsa*
Robert Quandt, *Illinois State University*
Karla Radke, *North Dakota State University*
Orlando Raola, *Santa Rosa College*
Casey Raymond, *State University of New York, Oswego*
Haley Redmond, *Texas Tech University*
Jimmy Reeves, *University of North Carolina, Wilmington*
Scott Reid, *Marquette University*
Paul Richardson, *Coastal Carolina University*
Alan Richardson, *Oregon State University*
Albert Rives, *Wake Forest University*
Mark Rockley, *Oklahoma State University*
Alan Rowe, *Norfolk State University*
Christopher Roy, *Duke University*
Erik Ruggles, *University of Vermont*
Beatriz Ruiz Silva, *University of California, Los Angeles*
Pam Runnels, *Germanna Community College*
Joel Russell, *Oakland University*
Jerry Sarquis, *Miami University, Ohio*
Nancy Savage, *University of New Haven*
Barbara Sawrey, *University of California, Santa Barbara*
Truman Schwartz, *Macalester College*
Louis Scudiero, *Washington State University*
Fatma Selampinar, *University of Connecticut*
Shawn Sendlinger, *North Carolina Central University*
Susan Shadle, *Boise State University*
George Shelton, *Austin Peay State University*
Peter Sheridan, *Colgate University*
Edwin Sibert, *University of Wisconsin, Madison*
Ernest Siew, *Hudson Valley Community College*
Roberta Silerova, *John Abbot College*
Virginia Smith, *United States Naval Academy*
Sally Solomon, *Drexel University*
Xianzhi Song, *University of Pittsburgh, Johnstown*
Brock Spencer, *Beloit College*
Clayton Spencer, *Illinois College*
Estel Sprague, *University of Cincinnati*
William Steel, *York College of Pennsylvania*
Wesley Stites, *University of Arkansas*
Meredith Storms, *University of North Carolina, Pembroke*
Steven Strauss, *Colorado State University*
Katherine Stumpo, *University of Tennessee, Martin*
Mark Sulkes, *Tulane University*
Luyi Sun, *Texas State University*
Duane Swank, *Pacific Lutheran University*
Keith Symcox, *University of Tulsa*
Agnes Tenney, *University of Portland*
Charles Thomas, *University of Tennessee, Martin*
Matthew Thompson, *Trent University*
Craig Thulin, *Utah Valley University*
Edmund Tisko, *University of Nebraska, Omaha*
Brian Tissue, *Virginia Polytechnic Institute and State University*
Laurie Tyler, *Union College*
Mike van Stipdonk, *Wichita State University*
Kris Varazo, *Francis Marion University*
William Vining, *University of Massachusetts*
Andrew Vruegdenhill, *Trent University*
Ed Walton, *California State Polytechnic University, Pomona*
Haobin Wang, *New Mexico State University*
Erik Wasinger, *California State University, Chico*
Rory Waterman, *University of Vermont*
Karen Wesenberg-Ward, *Montana Tech University*
Wayne Wesolowski, *University of Arizona*
Charles Wilkie, *Marquette University*
Ed Witten, *Northeastern University*
Karla Wohlers, *North Dakota State University*
Stephen Wood, *Brigham Young University*
Cynthia Woodbridge
Mingming Xu, *West Virginia University*
Tim Zauche, *University of Wisconsin, Platteville*
Noel Zaugg, *Brigham Young University, Idaho*
Corbin Zea, *Grand View University*
James Zimmerman, *Missouri State University*
Martin Zysmilich, *George Washington University*

Chemistry

The Science in Context

1

Particles of Matter

Measurement and the Tools of Science

ANCIENT UNIVERSE Light from the most distant galaxies in the universe, such as the one featured in this photo taken by NASA's Hubble Space Telescope, provide insights into the composition of the universe as it existed over 13 billion years ago.

PARTICULATE **REVIEW**

Atoms and Molecules: What's the Difference?

In Chapter 1 we explore how chemists classify different kinds of matter, from elements to compounds to mixtures. Hydrogen and helium were the first two elements formed after the universe began. Chemists use distinctively colored spheres to distinguish atoms of different elements in their drawings and models. For example, hydrogen is almost always depicted as white.

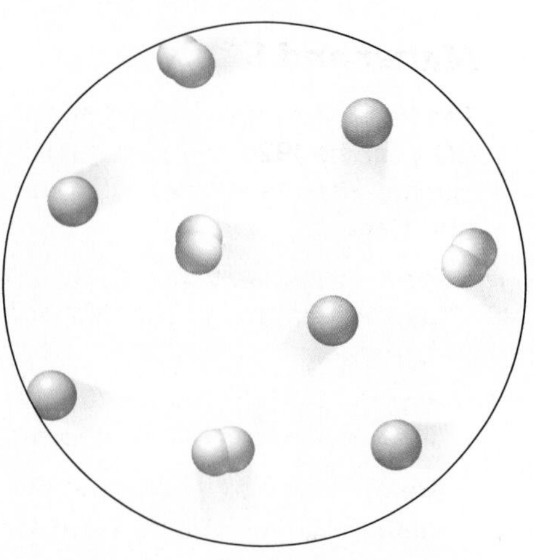

- How many of the following particles are shown in this image?
  - Hydrogen atoms
  - Hydrogen molecules
  - Helium atoms
- Are molecules composed of atoms, or are atoms composed of molecules?

(Answers to Particulate Review questions are in the back of the book.)

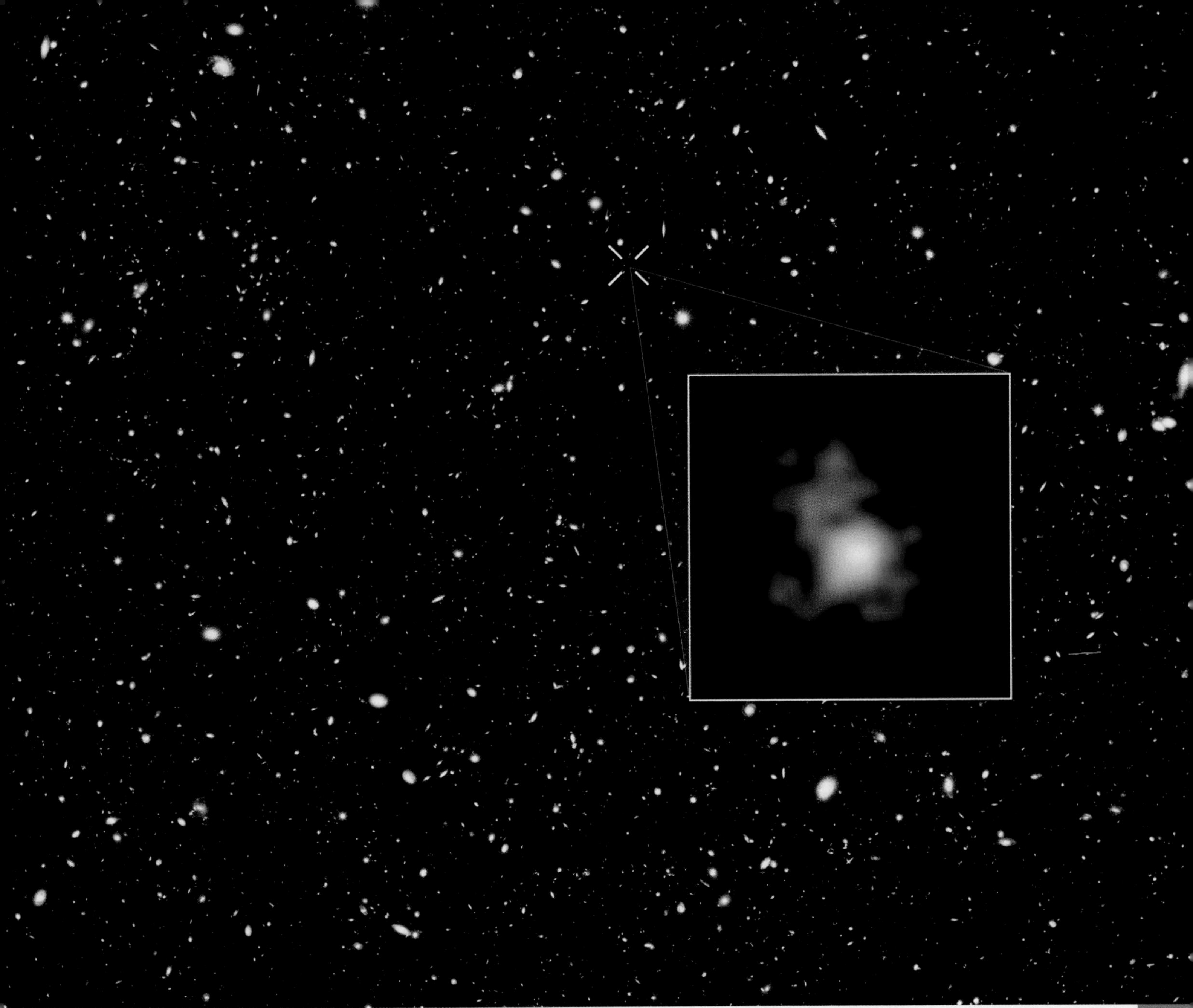

PARTICULATE **PREVIEW**

Matter and Energy

The temperature in outer space is 2.73 K. The temperature of dry ice (carbon dioxide, CO_2) can be 192 K warmer, but still cold enough to keep ice cream frozen on a hot summer day. As you read Chapter 1, look for ideas that will help you answer these questions:

- Particulate images of CO_2 as it sublimes are shown here. Which two phases of matter are involved in sublimation?
- What features of the images helped you decide which two phases were involved?
- What is the role of energy in this transformation of matter? Is energy added or is energy produced?

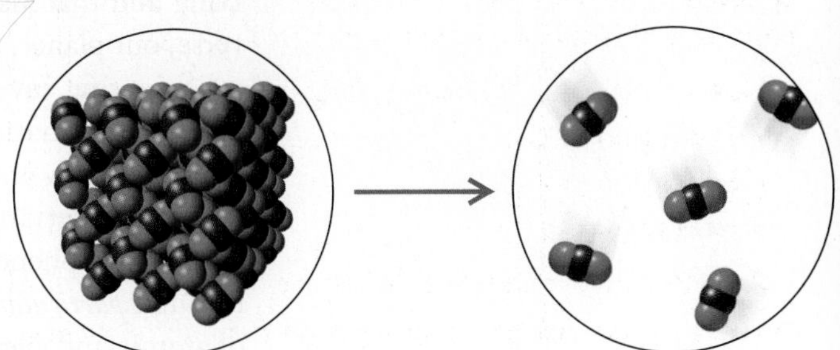

Learning Outcomes

LO1 Distinguish among pure substances, homogeneous mixtures, and heterogeneous mixtures, and between elements and compounds

LO2 Describe the particles that make up elements and compounds and how the particulate composition of substances is reflected in their chemical formulas and molecular structures

LO3 Distinguish between physical processes and chemical reactions, and between physical and chemical properties
Sample Exercise 1.1

LO4 Use a systematic approach (COAST) to problem solving

LO5 Describe the three states of matter and the transitions between them at the macroscopic and particulate levels
Sample Exercise 1.2

LO6 Describe the scientific method

LO7 Convert quantities from one system of units to another
Sample Exercises 1.3, 1.4, 1.9

LO8 Express uncertain values with the appropriate number of significant figures
Sample Exercise 1.5

LO9 Distinguish between exact and uncertain values, evaluate the precision and accuracy of experimental results, and identify outliers
Sample Exercises 1.6, 1.7, 1.8

1.1 How and Why

For thousands of years, we humans have sought to better understand the world around us. For most of that time we resorted to mythological explanations of natural phenomena. Many once believed, for example, that the Sun rose in the east and set in the west because it was carried across the sky by a god driving a chariot propelled by winged horses.

In recent times we have been able to move beyond such fanciful accounts of natural phenomena to explanations based on observation and scientific reasoning. Unfortunately, this movement toward rational explanations has not always been smooth. Consider, for example, the contributions of Galileo Galilei, the man Albert Einstein called the father of modern science. At the dawn of the 17th century, Galileo used advanced telescopes of his own design to observe the movement of the planets and their moons. He concluded that they, like Earth, revolved around the Sun. However, this view conflicted with a belief held by many religious leaders of his time that Earth was the center of the universe. In 1633 a religious tribunal forced Galileo to disavow his conclusion that Earth orbited the Sun and banned him (or anyone) from publishing the results of studies that called into question the Earth-centered view of the universe. The ban was not completely lifted until 1835—nearly 200 years after Galileo's death.

In the last century, advances in the design and performance of telescopes have led to the astounding discovery that we live in an expanding universe that probably began 13.8 billion years ago with an enormous release of energy. In this chapter and in later ones, we examine some of the data that led to the theory of the Big Bang and that also explain the formation of the elements that make up the universe, our planet, and ourselves.

Scientific investigations into the origin of the universe have stretched the human imagination and forced scientists to develop new models and new explanations of how and why things are the way they are. Frequently these efforts have involved observing and measuring large-scale phenomena, which we refer to as *macroscopic* phenomena. We seek to explain these macroscopic phenomena through *particulate* representations that show the structure of matter on the scale of atomic and even subatomic particles. Throughout this text, you will encounter many of these macroscopic–particulate connections. The authors of this book hope that your exploration of these connections will help you better understand how and why nature is the way it is.

mass the property that defines the quantity of matter in an object.

matter anything that has mass and occupies space.

chemistry the study of the composition, structure, and properties of matter, and of the energy consumed or given off during chemical reactions.

pure substance matter that has a constant composition and cannot be broken down into simpler matter by any physical process.

physical process a transformation of a sample of matter, such as a change in its physical state, that does not alter the chemical identity of any substance in the sample.

1.2 Macroscopic and Particulate Views of Matter

According to a formula widely used in medicine, the ideal weight for adult males 176 centimeters (cm, or 69 inches) tall (the average height of American and Canadian men) is 69 kilograms (kg, or 152 pounds). On average, this body mass consists of 12 kg of fat and 57 kg of lean body mass, including bones, organs, muscles, and blood. These values are measures of the total *mass* of all the *matter* in the body. In general, **mass** is the quantity of matter in any object. **Matter**, in turn, is a term that applies to everything in the body (and in the universe) that has mass and occupies space. **Chemistry** is the study of the composition, structure, and properties of matter.

Classes of Matter

The different forms of matter are classified according to the scheme shown in **Figure 1.1**. We begin on the left with **pure substances**, which have a constant composition that does not vary from one sample to another. The composition of pure water does not vary, no matter what its source or how much of it there is. Like all pure substances, water cannot be separated into simpler substances by any physical process. A **physical process** is a transformation of a sample of matter that does not alter the chemical identities of any of the substances in the sample, such as a change in physical state from solid to liquid.

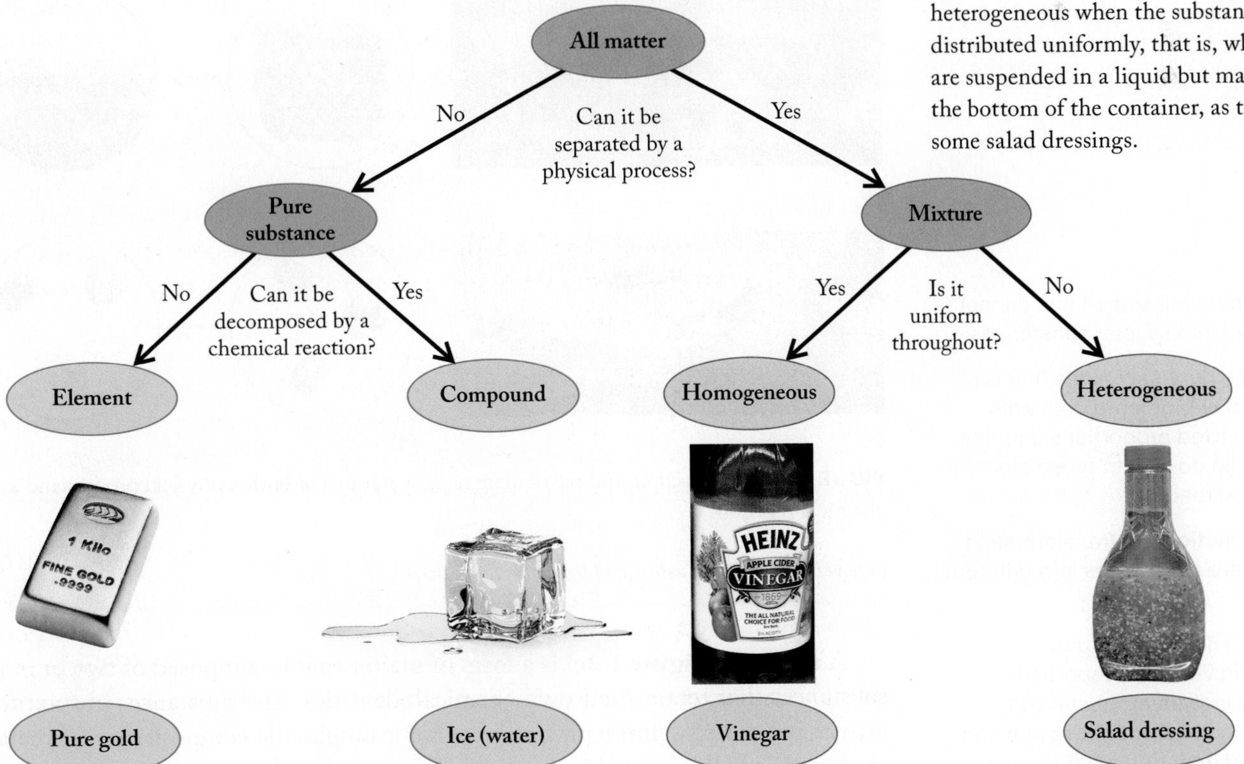

FIGURE 1.1 The two principal classes of matter are pure substances and mixtures. A pure substance may be a compound (such as water) or an element (such as gold). A mixture is homogeneous when the substances are distributed uniformly, as they are in vinegar (a mixture of mostly acetic acid and water). A mixture is heterogeneous when the substances are not distributed uniformly, that is, when solids are suspended in a liquid but may settle to the bottom of the container, as they do in some salad dressings.

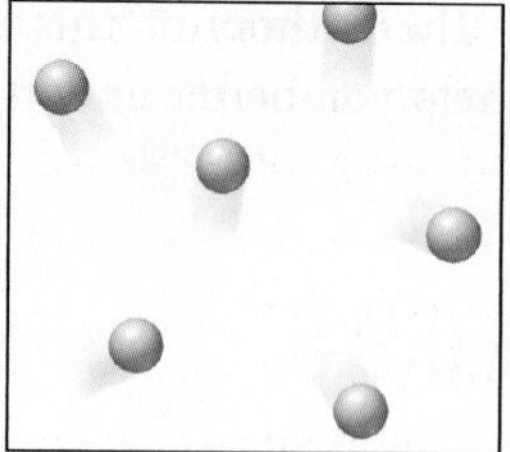
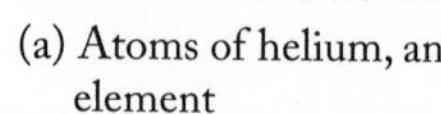
(a) Atoms of helium, an element

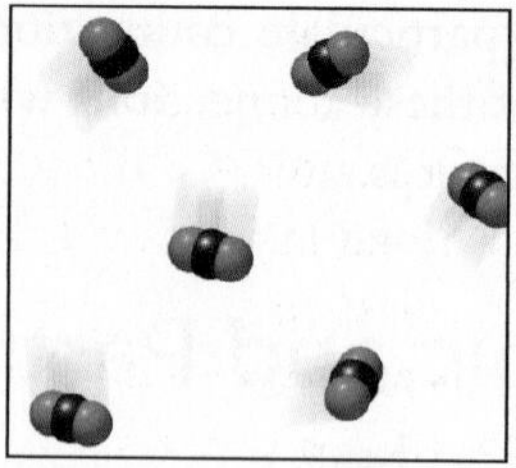
(b) Molecules of carbon dioxide, a compound

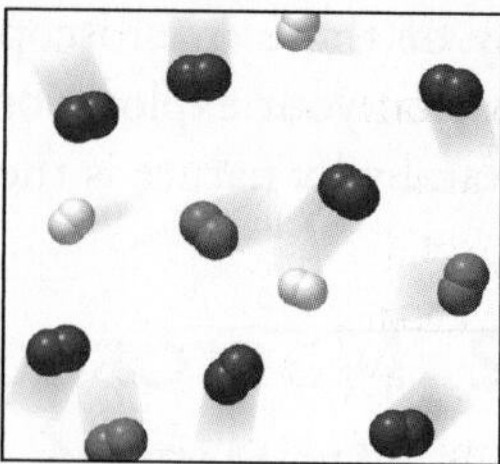
(c) Mixture of gases

FIGURE 1.2 All matter is made up of either pure substances (of which there are relatively few in nature) or mixtures. (a) The element helium (He), the second most abundant element in the universe, is a pure substance. (b) The compound carbon dioxide (CO_2), the gas used in many fire extinguishers, is also a pure substance. (c) This homogeneous mixture contains three substances: nitrogen (N_2, blue), hydrogen (H_2, white), and oxygen (O_2, red).

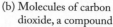

Pure substances are further classified as elements (**Figure 1.2a**) and compounds (**Figure 1.2b**). An **element** is a pure substance that cannot be broken down into simpler substances. The periodic table inside the front cover shows all 118 of the known elements. Only a few of them (including gold, silver, nitrogen, oxygen, and sulfur) occur in nature uncombined with other elements. Instead, most elements in nature are combined with other elements in the form of **compounds**, which are substances composed of two or more elements combined in fixed proportions. The elements in a compound can be separated from one another only by a **chemical reaction**: the transformation of one or more substances into one or more different substances. Compounds typically have properties that are very different from those of the elements of which they are composed. For example, common table salt (sodium chloride) has little in common with either sodium, which is a silver-gray metal that reacts violently when dropped in water, or chlorine, which is a toxic yellow-green gas.

CONCEPT TEST

Which photo in **Figure 1.3** depicts a physical process? Which photo depicts a chemical reaction? Use what you know about the difference between a physical process and a chemical reaction to match each photo to its corresponding particulate representation.

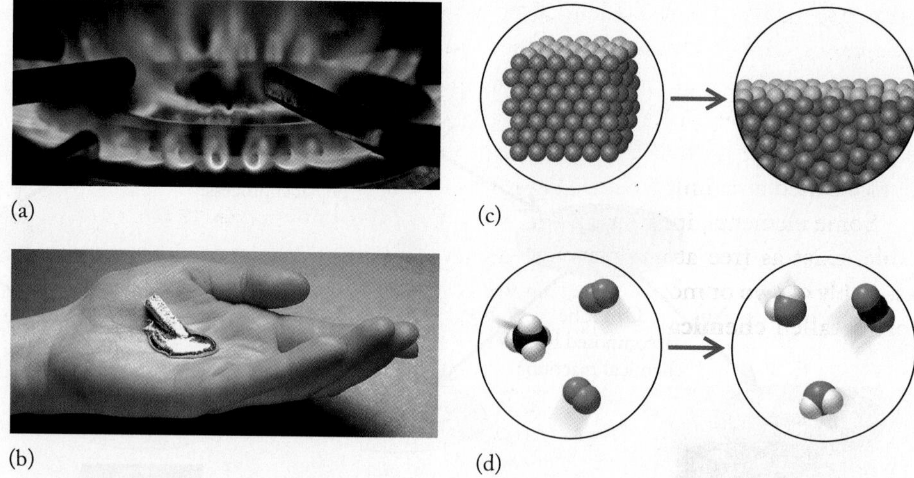

FIGURE 1.3 Macroscopic and particulate representations of both a physical process and a chemical reaction.

(Answers to Concept Tests are in the back of the book.)

element a pure substance that cannot be separated into simpler substances.

compound a pure substance that is composed of two or more elements combined in fixed proportions and that can be broken down into those elements by a chemical reaction.

chemical reaction the transformation of one or more substances into different substances.

mixture a combination of pure substances in variable proportions in which the individual substances retain their chemical identities and can be separated from one another by a physical process.

A **mixture** (**Figure 1.2c**) is a form of matter that is composed of two or more substances that retain their own chemical identities. The substances in mixtures are not present in definite proportions. For example, the composition of circulating blood in a human body is constantly changing as it delivers substances involved in energy production and cell growth to the cells and carries away the waste products of life's biochemical processes. Thus, blood contains more oxygen and less carbon dioxide when it leaves our lungs than it does when it enters them. The substances in mixtures can be separated from one another by physical processes (Figure 1.1), such as those described in Section 1.3.

Mixtures can be further classified as homogeneous and heterogeneous (Figure 1.1). In a **homogeneous mixture**, the substances making up the mixture are uniformly distributed. This means that the first sip you take from a bottle of water has the same composition as the last. (Bottled water contains small quantities of dissolved substances that either occur naturally in the water or are added prior to bottling to give it an agreeable taste. Bottled drinking water is not *pure* water.) Homogeneous mixtures are also called **solutions**, a term that chemists apply to homogeneous mixtures of gases and solids, too, not just liquids. For example, a sample of filtered air is a solution of nitrogen, oxygen, argon, carbon dioxide, and other atmospheric gases, and a "gold" ring is actually a solid solution of mostly gold plus other metals such as silver, copper, and zinc.

The substances in a **heterogeneous mixture** are *not* distributed uniformly. One way to tell that a liquid mixture is heterogeneous is to look for a boundary between the liquids in it (such as the oil and water layers in the bottle of salad dressing in Figure 1.1). Such a boundary indicates that the substances do not dissolve in one another. Another sign that a liquid may be a heterogeneous mixture is that it is not clear (transparent). Light cannot pass through such liquids because it is scattered by tiny solid particles or liquid drops that are *suspended*, but not dissolved, in the surrounding liquid. Human blood, for example, is opaque because the blood cells that are suspended in it absorb and scatter light.

homogeneous mixture a mixture in which the components are distributed uniformly throughout and have no visible boundaries or regions.

solution another name for a homogeneous mixture; solutions are often liquids, but they may also be solids or gases.

heterogeneous mixture a mixture in which the components are not distributed uniformly, so that the mixture contains distinct regions of different compositions.

atom the smallest particle of an element that retains the element's properties.

molecule a collection of atoms chemically bonded together in characteristic proportions.

chemical bond a force that holds two atoms or ions in a compound together.

A Particulate View

Given that compounds are formed from elements, do elements consist of yet smaller particles? The answer is yes. Each element consists of **atoms**: particles of one type unique to that element. Civilizations as old as the ancient Greeks believed in atoms, though people then had no evidence that atoms existed. Today, however, we have compelling evidence in the form of images of atoms, such as those on the surface of a disk made of high-purity silicon used to make computer chips (**Figure 1.4**) that has been magnified over 100 million times by using a device called a scanning tunneling microscope.

Some elements, including helium and the others in column 18 of the periodic table, exist as free atoms; other elements exist as molecules. A **molecule** is an assembly of two or more atoms that are held together in a characteristic pattern by forces called **chemical bonds**. A molecule of an element contains two or more

(a)

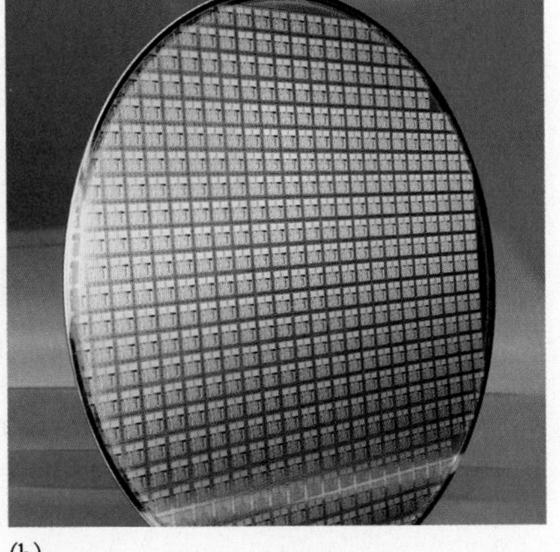

(b)

FIGURE 1.4 Since the 1980s, scientists have been able to image individual atoms by using an instrument called a scanning tunneling microscope (STM). (a) In this STM image of (b) a disk made of high-purity silicon, the fuzzy objects arranged in an array of hexagons are composed of six silicon atoms, which are colored blue to make them easier to see—keep in mind that atoms don't have colors!

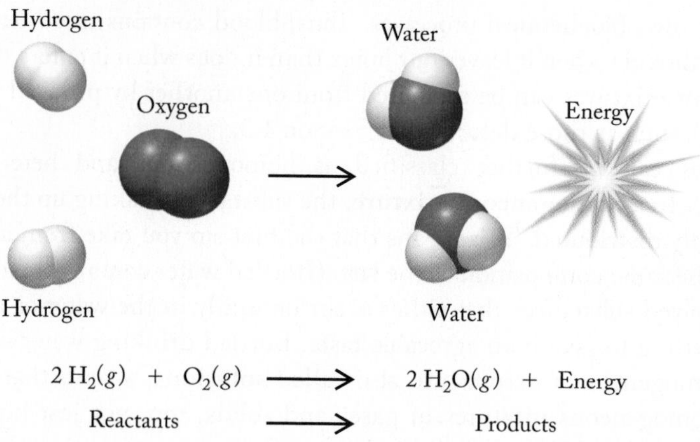

FIGURE 1.5 The reaction between hydrogen and oxygen is depicted with molecular models (white and red spheres) and in the form of a chemical equation. Energy is also a product of the reaction, meaning that energy is released when the water product is formed.

atoms of that element that are bonded together. For example, the air we breathe consists mostly of *diatomic* (two-atom) molecules of nitrogen gas, N_2, and oxygen gas, O_2. The subscripts in the **chemical formulas** of these two gases tell us that their molecules are each composed of two atoms of that element. Other elements also exist as diatomic molecules, including H_2 and elements in column 17 of the periodic table: F_2, Cl_2, Br_2, and I_2.

Most of the molecules in the universe, however, are compounds because they contain atoms of more than one element. The chemical formula of a molecular compound tells us the number of atoms of each element in one of its molecules. For example, the formula H_2O tells us that pure water is composed of molecules that each contain two hydrogen atoms and one oxygen atom (the subscript 1 is omitted), as shown in **Figure 1.5**.

The 2:1 ratio of hydrogen to oxygen atoms in molecules of H_2O also reflects the proportions of H_2 gas and O_2 gas that must react to form water. These gases *always* react with each other in the same proportion: two molecules of H_2 for every molecule of O_2 to form two molecules of H_2O. This relationship is illustrated in Figure 1.5 with models of the molecules involved and the chemical equation beneath them. In a **chemical equation**, chemical formulas represent the substances involved in a chemical reaction. The arrow in the middle of a chemical equation separates the formulas of the reactant(s) on the left from those of the product(s) on the right. In Figure 1.5, the *phase symbols* (g) and (ℓ) indicate that the reactants are gases and H_2O is a liquid, respectively. Solids are denoted by (s).

The reaction between hydrogen and oxygen also produces **energy**, which is generally defined as the capacity to do work. The reaction can be forced to run in reverse if enough energy is added to decompose water into hydrogen and oxygen (**Figure 1.6**). The reverse chemical reaction always produces two molecules of hydrogen gas for every one molecule of oxygen gas. This consistency illustrates the **law of constant composition**: every sample of a compound always contains the same elements combined in the same proportions.

chemical formula notation for representing elements and compounds; consists of the symbols of the constituent elements, each followed by subscripts that identify the number of atoms of each element present.

chemical equation notation in which chemical formulas express the identities and their coefficients express the quantities of substances involved in a chemical reaction.

energy the capacity to do work.

law of constant composition the principle that all samples of a compound contain the same elements combined in the same proportions.

ion a particle consisting of one or more atoms that has a net positive or negative electrical charge.

cation an ion with a positive charge.

anion an ion with a negative charge.

CONCEPT TEST

A compound with the formula NO is present in the exhaust gases leaving a car's engine. As NO travels through the car's exhaust system, some of it decomposes into nitrogen gas and oxygen gas. What is the ratio of nitrogen molecules to oxygen molecules formed from the decomposition of NO?

(Answers to Concept Tests are in the back of the book.)

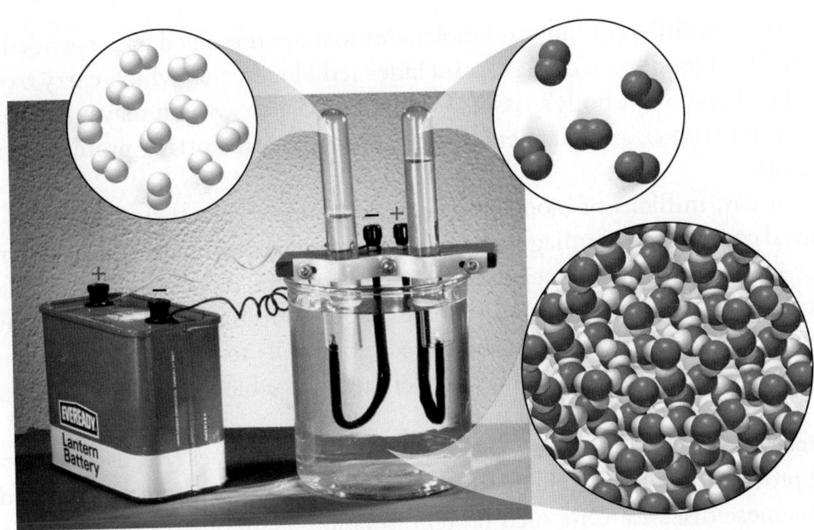

FIGURE 1.6 An electric current passed through water provides enough energy to decompose water into oxygen gas and hydrogen gas, both of which displace the water from the tops of their respective inverted test tubes. The ratio of the gases produced is always two molecules of hydrogen for every one molecule of oxygen, which is why the volume of H_2 gas produced is twice the volume of O_2 gas. These fixed proportions illustrate the law of constant composition.

Chemical formulas provide information about the ratios of the elements in molecular compounds, but formulas do not tell us how the atoms of each element are bonded to one another within each molecule, nor do they tell us anything about the shapes of molecules. To communicate information about bonding and shape, we need to draw *structural formulas* such as the one for ethanol (C_2H_5OH) in **Figure 1.7a**, which uses straight lines to represent the chemical bonds that connect the carbon (C), hydrogen (H), and oxygen (O) atoms within the molecule.

Although a structural formula shows which atoms are connected by bonds, it does not necessarily show how those atoms are arranged in three-dimensional space. *Molecular models* provide this 3-D perspective. *Ball-and-stick* molecular models (**Figure 1.7b**) use spheres to represent atoms and sticks to represent chemical bonds. The advantage of ball-and-stick models is that they show the correct angles between the bonds. However, the sizes of the spheres in these models may not be proportional to the sizes of the atoms they represent, and the atoms are spaced far enough apart to accommodate the stick bonds. (In real molecules, the atoms touch each other.) Both limitations are overcome with *space-filling* molecular models (**Figure 1.7c**), in which the spheres are drawn to scale and touch one another as atoms do in real molecules. One limitation of space-filling models is that the bond angles between atoms may be difficult to discern. An additional limitation of both the ball-and-stick and space-filling models is that atoms themselves do not have color. Representing oxygen atoms as red spheres and hydrogen atoms as white spheres is merely a convention used by chemists.

Not all compounds are composed of molecules. Many of them consist of positively and negatively charged particles called **ions** that are electrostatically attracted to one another. For example, the chemical formula of calcium chloride is $CaCl_2$, which means that this ionic compound contains two chloride (Cl^-) ions for every one calcium (Ca^{2+}) ion. Positive ions are called **cations**, whereas negative ions are called **anions**. Ions may consist of single atoms like Ca^{2+} and Cl^-, or they may contain two or more atoms bonded together that have an overall positive or negative charge, such as the hydroxide ion (OH^-).

Ethanol

```
     H   H
     |   |
  H—C—C—O—H
     |   |
     H   H
```

(a) Structural formula

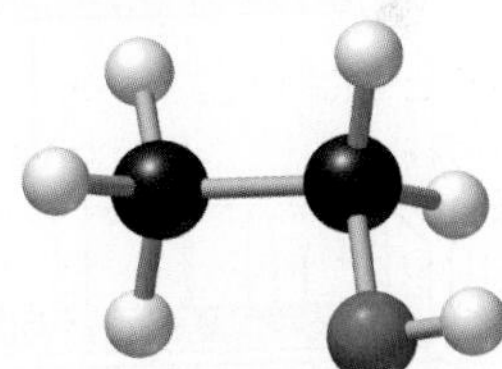

(b) Ball-and-stick model

(c) Space-filling model

FIGURE 1.7 Three ways to represent the arrangement of atoms in a molecule of ethanol: (a) structural formula, (b) ball-and-stick model, (c) space-filling model. In both molecular models, white spheres represent hydrogen atoms, black spheres represent carbon atoms, and red spheres represent oxygen atoms.

1.3 Mixtures and How to Separate Them

As noted in Section 1.2, mixtures can be separated into their component substances by physical processes. Consider, for example, how the components of human blood can be separated (**Figure 1.8**). Blood is a heterogeneous mixture

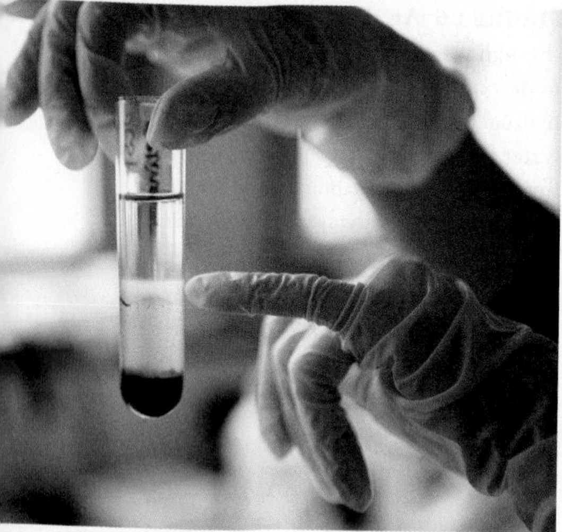

FIGURE 1.8 Separation of whole blood by centrifugation. The red blood cells at the bottom of the sample are separated from the upper layer of pale yellow plasma by a pale-red layer of white blood cells and platelets.

of hundreds of different ions and molecules that are dissolved or suspended in a liquid called blood *plasma*. Plasma includes red blood cells (which carry oxygen from the lungs to our body's tissues), white blood cells (which play a key role in fighting infections), and platelets (which combine with clotting proteins to stop bleeding).

Each day, millions of blood samples are separated into their suspended and dissolved components for diagnostic purposes or to recover materials used to treat sick or injured patients. To separate their components, blood samples are placed in centrifuge tubes and subjected to forces about 1000 times the force of gravity. Centrifugation causes the large, dense red blood cells to settle to the bottom of the tubes. Above them is a layer of platelets and white blood cells, and above that layer is clear, yellow-colored plasma.

Once the plasma has been separated, other techniques are used to isolate blood proteins dissolved in the plasma. Identification of these proteins helps doctors diagnose diseases. One such protein isolation technique is two-dimensional electrophoresis, in which the proteins are separated in two steps (**Figure 1.9**): first by charge, and then by size. The first step works because protein molecules have different charges because of the presence of positively and negatively charged ionic groups in their molecular structures. In the second step, fractions from the first step are further separated in a process in which smaller molecules move more rapidly and migrate farther than larger ones.

Different techniques are used to separate solid particles from liquid and gas mixtures. Consider, for example, the use of filters to trap particles of dust suspended in the air entering an automobile engine (**Figure 1.10a**) or to remove bacteria from the air in hospitals to control the spread of infectious diseases (**Figure 1.10b**). A key parameter in designing these filters is their pore size, which must be small enough to trap the target particles or microbes, but not so small that enormous pressure would be required to force an adequate flow of air through them. Filtration is also used to separate particles suspended in liquids—a key step in treating the water that many of us drink (**Figure 1.10c**).

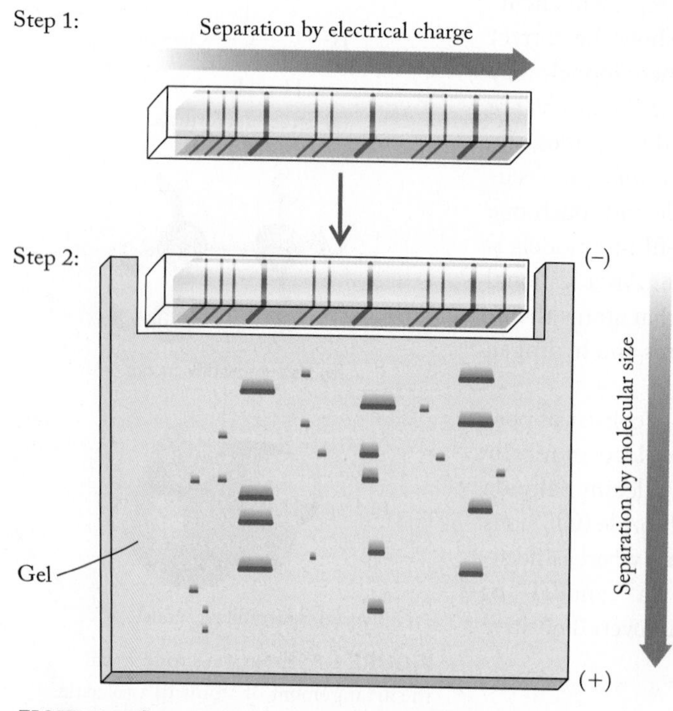

FIGURE 1.9 Separation of proteins by electrophoresis. In the first step, the proteins are separated based on whether they are positively or negatively charged—or have no net charge. In the second step, negatively charged proteins are separated mostly by molecular size: smaller molecules move faster than larger ones.

The components of a liquid mixture can also be separated by exploiting differences in their *volatilities*, that is, how easily they can be converted from liquids to gases. The most widely used of these techniques is distillation (**Figure 1.11**), which can be used to desalinate seawater because water is much more volatile than the sea salts dissolved in it. Distillation is also used to purify the water used in laboratories and other places where highly pure water is required. To prepare ultra high-purity water requires an additional separation step based on a technology called *ion exchange*, which can be used to remove essentially all the dissolved ions from water. During this process any dissolved cations in the water are replaced with H^+ ions, and any dissolved anions are replaced with OH^- ions. These two ions then combine to form additional molecules of water:

$$H^+(aq) + OH^-(aq) \rightarrow H_2O(\ell)$$

The phase symbol (*aq*) indicates that the ions are dissolved in water, that is, they are in an *aqueous* solution.

(a)

(b)

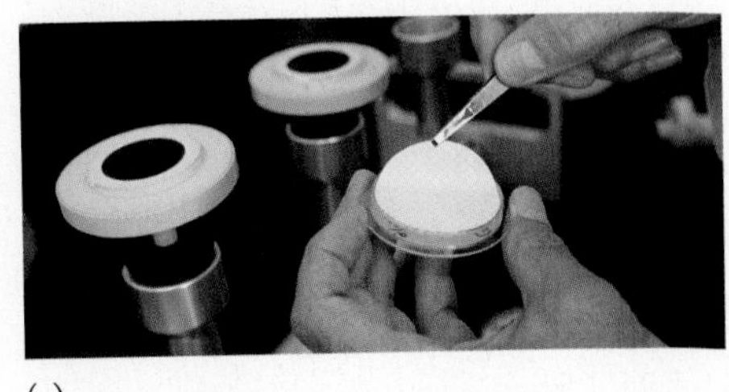
(c)

FIGURE 1.10 Filters are used to remove particles from gases such as (a) the air entering a car engine or (b) the air circulating through a hospital, and from liquids such as (c) the water being analyzed at a municipal water supply treatment plant. The pore size of the hospital filters must be very small to trap bacteria and viruses that are less than 1 micrometer (10^{-6} m) in diameter.

1.4 A Framework for Solving Problems

Throughout this book we include questions and problems that test your understanding of chemical principles and that develop your problem-solving skills. Solving chemistry problems is like playing a musical instrument: the more you practice, the better you become.

In this section we present a framework for solving problems that we follow throughout the book. Each Sample Exercise is followed by a Practice Exercise that can be solved using a similar approach. We strongly encourage you to hone your problem-solving skills by working all the Practice Exercises as you read. You should find the approach described here useful in solving the exercises in this book as well as problems you encounter in other courses and other contexts.

We use the acronym COAST (**C**ollect and **O**rganize, **A**nalyze, **S**olve, and **T**hink about the answer) to represent the four steps in this approach. As you read about it here and use it later, keep in mind that COAST is merely a *framework* for solving problems, not a recipe. Use it as a guide to develop your own approach to solving problems.

Collect and Organize The first step in solving a problem is to decide how to use the given information. You should be able to identify the key concept of the problem and the key terms used to express that concept. You may find it useful to restate the problem in your own words. Visualizing the problem may also be helpful—drawing a sketch based on molecular models or an experimental setup, or even just closing your eyes and picturing the situation. Once you have a clear understanding of what is being asked, sort through the information and separate what is relevant from what is not. Then assemble the relevant information, including chemical equations, molecular models, definitions, and the values of constants.

Analyze The next step is to analyze the information you have assembled to determine how to use it to answer the problem. Sometimes you can work backward to create these links: Consider the answer first and think about how you might reach it from the information you've assembled. This approach might mean finding some intermediate quantity to use in a later step. If the problem requires a numerical answer, frequently the units of the initial values and the final answer can help you identify how they are connected and which equations and conversion factors should be used. Also, when possible, try to estimate the value of your answer before solving the problem.

(a)

(b)

FIGURE 1.11 Distillation can be used to (a) purify compounds in the laboratory and (b) desalinate seawater for public consumption. This desalination plant is in Kuwait City.

Some Sample Exercises in this book involve performing a single-step calculation or answering a question based on a single concept. These exercises don't require assembling and sorting through a lot of information and equations, so we may combine the Collect, Organize, and Analyze steps before solving the problem.

Solve The solution to a problem that tests your understanding of a concept often flows directly from the Analyze step. To solve quantitative problems, you will need to set up and use the conversion factors and equations you have assembled to calculate the correct answer with appropriate units. Also, be sure that calculated answers contain the appropriate number of significant figures. (We discuss conversion factors in Section 1.9 and significant figures in Section 1.10.)

Think About It Does your answer make sense? Is this value about what you estimated it would be? Are the units correct? Is the number of significant figures appropriate? Finally, could you solve another problem, perhaps drawn from another context but based on the same concept?

The COAST approach should help you solve problems in a logical way and avoid pitfalls such as grabbing an equation that seems to have the right variables and just plugging in initial values or resorting to trial and error. As you study each step in a Sample Exercise, be sure to ask yourself these questions: *What* is done in this step? *How* is it done? *Why* is it done? With answers to these questions in mind, you will be ready to solve the Practice Exercises and end-of-chapter Questions and Problems in a systematic way.

1.5 Properties of Matter

Our senses provide a constant stream of information about the world around us. Subconsciously, we process this information to confirm that "all is well"—or, if it's not, to take appropriate action. For example, the normal components of the air we breathe are odorless, so when we smell something unusual as we enter a room or step outside, we know we are inhaling a gas not normally present in the air, and we respond accordingly.

Pure substances have distinctive properties. Some substances burn, whereas others put out fires. Some solid substances are shiny, whereas others are dull. Some are brittle, whereas others are malleable. These and other properties do not vary from one sample of a given pure substance to the next. Consider gold, for example: its color (very distinctive), its hardness (soft for a metal), its malleability (it can be hammered into very thin sheets called gold leaf), and its melting temperature (1064°C) all apply to any sample of pure gold. These characteristics are examples of **intensive properties**: properties that characterize matter independent of the quantity of it present in a sample. On the other hand, an object made of gold, such as an ingot, has a length, width, mass, and volume. These properties, which depend on how much of the substance is present, are **extensive properties**.

intensive property a property that is independent of the amount of substance present.

extensive property a property that varies with the quantity of the substance present.

CONCEPT TEST

Which of these properties of a sample of pure iron is intensive? (a) mass; (b) hardness; (c) volume

(Answers to Concept Tests are in the back of the book.)

Properties may be categorized as physical or chemical. The **physical properties** of a pure substance can be observed or measured without changing the substance into another substance. The intensive properties of gold are all physical properties, as is its **density (*d*)**—the ratio of the mass (*m*) of an object to its volume (*V*):

$$d = \frac{m}{V} \tag{1.1}$$

As we saw in Figure 1.5, hydrogen combines with oxygen in a chemical reaction that produces water and energy. This reaction is an example of combustion (burning), and H_2 is the fuel. High flammability is a **chemical property** of hydrogen. Like any other chemical property, flammability can be determined only by reacting one substance with another one and finding that one or more other substances are produced. A substance's chemical reactivity, including the rates of its reactions, the identities of other substances with which it reacts, and the products they form, define the chemical properties of the substance.

physical property a property of a substance that can be observed without changing it into another substance.

density (*d*) the ratio of the mass (*m*) of an object to its volume (*V*).

chemical property a property of a substance that can be observed only by reacting it to form another substance.

SAMPLE EXERCISE 1.1 Distinguishing Physical and Chemical Properties **LO3**

Which of the following properties of gold are chemical and which are physical?

a. Gold metal, which is insoluble in water, can be made soluble by reacting it with a mixture of nitric and hydrochloric acids known as aqua regia.
b. Gold melts at 1064°C.
c. Gold can be hammered into sheets so thin that light passes through them.
d. Gold metal can be recovered from gold ore by treating the ore with a solution containing cyanide, which reacts with and dissolves gold.

Collect, Organize, and Analyze Chemical properties describe how a substance reacts with other substances, whereas physical properties can be observed or measured without changing one substance into another. Properties (a) and (d) involve reactions that chemically change gold metal (which does not dissolve in water) into compounds of gold that do dissolve in water. Properties (b) and (c) describe physical changes that do not change the identity of the element: molten gold and gold leaf are still gold.

Solve Properties (a) and (d) are chemical properties, whereas (b) and (c) are physical properties.

Think About It When possible, we fall back on our experiences and observations. We know jewelry made of gold does not dissolve in water. Therefore, dissolving gold metal requires a change in its chemical identity: it can no longer be elemental gold. On the other hand, physical processes such as melting do not alter the chemical identity of the gold. Gold can be melted and then cooled to produce solid gold again.

Practice Exercise Which of the following properties of water are chemical and which are physical?

a. Water normally freezes at 0.0°C.
b. Water does not burn.
c. A cork floats on water, but a piece of copper sinks.
d. During digestion, starch reacts with water to form sugar.

(Answers to Practice Exercises are in the back of the book.)

solid a form of matter that has a definite volume and shape.

liquid a form of matter that occupies a definite volume but flows to assume the shape of its container.

gas a form of matter that has neither definite volume nor shape and that expands to fill its container; also called *vapor*.

1.6 States of Matter

Matter typically exists in one of three phases or physical states—namely, solid, liquid, or gas (**Figure 1.12**), which have the following characteristic properties:

- A **solid** has a definite volume and shape.
- A **liquid** has a definite volume but not a definite shape.
- A **gas** (or *vapor*) has neither a definite volume nor a definite shape. Rather, it expands to occupy the entire volume and shape of its container.
- Unlike solids, liquids and gases are fluids, which means they are free to flow.
- Unlike a solid or a liquid, a gas is compressible, which means it can be squeezed into a smaller volume if its container is not rigid and pressure is applied to it.

The differences between solids, liquids, and gases can be understood if we consider the three states on the particulate level. As shown in Figure 1.12a, each H_2O molecule in ice is surrounded by other H_2O molecules and is locked in place in a rigid three-dimensional lattice (array) of molecules. A molecule in this structure may vibrate, but it cannot move past the other molecules that surround it. In liquid water (Figure 1.12b), the H_2O molecules flow past one another, but they are still near one another, even though their nearest neighbors change over time. The H_2O molecules in water vapor (Figure 1.12c) are widely separated by empty space, so the volume of the molecules themselves is much smaller than the total volume occupied by the vapor. This separation accounts

FIGURE 1.12 Macroscopic and particulate views of the three states of water in this winter photograph of a hot spring in Yellowstone National Park: (a) solid (ice), (b) liquid (water: in the case of the hot spring, it is a solution of compounds in liquid water), and (c) gas (water vapor). Water vapor is invisible, but the mist above the spring consists of tiny drops of liquid water that form as water vapor condenses in the frigid air. The rich colors near the edges of the spring are produced by thermophiles, tiny organisms that thrive in extremely hot water.

for the compressibility of gases. Molecules in the gas phase move freely throughout the space occupied by the vapor.

We can transform water from one physical state to another by heating or cooling it (**Figure 1.13**). Ice forms on a pond when the temperature drops in the winter and the water freezes. This process is reversed when warmer temperatures return in the spring and the ice melts. The heat of the Sun may vaporize liquid water during the daytime, but colder temperatures at night may cause the water vapor to condense as dew.

Some solids change directly into gases with no intervening liquid phase. For example, snow may be converted to water vapor on a sunny but cold winter day even though the air temperature remains well below freezing. This transformation of solid directly to vapor is called **sublimation**. The reverse process, in which a gas (water vapor) is transformed directly into a solid (frost) without ever being a liquid, is called **deposition**.

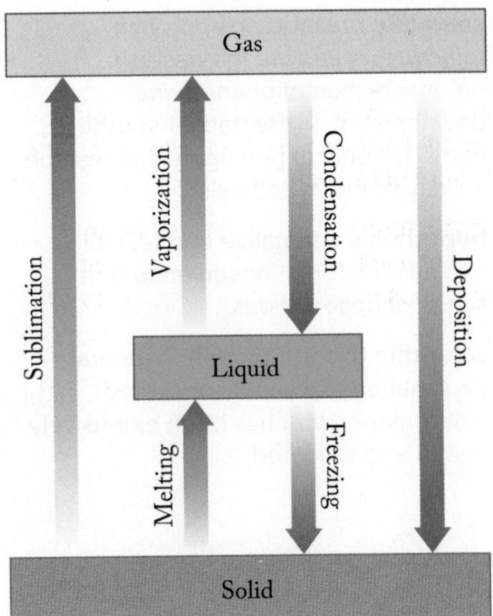

FIGURE 1.13 Matter changes from one state to another when energy is either added or removed. Arrows pointing up represent transformations that require the addition of energy *to* the matter; arrows pointing down represent transformations that release energy *from* the matter.

CONCEPT TEST

Is energy absorbed by a solid or released from it as it sublimes?

(Answers to Concept Tests are in the back of the book.)

SAMPLE EXERCISE 1.2 Distinguishing between Particulate Views of the Different States of Matter — LO5

Which physical state is represented in each box of **Figure 1.14**? (The particles could be atoms or molecules.) What change of state does each arrow indicate? What would the changes of state be if both arrows pointed in the opposite direction?

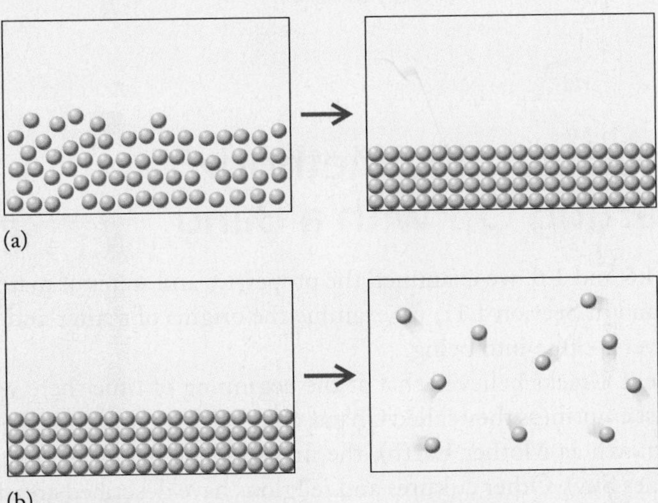

FIGURE 1.14 Changes in the physical state of matter.

Collect, Organize, and Analyze We need to look for patterns among the particles in each box:

- An ordered arrangement of particles with little space between the particles represents a solid.
- A less-ordered arrangement of particles with little space between the particles represents a liquid.
- Particles widely separated from one another represent a gas.

sublimation transformation of a solid directly into a vapor (gas).

deposition transformation of a vapor (gas) directly into a solid.

scientific method an approach to acquiring knowledge based on observation of phenomena, development of a testable hypothesis, and additional experiments that test the validity of the hypothesis.

hypothesis a tentative and testable explanation for an observation or a series of observations.

scientific theory (model) a general explanation of a widely observed phenomenon that has been extensively tested and validated.

Solve The particles in the left box in Figure 1.14a are close together but have a disordered arrangement, so they represent a liquid. The particles in the box on the right are ordered and are close together, so they represent a solid. The arrow represents a liquid turning into a solid, so the physical process is freezing. An arrow in the opposite direction would represent melting.

The particles in the box on the left in Figure 1.14b represent a solid because they are ordered, and are close together. Those to the right are widely separated from one another, so they represent a gas. The arrow represents a solid turning into a gas, so the physical process is sublimation. The reverse process—a gas becoming a solid—would be deposition.

Think About It The particles represented in Figure 1.14a are close together, which is consistent with how particles are distributed in solids and liquids such as ice and water (Figure 1.12). On the other hand, the particles in gases, as shown in the right panel of Figure 1.14b, are widely separated.

Practice Exercise (a) What physical state is represented by the particles in each box of **Figure 1.15**, and which change of state is represented? (b) Which change of state would be represented if the arrow pointed in the opposite direction?

FIGURE 1.15 A change in the physical state of matter.

(Answers to Practice Exercises are in the back of the book.)

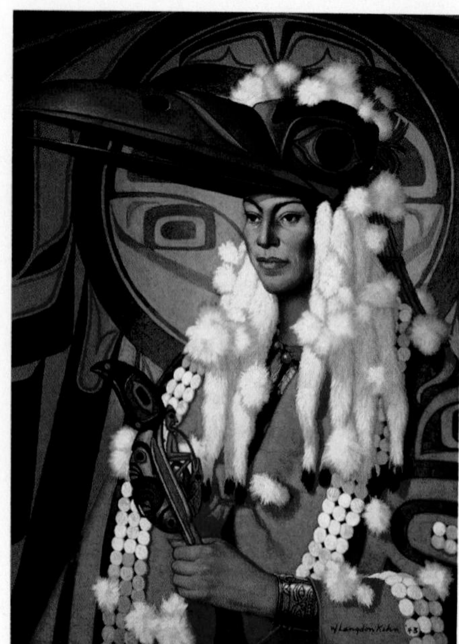

FIGURE 1.16 For centuries, cultures have passed on creation stories in which a supreme being is responsible for creating the Sun, the Moon, and Earth and its inhabitants. According to a creation myth of the indigenous peoples of coastal British Columbia, such as the Bellacoola woman depicted here, a deity in the form of a raven was responsible for releasing the Sun into the sky.

1.7 The Scientific Method: Starting Off with a Bang

In Sections 1.5 and 1.6, we examined the properties and states of matter. Here in Section 1.7 and in Section 1.11, we examine the origins of matter and a theory of how the universe came into being.

The ancient Greeks believed that at the beginning of time there was no matter, only a vast emptiness they called Chaos, and that from that emptiness emerged Gaia (also known as Mother Earth), the first supreme being, who gave birth to Uranus (Father Sky). Other cultures and religions have described creation in similar terms of supernatural beings creating ordered worlds out of vast emptiness. The opening verses of the book of Genesis, for example, describe darkness "without form and void" from which God created the heavens and Earth. Similar stories are part of indigenous cultures of Asia, Africa, and the Americas (**Figure 1.16**). Today we have the technology to take a different approach to explaining how the universe was formed and what physical forces control it. Our approach is based on the **scientific method** of inquiry (**Figure 1.17**).

This approach evolved in the late Renaissance, during a time when economic and social stability gave people an opportunity to study nature and question old

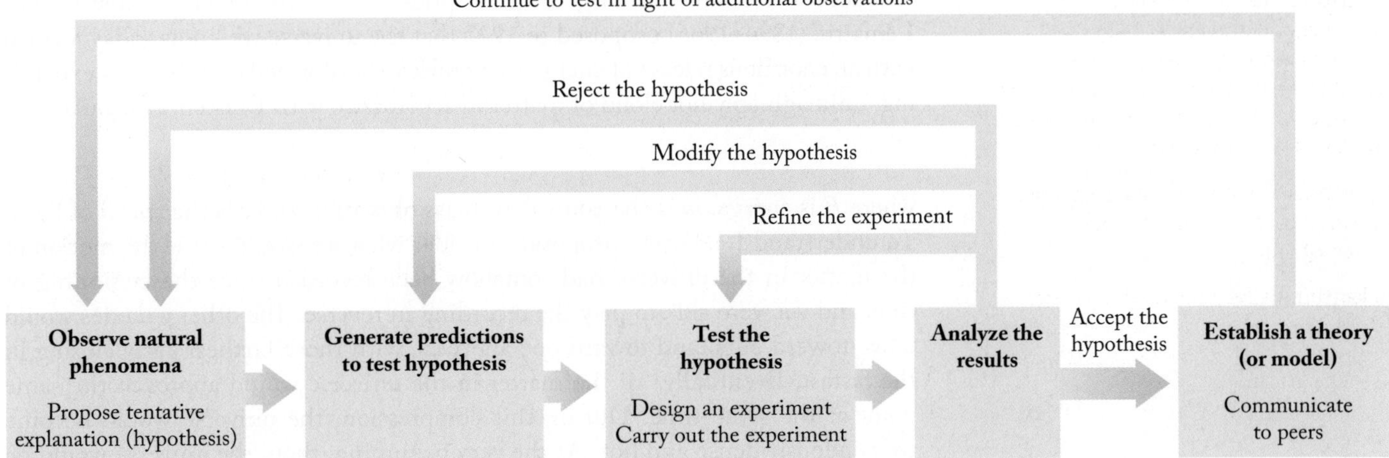

FIGURE 1.17 In the scientific method, observations lead to a tentative explanation, or hypothesis, which leads to more observations and testing, which may lead to the formulation of a succinct, comprehensive explanation called a theory. This process is rarely linear: it often involves looping back, as the results of one test lead to additional tests and a revised hypothesis. Science is a dynamic and self-correcting process.

beliefs. By the early 17th century, the English philosopher Francis Bacon (1561–1626) had published his *Novum Organum* (*New Organ* or *New Instrument*) in which he described how humans could acquire knowledge and understanding of the natural world through observation, experimentation, and reflection. In the process, he described how observations of a natural phenomenon could lead to a tentative explanation, or **hypothesis**, of what caused the phenomenon. Developing a hypothesis and devising experiments to test its validity are creative aspects of science. Further testing and observation may support a hypothesis, but unexpected results that don't support it are also valuable because they may result in modifications to the hypothesis that enable it to explain *all* experimental results. However, experimental results that do support a hypothesis can never "prove" it; the best that scientists can say is that no experiment or observation has yet to contradict the hypothesis.

One measure of the validity of a hypothesis is that it enables scientists to predict accurately the results of *future* experiments and observations. Ultimately, a hypothesis that withstands the tests of many experiments over time and explains the results of further observation and experimentation may be elevated to the rank of a **scientific theory** or **model**. Although *theory* in everyday conversation is often used as a synonym for "speculation," it is used by scientists to refer to an explanatory framework that can be used to test hypotheses empirically.

What are some of the ways the scientific method helped formulate the theory of the origin of the universe that has come to be called the Big Bang theory? In the early 20th century, American astronomer Edwin Hubble (1889–1953; **Figure 1.18**) and others used increasingly powerful telescopes to discover that the universe contains billions of galaxies, each containing many billions of stars. They also discovered that (1) these other galaxies were moving away from us and (2) the speeds with which these galaxies were receding was proportional to the distances from us—the farthest galaxies were moving away the fastest.

FIGURE 1.18 The Hubble Space Telescope is named in honor of American astronomer Edwin Hubble.

From these observations, Belgian priest and astronomer Georges-Henri Lemaître (1894–1966) proposed in 1927 that the universe we know today formed with an enormous release of energy that quickly transformed into a rapidly expanding, unimaginably hot cloud of matter in accordance with Einstein's formula,

$$E = mc^2$$

where E is energy, m is the equivalent mass of matter, and c is the speed of light. To understand Lemaître's proposal, consider what we would see if the motion of the matter in the universe had somehow been recorded since the beginning of time and we were able to play the recording in reverse. The other galaxies would move toward ours (and toward one another), with those farthest away closing in the fastest. Eventually, all the matter in the universe would approach the same point at the same time. During this compression, the universe would become tremendously dense and hot. At the very beginning, then, the universe would be squeezed into an infinitesimally small space with an inconceivably high temperature. At such a temperature, the distinction between matter and energy would become blurred. Indeed, at the very beginning there would be no matter, only energy. Now, if we were to play this recording in the forward direction, we would see an instantaneous release of an enormous quantity of energy in an event called the Big Bang.

Lemaître's explanation of how the universe began has been the subject of extensive testing—and controversy—since it was first proposed. Indeed, the term *Big Bang* was first used by British astronomer Sir Fred Hoyle (1915–2001) to poke fun at the idea. However, the results of many experiments and observations conducted since the 1920s support what is now described as the Big Bang theory. We use the term *theory* because this comprehensive explanation of natural phenomena has been validated by the results of extensive experimentation and observation and has also been used to accurately *predict* the results of *other* experiments and observations. In Section 1.11 we return to the Big Bang theory and examine a few of the experiments that have tested its validity.

CONCEPT **TEST**

If the volume of the universe is expanding and if the mass of the universe is not changing, is the density of the universe (a) increasing, (b) decreasing, or (c) remaining the same?

(Answers to Concept Tests are in the back of the book.)

1.8 SI Units

The rise of scientific inquiry in the 17th and 18th centuries brought about a heightened awareness of the need for accurate measurements and for expressing the results of those measurements in ways that others, including scientists in other countries, could understand. Standardization of the units of measurement was essential.

In 1791 French scientists proposed a standard unit of length, which they called the **meter** (m) after the Greek *metron*, which means "measure." They based the length of the meter on 1/10,000,000 of the distance along an imaginary line running from the North Pole to the equator. By 1794, hard work by teams of surveyors had established the length of the meter that is still in use today.

meter the standard unit of length, equivalent to 39.37 inches.

These French scientists also settled on a decimal-based system for designating lengths that are multiples or fractions of a meter (**Table 1.1**). They chose Greek

TABLE 1.1 Commonly Used Prefixes for SI Units

PREFIX		VALUE	
Name	Symbol	Numerical	Exponential
zetta	Z	1,000,000,000,000,000,000,000	10^{21}
exa	E	1,000,000,000,000,000,000	10^{18}
peta	P	1,000,000,000,000,000	10^{15}
tera	T	1,000,000,000,000	10^{12}
giga	G	1,000,000,000	10^{9}
mega	M	1,000,000	10^{6}
kilo	k	1000	10^{3}
hecto	h	100	10^{2}
deka	da	10	10^{1}
deci	d	0.1	10^{-1}
centi	c	0.01	10^{-2}
milli	m	0.001	10^{-3}
micro	μ	0.000001	10^{-6}
nano	n	0.000000001	10^{-9}
pico	p	0.000000000001	10^{-12}
femto	f	0.000000000000001	10^{-15}
atto	a	0.000000000000000001	10^{-18}
zepto	z	0.000000000000000000001	10^{-21}

prefixes for lengths much greater than 1 meter, such as *deka-* and *kilo-* for lengths of 10 meters (1 dekameter) and 1000 meters (1 kilometer). Then they chose Latin prefixes for lengths much smaller than a meter, such as *centi-* and *milli-* for lengths of 1/100 of a meter (1 centimeter) and 1/1000 of a meter (1 millimeter).

Eventually these decimal prefixes carried over to the names of standard units for other dimensions. In July 1799 a platinum rod 1 meter long and a platinum block having a mass of 1 kilogram (1000 grams) were placed in the French National Archives to serve as the legal standards for length and mass. These objects served as references for the metric system for expressing measured quantities.

Since 1960, scientists have, by international agreement, used a modern version of the French metric system: the *Système International d'Unités*, commonly abbreviated SI. **Table 1.2** lists the seven *SI base units*; many other SI units are derived from them. For example, a common SI unit for volume, the cubic meter (m^3), is derived from the base unit for length, the meter. The common SI unit for speed, meters per second (m/s), is derived from the base units for both length and time. **Table 1.3** lists some of these derived units and their equivalents in the U.S. Customary (USC) units. They include the volume corresponding to 1 cubic decimeter (a cube 1/10 meter on each side), which we call a liter (L).

Various combinations of base and derived SI units are used to express the densities of materials, that is, their mass-to-volume ratios. The selection of units depends on how dense the material is. For example, the densities of gases are often expressed in grams/liter (g/L) so that the density values are close to one. Liquids, on the other hand, are many hundreds of times denser than gases, so

TABLE 1.2 SI Base Units[a]

Quantity or Dimension	Unit Name	Unit Abbreviation
Mass	kilogram	kg
Length	meter	m
Temperature	kelvin	K
Time	second	s
Electrical current	ampere	A
Amount of a substance	mole	mol
Luminous intensity	candela	cd

[a]As of May 20, 2019, all SI units are based on constants of nature, such as the speed of light in a vacuum.

TABLE 1.3 SI Units and Their U.S. Customary Unit Equivalents

Quantity or Dimension	Equivalent Units
Mass	1 kg = 2.205 pounds (lb); 1 lb = 0.4536 kg = 453.6 g
	1 g = 0.03527 ounce (oz); 1 oz = 28.35 g
Length (distance)	1 m = 1.094 yards (yd); 1 yd = 0.9144 m (exactly)
	1 m = 39.37 inches (in); 1 foot (ft) = 0.3048 m (exactly)
	1 in = 2.54 cm (exactly)
	1 km = 0.6214 miles (mi); 1 mi = 1.609 km
Volume	1 m^3 = 35.31 ft^3; 1 ft^3 = 0.02832 m^3
	1 m^3 = 1000 liters (L) (exactly)
	1 L = 0.2642 gallon (gal); 1 gal = 3.785 L
	1 L = 1.057 quarts (qt); 1 qt = 0.9464 L

TABLE 1.4 Densities of Common Materials

ITEM	DENSITY (AT 20°C)
Solids	**g/cm^3**
Aluminum	2.70
Copper	8.96
Lead	11.34
Zinc	7.13
Liquids	**g/cm^3**
Gasoline	0.78
Olive oil	0.91
Diet soda	1.00
Milk	1.03
Soda	1.03
Seawater	1.03
Whole blood (human)	1.06
Gases	**g/L (at 20°C and 1 atm pressure)**
Air (dry)	1.204
Ammonia	0.717
Carbon dioxide	1.842
Propane gas	1.873
Water vapor	0.804

their densities are often expressed in g/cm^3, which is equivalent to g/mL. Solids have densities that are comparable to or greater than those of most liquids, so their values are usually expressed in g/cm^3, too (**Table 1.4**).

Modern science requires that the length of the meter, as well as the dimensions of other SI units, be defined by quantities that are much more constant than the length of a platinum rod in Paris. Two such quantities are the speed of light (c) and time. In 1983, 1 meter was redefined as the distance traveled in 1/299,792,458 of a second by the light emitted from a helium–neon laser. This modern definition of the meter is consistent with the one adopted in France in 1794.

1.9 Unit Conversions and Dimensional Analysis

Throughout this book we will often need to convert measurements and quantities from one unit to another. Sometimes this will require choosing the appropriate prefix, such as expressing the distance between Vancouver, BC, and Seattle, WA, in kilometers instead of meters. Other times we will need to convert a value from one unit system to another, such as converting a gasoline price per liter to an equivalent price per U.S. gallon.

To do these calculations and many others, we use an approach sometimes called the *unit factor method* but more often called *dimensional analysis*. The approach makes use of **conversion factors**, which are fractions in which the numerators and denominators have different units but represent equivalent quantities. This equivalency means that multiplying a quantity by a conversion factor is equal to multiplying by 1. The intrinsic value that the quantity represents does not change; it is simply expressed in different units.

The key to using conversion factors is to set them up correctly with the appropriate units in the numerator and denominator. For example, the odometer of a car traveling between the airports serving Vancouver and Seattle records the distance as 241 km. What is that distance in miles? To find out we turn the following equivalency from Table 1.3

$$1 \text{ km} = 0.6214 \text{ mi}$$

into the following conversion factor,

$$\frac{0.6214\ \text{mi}}{1\ \text{km}}$$

and multiply the distance in kilometers by it:

$$241\ \cancel{\text{km}} \times \frac{0.6214\ \text{mi}}{1\ \cancel{\text{km}}} = 150\ \text{mi}$$

We obtain a value with the desired units (in this case, miles) because those units are in the numerator of the conversion factor, whereas the units in the denominator (in this case, kilometers) of the conversion factor cancel out the units of the initial value itself. The general form of the equation for converting a value from one unit to another is

$$\cancel{\text{initial units}} \times \frac{\text{desired units}}{\cancel{\text{initial units}}} = \text{desired units}$$

Most conversion factors, such as the one for converting kilometers to miles, also have a "1" in the numerator or denominator. All these ones are *exactly* one; there is no uncertainty in their value, which is a topic we explore in Section 1.10.

conversion factor a fraction in which the numerator is equivalent to the denominator, even though they are expressed in different units, making the value of the fraction one.

CHEMTOUR
Dimensional Analysis

SAMPLE EXERCISE 1.3 Converting Units **LO7**

The masses of diamonds are usually expressed in carats (1 gram = 5 carats exactly). The Star of Africa (**Figure 1.19**) is one of the world's largest diamonds, having a mass of 530.02 carats. What is its mass in grams? In kilograms? In ounces?

Collect and Organize We have a mass value expressed in carats, and we want to convert it into grams (g), kilograms (kg), and ounces (oz). The problem states there are exactly 5 carats in 1 gram. From Table 1.1 we know that there are exactly 1000 grams in 1 kilogram, and from Table 1.3 that there are 0.03527 ounces in 1 gram.

Analyze We can use the above unit equivalencies to write conversion factors that enable us to convert carats into grams:

$$\frac{1\ \text{g}}{5\ \text{carat}}$$

grams into kilograms:

$$\frac{1\ \text{kg}}{1000\ \text{g}}$$

and grams into ounces:

$$\frac{0.03527\ \text{oz}}{1\ \text{g}}$$

Solve We use the above three conversion factors to convert the diamond's mass from carats into grams:

$$530.02\ \cancel{\text{carats}} \times \frac{1\ \text{g}}{5\ \cancel{\text{carat}}} = 106.00\ \text{g}$$

grams into kilograms:

$$106.00\ \cancel{\text{g}} \times \frac{1\ \text{kg}}{1000\ \cancel{\text{g}}} = 0.10600\ \text{kg}$$

and grams into ounces:

$$106.00\ \cancel{\text{g}} \times \frac{0.03527\ \text{oz}}{1\ \cancel{\text{g}}} = 3.739\ \text{oz}$$

Think About It The calculated masses may seem unusually large for a diamond, and they are in comparison to those of the ~1 carat diamonds in many engagement rings. The Star of Africa has an estimated value of U.S. \$1 billion.

FIGURE 1.19 The Star of Africa is one of the largest cut diamonds in the world. It is mounted in the handle of the Royal Sceptre in the British Crown Jewels displayed in the Tower of London.

Practice Exercise The Eiffel Tower in Paris is 324 m tall. What is this height, which includes a 24-m television antenna that did not exist when the tower was built in 1889, in kilometers and in centimeters?

(Answers to Practice Exercises are in the back of the book.)

SAMPLE EXERCISE 1.4 Multistep Conversions of U.S. Customary to SI Units **LO7**

On April 12, 1934, a wind gust of 231 miles per hour was recorded at the summit of Mount Washington in New Hampshire (**Figure 1.20**). What is this wind speed in meters per second?

FIGURE 1.20 A warning to Mount Washington hikers from the U.S. Forest Service.

Collect and Organize Speed is given in the USC units of miles per hour, and we need to convert it to meters per second. Table 1.3 contains the equivalency 1 km = 0.6214 mi, and we know from Table 1.1 that the *kilo-* prefix means 1000. We will also need two unit-of-time equivalencies: 1 hour = 60 minutes and 1 minute = 60 seconds.

Analyze The unit conversion patterns are

$$\text{mi} \rightarrow \text{km} \rightarrow \text{m}$$

in the numerator, and

$$\text{h} \rightarrow \text{min} \rightarrow \text{s}$$

in the denominator.

Solve We use the unit equivalencies described above as conversion factors corresponding to the conversion patterns we have identified. We arrange them so that their numerators and denominators cancel out the initial and intermediate units:

$$231 \frac{\text{mi}}{\text{h}} \times \frac{1\ \text{km}}{0.6214\ \cancel{\text{mi}}} \times \frac{1000\ \text{m}}{1\ \cancel{\text{km}}} \times \frac{1\ \cancel{\text{h}}}{60\ \cancel{\text{min}}} \times \frac{1\ \cancel{\text{min}}}{60\ \text{s}} = 103 \frac{\text{m}}{\text{s}}$$

Think About It The calculated value is reasonable because the product of the numerator values is 1000, and the product of the denominator values is (~2/3 × 60 × 60), or about 2400. Therefore, the calculated value should be near (1000/2400), or a little less than half the initial value, which it is.

Practice Exercise Light travels through exactly 1 meter of outer space in 1/299,792,458 of a second. How many kilometers does it travel in one year (a so-called light-year)?

(Answers to Practice Exercises are in the back of the book.)

1.10 Evaluating and Expressing Experimental Results

Although scientific measurements can be expressed in different units, they all have one thing in common: there is a limit to how accurate they can be. Nobody is perfect, and no analytical method is perfect either. The accuracy of every method has an inherent limit, so we need ways to express experimental results that reflect these limits in their certainty.

Significant Figures

significant figures all the certain digits in a measured value plus one estimated digit; the greater the number of significant figures, the greater the certainty with which the value is known.

One way to express how well we know a measured value is by using the appropriate number of **significant figures**, which are all the certain digits in a measured value *plus one estimated digit*. Suppose we determine the mass of a newly minted penny on the two laboratory scales shown in **Figure 1.21**. The upper one determines the masses of objects to the nearest 0.01 gram (g), whereas the lower one determines the mass of objects to the nearest 0.0001 g. According to the upper scale, our penny has a mass of 2.50 g; according to the lower scale, though, it has a mass of 2.4987 g. The mass obtained from the top scale has three significant figures: the 2, 5, and 0 are considered *significant*, which means we are confident in their values. The mass obtained using the bottom scale has five significant figures (2, 4, 9, 8, and 7). The mass of the penny can be determined with greater certainty with the bottom scale.

Now imagine that an aspirin tablet is placed on the lower scale in Figure 1.21, and the display reads 0.0180 g. How many significant figures are in this measurement? You might be tempted to say five because that is the number of digits displayed. However, the first two zeros are not considered significant because they serve only to set the location of the decimal point. These two zeros function the way exponents do when we use scientific notation to express values. Expressing 0.0180 g in scientific notation gives us 1.80×10^{-2} g. Only three of the digits (1, 8, and 0) in the decimal part indicate how precisely we know the value; the "−2" in the exponent does not. (See Appendix 1 for a review of how to express values using scientific notation.) The following guidelines will help you determine whether zeros (highlighted in green) are significant:

1. Zeros at the beginning of a value, as in 0.0592, are *never* significant. In this example, they just set the decimal place.
2. Zeros at the end of a value and after a decimal point, as in 3.00×10^8, are *always* significant.
3. Zeros at the end of a value that contains no decimal point, as in 96,500, may or may not be significant. If the zeros are there only to set the decimal place, then we should use scientific notation to indicate clearly that the value has only three significant figures: 9.65×10^4. Another approach to expressing the significance of ending zeroes in a value is to add a decimal point after them. For example, expressing the length of race sometimes called the "Olympic mile" as 1500. meters instead of 1500 meters makes it clear the value is known to more than two significant figures. In practice, this guideline is frequently not followed.
4. Zeros between nonzero digits, as in 101.3, are *always* significant.

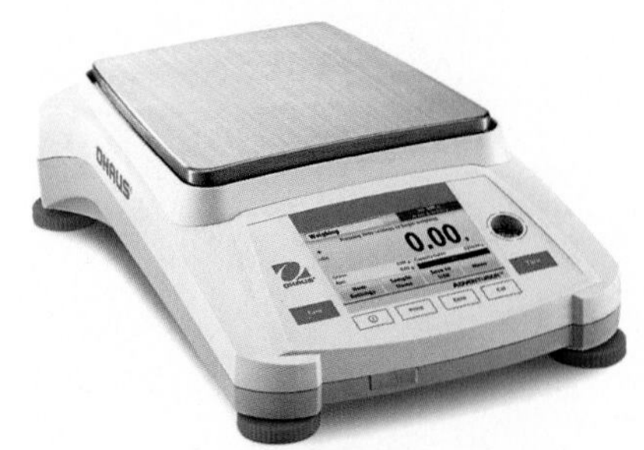

FIGURE 1.21 The mass of a penny can be measured to the nearest 0.01 g with the upper scale and to the nearest 0.0001 g with the scale below.

CHEMTOUR Significant Figures

CHEMTOUR Scientific Notation

CONCEPT TEST

How many significant figures are there in the values used as examples in guidelines 1, 2, and 4?

(Answers to Concept Tests are in the back of the book.)

Significant Figures in Calculations

How are significant figures used to express the results of calculations involving measured quantities? Suppose we suspect that a small nugget of yellow metal is pure gold. We could test our suspicion by determining the mass and volume of the nugget and then calculating its density. If the calculated density matches that of gold (19.3 g/mL), chances are good that the nugget is pure gold because few minerals are that dense. We find that the mass of the nugget is 4.72 g and its volume is 0.25 mL. What is the density of the nugget, expressed with the appropriate number of significant figures?

Using Equation 1.1 to calculate the density (d) from the mass (m) and volume (V) produces the following result:

$$d = \frac{m}{V} = \frac{4.72 \text{ g}}{0.25 \text{ mL}} = 18.88 \text{ g/mL}$$

This density value appears to be slightly less than that of pure gold, but how well do we know the result? The mass value is known to three significant figures, but the volume value is known only to two. At this point we need to apply the *weak-link principle*, which is based on the idea that a chain is only as strong as its weakest link. In calculations involving measured values, this principle means that we can know the answer to a calculation only as well as we know the least well-known measurement. In calculations involving multiplication or division, the weak link has the fewest significant figures. In this example, the weak link is the volume, 0.25 mL. Because it has only two significant figures, our final answer should be reported with only two significant figures. We must convert 18.88 to a number with two significant figures. We do this by a process called *rounding off.*

Rounding off a value means dropping the insignificant digits [all digits to the right of the last significant (first uncertain) digit] and then rounding that last significant digit either up or down. If the first digit in the string of insignificant digits dropped is greater than 5, we round up; if it is less than 5, we round down. If it is 5 and there is at least one nonzero digit to the right of it, we round up. For example, rounding 45.4501 to three significant figures makes it 45.5 (we rounded up because there was a nonzero digit to the right of the second 5).

If there are no nonzero digits to the right of the 5, then we need a tie-breaking rule for rounding down or up. A good rule to follow is to round the last significant digit to the nearest even number. For example, rounding 45.45 (or 45.450) to three significant figures makes either value 45.4, because the 4 in the tenths place is the nearest even number. However, we round off 45.55 (or 45.550) to 45.6, because the 6 in the tenths place is the nearest even number.

An important rule to remember is that you round off *only at the end of a calculation*—never on intermediate results. Otherwise, "rounding errors" can creep into calculations. Reasonable guidelines to follow are to carry at least one digit to the right of the last significant digit through all intermediate steps in a multistep calculation or simply to store all intermediate values in your calculator's memory until you have the final answer.

In the case of our gold nugget, the density value resulting from our calculation, 18.88 g/mL, must be rounded to two significant figures. Because the first insignificant digit to be dropped (shown in red) is greater than 5, we round up the blue 8 to 9:

$$18.88 \text{ g/mL} = 19 \text{ g/mL}$$

The density of pure gold (i.e., 19.3 g/mL) expressed to two significant figures is also 19 g/mL, so the nugget could be pure gold.

The weak-link principle for significant figures also applies to calculations involving addition and subtraction. To illustrate, consider how the volume of a gold nugget might be measured. One approach is to measure the volume of water the nugget displaces. Suppose we add 50.0 mL of water to a 100 mL graduated cylinder, as shown in **Figure 1.22**. The cylinder is graduated in milliliters, but the space between the graduations allows us to estimate the volumes of samples to the nearest tenth of a milliliter. For example, the meniscus of the water in Figure 1.22 is aligned exactly with the 50 mL graduation. We may record the volume of water as 50.0 mL because we can read the "50" part of this value directly from the

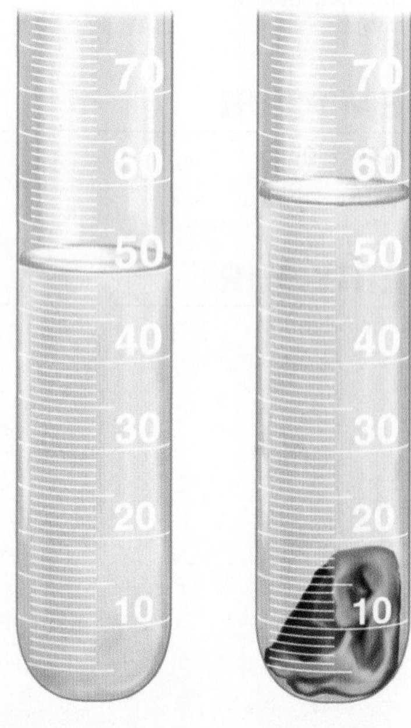

FIGURE 1.22 A nugget believed to be pure gold is placed in a graduated cylinder containing 50.0 mL of water. The volume rises to 58.5 mL, which means that the volume of the nugget is 8.5 mL.

cylinder, and we can estimate that the next digit is "0." An estimated final digit in a measured value is still considered significant.

Then we place the nugget in the graduated cylinder, being careful not to splash any water out. The level of the water is now about halfway between the 58 and 59 mL graduations, so we estimate the volume of the combined sample to be 58.5 mL. Taking the difference of the two estimated values, we have the volume of our nugget:

$$\begin{array}{r} 58.5\text{ mL} \\ -50.0\text{ mL} \\ \hline 8.5\text{ mL} \end{array}$$

The result has only two significant figures because the initial and final volumes are known to the nearest tenth of a milliliter. Therefore, we can report the difference between them to only the nearest tenth of a milliliter. When measured numbers are added or subtracted, the result is reported with the same number of significant digits to the right of the decimal as the measured number with the fewest digits to the right of the decimal.

Next, suppose the average mass of a recently minted penny is 2.50 g and a roll of pennies contains 50 of them. We use the upper scale from Figure 1.21 to determine the mass of a roll of pennies. Using the tare feature to correct for the mass of the wrapper, we find that the mass of pennies in the roll is 122.53 g. Dividing this value by the average mass of a penny and canceling out the grams units in the numerator and denominator, we get

$$\frac{122.53\text{ g}}{2.50\text{ g/penny}} = 49.012\text{ pennies}$$

If we were measuring uncountable substances, rounding off to 49.0 would be appropriate (because under these circumstances, the measured value of 2.50 g/penny would be the weak link in the calculation). However, because we use only whole numbers to count items, we round to 49 and conclude that the roll contains 49 pennies. We must add one penny to make the roll complete.

Suppose the penny we add is one we know has a mass of 2.5007 g because we determined its mass on the lower scale in Figure 1.21. What is the mass of all 50 pennies to the appropriate number of significant figures? Adding the mass of the 50th penny to the mass of the other 49, we have

$$\begin{array}{r} 122.53\text{ g} \\ +\ \ 2.5007\text{ g} \\ \hline 125.0307\text{ g} \end{array}$$

The mass of the first 49 pennies is known to two decimal places, whereas the mass of the 50th is known to four decimal places. By the weak-link principle, then, we can know the sum of the two numbers to only two decimal places. This makes the last two digits in the sum, shown in red, not significant, so we must round off the mass to 125.03 g.

SAMPLE EXERCISE 1.5 Using Significant Figures in Calculations LO8

The volume of a nugget of a shiny yellow mineral with a mass of 30.01 g is determined by the method of water displacement (see Figure 1.22). A 100 mL graduated cylinder contains 56.3 mL of water before the nugget is added and 62.6 mL after it is added. Is the nugget made of gold ($d = 19.3$ g/mL)?

Collect and Organize We are given the mass of a nugget and water displacement data that can be used to calculate its volume. Density is mass divided by volume.

Analyze The volume of the nugget is the difference between 62.6 mL and 56.3 mL, or about 6 mL. The ratio of the mass (30.01 g) to volume (~6 mL) is about 5 g/mL, which is much less than the density of gold.

Solve

$$d = \frac{m}{V}$$

$$= \frac{30.01 \text{ g}}{(62.6 - 56.3) \text{ mL}} = \frac{30.01 \text{ g}}{6.3 \text{ mL}} = \frac{4.7635 \text{ g}}{\text{mL}}$$

We know the volume of the nugget (6.3 mL) to two significant figures, so we can report its density to only two significant figures, too. As a result, we must round off the result to 4.8 g/mL. The density of gold is 19.3 g/mL, so the nugget cannot be pure gold.

Think About It The weak link in defining the certainty in this calculation is the volume of the irregularly shaped sample, which was calculated from the difference between two measured values that each had three significant figures, but the difference between them had only two. Our rough estimate in the Analyze step strongly suggested that the nugget was unlikely to be pure gold and our calculation in the Solve step confirmed that.

Practice Exercise Express the result of the following calculation to the appropriate number of significant figures:

$$\frac{0.391 \times 0.0821 \times (273 + 25)}{8.401}$$

Note: None of the starting values are exact numbers.

(Answers to Practice Exercises are in the back of the book.)

CONCEPT TEST

The distances between the opposite surfaces of a cube of metal were determined using a micrometer (**Figure 1.23**) and found to be 2.534, 2.537, and 2.529 cm. These values were multiplied together to calculate the volume of the cube, which was then divided into the mass of the cube, 135.080 g, to calculate its density. Was the volume of the cube or its mass the weak link in calculating its density?

(Answers to Concept Tests are in the back of the book.)

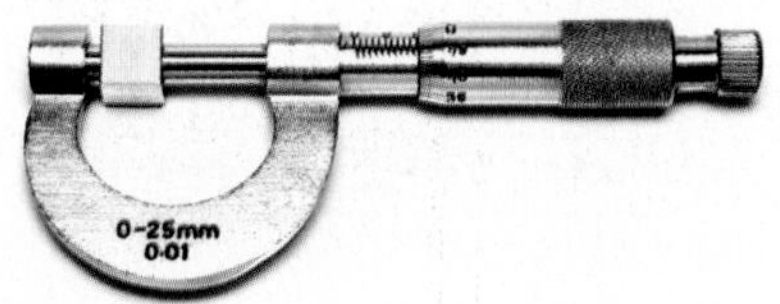

FIGURE 1.23 Micrometers are used to measure precisely the dimensions of objects such as this brass cube.

Measurements always have some degree of uncertainty, which limits the number of significant figures we can use to report any measurement. Some values are known exactly, however, such as 12 eggs in a dozen and 60 seconds in a minute. Because these values are definitions, there is no uncertainty in them and they are not considered when determining significant figures in the answer to a mathematical calculation in which they appear.

SAMPLE EXERCISE 1.6 Distinguishing Exact from Uncertain Values **LO9**

Which of the following data for the Washington Monument in Washington, DC, are exact numbers and which are not?

a. The monument is made of 36,491 white marble blocks.
b. The monument is 169 m tall.
c. There are 893 steps to the top.
d. The mass of its aluminum apex is 2.8 kg.
e. The area of its foundation is 1487 m^2.

precision agreement between the results of multiple measurements that were carried out in the same way.

accuracy agreement between one or more experimental values and the true (or accepted) value.

Collect, Organize, and Analyze One way to distinguish exact from inexact values is to answer the question, "Which quantities can be counted?" Values that can be counted are exact.

Solve (a) The number of marble blocks and (c) the number of stairs are quantities we can count, so they are exact numbers. The other three quantities are based on measurements of (b) length, (d) mass, and (e) area and therefore are not exact.

Think About It The "you can count them" property of exact numbers works in this exercise and in most others even when it takes a long time (such as counting the hairs on a dog) or if help is needed in visualizing the objects (such as using a microscope). These examples raise a good point: Is there uncertainty in the counting process itself? Can you think of a famous example in the history of American politics of an uncertain counting process?

Practice Exercise Which of the following statistics associated with the Golden Gate Bridge in San Francisco, CA, are exact numbers and which have some uncertainty?

a. The roadway is six lanes wide.
b. The width of the bridge is 27.4 m.
c. The bridge has a mass of 3.808×10^8 kg.
d. The length of the bridge is 2740 m.
e. The Pay-by-Plate toll for a car traveling south increased to $7.75 on July 7, 2017.

(Answers to Practice Exercises are in the back of the book.)

STEPWISE ANIMATION
Precision and Accuracy

Precision and Accuracy

Two terms—precision and accuracy—are used to describe how well we know a measured quantity or a value calculated from a measured quantity. **Precision** indicates how well the results of repeated measurements or analyses agree with each other, whereas **accuracy** reflects how close the results are to the true (or accepted) value of a quantity (**Figure 1.24**). The accuracy of a scale, for example, can be checked by measuring objects of known mass. Similarly, a thermometer can be calibrated by measuring the temperature at which a substance changes state. For example, ice melts to liquid water at 0.0°C, and an accurate thermometer dipped into a mixture of ice and liquid water reads 0.0°C. A measurement that is validated by calibration with an accepted standard material is considered accurate.

(a)

(b)

(c)

FIGURE 1.24 (a) The three dart throws meant to hit the center of the target are both accurate and precise. (b) The three throws meant to hit the center of the target are precise but not accurate. (c) This set of throws is neither precise nor accurate.

Another way to check the accuracy of measurements is to periodically analyze *control samples*, which contain known quantities of substances of interest. Control samples are used in clinical chemistry labs to ensure the accuracy of routine analyses of blood serum and urine samples. One substance that is included in comprehensive blood tests is creatinine, a product of protein metabolism that doctors use to monitor patients' kidney function. Suppose that a control sample known to contain 0.681 milligrams of creatinine per deciliter of sample is analyzed after every ten patient samples, and that the results of five control sample analyses are 0.685, 0.676, 0.669, 0.688, and 0.692 mg/dL. How precise (repeatable) are these results, and do they mean that the analyses of the control sample (and those subsequently conducted on the real samples) are accurate?

A widely used method for expressing the precision of data such as these involves calculating the average of all the values and then calculating how much

mean (arithmetic mean, $\bar{x}$) an average calculated by summing a set of related values and dividing the sum by the number of values in the set.

standard deviation (s) a measure of the amount of variation, or dispersion, in a set of related values. The lowercase Greek letter sigma (σ) is also used as the symbol for standard deviation.

confidence interval a range of values that has a specified probability of containing the true value of a measurement.

each value deviates from the average. To calculate the average, we sum the five control sample results and divide by the number of values (5):

$$\frac{(0.685 + 0.676 + 0.669 + 0.688 + 0.692)}{5}\text{ mg/dL} = \frac{3.410}{5}\text{ mg/dL} = 0.6820\text{ mg/dL}$$

We can write a general equation to represent this calculation for any number of measurements (n) of any parameter x. Individual results are identified by the generic symbol x_i, where i can be any integer from 1 to n, the number of measurements made. The capital Greek sigma ($\sum$) with an i subscript, $\sum_i(x_i)$, represents the *sum* of all the individual x_i values. This kind of average is called the **mean** or **arithmetic mean**, and it is represented by the symbol $\bar{x}$:

$$\bar{x} = \frac{\sum_i(x_i)}{n} \tag{1.2}$$

To evaluate the precision of the five analyses of the control sample, we first calculate how much each of the values, x_i, deviates from the mean value $\bar{x}$: $(x_i - \bar{x})$. Next, we square each deviation $(x_i - \bar{x})^2$ and then sum all the squared values. Next, we divide the sum by $(n - 1)$, and finally we take the square root of the quotient. The results of these calculations are shown in **Table 1.5**. The result is called the **standard deviation (s)** of the control sample values. Equation 1.3 combines the above steps for calculating s into a single equation:

$$s = \sqrt{\frac{\sum_i(x_i - \bar{x})^2}{n - 1}} \tag{1.3}$$

The standard deviation is sometimes reported with the mean value, using a $\pm$ sign to indicate that the uncertainty in the mean value spans the range from the mean plus the standard deviation to the mean minus the standard deviation. This expression is 0.6820 ± 0.0094 mg/dL for the five creatinine control samples, where the s value is rounded off to match the last decimal place of the mean. The smaller the value of s, the more tightly clustered the measurements are around the mean and the more precise the data are.

Now that we have evaluated the precision of the control sample data, we should evaluate how accurate they are—that is, how well they agree with 0.681 mg/dL, the actual creatinine value of the control sample. The mean of the five values is within 0.001 mg/dL of the actual value, so the results certainly seem accurate,

TABLE 1.5 Calculating the Standard Deviation of the Control Sample Creatinine Values

x_i (mg/dL)	$x_i - \bar{x}$ (mg/dL)	$(x_i - \bar{x})^2$	$\sum_i(x_i - \bar{x})^2$	$s = \sqrt{\frac{\sum_i(x_i - \bar{x})^2}{n - 1}}$
0.685	0.003	0.000009		
0.676	−0.006	0.000036		
0.669	−0.013	0.000169		
0.688	0.006	0.000036		
0.692	0.010	0.000100		
			0.000350	0.00935

but there is also a way to quantitatively express the certainty that they are accurate. It involves calculating the **confidence interval**, which is a range of values around a calculated mean (0.682 mg/dL in this case) that probably contains the true mean value, μ. Once we have calculated the size of this range, we can determine whether the actual creatinine value is within it. If it is, then the analyses are probably accurate.

To calculate the confidence interval, we use a statistical tool called the *t*-distribution and the following equation:

$$\mu = \bar{x} \pm \frac{t\,s}{\sqrt{n}} \tag{1.4}$$

A table of *t* values is in Appendix 1. **Table 1.6** contains a portion of it. The values are arranged based on two parameters: the number of values in a set of data (actually, $n - 1$) and the confidence level we wish to use in our decision making. A commonly used confidence level in chemical analysis is 95%. Using it means that the chances are 95% that the range we calculate using Equation 1.4 will contain the true mean value of, in this case, the amount of creatinine in our control sample. Using the 95% value for ($n - 1 = 5 - 1 = 4$) in Table 1.6, which is 2.776, and the mean and standard deviation values calculated above:

$$\mu = \bar{x} \pm \frac{t\,s}{\sqrt{n}} = \left(0.682 \pm \frac{2.776 \times 0.0094}{\sqrt{5}}\right)\text{ mg/dL} = (0.682 \pm 0.012)\text{ mg/dL}$$

Thus, we can say with 95% certainty that the true mean of our control sample data is between 0.670 and 0.694 mg/dL. Because this range includes the known concentration of the sample, 0.681 mg/dL, we may conclude that the results of analyses of the control samples (and, importantly, analyses of patient samples) are accurate.

TABLE 1.6 Values of *t*

	CONFIDENCE LEVEL (%)		
($n - 1$)	90	95	99
3	2.353	3.182	5.841
4	2.132	2.776	4.604
5	2.015	2.571	4.032
10	1.812	2.228	3.169
20	1.725	2.086	2.845
∞	1.645	1.960	2.576

SAMPLE EXERCISE 1.7 Evaluating the Precision of Analytical Results **LO9**

A group of students collects a sample of water from a river near their campus and shares it with five other groups of students. All six groups of students determine the concentration of dissolved oxygen in the sample. The results of their analyses are 9.2, 8.6, 9.0, 9.3, 9.1, and 8.9 mg O_2/L. What are the mean, standard deviation, and 95% confidence interval of these results?

Collect and Organize We are asked to calculate the mean, standard deviation, and 95% confidence interval of the results of six analyses of the same sample. Equations 1.2, 1.3, and 1.4 can be used to calculate the three statistical parameters we seek.

Analyze There are 6 values, so $n - 1 = 5$, and the appropriate *t* value (Table 1.6) to use in Equation 1.4 is 2.571. Mean and standard deviation functions are also included in many programmable calculators and in computer worksheet applications such as Microsoft Excel.

Solve

a. Calculating the mean:

$$\bar{x} = \frac{\Sigma_i(x_i)}{n} = \left(\frac{9.2 + 8.6 + 9.0 + 9.3 + 9.1 + 8.9}{6}\right)\text{ mg/L} = \frac{54.1}{6} = 9.02\text{ mg/L}$$

outlier a data point that is distant from the other observations.

Grubbs' test a statistical test used to detect an outlier in a set of data.

b. To calculate the standard deviation, we set up a data table like the one in Table 1.5:

x_i	$x_i - \bar{x}$	$(x_i - \bar{x})^2$
9.2	0.18	0.0324
8.6	−0.42	0.1764
9.0	−0.02	0.0004
9.3	0.28	0.0784
9.1	0.08	0.0064
8.9	−0.12	0.0144

The sum of the $(x_i - \bar{x})^2$ values is 0.3084 and

$$s = \sqrt{\frac{\sum_i (x_i - \bar{x})^2}{n-1}} = 0.25$$

c. Using Equation 1.4 to calculate the 95% confidence interval:

$$\mu = \bar{x} \pm \frac{ts}{\sqrt{n}} = \left(9.02 \pm \frac{2.571 \times 0.25}{\sqrt{6}}\right) \text{mg/L} = (9.02 \pm 0.26) \text{ mg/L}$$

Think About It Did you notice that the six data points used in these calculations each contained two significant figures (each was known to the nearest tenth of a mg/L), yet we expressed the mean with three significant figures (to the hundredths place)? This was appropriate because we knew the sum of the six results (54.1) to three significant figures and dividing this value by an exact number (six values) meant the quotient could be reported with three significant figures: 9.02. This increase in significant figures—and in the students' confidence in knowing the actual concentration of dissolved oxygen in the river—illustrates the importance of replicating analyses: doing so gives us more certainty about the true value of an experimental value than a single determination of it.

Practice Exercise Analyses of a sample of Dead Sea water produced these results for the concentration of sodium ions: 35.8, 36.6, 36.3, 36.8, and 36.4 mg/L. What are the mean, standard deviation, and 95% confidence interval of these results?

(Answers to Practice Exercises are in the back of the book.)

There is an assumption built into calculating means, standard deviations, and confidence intervals, which is that the variability in the data is random. *Random* means that data points are as likely to be above the mean value as below it and that there is a greater probability of values close to the mean than far away from it. This is called a *normal* distribution of the data points (it is also called a Gaussian distribution). Large numbers of such data produce a distribution profile called a bell curve. **Figure 1.25** shows such a curve, based on a study conducted by the U.S. Centers for Disease Control and Prevention on the average weights of Americans of different ages. The data in the figure are for the weights of 19-year-old males and have a mean of 175 pounds (79 kg) and a standard deviation of 41 pounds (19 kg). In randomly distributed data, 68% of the values—represented by the area under the curve highlighted in orange—are within 1 standard deviation of the mean.

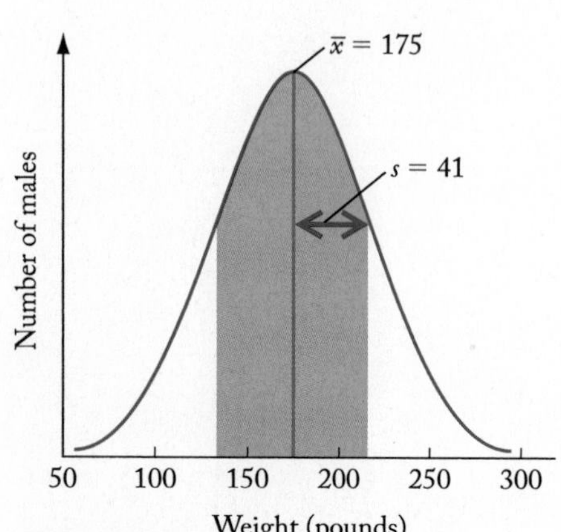

FIGURE 1.25 Weight distribution of 19-year-old American males. Source: U.S. Centers for Disease Control and Prevention (2012).

The concept of a normal distribution raises the question of how to handle **outliers**, that is, individual values that are much farther away from the mean than any of the other values. There may be a temptation to simply ignore such a value, but unless there is a valid reason for doing so, such as accidentally leaving out a step in an analytical procedure, it is unethical to disregard a value just because it is unexpected or not like the others.

Suppose a college freshman discovers a long-forgotten childhood piggy bank full of pennies and, being short of cash, decides to pack them into rolls of 50 to deposit them at a bank. To avoid having to count out hundreds of pennies, the student decides to weigh them out on a scale in a general chemistry lab that can weigh up to 300 grams to the nearest 0.001 g. The student separately weighs ten pennies from the piggy bank to determine their average mass, intending to multiply the average by 50 to calculate how many to weigh out and then pack into each roll. The results of the ten measurements are listed in **Table 1.7** from lowest value to highest value.

The results reveal that nine of the ten masses are quite close to 2.5 grams, but the tenth is considerably heavier. Was there an error in the measurement, or is there something unusual about that tenth penny? To answer questions such as these—and the broader one of whether an unusually high or low value is statistically different enough from the others in a set of data to be labeled an outlier—we analyze the data by using *Grubbs' test of a single outlier*, or **Grubbs' test**. In this test, the absolute difference between the suspected outlier and the mean of a set of data is divided by the standard deviation of the data set. The result is a statistical parameter that has the symbol Z:

$$Z = \frac{|x_i - \bar{x}|}{s} \tag{1.5}$$

If this calculated Z value is greater than the reference Z value (**Table 1.8**) for a given number of data points and a confidence level—usually 95%—then the suspect data point is an outlier and can be discarded.

To apply Grubbs' test to the mass of the tenth and heaviest penny, we first calculate the mean and standard deviation of the masses of all ten pennies. These values are 2.562 ± 0.1915 g. We use them in Equation 1.5 to calculate Z:

$$Z = \frac{|x_i - \bar{x}|}{s} = \frac{|3.107 - 2.562|}{0.1915} = 2.846$$

Next, we check the reference Z values in Table 1.8 for $n = 10$ data points, and we find that our calculated Z value is greater than both 2.290 and 2.482—the Z values above which we can conclude with 95% and 99% confidence, respectively, that the 3.107 g data point is an outlier. Stated another way, the probability that this data point *is not* an outlier is less than 1%. Note that Grubbs' test can be used *only once* to identify *only one* outlier in a set of data.

The tenth penny probably weighed much more than the others because U.S. pennies minted since 1983 weigh 2.50 grams new, but those minted before 1983 weighed 3.11 grams new. The older pennies are 95% copper and 5% zinc, whereas the newer pennies are 97.5% zinc and are coated with a thin layer of copper. (Copper is about 25% more dense than zinc.) Thus, a pre-1983 penny in a population of post-1983 pennies is an outlier based on its mass *and* based on its chemical composition.

TABLE 1.7 Masses in Grams of Ten Circulated Pennies

Mass (g)
2.486
2.495
2.500
2.502
2.502
2.505
2.506
2.507
2.515
3.107

TABLE 1.8 Reference Z Values for Grubbs' Test of a Single Outlier

	CONFIDENCE LEVEL (%)	
n	95	99
3	1.155	1.155
4	1.481	1.496
5	1.715	1.764
6	1.887	1.973
7	2.020	2.139
8	2.127	2.274
9	2.215	2.387
10	2.290	2.482
11	2.355	2.564
12	2.412	2.636

CONCEPT TEST

Calculate the mean mass of one penny in a roll of pennies that contains exactly half pre-1983 pennies. One synonym for *mean* is "average." Would the mass of any penny in this roll equal the average mass?

(Answers to Concept Tests are in the back of the book.)

SAMPLE EXERCISE 1.8 Testing Whether a Data Point Should Be Considered an Outlier LO9

Use Grubbs' test to determine whether the lowest value in the set of sodium ion concentration data from Practice Exercise 1.7 should be considered an outlier at the 95% confidence level.

Collect, Organize, and Analyze We want to determine whether the lowest value in the following set of five results is an outlier: 35.8, 36.6, 36.3, 36.8, and 36.4 mg/L. Grubbs' test (Equation 1.5) is used to determine whether a data point should be deemed an outlier. If the resulting Z value is equal to or greater than the reference Z values for $n = 5$ in Table 1.8, the suspect value is an outlier.

Solve Using the statistics functions of a programmable calculator, we find the mean and standard deviation of the data set to be 36.38 ± 0.38 mg/L. We calculate the value of Z by using Equation 1.5:

$$Z = \frac{|x_i - \bar{x}|}{s} = \frac{|35.8 - 36.38|}{0.38} = 1.5$$

This calculated Z value is less than 1.71, which is the reference Z value for $n = 5$ at the 95% confidence level (Table 1.8). Therefore, the lowest value is *not* an outlier, and it should be included with the other four values in any analysis of the data.

Think About It Although the lowest value may have appeared to be considerably lower than the others in the data set, Grubbs' test tells us that it is not *significantly* lower at the 95% confidence level.

Practice Exercise Duplicate determinations of the cholesterol concentration in a blood serum sample produce the following results: 181 mg/dL and 215 mg/dL. The patient's doctor is concerned about the difference between the two results because values above 200 mg/dL are considered "borderline high," so she orders a third cholesterol test. The result is 185 mg/dL. Should the doctor ignore the 215 mg/dL value?

(Answers to Practice Exercises are in the back of the book.)

1.11 Testing a Theory: The Big Bang Revisited

If all the matter in the universe started as a dense cloud of very hot gas accompanied by an enormous release of energy (the Big Bang), and if the universe has been expanding ever since, then the universe must have been cooling throughout time because gases cool as they expand. This is the principle behind the operation of refrigerators and air conditioners—and behind one of the most important experiments testing the validity of the Big Bang theory.

Temperature Scales

STEPWISE ANIMATION
Temperature Scale

If the universe is still expanding and cooling, some warmth must be left over from the Big Bang. By the 1950s, some scientists predicted that the leftover warmth should be enough to give interstellar space a temperature of 2.73 K, where K indicates a temperature value on the Kelvin scale.

Several temperature scales are in use today. In the United States the Fahrenheit scale is still the most popular. In the rest of the world and in science, temperatures are most often expressed in degrees Celsius or on the Kelvin scale. The Fahrenheit and Celsius scales differ in two ways, as shown in **Figure 1.26**. First, their zero points are different. Zero degrees Celsius (0°C) is the temperature at which water freezes under normal conditions, but that temperature is 32 degrees on the Fahrenheit scale (32°F). The other difference is the size of the temperature change corresponding to 1 degree. The difference between the freezing and

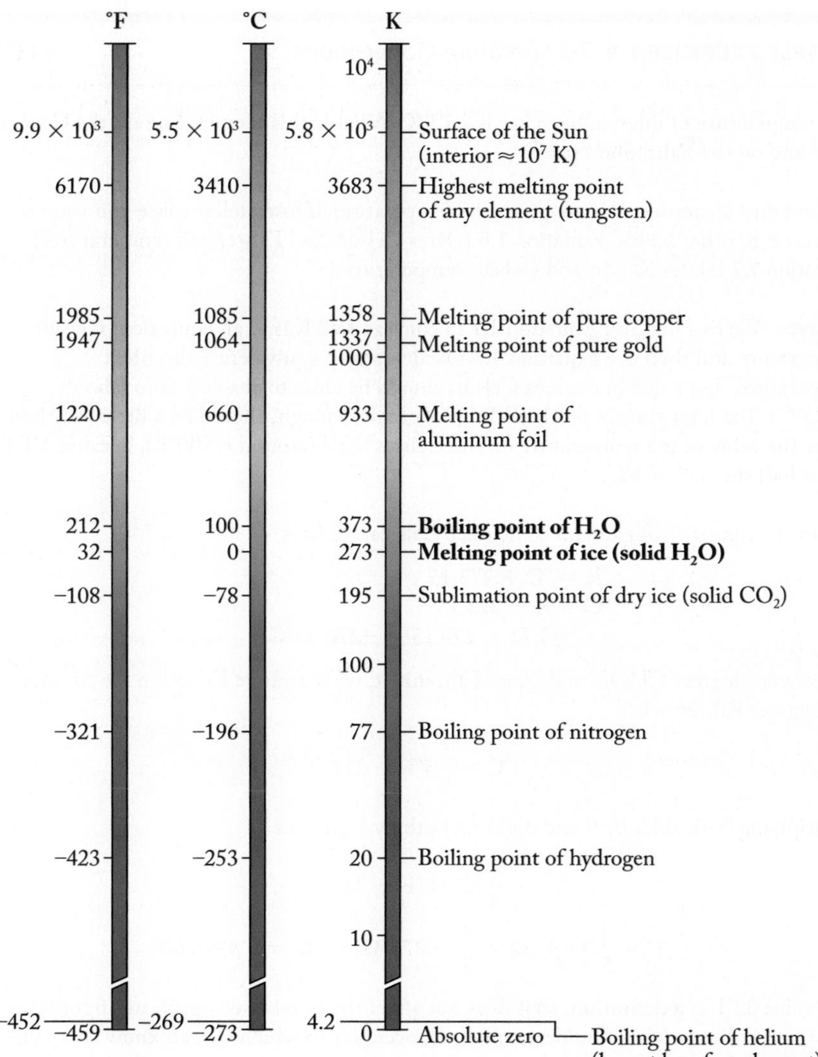

FIGURE 1.26 Three temperature scales are commonly used today, although the Fahrenheit scale is rarely used in scientific work. The freezing and boiling points of liquid water are the defining temperatures for each of the scales and their degree sizes.

CHEMTOUR Temperature Conversion

boiling points of water is 212 − 32 = 180 degrees on the Fahrenheit scale but only 100 − 0 = 100 degrees on the Celsius scale. This difference means that a Fahrenheit degree is 100/180, or 5/9, as large as a Celsius degree.

To convert temperatures from Fahrenheit into Celsius, we need to account for the differences in zero point and in degree size. Equation 1.6 does both:

$$°C = \frac{5}{9}(°F - 32) \tag{1.6}$$

The SI unit of temperature (Table 1.2) is the **kelvin (K)**. The zero point on the Kelvin scale is not related to the freezing of a substance; rather, it is the coldest temperature—called **absolute zero (0 K)**—that can theoretically exist. It is equivalent to −273.15°C. No one has ever been able to chill matter to absolute zero, but scientists have cooled samples to less than 10^{-9} K, so they have come very close.

The zero point on the Kelvin scale differs from that on the Celsius scale, but the size of 1 degree is the same on the two scales. For this reason, the conversion from a Celsius temperature to a Kelvin temperature is simply a matter of adding 273.15 to the Celsius value:

$$K = °C + 273.15 \tag{1.7}$$

kelvin (K) the SI unit of temperature.

absolute zero (0 K) the zero point on the Kelvin temperature scale; theoretically the lowest temperature possible.

FIGURE 1.27 Robert Dicke (1916–1997) predicted the existence of cosmic microwave background radiation. His prediction was confirmed by the serendipitous discovery of this radiation by Robert Wilson and Arno Penzias.

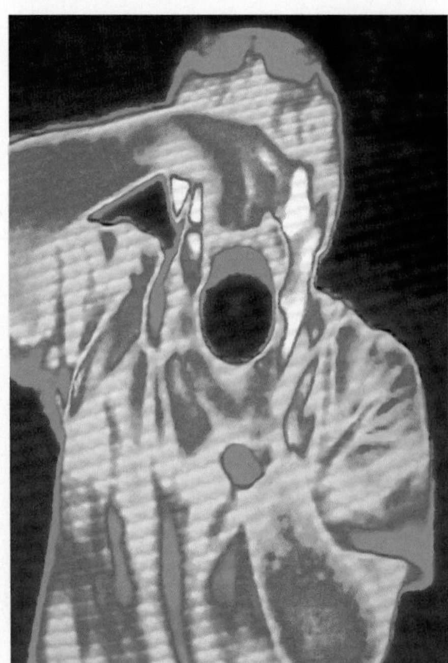

FIGURE 1.28 At normal body temperature, a human being emits radiation in the infrared region of the electromagnetic spectrum. In this thermal camera self-portrait, the red areas are the warmest and the blue regions are the coolest.

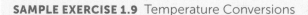

SAMPLE EXERCISE 1.9 Temperature Conversions **LO7**

The temperature of interstellar space is 2.73 K. What is this temperature on the Celsius scale and on the Fahrenheit scale?

Collect and Organize We are given the temperature of interstellar space and want to convert it to other scales. Equation 1.6 relates Celsius and Fahrenheit temperatures; Equation 1.7 relates Kelvin and Celsius temperatures.

Analyze We can first use Equation 1.7 to convert 2.73 K into an equivalent Celsius temperature and then use Equation 1.6 to calculate an equivalent Fahrenheit temperature. The value in degrees Celsius should be close to absolute zero (about −273°C). The temperature on the Fahrenheit scale, though, should be a little less than twice the value of the temperature on the Celsius scale (around −500°F), because 1°F is about half the size of 1°C.

Solve To convert from kelvins to degrees Celsius, we have

$$K = °C + 273.15$$
$$°C = K - 273.15$$
$$= 2.73 - 273.15 = -270.42°C$$

To convert degrees Celsius to degrees Fahrenheit, we rearrange Equation 1.6 to solve for degrees Fahrenheit:

$$°C = \frac{5}{9}(°F - 32)$$

Multiplying both sides by 9 and dividing both by 5 gives us

$$\frac{9}{5}°C = °F - 32$$

$$°F = \frac{9}{5}°C + 32 = \frac{9}{5}(-270.42) + 32 = -454.76°F$$

The value 32°F is a definition, so it does not affect the number of significant figures in the answer. The number that determines the accuracy to which we can know this value is −270.42°C.

Think About It The calculated Celsius value of −270.42°C makes sense because it represents a temperature only a few degrees above absolute zero, just as we estimated. The Fahrenheit value is within 10% of our estimate, so it is reasonable, too.

Practice Exercise The temperature of the Moon's surface varies from −233°C at night to 123°C during the day. What are these temperatures on the Kelvin and Fahrenheit scales?

(Answers to Practice Exercises are in the back of the book.)

An Echo of the Big Bang

By the early 1960s, Princeton University physicist Robert Dicke (1916–1997; **Figure 1.27**) had suggested the presence of residual energy left over from the Big Bang, and he was eager to test this hypothesis. He proposed building an antenna that could detect microwave energy reaching Earth from outer space. Why did he pick microwaves? Even matter as cold as 2.73 K emits a "glow" (an energy signature), but not a glow you can see or feel, like the infrared rays emitted by any warm object, including humans (**Figure 1.28**). Dicke wanted to build a microwave antenna because the glow from a 2.73 K object takes the form of microwaves.

Dicke's microwave detector was never built because of events that occurred just a short distance from Princeton. By the early 1960s, the United States had launched *Echo* and *Telstar*, the first communication satellites. These satellites were reflective spheres designed to bounce microwave signals to receivers on Earth. An antenna designed to receive these microwave signals had been built at Bell Laboratories in Holmdel, NJ (**Figure 1.29**). Two Bell Labs scientists, Robert W. Wilson (born 1936) and Arno A. Penzias (born 1933), were working to improve the antenna's reception when they encountered a problem. They found that no matter where they directed their antenna, it picked up a constant background microwave signal much like the hissing sound radios make when tuned between stations. They thought the signal was due to a flaw in the antenna or in one of the instruments connected to it, but extensive testing failed to detect such a flaw.

FIGURE 1.29 In 1965 Robert Wilson (left) and Arno Penzias discovered the microwave echo of the Big Bang while tuning this highly sensitive "horn" antenna at Bell Labs in Holmdel, NJ.

Then Wilson and Penzias learned of Dicke's prediction that a microwave echo of the Big Bang penetrated the universe, and they realized the cosmic significance of what they had thought was a problem with their satellite communication system. Others realized it, too, and in 1978 Wilson and Penzias shared the Nobel Prize in Physics for discovering the cosmic microwave background. Dicke did not share in the prize, even though he had accurately predicted what Wilson and Penzias discovered by accident.

Major discoveries in science almost always lead to new questions, new experiments, and new discoveries. The discovery of cosmic microwaves was no exception. It supported the Big Bang theory, but it also seemed to find a problem with it: the Bell Labs antenna picked up the same microwave signal no matter where in the sky it was pointed. In other words, the cosmic microwaves appeared to be uniformly distributed throughout the universe. This uniformity in the afterglow from the Big Bang troubled scientists because they knew there had to have been at least some heterogeneity in the density of the expanding universe to eventually allow clusters of matter to assemble and grow, leading to the formation of galaxies containing the first generation of stars.

Scientists doing work related to Dicke's proposed that, if such clustering had occurred, a record of it should exist as subtle differences in the cosmic background radiation. Unfortunately, a microwave antenna in New Jersey—or any other place on Earth—could not detect such slight variations in signals because too many other sources of microwaves interfere with such measurements. Detecting these tiny variations would require launching a radio antenna into space.

In late 1989 the United States launched such an antenna in the form of the *Cosmic Background Explorer* (*COBE*) satellite. After many months of collecting data and many more months of analyzing it, scientists released the results in 1992. The image in **Figure 1.30a** appeared on the front pages of newspapers and magazines around the world. This was a major news story because the predicted heterogeneity had been found, providing additional support for the Big Bang theory. At a news conference, the lead scientist on the *COBE* project called the map a "fossil of creation."

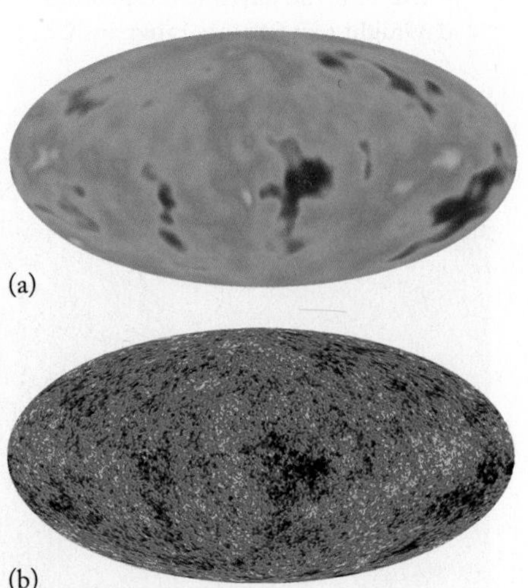

FIGURE 1.30 (a) Map of the cosmic microwave background released in 1992. It is a 360-degree image of the sky made by collecting microwave signals for a year from the microwave telescopes of the *COBE* satellite. (b) Higher-resolution image based on measurements made in 2012 by the *WMAP* satellite. Red regions are up to 200 μK warmer than the average interstellar temperature of 2.73 K, and blue regions are up to 200 μK colder than 2.73 K.

COBE's measurements, and those obtained over the next two decades by a satellite with even higher resolving power (**Figure 1.30b**), support the theory that the universe had undergone an uneven expansion and cooling. The blobs and ripples in the images in Figure 1.30 indicate that galaxy "seed clusters" formed early in the history of the universe. Cosmologists believe these ripples are a record of the next stage after the Big Bang in the creation of matter. This stage is examined in Chapter 2, where we discuss how some elements may have formed just after the Big Bang and how others continue to be formed by the nuclear reactions that fuel our Sun and all the stars in the universe.

CONCEPT TEST

When Lemaître first proposed how the universe might have formed following a cosmic release of energy, would it have been more appropriate to call his explanation a hypothesis or a theory? Why?

(Answers to Concept Tests are in the back of the book.)

SAMPLE EXERCISE 1.10 Integrating Concepts: Driving around Mars

In press conferences about the Martian rover *Curiosity* (**Figure 1.31**), scientists and engineers have expressed distances in SI and USC units and temperatures in K, °C, and °F in response to questions from journalists from different countries. After one early drive, they said *Curiosity* had stopped 8.0 ft away from an interesting football-shaped rock. *Curiosity*'s wheels are 50 cm in diameter, and the rover moves at a speed of 200 m/sol, where 1 sol = 1 Martian day = 24.65 h.

a. How many minutes would it take the rover to move to within 1.0 ft of the rock?
b. How many rotations of the wheels would be required to move that distance?
c. The rover landed in a valley about 15 miles away from the base of Mt. Sharp. How far is that in kilometers?
d. The distance from the base to the peak of Mt. Sharp is 5.5 km; the distance from the base to the peak of Mt. Everest on Earth is 15,000 ft. Which mountain is taller?
e. One night *Curiosity* recorded a temperature reported as −132°F. If the daytime temperature was 263 K, what was the day/night temperature range in °C?

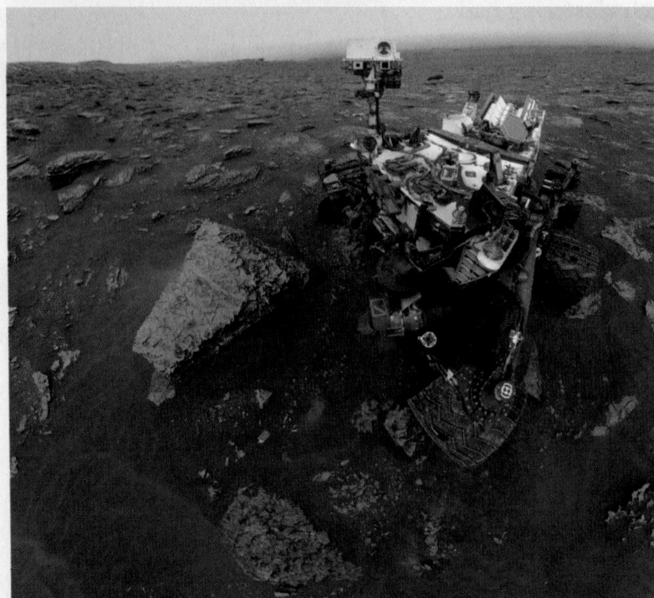

FIGURE 1.31 In late March 2018, NASA celebrated *Curiosity*'s 2000th Martian day (sol) of exploration by releasing this selfie taken by the rover with Mount Sharp in the background.

Collect, Organize, and Analyze We are given a series of measurements and are asked questions that require us to convert units of time, distance, speed, and temperature. The needed conversion factors are available on the inside back cover of the book or provided in the problem, including the rover's speed (200 m/sol) and the length of a Martian day (1 sol = 24.65 h).

Solve

a. The time to travel to within 1.0 ft of the rock is

$$(8.0 - 1.0)\ \cancel{\text{ft}} \times \frac{12\ \cancel{\text{in}}}{1\ \cancel{\text{ft}}} \times \frac{2.54\ \cancel{\text{cm}}}{1\ \cancel{\text{in}}} \times \frac{1\ \cancel{\text{m}}}{100\ \cancel{\text{cm}}} \times \frac{1\ \cancel{\text{sol}}}{200\ \cancel{\text{m}}} \times \frac{24.65\ \cancel{\text{h}}}{1\ \cancel{\text{sol}}} \times \frac{60\ \text{min}}{1\ \cancel{\text{h}}} = 16\ \text{min}$$

b. The diameter of the wheel is 50 cm, or

$$50\ \cancel{\text{cm}} \times \frac{1\ \cancel{\text{in}}}{2.54\ \cancel{\text{cm}}} \times \frac{1\ \text{ft}}{12\ \cancel{\text{in}}} = 1.64\ \text{ft}$$

and its circumference is

$$\pi \times d = 3.1416 \times 1.64\ \text{ft} = 5.15\ \text{ft}$$

The wheel has 7.0 ft to travel, so it requires

$$7.0\ \cancel{\text{ft}} \times \frac{1\ \text{revolution}}{5.15\ \cancel{\text{ft}}} = 1.4\ \text{revolutions}$$

c. $15\ \cancel{\text{mi}} \times \frac{1.6093\ \text{km}}{1\ \cancel{\text{mi}}} = 24\ \text{km}$

d. Mt. Sharp: $5.5\ \cancel{\text{km}} \times \frac{1\ \text{mi}}{1.6093\ \cancel{\text{km}}} = 3.4\ \text{mi}$

Mt. Everest: $15{,}000\ \cancel{\text{ft}} \times \frac{1\ \text{mi}}{5280\ \cancel{\text{ft}}} = 2.8\ \text{mi}$

Mt. Sharp is (3.4 − 2.8) = 0.6 miles higher than Mt. Everest.

e. $°\text{C} = \frac{5}{9}(-132°\text{F} - 32°\text{F}) = -91.1°\text{C}$

The high temperature that day was 263 K:

$$°\text{C} = \text{K} - 273.15 = 263 - 273.15 = -10.15 = -10°\text{C}$$

The temperature differential on the surface that day was

$$-10°\text{C} - (-91.1°\text{C}) = 81°\text{C}$$

Think About It Two of the reported starting values—the diameter (50 cm) of *Curiosity*'s wheels and its speed (200 m/sol)—ended with zeros and had no decimal points. This created ambiguity in how well these values are known and forced us to guess their certainties based on (1) the likely precision with which the components of *Curiosity* must have been fabricated and (2) the ability of the Jet Propulsion Laboratory engineers to control its speed. In both cases we assumed the values were known to two significant figures and used that many to express the results of the calculations in parts (a) and (b).

SUMMARY

LO1 **Matter** exists as **pure substances**, which may be either **elements** or **compounds**, and as **mixtures**. Mixtures may be **homogeneous** (these mixtures are also called **solutions**) or **heterogeneous**. (Section 1.2)

LO2 All matter consists of particles: **atoms, molecules** (combinations of atoms held together by **chemical bonds**), or **ions** (atoms or combinations of atoms that have a net negative or positive electrical charge). **Chemical formulas** are used to describe polyatomic forms of elements, such as H_2, and the elemental composition of compounds, such as H_2O. **Chemical equations** describe the proportions of the substances involved in a **chemical reaction**. Space-filling and ball-and-stick models are used to show the three-dimensional arrangement of the atoms in molecules. (Section 1.2)

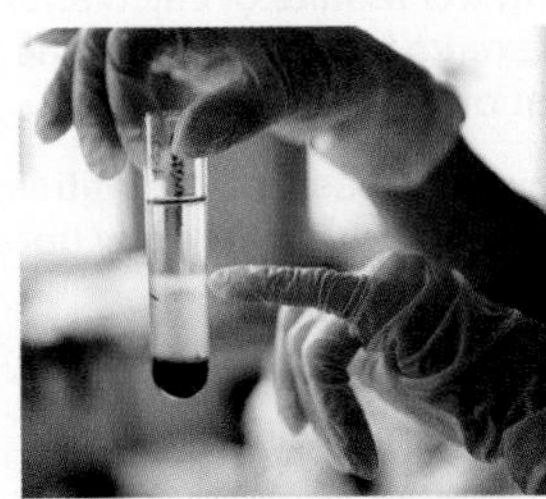

LO3 Matter undergoes **physical processes**, which do not change its chemical identity, and **chemical reactions**, which transform matter into different substances. Matter is described and defined in terms of its **physical properties** and **chemical properties**. Physical properties may be used to separate mixtures into pure substances. (Sections 1.1, 1.2, 1.3, and 1.5)

LO4 The COAST framework used in this book to solve problems has four components: Collect and Organize information and ideas; Analyze the information to determine how it can be used to obtain the answer; Solve the answer to the problem (often the math-intensive step); and Think About It, where you consider the answer's reasonableness, including its value and units. (Section 1.4)

LO5 The physical properties of the different states of matter are linked to the arrangement and mobility of the particles in them. For example, the particles in **solids** and **liquids** are packed close together, which make these states of matter nearly incompressible. **Gases**, however, are compressible because their particles are far apart. (Section 1.6)

LO6 The **scientific method** is the approach we use to acquire knowledge through observation, testable **hypotheses**, and experimentation. An extensively tested and well-validated hypothesis becomes accepted as a **scientific theory** or **model**. (Sections 1.7 and 1.11)

LO7 Dimensional analysis uses **conversion factors** (fractions in which the numerators and denominators have different units but represent the same quantity) to convert a value from one unit into another unit. (Section 1.9)

LO8 The limit of how well we know an uncertain value is expressed by the number of **significant figures** used to express the value. (Sections 1.8 and 1.10)

LO9 There are inherent uncertainties in all measured values, whereas other values, such as the number of eggs in a carton, are known exactly because they can be counted. The **precision** of measured values refers to how little variability there is between repeated measurements. The **accuracy** of a measurement, such as weighing an object on a scale, refers to how close the measured value is to the actual mass of the object. The **standard deviation** and **confidence interval** of a set of measurements provide measures of the uncertainty surrounding the average or **arithmetic mean** of the data set. An **outlier** in a data set may be identified by using the **Grubbs' test**. (Section 1.10)

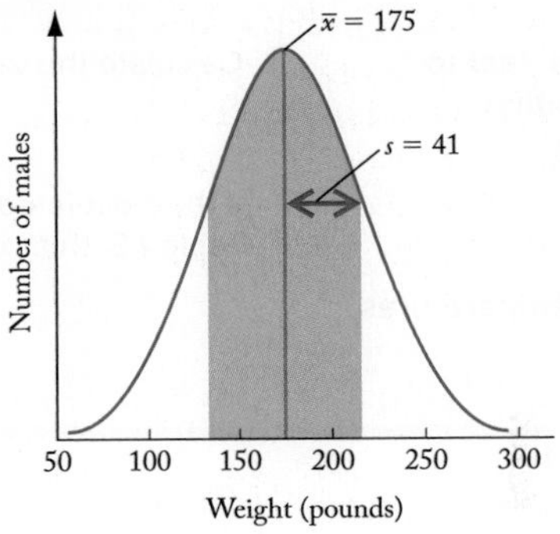

PARTICULATE PREVIEW WRAP-UP

Sublimation is the process by which a solid changes directly into a gas. The image on the left is a solid because it consists of tightly packed and ordered molecules, whereas the image on the right is a gas because there is space between the molecules and they are spread out to fill the circle. Energy must be added to convert a solid into a gas.

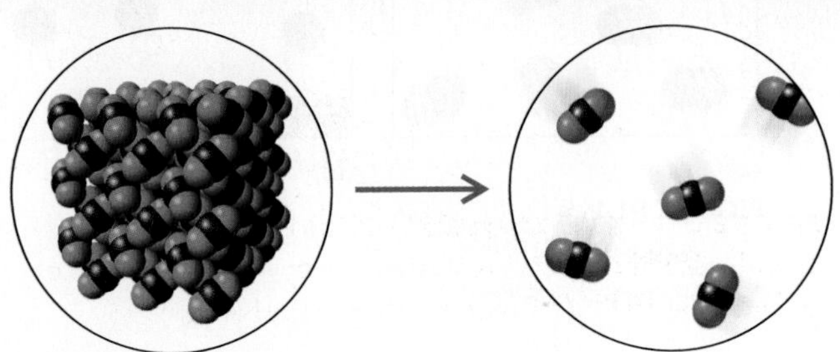

PROBLEM-SOLVING SUMMARY

Type of Problem	Concepts and Equations		Sample Exercises
Distinguishing physical properties from chemical properties	The chemical properties of a substance can be determined only by reacting it with another substance; physical properties can be determined without altering the substance's composition.		**1.1**
Recognizing physical states of matter	Particles in a solid are ordered; particles in a liquid are randomly arranged but close together; particles in a gas are separated by space and fill the volume of their container.		**1.2**
Using dimensional analysis to convert units	Convert values from one set of units to another by multiplying by conversion factors set up so that the original units cancel.		**1.3, 1.4**
Calculating density from mass and volume	$d = \frac{m}{V}$	(1.1)	**1.5**
Using significant figures in calculations	Apply the weak-link rule in expressing the results of calculations that involve uncertain values: the results are known only as well as the least well-known value used in the calculation.		**1.5**
Distinguishing exact from uncertain values	Quantities that can be counted are exact. Measured quantities or conversion factors that are not exact values are inherently uncertain.		**1.6**
Calculating mean, standard deviation, and confidence interval values	$\bar{x} = \frac{\sum_i(x_i)}{n}$ $s = \sqrt{\frac{\sum_i(x_i - \bar{x})^2}{n-1}}$ $\mu = \bar{x} \pm \frac{ts}{\sqrt{n}}$	(1.2) (1.3) (1.4)	**1.7**
Using Grubbs' test to identify an outlier	Calculate the value of Z for a suspected outlier x_i: $Z = \frac{\lvert x_i - \bar{x}\rvert}{s}$ If the calculated Z value is greater than the appropriate reference Z value in Table 1.8, then x_i is an outlier.	(1.5)	**1.8**
Converting temperatures	$°C = \frac{5}{9}(°F - 32)$ $K = °C + 273.15$	(1.6) (1.7)	**1.9**

VISUAL PROBLEMS

(Answers to boldface end-of-chapter questions and problems are in the back of the book.)

1.1. For each image in Figure P1.1, identify what class of matter is depicted (an element, a compound, a homogeneous mixture, or a heterogeneous mixture) and identify the physical state(s).

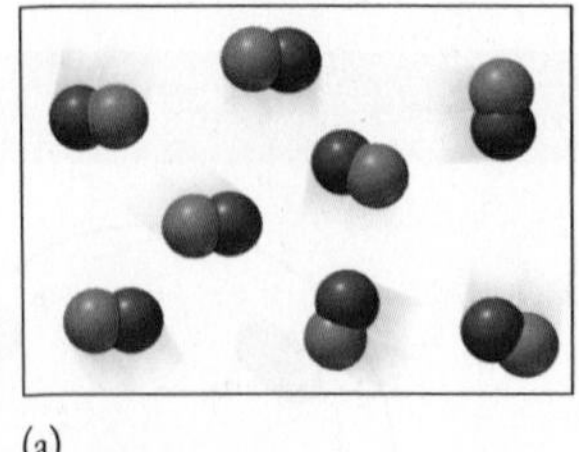

(a)

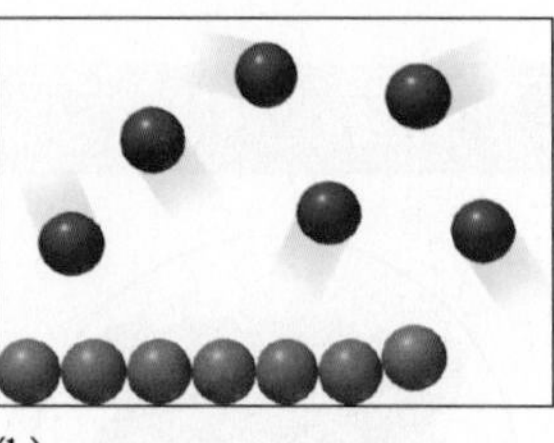

(b)

FIGURE P1.1

1.2. For each image in Figure P1.2, identify what class of matter is depicted (an element, a compound, a homogeneous mixture, or a heterogeneous mixture) and identify the physical state(s).

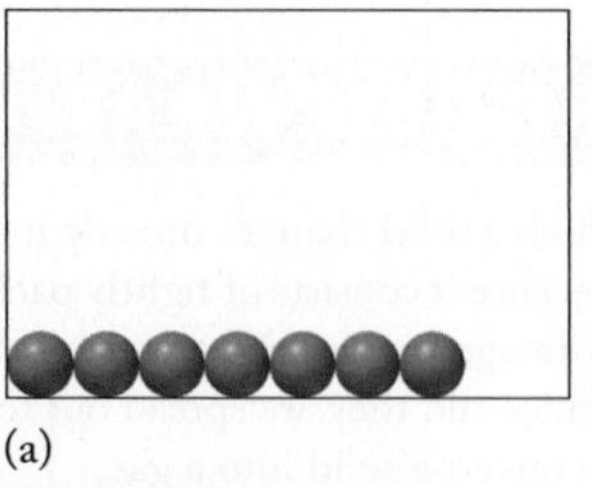

(a)

(b)

FIGURE P1.2

1.3. Which of the following statements best describes the change depicted in Figure P1.3?
a. A mixture of two gaseous elements undergoes a chemical reaction, forming a gaseous compound.
b. A mixture of two gaseous elements undergoes a chemical reaction, forming a solid compound.
c. A mixture of two gaseous elements undergoes deposition.
d. A mixture of two gaseous elements condenses.

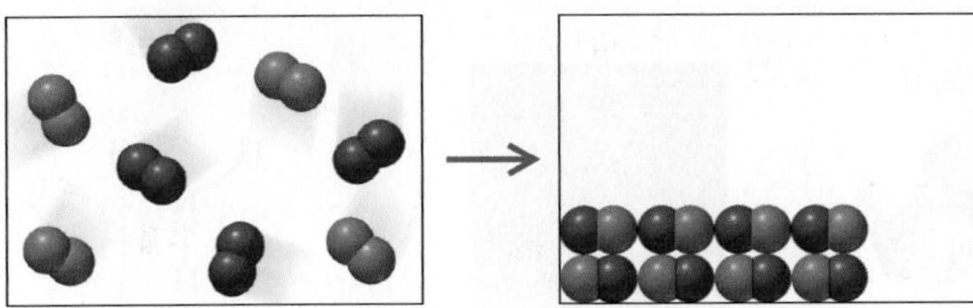

FIGURE P1.3

1.4. Which of the following statements best describes the change depicted in Figure P1.4?
a. A mixture of two gaseous elements is cooled to a temperature at which one of them condenses.
b. A mixture of two gaseous compounds is heated to a temperature at which one of them decomposes.
c. A mixture of two gaseous elements undergoes deposition.
d. A mixture of two gaseous elements reacts together to form two compounds, one of which is a liquid.

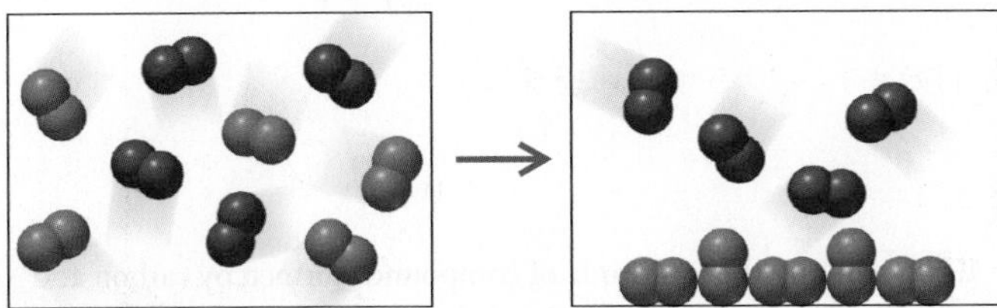

FIGURE P1.4

1.5. Suppose the CO_2 represented by the molecules in the box on the left in Figure P1.5 is cooled from room temperature to 180 K. Draw the particulate image of the CO_2 that would be present at that temperature in the box on the right. *Hint*: The normal sublimation point of CO_2 is 195 K.

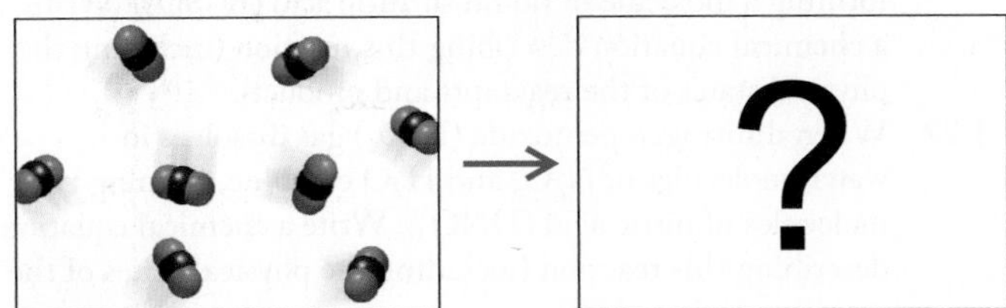

FIGURE P1.5

1.6. Suppose the temperature of the mixture of substances represented by the molecules in the box on the left in Figure P1.6 was reduced from 298 K to 50 K. Draw the particulate image of the mixture that would be present at 50 K in the box on the right. *Hint*: The colors used to represent the atoms of different elements are shown on the last page of the book.

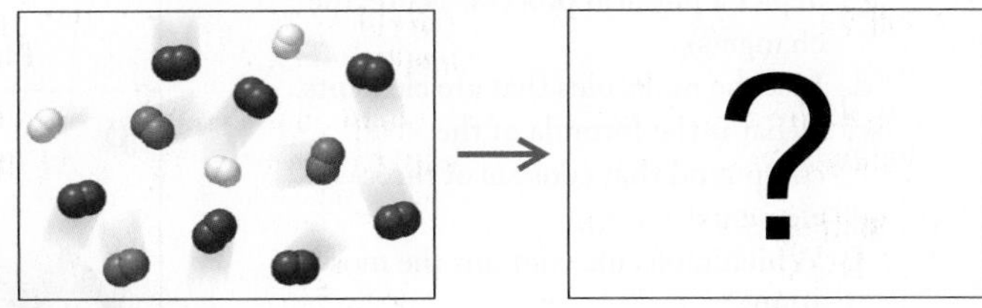

FIGURE P1.6

1.7. What are the chemical formulas of the three compounds with the molecular structures shown in Figure P1.7?

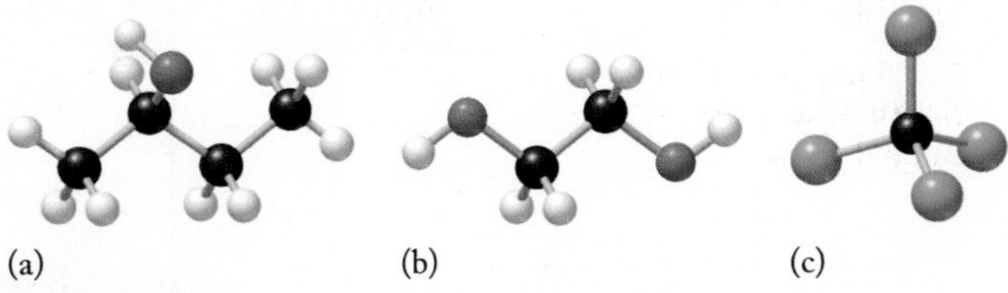

FIGURE P1.7

1.8. What are the chemical formulas of the three compounds with the molecular structures shown in Figure P1.8?

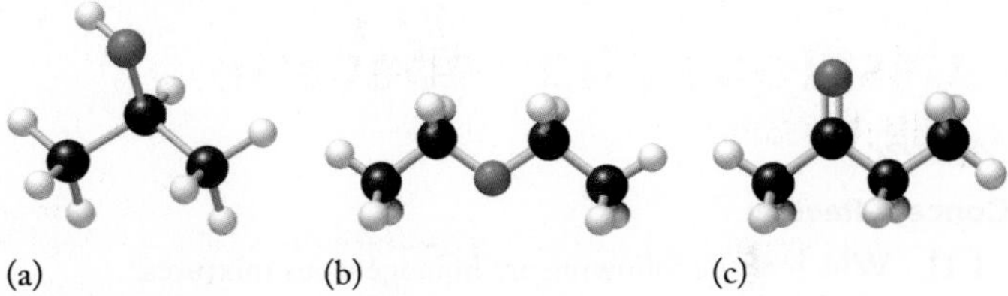

FIGURE P1.8

1.9. A pharmaceutical company checks the quality control process involved in manufacturing pills of one of its medicines by taking the mass of two samples of four pills each. Figure P1.9 shows graphs of the masses of the pills in each of the two samples. The pill is supposed to weigh 3.25 mg. Label each sample as both precise and accurate, precise but not accurate, accurate but not precise, or neither precise nor accurate.

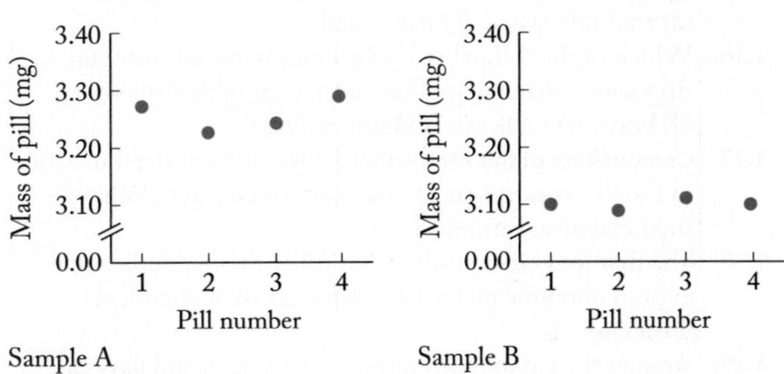

FIGURE P1.9

1.10. Use representations [A] through [I] in Figure P1.10 to answer questions (a)–(f).
   a. Which figures, if any, depict a chemical reaction?
   b. Which two representations depict the same compound?
   c. Which representations, if any, depict a physical process? Name the change(s).
   d. List the molecules that are elements.
   e. What is the formula of the compound that consists of three elements?
   f. Which molecule contains the most atoms?

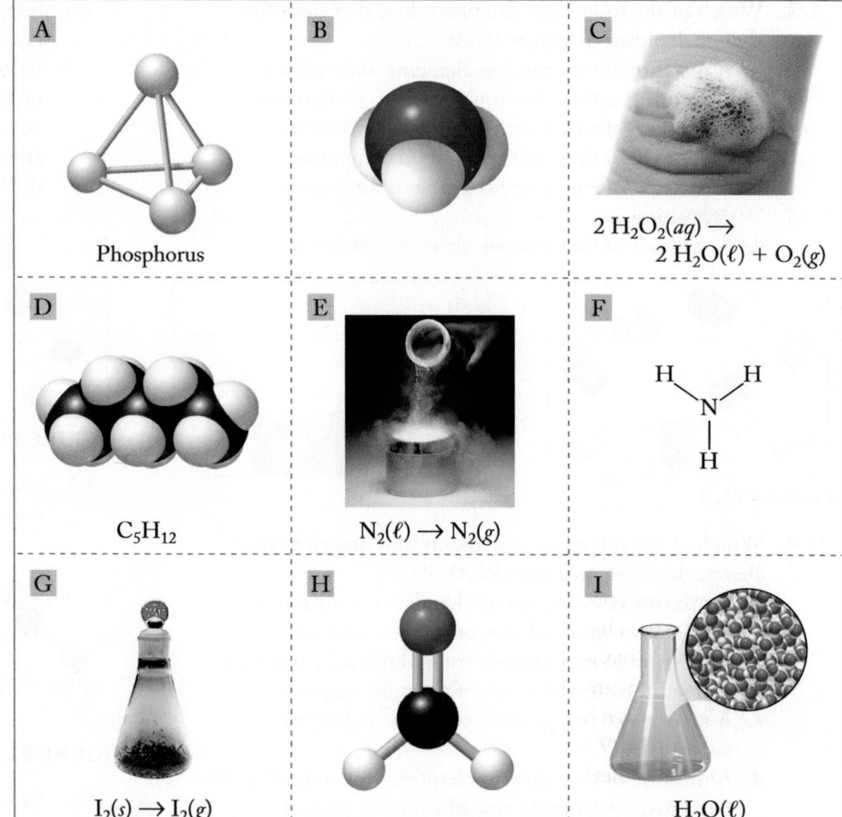

FIGURE P1.10

QUESTIONS AND PROBLEMS

Matter

Concept Review

1.11. Which of the following are homogeneous mixtures? (a) a gold wedding ring; (b) sweat; (c) bottled drinking water; (d) human blood; (e) compressed air in a scuba tank

1.12. Which of the following is a pure substance? (a) sweat; (b) blood; (c) air; (d) sucrose (table sugar); (e) milk

1.13. Which of the following is an element? (a) Cl_2; (b) H_2O; (c) HCl; (d) NaCl

1.14. Which of the following is *not* an element? (a) I_2; (b) O_3; (c) ClF; (d) S_8

1.15. Which of the following is a homogeneous mixture? (a) filtered water; (b) chicken noodle soup; (c) clouds; (d) trail mix snack; (e) fruit salad

1.16. Which of the following is a heterogeneous mixture? (a) air; (b) a sports drink (e.g., Gatorade); (c) muddy river water; (d) brass; (e) table salt (sodium chloride)

1.17. **Composition of the Atmosphere** Over 99% of the molecules in Earth's atmosphere are nitrogen and oxygen. What are their chemical formulas?

1.18. The first four elements in column 17 of the periodic table exist as diatomic molecules. What are their chemical formulas?

1.19. Among the compounds formed by nitrogen and oxygen are three that consist of molecules with (a) two atoms of nitrogen and one of oxygen, (b) two atoms of oxygen and one of nitrogen, and (c) one atom of each element. What are the chemical formulas of these compounds?

1.20. Among the multitude of compounds formed by carbon and hydrogen are three composed of molecules with (a) two atoms each of carbon and hydrogen, (b) six atoms each of carbon and hydrogen, and (c) one atom of carbon and four of hydrogen. What are the chemical formulas of these compounds?

1.21. **Acid Rain** In one of the reactions in the atmosphere that contributes to the formation of acid rain, molecules of water (H_2O) vapor and sulfur trioxide (SO_3) gas combine, forming a molecule of liquid sulfuric acid (H_2SO_4). Write a chemical equation describing this reaction (including the physical states of the reactants and product).

1.22. When dinitrogen pentoxide (N_2O_5) gas dissolves in water, molecules of N_2O_5 and H_2O combine, forming two molecules of nitric acid (HNO_3). Write a chemical equation describing this reaction (including the physical states of the reactants and product).

1.23. Which of the following describe physical properties and which describe chemical properties of the element sodium?
   a. Its density is greater than that of kerosene and less than that of water.
   b. It has a lower melting point than that of most metals.
   c. It is an excellent conductor of heat and electricity.
   d. It is soft and can be easily cut with a knife.
   e. Freshly cut sodium is shiny, but it rapidly tarnishes in contact with air.
   f. It reacts very vigorously with water to form hydrogen gas (H_2) and sodium hydroxide (NaOH).

1.24. Which of the following describe physical properties and which describe chemical properties of hydrogen gas (H_2)?
a. At room temperature, its density is less than that of any other gas.
b. It reacts vigorously with oxygen (O_2) to form water.
c. Liquefied H_2 boils at a very low temperature (−253°C).
d. H_2 gas does not conduct electricity.

1.25. Which of the following describe physical properties and which describe chemical properties of water?
a. Most, but not all, kinds of wood float in water.
b. Water has a remarkably high boiling point for a compound whose molecules consist of only three atoms, two of which are hydrogen.
c. Evaporation of the water in perspiration cools the skin.
d. A process called electrolysis can convert liquid water into hydrogen gas and oxygen gas.
e. Ice is less dense than liquid water.
f. Water is denser than gasoline.

1.26. Which of the following describe physical properties and which describe chemical properties of methane (the principal component of natural gas)?
a. Methane is lighter (less dense) than air.
b. Methane has a higher fuel value (energy produced per gram) than any other fossil fuel.
c. Methane can be liquefied for transport by ship if it is cooled to 109 K.
d. Liquefied methane is over 300 times denser than methane gas at its boiling point.
e. Leaking natural gas has caused explosions that destroyed peoples' homes.
f. Methane is colorless and odorless.

1.27. Which of the following are examples of chemical properties of formaldehyde (CH_2O)?
a. It has a characteristic acrid smell.
b. It is soluble in water.
c. It burns in air.
d. It is a gas at room temperature.
e. It is colorless.
f. Formaldehyde is used to synthesize many other compounds.

1.28. Which of the following is an example of a physical property of silver (Ag)?
a. Silverware slowly tarnishes over time.
b. Tarnished silverware can be restored to a shiny metallic finish if placed in a boiling hot aqueous solution of baking soda that also contains aluminum foil.
c. Silver coins in pirate treasure turn black when left in the sea for a few centuries.
d. Silver sinks in seawater.
e. Silver is an excellent conductor of electricity.
f. Silver is soft for a metal, though not as soft as gold.

1.29. Which of the following can be separated by filtration? (a) sugar dissolved in coffee; (b) sand and water; (c) octane from the other components in gasoline; (d) alcohol dissolved in water; (e) water vapor from the other gases in humid air

*1.30. Would filtration be a suitable way to separate dissolved proteins from blood plasma? Explain why or why not.

1.31. Examine Figure 1.12. In which physical state do particles have the greatest freedom of motion—solid, liquid, or gas? In which state do they have the least freedom of motion?

1.32. A pot of water on a stove is heated to a rapid boil. Identify the gas inside the bubbles that form in the boiling water.

1.33. A brief winter storm leaves a dusting of snow on the ground. During the sunny but very cold day after the storm, the snow disappears even though the air temperature never gets above freezing. If the snow didn't melt, where did it go?

1.34. What is in the space between the particles that make up a gas?

1.35. Can an extensive property be used to identify a substance? Explain why or why not.

1.36. Which of these properties of water are intensive and which are extensive?
a. The density of water at room temperature and pressure.
b. The temperature at which water freezes.
c. The mass of water in your body.
d. The mass of one molecule of water.
e. The rate at which water is flowing over Niagara Falls.

The Scientific Method: Starting Off with a Bang

Concept Review

1.37. What kinds of information are needed to formulate a hypothesis?

1.38. How does a hypothesis become a theory?

1.39. Is it possible to disprove a scientific hypothesis?

1.40. Is it possible to prove a theory?

1.41. Compare the meaning of the word *theory* as it is used in science with its use in normal conversation.

1.42. Why was the belief that matter consists of atoms considered a philosophy in ancient Greece, but a theory by the early 1800s?

Unit Conversions and Dimensional Analysis

Concept Review

1.43. Describe in general terms how the SI and USC systems of units differ.

1.44. Suggest two reasons why SI units are so widely used in science.

Problems

Note: The physical properties of the elements are listed in Appendix 3.

1.45. What is the mass of a magnesium block that measures 2.5 cm × 3.5 cm × 1.5 cm?

1.46. What is the mass of an osmium block that measures 6.5 cm × 9.0 cm × 3.25 cm? Do you think you could lift it with one hand?

1.47. Write the conversion factor(s) for each of the following unit conversions: (a) picoseconds (ps) to femtoseconds (fs); (b) kilograms (kg) to milligrams (mg); (c) the mass of a block of titanium in kilograms (kg) to its volume in cubic meters (m^3)

1.48. A single strand of natural silk may be as long as 4.0 km. What is this length in meters?

***1.49.** There are ten steps from the sidewalk up to the front door of a student's apartment. Each tread is 12.5 cm deep and 15.5 cm above the previous one. What is the distance diagonally from the bottom of the steps to the top in meters?

1.50. An Olympic-sized swimming pool is 50.00 m long and 25.0 m wide, and it must have a water depth of at least 2.0 m. What is the minimum volume of water in such a pool in cubic meters?

1.51. **Boston Marathon** To qualify to run in the 2019 Boston Marathon, a distance of 42.195 km, an 18-year-old woman had to have completed another marathon in 3 hours and 30 minutes or less. To qualify, what must this woman's average speed have been in meters per second?

1.52. **Olympic Mile** The length of a track race known as the "Olympic mile" is actually 1500. m. What percentage of a USC mile (5280. feet) is an Olympic mile?

***1.53.** If a wheelchair-marathon racer moving at 8.28 m/s expends energy at a rate of 665 Calories per hour, how much energy in Calories would be required to complete a marathon race (42.195 km) at this pace?

1.54. **Nearest Star** At a distance of 4.3 light-years, Proxima Centauri is the nearest star to our solar system. What is the distance to Proxima Centauri in kilometers?

1.55. What volume of gold would be equal in mass to a piece of copper with a volume of 125 cm^3?

*1.56. A small hot-air balloon is filled with 1.00×10^6 L of air ($d = 1.20$ g/L). As the air in the balloon is heated, it expands to 1.09×10^6 L. What is the density of the heated air in the balloon?

1.57. What is the volume of 1.00 kg of mercury?

1.58. A student wonders whether a piece of jewelry is made of pure silver. She determines that its mass is 3.17 g. She then drops it into a 10 mL graduated cylinder partially filled with water and determines that its volume is 0.3 mL. Could the jewelry be made of pure silver?

1.59. **The Density of Blood** The average density of human blood is 1.06 g/mL. What is the mass of blood in an adult with a blood volume of 5.5 L? Express your answer in (a) grams and (b) ounces.

1.60. **The Density of Earth** Earth has a mass of 5.98×10^{24} kg and a volume of 1.08×10^{12} km^3. What is the average density of our planet in units of grams per cubic centimeter?

Evaluating and Expressing Experimental Results

Concept Review

1.61. How many suspect data points can be identified from a data set by using Grubbs' test?

1.62. Which confidence interval is the largest for a given n value: 50%, 90%, or 95%?

1.63. The concentration of ammonia in an aquarium tank is determined each day for a week. When calculating the variability in the results, which will be larger: (a) standard deviation; (b) 95% confidence interval? Explain your selection.

1.64. During the analysis of a data set, if an outlier could not be found at the 95% confidence level, (a) could there be an outlier at the 90% confidence level? (b) Could there be an outlier at the 99% confidence level?

Problems

1.65. Which of these measured values has the fewest number of significant figures? (a) 545; (b) 6.4×10^{-3}; (c) 6.50; (d) 1.346×10^2

1.66. Which of these uncertain values has the greatest number of significant figures? (a) 545; (b) 6.4×10^{-3}; (c) 6.50; (d) 1.346×10^2

1.67. Which of these uncertain values has the fewest number of significant figures? (a) 1/545; (b) $1/(6.4 \times 10^{-3})$; (c) 1/6.50; (d) $1/(1.346 \times 10^2)$

1.68. Which of these uncertain values has the greatest number of significant figures? (a) 1/545; (b) $1/(6.4 \times 10^{-3})$; (c) 1/6.50; (d) $1/(1.346 \times 10^2)$

1.69. Which of these uncertain values has four significant figures? (a) 0.0592; (b) 0.08206; (c) 8.314; (d) 5420; (e) 5.4×10^3

1.70. Which of these uncertain values has only three significant figures? (a) 7.02; (b) 6.452; (c) 6.02×10^{23}; (d) 302; (e) 12.77

1.71. Perform each of the following calculations, and express the answer with the correct number of significant figures (only the highlighted values are exact):

a. $0.6274 \times 1.00 \times 10^3/[2.205 \times (2.54)^3] =$
b. $6 \times 10^{-18} \times (1.00 \times 10^3) \times 12 =$
c. $(4.00 \times 58.69)/(6.02 \times 10^{23} \times 6.84) =$
d. $[(26.0 \times 60.0)/43.53]/(1.000 \times 10^4) =$

1.72. Perform each of the following calculations, and express the answer with the correct number of significant figures (only the highlighted values are exact):

a. $[(12 \times 60.0) + 55.3]/(5.000 \times 10^3) =$
b. $(2.00 \times 183.9)/[(6.02 \times 10^{23}) \times (1.61 \times 10^{-8})^3] =$
c. $0.8161/[2.205 \times (2.54)^3] =$
d. $(9.00 \times 60.0) + (50.0 \times 60.0) + (3.00 \times 10^1) =$

1.73. Perform each of the following calculations, and express the answer with the correct number of significant figures and units (only the highlighted values are exact):

a. $[0.066\ \text{cm}^3 \times (6.67 \times 10^5\ \text{g/cm}^3)]/(2.838 - 2.211)\ \text{mL} =$

b. $(7.6\ \text{g} + 7.8\ \text{g} + 8.1\ \text{g} + 7.9\ \text{g} + 8.0\ \text{g})/5 =$

c. $(273 + 37.4)\ \text{K} \times (1.3806 \times 10^{-23}\ \text{J/K}) =$

d. $\sqrt{\dfrac{3(\frac{8.314\ \text{kgm}^2}{\text{s}^2\text{molK}})(273.15\ \text{K})}{0.0320\ \frac{\text{kg}}{\text{mol}}}} =$

1.74. Perform each of the following calculations, and express the answer with the correct number of significant figures and units (only the highlighted values are exact):

a. $(43\ \text{mg} + 38\ \text{mg} + 39\ \text{mg} + 37\ \text{mg} + 40\ \text{mg} + 41\ \text{mg})/6 =$

b. $(364.5\ \text{nm})\left(\dfrac{3^2}{3^2 - 2^2}\right) =$

c. $(2.998 \times 10^8\ \text{m/s})/[(656\ \text{nm}) \times \left(\dfrac{10^{-9}\ \text{m}}{1\ \text{nm}}\right)] =$

d. $(6.626 \times 10^{-34}\ \text{kg}\cdot\text{m}^2/\text{s})/[(9.109 \times 10^{-31}\ \text{kg}) \times (2.998 \times 10^8\ \text{m/s})] =$

***1.75.** The widths of copper lines in printed circuit boards must be close to a design value. Three manufacturers were asked to prepare circuit boards with copper lines that are 0.500 μm (micrometers) wide ($1\ \mu\text{m} = 1 \times 10^{-6}$ m). Each manufacturer's quality control department reported the following line widths on five sample circuit boards (given in micrometers):

Widths of Copper Lines (μm)

Manufacturer 1	Manufacturer 2	Manufacturer 3
0.512	0.514	0.500
0.508	0.513	0.501
0.516	0.514	0.502
0.504	0.514	0.502
0.513	0.512	0.501

a. What is the mean and standard deviation of the data provided by each manufacturer?

b. For which of the three sets of data does the 95% confidence interval include 0.500 μm?

c. Which of the data sets fit the description "precise and accurate," and which is "precise but not accurate"?

1.76. **Diabetes Test** Glucose concentrations in the blood above 110 mg/dL can be an early indication of several medical conditions, including diabetes. Suppose analyses of a series of blood samples from a patient at risk of diabetes produce these results: 106, 99, 109, 108, and 105 mg/dL.

a. What are the mean and the standard deviation of the data?

b. Patients with blood glucose levels above 120 mg/dL are considered diabetic. Is this value within the 95% confidence interval of these data?

1.77. Use Grubbs' test to decide whether the value 3.41 should be considered an outlier in the following data set from analyses of portions of the same sample conducted by six groups of students: 3.15, 3.03, 3.09, 3.11, 3.12, and 3.41.

1.78. Use Grubbs' test to decide whether any one of the values in this set of replicate measurements should be considered an outlier: 61, 75, 64, 65, 64, and 66.

Testing a Theory: The Big Bang Revisited

Concept Review

1.79. What is meant by an *absolute* temperature scale?

1.80. Can a temperature in °C ever have the same value in °F?

Problems

1.81. Silver and gold melt at 962°C and 1064°C, respectively. Convert these two temperatures to the Kelvin scale.

1.82. **Radiator Coolant** The coolant in an automobile radiator freezes at −39°C and boils at 110°C. What are these temperatures on the Fahrenheit scale?

1.83. **Critical Temperature** The discovery of new "high temperature" superconducting materials in the mid-1980s spurred a race to prepare the material with the highest superconducting temperature. The *critical temperatures* (T_c)—the temperatures at which the material becomes superconducting—of three such materials are −93.0 K, −250.0°C, and −231.1°F. Convert these temperatures into a single temperature scale, and determine which superconductor has the highest T_c value.

1.84. As air is cooled, which gas condenses first: N_2, He, or H_2O?

1.85. Liquid helium boils at 4.2 K. What is the boiling point of He in °C?

1.86. Liquid hydrogen boils at −253°C. What is the boiling point of H_2 on the Kelvin scale?

1.87. A person has a fever of 102.5°F. What is this temperature in °C?

1.88. Physiological temperature, or body temperature, is 37.0°C. What is this temperature in °F?

1.89. **Record Low** The lowest temperature measured on Earth is −128.6°F, recorded at Vostok, Antarctica, in July 1983. What is this temperature on the Celsius and Kelvin scales?

1.90. **Record High** The highest temperature ever recorded in the United States is 134°F at Greenland Ranch, Death Valley, CA, on July 13, 1913. What is this temperature on the Celsius and Kelvin scales?

Additional Problems

*1.91. Sodium chloride (NaCl) contains 1.54 g of Cl^- ions for every 1.00 g of Na^+ ions. Which of the following mixtures would react to produce sodium chloride with no sodium or chlorine left over?
a. 11.0 g of sodium and 17.0 g of chlorine
b. 6.5 g of sodium and 10.0 g of chlorine
c. 6.5 g of sodium and 12.0 g of chlorine
d. 6.5 g of sodium and 8.0 g of chlorine

1.92. Your laboratory instructor has given you two shiny, light gray metal cylinders. Your assignment is to determine which one is made of aluminum (d = 2.699 g/mL) and which one is made of titanium (d = 4.54 g/mL). The mass of each cylinder was determined on a scale to five significant figures. The volume was determined by immersing the cylinders in a graduated cylinder as shown in Figure P1.92. The initial volume of water was 25.0 mL in each graduated cylinder. The following data were collected:

	Mass (g)	Height (cm)	Diameter (cm)
Cylinder A	15.560	5.1	1.2
Cylinder B	35.536	5.9	1.3

a. Calculate the volume of each cylinder by using the dimensions of the cylinder only.
b. Calculate the volume from the water displacement method.
c. Which volume measurement allows for the greater number of significant figures in the calculated densities?
d. Express the density of each cylinder to the appropriate number of significant figures.

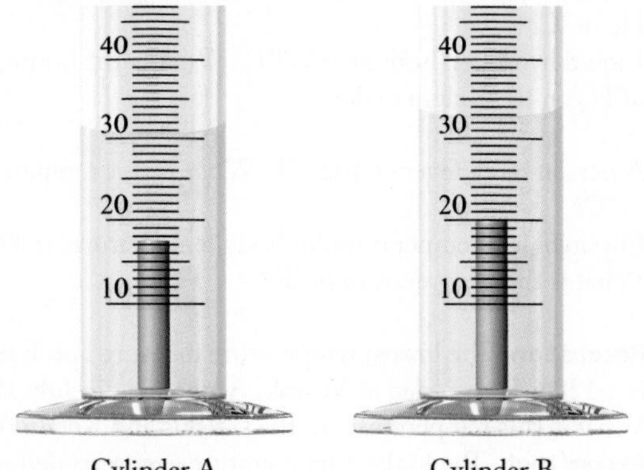

FIGURE P1.92

*1.93. Manufacturers of trail mix must control the distribution of the items in their products. Deviations of more than 2% outside specifications cause supply problems and downtime in the factory. A favorite trail mix is designed to contain 67% peanuts and 33% raisins. Bags of trail mix were sampled from the assembly line on different days. The bags were opened, and the contents counted, with the following results:

Day	Peanuts	Raisins
1	50	32
11	56	26
21	48	34
31	52	30

a. Calculate the mean and standard deviation in the percentage of peanuts and percentage of raisins in the four samples.
b. Do the 90% confidence intervals for these percentages include the target composition values: 67% peanuts and 33% raisins?

*1.94. Gasoline and water do not mix. Regular grade (87 octane) gasoline has a lower density (0.73 g/mL) than water (1.00 g/mL). A 100 mL graduated cylinder with an inside diameter of 3.2 cm contains 34.0 g of gasoline and 34.0 g of water. What is the combined height of the two liquid layers in the cylinder? The volume of a cylinder is $\pi r^2 h$, where r is the radius and h is the height.

1.95. At a natural history museum, a child was heard stating that dinosaur bones on display at the museum were 68,000,002 years old. When asked about this unusual age value, the child explained that during a visit two years earlier a tour guide had said the bones were 68 million years old. How would you explain to the child the meaning (and certainty) of a value such as 68 million years?

1.96. Ms. Goodson's geology classes are popular because of their end-of-the-year field trips. Some last several days, but all involve exactly eight hours of hiking per day. On one three-day trip the class's average hiking speeds were 1.6 mi/h, 1.4 mi/h, and 1.7 mi/h each day. What was the length of their trip in miles and kilometers?

*1.97. **Toothpaste Chemistry** Many popular brands of toothpaste contain about 1.00 mg of fluoride per gram of toothpaste. The fluoride compound that is most often used in toothpaste is sodium fluoride, NaF, which is 45% fluoride by mass. How many milligrams of NaF are in a 178-gram tube of toothpaste?

*1.98. **Test for HIV** Tests called ELISAs (enzyme-linked immunosorbent assays) detect and quantify substances such as HIV antibodies in biological samples. A "sandwich" assay traps the HIV antibody between two other molecules. The trapping event causes a detector molecule to change color. To make a sandwich assay for HIV, you need the following components: one plate to which the molecules are attached; a 0.550-mg sample of the recognition molecule that "recognizes" the HIV antibody; 1.200 mg of the capture molecule that "captures" the HIV antibody in a sandwich; and 0.450 mg of the detector molecule that produces a visible color when the HIV antibody is captured. You need to make 96 plates for an assay. You are given the following quantities of material: 100.00 mg of the recognition molecule; 100.00 mg of the capture molecule; and 50.00 mg of the detector molecule.

a. Do you have sufficient material to make 96 plates?
b. If you do, how much of each material is left after 96 sandwich assays are assembled? If you do not have sufficient material to make 96 assays, how many assays can you assemble?

1.99. Vitamin C Some people believe that large doses of vitamin C can cure the common cold. One commercial over-the-counter product consists of 500.0-mg tablets that are 20% by mass vitamin C. How many tablets are needed for a 1.00-g dose of vitamin C?

*1.100. **Patient Data** Measurements of a patient's temperature are routinely done several times a day in hospitals. Digital thermometers are used, and it is important to evaluate new thermometers and select the best ones. The accuracy of these thermometers is checked by immersing them in liquids of known temperature. Such liquids include an ice–water mixture at 0.0°C and boiling water at 100.0°C at exactly 1 atmosphere pressure (boiling point varies with atmospheric pressure). Suppose the data shown in the following table were obtained on three available thermometers and you were asked to choose the "best" one of the three. Which would you choose? Explain your choice.

Thermometer	Measured Temperature of Ice Water, °C	Measured Temperature of Boiling Water, °C
A	0.8	99.9
B	0.3	99.8
C	0.3	100.3

2

Atoms, Ions, and Molecules

Matter Starts Here

BABY TEETH A young child proudly holds a tooth that has just fallen out. A landmark study involving baby teeth showed that children living hundreds, even thousands, of miles away from atomic bomb test sites in the 1950s were exposed to and accumulated radioactive nuclides (radionuclides) produced during the tests.

PARTICULATE **REVIEW**

Elements versus Compounds

In Chapter 2 we begin to explore the structure of atoms and to discuss why they form the molecules and ions they do. All pure substances can be classified as either elements or compounds.

- Write the chemical formula for each substance shown here.
- Identify each substance as either an element or a compound.

(Review Section 1.2 if you need help.)

(Answers to Particulate Review questions are in the back of the book.)

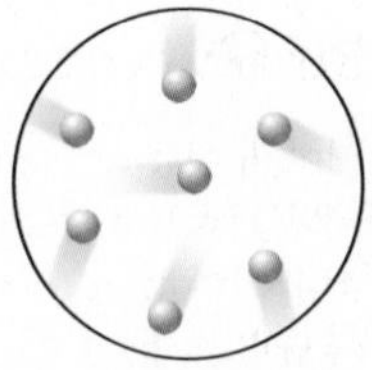

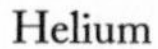

Helium

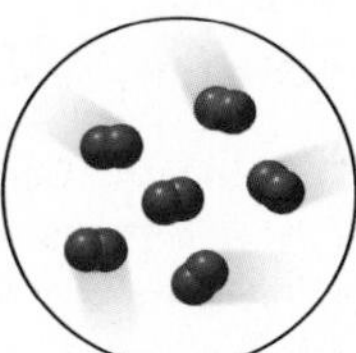

Nitrogen

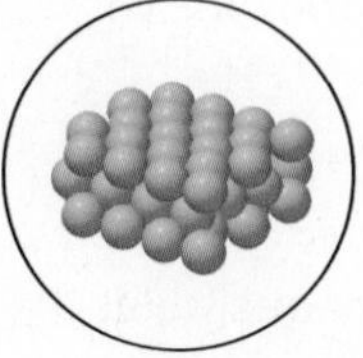

Gold

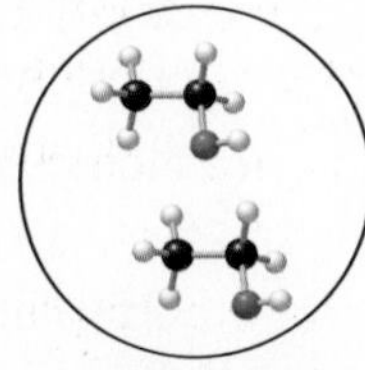

Ethanol

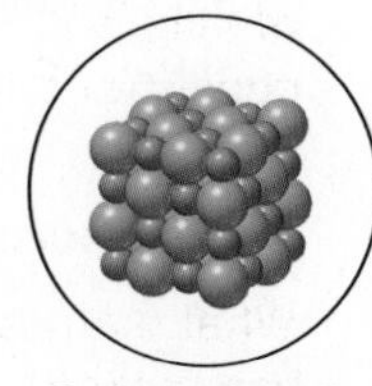

Sodium chloride

PARTICULATE **PREVIEW**

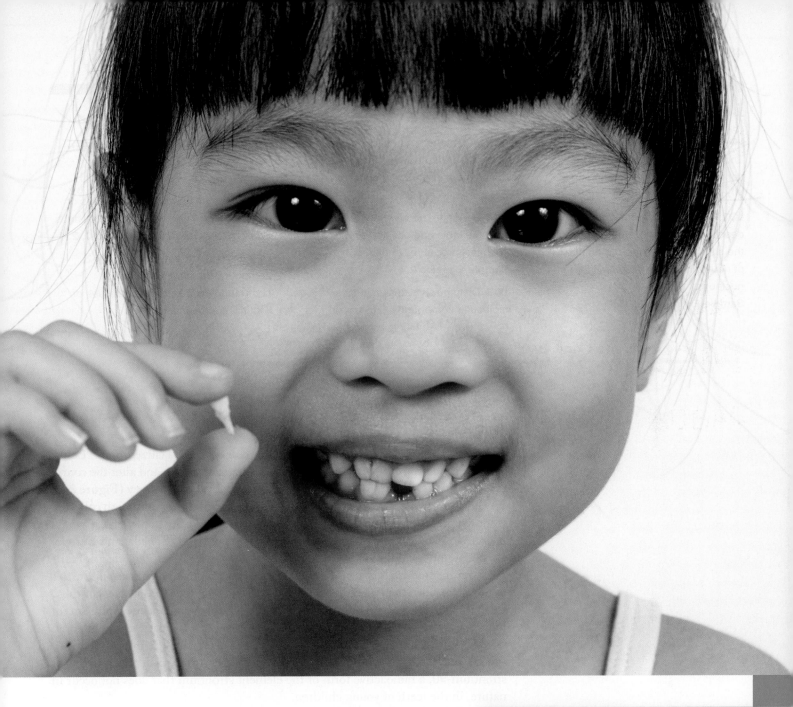

Ionic versus Molecular Compounds

Some compounds, such as NaCl, are composed of negatively and positively charged ions. Others are composed of molecules that contain the atoms of different elements held together by shared pairs of electrons called covalent bonds. As you read Chapter 2, look for ideas that will help you answer these questions:

- Which of the compounds represented here consist of different atoms bonded together and which consist of different ions bonded together?
- Which compound(s) are molecular? Which compound(s) are ionic?
- What features of these particulate images helped you make these distinctions?

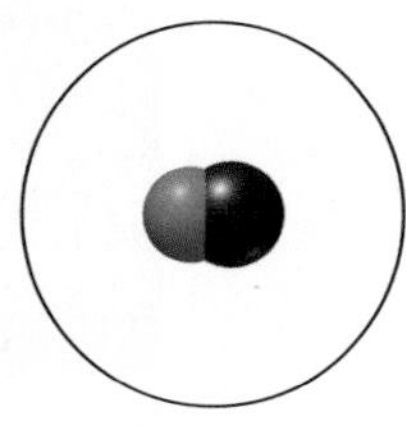

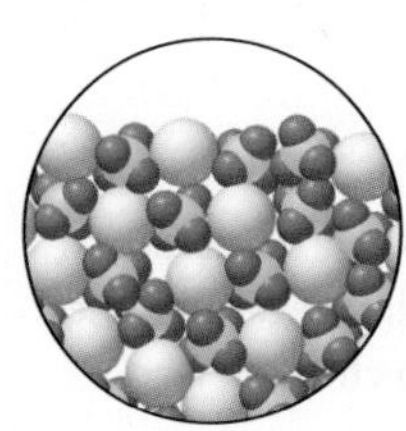

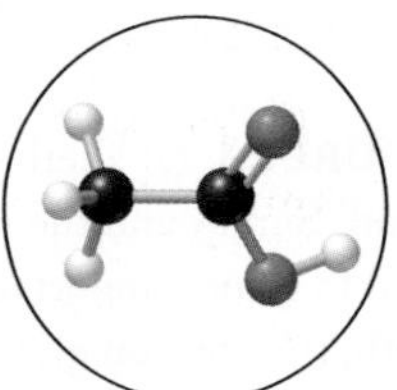

Learning Outcomes

LO1 Explain how the experiments of Thomson, Millikan, and Rutherford contributed to our understanding of atomic structure

LO2 Use the symbol of a nuclide to describe the composition of its nucleus: the number of protons, the number of neutrons, and the total number of nucleons
Sample Exercises 2.1, 2.2

LO3 Relate the natural abundance values of the stable isotopes of an element to the element's average atomic masses
Sample Exercise 2.3

LO4 Describe the development of the periodic table and use it to predict the chemical properties of elements
Sample Exercise 2.4

LO5 Distinguish between ionic and molecular compounds and explain how the law of multiple proportions is connected to the formulas of common molecular compounds
Sample Exercises 2.5 and 2.6

LO6 Relate the names and formulas of common molecular and ionic compounds
Sample Exercises 2.7, 2.8, 2.9, 2.10, 2.11

LO7 Distinguish between organic compounds and other molecular compounds and among the classes of organic compounds

LO8 Describe the origin of the elements

2.1 Atoms in Baby Teeth

In 1958 a group of physicians, scientists, and other concerned citizens in St. Louis, MO, began a project that would last for more than a decade—and alter the course of world events. Their project was known as the Baby Tooth Survey (**Figure 2.1**). It would eventually involve the analyses of more than 300,000 baby teeth collected from dentists' offices and young families in and around St. Louis, and it would inspire similar studies in other regions of the United States, Canada, and Europe.

The founders of the Baby Tooth Survey were worried about the effects of fallout from the hundreds of tests of nuclear weapons that had taken place in Earth's atmosphere since the end of World War II. They knew that the products of these tests included atoms exhibiting **radioactivity**, the spontaneous emission of high-energy radiation and particles. They also knew that radioactive fallout posed serious risks to human health. Their goal was to determine the impact of atmospheric nuclear testing on everyday citizens by measuring the concentration of strontium-90, a radioactive form of the element strontium that does not occur in nature, in the teeth of young children.

The Baby Tooth Survey focused on strontium-90 because of its radioactivity *and* because of its chemistry. Look at the periodic table of the elements on the inside front cover of this book and find Sr, the symbol for strontium. It is just below the one for calcium, Ca. Elements in the same column of the periodic table have similar chemical properties. Sometimes they can even substitute for each other in the compounds they form. This is the case for Sr^{2+} and Ca^{2+} ions, which have the same 2+ charge, as do all the ions formed by the elements in the second column. Thus, ions

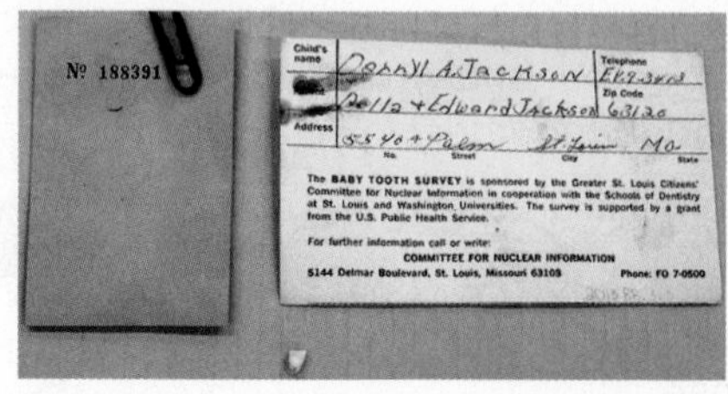

(a)

(b)

FIGURE 2.1 (a) Teeth collected in the Baby Tooth Survey and the information card used to record important data regarding the teeth; (b) a button given to children who donated their teeth.

of strontium-90 can accumulate in the rapidly growing, calcium-rich teeth and bones of young children. The baby teeth lost by the children of St. Louis provided convenient, noninvasive biomarkers of their exposure to radioactive fallout.

The results of the survey grabbed worldwide attention. They showed that radioactivity from strontium-90 in the teeth of children born in the 1950s and 60s was 100 times greater than in baby teeth that formed before the dawn of the nuclear age in 1945, and that the amount of strontium-90 correlated with the frequency of atomic bomb testing. Early results from the Baby Tooth Survey, coupled with other data on the incidence of childhood cancer, helped persuade U.S. President John F. Kennedy to negotiate a treaty with the Soviet Union and other nuclear powers to end atmospheric testing of nuclear weapons in 1963.

In this chapter we explore the structure and composition of atoms, including the subatomic particles that cause radioactivity and the subatomic particles that are responsible for ionic charges. We begin with the discoveries made in the late 19th and early 20th centuries that led to our understanding of atomic structure and that helped explain the chemical properties of the elements, including their tendencies to form ions. We end the chapter by exploring how hydrogen and helium formed just after the Big Bang, and we explain how other elements have continued to form both in the cores of giant stars and in celestial explosions called supernovas. These are the atoms that make up the entire universe, including all the matter in our world and in our bodies.

radioactivity the spontaneous emission of high-energy radiation and particles by materials.

cathode rays streams of electrons emitted by the cathode in a partially evacuated tube.

CHEMTOUR
Cathode-Ray Tube

2.2 Discovering the Structure of Atoms

By the end of the 19th century, many scientists had realized that atoms were not the smallest particles of matter, but rather consisted of even smaller *subatomic* particles. This realization came in part from the research of the British scientist Joseph John (J. J.) Thomson (1856–1940; **Figure 2.2**). However, not even Thomson himself recognized the fundamental importance of his work for the future of science and technology.

FIGURE 2.2 J. J. Thomson (1856–1940) discovered electrons in 1897 by using a cathode-ray tube, but he was not sure where electrons fit into the structure of atoms.

Electrons

Figure 2.3 shows the apparatus Thomson used in his experiments. It is called a cathode-ray tube (CRT), and it consists of a glass tube from which most of the air has been removed. Electrodes within the tube are attached to the poles of a high-voltage power supply. One electrode, called the cathode, is connected to the negative terminal of the power supply, whereas the other electrode, called the anode, is connected to the positive terminal. When these connections are made, electricity passes through the glass tube in the form of a beam of **cathode rays** emitted by the cathode. Cathode rays are invisible to the naked eye, but when the end of the CRT opposite the cathode is coated with a phosphorescent material, a glowing spot appears where the beam hits it.

STEPWISE ANIMATION
Parts of the Atom

Thomson found that cathode-ray beams were deflected by magnetic fields (Figure 2.3a) and by electric fields (Figure 2.3b). The directions of these deflections established that cathode rays were not rays of energy, but rather consisted of negatively charged particles. By adjusting the strengths of the electric and magnetic fields, Thomson was able to balance out the deflections (Figure 2.3c). From the strengths of the two opposing fields, he calculated the mass-to-charge (m/e) ratio of the particles. Thomson and others observed that these particles always

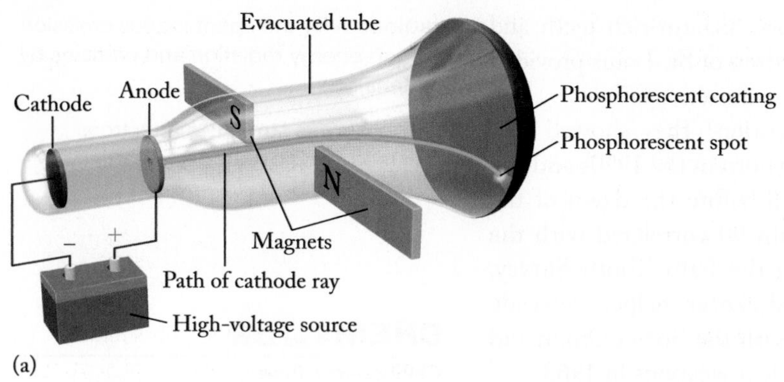

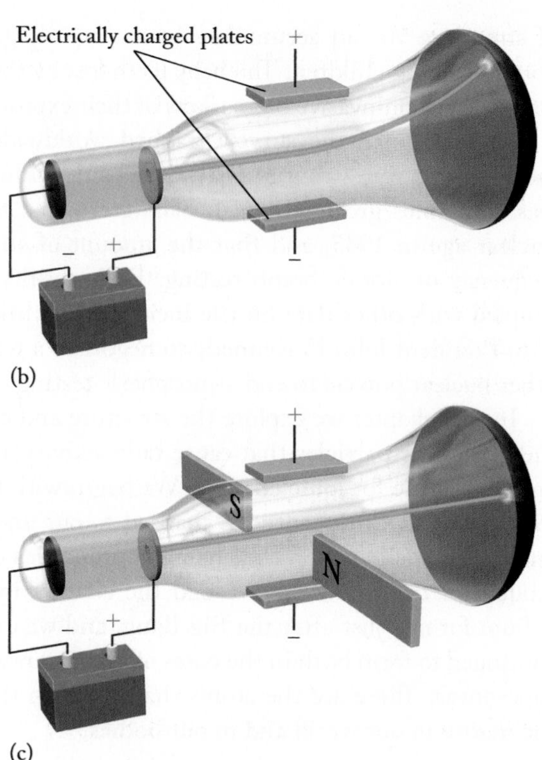

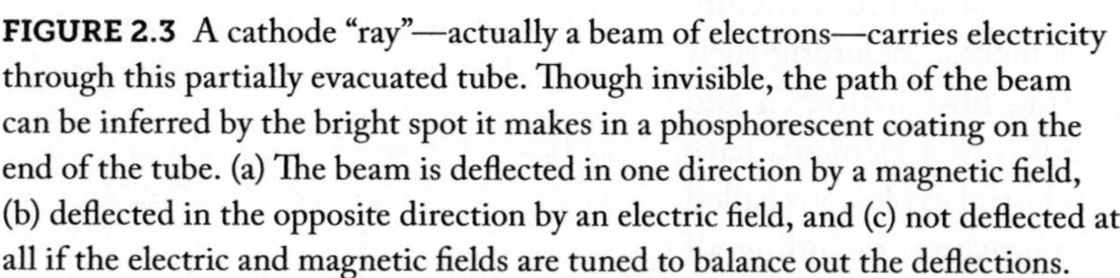

FIGURE 2.3 A cathode "ray"—actually a beam of electrons—carries electricity through this partially evacuated tube. Though invisible, the path of the beam can be inferred by the bright spot it makes in a phosphorescent coating on the end of the tube. (a) The beam is deflected in one direction by a magnetic field, (b) deflected in the opposite direction by an electric field, and (c) not deflected at all if the electric and magnetic fields are tuned to balance out the deflections.

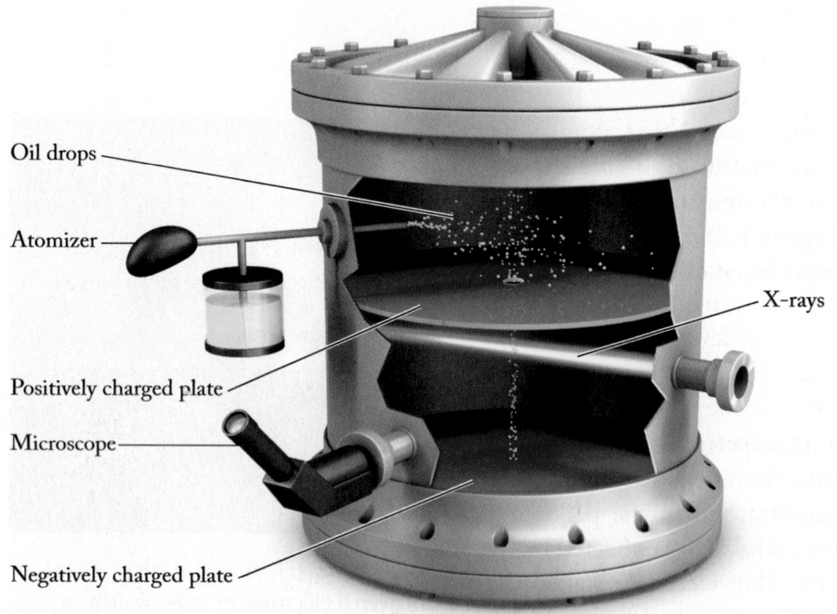

FIGURE 2.4 Millikan's oil-drop experiment.

CHEMTOUR
Millikan Oil-Drop Experiment

electron a subatomic particle that has a negative charge and little mass.

beta (β) particle a radioactive emission equivalent to a high-energy electron.

alpha (α) particle a radioactive emission with a charge of 2+ and a mass equivalent to that of a helium nucleus.

behaved the same way and always had the same mass-to-charge ratio, no matter what cathode material was used. This observation established that the particles in the cathode rays, which are now known as **electrons**, were fundamental particles present in all forms of matter.

In 1909 the American physicist Robert Millikan (1868–1953) advanced Thomson's work by quantifying the charge on the electron. Thomson had already calculated the mass-to-charge ratio of the electron; so, if Millikan could measure the charge of the electron, then he could determine its mass. **Figure 2.4** shows the components of Millikan's experimental setup. In it, high-energy X-rays remove electrons from molecules found in air in the lower of two connected chambers. As oil drops fall from the upper chamber into the lower one, they collide with the electrons removed from the air molecules and pick up their negative charge. Millikan measured the mass of the drops in the absence of an electric field, when their rate of fall was governed by gravity. He then repeated the experiment in the presence of an electric field, adjusting the field strength to make the drops fall at different rates and even suspending some of them in midair. From the strength of the electric field and the rate of fall, he calculated the charge on a drop. By measuring the charges on hundreds of drops, Millikan determined that the charge on each drop was a whole-number multiple of a minimum charge. He concluded that this minimum charge had to be the charge on one electron. Millikan's value was within 1% of the modern value: 1.602×10^{-19} C. (The coulomb, abbreviated C, is the SI unit for electric charge.) Knowing Thomson's value of m/e for the electron, Millikan calculated the mass of the electron: 9.109×10^{-28} g.

The discovery of the electron raised the possibility of other subatomic particles. Scientists in the 1890s knew that matter was electrically neutral, but they did not know how the electrons and the positive charges were arranged at the atomic level. Thomson proposed a *plum-pudding model* in which the atom was a diffuse sphere of positive charge with negatively charged electrons embedded in the sphere, like raisins in an English plum pudding (or blueberries in a muffin; **Figure 2.5**). Thomson's model did not last long. Its demise was linked to another scientific discovery in the 1890s: radioactivity.

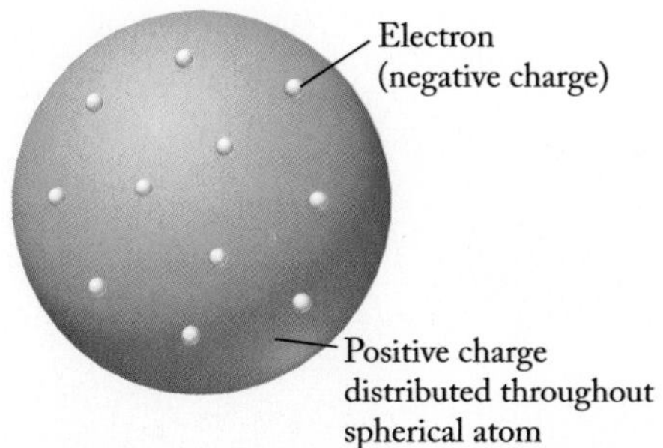

FIGURE 2.5 In J. J. Thomson's plum-pudding model, atoms consist of electrons distributed throughout a massive, positively charged but very diffuse sphere. The plum-pudding model was replaced after only a few years by a model based on experiments carried out under the direction of Ernest Rutherford, a former student of Thomson's.

CONCEPT **TEST**

Why did Thomson's discovery of the electron lead others to propose that a positive particle might exist within the atom?

(Answers to Concept Tests are in the back of the book.)

Radioactivity

In 1896 the French physicist Henri Becquerel (1852–1908) discovered that a mineral then called *pitchblende* (now known as *uraninite*; **Figure 2.6**), the principal natural source of the element uranium, produces radiation that can be detected on photographic plates. Becquerel and his contemporaries initially thought that this radiation consisted of the X-rays that had just been discovered by the German scientist Wilhelm Conrad Röntgen (1845–1923).[1] However, additional experiments by Becquerel, by the French wife-and-husband team of Marie Skłodowska Curie (1867–1934; **Figure 2.7**) and Pierre Curie (1859–1906), and by the British scientist Ernest Rutherford (1871–1937) showed that this radiation contained particles as well as rays. The scientists had discovered radioactivity.

FIGURE 2.6 The mineral now known as uraninite is the chief natural source of uranium.

In studying the particles emitted by pitchblende, Rutherford found that one type, which he named **beta (β) particles**, penetrated solid materials better than the second type, which he named **alpha (α) particles**. The degree to which β particles were deflected in a magnetic field allowed their mass-to-charge ratio to be calculated, and the results matched the electron mass-to-charge ratio determined by J. J. Thomson. These data established that β particles were simply high-energy electrons.

Rutherford discovered that α particles were deflected by an electric field or a magnetic field in the opposite direction from β particles. He concluded, therefore, that α particles were positively charged. The β particle (electron) was assigned a relative charge of 1−, whereas the corresponding charge of an α particle was 2+. In addition, α particles had nearly the same mass as an atom of helium, meaning they were over 10^3 times more massive than β particles.

The experiment that disproved Thomson's plum-pudding model was directed by Rutherford and carried out by two of his students at Manchester University: Hans Geiger (1882–1945; for whom the Geiger counter was named) and Ernest Marsden (1889–1970). Geiger and Marsden bombarded a thin foil of gold with a beam of α particles emitted from a radioactive source (**Figure 2.8a**). They then

FIGURE 2.7 The brilliant career of Marie Curie was honored by her being the only person to win Nobel prizes in two sciences: chemistry and physics.

[1]Röntgen discovered X-rays in experiments with a cathode-ray tube much like the apparatus used by J. J. Thomson. After encasing the tube in a black carton, Röntgen discovered that invisible rays escaped the carton and were detected by a photographic plate. Because he knew so little about these rays, he called them X-rays.

measured how many particles were deflected and to what extent. Rutherford's hypothesis was that if Thomson's model were correct, most of the α particles would pass straight through the diffuse positive spheres of gold atoms (each atom a "pudding" like the one shown in Figure 2.5), but a few of the particles would interact with the electrons (the "raisins") embedded in these spheres and be deflected slightly (**Figure 2.8b**).

Instead, Geiger and Marsden observed something quite unexpected. Although most of the α particles did indeed pass directly through the gold, about 1 in every 8000 particles was deflected from the foil through an average angle of 90 degrees. Even more surprising, a very few (about 1 out of 100,000) bounced back in the direction from which the α particles came (**Figure 2.8c**). Rutherford described his amazement at the result: "It is about as incredible as if you had fired a 15-inch shell at a piece of tissue paper and it came back and hit you." (The largest guns on British battleships in those days fired 15-inch-diameter shells.)

The results of the gold-foil experiments ended the short life of the plum-pudding model because the model could not account for the large angles of deflection. Rutherford concluded that those deflections occurred because the α particles occasionally encountered small, yet massive, regions of highly positive charge. Rutherford determined that the region of positive charge was only about 1/10,000 of the overall size of a gold atom. His model of the atom became the basis for our current understanding of atomic structure. It incorporates the assumption that an atom consists of a massive, but tiny, positively charged **nucleus** surrounded by a diffuse cloud of negatively charged electrons.

CHEMTOUR
Rutherford Experiment

Protons and Neutrons

In the decade following the gold-foil experiments, Rutherford and others observed that bombarding elements with α particles could change, or *transmute*, these elements into other elements. They also discovered that hydrogen nuclei were sometimes produced during transmutation reactions. By 1920, agreement was growing that hydrogen nuclei, which Rutherford called **protons** (from the Greek *protos*, meaning "first"), were part of all nuclei. For example, to account for both the mass and charge of an α particle, Rutherford assumed it was made of four protons, two of which had combined with two electrons to form two electrically neutral

Particles deflected by collisions with nucleus
Undeflected particles passing straight through
Beam of α particles
Thin gold foil
α emitter
Fluorescent screen
(a)

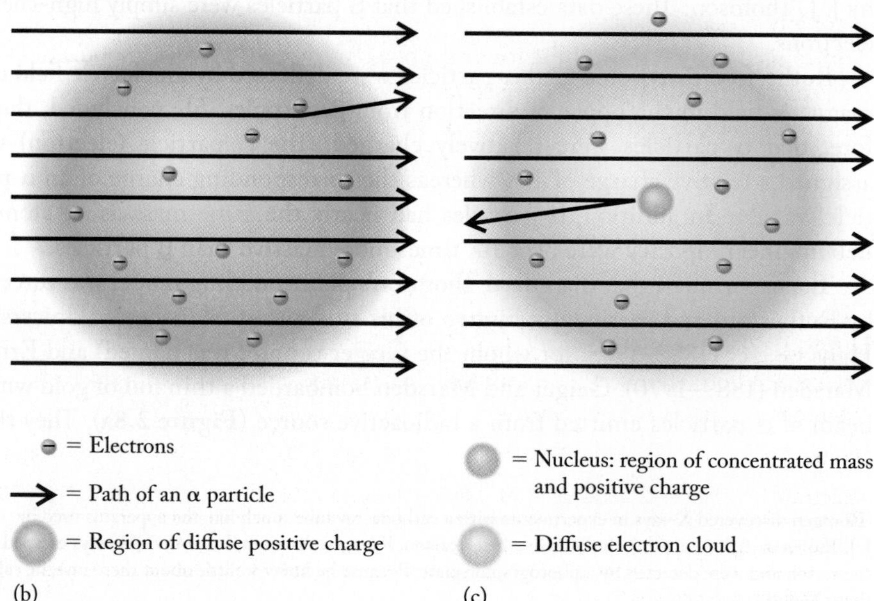

FIGURE 2.8 (a) In the Rutherford–Geiger–Marsden experiment, α particles from a radioactive source were allowed to strike a thin gold foil. A fluorescent screen surrounded the foil to detect any deflected particles. (b) If Thomson's plum-pudding model were correct, most of the α particles would pass through the gold foil and a few would be deflected slightly. (c) In fact, most particles passed straight through, but a few were scattered widely. This unexpected result led to the theory that an atom has a small, positively charged nucleus that contains most of the mass of the atom.

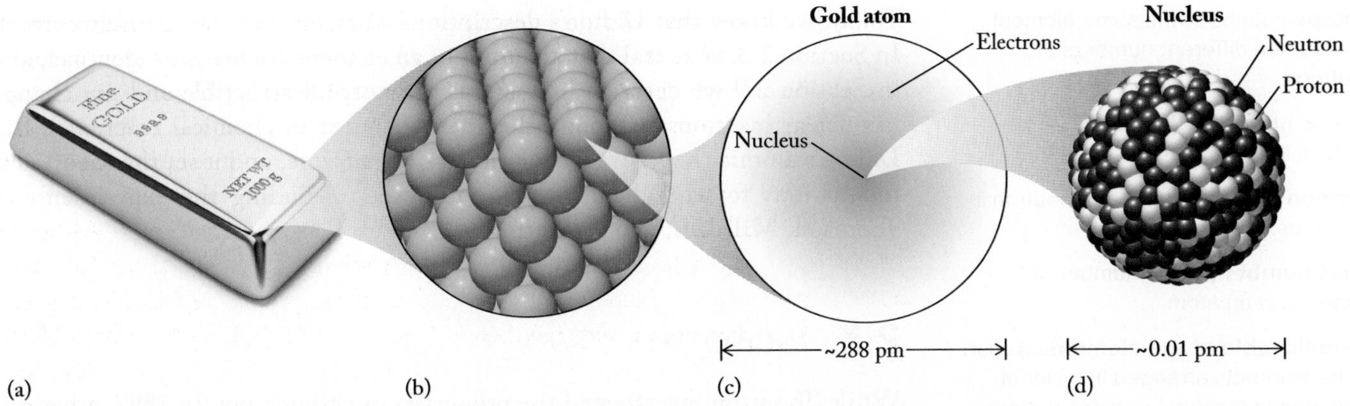

FIGURE 2.9 (a) A gold bar and (b) a particulate view of how the atoms are arranged within it. (c) The modern view of Rutherford's model of the gold atom includes electrons that surround (d) the nucleus, which is about 1/10,000 the overall size of the atom. The scales have units of picometers (1 pm = 10^{-12} m). The nucleus would be too small to see if drawn to scale in (c) and (d). If an atom were the size of the Rose Bowl (an oval stadium about 212 m across in the narrow direction), the nucleus of the atom would be the size of a dime at midfield.

particles that he called **neutrons**. Repeated attempts to produce neutrons by neutralizing protons with electrons were unsuccessful. In 1932, however, one of Rutherford's students, James Chadwick (1891–1974), discovered and characterized neutrons. With the discovery of neutrons, the current model of atomic structure was complete, as illustrated for the gold atom in **Figure 2.9**.

Table 2.1 summarizes the properties of neutrons, protons, and electrons. For convenience, the masses of these tiny particles are expressed in **unified atomic mass units (u)**. These units are also called **daltons (Da)**. One u is exactly 1/12 the mass of a carbon atom that has six protons and six neutrons in its nucleus. As you can see from the data in Table 2.1, the masses of the neutron and the proton are nearly the same and very close to 1 u. If you compare the u values in the table with the masses of the particles in grams, you will see that these particles are tiny indeed: 1 u is only 1.66054×10^{-24} g.

The unit name "dalton" honors the English chemist John Dalton (1766–1844), who published the first table of atomic masses in 1803. Dalton is also remembered for his *atomic theory*, published in 1808, in which he proposed that:

1. Each element consists of tiny, indestructible particles called atoms.
2. All atoms of an element are identical, and they are different from the atoms of any other element.
3. Atoms combine in simple ratios of whole numbers to form compounds. (We explore this point further in Section 2.6.)
4. Atoms are not changed, created, or destroyed in chemical reactions; a reaction only changes the arrangement in which the atoms are bound.

TABLE 2.1 Properties of Subatomic Particles

		MASS		CHARGE	
Particle	Symbol	In Atomic Mass Units (u)	In Grams (g)	Relative Value	Charge (C[a])
Neutron	$^{1}_{0}n$	1.00867 ≈ 1	1.67493×10^{-24}	0	0
Proton	$^{1}_{1}p$	1.00728 ≈ 1	1.67262×10^{-24}	1+	$+1.602 \times 10^{-19}$
Electron	$^{0}_{-1}e$	$5.485799 \times 10^{-4} \approx 0$	9.10939×10^{-28}	1−	-1.602×10^{-19}

[a]The *coulomb* (C) is the SI unit of electric charge. When a current of 1 ampere (see Table 1.2) passes through a conductor for 1 second, the quantity of electric charge that moves past any point in the conductor is 1 C.

nucleus the positively charged center of an atom that contains nearly all the atom's mass.

proton a positively charged subatomic particle present in the nucleus of an atom.

neutron an electrically neutral (uncharged) subatomic particle found in the nucleus of an atom.

unified atomic mass unit (u) the unit used to express the relative masses of atoms and subatomic particles; it is exactly 1/12 the mass of one atom of carbon with six protons and six neutrons in its nucleus.

dalton (Da) a unit of mass identical to 1 unified atomic mass unit (u); thus, 1 Da = 1 u.

isotopes atoms of the same element that contain different numbers of neutrons.

atomic number (*Z*) the number of protons in the nucleus of an atom.

nucleon either a proton or a neutron in a nucleus.

mass number (*A*) the number of nucleons in an atom.

periodic table of the elements a chart of the elements arranged in order of their atomic numbers and in a pattern based on their physical and chemical properties.

Today, we know that Dalton's descriptions of atoms were not entirely correct: In Section 2.3 we reveal that all atoms of an element are *not quite* identical, and in Section 2.9 we describe how atoms are not indestructible and *can* change in nuclear reactions (although they never change in chemical reactions). But Dalton's atomic theory was revolutionary for its time, and it set the theoretical foundations for the chemistry that followed, including the experiments of Thomson, Millikan, and Rutherford.

2.3 Isotopes

While Thomson investigated the properties of cathode rays in 1897, other scientists designed and built devices to produce beams of *positively* charged particles. One of Thomson's former students, Francis W. Aston (1877–1945), built modified cathode-ray tubes that were evacuated except for small quantities of *fill gases* such as neon. With these tubes he detected conventional beams of cathode rays, but he also detected secondary beams of positively charged particles. Charge was not the only thing different about the particles in these secondary beams. Whereas cathode rays are streams of electrons that all have the same mass and charge regardless of the cathode material or fill gas, the masses of the particles composing Aston's positive rays did depend on the identity of the fill gas. Aston's positively charged particles were not individual protons, but rather atoms of the fill gas that had lost electrons to form positively charged ions.

Aston used his positive-ray analyzer (**Figure 2.10**) to pass positively charged beams of particles through a magnetic field. Each particle in the beam was deflected along a path determined by the particle's mass: the greater the mass, the smaller the deflection. Using the purest sample of neon gas available, Aston determined that most of the particles had a mass of 20 u, but about 1 in 10 had a mass of 22 u.

STEPWISE ANIMATION
Simplified Mass Spectrometer

Since the time of John Dalton, scientists had believed that each element was composed of identical atoms, each having the same mass. Aston's research contradicted this long-held idea. To explain his data, Aston proposed that neon consists of two kinds of atoms, or **isotopes**. Both isotopes of neon had the same number of protons (10) in the nucleus, but one isotope had 10 neutrons in its nucleus, giving it a mass of 20 u, whereas the other isotope had 12 neutrons in its nucleus, giving it a mass of 22 u.

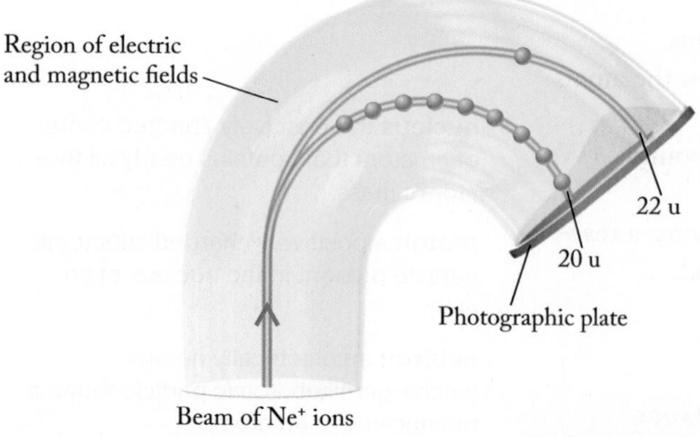

Aston's work showed that each element is in fact composed of atoms each having the *same number of protons* in its nucleus, but not necessarily the same number of neutrons, and therefore not necessarily the same mass. The number of protons is called the **atomic number (*Z*)** of an atom and identifies it as an atom of a particular element. For example, all atoms of neon have 10 protons in each of their nuclei. The total number of **nucleons** (neutrons and protons) in the nucleus of an atom defines its **mass number (*A*)**. Thus, isotopes of a given element all have the *same atomic number*, *Z*, but *different mass numbers*, *A*. A neutral atom has the same number of electrons as protons in its nucleus. The modern **periodic table of the elements** (inside the front cover of this book) displays the 118 known elements in order of atomic number.

FIGURE 2.10 Aston's positive-ray analyzer. A beam of positively charged ions of neon gas is passed through a focusing slit into a region of electric and magnetic fields. The ions are separated according to mass: those with a mass of 20 u—90% of the sample—hit the detector at one spot, and those with a mass of 22 u—the remaining 10%—hit the detector at a different spot. Aston's positive-ray analyzer was the forerunner of the modern mass spectrometer.

An atom with a specific combination of neutrons and protons is called a **nuclide**. The general symbol for identifying a nuclide is

$$^{A}_{Z}\text{X}$$

nuclide an atom with particular numbers of neutrons and protons in its nucleus.

where X represents the one- or two-letter symbol for the element. For example, the two isotopes of neon identified by Aston have the symbols:

$$^{20}_{10}\text{Ne} \quad ^{22}_{10}\text{Ne}$$

Because Z and X provide the same information—each by itself identifies the element—the subscript Z is frequently omitted. The isotope symbol is then simply written as ^{A}X (e.g., ^{20}Ne and ^{22}Ne for Aston's isotopes). This same information—mass number and element name—may also be spelled out. For example, the names of the two isotopes of neon that Aston discovered may be written as neon-20 and neon-22.

CONCEPT TEST

The radioactive atoms measured in the Baby Tooth Survey were strontium-90. Only one of the nuclides below is an isotope of strontium. Which one is it and why?

$$^{87}_{38}\text{X} \quad ^{90}_{40}\text{X} \quad ^{234}_{90}\text{X}$$

(Answers to Concept Tests are in the back of the book.)

SAMPLE EXERCISE 2.1 Writing Nuclide Symbols — LO2

Write symbols in the form $^{A}_{Z}$X for the nuclides that have (a) 6 protons and 6 neutrons, (b) 11 protons and 12 neutrons, and (c) 92 protons and 143 neutrons.

Collect, Organize, and Analyze We know the number of protons and neutrons in the nucleus of each nuclide. We need to write symbols in the $^{A}_{Z}$X form, where Z is the atomic number, A is the mass number, and X is the symbol of the element. The number of protons in the nucleus of an atom defines its atomic number (Z) and defines which element it is (X). The sum of the nucleons (protons plus neutrons) is the mass number (A).

Solve

a. This nuclide has six protons, so $Z = 6$. It must be an isotope of carbon. Six protons plus six neutrons give the isotope a mass number of 12, which makes it carbon-12, $^{12}_{6}$C.

b. This nuclide has 11 protons, which means $Z = 11$, so it must be an isotope of sodium. Eleven protons and 12 neutrons give the isotope a mass number of 23, so the isotope is sodium-23, $^{23}_{11}$Na.

c. This nuclide has 92 protons, so $Z = 92$, which makes it an isotope of uranium. The mass number is $92 + 143 = 235$. This isotope is uranium-235, $^{235}_{92}$U.

Think About It In working through this exercise, did you use the periodic table of the elements? Once you identify the number of protons in a nucleus (its atomic number), finding a symbol and identifying the element it represents is straightforward because the elements in the periodic table are arranged in order of increasing atomic number.

Practice Exercise Use the format ^{A}X to write the symbols of the nuclides having (a) 26 protons and 30 neutrons, (b) 7 protons and 8 neutrons, (c) 17 protons and 20 neutrons, and (d) 19 protons and 20 neutrons.

(Answers to Practice Exercises are in the back of the book.)

average atomic mass a weighted average of the masses of all the isotopes of an element, calculated by multiplying the natural abundance of each isotope by its mass in atomic mass units and then summing these products.

natural abundance the proportion of an isotope, usually expressed in percent, relative to all the isotopes of that element in a natural sample.

SAMPLE EXERCISE 2.2 Determining the Number of Neutrons in a Nuclide **LO2**

How many neutrons are in each of the following nuclides: (a) ^{14}N; (b) ^{32}P; (c) ^{157}Gd?

Collect, Organize, and Analyze We are given the symbols of three nuclides and asked to determine the number of neutrons in each of their nuclei. We know the value of Z from the element's symbol and we know the value of A from its superscripted number. Subtracting Z from A gives us the number of neutrons.

Solve

a. ^{14}N is a nuclide of nitrogen, whose atoms each have seven protons. The number of neutrons is $A - Z = 14 - 7 = 7$.

b. ^{32}P is a nuclide of phosphorus ($Z = 15$) with 32 nucleons per nucleus. The number of neutrons is $32 - 15 = 17$.

c. ^{157}Gd is a nuclide of gadolinium ($Z = 64$). The number of neutrons is $157 - 64 = 93$.

Think About It If we compare the ratios of neutrons to protons in these nuclei, we find that they increase as the atomic numbers of the nuclei increase. This is a reasonable trend that is reflected in the average atomic masses and atomic numbers of all the elements in the periodic table on the inside of the front cover.

Practice Exercise Determine the number of protons and neutrons in each of these radioactive nuclides: (a) ^{60}Co, used in cancer therapy; (b) ^{131}I, used in thyroid therapy; (c) ^{192}Ir, used to treat coronary disease.

(Answers to Practice Exercises are in the back of the book.)

2.4 Average Atomic Mass

Isotope	Mass (u)	Natural Abundance (%)
Neon-20	19.9924	90.4838
Neon-21	20.9940	0.2696
Neon-22	21.9914	9.2465

Each element in the periodic table is represented by its symbol. The number above this symbol is the element's atomic number (Z), and the number below the symbol is the element's **average atomic mass**. More precisely, the number below the symbol is the *weighted average* of the masses of all the isotopes of the element.

To understand the meaning of a weighted average, consider the masses and **natural abundances** of the three isotopes of neon in the table shown here. Natural abundances are usually expressed in percentages. Thus, 90.4838% of all neon atoms are neon-20, 9.2465% are neon-22, and only 0.2696% are neon-21. The abundance of neon-21 is so small that Aston could not detect it with his positive-ray analyzer. Modern mass spectrometers, which are the source of natural abundance data such as these, are vastly more sensitive and more precise than Aston's prototype.

To determine the average atomic mass of any element, we multiply the mass of each isotope by its natural abundance (in the language of mathematics, we *weight* the isotope's mass by using natural abundance as the *weighting factor*) and then sum the three weighted masses. To simplify the calculation for neon, we convert the percent abundance values into their decimal equivalents:

$$\begin{aligned}\text{Average atomic mass of neon} = \quad & (19.9924\ \text{u} \times 0.904838) = 18.08988\ \text{u}\\ + \quad & (20.9940\ \text{u} \times 0.002696) = 0.05660\ \text{u}\\ + \quad & (21.9914\ \text{u} \times 0.092465) = \underline{2.03344\ \text{u}}\\ & \qquad\qquad\qquad\qquad\qquad 20.17992\ \text{u}\end{aligned}$$

After accounting for significant figures, the average atomic mass of neon is 20.1799 u. No atom of neon has the average atomic mass; every atom of neon in the universe must have a mass equal to that of one of the three neon isotopes. The value we have calculated is simply the weighted average of these three isotopic masses.

This method of calculating average atomic mass works for every element. The general formula for these calculations is

$$m_X = a_1m_1 + a_2m_2 + a_3m_3 + \ldots \qquad (2.1)$$

where m_X is the average atomic mass of element X, which has isotopes with masses $m_1, m_2, m_3, \ldots$, the natural abundances of which, expressed in decimal form, are $a_1, a_2, a_3, \ldots$.

The term *natural abundance* may seem to mean that the abundances of the isotopes of an element are the same no matter where the element occurs in nature. This is not always the case. The isotopic abundances of some elements can vary slightly depending on the origins of samples and the presence of certain other elements in them.

One element with variable isotopic abundances is strontium. Sample Exercise 2.3 is based on experimentally determined isotopic abundances of strontium in a quantity of strontium carbonate, $SrCO_3$, that was carefully purified, blended, and analyzed by the U.S. National Institute of Standards and Technology (NIST). Portions of it are sold by NIST to scientists who wish to check the accuracy of isotopic analyses determined in their laboratories using mass spectrometry.

CONNECTION The use of reference materials with known composition to check the accuracy of chemical analyses is described in Section 1.9.

SAMPLE EXERCISE 2.3 Calculating Average Atomic Mass **LO3**

The following table contains the atomic masses and experimentally determined relative abundances of the isotopes of strontium in NIST Standard Reference Material Number 987:

Symbol	Mass (u)	Abundance (%)
^{84}Sr	83.9134	0.557
^{86}Sr	85.9093	9.857
^{87}Sr	86.9089	7.002
^{88}Sr	87.9056	82.584

What is the average atomic mass of strontium in the sample? Compare the calculated average to the atomic mass of Sr in the periodic table in the front of this book (87.62 u).

Collect, Organize, and Analyze We know the masses and the relative abundances of the four stable isotopes of strontium in a sample and are asked to calculate the average atomic mass of Sr in the sample. Using Equation 2.1, we multiply the mass of each isotope by its abundance (expressed as a decimal) and then add the products together.

Solve The weighted average atomic mass of Sr in the sample is:

$$\begin{aligned} {}^{84}\text{Sr:} &\quad 83.9134\text{ u} \times 0.00557 = 0.4674\text{ u} \\ {}^{86}\text{Sr:} &\quad 85.9093\text{ u} \times 0.09857 = 8.4681\text{ u} \\ {}^{87}\text{Sr:} &\quad 86.9089\text{ u} \times 0.07002 = 6.0854\text{ u} \\ {}^{88}\text{Sr:} &\quad 87.9056\text{ u} \times 0.82584 = 72.5960\text{ u} \\ \hline & \qquad\qquad\qquad\qquad\quad 87.6169\text{ u} \end{aligned}$$

We carried one more digit (highlighted in red) than allowed by the significant-figure rules in Chapter 1 to avoid rounding errors. Rounding the sum to the allowed number of significant figures gives us 87.617 u.

Think About It In this calculation the abundance values of the isotopes contain fewer significant digits (3, 4, 4, and 5 for ^{84}Sr, ^{86}Sr, ^{87}Sr, and ^{88}Sr, respectively) than do their

period a horizontal row in the periodic table.

group all the elements in the same column of the periodic table; also called *family*.

metals the elements on the left side of the periodic table that are typically shiny solids that conduct heat and electricity well and are malleable and ductile.

nonmetals the elements with properties opposite those of metals, including poor conductivity of heat and electricity.

metalloids (also called **semimetals**) the elements along the border between the metals and nonmetals in the periodic table; they have some metallic and some nonmetallic properties.

main group elements (also called **representative elements**) the elements in groups 1, 2, and 13 through 18 of the periodic table.

transition metals the elements in groups 3 through 12 of the periodic table.

masses (6 in every value). Therefore, isotopic abundance is the *weak link* in calculating this average atomic mass. The four abundance values should add up to 1.000, and they do. Sometimes this is not the case (check the neon abundances in the previous calculation). Finally, rounding the calculated value to the nearest 0.01 u yields the same average atomic mass of Sr (87.62 u) as in the periodic table inside the front cover of the book.

Practice Exercise Silver (Ag) has two stable isotopes: silver-107 (106.905 u) and silver-109 (108.905 u). If the average atomic mass of silver is 107.868 u, what is the natural abundance of each isotope? (*Hint*: Let x be the natural abundance of one of the isotopes. Then $1 - x$ is the natural abundance of the other.)

(Answers to Practice Exercises are in the back of the book.)

2.5 The Periodic Table of the Elements

Long before chemists knew about electrons, protons, and neutrons, they knew that groups of elements, such as Li, Na, and K, or F, Cl, and Br, had similar chemical (and sometimes physical) properties. When the elements were arranged by increasing atomic mass, repeating patterns of similar properties appeared among the elements. This *periodicity* in the chemical properties of the elements inspired several 19th-century scientists to create tables of the elements in which the elements were arranged in patterns based on similarities in their chemical properties.

The most successful of these scientists was the Russian chemist Dmitri Mendeleev (1834–1907). In 1872 he published a table (**Figure 2.11**) that was the forerunner of the modern periodic table (**Figure 2.12**). In addition to organizing all the elements that were known at the time, Mendeleev realized that there might be elements in nature that were yet to be discovered, so he left empty cells in his table for those unknown elements. Doing so allowed him to align the known elements so that those in each column had similar chemical properties. Based on the locations of the empty cells, Mendeleev predicted the chemical properties of the missing elements that ultimately were discovered. Mendeleev arranged the elements in his periodic table in order of increasing atomic mass, whereas the elements appear in order of increasing atomic number in modern periodic tables.

CONCEPT TEST

Why did Mendeleev skip cells in his periodic table?

(Answers to Concept Tests are in the back of the book.)

Group Number →

Row	I	II	III	IV	V	VI	VII	VIII
1	1 H							
2	7 Li	9.4 Be	11 B	12 C	14 N	16 O	19 F	
3	23 Na	24 Mg	27.3 Al	28 Si	31 P	32 S	35.5 Cl	
4	39 K	40 Ca	44 ?	48 Ti	51 V	52 Cr	55 Mn	56 Fe, 59 Co, 59 Ni
5	63 Cu	65 Zn	68 ?	72 ?	75 As	78 Se	80 Br	
6	85 Rb	87 Sr	88 ?Yt	90 Zr	94 Nb	96 Mo	100 ?	104 Ru, 104 Rh, 106 Pd
7	108 Ag	112 Cd	113 In	118 Sn	122 Sb	125 Te	127 J	
8	133 Cs	137 Ba	138 ?Di	140 ?Ce				
9								
10			178 ?Er	180 ?La	182 Ta	184 W		195 Os, 197 Ir, 198 Pt
11	199 Au	200 Hg	204 Tl	207 Pb	208 Bi			
12				231 Th		240 U		

FIGURE 2.11 Mendeleev organized his periodic table based on chemical and physical properties and atomic masses. He assigned three elements with similar properties to group VIII in rows 4, 6, and 10. Because he did this, the elements in the rows that followed lined up in columns with similar properties. In this way, rows 4 and 5 when combined contain spaces for 18 elements, corresponding to the 18 groups in the modern periodic table.

Atomic number / Symbol for element (key: 4 Be)

Period	1	2	3	4	5	6	7	8	9	10	11	12	13	14	15	16	17	18
1	1 H																	2 He
2	3 Li	4 Be											5 B	6 C	7 N	8 O	9 F	10 Ne
3	11 Na	12 Mg											13 Al	14 Si	15 P	16 S	17 Cl	18 Ar
4	19 K	20 Ca	21 Sc	22 Ti	23 V	24 Cr	25 Mn	26 Fe	27 Co	28 Ni	29 Cu	30 Zn	31 Ga	32 Ge	33 As	34 Se	35 Br	36 Kr
5	37 Rb	38 Sr	39 Y	40 Zr	41 Nb	42 Mo	43 Tc	44 Ru	45 Rh	46 Pd	47 Ag	48 Cd	49 In	50 Sn	51 Sb	52 Te	53 I	54 Xe
6	55 Cs	56 Ba	57 La	72 Hf	73 Ta	74 W	75 Re	76 Os	77 Ir	78 Pt	79 Au	80 Hg	81 Tl	82 Pb	83 Bi	84 Po	85 At	86 Rn
7	87 Fr	88 Ra	89 Ac	104 Rf	105 Db	106 Sg	107 Bh	108 Hs	109 Mt	110 Ds	111 Rg	112 Cn	113 Nh	114 Fl	115 Mc	116 Lv	117 Ts	118 Og

6 Lanthanides	58 Ce	59 Pr	60 Nd	61 Pm	62 Sm	63 Eu	64 Gd	65 Tb	66 Dy	67 Ho	68 Er	69 Tm	70 Yb	71 Lu
7 Actinides	90 Th	91 Pa	92 U	93 Np	94 Pu	95 Am	96 Cm	97 Bk	98 Cf	99 Es	100 Fm	101 Md	102 No	103 Lr

FIGURE 2.12 In the modern periodic table, the elements are arranged in order of atomic number (Z) and in a pattern related to their physical and chemical properties. The rows are called periods, and the columns contain groups (or families) of elements. The elements shown in tan are classified as metals; those shown in blue, nonmetals; and those shown in green, metalloids (also called semimetals).

Navigating the Modern Periodic Table

The modern periodic table (Figure 2.12) contains seven horizontal rows (also called **periods**) and 18 columns (known as **groups** or *families*) of elements. The periods are numbered at the far left of each row, and the group numbers appear at the top of each column. The periodic table inside the front cover shows a second set of column headings containing numbers followed by the letter A or B. These secondary headings were widely used in earlier versions of the table, and many scientists (and students) still find them useful.

The elements in the periodic table can be broadly categorized as metals, nonmetals, and metalloids (or semimetals). The elements highlighted in tan in Figure 2.12 are the **metals**. They tend to conduct heat and electricity well, they tend to be malleable (capable of being shaped by hammering) or ductile (capable of being drawn out in a wire), and they are shiny solids at room temperature, except for mercury (Hg), which is a liquid at room temperature (**Figure 2.13a**). The elements highlighted in blue are the **nonmetals**. They are poor conductors of heat and electricity, and the solids among them, including sulfur (**Figure 2.13b**), tend to be brittle. Several are gases; bromine is a volatile (easily vaporized) liquid, and iodine is a volatile solid. Lastly, the elements highlighted in green are the **metalloids** or **semimetals**, so named because they tend to have the physical properties of metals but the chemical properties of nonmetals (**Figure 2.13c**).

In the modern periodic table, groups 1, 2, and 13–18 are referred to collectively as **main group elements**, or **representative elements** (**Figure 2.14a**). These groups include the most abundant elements in the solar system and many of the most abundant elements on Earth. These are the "A" elements in the older group labeling system shown on the table inside the front cover. The elements in groups 3 through 12 are called **transition metals**; these are the "B" elements. Nearly all the elements in groups 3 through 12 exhibit the characteristic properties of metals—namely, they are hard, shiny, ductile, malleable, and excellent conductors of heat and electricity.

The first row contains only two elements—hydrogen and helium—and the second and third rows each contain only eight. Starting with the fourth row, all

(a) Metals

(b) Nonmetals

(c) Metalloids

FIGURE 2.13 (a) Metals: a spool of copper, gold that has been hammered into a thin foil, and mercury. (b) Nonmetals: sulfur, bromine, and iodine. (c) Metalloids: silicon, germanium, and antimony.

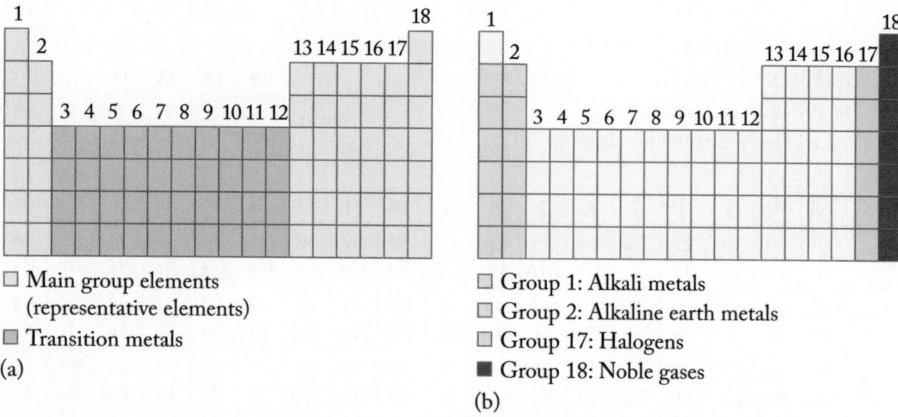

FIGURE 2.14 (a) The *main group* (or *representative*) elements are in groups 1, 2, and 13–18. In between are the *transition metals* in groups 3–12. (b) The commonly used names of groups 1, 2, 17, and 18.

STEPWISE ANIMATION

Lanthanides and Actinides in the Periodic Table

18 columns are full. The sixth and seventh rows contain additional elements, which appear in the two separate rows at the bottom of the main table. Elements in the row with atomic numbers from 58 to 71 are called the lanthanides (after element 57, lanthanum), and those with atomic numbers between 90 and 103 are called actinides (after element 89, actinium). All isotopes of the actinide elements are radioactive, and none of those with atomic numbers above 94 occur in nature. Therefore, they have no *natural* abundance, and their atomic mass values in the periodic table, which appear in brackets or parentheses, are based on the mass numbers of their isotopes that last the longest before undergoing radioactive decay.

Several of the groups have a name in addition to a number. The names are typically based on properties common to all the elements in that group (**Figure 2.14b**). For example, the elements in group 17 are called **halogens**. The word *halogen* is derived from the Greek for "salt former." Chlorine (Cl), for example, reacts with sodium (a metal in group 1) to form table salt. Other elements in group 1 (which are called **alkali metals**) and in group 2 (which are called **alkaline earth metals**) also react with members of the halogen family to form different salts. These reactivity patterns, and others, were the basis for Mendeleev's arrangement of the elements in his periodic table, and they illustrate what we mean by "similar chemical properties."

Compare Figures 2.11 and 2.12. First note the similarity in the arrangements of the lighter (smaller atomic number) elements through calcium ($Z = 20$). All the elements in groups 1, 2, and 13 through 17 of the modern table appear in the same order in Mendeleev's table, but group 18 (the **noble gases**) is missing from Mendeleev's table. There is a good reason for this. Helium was the first noble gas to be discovered, and that was not until 1895, many years after Mendeleev published his table. Noble gases have very limited chemical reactivity (indeed, helium and neon don't react at all) and so were elusive substances for early chemists to isolate and identify. Because Mendeleev arranged his table largely based on reactivity, he had no reason to predict the existence of the noble gases.

halogens the elements in group 17 of the periodic table.

alkali metals the elements in group 1 of the periodic table.

alkaline earth metals the elements in group 2 of the periodic table.

noble gases the elements in group 18 of the periodic table.

law of multiple proportions the principle that, when two masses of one element react with a given mass of another element to form two different compounds, the two masses of the first element have a ratio of two small whole numbers.

SAMPLE EXERCISE 2.4 Navigating the Periodic Table **LO4**

Use the periodic table on the inside front cover to determine the symbol and name of each of the following elements:

a. The third-row element in group 14
b. The fourth-row alkaline earth metal
c. The halogen with fewer than 16 protons in its nucleus

Collect and Organize We need to identify each element based on its: (a) row number and column number, (b) row number and the common name of its group, and (c) group name and atomic number.

Analyze (a) The target element is located at the intersection of row 3 and column 14, as shown in red in **Figure 2.15**. (b) The alkaline earth metals are in group 2, so the target element is located at the intersection of row 4 and column 2, as shown in blue in Figure 2.15. (c) The halogens are in group 17. As shown in green in Figure 2.15, only one of them has an atomic number (which is the number of protons per nucleus) that is less than 16.

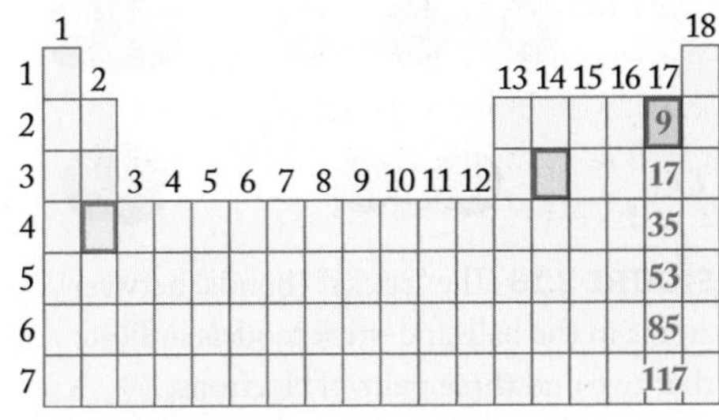

FIGURE 2.15

Solve (a) Si, silicon; (b) Ca, calcium; (c) F, fluorine.

Think About It Each element has a unique location in the periodic table determined by its atomic number, which defines the row it is in, and by its patterns of reactivity with other elements, which defines the group it is in.

Practice Exercise
Write the symbol and name of each element:

a. The metalloid in group 15 closest in mass to the noble gas krypton
b. The element in the fourth row that is an alkali metal
c. The transition metal in the fifth period with chemical properties most like those of zinc ($Z = 30$)
d. The nonmetal in the fourth period with chemical properties most like those of sulfur

(Answers to Practice Exercises are in the back of the book.)

2.6 Trends in Compound Formation

Mendeleev used patterns of reactivity to place elements in different groups in his early periodic table. Both Dalton and Mendeleev knew that when elements combine to form compounds, they do so in characteristic ratios. These ratios are reflected in the chemical formulas of compounds. For example, the formula of carbon dioxide, CO_2, indicates that every molecule of CO_2 consists of one atom of carbon combined with two atoms of oxygen.

Dalton's atomic view of compounds also explains why some elements (e.g., S and O) can form more than one compound (as in SO_2 and SO_3). Dalton determined that the ratio of the different masses of oxygen that react with a given mass of sulfur to form two distinct compounds could be expressed as a ratio of two small whole numbers. This principle was observed experimentally and is known as Dalton's **law of multiple proportions**.

To see what this principle means, consider SO_2 and SO_3. We determine in an experiment that under one set of conditions, 10 g of sulfur reacts with 10 g of oxygen to form SO_2. Under different conditions, however, 10 g of sulfur reacts with 15 g of oxygen to form SO_3. The ratio of the two masses of oxygen is 10:15, or 2:3, which is a ratio of two small whole numbers. This example illustrates the law of multiple proportions.

Similarly, we can confirm experimentally that the mass of oxygen that reacts with a given mass of nitrogen to form NO_2 (22.8 g of O for every 10.0 g of N) is twice as much as the mass of oxygen that reacts with the same mass of nitrogen to form NO (11.4 g of O for every 10.0 g of N). The ratio of the two oxygen masses is 22.8:11.4, or 2:1, again a ratio of small whole numbers. The law of multiple proportions and the law of constant composition (Section 1.2) were key ideas that formed the basis for Dalton's atomic theory.

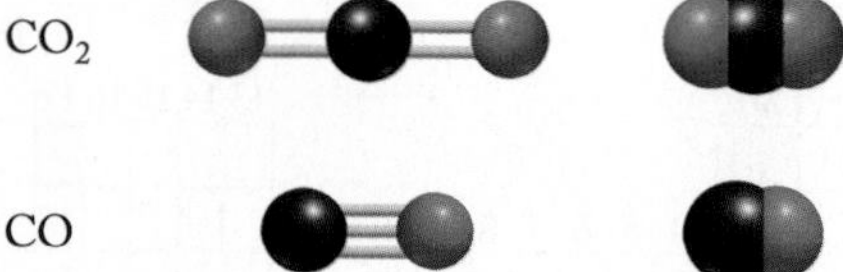

FIGURE 2.16 The "sticks" (bonds) between atoms in the ball-and-stick models indicate that two and three pairs of electrons, respectively, are shared. We discuss single (one bond, one pair of electrons shared), double (two bonds, two pairs), and triple (three bonds, three pairs) bonds in Chapter 8.

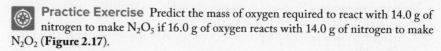

SAMPLE EXERCISE 2.5 Relating Chemical Formulas to the Law of Multiple Proportions **LO5**

Carbon can combine with oxygen to form either CO or CO_2 (**Figure 2.16**), depending on reaction conditions. If 26.6 g of O_2 reacts with 10.0 g of C to make CO_2, how many grams of O_2 react with 10.0 g of C to make CO?

Collect, Organize, and Analyze We know the mass ratio of oxygen to carbon in CO_2 and need to calculate the mass ratio of oxygen to carbon in CO. The ratio of O atoms to C atoms in CO_2 is 2:1. The ratio of the O atoms to C atoms in CO is 1:1. Therefore, half as much oxygen reacts with 10.0 g of carbon to make CO as reacts with 10.0 g of carbon to make CO_2.

Solve $(26.6\text{ g }O_2) \times \frac{1}{2} = 13.3\text{ g }O_2$

Think About It The result makes sense because it should take half as much O_2 (13.3 g instead of 26.6 g) to produce a compound that has one atom of oxygen for every atom of carbon instead of two atoms of oxygen for every atom of carbon.

Practice Exercise Predict the mass of oxygen required to react with 14.0 g of nitrogen to make N_2O_5 if 16.0 g of oxygen reacts with 14.0 g of nitrogen to make N_2O_2 (**Figure 2.17**).

(Answers to Practice Exercises are in the back of the book.)

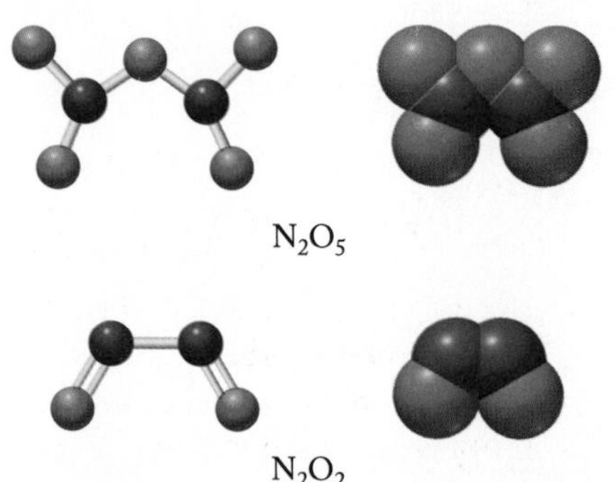

FIGURE 2.17 Two compounds of nitrogen and oxygen.

Molecular Compounds

We need to distinguish between compounds that are composed of ions and those that are composed of molecules. Compounds composed of pairs of metallic and nonmetallic elements tend to be ionic, whereas compounds composed of two nonmetals or metalloids tend to be molecular. The compounds we have examined so far in this section have been **molecular compounds**, each of which consisted of molecules composed of two different nonmetals. The molecular structures of several of these compounds are shown in **Figure 2.18**. Other molecular compounds are composed of molecules that contain the atoms of three, four, or more elements, which may be nonmetals or metalloids, and that are bound together by shared pairs of electrons called **covalent bonds**.

All the compounds in Figure 2.18 are present in the air we breathe. Each chemical formula specifies the number of atoms of each element in one molecule of the compound. Therefore, these chemical formulas are **molecular formulas**. The fact that the same two elements can form compounds with different molecular formulas (consistent with Dalton's law of multiple proportions) means that there are different ways to form covalent bonds between atoms of the same two elements.

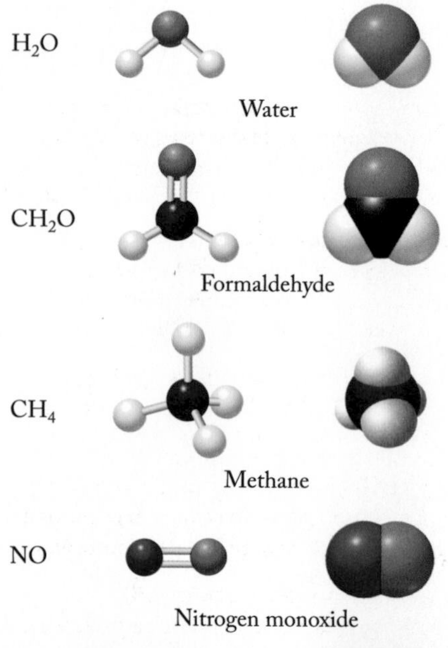

FIGURE 2.18 Ball-and-stick and space-filling models of some molecular compounds found in the air we breathe.

Ionic Compounds

Binary (two-element) **ionic compounds** contain atoms of metals (shown in tan in the periodic table inside the front cover) that have each lost one or more electrons, forming positively charged cations, and atoms of nonmetals (shown in blue) that have each gained one or more electrons, forming negatively charged anions. For example, table salt (**Figure 2.19**) is composed of crystals of sodium chloride,

NaCl, that each contain equal numbers of Na^+ ions (Na atoms that have each lost one electron),

$$Na \rightarrow Na^+ + e^-$$

and Cl^- ions (Cl atoms that have each gained one electron),

$$Cl + e^- \rightarrow Cl^-$$

Within NaCl crystals, each sodium ion is surrounded by six chloride ions, and each chloride ion is surrounded by six sodium ions. However, the smallest whole-number ratio of sodium ions to chloride ions is simply 1:1. Formulas based on the lowest whole-number ratio of the elements in a compound are called **empirical formulas**. The chemical formulas of ionic compounds, such as NaCl for sodium chloride, are empirical formulas. The empirical formula of an ionic compound describes a **formula unit**, the smallest electrically neutral unit within the crystal.

Ions such as Na^+ and Cl^- that are each composed of a single atom are called *monatomic* ions. The charges on the monatomic ions that different elements form strongly influence their chemical reactivity and, therefore, their placement among the different groups of the periodic table. **Figure 2.20** shows how the elements within a given group tend to form cations or anions with the same charge. It also shows that there is a pattern in how these charges vary among groups. For example, atoms of group 1 elements each form 1+ ions; atoms of group 2 elements each form 2+ ions (Figure 2.20). The charges on these monatomic cations match their group numbers. This correlation between group number and cation charge breaks down among the transition metals and the metallic elements on the right side of the periodic table, but there are still similarities in cation charge within groups. Figure 2.20 also shows that there is a pattern in the charges of the monatomic anions formed by the nonmetallic elements, namely, 3− in group 15, 2− in group 16, and 1− in group 17. We explore how these trends are related to the structure of the atoms of these elements, particularly the number of electrons surrounding their nuclei, in Chapter 7.

It is important to remember that ionic compounds, like molecular compounds, are electrically neutral. This means that the total positive charge on all the cations in an ionic compound must be equal in magnitude to the total negative charge on

FIGURE 2.19 Crystals of sodium chloride. The cubic shape of the crystals mirrors the cubic array of Na^+ and Cl^- ions that make up its structure. The empirical formula NaCl describes the smallest whole-number ratio of cations to anions in the structure, which is electrically neutral.

CHEMTOUR
NaCl Reaction

CONNECTION In Chapter 1 we saw that ions are particles with either a positive charge (cations) or a negative charge (anions).

1	2	3	4	5	6	7	8	9	10	11	12	13	14	15	16	17	18
H^+																	
Li^+														N^{3-}	O^{2-}	F^-	
Na^+	Mg^{2+}											Al^{3+}		P^{3-}	S^{2-}	Cl^-	
K^+	Ca^{2+}	Sc^{3+}	Ti^{3+} Ti^{4+}	V^{3+} V^{5+}	Cr^{2+} Cr^{3+}	Mn^{2+} Mn^{4+}	Fe^{2+} Fe^{3+}	Co^{2+} Co^{3+}	Ni^{2+} Ni^{3+}	Cu^+ Cu^{2+}	Zn^{2+}	Ga^+ Ga^{3+}			Se^{2-}	Br^-	
Rb^+	Sr^{2+}	Y^{3+}	Zr^{3+} Zr^{4+}							Ag^+	Cd^{2+}	In^+ In^{3+}	Sn^{2+} Sn^{4+}		Te^{2-}	I^-	
Cs^+	Ba^{2+}	La^{3+}									Hg_2^{2+} Hg^{2+}	Tl^+ Tl^{3+}	Pb^{2+} Pb^{4+}				

FIGURE 2.20 The charges of the most common monatomic ions (Hg_2^{2+} is diatomic). The ions formed by each highlighted element nearly always have the charge shown in the figure.

molecular compound a compound composed of molecules that contain the atoms of two or more elements covalently bonded together.

covalent bond a bond between two atoms created by sharing one or more pairs of electrons.

molecular formula a notation showing the number and type of atoms present in one molecule of a molecular compound.

ionic compound a compound composed of positively and negatively charged ions held together by electrostatic attraction.

empirical formula a formula showing the smallest whole-number ratio of the elements in a compound.

formula unit the smallest electrically neutral unit of an ionic compound.

all the anions in the compound. This neutrality requirement is the reason that the formula of the compound that contains only Na^+ ions and Cl^- ions is NaCl: the 1+ and 1− charges on these ions balance out only if the ratio of the ions is 1:1. On the other hand, the formula of the compound that contains only Mg^{2+} and F^- ions is MgF_2 because it takes two F^- ions to offset the 2+ charge on one Mg^{2+} ion.

CONCEPT TEST

Which of the following ionic formulas does *not* represent an electrically neutral compound? (*Hint*: Base your selection on the charges of the common ions in Figure 2.20.) (a) KBr; (b) CaF_2; (c) CsN; (d) TiO_2; (e) AgCl

(Answers to Concept Tests are in the back of the book.)

SAMPLE EXERCISE 2.6 Classifying Compounds as Molecular or Ionic LO5

Identify which of the following binary compounds are ionic and which are molecular: (a) sodium bromide (NaBr); (b) carbon dioxide (CO_2); (c) lithium iodide (LiI); (d) germanium tetrachloride ($GeCl_4$); (e) calcium chloride ($CaCl_2$).

Collect, Organize, and Analyze We must decide whether each of five binary compounds is ionic or covalent. A compound composed of a metallic and a nonmetallic element is usually ionic, whereas compounds composed of two nonmetals or metalloids tend to be molecular.

Solve NaBr, LiI, and $CaCl_2$ all contain a group 1 or group 2 metal and a group 17 nonmetal, so they are ionic compounds. CO_2 is composed of two nonmetals, and $GeCl_4$ is composed of a metalloid and a nonmetal, so both should be molecular compounds.

Think About It Labeling compounds as ionic or molecular is not as clear-cut as you might think based on this sample exercise. In later chapters we introduce binary molecular compounds that have some ionic "character," and we explore ways to predict how much ionic character they have.

Practice Exercise Identify the following compounds as molecular or ionic: (a) carbon disulfide (CS_2); (b) carbon monoxide (CO); (c) ammonia (NH_3); (d) water (H_2O); (e) sodium iodide (NaI).

(Answers to Practice Exercises are in the back of the book.)

2.7 Naming Inorganic Compounds and Writing Their Formulas

There are more than 140 million known chemical compounds, and new ones are being synthesized every day. Keeping track of all these compounds and assigning each a unique name that clearly conveys its chemical composition is an enormous undertaking. Today, the naming portion of this task is based on a widely used set of rules called **chemical nomenclature** from the Latin words *nomen* (name) and *calatus* (assigned).

Naming compounds is like naming life-forms in that it relies on dividing substances in categories, then subcategories, and so on. The two major categories in chemical nomenclature are *organic* and *inorganic* compounds. Inorganic compounds make up Earth's geosphere, including its crust (part of the lithosphere), the air we breathe (the atmosphere), and the water that covers most of Earth's

chemical nomenclature the rules that are followed in naming substances.

organic compound a compound composed of molecules containing carbon combined with hydrogen and other elements, such as nitrogen, oxygen, phosphorus, and sulfur.

surface (the hydrosphere). The biosphere, on the other hand, is rich in **organic compounds**, which are composed mostly of carbon and hydrogen, with lesser amounts of other elements such as nitrogen, oxygen, phosphorus, and sulfur.

For centuries, philosophers and scientists thought that organic compounds could be synthesized only by biological processes. In the 1820s, however, a German chemist named Friedrich Wöhler (1800–1882) showed that organic compounds could be synthesized in the laboratory from inorganic starting materials. Naming organic compounds and writing their molecular formulas is described in Section 2.8. The remainder of this section focuses on inorganic compounds, starting with those composed of only two elements.

Binary Molecular Compounds

Compounds formed by two nonmetals or by a metalloid and a nonmetal are usually composed of molecules, whereas most (though not all) compounds formed by a metal and a nonmetal are ionic. Let's begin by translating the formulas of molecular inorganic compounds into two-word compound names using the following three steps:

1. The first word is the name of the first element in the formula.
2. For the second word, change the ending of the name of the second element to *-ide.*
3. Use prefixes (**Table 2.2**) to indicate the number of atoms of each element in the molecule. (Exception: Do not use the prefix *mono-* with the first element in a name.)

TABLE 2.2 Prefixes for Naming Molecular Compounds

one	*mono-*
two	*di-*
three	*tri-*
four	*tetra-*
five	*penta-*
six	*hexa-*
seven	*hepta-*
eight	*octa-*
nine	*nona-*
ten	*deca-*

For example, the name of CO is carbon monoxide, SO_2 is sulfur dioxide, and PCl_3 is phosphorus trichloride. When prefixes ending in *o-* or *a-* (like *mono-* and *tetra-*) precede a name that begins with a vowel (such as *oxide*), the *o* or *a* at the end of the prefix is deleted to make the combination of prefix and name easier to pronounce. Thus, CO is carbon monoxide, not carbon *mono*oxide.

The order in which the symbols of the elements appear in the formulas of compounds depends on the relative positions of the elements in the periodic table: the one with the smaller group number nearly always appears first in the formulas of inorganic compounds. If a compound contains two elements from the same group—for example, sulfur and oxygen—the symbol of the element with the larger atomic number appears first.

Some common binary molecular compounds have widely used names that do not follow the naming guidelines just described. For example, H_2O is called water, not dihydrogen monoxide. Similarly, the common name for the compound with the formula NH_3 is ammonia, not nitrogen trihydride. There are other important molecular compounds with common names that predate current rules of chemical nomenclature. Many of them are organic compounds, and their names and formulas are discussed in Section 2.8.

SAMPLE EXERCISE 2.7 Relating the Formulas and Names of Molecular Compounds **LO6**

What are the names of the compounds with these molecular formulas: (a) N_2O; (b) N_2O_4; (c) N_2O_5? What are the molecular formulas of (d) sulfur trioxide, (e) sulfur monoxide, and (f) diphosphorus pentoxide?

Collect and Organize We are asked to translate the formulas of three molecular compounds into names and the names of three compounds into molecular formulas. The prefixes in Table 2.2 are used in the names of compounds to indicate the number of atoms of each element in a molecule: *mono-* means 1 atom, *di-* means 2 atoms, *tri-* means 3, *tetra-* means 4, and *penta-* means 5.

Analyze In the first question, the first element in all three compounds is nitrogen, so the first word in each name is *nitrogen* with the appropriate prefix. The second element in all three compounds is oxygen, so the second word in each name is *oxide* with the appropriate prefix. In the second question, all the molecules contain oxygen because their names end in *oxide*.

Solve (a) dinitrogen monoxide; (b) dinitrogen tetroxide (not dinitrogen *tetra*oxide); (c) dinitrogen pentoxide (not dinitrogen *penta*oxide); (d) SO_3; (e) SO; (f) P_2O_5

Think About It The prefixes *mono-*, *tetra-*, and *penta-* dropped their final letter when they combined with *oxide* to make the names of the compounds easier to pronounce.

 **Practice Exercise** Name (a) P_4O_{10}, (b) NO, and (c) NCl_3. Write the formulas for (d) sulfur hexafluoride, (e) iodine monochloride, and (f) dibromine monoxide.

(Answers to Practice Exercises are in the back of the book.)

Binary Ionic Compounds

Binary ionic compounds contain the monatomic cation of a metallic element and the monatomic anion of a nonmetal. Translating their formulas into names is a lot like the system we used for molecular compounds, but with fewer steps:

1. The first word is the name of the cation, which is simply the name of its parent element.
2. The second word is the name of the anion, which, for monatomic anions, is the name of its parent element, except that its ending is changed to *-ide*.

Prefixes are not needed in naming binary ionic compounds that each contain a metal and nonmetal whose ions nearly always have the same positive and negative charges (these elements are highlighted in Figure 2.20). Prefixes aren't needed because all ionic compounds are electrically neutral, which means the sums of the negative and positive charges of their ions must balance, and that dictates the number of each of the ions in each of their formulas. For example, a name like *magnesium fluoride* is unambiguous: it can mean only MgF_2 because all magnesium ions have 2+ charges and all fluoride ions have 1− charges.

FIGURE 2.21 Aluminum cans resist corrosion because of a thin layer of aluminum oxide that forms on the surfaces of the can. This layer protects the aluminum beneath the surface from combining with oxygen or other elements.

SAMPLE EXERCISE 2.8 Relating the Formulas and Names of Ionic Compounds **LO6**

Write the formulas of (a) potassium bromide, (b) calcium oxide, (c) sodium sulfide, (d) magnesium chloride, and (e) aluminum oxide (**Figure 2.21**).

Collect, Organize, and Analyze We need to write the formulas of five binary ionic compounds. Checking the names of the compounds against the positions of the elements in the periodic table and the charges on the monatomic ions they form (Figure 2.20), we see that all five are made up of main group elements that form these ions: (a) K^+ and Br^-, (b) Ca^{2+} and O^{2-}, (c) Na^+ and S^{2-}, (d) Mg^{2+} and Cl^-, and (e) Al^{3+} and O^{2-}.

Solve We must balance the positive and negative charges on the ions in each compound:

a. The 1+ and 1− charges of a K^+ ion and a Br^- ion add up to zero, so the formula is simply KBr.
b. The 2+ and 2− charges of a Ca^{2+} and a O^{2-} ion add up to zero, so the formula is CaO.
c. It takes two Na^+ ions to offset the 2− charge of one S^{2-} ion, so the formula of sodium sulfide is Na_2S.
d. It takes two Cl^- ions to offset the 2+ charge of one Mg^{2+} ion, so the formula of magnesium chloride is $MgCl_2$.
e. The simplest way to balance the 3+ and 2− charges on Al^{3+} and O^{2-} ions is with a ratio of 2 Al^{3+} ions for every 3 O^{2-} ions because $2 \times (3+) + 3 \times (2-) = 0$. Therefore, the formula of aluminum oxide is Al_2O_3.

Think About It The formulas of ionic compounds are based on the principle that the positive and negative charges on all the cations and anions in them must be in balance, that is, the sum of their charges must equal zero.

Practice Exercise Write the chemical formulas of (a) strontium chloride, (b) magnesium oxide, (c) sodium fluoride, and (d) calcium bromide.

(Answers to Practice Exercises are in the back of the book.)

Compounds of Metals That Form More than One Cation

Many of the transition metals and metallic elements in groups 13 through 16 form monatomic cations with more than one charge. For example, most of the copper found in nature is present as Cu^{2+}, but some copper compounds contain Cu^+ ions. Therefore, the name *copper chloride* does not distinguish between $CuCl_2$ and CuCl. To name these compounds, chemists historically used different names for Cu^+ and Cu^{2+} ions: they were called *cuprous* and *cupric* ions, respectively. Similarly, Fe^{2+} and Fe^{3+} ions were called *ferrous* and *ferric* ions, respectively. In both pairs of ions, the name with the lower charge ends in *-ous*, whereas the name with the higher charge ends in *-ic*.

In modern chemical nomenclature Roman numerals are used to indicate the charge on these metal ions. Thus, the modern name of $CuCl_2$ is copper(II) chloride (pronounced "copper-two chloride"), whereas CuCl is copper(I) chloride. Roman numerals are used to indicate the charge of metal cations unless the metal commonly forms only one. These metals include those in groups 1 through 3, which form ions with charges of 1+, 2+, and 3+, respectively, as well as Ag^+, Cd^{2+}, Zn^{2+}, and Al^{3+}.

FIGURE 2.22 Stainless steel pots and pans resist corrosion because the blend of metals in stainless steel includes chromium, which forms a thin layer of chromium(III) oxide on the surfaces of the cookware. This layer protects the metal atoms beneath the surface from combining with oxygen or other elements.

SAMPLE EXERCISE 2.9 Relating the Formulas and Names of Compounds of Metals That Form More than One Monatomic Cation **LO6**

What are the chemical formulas of (a) iron(II) sulfide and (b) chromium(III) oxide (**Figure 2.22**)? What are the names of (c) V_2O_5 and (d) $NiCl_2$?

Collect, Organize, and Analyze We are asked to write the formulas of two transition metal compounds from their names and to write the names of two others from their formulas. In the names of all four compounds, Roman numerals indicate the charges on the transition metal cations. The charges of the most common monatomic anions are shown in Figure 2.20. They include S^{2-}, O^{2-}, and Cl^-. Both vanadium and lead can form more than one monatomic cation, so the names of compounds (c) and (d) must include Roman numerals. As with all ionic compounds, the sum of the positive and negative charges of their ions must be zero.

Solve

a. The Roman numeral II in the compound's name means that the iron ions are Fe^{2+}. All sulfide ions are 2−. A 1:1 ratio of Fe^{2+} and S^{2-} ions balances their charges, making the formula of iron(II) sulfide FeS.
b. The Roman numeral III means that the chromium ions in the compound are Cr^{3+}. All oxide ions are 2−. To balance their positive and negative charges, we need two Cr^{3+} ions for every three O^{2-} ions, because $2 \times (3+) + 3 \times (2-) = 0$. Therefore, the formula of chromium(III) oxide is Cr_2O_3.
c. The compound with the formula V_2O_5 has five O^{2-} ions for every two vanadium ions. The negative charge from five O^{2-} ions is 10−, so the positive charge from two vanadium ions must be 10+. This means that each vanadium ion is 5+. Expressing this 5+ charge with the Roman numeral V gives us the compound name vanadium(V) oxide.
d. There are two Cl^- ions for every nickel ion in $NiCl_2$. The negative charge from two Cl^- ions is 2−, so the positive charge from the one nickel ion must be 2+. Expressing this 2+ charge with the Roman numeral II gives us the compound name nickel(II) chloride.

Think About It To check our work, we could try translating the formulas developed in parts (a) and (b) back into names and the names developed in parts (c) and (d) back into formulas to see if we could reproduce the starting names and formulas.

Practice Exercise Write the formulas of manganese(II) chloride and manganese(IV) oxide.

(Answers to Practice Exercises are in the back of the book.)

TABLE 2.3 Names, Formulas, and Charges of Some Common Polyatomic Ions

Name	Chemical Formula
Acetate	CH_3COO^-
Carbonate	CO_3^{2-}
Hydrogen carbonate or bicarbonate	HCO_3^-
Cyanide	CN^-
Hypochlorite	ClO^-
Chlorite	ClO_2^-
Chlorate	ClO_3^-
Perchlorate	ClO_4^-
Dichromate	$Cr_2O_7^{2-}$
Chromate	CrO_4^{2-}
Permanganate	MnO_4^-
Azide	N_3^-
Ammonium	NH_4^+
Nitrite	NO_2^-
Nitrate	NO_3^-
Hydroxide	OH^-
Peroxide	O_2^{2-}
Phosphate	PO_4^{3-}
Hydrogen phosphate	HPO_4^{2-}
Dihydrogen phosphate	$H_2PO_4^-$
Disulfide	S_2^{2-}
Sulfate	SO_4^{2-}
Hydrogen sulfate or bisulfate	HSO_4^-
Sulfite	SO_3^{2-}
Hydrogen sulfite or bisulfite	HSO_3^-
Thiocyanate	SCN^-

Polyatomic Ions

Ions that consist of two or more atoms joined by covalent bonds are called **polyatomic ions**. The names and formulas of some common polyatomic ions appear in **Table 2.3**. Only one of them, ammonium ion (NH_4^+), is a cation. When writing the formula of a compound with two or more of the same polyatomic ion per formula unit, we put parentheses around the formula of the polyatomic ion to make it clear that the subscript that follows applies to the entire ion. For example, the formula $Al_2(SO_4)_3$ means there are two Al^{3+} ions for every three SO_4^{2-} ions in each formula unit.

Polyatomic ions containing oxygen and one or more other elements are called **oxoanions**. Most oxoanions have a name based on the name of the element that appears first in the formula, with its ending changed to either *-ite* or *-ate*. The *-ate* oxoanion of an element has a greater number of oxygen atoms than its *-ite* counterpart; for example, SO_4^{2-} is the sulf*ate* ion, whereas SO_3^{2-} is the sulf*ite* ion, and NO_3^- is the nitr*ate* ion, whereas NO_2^- is the nitr*ite* ion.

If an element forms more than two oxoanions, as chlorine, bromine, and iodine do, prefixes are used to distinguish among them (see Table 2.3). The oxoanion with the greatest number of oxygen atoms may have the prefix *per-*, whereas the one with the fewest number of oxygen atoms may have the prefix *hypo-* in its name. Because these rules may not enable you to predict the chemical formula of an oxoanion from its name or its name from its formula, you need to memorize the formulas, charges, and names of the polyatomic ions in Table 2.3 that your instructor thinks are most important.

polyatomic ion a charged group of two or more atoms joined by covalent bonds.

oxoanion a polyatomic ion that contains oxygen in combination with one or more other elements.

SAMPLE EXERCISE 2.10 Relating the Formulas and Names of Compounds Containing Oxoanions **LO6**

What are the formulas of (a) sodium sulfite and (b) magnesium phosphate? What are the names of (c) $CaCO_3$ and (d) $KClO_4$?

Collect, Organize, and Analyze We are asked to write the formulas of two compounds whose names tell us they contain oxoanions (because of their -ite and -ate endings), and to write the names of two other compounds from formulas that contain oxoanions. Table 2.3 contains these pairs of names and formulas: sulfite is SO_3^{2-}, phosphate is PO_4^{3-}, carbonate is CO_3^{2-}, and perchlorate is ClO_4^-. The charges of the cations in all four compounds are shown in Figure 2.20: they are Na^+, Mg^{2+}, Ca^{2+}, and K^+. As with all ionic compounds, the sum of the positive and negative charges of the ions in these compounds must be zero.

Solve

a. To balance charges, we need twice as many Na^+ ions as SO_3^{2-} ions. Therefore, the formula of sodium sulfite is Na_2SO_3.
b. To balance charges, we need three Mg^{2+} ions for every two PO_4^{3-} ions. Therefore, the formula of magnesium phosphate is $Mg_3(PO_4)_2$. The phosphate ion is placed in parentheses because there is more than one of it and it is polyatomic. The magnesium ion is not placed in parentheses, even though there is more than one magnesium ion, because it is monatomic.
c. Combining the names of the Ca^{2+} ions and CO_3^{2-} ions, we have calcium carbonate.
d. Combining the names of the K^+ ions and ClO_4^- ions, we have potassium perchlorate.

Think About It To complete this exercise, we had to know the formulas and charges of several monatomic ions and oxoanions. The charges on the most common monatomic ions can be inferred from the positions of their parent elements in the periodic table, but knowing the names and formulas of common polyatomic ions requires at least some memorization. There are patterns, however, that can reduce how much you must memorize. For example, if you learn the formulas and charges of the nitrate and sulfate ions, then you can remember that removing one oxygen atom from each yields the formulas and charges of the nitrite and sulfite ions.

Practice Exercise What are the formulas of (a) strontium nitrate and (b) potassium sulfate, and the names of (c) NaClO and (d) $KMnO_4$?

(Answers to Practice Exercises are in the back of the book.)

Acids

Some compounds have special names that highlight their chemical properties. Among them are acids. We discuss acids in greater detail in later chapters, but for now it is sufficient to say that acids are compounds that release hydrogen ions (H^+) when they dissolve in water. The simplest acids are called binary acids because their molecules contain atoms of only two elements: hydrogen and a nonmetal such as one of the group 17 elements (**Table 2.4**). These compounds are all gases at room temperature, but when they dissolve in water, their molecules separate into hydrogen ions and halide ions. To name these binary acids:

1. Affix the prefix *hydro-* to the name of the second element in the formula.
2. Replace the last syllable in the second element's name with the suffix *-ic* and add *acid.*

A second important class of acids forms hydrogen ions and oxoanions when these acids dissolve in water. **Table 2.5** contains the formulas and names of

TABLE 2.4 Formulas and Names of Common Binary Acids

Molecular Formula	Molecular Name	Formula of Acid	Name of Acid
HF(*g*)	Hydrogen fluoride	HF(*aq*)	Hydrofluoric acid
HCl(*g*)	Hydrogen chloride	HCl(*aq*)	Hydrochloric acid
HBr(*g*)	Hydrogen bromide	HBr(*aq*)	Hydrobromic acid
HI(*g*)	Hydrogen iodide	HI(*aq*)	Hydroiodic acid

TABLE 2.5 Formulas and Names of Some Common Oxoacids

Chemical Formula	Name
H_2CO_3	Carbonic acid
HNO_3	Nitric acid
HNO_2	Nitrous acid
H_3PO_4	Phosphoric acid
H_2SO_4	Sulfuric acid
H_2SO_3	Sulfurous acid
HClO	Hypochlorous acid
$HClO_2$	Chlorous acid
$HClO_3$	Chloric acid
$HClO_4$	Perchloric acid

several of these *oxoacids*. Compare their names to the names of the oxoanions (see Table 2.3) that they form when they dissolve in water. Note that the oxoacids with names ending in *-ic* form oxoanions with names ending in *-ate*. For example, an aqueous solution of perchlor*ic* acid ($HClO_4$) contains hydrogen ions and perchlo*rate* (ClO_4^-) ions. On the other hand, all the oxoacids whose names end in *-ous* form oxoanions with names ending in *-ite*. Solutions of chlor*ous* acid ($HClO_2$), for example, contain hydrogen ions and chlor*ite* (ClO_2^-) ions.

SAMPLE EXERCISE 2.11 Relating the Formulas and Names of Oxoacids and Oxoanions **LO6**

What are the names and formulas of the oxoacids that release the following oxoanions when they ionize in aqueous solutions: (a) NO_2^-; (b) CrO_4^{2-}; (c) MnO_4^-?

Collect, Organize, and Analyze We are given the formulas of three oxoanions and asked to name the formulas of the oxoacids that form them when the acids ionize in water. According to Table 2.3, the names of the oxoanions are (a) nitrite, (b) chromate, and (c) permanganate. When the oxoanion name ends in *-ite*, the corresponding oxoacid name ends in *-ous*. When the anion name ends in *-ate*, the oxoacid name ends in *-ic*.

Solve Making the appropriate changes to the endings of the oxoanion names and adding the word *acid*, we get (a) nitrous acid (HNO_2), (b) chromic acid (H_2CrO_4), and (c) permanganic acid ($HMnO_4$).

Think About It You cannot tell the names of the oxoanions just by looking at them. You must remember the names associated with each family of polyatomic ions.

Practice Exercise Name the following acids, and name and write the formulas of the oxoanions their molecules form when each of them releases a H^+ ion in an aqueous solution: (a) HBrO; (b) HIO_2; (c) H_2SeO_3.

(Answers to Practice Exercises are in the back of the book.)

2.8 Organic Compounds: A First Look

All the compounds we have discussed so far in this chapter—those composed of molecules and those made of ions—have been *inorganic* compounds. These compounds make up most of the world *around* us, but, except for water, these compounds constitute very little of the matter *inside* us.

We are living organisms, which means our bodies are mostly water and *organic* compounds composed of molecules that always contain carbon atoms, almost always contain hydrogen atoms, and frequently contain **heteroatoms**—typically oxygen, nitrogen, sulfur, phosphorus, and the halogens. The study of these compounds is called **organic chemistry**. Although *organic* has connotations in everyday conversation that often refer to farming and foods grown without the use of pesticides or synthetic fertilizers, the word *organic* as used by chemists refers to a compound's chemical composition and molecular structure. Still, chemists tend to distinguish between organic compounds derived from *natural products*, which are produced by living organisms, and *synthetic* organic compounds produced in the laboratory.

Hydrocarbons

Organic compounds that contain no heteroatoms are called **hydrocarbons** because their molecules contain only hydrogen and carbon atoms. One class of hydrocarbons, called **alkanes**, is composed of molecules in which each carbon atom is bonded to four other atoms by sharing a pair of electrons with each of them, that is, by forming four covalent bonds. These bonds are also called single bonds because they each involve the sharing of a single pair of electrons.

The names and structural formulas of the two simplest alkanes are

$$
\begin{array}{ccccc}
 & H & \\
 & | & \\
H- & C & -H \\
 & | & \\
 & H &
\end{array}
\qquad
\begin{array}{cccc}
 & H & H & \\
 & | & | & \\
H- & C- & C & -H \\
 & | & | & \\
 & H & H &
\end{array}
$$

Methane Ethane

The name of each of these alkanes ends in *-ane*, as do the names of all alkanes. Methane and ethane are important components of natural gas. Alkanes made of larger molecules make up other important fuels, including gasoline, diesel fuel, jet fuel, and heating oil. Because of their importance as fuels, we explore their molecular structures and chemical properties in detail in Chapter 6, which focuses on combustion and the energetics of chemical reactions.

Another class of hydrocarbons contains carbon atoms bonded to fewer than four other atoms because they contain pairs of carbon atoms that bond together by sharing *two* pairs or double bonds. These compounds are called **alkenes**. Two of the most common have these molecular structures:

$$
\begin{array}{cccc}
H- & C= & C & -H \\
 & | & | & \\
 & H & H &
\end{array}
\qquad
\begin{array}{ccccc}
 & & & H & \\
 & & & | & \\
H- & C= & C- & C & -H \\
 & | & | & | & \\
 & H & H & H &
\end{array}
$$

Ethylene (ethene) Propylene (propene)

Like all alkenes, the names of these compounds end in *-ene*. The names in parentheses, ethene and propene, follow the naming rules established by the International Union of Pure and Applied Chemistry (IUPAC). (We followed IUPAC rules in naming the inorganic compounds in Section 2.7.) The beginning letters of ethene and propene correspond to the names of the alkanes with the same number of carbon atoms (ethane and propane). This pattern is repeated for all

heteroatom an atom of an element other than carbon and hydrogen within a molecule of an organic compound.

organic chemistry the study of organic compounds.

hydrocarbon an organic compound whose molecules are composed only of carbon and hydrogen atoms.

alkane a hydrocarbon in which all the bonds are single bonds.

alkene a hydrocarbon containing one or more carbon–carbon double bonds.

alkyne a hydrocarbon containing one or more carbon–carbon triple bonds.

alcohol an organic compound containing the OH functional group.

functional group a group of atoms in the molecular structure of an organic compound that imparts characteristic chemical and physical properties.

alkenes. However, many organic compounds are better known by their common names, such as ethylene and propylene, which predate the current IUPAC naming rules. We use these common names throughout this book, providing the IUPAC names in parentheses.

A third class of hydrocarbons, called **alkynes**, contain pairs of carbon atoms that share three pairs of electrons, forming C≡C triple bonds:

$$\mathrm{H{-}C{\equiv}C{-}H} \qquad \mathrm{H{-}C{\equiv}C{-}CH_3}$$

Acetylene (ethyne) Propyne

As with alkenes, the names of alkynes are linked to the names of the alkanes with the same number of carbon atoms. All their IUPAC names end in *-yne*.

Heteroatoms and Functional Groups

Incorporating heteroatoms into the molecular structure of hydrocarbons produces organic compounds with physical and chemical properties that are very different from those of their parent hydrocarbons. For example, replacing a hydrogen atom with an OH group in the molecular structures of methane and ethane produces molecules with these simplified structural formulas (the O–H bonds are not shown):

$$\mathrm{H{-}CH_2{-}OH} \qquad \mathrm{H{-}CH_2{-}CH_2{-}OH}$$

Methanol Ethanol

The names of these compounds use the parent alkane root (i.e., *methan-* and *ethan-*, respectively), but they each end in *-ol.* The presence of the OH groups, highlighted in red, means that these compounds are **alcohols**. Unlike their parent alkanes, which are gases at room temperature and pressure, methanol and ethanol are colorless liquids. Alkanes don't dissolve in water (at least, not much), but these two alcohols dissolve freely in water (and water dissolves freely in them).

The OH group in alcohols is an example of a **functional group**, that is, a group of atoms in the molecular structure of a compound that has a significant impact on the physical and chemical properties of the compound. The common functional groups, whose chemical reactivity we explore in the chapters ahead, are listed in **Table 2.6**. All the functional groups contain an atom of oxygen or nitrogen or both. The symbols R and R′ are used to represent portions of the molecule that contain only carbon and hydrogen atoms.

There are many other rules that apply to naming compounds such as those in Table 2.6. See Appendix 7 for a comprehensive summary. Rather than dwelling on all of them right now, try instead to relate the names and structures of the sample compounds by using the names of methane and ethane and the class of compounds to which each belongs. For example, the name of the sample aldehyde in Table 2.6, *ethanal,* is derived from *ethane*, the alkane that also has two carbon atoms per molecule, and the functional group *al*dehyde.

TABLE 2.6 Classes of Organic Compounds and Their Functional Groups

Class of Compounds	Structural Formula of Functional Group	Sample Compound	Name of Compound	Common Use of Compound
Alcohols	R—OH	CH_3CH_2—OH	Ethanol	Alcoholic beverages
Ethers	R—O—R'	H_3C—O—CH_3	Dimethyl ether (methoxymethane)	Organic synthesis
Aldehydes	R—C(=O)—H	H_3C—C(=O)—H	Ethanal (acetaldehyde)	Making plastic resin
Ketones	R—C(=O)—R'	H_3C—C(=O)—CH_3	Acetone (propanone)	Nail polish remover
Carboxylic acids	R—C(=O)—OH	H_3C—C(=O)—OH	Acetic acid (ethanoic acid)	Vinegar
Esters	R—C(=O)—O—R'	H_3C—C(=O)—O—CH_3	Methyl acetate (methyl ethanoate)	Nail polish remover, glue
Amines	R—NH_2	H_3C—NH_2	Methylamine (aminomethane)	Synthesis of pharmaceuticals and industrial chemicals
Amides	R—C(=O)—NH_2	H_3C—C(=O)—NH_2	Acetamide (ethanamide)	Solvent, plastics additive

CONCEPT TEST

What functional group is present in each of the three organic compounds shown in **Figure 2.23**? These molecular models use different colors to represent atoms (see Section 1.2), namely, black is carbon, white is hydrogen, red is oxygen, and blue is nitrogen.

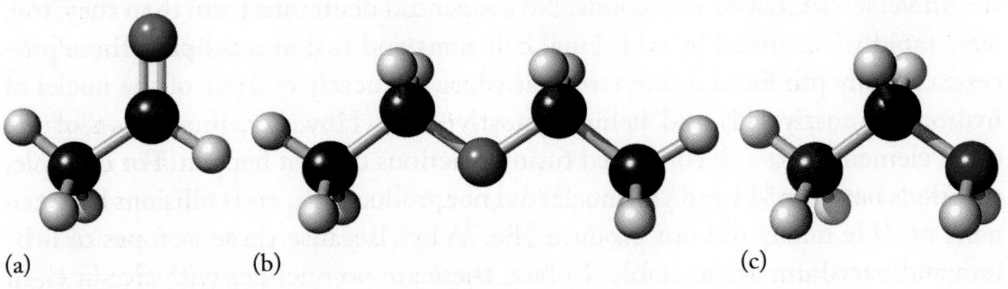

FIGURE 2.23 Organic compounds.

(Answers to Concept Tests are in the back of the book.)

2.9 Nucleosynthesis: The Origin of the Elements

We began this chapter by discussing remarkable advances in the early 20th century that provided insights into the structure of atoms and helped to explain why elements react the way they do. Later in the century, scientists increased our understanding of how the elements may have originally formed by determining the composition of the very early universe—the particles that appeared in the moments immediately following the Big Bang.

Scientists believe that much of the energy released at the instant of the Big Bang transformed into matter within a few microseconds (**Figure 2.24**). This matter consisted of electrons and **quarks**. Less than a millisecond later, the universe had expanded and "cooled" to a mere 10^{12} K, and quarks combined with one another to form neutrons and protons. Thus, in less than a second, the matter in the universe consisted of the three types of subatomic particles that would eventually make up all atoms.

Primordial Nucleosynthesis

By about four minutes after the Big Bang, the universe had expanded and cooled to 10^9 K. In this hot, dense subatomic "soup," neutrons and protons that collided with one another began to fuse in a process called primordial **nucleosynthesis**. (We discuss nuclear fusion in more detail in Chapter 19.) In one step, protons (p) and neutrons (n) fused to form *deuterons* (d), which are nuclei of the deuterium ($^2_1$H) isotope of hydrogen:

$$^1_1\text{p} + {}^1_0\text{n} \rightarrow {}^2_1\text{d} \qquad (2.2)$$

In writing this equation, we follow the rules presented in Section 2.3 for writing nuclide symbols: the superscript is a mass number and the subscript is an atomic number. We use a similar convention for writing the symbols of subatomic particles, except that subscripts represent the charges on particles. For example, the symbol of the neutron is $^1_0$n because a neutron has a mass number of 1 and a charge of 0.

Equation 2.2 is an example of a *nuclear equation*. It is related to the chemical equations we begin working with in Chapter 3 in that it is *balanced*. This means that the sum of the masses (superscripts) of the particles to the left of the arrow is equal to the sum of the masses of the particles to the right of the arrow. Similarly, the sum of the charges (subscripts) of the particles on the left side is equal to the sum of the charges of the particles on the right side.

Deuteron formation proceeded rapidly, consuming most of the neutrons in the universe in a matter of seconds. No sooner did deuterons form than they, too, were rapidly consumed by additional collisions and fusion reactions. These processes rapidly produced a universe that consisted nearly entirely of the nuclei of hydrogen (mostly $^1_1$H) and helium (mostly $^4_2$He). However, production of the other elements through continued fusion reactions did not happen. For example, collisions between $^1_1$H and $^4_2$He nuclei did not produce $^5_3$Li, and collisions between pairs of $^4_2$He nuclei did not produce $^8_4$Be. Why? Because these isotopes of lithium and beryllium are unstable. In fact, there are no nuclides with five or eight nucleons.

Within minutes of the Big Bang, therefore, primordial nucleosynthesis shut down, leaving an expanding and cooling universe that consisted nearly

quarks elementary particles that combine to form neutrons and protons.

nucleosynthesis the natural formation of nuclei because of fusion and other nuclear processes.

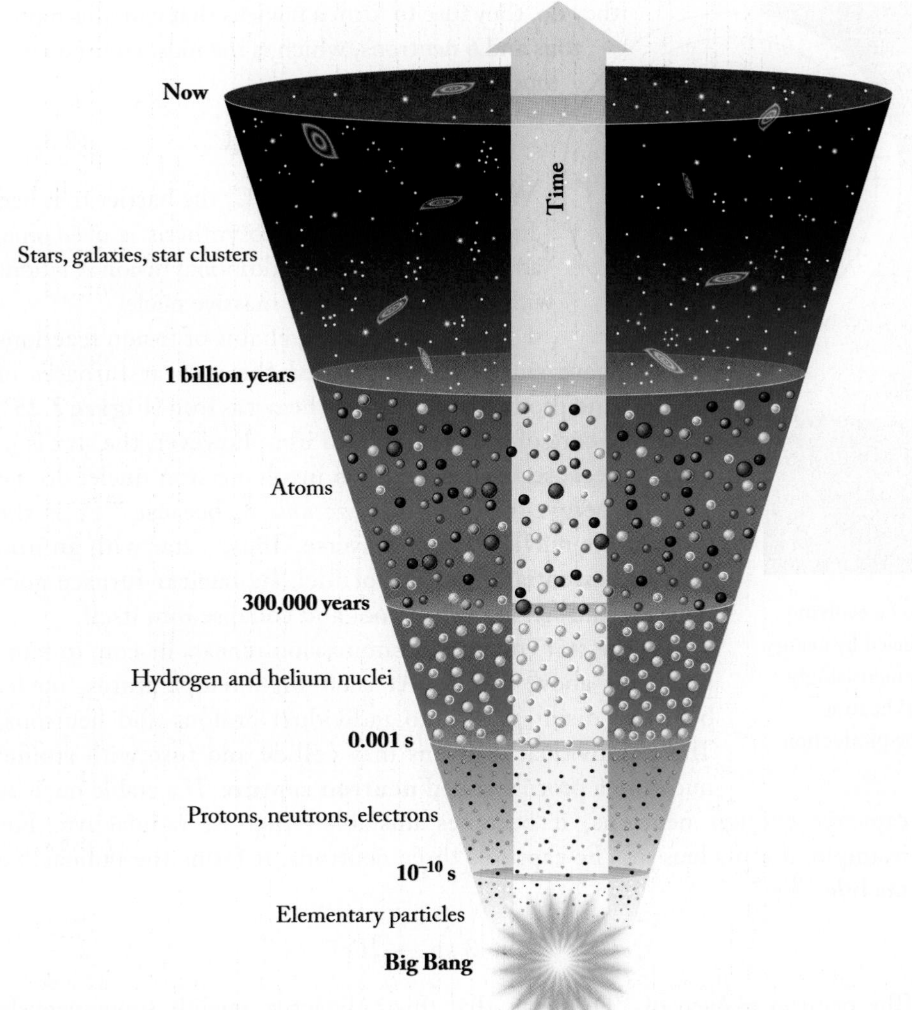

FIGURE 2.24 Timeline for energy and matter transformations believed to have occurred since the universe began. In this model, protons and neutrons were formed from quarks in the first millisecond after the Big Bang, followed by hydrogen and helium nuclei. Whole atoms of H and He did not form until after 300,000 years of expansion and cooling, and other elements did not form until the first galaxies appeared, around 1 billion years after the Big Bang. According to this model, our solar system, our planet, and all life-forms on it are composed of elements synthesized in stars that were born, burned brightly, and then disappeared millions to billions of years after the Big Bang.

entirely of hydrogen and helium. The fact that the universe today is still 99% hydrogen and helium and that these elements occur in the same proportion predicted by physicists' models of primordial nucleosynthesis (about 3 parts hydrogen to 1 part helium, by mass) is strong evidence in support of the theory of the Big Bang.

Stellar Nucleosynthesis

CHEMTOUR
The Synthesis of Elements

How, then, did the other elements in the periodic table form, including those that make up most of our planet? Scientists theorize that the synthesis of elements more massive than helium had to wait until nuclear fusion resumed in the first generation of stars. Inside the coalescing masses of hydrogen and helium that would turn into the first stars, the gases underwent enormous compressional heating. The nuclear furnaces that are the source of the energy in all stars were ignited as hydrogen nuclei began to fuse, making more helium nuclei.

Once nuclear fusion begins in stars, it may continue in stages. When the hydrogen at the star's core has all fused into helium, gravity causes the core to contract, creating even higher temperatures and pressures. In the cores of large stars, helium nuclei are so densely packed that they collide three at a time. When

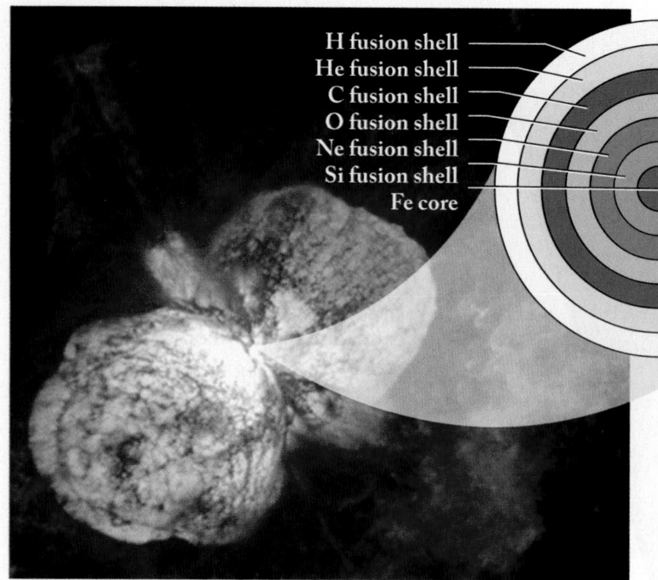

FIGURE 2.25 The star Eta Carinae is believed to be evolving toward an explosion. The outer regions are still fueled by energy released as hydrogen isotopes fuse, but the star is increasingly hotter and denser closer to the center. This central heating allows the fusion of larger nuclei and results in the production of ^{56}Fe in the core.

they do, they fuse to form a nucleus that contains 6 protons and 6 neutrons, which is the most common isotope of carbon:

$$3\,{}^{4}_{2}\text{He} \rightarrow {}^{12}_{6}\text{C} \tag{2.3}$$

With the formation of ^{12}C, the barrier that had halted primordial nucleosynthesis is overcome, and the stage is set for additional fusion reactions with progressively more massive nuclei.

For billions of years, chains of fusion reactions have simultaneously fueled the nuclear furnaces of stars and produced elements as heavy as iron (**Figure 2.25**). Once the core of a star turns into iron, however, the star is in trouble because fusion reactions involving iron nuclei do *not* release energy. Instead, they *consume* it, because ^{56}Fe is the most stable nuclide in the universe. Thus, a star with an iron core has essentially run out of fuel. Its nuclear furnace goes out, and the star begins to cool and collapse into itself.

As the star collapses, compression reheats its core to temperatures above 10^9 K. At such high temperatures, nuclei begin to disintegrate into individual protons and neutrons. These individual neutrons may collide and fuse with atomic nuclei in a process called **neutron capture**. If a stable nucleus captures enough neutrons, it becomes unstable—that is, radioactive. For example, if a nucleus of ^{56}Fe captures three neutrons, it forms the radioactive nuclide ^{59}Fe:

$$^{56}_{26}\text{Fe} + 3\,^{1}_{0}\text{n} \rightarrow {}^{59}_{26}\text{Fe}$$

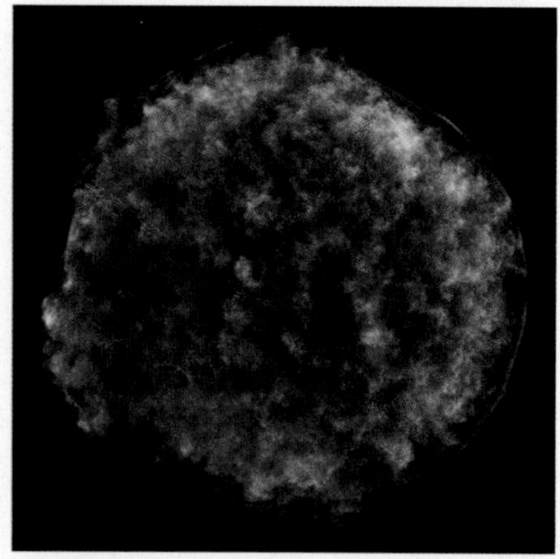

FIGURE 2.26 This colorized picture from the Chandra X-ray Observatory shows the remains of SN 1572, called "Tycho's supernova," in the constellation Cassiopeia. The expanding bubble of debris colored red and green is a cloud of hot ionized gas inside a more rapidly moving shell of extremely high-energy electrons in blue.

The neutron *richness* of ^{59}Fe means that this radioactive nuclide spontaneously undergoes a kind of radioactive decay that reduces the ratio of neutrons to protons in its nucleus. This is called **beta (β) decay** because it involves the ejection of a high-energy electron, or β particle. The decay process is summarized in the following nuclear equation:

$$^{59}_{26}\text{Fe} \rightarrow {}^{59}_{27}\text{Co} + {}^{\;0}_{-1}\beta \tag{2.4}$$

The nuclide formed in the reaction has an atomic number of 27, or one more than the radioactive isotope that produced it. This increase is the result of the decomposition of a neutron into a proton, which remains in the nucleus, and an electron (β particle) that is ejected from it. The additional proton results in an increase in atomic number. The combination of repeated neutron capture and β decay events in the cores of collapsing stars produces the most massive stable nuclides in the periodic table.

Eventually, the enormous heating that occurs when a massive star collapses produces a gigantic explosion. Cosmologists call such an event a *supernova*. In addition to finishing the job of synthesizing the elemental building blocks found in the universe—up to and including the isotopes of uranium—a supernova serves as its own element-distribution system, blasting its inventory of elements throughout its galaxy (**Figure 2.26**). The legacies of supernovas are found in the elemental composition of later-generation stars like our Sun and in the planets that orbit these stars. Indeed, all the matter in our solar system—and in us—is essentially demolition debris from ancient exploding stars.

neutron capture the absorption of a neutron by a nucleus.

beta (β) decay a spontaneous process by which a neutron in a radioactive nuclide is transformed into a proton and emits a high-energy electron (β particle).

SAMPLE EXERCISE 2.12 Integrating Concepts: Radioactive Isotopes in Medicine

Dmitri Mendeleev predicted the existence of technetium ($Z = 43$), one of the elements that was unknown in his time. Technetium has more than 30 isotopes, with mass numbers ranging from 85 to 118. Every isotope of technetium is radioactive. There are medical procedures that take advantage of this radioactivity, using ^{99}Tc to image many parts of the human body, including the brain, heart, thyroid, lungs, and kidneys (**Figure 2.27**). More than 20 million medical procedures using ^{99}Tc are performed each year. One challenge, though, is how to make ^{99}Tc available to doctors precisely where and when it is needed because it rapidly undergoes radioactive decay. The solution to this problem is to transport a more stable substance, in this case, molybdenum-99 ($Z = 42$), from which technetium-99 can be isolated. Most of the molybdenum-99 on Earth is produced from ^{235}U in nuclear reactors.

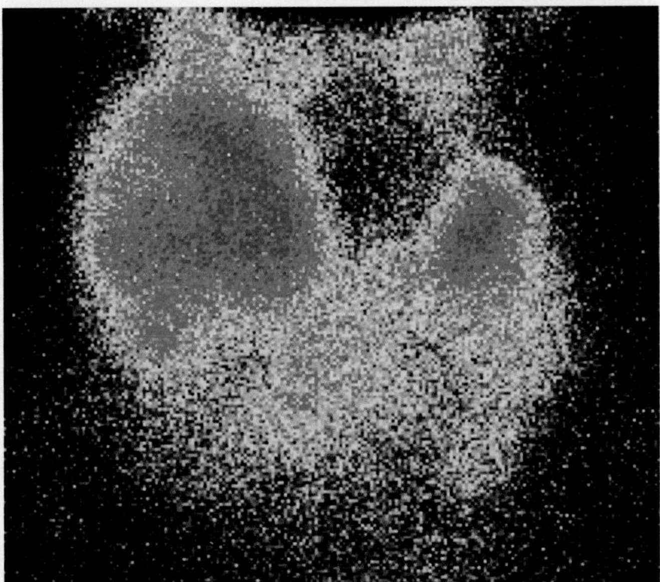

FIGURE 2.27 Technetium is used to study the function of organs such as the thyroid gland. Typically, an image is taken soon after injection of a radioactive Tc compound. After a time, the thyroid is re-imaged to see how much Tc remains. The distribution of Tc should be symmetrical. Here the lobes are not the same, indicating abnormal thyroid function.

a. What are the nuclide symbols for the lightest and the heaviest isotopes of technetium?
b. How many protons, neutrons, and electrons are in a neutral atom of ^{99}Tc?
c. When doctors need to administer ^{99}Tc to a patient, they obtain a sample of radioactive molybdenum-99 that then undergoes beta decay, forming ^{99}Tc. Write a balanced nuclear equation to show the beta decay of molybdenum-99 to form technetium-99.

Collect and Organize We are given the mass numbers of two technetium isotopes and are asked to write their complete nuclide symbols, and we need to translate the nuclide symbol ^{99}Tc into the numbers of protons, neutrons, and electrons per atom. We are also given the molybdenum isotope that decays to ^{99}Tc and are asked to write a balanced chemical equation describing the reaction.

Analyze The atomic number (Z) of an element is the same as the number of protons and the number of electrons in each of its atoms. The mass number (A) of a nuclide is the sum of the numbers of protons and neutrons in each of its nuclei. In the symbol of a nuclide, Z is the subscript and A is the superscript that precede the element symbol. Beta decay involves the ejection of high-energy electrons, $_{-1}^{0}\beta$, and produces nuclides with an atomic number one greater than the radionuclides undergoing decay.

Solve

a. For the lightest isotope of Tc, $Z = 43$ and $A = 85$; for the heaviest isotope, $Z = 43$ and $A = 118$. Therefore, the two symbols are $_{43}^{85}Tc$ and $_{43}^{118}Tc$, respectively.
b. For ^{99}Tc: 43 protons, 56 neutrons, 43 electrons
c. $_{42}^{99}Mo \rightarrow _{43}^{99}Tc + _{-1}^{0}\beta$

Think About It Because one of its isotopes emits radiation that can be easily detected, technetium incorporated into soluble molecules can be used to observe the flow of fluids in organisms. The thyroid is a symmetrical, butterfly-shaped gland that surrounds the trachea. It appears in Figure 2.27 as the two lobes imaged after the administration of a technetium-containing agent. The gland should be symmetrical, but one lobe is smaller than the other. The irregular uptake of the agent indicates a malfunctioning thyroid.

SUMMARY

LO1 Thomson's experiments with cathode rays, Millikan's oil-drop experiments, Rutherford's gold-foil experiments, and studies of **radioactivity** by Henri Becquerel and Marie and Pierre Curie led to our understanding of atomic structure. (Sections 2.1 and 2.2)

LO2 Atoms consist of **nuclei** that each contain one or more **nucleons**, that is, **protons** and **neutrons**. The nucleus accounts for nearly all the mass of an atom; it is surrounded by much less massive **electrons**, whose negative charges balance the positive charges of the protons. Unique symbols for each **isotope** of every element indicate the identity of an atom by showing its **atomic number** (the number of protons in its nucleus) and its **mass number** (the number of nucleons in the nucleus). (Sections 2.2 and 2.3)

LO3 The **average atomic mass** of an element is the weighted average of all the isotopes of that element. It is calculated by multiplying the mass of each stable isotope by the isotope's **natural abundance** and then summing these products. (Section 2.4)

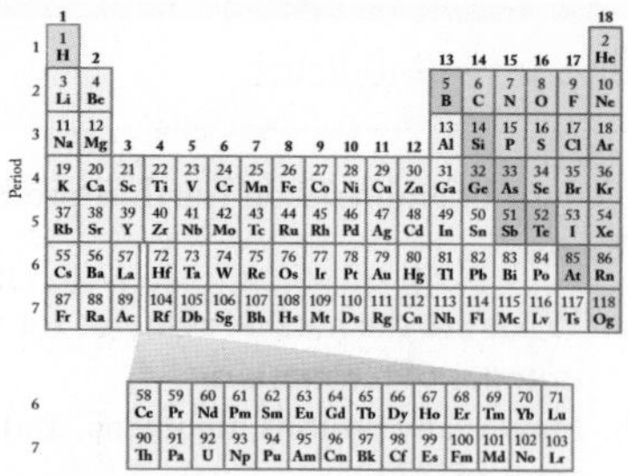

LO4 Russian chemist Dmitri Mendeleev developed a forerunner of the modern **periodic table of the elements** in which the elements were arranged in patterns based on their physical and chemical properties. Elements in the same vertical column are said to be in the same **group**. The **main group** (or **representative**) **elements** are in groups 1, 2, and 13–18, and the **transition metals** are in groups 3–12. Within the periodic table, three regions contain **metals, metalloids**, and **nonmetals**. (Sections 2.3 and 2.5)

LO5 The **law of multiple proportions** describes the ratio of the masses of elements in simple compounds in terms of small whole numbers. Compounds composed of pairs of metallic and nonmetallic elements tend to be **ionic compounds**, whereas those composed of two nonmetals or metalloids tend to be **molecular compounds**. The **empirical formula** of a molecular or ionic compound gives the smallest whole-number ratio of the atoms (or ions) in it. (Section 2.6)

LO6 Standardized naming conventions allow us to translate the names of molecular and ionic compounds into their chemical formulas, and vice versa. (Section 2.7)

LO7 **Organic compounds** always contain carbon atoms and almost always contain hydrogen atoms. Different classes of organic compounds are composed of molecules with characteristic **functional groups**. (Section 2.8)

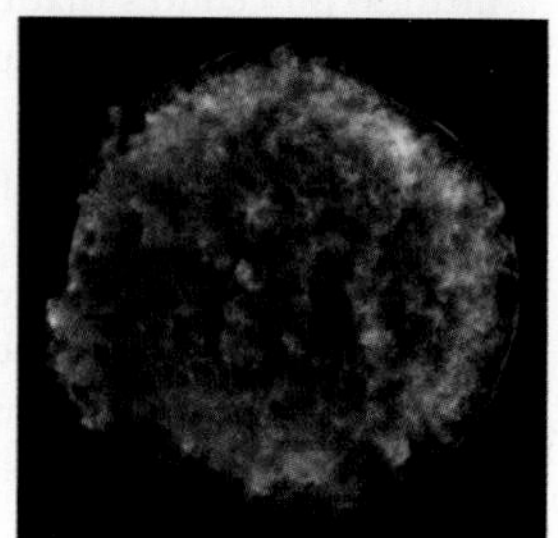

LO8 During primordial **nucleosynthesis**, protons and neutrons fused, producing nuclei of helium. Nuclei of atoms as massive as ^{56}Fe formed when the nuclei of lighter elements fused in the cores of giant stars in a process called stellar nucleosynthesis, which continues today. Even more massive nuclei are formed by a combination of other nuclear reactions, leading to supernovas (explosions of giant stars). (Section 2.9)

PARTICULATE PREVIEW WRAP-UP

The middle compound is ionic because it contains two types of particles (one is a polyatomic ion) arranged in an extended, ordered array. The other two are molecular compounds. The images show single molecules containing atoms of two (on the left) and three (on the right) elements bonded together.

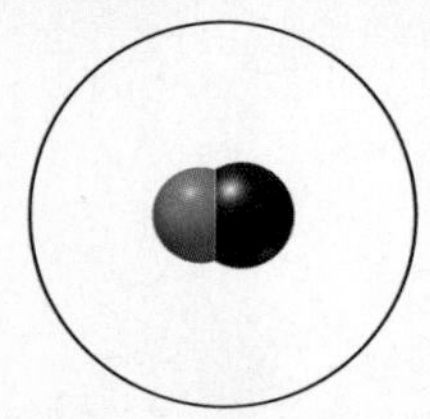

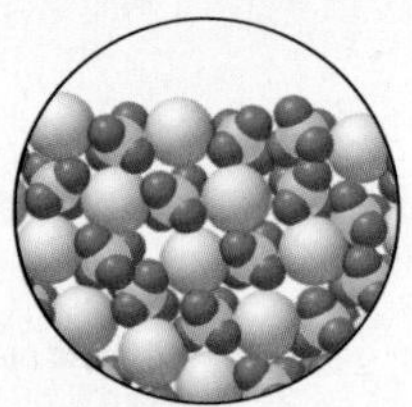

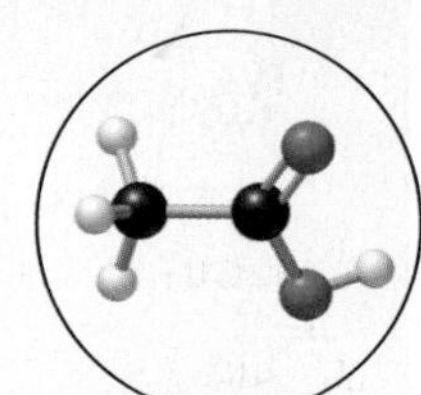

PROBLEM-SOLVING SUMMARY

Type of Problem	Concepts and Equations	Sample Exercises
Writing symbols of isotopes	To the left of the element symbol, place a superscript for the mass number (A) and (if needed) a subscript for the atomic number (Z).	**2.1, 2.2**
Calculating the average atomic mass of an element	Multiply the mass (m) of each stable isotope of the element times the natural abundance (a) of that isotope converted to a decimal; then sum these products: $m_X = a_1m_1 + a_2m_2 + a_3m_3 + \ldots$ (2.1)	**2.3**
Locating elements in the periodic table	Use the row number, group designation, and chemical properties to locate elements.	**2.4**
Predicting the composition of different compounds formed by the same two elements	Use the chemical formulas of the two compounds and the law of multiple proportions.	**2.5**
Classifying compounds as ionic or molecular	Ionic compounds contain metallic and nonmetallic elements, whereas molecular compounds contain nonmetals or metalloids.	**2.6**
Naming binary compounds and writing their formulas	Apply the naming rules in Section 2.7.	**2.7, 2.8**
Naming transition metal compounds and writing their formulas	Use a Roman numeral to indicate the charge on the transition metal cation.	**2.9**
Naming oxoacids and compounds containing oxoanions and writing their formulas	Apply the naming rules in Section 2.7.	**2.10, 2.11**

VISUAL PROBLEMS

(Answers to boldface end-of-chapter questions and problems are in the back of the book.)

2.1. Alpha and beta particles emitted by a sample of pitchblende escape through a narrow channel in the shielding surrounding the sample and into an electrical field as shown in Figure P2.1. Which path in the figure corresponds to each form of radiation?

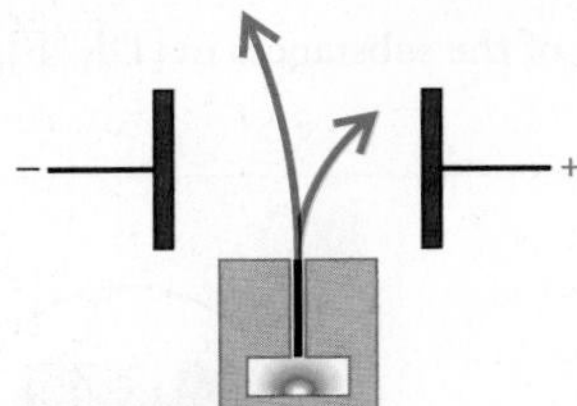

FIGURE P2.1

2.2. Does the particle that follows the red path in Figure P2.1 penetrate solid objects better than the radiation following the green path? Explain your answer.

2.3. The images in Figure P2.3 represent the nuclei of three nuclides. Assuming all the nucleons are visible, what are the symbols of these nuclides (include both mass number and atomic number).

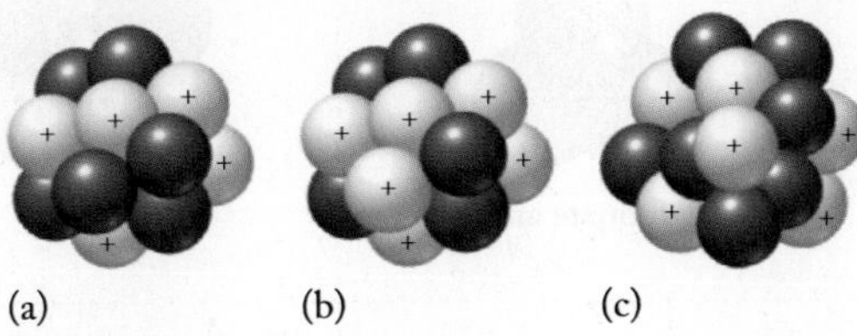

FIGURE P2.3

2.4. The images in Figure P2.4 represent the nuclei of three nuclides. Assuming all the nucleons are visible, what are the symbols of these nuclides (include both mass number and atomic number).

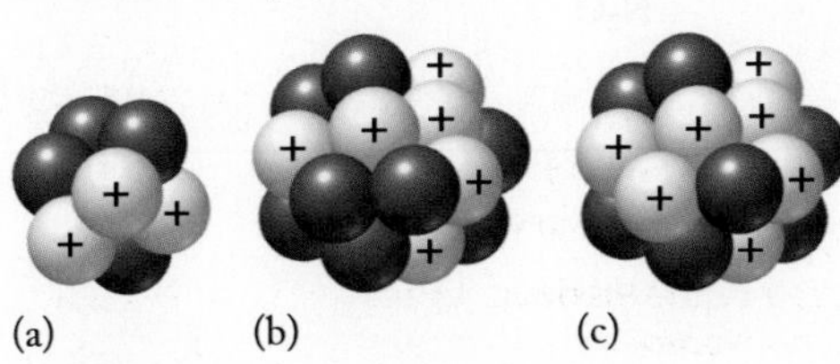

FIGURE P2.4

2.5. In Figure P2.5 the blue spheres represent nitrogen atoms and the red spheres represent oxygen atoms. The figure represents which of the following gases? (a) N_2O_3; (b) N_7O_{11}; (c) a mixture of NO_2 and NO; (d) a mixture of N_2 and O_3

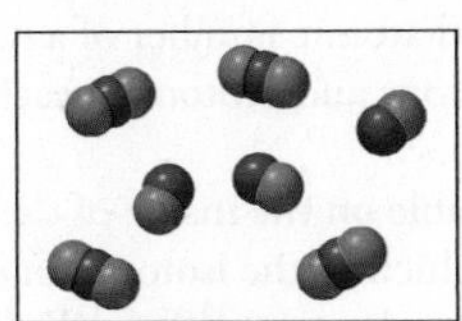

FIGURE P2.5

2.6. In Figure P2.6 the black spheres represent carbon atoms and the red spheres represent oxygen atoms. Which of the following statements about the two equal-volume compartments is or are true?

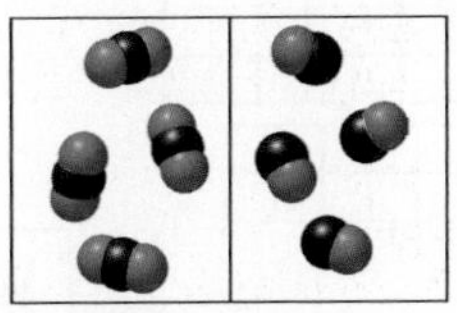

FIGURE P2.6

a. The compartment on the left contains CO_2; the one on the right contains CO.
b. The compartments contain the same mass of carbon.
c. The ratio of oxygen to carbon in the gas in the left compartment is twice that of the gas in the right compartment.
d. The pressures inside the two compartments are equal. (Assume that the pressures are proportional to the number of molecules in each compartment.)

2.7. Which of the highlighted elements in Figure P2.7 is (a) a reactive nonmetal, (b) a chemically inert gas, (c) a reactive metal?

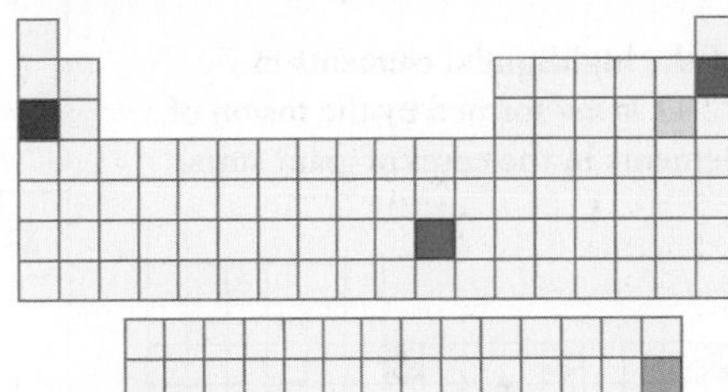

FIGURE P2.7

2.8. Which of the highlighted elements in Figure P2.8 forms monatomic ions with a charge of (a) 1+, (b) 2+, (c) 3+, (d) 1−, (e) 2−?

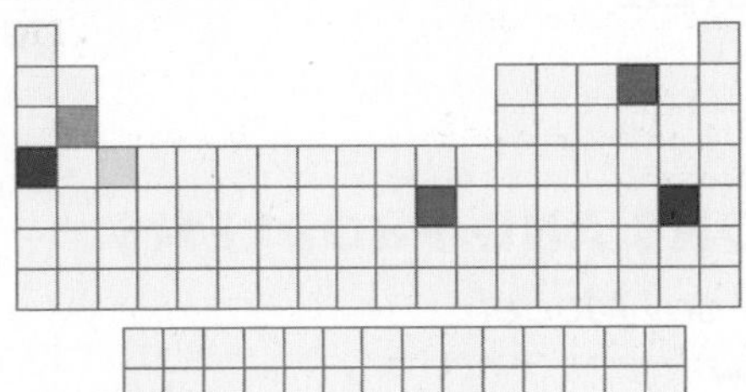

FIGURE P2.8

2.9. Which of the highlighted elements in Figure P2.9 forms an oxide with the following formula: (a) XO; (b) X_2O; (c) XO_2; (d) X_2O_3?

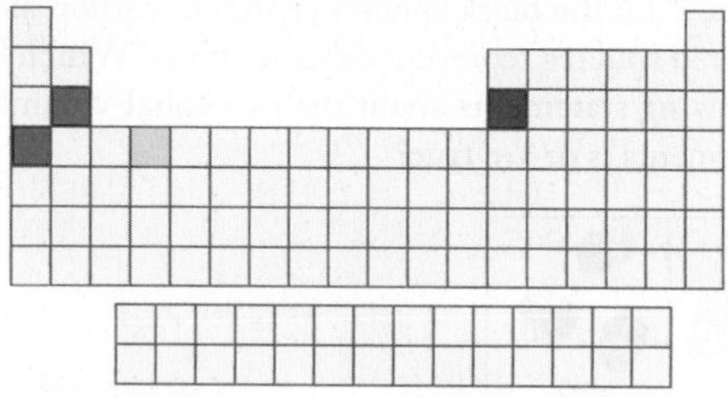

FIGURE P2.9

2.10. Which of the highlighted elements in Figure P2.10 forms an oxoanion with the following generic formula: (a) XO_4^-; (b) XO_4^{2-}; (c) XO_4^{3-}; (d) XO_3^-?

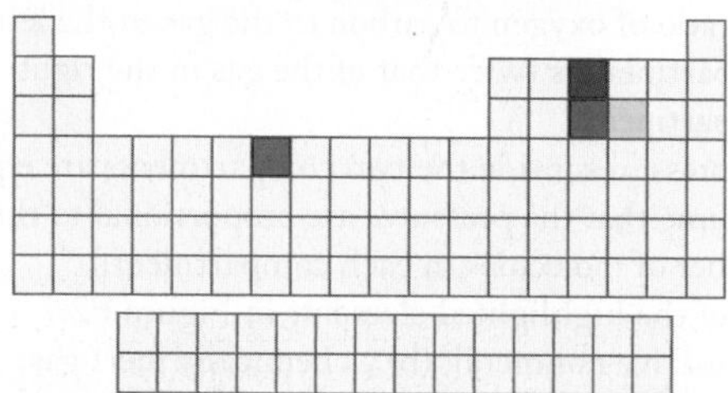

FIGURE P2.10

2.11. Which of the highlighted elements in Figure P2.11 is *not* formed by the fusion of lighter elements in the cores of giant stars?

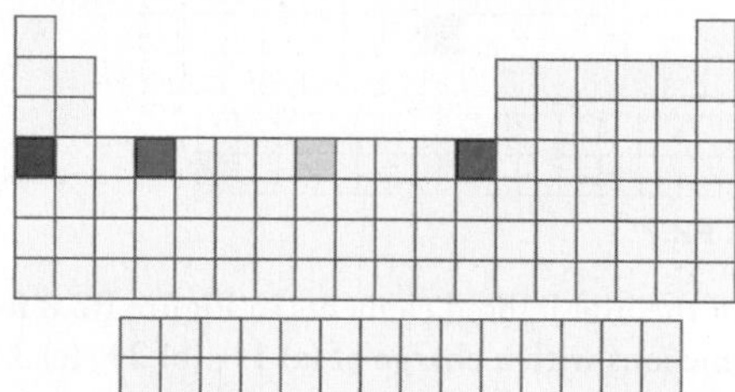

FIGURE P2.11

2.12. Use representations [A] through [I] in Figure P2.12 to answer questions (a)–(f).

a. Which are molecular compounds?
b. Which are ionic compounds?
c. Which incorporates both covalent and ionic bonding?
d. Which two representations demonstrate the law of multiple proportions?
e. Which of the representations, if any, have the same empirical formula?
f. What are the names of the substances in [D], [F], and [H]?

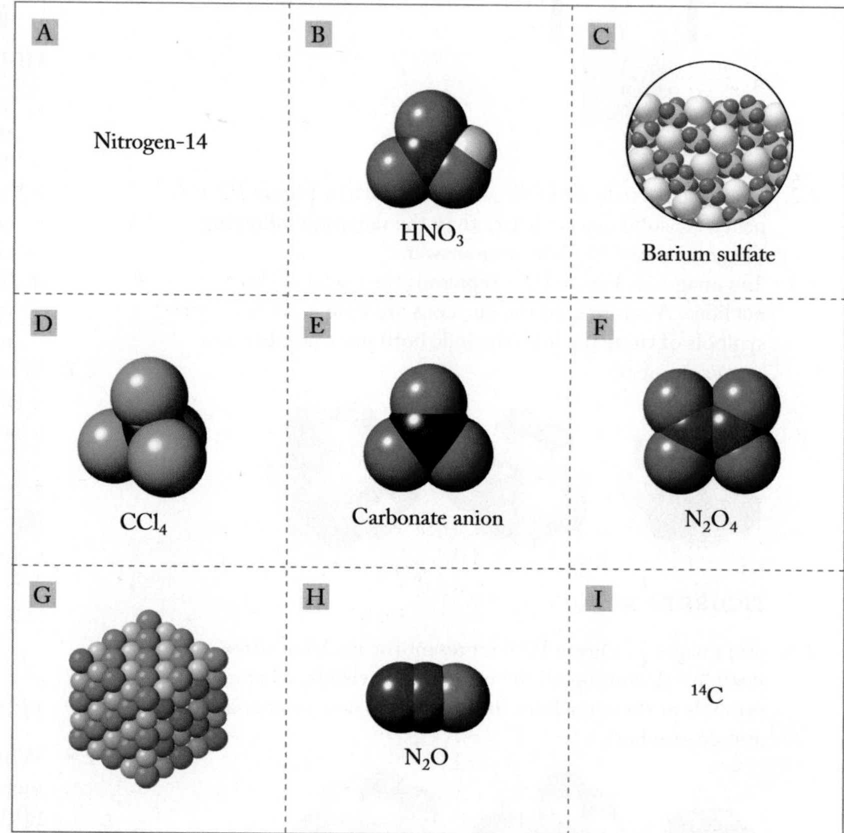

FIGURE P2.12

QUESTIONS AND PROBLEMS

The Rutherford Model

Concept Review

2.13. Explain how the results of the gold-foil experiment led Rutherford to dismiss the plum-pudding model of the atom and create his own model based on a nucleus surrounded by electrons.

2.14. Had the plum-pudding model been valid, how would the results of the gold-foil experiment have differed from what Geiger and Marsden observed?

2.15. What properties of cathode rays led Thomson to conclude that they were not pure energy, but rather particles with an electric charge?

*2.16. What would be observed if the charges on the plates of Millikan's apparatus (Figure 2.4) were reversed?

Isotopes and Average Atomic Mass

Concept Review

2.17. What is meant by a *weighted average*?

2.18. Explain how natural abundance percentages are related to average atomic masses.

2.19. Explain the inherent redundancy in the nuclide symbol $^A_Z X$.

2.20. How are the mass number and atomic number of a nuclide related to the number of neutrons and protons in each of its nuclei?

2.21. Use the data in the periodic table on the inside of the front cover of this book to decide which of the isotopes in each of the following pairs is more abundant. (a) ^{10}B or ^{11}B; (b) ^6Li or ^7Li; (c) ^{14}N or ^{15}N; (d) ^{20}Ne or ^{22}Ne

2.22. If the nucleus of an atom contains an equal number of neutrons and protons, is the ratio of its mass number to its atomic number greater than, less than, or equal to 1?

Problems

2.23. How many protons, neutrons, and electrons are there in the following atoms? (a) ^{14}C; (b) ^{59}Fe; (c) ^{90}Sr; (d) ^{210}Pb

2.24. How many protons, neutrons, and electrons are there in the following atoms? (a) ^{11}B; (b) ^{19}Fe; (c) ^{131}I; (d) ^{222}Rn

2.25. Fill in the missing information about atoms of the four nuclides in the following table.

Symbol	^{16}O	?	?	?
Number of Protons	?	26	?	79
Number of Neutrons	?	30	?	?
Number of Electrons	?	?	50	?
Mass Number	?	?	118	197

2.26. Fill in the missing information about atoms of the four nuclides in the following table.

Symbol	^{27}Al	?	?	?
Number of Protons	?	42	?	92
Number of Neutrons	?	56	?	?
Number of Electrons	?	?	60	?
Mass Number	?	?	143	238

2.27. The natural abundances of the two stable isotopes of boron can vary. If a sample collected from a volcanic crater lake in southeastern Australia contained 81.07% ^{11}B (mass = 11.0093 u) and 18.93% ^{10}B (mass = 10.0129 u), what is the average atomic mass of the boron in the sample? Compare your calculated result to the average atomic mass of boron listed on the inside of the front cover of this book.

2.28. The natural abundances of the two stable isotopes of lithium can vary a little. If a sample of Li is 7.714% ^{6}Li (mass = 6.015123 u) and 92.286% ^{7}Li (mass = 7.016003 u), what is the average atomic mass of the lithium in the sample? Compare your calculated result to the average atomic mass of lithium listed on the inside of the front cover of this book.

2.29. **Chemistry of Mars** NASA's 1997 mission to Mars included a small robot, the *Sojourner*, that analyzed the composition of Martian rocks. Magnesium oxide from a boulder dubbed "Barnacle Bill" was analyzed and found to have the following isotopic composition:

Mass (u)	Natural Abundance (%)
39.9872	78.70
40.9886	10.13
41.9846	11.17

If essentially all the oxygen in the Martian MgO sample is oxygen-16 (which has a mass of 15.9948 u), is the average atomic mass of magnesium on Mars the same as on Earth (24.31 u)?

2.30. Platinum has six isotopes: ^{190}Pt, ^{192}Pt, ^{194}Pt, ^{195}Pt, ^{196}Pt, and ^{198}Pt.

a. How many neutrons are there in each isotope?
b. The natural abundances of the six isotopes are 0.014% ^{190}Pt (189.96 u); 0.782% ^{192}Pt (191.96 u); 32.967% ^{194}Pt (193.96 u); 33.832% ^{195}Pt (194.97 u); 25.242% ^{196}Pt (195.97 u); and 7.163% ^{198}Pt (197.97 u). Calculate the average atomic mass of platinum and compare it with the value in the periodic table on the inside front cover.

2.31. Use the following table of abundances and masses of five naturally occurring titanium isotopes to calculate the mass of ^{48}Ti.

Symbol	Mass (u)	Natural Abundance (%)
^{46}Ti	45.95263	8.25
^{47}Ti	46.9518	7.44
^{48}Ti	?	73.72
^{49}Ti	48.94787	5.41
^{50}Ti	49.9448	5.18
Average	47.87	

2.32. Use the following table of abundances and masses of the three naturally occurring argon isotopes to calculate the mass of ^{40}Ar.

Symbol	Mass (u)	Natural Abundance (%)
^{36}Ar	35.96755	0.337
^{38}Ar	37.96272	0.063
^{40}Ar	?	99.60
Average	39.948	

The Periodic Table of the Elements

Concept Review

2.33. Mendeleev ordered the elements in his version of the periodic table according to their atomic masses instead of their atomic numbers. Why?

2.34. Why didn't Mendeleev include the noble gases in his version of the periodic table?

2.35. Mendeleev arranged the elements on the left side of his periodic table according to the formulas of the binary compounds they form with oxygen, and he used those formulas as column labels. For example, group 1 of the modern periodic table was labeled "R_2O" in Mendeleev's table, where "R" represented one of the elements in the group. What labels did Mendeleev use for groups 2, 3, and 4 of the modern periodic table?

2.36. Mendeleev arranged the elements on the right side of his periodic table according to the formulas of the binary compounds they form with hydrogen, and he used those formulas as column labels. Which groups of the modern periodic table were labeled "HR," "H_2R," and "H_3R" in Mendeleev's table, where "R" represented one of the elements in the group?

Problems

2.37. Which element is most likely to form a cation with a 3+ charge? (a) S; (b) P; (c) Be; (d) Al

2.38. Which element is most likely to form a monatomic ion with a 3− charge? (a) S; (b) P; (c) Be; (d) Al

2.39. Which ions have the same number of electrons as an atom of argon? (a) S^{2-}; (b) P^{3-}; (c) Be^{2+}; (d) Ca^{2+}

2.40. Which ions have the same number of electrons as an atom of krypton? (a) Se^{2-}; (b) As^{3-}; (c) Ca^{2+}; (d) K^{+}

2.41. Classify each of the following fifth-row elements as a metal, a metalloid, or a nonmetal: (a) Y; (b) Ag; (c) Sn; (d) Te; (e) I.

2.42. Classify each of the following fourth-row elements as a metal, a metalloid, or a nonmetal: (a) Ca; (b) Cr; (c) Cu; (d) As; (e) Br.

2.43. Classify each of the following fourth-row elements as an alkali metal, an alkaline earth metal, a transition metal, a halogen, or a noble gas: (a) Br; (b) Ca; (c) K; (d) Kr; (e) V.

2.44. Which element in the second row of the periodic table is (a) a halogen; (b) an alkali metal; (c) an alkaline earth metal; (d) a noble gas?

2.45. **Elements in TNT** Molecules of the explosive called TNT contain atoms of hydrogen and of the second-row elements in groups 14, 15, and 16. Which three elements are these?

2.46. **Chemical Weapons** Phosgene, a colorless, poisonous gas, was used as a chemical weapon during World War I. Despite its name, molecules of phosgene contain no atoms of phosphorus. Instead, they contain atoms of carbon, of the group 16 element in the second row of the periodic table, and of the group 17 element in the third row. What are the names and the atomic numbers of these last two elements?

2.47. **Catalytic Converters** The catalytic converters used to remove pollutants from automobile exhaust contain the compounds of several expensive elements, including those described below. Which elements are they?

a. The group 10 transition metal in the fifth row of the periodic table.
b. The transition metal whose atoms contain one less proton than the atoms of your answer to part (a).
c. The name of the transition metal with an average atomic mass of 195.08 u.

2.48. **Swimming Pool Chemistry** Compounds containing chlorine have long been used to disinfect the water in swimming pools, but in recent years a compound of a less corrosive halogen has become a popular alternative disinfectant. What is the name of this fourth-row element?

Trends in Compound Formation

Concept Review

2.49. How does Dalton's atomic theory of matter explain the fact that when water is decomposed into hydrogen and oxygen gas, the ratio of the volumes of the two gases is 2:1?

2.50. **Pollutants in Automobile Exhaust** In the internal combustion engines that power most automobiles, nitrogen and oxygen may combine to form NO. When NO in automobile exhaust is released into the atmosphere, it reacts with more oxygen, forming NO_2, a key ingredient in smog. How do these reactions illustrate Dalton's law of multiple proportions?

2.51. Describe the types of elements that combine to form molecular compounds and the types that combine to form ionic compounds.

2.52. How do the properties of ionic compounds differ from those of molecular compounds?

Problems

2.53. Cobalt forms two sulfides: CoS and Co_2S_3. Predict the ratio of the two masses of sulfur that combine with a fixed mass of cobalt to form CoS and Co_2S_3.

2.54. Lead forms two oxides: PbO and PbO_2. Predict the ratio of the two masses of oxygen that combine with a fixed mass of lead to form PbO and PbO_2.

2.55. When 5.0 g of sulfur is combined with 5.0 g of oxygen, 10.0 g of sulfur dioxide (SO_2) is formed. What mass of oxygen would be required to convert 5.0 g of sulfur into sulfur trioxide (SO_3)?

*2.56. Nitrogen monoxide (NO) is 46.7% nitrogen by mass. Use the law of multiple proportions to calculate the mass percentage of nitrogen in nitrogen dioxide (NO_2).

2.57. Fill in the missing information in the following table of monatomic ions.

Symbol	$^{34}S^{2-}$	?	?	?
Number of Protons	?	9	?	50
Number of Neutrons	?	10	52	?
Number of Electrons	?	10	36	46
Mass Number	?	?	90	120

2.58. Fill in the missing information in the following table of monatomic ions.

Symbol	$^{40}Ca^{2+}$	?	?	?
Number of Protons	?	17	?	56
Number of Neutrons	?	18	36	?
Number of Electrons	?	18	28	54
Mass Number	?	?	66	138

2.59. Which of these compounds consist of molecules, and which consist of ions? (a) P_4O_{10}; (b) $SrCl_2$; (c) MgF_2; (d) SO_3

2.60. Which of these compounds consist of molecules, and which consist of ions? (a) Mg_3N_2; (b) BaS; (c) AgCl; (d) NCl_3

2.61. Would compounds formed from each pair of elements contain covalent bonds or ionic bonds? (a) cesium and fluorine; (b) nitrogen and chlorine; (c) carbon and oxygen; (d) magnesium and oxygen

2.62. Would compounds formed from each pair of elements contain covalent bonds or ionic bonds? (a) silver and oxygen; (b) sulfur and oxygen; (c) carbon and hydrogen; (d) barium and nitrogen

2.63. Give the total number of atoms in a formula unit of these compounds: (a) $Mg(NO_3)_2$; (b) Al_2O_3; (c) $(NH_4)_2CO_3$; (d) $Ca_3(PO_4)_2$.

2.64. Give the total number of atoms in a formula unit of these compounds: (a) K_2SO_4; (b) $Sr_3(AsO_4)_2$; (C) $Fe(NO_3)_3$; (d) $In_2(SO_4)_3$.

Naming Compounds and Writing Formulas

Concept Review

2.65. Consider a mythical element X, which forms only two oxoanions: XO_2^{2-} and XO_3^{2-}. Which of the two has a name that ends in *-ite*?

2.66. Concerning the oxoanions in Problem 2.65, would the name of either of them require a prefix such as *hypo-* or *per-*? Explain why or why not.

2.67. What is the role of Roman numerals in the names of the compounds formed by transition metals?

2.68. Why do the names of the ionic compounds formed by the alkali metals and by the alkaline earth metals not include Roman numerals?

Problems

2.69. What are the names of these compounds of phosphorus and oxygen? (a) PO; (b) P_2O_3; (c) P_2O_4; (d) P_2O_5

2.70. What are the names of these compounds of chlorine and oxygen? (a) ClO_3; (b) ClO; (c) ClO_2; (d) Cl_2O_7

2.71. What are the formulas and names of the ionic compounds containing the following pairs of elements? (a) sodium and sulfur; (b) strontium and chlorine; (c) aluminum and oxygen; (d) lithium and hydrogen

2.72. What are the formulas and names of the ionic compounds containing the following pairs of elements? (a) potassium and bromine; (b) barium and oxygen; (c) lithium and nitrogen; (d) aluminum and chlorine

2.73. **Milk of Magnesia** Milk of magnesia is a slurry of $Mg(OH)_2$ in water. What is the chemical name of this compound?

2.74. **Smelling Salts** Smelling salts are an antidote for fainting, made of $(NH_4)_2CO_3$, which smells of ammonia. What is the chemical name of $(NH_4)_2CO_3$?

2.75. What are the names of these sodium compounds? (a) Na_2O; (b) Na_2S; (c) Na_2SO_4; (d) $NaNO_3$; (e) $NaNO_2$

2.76. What are the names of these potassium compounds? (a) K_3PO_4; (b) K_2O; (c) K_2SO_3; (d) KNO_3; (e) KNO_2

2.77. What are the formulas of these compounds? (a) potassium sulfide; (b) potassium selenide; (c) rubidium sulfate; (d) rubidium nitrite; (e) magnesium sulfate

2.78. What are the formulas of these compounds? (a) rubidium nitride; (b) potassium selenite; (c) rubidium sulfite; (d) rubidium nitrate; (e) magnesium sulfite

2.79. What are the formulas of these compounds? (a) sodium hypobromite; (b) potassium sulfate; (c) lithium iodate; (d) magnesium nitrite

*2.80. What are the formulas of these compounds? (a) potassium tellurite; (b) sodium arsenate; (c) calcium selenite; (d) potassium chlorate

2.81. Which compound is magnesium nitrite? (a) Mg_3N_2; (b) $Mg(NO_2)_2$; (c) $Mg(NO_3)_2$; (d) $Mg(NO)_2$

2.82. Which compound is aluminum phosphate? (a) Al_3P_3; (b) $AlPO_3$; (c) $AlPO_4$; (d) $Al_2(PO_4)_3$

2.83. Which of these chemical names is followed by a chemical formula that does not fit the name? (a) calcium oxide, CaO; (b) lithium sulfate, $LiSO_4$; (c) barium sulfide, BaS; (d) potassium oxide, K_2O

2.84. Which of these chemical names is followed by a chemical formula that does not fit the name? (a) aluminum nitride, AlN; (b) aluminum sulfate, $Al_2(SO_4)_3$; (c) potassium chloride, KCl_2; (d) cesium sulfate, Cs_2SO_4

2.85. **Ions in Blood** The most abundant cations in blood plasma are Na^+, K^+, Mg^{2+}, and Ca^{2+}. Two of the principal anions are Cl^- and $H_2PO_4^-$. Write the formulas of the eight ionic compounds these cations and anions form.

2.86. Evaporation of seawater gives a mixture of sodium compounds that also contain chloride, sulfate, carbonate, and fluoride ions. Write the chemical formulas of all these compounds.

2.87. What are the names of these compounds? (a) Cr_2Te_3; (b) $V_2(SO_4)_3$; (c) $Fe_2(CrO_4)_3$; (d) MnO

2.88. What are the names of these compounds? (a) $FePO_4$; (b) $CuSO_4$; (c) Ag_2CO_3; (d) $Zn(NO_2)_2$

2.89. What are the formulas for these transition metal compounds? (a) zinc dichromate; (b) iron(III) acetate; (c) mercury(I) peroxide; (d) scandium thiocyanate

2.90. What are the formulas for these transition metal compounds? (a) mercury(II) hydroxide; (b) silver perchlorate; (c) manganese(IV) nitrate; (d) vanadium(IV) oxide

2.91. What are the formulas of the following acids? (a) hydroiodic acid; (b) hypobromous acid; (c) iodic acid; (d) hydrobromic acid

2.92. What are the formulas of the following acids? (a) selenous acid; (b) hydrocyanic acid; (c) phosphoric acid; (d) nitrous acid

2.93. What are the names of the acids with the following formulas? (a) HF; (b) $HBrO_3$; (c) HBr; (d) HIO_4

2.94. What are the names of the acids with the following formulas? (a) HF; (b) HClO; (c) $HBrO_2$; (d) $HClO_2$

Organic Compounds: A First Look

Concept Review

2.95. What classes of organic compounds contain no heteroatoms?

2.96. What classes of organic compounds contain oxygen atoms?

2.97. What classes of organic compounds contain nitrogen atoms? Does any class of organic compounds contain both oxygen and nitrogen atoms?

2.98. Are most organic compounds molecular or ionic?

Problems

2.99. To which class of organic compounds does each of the following compounds belong?
a. $CH_3CH_2CH_2CH_2CH_2CH_2CH_2CH_3$ (octane—a principal component of gasoline)
b. $HC{\equiv}CH$ (ethyne, also known as acetylene—used in welding torches)

2.100. To which class of organic compounds does each of the following compounds belong?
a. $CH_3CH_2{-}O{-}CH_2CH_3$ (once used as anesthesia for surgery)
b. $CH_3CH_2CH_2CH_2OH$ (a component of brake fluids and perfumes)

2.101. **Scent of Pears** What is the functional group present in propyl acetate (the substance responsible for the smell of pears; Figure P2.101)?

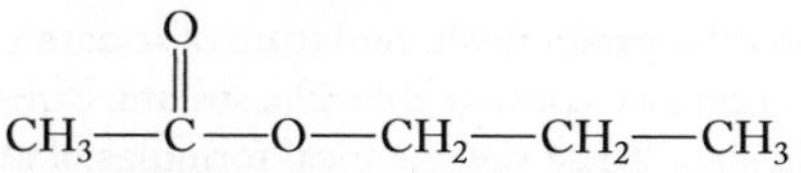

$$CH_3{-}\overset{\overset{\displaystyle O}{\|}}{C}{-}O{-}CH_2{-}CH_2{-}CH_3$$

FIGURE P2.101

*2.102. What two functional groups are present in glycine (the simplest amino acid; Figure P2.102)?

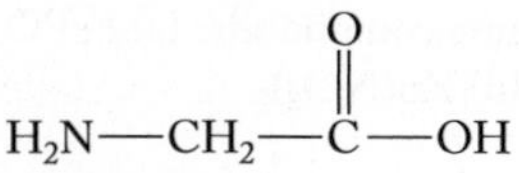

$$H_2N{-}CH_2{-}\overset{\overset{\displaystyle O}{\|}}{C}{-}OH$$

FIGURE P2.102

Nucleosynthesis: The Origin of the Elements

Concept Review

2.103. In the history of the universe, which of these particles formed first, and which formed last? (a) deuteron; (b) neutron; (c) proton; (d) electron

2.104. Why did early nucleosynthesis last such a short time?

2.105. In the current cosmological model, the volume of the universe is increasing with time. How might this expansion affect the density of the universe?

2.106. **Components of Solar Wind** Most of the ions in the solar wind emitted by the Sun are hydrogen ions. The ions of which element should be next most abundant?

Additional Problems

2.107. The molecular compounds HClO, $HClO_2$, and $HClO_4$ are three of the four common oxoacids of chlorine. Give the formula and chemical name of the fourth oxoacid of chlorine.

2.108. **Fusion in Stars** One reaction in the process of carbon fusion in massive stars involves the combination of two carbon-12 nuclei to form the nucleus of a new element and an alpha particle. Write an equation that describes this process.

2.109. A process called neon fusion takes place in massive stars. In one of the reactions in this process, an alpha particle combines with a neon-21 nucleus to produce another element and a neutron. Write an equation that describes this process.

2.110. In April 1897, J. J. Thomson presented the results of his experiment with cathode-ray tubes (Figure P2.110) in which he proposed that the rays were beams of negatively charged particles, which he called "corpuscles."

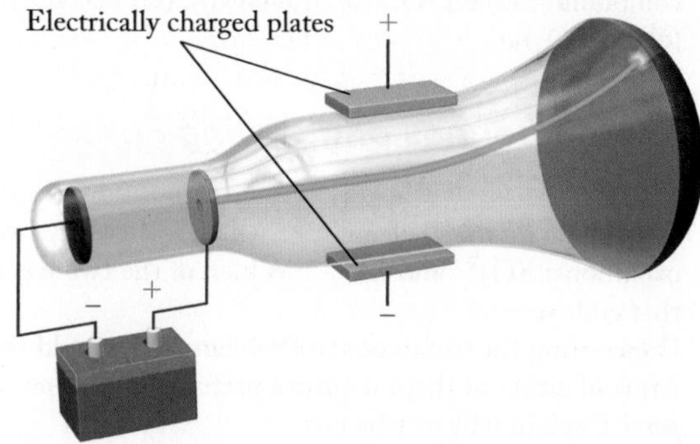

FIGURE P2.110

a. What is the name we use for these particles today?
b. Why did the beam deflect when passed between electrically charged plates, as shown in Figure P2.110?
c. If the polarity of the plates was switched, how would the position of the light spot on the phosphorescent screen change?
d. If the voltage on the plates was reduced by half, how would the position of the light spot change?

***2.111.** Suppose the electrically charged disks at the end of the cathode-ray tube were replaced with a radioactive source, as shown in Figure P2.111. Also suppose the radioactive material inside the source emits α and β particles. The only way for either kind of particle to escape the source is through a narrow channel drilled through a block of lead.
a. How many light spots do you expect to see on the phosphorescent screen?
b. What are their positions relative to the electrical plates, and which particle produces which spot?

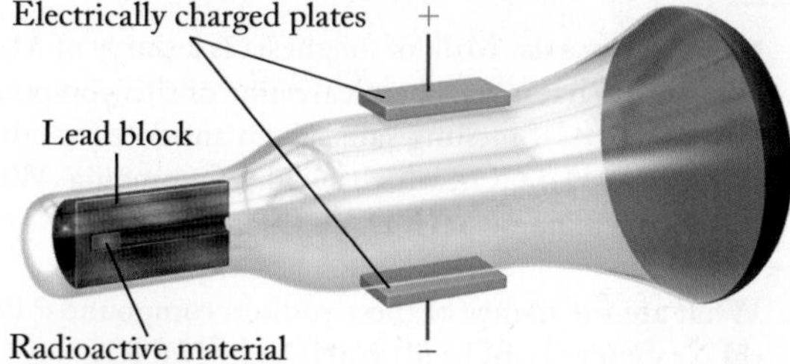

FIGURE P2.111

*2.112. Suppose the radioactive material inside the source in the apparatus shown in Figure P2.111 emits protons and α particles and suppose both kinds of particles have the same velocities.
a. How many light spots do you expect to see on the phosphorescent screen?
b. What are their positions on the screen (above, at, or below the center)? Which particle produces which spot?

2.113. **Early Universe** Cosmologists estimate that the matter in the early universe was 75% hydrogen-1 and 25% helium-4, by mass, when atoms first formed.

a. Assuming these proportions are correct, what was the ratio of hydrogen to helium *atoms* in the early universe?
b. The ratio of hydrogen to helium atoms in our solar system is slightly less than 10:1. Compare this value with the value you calculated in part (a).
c. Propose a hypothesis that accounts for the difference in composition between the solar system and the early universe.
d. Describe an experiment that would test your hypothesis.

2.114. **Emergency Air Supply** Potassium forms three compounds with oxygen: K_2O (potassium oxide), K_2O_2 (potassium peroxide), and KO_2 (potassium superoxide). Elemental potassium is rarely encountered; it reacts violently with water and is very corrosive to human tissue. Potassium superoxide is used in self-contained breathing apparatuses as a source of oxygen for use in mines, submarines, and spacecraft. Potassium peroxide binds carbon dioxide and is used to scrub (remove) toxic CO_2 from the air in submarines. Predict the ratio of the masses of oxygen that combine with a fixed mass of potassium in K_2O, K_2O_2, and KO_2.

***2.115.** **Bronze Age** Historians and archaeologists often apply the term "Bronze Age" to the period in Mediterranean and Middle Eastern history when bronze was the preferred material for making weapons, tools, and other metal objects. Ancient bronze was prepared by blending molten copper (88% by mass) and tin (12% by mass) into a mixture called an alloy. What is the ratio of copper to tin atoms in a piece of bronze with this composition?

*2.116. In his version of the periodic table, Mendeleev arranged elements according to the formulas of the compounds they formed with hydrogen and oxygen. The elements in one of his eight groups formed compounds with the generic formulas MH_3 and M_2O_5, where M was the symbol of an element in the group. Which Roman numeral did Mendeleev assign to this group?

2.117. In the Mendeleev table in Figure 2.11, there are no symbols for elements with predicted atomic masses of 44, 68, and 72.

a. Which elements are these?
b. Mendeleev anticipated the later discovery of these three elements and gave them the tentative names ekaaluminum, ekaboron, and ekasilicon, reflecting the probability that their properties would resemble those of aluminum, boron, and silicon, respectively. What are the modern names of ekaaluminum, ekaboron, and ekasilicon?
c. When were these elements finally discovered? (*Hint*: Consult a reference such as webelements.com.)

2.118. In chemical nomenclature, the prefix *thio-* is used to indicate that a sulfur atom has replaced an oxygen atom in the structure of a molecule or a polyatomic ion.

a. With this rule in mind, write the formula for the thiosulfate ion.
b. What is the formula of sodium thiosulfate?

2.119. There are two stable isotopes of gallium. Their masses are 68.92558 u and 70.9247050 u. If the average atomic mass of gallium is 69.7231 u, what is the natural abundance of the lighter isotope?

2.120. There are two stable isotopes of bromine. Their masses are 78.9183 u and 80.9163 u. If the average atomic mass of bromine is 79.9091 u, what is the natural abundance of the heavier isotope?

***2.121.** Using the information in the previous question:

a. Predict the possible masses of individual molecules of Br_2.
b. Calculate the natural abundance of molecules with each of the masses predicted in part (a) in a sample of Br_2.

*2.122. There are three stable isotopes of magnesium. Their masses are 23.9850 u, 24.9858 u, and 25.9826 u. If the average atomic mass of magnesium is 24.3050 u and the natural abundance of the lightest isotope is 78.99%, what are the natural abundances of the other two isotopes?

2.123. Write the names and formulas of the following compounds from this list of elements: Li, Fe, Al, O, C, and N.

a. A molecular substance AB_2, where A is a group 14 element and B is a group 16 element
b. An ionic compound C_3D, where C is a group 1 element and D is a group 15 element

2.124. Write the chemical symbol of each of the following species: (a) a cation with a mass number of 24, an atomic number of 12, and a charge of 2+; (b) a member of group 15 that has a charge of 3+, 48 electrons, and 70 neutrons; (c) a noble gas atom with 48 neutrons.

***2.125.** Predict some physical and chemical properties of radium (Ra). Predict the melting points of $RaCl_2$ and RaO. (*Hint*: Research the melting points of the other alkaline earth element chlorides and oxides.)

2.126. From their positions in the periodic table, predict which of the following elements would be good electrical conductors: Ti, Ne, N, Ag, Tb, Br, and Mo.

2.127. Argon has a larger average atomic mass than potassium, yet it is placed before potassium in the modern periodic table. Explain.

*2.128. It takes nearly twice the energy to remove an electron from a helium atom as it does to remove an electron from a hydrogen atom. Propose an explanation for this.

3

Stoichiometry

Mass, Formulas, and Reactions

WILDFIRE: THE RESULT OF UNCONTROLLED COMBUSTION The principal products of combustion of wood and fossil fuels are carbon dioxide and water.

PARTICULATE **REVIEW**

Molecules and Names

In Chapter 3 we examine the composition of pure substances and explore how the quantities of reactants and products in chemical reactions are related. Unlike today's atmosphere, which is mostly elemental nitrogen and oxygen, Earth's early atmosphere consisted predominantly of the compounds shown here.

- Name these compounds.

(Review Section 2.7 if you need help.)

(Answers to Particulate Review questions are in the back of the book.)

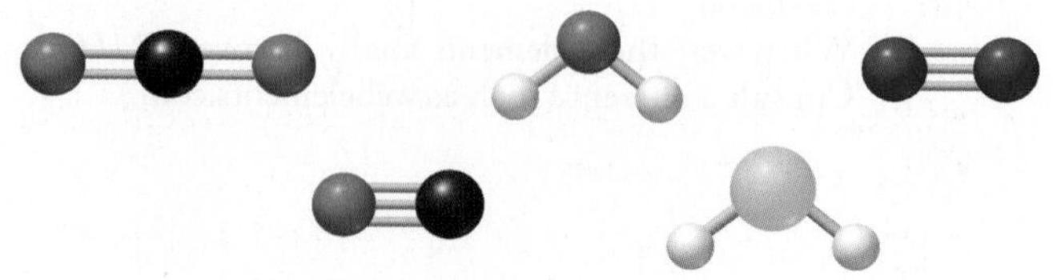

PARTICULATE **PREVIEW**

How Many Particles?

Combustion can be uncontrolled, as in the fire above, or a controlled reaction used for heating and cooking. Methane, the principal component of natural gas, undergoes a combustion reaction with oxygen. As you read Chapter 3, look for ideas that will help you answer these questions:

- How many oxygen molecules are necessary to completely react with these methane molecules?
- Identify the products of this reaction.
- Draw all the molecules of product formed in this reaction.

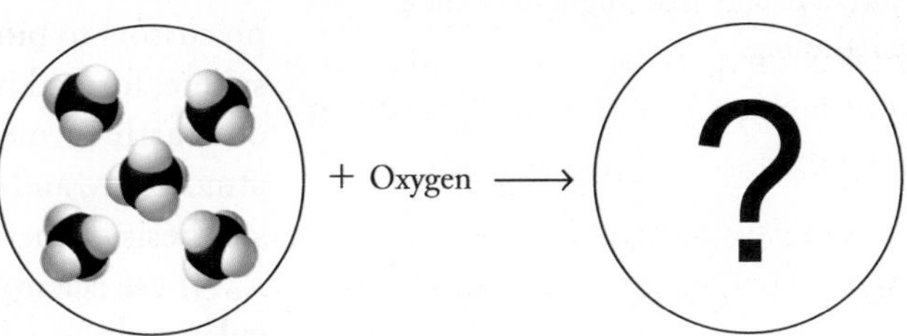

Learning Outcomes

LO1 Use molar mass and the Avogadro constant to make conversions among mass, number of particles, and number of moles
Sample Exercises 3.1, 3.2, 3.3, 3.4, 3.5, 3.6, 3.7

LO2 Write balanced chemical equations to describe chemical reactions
Sample Exercises 3.8, 3.9, 3.10, 3.11

LO3 Use balanced chemical equations and molar masses to relate the quantities of reactants consumed and the quantities of products formed
Sample Exercises 3.12, 3.13

LO4 Identify the limiting reactant in a chemical reaction
Sample Exercises 3.14

LO5 Calculate the theoretical and percent yields in a chemical reaction
Sample Exercises 3.15, 3.16, 3.17

LO6 Convert between the chemical formula and percent composition of a compound
Sample Exercises 3.18, 3.19, 3.20

LO7 Determine the molecular formula of a compound from its empirical formula and molar mass
Sample Exercise 3.21

LO8 Use data from combustion analysis to determine the empirical formula of a substance
Sample Exercises 3.22, 3.23

3.1 Air, Life, and Molecules

FIGURE 3.1 The apparatus used by Miller (shown) and Urey to simulate the synthesis of amino acids in the atmosphere of early (prebiotic) Earth. Discharges between the tungsten electrodes were meant to provide the sort of energy that might have come from lightning.

In Chapter 2 we described how elements are synthesized in the cores of giant stars and dispersed throughout galaxies when the stars explode as supernovas. The dispersed elements become the building blocks of other stars and planets. As our solar system formed, the inner planets—Mercury, Venus, Earth, and Mars—were rich in nonvolatile elements such as iron, silicon, magnesium, and aluminum. Earth was also rich in oxygen—in the form of stable compounds with these and other elements. These compounds formed the rocks and minerals of Earth's crust and an early atmosphere. As Earth cooled, atmospheric water vapor condensed into torrential rains that filled up depressions in the crust, forming the first oceans.

How could the substances in lifeless rocks and air or dissolved in primordial seas have combined to become the organic building blocks of life—compounds like simple sugars, amino acids, and the molecules that form DNA? No one knows for sure, but in the early 1950s Nobel Prize winner Harold Urey (1893–1981) and his student Stanley Miller (1930–2007) tested this hypothesis by assembling a mixture of gases believed to have been present in Earth's early (prebiotic) atmosphere. Urey and Miller subjected that mixture of gases to an electric current, simulating the lightning that would have been prevalent on early Earth (**Figure 3.1**). The resulting chemical reactions produced several amino acids, the building blocks of proteins. Samples from additional experiments by Miller in an apparatus simulating the composition of gases from a volcanic eruption were reanalyzed in 2007, and more than 20 amino acids were detected.

Amino acids have also been found in meteorites, which have bombarded Earth since it formed. Extensive analyses of the Murchison meteorite, which landed in Australia in 1969, showed that it contains many of the same amino acids synthesized in the Miller–Urey experiments. Regardless of whether important precursors to biological compounds were synthesized throughout the early atmosphere, formed in the localized environment around volcanic eruptions, or arrived on Earth in meteors, current thinking favors an early Earth with a primitive atmosphere and an ocean of water that provided an environment conducive to the synthesis of the molecules of life.

These scientific theories about the origin of life from simple inorganic molecules and the evidence that supports them are the product of research carried out in the 20th and 21st centuries. In the early 19th century, when molecular science was in its infancy, scientists believed that organic compounds could only be made by living organisms. In addition, the prevailing understanding held that chemical compounds were different only because they had different elemental compositions. A serendipitous experiment by Friedrich Wöhler (1800–1882) in 1828 led to the demise of both ideas and to their replacement with a view of matter that is a cornerstone of modern molecular science.

isomers compounds with the same molecular formula but different arrangements of the atoms in their molecules.

Wöhler attempted to synthesize the compound ammonium cyanate (NH_4NCO; **Figure 3.2a, b**) by using silver cyanate ($AgNCO$) and ammonium chloride (NH_4Cl). The product of his reaction demonstrated none of the chemical or physical properties of cyanates. After several other attempts at synthesis, he established that the white crystals produced by the reaction were urea (H_2NCONH_2; **Figure 3.2c**). At that time, urea was known only as a material isolated from urine—that is, the product of a living being. Wöhler famously reported to his scientific mentor Jöns Jacob Berzelius (1779–1848) that he had succeeded in making urea "without the intervention of a kidney." This result and others like it ultimately led to the modern definition of organic compounds (Chapter 2)—namely, compounds of carbon and hydrogen, often including other elements—and to the understanding that organic compounds identical to those found in living systems can be made in the laboratory.

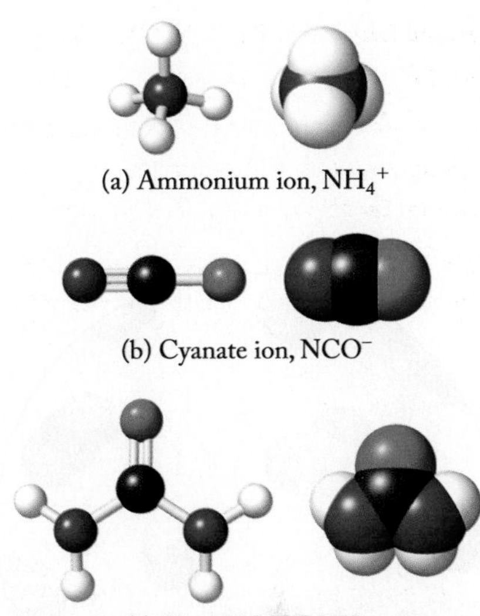

FIGURE 3.2 (a, b) Ball-and-stick and space-filling models of the ions in ammonium cyanate and (c) a molecule of urea.

Wöhler's result had another very important consequence. The formulas for ammonium cyanate and urea, when written to simply illustrate their composition, are both CH_4N_2O. Thus, two different substances can have the same atomic composition. Scientists ultimately reasoned that the *arrangement* of atoms within the molecules of each compound must be different. As a result, chemical substances must be defined not only by the number and kind of atoms in their molecules but also by the arrangement of those atoms. Composition and arrangement both matter. Berzelius called urea and ammonium cyanate **isomers** (literally, "same units"), which is the term still used to describe two or more different compounds with the same molecular formula.

Here in Chapter 3 we begin with a molecular view of the physical processes and chemical reactions that may have occurred on early Earth. We explore how the quantities of substances produced and consumed in chemical reactions are related. Then we examine the analytical methods that reveal the composition of pure substances. Finally, we consider a class of chemical reactions that consume oxygen and a fuel, liberating energy—including the energy necessary for life. We concentrate on the quantitative aspects of identifying chemical substances and their reactions but keep Wöhler's result in mind: how the atoms within molecules are arranged matters, too. We address the important issue of how atoms connect to each other and the resulting shapes of molecules in detail in Chapters 8 and 9.

Chemical Reactions and Earth's Early Atmosphere

The Earth that formed 4.6 billion years ago was a hot, molten sphere that gradually separated into distinct regions based on differences in density and melting point. The densest elements, notably iron and nickel, sank to the center of the planet. A less dense mantle, rich in compounds containing aluminum, magnesium, silicon, and oxygen, formed around the core. As time passed and Earth cooled, the mantle fractionated further, allowing a solid crust to form from the components of the mantle that were the least dense and had the highest melting points. The core also separated into a solid inner core and a molten outer core. **Figure 3.3** shows the elemental compositions of these layers.

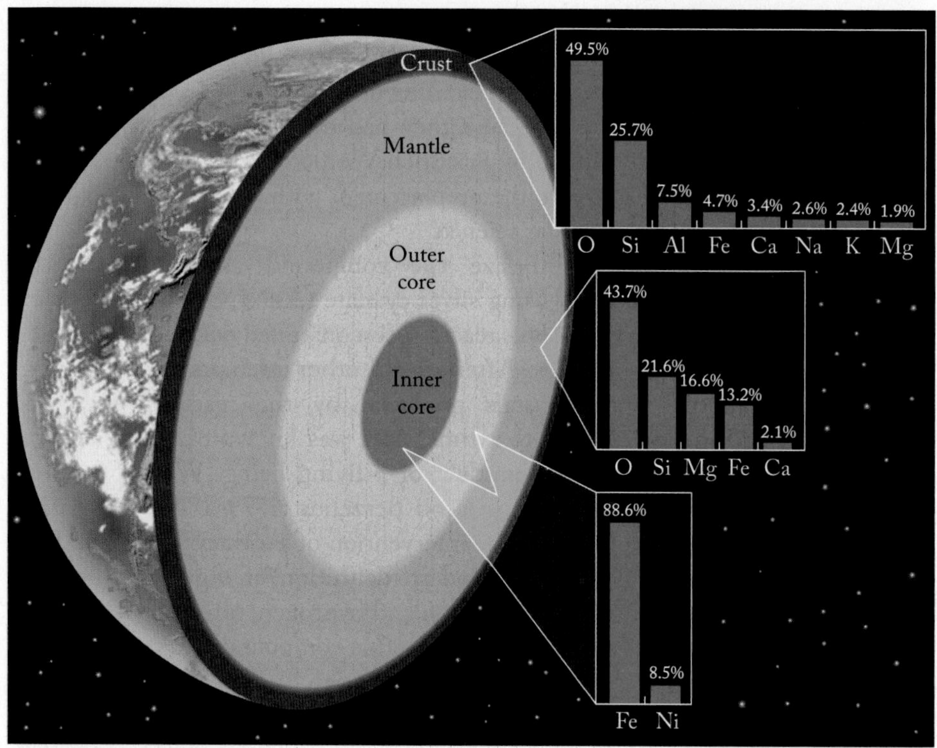

FIGURE 3.3 Earth is composed of a solid inner core, consisting mostly of nickel and iron, surrounded by a molten outer core of similar composition. A rocky mantle, composed mostly of oxygen, silicon, magnesium, and iron, lies between the outer core and a relatively thin solid crust.

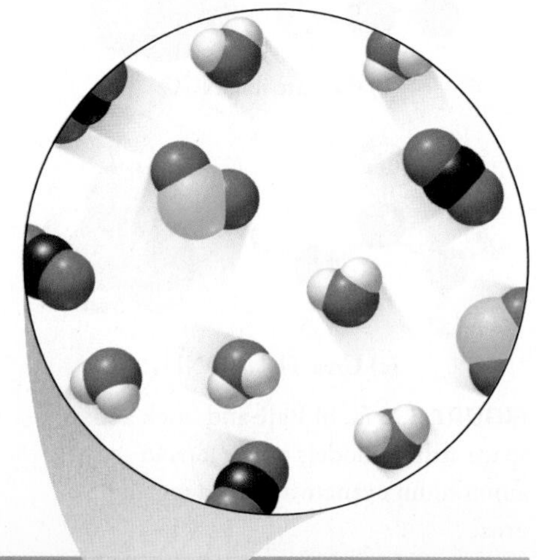

FIGURE 3.4 The Shiveluch volcano on the Russian Kamchatka Peninsula erupted on December 5, 2017.

Earth's early crust was torn by the impact of asteroids and widespread volcanic activity. The gases released by these impacts and eruptions generated a primitive atmosphere with a chemical composition different from that of the air we breathe now. Current research suggests that the early atmosphere of Earth was nearly devoid of molecular oxygen (O_2) yet was rich in oxygen-containing compounds, including carbon dioxide (CO_2), carbon monoxide (CO), and water vapor (H_2O). Minor components of the atmosphere included hydrogen (H_2), methane (CH_4), and hydrogen sulfide (H_2S). Oxides of nitrogen and additional CO may have arisen when the atmosphere was heated during bombardment by large meteorites, whereas traces of other volatile oxides, including sulfur dioxide (SO_2) and sulfur trioxide (SO_3), may have resulted from volcanic activity. Today, the most abundant gases released by volcanoes like Mount Shiveluch in Russia are water vapor, CO_2, and SO_2 (**Figure 3.4**).

Sometimes these compounds in Earth's prebiotic atmosphere combined to make substances with more elaborate molecular structures. For example, sulfur trioxide gas and water vapor are the **reactants** that combine to form liquid sulfuric acid, H_2SO_4, as a **product** (**Figure 3.5**). The reaction between SO_3 and H_2O is an example of a **combination reaction**, a reaction in which two (or more) substances combine to form one product. We use the formulas of the reactants and products in a chemical equation to describe the reaction:

$$SO_3(g) + H_2O(g) \rightarrow H_2SO_4(\ell) \tag{3.1}$$

Sulfur trioxide + Water → Sulfuric acid

Reactants → Product

FIGURE 3.5 In this combination reaction, a molecule of SO_3 and a molecule of H_2O form a molecule of H_2SO_4.

An important feature of any chemical equation is that it is *balanced*: every atom that is present in the reactants is also present in the products. This conservation of atoms means that there is also a conservation of mass: the sum of the masses of the reactants always equals the sum of the masses of the products. Keep in mind that chemical equations do not explain *how* or *why* reactions occur. They just tell us how the quantities of reactants and products are related.

CONNECTION In Section 1.2 we defined a *chemical reaction* as the transformation of one or more substances into different substances, and we used *chemical formulas* in a *chemical equation* to describe this transformation.

The sulfuric acid that formed in Earth's early atmosphere eventually fell to the planet's surface as highly acidic rain. This rain landed on a crust made up mostly of metal oxides and metalloid oxides, including the mineral hematite, Fe_2O_3. When that sulfuric acid mixed with hematite, another chemical reaction took place—one that produced slightly water-soluble iron(III) sulfate, $Fe_2(SO_4)_3$, and liquid water. This reaction is described by the following chemical equation:

CONNECTION In Section 2.9 we wrote balanced *nuclear equations* in terms of both mass and charge.

$$Fe_2O_3(s) + 3\,H_2SO_4(aq) \rightarrow 3\,H_2O(\ell) + Fe_2(SO_4)_3(aq) \qquad (3.2)$$

The "3"s in front of both H_2SO_4 and H_2O are coefficients that balance the number of atoms of each element on both sides of the reaction arrow—something we explore in Section 3.3. The absence of a coefficient in front of Fe_2O_3 and $Fe_2(SO_4)_3$ is the same as having a coefficient of "1" in front of them. The chemical formula itself indicates the composition of one molecule (or one formula unit) of the substance. As we explain in Section 3.2, the chemical formula also indicates the number of *moles* of each element in one *mole* of a compound.

3.2 The Mole

Equation 3.1 describes a chemical reaction between one molecule of sulfur trioxide and one molecule of water. In our macroscopic world, chemists rarely work with individual atoms or single molecules because they are too small to manipulate easily. Instead, we usually deal with quantities of reactants and products large enough to see and work with comfortably. Such measurable quantities contain enormous numbers of particles—atoms, molecules, or ions—so we need a unit that can relate measurable quantities of substances to the quantities of particles they contain. That unit is the **mole (mol)**, the SI base unit for expressing quantities of substances (see Table 1.2).

CHEMTOUR
Avogadro Constant

One mole of any pure substance is defined as the amount of it that contains exactly $6.02214076 \times 10^{23}$ characteristic particles of the substance, for example, atoms of an element, molecules of a molecular substance, formula units of an ionic compound. When used in calculations this very large value is called the **Avogadro constant (N_A)** after the Italian scientist Amedeo Avogadro (1776–1856), whose research enabled other scientists to accurately determine the atomic masses of the elements. We round off the Avogadro constant to four significant figures, 6.022×10^{23} particles/mol throughout this book. One mole of water contains 6.022×10^{23} particles—in this case, molecules of water. To put this quantity in perspective, 1 mole of basketballs would fill an empty sphere the size of Earth, but 1 mole of water molecules occupies a volume of roughly 15 mL (i.e., one tablespoon).

reactant a substance consumed during a chemical reaction.

product a substance formed during a chemical reaction.

combination reaction a reaction in which two (or more) substances combine to form one product.

mole (mol) the SI base unit for expressing quantities of substances. One mole contains $6.02214076 \times 10^{23}$ particles: atoms, molecules, or formula units.

Avogadro constant (N_A) the fundamental constant representing the number of characteristic particles in 1 mole of a substance: $6.02214076 \times 10^{23}$ particles/mol.

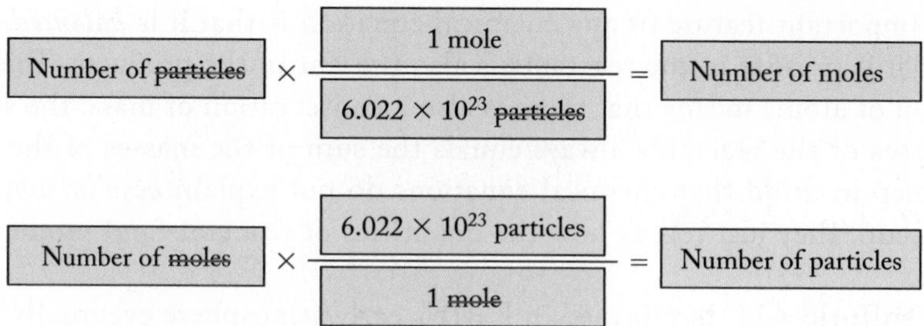

FIGURE 3.6 Converting between a number of particles and an equivalent number of moles (or vice versa) is a matter of dividing (or multiplying) by the Avogadro constant. Note how the units cancel, leaving only the units we sought.

FIGURE 3.7 The quantities shown are equivalent to 1 mole of each material: 4.003 g of helium gas in the balloon and, left to right in front of the balloon, 63.55 g of copper metal, 55.84 g of solid iron, 32.06 g of solid sulfur, and 200.59 g of liquid mercury.

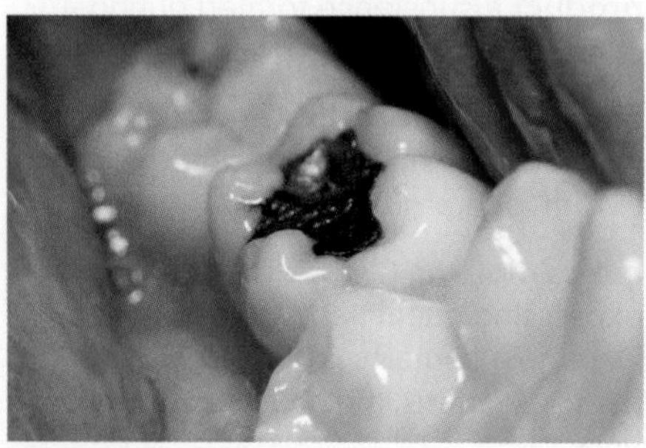

FIGURE 3.8 One of the materials used to fill dental caries is a mixture of three compounds: Ag_2Hg_3, Ag_3Sn, and Sn_8Hg. This material is one of a class of metallic mixtures known as alloys.

Multiplying a number of moles of a substance by the Avogadro constant produces the number of particles in that many moles. For example, a 1 L bottle of seltzer water contains 55 moles of H_2O. The number of molecules of water in that 1 L bottle is

$$\frac{55\ \cancel{\text{mol H}_2\text{O}}}{1\text{ L}} \times \frac{6.022 \times 10^{23}\text{ molecules H}_2\text{O}}{1\ \cancel{\text{mol H}_2\text{O}}} = \frac{3.3 \times 10^{25}\text{ molecules H}_2\text{O}}{1\text{ L}}$$

The use of this conversion factor is illustrated in **Figure 3.6**, and one-mole quantities of some common elements are shown in **Figure 3.7**.

SAMPLE EXERCISE 3.1 Converting Number of Moles into Number of Particles — LO1

It's not unusual for the polluted air above a large metropolitan area to contain as much as 5×10^{-10} moles of SO_2 per liter of air. How many molecules of SO_2 are there per liter of air?

Collect and Organize The problem gives the number of moles of SO_2 per liter of air. We want to find the number of molecules of SO_2 in 1 L. There are 6.022×10^{23} particles in 1 mole.

Analyze We convert the number of moles/liter into the number of molecules/liter by multiplying by 6.022×10^{23} molecules/mole:

$$\boxed{\frac{\text{mole}}{\text{L}}} \xrightarrow{\frac{6.022 \times 10^{23}\text{ molecules}}{1\text{ mole}}} \boxed{\frac{\text{molecules}}{\text{L}}}$$

We can estimate the answer using an approximate value of the Avogadro constant (about 6×10^{23}) and the number of moles of SO_2 (5×10^{-10}). The product of the two is 3×10^{14}.

Solve The number of SO_2 molecules per liter of air is

$$\frac{5 \times 10^{-10}\ \cancel{\text{mol SO}_2}}{1\text{ L air}} \times \frac{6.022 \times 10^{23}\text{ molecules SO}_2}{1\ \cancel{\text{mol SO}_2}} = \frac{3 \times 10^{14}\text{ molecules SO}_2}{1\text{ L air}}$$

Think About It The result tells us that a tiny fraction of 1 mole of a molecular substance (10^{-10}) is equivalent to a very large number of molecules (10^{14}), which makes sense given the immensity of the Avogadro constant (between 10^{23} and 10^{24}).

 **Practice Exercise** If 1.0 mL of seawater contains about 2.5×10^{-14} moles of dissolved gold, how many atoms of gold are in that volume of seawater?

(Answers to Practice Exercises are in the back of the book.)

SAMPLE EXERCISE 3.2 Relating Formulas, Moles, and Particles **LO1**

Dental fillings contain compounds with the formulas Ag_2Hg_3, Ag_3Sn, and Sn_8Hg (**Figure 3.8**). Which compound contains the most atoms per mole? How many atoms of silver are there in 1 mole of Ag_3Sn?

Collect, Organize, and Analyze We are given the formulas of three compounds: Ag_2Hg_3, Ag_3Sn, and Sn_8Hg. A chemical formula indicates the number of moles of each element in 1 mole of that compound. The number of atoms in a mole of *any* element is defined by the Avogadro constant: $N_A = 6.022 \times 10^{23}$ atoms/mol.

Solve From the formulas of the three compounds we have:

- 1 mole of Ag_2Hg_3 contains 2 moles of silver and 3 moles of mercury, or $2 + 3 = 5$ total moles of atoms.
- 1 mole of Ag_3Sn contains 3 moles of silver and 1 mole of tin, or $3 + 1 = 4$ total moles of atoms.
- 1 mole of Sn_8Hg contains 8 moles of tin and 1 mole of mercury, or $8 + 1 = 9$ total moles of atoms.

Thus, 1 mole of Sn_8Hg contains the greatest number of atoms:

$$\frac{9\ \cancel{\text{total mol}}}{1\ \text{mol Sn}_8\text{Hg}} \times \frac{6.022 \times 10^{23}\ \text{atoms}}{1\ \cancel{\text{mol}}} = \frac{5.420 \times 10^{24}\ \text{atoms}}{1\ \text{mol Sn}_8\text{Hg}}$$

The number of atoms of Ag per mole of Ag_3Sn is:

$$\frac{3\ \cancel{\text{mol Ag}}}{1\ \text{mol Ag}_3\text{Sn}} \times \frac{6.022 \times 10^{23}\ \text{atoms Ag}}{1\ \cancel{\text{mol Ag}}} = \frac{1.807 \times 10^{24}\ \text{atoms Ag}}{1\ \text{mol Ag}_3\text{Sn}}$$

Think About It Chemical formulas reveal the number of moles of each element per mole of that compound. When we compare two compounds, their chemical formulas allow us to quickly determine the relative numbers of each type of atom.

Practice Exercise How many electrons are in a 1-ounce gold coin (**Figure 3.9**) that contains 0.143 mol of pure gold?

(Answers to Practice Exercises are in the back of the book.)

(a)

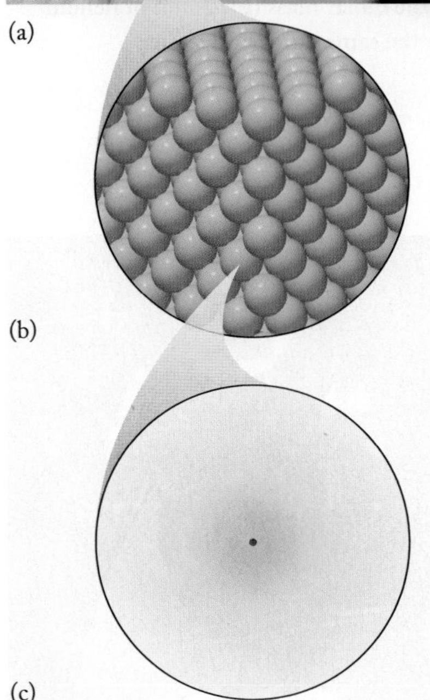

(b)

(c)

FIGURE 3.9 (a) Five 1-ounce gold coins, each containing about 0.143 mol of pure gold. (b) A lattice of gold atoms. (c) The nucleus of one gold atom, surrounded by electrons.

The mole enables us to calculate the number of particles in any sample of a given substance from the known (or measured) mass of the sample. We can do this because the mole represents a fixed number of particles (the Avogadro constant). One useful analogy is to compare quantities of different fruits. A dozen grapes and a dozen apples have very different masses, but each contains 12 pieces of fruit (**Figure 3.10**). Indeed, the mole is sometimes referred to as "the chemist's dozen."

FIGURE 3.10 The mass of a dozen grapes is less than that of a dozen apples, even though both contain the same number of pieces of fruit.

CONCEPT **TEST**

How does a unit of measure such as a gross (144 units) relate to the concept of the mole?

(Answers to Concept Tests are in the back of the book.)

Molar Mass

The mole provides an important link between the values of atomic mass in the periodic table and the masses of macroscopic quantities of elements and compounds. Recall from Section 2.4 that the mass value listed for each element in the periodic table is the average atomic mass of the atoms of the element, expressed in unified atomic mass units (u) or daltons (Da). That same value is also the mass of 1 mole

2 He 4.003	
Average atomic mass of He	4.003 u/atom
Mass of 1 mol of He	4.003 g
Molar mass of He	4.003 g/mol

FIGURE 3.11 The atomic mass (in u/atom) and the molar mass (in g/mol) of helium have the same numerical value.

of atoms of that element expressed in grams. Thus, the average mass of one atom of helium is 4.003 u, and the mass of 1 mole of helium is 4.003 g. The mass in grams of 1 mole of any pure substance is called its **molar mass ($\mathcal{M}$)**. Thus, the molar mass of helium is 4.003 g/mol (**Figure 3.11**). This numerical equality is not a coincidence: the value of N_A was defined to create an equality between the average atomic mass (u) of any element in the periodic table and its molar mass in g/mol.

Because the mass of electrons is insignificant compared with that of protons and neutrons (Table 2.1), the mass of an ion (e.g., Na^+ or Cl^-) is essentially the same as the mass of the corresponding atom, even though cations have fewer electrons than neutral atoms and anions have more.

Molar mass is an extremely useful conversion factor that allows us to convert between the mass of a sample of any substance and the number of moles in that sample or to convert a quantity in moles to an equivalent mass in grams. Although chemical reactions take place between particles (i.e., atoms, molecules, and ions), in the laboratory we almost always measure the *masses* of products and reactants.

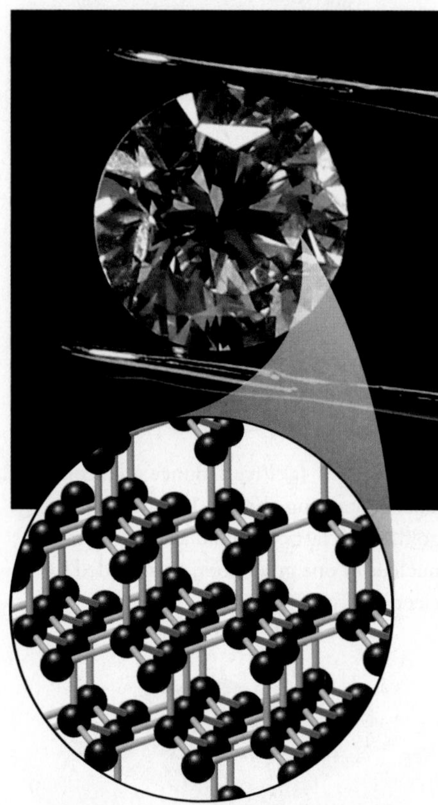

FIGURE 3.12 A 3.25-carat diamond.

SAMPLE EXERCISE 3.3 Converting Mass into Number of Moles LO1

The active ingredient in many antacid tablets is calcium carbonate. Some of these tablets contain 0.400 g of calcium (as Ca^{2+} ions). How many moles of calcium are in each tablet?

Collect, Organize, and Analyze We are given the mass of Ca^{2+} ions in each tablet, and asked to convert this mass into an equivalent number of moles of calcium. We can convert a mass of Ca^{2+} in grams into an equivalent quantity in moles by dividing by the molar mass of Ca, which is 40.08 g/mol to four significant figures.[1]

$$\boxed{\text{g Ca}^{2+}} \xrightarrow[\text{molar mass}]{1} \boxed{\text{mol Ca}^{2+}}$$

As noted in the text, the tiny mass of an electron means that the mass of a monatomic ion is not significantly different from the mass of its parent atom. The ratio of the mass of Ca^{2+} ions in the sample (0.400 g) to the molar mass of Ca (about 40.1 g/mol) is about 0.01, so our answer should be about 0.01 mol.

Solve

$$0.400\ \cancel{\text{g Ca}^{2+}} \times \frac{1\ \text{mol Ca}^{2+}}{40.08\ \cancel{\text{g Ca}^{2+}}} = 0.00998\ \text{mol Ca}^{2+} = 9.98 \times 10^{-3}\ \text{mol Ca}^{2+}$$

Think About It We predicted that the antacid tablet contained about 0.01 mol of calcium, and that prediction was very close to the calculated value.

Practice Exercise The mass of the diamond in **Figure 3.12** is 3.25 carats (1 carat = 0.200 g). Diamonds are nearly pure carbon. How many moles of carbon are in the diamond?

(Answers to Practice Exercises are in the back of the book.)

SAMPLE EXERCISE 3.4 Converting Number of Moles into Mass LO1

A helium balloon sold at an amusement park contains 0.462 mol He. How many grams of He are in the balloon?

molar mass ($\mathcal{M}$) the mass of 1 mole of a substance; the molar mass of an element in grams per mole is numerically equal to that element's average atomic mass in unified atomic mass units.

molecular mass the mass of one molecule of a molecular compound.

formula mass the mass of one formula unit of an ionic compound.

[1]In sample exercises involving molar masses, we usually round off the values given in the periodic table to four significant figures, except for elements with molar masses greater than 100.

Collect, Organize, and Analyze We are asked to convert 0.462 mol He into g He. We will need to use the molar mass of helium, 4.003 g/mol, to convert from moles to grams:

$$\boxed{\text{mol He}} \xrightarrow{\text{molar mass}} \boxed{\text{g He}}$$

Because the balloon contains nearly $\frac{1}{2}$ mole of helium, the answer should be a little less than half of 4, or about 2 g.

Solve

$$0.462 \text{ }\cancel{\text{mol He}} \times \frac{4.003 \text{ g He}}{1 \text{ }\cancel{\text{mol He}}} = 1.85 \text{ g He}$$

Think About It We predicted that the balloon contained a little less than 2 g He, and that prediction turned out to be correct.

Practice Exercise
How many grams of gold are there in 0.250 mol of gold?

(Answers to Practice Exercises are in the back of the book.)

CONCEPT TEST

Which contains more atoms: 1 gram of gold (Au) or 1 gram of silver (Ag)?

(Answers to Concept Tests are in the back of the book.)

Molecular mass of SO_3	80.06 u/molecule
Mass of 1 mol of SO_3	80.06 g
Molar mass of SO_3	80.06 g/mol

FIGURE 3.13 The molecular mass (in u/molecule) and the molar mass (in g/mol) of sulfur trioxide have the same numerical value.

Molecular Masses and Formula Masses

The **molecular mass** of a molecular compound is the sum of the atomic masses of the atoms in that molecule. Because atomic masses are given in unified atomic mass units, molecular masses are also reported in u. Thus, the molecular mass of sulfur trioxide, SO_3, is the sum of the masses of 1 sulfur atom and 3 oxygen atoms (**Figure 3.13**):

$$32.06 \text{ u} + (3 \times 16.00 \text{ u}) = 80.06 \text{ u}$$

Just as the molar mass of an element in grams per mole is numerically the same as the atomic mass of one atom of the element in unified atomic mass units (Figure 3.11), the molar mass of a molecular compound in grams per mole is numerically the same as its molecular mass in unified atomic mass units. Thus, the molar mass of SO_3 is 80.06 g/mol (Figure 3.13).

As with atomic masses, we can use the concept of moles to scale from molecular masses to molar masses, which are quantities that are large enough to measure and manipulate in the laboratory. For example, the molar mass of SO_3 in grams per mole is the sum of the masses of 1 mole of sulfur atoms and 3 moles of oxygen atoms:

$$\mathcal{M}_{SO_3} = 32.06 \text{ g/mol} + 3(16.00 \text{ g/mol}) = 80.06 \text{ g/mol}$$

The concept of molar mass applies not only to molecular compounds but also to ionic compounds. For example, 1 mole of BaS contains 1 mole (137.33 g) of Ba^{2+} ions and 1 mole (32.06 g) of S^{2-} ions. Therefore, the molar mass of BaS is

$$\mathcal{M}_{BaS} = 137.33 \text{ g/mol} + 32.06 \text{ g/mol} = 169.39 \text{ g/mol}$$

There are no discrete molecules of BaS. It is an ionic compound, so its crystals consist of ordered arrays of cations and anions (**Figure 3.14**). The formula unit defines the smallest integer ratio of positive and negative ions (1:1, in this case) that describes the composition of an ionic compound. The mass of one formula unit of an ionic compound is called its **formula mass**, which is analogous to the

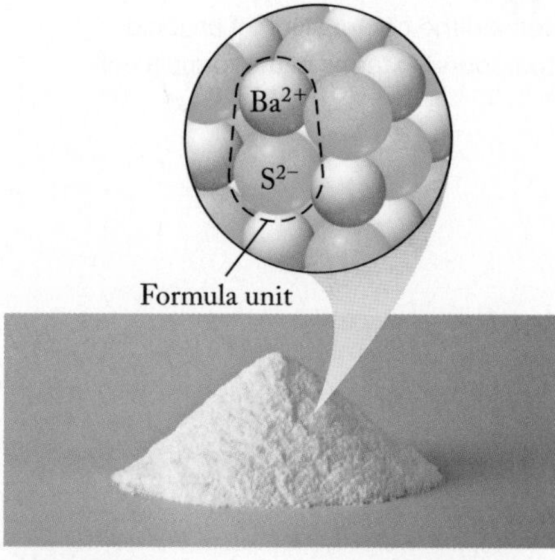

FIGURE 3.14 Barium sulfide is an ionic compound used in white paint. A crystal of barium sulfide consists of a three-dimensional array of Ba^{2+} ions and S^{2-} ions. The 1:1 ratio of Ba^{2+} ions to S^{2-} ions is represented in the formula unit of the crystal (enclosed in a dashed oval) and in the empirical formula of the compound (BaS).

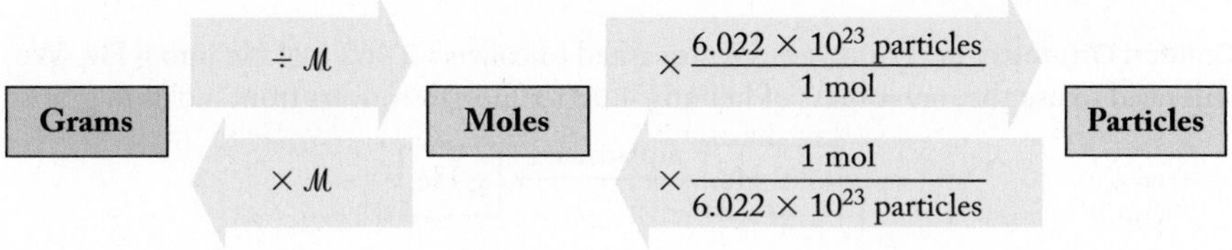

FIGURE 3.15 The mass of a sample of a pure substance, the number of moles in it, and the number of particles in it are related by the molar mass ($\mathcal{M}$) of the substance and the Avogadro constant (N_A).

molecular mass of a molecular compound. **Figure 3.15** summarizes how to use the Avogadro constant and molar mass to interconvert quantities of elements or compounds expressed in grams, moles, or numbers of particles.

CONNECTION In Section 2.6 we defined the composition of an ionic compound in terms of the formula unit.

SAMPLE EXERCISE 3.5 Calculating the Molar Mass of a Compound — LO1

Calculate the molar masses of (a) H_2O and (b) H_2SO_4.

Collect and Organize We are given the formulas of two compounds and want to find their molar masses. The molar mass of a molecular compound is the sum of the molar masses of the elements in the molecular formula of the compound, each multiplied by the number of atoms of that element in the molecular formula. The molar mass of an element in grams per mole can be found in the periodic table.

Analyze

a. One mole of H_2O contains 2 moles of H atoms and 1 mole of O atoms. Therefore,

$$\mathcal{M}_{H_2O} = (2 \times \mathcal{M}_H) + \mathcal{M}_O$$

b. One mole of H_2SO_4 contains 2 moles of H atoms, 1 mole of S atoms, and 4 moles of O atoms. Therefore,

$$\mathcal{M}_{H_2SO_4} = (2 \times \mathcal{M}_H) + \mathcal{M}_S + (4 \times \mathcal{M}_O)$$

Solve

a. The molar mass of H_2O is

$$2(1.008 \text{ g/mol}) + 16.00 \text{ g/mol} = 18.02 \text{ g/mol } H_2O$$

b. The molar mass of H_2SO_4 is

$$2(1.008 \text{ g/mol}) + 32.06 \text{ g/mol} + 4(16.00 \text{ g/mol}) = 98.08 \text{ g/mol } H_2SO_4$$

Think About It The molar mass of a compound is the sum of the masses of the elements in that compound. Because H_2SO_4 contains more atoms and heavier atoms than H_2O does, we expect the molar mass of sulfuric acid to be larger than that of water.

Practice Exercise When green plants take in water (H_2O) and carbon dioxide (CO_2), they produce glucose ($C_6H_{12}O_6$) and oxygen (O_2). Calculate the molar masses of carbon dioxide, oxygen, and glucose.

(Answers to Practice Exercises are in the back of the book.)

SAMPLE EXERCISE 3.6 Converting Between Grams, Moles, and Formula Units — LO1

Lithium carbonate is used in the treatment of bipolar disorder (**Figure 3.16**). How many moles and how many formula units of lithium carbonate are in a capsule containing 0.300 g Li_2CO_3?

Collect and Organize We are given the mass of Li_2CO_3 in a capsule and asked to express that mass in moles and in formula units of Li_2CO_3. The number of formula units of an ionic compound is analogous to the number of molecules of a molecular compound. The formula unit of lithium carbonate is Li_2CO_3. The Avogadro constant relates the number of formula units of an ionic compound to the number of moles of an ionic compound.

Analyze Figure 3.15 illustrates how to convert the mass of a substance in grams into an equivalent number of moles by dividing by the substance's molar mass and then into an equivalent number of particles (formula units in this case) by multiplying the number of moles by the Avogadro constant:

$$\boxed{\text{g } Li_2CO_3} \xrightarrow{\frac{1}{\text{molar mass}}} \boxed{\text{mol } Li_2CO_3} \xrightarrow{N_A} \boxed{\text{formula units } Li_2CO_3}$$

The capsule contains 0.300 g of Li_2CO_3, which probably has a molar mass of ~100 g/mol, so the number of moles of Li_2CO_3 should be about 0.003, and the number of formula units should be about 2×10^{21}.

Solve First we need to calculate the molar mass of Li_2CO_3 by adding twice the molar mass of Li to the molar mass of C, and three times the molar mass of O:

$$2(6.941 \text{ g/mol}) + 12.01 \text{ g/mol} + 3(16.00 \text{ g/mol}) = 73.89 \text{ g/mol } Li_2CO_3$$

Using this molar mass to convert g Li_2CO_3 into mol Li_2CO_3:

$$0.300 \cancel{\text{g } Li_2CO_3} \times \frac{1 \text{ mol } Li_2CO_3}{73.89 \cancel{\text{g } Li_2CO_3}} = 0.004060 \text{ mol } Li_2CO_3 = 4.060 \times 10^{-3} \text{ mol } Li_2CO_3$$

This value should be expressed with three significant figures, 4.060×10^{-3} mol, because the mass of Li_2CO_3 has only three significant figures.

Converting the calculated moles of Li_2CO_3 into formula units of Li_2CO_3:

$$4.060 \times 10^{-3} \cancel{\text{mol } Li_2CO_3} \times \frac{6.022 \times 10^{23} \text{ formula units } Li_2CO_3}{1 \cancel{\text{mol } Li_2CO_3}}$$

$$= 2.44 \times 10^{21} \text{ formula units } Li_2CO_3$$

Think About It The calculated number of moles of Li_2CO_3 is about the predicted value, as is the number of formula units, so both values are reasonable. We could never count out such an enormous number of formula units, but calculations such as this one allow us to know how many there are in a sample.

Practice Exercise How many moles of $CaCO_3$ are in an antacid tablet containing 0.500 g of $CaCO_3$? How many formula units of $CaCO_3$ are in this tablet?

(Answers to Practice Exercises are in the back of the book.)

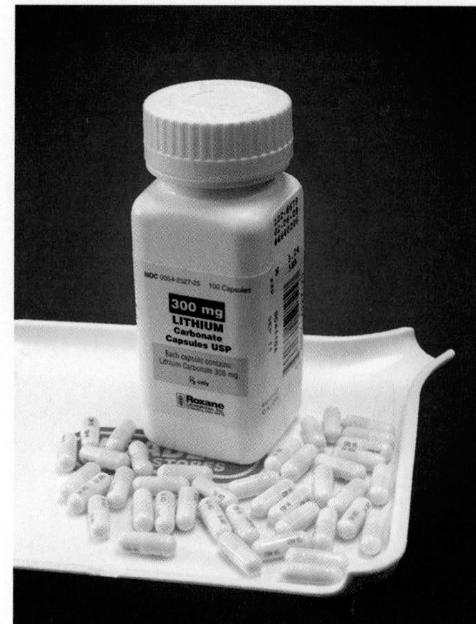

FIGURE 3.16 Lithium carbonate capsules can be prescribed for the treatment of bipolar disorder.

SAMPLE EXERCISE 3.7 Calculating the Masses of Elements in Their Compounds LO1

The discovery that iron metal could be extracted from iron-containing minerals such as hematite (Fe_2O_3) and magnetite (Fe_3O_4) ushered in the Iron Age around 1200 BCE (**Figure 3.17**). If a metalworker were able to extract all the iron in 250.0-g samples of Fe_2O_3 and Fe_3O_4, what masses of iron would each sample produce?

Collect and Organize We are given the masses of two iron oxide samples and asked to calculate the mass of iron in each of them. The chemical formulas of the oxides tell us the number of moles of Fe per mole of each oxide. The molar mass of a compound or element can be used to convert its mass into the equivalent number of moles as shown in Figure 3.15.

FIGURE 3.17 A knife and ax head excavated in Jerusalem were forged from iron between the 10th and 8th centuries BCE.

Analyze To begin, we need to calculate the molar masses of Fe_2O_3 and Fe_3O_4, and then use these values to convert the masses of Fe_2O_3 and Fe_3O_4 into moles of Fe_2O_3 and Fe_3O_4 as shown in Figure 3.15. Next, we convert these mole values into moles of Fe: there are 2 mol Fe/1 mol Fe_2O_3 and 3 mol Fe/1 mol Fe_3O_4. Then, we convert the two mol Fe values into mass (g) Fe values by multiplying by the molar mass of Fe, 55.84 g/mol. Summarizing the conversion steps for Fe_2O_3:

$$\boxed{\text{g Fe}_2\text{O}_3} \xrightarrow[\text{molar mass}]{1} \boxed{\text{mol Fe}_2\text{O}_3} \xrightarrow[\text{mol Fe}_2\text{O}_3]{\text{2 mol Fe}} \boxed{\text{mol Fe}} \xrightarrow{\text{molar mass}} \boxed{\text{g Fe}}$$

Solve Calculating the molar masses of Fe_2O_3 and Fe_3O_4:

$$Fe_2O_3\text{: } 2(55.84 \text{ g/mol}) + 3(16.00 \text{ g/mol}) = 159.68 \text{ g/mol } Fe_2O_3$$
$$Fe_3O_4\text{: } 3(55.84 \text{ g/mol}) + 4(16.00 \text{ g/mol}) = 231.52 \text{ g/mol } Fe_3O_4$$

Converting grams of Fe_2O_3 into grams of Fe:

$$250.0 \text{ g } \cancel{Fe_2O_3} \times \frac{1 \text{ mol } \cancel{Fe_2O_3}}{159.68 \text{ g } \cancel{Fe_2O_3}} \times \frac{2 \text{ mol Fe}}{1 \text{ mol } \cancel{Fe_2O_3}} \times \frac{55.84 \text{ g Fe}}{1 \text{ mol Fe}} = 174.8 \text{ g Fe}$$

Repeating the calculation for Fe_3O_4:

$$250.0 \text{ g } \cancel{Fe_3O_4} \times \frac{1 \text{ mol } \cancel{Fe_3O_4}}{231.52 \text{ g } \cancel{Fe_3O_4}} \times \frac{3 \text{ mol Fe}}{1 \text{ mol } \cancel{Fe_3O_4}} \times \frac{55.84 \text{ g Fe}}{1 \text{ mol Fe}} = 180.9 \text{ g Fe}$$

Think About It The mass of extractable iron in 250.0 g Fe_3O_4 is greater than that in the same mass of Fe_2O_3. This makes sense because the mole ratio of Fe:O in Fe_3O_4 is 3:4, which is greater than the Fe:O mole ratio in Fe_2O_3 (2:3). Its greater Fe:O ratio means that more of the mass of Fe_3O_4 is iron and less of it is oxygen.

Practice Exercise How many grams of chromium are there in 225 g sodium chromate (Na_2CrO_4) and in 225 g sodium dichromate ($Na_2Cr_2O_7$)?

(Answers to Practice Exercises are in the back of the book.)

Moles and Chemical Equations

The concepts of mole and molar mass offer three new interpretations of Equation 3.1:

$$SO_3(g) + H_2O(g) \rightarrow H_2SO_4(\ell)$$

1. Based on the coefficients in the chemical equation, 1 *mole* of SO_3 reacts with 1 *mole* of H_2O, producing 1 *mole* of H_2SO_4.
2. Based on the molar masses of the reactants and products, 80.06 *grams* of SO_3 react with 18.02 *grams* of H_2O, producing 98.08 *grams* of H_2SO_4.
3. Based on the Avogadro constant, 6.022×10^{23} *molecules* of SO_3 react with 6.022×10^{23} *molecules* of H_2O, forming 6.022×10^{23} *molecules* of H_2SO_4. In terms of lowest whole-number ratios, 1 *molecule* of SO_3 reacts with 1 *molecule* of H_2O, forming 1 *molecule* of H_2SO_4.

These interpretations of this equation, however, do not limit the amount of SO_3 that can react to 80.06 g. Reacting twice that amount, or 160.12 g of SO_3, is possible if twice the molar mass of water, 36.04 g of H_2O, is available. That is, 2 moles of SO_3 can react with 2 moles of H_2O to produce 2 moles of H_2SO_4. In fact, the equation means that *any* number of moles (*x* mol) of SO_3 react with an equal number of moles (*x* mol) of H_2O to produce the same quantity (*x* mol) of H_2SO_4 as shown in the table that follows.

	$SO_3(g)$	+	$H_2O(g)$	→	$H_2SO_4(\ell)$
Particle Ratios	1 molecule	+	1 molecule	→	1 molecule
Mole Ratios	1 mol	+	1 mol	→	1 mol
Mass Ratios	80.06 g	+	18.02 g	→	98.08 g
General Case (moles)	*x* mol	+	*x* mol	→	*x* mol
General Case (masses)	*x*(80.06 g/mol)	+	*x*(18.02 g/mol)	→	*x*(98.08 g/mol)

The mole ratio differs when different substances combine and is therefore specific for a given reaction. For example, in the reaction in Equation 3.2 between sulfuric acid and Fe_2O_3,

$$Fe_2O_3(s) + 3\ H_2SO_4(aq) \rightarrow 3\ H_2O(\ell) + Fe_2(SO_4)_3(aq)$$

the mole ratio of Fe_2O_3 to H_2SO_4 is 1:3, and an abbreviated version of the table above with just these two reactants looks like this.

	$Fe_2O_3(s)$	+	$3\ H_2SO_4(\ell)$	→
Particle Ratios	1 formula unit	+	3 molecules	→
Mole Ratios	1 formula mass	+	3 mol	→
Mass Ratios	159.68 g	+	98.08 g	→
General Case (moles)	*x* mol	+	3*x* mol	→
General Case (masses)	*x*(159.68 g/mol)	+	3*x*(98.08 g/mol)	→

This quantitative relationship between the reactants and products involved in a chemical reaction is called the **stoichiometry** of the reaction. In a reaction where all reactants are completely converted into products, the sum of the masses of the reactants equals the sum of the masses of the products. This fact illustrates a fundamental relation known as the **law of conservation of mass**, which applies to all chemical reactions. This law works in an equation representing any chemical reaction for two reasons: (1) the total number of atoms (and hence moles) of each element to the left of the reaction arrow must equal the total number of atoms (and moles) of that element to the right of the reaction arrow, and (2) the identity of atoms does not change in a chemical reaction.

stoichiometry the quantitative relationship between reactants and products in a chemical reaction.

law of conservation of mass the principle that the sum of the masses of the reactants in a chemical reaction is equal to the sum of the masses of the products.

SAMPLE EXERCISE 3.8 Visualizing Chemical Reactions **LO2**

Using blue spheres to represent nitrogen atoms and red spheres to represent oxygen atoms, sketch a picture describing the reaction between two molecules of N_2 and four molecules of O_2 to form NO_2.

Collect, Organize, and Analyze We are given the chemical formulas for two reactants and will use colored spheres to sketch their conversion to a product. We choose blue spheres to represent nitrogen atoms and red spheres to represent oxygen atoms. We use the atoms contained in two molecules of nitrogen and four molecules of oxygen to construct the appropriate number of NO_2 molecules.

Solve From a total of 4 nitrogen and 8 oxygen atoms, we can make four molecules of NO_2:

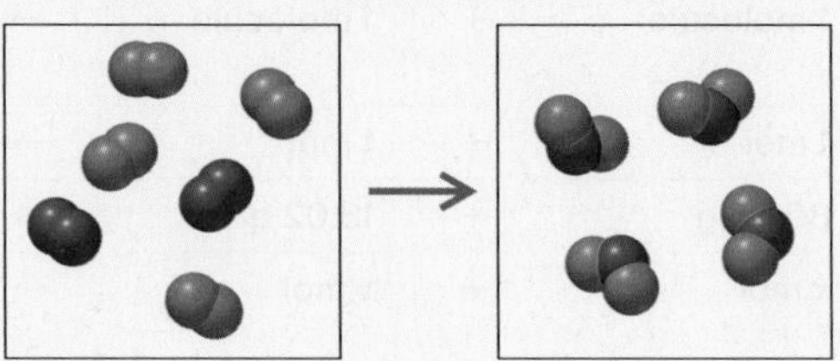

Think About It A chemical reaction involves rearrangements of atoms and molecules, resulting in new molecules. The number of each type of atom present before the reaction equals the number of each type of atom present after the reaction.

Practice Exercise Using colored spheres to represent nitrogen atoms (blue) and hydrogen atoms (white), sketch a picture describing the reaction between two molecules of N_2 and six molecules of H_2 to form ammonia, NH_3.

(Answers to Practice Exercises are in the back of the book.)

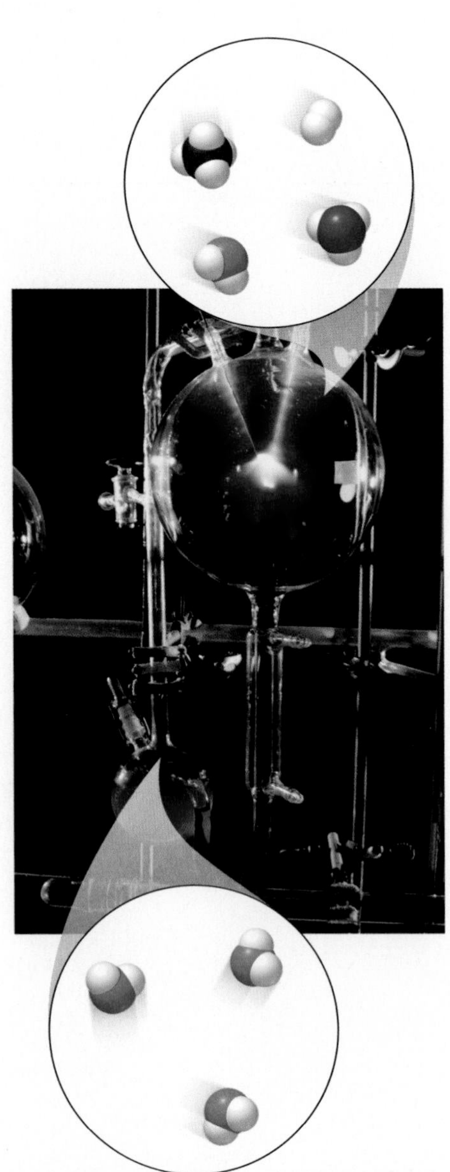

FIGURE 3.18 A molecular view of the reaction mixture in the Miller–Urey experiment. The large flask contained water, ammonia, hydrogen, and methane in ratios believed at that time to represent the composition of Earth's early atmosphere. The smaller flask was heated to add water vapor to the gas mixture in the large flask.

3.3 Writing Balanced Chemical Equations

A *balanced* chemical equation has the same number of atoms of each element on both sides of the equation. Here in Section 3.3, we practice writing balanced chemical equations describing chemical reactions that may account for the products observed by Miller and Urey in their experiments on the origins of life on Earth mentioned at the beginning of this chapter.

Figure 3.18 revisits the apparatus we saw in Figure 3.1. A key molecule in the synthesis of glycine, one of the amino acids produced in the Miller–Urey experiment, is hydrogen cyanide (HCN). Hydrogen cyanide can be formed from methane (CH_4) and ammonia (NH_3) in a reaction that also produces hydrogen (H_2). Let's write a balanced chemical equation describing this reaction. We know the formulas of the reactants and products, but we don't know the stoichiometry of the reaction—that is, we need to determine the coefficients that precede their formulas in a balanced chemical equation. There are several approaches to writing chemical equations. Let's apply the following four-step method, which works well if we have correctly identified all the reactants and products.

1. *Write a preliminary expression containing a single particle (atom, molecule, or formula unit) of each reactant and product with a reaction arrow separating reactants from products. Include phase symbols indicating physical states.*

The following expression represents the reaction of interest (particulate models of the molecules involved are included):

$$CH_4(g) + NH_3(g) \rightarrow HCN(g) + H_2(g)$$
$$\text{Reactants} \rightarrow \text{Products}$$

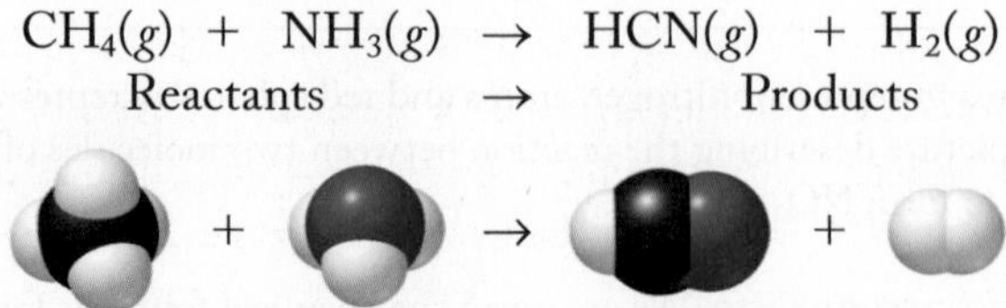

2. *Check whether the expression is balanced by counting the atoms of each element on each side of the reaction arrow.*

Adding up the atoms of each element in the reactants and products, we find that the numbers of C atoms and N atoms are balanced, but the number of H atoms is not: there are 7 on the reactant side but only 3 on the product side.

Element	Reactant Side	Product Side	Balanced?
C	1	1	✓
N	1	1	✓
H	4 + 3 = 7	1 + 2 = 3	✗

3. *Choose an element that appears in only one reactant and one product and balance it first.*

In this example, the two elements that occur only once on each side of the reaction arrow are C and N, but they are already balanced, so we can skip this step.

4. *Choose coefficients for the other substances so that the number of atoms for each element is the same on both sides of the reaction arrow.*

We need 4 more H atoms on the product side to have a balanced equation. Changing the coefficient in front of either HCN or H_2 could increase the number of H atoms but changing the coefficient of HCN would also increase the numbers of C and N atoms, which are already balanced. Therefore, the simplest approach is to increase the coefficient of H_2 from 1 to 3, which increases the number of H atoms from 2 for one molecule of H_2 to 6 for three molecules of H_2.

CHEMTOUR
Balancing Equations

$$CH_4(g) + NH_3(g) \rightarrow HCN(g) + \underline{3}\,H_2(g)$$

A final check of the number of atoms of each element on both sides of the reaction arrow confirms that the equation is balanced:

Element	Reactant Side	Product Side	Balanced?
C	1	1	✓
N	1	1	✓
H	4 + 3 = 7	1 + (3 × 2) = 7	✓

Note that balance was achieved by changing a coefficient, *not* by changing any subscripts in the chemical formulas. Changing subscripts would change the identities of the substances, and that would change the reaction itself. Consequently, only coefficients may be changed when balancing chemical equations.

Let's get more practice writing balanced chemical equations by using another important reaction in the Miller–Urey experiment. Analysis of the vapor above the solution in the large flask revealed the presence of both carbon dioxide and carbon monoxide gas. The reaction between methane and water yields carbon dioxide and hydrogen.

1. *Write a preliminary expression containing a single particle of each reactant and product.* Methane has the formula CH_4, water is H_2O, carbon dioxide is CO_2, and hydrogen is H_2. Recall from Chapter 1 that all the gaseous nonmetallic elements—except for the noble gases—exist in nature as diatomic molecules: H_2, N_2, O_2, F_2, and Cl_2. Bromine, a liquid at room temperature,

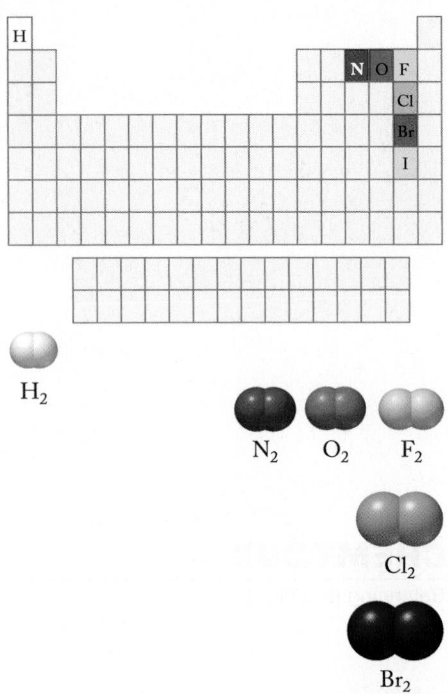

FIGURE 3.19 The diatomic elements are hydrogen, nitrogen, oxygen, and the group 17 halogens. The halogen astatine (At) near the bottom of group 17 is the rarest terrestrial element, and virtually none of its bulk physical properties are known.

and I_2, a solid, are also diatomic (**Figure 3.19**). Whenever these substances are involved in chemical reactions, they must be written as diatomic molecules.

$$CH_4(g) + H_2O(g) \rightarrow CO_2(g) + H_2(g)$$

2. *Check whether the expression is balanced.*

Element	Reactant Side	Product Side	Balanced?
C	1	1	✓
O	1	2	✗
H	4 + 2 = 6	2	✗

The expression is not yet balanced.

3. *Choose an element that appears in only one reactant and one product to balance first.* In this case we could choose either carbon or oxygen. Carbon is already balanced, so we focus on oxygen. With 1 O atom on the reactant side and 2 O atoms on the product side, we can make the number of O atoms equal on both sides of the equation by placing a coefficient of 2 in front of H_2O.

$$CH_4(g) + \underline{2}\,H_2O(g) \rightarrow CO_2(g) + \underline{\quad}\,H_2(g)$$

Assigning this coefficient balances the number of oxygen atoms but not the number of hydrogen atoms.

Element	Reactant Side	Product Side	Balanced?
C	1	1	✓
O	2 × 1 = 2	2	✓
H	4 + (2 × 2) = 8	2	✗

4. *Choose coefficients for the other substances so that the numbers of atoms of each element are the same on both sides of the equation.*

We can balance the H atoms by placing a coefficient of 4 in front of H_2, which completes the balancing process:

$$CH_4(g) + \underline{2}\,H_2O(g) \rightarrow CO_2(g) + \underline{4}\,H_2(g)$$

Element	Reactant Side	Product Side	Balanced?
C	1	1	✓
O	2 × 1 = 2	2	✓
H	4 + (2 × 2) = 8	4 × 2 = 8	✓

SAMPLE EXERCISE 3.9 Writing a Balanced Chemical Equation **LO2**

Write a balanced chemical equation for the gas-phase reaction between SO_2 and O_2 that forms SO_3. This reaction, one of the causes of acid rain, occurs when fuels containing sulfur are burned.

Collect, Organize, and Analyze We are given the chemical formulas and physical states of the reactants and products. We need to write a balanced chemical equation in which the same number of atoms of each element appear on both sides of the equation. To do so, we assign coefficients as needed to make the numbers of atoms of each element on the reactant (left) and product (right) sides of the reaction arrow equal.

Solve

1. Starting with a reaction expression that has single molecules of the reactants and product:

$$SO_2(g) + O_2(g) \rightarrow SO_3(g)$$

2. Checking whether the expression is already balanced:

Element	Reactant Side	Product Side	Balanced?
S	1	1	✔
O	2 + 2 = 4	3	✗

The sulfur atoms are balanced but the oxygen atoms are not.

3. Balancing the O atoms presents a challenge because increasing their number on the product side by increasing the SO_3 coefficient from 1 to 2 unbalances the sulfur atoms:

$$SO_2(g) + O_2(g) \rightarrow \underline{2}\ SO_3(g)$$

There are now two more O atoms and one more S atom on the product side than on the reactant side. Fortunately, both imbalances can be solved simultaneously by increasing the SO_2 coefficient from 1 to 2:

$$\underline{2}\ SO_2(g) + O_2(g) \rightarrow \underline{2}\ SO_3(g)$$

$$+ \quad \rightarrow$$

A check of the atom inventory on both sides confirms that the equation is now balanced:

Element	Reactant Side	Product Side	Balanced?
S	2 × 1 = 2	2 × 1 = 2	✔
O	(2 × 2) + 2 = 6	2 × 3 = 6	✔

Think About It Instead of increasing the number of O atoms on the product side to bring the equation into balance, we might have tried *decreasing* their number on the reactant side by reducing the coefficient in front of O_2 from 1 to $\frac{1}{2}$:

$$SO_2(g) + \tfrac{1}{2}O_2(g) \rightarrow SO_3(g)$$

This fix has the advantage of bringing the number of O atoms into balance without unbalancing the number of S atoms. It has the disadvantage of creating a fractional coefficient, but this problem can be solved by multiplying all the coefficients by 2, which gives us the same balanced equation we obtained in the Solve section:

$$2\,SO_2(g) + O_2(g) \rightarrow 2\,SO_3(g)$$

We use this strategy again in Sample Exercise 3.10 and in writing chemical equations for some of the combustion reactions in Section 3.4.

Practice Exercise Write balanced chemical equations for (a) the reaction between elemental phosphorus, $P_4(s)$, and oxygen to make $P_4O_{10}(s)$ and (b) the reaction of water with $P_4O_{10}(s)$ to produce phosphoric acid, $H_3PO_4(aq)$.

(Answers to Practice Exercises are in the back of the book.)

CONCEPT **TEST**

Why can't we write a balanced equation for the reaction between SO_2 and O_2 in Sample Exercise 3.9 this way?

$$SO_2(g) + O(g) \rightarrow SO_3(g)$$

(Answers to Concept Tests are in the back of the book.)

A chemical equation is analogous to a mathematical equation, because the equation in Sample Exercise 3.9 is still balanced if we multiply all the coefficients by the same number, such as 2:

$$4\,SO_2(g) + 2\,O_2(g) \rightarrow 4\,SO_3(g)$$

It is also still balanced if we multiply all the coefficients by a fraction, such as $\frac{1}{2}$. It is conventional, however, to write a balanced equation with the smallest whole-number coefficients.

SAMPLE EXERCISE 3.10 Writing Balanced Chemical Equations Using Fractional Coefficients LO2

Write a balanced chemical equation for the gas-phase reaction in which dinitrogen pentoxide is formed from nitrogen and oxygen.

Collect, Organize, and Analyze We are asked to write a balanced chemical equation given the names and physical states of the reactants and products. Nitrogen and oxygen gases exist in nature as diatomic molecules, so their molecular formulas are N_2 and O_2. The formula for dinitrogen pentoxide is N_2O_5. To write a balanced chemical equation, we must assign coefficients as needed to make the numbers of atoms of each element on the reactant (left) and product (right) sides of the reaction arrow equal.

Solve

1. Write a preliminary expression using correct formulas:

$$N_2(g) + O_2(g) \rightarrow N_2O_5(g)$$

2. The expression is not balanced as written:

Element	Reactant Side	Product Side	Balanced?
N	2	2	✔
O	2	5	✗

3. One way to balance the O atoms without unbalancing the N atoms is to place the fractional coefficient $\frac{5}{2}$ in front of O_2:

$$N_2(g) + \tfrac{5}{2}O_2(g) \rightarrow N_2O_5(g)$$

which gives us five O atoms on both sides. To obtain whole-number coefficients we multiply all three coefficients by 2:

$$\underline{2}\,N_2(g) \quad + \quad \underline{5}\,O_2(g) \rightarrow \underline{2}\,N_2O_5(g)$$

The equation is now balanced.

Think About It The use of a fractional coefficient in front of O_2 allowed us to balance the number of O atoms without unbalancing the number of N atoms (see the solution to Sample Exercise 3.9), thereby completing the exercise in fewer steps. In balancing a chemical equation, always check the equation after you are done to make sure that an element balanced in a previous step is still balanced at the end.

Practice Exercise Write a balanced chemical equation for the reaction between carbon monoxide, $CO(g)$, and oxygen to form carbon dioxide, $CO_2(g)$.

(Answers to Practice Exercises are in the back of the book.)

3.4 Combustion Reactions

Both methane and carbon monoxide are still present in our atmosphere, although at much lower concentrations than on early Earth. Atmospheric methane (natural gas) can be traced to several sources, including wetlands, cattle ranching, rice production, and oil drilling and fracking operations. Carbon dioxide in the atmosphere arises from burning carbon-containing compounds. Natural sources of CO_2 include volcanic activity and forest fires. In our industrialized world, the greatest use of methane and petroleum-based fuels is energy production.

Natural gas and other hydrocarbon fuels are burned to heat buildings and to warm water. Burning CH_4 is an example of an important class of chemical reactions known as **combustion reactions**, or simply *combustion* (**Figure 3.20**). Combustion refers to the reaction of a fuel with oxygen, such as the burning of a substance in air, which contains 21% oxygen.

When the combustion of a hydrocarbon is complete, the only products are carbon dioxide and water. In the presence of O_2, the more stable form of carbon is CO_2, not CO, so carbon monoxide also reacts with oxygen to form carbon dioxide. In the combustion reaction between methane and oxygen, the carbon becomes carbon dioxide when a sufficient supply of oxygen is available. The hydrogen in the fuel combines with oxygen to form water vapor. Let's see how to write a balanced equation describing the combustion of methane.

FIGURE 3.20 The combustion of methane in a gas stove produces carbon dioxide and water.

CONNECTION Molecules like methane (CH_4) are members of a class of organic compounds known as hydrocarbons because they are composed of only hydrogen and carbon (see Chapter 2).

combustion reaction a heat-producing reaction between oxygen and another element or compound.

1. Our first step is to write the formulas of the reactants on the left and products on the right.

$$CH_4(g) + O_2(g) \rightarrow CO_2(g) + H_2O(g)$$

Element	Reactant Side	Product Side	Balanced?
C	1	1	✔
H	4	2	✗
O	2	2 + 1 = 3	✗

The carbon atoms are balanced but the hydrogen and oxygen atoms are not.

2. Like carbon, H appears in only one reactant and one product. Adding a coefficient of 2 in front of H_2O makes the number of H atoms equal on both sides of the equation but increases the number of O atoms on the product side:

$$CH_4(g) + \_\ O_2(g) \rightarrow CO_2(g) + \underline{2}\ H_2O(g)$$

Element	Reactant Side	Product Side	Balanced?
C	1	1	✔
H	4	2 × 2 = 4	✔
O	2	2 + (2 × 1) = 4	✗

3. This leaves only the O atoms to balance. A coefficient of 2 in front of O_2 makes the atoms of O equal on both sides of the equation:

$$CH_4(g) + \underline{2}\ O_2(g) \rightarrow CO_2(g) + \underline{2}\ H_2O(g)$$

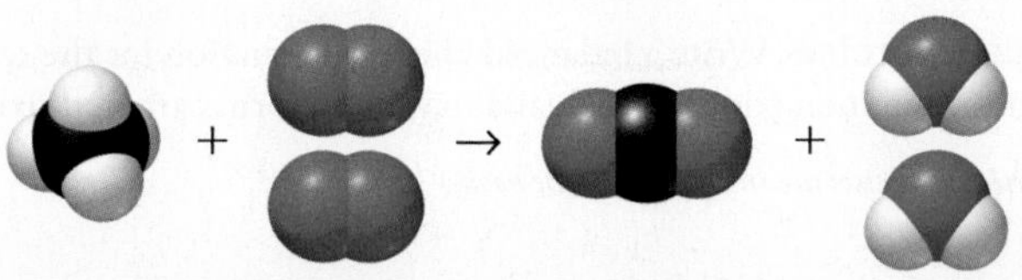

Element	Reactant Side	Product Side	Balanced?
C	1	1	✔
H	4	2 × 2 = 4	✔
O	2 × 2 = 4	2 + (2 × 1) = 4	✔

Balancing first C, then H, then O is a useful strategy when writing the equations for many combustion reactions involving hydrocarbons.

SAMPLE EXERCISE 3.11 Writing a Balanced Chemical Equation for a Combustion Reaction **LO2**

Methane (CH_4) is the principal ingredient in natural gas, but significant amounts of the hydrocarbon gases ethane (C_2H_6) and propane (C_3H_8) are also present in most natural gas samples. Write a balanced chemical equation describing the complete combustion of C_2H_6.

Collect, Organize, and Analyze We need to write a balanced chemical equation for the combustion of C_2H_6. The products of complete combustion of hydrocarbons are carbon dioxide and water.

Solve The preliminary expression with single particles of all known reactants and products describing the combustion of ethane is

$$C_2H_6(g) + O_2(g) \rightarrow CO_2(g) + H_2O(g)$$

Element	Reactant Side	Product Side	Balanced?
C	2	1	✗
H	6	2	✗
O	2	2 + 1 = 3	✗

None of the elements is currently balanced. Because this reaction is between oxygen and a hydrocarbon, we balance C first, then H, then O. Balance the carbon atoms first by giving CO_2 in the products a coefficient of 2:

$$C_2H_6(g) + \underline{\ \ } O_2(g) \rightarrow \underline{2}\, CO_2(g) + \underline{\ \ } H_2O(g)$$

Element	Reactant Side	Product Side	Balanced?
C	2	2 × 1 = 2	✓
H	6	2	✗
O	2	(2 × 2) + 1 = 5	✗

Then balance the hydrogen atoms by giving H_2O (a product) a coefficient of 3:

$$C_2H_6(g) + \underline{\ \ } O_2(g) \rightarrow \underline{2}\, CO_2(g) + \underline{3}\, H_2O(g)$$

Element	Reactant Side	Product Side	Balanced?
C	2	2 × 1 = 2	✓
H	6	3 × 2 = 6	✓
O	2	(2 × 2) + (3 × 1) = 7	✗

At this stage, the oxygen atoms (and the entire equation) can be balanced by giving O_2 a coefficient of $\frac{7}{2}$:

$$C_2H_6(g) + \tfrac{7}{2} O_2(g) \rightarrow \underline{2}\, CO_2(g) + \underline{3}\, H_2O(g)$$

Element	Reactant Side	Product Side	Balanced?
C	2	2 × 1 = 2	✓
H	6	3 × 2 = 6	✓
O	7	(2 × 2) + (3 × 1) = 7	✓

To remove the fraction, we multiply all of the coefficients by 2:

$$\underline{2}\, C_2H_6(g) + \underline{7}\, O_2(g) \rightarrow \underline{4}\, CO_2(g) + \underline{6}\, H_2O(g)$$

Think About It Balancing chemical equations for combustion reactions of hydrocarbons often requires several iterations of the coefficients, but, as in Sample Exercise 3.10, all the elements were balanced in fewer steps by giving O_2 a temporary fractional coefficient.

Practice Exercise Write a balanced chemical equation describing the complete combustion of propane (C_3H_8).

(Answers to Practice Exercises are in the back of the book.)

3.5 Stoichiometric Calculations and the Carbon Cycle

Earth's atmosphere underwent a major change beginning about 2.5 billion years ago with the evolution of bacteria capable of *photosynthesis*. Photosynthesis is driven by the energy in sunlight and involves several steps, but the overall reaction is

$$6\,CO_2(g) + 6\,H_2O(\ell) \rightarrow \underset{\text{Glucose}}{C_6H_{12}O_6(aq)} + 6\,O_2(g)$$

Although photosynthetic bacteria are believed to have been the initial source of oxygen in our atmosphere, today O_2 comes mostly from green plants. The reverse reaction, called *respiration*, is the major source of energy for all living things on Earth:

$$\underset{\text{Glucose}}{C_6H_{12}O_6(aq)} + 6\,O_2(g) \rightarrow 6\,CO_2(g) + 6\,H_2O(\ell)$$

Photosynthesis and respiration are key reactions in the *carbon cycle* (**Figure 3.21**). The two processes are nearly, but not exactly, in balance in Earth's biosphere. If they were exactly in balance, no net change would have taken place in the concentrations of atmospheric carbon dioxide or oxygen in the past 2.5 billion years. However, about 0.01% of the decaying mass of plants and animals (called *detritus*) is incorporated into sediments and soil when organisms die. Shielded in this way from exposure to oxygen, the carbon in this mass is not converted back into CO_2. Although 0.01% may not seem like much, over hundreds of millions of years it has added up to the removal of about 10^{20} kg of carbon dioxide from the atmosphere. About 10^{15} kg of this buried carbon is in the form of fossil fuels: coal, petroleum, and natural gas.

FIGURE 3.21 The carbon cycle. ① Green plants and marine phytoplankton incorporate CO_2 into their biomass. ② Some of this biomass becomes the biomass of animals. ③ As photosynthetic organisms and animals respire, they release CO_2 back into the environment. ④ When they die, the decay of their tissues releases most of their carbon content as CO_2, but about 0.01% is incorporated into carbonate minerals and deposits of coal, petroleum, and natural gas (fossil fuels, ⑤). ⑥ Mining and the combustion of fossil fuels for human use are shifting the natural equilibrium that has controlled the concentration of CO_2 in the atmosphere.

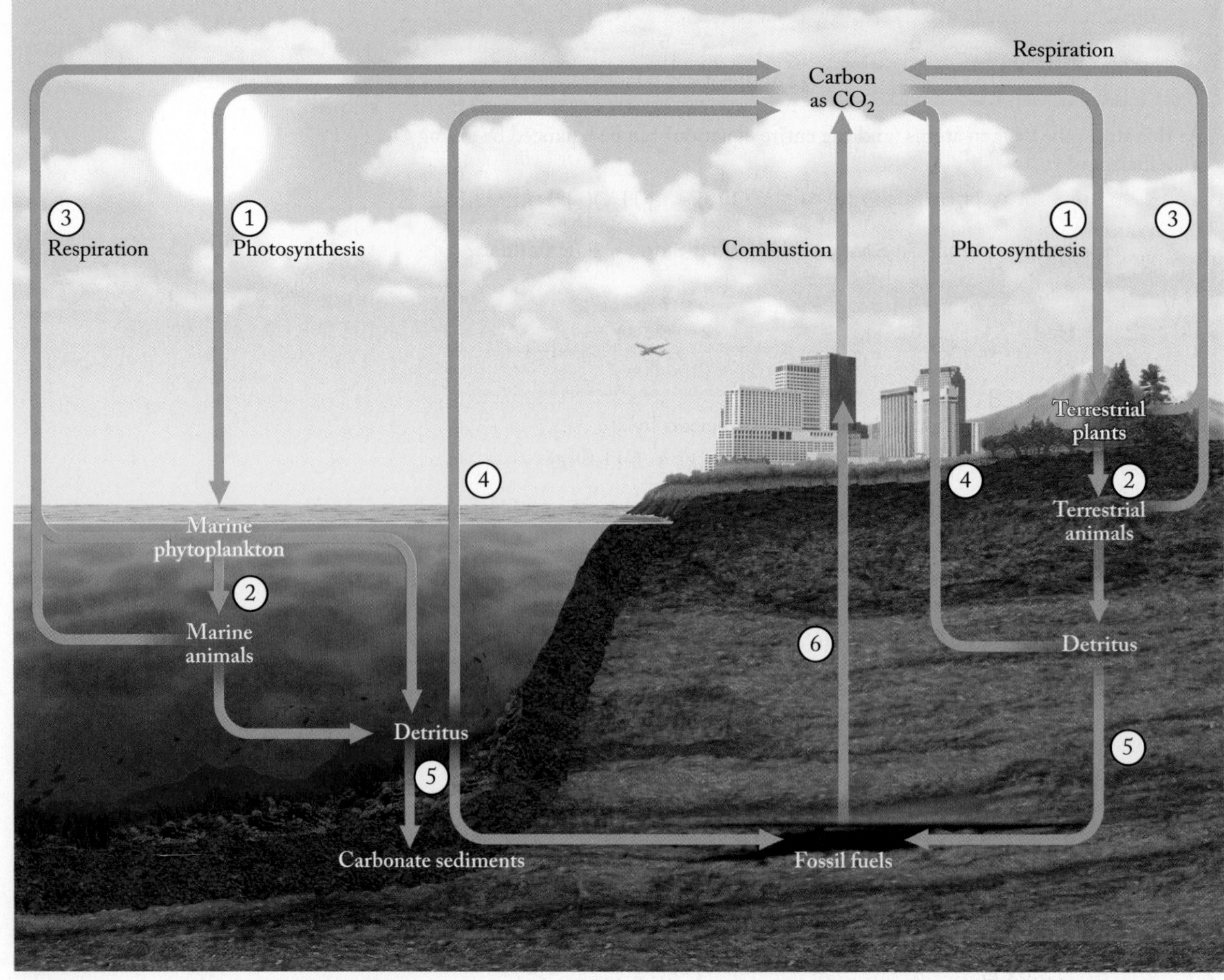

Fossil fuels have served as the world's primary energy resoure for more than a century with a rapid increase in the rate of consumption of these fuels since 1950. In 2018, the combustion of fossil fuels and a variety of industrial processes resulted in the conversion of 10.1 trillion (10.1×10^{12}) kilograms of fossil carbon into carbon dioxide gas, which was released into the atmosphere. Assessing the environmental impacts of these additions of CO_2 to Earth's atmosphere has been the subject of intense research in recent years for reasons we discuss in Chapters 8 and 16. For now, let's focus on how we can use balanced chemical equations to calculate how much fossil CO_2 is produced by combustion of a widely used fossil fuel: gasoline.

In many industrial countries, including the United States and Canada, the most heavily used fossil fuel is petroleum. In the United States nearly half of the petroleum is refined into gasoline: 1.43×10^{11} U.S. gallons of it in 2017. To evaluate the environmental impact of converting this much gasoline into atmospheric CO_2, let's calculate how much is produced by the complete combustion of just one gallon (3.785 liters) of it. Gasolines are complex mixtures of hydrocarbons of variable composition, so we will make two simplifying assumptions: (1) a typical density of gasoline is 0.73 g/mL and (2) the average number of carbon atoms per molecule in gasoline is 7. Therefore, we will use heptane, C_7H_{16}, as a model gasoline hydrocarbon.

To begin, let's write a chemical equation describing the combustion of heptane. We start with a reaction expression that contains single molecules (or moles) of the known reactants and products:

$$C_7H_{16}(\ell) + O_2(g) \rightarrow CO_2(g) + H_2O(g)$$

To balance the expression, we begin by balancing the number of atoms of carbon on both sides of the reaction arrow:

$$C_7H_{16}(\ell) + O_2(g) \rightarrow 7\ CO_2(g) + H_2O(g)$$

then balancing the number of atoms of hydrogen:

$$C_7H_{16}(\ell) + O_2(g) \rightarrow 7\ CO_2(g) + 8\ H_2O(g)$$

Carbon Cycle

and finally, balancing the number of atoms of oxygen:

$$C_7H_{16}(\ell) + 11\ O_2(g) \rightarrow 7\ CO_2(g) + 8\ H_2O(g)$$

This yields a balanced chemical equation.

Next, we need to calculate the mass of heptane in a gallon and then convert that mass into an equivalent quantity in moles by dividing by the molar mass of heptane. The last step requires our first calculating the molar mass of a compound that has 7 atoms of carbon and 16 atoms of hydrogen per molecule:

$$\mathcal{M}\ C_7H_{16} = 7(12.01\ \text{g/mmol}) + 16(1.008\ \text{g/mol}) = 100.20\ \text{g/mol}$$

The starting quantity of gasoline is a volume of exactly one U.S. gallon or 3.785 liters. Multiplying volume by density with the appropriate unit conversions will yield a mass in grams, which can be converted to a quantity of moles using the calculated molar mass:

$$3.875\ \cancel{\text{L}} \times \frac{1000\ \cancel{\text{mL}}}{\cancel{\text{L}}} \times \frac{0.73\ \text{g}}{\cancel{\text{mL}}} \times \frac{1\ \text{mol}}{100.20\ \text{g}} = 28.23\ \text{mol}\ C_7H_{16}$$

The mole ratio of CO_2 to C_7H_{16} in the balanced chemical equation is 7:1 and the molar mass of CO_2 is

$$12.01\ \text{g/mol} + 2(16.00\ \text{g/mol}) = 44.01\ \text{g/mol}\ CO_2$$

Using these two values as the final conversion factors:

$$28.23 \text{ mol } C_7H_{16} \times \frac{7 \text{ mol } CO_2}{1 \text{ mol } C_7H_{16}} \times \frac{44.01 \text{ g } CO_2}{1 \text{ mol } CO_2} = 8.7 \times 10^3 \text{ g or } 8.7 \text{ kg } CO_2$$

The mass of CO_2 is expressed to two significant figures because that was the number in the density value used in its calculation. If 8.7 kg CO_2 (about 19 pounds) per gallon of gasoline seems like a lot, remember that one mole (12.01 g) of carbon produces one mole (44.01 g) of CO_2, which represents a mass gain of 44.01/12.01 or nearly 4 times. Then consider the mass of CO_2 produced by all the gasoline engines in the United States in 2017:

$$1.43 \times 10^{11} \text{ gallons gasoline} \times \frac{8.7 \text{ kg } CO_2}{\text{gallon gasoline}} = 1.2 \times 10^{12} \text{ kg } CO_2$$

This procedure can be applied to determining the mass of any substance (reactant or product) involved in any chemical reaction if we know (1) the mass of another substance in the reaction and (2) the *stoichiometric relation* between the two substances, that is, their mole ratio in the balanced chemical equation.

SAMPLE EXERCISE 3.12 Calculating a Product Mass from a Reactant Mass **LO3**

In 2014 electric power plants in the United States consumed about 1.57×10^{11} kg of natural gas. Natural gas is mostly methane (CH_4), so we can approximate the combustion reaction generating the energy by the equation:

$$CH_4(g) + 2\,O_2(g) \rightarrow CO_2(g) + 2\,H_2O(g)$$

How many kilograms of CO_2 were released into the atmosphere from these power plants in 2014?

Collect, Organize, and Analyze We are asked to calculate the mass of CO_2 produced from the combustion of 1.57×10^{11} kg of CH_4. We can change the mass of CH_4 in kilograms to grams and then convert into moles of CH_4 by multiplying by the molar mass of CH_4. According to the balanced equation, 1 mole of carbon dioxide is produced for every 1 mole of methane consumed (i.e., the mole ratio of CO_2:CH_4 is 1:1), so moles of CH_4 = moles of CO_2. Finally, we convert moles of CO_2 into kilograms of CO_2. We can combine all these steps into a single calculation:

$$\boxed{\text{kg } CH_4} \xrightarrow{\frac{10^3\text{ g}}{1\text{ kg}}} \boxed{\text{g } CH_4} \xrightarrow{\frac{1}{\text{molar mass}}} \boxed{\text{mol } CH_4} \xrightarrow{\frac{1\text{ mol } CO_2}{1\text{ mol } CH_4}}$$

$$\boxed{\text{mol } CO_2} \xrightarrow{\text{molar mass}} \boxed{\text{g } CO_2} \xrightarrow{\frac{1\text{ kg}}{10^3\text{ g}}} \boxed{\text{kg } CO_2}$$

Solve The chemical equation is balanced as written (you should always check to be sure). To convert between grams and moles of CH_4 and CO_2, we need to calculate the molar masses of these compounds. The molar mass of CH_4 is

$$12.01 \text{ g/mol} + 4(1.008 \text{ g/mol}) = 16.04 \text{ g/mol } CH_4$$

and the molar mass of CO_2 is

$$12.01 \text{ g/mol} + 2(16.00 \text{ g/mol}) = 44.01 \text{ g/mol } CO_2$$

Convert the mass of CH_4 into moles:

$$1.57 \times 10^{11}\ \text{kg CH}_4 \times \frac{10^3\ \text{g}}{1\ \text{kg}} \times \frac{1\ \text{mol CH}_4}{16.04\ \text{g CH}_4} = 9.788 \times 10^{12}\ \text{mol CH}_4$$

Convert moles of CH_4 into moles of CO_2 by using the mole ratio from the balanced chemical equation:

$$9.788 \times 10^{12}\ \text{mol CH}_4 \times \frac{1\ \text{mol CO}_2}{1\ \text{mol CH}_4} = 9.788 \times 10^{12}\ \text{mol CO}_2$$

Convert moles of CO_2 into mass of CO_2:

$$9.788 \times 10^{12}\ \text{mol CO}_2 \times \frac{44.01\ \text{g CO}_2}{1\ \text{mol CO}_2} \times \frac{1\ \text{kg}}{10^3\ \text{g}} = 4.31 \times 10^{11}\ \text{kg CO}_2$$

We can combine the three separate calculations into a single calculation, and in subsequent problems we may not show the individual steps:

$$1.57 \times 10^{11}\ \text{kg CH}_4 \times \frac{10^3\ \text{g}}{1\ \text{kg}} \times \frac{1\ \text{mol CH}_4}{16.04\ \text{g CH}_4} \times \frac{1\ \text{mol CO}_2}{1\ \text{mol CH}_4} \times \frac{44.01\ \text{g CO}_2}{1\ \text{mol CO}_2} \times \frac{1\ \text{kg}}{10^3\ \text{g}}$$

$$= 4.31 \times 10^{11}\ \text{kg CO}_2$$

Think About It Complete combustion of 1 mole (16.04 g) of CH_4 produces 1 mole (44.01 g) of CO_2. Therefore, there will be a reactant-to-product gain in mass of 44.01/16.04 = 2.744 times when the 4 moles of H atoms (4.03 g) in each mole of CH_4 are replaced by 2 moles of O atoms (32.00 g) in each mole of CO_2. Therefore, the calculated mass of CO_2 should be 1.57×10^{11} kg $\times$ 2.744 = 4.31×10^{11} kg of CO_2, and it is.

Practice Exercise Disposable lighters burn butane (C_4H_{10}) and produce CO_2 and H_2O. Write a balanced chemical equation for this combustion reaction and determine how many grams of CO_2 are produced by burning 1.00 g of C_4H_{10}.

(Answers to Practice Exercises are in the back of the book.)

SAMPLE EXERCISE 3.13 Calculating the Mass of a Reactant Needed to Produce a Mass of Product LO3

Copper was among the first metals to be refined from minerals collected by ancient metalworkers. The production of copper was already an industry by 3500 BCE. Cuprite is a copper mineral commonly found near Earth's surface, making it a likely resource for Bronze Age coppersmiths. Cuprite has the formula Cu_2O and can be converted to copper metal by reacting it with charcoal:

$$2\ \text{Cu}_2\text{O}(s) + \text{C}(s) \rightarrow 4\ \text{Cu}(s) + \text{CO}_2(g)$$

How much cuprite must be consumed in the above reaction to prepare 256 g of copper for a bracelet such as the one shown in **Figure 3.22**?

Collect and Organize We are given a balanced chemical equation and the mass of a product, Cu, and are asked to find the mass of one of the reactants, Cu_2O. The mole ratio is 4 mol Cu:2 mol Cu_2O, which simplifies to 2:1.

Analyze To use this mole ratio, we first need to convert g Cu to mol Cu, and then use the 2:1 ratio to calculate an equivalent mol Cu_2O value. Finally, we use the molar mass of Cu_2O to calculate the mass of it required for the reaction.

$$\boxed{\text{g Cu}} \xrightarrow{\frac{1}{\text{molar mass}}} \boxed{\text{mol Cu}} \xrightarrow{\frac{2\ \text{mol Cu}_2\text{O}}{4\ \text{mol Cu}}} \boxed{\text{mol Cu}_2\text{O}} \xrightarrow{\text{molar mass}} \boxed{\text{g Cu}_2\text{O}}$$

FIGURE 3.22 Copper bracelet.

Solve Carrying out the steps described in the analysis of the problem:

$$256 \text{ g Cu} \times \frac{1 \text{ mol Cu}}{63.55 \text{ g Cu}} \times \frac{1 \text{ mol } Cu_2O}{2 \text{ mol Cu}} \times \frac{143.10 \text{ g } Cu_2O}{1 \text{ mol } Cu_2O} = 288 \text{ g } Cu_2O$$

Think About It There are only three significant figures in our answer because there are three significant figures in 256 g of Cu. The 2:1 mole ratio in the stoichiometry of the reaction makes sense because there are 2 moles of Cu in 1 mole of Cu_2O. The presence of oxygen in Cu_2O explains the greater mass of it (288 g) that was needed to produce 256 g of copper.

Practice Exercise The copper mineral chalcocite (Cu_2S) can be converted to copper simply by heating in air: $Cu_2S(s) + O_2(g) \rightarrow 2\ Cu(s) + SO_2(g)$. How much Cu_2S is needed to make 256 g of Cu? How many grams of SO_2 are also produced?

(Answers to Practice Exercises are in the back of the book.)

3.6 Limiting Reactants and Percent Yield

Let's return to photosynthesis, the process responsible for the O_2 in our present-day atmosphere and for the energy that sustains life on Earth's surface. The stoichiometry of the reaction calls for equal moles, referred to as *equimolar amounts*, of CO_2 and H_2O:

$$6\ CO_2(g) + 6\ H_2O(\ell) \rightarrow \underset{\text{Glucose}}{C_6H_{12}O_6(aq)} + 6\ O_2(g)$$

In nature there is little likelihood of having the exact mole ratio of reactants at a reaction site, so let's consider what happens when more than six molecules of water are available for every six molecules of CO_2 (this is not a theoretical consideration because having water in excess is a common occurrence in biological systems). The photosynthetic production of glucose will continue until all the CO_2 at the reaction site is consumed, leaving surplus water unreacted. In this example, carbon dioxide is the **limiting reactant**, meaning that the maximum amount of product is determined by the quantity of CO_2 available and not by the quantity of H_2O, which is in excess. **Figure 3.23** provides another example of a reaction mixture with a limiting reactant.

The concept of the limiting reactant applies to everyday tasks, too. Suppose you are assembling bicycles that consist of one bike frame, one bike seat, and two wheels:

$$\text{frame} + \text{bike seat} + 2 \text{ wheels} \rightarrow \text{bicycle}$$

You have 100 bike frames, 100 seats, and 100 wheels. If you used all the bike frames and all the seats, you could build 100 bicycles. But you only have enough wheels to build 50 bicycles, because each bicycle needs two wheels. The wheels, then, are the limiting reactant in this scenario.

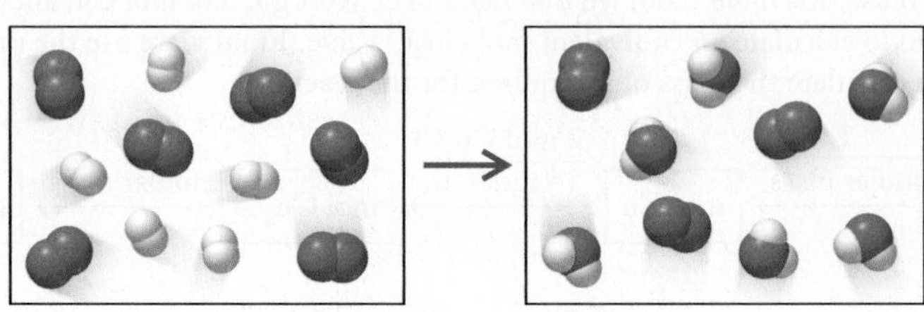

FIGURE 3.23 The reaction of hydrogen with oxygen produces water. A mixture containing equal numbers of hydrogen and oxygen molecules produces only as many water molecules as the number of H_2 molecules available. In this case, H_2 is the limiting reactant and O_2 is in excess.

Calculations Involving Limiting Reactants

limiting reactant a reactant that is consumed completely in a chemical reaction; the amount of product formed depends on the amount of the limiting reactant available.

Suppose you are asked to calculate the mass of a product formed from given masses of two or more reactants. How do you know whether one reactant is limiting? And how do you know which one is limiting? It is tempting to select the reactant present in the smaller amount by mass. Avoid this temptation and instead take a systematic approach based on the stoichiometry of the reaction. Several approaches are possible, two of which we present here—namely, one that uses stoichiometric calculations and one that uses mole ratios of reactants.

In our first method, we calculate how much product would be formed if reactant A were completely consumed. Then we repeat the calculation for reactant B (and so on for any additional reactants). Let's try this approach with the earlier reaction of Equation 3.1 between sulfur trioxide gas and water vapor to form sulfuric acid:

CHEMTOUR
Limiting Reactant

$$SO_3(g) + H_2O(g) \rightarrow H_2SO_4(\ell) \qquad (3.1)$$

This reaction contributed to the acidity of Earth's early atmosphere (see Section 3.1). Suppose we carry out this reaction in the laboratory by using 20.00 g of $SO_3(g)$ and 10.00 g of $H_2O(g)$. Which is the limiting reactant, and how many grams of H_2SO_4 are produced from these masses of reactants?

Calculating the numbers of moles of SO_3 and H_2O from the two masses:

$$20.00\ \cancel{\text{g SO}_3} \times \frac{1\ \text{mol SO}_3}{80.06\ \cancel{\text{g SO}_3}} = 0.2498\ \text{mol SO}_3$$

$$10.00\ \cancel{\text{g H}_2\text{O}} \times \frac{1\ \text{mol H}_2\text{O}}{18.02\ \cancel{\text{g H}_2\text{O}}} = 0.5549\ \text{mol H}_2\text{O}$$

Then we calculate the mass of sulfuric acid produced if all the SO_3 were consumed:

$$0.2498\ \cancel{\text{mol SO}_3} \times \frac{1\ \cancel{\text{mol H}_2\text{SO}_4}}{1\ \cancel{\text{mol SO}_3}} \times \frac{98.08\ \text{g H}_2\text{SO}_4}{1\ \cancel{\text{mol H}_2\text{SO}_4}} = 24.50\ \text{g H}_2\text{SO}_4$$

Next, we carry out the same calculation to determine how much sulfuric acid is produced if the 10.00 g of H_2O were to react completely:

$$0.5549\ \cancel{\text{mol H}_2\text{O}} \times \frac{1\ \cancel{\text{mol H}_2\text{SO}_4}}{1\ \cancel{\text{mol H}_2\text{O}}} \times \frac{98.08\ \text{g H}_2\text{SO}_4}{1\ \cancel{\text{mol H}_2\text{SO}_4}} = 54.42\ \text{g H}_2\text{SO}_4$$

The sulfur trioxide is the limiting reactant because, when it is completely consumed, a smaller amount of product is formed. Thus, when 20.00 g of SO_3 and 10.00 g of H_2O react, the maximum amount of product that can form is 24.50 g of H_2SO_4. Making that amount of sulfuric acid consumes all the available SO_3 but not all the available H_2O.

The second method compares the mole ratio of the reactants to the mole ratio required by the balanced chemical equation. To use this approach, we take the following steps:

1. Convert masses of reactants A and B into moles.
2. Calculate the mole ratio of A to B.
3. Compare this mole ratio with the stoichiometric mole ratio from the balanced chemical equation. If

$$\left(\frac{\text{mol A}}{\text{mol B}}\right)_{\text{calculated}} > \left(\frac{\text{mol A}}{\text{mol B}}\right)_{\text{stoichiometric}} \qquad (3.3)$$

theoretical yield the maximum amount of product possible in a chemical reaction for given quantities of reactants; also called *stoichiometric yield.*

then the reaction mixture has more A than is needed to consume all of B, and B is the limiting reactant. If

$$\left(\frac{\text{mol A}}{\text{mol B}}\right)_{\text{calculated}} < \left(\frac{\text{mol A}}{\text{mol B}}\right)_{\text{stoichiometric}} \tag{3.4}$$

then there is not enough A in the reaction mixture to consume all of B, and A is the limiting reactant. If the two ratios are equal, the masses are the stoichiometric amounts and both reactants are consumed completely.

For the same example using 20.00 g of SO_3 and 10.00 g of H_2O, we calculate the mole ratio of SO_3 to H_2O after calculating the moles of SO_3 and H_2O available:

$$\frac{\text{mol SO}_3}{\text{mol H}_2\text{O}} = \frac{20.00\ \cancel{\text{g SO}_3} \times \dfrac{1\ \text{mol SO}_3}{80.06\ \cancel{\text{g SO}_3}}}{10.00\ \cancel{\text{g H}_2\text{O}} \times \dfrac{1\ \text{mol H}_2\text{O}}{18.02\ \cancel{\text{g H}_2\text{O}}}} = \frac{0.2498\ \text{mol SO}_3}{0.5549\ \text{mol H}_2\text{O}} = 0.4502$$

According to Equation 3.1, the stoichiometric ratio of SO_3 to H_2O is 1/1 = 1. The SO_3/H_2O mole ratio for our reaction conditions is 0.2498/0.5549 = 0.4502. Expressing this outcome as a mathematical relationship, we get

$$\left(\frac{\text{mol SO}_3}{\text{mol H}_2\text{O}}\right)_{\text{calculated}} < \left(\frac{\text{mol SO}_3}{\text{mol H}_2\text{O}}\right)_{\text{stoichiometric}}$$

This inequality matches Equation 3.4, which tells us that SO_3 (reactant A in Equation 3.4) must be the limiting reactant.

These two approaches to determining a limiting reactant lead to the same conclusion and are of similar complexity. In both cases, the masses of the reactants are given in grams, so we must convert those masses into moles. The advantage of the first method is that it not only determines the identity of the limiting reactant, but it also determines the **theoretical yield**, the maximum amount of product that would be produced by the given mixture if the limiting reactant were completely converted into product. The theoretical yield is sometimes called the *stoichiometric yield* or *100% yield.*

SAMPLE EXERCISE 3.14 Identifying Limiting Reactants LO4

During strenuous exercise, respiration (the reaction between glucose and oxygen that produces CO_2 and water; Section 3.5) is limited by the availability of oxygen to the muscles. We can model this situation by considering the reaction between 1.32 g $C_6H_{12}O_6$ (**Figure 3.24**) and 1.32 g O_2. Is one of these reactants a limiting reactant, or is the mixture stoichiometric? If one reactant is limiting, which one is it?

Collect and Organize We are given quantities of two reactants ($C_6H_{12}O_6$ and O_2) and we want to determine which, if either, is limiting. Respiration is described (Section 3.5) by this chemical equation:

$$C_6H_{12}O_6(aq) + 6\ O_2(g) \rightarrow 6\ CO_2(g) + 6\ H_2O(\ell)$$

Analyze We can determine the limiting reactant from the ratio of O_2 to $C_6H_{12}O_6$ in the chemical equation for the reaction. That means we will use the second method described in the preceding discussion in which we compare the ratio of the amounts of reactants with the stoichiometric ratio in the balanced equation to determine whether one of the reactants is limiting.

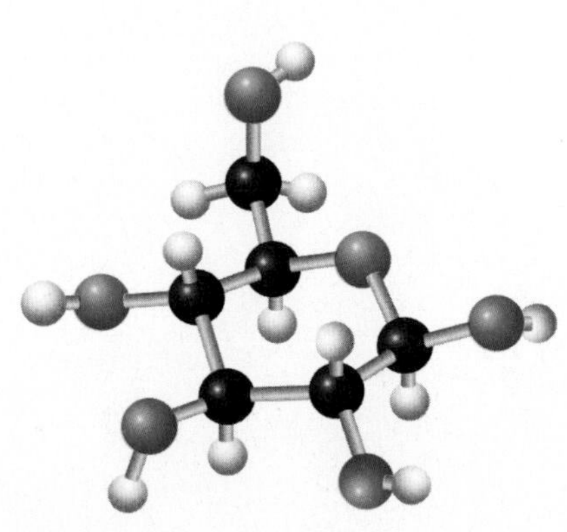

FIGURE 3.24 The molecular structure of glucose, which is consumed during respiration.

Solve Convert the given masses of reactants to moles:

$$1.32 \text{ g O}_2 \times \frac{1 \text{ mol O}_2}{32.00 \text{ g O}_2} = 4.13 \times 10^{-2} \text{ mol O}_2$$

$$1.32 \text{ g C}_6\text{H}_{12}\text{O}_6 \times \frac{1 \text{ mol C}_6\text{H}_{12}\text{O}_6}{180.16 \text{ g C}_6\text{H}_{12}\text{O}_6} = 7.33 \times 10^{-3} \text{ mol C}_6\text{H}_{12}\text{O}_6$$

and calculate the mole ratio:

$$\frac{4.13 \times 10^{-2} \text{ mol O}_2}{7.33 \times 10^{-3} \text{ mol C}_6\text{H}_{12}\text{O}_6} = 5.6$$

This ratio is less than the stoichiometric ratio of 6 moles O_2 to 1 mole $C_6H_{12}O_6$, so oxygen is the limiting reactant, and there is a slight excess of glucose.

Think About It The calculation indicates that when equal masses of glucose and oxygen react, oxygen is the limiting reactant. This makes sense because the mass of 6 moles O_2 is 6 mol × 32 g/mol = 192 g, whereas the mass of 1 mole of glucose is 180 g. Therefore, a larger mass of O_2 is needed to react with a given mass of glucose.

Practice Exercise Any fuel–oxygen mixture that contains more oxygen than is needed to burn the fuel completely is called a *lean mixture*, whereas a mixture containing too little oxygen to allow complete combustion of the fuel is called a *rich mixture*. A high-performance heater that burns propane, $C_3H_8(g)$, is adjusted so that 100.0 g of $O_2(g)$ enters the system for every 26.0 g of propane. Is this mixture lean or rich?

(Answers to Practice Exercises are in the back of the book.)

SAMPLE EXERCISE 3.15 Calculating Theoretical Yield **LO5**

The results of the Miller–Urey experiments required a chemical reaction to account for the formation of glycine (**Figure 3.25**). The reaction of carbon dioxide, ammonia, and methane that produces glycine, water, and carbon monoxide is described by the following chemical equation:

$$2\,CO_2(g) + NH_3(g) + CH_4(g) \rightarrow C_2H_5NO_2(s) + H_2O(\ell) + CO(g)$$

How much glycine could be expected from the reaction of 21.3 g of CO_2, 4.53 g of NH_3, and 4.27 g of CH_4? How much of which reactant(s) would remain unreacted?

Collect and Organize We are given a balanced chemical equation and the quantities of three reactants (CH_4, NH_3, and CO_2). We are asked how much glycine can be produced, which will require us to identify which one of the reactants is the limiting reactant and which reactants will be in excess. Because there are three reactants, we must use the first method discussed previously.

Analyze First we calculate the number of moles of each reactant we have available. Then, we calculate the theoretical quantity of glycine that would be produced from completely consuming each reactant. The reactant yielding the smallest amount of glycine is the limiting reactant and determines the theoretical amount of glycine that we can expect from the reaction. To calculate the quantity of the reactants in excess that will be left over, we use the stoichiometry of the reaction to calculate how much of each reactant is consumed during the reaction and then subtract that value from the initial amount present in the reaction mixture.

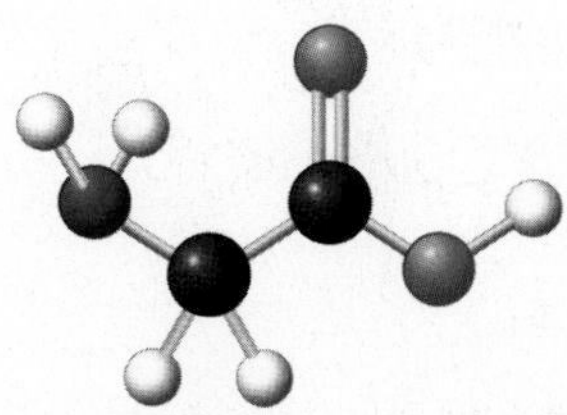

FIGURE 3.25 The molecular structure of glycine, the simplest amino acid.

Solve

1. Using the molar masses of CO_2, NH_3, and CH_4, we calculate the number of moles of each that are available:

$$21.3\ \text{g CO}_2 \times \frac{1\ \text{mol CO}_2}{44.01\ \text{g CO}_2} = 0.4840\ \text{mol CO}_2$$

$$4.53\ \text{g NH}_3 \times \frac{1\ \text{mol NH}_3}{17.03\ \text{g NH}_3} = 0.2660\ \text{mol NH}_3$$

$$4.27\ \text{g CH}_4 \times \frac{1\ \text{mol CH}_4}{16.04\ \text{g CH}_4} = 0.2662\ \text{mol CH}_4$$

Using the mole ratios of $C_2H_5NO_2$ to each reactant and the molar mass of $C_2H_5NO_2$, we calculate the amount of glycine that could be produced from each reactant:

$$0.4840\ \text{mol CO}_2 \times \frac{1\ \text{mol C}_2\text{H}_5\text{NO}_2}{2\ \text{mol CO}_2} \times \frac{75.07\ \text{g C}_2\text{H}_5\text{NO}_2}{1\ \text{mol C}_2\text{H}_5\text{NO}_2} = 18.2\ \text{g C}_2\text{H}_5\text{NO}_2$$

$$0.2660\ \text{mol NH}_3 \times \frac{1\ \text{mol C}_2\text{H}_5\text{NO}_2}{1\ \text{mol NH}_3} \times \frac{75.07\ \text{g C}_2\text{H}_5\text{NO}_2}{1\ \text{mol C}_2\text{H}_5\text{NO}_2} = 20.0\ \text{g C}_2\text{H}_5\text{NO}_2$$

$$0.2662\ \text{mol CH}_4 \times \frac{1\ \text{mol C}_2\text{H}_5\text{NO}_2}{1\ \text{mol CH}_4} \times \frac{75.07\ \text{g C}_2\text{H}_5\text{NO}_2}{1\ \text{mol C}_2\text{H}_5\text{NO}_2} = 20.0\ \text{g C}_2\text{H}_5\text{NO}_2$$

The smallest of the theoretical yields, 18.2 g, represents the maximum quantity of $C_2H_5NO_2$ that could be produced and identifies CO_2 as the limiting reactant.

2. The masses of NH_3 and CH_4 that would be consumed in a reaction in which 0.4840 mol CO_2 is completely consumed are

$$0.4840\ \text{mol CO}_2 \times \frac{1\ \text{mol NH}_3}{2\ \text{mol CO}_2} \times \frac{17.03\ \text{g NH}_3}{1\ \text{mol NH}_3} = 4.12\ \text{g NH}_3$$

$$0.4840\ \text{mol CO}_2 \times \frac{1\ \text{mol CH}_4}{2\ \text{mol CO}_2} \times \frac{16.04\ \text{g CH}_4}{1\ \text{mol CH}_4} = 3.88\ \text{g CH}_4$$

Therefore, the masses of unreacted NH_3 and CH_4 would be

$$(4.53 - 4.12)\ \text{g NH}_3 = 0.41\ \text{g NH}_3$$

$$(4.27 - 3.88)\ \text{g CH}_4 = 0.39\ \text{g CH}_4$$

Think About It The mass of CO_2 in the original mixture is nearly 5 times the mass of each of the other two reactants, yet CO_2 is the limiting reactant. This may not seem reasonable, but it is for two reasons: the molar mass of CO_2 is over 2.5 times the molar masses of NH_3 and CH_4, and 2 moles of CO_2 are consumed for every 1 mole of NH_3 and 1 mole of CH_4.

Practice Exercise One of the intermediates in the synthesis of glycine from ammonia, carbon dioxide, and methane is $C_2H_4N_2$, produced by this reaction:

$$3\ \text{CH}_4(g) + 5\ \text{CO}_2(g) + 8\ \text{NH}_3(g) \rightarrow 4\ \text{C}_2\text{H}_4\text{N}_2(g) + 10\ \text{H}_2\text{O}(\ell)$$

How many grams of $C_2H_4N_2$ could be expected from the reaction of 14.2 g of CO_2, 2.27 g of NH_3, and 2.14 g of CH_4?

(Answers to Practice Exercises are in the back of the book.)

Actual Yields versus Theoretical Yields

Recall from the beginning of this section that the theoretical yield is the maximum amount of product possible from the given quantities of reactants, provided the reaction works perfectly. In nature, industry, or the laboratory, the **actual yield** may be less than the theoretical yield for several reasons. Sometimes reactants combine to form products other than the ones desired. In others, fractions of the reactants remain unreacted no matter how long the reaction runs. For these and other reasons, it is useful to distinguish between the theoretical yield and the actual yield of a chemical reaction and to calculate the **percent yield**:

$$\text{percent yield} = \frac{\text{actual yield}}{\text{theoretical yield}} \times 100\% \quad (3.5)$$

Percent yield is a measure of a reaction's efficiency at forming the desired product.

SAMPLE EXERCISE 3.16 Calculating Percent Yield **LO5**

The industrial process for making the ammonia used in fertilizer, explosives, and many other products is based on the reaction between nitrogen and hydrogen at high temperature and pressure:

$$N_2(g) + 3\,H_2(g) \rightarrow 2\,NH_3(g)$$

If 18.20 kg of NH_3 is produced by a reaction mixture that initially contains 6.00 kg of H_2 and an excess of N_2, what is the percent yield of the reaction?

Collect and Organize We know that the actual yield of NH_3 is 18.20 kg. We also know that the reaction mixture initially contained 6.00 kg of H_2 and that H_2 must be the limiting reactant because the problem specifies the presence of excess N_2. We need to find the percent yield.

Analyze We need to use the mass of H_2 to calculate how much NH_3 could have been produced—the theoretical yield. We can then use the theoretical yield and the actual yield to calculate the percent yield. We need the molar masses of H_2 and NH_3 and the stoichiometry of the reaction, which specifies that 2 moles of NH_3 are produced for every 3 moles of H_2 consumed.

$$\boxed{\text{g } H_2} \xrightarrow{\frac{1}{\text{molar mass}}} \boxed{\text{mol } H_2} \xrightarrow{\frac{2 \text{ mol } NH_3}{3 \text{ mol } H_2}} \boxed{\text{mol } NH_3} \xrightarrow{\text{molar mass}} \boxed{\text{g } NH_3} = \boxed{\text{theoretical yield}}$$

$$\boxed{\%\text{ yield}} = \boxed{\frac{\text{actual yield}}{\text{theoretical yield}}} \times 100\%$$

Solve The molar masses we need are:

$$\mathcal{M}_{H_2} = 2(1.008 \text{ g/mol}) = 2.016 \text{ g/mol}$$

$$\mathcal{M}_{NH_3} = 14.01 \text{ g/mol} + 3(1.008 \text{ g/mol}) = 17.03 \text{ g/mol}$$

Calculating the theoretical yield of NH_3:

$$6.00 \text{ kg } H_2 \times \frac{10^3 \text{ g } H_2}{1 \text{ kg } H_2} \times \frac{1 \text{ mol } H_2}{2.016 \text{ g } H_2} \times \frac{2 \text{ mol } NH_3}{3 \text{ mol } H_2} \times \frac{17.03 \text{ g } NH_3}{1 \text{ mol } NH_3} \times \frac{1 \text{ kg } NH_3}{10^3 \text{ g } NH_3} = 33.8 \text{ kg } NH_3$$

Dividing the actual yield by this theoretical yield to determine percent yield:

$$\frac{18.20 \text{ kg } NH_3}{33.8 \text{ kg } NH_3} \times 100\% = 53.8\%$$

Think About It In the text preceding this sample exercise we noted that some reactions proceed only partway in converting reactants into products, but no further, leaving reactants unreacted no matter how long the reaction runs. The synthesis of ammonia is such a reaction, with percent yields that are usually much less than 100% and making the calculated value a reasonable one.

actual yield the amount of product obtained from a chemical reaction, which may be less than the theoretical yield.

percent yield the ratio, expressed as a percentage, of the actual yield of a chemical reaction to the theoretical yield.

Practice Exercise Most of the hydrogen gas used in industry and as fuel is produced in a reaction between methane and high-temperature steam: $CH_4(g) + H_2O(g) \rightarrow 3\,H_2(g) + CO(g)$. If 1.20 kg CH_4 and an excess of steam produces 318 g H_2, what is the percent yield of the reaction?

(Answers to Practice Exercises are in the back of the book.)

SAMPLE EXERCISE 3.17 Effect of Percent Yield on Calculating the Mass of Reactant Needed **LO5**

Another way of producing hydrogen gas (see Practice Exercise 3.16) is the water–gas shift reaction:

$$H_2O(g) + CO(g) \rightarrow H_2(g) + CO_2(g)$$

At 200°C, the reaction produces a 96% yield. How many grams of H_2O and CO are needed to generate 176 g of H_2 under these conditions?

Collect and Organize This is a stoichiometry problem like Sample Exercises 3.12 and 3.13, except that the reaction yield is only 96%.

Analyze We need to convert the mass of H_2 we want to make into moles of H_2, and then we use the mole ratios of H_2:CO and H_2:H_2O to find the quantities of reactants we need. To produce 176 g of H_2 (actual yield), we know the percent yield (96%) and must calculate the theoretical yield. Then by analogy to Sample Exercise 3.13, we can calculate the amount of both reactants needed.

Solve

1. The number of moles of H_2 we want to produce is:

$$176 \text{ g } H_2 \times \frac{1 \text{ mol } H_2}{2.016 \text{ g } H_2} = 87.30 \text{ mol } H_2$$

This value represents an actual yield value that is only 96% of the theoretical yield. Solving for the theoretical yield using Equation 3.5:

$$\text{percent yield} = \frac{\text{actual yield}}{\text{theoretical yield}} \times 100\%$$

$$= 96\% = \frac{87.30 \text{ mol } H_2}{x} \times 100\%$$

$$x = 90.9 \text{ mol } H_2$$

2. Using the mole ratios of H_2:H_2O and H_2:CO, we find the number of moles of H_2O and CO:

$$90.9 \text{ mol } H_2 \times \frac{1 \text{ mol } H_2O}{1 \text{ mol } H_2} = 90.9 \text{ mol } H_2O$$

$$90.9 \text{ mol } H_2 \times \frac{1 \text{ mol CO}}{1 \text{ mol } H_2} = 90.9 \text{ mol CO}$$

3. To convert from moles of reactants to the mass of the reactants, we multiply by the molar mass:

$$90.9 \text{ mol } H_2O \times \frac{18.02 \text{ g } H_2O}{1 \text{ mol } H_2O} = 1.64 \times 10^3 \text{ g } H_2O$$

$$90.9 \text{ mol CO} \times \frac{28.01 \text{ g CO}}{1 \text{ mol CO}} = 2.55 \times 10^3 \text{ g CO}$$

Think About It As a check of our answers, let's calculate how much H_2 would be produced from 90.9 moles of CO if the reaction proceeded in 100% yield:

$$90.9 \text{ mol CO} \times \frac{1 \text{ mol } H_2}{1 \text{ mol CO}} \times \frac{2.016 \text{ g } H_2}{1 \text{ mol } H_2} = 183 \text{ g } H_2$$

The percent yield is

$$\frac{176 \text{ g } H_2}{183 \text{ g } H_2} \times 100\% = 96\%$$

Practice Exercise How much aluminum oxide and how much carbon are needed to prepare 454 g (1 pound) of aluminum by the balanced chemical equation:

$$2\ Al_2O_3(s) + 3\ C(s) \rightarrow 4\ Al(s) + 3\ CO_2(g)$$

if the reaction proceeds to 78% yield?

(Answers to Practice Exercises are in the back of the book.)

percent composition the composition of a compound expressed in terms of the percentage by mass of each element in the compound.

3.7 Determining Empirical Formulas from Percent Composition

Natural sources of methane such as swamps, bogs, rice paddies, and animals introduce about 570 million kg CH_4 into the atmosphere every day. Other volatile hydrocarbons are emitted by other natural sources, such as conifers (**Figure 3.26**), which emit a variety of hydrocarbons including pinene, $C_{10}H_{16}$. One way to distinguish between CH_4 and $C_{10}H_{16}$ is in terms of their **percent composition**: the composition of a compound expressed in terms of the percentages of the masses of the constituent elements with respect to the total mass of the compound. Putting this definition in equation form:

$$\text{mass \% of an element} = \frac{\text{mass of element in 1 mole of compound}}{\text{molar mass of compound}} \times 100\% \quad (3.6)$$

To calculate the percent composition of CH_4 and $C_{10}H_{16}$, we must calculate the carbon content of both compounds. Suppose we have 1 mole of CH_4, which has a molar mass of

$$4(1.008 \text{ g H/mol}) + 12.011 \text{ g C/mol} = 16.043 \text{ g/mol}$$

Of this 16.043 g, carbon accounts for 12.011 g. Therefore, the mass % of carbon in CH_4 is

$$\text{mass \% C} = \frac{\text{mass of C}}{\text{molar mass } CH_4} \times 100\% = \frac{12.011 \text{ g C}}{16.043 \text{ g } CH_4} \times 100\% = 74.87\%$$

Similarly, for hydrogen content:

$$\text{mass \% H} = \frac{\text{mass of H}}{\text{molar mass } CH_4} \times 100\% = \frac{4(1.008 \text{ g H})}{16.043 \text{ g } CH_4} \times 100\% = 25.13\%$$

Because CH_4 contains only C and H, we could also determine the percent H by subtracting the percent C from 100%:

$$100.00\% - 74.87\% = 25.13\%$$

The benefit of calculating the percentages separately, instead of by subtraction, is that you can check to make sure your answers sum to 100%.

FIGURE 3.26 The odor of the longleaf pine (*Pinus palustris*) comes from the volatile hydrocarbon pinene.

How does this compare with the percent C in pinene? If we repeat the calculation for $C_{10}H_{16}$, the molar mass is

$$16(1.008 \text{ g H/mol}) + 10(12.01 \text{ g C/mol}) = 136.23 \text{ g/mol}$$

The carbon content of $C_{10}H_{16}$ is

CHEMTOUR Percent Composition

$$\text{mass \% C} = \frac{\text{mass of C}}{\text{total mass}} = \frac{10(12.01 \text{ g C})}{136.23 \text{ g } C_{10}H_{16}} = 0.8816 = 88.16\%$$

and the hydrogen content is

$$\text{mass \% H} = 100.00\% - 88.16\% = 11.84\%$$

Pinene contains a higher percent carbon by mass and a correspondingly lower percent hydrogen than those of methane. Methane, in fact, has the highest percent hydrogen of all hydrocarbons.

SAMPLE EXERCISE 3.18 Calculating Percent Composition from a Chemical Formula **LO6**

A compound with the formula $C_2H_2F_4$ is a propellant in the inhalers used by asthma sufferers (**Figure 3.27**). What is the percent composition of $C_2H_2F_4$?

FIGURE 3.27 Patients suffering from asthma and other respiratory ailments use inhalers to relieve their discomfort. The medication is delivered using compounds such as $C_2H_2F_4$ as the propellant.

Collect, Organize, and Analyze We are asked to calculate the percent composition of $C_2H_2F_4$. We need the molar masses of each of the elements (C, H, and F) to calculate the molar mass of $C_2H_2F_4$ and the relative contribution of each element to the compound's molar mass. To calculate the percent composition of $C_2H_2F_4$, we must determine the percentage by mass of each element in 1 mole of $C_2H_2F_4$. These percentages can be calculated by dividing the mass of each element in 1 mole of $C_2H_2F_4$ by the molar mass of $C_2H_2F_4$.

Solve The molar mass of $C_2H_2F_4$ is

$$2(12.01 \text{ g/mol}) + 2(1.008 \text{ g/mol}) + 4(19.00 \text{ g/mol}) = 102.04 \text{ g/mol}$$

Thus, the percent composition of this compound is

$$\text{mass \% C} = \frac{24.02 \text{ g C}}{102.04 \text{ g}} \times 100\% = 23.54\%$$

$$\text{mass \% H} = \frac{2.016 \text{ g H}}{102.04 \text{ g}} \times 100\% = 1.98\%$$

$$\text{mass \% F} = \frac{76.00 \text{ g F}}{102.04 \text{ g}} \times 100\% = 74.48\%$$

Think About It Although C and F have similar molar masses, the observation that the percentage of F is nearly three times the percentage of C makes sense because there are 2 moles of F for every 1 mole of C in $C_2H_2F_4$. It also makes sense that the percentage of the lightest element, hydrogen, is very small.

Practice Exercise Determine the percent composition of tetradecafluorohexane (C_6F_{14}; **Figure 3.28**), which is used to treat patients suffering from smoke inhalation.

(Answers to Practice Exercises are in the back of the book.)

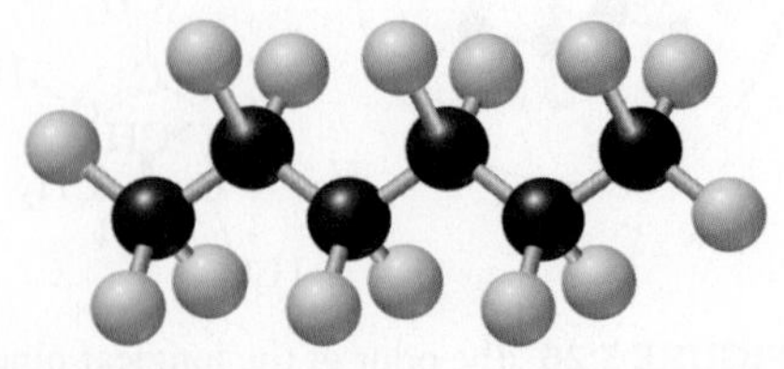

FIGURE 3.28 Molecular struture of tetradecafluorohexane.

The three mass percentages in Sample Exercise 3.18 sum to 100.00%. Percent composition values should always sum to 100%, or very close to 100%, if we have accounted for all the elements that make up the total mass of the compound. The total may deviate slightly from 100% because of rounding.

We typically determine the percent composition of a substance in the laboratory by measuring the amount of each element in a given mass of the substance. We can use these data to determine the *empirical formula* of the substance. There are four steps to deriving an empirical formula from percent composition data obtained in the laboratory:

1. Assume you have exactly 100 g of the substance, so that the percent composition values are equivalent to the values of the masses of the elements expressed in grams. As an initial check, add the mass percentages given in the problem to make sure they total 100%.
2. Convert the mass of each element into moles.
3. Compute a simple mole ratio by dividing all the mole values by the smallest one.
4. If necessary, convert the ratio from step 3 into a ratio of whole numbers, which will serve as subscripts following the symbols of the elements in the empirical formula of the compound.

CONNECTION We discussed empirical formulas in Chapter 2. *Empirical*, which is related to the word *experiment*, means "derived from experimental data," such as the data from experiments to determine percent composition.

The best way to learn this process is by using it. Let's start with carbon tetrachloride, a compound once widely used as a fire-extinguishing agent and as a solvent in dry cleaning. It is 7.81% C and 92.19% Cl, which means that in exactly 100 g of this substance there are 7.81 g C and 92.19 g Cl. We use these masses to calculate the number of moles of C and Cl in 100 g of the compound:

$$7.81\ \cancel{\text{g C}} \times \frac{1\ \text{mol C}}{12.01\ \cancel{\text{g C}}} = 0.650\ \text{mol C}$$

$$92.19\ \cancel{\text{g Cl}} \times \frac{1\ \text{mol Cl}}{35.45\ \cancel{\text{g Cl}}} = 2.601\ \text{mol Cl}$$

To calculate a simple ratio of these mole values, we divide both numbers by the smaller of the two:

$$\frac{0.650\ \text{mol C}}{0.650} = 1.000\ \text{or}\ 1\ \text{mol C} \qquad \frac{2.601\ \text{mol Cl}}{0.650} = 4.00\ \text{or}\ 4\ \text{mol Cl}$$

Both values are whole numbers, so the empirical formula of carbon tetrachloride is CCl_4.

We can use these same steps for substances that contain more than two elements. For example, elemental analysis of a commonly used refrigeration gas gives the following percent composition of a sample: 28.58% C, 3.60% H, and 67.82% F. The sum of these percentages totals 100%, so all elements are accounted for in our analysis. Our calculations are then

$$28.58\%\ \text{C} = 28.58\ \text{g C in 100 g}$$

$$3.60\%\ \text{H} = 3.60\ \text{g H in 100 g}$$

$$67.82\%\ \text{F} = 67.82\ \text{g F in 100 g}$$

and the corresponding numbers of moles are

$$28.58\ \cancel{\text{g C}} \times \frac{\text{mol C}}{12.01\ \cancel{\text{g C}}} = 2.380\ \text{mol C}$$

$$67.82\ \cancel{\text{g F}} \times \frac{\text{mol F}}{19.00\ \cancel{\text{g F}}} = 3.569\ \text{mol F}$$

$$3.60\ \cancel{\text{g H}} \times \frac{\text{mol H}}{1.008\ \cancel{\text{g H}}} = 3.57\ \text{mol H}$$

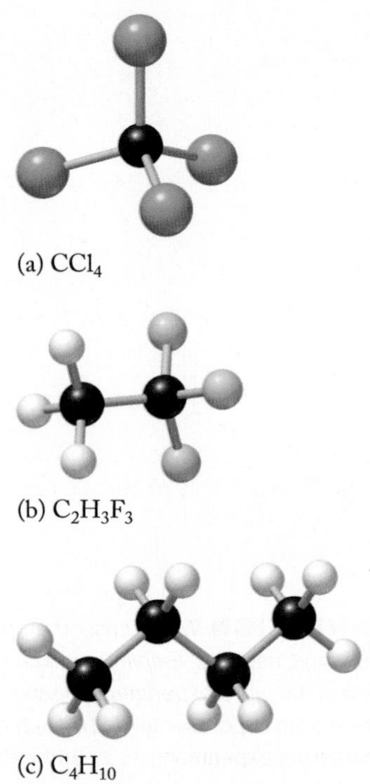

FIGURE 3.29 The molecular structures of (a) carbon tetrachloride, CCl_4; (b) the refrigerant gas, $C_2H_3F_3$; and (c) the fuel gas butane, C_4H_{10}.

The mole ratio for C:H:F is 2.380:3.57:3.569. To convert this ratio to small whole numbers, we divide all three values by 2.380 mol, which is the smallest value:

$$\frac{2.380 \text{ mol C}}{2.380} = 1.000 \text{ or } 1 \text{ mol C} \qquad \frac{3.569 \text{ mol F}}{2.380} = 1.500 \text{ or } 1.5 \text{ mol F}$$

$$\frac{3.57 \text{ mol H}}{2.380} = 1.50 \text{ or } 1.5 \text{ mol H}$$

To change the mole ratio 1:1.5:1.5 to a ratio of whole numbers, we multiply the numbers by a factor that results in whole numbers. Multiplying 1.5 (or $\frac{3}{2}$) by 2 yields 3, so we multiply all three terms in this ratio by 2 to obtain a C:H:F ratio of 2:3:3. Thus, the empirical formula is $C_2H_3F_3$.

Both CCl_4 and $C_2H_3F_3$ are molecular compounds whose empirical formulas match their molecular formulas (**Figure 3.29a, b**). This is not always the case. Butane, another hydrocarbon found in natural gas and used as a fuel, has a percent composition of 82.66% C and 17.34% H. We can calculate its empirical formula as follows:

$$82.66\% \text{ C} = 82.66 \cancel{\text{g C}} \times \frac{1 \text{ mol C}}{12.01 \cancel{\text{g C}}} = 6.883 \text{ mol C}$$

$$17.34\% \text{ H} = 17.34 \cancel{\text{g H}} \times \frac{1 \text{ mol H}}{1.008 \cancel{\text{g H}}} = 17.20 \text{ mol H}$$

The C:H mole ratio is 6.883:17.20, so dividing by 6.883 mol gives

$$\frac{6.883}{6.883} : \frac{17.20}{6.883} = 1:2.499$$

When we multiply both numbers by 2, the whole-number ratio becomes 2:5, so the empirical formula for butane is C_2H_5. The structure of a molecule of butane is shown in **Figure 3.29c**; it has the *molecular* formula C_4H_{10}. It is important to keep in mind that percent composition data are used to determine empirical formulas—the lowest whole-number ratio of atoms in a molecule. Empirical formulas are not necessarily molecular formulas that tell us the actual numbers of atoms in the molecule.

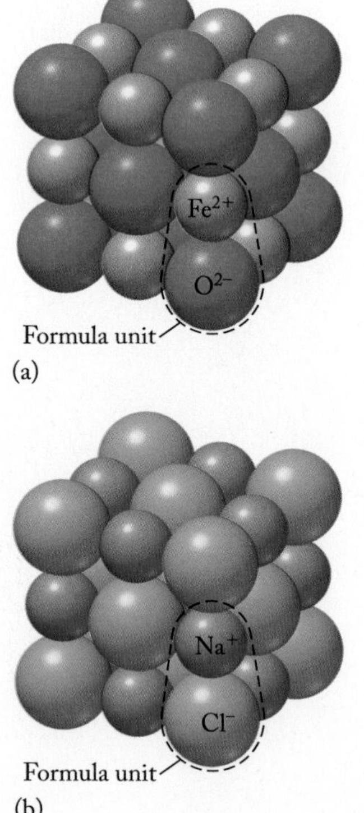

FIGURE 3.30 (a) The Fe^{2+} and O^{2-} ions in FeO adopt the halite structure, which is the same three-dimensional pattern as (b) the Na^+ and Cl^- ions in NaCl.

For ionic compounds such as FeO and Fe_2O_3, chemical formulas *are* empirical formulas because the chemical formulas of ionic compounds represent the ratio of the ions in one formula unit of the compound. Samples of ionic compounds contain vast three-dimensional arrays of their component ions: Fe^{2+} and O^{2-} ions in FeO (**Figure 3.30a**), and Fe^{3+} and O^{2-} ions in Fe_2O_3. The arrangement of the Fe^{2+} and O^{2-} ions in FeO is called a ***halite structure*** because it is also the structure of the mineral halite (NaCl; **Figure 3.30b**). In an ionic compound the chemical formula, the empirical formula, and the formula unit are all the same.

CONCEPT **TEST**

Which pairs of the following compounds have the same empirical formula and same percent composition?

a. Ethylene (C_2H_4), a gas used to ripen bananas

b. Acetylene (C_2H_2), a gas used in welding

c. Eicosene ($C_{20}H_{40}$), a compound used to attract Japanese beetles to traps

d. Benzene (C_6H_6), a known carcinogen found in gasoline

(Answers to Concept Tests are in the back of the book.)

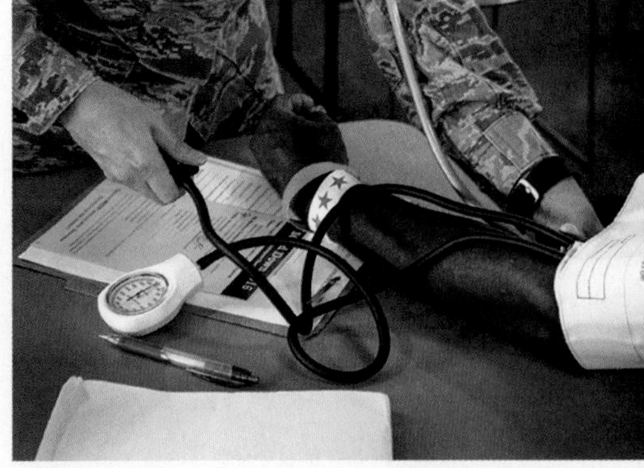

FIGURE 3.31 A stethoscope, cuff, and gauge used to measure blood pressure.

SAMPLE EXERCISE 3.19 Deriving an Empirical Formula from Percent Composition **LO6**

Nitrous oxide, the laughing gas used in dentists' offices, is 63.65% N. What is the empirical formula of nitrous oxide?

Collect and Organize We are given the percent composition by mass of N (63.65%) and asked to determine the simplest whole-number ratio of nitrogen and oxygen in the chemical formula.

Analyze We can obtain the percent composition by mass of oxygen in nitrous oxide by subtracting the percent N from 100%. We can use the four steps described previously to determine the empirical formula.

Solve The percent O in the sample is:

$$\%O = 100.00\% - \%N = 100.00\% - 63.65\% = 36.35\%$$

Now we have the data we need to determine the empirical formula of nitrous oxide.

1. Convert percent composition values into masses by assuming a sample size of exactly 100 g:

$$63.65\% \text{ N} = 63.65 \text{ g N in } 100 \text{ g}$$

$$36.35\% \text{ O} = 36.35 \text{ g O in } 100 \text{ g}$$

2. Convert masses into moles:

$$63.65 \cancel{\text{g N}} \times \frac{1 \text{ mol N}}{14.01 \cancel{\text{g N}}} = 4.543 \text{ mol N}$$

$$36.35 \cancel{\text{g O}} \times \frac{1 \text{ mol O}}{16.00 \cancel{\text{g O}}} = 2.272 \text{ mol O}$$

3. Simplify the mole ratio 4.543:2.272 by dividing by the smaller value:

$$\frac{4.543 \text{ mol N}}{2.272} = 2.000 \text{ mol N} \qquad \frac{2.272 \text{ mol O}}{2.272} = 1.000 \text{ mol O}$$

   The N:O mole ratio is 2:1.

4. Since the mole ratio of N to O is already a ratio of whole numbers, the empirical formula of nitrous oxide is N_2O.

Think About It The molar masses of N and O are similar, so a ≈2:1 ratio of the mass of N to the mass of O is consistent with an empirical formula that is 2:1 nitrogen to oxygen. "Nitrous oxide"—whose molecular formula is also N_2O—is formally named "dinitrogen monoxide" under the rules introduced in Chapter 2.

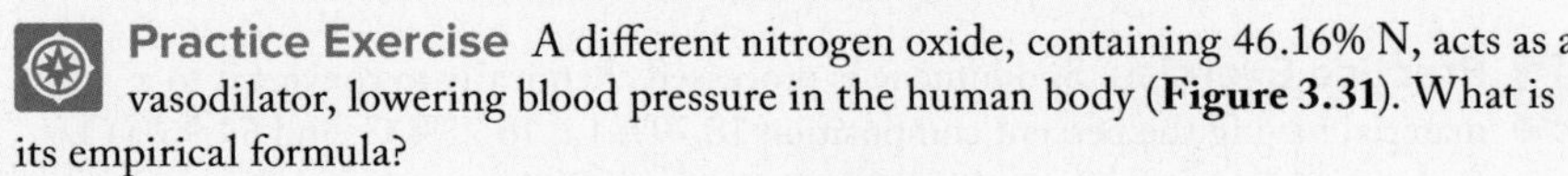

Practice Exercise A different nitrogen oxide, containing 46.16% N, acts as a vasodilator, lowering blood pressure in the human body (**Figure 3.31**). What is its empirical formula?

(Answers to Practice Exercises are in the back of the book.)

(a)

(b)

(c)

FIGURE 3.32 (a) The mineral spodumene is the most important source of lithium metal for use in (b) the lithium batteries for portable electronic devices and (c) automobiles.

SAMPLE EXERCISE 3.20 Deriving an Empirical Formula of a Compound Containing More than Two Elements **LO6**

Rechargeable lithium batteries are essential to portable electronic devices and are increasingly important to the automotive industry in gas–electric hybrid and all-electric vehicles (**Figure 3.32**). Analysis of the most common lithium-containing mineral, spodumene, shows it is composed of 3.73% Li, 14.50% Al, 30.18% Si, and 51.59% O. What is the empirical formula of spodumene?

Collect and Organize We are given the percent composition by mass of spodumene and asked to determine the simplest whole-number ratio of the four elements (Li, Al, Si, and O), which defines the empirical formula of spodumene.

Analyze We follow the same steps as in Sample Exercise 3.19, except that here we must calculate more than one mole ratio.

Solve

1. Assuming an exactly 100-g sample of spodumene, we have 3.73 g of Li, 14.50 g of Al, 30.18 g of Si, and 51.59 g of O.
2. Converting the masses into moles, we have

$$3.73\ \cancel{\text{g Li}} \times \frac{1\ \text{mol Li}}{6.941\ \cancel{\text{g Li}}} = 0.5374\ \text{mol Li}$$

$$14.50\ \cancel{\text{g Al}} \times \frac{1\ \text{mol Al}}{26.98\ \cancel{\text{g Al}}} = 0.5374\ \text{mol Al}$$

$$30.18\ \cancel{\text{g Si}} \times \frac{1\ \text{mol Si}}{28.09\ \cancel{\text{g Si}}} = 1.074\ \text{mol Si}$$

$$51.59\ \cancel{\text{g O}} \times \frac{1\ \text{mol O}}{16.00\ \cancel{\text{g O}}} = 3.224\ \text{mol O}$$

3. Dividing each mole value by the smallest value yields a simple ratio of the four elements:

$$\frac{0.5374\ \text{mol Li}}{0.5374} = 1.000\ \text{or 1 mol Li} \qquad \frac{0.5374\ \text{mol Al}}{0.5374} = 1.000\ \text{or 1 mol Al}$$

$$\frac{1.074\ \text{mol Si}}{0.5374} = 1.999\ \text{or 2 mol Si} \qquad \frac{3.224\ \text{mol O}}{0.5374} = 5.999\ \text{or 6 mol O}$$

4. All these results are either whole numbers or very close to whole numbers, so no further simplification of terms is needed. Our mole ratio is Li:Al:Si:O = 1:1:2:6, which means the empirical formula of spodumene is $LiAlSi_2O_6$.

Think About It The sum of the mass percentages should be very close to 100%; in this case, 3.73% + 14.50% + 30.18% + 51.59% = 100.00%, so we have accounted for all the elements. Lithium is present in the lowest percentage by mass of all four elements in spodumene, but spodumene remains the most commercially important source of lithium for batteries.

Practice Exercise Spodumene is processed chemically to convert it to a material having the percent composition 18.79% Li, 16.25% C, and 64.96% O. What is the empirical formula of this lithium compound?

(Answers to Practice Exercises are in the back of the book.)

3.8 Comparing Empirical and Molecular Formulas

An empirical formula represents the simplest whole-number ratio of the elements contained in a compound, but it is not necessarily the same as the molecular formula of a molecular compound. To further examine the difference between empirical and molecular formulas, let's consider the organic molecule glycolaldehyde

(**Figure 3.33**). About one in 20 of the meteorites that fall to Earth today contains a variety of organic compounds, including glycolaldehyde. Some scientists view the presence of these compounds as evidence that the molecular building blocks of life on Earth may have come from space. Glycolaldehyde is also present in our bodies as a product of the metabolism of sugars and proteins.

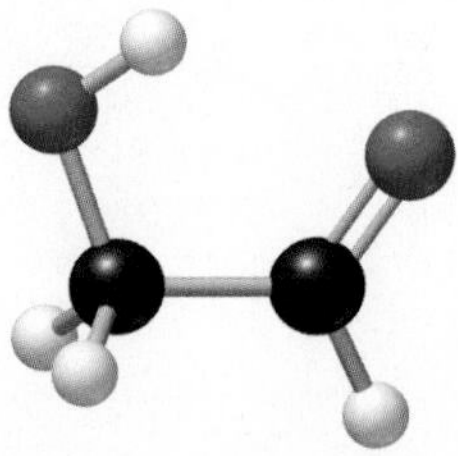

FIGURE 3.33 Glycolaldehyde is one of over 100 organic compounds detected in interstellar gases. It is also the smallest molecule among those identified as sugars.

Elemental analysis of glycolaldehyde provides the following percent composition data: 40.00% C, 6.71% H, and 53.28% O. We can use this information to calculate the empirical formula of the compound and then compare our result with the molecular structure in Figure 3.33. Using the method developed in Section 3.7, we obtain these results:

$$40.00\ \cancel{\text{g C}} \times \frac{1\ \text{mol C}}{12.01\ \cancel{\text{g C}}} = 3.331\ \text{mol C}$$

$$6.71\ \cancel{\text{g H}} \times \frac{1\ \text{mol H}}{1.008\ \cancel{\text{g H}}} = 6.66\ \text{mol H}$$

$$53.28\ \cancel{\text{g O}} \times \frac{1\ \text{mol O}}{16.00\ \cancel{\text{g O}}} = 3.33\ \text{mol O}$$

The mole ratio of C:H:O is 3.331:6.66:3.330 = 1:2:1, so the empirical formula is CH_2O.

The molecular formula of the molecule in Figure 3.33 is $C_2H_4O_2$, which represents the actual numbers of C, H, and O atoms in one molecule of glycolaldehyde. This molecular formula is a multiple of the empirical formula, which is consistent with the idea that the empirical formula represents a ratio of the atoms of the elements. In this case, we would multiply each subscript in the empirical formula by 2 to obtain the molecular formula.

Occasionally, empirical and molecular formulas are identical. For example, formaldehyde—which, like glycolaldehyde, is 40.00% C, 6.71% H, and 53.28% O—has the molecular formula CH_2O, which is the same as its empirical formula (**Figure 3.34**). Many other compounds also have the empirical formula CH_2O. Each subscript in these molecular formulas is a multiple of the corresponding subscript in the empirical formula. For example, a multiplier (*n*) of 6 converts the empirical formula, CH_2O, into the molecular formula of many sugars, including $C_6H_{12}O_6$:

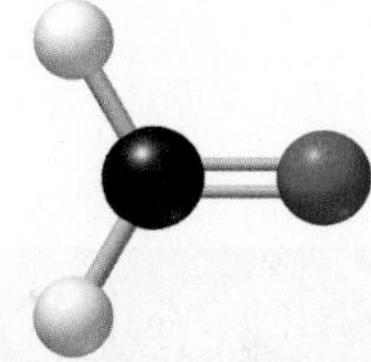

FIGURE 3.34 The empirical formula of the molecular compound formaldehyde is identical to its molecular formula, CH_2O.

$$(CH_2O)_n = (CH_2O)_6 = \underset{\text{Glucose}}{C_6H_{12}O_6}$$

The key to translating empirical formulas into molecular formulas is to determine the value of *n*, and the key to determining the value of *n* is knowing the molar mass. For example, suppose elemental analysis determines that the

CONCEPT TEST

Which of the following compounds have empirical formulas that are identical to their molecular formulas?

a. $C_2H_6O_2$ (ethylene glycol), used in antifreeze

b. C_3H_8O (isopropanol), also known as rubbing alcohol

c. $C_6H_{12}O_6$ (glucose), also known as "blood sugar" and the primary source of energy for our body's cells

(Answers to Concept Tests are in the back of the book.)

empirical formula of a liquid hydrocarbon is C_4H_5. Using the mole ratio given by this formula and the average atomic masses of the two elements, we can calculate the molar mass corresponding to the empirical formula:

$$4(12.01 \text{ g/mol}) + 5(1.008 \text{ g/mol}) = 53.08 \text{ g/mol}$$

Each molecule of this hydrocarbon consists of either the number of atoms in the empirical formula or a multiple of those atoms. Additional analysis of the compound reveals its molar mass to be 106 g/mol. What multiple of 53.08 g/mol (the mass of 4 moles of carbon atoms and 5 moles of hydrogen atoms) has a mass of 106 g/mol? Do the following calculation to find out:

$$n = \frac{\text{molar mass}}{\text{empirical formula mass}} = \frac{106 \text{ g/}\cancel{\text{mol}}}{53 \text{ g/}\cancel{\text{mol}}} = 2.00 = 2$$

Thus, 2 is the value of the multiplier, n, and the molecular formula of the compound is

$$(C_4H_5)_n = (C_4H_5)_2 = C_8H_{10}$$

SAMPLE EXERCISE 3.21 Deriving a Molecular Formula from an Empirical Formula **LO7**

Lycopene (molar mass 536.88 g/mol) and cymene (134.22 g/mol) are natural products found in tomatoes and cumin (a common spice), respectively (**Figure 3.35**). Lycopene in the diet has been shown to reduce the risk of prostate cancer, whereas cumin is a home remedy for stomach ailments. Both compounds are 89.49% C and 10.51% H by mass. What are the empirical and molecular formulas of lycopene and cymene?

Collect, Organize, and Analyze The problem gives the percent composition and molar masses of two compounds. We want to determine both the empirical and molecular formulas of lycopene and cymene. Percent composition data can be used to derive empirical formulas as described in Section 3.7 and illustrated in Sample Exercises 3.19 and 3.20.

Solve Assuming a 100-g sample, we have 89.49 g of C and 10.51 g of H, which we can convert to moles:

$$89.49 \text{ g}\cancel{\text{C}} \times \frac{1 \text{ mol C}}{12.01 \text{ g}\cancel{\text{C}}} = 7.451 \text{ mol C}$$

$$10.51 \text{ g}\cancel{\text{H}} \times \frac{1 \text{ mol H}}{1.008 \text{ g}\cancel{\text{H}}} = 10.43 \text{ mol H}$$

The mole ratio of C to H is 7.451:10.43, or 1.000:1.400. We can convert the fractional mole ratio to a ratio of whole numbers by multiplying by 5:

$$\text{C:H} = 5 \times (1:1.4) = 5:7$$

which means the empirical formula is C_5H_7.

To determine the molecular formula, we first determine the mass of the empirical formula, C_5H_7:

$$5(12.01 \text{ g/mol}) + 7(1.008 \text{ g/mol}) = 67.11 \text{ g/mol}$$

Next, we divide the molar masses for lycopene and cymene by this empirical formula mass to determine the values of the multiplier, n:

$$n = \frac{\text{molar mass lycopene}}{\text{empirical formula mass}} = \frac{536.88 \text{ g/}\cancel{\text{mol}}}{67.11 \text{ g/}\cancel{\text{mol}}} = 8$$

$$n = \frac{\text{molar mass cymene}}{\text{empirical formula mass}} = \frac{134.22 \text{ g/}\cancel{\text{mol}}}{67.11 \text{ g/}\cancel{\text{mol}}} = 2$$

(a)

(b)

FIGURE 3.35 (a) Tomatoes and (b) ground cumin (a cooking spice) both contain hydrocarbons reputed to offer therapeutic benefits.

The molecular formula of lycopene is

$$(C_5H_7)_n = (C_5H_7)_8 = C_{40}H_{56}$$

and the molecular formula of cymene is

$$(C_5H_7)_n = (C_5H_7)_2 = C_{10}H_{14}$$

Think About It The percent composition of a compound reveals only the empirical formula. All compounds with the same empirical formula *must* have the same percent composition.

Practice Exercise *Pheromones* are chemical substances secreted by members of a species to stimulate a response in other individuals of the same species. For example, females of a species secrete certain pheromones to attract males for mating. The percent composition of eicosene (280 g/mol), a compound like the Japanese beetle mating pheromone, is 85.63% C and 14.37% H. Determine its molecular formula.

(Answers to Practice Exercises are in the back of the book.)

mass spectrometer an instrument that separates and counts ions according to their mass-to-charge ratios.

molecular ion (M^+) the peak of highest mass in a mass spectrum; it has the same mass as the molecule from which it came.

mass spectrum a graph of the data from a mass spectrometer.

CONCEPT TEST

Lycopene and cymene (Sample Exercise 3.21) have the same percent composition. Are they isomers?

(Answers to Concept Tests are in the back of the book.)

Molecular Mass and Mass Spectrometry

How do chemists who isolate an unknown molecular compound from reaction mixtures or biological systems determine the molecular mass of the compound? They often use a powerful analytical technique called *mass spectrometry*.

Mass spectrometers ionize molecules and then separate the ions by the ratio of their masses (m) to their electric charges (z). In many mass spectrometers, samples are vaporized and then bombarded with a beam of high-energy electrons (**Figure 3.36**). These electrons smash into gas-phase molecules with such force that they knock electrons off the molecules, forming positively charged ions known as the **molecular ions (M^+)**:

$$M + e^- \rightarrow M^+ + 2\,e^-$$

Sometimes the collisions are so forceful that they break molecules into fragments and ionize the fragments. The charged fragments and molecular ions are ejected into a second region of the mass spectrometer where they are separated based on their m/z ratios. Then they pass into a detector and are counted. The resulting data are displayed as a **mass spectrum** (plural *spectra*) in which the horizontal axis is m/z and the spectrum is a displayed as a series of vertical bars at various m/z values. The heights of the bars indicate the number of ions reaching the detector with that m/z value. The bar with the highest m/z value in the spectrum is usually the molecular ion (M^+) peak.

CONNECTION The positive-ray analyzer that Francis Aston built and used to separate two of the isotopes of neon (see Section 2.3) was the forerunner of today's mass spectrometers.

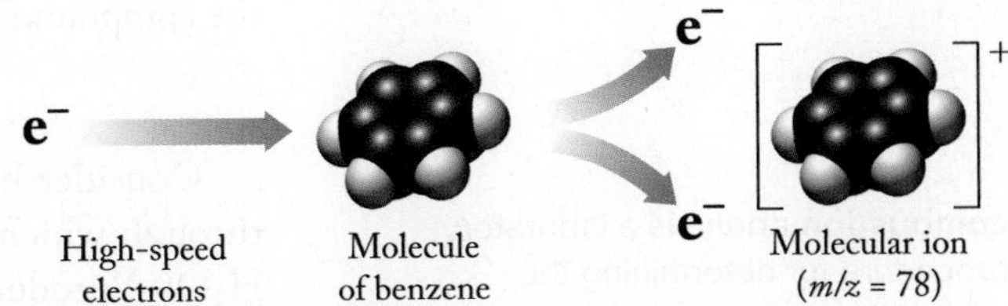

FIGURE 3.36 In many mass spectrometers, sample molecules are bombarded with beams of high-energy electrons, producing molecular ions with positive charges, as shown here for benzene, C_6H_6.

FIGURE 3.37 Mass spectra of (a) acetylene and (b) benzene. The molecular-ion bar in both spectra is labeled M^+. The bars at much smaller m/z values are produced when the ionizing beam also breaks apart the molecules into fragments. The molar masses of C_2H_2 and C_6H_6 are 26 and 78 g/mol, respectively.

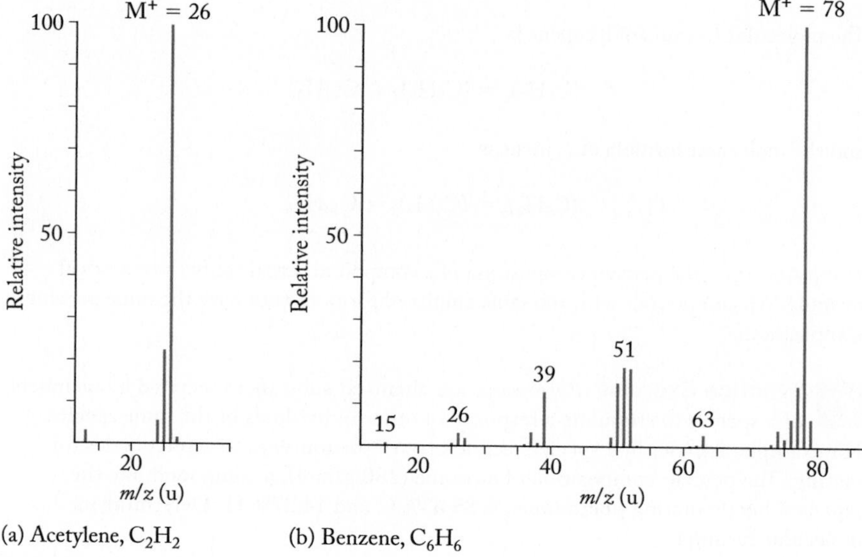

If the ions are singly charged ions ($z = 1+$), then the m/z value of that bar is simply m, the molecular mass of the compound.

Figure 3.37 shows the mass spectra of acetylene and benzene. In both spectra, the tall bar with the greatest m/z value (at 26 u in the acetylene spectrum and at 78 u in the benzene spectrum) is the molecular ion. The bars at lower m/z values represent fragments. Distinctive fragmentation patterns help scientists confirm molecular structures.

CONCEPT TEST

Does the molecular ion in a mass spectrum correspond to the mass associated with the empirical formula or the molecular formula of a compound?

(Answers to Concept Tests are in the back of the book.)

3.9 Combustion Analysis

In **combustion analysis**, the complete combustion (defined in Section 3.4) of a compound, followed by an analysis of the products, enables chemists to determine the chemical composition of that compound. To ensure that combustion is complete, the process is carried out in excess oxygen, which means more oxygen is present than the stoichiometric amount. Combusting an organic compound means converting all the carbon in the compound to CO_2 and all the hydrogen in the compound to H_2O:

$$C_aH_b + \text{excess } O_2(g) \rightarrow a\, CO_2(g) + b/2\, H_2O(g)$$

combustion analysis a laboratory procedure for determining the composition of a substance by burning it completely in excess oxygen to produce known compounds whose masses are used to determine the composition of the original material.

Consider burning a hydrocarbon of unknown composition in a chamber through which a stream of pure oxygen flows (**Figure 3.38**). The $CO_2(g)$ and $H_2O(g)$ produced flow first through a tube packed with $Mg(ClO_4)_2(s)$, which selectively absorbs the $H_2O(g)$, and then through a tube containing $NaOH(s)$, which absorbs the $CO_2(g)$. The masses of these tubes are measured before and after combustion. Suppose the mass of the tube that traps $CO_2(g)$ increases by 1.320 g, and the mass of the tube that traps $H_2O(g)$ increases by 0.540 g.

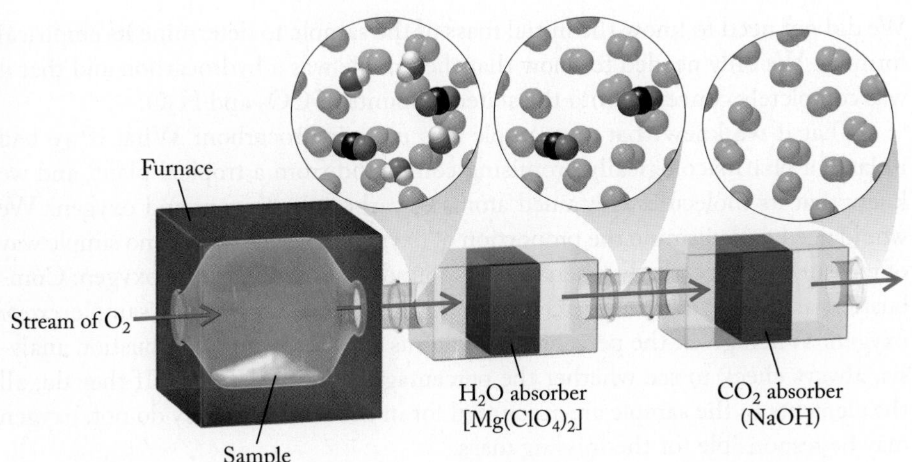

FIGURE 3.38 A carbon/hydrogen elemental analyzer relies on the complete combustion of organic compounds in excess oxygen. The products are H_2O vapor and CO_2. Water vapor is absorbed by a $Mg(ClO_4)_2$ filter, whereas carbon dioxide is absorbed by a NaOH filter. The empirical formula of the compound is calculated from the masses of H_2O and CO_2 absorbed.

How can we use these results to determine the empirical formula of the hydrocarbon?

First, let's establish what we know about the hydrocarbon in this example:

1. Hydrocarbons contain only carbon and hydrogen.
2. Complete conversion of its carbon into CO_2 produces 1.320 g of CO_2.
3. Complete conversion of its hydrogen into H_2O produces 0.540 g of H_2O.

To derive an empirical formula for the hydrocarbon, we must determine the number of moles of carbon in 1.320 g of carbon dioxide and the number of moles of hydrogen in 0.540 g of water vapor, because these quantities are equal to the number of moles of carbon and hydrogen in the sample that was combusted. Converting the mass of carbon dioxide to moles of carbon and the mass of water to moles of hydrogen:

STEPWISE ANIMATION
Combustion Analysis

$$1.320\ \cancel{\text{g CO}_2} \times \frac{1\ \cancel{\text{mol CO}_2}}{44.01\ \cancel{\text{g CO}_2}} \times \frac{1\ \text{mol C}}{1\ \cancel{\text{mol CO}_2}} = 0.02999\ \text{mol C} \approx 0.0300\ \text{mol C}$$

$$0.540\ \cancel{\text{g H}_2\text{O}} \times \frac{1\ \cancel{\text{mol H}_2\text{O}}}{18.02\ \cancel{\text{g H}_2\text{O}}} \times \frac{2\ \text{mol H}}{1\ \cancel{\text{mol H}_2\text{O}}} = 0.05993\ \text{mol H} \approx 0.0600\ \text{mol H}$$

The empirical formula is based on the mole ratio of C and H. If we divide both mole values by the smaller of the two, we get

$$\frac{0.0300\ \text{mol C}}{0.0300} = 1\ \text{mol C} \qquad \frac{0.0600\ \text{mol H}}{0.0300} = 2\ \text{mol H}$$

We obtain CH_2 for the empirical formula of the hydrocarbon.

If we want to extend this analysis to determine a molecular formula, we need to know the molecular mass of the hydrocarbon. Suppose its mass spectrum has a molecular ion with a mass of 84 u. This means the molar mass of the compound is 84 g/mol. The empirical formula mass of CH_2 is 14 g/mol. Dividing the molar mass by the empirical formula mass, we get the multiplier, n:

$$n = \frac{84\ \text{g/mol}}{14\ \text{g/mol}} = 6$$

The molecular formula is therefore

$$(CH_2)_6 = C_6H_{12}$$

We did not need to know the initial mass of the sample to determine its empirical formula. We only needed to know that the sample was a hydrocarbon and that it was completely converted into the stated amounts of CO_2 and H_2O.

What if we knew that our sample was *not* a hydrocarbon? What if we had isolated a pharmacologically promising compound from a tropical plant, and we knew that its molecules contained atoms of carbon, hydrogen, and oxygen? We would need to determine the proportion of oxygen in it, but there is no simple way of measuring that directly when the compound is burned in excess oxygen. Combustion analyses yield the percentages by mass of all atoms in a sample *except* oxygen. When given the percentages of atoms by mass from a combustion analysis, always check to see whether the percentages add up to 100%. If they do, all the elements in the sample are accounted for in the results. If they do not, oxygen may be responsible for the missing mass.

FIGURE 3.39 Citrus fruits such as lemons, limes, oranges, and grapefruit get their characteristic odor from a hydrocarbon compound called limonene.

SAMPLE EXERCISE 3.22 Combustion Analysis of a Hydrocarbon LO8

Limonene is a hydrocarbon that contributes to the odor of citrus fruits, including lemons (**Figure 3.39**). Combustion of 0.671 g of limonene yielded 2.168 g of CO_2 and 0.710 g of H_2O. What is the empirical formula of limonene? If the molar mass of limonene is 136.24 g/mol, then what is its molecular formula?

Collect and Organize We know the masses of CO_2 and H_2O produced during the combustion of a hydrocarbon sample, and we are asked to determine its empirical formula and then, from its molecular mass, its molecular formula.

Analyze First we determine the number of moles of C and H in the CO_2 and H_2O produced during combustion. These values are equal to the number of moles of C and H in the combusted sample. Then we calculate the C:H mole ratio and convert it into a ratio of small whole numbers to obtain the empirical formula. Next, we calculate the mass of the empirical formula and divide this mass into the molar mass to obtain the multiplier, n, which allows us to convert the empirical formula to a molecular formula.

$$\boxed{\text{g } CO_2} \xrightarrow{\frac{1}{\text{molar mass}}} \boxed{\text{mol } CO_2} \xrightarrow{\frac{1 \text{ mol C}}{1 \text{ mol } CO_2}} \boxed{\text{mol C}}$$

$$\boxed{\text{g } H_2O} \xrightarrow{\frac{1}{\text{molar mass}}} \boxed{\text{mol } H_2O} \xrightarrow{\frac{2 \text{ mol H}}{1 \text{ mol } H_2O}} \boxed{\text{mol H}}$$

Solve The moles of C and H in the CO_2 and H_2O collected during combustion are

$$2.168 \text{ g } CO_2 \times \frac{1 \text{ mol } CO_2}{44.01 \text{ g } CO_2} \times \frac{1 \text{ mol C}}{1 \text{ mol } CO_2} = 0.04926 \text{ mol C}$$

$$0.710 \text{ g } H_2O \times \frac{1 \text{ mol } H_2O}{18.02 \text{ g } H_2O} \times \frac{2 \text{ mol H}}{1 \text{ mol } H_2O} = 0.07880 \text{ mol H}$$

The mole ratio of the two elements in the sample is

$$0.04926 \text{ mol C} : 0.07880 \text{ mol H}$$

Dividing through by the smallest value (0.04926 mol) gives a mole ratio of 1.000:1.600. We can convert this ratio to whole numbers by multiplying by 5, making the empirical formula of the sample C_5H_8.

The molar mass of limonene is 136.24 g/mol. The mass of the empirical formula of limonene is 5(12.01 g/mol) + 8(1.008 g/mol) = 68.11 g/mol, so the multiplier n is

$$n = \frac{\text{molar mass}}{\text{empirical formula mass}} = \frac{136.24 \text{ g/mol}}{68.11 \text{ g/mol}} = 2$$

and the molecular formula is

$$(C_5H_8)_n = (C_5H_8)_2 = C_{10}H_{16}$$

Think About It One way to check the accuracy of the empirical formula determination is to convert the moles of C and H into grams of C and H, add the masses together and compare their sum to the mass of the original sample: 0.04926 mol C × 12.01 g C/mol C = 0.5916 g C; 0.07880 mol H × 1.008 g H/mol H = 0.07943 g H. Summing these two mass: 0.5916 g C + 0.07943 g H = 0.671 g, which exactly matches the mass of the sample that was analyzed.

Practice Exercise Cembrene A ($\mathcal{M}$ = 272.47 g/mol) is another naturally occurring hydrocarbon, extracted from coral (**Figure 3.40**). Combustion of 0.0341 g of cembrene A yields 0.1101 g of CO_2 and 0.0360 g of H_2O. What are the empirical and molecular formulas of cembrene A?

(Answers to Practice Exercises are in the back of the book.)

FIGURE 3.40 Cembrene A is a colorless oil that can be isolated from some species of coral.

SAMPLE EXERCISE 3.23 Combustion Analysis for Compounds Containing Oxygen — LO8

Eugenol is an ingredient in several spices, including bay leaves and cloves (**Figure 3.41**). Eugenol contains carbon, hydrogen, and oxygen. Combustion of 21.80 mg of eugenol yields 58.5 mg of CO_2 and 14.4 mg of H_2O. What is the empirical formula of eugenol? If the mass spectrum of eugenol shows a molecular ion at 164 u, what is the molecular formula of eugenol?

Collect and Organize We are given the masses of CO_2 (58.5 mg) and H_2O (14.4 mg) produced by the combustion of 21.80 mg of eugenol and are asked to determine its empirical formula. We also know its molar mass, and we need to determine its molecular formula.

Analyze First we determine the number of moles of C and H in the CO_2 and H_2O produced during combustion. These values are equal to the number of moles of C and H in the combusted sample. Multiplying these mole values by the atomic masses of C and H yields the mass of these elements in the original sample, which we then subtract from the total mass of the sample to obtain the mass of O in the sample. This mass can then be converted into moles of O in the sample:

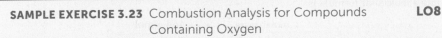

$$\boxed{\text{mg sample}} - \boxed{\text{mg C}} - \boxed{\text{mg H}} = \boxed{\text{mg O}} \xrightarrow{\frac{1\text{ g}}{10^3\text{ mg}}} \boxed{\text{g O}} \xrightarrow{\frac{1}{\text{molar mass}}} \boxed{\text{mol O}}$$

The mole ratio of C:H:O is then calculated and reduced to a ratio of small whole numbers to obtain the empirical formula. We then divide the mass of the empirical formula into the molar mass to obtain the multiplier, n, which makes it possible to convert the empirical formula to a molecular formula.

Solve The moles of C and H in the CO_2 and H_2O collected during combustion are

$$58.5\text{ mg }CO_2 \times \frac{1\text{ g }CO_2}{10^3\text{ mg }CO_2} \times \frac{1\text{ mol }CO_2}{44.01\text{ g }CO_2} \times \frac{1\text{ mol C}}{1\text{ mol }CO_2} = 1.329 \times 10^{-3}\text{ mol C}$$

$$14.4\text{ mg }H_2O \times \frac{1\text{ g }H_2O}{10^3\text{ mg }H_2O} \times \frac{1\text{ mol }H_2O}{18.02\text{ g }H_2O} \times \frac{2\text{ mol H}}{1\text{ mol }H_2O} = 1.598 \times 10^{-3}\text{ mol H}$$

The masses of C and H are calculated in milligrams to match the original sample:

$$1.329 \times 10^{-3}\text{ mol C} \times \frac{12.01\text{ g C}}{1\text{ mol C}} \times \frac{10^3\text{ mg C}}{1\text{ g C}} = 15.96\text{ mg C}$$

$$1.598 \times 10^{-3}\text{ mol H} \times \frac{1.008\text{ g H}}{1\text{ mol H}} \times \frac{10^3\text{ mg H}}{1\text{ g H}} = 1.611\text{ mg H}$$

(a)

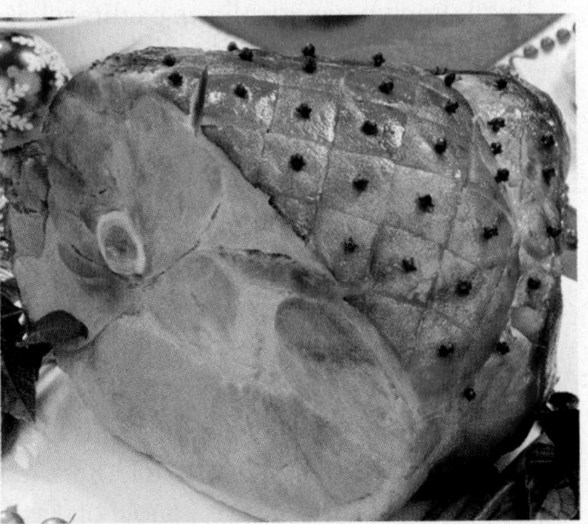

(b)

FIGURE 3.41 (a) The bay leaf is commonly used to season soups and tomato sauces, whereas (b) cloves are often added as a spice to baked hams. Both cooking staples contain the compound eugenol.

The sum of these two masses (15.96 mg C + 1.611 mg H = 17.57 mg) is less than the mass of the sample (21.80 mg). The difference must be the mass of oxygen in the sample:

$$(21.80 - 17.57)\text{ mg O} = 4.23\text{ mg O}$$

The number of moles of O in the sample is

$$4.23\text{ mg O} \times \frac{1\text{ g O}}{10^3\text{ mg O}} \times \frac{1\text{ mol O}}{16.00\text{ g O}} = 2.644 \times 10^{-4}\text{ mol O}$$

The mole ratio of the three elements in the sample is

$$1.329 \times 10^{-3}\text{ mol C}:1.598 \times 10^{-3}\text{ mol H}:2.644 \times 10^{-4}\text{ mol O}$$

$$1.329 \times 10^{-3}\text{ mol C}:1.598 \times 10^{-3}\text{ mol H}:2.64 \times 10^{-4}\text{ mol O}$$

Dividing through by the smallest value (2.64×10^{-4} mol) gives a mole ratio of 5.03:6.04:1, or 5:6:1, making the empirical formula of the sample C_5H_6O. The empirical formula mass is

$$5(12.01\text{ g/mol}) + 6(1.008\text{ g/mol}) + 16.00\text{ g/mol} = 82.10\text{ g/mol}$$

The molecular ion identifies the molecular mass of eugenol as 164 u, so the molar mass of eugenol is 164 g/mol. Therefore, the multiplier, n, to convert the empirical formula into a molecular formula is

$$n = \frac{\text{molar mass}}{\text{empirical formula mass}} = \frac{164\text{ g/mol}}{82.10\text{ g/mol}} = 2$$

and the molecular formula is

$$(C_5H_6O)_n = (C_5H_6O)_2 = C_{10}H_{12}O_2$$

Think About It Our answer makes sense because we have a simple whole-number ratio of the elements.

Practice Exercise Vanillin is a compound containing carbon, hydrogen, and oxygen that gives vanilla beans their distinctive flavor (**Figure 3.42**). The combustion of 30.4 mg of vanillin produces 70.4 mg of CO_2 and 14.4 mg of H_2O. The mass spectrum of vanillin shows a molecular ion at 152 u. Use this information to determine the molecular formula of vanillin.

(Answers to Practice Exercises are in the back of the book.)

FIGURE 3.42 The distinctive flavor of vanilla ice cream comes from the seeds of the vanilla bean and the compound vanillin.

SAMPLE EXERCISE 3.24 Integrating Concepts: Taxol

Taxol, known generically as paclitaxel, is a molecular compound used in the treatment of cancer. It was originally isolated from the bark of the Pacific yew tree (**Figure 3.43**), but subsequently has been synthesized in the laboratory.

a. The yew tree provides about 100 mg of Taxol for every 1.00 kg of bark. If 1 tree has about 3 kg of bark and 9000 trees were needed to isolate enough Taxol for its first clinical trial, how many grams of Taxol were used for those initial tests?
b. One of the laboratory syntheses of Taxol begins with camphor (78.89% C; 10.59% H; $\mathcal{M}$ = 152.23 g/mol) and requires 23 different chemical reactions to convert it into Taxol (66.11% C; 6.02% H; 1.64% N; $\mathcal{M}$ = 853.88 g/mol). The overall yield of the process to produce 1 mole of Taxol, starting with 1 mole of camphor, is 0.10%. (i) What are the molecular formulas of camphor and Taxol? (ii) How much camphor must be used to make the amount of Taxol extracted from 9000 trees?

Collect and Organize We are given information about the amount of Taxol in bark, the percent composition and molar mass of camphor and Taxol, and the yield of the laboratory synthesis. Using dimensional analysis and other methods from this chapter, we can answer the questions posed.

Analyze The percent compositions given for both substances do not add to 100%. We may assume that both compounds contain oxygen.

Solve

a. We begin by calculating the amount of Taxol in 9000 trees:

$$9000\text{ trees} \times \frac{3\text{ kg bark}}{1\text{ tree}} \times \frac{0.100\text{ g Taxol}}{1.00\text{ kg bark}} = 3 \times 10^3\text{ g Taxol}$$

b. (i) The percent carbon in camphor added to the percent hydrogen is less than 100%:

$$78.89\% + 10.59\% = 89.48\%$$

Because the reported percentages do not add up to 100%, the camphor most likely contains oxygen, the percent of which we cannot find in a combustion analysis. We can assume the percent oxygen in camphor is 100% − 89.48% = 10.52%. We can now determine the empirical formula of camphor by assuming we have 100.00 g of camphor, which would contain 78.89 g of C, 10.59 g of H, and 10.52 g of O. The first step is to determine how many moles of C, H, and O that is:

$$78.89\ \text{g C} \times \frac{1\ \text{mol C}}{12.01\ \text{g C}} = 6.569\ \text{mol C}$$

$$10.59\ \text{g H} \times \frac{1\ \text{mol H}}{1.008\ \text{g H}} = 10.51\ \text{mol H}$$

$$10.52\ \text{g O} \times \frac{1\ \text{mol O}}{16.00\ \text{g O}} = 0.6575\ \text{mol O}$$

Dividing by the smallest number of moles:

$$\frac{6.569\ \text{mol C}}{0.6575} = 10\ \text{mol C}$$

$$\frac{10.51\ \text{mol H}}{0.6575} = 16\ \text{mol H}$$

$$\frac{0.6575\ \text{mol O}}{0.6575} = 1\ \text{mol O}$$

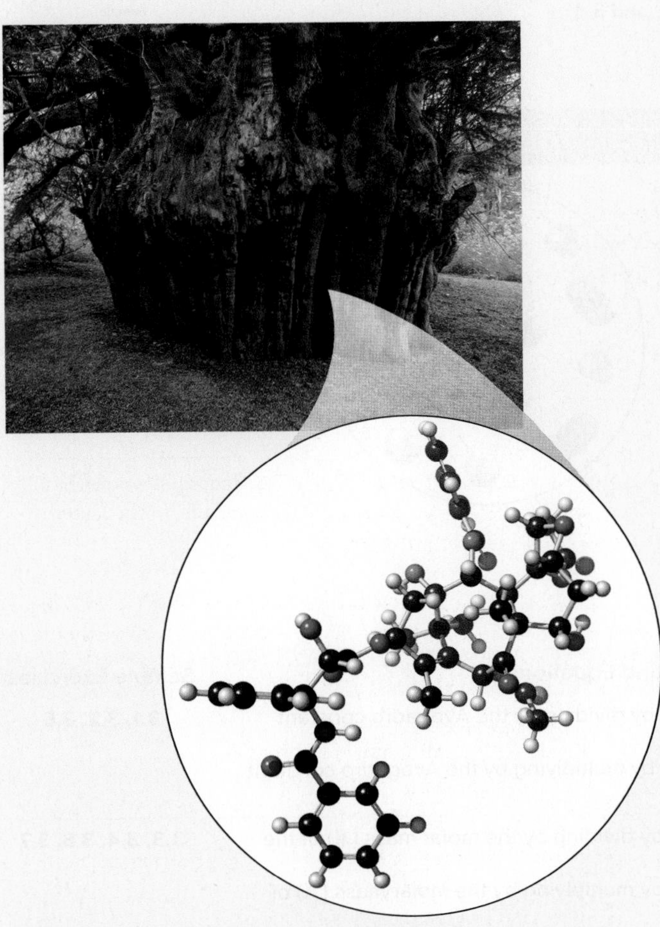

FIGURE 3.43 The bark of the ancient yew tree contains paclitaxel, a powerful cancer-fighting molecule that is marketed under the commercial name Taxol.

The empirical formula of camphor is $C_{10}H_{16}O$. The corresponding mass is [(10 × 12.01 g/mol) + (16 × 1.008 g/mol) + (16.00 g/mol)] = 152.23 g/mol, which matches the molar mass of camphor. Therefore, the molecular formula of camphor is the same as its empirical formula.

Carrying out the same procedure for Taxol:

$$66.11\% + 6.02\% + 1.64\% = 73.77\%$$

$$100\% - 73.77\% = 26.23\%\ \text{oxygen}$$

Calculating the empirical formula:

$$66.11\ \text{g C} \times \frac{1\ \text{mol C}}{12.01\ \text{g C}} = 5.505\ \text{mol C}$$

$$6.02\ \text{g H} \times \frac{1\ \text{mol H}}{1.008\ \text{g H}} = 5.97\ \text{mol H}$$

$$1.64\ \text{g N} \times \frac{1\ \text{mol N}}{14.01\ \text{g N}} = 0.117\ \text{mol N}$$

$$26.23\ \text{g O} \times \frac{1\ \text{mol O}}{16.00\ \text{g O}} = 1.639\ \text{mol O}$$

Dividing by the smallest number of moles:

$$\frac{5.505\ \text{mol C}}{0.117} = 47\ \text{mol C}$$

$$\frac{5.97\ \text{mol H}}{0.117} = 51\ \text{mol H}$$

$$\frac{0.117\ \text{mol N}}{0.117} = 1\ \text{mol N}$$

$$\frac{1.639\ \text{mol O}}{0.117} = 14\ \text{mol O}$$

The empirical formula of Taxol is $C_{47}H_{51}NO_{14}$, which has a mass of [(47 × 12.01 g/mol) + (51 × 1.008 g/mol) + (1 × 14.01 g/mol) + (14 × 16.00 g/mol)] = 853.88 g/mol. This value is the same as the molar mass of Taxol, so $C_{47}H_{51}NO_{14}$ is also the molecular formula.

(ii) We need 3×10^3 g of Taxol from a process that has a yield of 0.10%. The molar ratio in the process is 1 mole of camphor produces 1 mole of Taxol.

$$3 \times 10^3\ \text{g} = \text{Taxol} = 0.10\%\ \text{of the theoretical yield}$$

The theoretical yield for the process is

$$3 \times 10^3\ \text{g} = 0.0010x \qquad x = 3 \times 10^6\ \text{g}$$

Starting with that amount of Taxol as the theoretical yield:

$$3 \times 10^6\ \text{g Taxol} \times \frac{1\ \text{mol Taxol}}{853.88\ \text{g Taxol}} \times \frac{1\ \text{mol camphor}}{1\ \text{mol Taxol}} \times \frac{152.23\ \text{g camphor}}{1\ \text{mol camphor}} = 5 \times 10^5\ \text{g camphor}$$

Thus, we need 500 kg of camphor to synthesize as much Taxol as there is in 9000 Pacific yew trees.

Think About It The 0.10% yield of the laboratory synthesis of Taxol probably seems strangely low but remember that there were 23 steps in the synthesis. If the average yield of each of them was 74%, the overall yield of the synthesis would be $(0.74)^{23}$ or 0.1%.

SUMMARY

LO1 The **Avogadro constant** ($N_A = 6.022 \times 10^{23}$ particles) and the **mole** (the amount of a material that contains this many particles) can be used to convert among grams of a substance, moles of a substance, and number of particles. One mole of any substance has a mass equal to the sum of the molar masses of the elements in its formula. (Section 3.2)

LO2 Balanced chemical equations are essential tools to describe chemical reactions. Correct equations indicate the quantitative relationship in terms of moles between **reactants**, whose formulas appear first in the equation, and **products**, whose formulas appear after the reaction arrow. In a balanced chemical equation, the number of atoms of each element is the same on the reactant side and on the product side. (Sections 3.1–3.4)

LO3 Mole ratios from balanced chemical equations can be used to relate the masses of reactants consumed and the masses of products formed in a chemical reaction. (Section 3.5)

LO4 Balanced equations and stoichiometric calculations can be used to determine which substances are **limiting reactants**. (Section 3.6)

LO5 The **actual yield** (often expressed as **percent yield**) of a chemical reaction is frequently less than the **theoretical yield** predicted by the stoichiometry of the balanced chemical equation. If we know the amounts of reactants consumed and products formed in a chemical reaction, we can calculate the percent yield as the ratio of actual yield to theoretical yield, expressed as a percentage. (Section 3.6)

LO6 Writing balanced chemical equations requires knowing the correct formulas for substances. Empirical formulas for substances can be determined from their **percent composition** by mass. (Section 3.7)

LO7 The molecular formula of a compound may or may not be the same as the empirical formula. The molecular formula can be determined from the empirical formula and the molecular mass of the compound. (Section 3.8)

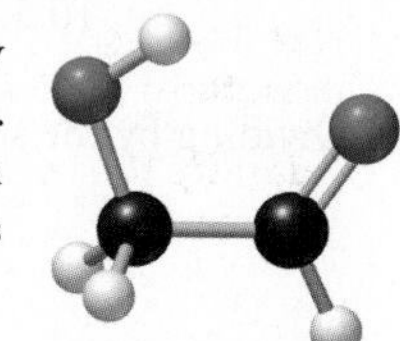

LO8 Empirical formulas of organic compounds can be determined from the **combustion analysis**. Production of CO_2, H_2O, and other oxides can provide the percent by mass of all elements in the compound except oxygen, which can be determined by applying the law of conservation of mass. (Sections 3.8 and 3.9)

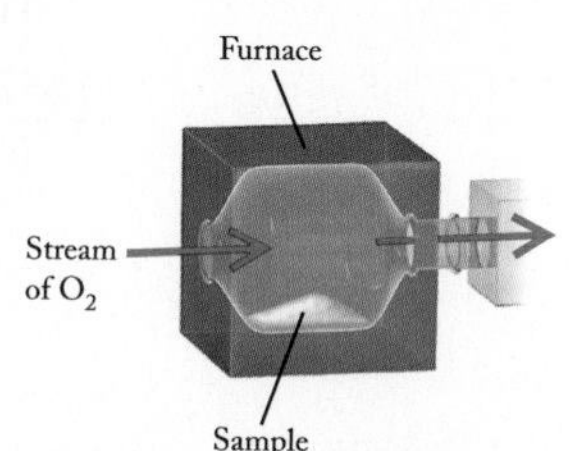

PARTICULATE PREVIEW WRAP-UP

Ten oxygen molecules will completely combust the five methane molecules to produce five carbon dioxide molecules and ten water molecules.

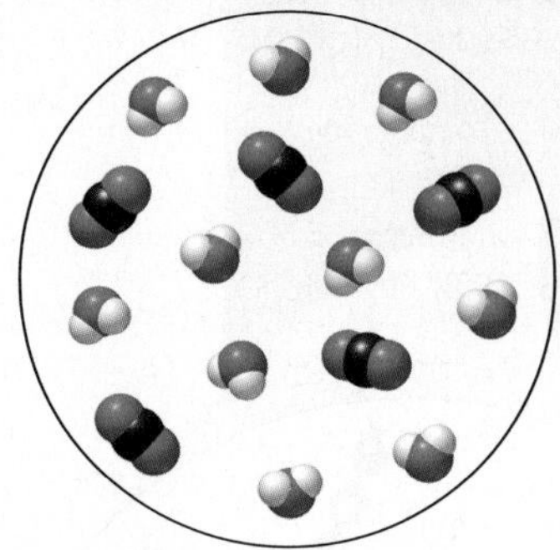

PROBLEM-SOLVING SUMMARY

Type of Problem	Concepts and Equations	Sample Exercises
Convert number of particles into number of moles (or vice versa)	Convert number of particles to moles by dividing by the Avogadro constant ($N_A = 6.022 \times 10^{23}$ particles/mol). Convert number of moles to particles by multiplying by the Avogadro constant ($N_A = 6.022 \times 10^{23}$ particles/mol).	**3.1, 3.2, 3.6**
Convert mass of a substance into number of moles (or vice versa)	Convert mass of substance to moles by dividing by the molar mass ($\mathcal{M}$) of the substance. Convert moles of substance to mass by multiplying by the molar mass ($\mathcal{M}$) of the substance.	**3.3, 3.4, 3.6, 3.7**

Type of Problem	Concepts and Equations	Sample Exercises
Calculate molar mass of a compound	Multiply the molar mass of each element by its subscript in the compound's formula, then sum the resulting values.	**3.5, 3.6, 3.7**
Write a balanced chemical equation	Change the coefficients in the equation so that the numbers of atoms of each element are the same on both sides of the reaction arrow.	**3.8, 3.9, 3.10**
Write a balanced chemical equation for a combustion reaction	The C and H in organic compounds react with O_2 to form CO_2 and H_2O, for example: $$CH_4(g) + 2\ O_2(g) \rightarrow CO_2(g) + 2\ H_2O(g)$$ Balance moles of C first, then H, then O.	**3.11**
Calculate mass of a product from mass of a reactant	Use $\mathcal{M}_{\text{reactant}}$ to convert mass of reactant into moles of reactant, use reaction stoichiometry to calculate moles of product, and then use $\mathcal{M}_{\text{product}}$ to calculate mass of product.	**3.12, 3.13**
Identify the limiting reactant	Method 1: Calculate how much product each reactant could make; the reactant making the *least* amount of product is the limiting reactant. Method 2: Convert given masses of reactants into moles. Compare the mole ratio of reactants to the corresponding mole ratio in the stoichiometric equation. If $$\left(\frac{\text{mol A}}{\text{mol B}}\right)_{\text{calculated}} > \left(\frac{\text{mol A}}{\text{mol B}}\right)_{\text{stoichiometric}} \quad (3.3)$$ then B is the limiting reactant. If $$\left(\frac{\text{mol A}}{\text{mol B}}\right)_{\text{calculated}} < \left(\frac{\text{mol A}}{\text{mol B}}\right)_{\text{stoichiometric}} \quad (3.4)$$ then A is the limiting reactant.	**3.14, 3.15**
Calculate percent yield	Calculate the theoretical yield of product using the mass of limiting reactant. Divide actual yield (given) by theoretical yield: $$\text{percent yield} = \frac{\text{actual yield}}{\text{theoretical yield}} \times 100\% \quad (3.5)$$	**3.16, 3.17**
Calculate percent composition from a chemical formula	Calculate the molar mass of the compound represented by the chemical formula and divide it into the mass of each element in 1 mole of the compound: $$\text{mass \% of an element} = \frac{\text{mass of element in 1 mole of compound}}{\text{molar mass of compound}} \quad (3.6)$$	**3.18**
Determine an empirical formula from percent composition	Assuming a 100.00-g sample, assign the mass (g) of each element to equal its percentage. Divide the mass of each element by its molar mass to get moles. Simplify mole ratios to lowest whole numbers and use those numbers as subscripts in the empirical formula of the compound.	**3.19, 3.20**
Relate empirical and molecular formulas	Calculate the multiplier *n* by dividing the compound's molecular mass by its empirical formula mass. Multiply the subscripts in the empirical formula by this conversion factor.	**3.21**
Determine an empirical formula from combustion analysis	For hydrocarbons, convert given masses of CO_2 and H_2O into moles of CO_2 and H_2O, and then to moles of C and H. For compounds containing O, convert moles of C and H into masses of C and H and subtract the sum of these values from the sample mass to calculate mass of O. Convert mass of O to moles of O. Simplify the mole ratio of C to H to O.	**3.22, 3.23**

VISUAL PROBLEMS

(Answers to boldface end-of-chapter questions and problems are in the back of the book.)

3.1. Each of the pairs of containers pictured in Figure P3.1 contains substances composed of elements X (red spheres) and Y (blue spheres). For each pair, write a balanced chemical equation describing the reaction that takes place. Be sure to indicate the physical states of the reactants and products, using the appropriate symbols in parentheses.

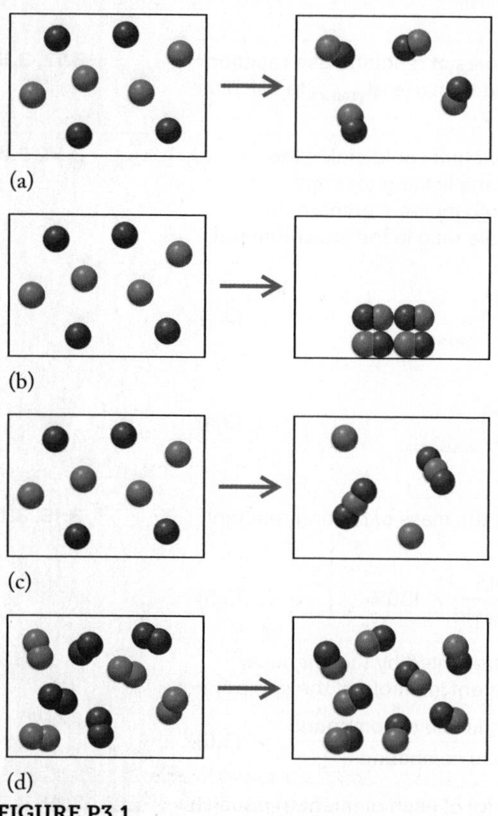

FIGURE P3.1

3.2. Identify the limiting reactant in each of the pairs of containers pictured in Figure P3.2. The red spheres represent atoms of element X, whereas the blue spheres represent atoms of element Y. Each question mark means that there is unreacted reactant left over. Identify which reactant is represented by the question mark.

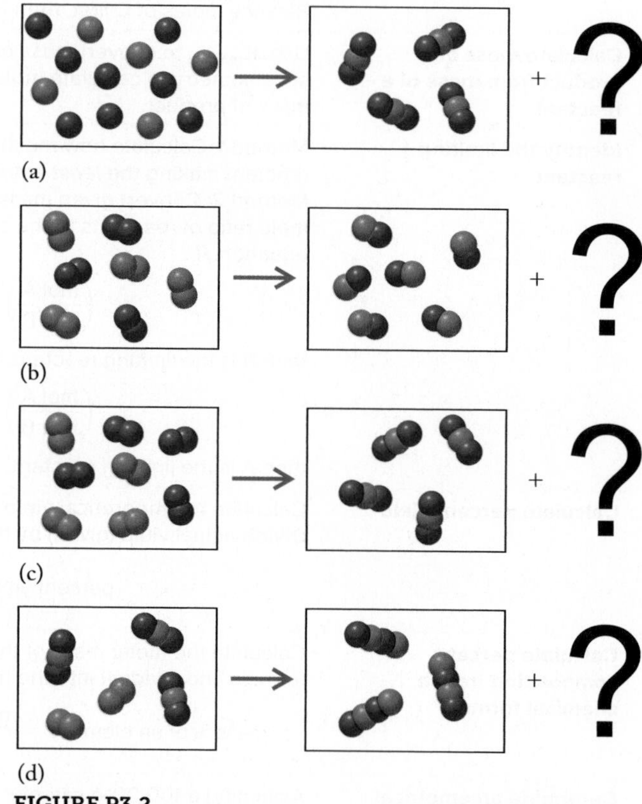

FIGURE P3.2

3.3. Which of the drawings in Figure P3.3 best illustrates the 100% reaction between N_2 and O_2 to produce N_2O? The red spheres represent atoms of oxygen, whereas the blue spheres represent atoms of nitrogen.

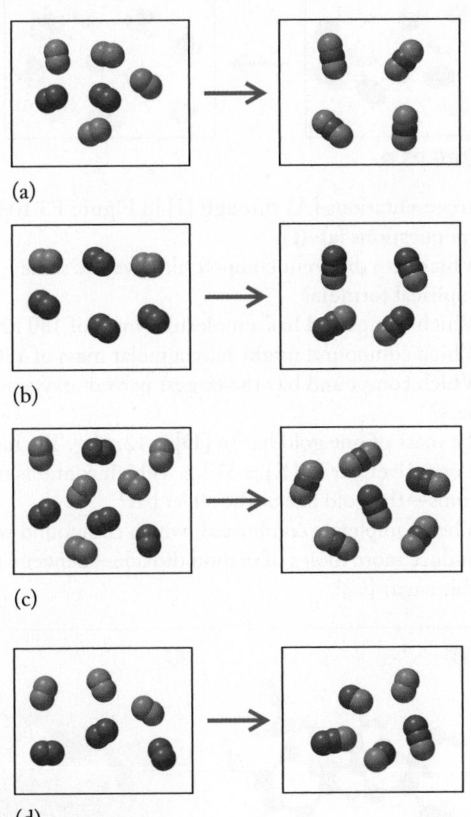

FIGURE P3.3

3.4. Is there a limiting reactant in any of the reactions depicted in Figure P3.3? If so, what is it, and how much of the excess reactant remains?

3.5. Which of the molecules in Figure P3.5 have the same empirical formulas?

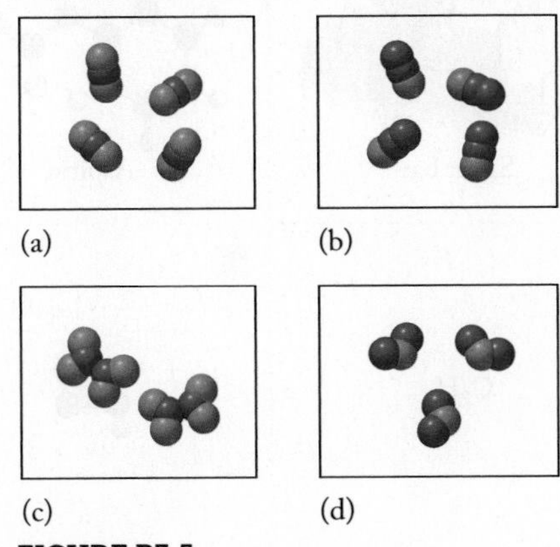

FIGURE P3.5

3.6. Which of the pairs of particulate images in Figure P3.6 represent balanced chemical equations? Write balanced chemical equations for any pairs of particulate images that are unbalanced as drawn. The red spheres represent oxygen, black represent carbon, blue represent nitrogen, and yellow represent sulfur.

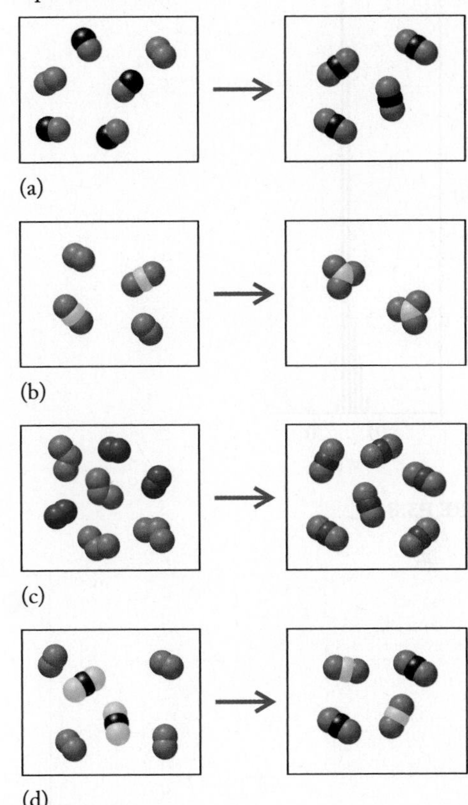

FIGURE P3.6

3.7. The two major products of combustion are CO_2 and H_2O. Figure P3.7 shows two mass spectra. Which belongs to water, and which belongs to carbon dioxide?

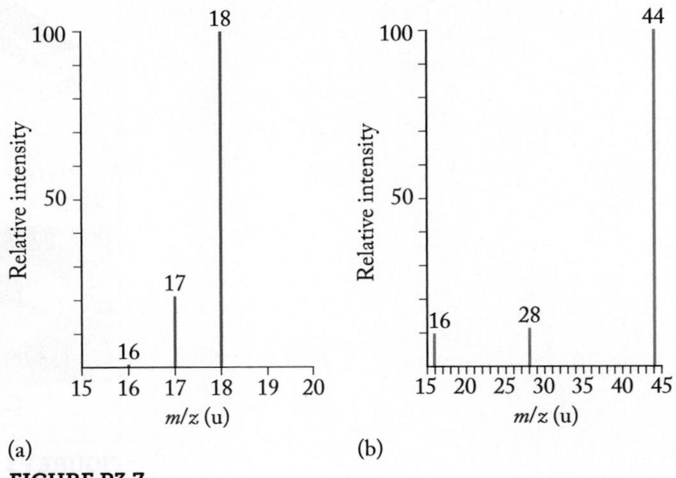

FIGURE P3.7

3.8. Figure P3.8 shows the mass spectrum of a simple hydrocarbon that is a widely used fuel for heating and electricity generation. What is the molecular formula of this hydrocarbon?

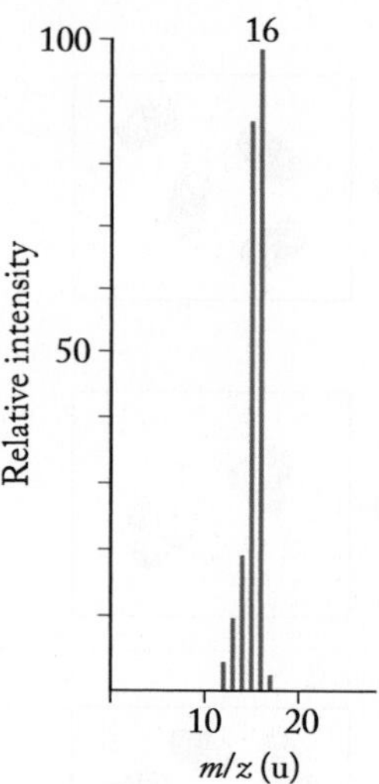

FIGURE P3.8

3.9. What is the percent yield of NH_3 for the reaction depicted in Figure P3.9? The blue spheres represent nitrogen, whereas the white spheres are hydrogen.

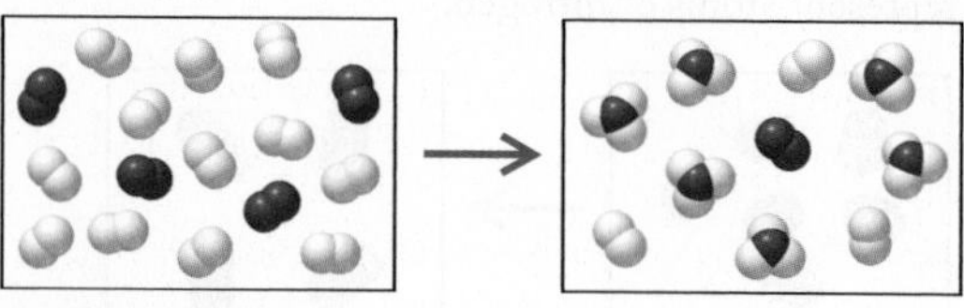

FIGURE P3.9

3.10. Use representations [A] through [I] in Figure P3.10 to answer questions (a)–(f).

a. Which two different compounds have the same empirical formula?
b. Which compound has a molecular mass of 180 u?
c. Which compound might have a molar mass of 180 g?
d. Which compound has the largest percent oxygen by mass?
e. The mass of one gold bar in [D] is 12.4 kg. The mass of one silver bar in [E] is 31 kg. Which contains more atoms—the gold bar or the silver bar?
f. When completely combusted, which compound will produce more moles of carbon dioxide—benzene [A] or table sugar [C]?

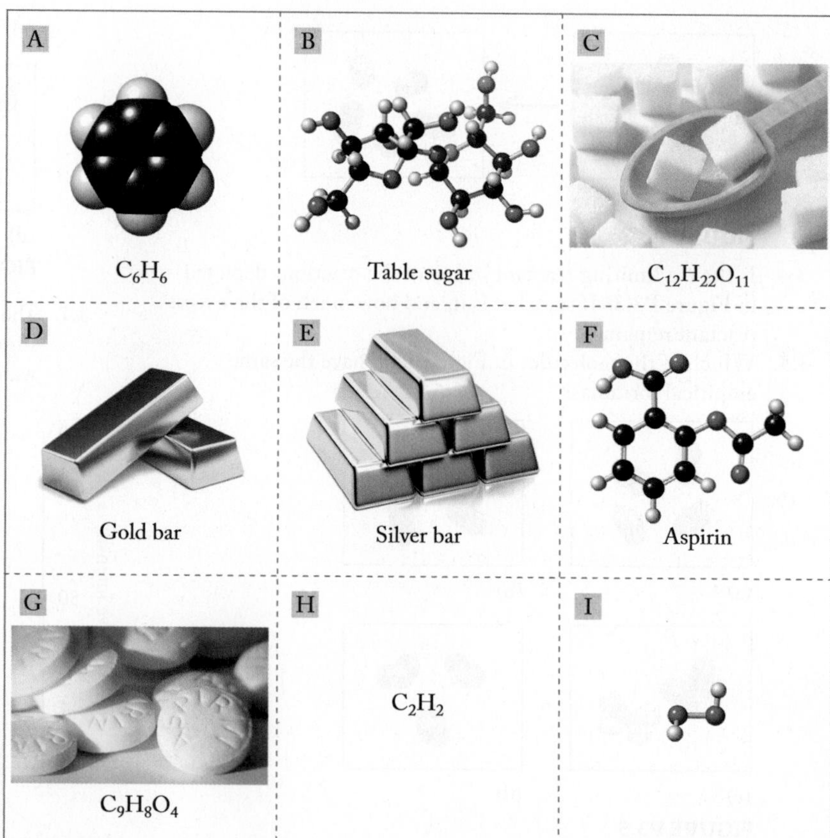

FIGURE P3.10

QUESTIONS AND PROBLEMS

Air, Life, and Molecules

Concept Review

3.11. In a combination reaction, can the reactants be two elements? Two compounds? One compound and one element?

3.12. Based on the compositions and physical states of Earth's various layers (Figure 3.3), does Al_2O_3 or Ni have the higher melting point?

3.13. Based on the distribution of the elements in Earth's layers (see Figure 3.3), which of the following substances should be the densest? (a) $SiO_2(s)$; (b) $Al_2O_3(s)$; (c) $Fe(\ell)$

*3.14. The proportions of the elements that make up the asteroid 433 Eros are like those that make up Earth. Scientists believe that this similarity means that 433 Eros and Earth formed around the same time. If another asteroid formed *after* Earth formed a solid crust, and if this asteroid was the product of a collision between Earth and an even larger asteroid, how would the new asteroid's composition differ from that of 433 Eros?

The Mole

Concept Review

3.15. In principle we could use the more familiar unit *dozen* in place of mole when expressing the quantities of particles (atoms, ions, or molecules) in chemical reactions. What would be the disadvantage in doing so?

3.16. In what way are the molar mass of an ionic compound and its formula mass the same, and in what ways are they different?

3.17. Do molecular compounds containing three atoms per molecule always have a molar mass greater than that of molecular compounds containing two atoms per molecule? Explain.

3.18. Without calculating their molar masses (though you may consult the periodic table), predict whether NO_2 or N_2O has the larger molar mass.

Problems

3.19. Earth's atmosphere contains many volatile substances that are present in trace amounts. The following quantities of these trace gases were found in a 1.0 mL sample of air. Calculate the number of moles of each gas in the sample.

a. 4.4×10^{14} atoms of $Ne(g)$
b. 4.2×10^{13} molecules of $CH_4(g)$
c. 2.5×10^{12} molecules of $O_3(g)$
d. 4.9×10^{9} molecules of $NO_2(g)$

3.20. The following quantities of trace gases were found in a 1.0 mL sample of air. Calculate the number of moles of each compound in the sample.

a. 1.4×10^{13} molecules of $H_2(g)$
b. 1.5×10^{14} atoms of $He(g)$
c. 7.7×10^{12} molecules of $N_2O(g)$
d. 3.0×10^{12} molecules of $CO(g)$

3.21. How many atoms of titanium are there in 0.125 mol of each of the following?

a. ilmenite, $FeTiO_3$
b. titanium(IV) chloride
c. Ti_2O_3
d. Ti_3O_5

3.22. How many atoms of iron are there in 2.5 moles of each of the following?

a. wolframite, $FeWO_4$
b. pyrite, FeS_2
c. magnetite, Fe_3O_4
d. hematite, Fe_2O_3

3.23. Which substance in each of the following pairs of quantities contains more moles of sulfur?

a. 3 moles of Al_2S_3 or 4 moles of Fe_2S_3
b. 3 moles of Li_2SO_4 or 4 moles of CaS
c. 2 moles of SO_3 or 2 moles of SO_2

3.24. Which substance in each of the following pairs of quantities contains more moles of nitrogen?

a. 2 moles of Li_3N or 0.5 mole of $Ca(NO_3)_2$
b. 1 mole of NO or 0.4 mole of NO_2
c. 3 moles of NF_3 or 1 mole of N_2O_5

3.25. Elemental Composition of Amino Acids Carbon, hydrogen, nitrogen, and oxygen are the most prevalent components of amino acids, the fundamental building blocks of proteins in living systems. How many moles of oxygen atoms are in 1.50 moles of the following compounds?

a. Glycine, the smallest amino acid: $C_2H_5NO_2$
b. Lysine, essential in the diet of humans: $C_6H_{14}N_2O_2$
c. Asparagine, produced in our bodies: $C_4H_8N_2O_3$

3.26. **Composition of Uranium Ores** The uranium used for nuclear fuel exists in nature in several minerals. Calculate how many moles of uranium are in 1 mole of the following:

a. carnotite, $K_2(UO_2)_2(VO_4)_2$
b. uranophane, $CaU_2Si_2O_{11}$
c. autunite, $Ca(UO_2)_2(PO_4)_2$

3.27. How many moles of carbon are there in 500.0 g of carbon?

3.28. How many moles of gold are there in 2.00 ounces of gold?

3.29. Cancer Therapy with Iridium Metal When iridium-192 is used in cancer treatment, a small cylindrical piece of ^{192}Ir, 0.6 mm in diameter and 3.5 mm long, is surgically inserted into the tumor. If the density of iridium is 22.42 g/cm^3, how many iridium atoms are present in the sample?

*3.30. **Gold Nanoparticles** The product shown in Figure P3.30 contains gold nanoparticles in water. The manufacturer claims that drinking this beverage improves human health. How many gold atoms are in a gold nanoparticle with a diameter of 2.00 nm (d = 19.3 g/mL, 1 u = 1.66054×10^{-24} g)?

FIGURE P3.30

3.31. Calculate the molar masses of the following molecules found in Earth's atmosphere: (a) SO_2; (b) O_3; (c) CO_2; (d) N_2O_5.

3.32. Determine the molar masses of the following minerals:
a. rhodonite, $MnSiO_3$
b. scheelite, $CaWO_4$
c. ilmenite, $FeTiO_3$
d. magnesite, $MgCO_3$

3.33. Flavoring Additives Calculate the molar masses of the following common flavors in food:
a. vanillin, $C_8H_8O_3$
b. oil of cloves, $C_{10}H_{12}O_2$
c. anise oil, $C_{10}H_{12}O$
d. oil of cinnamon, C_9H_8O

3.34. **Sweeteners** Calculate the molar masses of the following common sweeteners:
a. sucrose, $C_{12}H_{22}O_{11}$
b. saccharin, $C_7H_5NO_3S$
c. aspartame, $C_{14}H_{18}N_2O_5$
d. fructose, $C_6H_{12}O_6$

3.35. How many grams of iron are there in 1 mole of the following compounds? (a) FeO; (b) Fe_2O_3; (c) $Fe(OH)_3$; (d) Fe_3O_4

3.36. How many grams of sodium are there in 1 mole of the following compounds? (a) NaCl; (b) Na_2SO_4; (c) Na_3PO_4; (d) $NaNO_3$

3.37. How many moles of SiO_2 are there in a quartz crystal (SiO_2) that has a mass of 45.2 g?

3.38. How many moles of NaCl are there in a crystal of halite that has a mass of 6.82 g?

3.39. Early Atmosphere I In Urey and Miller's experiments, some of the gases they reacted were H_2O, H_2S, SO_2, and SO_3. Suppose four balloons are each filled with 10.0 g of one of these gases. Which balloon contains the most molecules?

3.40. **Early Atmosphere II** Which contains more atoms? (a) 0.1 mol H_2O or 0.1 mol H_2S; (b) 0.1 mol SO_3 or 0.2 mol SO_2

3.41. What is the mass of 0.122 mol $MgCO_3$?

3.42. What is the volume of 1.00 mol benzene (C_6H_6) at 20°C? The density of benzene is 0.879 g/mL.

***3.43.** The density of uranium (U; 19.05 g/cm^3) is more than five times greater than that of diamond (C; 3.514 g/cm^3). If you have a cube (1 cm on a side) of each element, which cube contains more atoms?

*3.44. Aluminum (d = 2.70 g/mL) and strontium (d = 2.64 g/mL) have nearly the same density. If we manufacture two cubes, each containing 1 mole of one element or the other, which cube will be smaller? What are the dimensions of this cube?

Writing Balanced Chemical Equations

Concept Review

3.45. In a balanced chemical equation, does the number of moles of reactants always equal the number of moles of products?

3.46. In a balanced chemical equation, does the sum of the coefficients for the reactants always equal the sum of the coefficients for the products?

3.47. In a balanced chemical equation, must the sum of the masses of all the gaseous reactants always equal the sum of the masses of the gaseous products?

3.48. In an unbalanced chemical equation, can the equation be balanced by changing subscripts in formulas rather than changing coefficients?

Problems

3.49. Using different-colored spheres to represent C and O, sketch the reaction between four C atoms and the necessary number of O_2 molecules that forms an equal number of CO and CO_2 molecules.

3.50. Using different-colored spheres to represent N and O, sketch the reaction between three molecules of N_2 and sufficient O_2 to produce a mixture in which there are an equal number of NO_2 and N_2O_4 molecules.

3.51. Ammonia (NH_3) can be formed by the reaction of two elements. What is the coefficient of $NH_3(g)$ in the balanced chemical equation?

3.52. Aluminum reacts with elemental oxygen at high temperatures to give pure aluminum oxide. What is the coefficient of $O_2(g)$ in the balanced chemical equation?

***3.53. Chemical Weathering of Rocks and Minerals** Write balanced chemical equations for the following chemical reactions, which contribute to weathering of the iron–silicate minerals ferrosilite ($FeSiO_3$), fayalite (Fe_2SiO_4), and greenalite [$Fe_3Si_2O_5(OH)_4$]:
a. $FeSiO_3(s) + H_2O(\ell) \rightarrow Fe_3Si_2O_5(OH)_4(s) + H_4SiO_4(aq)$
b. $Fe_2SiO_4(s) + CO_2(g) + H_2O(\ell) \rightarrow FeCO_3(s) + H_4SiO_4(aq)$
c. $Fe_3Si_2O_5(OH)_4(s) + CO_2(g) + H_2O(\ell) \rightarrow FeCO_3(s) + H_4SiO_4(aq)$

3.54. **Chemistry of Geothermal Vents** Some scientists believe that life on Earth may have originated near deep-ocean vents. Write balanced chemical equations describing the following reactions, which are among those taking place near such vents:
a. $CH_3SH(aq) + CO(aq) \rightarrow CH_3COSCH_3(aq) + H_2S(aq)$
b. $H_2S(aq) + CO(aq) \rightarrow CH_3CO_2H(aq) + S_8(s)$

3.55. Physiologically Active Nitrogen Oxides The oxides of nitrogen are biologically reactive substances. NO is a powerful agent for dilating blood vessels; N_2O is the anesthetic known as laughing gas; NO_2 has an acrid odor and is corrosive to lung tissue. Write balanced chemical equations describing the following reactions for the formation of nitrogen oxides:
a. $N_2(g) + O_2(g) \rightarrow NO(g)$
b. $NO(g) + O_2(g) \rightarrow NO_2(g)$
c. $NO(g) + NO_3(g) \rightarrow NO_2(g)$
d. $N_2(g) + O_2(g) \rightarrow N_2O(g)$

3.56. **Purifying Natural Gas** If natural gas contains significant amounts of sulfur as H_2S, it is called sour natural gas. For the gas to be commercially useful as a fuel, the H_2S must be removed. Once it is separated from the natural gas, it is reacted with oxygen in two different processes to yield either elemental sulfur (S_8), a commercial material that can be sold, or sulfur dioxide (SO_2). This sulfur dioxide product can be reacted with more H_2S to make additional elemental sulfur. Write balanced chemical equations describing the following reactions linked to the production of elemental sulfur.

a. $H_2S(g) + O_2(g) \rightarrow S_8(s) + H_2O(g)$
b. $H_2S(g) + O_2(g) \rightarrow SO_2(g) + H_2O(g)$
*c. $H_2S(g) + SO_2(g) \rightarrow S_8(s) + H_2O(g)$

***3.57.** Write a balanced chemical equation for each of the following reactions:

a. Dinitrogen pentoxide reacts with sodium metal to produce sodium nitrate and nitrogen dioxide.
b. A mixture of nitric acid and nitrous acid is formed when water reacts with dinitrogen tetroxide.
c. At high pressure, nitrogen monoxide reacts to form dinitrogen monoxide and nitrogen dioxide.
d. Acetylene (C_2H_2) burns and becomes carbon dioxide and water vapor.

3.58. Write a balanced chemical equation for each of the following reactions:

a. Carbon dioxide reacts with carbon to form carbon monoxide.
b. Potassium reacts with water to give potassium hydroxide and hydrogen gas.
c. Phosphorus (P_4) burns in air to give diphosphorus pentoxide.
d. Octane (C_8H_{18}) burns and becomes carbon dioxide and water vapor.

Combustion Reactions; Stoichiometric Calculations and the Carbon Cycle

Concept Review

3.59. Does the sum of the masses of the products always equal the sum of the masses of the reactants in a balanced chemical equation?

*3.60. There are two ways to write the equation for the combustion of ethane:

$$C_2H_6(g) + \tfrac{7}{2}O_2(g) \rightarrow 3\,H_2O(g) + 2\,CO_2(g)$$
$$2\,C_2H_6(g) + 7\,O_2(g) \rightarrow 6\,H_2O(g) + 4\,CO_2(g)$$

Do these two different ways of writing the equation affect the calculation of how much CO_2 is produced from a known quantity of C_2H_6?

Problems

3.61. The manufacture of aluminum includes the production of cryolite (Na_3AlF_6) from the following reaction:

$$6\,HF(g) + 3\,NaAlO_2(s) \rightarrow Na_3AlF_6(s) + 3\,H_2O(\ell) + Al_2O_3(s)$$

How much $NaAlO_2$ (sodium aluminate) is required to produce 1.00 kg of Na_3AlF_6?

3.62. Chromium metal can be produced from the high-temperature reaction of Cr_2O_3 [chromium(III) oxide] with silicon or aluminum by each of the following reactions:

$$Cr_2O_3(s) + 2\,Al(\ell) \rightarrow 2\,Cr(\ell) + Al_2O_3(s)$$
$$2\,Cr_2O_3(s) + 3\,Si(\ell) \rightarrow 4\,Cr(\ell) + 3\,SiO_2(s)$$

a. Calculate the number of grams of aluminum required to prepare 400.0 g of chromium metal by the first reaction.
b. Calculate the number of grams of silicon required to prepare 400.0 g of chromium metal by the second reaction.

3.63. When $NaHCO_3$ is heated above 80°C, it decomposes as described by this chemical equation:

$$2\,NaHCO_3(s) \rightarrow Na_2CO_3(s) + H_2O(g) + CO_2(g)$$

How many grams of Na_2CO_3 are produced from the decomposition of 50.0 g $NaHCO_3$?

3.64. **Egyptian Cosmetics** Pb(OH)Cl, one of the lead compounds used in ancient Egyptian cosmetics, was prepared from PbO according to the following recipe:

$$PbO(s) + NaCl(aq) + H_2O(\ell) \rightarrow Pb(OH)Cl(s) + NaOH(aq)$$

How many grams of PbO and how many grams of NaCl would be required to produce 10.0 g of Pb(OH)Cl?

***3.65.** Suppose the fuel for an electric power plant is coal that contains 3.0% sulfur by mass. During combustion, the sulfur is converted into sulfur dioxide. How many grams of sulfur dioxide are produced for every kilogram of coal that is burned to produce electricity?

*3.66. The uranium minerals found in nature must be refined and enriched in ^{235}U before the uranium can be used as a fuel in nuclear reactors. One procedure for enriching uranium relies on the reaction of UO_2 with HF to form UF_4, which is then converted into UF_6 by reaction with fluorine:

$$UO_2(g) + 4\,HF(aq) \rightarrow UF_4(g) + 2\,H_2O(\ell)$$
$$UF_4(g) + F_2(g) \rightarrow UF_6(g)$$

a. How many kilograms of HF are needed to completely react with 5.00 kg of UO_2?
b. How much UF_6 can be produced from 850.0 g of UO_2?

3.67. In Brazil automobiles use ethanol (C_2H_6O) as fuel, whereas in the United States we rely on gasoline. Using C_8H_{18} (octane) to represent gasoline, write balanced chemical equations for the complete combustion of ethanol and octane. Which fuel produces more CO_2 per gram of fuel?

3.68. Driving 1000 miles a month is not unusual for a short-distance commuter. If your vehicle gets 25 mpg, you would use 40 gallons ($\approx$150 L) of gasoline every month. If gasoline is approximated as C_8H_{18} ($d = 0.703$ g/mL), how much carbon dioxide does your vehicle emit every month?

3.69. Chalcopyrite ($CuFeS_2$) is an abundant copper mineral that can be converted into elemental copper. How much Cu could be produced from 1.00 kg of $CuFeS_2$?

*3.70. **Mining for Gold** Unlike most metals, gold is found in nature as the pure element. Miners in California in 1849 searched for gold nuggets and gold dust in streambeds, where the denser gold could be easily separated from sand and gravel. However, larger deposits of gold are found in veins of rock and can be separated chemically in a two-step process:

(1) $4\ Au(s) + 8\ NaCN(aq) + O_2(g) + 2\ H_2O(\ell) \rightarrow 4\ NaAu(CN)_2(aq) + 4\ NaOH(aq)$

(2) $2\ NaAu(CN)_2(aq) + Zn(s) \rightarrow 2\ Au(s) + Na_2[Zn(CN)_4](aq)$

If a 1.0×10^3 kg sample of rock is 0.019% gold by mass, how much Zn is needed to react with the gold extracted from the rock? Assume that reactions (1) and (2) are 100% efficient.

Limiting Reactants and Percent Yield

Concept Review

3.71. If a reaction vessel contains equal masses of Fe and S, a mass of FeS corresponding to which of the following could theoretically be produced?
a. the sum of the masses of Fe and S
b. more than the sum of the masses of Fe and S
c. less than the sum of the masses of Fe and S

3.72. Can the percent yield of a chemical reaction ever exceed 100%?

3.73. Give two reasons that the actual yield from a chemical reaction is usually less than the theoretical yield.

3.74. A chemical reaction produces less than the expected amount of product. Is this result a violation of the law of conservation of mass?

3.75. **Laboratory Errors** A student conducting an experiment based on the thermal decomposition of $NaHCO_3$ (see Problem 3.63) accidentally spills some of the reactant after determining its mass and before placing it in an oven set at 110°C. The student decides to carry out the decomposition reaction on what is left. How will the lost $NaHCO_3$ affect the results of the experiment, including the mass of the decomposed sample and the calculated values of the actual yield and percent yield?

3.76. Consider the reaction of 10 g A with 10 g B in the combination reaction $A + B \rightarrow C$. How do the theoretical yield, actual yield, and percent yield change (or not change) if 20 g A are reacted with 20 g B?

Problems

3.77. **Making Hollandaise Sauce** A recipe for 1 cup of hollandaise sauce calls for $\frac{1}{2}$ cup of butter, $\frac{1}{4}$ cup of hot water, 4 egg yolks, and the juice of a medium-sized lemon. How many cups of this sauce can be made from a pound (2 cups) of butter, a dozen eggs, 4 medium lemons, and an unlimited supply of hot water?

3.78. A factory making toy wagons has 13,466 wheels, 3360 handles, and 2400 wagon beds in stock. What is the maximum number of wagons the factory can make?

3.79. Given the amounts of reactants shown, calculate the theoretical yield in grams of the product of each of these *unbalanced* descriptions of chemical reactions.
a. $Li(s) + N_2(g) \rightarrow Li_3N(s)$
5.0 g 2.0 g ? g
b. $P_2O_5(s) + H_2O(\ell) \rightarrow H_3PO_4(aq)$
25.0 g 36.0 g ? g
c. $SO_2(g) + O_2(g) \rightarrow SO_3(g)$
6.4 g 4.0 g ? g

3.80. Given the amounts of reactants shown, calculate the theoretical yield in grams of the product indicated by the question mark for each of these *unbalanced* descriptions of chemical reactions.
a. $Cu_2O(s) + H_2(g) \rightarrow Cu(s) + H_2O(\ell)$
30.0 g 12.0 g ? g
b. $Mg(s) + HCl(g) \rightarrow MgCl_2(s) + H_2(g)$
24.3 g 10.0 g ? g
c. $CuCl_2(aq) + Zn(s) \rightarrow Cu(s) + ZnCl_2(aq)$
11.6 g 10.0 g ? g

3.81. Ammonia rapidly reacts with hydrogen chloride, making ammonium chloride. Write a balanced chemical equation for the reaction and calculate the number of grams of excess reactant when 3.0 g of NH_3 reacts with 5.0 g of HCl.

3.82. Sulfur trioxide dissolves in water, producing H_2SO_4. How much sulfuric acid can be produced from 10.0 mL of water ($d = 1.00$ g/mL) and 25.6 g of SO_3?

***3.83.** Phosgenite, a lead compound with the formula $Pb_2Cl_2CO_3$, is found in ancient Egyptian cosmetics. Phosgenite was prepared by the reaction of PbO, NaCl, H_2O, and CO_2. An unbalanced description of the reaction is

$PbO(s) + NaCl(aq) + H_2O(\ell) + CO_2(g) \rightarrow Pb_2Cl_2CO_3(s) + NaOH(aq)$

a. Write a balanced chemical equation describing the reaction.
b. How many grams of phosgenite can be obtained from 10.0 g of PbO and 10.0 g of NaCl in the presence of excess water and CO_2?
c. If 2.72 g of phosgenite is produced in the laboratory from the amounts of starting materials stated in part (b), what is the percent yield of the reaction?

3.84. Potassium superoxide (KO_2) reacts with carbon dioxide to form potassium carbonate and oxygen:

$4\ KO_2(s) + 2\ CO_2(g) \rightarrow 2\ K_2CO_3(s) + 3\ O_2(g)$

This reaction makes potassium superoxide useful in a self-contained breathing apparatus. How much O_2 could be produced from 2.50 g of KO_2 and 4.50 g of CO_2?

3.85. The reaction of 5.0 g of pentane (C_5H_{12}) with 5.0 g of oxygen gas produces 20.4 g of CO_2. What is the percent yield of this reaction?

3.86. Baking soda ($NaHCO_3$) can be made in large quantities by the following reaction:

$NaCl(aq) + NH_3(aq) + CO_2(aq) + H_2O(\ell) \rightarrow NaHCO_3(s) + NH_4Cl(aq)$

If 10.0 g of NaCl reacts with excesses of the other reactants and 4.2 g of $NaHCO_3$ is isolated, what is the percent yield of the reaction?

3.87. **Chemistry of Fermentation** Yeast converts glucose ($C_6H_{12}O_6$) into ethanol ($d = 0.789$ g/mL) in a process called fermentation. An equation for the reaction can be written as follows:

$$C_6H_{12}O_6(aq) \rightarrow C_2H_5OH(\ell) + CO_2(g)$$

a. Write a balanced chemical equation for this fermentation reaction.
b. If 100.0 g of glucose yields 50.0 mL of ethanol, what is the percent yield for the reaction?

*3.88. **Composition of Seawater** A 1 liter sample of seawater contains 19.4 g of Cl^-, 10.8 g of Na^+, and 1.29 g of Mg^{2+}.
a. How many moles of each ion are present?
b. If we evaporated the seawater, would there be enough Cl^- present to form the chloride salts of all the sodium and magnesium present?

Determining Empirical Formulas from Percent Composition; Comparing Empirical and Molecular Formulas

Concept Review

3.89. What is the difference between an empirical formula and a molecular formula?

3.90. Why is the empirical formula of an ionic compound represented by a formula unit? Why is the term molecular formula inappropriate to use when describing ionic compounds?

3.91. Do the empirical and molecular formulas of a compound always have the same percent composition values? Explain your answer.

3.92. Is the element with the largest atomic mass always the element present in the highest percentage by mass in a compound? Explain your answer.

3.93. Among the naturally occurring hydrocarbons emitted by plants are three compounds named camphene, carene, and thujene. If all three of these compounds have the same percent composition and the same molar mass, do they have the same empirical formula? Do they have the same molecular formula? Are they isomers?

*3.94. How might the compounds in Problem 3.93 differ from each other if they have the same molar mass and percent composition?

Problems

3.95. Gasoline consists primarily of a mixture of the hydrocarbons C_6H_{14}, C_7H_{16}, C_8H_{18}, and C_9H_{20}. What is the empirical formula of each compound?

3.96. The biosynthesis of carbohydrate includes the molecules shown below. What is the empirical formula of each compound shown in Figure P3.96?

Glyceraldehyde	Dihydroxyacetone	Erythrulose
		CH_2OH
HC=O	CH_2OH	C=O
HC—OH	C=O	HC—OH
CH_2OH	CH_2OH	CH_2OH

FIGURE P3.96

3.97. Calculate the percent composition of (a) Na_2O, (b) NaOH, (c) $NaHCO_3$, and (d) Na_2CO_3.

3.98. Calculate the percent composition of (a) sodium sulfate, (b) dinitrogen tetroxide, (c) strontium nitrate, and (d) aluminum sulfide.

3.99. **Organic Compounds in Space** The following compounds have been detected in space. Which of them contains the greatest percentage of carbon by mass? Do any two of the following compounds have the same empirical formula?
a. naphthalene, $C_{10}H_8$
b. chrysene, $C_{18}H_{12}$
c. pentacene, $C_{22}H_{14}$
d. pyrene, $C_{16}H_{10}$

3.100. Of the nitrogen oxides—N_2O, NO, N_2O_3, N_2O_2, NO_2, and N_2O_4—which are more than 50% oxygen by mass? Which, if any, have the same empirical formula?

3.101. Methane (CH_4) and tetrafluoromethane (CF_4) both contain 20% carbon per mole. Which one has the greater percent C by mass?

3.102. Silane (SiH_4) is used in the electronics industry to manufacture thin films of silicon. What is the percent Si by mass in SiH_4? Does silane have the same percent Si by mass as disilane (Si_2H_6)?

3.103. **Surgical-Grade Titanium** Medical implants and high-quality jewelry items for body piercings are frequently made of a material known as G23Ti, or surgical-grade titanium. The percent composition of the material is 64.39% titanium, 24.19% aluminum, and 11.42% vanadium. What is the empirical formula for surgical-grade titanium?

3.104. A sample of an iron-containing compound is 22.0% iron, 50.2% oxygen, and 27.8% chlorine by mass. What is the empirical formula of this compound?

3.105. **Sour Candy** Tartaric acid ($C_4H_6O_6$) and citric acid ($C_6H_8O_7$) are both used commercially to give sour candies (Figure P3.105) their characteristic sour taste. Which compound has the larger percent C by mass?

FIGURE P3.105

3.106. **Chlorofluorocarbons** CFCs (chlorofluorocarbons) are molecules used as refrigerants, but they also contribute to the destruction of the ozone layer. One CFC consists of two carbon atoms, two fluorine atoms, and four chlorine atoms. What is the empirical formula of this CFC? What is its molecular formula?

3.107. **Asbestosis** Asbestosis is a lung disease caused by inhaling asbestos fibers. In addition, fiber from a form of asbestos called chrysotile is considered to be a human carcinogen by the U.S. Department of Health and Human Services. Chrysotile's composition is 26.31% magnesium, 20.20% silicon, and 1.45% hydrogen, with the remainder of the mass as oxygen. Determine the empirical formula of chrysotile.

3.108. **Chemistry of Soot** A candle flame produces easily seen specks of soot near the edges of the flame, especially when the candle is moved. A piece of glass held over a candle flame will become coated with soot, which is the result of the incomplete combustion of candle wax. Elemental analysis of a compound extracted from a sample of this soot gave these results: 92.26% C and 7.74% H by mass. Calculate the empirical formula of the compound.

3.109. **Making the DNA Bases** Adenine (135.14 g/mol; 44.44% C, 3.73% H, and 51.84% N) was detected in mixtures of HCN, ammonia, and water under conditions that simulate early Earth. This observation suggests a possible origin for one of the bases found in DNA. What are the empirical and molecular formulas for adenine?

3.110. **Making Sugars for RNA** Ribose, the sugar found in RNA, has been detected in experiments designed to mimic the conditions of early Earth. If ribose contains 40.00% C, 6.71% H, and 53.28% O, with a molar mass of 150.13 g/mol, what are the empirical and molecular formulas for ribose?

Combustion Analysis

Concept Review

3.111. Explain why it is important for combustion analysis to be carried out in an excess of oxygen.

3.112. Why is the quantity of CO_2 obtained in a combustion analysis not a direct measure of the oxygen content of the starting compound?

3.113. What additional information is needed to determine a molecular formula from the results of an elemental analysis of an organic compound?

*3.114. If a compound containing sulfur is subjected to combustion analysis in excess oxygen, what is the most likely molecular formula for the sulfur-containing product?

***3.115.** If a compound containing nitrogen is subjected to combustion analysis in excess oxygen, what is the most likely molecular formula for the nitrogen-containing product?

*3.116. Suppose an insufficient amount of oxygen is used for a combustion analysis of a hydrocarbon, and some of the carbon is converted to CO rather than to CO_2. Will the empirical formula determined from the results of this analysis be too low or too high in carbon?

Problems

3.117. The combustion of 135.0 mg of a hydrocarbon produces 440.0 mg of CO_2 and 135.0 mg of H_2O. The molar mass of the hydrocarbon is 270 g/mol. Determine the empirical and molecular formulas of this compound.

3.118. A 0.100-g sample of a compound containing C, H, and O is burned in oxygen, producing 0.1783 g of CO_2 and 0.0734 g of H_2O. Determine the empirical formula of the compound.

3.119. **GRAS List for Food Additives** The compound geraniol ($\mathcal{M}$ = 154.25 g/mol) is on the Food and Drug Administration's GRAS (generally recognized as safe) list and can be used in foods and personal care products. By itself, geraniol smells like roses, but it is frequently blended with other fragrances on the GRAS list and then added to products to produce a pleasant peachlike or lemonlike aroma. In an analysis, the complete combustion of 175 mg of geraniol produced 499 mg of CO_2 and 184 mg of H_2O. What are the empirical and molecular formulas of geraniol?

*3.120. The combustion of 40.5 mg of a compound containing C, H, and O, and extracted from the bark of the sassafras tree, produces 110.0 mg of CO_2 and 22.5 mg of H_2O. The molar mass of the compound is 162 g/mol. Determine its empirical and molecular formulas.

3.121. **Ethnobotany** One of the ingredients in the Native American stomachache remedy derived from common chokecherry is caffeic acid ($\mathcal{M}$ = 180.16 g/mol). Combustion of 1.00×10^2 mg of caffeic acid yielded 220 mg of CO_2 and 40.3 mg of H_2O. What are the empirical and molecular formulas of caffeic acid?

3.122. Coniine ($\mathcal{M}$ = 127.23 g/mol), a substance isolated from poison hemlock, contains only carbon, hydrogen, and nitrogen. Combustion of 5.024 mg of coniine yields 13.90 mg CO_2 and 6.048 mg of H_2O. What are the empirical and molecular formulas of coniine?

Additional Problems

***3.123.** As a solution of copper sulfate slowly evaporates, beautiful blue crystals made of copper(II) and sulfate ions form such that water molecules are trapped inside the crystals. The overall formula of the compound is $CuSO_4 \cdot 5H_2O$.

a. What is the percent by mass of water in this compound?
b. At high temperatures, the water in the compound is driven off as steam. What mass percentage of the original sample of the blue solid is lost as a result?

3.124. **Production of Aluminum** Aluminum is mined as the mineral bauxite, which consists primarily of Al_2O_3 (alumina).

a. How much aluminum is produced from 1 metric ton (1 metric ton = 10^3 kg) of Al_2O_3?
$$2\,Al_2O_3(s) \rightarrow 4\,Al(s) + 3\,O_2(g)$$
b. The oxygen produced in part (a) can react with carbon to produce carbon monoxide:
$$O_2(g) + 2\,C(s) \rightarrow 2\,CO(g)$$
Write a balanced chemical equation describing the reaction involving the following reactants and products:
$$Al_2O_3(s) + C(s) \rightarrow Al(s) + CO(g)$$
c. How much CO can be produced from the O_2 made in part (a)?

***3.125. Chemistry of Copper Production** "Native," or elemental, copper can be found in nature, but most copper is mined as oxide or sulfide minerals. Chalcopyrite ($CuFeS_2$) is one copper mineral that can be converted to elemental copper in a series of chemical steps. Reacting chalcopyrite with oxygen at high temperature produces a mixture of copper sulfide and iron oxide. The iron oxide is separated from CuS by reaction with sand. CuS is converted to Cu_2S, and the Cu_2S is then burned in air to produce Cu and SO_2:

(1) $2\ CuFeS_2(s) + 3\ O_2(g) \rightarrow 2\ CuS(s) + 2\ FeO(s) + 2\ SO_2(g)$

(2) $FeO(s) + SiO_2(s) \rightarrow FeSiO_3(s)$

(3) $2\ CuS(s) \rightarrow Cu_2S(s) + \frac{1}{8}\ S_8(s)$

(4) $Cu_2S(s) + O_2(g) \rightarrow 2\ Cu(s) + SO_2(g)$

An average copper penny minted in the 1960s has a mass of about 3.0 g.

a. How much chalcopyrite had to be mined to produce one dollar's worth of pennies?
b. How much chalcopyrite had to be mined to produce one dollar's worth of pennies if reaction 1 above had a percent yield of 85% and reactions 2, 3, and 4 had percent yields of essentially 100%?
c. How much chalcopyrite had to be mined to produce one dollar's worth of pennies if each reaction involving copper proceeded with an 85% yield?

***3.126. Mining for Gold** Gold can be extracted from the surrounding rock by using a solution of sodium cyanide. Although effective for isolating gold, toxic cyanide finds its way into watersheds, causing environmental damage and harming human health.

$$4\ Au(s) + 8\ NaCN(aq) + O_2(g) + 2\ H_2O(\ell) \rightarrow 4\ NaAu(CN)_2(aq) + 4\ NaOH(aq)$$

$$2\ NaAu(CN)_2(aq) + Zn(s) \rightarrow 2\ Au(s) + Na_2[Zn(CN)_4](aq)$$

a. If a sample of rock contains 0.009% gold by mass, how much NaCN is needed to extract the gold from 1 metric ton (1 metric ton = 10^3 kg) of rock as $NaAu(CN)_2$?
b. How much zinc is needed to convert the $NaAu(CN)_2$ from part (a) to metallic gold?
c. The gold recovered in part (b) is manufactured into a gold ingot in the shape of a cube. The density of gold is 19.3 g/cm^3. How big is the cube of gold?

***3.127. Preparing Nuclear Reactor Fuel** Uranium oxides used in the preparation of fuel for nuclear reactors are separated from other metals in minerals by converting the uranium to $UO_x(NO_3)_y(H_2O)_z$, where uranium has a positive charge ranging from 3+ to 6+.

a. Roasting $UO_x(NO_3)_y(H_2O)_z$ at 400°C leads to loss of water and decomposition of the nitrate ion to nitrogen oxides, leaving behind a product with the formula U_aO_b that is 83.22% U by mass. What are the values of *a* and *b*? What is the charge on U in U_aO_b?
b. Higher temperatures produce a different uranium oxide, U_cO_d, with a higher uranium content, 84.8% U. What are the values of *c* and *d*? What is the charge on U in U_cO_d?
c. The values of *x*, *y*, and *z* in $UO_x(NO_3)_y(H_2O)_z$ are found by gently heating the compound to remove all the water. In a laboratory experiment, 1.328 g of $UO_x(NO_3)_y(H_2O)_z$ produced 1.042 g of $UO_x(NO_3)_y$. Continued heating generated 0.742 g of U_nO_m. Using the information in parts (a) and (b), calculate *x*, *y*, and *z*.

*3.128. **Fate of Fertilizer** Large quantities of fertilizer are washed into the Mississippi River from agricultural land in the Midwest. The excess nutrients collect in the Gulf of Mexico, promoting the growth of algae and endangering other aquatic life.

a. One commonly used fertilizer is ammonium nitrate. What is the chemical formula of ammonium nitrate?
b. Corn farmers typically use 5.0×10^3 kg of ammonium nitrate per square kilometer of cornfield per year. Ammonium nitrate can be prepared by the following reaction:

$$NH_3(aq) + HNO_3(aq) \rightarrow NH_4NO_3(aq)$$

How much nitric acid would be required to make the fertilizer needed for 1 km^2 of cornfield per year?

c. The ammonium ions can be converted into NO_3^- by bacterial action.

$$NH_4^+(aq) + 2\ O_2(g) \rightarrow NO_3^-(aq) + H_2O(\ell) + 2\ H^+(aq)$$

If 10% of the ammonium component of 5.0×10^2 kg of fertilizer ends up as nitrate, how much oxygen would be consumed?

3.129. Composition of Over-the-Counter Medicines Calculate the number of molecules or formula units of compound in each of the following common, over-the-counter medications:

a. ibuprofen, a pain reliever and fever reducer that contains 200.0 mg of the active ingredient, $C_{13}H_{18}O_2$
b. an antacid containing 500.0 mg of calcium carbonate
c. an allergy tablet containing 4 mg of chlorpheniramine ($C_{16}H_{19}ClN_2$)

3.130. **Chemistry of Pain Relievers** The common pain relievers aspirin ($C_9H_8O_4$), acetaminophen ($C_8H_9NO_2$), and naproxen sodium ($C_{14}H_{13}O_3Na$) are all available in tablets containing 200.0 mg of the active ingredient. Which compound contains the greatest number of molecules per tablet? How many molecules of the active ingredient are present in each tablet?

3.131. Some catalytic converters in automobiles contain the manganese oxides Mn_2O_3 and MnO_2.

a. What are the names of Mn_2O_3 and MnO_2?
b. Calculate the percent manganese by mass in Mn_2O_3 and MnO_2.
c. Explain how Mn_2O_3 and MnO_2 are consistent with the law of multiple proportions.

*3.132. Several chemical reactions have been proposed for the formation of organic compounds from inorganic precursors. Here is one of them:

$$H_2S(g) + FeS(s) + CO_2(g) \rightarrow FeS_2(s) + HCO_2H(\ell)$$

a. Identify the ions in FeS and FeS_2. Give correct names for each compound.
*b. How much HCO_2H is obtained by reacting 1.00 g of FeS, 0.50 g of H_2S, and 0.50 g of CO_2 if the reaction results in a 50.0% yield?

***3.133.** The formation of organic compounds by the reaction of iron(II) sulfide with carbonic acid is described by the following chemical equation:

$$2\ FeS(s) + H_2CO_3(aq) \rightarrow 2\ FeO(s) + 1/n\ (CH_2O)_n(s) + 2\ S(s)$$

a. How much FeO is produced starting with 1.50 g of FeS and 0.525 mol of H_2CO_3 if the reaction results in a 78.5% yield?

*b. If the carbon-containing product has a molar mass of 3.00×10^2 g/mol, what is the chemical formula of the product?

*3.134. **Marine Chemistry of Iron** On the seafloor, iron(II) oxide reacts with water to form Fe_3O_4 and hydrogen in a process called serpentization.

a. Write a balanced chemical equation describing serpentization using these reactants and products:

$$FeO(s) + H_2O(\ell) \rightarrow Fe_3O_4(s) + H_2(g)$$

b. When CO_2 is present, the product is methane, not hydrogen. Write a balanced chemical equation based on these reactants and products:

$$FeO(s) + H_2O(\ell) + CO_2(g) \rightarrow Fe_3O_4(s) + CH_4(g)$$

3.135. Composition of Solar Wind The solar wind is made up of ions, mostly protons, flowing out from the Sun at about 400 km/s. Near Earth, each cubic kilometer of interplanetary space contains, on average, 6×10^{15} solar-wind ions. How many moles of ions are in a cubic kilometer of near-Earth space?

3.136. The Hope Diamond at the Smithsonian National Museum of Natural History has a mass of 45.52 carats (Figure P3.136). Diamond is a crystalline form of carbon.

a. How many moles of carbon are in the Hope Diamond (1 carat = 200.0 mg)?

b. How many carbon atoms are in the diamond?

FIGURE P3.136

***3.137.** E-85 is an alternative fuel for automobiles and light trucks that consists of 85% (by volume) ethanol (CH_3CH_2OH) and 15% gasoline. The density of ethanol is 0.79 g/mL. How many moles of ethanol are in a gallon of E-85?

*3.138. A 100.00-g sample of white powder A is heated to 550°C. At that temperature the powder decomposes, giving off colorless gas B, which is denser than air and is neither flammable nor does it support combustion. The products also include 56 g of a second white powder C. When gas B is bubbled through a solution of calcium hydroxide, substance A reforms. What are the identities of substances A, B, and C?

***3.139.** You are given a 0.6240-g sample of a substance with the generic formula $MCl_2(H_2O)_2$. After complete drying of the sample (which means removing the 2 moles of H_2O per mole of MCl_2), the sample has a mass of 0.5471 g. What is the identity of element M?

3.140. A compound found in crude oil consists of 93.71% C and 6.29% H by mass. The molar mass of the compound is 128 g/mol. What is its molecular formula?

3.141. A reaction vessel for synthesizing ammonia by reacting nitrogen and hydrogen is charged with 6.04 kg of H_2 and excess N_2. A total of 28.0 kg of NH_3 is produced. What is the percent yield of the reaction?

3.142. If a cube of table sugar, which is made of sucrose ($C_{12}H_{22}O_{11}$) is added to concentrated sulfuric acid, the acid "dehydrates" the sugar, removing the hydrogen and oxygen from it and leaving behind a lump of carbon. What percentage of the initial mass of sugar is carbon?

***3.143.** A power plant burns 1.0×10^2 metric tons of coal that contains 3.0% (by mass) sulfur (1 metric ton = 10^3 kg). The sulfur is converted to SO_2 during combustion.

a. How many metric tons of SO_2 are produced?

b. When SO_2 escapes into the atmosphere, it may combine with O_2 and H_2O, forming sulfuric acid (H_2SO_4). Write a balanced chemical equation describing this reaction.

c. How many metric tons of sulfuric acid, a component of acid rain, could be produced from the quantity of SO_2 calculated in part (a)?

3.144. **Reducing SO_2 Emissions** With respect to the previous question, one way to reduce the formation of acid rain involves trapping the SO_2 by passing smokestack gases through a spray of calcium oxide and O_2. The product of this reaction is calcium sulfate.

a. Write a balanced chemical equation describing this reaction.

b. How many metric tons of calcium sulfate would be produced from each ton of SO_2 that is trapped?

3.145. In the early 20th century, Londoners suffered from severe air pollution caused by burning high-sulfur coal. The sulfur dioxide that was emitted into the air mixed with London fog, forming sulfuric acid. For every gram of sulfur that was burned, how many grams of sulfuric acid could have formed?

*3.146. **Gas Grill Reaction** The burner in a gas grill mixes 24 volumes of air for every one volume of propane (C_3H_8) fuel. Like all gases, the volume that propane occupies is directly proportional to the number of moles of it at a given temperature and pressure. Air is 21% (by volume) O_2. Is the flame produced by the burner fuel-rich (excess propane in the reaction mixture), fuel-lean (not enough propane), or stoichiometric (just right)?

3.147. If you had equal masses of the substances in the following pairs of compounds, which of the two would contain the greater number of ions? (a) NaBr or KCl; (b) NaCl or $MgCl_2$; (c) $BaCl_2$ or Li_2CO_3

*3.148. **Ozone Generators** Some indoor air-purification systems work by converting a little of the oxygen in the air to ozone, which oxidizes mold and mildew spores and other biological air pollutants. The chemical equation for the ozone generation reaction is

$$3\ O_2(g) \rightarrow 2\ O_3(g)$$

It is claimed that one such system generates 4.0 g of O_3 per hour from dry air passing through the purifier at a flow of 5.0 L/min. If 1 liter of indoor air contains 0.28 g of O_2,

a. what fraction of the molecules of O_2 is converted to O_3 by the air purifier?
b. what is the percent yield of the ozone generation reaction?

***3.149.** **Rebreathing Devices** In the first episode of George Lucas's *Star Wars* series, Qui-Gon Jinn and Obi-Wan Kenobi can visit the underwater world of the Gungans only by using A99 Aquata Breathers, which allow them to survive underwater for up to two hours. Although the tiny devices may be from the farfetched world of science fiction, current technology exists for transforming carbon dioxide to oxygen. These self-contained rebreathers are used by a select group of underwater cave explorers and can act as self-rescue devices. The chemistry is based on the following chemical reactions, using either potassium superoxide or sodium peroxide:

$$4\ KO_2(s) + 2\ CO_2(g) \rightarrow 2\ K_2CO_3(s) + 3\ O_2(g)$$

$$2\ Na_2O_2(s) + 2\ CO_2(g) \rightarrow 2\ Na_2CO_3(s) + O_2(g)$$

a. The respiratory rate at rest for an average, healthy adult is 12 breaths per minute. If the average breath takes in 0.500 L of O_2 ($d = 1.429$ g/L) into the lungs, how many grams of KO_2 are needed to produce enough oxygen for 2 hours underwater?
b. Would you need more or less Na_2O_2 to produce an equivalent amount of oxygen?
c. Given the densities of KO_2 ($d = 2.14$ g/mL) and Na_2O_2 ($d = 2.805$ g/mL), which solid material would occupy less volume in a rebreather device?

3.150. **Smelly Socks** Socks containing silver nanoparticles embedded in the fabric are currently marketed as an antidote to smelly socks. Silver is known to have antimicrobial properties, and silver ions are toxic to aquatic life. A study at Arizona State University found that much of the silver particles are lost upon laundering the socks in mild acid.

a. Each sock in the study began with 1360 μg of silver. How many moles of silver are contained in each sock?
b. As much as 650 μg of silver was lost after four washings. What percent of the silver was lost?

4

Reactions in Solution

Aqueous Chemistry in Nature

SURVIVING IN SEAWATER Marine mammals such as this humpback whale have the ability to regulate the concentrations of dissolved salts in their bodies to about one-third that of the seawater they live in.

PARTICULATE **REVIEW**

Ionic or Covalent in Solution?

In Chapter 4 we examine what happens to ionic and molecular compounds when they dissolve in water and the reactions that result from collisions of the dissolved particles. The beaker on the left contains 0.1 moles of ethylene glycol dissolved in water; the beaker on the right contains 0.1 moles of sodium sulfate dissolved in water. Answer the following questions about these two aqueous solutions.

- Is ethylene glycol a molecular compound or an ionic compound? Are its bonds covalent or ionic?
- Is sodium sulfate a molecular compound or an ionic compound? Are its bonds covalent or ionic?
- What features of the particles help you answer these questions?

(Review Figures 1.7, 2.18, and 2.19 if you need help.)

(Answers to Particulate Review questions are in the back of the book.)

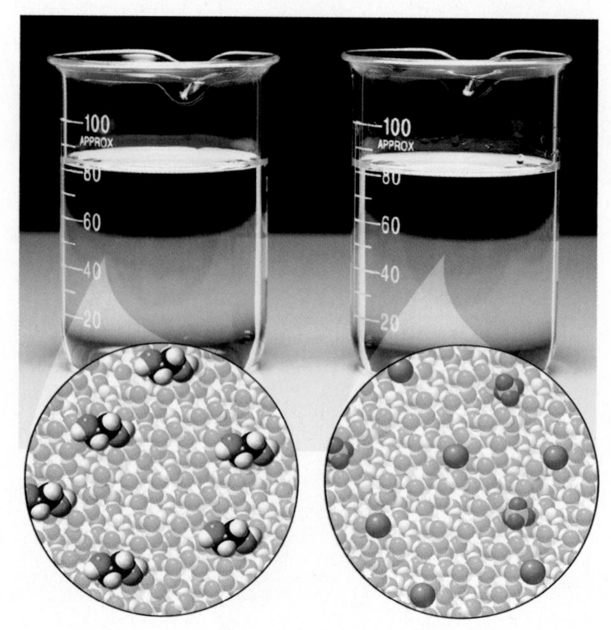

PARTICULATE **PREVIEW**

Products and Leftover Reactants

This picture shows 25 mL of 0.5 M NH_4Cl about to be poured into 75 mL of 0.1 M $AgNO_3$. As you read Chapter 4, look for ideas that will help you answer these questions:

- What are the possible products when these solutions mix? Will the ions remain dissolved in water or form a precipitate?
- Which is the limiting reactant? Will any nitrate ions remain in solution? Why or why not?
- Sketch a drawing of the particulate contents after pouring all the aqueous NH_4Cl solution into the aqueous $AgNO_3$ solution.

Learning Outcomes

LO1 Express the concentrations of solutions in different units, including molarity
Sample Exercises 4.1, 4.2, 4.3

LO2 Calculate the mass of a solute or the volume of a highly concentrated solution required to make a solution of specified volume and concentration
Sample Exercises 4.4, 4.5

LO3 Use Beer's law and absorbance data to determine the concentration of an unknown solution
Sample Exercise 4.6

LO4 Write molecular, overall ionic, and net ionic equations for reactions in aqueous solution
Sample Exercises 4.7, 4.10, 4.11, 4.12

LO5 Classify substances as soluble or only slightly soluble in aqueous solution and identify strong electrolytes, weak electrolytes, nonelectrolytes, and Brønsted–Lowry acids and bases based on their properties
Sample Exercise 4.8

LO6 Calculate the concentration of a solution from titration data
Sample Exercises 4.9, 4.10

LO7 Use solubility rules to predict the products of precipitation reactions and to quantify results from precipitation reactions
Sample Exercises 4.11, 4.12, 4.13, 4.14, 4.15

LO8 Determine oxidation numbers and use them to identify redox reactions, oxidizing agents, and reducing agents. Use the activity series to explain redox reactions between metals and metal ions
Sample Exercises 4.16, 4.17, 4.19, 4.20

LO9 Use half-reactions to write balanced equations for redox reactions
Sample Exercises 4.18, 4.21

4.1 Ions and Molecules in Oceans and Cells

Chapter 3 began with a description of how Earth formed about 4.6 billion years ago: as a hot, molten sphere that gradually cooled and formed a solid crust surrounded by an atmosphere of gases. Further cooling allowed one of the atmosphere's principal components, water vapor, to condense. The resultant torrential rain ran down from the crust's highlands and collected in its depressions, forming the first oceans. Life on Earth may have begun in these oceans. According to one theory, life originated deep beneath the surface of the oceans at hydrothermal vents, where hot, mineral-laden seawater emerging from Earth's crust mixed with cold seawater, forming mineral-rich deposits. Others have argued that only shallow ponds, not the marine environment, provided the appropriate concentrations of ions for cells to arise. Whether either theory is correct or not, water is essential for life as we know it on Earth. On early Earth, water provided the medium in which dissolved molecules and ions moved about, collided with one another, and participated in chemical reactions that led to the formation of more complex molecules, and eventually to assemblies of molecules capable of reproduction.

CONNECTION Recall from Chapter 1 that solutions are mixtures that are homogeneous: the substances making up the mixture are distributed uniformly.

Debate over the existence of life elsewhere in the universe hinges on the prospect of liquid water on other planets. The controversial claim that Martian meteorites collected in Antarctica contain fossilized life-forms (**Figure 4.1**) is consistent with evidence that water flowed on the surface of Mars in the distant past. Some Martian rocks and pebbles photographed by the *Curiosity* rover are round and smooth like streambed deposits on Earth, suggesting that they were formed by similar processes (**Figure 4.2**). Furthermore, analysis by *Curiosity* has revealed that some samples of Martian surface soil contain 2% water by weight. These and other observations support the idea that liquid water may now exist just under the surface of Mars, and where there is water, there could be life.

solvent the component of a solution that is present in the largest number of moles.

solute any component in a solution other than the solvent; a solution may contain one or more solutes.

(a)

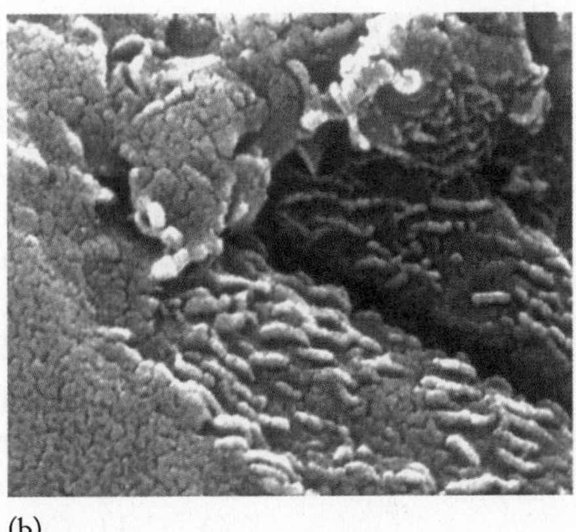

(b)

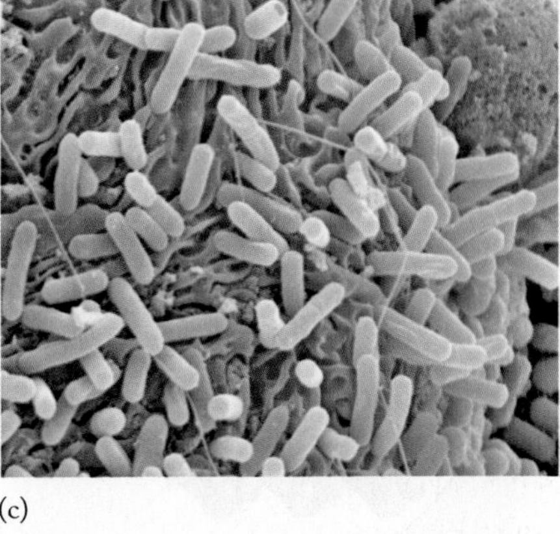

(c)

FIGURE 4.1 (a) Meteorite ALH84001, thought to have originated from Mars, contains (b) microscopic features that some scientists believe to be fossilized bacteria. (c) An Earth-grown colony of the bacterium *Escherichia coli*, for comparison.

Understanding the chemistry of the biosphere requires understanding the principles of chemical reactions between substances dissolved in water. When one element or compound dissolves in another, a *solution* forms—a homogeneous mixture of two or more substances (**Figure 4.3**). The substance present in the greatest number of moles is called the **solvent**, and the other substances in the solution are called **solutes**. When the solvent is water, the solution is an *aqueous solution*, and here in Chapter 4 you may assume all the solutions we discuss are aqueous. (It bears noting, however, that in the most general case, a solvent does not have to be a liquid. Many materials in Earth's crust, for example, are *solid solutions*, which are uniform mixtures of solid substances. Moreover, Earth's atmosphere can be thought of as a *gaseous solution*; gases are the subject of Chapter 5.)

CONCEPT TEST

Which, if any, of the following are solutions? (a) muddy river water; (b) helium gas; (c) clear cough syrup; (d) filtered dry air

(Answers to Concept Tests are in the back of the book.)

(a) Mars

(b) Earth

FIGURE 4.2 (a) Rocks on the Martian surface are similar in size and shape to (b) rocks on Earth in an area where a fast-moving stream of water once flowed.

FIGURE 4.3 Adding table sugar (the solute) to water (the solvent) produces a solution of sugar molecules evenly distributed among water molecules. The covalent bonds in the sugar molecules (shown here as space-filling models) do not break upon dissolving, and the particles of solute in the solution are neutral molecules.

Dissolved ionic and molecular compounds are present in all the waters on Earth's surface. The amounts of these dissolved compounds in a given volume range from very low (in the water from melting glaciers) to very high (in seawater). The liquids in the cells of all living things are also saline solutions, though not as salty as seawater. For example, the most abundant ionic solutes in our cells and in our blood plasma—Na^+, Cl^-, HCO_3^-, K^+, Ca^{2+}, HPO_4^{2-}, and Mg^{2+} ions—are even more abundant in seawater. The proportions of these ions present in the blood of healthy people are remarkably constant, as are the principal molecular solutes: CO_2, $C_6H_{12}O_6$ (glucose), and hemoglobin. Therefore, clinical assessments of a patient's well-being often include tests of whether these solutes are within their normal quantitative ranges. Departures from normal ranges may indicate problems, such as kidney disease or drug abuse.

Dissolved particles are involved in several major classes of chemical reactions that are essential to the proper function of all living cells. Molecules of CO_2, produced when our bodies metabolize food to produce energy, combine with molecules of H_2O, forming hydrogen ions and bicarbonate ions:

$$CO_2(aq) + H_2O(\ell) \rightarrow H^+(aq) + HCO_3^-(aq)$$

The H^+ ions produced in this reaction combine with additional molecules of water, forming **hydronium ions, H_3O^+**:

$$H^+(aq) + H_2O(\ell) \rightarrow H_3O^+(aq)$$

Reactions such as these that involve the transfer of H^+ ions are called *acid–base reactions*. Acid–base reactions are a major class of aqueous-phase reactions that we explore in this chapter.

A second class of reactions involves pairs of dissolved cations and anions that collide to form only slightly soluble ionic compounds that precipitate from (settle out of) solution. Such a *precipitation reaction* occurs when Ca^{2+} and HPO_4^{2-} ions combine with water molecules to form a mineral that has the common name hydroxyapatite, which is the principal component in tooth enamel:

$$10\ Ca^{2+}(aq) + 6\ HPO_4^{2-}(aq) + 2\ H_2O(\ell) \rightarrow Ca_{10}(PO_4)_6(OH)_2(s) + 8\ H^+(aq)$$

In this chapter we examine trends in the solubility of ionic compounds that enable you to predict whether precipitates form when solutions of different ionic compounds are mixed.

Much of a cell's energy comes from a process we discussed in Chapter 3, in which glucose reacts with O_2 (conveyed in the blood by hemoglobin), forming carbon dioxide and water:

$$C_6H_{12}O_6(aq) + 6\ O_2(g) \rightarrow 6\ CO_2(g) + 6\ H_2O(\ell)$$

This reaction is an example of a third important class of reactions, called *redox* reactions, in which elements (oxygen in this case) are *red*uced, which means that their atoms gain electrons, while others (the carbon in glucose) are *ox*idized, which means that their atoms lose electrons. We explore redox reactions in Section 4.8.

4.2 Expressing Concentrations

In Chapter 3 we used the ideas of the mole and the stoichiometric relationship between reactants and products to deal with macroscopic quantities of materials involved in chemical reactions. In this section we develop additional tools that enable us to apply the concepts of stoichiometry to reactions taking place in solutions. **Figure 4.4** shows a common acid–base reaction: an antacid tablet containing sodium bicarbonate is added to stomach acid, releasing bubbles of $CO_2(g)$ and neutralizing some of the $HCl(aq)$ in gastric fluid. How does a manufacturer of antacid tablets determine the quantity of sodium bicarbonate needed to neutralize a portion of stomach acid? The answer lies in part in knowing both the balanced equation that describes the reaction and the *concentration* of the solution.

hydronium ion (H_3O^+) a H^+ ion bonded to a molecule of water, H_2O; the chemical form of a hydrogen ion in an aqueous solution.

concentration the amount of a solute in a given mass or volume of solvent or solution.

Concentration Units

The **concentration** of any solution, or the amount of solute in a given amount of solvent or solution, can be expressed in a variety of ways. Qualitatively, solutions are described as *concentrated* or *dilute* if they contain relatively high or low ratios of solute to solvent.

To be more specific, we need to be able to quantify the concentration of a solution. Some quantitative measures of concentration are based on *mass-to-volume ratios*, such as milligrams of solute per liter of solution. When clinical laboratories report the concentration of substances in blood plasma, a common unit is milligrams per deciliter (dL; 1 dL = 0.10 L = 100 mL) of plasma (solution). For example, the reference concentration for total cholesterol is <200 mg/dL, which is the same as <2000 mg/L:

$$\frac{200\text{ mg}}{\cancel{\text{dL}}} \times \frac{1.0\ \cancel{\text{dL}}}{100\ \cancel{\text{mL}}} \times \frac{1000\ \cancel{\text{mL}}}{1\text{ L}} = \frac{2000\text{ mg}}{\text{L}}$$

Most scientists interested in studying chemical processes in natural waters or laboratory solutions prefer to work with concentrations based on moles of solute.

$$NaHCO_3(s) + H_3O^+(aq) + Cl^-(aq) \rightarrow Na^+(aq) + Cl^-(aq) + 2\,H_2O(\ell) + CO_2(g)$$

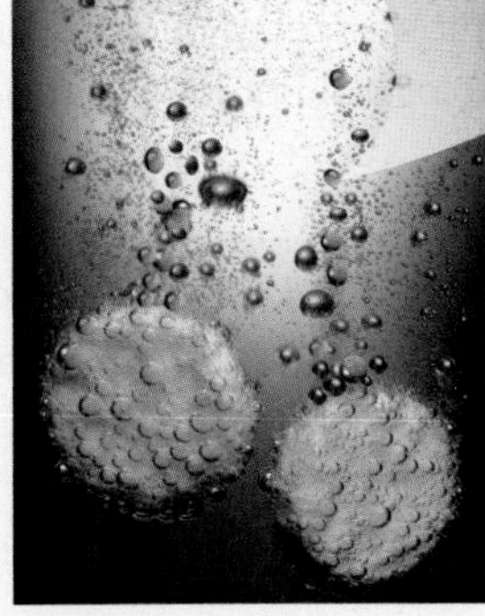

FIGURE 4.4 The reaction between a base (sodium bicarbonate) and stomach acid generates bubbles of carbon dioxide and provides relief for an upset stomach. (Solvent water molecules have been omitted in the particle representation for clarity.)

molarity (M) the concentration of a solution expressed in moles of solute per liter of solution ($M = n/V$).

CHEMTOUR
Molarity

For these scientists, the preferred concentration unit is moles of solute per liter of solution. This ratio is called **molarity (M)**:[1]

$$\text{molarity } (M) = \frac{\text{moles } (n) \text{ of solute}}{\text{volume } (V) \text{ of solution in liters}}$$

or

$$M = \frac{n}{V} \tag{4.1}$$

Molarity

If we know the volume and molarity for any solution, we can readily calculate the mass of solute in the solution. First, we rearrange Equation 4.1:

$$n = V \times M \tag{4.2}$$

Remember from Chapter 3 that the moles (n) of a substance can also be calculated from its mass (m) and molar mass ($\mathcal{M}$):

$$\text{moles} = \frac{\text{mass (g)}}{\text{molar mass (g/mol)}}$$

Setting these two expressions for calculating the number of moles of solute equal to each other:

$$\frac{\text{mass (g)}}{\text{molar mass (g/mol)}} = V \times M$$

Solving for the mass of solute:

$$m_{\text{solute}} = (V \times M)\, \mathcal{M} \tag{4.3}$$

To check that Equation 4.3 is correct, note how the units cancel out:

$$\text{g} = \left(\cancel{\text{L}} \times \frac{\cancel{\text{mol}}}{\cancel{\text{L}}}\right) \times \frac{\text{g}}{\cancel{\text{mol}}}$$

Equation 4.3 is useful when we want to calculate the mass of a solute needed to prepare a solution of a desired volume and molarity or when we have a solution of known molarity and want to know the mass of solute in a given volume of the solution.

In many environmental and biological systems, solute concentrations are often much less than 1.0 M. **Table 4.1** lists the major ions in seawater and in human serum and their average concentrations, using three common units. For instance, the concentration units in column 4 are not moles per liter but *millimoles* per liter, which means we are describing concentration in terms of *millimolarity* (mM; 1 mM = 10^{-3} M) rather than molarity. On an even smaller scale, the concentrations of minor and trace elements in seawater are often expressed in terms of *micromolarity* (μM; 1 μM = 10^{-6} M), *nanomolarity* (nM; 1 nM = 10^{-9} M), and even *picomolarity* (pM; 1 pM = 10^{-12} M). The concentration ranges of many biologically active substances in blood, urine, and other biological liquids are also so small that they are often expressed in units such as these.

CONCEPT TEST

Which of the following aqueous sodium chloride solutions is the least concentrated? (a) 0.0053 M NaCl; (b) 54 mM NaCl; (c) 550 μM NaCl; (d) 56,000 nM NaCl

(Answers to Concept Tests are in the back of the book.)

[1]Note that molarity is symbolized by M, whereas molar mass is $\mathcal{M}$ throughout this text.

TABLE 4.1 Average Concentrations of 11 Major Constituents of Seawater and Their Normal Ranges in Human Serum

	SEAWATER			HUMAN SERUM
Constituent	g/kg[a]	mmol/kg[b]	mmol/L = m*M*[c]	mmol/L = m*M*[c]
Na^+	10.781	468.96	480.57	135–145
K^+	0.399	10.21	10.46	3.5–5.0
Mg^{2+}	1.284	52.83	54.14	0.08–0.12
Ca^{2+}	0.4119	10.28	10.53	0.2–0.3
Sr^{2+}	0.00794	0.0906	0.0928	$<3 \times 10^{-4}$
Cl^-	19.353	545.88	559.40	98–108
SO_4^{2-}	2.712	28.23	28.93	0.3
HCO_3^-	0.126	2.06	2.11	22–30
Br^-	0.0673	0.844	0.865	0.04–0.06
$B(OH)_3$	0.0257	0.416	0.426	$<8 \times 10^{-4}$
F^-	0.00130	0.068	0.070	5–6

[a]g/kg = grams of solute per kilogram of solution.
[b]mmol/kg = millimoles of solute per kilogram of solution. (Oceanographers prefer this unit to one based on solution volume because the volume of a given mass of water varies with changing temperature and pressure, whereas its mass remains constant.)
[c]mmol/L = m*M* = millimoles of solute per liter of solution.

Some concentration units are not mass-to-volume ratios like molarity but rather are *mass-to-mass ratios.* Two such mass-to-mass concentration units are *parts per million* (ppm) and *parts per billion* (ppb). These units are convenient for expressing the very small concentrations frequently encountered in environmental samples: the Safe Drinking Water Act passed by the U.S. Congress in 1974 uses these units to identify safe maximum levels for contaminants in our drinking water, from harmful substances such as lead to helpful substances such as fluoride, which prevents dental cavities. A solution with a concentration of 1 ppm contains 1 part solute for every million parts of solution, which means 1 g of solute for every million (10^6) grams of solution. Put another way, each gram of a 1 ppm solution contains one-millionth of a gram (10^{-6} g = 1 μg) of solute. A 1 ppm solution is also the same as 1 mg of solute per kilogram of solution:

$$1\text{ ppm} = \frac{1\ \cancel{\mu\text{g solute}}}{1\ \cancel{\text{g solution}}} \times \frac{1\text{ mg solute}}{10^3\ \cancel{\mu\text{g solute}}} \times \frac{10^3\ \cancel{\text{g solution}}}{1\text{ kg solution}} = \frac{1\text{ mg solute}}{1\text{ kg solution}}$$

Let's express the smallest seawater concentration in Table 4.1 (0.00130 g F^-/kg seawater) in parts per million. To do so, we convert grams of F^- into milligrams of F^- because 1 mg of solute per kilogram of solution is the same as 1 ppm:

$$\frac{0.00130\ \cancel{\text{g F}^-}}{1\text{ kg seawater}} \times \frac{10^3\text{ mg F}^-}{1\ \cancel{\text{g F}^-}} = \frac{1.30\text{ mg F}^-}{1\text{ kg seawater}} = 1.30\text{ ppm F}^-$$

Thus, parts per million is a particularly convenient unit for expressing the concentration of F^- ions in seawater because it avoids the use of exponents or lots of zeroes to set the decimal place.

Some substances are so harmful that a concentration of less than 1 ppm can still be cause for concern. For example, the maximum concentration of copper ions in drinking water in the United States is 1.3 mg/L or 1.3 ppm (assuming the density of drinking water is 1.00 kg/L). However, the maximum concentration of cadmium ions in drinking water is 0.005 mg/L. For these substances, ppb is a more appropriate unit. A concentration of 1 ppb is 1/1000 as concentrated as 1 ppm. Because 1 ppm is the same as 1 mg of solute per kilogram of solution, 1 ppb is equivalent to 1 μg of solute per kilogram of solution:

$$1\ \text{ppb} = \frac{1\ \text{ppm}}{1000} = \frac{0.001\ \mu\text{g solute}}{1\ \cancel{\text{g solution}}} \times \frac{10^3\ \cancel{\text{g solution}}}{1\ \text{kg solution}} = \frac{1\ \mu\text{g solute}}{1\ \text{kg solution}}$$

Thus, a 0.005 mg/L solution of Cd^{2+} is equivalent to 5 ppb.

SAMPLE EXERCISE 4.1 Comparing Ion Concentrations in Aqueous Solutions **LO1**

The average concentration of chloride ion in seawater is 19.353 g/kg solution. The World Health Organization (WHO) recommends that the concentration of Cl^- ions in drinking water not exceed 250 ppm (2.50×10^2 ppm). How many times greater is the Cl^- ion concentration in seawater than the WHO limit for Cl^- ions in drinking water?

Collect, Organize, and Analyze We are asked to compare concentrations that are expressed in two sets of units: g/kg and ppm. To facilitate the comparison, let's convert the drinking water standard from ppm (or mg/kg) into g/kg using a conversion factor based on the equality 1 g = 1000 mg.

Solve

The maximum chloride ion concentration in drinking water is

$$250\ \text{ppm} = 250\ \frac{\text{mg}}{\text{kg}} \times \frac{1\ \text{g}}{1000\ \cancel{\text{mg}}} = 0.250\ \frac{\text{g}}{\text{kg}}$$

Dividing this drinking water maximum into the concentration of Cl^- ions in seawater:

$$\frac{19.353\ \frac{\text{g}}{\text{kg}}}{0.250\ \frac{\text{g}}{\text{kg}}} = 77.4$$

Therefore, the concentration of Cl^- ions in seawater is 77.4 times greater than the maximum acceptable concentration of these ions in drinking water.

CONNECTION In Chapter 1 we discussed distilling seawater to separate the sea salts from the water, thereby generating fresh drinking water.

Think About It Common sense says that seawater is much saltier than drinking water. Drinking seawater induces nausea and vomiting, and it may even cause death. We could also have converted the seawater concentration into ppm and compared that value to the drinking water standard. Doing so:

$$19.353\ \frac{\text{g}}{\text{kg}} \times \frac{1000\ \text{mg}}{1\ \text{g}} = 19{,}353\ \frac{\text{mg}}{\text{kg}} = 19{,}353\ \text{ppm} \quad \text{and} \quad \frac{19{,}353\ \text{ppm}}{250\ \text{ppm}} = 77.4$$

gives the same concentration ratio, confirming the results of the first calculation.

Practice Exercise The WHO drinking water standard for arsenic is <10.0 μg/L. Arsenic concentrations in water from some wells in Bangladesh are as high as 1.2 mg/L. How many times above the WHO standard is this level?

(Answers to Practice Exercises are in the back of the book.)

SAMPLE EXERCISE 4.2 Calculating Molarity from Mass and Volume **LO1**

Pipes made of polyvinyl chloride (**Figure 4.5**) are widely used in homes and office buildings, though their use is restricted because vinyl chloride (CH_2CHCl, $\mathcal{M}$ = 62.49 g/mol) may leach from them. The maximum concentration of vinyl chloride allowed in drinking water in the United States is 0.002 mg/L. What molarity is this?

Collect and Organize We are given a concentration in milligrams of solute per liter of solution, and we are asked to convert it to units of molarity (moles of solute per liter of solution). The molar mass (g/mol) of a substance relates the mass and number of moles of a given quantity of the substance.

Analyze The solute mass is given in milligrams, so we need to convert this mass first to grams and then to moles to calculate molarity. The mass/volume concentration value is small; the molarity value should be smaller still.

$$\boxed{\frac{\text{mg VC}}{\text{L}}} \xrightarrow{\frac{\text{g}}{10^3\text{ mg}}} \boxed{\frac{\text{g VC}}{\text{L}}} \xrightarrow{\frac{1}{\text{molar mass}}} \boxed{\frac{\text{mol VC}}{\text{L}}}$$

Solve The molarity of vinyl chloride in a 0.002 mg/L solution is

$$\frac{0.002\ \cancel{\text{mg}}}{\text{L}} \times \frac{1\ \cancel{\text{g}}}{10^3\ \cancel{\text{mg}}} \times \frac{1\text{ mol}}{62.49\ \cancel{\text{g}}} = \frac{3.2 \times 10^{-8}\text{ mol}}{\text{L}} = 3 \times 10^{-8}\ M\text{, or } 0.03\ \mu M$$

Think About It As predicted, the molar concentration equivalent to the drinking water standard is very low, reflecting the danger that exposure to vinyl chloride poses to human health.

Practice Exercise When 1.00 L of water from the surface of the Dead Sea is evaporated, 179 g of $MgCl_2$ is recovered. What was the molarity of $MgCl_2$ in the original sample?

(Answers to Practice Exercises are in the back of the book.)

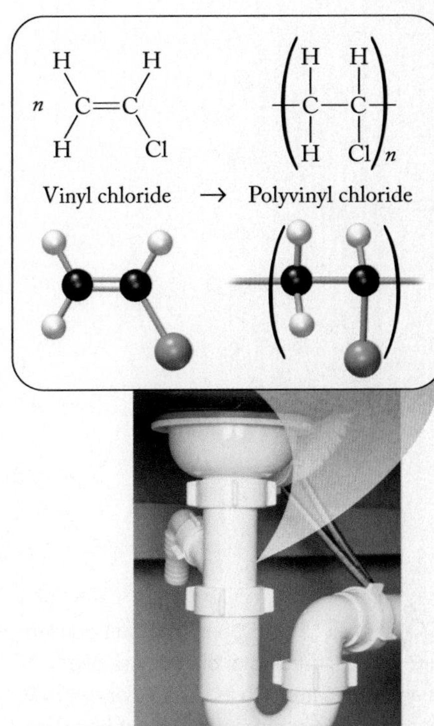

FIGURE 4.5 Vinyl chloride, a suspected carcinogen, leaches from pipes made of polyvinyl chloride (PVC). Vinyl chloride is the common name of a small molecule, called a monomer (Greek for "one unit"), from which the very large molecule called a polymer ("many units") is formed. The PVC polymer usually consists of hundreds of molecules of vinyl chloride bonded together.

SAMPLE EXERCISE 4.3 Calculating Molarity from Density — LO1

The concentration of Na^+ ions in a sample of water from the Great Salt Lake in Utah is 83.6 mg/g. What is the molarity of Na^+ ions if the density of the water is 1.160 g/mL solution?

Collect and Organize We are asked to convert a concentration value expressed in mg/g to one expressed in mol/L. We are given the density of the sample, and the molar mass of Na is 22.99 g/mol.

Analyze We need to convert mg Na^+ ions to an equivalent number of moles:

$$\boxed{\text{mg Na}^+} \xrightarrow{\frac{\text{g}}{10^3\text{ mg}}} \boxed{\text{g Na}^+} \xrightarrow{\frac{1}{\text{molar mass}}} \boxed{\text{mol Na}^+}$$

and use the density of the sample to convert milligrams of it to liters:

$$\boxed{\text{g sample}} \xrightarrow{\frac{1}{\text{density}}} \boxed{\text{mL sample}} \xrightarrow{\frac{\text{L}}{10^3\text{ mL}}} \boxed{\text{L sample}}$$

Dividing the results of the first calculation by the results of the second will give us the moles of Na^+ ions in one liter of sample, that is, molarity.

Solve Calculating moles of Na^+ ions in 1 g of lake water:

$$83.6\ \cancel{\text{mg Na}^+} \times \frac{1\text{ g}}{10^3\ \cancel{\text{mg}}} \times \frac{1\text{ mol Na}^+}{22.99\ \cancel{\text{g Na}^+}} = 3.64 \times 10^{-3}\text{ mol Na}^+$$

The volume of exactly 1 g of lake water is

$$1.000\ \text{g} \times \frac{1\ \text{mL}}{1.160\ \text{g}} \times \frac{1\ \text{L}}{10^3\ \text{mL}} = 8.621 \times 10^{-4}\ \text{L}$$

The molarity of Na^+ ions in the sample is

$$\frac{3.64 \times 10^{-3}\ \text{mol Na}^+}{8.621 \times 10^{-4}\ \text{L}} = 4.22\ M$$

Think About It Let's compare the initial (83.3 mg/g = 83.3 g/kg) and calculated (4.22 *M*) concentrations of Na^+ ions in water from the Great Salt Lake with Na^+ ion concentrations in seawater in Table 4.1, column 2 (10.781 g/kg) and column 4 (480.57 m*M* = 0.48057 *M*). Both the lake water samples are much higher than the comparable seawater values, though the ratio is greater for the molarity concentrations, 4.22 *M*/0.48057 *M* = 8.78, than the mass-based concentrations, (83.3 g/kg)/(10.781 g/kg) = 7.73. This difference makes sense because the density of the lake water sample (1.160 g/mL) is over 13% higher than the average density of seawater (1.025 g/mL). Therefore, concentrations based on sample volume, such as one liter, will be proportionately greater than those based on sample mass, such as one kilogram. Compare the values in columns 3 and 4 of Table 4.1 to gain another perspective on this difference.

CONNECTION Density, first introduced in Chapter 1, is the ratio of the mass of a quantity of material to its volume, or $d = m/V$. Density is an intensive physical property, so its value is independent of the quantity of material present.

Practice Exercise If the density of ocean water at a depth of 10,000 m is 1.071 g/mL and if 25.0 g of water at that depth contains 190 mg of potassium chloride, what is the molarity of potassium chloride in the water?

(Answers to Practice Exercises are in the back of the book.)

SAMPLE EXERCISE 4.4 Calculating the Quantity of Solute Needed to Prepare a Solution of Known Concentration **LO2**

An aqueous solution called *Kalkwasser* [German for "limewater"; 0.0225 *M* $Ca(OH)_2$] may be added to saltwater aquaria to reduce acidity and add Ca^{2+} ions. How many grams of $Ca(OH)_2$ do you need to make 500.0 mL of Kalkwasser?

Collect, Organize, and Analyze We know the volume and molarity of the solution we must prepare. We also know the identity of the solute, which enables us to calculate its molar mass. We can use Equation 4.3 to calculate the mass of $Ca(OH)_2$ needed. We also must convert the volume from milliliters to liters to obtain the mass, in grams, of solute required.

$$\boxed{\text{mL solution}} \xrightarrow{\frac{\text{L}}{10^3\ \text{mL}}} \boxed{\text{L solution}} \xrightarrow{\frac{\text{mol solute}}{\text{L solution}}} \boxed{\text{mol solute}} \xrightarrow{\text{molar mass}} \boxed{\text{g solute}}$$

The molar mass of $Ca(OH)_2$ is

$$40.08\ \text{g/mol} + 2(16.00\ \text{g/mol}) + 2(1.008\ \text{g/mol}) = 74.10\ \text{g/mol}$$

Solve Equation 4.3 can be used to calculate the mass of $Ca(OH)_2$ needed:

$$m_{\text{solute}} = (V \times M)\,\mathcal{M}$$

$$= 500.0\ \text{mL} \times \frac{1\ \text{L}}{10^3\ \text{mL}} \times \frac{0.0225\ \text{mol}}{1\ \text{L}} \times \frac{74.10\ \text{g}}{1\ \text{mol}} = 0.834\ \text{g}$$

Figure 4.6 shows the technique for making this solution. A key feature in the preparation is that we suspend 0.834 g of solute in 200–300 mL of water and then add additional water until the final solution has the volume specified.

(a)

(b)

(c)

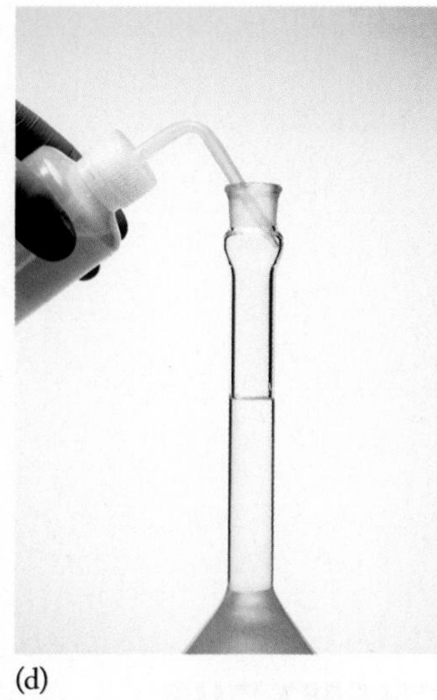
(d)

FIGURE 4.6 Preparing 500.0 mL of 0.0225 M $Ca(OH)_2$. (a) Weigh out the desired quantity of solid $Ca(OH)_2$; (b) transfer the $Ca(OH)_2$ to a 500 mL volumetric flask and add 200 to 300 mL of water; (c) swirl to suspend the solute; (d) dilute to exactly 500 mL. Stopper the flask and invert it several times to thoroughly mix the solution.

Think About It To complete an exercise like this one requires that we remember the definition of molarity: moles of solute per liter of solution. The solution we want has a concentration of 0.0225 M, or 0.0225 mol/L. Because we need only 0.5000 L of this solution, we do not need 0.0225 mol of $Ca(OH)_2$, but only half that amount (0.0112 mol). Finally, we calculate the number of grams of calcium hydroxide in 0.0112 mol of calcium hydroxide (0.834 g). The amount of $Ca(OH)_2$ needed is less than a teaspoon.

Practice Exercise An aqueous solution known as Ringer's lactate is administered intravenously to trauma victims suffering from blood loss or severe burns. The solution contains the chloride salts of sodium, potassium, and calcium and is 4.00 mM in sodium lactate ($NaC_3H_5O_3$; **Figure 4.7**). How many grams of sodium lactate are needed to prepare 10.0 L of Ringer's lactate?

(Answers to Practice Exercises are in the back of the book.)

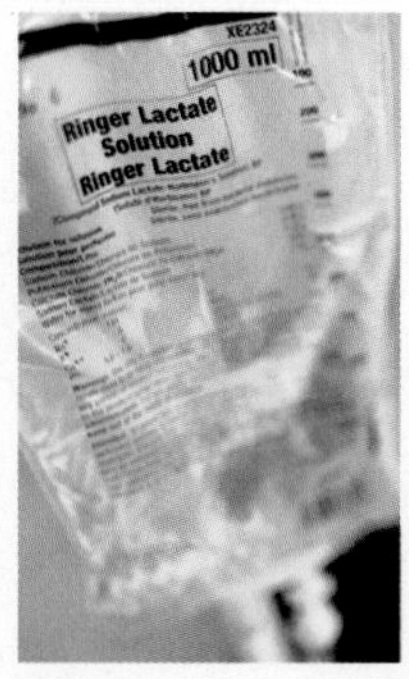

(a)

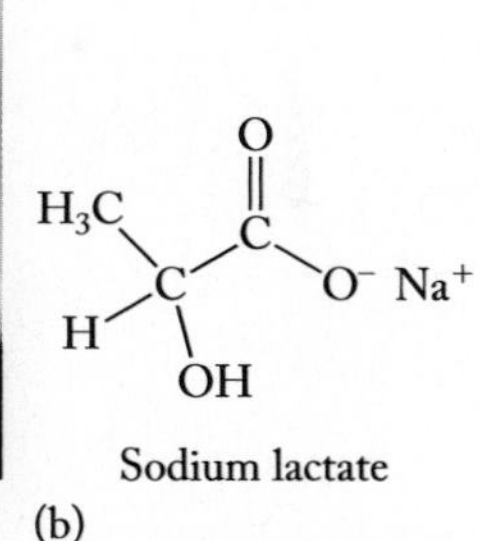

(b)

FIGURE 4.7 (a) 1000 mL of Ringer's lactate solution. (b) Structure of sodium lactate, $NaC_3H_5O_3$.

4.3 Dilutions

In laboratories such as those that test drinking water quality, scientists purchase **stock solutions** of substances such as pesticides and toxic metals from chemical manufacturers. To make their distribution more efficient, these solutions are packaged and sold in high concentrations—too high to be used directly in analytical procedures. So, to prepare a solution with a solute concentration that is near the substance's expected concentration in the sample of drinking water being tested, the laboratory will *dilute* the stock solution. **Dilution** is the process of lowering the concentration of a solution by adding solvent to a known volume of the initial solution. The stock solution is concentrated—it contains a much larger solute-to-solvent ratio than does the dilute solution.

Suppose we need to prepare 250.0 mL of an aqueous solution that is 0.0100 M Cu^{2+}, starting with a stock solution that is 0.1000 M Cu^{2+}. We can use Equation 4.2 to calculate the number of moles of Cu^{2+} ions in the diluted solution:

$$n_{\text{diluted}} = V_{\text{diluted}} \times M_{\text{diluted}}$$
$$= (0.2500\ \cancel{\text{L}})(0.0100\ \text{mol}/\cancel{\text{L}}) = 2.50 \times 10^{-3}\ \text{mol Cu}^{2+}$$

stock solution a concentrated solution of a substance used to prepare solutions of lower concentration.

dilution the process of lowering the concentration of a solution by adding more solvent.

The subscript "diluted" refers to the solution we are preparing, and V refers to the final volume of the solution in liters. Using the same equation, we can calculate the *initial* quantities, that is, the volume of the stock solution we need to deliver 2.50×10^{-3} of Cu^{2+}:

$$n_{\text{initial}} = V_{\text{initial}} \times M_{\text{initial}}$$

$$V_{\text{initial}} = \frac{n_{\text{initial}}}{M_{\text{initial}}} = \frac{2.50 \times 10^{-3}\ \cancel{\text{mol}}}{0.1000\ \cancel{\text{mol}}/\text{L}} = 2.50 \times 10^{-2}\ \text{L} = 25.0\ \text{mL}$$

To make the diluted solution, we add water to 25.0 mL of the stock solution to make a final volume of 250.0 mL (**Figure 4.8**).

Because the number of moles of solute (Cu^{2+} ions) is the same in the diluted solution and in the portion of stock solution required ($n_{\text{diluted}} = n_{\text{initial}}$), we can combine these two equations into one that applies to all situations in which we know any three of these four variables: the initial volume (V_{initial}) of stock solution,

CHEMTOUR
Dilutions

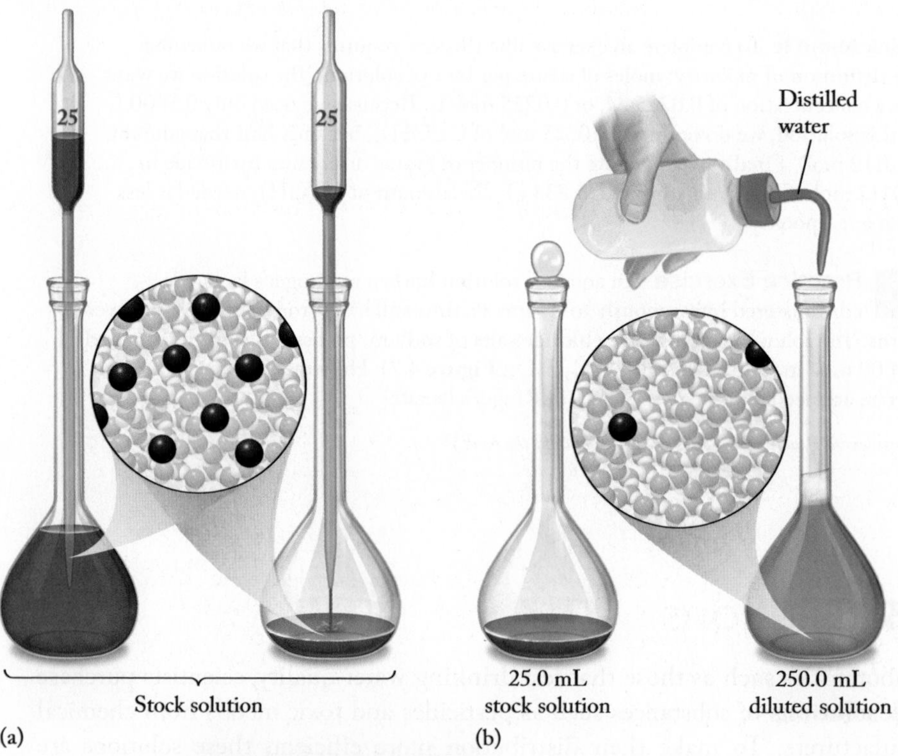

Volume (V_{initial}) of stock solution

↓ Multiply by molarity of stock solution (M_{initial})

Moles of solute in stock solution = **Moles of solute in diluted solution**

↓ Divide by total volume of diluted solution (V_{diluted})

Molarity of diluted solution (M_{diluted})

FIGURE 4.8 To prepare 250.0 mL of a solution that is 0.0100 M in Cu^{2+}, (a) a pipet is used to withdraw 25.0 mL of a 0.1000 M stock solution. This volume of stock solution is transferred to a 250.0 mL volumetric flask. (b) Distilled water is added to bring the volume of the transferred stock solution to 250.0 mL, and the resulting diluted solution is thoroughly mixed. Note that the color of the diluted solution is lighter than that of the stock solution.

the initial solute concentration ($M_{initial}$), the volume ($V_{diluted}$) of the diluted solution, and the solute concentration ($M_{diluted}$) in the diluted solution:

$$V_{initial} \times M_{initial} = n_{initial} = n_{diluted} = V_{diluted} \times M_{diluted}$$

or simply

$$V_{initial} \times M_{initial} = V_{diluted} \times M_{diluted} \qquad (4.4)$$

Equation 4.4 works because each side of the equation represents a quantity of solute that does not change because of dilution. Furthermore, this equation can be used for *any* units of volume and concentration if the units used to express the initial and diluted volumes are the same and the units for the initial and diluted concentrations are the same.

Equation 4.4 applies to all systems, even those much larger than a laboratory stock solution to be diluted. In estuaries, for instance, where rivers flow into the sea, seawater is diluted by the freshwater that enters it. Suppose a coastal bay contains 3.8×10^{12} L of ocean water mixed with 1.2×10^{12} L of river water for a total volume of 5.0×10^{12} L. We can calculate the Na^+ concentration in the bay if the Na^+ concentration in the open ocean is 0.48 *M* and if we assume that the river water does not contribute significantly to the Na^+ content in the bay. Using these values to solve Equation 4.4 for $M_{diluted}$, the Na^+ concentration in the bay, we have

$$M_{diluted} = \frac{V_{initial} \times M_{initial}}{V_{diluted}} = \frac{(3.8 \times 10^{12}\ \cancel{L}) \times 0.48\ M}{5.0 \times 10^{12}\ \cancel{L}} = 0.36\ M$$

The bay is less salty than the ocean because it is diluted by the freshwater from the river.

SAMPLE EXERCISE 4.5 Preparing a Dilution **LO2**

The solution used in hospitals for intravenous infusion—called *physiological saline* or *saline solution*—is 0.155 *M* in NaCl. It is typically prepared by diluting a stock solution, the concentration of which is 1.76 *M*, with water. What volume of stock solution and what volume of water are required to prepare 2.00 L of physiological saline?

Collect, Organize, and Analyze We know the volume ($V_{diluted} = 2.00$ L) and concentration ($M_{diluted} = 0.155\ M$) of the diluted solution and the concentration of stock solution available ($M_{initial} = 1.76\ M$). That is, we know three of the four variables in Equation 4.4. We must find $V_{initial}$, the volume of the stock solution required to make 2.00 L of physiological saline.

Solve Solving Equation 4.4 for $V_{initial}$ and using the values given:

$$V_{initial} = \frac{V_{diluted} \times M_{diluted}}{M_{initial}} = \frac{2.00\ L \times 0.155\ M}{1.76\ M} = 0.176\ L$$

To make 2.00 L of the diluted solution, we need to add: 2.00 L − 0.176 L = 1.82 L of water. If we prepare the dilute solution in a volumetric flask like that in Figure 4.8, rather than add 1.82 L of water, we would add enough water to fill the flask to the mark.

Think About It Because the concentration of the stock solution is about ten times the concentration of the diluted solution, the volume of stock solution required should be about one-tenth the final volume. Our result of 0.176 L of stock solution is reasonable.

Practice Exercise The concentration of Pb^{2+} in a stock solution is 1.000 mg/mL. What volume of this solution should be diluted to 500.0 mL to produce a solution in which the Pb^{2+} concentration is 25.0 mg/L?

(Answers to Practice Exercises are in the back of the book.)

(a)

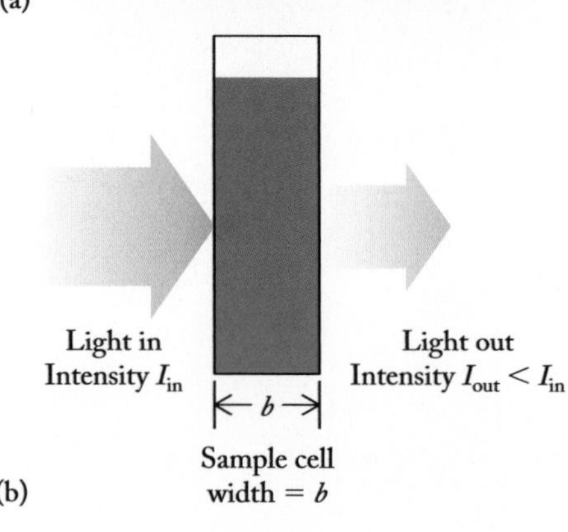

(b)

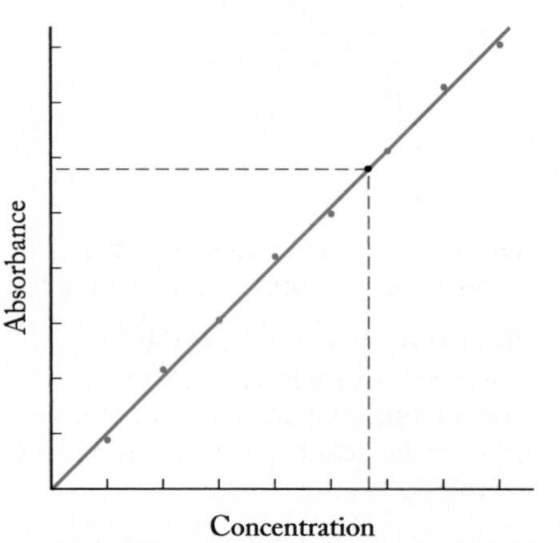

(c)

FIGURE 4.9 (a) A spectrophotometer, used to measure the absorbance of solutions. (b) The intensity of light leaving the solution (I_{out}) is less than the intensity of light entering the solution (I_{in}). (c) The red line represents a calibration curve for Beer's law, plotting absorbance versus concentration for a series of standard samples. The horizontal dashed blue line represents the absorbance of a solution of unknown concentration. By extending the line to where it meets the calibration curve and following the vertical dashed blue line down to the x-axis, we can determine the concentration of the unknown solution.

CONCEPT **TEST**

Why is a stock solution always more concentrated than the solutions made from it?

(Answers to Concept Tests are in the back of the book.)

Determining Concentration

The intensity of the blue color of the Cu^{2+} solution in Figure 4.8 decreases as we dilute the solution. Is there a way of using the intensity of the color to quantify the concentration? The color of the Cu^{2+} solution arises from the absorption of visible light, a topic we explore in more detail in Chapters 7 and 22. For now, it is sufficient to know that the **absorbance (A)** of a solution is a measure of the intensity of the color. **Beer's law** (Equation 4.5) relates absorbance to three quantities: concentration (c), the path length that the light travels through the solution (b), and a constant (called the **molar absorptivity**, or **ε**) characteristic of the dissolved solute.

$$A = \varepsilon bc \tag{4.5}$$

Absorbance is measured with a spectrophotometer like that shown in **Figure 4.9a**, in which a solution is placed in a sample cell of fixed length b, typically 1.00 cm. The cell is placed in a beam of light (**Figure 4.9b**). The value of A represents how much light is absorbed by the solution. If we express the concentration of the solution, c, in molarity (M), we can calculate ε. A graph of A versus c for solutions of different Cu^{2+} concentration yields a straight line because the amount of light absorbed by a sample is *directly proportional* to the concentration of that sample. Such a graph (**Figure 4.9c**) is called a **calibration curve**. Once a calibration curve is generated, the concentration of any solution of Cu^{2+} can be determined by measuring its absorbance, finding that value on the y-axis, reading across to where it intersects the calibration curve, and then reading the value on the x-axis below, which is the concentration. We can also use a calibration curve to determine molar absorptivity. If $b = 1.00$ cm, then the slope of the graph is equal to ε in units of M^{-1} cm^{-1}. This method of analysis is called *spectrophotometry*. Sample Exercise 4.6 shows how we can apply Beer's law to find the concentration of a solution of iron ions.

SAMPLE EXERCISE 4.6 Applying Beer's Law **LO3**

One way to determine the concentration of Fe^{2+} in an environmental sample is to add a solution of phenanthroline (phen) molecules that will bond to the metal ions, creating an intensely orange-red solution (**Figure 4.10**). The absorbance of a 5.00×10^{-5} M solution is measured as 0.55 in a 1.00 cm cell. (a) Calculate the molar absorptivity, ε, and (b) determine the concentration of such a solution whose absorbance is 0.36.

Collect and Organize Given the concentration and absorbance of a solution, we are asked to calculate the molar absorptivity (ε).

Analyze (a) We can rearrange Beer's law (Equation 4.5) to solve for ε and then use this value to find the concentration of the unknown solution. (b) Concentration and absorbance are linearly related by Beer's law, so we predict that a smaller absorbance for the unknown solution should correspond to a lower concentration. (In fact, a concentration of zero would correspond to zero absorbance.)

Solve

a. Rearranging Equation 4.5 and substituting for A, b, and c:

$$\varepsilon = \frac{A}{bc} = \frac{0.55}{(1.00\ \text{cm})(5.00 \times 10^{-5}\ M)} = 1.1 \times 10^{4}\ M^{-1}\ \text{cm}^{-1}$$

b. Solving Beer's law for concentration and substituting for A, b, and ε:

$$c = \frac{A}{\varepsilon b} = \frac{0.36}{(1.1 \times 10^{4}\ M^{-1} \cdot \cancel{\text{cm}^{-1}})(1.00\ \cancel{\text{cm}})} = 3.3 \times 10^{-5}\ M$$

Think About It As predicted, the concentration of the unknown solution is lower than $5.00 \times 10^{-5}\ M$.

Practice Exercise One analytical method for determining the concentration of copper ions in a sample uses a solution of cuproine molecules to bond to the metal, like the bonds between Fe^{2+} ions and phenanthroline molecules. The following data were collected for four standard solutions of Cu^{2+} with cuproine:

Concentration of Cu^{2+} (M)	Absorbance
1.00×10^{-4}	0.64
1.25×10^{-4}	0.80
1.50×10^{-4}	0.96
1.75×10^{-4}	1.12

Construct a calibration curve based on these data and use it to determine the concentration of a copper solution with an absorbance $A = 0.74$.

(Answers to Practice Exercises are in the back of the book.)

FIGURE 4.10 Adding a colorless solution of phenanthroline to a colorless solution of Fe^{2+} forms an orange-red solution when the phenanthroline molecules bond to the Fe^{2+} ions.

4.4 Electrolytes and Nonelectrolytes

The high concentrations of NaCl and other salts in seawater (Table 4.1) make it a good conductor of electricity. Suppose we immerse two *electrodes* (solid conductors) in distilled water (**Figure 4.11a**) and connect them to a battery and a light bulb. For electricity to flow and the bulb to light up, the circuit must be completed by mobile charge carriers (ions) in the solution. The light bulb does not light when the electrodes are placed in distilled water because there are very few ions in distilled water. When the electrodes are pressed into solid NaCl (**Figure 4.11b**), no electric current flows and the bulb does not light because the Na^+ and Cl^- ions in solid NaCl are fixed in position. However, when the electrodes are immersed in aqueous 0.50 M NaCl (**Figure 4.11c**), the bulb lights because the many mobile ions in the solution act as charge carriers between the two electrodes. The saline solution conducts electricity because Na^+ ions are attracted to and migrate toward the electrode connected to the negative terminal of the battery, and Cl^- ions are attracted to and migrate toward the positive electrode, completing an electric circuit.

Ions in Solution

Any solute that imparts electrical conductivity to an aqueous solution is called an **electrolyte**. Sodium chloride is categorized as a **strong electrolyte** because it *dissociates* completely in water, which means it completely breaks up into its component ions when it dissolves, releasing charge-carrying Na^+ cations and Cl^- anions.

absorbance (A) a measure of the quantity of light absorbed by a sample.

Beer's law the relation of the absorbance of a solution (A) to concentration (c), the light's path length (b), and the solute's molar absorptivity (ε) by the equation $A = \varepsilon bc$.

molar absorptivity (ε) a measure of how well a compound or ion absorbs light.

calibration curve a graph showing how a measurable property, such as absorbance, varies for a set of standard samples of known concentration that can later be used to determine the unknown concentration of a sample from its absorbance.

electrolyte a substance that dissociates into ions when it dissolves in water, enhancing the conductivity of the solvent.

strong electrolyte a substance that dissociates completely into ions when it dissolves in water.

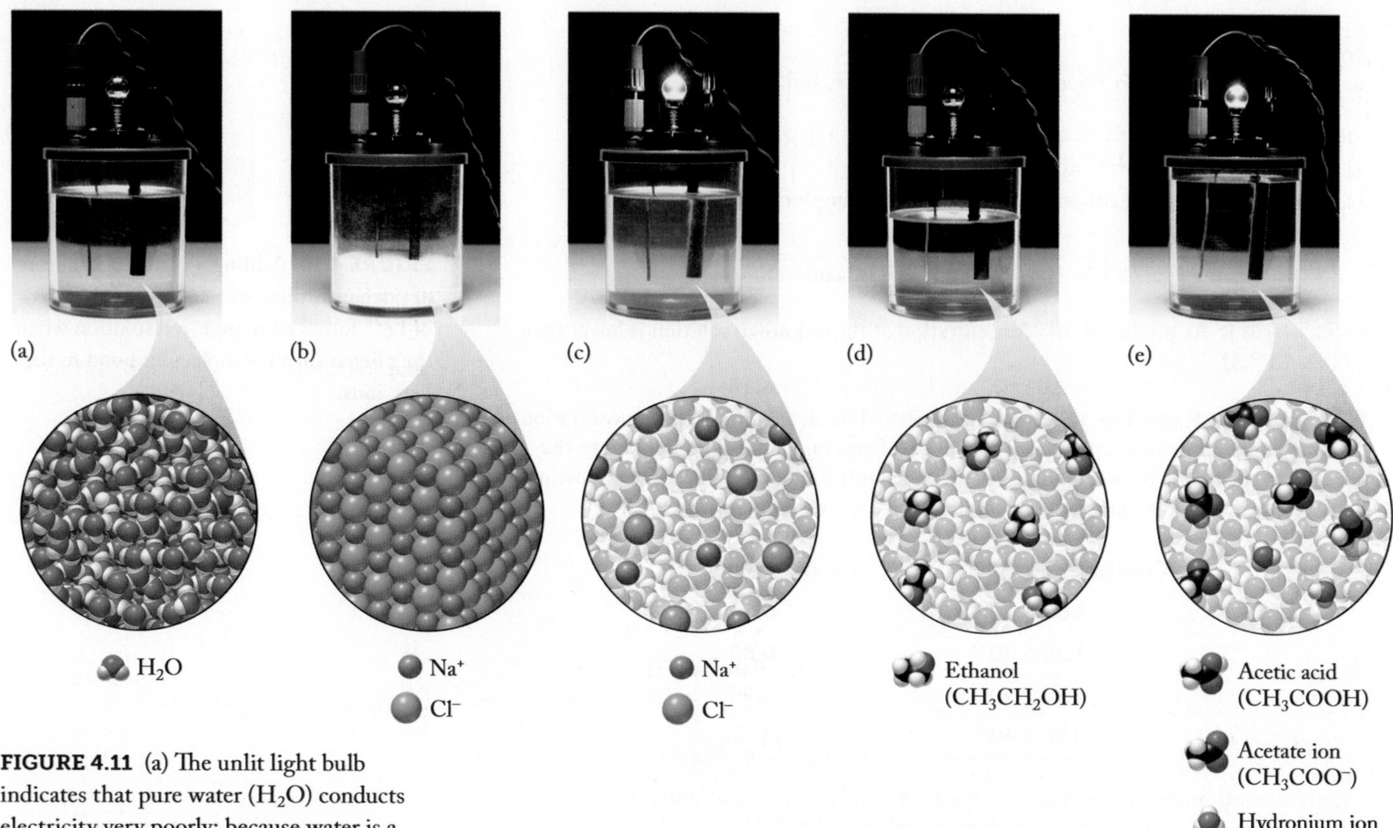

FIGURE 4.11 (a) The unlit light bulb indicates that pure water (H_2O) conducts electricity very poorly; because water is a molecular material, it contains very few ions. (b) The unlit light bulb indicates that solid sodium chloride (NaCl) does not conduct electricity. (c) The brightly lit bulb indicates that a 0.50 *M* solution of NaCl conducts electricity very well. (d) The unlit bulb indicates that a 0.50 *M* solution of ethanol (CH_3CH_2OH) does not conduct electricity any better than pure water. (e) The dimly lit bulb indicates that a 0.50 *M* solution of acetic acid (CH_3COOH) conducts electricity better than pure water or the ethanol solution, but not as well as the NaCl solution.

CHEMTOUR
Ions in Solution

Other substances, however, do *not* dissociate when dissolved in water. For instance, a 0.50 *M* solution of ethanol (CH_3CH_2OH) conducts electricity no better than pure water because each CH_3CH_2OH molecule remains intact in the solution and no ions are formed (**Figure 4.11d**). Solutes that do not form ions when dissolved in water are called **nonelectrolytes**.

In addition to strong electrolytes (complete dissociation into ions) and nonelectrolytes (no dissociation), there are **weak electrolytes**. Only a small fraction of a weak electrolyte's molecules dissociates into ions. Instead, most of them remain intact as whole, neutral molecules. One example of a weak electrolyte is acetic acid (CH_3COOH). In **Figure 4.11e**, the beaker contains a 0.50 *M* aqueous solution of acetic acid. The molarity of the solution in part (e) is equivalent to that in parts (c) and (d) of the figure, meaning that the same number of moles of acetic acid were dissolved as the number of moles of NaCl or moles of CH_3CH_2OH. Because the bulb connected to the acetic acid solution glows, but not very brightly, we can conclude that there are some ions present, but not as many as are present in the 0.50 *M* NaCl solution (Figure 4.11c). The acetic acid molecules dissociate to some extent, forming hydronium ions and acetate ions, but the process does not go to completion:

$$CH_3COOH(aq) + H_2O(\ell) \rightleftharpoons CH_3COO^-(aq) + H_3O^+(aq)$$

In this case, the arrow with the top half pointing right and the bottom half pointing left indicates that all species (the reactant and both products) are simultaneously present in solution: undissociated CH_3COOH molecules, acetate ions, and hydrogen ions. This type of arrow is used to indicate that the dissociation process is incomplete. It is important to understand that dissolving and dissociating are two different processes. All the acetic acid molecules dissolve in water, but only a tiny fraction of them dissociate into ions. A balance, called a *dynamic equilibrium*, is achieved in solutions of weak electrolytes, at which point the net concentrations of the reactants and the products in the dissociation reaction do not change. Solutions of weak electrolytes, such as weak acids, consist of dissociated ions and undissociated molecules existing together in dynamic equilibrium. For this reason, the arrow used above is called an *equilibrium arrow*.

nonelectrolyte a substance that does *not* dissociate into ions and therefore does *not* result in conductivity when dissolved in water.

weak electrolyte a substance that dissolves in water with most molecules staying intact, but with a small fraction dissociating into ions.

acid (Brønsted–Lowry acid) a proton donor.

base (Brønsted–Lowry base) a proton acceptor.

CONCEPT **TEST**

Why were equivalent molar concentrations used in the solutions in Figure 4.11?

(Answers to Concept Tests are in the back of the book.)

4.5 Acid–Base Reactions: Proton Transfer

Although the behavior of acetic acid molecules when dissolved in water is best described as a dynamic equilibrium, not all acids are weak electrolytes. Pure HCl is a molecular compound (called hydrogen chloride) and a gas. When HCl dissolves in water, however, it ionizes completely, producing $H_3O^+(aq)$ and $Cl^-(aq)$. This solution is called hydrochloric acid.

CONNECTION We discussed the naming of acids in Section 2.7.

CONCEPT **TEST**

Which drawing in **Figure 4.12** depicts the particles in an aqueous solution of an acid that is a weak electrolyte, and which drawing depicts the particles in an aqueous solution of an acid that is a strong electrolyte? [The solvent molecules (water) have been omitted.]

(Answers to Concept Tests are in the back of the book.)

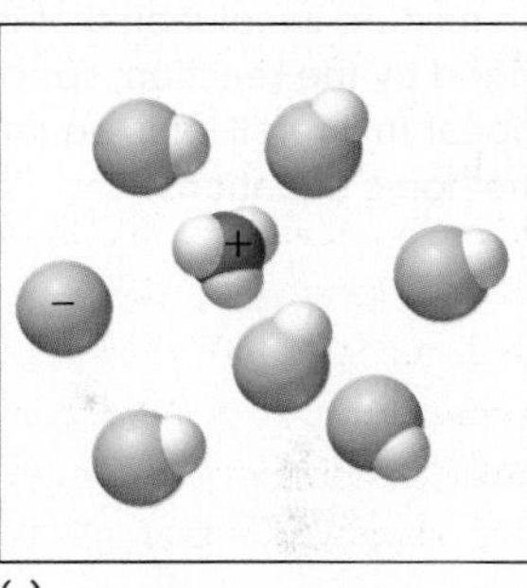

(a)

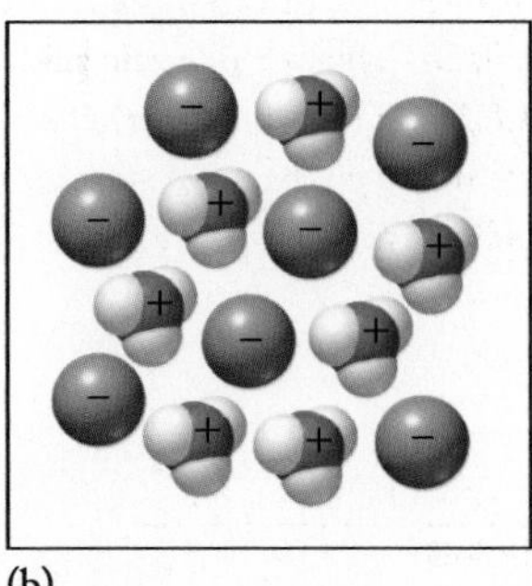

(b)

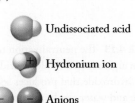

Undissociated acid

Hydronium ion

Anions

FIGURE 4.12 Aqueous solutions of acids. (Solvent water molecules have been omitted for clarity.)

As a molecule of HCl dissolves, it donates a proton (H^+ ion) to a water molecule:

$$\underset{\substack{\text{proton donor}\\\text{(acid)}}}{HCl(g)} + \underset{\substack{\text{proton acceptor}\\\text{(base)}}}{H_2O(\ell)} \rightarrow \underset{\substack{\text{hydronium}\\\text{ion}}}{H_3O^+(aq)} + Cl^-(aq) \quad (4.6)$$

This behavior is consistent with our definition of an **acid** as a *proton donor*. The water molecule accepts the proton, so water in this case is a *proton acceptor*, which is the corresponding definition of a **base**. These definitions were originally proposed by chemists Johannes Brønsted (1879–1947) and Thomas Lowry (1874–1936), so they are called **Brønsted–Lowry acids** and **Brønsted–Lowry bases**. We develop this concept of proton transfer more completely in Chapter 15, but for now it provides a convenient way to define acids and bases in aqueous solutions.

We often use $H^+(aq)$ to represent the cation produced in aqueous solution by an acid. Hydrogen ions do not have an independent existence in water, however, because each proton combines with a water molecule to form a hydronium ion (H_3O^+):

$$H^+(aq) + H_2O(\ell) \rightarrow H_3O^+(aq)$$

neutralization reaction a reaction that takes place when an acid reacts with a base and produces a solution of a salt in water.

salt the product of a neutralization reaction; it is made up of the cation of the base in the reaction plus the anion of the acid.

molecular equation an equation describing a reaction in solution in which the reactants and products are written as neutral compounds.

overall ionic equation an equation describing a reaction in solution that shows all the species, both ionic and molecular, present in the reaction mixture.

net ionic equation an equation describing a reaction in solution that contains only the species changed by the reaction; it is obtained by eliminating the spectator ions from the overall ionic equation.

spectator ion an ion present in a reaction mixture in solution that is unchanged by the reaction; spectator ions appear in overall ionic equations but not in net ionic equations.

In aqueous acid–base chemistry, the terms hydrogen ion, proton, and hydronium ion all refer to the same species.

In the reaction between aqueous solutions of HCl and NaOH,

$$HCl(aq) + NaOH(aq) \rightarrow NaCl(aq) + H_2O(\ell) \qquad (4.7)$$

the HCl functions as an acid by donating a proton to the hydroxide ion, and the hydroxide ion functions as a base by accepting the proton; the H^+ and OH^- ions combine to form a molecule of water. The remaining two ions—the anion Cl^- and the cation Na^+—form a salt, sodium chloride, which remains dissolved and dissociated in the aqueous solution. This reaction is called a **neutralization reaction** because the acid and the base no longer exist as a result of the reaction. The products of the neutralization of an acid like HCl with a base like NaOH in solution are always water and a salt. A **salt** is a substance formed along with water as a product of a neutralization reaction.

Equation 4.7 is written as a **molecular equation**, meaning each reactant and product is written as a neutral compound. Balanced molecular equations are sometimes the easiest equations to write, so reactions involving substances that dissociate in solution are often written first as molecular equations.

Another equation representing the reaction in Equation 4.7 is one that shows the reactants and products as the particles that exist in solution, which means that any substance that dissociates completely is written as individual ions:

$$\overbrace{H^+(aq) + Cl^-(aq)}^{HCl(aq)} + \overbrace{Na^+(aq) + OH^-(aq)}^{NaOH(aq)} \rightarrow \overbrace{Na^+(aq) + Cl^-(aq)}^{NaCl(aq)} + H_2O(\ell) \qquad (4.8)$$

This form, called an **overall ionic equation**, distinguishes soluble ionic substances from molecular substances in a chemical reaction taking place in solution. Any nonelectrolytes or weak electrolytes are written as neutral molecules.

Notice in Equation 4.8 that chloride ions and sodium ions appear on both sides of the reaction arrow. If this were an algebraic equation, we would remove the terms that are the same on both sides. When we do the same thing in chemistry, the result is a **net ionic equation** (Equation 4.9). The ions that are removed (in this case, sodium ion and chloride ion) are called **spectator ions**, and the resulting equation shows only the species taking part in the reaction:

$$H^+(aq) + OH^-(aq) \rightarrow H_2O(\ell) \qquad (4.9)$$

The spectator ions are unchanged by the reaction and remain in solution. The chemical change in a neutralization reaction occurs when a hydrogen ion reacts

(a) Molecular equation $HCl(aq) + NaOH(aq) \rightarrow NaCl(aq) + H_2O(\ell)$

(b) Overall ionic equation

$$H^+(aq) + Cl^-(aq) + Na^+(aq) + OH^-(aq) \rightarrow Na^+(aq) + Cl^-(aq) + H_2O(\ell)$$

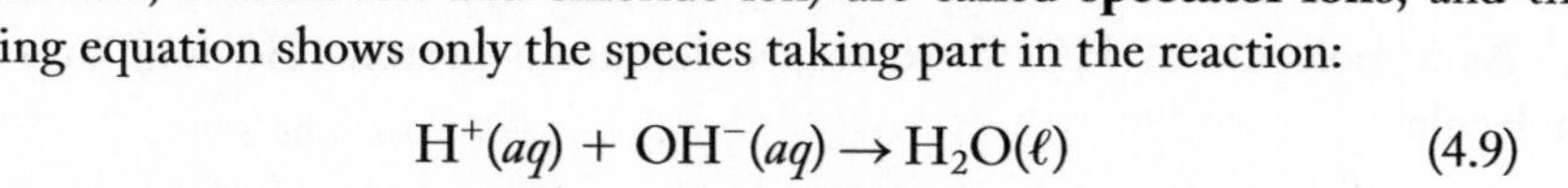

(c) Net ionic equation = overall ionic equation − spectator ions

$$H^+(aq) + \cancel{Cl^-(aq)} + \cancel{Na^+(aq)} + OH^-(aq) \rightarrow \cancel{Na^+(aq)} + \cancel{Cl^-(aq)} + H_2O(\ell)$$

$$H^+(aq) + OH^-(aq) \rightarrow H_2O(\ell)$$

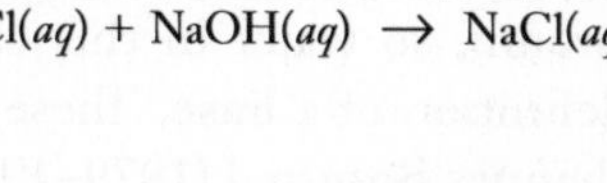

FIGURE 4.13 The neutralization reaction between aqueous hydrochloric acid and sodium hydroxide that produces sodium chloride and water can be written in three ways: (a) a molecular equation, (b) an overall ionic equation, and (c) a net ionic equation. (Solvent water molecules have been omitted for clarity.)

with a hydroxide ion to produce a molecule of water. The net ionic equation enables us to focus on those species that participate in the reaction. The three ways of depicting the reaction between HCl and NaOH—molecular, overall ionic, and net ionic equations—are summarized in **Figure 4.13.**

SAMPLE EXERCISE 4.7 Writing Neutralization Reaction Equations **LO4**

Write (a) molecular, (b) overall ionic, and (c) net ionic equations that describe the reaction that takes place when an aqueous solution of nitric acid is neutralized by an aqueous solution of calcium hydroxide.

Collect and Organize Solutions of nitric acid and calcium hydroxide take part in a neutralization reaction. Our task is to write three different equations describing the neutralization reaction.

Analyze We know that calcium hydroxide [$Ca(OH)_2$] is the base and nitric acid (HNO_3) is the acid. The products of a neutralization reaction are water and a salt. (a) All species are written as neutral compounds in the molecular equation. The products are water and calcium nitrate, the salt made from the cation of the base and the anion of the acid. (b) Writing each substance in the balanced molecular equation in terms of the ions it forms when it dissolves in water produces the overall ionic equation. (c) Removing spectator ions from the overall ionic equation yields the net ionic equation.

Solve

a. The molecular equation is

$$HNO_3(aq) + Ca(OH)_2(aq) \rightarrow Ca(NO_3)_2(aq) + H_2O(\ell) \quad \text{(unbalanced)}$$

$$2\,HNO_3(aq) + Ca(OH)_2(aq) \rightarrow Ca(NO_3)_2(aq) + 2\,H_2O(\ell) \quad \text{(balanced)}$$

b. The overall ionic equation is

$$2\,H^+(aq) + 2\,NO_3^-(aq) + Ca^{2+}(aq) + 2\,OH^-(aq) \rightarrow Ca^{2+}(aq) + 2\,NO_3^-(aq) + 2\,H_2O(\ell)$$

c. The spectator ions are Ca^{2+} and NO_3^-. Removing them gives us the net ionic equation:

$$2\,H^+(aq) + 2\,OH^-(aq) \rightarrow 2\,H_2O(\ell)$$

or

$$H^+(aq) + OH^-(aq) \rightarrow H_2O(\ell)$$

Think About It Both reactants are soluble compounds that form ions in aqueous solutions. Of these ions, only the H^+ ions formed by HNO_3 and the OH^- ions from $Ca(OH)_2$ are no longer ions on the product side, but rather molecules of H_2O. Therefore, it is reasonable that these are the only species in the net ionic equation describing the reaction.

Practice Exercise Write (a) molecular, (b) overall ionic, and (c) net ionic equations for the reaction between an aqueous solution of acetic acid, $CH_3COOH(aq)$, and an aqueous solution of sodium hydroxide. The products are sodium acetate and water. *Note*: Acetic acid is a weak electrolyte.

(Answers to Practice Exercises are in the back of the book.)

For billions of years, neutralization reactions have played key roles in the chemical transformations of Earth's crust that geologists call *chemical weathering.* One type of chemical weathering occurs when carbon dioxide in the atmosphere dissolves in rainwater to create a weakly acidic solution of carbonic acid, $H_2CO_3(aq)$. This solution reacts with calcium carbonate, which occurs as chalk, limestone, and marble, and is only slightly soluble in pure water. The reaction between carbonic acid and calcium carbonate is responsible for the formation of stalactites and stalagmites (**Figure 4.14**). Chemical weathering caused by even more acidic rain is responsible for the degradation of statues and the exteriors of buildings made of marble (**Figure 4.15**). The rain that falls in many regions of North America and Europe contains dilute sulfuric acid, which reacts with structures made of marble and limestone (forms of calcium carbonate, $CaCO_3$) forming calcium sulfate $CaSO_4$, which is more soluble in water than calcium carbonate.

$$CaCO_3(s) + H_2SO_4(aq) \rightarrow CaSO_4(aq) + H_2O(\ell) + CO_2(g) \quad (4.10)$$

Table 4.2 lists some common nonmetal oxides that are classified as atmospheric pollutants because they produce acid rain.

Nonmetal oxides other than carbon dioxide form solutions that are much more acidic than carbonic acid. Dinitrogen pentoxide, for example, forms nitric acid in water in a **hydrolysis** reaction.

$$N_2O_5(g) + H_2O(\ell) \rightarrow 2\ HNO_3(aq)$$

Similarly, sulfur trioxide forms sulfuric acid:

$$SO_3(g) + H_2O(\ell) \rightarrow H_2SO_4(aq)$$

Nitric acid and sulfuric acid are categorized as **strong acids** because they dissociate completely in water.

$$H_2O(\ell) + HNO_3(aq) \rightarrow H_3O^+(aq) + NO_3^-(aq)$$
$$H_2O(\ell) + H_2SO_4(aq) \rightarrow H_3O^+(aq) + HSO_4^-(aq)$$

They are listed in **Table 4.3** along with other common strong acids. Strong acids are strong electrolytes. The arrows in the ball-and-stick figures of the acids point to the bonds that ionize in solution to produce protons. Because one molecule of nitric acid donates one proton, nitric acid is called a *monoprotic acid*. Hydrochloric acid is another strong monoprotic acid. Sulfuric acid can donate up to two protons per molecule, which makes it a *diprotic acid*. In some H_2SO_4 solutions, however, only one of the two H atoms in each molecule ionizes. Sulfuric acid, then, is a strong acid when it comes to donating the first proton. This dissociation

hydrolysis the reaction of water with another material. The hydrolysis of nonmetal oxides produces acids.

strong acid an acid that completely dissociates into ions in aqueous solution.

weak acid an acid that is a weak electrolyte and so has a limited capacity to donate protons to the medium.

carboxylic acid an organic compound containing the –COOH group.

strong base a base that completely dissociates into ions in aqueous solution.

weak base a base that is a weak electrolyte and so has a limited capacity to accept protons.

FIGURE 4.14 In Carlsbad Caverns in New Mexico, stalagmites of limestone grow up from the cavern floor and stalactites grow downward from the ceiling.

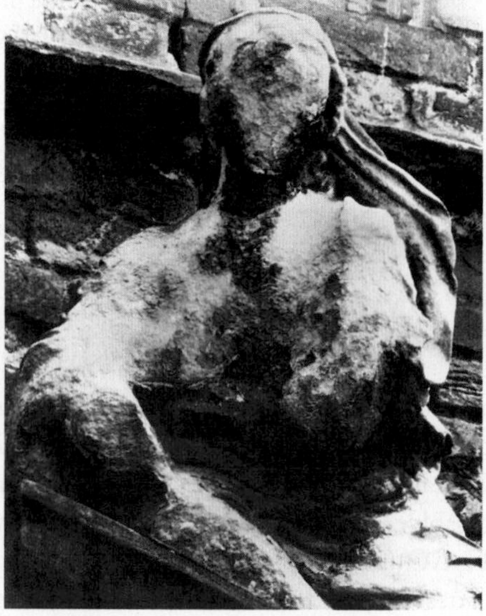

FIGURE 4.15 Atmospheric sulfuric acid, made when $SO_3(g)$ dissolves in rainwater, attacks marble statues by converting the calcium carbonate to calcium sulfate, which is slowly washed away by rain and melting snow. The picture on the left was taken in 1908; the one on the right, in 1968.

results in the formation of the HSO_4^- ion (hydrogen sulfate), which is also an acid because HSO_4^- ions can dissociate in solution into H^+ and SO_4^{2-} ions. Not all HSO_4^- ions may dissociate, however, so the hydrogen sulfate anion is a **weak acid**—that is, it is an acid in which only a small fraction of all the ions dissociate in water. Acetic acid, the weak electrolyte shown previously in Figure 4.11e, is a weak acid, too.

The two equations for the dissociation of the two protons of sulfuric acid are written differently to reflect these different dissociation behaviors. First, the complete dissociation of the strong acid H_2SO_4:

$$H_2O(\ell) + H_2SO_4(aq) \rightarrow H_3O^+(aq) + HSO_4^-(aq)$$

Second, the dynamic equilibrium and partial dissociation of the weak acid HSO_4^-:

$$H_2O(\ell) + HSO_4^-(aq) \rightleftharpoons H_3O^+(aq) + SO_4^{2-}(aq)$$

Beyond the strong acids listed in Table 4.3, nearly all other acids are weak acids, even if they are *polyprotic acids* such as carbonic acid (H_2CO_3) or phosphoric acid (H_3PO_4). For any diprotic acid, the HX^- anion formed when the first proton dissociates is a weaker acid than the original acid H_2X. This pattern applies to triprotic acids (H_3X) as well. Thus, H_3X is a stronger acid than H_2X^-, and H_2X^- is a stronger acid than HX^{2-}.

Acetic acid (CH_3COOH) is a member of a large group of organic compounds called **carboxylic acids** because they all contain the –COOH group in their structures. All carboxylic acids, not just acetic acid, are weak acids and hence dissociate only partially in water.

CONCEPT TEST

A molecule called EDTA has four carboxylic acid groups and hence four acidic protons. We can symbolize it as H_4E to highlight this property. Pick the more acidic member of each pair listed: (a) H_2E^{2-} or HE^{3-}; (b) H_4E or H_3E^-.

(Answers to Concept Tests are in the back of the book.)

As with acids, bases are classified as strong or weak depending on the extent to which they dissociate or hydrolyze in aqueous solution. **Strong bases** are strong electrolytes that dissociate completely, whereas **weak bases** are weak electrolytes. Strong bases include the hydroxides of group 1 and group 2 metals, which dissociate completely when dissolved in water.

Although substances such as $NaOH(aq)$ and $Ca(OH)_2(aq)$ can be considered bases because they produce hydroxide ions in aqueous solution, Brønsted–Lowry bases are proton acceptors. Hydroxide ions are bases because they can accept protons. This definition permits other substances that do not consist of hydroxide ions to be categorized as bases as well.

Ammonia (NH_3), for example, is a weak base, even though it does not contain hydroxide ions. Ammonia in its

TABLE 4.2 Volatile Nonmetal Oxides and Their Acids

Oxide	Acid
SO_2	H_2SO_3
SO_3	H_2SO_4
NO_2	HNO_2, HNO_3
N_2O_5	HNO_3
CO_2	H_2CO_3

TABLE 4.3 Strong Acids

Acid	Molecular Formula	Bond(s) Ionizing in Solution
Hydrochloric acid	HCl	
Hydrobromic acid	HBr	
Hydroiodic acid	HI	
Nitric acid[a]	HNO_3	
Sulfuric acid	H_2SO_4	
Perchloric acid	$HClO_4$	

[a]The dashed bonds in the model of nitric acid indicate bonds that are part single bond and part double bond in character. We discuss that phenomenon in detail in Chapter 8.

STEPWISE
ANIMATION
Strong vs. Weak Acids

molecular form is a gas at room temperature. When NH_3 dissolves in water it produces a solution that conducts electricity weakly, which means that ammonia is a weak electrolyte:

$$\underset{\substack{\text{base} \\ \text{(proton acceptor)}}}{NH_3(aq)} + \underset{\substack{\text{acid} \\ \text{(proton donor)}}}{H_2O(\ell)} \rightleftharpoons NH_4^+(aq) + OH^-(aq) \quad (4.11)$$

Ammonia is an inorganic molecule, but it can be viewed as the parent in a family of organic molecules that are typically weak bases. These compounds are called **amines**, and they are characterized by having one or more of the hydrogen atoms in NH_3 replaced by an organic group. We deal with the structure of ammonia in Chapters 8 and 9, but for now you need only to recognize amines as weak bases. They all behave similarly to ammonia in water by accepting a proton from the water molecule and producing a hydroxide ion in solution. The simplest member of this family is methylamine: CH_3NH_2. Because amines are weak bases, only a small fraction of the molecules accept protons in an aqueous solution; this behavior is indicated by an equilibrium arrow, just as we used with ammonia:

$$CH_3NH_2(g) + H_2O(\ell) \rightleftharpoons CH_3NH_3^+(aq) + OH^-(aq)$$

If a second or third hydrogen atom in NH_3 is replaced, whether by the same or different organic group, the compound still behaves as a weak base. For example, dimethyl amine, $(CH_3)_2NH$, contains one nitrogen atom with two CH_3 groups and one hydrogen atom bonded to it. Trimethyl amine has three CH_3 groups (and no hydrogen atoms) bonded to the central nitrogen atom: $(CH_3)_3N$. The most general way of describing amines is to use the symbol R for any organic group. Then the three general types of amines may be symbolized as RNH_2, R_2NH, and R_3N.

Finally, some substances, such as water, can behave as either acids or bases, depending on what other substances are in solution. Water is called an **amphiprotic** substance because it can function as a proton acceptor (Brønsted–Lowry base) in solutions containing acid solutes or as a proton donor (Brønsted–Lowry acid) in solutions containing basic solutes. In the hydrolysis of HCl (Equation 4.6), water functions as a base; in the hydrolysis of NH_3 (Equation 4.11), water functions as an acid.

amine an organic compound that functions as a base and has the general formula RNH_2, R_2NH, or R_3N, where R is any organic group.

amphiprotic describes a substance that can behave as either a proton acceptor or a proton donor.

titration an analytical method for determining the concentration of a solute in a sample by reacting the solute with a standard solution of known concentration.

CONCEPT **TEST**

Identify the following compounds as acids or bases or neither in water. If the compound is an acid or base, identify it as weak or strong. (a) CH_3CH_2COOH; (b) CH_3OH; (c) $(CH_3CH_2)_2NH$; (d) HNO_3

(Answers to Concept Tests are in the back of the book.)

SAMPLE EXERCISE 4.8 Comparing Electrolytes, Acids, and Bases — LO5

Classify each of the following compounds in aqueous solution as a strong electrolyte, a weak electrolyte, or a nonelectrolyte; and then again as an acid, a base, or neither an acid nor a base: (a) NaCl; (b) HCl; (c) NaOH; (d) CH_3COOH; (e) $CH_3CH_2NH_2$.

Collect, Organize, and Analyze We are given the formulas of five compounds and asked to classify them based on their behavior in an aqueous solution.

Strong electrolytes are substances that dissociate completely into ions when dissolved in water. Weak electrolytes dissociate only partially in water. Nonelectrolytes do not dissociate at all. A Brønsted–Lowry acid is defined as a proton donor, whereas a Brønsted–Lowry base is a proton acceptor. Structures with a –COOH group are weak acids, whereas compounds containing an amine group, $-NH_2$, are weak bases.

Solve

a. NaCl is an ionic compound that dissociates completely into ions in aqueous solution, as shown in Figure 4.11c. Therefore, NaCl is a strong electrolyte. Sodium chloride has no protons to donate, and neither Na^+ nor Cl^- ions readily accept protons, so NaCl is neither an acid nor a base.

b. HCl dissociates completely into H^+ and Cl^- ions when dissolved in water, making HCl a strong electrolyte. Aqueous HCl is a proton donor, so it is also a Brønsted–Lowry acid.

c. NaOH is an ionic compound that dissociates completely into Na^+ and OH^- ions in aqueous solution. NaOH is a strong electrolyte. The OH^- ion formed upon dissolution of NaOH readily accepts a proton to form water, making NaOH a Brønsted–Lowry base.

d. Some molecules of acetic acid, CH_3COOH, when dissolved in water, dissociate into H^+ and CH_3COO^- ions (Figure 4.11e), creating a weakly conducting solution; thus, CH_3COOH is a weak electrolyte. Since acetic acid is a proton donor, it is a Brønsted–Lowry acid.

e. Some molecules of $CH_3CH_2NH_2$ function as proton acceptors when dissolved in water (which functions as a proton donor) to form $CH_3CH_2NH_2^+$ ions, creating a weakly conducting solution. Therefore, $CH_3CH_2NH_2$ is both a weak electrolyte and a Brønsted–Lowry base.

Think About It Strong acids and bases are also strong electrolytes, and weak acids and bases are weak electrolytes. However some strong electrolytes are neither acidic nor basic, but instead are neutral compounds.

Practice Exercise Classify aqueous solutions of each of the following compounds as a strong electrolyte, weak electrolyte, or nonelectrolyte; and as an acid, a base, or neither: (a) potassium sulfate; (b) HBr; (c) NH_3; (d) ethanol (CH_3CH_2OH).

(Answers to Practice Exercises are in the back of the book.)

4.6 Titrations

We can use the reactions of acids and bases to analyze solutions and determine the concentrations of dissolved substances. A technique called **titration** is a common analytical method based on measured volumes of reactants. Acid–base neutralization reactions are frequently the basis of analyses done by titration.

One use of titrations is to analyze water draining from abandoned coal mines. Sulfide-containing minerals called pyrites may be exposed when coal is removed from a site. The ensuing reaction between the minerals and microorganisms in the presence of oxygen produces very acidic solutions of sulfuric acid. To manage these hazardous wastes appropriately, we must know the concentrations of acid present in them. Titrations give us this information.

Suppose we have 100.0 mL of drainage water containing an unknown concentration of sulfuric acid. We can use an acid–base titration in which the sulfuric acid in the water sample is neutralized by reaction with a standard solution of 0.00100 *M* NaOH. The NaOH solution in this case is called the **titrant**; it is a **standard solution**, meaning its concentration is known accurately. The neutralization reaction is

$$H_2SO_4(aq) + 2\ NaOH(aq) \rightarrow Na_2SO_4(aq) + 2\ H_2O(\ell)$$

To determine the concentration of sulfuric acid in the sample, we determine the volume of titrant needed to neutralize a known volume of the sample.

Suppose 22.40 mL of the NaOH solution is required to react completely with the H_2SO_4 in the sample. Because we know the volume and molarity of the NaOH solution, we can calculate the number of moles of NaOH reacted:

$$\begin{aligned} n &= V \times M \\ &= 22.40\ \cancel{\text{mL}} \times \frac{1\ \cancel{\text{L}}}{10^3\ \cancel{\text{mL}}} \times \frac{1.00 \times 10^{-3}\ \text{mol NaOH}}{\cancel{\text{L}}} \\ &= 2.24 \times 10^{-5}\ \text{mol NaOH} \end{aligned}$$

We know from the stoichiometry of the reaction that 2 moles of NaOH are required to neutralize 1 mole of H_2SO_4, so the number of moles of H_2SO_4 in the 100.0 mL sample must be

$$n_{H_2SO_4} = 2.24 \times 10^{-5}\ \cancel{\text{mol NaOH}} \times \frac{1\ \text{mol } H_2SO_4}{2\ \cancel{\text{mol NaOH}}} = 1.12 \times 10^{-5}\ \text{mol } H_2SO_4$$

We can combine these two steps into one mathematical expression and do so throughout the rest of the text:

$$\begin{aligned} n_{H_2SO_4} &= 0.02240\ \cancel{\text{L}} \times \frac{1.00 \times 10^{-3}\ \cancel{\text{mol NaOH}}}{\cancel{\text{L}}} \times \frac{1\ \text{mol } H_2SO_4}{2\ \cancel{\text{mol NaOH}}} \\ &= 1.12 \times 10^{-5}\ \text{mol } H_2SO_4 \end{aligned}$$

Therefore, the concentration of H_2SO_4 is

$$\begin{aligned} [H_2SO_4] &= \frac{1.12 \times 10^{-5}\ \text{mol } H_2SO_4}{0.1000\ \text{L}} = \frac{1.12 \times 10^{-4}\ \text{mol } H_2SO_4}{\text{L}} \\ &= 1.12 \times 10^{-4}\ M\ H_2SO_4 \end{aligned}$$

Couldn't we have used Equation 4.4 (i.e., $V_{\text{initial}}M_{\text{initial}} = V_{\text{diluted}}M_{\text{diluted}}$) to solve for the molarity of sulfuric acid given that we knew both the concentration and the volume of the sodium hydroxide solution in addition to the volume of sulfuric acid solution? That may seem possible because Equation 4.4 can be rearranged to solve for a volume when given another volume and two concentrations, but that equation applies only to situations in which the moles of solute remain unchanged as more solvent is added to dilute the solution. In the problem we just worked, we do not have one solution that is diluted but rather a reaction that occurs when two aqueous solutions are mixed together.

How did we determine in the first place that exactly 22.40 mL of the NaOH solution was needed to react with the sulfuric acid in the sample? The glassware required to perform an acid–base titration is illustrated in **Figure 4.16**. The standard solution of NaOH is poured into a *buret*, a narrow glass cylinder with volume markings. The solution is gradually added to the sample until the point in the titration when just enough standard solution has been added to completely react with all the solute in the sample. This is called the **equivalence point** of the titration.

titrant the standard solution added to the sample in a titration.

standard solution a solution of known concentration used in titrations.

equivalence point the point in a titration at which just enough titrant has been added to completely react the substance being analyzed.

end point the point in a titration that is reached when just enough standard solution has been added to cause the indicator to change color.

How do we know when the equivalence point has been reached? Chemists add a small amount of indicator solution to the sample before it is titrated. Indicator solutions contain molecules that change color as the concentration of $H^+(aq)$ ions changes. To detect the end point in our sulfuric acid–sodium hydroxide titration, we might use the indicator *phenolphthalein*, which is colorless in acidic solutions but pink in basic solutions. If the correct indicator has been chosen, the equivalence point is very close to the **end point**, the point at which the indicator changes color. The part of the titration that requires the most skill is adding just enough NaOH solution to reach the end point, the point at which a pink color first persists in the solution being titrated. To catch this end point precisely, one must add the standard solution no faster than one drop at a time with thorough mixing between drops.

Titration is an effective method for determining solution concentrations; however, it is not applicable in all situations. We have already seen in Section 4.3 how spectrophotometry can be used to determine concentration. We explore yet another method in Section 4.7.

(a)

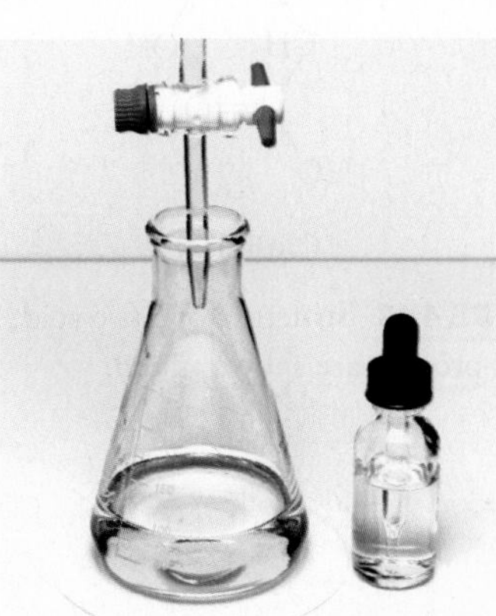
(b)

(c)

FIGURE 4.16 Determining a sulfuric acid concentration. (a) A known volume of a sample containing H_2SO_4 is placed in the flask. The buret is filled with the titrant, which is an aqueous NaOH solution of known concentration. A few drops of phenolphthalein indicator solution are added to the flask. (b) Titrant is carefully added to the flask until the indicator changes from colorless to faint pink, signaling that the acid has been neutralized and the end point has been reached. (c) If more titrant is added, the color becomes darker, signaling that more titrant than was needed to reach the end point has been added.

SAMPLE EXERCISE 4.9 Calculating Molarity from Titration Data **LO6**

Vinegar is an aqueous solution of acetic acid (CH_3COOH) that can be made from any source containing starch or sugar. Apple cider vinegar is made from apple juice that is fermented to produce alcohol, which then reacts with oxygen from the air in the presence of certain bacteria to produce vinegar. Commercial vinegar must contain no less than 4% acetic acid—that is, no fewer than 4 grams of acetic acid per 100 mL of vinegar. Suppose the titration of a 25.00 mL sample of vinegar requires 11.20 mL of a 5.95 *M* solution of NaOH. What is the molarity of the vinegar? Could this be a commercial sample of vinegar?

Collect, Organize, and Analyze We are given the volume and concentration of the titrant and the volume of the sample, and we are asked to calculate the concentration of the sample. We start with the chemical equation for the neutralization reaction:

$$CH_3COOH(aq) + NaOH(aq) \rightarrow CH_3COONa(aq) + H_2O(\ell)$$

The following sequence allows us to calculate the molarity of the acetic acid in the sample, from which we can calculate the grams of acetic acid per 100 mL and determine if the sample could be commercial-grade vinegar.

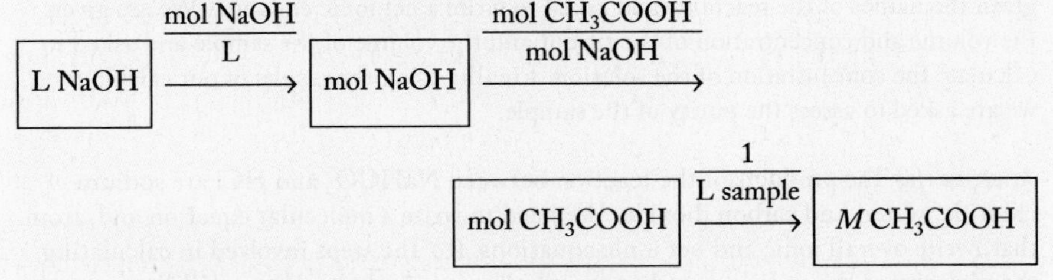

Solve The stoichiometry of the titration reaction is 1 mol NaOH per mol CH_3COOH. Therefore, number of moles of acetic acid in the sample is

$$n_{CH_3COOH} = 0.01120\ \text{L NaOH} \times \frac{5.95\ \text{mol NaOH}}{1\ \text{L NaOH}} \times \frac{1\ \text{mol CH}_3\text{COOH}}{1\ \text{mol NaOH}}$$

$$= 0.0666\ \text{mol CH}_3\text{COOH}$$

and its concentration is

$$[CH_3COOH] = \frac{0.0666\ \text{mol CH}_3\text{COOH}}{0.02500\ \text{L vinegar}} = 2.66\ M$$

The number of grams of acetic acid in the sample is

$$m_{CH_3COOH} = 0.0666 \text{ mol } CH_3COOH \times \frac{60.05 \text{ g } CH_3COOH}{1 \text{ mol } CH_3COOH} = 4.00 \text{ g } CH_3COOH$$

The original sample had a volume of 25.00 mL, so the mass of acetic acid per 100 mL of sample is

$$m_{CH_3COOH} = \frac{4.00 \text{ g } CH_3COOH}{25.00 \text{ mL vinegar}} \times 100 \text{ mL vinegar} = 16.0 \text{ g}$$

At 16.0 g/100 mL, or 16%, acetic acid, the sample has more than enough of the acid to qualify as a commercial vinegar.

Think About It About 11 mL of titrant were needed to neutralize the sample. That is about half the volume of the sample, so the calculated molarity of the vinegar should be about half that of the titrant, which it is. Vinegar this concentrated is typically used for pickling. Table vinegar is usually between 4% and 8% acetic acid.

Practice Exercise Citric acid ($C_6H_8O_7$; **Figure 4.17**), a weak triprotic acid found in lemon juice, is widely used as a flavoring in beverages. What is the molarity of citric acid in commercially available lemon juice if 14.26 mL of 1.751 *M* NaOH is required in a titration to neutralize 20.00 mL of the juice? How many grams of citric acid are in 100.0 mL of the juice?

(Answers to Practice Exercises are in the back of the book.)

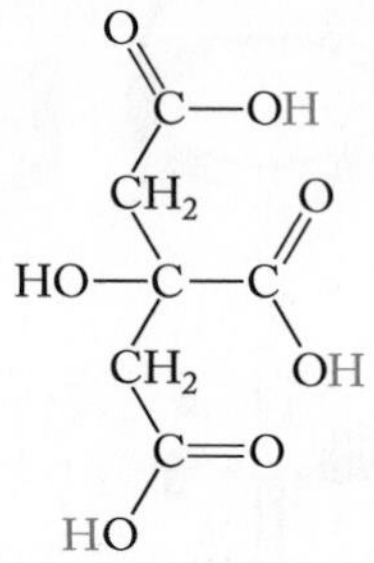

Citric acid

FIGURE 4.17 Structure of citric acid; acidic protons are shown in red.

SAMPLE EXERCISE 4.10 Using Titration Data to Calculate the Purity of a Sample — LO4, LO6

The over-the-counter antacid shown in **Figure 4.18**, contains sodium bicarbonate, which neutralizes excess stomach acid (aqueous HCl). A sample that contains a 650 mg tablet of the antacid dissolved in 25.00 mL of water required 25.27 mL of a 0.3000 *M* solution of HCl to be neutralized. (a) Write a net ionic equation for the reaction between $NaHCO_3$ and HCl. (b) What is the molarity of the antacid solution? (c) Is the tablet pure $NaHCO_3$?

Collect and Organize We are assigned three tasks in this sample exercise. We are given the names of the reactants and asked to write a net ionic equation. We are given the volume and concentration of the titrant and the volume of the sample and asked to calculate the concentration of the solution. Finally, using the results of our calculations, we are asked to assess the purity of the sample.

Analyze (a) The products of the reaction between $NaHCO_3$ and HCl are sodium chloride, water, and carbon dioxide. We need to write a molecular equation and, from that, write overall ionic and net ionic equations. (b) The steps involved in calculating the molarity of the sodium bicarbonate solution (not including liter–milliliter conversions) are

$$\boxed{\text{L HCl}} \xrightarrow{\frac{\text{mol HCl}}{\text{L}}} \boxed{\text{mol HCl}} \xrightarrow{\frac{\text{mol NaHCO}_3}{\text{mol HCl}}}$$

$$\boxed{\text{mol NaHCO}_3} \xrightarrow{\frac{1}{\text{L sample}}} \boxed{M\ \text{NaHCO}_3}$$

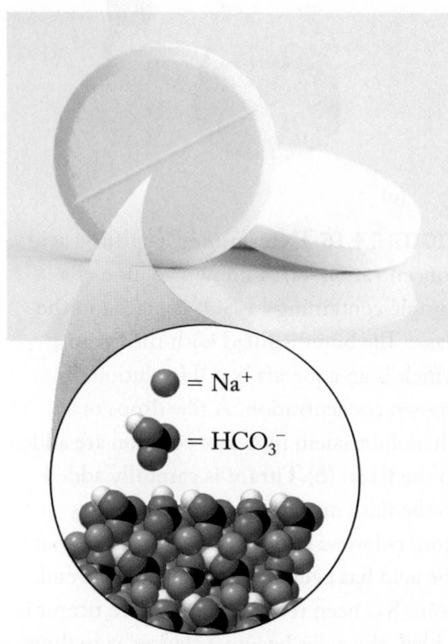

FIGURE 4.18 Sodium bicarbonate antacid tablets.

(c) The purity of the tablet can be determined by converting the number of moles of $NaHCO_3$ calculated in part (b) into an equivalent number of grams,

$$\text{mol NaHCO}_3 \xrightarrow{\text{molar mass}} \text{g NaHCO}_3$$

and comparing the result with the mass of the tablet (650 mg).

Solve

a. The molecular equation for the neutralization reaction is

$$NaHCO_3(aq) + HCl(aq) \rightarrow H_2O(\ell) + CO_2(g) + NaCl(aq)$$

The ionic equation is

$$Na^+(aq) + HCO_3^-(aq) + H^+(aq) + Cl^-(aq) \rightarrow H_2O(\ell) + CO_2(g) + Na^+(aq) + Cl^-(aq)$$

and the net ionic equation is

$$\cancel{Na^+(aq)} + HCO_3^-(aq) + H^+(aq) + \cancel{Cl^-(aq)} \rightarrow H_2O(\ell) + CO_2(g) + \cancel{Na^+(aq)} + \cancel{Cl^-(aq)}$$

$$HCO_3^-(aq) + H^+(aq) \rightarrow H_2O(\ell) + CO_2(g)$$

b. The number of moles of $NaHCO_3$ in the sample is

$$n_{NaHCO_3} = 0.02527\ \cancel{\text{L HCl}} \times \frac{0.3000\ \cancel{\text{mol HCl}}}{1\ \cancel{\text{L HCl}}} \times \frac{1\ \text{mol NaHCO}_3}{1\ \cancel{\text{mol HCl}}}$$

$$= 0.007581\ \text{mol NaHCO}_3$$

and the concentration of the HCO_3^- ion is

$$[HCO_3^-] = \frac{0.007581\ \text{mol NaHCO}_3}{0.02500\ \text{L solution}} = 0.3032\ M\ \text{NaHCO}_3$$

c. To calculate the mass of $NaHCO_3$ in the sample, we first must calculate its molar mass:

$$\mathcal{M}_{NaHCO_3} = 22.99\ \text{g/mol} + 1.008\ \text{g/mol} + 12.01\ \text{g/mol} + 3(16.00\ \text{g/mol})$$

$$= 84.01\ \text{g/mol NaHCO}_3$$

and then the mass of sodium bicarbonate in the tablet:

$$m_{NaHCO_3} = 0.007581\ \cancel{\text{mol NaHCO}_3} \times \frac{84.01\ \text{g NaHCO}_3}{1\ \cancel{\text{mol NaHCO}_3}} = 0.637\ \text{g} = 637\ \text{mg NaHCO}_3$$

This mass is slightly less than the mass of the tablet (650 mg), which indicates that the tablet was not pure $NaHCO_3$.

Think About It The experimentally determined mass of $NaHCO_3$ in the tablet corresponded to 637/650 or 98% of the mass of the tablet. The experimental results may mean that the tablet was not pure $NaHCO_3$, though analytical error is another possibility. Additional titrations are needed to assess the variability in the results, and carefully prepared standard solutions should be titrated to assess accuracy.

Practice Exercise What is the molarity of calcium bicarbonate if 9.870 mL of 1.000 M HNO_3 is required in a titration to neutralize 50.00 mL of a solution of $Ca(HCO_3)_2$?

(Answers to Practice Exercises are in the back of the book.)

4.7 Precipitation Reactions

Some of the most abundant elements in Earth's crust, including silicon (Si), aluminum (Al), and iron (Fe), are not abundant in seawater for the simple reason that most compounds containing these elements are only slightly soluble in water. Earth's crust must be made of compounds with limited water solubility, or else they would never survive as solids in the presence of all the water on our planet.

TABLE 4.4 Solubility Guidelines for Ionic Compounds in Water

All compounds containing the following ions are soluble:
- Cations: Group 1 ions (alkali metals) and NH_4^+
- Anions: NO_3^- and CH_3COO^- (acetate)

Compounds containing the following anions are soluble except as noted:
- Cl^-, Br^-, and I^-, except those of Ag^+, Cu^+, Hg_2^{2+}, and Pb^{2+}
- SO_4^{2-}, except those of Ba^{2+}, Ca^{2+}, Hg_2^{2+}, Pb^{2+}, and Sr^{2+}

The following compounds are only slightly soluble:
- All hydroxides except those of group 1 cations and $Ca(OH)_2$, $Sr(OH)_2$, and $Ba(OH)_2$
- All sulfides except those of group 1 cations and NH_4^+, CaS, SrS, and BaS
- All carbonates except those of group 1 cations and NH_4^+
- All phosphates except those of group 1 cations and NH_4^+
- Most fluorides, though not those of group 1 cations and NH_4^+

The reasons why some compounds don't readily dissolve in water whereas others do involve many factors. For now, we will rely on the solubility guidelines in **Table 4.4**, which allow us to predict the solubility of common ionic compounds. *Soluble* and *slightly soluble* are qualitative terms. In principle, all ionic compounds dissolve in water to some extent. When ionic compounds dissolve in water, they do so by dissociating into ions. In practice, we consider a compound to be only slightly soluble in water if the maximum amount that dissolves gives a concentration of less than 0.01 *M*.

Precipitate Formation

In Section 4.5, we examined what happens in neutralization reactions such as the addition of aqueous sulfuric acid to aqueous barium hydroxide. When this reaction occurs (**Figure 4.19**), however, a white solid called a **precipitate** forms at the bottom of the beaker. Such reactions are called **precipitation reactions**.

We can use Table 4.4 to determine whether a precipitate forms when solutions of ionic solutes are mixed together. For example, does a precipitate form when aqueous solutions of potassium nitrate (KNO_3) and sodium iodide (NaI) are mixed? To answer this question we need to follow two steps:

1. *Recognize that both salts are in solution, which means that they are both soluble in water and therefore have dissociated into their respective cations and anions.* Table 4.4 specifies that all compounds containing cations from group 1 in the periodic table are soluble in water. Potassium and sodium are in group 1, so salts containing them are soluble and dissociate in water into their ions:
   - Solution 1 consists of potassium ions and nitrate ions: $K^+(aq)$ and $NO_3^-(aq)$
   - Solution 2 consists of sodium ions and iodide ions: $Na^+(aq)$ and $I^-(aq)$

   From this information we can set up the reactant side of the overall ionic equation:

$$K^+(aq) + NO_3^-(aq) + Na^+(aq) + I^-(aq) \rightarrow ?$$

2. *Determine whether any of the possible combinations of ions produces a slightly soluble product.* In step 1 we determined that there are four kinds of ions present when the two solutions are mixed. When these ions collide with one another, do any of them form a slightly soluble ionic compound, that is, a precipitate? KI and $NaNO_3$ are soluble according to Table 4.4. (The other possible combinations of ions are KNO_3 and NaI, but those are the compounds we started with and we already determined them to be soluble and dissociate.) As a result, no precipitate forms when the two solutions are mixed:

$$K^+(aq) + NO_3^-(aq) + Na^+(aq) + I^-(aq) \rightarrow \text{no reaction}$$

$Ba^{2+}(aq) + SO_4^{2-}(aq) \rightarrow BaSO_4(s)$

FIGURE 4.19 Addition of aqueous sulfuric acid to a solution of barium hydroxide precipitates white barium sulfate.

Table 4.5 conveniently summarizes these solubility rules. It contains the same information as Table 4.4, but it is organized to show trends across common cations and anions. Does a precipitate form when an aqueous solution of lead(II) nitrate is added to an aqueous solution of sodium iodide? As before, NaI(*aq*) exists in solution as $Na^+(aq)$ and $I^-(aq)$ ions. Because Table 4.4 (and Table 4.5) indicate

TABLE 4.5 Solubility Matrix for Ionic Compounds in Water

ANION	CATION: Group 1 or NH_4^+	Ca^{2+}, Sr^{2+}, or Ba^{2+}	Cu^+ or Ag^+	Hg_2^{2+} or Pb^{2+}
NO_3^- or CH_3COO^-	Soluble	Soluble	Soluble	Soluble
Cl^-, Br^-, or I^-	Soluble	Soluble	Slightly soluble	Slightly soluble
OH^- or S^{2-}	Soluble	Soluble	Slightly soluble	Slightly soluble
SO_4^{2-}	Soluble	Slightly soluble	Soluble	Slightly soluble
F^-, CO_3^{2-}, or PO_4^{3-}	Soluble	Slightly soluble	Slightly soluble	Slightly soluble

precipitate a solid product formed from a reaction in solution.

precipitation reaction a reaction that produces a slightly soluble product upon mixing two solutions.

that all nitrate salts are soluble, $Pb(NO_3)_2$ dissolves in water and forms ions. The ions present in the mixed solutions are

$$Pb^{2+}(aq) + 2\,NO_3^-(aq) + Na^+(aq) + I^-(aq) \rightarrow ?$$

[There is a coefficient of 2 present in front of nitrate because there are 2 moles of nitrate for every mole of $Pb^{2+}(aq)$ ions.] When these ions collide, do ionic bonds form and produce a slightly soluble ionic compound? The solubility guidelines in Table 4.4 indicate that PbI_2 is only slightly soluble, so we predict that $PbI_2(s)$ will precipitate (**Figure 4.20**).

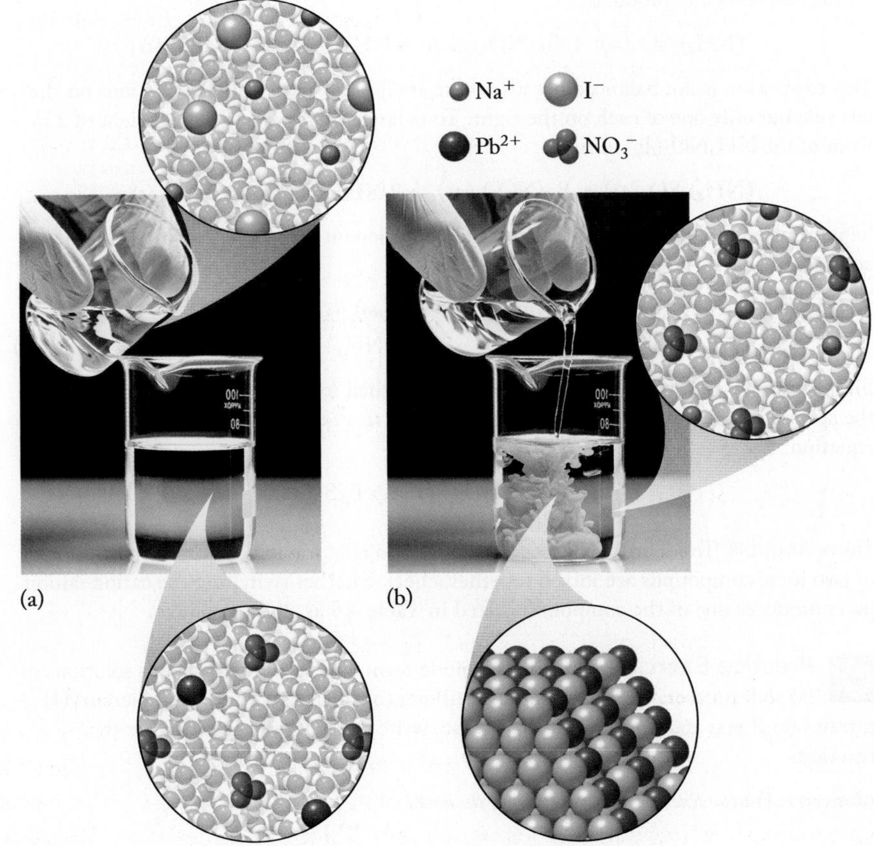

FIGURE 4.20 (a) One beaker contains a 0.1 *M* solution of $Pb(NO_3)_2$, and the other contains a 0.1 *M* solution of NaI. Both solutions are colorless. (b) As the NaI solution is poured into the $Pb(NO_3)_2$ solution, a yellow precipitate of PbI_2 forms.

Because a reaction takes place, we can write an equation that describes the process, starting with the molecular equation:

Unbalanced: $Pb(NO_3)_2(aq) + NaI(aq) \rightarrow PbI_2(s) + NaNO_3(aq)$

Balanced: $Pb(NO_3)_2(aq) + 2\ NaI(aq) \rightarrow PbI_2(s) + 2\ NaNO_3(aq)$

Converting the molecular equation into an overall ionic equation:

$$Pb^{2+}(aq) + 2\ NO_3^-(aq) + 2\ Na^+(aq) + 2\ I^-(aq) \rightarrow PbI_2(s) + 2\ Na^+(aq) + 2\ NO_3^-(aq)$$

Removing the spectator ions NO_3^- and Na^+ gives the net ionic equation:

$$Pb^{2+}(aq) + 2\ I^-(aq) \rightarrow PbI_2(s)$$

SAMPLE EXERCISE 4.11 Writing Net Ionic Equations for Precipitation Reactions I **LO4, LO7**

A precipitate forms when aqueous solutions of ammonium sulfate and barium nitrate are mixed. Write the net ionic equation for this reaction.

Collect, Organize, and Analyze We are to write the net ionic equation describing the precipitation reaction that occurs when solutions of ammonium sulfate and barium nitrate are mixed together. The formulas of the reactants are $(NH_4)_2SO_4$ and $Ba(NO_3)_2$. According to Table 4.4 all ammonium salts and all nitrate salts are soluble in water. Therefore, the precipitate must be $BaSO_4$ and the only other product is soluble NH_4NO_3.

Solve Let's start by writing a preliminary expression containing single formula units of all the reactants and products:

$$(NH_4)_2SO_4(aq) + Ba(NO_3)_2(aq) \rightarrow NH_4NO_3(aq) + BaSO_4(s)$$

This expression is not balanced because there are 2 NH_4^+ ions and 2 NO_3^- ions on the left side but only one of each on the right. To balance them we add a coefficient of 2 in front of the $NH_4NO_3(aq)$ term:

$$(NH_4)_2SO_4(aq) + Ba(NO_3)_2(aq) \rightarrow 2\ NH_4NO_3(aq) + BaSO_4(s)$$

Separating the soluble compounds into their component ions yields the overall ionic equation:

$$\cancel{2\ NH_4^+(aq)} + SO_4^{2-}(aq) + Ba^{2+}(aq) + \cancel{2\ NO_3^-(aq)} \rightarrow \cancel{2\ NH_4^+(aq)} + \cancel{2\ NO_3^-(aq)} + BaSO_4(s)$$

Simplifying by crossing out (equation above) and then removing (equation below) all the spectator ions that appear on both sides of the reaction arrow yields the net ionic equation:

$$Ba^{2+}(aq) + SO_4^{2-}(aq) \rightarrow BaSO_4(s)$$

Think About It To identify whether a precipitation reaction will occur when solutions of two ionic compounds are mixed together, check whether switching the cation–anion pairs produces one of the compounds listed in Table 4.5 as slightly soluble.

Practice Exercise Does a precipitate form when you mix aqueous solutions of (a) sodium acetate and ammonium sulfate; (b) calcium chloride and mercury(I) nitrate? (c) If you answered yes in either case, write the net ionic equation for the reaction.

(Answers to Practice Exercises are in the back of the book.)

Precipitation reactions can be used to synthesize slightly soluble salts. For example, barium sulfate, used in medical procedures to image the gastrointestinal tract, can be made by mixing an aqueous solution of any water-soluble barium salt with a solution of a soluble sulfate salt. The precipitate from such a reaction can be collected by filtration, dried, and used for whatever purpose we may have. The soluble compound remaining in solution can also be collected. The ammonium chloride left in solution in Sample Exercise 4.11, for instance, can be isolated by evaporating the water to leave a residue of solid NH_4NO_3.

SAMPLE EXERCISE 4.12 Writing Net Ionic Equations for Precipitation Reactions II — LO4, LO7

Write a net ionic equation for the reaction between aqueous sulfuric acid and barium hydroxide as pictured in Figure 4.19. Is this the same net ionic equation as in Sample Exercise 4.11?

Collect and Organize We are asked to write a net ionic equation given the names of the reactants. Figure 4.19 shows that the reaction produces a white precipitate. We then need to write the net ionic equation and compare it with the one from Sample Exercise 4.11.

Analyze First, we need to write a molecular equation for the reaction. Next, we must identify which reactants and products are soluble in water. We write any soluble ionic compounds as ions, leaving any precipitates and molecular compounds unchanged. Finally, we will simplify the overall ionic equation by removing any spectator ions that appear on both sides of the equation to obtain the net ionic equation.

Sulfuric acid is a strong diprotic acid. As mentioned previously in this chapter, its first proton completely dissociates but its second proton does not:

$$H_2SO_4(aq) \rightarrow H^+(aq) + HSO_4^-(aq) \qquad HSO_4^-(aq) \rightleftharpoons H^+(aq) + SO_4^{2-}(aq)$$

Therefore, the correct way to write the ionic species present in solution in an overall ionic equation when sulfuric acid is a reactant is $H^+(aq) + HSO_4^-(aq)$, *not* $2\,H^+(aq) + SO_4^{2-}(aq)$.

Solve

1. Molecular equation:

$$Ba(OH)_2(aq) + H_2SO_4(aq) \rightarrow BaSO_4(s) + 2\,H_2O(\ell)$$

2. Overall ionic equation:

$$Ba^{2+}(aq) + 2\,OH^-(aq) + H^+(aq) + HSO_4^-(aq) \rightarrow BaSO_4(s) + 2\,H_2O(\ell)$$

3. The equation cannot be simplified any further, so the net ionic equation is the same as the overall ionic equation in this case.

This net ionic equation is *not* the same as for the reaction between ammonium sulfate and barium nitrate in Sample Exercise 4.11, even though both reactions form the same precipitate.

Think About It The reaction between barium hydroxide and sulfuric acid is *both* a neutralization reaction *and* a precipitation reaction. Formation of a slightly soluble salt and water means that all the ions on the reactant side are changed by the reaction, which makes the net ionic equation the same as the overall ionic equation.

Practice Exercise Write a net ionic equation for the reaction between barium hydroxide and phosphoric acid (H_3PO_4) in water.

(Answers to Practice Exercises are in the back of the book.)

CONCEPT TEST

Lead(II) chromate ($PbCrO_4$), is a paint pigment called *chrome yellow* that was used for decades to paint school buses and the lines on highways. Design a synthesis of chrome yellow that uses a precipitation reaction.

(Answers to Concept Tests are in the back of the book.)

Using Precipitation in Analysis

Chemists use the insolubility of ionic compounds to determine concentrations of ions in solution. For example, NaCl (in the form of *rock salt*) is used to melt ice and snow on roads during the winter. However, the resulting NaCl solutions may run off into nearby water supplies, and that water can become contaminated with high levels of sodium and chloride ions. To determine whether sources of drinking water have become contaminated with such runoff, analytical chemists can determine the concentration of chloride ions by reacting a sample of the water with a solution of silver nitrate, $AgNO_3(aq)$. Any chloride ions in the water combine with Ag^+ ions to form a precipitate of AgCl:

$$NaCl(aq) + AgNO_3(aq) \rightarrow AgCl(s) + NaNO_3(aq) \qquad (4.12)$$

This precipitate can be filtered and dried, and its mass determined. From its mass, we can calculate the concentration of chloride in the original water sample.

Equation 4.12 is the molecular equation describing the formation of AgCl. Because both reactants are soluble ionic compounds and completely dissociate in solution, the overall ionic equation is

$$Na^+(aq) + Cl^-(aq) + Ag^+(aq) + NO_3^-(aq) \rightarrow AgCl(s) + Na^+(aq) + NO_3^-(aq)$$

Eliminating the spectator ions $Na^+(aq)$ and $NO_3^-(aq)$ yields the net ionic equation:

$$Ag^+(aq) + Cl^-(aq) \rightarrow AgCl(s)$$

When performing this analysis, we must add enough Ag^+ ions to ensure that all the Cl^- ions precipitate. Specifically, if we use a molar ratio of $Ag^+(aq)/Cl^-(aq)$ greater than 1, then the $Ag^+(aq)$ is present *in excess*. Precipitation reactions such as this may also include indicators that produce a visual signal (like a color change) that defines when the reaction is complete.

CONNECTION To review limiting reactants and reactants in excess when calculating the yield of reactions, see Chapter 3.

SAMPLE EXERCISE 4.13 Calculating the Mass of a Precipitate **LO7**

When hydrofluoric acid, $HF(aq)$, gets on the skin, it migrates away from the site of exposure and shuts down the microcapillaries that carry blood. The clinical management of burns on the skin from exposure to hydrofluoric acid may include the injection of a soluble calcium salt in the skin near the site of contact to stop the migration of $HF(aq)$. As F^- ions from the acid move through the skin, they combine with Ca^{2+} ions and form deposits of only slightly soluble CaF_2. If 1.00 mL of a 2.24 *M* aqueous solution of Ca^{2+} is injected, what is the mass of CaF_2 produced if all the calcium ions react with fluoride ions?

Collect and Organize We are given the volume and concentration of a solution of Ca^{2+} ions and asked to determine the maximum amount of CaF_2 precipitate that could be formed upon reaction with F^- ions.

Analyze The net ionic reaction for the process involved is

$$Ca^{2+}(aq) + 2\,F^-(aq) \rightarrow CaF_2(s)$$

The problem states that all the calcium ions react, so we can calculate the mass of product assuming Ca^{2+} ions are the limiting reactant by (1) converting the volume × concentration of the Ca^{2+} solution into mol Ca^{2+}, (2) using the stoichiometry of the equation to calculate mol CaF_2, and (3) using $\mathcal{M}_{CaF_2}$ to calculate g CaF_2.

Solve First we need to calculate the molar mass of CaF_2:

$$\mathcal{M}_{CaF_2} = 40.08\text{ g/mol} + 2(19.00\text{ g/mol}) = 78.08\text{ g/mol CaF}_2$$

Following the steps described in the Analyze section:

$$m_{CaF_2} = 1.00 \times 10^{-3}\ \cancel{\text{L solution}} \times \frac{2.24\ \cancel{\text{mol Ca}^{2+}}}{1\ \cancel{\text{L solution}}} \times \frac{1\ \cancel{\text{mol CaF}_2}}{1\ \cancel{\text{mol Ca}^{2+}}} \times \frac{78.08\text{ g CaF}_2}{1\ \cancel{\text{mol CaF}_2}}$$

$$= 0.175\text{ g CaF}_2$$

Think About It As a check on the solution to this exercise, consider that 1.0 mL of 2.24 *M* Ca^{2+} ions contains 1.0 $\cancel{\text{mL}}$ × 2.24 mol/$\cancel{\text{L}}$ = 2.24 mmoles of Ca^{2+} ions (1 mmol = 0.001 mol). This quantity of Ca^{2+} ions produces the same quantity of CaF_2 whose mass is 2.24 m$\cancel{\text{mol}}$ CaF_2 × 78.08 g/$\cancel{\text{mol}}$ = 175 mg or 0.175 g CaF_2, which is that answer we obtained above.

Practice Exercise A deep red pigment known as vermilion has been used by artists and craftsmen for thousands of years. The pigment is prepared from the mineral cinnabar, which is a form of mercury(II) sulfide (HgS). Synthetic cinnabar can be prepared by precipitating HgS from mixing solutions containing Hg^{2+} ions and S^{2-} ions. How many grams of cinnabar could be made in the precipitation reaction that occurs when 50.00 mL of 0.0150 *M* $Hg(NO_3)_2$ is mixed with excess $Na_2S(aq)$?

(Answers to Practice Exercises are in the back of the book.)

SAMPLE EXERCISE 4.14 Calculating Solute Concentration from a Mass of Precipitate — LO7

To determine the concentration of chloride ions in a 100.0 mL sample of groundwater, a chemist adds a large enough volume of a solution of $AgNO_3$ to the sample to precipitate the Cl^- ions as AgCl ($\mathcal{M}$ = 143.32 g/mol). The mass of the resulting AgCl precipitate is 71.7 mg. What is the Cl^- ion concentration in the sample in mg/L?

Collect and Organize We are given the mass (71.7 mg) of AgCl formed by precipitating the Cl^- ions from a 100.0 mL sample and asked to calculate the concentration of Cl^- ions in the sample in mg/L. Each mole of AgCl ($\mathcal{M}$ = 143.32 g/mol) precipitate contains 1 mole of Cl^- ions.

Analyze The diagram below outlines a strategy for converting a mass of AgCl into a concentration of Cl^- ions in mg/L. The initial mass of AgCl and final mass of Cl^- ions are both in mg, so the "milli" prefix is carried through the calculation to avoid unnecessary milligram-to-gram and gram-to-milligram conversion steps.

$$\boxed{\text{mg AgCl}} \xrightarrow{\frac{1}{\text{molar mass}}} \boxed{\text{mmol AgCl}} \xrightarrow{\frac{1\text{ mol Cl}^-}{1\text{ mol AgCl}}} \boxed{\text{mmol Cl}^-} \xrightarrow{\text{molar mass}}$$

$$\boxed{\text{mg Cl}^-} \xrightarrow{\frac{1}{100.0\text{ mL}}} \boxed{\frac{\text{mg Cl}^-}{\text{mL}}} \xrightarrow{\frac{10^3\text{ mL}}{\text{L}}} \boxed{\frac{\text{mg Cl}^-}{\text{L}}}$$

saturated solution a solution that contains the maximum concentration of a solute possible at a given temperature.

solubility the maximum amount of a substance that dissolves in a given volume of solution at a given temperature.

supersaturated solution a solution that contains more than the maximum quantity of solute predicted to be soluble in a given volume of solution at a given temperature.

Solve Converting the mass of AgCl precipitate into mg Cl^- ions /L:

$$71.7\ \text{mg AgCl}(s) \times \frac{1\ \text{g AgCl}(s)}{10^3\ \text{mg AgCl}(s)} \times \frac{1\ \text{mol AgCl}(s)}{143.32\ \text{g AgCl}(s)} \times \frac{1\ \text{mol Cl}^-(aq)}{1\ \text{mol AgCl}(s)} \times \frac{35.45\ \text{g Cl}^-(aq)}{1\ \text{mol Cl}^-(aq)}$$

$$\times \frac{10^3\ \text{mg Cl}^-}{1\ \text{g Cl}^-(aq)} \times \frac{1}{100.0\ \text{mL}} \times \frac{1000\ \text{mL}}{\text{L}} = 177\ \text{mg Cl}^-/\text{L}$$

Think About It Another way to solve this problem starts by expressing the mass of AgCl precipitated from a 100.0 mL sample as if it were a concentration and then using a ratio of molar masses to calculate an equivalent concentration of Cl^- ions in mg/L:

$$\frac{71.7\ \text{mg AgCl}(s)}{100.0\ \text{mL}} \times \frac{1\ \text{g AgCl}(s)}{10^3\ \text{mg AgCl}^-(s)} \times \frac{1\ \text{mol Cl}^-(aq)}{1\ \text{mol AgCl}(s)} \times \frac{35.45\ \text{g Cl}^-(aq)/1\ \text{mol Cl}^-(aq)}{143.32\ \text{g AgCl}(s)/1\ \text{mol AgCl}(s)}$$

$$\times \frac{10^3\ \text{mg Cl}^-}{1\ \text{g Cl}^-(aq)} \times \frac{1000\ \text{mL}}{\text{L}} = 177\ \text{mg Cl}^-/\text{L}$$

Practice Exercise The concentration of $SO_4^{2-}(aq)$ in a 50.0 mL sample of river water is determined using a precipitation titration in which $Ba^{2+}(aq)$ is added to precipitate $BaSO_4(s)$. What is the $SO_4^{2-}(aq)$ concentration in molarity if 6.55 mL of a 0.00100 M $Ba^{2+}(aq)$ solution is consumed?

(Answers to Practice Exercises are in the back of the book.)

Saturated Solutions and Supersaturation

A solution containing the maximum amount of solute that can dissolve is a **saturated solution** (**Figure 4.21**). The maximum amount of solute that can dissolve in a given quantity of solvent at a given temperature is called the **solubility** of that solute in the solvent. Precipitates form when the solubility of a solute is exceeded.

Solubility depends on temperature; often, the higher the temperature, the greater the solubility of solids in water. The makers of rock candy (**Figure 4.22**), for example, exploit the property that more table sugar dissolves in hot water than in cold. After a large amount of sugar is dissolved in hot water, the solution is slowly cooled. An object with a slightly rough surface (such as a string or a wooden stirrer) suspended in the solution serves as a site for crystallization. As the solution cools below the temperature at which the solubility of sugar is exceeded, crystals of sugar grow on the rough surface. The formation of crystals as the temperature of a saturated solution is lowered is a form of precipitation.

FIGURE 4.21 (a) Although seawater is a concentrated aqueous solution of NaCl, it is far from saturated. (b) This solution is in equilibrium with solid NaCl (the white mass at the bottom of the beaker) at 20°C, so the liquid is a saturated solution of NaCl. Crystals in the solid are constantly dissolving, while ions in the solution are constantly precipitating. Because of this dynamic equilibrium, the concentration of solute [$Na^+(aq)$ and $Cl^-(aq)$ ions] in the solvent remains constant.

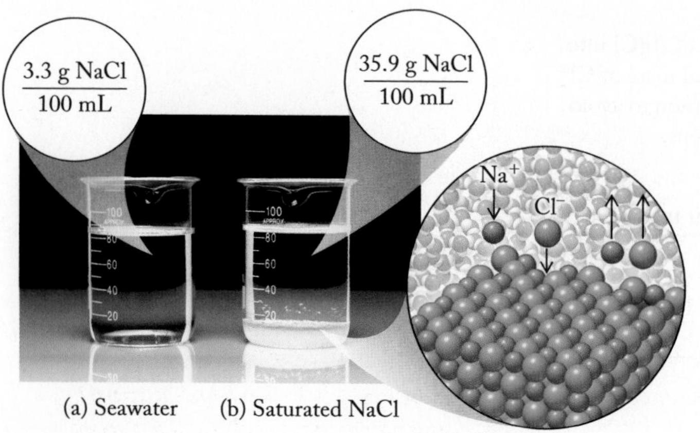

Sometimes, more solute dissolves in a volume of liquid than the amount predicted by the solute's solubility in that liquid, creating a **supersaturated solution**. This can happen when the temperature of a saturated solution drops slowly or when the volume of an unsaturated solution is reduced through slow evaporation. Slow evaporation of the solvent is partly responsible for the formation of stalactites and stalagmites, as well as of the salt flats in the American Southwest. Sooner or later, usually in response to a disruption such as a change in temperature, a mechanical shock, or the addition of a *seed crystal* (a small crystal of the solute that provides a site for crystallization), the solute in a supersaturated solution rapidly comes out of solution (**Figure 4.23**).

SAMPLE EXERCISE 4.15 Limiting Reactants in Aqueous Phase Reactions LO7

Acid rain in the form of aqueous H_2SO_4 slowly erodes marble statues, as shown in Figure 4.15. Can 16.75 mL of 1.00×10^{-2} M H_2SO_4 completely dissolve 15.0 mg of $CaCO_3$?

Collect and Organize We are given the quantities of two reactants: the volume and concentration of one and the mass of the other and are essentially being asked to determine which one is the limiting reactant. Equation 4.10 is the molecular equation that describes the reaction between dilute sulfuric acid and solid calcium carbonate.

$$CaCO_3(s) + H_2SO_4(aq) \rightarrow CaSO_4(aq) + H_2O(\ell) + CO_2(g) \qquad (4.10)$$

One mole of $H_2SO_4(aq)$ reacts with 1 mole of $CaCO_3(s)$.

Analyze One way to answer the question is to calculate the mass of $CaCO_3$ that could be consumed by all the H_2SO_4 available. The steps for doing that are:

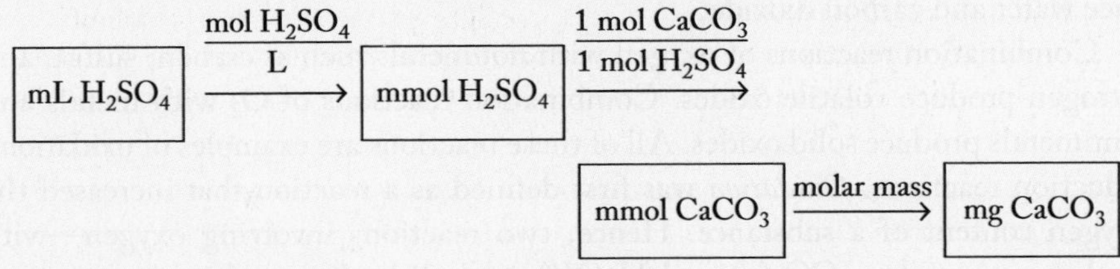

Solve Calculate the mass of calcium carbonate that could be consumed by 16.75 mL of the sulfuric acid solution:

$$16.75 \text{ mL } H_2SO_4 \times \frac{1.00 \times 10^{-2} \text{ mol } H_2SO_4}{1 \text{ L } H_2SO_4} \times \frac{1 \text{ mol } CaCO_3}{1 \text{ mol } H_2SO_4} \times \frac{100.09 \text{ g } CaCO_3}{1 \text{ mol } CaCO_3}$$

$$= 0.0168 \text{ g} = 16.8 \text{ mg } CaCO_3$$

There is more than enough sulfuric acid to completely dissolve 15.0 mg $CaCO_3$.

Think About It According to Table 4.5, $CaSO_4$ is slightly soluble in water, so perhaps you are wondering why there is an (aq) symbol after $CaSO_4$ in Equation 4.10. Actually, the concentration of a saturated solution of $CaSO_4$ is 0.015 M at 20°C. In this case, 0.01 M H_2SO_4 produces a solution that contains ~0.01 M $CaSO_4$, which is <0.015 M and should not precipitate.

Practice Exercise Heavy metal ions such as lead(II) can be precipitated from laboratory wastewater by adding sodium sulfide, Na_2S. Will all of the lead be removed from 15.00 mL of 7.52×10^{-3} M $Pb(NO_3)_2$ upon addition of 12.50 mL of 0.0115 M Na_2S?

(Answers to Practice Exercises are in the back of the book.)

(a) T = close to boiling point

(b) Put in a wooden stirrer

(c) Cool to room temperature

(d) Rock candy

FIGURE 4.22 Making rock candy. (a) A large quantity of table sugar is dissolved in hot water. The solution is not saturated at this high temperature, however. (b) A wooden stirrer is added and the solution is allowed to cool. (c) As the solution cools, it reaches the temperature at which the concentration of sugar exceeds the maximum that can remain in solution, and crystals ("rocks") of solid sugar ("candy") precipitate and attach to the stirrer. (d) Rock candy.

(a) (b) (c)

FIGURE 4.23 (a) A seed crystal is added to a supersaturated solution of sodium acetate. (b) The seed crystal becomes a site for rapid growth of sodium acetate crystals. (c) Crystal growth continues until the solution is no longer supersaturated but merely saturated with sodium acetate.

4.8 Oxidation–Reduction Reactions: Electron Transfer

CONNECTION In Chapter 3 we defined combustion as a heat-producing reaction between oxygen and another element or compound.

In the 15th century, Leonardo da Vinci observed: "Where flame cannot live, no animal that draws breath can live." Although oxygen was not isolated and recognized as an element until the 18th century, da Vinci captured an important property of the substance as a requirement for combustion and metabolism. Oxygen makes up about 50% by mass of Earth's crust, it is almost 89% by mass of the water that covers 70% of Earth's surface, and it is about 20% of the volume of Earth's atmosphere. It is essential for the metabolism of most creatures, just as it is essential for combustion reactions. What type of reaction is involved in metabolism and in the combustion reaction in **Figure 4.24**? Methane and oxygen are neither acids nor bases. No precipitates are formed in the reactions, which produce water and carbon dioxide.

FIGURE 4.24 The combustion of methane in this rocket engine being test fired at NASA is an oxidation–reduction reaction.

Combination reactions of oxygen with nonmetals such as carbon, sulfur, and nitrogen produce volatile oxides. Combination reactions of O_2 with metals and semimetals produce solid oxides. All of these reactions are examples of oxidation–reduction reactions. *Oxidation* was first defined as a reaction that increased the oxygen content of a substance. Hence, two reactions involving oxygen—with methane to produce $CO_2(g)$ and $H_2O(\ell)$, and with elemental iron to produce $Fe_2O_3(s)$—are both oxidations from the point of view of the CH_4 and the Fe because the products contain more oxygen than the starting materials:

$$CH_4(g) + 2\,O_2(g) \rightarrow CO_2(g) + 2\,H_2O(g)$$
$$4\,Fe(s) + 3\,O_2(g) \rightarrow 2\,Fe_2O_3(s)$$

Reduction reactions were originally defined in a similar fashion—namely, reactions in which the oxygen content of a substance is reduced. A classic example is the reduction of iron ore (mostly Fe_2O_3) to metallic iron:

$$Fe_2O_3(s) + 3\,CO(g) \rightarrow 2\,Fe(s) + 3\,CO_2(g)$$

The oxygen content of the iron ore is reduced—Fe_2O_3 is converted into Fe—as a result of its reaction with $CO(g)$. At the same time, though, the carbon in CO is oxidized because the oxygen content of CO increases as it is converted into CO_2.

CONCEPT **TEST**

The following equation describes the conversion of one iron mineral, called magnetite (Fe_3O_4), into another, called hematite (Fe_2O_3):

$$4\,Fe_3O_4(s) + O_2(g) \rightarrow 6\,Fe_2O_3(s)$$

Identify which element is oxidized and which is reduced and explain your choices.

(Answers to Concept Tests are in the back of the book.)

Oxidation Numbers

The processes of oxidation and reduction always occur together, which gives rise to the common way of referring to them as *redox reactions*. To distinguish between oxidation and reduction in redox reactions, chemists have developed a system of oxidation numbers to assign to atoms in reactants and products. The **oxidation number (O.N.)**, or **oxidation state**, of an atom in a molecule or ion may be a positive or negative number or zero. The oxidation number represents the positive or negative character an atom has or appears to have as determined by the following rules:

1. The oxidation numbers of the atoms in a neutral molecule sum to zero; those of the atoms in an ion sum to the charge on the ion.
2. Each atom in a pure element has an oxidation number of zero:

   F_2 O.N. = 0 for each F O_2 O.N. = 0 for each O
   Fe O.N. = 0   Na O.N. = 0

3. In monatomic ions, the oxidation number is the charge on the ion:

   F^- O.N. = −1 O^{2-} O.N. = −2
   Fe^{3+} O.N. = +3 Na^+ O.N. = +1

   Note that the number comes first when we write the symbol of an ion, Fe^{3+}, but the sign comes first when we write an oxidation number, +3.
4. In compounds containing fluorine and one or more other elements, the oxidation number of the fluorine is *always* −1:

   KF O.N. = −1 for F, which means O.N. = +1 for K (rule 1)
   OF_2 O.N. = −1 for each F, which means O.N. = +2 for O (rule 1)
   CF_4 O.N. = −1 for each F, which means O.N. = +4 for C (rule 1)

5. In most compounds, the oxidation number of hydrogen is +1 and that of oxygen is −2. Exceptions include hydrogen in metal hydrides (e.g., LiH), where O.N. = −1 for H, and oxygen in peroxides such as H_2O_2, where O.N. = −1 for each O atom.
6. Unless combined with oxygen or fluorine, chlorine and bromine have an oxidation number of −1:

   $CaCl_2$ O.N. = −1 for Cl, which means O.N. = +2 for Ca (rule 1)
   $AlBr_3$ O.N. = −1 for Br, which means O.N. = +3 for Al (rule 1), *but*
   ClO_4^- O.N. = −2 for O (rule 5), which means O.N. = +7 for Cl (rule 1)

SAMPLE EXERCISE 4.16 Determining Oxidation Numbers **LO8**

What is the oxidation number of sulfur in (a) SO_2, (b) Na_2S, and (c) $CaSO_4$?

Collect, Organize, and Analyze To assign the oxidation number of sulfur in three of its compounds, we apply the rules for determining the oxidation numbers of the elements in each compound. SO_2 is the molecule sulfur dioxide. Na_2S is an ionic compound, and the oxidation number of an ion in an ionic compound equals its charge. Sulfur in $CaSO_4$ is part of the sulfate ion, which means its oxidation number added to those of the four O atoms must add up to the charge of the ion; we determine the oxidation number for calcium from rule 1.

Solve

a. From rule 5, O.N. = −2 for the O in SO_2. Rule 1 says that the sum of the oxidation numbers for S and the two O in the neutral molecule must be zero. Letting O.N. for S be x:

$$x + 2(-2) = 0$$
$$x = +4 \qquad \text{O.N. for S in } SO_2 = +4$$

oxidation number (O.N.) (also called **oxidation state**) a positive or negative number based on the number of electrons an atom gains or loses when it forms an ion, or that it shares when it forms a covalent bond with another element; pure elements have an oxidation number of zero.

b. Sodium forms only one ion, Na^+, which means that, according to rule 3, O.N. = +1. To balance the O.N. values in Na_2S (rule 1), we let y stand for the O.N. of sulfur:

$$2(+1) + y = 0$$
$$y = -2 \qquad \text{O.N. for S in } Na_2S = -2$$

c. The charge on the calcium ion is always 2+. This means that the charge on the sulfate ion must be 2−. Assigning O.N. = −2 for oxygen (rule 5) and z for the O.N. of sulfur:

$$z + 4(-2) = -2$$
$$z = +6 \qquad \text{O.N. for S in } CaSO_4 = +6$$

Think About It We found oxidation numbers for sulfur ranging from −2 for its monatomic ion to +4 and +6 in its oxoanions. This pattern applies to other nonmetals (except O and F), though their maximum positive values vary with group number: from +4 in group 14 to +7 in group 17.

Practice Exercise Determine the oxidation number of chromium in (a) $Cr(OH)_3$, (b) $BaCrO_4$, and (c) $K_2Cr_2O_7$.

(Answers to Practice Exercises are in the back of the book.)

Changes in Oxidation Numbers in Redox Reactions

Now that we know how to assign oxidation numbers to atoms in molecules and ions, we can examine how oxidation numbers change during a redox reaction. Let's look at the formation of sodium chloride from its elements:

$$2\,Na(s) + Cl_2(g) \rightarrow 2\,NaCl(s)$$

First, we assign oxidation numbers to all the atoms in the reactants and the ions in the product:

OXIDATION NUMBERS			
Atoms in Reactants		**Ions in Product**	
Na:	0	Na^+ in NaCl:	+1
Cl in Cl_2:	0	Cl^- in NaCl:	−1

The oxidation numbers of the sodium atoms and chlorine atoms change; such a change for any element defines a reaction as redox. The sodium atom on the left changes from O.N. = 0 in Na(s) to +1 in NaCl(s), and each chlorine atom changes from O.N. = 0 in $Cl_2(g)$ to −1 in NaCl(s). Because the oxidation number of sodium *increases*, sodium is said to be *oxidized* in this reaction. Likewise, because the oxidation number of chlorine *decreases*, chlorine is said to be *reduced*. In redox reactions, we can keep track of the oxidation numbers and the electrons transferred by annotating the equation in the following way:

Sodium: 0 oxidation +1

$$2\,Na(s) + Cl_2(g) \rightarrow 2\,NaCl(s)$$

Chlorine: 0 reduction −1

oxidation a chemical change in which a species loses electrons; the oxidation number of the species increases.

reduction a chemical change in which a species gains electrons; the oxidation number of the species decreases.

oxidizing agent a substance in a redox reaction that contains the element being reduced.

CONCEPT **TEST**

Determine whether the following are redox reactions. For the redox reactions, identify the atoms that undergo changes in oxidation number.

a. $Sn^{2+}(aq) + Br_2(aq) \rightarrow Sn^{4+}(aq) + 2\ Br^-(aq)$

b. $2\ F_2(g) + 2\ H_2O(\ell) \rightarrow 4\ HF(aq) + O_2(g)$

c. $NaHCO_3(aq) + HCl(aq) \rightarrow NaCl(aq) + CO_2(g) + H_2O(\ell)$

(Answers to Concept Tests are in the back of the book.)

Electron Transfer in Redox Reactions

Ultimately, the meanings of oxidation and reduction were expanded, so that **oxidation** now refers to any chemical reaction in which a substance *loses electrons*, whereas **reduction** refers to any reaction in which a substance *gains electrons*. If one substance loses electrons, another substance must gain them; that is, if one species is oxidized, another must be reduced. The defining event in oxidation–reduction reactions is the *transfer* of electrons from one substance to another.

For example, when a piece of copper wire [elemental copper, Cu(*s*)] is placed in a colorless solution of silver nitrate [$Ag^+(aq) + NO_3^-(aq)$], the solution gradually turns blue, the color of a solution of $Cu^{2+}(aq)$, and branchlike structures of Ag(*s*) form in the medium (**Figure 4.25**). As silver ions (Ag^+) collide with the surface of the solid Cu, electrons transfer from the copper atoms to the silver ions, resulting in a layer of silver atoms depositing on the copper solid. Two electrons transfer, one to each Ag^+ ion, from one copper atom. Simultaneously, for each two Ag^+ ions that deposit as silver atoms, one copper atom enters solution as a Cu^{2+} ion. The process can be summarized as

Silver: +1 reduction 0

$$Ag^+(aq) + Cu(s) \rightarrow Ag(s) + Cu^{2+}(aq)$$

Copper: 0 oxidation +2

Because the silver ion is reduced as the copper atom is oxidized, the silver ion is the **oxidizing agent** in the redox reaction. We say that "silver ion oxidizes copper." As an oxidizing agent, the silver ion is reduced. This is always the case: the *oxidizing*

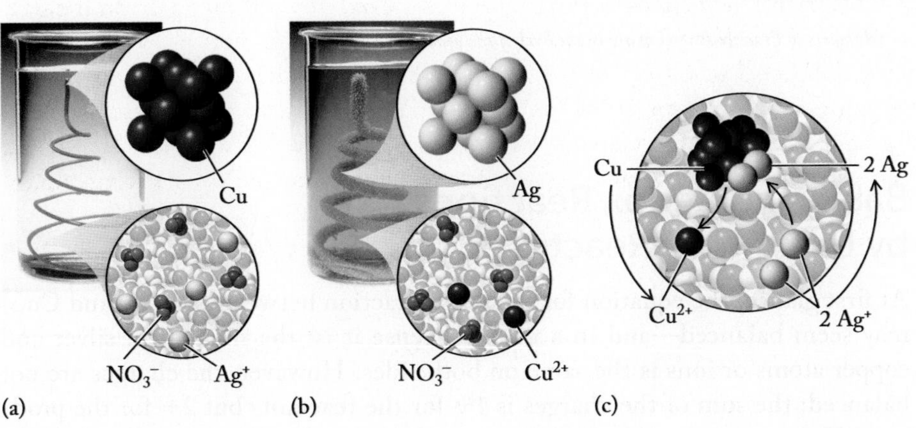

FIGURE 4.25 (a) When a Cu wire is immersed in a solution of $AgNO_3$, Cu metal oxidizes to Cu^{2+} ions as Ag^+ ions are reduced to Ag metal. (b) A day later, the solution has the blue color of a solution of $Cu(NO_3)_2$, and the wire is coated with Ag metal. (c) Copper atoms donate two electrons to silver ions, as silver atoms form a layer on the surface of the copper wire and copper ions enter solution.

reducing agent a substance in a redox reaction that contains the element being oxidized.

agent in any redox reaction is always *reduced* in the process. Any species that is reduced experiences a *reduction in oxidation number,* as is the case for the silver ion, which goes from O.N. = +1 to O.N. = 0. Likewise, as the copper atom is oxidized, the silver ion is reduced. Copper in this reaction is therefore the **reducing agent.** The *reducing* agent is always *oxidized* and experiences an *increase in oxidation number;* in this reaction, the copper atom goes from O.N. = 0 to O.N. = +2. Every redox reaction has both an oxidizing agent and a reducing agent.

SAMPLE EXERCISE 4.17 Identifying Oxidizing Agents and Reducing Agents **LO8**

Hydrazine is used as rocket fuel, but it is also used to remove dissolved oxygen gas from aqueous solutions:

$$O_2(aq) + N_2H_4(aq) \rightarrow 2\ H_2O(\ell) + N_2(g)$$

Identify the species oxidized, the species reduced, the oxidizing agent, and the reducing agent.

Collect and Organize The reactants are the element oxygen and the compound hydrazine. The products are water and nitrogen.

Analyze The O.N. of the oxygen atoms in O_2 changes from zero to −2 in H_2O. The O.N. of nitrogen changes from −2 in N_2H_4 to zero in N_2. The O.N. = +1 of hydrogen is unchanged.

Solve

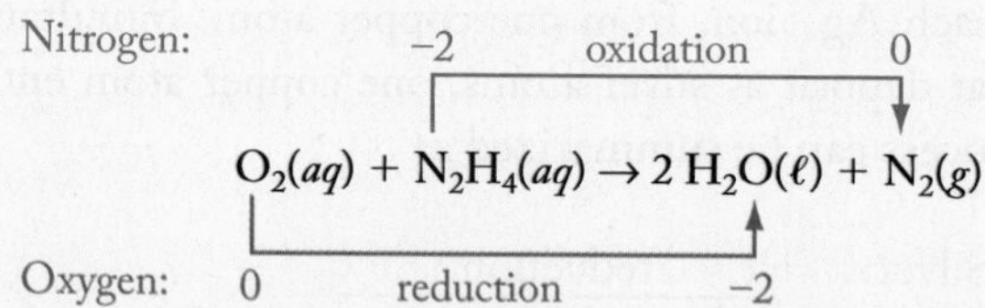

Oxygen is the oxidizing agent because the oxidation number of nitrogen increases. This means that O_2 is reduced and N_2H_4 is the reducing agent.

Think About It We tend to identify the specific *atoms* that are oxidized or reduced, whereas whole *molecules or ions* are identified as the oxidizing agents or reducing agents. In the equation in this exercise, nitrogen is oxidized but N_2H_4 is the reducing agent.

Practice Exercise In the reaction between $O_2(g)$ and $SO_2(g)$ to make $SO_3(g)$, identify the species oxidized, the species reduced, the oxidizing agent, and the reducing agent.

(Answers to Practice Exercises are in the back of the book.)

Balancing Redox Reactions by Using Half-Reactions

At first glance, the equation for the redox reaction between $Ag^+(aq)$ and Cu(*s*) may seem balanced—and in a material sense it is: the number of silver and copper atoms or ions is the same on both sides. However, the charges are not balanced: the sum of the charges is 1+ for the reactants but 2+ for the products. The reason for this imbalance is that one electron is involved in the reduction reaction but two are involved in the oxidation reaction. Equal numbers of electrons must be lost and gained, so to balance equations describing redox reactions like this one, we must develop a method that accounts for electron transfer.

The change in oxidation number reflects the change in the number of electrons associated with an atom or ion. This change in the number of electrons happens because electrons are transferred from one atom or ion to another during a redox reaction. By tracking changes in oxidation numbers, we can gain additional insight into the transfer of electrons as the redox reaction takes place.

Think of this reaction between Cu(*s*) and $AgNO_3(aq)$ as consisting of one oxidation and one reduction. Each of these reactions represents *half* of the overall reaction; hence, each is called a **half-reaction**. We can write a balanced chemical equation by combining half-reactions following these five steps:

1. *Write one expression for the oxidation half-reaction and a separate one for the reduction half-reaction*:

   Oxidation: $Cu(s) \rightarrow Cu^{2+}(aq)$
   Reduction: $Ag^{+}(aq) \rightarrow Ag(s)$

2. *Balance the number of atoms of each element in each half-reaction.* This example requires no changes at this point because both contain single atoms of one element on both sides of the reaction arrow.
3. *Balance the charges* by adding electrons to the appropriate side. Always *add* electrons; never subtract them. We add 2 electrons to the product side of the Cu half-reaction:

   Oxidation: $Cu(s) \rightarrow Cu^{2+}(aq) + 2\ e^{-}$

   and 1 electron to the reactant side of the Ag half-reaction:

   Reduction: $1\ e^{-} + Ag^{+}(aq) \rightarrow Ag(s)$

   This gives us a total charge of 0 on both sides of both half-reactions. Note that electrons are added on the reactant side of the reduction equation (reduction is the gain of electrons) and on the product side of the oxidation equation (oxidation is the loss of electrons). This is always the case.
4. *Multiply each equation by the appropriate whole number* to make the number of electrons lost in the oxidation half-reaction equal the number of electrons gained in the reduction half-reaction:

   Oxidation: $Cu(s) \rightarrow Cu^{2+}(aq) + 2\ e^{-}$
   Reduction: $2 \times [1\ e^{-} + Ag^{+}(aq) \rightarrow Ag(s)]$

   As a result we have the following half-reactions:

   Oxidation: $Cu(s) \rightarrow Cu^{2+}(aq) + 2\ e^{-}$
   Reduction: $2\ e^{-} + 2\ Ag^{+}(aq) \rightarrow 2\ Ag(s)$

5. *Add the two equations to generate one that represents the overall redox reaction*:

   Oxidation: $Cu(s) \rightarrow Cu^{2+}(aq) + \cancel{2\ e^{-}}$
   Reduction: $\cancel{2\ e^{-}} + 2\ Ag^{+}(aq) \rightarrow 2\ Ag(s)$

   Overall equation: $2\ Ag^{+}(aq) + Cu(s) \rightarrow 2\ Ag(s) + Cu^{2+}(aq)$

If we have carried out step 4 correctly, the number of electrons in the oxidation half-reaction is the same as the number of electrons in the reduction half-reaction, and they cancel out when we write the overall equation. In redox reactions, we can

CONNECTION In Chapter 2 we introduced nuclear equations, which must also be balanced in both mass and charge.

STEPWISE ANIMATION
Balancing Redox Reactions

half-reaction one of the two halves of an oxidation–reduction reaction; one half-reaction is the oxidation component, and the other is the reduction component.

keep track of the oxidation numbers and the electrons transferred by annotating the equation in the following way:

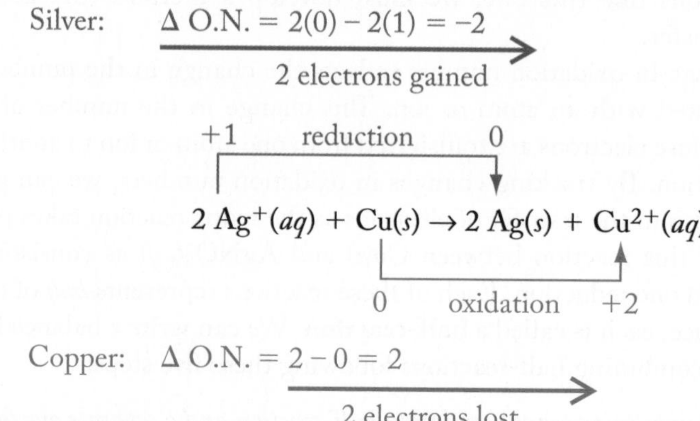

The equation is now balanced in terms of number of atoms and charge. The equation is a net ionic equation because only species changed by the redox reaction appear in it. The nitrate ions from the silver nitrate solution (Figure 4.25) are spectator ions and are not shown in the Ag half-reaction or the net ionic equation.

We review the methods for balancing chemical equations as described here when we discuss applications of oxidation–reduction reactions in Chapter 18.

(a)

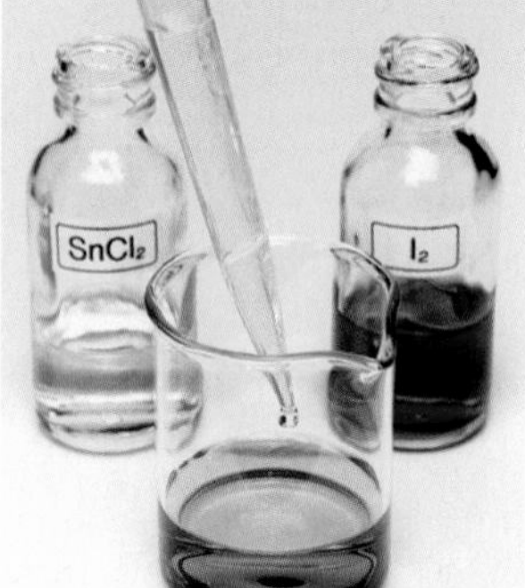

(b)

(c)

FIGURE 4.26 (a) The beaker contains an aqueous solution of iodine. (b) When drops of a solution of $SnCl_2$ are added to the beaker, the yellow-brown color of the iodine solution begins to fade as Sn^{2+} ions reduce I_2 to colorless I^- ions. (c) The color disappears when enough Sn^{2+} has been added to reduce all the I_2.

SAMPLE EXERCISE 4.18 Balancing Redox Reactions with Half-Reactions LO9

Iodine is slightly soluble in water forming yellow-brown aqueous solutions. When colorless $Sn^{2+}(aq)$ is dissolved in it (**Figure 4.26**), the solution turns colorless as $I^-(aq)$ forms and $Sn^{2+}(aq)$ is converted into $Sn^{4+}(aq)$. (a) Is this a redox reaction? (b) Write the net ionic equation describing the reaction.

Collect and Organize The reactants in this process are molecules of I_2 and Sn^{2+} ions; the products are I^- and Sn^{4+} ions.

Analyze The oxidation number of tin increases from +2 in Sn^{2+} ions to +4 in Sn^{4+} ions. The oxidation number of iodine is reduced from 0 in molecules of I_2 to −1 in I^- ions. Combining half-reactions is one way to write a net ionic equation for a redox reaction, so we need to write half-reactions based on the oxidation of Sn^{2+} ions to Sn^{4+} ions and on the reduction of I_2 to I^- ions. Tin and iodine are in separate equations when we write the respective half-reactions.

Solve

a. The oxidation numbers of the atoms of two elements change, so this is a redox reaction.
b. Writing the symbols of all reactants and products:

$$Sn^{2+}(aq) + I_2(aq) \rightarrow I^-(aq) + Sn^{4+}(aq)$$

Balancing this expression using the five-step process.

1. Separate the half-reactions.

Oxidation: $Sn^{2+}(aq) \rightarrow Sn^{4+}(aq)$
Reduction: $I_2(aq) \rightarrow I^-(aq)$

2. Balance the number of atoms of each element.

Oxidation: $Sn^{2+}(aq) \rightarrow Sn^{4+}(aq)$
Reduction: $I_2(aq) \rightarrow 2\,I^-(aq)$

3. Balance the charges by adding electrons.

Oxidation: $Sn^{2+}(aq) \rightarrow Sn^{4}+(aq) + 2\,e^-$
Reduction: $2\,e^- + I_2(aq) \rightarrow 2\,I^-(aq)$

4. Balance the numbers of electrons gained and lost. This step is not needed because they're already equal.
5. Combine the two half-reactions.

$$\begin{array}{ll} \text{Oxidation:} & Sn^{2+}(aq) \rightarrow Sn^{4+}(aq) + \cancel{2\,e^-} \\ \text{Reduction:} & \cancel{2\,e^-} + I_2(aq) \rightarrow 2\,I^-(aq) \\ \hline \text{Overall equation:} & Sn^{2+}(aq) + I_2(aq) \rightarrow 2\,I^-(aq) + Sn^{4+}(aq) \end{array}$$

Think About It To assess the accuracy of the overall chemical equation,

1. check for atom balance: 1 Sn + 2 I = 2 I + 1 Sn ✓
2. check for charge balance: reactant side 2+, product side 2(1−) + (4+) = 2+ ✓.

Practice Exercise When a nail made of Fe(*s*) is placed in an aqueous solution of a soluble palladium(II) salt [containing $Pd^{2+}(aq)$], it gradually disappears because the iron enters the solution as $Fe^{3+}(aq)$, and palladium metal [Pd(*s*)] forms. (a) Is this a redox reaction? (b) Balance the equation that describes this reaction.

(Answers to Practice Exercises are in the back of the book.)

The Activity Series of Metals

If we replaced the $AgNO_3(aq)$ solution in Figure 4.25a with a $Zn(NO_3)_2$ solution, no reaction between it and Cu metal would be observed. By immersing copper wire in solutions of other metal cations, we would find, for example, that Cu metal reduces Ag^+ and Au^{3+} ions to Ag and Au metal, but it does not reduce Zn^{2+}, Ni^{2+}, and Al^{3+} ions to their metals. Similarly, we could explore the ability of zinc metal to reduce these same cations and find that all except Al^{3+} (and Zn^{2+}) are reduced by Zn metal.

Why do particular metals reduce some metal cations but not others? A detailed explanation awaits in Chapter 18, but for now we can use the **activity series** in **Table 4.6** to rank the strengths of metals as reducing agents: from the highly reactive group 1 and group 2 metals at the top of the table to the largely inert metals, such as platinum and gold, at the bottom. Metals at the top of the table can reduce the cations of metals below them as the top metal's oxidation half-reaction proceeds as written in the table, whereas the lower metal's cations are reduced to metal atoms as its half-reaction proceeds in reverse—as a reduction half-reaction.

SAMPLE EXERCISE 4.19 Using the Activity Series I **LO8**

An aluminum wire is dipped into a solution containing Na^+, Sn^{2+}, and Cu^{2+} ions. Which ions will be present in the solution after the reactions are complete?

Collect and Organize We are asked to predict which metal ions in solution will be reduced by aluminum. We use the activity series in Table 4.6 to guide our answer.

Analyze The activity series is arranged based on metals' decreasing strengths as reducing agents. Alternatively, the ease of oxidation of the corresponding metals at the top of the series can reduce the cations of the metals below them.

activity series a high-to-low ranking of metals based on their strengths as reducing agents.

Solve Aluminum is above copper and tin but below sodium in Table 4.6. Therefore, Al reduces both Cu^{2+} and Sn^{2+} ions, removing them from the solution, but Na^+ ions remain. Aluminum is oxidized as Cu^{2+} and Sn^{2+} ions are reduced, so Al^{3+} ions are also present in the solution when the reaction is over.

Think About It The arrangement of metals in the activity series makes sense from what we know about the chemical reactivity of groups of metals: the highly reactive group 1 and group 2 metals are clustered at the top and many of those that are stable as free metals such as gold, silver, and copper, which are used to make coins, are near the bottom.

Practice Exercise Which substance listed in Table 4.6 is most easily oxidized? Which substance is most easily reduced?

(Answers to Practice Exercises are in the back of the book.)

Why is hydrogen gas listed in Table 4.6? What does its position in the activity series tell us? Table 4.6 allows us to predict which metals will react with strong aqueous acids like HCl. Any metal below H_2 in Table 4.6 will not react with H^+ and hence will be resistant to most strong acids. One reason coins and jewelry made from silver, gold, and platinum last is that they do not react with most acids.

TABLE 4.6 An Activity Series for Metals in Aqueous Solution

Metal	Oxidation Reaction
Lithium	$Li(s) \rightarrow Li^+(aq) + e^-$
Potassium	$K(s) \rightarrow K^+(aq) + e^-$
Barium	$Ba(s) \rightarrow Ba^{2+}(aq) + 2\ e^-$
Calcium	$Ca(s) \rightarrow Ca^{2+}(aq) + 2\ e^-$
Sodium	$Na(s) \rightarrow Na^+(aq) + e^-$
Magnesium	$Mg(s) \rightarrow Mg^{2+}(aq) + 2\ e^-$
Aluminum	$Al(s) \rightarrow Al^{3+}(aq) + 3\ e^-$
Manganese	$Mn(s) \rightarrow Mn^{2+}(aq) + 2\ e^-$
Zinc	$Zn(s) \rightarrow Zn^{2+}(aq) + 2\ e^-$
Chromium	$Cr(s) \rightarrow Cr^{3+}(aq) + 3\ e^-$
Iron	$Fe(s) \rightarrow Fe^{2+}(aq) + 2\ e^-$
Cobalt	$Co(s) \rightarrow Co^{2+}(aq) + 2\ e^-$
Nickel	$Ni(s) \rightarrow Ni^{2+}(aq) + 2\ e^-$
Tin	$Sn(s) \rightarrow Sn^{2+}(aq) + 2\ e^-$
Lead	$Pb(s) \rightarrow Pb^{2+}(aq) + 2\ e^-$
Hydrogen	$H_2(g) \rightarrow 2\ H^+(aq) + 2\ e^-$
Copper	$Cu(s) \rightarrow Cu^{2+}(aq) + 2\ e^-$
Silver	$Ag(s) \rightarrow Ag^+(aq) + e^-$
Mercury	$Hg(\ell) \rightarrow Hg^{2+}(aq) + 2\ e^-$
Platinum	$Pt(s) \rightarrow Pt^{2+}(aq) + 2\ e^-$
Gold	$Au(s) \rightarrow Au^{3+}(aq) + 3\ e^-$

SAMPLE EXERCISE 4.20 Using the Activity Series II **LO8**

Use the activity series to predict the products of the reaction between zinc metal and aqueous hydrochloric acid and write a net ionic equation describing the reaction.

Collect, Organize, and Analyze The reactants in this process are $Zn(s)$ and $HCl(aq)$, which is a strong acid that forms $H^+(aq)$ and $Cl^-(aq)$ ions in aqueous solutions. The relevant half-reactions in Table 4.6 are

$$Zn(s) \rightarrow Zn^{2+}(aq) + 2\,e^- \quad \text{and} \quad H_2(g) \rightarrow 2\,H^+(aq) + 2\,e^-$$

Zinc is listed above hydrogen in Table 4.6.

Solve Placement of the half-reactions in the table tells us that zinc metal reduces H^+ ions, producing Zn^{2+} ions (as the Zn half-reaction runs in the forward direction) and H_2 gas (as the H_2 half-reaction runs in reverse).

Writing the two half-reactions in the appropriate directions and combining them:

$$\begin{aligned} Zn(s) &\rightarrow Zn^{2+}(aq) + \cancel{2\,e^-} \\ 2\,H^+(aq) + \cancel{2\,e^-} &\rightarrow H_2(g) \\ \hline Zn(s) + 2\,H^+(aq) &\rightarrow Zn^{2+}(aq) + H_2(g) \end{aligned}$$

Think About It The activity series is a list of net ionic equations describing half-reactions, so selecting the two that are related to a redox reaction of interest—and reversing the one that is lower in the series before combining it with the other—is a convenient way to write a net ionic equation for the overall reaction. Knowing that H^+ ions were a reactant was another clue that the H_2 half-reaction ran in reverse as a reduction. Combining the two was made easy because two electrons were gained in one half-reaction and lost in the other.

Practice Exercise Predict the products of the reaction between aluminum metal and silver nitrate. Write a net ionic equation describing the reaction.

(Answers to Practice Exercises are in the back of the book.)

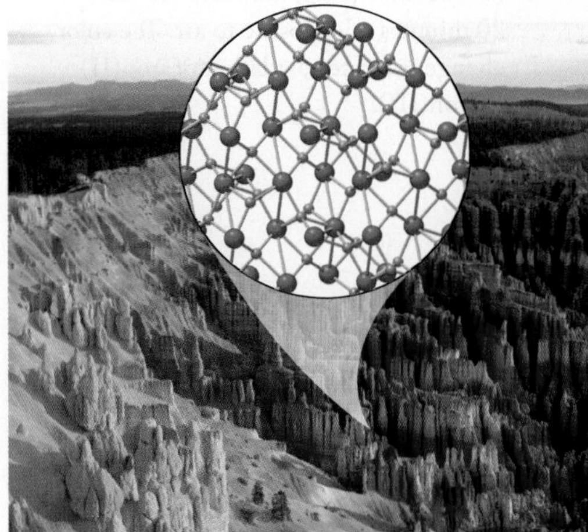

FIGURE 4.27 Red-orange rock formations called hoodoos in Bryce Canyon, Utah, owe their color to high concentrations of iron(III) oxide that result from weathering of the rock.

Redox in Nature

Redox processes play a major role in determining the character of Earth's rocks and soils. For example, the orange-red rocks of Bryce Canyon National Park in Utah contain the iron(III) oxide mineral called hematite (**Figure 4.27**). Rocks containing hematite are relatively weak and easily broken, causing the fascinating shapes of such deposits. Iron(III) oxide is also the form of iron known as rust—the crumbly, red-brown solid that forms on iron objects exposed to water and oxygen.

Soil color and mineral content are influenced by moisture content. Wetlands, which are areas that are either saturated or flooded with water during much of the year, are protected environments in many areas of the United States. Defining a given area as a wetland relies in part on the characteristics of the soils, especially their color (**Figure 4.28**), which is strongly influenced by the redox chemistry of iron.

Figure 4.29a shows a solution of iron(II) ammonium sulfate, $(NH_4)_2Fe(SO_4)_2(aq)$, upon addition of $NaOH(aq)$. The pale greenish color of the liquid is characteristic of $Fe^{2+}(aq)$. Addition of $NaOH(aq)$ causes $Fe(OH)_2(s)$ to precipitate as a blue-gray solid, similar in color to the wetland soils shown in Figure 4.28. When this mixture is filtered, the iron(II) hydroxide residue immediately starts to darken (**Figure 4.29b**). After about 20 minutes of exposure to

FIGURE 4.28 The soil in this wetland has a mostly blue-gray color because of the presence of iron(II) compounds. However, the orange-red mottling in the soil indicates that atmospheric O_2 penetrated into the soil and oxidized some of the Fe(II) compounds to Fe(III) oxide.

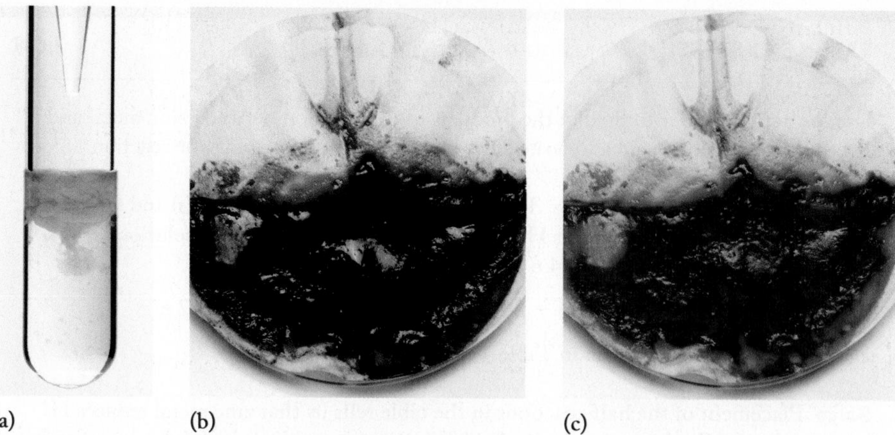

FIGURE 4.29 (a) A solution of iron(II) ammonium sulfate upon addition of NaOH(*aq*). The precipitate is iron(II) hydroxide. (b) The iron(II) hydroxide precipitate immediately after filtration. (c) The same precipitate after about 20 minutes of exposure to air. The color change indicates oxidation of iron(II) to iron(III).

oxygen in the air (**Figure 4.29c**), the precipitate turns the orange-red color of iron(III) hydroxide. The iron(II) has been oxidized—just as the iron(II) compounds in a wetland soil are oxidized when exposed to oxygen.

Redox reactions take place in acidic, neutral, or basic solutions, and H^+ ions, OH^- ions, or even molecules of water may play roles in these reactions. For example, a method for determining the concentration of $Fe^{2+}(aq)$ involves its oxidation to $Fe^{3+}(aq)$ by intensely purple permanganate, MnO_4^- ions (**Figure 4.30**), which are reduced to $Mn^{2+}(aq)$. Thus, the key components of the redox reaction are

$$\underset{\text{nearly colorless}}{Fe^{2+}(aq)} + \underset{\text{deep purple}}{MnO_4^-(aq)} \rightarrow \underset{\text{yellow-orange}}{Fe^{3+}(aq)} + \underset{\text{pale pink}}{Mn^{2+}(aq)}$$

A chemical equation describing the reaction will be balanced when (a) the number of atoms of each element on both sides of the reaction is the same and (b) the total charges on each side of the reaction arrow are the same. We proceed in the same way as for the previous examples in neutral solutions.

1. Separate iron and manganese into their respective half-reactions.

   Oxidation: $Fe^{2+}(aq) \rightarrow Fe^{3+}(aq)$
   Reduction: $MnO_4^-(aq) \rightarrow Mn^{2+}(aq)$

2. Balance the atoms of the elements being oxidized or reduced.

   a. Oxidation: $Fe^{2+}(aq) \rightarrow Fe^{3+}(aq)$ (Fe atoms balanced)
   Reduction: $MnO_4^-(aq) \rightarrow Mn^{2+}(aq)$ (Mn atoms balanced, but not O atoms)

   Balance O atoms by adding four molecules of H_2O.

   b. Oxidation: $Fe^{2+}(aq) \rightarrow Fe^{3+}(aq)$
   Reduction: $MnO_4^-(aq) \rightarrow Mn^{2+}(aq) + 4\,H_2O(\ell)$
   (O atoms balanced, H atoms are not)

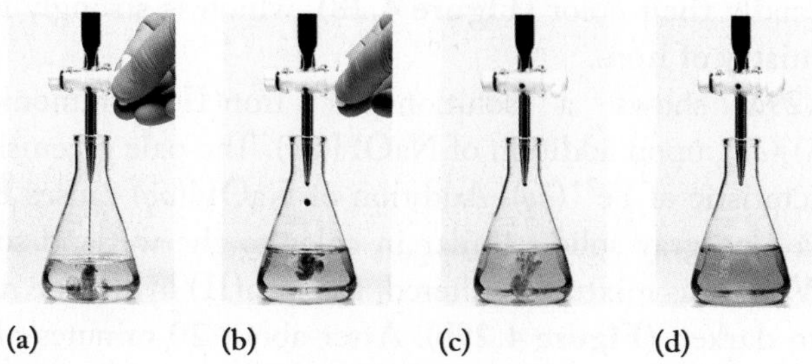

FIGURE 4.30 In this colorful redox titration, (a) a dark purple standard solution of permanganate, MnO_4^-, ions is added to an acidic sample containing nearly colorless Fe^{2+} ions. (b) As the titration reaction proceeds, Fe^{2+} ions are oxided to Fe^{3+} ions while MnO_4^- ions are reduced to Mn^{2+} ions. (c) Increasing concentrations of pale yellow Fe^{3+} ions and faintly pink Mn^{2+} ions combine to give the sample a pale amber color. (d) When all the Fe^{2+} ions have been consumed, the next drop of unreacted purple MnO_4^- ions gives the sample a peach color that signals the end point of the titration.

In neutral or acidic solutions, balance H atoms by adding H^+ ions:

c. Oxidation: $Fe^{2+}(aq) \rightarrow Fe^{3+}(aq)$

Reduction: $8\ H^+(aq) + MnO_4^-(aq) \rightarrow Mn^{2+}(aq) + 4\ H_2O(\ell)$

(all atoms balanced)

3. Balance charges by adding electrons.

Oxidation: $Fe^{2+}(aq) \rightarrow Fe^{3+}(aq) + 1\ e^-$

Reduction: $5\ e^- + 8\ H^+(aq) + MnO_4^-(aq) \rightarrow Mn^{2+}(aq) + 4\ H_2O(\ell)$

Note that the number of electrons added to the reduction half-reaction also accounts for the oxidation state of Mn changing from +7 to +2.

4. Balance numbers of electrons lost in the oxidation half-reaction and gained in the reduction half-reaction.

Oxidation: $5 \times [Fe^{2+}(aq) \rightarrow Fe^{3+}(aq) + 1\ e^-]$

Reduction: $1 \times [5\ e^- + 8\ H^+(aq) + MnO_4^-(aq) \rightarrow Mn^{2+}(aq) + 4\ H_2O(\ell)]$

5. Add the two equations and cancel species that appear on both sides of the reaction arrow.

Oxidation: $5\ Fe^{2+}(aq) \rightarrow 5\ Fe^{3+}(aq) + \cancel{5\ e^-}$

Reduction: $\cancel{5\ e^-} + 8\ H^+(aq) + MnO_4^-(aq) \rightarrow Mn^{2+}(aq) + 4\ H_2O(\ell)]$

$$8\ H^+(aq) + MnO_4^-(aq) + 5\ Fe^{2+}(aq) \rightarrow 5\ Fe^{3+}(aq) + Mn^{2+}(aq) + 4\ H_2O(\ell)$$

Always check the atom and charge balance of the final net ionic equation. The reactant side contains 8 H, 1 Mn, 4 O, and 5 Fe atoms and the product side contains 8 H, 1 Mn, 4 O, and 5 Fe atoms, so the atoms are balanced. On the reactant side, the charge is (8+) + (1−) + (10+) = 17+, and on the product side it is (15+) + (2+) = 17+, so the charges are balanced, too.

This method works for writing the net ionic equation for any redox reaction taking place in neutral or acidic aqueous solutions. The oxidation in Figure 4.29, however, was carried out in a basic medium, so how do we write a chemical equation for that reaction? The procedure is almost the same as for acidic solutions: in fact, we carry out the same steps as in the previous exercise, but we add one more in which hydroxide ions are added to neutralize any H^+ ions, forming molecules of H_2O.

SAMPLE EXERCISE 4.21 Balancing Redox Equations That Involve Hydroxide Ions — LO9

Wetland soil is blue-gray because it contains $Fe(OH)_2(s)$, whereas well-aerated soils are often orange-red because of the presence of $Fe(OH)_3(s)$ (Figure 4.28). Write the balanced equation for the reaction of $O_2(g)$ with $Fe(OH)_2(s)$ in soil that produces $Fe(OH)_3(s)$ in basic solution.

Collect and Organize We know the reactants and product and can write the unbalanced equation:

$$Fe(OH)_2(s) + O_2(g) \rightarrow Fe(OH)_3(s)$$

Our task is to balance the equation with respect to both mass and charge.

Analyze This reaction takes place with both iron compounds in the solid phase in the presence of hydroxide ions in water. For the oxidation half-reaction, iron(II) hydroxide is oxidized to iron(III) hydroxide. For the reduction half-reaction, O_2 is converted into the additional OH^- ion in the iron(III) hydroxide product.

Solve

1. Separate the equations.

   Oxidation: $Fe(OH)_2(s) \rightarrow Fe(OH)_3(s)$

   Reduction: $O_2(g) \rightarrow OH^-(aq)$

2. Balance the atoms.

   a. Balance all atoms except H and O. This step is not needed because Fe is already balanced.

   b. Balance O atoms by adding water as needed.

   Oxidation: $H_2O(\ell) + Fe(OH)_2(s) \rightarrow Fe(OH)_3(s)$

   Reduction: $O_2(g) \rightarrow OH^-(aq) + H_2O(\ell)$

   c. Balance H atoms by adding H^+ ions as needed.

   $$H_2O(\ell) + Fe(OH)_2(s) \rightarrow Fe(OH)_3(s) + H^+(aq)$$

   $$3\ H^+(aq) + O_2(g) \rightarrow OH^-(aq) + H_2O(\ell)$$

3. Balance the charges.

   $$H_2O(\ell) + Fe(OH)_2(s) \rightarrow Fe(OH)_3(s) + H^+(aq) + 1\ e^-$$

   $$4\ e^- + 3\ H^+(aq) + O_2(g) \rightarrow OH^-(aq) + H_2O(\ell)$$

4. Balance the numbers of electrons lost and gained.

   $$4 \times [H_2O(\ell) + Fe(OH)_2(s) \rightarrow Fe(OH)_3(s) + H^+(aq) + 1\ e^-]$$

   $$1 \times [4\ e^- + 3\ H^+(aq) + O_2(g) \rightarrow OH^-(aq) + H_2O(\ell)]$$

5. Add the two equations.

   $$3\ \cancel{4}\ H_2O(\ell) + 4\ Fe(OH)_2(s) + \cancel{4\ e^-} + \cancel{3\ H^+(aq)} + O_2(g) \rightarrow$$

   $$4\ Fe(OH)_3(s) + \cancel{3}\ 4\ H^+(aq) + \cancel{4\ e^-} + OH^-(aq) + \cancel{H_2O(\ell)}$$

   This gives us the balanced equation:

   $$3\ H_2O(\ell) + 4\ Fe(OH)_2(s) + O_2(g) \rightarrow 4\ Fe(OH)_3(s) + OH^-(aq) + H^+(aq)$$

   This equation can be simplified further because $OH^- + H^+$ form $H_2O(\ell)$:

   $$3\ H_2O(\ell) + 4\ Fe(OH)_2(s) + O_2(g) \rightarrow 4\ Fe(OH)_3(s) + \underbrace{OH^-(aq) + H^+(aq)}_{H_2O(\ell)}$$

   Canceling the molecule of H_2O common to both sides produces a balanced equation:

   $$2\ H_2O(\ell) + 4\ Fe(OH)_2(s) + O_2(g) \rightarrow 4\ Fe(OH)_3(s)$$

Think About It A check of the equation reveals that it is balanced in terms of both the number of atoms of each element and electrical charges (there aren't any). Another approach to writing chemical equations for redox reactions balances the changes in oxidation number in the first step instead of balancing the losses and gains of electrons in the last step.

The reaction above involves the oxidation of the Fe^{2+} ions $Fe(OH)_2(s)$ to Fe^{3+} ions in $Fe(OH)_3(s)$. This means the change in oxidation number (ΔO.N.) for each iron ion is +1. The oxidation number for each of the 2 O atoms in a molecule of O_2 changes from 0 to −2 as it becomes part of an OH^- ion on the product side. The overall ΔO.N. per molecule of O_2 is 2 atoms × (−2/atom) = −4. To balance the +1 and −4 changes in oxidation number, we need 4 iron ions for every 1 molecule of O_2 so that $4 \times (+1) - 4 = 0$. A preliminary expression based on the 4:1 mole ratio looks like this

$$4\ Fe(OH)_2(s) + O_2(g) \rightarrow 4\ Fe(OH)_3(s)$$

The Fe atoms are balanced, but the right side has 12 H atoms and 12 O atoms whereas the left side has only 8 H atoms and 10 O atoms. To correct this imbalance, we add 2 molecules of H_2O to the left side, producing a balanced chemical equation:

$$4\ Fe(OH)_2(s) + O_2(g) + 2\ H_2O(\ell) \rightarrow 4\ Fe(OH)_3(s)$$

Practice Exercise Hydroperoxide ions, $HO_2^-(aq)$, react with permanganate ions, $MnO_4^-(aq)$, producing $MnO_2(s)$ and oxygen gas. Balance the equation for the oxidation of hydroperoxide ion to $O_2(g)$ by a permanganate ion in a basic solution.

(Answers to Practice Exercises are in the back of the book.)

Reactions in aqueous solutions are an integral part of our daily lives. On a large scale, in oceans, rivers, and rain, they shape our physical world. Many reactions that produce the substances that are part of modern life—from paint pigments to drugs—are run in water, and many analytical procedures rely on reactions in water to determine the content of aqueous solutions that we drink, swim in, and use in countless consumer products such as car batteries and shampoos. On a small scale, reactions in the water within the cells of our bodies and in all living organisms make the chemical processes that are essential to life possible. On Earth, there may be water without life, but there is no life without water.

SAMPLE EXERCISE 4.22 Integrating Concepts: Shelf-Stability of Drugs

Commercial pharmaceutical agents undergo extensive analysis to establish how long they may be stored without degradation and loss of potency. A candidate drug, a diprotic acid, has the following percent composition from combustion analysis: C, 62.50%; H, 4.20%. A standard tablet containing 325 mg of the drug is dissolved in 100.00 mL of water and titrated to the equivalence point with 16.45 mL of 0.2056 *M* NaOH(*aq*). After storage at 50°C under high humidity for one month, a second tablet, when dissolved in 50.00 mL of water, requires 10.10 mL of 0.1755 *M* NaOH(*aq*) to be completely neutralized.

a. What is the empirical formula of the drug?
b. What is its molar mass?
c. What is its molecular formula?
d. What is the percent of active drug substance remaining in the tablet after storage?

Collect and Organize We are given the percent composition of the drug and information from titration experiments. From these data we can find the empirical formula, the molar mass, and the amount of active drug in the stored tablet.

Analyze The percent composition data from the combustion analysis do not add to 100%. We may assume that the amount missing results from oxygen in the molecule. The drug is a diprotic acid, so neutralizing 1 mole of the drug requires 2 moles of NaOH. The titration of the drug requires about 20 mL of 0.2 *M* NaOH, which is about 0.004 mol OH^-. That means the pure sample is about 0.002 mol of the drug; if about 0.3 g contains 0.002 mol, the drug must have a molar mass of about 150 g. If any of the drug decomposes upon storage, we should have less than 325 mg in the second sample.

Solve

a. The percent oxygen in the drug molecule is

$$100.00\% - (62.50\% + 4.20\%) = 33.30\%$$

Calculating the moles of C, H, and O in exactly 100 g of the drug:

$$\frac{62.50 \text{ g C}}{12.01 \text{ g/mol}} = 5.20 \text{ mol C}$$

$$\frac{4.20 \text{ g H}}{1.008 \text{ g/mol}} = 4.17 \text{ mol H}$$

$$\frac{33.30 \text{ g O}}{16.00 \text{ g/mol}} = 2.08 \text{ mol O}$$

Reducing these quantities to ratios of small whole numbers,

$$\frac{5.20 \text{ g C}}{2.08} = 2.50 \text{ mol C}$$

$$\frac{4.17 \text{ mol H}}{2.08} = 2.00 \text{ mol H}$$

$$\frac{2.08 \text{ mol O}}{2.08} = 1.00 \text{ mol O}$$

requires multiplying by 2, which yields

$$5 \text{ mol C} : 4 \text{ mol H} : 2 \text{ mol O}$$

The empirical formula is $C_5H_4O_2$.

b. To find the molar mass of the drug from the titration data, we first find the number of moles of drug in the sample:

$$0.01645 \text{ }\cancel{\text{L NaOH}} \times \frac{0.2056 \cancel{\text{ mol NaOH}}}{1 \cancel{\text{ L NaOH}}} \times \frac{1 \text{ mol drug}}{2 \cancel{\text{ mol NaOH}}}$$

$$= 0.001691 \text{ mol drug}$$

If 0.325 g of drug is 0.001691 mol of drug, then the mass of 1 mole is

$$\mathcal{M}_{drug} = \frac{0.325 \text{ g}}{0.001691 \text{ mol}} = 192.2 \text{ g/mol}$$

c. The mass of 1 mole of empirical formula units ($C_5H_4O_2$) is

$$(5 \times 12.01 \text{ g/mol}) + (4 \times 1.008 \text{ g/mol}) + (2 \times 16.00 \text{ g/mol}) = 96.08 \text{ g/mol}$$

The number of moles of formula units per mole of molecules is

$$\frac{192.16 \frac{\text{g}}{\text{mol}}}{96.08 \frac{\text{g}}{\text{mol}}} = 2$$

The molecular formula of the drug, then, is

$$(C_5H_4O_2)_2 = C_{10}H_8O_4$$

d. The stored tablet contains

$$0.01010 \text{ L NaOH} \times \frac{0.1755 \text{ mol NaOH}}{1 \text{ L NaOH}} \times \frac{1 \text{ mol drug}}{2 \text{ mol NaOH}} = 0.0008863 \text{ mol drug} = 8.863 \times 10^{-4} \text{ mol}$$

Comparing this value with the original amount yields:

$$\frac{8.863 \times 10^{-4} \text{ mol}}{1.691 \times 10^{-3} \text{ mol}} \times 100\% = 52.41\%$$

Thus, 52.41% of the active drug is left in the tablet.

Think About It The molar mass we calculated is close to our estimated value, so it seems reasonable. If only half the active agent is present in the tablet after storage under these conditions, the manufacturer may have to reformulate the drug or package it differently to protect it from its surroundings.

SUMMARY

LO1 The **concentration** of **solute** in a solution can be expressed many different ways: as mass of solute per mass of solution (such as grams of solute per kilogram of solution), parts per million (1 ppm = 1 μg solute/g solution = 1 mg solute/kg solution), or parts per billion (1 ppb = 1 μg solute/kg solution). Solute concentration can also be expressed as mass of solute per volume of **solvent** and as moles of solute per liter of solution, or **molarity (*M*)**. One set of units can be converted into another by using dimensional analysis. (Sections 4.1 and 4.2)

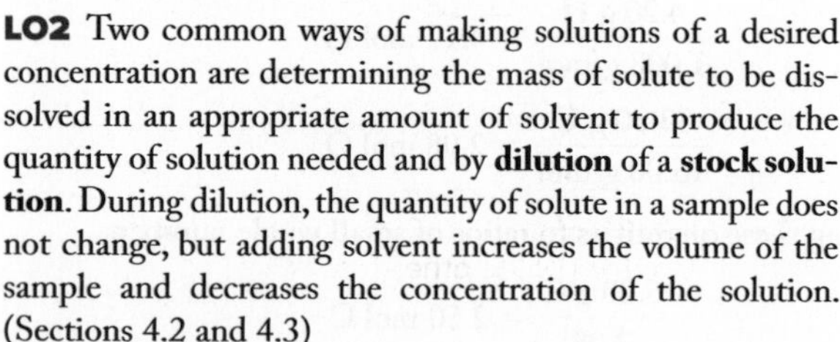

LO2 Two common ways of making solutions of a desired concentration are determining the mass of solute to be dissolved in an appropriate amount of solvent to produce the quantity of solution needed and by **dilution** of a **stock solution**. During dilution, the quantity of solute in a sample does not change, but adding solvent increases the volume of the sample and decreases the concentration of the solution. (Sections 4.2 and 4.3)

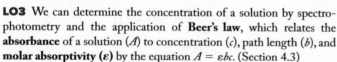

LO3 We can determine the concentration of a solution by spectrophotometry and the application of **Beer's law**, which relates the **absorbance** of a solution (A) to concentration (c), path length (b), and **molar absorptivity (ε)** by the equation $A = \varepsilon bc$. (Section 4.3)

LO4 In **molecular equations** reactants and products are written as neutral compounds. **Overall ionic equations** include all species in the reaction mixture, including the ions that exist in solution. **Net ionic equations** eliminate **spectator ions** and show only the species that are involved in the reaction. (Sections 4.5, 4.6, 4.7, and 4.8)

LO5 Measuring the conductivity of a solution enables us to identify solutes as **strong electrolytes, weak electrolytes,** or **nonelectrolytes.** In a **neutralization reaction, Brønsted–Lowry acids** are proton donors and **Brønsted–Lowry bases** are proton acceptors; they are also strong or weak electrolytes in aqueous solution. (Sections 4.4 and 4.5)

LO6 **Titrations** are volumetric methods of chemical analysis in which the concentrations of solutions are determined by reacting them with known volumes and concentrations of standard solutions called **titrants.** When the amount of titrant is enough to completely react the dissolved solute in the sample, the titration has reached the **equivalence point.** (Section 4.6)

LO7 **Solubility** rules define substances that dissolve readily in water and those that have very limited solubility. We can remove ions from solution by **precipitation,** and we can quantify the amount of precipitate formed and the concentration of ions left in solution after reaction. (Section 4.7)

LO8 In a redox reaction, substances either gain electrons (and thereby undergo **reduction**) or lose electrons (undergo **oxidation**). A reaction is a redox reaction if the **oxidation numbers (O.N.),** or **oxidation states,** of the atoms in the reactants change during the reaction. The **activity series** allows us to predict whether a particular metal will reduce an ion of a different metal. (Section 4.8)

LO9 Oxidation and reduction are complementary processes. When balancing equations, the number of electrons lost by the substance being oxidized must match the number of electrons gained by the substance being reduced. (Section 4.8)

PARTICULATE PREVIEW WRAP-UP

The balanced equation when the contents of the two beakers are mixed is

$$NH_4Cl(aq) + AgNO_3(aq) \rightarrow NH_4NO_3(aq) + AgCl(s)$$

Therefore, ammonium nitrate remains as dissociated ions dissolved in water, and silver chloride precipitates as a solid. $AgNO_3(aq)$ is the limiting reagent, and all the nitrate ions remain in solution because they are spectator ions. The starting materials are drawn showing four cations and four anions for each reactant, so they form the products drawn here.

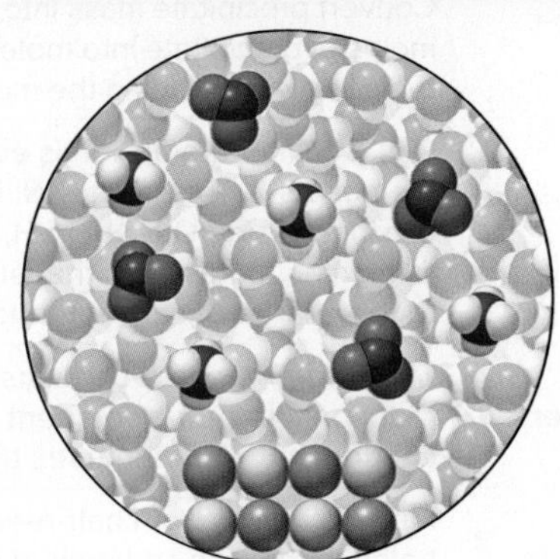

PROBLEM-SOLVING SUMMARY

Type of Problem	Concepts and Equations	Sample Exercises
Comparing the concentrations of aqueous solutions	Use conversion factors to express concentrations in different units.	**4.1**
Calculating molarity from solute mass and solution volume or from solute mass, solution mass, and density	Convert the solute mass into grams and then into moles. Convert the solution mass into volume by dividing by the density of the solvent. Divide moles of solute by liters of solution: $M = \frac{n}{V}$ (4.1)	**4.2, 4.3**
Calculating the mass of solute or volume of stock solution to prepare a solution	Multiply the known concentration (in mol/L) by the target volume (in L) to obtain the moles of solute needed: $n = V \times M$ (4.2) Then multiply moles of solute by solute molar mass, $\mathcal{M}$, to get mass of solute needed: $m_{solute} = (V \times M)\mathcal{M}$ (4.3) Given three of the four variables, use $V_{initial} \times M_{initial} = V_{diluted} \times M_{diluted}$ (4.4) to solve for the fourth.	**4.4, 4.5**
Applying Beer's law	Calculate the value of the absorbance (A), concentration (c), or molar absorptivity (ε) of a sample using Equation 4.5 and the values of the other variables in the equation. $A = \varepsilon bc$ (4.5)	**4.6**
Writing molecular, overall ionic, and net ionic equations for neutralization reactions	Balance the molecular equation by balancing the moles of H^+ ions donated by the acid and accepted by the base. Next, create the overall ionic equation that incorporates the formulas of the ions in strong electrolytes. Finally, create the net ionic equation by eliminating spectator ions from the overall ionic equation.	**4.7, 4.11**
Comparing electrolytes, acids, and bases	Use the definitions of a strong electrolyte, a weak electrolyte, a nonelectrolyte, an acid, and a base to classify compounds into one or more categories.	**4.8**
Calculating molarity from titration data	Use the volume and concentration of titrant used to neutralize a sample along with a balanced chemical equation to calculate the number of moles in a sample of known volume and determine its molarity.	**4.9, 4.10**
Predicting precipitation reactions	Write all the ions present in the solutions being mixed. If any cation/anion pair forms a slightly soluble compound, assume that compound will precipitate.	**4.11, 4.12**
Calculating the mass of a precipitate	Use the stoichiometry of the net ionic equation to calculate moles of precipitate from moles of limiting reactant. Use the molar mass of the precipitate to convert moles of it into grams.	**4.13, 4.15**

Type of Problem	Concepts and Equations	Sample Exercises
Calculating a solute concentration from a precipitate mass	Convert precipitate mass into moles by dividing by its molar mass. Convert moles of precipitate into moles of solute. Calculate molarity of the solute in the sample by dividing the moles of solute by the volume of sample in liters.	**4.14**
Determining oxidation numbers (O.N.)	O.N. of a monatomic ion is equal to the ion's charge. O.N. of a pure element is 0. To assign O.N. in a molecule containing more than one type of atom, assign O.N. +1 to H, −2 to O, and then calculate O.N. for any remaining atoms such that all the O.N. values sum to 0 for compounds or to the charge of a polyatomic ion.	**4.16**
Identifying oxidizing and reducing agents and number of electrons transferred	The *oxidizing* agent contains an atom whose O.N. *decreases* during the reaction; the *reducing* agent contains an atom whose O.N. *increases*; the change in O.N. determines the number of electrons transferred.	**4.17**
Balancing redox reactions with half-reactions	Multiply one or both half-reactions by the appropriate coefficient(s) to balance the loss and gain of electrons. Combine the two half-reactions and simplify.	**4.18**
Using the activity series	Any metal in Table 4.6 will reduce a cation listed *below* it in the activity series.	**4.19, 4.20**
Balancing redox reactions that involve acidic or basic conditions	Follow the steps described in Section 4.8 on balancing redox chemical equations.	**4.21**

VISUAL PROBLEMS

(Answers to boldface end-of-chapter questions and problems are in the back of the book.)

4.1. In Figure P4.1, which shows a solution containing three binary acids, one of the three is a weak acid and the other two are strong acids. Which color sphere represents the anion of the dissociated weak acid?

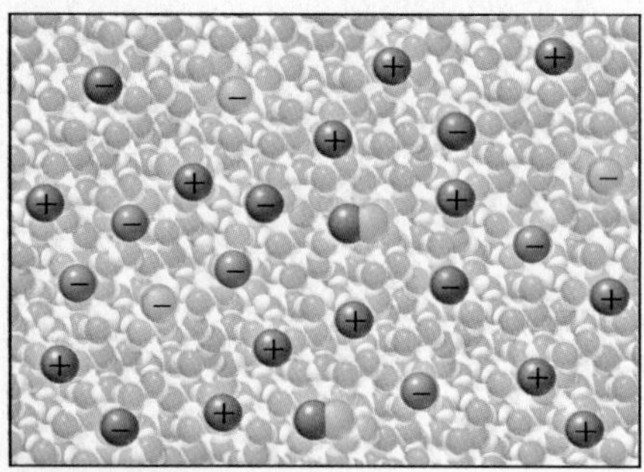

FIGURE P4.1

4.2. Solutions of sodium chloride and silver iodide are mixed together and shaken vigorously. Which colored spheres in Figure P4.2 represent the following ions? (a) Na^+; (b) Cl^-; (c) I^-

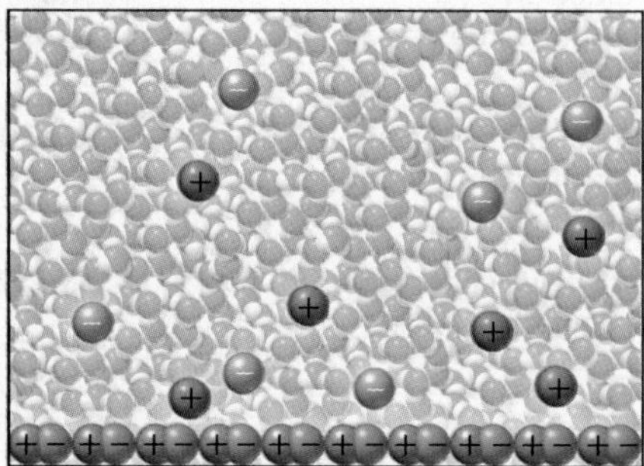

FIGURE P4.2

4.3. Which of the highlighted elements in Figure P4.3 forms an acid with the following generic formula? (a) HX; (b) H_2XO_4; (c) HXO_3; (d) H_3XO_4

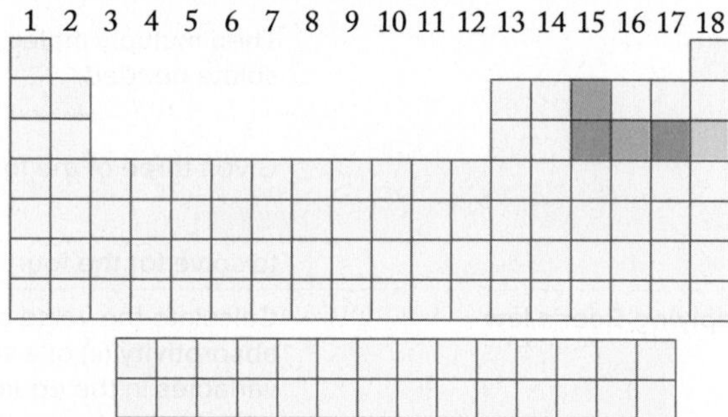

FIGURE P4.3

4.4. In which of the highlighted groups of elements in Figure P4.4 will you find an element that forms the following? (a) slightly soluble halides; (b) slightly soluble hydroxides; (c) hydroxides that are soluble; (d) binary compounds with hydrogen that are strong acids

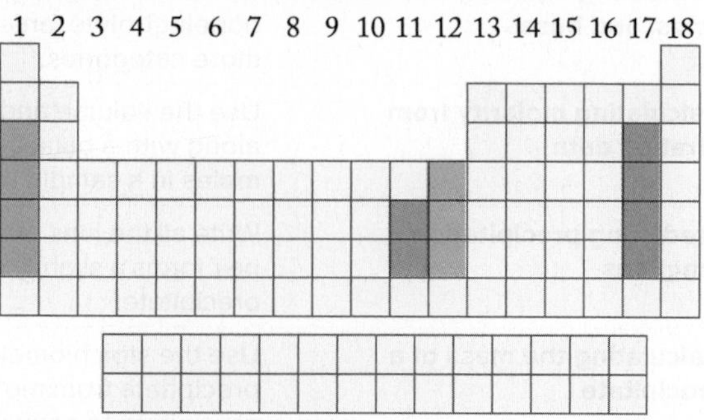

FIGURE P4.4

4.5. Which of the drawings in Figure P4.5 depicts a strong electrolyte? A weak electrolyte? A strong acid? A weak acid? A nonelectrolyte? Each drawing may fit more than one category. (Solvent water molecules have been omitted for clarity.)

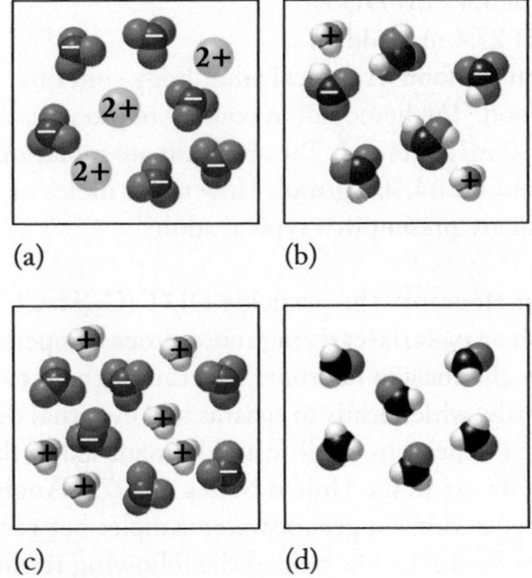

FIGURE P4.5

4.6. Which of the three half-reactions shown in Figure P4.6 depicts an oxidation? Which depicts a reduction? (Solvent water molecules have been omitted for clarity.)

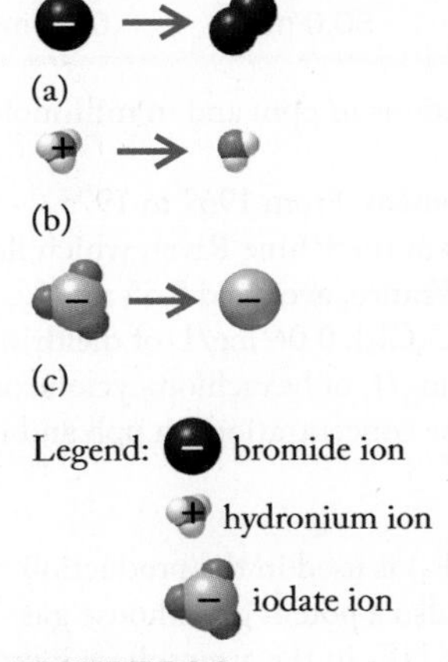

FIGURE P4.6

4.7. Which ions in Figure P4.7 will remain in solution? (Solvent water molecules have been omitted for clarity.)

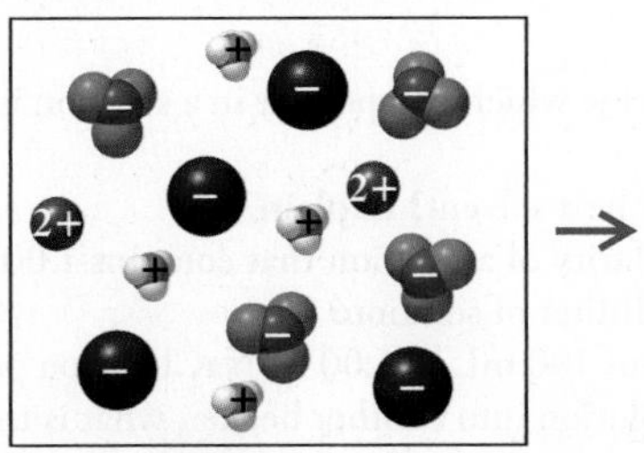

Legend: bromide ions; lead(II) ions

FIGURE P4.7

4.8. Use representations [A] through [I] in Figure P4.8 to answer questions (a)–(f).

a. Which solutes form aqueous solutions that conduct electricity?
b. Solutions of which solutes will produce a precipitate when mixed?
c. Which solutes are nonelectrolytes?
d. Which, if any, depict(s) precipitation reaction(s)?
e. Which, if any, depict(s) redox reaction(s)?
f. Which, if any, depict(s) acid–base reaction(s)?

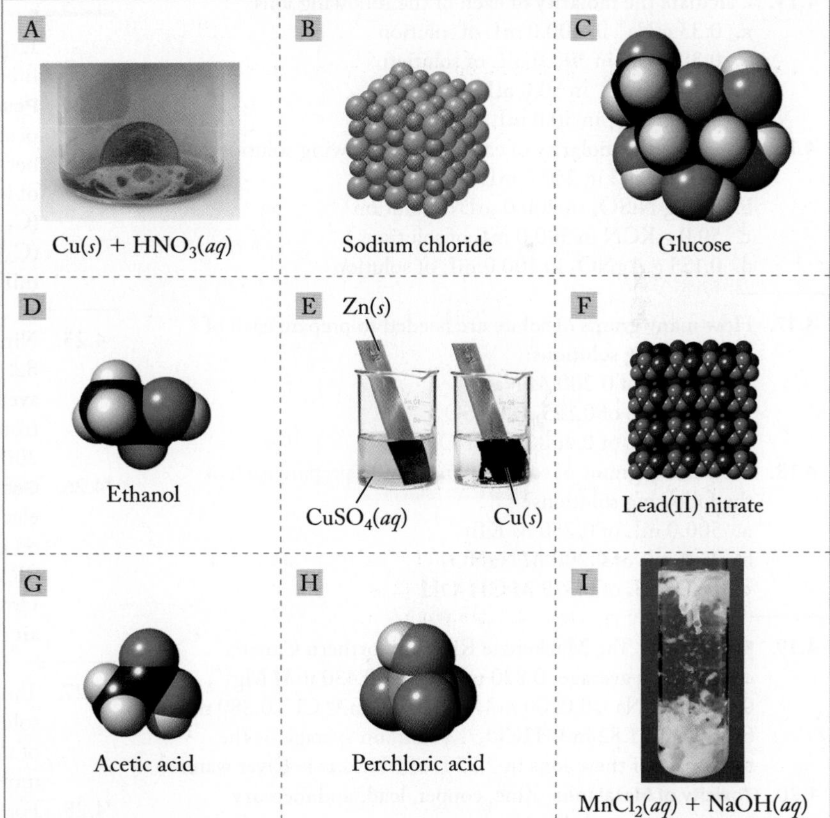

FIGURE P4.8

QUESTIONS AND PROBLEMS

Quantifying Particles in Solution

Concept Review

4.9. How do you decide which component in a solution is the solvent?

4.10. Can a solid ever be a solvent? Explain.

4.11. What is the molarity of a solution that contains 1.00 mmol of solute per milliliter of solution?

4.12. A beaker contains 100 mL of 1.00 *M* NaCl. If you pour 50 mL of the solution into another beaker, what is the molarity of the solution in each of the two beakers?

Problems

4.13. Calculate the molarity of each of the following solutions:
a. 0.56 mol of $BaCl_2$ in 100.0 mL of solution
b. 0.200 mol of Na_2CO_3 in 200.0 mL of solution
c. 0.325 mol of $C_6H_{12}O_6$ in 250.0 mL of solution
d. 1.48 mol of KNO_3 in 250.0 mL of solution

4.14. Calculate the molarity of each of the following solutions:
a. 0.150 mol of urea (CH_4N_2O) in 250.0 mL of solution
b. 1.46 mol of $NaC_2H_3O_2$ in 1.000 L of solution
c. 1.94 mol of methanol (CH_3OH) in 5.000 L of solution
d. 0.045 mol of sucrose ($C_{12}H_{22}O_{11}$) in 50.0 mL of solution

4.15. Calculate the molarity of each of the following ions:
a. 0.33 g Na^+ in 100.0 mL of solution
b. 0.38 g Cl^- in 100.0 mL of solution
c. 0.46 g SO_4^{2-} in 50.0 mL of solution
d. 0.40 g Ca^{2+} in 50.0 mL of solution

4.16. Calculate the molarity of each of the following solutions:
a. 64.7 g LiCl in 250.0 mL of solution
b. 29.3 g $NiSO_4$ in 200.0 mL of solution
c. 50.0 g KCN in 500.0 mL of solution
d. 0.155 g $AgNO_3$ in 100.0 mL of solution

4.17. How many grams of solute are needed to prepare each of the following solutions?
a. 1.000 L of 0.200 *M* NaCl
b. 250.0 mL of 0.125 *M* $CuSO_4$
c. 500.0 mL of 0.400 *M* CH_3OH

4.18. How many grams of solute are needed to prepare each of the following solutions?
a. 500.0 mL of 0.250 *M* KBr
b. 25.0 mL of 0.200 *M* $NaNO_3$
c. 100.0 mL of 0.375 *M* CH_3OH

4.19. **River Water** The Mackenzie River in northern Canada contains, on average, 0.820 m*M* Ca^{2+}, 0.430 m*M* Mg^{2+}, 0.300 m*M* Na^+, 0.0200 m*M* K^+, 0.250 m*M* Cl^-, 0.380 m*M* SO_4^{2-}, and 1.82 m*M* HCO_3^-. What, on average, is the total mass of these ions in 2.75 L of Mackenzie River water?

4.20. **Toxicity of Metal Ions** Zinc, copper, lead, and mercury ions are toxic to Atlantic salmon at concentrations of 6.42×10^{-2} m*M*, 7.16×10^{-3} m*M*, 0.965 m*M*, and 5.00×10^{-2} m*M*, respectively. What are the corresponding concentrations in milligrams per liter?

4.21. Calculate the number of moles of solute contained in the following volumes of aqueous solutions of four pesticides:
a. 0.400 L of 0.024 *M* lindane
b. 1.65 L of 0.473 m*M* dieldrin
c. 25.8 L of 3.4 m*M* DDT
d. 154 L of 27.4 m*M* aldrin

4.22. **Hemoglobin in Blood** A typical adult body contains 6.0 L of blood. The hemoglobin content of blood is about 15.5 g/100.0 mL of blood. The approximate molar mass of hemoglobin is 64,500 g/mol. How many moles of hemoglobin are present in a typical adult?

4.23. **DDT Affects Neurons** The pesticide DDT ($C_{14}H_9Cl_5$) kills insects such as malaria-carrying mosquitoes by opening sodium ion channels in neurons. This causes them to fire spontaneously, which leads to spasms and eventual death. However, its toxicity in wildlife and humans led to the banning of its use in the United States in 1972. Analysis of DDT concentrations in groundwater samples between 1969 and 1971 in Pennsylvania yielded the following results:

Location	Sample Size	Mass of DDT
Orchard	250.0 mL	0.030 mg
Residential	1.750 L	0.035 mg
Residential after a storm	50.0 mL	0.57 mg

Express these concentrations in ppm and in millimoles per liter.

4.24. **Pesticides in the Environment** From 1969 to 1975, pesticide concentrations in the Rhine River, which flows between Germany and France, averaged 0.55 mg/L of hexachlorobenzene (C_6Cl_6), 0.06 mg/L of dieldrin ($C_{12}H_8Cl_6O$), and 1.02 mg/L of hexachlorocyclohexane ($C_6H_6Cl_6$). Express these concentrations in ppb and in millimoles per liter.

4.25. Nitrogen trifluoride (NF_3) is used in the production of flat panel displays. It is also a potent greenhouse gas. The average concentration of NF_3 in the atmosphere increased from 0.02 parts per trillion (ppt) in 1978 to 0.454 ppt in 2008. What is the concentration of NF_3 in mg per kg of air?

*4.26. **Gases Found in Air** Sulfur hexafluoride (SF_6) is used in electrical transformers. Like NF_3, it has a potential impact on climate. Between 1978 and 2012, the concentration of SF_6 increased from 0.51 parts per trillion (ppt) to 7.48 ppt. How many *more* molecules of SF_6 were found in one liter of air in 2012 than in 1978? (1 mole of gas = 22.4 L of gas.)

***4.27.** The concentration of copper(II) sulfate in one brand of soluble plant fertilizer is 0.07% by mass. If a 20 g sample of this fertilizer is dissolved in 2.0 L of solution, what is the molarity of Cu^{2+}?

*4.28. For which of the following compounds is it possible to make a 1.0 *M* solution at 20°C?
a. $CuSO_4$, solubility = 32.0 g/100 mL
b. $Ba(OH)_2$, solubility = 3.9 g/100 mL
c. $FeCl_2$, solubility = 68.5 g/100 mL
d. $Ca(OH)_2$, solubility = 0.173 g/100 mL

Dilutions

Problems

4.29. Calculate the final concentrations of the following aqueous solutions after each has been diluted to a final volume of 25.0 mL:
a. 3.00 mL of 0.175 M K^+
b. 2.50 mL of 10.6 mM LiCl
c. 15.00 mL of 7.24 × 10^{-2} mM Zn^{2+}

4.30. **Dilution of Adult-Strength Cough Syrup** A standard dose of an over-the-counter cough suppressant for adults is 20.0 mL. A portion this size contains 35 mg of the active pharmaceutical ingredient (API). Your pediatrician says you may give this medication to your 6-year-old child, but the child may take only 10.0 mL at a time and receive a maximum of 4.00 mg of the API. What is the concentration in mg/mL of the adult-strength medication, and how many millimeters of it would you need to dilute to make 100.0 mL of child-strength cough syrup?

4.31. The concentration of Na^+ ions in seawater, 0.481 M, is higher than in the cytosol, the fluid inside human cells (12 mM). How much water must be added to 1.50 mL of seawater to make the Na^+ ion concentration equal to that found in the cytosol? Assume the volumes are additive.

4.32. The concentration of chloride ion in blood, 116 mM, is less than that in the ocean, 0.559 M. Describe how you would prepare 2.50 mL of a solution of 116 mM chloride ion from seawater.

4.33. Water is allowed to evaporate from 100.0 mL of 0.24 M Na_2SO_4 until the solution volume is 60.0 mL. What is the molar concentration of the evaporated solution?

4.34. **Mixing Fertilizer** The label on a bottle of "organic" liquid fertilizer concentrate states that it contains 8 g of phosphate per 100.0 mL and that 15.8 mL should be diluted with water to make 4.0 L of fertilizer to be applied to growing plants. What is the phosphate concentration in grams per liter in the diluted fertilizer?

4.35. If the absorbance of a solution of copper ion decreases by 45% upon dilution, how much water was added to 15.0 mL of a 1.00 M solution of Cu^{2+}?

4.36. By what percentage does the absorbance decrease if 12.25 mL of water is added to a 16.75 mL sample of 0.500 M Cr^{3+}?

***4.37.** The reaction of $SnCl_2(aq)$ with $Pt^{4+}(aq)$ in aqueous HCl yields a yellow-orange solution of a 1:1 Pt–Sn compound with a molar absorptivity (ε) of 1.3 × 10^4 M^{-1} cm^{-1}. What is the absorbance in a cell with a path length of 1.00 cm of a solution prepared by adding 100 mL of an aqueous solution of 5.2 mg $(NH_4)_2PtCl_6$ to 100 mL of an aqueous solution of 2.2 mg $SnCl_2$?

*4.38. The reaction of $SnCl_2(aq)$ with $RhCl_3(aq)$ in aqueous HCl yields a red solution of a 1:1 Rh–Sn compound. If a solution prepared by adding 150 mL of a 0.272 mM aqueous solution of $SnCl_2$ to 50 mL of an aqueous solution of 8.5 mg $RhCl_3$ has an absorbance of 0.85, as measured in a 1.00 cm cell, what is the molar absorptivity of the red compound?

Electrolytes and Nonelectrolytes

Concept Review

4.39. A solution of table salt is a good conductor of electricity, but a solution containing an equal molar concentration of table sugar is not. Why?

*4.40. **Corrosion at Sea** Metallic fixtures on the bottom of a ship corrode more quickly in seawater than in freshwater. Why?

4.41. Explain why liquid methanol (CH_3OH) cannot conduct electricity, whereas molten NaOH can.

4.42. **Fuel Cells** The electrolyte in an electricity-generating device called a *fuel cell* consists of a mixture of Li_2CO_3 and K_2CO_3 heated to 650°C. These two ionic solids melt at this temperature. Explain how this mixture of molten carbonates can conduct electricity.

***4.43.** A beaker containing an aqueous solution of potassium chloride sits uncovered on a lab bench overnight. Half of the solvent evaporates. How does the conductivity of the more concentrated solution compare to the original, more dilute solution?

*4.44. Two 0.1 M aqueous solutions of sucrose are prepared, one using distilled water and one using tap water. The distilled water solution does not conduct electricity, but the tap water solution does. Suggest a reason why these two solutions have different conductivities.

Problems

4.45. Rank the following solutions on the basis of their ability to conduct electricity, starting with the most conductive: (a) 1.0 M NaCl; (b) 1.2 M KCl; (c) 1.0 M Na_2SO_4; (d) 0.75 M LiCl.

4.46. Rank the conductivities of 1 M aqueous solutions of each of the following solutes, starting with the most conductive: (a) acetic acid; (b) methanol; (c) sucrose (table sugar); (d) hydrochloric acid.

4.47. Calculate the molarity of Na^+ ions in a 0.025 M aqueous solution of: (a) NaBr; (b)Na_2SO_4; (c) Na_3PO_4.

4.48. Calculate the molarity of each ion in a 0.025 M aqueous solution of: (a) KCl; (b)$CuSO_4$; (c) $CaCl_2$.

4.49. Which of the following solutions has the greatest number of particles (atoms or ions) of solute per liter? (a) 1 M NaCl; (b) 1 M $CaCl_2$; (c) 1 M ethanol; (d) 1 M acetic acid

4.50. Which of the following solutions contains the most solute particles per liter? (a) 1 M KBr; (b) 1 M $Mg(NO_3)_2$; (c) 4 M ethanol; (d) 4 M acetic acid

---

4.51. Rank these aqueous solutions in order of increasing concentration of potassium ions: (a) 0.1 M potassium chloride; (b) 0.2 M potassium sulfate; (c) 0.3 M potassium acetate.

4.52. Rank these aqueous solutions in order of increasing conductivity: (a) 0.1 M calcium chloride; (b) 0.2 M ammonium chloride; (c) 0.3 M sodium chloride.

Acid–Base Reactions: Proton Transfer

Concept Review

4.53. What name is given to a proton donor?

4.54. What is the difference between a strong acid and a weak acid?

4.55. Identify each compound as either a weak acid or a strong acid in aqueous solution: (a) HNO_3; (b) HNO_2; (c) $CH_3CH_2CH_2COOH$; (d) H_2SO_4.

4.56. An aqueous solution of acetic acid consists of water molecules, molecules of CH_3COOH, acetate ions, and protons. Which species is present in the largest amount? Which substances are present in the smallest amount?

4.57. What name is given to a proton acceptor?

4.58. What is the difference between a strong base and a weak base?

4.59. Identify each compound as either a weak base or a strong base in aqueous solution: (a) $Ca(OH)_2$; (b) NH_3; (c) $CH_3CH_2NH_2$; (d) NaOH.

4.60. Write the net ionic equation for the neutralization of a strong acid by a strong base.

Problems

4.61. For each of the following acid–base reactions, identify the acid and the base, and then write the overall ionic and net ionic equations.

a. $H_2SO_4(aq) + Ca(OH)_2(aq) \rightarrow CaSO_4(s) + 2\ H_2O(\ell)$

b. $PbCO_3(s) + H_2SO_4(aq) \rightarrow$
$PbSO_4(s) + CO_2(g) + H_2O(\ell)$

c. $Ca(OH)_2(s) + 2\ CH_3COOH(aq) \rightarrow$
$Ca(CH_3COO)_2(aq) + 2\ H_2O(aq)$

4.62. Complete and balance each of the following neutralization reactions, name the products, and write the overall ionic and net ionic equations.

a. $HBr(aq) + KOH(aq) \rightarrow$

b. $H_3PO_4(aq) + Ba(OH)_2 + (aq) \rightarrow$

c. $Al(OH)_3(s) + HCl(aq) \rightarrow$

d. $CH_3COOH(aq) + Sr(OH)_2(aq) \rightarrow$

4.63. Write a balanced molecular equation and a net ionic equation for the following reactions:

a. Solid magnesium hydroxide reacts with a solution of sulfuric acid.

b. Solid magnesium carbonate reacts with a solution of hydrochloric acid.

c. Ammonia gas reacts with hydrogen chloride gas.

d. Gaseous sulfur trioxide is dissolved in water and reacts with a solution of sodium hydroxide.

4.64. Write a balanced molecular equation and a net ionic equation for the following reactions:

a. Solid aluminum hydroxide reacts with a solution of hydrobromic acid.

b. A solution of sulfuric acid reacts with solid sodium carbonate.

c. A solution of calcium hydroxide reacts with a solution of nitric acid.

d. Solid potassium oxide is dissolved in water and reacts with a solution of sulfuric acid.

4.65. **Toxicity of Lead Pigments** The use of lead(II) carbonate and lead(II) hydroxide as white pigments in paint was discontinued because children have been known to eat paint chips. The pigments dissolve in stomach acid, and lead ions enter the nervous system and interfere with neurotransmissions in the brain, causing neurological disorders. Using net ionic equations, show why lead(II) carbonate and lead(II) hydroxide dissolve in acidic solutions.

4.66. **Lawn Care** Many homeowners treat their lawns with $CaCO_3(s)$ to reduce the acidity of the soil. Write a net ionic equation for the reaction of $CaCO_3(s)$ with a strong acid.

Titrations

Problems

4.67. How many milliliters of 0.250 *M* NaOH are required to neutralize the following solutions?

a. 60.0 mL of 0.0750 *M* HCl

b. 35.0 mL of 0.226 *M* HNO_3

c. 75.0 mL of 0.190 *M* H_2SO_4

4.68. How many milliliters of 0.250 *M* HNO_3 are needed to neutralize the following solutions?

a. 25.0 mL of 0.395 *M* KOH

b. 78.6 mL of 0.0100 *M* $Al(OH)_3$

c. 65.9 mL of 0.475 *M* NaOH

4.69. If 22.4 mL of 0.25 *M* sodium hydroxide are required to neutralize 15.0 mL of a hydroiodic acid solution, how many grams of hydrogen iodide were dissolved in the solution?

4.70. If 12.5 mL of 0.10 *M* hydrobromic acid are required to neutralize 25.0 mL of a calcium hydroxide solution, how many grams of calcium hydroxide were dissolved in the solution?

***4.71.** The solubility of slaked lime, $Ca(OH)_2$, in water at 20°C is 0.185 g/100.0 mL. What volume of 0.00100 *M* HCl is needed to neutralize 10.0 mL of a saturated $Ca(OH)_2$ solution?

*4.72. The solubility of magnesium hydroxide, $Mg(OH)_2$, in water is 9.0×10^{-4} g/100.0 mL at 20°C. What volume of 0.00100 *M* HNO_3 is required to neutralize 1.00 L of saturated $Mg(OH)_2$ solution?

4.73. A 10.0 mL dose of the antacid in Figure P4.73 contains 830 mg of magnesium hydroxide. What volume of 0.10 *M* stomach acid (HCl) could one dose neutralize?

FIGURE P4.73

*4.74. **Exercise Physiology** The ache, or "burn," you feel in your muscles during strenuous exercise is related to the accumulation of lactic acid, which has the structure shown in Figure P4.74. Only the hydrogen atom in the −COOH group is acidic, that is, can be released as an H^+ ion in aqueous solutions. To determine the concentration of a solution of lactic acid, a chemist titrates a 20.00 mL sample of it with 0.1010 *M* NaOH and finds that 12.77 mL of titrant is required to reach the equivalence point. What is the concentration of the lactic acid solution in moles per liter?

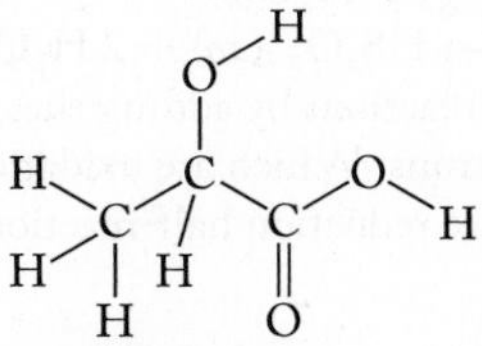

FIGURE P4.74

Precipitation Reactions

Concept Review

4.75. What is the difference between a saturated solution and a supersaturated solution?

4.76. When two aqueous solutions of ionic compounds are mixed but no precipitate forms, what does this suggest about the overall ionic equation?

4.77. An aqueous solution containing Ca^{2+}, Cl^-, CO_3^{2-}, and NO_3^- is allowed to evaporate. Which compound will precipitate first?

4.78. A precipitate may appear when two completely clear aqueous solutions are mixed. What circumstances are responsible for this event?

4.79. Is a saturated solution always a concentrated solution? Explain.

4.80. **Behavior of Honey** Honey is a concentrated solution of sugar molecules in water. Clear, viscous honey becomes cloudy after being stored for long periods. Explain how this transition illustrates supersaturation.

Problems

4.81. According to the solubility rules in Table 4.4 and Table 4.5, which of the following compounds have limited solubility in water? (a) barium sulfate; (b) barium hydroxide; (c) lanthanum nitrate; (d) sodium acetate; (e) lead hydroxide; (f) calcium phosphate

4.82. **Ocean Vents** The black "smoke" that flows out of deep ocean hydrothermal vents (Figure P4.82) consists of metal sulfides suspended in seawater. Of the following cations that are present in the water flowing up through these vents, which ones could contribute to the formation of the black smoke? Na^+, Li^+, Mn^{2+}, Fe^{2+}, Ca^{2+}, Mg^{2+}, Zn^{2+}, Pb^{2+}, Cu^{2+}

FIGURE P4.82

4.83. Complete and balance the molecular equations for the precipitation reactions, if any, between the following pairs of reactants, and write the overall and net ionic equations.

a. $Pb(NO_3)_2(aq) + Na_2SO_4(aq) \rightarrow$

b. $NiCl_2(aq) + NH_4NO_3(aq) \rightarrow$

c. $FeCl_2(aq) + Na_2S(aq) \rightarrow$

d. $MgSO_4(aq) + BaCl_2(aq) \rightarrow$

***4.84.** Complete and balance the molecular equations for the precipitation reactions, if any, between aqueous solutions of the following pairs of reactants, and write the overall and net ionic equations.

a. ammonium acetate and potassium sulfide

b. strontium hydroxide and lithium phosphate

c. silver nitrate and hydrochloric acid

d. lithium fluoride and potassium chloride

4.85. Calculate the mass of $MgCO_3$ precipitated by mixing 10.0 mL of a 0.200 *M* Na_2CO_3 solution with 5.00 mL of 0.0500 *M* $Mg(NO_3)_2$ solution.

4.86. Toxic chromate can be precipitated from an aqueous solution by bubbling SO_2 through the solution. How many grams of SO_2 are required to treat 3.0×10^8 L of 0.050 m*M* CrO_4^-?

$$2\,CrO_4^{2-}(aq) + 3\,SO_2(g) + 4\,H^+(aq) \rightarrow Cr_2(SO_4)_3(s) + 2\,H_2O(\ell)$$

4.87. Iron(II) can be precipitated from a slightly basic aqueous solution by bubbling oxygen through the solution, which converts soluble $Fe(OH)^+$ to practically insoluble $Fe(OH)_3$. How many grams of O_2 are consumed to precipitate all of the iron in 75 mL of 0.090 *M* iron(II)?

$$4\,Fe(OH)^+(aq) + 4\,OH^-(aq) + O_2(g) + 2\,H_2O(\ell) \rightarrow 4\,Fe(OH)_3(s)$$

4.88. Given the following equation, how many grams of $PbCO_3$ will dissolve when 1.00 L of 1.00 *M* H^+ is added to 5.00 g of $PbCO_3$?

$$PbCO_3(s) + 2\,H^+(aq) \rightarrow Pb^{2+}(aq) + H_2O(\ell) + CO_2(g)$$

***4.89.** **Treating Drinking Water** Phosphate can be removed from drinking-water supplies by treating the water with $Ca(OH)_2$. How much $Ca(OH)_2$ is required to remove 90% of the PO_4^{3-} from 4.5×10^6 L of drinking water containing 25 mg/L of PO_4^{3-}?

$$5\,Ca(OH)_2(aq) + 3\,PO_4^{3-}(aq) \rightarrow Ca_5OH(PO_4)_3(s) + 9\,OH^-(aq)$$

4.90. Toxic cyanide ions can be removed from wastewater by adding hypochlorite.

$$2\,CN^-(aq) + 5\,OCl^-(aq) + H_2O(\ell) \rightarrow N_2(g) + 2\,HCO_3^-(aq) + 5\,Cl^-(aq)$$

a. If 1.50×10^3 L of 0.125 *M* OCl^- is required to remove the CN^- in 3.4×10^6 L of wastewater, what is the CN^- concentration in the water in mg/L?

*b. How many milliliters of 0.575 *M* $AgNO_3$ would you need to add to a 50.00 mL aliquot of the final solution (consider the volumes simply additive) to precipitate the chloride ions formed in the reaction?

4.91. For each of the following aqueous mixtures, determine which ionic concentrations decrease and which remain the same.
a. Sodium chloride and silver nitrate are dissolved in 100 mL of water.
b. Equimolar amounts of sodium hydroxide and hydrochloric acid react.
c. Ammonium sulfate and potassium bromide are dissolved in 100 mL of water.

4.92. For each of the following aqueous mixtures, determine which ionic concentrations decrease and which remain the same.
a. Sodium chloride and iron(II) chloride are dissolved in 100 mL of water.
b. Equimolar amounts of sodium carbonate and sulfuric acid react.
c. Potassium sulfate and barium nitrate are dissolved in 100 mL of water.

Oxidation–Reduction Reactions: Electron Transfer

Concept Review

4.93. How are the gains or losses of electrons related to changes in oxidation numbers?

4.94. What is the sum of the oxidation numbers of the atoms in a molecule?

4.95. What is the sum of the oxidation numbers of all the atoms in each of the following polyatomic ions? (a) OH^-; (b) NH_4^+; (c) SO_4^{2-}; (d) PO_4^{3-}

4.96. Gold does not dissolve in concentrated H_2SO_4 but readily dissolves in H_2SeO_4 (selenic acid). Which acid is the stronger oxidizing agent?

4.97. Silver dissolves in sulfuric acid to form silver sulfate and H_2, but gold does not dissolve in sulfuric acid to form gold sulfate. Which of the two metals is the better reducing agent?

4.98. What is meant by a half-reaction?

***4.99.** When an electric current is passed through molten NaCl, the salt decomposes in much the way that water decomposes when current is passed through it (see Figure 1.6). Decomposition occurs via two half-reactions: reduction takes place at the negative electrode and oxidation takes place at the positive electrode. Write the two-half reactions that describe the decomposition of NaCl.

4.100. Electron gain is associated with ______ half-reactions, whereas electron loss is associated with ______ half-reactions.

Problems

4.101. Give the oxidation number of boron in each of the following: (a) HBO_2 (metaboric acid); (b) H_3BO_3 (boric acid); (c) $Na_2B_4O_7$ (sodium borate).

4.102. Give the oxidation number of nitrogen in each of the following: (a) elemental nitrogen (N_2); (b) hydrazine (N_2H_4); (c) ammonium ion (NH_4^+).

4.103. Balance the following half-reactions by adding the appropriate number of electrons. Identify the oxidation half-reactions and the reduction half-reactions.
a. $Br_2(\ell) \rightarrow 2\,Br^-(aq)$
b. $Pb(s) + 2\,Cl^-(aq) \rightarrow PbCl_2(s)$
c. $O_3(g) + 2\,H^+(aq) \rightarrow O_2(g) + H_2O(\ell)$
d. $2\,H_2SO_3(aq) + H^+(aq) \rightarrow HS_2O_4^-(aq) + 2\,H_2O(\ell)$

4.104. Balance the following half-reactions by adding the appropriate number of electrons. Which are oxidation half-reactions, and which are reduction half-reactions?
a. $Fe^{2+}(aq) \rightarrow Fe^{3+}(aq)$
b. $AgI(s) \rightarrow Ag(s) + I^-(aq)$
c. $VO_2^+(aq) + 2\,H^+(aq) \rightarrow VO^{2+}(aq) + H_2O(\ell)$
d. $I_2(s) + 6\,H_2O(\ell) \rightarrow 2\,IO_3^-(aq) + 12\,H^+(aq)$

4.105. Balance the following net ionic reactions. Identify the oxidizing agent and the reducing agent. Identify which elements are oxidized and which are reduced:
a. $MnO_2(s) + HCl(aq) \rightarrow Mn^{2+}(aq) + Cl_2(g)$
b. $I_2(s) + S_2O_3^{2-}(aq) \rightarrow S_4O_6^{2-}(aq) + I^-(aq)$
c. $MnO_4^-(aq) + Fe^{2+}(aq) \rightarrow Mn^{2+}(aq) + Fe^{3+}(aq)$

4.106. Balance the following net ionic reactions. Identify the oxidizing agent and the reducing agent. Identify which elements are oxidized and which are reduced:
a. $MnO_4^-(aq) + S^{2-}(aq) \rightarrow MnO_2(s) + S(s)$
b. $IO_3^-(aq) + I^-(aq) \rightarrow I_2(s)$
c. $Mn^{2+}(aq) + BiO_3^-(aq) \rightarrow MnO_4^-(aq) + Bi^{3+}(aq)$

4.107. **Earth's Crust** The following chemical reactions have helped to shape Earth's crust. Determine the oxidation numbers of all the elements in the reactants and products, and identify which elements are oxidized and which are reduced.
a. $3\,SiO_2(s) + 2\,Fe_3O_4(s) \rightarrow 3\,Fe_2SiO_4(s) + O_2(g)$
b. $SiO_2(s) + 2\,Fe(s) + O_2(g) \rightarrow Fe_2SiO_4(s)$
c. $4\,FeO(s) + O_2(g) + 6\,H_2O(\ell) \rightarrow 4\,Fe(OH)_3(s)$

4.108. Determine the oxidation numbers of each of the elements in the following reactions, and identify which of them are oxidized or reduced, if any.
a. $SiO_2(s) + 2\,H_2O(\ell) \rightarrow H_4SiO_4(aq)$
b. $2\,MnCO_3(s) + O_2(g) \rightarrow 2\,MnO_2(s) + 2\,CO_2(g)$
c. $3\,NO_2(g) + H_2O(\ell) \rightarrow$
$2\,NO_3^-(aq) + NO(g) + 2\,H^+(aq)$

***4.109.** Combine the half-reaction for the reduction of O_2

$$O_2(aq) + 4\,H^+(aq) + 4\,e^- \rightarrow 2\,H_2O(\ell)$$

with the following oxidation half-reactions (which are based on common iron minerals) to develop complete redox reactions:
a. $2\,FeCO_3(s) + H_2O(\ell) \rightarrow$
$Fe_2O_3(s) + 2\,CO_2(g) + 2\,H^+(aq) + 2\,e^-$
b. $3\,FeCO_3(s) + H_2O(\ell) \rightarrow$
$Fe_3O_4(s) + 3\,CO_2(g) + 2\,H^+(aq) + 2\,e^-$
c. $2\,Fe_3O_4(s) + H_2O(\ell) \rightarrow 3\,Fe_2O_3(s) + 2\,H^+(aq) + 2\,e^-$

4.110. Uranium is found in Earth's crust as UO_2 and an assortment of compounds containing UO_2^{n+} cations. Add the following pairs of reduction and oxidation equations to develop overall equations for converting soluble uranium polyatomic ions into practically insoluble UO_2.

a. $6\,H^+(aq) + UO_2(CO_3)_3^{4-}(aq) + 2\,e^- \rightarrow UO_2(s) + 3\,CO_2(g) + 3\,H_2O(\ell)$
$Fe^{2+}(aq) + 3\,H_2O(\ell) \rightarrow Fe(OH)_3(s) + 3\,H^+(aq) + e^-$

b. $6\,H^+(aq) + UO_2(CO_3)_3^{4-}(aq) + 2\,e^- \rightarrow UO_2(s) + 3\,CO_2(g) + 3\,H_2O(\ell)$
$HS^-(aq) + 4\,H_2O(\ell) \rightarrow SO_4^{2-}(aq) + 9\,H^+(aq) + 8\,e^-$

c. $2\,e^- + UO_2(HPO_4)_2^{2-}(aq) \rightarrow UO_2(s) + 2\,HPO_4^{2-}(aq)$
$3\,OH^-(aq) \rightarrow H_2O(\ell) + HO_2^-(aq) + 2\,e^-$

4.111. Nitrogen in the hydrosphere is found primarily as ammonium ions and nitrate ions. Complete and balance the following chemical equation describing the oxidation of ammonium ions to nitrate ions in acid solution:

$$NH_4^+(aq) + O_2(g) \rightarrow NO_3^-(aq)$$

4.112. **When Soil Smells Bad** In sediments and waterlogged soil, dissolved O_2 concentrations are so low that the microorganisms living there must rely on other sources of oxygen for respiration. Some bacteria can extract the oxygen from sulfate ions, reducing the sulfur in them to hydrogen sulfide gas and giving the sediments or soil a distinctive rotten-egg odor.

a. What is the change in oxidation state of sulfur as a result of this reaction?
b. Write the balanced net ionic equation for the reaction, under acidic conditions, that releases O_2 from sulfate and forms hydrogen sulfide gas.

4.113. Chromium is more toxic and more soluble in natural waters as $HCrO_4^-$ than as chromium(III) ion. In the presence of H_2S, the following reaction takes place in neutral solution:

$$HCrO_4^-(aq) + H_2S(aq) \rightarrow Cr_2O_3(s) + SO_4^{2-}(aq)$$

a. Assign oxidation numbers to the reactants and products.
b. Balance the equation.
c. How many electrons are transferred for each atom of chromium that reacts?

4.114. The water-soluble uranyl cation (UO_2^+) can be removed by reaction with methane gas:

$$UO_2^+(aq) + CH_4(g) \rightarrow UO_2(s) + HCO_3^-(aq)$$

a. Assign oxidation numbers to the reactants and products.
b. Balance the equation in acidic solution.
c. How many electrons are transferred for each atom of uranium that reacts?

4.115. The solubilities of Fe and Mn in freshwater streams are affected by changes in their oxidation states. Complete and balance the following redox equation in which soluble Mn^{2+} becomes solid MnO_2:

$$Fe(OH)_2^+(aq) + Mn^{2+}(aq) \rightarrow MnO_2(s) + Fe^{2+}(aq)$$

4.116. **Bactericide and Virucide** The water-soluble gas ClO_2 is known as an oxidative biocide. It destroys bacteria by oxidizing their cell walls and destroys viruses by attacking their viral envelopes. ClO_2 may be prepared for use as a decontaminating agent from several different starting materials in slightly acidic solutions. Complete and balance the following chemical equations for the synthesis of ClO_2.

a. $ClO_3^-(aq) + SO_2(g) \rightarrow ClO_2(g) + SO_4^{2-}(aq)$
b. $ClO_3^-(aq) + Cl^-(aq) \rightarrow ClO_2(g) + Cl_2(g)$
c. $ClO_3^-(aq) + Cl_2(g) \rightarrow ClO_2(g) + O_2(g)$

4.117. Refer to Table 4.6 to determine which of the following metals will reduce aqueous Fe^{2+} to iron metal: lead, copper, zinc, or aluminum.

4.118. Which ions will oxidize aluminum? Li^+; Ca^{2+}; Ag^+; Sn^{2+}

4.119. Through appropriate experiments, we could expand the activity series in Table 4.6 to include additional metals. If aluminum is oxidized by V^{3+} but aluminum does not reduce Sc^{3+}, where would you place vanadium and scandium in the activity series? Which metal would you test to firmly establish scandium's position?

4.120. If iron is oxidized by Cd^{2+} but iron does not reduce Ga^{3+}, where would you place cadmium and gallium in the activity series? Which metal would you test to firmly establish gallium's position?

4.121. Dichromate ion oxidizes Fe^{2+} ion in aqueous, acidic solution, producing Fe^{3+} and Cr^{3+} by the unbalanced chemical equation:

$$Cr_2O_7^{2-}(aq) + Fe^{2+}(aq) \rightarrow Fe^{3+} + 2\,Cr^{3+}(aq)$$

a. Balance the equation.
b. If 15.2 mL of 0.135 M $Cr_2O_7^{2-}$ is required to completely react with 100.0 mL of Fe^{2+}, what is the concentration of the Fe^{2+} solution?

*4.122. Ozone (O_3) reacts with iodide ion (I^-) in basic solution to form O_2 and I_2 by the following unbalanced chemical equation:

$$O_3(aq) + I^-(aq) \rightarrow O_2(g) + I_2(aq)$$

a. Balance the equation.
b. A saturated solution of ozone in 125 mL of water at 0°C is treated with 10 mL of 2.0 M KI. After the reaction is complete, the solution is titrated with 0.100 M H^+. If 54.7 mL of acid is needed, what is the concentration of O_3 in a saturated solution?

Additional Problems

4.123. A puddle of coastal seawater, caught in a depression formed by some coastal rocks at high tide, begins to evaporate on a hot summer day as the tide goes out. If the volume of the puddle decreases to 23% of its initial volume, what is the concentration of Na^+ after evaporation if initially it was 0.449 M?

4.124. **Antifreeze** Ethylene glycol is the common name for the liquid used to keep the coolant in automobile cooling systems from freezing. It is 38.7% carbon, 9.7% hydrogen, and 51.6% oxygen by mass. Its molar mass is 62.07 g/mol, and its density is 1.106 g/mL at 20°C.
a. What is the empirical formula of ethylene glycol?
b. What is the molecular formula of ethylene glycol?
c. In a solution prepared by mixing equal volumes of water and ethylene glycol, which ingredient is the solute, and which is the solvent?

4.125. According to the label on a bottle of concentrated hydrochloric acid, the contents are 36.0% HCl by mass and have a density of 1.18 g/mL.
a. What is the molarity of concentrated HCl?
b. What volume of it would you need to prepare 0.250 L of 2.00 *M* HCl?
c. What mass of sodium hydrogen carbonate would be needed to neutralize the spill if a bottle containing 1.75 L of concentrated HCl dropped on a lab floor and broke open?

*4.126. Why is $HSO_4^-(aq)$ a weaker acid than $H_2SO_4(aq)$?

4.127. **Synthesis and Toxicity of Chlorine** Chlorine was first prepared in 1774 by heating a mixture of NaCl and MnO_2 in sulfuric acid:

$$NaCl(aq) + H_2SO_4(aq) + MnO_2(s) \rightarrow Na_2SO_4(aq) + MnCl_2(aq) + H_2O(\ell) + Cl_2(g)$$

a. Assign oxidation numbers to the elements in each compound and balance the redox reaction in acid solution.
b. Write a net ionic equation describing the reaction for the formation of chlorine.
c. If chlorine gas is inhaled, it causes pulmonary edema (fluid in the lungs) because it reacts with water in the alveolar sacs of the lungs to produce the strong acid HCl and the weaker acid HOCl. Balance the equation for the conversion of Cl_2 to HCl and HOCl.

*4.128. When a solution of dithionate ions ($S_2O_4^{2-}$) is added to a solution of chromate ions (CrO_4^{2-}), the products of the reaction under basic conditions include soluble sulfite ions and solid chromium(III) hydroxide. This reaction is used to remove chromium(VI) from wastewater generated by factories that make chrome-plated metals.
a. Write the net ionic equation for this redox reaction.
b. Which element is oxidized and which is reduced?
c. Identify the oxidizing and reducing agents in this reaction.
d. How many grams of sodium dithionate would be needed to remove the chromium(VI) in 100.0 L of wastewater that contains 0.00148 *M* chromate ion?

4.129. **An Iron Battery** A prototype battery based on iron compounds with large, positive oxidation numbers was developed in 1999. In the following reactions, assign oxidation numbers to the elements in each compound and balance the redox reactions in basic solution.
a. $FeO_4^{2-}(aq) + H_2O(\ell) \rightarrow FeO(OH)(s) + O_2(g) + OH^-(aq)$
b. $FeO_4^{2-}(aq) + H_2O(\ell) \rightarrow Fe_2O_3(s) + O_2(g) + OH^-(aq)$

4.130. **Polishing Silver** Silver tarnish is the result of silver metal reacting with sulfur compounds, such as H_2S, in the air. The tarnish on silverware (Ag_2S) can be removed by soaking in a solution of $NaHCO_3$ (baking soda) in a basin lined with aluminum foil.
a. Write a balanced equation for the tarnishing of Ag to Ag_2S and assign oxidation numbers to the reactants and products. How many electrons are transferred per mole of silver?
b. Write a balanced equation for the reaction of Ag_2S with Al metal, $NaHCO_3$, and water to produce $Al(OH)_3$, H_2S, H_2, and Ag metal.

4.131. Many nonmetal oxides react with water to form acidic solutions. Give the formula and name for the acids produced from the following reactions:
a. $P_4O_{10}(s) + 6\ H_2O(\ell) \rightarrow$
b. $SeO_2(s) + H_2O(\ell) \rightarrow$
c. $B_2O_3(s) + 3\ H_2O(\ell) \rightarrow$

4.132. Write overall and net ionic equations for the reactions that occur when
a. a sample of acetic acid is titrated with a solution of KOH.
b. a solution of sodium carbonate is mixed with a solution of calcium chloride.
c. calcium oxide dissolves in water.

***4.133.** Write both the molecular equation and the net ionic equation for each of the following reactions.
a. a neutralization reaction that results in the formation of sodium fluoride
b. a precipitation reaction that results in the formation of strontium sulfate
c. a redox reaction that results in the formation of potassium chloride

*4.134. **Wastewater Treatment** Show with appropriate net ionic equations how Cr^{3+} and Cd^{2+} can be removed from wastewater by treatment with solutions of sodium hydroxide.

***4.135.** **Fluoride Ion in Drinking Water** Sodium fluoride is added to drinking water in many municipalities to protect teeth against cavities. The target of the fluoridation is hydroxyapatite, $Ca_{10}(PO_4)_6(OH)_2$, a compound in tooth enamel. There is concern, however, that fluoride ions in water may contribute to skeletal fluorosis, an arthritis-like disease.
a. Write a net ionic equation for the reaction between hydroxyapatite and sodium fluoride that produces fluorapatite, $Ca_{10}(PO_4)_6F_2$.
b. The EPA currently restricts the concentration of F^- in drinking water to 4 mg/L. Express this concentration of F^- in molarity.
c. One study of skeletal fluorosis suggests that drinking water with a fluoride concentration of 4 mg/L for 20 years raises the fluoride content in bone to 6 mg/g, a level at which a patient may experience stiff joints and other symptoms. How much fluoride (in milligrams) is present in a 100 mg sample of bone with this fluoride concentration?

*4.136. **Rocket Fuel in Drinking Water** Near Las Vegas, NV, improper disposal of perchlorates used to manufacture rocket fuel has contaminated a stream that flows into Lake Mead, the largest artificial lake in the United States and a major supply of drinking and irrigation water for the American Southwest. The EPA has proposed an advisory range for perchlorate concentrations in drinking water of 4 to 18 μg/L. The perchlorate concentration in the stream averages 700.0 μg/L, and the stream flows at an average rate of 609 million liters per day.

a. What are the formulas of sodium perchlorate and ammonium perchlorate?
b. How many kilograms of perchlorate flow from the Las Vegas stream into Lake Mead each day?
c. What volume of perchlorate-free lake water would have to mix with the stream water each day to dilute the stream's perchlorate concentration from 700.0 to 4 μg/L?
d. Since 2003, Maryland (MD), Massachusetts (MA), and New Mexico (NM) have limited perchlorate concentrations in drinking water to 0.1 μg/L. Five replicate samples were analyzed for perchlorates by laboratories in each state, and the following data (μg/L) were collected:

MD	MA	NM
1.1	0.90	1.2
1.1	0.95	1.2
1.4	0.92	1.3
1.3	0.90	1.4
0.9	0.93	1.1

Which of the labs produced the most precise analytical results?

***4.137. Making Apple Cider Vinegar** Some people who prefer natural foods make their own apple cider vinegar. They start with freshly squeezed apple juice that contains about 6% natural sugars. These sugars, which all have nearly the same empirical formula, CH_2O, are fermented with yeast in a chemical reaction that produces equal numbers of moles of ethanol (Figure P4.137a) and carbon dioxide. The product of this fermentation, called hard cider, undergoes an acid fermentation step in which ethanol and dissolved oxygen gas react to form acetic acid (Figure P4.137b) and water. This acetic acid is the principal solute in vinegar.

Ethanol
$CH_3—CH_2—OH$
(a)

Acetic acid
$CH_3—COOH$
(b)

FIGURE P4.137

a. Write a balanced chemical equation describing the fermentation of natural sugars to ethanol and carbon dioxide. You may use the empirical formula, CH_2O.
b. Write a balanced chemical equation describing the acid fermentation of ethanol to acetic acid.
c. What are the oxidation states of carbon in the reactants and products of the two fermentation reactions?
d. If a sample of apple juice contains 1.00×10^2 g of natural sugar, what is the maximum quantity of acetic acid that could be produced by the two fermentation reactions?

*4.138. **Acidic Mine Drainage** Water draining from abandoned mines on Iron Mountain in California is extremely acidic and leaches iron, zinc, and other metals from the underlying rock (Figure P4.138). One liter of drainage contains as much as 80.0 g of dissolved iron and 6 g of zinc.

FIGURE P4.138

a. Calculate the molarity of iron and of zinc in the drainage.
b. One source of the dissolved iron is the reaction between water containing H_2SO_4 and solid $Fe(OH)_3$. Complete the following chemical equation and write a net ionic equation for the process.

$$2\ Fe(OH)_3(s) + 3\ H_2SO_4(aq) \rightarrow$$

c. Sources of zinc include the mineral smithsonite, $ZnCO_3$. Write a balanced net ionic equation for the reaction between smithsonite and H_2SO_4 that produces $Zn^{2+}(aq)$.
d. One member of a class of minerals called ferrites is found to contain a mixture of zinc(II), iron(II), and iron(III) oxides. The generic formula for the mineral is $Zn_xFe_{1-x}O \cdot Fe_2O_3$. If acidic mine waste flowing through a deposit of this mineral contains 80 g of Fe and 6 g of Zn as a result of dissolution of the mineral, what is the value of x in the formula of the mineral in the deposit?

***4.139.** A food chemist determines the concentration of acetic acid in a sample of apple cider vinegar (see Problem 4.137) by acid–base titration. The density of the sample is 1.01 g/mL. The titrant is 1.002 *M* NaOH. The average volume of titrant required to titrate 25.00 mL aliquots of the vinegar is 20.78 mL. What is the concentration of acetic acid in the vinegar? Express your answer the way a food chemist probably would: as percent by mass.

*4.140. One way to follow the progress of a titration and detect its equivalence point is by monitoring the conductivity of the titration reaction mixture. For example, consider the way the conductivity of a sample of sulfuric acid changes as it is titrated with a standard solution of barium hydroxide before and then after the equivalence point.

a. Write the overall ionic equation for the titration reaction.
b. Which of the four graphs in Figure P4.140 comes closest to representing the changes in conductivity during the titration? (The zero point on the y-axis of these graphs represents the conductivity of pure water; the break points on the x-axis represent the equivalence point.)

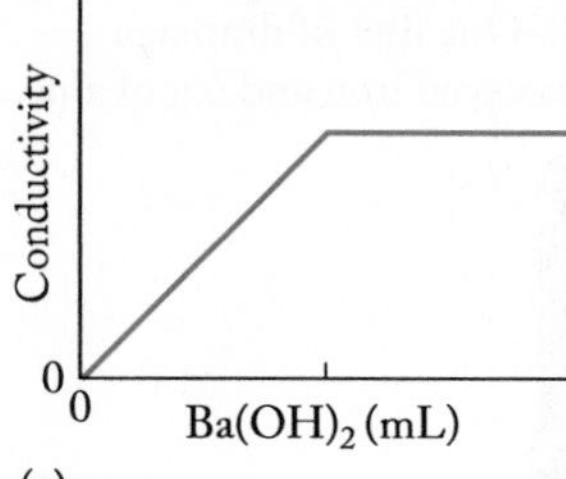

(a)

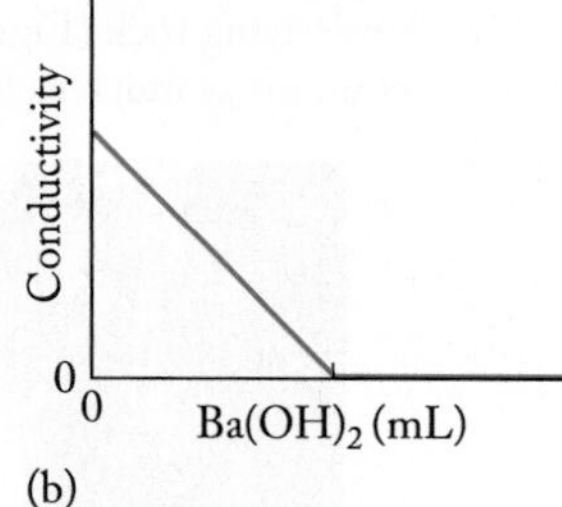

(b)

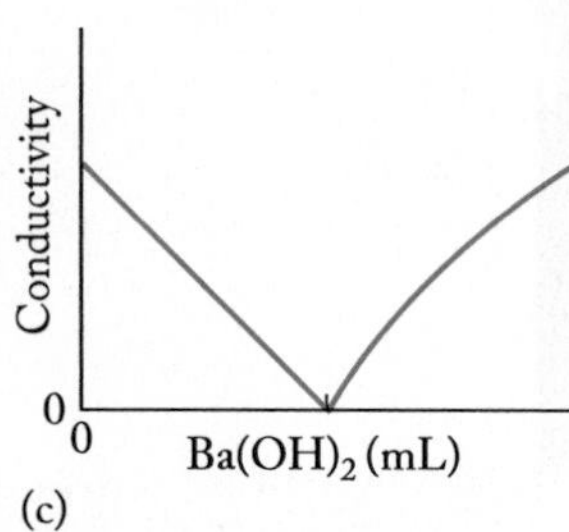

(c)

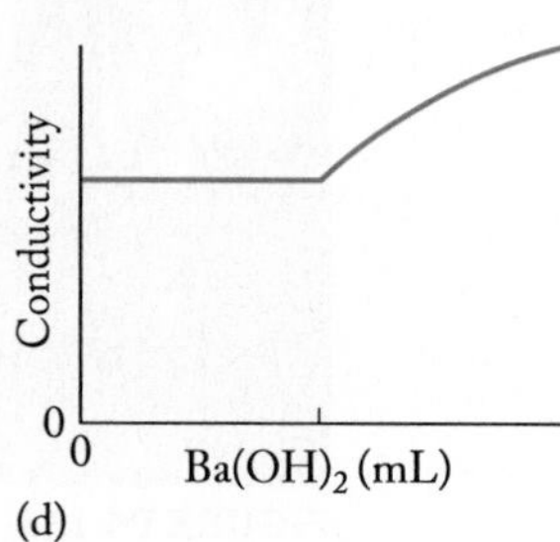

(d)

FIGURE P4.140

*4.141. Which of the graphs in Figure P4.141 best represents the changes in conductivity that occur before and after the equivalence point in each of the following titrations:

a. sample = $AgNO_3(aq)$; titrant = $KCl(aq)$
b. sample = $HCl(aq)$; titrant = $LiOH(aq)$
c. sample = $CH_3COOH(aq)$; titrant = $NaOH(aq)$

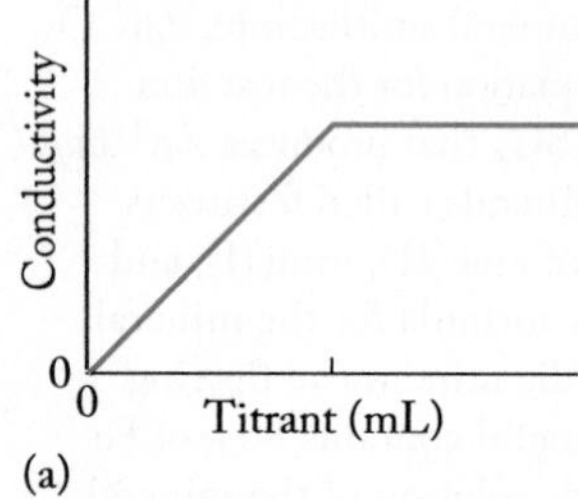

(a)

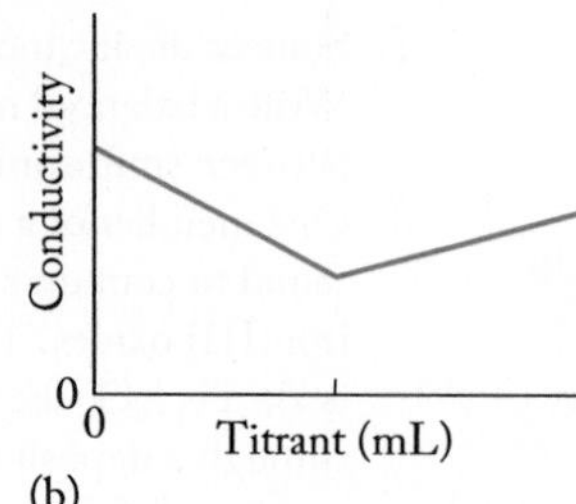

(b)

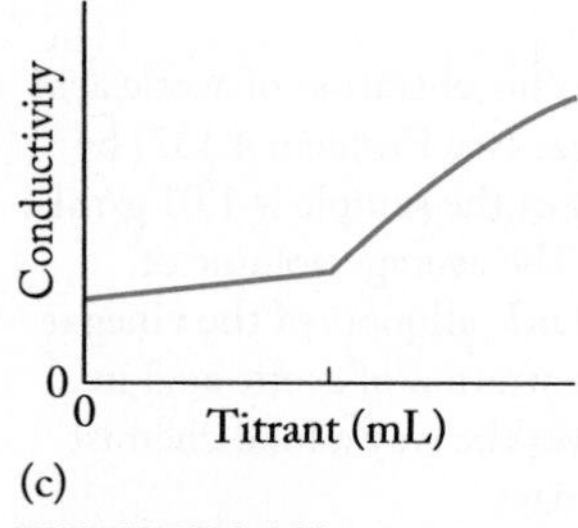

(c)

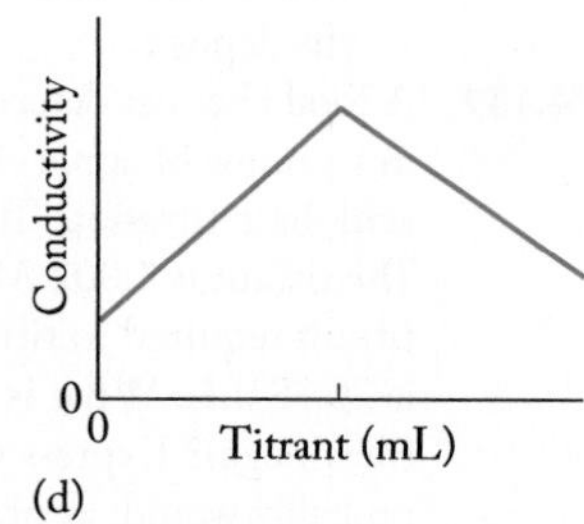

(d)

FIGURE P4.141

4.142. When electrodes connected to a light bulb are inserted into an aqueous solution of acetic acid, the bulb glows dimly. Will the bulb become brighter, remain the same, or turn off after one equivalent of aqueous NaOH is added to the solution? Write a balanced net ionic equation that supports your answer.

*4.143. When electrodes connected to a light bulb are inserted into a beaker containing silver carbonate and water, will the bulb not glow, glow dimly, or glow brightly? What do you think will happen after addition of one equivalent of aqueous HCl? Write a balanced net ionic equation that supports your answer.

4.144. **Superoxide Dismutases** Oxygen in the form of superoxide ions (O_2^-) is quite hazardous to human health. Superoxide dismutases are enzymes that convert superoxide ions to hydrogen peroxide and oxygen by the unbalanced chemical equation:

$$O_2^-(aq) + H^+(aq) \rightarrow H_2O_2(aq) + O_2(aq)$$

a. Identify the oxidation and reduction half-reactions.
b. Balance the equation.

*4.145. **Nitrogen-Fixing Bacteria** Bacteria found among the roots of legumes perform an important biological function, converting nitrogen to ammonia in a process known as nitrogen fixation. The electrons required for this redox reaction are supplied by enzymes called nitrogenases that contain transition metals. The unbalanced chemical equation for this process is

$$N_2(g) + H^+(aq) + M^{2+}(aq) \rightarrow NH_3(aq) + H_2(g) + M^{3+}(aq)$$

where M represents a transition metal such as iron. Nitrification is a multistep process in which the nitrogen in organic and inorganic compounds is biochemically oxidized. Bacteria and fungi are responsible for a part of the *nitrification process* described by the reaction:

$$NH_4^+(aq) + M^{3+}(aq) \rightarrow NO_2^-(aq) + M^{2+}(aq)$$

a. What are the oxidation numbers of nitrogen in the reactants and products of each reaction?
b. Which compounds or ions are being reduced in each reaction?
c. Balance the equations in acidic solution.

4.146. **Formations in Caves** The stalactites and stalagmites in most caves are made of limestone (calcium carbonate; see Figure 4.14). In the Lower Kane Cave in Wyoming, however, they are made of gypsum (calcium sulfate). The presence of $CaSO_4$ is explained by the following sequence of reactions:

$$H_2S(aq) + 2\,O_2(g) \rightarrow H_2SO_4(aq)$$

$$H_2SO_4(aq) + CaCO_3(s) \rightarrow CaSO_4(s) + H_2O(\ell) + CO_2(g)$$

a. Which (if either) of these reactions is a redox reaction? How many electrons are transferred?
b. Write a net ionic equation for the reaction of H_2SO_4 with $CaCO_3$.
c. How would the net ionic equation differ if the reaction were written as follows?

$$H_2SO_4(aq) + CaCO_3(s) \rightarrow CaSO_4(s) + H_2CO_3(aq)$$

4.147. Balance this net ionic reaction and answer the questions that follow:

$$BrO_3^-(aq) + Br^-(aq) \rightarrow Br_2(aq)$$

a. Is $BrO_3^-(aq)$ reduced?
b. What is the product of $BrO_3^-(aq)$ reduction in this reaction?
c. Is $Br^-(aq)$ oxidized?
d. What is the product of $Br^-(aq)$ oxidation in this reaction?

4.148. Which of the following reactions of calcium compounds is/are redox reactions?

a. $CaCO_3(s) \rightarrow CaO(s) + CO_2(g)$
b. $CaO(s) + SO_2(g) \rightarrow CaSO_3(s)$
c. $CaCl_2(s) \rightarrow Ca(s) + Cl_2(g)$
d. $3\,Ca(s) + N_2(g) \rightarrow Ca_3N_2(s)$

4.149. Preparation of Fluorine Gas HF is prepared by reacting CaF_2 with H_2SO_4:

$$CaF_2(s) + H_2SO_4(\ell) \rightarrow 2\,HF(g) + CaSO_4(s)$$

HF can in turn be electrolyzed when dissolved in molten KF to produce fluorine gas:

$$2\,HF(\ell) \rightarrow F_2(g) + H_2(g)$$

Fluorine is extremely reactive, so it is typically sold as a 5% mixture by volume in an inert gas such as helium. How much CaF_2 is required to produce 500.0 L of 5% F_2 in helium? Assume the density of F_2 gas is 1.70 g/L.

5

Properties of Gases

The Air We Breathe

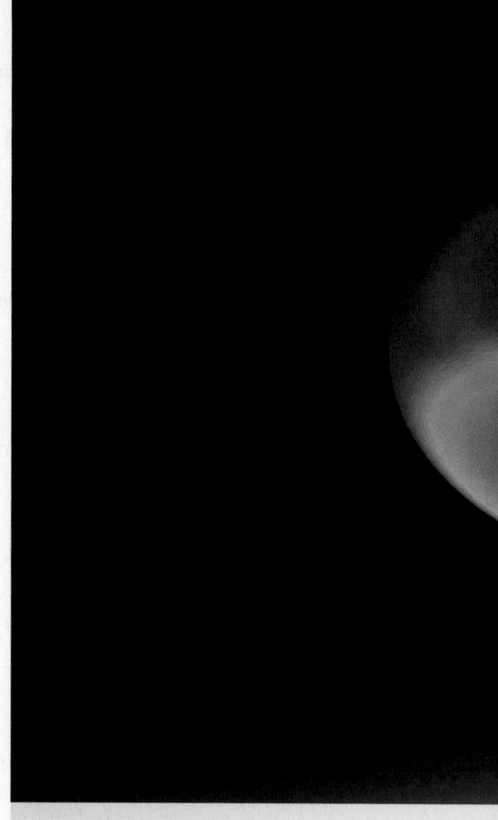

THE AIR WE BREATHE An enormous balloon like the one in this photograph is scheduled to lift NASA's Galactic/Extragalactic Spectroscopic Terahertz Observatory into the sky above Antarctica in December 2021. Its mission is to study the composition of trace gases and dust that are found between stars.

PARTICULATE **REVIEW**

Particles in the Gas Phase

In Chapter 5 we focus on the properties of gases, including those that serve as fuels in combustion reactions, as described in Chapter 3, and in other forms of energy production. One such fuel is hydrogen gas, which can be produced by passing an electric current through water, causing molecules of liquid H_2O to decompose into molecules of H_2 and O_2 gas.

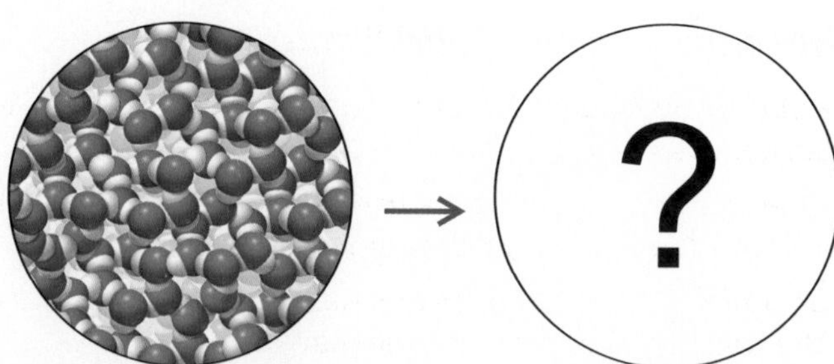

- Write a balanced chemical equation describing this decomposition reaction.
- In the circle on the right, draw the products that would be produced by decomposition of six of the water molecules on the left.
- Classify the products as elements, compounds, or a mixture. Choose all that apply.

(Review Sections 1.1, 1.2, and 3.3 if you need help.)

(Answers to Particulate Review questions are in the back of the book.)

PARTICULATE **PREVIEW**

Pressure, Volume, and Temperature

As you read Chapter 5, look for ideas that will help you answer these questions.

- Draw particulate images of the helium in the tank and in the balloon. How do these drawings differ?
- Suppose the tank and several helium-filled balloons are placed in the trunk of a car on a hot summer day.
  - How would your particulate image for the helium in the balloon change?
  - How would your particulate image for the helium in the tank change?

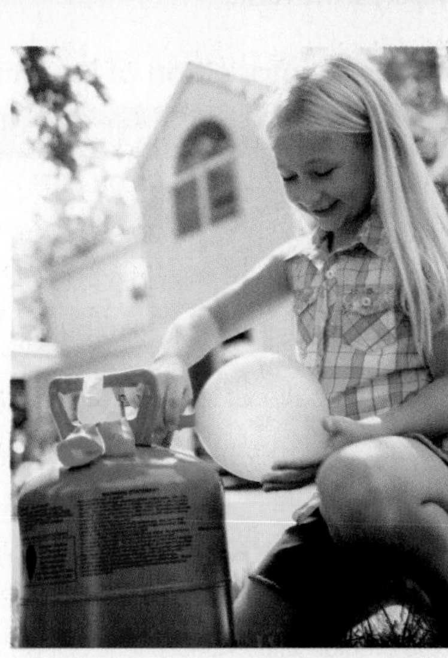

Learning Outcomes

LO1 Describe how gas pressure is measured and convert between the different units used to express gas pressure
Sample Exercises 5.1, 5.2

LO2 Use the individual, combined, and ideal gas laws to relate changes in the volume, temperature, pressure, and number of moles of a gas and express these changes quantitatively
Sample Exercises 5.3, 5.4, 5.5, 5.6, 5.7

LO3 Relate the volumes of gas phase reactants and products to the quantities of other reactants and products by using the stoichiometry of a reaction and the ideal gas law
Sample Exercises 5.8, 5.9

LO4 Relate the density of a gas to its molar mass, temperature, and pressure
Sample Exercises 5.10, 5.11

LO5 Determine the mole fraction and partial pressure of a gas in a mixture of gases
Sample Exercises 5.12, 5.13, 5.14

LO6 Use kinetic molecular theory to explain the physical properties of gases

LO7 Relate the root-mean-square speed of a gas and its rate of effusion and diffusion to its temperature and molar mass
Sample Exercises 5.15, 5.16

LO8 Explain the molecular origins of nonideal behavior of real gases at high temperatures and low pressures and relate such behavior to the terms in the van der Waals equation
Sample Exercise 5.17

5.1 Air: An Invisible Necessity

How often do you think about air? You probably remember how important air is when you don't have enough of it, such as when you dive into a pool of water or hike to the top of a tall mountain. Otherwise you probably don't give much thought to breathing or to the mixture of gases that make up the air you breathe. You may not think much about the oxygen needed to stay alive because it is colorless, tasteless, odorless—and free. But the exchange of gases between your lungs and the surrounding atmosphere is crucial to life. Air is an invisible necessity.

FIGURE 5.1 The space above crystals of solid iodine fill with purple iodine vapor, and the space above a puddle of liquid bromine fills with orange-brown bromine vapor. The solid and liquid phases of these elements are hundreds of times denser than the vapors above them.

Having the right mixture of gases in our bodies is critical during surgery, which is why anesthesiologists in a hospital operating room constantly monitor levels of oxygen and carbon dioxide in the blood. The management of the delicate balance of gases entering and leaving a patient can mean the difference between a normal recovery and an irreversible coma.

We have seen how dissolved compounds react in aqueous solution. Chemical reactions also take place in the gas phase, and gases are intimately involved in chemical reactions in living systems and the material world. Most life in our biosphere requires oxygen. Insects, birds, mammals, plants, and even underwater organisms need O_2 to metabolize nutrients.

TABLE 5.1 Composition of Dry Air[a]

Component	% (by volume)
Nitrogen	78.08
Oxygen	20.95
Argon	0.934
Carbon dioxide	0.0409[b]
Neon	0.0018
Helium	0.00052
Methane	0.00018
Hydrogen	0.00011

[a]Includes major and minor gases (with concentrations >1 ppm by volume).
[b]2018 average value. Atmospheric CO_2 is increasing by about 2 ppm each year.

How do gases differ from solids and liquids? Gases have neither definite volumes nor definite shapes; they expand to occupy the entire volume of their container and assume the container's shape. Under everyday conditions, other properties also distinguish gases from liquids and solids:

1. Unlike the volume occupied by a liquid or solid, the volume occupied by a gas changes significantly with pressure. If we carry an inflated balloon from sea level (0 m) to the top of a 1600-m mountain, the balloon volume increases by about 20%. The volume of a liquid or solid is unchanged under these conditions.
2. The volume of a gas changes with temperature. For example, the volume of a balloon filled with room-temperature air (~20°C) decreases about 7% when the balloon is taken outside on a cold winter's day (~0°C), whereas

the volume of a liquid or solid remains practically unchanged by this modest temperature change.

3. Gases are **miscible**, which means they can be mixed in any proportion (unless they chemically react with one another). A hospital patient experiencing respiratory difficulties may be given a mixture of nitrogen and oxygen in which the proportion of oxygen is much higher than its proportion in air. Alternatively, a scuba diver may leave the ocean surface with a tank of air containing a homogeneous mixture of 17% oxygen, 34% nitrogen, and 49% helium. In contrast, many liquids are immiscible, such as oil and water.
4. Gases are typically much less dense than liquids or solids (**Figure 5.1**). One indicator of this large difference is that gas densities are expressed in grams per *liter*, whereas liquid densities are expressed in grams per *milliliter*. The density of dry air at 20°C and typical atmospheric pressure is 1.20 g/L, for example, whereas the density of liquid water under the same conditions is 1.00 g/mL—more than 800 times greater than the density of dry air.

These four observations about gases are consistent with the idea that the particles of a gas (be they molecules or atoms) are farther apart than the particles in solids and liquids. The larger spaces between the molecules in air, for example, make it possible to compress air into scuba tanks. Greater distances between molecules also account for both the lower densities and the miscibility of gases.

miscible capable of being mixed in any proportion.

pressure (*P*) the ratio of a force to the surface area over which the force is applied.

atmospheric pressure (P_{atm}) the force exerted by the gases surrounding Earth on Earth's surface and on all surfaces of all objects.

barometer an instrument that measures atmospheric pressure.

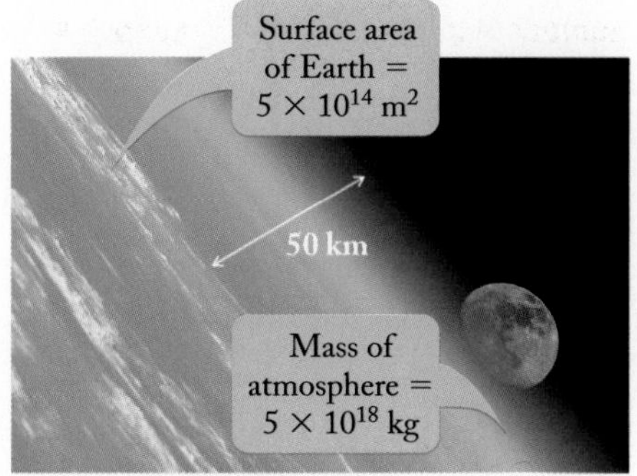

(a)

(b)

FIGURE 5.2 (a) Atmospheric pressure results from the force exerted by the atmosphere on Earth's surface. (b) If you stretch out your hand palm upward, the mass of the column of air above your palm is about 100 kg. This textbook has a mass of about 2.5 kg, so the mass of the atmosphere on your palm is equivalent to the mass of about 40 such textbooks.

5.2 Atmospheric Pressure and Collisions

Earth is surrounded by a layer of gases 50 km thick. We call this mixture of gases either *air* or *the atmosphere*. By volume it is composed primarily of nitrogen (78%) and oxygen (21%), with lesser amounts of other gases (**Table 5.1**). To put the thickness of the atmosphere in perspective: if Earth were the size of an apple, the atmosphere would be about as thick as the apple's skin.

Earth's atmosphere is pulled toward Earth by gravity and exerts a force that is spread across the entire surface of the planet (**Figure 5.2**). The molecules of N_2 and O_2 (along with the other elements and compounds in Table 5.1) are in constant motion in the atmosphere. As these molecules collide with one another and with the surface of Earth, the force of each collision creates **pressure (*P*)**. Pressure is defined as the ratio of force (*F*) to surface area (*A*):

$$P = \frac{F}{A} \tag{5.1}$$

The force exerted by the atmosphere of colliding atoms and molecules on Earth's surface is called **atmospheric pressure (P_{atm})**.

Atmospheric pressure is measured with an instrument called a **barometer**. A simple but effective barometer design consists of a tube nearly 1 m long, filled with mercury, and closed at one end (**Figure 5.3**). The tube is inverted with its open end immersed in a pool of mercury that is open to the atmosphere (which means that molecules of N_2 and O_2 collide against the surface of the liquid mercury). Gravity pulls the mercury in the tube downward, creating a vacuum at the top of the tube, while atmospheric pressure pushes the mercury in the pool up into the tube (from the force of the collisions). The net effect of these opposing forces is indicated by the height of the mercury in the tube, which provides a measure of atmospheric pressure.

atmosphere (atm) a unit of pressure equal to Earth's mean atmospheric pressure at sea level.

torr a unit of pressure. There are exactly 760 torr in 1 atm of pressure.

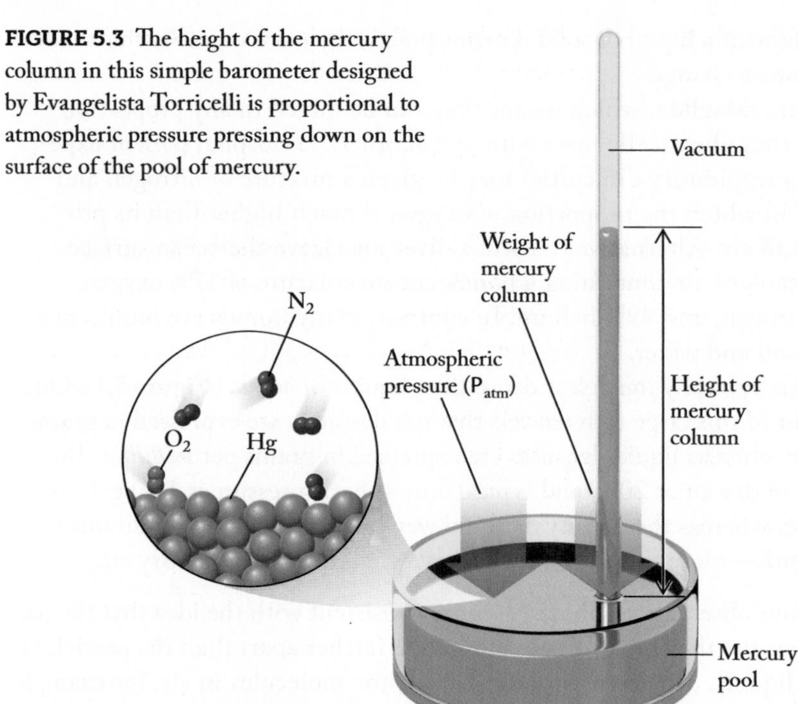

FIGURE 5.3 The height of the mercury column in this simple barometer designed by Evangelista Torricelli is proportional to atmospheric pressure pressing down on the surface of the pool of mercury.

STEPWISE ANIMATION
Measuring Gas Pressure

Atmospheric pressure varies from place to place and with changing weather conditions. Several units are used to express pressure. An **atmosphere (atm)** of pressure, or 1 atm, is the pressure capable of supporting a column of mercury 760 mm high in a barometer. This column height is the average height of mercury in a barometer at sea level and is the physical basis for another unit of pressure: *millimeters of mercury (mmHg)*. Pressure is also expressed in a unit called the **torr**

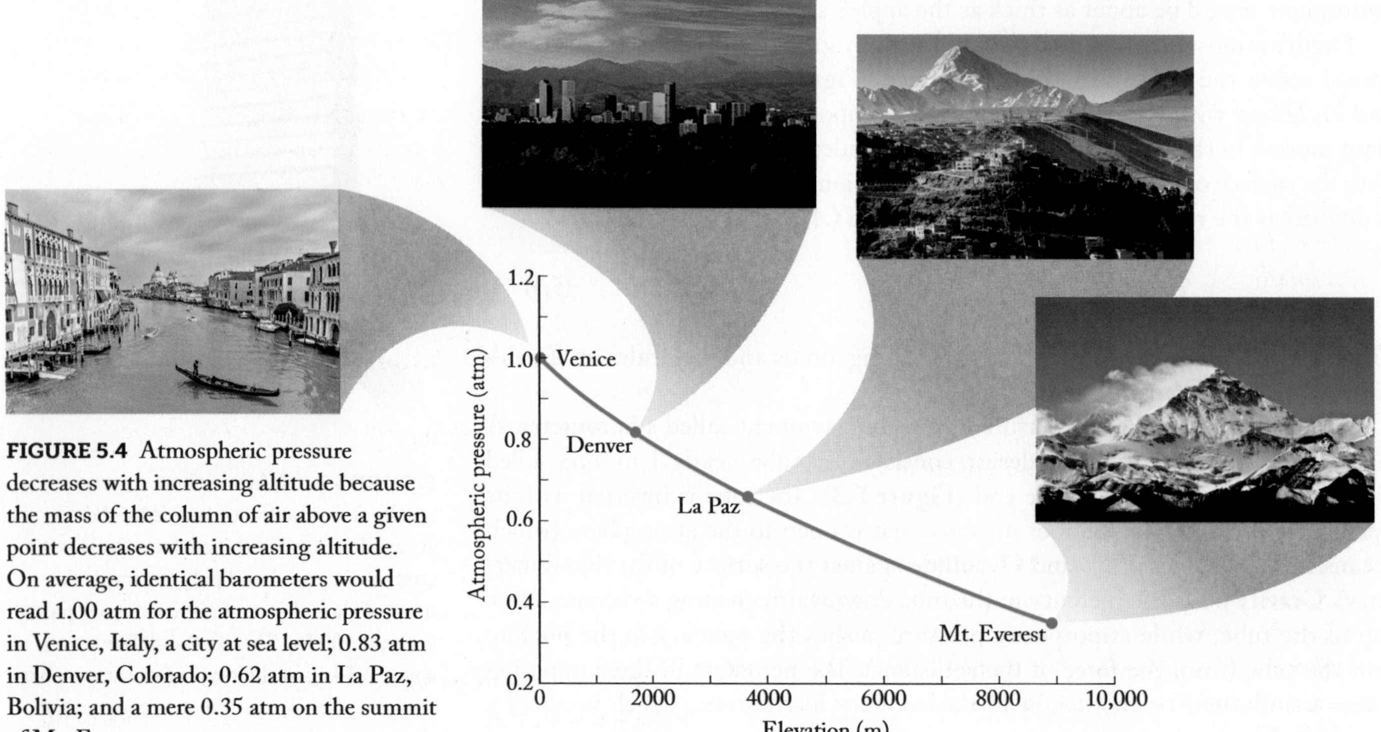

FIGURE 5.4 Atmospheric pressure decreases with increasing altitude because the mass of the column of air above a given point decreases with increasing altitude. On average, identical barometers would read 1.00 atm for the atmospheric pressure in Venice, Italy, a city at sea level; 0.83 atm in Denver, Colorado; 0.62 atm in La Paz, Bolivia; and a mere 0.35 atm on the summit of Mt. Everest.

in honor of Evangelista Torricelli (1608–1647), the Italian mathematician and physicist who invented the barometer. There are exactly 760 torr in 1 atm, which leads to these unit equivalents:

$$1 \text{ atm} = 760 \text{ torr} = 760 \text{ mmHg}$$

The SI unit of pressure is the *pascal* (Pa), named in honor of the French mathematician and physicist Blaise Pascal (1623–1662), who was the first to propose that atmospheric pressure decreases with increasing altitude. The atmospheric pressure at any given location on Earth's surface is related to the mass of the column of air *above* that location (**Figure 5.4**). As altitude increases, the mass of air in the column decreases, which means the mass of the gases in the column decreases. Less mass means fewer collisions among molecules in the atmosphere and therefore a smaller force exerted by the air. According to Equation 5.1, a smaller force means less pressure.

The pascal is a derived SI unit; it is defined using the SI base units kilogram, meter, and second:

$$1 \text{ Pa} = \frac{1 \text{ kg}}{\text{m} \cdot \text{s}^2}$$

To understand the logic of this combination of units, consider the relationship in physics that states that pushing an object of mass m with a force F causes the object to accelerate at the rate a:

$$F = ma \tag{5.2}$$

Combining Equations 5.1 and 5.2, we have

$$P = \frac{F}{A} = \frac{ma}{A} \tag{5.3}$$

If m is in kilograms, a is in meters per second-squared (m/s^2), and A is in square meters (m^2), the units for P are

$$\frac{\text{kg}\,\dfrac{\text{m}}{\text{s}^2}}{\text{m}^2} = \text{kg}\,\frac{\text{m}}{\text{s}^2} \times \frac{1}{\text{m}^2} = \frac{\text{kg}}{\text{m} \cdot \text{s}^2}$$

The relationship between atmospheres and pascals is

$$1 \text{ atm} = 101{,}325 \text{ Pa}$$

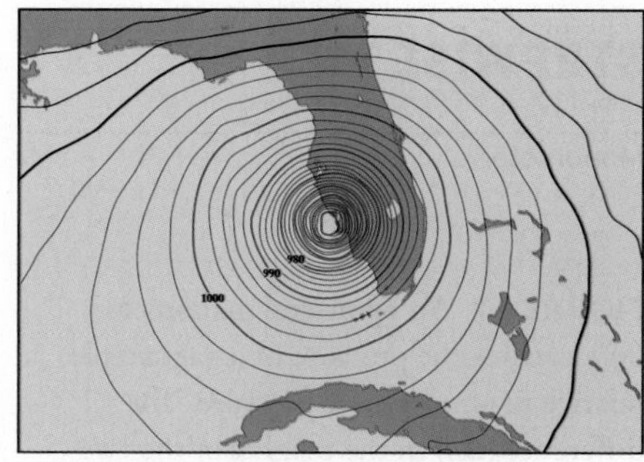

FIGURE 5.5 Changes in atmospheric pressure are associated with major weather events such as this hurricane.

Thus, 1 Pa is a tiny quantity of pressure. In many applications it is more convenient to express pressure in kilopascals.

For many years, meteorologists have expressed atmospheric pressure in *millibars* (mbar). Weather maps show changes in atmospheric pressure by constant-pressure contour lines, called *isobars*, spaced 4 mbar apart (**Figure 5.5**). There are exactly 10 mbar in 1 kPa; thus,

$$\begin{aligned} 1 \text{ atm} &= (101.325 \cancel{\text{kPa}})(10 \text{ mbar}/\cancel{\text{kPa}}) \\ &= 1013.25 \text{ mbar} \end{aligned}$$

Other units of pressure are derived from masses and areas in the U.S. Customary System, such as pounds per square inch (lb/in^2, psi) for tire pressures and inches of mercury for atmospheric pressure in weather reports. The relationships between different units for pressure and 1 atm are summarized in **Table 5.2**.

TABLE 5.2 Units for Expressing Pressure

Unit	Value
Atmosphere (atm)	1 atm
Millimeter of mercury (mmHg)	1 atm = 760 mmHg
Torr	1 atm = 760 torr
Pascal (Pa)	1 atm = 1.01325×10^5 Pa
Kilopascal (kPa)	1 atm = 101.325 kPa
Bar	1 atm = 1.01325 bar
Millibar (mbar)	1 atm = 1013.25 mbar
Pounds per square inch (psi)	1 atm = 14.7 psi
Inches of mercury	1 atm = 29.92 inches of Hg

manometer an instrument for measuring the pressure exerted by a gas.

gauge pressure the pressure of a gas that is above atmospheric pressure.

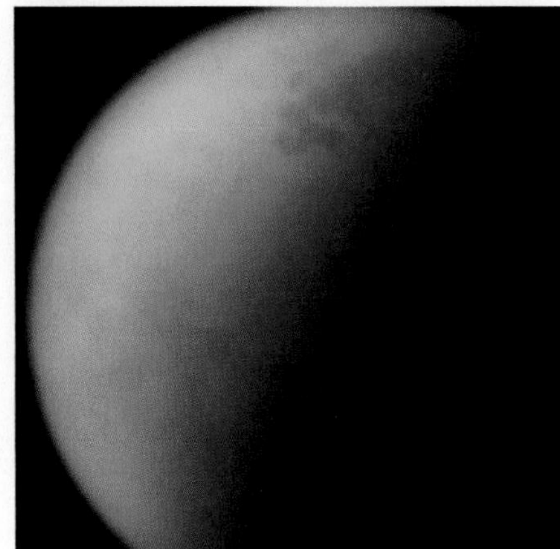

FIGURE 5.6 Saturn's moon Titan as recorded by the NASA spacecraft *Cassini* during its nearly 20-year exploration of the solar system that ended on September 15, 2017.

SAMPLE EXERCISE 5.1 Calculating Atmospheric Pressure **LO1**

The atmosphere on Titan (**Figure 5.6**), one of Saturn's moons, is composed almost entirely of nitrogen (98%). Its oceans contain substances thought to be important in the evolution of life, however, making Titan interesting for planetary scientists. The mass of Titan's atmosphere is estimated to be 9.0×10^{18} kg. The surface area of Titan is 8.3×10^{13} m^2. The acceleration due to gravity on Titan's atmosphere is 1.35 m/s^2. From these values, calculate an average atmospheric pressure in kilopascals.

Collect, Organize, and Analyze Equation 5.3 allows us to use the mass of Titan's atmosphere, its surface area, and acceleration due to its gravity to calculate its atmospheric pressure.

Solve Substituting the given values into Equation 5.3:

$$P = \frac{ma}{A}$$

$$= \frac{(9.0 \times 10^{18}\ \text{kg})(1.35\ \text{m/s}^2)}{8.3 \times 10^{13}\ \text{m}^2}$$

$$= \frac{1.46 \times 10^5\ \text{kg}}{\text{m} \cdot \text{s}^2} = 1.5 \times 10^5\ \text{Pa}$$

$$= 1.5 \times 10^2\ \text{kPa}$$

Think About It The atmospheric pressure on Titan is similar to Earth's (1.01×10^2 kPa). The greater mass of Titan's atmosphere and its smaller surface area compensate for its weaker acceleration due to gravity.

Practice Exercise Calculate the pressure in pascals exerted on a tabletop by a sugar cube that is 1.00 cm on each side and has a mass of 4.15 g. The acceleration due to gravity on Earth is 9.8 m/s^2.

(Answers to Practice Exercises are in the back of the book.)

STEPWISE ANIMATION
Manometer

Scientists conducting experiments with gases may monitor the pressures exerted by the gases by using a **manometer**. One type of manometer is illustrated in **Figure 5.7**. It consists of a U-shaped tube filled with mercury (or another dense liquid) that is connected to a container holding the gas sample

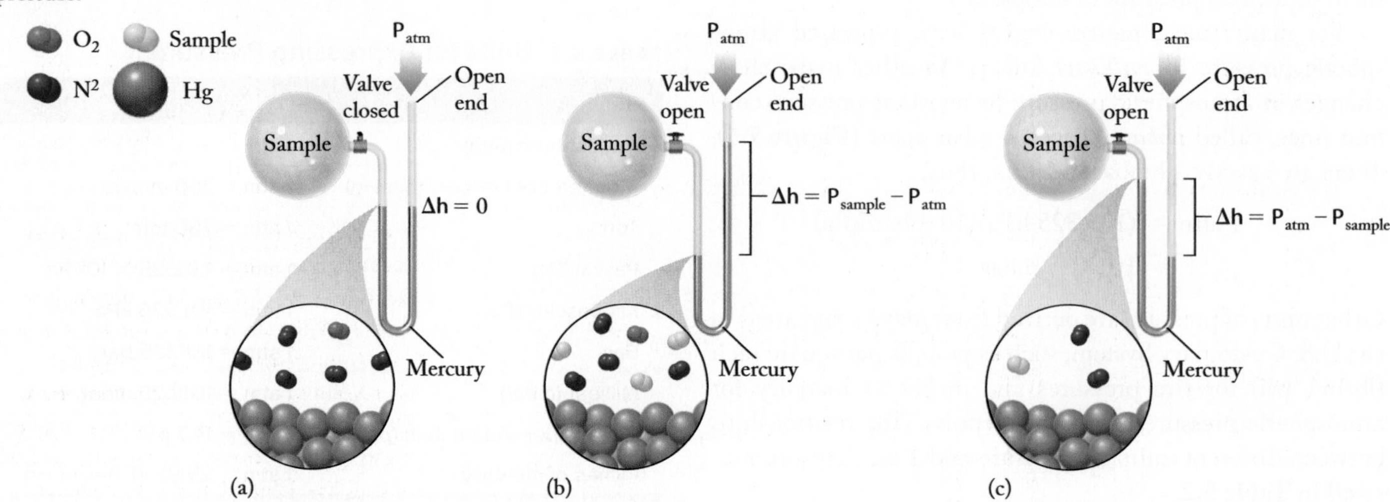

FIGURE 5.7 An open-end manometer (a) measures the pressure of a gas sample relative to atmospheric pressure. The difference (Δh) in the heights of the two mercury columns in (b) is caused by a sample pressure that is greater than the atmospheric pressure. The pressure of the sample in (c) is less than atmospheric pressure.

whose pressure is being measured. The other end of the manometer is open to the atmosphere. When the valve is opened, the gas exerts pressure on the mercury.

The difference in heights of the mercury column represents the difference between the pressure in the flask and atmospheric pressure. When the pressure in the flask is *greater* than atmospheric pressure (Figure 5.7b), the level in the arm open to the atmosphere is higher than the level in the arm attached to the flask; when the pressure in the flask is *lower* than atmospheric pressure (Figure 5.7c), the level in the arm attached to the flask is higher than the level in the arm open to the atmosphere.

Manometers have been largely displaced by pressure sensors based on flexible metallic or ceramic diaphragms. As the pressure on one side of the diaphragm increases, it distorts away from that side. This is the mechanism used to sense changes in atmospheric pressure in barometers, including the recording barometer, or *barograph*, shown in **Figure 5.8**.

Though manometers are not as widely used as they once were, many other devices are used, which, like the manometer in Figure 5.7, measure the pressure of a gas relative to atmospheric pressure. These devices are said to measure **gauge pressure**. Familiar examples include the gauges used to measure the pressure in automobile or bicycle tires. The values these devices display represent how much air pressure there is inside a tire ***above atmospheric pressure***. Thus, a gauge pressure of 2.0 atm represents a total gas pressure of $2.0 + 1.0 = 3.0$ atm.

(a)

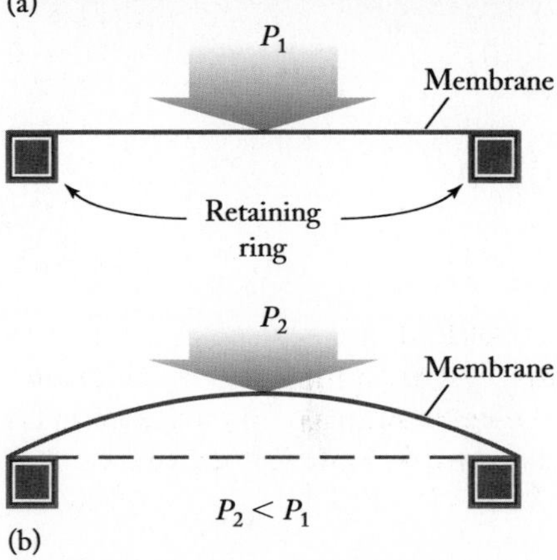

(b)

FIGURE 5.8 (a) The pressure sensor in this classic barograph is a partially evacuated corrugated metal can. (b) As atmospheric pressure decreases, the lid of the can distorts outward. This motion is amplified by a series of levers and transmitted via the horizontal arm to a pen tip that records the pressure on the graph paper as the drum slowly turns. A week's worth of barometric data can be recorded in this way.

SAMPLE EXERCISE 5.2 Measuring Gas Pressure with a Manometer **LO1**

Oyster shells are composed of calcium carbonate ($CaCO_3$). When the shells are roasted, they produce solid calcium oxide (CaO) and CO_2. A chemist roasts some oyster shells in an evacuated flask that is attached to a manometer (**Figure 5.9a**). When the roasting is complete, and the system has cooled to room temperature, the difference in levels of mercury (Δh) in the arms of the manometer is 144 mm (**Figure 5.9b**). Calculate the pressure created by the $CO_2(g)$ in (a) torr, (b) atmospheres, and (c) kilopascals on a day when the atmospheric pressure is 758 torr.

Collect, Organize, and Analyze The difference in the heights of the mercury levels in the arms of the manometer is equal to $P_{sample} - P_{atmosphere}$ (see Figure 5.7b) in mmHg. The conversion factors from Table 5.2 that we need are 760 mmHg = 760 torr = 1 atm = 101.325 kPa. The pressures expressed in mmHg and torr should be the same, and they should be about 800 times greater than the pressure expressed in atmospheres and about 8 times greater than the pressure in kPa.

Solve

a. Calculating the partial pressure of CO_2 (P_{sample}) in mmHg and torr:

$$\Delta h = P_{sample} - P_{atmosphere}$$

or

$$P_{sample} = P_{atmosphere} + \Delta h$$
$$= 758 + 144 = 902.3 \text{ mmHg} = 902 \text{ torr}$$

b. Converting torr to atmospheres:

$$902 \text{ torr} \times \frac{1 \text{ atm}}{760 \text{ torr}} = 1.187 \text{ atm} = 1.19 \text{ atm}$$

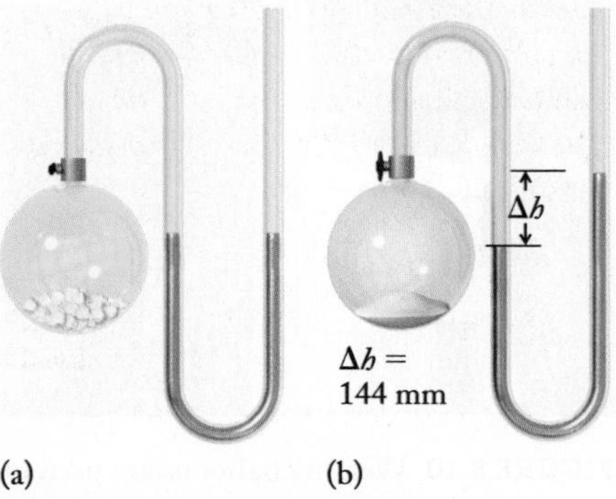

(a) (b)

FIGURE 5.9 Roasting $CaCO_3$ in a flask connected to an open-end manometer. (a) An evacuated flask containing oyster shells. (b) The same setup after the shells have been roasted and the sample has been allowed to cool to room temperature. Decomposition of $CaCO_3$ during sample roasting produces CO_2, the partial pressure of which is detected by the manometer.

c. Converting atmospheres to kilopascals:

$$1.19 \text{ atm} \times \frac{101.325 \text{ kPa}}{1 \text{ atm}} = 120 \text{ kPa}$$

Think About It The calculated pressure values make sense given our estimates and the relative heights of the mercury levels in the barometer, which indicate that $P_{\text{sample}} > P_{\text{atmosphere}}$. Therefore, the P_{sample} values should be greater than 760 torr, 1 atm, and 101.35 kPa.

Practice Exercise What if exactly half the quantity of oyster shells were roasted and P_{CO_2} was determined as described in this sample exercise. What would be the value of Δh in millimeters? On which side of the manometer, sample or atmosphere, would the mercury level be higher?

(Answers to Practice Exercises are in the back of the book.)

5.3 The Gas Laws

In Section 5.1 we summarized some of the properties of gases and described the effect of pressure and temperature on volume, mostly in qualitative terms. Our knowledge of the quantitative relationships among *P*, *T*, and *V* goes back more than three centuries, to a time before the field of chemistry as we now know it even existed. Some of the experiments that led to our understanding of how gases behave were driven by human interest in hot-air balloons, and today balloons are still used to study weather and atmospheric phenomena (**Figure 5.10**). Their successful and safe use requires an understanding of gas properties that was first gained in the 17th and 18th centuries.

FIGURE 5.10 Weather balloons are used to carry meteorological instruments into the upper atmosphere.

Boyle's Law: Relating Pressure and Volume

When a diver exhales underwater, the bubbles increase in size as they rise toward the surface. As the bubbles rise, the pressure exerted on them decreases—and as the pressure exerted on a gas decreases, its volume increases. The inverse is also true: as the pressure on a gas increases, its volume decreases. In other words, gases are compressible, a property that allows us to store gases (which would typically occupy large volumes under atmospheric pressure) in relatively small metal cylinders at high pressure.

Boyle's law the principle that the volume of a given amount of gas at constant temperature is inversely proportional to its pressure.

The relationship between the pressure and volume of a fixed quantity of gas (constant value of *n*, where *n* is the number of moles) at constant temperature was investigated by the British chemist Robert Boyle (1627–1691), who conducted experiments with a J-shaped tube open at one end and closed at the other end (**Figure 5.11**). When a small amount of mercury was poured into the tube, a column of air became trapped at the closed end. The pressure on the trapped air was changed by varying the amount of mercury poured into the open arm. The more mercury added, the greater the force exerted by the mercury ($F = ma$), and hence the greater the force per unit area ($P = F/A$) on the column of trapped air in the closed arm. The pressure exerted on this trapped air depended on the difference in the height of the mercury in the two sides of the tube and the atmospheric pressure.

As Boyle added mercury to the open arm, the *amount* (number of moles) of trapped air remained constant, but the volume it occupied decreased. At the time

Boyle discovered the relationship between P and V, scientists lacked a clear understanding of atoms or molecules. We now know that because the molecules in the trapped gas were in constant random motion within a smaller, confined space, the number of collisions with one another and with the surface of the mercury *increased*. The more frequent the collisions, the greater the force exerted by the molecules against the surface of the mercury and thus the greater the pressure of the gas. The mathematical relationship describing the inverse relationship between volume and pressure (under conditions of constant moles and constant temperature) is known as **Boyle's law (Figure 5.12)**:

$$P \propto \frac{1}{V} \quad (T \text{ and } n \text{ constant}) \tag{5.4}$$

where the symbol $\propto$ means "is proportional to."

Mathematically, we can replace the proportionality symbol with an equals sign and a constant:

$$P = (\text{constant})\frac{1}{V}$$

$$PV = \text{constant} \tag{5.5}$$

The value of the constant depends on the quantity of trapped air and its temperature.

Because the value of the product PV in Equation 5.5 does not change for a given quantity of trapped air at constant temperature, any two combinations of pressure and volume for a given sample of gas at constant temperature are related as follows:

$$P_1V_1 = P_2V_2 \tag{5.6}$$

This relationship, which applies to all gases, is illustrated by the dashed lines in Figure 5.12c, d. For example, when 44.8 L (V_1) of gas has a pressure of 0.500 atm (P_1),

$$P_1V_1 = (0.500 \text{ atm})(44.8 \text{ L}) = 22.4 \text{ L} \cdot \text{atm}$$

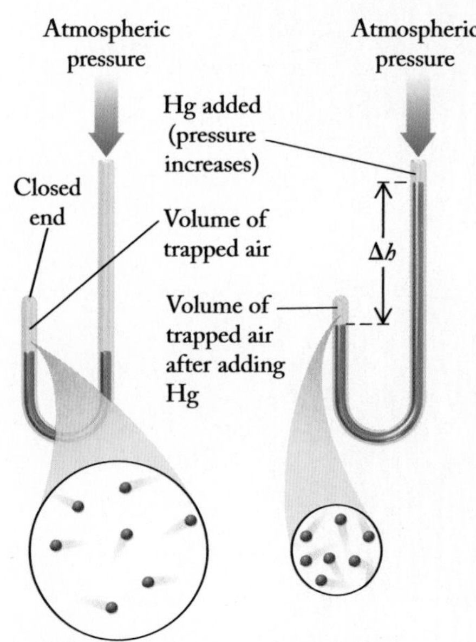

FIGURE 5.11 Boyle used a J-shaped tube for his experiments on the relationship between P and V. The volume of air trapped in the closed end of the tube decreases as the difference in height of mercury in the tube increases. The total pressure on the gas is equal to the pressure exerted by the added mercury (Δh) plus atmospheric pressure.

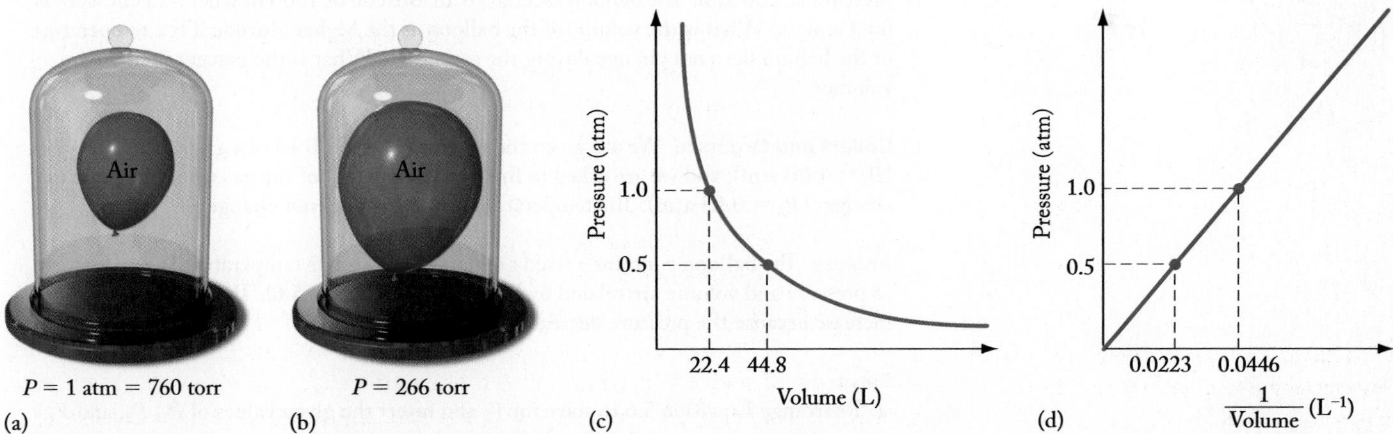

FIGURE 5.12 The sizes of the balloon in these images demonstrates the inverse relationship between pressure and volume. (a) The balloon inside the bell jar is surrounded by a pressure of 760 torr (1 atm). (b) After some of the air is pumped out of the bell jar, the pressure of the air inside it is only 266 torr. The decrease in pressure allows the balloon to expand until the pressure inside and outside the balloon is equal. (c) At constant temperature, the pressure of a given quantity of gas is inversely proportional to the volume it occupies; the graph of an inverse proportion is a hyperbola. (d) The inverse proportion between P and V means that a plot of P versus $1/V$ is a straight line.

If this quantity of the gas is compressed into a container half the size, then V_2 is 22.4 L, and P_2, according to Equation 5.6, is

$$P_1V_1 = P_2V_2$$

$$P_2 = \frac{P_1V_1}{V_2} = \frac{(0.500\ \text{atm})(44.8\ \cancel{\text{L}})}{22.4\ \cancel{\text{L}}}$$

$$= 1.00\ \text{atm}$$

Thus, when the volume is cut in half, Boyle's law shows that the pressure doubles.

CONCEPT TEST

Which of the graphs in **Figure 5.13** correctly depicts the relationship between the product of pressure and volume (*PV*) as a function of pressure (*P*) for a given quantity of gas at constant temperature?

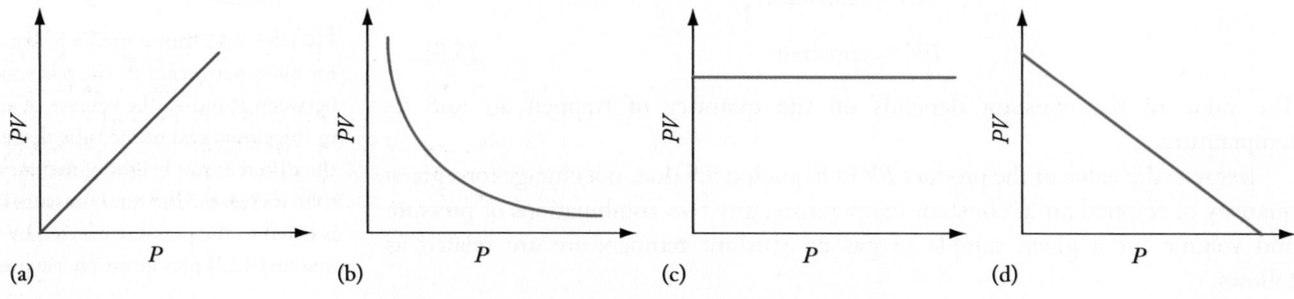

FIGURE 5.13

(Answers to Concept Tests are in the back of the book.)

SAMPLE EXERCISE 5.3 Using Boyle's Law **LO2**

A balloon is partially inflated with 5.00 L of helium at sea level where the atmospheric pressure is 1.00 atm. The balloon ascends to an altitude of 1600 m where the pressure is 0.83 atm. (a) What is the volume of the balloon at the higher altitude if the temperature of the helium does not change during the ascent? (b) What is the percent increase in volume?

Collect and Organize We are given the volume ($V_1 = 5.00$ L) of a gas and its pressure ($P_1 = 1.00$ atm), and we are asked to find the volume (V_2) of the gas when the pressure changes ($P_2 = 0.83$ atm). The temperature of the gas does not change.

Analyze The balloon contains a fixed amount of gas and its temperature is constant, so pressure and volume are related by Boyle's law (Equation 5.6). The volume should increase because the pressure decreases.

Solve

a. Rearrange Equation 5.6 to solve for V_2 and insert the given values of P_1, P_2, and V_1:

$$V_2 = \frac{P_1V_1}{P_2}$$

$$V_2 = \frac{(1.00\ \cancel{\text{atm}})(5.00\ \text{L})}{0.83\ \cancel{\text{atm}}} = 6.0\ \text{L}$$

b. To calculate the percent increase in the volume, we must determine the difference between V_1 and V_2 and compare the result with V_1:

$$\% \text{ increase in volume} = \frac{V_2 - V_1}{V_1} \times 100\%$$
$$= \frac{6.0 \text{ L} - 5.0 \text{ L}}{5.0 \text{ L}} \times 100\% = 20\%$$

Think About It The volume did, in fact, increase when the pressure decreased. The lower atmospheric pressure means fewer collisions by molecules in the atmosphere on the outside of the helium balloon. The helium balloon will rise until it reaches an altitude where the pressure inside the balloon (i.e., the number of collisions by the helium atoms) equals the atmospheric pressure. Equation 5.6 can thus be used to calculate the change in pressure that takes place at constant temperature when the volume of a quantity of gas changes.

Practice Exercise A scuba diver exhales 3.50 L of air while swimming at a depth of 20.0 m, where the sum of atmospheric pressure and water pressure is 3.00 atm. By the time the exhaled air rises to the surface, where the pressure is 1.00 atm, what is its volume?

(Answers to Practice Exercises are in the back of the book.)

Charles's law the principle that the volume of a fixed quantity of gas at constant pressure is directly proportional to its absolute temperature.

absolute temperature temperature expressed in kelvins on the absolute (Kelvin) temperature scale, on which 0 K is the lowest possible temperature.

Charles's Law: Relating Volume and Temperature

Nearly a century after Boyle's discovery of the inverse relationship between the pressure exerted by a gas and its volume, the French scientist Jacques Charles (1746–1823) documented the linear relationship between the volume and temperature of a fixed quantity of gas at constant pressure. Now known as **Charles's law**, this relationship states that, when the pressure exerted by a gas is held constant, the volume of a fixed quantity of gas is directly proportional to the **absolute temperature** (the temperature on the Kelvin scale) of the gas:

$$V \propto T \quad (P \text{ and } n \text{ constant}) \tag{5.7}$$

Because the lowest temperature on the absolute temperature scale is 0 K, temperatures expressed in kelvin are always positive numbers.

The effects of Charles's law can be seen in **Figure 5.14**, in which a balloon has been attached to a flask, trapping a fixed amount of gas in the apparatus. Heating the flask causes the gas to expand, inflating the balloon. The higher the temperature, the larger the volume occupied by the gas and the bigger the balloon.

As with Boyle's law, we can replace the proportionality symbol in Equation 5.7 with an equals sign if we include a proportionality constant:

$$V = (\text{constant})\,T$$
$$\frac{V}{T} = \text{constant} \tag{5.8}$$

The value of the constant depends on the number of moles of gas in the sample (n) and on the pressure of the gas (P). When those two parameters are held constant, the ratio V/T does not change, and any two combinations of volume and temperature are related as follows:

$$\frac{V_1}{T_1} = \frac{V_2}{T_2} \tag{5.9}$$

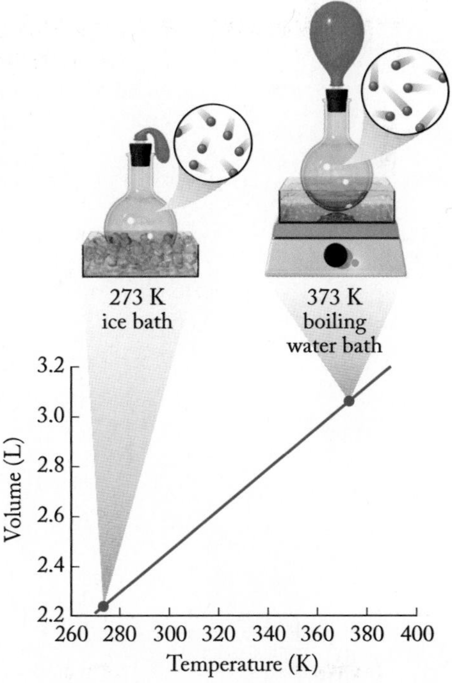

FIGURE 5.14 A balloon attached to a flask inflates as the temperature of the gas inside the flask increases from 273 K to 373 K at constant atmospheric pressure. This behavior is described by Charles's law.

What happens when a balloon containing 2.00 L of air at 20°C is taken outside on a day when the temperature is 0°C? The amount of gas in the balloon is fixed, and the atmospheric pressure is constant, so Charles's law applies. Experience tells us that the volume of the balloon decreases. We can calculate the final volume with Equation 5.9, provided that we express T_1 and T_2 in kelvin, not degrees Celsius. Solving Equation 5.9 for V_2:

$$V_2 = \frac{V_1 T_2}{T_1}$$

and substituting the known volume and temperature values:

$$V_2 = \frac{2.00 \text{ L} \times 273 \cancel{\text{K}}}{293 \cancel{\text{K}}} = 1.86 \text{ L}$$

CONNECTION We learned in Chapter 1 that Kelvin and Celsius temperatures are related by the equation K = °C + 273.15.

As predicted, the volume of the gas decreases when the temperature of the gas decreases. The percent change in the volume of the balloon is

$$\% \text{ decrease} = \frac{V_1 - V_2}{V_1} \times 100\% = \frac{2.00 \cancel{\text{L}} - 1.86 \cancel{\text{L}}}{2.00 \cancel{\text{L}}} \times 100\% = 7.0\%$$

CONCEPT TEST

Suppose you have two graphs: graph 1 plots volume (*V*) as a function of temperature (*T*) for 1 mole of a gas at a pressure of 1.00 atm, and graph 2 plots volume (*V*) as a function of *T* for 1 mole of a gas at 2.00 atm pressure. How do the slopes of the two graphs differ?

(Answers to Concept Tests are in the back of the book.)

SAMPLE EXERCISE 5.4 Using Charles's Law LO2

Charles was drawn to the study of gases because of his interest in hot-air balloons. What temperature (in °C) is required to increase the volume of a sealed balloon from 2.00 L to 3.00 L if the initial temperature is 15°C and the atmospheric pressure is constant?

Collect and Organize We are given the volume (V_1 = 2.00 L) of a gas at an initial Celsius temperature (T_1 = 15°C), and we are asked to calculate the Celsius temperature needed to increase the volume of the gas (V_2 = 3.00 L). The container (a balloon) is sealed, so n is constant, as is atmospheric pressure.

Analyze The relationship between volume and temperature is given by Charles's law (Equation 5.9). According to Charles's law, the temperature increases when the volume increases. If the volume increases by 50% (i.e., from 2 L to 3 L), then the absolute temperature must also increase by 50%.

Solve First, we rearrange Equation 5.9 to solve for T_2:

$$T_2 = \frac{V_2 T_1}{V_1}$$

We then substitute for V_1, V_2, and T_1, remembering to convert temperature to kelvins:

$$T_2 = \frac{V_2 T_1}{V_1} = \frac{(3.00 \cancel{\text{L}})(273 + 15 \text{ K})}{2.00 \cancel{\text{L}}} = 432 \text{ K}$$

The final step is to convert T_2 to degrees Celsius:

$$T_2 = 432 \text{ K} - 273 = 159°\text{C}$$

Think About It To increase the volume by 1.5 times [(1.5)(2.00 L) = 3.00 L], *absolute* temperature must increase 1.5 times [(1.5)(288 K) = 432 K].

Practice Exercise Hot expanding gases can be used to perform useful work in a cylinder fitted with a movable piston, as in **Figure 5.15**. If the temperature of a gas confined to such a cylinder is increased from 245°C to 560°C, what is the ratio of the initial volume to the final volume if the pressure exerted on the gas remains constant?

(Answers to Practice Exercises are in the back of the book.)

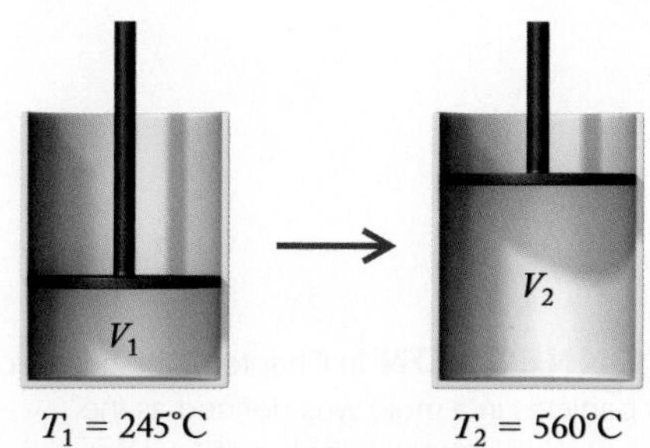

FIGURE 5.15 The volume of a gas in a cylinder changes as the temperature of the gas increases from 245°C to 560°C.

Charles's law also made it possible to determine that absolute zero (0 K) is equal to −273.15°C. Suppose the volume of a fixed quantity of gas at constant pressure is plotted as a function of temperature. Some typical results for three gases are graphed in **Figure 5.16**, showing that the decrease in volume is linear as temperature decreases. There is a limit to how much the temperature of a gas can decrease because at some temperature the gas liquefies, at which point the gas laws no longer apply. However, we can *extrapolate* from our measured data to the point where the volume of a gas would reach zero if condensation did not occur. The dashed lines in Figure 5.16 cross the temperature axis at −273.15°C, the temperature defined as 0 K, or absolute zero.

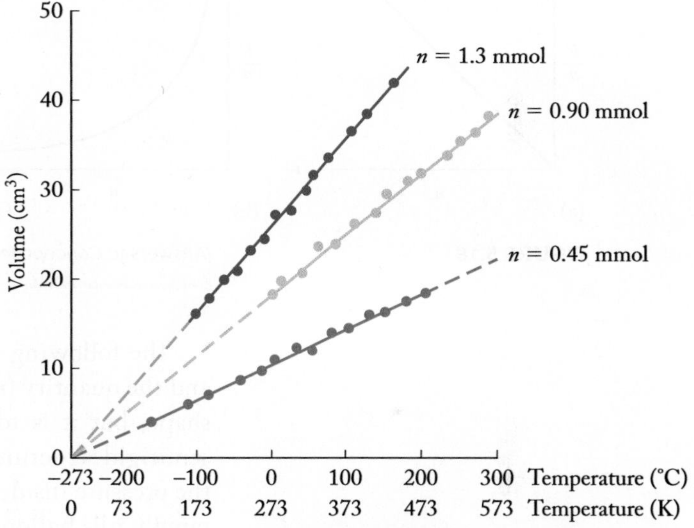

FIGURE 5.16 The volumes of three quantities of gas at the same pressure plotted against temperature on both the Celsius and Kelvin scales. As predicted by Charles's law, volume decreases as the temperature decreases; the relationship is linear on both scales. The dashed lines show an extrapolation from the experimental data to a point corresponding to zero volume. This temperature is known as absolute zero: −273.15°C on the Celsius scale and 0 K on the Kelvin scale.

Avogadro's Law: Relating Volume and Quantity of Gas

Boyle's and Charles's experiments involved constant quantities of gas. What happens when the amount of gas is *not* held constant, such as when a balloon is inflated (**Figure 5.17**)? As you probably know from

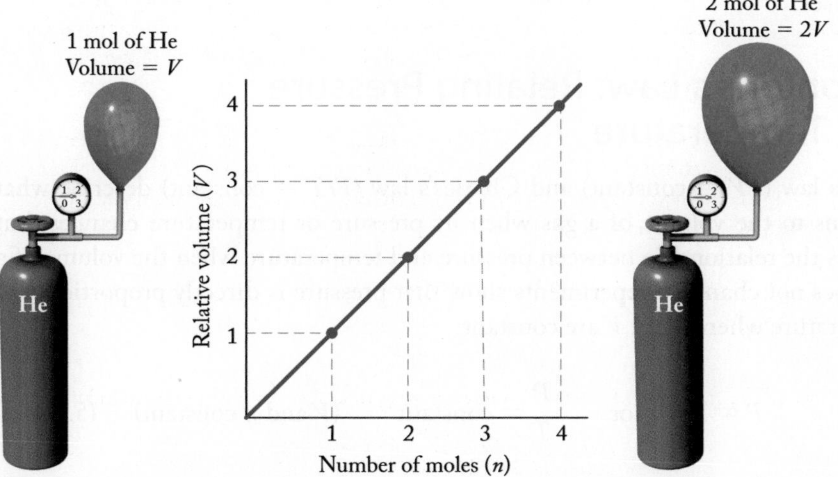

FIGURE 5.17 The balloon on the right has twice the volume of the balloon on the left because it is filled with twice the quantity of helium. This direct relationship between volume and number of moles of a gas at constant temperature and pressure demonstrates Avogadro's law.

personal experience, adding more gas to a balloon causes its volume to increase. If some of this gas escapes, balloon volume decreases. From these observations we may conclude that the volume of a sample of gas is proportional to the quantity (number of moles) of gas in the sample. This relationship between V and n is known as **Avogadro's law** to honor Amedeo Avogadro (1776–1856), who articulated that the volume of a gas at a given temperature and pressure is directly proportional to the quantity of the gas in moles:

CONNECTION In Chapter 3 the number of particles in a mole was defined as the Avogadro constant, in honor of Amedeo Avogadro's early work with gases that led to determining atomic masses.

$$V \propto n \quad \text{or} \quad \frac{V}{n} = \text{constant} \qquad (P \text{ and } T \text{ constant}) \qquad (5.10)$$

CONCEPT TEST

Which graph in **Figure 5.18** correctly describes the relationship between the value of *V*/*n* as *n* is increased at constant *P* and *T*?

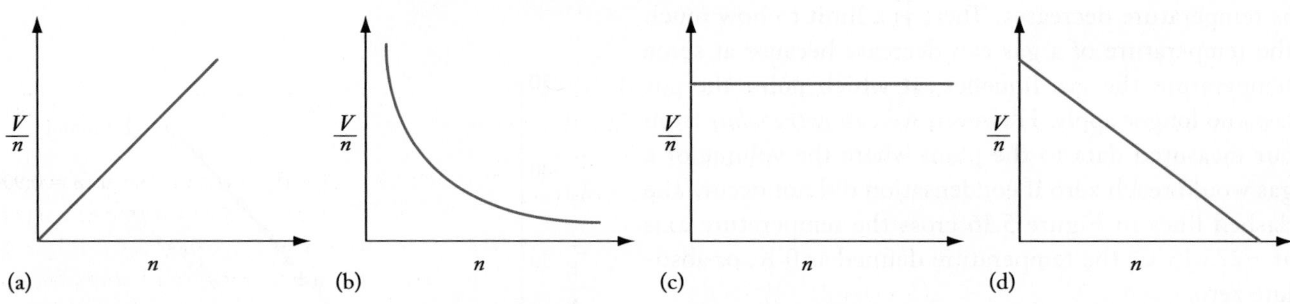

FIGURE 5.18

(Answers to Concept Tests are in the back of the book.)

The following observations illustrate the relationship between pressure (P) and the quantity (n) of a gas. Suppose a bicycle tire is inflated enough to hold its shape, but it is too soft to ride on. If more air is pumped into the tire, its semirigid structure limits any increase in volume, and adding more air increases the pressure inside the tire. As another example, consider a carnival vendor who rapidly fills balloons from a helium tank such as the one in Figure 5.17. If balloon sales are brisk, the pressure reading on the tank gauge will decrease as more and more balloons are filled and the quantity of helium inside the cylinder decreases.

Amontons's Law: Relating Pressure and Temperature

Boyle's law (PV = constant) and Charles's law (V/T = constant) describe what happens to the volume of a gas when its pressure or temperature changes, but what is the relationship between pressure and temperature when the volume of a gas does not change? Experiments show that pressure is directly proportional to temperature when n and V are constant:

$$P \propto T \quad \text{or} \quad \frac{P}{T} = \text{constant} \qquad (V \text{ and } n \text{ constant}) \qquad (5.11)$$

The relationship in Equation 5.11 means that as the absolute temperature of a fixed amount of gas held at a constant volume increases, the pressure of the gas increases (**Figure 5.19**). This statement is referred to as **Amontons's law** in honor

Avogadro's law the principle that the volume of a gas at a given temperature and pressure is proportional to the quantity of the gas.

Amontons's law the principle that the pressure of a quantity of gas at constant volume is directly proportional to its absolute temperature.

of the French physicist Guillaume Amontons (1663–1705), a contemporary of Robert Boyle, who constructed a thermometer based on the observation that the pressure of a gas is directly proportional to its temperature.

Where do we see evidence of Amontons's law? Suppose a bicycle tire is inflated to a recommended gauge pressure of 85 psi (5.8 atm) on an afternoon in late autumn when the temperature is 25°C. The next morning, after the temperature dropped to 0°C, the pressure in the tire dropped to 76 psi (5.2 atm). Did the tire leak? Perhaps, but the decrease in pressure could also be explained by Amontons's law.

We can confirm that the decrease in tire pressure was consistent with the decrease in temperature by relating the initial and final *total* pressures in the tire to the initial (T_1) and final (T_2) absolute temperatures.

$$\frac{P_1}{T_1} = \frac{P_2}{T_2} \quad (5.12)$$

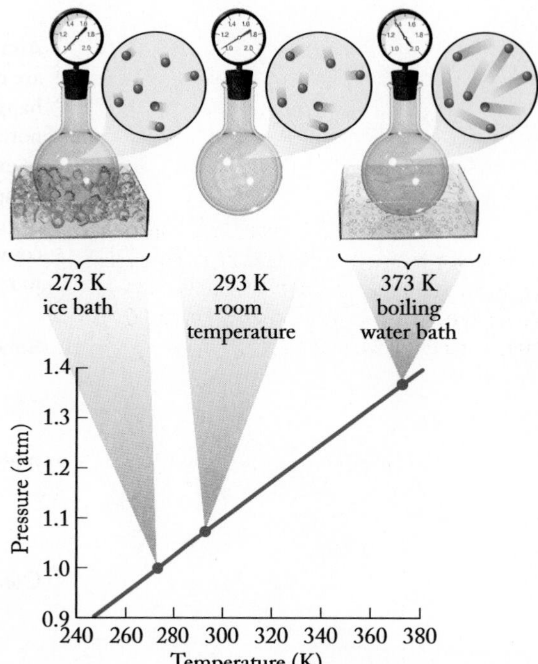

FIGURE 5.19 The pressure of a given quantity of gas is directly proportional to its absolute temperature at constant volume. The average speeds of gas molecules increase with increasing temperature, causing more frequent and more forceful collisions with the walls of the flasks and higher pressures.

The pressure measured by a tire gauge reflects how much additional pressure above atmospheric pressure is in the tire. To calculate the total pressure initially in the tire, we add 1.0 atm to the gauge pressure so that $P_1 = 5.8 + 1.0 = 6.8$ atm. The predicted P_2 caused by the temperature drop is

$$P_2 = \frac{P_1T_2}{T_1} = \frac{(6.8 \text{ atm})(273 \cancel{\text{K}})}{298 \cancel{\text{K}}} = 6.2 \text{ atm}$$

This total pressure is equivalent to a gauge pressure of $6.2 - 1.0 = 5.2$ atm, which is the value that was measured at 273 K. Therefore, the decrease in tire pressure resulted from a decrease in temperature, not a leak.

Why is pressure proportional to absolute temperature? Like pressure, temperature is directly related to molecular motion. The average speed at which a population of molecules moves increases with increasing temperature (see Figure 5.19). For a given number of gas molecules, increasing the temperature increases the average speed and therefore increases both the frequency and force with which the molecules collide with the walls of their container. Because pressure is related to the frequency and force of the collisions, it follows that higher temperatures produce higher pressures if other factors such as volume and quantity of gas are held constant.

SAMPLE EXERCISE 5.5 Using Amontons's Law **LO2**

Labels on aerosol cans caution against incineration because the cans may explode when their internal pressure exceeds 3.00 atm. At what temperature (in °C) will an aerosol can burst if the initial pressure inside the can is 2.20 atm at 25°C?

Collect and Organize We are given the temperature ($T_1 = 25$°C) and pressure ($P_1 = 2.20$ atm) of a gas, and we are asked to determine the temperature (T_2) at which the pressure ($P_2 = 3.00$ atm) will cause the can to explode.

Analyze Because the gas is enclosed in an aerosol can, the volume and quantity of gas are constant. We can use Amontons's law (Equation 5.12) because it describes what happens to pressure as the temperature of a gas is changed. To estimate the answer, notice that the pressure in the can must increase by about 50% in order for the pressure to exceed 3 atm. Pressure is directly proportional to temperature, so the *absolute* temperature must also increase by about 50%. The initial temperature is given in degrees Celsius. Because Equation 5.12 works only with absolute temperatures, we must convert T_1 to kelvins. After solving for T_2, we need to convert it back to degrees Celsius to report our answer.

Solve Rearrange Equation 5.12 to solve for T_2:

$$T_2 = \frac{T_1 P_2}{P_1}$$

After converting T_1 from degrees Celsius to kelvins, we have

$$T_2 = \frac{(25 + 273)\text{ K} \times 3.00 \cancel{\text{ atm}}}{2.20 \cancel{\text{ atm}}} = 406\text{ K}$$

Convert T_2 to degrees Celsius:

$$T_2 = 406\text{ K} - 273 = 133^\circ\text{C}$$

Think About It This temperature is certainly higher than the original temperature. In fact, 406 K is about 1.5 times (or 50% higher than) the initial temperature, 298 K, so the answer makes sense.

Practice Exercise The air pressure in the tires of an automobile is adjusted to 2.3 atm at a gas station in San Diego, where the air temperature is 20°C (68°F) and atmospheric pressure is 0.987 atm. After a 3-hour drive along Interstate 8 (I-8), the car and driver are in Yuma, Arizona, where the temperature is 43°C (110°F). What is the pressure in the tires, if we assume that atmospheric pressure is still 0.987 atm? *Note*: The measured tire pressures are *gauge* pressures, meaning that they indicate the pressures in the tires *above atmospheric pressure.*

(Answers to Practice Exercises are in the back of the book.)

ideal gas a gas whose behavior is predicted by the linear relationships defined by Boyle's, Charles's, Avogadro's, and Amontons's laws.

ideal gas equation (also called **ideal gas law**) the principle relating the pressure, volume, number of moles, and temperature of an ideal gas; expressed as $PV = nRT$, where R is the universal gas constant.

universal gas constant the constant R in the ideal gas equation; its value and units depend on the units used for the variables in the equation.

standard temperature and pressure (STP) 0°C and 1 bar as defined by IUPAC; 0°C and 1 atm are commonly used in the United States and in this textbook.

molar volume the volume occupied by 1 mole of an ideal gas at STP; 22.4 L.

Gases that behave according to the relationships discovered by Boyle, Charles, Avogadro, and Amontons are called **ideal gases**. In an ideal gas, the atoms or molecules are assumed *not* to interact with one another; rather, they move independently with speeds that are related to their masses and to the temperature of the gas. Most gases exhibit ideal behavior at the pressures and temperatures typically encountered in the atmosphere. Under these conditions, the volumes occupied by gas molecules or atoms are insignificant compared with the overall volume occupied by the gas.

5.4 The Ideal Gas Law

What would happen to the volume of a weather balloon if it is launched from the surface of Earth and allowed to drift to an elevation of 10,000 m? The volume of the balloon would expand as the atmospheric pressure decreased, but the air temperature would simultaneously decrease as the balloon ascended, tending to make the volume smaller. How do we determine the final volume of the balloon when both pressure and temperature change at the same time for a given quantity of gas? Taken individually, none of the four gas laws and their accompanying mathematical relations apply. Nevertheless, we can derive a relationship that describes this situation.

Boyle's law (Equation 5.5) states that volume and pressure are inversely proportional, whereas Charles's law (Equation 5.8) states that volume is directly proportional to temperature. Putting these two laws together and combining their constants gives Equation 5.13, which relates P, V, and T:

$$PV = \text{constant} \qquad \text{Boyle's law}$$

$$\frac{V}{T} = \text{constant} \qquad \text{Charles's law}$$

$$\frac{PV}{T} = \text{combined constant} \tag{5.13}$$

Avogadro's law (Equation 5.10), which states that volume is directly proportional to the number of moles of gas when T and P are constant, can also be included in Equation 5.13. After combining the constant in Equation 5.10 with the combined constant in Equation 5.13,

$$\frac{PV}{nT} = \text{combined constant} \quad \text{or} \quad PV = (\text{combined constant}) \times nT \tag{5.14}$$

Equation 5.14 can be turned into the **ideal gas equation** or **ideal gas law** (Equation 5.15) by inserting an appropriate constant, R, which is called the **universal gas constant**:

$$PV = nRT \tag{5.15}$$

As **Table 5.3** shows, R has different values depending on the units used. When $R = 0.08206\ \text{L} \cdot \text{atm/(mol} \cdot \text{K)}$, for example, the quantity of gas should be expressed in moles, the volume in liters, the pressure in atmospheres, and the temperature in kelvins.

TABLE 5.3 Values for the Universal Gas Constant (R)

Value of R	Units
0.08206	L · atm/(mol · K)
8.314	kg · m^2/(s^2 · mol · K)
8.314	J/(mol · K)
8.314	m^3 · Pa/(mol · K)
62.37	L · torr/(mol · K)

A useful reference point when studying the properties of gases is **standard temperature and pressure (STP)**, defined by the International Union of Pure and Applied Chemistry (IUPAC) as 0°C and 1 bar. The pressure unit of 1 atm is very close to 1 bar, so we consider STP to be 0°C and 1 atm in calculations in this textbook. The volume of 1 mole of an ideal gas at STP is known as the **molar volume** (**Figure 5.20**). We can calculate the molar volume from the ideal gas equation by solving the equation for V and inserting the values of n, P, and T at STP:

$$V = \frac{(1\ \cancel{\text{mol}})\left(0.08206\ \frac{\text{L} \cdot \cancel{\text{atm}}}{\cancel{\text{mol}} \cdot \cancel{\text{K}}}\right)(273\ \cancel{\text{K}})}{1\ \cancel{\text{atm}}} = 22.4\ \text{L}$$

Many chemical and biochemical processes take place at pressures near 1 atm and at temperatures between 0°C and 40°C. Within this range, the volume that 1 mole of gaseous reactant or product occupies is no more than about 15% greater than the molar volume. Therefore, volumes can be estimated easily if molar amounts are known. An important feature of molar volume is that it applies to any ideal gas, independent of its chemical composition. In other words, at STP, 1 mole of helium occupies the same volume—22.4 L—as 1 mole of methane (CH_4) or carbon dioxide (CO_2).

FIGURE 5.20 The box contains 1 mole of gas; the molar volume of an ideal gas is 22.4 L at 0°C and 1 atm of pressure. A basketball fits loosely into a box having this volume.

combined gas law (also called **general gas equation**) the principle relating the pressure, volume, and temperature of a quantity of an ideal gas:

$$\frac{P_1V_1}{T_1} = \frac{P_2V_2}{T_2}$$

We can derive another relationship that is useful when a system starts in an initial state (P_1, V_1, T_1, n_1) and moves to a final state (P_2, V_2, T_2, n_2):

$$P_1V_1 = n_1RT_1 \quad \text{so} \quad \frac{P_1V_1}{n_1T_1} = R$$

and

$$P_2V_2 = n_2RT_2 \quad \text{so} \quad \frac{P_2V_2}{n_2T_2} = R \tag{5.16}$$

Because both the initial-state and final-state expressions are equal to R, they are also equal to each other:

CHEMTOUR
Ideal Gas Law

$$\frac{P_1V_1}{n_1T_1} = \frac{P_2V_2}{n_2T_2} \tag{5.17}$$

Some processes, such as breathing (**Figure 5.21**), illustrate the relationship between P, V, and n in Equation 5.17. In closed systems, such as weather balloons, the amount of gas is constant ($n_1 = n_2$), but pressure, temperature, and volume vary, which simplifies Equation 5.17 to

$$\frac{P_1V_1}{T_1} = \frac{P_2V_2}{T_2} \tag{5.18}$$

Equation 5.18 is known as the **combined gas law**, or the **general gas equation**. Sample Exercise 5.6 shows how to use this equation for determining the effect of changes in P and T on the volume (V) of a weather balloon. Other simplified versions of Equation 5.17 are certainly possible when variables in addition to n are fixed, but only Equation 5.18 is known as the combined gas law.

CONCEPT TEST

Which are correct variations of Equation 5.17?

a. $\frac{n_2T_2}{P_2} = \frac{n_1T_1}{P_1}$ at constant V

b. $\frac{n_2V_2}{P_2} = \frac{n_1V_1}{P_1}$ at constant T

c. $\frac{n_2T_2}{V_2} = \frac{n_1T_1}{V_1}$ at constant P

d. $\frac{T_1}{n_1V_1} = \frac{T_2}{n_2V_2}$ at constant P

(Answers to Concept Tests are in the back of the book.)

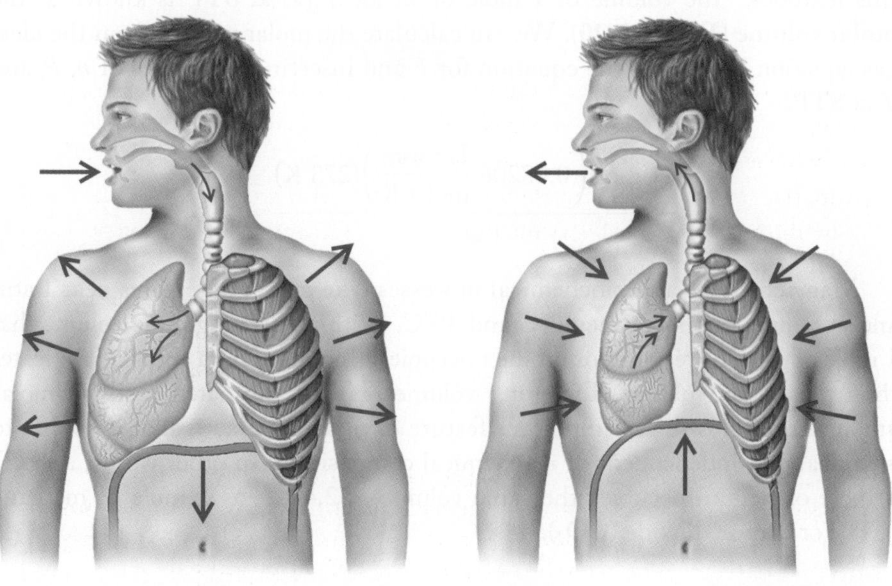

FIGURE 5.21 Breathing illustrates the relationship between P, V, and n. (a) When you inhale, your rib cage expands, and your diaphragm moves down, increasing the volume of your lungs. Increased volume decreases the pressure inside your lungs, which allows more air to flow into them. (b) As you exhale, increased pressure decreases lung volume, forcing air out.

SAMPLE EXERCISE 5.6 Calculations Involving Changes in *P*, *V*, and *T* **LO2**

A weather balloon filled with 100.0 L of He is launched from ground level ($T = 20°C$, $P = 755$ torr). No gas is added or removed from the balloon during its flight. Calculate its volume at an altitude of 10 km, where the temperature and pressure in the balloon are −52°C and 195 torr, respectively.

Collect, Organize, and Analyze We are given the initial temperature, pressure, and volume of a gas, and we are asked to determine the final volume after the pressure and temperature have changed. The decrease in temperature leads to a *decrease* in volume, but the decrease in pressure leads to an *increase* in volume. We need to express the temperatures in kelvins and estimate which variable dominates the change. Because the quantity of gas does not change (n is constant), we can solve for V_2 in the combined gas law (Equation 5.18).

Solve First, convert the given Celsius temperatures to kelvins:

$$T_1 = 20°C + 273 = 293 \text{ K}$$
$$T_2 = -52°C + 273 = 221 \text{ K}$$

Then, use the general gas equation (Equation 5.18) to solve for V_2:

$$\frac{P_1V_1}{T_1} = \frac{P_2V_2}{T_2}$$

$$V_2 = V_1 \times \frac{P_1}{P_2} \times \frac{T_2}{T_1}$$

$$V_2 = (100.0 \text{ L}) \times \frac{755 \cancel{\text{torr}}}{195 \cancel{\text{torr}}} \times \frac{221 \cancel{\text{K}}}{293 \cancel{\text{K}}} = 292 \text{ L}$$

Think About It The volume increases nearly threefold as the balloon ascends to 10 km. This result makes sense because atmospheric pressure decreases to about one-quarter of its ground-level value during the ascent. The volume would have increased more if not for the counteracting effect of the lower temperature at 10 km.

Practice Exercise The balloon in Sample Exercise 5.6 is designed to continue its ascent to an altitude of 30 km, where it bursts, releasing a package of meteorological instruments that parachute back to Earth. If the atmospheric pressure at 30 km is 28.0 torr and the temperature is −45°C, what is the volume of the balloon when it bursts?

(Answers to Practice Exercises are in the back of the book.)

The ideal gas law describes the relations among number of moles, pressure, volume, and temperature for any gas, provided it behaves ideally. Because many gases behave ideally at typical atmospheric pressures, we can apply the ideal gas equation to many situations outside the laboratory. In Sample Exercise 5.7, we calculate the mass of oxygen in an alpine climber's compressed-oxygen cylinder.

SAMPLE EXERCISE 5.7 Applying the Ideal Gas Law **LO2**

Bottles of compressed O_2 carried by climbers ascending Mt. Everest have an internal volume of 5.90 L. Assume that one such bottle has been filled with O_2 to a pressure of 2025 psi at 25°C. Also assume that O_2 behaves as an ideal gas. (a) How many moles of O_2 are in the bottle? (b) What is the mass in grams of O_2 in the bottle?

Collect, Organize, and Analyze The ideal gas equation enables us to use the given values for P, V, and T to calculate n, the number of moles of O_2. Then we can use its molar mass (32.00 g/mol) to calculate the mass of O_2 in the bottle. We can estimate the answer by remembering that 1 mole of gas at STP occupies 22.4 L. Our oxygen bottle has a volume of about 6 L, which is about one-quarter of the molar volume at STP, but the pressure (2025 psi) is more than 100 times greater than 1 atm (see Table 5.2). Thus, we predict that the bottle contains more than 25 moles of O_2.

Solve

a. First, rearrange the ideal gas equation to solve for n:

$$PV = nRT$$

$$n = \frac{PV}{RT}$$

Before using this expression for n, we need to convert pressure into atmospheres and temperature into kelvins:

$$P = (2025 \text{ psi})\left(\frac{1 \text{ atm}}{14.7 \text{ psi}}\right) = 138 \text{ atm} \qquad T = 25^\circ\text{C} + 273 = 298 \text{ K}$$

$$n = \frac{(138 \text{ atm})(5.90 \text{ L})}{\left(0.08206 \frac{\text{L} \cdot \text{atm}}{\text{mol} \cdot \text{K}}\right)(298 \text{ K})} = 33.3 \text{ mol}$$

b. To convert moles into grams, multiply moles by the molar mass:

$$(33.3 \text{ mol}) \frac{(32.00 \text{ g})}{1 \text{ mol}} = 1.07 \times 10^3 \text{ g}$$

Think About It Our answer in part (a) is certainly reasonable based on our estimate. Most climbers require several bottles to climb Mt. Everest and return.

Practice Exercise Starting with the moles of O_2 calculated in Sample Exercise 5.7, calculate the volume of O_2 the bottle could deliver to a climber at an altitude where the temperature is −38°C and the atmospheric pressure is 0.35 atm.

(Answers to Practice Exercises are in the back of the book.)

5.5 Gases in Chemical Reactions

Gases are reactants or products in many important reactions. For example, if a commercial airliner flying at a high altitude loses cabin pressure, oxygen masks are deployed automatically for the passengers to breathe until the aircraft can descend or the cabin can be pressurized again. The oxygen gas that flows in the masks is produced by the decomposition of solid sodium chlorate, $NaClO_3$. Similarly, when an air bag deploys during an automobile accident, the nitrogen gas that rapidly inflates the protective air bag is the product of the decomposition of solid sodium azide, NaN_3. Gases are also reactants, as in the combustion of charcoal in a backyard grill. Solid carbon reacts with oxygen gas in the air to produce carbon dioxide and the heat used to cook our food:

$$C(s) + O_2(g) \rightarrow CO_2(g) + \text{heat} \qquad (5.19)$$

In any chemical reaction that involves a gas as either a reactant or a product, the volume of the gas indirectly defines the amount of it in the reaction. If T and P are known, we can use the ideal gas equation to relate volume to the number of moles of gas in the system. Once we know that, we can use stoichiometric calculations to relate quantities of gas to quantities of other reactants and products, including heat. For example, what volume of oxygen is needed to completely burn 1.00 kg (about 2 lb) of charcoal at 1.00 atm of pressure on an average summer day (25°C)? Before starting the calculation, we must first write a balanced chemical equation for the reaction. Equation 5.19 is already balanced, so 1 mole of C reacting with 1 mole of O_2 should yield 1 mole of CO_2. If we start with 1.00 kg of C, we have

$$1.00 \text{ kg C} \times \frac{10^3 \text{ g}}{1 \text{ kg}} \times \frac{1 \text{ mol C}}{12.01 \text{ g C}} = 83.3 \text{ mol C}$$

According to the stoichiometry of the reaction, we need 83.3 moles of O_2 to completely react with the given amount of C. The volume of O_2 that corresponds to 83.3 moles of O_2 is calculated using the ideal gas equation by first rearranging the equation to solve for V and then substituting the values of n, R, T (in kelvins), and P:

$$V = \frac{nRT}{P} = \frac{(83.3 \text{ mol})\left(0.08206 \dfrac{\text{L} \cdot \text{atm}}{\text{mol} \cdot \text{K}}\right)(298 \text{ K})}{1.00 \text{ atm}} = 2.04 \times 10^3 \text{ L}$$

SAMPLE EXERCISE 5.8 Combining Stoichiometry and the Ideal Gas Law **LO3**

Oxygen generators in some airplanes (**Figure 5.22**) are based on the chemical reaction between solid sodium chlorate ($\mathcal{M}$ = 106.44 g/mol) and iron:

$$NaClO_3(s) + Fe(s) \rightarrow O_2(g) + NaCl(s) + FeO(s)$$

The resultant O_2 is blended with cabin air to provide 10–15 minutes of breathable air for passengers. How many grams of $NaClO_3$ are needed in a typical generator to produce 125 L of O_2 gas at 1.00 atm and 20.0°C?

Collect and Organize We are given the volume of O_2 to be generated at a particular pressure and temperature. Using this information, we can determine the mass of $NaClO_3$ needed based on the stoichiometric relations in the balanced chemical equation.

Analyze The solution requires two calculations. (1) According to the balanced chemical equation, 1 mole of $NaClO_3$ is needed to produce 1 mole of O_2. If we can determine how many moles of O_2 occupy a volume of 125 L at 1.00 atm pressure and 20.0°C, we can determine the number of moles of $NaClO_3$ we need. To determine the moles of O_2 (n_{O_2}), we can use the ideal gas law (Equation 5.15). (2) Then we use our calculated value of n_{O_2} and the balanced chemical equation to determine the number of moles and number of grams of $NaClO_3(s)$ required.

To estimate the answer, notice that the volume of gas we wish to make (125 L) is about five times the molar volume of an ideal gas at STP (22.4 L). The difference between the temperature at STP (0°C = 273 K) and the temperature in this problem (20°C = 293 K) is relatively small, so we need about 5 moles of O_2, which requires about 5 moles or about 500 g $NaClO_3$.

Solve Use the rearranged ideal gas equation to solve for moles of O_2:

$$n = \frac{PV}{RT}$$

$$n = \frac{(1.00 \text{ atm})(125 \text{ L})}{\left(0.08206 \dfrac{\text{L} \cdot \text{atm}}{\text{mol} \cdot \text{K}}\right)(273 + 20.0) \text{ K}} = 5.199 \text{ mol} = 5.20 \text{ mol } O_2$$

FIGURE 5.22 A flight attendant demonstrates how to use the oxygen mask on an airplane.

To convert moles of O_2 into an equivalent mass of $NaClO_3$, we use the stoichiometry of the reaction to calculate the equivalent number of moles of $NaClO_3$, and we use molar mass to determine the number of grams of $NaClO_3$ required:

$$5.20 \text{ mol } O_2 \times \frac{1 \text{ mol NaClO}_3}{1 \text{ mol } O_2} \times \frac{106.44 \text{ g NaClO}_3}{1 \text{ mol NaClO}_3} = 553 \text{ g NaClO}_3$$

Think About It We predicted that about 500 g $NaClO_3$ would be needed, which is within about 10% of the calculated value.

Practice Exercise Automobile air bags (**Figure 5.23**) inflate during a crash or sudden stop as a result of a series of rapid chemical reactions that include the generation of nitrogen gas from solid sodium azide:

$$2\ NaN_3(s) \rightarrow 2\ Na(s) + 3\ N_2(g)$$

How many grams of sodium azide are needed to produce 50.0 L of N_2 at a pressure of 1.20 atm at 15°C?

(Answers to Practice Exercises are in the back of the book.)

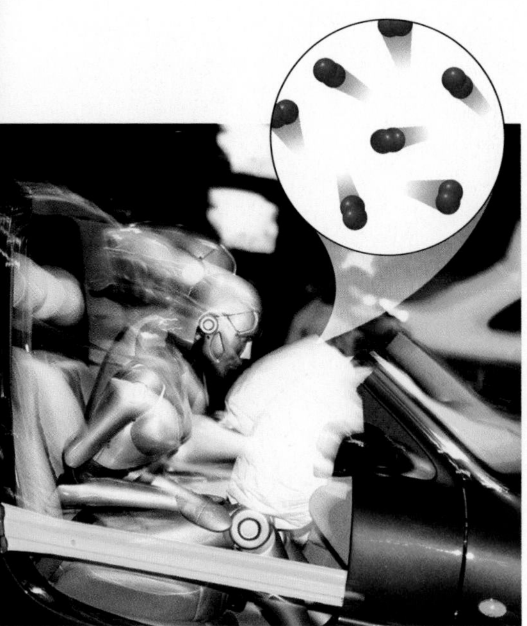

FIGURE 5.23 An automobile air bag inflates when solid NaN_3 decomposes, rapidly producing N_2 gas.

SAMPLE EXERCISE 5.9 Reactant Volumes and the Ideal Gas Law — LO3

Scientists exploring nonbiological reactions that lead to amino acid formation are interested in reactions between methane and ammonia. One such reaction produces cyanamide, H_2NCN:

$$CH_4(g) + 2\ NH_3(g) \rightarrow H_2NCN(g) + 4\ H_2(g)$$

If a reaction mixture contains 1.25 L of methane and 3.00 L of ammonia, what volume of hydrogen could theoretically be produced at constant temperature and pressure?

Collect and Organize We are given the chemical equation and the volumes of the two reactant gases, and we are asked to calculate the volume of product gas. Temperature and pressure are constant, but their values are not specified. According to Avogadro's law (Equation 5.10), the volume of a gas is proportional to the number of moles of gas in it at constant temperature and pressure.

Analyze When the quantities of two reactants are given, one is likely to be the limiting reactant, so solving this exercise will involve identifying that reactant. The stoichiometric ratio is 4 moles of H_2 produced for every 1 mole of CH_4 and 2 moles of NH_3 consumed, and it is equivalent to the same ratio of volumes at constant T and P. For example, 4 L H_2 : 1 L CH_4 : 2 L NH_3.

Solve Identifying the limiting reactant by calculating the volume of H_2 it could produce:

$$\text{Methane:} \quad 1.25 \text{ L CH}_4 \times \frac{4 \text{ L H}_2}{1 \text{ L CH}_4} = 5.00 \text{ L H}_2$$

$$\text{Ammonia:} \quad 3.00 \text{ L NH}_3 \times \frac{4 \text{ L H}_2}{2 \text{ L NH}_3} = 6.00 \text{ L H}_2$$

Methane is the limiting reactant, and the theoretical yield is 5.00 L H_2.

Think About It Had we known the temperature and pressure of the reaction system, we could have used the ideal gas equation to calculate the moles of the two reactants and then calculate the moles of H_2 each could theoretically produce. We then could convert the lesser of those two values into a volume of H_2. This approach would have entailed a lot more arithmetic and yielded the same result: 5.00 L H_2.

Practice Exercise Dinitrogen monoxide, N_2O, commonly known as nitrous oxide, is used as an anesthetic or "laughing gas" in many dental procedures (**Figure 5.24**). If 316 mL of nitrogen is combined with 178 mL of oxygen, what volume of N_2O is produced at constant temperature and pressure if the reaction proceeds to 82% yield?

$$2\ N_2(g) + O_2(g) \rightarrow 2\ N_2O(g)$$

(Answers to Practice Exercises are in the back of the book.)

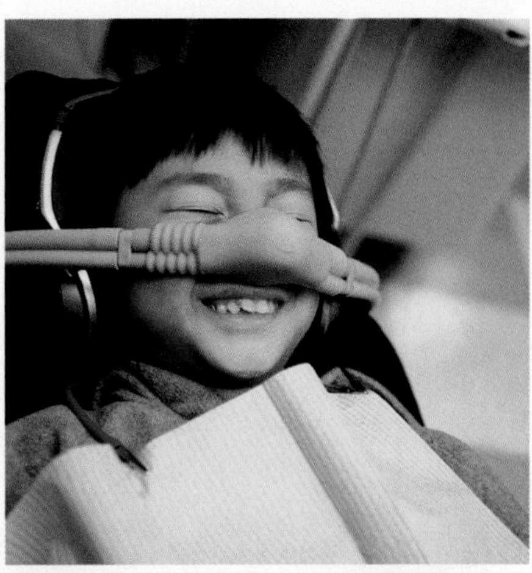

FIGURE 5.24 A child is administered "laughing gas" before a dental procedure.

5.6 Gas Density

Releases of large quantities of $CO_2(g)$ from volcanic activity contribute to growing atmospheric concentrations of the gas, but volcanic emissions can be life-threatening, too. On the Dieng Plateau in Indonesia in 1979, 149 people died from asphyxiation in a valley after a sudden, massive release of carbon dioxide from a nearby volcano (**Figure 5.25**). The deaths occurred because the CO_2 settled into the bottom of the valley, displacing the air (and oxygen) there. This happened because CO_2 is denser than air.

The density of a gas at STP can be calculated by dividing its molar mass by its molar volume. Carbon dioxide has a molar mass of 44.01 g/mol, so its density at STP is

$$\frac{44.01\ \text{g/}\cancel{\text{mol}}}{22.4\ \text{L/}\cancel{\text{mol}}} = 1.96\ \text{g/L}$$

The density of air, which is mostly a mixture of N_2 and O_2, is only about 1.3 g/L at STP. The greater density of CO_2 also explains why it so effectively fights fires: it blankets burning fuel, separating it from the O_2 it needs to burn, thus extinguishing the fire.

CONCEPT **TEST**

Which gas has the highest density at STP: CH_4, Cl_2, Kr, or C_3H_8?

(Answers to Concept Tests are in the back of the book.)

We can calculate the density of an ideal gas at any temperature and pressure by using the ideal gas equation. Because density is the mass of a sample divided by its volume, $d = m/V$, we need to identify the variables in the ideal gas equation that represent the density. Mass can be determined from the number of moles, and the V in the ideal gas equation is the volume. We can rearrange Equation 5.15:

$$\frac{P}{RT} = \frac{n}{V} \tag{5.20}$$

If we multiply both sides of the equation by the molar mass ($\mathcal{M}$) of the gas, then the numerator on the right side of the equation is the product of $\mathcal{M} \times n$, which has units of [(grams/~~mole~~) × ~~mole~~] or grams. In other words, the product is mass (m):

$$\frac{P\mathcal{M}}{RT} = \frac{n\mathcal{M}}{V} = \frac{m}{V}$$

The right-most fraction (mass/volume) is the definition of density (d). Therefore:

$$d = \frac{P\mathcal{M}}{RT} \tag{5.21}$$

FIGURE 5.25 The release of CO_2 from volcanic centers on Indonesia's Dieng Plateau in 1979 killed many people and animals as the dense gas flowed over the valley floor, effectively displacing the air (i.e., oxygen) that the inhabitants and their livestock needed to survive.

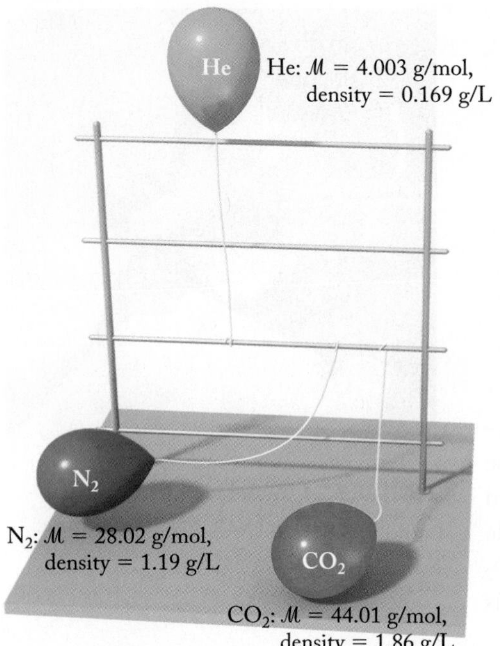

FIGURE 5.26 At 15°C and 1 atm, the density of the gas inside each balloon determines whether the balloon floats (helium), hovers just slightly above the benchtop (nitrogen), or sinks (carbon dioxide). Note the correlation between molar mass and density: the larger the molar mass of a gas, the greater the density.

Figure 5.26 illustrates the relationship between density and molar mass of three pure gases and air. Note how the balloon containing $CO_2(g)$ does not float at all but sinks to the benchtop because the density of this pure gas is greater than the density of air. In the absence of any wind currents or mixing, carbon dioxide will always sink through air, just as it filled the valley in the Dieng Plateau disaster.

Equation 5.21 can also be used to explain why a balloon inflated with hot air rises (**Figure 5.27**). Like a bicycle tire, the volume of the balloon is essentially constant once inflated. As the temperature of the air inside the balloon increases, the balloon can no longer expand. However, the balloons in Figure 5.27 represent open systems—gas can escape from the bottom of the balloon, so n decreases. This effect can be seen in Equation 5.20, where increasing T at constant P and V means that n must decrease. If n decreases, d will also decrease. Put another way, the density of a given quantity of gas decreases with increasing temperature, giving the balloon buoyancy.

FIGURE 5.27 Increasing the temperature of the air in the balloon leads to a decrease in density as the air expands. Hot air in a balloon is less dense than cooler air, making the balloon buoyant.

CONCEPT TEST

Which graph in **Figure 5.28** best approximates (a) the relationship between density and pressure (n and T constant) and (b) the relationship between density and temperature (n and P constant) for an ideal gas?

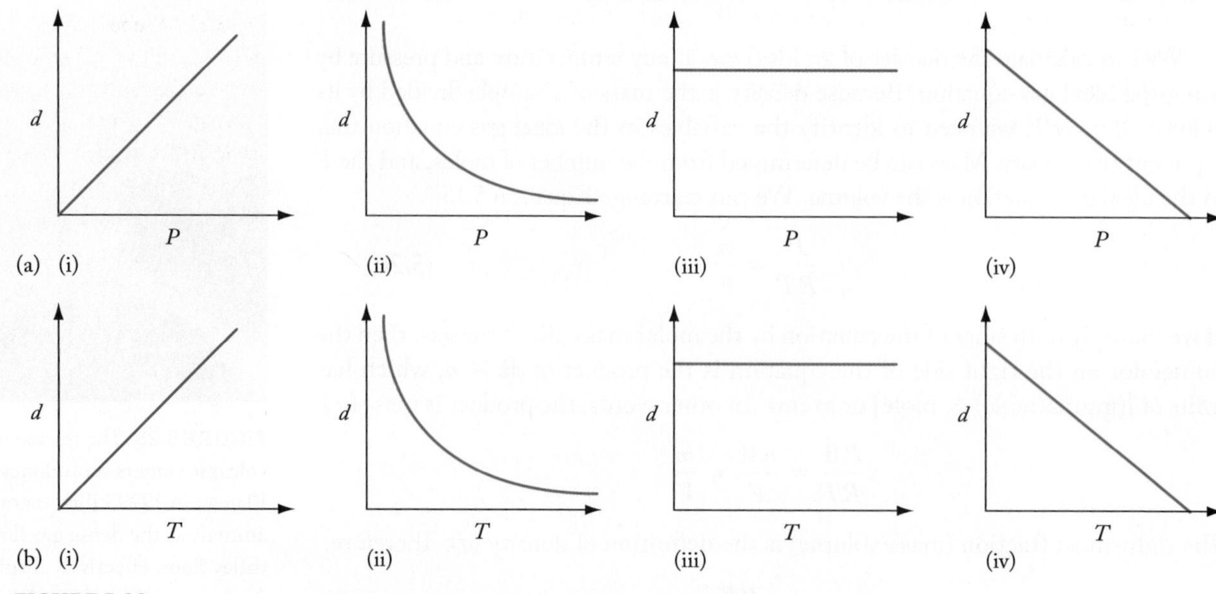

FIGURE 5.28

(Answers to Concept Tests are in the back of the book.)

SAMPLE EXERCISE 5.10 Calculating the Density of a Gas LO4

Calculate the density of air at 1.00 atm and 302 K and compare your answer with the density of air at STP (1.29 g/L). Assume that air has an average molar mass of 28.8 g/mol. (This average molar mass is the weighted average of the molar masses of the gases in air as listed in Table 5.1.)

Collect, Organize, and Analyze We are provided with the average molar mass, temperature, and atmospheric pressure of air, and we can use Equation 5.21 to calculate its density. The density of a gas is inversely proportional to temperature, and the temperature in this exercise (302 K) is higher than the temperature at STP, so the density of the air should be less than 1.29 g/L.

Solve Inserting the values of P, T, and $\mathcal{M}$ into Equation 5.21, we have

$$d = \frac{P\mathcal{M}}{RT} = \frac{(1.00\ \cancel{\text{atm}})\left(28.8\ \frac{\text{g}}{\cancel{\text{mol}}}\right)}{[0.08206\ \text{L} \cdot \cancel{\text{atm}}/(\cancel{\text{mol}}) \cdot \cancel{\text{K}}](302\ \cancel{\text{K}})} = 1.16\ \text{g/L}$$

Think About It The calculated density of air at 302 K is less than the value at 273 K (1.29 g/L) as we expected. The density of air (and most gases) is considerably less than that of liquids. For example, the density of liquid water is 1.00 g/mL or 1.00 kg/L, about 1000 times greater than the density of air.

Practice Exercise Air is a mixture of mostly nitrogen and oxygen. A balloon filled with only oxygen is released in a room full of air. Will it sink to the floor or float to the ceiling?

(Answers to Practice Exercises are in the back of the book.)

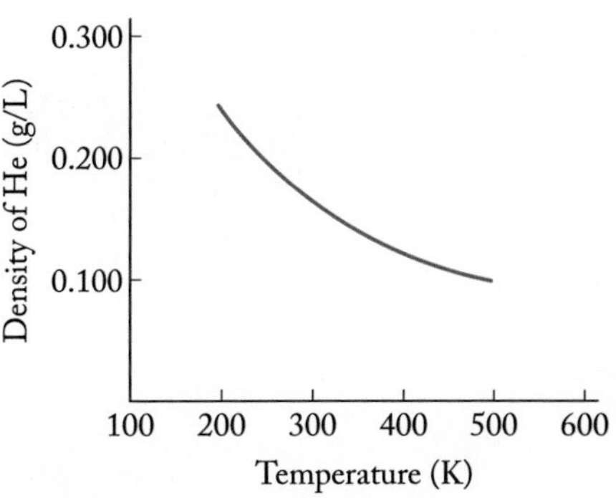

FIGURE 5.29 At $P = 1$ atm, the density of helium decreases with increasing temperature. The density of any ideal gas at constant pressure is inversely proportional to its absolute temperature.

Equation 5.21, which shows that density decreases as temperature increases (**Figure 5.29**), can be rearranged so that the molar mass of a gas can be calculated from its density at any temperature and pressure:

$$\mathcal{M} = \frac{dRT}{P} \tag{5.22}$$

Gas density can be determined with a glass tube of known volume attached to a vacuum pump (**Figure 5.30**). The mass of the tube is determined when it has essentially no gas in it and again when it is filled with the test gas. The difference in the masses divided by the volume of the tube is the density of the gas.

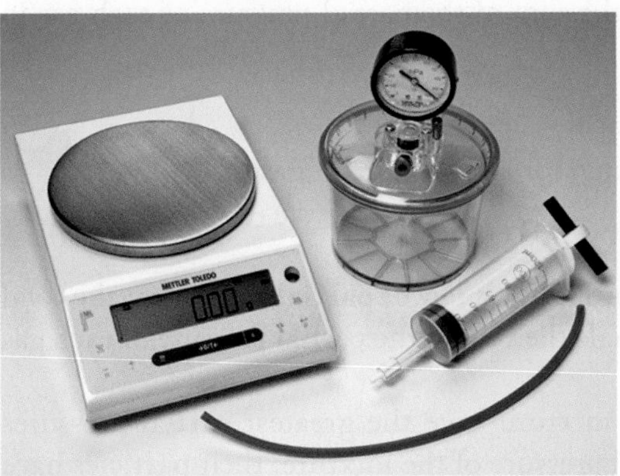

(a)

(b)

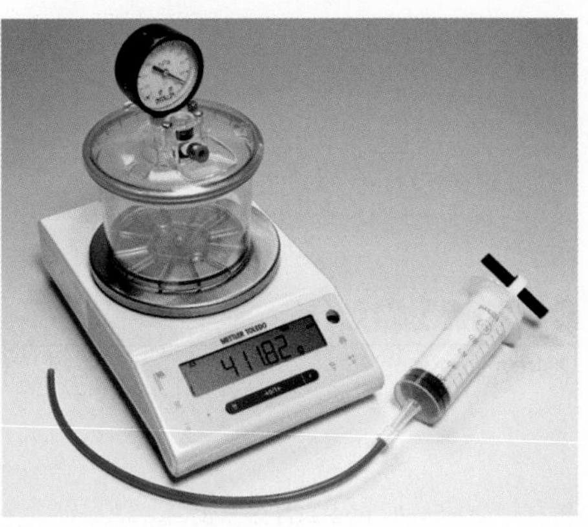

(c)

FIGURE 5.30 (a) A gas collection tube with an internal volume of 235 mL is evacuated by connecting it to a vacuum pump. (b) The mass of the evacuated tube is measured. (c) Then the tube is opened to the atmosphere and refills with air. The difference in mass between the filled and evacuated tubes (0.273 g, or 273 mg) is the mass of the air inside it. The density of the air sample is 273 mg/235 mL, or 1.16 mg/mL.

SAMPLE EXERCISE 5.11 Calculating Molar Mass from Density **LO4**

Vent pipes at solid-waste landfills often emit foul-smelling gases that may be either relatively pure substances or mixtures of several gases. A sample of such an emission has a density of 0.650 g/L at 25.0°C and 757 mmHg. What is the molar mass of the gas emitted? (If the sample is a mixture, then the answer will be the weighted average of the molar masses of the individual gases.)

Collect, Organize, and Analyze We are given the density, temperature, and pressure of a gaseous sample. We can use Equation 5.22 to calculate the molar mass. Don't forget to convert the temperature to kelvins and the pressure to atmospheres before using these values in the equation.

$$T = 273 + 25.0 = 298\text{ K}$$

$$P = 757\text{ mmHg} \times \frac{1}{760\text{ mmHg}} = 0.996\text{ atm}$$

Many gases have relatively low molar masses, so we expect that the molar mass should be similar to that of typical gases we have discussed in this chapter.

Solve

$$\mathcal{M} = \frac{dRT}{P} = \left(\frac{0.650\text{ g}}{\text{L}}\right)\left(\frac{0.08206\text{ L}\cdot\text{atm}}{\text{mol}\cdot\text{K}}\right)\left(\frac{298\text{ K}}{0.996\text{ atm}}\right)$$

$$= 16.0\text{ g/mol}$$

Think About It The density of the landfill gas (0.650 g/L) is a little more than half the density of air (1.16 g/L) at 302 K, which we calculated in Sample Exercise 5.10, so the calculated molar mass of the gas should be a little more than half that of air (28.8 g/mol), which it is. The calculated value is also consistent with that of methane, a principal component in the mixture of gases emitted by decomposing solid waste.

Practice Exercise When HCl(*aq*) and $NaHCO_3$(*aq*) are mixed, a chemical reaction takes place in which a gas is one of the products. A sample of the gas has a density of 1.81 g/L at 1.00 atm and 23.0°C. What is the molar mass of the gas? Write a balanced chemical equation to identify the gas.

(Answers to Practice Exercises are in the back of the book.)

5.7 Dalton's Law and Mixtures of Gases

CHEMTOUR Dalton's Law

Air is mostly N_2 and O_2 (Table 5.1) with smaller quantities of other gases. Each gas in air, and in any other gas mixture, exerts its own pressure, called a **partial pressure**. Atmospheric pressure is the sum of the partial pressure of each gas in the air:

$$P_{\text{atm}} = P_{\text{N}_2} + P_{\text{O}_2} + P_{\text{Ar}} + P_{\text{CO}_2} + \ldots$$

A similar expression can be written for any mixture of gases:

$$P_{\text{total}} = P_1 + P_2 + P_3 + P_4 + \ldots \quad (5.23)$$

This is a mathematical expression of **Dalton's law of partial pressures**: the total pressure of any mixture of gases equals the sum of the partial pressure of each gas in the mixture.

The most abundant gases in a mixture have the greatest partial pressures and contribute the most to the total pressure of the mixture: their particles have the largest number of collisions simply because there are more of them. The mathematical term used to express the abundance of each component, x, is its **mole fraction (X_x)**,

$$X_x = \frac{n_x}{n_{\text{total}}} \quad (5.24)$$

where n_x is the number of moles of component x and n_{total} is the sum of the number of moles of all the components of the mixture.

The mole fraction as an expression of concentration has three characteristics worth noting: (1) unlike molarity, mole fractions have no units; (2) unlike molarity, mole fractions are based on numbers of moles and can be used for any mixture or solution (solid, liquid, or gas); and (3) the mole fractions of all the components in a mixture must sum to 1.

To see how mole fractions work, consider a sample of the atmosphere (air) that contains 100.0 moles of atmospheric gases. This sample consists of 21.0 moles of O_2 and 78.1 moles of N_2. The mole fraction of O_2 in the air sample is

$$X_{O_2} = \frac{n_{O_2}}{n_{\text{total}}} = \frac{21.0}{100.0} = 0.210$$

and the mole fraction of N_2 is

$$X_{N_2} = \frac{n_{N_2}}{n_{\text{total}}} = \frac{78.1}{100.0} = 0.781$$

Mole fractions express concentrations based on numbers of moles (and molecules) of substances in mixtures. In a sample of air, for example, the greater mole fraction of N_2 also means there are more moles (and molecules) of nitrogen than there are moles of oxygen:

$$N_2 \quad 78.1\ \cancel{\text{mol}} \times (6.022 \times 10^{23}\ \text{molecules}/\cancel{\text{mol}}) = 4.70 \times 10^{25}\ \text{molecules}$$

$$O_2 \quad 21.0\ \cancel{\text{mol}} \times (6.022 \times 10^{23}\ \text{molecules}/\cancel{\text{mol}}) = 1.26 \times 10^{25}\ \text{molecules}$$

CONCEPT **TEST**

The partial pressure of one component of a gas mixture cannot be measured directly. Why?

(Answers to Concept Tests are in the back of the book.)

The partial pressure of any gas in air can be calculated by multiplying the mole fraction of the gas times the total atmospheric pressure. If the total pressure is 1.00 atm, for instance, as it often is at sea level, the partial pressures of O_2 and N_2 in the example above are

$$P_{O_2} = X_{O_2}P_{\text{total}} = (0.210)(1.00\ \text{atm}) = 0.210\ \text{atm}$$

$$P_{N_2} = X_{N_2}P_{\text{total}} = (0.781)(1.00\ \text{atm}) = 0.781\ \text{atm}$$

These two equations are specific instances of the general equation for the partial pressure of one gas, x, in a mixture of gases:

$$P_x = X_xP_{\text{total}} \quad (5.25)$$

The pressure of a gas is directly proportional to the *quantity* of the gas in a given volume and does *not* depend on the *identity* of the gas or whether the gas

partial pressure the contribution to the total pressure made by one gas in a mixture of gases.

Dalton's law of partial pressures the principle that the total pressure of any mixture of gases equals the sum of the partial pressure of each gas in the mixture.

mole fraction (X_x) the ratio of the number of moles of a component in a mixture to the total number of moles in the mixture.

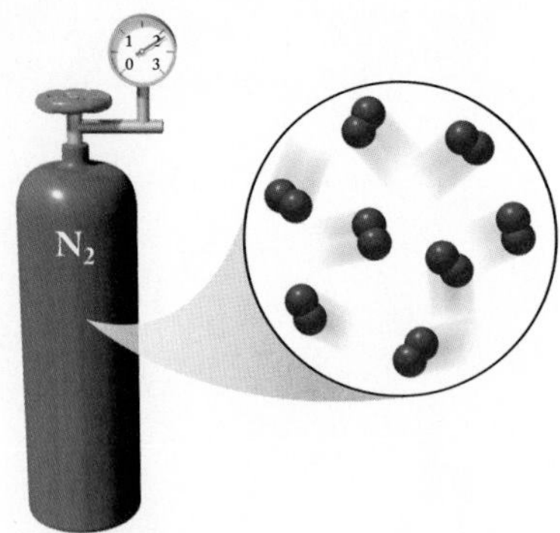

8 mol N_2

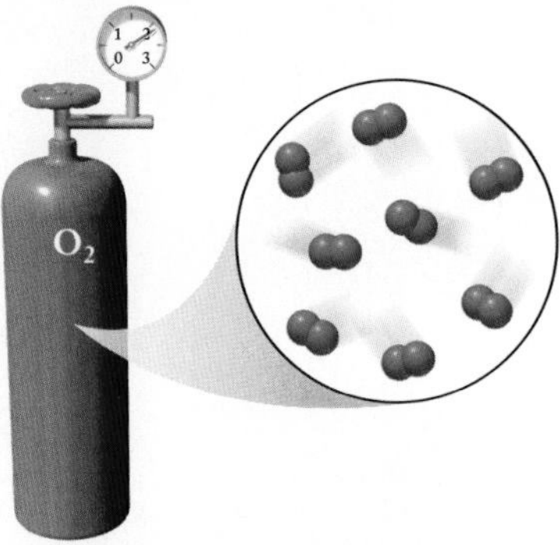

8 mol O_2

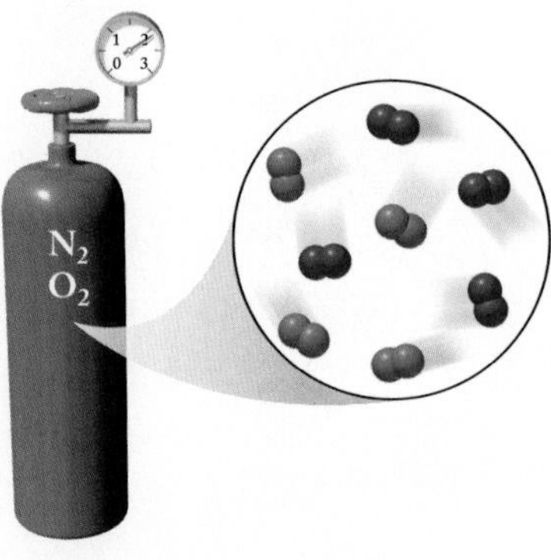

8 mol gas

FIGURE 5.31 Pressure is directly proportional to number of moles of an ideal gas, independent of the identity of the gas and independent of whether the sample is a pure gas or a mixture. In three containers that have the same volume and temperature, 8 moles of nitrogen, 8 moles of oxygen, and 8 moles of a 50:50 mixture of nitrogen and oxygen all exert the same pressure, as indicated by the gauges.

is pure or a mixture (**Figure 5.31**). Because P_{total} is a measure of the total number of collisions within a mixture, the partial pressure of one gas is due to some fraction of the total collisions. That fraction is the mole fraction of that gas.

SAMPLE EXERCISE 5.12 Calculating Mole Fractions — LO5

Scuba divers who descend more than 45 m below the surface may use a gas mixture called Trimix to support their breathing. Trimix is available in several different concentrations, including 11.7% He, 56.2% N_2, and 32.1% O_2, by mass. This mixture of gases avoids high concentrations of dissolved nitrogen in the blood at great depth, which can lead to *nitrogen narcosis*, a potentially deadly condition for divers. Calculate the mole fraction of each gas in this mixture.

Collect, Organize, and Analyze If we assume a 100.0 g sample of Trimix with a specified composition, then we can calculate the mass of each gas from the mass percentages. Next, we can convert each mass to moles, from which we can determine the total number of moles in the sample and the mole fraction of each gas. Mole fraction is defined in Equation 5.24, and to calculate it we need to calculate both the total number of moles of gas in the mixture and the number of moles of each component.

We can estimate the answer by considering the relative molar masses of the three gases. The molar masses of nitrogen and oxygen are similar and seven to eight times larger, respectively, than the molar mass of helium. Because N_2 constitutes a higher percentage of the mass of the mixture than does O_2, the mole fraction of nitrogen should be larger than that of oxygen. Furthermore, the mole fractions of He, N_2, and O_2 should be quite different from the mass percentages of the three gases because the molar mass of He is much less than the molar masses of N_2 and O_2.

Solve In 100.0 g of this mixture we would have 11.7 g of He, 56.2 g of N_2, and 32.1 g of O_2. Using molar masses to convert these masses into moles yields

$$(11.7\ \cancel{\text{g He}})\left(\frac{1\text{ mol He}}{4.003\ \cancel{\text{g He}}}\right) = 2.92\text{ mol He}$$

$$(56.2\ \cancel{\text{g N}_2})\left(\frac{1\text{ mol N}_2}{28.02\ \cancel{\text{g N}_2}}\right) = 2.01\text{ mol N}_2$$

$$(32.1\ \cancel{\text{g O}_2})\left(\frac{1\text{ mol O}_2}{32.00\ \cancel{\text{g O}_2}}\right) = 1.00\text{ mol O}_2$$

Mole fractions are calculated with Equation 5.24. The total number of moles in 100.0 g of Trimix is the sum of the number of moles of the constituent gases: He, N_2, and O_2.

$$n_{total} = 2.92 + 2.01 + 1.00 = 5.93\text{ mol}$$

Thus, X_{He}, X_{N_2}, and X_{O_2} are

$$X_{He} = \frac{2.92\ \cancel{\text{mol}}\text{ He}}{5.93\ \cancel{\text{mol}}} = 0.492$$

$$X_{N_2} = \frac{2.01\ \cancel{\text{mol}}\text{ N}_2}{5.93\ \cancel{\text{mol}}} = 0.339$$

$$X_{O_2} = \frac{1.00\ \cancel{\text{mol}}\text{ O}_2}{5.93\ \cancel{\text{mol}}} = 0.169$$

Alternatively, we could use Equation 5.24 for all but one of the gases in the sample and then calculate the mole fraction of the final component by subtracting the sum of the calculated mole fractions from 1.00. For example, we could use the equation to calculate the mole fractions of He and N_2 and then determine the O_2 mole fraction by difference:

$$X_{O_2} = 1 - (X_{He} + X_{N_2}) = 1 - (0.492 + 0.339) = 0.169$$

Think About It We can check the answer by summing the mole fractions; they should add up to 1.

$$X_{He} + X_{N_2} + X_{O_2} = 1$$

$$0.492 + 0.339 + 0.169 = 1.00$$

Finally, as we predicted, the mole fractions of the three gases do not reflect their mass percentages. Although nitrogen is present in the greatest amount by mass in Trimix, helium has the largest mole fraction.

Practice Exercise A gas mixture called Heliox that is 52.17% O_2 and 47.83% He by mass is used in scuba tanks for descents more than 65 m below the surface. Calculate the mole fractions of He and O_2 in this mixture.

(Answers to Practice Exercises are in the back of the book.)

CONCEPT TEST

Does each of the gases in an equimolar mixture of four gases have the same mole fraction?

(Answers to Concept Tests are in the back of the book.)

Let's apply Equation 5.25 to the "thin" air at high altitudes. The mole fraction of oxygen in air is 0.210 and does not change significantly with increasing altitude. The total pressure of the atmosphere, however, and therefore the partial pressure of oxygen, decreases with increasing altitude, as illustrated in Sample Exercise 5.13.

SAMPLE EXERCISE 5.13 Calculating Partial Pressure — LO5

Calculate the partial pressure of O_2 (in atm) in the air outside an airplane cruising at an altitude of 10 km, where the atmospheric pressure is 190.0 torr. The mole fraction of O_2 in the air is 0.210.

Collect, Organize, and Analyze We are given both the mole fraction of oxygen and atmospheric pressure, and we are asked to calculate the partial pressure of oxygen. Equation 5.25 relates the partial pressure of a gas in a gas mixture to its mole fraction in the mixture and to the total gas pressure.

Solve

$$P_{O_2} = X_{O_2}P_{total} = (0.210)(190.0\ \cancel{torr})\left(\frac{1\ atm}{760\ \cancel{torr}}\right)$$

$$= 0.0525\ atm$$

Think About It The partial pressure of oxygen in air at sea level is 0.210 atm. At higher altitudes, the atmosphere becomes thinner (less dense). Our answer of 0.0525 atm makes sense because, at an altitude of 10 km, the air is much less dense than at sea level.

Practice Exercise Assume a scuba diver is working at a depth where the total pressure is 5.0 atm (about 50 m below the surface). What mole fraction of oxygen is necessary in the gas mixture the diver breathes for the partial pressure of oxygen to be 0.21 atm?

(Answers to Practice Exercises are in the back of the book.)

Sample Exercises 5.12 and 5.13 illustrate how much humans depend on the atmosphere and the properties of gases that compose it. If we venture very far from Earth's surface into the mountains or into the oceans, we have to take the right blend of gases with us to sustain life. The scuba diver can't breathe the same mixture of gases as we do on the surface of Earth because the increased partial pressure of O_2 can be toxic below 30 m. Similarly, a higher partial pressure of N_2 could lead to nitrogen narcosis, a dangerous condition caused by a high concentration of nitrogen in the blood that leads to hallucinations. Alpine climbers on the tallest peaks on Earth must be vigilant against the opposite problem. For them, the lower atmospheric pressure leads to a lower partial pressure of oxygen, making it hard to get enough oxygen into the blood. Most mountaineers in these locations carry bottled oxygen. The lower partial pressure of oxygen at high elevations is also the reason that aircraft cabins are pressurized.

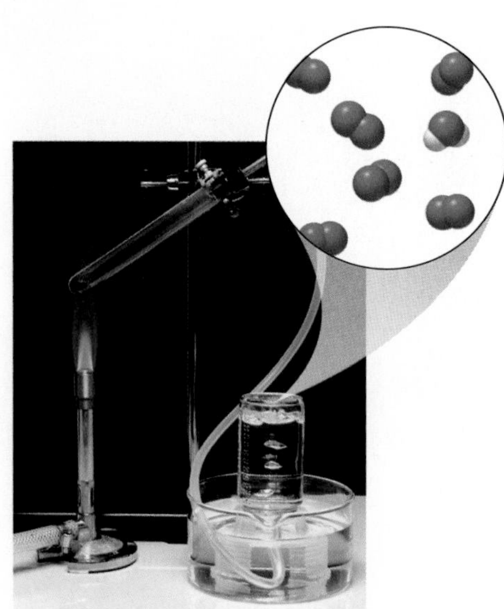

FIGURE 5.32 Oxygen produced by the thermal decomposition of $KClO_3$ is collected by water displacement.

Dalton's law of partial pressures is also useful in the laboratory when we want to measure the pressure exerted by a gaseous product in a chemical reaction. For example, heating potassium chlorate ($KClO_3$) in the presence of MnO_2 causes it to decompose into $KCl(s)$ and $O_2(g)$. The oxygen gas produced by this reaction can be collected by bubbling the gas into an inverted bottle that is initially filled with water (**Figure 5.32**). As the reaction proceeds, $O_2(g)$ displaces the water in the bottle. When the reaction is complete, the volume of water displaced provides a measure of the volume of O_2 produced. If you know the temperature of the water and the barometric pressure at the time of the reaction, then you can use the ideal gas law to calculate the number of moles of O_2 produced.

This procedure works for any gas that neither reacts with nor dissolves appreciably in water. However, one additional step is needed to calculate the number of moles of gas produced from the ideal gas law. At room temperature (nominally 20°C), any enclosed space above a pool of liquid water contains some $H_2O(g)$, meaning some water vapor is in the collection flask in addition to the gas produced by the reaction. In the $KClO_3$ reaction in Figure 5.32, therefore, the gas collected is a mixture of both $O_2(g)$ and $H_2O(g)$. A mixture of a gas and water vapor is said to be *wet*, in contrast to a gas that contains no water vapor, which is said to be *dry*. The amount of water vapor increases as temperature increases (**Table 5.4**). Dalton's law of partial pressures gives us the total pressure of the mixture at the point where the water level inside the collection vessel matches the water level outside the vessel:

TABLE 5.4 Partial Pressure of $H_2O(g)$ at Selected Temperatures

Temperature (°C)	Pressure (torr)
5	6.5
10	9.2
15	12.8
20	17.5
25	23.8
30	31.8
35	42.2
40	55.3
45	71.9
50	92.5

$$P_{\text{total}} = P_{\text{atm}} = P_{O_2} + P_{H_2O} \quad (5.26)$$

To calculate the quantity of oxygen produced using the ideal gas law, we must know P_{O_2}, which we get by subtracting P_{H_2O} at 20°C (see Table 5.4) from P_{atm}. If we know the values of T and V, we can calculate the number of moles (or subsequently, the number of grams) of oxygen produced.

SAMPLE EXERCISE 5.14 Determining the Quantity of a Gas Collected over Water **LO5**

During the decomposition of $KClO_3$, 92.0 mL of gas was collected by the displacement of water at 25.0°C. If atmospheric pressure is 756 torr, what mass of O_2 was collected?

Collect and Organize We are given the atmospheric pressure, the volume of oxygen, and the temperature, and we are asked to calculate the mass of oxygen collected by water displacement. According to Table 5.4, the vapor pressure of water at 25°C is 23.8 torr.

Analyze Because O_2 is collected by water displacement, the total pressure inside the collection vessel, which is 756 torr, is the sum of the partial pressure of O_2 and the vapor pressure of water (23.8 torr). At 25°C the molar volume of an ideal gas should be about 10% more than 22.4 L, or about 25 L. The volume of gas collected is about 0.1 L/25 L = 0.004 molar volumes. The gas is mostly O_2, so multiplying 0.004 × 32.00 g O_2/mol gives an estimate of 0.12 g O_2 collected.

Solve Calculating P_{O_2} in the collected gas

$$P_{O_2} = P_{total} - P_{H_2O} = 756 \text{ torr} - 23.8 \text{ torr} = 732 \text{ torr}$$

Using the ideal gas equation to calculate the moles of O_2 requires first converting P_{O_2} to atmospheres, V to liters, and T to kelvins:

$$732 \text{ torr} \times \frac{1 \text{ atm}}{760 \text{ torr}} = 0.963 \text{ atm}$$

$$92.0 \text{ mL} \times \frac{10^{-3} \text{ L}}{1 \text{ mL}} = 0.0920 \text{ L}$$

$$25°\text{C} = 25 + 273 = 298 \text{ K}$$

$$n = \frac{PV}{RT} = \frac{(0.963 \text{ atm})(0.0920 \text{ L})}{[0.08206 \text{ L} \cdot \text{atm/(mol} \cdot \text{K)}](298 \text{ K})} = 0.00362 \text{ mol}$$

The mass of O_2 collected is

$$m_{O_2} = 0.00362 \text{ mol} \times \frac{32.00 \text{ g}}{1 \text{ mol}} = 0.116 \text{ g}$$

Think About It The calculated mass of O_2 matches the preliminary estimate even though we ignored the presence of water vapor in developing the estimate. However, the estimate is based on calculations that often use values expressed with only one significant figure. The contribution of water vapor to total pressure becomes significant when the goal is a quantity expressed with two or more significant figures.

Practice Exercise Electrical energy can be used to separate water into $O_2(g)$ and $H_2(g)$—a process called electrolysis. The gases are collected separately. In one demonstration of this reaction, 27 mL of H_2 was collected over water at 25°C. If the atmospheric pressure was 761 torr, how many milligrams of H_2 were collected?

(Answers to Practice Exercises are in the back of the book.)

5.8 The Kinetic Molecular Theory of Gases

The fundamental discoveries of Boyle, Charles, and Amontons all took place before scientists recognized the existence of molecules. Only Avogadro, because he lived after Dalton, had the advantage of knowing Dalton's proposal that matter was composed of tiny particles called atoms. Thus, Avogadro recognized that the volume of a gas was directly proportional to the number of particles (moles) of gas. Still, Avogadro's law, V/n = constant (Equation 5.10), does not explain why 1 mole of relatively small He(g) atoms at STP has the same volume as 1 mole of much larger $SF_6(g)$ molecules. Dalton's law of partial pressures (Equation 5.23) does not explain *why* each gas in a mixture contributes a partial pressure based on its mole fraction. Similarly, Boyle's, Charles's, and Amontons's laws demand explanations for why pressure and volume are inversely proportional to each other, but both are directly proportional to temperature.

In this section we examine the **kinetic molecular theory** of gases, a unifying theory developed in the late 19th century that explains the relationships described by the ideal gas law and its predecessors, the laws of Boyle, Charles, Dalton, Amontons, and Avogadro.

The main assumptions of the kinetic molecular theory are:

1. Gas molecules have tiny volumes compared with the collective volume they occupy. Their individual volumes are so small as to be considered negligible, allowing particles in a gas to be treated as *point masses*—masses with essentially no volume. Gas molecules are separated by large distances; hence, a gas is mostly empty space.
2. Gas molecules move constantly and randomly throughout the volume they collectively occupy.
3. The motion of these molecules is associated with an average kinetic energy that is proportional to the absolute temperature of the gas. All populations of gas molecules at the same temperature have the same average kinetic energy.
4. Gas molecules continually collide with one another and with their container walls. These collisions are *elastic*; that is, they result in no net transfer of energy to the walls. Therefore, the average kinetic energy of gas molecules is not affected by these collisions and remains constant as long as there is no change in temperature.
5. Each gas molecule acts independently of all other molecules in a sample. We assume there are no forces of attraction or repulsion between the molecules.

Explaining Boyle's, Dalton's, and Avogadro's Laws

Every collision between a gas molecule and a wall of its container generates a force. The more frequent the collisions, the greater the force and thus the greater the pressure. Compressing molecules into a smaller space by reducing the volume of the container means more collisions take place per unit time, and the pressure increases (Boyle's law: $P \propto 1/V$). Because $P \propto n$, we can also increase the number of collisions by increasing the number of gas molecules (number of moles, n) in a container of fixed volume (**Figure 5.33a and b**). We can increase n either by adding more of the same gas to a container of fixed volume or by adding some quantity of a different gas to the container. If we add two gases, such as N_2 and O_2, to a container in a 4:1 mole ratio, then the number of collisions involving nitrogen molecules and a container wall will be greater than the number involving oxygen molecules. Therefore, the pressure due to nitrogen (P_{N_2}) will be greater than that due to oxygen (P_{O_2}), and the total pressure will be $P_{total} = P_{O_2} + P_{N_2}$. This statement is precisely what Dalton proposed in his law of partial pressures.

Kinetic molecular theory explains Avogadro's law ($V \propto n$), too. Avogadro observed that the volume of a gas *at constant pressure* is directly proportional to the number of moles of the gas. Because additional gas molecules in a given volume

kinetic molecular theory a model that describes the behavior of ideal gases; all equations defining relationships between pressure, volume, temperature, and number of moles of gases can be derived from the theory.

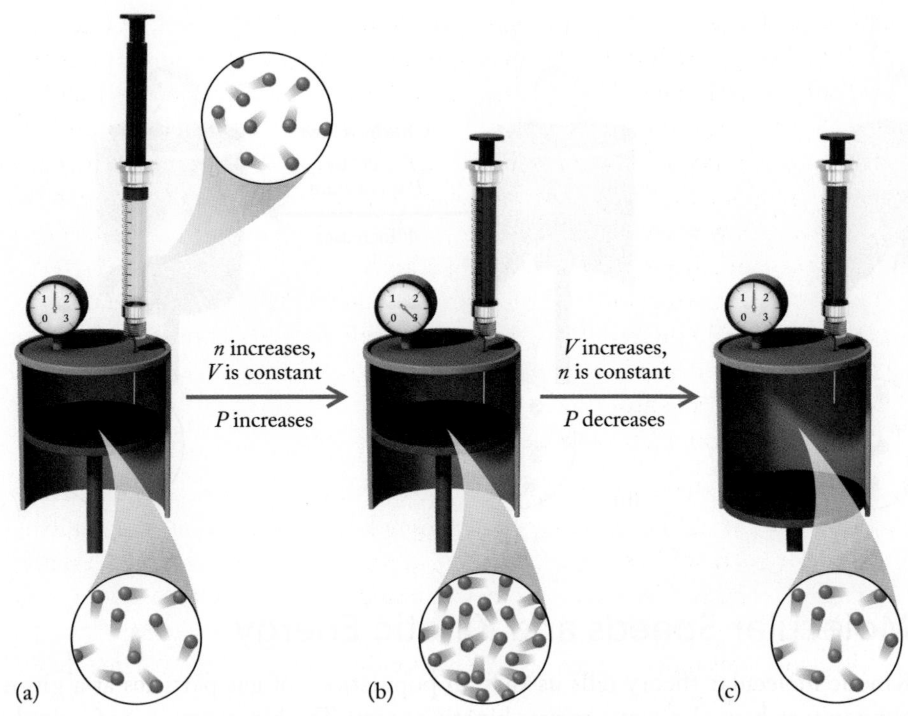

FIGURE 5.33 (a) The pressure inside a cylinder fitted with a movable piston is 1.5 atm. (b) More gas is added to the cylinder, doubling the moles of gas with no change in volume. The pressure in the cylinder doubles to 3.0 atm. (c) Downward movement of the piston doubles the volume of the gas in the cylinder. Its pressure returns to 1.5 atm.

lead to increased numbers of collisions and an increase in pressure, one way to reduce the pressure is to allow the gas to expand (**Figure 5.33c**).

CONCEPT **TEST**

If the collisions between molecules and the walls of the container were *not* elastic and energy was lost to the walls, would the pressure of the gas be higher or lower than that predicted by the ideal gas law?

(Answers to Concept Tests are in the back of the book.)

Explaining Amontons's and Charles's Laws

Amontons's law ($P \propto T$) and Charles's law ($V \propto T$) both depend on temperature, and both can be explained in terms of kinetic molecular theory. Why is pressure directly proportional to temperature? Like pressure, temperature is directly related to molecular motion. The average speed at which a population of molecules moves increases with increasing temperature. For a given number of gas molecules, increasing the temperature increases the velocity of the molecules and therefore increases the frequency and average kinetic energy with which the molecules collide with one another and with the walls of their container. Because pressure is related to the frequency and force of these collisions, it follows that higher temperatures produce higher pressures, as long as volume and quantity of gas are constant. **Figure 5.34a** illustrates Amontons's law on a molecular level.

How do increased frequency of collisions and higher average kinetic energy explain Charles's law, which states that the volume of a quantity of an ideal gas is proportional to its absolute temperature at constant pressure? As temperature increases, the system must expand—increase its volume—to maintain a constant pressure (**Figure 5.34b**).

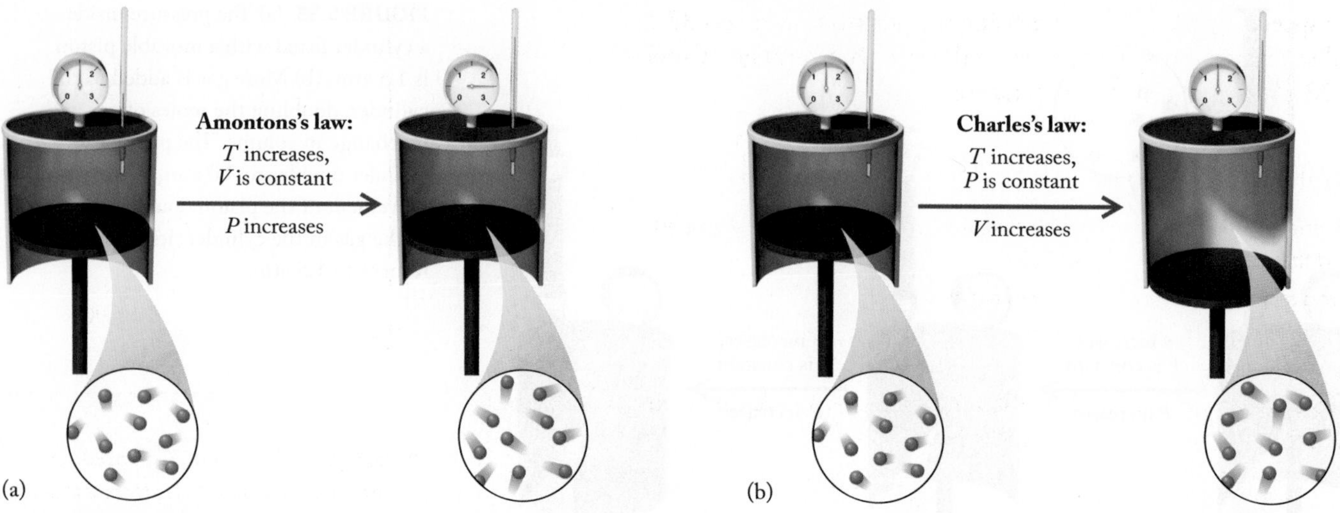

FIGURE 5.34 (a) If the temperature of a quantity of gas increases while the volume is held constant, the pressure of the gas increases in accordance with Amontons's law. (b) If the temperature of the gas increases, maintaining a constant pressure requires that its volume increase, in accordance with Charles's law.

Molecular Speeds and Kinetic Energy

Kinetic molecular theory tells us that all populations of gas particles at a given temperature have the same *average* kinetic energy. The kinetic energy of a single molecule or atom of a gas is given by the following equation

$$\text{KE} = \tfrac{1}{2}mu^2$$

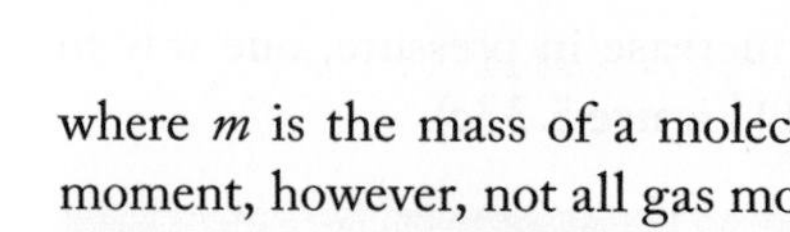

where m is the mass of a molecule of the gas and u is its speed. At any given moment, however, not all gas molecules in a population are traveling at the same speed. Even elastic collisions between two molecules may result in one of them moving with a greater speed than the other after the collision. One molecule might even stop moving. Thus, collisions between gas molecules cause the molecules in any sample to have a range of speeds.

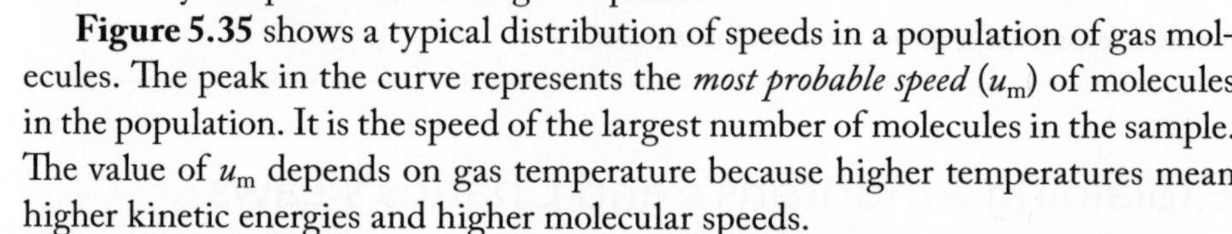

Figure 5.35 shows a typical distribution of speeds in a population of gas molecules. The peak in the curve represents the *most probable speed* (u_m) of molecules in the population. It is the speed of the largest number of molecules in the sample. The value of u_m depends on gas temperature because higher temperatures mean higher kinetic energies and higher molecular speeds.

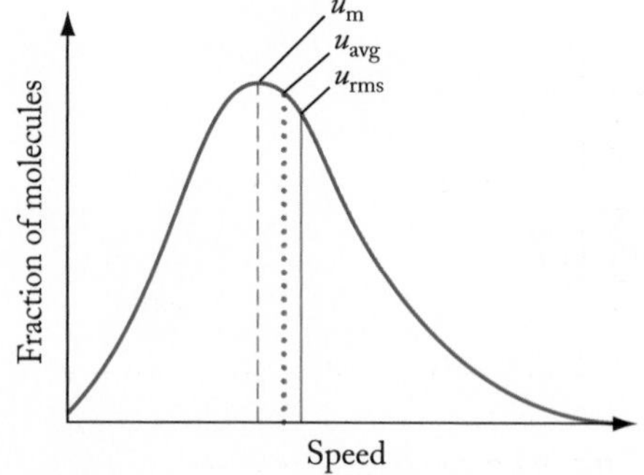

FIGURE 5.35 At any given temperature, the speeds of gas molecules cover a range of values. The most probable speed (u_m, dashed line) corresponds to the highest point on the curve. More molecules have the most probable speed than any other speed. The average speed (u_{avg}, dotted line), which is a little greater than the most probable speed, is the arithmetic mean of all the molecules' speeds. The root-mean-square speed (u_{rms}, solid line) is directly proportional to the square root of the absolute temperature of the gas and inversely proportional to the square root of its molar mass.

Because the distribution of speeds is not symmetrical, the *average speed* (u_{avg}), which is the simple arithmetic mean of all the speeds of all the molecules in the population, is a little higher than the most probable speed. The **root-mean-square speed (u_{rms})** is the speed of a molecule that has the average kinetic energy of a population of gas molecules. Average kinetic energy and u_{rms} are related by an equation with a familiar appearance:

$$\text{KE}_{avg} = \tfrac{1}{2}m(u_{rms})^2 \tag{5.27}$$

The value of the u_{rms} of a gas can be calculated from its molar mass and temperature:

$$u_{rms} = \sqrt{\frac{3RT}{\mathcal{M}}} \tag{5.28}$$

Equation 5.28 conveys the message that the velocities of gas particles are proportional to the square root of their absolute temperature and inversely proportional to the square root of their masses: the smaller the particles, the faster they move. **Figure 5.36** shows how the most probable molecular speed (u_m, dashed lines) of a gas increases with increasing temperature and how the distribution of these speeds broadens with increasing temperature. **Figure 5.37** shows the different distributions of speeds for several gases with a range of molar masses from 2 to 32 g/mol at the same temperature.

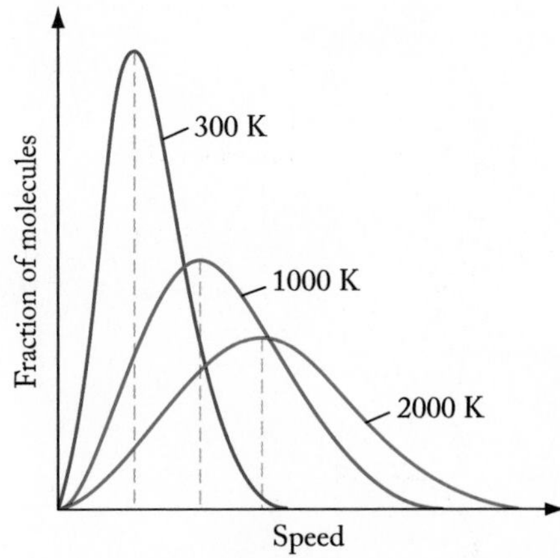

FIGURE 5.36 The molecular speeds of a gas increase with increasing temperature.

CONCEPT TEST

Rank these gases in order of increasing root-mean-square speed at 20°C, lowest speed first: H_2, CO_2, Ar, SF_6, UF_6, Kr.

(Answers to Concept Tests are in the back of the book.)

To calculate u_{rms} using Equation 5.28 we need an R value with velocity units in it. Table 5.3 contains such an R value:

$$R = 8.314 \frac{\text{kg} \cdot \left(\frac{\text{m}}{\text{s}}\right)^2}{\text{mol} \cdot \text{K}}$$

Using this value for R requires that we express molar mass in kilograms per mole instead of grams per mole, as illustrated in the following sample exercise.

CHEMTOUR
Molecular Speed

SAMPLE EXERCISE 5.15 Calculating Root-Mean-Square Speeds LO7

Calculate the root-mean-square speed of nitrogen molecules at 300.0 K in meters per second.

Collect, Organize, and Analyze We are asked to calculate the value of u_{rms} of N_2 molecules at 300.0 K. The molar mass of N_2 is 28.01 g/mol. Equation 5.28 relates T, $\mathcal{M}$, and R, but first the value of $\mathcal{M}$ must be converted into kg/mol.

Solve Convert $\mathcal{M}_{N_2}$ to kg/mol

$$28.01 \frac{\text{g}}{\text{mol}} \times \frac{1 \text{ kg}}{1000 \text{ g}} = 0.02801 \text{ kg/mol}$$

$$u_{rms} = \sqrt{\frac{3RT}{\mathcal{M}}} = \sqrt{\frac{3\left(\frac{8.314 \text{ kg} \cdot (\text{m/s})^2}{\text{mol} \cdot \text{K}}\right)(300.0 \text{ K})}{0.02801 \frac{\text{kg}}{\text{mol}}}}$$

$$= 516.9 \text{ m/s}$$

Think About It The calculated root-mean-square speed of an N_2 molecule is very high but makes sense because the value of $3RT$ is about 8000 and the value of $\mathcal{M}_{N_2}$ in kg/mol is about 0.03, making the value of the fraction about $(250{,}000)^{1/2}$ or 500.

Practice Exercise Calculate the root-mean-square speed of helium at 300.0 K in meters per second and compare your result with the root-mean-square speed of nitrogen calculated in Sample Exercise 5.15.

(Answers to Practice Exercises are in the back of the book.)

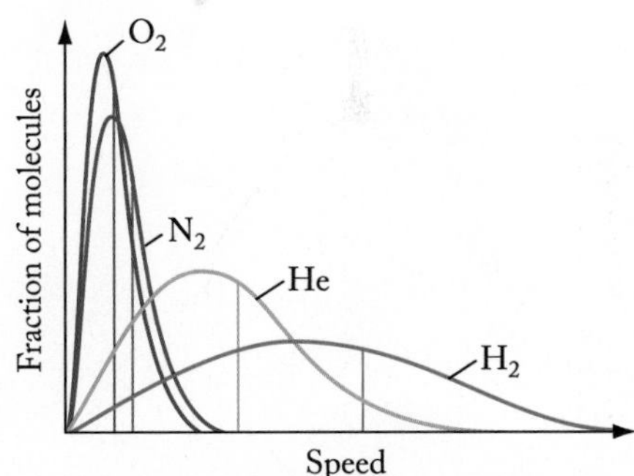

FIGURE 5.37 Molecular speed distributions and u_{rms} values for samples of oxygen, nitrogen, helium, and hydrogen gas at the same temperature.

Why are the pressures exerted by two different gases the same at a given temperature when their root-mean-square speeds are quite different? The answer lies in the assumption in the kinetic molecular theory that all populations of gas molecules have the same average kinetic energy at a given temperature, independent of their molar mass. To better understand this, let's calculate the average kinetic

root-mean-square speed (u_{rms}) the square root of the average of the squared speeds of all the molecules in a population of gas molecules; a molecule possessing the average kinetic energy moves at this speed.

energy of 1 mole of N_2 and 1 mole of He at 300 K. According to the solution to Sample Exercise 5.15, the u_{rms} of N_2 molecules at 300 K is 517 m/s. The root-mean-square speed of He atoms is 1367 m/s (the result for the practice exercise of Sample Exercise 5.15). Using these values of speed and the mass of 1 mole of each gas ($m = \mathcal{M}$) in Equation 5.27 gives

$$\begin{aligned} KE_{N_2} &= \tfrac{1}{2}\mathcal{M}_{N_2}\,(u_{rms,N_2})^2 = \tfrac{1}{2}(2.801 \times 10^{-2}\text{ kg}) \cdot (516.9\text{ m/s})^2 \\ &= 3.74 \times 10^3\text{ kg} \cdot (\text{m/s})^2 = 3.74 \times 10^3\text{ J} = 3.74\text{ kJ} \\ KE_{He} &= \tfrac{1}{2}\mathcal{M}_{He}\,(u_{rms,He})^2 = \tfrac{1}{2}(4.003 \times 10^{-3}\text{ kg}) \cdot (1367\text{ m/s})^2 \\ &= 3.74 \times 10^3\text{ kg} \cdot (\text{m/s})^2 = 3.74 \times 10^3\text{ J} = 3.74\text{ kJ} \end{aligned}$$

This calculation demonstrates that the same quantities of two *different* gases have the same average kinetic energy at the same temperature, which means they should exert the same pressure. The slower N_2 molecules collide with the container walls less often, but because they are more massive than He atoms, the N_2 molecules exert a greater force during each collision.

Equation 5.28 enables us to compare the relative root-mean-square speeds of two gases at the same temperature. Consider an equimolar mixture of $N_2(g)$ and He(g). We have just demonstrated that at any given temperature their average kinetic energies are the same:

$$KE_{N_2} = \tfrac{1}{2}m_{N_2}(u_{rms,N_2})^2 = \tfrac{1}{2}m_{He}(u_{rms,He})^2 = KE_{He}$$

or

$$m_{N_2}(u_{rms,N_2})^2 = m_{He}(u_{rms,He})^2$$

Rearranging this equation to express the ratio of the root-mean-square speeds in terms of the ratio of the molar masses and then taking the square root of each side, we get

$$\frac{(u_{rms,He})^2}{(u_{rms,N_2})^2} = \frac{\mathcal{M}_{N_2}}{\mathcal{M}_{He}}$$

$$\frac{u_{rms,He}}{u_{rms,N_2}} = \sqrt{\frac{\mathcal{M}_{N_2}}{\mathcal{M}_{He}}} = \sqrt{\frac{28.01\text{ g/mol}}{4.003\text{ g/mol}}} = 2.645$$

The root-mean-square speed of helium atoms should be 2.645 times higher than that of nitrogen molecules. We can test this by using the values for u_{rms} used above to calculate KE_{N_2} and KE_{He}:

$$\frac{u_{rms,He}}{u_{rms,N_2}} = \frac{1367\text{ m/s}}{516.9\text{ m/s}} = 2.645$$

The relationship between root-mean-square speeds and molar masses applies to any pair of gases x and y:

$$\frac{u_{rms,x}}{u_{rms,y}} = \sqrt{\frac{\mathcal{M}_y}{\mathcal{M}_x}} \tag{5.29}$$

This equation leads to the conclusion that the root-mean-square speed of more massive particles is *lower* than the root-mean-square speed of lighter particles.

CONCEPT **TEST**

Because root-mean-square speed depends on temperature, why doesn't the ratio of the u_{rms} of two gases change with temperature?

(Answers to Concept Tests are in the back of the book.)

Graham's Law: Effusion and Diffusion

Figure 5.38 shows two balloons—one filled with nitrogen and the other with helium. The volume, temperature, and pressure of the gases are identical in the two balloons, and the pressure inside each balloon is greater than atmospheric pressure. Over time, the volume of the helium balloon decreases significantly, but the volume of the nitrogen balloon does not. Why?

The skin of any balloon is slightly permeable; that is, it has microscopic holes that allow gas to escape, reducing the pressure inside the balloon. Why does the helium leak more quickly than the nitrogen? We now know the relationship between root-mean-square speeds and molar masses of two gases, and we have calculated that helium atoms move about 2.65 times faster than nitrogen atoms at the same temperature. The gas with the greater root-mean-square speed (He) leaks out of the balloon at a higher rate. The process of moving through a small opening from a higher-pressure region to a lower-pressure region is called **effusion**.

In the 19th century, Scottish chemist Thomas Graham (1805–1869) recognized that the effusion rate of a gas is related to its molar mass. Today this relation is known as **Graham's law of effusion**, and it states that the effusion rate of any gas is inversely proportional to the square root of its molar mass. We can derive a mathematical representation of Graham's law starting from Equation 5.28 for two gases x and y:

$$u_{\text{rms},x} = \sqrt{\frac{3RT}{\mathcal{M}_x}}$$

$$u_{\text{rms},y} = \sqrt{\frac{3RT}{\mathcal{M}_y}}$$

We take the ratio of the root-mean-square speeds, which we designate as the effusion rates r_x and r_y:

$$\frac{r_x}{r_y} = \frac{u_{\text{rms},x}}{u_{\text{rms},y}} = \frac{\sqrt{\dfrac{3RT}{\mathcal{M}_x}}}{\sqrt{\dfrac{3RT}{\mathcal{M}_y}}} \tag{5.30}$$

Equation 5.30 simplifies to Equation 5.31, the usual form of Graham's law:

$$\frac{r_x}{r_y} = \sqrt{\frac{\mathcal{M}_y}{\mathcal{M}_x}} \tag{5.31}$$

Using Equation 5.31 for the helium and nitrogen balloons, we see that the effusion rate of helium compared to nitrogen is:

$$\frac{r_{\text{He}}}{r_{\text{N}_2}} = \sqrt{\frac{\mathcal{M}_{\text{N}_2}}{\mathcal{M}_{\text{He}}}} = \sqrt{\frac{28.01\ \cancel{\text{g/mol}}}{4.003\ \cancel{\text{g/mol}}}} = 2.645$$

This expression has the same form—and hence the ratio has the same value—as the ratio of the root-mean-square speeds of the N_2 and He molecules.

Two things about Equation 5.31 are worth noting: (1) Like the root-mean-square speed in Equation 5.29, Graham's law involves a square root. (2) With labels x and y for two gases in a mixture, the left side of Equation 5.31 has the x term in the numerator and the y term in the denominator, whereas the right side of the equation has the opposite: the y term is in the numerator and x term in the denominator. Keeping the labels straight is the key to solving problems involving Graham's law.

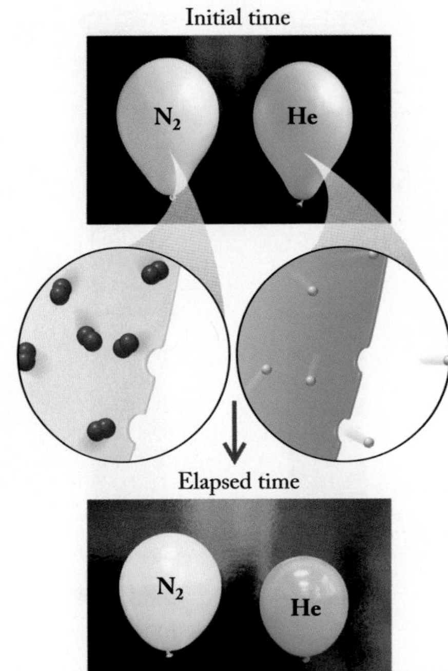

FIGURE 5.38 Two balloons at the same volume, temperature, and pressure—one filled with nitrogen gas and the other filled with helium gas. Over time, the volume of the helium balloon decreases more quickly than does the volume of the nitrogen balloon. The helium atoms are lighter than the nitrogen molecules and therefore have higher root-mean-square speeds. As a consequence, helium atoms encounter holes more frequently and escape from the balloon more rapidly, causing the balloon containing helium to shrink more quickly than the balloon containing nitrogen.

effusion the process by which a gas escapes from its container through a tiny hole into a region of lower pressure.

Graham's law of effusion the principle that the rate of effusion of a gas is inversely proportional to the square root of its molar mass.

FIGURE 5.39 Bromine vapor in the lower part (a) of this tube diffuses upward over time (b). Notice how the top contains some vapor but is not uniformly filled.

How does the kinetic molecular theory explain Graham's law of effusion? The escape of a gas molecule from a balloon requires that the molecules encounter one of the microscopic holes in the balloon. The faster a gas molecule moves, the more likely it is to find one of these holes.

Effusion of gas molecules is related to **diffusion**, which is the spread of one substance through another. If we focus our attention on gases, then diffusion contributes to the passage of odors such as a perfume throughout a room and the smell of baking bread throughout a house. **Figure 5.39** illustrates the diffusion of orange-brown bromine vapor through an apparatus. The rates of diffusion of gases depend on the average speeds of the gas molecules, which depend on their molar masses. Like effusion, the diffusion of gases is described by Graham's law.

SAMPLE EXERCISE 5.16 Applying Graham's Law to Rates of Diffusion of Gases LO7

An odorous gas emitted by a hot spring was found to diffuse 2.92 times slower than helium. What is the molar mass of the emitted gas?

Collect and Organize We are asked to determine the molar mass of an unknown gas on the basis of its rate of diffusion relative to the rate for helium.

Analyze Graham's law (Equation 5.31) relates the relative rate of diffusion of two gases to their molar masses. We know that helium ($\mathcal{M}_{He} = 4.003$ g/mol) diffuses 2.92 times faster than the unidentified gas (y). Because helium diffuses faster than the unknown gas, we predict that the unidentified gas has a larger molar mass. In fact, it should be about nine (or 3^2) times the mass of He according to Graham's law, or about 36 g/mol, because the ratio of the molar masses will equal the square of the ratio of the diffusion rates.

Solve Beginning with Equation 5.31, and rearranging it to obtain an expression for $\mathcal{M}_y$:

$$\frac{r_x}{r_y} = \sqrt{\frac{\mathcal{M}_y}{\mathcal{M}_x}} \quad \text{or} \quad \mathcal{M}_{He}\left(\frac{r_{He}}{r_y}\right)^2 = \mathcal{M}_y$$

Substituting for the ratio of the diffusion rates and the mass of 1 mole of helium gives

$$\mathcal{M}_y = \mathcal{M}_{He}\left(\frac{r_{He}}{r_y}\right)^2 = (4.003\ \text{g/mol})(2.92)^2 = 34.1\ \text{g/mol}$$

Think About It The molar mass of the unidentified gas is 34.1 g/mol, which is consistent with our prediction. One possibility for the identity of this gas is $H_2S(g)$, $\mathcal{M} = 34.08$ g/mol, a foul-smelling and toxic gas frequently emitted from volcanoes that is also responsible for the odor of rotten eggs.

Practice Exercise

Helium effuses 3.16 times as fast as which other noble gas?

(Answers to Practice Exercises are in the back of the book.)

5.9 Real Gases

Up to now we have treated all gases as ideal. This is acceptable because, under typical atmospheric pressures and temperatures, most gases *do* behave ideally. We have also assumed, according to kinetic molecular theory, that the volume occupied by individual gas molecules is negligible compared with the total volume occupied by the gas. In addition, we have assumed that no interactions occur between gas molecules other than random elastic collisions. These assumptions are not valid, however, when we begin to compress gases into increasingly smaller volumes.

diffusion the spread of one substance (usually a gas or liquid) through another.

Deviations from Ideality

How does 1.0 mole of a gas behave as we increase the pressure on it? From the ideal gas law, we know that $PV/RT = n$, so for 1 mole of gas, PV/RT should remain equal to 1.0 regardless of how we change the pressure. The relationship between PV/RT and P for an ideal gas is shown by the purple line in **Figure 5.40**. However, the curves for PV/RT versus P for CH_4, H_2, and CO_2 at pressures above 10 atm are not horizontal, straight lines like this ideal curve. Not only do the curves diverge from the ideal, but the shapes of the curves also differ for each gas, indicating that when we are dealing with real gases, the identity of the gas does matter.

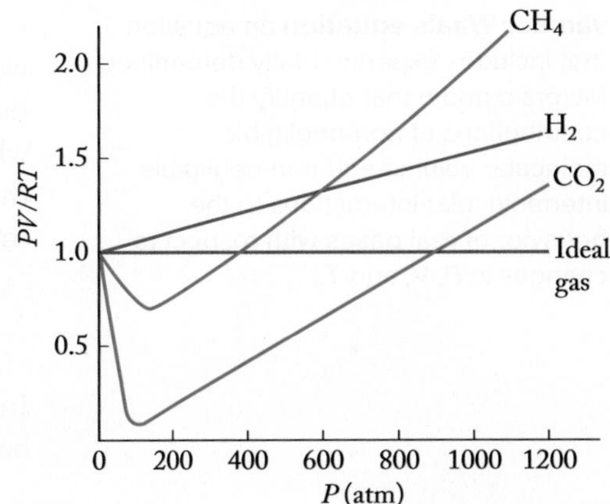

FIGURE 5.40 The effect of very high pressure on the behavior of real and ideal gases. The curves for real gases drop below the horizontal ideal-gas line because of interactions among their molecules. Those curves go above the line as gas molecules take up a larger fraction of the total volume that the gas occupies.

Why don't real gases behave like ideal gases at high pressure? One reason is that the ideal gas law considers gas molecules to have so little volume compared with the volume of their container that they are assumed to have no volume at all. At high pressures, however, more molecules are squeezed into a given volume (**Figure 5.41**). Under these conditions the volume occupied by the molecules can become significant. What is the impact of this on the graph of PV/RT versus P?

In PV/RT, the V actually refers to the *free volume* ($V_{\text{free volume}}$), the empty space not occupied by gas molecules. Because $V_{\text{free volume}}$ is difficult to measure, we instead measure V_{total}, the total volume of the container holding the gas:

$$V_{\text{total}} = V_{\text{free volume}} + V_{\text{molecules}}$$

In an ideal gas and in a real gas at low pressure, $V_{\text{total}} \approx V_{\text{free volume}}$ because $V_{\text{molecules}}$ is negligible (≈ 0) in relation to the volume of the container (V_{total}). When this condition applies, we can use the ideal gas equation with confidence. Consider, however, what happens to a real gas in a closed but flexible container as we increase the external pressure, causing the container—and hence the volume occupied by the gas—to shrink. As the external pressure increases, the assumption that $V_{\text{molecules}} \approx 0$ becomes less valid because the portion of the container's volume occupied by the molecules themselves increases. The relationship $V_{\text{total}} = V_{\text{free volume}} + V_{\text{molecules}}$ still applies, but now, $V_{\text{molecules}} \neq 0$, which means $V_{\text{free volume}}$ can no longer be approximated by V_{total}. Because $V_{\text{total}} > V_{\text{free volume}}$, the ratio PV/RT is also larger for a real gas than for an ideal gas:

$$\left(\frac{PV_{\text{total}}}{RT}\right)_{\text{real gas}} > \left(\frac{PV_{\text{free volume}}}{RT}\right)_{\text{ideal gas}}$$

As a consequence, the curve for a real gas in Figure 5.40 diverges upward from the line for an ideal gas.

A second factor also causes the ratio PV/RT to diverge from 1. Kinetic molecular theory assumes that molecules do not interact, but real molecules do attract one another. These attractive forces function over short distances, so the assumption that molecules behave independently is a good one as long as the molecules are far apart. However, as pressure increases on a population of gas molecules and they are pushed closer together, intermolecular attractive forces can become significant. Likewise, lowering the temperature causes molecules to move more slowly,

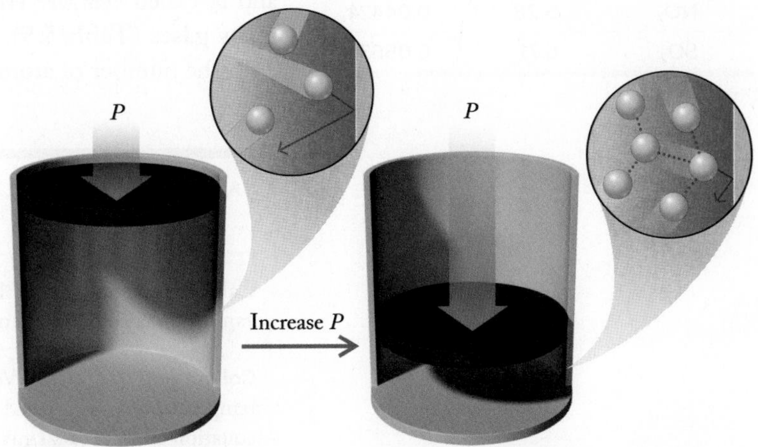

FIGURE 5.41 At high pressures, a larger fraction of the volume of a gas is occupied by the gas molecules. The larger density of the molecules also results in more interactions between them (broad red arrows in expanded view). These interactions reduce the frequency and force of collisions with the walls of the container (red arrow), thereby reducing the pressure.

van der Waals equation an equation that includes experimentally determined factors *a* and *b* that quantify the contributions of non-negligible molecular volume and non-negligible intermolecular interactions to the behavior of real gases with respect to changes in *P*, *V*, and *T*.

resulting in more attractions and more deviations from ideal behavior. In fact, as we see in Chapter 10, a sufficiently cooled gas will condense because of such attractions. These attractive forces cause the molecules to associate with one another, which decreases the force of their collisions with the walls of their container, thereby decreasing the pressure exerted by the gas. If the value of P in PV/RT is smaller in the real gas than the ideal value, the value of the ratio decreases:

$$\frac{P_{\text{real}}V}{RT} < \frac{P_{\text{ideal}}V}{RT}$$

In this case, the curve for a real gas diverges from the ideal gas line by moving below it on the graph.

CONCEPT TEST

For which gases in Figure 5.40 does the impact of molecular interactions outweigh the impact of molecular volume on *PV/RT* at 200 atm?

(Answers to Concept Tests are in the back of the book.)

TABLE 5.5 Van der Waals Constants of Selected Gases

Substance	*a* ($L^2 \cdot$ atm/mol^2)	*b* (L/mol)
H_2	0.244	0.0266
He	0.0341	0.02370
N_2	1.39	0.0391
O_2	1.36	0.0318
Ar	1.34	0.0322
CH_4	2.25	0.0428
CO	1.45	0.0395
CO_2	3.59	0.0427
H_2O	5.46	0.0305
HCl	3.67	0.04081
NO	1.34	0.02789
NO_2	5.28	0.04424
SO_2	6.71	0.05636

The van der Waals Equation for Real Gases

Because the ideal gas equation does not hold at high pressures or low temperatures, we need another equation that can be used under these conditions—one that accounts for the following facts:

1. The free volume of a real gas is less than the total volume because its molecules occupy significant space.
2. The observed pressure is less than the pressure of an ideal gas because of intermolecular attractions.

The **van der Waals equation,**

$$\left(P + \frac{n^2a}{V^2}\right)(V - nb) = nRT \tag{5.32}$$

includes terms to correct for pressure (n^2a/V^2) and volume (nb). The values of a and b, called *van der Waals constants*, have been determined experimentally for many gases (**Table 5.5**). Both a and b increase with increasing molar mass and with the number of atoms in each molecule of a gas.

SAMPLE EXERCISE 5.17 Calculating Pressure with the van der Waals Equation — LO8

Calculate the pressure of 1.00 mol of N_2 in a 1.00 L container at 300.0 K, using first the van der Waals equation and then the ideal gas equation.

Collect and Organize We are given the amount of nitrogen, its volume, and its temperature, and we need to calculate its pressure by using both the van der Waals equation and the ideal gas equation. Using the van der Waals equation means we need to know the values of the van der Waals constants for nitrogen.

Analyze The van der Waals constants for nitrogen gas are $a = 1.39\ L^2 \cdot atm/mol^2$ and $b = 0.0391$ L/mol in Table 5.5. The values of a and b represent experimentally determined corrections for the interactions between molecules and the portion of the total volume occupied by the gas molecules. One mole of ideal gas at STP occupies

22.4 L. Compressing the gas to a volume of about 1/20 of the initial volume requires a pressure of about 20 atm. Heating it to 300 K at constant volume causes a relatively minor change in the pressure. We can estimate the effect of a pressure of about 20 atm on the ideality of the nitrogen from the curves in Figure 5.40. Although we do not know which curve best describes the behavior of N_2, the scale of the x-axis is rather large. At a pressure of 20 atm, none of the curves deviates substantially from ideality, so we expect a relatively small difference in the pressure calculated for N_2 as an ideal gas or by using the van der Waals equation.

Solve Begin with Equation 5.32 then solve it for pressure:

$$\left(P + \frac{n^2a}{V^2}\right)(V - nb) = nRT \quad \text{or} \quad P = \frac{nRT}{V - nb} - \frac{n^2a}{V^2}$$

$$P_{N_2} = \frac{(1.00\ \text{mol})\left(0.08206\ \frac{\text{L}\cdot\text{atm}}{\text{mol}\cdot\text{K}}\right)(300.0\ \text{K})}{1.00\ \text{L} - (1.00\ \text{mol})\left(0.0391\ \frac{\text{L}}{\text{mol}}\right)} - \frac{(1.00\ \text{mol})^2\left(1.39\ \frac{\text{L}^2\cdot\text{atm}}{\text{mol}^2}\right)}{(1.00\ \text{L})^2}$$

$$= 24.2\ \text{atm}$$

If nitrogen behaved as an ideal gas, we would have

$$P = \frac{nRT}{V}$$

$$P_{N_2} = \frac{(1.00\ \text{mol})\left(0.08206\ \frac{\text{L}\cdot\text{atm}}{\text{mol}\cdot\text{K}}\right)(300.0\ \text{K})}{1.00\ \text{L}}$$

$$= 24.6\ \text{atm}$$

Think About It This pressure is about 25 times greater than normal atmospheric pressure, so we expect some difference in the calculated pressures. However, the small deviation from ideality, $0.4/24.6 \times 100 = 2\%$, supports our prediction. Furthermore, because of the likely deviation from ideal behavior, we expect the pressure calculated using the ideal gas equation to be too high, which turns out to be true: $P_{\text{ideal}} > P_{\text{real}}$.

Practice Exercise Assuming the conditions stated in Sample Exercise 5.17, use the van der Waals equation to calculate the pressure for 1.00 mol He gas. Compare the real-gas pressures of He and N_2 and suggest why they differ.

(Answers to Practice Exercises are in the back of the book.)

We began this chapter with a description of our atmosphere as an invisible necessity. Looking back, we can now answer the questions we raised about gases and our atmosphere. We now understand how elevation affects the quantity of gas available, making pressurized cabins on commercial airliners a necessity for safe, comfortable travel. We have seen the chemical reactions used to supply oxygen in aircraft under emergency conditions and applied the principles of stoichiometry to calculate the quantity of reactant needed to generate a given quantity of emergency oxygen gas. Undersea explorers carry a different mixture of gases in a scuba tank from what we breathe under normal conditions because the high pressures under water increase the partial pressures of oxygen and nitrogen to unhealthy levels. Hot-air balloons rise because volume is directly proportional to temperature and because the density of air decreases as temperature rises. Understanding the physical properties of gases, especially their compressibility and their responses to changing conditions, makes their use in many everyday items, from automobile tires to weather balloons and fire extinguishers, possible because their behavior is reliable and predictable.

SAMPLE EXERCISE 5.18 Integrating Concepts: Scuba Diving

The air bubbles exhaled by a scuba diver (**Figure 5.42**) expand as they rise to the surface. The lungs of a diver behave similarly unless certain precautions are taken, and serious medical consequences may result if divers deviate from these precautions. In the following calculations, assume that the pressure at sea level is 1.00 atm and that the pressure increases by 1.0 atm for every 10 m increase in depth beneath the surface. Assume a diver's lungs have an air volume of 5.1 L and that the dry air in a scuba tank is 78.1% N_2, 20.9% O_2, and 1.00% Ar by volume. (a) What is the partial pressure of each gas in the diver's lungs as she begins her dive? (b) What is the partial pressure of each gas in the diver's lungs at a depth of 30.0 m? (c) Suppose the scuba diver is accompanied during her descent by a diver without a supplemental air supply. This breath-hold diver also has a lung air capacity of 5.1 L and dives to 30.0 m and ascends back to the surface, all without exhaling. What is the volume of air in the breath-hold diver's lungs at a depth of 30.0 m and upon the return to the surface? (d) What would happen to the volume of air in the scuba diver's lungs if she held her breath while ascending from 30.0 m to the surface?

Air may be used in scuba tanks for shallow dives. However, nitrogen narcosis (the symptoms of which include dulling of the senses, inactivity, and even unconsciousness) can result during deeper dives because of the anesthetic qualities of nitrogen under pressure. Breathing pure oxygen is not a solution because oxygen at high levels is quite toxic to the central nervous system and other parts of the body. Therefore, mixtures of N_2, O_2, and He, generically called Trimix, are used to minimize nitrogen narcosis and oxygen toxicity. For most dives, the oxygen concentration in the tank is adjusted to a maximum partial pressure of 1.5 atm at the working depth; the "equivalent narcotic depth" (END) for N_2 pressure is normally set at 4.0 atm. Helium is added as needed to achieve these values. (e) What Trimix composition (in percent by volume) achieves these values for a dive to 70.0 m below the surface?

FIGURE 5.42

Collect and Organize We are given initial and final conditions for mixtures of gases and desired partial pressures for gases at known total pressures. Boyle's law relates the volume and pressure of a quantity of a gas at constant temperature, and Dalton's law of partial pressures relates the partial pressures of the gases in a mixture to its total pressure and the gases' mole fractions in the mixture.

Analyze We can convert percent composition data into partial pressures at 1 atm and then into partial pressures under other conditions. The pressure at 30.0 m is about 4 atm, so the partial pressures of gases in the scuba diver's lungs will increase by a factor of 4. The air in the scuba diver's lungs will have the same volume at 30 m; the breath-holding diver's lungs will be compressed to about $\frac{1}{4}$ their volume on the surface. As the two ascend, both divers' lungs will expand as pressure drops, but the scuba diver's lungs start at a much higher volume than the breath-hold diver's. A dive to 70.0 m means experiencing an ambient pressure of 8.0 atm. Using that as the total pressure:

$$P_{total} = 8.0 \text{ atm} = P_{N_2} + P_{O_2} + P_{He} = 4.0 \text{ atm} + 1.5 \text{ atm} + P_{He}$$

Therefore, P_{He} should be 2.5 atm.

Solve

a. The percent composition by volume can be converted into mole fractions by recognizing, for example, that 78.1% nitrogen means 0.781 moles of nitrogen for every 1 mole of gas:

$$X_{N_2} = \frac{n_{N_2}}{n_{total}} = \frac{0.781}{1} = 0.781$$

The mole fraction of oxygen in the scuba diver's lungs is 0.209, and that of argon is 0.010.

b. For every 10 m of descent below the surface, the pressure increases by 1.0 atm. At 30.0 m, the pressure would increase by 3.0 atm, so the total pressure is 3.0 atm + 1.0 atm = 4.0 atm. The partial pressure of each gas at 4.0 atm total pressure is:

$$P_{N_2} = X_{N_2}P_{total} = 0.781 \times 4.0 \text{ atm} = 3.1 \text{ atm}$$

$$P_{O_2} = X_{O_2}P_{total} = 0.209 \times 4.0 \text{ atm} = 0.84 \text{ atm}$$

$$P_{Ar} = X_{Ar}P_{total} = 0.010 \times 4.0 \text{ atm} = 0.040 \text{ atm}$$

c. The breath-hold diver does not breathe in more air, so the quantity of gas in her lungs remains constant. As the pressure increases, Boyle's law predicts her lungs will decrease in volume:

$$V_{final} = V_{initial} \times \frac{P_{initial}}{P_{final}} = 5.1 \text{ L} \times \frac{1.0 \text{ atm}}{4.0 \text{ atm}} = 1.3 \text{ L}$$

Upon return to the surface, her lungs will re-expand to their original volume of 5.1 L.

d. Because the scuba diver has been breathing air supplied by her scuba tank, her lungs retain their volume of 5.1 L at a depth of 30 m. If she holds her breath as she ascends to the surface, her lungs will expand (or try to):

$$5.1\ \text{L} \times \frac{4.0\ \text{atm}}{1.0\ \text{atm}} = 20\ \text{L}$$

e. The pressure change during a dive to 70.0 m would be

$$70.0\ \text{m} \times \frac{1.0\ \text{atm}}{10\ \text{m}} = 7.0\ \text{atm}$$

The total pressure at 70.0 m, then, would be 7.0 atm + 1.0 atm = 8.0 atm.

The partial pressure of helium in the tank must be

$$8.0\ \text{atm} = 4.0\ \text{atm} + 1.5\ \text{atm} + x$$

$$x = 2.5\ \text{atm}$$

To achieve these partial pressures at a depth of 70.0 m, the percent composition of the gas mixture in the scuba tank must be:

$$N_2:\quad \frac{P_{N_2}}{P_{\text{total}}} \times 100\% = \frac{4.0\ \text{atm}}{8.0\ \text{atm}} \times 100\% = 50 = 50\%$$

$$O_2:\quad \frac{P_{O_2}}{P_{\text{total}}} \times 100\% = \frac{1.5\ \text{atm}}{8.0\ \text{atm}} \times 100\% = 18.8 = 19\%$$

$$\text{He}:\quad \frac{P_{\text{He}}}{P_{\text{total}}} \times 100\% = \frac{2.5\ \text{atm}}{8.0\ \text{atm}} \times 100\% = 31.3 = 31\%$$

Think About It An important rule in scuba diving is "never hold your breath." If the scuba diver held her breath on an ascent from 30 m, the resulting expansion of air in her lungs could rupture them or force air into blood vessels, causing embolisms. A breath-hold diver is not immune to damage from pressure changes either. During deep dives, a condition called thoracic squeeze may develop, which could also result in hemorrhage of lung tissues.

SUMMARY

LO1 Gases diffuse to occupy the entire volume of their container. The volume occupied by a gas changes significantly with pressure and temperature. Gases are **miscible**, mixing in any proportion. They are much less dense than liquids or solids. **Pressure** is defined as the ratio of force to surface area, and gas pressure is measured with **barometers** and **manometers**. Pressure can be expressed in a variety of units, and we can convert between them by using dimensional analysis. (Sections 5.1 and 5.2)

LO2 **Boyle's law, Charles's law, Avogadro's law,** and **Amontons's law** describe the behavior of gases under different conditions. They may be used individually or in a combined form called the **ideal gas law** ($PV = nRT$) to calculate the volume, temperature, pressure, or number of moles of a gas. (Sections 5.3 and 5.4)

LO3 The ideal gas equation and the stoichiometry of a chemical reaction can be used to calculate the volumes of gases required or produced in the reaction. (Section 5.5)

LO4 The density of a gas can be calculated for a given set of conditions of volume, temperature, and pressure. The density of a gas can also be used to calculate its molar mass. (Section 5.6)

LO5 In a gas mixture, the contribution each component gas makes to the total gas pressure is called the **partial pressure** of that gas. **Dalton's law of partial pressures** allows us to calculate the partial pressure (P_x) of any constituent gas x in a gas mixture if we know its **mole fraction** (X_x) and the total pressure. (Section 5.7)

LO6 **Kinetic molecular theory** describes gases as particles in constant random motion. This motion is associated with an average kinetic energy that is proportional to the absolute temperature of the gas. Pressure arises from elastic collisions between gases and the walls of their container. All relationships between P, V, T, and n can be derived from kinetic molecular theory. (Section 5.8)

LO7 Particles move at **root-mean-square speeds** (u_{rms}) that are inversely proportional to the square root of their molar masses and directly proportional to the square root of their temperature. **Graham's law of effusion** states that the rate of **effusion** (escape through a pinhole) or **diffusion** (spreading) of a gas at a fixed temperature is inversely proportional to the square root of its molar mass. (Section 5.8)

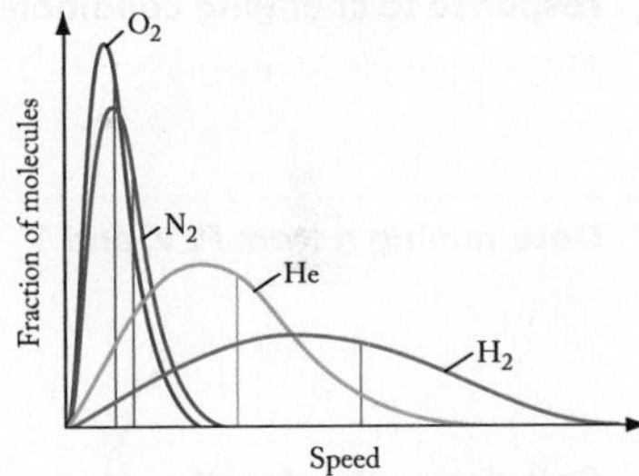

LO8 At high pressures, the behavior of real gases deviates from the predictions of the ideal gas law. The **van der Waals equation**, a modified form of the ideal gas equation, accounts for real gas properties. (Section 5.9)

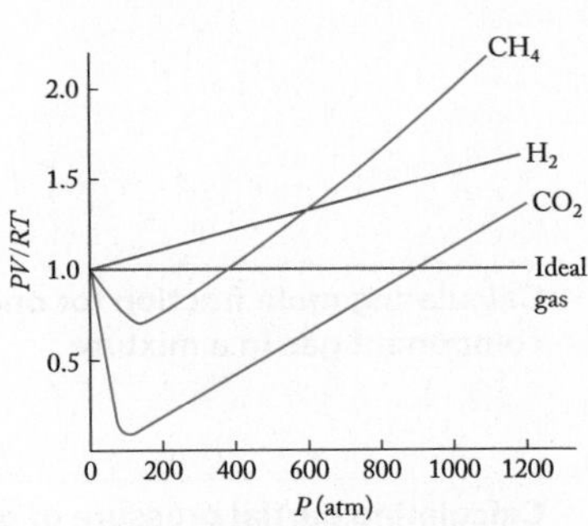

PARTICULATE PREVIEW WRAP-UP

Helium atoms fill both the tank and the balloon, but there are more atoms in the same volume of gas in the tank than in a single balloon because the tank contains enough helium to fill many balloons. The drawing for the gas in the tank in a car trunk on a hot day does not change because the quantity and volume of the gas does not change significantly. However, the helium-filled balloons in the trunk will expand because higher temperatures increase the kinetic energies of the helium atoms. More collisions between the atoms and the walls of the balloons cause the balloons to expand. The same number of He atoms would occupy a larger volume, or a particulate view of the same balloon volume would contain fewer helium atoms.

PROBLEM-SOLVING SUMMARY

Type of Problem	Concepts and Equations	Sample Exercises
Calculating pressure of any gas; calculating atmospheric pressure	Divide the force by the area over which the force is applied, using the equation $$P = \frac{F}{A} \quad (5.1)$$	**5.1, 5.2**
Calculating changes in *P*, *V*, or *T* in response to changing conditions	Rearrange $$\frac{P_1V_1}{T_1} = \frac{P_2V_2}{T_2} \quad (5.18)$$ for whichever variable is sought and then substitute given values. (*T* must be in kelvins, and *n* must be constant.)	**5.3, 5.4, 5.5, 5.6**
Determining *n* from *P*, *V*, and *T*	Rearrange $$PV = nRT \quad (5.15)$$ for *n* and then substitute given values of *P*, *T*, and *V*. (*T* must be in kelvins.)	**5.7, 5.8, 5.9**
Calculating the density of a gas and calculating molar mass from density	Substitute values for pressure, absolute temperature, and molar mass into the equation $$d = \frac{P\mathcal{M}}{RT} \quad (5.21)$$ Substitute values for pressure, absolute temperature, and density into the equation $$\mathcal{M} = \frac{dRT}{P} \quad (5.22)$$	**5.10, 5.11**
Calculating mole fraction for one component gas in a mixture	Divide the number of moles of the component gas by the total number of moles in the mixture: $$X_x = \frac{n_x}{n_{total}} \quad (5.24)$$	**5.12**
Calculating partial pressure of one component gas in a mixture and total pressure in the mixture	Substitute the mole fraction of the component gas and the total pressure in the equation $$P_x = X_xP_{\text{total}} \quad (5.25)$$ Solve the equation $$P_{\text{total}} = P_1 + P_2 + P_3 + P_4 + \ldots \quad (5.23)$$ for the partial pressure of the component gas and then substitute given values for other partial pressures and total pressure.	**5.13, 5.14**

Type of Problem	Concepts and Equations	Sample Exercises
Calculating root-mean-square speeds	Substitute absolute temperature, molar mass, and the value 8.314 kg · m²/(s² · mol · K) for *R* in the equation $$u_{\text{rms}} = \sqrt{\frac{3RT}{\mathcal{M}}} \quad (5.28)$$	**5.15**
Calculating relative rate of effusion or diffusion from molar masses	Substitute values for molar masses into the equation $$\frac{r_x}{r_y} = \sqrt{\frac{\mathcal{M}_y}{\mathcal{M}_x}} \quad (5.31)$$	**5.16**
Calculating *P* for a real gas	Solve the van der Waals equation $$\left(P + \frac{n^2a}{V^2}\right)(V - nb) = nRT \quad (5.32)$$ for *P* and substitute given values of *n*, *V*, and *T* (in kelvins), plus values of *a* and *b* from Table 5.5.	**5.17**

VISUAL PROBLEMS

(Answers to boldface end-of-chapter questions and problems are in the back of the book.)

5.1. Shown in Figure P5.1 are three barometers. The one in the center is located at sea level. Which barometer is most likely to reflect the atmospheric pressure in Denver, CO, where the elevation is approximately 1500 m? Explain your answer.

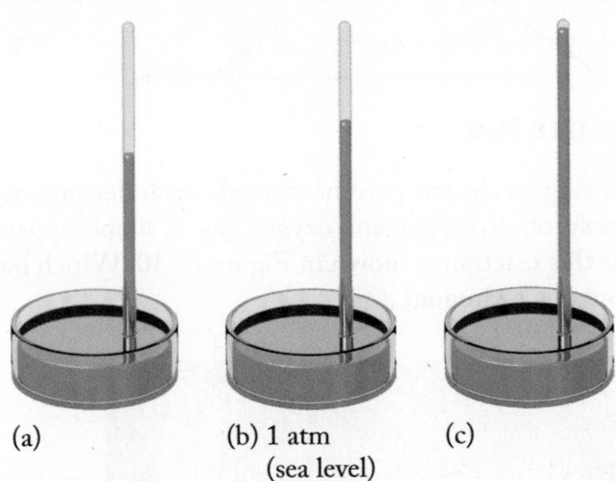

FIGURE P5.1

5.2. A flask is filled with purple iodine vapor. Which of the drawings in Figure P5.2 most accurately reflects the gas in the flask?

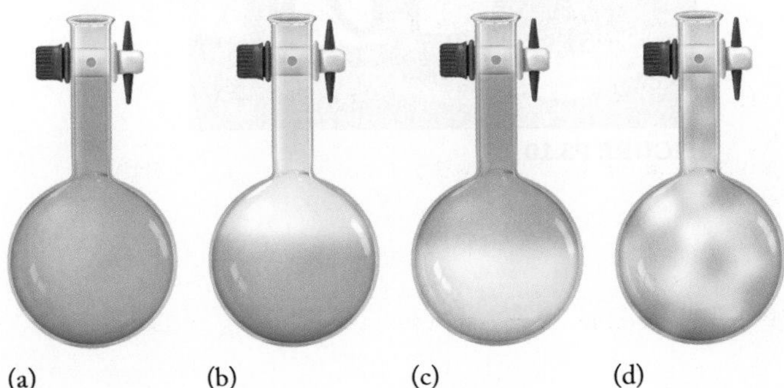

FIGURE P5.2

5.3. Which of the three changes shown in Figure P5.3 best illustrates what happens when (1) the atmospheric pressure on a helium-filled balloon is increased at constant temperature and (2) the temperature of a helium-filled balloon is increased at constant pressure?

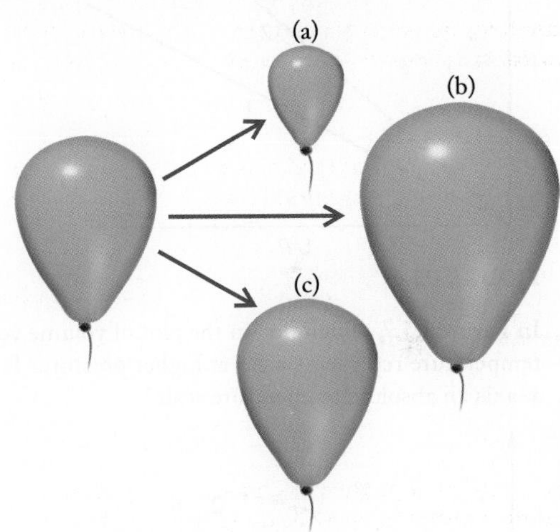

FIGURE P5.3

5.4. Figure P5.4 illustrates the relationship between density and pressure for two gases: methane and nitrogen.

a. Which line should be labeled CH_4?

b. Add lines for the density of He and NO.

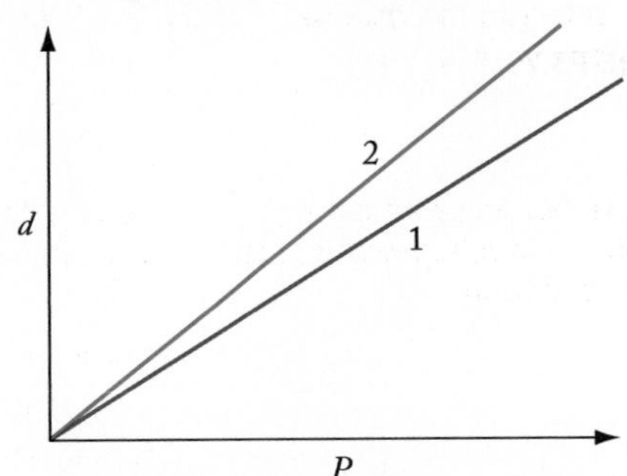

FIGURE P5.4

5.5. Which of the three changes shown in Figure P5.5 best illustrates what happens when the amount of gas in a helium-filled rubber balloon is increased at constant temperature and pressure?

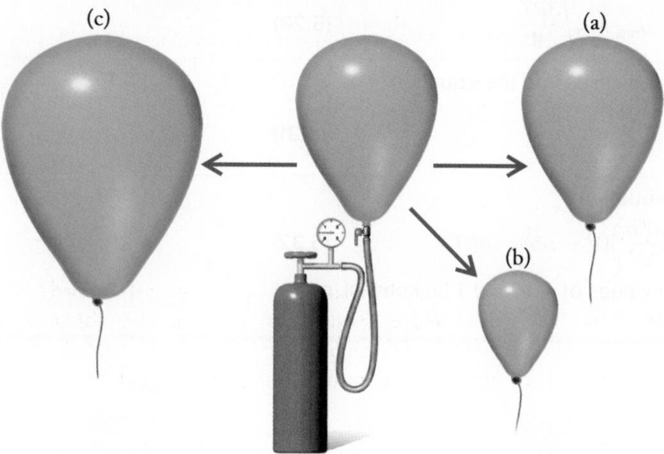

FIGURE P5.5

5.6. Which line in the plot of volume versus reciprocal pressure in **Figure P5.6** corresponds to the higher temperature?

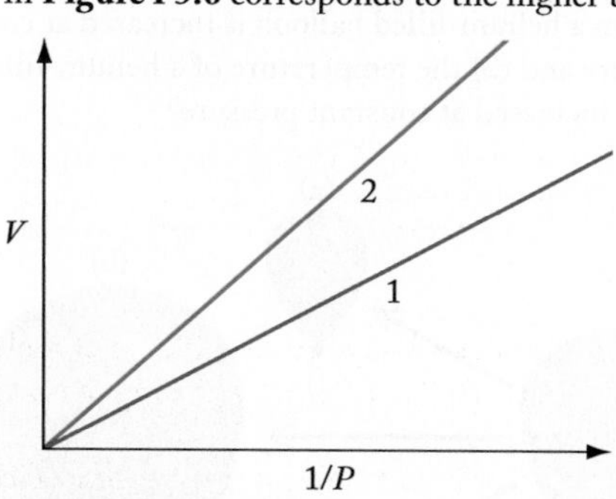

FIGURE P5.6

5.7. In Figure P5.7, which line in the plot of volume versus temperature represents a gas at higher pressure? Is the x-axis an absolute temperature scale?

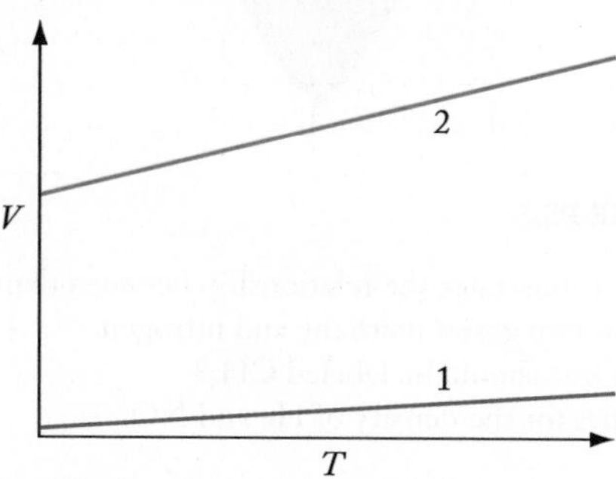

FIGURE P5.7

5.8. In Figure P5.8, which of the two plots of volume versus pressure at constant temperature is *not* consistent with the ideal gas law?

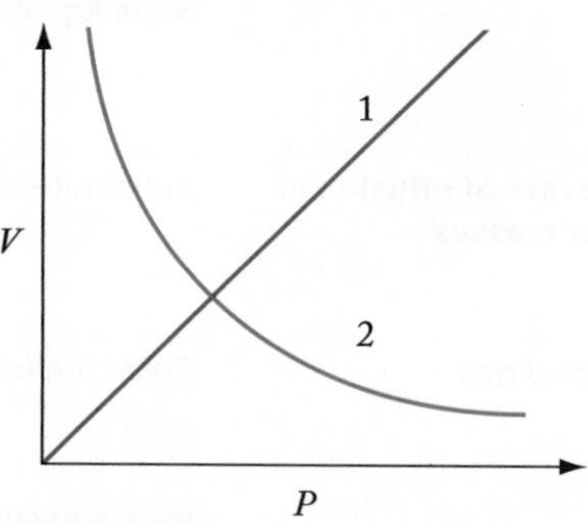

FIGURE P5.8

5.9. In Figure P5.9, which of the two plots of volume versus temperature at constant pressure is *not* consistent with the ideal gas law?

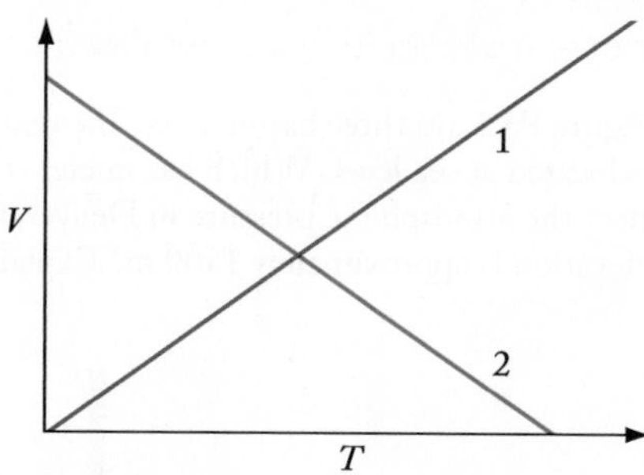

FIGURE P5.9

5.10. Passing an electric current through water decomposes the water into hydrogen and oxygen gas. A simple apparatus for this reaction is shown in Figure P5.10. Which buret contains hydrogen?

FIGURE P5.10

5.11. Passing an electric current through cold (−37°C) liquid ammonia decomposes the ammonia into nitrogen and hydrogen gas. Which tube in Figure P5.11 contains nitrogen?

FIGURE P5.11

5.12. Which of the graphs in Figure P5.12 best represents the relationship between number of collisions with the container and pressure?

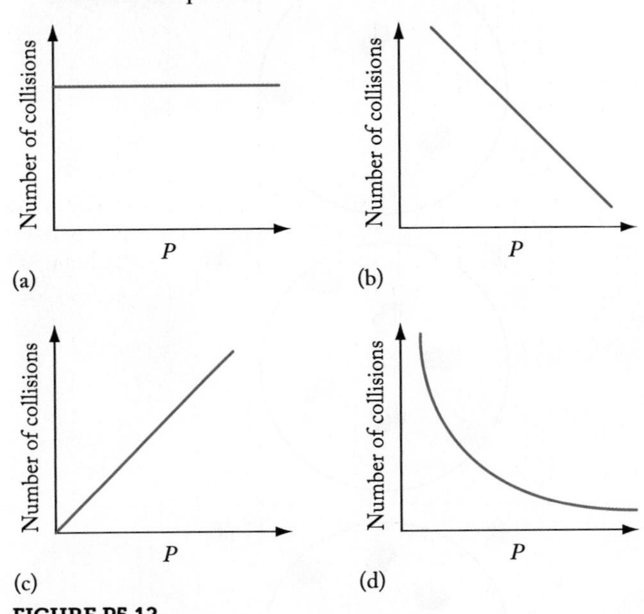

FIGURE P5.12

5.13. Which of the particulate views in Figure P5.13 best depicts the arrangement of molecules in a mixture of gases? The blue and pink spheres represent atoms of helium and neon, respectively.

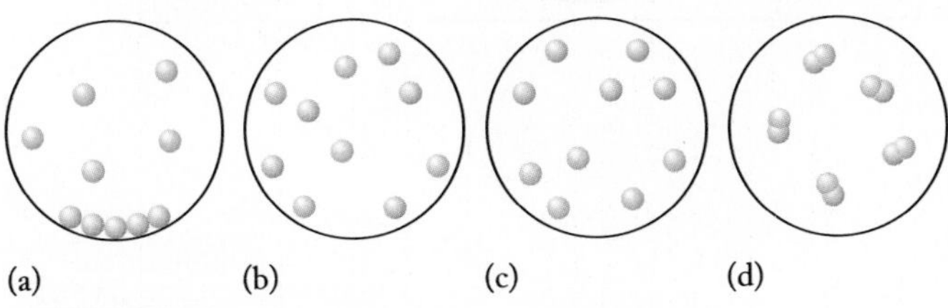

FIGURE P5.13

5.14. The particulate views of Figure P5.14 represent equal volumes of four mixtures of N_2 and O_2 at the same temperature. Is the total pressure in the samples the same? Which has the highest partial pressure of N_2?

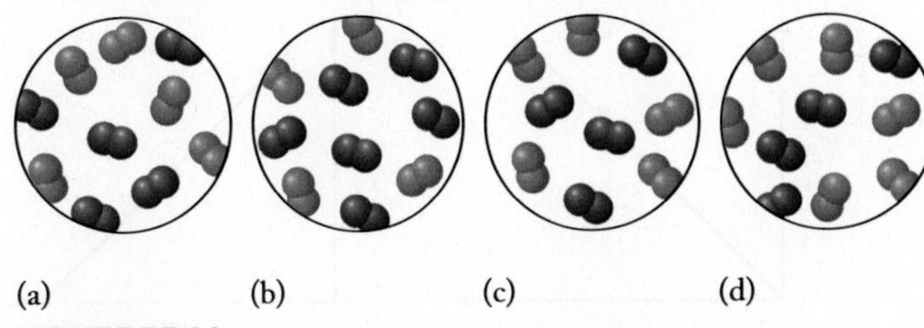

FIGURE P5.14

5.15. Figure P5.15 shows the distribution of molecular speeds of CO_2 and SO_2 molecules at 25°C. Which curve is the profile for SO_2? Which of these profiles should match that of propane (C_3H_8), a common fuel in portable grills?

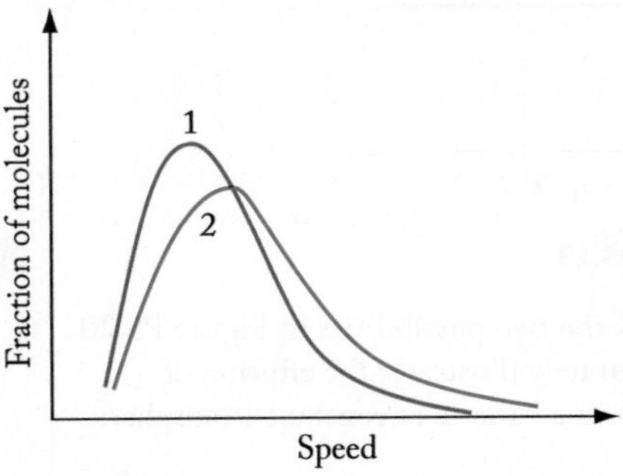

FIGURE P5.15

5.16. How would a graph showing the distribution of molecular speeds of CO_2 at −100°C differ from the graph for CO_2 shown in Figure P5.15?

5.17. A container with a pinhole leak contains a mixture of the elements highlighted in Figure P5.17. Which element leaks the most slowly from the container?

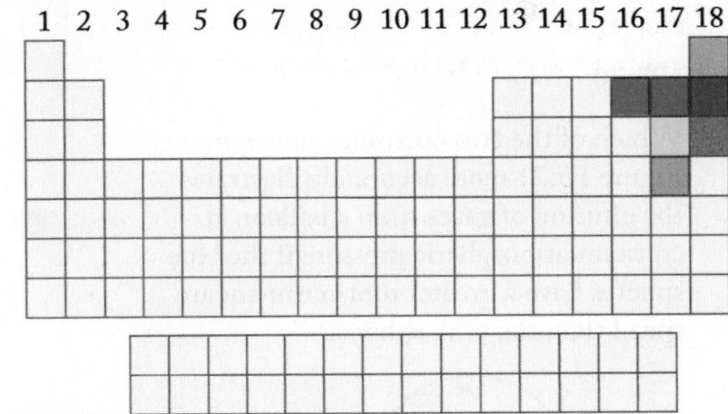

FIGURE P5.17

5.18. A container with a pinhole leak contains a mixture of the elements highlighted in Figure P5.17. Which element has the smallest root-mean-square speed?

5.19. The rate of effusion of a gas increases with temperature. Which graph in Figure P5.19 best describes the ratio of the rates of effusion of two gases (*x* and *y*) as a function of temperature?

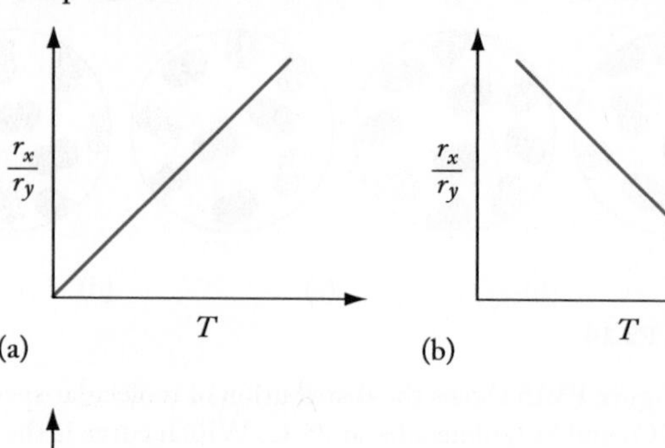

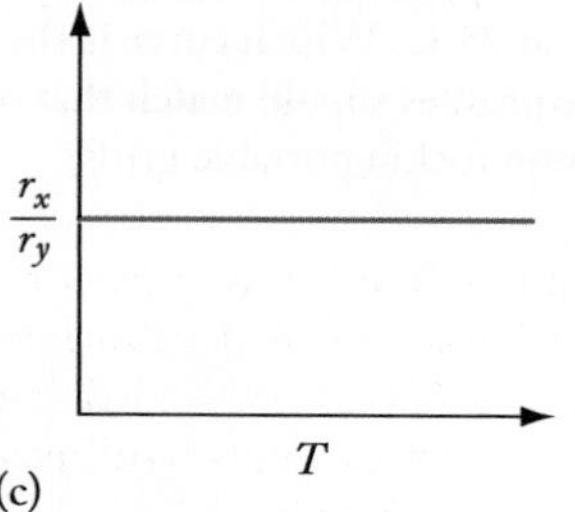

FIGURE P5.19

5.20. Which of the two possibilities in Figure P5.20 more accurately illustrates the effusion of helium from a balloon at constant atmospheric pressure?

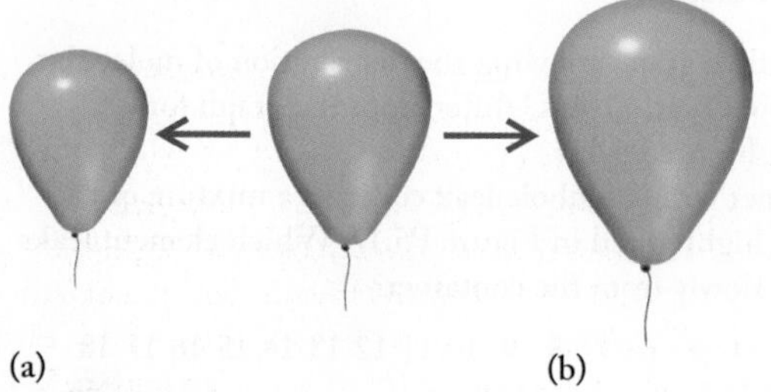

FIGURE P5.20

5.21. Which of the two outcomes shown in Figure P5.21 more accurately illustrates the effusion of gases from a balloon at constant atmospheric pressure if the blue spheres have a greater root-mean-square speed than the pink spheres?

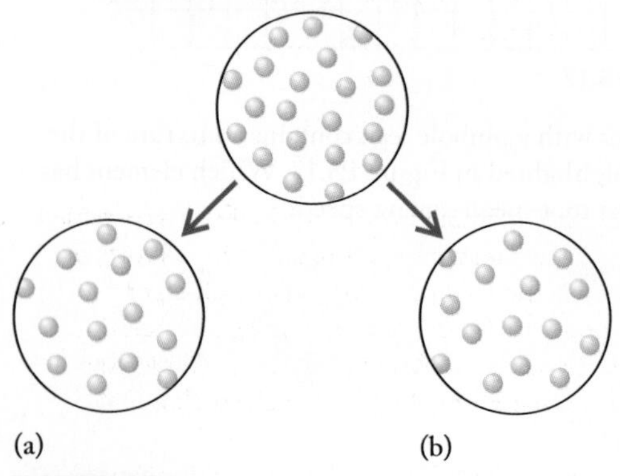

FIGURE P5.21

5.22. Use representations [A] through [I] in Figure P5.22 to answer questions (a)–(f). The pink balloons contain hydrogen, the yellow balloons contain nitrogen, and the gray balloons contain oxygen.

a. Identify three different changes that could be responsible for the change in size of the pink balloon from [A] to [C].
b. If the smaller pink balloon in [A] corresponds to the particulate view in [B], which of the changes identified in part (a) would result in the larger pink balloon in [C] also corresponding to the particulate view in [B]?
c. If the smaller yellow balloon in [D] corresponds to the particulate view in [E], which particulate view corresponds to the larger yellow balloon in [F] if no additional nitrogen has been added?
d. If the larger gray balloon in [I] corresponds to the particulate view in [E], which particulate view represents the gas at a lower temperature?
e. If each gray balloon contains 1 mol of gas at 25°C, in which balloon are the collisions between the oxygen molecules and the inside of the balloon more frequent?
f. Which balloon contains the gas with the shortest mean free path?

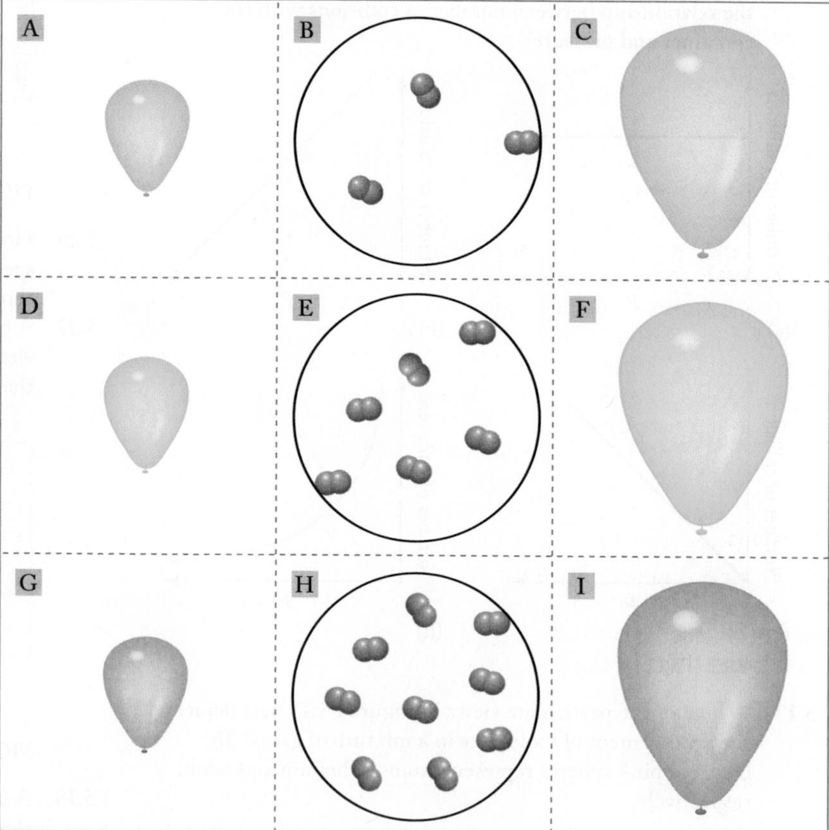

FIGURE P5.22

QUESTIONS AND PROBLEMS

Atmospheric Pressure and Collisions

Concept Review

5.23. Describe the difference between force and pressure.

5.24. Describe the role of collisions in generating pressure.

5.25. How does Torricelli's barometer measure atmospheric pressure?

5.26. What is the relationship between *torr* and *atmospheres* of pressure?

5.27. Which quantity represents the larger pressure, 10 torr or 10 atm?

5.28. In which balloon do He atoms experience more collisions at room temperature: a 1 L balloon at 0.1 atm of pressure or a 0.5 L balloon at 0.1 atm of pressure?

5.29. What is the relationship between *millibars* and *pascals* of pressure?

5.30. Three barometers based on Torricelli's design are constructed using water ($d = 1.00$ g/mL), ethanol ($d = 0.789$ g/mL), and mercury ($d = 13.546$ g/mL). Which barometer contains the tallest column of liquid?

5.31. Why does an ice skater exert more pressure on ice when wearing newly sharpened skates than when wearing skates with dull blades?

5.32. Why is it easier to travel over deep snow when wearing snowshoes over your boots (Figure P5.32) rather than just your boots?

5.33. Why does atmospheric pressure decrease with increasing elevation?

*5.34. Pieces of different metals have the same mass but different densities. Could these objects ever exert the same pressure?

FIGURE P5.32

Problems

5.35. Air Pressure and the Eiffel Tower In his book *Caesar's Last Breath*, author Sam Kean states that a column of air around the Eiffel Tower in Paris (Figure P5.35) weighs more than the tower itself. The tower reaches a height of 324 m and occupies a square base 125 m on a side. The pressure exerted by the Eiffel Tower is estimated to be 4.5 kg/cm^2. If the density of air is taken to be 1.19 g/L, what is the pressure exerted by a column of air, 324 × 125 × 125 m?

FIGURE P5.35

5.36. The gold block represented in Figure P5.36 has a mass of 38.6 g. Calculate the pressure exerted by the block when it is on (a) a square face and (b) a rectangular face.

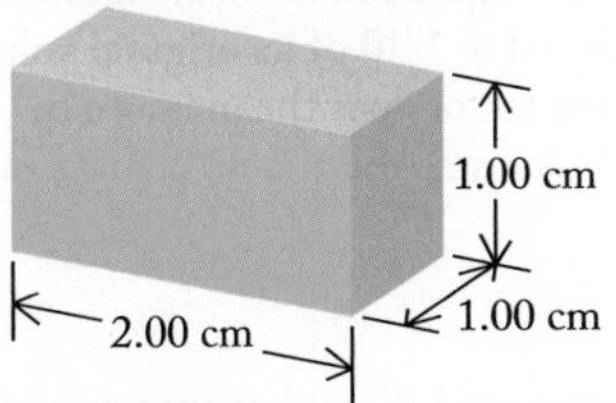

FIGURE P5.36

5.37. Convert the following pressures into atmospheres: (a) 2.0 kPa; (b) 562 torr.

5.38. Convert the following pressures into millimeters of mercury: (a) 0.541 atm; (b) 2.8 kPa.

5.39. Record High Atmospheric Pressure The highest atmospheric pressure recorded on Earth was measured at Tosontsengel, Mongolia, on December 19, 2001, when the barometer read 108.6 kPa. Express this pressure in (a) millimeters of mercury, (b) atmospheres, and (c) millibars.

5.40. **Record Low Atmospheric Pressure** Hurricane Irene registered an atmospheric pressure of 982 mbar in August 2011. Hurricane Katrina registered an atmospheric pressure of 86.2 kPa in September 2005. What was the *difference* in pressure between the two hurricanes in (a) millimeters of mercury, (b) atmospheres, and (c) millibars?

***5.41. Pressure on Venus** Venus is similar to Earth in size and distance from the Sun but has an atmosphere inhospitable for human life that consists of 96.5% CO_2 and 3.5% N_2 by mass.

a. Calculate the atmospheric pressure on Venus given the mass of the Venusian atmosphere (4.8×10^{20} kg), the surface area (4.6×10^{14} m^2), and the acceleration due to gravity (8.87 m/s^2).

*b. Given the similarity between Venus and Earth in surface area, why do you suppose atmospheric pressure on Venus is so much higher than on Earth?

*5.42. **Pressure on Uranus** Uranus is much larger than Earth, with a surface area of 8.1×10^{15} m^2. The atmosphere on Uranus is composed of low-density gases: 82.5% H_2, 15.2% He, and 2.3% CH_4. The acceleration due to gravity is similar to that on Earth, 8.7 m/s^2, yet the atmospheric pressure on Uranus is estimated to exceed 10^8 Pa. What is the mass of the atmosphere on Uranus?

The Gas Laws

Concept Review

5.43. If the quantity of helium in a weather balloon is increased by a factor of 5 at constant pressure and temperature, by how much does the volume of the balloon change?

5.44. How do we explain Boyle's law on a molecular basis?

5.45. A balloonist is rising too fast for her taste. Should she increase the temperature of the gas in the balloon or decrease it?

5.46. Could the pilot of the balloon in Problem 5.45 reduce her rate of ascent by allowing some gas to leak out of the balloon? Explain your answer.

5.47. The volume of a quantity of gas decreases by 50% when it cools from 20°C to 10°C. Does the pressure of the gas increase, decrease, or remain the same?

5.48. If the volume of gasoline vapor and air in an automobile engine cylinder is reduced to 1/10 of its original volume before ignition, by what factor does the pressure in the cylinder increase? (Assume there is no change in temperature.)

Problems

5.49. What is the final pressure of 1.00 mol of ammonia gas, initially at 1.00 atm, if the volume is
a. gradually decreased from 78.0 mL to 39.0 mL at constant temperature?
b. increased from 43.5 mL to 65.5 mL at constant temperature?
c. decreased by 40% at constant temperature?

5.50. The behavior of 485 mL of an ideal gas in response to pressure is studied in a vessel with a movable piston. What is the final volume of the gas if the pressure on the sample is
a. increased from 715 torr to 3.55 atm at constant temperature?
b. decreased from 1.15 atm to 520 torr at constant temperature?
c. increased by 26% at constant temperature?

5.51. **Underwater Archeology** A scuba diver releases a balloon containing 153 L of helium attached to a tray of artifacts at an underwater archeological site (Figure P5.51). When the balloon reaches the surface, it has expanded to a volume of 352 L. The pressure at the surface is 1.00 atm; what is the pressure at the underwater site? Pressure increases by 1.0 atm for every 10 m of depth; at what depth was the diver working? Assume the temperature remains constant.

FIGURE P5.51

5.52. **Breath-Hold Diving** The world record for diving without supplemental air tanks is about 125 m, a depth at which the pressure is about 12.5 atm. If a diver's lungs contain 6.0 L of air at the surface of the water, what is this volume at a depth of 125 m?

5.53. Use the data in the first two columns of the table that follows to plot a graph of the volume (V) of 1.00 mol H_2 as a function of the reciprocal of its pressure ($1/P$) at 298 K; use the data in the second two columns to plot a graph of the volume (V) of 1.00 mol H_2 as a function of its absolute temperature (T) at $P = 1.00$ atm.

P (torr)	*V* (L)	*V* (L)	*T* (K)
100	186	7.88	96
120	155	3.94	48
240	77.5	1.97	24
380	48.9	0.79	9.6
500	37.2	0.39	4.8

a. How would these graphs differ if 1.00 mol Ar had been used instead of 1.00 mol H_2?
b. How would these graphs differ if 0.50 mol H_2 had been used instead of 1.00 mol H_2?
c. How would the second graph differ if the pressure of the gas had been 3.00 atm?

5.54. Use the data in the first two columns of the table below to plot a graph of the volume (V) of 1.00 mol He as a function of the reciprocal of its pressure ($1/P$) at 298 K; use the data in the second two columns to plot a graph of the volume (V) of 0.50 mol He as a function of its absolute temperature (T) at 1.00 atm.

P (atm)	*V* (L)	*V* (L)	*T* (K)
0.10	246.3	3.94	96
0.25	98.5	1.97	48
0.50	49.3	0.79	24
0.75	32.8	0.39	9.6
1.0	24.6	0.20	4.8

a. How would the first graph differ if 1.00 mol Ar had been used instead of 1.00 mol He?
b. How would the second graph differ if 1.00 mol He had been used instead of 0.50 mol He?
c. How would the second graph differ if the pressure of the gas had been 0.50 atm?

5.55. A cylinder with a piston (Figure P5.55) contains a sample of gas at 25°C. The piston moves to keep the pressure constant inside the cylinder. At what gas temperature would the piston move so that the volume inside the cylinder doubled?

FIGURE P5.55

5.56. The temperature of the gas in Problem 5.55 is reduced to a temperature at which the volume inside the cylinder has decreased by 25% from its initial volume at 25.0°C. What is the new temperature?

5.57. Calculate the volume of a 2.68 L sample of gas in the cylinder shown in Figure P5.55 after it is subjected to the following changes in conditions:
a. The gas is warmed from 250 K to a final temperature of 398 K at constant pressure.

b. The pressure is increased by 33% at constant temperature.
c. Ten percent of the gas leaks from the cylinder at constant temperature and pressure.

5.58. Calculate the temperature of a 5.6 L sample of gas in the cylinder in Figure P5.55 after it is subjected to the following changes in conditions:
a. The gas, at constant pressure, is cooled from 78°C to a temperature at which its volume is 4.3 L.
b. The external pressure is doubled at constant volume.
c. The number of molecules in the cylinder is increased by 15% at constant pressure and volume.

5.59. Which of the following actions would produce the greater increase in the volume of a gas sample: (a) lowering the pressure from 760 torr to 720 torr at constant temperature or (b) raising the temperature from 10°C to 40°C at constant pressure?

5.60. Which of the following actions would produce the greater increase in the volume of a gas sample: (a) doubling the amount of gas in the sample at constant temperature and pressure or (b) raising the temperature from 244°C to 1100°C at constant pressure?

***5.61.** What happens to the volume of gas in a cylinder with a movable piston under the following conditions?
a. Both the absolute temperature and the external pressure on the piston double.
b. The absolute temperature is halved and the external pressure on the piston doubles.
c. The absolute temperature increases by 75% and the external pressure on the piston increases by 50%.

*5.62. What happens to the pressure of a gas under the following conditions?
a. The absolute temperature is halved and the volume doubles.
b. Both the absolute temperature and the volume double.

5.63. Liquid Nitrogen–Powered Car Students at the University of North Texas and the University of Washington built a car propelled by compressed nitrogen gas. The gas was obtained by boiling liquid nitrogen stored in a 182 L tank. What volume of nitrogen is released at 0.927 atm of pressure and 25°C from a tank full of liquid nitrogen ($d = 0.808$ g/mL)?

5.64. **Fuel for Camp Stoves** The fuel canister for a portable camping stove (Figure P5.64) contains 190 g of pressurized butane (C_4H_{10}). What volume of butane does this represent at 25°C if it escapes into the air at 750 torr?

FIGURE P5.64

5.65. Temperature Effects on Bicycle Tires A bicycle racer inflates his tires to 7.1 atm on a warm autumn afternoon when temperatures reached 27°C. By morning the temperature has dropped to 5.0°C. What is the pressure in the tires if we assume that the volume of each tire does not change significantly?

*5.66. A balloon vendor at a street fair is using a tank of helium to fill her balloons. The tank has a volume of 145 L and a pressure of 136 atm at 25°C. After a while, she notices that the valve has not been closed properly and the pressure has dropped to 94 atm. How many moles of gas have been lost?

The Ideal Gas Law; Gases in Chemical Reactions

Concept Review

5.67. What is meant by standard temperature and pressure (STP)? What is the volume of 1 mole of an ideal gas at STP?

5.68. Which of the following are *not* characteristics of an ideal gas?
a. The molecules of gas have little volume compared with the volume that they occupy.
b. Its volume is independent of temperature.
c. The density of all ideal gases is the same.
d. Gas atoms or molecules do not interact with one another.

***5.69.** What does the slope represent in a graph of pressure as a function of $1/V$ at constant temperature for an ideal gas?

*5.70. How would the graph in Problem 5.69 change if we increased the temperature to a larger but constant value?

Problems

5.71. How many moles of air are there in a bicycle tire with a volume of 2.36 L if it has an internal pressure of 6.8 atm at 17.0°C?

5.72. At what temperature will 1.00 mole of an ideal gas in a 1.00 L container exert a pressure of
a. 1.00 atm?
b. 2.00 atm?
c. 0.75 atm?

5.73. Hyperbaric Oxygen Therapy Hyperbaric oxygen chambers are used to treat divers suffering from decompression sickness (the "bends") with pure oxygen at greater than atmospheric pressure. Other clinical uses include treatment of patients with thermal burns, necrotizing fasciitis, and CO poisoning. What is the pressure in a chamber with a volume of 2.36×10^3 L that contains 4635 g of $O_2(g)$ at a temperature of 298 K?

5.74. Hydrogen holds promise as an "environment friendly" fuel. How many grams of $H_2(g)$ are present in a 50.0 L fuel tank at $P = 25.0$ bar and $T = 20$°C?

5.75. A weather balloon with a volume of 200.0 L is launched at 20°C at sea level, where the atmospheric pressure is 1.00 atm.
a. What is the volume of the balloon at 20,000 m where atmospheric pressure is 63 torr and the temperature is 220 K?
b. What is the percent change in volume for the balloon in part (a)?

c. At even higher elevations (the stratosphere, up to 50,000 m), the temperature of the atmosphere increases to 270 K but the pressure decreases to 0.80 torr. What is the volume of the balloon under these conditions?
d. If the balloon is designed to rupture when the volume exceeds 400.0 L, will the balloon break under either of these sets of conditions?

5.76. A sealed, flexible foil bag of potato chips containing 0.500 L of air is carried from Boston (P_{atm} = 1.0 atm) to Denver (P_{atm}= 0.83 atm).
a. What would the volume of the bag be upon arrival in Denver?
b. What is the percent change in volume for the balloon in part (a)?
c. If the structural limitations of the bag allow for only a 10% expansion in the volume of the bag, what is the pressure in the bag?
d. If the bag is placed in checked luggage, the pressure and temperature in the hold will decrease considerably during flight to T = 210 K and P = 126 torr. Calculate the pressure in the bag during flight.

5.77. **Miners' Lamps** Before the development of reliable batteries, miners' lamps burned acetylene (C_2H_2) produced by the reaction of calcium carbide with water:

$$CaC_2(s) + H_2O(\ell) \rightarrow C_2H_2(g) + CaO(s)$$

A lamp uses 1.00 L of acetylene per hour at 1.00 atm pressure and 18°C.
a. How many moles of C_2H_2 are used per hour?
b. How many grams of calcium carbide must be in the lamp for a 4-h shift?

5.78. Acid precipitation dripping on limestone produces carbon dioxide by the following reaction:

$$CaCO_3(s) + 2\,H_3O^+(aq) \rightarrow Ca^{2+}(aq) + CO_2(g) + 3\,H_2O(\ell)$$

If 15.0 mL of CO_2 was produced at 25°C and 760 torr, then
a. how many moles of CO_2 were produced?
b. how many milligrams of $CaCO_3$ were consumed?

***5.79.** Air is about 78% nitrogen by volume and 21% oxygen. Pure nitrogen can be produced by the decomposition of ammonium dichromate:

$$(NH_4)_2Cr_2O_7(s) \rightarrow N_2(g) + Cr_2O_3(s) + 4\,H_2O(g)$$

Oxygen can be generated by the thermal decomposition of potassium chlorate:

$$2\,KClO_3(s) \rightarrow 2\,KCl(s) + 3\,O_2(g)$$

How many grams of ammonium dichromate and how many grams of potassium chlorate would be needed to make 200.0 L of "air" at 0.85 atm and 273 K?

*5.80. Nitrogen can be produced from sodium metal and potassium nitrate by the reaction:

$$10\,Na(s) + 2\,KNO_3(s) \rightarrow K_2O(s) + 5\,Na_2O(s) + N_2(g)$$

If we generate oxygen by the thermal decomposition of potassium chlorate,

$$2\,KClO_3(s) \rightarrow 2\,KCl(s) + 3\,O_2(g)$$

how many grams of potassium nitrate and how many grams of potassium chlorate would be needed to make 200.0 L of gas containing 160.0 L of N_2 and 40.0 L of O_2 at 1.00 atm and 290 K?

5.81. **Healthy Air for Sailors** The CO_2 that builds up in the air of a submerged submarine can be removed by reacting it with sodium peroxide:

$$2\,Na_2O_2(s) + 2\,CO_2(g) \rightarrow 2\,Na_2CO_3(s) + O_2(g)$$

If a sailor exhales 150.0 mL of CO_2 per minute at 20°C and 1.02 atm, how much sodium peroxide is needed per sailor in a 24-h period?

5.82. **Rescue Breathing Devices** Self-contained, self-rescue breathing devices, like the one shown in Figure P5.82, convert CO_2 into O_2 according to the following reaction:

$$4\,KO_2(s) + 2\,CO_2(g) \rightarrow 2\,K_2CO_3(s) + 3\,O_2(g)$$

How many grams of KO_2 are needed to produce 100.0 L of O_2 at 20°C and 1.00 atm?

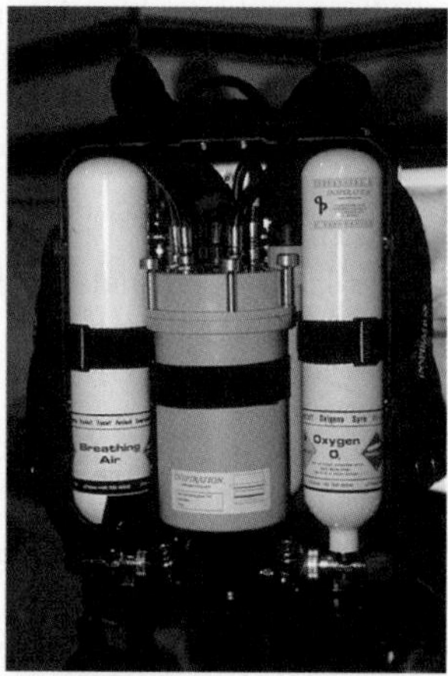

FIGURE P5.82

Gas Density

Concept Review

5.83. Do all gases at the same pressure and temperature have the same density? Explain your answer.

5.84. Birds and sailplanes take advantage of thermals (rising columns of warm air) to gain altitude with less effort than usual. Why does warm air rise?

5.85. How does the density of a gas sample change when (a) its pressure is increased and (b) its temperature is decreased?

5.86. Which will increase the density of a gas: doubling its temperature or doubling its pressure?

Problems

5.87. **Biological Effects of Radon Exposure** Radon is a naturally occurring radioactive gas found in the ground and in building materials. It is easily inhaled and emits α particles when it decays. Cumulative radon exposure is a significant risk factor for lung cancer.
a. Calculate the density of radon at 298 K and 1.00 atm of pressure.
b. Are radon concentrations likely to be greater in the basement or on the top floor of a building?

*5.88. Four empty balloons, each with a mass of 10.0 g, are inflated to a volume of 20.0 L. The first balloon contains He; the second, Ne; the third, CO_2; and the fourth, CO. If the density of air at 25°C and 1.00 atm is 0.00117 g/mL, which of the balloons float in this air?

5.89. A 150.0 mL flask contains 0.391 g of a volatile oxide of sulfur. The pressure in the flask is 750 torr and the temperature is 22°C. Is the gas SO_2 or SO_3?

5.90. A 100.0 mL flask contains 0.193 g of a volatile oxide of nitrogen. The pressure in the flask is 760 torr at 17°C. Is the gas NO, NO_2, or N_2O_5?

5.91. A 0.375 g sample of benzene vapor has a volume of 149 mL measured at 95.0°C and 740.0 torr. Calculate the molar mass of benzene.

5.92. Calculate the density of toluene vapor (molar mass 92 g/mol) at 1.00 atm pressure and 227.0°C.

Dalton's Law and Mixtures of Gases

Concept Review

5.93. What is meant by the *partial pressure* of a gas?

5.94. Can a barometer be used to measure just the partial pressure of oxygen in the atmosphere? Why or why not?

5.95. Which gas sample has the largest volume at 25°C and 1.00 atm pressure? (a) 0.500 mol of dry H_2; (b) 0.500 mol of dry N_2; (c) 0.500 mol of wet H_2 (i.e., H_2 collected over water)

5.96. Two identical balloons are filled to the same volume at the same pressure and temperature. One balloon is filled with air and the other with helium. Which balloon contains more particles (atoms and molecules)?

5.97. A mixture of gases has a partial pressure of $N_2 = 0.5$ atm and a partial pressure of $O_2 = 1.0$ atm. Which molecules experience more collisions: the molecules of nitrogen or the molecules of oxygen?

5.98. Why does the partial pressure of water increase as temperature increases?

Problems

5.99. What pressure is exerted by a gas mixture containing 2.00 g of H_2 and 7.00 g of N_2 at 273°C in a 10.0 L container? What is the contribution of N_2 to the total pressure?

5.100. A gas mixture contains 7.0 g of N_2, 2.0 g of H_2, and 16.0 g of CH_4. What is the mole fraction of H_2 in the mixture? Calculate the pressure of the gas mixture and the partial pressure of each constituent gas if the mixture is in a 1.00 L vessel at 0°C.

5.101. **The X-15 and Mach 6** On November 9, 1961, the Bell X-15 test aircraft exceeded Mach 6, or six times the speed of sound. A few weeks later it reached an elevation above 300,000 feet, effectively putting it in space. The combustion of ammonia and oxygen provided the fuel for the X-15.

a. Write a balanced chemical equation for the combustion of ammonia given that nitrogen ends up as nitrogen dioxide.

b. What ratio of partial pressures of ammonia and oxygen is needed for this reaction?

5.102. **Rocket Fuels** Before settling on hydrogen as the fuel of choice for space vehicles, several other fuels were explored, including hydrazine (N_2H_4) and pentaborane (B_5H_9). Both are gases under conditions in space.

a. Write balanced chemical equations for the combustion of hydrazine and pentaborane given that the products in addition to water are $NO_2(g)$ and $B_2O_3(s)$, respectively.

b. What ratios of partial pressures of hydrazine to oxygen and pentaborane to oxygen are needed for these reactions?

5.103. A sample of oxygen was collected over water at 25°C and 1.00 atm.

a. If the total sample volume was 0.480 L, how many moles of O_2 were collected?

*b. If the same volume of oxygen is collected over ethanol instead of water, does it contain the same number of moles of O_2?

5.104. Water and ethanol were removed from the O_2 samples in Problem 5.103.

a. What is the volume of the dry O_2 gas sample at 25°C and 1.00 atm?

b. What is the volume of the dry O_2 gas sample at 25°C and 1.00 atm if $P_{ethanol} = 49$ torr at 25°C?

5.105. The following reactions were carried out in sealed containers. Will the total pressure after each reaction is complete be greater than, less than, or equal to the total pressure before the reaction? Assume all reactants and products are gases at the same temperature.

a. $N_2O_5(g) + NO_2(g) \rightarrow 3\,NO(g) + 2\,O_2(g)$

b. $2\,SO_2(g) + O_2(g) \rightarrow 2\,SO_3(g)$

c. $C_3H_8(g) + 5\,O_2(g) \rightarrow 3\,CO_2(g) + 4\,H_2O(g)$

d. $4\,NH_3(g) + 5\,O_2(g) \rightarrow 4\,NO(g) + 6\,H_2O(g)$

5.106. The following reactions were carried out in a cylinder with a piston. If the external pressure is constant, in which reactions will the volume of the cylinder increase? Assume all reactants and products are gases at the same temperature.

a. $CH_4(g) + NH_3(g) \rightarrow HCN(g) + 3\,H_2(g)$

b. $H_2S(g) + 2\,O_2(g) \rightarrow H_2O(g) + SO_3(g)$

c. $H_2(g) + Cl_2(g) \rightarrow 2\,HCl(g)$

d. $2\,NO_2(g) \rightarrow 2\,NO(g) + O_2(g)$

5.107. **High-Altitude Mountaineering** Climbers use pure oxygen (Figure P5.107) near the summits of 8000-m peaks, where $P_{atm} = 0.35$ atm. How much more O_2 is there in a lung full of pure O_2 at this elevation than in a lung full of air at sea level?

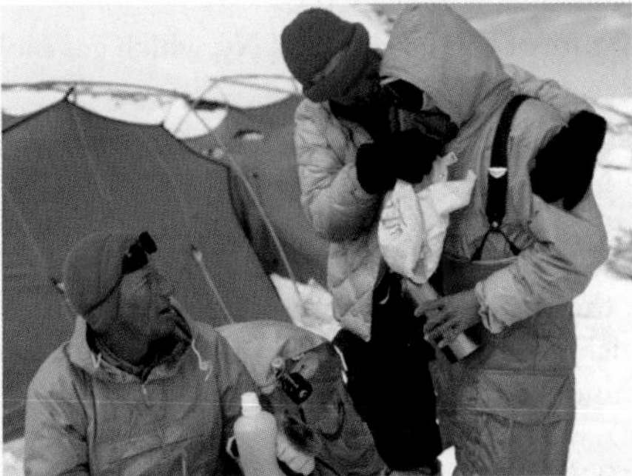

FIGURE P5.107

5.108. **Scuba Diving** A scuba diver is at a depth of 50 m, where the pressure is 5.0 atm. What should be the mole fraction of O_2 in the gas mixture the diver breathes to produce the same partial pressure of oxygen as the gas mixture at sea level?

5.109. Carbon monoxide at a pressure of 680 torr reacts completely with O_2 at a pressure of 340 torr in a sealed vessel to produce CO_2. What is the final pressure in the flask?

5.110. Ozone reacts completely with NO, producing NO_2 and O_2. A 10.0 L vessel is filled with 0.280 mol of NO and 0.280 mol of O_3 at 350 K. Find the partial pressure of each product and the total pressure in the flask at the end of the reaction.

***5.111.** **Ammonia Production** Ammonia is produced industrially from the reaction of hydrogen with nitrogen under pressure in a sealed reactor. What is the percent decrease in pressure of a sealed reaction vessel during the reaction between 3.60×10^3 mol of H_2 and 1.20×10^3 mol of N_2 if half of the N_2 is consumed?

*5.112. A mixture of 0.156 mol of C is reacted with 0.117 mol of O_2 in a sealed 10.0 L vessel at 500 K, producing a mixture of CO and CO_2. The total pressure is 0.640 atm. What is the partial pressure of CO?

The Kinetic Molecular Theory of Gases

Concept Review

5.113. What is meant by the *root-mean-square speed* of gas molecules?

5.114. Why don't all molecules in a sample of air move at exactly the same speed?

5.115. How does the root-mean-square speed of the molecules in a gas vary with (a) molar mass and (b) temperature?

5.116. Does pressure affect the root-mean-square speed of the molecules in a gas? Explain your answer.

5.117. How can Graham's law of effusion be used to determine the molar mass of an unknown gas?

5.118. Is the ratio of the rates of effusion of two gases the same as the ratio of their root-mean-square speeds?

5.119. Why is it possible to readily separate the group 18 elements by diffusion but not the group 15 elements?

5.120. If gas X diffuses faster in air than gas Y, is gas X also likely to effuse faster than gas Y?

Problems

5.121. Rank the gases SO_2, CO_2, and NO_2 in order of increasing root-mean-square speed at 0°C.

5.122. In a mixture of CH_4, NH_3, and N_2, which gas molecules are, on average, moving fastest?

5.123. At 286 K, three gases, A, B, and C, have root-mean-square speeds of 351 m/s, 433 m/s, and 472 m/s, respectively.
a. Which gas is O_2?
b. Are the other gases heavier or lighter than O_2?
c. Which one could be a different diatomic element?

5.124. Determine the root-mean-square speeds of CO_2, SO_2, and NO_2 molecules that have an average kinetic energy of 4.2×10^{-21} J per molecule.

***5.125.** The root-mean-square speed of helium gas at 300 K is 1.370×10^3 m/s. Sketch a graph of $u_{rms,He}$ versus T for $T = 300, 450, 600, 750$, and 900 K. Is the graph linear?

*5.126. The root-mean-square speed of N_2 gas at 300 K is 516.8 m/s. Why doesn't doubling the temperature also double u_{rms,N_2}?

5.127. Molecules made out of two atoms of deuterium (^2H, an isotope of hydrogen) have the formula D_2. What is the ratio of the root-mean-square speed of D_2 to that of H_2 at constant temperature?

5.128. **Enriching Uranium** The two isotopes of uranium, ^{238}U and ^{235}U, can be separated by diffusion of the corresponding UF_6 gases. What is the ratio of the root-mean-square speed of $^{238}UF_6$ to that of $^{235}UF_6$ at constant temperature?

5.129. A student measured the relative rates of effusion of carbon dioxide and propane (C_3H_8) and found that they effuse at the same rate. Did the student make a mistake?

*5.130. The density of Ne at 760 torr is exactly half its density at 2.00 atm at constant temperature. Is the root-mean-square speed of Ne at 2.00 atm half, twice, or the same as $u_{rms,Ne}$ at 760 torr?

5.131. An unknown pure gas X effuses at half the rate of effusion of O_2 at the same temperature and pressure. What is the molar mass of gas X?

5.132. **Decay Products of Uranium Minerals** Radon and helium are both by-products of the radioactive decay of uranium minerals. A fresh sample of carnotite, $K_2(UO_2)_2(VO_4)_2 \cdot 3\,H_2O$, is put on display in a museum. Calculate the relative rates of diffusion of helium and radon under fixed conditions of pressure and temperature. Which gas diffuses more rapidly through the display case?

5.133. **Isotope Use by Plants** During photosynthesis, green plants preferentially use $^{12}CO_2$ over $^{13}CO_2$ in making sugars, and food scientists can frequently determine the source of sugars used in foods on the basis of the ratio of ^{12}C to ^{13}C in a sample.
a. Calculate the relative rates of diffusion of $^{13}CO_2$ and $^{12}CO_2$.
b. Specify which gas diffuses faster.

5.134. At a fixed temperature, how much faster does NO effuse than NO_2?

5.135. One balloon was filled with H_2, another with He. The person responsible for filling them neglected to label them. After 24 h the volumes of both balloons had decreased but by different amounts. Which balloon contained hydrogen?

5.136. Compounds sensitive to oxygen are often manipulated in *glove boxes* that contain a pure nitrogen or pure argon atmosphere. A balloon filled with carbon monoxide was placed in a glove box. After 24 h, the volume of the balloon was unchanged. Did the glove box contain N_2 or Ar?

Real Gases

Concept Review

5.137. Rearrange the van der Waals equation to solve for *P*. Why is the pressure exerted by a real gas lower than the pressure for an ideal gas at the same temperature and volume?

5.138. Under what conditions is the pressure exerted by a real gas *less* than that predicted for an ideal gas?

5.139. The van der Waals equation contains two constants, *a* and *b*, that depend on the identity of the gas. Which gas, Ne or Kr, has a higher value of constant *a*?

5.140. Explain why the constant *a* in the van der Waals equation generally increases with the molar mass of the gas.

Problems

5.141. The graphs of PV/RT versus P (see Figure 5.40) for 1 mole of CH_4 and 1 mole of H_2 differ in how they deviate from ideal behavior. For which gas is the effect of the volume occupied by the gas molecules more important than the attractive forces between molecules?

5.142. Which noble gas is expected to deviate the most from ideal behavior in a graph of PV/RT versus P?

5.143. At high pressures, real gases do not behave ideally. (a) Use the van der Waals equation and data in the text to calculate the pressure exerted by 40.0 g of H_2 at 20°C in a 1.00 L container. (b) Repeat the calculation, assuming that the gas behaves like an ideal gas. (c) Explain the difference between your answers to parts (a) and (b) using kinetic molecular theory.

5.144. (a) Calculate the pressure exerted by 1.00 mol of CO_2 in a 1.00 L vessel at 300 K, assuming that the gas behaves ideally. (b) Repeat the calculation using the van der Waals equation. (c) Explain the difference between your answers to parts (a) and (b) using kinetic molecular theory.

Additional Problems

5.145. The volume of a sample of propane gas at 12.5 atm is 10.6 L. What volume does the gas occupy if the pressure is reduced to 1.05 atm and the temperature remains constant?

5.146. A 22.4 L sample of hydrogen chloride gas is heated from 15°C to 78°C. What volume does it occupy at the higher temperature if the pressure remains constant?

5.147. A gas cylinder in the back of an open truck experiences a temperature change from 12°F to 143°F as it is driven from high in the Rocky Mountains of Colorado to Death Valley, CA. If the pressure gauge on the tank reads 2200 psi in Colorado, what will the gauge read in Death Valley to the nearest psi?

5.148. The temperature of a quantity of methane gas at a pressure of 761 torr is 18.6°C. Predict the temperature of the gas if the pressure is reduced to 355 torr while the volume remains constant.

5.149. A souvenir soccer ball is partially deflated and then put into a suitcase. At a pressure of 0.947 atm and a temperature of 27°C, the ball has a volume of 1.034 L. What volume does it occupy during an airplane flight if the pressure in the baggage compartment is 0.235 atm and the temperature is −35°C?

5.150. **Asthma Therapy** A gas mixture used experimentally for asthma treatments contains 17.5 mol of helium for every 0.938 mol of oxygen. What is the mole fraction of oxygen in the mixture?

5.151. **Blood Pressure** A typical blood pressure in a resting adult is "120 over 80," meaning 120 mmHg with each beat of the heart and 80 mmHg of pressure between heartbeats. Express these pressures in the following units: (a) torr; (b) atm; (c) bar; (d) kPa.

*5.152. **Scuba Tanks** A popular scuba tank is the "aluminum 80," so named because it can deliver 80 cubic feet of air at "normal" temperature (72°F) and pressure (1.00 atm, 14.7 psi) when filled with air at a pressure of 3000 psi. A particular aluminum 80 tank has a mass of 15 kg when empty. What is its mass when filled with air at 3000 psi?

5.153. The flame produced by the burner of a gas (propane, C_3H_8) grill is a pale blue color when enough air mixes with the propane to burn it completely. For every gram of propane that flows through the burner, what volume of air is needed to burn it completely? Assume that the temperature of the burner is 200°C, the pressure is 1.00 atm, and the mole fraction of O_2 in air is 0.21.

5.154. Which noble gas effuses at about half the effusion rate of O_2? Are there any that effuse at twice the effusion rate of O_2?

5.155. **Anesthesia** A common anesthesia gas is halothane, with the structure shown in Figure P5.155. Liquid halothane boils at 50.2°C and 1.00 atm. If halothane behaved as an ideal gas, what volume would 10.0 mL of liquid halothane (d = 1.87 g/mL) occupy at 60°C and 1.00 atm of pressure? What is the density of halothane vapor at 55°C and 1.00 atm of pressure?

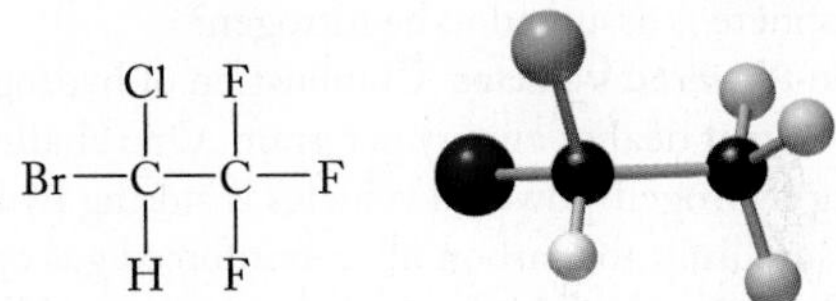

FIGURE P5.155

5.156. One cotton ball soaked in ammonia and another soaked in hydrochloric acid were placed at opposite ends of a 1.00 m glass tube (Figure P5.156). The vapors diffused toward the middle of the tube and formed a white ring of ammonium chloride where they met.

a. Write the chemical equation for this reaction.
b. Will the ammonium chloride ring form closer to the end of the tube with ammonia or the end with hydrochloric acid? Explain your answer.
c. Calculate the distance from the ammonia end to the position of the ammonium chloride ring.

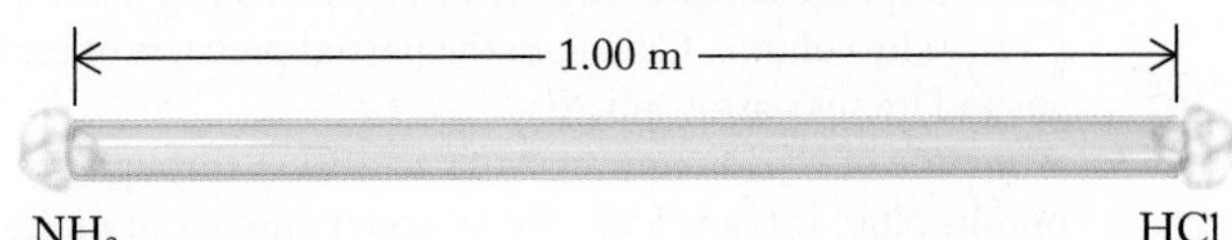

FIGURE P5.156

***5.157.** The same apparatus described in Problem 5.156 was used in another series of experiments. A cotton ball soaked in either hydrochloric acid (HCl) or acetic acid (CH_3COOH) was placed at one end. Another cotton ball soaked in one of three amines—CH_3NH_2, $(CH_3)_2NH$, or $(CH_3)_3N$—was placed in the other end (Figure P5.157).

a. In one combination of acid and amine, a white ring was observed almost exactly halfway between the two ends. Which acid and which amine were used?
b. Which combination of acid and amine would produce a ring closest to the amine end of the tube?
c. Do any two of the six combinations result in the formation of product at the same position in the ring? Assume measurements can be made to the nearest centimeter.

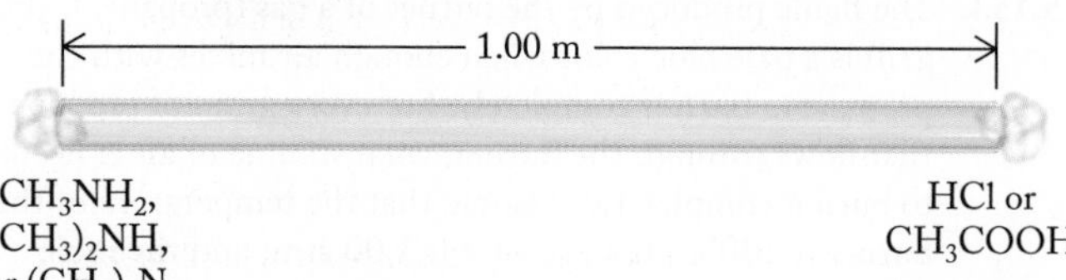

FIGURE P5.157

5.158. **Early Earth Atmosphere** Life as we know it on Earth depends on a constant atmospheric composition of 21% O_2 and 79% N_2 by volume. Over the past ~3 billion years, the amount of oxygen in the atmosphere has ranged from as low as 15% to as high as 35% by volume.

a. What partial pressures of oxygen correspond to this range?
b. How much does the density of air change over this range of oxygen concentrations if the balance of the atmosphere is assumed to be nitrogen?

5.159. Hydrogen-Powered Vehicles Combustion of hydrogen supplies a great deal of energy per gram. One challenge in designing hydrogen-powered vehicles is storing hydrogen. The pressure limit for carbon fiber–reinforced gas cylinders is 10,000 psi (1 atm = 14.7 psi). A fuel tank capable of holding up to 10 kg of H_2 is required to meet a reasonable driving range for a vehicle. If we limit the volume of the fuel tank to 400 L, what pressure must the hydrogen be under at 300 K? Does your answer conform to the safety requirements?

5.160. A sample of 11.4 L of an ideal gas at 25.0°C and 735 torr is compressed and heated so that the volume is 7.9 L and the temperature is 72.0°C. What is the pressure in the container?

5.161. A sample of a gas has a mass of 2.889 g and a volume of 940 mL at 735 torr and 31°C. What is its molar mass?

5.162. Uranus has a total atmospheric pressure of 130 kPa and its atmosphere consists of 83% $H_2(g)$, 15% $He(g)$, and 2% $CH_4(g)$ by volume. Calculate the partial pressure of each gas in Uranus's atmosphere.

5.163. A sample of $N_2(g)$ requires 240 s to diffuse through a porous plug. It takes 530 s for an equal number of moles of an unknown gas X to diffuse through the plug under the same conditions of temperature and pressure. What is the molar mass of gas X?

5.164. The rate of effusion of an unknown gas is 0.10 m/s and the rate of effusion of $SO_3(g)$ is 0.052 m/s under identical experimental conditions. What is the molar mass of the unknown gas?

5.165. Derive an equation that describes the relationship between root-mean-square speed, u_{rms}, of a gas and its density.

5.166. Derive an equation that expresses the ratio of the densities (d_1 and d_2) of a gas under two different combinations of temperature and pressure, (T_1, P_1) and (T_2, P_2).

5.167. Denitrification in the Environment In some aquatic ecosystems, nitrate (NO_3^-) is converted to nitrite (NO_2^-), which then decomposes to nitrogen and water. As an example of this second reaction, consider the decomposition of ammonium nitrite:

$$NH_4NO_2(aq) \rightarrow N_2(g) + 2\,H_2O(\ell)$$

What would be the change in pressure in a sealed 10.0 L vessel caused by the formation of N_2 gas when the ammonium nitrite in 1.00 L of 1.0 *M* NH_4NO_2 decomposes at 25°C?

*5.168. When sulfur dioxide bubbles through a solution containing nitrite, chemical reactions that produce gaseous N_2O and NO may occur.

a. How much faster, on average, would NO molecules be moving than N_2O molecules in such a reaction mixture?
b. If these two nitrogen oxides were to be separated based on differences in their rates of effusion, would unreacted SO_2 interfere with the separation? Explain your answer.

***5.169. Using Wetlands to Treat Agricultural Waste** Wetlands can play a significant role in removing fertilizer residues from rain runoff and groundwater; one way they do this is through denitrification, which converts nitrate ions to nitrogen gas:

$$2\,NO_3^-(aq) + 5\,CO(g) + 2\,H^+(aq) \rightarrow N_2(g) + H_2O(\ell) + 5\,CO_2(g)$$

Suppose 200.0 g of NO_3^- flows into a swamp each day. What volume of N_2 would be produced at 17°C and 1.00 atm if the denitrification process were complete? What volume of CO_2 would be produced? Suppose the gas mixture produced by the decomposition reaction is trapped in a container at 17°C; what is the density of the mixture if we assume that P_{total} = 1.00 atm?

5.170. Ammonium nitrate decomposes on heating. The products depend on the reaction temperature:

$$NH_4NO_3(s) \xrightarrow{>300°C} N_2(g) + \tfrac{1}{2}\,O_2(g) + 2\,H_2O(g)$$

$$NH_4NO_3(s) \xrightarrow{200°C\text{–}260°C} N_2O(g) + 2\,H_2O(g)$$

A sample of NH_4NO_3 decomposes at an unspecified temperature, and the resulting gases are collected over water at 20°C.

a. Without completing a calculation, predict whether the volume of gases collected can be used to distinguish between the two reaction pathways. Explain your answer.
*b. The gas produced during the thermal decomposition of 0.256 g of NH_4NO_3 displaces 79 mL of water at 20°C and 760 torr of atmospheric pressure. Is the gas N_2O or a mixture of N_2 and O_2?

5.171. **Hydrogen-Producing Enzymes** Generating hydrogen from water or methane is energy intensive. A non-natural enzymatic process has been developed that produces 12 moles of hydrogen per mole of glucose by the reaction:

$$C_6H_{12}O_6(aq) + 6\ H_2O(\ell) \rightarrow 12\ H_2(g) + 6\ CO_2(g)$$

What volume of hydrogen could be produced from 256 g of glucose at STP?

*5.172. **Oxygen Generators** Several devices are available for generating oxygen. Breathing pure oxygen for any length of time, however, can be dangerous to the human body, so a better rebreathing apparatus would be one that also produces nitrogen. One possible reaction that produces both N_2 and O_2 is the thermal decomposition of ammonium nitrate:

$$2\ NH_4NO_3(s) \rightarrow 2\ N_2(g) + O_2(g) + 4\ H_2O(g)$$

a. The respiratory rate at rest for an average, healthy adult is 12 breaths per minute. If the average breath takes in 500 mL of gas into our lungs, what mass in grams of NH_4NO_3 is needed to satisfy these needs for 1 h if $P = 1.00$ atm and $T = 37°C$?

b. What are the partial pressures of N_2 and O_2 in this mixture?

5.173. **Tropical Storms** The severity of a tropical storm is related to the depressed atmospheric pressure at its center. In August 1985, Typhoon Odessa reached maximum winds of about 90 mi/h and the pressure was 40 mbar lower at the center than normal atmospheric pressure. In contrast, the central pressure of Hurricane Andrew (Figure P5.173) was 90 mbar lower than its surroundings when it hit southern Florida with winds as high as 266 km/h (165 mi/h). If a small weather balloon with a volume of 50.0 L at a pressure of 1.0 atm was deployed above the center of Andrew, what was the volume of the balloon when it reached the surface of the ocean?

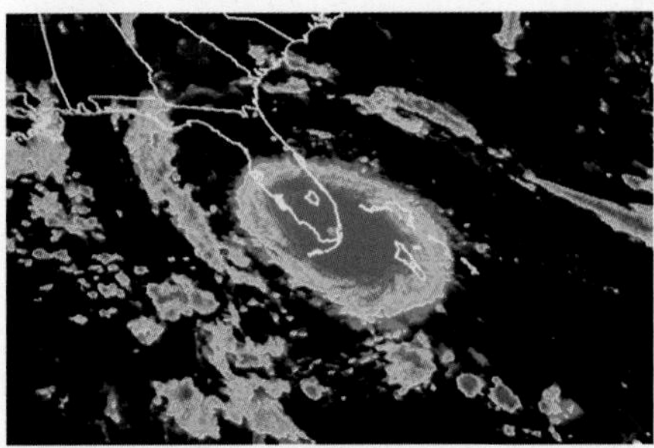

FIGURE P5.173

5.174. **Manufacturing a Proper Air Bag** The overall reaction in an automobile air bag is

$$20\ NaN_3(s) + 6\ SiO_2(s) + 4\ KNO_3(s) \rightarrow 32\ N_2(g) + 5\ Na_4SiO_4(s) + K_4SiO_4(s)$$

Calculate how many grams of sodium azide (NaN_3) are needed to inflate a 40 × 40 × 20 cm bag to a pressure of 1.25 atm at a temperature of 20°C. How much more sodium azide is needed if the air bag must produce the same pressure at 10°C?

5.175. Suppose 0.200 L of O_2 is collected over water at 25.0°C and the atmospheric pressure is 750.0 torr. The vapor pressure of water at 25.0°C is 24.0 torr. How many moles of O_2 have been collected?

5.176. Use Dalton's law of partial pressures to calculate the mole fraction of water vapor in the gas sample collected over water in Problem 5.175.

6

Thermochemistry

Energy Changes in Chemical Reactions

ENERGY PRODUCTION AND CLIMATE CHANGE Melting glaciers in polar regions are among the environmental impacts of climate change linked to the combustion of fossil fuels.

PARTICULATE **REVIEW**

Neutralization Reactions Revisited

In Chapter 6 we consider the energy changes that occur during reactions such as the combustion reactions from Chapter 3 and neutralization reactions from Chapter 4. Here we see the key molecules and ions involved in the titration of a sample containing hydrochloric acid with a standard solution of sodium hydroxide.

- Which of the illustrated particles are present in the buret?
- Which of the particles are present in the flask before any sodium hydroxide has been added?
- Which particles are present after all the acid has been neutralized?
- Write a balanced net ionic equation for the reaction.

(Review Sections 4.5 and 4.6 if you need help.)

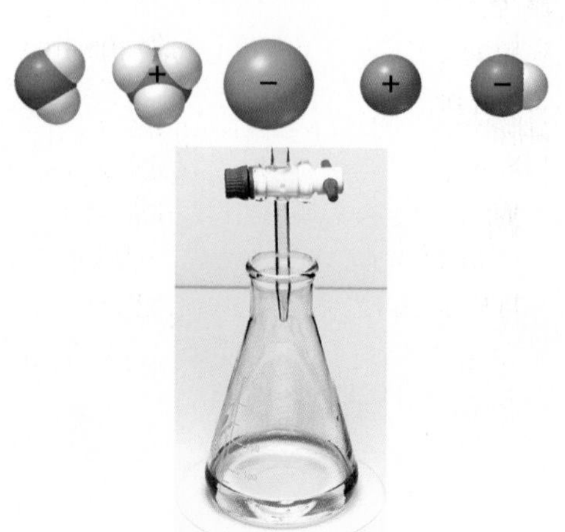

(Answers to Particulate Review questions are in the back of the book.)

PARTICULATE **PREVIEW**

Breaking Bonds and Energy

Ultraviolet rays (UV rays) from the Sun can cause molecules of ozone to break apart into oxygen molecules and oxygen atoms according to the chemical reaction depicted here. As you read Chapter 6, look for ideas that will help you answer these questions:

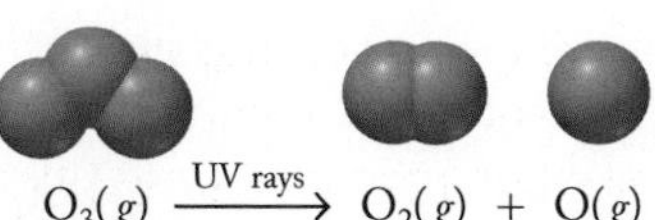

- When the bonds in O_3 break, is this (a) accompanied by a release of energy or (b) due to the absorption of energy?
- Is energy absorbed or released during this reaction?
- Would energy be absorbed or released if the reaction ran in reverse: $O_2(g) + O(g) \rightarrow O_3(g)$?

Learning Outcomes

LO1 Explain kinetic energy and potential energy at the atomic and molecular level
Sample Exercise 6.1

LO2 Differentiate between a system and its surroundings and identify familiar endothermic and exothermic processes
Sample Exercise 6.2

LO3 Calculate changes in the internal energy of a system
Sample Exercises 6.3, 6.4

LO4 Calculate the energy changes associated with physical and chemical processes
Sample Exercises 6.5, 6.6, 6.7, 6.8, 6.9

LO5 Calculate thermochemical values by using data from calorimetry experiments
Sample Exercises 6.10, 6.11

LO6 Calculate enthalpies of reaction
Sample Exercises 6.12, 6.13, 6.15

LO7 Recognize and write equations for formation reactions
Sample Exercise 6.14

LO8 Calculate and compare fuel and food values and fuel densities
Sample Exercises 6.16, 6.17

6.1 Sunlight Unwinding

All physical changes of matter, such as ponds freezing in winter and thawing in spring, involve changes in energy, as do the chemical reactions we studied in Chapter 4. Energy—to power an automobile, heat a home, or support life—may seem like an abstract idea because energy has no mass and no volume. However, we see its effects very clearly when matter changes from one state to another—as sunlight melts snow, or a gas flame boils water and converts it into steam—and when energy itself is converted from one form into another, as when the chemical energy of gasoline is converted into mechanical energy to move a car. Part of the energy in the gasoline contributes nothing to moving the vehicle and is lost to the surroundings as heat. The sum of the energy used to move the vehicle and the energy lost as heat equals the energy released during the combustion of the gasoline. That is, energy is neither created nor destroyed during chemical reactions.

We can roast marshmallows by using energy from a campfire, but where does that energy come from? R. Buckminster Fuller (1895–1983), a 20th-century architect, inventor, and futurist, described a burning log like this: trees gather the energy in sunlight and combine it with water and carbon dioxide to make the molecules that compose wood. When the wood is burned, the chemical products are carbon dioxide and water, and the fire is, as Fuller said, "All that sunlight unwinding." The sunlight unwinding is the net release of chemical energy as bonds break between atoms in the molecules of the wood and bonds form between atoms in the molecules of carbon dioxide and water. Through the transforming power of green plants, sunlight is the source of the chemical energy stored in all the substances we consume as food and fuel.

CONNECTION We introduced energy as the capacity to do work in Chapter 1.

Nearly every chemical reaction, from the combustion of fuels to neutralization reactions to the dissolution of salts, involves energy as either a product or a reactant. For example, the reaction in which hydrogen combines with oxygen to form water releases energy:

$$2\ H_2(g) + O_2(g) \rightarrow 2\ H_2O(g) + \text{energy}$$

The energy derived from the reaction between hydrogen and oxygen may be converted into motion, as when a spacecraft fueled by hydrogen lifts off. The study of energy and its transformations from one form to another is called **thermodynamics**. The part of thermodynamics that deals with changes in energy accompanying chemical reactions is known as **thermochemistry**.

When we put an ice cube, initially at −18°C—the typical temperature of a freezer—into room-temperature water (25°C), the ice cube melts and the water cools because energy moves from the room-temperature water into the colder ice cube and the 0°C water it produces as it melts. The process by which energy moves from warmer to cooler objects is called *heat transfer*, or more generally *energy transfer*. The difference in temperature defines the direction of energy flow when two objects come into contact: energy transfer in the form of **heat** always flows from the hotter object to the colder one (**Figure 6.1**). The ice cube changes state from solid to liquid as energy is transferred to it from the liquid water. The warmer water remains in the liquid state, but its temperature decreases as energy from it is transferred to the ice cube and its melt water. Ultimately, the warmer and colder water reach the same temperature, which is higher than the initial temperature of the ice cube but lower than the initial temperature of the water. At this point, **thermal equilibrium** has been reached, and no further energy transfer occurs.

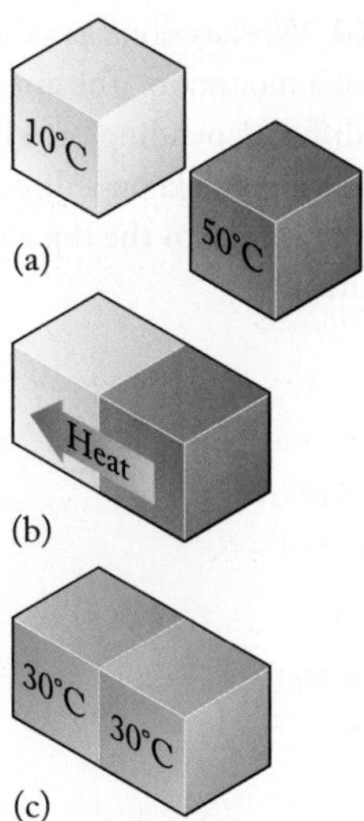

FIGURE 6.1 (a) Two identical blocks at different temperatures are brought into contact. (b) Heat is transferred from the block at higher temperature to the block at lower temperature until thermal equilibrium (equal temperature) (c) is reached.

How do we measure the amount of energy involved in physical and chemical processes? We cannot directly measure the amount of energy in a chemical reaction, but by measuring changes in the temperature, we can relate *the change* in energy to the identities and amounts of reactants and products involved in changes of state and in chemical reactions. Studying energy changes gives us important insights into the way nature works and how human activities affect our world.

6.2 Forms of Energy

The energy produced by a chemical reaction can be used to do **work**, can be transferred to an object to raise its temperature, or both. Energy that does work includes electrical, mechanical, light, and sound energy. Whatever the form, energy used to do work causes motion: the location or shape of an object changes when an energy source does work on the object. For example, the energy derived from the reaction between hydrogen and oxygen is converted into motion when a hydrogen-fueled spacecraft blasts off.

Work, Potential Energy, and Kinetic Energy

How does the combustion of hydrogen lead to the work done in lifting a spacecraft into orbit? Let's start with the classical view of work and energy from physics and consider a skier poised on top of a steep slope. In the physical sciences, work (w) is done whenever a force (F) moves an object through a distance (d). The amount of mechanical work done is

$$w = F \times d \tag{6.1}$$

How does Equation 6.1 relate to skiers ascending a mountain (**Figure 6.2**)? The work (w) done by the gondola on a skier equals the distance of the ride (d) times the force (F) needed to overcome gravity and transport the skier up the mountain. Some of the work done is stored in the skier as **potential energy (PE)**, which is the energy an object has because of its position or composition. The skier's potential energy depends on her mass (m), her height (h) above a zero-energy

thermodynamics the study of energy and its transformations.

thermochemistry the study of the energy changes that occur during chemical reactions.

heat the energy transferred between objects because of a difference in their temperatures.

thermal equilibrium a condition in which temperature is uniform throughout a material and no energy flows from one point to another.

work a form of energy—specifically, the energy required to move an object through a given distance.

potential energy (PE) the energy stored in an object because of its position or composition.

FIGURE 6.2 Work is done as skiers ascend to the top of a mountain. The amount of work may differ, depending on whether the skiers (1) ride a gondola on a direct route to the top or (2) hike to the top along a winding path.

reference point—in this case the bottom of the ski slope—and acceleration due to the force of gravity (g):

$$PE = m \times g \times h \quad (6.2)$$

The kind of potential energy in Equation 6.2 is called *gravitational* potential energy, which is the energy an object has because of its position in a gravitational field. The equation tells us that gravitational PE depends on the position of the skier and not on the path (**Figure 6.3**) she took to reach that position, nor does it depend on the quantity of work done to get her there. This means that gravitational PE, like all forms of potential energy, is a **state function**, which means it is independent of the pathway followed to acquire the potential energy.

Now let's consider the gravitational potential energy of a skier standing at the top of the ski jump in **Figure 6.4a** and how that energy is converted into **kinetic energy (KE; Figure 6.4b)** from the start of the run and until he stops at the bottom of the hill. The ski jumper's total energy at any position on the hill

state function a property of an entity based solely on its chemical or physical state or both, but not on how it achieved that state.

kinetic energy the energy of an object in motion due to its mass (m) and its speed (u): $KE = \frac{1}{2}mu^2$.

law of conservation of energy the principle that energy cannot be created or destroyed but can be converted from one form into another.

CHEMTOUR

State Functions and Path Functions

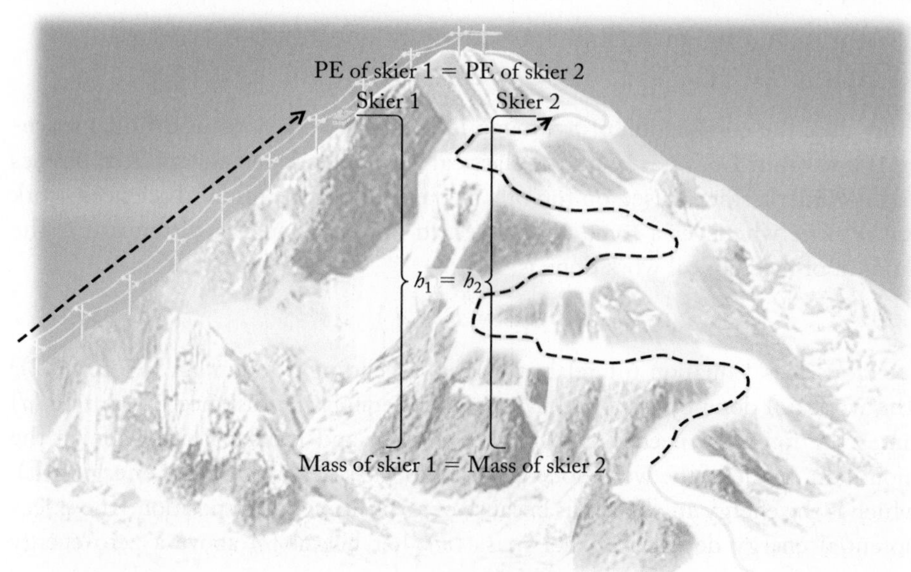

FIGURE 6.3 The potential energy of a skier depends only on the skier's mass and height above the base of the slope. If two skiers are at the same height ($h_1 = h_2$) and both skiers have the same mass, then they have the same potential energy, no matter how each skier got to that height.

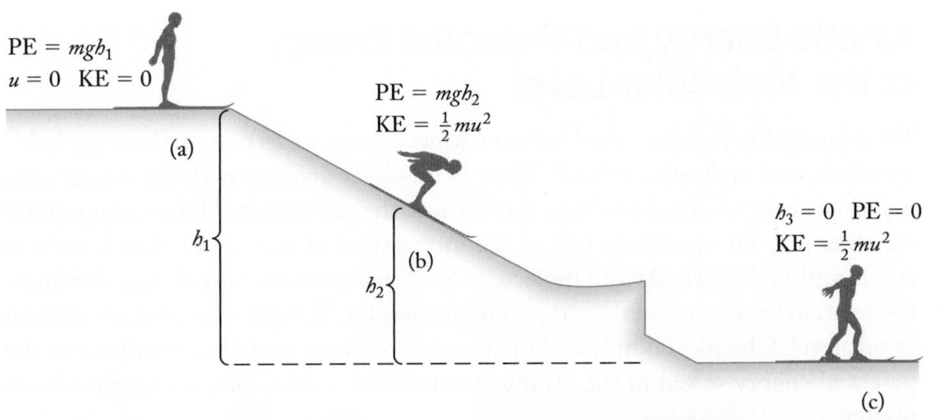

FIGURE 6.4 (a) A ski jumper at the top of a ski jump has gravitational potential energy (PE) due to his position (h_1) above the bottom of the slope, his mass (m), and the acceleration due to gravity (g): PE = mgh_1. (b) During his run, the ski jumper's potential energy is converted into kinetic energy: KE = $\frac{1}{2}mu^2$. While he moves down the slope, he has both KE and PE. (c) At the end of the run, the ski jumper's gravitational PE is 0. His KE decreases from its maximum value to 0 as he slows to a stop at the bottom of the hill.

is the sum of his potential and kinetic energies. The jumper's kinetic energy is proportional to his mass (m) times the square of his speed (u) as described by Equation 5.27:

$$\text{KE} = \tfrac{1}{2}mu^2 \tag{5.27}$$

A 100 kg (220 lb) jumper has twice the gravitational PE as a 50 kg (110 lb) jumper at the top of the run, and, if moving at the same speed as a lighter jumper at the bottom of the jump, has twice the kinetic energy. According to the **law of conservation of energy**, energy cannot be created or destroyed. However, it can be converted from one form to another, as the example of the two ski jumpers illustrates.

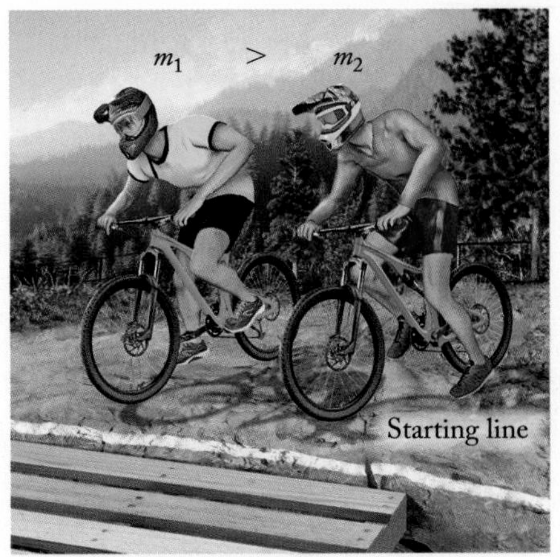

FIGURE 6.5 Two mountain bikers of different mass are at the same position at the start of a race with respect to the bottom of the hill.

CONCEPT TEST

Sports competitions such as the X-Games include downhill mountain bike races. Two mountain bikers with masses m_1 and m_2 ($m_1 > m_2$) are poised at the starting gate of a race course (**Figure 6.5**). Is the potential energy of rider 1 the same as that of rider 2? If the energies are different, which rider has more potential energy?

(Answers to Concept Tests are in the back of the book.)

CONCEPT TEST

Two mountain bikers with masses m_1 and m_2, where $m_1 > m_2$, go past the same elevation on parallel race courses at the same time (**Figure 6.6**). At that moment, is the potential energy of rider 1 more than, less than, or equal to the PE of rider 2? If the two are moving at the same speed, which has the greater kinetic energy?

FIGURE 6.6 The same two mountain bikers in Figure 6.5 reach the finish line at the same time.

Kinetic Energy and Potential Energy at the Molecular Level

The relationship just described between kinetic energy and potential energy holds for atoms and molecules as well. There is no direct analogy with the kinetic and potential energies of a skier, however, because the gravitational forces that control the skier play no significant role in the interactions of very small objects such as atoms and molecules. At the molecular level, temperature and charge dominate the relation between kinetic and potential energies. Temperature governs motion at this level. Chemical bonds and interactions between particles contribute to the potential energy stored in the arrangements of the atoms, ions, and molecules in matter.

The kinetic energy of an atomic particle depends on its mass and speed, just as with macroscopic objects. However, because the particle's speed depends on temperature, its kinetic energy does, too. As the temperature of a population of particles increases, the average kinetic energy of the particles also increases. Consider, for example, how the molecules in the vapor phase above a liquid behave at different temperatures. We pick the gas phase because the molecules in a gas at normal pressures are widely separated and behave essentially independently of one another, and they all behave the same way regardless of their identities. If we have two samples of water vapor (molecular mass 18.02 u) at room temperature, the two populations of H_2O molecules have the same average kinetic energy. The average speeds of the molecules are the same because their masses are identical. If we increase the temperature of one sample, that population of H_2O molecules acquires

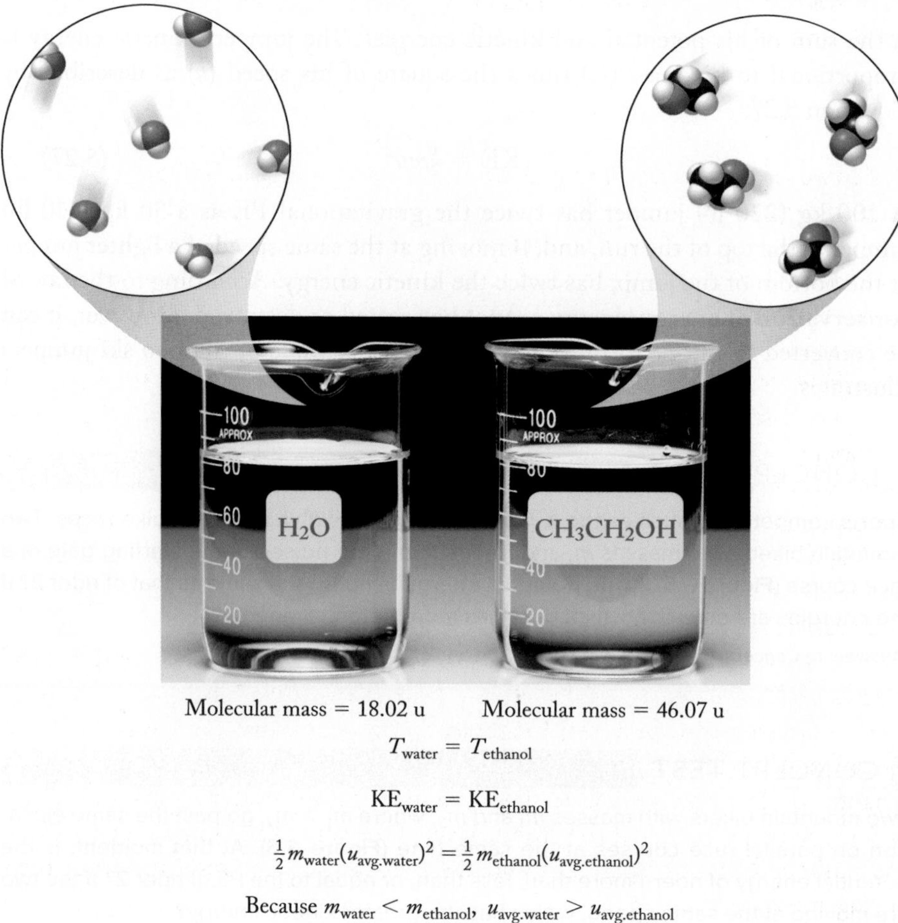

$$T_{\text{water}} = T_{\text{ethanol}}$$

$$\text{KE}_{\text{water}} = \text{KE}_{\text{ethanol}}$$

$$\tfrac{1}{2} m_{\text{water}}(u_{\text{avg.water}})^2 = \tfrac{1}{2} m_{\text{ethanol}}(u_{\text{avg.ethanol}})^2$$

$$\text{Because } m_{\text{water}} < m_{\text{ethanol}},\ u_{\text{avg.water}} > u_{\text{avg.ethanol}}$$

FIGURE 6.7 Two populations of gas-phase molecules have the same temperature and therefore the same average kinetic energy. Because ethanol molecules have a greater mass than water molecules, the average speed of the water molecules in the water vapor above the liquid water is greater than the average speed of the gas-phase ethanol molecules above the liquid ethanol.

a higher average kinetic energy and the average speed of its molecules increases. An equivalent population of ethanol molecules (molecular mass 46.07 u) in the gas phase at room temperature has the same average kinetic energy as the water molecules at room temperature, but the average speed of the ethanol molecules is lower because their mass is higher (**Figure 6.7**).

The kinetic energy associated with the total random motion of molecules is called **thermal energy**, and the thermal energy of a given sample of matter is proportional to the temperature of the sample. It also depends on the number of particles in the sample. The water in a swimming pool and in a cup of water taken from the pool have the same temperature, so their molecules have the same average kinetic energy. However, the water in the pool has considerably more thermal energy than the water in the cup simply because the pool contains many more molecules of water.

thermal energy the kinetic energy of atoms, ions, and molecules.

electrostatic potential energy (E_{el}) the energy a particle has because of its electrostatic charge and its position with respect to another particle; it is directly proportional to the product of the charges of the particles and inversely proportional to the distance between them.

CONCEPT TEST

If we heat a cup of water from a swimming pool almost to the boiling point, will its thermal energy be more than, less than, or the same as the thermal energy of all the water in the pool?

An important form of potential energy at the atomic level arises from electrostatic interactions between charged particles. Just as the potential energy of skiers is determined by their distances above the bottom of the slope, the **electrostatic potential energy (E_{el})** of charged particles is related to the distance between them. The magnitude of this electrostatic potential energy, also known as *coulombic interaction*, is directly proportional to the charges (Q_1 and Q_2) on the particles and is inversely proportional to the distance (d) between them:

$$E_{el} \propto \frac{Q_1 \times Q_2}{d} \qquad (6.3)$$

If the two charges in Equation 6.3 are either both positive or both negative, their product is positive, E_{el} is positive, and the particles repel each other. If one particle is positive and the other negative, E_{el} is negative, and the particles attract each other. Equation 6.3 is called Coulomb's law because it traces back to the work of French engineer Charles Augustin de Coulomb (1736–1806), who first measured the interactions between charged particles. The change in E_{el} that two oppositely charged particles experience as they approach each other is illustrated in **Figure 6.8**. When the ions are far apart (the right end of the curve), there is no attraction between them and $E_{el} = 0$. As the ions approach each other, E_{el} decreases as electrostatic attractions increase, reaching an energy minimum that corresponds to maximum stability. If the ions are pushed even closer together (to the left of the energy minimum), E_{el} rises and the ion pair becomes less stable as decreasing distance between their positively charged nuclei creates a growing electrostatic repulsion between the ions.

Ions are not the only particles that experience coulombic interactions. Neutral species such as water molecules attract each other as well, because there is an uneven electron distribution throughout the molecule that produces regions of partial positive and negative charges. We will explore this topic in detail in Chapter 10. Whether we are dealing with matter composed of atoms, molecules, or ions, the total energy at the atomic level is the sum of the kinetic energy from the random motion of particles and the potential energy resulting from their arrangement.

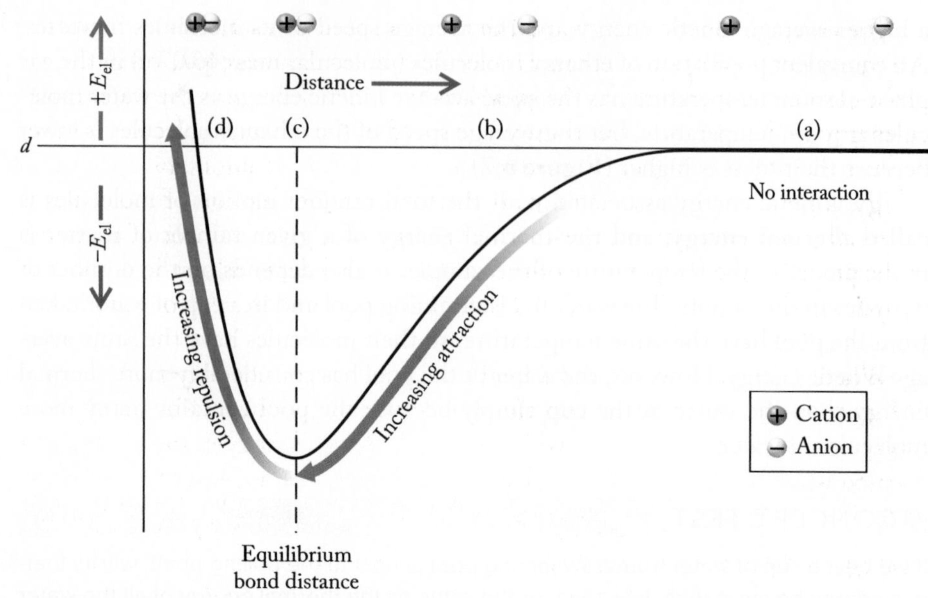

FIGURE 6.8 Ionic interaction. (a) A positive ion and a negative ion are so far apart (*d* is large) that they do not interact at all. (b) As the ions move closer together (*d* decreases), the electrostatic potential energy between them becomes more negative. (c) At this distance, the attraction between them produces an arrangement that is the most favorable energetically because the ions have the lowest electrostatic potential energy. This is also the equilibrium bond distance. (d) If the ions are forced even closer together, the electrostatic potential energy increases because of increasing repulsions.

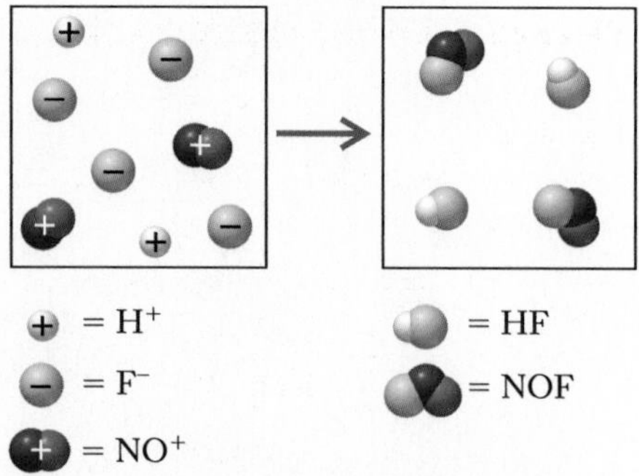

FIGURE 6.9 Protons and nitrosonium ions react with fluoride ions to form HF and NOF, respectively.

SAMPLE EXERCISE 6.1 Kinetic Energy of Gas-Phase Ions **LO1**

Protons (H^+) and nitrosonium ions (NO^+) react with fluoride ions (F^-) in the gas phase under laboratory conditions (constant temperature and pressure), forming HF and NOF, respectively, as shown in **Figure 6.9**. (a) At a given temperature, which ion has the greatest kinetic energy? (b) Which cation is moving at the higher average speed?

Collect and Organize We are asked to predict which of three gas-phase particles has the greatest kinetic energy, and which of two positively charged particles has the higher average speed.

Analyze Particles of gas at the same temperature have the same kinetic energy (KE), which is related to their average speeds (u) by Equation 5.27, $KE = \frac{1}{2}mu^2$. For particles of equal kinetic energy, this relationship indicates that a heavier particle will move more slowly than a lighter particle.

Solve

a. Because the temperature is constant, both H^+ and NO^+ ions have the same average kinetic energy.
b. Expressing this equality using Equation 5.27:

$$\tfrac{1}{2}m_{H^+} \times u_{H^+}^2 = \tfrac{1}{2}m_{NO^+} \times u_{NO^+}^2$$

Rearranging the terms and simplifying:

$$\frac{u_{H^+}}{u_{NO^+}} = \sqrt{\frac{m_{NO^+}}{m_{H^+}}}$$

The mass of a NO^+ ion is much greater than the mass of a proton; therefore, the protons in the reaction mixture have a higher average speed.

Think About It We have confirmed that heavier particles move more slowly than lighter particles with the same kinetic energy.

Practice Exercise Consider two positively charged particles, A^+ and B^+. If A^+ is 33% heavier than B^+, how much slower must it move to equal the kinetic energy of B^+?

(Answers to Practice Exercises are in the back of the book.)

The energy given off or absorbed during a chemical reaction is equal to the difference in the energy of the reactants and products. For example, when hydrogen molecules burn in oxygen, the products are water and a considerable amount of energy (**Figure 6.10**). The energy given off by this reaction can be used to power rockets and is now being used to run some buses and automobiles. Because energy is given off in the reaction, the product molecules must be at a lower potential energy than the reactant molecules. The difference between the energy of the products and the energy of the reactants is the energy released. In the case of hydrogen combustion (**Figure 6.11a**), this energy is now used to fuel passenger cars (**Figure 6.11b**).

FIGURE 6.10 Energy from the combustion of hydrogen can be used to launch rockets.

Where does the energy released by the reaction between hydrogen and oxygen in Figure 6.11 come from? During the combustion of H_2, the bonds in molecules of H_2 and O_2 must be broken, which requires energy, but even more energy is released when the O–H bonds in H_2O form. The result is a net release of energy.

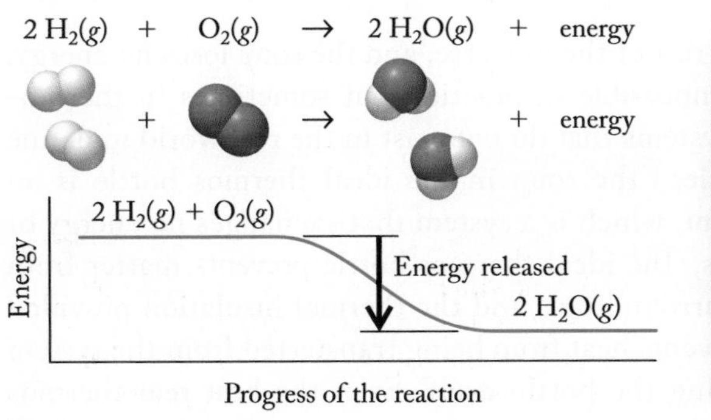

FIGURE 6.11 (a) Hydrogen reacts with oxygen to produce water. Because this reaction releases energy, the product molecules are at a lower energy than those of the reactants. (b) Fueling a hydrogen-powered vehicle.

6.3 Systems, Surroundings, and Energy Transfer

In both thermochemistry and thermodynamics, the specific part of the universe we are studying is called the **system** and everything else is called the **surroundings**. Although a system can be as large as a galaxy or as small as a living cell, most of the systems we examine in this chapter fit on a laboratory bench. Typically, we limit our concern about surroundings to that part of the universe that can exchange energy and matter with the system. When evaluating the energy gained or lost in a chemical reaction, the system may be just the particles involved in the reaction, or it may also include the vessel in which the reaction occurs.

Isolated, Closed, and Open Systems

Thermodynamic systems may be isolated, closed, or open (**Figure 6.12**). These designations are important because they define the system we are dealing with and the part of the universe that the system interacts with. In the following discussion, hot soup is the system.

system the part of the universe that is the focus of a thermochemical study.

surroundings everything that is not part of the system.

What kind of system is hot soup in an ideal, closed thermos bottle? An ideal thermos bottle takes no energy away from the soup, thereby completely

FIGURE 6.12 Transfer of energy and matter in isolated, closed, and open systems. (a) Hot soup in a tightly sealed thermos bottle approximates an isolated system: no vapor escapes, no matter is added or removed, and no energy escapes to the surroundings. (b) Hot soup in a cup with a lid is a closed system: the soup transfers energy to the surroundings as it cools, but no matter escapes and none is added. (c) Hot soup in a cup with no lid is an open system: it transfers both matter (steam) and energy (heat) to the surroundings as it cools. Matter in the form of pepper, grated cheese, and crackers, as well as other matter from the surroundings, may be added to the soup.

(a) **Isolated system:** A thermos bottle containing hot soup with the lid screwed on tightly

(b) **Closed system:** A cup of hot soup with a lid

(c) **Open system:** An open cup of hot soup

insulating the soup from the rest of the universe, and the soup loses no energy. Such perfect insulation is impossible in practice, but sometimes in thermochemistry we must discuss systems that do not exist in the real world to define the total range of possibilities. The soup in the ideal thermos bottle is an example of an **isolated system**, which is a system that exchanges no energy or matter with its surroundings. The ideal thermos bottle prevents matter from being exchanged with the surroundings, and the thermal insulation provided by the ideal thermos also prevents heat from being transferred from the system to the surroundings, including the bottle itself. Even the best real thermos bottle cannot maintain the soup as an isolated system over time because heat will leak out, and the contents will cool. But for short periods, the soup in a good thermos bottle *approximates* an isolated system.

Hot soup in a cup with a lid is an example of a **closed system**, which is a system that exchanges energy but not matter with its surroundings. Because the cup has a lid, no vapor (which is matter) escapes from the soup and no matter can be added. Only energy is exchanged between the soup and its surroundings. Energy from the soup is transferred—first into the cup walls, then into the air and the tabletop—and the soup gradually cools. Many real systems are closed systems.

Soup in an open cup is an **open system**, one that can exchange both energy and matter with its surroundings. Energy, in the form of heat, from the soup is transferred to the surroundings (cup, air, tabletop), and matter, in the form of water vapor, leaves the system and enters the air. We may add matter to the system from the surroundings by sprinkling on a little grated cheese, some ground pepper, or a few crumbled crackers.

Most of the real systems we deal with are closed, and we may treat some of them as isolated, at which point we assume behavior that is more ideal than real. We do this because isolated and closed systems are easier to model quantitatively than open systems. However, many important systems are open—including cells, organisms, and Earth itself.

isolated system a system that exchanges neither energy nor matter with the surroundings.

closed system a system that exchanges energy but not matter with the surroundings.

open system a system that exchanges both energy and matter with the surroundings.

exothermic process a thermochemical or physical process in which energy flows from a system into its surroundings.

endothermic process a thermochemical or physical process in which energy flows from the surroundings into the system.

CONCEPT TEST

Identify the following systems as isolated, closed, or open: (a) the water in a pond; (b) a carbonated beverage in a sealed bottle; (c) a sandwich wrapped in thermally conducting plastic wrap; (d) a live chicken.

Exothermic and Endothermic Processes

Chemists classify chemical reactions and changes of state based on whether they give off or absorb energy. For example, condensation of a gas is accompanied by a transfer of energy from the gas (the system) to its surroundings (**Figure 6.13a**), which makes condensation **exothermic** from the point of view of the system. This flow of energy can be detected because it causes the temperature of the surroundings to increase. Combustion reactions are also exothermic because energy flows from hot reaction mixtures (the system) to their surroundings. In contrast, processes that absorb energy from their surroundings are **endothermic** (**Figure 6.13b**). For example, when ice cubes (the system) are added to a glass of room-temperature water (the surroundings), energy flows spontaneously from the warmer water to the cooler ice cubes, causing them to melt. The melting process is endothermic.

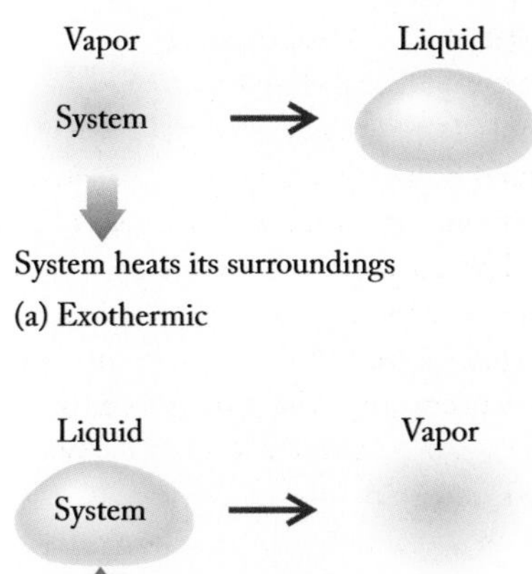

FIGURE 6.13 A process that is exothermic in one direction, such as (a) the condensation of a vapor, is endothermic in the reverse direction, such as (b) the evaporation of a liquid. Reversing a process changes the direction in which energy is transferred but not the quantity of energy transferred.

CONNECTION In Chapter 3 we defined combustion as the reaction of oxygen with another element.

In the condensation of water vapor (the system) from humid air, droplets of liquid water form on the outside of a glass containing an ice-cold drink (part of the surroundings) because energy flows from the system to the surroundings. From the point of view of the system, this process is exothermic. If we pour out the cold drink and pour hot coffee into the glass, the water droplets (the system) on the outside surface of the glass absorb energy from the coffee and vaporize in a process that is endothermic from the point of view of the system. This illustrates an important concept: a process that is exothermic in one direction (vapor → liquid; releases energy) is endothermic in the reverse direction (liquid → vapor; absorbs energy).

We use the symbol q to represent the *quantity* of energy transferred during a chemical reaction or a change of state. For historical reasons, q is called heat and represents energy transferred directly because of a difference in temperature. If the reaction or process is endothermic, q is positive, indicating that energy is *gained* by the system. If the reaction or process is exothermic, q is negative, meaning that the system *loses* energy to its surroundings. In **Figure 6.14**, endothermic changes of state are represented by arrows pointing upward: solid → liquid, liquid → gas, and solid → gas. The opposite (i.e., exothermic) changes of state—liquid → solid, gas → liquid, and gas → solid—are represented by arrows pointing downward. To summarize:

$$\text{Exothermic: } q < 0 \qquad \text{Endothermic: } q > 0$$

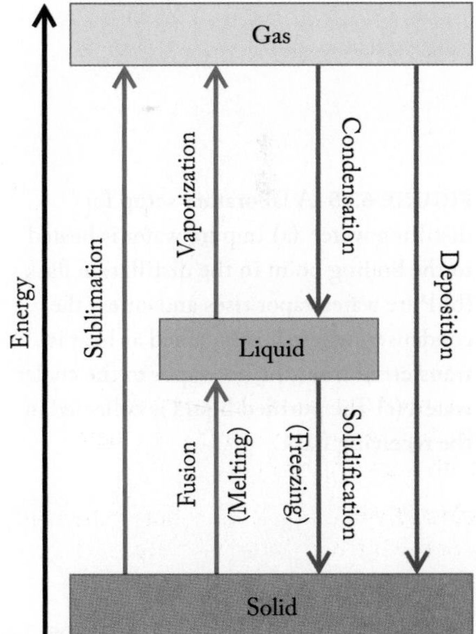

FIGURE 6.14 Matter can be transformed from one physical state to another through adding or removing energy. Red, upward-pointing arrows represent endothermic processes (energy enters the system from the surroundings; $q > 0$). Blue, downward-pointing arrows represent exothermic processes (energy leaves the system and enters the surroundings; $q < 0$).

SAMPLE EXERCISE 6.2 Identifying Exothermic and Endothermic Processes **LO2**

Describe the flow of energy during the purification of water by distillation (**Figure 6.15**), identify the steps in the process as either endothermic or exothermic, and give the sign of q associated with each step. Consider the water being purified to be the system.

Collect and Organize The water is the system, so we must evaluate how the water gains or loses energy during distillation.

Analyze In distillation, energy is transferred during three steps: (1) Liquid water is heated to the boiling point and (2) vaporizes. (3) The vapors are cooled and condense back to a liquid as they pass through the condenser.

Solve Energy flows from the surroundings (the hot plate) to heat the impure water (the system) to its boiling point and then to vaporize it (Figure 6.15a). Therefore, steps 1 and 2 are endothermic. The sign of q is positive for both processes. In step 3, energy flows from the system (the water vapor) into the surroundings (the condenser walls, Figure 6.15b), so this process is exothermic, and the sign of q is negative.

Think About It *Endothermic* means that energy is transferred from the surroundings into the system—the water in the distillation flask. When the water vapor is cooled in the condenser, energy flows from the vapor as it is converted from a gas to a liquid (Figure 6.15b); the process is exothermic.

Practice Exercise What is the sign of q as (a) a match burns, (b) drops of molten candle wax solidify, and (c) perspiration evaporates from skin? In each case, define the system and indicate whether the process is endothermic or exothermic.

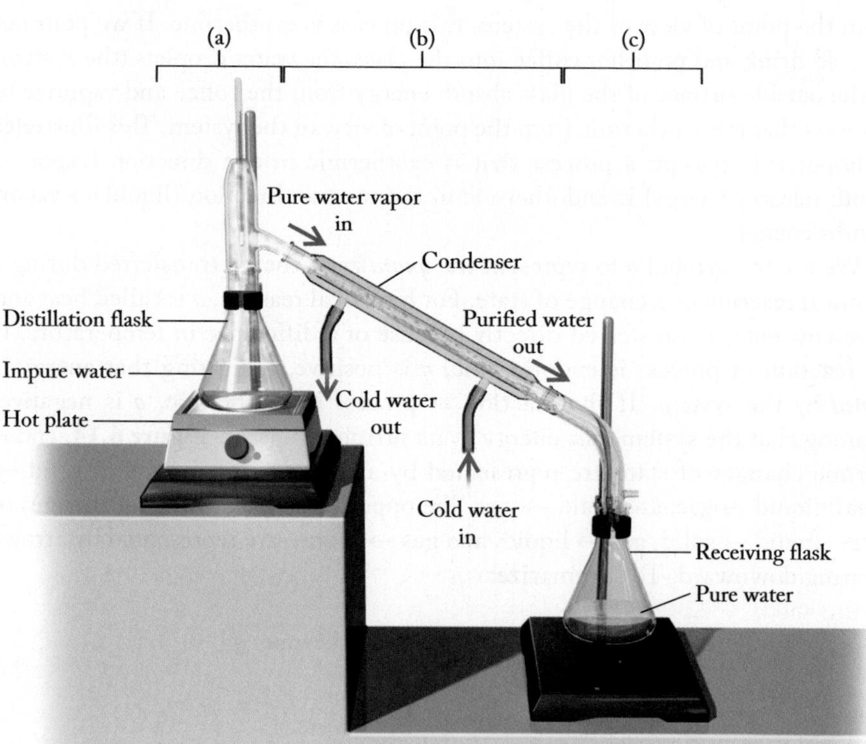

FIGURE 6.15 A laboratory setup for distilling water. (a) Impure water is heated to the boiling point in the distillation flask. (b) Pure water vapor rises and enters the condenser, where it is liquefied as heat is transferred from the hot vapor to the cooler water. (c) The purified liquid is collected in the receiving flask.

CONNECTION In Chapter 1 we discussed the arrangement of molecules in ice, water, and water vapor.

internal energy (*E*) the sum of all the kinetic and potential energies of all the components of a system.

Now let's consider the flow of energy when an ice cube is left on a kitchen counter (**Figure 6.16**). Qualitatively, as the cube (the system) absorbs energy from the air and the counter (the surroundings) and starts to melt, the attractive forces that hold the water molecules in place in solid ice are overcome. The water molecules now have more freedom of motion and more kinetic energy. After all the ice has melted into liquid water at 0°C, the temperature of the water (the system) slowly rises to room temperature. As the temperature increases, the average kinetic energy of the molecules increases, and they move more rapidly. (In Section 6.4 we present a quantitative view of this process.)

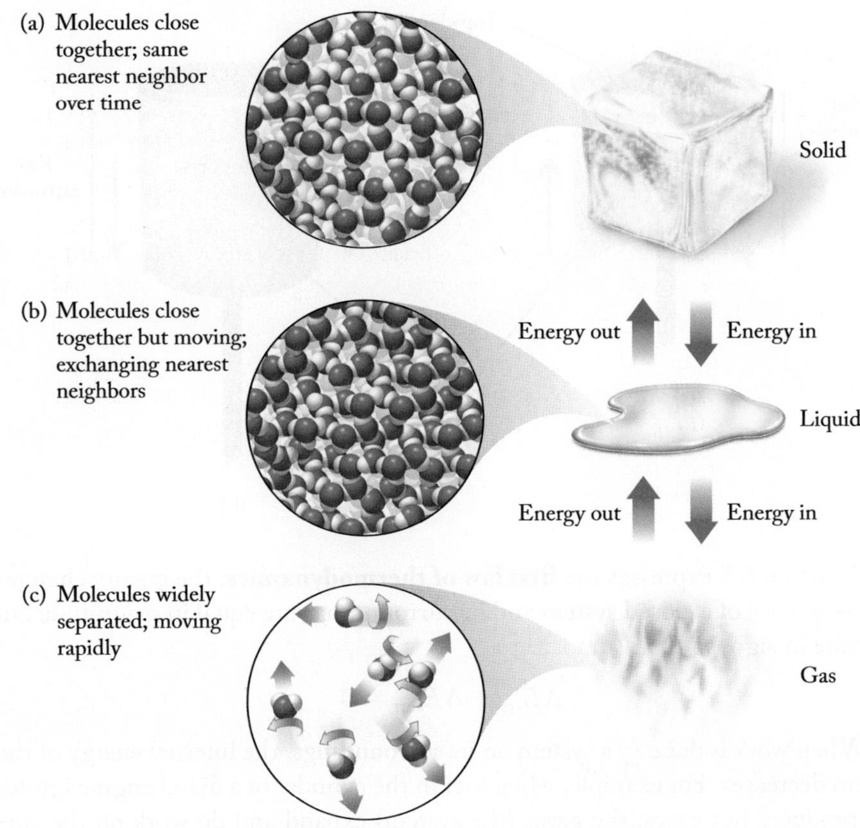

FIGURE 6.16 Changes of state. (a) The molecules in solid ice are held together in a rigid three-dimensional arrangement; they have the same nearest neighbors over time. Solid ice absorbs energy and is converted to liquid water. (b) As the solid melts, the molecules in the liquid state exchange nearest neighbors and occupy many more positions relative to one another than were possible in the solid. The liquid absorbs energy and is converted to a gas. (c) The molecules in a gas are widely separated from one another and move rapidly. The reverse of these processes occurs when water in the gas state loses energy and condenses to a liquid. The liquid also loses energy when it is converted to a solid.

CHEMTOUR
Internal Energy

The **internal energy (*E*)** of a system is the sum of the kinetic and potential energies of all the components of the system (**Figure 6.17**). It is not possible to determine the absolute values of kinetic and potential energies, but *changes* in internal energy (ΔE) can be determined because a change in a system's physical state or temperature is a measure of the change in its internal energy. (The capital Greek delta, Δ, is the standard way scientists symbolize change in a quantity.) The change in internal energy is the difference between the final internal energy of the system and its initial internal energy:

$$\Delta E = E_{\text{final}} - E_{\text{initial}} \tag{6.4}$$

Internal energy is a state function because ΔE depends only on the initial and final states. How the change occurs in the system does not matter.

The law of conservation of energy (Section 6.2) applies to the transfer of energy in materials. The total energy change experienced by a system must be balanced by the total energy change experienced by its surroundings.

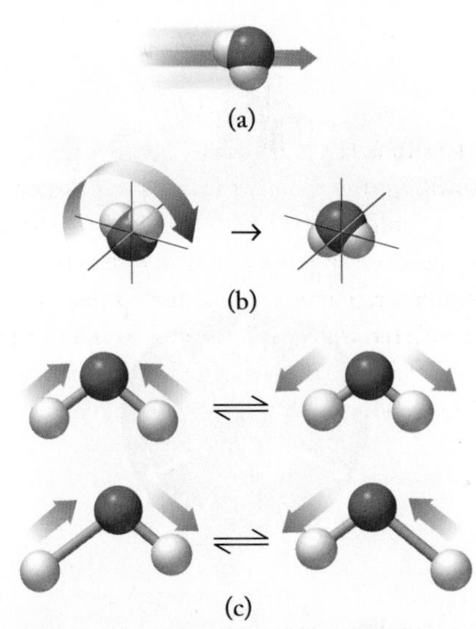

FIGURE 6.17 Some of the types of molecular motion that contribute to the overall internal energy of a system: (a) translational motion, motion from place to place along a path; (b) rotational motion, motion about a fixed axis; and (c) vibrational motion, movement back and forth from some central position.

P–V Work and Energy Units

Doing work *on* a system is a way to *add* to its internal energy. For example, compressing a quantity of gas (the system) into a smaller volume does work on the gas, and that work causes the temperature of the gas to rise, meaning its internal energy increases. The total increase in the internal energy of a closed system is the sum of the work done on it (w) and any other energy (q) gained:

$$\Delta E = q + w \tag{6.5}$$

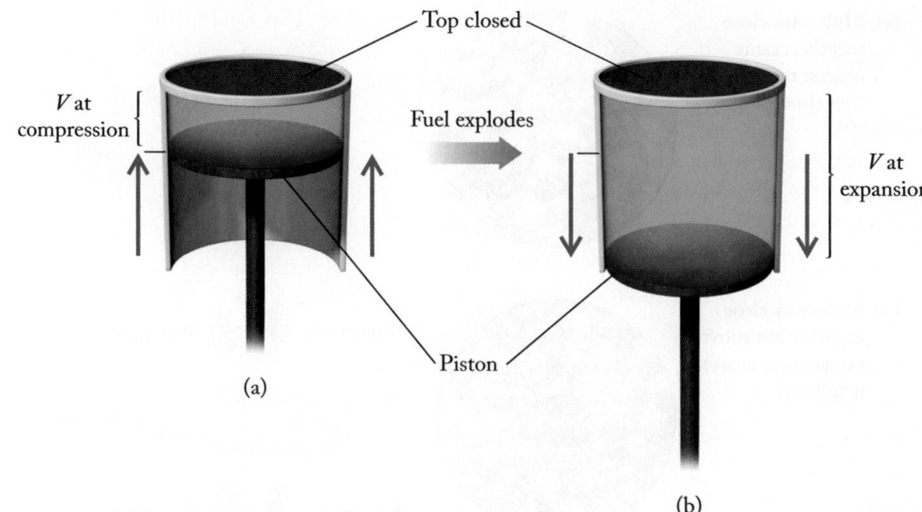

FIGURE 6.18 Work performed by changing the volume of a gas. (a) Highest position of the piston after it compresses the gas in the cylinder, causing the fuel to ignite in a diesel engine. The piston does work on the gas in decreasing the gas volume. (b) The exploding fuel releases energy, causing the gas in the cylinder to expand and push the piston down. The gas has done work on the piston.

CHEMTOUR
Pressure–Volume Work

Equation 6.5 expresses the **first law of thermodynamics**: the energy changes ($\Delta E = q + w$) of a closed system and its surroundings are equal in magnitude but opposite in sign, so their sum is zero:

$$\Delta E_{\text{sys}} + \Delta E_{\text{surr}} = 0$$

When work is done *by* a system on its surroundings, the internal energy of the system decreases. For example, when fuel in the cylinder of a diesel engine ignites and produces hot gases, the gases (the system) expand and do work on the surroundings by pushing on the piston (**Figure 6.18**).

The hot-air balloon depicted in **Figure 6.19** provides another example of work being done by a system on its surroundings. The air in the balloon, which we define as the system, is heated by a burner located at the balloon's bottom, which is open. Heating this air causes the balloon to expand. As the volume of the balloon increases, the balloon presses against the air outside the balloon, thus doing work on that outside air (the surroundings). This type of work, in which the pressure on a system remains constant but the volume of the system changes, is called **pressure–volume (*P–V*) work**. (In this example, we are ignoring the work to lift the mass of the collapsed balloon to the height of the filled balloon.)

FIGURE 6.19 In a completely deflated hot-air balloon, $V = 0$. Inflating the balloon causes it to do work on the atmosphere by pushing against the air. Because this work involves a change in volume, it is called *P–V* work.

What happens to the balloon in terms of heat transferred and work done under constant pressure? First, adding hot air to the balloon increases its internal energy and causes the balloon to expand. Second, the expansion of the balloon against the pressure of its surroundings is *P–V* work done by the system on its surroundings. The internal energy of the system decreases as it performs this *P–V* work. We can relate this change in internal energy to the energy gained by the system by heating (q) and by the work done by the system ($P\Delta V$) by writing Equation 6.5 in the form

$$\Delta E = q + (-P\Delta V) = q - P\Delta V \quad (6.6)$$

The negative sign in front of $P\Delta V$ is appropriate in this case because, when the system expands (positive change in volume ΔV), it loses energy (negative change in internal energy ΔE) as it does work on its surroundings. Correspondingly, when the surroundings do work on the system (e.g., when a gas is compressed), the quantity ($-P\Delta V$) has a positive value.

The sign of q may also be either positive or negative (**Figure 6.20**). If the system is heated by its surroundings, then q is positive ($q > 0$). Energy is added to the balloon, for instance, when it is being inflated, because energy flows from the surroundings (the burner) into the system (the air in the balloon). When the balloon expands, work is done by the system, w is negative ($w < 0$), and the internal energy decreases. If energy is transferred from the system into the surroundings due to a temperature difference, q is negative ($q < 0$), and the internal energy decreases. According to Equation 6.6, the change in internal energy of a system is positive when more energy enters the system than leaves it and negative when more energy leaves the system than enters it. Figure 6.20 also illustrates the first law of thermodynamics: the energy changes ($\Delta E = q + w$) of the system and the surroundings are equal in magnitude but opposite in sign, so their sum is zero.

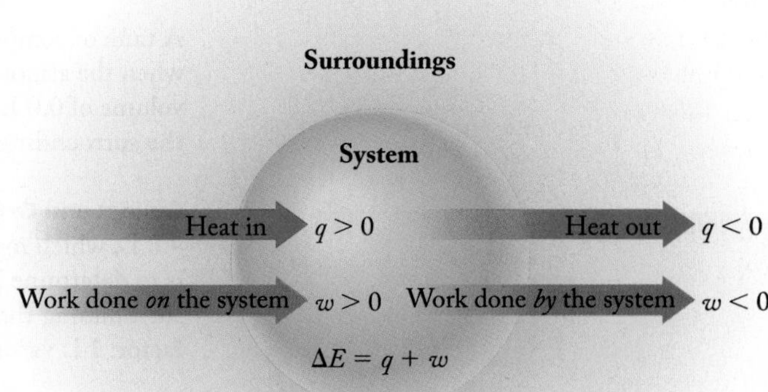

FIGURE 6.20 Energy entering a system by heating and work done on the system by the surroundings are both positive quantities because both increase the internal energy of the system. Energy released by a system to the surroundings and work done by a system on the surroundings are both negative quantities because both decrease the internal energy of a system.

What units are needed to calculate ΔE by using Equation 6.6? Energy changes that accompany chemical reactions and changes in physical state are sometimes expressed in calories. A **calorie (cal)** is the quantity of energy required to raise the temperature of 1 g of water from 14.5°C to 15.5°C. The SI unit of energy, used throughout this text, is the **joule (J)**; 1 cal = 4.184 J. The *Calorie* (Cal; note the capital *C*) in nutrition is one kilocalorie (kcal): 1 Cal = 1000 cal = 1 kcal. If the SI unit for energy is a joule, but pressure is measured in atmospheres and volume in liters, how can we calculate work? A conversion factor, 1 L · atm = 101.32 J, connects joules to liter-atmospheres.

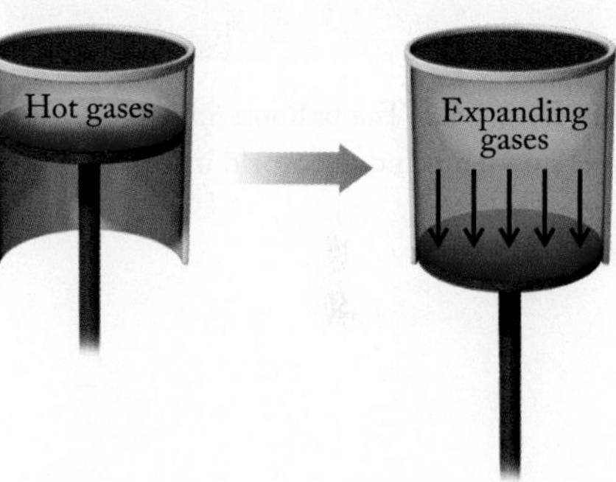

FIGURE 6.21 Hot gases expand, pushing down on the piston.

SAMPLE EXERCISE 6.3 Calculating Changes in Internal Energy LO3

Figure 6.21 shows a simplified version of a piston and cylinder in an engine. Suppose combustion of fuel injected into the cylinder produces 155 J of energy. The hot gases in the cylinder expand, pushing the piston down and doing 93 J of *P–V* work on the piston. If the system is the gases in the cylinder, what is the change in internal energy of the system?

Collect and Organize We are given q and w and asked to calculate ΔE. Are q and w positive or negative according to the sign convention in Figure 6.20?

Analyze The change in internal energy is related to the work done by or on a system and the energy gained or lost by the system (Equation 6.5). The system (gases in the cylinder) absorbs energy by heating, so $q > 0$, and the system does work on the surroundings (the piston), so $w < 0$.

Solve When we substitute values into Equation 6.5, the change in internal energy for the system is

$$\Delta E = q + w = (155\ \text{J}) + (-93\ \text{J}) = 62\ \text{J}$$

Think About It More energy enters the system (155 J) than leaves it (93 J), so a positive value of ΔE is reasonable.

Practice Exercise The piston in Figure 6.21 compresses the air in the cylinder by doing 64 J of work on the gas. As a result, the air gives off 32 J of energy to the surroundings. If the system is the air in the cylinder, what is the change in its internal energy?

first law of thermodynamics the principle that the energy gained or lost by a system must equal the energy lost or gained by its surroundings.

pressure–volume (*P–V*) work the work associated with the expansion or compression of a gas.

calorie (cal) the amount of energy necessary to raise the temperature of 1 g of water by 1°C; for example, from 14.5°C to 15.5°C.

joule (J) the SI unit of energy; 4.184 J = 1 cal.

SAMPLE EXERCISE 6.4 Calculating *P–V* Work — LO3

A tank of compressed helium is used to inflate balloons for sale at a carnival on a day when the atmospheric pressure is 0.99 atm. If each balloon is inflated from an initial volume of 0.0 L to a final volume of 4.8 L, how much *P–V* work, in joules, is done on the surrounding atmosphere by 100 balloons when they are inflated?

Collect and Organize Each of the 100 balloons goes from empty ($V = 0.0$ L) to 4.8 L, which means $\Delta V = 4.8$ L, and the atmospheric pressure P is 0.99 atm. Our task is to determine how much *P–V* work is done by the 100 balloons on the atmosphere surrounding them. To express our answer in units of joules, we will need the conversion factor: 1 L · atm = 101.32 J.

Analyze We focus on the work done on the atmosphere (the surroundings) by the balloons and the helium they contain (the system).

Solve The volume change (ΔV) as all the balloons are inflated is

$$100 \text{ balloons} \times 4.8 \text{ L/balloon} = 4.8 \times 10^2 \text{ L}$$

The work (w) done by the balloons is

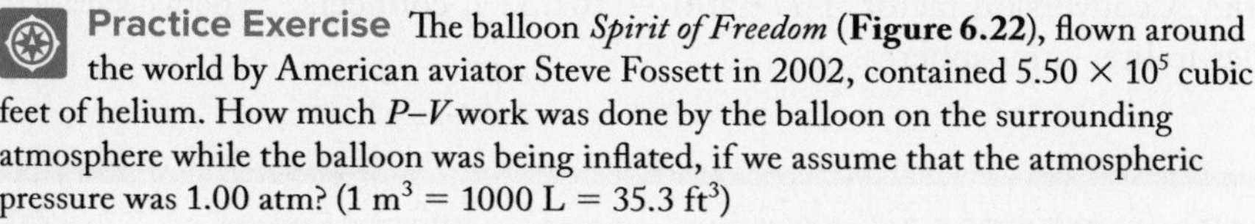

$$w = -P\Delta V = -0.99 \text{ atm} \times (4.8 \times 10^2) \cancel{\text{L}} \times \frac{101.32 \text{ J}}{\cancel{\text{L}} \cdot \text{atm}} = -4.8 \times 10^4 \text{ J} = -48 \text{ kJ}$$

Think About It Because work is done *by* the system on its surroundings, the work is negative from the point of view of the system (see Figure 6.20).

Practice Exercise The balloon *Spirit of Freedom* (**Figure 6.22**), flown around the world by American aviator Steve Fossett in 2002, contained 5.50×10^5 cubic feet of helium. How much *P–V* work was done by the balloon on the surrounding atmosphere while the balloon was being inflated, if we assume that the atmospheric pressure was 1.00 atm? (1 m^3 = 1000 L = 35.3 ft^3)

FIGURE 6.22 The balloon *Spirit of Freedom* was flown around the world in 2002.

6.4 Enthalpy and Enthalpy Changes

Many physical and chemical changes occur at constant atmospheric pressure (P) because they are done in vessels open to the air. Examples include boiling water for tea or any of the chemical reactions described in Chapter 4. The thermodynamic parameter that relates the flow of energy into or out of a system during chemical reactions or physical changes at constant pressure is called the **enthalpy change (ΔH)**. We symbolize this enthalpy change at constant pressure as q_P, where the subscript P specifies a process taking place at constant pressure. We then rearrange Equation 6.6 to define ΔH as

$$\Delta H = q_P = \Delta E + P\Delta V \tag{6.7}$$

Thus, the change in enthalpy is the change in the internal energy of the system at constant pressure plus the *P–V* work done by the system on its surroundings.

The **enthalpy (H)** of a thermodynamic system is the sum of the internal energy and the pressure–volume product ($H = E + PV$). However, as we saw with internal energy, determining the absolute values of these parameters is difficult, whereas determining *changes* is fairly easy. We therefore concentrate on the *change* in enthalpy (ΔH) of a system or its surroundings. According to Equation 6.7, the enthalpy change for a reaction run at constant pressure is equal to q_P,

enthalpy change (ΔH) the heat absorbed by an endothermic process or given off by an exothermic process occurring at constant pressure.

enthalpy (H) the sum of the internal energy and the pressure–volume product of a system; $H = E + PV$.

which is the heat gained or lost by the system during the reaction. This statement also means that the units for ΔH are the same as those for q—namely, enthalpy has units of joules (J), and, if it is reported with respect to the quantity of a substance, J/g or J/mol.

Both ΔH and ΔE represent changes in a state function of a system. These two terms are similar, but the difference between them is important. ΔE includes *all* the energy (heat and work) exchanged by the system with the surroundings: $\Delta E = q + w$. ΔH, on the other hand, is only q, the heat, exchanged at constant pressure: $\Delta H = q_P$. If a chemical reaction does not involve changes in volume, then ΔE and ΔH have very similar values. But if, for example, a reaction consumes or produces gas and the system experiences large changes in volume as a result, then ΔE and ΔH can be quite different.

When energy flows out of a system, q is negative according to our sign convention (Figure 6.20), so the enthalpy change is negative: $\Delta H < 0$. When energy flows into a system, q is positive, so $\Delta H > 0$. As we saw in Section 6.3, positive q values indicate endothermic processes, whereas negative q values indicate exothermic processes. Knowing that, we can relate the terms *exothermic* and *endothermic* to enthalpy changes as well.

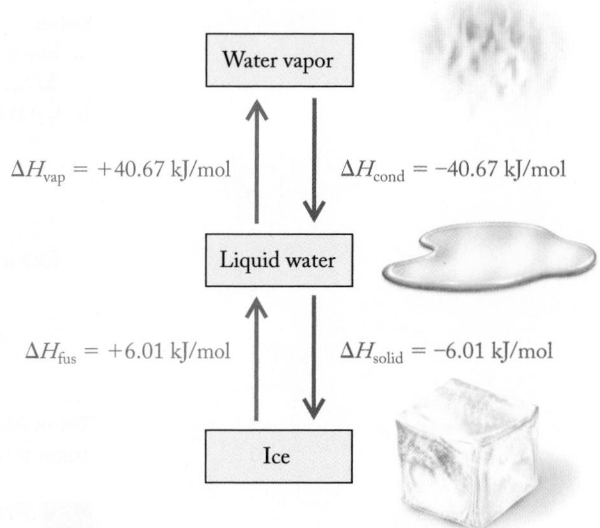

FIGURE 6.23 The enthalpy changes at 100°C for vaporization of liquid water, ΔH_{vap} = 40.67 kJ/mol, and condensation of water vapor, ΔH_{cond} = −40.67 kJ/mol, are equal in magnitude but opposite in sign. So, too, are the enthalpy changes at 0°C for melting ice, ΔH_{fus} = 6.01 kJ/mol, and freezing water, ΔH_{solid} = −6.01 kJ/mol.

For example, the flow of energy into a melting ice cube (the system; **Figure 6.23**) from its surroundings is an endothermic process, meaning that q—and therefore ΔH—are positive. To make ice cubes in a freezer, however, the water (the system) must lose energy, which means the process is exothermic and both q and ΔH are negative. The enthalpy changes for the two processes—melting (fusion) and freezing (solidification)—have different signs but the same absolute value: the enthalpy of fusion (ΔH_{fus}) = 6.01 kJ/mol of ice, but the enthalpy of solidification (ΔH_{solid}) = −6.01 kJ/mol of water. The enthalpies of vaporization and condensation (Figure 6.23) and the enthalpies of sublimation and deposition are similarly paired: equal in magnitude but opposite in sign.

In terms of symbols, we put subscripts on ΔH to clarify not only the specific process but also the part of the universe to which the value applies. If we wish to indicate the enthalpy change associated with the system, we may write ΔH_{sys}.

SAMPLE EXERCISE 6.5 Determining the Value and Sign of ΔH **LO4**

Between periods of a hockey game, an ice-resurfacing machine (**Figure 6.24**) spreads 855 L of water across the surface of a hockey rink. (a) If the system is the water, what is the sign of ΔH_{sys} as the water freezes? (b) To freeze this volume of water at 0°C, what is the value of ΔH_{sys}? The density of water is 1.00 g/mL. Assume atmospheric pressure is constant.

Collect, Organize, and Analyze We are given the volume and density of the water, and we need to determine the sign and value of ΔH_{sys}. The water from the ice-resurfacing machine is the system, so we can determine the sign of ΔH by determining whether energy transfer is into or out of the water. (a) The freezing takes place at constant pressure, so $\Delta H_{sys} = q_P$. (b) To calculate the amount of energy lost from the water as it freezes, we must convert 855 L of water into moles of water because the conversion factor between the quantity of energy removed and the quantity of water that freezes is the enthalpy of solidification of water, ΔH_{solid} = −6.01 kJ/mol.

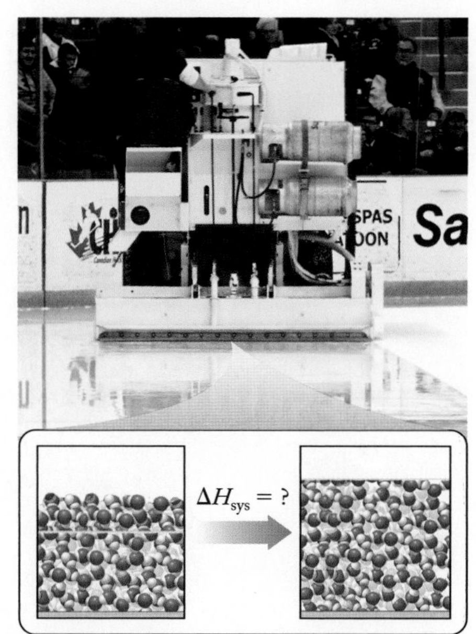

FIGURE 6.24 An ice-resurfacing machine spreads a thin layer of liquid water on top of the existing ice. The liquid freezes, providing a fresh, smooth surface for skaters. If the water represents the system, the energy change for the freezing of the water is described by ΔH_{sys}.

Solve

a. For the water (the system) to freeze, energy must be removed from it. Therefore, ΔH_{sys} must be a negative value.

b. Converting the volume of water into moles:

$$855\ \text{L} \times \frac{1000\ \text{mL}}{1\ \text{L}} \times \frac{1.00\ \text{g}}{1\ \text{mL}} \times \frac{1\ \text{mol}}{18.02\ \text{g}} = 4.745 \times 10^4\ \text{mol}$$

Calculating ΔH_{sys}:

$$\Delta H_{sys} = 4.745 \times 10^4\ \text{mol} \times \frac{-6.01\ \text{kJ}}{1\ \text{mol}} = -2.85 \times 10^5\ \text{kJ}$$

Think About It The magnitude of the answer to part (b) is reasonable given the large amount of water that is frozen to refinish the rink's surface.

Practice Exercise The flame in a torch used to cut metal is produced by burning acetylene (C_2H_2) in pure oxygen. If we assume that the combustion of 1 mole of acetylene releases 1251 kJ of energy, what mass of acetylene is needed to cut through a piece of steel if the process requires 5.42×10^4 kJ of energy?

SAMPLE EXERCISE 6.6 The Sign of ΔH in a Chemical Reaction LO4

The chemical cold packs (**Figure 6.25**) used to treat minor sports injuries contain a small pouch of water inside a bag of solid ammonium nitrate. To activate the pack, you press on it to rupture the pouch, allowing the ammonium nitrate to dissolve in the water. The result is a cold, aqueous solution of ammonium nitrate. Write a balanced chemical equation describing the dissolution of solid NH_4NO_3. If we define the system as the cold pack and the surroundings as an injured player's sprained ankle (for example), what are the signs of ΔH_{sys} and q_{surr}?

Collect, Organize, and Analyze The solution's temperature drops when we dissolve NH_4NO_3 in water. We want to write a balanced chemical equation for the dissolution of ammonium nitrate and determine the sign of ΔH for the solution (the system) and q_{surr}. The first law of thermodynamics says energy must be conserved, so all the energy must be accounted for between system and surroundings.

Solve According to the solubility rules in Tables 4.4 and 4.5, all ammonium salts are soluble. We can write a balanced chemical equation for the dissolution of NH_4NO_3 as

$$NH_4NO_3(s) \rightarrow NH_4^+(aq) + NO_3^-(aq)$$

The low temperature of the cold pack (the system) means that heat flows into it from the injured player's sprained ankle). Therefore, the sign of ΔH_{sys} is positive ($\Delta H_{sys} > 0$), and the sign of q_{surr} is negative ($q_{surr} < 0$).

Think About It The first law of thermodynamics dictates that the signs of ΔH_{sys} and q_{surr} are opposite.

Practice Exercise Potassium hydroxide can be used to unclog sink drains. The reaction between potassium hydroxide and water is quite exothermic. What are the signs of ΔH_{sys} and q_{surr}?

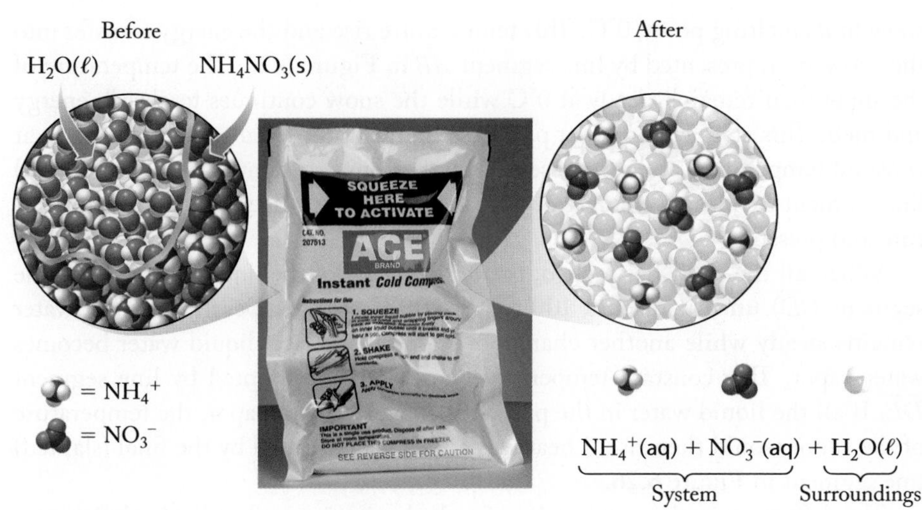

FIGURE 6.25 A chemical cold pack contains a bag of water inside a larger bag of ammonium nitrate. Breaking the inner bag and allowing the NH_4NO_3 to dissolve leads to a decrease in temperature.

6.5 Heating Curves, Molar Heat Capacity, and Specific Heat

Many winter hikers use portable stoves fueled by propane to prepare hot meals. Ice or snow may be their only source of water. In this section, we use this scenario to examine the transfer of energy into water that begins as snow and ends up as vapor.

Hot Soup on a Cold Day

What changes of temperature and state does water undergo when hikers melt snow as a first step in preparing soup from a dry soup mix? Suppose they start with a saucepan filled with snow at −18°C at constant pressure. They place the pan above the flame of a portable stove, and energy begins to flow into the snow. The temperature of the snow immediately begins to rise. If the flame is steady, so that energy flow is constant, the temperature of the snow changes as indicated on the graph shown in **Figure 6.26**. First, heating increases the temperature of the

FIGURE 6.26 The energy required to melt snow and boil the resultant water is illustrated by the four line segments on the heating curve of water: heating snow to its melting point, $\overline{AB}$; melting the snow to form liquid water, $\overline{BC}$; heating the water to its boiling point, $\overline{CD}$; and boiling the water to convert it to vapor, $\overline{DE}$.

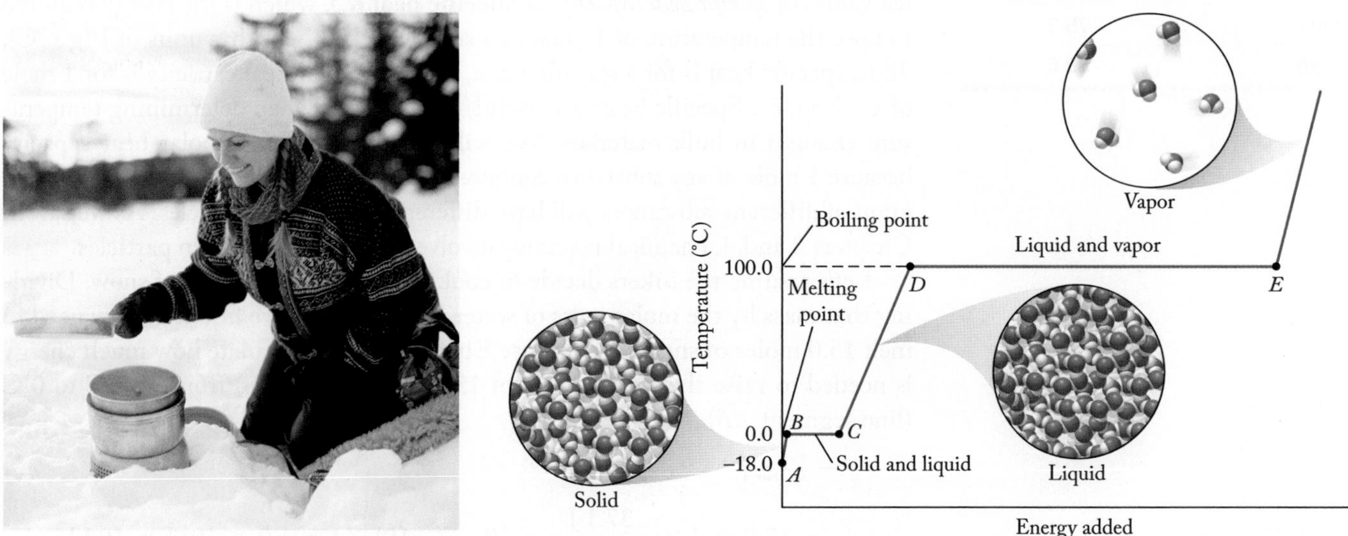

snow to its melting point, 0°C. This temperature rise and the energy transfer into the snow are represented by line segment $\overline{AB}$ in Figure 6.26. The temperature of the snow then remains steady at 0°C while the snow continues to absorb energy and melt. This state change that produces liquid water from snow takes place at constant temperature and is represented by the constant-temperature (horizontal) line segment $\overline{BC}$. Phase changes of pure materials take place at constant temperature and pressure.

CHEMTOUR
Heating Curves

When all the snow has melted, the temperature of the liquid water rises (line segment $\overline{CD}$) until it reaches 100°C. At 100°C, the temperature of the water remains steady while another change of state takes place: liquid water becomes water vapor. This constant-temperature process is represented by line segment $\overline{DE}$. If all the liquid water in the pan were converted into vapor, the temperature of the vapor would then rise if heat was added, as indicated by the final (slanted) line segment in Figure 6.26.

The difference in the x-axis values for the beginning and end of each line segment in Figure 6.26 indicates how much energy is required in each step in this process. In the first step (line segment $\overline{AB}$), the energy required to raise the temperature of the snow from −18°C to 0°C can be calculated if we know how many moles of snow we have and how much energy is required to change the temperature of 1 mole of snow.

Molar heat capacity (c_P) is the quantity of energy required to raise the temperature of 1 *mole* of a substance by 1°C. The symbol we use for molar heat capacity is c_P, where the subscript P again indicates the value for a process taking place at *constant pressure*. Molar heat capacities for several substances are listed in **Table 6.1**. Ice has a molar heat capacity of 37.1 J/(mol · °C) at constant pressure. These units indicate that if we know the number of moles of snow and the temperature change it experiences as we bring it to its melting point, then we can calculate q for the process:

TABLE 6.1 Molar Heat Capacities of Selected Substances at 25°C

Substance	c_P [J/(mol · °C)]
Al(s)	24.4
Cu(s)	24.5
Fe(s)	25.1
C(s), graphite	8.54
$CH_3CH_2OH(\ell)$, ethanol	113.1
$H_2O(s)$	37.1
$H_2O(\ell)$	75.3
$H_2O(g)$	33.6

$$q = nc_P\Delta T \tag{6.8}$$

where n is the number of moles of the substance absorbing or releasing energy and ΔT is the temperature change in degrees Celsius.

In addition to molar heat capacity, other measures exist that specify how much energy is required to increase the temperature of a substance, each referring to a specific amount of the substance. For example, some tables of thermodynamic data list values of *specific heat capacity*, or **specific heat (c_s)**, which is the energy required to raise the temperature of 1 *gram* of a substance by 1°C; c_s has units of J/(g · °C). Thus, specific heat is for a specific mass, whereas molar heat capacity is for 1 mole of a substance. Specific heats are useful, for example, when determining temperature changes in bulk materials. We will mostly work with molar heat capacity because 1 mole of any substance contains the same number of particles, whereas 1 gram of different substances will have different numbers of particles. As we saw in Chapters 3 and 4, chemical reactions involve interactions between particles.

Let's assume the hikers decide to cook their meal with 270 g of snow. Dividing that mass by the molar mass of water (18.02 g/mol), we find that they need to melt 15.0 moles of snow. We can use Equation 6.8 to calculate how much energy is needed to raise the temperature of 15.0 moles of $H_2O(s)$ from −18°C to 0°C (line segment $\overline{AB}$):

$$q = nc_P\Delta T$$

$$= 15.0\ \cancel{\text{mol}} \times \frac{37.1\ \text{J}}{\cancel{\text{mol}} \cdot \cancel{^\circ\text{C}}} \times [0 - (-18)]\cancel{^\circ\text{C}} = 1.0 \times 10^4\ \text{J} = 10\ \text{kJ}$$

This value is positive, which means that the system (the snow) gains energy as it warms up.

The energy absorbed as the snow melts, or *fuses* (line segment $\overline{BC}$ in Figure 6.26), can be calculated using the enthalpy change that takes place as 1 mole of snow melts. This enthalpy change is called the **molar enthalpy of fusion (ΔH_{fus})**, and the energy absorbed as n moles of a substance melts is given by

$$q = n\Delta H_{fus} \quad (6.9)$$

The molar enthalpy of fusion for water is 6.01 kJ/mol. Using this value and the known number of moles of snow, we get

$$q = 15.0 \cancel{\text{mol}} \times \frac{6.01 \text{ kJ}}{\cancel{\text{mol}}} = 90.2 \text{ kJ}$$

This value is positive because energy enters the system. It represents the energy needed to overcome the attractive forces between water molecules in the solid as the snow becomes a liquid. No factor for temperature appears in this calculation because state changes of pure substances take place at constant temperature, as the two horizontal line segments in Figure 6.26 indicate. Snow at 0°C becomes liquid water at 0°C.

While the water temperature increases from 0°C to 100°C (line segment $\overline{CD}$), the relation between temperature and energy absorbed is again defined by Equation 6.8, but this time c_P represents the molar heat capacity of $H_2O(\ell)$, 75.3 J/(mol · °C) (see Table 6.1):

$$\begin{aligned} q &= nc_P\Delta T \\ &= 15.0 \cancel{\text{mol}} \times \frac{75.3 \text{ J}}{\cancel{\text{mol}} \cdot \cancel{°\text{C}}} \times 100\cancel{°\text{C}} = 1.13 \times 10^5 \text{ J} = 113 \text{ kJ} \end{aligned}$$

This value is positive because the system takes in energy as its temperature rises.

At this point in our story, our hiker–chefs are ready to make their soup and enjoy a hot meal. However, if they accidentally leave the boiling water unattended, it will eventually vaporize completely. How much heat is needed to boil away all the water in the pot? As line segment $\overline{DE}$ in Figure 6.26 shows, the temperature of the water remains at 100°C until all of it is vaporized. The enthalpy change associated with changing 1 mole of a liquid to a gas is the **molar enthalpy of vaporization (ΔH_{vap})**, and the quantity of energy absorbed as n moles of a substance vaporizes is given by

$$q = n\Delta H_{vap} \quad (6.10)$$

The molar enthalpy of vaporization for water is 40.67 kJ/mol, so to completely vaporize the water we need an additional 610 kJ of energy:

$$q = (15.0 \cancel{\text{mol}})(40.67 \text{ kJ}/\cancel{\text{mol}}) = 610 \text{ kJ}$$

This value is positive because the system must take in energy to overcome the attractive forces between liquid water molecules and separate them into vapor. Again, no factor for temperature appears because the phase change takes place at constant temperature. Only after all the water has vaporized does its temperature increase above 100°C, along the line above point E in Figure 6.26. The molar heat capacity of steam, 33.6 J/(mol · °C) (Table 6.1), is used to calculate the energy required to heat the vapor to any temperature above 100°C.

The c_P values for $H_2O(s)$, $H_2O(\ell)$, and $H_2O(g)$ are different. Nearly all substances have different molar heat capacities in their different physical states.

molar heat capacity (c_P) the quantity of energy required to raise the temperature of 1 mole of a substance by 1°C at constant pressure.

specific heat (c_s) the quantity of energy required to raise the temperature of 1 g of a substance by 1°C at constant pressure.

molar enthalpy of fusion (ΔH_{fus}) the energy required to convert 1 mole of a solid substance at its melting point into the liquid state.

molar enthalpy of vaporization (ΔH_{vap}) the energy required to convert 1 mole of a liquid substance at its boiling point to the vapor state.

Notice in Figure 6.26 that line segment $\overline{DE}$, the phase change from liquid water to water vapor, is much longer than the line segment $\overline{BC}$, which represents the phase change from solid snow to liquid water. The relative lengths of these lines indicate that the molar enthalpy of vaporization of water (40.67 kJ/mol) is much larger than the molar enthalpy of fusion of snow (6.01 kJ/mol). Why does it take more energy to boil 1 mole of water than to melt 1 mole of snow? The answer is related to the extent to which attractive forces between molecules must be overcome in each process and to the change in the internal energy of the system as energy is added (**Figure 6.27**).

Attractive forces determine both the organization of the molecules in any substance and their relation to their nearest neighbors. In forms of frozen water, like ice and snow, these attractive forces are strong enough to hold the water molecules in place relative to one another. Melting the snow requires overcoming these attractive forces. The energy added to the system at the melting point is sufficient to overcome the attractive forces and change the arrangement (and hence the potential energy).

Intermolecular attractive forces still exist in liquid water, but the energy added at the melting point increases the energy of the molecules and enables them to move with respect to their nearest neighbors. The molecules are closer together in the liquid than they are in the solid, but their relative positions constantly change; they have different nearest neighbors over time.

Once the snow has completely melted, added energy causes the temperature of the liquid water to rise, increasing its internal energy. When water vaporizes at the boiling point, the attractive forces between water molecules must be overcome to separate the molecules from one another as they enter the gas phase. Separating the molecules widely in space requires work (i.e., energy), and the amount required is much larger than that for the solid-to-liquid state change. This difference is consistent with the relative lengths of $\overline{BC}$ and $\overline{DE}$ in Figure 6.26.

FIGURE 6.27 Macroscopic and molecule-level views of (a) ice, (b) water, and (c) water vapor.

SAMPLE EXERCISE 6.7 Calculating the Energy Required to Raise the Temperature of Water **LO4**

Calculate the amount of energy required to convert 237 g of solid ice at 0.0°C to hot water at 80.0°C. The molar enthalpy of fusion (ΔH_{fus}) of ice is 6.01 kJ/mol. The molar heat capacity of liquid water is 75.3 J/(mol · °C).

Collect, Organize, and Analyze This exercise refers to the process symbolized by segments $\overline{BC}$ and $\overline{CD}$ in **Figure 6.28**. We can calculate the temperature change (ΔT) of the water from the initial and final temperatures. Four mathematical steps are required: (1) calculate n, the number of moles of ice; (2) determine the amount of energy required to melt the ice, using ΔH_{fus} = 6.01 kJ/mol in Equation 6.9; (3) calculate the amount of energy required to raise the liquid water temperature from 0.0°C to 80.0°C, using c_P = 75.3 J/(mol · °C) in Equation 6.8; and (4) add the results of steps (2) and (3). We can calculate the number of moles in 237 g of water by using the molar mass of water ($\mathcal{M}$ = 18.02 g/mol).

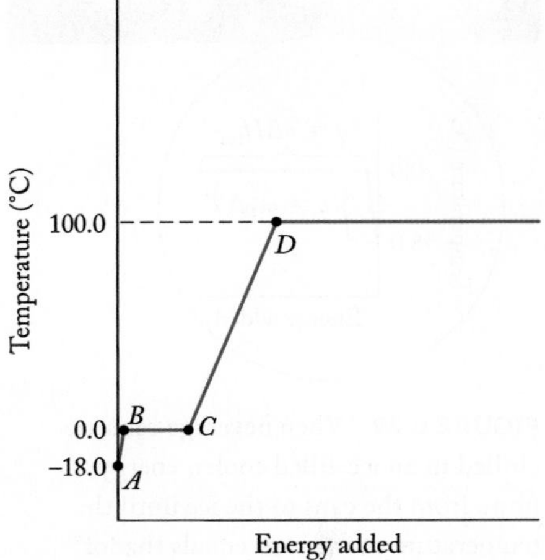

FIGURE 6.28 A portion of the heating curve for water.

Solve

1. Calculate the number of moles of water:

$$n = 237 \text{ g H}_2\text{O} \times \frac{1 \text{ mol H}_2\text{O}}{18.02 \text{ g H}_2\text{O}} = 13.2 \text{ mol H}_2\text{O}$$

2. Determine the amount of energy needed to melt the ice (Equation 6.9):

$$q_1 = n\Delta H_{fus} = 13.2 \text{ mol} \times 6.01 \text{ kJ/mol} = 79.3 \text{ kJ}$$

3. Determine the amount of energy needed to warm the water (Equation 6.8):

$$q_2 = nc_P\Delta T$$

$$= 13.2 \text{ mol} \times \frac{75.3 \text{ J}}{\text{mol} \cdot {}^\circ\text{C}} \times (80.0 - 0.0){}^\circ\text{C} = 79{,}517 \text{ J} = 79.5 \text{ kJ}$$

4. Add the results of parts 2 and 3:

$$q_1 + q_2 = 79.3 \text{ kJ} + 79.5 \text{ kJ} = 158.8 \text{ kJ}$$

Think About It Using the definitions of molar enthalpy of fusion and molar heat capacity, we can also think our way through the solution to this exercise without referring to Equations 6.8 and 6.9. Molar enthalpy of fusion defines the amount of energy needed to melt 1 mole of ice. Multiplying that value (6.01 kJ/mol) by the number of moles (13.2 mol) gives us the energy required to melt the given amount of ice. By the same token, the molar heat capacity of water [75.3 J/ (mol · °C)] defines the amount of energy needed to raise the temperature of 1 mole of liquid water by 1°C. We know the number of moles (13.2 mol), and we know the number of degrees by which we want to raise the temperature (80.0°C − 0.00°C = 80.0°C); multiplying those factors together gives us the energy needed to raise the temperature of the water.

Practice Exercise Calculate the change in energy when 125 g of water vapor at 100.0°C condenses to liquid water and then cools to 25.0°C.

Water is an extraordinary substance for many reasons, one of which is its high molar heat capacity. The ability of water to absorb large quantities of thermal energy is one reason it is used as a *heat sink* both in automobile radiators and in our bodies. The term "heat sink" is often used to identify matter that can absorb energy without changing phase or significantly changing its temperature. Weather and climate changes are largely driven and regulated by cycles involving retention of energy by our planet's oceans, which serve as giant heat sinks for solar energy.

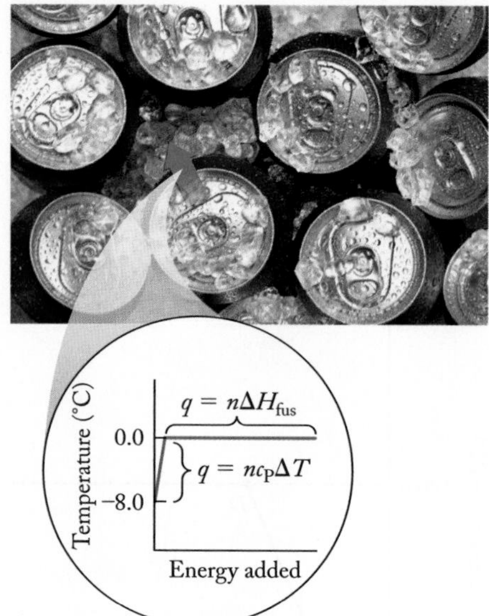

FIGURE 6.29 When beverage cans are chilled in an ice-filled cooler, energy flows from the cans to the ice until the temperature of the cans equals that of the ice.

Cold Drinks on a Hot Day

Suppose we throw a party and plan to chill three cases (72 aluminum cans, each containing 355 mL) of beverages by placing the cans in an insulated cooler and covering them with ice cubes (**Figure 6.29**). If the temperature of the ice (sold in 10-pound bags) is −8.0°C and the temperature of the beverages is initially 25.0°C, how many bags of ice do we need to chill the cans and their contents to 0.0°C (as in "ice cold")?

You may already have an idea that more than one bag, but probably fewer than ten, will be needed. We can use the energy transfer relationships we have defined to predict more accurately how much ice is required. In doing so, we assume that whatever energy is absorbed by the ice is lost by the cans and the beverages in them. As the ice absorbs energy from the cans and the beverages, the temperature of the ice increases from −8.0°C to 0.0°C, and, as we saw in analyzing Figure 6.26, the resulting liquid water remains at 0.0°C until all the ice has melted. We need enough ice so that the last of it melts just as the temperature of the beverages and the cans reaches 0.0°C. Our analysis of this cooling process is a little simpler than our snow/liquid water/water vapor analysis because here there is no change of state. All the energy transferred goes only into cooling the cans and beverages to 0.0°C. The cans stay in the solid state, and the beverages stay in the liquid state.

First let's consider the energy lost in the cooling process. Two materials are to be chilled: 72 aluminum cans and 72 × 355 mL = 25,600 mL of beverages. The beverages are mostly water. The other ingredients in the beverages are present in such small concentrations that they will not affect our calculations, so we can assume that we need to reduce the temperature of 25,600 mL of water by 25.0°C. We can calculate the amount of energy lost with Equation 6.8 if we first calculate the number of moles of water in 25,600 mL, assuming a density of 1.000 g/mL:

$$25{,}600 \ \cancel{\text{mL H}_2\text{O}} \times 1.000 \frac{\text{g}}{\cancel{\text{mL}}} \times \frac{1 \text{ mol H}_2\text{O}}{18.02 \ \cancel{\text{g H}_2\text{O}}} = 1420 \text{ mol H}_2\text{O}$$

We can use the molar heat capacity of water and Equation 6.8 to calculate the energy lost by 1420 moles of water as its temperature decreases from 25.0°C to 0.0°C:

$$\begin{aligned} q_{\text{beverage}} &= nc_P\Delta T \\ &= 1420 \ \cancel{\text{mol}} \times \frac{75.3 \text{ J}}{\cancel{\text{mol}} \cdot \cancel{°\text{C}}} \times (-25.0 \cancel{°\text{C}}) \\ &= -2.67 \times 10^6 \text{ J} \end{aligned}$$

We must also consider the energy released in lowering the temperature of 72 aluminum cans by 25.0°C. The typical mass of a soda can is 12.5 g. The molar heat capacity of solid aluminum (Table 6.1) is 24.4 J/(mol · °C), and the molar mass of aluminum is 26.98 g/mol. Thus,

$$\begin{aligned} q_{\text{cans}} &= nc_P\Delta T \\ &= 72 \ \cancel{\text{cans}} \times \frac{12.5 \ \cancel{\text{g Al}}}{\cancel{\text{can}}} \times \frac{1 \ \cancel{\text{mol}}}{26.98 \ \cancel{\text{g Al}}} \times \frac{24.4 \text{ J}}{\cancel{\text{mol}} \cdot \cancel{°\text{C}}} \times (-25.0 \cancel{°\text{C}}) \\ &= -2.03 \times 10^4 \text{ J} \end{aligned}$$

The total quantity of energy that must be removed from the cans and beverages is

$$\begin{aligned} q_{\text{total lost}} = q_{\text{beverage}} + q_{\text{cans}} &= [(-2.67 \times 10^6) + (-2.03 \times 10^4)] \text{ J} \\ &= (-2.67 - 0.0203) \times 10^6 \text{ J} = -2.69 \times 10^6 \text{ J} \\ &= -2.69 \times 10^3 \text{ kJ} \end{aligned}$$

This quantity of energy must be absorbed by the ice as it warms to its melting point and then melts. The calculation of the amount of ice needed is an algebra problem. Let n be the number of moles of ice needed. The energy absorbed is the sum of (1) the energy needed to raise the temperature of n moles of ice from −8.0° to 0.0°C and (2) the energy needed to melt n moles of ice. These quantities can be calculated with Equation 6.8 for step 1 and Equation 6.9 for step 2. The c_P values in Table 6.1 have units of joules, whereas ΔH_{fus} values given earlier have units of kilojoules. We need to have both terms in the same units, so we use 0.0371 kJ/ (mol · °C) for c_P:

$$\begin{aligned} q_{\text{total gained}} &= q_1 + q_2 \\ &= nc_P\Delta T + n\Delta H_{fus} \\ &= n\left(\frac{0.0371\text{ kJ}}{\text{mol}\cdot\cancel{°\text{C}}}\right)(8.0\cancel{°\text{C}}) + n(6.01\text{ kJ/mol}) \\ &= n(6.31\text{ kJ/mol}) \end{aligned}$$

The energy lost by the cans and the beverages balances the energy gained by the ice:

$$\begin{aligned} -q_{\text{total lost}} &= +q_{\text{total gained}} \\ 2.69 \times 10^3\ \cancel{\text{kJ}} &= n(6.31\ \cancel{\text{kJ}}/\text{mol}) \\ n &= 4.26 \times 10^2\text{ mol ice} \end{aligned}$$

Converting 4.26×10^2 mol of ice into pounds gives us

$$(4.26 \times 10^2\ \cancel{\text{mol}}) \times \frac{18.02\ \cancel{\text{g}}}{\cancel{\text{mol}}} \times \frac{1\text{ lb}}{453.6\ \cancel{\text{g}}} = 16.9\text{ lb of ice}$$

Thus, we need at least two 10-pound bags of ice to chill three cases of our favorite beverages.

CONCEPT TEST

The energy lost by the beverages inside the 72 cans in the preceding discussion was more than 100 times the energy lost by the aluminum cans themselves. What factors contributed to this large difference between the energy lost by the cans and the energy lost by their contents?

SAMPLE EXERCISE 6.8 Calculating the Temperature of Iced Tea LO4

If you add 250.0 g of ice, initially at −18.0°C, to a cup (237 g) of freshly brewed tea, initially at 100.0°C, and all the ice melts, what is the final temperature of the tea? Assume that the mixture is an isolated system (in an ideal insulated container) and that tea has the same molar heat capacity, density, and molar mass as water.

Collect and Organize We know the mass of the tea, its initial temperature, and the molar heat capacity of tea, for which we use the c_P value for water. We know the amount of ice, its initial temperature, and the molar enthalpy of fusion of ice. Our task is to find the final temperature of the tea–ice-water mixture.

Analyze The amount of energy released when the tea is cooled will be the same as the amount of energy gained by the ice and, once it is melted, the amount gained by the water that is formed as it warms to the final temperature. If we assume that the ice melts completely, the tea loses energy through three processes:

1. q_1: raising the temperature of the ice to 0.0°C
2. q_2: melting the ice
3. q_3: bringing the mixture to the final temperature; the temperature of the water from the melted ice rises and the temperature of the tea (T_{initial} = 100.0°C) falls to the final temperature of the mixture (T_{final} = ?)

The energy gained by the ice (and ice melt) equals the energy lost by the tea:

$$q_{\text{ice}} = -q_{\text{tea}}$$

From our analysis, we know that

$$q_{\text{ice}} = q_1 + q_2 + q_3$$

Solve The energy lost by the tea as it cools from 100.0°C to T_{final} is, from Equation 6.8,

$$\begin{aligned} q_{\text{tea}} &= nc_P\Delta T_{\text{tea}} \\ &= 237\text{ g} \times \frac{1\text{ mol}}{18.02\text{ g}} \times \frac{75.3\text{ J}}{\text{mol}\cdot{}^\circ\text{C}} \times (T_{\text{final}} - 100.0^\circ\text{C}) \\ &= (990\text{ J/}^\circ\text{C})(T_{\text{final}} - 100.0^\circ\text{C}) = (0.990\text{ kJ/}^\circ\text{C})(T_{\text{final}} - 100.0^\circ\text{C}) \end{aligned}$$

The transfer of this energy is responsible for the changes in the ice. We can treat the energy transfer from the hot tea to the ice in terms of the three processes we identified in the Analyze step. In step 1, the ice is warmed from −18.0°C (T_{initial}) to 0.0°C (T_{final} for this step):

$$\begin{aligned} q_1 &= n_{\text{ice}}c_{P,\text{ice}}\Delta T_{\text{ice}} \\ &= 250.0\text{ g} \times \frac{1\text{ mol}}{18.02\text{ g}} \times \frac{37.1\text{ J}}{\text{mol}\cdot{}^\circ\text{C}} \times [(0.0) - (-18.0)]^\circ\text{C} \\ &= 9.26 \times 10^3\text{ J} = 9.26\text{ kJ} \end{aligned}$$

In step 2, the ice melts, requiring the absorption of energy:

$$\begin{aligned} q_2 &= n_{\text{ice}}\Delta H_{\text{fus,ice}} \\ &= 250.0\text{ g} \times \frac{1\text{ mol}}{18.02\text{ g}} \times \frac{6.01\text{ kJ}}{\text{mol}} \\ &= 83.4\text{ kJ} \end{aligned}$$

In step 3, the water from the ice, initially at 0.0°C, warms to the final temperature (where $\Delta T = T_{\text{final}} - T_{\text{initial}} = T_{\text{final}} - 0.0^\circ\text{C}$):

$$\begin{aligned} q_3 &= n_{\text{water}}c_{P,\text{water}}\Delta T_{\text{water}} \\ &= 250.0\text{ g} \times \frac{1\text{ mol}}{18.02\text{ g}} \times \frac{75.3\text{ J}}{\text{mol}\cdot{}^\circ\text{C}} \times (T_{\text{final}} - 0.0^\circ\text{C}) \\ &= (1045\text{ J/}^\circ\text{C})(T_{\text{final}}) = (1.045\text{ kJ/}^\circ\text{C})(T_{\text{final}}) \end{aligned}$$

The sum of the quantities of energy absorbed by the ice and the water from it during steps 1 through 3 must balance the energy lost by the tea:

$$q_{\text{ice}} = q_1 + q_2 + q_3 = -q_{\text{tea}}$$

$$9.26\text{ kJ} + 83.4\text{ kJ} + (1.045\text{ kJ/}^\circ\text{C})(T_{\text{final}}) = -[(0.990\text{ kJ/}^\circ\text{C})(T_{\text{final}} - 100.0^\circ\text{C})]$$

Rearranging the terms to solve for T_{final}, we have

$$(2.04\text{ kJ/}^\circ\text{C})T_{\text{final}} = -9.26\text{ kJ} - 83.4\text{ kJ} + 99.0\text{ kJ} = 6.34\text{ kJ}$$

$$T_{\text{final}} = 3.1^\circ\text{C}$$

calorimetry the measurement of the quantity of energy transferred during a physical change or chemical process.

calorimeter a device used to measure the absorption or release of energy by a physical change or chemical process.

Think About It This calculation was carried out under the assumption that the system (the tea plus the ice) was isolated and the vessel was a perfect insulator. Our answer, therefore, is an "ideal" answer and reflects the coldest temperature we can expect the tea to reach. In the real world, the ice would absorb some energy from the surroundings (the container and the air) and the tea would lose some energy to the surroundings, so the final temperature of the beverage could be different from the ideal value we calculated.

Practice Exercise Calculate the final temperature of a mixture of 350 g of ice, initially at −18.0°C, and 437 g of water, initially at 100.0°C.

6.6 Calorimetry: Measuring Heat Capacity and Enthalpies of Reaction

Up to this point we have discussed molar heat capacities (c_P) and the enthalpy changes (ΔH_{fus} and ΔH_{vap}) associated with phase changes, but we have not broached the issue of how we know the values of these parameters. The experimental method of measuring the quantities of energy associated with chemical reactions and physical changes is called **calorimetry**. The device used to measure the energy released or absorbed during a process is a **calorimeter**.

Determining Molar Heat Capacity and Specific Heat

When we determined the amount of ice needed to cool 72 aluminum cans of beverage in Section 6.5, we used the molar heat capacity of aluminum to determine how much energy was lost by the cans. How are molar heat capacities determined?

We can apply the first law of thermodynamics to determine the specific heat of aluminum. Recall from Section 6.5 that specific heat c_s is defined as the quantity of energy required to raise the temperature of 1 g of a substance by 1°C. The units of specific heat, J/(g · °C), tell us that we need to determine how much energy it takes to raise or lower the temperature of a mass of aluminum from initial to final temperatures that we can also measure accurately. Suppose we employ the strategy of heating a quantity of Al beads and then using the hot beads to raise the temperature of a known quantity of water from initial to final temperatures that we can measure accurately.

We start by transferring 23.5 g of Al beads to a glass test tube and placing the test tube in a bath of boiling water (**Figure 6.30a**). After several minutes the beads have been warmed to the temperature of the boiling water: 100.0°C. While the metal is warming, 130.0 g of water in a beaker are placed in a Styrofoam box (**Figure 6.30b**), which we will assume effectively insulates its contents from their surroundings. A thermometer allows us to monitor the temperature of the water, which initially is 23.1°C. Next, we remove the test tube containing the Al beads from the boiling water, remove the lid of the insulated box, pour the hot beads into the water in the beaker, and quickly close the lid (**Figure 6.30c**). The hot beads warm the temperature of the water to 26.0°C, producing a change in the temperature of the water in the beaker of ΔT_{water} = 26.0°C − 23.1°C = 2.9°C,

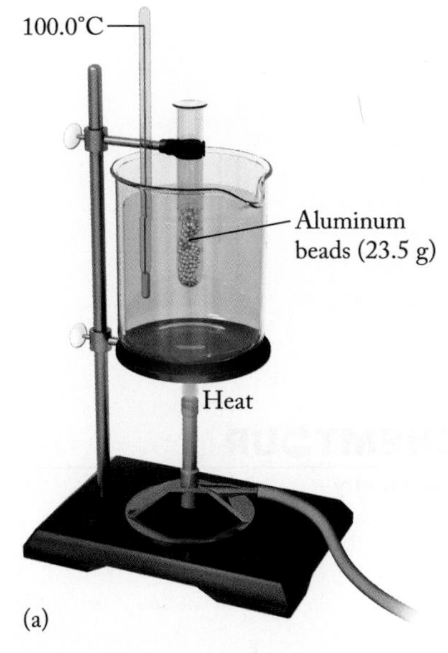

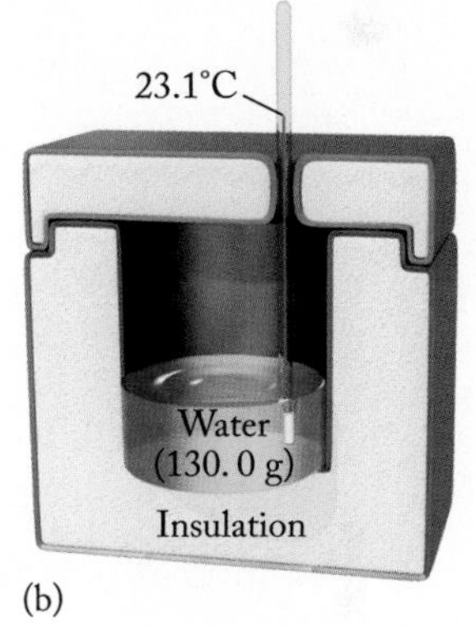

FIGURE 6.30 Experimental setup to determine the molar heat capacity of a metal. (a) Pure aluminum beads having a combined mass of 23.5 g are heated to 100.0°C in boiling water; (b) 130.0 g of water at 23.1°C is placed in a Styrofoam box; (c) the hot Al beads are dropped into the water, and the temperature at thermal equilibrium is 26.0°C.

while the temperature of the aluminum beads changes by $\Delta T_{Al} = 26.0°C - 100.0°C = -74.0°C$.

The measured temperatures were the result of energy transferred from the aluminum beads to the water in the beaker. According to the first law of thermodynamics, the energy lost by the beads and the energy gained by the water are equal in magnitude but opposite in sign:

$$-q_{Al} = q_{water}$$

To determine how much energy was lost by the aluminum, we first calculate how much energy was gained by the water. We know the mass of water (130.0 g), its temperature change (2.9°C), and the specific heat of water (see Section 6.3):

$$c_{s,water} = 4.184 \frac{J}{g \cdot °C}$$

CHEMTOUR
Calorimetry

We can combine these three values using an equation like Equation 6.8 but based on mass rather than moles:

$$q = mc_s\Delta T \tag{6.11}$$

Inserting the measured values and $c_{s,water}$:

$$q_{water} = (130.0 \text{ g } H_2O)\left(\frac{4.184 \text{ J}}{\text{g } H_2O \cdot °C}\right)(2.9°C)$$
$$= 1.57 \times 10^3 \text{ J}$$

To find the specific heat of aluminum, we use Equation 6.11 for aluminum and the fact that the energy that increased the temperature of the water came from the aluminum beads. Therefore,

$$q_{Al} = -q_{water} = -1.57 \times 10^3 \text{ J} = mc_{s,Al}\Delta T_{Al} = (23.5 \text{ g Al})(c_{s,Al})(-74.0°C)$$

Solving for $c_{s,Al}$:

$$c_{s,Al} = \frac{-1.57 \times 10^3 \text{ J}}{(23.5 \text{ g})(-74.0°C)} = \frac{0.90 \text{ J}}{g \cdot °C}$$

From this specific heat value, we can also calculate the molar heat capacity of aluminum, $c_{P,Al}$, by multiplying $c_{s,Al}$ by the molar mass of aluminum:

$$c_{P,Al} = c_s \times \mathcal{M} = \frac{0.90 \text{ J}}{g \cdot °C} \times 26.98 \frac{g}{mol} = 24 \frac{J}{mol \cdot °C}$$

SAMPLE EXERCISE 6.9 Determining the Specific Heat of a Metal **LO4**

Brass is the name of an alloy (homogenous mixture of metals) that contains mostly copper and zinc and only minor concentrations of other metals. A 50.0 g sample of brass is heated to 115°C in a labware drying oven and then immersed in 50.0 g H_2O ($T_{initial} = 21.4°C$) in a beaker housed in an insulated apparatus like that shown in Figure 6.30. The temperature of the H_2O increases to 29.2°C. What is the specific heat of the brass sample?

Collect and Organize We are given the masses and initial and final temperatures of samples of brass and water. We are asked to use these data to calculate the specific heat of the brass sample. The specific heat of water is 4.184 J/g · °C. Equation 6.11,

$q = mc_s\Delta T$, relates energy that must be transferred to an object of mass m to raise its temperature by ΔT to the specific heat of the object.

Analyze Following the approach described in the prior discussion to calculate the specific heat of aluminum, we should first calculate the quantity of energy absorbed by the water (q_{water}) using Equation 6.11. This q_{water} value is equal in magnitude but opposite in sign to the energy lost by the brass sample: $q_{water} = -q_{brass}$. We can use m_{brass} and the calculated values of ΔT_{brass} and q_{brass} to calculate $c_{s,brass}$ using Equation 6.11. The masses of the brass and water samples are the same, but the final temperature of the two samples in thermal equilibrium with each other is only about 8°C from the initial water temperature and about 86°C from the initial brass temperature, which means $c_{s,brass}$ is likely to be about $0.1 \times c_{s,water}$ or about 0.4 J/g · °C.

Solve Calculating the energy absorbed by the water sample:

$$\begin{aligned} q_{water} &= m_{water}c_{s,water}\Delta T_{water} \\ &= 50.0\text{ g} \times 4.184\,\frac{\text{J}}{\text{g}\cdot\text{°C}} \times (29.2 - 21.4)\text{°C} \\ &= 1632\text{ J} \end{aligned}$$

Relating q_{water} to q_{brass} and solving for $c_{s,brass}$

$$\begin{aligned} -q_{brass} &= q_{water} = 1632\text{ J} \\ q_{brass} &= -1632\text{ J} = m_{brass}c_{s,brass}\Delta T_{brass} = 50.0\text{ g} \times c_{s,brass} \times (29.2 - 115)\text{°C} \\ c_{s,brass} &= 0.38\,\frac{\text{J}}{\text{g}\cdot\text{°C}} \end{aligned}$$

Think About It As predicted, the specific heat of brass is about one-tenth the specific heat of water. This value also makes sense because the molar heat capacity of water in Table 6.1 is about 3.1 times that of copper (75.3/24.5), which is the principal component of brass. This factor coupled with the ratio of their molar masses, 64 g/mol for Cu and 18 g/mol for H_2O, or 3.6, means that the ratio of heat capacities expressed per gram, that is, specific heat values of water and of an alloy that is mostly Cu, should be about 3.1×3.6 or about 11, which matches the ratio obtained using the calculated $c_{s,brass}$ value: $4.184/0.38 = 11$. Note that only two digits are used to express the value of $c_{s,brass}$ because that is the number of significant figures in the value of ΔT_{water} that was used in its calculation.

Practice Exercise A student adds 2.55 g of glass beads to a round bottom flask to promote the even boiling of a liquid in the flask. If the specific heat of the glass is 0.83 J/g · °C, how much additional energy is required to raise the temperature of the flask and its contents from 22.5°C to the boiling point of the liquid at 78.4°C?

Enthalpies of Reaction

The energy transfer accompanying any chemical reaction is defined by a quantity known as the **enthalpy of reaction (ΔH_{rxn})**, also called the *heat of reaction*. The "rxn" subscript may be changed to reflect a specific type of reaction being studied, such as ΔH_{comb} for the enthalpy of a combustion reaction.

Much of the energy we use every day can be traced to combustion of a fuel such as natural gas (mostly methane, CH_4), propane (C_3H_8), or gasoline (a mixture of hydrocarbons). Our bodies obtain energy from respiration, which is essentially the combustion of glucose ($C_6H_{12}O_6$). We have already written balanced chemical equations for the combustion of CH_4, C_3H_8, and $C_6H_{12}O_6$ in

enthalpy of reaction (ΔH_{rxn}) the energy absorbed or given off by a chemical reaction under conditions of constant pressure; also called *heat of reaction*.

thermochemical equation the chemical equation of a reaction that includes the change in enthalpy that accompanies that reaction.

bomb calorimeter a constant-volume device used to measure the energy released during a combustion reaction.

heat capacity (C_P) the quantity of energy needed to raise the temperature of an object by 1°C at constant pressure.

calorimeter constant ($C_{calorimeter}$) the heat capacity of a calorimeter.

Chapter 3. Now let's consider **thermochemical equations** for these processes by adding to them their ΔH_{comb} values:

$$CH_4(g) + 2\,O_2(g) \rightarrow CO_2(g) + 2\,H_2O(g) \quad \Delta H_{comb} = -802 \text{ kJ/mol } CH_4$$

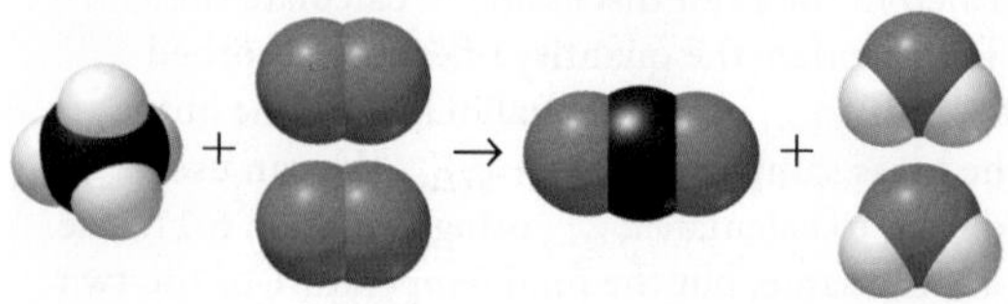

$$C_3H_8(g) + 5\,O_2(g) \rightarrow 3\,CO_2(g) + 4\,H_2O(g) \quad \Delta H_{comb} = -2220 \text{ kJ/mol } C_3H_8$$

$$C_6H_{12}O_6(s) + 6\,O_2(g) \rightarrow 6\,CO_2(g) + 6\,H_2O(g) \quad \Delta H_{comb} = -2803 \text{ kJ/mol } C_6H_{12}O_6$$

All three of these reactions are *exothermic* in the forward direction, so the enthalpy change for each reaction is less than zero: $\Delta H_{rxn} < 0$.

Knowing ΔH_{comb} values allows us to compare fuels and to calculate how much fuel is needed to supply a desired quantity of energy. For example, we can calculate how much propane we need to melt the 270 g (15.0 mol) of snow in Figure 6.26 and raise its temperature to 100°C. In the beginning of Section 6.5, we calculated that the energy required to melt this much snow is 113 kJ. If we get 2220 kJ/mol C_3H_8, then we need to burn at least 2.24 g of C_3H_8:

$$113 \cancel{\text{kJ}} \times \frac{1 \cancel{\text{mol } C_3H_8}}{2220 \cancel{\text{kJ}}} \times \frac{44.09 \text{ g } C_3H_8}{1 \cancel{\text{mol } C_3H_8}} = 2.24 \text{ g } C_3H_8$$

How much energy would we get from burning 2.24 g of glucose? Converting the mass of glucose to moles and multiplying by ΔH_{comb}:

$$2.24 \cancel{\text{g } C_6H_{12}O_6} \times \frac{1 \cancel{\text{mol } C_6H_{12}O_6}}{180.16 \cancel{\text{g } C_6H_{12}O_6}} \times \frac{2803 \text{ kJ}}{1 \cancel{\text{mol } C_6H_{12}O_6}} = 34.9 \text{ kJ}$$

Enthalpies of combustion are best measured with a device called a **bomb calorimeter (Figure 6.31)**. To measure the enthalpy of a combustion reaction, the sample is placed in a sealed vessel (called a *bomb*) capable of withstanding high pressures and submerged in a large volume of water in a heavily insulated container. Oxygen is introduced into the bomb, and the mixture is ignited with an electric spark. As combustion occurs, energy generated by the reaction flows into the walls of the bomb and then into the water surrounding the bomb. A good bomb calorimeter keeps the system contained within the bomb and ensures that all energy generated by the reaction stays in the calorimeter. The surroundings consist of the bomb, the water, the insulated container, and minor components (stirrer, thermometer, and any other materials). A bomb calorimeter is a more sophisticated version of the calorimeter shown in Figure 6.30.

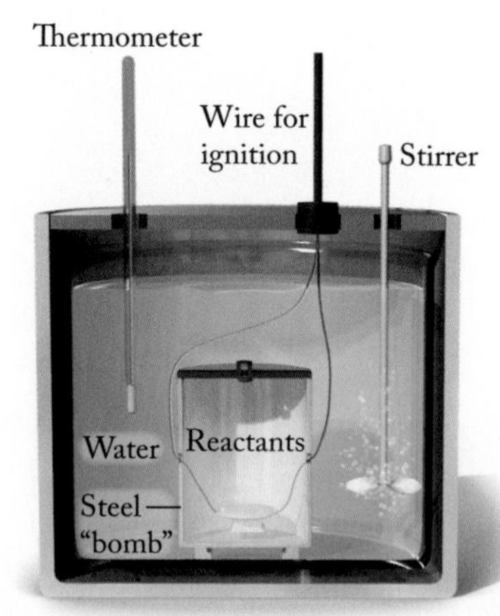

FIGURE 6.31 A bomb calorimeter.

The energy produced by the reaction is determined by measuring the temperature of the water before and after the reaction. The water is at the same temperature as the parts of the calorimeter it contacts—the walls of the bomb, the thermometer, and the stirrer—so the temperature change of the water takes the entire calorimeter into account.

Measuring the change in temperature of the water is not the whole story, however. We also need to know the **heat capacity (C_P)** of the calorimeter, which is the quantity of energy required to increase its temperature by 1°C at constant pressure. There is an important difference between heat capacity and specific heat (c_s) and molar heat capacity (c_P). *Specific heat* [J/(g · °C)] is the energy required to raise the temperature of 1 *gram* of a substance by 1°C, and *molar heat capacity* [J/(mol · °C)] is the energy required to raise the temperature of 1 *mole* of a substance by 1°C. Thus, specific heat and molar heat capacity are intensive properties of substances. However, heat capacity is an extensive property of an object, such as a calorimeter. The heat capacity of a calorimeter is also referred to as a **calorimeter constant** **($C_{\text{calorimeter}}$)** and has units of J/°C. If we know a calorimeter's $C_{\text{calorimeter}}$ value and if we can measure its change in temperature, we can calculate the quantity of energy that flowed from the reactants into the calorimeter ($q_{\text{calorimeter}}$) and caused the temperature change:

$$q_{\text{calorimeter}} = C_{\text{calorimeter}}\,\Delta T \tag{6.12}$$

Rearranging terms:

$$C_{\text{calorimeter}} = \frac{q_{\text{calorimeter}}}{\Delta T} \tag{6.13}$$

Thus, heat capacity is expressed in units of energy divided by temperature—usually kilojoules per degree Celsius (kJ/°C).

Equation 6.13 can be used to determine $C_{\text{calorimeter}}$ for a bomb calorimeter. To do this, we burn a quantity of material in the calorimeter that produces a known quantity of energy when it burns—in other words, a material whose ΔH_{comb} value is already known. Benzoic acid ($C_7H_6O_2$) is often used for this purpose because it can be obtained in a very pure form. Once $C_{\text{calorimeter}}$ has been determined, the calorimeter can be used to determine ΔH_{comb} for other substances. The observed increases in water temperature can be used to calculate the quantities of energy produced by combustion reactions on a per-gram or per-mole basis.

Because there is no change in the volume of the reaction mixture in a bomb calorimeter, this technique is referred to as *constant-volume calorimetry*. No P–V work is done, so Equation 6.6:

$$\Delta E_{\text{comb}} = q_{\text{comb}} - P\Delta V_{\text{comb}}$$

is simply

$$\Delta E_{\text{comb}} = q_{\text{comb}}$$

The energy gained by the calorimeter, $q_{\text{calorimeter}}$, equals the energy lost by the combustion reaction system: $-q_{\text{ccomb}}$. Therefore,

$$q_{\text{calorimeter}} = -\Delta E_{\text{comb}} \tag{6.14}$$

The pressure inside a bomb calorimeter can change during combustion reactions, and in such cases ΔE_{comb} is not *exactly* the same as ΔH_{comb}. However, the overall pressure change after thermal equilibrium has been achieved is usually so small that ΔE_{comb} is *nearly* the same as ΔH_{comb}. Hence, we discuss enthalpies of reactions (ΔH_{rxn}) throughout and apply the approximate relation:

$$q_{\text{calorimeter}} = -\Delta H_{\text{comb}}$$

or more generally

$$q_{\text{calorimeter}} = -\Delta H_{\text{rxn}} \tag{6.15}$$

Other types of calorimeters allow the volume to change while the pressure remains constant, so $q_{\text{calorimeter}}$ in those cases *is* the same as the enthalpy of reaction (ΔH_{rxn}).

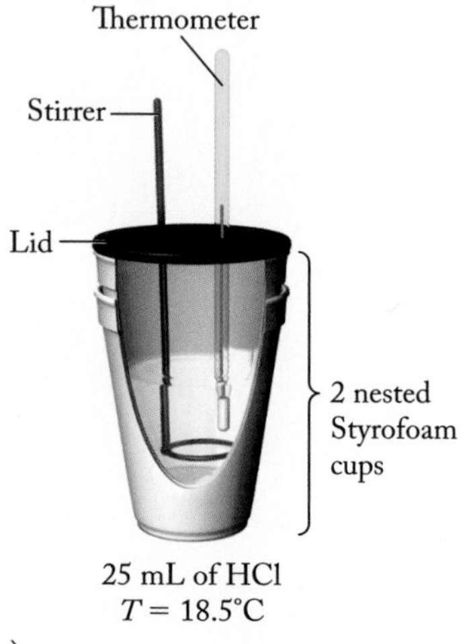

(a)

(b)

FIGURE 6.32 A cutaway image of a coffee-cup calorimeter made of two nested Styrofoam cups. A thermometer enables us to observe the temperature of the contents, which we can agitate using a simple stirrer to ensure mixing of solutions. (a) The inside cup contains 25 mL of 1.0 M HCl. (b) The rapid addition of 25 mL of 1.0 M NaOH causes the temperature to rise because energy is given off in the ensuing neutralization reaction.

SAMPLE EXERCISE 6.10 Determining a Calorimeter Constant and the Value of ΔH_{rxn} LO5

a. What is the calorimeter constant of a bomb calorimeter if burning 1.000 g of benzoic acid in it causes the temperature of the calorimeter to rise by 7.248°C? The enthalpy of combustion of benzoic acid is $\Delta H_{comb} = -26.38$ kJ/g.
b. When 0.500 g of a mixture of hydrocarbons is burned in this bomb calorimeter, the calorimeter's temperature increases by 6.76°C. What change in enthalpy occurs during the combustion reaction?
c. What is ΔH_{comb} for this reaction in kJ/g?

Collect and Organize We are asked to find the heat capacity (calorimeter constant) of a calorimeter based on how much the temperature of the calorimeter increases when a known amount of benzoic acid is burned in it, and then use the calorimeter constant to determine the enthalpy change that accompanies a combustion reaction.

Analyze (a) We need to determine the amount of energy required to raise the temperature of the calorimeter by 1°C. The heat capacity of a calorimeter can be calculated using Equation 6.13 and the knowledge that the combustion of 1.000 g of benzoic acid produces 26.38 kJ of heat. (b) Once we have determined the calorimeter constant, we can combine Equations 6.12 and 6.15 to calculate the enthalpy change that occurs during the combustion reaction. (c) The value of the enthalpy of reaction expressed in kJ/g should be twice the enthalpy change that accompanied the combustion of 0.500 g in part (b).

Solve

a. $C_{calorimeter} = \dfrac{q_{calorimeter}}{\Delta T} = \dfrac{26.38 \text{ kJ}}{7.248°\text{C}} = 3.640 \text{ kJ/°C}$

b. Determining the enthalpy change during the combustion of the hydrocarbon mixture:

$$\begin{aligned}\Delta H_{rxn} &= -q_{calorimeter} = -C_{calorimeter}\Delta T \\ &= -\frac{3.640 \text{ kJ}}{\cancel{°\text{C}}}(6.76\ \cancel{°\text{C}}) \\ &= -24.6 \text{ kJ}\end{aligned}$$

c. Doubling the ΔH value from part (b):

$$\Delta H_{comb} \text{ (kJ/g)} = 2 \times \Delta H_{rxn} = 2 \times -24.6 \text{ kJ} = -49.2 \text{ kJ/g}$$

Think About It The calorimeter constant is determined for a specific calorimeter. Using the calorimeter constant to determine ΔH_{rxn} values assumes that the constant does not change as the calorimeter is used. However, it makes sense that any alteration that could change its thermal mass, such as replacing a broken temperature sensor or refilling the water jacket surrounding the bomb, would require recalibrating the instrument, that is, determining a new calorimeter constant value (or confirming that it has not changed.)

Practice Exercise After two semesters of use in a university laboratory, a laboratory instructor replaces the water in a bomb calorimeter. Burning 1.000 g of benzoic acid in the calorimeter causes its temperature to rise by 6.009°C. What is the value of $C_{calorimeter}$? To check the accuracy of $C_{calorimeter}$, 0.500 g of graphite is burned in the calorimeter and its temperature rises 3.73°C. How much energy (in kilojoules) is released during the combustion reaction?

Calorimetry can be used to measure ΔH_{rxn} for other chemical reactions, such as the neutralization reactions discussed in Chapter 4. A simple laboratory apparatus for measuring the heat of reaction between a strong acid and a strong base is shown in **Figure 6.32**. A nested pair of Styrofoam coffee cups acts as our calorimeter, much like the apparatus in Figure 6.30. The pressure remains constant in this kind of coffee-cup calorimeter, so q_{rxn} will be the same as the enthalpy of reaction (ΔH_{rxn}).

Suppose we place 25.0 mL of 1.0 *M* aqueous HCl in the coffee cup and its temperature is 18.5°C. Rapid addition of 25.0 mL of 1.0 *M* NaOH causes the temperature to rise to 25.0°C. What is ΔH_{rxn} for the reaction? (This is the reaction depicted in the Particulate Review on the first page of this chapter.)

$$HCl(aq) + NaOH(aq) \rightarrow NaCl(aq) + H_2O(\ell)$$

The rise in temperature is the result of energy flowing from the reaction (the system) to the water (the surroundings). The amount of heat, q, gained by the surroundings is

$$q_{H_2O} = n_{H_2O} c_P \Delta T$$

First, we need to calculate the moles of water, using the density of H_2O as a reasonable approximation of the density of the solution:

$$n_{H_2O} = 50.0\ \cancel{mL\ H_2O} \times \frac{1.00\ \cancel{g\ H_2O}}{1\ \cancel{mL\ H_2O}} \times \frac{1.00\ mol\ H_2O}{18.02\ \cancel{g\ H_2O}} = 2.77\ mol\ H_2O$$

The total volume of the solution is 50.0 mL because we have mixed two 25.0 mL solutions in our calorimeter. Substituting n_{H_2O} and the values of c_P and ΔT,

$$q_{H_2O} = n_{H_2O} c_P \Delta T = (2.77\ \cancel{mol\ H_2O}) \times \left(\frac{75.3\ J}{1\ \cancel{mol\ H_2O} \cdot \cancel{°C}}\right) \times (25.0\cancel{°C} - 18.5\cancel{°C})$$

$$= 1356\ J$$

The heat gained by the surroundings must equal the heat lost by the system,

$$q_{H_2O} = -q_{rxn} = -1356\ J = -1.356\ kJ$$

Because 25.0 mL of 1.0 *M* HCl contains 0.025 mol of HCl,

$$\Delta H_{rxn} = \frac{-1.356\ kJ}{0.025\ mol\ HCl} = -54\ kJ/mol$$

In this experiment we made three important assumptions:

1. The coffee cup does not absorb any of the heat generated by the reaction.
2. The densities of the relatively dilute solutions are the same as that of pure water.
3. The molar heat capacity of the solution is the same as that of pure water.

These assumptions are reasonably valid for our example, but they would not be in cases where the concentrations of the solutions are substantially higher. Calorimeters suitable for accurate determination of ΔH_{rxn} are commercially available.

SAMPLE EXERCISE 6.11 Coffee-Cup Calorimetry: ΔH_{soln} of NH_4NO_3 **LO5**

Dissolving 80.0 g of NH_4NO_3 in 505 g of water in a coffee-cup calorimeter causes the temperature of the water to decrease by 13.3°C. What is the enthalpy change that accompanies the dissolution process, ΔH_{soln}, expressed in kJ/mol?

Collect and Organize We are given the masses of solute and solvent used to prepare a solution of ammonium nitrate. The temperature decreases by 13.3°C. We need to determine the value of ΔH_{soln}.

Analyze Ammonium nitrate is the system, and water represents its surroundings.

a. First, we use $q = nc_P\Delta T$ (Equation 6.8) to find q_{surr}.
b. Then we use the relationship $q_{sys} = -q_{surr}$ to obtain q_{sys}.
c. Finally, we calculate ΔH_{soln} by dividing q_{sys} by the number of moles of NH_4NO_3.

Solve

a. The number of moles of water is

$$505\ \text{g H}_2\text{O} \times \frac{1\ \text{mol H}_2\text{O}}{18.02\ \text{g H}_2\text{O}} = 28.0\ \text{mol H}_2\text{O}$$

The heat lost by the water, q_{surr}, is

$$q_{surr} = n_{H_2O}c_{P,H_2O}\Delta T = 28.0\ \text{mol H}_2\text{O} \times \frac{75.3\ \text{J}}{\text{mol H}_2\text{O} \cdot {}^\circ\text{C}} \times (-13.3^\circ\text{C}) = -2.80 \times 10^4\ \text{J}$$

b. The heat gained by the system, q_{sys}, is

$$q_{sys} = -q_{surr} = -(-2.80 \times 10^4\ \text{J}) = 2.80 \times 10^4\ \text{J}$$

c. The number of moles of NH_4NO_3 (the system) is

$$80.0\ \text{g NH}_4\text{NO}_3 \times \frac{1\ \text{mol NH}_4\text{NO}_3}{80.05\ \text{g NH}_4\text{NO}_3} = 0.999\ \text{mol NH}_4\text{NO}_3$$

and the enthalpy of solution of NH_4NO_3 is

$$\Delta H_{soln} = \frac{q_{sys}}{n_{NH_4NO_3}} = \frac{2.80 \times 10^4\ \text{J}}{0.999\ \text{mol NH}_4\text{NO}_3} = 2.81 \times 10^4\ \text{J/mol NH}_4\text{NO}_3$$

$$= 28.1\ \text{kJ/mol NH}_4\text{NO}_3$$

Think About It The value for ΔH_{soln}, 28.1 kJ/mol, is positive, which is consistent with the use of ammonium nitrate in chemical cold packs.

Practice Exercise Addition of 114 g of potassium fluoride to 0.600 L of water ($d = 1.00$ g/mL) causes the temperature to rise by 3.6°C. What is ΔH_{soln} for KF?

6.7 Hess's Law

Most organisms rely on glucose as a fuel for their bodies, but not all of them produce CO_2 and water as the only products. The natural world abounds in organisms that metabolize glucose in different ways. For example, some anaerobic bacteria convert glucose to carbon dioxide, acetic acid, and hydrogen in a reaction that has an enthalpy of reaction of +90 kJ:

$$C_6H_{12}O_6(s) + 2\ H_2O(\ell) \rightarrow 2\ CH_3CO_2H(\ell) + 4\ H_2(g) + 2\ CO_2(g) \qquad \Delta H_1 = +90\ \text{kJ} \quad (6.16)$$

Methanosarcina, a strain of bacteria, converts acetic acid to methane (CH_4) in a reaction that has an enthalpy of reaction of +16 kJ:

$$CH_3CO_2H(\ell) \rightarrow CH_4(g) + CO_2(g) \qquad \Delta H_2 = +16\ \text{kJ} \quad (6.17)$$

In a world where fuel supplies are short, could we use these two reactions to prepare methane and hydrogen, two useful fuels? In principle, we can write an overall reaction equation that results from adding the reactions in Equations 6.16 (step 1) and 6.17 (step 2). However, doing so requires that we multiply Equation 6.17 by 2 to balance the production of acetic acid in step 1 with its consumption in step 2.

$$\begin{array}{ll} (1) & C_6H_{12}O_6(s) + 2\ H_2O(\ell) \rightarrow 2\ CH_3CO_2H(\ell) + 4\ H_2(g) + 2\ CO_2(g) \\ (2) & 2\ [CH_3CO_2H(\ell) \rightarrow CH_4(g) + CO_2(g)] \\ \hline (3) & C_6H_{12}O_6(s) + 2\ H_2O(\ell) \rightarrow 4\ H_2(g) + 4\ CO_2(g) + 2\ CH_4(g) \quad (6.18) \end{array}$$

After simplifying, we obtain the reaction in Equation 6.18: the conversion of glucose to methane, hydrogen, and carbon dioxide (step 3). Just as we obtain the overall chemical equation 3 by adding equations 1 and 2, we obtain the enthalpy of reaction for the overall reaction by adding the ΔH values for reactions 1 and 2. Since we multiplied Equation 6.17 by 2 before adding it to Equation 6.16, we also need to multiply ΔH_2 by 2. The thermochemical equation for the overall reaction is therefore

$$\Delta H_1 + 2\Delta H_2 = \Delta H_3$$
$$+90 \text{ kJ} + 2(16 \text{ kJ}) = +122 \text{ kJ}$$

Hess's law the principle that the enthalpy of reaction ΔH_{rxn} for a reaction that is the sum of two or more reactions is equal to the sum of the ΔH_{rxn} values of the constituent reactions; also called *Hess's law of constant heat of summation*.

This calculation for ΔH_{rxn} is an application of **Hess's law**. Also known as *Hess's law of constant heat of summation*, it states that the enthalpy of reaction ΔH_{rxn} for a process that is the sum of two or more other reactions is equal to the sum of the ΔH_{rxn} values of the constituent reactions.

Hess's law is especially useful for calculating enthalpy changes that are difficult to measure directly. For example, CO_2 is the principal product of the combustion of carbon in the form of charcoal:

$$\text{Reaction A:} \quad C(s) + O_2(g) \rightarrow CO_2(g) \qquad \Delta H_{comb} = -393.5 \text{ kJ}$$

CHEMTOUR Hess's Law

When the oxygen supply is limited, however, the products include carbon monoxide:

$$\text{Reaction B:} \quad C(s) + \tfrac{1}{2}O_2(g) \rightarrow CO(g)$$

It is difficult to measure the enthalpy of combustion of this reaction directly because, as long as any oxygen is present, some of the $CO(g)$ formed reacts with the O_2 to form $CO_2(g)$, yielding a mixture of CO and CO_2 as the product. However, we can use Hess's law to obtain this value *indirectly* by working with ΔH_{comb} values that we can measure.

Because we can run reaction A with excess oxygen and thereby force it to completion, we can measure the enthalpy of combustion, which is −393.5 kJ. We can also react a sample of pure $CO(g)$ with oxygen and measure ΔH_{comb} for that reaction:

$$\text{Reaction C:} \quad CO(g) + \tfrac{1}{2}O_2(g) \rightarrow CO_2(g) \qquad \Delta H_{comb} = -283.0 \text{ kJ}$$

Hess's law gives us a way to calculate ΔH_{comb} for reaction B from the measured values for reactions A and C. To do this, we must find a way to combine the equations for reactions A and C so that the sum equals reaction B. Once we have that combination, we can calculate the ΔH_{comb} value we do not know from the two that we do.

One approach to this analysis is to focus on the reactants and products in the reaction whose ΔH_{comb} value is unknown—reaction B in this example. This reaction has carbon and oxygen as reactants and carbon monoxide as a product. Reaction C has carbon monoxide as a reactant. If we add reaction C to reaction B, the carbon monoxide cancels out, and we end up with reaction A as the sum of C and B:

$$\begin{array}{lll} \text{Reaction B:} & C(s) + \tfrac{1}{2}O_2(g) \rightarrow \cancel{CO(g)} & \Delta H_{comb} = ? \\ \text{Reaction C:} & \cancel{CO(g)} + \tfrac{1}{2}O_2(g) \rightarrow CO_2(g) & \Delta H_{comb} = -283.0 \text{ kJ} \\ \hline \text{Reaction A:} & C(s) + O_2(g) \rightarrow CO_2(g) & \Delta H_{comb} = -393.5 \text{ kJ} \end{array}$$

We now use algebra to find ΔH_{comb} for reaction B:

$$\Delta H_B + \Delta H_C = \Delta H_A$$
$$\Delta H_B = \Delta H_A - \Delta H_C$$
$$= -393.5 \text{ kJ} - (-283.0 \text{ kJ}) = -110.5 \text{ kJ}$$

ΔH is a state function, so we can manipulate equations in two important ways when applying Hess's law, should the need arise. (1) We can multiply the coefficients in a balanced equation and the ΔH for the reaction by the same factor to change the quantity of material we are dealing with, as we did with Equation 6.17. (2) We can reverse a reaction (i.e., make the reactants the products and the products the reactants) if we also change the sign of ΔH.

Hess's law is a direct consequence of the fact that enthalpy is a state function. In other words, for a particular set of reactants and products, the enthalpy change of the reaction is the same whether the reaction takes place in one step or in a series of steps. The concept of enthalpy and its expression in Hess's law are very useful because the enthalpy changes associated with many reactions can be calculated from a few that have been measured.

SAMPLE EXERCISE 6.12 Applying Hess's Law in Biology **LO6**

One source of energy in our bodies is the conversion of sugars to CO_2 and H_2O (called respiration; see Section 3.5). Maltose is one of the sugars found in foods. Show how we could combine reactions A and B below to get reaction C. How is ΔH_C related to the values of ΔH_A and ΔH_B?

Reaction A: maltose(s) + $H_2O(\ell) \rightarrow$ 2 glucose(s) $\quad \Delta H_A$

Reaction B: 6 $CO_2(g)$ + 6 $H_2O(\ell) \rightarrow$ glucose(s)+ 6 $O_2(g)$ $\quad \Delta H_B$

Reaction C: maltose(s) + 12 $O_2(g) \rightarrow$ 12 $CO_2(g)$ + 11 $H_2O(\ell)$ $\quad \Delta H_C$

Collect and Organize We are asked to combine two chemical equations to write the equation of an overall reaction and to combine two ΔH_{rxn} values to obtain an overall ΔH_{rxn} value. Hess's law allows us to combine the enthalpy changes that accompany the steps in an overall chemical reaction to calculate the value of ΔH_{rxn} of the overall reaction.

Analyze Maltose is a reactant in reactions A and C, so we can leave reaction A unchanged. The products of reaction C appear as reactants in reaction B, so we need to reverse B to get CO_2 and H_2O on the product side. To balance the moles of glucose produced in reaction A and consumed in reaction B, we also need to multiply the reversed reaction B by 2 before adding it to reaction A. Reversing reaction B and multiplying its coefficients by 2 means that we must also change the sign of ΔH_B and multiply its value by 2.

Solve We start with reaction A as written, and reverse and double reaction B:

maltose(s) + $H_2O(\ell) \rightarrow$ 2 glucose(s) $\quad \Delta H_A$

2 × [glucose(s) + 6 $O_2(g) \rightarrow$ 6 $CO_2(g)$ + 6 $H_2O(\ell)$] $\quad 2 \times [-\Delta H_B]$

Adding A + 2(−B):

$$
\begin{array}{rll}
\text{A:} & \text{maltose}(s) + \cancel{H_2O(\ell)} \rightarrow \cancel{2\ \text{glucose}(s)} & \Delta H_A \\
-(2 \times \text{B}): & \cancel{2\ \text{glucose}(s)} + 12\ O_2(g) \rightarrow 12\ CO_2(g) + \overset{11}{\cancel{12}}\ H_2O(\ell) & -2\Delta H_B \\
\hline
\text{C:} & \text{maltose}(s) + 12\ O_2(g) \rightarrow 12\ CO_2(g) + 11\ H_2O(\ell) & \Delta H_C
\end{array}
$$

Hess's law allows us to add the respective enthalpies: $\Delta H_C = \Delta H_A - 2\Delta H_B$.

Think About It Chemical equations can be added just like algebraic equations, multiplying them by coefficients if necessary, to obtain an equation for a new chemical reaction. A key to applying Hess's law is to treat the values of ΔH the same way that you treat the coefficients in the chemical equations and to change the sign of ΔH if an equation is reversed.

Practice Exercise How would you combine reactions A–C, shown below, to obtain reaction D?

A: $2\,H_2(g) + O_2(g) \rightarrow 2\,H_2O(g)$

B: $H_3BNH_3(s) \rightarrow NH_3(g) + BH_3(g)$

C: $H_3BNH_3(s) \rightarrow 2\,H_2(g) + HBNH(s)$

D: $NH_3(g) + BH_3(g) + O_2(g) \rightarrow 2\,H_2O(g) + HBNH(s)$

SAMPLE EXERCISE 6.13 Calculating Enthalpies of Reaction by Using Hess's Law — LO6

Hydrocarbons burned in a limited supply of air may not burn completely, and $CO(g)$ may be generated. One reason furnaces and hot-water heaters fueled by natural gas need to be vented is that incomplete combustion can produce toxic carbon monoxide:

Reaction A: $2\,CH_4(g) + 3\,O_2(g) \rightarrow 2\,CO(g) + 4\,H_2O(g)$ $\quad \Delta H_A = ?$

Use reactions B and C to calculate the ΔH_{comb} for reaction A.

Reaction B: $CH_4(g) + 2\,O_2(g) \rightarrow CO_2(g) + 2\,H_2O(g)$ $\quad \Delta H_B = -802\ kJ$

Reaction C: $2\,CO(g) + O_2(g) \rightarrow 2\,CO_2(g)$ $\quad \Delta H_C = -566\ kJ$

Collect and Organize We are given two reactions (B and C) with thermochemical data and a third (A) for which we are asked to find ΔH_A. All the reactants and products of reaction A are present in reaction B or C or both.

Analyze We can manipulate the equations for reactions B and C so that they sum to give the equation for which ΔH_{comb} is unknown. Then we can calculate this unknown value by applying Hess's law.

The reaction of interest (A) has methane on the reactant side. Because reaction B also has methane as a reactant, we can use B as written. Reaction A has CO as a product. Reaction C involves CO as a reactant, so we must reverse C to get CO on the product side. Once we reverse C, we must change the sign of ΔH_C. If the coefficients as given do not allow us to sum the two reactions to yield reaction A, we can multiply one or both reactions by other factors.

Solve We start with reaction B as written and add the reverse of reaction C, remembering to change the sign of ΔH_C:

B: $CH_4(g) + 2\,O_2(g) \rightarrow CO_2(g) + 2\,H_2O(g)$ $\quad \Delta H_B = -802\ kJ$

C (reversed): $2\,CO_2(g) \rightarrow 2\,CO(g) + O_2(g)$ $\quad -\Delta H_C = 566\ kJ$

Because methane has a coefficient of 2 in reaction A, we multiply all the terms in reaction B, including ΔH_{comb}, by 2:

$$2 \times [CH_4(g) + 2\,O_2(g) \rightarrow CO_2(g) + 2\,H_2O(g)] \qquad 2 \times [\Delta H_B = -802\ kJ]$$

Because the carbon monoxide in reaction A has a coefficient of 2, we do not need to multiply reaction C by any factor. Now we add (2 × B) to the reverse of C and cancel out common terms:

2 × B: $2\,CH_4(g) + \overset{3}{\cancel{4}}\,O_2(g) \rightarrow \cancel{2\,CO_2(g)} + 4\,H_2O(g)$ $\quad 2\Delta H_B = -1604\ kJ$

C (reversed): $\cancel{2\,CO_2(g)} \rightarrow 2\,CO(g) + \cancel{O_2(g)}$ $\quad -\Delta H_C = 566\ kJ$

A: $2\,CH_4(g) + 3\,O_2(g) \rightarrow 2\,CO(g) + 4\,H_2O(g)$ $\quad \Delta H_A = -1038\ kJ$

Think About It We used Hess's law to calculate the enthalpy of combustion of methane to make CO. This is impossible to achieve in an experiment because any CO produced will react with O_2 to give CO_2, resulting in a product mixture of CO and CO_2. The answer of −1038 kJ is less negative than the enthalpy change accompanying the complete combustion of 2 moles of CH_4 [2 × (−802 kJ)], so the answer is reasonable.

Practice Exercise It does not matter how you assemble the equations in a Hess's law problem. Show that reactions A and C can be summed to give reaction B and result in the same value for ΔH_B.

6.8 Standard Enthalpies of Formation and Reaction

As noted in Section 6.3, it is impossible to measure the *absolute* value of the internal energy of a substance. The same is true for the enthalpy of a substance. However, we can establish *relative* enthalpy values that are referenced to a convenient standard. This approach is like using the freezing point of water as the zero point on the Celsius temperature scale or to using sea level as the zero point for expressing altitude. The enthalpy value referenced to this zero point is a substance's **standard enthalpy of formation (ΔH_f°)**, or *standard heat of formation*, defined as the enthalpy change that takes place at constant pressure when 1 mole of a substance is formed from its constituent elements in their standard states. A reaction that fits this description is known as a **formation reaction.**

The adjective *standard*, indicated by the symbol °, as seen in ΔH_f°, indicates that the associated reaction occurs under **standard conditions**, which means at a constant pressure of 1 bar and at some specified temperature. The value of 1 bar of pressure is very close to 1 atm; for the level of precision used in this book, a standard pressure of 1 bar will be considered equivalent to 1 atm. There is no universal standard temperature, though many tables of thermodynamic data, including those in the appendix of this book, apply to processes occurring at 25°C.

We also use the term **standard state** to describe the most stable physical state of a substance under standard conditions. In their standard states, oxygen is a gas, water is a liquid, and carbon is solid graphite. By definition, for a pure element this is its most stable form under standard conditions. It is the zero point of enthalpy values. **Table 6.2** lists standard enthalpies of formation for several substances. (A more complete list can be found in Appendix 4.[1])

The general symbol for the enthalpy change associated with a reaction that takes place under standard conditions is ΔH°_{rxn}, and the value is called either a **standard enthalpy of reaction** or a *standard heat of reaction*. Implied in our notion of standard states and standard conditions is the assumption that parameters such as ΔH change with temperature and pressure. That assumption is correct, although the changes are so small that we ignore them in the calculations in this textbook.

Because the definition of a formation reaction specifies 1 mole of product, writing balanced equations for formation reactions may require the use of something we typically avoided in Chapter 3—namely, fractional coefficients in the final form of our balanced equations. For example, the thermochemical equation for the synthesis of ammonia from nitrogen and hydrogen is usually written

$$N_2(g) + 3\ H_2(g) \rightarrow 2\ NH_3(g) \qquad \Delta H^\circ_{rxn} = -92.2\ \text{kJ}$$

TABLE 6.2 Standard Enthalpies of Formation for Selected Substances at 25°C

Substance	ΔH_f° (kJ/mol)
$Br_2(\ell)$	0
C(s, graphite)	0
$CH_4(g)$, methane	−74.8
$C_2H_2(g)$, acetylene	226.7
$C_2H_4(g)$, ethylene	52.26
$C_2H_6(g)$, ethane	−84.68
$C_3H_8(g)$, propane	−103.8
$C_4H_{10}(g)$, butane	−125.6
$CH_3OH(\ell)$, methanol	−238.7
$CH_3CH_2OH(\ell)$, ethanol	−277.7
$CH_3COOH(\ell)$, acetic acid	−484.5
CO(g)	−110.5
$CO_2(g)$	−393.5
$H_2(g)$	0
$H_2O(g)$	−241.8
$H_2O(\ell)$	−285.8
$N_2(g)$	0
$NH_3(g)$, ammonia	−46.1
$N_2H_4(g)$, hydrazine	95.4
$N_2H_4(\ell)$	50.63
NO(g)	90.3
$O_2(g)$	0

[1]The website of the National Institute of Standards and Technology contains much additional data: http://webbook.nist.gov.

Although all reactants and products in this equation are in their standard states, it is not a formation reaction because 2 moles of product are formed, which is why we denote the enthalpy change of this reaction by ΔH°_{rxn} rather than ΔH°_f. The formation reaction for ammonia must, by definition, show 1 mole of product being formed, so we divide each coefficient in the above equation by 2. Because energy is a stoichiometric quantity, the enthalpy of reaction is divided by 2 as well. Thus, the thermochemical equation representing the formation reaction of ammonia is

$$\tfrac{1}{2}\,N_2(g) + \tfrac{3}{2}\,H_2(g) \rightarrow NH_3(g) \qquad \Delta H^\circ_f = -46.1\text{ kJ}$$

standard enthalpy of formation (ΔH°_f) the enthalpy change of a formation reaction; also called *standard heat of formation* or *heat of formation*.

formation reaction a reaction in which 1 mole of a substance is formed from its component elements in their standard states.

standard conditions in thermodynamics: a pressure of 1 bar (~1 atm) and some specified temperature, assumed to be 25°C unless otherwise stated; for solutions, a concentration of 1 *M* is specified.

standard state the most stable form of a substance under 1 bar pressure and some specified temperature (25°C unless otherwise stated).

standard enthalpy of reaction (ΔH°_{rxn}) the energy associated with a reaction that takes place under standard conditions; also called *standard heat of reaction*.

SAMPLE EXERCISE 6.14 Recognizing Formation Reactions LO7

Which of the following reactions are formation reactions at 25°C? For those that are not, explain why not.

a. $H_2(g) + \tfrac{1}{2}\,O_2(g) \rightarrow H_2O(g)$
b. $C(s\text{, graphite}) + 2\,H_2(g) + \tfrac{1}{2}\,O_2(g) \rightarrow CH_3OH(\ell)$
(CH$_3$OH is methanol, a liquid in its standard state.)
c. $CH_4(g) + 2\,O_2(g) \rightarrow CO_2(g) + 2\,H_2O(\ell)$
d. $P_4(s) + 2\,O_2(g) + 6\,Cl_2(g) \rightarrow 4\,POCl_3(\ell)$
(In their standard states, P_4 is a solid, Cl_2 is a gas, and $POCl_3$ is a liquid.)

Collect and Organize We are given four balanced chemical equations and information about the standard states of specific reactants and products. We want to determine which of these are formation reactions.

Analyze For a reaction to be a formation reaction, it must produce 1 mole of a substance from its component elements in their standard states. Therefore, we must evaluate each reaction for the quantity of product and for the state of each reactant.

Solve

a. The reaction shows 1 mole of water vapor formed from its constituent elements in their standard states. This, then, is the formation reaction for $H_2O(g)$, and its heat of reaction is ΔH°_f.
b. The reaction shows 1 mole of liquid methanol formed from its constituent elements in their standard states. This reaction is a formation reaction.
c. This is *not* a formation reaction because the reactants are not elements in their standard states and because more than one product is formed.
d. This is *not* a formation reaction because the product is 4 moles of $POCl_3$. Note, however, that all the constituent elements are in their standard states and that only one compound is formed. We could convert this into a formation reaction by dividing all the coefficients by 4.

Think About It Most chemical equations that describe formation reactions, including the (a) and (b) equations above, contain fractional coefficients on the reactant side. They are needed so that the coefficients for the single products can be 1. Also, these equations sometimes describe reactions that don't happen on their own. Rather, they provide reference values against which the enthalpy changes in reactions that do happen can be evaluated.

 **Practice Exercise** Write formation reactions for (a) $CaCO_3(s)$, (b) $CH_3COOH(\ell)$ (acetic acid), and (c) $KMnO_4(s)$.

The standard enthalpies of formation of acetylene and ethylene are positive, which means that the formation reactions for these compounds are endothermic. The other hydrocarbon fuels in Table 6.2 have negative enthalpies of formation, but the values for all of them are less negative than the values for water and carbon dioxide. Recall that the products of the complete combustion of hydrocarbons are carbon dioxide and water. The more negative the enthalpy of formation (ΔH_f°) of a substance, the more stable it is. The implication of this, as stated earlier, is that the reaction of fuels with oxygen to produce CO_2 and H_2O is exothermic. The reactions produce energy, which is why hydrocarbons are useful as fuels.

Standard enthalpies of formation ΔH_f° are used to predict standard enthalpies of reaction ΔH_{rxn}°. We can calculate the standard heat of reaction for any reaction by determining the difference between the ΔH_f° values of the products and the ΔH_f° values of the reactants. To see how this approach works, consider the reaction we saw earlier as Equation 6.18:

$$C_6H_{12}O_6(s) + 2\,H_2O(\ell) \rightarrow 4\,H_2(g) + 4\,CO_2(g) + 2\,CH_4(g)$$

In this calculation, we use data from Table 6.2 in the following equation:

$$\Delta H_{rxn}^\circ = \sum n_{products}\,\Delta H_{f,products}^\circ - \sum n_{reactants}\,\Delta H_{f,reactants}^\circ \qquad (6.19)$$

where $n_{products}$ is the number of moles of each product in the balanced equation and $n_{reactants}$ is the number of moles of each reactant. Equation 6.19 states that the value of ΔH_{rxn}° for any chemical reaction equals the sum ($\sum$) of the ΔH_f° value for each product times the number of moles of that product in the balanced equation, minus the sum of the ΔH_f° value for each reactant times the number of moles of that reactant in the balanced chemical equation. Thus, we multiply the value of ΔH_f° for H_2O in Equation 6.18 by 2 before summing with the value of ΔH_f° for $C_6H_{12}O_6$ because the H_2O coefficient is 2 in the balanced equation. We do the same with the ΔH_f° values of the products, multiplying them by 4, 4, and 2, respectively.

Once we insert ΔH_f° values from Table 6.2 for the reactants and the products in Equation 6.18, remembering that $\Delta H_f^\circ = 0$ for $H_2(g)$ because it is a pure element in its most stable form, Equation 6.19 becomes

$$\begin{aligned}\Delta H_{rxn}^\circ &= [(4\text{ mol }H_2)(0.0\text{ kJ/mol}) + (4\text{ mol }CO_2)(-393.5\text{ kJ/mol})\\ &\quad + (2\text{ mol }CH_4)(-74.8\text{ kJ/mol})] - [(1\text{ mol }C_6H_{12}O_6)(-1274.4\text{ kJ/mol})\\ &\quad + (2\text{ mol }H_2O)(-285.8\text{ kJ/mol})]\\ &= [(0.0\text{ kJ}) + (-1574\text{ kJ}) + (-149.6\text{ kJ})] - [(-1274.4\text{ kJ}) + (-571.6\text{ kJ})]\\ &= 122.4\text{ kJ}\end{aligned}$$

Now, let's consider the reaction between methane and steam that yields carbon dioxide and hydrogen gas:

$$CH_4(g) + 2\,H_2O(g) \rightarrow CO_2(g) + 4\,H_2(g)$$

This reaction is used to synthesize the hydrogen used in the steel industry to remove impurities from molten iron and in the chemical industry to make hundreds of compounds, including ammonia (NH_3) for use in fertilizers and as a refrigerant. The reaction is also important in the manufacture of hydrogen fuel for fuel cells, which are used to generate electricity directly from a chemical reaction.

Inserting the appropriate values from Table 6.2 into Equation 6.19, along with the coefficients 2 for H_2O and 4 for H_2,

$$\begin{aligned}\Delta H_{rxn}^\circ &= [(1\text{ mol }CO_2)(-393.5\text{ kJ/mol}) + (4\text{ mol }H_2)(0.0\text{ kJ/mol})]\\ &\quad - [(1\text{ mol }CH_4)(-74.8\text{ kJ/mol}) + (2\text{ mol }H_2O)(-241.8\text{ kJ/mol})]\\ &= +165\text{ kJ}\end{aligned}$$

The positive standard enthalpy of reaction tells us that this reaction between water vapor and methane is endothermic. In other words, energy must be added to make the reaction take place (typically conducted at temperatures near 1000°C).

Enthalpy is a state function, and, as noted in Section 6.2, the value of a state function is independent of the path taken to achieve that state. Thus, values of ΔH_{rxn} are independent of pathway. It does not matter what path we take to get from the reactants to the products; the enthalpy difference between them is always the same. In fact, on an industrial scale, the reaction to produce hydrogen gas from methane is carried out in two steps. In the first step, methane reacts with a limited supply of steam to produce carbon monoxide and hydrogen gas (which is called the *steam reforming reaction*). In the second step, the carbon monoxide reacts with additional steam and more hydrogen gas:

Step 1:	$CH_4(g) + H_2O(g) \rightarrow \cancel{CO(g)} + 3\ H_2(g)$	$\Delta H_1 = +206$ kJ
Step 2:	$\cancel{CO(g)} + H_2O(g) \rightarrow CO_2(g) + H_2(g)$	$\Delta H_2 = -41$ kJ
Overall reaction:	$CH_4(g) + 2\ H_2O(g) \rightarrow CO_2(g) + 4\ H_2(g)$	$\Delta H_3 = +165$ kJ

Figure 6.33 depicts the enthalpy changes involved in these two steps, as well as the overall reaction. The enthalpy change for the overall reaction has a value of +165 kJ, the same value we calculated with Equation 6.19 by using enthalpies of formation.

Just as we saw in Section 6.3 with state changes, once we know the value of the energy associated with running a reaction in the forward direction, we also know the value of the energy change when the reaction runs in reverse: it has the same magnitude but the opposite sign. This means that, having calculated a value of $\Delta H^\circ_{rxn} = +165$ kJ for the steam reforming reaction, we can write

$$CO(g) + 3\ H_2(g) \rightarrow CH_4(g) + H_2O(g) = \Delta H^\circ_{rxn} -206 \text{ kJ}$$

for the exothermic reverse reaction, as illustrated in **Figure 6.34**.

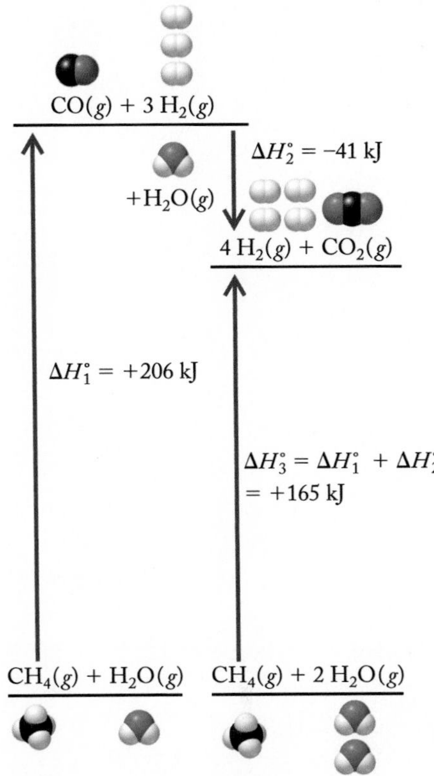

FIGURE 6.33 Hess's law predicts that the enthalpy change for the production of 4 moles of $H_2(g)$ and 1 mole of $CO_2(g)$ from 1 mole of $CH_4(g)$ and 2 moles of $H_2O(g)$, ΔH°_3, is the sum of the enthalpies of two reactions: $\Delta H^\circ_3 = \Delta H^\circ_1 + \Delta H^\circ_2$.

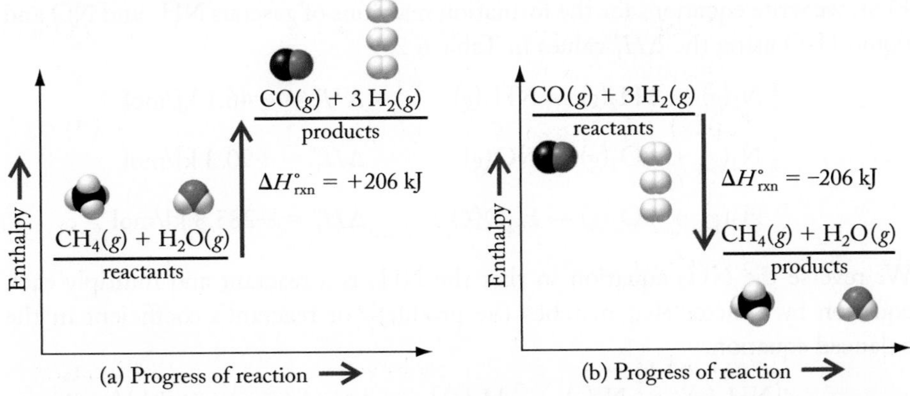

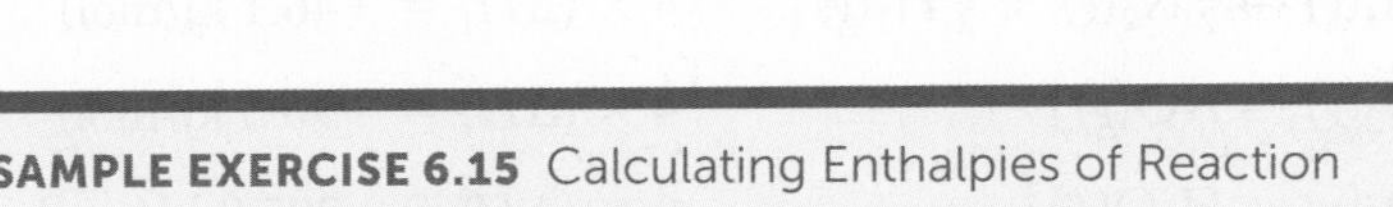

FIGURE 6.34 (a) The reaction of methane with water to produce carbon monoxide and hydrogen is endothermic. (b) The reverse reaction, carbon monoxide plus hydrogen producing methane and water, is exothermic. The value of the enthalpy change of the two reactions has the same magnitude but is positive for the endothermic reaction and negative for the exothermic reaction.

SAMPLE EXERCISE 6.15 Calculating Enthalpies of Reaction **LO6**

Using the appropriate values from Table 6.2, calculate ΔH°_{rxn} for the combustion of the fuel propane (C_3H_8) in air.

Collect and Organize The reactants are propane and elemental oxygen, O_2, and the products are carbon dioxide and water. Even though combustion is an exothermic reaction, we assume that water is produced as liquid, $H_2O(\ell)$, so that all products and reactants are in their standard states. The heats of formation of propane, carbon dioxide, and $H_2O(\ell)$ are given in Table 6.2. The heat of formation of the element O_2 in its standard state is zero. Our task is to use the balanced equation and the data from Table 6.2 to calculate the enthalpy of combustion.

Analyze Equation 6.19 defines the relation between heats of formation of reactants and products and the standard enthalpy of reaction. We also need the balanced equation for the combustion of propane:

$$C_3H_8(g) + 5\,O_2(g) \rightarrow 3\,CO_2(g) + 4\,H_2O(\ell)$$

Solve Inserting ΔH°_f values for the products [$CO_2(g)$ and $H_2O(\ell)$] and reactants [$C_3H_8(g)$ and $O_2(g)$] from Table 6.2 and the coefficients in the balanced chemical equation into Equation 6.19, we get

$$\begin{aligned}\Delta H^\circ_{rxn} &= [(3\text{ mol }CO_2)(-393.5\text{ kJ/mol}) + (4\text{ mol }H_2O)(-285.8\text{ kJ/mol})] \\ &\quad -[(1\text{ mol }C_3H_8)(-103.8\text{ kJ/mol}) + (5\text{ mol }O_2)(0.0\text{ kJ/mol})] \\ &= -2219.9\text{ kJ}\end{aligned}$$

Think About It The result of the calculation has a large negative value, which means that the combustion reaction is highly exothermic, as expected for a hydrocarbon fuel.

 **Practice Exercise** Calculate ΔH°_{rxn} for the *water–gas shift reaction*:

$$CO(g) + H_2O(g) \rightarrow CO_2(g) + H_2(g)$$

When we use standard enthalpies of formation (ΔH°_f) to determine standard enthalpies of reaction ΔH°_{rxn}, we are applying Hess's law. Consider, for example, the reaction of ammonia with oxygen to make $NO(g)$ and liquid water (this is the first step of the industrial synthesis of nitric acid):

$$4\,NH_3(g) + 5\,O_2(g) \rightarrow 4\,NO(g) + 6\,H_2O(\ell)$$

First, we write equations for the formation reactions of gaseous NH_3 and NO and liquid H_2O using the ΔH°_f values in Table 6.2:

$$\tfrac{1}{2}N_2(g) + \tfrac{3}{2}H_2(g) \rightarrow NH_3(g) \qquad \Delta H^\circ_f = -46.1\text{ kJ/mol}$$

$$\tfrac{1}{2}N_2(g) + \tfrac{1}{2}O_2(g) \rightarrow NO(g) \qquad \Delta H^\circ_f = +90.3\text{ kJ/mol}$$

$$H_2(g) + \tfrac{1}{2}O_2(g) \rightarrow H_2O(\ell) \qquad \Delta H^\circ_f = -285.8\text{ kJ/mol}$$

We reverse the NH_3 equation so that the NH_3 is a reactant and multiply each equation by a factor that matches the product's or reactant's coefficient in the balanced equation.

$$4[NH_3(g) \rightarrow \tfrac{1}{2}N_2(g) + \tfrac{3}{2}H_2(g)] \qquad 4 \times (\Delta H^\circ_f = +46.1\text{ kJ/mol})$$

$$4[\tfrac{1}{2}N_2(g) + \tfrac{1}{2}O_2(g) \rightarrow NO(g)] \qquad 4 \times (\Delta H^\circ_f = +90.3\text{ kJ/mol})$$

$$6[H_2(g) + \tfrac{1}{2}O_2(g) \rightarrow H_2O(\ell)] \qquad 6 \times (\Delta H^\circ_f = -285.8\text{ kJ/mol})$$

Their sum yields the reaction of interest:

$$4\,NH_3(g) \rightarrow \cancel{2\,N_2(g)} + \cancel{6\,H_2(g)} \qquad \Delta H^\circ_{rxn} = +184.4\text{ kJ}$$

$$\cancel{2\,N_2(g)} + 2\,O_2(g) \rightarrow 4\,NO(g) \qquad \Delta H^\circ_{rxn} = +361.2\text{ kJ}$$

$$\cancel{6\,H_2(g)} + 3\,O_2(g) \rightarrow 6\,H_2O(\ell) \qquad \Delta H^\circ_{rxn} = -1714.8\text{ kJ}$$

$$4\,NH_3(g) + 5\,O_2(g) \rightarrow 4\,NO(g) + 6\,H_2O(\ell) \qquad \Delta H^\circ_{rxn} = -1169.2\text{ kJ}$$

This is the same mathematical result that we get when we use Equation 6.19:

$$\Delta H^\circ_{rxn} = \sum n_{products}\,\Delta H^\circ_{f,products} - \sum n_{reactants}\,\Delta H^\circ_{f,reactants}$$

$$= [4\ \cancel{mol}\ NO(+90.3\ kJ/\cancel{mol}) + 6\ \cancel{mol}\ H_2O(-285.8\ kJ/\cancel{mol})]$$

$$-[4\ \cancel{mol}\ NH_3(-46.1\ kJ/\cancel{mol}) + 5\ \cancel{mol}\ O_2(0\ kJ/\cancel{mol})] = -1169.2\ kJ$$

saturated hydrocarbon an alkane; compounds containing the maximum ratio of hydrogen atoms to carbon atoms.

methylene group ($-CH_2-$) a structural unit that can make two bonds.

methyl group ($-CH_3$) a structural unit that can make only one bond.

6.9 Fuels, Fuel Values, and Food Values

In Section 6.6 we saw that the enthalpy of combustion of propane (−2220 kJ/mol) is much greater than that of methane (−802 kJ/mol)—that is, much more energy is released in the combustion of 1 mole of propane. Does this make propane an inherently more valuable fuel? Not necessarily because we do not purchase fuels, or anything else for that matter, in units of moles. Depending on the fuel, we buy it either by mass (coal) or by volume (gasoline and diesel oil). Before addressing the question of which fuel is better, let's take a quick look at the molecular structures of methane, propane, and other related hydrocarbons.

CONNECTION In Chapter 2 we defined hydrocarbons as compounds composed of only carbon and hydrogen atoms, and we saw that alkanes are hydrocarbons in which each carbon atom is bonded to four other atoms.

Alkanes

Methane and propane belong to the family of organic compounds called *alkanes*—hydrocarbons distinguished by the fact that each carbon atom is bonded to four other atoms. Alkanes are classified as **saturated hydrocarbons** because they contain the maximum ratio of hydrogen atoms to carbon atoms.

Alkanes are also known as *paraffins*, a name derived from Latin meaning "little affinity." This is a perfect description of the alkanes, which tend to be much less reactive than the other hydrocarbon families. *Unreactive* may not seem like the correct term to apply to compounds that are fuels, but alkanes do not react readily, even with oxygen. As evidence of this, most fuels need some source of energy, such as a spark from a spark plug in an engine, to initiate combustion.

The first four members of the alkane family (**Figure 6.35**) are methane (1 carbon atom: a C_1 alkane), ethane (2 carbon atoms: a C_2 alkane), propane (a C_3 alkane), and butane (a C_4 alkane). The names of alkanes with more than 4 carbon atoms are derived from the IUPAC prefix for the number of carbon atoms per molecule, followed by *-ane*. **Table 6.3** lists the prefixes for the C_1 through C_{10} alkanes.

Alkanes all have the general formula C_nH_{2n+2}, and each alkane differs from the previous one by one $-CH_2-$ unit, which is called a **methylene group**. The terminal $-CH_3$ groups are called **methyl groups**. All the alkanes in the table are

TABLE 6.3 Prefixes for Naming Alkanes

Prefix	Condensed Structure	Name
Meth-	CH_4	Methane
Eth-	CH_3CH_3	Ethane
Prop-	$CH_3CH_2CH_3$	Propane
But-	$CH_3(CH_2)_2CH_3$	Butane
Pent-	$CH_3(CH_2)_3CH_3$	Pentane
Hex-	$CH_3(CH_2)_4CH_3$	Hexane
Hept-	$CH_3(CH_2)_5CH_3$	Heptane
Oct-	$CH_3(CH_2)_6CH_3$	Octane
Non-	$CH_3(CH_2)_7CH_3$	Nonane
Dec-	$CH_3(CH_2)_8CH_3$	Decane

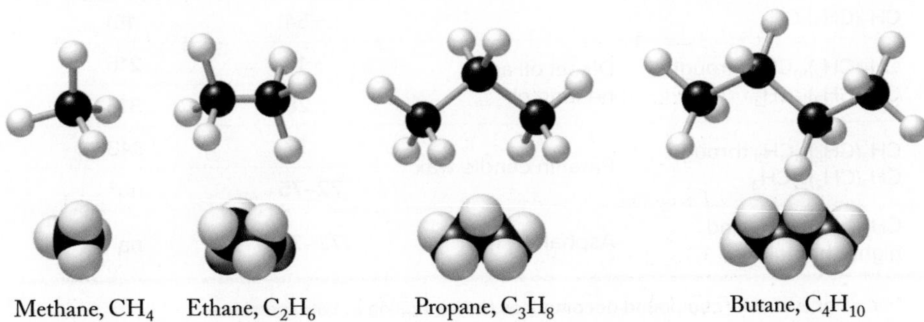

FIGURE 6.35 Ball-and-stick and space-filling models of the first four members of the alkane family: methane, ethane, propane, and butane.

straight-chain hydrocarbon a hydrocarbon in which the carbon atoms are bonded together in one continuous carbon chain.

branched-chain hydrocarbon a hydrocarbon in which the chain of carbon atoms is not linear.

straight-chain hydrocarbons, which means a continuous sequence of carbon atoms with no branching. Straight-chain alkanes have a methyl group at each end with methylene groups connecting them. **Table 6.4** lists the melting points and boiling points for some straight-chain alkanes. These data show similar trends—namely, as straight-chain alkanes increase in molar mass, their melting and boiling points increase.

CONCEPT TEST

Alkanes have the general formula C_nH_{2n+2}. If an alkane has a molar mass of 114 g/mol, what is the value of *n*?

The alkane family would be huge even if it consisted of only straight-chain hydrocarbons, but another structural possibility makes the family even larger. For example, alkanes with four or more carbon atoms can also be **branched-chain hydrocarbons**, as shown for the C_4 alkanes in **Table 6.5**. A *branch* is a side chain attached to the main carbon chain. The molecular formula of both structures in Table 6.5 is C_4H_{10}, but they represent different compounds with different properties. In recognition of these differences, the two compounds have different formal names (butane and 2-methylpropane).

The principal source of liquid alkanes is crude oil. Natural gas is the major source for the simplest alkane, methane, and also smaller quantities of low-molar-mass alkanes, including ethane, propane, and butane. Methane is also produced during bacterial decomposition of vegetable matter in the absence of air, a condition that frequently arises in swamps. Hence methane's common name is swamp gas or marsh gas (**Figure 6.36**).

Combustion reactions between hydrocarbons and oxygen power our vehicles, warm our homes, cook our meals, and generate electricity. Gasoline, kerosene, and diesel oils are complex mixtures of alkanes and other hydrocarbons containing up

TABLE 6.4 Melting Points and Boiling Points for Selected Straight-Chain Alkanes

Condensed Structure	Use	Melting Point (°C)	Normal Boiling Point (°C)
$CH_3CH_2CH_3$	Gaseous fuels	−190	−42
$CH_3(CH_2)_2CH_3$		−138	−0.5
$CH_3(CH_2)_3CH_3$	Gasoline	−130	36
$CH_3(CH_2)_4CH_3$		−95	69
$CH_3(CH_2)_5CH_3$		−91	98
$CH_3(CH_2)_6CH_3$		−57	126
$CH_3(CH_2)_7CH_3$		−54	151
$CH_3(CH_2)_{10}CH_3$ through $CH_3(CH_2)_{16}CH_3$	Diesel oil and heating oil	−10 28	216 316
$CH_3(CH_2)_{18}CH_3$ through $CH_3(CH_2)_{32}CH_3$	Paraffin candle wax	37 72–75	343 na[a]
$CH_3(CH_2)_{34}CH_3$ and higher homologs	Asphalt	72–76	na

[a]na = not available; compound decomposes before boiling at 1 atm pressure.

TABLE 6.5 Comparing Straight-Chain and Branched Hydrocarbons

	Butane	2-Methylpropane
Condensed Structure	$CH_3CH_2CH_2CH_3$	$CH_3CH(CH_3)CH_3$ or CH_3 / CH_3CHCH_3
Ball-and-Stick Model	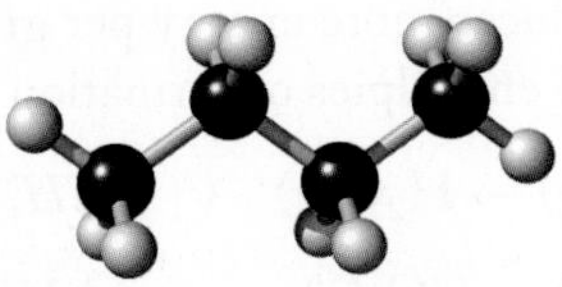	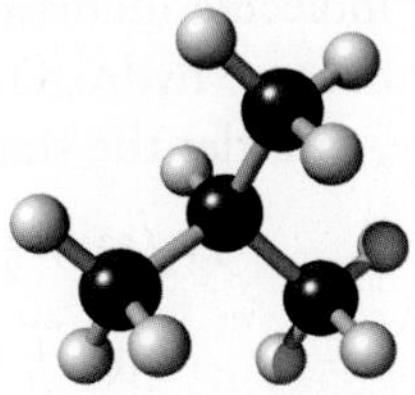
Space-Filling Model		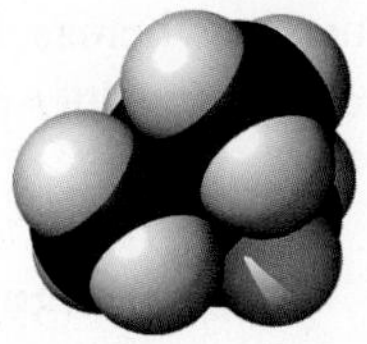
Melting Point (°C)	−138	−160
Normal Boiling Point (°C)	0	−12
Density of Gas (g/L)	0.5788	0.5934

to 18 carbon atoms per molecule, as shown in Table 6.4. The energies released by the combustion of these fuels and several individual alkanes are listed in **Table 6.6**. Alkanes with higher molar masses are viscous liquids that are used as lubricating oils. Low-melting solid alkanes (C_{20}–C_{40}) are used in candles and in manufacturing matches. Very heavy hydrocarbon gums and solid residues (C_{36} and up) are used for paving roads.

TABLE 6.6 Standard Enthalpies of Combustion and Fuel Values of H_2 and Hydrocarbon Fuels

Substance	ΔH°_{comb} (kJ/mol)	Fuel Value (kJ/g)
$H_2(g)$	−241.8	119.9
$CH_4(g)$, methane	−802.3	50.01
$C_3H_8(g)$, propane	−2044.2	46.36
$C_5H_{12}(\ell)$, pentane	−3272.6	45.38
$C_8H_{16}(\ell)$, octane	−5074.9	44.43
Gasoline (d = 0.737 g/mL)		43.4
Kerosene (d = 0.821 g/mL)		43.0
Diesel oil (d = 0.846 g/mL)		42.6

(a)

(b)

FIGURE 6.36 Methane (CH_4) is a renewable source of energy because it can be produced by the degradation of organic matter by methanogenic bacteria. Methane is produced in swamps, so it is commonly called swamp gas. (a) Methane bubbles trapped in a frozen pond. Methane is produced by rotting organic matter at the bottom of the pond. (b) Experiments in the bulk production of methane from natural sources, like the one shown here of a plastic tarp covering a lagoon of animal waste, are being carried out worldwide to augment the fuel supply.

Fuel Value

To calculate the energy released when 1 g of a hydrocarbon fuel burns in air, producing CO_2 and water vapor, we divide the absolute value of its standard enthalpy of combustion ($\Delta H^\circ_{\text{comb}}$, kJ/mol) by its molar mass (g/mol). The result is an energy value expressed in kJ/g that is called **fuel value**.

Fuel values for several alkanes are listed in Table 6.6. Note how they decrease with increasing molar mass. Why is that the case? The answer lies in the hydrogen-to-carbon ratios in these compounds. As the number of carbon atoms per molecule increases, the hydrogen-to-carbon ratio decreases. This is important because hydrogen produces more energy per gram than carbon does. To see why, consider the standard enthalpies of formation of $H_2O(g)$ and $CO_2(g)$:

$$H_2(g) + \tfrac{1}{2}O_2(g) \rightarrow H_2O(g) \qquad \Delta H^\circ_f = -241.8 \text{ kJ/mol}$$

$$C(s) + O_2(g) \rightarrow CO_2(g) \qquad \Delta H^\circ_f = -393.5 \text{ kJ/mol}$$

These equations tell us that the combustion of 1 mol (2.016 g) H_2 releases 241.8 kJ of energy, whereas 1 mol (12.011 g) C releases 393.5 kJ of energy. Expressing these energy values per gram of fuel we get:

$$241.8 \text{ kJ}/2.016 \text{ g} = 119.9 \text{ kJ/g for hydrogen}$$

$$393.5 \text{ kJ}/12.011 \text{ g} = 32.76 \text{ kJ/g for carbon}$$

Thus, gram for gram, hydrogen has nearly four times the fuel value of carbon (which is why H_2 appears at the top of the list of fuels in Table 6.6).

Table 6.6 also contains fuel values for three common fuels: gasoline, kerosene, and diesel oil. Each is a complex mixture of hydrocarbons whose composition can vary from sample to sample, depending on the source of the crude oil from which it was obtained. This variability can lead to differences in fuel values. Those in Table 6.6 apply to samples of these fuels that have the physical densities (expressed in g/mL) also listed in the table.

Another term frequently used to compare the energy content of liquid fuels is **fuel density**. Fuel density describes the amount of energy available per unit volume of fuel. They are typically expressed in kJ/mL for liquid fuels and kJ/L for gaseous fuels. Both fuel value and fuel density are reported as positive numbers, but it is understood that these values refer to the energy released from the fuel when it is burned.

CONCEPT TEST

Without doing any calculations, predict which compound in each pair releases more energy during combustion in air: (a) 1 mole of CH_4 or 1 mole of H_2; (b) 1 g of CH_4 or 1 g of H_2.

fuel value the quantity of energy released during the complete combustion of 1 g of a substance.

fuel density the quantity of energy released during the complete combustion of 1 L of a liquid fuel.

food value the quantity of energy produced when a material consumed by an organism for sustenance is burned completely; it is typically reported in Calories (kilocalories) per gram of food.

SAMPLE EXERCISE 6.16 Comparing Fuel Values and Fuel Densities LO8

Most automobiles run on either gasoline or diesel oil. Use the data in Table 6.6 to compare the fuel values (kJ/g) and fuel densities (kJ/mL) of these widely used fuels.

Collect, Organize, and Analyze We are to compare the fuel values and fuel densities of gasoline and diesel oil. Table 6.6 contains the fuel values of both fuels and their physical densities, which we can use to convert their fuel values to fuel densities.

Solve The fuel values of gasoline and diesel oil in Table 6.6 are 43.4 and 42.6 kJ/g, respectively. The ratio of these two values: 43.4/42.6 = 1.019, tell us that the fuel value of gasoline is about 2% higher than the fuel value of diesel oil. Using the physical densities of the fuels to calculate their fuel densities:

$$\text{Gasoline:} \qquad 43.4\,\frac{\text{kJ}}{\text{g}} \times 0.737\,\frac{\text{g}}{\text{mL}} = 32.0\ \text{kJ/mL}$$

$$\text{Diesel oil:} \qquad 42.6\,\frac{\text{kJ}}{\text{g}} \times 0.846\,\frac{\text{g}}{\text{mL}} = 36.0\ \text{kJ/mL}$$

The ratio of these two values is 36.0/32.0 = 1.125. Thus, the fuel density of diesel oil is about 12% higher than the fuel density of gasoline.

Think About It The trend toward lower fuel values with increasing molar mass among the hydrocarbons in Table 6.6 is reflected in the fuel values of gasoline and diesel oil: the hydrocarbons in gasoline have, on average, smaller molar masses and greater hydrogen-to-carbon ratios (see Table 6.4) than those in diesel oil. However, the difference in the fuel values is small compared to the difference in the physical densities of the two fuels, which leads to the much higher fuel density of diesel oil compared to gasoline. Because both fuels are sold by volume, the higher fuel density of diesel contributes to its overall value as a fuel.

Practice Exercise High purity kerosene (see Table 6.6) is used to fuel jet engines; less pure grades are used in heating systems. Use the data in Table 6.6 to compare the fuel value and fuel density of kerosene to those of gasoline and diesel oil.

Food Value

Food serves the same purpose in living systems as fuel does in mechanical systems. The chemical reactions that convert food into energy resemble combustion but consist of many more steps that are much more highly controlled. Carbon dioxide and water are the ultimate products, however, and in a fundamental way, metabolism of food by a living system and combustion of fuel in an engine are the same process. The **food value** of what we eat—the amount of energy produced when food is burned completely—can be determined using the same equipment and by applying the same concepts of thermochemistry we have developed to evaluate fuels for vehicles. That is, we can analyze the relative food value of the items we consume by burning material in a bomb calorimeter and measuring the quantity of energy released.

What, for example, is the food value of a serving of peanuts (about 28 g, or 40 peanuts)? To determine its energy content per unit of mass, we burn the peanuts in a calorimeter. A portion of the mass of the peanuts (about 1.5 g) is water, so we first prepare the sample by drying it. This is a necessary step to get an accurate value for the enthalpy of reaction, ΔH_{rxn}, because we need to make sure that all the sample mass is due to the elemental composition of the peanuts alone.

Once the peanuts are dry, suppose they have a mass of 26.5 g. We put them in a calorimeter for which $C_{\text{calorimeter}} = 41.8$ kJ/°C and burn them completely in excess oxygen. If the temperature of the calorimeter rises by 16.0°C, what is the food value of the peanuts?

To answer this question, we use Equation 6.12 to determine the quantity of energy that flowed from the peanuts to the calorimeter:

$$q_{\text{calorimeter}} = C_{\text{calorimeter}}\Delta T = (41.8\ \text{kJ/°C})(16.0\text{°C}) = +669\ \text{kJ}$$

(a)

(b)

(c)

FIGURE 6.37 (a) A can of soda, (b) a protein bar, and (c) a stack of saltines all contain between 120 and 160 Calories, or about 6% to 8% of our daily energy needs.

where the plus sign indicates that energy flowed *into* the calorimeter. The energy lost by the peanuts as they burned has the same value but opposite sign:

$$\begin{aligned} -q_{\text{peanuts}} &= q_{\text{calorimeter}} \\ &= -669 \text{ kJ} \end{aligned}$$

Just as with fuel values, however, food value is reported as a positive number, and it is understood that the energy contained in the food is released when the food is metabolized. The peanuts' food value is therefore

$$\frac{669 \text{ kJ}}{1 \text{ serving of peanuts}} \times \frac{1 \text{ serving of peanuts}}{26.5 \text{ g}} = 25.2 \text{ kJ/g}$$

What is the food value of these peanuts in nutritional Calories (kilocalories)? We can use the definition of Calorie from Section 6.3 to convert energy in kilojoules into the familiar unit used by nutritionists:

$$\frac{669 \text{ kJ}}{\text{serving of peanuts}} \times \frac{1 \text{ Cal}}{4.184 \text{ kJ}} = \frac{160 \text{ Cal}}{\text{serving of peanuts}}$$

A single serving of peanuts represents 6% to 8% of the Calories needed by an adult with normal activity in one day. **Figure 6.37** shows some other foods that have a similar food value.

SAMPLE EXERCISE 6.17 Calculating Food Value **LO8**

Glucose ($C_6H_{12}O_6$) is a simple sugar formed by photosynthesis in plants. The complete combustion of 0.5763 g of glucose in a calorimeter ($C_{\text{calorimeter}}$ = 6.20 kJ/°C) raises the temperature of the calorimeter by 1.45°C. What is the food value of glucose in Calories per gram?

Collect and Organize We are asked to determine the food value of glucose, which means the energy given off when 1 g is burned. We have the mass of glucose burned and the calorimeter constant. To convert kilojoules to Calories we use the conversion factor 1 Cal = 4.184 kJ. Equation 6.12 ($q_{\text{calorimeter}} = C_{\text{calorimeter}}\Delta T$) can be used to calculate $q_{\text{calorimeter}}$ from its temperature change and heat capacity (calorimeter constant).

Analyze The quantity of energy absorbed by the calorimeter is equal in magnitude but opposite in sign to the enthalpy change in the reaction mixture as described by Equation 6.15 ($q_{\text{calorimeter}} = -\Delta H_{\text{rxn}}$). However, this sign change is not needed in this case because food (and fuel) values are expressed as positive "energy released" values.

Solve

$$q_{\text{calorimeter}} = C_{\text{calorimeter}}\Delta T = (6.20 \text{ kJ/°C})(1.45\text{°C}) = 8.99 \text{ kJ}$$

This positive quantity of energy absorbed by the calorimeter can be converted directly to a food value by dividing by the sample mass:

$$\frac{8.99 \text{ kJ}}{0.5763 \text{ g}} = 15.6 \text{ kJ/g}$$

Converting this value into Calories:

$$(15.6 \text{ kJ/g})\left(\frac{1 \text{ Cal}}{4.184 \text{ kJ}}\right) = 3.73 \text{ Cal/g}$$

Think About It Using calorimetry to determine food or fuel values is simplified by the fact that the energy transferred to the calorimeter can be used directly to calculate positive food or fuel values. If a 1-gram sample had been combusted, its food (or fuel) value would have been equal to ($C_{\text{calorimeter}} \cdot \Delta T$).

Practice Exercise Sucrose (table sugar) has the formula $C_{12}H_{22}O_{11}$ ($\mathcal{M}$ = 342.30 g/mol) and a food value of 16.4 kJ/g. Determine the calorimeter constant of the calorimeter in which the combustion of 1.337 g of sucrose raises the temperature by 1.96°C.

In this chapter we have considered the flow of energy and its role in defining the behavior of physical, chemical, and biological processes. The interaction of energy with matter—the transfer of energy into and out of materials—causes phase changes and alters the temperature of matter. The magnitude of these changes and the temperature ranges over which they occur are characteristic of the quantity and identity of the matter involved. Energy is also a product or a reactant in virtually all chemical reactions and, as such, behaves stoichiometrically, which means the chemical changes that a specific quantity of material undergoes are characterized by the release or consumption of a specific amount of energy. Understanding the energy contained in fuels is particularly important because fuels provide the bulk of the power needed by the equipment and devices of modern life—from cars and airplanes to computers, air conditioners, and lightbulbs. Just as significant is the energy contained in foods that sustain life. How we deal with the needs for energy in the near future—in terms of both fuel and food—will determine the quality of all our lives and the health of our planet.

SAMPLE EXERCISE 6.18 Integrating Concepts: Recycling Aluminum

Chemical engineers frequently analyze the energy required to carry out industrial procedures to assess costs of operations and support new approaches to producing materials. Over the last century, aluminum—both alone and in combination with other metals—has replaced steel for use where high strength-to-weight ratios and corrosion resistance are paramount. The industrial process for converting Al_2O_3 (alumina) into aluminum, called the Hall–Héroult process after Charles Martin Hall (1863–1914) and Paul Héroult (1863–1914), is based on passing an electric current through a solution of alumina dissolved in molten cryolite (Na_3AlF_6). As electricity passes through the solution, aluminum ions are reduced to aluminum metal at negatively charged electrodes, while positively charged carbon electrodes are oxidized to carbon dioxide. The process is described by the following overall reaction equation:

$$2\,Al_2O_3\text{ (in molten }Na_3AlF_6) + 3\,C(s\text{, graphite}) \rightarrow 4\,Al(\ell) + 3\,CO_2(g)$$

The principal energy cost of the Hall–Héroult process is the electricity needed to reduce Al_2O_3; the major cost in recycling is the energy required to melt aluminum metal. We can use thermochemistry principles to estimate the energy requirements for the Hall–Héroult process and compare it with the energy needed for recycling.

a. Assign oxidation numbers to the reactants and products in the reaction and determine how many moles of electrons are needed per mole of aluminum produced.
b. Calculate the standard enthalpy of reaction for the reduction reaction of alumina from the standard enthalpies of formation. Use Table A4.3 in Appendix 4 as needed.
c. Estimate the energy required to recycle 1.00 mole of aluminum by heating it from 25°C to its melting point (660°C) until all the aluminum melts.

Collect and Organize We are given a balanced chemical equation. We will need the melting point of aluminum, 660°C; the molar enthalpy of fusion, $\Delta H_{\text{fus,Al}} = 10.79$ kJ/mol; and the molar specific heat of aluminum, $c_{\text{P,Al}}$, 24.4 J/(mol · K). Additional thermodynamic data are available in Appendix 4 to calculate $\Delta H^\circ_{\text{rxn}}$. We are asked to compare the energy savings for recycling aluminum with reducing it from Al_2O_3.

Analyze The energy needed to make 1 mole of aluminum—that is, the standard enthalpy change for the reaction—is calculated using standard enthalpies of formation and Equation 6.19. The energy required to recycle 1 mole of aluminum represents the sum of the energy needed to melt a mole of aluminum (the enthalpy of fusion) and the energy needed to heat the aluminum to its melting point, which can be calculated using its molar heat capacity and Equation 6.8.

Solve

a. The balanced chemical equation is given as:

$$2\,Al_2O_3(s) + 3\,C(s\text{, graphite}) \rightarrow 4\,Al(\ell) + 3\,CO_2(g)$$

where we have omitted the "in molten Na_3AlF_6" because the latter is the solvent for the reaction. According to the rules for assigning oxidation numbers (Chapter 4), both liquid Al and solid C have an oxidation number of zero because they are pure elements. Assigning the oxidation number for oxygen as −2 in both Al_2O_3 and CO_2, we can calculate the oxidation numbers for Al and C as follows:

$$\text{Al:}\quad 2x + 3(-2) = 0 \qquad x = +3 \qquad\qquad \text{C:}\quad x + 2(-2) = 0 \qquad x = +4$$

To determine how many moles of electrons are needed per mole of aluminum produced, we need to determine the number of electrons transferred in this process:

Carbon: $\Delta\,\text{O.N.} = 3(+4) - 3(0) = +12$ → 12 electrons lost

0 oxidation +4

$$2\,Al_2O_3 + 3\,C(s,\text{ graphite}) \rightarrow 4\,Al(\ell) + 3\,CO_2(g)$$

+3 reduction 0

Aluminum: $\Delta\,\text{O.N.} = 4(0) - 4(+3) = -12$ → 12 electrons gained

A total of 12 moles of electrons are transferred for every 4 moles of aluminum produced, so 3 moles of electrons are required to produce each mole of aluminum.

b. Using Equation 6.19, we can calculate the standard enthalpy of reaction for the reduction reaction of alumina from the standard enthalpies of formation of the reactants and products:

$$\Delta H^\circ_{\text{rxn}} = [3(\Delta H^\circ_{\text{f},CO_2(g)}) + 4(\Delta H^\circ_{\text{f},Al(\ell)})] - [2(\Delta H^\circ_{\text{f},Al_2O_3(s)}) + 3(\Delta H^\circ_{\text{f},C(s)})]$$

$$= \left[(3\text{ mol }CO_2)\left(\frac{-393.5\text{ kJ}}{1\text{ mol }CO_2}\right) + (4\text{ mol Al})\left(\frac{10.6\text{ kJ}}{1\text{ mol Al}}\right)\right] - \left[(2\text{ mol }Al_2O_3)\left(\frac{-1675.7\text{ kJ}}{1\text{ mol }Al_2O_3}\right) + (3\text{ mol C})\left(\frac{0.0\text{ kJ}}{1\text{ mol C}}\right)\right]$$

$$= +2213.3\text{ kJ}$$

Dividing this value by the 4 moles of aluminum produced in the reaction as written, we get

$$\frac{2213.3\text{ kJ}}{4\text{ mol Al}} = 553.3\text{ kJ/mol}$$

This reaction is endothermic, requiring 553 kJ for every mole of aluminum produced.

c. The energy needed to heat 1 mole of aluminum from 25°C to 660°C can be calculated from its molar heat capacity [24.4 J/(mol · °C), per Table 6.1] and Equation 6.8:

$$q = nc_P\Delta T$$

$$= 1.00\text{ mol} \times 24.4\frac{\text{J}}{\text{mol}\cdot{}^\circ\text{C}} \times (660 - 25)^\circ\text{C} \times \frac{1\text{ kJ}}{1000\text{ J}}$$

$$= 15.5\text{ kJ}$$

Once the aluminum is at its melting point, the energy required to melt 1 mole of Al is its enthalpy of fusion ($\Delta H^\circ_{\text{fus}} = 10.6$ kJ/mol):

$$10.6\text{ kJ/mol} \times 1.00\text{ mol} = 10.6\text{ kJ}$$

The estimated total energy to heat and melt 1.00 mole of aluminum is

$$15.5\text{ kJ} + 10.6\text{ kJ} = 26.1\text{ kJ}$$

This value represents

$$\frac{26.1\text{ kJ}}{553\text{ kJ}} \times 100\% = 4.75\%$$

of the energy needed electrically to produce 1 mole of aluminum from its ore. The high cost of electricity makes recycling aluminum economically attractive as well as environmentally sound.

Think About It Other energy costs arise in both the production and recycling of aluminum, and the numbers calculated here should be viewed as estimates based on ideal situations; overall, however, recycling saves aluminum manufacturers about 95% of the energy required to produce the metal from the ore. This energy savings has inspired the rapid growth of a global aluminum recycling industry, and in the United States alone, aluminum recycling is a $1 billion-per-year business. Junkyards in the United States currently recycle 85% of the aluminum in cars and more than 50% of the aluminum in food and beverage containers.

SUMMARY

LO1 **Potential energy (PE)** is the energy stored in an object because of its position and is a **state function. Kinetic energy (KE)** is the energy of motion. Heating a sample increases the average kinetic energy of the atoms in the sample. Energy is stored in compounds, and energy is absorbed or released when they are transformed into different compounds or when a change of state occurs. (Sections 6.1 and 6.2)

LO2 The focus of a thermochemical study defines a **system**; everything other than the system is considered **surroundings**. In an **exothermic process**, the system loses energy to its surroundings ($q < 0$); in an **endothermic process**, the system absorbs energy ($q > 0$) from its surroundings. (Section 6.3)

LO3 The sum of the kinetic and potential energies of a system is called its **internal energy (*E*)**. The internal energy of a system is increased ($\Delta E = E_{\text{final}} - E_{\text{initial}}$ is positive) when it is heated ($q > 0$) or if work is done on it ($w > 0$). When the pressure on a system is constant but the volume of the system changes, the work done is called **pressure–volume (*P*–*V*) work.** (Section 6.3)

LO4 The **enthalpy (*H*)** of a system is given by $H = E + PV$. The **enthalpy change (ΔH)** of a system is equal to the energy in the form of heat (q_P) added to or removed from the system at constant pressure: $\Delta H > 0$ for endothermic reactions and $\Delta H < 0$ for exothermic reactions. Thermodynamic values like **molar heat capacity (c_P)** and **specific heat (c_s)** can be used to quantify physical and chemical changes in systems. (Sections 6.4 and 6.5)

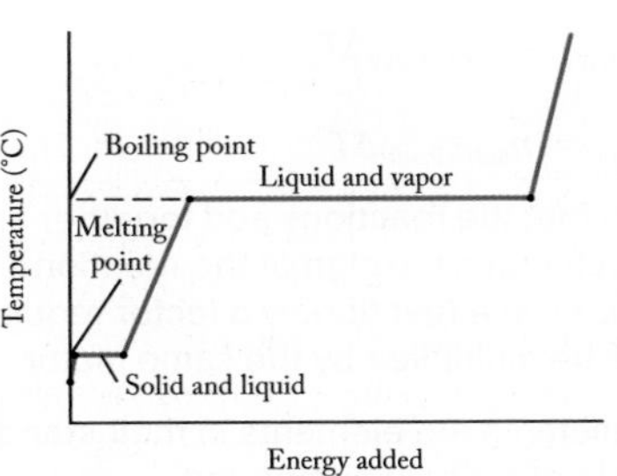

LO5 A **calorimeter**, characterized by its **calorimeter constant** (its characteristic **heat capacity**), is a device used to measure the amount of energy involved in physical and chemical processes. The enthalpy change associated with a reaction is defined by the **enthalpy of reaction (ΔH_{rxn})**. (Section 6.6)

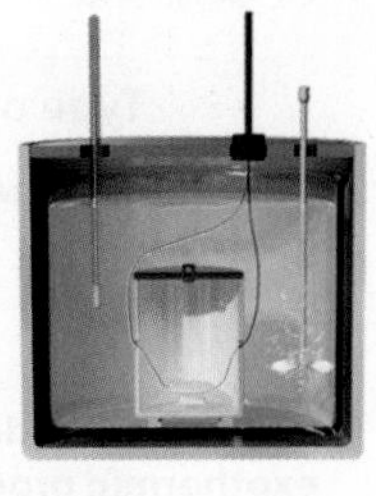

LO6 **Hess's law** states that the enthalpy of a reaction (ΔH_{rxn}) that is the sum of two or more other reactions is equal to the sum of the ΔH_{rxn} values of the constituent reactions. It can be used to calculate enthalpy changes in reactions that are hard or impossible to measure directly. (Section 6.7)

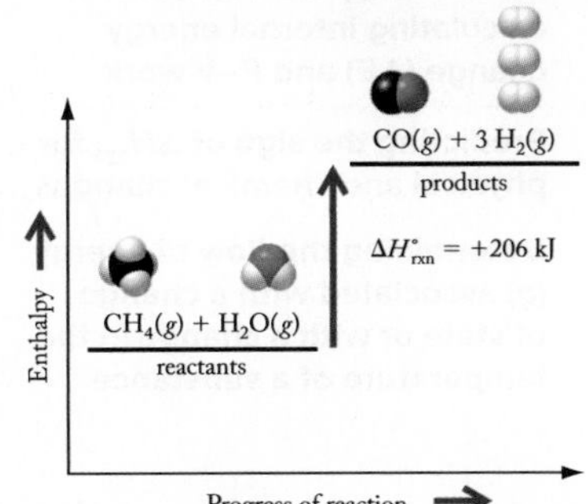

LO7 The **standard enthalpy of formation (ΔH_f°)** of a substance is the amount of energy involved in a **formation reaction**, in which 1 mole of the substance is made from its constituent elements in their **standard states** (under **standard conditions**). Enthalpy changes for physical changes and chemical reactions can be calculated from the enthalpies of formation of the reactants and products. (Section 6.8)

LO8 **Fuel value** is the amount of energy released on complete combustion of 1 g of a fuel. **Food value** is the amount of energy released when a material consumed by an organism for sustenance is burned completely; nutritionists often express food values in Calories (kilocalories) rather than in the SI unit kilojoules. (Section 6.9)

PARTICULATE **PREVIEW WRAP-UP**

Energy is absorbed by ozone to break its bonds. Bond breaking is endothermic; bond formation is exothermic. Energy is absorbed during this reaction, but energy would be released if the reaction ran in reverse.

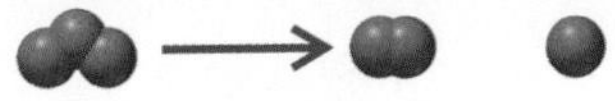

PROBLEM-SOLVING SUMMARY

Type of Problem	Concepts and Equations	Sample Exercises
Calculating kinetic and potential energy	$KE = \frac{1}{2}mu^2$ (5.27) $E_{el} \propto \frac{Q_1 \times Q_2}{d}$ (6.3)	**6.1**
Identifying endothermic and exothermic processes, and calculating internal energy change (ΔE) and P–V work	For the system: $\Delta E = q + w$ (6.5) where $w = -P\Delta V$.	**6.2, 6.3, 6.4**
Predicting the sign of ΔH_{sys} for physical and chemical changes	Exothermic: $\Delta H_{sys} < 0$ Endothermic: $\Delta H_{sys} > 0$	**6.5, 6.6**
Determining the flow of energy (q) associated with a change of state or with a change in the temperature of a substance	Heating a substance: $q = nc_P\Delta T$ (6.8) or melting a solid at its melting point: $q = n\Delta H_{fus}$ (6.9) or vaporizing a liquid at its boiling point: $q = n\Delta H_{vap}$ (6.10)	**6.7, 6.8, 6.9**
Measuring the heat capacity (calorimeter constant) of a calorimeter and ΔH_{rxn}	$C_{calorimeter} = q_{calorimeter}/\Delta T$ (6.13) where $C_{calorimeter}$ is the heat capacity of the calorimeter, $q_{calorimeter}$ is the heat released by a standard combustion reaction, and ΔT is the temperature change of the calorimeter.	**6.10**
Measuring the enthalpy of reaction	The energy change of the system is equal in magnitude, but opposite in sign, to the energy change of the surroundings: $q_{gained,sys} = -q_{lost,surr}$ where the energy depends upon the amount of substance (moles), molar heat capacity, and change in temperature: $q_{gained,sys} = n_{sys}c_{P,sys}\Delta T$ $q_{lost,surr} = n_{surr}c_{P,surr}\Delta T$	**6.11**
Using Hess's law	Reorganize the information so that the reactions add together as desired. Reversing a reaction changes the sign of the reaction's ΔH_{rxn} value. Multiplying the coefficients in a reaction by a factor requires that the reaction's ΔH_{rxn} value be multiplied by the same factor.	**6.12, 6.13**
Recognizing and writing formation reactions	In a formation reaction, the reactants are elements in their standard states and the product is 1 mole of a single compound.	**6.14**
Calculating standard enthalpies of reaction from heats of formation	$\Delta H^\circ_{rxn} = \sum n_{products}\Delta H^\circ_{f,products} - \sum n_{reactants}\Delta H^\circ_{f,reactants}$ (6.19)	**6.15**
Calculating fuel value and food value	The fuel value or food value of a substance is the energy released by the complete combustion of 1 g of the substance.	**6.16, 6.17**

VISUAL PROBLEMS

(Answers to boldface end-of-chapter questions and problems are in the back of the book.)

6.1. A brick lies perilously close to the edge of the flat roof of a building (Figure P6.1). The roof edge is 50 ft above street level, and the brick has 500 J of potential energy with respect to street level. Someone edges the brick off the roof, and it begins to fall. What is the brick's kinetic energy when it is 35 ft above street level? What is its kinetic energy the instant before it hits the street surface?

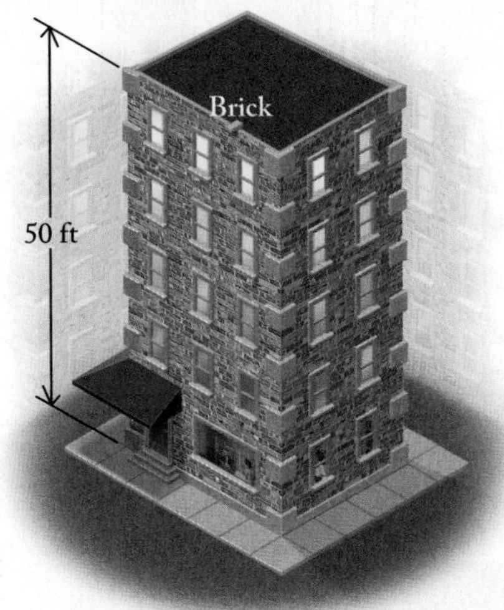

FIGURE P6.1

6.2. Figure P6.2 shows pairs of cations and anions in contact. The energy of each interaction is proportional to Q_1Q_2/d, where Q_1 and Q_2 are the charges on the cation and anion, and d is the distance between their nuclei. Which pair experiences the greatest attraction, and which experiences the smallest?

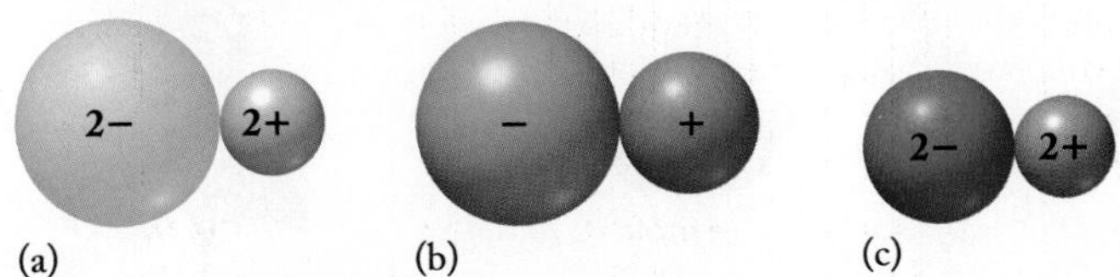

FIGURE P6.2

6.3. Figure P6.3 shows a sectional view of an assembly, consisting of a gas trapped inside a stainless-steel cylinder with a stainless-steel piston and a block of iron on top of the piston to keep it in place.

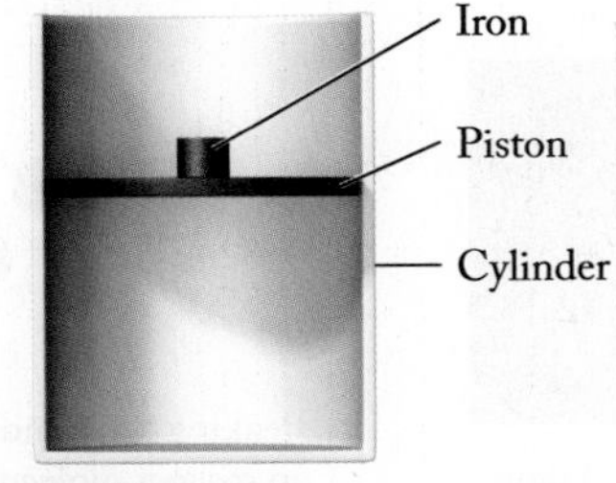

FIGURE P6.3

a. If the cylinder is heated for a few seconds with a blowtorch, is the piston higher or lower in the cylinder?
b. Has heat (q) been added to the system?
c. Has the system done work (w) on its surroundings, or have the surroundings done work on the system?

6.4. The closed, rigid, metal box lying on the wooden kitchen table in Figure P6.4 is about to be involved in an accident.
a. Are the contents of the box an isolated system, a closed system, or an open system?
b. What will happen to the internal energy of the system if the table catches fire and burns?
c. Will the system do any work on the surroundings while the fire is burning or after the table has burned away and collapsed?

FIGURE P6.4

6.5. The diagram in Figure P6.5 shows how a chemical reaction in a cylinder with a piston affects the volume of the system.
a. In this reaction, does the system do work on the surroundings?
b. If the reaction is endothermic, does the internal energy of the system increase or decrease when the reaction is proceeding?

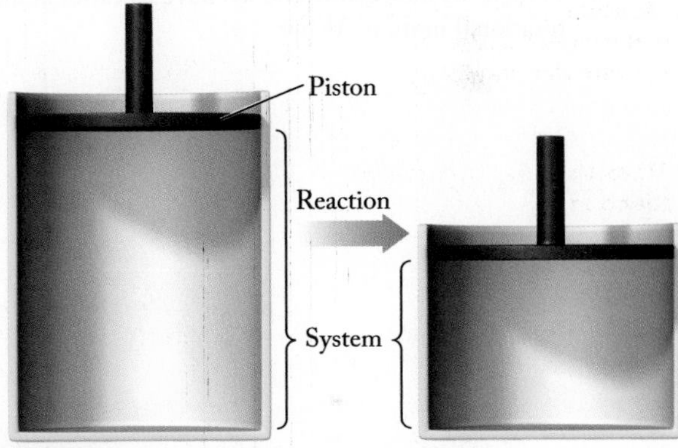

FIGURE P6.5

6.6. The enthalpy diagram in Figure P6.6 indicates the enthalpies of formation of four compounds made from the elements listed on the "zero" line of the vertical axis.

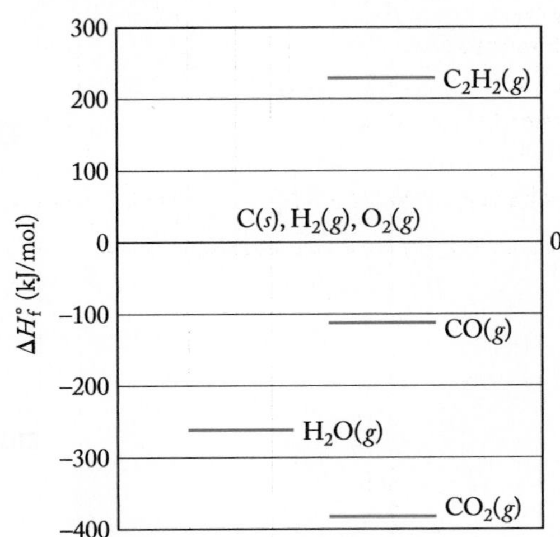

FIGURE P6.6

a. Why are the elements all put on the same horizontal line?
b. Why is $C_2H_2(g)$ sometimes called an "endothermic" compound?
c. How could the data be used to calculate the heat of the reaction that converts a stoichiometric mixture of $C_2H_2(g)$ and $O_2(g)$ to $CO(g)$ and $H_2O(g)$?

6.7. The process illustrated in Figure P6.7 takes place at constant pressure.
a. Write a balanced equation for the process.
b. Is w positive, negative, or zero for this reaction?
c. Using data from Appendix 4, calculate ΔH°_{rxn} for the formation of 1 mole of the product.

6.8. Use representations [A] through [I] in Figure P6.8 to answer questions (a)–(f).
a. Which processes are exothermic?
b. Which processes have a positive ΔH?
c. In which processes does the system gain energy?
d. In which processes do the surroundings lose energy?
e. Compare a flame of methane [D] at 1000°C to a flame of propane [E] at 1000°C in terms of (i) average kinetic energy and (ii) the average speed of the molecules.
f. Which substance(s) would *not* have vibrational motion or rotational motion? Why?

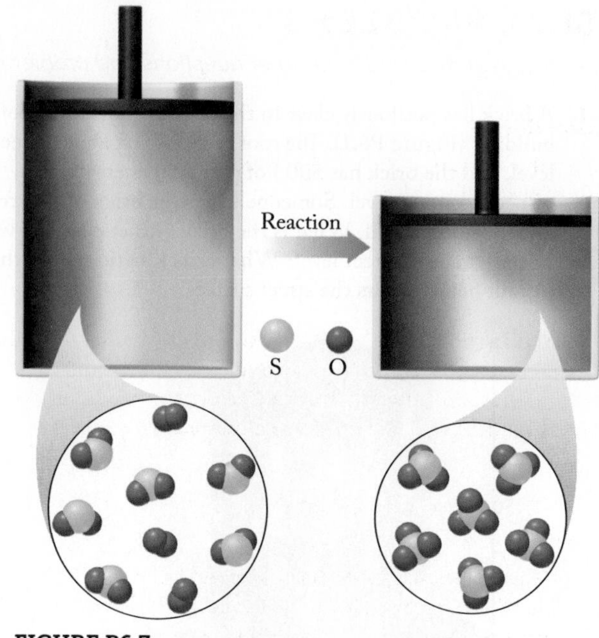

FIGURE P6.7

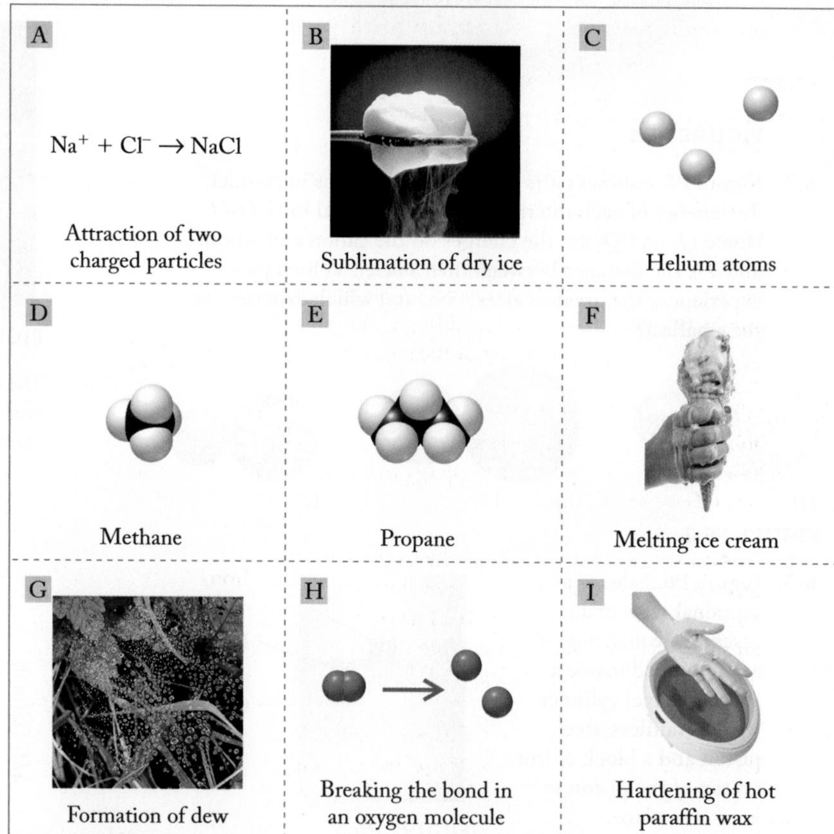

FIGURE P6.8

QUESTIONS AND PROBLEMS

Forms of Energy

Concept Review

6.9. How are energy and work related?

6.10. Explain the difference between potential energy and kinetic energy.

6.11. Explain what is meant by a state function.

6.12. Are kinetic energy and potential energy both state functions?

6.13. Explain the nature of the potential energy in the following: (a) the new battery for your remote control; (b) a gallon of gasoline; (c) the crest of a wave before it crashes onto shore.

6.14. Explain the kinetic energy in a stationary ice cube.

Systems, Surroundings, and Energy Transfer

Concept Review

6.15. What is meant by the terms *system* and *surroundings*?

6.16. Do all exothermic processes that do work on the surroundings have the same sign for ΔE?

6.17. Why don't all endothermic processes that do work on the surroundings have the same sign for ΔE?

*6.18. Does the amount of P–V work done by a gas on its surroundings depend on the identity of the gas?

Problems

6.19. Which of the following processes are exothermic, and which are endothermic? (a) a candle burns; (b) rubbing alcohol feels cold on the skin; (c) a supersaturated solution crystallizes, causing the temperature of the solution to rise

6.20. Which of the following processes are exothermic, and which are endothermic? (a) frost forms on a car window in the winter; (b) water condenses on a glass of ice water on a humid summer afternoon; (c) adding ammonium nitrate to water causes the temperature of the solution to decrease

---

6.21. What happens to the internal energy of a liquid at its boiling point when it vaporizes?

6.22. What happens to the internal energy of a gas when it expands (with no heat flow)?

6.23. Atmospheric Research Atmospheric scientists often use balloons to carry their instruments aloft. How much P–V work does a gas in a balloon do on its surroundings at a constant pressure of 1.00 atm if the volume of gas increases from 1.250×10^5 L to 1.050×10^6 L? Express your answer in L · atm and joules (J).

6.24. An expanding gas does 150.0 J of work on its surroundings at a constant pressure of 1.01 atm. If the gas initially occupied 68 mL, what is the final volume of the gas?

6.25. Calculate ΔE for the following situations:
a. $q = 120.0$ J; $w = -40.0$ J
b. $q = 9.2$ kJ; $w = 0.70$ J
c. $q = -625$ J; $w = -315$ J

6.26. Calculate ΔE for
a. the combustion of a gas that releases 210.0 kJ of heat to its surroundings and does 65.5 kJ of work on its surroundings.
b. a chemical reaction that produces 90.7 kJ of heat but does no work on its surroundings.

***6.27.** The following reactions take place in a cylinder equipped with a movable piston at atmospheric pressure (Figure P6.27). Which reactions will result in work being done on the surroundings? Assume the system returns to an initial temperature of 110°C. (*Hint*: The volume of a gas is proportional to number of moles, n, at constant temperature and pressure.)
a. $CH_4(g) + 2\,O_2(g) \rightarrow CO_2(g) + 2\,H_2O(g)$
b. $C_3H_8(g) + 5\,O_2(g) \rightarrow 3\,CO_2(g) + 4\,H_2O(g)$
c. $N_2(g) + 2\,O_2(g) \rightarrow 2\,NO_2(g)$

FIGURE P6.27

*6.28. In which direction will the piston shown in Figure P6.27 move when the following reactions are carried out at atmospheric pressure inside the cylinder and after the system has returned to its initial temperature of 110°C? (*Hint*: The volume of a gas is proportional to number of moles, n, at constant temperature and pressure.)
a. $N_2(g) + 3\,H_2(g) \rightarrow 2\,NH_3(g)$
b. $C(s) + O_2(g) \rightarrow CO_2(g)$
c. $CH_3CH_2OH(g) + 3\,O_2(g) \rightarrow 2\,CO_2(g) + 3\,H_2O(g)$

***6.29. Filling Air Bags** Automobile air bags produce nitrogen gas from the reaction:

$$2\,NaN_3(s) \rightarrow 2\,Na(s) + 3\,N_2(g)$$

a. If 2.25 g of NaN_3 reacts to fill an air bag, how much P–V work will the N_2 do against an external pressure of 1.00 atm given that the density of nitrogen is 1.165 g/L at 20°C?
b. If the process releases 2.34 kJ of heat, what is ΔE for the system?

*6.30. **Black Powder** Igniting gunpowder produces nitrogen and carbon dioxide gas that propels the bullet by the reaction:

$$2\,KNO_3(s) + \tfrac{1}{8}\,S_8(s) + 3\,C(s) \rightarrow K_2S(s) + N_2(g) + 3\,CO_2(g)$$

a. If 1.00 g of KNO_3 reacts, how much P–V work will the gases do against an external pressure of 1.00 atm given that the densities of nitrogen and CO_2 are 1.165 g/L and 1.830 g/L, respectively, at 20°C?
b. If the reaction produces 21.6 kJ of heat, what is ΔE for the system?

Enthalpy and Enthalpy Changes

Concept Review

6.31. What is meant by an *enthalpy change*?

6.32. Describe the difference between an internal energy change (ΔE) and an enthalpy change (ΔH).

6.33. Which symbol, ΔH_{comb} or ΔH_{fus}, refers to a physical change?

6.34. Which symbol, ΔH_f or ΔH_{fus}, refers to a chemical change?

Problems

6.35. **A Clogged Sink** Adding Drano to a clogged sink causes the drainpipe to get warm. What is the sign of ΔH for this process?

6.36. **Hot Packs for Cold Fingers** Skiers can purchase small packets of finely divided iron that, after reaction with water, provide heat to frozen fingers and toes (Figure P6.36). Define the system and the surroundings in this example and indicate the sign of ΔH_{sys}.

(a)

(b)

FIGURE P6.36

6.37. **Diatomic Elements** The stable forms of hydrogen and oxygen at room temperature and pressure are gas phase, diatomic molecules H_2 and O_2. What is the sign of ΔH for the following processes?

a. A solid with metallic properties is formed when hydrogen gas is compressed under extremely high pressures:

$$H_2(g) \rightarrow H_2(s)$$

b. High-energy light shines on oxygen gas in the upper reaches of the atmosphere, converting oxygen gas to oxygen atoms:

$$O_2(g) \rightarrow 2\ O(g)$$

6.38. **White Phosphorus** Elemental phosphorus is found in several allotropes, including white phosphorous, $P_4(s)$. Predict the sign of the enthalpy change for the following reactions:

$$P_4(s) \rightarrow P_4(g)$$

$$P_4(g) \rightarrow 2\ P_2(g)$$

$$4\ P(g) \rightarrow P_4(s)$$

6.39. **Plaster of Paris** Gypsum is the common name of calcium sulfate dihydrate ($CaSO_4 \cdot 2\ H_2O$). When gypsum is heated to 150°C, it loses most of the water in its formula and forms plaster of Paris ($CaSO_4 \cdot 0.5\ H_2O$):

$$2\ (CaSO_4 \cdot 2\ H_2O)(s) \rightarrow 2\ (CaSO_4 \cdot 0.5\ H_2O)(s) + 3\ H_2O(g)$$

What is the sign of ΔH for making plaster of Paris from gypsum?

6.40. **Kitchen Chemistry** A simple "kitchen chemistry" experiment requires placing some vinegar in a soda bottle. A deflated balloon containing baking soda is stretched over the mouth of the bottle. Adding the baking soda to the vinegar starts the following reaction and inflates the balloon:

$$NaHCO_3(aq) + CH_3COOH(aq) \rightarrow CH_3COONa(aq) + CO_2(g) + H_2O(\ell)$$

If the contents of the bottle are considered the system, is work being done on the surroundings or on the system?

Heating Curves, Molar Heat Capacity, and Specific Heat

Concept Review

6.41. What is the difference between *specific heat* and *molar heat capacity*?

6.42. What happens to the molar heat capacity of a material if its mass is doubled? Is the same true for the specific heat?

6.43. Are the enthalpies of fusion and vaporization of a given substance usually the same?

6.44. An equal amount of energy is added to pieces of metal A and metal B that have the same mass. Why does the metal with the smaller specific heat reach the higher temperature?

***6.45.** **Cooling an Automobile Engine** Most automobile engines are cooled by water circulating through them and a radiator, but the original Volkswagen Beetle had an air-cooled engine. Why might car designers choose water cooling over air cooling?

*6.46. **Nuclear Reactor Coolants** The reactor-core cooling systems in some nuclear power plants use liquid sodium as the coolant. Sodium has a thermal conductivity of 1.42 J/(cm · s · K), which is quite high compared with that of water [6.1×10^{-3} J/(cm · s · K)]. The respective molar heat capacities are 28.28 J/(mol · K) and 75.31 J/(mol · K). What is the advantage of using liquid sodium over water in this application?

Problems

6.47. How much energy is required to raise the temperature of 100.0 g of water from 30.0°C to 100.0°C?

6.48. **Cooking in the Mountains** At an elevation where the boiling point of water is 93°C, 100.0 g of water at 30°C absorbs 290.0 kJ of energy from a mountain climber's stove. Is this amount of energy sufficient to heat the water to its boiling point?

6.49. Use the following data to sketch a heating curve for 1 mole of methanol. Start the curve at −100°C and end it at 100°C.

Boiling point	65°C
Melting point	−94°C
ΔH_{vap}	37 kJ/mol
ΔH_{fus}	3.18 kJ/mol
Molar heat capacity (s)	48.7 J/(mol · °C)
(ℓ)	81.1 J/(mol · °C)
(g)	43.9 J/(mol · °C)

6.50. Use the following data to sketch a heating curve for 1 mole of octane. Start the curve at −57°C and end it at 150°C.

Boiling point	125.7°C
Melting point	−56.8°C
ΔH_{vap}	41.5 kJ/mol
ΔH_{fus}	20.7 kJ/mol
Molar heat capacity (ℓ)	254.6 J/(mol · °C)
Molar heat capacity (*g*)	316.9 J/(mol · °C)

6.51. Keeping an Athlete Cool During a strenuous workout, an athlete generates 2000.0 kJ of energy. What mass of water would have to evaporate from the athlete's skin to dissipate this much heat?

6.52. **Hypothermia** Damp clothes can prove fatal when outdoor temperatures drop (death by hypothermia).

a. If the clothes you are wearing absorb 1.00 kg of water and then dry in a cold wind on Mount Washington, how much heat would your body lose during this process? $c_{P,H_2O} = 75.3$ J/(mol · °C), and $\Delta H_{vap,H_2O} = 40.67$ kJ/mol.

*b. If the heat lost by your body is not replaced, what will the final temperature of your body be after the 1.00 kg of water has evaporated? Use your own body weight and assume that the specific heat of your body is 4.25 J/g. (The specific heat is given in units of J/g.)

6.53. The same quantity of energy is added to 10.00 g pieces of gold, magnesium, and platinum, all initially at 25°C. The molar heat capacities of these three metals are 25.41 J/(mol · °C), 24.79 J/(mol · °C), and 25.95 J/(mol · °C), respectively. Which piece of metal has the highest final temperature?

6.54. Which of the following would reach the higher temperature after 10.00 g of iron [$c_P = 25.1$ J/(mol · °C)] at 150°C is added: 100 mL of water [$d = 1.00$ g/mL, $c_P = 75.3$ J/(mol · °C)] or 200 mL of ethanol [CH_3CH_2OH, $d = 0.789$ g/mL, $c_P = 113.1$ J/(mol · °C)]?

***6.55.** Exactly 10.0 mL of water at 25.0°C is added to a hot iron skillet. All the water is converted into steam at 100.0°C. The mass of the pan is 1.20 kg and the molar heat capacity of iron is 25.19 J/(mol · °C). What is the temperature change of the skillet?

*6.56. A 20.0 g piece of iron and a 20.0 g piece of gold, both at 100.0°C, were dropped into the same 1.00 L of water at 20.0°C. The molar heat capacities of iron and gold are 25.19 J/(mol · °C) and 25.41 J/(mol · °C), respectively. What is the final temperature of the water and the pieces of metal?

Calorimetry: Measuring Heat Capacity and Enthalpies of Reaction

Concept Review

6.57. Why is it necessary to know the heat capacity of a calorimeter?

6.58. Could an endothermic reaction be used to measure the heat capacity of a calorimeter?

6.59. If we replace the water in a bomb calorimeter with another liquid, why do we need to redetermine the heat capacity of the calorimeter?

6.60. When measuring the enthalpy of combustion of a very small amount of material, would you prefer to use a calorimeter having a heat capacity that is small or large? Explain your reasoning.

Problems

6.61. Designing Cold Packs If you were designing a chemical cold pack, which of the following salts would you choose to provide the greatest drop in temperature per gram: NH_4Cl ($\Delta H_{soln} = 14.6$ kJ/mol), NH_4NO_3 ($\Delta H_{soln} = 25.7$ kJ/mol), or $NaNO_3$ ($\Delta H_{soln} = 20.4$ kJ/mol)?

6.62. Dissolving calcium hydroxide ($\Delta H_{soln} = -16.2$ kJ/mol) in water is an exothermic process. How much lithium hydroxide ($\Delta H_{soln} = -23.6$ kJ/mol) would be required to produce the same amount of energy as dissolving 15.00 g of $Ca(OH)_2$?

6.63. The standard enthalpy of combustion of benzoic acid (molar mass 122 g/mol) is −3225 kJ/mol. Calculate the heat capacity of a bomb calorimeter if a temperature increase of 2.16°C occurs on combusting 0.500 g of benzoic acid in the presence of excess O_2.

6.64. Burning 1.43 g of methane [$CH_4(g)$, molar mass 16.0 g/mol] with excess O_2 raised the temperature of 250 g of water in a bomb calorimeter from 22°C to 98°C. Calculate the molar enthalpy of combustion of methane.

6.65. The complete combustion of 1.200 g of cinnamaldehyde (C_9H_8O, one of the compounds in cinnamon) in a bomb calorimeter ($C_{calorimeter} = 3.640$ kJ/°C) produced an increase in temperature of 12.79°C. Calculate the molar enthalpy of combustion of cinnamaldehyde (ΔH_{comb}) in kilojoules per mole of cinnamaldehyde.

6.66. **Aromatic Spice** The aromatic hydrocarbon cymene ($C_{10}H_{14}$) is found in nearly 100 spices and fragrances, including coriander, anise, and thyme. The complete combustion of 1.608 g of cymene in a bomb calorimeter ($C_{calorimeter} = 3.640$ kJ/°C) produced an increase in temperature of 19.35°C. Calculate the molar enthalpy of combustion of cymene (ΔH_{comb}) in kilojoules per mole of cymene.

6.67. Hormone Mimics Phthalates, used to make plastics flexible, are among the most abundant industrial contaminants in the environment. Several have been shown to act as hormone mimics in humans by activating the receptors for estrogen, a female sex hormone. In characterizing the compounds completely, the value of ΔH_{comb} for dimethyl phthalate ($C_{10}H_{10}O_4$) was determined to be −4685 kJ/mol. Assume that 1.00 g of dimethyl phthalate is combusted in a calorimeter whose heat capacity ($C_{calorimeter}$) is 7.854 kJ/°C at 20.215°C. What is the final temperature of the calorimeter?

6.68. Diborane, B_2H_6, was once considered for use as a rocket fuel. It is a highly reactive gas, burning with a green flame in air.

$$B_2H_6(g) + 3\ O_2(g) \rightarrow B_2O_3(s) + 3\ H_2O(g)$$

The ΔH_{comb} value for diborane is −1958 kJ/mol. Assume 0.368 g of diborane is combusted in a calorimeter whose heat capacity ($C_{calorimeter}$) is 7.854 kJ/°C at 20.611°C. What is the final temperature of the calorimeter?

6.69. Adding 2.00 g of Mg metal to 95.0 mL of 1.00 M HCl in a coffee-cup calorimeter leads to a temperature increase of 9.2°C.

a. Write a balanced net ionic equation for the reaction.
b. If the molar heat capacity of 1.00 M HCl is the same as that for water [c_P = 75.3 J/(mol · °C)], what is ΔH_{rxn}?

6.70. What is ΔH_{rxn} for the precipitation of AgCl if adding 125 mL of 1.00 M $AgNO_3$ to 125 mL of 1.00 M NaCl at 18.6°C causes the temperature to increase to 26.4°C? (Assume $c_{P,soln} = c_{P,water}$.)

***6.71.** How much glucose must be metabolized to completely evaporate 1.00 g of water at 37°C, given $\Delta H_{comb,glucose}$ = −2803 kJ/mol?

6.72. If all the energy obtained from burning 275 g of propane ($\Delta H_{comb,C_3H_8}$ = −2220 kJ/mol) is used to heat water, how many liters of water can be heated from 20.0°C to 100.0°C?

Hess's Law

Concept Review

6.73. How is Hess's law consistent with the law of conservation of energy?

6.74. Why is it important for Hess's law that enthalpy is a state function?

Problems

6.75. How can the first two of the following reactions be combined to obtain the third reaction?

a. $CO(g) + NH_3(g) \rightarrow HCN(g) + H_2O(g)$
b. $CO(g) + 3\ H_2(g) \rightarrow CH_4(g) + H_2O(g)$
c. $CH_4(g) + NH_3(g) \rightarrow HCN(g) + 3\ H_2(g)$

6.76. **Cleansing the Atmosphere** The atmosphere contains the highly reactive molecule OH, which acts to remove selected pollutants. Use the values for ΔH_{rxn} given below to find the ΔH_{rxn} for the formation of OH and H from water.

$\frac{1}{2}H_2(g) + \frac{1}{2}O_2(g) \rightarrow OH(g)$ $\quad \Delta H_{rxn} = 42.1$ kJ

$H_2(g) \rightarrow 2\ H(g)$ $\quad \Delta H_{rxn} = 435.9$ kJ

$H_2(g) + \frac{1}{2}O_2(g) \rightarrow H_2O(g)$ $\quad \Delta H_{rxn} = -241.8$ kJ

$H_2O(g) \rightarrow H(g) + OH(g)$ $\quad \Delta H_{rxn} = ?$

6.77. Ozone Layer The destruction of the ozone layer by chlorofluorocarbons (CFCs) can be described by the following reactions:

$ClO(g) + O_3(g) \rightarrow Cl(g) + 2\ O_2(g)$ $\quad \Delta H_{rxn} = -29.90$ kJ

$2\ O_3(g) \rightarrow 3\ O_2(g)$ $\quad \Delta H_{rxn} = 24.18$ kJ

Determine the value of the heat of reaction for the following:

$Cl(g) + O_3(g) \rightarrow ClO(g) + O_2(g)$ $\quad \Delta H_{rxn} = ?$

6.78. **Low-Emission Vehicles** Adding ammonia to the exhaust gases from motor vehicles can reduce the emissions of nitrogen oxide pollutants through selective catalytic reduction (SCR):

$6\ NO(g) + 4\ NH_3(g) \rightarrow 5\ N_2(g) + 6\ H_2O(g)$ $\quad \Delta H^\circ_{rxn} = -1807$ kJ

$6\ NO_2(g) + 8\ NH_3(g) \rightarrow 7\ N_2(g) + 12\ H_2O(g)$ $\quad \Delta H^\circ_{rxn} = -2730$ kJ

Determine the value of the heat of reaction for the following:

$3\ NO(g) + N_2(g) + 3\ H_2O(g) \rightarrow 3\ NO_2(g) + 2\ NH_3(g)$ $\quad \Delta H^\circ_{rxn} = ?$ kJ

Standard Enthalpies of Formation and Reaction

Concept Review

6.79. Explain how the use of ΔH°_f to calculate ΔH°_{rxn} is an example of Hess's law.

6.80. Why is the standard enthalpy of formation of CO(g) difficult to measure experimentally?

***6.81.** Are the standard enthalpies of formation of all allotropes of an element the same? Explain.

6.82. Explain why the heats of formation of elements in their standard states are zero.

Problems

6.83. For which of the following reactions does ΔH°_{rxn} represent an enthalpy of formation?

a. $C(s) + O_2(g) \rightarrow CO_2(g)$
b. $CO_2(g) + C(s) \rightarrow 2\ CO(g)$
c. $CO_2(g) + H_2(s) \rightarrow H_2O(g) + CO(g)$
d. $2\ H_2(g) + C(s) \rightarrow CH_4(g)$

6.84. For which of the following reactions does ΔH°_{rxn} also represent an enthalpy of formation?

a. $2\ N_2(g) + 3\ O_2(g) \rightarrow 2\ NO_2(g) + 2\ NO(g)$
b. $N_2(g) + O_2(g) \rightarrow 2\ NO(g)$
c. $2\ NO_2(g) \rightarrow N_2O_4(g)$
d. $N_2(g) + 2\ O_2(g) \rightarrow 2\ NO_2(g)$

6.85. Use the following standard heats of formation to calculate the molar enthalpy of vaporization of liquid hydrogen peroxide: ΔH°_f of $H_2O_2(\ell)$ is −188 kJ/mol and ΔH°_f of $H_2O_2(g)$ is −136 kJ/mol.

6.86. Use the following standard heats of formation to calculate the molar enthalpy of vaporization of acetic acid: ΔH°_f of $CH_3COOH(\ell)$ is −484.5 kJ/mol and ΔH°_f of $CH_3COOH(g)$ is −432.8 kJ/mol.

6.87. Methanogenesis Some methanogenic bacteria use acetic acid (CH_3COOH) and hydrogen to make methane. Write a balanced chemical reaction for the synthesis of methane and water from hydrogen and acetic acid and calculate the standard enthalpy of the reaction by using the appropriate enthalpies of formation from Appendix 4.

6.88. **Military Explosives** Explosives called amatols are mixtures of ammonium nitrate and trinitrotoluene (TNT). They were introduced during World War I when TNT was in short supply. The mixtures can provide 30% more explosive power than TNT alone. Above 300°C, ammonium nitrate decomposes to N_2, O_2, and H_2O. Write a balanced chemical equation describing the decomposition of ammonium nitrate and determine the standard enthalpy of reaction by using the appropriate standard enthalpies of formation from Appendix 4.

6.89. **Improvised Explosives** Mixtures of fertilizer (ammonium nitrate) and fuel oil (a mixture of long-chain hydrocarbons such as decane, $C_{10}H_{22}$) can produce powerful explosions. Determine the enthalpy change of the following explosive reaction by using the appropriate enthalpies of formation ($\Delta H^\circ_{f,C_{10}H_{22}} = 249.7$ kJ/mol):

$$3\ NH_4NO_3(s) + C_{10}H_{22}(\ell) + 14\ O_2(g) \rightarrow 3\ N_2(g) + 17\ H_2O(g) + 10\ CO_2(g)$$

*6.90. **A Little TNT** Trinitrotoluene (TNT) is a highly explosive compound. The thermal decomposition of TNT is described by the following chemical equation:

$$2\ C_7H_5N_3O_6(s) \rightarrow 12\ CO(g) + 5\ H_2(g) + 3\ N_2(g) + 2\ C(s)$$

If ΔH_{rxn} for this reaction is −10,153 kJ/mol, how much TNT is needed to equal the explosive power of 1 mole of ammonium nitrate (Problem 6.89)?

6.91. How can the standard enthalpy of formation ΔH°_f of $CO(g)$ be calculated from the standard enthalpy of formation ΔH°_f of $CO_2(g)$ and the standard enthalpy of combustion ΔH°_{comb} of $CO(g)$?

6.92. **The Martian** In the motion picture *The Martian*, a stranded astronaut uses hydrazine, N_2H_4, to produce hydrogen, which he then reacts with oxygen to make water. The enthalpy of the reaction between N_2H_4 and O_2 is $\Delta H^\circ_{rxn} = -622$ kJ/mol. Calculate the standard enthalpy of formation of $N_2H_4(\ell)$ from the standard enthalpy changes of the following reactions:

$$N_2H_4(\ell) + O_2(g) \rightarrow N_2(g) + 2\ H_2O(g) \qquad \Delta H^\circ_{rxn} = -622\ \text{kJ}$$

$$H_2(g) + \tfrac{1}{2}\ O_2(g) \rightarrow H_2O(g) \qquad \Delta H^\circ_{rxn} = -285.8\ \text{kJ}$$

$$N_2(g) + 2\ H_2(g) \rightarrow N_2H_4(\ell) \qquad \Delta H^\circ_{rxn} = ?$$

6.93. **Burning Phosphorus** White phosphorus, P_4, ignites in air to form P_4O_{10} in a rapid reaction used in incendiary devices. Red phosphorus, an allotrope of element 15, also reacts with O_2 in air to form the same product, but at a much slower rate. What is ΔH_f for red phosphorus,

$$P_4(white,\ s) \rightarrow 4\ P(red,\ s) \qquad \Delta H_{rxn} = ?$$

given ΔH_{rxn} for the following reactions?

$$P_4(white,\ g) + 5\ O_2(g) \rightarrow P_4O_{10}(g) \qquad \Delta H_{rxn} = -3036\ \text{kJ}$$

$$4\ P(red,\ g) + 5\ O_2(g) \rightarrow P_4O_{10}(g) \qquad \Delta H_{rxn} = -3018\ \text{kJ}$$

6.94. Baking soda decomposes on heating as follows:

$$2\ NaHCO_3(s) \rightarrow Na_2CO_3(s) + CO_2(g) + H_2O(\ell) \qquad \Delta H^\circ_{rxn} = -129.3\ \text{kJ}$$

Calculate the standard enthalpy of formation of $NaHCO_3(s)$ using the following information:

$$\Delta H^\circ_f[CO_2(g)] = -394\ \text{kJ/mol}$$

$$\Delta H^\circ_f[Na_2CO_3(s)] = -1131\ \text{kJ/mol}$$

$$\Delta H^\circ_f[H_2O(\ell)] = -286\ \text{kJ/mol}$$

Fuels, Fuel Values, and Food Values

Concept Review

6.95. What is meant by *fuel value*?

6.96. What are the units of fuel values?

6.97. How are fuel values calculated from molar enthalpies of combustion?

6.98. Is the fuel value of liquid propane the same as that of propane gas?

Problems

6.99. **Flex Fuel Vehicles** An increasing number of vehicles in the United States can run on either gasoline [$C_9H_{20}(\ell)$, $\Delta H_{comb,gasoline} = -6160$ kJ/mol] or ethanol [$CH_3CH_2OH(\ell)$, $\Delta H_{comb,ethanol} = -1367$ kJ/mol].

a. Which fuel has the greater fuel value?

b. The densities of C_9H_{20} and CH_3CH_2OH are 0.718 g/mL and 0.794 g/mL, respectively. Which has the greater fuel value expressed as kJ/L?

6.100. **Arctic Exploration** Food contains three main categories of compounds: carbohydrate, protein, and fat. Arctic explorers often eat a high-fat diet because fats have a high food value. The average food values for carbohydrate, protein, and fat are 4 Cal/g, 4 Cal/g, and 9 Cal/g, respectively. A typical fat has the chemical formula $C_{18}H_{36}O_2$, glucose ($C_6H_{12}O_6$) is a representative carbohydrate, and the amino acid alanine ($C_3H_6NO_2$) is a building block of proteins. Express the food values given above as ΔH_{comb} in kJ/mol.

6.101. **Biofuels** The microorganism *C. acetobutylicum* can convert glucose into butanol ($C_4H_{10}O$). Butanol is touted as a better renewable biofuel than ethanol.

a. Calculate the fuel value of $C_4H_{10}O$, given that $\Delta H^\circ_{comb} = -2676$ kJ/mol.

b. How much energy is released during the combustion of 1.00 kg of $C_4H_{10}O$?

c. How many grams of $C_4H_{10}O$ must be burned to heat 1.00 kg of water from 25.0°C to 85.0°C? Assume that all the energy released during combustion is used to heat the water.

6.102. **Biofuels in Brazil** In response to increasing need for energy self-sufficiency, Brazil has managed to convert almost all its automobiles to burn ethanol (CH_3CH_2OH) as fuel.

a. Use the values for ΔH°_f from Appendix 4 to calculate the fuel value of ethanol.

b. How much energy is released during the combustion of 1.00 kg of CH_3CH_2OH?

c. How many grams of CH_3CH_2OH are needed to heat 1.00 kg of water from 25.0°C to 85.0°C? Assume that all the energy released during combustion is used to heat the water.

d. Is the amount of CH_3CH_2OH needed greater than or less than the amount of butanol ($C_4H_{10}O$) needed to heat 1.00 kg of water from 25.0°C to 85.0°C (see Problem 6.101)?

Additional Problems

6.103. **Diagnosing Car Trouble** Mechanics sometimes use diethyl ether ($CH_3CH_2OCH_2CH_3$) to diagnose starting problems in vehicles because it has a low ignition temperature. If sprayed into the air intake, it can help determine whether the spark and ignition system of the car are functioning, and the fuel delivery system is not. If the normal fuel is not entering the engine, the engine will run only until the diethyl ether vapors are completely burned. Compare the fuel value and fuel density of diethyl ether ($\Delta H^\circ_{comb} = -2726.3$ kJ/mol; density = 0.7134 g/mL) to that of diesel oil (determined in Sample Exercise 6.16).

6.104. The standard enthalpies of combustion of ethyne [$C_2H_2(g)$], $C(s)$, and $H_2(g)$ are −1299.6, −393.5, and −285.9 kJ/mol, respectively. Use this information to calculate the standard enthalpy of formation of ethyne.

6.105. Carbon tetrachloride (CCl_4) was at one time used as a fire-extinguishing agent. It has a molar heat capacity of 131.3 J/(mol · °C). How much energy is required to raise the temperature of 275 g of CCl_4 from room temperature (22°C) to its boiling point (77°C)?

6.106. Ethylene glycol ($HOCH_2CH_2OH$) is mixed with the water in radiators to cool car engines. How much heat will 725 g of pure ethylene glycol remove from an engine as it is warmed from 0°C to its boiling point of 196°C? The c_P of ethylene glycol is 149.5 J/(mol · °C).

6.107. Sodium may be used as a heat-storage material in some devices. The specific heat (c_s) of sodium metal is 1.23 J/(g · °C). How many moles of sodium metal are required to absorb 1.00×10^3 kJ of energy?

6.108. The water in a bomb calorimeter was replaced with the organic compound methylene chloride (CH_2Cl_2). Burning 2.23 g of glucose ($C_6H_{12}O_6$; $\Delta H_{comb} = -2801$ kJ/mol) in the calorimeter causes its temperature to rise 9.64°C. What is the heat capacity of the calorimeter?

6.109. The standard enthalpy of formation of NH_3 is −46.1 kJ/mol. What is ΔH°_{rxn} for the following reactions?
a. $N_2(g) + 3\,H_2(g) \rightarrow 2\,NH_3(g)$
b. $NH_3(g) \rightarrow \frac{3}{2}\,H_2(g) + \frac{1}{2}\,N_2(g)$

*6.110. **Hung Out to Dry** Laundry left outside to dry on a clothesline in the winter slowly dries by sublimation. The increase in internal energy of water vapor produced by sublimation is less than the amount of heat absorbed. Explain.

6.111. Chlorofluorocarbons (CFCs) such as CF_2Cl_2 are refrigerants whose use has been phased out because of their destructive effect on Earth's ozone layer. The standard enthalpy of vaporization of CF_2Cl_2 is 17.4 kJ/mol, compared with $\Delta H^\circ_{vap} = 40.67$ kJ/mol for liquid water. How many grams of liquid CF_2Cl_2 are needed to cool 200.0 g of water from 50.0°C to 40.0°C? The specific heat of water is 4.184 J/(g · °C).

6.112. A 100.0 mL sample of 1.0 *M* NaOH is mixed with 50.0 mL of 1.0 *M* H_2SO_4 in a large Styrofoam coffee cup; the cup is fitted with a lid through which a calibrated thermometer passes. The temperature of each solution before mixing is 22.3°C. After the NaOH solution is added to the coffee cup and the mixed solutions are stirred with the thermometer, the maximum temperature measured is 31.4°C. Assume that the density of the mixed solutions is 1.00 g/mL, the specific heat of the mixed solutions is 4.18 J/(g · °C), and no heat is lost to the surroundings.
a. Write a balanced chemical equation for the reaction that takes place in the Styrofoam cup.
b. Is any NaOH or H_2SO_4 left in the Styrofoam cup when the reaction is over?
c. Calculate the enthalpy change per mole of H_2SO_4 in the reaction.

6.113. Varying the scenario in Problem 6.112, assume this time that 65.0 mL of 1.0 *M* H_2SO_4 is mixed with 100.0 mL of 1.0 *M* NaOH and that both solutions are initially at 25.0°C. Assume that the mixed solutions in the Styrofoam cup have the same density and specific heat as in Problem 6.112 and no heat is lost to the surroundings. What is the maximum measured temperature in the Styrofoam cup?

*6.114. An insulated container is used to hold 50.0 g of water at 25.0°C. A 7.25 g sample of copper is placed in a dry test tube and heated for 30 min in a boiling water bath at 100.1°C. The heated test tube is carefully removed from the water bath with laboratory tongs and inclined so that the copper slides into the water in the insulated container. Given that the specific heat of solid copper is 0.385 J/(g · °C), calculate the maximum temperature of the water in the insulated container after the copper metal is added.

6.115. The mineral magnetite (Fe_3O_4) is magnetic, whereas iron(II) oxide is not.
a. Write and balance the chemical equation for the formation of magnetite from iron(II) oxide and oxygen.
b. Given that 318 kJ of heat is released for each mole of Fe_3O_4 formed, what is the enthalpy change of the balanced reaction of formation of Fe_3O_4 from iron(II) oxide and oxygen?

6.116. Which of the following substances has a standard enthalpy of formation equal to zero? (a) Pb at 1000°C; (b) $C_3H_8(g)$ at 25.0°C and 1 atm pressure; (c) solid glucose at room temperature; (d) $N_2(g)$ at 25.0°C and 1 atm pressure

***6.117.** The standard enthalpy of formation of liquid water is −285.8 kJ/mol.
a. What is the significance of the negative sign associated with this value?
b. Why is the magnitude of this value so much larger than the enthalpy of vaporization of water ($\Delta H^\circ_{vap} =$ 40.67 kJ/mol)?
c. Calculate the amount of heat produced in making 50.0 mL of water from its elements under standard conditions.

6.118. Acetylene, C_2H_2 ($\Delta H^\circ_f = 226.7$ kJ/mol), and benzene, C_6H_6 ($\Delta H^\circ_f = 49.0$ kJ/mol), are sometimes referred to as endothermic compounds.
a. Why are C_2H_2 and C_6H_6 called endothermic compounds?
b. Calculate the standard molar enthalpy of combustion of acetylene and benzene.

***6.119.** Balance the following chemical equation, name the reactants and products, and calculate the standard enthalpy change by using the data in Appendix 4.

$$FeO(s) + O_2(g) \rightarrow Fe_2O_3(s)$$

*6.120. Add reactions 1, 2, and 3, and label the resulting reaction 4. Consult Appendix 4 to find the standard enthalpy change for the balanced reaction 4.

(1) $Zn(s) + \frac{1}{8}S_8(s) \rightarrow ZnS(s)$

(2) $ZnS(s) + 2\ O_2(g) \rightarrow ZnSO_4(s)$

(3) $\frac{1}{8}S_8(s) + O_2(g) \rightarrow SO_2(g)$

6.121. Rocket Plane Fuel The X37B surveillance plane under development uses a mixture of hydrazine (N_2H_4) and dinitrogen tetroxide (N_2O_4) as fuel.

a. The standard enthalpies of formation of $N_2H_4(g)$, $N_2O_4(g)$, and $H_2O(g)$ are 95.35, 9.2, and −241.8 kJ/mol, respectively. Use this information to calculate the standard enthalpy of the following reaction:

$$2\ N_2H_4(g) + N_2O_4(g) \rightarrow 3\ N_2(g) + 4\ H_2O(g)$$

b. Would more energy be released if the hydrazine was in the liquid state if $\Delta H^\circ_{f,N_2H_4,\ell} = 50.63$ kJ/mol?

6.122. The specific heat of solid copper is 0.385 J/(g · °C). What thermal energy change occurs when a 35.3 g sample of copper is cooled from 35.0°C to 15.0°C? Be sure to give your answer the proper sign. This amount of energy is used to melt solid ice at 0.0°C. The molar enthalpy of fusion of ice is 6.01 kJ/mol. How many moles of ice are melted?

***6.123. Metabolism of Methanol** Methanol is toxic because it is metabolized in a two-step process *in vivo* to formic acid (HCOOH). Consider the following overall reaction under standard conditions:

$$O_2(g) + 2\ CH_3OH(\ell) \rightarrow 2\ HCOOH(\ell) + 2\ H_2O(\ell)$$

a. Is this reaction endothermic or exothermic?

b. What is the value of ΔH°_{rxn} for this reaction?

c. How much energy would be absorbed or released if 60.0 g of methanol were metabolized in this reaction?

d. In the first step of metabolism, methanol is converted into formaldehyde (CH_2O), which is then converted into formic acid. Would you expect ΔH°_{rxn} for the metabolism of 1 mole of $CH_3OH(\ell)$ producing 1 mole of formaldehyde to be larger or smaller than 509.8 kJ?

6.124. Use Hess's law and the following data to calculate the standard enthalpy of formation of ethanol, $CH_3CH_2OH(\ell)$.

$CH_3CH_2OH(\ell) + 3\ O_2(g) \rightarrow 2\ CO_2(g) + 3\ H_2O(\ell)$ $\quad \Delta H^\circ_{rxn} = -1368.2$ kJ/mol

$C(s) + O_2(g) \rightarrow CO_2(g)$ $\quad \Delta H^\circ_f = -393.5$ kJ/mol

$H_2(g) + \frac{1}{2}O_2(g) \rightarrow H_2O(\ell)$ $\quad \Delta H^\circ_f = -285.9$ kJ/mol

6.125. Use Hess's law and the following data to calculate the standard enthalpy of formation of $CH_4(g)$.

$C(s) + O_2(g) \rightarrow CO_2(g)$ $\quad \Delta H^\circ_f = -393.5$ kJ/mol

$H_2(g) + \frac{1}{2}O_2(g) \rightarrow H_2O(\ell)$ $\quad \Delta H^\circ_f = -285.9$ kJ/mol

$CO_2(g) + 2\ H_2O(\ell) \rightarrow CH_4(g) + 2\ O_2(g)$ $\quad \Delta H^\circ_{rxn} = -890.4$ kJ/mol

6.126. The reaction of $CH_3OH(g)$ with $N_2(g)$ to give $HCN(g)$ and $NH_3(g)$ requires 164 kJ/mol of heat.

a. Write a balanced chemical equation for this reaction.

b. Should the thermal energy involved be written as a reactant or as a product?

c. What is the enthalpy change in the reaction of 60.0 g of $CH_3OH(g)$ with excess $N_2(g)$ to give $HCN(g)$ and $NH_3(g)$ in this reaction?

6.127. Calculate ΔH°_{rxn} for the reaction

$$2\ Ni(s) + \frac{1}{4}S_8(s) + 3\ O_2(g) \rightarrow 2\ NiSO_3(s)$$

from the following information:

(1) $NiSO_3(s) \rightarrow NiO(s) + SO_2(g)$ $\quad \Delta H^\circ_{rxn} = 156$ kJ

(2) $\frac{1}{8}S_8(s) + O_2(s) \rightarrow SO_2(g)$ $\quad \Delta H^\circ_f = -297$ kJ

(3) $Ni(s) + \frac{1}{2}O_2(g) \rightarrow NiO(s)$ $\quad \Delta H^\circ_f = -241$ kJ

6.128. Use the following information to calculate the enthalpy change involved in the complete reaction of 3.0 g of carbon to form $PbCO_3(s)$ in reaction 4. Be sure to give the proper sign (positive or negative) with your answer.

(1) $Pb(s) + \frac{1}{2}O_2(g) \rightarrow PbO(s)$ $\quad \Delta H^\circ_{rxn} = -219$ kJ

(2) $C(s) + O_2(g) \rightarrow CO_2(g)$ $\quad \Delta H^\circ_{rxn} = -394$ kJ

(3) $PbCO_3(s) \rightarrow PbO(s) + CO_2(g)$ $\quad \Delta H^\circ_{rxn} = 86$ kJ

(4) $Pb(s) + C(s) + \frac{3}{2}O_2(g) \rightarrow PbCO_3(s)$ $\quad \Delta H^\circ_{rxn} = ?$

***6.129. Butanol as Automobile Fuel** The use of butanol [$CH_3(CH_2)_2CH_2OH$] as a biofuel has been proposed. Calculate the standard molar enthalpy for the complete combustion of liquid butanol by using the standard enthalpies of formation of the reactants and gaseous products as given in Appendix 4. Which has a greater fuel value, butanol or ethanol?

6.130. Adding 1.56 g of K_2SO_4 to 6.00 mL of water at 16.2°C causes the temperature of the solution to drop by 7.70°C. How many grams of NaOH ($\Delta H_{soln} = -44.3$ kJ/mol) would you need to add to raise the temperature back to 16.2°C?

6.131. Two solids, 5.00 g of NaOH and 4.20 g of KOH, are added to 150 mL of water [$c_P = 75.3$ J/(mol · °C); $T = 23$°C] in a calorimeter. Given that $\Delta H_{soln,NaOH} = -44.3$ kJ/mol and $\Delta H_{soln,KOH} = -56.0$ kJ/mol, what is the final temperature of the solution?

6.132. **The 100-Meter Dash** A 1995 article in *Discover* magazine on world-class sprinters contained the following statement: "In one race, a field of eight runners releases enough energy to boil a gallon jug of ice at 0.0°C in ten seconds!" How much "energy" do the runners release in 10 seconds? Assume that the ice has a mass of 128 ounces.

***6.133. Specific Heats of Metals** In 1819, Pierre Dulong and Alexis Petit reported that the product of the atomic mass of a metal times its specific heat is approximately constant, an observation called the *law of Dulong and Petit*.

Use the data in the table below to answer the following questions.

Element	$\mathcal{M}$ (g/mol)	c_s [J/(g · °C)]	$\mathcal{M} \times c_s$
Bismuth		0.120	
Lead	207.2	0.123	25.5
Gold	197.0	0.125	
Platinum	195.1		
Tin	118.7	0.215	
		0.233	
Zinc	65.4	0.388	
Copper	63.5	0.397	
		0.433	
Iron	55.8	0.460	
Sulfur	32.1		
		Average value:	

a. Complete each row in the table by multiplying each given molar mass and specific heat pair (one result has been entered in the table). What are the units of the resulting values in column 4?
b. Next, calculate the average of the values in column 4.
c. Use the mean value from part (b) to calculate the missing atomic masses in the table. Do you feel confident in identifying the elements from the calculated atomic masses?
d. Use the average value from part (b) to predict the missing specific heat values in the table.

*6.134. **Odor of Urine** Urine odor gets worse with time because urine contains the metabolic product urea [$CO(NH_2)_2$]. Urea is slowly converted to ammonia, which has a sharp, unpleasant odor, and carbon dioxide:

$$CO(NH_2)_2(aq) + H_2O(\ell) \rightarrow CO_2(aq) + 2\ NH_3(aq)$$

This reaction is much too slow for the enthalpy change to be measured directly by using a temperature change. Instead, the enthalpy change for the reaction may be calculated from the following data:

Compound	ΔH_f° (kJ/mol)
Urea(aq)	−319.2
$CO_2(aq)$	−412.9
$H_2O(\ell)$	−285.8
$NH_3(aq)$	−80.3

Calculate the standard molar enthalpy change for the reaction.

6.135. Use the following data to determine whether the conversion of diamond into graphite is exothermic or endothermic:

$$C(s,\text{ diamond}) + O_2(g) \rightarrow CO_2(g) \qquad \Delta H^\circ = -395.4\text{ kJ}$$
$$2\ CO_2(g) \rightarrow 2\ CO(g) + O_2(g) \qquad \Delta H^\circ = 566.0\text{ kJ}$$
$$2\ CO(g) \rightarrow C(s,\text{ graphite}) + CO_2(g) \qquad \Delta H^\circ = -172.5\text{ kJ}$$
$$C(s,\text{ diamond}) \rightarrow C(s,\text{ graphite}) \qquad \Delta H^\circ = ?$$

*6.136. **Rocket Fuels** The payload of a rocket includes a fuel and oxygen for combustion of the fuel. Reactions 1 and 2 describe the combustion of dimethylhydrazine and hydrogen, respectively. Kilogram for kilogram, which is the better rocket fuel, dimethylhydrazine or hydrogen?

(1) $(CH_3)_2NNH_2(\ell) + 4\ O_2(g) \rightarrow N_2(g) + 4\ H_2O(g) + 2\ CO_2(g) \qquad \Delta H^\circ_{rxn} = -1694\text{ kJ}$

(2) $H_2(g) + \frac{1}{2}O_2(g) \rightarrow H_2O(g) \qquad \Delta H^\circ_{rxn} = -242\text{ kJ}$

6.137. At high temperatures, such as those in the combustion chambers of automobile engines, nitrogen and oxygen form nitrogen monoxide:

$$N_2(g) + O_2(g) \rightarrow 2\ NO(g) \qquad \Delta H^\circ_{comb} = +180\text{ kJ}$$

Any NO released into the environment is oxidized to NO_2:

$$2\ NO(g) + O_2(g) \rightarrow 2\ NO_2(g) \qquad \Delta H^\circ_{comb} = -112\text{ kJ}$$

Is the overall reaction,

$$N_2(g) + 2\ O_2(g) \rightarrow 2\ NO_2(g)$$

exothermic or endothermic? What is ΔH°_{comb} for this reaction?

6.138. You are given the following data:

$$\tfrac{1}{2}N_2(g) + \tfrac{1}{2}O_2(g) \rightarrow NO(g) \qquad \Delta H^\circ_{rxn} = +90.3\text{ kJ}$$
$$NO(g) + \tfrac{1}{2}Cl_2(g) \rightarrow NOCl(g) \qquad \Delta H^\circ_{rxn} = -38.6\text{ kJ}$$
$$2\ NOCl(g) \rightarrow N_2(g) + O_2(g) + Cl_2(g) \qquad \Delta H^\circ_{rxn} = ?$$

a. Which of the ΔH°_{rxn} values represent enthalpies of formation?
b. Determine ΔH°_{rxn} for the decomposition of NOCl.

6.139. **Hydrogen as Fuel** Hydrogen is attractive as a fuel because it has a high fuel value and produces no CO_2.

$$H_2(g) + \tfrac{1}{2}O_2(g) \rightarrow H_2O(g)$$

Unfortunately, the production and storage of hydrogen fuel remain problematic.

a. One problem with hydrogen is its low fuel density. What is the fuel density (kJ/L) for H_2 given that the density of hydrogen gas is 0.0899 g/L? If we could liquefy hydrogen, what would the fuel density of hydrogen be, given that the density of liquid hydrogen is 70.8 g/L?
b. One solution to the problem of hydrogen storage is to use solid, hydrogen-containing compounds that release hydrogen upon heating at low temperature. One such compound is ammonia borane (H_3NBH_3). Calculate the enthalpy change (ΔH) for the following reactions given that $\Delta H^\circ_{f,H_3NBH_3} = -38.1$ kJ/mol, $\Delta H^\circ_{f,BH_3} = +110.2$ kJ/mol, $\Delta H^\circ_{f,NH_3} = -46.1$ kJ/mol, $\Delta H^\circ_{f,H_2NBH_2} = -66.5$ kJ/mol, and $\Delta H^\circ_{f,HNBH} = +56.9$ kJ/mol.

$$H_3NBH_3(g) \rightarrow NH_3(g) + BH_3(g)$$
$$H_3NBH_3(g) \rightarrow H_2(g) + H_2NBH_2(g)$$
$$H_2NBH_2(g) \rightarrow H_2(g) + HNBH(g)$$

c. How many kilograms of ammonia borane are needed to supply 10.0 kg of hydrogen?

*6.140. **Industrial Use of Cellulose** Research is being carried out on cellulose as a source of chemicals for the production of fibers, coatings, and plastics. Cellulose consists of long chains of glucose molecules ($C_6H_{12}O_6$), so for the purposes of modeling the reaction, we can consider the conversion of glucose to formaldehyde (CH_2O).

a. Is the reaction to convert glucose into formaldehyde an oxidation or a reduction?

b. Calculate the heat of reaction for the conversion of 1 mole of glucose into formaldehyde, given the following thermochemical data:

ΔH°_{comb} of formaldehyde gas = −572.9 kJ/mol

ΔH°_f of solid glucose = −1274.4 kJ/mol

$$C_6H_{12}O_6(s) \rightarrow 6\ CH_2O(g)$$

Glucose   Formaldehyde

6.141. Smelting Iron Iron metal is obtained by reducing iron oxide with carbon. The balanced chemical equation for making iron from Fe_2O_3 is

$$2\ Fe_2O_3(s) + 3\ C(s) \rightarrow 4\ Fe(s) + 3\ CO_2(g)$$

Iron melts at 1538°C with ΔH_{fus} = 19.4 kJ/mol. The molar heat capacity of iron is $c_{P,Fe}$ = 25.1 J/(mol · °C). Is the energy required to melt recycled iron less than that needed to reduce the iron in Fe_2O_3 to the free metal?

6.142. **Recycling Copper** How does the energy required to recycle 1.00 mole of copper compare with that required to recover copper from CuO? The balanced chemical equation for the smelting of copper is $CuO(s) + CO(g) \rightarrow Cu(s) + CO_2(g)$. Copper melts at 1084.5°C with ΔH°_{fus} = 13.0 kJ/mol and a molar heat capacity $c_{P,Cu}$ = 24.5 J/(mol · °C). In addition, $\Delta H^\circ_{f,CuO}$ = −155 kJ/mol.

6.143. Converting Diamond to Graphite The standard state of carbon is graphite. $\Delta H^\circ_{f,diamond}$ is +1.896 kJ/mol. Diamond masses are normally given in carats, where 1 carat = 0.20 g. Determine the standard enthalpy of the reaction for the conversion of a 4-carat diamond into an equivalent mass of graphite. Is this reaction endothermic or exothermic?

7

A Quantum Model of Atoms

Waves, Particles, and Periodic Properties

RAINBOWS When sunlight passes through water droplets in the sky, a rainbow is often the result. Scientists' efforts to explain why energy (such as sunlight) interacted with matter (such as raindrops) to produce the colors of a rainbow led to the development of quantum theory and an entirely new way of viewing matter at the atomic level.

PARTICULATE **REVIEW**

Atoms, Ions, and Their Electrons

In Chapter 7 we explore periodic trends in the sizes of atoms and monatomic ions, along with the energy changes involved with forming ions from neutral parent atoms.

- The spheres shown here represent the charges and relative sizes of the most common monatomic ions formed by potassium, aluminum, fluorine, and sulfur. Match each element to its ion.
- How many electrons are in a neutral atom of each of these elements?
- How many electrons are in each ion shown here?

  (Review Sections 2.3 and 2.6 if you need help.)

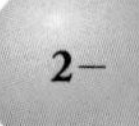

(Answers to Particulate Review questions are in the back of the book.)

PARTICULATE **PREVIEW**

Energy Changes and Ion Formation

In Section 6.2 we learned that ions of opposite charge are attracted to each other. Separating ions that are bonded to one another is an endothermic process. As you read Chapter 7, look for ideas that will help you answer these questions:

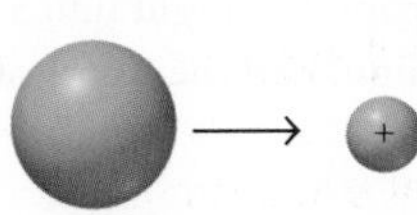

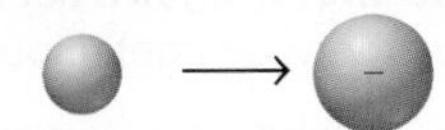

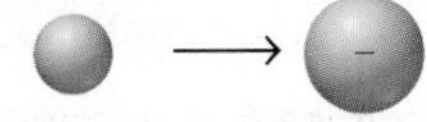

- Is an electron gained or lost in each of the two processes depicted here?
- Which process usually emits energy?
- Which process requires the absorption of energy?

Learning Outcomes

LO1 Interconvert the energies, wavelengths, and frequencies of electromagnetic radiation and correlate their values to the appropriate regions of the electromagnetic spectrum
Sample Exercises 7.1, 7.2

LO2 Describe quantum theory and use it to explain the photoelectric effect
Sample Exercise 7.3

LO3 Calculate the energies and wavelengths of photons absorbed and emitted by atoms in their electron transitions between atomic energy levels
Sample Exercises 7.4, 7.5

LO4 Apply the Heisenberg uncertainty principle to particles in motion and calculate their de Broglie wavelengths
Sample Exercises 7.6, 7.7

LO5 Assign quantum numbers to orbitals and use their values to describe the sizes, energies, and orientations of orbitals
Sample Exercises 7.8, 7.9

LO6 Use the aufbau principle and Hund's rule to write electron configurations and draw orbital diagrams of atoms and monatomic ions
Sample Exercises 7.10, 7.11, 7.12, 7.13

LO7 Use atomic orbitals and the concept of effective nuclear charge to explain the ionization energies and relative sizes of atoms and monatomic ions
Sample Exercises 7.14, 7.15

LO8 Relate the electron affinities of the elements to their positions in the periodic table

FIGURE 7.1 Isaac Newton used a prism to separate sunlight into a spectrum containing all the colors of the rainbow.

7.1 Rainbows of Light

Rainbows can put visual exclamation points on passing summer showers. They often appear just after the clouds that brought the shower begin to clear, allowing sunlight to illuminate the trailing edge of the receding rain. A rainbow's colors are nearly always arranged in the same pattern with red on top, violet on the bottom, and the other colors in between. Have you ever wondered why?

The answer has to do with the ways that radiant energy (sunlight) interacts with matter (raindrops). Such interactions are a recurring theme in this chapter. Although humans have enjoyed the colors of rainbows ever since we evolved as an intelligent species, it was only a few hundred years ago that a brilliant Englishman named Isaac Newton (1642–1727) was able to explain how and why rainbows appear. He discovered that passing sunlight through a glass prism (**Figure 7.1**) separated the light into its component colors in much the way raindrops do. Then he showed how the colors could be recombined into white sunlight with a second prism, thereby proving that white light is a blend of all the colors in a rainbow.

In the decades following Newton's discoveries, the quality of prisms improved dramatically, enabling scientists in Europe and Great Britain to separate sunlight into its component colors with increasing resolution. Then, in 1800, another

Fraunhofer lines a set of dark lines in the otherwise continuous solar spectrum.

atomic emission spectrum a characteristic series of bright lines produced by high-temperature atoms.

atomic absorption spectrum a characteristic series of dark lines produced when free, gaseous atoms are illuminated by a continuous source of radiation.

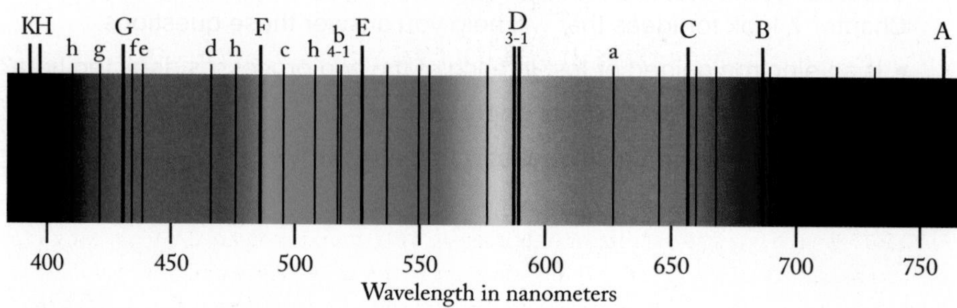

FIGURE 7.2 The spectrum of sunlight contains many narrow gaps, which appear as dark lines called Fraunhofer lines.

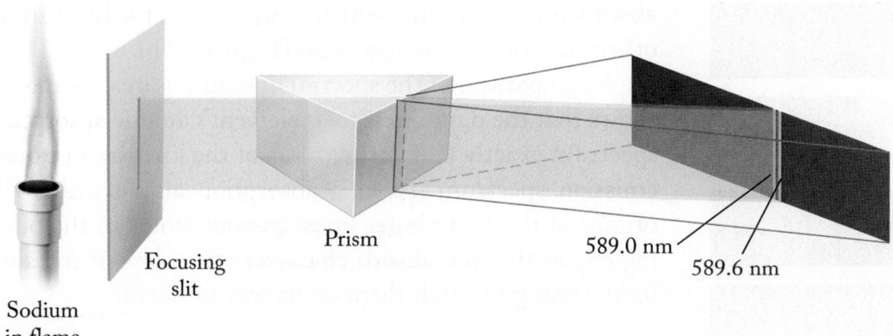

FIGURE 7.3 Sodium atoms heated in a flame produce a bright yellow-orange light that is caused by two closely spaced emission lines with wavelengths 589.0 nm and 589.6 nm (see Section 7.2).

English scientist named William Hyde Wollaston (1766–1828) made a second discovery about the sun's rainbow of colors: it was not continuous. Rather, it contained dark, narrow lines. Using even better prisms, the German physicist Joseph von Fraunhofer (1787–1826) resolved more than 500 of these lines, which are now called **Fraunhofer lines** (**Figure 7.2**). He labeled the darkest lines with the letters of the alphabet, starting with *A* at the red end of the Sun's spectrum.

Labeling his lines, however, did not mean that Fraunhofer, or anyone else, knew why they existed. Discovering why would come later in the 19th century with the work of two more German scientists: the chemist Robert Wilhelm Bunsen (1811–1899) and physicist Gustav Robert Kirchhoff (1824–1887). They collaborated on extensive studies of the light emitted (given off) by elements, particularly group 1 and 2 elements, when they are vaporized, and their atoms are heated by the transparent flames produced by a burner designed by Bunsen. Unlike the spectrum of sunlight, which displays narrow gaps in an otherwise continuous spectrum of all colors, the spectra produced by hot atoms in flames consist of only a few bright lines on a dark background, as shown in **Figure 7.3** and **Figure 7.4**. Bunsen and Kirchhoff discovered that many of the lines in these **atomic emission spectra** exactly matched colors that were missing in the Sun's visible spectrum. For example, the Fraunhofer *D* line corresponded to the yellow-orange light produced by hot sodium vapor (Figure 7.3).

These experiments, and others employing light sources called gas discharge tubes, showed that each element emits a characteristic line spectrum when its atoms are heated to a sufficiently high temperature (Figure 7.4). Other experiments showed that when free atoms of an element are illuminated by a continuous source of light, such as that emitted by a light bulb (**Figure 7.5a**) or the surface of the Sun, the atoms may absorb particular colors of light, producing an **atomic**

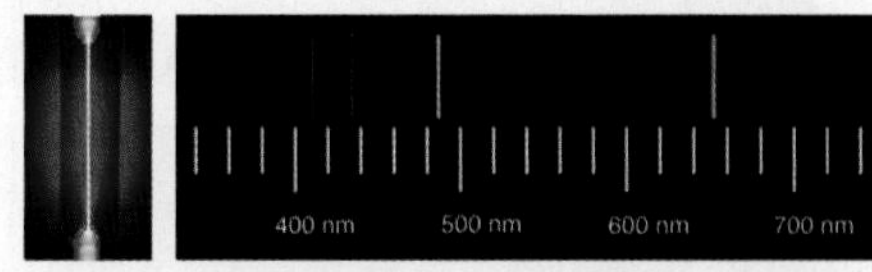

(a) Visible emission spectrum of hydrogen

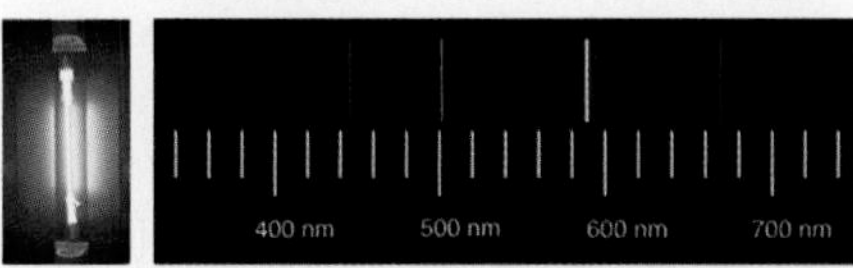

(b) Visible emission spectrum of helium

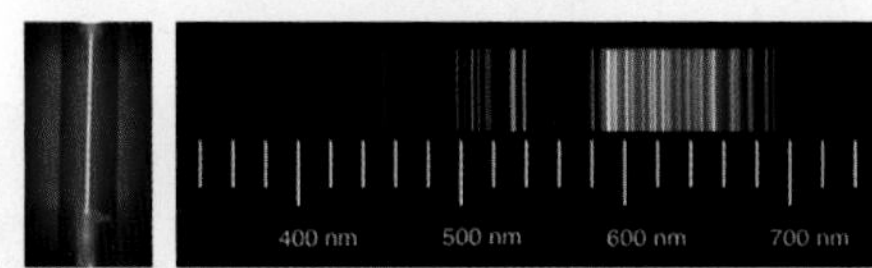

(c) Visible emission spectrum of neon

FIGURE 7.4 Atoms of elements such as (a) hydrogen, (b) helium, and (c) neon emit characteristic line spectra in gas discharge tubes.

CHEMTOUR
Light Diffraction

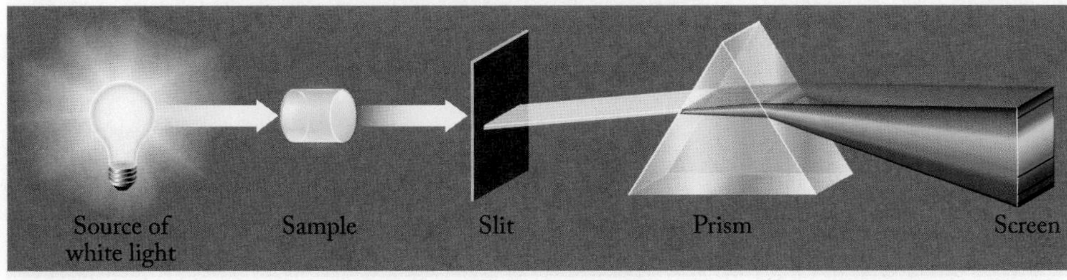

(a)

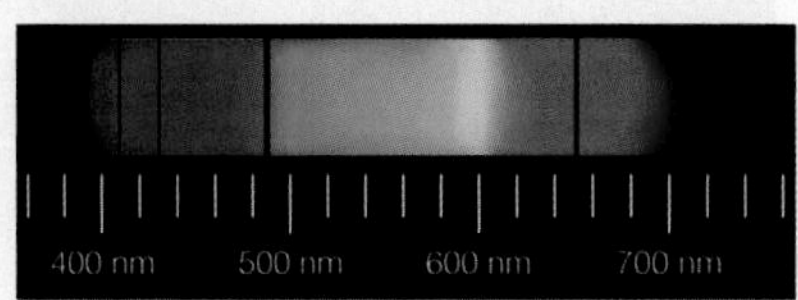

Absorption spectrum of hydrogen

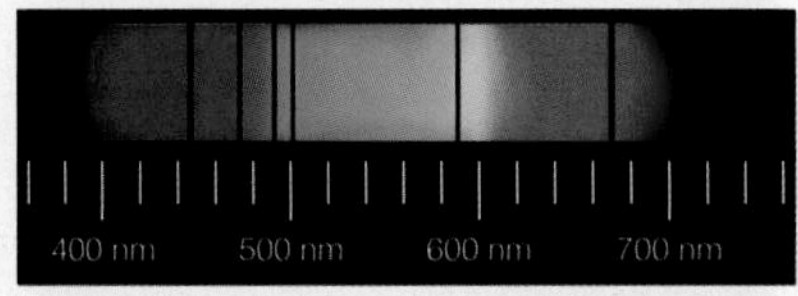

Absorption spectrum of helium

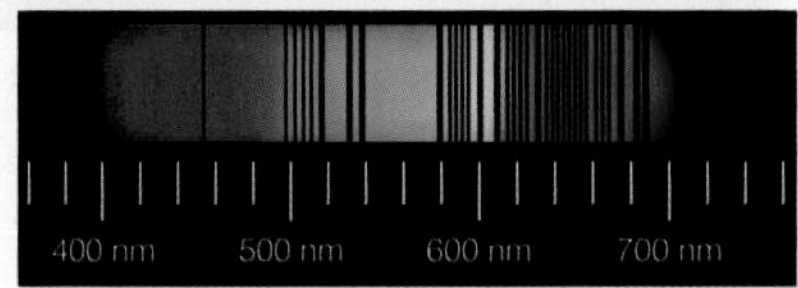

Absorption spectrum of neon

(b)

FIGURE 7.5 (a) When gaseous atoms of hydrogen, helium, and neon are illuminated by an external source of white light (containing all colors of the visible spectrum), the resultant atomic absorption spectra contain dark lines that are characteristic of each element. (b) The dark lines in the atomic absorption spectra of these elements match the bright lines in their atomic emission spectra, shown in Figure 7.4.

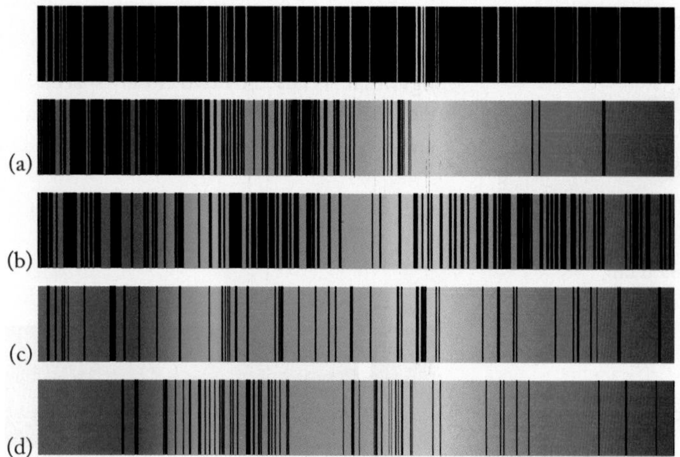

FIGURE 7.6 Atomic emission spectrum of mercury vapor (top) and four atomic absorption spectra (a–d).

absorption spectrum, which consists of dark lines in an otherwise continuous spectrum (**Figure 7.5b**).

A comparison of the spectral lines in Figures 7.4 and 7.5 shows that the dark lines in an element's atomic absorption spectrum exactly match the colors of the lines in its atomic emission spectrum. Atomic absorption also explains the origins of the Fraunhofer lines: gaseous atoms in the outer regions of the Sun absorb characteristic colors of the sunlight passing through them on its way to Earth.

CONCEPT TEST

The top spectrum in **Figure 7.6** is the emission spectrum of mercury vapor. Select the absorption spectrum of mercury vapor from the other four spectra.

(Answers to Concept Tests are in the back of the book.)

In this chapter we explore why the atoms of the elements have unique and complementary atomic emission and absorption spectra. We also link the

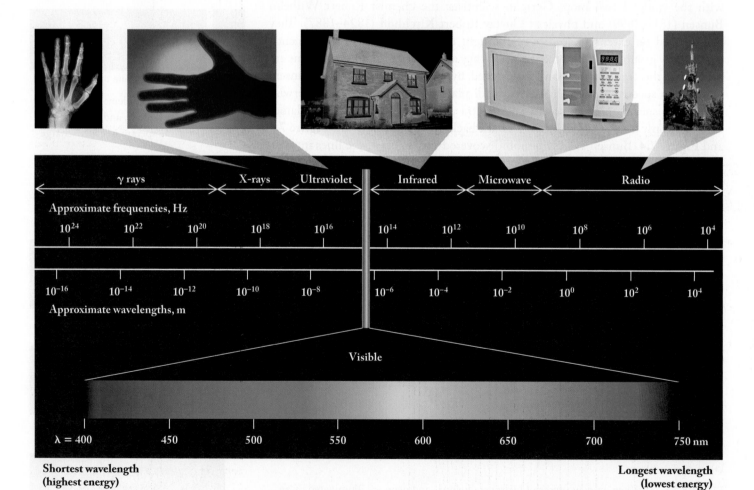

FIGURE 7.7 Visible light occupies a tiny fraction of the electromagnetic spectrum, which ranges from ultrashort-wavelength, high-frequency gamma (γ) rays to long-wavelength, low-frequency radio waves. Note that frequencies increase from right to left, and wavelengths increase from left to right.

positions of the lines in these spectra to the internal structure of atoms and to the arrangement and motion of the electrons inside atoms. We begin this exploration of atomic structure by addressing the properties of sunlight and other invisible forms of radiant energy, and how the nature of radiant energy allows it to interact with matter.

STEPWISE ANIMATION
Absorption of Light

7.2 Waves of Energy

Visible light is a small part of the **electromagnetic spectrum** (**Figure 7.7**). This spectrum is a continuous range of radiant energy that extends from high-energy gamma rays to low-energy radio waves. All forms of radiant energy are examples of **electromagnetic radiation**.

CONNECTION In Chapter 1 we discussed the background microwave radiation of the universe, discovered by Penzias and Wilson, which provided evidence for the Big Bang.

The term *electromagnetic* comes from a theory, proposed by the Scottish scientist James Clerk Maxwell (1831–1879), that radiant energy moves through space (or any transparent medium) in a way that resembles waves flowing across a body of water. Unlike water waves, which oscillate only up and down, Maxwell's waves of radiant energy have two components: an oscillating electric field and an oscillating magnetic field (**Figure 7.8**). These two fields are perpendicular to each other and travel together through space. Maxwell derived a set of equations based on his oscillating-wave model that accurately describes nearly all the observed properties of light.

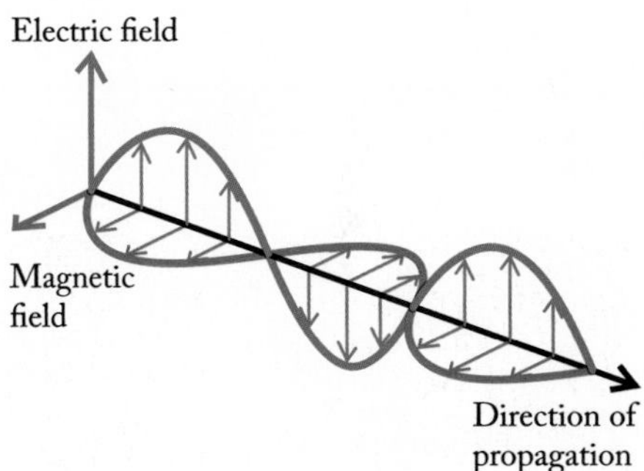

FIGURE 7.8 Electromagnetic waves consist of electric and magnetic fields that oscillate in planes oriented at right angles to each other.

A wave of electromagnetic radiation, like any wave traveling through any medium, has a characteristic **wavelength** (**λ**, the distance from crest to crest) and **frequency** (**ν**, the number of crests that pass a stationary point of reference per second), as shown in **Figure 7.9a**. Frequencies have units of **hertz (Hz)**, also called *cycles per second* (cps): 1 Hz = 1 cps = 1/s. The product of the wavelength and frequency of any electromagnetic radiation is the universal constant c, which is the symbol for the speed of light in a vacuum (2.998×10^8 m/s):

$$c = \lambda\nu \qquad (7.1)$$

Thus, wavelength and frequency have a reciprocal relationship: as wavelength decreases, frequency increases. Another characteristic of a wave is its **amplitude**, the height of the crest or the depth of the trough with respect to the center line of the wave (**Figure 7.9b**).

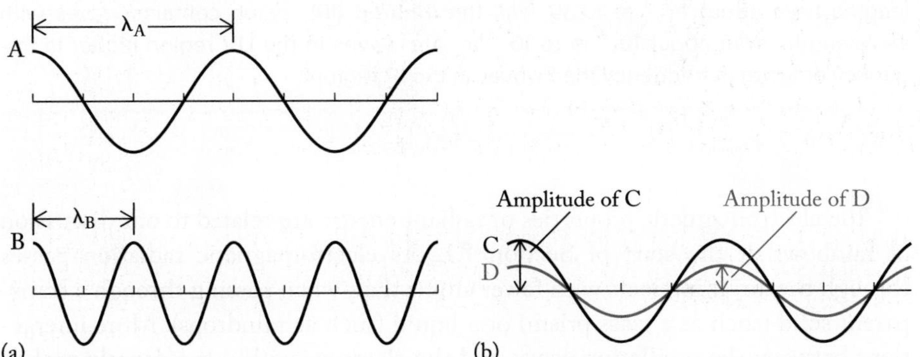

FIGURE 7.9 Every wave has a characteristic wavelength (λ), frequency (ν), and amplitude (intensity). (a) Wave A has a longer wavelength (and lower frequency) than wave B. (b) Waves C and D have the same wavelength and frequency, but the amplitude of C is greater than that of D.

electromagnetic spectrum a continuous range of radiant energy that includes radio waves, microwaves, infrared radiation, visible light, ultraviolet radiation, X-rays, and gamma rays.

electromagnetic radiation any form of radiant energy in the electromagnetic spectrum.

wavelength (λ) the distance from crest to crest or trough to trough on a wave.

frequency (ν) the number of crests of a wave that pass a stationary point of reference per second.

hertz (Hz) the SI unit of frequency, equivalent to 1 cycle per second, or simply 1/s.

amplitude the height of the crest or depth of the trough of a wave with respect to the center line of the wave.

SAMPLE EXERCISE 7.1 Calculating Frequency from Wavelength **LO1**

What is the frequency of the yellow-orange light ($\lambda = 589$ nm) produced by sodium-vapor streetlights?

Collect and Organize We are asked to find the frequency of light that has a wavelength of 589 nm. Frequency and wavelength are related by Equation 7.1 ($c = \lambda\nu$), where c is the speed of light $= 2.998 \times 10^8$ m/s.

CHEMTOUR
Electromagnetic Radiation

Analyze Before using Equation 7.1, we need to convert the wavelength of the light from nanometers to meters (1 nm $= 10^{-9}$ m). The unit labels in Figure 7.7 indicate that frequencies of visible light are in the 10^{14} Hz range, so a correct answer should have a value in this range.

Solve Rearrange Equation 7.1 to solve for frequency:

$$\nu = \frac{c}{\lambda}$$

Convert wavelength from nanometers to meters:

$$589\ \text{nm} \times \frac{10^{-9}\ \text{m}}{1\ \text{nm}} = 589 \times 10^{-9}\ \text{m} = 5.89 \times 10^{-7}\ \text{m}$$

Use this wavelength value to calculate frequency:

$$\nu = \frac{2.998 \times 10^8\ \text{m/s}}{5.89 \times 10^{-7}\ \text{m}}$$
$$= 5.09 \times 10^{14}\ \text{s}^{-1} = 5.09 \times 10^{14}\ \text{Hz}$$

Think About It Our calculated frequency is in the range we expected. Remember that wavelengths and frequencies are inversely proportional to each other: as one increases, the other decreases.

Practice Exercise The radio waves transmitted by a radio station have a frequency of 90.9 MHz. What is the wavelength of these waves in meters?

(Answers to Practice Exercises are in the back of the book.)

CONCEPT TEST

The ultraviolet (UV) region of the electromagnetic spectrum contains waves with wavelengths from about 10^{-9} m to 10^{-7} m; the infrared (IR) region contains waves with wavelengths from about 10^{-6} m to 10^{-4} m. Are waves in the UV region higher in frequency or lower in frequency than waves in the IR region?

The electromagnetic properties of radiant energy are related to our discussion of rainbows at the start of Section 7.1. As electromagnetic radiation passes through the air, it interacts with fewer atoms than when passing through a transparent solid (such as a glass prism) or a liquid (such as raindrops). More interactions between the oscillating waves and the electrons within the densely packed atoms and molecules in solids and liquids cause the waves to slow and their path to bend. Shorter wavelengths are bent more than longer wavelengths as they pass between the raindrops, causing the violet end of the visible spectrum to appear on the bottom of rainbows, while the red end appears at the top.

7.3 Particles of Energy and Quantum Theory

quantum (plural *quanta*) the smallest discrete quantity of a form of energy.

Planck constant (*h*) the proportionality constant between the energy and frequency of electromagnetic radiation: $E = h\nu$; $h = 6.626 \times 10^{-34}$ J · s.

As studies of electromagnetic radiation progressed in the 19th century, scientists discovered limitations to Maxwell's wave model. One limitation was the inability to account for radiation given off by very hot objects. Consider what happens, for example, when a metal rod is heated in a flame. At first the rod gives off only heat in the form of infrared radiation, which we can feel but not see. As the temperature of the metal is increased from 500 to 1000 K (**Figure 7.10a**), the most intense radiation is still in the infrared region of the spectrum, but a small fraction of the emitted light is in the red region of the visible spectrum, as shown in **Figure 7.10b**. With still more heating, the intensity of the red glow increases, and the color of the rod begins to shift to red-orange, orange (**Figure 7.10c**), yellow, and eventually to white (**Figure 7.10d**) as the metal emits all wavelengths of visible light. Even white-hot metal emits little radiation in the ultraviolet region and none at even shorter wavelengths.

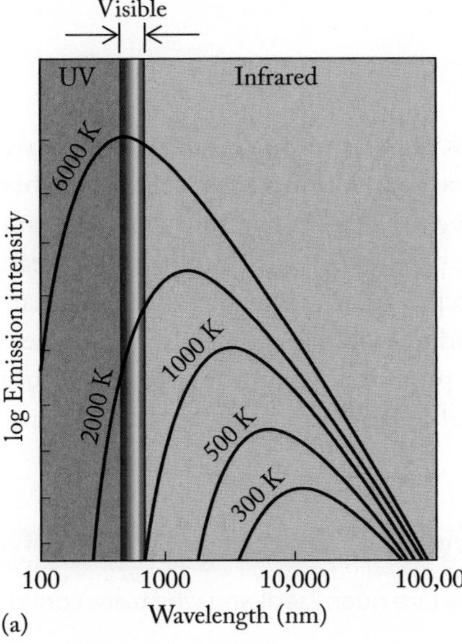

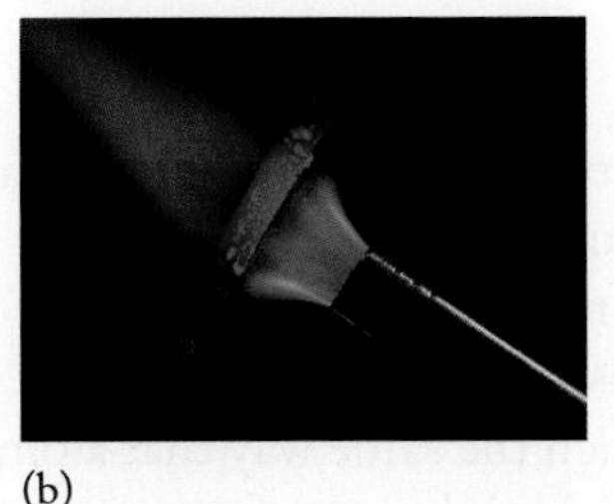
(b)

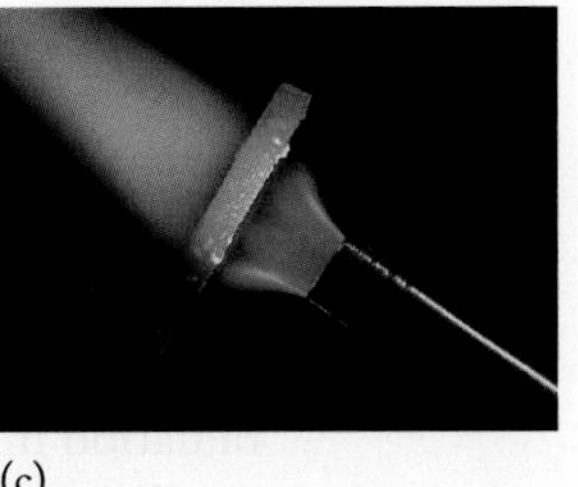
(c)

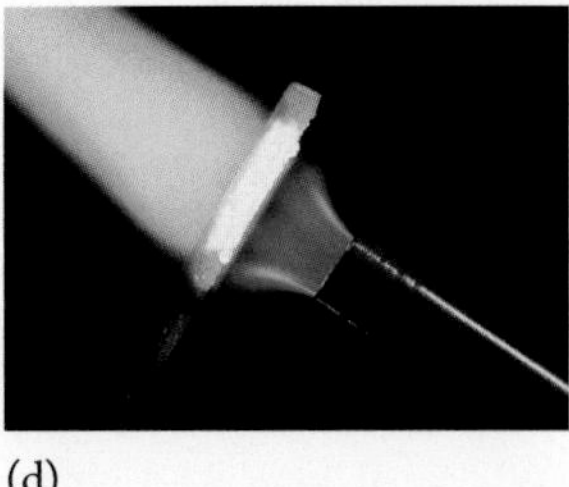
(d)

FIGURE 7.10 (a) Plots of emission intensity versus wavelength show that as the temperature of a solid increases, the intensity of the radiation that it emits increases and the wavelength of maximum intensity decreases. As a metal rod is heated, (b) it first glows red, (c) then turns orange, and (d) finally becomes white-hot as light of all colors is emitted.

This phenomenon, which is called *blackbody* radiation, was well known at the end of the 19th century. However, none of Maxwell's equations for electromagnetic radiation could account for the emission spectrum of a heated metal filament. Another explanation—indeed, another model of radiation behavior—was required.

Quantum Theory

In 1900 the German scientist Max Planck (1858–1947; **Figure 7.11**) proposed such a model in which he discarded classical physics and made a bold assumption about radiant energy: no matter what its source, it can never be truly continuous. Instead, Planck proposed that radiant objects emit energy only in integral multiples of an elementary unit, or **quantum**, of energy defined by the equation

$$E = h\nu \qquad (7.2)$$

where ν is the frequency of the radiation that such an object emits and h is the **Planck constant**, 6.626×10^{-34} J · s.

FIGURE 7.11 German scientist Max Planck is considered the father of quantum physics. He won the 1918 Nobel Prize in Physics for his pioneering work on the quantized nature of electromagnetic radiation. Planck was revered by his colleagues for his personal qualities as well as his scientific accomplishments.

FIGURE 7.12 Quantized and unquantized heights. A flight of stairs exemplifies quantization because each step rises by a discrete height to the next step. The rise on a ramp, on the other hand, is continuous, not quantized.

To visualize the meaning of Planck's quanta, consider two ways you might get from the sidewalk to the entrance of a building (**Figure 7.12**). If you walked up the steps, you would be able to stand at only discrete heights above the sidewalk—each height equal to the rise of a single step. You could not stand at a height between two adjacent steps because there would be nothing to stand on at that height. If you walked up the ramp, however, you could stop at any height between the sidewalk and the entrance. The discrete height changes represented by the steps are also a model of Planck's hypothesis that energy is **quantized**; that is, it is released (analogous to walking down the steps) or absorbed (walking up the steps) in discrete packets, or quanta, of energy.

Combining Equations 7.1 and 7.2 yields an equation that relates the energy of a quantum of radiant energy to its wavelength:

$$E = \frac{hc}{\lambda} \tag{7.3}$$

Given the extremely tiny value of the Planck constant (h), the energy values we obtain using Equation 7.3 tend to be very small because E represents the number of joules in a single quantum of energy. Today we call the tiny packets of radiant energy **photons**. They represent elementary building blocks of electromagnetic radiation in much the same way that atoms represent the building blocks of matter. The observed brightness of a source of radiant energy is the sum of the energies of the enormous number of photons it produces per unit of time. Because Planck's energy model is characterized by *quantum* building blocks, it has become known as **quantum theory**.

quantized having values restricted to whole-number multiples of a specific base value.

photon a quantum of electromagnetic radiation.

quantum theory a model based on the idea that energy is absorbed and emitted in discrete quantities of energy called quanta.

CONCEPT TEST

Which of these quantities vary by discrete values (are quantized) and which are continuous (not quantized)?

a. the volume of water that evaporates from a lake each day during a summer heat wave
b. the number of eggs remaining in a carton
c. the time it takes you to get ready for class in the morning
d. the number of red lights encountered when driving the length of Fifth Avenue in New York City

SAMPLE EXERCISE 7.2 Calculating the Energy of a Single Photon LO1

Chlorophyll *b*, a pigment found in plants, absorbs light with wavelengths of 462 nm and 647 nm. What are the energies of a photon associated with each of these wavelengths? What colors of light does chlorophyll *b* absorb at each wavelength?

Collect, Organize, and Analyze We are asked to calculate the energies of photons with two different wavelengths. To use Equation 7.3, which relates these quantities, we need to use the value of the Planck constant (h), which is 6.626×10^{-34} J · s, and the speed of light (c), 2.998×10^{8} m/s. The photons' wavelengths are given in nanometers, but this value of c has units of meters per second, so we need to convert nanometers to meters for the distance units to cancel. The value of h is extremely small, so the energy of one photon, even factoring in the speed of light, should be very low.

Solve

$$E = \frac{hc}{\lambda} = \frac{(6.626 \times 10^{-34}\,\text{J}\cdot\text{s})\left(2.998 \times 10^{8}\,\frac{\text{m}}{\text{s}}\right)}{462\,\text{nm} \times \frac{10^{-9}\,\text{m}}{1\,\text{nm}}}$$
$$= 4.30 \times 10^{-19}\,\text{J}$$

$$E = \frac{hc}{\lambda} = \frac{(6.626 \times 10^{-34}\,\text{J}\cdot\text{s})\left(2.998 \times 10^{8}\,\frac{\text{m}}{\text{s}}\right)}{647\,\text{nm} \times \frac{10^{-9}\,\text{m}}{1\,\text{nm}}}$$
$$= 3.07 \times 10^{-19}\,\text{J}$$

According to Figure 7.7, a wavelength of 462 nm is blue light, and a wavelength of 647 nm is orange light.

Think About It These quantities of energy are extremely small, as they should be, because a single photon is an atomic-level particle of radiant energy. The results of the calculations confirm that a photon of orange light has less energy than a photon of blue light.

Practice Exercise Leaves contain an array of different light-absorbing molecules to harvest the full spectrum of visible light. Some of these include β-carotene (the compound that makes carrots orange), which absorbs at $\lambda = 453$ nm and 482 nm. How much energy do single photons of 453 nm and 482 nm light have?

The Photoelectric Effect

STEPWISE ANIMATION
The Photoelectric Effect

Although Planck's quantum model explained the emission spectra of hot objects, there was no experimental evidence in 1900 to support the existence of quanta of energy. In 1905 Albert Einstein (1879–1955) supplied that evidence. It came from his studies of a phenomenon called the **photoelectric effect**, in which electrons are emitted from metals when they are illuminated by and absorb electromagnetic radiation. Because light releases these electrons, they are called *photoelectrons.*

Photoelectrons are emitted by a material when the frequency of incident radiation is above a minimum **threshold frequency (ν_0)** characteristic of that material (**Figure 7.13**). Radiation at frequencies less than the threshold value produces no photoelectrons, no matter how intense the radiation is. On the other hand, even a dim source of radiant energy produces at least a few photoelectrons when the frequencies it emits are equal to, or greater than, the threshold frequency.

Einstein used Planck's quantum theory to explain this behavior. He proposed that the threshold frequency is the frequency of the minimum quantum of absorbed energy needed to remove a single electron from the surface of a material. This minimum quantity of energy is related to the strength of the attraction between the nuclei of surface atoms and the electrons surrounding them, and it is called the material's **work function (ϕ)**:

$$\phi = h\nu_0 \tag{7.4}$$

If a photoelectric material is illuminated with radiation frequencies above the threshold frequency ($\nu > \nu_0$), any energy in excess of ϕ is imparted to each ejected electron as kinetic energy:

$$\text{KE}_{\text{electron}} = h\nu - h\nu_0 = h\nu - \phi \tag{7.5}$$

The higher the frequency of the incident light above the threshold frequency, the higher the kinetic energy, and hence the velocity, of the ejected electrons.

photoelectric effect the phenomenon of light striking a metal surface and producing an electric current (a flow of electrons).

threshold frequency (ν_0) the minimum frequency of light required to produce the photoelectric effect.

work function (ϕ) the amount of energy needed to remove an electron from the surface of a metal.

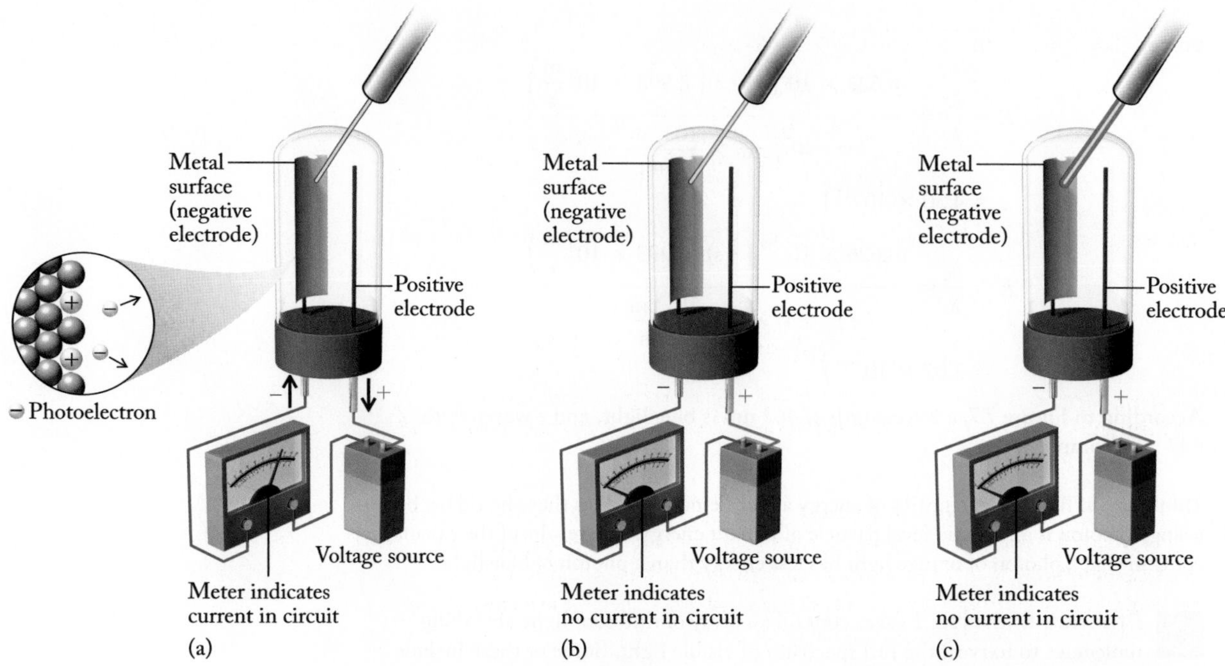

FIGURE 7.13 A phototube includes a positive electrode and a negative metal electrode. (a) If radiation of high enough frequency and energy (violet light in this illustration) illuminates the negative electrode, electrons are dislodged from the surface and flow toward the positive electrode. This flow of electrons produces an electric current that is detected by the meter. The size of the current is proportional to the intensity of the radiation—to the number of photons per unit time striking the negative electrode. (b) The frequency of the red light is below the threshold frequency, so it cannot dislodge electrons and does not produce the photoelectric effect. (c) Even if many low-frequency photons bombard the surface of the metal, no electrons are emitted, and no electrical current flows.

Wave–Particle Duality

Einstein's explanation of the photoelectric effect was of such profound significance that he was awarded the Nobel Prize in Physics for it in 1921. Why does the concept of photons have such far-reaching significance in science? Classical physics had identified the wave nature of light but could not account for light behaving as a particle. Only quantum physics recognizes the **wave–particle duality** of light: light behaves as both a wave *and* a particle.

wave–particle duality the behavior of an object that exhibits the properties of both a wave and a particle.

SAMPLE EXERCISE 7.3 Applying the Photoelectric Effect — LO2

Can germanium be used to detect infrared radiation ($\lambda = 902$ nm) emitted by a remote-control device? The work function of germanium is 7.61×10^{-19} J.

Collect, Organize, and Analyze We know the work function of germanium and a wavelength of infrared radiation, and we are asked to determine whether the radiation will dislodge photoelectrons from germanium. Equation 7.4 relates a material's work function (ϕ) to its threshold frequency (ν_0). To use Equation 7.4 to solve this problem, we need to convert wavelength to frequency, which we can do using Equation 7.1.

Solve Calculate the frequency of a 902 nm photon:

$$\lambda\nu = c = 2.998 \times 10^8 \text{ m/s}$$

$$\nu = \frac{c}{\lambda} = \frac{2.998 \times 10^8 \text{ m/s}}{902 \text{ nm} \times \dfrac{10^{-9} \text{ m}}{1 \text{ nm}}} = 3.33 \times 10^{14} \text{ s}^{-1}$$

Rearrange Equation 7.4 to solve for the threshold frequency for germanium:

$$\phi = h\nu_0$$

$$\nu_0 = \frac{\phi}{h} = \frac{7.61 \times 10^{-19} \text{ J}}{6.626 \times 10^{-34} \text{ J} \cdot \text{s}} = 1.15 \times 10^{15} \text{ s}^{-1}$$

The frequency of the infrared radiation is lower than the threshold frequency for germanium. Therefore, the infrared radiation will *not* liberate photoelectrons, and germanium *cannot* be used to detect it.

Think About It Another way to solve this problem would have been to use Equation 7.3 (i.e., $E = hc/\lambda$) to determine the energy of a photon of 902 nm radiation. If this energy were equal to or greater than the work function, germanium could be used to detect the radiation. According to the results of the calculation above, the energy of a 902 nm photon is less than the work function of germanium.

Practice Exercise The work function of silver is 7.59×10^{-19} J. What is the longest wavelength (in nanometers) of electromagnetic radiation that can eject an electron from the surface of a piece of silver?

CONCEPT **TEST**

Use the electromagnetic spectrum in Figure 7.7 to answer this question without making a calculation. If a photon of orange light has sufficient energy to eject a photoelectron from the surface of a metal, does a photon of green light have enough energy to do so? If your answer is yes, what happens to the excess energy?

7.4 The Hydrogen Spectrum and the Bohr Model

In formulating his quantum theory, Planck was influenced by the results of investigating the emission spectra produced by free (gas-phase) atoms, which led him to question whether any spectrum, even that of an incandescent light bulb, was truly continuous. Among these earlier results was a discovery made in 1885 by a Swiss mathematician and schoolteacher named Johann Balmer (1825–1898).

The Hydrogen Emission Spectrum

Balmer determined that the wavelengths of the four brightest lines in the visible region of the emission spectrum of hydrogen (Figure 7.4a) fit the simple equation

$$\lambda = 364.5 \text{ nm}\left(\frac{m^2}{m^2 - n^2}\right) \tag{7.6}$$

where n is 2 and m is a whole number greater than 2—specifically, 3 for the red line, 4 for the green, 5 for the blue, and 6 for the violet. Without having seen any other lines in hydrogen's visible emission spectrum, Balmer predicted there should be at least one more ($n = 7$) at the edge of the violet region, and indeed such a line was later discovered.

Balmer also predicted that a series of hydrogen emission lines should exist in regions outside the visible range, lines with wavelengths calculated by replacing $n = 2$ in Equation 7.6 with $n = 1, 3, 4$, and so forth. He was right. In 1908 the German physicist Friedrich Paschen (1865–1947) discovered hydrogen emission lines in the infrared region, corresponding to $n = 3$ in Balmer's equation. A few years later, Theodore Lyman (1874–1954) at Harvard University discovered hydrogen emission lines in the UV region corresponding to $n = 1$. By the 1920s, the $n = 4$ and $n = 5$ series of emission lines also had been discovered. Like the $n = 3$ lines, they are in the infrared region.

CHEMTOUR

Emission Spectra and the Bohr Model of the Atom

In 1888 the Swedish physicist Johannes Robert Rydberg (1854–1919) revised Balmer's equation, changing wavelength to *wave number* ($1/\lambda$), which is the number of wavelengths per unit of distance. Rydberg's equation is

$$\frac{1}{\lambda} = [1.097 \times 10^{-2}\ \text{nm}^{-1}]\left(\frac{1}{n_1^{\,2}} - \frac{1}{n_2^{\,2}}\right) \tag{7.7}$$

where n_1 is a positive whole number that remains fixed for a series of emission lines and n_2 is a whole number equal to $n_1 + 1, n_1 + 2, \ldots$, for successive lines in the series. Values of n_1 in Rydberg's equation correspond to values of n in Balmer's equation, so setting $n_1 = 2$ and $n_2 = 3, 4, 5$, or 6 produces the wave numbers of the visible hydrogen emission lines.

SAMPLE EXERCISE 7.4 Calculating the Wavelength of a Line in the Hydrogen Emission Spectrum **LO3**

What is the wavelength of the visible hydrogen atomic emission line corresponding to $n_2 = 3$ in Equation 7.7?

Collect, Organize, and Analyze We are asked to calculate the wavelength of a line in the hydrogen emission spectrum given an n_2 value of 3. The n_1 value of all visible hydrogen emission lines is 2. Wavelengths of visible radiation are between 400 nm and 750 nm.

Solve

$$\frac{1}{\lambda} = [1.097 \times 10^{-2}\ \text{nm}^{-1}]\left(\frac{1}{2^2} - \frac{1}{3^2}\right) = [1.097 \times 10^{-2}\ \text{nm}^{-1}]\left(\frac{1}{4} - \frac{1}{9}\right)$$

$$= [1.097 \times 10^{-2}\ \text{nm}^{-1}](0.1389) = 1.524 \times 10^{-3}\ \text{nm}^{-1}$$

$$\lambda = 656\ \text{nm}$$

Think About It The calculated wavelength is in the visible region of the electromagnetic spectrum, so the answer is reasonable.

Practice Exercise What is the wavelength, in nanometers, of the line in the hydrogen spectrum corresponding to $n = 2$ and $m = 4$ in Equation 7.6?

When Balmer and Rydberg derived their equations describing the hydrogen spectrum, they did not know why the equations worked. The discrete frequencies of hydrogen's emission lines indicated that only certain levels of internal energy were available in hydrogen atoms. However, classical physics could not explain the existence of these internal energy levels. A new model that could explain these observations at the atomic level was needed.

The Bohr Model of Hydrogen

In the early 20th century, Ernest Rutherford established that atoms are mostly empty space occupied by negatively charged electrons surrounding a tiny nucleus containing most of the atom's mass and all its positive charge. What, then, keeps the electrons from falling into the nucleus? Rutherford suggested that the electrons might orbit the nucleus the way planets orbit the Sun. However, classical physics predicts that negative electrons orbiting a positive nucleus should emit energy in the form of electromagnetic radiation and eventually spiral into the nucleus. If this happened, no atom would be stable.

CONNECTION We discussed Rutherford's gold-foil experiment and the development of the idea of the nuclear atom in Section 2.2.

The Danish physicist Niels Bohr (1885–1962) was familiar with the challenge of trying to explain atom stability based on the Rutherford model because he had studied with Rutherford. Bohr proposed a theoretical model for the hydrogen atom that assumed its one electron travels around the nucleus in one of an array of concentric orbits. Each orbit represents an allowed energy level and is designated by the value of n:

$$E = -2.178 \times 10^{-18}\,\text{J}\left(\frac{1}{n^2}\right) \tag{7.8}$$

where $n = 1, 2, 3, \ldots, \infty$. In the Bohr model, an electron in the orbit closest to the nucleus ($n = 1$) has the lowest energy:

$$E = -2.178 \times 10^{-18}\,\text{J}\left(\frac{1}{1^2}\right) = -2.178 \times 10^{-18}\,\text{J}$$

The next-closest orbit has an n value of 2 and the energy of an electron in it is

$$E = -2.178 \times 10^{-18}\,\text{J}\left(\frac{1}{2^2}\right) = -5.445 \times 10^{-19}\,\text{J}$$

This value is *less* negative than the value of the electron in the $n = 1$ orbit. Thus, as the value of n increases, the radius of the orbit increases and so, too, does the energy of an electron in the orbit—that is, its value becomes less negative. As n approaches ∞, E approaches zero:

$$E = -2.178 \times 10^{-18}\,\text{J}\left(\frac{1}{\infty^2}\right) = 0$$

Zero energy means that there is no longer any electrostatic attraction between the electron and the positively charged nucleus, in which case the electron is no longer part of the atom. The H atom has become two separate particles: an H^+ ion and a free electron.

CONNECTION We discussed the potential energy of attraction between charged particles in Section 6.2.

An important feature of the Bohr model is that it provides a theoretical framework for explaining the experimental observations and equations developed by Balmer and by Rydberg. To see how, consider what happens when an electron moves between two allowed energy levels in Bohr's model. If we label the initial energy level (the level where the electron starts) n_{initial}, and if we label the second level (the level where the electron ends up) n_{final}, then the change in energy of the electron is

$$\Delta E = -2.178 \times 10^{-18}\,\text{J}\left(\frac{1}{n_{\text{final}}^{\;2}} - \frac{1}{n_{\text{initial}}^{\;2}}\right) \tag{7.9}$$

If the electron moves to an orbit farther from the nucleus, then $n_{\text{final}} > n_{\text{initial}}$, and the value of the terms inside the parentheses in Equation 7.9 is negative because

$$\frac{1}{n_{\text{final}}^{\;2}} < \frac{1}{n_{\text{initial}}^{\;2}}$$

This negative value multiplied by the negative coefficient would give us a positive ΔE and represent an increase in electron energy. On the other hand, if the electron moves from an outer orbit to one closer to the nucleus, then $n_{\text{final}} < n_{\text{initial}}$, and the sign of ΔE is negative. This means the electron has lost energy. Equation 7.9 demonstrates that energy in the hydrogen atom is *quantized* because ΔE can have only certain values determined by n_{final} and n_{initial}.

When the electron in a hydrogen atom is in the lowest ($n = 1$) energy level, the atom is said to be in its **ground state**. If the electron in a hydrogen atom is in an energy level above $n = 1$, then the atom is said to be in an **excited state**. According to the Bohr model, a hydrogen atom's electron can move from the ground state ($n = 1$) energy level to an excited state (e.g., $n = 3$) by absorbing a quantity of energy (ΔE) that exactly matches the energy difference between the two states. Similarly, an electron in an excited state can move to an even higher energy level by absorbing a quantity of energy that exactly matches the energy difference between the two excited states. An electron in an excited state can also move to a lower-energy excited state, or to the ground state, by emitting a quantity of energy that exactly matches the energy difference between those two states. Any change in electron energy that occurs by absorption of energy (i.e., $\Delta E > 0$) or emission of energy (i.e., $\Delta E < 0$) is called an **electron transition.**

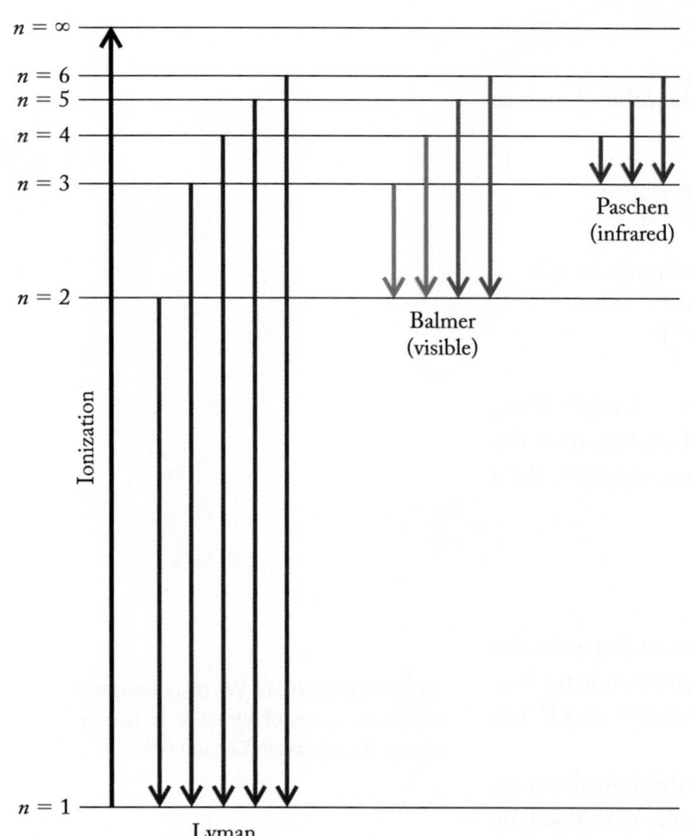

FIGURE 7.14 An energy-level diagram showing possible electron transitions for the electron in the hydrogen atom. The black arrow pointing up represents ionization. Arrows pointing down represent the emission of energy that accompanies an electron falling to a lower energy level. This diagram shows several possible transitions for the single electron in hydrogen; each arrow does *not* represent a different electron in the atom.

Energy-level diagrams, such as the one depicted in **Figure 7.14** for the hydrogen atom, show some of the transitions that electrons can make in these atoms from one energy level to another. Note how the differences in energy between levels decrease as the value of n increases. This means that transitions to and from the ground state ($n = 1$) involve the greatest losses and gains of energy, including the gain represented by the black arrow pointing upward, which depicts the absorption of enough energy to completely remove the electron from a hydrogen atom (ionization).

The downward-pointing arrows in Figure 7.14 represent decreases in a hydrogen atom's internal energy as it loses quanta of energy that exactly match the difference in energy (ΔE) between the initial and final states that define the length of the arrow. Often these quanta are the energies of photons that are emitted as the electron moves from a higher energy level to a lower one. Note how the longest of these arrows point to the ground state; these arrows represent the energies of ultraviolet radiation discovered by Theodore Lyman. The shorter arrows pointing to the $n = 2$ level represent the energies of the visible emission lines discovered by Johannes Balmer, and the even shorter ones pointing to the $n = 3$ level represent the even lower energies of the infrared radiation discovered by Friedrich Paschen. If the colored arrows had pointed up, they would represent absorption of quanta (including photons) leading to increases in the internal energy of the atom. Keep in mind that downward and upward arrows between the same two energy levels represent loss or gain of the same quantum of energy and emission or absorption of photons with same wavelength.

Equations 7.9 and 7.7 are very much alike. Their coefficients differ only because different units are used to express wave number and energy. In addition, Bohr was able to derive his coefficient from a combination of physical constants, including the Planck constant and the charge of an electron. It is not an arbitrary value that just happens to work; it is part of atomic structure. The key point is that Equation 7.7, which was developed to fit the absorption and emission spectra of hydrogen, has the same form as Equation 7.9, the theoretical equation developed by Bohr to explain the internal structure of the hydrogen atom. Thus, atomic emission and absorption spectra reveal the energies of electrons inside atoms.

ground state the most stable, lowest-energy state available to an atom, ion, or molecule.

excited state any energy state above the ground state in an atom, ion, or molecule.

electron transition the movement of an electron between energy levels.

CONCEPT TEST

Based on the lengths of the arrows in Figure 7.14, rank the following electron transitions in order of greatest change in energy to smallest change:

a. $n = 4 \rightarrow n = 2$
b. $n = 3 \rightarrow n = 2$
c. $n = 2 \rightarrow n = 1$
d. $n = 4 \rightarrow n = 3$

SAMPLE EXERCISE 7.5 Calculating the Energy Change of an Electron Transition in the Hydrogen Atom — LO3

How much energy is required to ionize a ground-state hydrogen atom? Express your answer in energy per atom and energy per mole.

Collect and Organize We are asked to determine the energy required to remove an electron from a hydrogen atom in its ground state. Equation 7.9 enables us to calculate the energy change associated with any electron transition in a hydrogen atom. There are $N_A = 6.022 \times 10^{23}$ atoms per mole.

Analyze To use Equation 7.9, we need to identify the initial (n_{initial}) and final (n_{final}) energy levels. The ground state of a hydrogen atom corresponds to the $n = 1$ energy level. If the atom is ionized, $n = \infty$ and there is no longer any electrostatic attraction between the positively charged nucleus and the electron.

Solve

$$\Delta E = -2.178 \times 10^{-18}\ \text{J}\left(\frac{1}{n_{\text{final}}^2} - \frac{1}{n_{\text{initial}}^2}\right)$$

$$= -2.178 \times 10^{-18}\ \text{J}\left(\frac{1}{\infty^2} - \frac{1}{1^2}\right)$$

Dividing by ∞^2 yields zero, so the difference in parentheses simplifies to -1. Therefore:

$$\Delta E = 2.178 \times 10^{-18}\ \text{J}$$

This value represents the energy needed to ionize one H atom. To ionize 1 mole of them requires

$$2.178 \times 10^{-18}\ \text{J/atom} \times 6.022 \times 10^{23}\ \text{atoms/mol} \times 1\ \text{kJ}/10^3\ \text{J} = 1312\ \text{kJ/mol}$$

Think About It The sign of ΔE is positive because energy must be added to overcome the electrostatic attraction between the negatively charged electron and the proton in the nucleus of an H atom.

Practice Exercise Calculate the energy required to ionize an excited-state hydrogen atom with an electron initially in the $n = 3$ energy level. Before doing the calculation, predict whether this energy is greater than or less than the 2.178×10^{-18} J needed to ionize a ground-state hydrogen atom.

One of the strengths of the Bohr model of the hydrogen atom is that it accurately predicts the energy needed to remove the electron. This energy is called the ionization energy of the hydrogen atom. We examine the ionization energies of

matter wave the wave associated with any particle.

other elements in Section 7.11. However, the Bohr model applies only to hydrogen atoms and to ions that have only a single electron. The model does not agree with the observed spectra of multielectron atoms and ions because it does not account for the way electrons interact with each other. Thus, the picture of the atom provided by Bohr's model is extremely limited. Nevertheless, it enabled scientists to begin using quantum theory to explain the behavior of matter at the atomic level.

CONCEPT **TEST**

Can the Bohr model be used to explain the emission spectrum of He atoms? Why or why not?

7.5 Electron Waves

A decade after Bohr published his model of the hydrogen atom, a French graduate student named Louis de Broglie (1892–1987) provided a theoretical basis for the stability of electron orbits. His approach incorporated yet another significant advance in the way early-20th-century scientists conceived of atoms and subatomic particles—namely, to think of them not only as particles of matter but also as waves. His work provided an answer to the question of why an electron in the hydrogen atom does not spiral into the nucleus.

de Broglie Wavelengths

CHEMTOUR
de Broglie Wavelength

de Broglie proposed that if light, which we normally think of as a wave, has particle properties, then perhaps the electron, which we normally think of as a particle, has wave properties. If that were true, an electron moving around in an atom should have a characteristic wavelength. de Broglie calculated electron wavelengths from Einstein's equations relating energy and mass, $E = mc^2$, and the energy and wavelength of a photon, $E = hc/\lambda$:

$$\lambda = \frac{hc}{E} = \frac{hc}{mc^2} = \frac{h}{mc} \tag{7.10}$$

To apply Equation 7.10 to electrons, de Broglie replaced c (the speed of light) with u, the speed of an orbiting electron in an atom:

$$\lambda = \frac{h}{mu} \tag{7.11}$$

where h is the Planck constant, m is the mass of the electron in kilograms, and u is its velocity in meters per second. The wavelength of an electron calculated in this way is often called the *de Broglie wavelength* of the electron.

de Broglie's equation is not restricted to electrons—any moving particle has wavelike properties. In other words, the particle behaves as a **matter wave**. de Broglie predicted that moving particles much bigger than electrons, such as atomic nuclei, molecules, and even macroscopic objects such as tennis balls and airplanes, should have characteristic wavelengths described by Equation 7.11. The wavelengths of such large objects are extremely small, though, given the tiny size of the Planck constant in the numerator and the large size of the mass in the denominator of Equation 7.11. As a result, we never notice the wave nature of large objects in motion.

SAMPLE EXERCISE 7.6 Calculating the Wavelength of a Particle in Motion **LO4**

(a) Compare the wavelength of a 142-g baseball thrown at 44 m/s (98 mi/h) with the size of the ball, which has a diameter of 7.5 cm. (b) Compare the wavelength of an electron ($m_e = 9.109 \times 10^{-31}$ kg) moving at one-tenth the speed of light in a hydrogen atom with the size of the atom (diameter = 1.06×10^{-10} m).

Collect, Organize, and Analyze Given the masses and velocities of these two moving objects, Equation 7.11 may be used to calculate their wavelengths. Given the small value of *h*, it is likely that the wavelength of a pitched baseball is only a tiny fraction of the size of the baseball. However, the wavelength of an electron moving at one-tenth the speed of light may be a much greater fraction of the size of the atom that it orbits. The fraction on the right side of Equation 7.11 has units of joule-seconds in the numerator and mass and velocity in the denominator. To combine these units in a way that gives us a unit of length, we need to use the following conversion factor:

$$1 \text{ J} = 1 \text{ kg} \cdot \text{m}^2/\text{s}^2$$

To use this equality, we must express the mass of the baseball in kilograms:

$$142 \text{ g} = 0.142 \text{ kg}$$

Solve

a. For the baseball:

$$\lambda = \frac{h}{mu} = \frac{6.626 \times 10^{-34} \text{ J} \cdot \text{s}}{(0.142 \text{ kg})(44 \text{ m/s})} \times \frac{1 \text{ kg} \cdot \text{m}^2/\text{s}^2}{1 \text{ J}}$$

$$= 1.06 \times 10^{-34} \text{ m}$$

The wavelength of the baseball is

$$\frac{1.06 \times 10^{-34} \text{ m}}{0.075 \text{ m}} \times 100\% = 1.4 \times 10^{-31}\%$$

of the ball's diameter.

b. The wavelength of an electron moving at one-tenth the speed of light is

$$\lambda = \frac{h}{mu} = \frac{6.626 \times 10^{-34} \text{ J} \cdot \text{s}}{(9.109 \times 10^{-31} \text{ kg})(2.998 \times 10^7 \text{ m/s})} \times \frac{1 \text{ kg} \cdot \text{m}^2/\text{s}^2}{1 \text{ J}}$$

$$= 2.43 \times 10^{-11} \text{ m}$$

and

$$\frac{2.43 \times 10^{-11} \text{ m}}{1.06 \times 10^{-10} \text{ m}} \times 100\% = 23\%$$

of the diameter of a hydrogen atom.

Think About It The matter wave of the baseball is much too small to be observed, so its wavelike character contributes nothing to the behavior of the baseball. We expected that. Our experience with objects in the world is that they behave like particles, not like waves. For the electron, however, the wavelength is a significant percentage of the size of the hydrogen atom. This is *not* a familiar experience to us—a particle that also has properties of waves such as frequency and wavelength. The implications of this for our understanding of the structure of the atom are discussed in the rest of this section.

 **Practice Exercise** The velocity of the electron in the ground state of a hydrogen atom is 2.2×10^6 m/s. What is the wavelength of this electron in meters?

de Broglie explained the stability of the electron levels in Bohr's model of the hydrogen atom by proposing that the electron in a hydrogen atom behaves like a circular wave oscillating around the nucleus. To understand the implications of this statement, we need to examine what is required to make a stable, circular wave. Consider the motion of a vibrating violin string of length L (**Figure 7.15a**). Because the string is fixed at both ends, there is no vibration at the ends and the vibration is at maximum in the middle. The wave created by this combination of fixed ends and maximum vibration in the middle is a **standing wave**: a wave that oscillates back and forth within a fixed space rather than moving through space the way waves of light travel through space. On a standing wave, any points that have zero displacement, such as the two ends of the string, are called **nodes**. The sound wave produced in this way on a violin string is called the *fundamental* of the string. The wavelength of the fundamental is $2L$. It has the lowest frequency and longest wavelength possible for that string. The length of the string is one-half the wavelength of the fundamental ($L = \lambda/2$).

If the string is held down in the middle (creating a third node) and plucked halfway between the middle and one end, a new, higher-frequency wave called the *first harmonic* is produced. The wavelength of this harmonic is equal to L because $L = 2(\lambda/2) = \lambda$. We can continue generating higher frequencies with wavelengths that are related to the length of the string by the equation

$$L = \frac{n\lambda}{2}$$

where n is a whole number such that $L = 3(\lambda/2)$, $4(\lambda/2)$, $5(\lambda/2)$, and so on.

The standing-wave pattern for a circular wave generated by an electron differs slightly from that of a vibrating violin string in that the circular wave has no defined stationary ends. Instead, the electron vibrates in an endless series of waves. However, standing waves are only produced if, as shown in **Figure 7.15b**, the circumference of the circle equals a whole-number multiple of the electron's wavelength:

$$\text{Circumference} = n\lambda \quad (7.12)$$

Equation 7.12 offers a new interpretation of Bohr's number n: it represents the number of matter waves in a given energy level. de Broglie's work also solved the problem of a negatively charged electron spiraling into the positively charged nucleus. Because $n = 1$ represents the minimum circumference of the circular wave of a moving electron, the electron must be a minimum distance from the nucleus. Furthermore, the lowest energy possible for an electron in a hydrogen atom occurs for a standing wave with circumference λ, so the electron cannot get any closer to the nucleus than that.

de Broglie's research created a quandary for the graduate faculty at the University of Paris, where he studied. Bohr's model of electrons moving between allowed energy levels had been widely criticized as an arbitrary suspension of well-tested physical laws. de Broglie's rationalization of Bohr's model seemed even more outrageous to many scientists. Before the faculty would accept his thesis, they wanted another opinion, so they sent it to Albert Einstein for review. Einstein wrote back that he found the young man's work "quite interesting." That endorsement was good enough for the faculty: de Broglie's thesis was accepted in 1924 and immediately submitted for publication. Five years later, he was awarded the Nobel Prize in Physics.

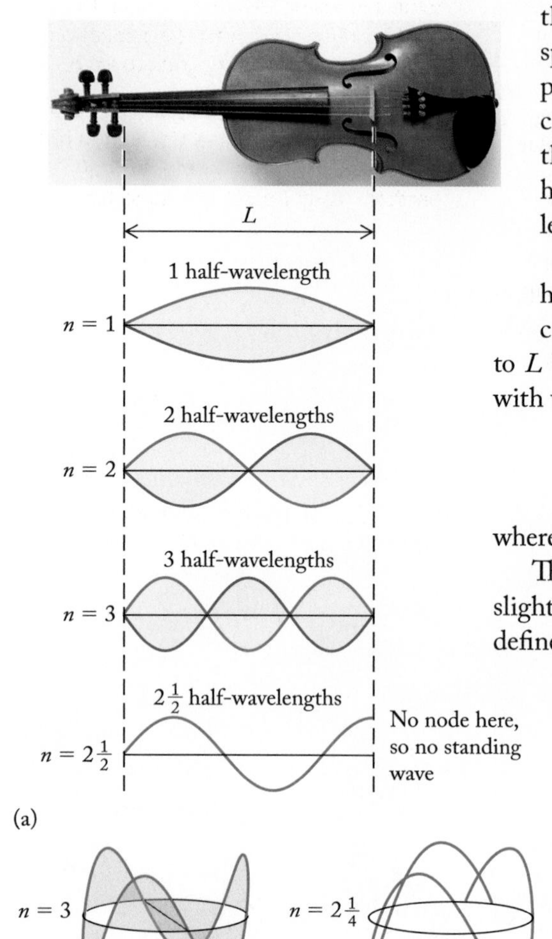

FIGURE 7.15 Linear and circular standing waves. (a) The wavelength of a standing wave in a violin string fixed at both ends is related to the distance L between the ends of the string by the equation $L = n(\lambda/2)$. In the standing waves shown, $n = 1, 2$, and 3. An n value of $2\frac{1}{2}$ does not produce a standing wave because a standing wave can have no string motion at either end. (b) The circular standing waves proposed by de Broglie account for the stability of the energy levels in Bohr's model of the hydrogen atom. Each stable wave must have a circumference equal to $n\lambda$ with n restricted to being an integer, such as the $n = 3$ circular standing wave shown here. If the circumference is not an exact multiple of λ, as shown in the $2\frac{1}{4}$ image, no standing wave occurs.

The Heisenberg Uncertainty Principle

If electrons exhibit both particle-like and wavelike behavior, then what impact does the wave behavior have on our ability to locate the electron? A wave, by its very nature, is spread out in space. The question "Where is the electron?" has one answer if we treat the electron as a particle and a very different answer if we treat it as a wave. This issue was addressed by the German physicist Werner Heisenberg (1901–1976), who proposed the following thought experiment: Watch an electron with a microscope to "see" the electron's path around an atom. The microscope (if it existed) would need to use gamma rays for illumination (rather than visible light) because gamma rays are the only part of the electromagnetic spectrum with wavelengths short enough to match the diminutive size of electrons. According to Equations 7.2 and 7.3, however, the short wavelengths and high frequencies of gamma rays mean that they have enormous energies—so large that any gamma ray striking an electron would knock the electron off course. The only way to *not* affect the electron's motion would be to use a much lower-energy, longer-wavelength source of radiation to illuminate it—but then we would be unable to see the tiny electron clearly.

This situation presents a quantum mechanical dilemma. The only means for clearly observing an electron make it impossible to know the electron's motion or, more precisely, its momentum, which is defined as an object's velocity times its mass. Therefore, we can never know exactly both the position and the momentum of the electron simultaneously. This conclusion is known as the **Heisenberg uncertainty principle**, and it is expressed mathematically by Equation 7.13:

$$\Delta x \cdot m\Delta u \geq \frac{h}{4\pi} \qquad (7.13)$$

where Δx is the uncertainty in the position of the electron, m is its mass, Δu is the uncertainty in its velocity, and h is the Planck constant. To Heisenberg, this uncertainty was the essence of quantum mechanics. Its message for us is that there are limits to what we can observe, measure, and therefore know.

standing wave a wave confined to a given space, with a wavelength (λ) related to the length L of the space by $L = n(\lambda/2)$, where n is a whole number.

node a location in a standing wave that experiences no displacement; in the context of orbitals, nodes are locations at which electron density goes to zero.

Heisenberg uncertainty principle the principle that we cannot determine both the position and the momentum of a particle in an atom at the same time.

SAMPLE EXERCISE 7.7 Calculating (Heisenberg) Uncertainty **LO4**

Use the data in Sample Exercise 7.6 to compare the uncertainty in the velocity of the baseball with the uncertainty in the velocity of the electron. Assume that the position of the baseball is known to within one wavelength of red light ($\Delta x_{\text{baseball}} = 680$ nm) and that the position of the electron is known to within the radius of the hydrogen atom ($\Delta x_{\text{electron}} = 5.29 \times 10^{-11}$ m).

Collect and Organize We are asked to calculate the uncertainty in the velocities of two particles given the uncertainties in their positions. From Sample Exercise 7.6 we know that the mass of a baseball is 0.142 kg and the mass of an electron is 9.109×10^{-31} kg.

Analyze Equation 7.13 provides a mathematical connection between these variables. According to Equation 7.13, the uncertainty in the velocity and position of a particle is inversely proportional to its mass. Therefore, we can expect little uncertainty in the velocity of the baseball, but much greater uncertainty in the velocity of the electron. We need to rearrange the terms in the equation to solve for the uncertainty in velocity (Δu):

$$\Delta u \geq \frac{h}{4\pi\Delta x m}$$

Solve For the baseball:

$$\Delta u \geq \frac{6.626 \times 10^{-34}\ \text{J} \cdot \text{s}}{4\pi(6.8 \times 10^{-7}\ \text{m})(0.142\ \text{kg})} \times \frac{1\ \text{kg} \cdot \text{m}^2/\text{s}^2}{1\ \text{J}}$$

$$\geq 5.5 \times 10^{-28}\ \text{m/s}$$

For the electron:

$$\Delta u \geq \frac{6.626 \times 10^{-34}\ \text{J}\cdot\text{s}}{4\pi(5.29 \times 10^{-11}\ \text{m})(9.109 \times 10^{-31}\ \text{kg})} \times \frac{1\ \text{kg}\cdot\text{m}^2/\text{s}^2}{1\ \text{J}}$$
$$\geq 1.09 \times 10^6\ \text{m/s}$$

Comparing the two values, we see that the uncertainty in the velocity of the baseball is extremely small (about 10^{-28} m/s), whereas that of the electron is huge (greater than 1 million m/s).

Think About It The uncertainty in the measurement of the baseball's velocity is so minuscule that it is insignificant. This result is expected for objects in the macroscopic world. However, the uncertainty in the measurement of the velocity of the electron is huge, which is what we must expect at the atomic level, where "particles" such as the electron also behave like waves.

Practice Exercise What is the uncertainty, in meters, in the position of an electron moving near a nucleus at a speed of 8×10^7 m/s? Assume the uncertainty in the velocity of the electron is 1% of its value—that is, $\Delta u = (0.01)(8 \times 10^7\ \text{m/s})$.

When Heisenberg proposed his uncertainty principle, he was working with Bohr at the University of Copenhagen. The two scientists had widely different views about the significance of the uncertainty principle and the idea that particles could behave like waves. To Heisenberg, uncertainty was a fundamental characteristic of nature. To Bohr, it was merely a mathematical consequence of the wave–particle duality of electrons; there was no physical meaning to an electron's position and path. The debate between these two gifted scientists was heated at times. Heisenberg later wrote about one particularly emotional debate:

> [A]t the end of the discussion I went alone for a walk in the neighboring park [and] repeated to myself again and again the question: "Can nature possibly be as absurd as it seems?"[1]

The Heisenberg uncertainty principle is fundamental to our present understanding of the atom. If we cannot know both the position and momentum of an electron in a hydrogen atom, then the electron cannot be moving in circular orbits as implied by Bohr's original model. As we see in Section 7.6, the Heisenberg uncertainty principle limits us to knowing only the *probability* of finding an electron at a location in an atom.

wave mechanics (also called **quantum mechanics**) a mathematical description of the wavelike behavior of particles on the atomic level.

Schrödinger wave equation a description of how the electron matter wave varies with location and time around the nucleus of a hydrogen atom.

wave function (ψ) a solution to the Schrödinger wave equation.

orbital a region around the nucleus of an atom where the probability of finding an electron is high; each orbital is defined by the square of the wave function (ψ^2) and is identified by a unique combination of three quantum numbers.

quantum number a number that specifies the energy, the probable location or orientation of an orbital, or the spin of an electron within an orbital.

principal quantum number (n) a positive integer describing the relative size and energy of an atomic orbital or group of orbitals in an atom.

angular momentum quantum number (ℓ) an integer having any value from 0 to $n - 1$ that defines the shape of an orbital.

magnetic quantum number (m_ℓ) an integer that may have any value from $-\ell$ to $+\ell$, where ℓ is the angular momentum quantum number; it defines the orientation of an orbital in space.

7.6 Quantum Numbers and Electron Spin

Many of the leading scientists of the 1920s were unwilling to accept the dual wave–particle nature of electrons proposed by de Broglie unless the model could be used to predict the features of the hydrogen emission spectrum. This kind of application required the development of equations describing the behavior of electron waves. During his Christmas vacation in 1925, the Austrian physicist Erwin Schrödinger (1887–1961) did just that, developing in a few weeks the mathematical foundation for what came to be called **wave mechanics** or **quantum mechanics**.

Schrödinger's mathematical description of electron waves is called the **Schrödinger wave equation**. It is not discussed in detail in this book, but you should know that solutions to the wave equation are called **wave functions (ψ)**:

[1]Heisenberg, W. *Physics and Philosophy: The Revolution in Modern Science* (Harper & Row, 1958), p. 42.

mathematical expressions that describe how the matter wave of an electron in an atom varies both with time and with the location of the electron in the atom. Wave functions define the energy levels in the hydrogen atom. They can be simple trigonometric functions, such as sine or cosine waves, or they can be very complex.

What is the physical significance of a wave function? Actually, there is none. However, the *square of a wave function* (ψ^2) does have physical meaning. Initially, Schrödinger believed that a wave function depicted the "smearing" of an electron through three-dimensional space. This notion of subdividing a discrete particle was later rejected in favor of the model developed by German physicist Max Born (1882–1970), who proposed that ψ^2 defines an **orbital**: the space around the nucleus of an atom where the probability of finding an electron is high. Born later showed that his interpretation could be used to calculate the probability of a transition between two orbitals, as happens when an atom absorbs or emits a photon.

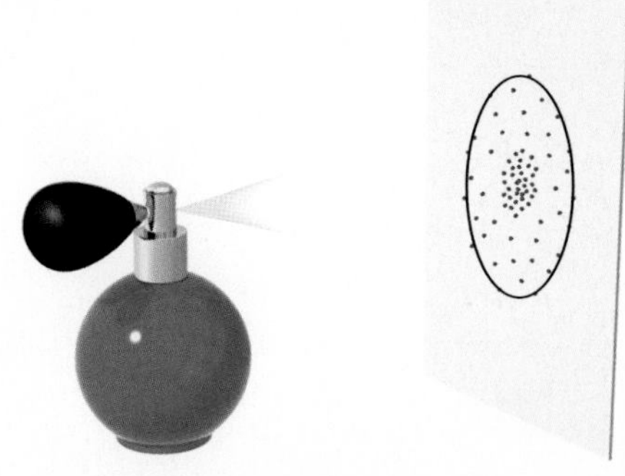

FIGURE 7.16 The probability of encountering an ink spot in the pattern produced by an ink spray decreases with increasing distance from the center of the pattern.

To help visualize the probabilistic meaning of ψ^2, consider what happens when we spray ink onto a flat surface (**Figure 7.16**). If we then draw a circle encompassing most of the ink spots, we are identifying the region of maximum probability for finding the spots.

Quantum mechanical orbitals in an atom are not two-dimensional concentric orbits, as in Bohr's model of the hydrogen atom, or even two-dimensional circles, as in Figure 7.16. Instead, as we see in detail in Section 7.7, orbitals are three-dimensional regions of space with distinctive shapes, orientations, and average distances from the nucleus. Each orbital is a solution to Schrödinger's wave equation and is identified by a unique combination of three integers, or **quantum numbers**, whose values flow directly from the mathematical solutions to the wave equation. The quantum numbers are n, ℓ, and m_ℓ:

- The **principal quantum number n** is like Bohr's number n for the hydrogen atom in that it is a positive integer that indicates the relative size and energy of an orbital or group of orbitals in an atom. Orbitals with the same value of n are in the same *shell*. Orbitals with larger values of n are farther from the nucleus and, in the hydrogen atom, represent higher energy levels, consistent with Bohr's model of the hydrogen atom. In multielectron atoms, the relationship between energy levels and orbitals is more complex but increasing values of n generally represent higher energy levels.
- The **angular momentum quantum number ℓ** is an integer with a value ranging from zero to $n - 1$ that defines the shape of an orbital. Orbitals with the same value of n and ℓ are in the same *subshell* and represent equal energy levels. Orbitals with a given value of ℓ are identified with a letter according to the following scheme:

Value of ℓ	0	1	2	3
Type of Orbital	*s*	*p*	*d*	*f*

- The choice of letters to designate the values of ℓ (i.e., *s*, *p*, *d*, and *f*) may seem a bit odd. Before quantum mechanics was developed, scientists recording line spectra of the elements described the lines they observed as ***sharp***, ***principal***, ***diffuse***, and ***fundamental***. Designating orbitals with the letters *s*, *p*, *d*, and *f* recognizes this historical convention.
- The **magnetic quantum number m_ℓ** is an integer with a value from $-\ell$ to $+\ell$. It defines the orientation of an orbital in the space around the nucleus of an atom.

Each subshell in an atom has a two-part designation containing the appropriate value of n and a letter designation for ℓ. For example, orbitals with $n = 3$ and $\ell = 1$ are called 3*p* orbitals, and electrons in 3*p* orbitals are called 3*p* electrons.

TABLE 7.1 Quantum Numbers of the Orbitals in the First Four Shells

Value of n	Allowed Value of ℓ	Subshell Label	Allowed Values of m_ℓ	NUMBER OF ORBITALS In Subshell	NUMBER OF ORBITALS In Shell
1	0	*s*	0	1	1
2	0	*s*	0	1	4
	1	*p*	−1, 0, +1	3	
3	0	*s*	0	1	9
	1	*p*	−1, 0, +1	3	
	2	*d*	−2, −1, 0, +1, +2	5	
4	0	*s*	0	1	16
	1	*p*	−1, 0, +1	3	
	2	*d*	−2, −1, 0, +1, +2	5	
	3	*f*	−3, −2, −1, 0, +1, +2, +3	7	

CHEMTOUR
Quantum Numbers

How many 3*p* orbitals are there? The number depends on how many m_ℓ values can an $\ell = 1$ (*p*) subshell have, and the answer is three: −1, 0, and +1. These three m_ℓ values mean that there are three 3*p* orbitals, each with a unique combination of *n*, ℓ, and m_ℓ values. **Table 7.1** lists all the possible combinations of these three quantum numbers for the orbitals of the first four shells.

SAMPLE EXERCISE 7.8 Identifying the Subshells and Orbitals in an Energy Level — LO5

(a) What are the designations of all the subshells in the $n = 3$ shell? (b) How many orbitals are in these subshells?

Collect and Organize We are asked to describe the subshells in the third shell and to determine how many orbitals are in all these subshells. Table 7.1 lists all the subshells in the first four shells.

Analyze The designations of subshells are based on the possible values of the quantum numbers *n* and ℓ. The allowed values of ℓ depend on the value of *n*, because ℓ is an integer from 0 up to $n - 1$. The number of orbitals in a subshell depends on the number of possible values of m_ℓ (i.e., integers that range from $-\ell$ to $+\ell$).

Solve

a. The allowed values of ℓ for $n = 3$ range from 0 through $(n - 1)$, so they are 0, 1, and 2. These ℓ values correspond to the subshell designations *s*, *p*, and *d*, respectively. The appropriate subshell names are thus 3*s*, 3*p*, and 3*d*.

b. The possible values of m_ℓ from $-\ell$ to $+\ell$ are as follows:

$\ell = 0$; $m_\ell = 0$: This combination of ℓ and m_ℓ values for the $n = 3$ shell represents a single 3*s* orbital.

$\ell = 1$; $m_\ell = -1, 0,$ or $+1$: These three combinations of ℓ and m_ℓ values for the $n = 3$ shell represent the three 3*p* orbitals.

$\ell = 2$; $m_\ell = -2, -1, 0, +1,$ or $+2$: These five combinations of ℓ and m_ℓ values represent the five 3*d* orbitals.

Think About It There are a total of $1 + 3 + 5 = 9$ orbitals in the $n = 3$ shell. The total number of orbitals in a shell increases with increasing values of *n* because as *n* increases the allowed values ℓ and m_ℓ also increase. Thus, there is only 1 orbital in the $n = 1$ shell, there are 4 in $n = 2$, and 16 in $n = 4$.

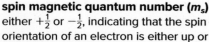

Practice Exercise How many orbitals are there in the $n = 5$ shell?

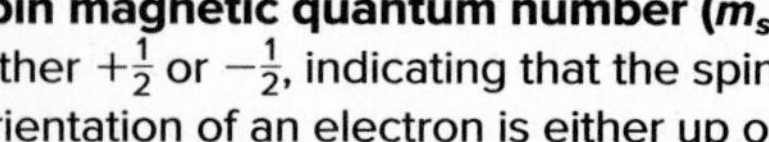

spin magnetic quantum number (m_s) either $+\frac{1}{2}$ or $-\frac{1}{2}$, indicating that the spin orientation of an electron is either up or down, respectively.

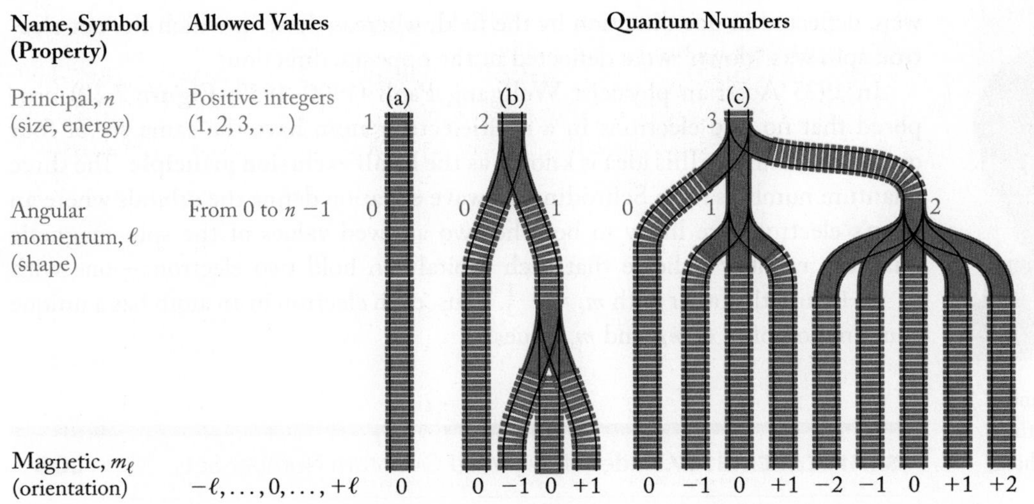

FIGURE 7.17 These three sets of train tracks provide a visual metaphor for the orbitals in the first three energy shells in an atom. The single track labeled (a) represents the first shell and its 1*s* orbital, which has ℓ and m_ℓ values of 0. The set labeled (b) has two ℓ branches, labeled 0 and 1, representing the 2*s* and 2*p* orbitals. The 1 branch further divides into branches with m_ℓ values of −1, 0, and +1, representing the three 2*p* orbitals. The set labeled (c) divides into three main branches with ℓ values of 0, 1, and 2, representing the 3*s*, 3*p*, and 3*d* subshells, respectively.

Figure 7.17 visually summarizes the quantum number rules. This system of quantum numbers provides a useful "shorthand" to refer to the orbitals that are solutions to Schrödinger's wave equation. Even so, the Schrödinger wave equation could account for most, but not all, aspects of atomic spectra. The emission spectrum of hydrogen, for example, has a pair of red lines at 656 nm where Balmer thought there was only one line (**Figure 7.18**). There are also pairs of lines in the spectra of multielectron atoms that have a single electron in their outermost shells (see Figure 7.3). The Schrödinger equation cannot explain these pairs of lines.

In 1925, two students at the University of Leiden in the Netherlands, Samuel Goudsmit (1902–1978) and George Uhlenbeck (1900–1988), proposed that the pairs of lines, called *doublets*, were caused by a property they called electron spin. In their model, electrons spin in one of two directions, designated "spin up" and "spin down." A moving electron (or any charged particle) creates a magnetic field by its motion. The spinning motion produces a second magnetic field oriented up or down. To account for these two spin orientations, Goudsmit and Uhlenbeck proposed a fourth quantum number, the **spin magnetic quantum number (m_s)**. The values of m_s are $+\frac{1}{2}$ for spin up and $-\frac{1}{2}$ for spin down.

Even before Goudsmit and Uhlenbeck proposed the electron-spin hypothesis, two other scientists, Otto Stern (1888–1969) and Walther Gerlach (1889–1979), had observed the effect of electron spin when they shot a beam of silver ($Z = 47$) atoms through a magnetic field (**Figure 7.19**). Those atoms in which the net electron spin was "up"

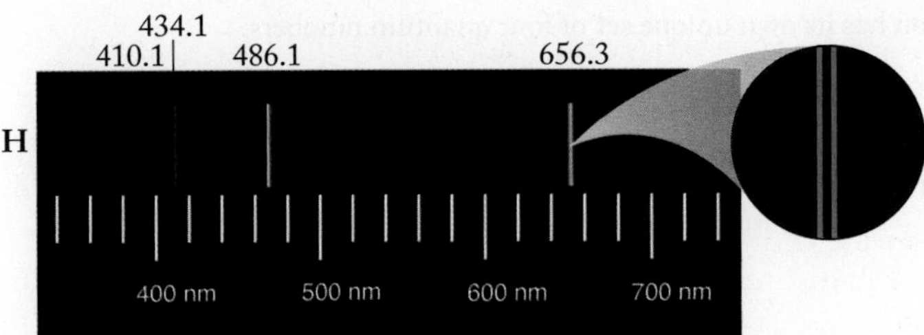

FIGURE 7.18 The Schrödinger equation does not account for the appearance of closely spaced pairs of bright lines in the emission spectra of atoms, such as the red lines at 656.272 nm and 656.285 nm in the spectrum of hydrogen.

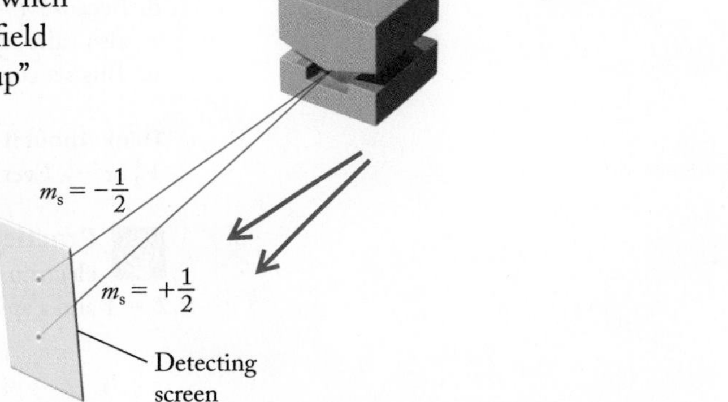

FIGURE 7.19 A narrow beam of silver atoms passed through a magnetic field is split into two beams because of the interactions between the field and the spinning electrons in the atoms. This observation led to proposing the fourth quantum number, m_s. (The blue arrows indicate the direction of the beam.)

FIGURE 7.20 Wolfgang Pauli (left) and Niels Bohr are apparently amused by the behavior of a toy called a tippe top, which, when spun on its base, tips itself over and spins on its stem. Although the toy's behavior is caused by a combination of friction and the top's angular momentum and is not quantum mechanical in origin, it provides a visual metaphor for the two spin orientations of an electron in an orbital.

were deflected in one direction by the field, whereas those in which the net electron spin was "down" were deflected in the opposite direction.

In 1925 Austrian physicist Wolfgang Pauli (1900–1958; **Figure 7.20**) proposed that no two electrons in a multielectron atom have the same set of four quantum numbers. This idea is known as the **Pauli exclusion principle**. The three quantum numbers from Schrödinger's wave equation define the orbitals where an atom's electrons are likely to be. The two allowed values of the spin magnetic quantum number indicate that each orbital can hold two electrons—one with $m_s = +\frac{1}{2}$ and the other with $m_s = -\frac{1}{2}$. Thus, each electron in an atom has a unique combination of n, ℓ, m_ℓ, and m_s values.

SAMPLE EXERCISE 7.9 Identifying Valid Quantum Number Sets **LO5**

Which of these combinations of quantum numbers are valid?

	n	ℓ	m_ℓ	m_s
(a)	1	0	−1	$+\frac{1}{2}$
(b)	3	2	−2	$+\frac{1}{2}$
(c)	2	2	0	0
(d)	2	0	0	$-\frac{1}{2}$
(e)	−3	−2	−1	$-\frac{1}{2}$

Collect, Organize, and Analyze We are asked to validate five sets of quantum numbers. These rules apply: A principal quantum number (n) can be any positive integer. The values of ℓ in a given shell are integers from 0 to $(n - 1)$; the values of m_ℓ in a given subshell are all integers from $-\ell$ to $+\ell$, including 0. The only two options for m_s are $+\frac{1}{2}$ or $-\frac{1}{2}$.

Solve

a. Because $n = 1$, the maximum (and only) value of ℓ is $(n - 1) = 1 - 1 = 0$. Therefore, the values of n and ℓ are valid. However, if $\ell = 0$, then m_ℓ must be 0; it cannot be −1. Therefore, this set is not valid. The spin quantum number is a possible value.
b. Because $n = 3$, ℓ can be 2 and m_ℓ can be −2. Also, $m_s = +\frac{1}{2}$ is a valid choice for the spin magnetic quantum number. This set is valid.
c. Because $n = 2$, ℓ cannot be 2, making this set invalid. In addition, m_s has an invalid value (0).
d. Because $n = 2$, ℓ can be 0, and for that value of ℓ, m_ℓ must be 0. The value of m_s is also valid, and so is the set.
e. This set contains two impossible values, $n = -3$ and $\ell = -2$, so it is invalid.

Think About It The values of n, ℓ, and m_ℓ are related mathematically, and m_s can be either $+\frac{1}{2}$ or $-\frac{1}{2}$. Every electron has its own unique set of four quantum numbers.

 **Practice Exercise** Write all the possible sets of quantum numbers for an electron in the $n = 3$ shell that has an angular momentum quantum number $\ell = 1$ and a spin quantum number $m_s = +\frac{1}{2}$.

Pauli exclusion principle the principle that no two electrons in an atom can have the same set of four quantum numbers.

Momentous advances in chemistry and physics were made in the first three decades of the 20th century because of the brilliant minds of Einstein, Planck, Rutherford, Bohr, de Broglie, Schrödinger, and others, who have forever changed scientists' view of the fundamental structure of matter and the universe.

Figure 7.21 summarizes and connects some of these advances.

How different is the view of the interaction of matter and energy provided by quantum mechanics from the laws governing the behavior of large objects? A pebble picked up and dropped immediately falls to the ground. Electrons, however, remain in excited states for indeterminate (though usually short) times before falling to their ground states. This lack of determinacy bothered Einstein and many of his colleagues. Had they discovered an underlying theme of nature—that some processes cannot be described or known with certainty? Are there fundamental limits to how well we can know and understand our world and the events that change it?

The variable lifetimes of excited states are the basis for the lasers used in bar-code scanners at grocery stores, laser pointers, DVD players, and in surgery (**Figure 7.22**). The word *laser* is an acronym for **L**ight **A**mplification by **S**timulated **E**mission of **R**adiation. The phrase *stimulated emission* is linked to the fact that the atoms in lasers remain in excited states for unusually long times, which increases the populations of these excited states. Then, if a photon with just the right energy encounters one of these excited-state atoms, it stimulates the decay of the excited state, producing a second photon that matches the first: same wavelength, same phase, and going in the same direction. When these two photons encounter two more excited-state atoms, they could produce four in-phase photons, which could produce four more, and so on. In this way, one incident photon is rapidly amplified into an intense pulse of monochromatic light. To produce a steady beam of this light, electrical energy continuously pumps up the lasing material, repopulating the excited state. By manipulating the composition of lasers, scientists can adjust the electronic energy levels inside them and the color of the light they produce.

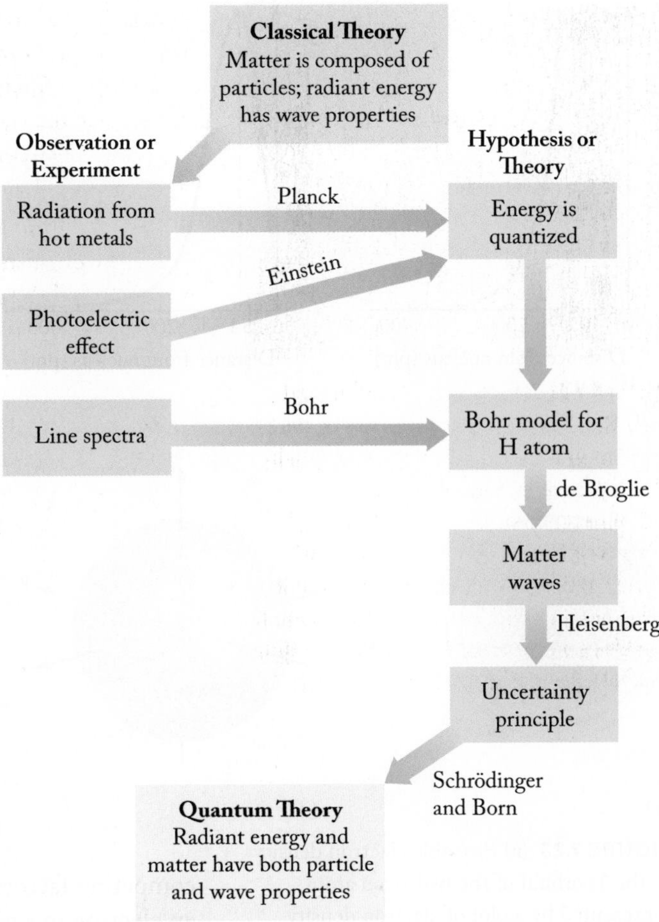

FIGURE 7.21 During the first three decades of the 20th century, quantum theory evolved from classical 19th-century theories of the nature of matter and energy. The arrows trace the development of modern quantum theory, which incorporates the assumption that radiant energy has both wavelike and particle-like properties and that mass (matter) also has both wavelike and particle-like properties.

7.7 The Sizes and Shapes of Atomic Orbitals

As noted in Section 7.6, the orbitals that are solutions to Schrödinger's wave equation have three-dimensional shapes that are graphical representations of ψ^2. In this section, we examine the shapes of atomic orbitals and explain how those shapes affect the energies of the electrons in them.

s Orbitals

Figure 7.23 shows several representations of the 1*s* orbital of the hydrogen atom. In Figure 7.23a, electron density is plotted against distance from the nucleus and shows that density decreases with increasing distance. However, Figure 7.23b provides a more useful profile of electron distribution. To understand why, imagine if the hydrogen atom were like an onion, made of many concentric spherical

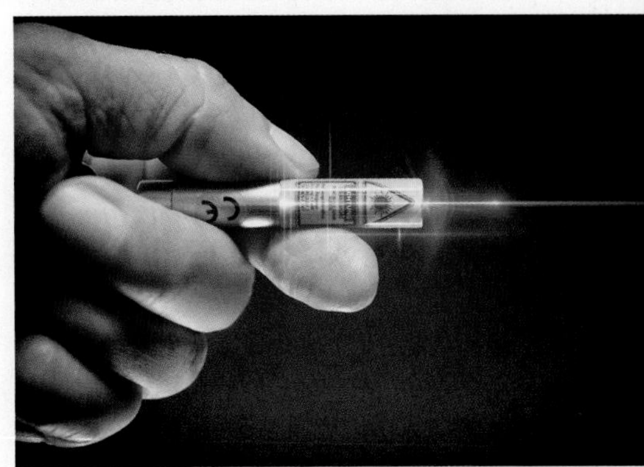

FIGURE 7.22 Laser pointers emitting different colors are commonplace.

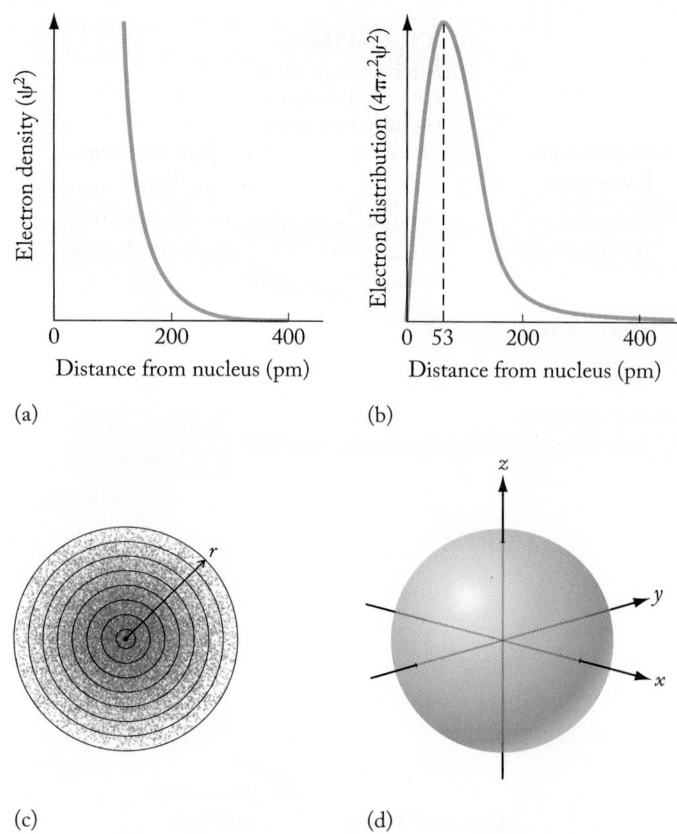

FIGURE 7.23 (a) Probable electron density in the 1*s* orbital of the hydrogen atom represented by a plot of electron density (ψ^2) versus distance from the nucleus. (b) Electron distribution in the 1*s* orbital versus distance from the nucleus. The distribution is essentially zero for both very short distances from the nucleus and very long distances from the nucleus. The maximum probability occurs at $r = 53$ pm. (c) Cross section through the hydrogen atom, with the space surrounding the nucleus divided into an arbitrary number of thin, concentric, hollow layers. Each layer has a unique value for radius *r*. The probability of finding an electron in a layer of radius *r* depends on the volume of the layer and the density of electrons in the layer. (d) Boundary–surface representation of a sphere within which the probability of finding a 1*s* electron is 90%.

layers, all the same thickness. A cross section of this image of the atom is shown in Figure 7.23c. What would be the probability of finding the electron in one of these spherical layers? A layer very close to the nucleus has a very small radius, so it accounts for only a small fraction of the total volume of the atom. A layer with a larger radius makes up a much larger fraction of the volume of the atom because the volume of the layers increases as a function of r^2. (The volume of a sphere depends on r^3, but here we are discussing a spherical shell, the volume of which depends on r^2.) Even though electron densities are higher closer to the nucleus (as Figure 7.23a shows), the volumes of the spherical layers closest to the nucleus are so small that the chances of the electron being near the center of an atom are extremely low; this is shown in Figure 7.23b, where the curve starts off at essentially zero for electron distribution values at distances very close to the nucleus. Farther from the nucleus, electron densities are lower, but the volumes of the layers are much larger, so the probability of the electron being in one of these layers is relatively high, represented by the peak in the curve of Figure 7.23b. At greater distances, volumes of the layers are very large but ψ^2 drops to nearly zero (see Figure 7.23a), so the chances of finding an electron in layers far from the nucleus are very small.

Figure 7.23b, then, represents a combination of two competing factors: increasing layer volume and decreasing probability of finding an electron in a given layer. This combination produces a *radial distribution profile* for the electron. Figure 7.23b is not a plot of ψ^2 versus distance from the nucleus as in Figure 7.23a, but rather a plot of $4\pi r^2\psi^2$ versus distance from the nucleus. In geometry, $4\pi r^2$ is the formula for the surface area of a sphere, but here it represents the volume of one of the thin spherical layers in Figure 7.23c.

A significant feature of the curve in Figure 7.23b is that its maximum value corresponds to the most likely radial distance of the electron from the nucleus. The value of *r* corresponding to this maximum for the 1*s* orbital of hydrogen is 53 picometers (pm).

Figure 7.23d provides a view of the spherical shape of this (or any other) *s* orbital. The surface of the sphere encloses the volume within which the probability of finding a 1*s* electron is 90%. This type of depiction, called a *boundary–surface representation*, is one of the most useful ways to view the relative sizes, shapes, and orientations of orbitals. All *s* orbitals are spheres, which have only one orientation and in which electron density depends only on distance from the nucleus. Boundary surfaces are a useful way to depict the shape and relative size of an orbital.

The relative sizes of 1*s*, 2*s*, and 3*s* orbitals are shown in **Figure 7.24**. Note that orbital size increases with increasing values of the principal quantum number *n*. Note also that the sections of the spheres above the profile curves show bands in which the density of dots is high. The dots represent the probability of an electron being in these regions of three-dimensional space, and each band is called a *local maximum* of electron density. In all three profiles, a local maximum occurs close to the nucleus. This means that electrons in *s* orbitals—even *s* orbitals with high values of *n*—have some probability of being close to the nucleus.

The local maxima in any *s* orbital are separated from other local maxima by nodes. The number of nodes in any *s* orbital is equal to $n - 1$. Nodes have the

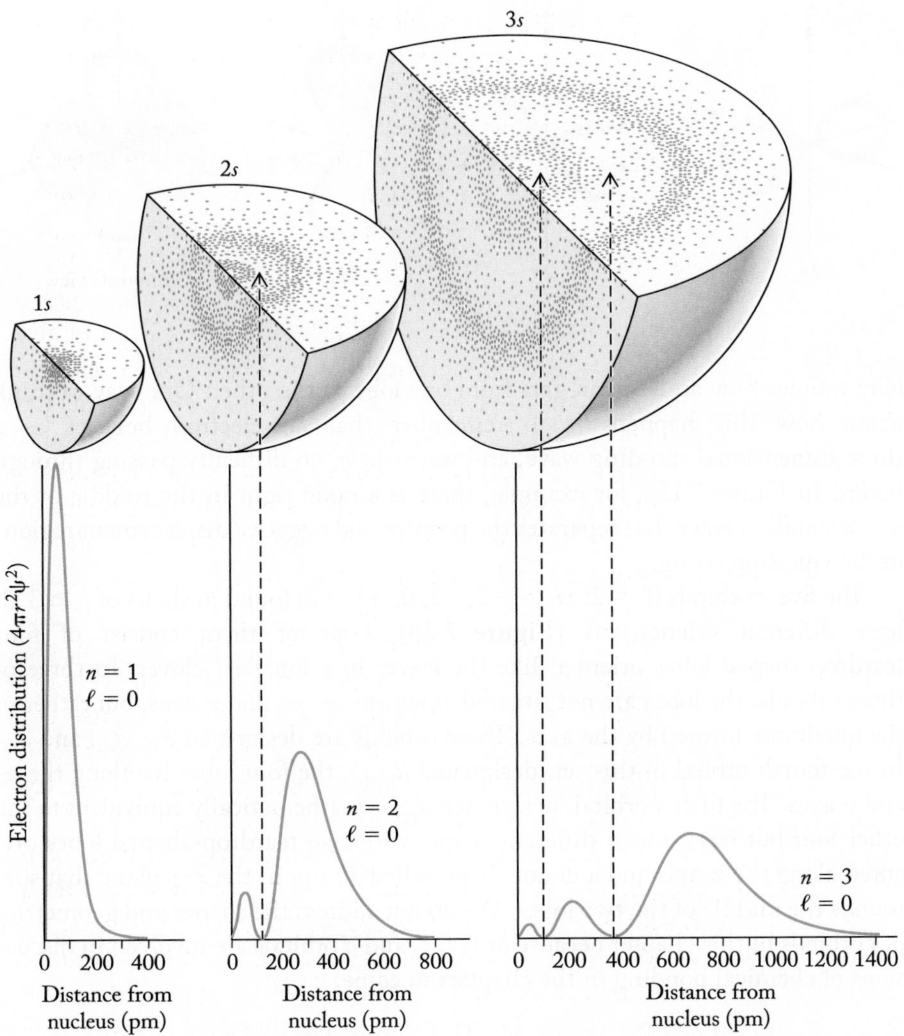

FIGURE 7.24 These radial distribution profiles of 1*s*, 2*s*, and 3*s* orbitals have 0, 1, and 2 nodes, respectively, identifying (with dashed arrows) locations of zero electron density. Electrons in all these *s* orbitals have some probability of being close to the nucleus, but 3*s* electrons are more likely to be farther away from the nucleus than 2*s* electrons, which are more likely to be farther away than 1*s* electrons.

same meaning here as they do in one-dimensional standing waves (recall Figure 7.15a): they are places where the wave has zero amplitude. In the context of electrons as three-dimensional matter waves, nodes are locations at which electron density goes to zero.

CONCEPT **TEST**

How many nodes are there in the electron distribution profile of the 6s orbital?

p and *d* Orbitals

All shells with $n \geq 2$ have a subshell containing three *p* orbitals ($\ell = 1$; $m_\ell = -1$, 0, +1). Each of these orbitals has two teardrop-shaped lobes, oriented opposite each other, along one of the three perpendicular Cartesian axes *x*, *y*, and *z* (**Figure 7.25**). These orbitals are designated p_x, p_y, and p_z, depending on the axis along which the lobes are situated. The two lobes of a *p* orbital may be labeled with plus and minus signs (not in this text, but in other sources), indicating that the sign of the wave function defining them is either $+\psi$ or $-\psi$. (These signs are not in any way connected with electric charges.) An electron in a *p* orbital occupies *both* lobes. Because a node of zero probability separates the two lobes, you

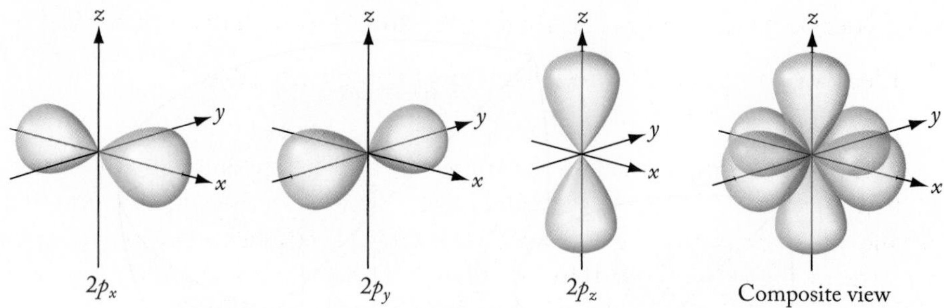

FIGURE 7.25 Boundary–surface views of the three p orbitals, showing their orientation along the x-, y-, and z-axes. The nucleus of the atom containing these orbitals is located at the origin. These shapes are an elongated version of the theoretical shapes of the orbitals. We use this version throughout the book to make it easier to see the orientation of the lobes.

may wonder how an electron gets from one lobe to the other. One way to think about how this happens is to remember that an electron behaves as a three-dimensional standing wave, and waves have no difficulty passing through nodes. In Figure 7.15a, for example, there is a node right in the middle of the $n = 2$ standing wave that separates the positive and negative displacement regions in the vibrating string.

The five d orbitals ($\ell = 2$; $m_\ell = -2, -1, 0, +1, +2$) found in shells of $n \geq 3$ all have different orientations (**Figure 7.26**). Four of them consist of four teardrop-shaped lobes oriented like the leaves in a four-leaf clover. In three of these orbitals, the lobes are not situated along the x-, y-, and z-axes, but rather in the quadrants formed by the axes. These orbitals are designated d_{xy}, d_{xz}, and d_{yz}. In the fourth orbital in this set, designated $d_{x^2-y^2}$, the four lobes lie along the x- and y-axes. The fifth d orbital, designated d_{z^2}, is mathematically equivalent to the other four but has a much different shape, with two teardrop-shaped lobes oriented along the z-axis and a donut shape called a *torus* in the x–y plane that surrounds the middle of the two lobes. We do not address the shapes and geometries of f orbitals in this chapter because only s, p, and d orbitals are involved in discussions of chemical bonding in the chapters to come.

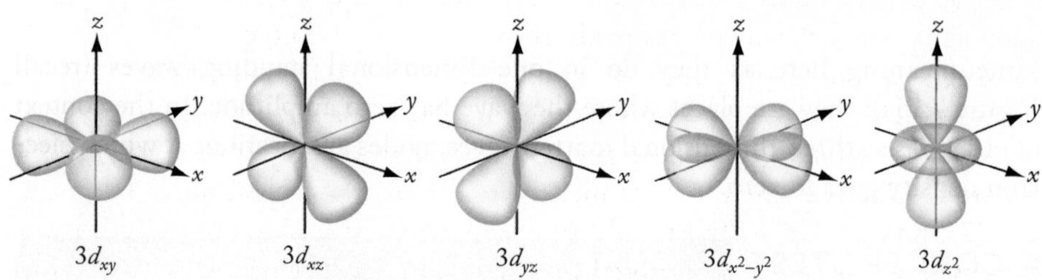

FIGURE 7.26 Boundary–surface views of the five d orbitals, showing their orientation relative to the x-, y-, and z-axes. The d_{xy}, d_{xz}, and d_{yz} orbitals are *not* aligned along any axis; the $d_{x^2-y^2}$ orbital lies along the x- and y-axes; the d_{z^2} orbital consists of two teardrop-shaped lobes along the z-axis with a donut-shaped torus ringing the point where the two lobes meet.

7.8 The Periodic Table and Filling the Orbitals of Multielectron Atoms

CONNECTION We described in Section 2.5 how Mendeleev developed the first useful periodic table of the elements, not only decades before de Broglie gave us electron waves and Schrödinger pioneered quantum mechanics, but even before atoms were known to consist of electrons, protons, and neutrons.

Now that we have developed a model to describe the atomic orbitals, we can use that model to describe atoms containing more than one electron. To explore the orbital-filling sequence, let's start at the beginning of the periodic table with hydrogen and put each successive electron into the lowest-energy orbital available as we move through the table element by element. This method is based on the **aufbau principle** (German *aufbauen*, "to build up"), which states that the most stable atomic structures are those in which the electrons are in the lowest-energy orbitals available.

To decide which orbitals contain electrons, we start with two rules:

1. Electrons always go into the lowest-energy orbital available.
2. Each orbital can hold up to two electrons.

Using these rules, let's assign the single electron in a hydrogen ($Z = 1$) atom to the appropriate orbital. We have already discussed how the single electron in a ground-state atom of hydrogen is in the 1*s* orbital. We represent this arrangement with the **electron configuration** $1s^1$, where the first "1" indicates the principal quantum number (*n*) of the orbital, "*s*" indicates the type of orbital, and the superscript "1" indicates that there is *one* electron in this 1*s* orbital. Because the hydrogen atom has only one electron, $1s^1$ is the complete electron configuration for a hydrogen atom in its ground state.

To be unambiguous about the electron configuration, we use **orbital diagrams** to show how electrons, represented by single-headed arrows, are distributed among orbitals, represented by boxes. A single-headed arrow pointing upward represents an electron with spin up ($m_s = +\frac{1}{2}$), and a downward-pointing single-headed arrow represents an electron with spin down ($m_s = -\frac{1}{2}$). Because hydrogen has only one electron, its orbital diagram contains one single-headed arrow in a box labeled 1*s* (**Figure 7.27a**).

The atomic number of helium is 2, which tells us there are two protons and two electrons in the neutral atom. Using the aufbau principle, we simply add another electron to the 1*s* orbital. We know that each orbital can contain two electrons, but the spin quantum numbers for the two electrons cannot be the same. One must be $+\frac{1}{2}$ and the other must be $-\frac{1}{2}$. These two electrons are said to be *spin-paired*. Their presence gives helium a ground-state electron configuration of $1s^2$. With two electrons, the 1*s* orbital is now full and so is the $n = 1$ shell (**Figure 7.27b**).

The concept of a *filled shell* is key to understanding chemical properties: elements composed of atoms that have filled *s* and *p* subshells in their outermost shells are chemically stable and generally unreactive. Helium is such an element, as are all the other elements in group 18.

The location of lithium ($Z = 3$) in the periodic table—the first element in the second period—is a signal that lithium has one electron in its $n = 2$ shell. The row numbers in the periodic table correspond to the *n* values of the outermost shells of the elements in the rows. The second shell has four orbitals (one 2*s* and three 2*p*), so it can hold up to eight electrons. Lithium's third electron occupies the lowest-energy orbital in the second shell, which is the 2*s* orbital, making its electron configuration $1s^2 2s^1$.

Now that we've begun to place electrons in the second shell, let's consider the forces that an electron in the second shell experiences. An electron in a 2*s* or 2*p* orbital has an electrostatic attraction to the positively charged nucleus, but that electron is also partially shielded from the nucleus by the two negative charges on the 1*s* electrons (**Figure 7.28**). This shielding essentially reduces the nuclear charge experienced by a lithium atom's 2*s* electron, that is, its **effective nuclear charge (Z_{eff})**, from +3 to about +1.

Returning to the ground-state lithium atom, we saw that its third electron occupies the 2*s* orbital. Why is the 2*s* orbital lower in energy than any of the 2*p* orbitals? We can best explain this difference in energies by comparing the radial distribution profiles of these orbitals (**Figure 7.29**). The small peak on the 2*s* curve near the nucleus indicates that an electron in the 2*s* orbital is closer to the nucleus more of the time than an electron in a 2*p* orbital, which has no such

aufbau principle the method of building electron configurations of atoms by adding one electron at a time as atomic number increases across the rows of the periodic table; each electron goes into the lowest-energy orbital available.

electron configuration the distribution of electrons among the orbitals of an atom or ion.

orbital diagram depiction of the arrangement of electrons in an atom or ion, using boxes to represent orbitals.

effective nuclear charge (Z_{eff}) the attraction toward the nucleus experienced by a valence-shell electron, which is approximately equal to the sum of the positive charge on the nucleus and the total negative charge of all inner-shell electrons.

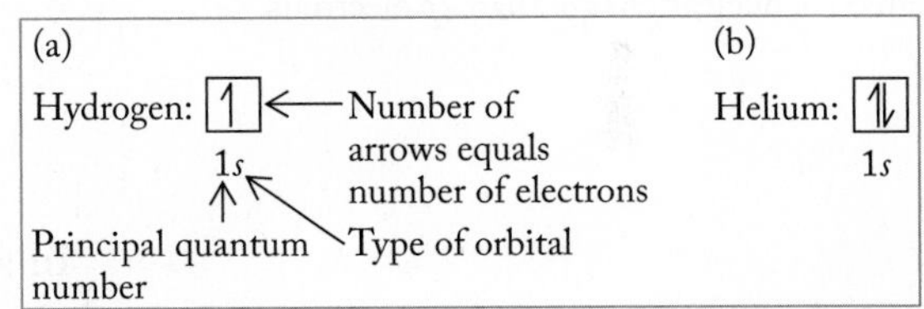

FIGURE 7.27 Orbital diagrams for hydrogen and helium indicate the orbitals in which the electrons are found and the number of electrons in each orbital. The label on the orbital, 1*s*, indicates the value of the principal quantum number for the orbital ($n = 1$) and the type of orbital (*s*). (a) Hydrogen atoms have one electron in the 1*s* orbital, whereas (b) helium atoms have two.

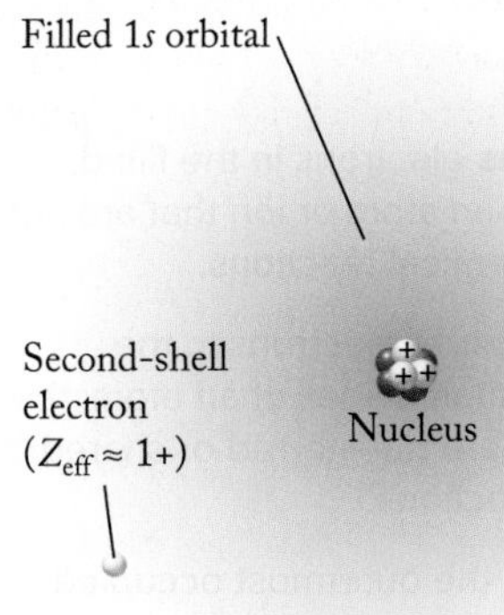

FIGURE 7.28 The effective nuclear charge (Z_{eff}) experienced by an outer-shell electron in Li equals the actual nuclear charge (3+) less the shielding effect of the negative charges of the two 1*s* electrons (2−). This shielding produces a net Z_{eff} of about 1+.

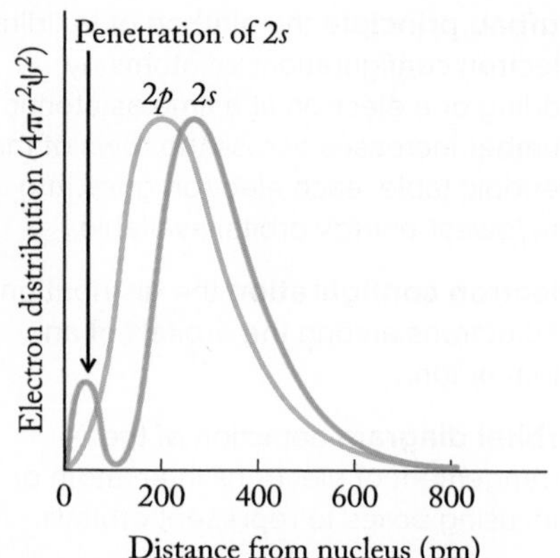

FIGURE 7.29 Radial distribution profiles of electrons in 2*s* and 2*p* orbitals. The 2*s* orbital is lower in energy because electrons in it penetrate more closely to the nucleus, as indicated by the local maximum in electron distribution about 50 pm from the nucleus. As a result, 2*s* electrons experience a greater effective nuclear charge than 2*p* electrons.

secondary peak. This closer proximity means that a 2*s* electron experiences a greater Z_{eff} than an electron in a 2*p* orbital, even though both are shielded by the 1*s* electrons. Therefore, a 2*s* electron is lower in energy than a 2*p* electron, which is why the 2*s* orbital fills first.

We can simplify the electron configuration of Li and all the elements that follow it in the periodic table by using a *condensed electron configuration*. For Li, this is [He]$2s^1$. In this form, the symbols representing all the electrons in orbitals that were filled in the rows above the element of interest are replaced by the symbol of the group 18 element at the end of the row above the element. For Li, this is [He]. Condensed electron configurations are useful because they replace the orbital notation for the **core electrons** in filled shells and subshells with the symbol of a single noble gas. Only the orbital notation for the electrons in the outermost shell is written out. These outermost electrons are called **valence electrons.** The distinction between core and valence electrons is used in later chapters to explain reactions and bond formation.

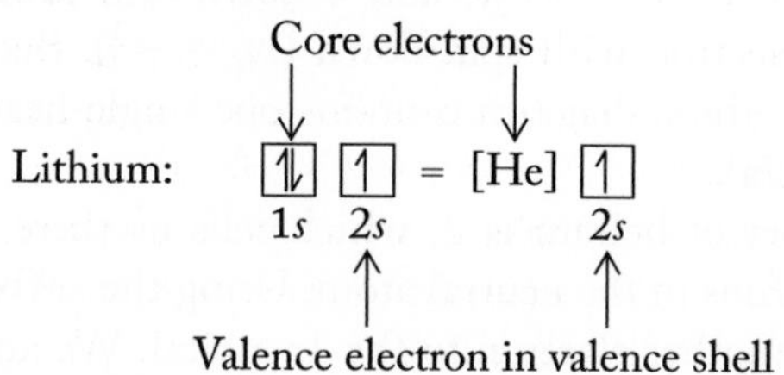

In any atom, the shell containing the valence electrons is referred to as the **valence shell**. Lithium has a single electron in its valence-shell *s* orbital, as does hydrogen, the element directly above it in the periodic table; therefore, both atoms have the valence-shell configuration ns^1. Here *n* represents both the number of the row in which the atom is located on the table and the principal quantum number of the valence shell in the atom.

Beryllium ($Z = 4$) is the fourth element in the periodic table and the first in group 2. The configuration for its four electrons is $1s^22s^2$ or [He]$2s^2$. The other elements in group 2 also have two spin-paired electrons in the *s* orbital of their valence shell. The second shell is not filled at this point because it also has three empty *p* orbitals that will fill as we move to the next elements in the periodic table.

Boron ($Z = 5$) is the next element and the first element in group 13. Its fifth electron is in one of its three 2*p* orbitals, resulting in the condensed electron configuration [He]$2s^22p^1$. Designating the 2*p* orbital that contains the fifth electron as $2p_x$, $2p_y$, or $2p_z$ is not important because these three orbitals all have the same energy; we say they are **degenerate orbitals**.

The next element is carbon ($Z = 6$). It has four electrons in its valence shell (the $n = 2$ shell), so its condensed electron configuration is [He]$2s^22p^2$. Are the two electrons in the 2*p* orbitals both in the same orbital? All electrons have a negative charge and repel one another. Thus, they occupy orbitals that are as far away from each other as possible, which means the two 2*p* electrons in carbon occupy separate 2*p* orbitals. This distribution pattern follows **Hund's rule**, named after the German physicist Friedrich Hund (1896–1997), which states that the lowest-energy electron configuration for degenerate orbitals (such as the $2p_x$, $2p_y$, and $2p_z$ orbitals in the 2*p* subshell) is the one with the maximum number of unpaired parallel valence electrons (i.e., all their spins have the same direction).

Hund's rule is the third rule we apply when determining electron configurations. By convention, the first electron placed in an orbital has a positive spin. When electrons fill a degenerate set of orbitals, Hund's rule requires that an

core electrons electrons in the filled, inner shells of an atom or ion that are not involved in chemical reactions.

valence electrons electrons in the outermost occupied shell of an atom; the electrons that are transferred or shared in chemical reactions.

valence shell the outermost occupied shell of an atom.

degenerate orbitals describes orbitals of the same energy.

Hund's rule the lowest-energy electron configuration of an atom has the maximum number of unpaired electrons, all of which have the same spin, in degenerate orbitals.

electron with a positive spin occupy each valence orbital before any electron with a negative spin enters any orbital. Electrons do not pair in degenerate orbitals until each orbital is occupied by a single electron. To obey Hund's rule, the orbital diagram for carbon must be

Carbon: [↑↓] [↑↓] [↑|↑|]
$1s$ $2s$ $2p$

which shows that the two $2p$ electrons are unpaired and have the same spin.

The next element is nitrogen ($Z = 7$), represented by either $1s^22s^22p^3$ or $[He]2s^22p^3$. According to Hund's rule, the third $2p$ electron is assigned to the third $2p$ orbital, so that the electron distribution is

Nitrogen: [↑↓] [↑↓] [↑|↑|↑]
$1s$ $2s$ $2p$

As we proceed across the second row to neon ($Z = 10$), we fill the $2p$ orbitals as shown in **Figure 7.30**. The last three $2p$ electrons added (in oxygen, fluorine, and neon) spin pair with the first three, so that in neon, the three $2p$ orbitals are all filled. At this point the $n = 2$ shell is full. Helium, neon, and all the other noble gases in group 18 have filled s and p orbitals in their valence shells. This same trend is seen throughout the periodic table: *main group elements in the same column of the periodic table have the same valence-shell electron configuration.* Argon, krypton, xenon, and radon also have filled s and p orbitals in their outermost occupied shells, and they also are chemically inert gases at room temperature.

CHEMTOUR
Electron Configuration

Sodium ($Z = 11$) follows neon in the periodic table. It is the third element in group 1 and the first element in the third period. Ten of its electrons are distributed as in neon. The 11th electron is in the lowest-energy orbital available after $2p$ has been filled, which is $3s$. The condensed electron configuration of Na is $[Ne]3s^1$. Just as we write condensed electron configurations that provide orbital notation for only the outermost occupied shell, we can also condense orbital diagrams in

	Orbital diagram 1s	2s	2p	Electron configuration	Condensed configuration
H	[↑]	[ ]	[| |]	$1s^1$	
He	[↑↓]	[ ]	[| |]	$1s^2$	
Li	[↑↓]	[↑]	[| |]	$1s^22s^1$	$[He]2s^1$
Be	[↑↓]	[↑↓]	[| |]	$1s^22s^2$	$[He]2s^2$
B	[↑↓]	[↑↓]	[↑| |]	$1s^22s^22p^1$	$[He]2s^22p^1$
C	[↑↓]	[↑↓]	[↑|↑|]	$1s^22s^22p^2$	$[He]2s^22p^2$
N	[↑↓]	[↑↓]	[↑|↑|↑]	$1s^22s^22p^3$	$[He]2s^22p^3$
O	[↑↓]	[↑↓]	[↑↓|↑|↑]	$1s^22s^22p^4$	$[He]2s^22p^4$
F	[↑↓]	[↑↓]	[↑↓|↑↓|↑]	$1s^22s^22p^5$	$[He]2s^22p^5$
Ne	[↑↓]	[↑↓]	[↑↓|↑↓|↑↓]	$1s^22s^22p^6$	$[He]2s^22p^6 = [Ne]$

FIGURE 7.30 Orbital diagrams and condensed electron configurations for the first ten elements show that each orbital (indicated by a square in the orbital diagrams) holds at most two electrons and the two electrons must be of opposite spin. The orbitals are filled in order of increasing quantum numbers n and ℓ. Condensed electron configurations for all elements are given in Appendix 3.

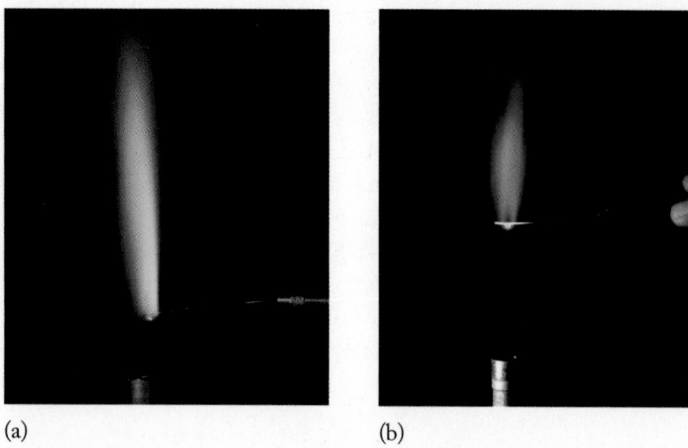

(a) (b)

FIGURE 7.31 The alkali metals produce characteristic colors in Bunsen burner flames because the high flame temperatures produce excited-state atoms of these elements. (a) The yellow-orange glow of Na atoms. (b) The lavender color of potassium atoms.

this same way, so that the condensed orbital diagram for sodium is

Sodium: [Ne] [1] 3*s*

This diagram reinforces the message that the electron configuration of a sodium atom consists of a neon core plus a single electron in the 3*s* orbital of the valence shell. The sodium atom has the same generic valence-shell configuration as lithium and hydrogen—namely, ns^1, where n is the period number. This pattern for main group elements in the same group continues throughout the periodic table. The electron configuration of magnesium ($Z = 12$), for instance, is [Ne]$3s^2$, and the electron configuration of every other element in group 2 consists of the immediately preceding noble gas core followed by ns^2.

The next six elements in the periodic table—aluminum, [Ne]$3s^23p^1$, to argon, [Ne]$3s^23p^6$—show a pattern of increasing numbers of 3*p* electrons, a trend that continues until all three 3*p* orbitals are filled (six electrons), which means the *s* and *p* orbitals of the $n = 3$ shell are filled (eight electrons). Thus, argon is chemically inert, as predicted by its position in group 18.

Before leaving the third row, let's revisit the condensed electron configuration of sodium, [Ne]$3s^1$. This configuration represents a ground-state sodium atom because all the electrons, and most importantly its valence electron, occupy the lowest-energy orbitals available. Now think back to the discussion about atomic emission spectra in Section 7.1 and the distinctive yellow-orange glow that sodium makes in the flames of Bunsen burners, as shown back in Figure 7.3 and here in **Figure 7.31a**. Each sodium atom absorbs a quantum of energy from the Bunsen burner that raises the valence electron from the ground state to an excited state (a transition represented by the orange arrow pointing to the right in **Figure 7.32**). The easiest excited state to populate is the one with the smallest energy above the ground state. We have seen that the 3*p* orbitals fill after the 3*s* orbital because they have the next lowest energy. Therefore, the lowest-energy (or *first*) excited state of sodium is one in which its 3*s* electron has moved up to a 3*p* orbital. This excited state has the electron configuration [Ne]$3p^1$ (see Figure 7.32) and a very short lifetime. The electron typically takes less than a nanosecond to fall back to the ground state in a transition represented by the yellow-orange arrow pointing to the left in Figure 7.32. This transition releases a quantum of energy ($h\nu$) equal to the difference in energy between the 3*p* and 3*s* orbitals in a Na atom—the energy of a photon of yellow-orange light.

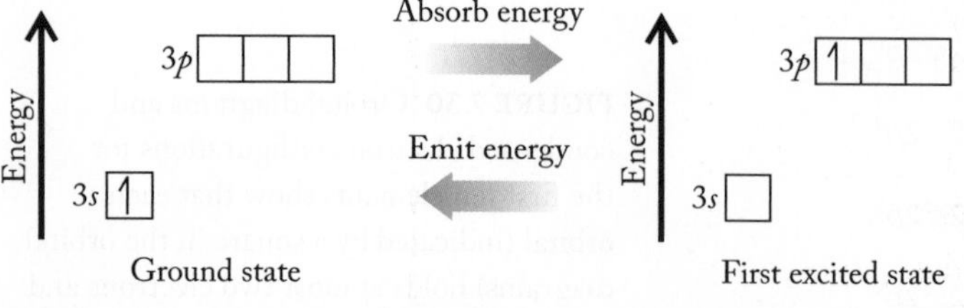

FIGURE 7.32 A ground-state Na atom absorbs a quantum of energy as its valence electron moves from the 3*s* orbital to a 3*p* orbital. This 3*p* electron in the excited-state atom spontaneously falls back to the empty 3*s* orbital, emitting a photon of yellow-orange light. The energy of the photon exactly matches the difference in energy between the 3*p* and 3*s* orbitals of Na atoms.

After argon comes potassium ($Z = 19$) in the fourth row of group 1 ([Ar]$4s^1$). Potassium atoms, like those of sodium, produce a distinctive emission spectrum when heated in a Bunsen burner flame (**Figure 7.31b**). Potassium is followed by calcium ($Z = 20$) in group 2 ([Ar]$4s^2$), at which point the 4*s* orbital is filled but the 3*d* orbitals are still empty. (Review Table 7.1 if you need help in recalling that the $n = 3$ shell contains a *d* subshell.) Why were the 3*d* orbitals not filled before 4*s*?

Applying the first aufbau principle—in building atoms, each electron goes in the lowest-energy orbital available—would be straightforward were it not for the fact that the differences in energy between shells get smaller as n gets larger (**Figure 7.33**). These smaller differences result in orbitals with large ℓ values in one shell having energies similar to those of orbitals with small ℓ values in the next higher shell. Note in Figure 7.33 that the energy of the 4*s* orbital is slightly lower than that of the 3*d* orbitals. The 4*s* orbitals in potassium and calcium are the lowest-energy orbitals available and are filled before any electrons go into a 3*d* orbital.

The element after calcium is scandium ($Z = 21$). It is the first element in the central region of the periodic table, the region populated by transition metals. Scandium has the condensed electron configuration $[Ar]3d^14s^2$. Note that the orbitals are listed in order of increasing principal quantum number, not necessarily in the order in which they were filled. The advantage of writing configurations based on n is that it helps us predict the order in which elements lose valence electrons when they form monatomic cations (as discussed in next section). The 3*d* orbitals are filled in the transition metals from scandium to zinc ($Z = 30$). This pattern of filling the *d* orbitals of the shell whose principal quantum number is 1 less than the period number, $(n - 1)d$, is followed throughout the periodic table: the 4*d* orbitals are filled in the transition metals of the fifth period, and so on (**Figure 7.34**).

The element after scandium is titanium ($Z = 22$), which has one more *d* electron than scandium, so its condensed electron configuration is $[Ar]3d^24s^2$. Vanadium ($Z = 23$) has the configuration $[Ar]3d^34s^2$. At this point, you may feel you can accurately predict the electron configurations of the remaining transition metals in the fourth period. However, because the energies of the 3*d* and 4*s* orbitals are similar, the sequence of *d*-orbital filling deviates in two spots from the pattern you might expect. The next element, chromium ($Z = 24$), has the configuration $[Ar]3d^54s^1$:

Chromium: [Ar] [1][1][1][1][1] [1]
$3d^5$ $4s^1$

This half-filled set of *d* orbitals is an energetically favored configuration. Apparently, the stability of having five half-filled 3*d* orbitals compensates for the energy needed to raise a 4*s* electron to a 3*d* orbital. As a result, $[Ar]3d^54s^1$ is a lower-energy electron configuration than $[Ar]3d^44s^2$.

Another deviation from the expected filling pattern is observed near the end of a row of transition metals. Copper ($Z = 29$) has the electron configuration $[Ar]3d^{10}4s^1$ instead of the expected $[Ar]3d^94s^2$ because a filled set of *d* orbitals also represents a lower-energy electron configuration.

Figure 7.34 illustrates the overall orbital-filling pattern described above. It also shows how the periodic table can be used to predict the electron configurations of the elements. The color patterns and labels in Figure 7.34a indicate which type of orbital is filled going across each row from left to right. For example, groups 1 and 2 are called *s* block elements because their outermost electrons are in *s* orbitals. Similarly, groups 13 through 18 (except for helium) are called the *p* block elements because their outermost electrons are in *p* orbitals. Note how the principal quantum numbers (n) of the outermost orbitals in the *s* and *p* blocks match their row numbers: 2*s* and 2*p* in row 2, 3*s* and 3*p* in row 3, and so on. This means that the periodic table is a very useful reference for writing the electron configurations of the ground states of atoms. For example, barium is the group 2 element in the sixth row. This location means that a ground-state Ba atom has 2 electrons in its 6*s* orbital, so its condensed electron configuration is

$$\text{Ba:} \quad [Xe]6s^2$$

Between the *s* and *p* blocks are the transition metals in groups 3 through 12, which make up the *d* block, and the two rows of elements at the bottom of the

FIGURE 7.33 The energy levels in multielectron atoms increase with increasing values of n and with increasing values of ℓ within a shell. The difference in energy between adjacent shells decreases with increasing values of n, which may result in the energies of subshells in two adjacent shells overlapping. For example, electrons in 3*d* orbitals have slightly higher energy than those in the 4*s* orbital, resulting in the order of subshell filling $4s \rightarrow 3d \rightarrow 4p$.

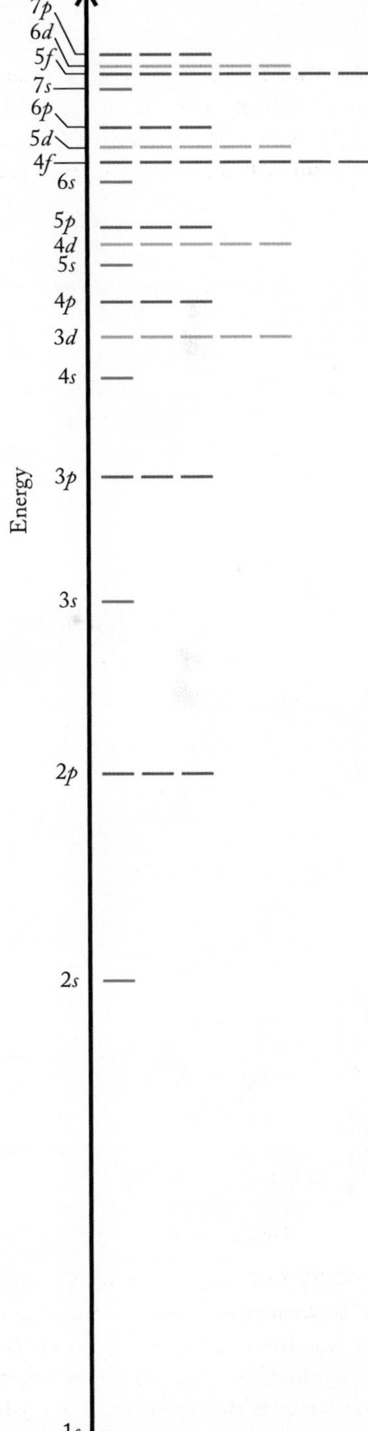

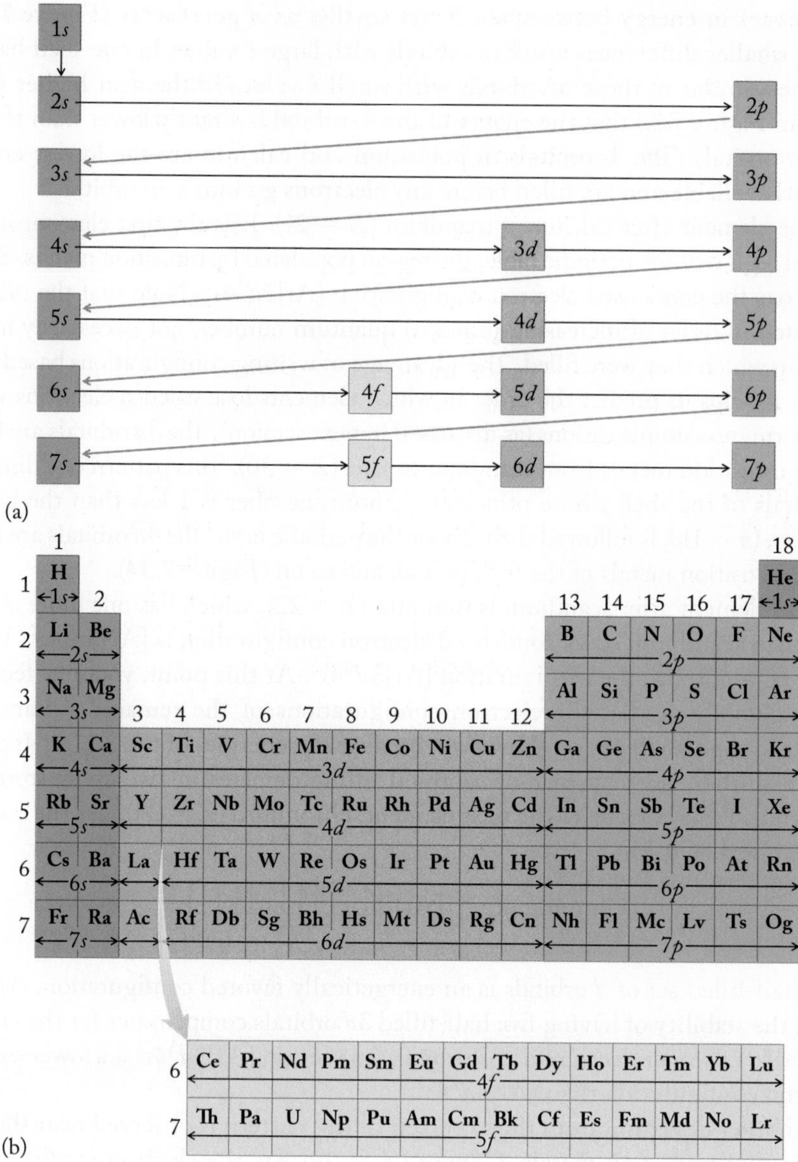

FIGURE 7.34 (a) This diagram shows the sequence in which atomic orbitals fill. (b) The same color coding in this version of the periodic table highlights the four "blocks" of elements in which valence-shell *s* (green), *p* (blue), *d* (orange), and *f* (purple) orbitals are filled as atomic number increases across a row of the table.

periodic table, called the lanthanides and actinides, which make up the *f* block. Both of the *f* block series are 14 elements long. The *n* value of the *d* orbitals being filled in a row is always one less than the row number, and the *n* value of the *f* orbitals being filled is two less than the row number.

Several relationships are worth noting between the quantum numbering system (Figure 7.17) and the organization of the periodic table into rows and *s*, *p*, *d*, and *f* blocks (Figure 7.34b):

- Each row number corresponds to a value of n. The first row begins the $n = 1$ shell, the second row begins the $n = 2$ shell, and so on.
- There are n subshells in the nth shell. There is one subshell (1*s*) in the $n = 1$ shell, there are two subshells (2*s* and 2*p*) in the $n = 3$ shell, and so on.
- There are n^2 orbitals in the nth shell and a maximum of $2n^2$ electrons. The first row has $n = 1$, so $1^2 = 1$ orbital (1*s*). The second row has $n = 2$, so $2^2 = 4$ orbitals (2*s*, $2p_x$, $2p_y$, and $2p_z$), and so on.

- Each block represents a subshell and there are $(2\ell + 1)$ orbitals in each subshell. The *s* block (subshell) consists of $(2 \times 0 + 1 = 1)$ one *s* orbital in each row. The *p* block consists of $(2 \times 1 + 1 = 3)$ three *p* orbitals in each row. The *d* block consists of $(2 \times 2 + 1 = 5)$ five *d* orbitals in each row. Lastly, the *f* block consists of $(2 \times 3 + 1 = 7)$ seven *f* orbitals in each row.
- Because the *s*, *p*, *d*, and *f* subshells always contain 1, 3, 5, and 7 orbitals, respectively, and each orbital contains 2 electrons, the *s* block consists of 2 elements in one row of the periodic table, the *p* block consists of 6 elements, the *d* block consists of 10 elements, and the *f* block consists of 14 elements. In the $n = 4$ shell, for example, there are 2 elements in the *s* block (K and Ca), 6 elements in the *p* block (Ga through Kr), 10 elements in the *d* block (Y through Cd), and 14 elements in the *f* block (Ce through Lu).

With these patterns in mind, let's write the condensed electron configuration of lead ($Z = 82$), which is the group 14 element in the sixth row. The noble gas that most closely precedes it is Xe ($Z = 54$). The difference in atomic numbers means that we need to account for $82 - 54 = 28$ electrons in the electron configuration symbols. The location of Pb and the block labels in Figure 7.34b indicates that these 28 electrons are distributed as follows:

2 electrons in 6*s*
14 electrons in 4*f*
10 electrons in 5*d*
2 electrons in 6*p*

Therefore, the condensed electron configuration of ground-state lead atoms reflects this distribution:

$$\text{Pb:} \quad [\text{Xe}]4f^{14}5d^{10}6s^26p^2$$

SAMPLE EXERCISE 7.10 Writing Electron Configurations **LO6**

Strontium salts give off a bright red light at the high temperature of fireworks, explosions, and signal flares (**Figure 7.35**). What is the ground-state electron configuration of strontium atoms?

Collect, Organize, and Analyze Strontium is in the fifth row of group 2 of the periodic table. This means that the last orbital to be filled in an atom of Sr is 5*s* (see Figure 7.34b). The sequence of filling the atom's inner-shell orbitals (Figure 7.34a) is 1*s*, 2*s*, 2*p*, 3*s*, 3*p*, 4*s*, 3*d*, 4*p*, and then 5*s*. The atomic number of strontium, $Z = 38$, means that each strontium atom has 38 electrons.

Solve An *s* orbital can hold up to 2 electrons, a set of three *p* orbitals can hold 6 electrons, and a set of five *d* orbitals can hold 10. Therefore, the distribution of the 38 electrons in an atom of Sr is $1s^22s^22p^63s^23p^63d^{10}4s^24p^65s^2$.

Think About It The location of Sr in Figure 7.34 also indicates that the condensed electron configuration of Sr is $[\text{Kr}]5s^2$.

Practice Exercise Gallium arsenide (GaAs) is used in the red lasers in bar-code readers. Write the electron configuration of a ground-state atom of gallium ($Z = 31$) and a ground-state arsenic atom ($Z = 33$).

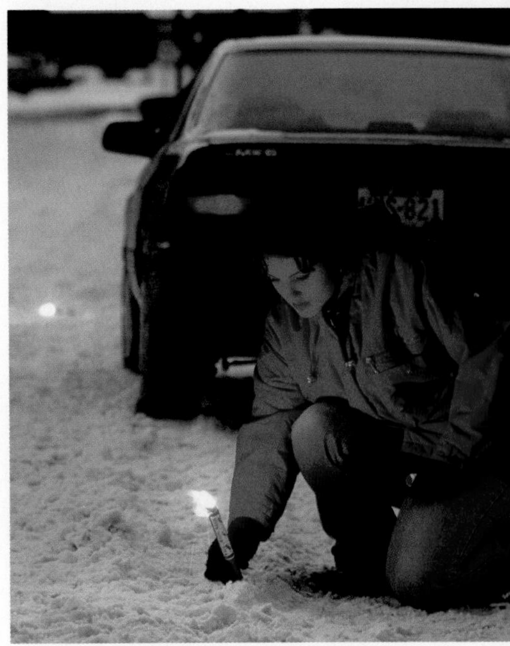

FIGURE 7.35 Strontium nitrate is a common ingredient in road flares that give off red light.

SAMPLE EXERCISE 7.11 Writing Condensed Electron Configurations of Transition Metal Atoms **LO6**

Write the condensed electron configuration of an atom of silver.

Collect and Organize We need to write an element's condensed electron configuration, in which filled inner-shell orbitals are represented by the atomic symbol of the noble gas immediately preceding the element of interest in the periodic table. Silver ($Z = 47$) is the group 11 element in the fifth row of the periodic table.

Analyze Krypton ($Z = 36$) is the noble gas at the end of the fourth row. The difference between their atomic numbers ($47 - 36$) means that we need to assign 11 electrons to the orbitals after [Kr] in the condensed electron configuration of silver.

Solve According to the orbital-filling pattern in Figure 7.34a, 2 of the 11 electrons would be in the 5*s* orbital and the other 9 would be in 4*d* orbitals. However, a filled set of *d* orbitals is more stable than a partially filled set. Therefore silver, like copper just above it in the periodic table, has 10 electrons in its outermost *d* orbital and only 1 electron in its outermost *s* orbital. The condensed electron configuration of an Ag atom is $[Kr]4d^{10}5s^1$.

Think About It We can generate a tentative electron configuration by simply moving across a row in Figure 7.34b until we come to the element of interest. However, we may need to move an *s* electron to a *d* orbital to take into account the stability of half-filled and filled *d* orbitals.

Practice Exercise Write the condensed electron configuration of a ground-state atom of cobalt ($Z = 27$).

7.9 Electron Configurations of Ions

In Section 2.6 we learned that metals and nonmetals can form cations or anions (Figure 2.20) by losing or gaining one or more electrons. How does the electron configuration of an ion compare with that of the neutral parent atom? To write the electron configuration of an ion, we begin with the electron configuration of its neutral parent atom. For a cation, we remove the appropriate number of electrons from the orbital(s) with the highest principal quantum number. For an anion, we add the appropriate number of electrons to one or more partially filled outer-shell orbitals.

Ions of the Main Group Elements

The *s* block elements (see Figure 7.34) form monatomic cations by losing all the outer-shell electrons, producing ions with the electron configurations of the noble gases immediately preceding them in the periodic table. For example, an atom of sodium forms a Na^+ ion by losing the single 3*s* electron:

$$Na \rightarrow Na^+ + e^-$$
$$[Ne]3s^1 \rightarrow [Ne] + e^-$$

The nonmetal elements of the *p* block that form monatomic anions do so by gaining enough electrons to fill the outer-shell *p* orbitals, producing ions with the electron configurations of the noble gases at the right ends of the corresponding rows in the periodic table. For example, an atom of fluorine forms a F^- ion by gaining one electron, which fills the set of three $2p$ orbitals and gives it the electron configuration of neon:

$$F + e^- \rightarrow F^-$$
$$[He]2s^22p^5 + e^- \rightarrow [He]2s^22p^6 = [Ne]$$

A Na^+ ion and a F^- ion both have the same electron configuration as the neutral atom of Ne. We say that Na^+, F^-, and Ne are **isoelectronic**, meaning they have the same electron configuration.

CONNECTION In previous chapters we learned the charges on the ions of common elements. Electron configurations help us understand why these ions have the charges they do.

SAMPLE EXERCISE 7.12 Determining Isoelectronic Species **LO6**

(a) Write the condensed electron configurations of the ions in CsF, $MgCl_2$, CaO, and KBr. (b) Which ions in part (a) are isoelectronic?

Collect and Organize In part (a), we must determine the electron configuration of each ion in four binary ionic compounds. In part (b), we must identify the configurations from part (a) that are equivalent. Figure 7.34a provides information on the order in which orbitals are filled.

Analyze The elements in the compounds include

- two from group 1, Cs and K, which are present as 1+ cations
- two from group 2, Mg and Ca, which are present as 2+ cations
- one from group 16, O, which is present as a 2− anion
- three from group 17, F, Cl, and Br, which are present as 1− anions

Solve

a. Let's arrange these atoms and ions in a table in which we list the condensed electron configurations of the parent atoms, the formulas of the ions (derived from the parent elements' positions in the periodic table—see Figure 2.20), followed by the number of electrons lost or gained to form each ion, and finally the condensed electron configuration of the ion that is produced.

Element	Electron Configuration of Atom	Formula of Ion	Number of Electrons Lost/Gained	Electron Configuration of Ion
Cs	$[Xe]6s^1$	Cs^+	−1	[Xe]
K	$[Ar]4s^1$	K^+	−1	[Ar]
Mg	$[Ne]3s^2$	Mg^{2+}	−2	[Ne]
Ca	$[Ar]4s^2$	Ca^{2+}	−2	[Ar]
O	$[He]2s^22p^4$	O^{2-}	+2	$[He]2s^22p^6 = [Ne]$
F	$[He]2s^22p^5$	F^-	+1	$[He]2s^22p^6 = [Ne]$
Cl	$[Ne]3s^23p^5$	Cl^-	+1	$[Ne]3s^23p^6 = [Ar]$
Br	$[Ar]3d^{10}4s^24p^5$	Br^-	+1	$[Ar]3d^{10}4s^24p^6 = [Kr]$

b. Based on the ion electron configurations in the last column, we have three ions that are isoelectronic with Ne (Mg^{2+}, O^{2-}, and F^-) and three that are isoelectronic with Ar (K^+, Ca^{2+}, and Cl^-).

Think About It The electron configurations of all the ions make sense because each cation has the stable electron configuration of the noble gas in the preceding row of the periodic table, and each anion is isoelectronic with the noble gas element at the end of its row.

 **Practice Exercise** Write the electron configurations of the ions in KI, BaO, Rb_2O, and $Al(NO_3)_3$. Which of these ions are isoelectronic with Ar?

isoelectronic atoms or ions having identical electron configurations.

CONCEPT TEST

Potassium has one valence electron in the 4s orbital. Figure 7.31b shows the color produced from excited-state potassium atoms. Write a condensed electron configuration for an excited-state potassium atom. How does that electron configuration differ from that of a potassium ion?

Transition Metal Cations

As with the main group elements, writing the electron configurations of transition metal cations begins by considering the metal atoms from which the cations form. Zinc atoms, like those of many transition metals, form ions with 2+ charges by losing both electrons from the *s* orbital in the outermost shell:

$$\mathrm{Zn} \rightarrow \mathrm{Zn}^{2+} + 2\ \mathrm{e}^-$$

$$[\mathrm{Ar}]3d^{10}4s^2 \rightarrow [\mathrm{Ar}]3d^{10} + 2\ \mathrm{e}^-$$

A few transition metals, including silver, have only one outer-shell *s* electron and may form singly charged ions:

$$\mathrm{Ag} \rightarrow \mathrm{Ag}^{+} + \mathrm{e}^-$$

$$[\mathrm{Kr}]4d^{10}5s^1 \rightarrow [\mathrm{Kr}]4d^{10} + \mathrm{e}^-$$

We might have expected Zn and Ag atoms to lose the 3*d* and 4*d* electrons, respectively, reasoning that the last orbitals to be filled should be the first to lose electrons when an atom forms a positive ion. Instead, the rule that the electrons in orbitals with the highest *n* value ionize first applies to all atoms, including transition metals. Preferential loss of outer-shell *s* electrons explains why the most frequently encountered charge on transition metal ions is 2+.

Transition metal atoms may also lose one or more *d* electrons in addition to the two valence *s* electrons when forming ions with charges >2+. For example, an atom of scandium, $[\mathrm{Ar}]3d^14s^2$, loses both 4*s* electrons *and* the 3*d* electron when it forms a Sc^{3+} ion. The chemistry of titanium, $[\mathrm{Ar}]3d^24s^2$, is dominated by its atoms losing all four of the 4*s* and 3*d* electrons to form Ti^{4+} ions. In other cases, some—but not all—of the outermost *d* electrons are lost when a transition metal atom forms an ion. For example, when an atom of copper, $[\mathrm{Ar}]3d^{10}4s^1$, forms a Cu^{2+} ion, it loses its 4*s* electron and one 3*d* electron, giving it a $[\mathrm{Ar}]3d^9$ electron configuration.

SAMPLE EXERCISE 7.13 Writing Condensed Electron Configurations of Transition Metal Ions **LO6**

What are the condensed electron configurations of Fe^{3+} and Ni^{2+}?

Collect and Organize We are asked to write the electron configurations of ions formed by two transition metals: iron (Z = 26) and nickel (Z = 28). We can use Figure 7.34a, which shows the order in which orbitals are filled, to determine the electron configurations of the neutral atoms. Then, electrons in orbitals with the highest *n* value ionize first. For transition metal atoms that means their outermost *s* electrons, not their outermost *d* electrons.

Analyze The location of iron in group 8 of the periodic table indicates that its atoms have two 4*s* electrons and six 3*d* electrons built on an argon core. Nickel is in group 10, so its atoms have two more 3*d* electrons than iron. As a result, the electron configurations of Fe and Ni *atoms* are

$$\mathrm{Fe} = [\mathrm{Ar}]3d^64s^2 \text{ and } \mathrm{Ni} = [\mathrm{Ar}]3d^84s^2$$

Solve Fe atoms lose their 4*s* electrons *and* one 3*d* electron when forming Fe^{3+} ions, whereas Ni atoms lose two 4*s* electrons to form Ni^{2+} ions, producing these condensed electron configurations:

$$Fe \rightarrow Fe^{3+} + 3\ e^-$$
$$[Ar]3d^64s^2 \rightarrow [Ar]3d^5 + 3\ e^-$$
$$Ni \rightarrow Ni^{2+} + 2\ e^-$$
$$[Ar]3d^84s^2 \rightarrow [Ar]3d^8 + 2\ e^-$$

Think About It Both atoms lose both of their valence-shell *s* electrons, which is what usually happens when transition metals form monatomic cations. By also losing one of its 3*d* electrons, each Fe atom forms an Fe^{3+} ion with a stable, half-filled set of 3*d* orbitals.

Practice Exercise Write the electron configurations for the manganese atom and the ions Mn^{3+} and Mn^{4+}.

We have yet to consider the lanthanides (elements 58 through 71) and actinides (elements 90 through 103), represented by the two purple rows in Figure 7.34b. The lanthanides have partly filled 4*f* orbitals, whereas the actinides have partly filled 5*f* orbitals. There are 14 elements in each group, reflecting the capacity of the seven orbitals in each *f* subshell ($\ell = 3$; $m_\ell = -3, -2, -1, 0, +1, +2,$ and $+3$). According to Figure 7.34a, the 4*f* orbitals are not filled until after the 6*s* orbital has been filled. This order of filling occurs because 6*s* and 4*f* orbitals have similar energies (Figure 7.33). Similarly, the 5*f* orbitals are filled after the 7*s* orbital is filled.

The periodic table is a useful reference for predicting the physical and chemical properties of elements. Our quantum mechanical perspectives on atomic structure provide us with a theoretical basis for explaining why particular families of elements behave similarly. The original table was based on periodic trends in observable chemical properties. Now we know that the chemical properties of an element are closely linked to the electron configurations of the atoms of the element.

CONCEPT **TEST**

The electron configuration of Eu ($Z = 63$) is $[Xe]4f^76s^2$, but the electron configuration of Gd ($Z = 64$) is $[Xe]4f^75d^16s^2$. Suggest a reason why the additional electron in Gd goes into a 5*d* orbital instead of a 4*f* orbital, which would result in the electron configuration $[Xe]4f^86s^2$.

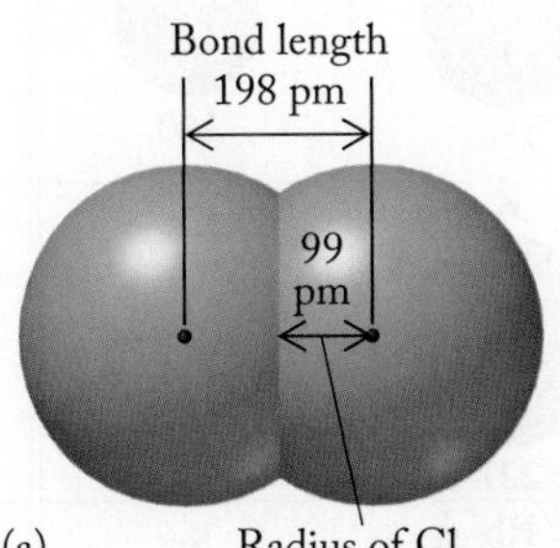

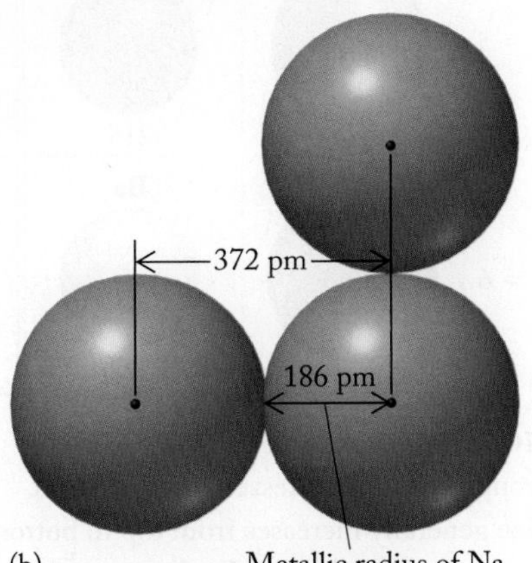

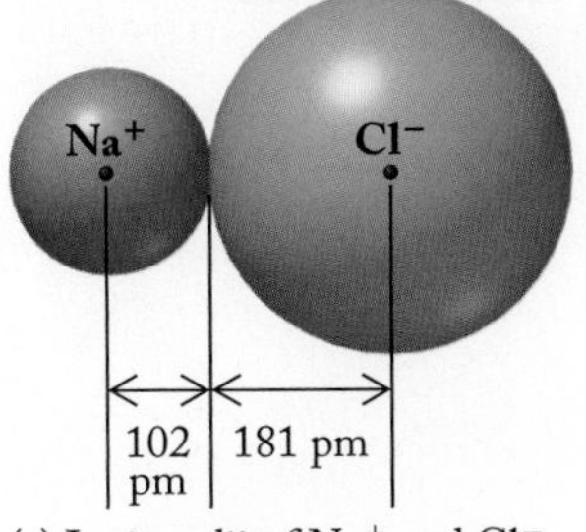

(c) Ionic radii of Na^+ and Cl^-

FIGURE 7.36 A comparison of covalent, metallic, and ionic radii. (a) A covalent radius is half the length of the bond between identical atoms in a molecule, such as the bond in Cl_2. (b) A metallic radius is based on the distance of closest approach of adjacent atoms in a crystalline metal. (c) Ionic radii are based on the distances between the centers of adjacent ions in the crystals of many ionic compounds.

7.10 The Sizes of Atoms and Ions

The sizes of atoms are usually expressed in terms of their radii. The atomic radius of an element that occurs in nature as a diatomic molecule, such as N_2 and O_2, is simply half the distance between the nuclear centers in the molecule (**Figure 7.36a**). The atomic radius of a metal, also called its *metallic* radius, is half the distance between the nuclear centers in the solid metal (**Figure 7.36b**). The values of ionic radii are derived from the distances between nuclear centers in solid ionic compounds (**Figure 7.36c**). The periodic trends in the relative sizes of atoms are shown in **Figure 7.37**. Several factors contribute to these trends.

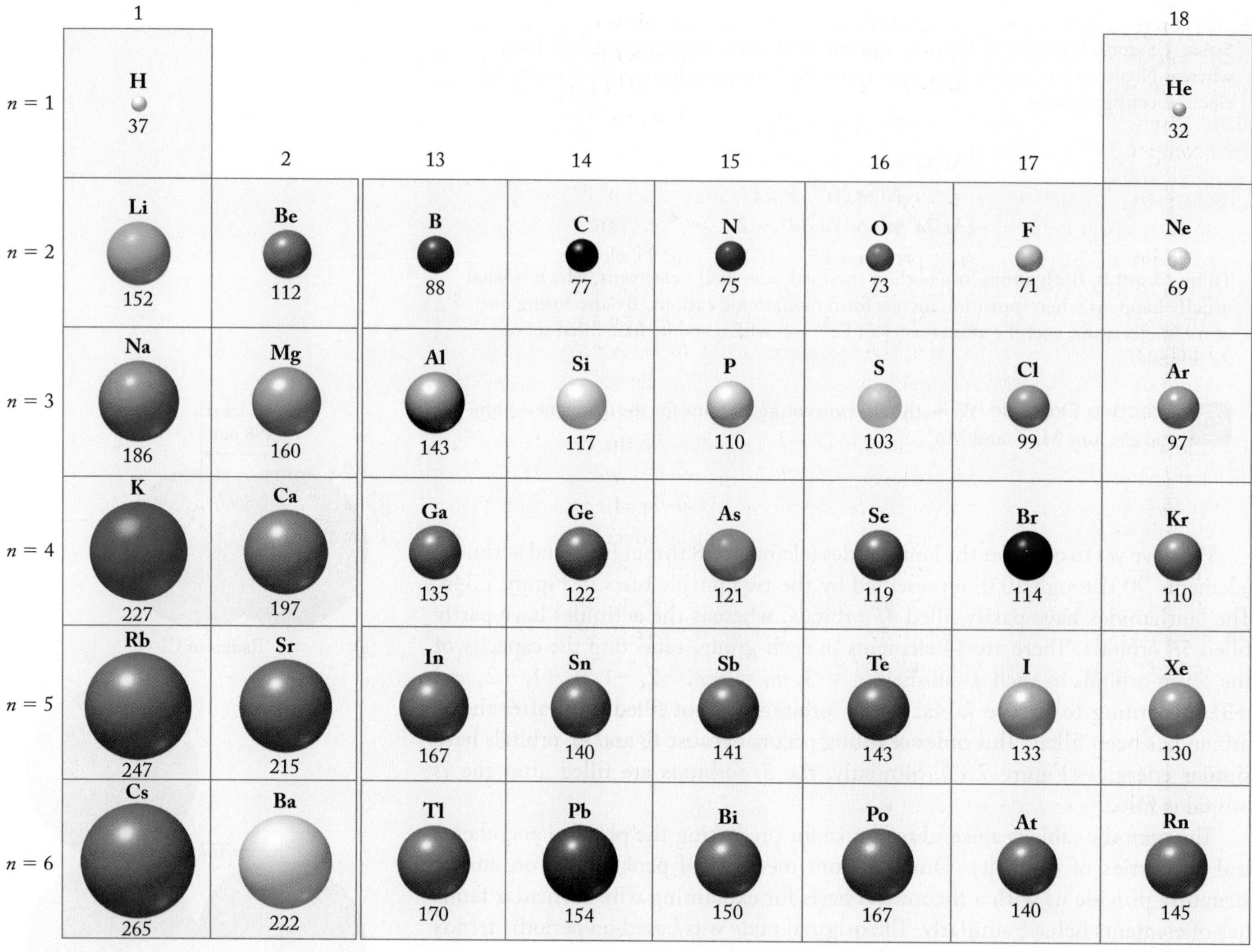

FIGURE 7.37 Atomic radii of the main group elements expressed in picometers. Size generally increases from top to bottom in any group and generally decreases from left to right across any period.

Trends in Atom and Ion Sizes

The radial distribution profiles in Figure 7.24 show how the distance from the nucleus to the *s* orbitals increases as the principal quantum number *n* increases. This same trend is observed for *p, d,* and *f* orbitals as well. Therefore, we expect the atomic radii of elements in the same group of the periodic table to increase with increasing atomic number. For example, the atomic radii of the halogens increase as we move from the top to the bottom of group 17 of the periodic table:

Element	F	Cl	Br	I	At
Atomic Number	9	17	35	53	85
n Value of Valence Shell	2	3	4	5	6
Atomic Radius (pm)	71	99	114	133	140

This pattern holds for other groups of elements, as shown in Figure 7.37.

As we move across a row (period) in the periodic table, each successive element has one additional electron. We might expect that the addition of electrons across a row would mean a corresponding increase in the size of the atom.

Surprisingly, however, experimental data do not support this prediction. Atomic radii tend to *decrease* as the atomic number increases across a row of the periodic table. This pattern is particularly evident as we move from left to right across the main group elements (Figure 7.37). To explain these data, we need to consider two competing interactions:

1. *Increasing effective nuclear charge.* Each time the atomic number increases, so does the positive charge of the nucleus. Consider, for example, an atom of sodium-23, which has 11 protons, 12 neutrons, and 11 electrons, and an atom of magnesium-24, which has 12 protons, 12 neutrons, and 12 electrons. The one valence electron of Na and the two valence electrons of Mg are both in the 3*s* orbital, meaning that the valence electron in Na is probably at the same distance from the Na nucleus as the two valence electrons in Mg are from the Mg nucleus. The two valence electrons in Mg experience a larger effective nuclear charge than the valence electron in Na: the electrostatic attraction of electrons (regardless of whether it is one or two) in a 3*s* orbital to 12 protons is stronger than the attraction to 11 protons because the distance is unchanged. As Z_{eff} increases, the size of atoms decreases.
2. *Increasing repulsion between valence electrons.* As atomic number increases, so does the number of valence electrons. More electrons mean more electron–electron repulsions (because two negatively charged particles repel one another), which tends to increase the size of the atoms.

CHEMTOUR
Periodic Trends

The experimental data tabulated in Figure 7.37 show that atomic size decreases with increasing *Z* across each row of main group elements. This means that interaction 1, effective nuclear charge, more than offsets interaction 2, repulsions between the increasing number of valence electrons.

The cations of the main group elements are much smaller than their parent atoms, whereas the anions are much larger (**Figure 7.38**). To understand these trends, let's revisit what happens when a Na atom forms a Na^+ ion. The atom loses its only valence-shell (3*s*) electron, forming an ion with a much smaller neonlike

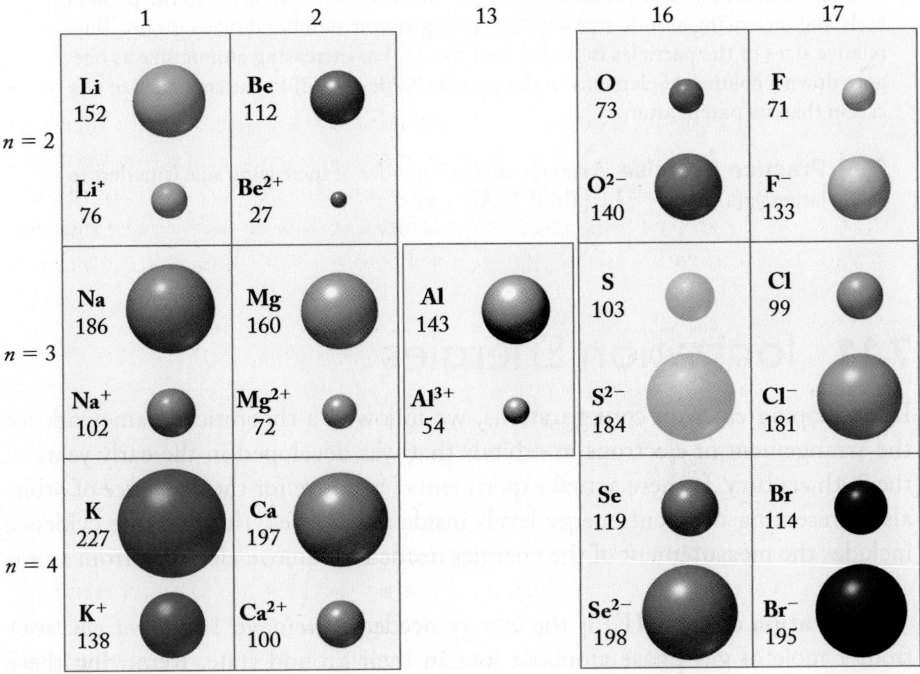

FIGURE 7.38 Comparison of atomic and ionic radii. Values are in picometers.

core of 1*s*, 2*s*, and 2*p* electrons. On the other hand, when a Cl atom gains an electron and forms a Cl^- ion, it contains one more electron (18) than a Cl atom (17). The additional electron means more electron–electron repulsion, which increases the size of the valence shell and results in a Cl^- ion that is bigger than a Cl atom. This phenomenon causes all monatomic anions to be larger than the atoms from which they form.

CONCEPT TEST

Rank the following orbitals in an atom of silver in order of decreasing effective nuclear charge experienced by the electrons in them: 1*s*, 2*s*, 3*s*, 4*s*, 2*p*, 3*p*, 4*p*.

SAMPLE EXERCISE 7.14 Ordering Atoms and Ions by Size **LO7**

Arrange each set by size, largest to smallest: (a) O, P, S; (b) Na^+, Na, K.

Collect and Organize We are asked to rank a set of three atoms according to their size and a set of two atoms and one cation of one of the atoms according to their size. The location of elements in the periodic table can be used to determine the relative sizes of the atoms.

Analyze Sizes decrease as we move from left to right across a period and increase as we move down a column. In addition, a cation is always smaller than the atom from which it is made.

Solve

a. S is below O in group 16, so in terms of atomic size, S > O. S is to the right of P, so P > S. The size order is thus P > S > O.
b. Cations are smaller than their atoms, so Na > Na^+. K is below Na in the alkali metals group, and size increases down a group, so K > Na. Therefore, the size order is K > Na > Na^+.

Think About It The trend in the sizes of the atoms in set (a) reflects decreasing atomic size with increasing effective nuclear charge within a row of elements in the periodic table and increasing atomic size with increasing atomic number down a group. The relative sizes of the particles in set (b) are linked (1) to increasing atomic size as one goes down a column of elements in the periodic table and (2) to the smaller size of a cation than its parent atom.

Practice Exercise Arrange each set in order of increasing size (smallest to largest): (a) Cl^-, F^-, Li^+; (b) P^{3-}, Al^{3+}, Mg^{2+}.

7.11 Ionization Energies

In developing electron configurations, we followed a theoretical framework for the arrangement of electrons in orbitals that was developed in the early years of the 20th century. Is there actual experimental evidence for the existence of orbitals representing different energy levels inside atoms? Yes, there is. The evidence includes the measurement of the energies needed to remove electrons from atoms and ions.

ionization energy (IE) the quantity of energy needed to remove 1 mole of electrons from 1 mole of ground-state atoms or ions in the gas phase.

Ionization energy (IE) is the energy needed to remove 1 mole of electrons from 1 mole of gas-phase atoms or ions in their ground state. Removing these electrons always requires an addition of energy to the system because a negatively charged electron is attracted to a positively charged nucleus, and overcoming that attractive force requires energy. The amount of energy needed to remove 1 mole of electrons from 1 mole of atoms to make 1 mole of cations with a 1+ charge is called the *first ionization energy* (IE_1); the energy needed to remove 1 mole of electrons from 1 mole of 1+ cations to make 1 mole of cations with a 2+ charge is the *second ionization energy* (IE_2); and so forth. For example,

$$Mg(g) \rightarrow Mg^{+}(g) + 1\ e^{-} \qquad (IE_1 = 738\ \text{kJ/mol})$$

$$Mg^{+}(g) \rightarrow Mg^{2+}(g) + 1\ e^{-} \qquad (IE_2 = 1451\ \text{kJ/mol})$$

The total energy required to make 1 mole of $Mg^{2+}(g)$ cations from 1 mole of $Mg(g)$ atoms is the sum of these two ionization energies:

$$Mg(g) \rightarrow Mg^{2+}(g) + 2\ e^{-}$$

$$\text{Total IE} = (738 + 1451)\ \text{kJ/mol} = 2189\ \text{kJ/mol}$$

Figure 7.39 shows how the first ionization energies of the main group elements vary. The IE_1 of hydrogen is 1312 kJ/mol, whereas the IE_1 of helium is nearly twice as big: 2372 kJ/mol. This difference is reasonable because He atoms have two protons per nucleus, whereas H atoms have only one. Twice the nuclear charge increases Z_{eff}, so nearly twice the ionization energy is needed to pull an electron away from the nucleus. In general, first ionization energies increase from left to right across a period. Thus, the easiest element to ionize in a period is the group 1 element, and the hardest is the group 18 element. This pattern makes sense because as the charge of the nucleus increases across a row, so do Z_{eff} and the attractions between the nucleus and the valence electrons.

Two anomalies occur in the general trend of increasing IE_1 with increasing Z across a row. One shows up in the decrease in IE_1 values between the group 2 and group 13 elements in the second and third rows. Note that B and Al lose a *p* electron when they ionize, whereas Be and Mg lose an *s* electron. It is easier to remove a *p* electron because it experiences a smaller effective nuclear charge than an

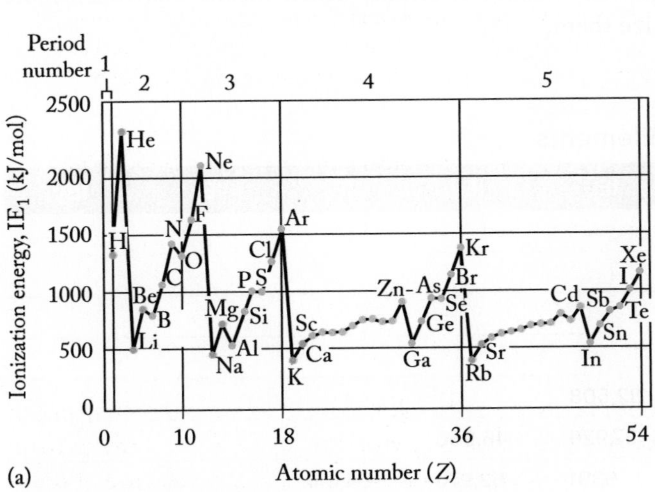

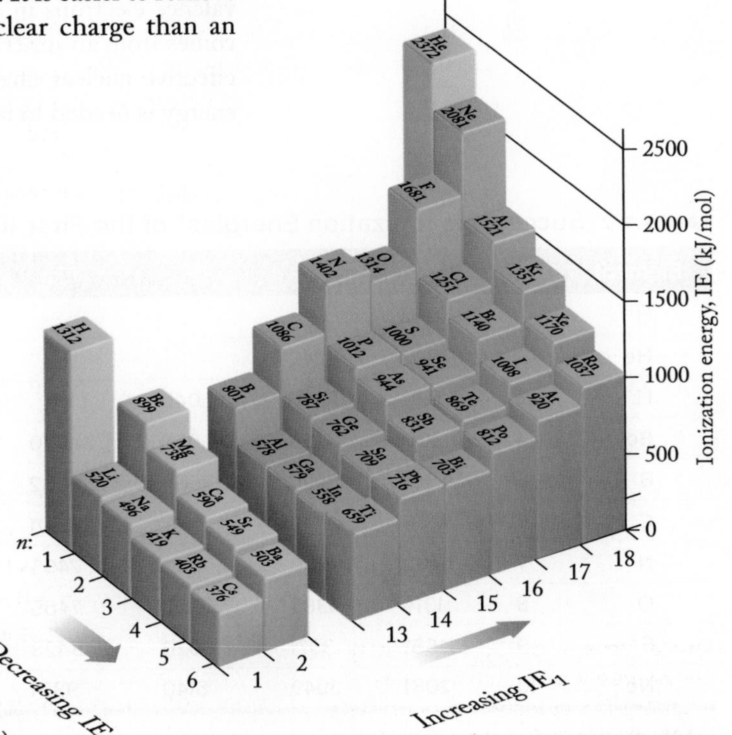

FIGURE 7.39 The first ionization energies (IE_1) of the elements generally increase from left to right in a period and decrease from top to bottom in a group: (a) IE_1 values for the elements in the first five periods; (b) IE_1 values for all the main group elements.

s electron in the same shell. A second anomaly occurs between the group 15 and 16 elements in each row in which there is a decrease in IE_1 values because ionization of group 16 atoms involves removal of an electron from a filled *p* orbital (see the orbital diagram below). This takes slightly less energy because removal of the electron reduces electron–electron repulsion.

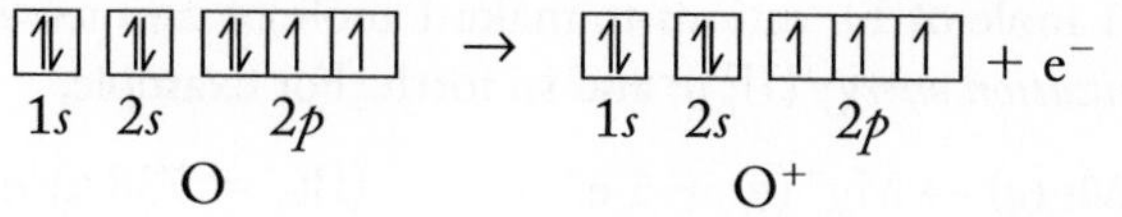

$$\underset{1s}{[\uparrow\downarrow]}\ \underset{2s}{[\uparrow\downarrow]}\ \underset{2p}{[\uparrow\downarrow|\uparrow|\uparrow]} \rightarrow \underset{1s}{[\uparrow\downarrow]}\ \underset{2s}{[\uparrow\downarrow]}\ \underset{2p}{[\uparrow|\uparrow|\uparrow]} + e^-$$

O $\qquad\qquad$ O^+

Figure 7.39 also shows how IE_1 values decrease with increasing atomic number down in each group. There are two reasons for this. First, the increasing size of the atoms (see Figure 7.37) means that the valence-shell electrons in atoms of elements farther down each group are farther from the nucleus, and so they experience less electrostatic attraction to the positive charge of the nucleus. In addition, the greater number of protons and greater nuclear charges of atoms farther down each group are offset by more inner shells of electrons that effectively shield valence-shell electrons from the greater nuclear charge. As a result, the effective nuclear charge experienced by, for example, the outer-shell *s* electron in an atom of any group 1 element is essentially the same: +1.

Another perspective on energy levels in atoms is provided by *successive* ionization energies. Consider the trend in the energies needed to remove all the electrons in the atoms of the first 10 elements (**Table 7.2**). For multielectron atoms, the second ionization energy (IE_2) is always greater than IE_1 because the second electron is being removed from an ion that already has a positive charge. Because electrons carry a negative charge, they are held more strongly in a cation (have stronger electrostatic attractions to the nucleus) than in the atom from which the cation was formed.

The energy needed to remove a third electron is greater still because it is being removed from a 2+ ion. Superimposed on this trend are much more dramatic increases in ionization energy (separated by the red line in Table 7.2) when all the valence electrons in an ionizing atom have been removed and the next electron comes from an inner shell. Core electrons experience less shielding and a greater effective nuclear charge than valence electrons, which means that much more energy is needed to ionize them.

TABLE 7.2 Successive Ionization Energies[a] of the First 10 Elements

Element	Z	IE_1	IE_2	IE_3	IE_4	IE_5	IE_6	IE_7	IE_8	IE_9
H	1	1312								
He	2	2372	5249							
Li	3	520	7296	12,040						
Be	4	900	1758	15,050	21,070					
B	5	801	2426	3660	24,682	32,508				
C	6	1086	2348	4617	6201	37,926	46,956			
N	7	1402	2860	4581	7465	9391	52,976	64,414		
O	8	1314	3383	5298	7465	10,956	13,304	71,036	84,280	
F	9	1681	3371	6020	8428	11,017	15,170	17,879	92,106	106,554
Ne	10	2081	3949	6140	9391	12,160	15,231	19,986	23,057	115,584

[a]All values in kilojoules per mole of atoms.

electron affinity (EA) the energy change that occurs when 1 mole of electrons combines with 1 mole of atoms or monatomic cations in the gas phase.

SAMPLE EXERCISE 7.15 Ranking Ionization Energies **LO7**

Arrange argon, magnesium, and phosphorus in order of increasing first ionization energy.

Collect, Organize, and Analyze We are asked to order three elements based on their IE_1 values. All three are in the third row of the periodic table. First ionization energies generally increase from left to right across a row because of increasing effective nuclear charge.

Solve Ranking the elements in order of increasing first ionization energies, according to their increasing group numbers:

$$Mg < P < Ar$$

Think About It Magnesium forms stable 2+ cations, so we expect its ionization energy to be smaller than the values for phosphorus and argon, which do not form cations. Argon is a noble gas with a stable valence electron configuration, so its first ionization energy should be the highest of the three. We can check our prediction against Figure 7.39b: Mg, 738 kJ/mol; P, 1012 kJ/mol; Ar, 1521 kJ/mol.

Practice Exercise Arrange cesium, calcium, and neon in order of decreasing first ionization energy.

CONCEPT **TEST**

Why does ionization energy decrease with increasing atomic number from helium to neon to argon?

7.12 Electron Affinities

In Section 7.11 we examined the periodic nature of the energy required to ionize atoms. Now we look at a complementary process, called **electron affinity (EA)**, which is the change in energy when 1 mole of electrons is added to 1 mole of gas-phase atoms or monatomic cations.

The energy associated with adding 1 mole of electrons to 1 mole of chlorine atoms in the gas phase is

$$Cl(g) + e^- \rightarrow Cl^-(g) \qquad EA_1 = -349 \text{ kJ/mol}$$

The electron affinities of many elements are negative, meaning the formation of anions with a 1− charge releases energy (**Figure 7.40**). The values in Figure 7.40, however, reveal that the trends in EA are not as regular as the trends in size and IE. Electron affinity increases down a column only for the group 1 metals, whereas other groups do not display a clear trend. In general, electron affinity becomes more negative with increasing atomic number across a row, but there are exceptions to that trend, too. The halogens (group 17) have the most negative EA values, which seems logical given that each halogen atom becomes isoelectronic with a noble gas electron configuration upon gaining one more electron.

The addition of an electron to an atom of a noble gas consumes energy, which makes sense because these atoms

1	2	13	14	15	16	17	18
H −72.6							He (0.0)[a]
Li −59.6	Be >0	B −26.7	C −122	N +7	O −141	F −328	Ne (+29)[a]
Na −52.9	Mg >0	Al −42.5	Si −134	P −72.0	S −200	Cl −349	Ar (+35)[a]
K −48.4	Ca −2.4	Ga −28.9	Ge −119	As −78.2	Se −195	Br −325	Kr (+39)[a]
Rb −46.9	Sr −5.0	In −28.9	Sn −107	Sb −103	Te −190	I −295	Xe (+41)[a]
Cs −45.5	Ba −14	Tl −19.2	Pb −35.2	Bi −91.3	Po −183.3	At −270[a]	Rn (+41)[a]

[a]Calculated values.

FIGURE 7.40 Electron affinity (EA) values of main group elements are expressed in kilojoules per mole. The more negative the value, the more energy is released when 1 mole of atoms combines with 1 mole of electrons to form 1 mole of anions with a 1− charge. A greater release of energy reflects stronger electrostatic attractions between the nuclei of the elements and free electrons.

already have stable electron configurations. Beryllium and magnesium have very small, but positive, electron affinities because an additional electron occupies a valence-shell *p* orbital, which is significantly higher in energy than a valence-shell *s* orbital.

CONCEPT TEST

Describe at least one similarity and one difference in the periodic trends for first ionization energies and electron affinities among the main group elements.

SAMPLE EXERCISE 7.16 Integrating Concepts: Red Fireworks

The brilliant red color of some fireworks (**Figure 7.41**) is produced by the presence of lithium carbonate (Li_2CO_3) in the shells that are launched into the night sky and explode at just the right time. The explosions produce enough energy to vaporize the lithium compound and produce excited-state lithium atoms. These atoms quickly lose energy as they transition from their excited states to their ground states, emitting photons of red light that have a wavelength of 670.8 nm.

a. How much energy is in each photon of red light emitted by an excited-state lithium atom?
b. What is the frequency of the red light?
c. Lithium atoms in the lowest-energy state above the ground state emit these red photons. What is the difference in energy between the ground state of a Li atom and its lowest-energy excited state?
d. What is the electron configuration of a ground-state lithium atom?
e. What is the electron configuration of a lithium atom in its lowest-energy excited state?

FIGURE 7.41 Lithium carbonate is sometimes used to produce vivid red colors in fireworks.

Collect and Organize We are asked to calculate the energy of photons of red light from their wavelength. We are also asked to relate this energy to the difference in energies between a lithium atom's lowest-energy excited state and its ground state and to write the electron configurations of these states. The energy of a photon is related to its wavelength (λ) by Equation 7.3:

$$E = \frac{hc}{\lambda}$$

Lithium is the group 1 element in the second row of the periodic table.

Analyze The energy of a photon (E) emitted by an excited-state atom is the same as the difference in energies (ΔE) between the excited state and the lower-energy state (in this case, the ground state) of the transition that produced the photon. All the group 1 elements have one electron in their valence-shell *s* orbital. In lithium this electron is in the 2*s* orbital, which is occupied after the 1*s* has filled. The next higher energy orbitals are the 2*p* orbitals.

Solve

a. The energy of a photon with a wavelength of 670.8 nm is

$$E = \frac{hc}{\lambda} = \frac{(6.626 \times 10^{-34}\ \text{J} \cdot \text{s})\left(2.998 \times 10^{8}\ \frac{\text{m}}{\text{s}}\right)}{670.8\ \text{nm}\left(\frac{10^{-9}\ \text{m}}{\text{nm}}\right)} = 2.961 \times 10^{-19}\ \text{J}$$

b. Frequency and wavelength are related:

$$c = \nu\lambda \quad \text{so} \quad \nu = \frac{c}{\lambda}$$

The frequency of red light with a wavelength of 670.8 nm is

$$\nu = \frac{2.998 \times 10^{8}\ \frac{\text{m}}{\text{s}}}{670.8\ \text{nm}\left(\frac{10^{-9}\ \text{m}}{\text{nm}}\right)} = 4.469 \times 10^{14}\ \text{s}^{-1}$$

c. The difference in energies between the ground state of a Li atom and its lowest-energy excited state must match exactly the energy of the photon emitted; therefore, $\Delta E = 2.961 \times 10^{-19}$ J.
d. The electron configuration of a ground-state Li atom is $1s^22s^1$.
e. In the lowest-energy excited state, the valence electron has moved to the next higher energy orbital above the 2*s* orbital, which is one of the three 2*p* orbitals. Therefore, the electron configuration of the excited state is $1s^22p^1$.

Think About It You might expect that the colors that lithium and other elements make in fireworks explosions are the same colors they produce in Bunsen burner flames. You would be right. Many of the colors are produced by transitions from lowest-energy excited states to ground states because the lowest-energy excited state is the one most easily populated by the thermal energy available in a gas flame or exploding firework shell.

SUMMARY

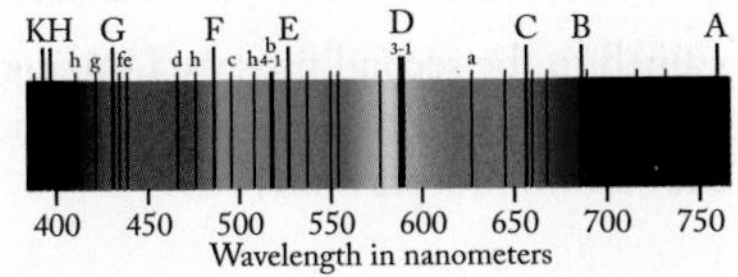

LO1 Visible light and other forms of **electromagnetic radiation** have wavelike properties described by characteristic **wavelengths** and **frequencies**. The speed of light is a constant in a vacuum, $c = 2.998 \times 10^8$ m/s, but slows as it passes through liquids and solids. (Sections 7.1 and 7.2)

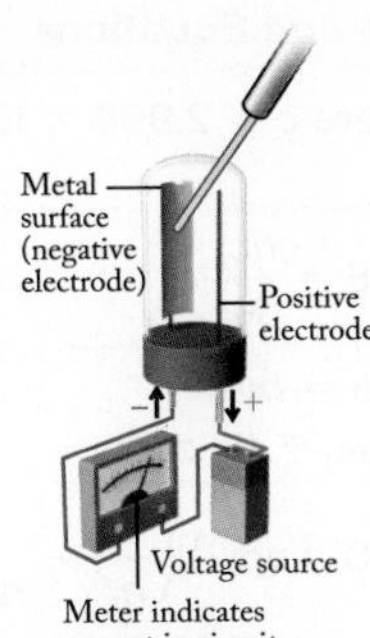

LO2 Max Planck used **quantum theory**, which incorporates the assumption that all electromagnetic radiation consists of discrete particles of energy called **quanta**, to explain blackbody radiation. The **Planck constant** relates the energy of a quantum of light, called a **photon**, to its frequency. Einstein also used quantum theory to explain the **photoelectric effect**: a process in which a metal surface emits electrons when illuminated by electromagnetic radiation that is at or above a **threshold frequency**. (Section 7.3)

LO3 Free atoms in flames and in gas-discharge tubes produce **atomic emission spectra** consisting of narrow bright lines at characteristic wavelengths. When continuous radiation passes through atomic gases, absorption produces the dark lines of **atomic absorption spectra**. The bright lines of an element's emission spectrum and the dark lines of its absorption spectrum are at the same wavelengths. A **ground-state** atom or ion has all its electrons in the lowest possible energy levels. Other arrangements are called **excited states. Electron transitions** from higher to lower energy levels in an atom cause the atom to emit particular frequencies of radiation; absorption of radiation of the same frequencies accompanies electron transitions from the same lower to higher energy levels. (Sections 7.1 and 7.4)

LO4 de Broglie proposed that electrons in atoms, as well as all other moving particles, have wave properties and can be treated as **matter waves.** He explained the stability of the electron orbits in the Bohr hydrogen atom in terms of **standing waves**: the circumferences of the allowed orbits had to be whole-number multiples of the hydrogen electron's characteristic wavelength. The **Heisenberg uncertainty principle** states that both the position and momentum of an electron cannot be precisely known at the same time. (Section 7.5)

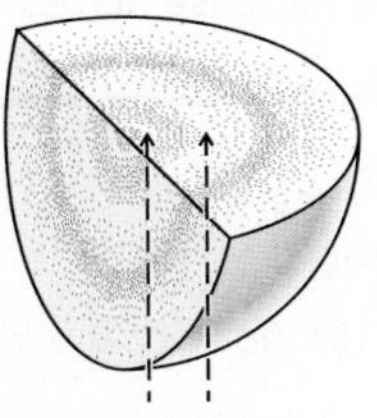

LO5 The solutions to **Schrödinger's wave equation** are mathematical expressions called **wave functions (ψ)**, where ψ^2 defines the regions within an atom, called **orbitals**, that describe the probability of finding an electron at a given distance from the nucleus. Orbitals have characteristic three-dimensional sizes, shapes, and orientations that are depicted by boundary–surface representations. Each orbital has a unique set of three **quantum numbers** that come directly from the solution to the Schrödinger wave equation: the **principal quantum number n**, which defines orbital size and energy level; the **angular momentum quantum number ℓ**, which defines orbital shape; and the **magnetic quantum number m_ℓ**, which defines orbital orientation in space. A fourth quantum number, the **spin magnetic quantum number m_s**, is necessary to explain certain characteristics of emission spectra. (Sections 7.6 and 7.7)

LO6 According to the **aufbau principle**, electrons fill the lowest-energy atomic orbitals of a ground-state atom first. Two electrons in the same orbital have opposite-spin quantum numbers m_s: $+\frac{1}{2}$ and $-\frac{1}{2}$. The **Pauli exclusion principle** states that no two electrons in an atom can have the same four values of n, ℓ, m_ℓ, and m_s. An **electron configuration** is a set of numbers and letters expressing the number of electrons that occupy each orbital in an atom or ion. The electrons in the outermost occupied shell of an atom are **valence electrons.** All the orbitals of a given subshell are **degenerate**; they all have the same energy. **Hund's rule** states that in any set of degenerate orbitals, one electron must occupy each orbital before a second electron occupies any orbital in the set. We use **orbital diagrams** to show in detail how electrons are distributed among orbitals, which are represented by boxes. (Sections 7.8 and 7.9)

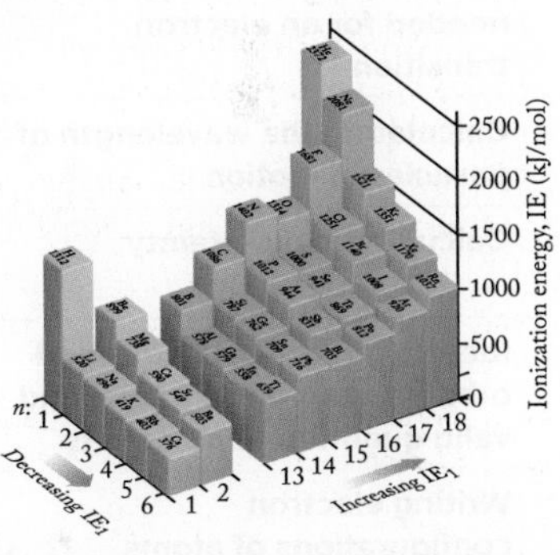

LO7 Effective nuclear charge (Z_{eff}) is the net nuclear charge experienced by outer-shell electrons when they are shielded from the full nuclear charge by inner-shell electrons. Greater Z_{eff} means lower energy. The sizes of atoms increase with increasing atomic number in a group of elements because valence-shell electrons with higher n values are, on average, farther from the nucleus. However, the sizes of atoms decrease with increasing atomic number across a row of elements because the valence electrons experience higher effective nuclear charges. Anions are larger than their parent atoms because of additional electron–electron repulsion, but cations are smaller than their parent atoms—sometimes much smaller when all the electrons in the valence shell are lost. **Ionization energy (IE)** is the amount of energy needed to remove 1 mole of electrons from 1 mole of atoms or ions in the gas phase. IE values generally increase with increasing effective nuclear charge across a row and decrease with increasing atomic number down a group in the periodic table. The energy differences between the different shells and subshells of atoms are reflected in the values of successive ionization energies (IE_1, IE_2, IE_3, . . .). (Sections 7.8, 7.10, 7.11)

LO8 Electron affinity (EA) is the energy change that occurs when 1 mole of electrons combines with 1 mole of atoms or ions in the gas phase. Electron affinity values of many main group elements are negative (energy is released when they acquire electrons) but positive for nitrogen and the noble gases. (Section 7.12)

PARTICULATE PREVIEW WRAP-UP

Each process forms an ion from a neutral parent atom. Because cations are positively charged and smaller than their neutral parent atoms, an electron is lost in the first process. Because anions are negatively charged and larger than their neutral parent atoms, an electron is gained in the second process. Ionizing an electron from a neutral atom to form a cation requires the absorption of energy. For most elements, the second process emits energy.

PROBLEM-SOLVING SUMMARY

Type of Problem	Concepts and Equations		Sample Exercises
Calculating frequency from wavelength	$\lambda\nu = c$ where $c = 2.998 \times 10^8$ m/s	(7.1)	**7.1**
Calculating the energy of a photon	$E = \frac{hc}{\lambda}$	(7.3)	**7.2**
Applying the photoelectric effect	$\phi = h\nu_0$ $KE_{electron} = h\nu - \phi$	(7.4) (7.5)	**7.3**
Calculating the wavelength of a line in the hydrogen spectrum	$\frac{1}{\lambda} = [1.097 \times 10^{-2}\ \text{nm}^{-1}]\left(\frac{1}{n_1^2} - \frac{1}{n_2^2}\right)$	(7.7)	**7.4**
Calculating the energy needed for an electron transition	$\Delta E = -2.178 \times 10^{-18}\ \text{J}\left(\frac{1}{n_{final}^2} - \frac{1}{n_{initial}^2}\right)$	(7.9)	**7.5**
Calculating the wavelength of particles in motion	$\lambda = \frac{h}{mu}$	(7.11)	**7.6**
Calculating uncertainty	$\Delta x \cdot m\Delta u \geq \frac{h}{4\pi}$	(7.13)	**7.7**
Identifying the subshells and orbitals in an energy level and valid quantum number sets	n is the shell number, ℓ defines both the subshell and the type of orbital; an orbital has a unique combination of allowed n, ℓ, and m_ℓ values. ℓ is any integer from 0 to $n - 1$; m_ℓ is any integer from $-\ell$ to $+\ell$, including zero.		**7.8, 7.9**
Writing electron configurations of atoms and ions; determining isoelectronic species	Orbitals fill up in the following sequence: $1s^2$, $2s^2$, $2p^6$, $3s^2$, $3p^6$, $4s^2$, $3d^{10}$, $4p^6$, $5s^2$, $4d^{10}$, $5p^6$, $6s^2$, $4f^{14}$, $5d^{10}$, $6p^6$ (superscripts represent maximum numbers of electrons). Arrange orbitals in electron configuration based on the increasing value of n. In forming transition metal ions, electrons are removed to maximize the number of d electrons; there is enhanced stability in half-filled and filled d subshells.		**7.10, 7.11, 7.12, 7.13**
Ordering atoms and ions by size	Effective nuclear charge and shielding explain observable periodic trends. In general, sizes decrease from left to right across a row and increase down a column of the periodic table. Cations are smaller, and anions are larger, than their parent atoms.		**7.14**
Ranking ionization energies	First ionization energies generally increase across a row and decrease down a column.		**7.15**

VISUAL PROBLEMS

(Answers to boldface end-of-chapter questions and problems are in the back of the book.)

7.1. Which of the elements highlighted in Figure P7.1 consist of ground-state atoms with:

a. a single s electron in the valence shell? (More than one answer is possible.)

b. filled sets of s and p orbitals in the valence shell?

c. filled sets of d orbitals?

d. half-filled sets of d orbitals?

e. two s electrons in the valence shell?

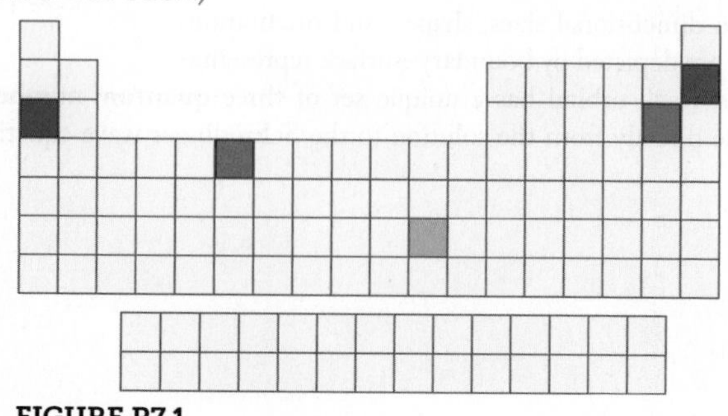

FIGURE P7.1

7.2. Which of the highlighted elements in Figure P7.1:
  a. forms a common monatomic ion that is larger than its parent atom?
  b. has the most unpaired electrons per ground-state atom?
  c. has a complete octet of valence electrons?
  d. forms a common monatomic ion that is isoelectronic with another highlighted element?

7.3. Which of the elements highlighted in Figure P7.3 forms monatomic ions by
  a. losing an *s* electron?
  b. losing two *s* electrons?
  c. losing two *s* electrons and a *d* electron?
  d. adding an electron to a *p* orbital?
  e. adding electrons to two *p* orbitals?

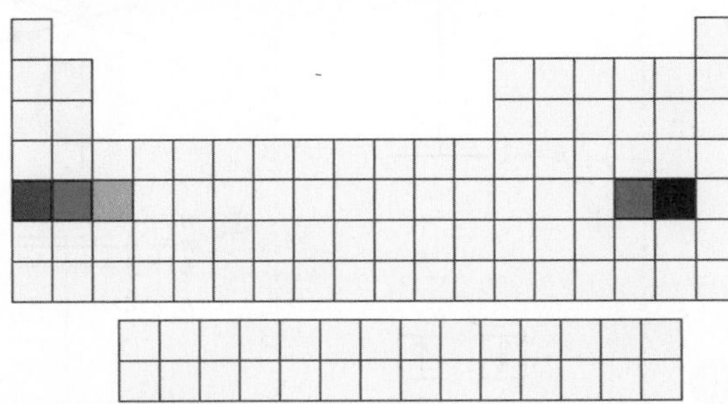

FIGURE P7.3

7.4. Which of the highlighted elements in Figure P7.3:
  a. forms common monatomic ions smaller than the parent atoms? (More than one answer is possible.)
  b. has the largest first ionization energy, IE_1?
  c. has the largest second ionization energy, IE_2?

7.5. Rank the elements highlighted in Figure P7.3 by:
  a. increasing atomic size.
  b. increasing size of the most common monatomic ions.

7.6. Which arrow in Figure P7.6 represents:
  a. emission of light with the longest wavelength?
  b. absorption of light of an atom in an excited state?
  c. a transition requiring absorption of the most energy?

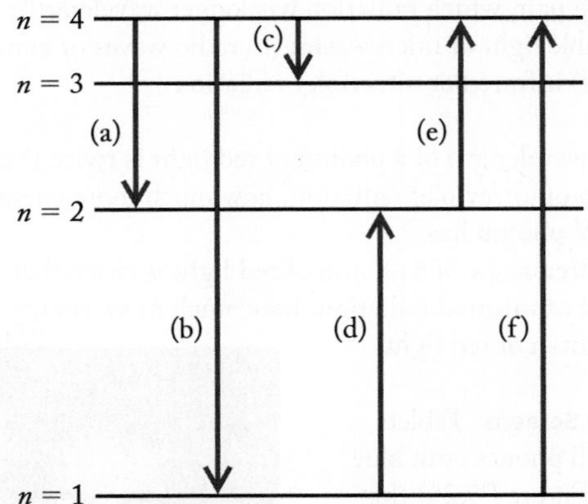

FIGURE P7.6

7.7. In Figure P7.7, which sphere represents a Na atom, a Na^+ ion, and a K atom?

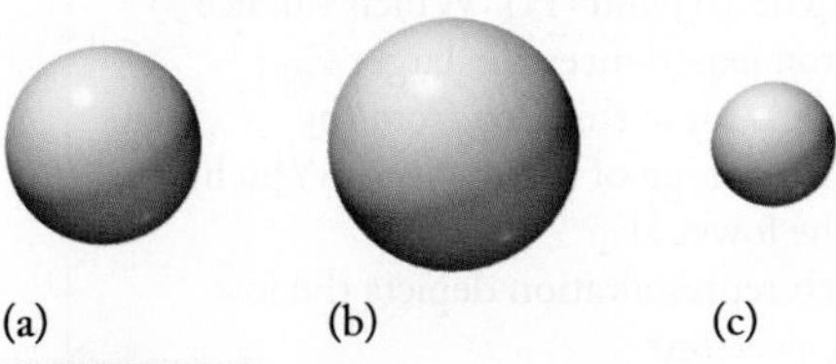

FIGURE P7.7

7.8. Which orbital diagram in Figure P7.8 represents:
  a. the ground state of an oxygen cation, O^+?
  b. an excited state of a nitrogen atom?
  c. a violation of Hund's rule?
  d. a violation of the Pauli exclusion principle?

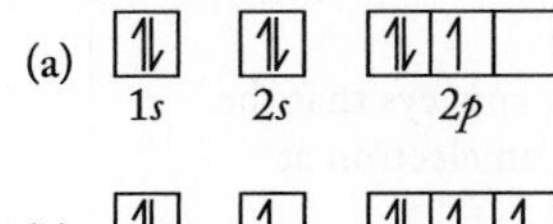

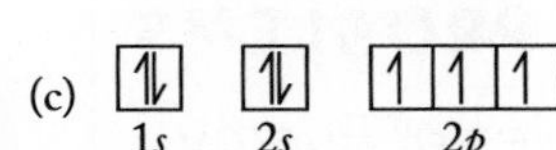

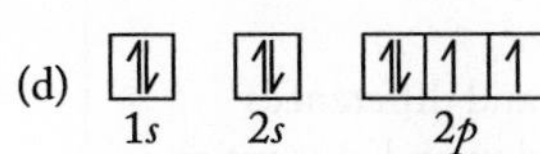

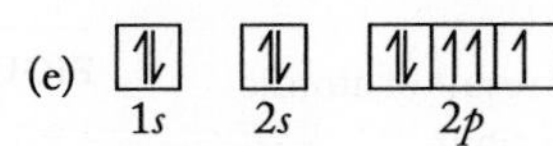

(e) 1s 2s 2p

FIGURE P7.8

7.9. Which light represented by the waves in Figure P7.9 has:
  a. the highest frequency?
  b. the lowest energy?
  c. the highest energy?

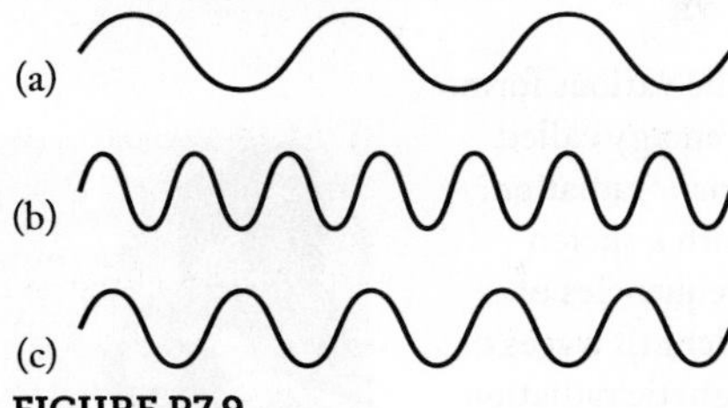

FIGURE P7.9

7.10. Use representations [A] through [I] in Figure P7.10 to answer questions (a)–(f).
   a. Compare [A] and [H]. Which valence electron experiences the larger Z_{eff}? Which is more shielded from the positive charge of the nucleus? Which has the lower IE_1?
   b. Which representation depicts the loss of an electron?
   c. Which representations can be discussed using only a quantized model of light?
   d. Which representations are consistent with the model of an atom that uses probability to locate electrons?
   e. What do the arrows represent in [E] and [I]?
   f. Which representation conveys that the probability of finding an electron at the nucleus is zero?

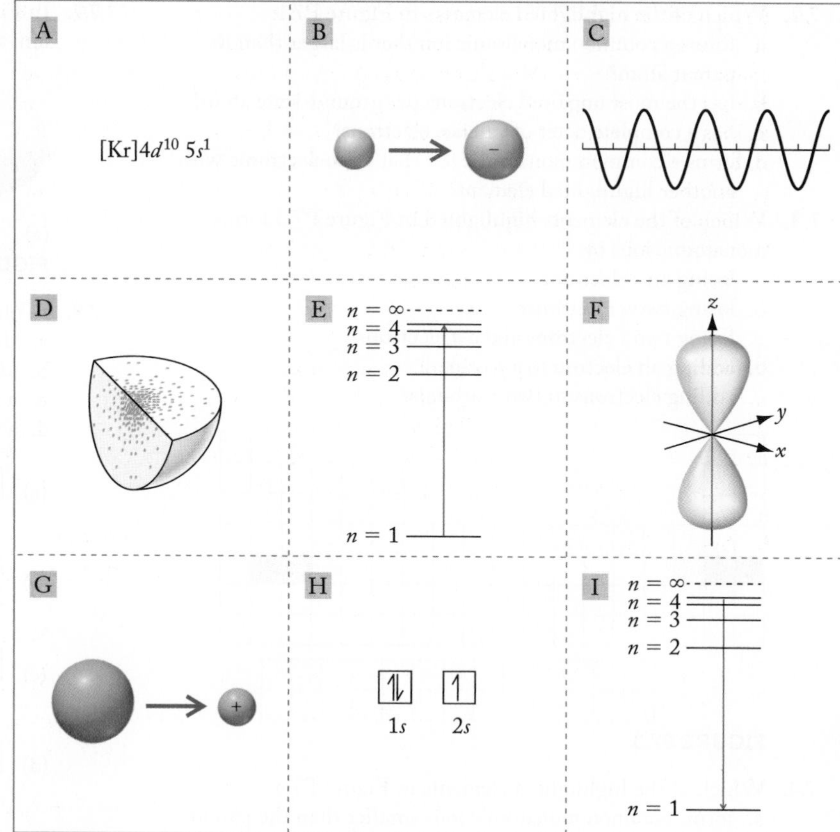

FIGURE P7.10

QUESTIONS AND PROBLEMS

Rainbows of Light; Waves of Energy

Concept Review

7.11. Describe the similarities and differences between the atomic emission and absorption spectra of hydrogen.

7.12. Are Fraunhofer lines the result of atomic emission or atomic absorption?

7.13. How did the study of the atomic emission spectra of elements lead to the identification of the Fraunhofer lines in sunlight?

*7.14. What would happen to the appearance of the Fraunhofer lines in the solar spectrum if sunlight were passed through a flame containing high-temperature calcium atoms and then analyzed?

7.15. Why are the various forms of radiant energy called *electromagnetic* radiation?

7.16. Explain with a sketch why the frequencies of long-wavelength waves of electromagnetic radiation are lower than those of short-wavelength waves.

7.17. **Dental X-rays** When X-ray images are taken of your teeth and gums in the dentist's office, your body is covered with a lead shield (Figure P7.17). Explain the need for this precaution.

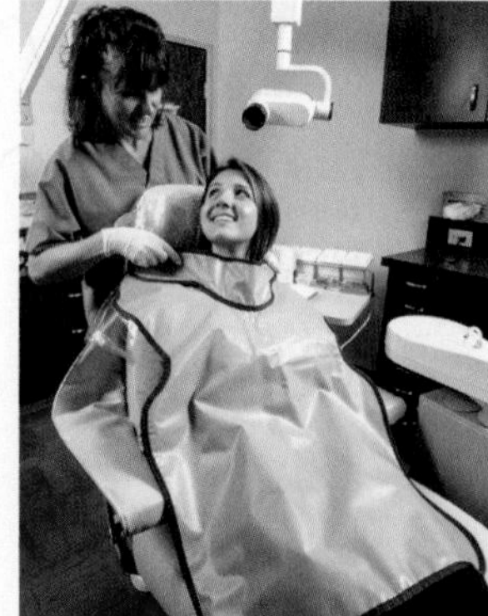

FIGURE P7.17

7.18. **UV Radiation and Skin Cancer** Ultraviolet radiation causes skin damage that may lead to cancer, but exposure to infrared radiation does not seem to cause skin cancer. Why do you think this is so?

7.19. Gamma rays are an example of "ionizing" radiation because they have the energy to break apart molecules into molecular ions and free electrons. What other forms of electromagnetic radiation could also be ionizing radiation?

*7.20. If light consists of waves, why don't things look "wavy" to us?

Problems

7.21. In each pair, which radiation has higher frequency? (a) ultraviolet or infrared radiation; (b) visible light or gamma rays; (c) microwaves or radio waves

7.22. In each pair, which radiation has longer wavelength? (a) visible light or microwaves; (b) radio waves or gamma rays; (c) infrared or ultraviolet radiation

---

7.23. If the wavelength of a photon of red light is twice that of a photon of ultraviolet radiation, how much more energy does the UV photon have?

7.24. If the frequency of a photon of red light is twice that of a photon of infrared radiation, how much more energy does the photon of red light have?

7.25. **Digital Screens** Tablets and cell phones emit blue light (Figure P7.25) that is known to disrupt sleep patterns when viewed before going to bed. The longest wavelength that corresponds to blue light is 495 nm. What is the frequency of this radiation?

FIGURE P7.25

7.26. **Submarine Communications** In the 1990s the Russian and American navies developed extremely low-frequency communications networks to send messages to submerged submarines. The frequency of the carrier wave of the Russian network was 82 Hz, whereas the Americans used 76 Hz.

a. What was the ratio of the wavelengths of the Russian network to the American network?

b. To calculate the actual underwater wavelength of the transmissions in either network, what additional information would you need?

7.27. **Drones and Phones** Remote control aerial vehicles such as the drone in Figure P7.27 typically operate at gigahertz (GHz) frequencies, whereas cell phones operate using megahertz (MHz) frequencies. Calculate the wavelengths corresponding to the following frequencies:

a. quadcopter drone, 2.4 GHz

b. Apple iPhone, 848 MHz

FIGURE P7.27

7.28. Which radiation has the longer wavelength: (a) radio waves from an AM radio station broadcasting at 680 kHz or (b) infrared radiation emitted by the surface of Earth ($\lambda = 15\ \mu m$)?

7.29. Which radiation has the lower frequency: (a) radio waves from an AM radio station broadcasting at 1030 kHz or (b) the red light ($\lambda = 633$ nm) from a helium–neon laser?

7.30. Which radiation has the higher frequency: (a) the red light on a bar-code reader at a grocery store or (b) the green light on the battery charger for a laptop computer?

7.31. **Speed of Light** How long does it take moonlight to reach Earth when the distance to the moon is 384,000 km?

7.32. **Exploration of the Solar System** How long would it take an instruction to a Martian rover to travel to Mars from a NASA site on Earth when Earth and Mars are 68.4 million km apart?

Particles of Energy and Quantum Theory

Concept Review

7.33. What is the difference between a quantum and a photon?

7.34. A variable power supply is connected to an incandescent light bulb. At the lowest power setting, the bulb feels warm to the touch but produces no light. At medium power, the light bulb filament emits a red glow. At the highest power, the light bulb emits white light. Explain this emission pattern.

***7.35.** What effect does the intensity (amplitude) of a wave have on the emission of electrons from a surface, if we assume that the frequency of incident radiation is above the threshold frequency?

7.36. Is the kinetic energy of a photoelectron quantized?

7.37. Which of the following have quantized values? Explain your selections.

a. the elevation of the treads of a moving escalator

b. the elevations at which the doors of an elevator open

c. the speed of an automobile

7.38. Which of the following have quantized values? Explain your selections.

a. the pitch of a note played on a slide trombone

b. the pitch of a note played on a flute

c. the wavelengths of light produced by the heating elements in a toaster

d. the wind speed at the top of Mt. Everest

Problems

7.39. **Tanning Booths** Prolonged exposure to ultraviolet radiation in tanning booths significantly increases the risk of skin cancer. What is the energy of a photon of UV light with a wavelength of 305 nm? Is this enough energy to break a C–C bond with a bond energy of 345 kJ/mol?

7.40. **Remote Controls** Most remote controllers for AV equipment emit radiation with the wavelength profile shown in Figure P7.40. What type of radiation is shown and what is the energy of a photon at the peak wavelength?

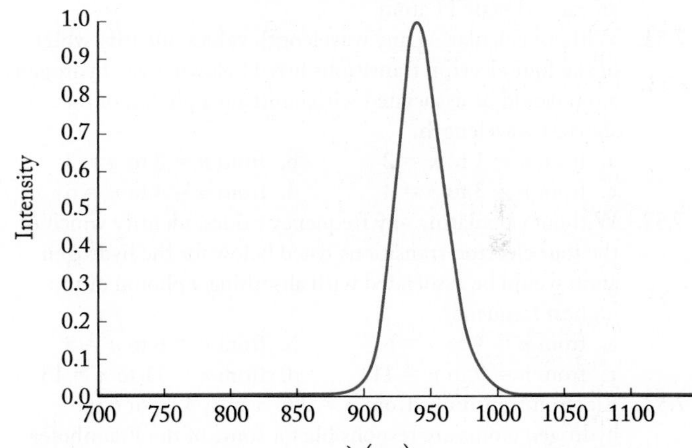

FIGURE P7.40

***7.41.** Thin layers of potassium ($\phi = 3.68 \times 10^{-19}$ J) and sodium ($\phi = 4.41 \times 10^{-19}$ J) are exposed to radiation of wavelength 300 nm. Which metal emits electrons with the greater velocity? What is the velocity of these electrons?

7.42. Titanium ($\phi = 6.94 \times 10^{-19}$ J) and silicon ($\phi = 7.24 \times 10^{-19}$ J) surfaces are irradiated with UV radiation with a wavelength of 250 nm. Which surface emits electrons with the longer wavelength? What is the wavelength of the electrons emitted by the titanium surface?

7.43. **Solar Power** Photovoltaic cells convert solar energy into electricity. Could tantalum ($\phi = 6.81 \times 10^{-19}$ J) be used to convert visible light to electricity? Assume that most of the electromagnetic energy from the Sun is in the visible region near 500 nm.

7.44. With reference to Problem 7.43, could tungsten ($\phi = 7.20 \times 10^{-19}$ J) be used to construct solar cells?

7.45. The power of a red laser ($\lambda = 630$ nm) is 1.00 watt (abbreviated W, where 1 W = 1 J/s). How many photons per second does the laser emit?

*7.46. The energy density of starlight in interstellar space is 10^{-15} J/m^3. If the average wavelength of starlight is 500 nm, what is the corresponding density of photons per cubic meter of space?

The Hydrogen Spectrum and the Bohr Model

Concept Review

7.47. Why should hydrogen have the simplest atomic spectrum of all the elements?

7.48. For an electron in a hydrogen atom, how is the value of n of its orbit related to its energy?

7.49. Does the electromagnetic energy emitted by an excited-state H atom depend on the individual values of n_1 and n_2, or only on the difference between them ($n_1 - n_2$)?

7.50. Explain the difference between a ground-state H atom and an excited-state H atom.

7.51. Without calculating any wavelength values, identify which of the four electron transitions listed below for the hydrogen atom would be associated with emitting a photon of the shortest wavelength.

a. from $n = 1$ to $n = 2$
b. from $n = 2$ to $n = 3$
c. from $n = 3$ to $n = 4$
d. from $n = 4$ to $n = 5$

7.52. Without calculating any frequency values, identify which of the four electron transitions listed below for the hydrogen atom would be associated with absorbing a photon of the highest frequency.

a. from $n = 4$ to $n = 6$
b. from $n = 6$ to $n = 8$
c. from $n = 9$ to $n = 11$
d. from $n = 11$ to $n = 13$

7.53. Electron transitions from $n = 2$ to $n = 3$, 4, 5, or 6 in hydrogen atoms are responsible for some of the Fraunhofer lines in the Sun's spectrum. Are there any Fraunhofer lines caused by transitions that start from $n = 3$ in hydrogen atoms?

7.54. In the visible portion of the atomic emission spectrum of hydrogen (Figure 7.4), are there any bright lines caused by electron transitions to the $n = 1$ state?

7.55. Balmer observed a hydrogen emission line for the transition from $n = 6$ to $n = 2$, but not for the transition from $n = 7$ to $n = 2$. Why?

*7.56. In what ways should the emission spectra of H and He$^+$ be alike, and in what ways should they be different?

Problems

7.57. Calculate the wavelength of the photons emitted by hydrogen atoms when they undergo transitions from $n = 4$ to $n = 3$. In which region of the electromagnetic spectrum does this radiation occur?

7.58. Calculate the frequency of the photons emitted by hydrogen atoms when they undergo transitions from $n = 5$ to $n = 3$. In which region of the electromagnetic spectrum does this radiation occur?

***7.59.** The energies of the photons emitted by one-electron atoms and ions fit the equation

$$E = (2.178 \times 10^{-18}\ \text{J})Z^2\left(\frac{1}{n_1^{\,2}} - \frac{1}{n_2^{\,2}}\right)$$

where Z is the atomic number, n_2 and n_1 are positive integers, and $n_2 > n_1$.

a. As the value of Z increases, does the wavelength of the photon associated with the transition from $n = 2$ to $n = 1$ increase or decrease?
b. Can the wavelength associated with the transition from $n = 2$ to $n = 1$ ever be observed in the visible region of the spectrum?

*7.60. Can transitions from higher energy states to the $n = 2$ level in He$^+$ ever produce visible light? If so, for what values of n_2? (*Hint*: The equation in Problem 7.59 may be useful.)

7.61. The transition from $n = 3$ to $n = 2$ in a hydrogen atom produces a photon with a wavelength of 656 nm. What is the wavelength of the transition from $n = 3$ to $n = 2$ in a Li^{2+} ion? (Use the equation in Problem 7.59.)

*7.62. The hydrogen atomic emission spectrum includes a UV line with a wavelength of 92.3 nm.

a. Is this line associated with a transition between different excited states or between an excited state and the ground state?
b. What is the value of n_1 of this transition?
c. What is the energy of the longest wavelength photon that a ground-state hydrogen atom can absorb?

Electron Waves

Concept Review

7.63. Identify the symbols in the de Broglie equation $\lambda = h/mu$, and explain how the relation links the properties of a particle to those of a wave.

7.64. How does de Broglie's hypothesis that electrons behave like waves explain the stability of the electron orbits in the Bohr model of the hydrogen atom?

7.65. Would the density of an object influence its de Broglie wavelength?

7.66. Would the shape of an object influence its de Broglie wavelength?

Problems

7.67. Calculate the wavelengths of the following objects:

a. a muon (a subatomic particle with a mass of 1.884×10^{-25} g) traveling at 325 m/s
b. electrons ($m_e = 9.10938 \times 10^{-28}$ g) moving at 4.05×10^6 m/s in an electron microscope
c. a 75-kg athlete running a 4-minute mile
d. Earth (mass = 6.0×10^{27} g) moving through space at 3.0×10^4 m/s

7.68. Two objects are moving at the same velocity. Which (if any) of the following statements about them is true?
a. The de Broglie wavelength of the heavier object is longer than that of the lighter one.
b. If one object has twice as much mass as the other, its wavelength is one-half the wavelength of the other.
c. Doubling the velocity of one of the objects will have the same effect on its wavelength as doubling its mass.

7.69. Which (if any) of the following statements about the frequency of a particle is true?
a. Heavy, fast-moving objects have lower frequencies than those of lighter, faster-moving objects.
b. Only very light particles can have high frequencies.
c. Doubling the mass of an object and halving its velocity results in no change in its frequency.

7.70. How rapidly would each of the following particles be moving if they all had the same wavelength as a photon of red light ($\lambda = 750$ nm)?
a. an electron of mass 9.10938×10^{-28} g
b. a proton of mass 1.67262×10^{-24} g
c. a neutron of mass 1.67493×10^{-24} g
d. an α particle of mass 6.64×10^{-24} g

***7.71.** Kinetic molecular theory tells us that helium atoms at 500 K are in constant random motion.
a. Calculate the root-mean-square speed of helium atoms at 500 K.
b. What is the wavelength of a helium atom at 500 K?

*7.72. Neon atoms are expected to move more slowly than helium atoms at the same temperature.
a. Using Graham's law of effusion and your answer from Problem 7.71, calculate the root-mean-square speed of neon atoms at 500 K.
b. What is the wavelength of a neon atom at 500 K?

7.73. **Particles in a Cyclotron** The first cyclotron was built in 1930 at the University of California, Berkeley, and was used to accelerate molecular ions of hydrogen, H_2^+, to a velocity of 4×10^6 m/s. (Modern cyclotrons can accelerate particles to nearly the speed of light.) If the uncertainty in the velocity of the H_2^+ ion was 3%, what was the uncertainty of its position?

7.74. **Radiation Therapy** An effective treatment for some cancerous tumors involves irradiation with "fast" neutrons. The neutrons from one treatment source have an average velocity of 3.1×10^7 m/s. If the velocities of individual neutrons are known to within 2% of this value, what is the uncertainty in the position of one of them?

Quantum Numbers and Electron Spin; The Sizes and Shapes of Atomic Orbitals

Concept Review

7.75. How does the concept of an orbit in the Bohr model of the hydrogen atom differ from the concept of an orbital in quantum theory?

7.76. What properties of an orbital are defined by each of the three quantum numbers n, ℓ, and m_ℓ?

7.77. Why does the size of an s orbital increase as the value of quantum number n increases?

7.78. Why do we need two quantum numbers to define a type of orbital when four are needed to describe an electron in an orbital?

Problems

7.79. How many orbitals are there in an atom with each of the following principal quantum numbers? (a) 1; (b) 2; (c) 3; (d) 4; (e) 5

7.80. How many orbitals are there in an atom with the following combinations of quantum numbers?
a. $n = 3, \ell = 2$
b. $n = 3, \ell = 1$
c. $n = 4, \ell = 2, m_\ell = 2$

7.81. What is the maximum number of electrons in the $n = 2$ shell of an atom?

7.82. What is the maximum number of electrons in an atom for which $n = 3$ and $\ell = 2$?

7.83. What set of orbitals corresponds to each of the following sets of quantum numbers? How many electrons could occupy these orbitals?
a. $n = 2, \ell = 0$
b. $n = 3, \ell = 1$
c. $n = 4, \ell = 2$
d. $n = 1, \ell = 0$

7.84. What set of orbitals corresponds to each of the following sets of quantum numbers? How many electrons could occupy these orbitals?
a. $n = 2, \ell = 1$
b. $n = 5, \ell = 3$
c. $n = 3, \ell = 2$
d. $n = 4, \ell = 3$

7.85. Which of the following combinations of quantum numbers are allowed?
a. $n = 1, \ell = 1, m_\ell = 0, m_s = +\frac{1}{2}$
b. $n = 3, \ell = 0, m_\ell = 0, m_s = -\frac{1}{2}$
c. $n = 1, \ell = 0, m_\ell = 1, m_s = -\frac{1}{2}$
d. $n = 2, \ell = 1, m_\ell = 2, m_s = +\frac{1}{2}$

7.86. Which of the following combinations of quantum numbers are allowed?
a. $n = 3, \ell = 2, m_\ell = 0, m_s = -\frac{1}{2}$
b. $n = 5, \ell = 4, m_\ell = 4, m_s = +\frac{1}{2}$
c. $n = 3, \ell = 0, m_\ell = 1, m_s = +\frac{1}{2}$
d. $n = 4, \ell = 4, m_\ell = 1, m_s = -\frac{1}{2}$

The Periodic Table and Filling the Orbitals of Multielectron Atoms; Electron Configurations of Ions

Concept Review

7.87. What is meant when two or more orbitals are said to be degenerate?

7.88. Explain how the electron configurations of the group 2 elements are linked to their location in the periodic table developed by Mendeleev (see Figure 2.11).

7.89. How do we know from examining the structure of the periodic table that the 4s orbital is filled before the 3d orbital?

7.90. Explain why so many transition metals form ions with a 2+ charge.

7.91. Why is there only one ground-state electron configuration for an atom but many excited-state electron configurations?

7.92. Can the ground-state electron configuration for an atom ever be an excited-state electron configuration for a different atom?

Problems

7.93. List the following orbitals in order of increasing energy in a multielectron atom:

a. $n = 3, \ell = 2$ b. $n = 5, \ell = 1$
c. $n = 3, \ell = 0$ d. $n = 4, \ell = 1$

7.94. Place the following orbitals in order of increasing energy in a multielectron atom:

a. $n = 2, \ell = 1$ b. $n = 5, \ell = 3$
c. $n = 3, \ell = 2$ d. $n = 4, \ell = 3$

7.95. What are the electron configurations of Li, Li^+, Ca, F^-, Na^+, Mg^{2+}, and Al^{3+}?

7.96. Which species listed in Problem 7.95 are isoelectronic with Ne?

7.97. In what way are the electron configurations of H, Li, Na, K, Rb, and Cs similar?

7.98. In what way are the electron configurations of C, Si, and Ge similar?

7.99. What are the condensed electron configurations of the following ground-state atoms and ions? How many unpaired electrons are there in each? (a) K; (b) K^+; (c) S^{2-}; (d) N; (e) Ba; (f) Ti^{4+}; (g) Al

7.100. What are the condensed electron configurations of the following ground-state atoms and ions? How many unpaired electrons are there in each? (a) O; (b) P^{3+}; (c) Na^+; (d) Sc; (e) Ag^+; (f) Cd^{2+}; (g) Zr^{2+}

7.101. Identify the element and the number of unpaired electrons in each of the following ground-state, condensed electron configurations of gas-phase atoms:

a. $[Ar]3d^24s^2$ b. $[Ar]3d^54s^1$
c. $[Ar]3d^{10}4s^1$ *d. $[Kr]4d^{10}$

7.102. Identify the ion and the number of unpaired electrons in each of the following ground-state, condensed electron configurations of monatomic cations:

a. [Ne] b. $[Kr]4d^{10}5s^2$
c. $[Ar]3d^{10}$ d. $1s^22s^22p^6$

7.103. Identify the element with the lowest atomic number that has two unpaired electrons in its ground state.

7.104. Identify the first-row transition metal that has no unpaired electrons in its ground state.

7.105. Does the ground-state electron configuration of Na^+ represent an excited-state electron configuration of Ne?

7.106. Which excited state of the hydrogen atom is likely to be of higher energy, $1s^12s^1$ or $1s^13s^1$?

7.107. Predict the charge of the monatomic ions formed by Al, N, Mg, and Cs.

7.108. Predict the charge of the monatomic ions formed by S, P, Zn, and I.

7.109. Which of the following electron configurations represent an excited state?

a. $[He]2s^12p^5$ b. $[Kr]4d^{10}5s^25p^1$
c. $[Ar]3d^{10}4s^24p^5$ d. $[Ne]3s^23p^24s^1$

7.110. Which of the following electron configurations represent an excited state?

a. $[Ne]3s^23p^1$ b. $[Ar]3d^{10}4s^14p^2$
c. $[Kr]4d^{10}5s^15p^1$ d. $[Ne]3s^23p^64s^1$

7.111. Boric acid, H_3BO_3, gives off a green color (Figure P7.111) when ignited.

a. Write ground-state electron configurations for B and O.
b. Assign oxidation numbers to each of the elements in boric acid. If the oxidation number reflects the ionic charge on the element, write ground-state electron configurations for each ion.

FIGURE P7.111

7.112. Introducing calcium chloride into a flame imparts an intense orange color (Figure P7.112).

a. Write ground-state electron configurations for Ca and Cl.
b. Calcium chloride contains calcium and chloride ions. Write ground-state electron configurations for each ion.

FIGURE P7.112

7.113. Magnesium metal burns in air forming solid magnesium oxide.

a. Write a balanced chemical equation describing this reaction.
b. Write the electron configurations of the atoms in magnesium metal and the ions in magnesium oxide.
c. Which element is oxidized, and which is reduced in this reaction?

7.114. Sodium metal reacts (sometimes explosively) with chlorine gas, forming solid sodium chloride.

a. Write a balanced chemical equation describing this reaction.
b. Write the electron configurations of the atoms in sodium metal and the ions in sodium chloride.
c. Which element is oxidized, and which is reduced in this reaction?

***7.115.** Appendix 3 lists the ground state of gas-phase palladium atoms as [Kr]$4d^{10}$. Which of the following is the *best* explanation for this observation?
a. The aufbau principle only applies to transition metal atoms in period 4.
b. Gas-phase palladium atoms are always in an excited state.
c. The energy of the 5*s* orbital must be higher than that of the 4*d* orbital for Pd.
d. Palladium is always present as a Pd^{2+} cation because of the photoelectric effect.

*7.116. The ground-state electron configuration in Appendix 3 for gas-phase silver ions is [Kr]$4d^{10}5s^1$. Which of the following statements best explains the electron configuration of silver *and* is consistent with the answer to Problem 7.115?
a. *nd* orbitals always fill before $(n + 1)s$ orbitals.
b. The energy of the 4*d* orbital is still lower in energy than the 5*s* orbital in Ag atoms.
c. The repulsion between electrons in the 4*d* orbital are greater than in the 5*s* orbital.
d. Gas-phase silver atoms are always in an excited state.

***7.117.** Removing a 1*s* electron from beryllium requires X-rays with wavelength 110.7 nm. Table 7.2 lists the first ionization energy of Be as 900 kJ/mol.
a. Why don't these data represent the same process?
b. Does the energy to remove the 1*s* electron correspond to the second ionization energy of Be, which is 1758 kJ/mol?

*7.118. The first ionization energies of group 1 elements decrease from top to bottom in the group. The energies to move an electron from $n = 1$ to $n = \infty$, however, increase from Li to Cs. Why are the trends opposite?

The Sizes of Atoms and Ions

Concept Review

7.119. Sodium atoms are much larger than chlorine atoms, but in NaCl sodium ions are much smaller than chloride ions. Why?

7.120. Why does atomic size tend to decrease with increasing atomic number across a row of the periodic table?

7.121. Which of the following group 1 elements has the largest atoms: Li, Na, K, or Rb? Explain your selection.

7.122. Which of the following group 17 elements has the largest monatomic ions: F, Cl, Br, or I? Explain your selection.

Problems

7.123. Using only the periodic table as a guide, arrange each set of particles by size, from largest to smallest:
a. Al, P, Cl, Ar
b. C, Si, Ge, Sn
c. Li^+, Li, Na, K
d. F, Ne, Cl, Cl^-

7.124. Using only the periodic table, arrange each set of particles by size, from largest to smallest:
a. Li, B, N, Ne
b. Mg, K, Ca, Sr
c. Rb^+, Sr^{2+}, Cs, Fr
d. S^{2-}, Cl^-, Ar, K^+

Ionization Energies

Concept Review

7.125. How do ionization energies change with increasing atomic number (a) down a group of elements in the periodic table and (b) from left to right across a period of elements?

7.126. The first ionization energies of the main group elements are given in Figure 7.39. Explain the differences in the first ionization energy between (a) He and Li; (b) Li and Be; (c) Be and B; (d) N and O.

7.127. Explain why it is more difficult to ionize a fluorine atom than a boron atom.

7.128. Do you expect the ionization energies of anions of group 17 elements to be lower or higher than for neutral atoms of the same group?

Problems

7.129. Which of the following elements should have the smallest *second* ionization energy? Br, Kr, Rb, Sr, Y

7.130. Why is the first ionization energy (IE_1) of Al ($Z = 13$) less than the IE_1 of Mg ($Z = 12$) *and* less than the IE_1 of Si ($Z = 14$)?

7.131. Without referring to Figure 7.39, arrange the following groups of elements in order of increasing first ionization energy.
a. F, Cl, Br, I
b. Li, Be, Na, Mg
c. N, O, F, Ne

7.132. Without referring to Figure 7.39, arrange the following groups of elements in order of increasing first ionization energy.
a. Mg, Ca, Sr, Ba
b. He, Ne, Ar, Kr
c. P, S, Cl, Ar

Electron Affinities

Concept Review

7.133. An electron affinity (EA) value that is negative indicates that the free atoms of an element are less stable than the 1− anions they form by acquiring electrons. Does this mean that all the elements with negative EA values exist in nature as anions? Give some examples to support your answer.

7.134. The electron affinities of the group 17 elements are all negative values, but the EA values of the group 18 noble gases are all positive. Explain this difference.

7.135. The electron affinities of the group 17 elements increase with increasing atomic number. Suggest a reason for this trend.

7.136. Ionization energies generally increase with increasing atomic number across the second row of the periodic table, but electron affinities generally decrease. Explain the opposing trends.

Additional Problems

7.137. Searching for Extraterrestrial Life Project Ozma was an early attempt to find any advanced civilizations elsewhere in our universe. Launched in 1959, scientists involved with the project decided to search space for a signal emitted by H atoms at 1240 MHz. They reasoned that other intelligent life might come to the same conclusion about a "universal" communication frequency.

a. Calculate the wavelength and energy of this light.
b. Does this frequency correspond to Balmer, Lyman, or Paschen lines in the emission spectrum of H?

7.138. **X-ray Absorption Spectroscopy** Electrons in multielectron atoms absorb X-rays at characteristic energies, leading to ionization. The characteristic energies for each element allow scientists to identify the element. For magnesium, X-rays with λ= 952 pm are required to selectively eject an electron from $n = 1$, but $\lambda = 197$ nm removes an electron from $n = 2$.

a. Calculate the frequency and energy of these X-rays.
b. Why are the wavelengths and energies different for the two electrons?
c. In terms of Bohr's values of n, what transitions do these represent?

7.139. Satellite Radio and Weather Radio Sirius XM radio transmits at 2.3 GHz using a network of satellites. The National Oceanic and Atmospheric Administration (NOAA) transmits warnings for tornados over weather radios at 162 MHz. Which of these signals uses photons of higher energy? Of shorter wavelength? Of higher frequency?

7.140. **Interstellar Hydrogen** Astronomers have detected hydrogen atoms in interstellar space in the $n = 732$ excited state. Suppose an atom in this excited state undergoes a transition from $n = 732$ to $n = 731$.

a. How much energy does the atom lose because of this transition?
b. What is the wavelength of radiation corresponding to this transition?
c. What kind of telescope would astronomers need to detect radiation of this wavelength? (*Hint*: It would not be one designed to capture visible light.)

***7.141.** When an atom absorbs an X-ray of sufficient energy, one of its 2*s* electrons may be emitted, creating a hole that can be spontaneously filled when an electron in a higher-energy orbital—a 2*p*, for example—falls into it. A photon of electromagnetic radiation with an energy that matches the energy lost in the $2p \rightarrow 2s$ transition is emitted. Predict how the wavelengths of $2p \rightarrow 2s$ photons would differ between (a) different elements in the fourth row of the periodic table and (b) different elements in the same column (e.g., between the noble gases from Ne to Rn).

7.142. Compare the orbital representations labeled (i), (ii), and (iii) in Figure P7.142.

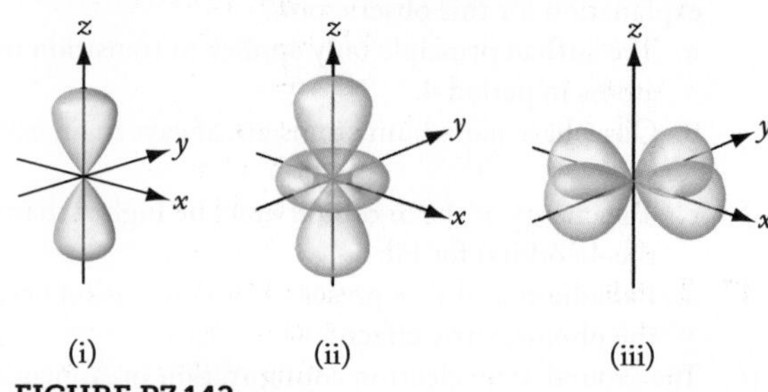

FIGURE P7.142

a. Which orbital has the lowest energy?
b. Which two orbitals have the same value for the quantum number ℓ?
c. What is the maximum number of electrons that can occupy (i)? (ii)? (iii)?

***7.143.** The spin quantum number, m_s, was first detected by passing a beam of silver atoms through a magnetic field. Why doesn't this experiment allow scientists to distinguish between the two most likely ground-state electron configurations for silver: $[Kr]4d^{10}5s^1$ and $[Kr]4d^9 5s^2$?

7.144. A green flame is observed when copper(II) chloride is heated in a flame (Figure P7.144).

a. Write the ground-state electron configuration for copper atoms.
b. There are two common forms of copper chloride, copper(I) chloride and copper(II) chloride. What is the difference in the ground-state electron configurations of copper in these two compounds?

FIGURE P7.144

***7.145.** Two helium ions (He^+) in the $n = 3$ excited state emit photons of radiation as they return to the ground state. One ion does so in a single transition from $n = 3$ to $n = 1$. The other does so in two steps: $n = 3$ to $n = 2$ and then $n = 2$ to $n = 1$. Which of the following statements about these two pathways is true?

a. The sum of the energies lost in the two-step process is the same as the energy lost in the single transition from $n = 3$ to $n = 1$.
b. The sum of the wavelengths of the two photons emitted in the two-step process is equal to the wavelength of the single photon emitted in the transition from $n = 3$ to $n = 1$.
c. The sum of the frequencies of the two photons emitted in the two-step process is equal to the frequency of the single photon emitted in the transition from $n = 3$ to $n = 1$.
d. The wavelength of the photon emitted by the He^+ ion in the $n = 3$ to $n = 1$ transition is shorter than the wavelength of a photon emitted by an H atom in an $n = 3$ to $n = 1$ transition.

*7.146. Use your knowledge of electron configurations to explain the following observations:
a. Silver tends to form ions with a charge of 1+, but the element to the right of silver in the periodic table tends to form ions with 2+ charges.
b. The heavier group 13 elements (Ga, In, Tl) tend to form ions with charges of 1+ or 3+ but not 2+.
c. The heavier elements of group 14 (Sn, Pb) and group 4 (Ti, Zr, Hf) tend to form ions with charges of 2+ or 4+.

7.147. Trends in ionization energies of the elements as a function of the position of the elements in the periodic table are a useful test of our understanding of electronic structure.
a. Should the same trend in the first ionization energies for elements with atomic numbers $Z = 31$ through $Z = 36$ be observed for the second ionization energies of the same elements? Explain why or why not.
b. Which element should have the greater second ionization energy: Rb ($Z = 37$) or Kr ($Z = 36$)? Why?

7.148. **Chemistry of Photo-Gray Glasses** "Photo-gray" lenses for eyeglasses darken in bright sunshine because the lenses contain tiny, transparent AgCl crystals. Exposure to light removes electrons from Cl^- ions, forming a chlorine atom in an excited state (indicated below by the asterisk):

$$Cl^- + h\nu \rightarrow Cl^* + e^-$$

The electrons are transferred to Ag^+ ions, forming silver metal:

$$Ag^+ + e^- \rightarrow Ag$$

Silver metal is reflective, giving rise to the photo-gray color.
a. Write a balanced chemical equation describing the overall chemical reaction.
b. Which elements are oxidized, and which are reduced?
c. Write condensed electron configurations of Cl^-, Cl, Ag, and Ag^+.
d. Would more energy be needed to remove an electron from a Br^- ion or from a Cl^- ion? Explain your answer.
*e. How might substitution of AgBr for AgCl affect the light sensitivity of photo-gray lenses?

7.149. The first ionization energy of a gas-phase atom of a particular element is 6.24×10^{-19} J. What is the maximum wavelength of electromagnetic radiation that could ionize this atom?

7.150. Tin (in group 14) forms both Sn^{2+} and Sn^{4+} ions, but magnesium (in group 2) forms only Mg^{2+} ions.
a. Write condensed ground-state electron configurations for the ions Sn^{2+}, Sn^{4+}, and Mg^{2+}.
b. Which neutral atoms have ground-state electron configurations identical to Sn^{2+} and Mg^{2+}?
c. Which 2^+ ion is isoelectronic with Sn^{4+}?

7.151. **Rare Earth Magnets** The rare earth elements, $Z = 57$–71, are used in the strong magnets found in many automobiles.
a. Predict the ground-state electron configurations for these 15 elements.
b. Using Table A3.1 in the appendix, which of these elements have electron configurations different from those predicted in part (a)?
*c. How might one rationalize the differences between the observed and predicted electron configurations for the elements identified in part (b)?
d. Can we distinguish between the observed and predicted electron configurations for the elements in part (b) by determining the number of unpaired electrons?

7.152. **Oxygen Ions in Space** Between 1999 and 2007 the Far Ultraviolet Spectroscopic Explorer satellite analyzed the spectra of emission sources within the Milky Way. Among the satellite's findings were interplanetary clouds containing oxygen atoms that have lost five electrons.
a. Write an electron configuration for these highly ionized oxygen atoms.
b. Which electrons have been removed from the neutral atoms?
c. The ionization energies corresponding to removal of the third, fourth, and fifth electrons are 4581 kJ/mol, 7465 kJ/mol, and 9391 kJ/mol, respectively. Explain why removal of each additional electron requires more energy than removal of the previous one.
d. What is the maximum wavelength of radiation that will remove the fifth electron from an O atom?

***7.153.** Effective nuclear charge (Z_{eff}) is related to atomic number (Z) by the shielding parameter (σ) according to the equation $Z_{eff} = Z - \sigma$.
a. Calculate Z_{eff} for the outermost s electrons of Ne and Ar given $\sigma = 4.24$ (for Ne) and 11.24 (for Ar).
b. Explain why the shielding parameter is much greater for Ar than for Ne.

7.154. **Fog Lamp Technology** Sodium fog lamps and street lamps contain gas-phase sodium atoms and sodium ions. Sodium atoms emit yellow-orange light at 589 nm. Do sodium ions emit the same yellow-orange light? Explain why or why not.

7.155. How can an electron get from the (+) lobe of a p orbital to the (−) lobe without going through the node between the lobes?

7.156. Einstein did not fully accept the uncertainty principle, remarking that "He [God] does not play dice." What do you think Einstein meant? Niels Bohr allegedly responded by saying, "Albert, stop telling God what to do." What do you think Bohr meant?

***7.157.** The wavelengths of Fraunhofer lines in galactic spectra are not the same as those in sunlight: they tend to be shifted to longer wavelengths (*redshifted*), in part because of the Doppler effect. The Doppler effect is described by the equation

$$\frac{(\nu - \nu')}{\nu} = \frac{u}{c}$$

where ν is the unshifted frequency, ν' is the perceived frequency, c is the speed of light, and u is the speed at which the object is moving. If hydrogen in a galaxy that is receding from Earth at half the speed of light emits radiation with a wavelength of 656 nm, will the radiation still be in the visible part of the electromagnetic spectrum when it reaches Earth?

7.158. The work function of mercury is 7.22×10^{-19} J.
a. What is the minimum frequency of radiation required to eject photoelectrons from a mercury surface?
b. Could visible light produce the photoelectric effect in mercury?

8

Chemical Bonds

What Makes a Gas a Greenhouse Gas?

RICE PADDIES AND CLIMATE CHANGE Microorganisms growing in flooded rice paddies release methane gas, which is a more potent greenhouse gas than carbon dioxide.

PARTICULATE **REVIEW**

Breaking and Making Bonds

In Chapter 8 we learn how chemists model and represent the formation of chemical bonds between atoms.

- When methane burns in air, oxygen is consumed, and carbon dioxide and water vapor are produced. Molecular models of the reactants and products are shown here. Write a balanced chemical equation describing this reaction.
- Which bonds and how many of them are broken during the combustion of 1 mole of methane?
- Which bonds and how many of them are formed during the combustion of 1 mole of methane?

(Review Section 3.3 if you need help.)

(Answers to Particulate Review questions are in the back of the book.)

PARTICULATE **PREVIEW**

Single, Double, or Triple Bond?

These space-filling models show an oxygen molecule, a fluorine molecule, and a nitrogen molecule. As you read Chapter 8, look for ideas that will help you answer these questions:

- One of the three molecules contains a single bond, one contains a double bond, and one contains a triple bond. Does a space-filling model tell you what kind of bond exists within a substance?
- What is the molecular formula for each substance? Does a molecular formula tell you what kind of bond exists within a substance?
- What information does a ball-and-stick model contain that a space-filling model does not?

Learning Outcomes

LO1 Describe ways in which covalent, ionic, and metallic bonds are alike and ways in which they differ
Sample Exercise 8.1

LO2 Draw Lewis structures of molecular compounds and ionic compounds
Sample Exercises 8.2, 8.3, 8.4, 8.5

LO3 Predict the polarity of covalent bonds based on differences in electronegativity between the bonded atoms
Sample Exercise 8.6

LO4 Explain how molecules of some atmospheric gases absorb infrared radiation and contribute to the greenhouse effect

LO5 Draw resonance structures and evaluate their contribution to the bonding in molecules and polyatomic ions using formal charges
Sample Exercises 8.7, 8.8, 8.9, 8.10, 8.11, 8.12

LO6 Describe how bond order, bond energy, and bond length are related, and estimate the enthalpy of a reaction (ΔH_{rxn}) from average bond energies
Sample Exercise 8.13

8.1 Types of Chemical Bonds and the Greenhouse Effect

In Chapter 7 we saw that quantum theory describes matter's interaction with energy at the atomic level. In this chapter we continue our exploration of matter and energy as we step from considering individual atoms to explaining how atoms are bonded to one another in molecules and how molecules interact with energy.

We begin with molecules of carbon dioxide in Earth's atmosphere. Most scientists agree that recent increases in the average temperature of Earth's surface (about half a Celsius degree in the last half century) are linked to increases in the concentration of carbon dioxide and other *greenhouse gases* in the atmosphere. These gases trap Earth's heat in its atmosphere, much like panes of glass trap heat in a greenhouse or in a car on a hot summer day (**Figure 8.1**). The amount of some greenhouse gases in the atmosphere may surprise you. Energy production, livestock animals, landfills, and wetlands introduce more than 0.5 teragram CH_4 (1 teragram = 1 Tg = 1 million metric tons = 10^{12} g) into the atmosphere *every* day. Increases in atmospheric concentrations of other greenhouse gases have also been linked to human activity, particularly to increasing rates of fossil fuel combustion and the destruction of forests that would otherwise consume CO_2 during photosynthesis.

FIGURE 8.1 The glass windows of a greenhouse allow sunlight to enter but trap the warm air inside. Greenhouse gases, such as CH_4 (methane) and CO_2 (carbon dioxide), behave in a similar fashion. Recent increases in atmospheric CO_2 concentrations have been linked to rising global temperatures and the threat of global climate change.

Were it not for the presence of atmospheric CO_2 and other greenhouse gases, Earth would be much colder than it is. Indeed, it would be too cold to be habitable. The *greenhouse effect* resulting from the presence of these gases has moderated Earth's average global temperature to within a narrow range, 15°C–17°C, for many thousands of years. However, during the last 60 years, atmospheric concentrations of CO_2 have increased from about 315 ppm to more than 400 ppm. The atmosphere has not contained this much CO_2 since more than half a million years ago, long before humans evolved.

Carbon dioxide allows solar radiation to pass through the atmosphere and warm Earth's surface, but it traps heat that would otherwise radiate from the warm surface back into space. Why is CO_2 so good at trapping heat? Heat is associated with the infrared region of the electromagnetic spectrum (Chapter 7). Carbon dioxide traps heat because it absorbs infrared radiation. To understand how CO_2 does this, whereas O_2 and N_2 do not, we need to learn more about the chemical bonds that hold atoms together in molecules.

Forming Bonds from Atoms

Fewer than 100 stable (nonradioactive) elements make up all the matter in our world. These elements combine to form more than 60 million compounds, and the number grows daily as chemists synthesize new ones. This multitude of compounds exists because they are more stable than their constituent elements. In other words, two atoms that are linked by a chemical bond tend to be lower in chemical energy than those same two atoms without a bond connecting them.

Why are bonded atoms more stable? The principal reason stems from the concept of electrostatic potential energy (E_{el}) between charged particles. Recall from Chapter 6 that E_{el} is directly proportional to the product of the charges (Q_1 and Q_2) on pairs of ions (or subatomic particles) and is inversely proportional to the distance (d) between them:

$$E_{el} \propto \frac{Q_1 \times Q_2}{d} \qquad (6.3)$$

When we apply Equation 6.3 to particles of opposite charge, such as a cation–anion pair, the product of Q_1 and Q_2 is negative and so is the value of E_{el}. The value of E_{el} becomes more negative as the particles approach each other and the distance (d) between them decreases (review Figure 6.8). As the oppositely charged pair of ions moves very close together, however, their attraction to each other is offset by electrostatic repulsions between the ion's outer-shell electrons and between their positive nuclei. At the energy minimum in Figure 6.8, the forces of attraction and repulsion are in balance, and the ions are held together by a stable **ionic bond**.

ionic bond a chemical bond that results from the electrostatic attraction between a cation and an anion.

CONNECTION Negative values of E_{el} quantify the attraction between particles of opposite charge; positive values relate to the repulsion of particles with the same charge, as described in Chapter 6.

This concept of electrostatic potential energy can also explain why two neutral atoms may come together to form a stable molecule. Let's consider the case of two hydrogen atoms (**Figure 8.2a**) that are sufficiently far apart that they do not interact with each other and have zero potential energy with respect to each other. As they approach each other (moving from right to left along the x axis of the graph), the proton in the nucleus of one H atom is attracted to the electron of the other, and vice versa. This mutual attraction produces a negative E_{el} in the (b) region of Figure 8.2. As the distance between the atoms decreases, the value of E_{el} continues

CHEMTOUR
Bonding

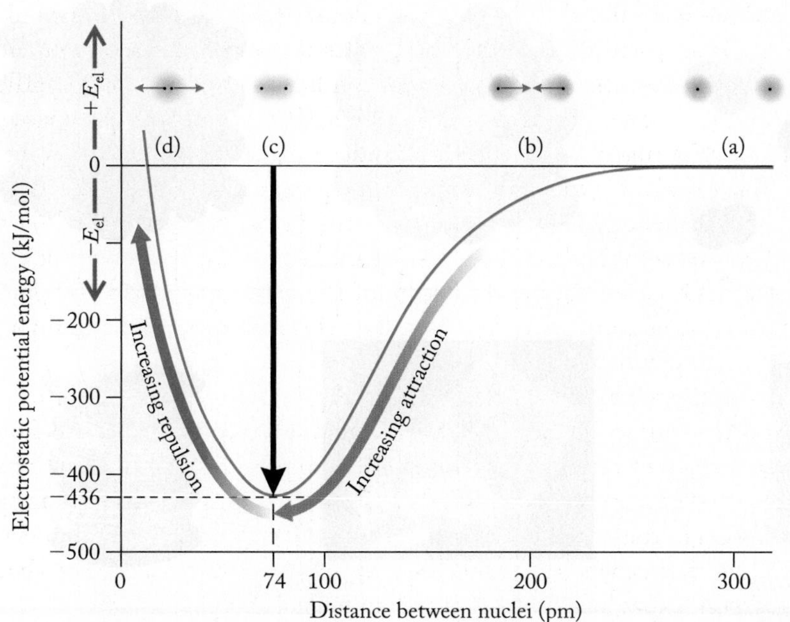

FIGURE 8.2 The electrostatic potential energy (E_{el}) profile of a H—H bond: (a) Two H atoms that are far apart do not interact, so $E_{el} = 0$. (b) As the atoms approach each other, mutual attraction between their positive nuclei and negative electrons causes E_{el} values to decrease. (c) At 74 pm, E_{el} reaches a minimum as the two atoms form a H—H bond. (d) If the atoms are even closer together, repulsion between their nuclei causes E_{el} to increase and destabilizes the bond.

bond length the distance between the nuclear centers of two atoms joined in a bond.

metallic bond a chemical bond consisting of metal atoms sharing a "sea" of electrons.

delocalized electrons electrons that are shared among more than two atoms.

to decrease, eventually reaching a minimum when the nuclei are 74 pm apart at point (c). At this optimal distance, the strengths of the attractions between the two nuclei and the two electrons are equal; they have formed a single covalent bond. They are no longer two separate H atoms but rather a more stable, single molecule of H_2. If the atoms move any closer together, the repulsion between their two positive nuclei more than offsets the mutual proton–electron attractions, so E_{el} rises rapidly with any further decrease in d in region (d) of Figure 8.2.

The distance between the atoms when E_{el} is at a minimum corresponds to the length of a H—H covalent bond; in this case, the H—H **bond length** is 74 pm. The value of E_{el} at this distance, −436 kJ/mol, is the energy *released* when 2 moles of hydrogen atoms bond together, forming 1 mole of H_2 molecules. Breaking all these bonds would require adding 436 kJ of energy to the molecules. This energy requirement is a measure of H—H bond energy, or *bond strength*. We explore the significance of bond length and bond strength in more detail in Section 8.7.

We can also use electrostatic potential energy to explain the interactions between the atoms in a solid piece of metal. As with atoms in molecules, the positive nucleus of each atom in a metallic solid is attracted to the electrons of the atoms that surround it. These attractions result in the formation of **metallic bonds**, which are unlike covalent bonds in that they are *not* a pair of electrons shared by two atoms. Instead, the shared electrons in metallic bonds are highly **delocalized**, forming a "sea" of mobile electrons that move freely among all the atoms in a metallic solid. This electron mobility makes metals good conductors of heat and electricity. We discuss metallic bonding in greater detail in Chapter 9.

TABLE 8.1 Types of Chemical Bonds

	Covalent	Ionic	Metallic
Elements Involved	Nonmetals, metalloids	Metals + nonmetals	Metals
Electron Distribution	Shared	Transferred	Delocalized
Particulate View	 Br Br	 Na^+ Cl^-	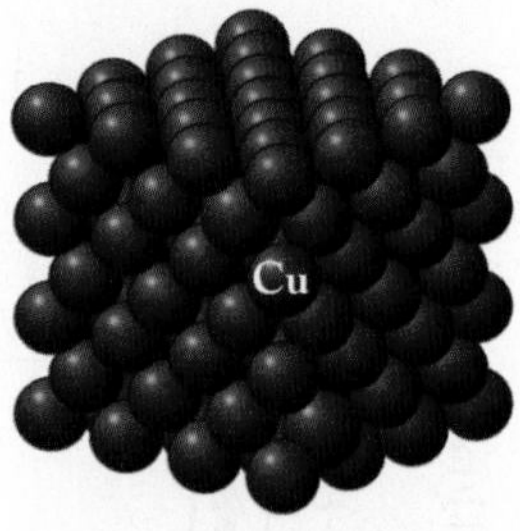 Cu
Macroscopic View			

Table 8.1 summarizes some of the differences among covalent, ionic, and metallic bonds. As you consider these properties, keep in mind that many bonds do not fall exclusively into just one category, but rather have some covalent, ionic, and even metallic character.

CONNECTION In Chapter 1 we described a chemical bond as the energy that holds two atoms in a molecule together. A covalent bond was defined in Chapter 2 as a bond between two atoms created by sharing pairs of electrons.

SAMPLE EXERCISE 8.1 Describing Similarities and Differences in Bonding LO1

Figure 8.3 shows particulate representations of barium sulfate, carbon dioxide, and potassium chloride. Identify each representation, name the type of bonding present in each compound, and determine what particles are attracted to one another when the electrostatic potential energy is at a minimum.

Collect and Organize We are given the names and representations of three compounds. We need to determine what kind of particles (atoms, molecules, or ions) are present in each compound and each representation to identify the type of bonding and describe the electrostatic attractions. Table 8.1 summarizes the three types of bonds.

Analyze Barium sulfate is an ionic compound: the *-ate* ending of its name indicates that it contains oxoanions of sulfur bonded to Ba^{2+} cations. Carbon dioxide is a binary compound of two nonmetals. The *di-* prefix means that it is composed of molecules that each contain one atom of carbon bonded to two atoms of oxygen. The *-ide* ending of potassium chloride conveys that it is an ionic compound of K^+ cations (potassium is a group 1 metallic element) and Cl^- anions (chlorine is a group 17 nonmetal).

Solve Representations (i) and (iii) both depict ionic compounds with 1:1 ratios of cations to anions that are electrostatically attracted to form ionic bonds. The polyatomic ions in (i) each contain one S atom bonded to 4 O atoms and are held together by covalent bonds. Their presence means that (i) must represent $BaSO_4$, and (iii) must represent KCl. Each of the molecules represented in (ii) consist of covalent bonds that are formed by electrostatic attractions between one central carbon atom and two oxygen atoms.

Think About It In addition to covalent electrostatic attractions between atoms and ionic electrostatic attractions between ions, ionic compounds that contain polyatomic ions have both covalent and ionic electrostatic interactions.

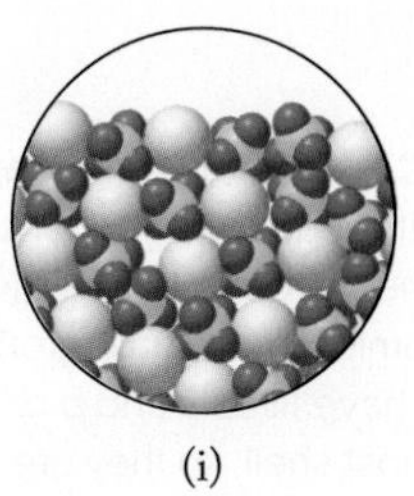

(i)

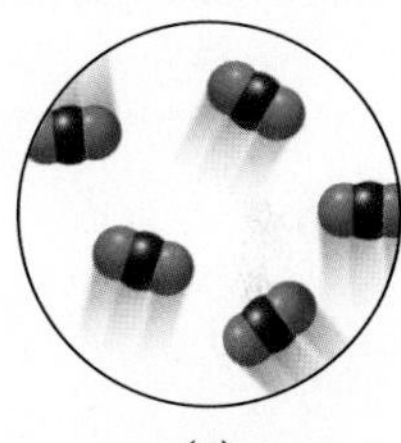

(ii)

(iii)

FIGURE 8.3 Particulate representations for Sample Exercise 8.1.

Practice Exercise **Figure 8.4** shows particulate representations of gold, calcium carbonate, and phosphorus. Identify each representation, name the type of bonding present in each substance, and determine which particles are attracted to one another.

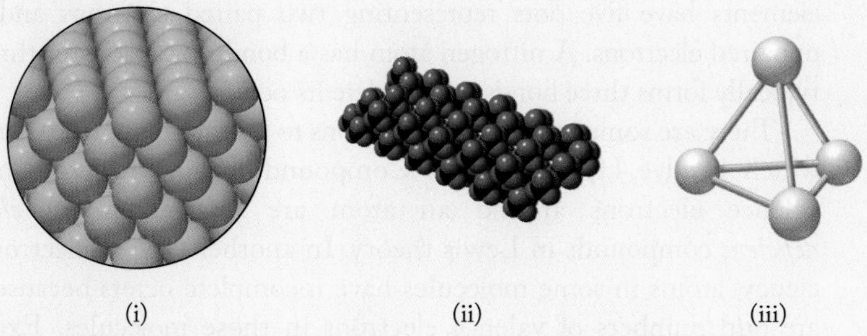

(i) (ii) (iii)

FIGURE 8.4 Particulate representations for Practice Exercise 8.1.

(Answers to Practice Exercises are in the back of the book.)

8.2 Lewis Structures

In 1916 American chemist Gilbert N. Lewis (1875–1946) proposed that atoms form chemical bonds by sharing electrons. He further suggested that through this sharing, each atom is surrounded by enough electrons to mimic the electron configuration of a noble gas. Today we associate such an electron configuration with filled *s* and *p* orbitals in the valence shells of atoms. Lewis's view of chemical bonding predated quantum mechanics and the notion of atomic orbitals, but it was consistent with what he called the **octet rule**: atoms of most main group elements lose, gain, or share electrons so that each atom has eight valence electrons (i.e., an *octet* of electrons). Hydrogen is an important exception to the octet rule—each of its atoms shares only a single pair of bonding electrons, thereby acquiring a "duet" instead of an octet. The space-filling models in Table 8.1 show the atoms involved in different types of bonding, but they do not show how their electrons are distributed. Lewis devised a way of representing individual valence electrons in atoms, to show how these electrons are distributed in various types of bonds.

CONNECTION The elements in groups 1, 2, and 13 through 18 in the periodic table are called main group elements (Chapter 2). The monatomic ions of main group elements (except H^+) have filled *s* and *p* orbitals in their outermost shell, so they are isoelectronic with a noble gas (Chapter 7).

Lewis Symbols

Lewis developed a system of symbols, called **Lewis symbols** or **Lewis dot symbols**, to represent the number of chemical bonds an atom typically forms to complete its octet. In other words, the Lewis symbol of an atom depicts its **bonding capacity**. A Lewis symbol consists of the symbol of an element surrounded by dots representing the valence electrons in one of its atoms. The dots are placed on the four sides of the symbol (top, bottom, right, and left). The order in which they are placed does not matter *if one dot is placed on each side before any dots are paired.* The number of *unpaired* dots in a Lewis symbol indicates the typical bonding capacity of the atom.

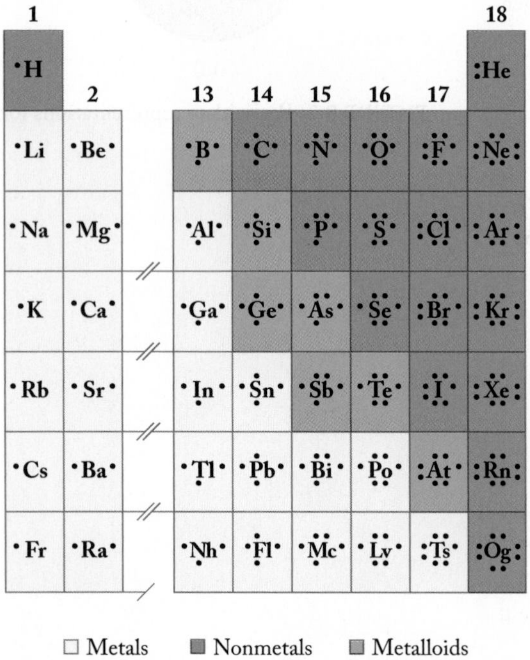

1	2	13	14	15	16	17	18
H							He
Li	Be	B	C	N	O	F	Ne
Na	Mg	Al	Si	P	S	Cl	Ar
K	Ca	Ga	Ge	As	Se	Br	Kr
Rb	Sr	In	Sn	Sb	Te	I	Xe
Cs	Ba	Tl	Pb	Bi	Po	At	Rn
Fr	Ra	Nh	Fl	Mc	Lv	Ts	Og

□ Metals ■ Nonmetals ■ Metalloids

FIGURE 8.5 Lewis symbols of the main group elements of the periodic table. Because elements in a family have similar outer-shell electron configurations, they have the same number of dots in their Lewis symbols, which represent valence electrons.

Figure 8.5 shows the Lewis symbols of the main group elements. Because all elements in a family have the same number of valence electrons, they all have the same arrangement of dots in their Lewis symbols. For example, the Lewis symbols of carbon and all other group 14 elements have four unpaired dots representing four unpaired electrons. Four unpaired electrons mean that a carbon atom has a bonding capacity of four: it typically forms four chemical bonds. In doing so, it completes its octet because these four bonds contain the carbon atom's original four valence electrons plus four more from the atoms with which it forms bonds. Similarly, the Lewis symbols of nitrogen and all other group 15 elements have five dots representing two paired electrons and three unpaired electrons. A nitrogen atom has a bonding capacity of three and typically forms three bonds to complete its octet.

There are some important exceptions to Lewis's octet rule, several of which involve H, Be, and B. Compounds having fewer than eight valence electrons around an atom are referred to as *electron-deficient* compounds in Lewis theory. In another type of electron deficiency, atoms in some molecules have incomplete octets because there are odd numbers of valence electrons in those molecules. Examples include NO and NO_2. We explore electron-deficient compounds and the chemical implications of the unpaired electrons in molecules in Section 8.6.

Before we start to use Lewis symbols to explore how elements form compounds, we need to keep in mind that the number of unpaired dots in the symbols indicates the number of bonds that the atoms of an element *typically* form. We see later in this chapter that atoms may exceed their bonding capacities in some molecules and polyatomic ions, and atoms fail to reach them in others. Bonding capacity is a useful concept, but treat it more as a guideline than a strict requirement.

Drawing Lewis Structures

The properties of molecular substances depend on which elements are included, the number of atoms of each element in each of their molecules, and how those atoms are bonded together. A **Lewis structure** is a two-dimensional representation of a molecule showing how the atoms in the molecule are connected. Because covalent bonds are *pairs* of shared valence electrons, Lewis structures focus on how these electron pairs, called **bonding pairs**, are distributed among the atoms in the molecule. In a Lewis structure, the pairs of electrons that are shared in chemical bonds are drawn with lines, as in the molecular structures we first saw in Chapter 1. A pair of electrons shared between two atoms is called a **single bond**. Electron pairs that are *not* involved in bond formation appear as pairs of dots on one atom. These unshared electron pairs are called **lone pairs**. Atoms with one or more lone pairs in their Lewis symbols frequently have that same number of lone pairs of electrons in the Lewis structures of the molecules they form.

The following guidelines describe a five-step approach to drawing Lewis structures. These initial guidelines are just a starting point in the discussion of molecular structure. We will continue to refine these guidelines as we draw structures for increasingly more challenging molecules and learn more about the distribution of electrons in bonds.

1. *Determine the number of valence electrons.* For a neutral molecule, count the valence electrons in all the atoms in the molecule. For a polyatomic ion, count the valence electrons and then add (for anions) or subtract (for cations) the number of electrons needed to account for the charge on the ion. This total number of valence electrons is used to form bonds and lone pairs in the Lewis structure.
2. *Arrange the symbols of the elements to show how the atoms are bonded and then connect them with single bonds (single pairs of bonding electrons).* Put the atom with the greatest bonding capacity in the center. (If two elements have the same bonding capacity, choose the one that is less electronegative, an atomic property we examine in Section 8.3.) Place the remaining atoms around the central atom and those that form the fewest bonds (such as hydrogen) around the periphery. Connect the atoms with single bonds. The resulting representation is often called a *skeletal structure*.
3. *Complete the octets of the atoms bonded to the central atom by adding lone pairs of electrons.* Place lone pairs on each of the outer atoms until each outer atom has an octet (including the two electrons used to connect it to the central atom in the skeletal structure). Remember that the same bonding electrons are used to complete the octets of both atoms they connect; in other words, bonding electrons count twice when completing octets, but only once when calculating the total number of valence electrons in a Lewis structure.
4. *Compare the number of valence electrons in the Lewis structure with the number determined in step 1.* If valence electrons remain unused in the structure, place lone pairs on the central atom (even if doing so means giving it more than an octet of valence electrons), until all the valence electrons counted in step 1 have been included in the Lewis structure.

octet rule the tendency of atoms of main group elements to make bonds by gaining, losing, or sharing electrons to achieve a valence shell containing eight electrons (i.e., four electron pairs).

Lewis symbol (also called **Lewis dot symbol**) a system in which the chemical symbol for an element is surrounded by one or more dots representing each of its atom's valence electrons.

bonding capacity the number of covalent bonds an atom forms to have an octet of electrons in its valence shell.

Lewis structure a two-dimensional representation of the bonds and lone pairs of valence electrons in a molecule or polyatomic ion.

bonding pair a pair of electrons shared between two atoms.

single bond a chemical bond that results when two atoms share one pair of electrons.

lone pair a pair of valence electrons that is not shared.

5. *Complete the octet on the central atom.* If there is an octet on the central atom, the structure is complete. If there is less than an octet on the central atom, create additional bonds to it by converting one or more lone pairs of electrons on outer atoms into bonding pairs.

CHEMTOUR
Lewis Structures

SAMPLE EXERCISE 8.2 Drawing the Lewis Structure of Chloroform LO2

Chloroform ($CHCl_3$) is a liquid with a low boiling point that was once used as an anesthetic in surgery. Draw its Lewis structure.

Collect and Organize We are given the molecular formula of chloroform, $CHCl_3$, and we can use the five-step approach described previously to generate the Lewis structure.

Analyze The formula $CHCl_3$ indicates that a chloroform molecule contains one carbon atom, one hydrogen atom, and three chlorine atoms. Because carbon is a group 14 element, it has four valence electrons and needs four more to complete its octet. Hydrogen, in group 1, has one valence electron in the 1*s* orbital and needs one more to complete its duet. Chlorine, in group 17, has seven valence electrons (in the 3*s* and 3*p* atomic orbitals) and needs one more to complete its octet.

Solve

1. The number of valence electrons in the $CHCl_3$ molecule is

Element:	C	+	H	+	3 Cl
Valence electrons:	4	+	1	+	(3 × 7) = 26

2. The carbon atom has the most (four) unpaired electrons in its Lewis symbol and hence has the greatest bonding capacity. Therefore, C is the central atom in the molecule. Each H atom and each Cl atom needs one more electron to achieve the electron configuration of a noble gas, and so each forms one bond to the C atom:

H
|
Cl—C—Cl
|
Cl

3. In this structure the hydrogen atom has a duet because it shares a bonding pair of electrons with the central carbon atom. Each chlorine atom has a duet, too, but each needs an octet, so add three lone pairs of electrons to each of the three chlorine atoms:

H
|
:C̤̈l—C—C̤̈l:
|
:C̤̈l:

4. This structure contains four pairs of bonding electrons and nine lone pairs, for a total of

$$(4 \times 2) + (9 \times 2) = 26 \text{ electrons}$$

which is the number of electrons determined in step 1.

5. The carbon atom is surrounded by four bonds, which means eight electrons, so it has a full octet. The structure is complete.

Think About It In this example there was no need to change the structure in steps 4 and 5 because all the valence electrons were already included and all the atoms had an octet (or duet in the case of the hydrogen atom).

Practice Exercise
Draw the Lewis structure of methane, CH_4.

SAMPLE EXERCISE 8.3 Drawing the Lewis Structure of Ammonia **LO2**

Draw the Lewis structure of ammonia, NH_3.

Collect, Organize, and Analyze The formula NH_3 indicates that each molecule contains one atom of nitrogen and three atoms of hydrogen. Nitrogen is a group 15 element with five valence electrons, three of which are unpaired, for a bonding capacity of 3. Each hydrogen atom has one valence electron in the 1*s* orbital and a bonding capacity of 1.

Solve

1. The number of valence electrons in the NH_3 molecule is

Element:	N	+	3 H
Valence electrons:	5	+	$(3 \times 1) = 8$

2. The nitrogen atom has the greater bonding capacity and is the central atom. (With a bonding capacity of 1, hydrogen can never be a central atom.) Connecting each H atom to the nitrogen atom with a covalent bond yields

$$\begin{array}{c} H—N—H \\ | \\ H \end{array}$$

3. Each bonded H atom has a complete duet of electrons.
4. The three bonds in the structure represent $3 \times 2 = 6$ valence electrons, but we need eight to match the number determined in step 1. We add the two remaining valence electrons as a lone pair on nitrogen:

$$\begin{array}{c} H—\ddot{N}—H \\ | \\ H \end{array}$$

5. The N atom now has an octet of electrons, so the Lewis structure is complete.

Think About It The Lewis symbol of the nitrogen atom contains two electrons that are paired and three that are unpaired. It is reasonable, therefore, that the nitrogen atom in NH_3 has one lone pair and three bonding pairs of electrons.

Practice Exercise
Draw the Lewis structure of phosphorus trichloride.

Lewis Structures of Molecules with Double and Triple Bonds

We can use Lewis structures to show the bonding in molecules in which two atoms share more than one pair of bonding electrons. A bond in which two atoms share *two* pairs of electrons is called a **double bond**. For example, when the two oxygen atoms in a molecule of O_2 share two pairs of electrons, they form an O═O double bond. Similarly, when the two nitrogen atoms in a molecule of N_2 share *three* pairs of electrons, they form a N≡N **triple bond**. In Section 8.7 we discuss the characteristics of these multiple bonds and compare them with those of single bonds.

How do we know when a Lewis structure has a double or triple bond? Typically, we find out when we apply steps 3 through 5 in the guidelines. Suppose we fill the octets of all the atoms attached to the central atom in step 3, and in doing so we use all the valence electrons available. If we discover that the central atom does *not* have an octet, we can provide the needed electrons in step 5 by converting

double bond a chemical bond in which two atoms share two pairs of electrons.

triple bond a chemical bond in which two atoms share three pairs of electrons.

aldehyde an organic compound having a carbonyl group with a single bond to a hydrogen atom and a single bond to another atom or group of atoms, designated as R– in the general formula RCHO.

carbonyl group a carbon atom with a double bond to an oxygen atom.

one or more lone pairs of electrons from one of the other atoms into a bonding pair. Drawing the Lewis structure for the organic molecule formaldehyde (H_2CO) illustrates one such application of the guidelines.

1. The total number of valence electrons is

Element:	C	+	2 H	+	O
Valence electrons:	4	+	(2 × 1)	+	6 = 12

2. Of the three elements, carbon has the greatest bonding capacity (4), so it is the central atom. Connecting it with single bonds to the other three atoms yields:

H—C—H
 |
 O

3. Each H atom has a single covalent bond (2 electrons) completing its valence shell. Oxygen needs three lone pairs of electrons to complete its octet:

H—C—H
 |
 :Ö:

4. There are 12 valence electrons in this structure, which matches the number determined in step 1.
5. The central C atom has only 6 electrons. To provide the carbon atom with the 2 additional electrons it needs to have an octet—without removing any electrons from oxygen, which already has an octet—we convert one of the lone pairs on the oxygen atom into a bonding pair between C and O:

H—C—H (with :Ö:) → $H_2C{=}O$ (with two lone pairs on O)

CONNECTION We have already seen a few families of organic molecules. We introduced carboxylic acids (RCOOH) and amines (RNH_2, R_2NH, R_3N) in Chapter 4. We introduced alkanes (compounds of carbon atoms and hydrogen atoms linked by single bonds) in Chapter 2 and explored them further in Chapter 6. Remember that R– stands for any organic group. Table 2.6 lists many of the functional groups in organic compounds.

It does not matter which of the three lone pairs we move because they are equivalent. The central carbon atom now has a complete octet, and the oxygen atom still does. This structure makes sense because the four covalent bonds around carbon—two single bonds and one double bond—match its bonding capacity. Similarly, the double bond to oxygen makes sense because oxygen is a group 16 element with a bonding capacity of 2, and it has two bonds in this structure.

Notice that we have drawn the double bond and the two single bonds on the central carbon atom at an angle of about 120° from each other in the final structure; we have done the same thing with the lone pairs of electrons on oxygen with respect to the double bond. Electrons are negatively charged and repel each other, so we draw them as far apart from each other as possible. This logic is an important part of drawing three-dimensional structures of molecules in Chapter 9.

Formaldehyde is the smallest member of a family of organic compounds known as **aldehydes**. All aldehydes contain the functional group that consists of a carbon atom with a single bond to a hydrogen atom and a double bond to an oxygen atom (**Figure 8.6**). The C=O group is called a **carbonyl group**. The fourth bond on the carbon atom connects to another hydrogen atom in formaldehyde, but in all other aldehydes it connects to another carbon-containing group, giving aldehydes a general formula of RCHO. It is not necessary for you to know how to name these compounds, but it is important for you to recognize the aldehyde functional group.

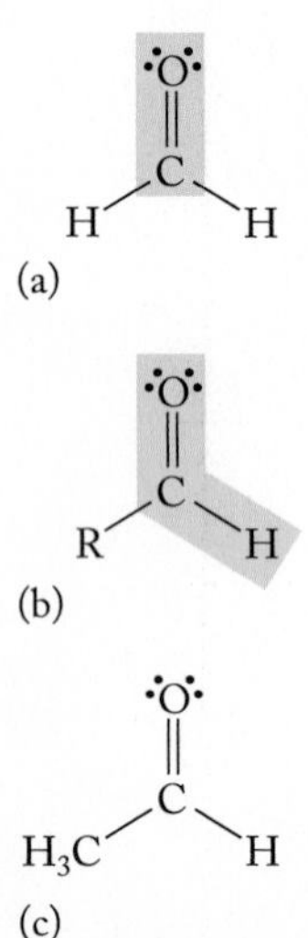

FIGURE 8.6 (a) Formaldehyde is the smallest member of the aldehyde family. The carbonyl group is highlighted. (b) All aldehydes contain a carbonyl group with a single bond to a hydrogen atom; R– represents any other organic group. The aldehyde group is highlighted. (c) Acetaldehyde is the next largest aldehyde; in acetaldehyde, R is CH_3.

The five-step guidelines for drawing Lewis structures are particularly useful when molecules or polyatomic ions have a central atom bonded to atoms with lower bonding capacity. Many inorganic molecules, polyatomic ions, and small organic molecules (fewer than four carbon atoms) fit this description. To draw some larger Lewis structures, however, we may sometimes need to break the structure into subunits and consider more than one "central" atom, as in hydrogen peroxide:

$$\mathrm{H{-}\ddot{\underset{\cdot\cdot}{O}}{-}\ddot{\underset{\cdot\cdot}{O}}{-}H}$$

In these cases, we treat each atom that is bonded to two other atoms as the "central atom" within a three-atom subunit, as shown by the oxygen atoms highlighted in red:

$$\mathrm{{-}\ddot{\underset{\cdot\cdot}{O}}{-}\ddot{\underset{\cdot\cdot}{O}}{-}H}$$

$$\mathrm{H{-}\ddot{\underset{\cdot\cdot}{O}}{-}\ddot{\underset{\cdot\cdot}{O}}{-}}$$

SAMPLE EXERCISE 8.4 Drawing the Lewis Structure of Acetylene **LO2**

Draw the Lewis structure of acetylene, C_2H_2, the hydrocarbon fuel used in oxyacetylene torches for welding and cutting metal.

Collect, Organize, and Analyze We are asked to draw the Lewis structure of acetylene, which has the molecular formula C_2H_2. We follow the five steps in the guidelines. Each molecule contains two atoms of carbon and two atoms of hydrogen. Carbon is a group 14 element with a bonding capacity of 4.

Solve

1. The total number of valence electrons is

Element:	2 C	+	2 H
Valence electrons:	(2×4)	+	$(2 \times 1) = 10$

2. Of the two elements, carbon has the greater bonding capacity, so the two carbon atoms are placed in the center of the molecule. Connecting each carbon atom with single bonds to the other atoms, we have:

$$\mathrm{H{-}C{-}C{-}H}$$

3. Each H atom has a single covalent bond and a complete valence shell.
4. So far, the current structure has 6 valence electrons, but it should have 10. This means we must add two pairs of electrons. We could add them as a lone pair on each of the two carbon atoms, or we could add them as a two bonding pairs between the two carbon atoms. The second option has the advantage of completing the octets of both carbon atoms (step 5), so that is the one we should use:

$$\mathrm{H{-}C{\equiv}C{-}H}$$

Think About It This structure makes sense because carbon has a bonding capacity of 4, and there are four covalent bonds around each carbon atom—one single bond and one triple bond—giving both atoms complete octets. Acetylene's triple bond makes it the smallest member of the *alkyne* family (see Section 2.8).

 **Practice Exercise**

Determine the Lewis structure for carbon dioxide.

Lewis Structures of Ionic Compounds

Crystals of sodium chloride are held together by the attraction between oppositely charged Na^+ and Cl^- ions. As described in Chapter 7, atoms of sodium and the other group 1 elements achieve noble gas electron configurations by losing their valence-shell *s* electron, forming 1+ cations, whereas chlorine and the other group 17 nonmetals gain one electron as they form anions with 1− charges, as shown below. Therefore, the ions of group 1 (and group 2) elements have no valence-shell electrons, and their Lewis structures have no dots around them. On the other hand, nonmetals such as chlorine, which acquire electrons to achieve noble gas electron configurations, have filled *s* and *p* orbitals in their valence shells and complete octets:

$$\text{Na}\cdot + \cdot\ddot{\underset{\cdot\cdot}{\text{Cl}}}\text{:} \rightarrow \text{Na}^+\left[\text{:}\ddot{\underset{\cdot\cdot}{\text{Cl}}}\text{:}\right]^-$$

The brackets indicate that all eight valence electrons are associated with the anion, and the charge of the ion is placed outside the bracket to indicate the overall charge of everything inside.

CONNECTION The ionization energy of an element (Section 7.11) is the amount of energy required to remove 1 mole of electrons from 1 mole of atoms in the gas phase.

All alkali metal elements and alkaline earth elements tend to lose rather than share their valence electrons when bonding with nonmetals, in part because these two families of metals have low ionization energies. The cations they form by losing these electrons obey the octet rule because they have the electron configurations of the noble gases that precede the parent elements in the periodic table.

SAMPLE EXERCISE 8.5 Drawing Lewis Structures of Binary Ionic Compounds — LO2

Draw the Lewis structure of calcium oxide.

Collect, Organize, and Analyze The *-ide* ending of calcium oxide tells us that it is a binary compound containing a metallic element and a nonmetal, which means it is an ionic compound. We learned in Chapter 7 that atoms of Ca and the other group 2 elements tend to form 2+ cations by losing their valence shell *s* electrons and that O atoms, like those of other group 16 elements, form monatomic ions by gaining 2 electrons.

Solve Formation of the calcium and oxide ions from their parent atoms leads to these Lewis symbols for the ions:

$$\cdot\text{Ca}\cdot \xrightarrow{-2\,e^-} \text{Ca}^{2+}$$

$$\text{:}\ddot{\underset{\cdot}{\text{O}}}\cdot \xrightarrow{+2\,e^-} \left[\text{:}\ddot{\underset{\cdot\cdot}{\text{O}}}\text{:}\right]^{2-}$$

Combining the Lewis symbols of Ca^{2+} and O^{2-} ions yields this Lewis structure of CaO:

$$\text{Ca}^{2+}\left[\text{:}\ddot{\underset{\cdot\cdot}{\text{O}}}\text{:}\right]^{2-}$$

Think About It The lack of dots in the symbol of the Ca^{2+} ion reinforces the fact that atoms of Ca, like all main group metal atoms, lose all the electrons in their valence shells when they form monatomic cations. In contrast, atoms of nonmetals form monatomic anions with complete octets.

Practice Exercise Draw the Lewis structure of magnesium fluoride.

polar covalent bond a bond characterized by unequal sharing of bonding pairs of electrons between atoms.

bond polarity a measure of the extent to which bonding electrons are unequally shared.

electronegativity a relative measure of the ability of an atom to attract electrons in a bond to itself.

8.3 Polar Covalent Bonds

When Lewis proposed that atoms form chemical bonds by sharing electrons, he knew that electron sharing in covalent bonds did not necessarily mean *equal* sharing. For example, Lewis knew that molecules of HCl ionized to form H^+ ions and Cl^- ions when HCl dissolves in water. To explain this phenomenon, Lewis proposed that the bonding pair of electrons in a molecule of HCl is closer to the chlorine end of the bond than to the hydrogen end. This unequal sharing makes the H—Cl bond a **polar covalent bond**. As a result, when the H—Cl bond breaks, the one shared pair of electrons remains with the Cl atom to form a Cl^- ion, simultaneously changing the H atom into an H^+ ion with no electrons.

This **bond polarity** in HCl means that the bond functions as a tiny electric *dipole*—that is, there is a slightly positive pole at the H end of the bond and a slightly negative pole at the Cl end, analogous to the positive and negative ends of a battery (**Figure 8.7**). Figure 8.7 also shows two representations we use to depict unequal sharing of bonding pairs of electrons. The arrow with a plus sign embedded in its tail (+→) indicates the *direction of polarity*: the arrow points toward the more negative, electron-rich atom in the bond, and the position of the plus sign indicates the more positive, electron-poor atom. The other way to represent unequal sharing makes use of the lowercase Greek delta (δ) followed by a + or − sign. The deltas represent *partial* electrical charges, as opposed to the full electrical charges that accompany the complete transfer of one or more electrons when atoms become ions.

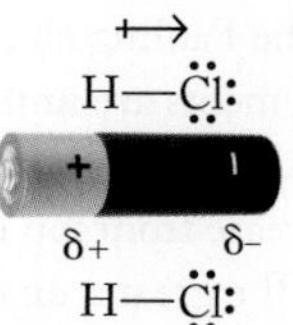

FIGURE 8.7 Just as a battery has positive and negative terminals, a polar bond such as the one in a molecule of HCl has positive and negative ends. The polarity of a bond is indicated by the arrow with a positive tail above the bond or by delta symbols (δ+ and δ−).

Figure 8.8 shows examples of the equal and unequal sharing of bonding pairs of electrons. In part (a), the two Cl atoms of Cl_2 share a pair of electrons equally in a nonpolar covalent bond; in part (b), HCl has a polar covalent bond; and in part (c), the ionic bond in NaCl results from electron transfer. In these images, the degree of charge separation is represented using color: yellow-green represents no charge separation (nonpolar covalent bonding), whereas colors near the violet end of the visible spectrum represent low electron density and those toward the red end represent high electron densities.

The American chemist Linus Pauling (1901–1994) developed the concept of **electronegativity** to explain bond polarity. Pauling assigned electronegativity

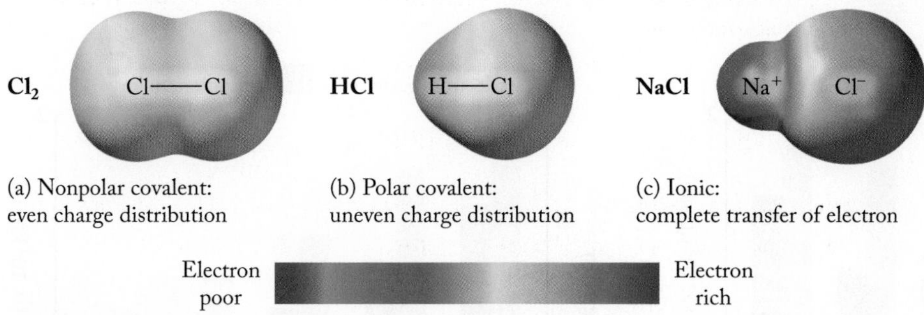

FIGURE 8.8 Variations in valence electron distribution are represented using colored surfaces in these molecular models. (a) In the covalent bond in Cl_2, the same color pattern on both ends of the bond means that the two atoms share their bonding pair of electrons equally. (b) Unequal sharing of the bonding pair of electrons in HCl is shown by the orange-red color of the Cl atom and the blue-green color around the H atom. (c) In NaCl, which is ionic, the violet color on the surface of the sodium ion indicates that it has a 1+ charge, whereas the deep red of the chloride ion reflects its 1− charge.

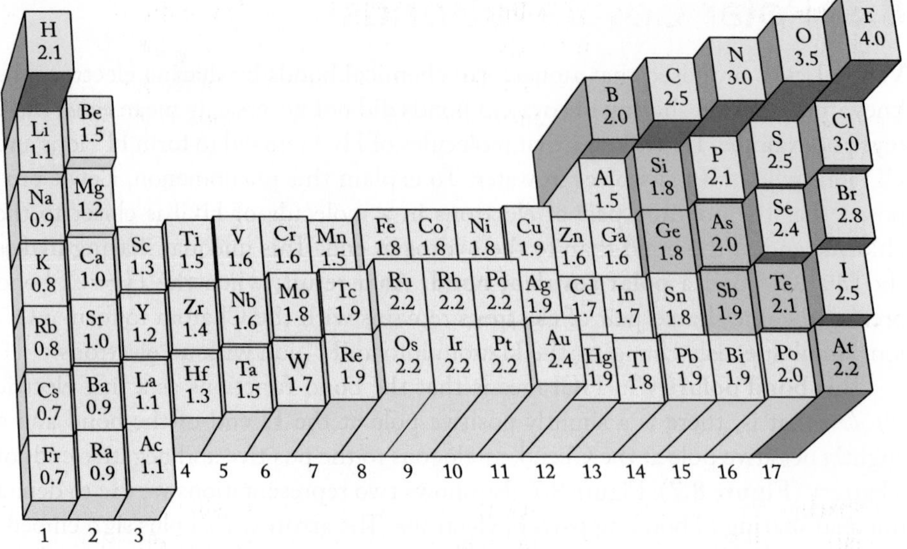

FIGURE 8.9 The Pauling electronegativity values of the elements are unitless numbers that increase from left to right across a period and decrease from top to bottom down a group. The greater an element's electronegativity, the greater the ability of an atom to attract electrons toward itself within a chemical bond.

values to the elements (**Figure 8.9**) based on the idea that the bonds between atoms of different elements are neither 100% covalent nor 100% ionic, but somewhere in between. The degree of **ionic character** of a bond depends on the differences in the abilities of the two atoms to attract the electrons they share: the greater the difference, the more ionic is the bond between them.

Electronegativity is a periodic property, with values generally increasing from left to right across a row in the periodic table and decreasing from top to bottom down a group. The reasons for these trends are essentially the same ones that produce similar trends in first ionization energies (see Figure 7.39). Increasing attraction between the nuclei of atoms and their outer-shell electrons as the atomic number increases across a row produces both higher ionization energies and greater electronegativities (**Figure 8.10a**). Within a group of elements, the weaker attraction between nuclei and outer-shell electrons with increasing atomic number leads to lower ionization energies and smaller electronegativities (**Figure 8.10b**). For these two reasons, the most electronegative elements—fluorine, oxygen, and nitrogen—are in the upper right corner of the periodic table, whereas the least electronegative elements—francium, cesium, and rubidium—are in the lower left corner. Other electronegativity scales have been developed with slightly different

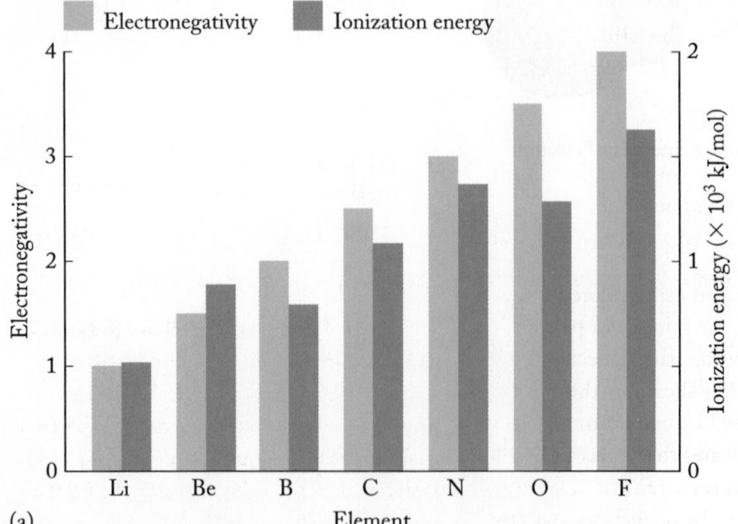

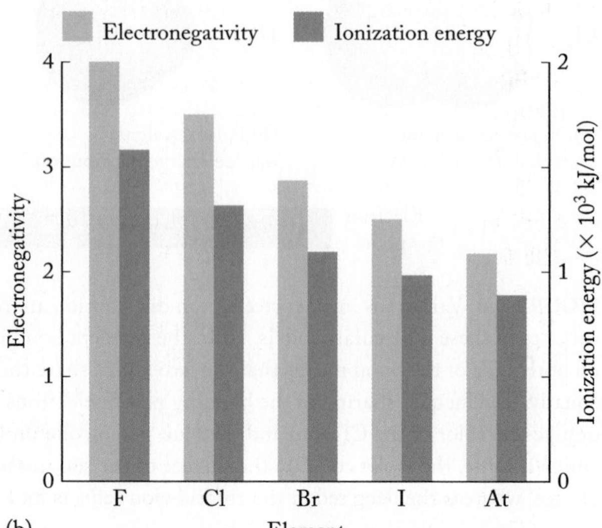

FIGURE 8.10 The trends in the electronegativities of the main group elements follow those of first ionization energies: both tend to (a) increase with increasing atomic number across a row and (b) decrease with increasing atomic number within a group.

values from those proposed by Pauling; regardless of the scale you use, though, the trends in electronegativity remain the same as those shown in Figure 8.10.

Having defined electronegativity, let's elaborate on step 2 of the guidelines for drawing Lewis structures: *The central atom in a Lewis structure is often the atom with the lowest electronegativity.*

ionic character an estimate of the magnitude of charge separation in a covalent bond.

CONCEPT **TEST**

Draw arrows on the Lewis structure of CO_2 to indicate the polarity of the bonds.

(Answers to Concept Tests are in the back of the book.)

Polarity and Type of Bond

Comparing electronegativity values allows us to determine which end of a bond is relatively electron-rich and which end is relatively electron-poor. The greater the difference in electronegativity (ΔEN), the more uneven the distribution of electrons and the more polar the covalent bond. **Figure 8.11** shows this trend for compounds formed between hydrogen and the halogens. For example, a H—F bond is more polar than a H—Cl bond because the ΔEN between H (2.1) and F (4.0) is 1.9, whereas the ΔEN between H (2.1) and Cl (3.0) is 0.9.

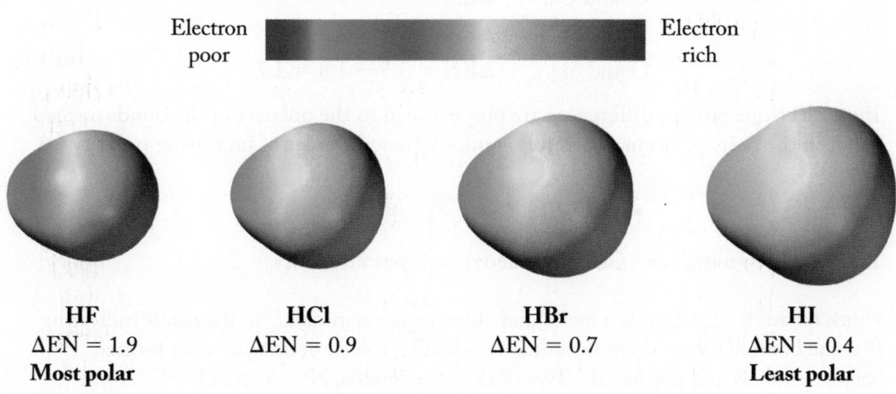

FIGURE 8.11 The greater the electronegativity difference (ΔEN) between any two atoms, the more polar the bond. Among the hydrogen halides, HF has the largest ΔEN and HI has the smallest. Therefore, the HF bond is the most polar in this group, and the HI bond is the least polar.

Figure 8.12 provides an approximate scale for judging the degree of polarity in a chemical bond based on electronegativity differences. We saw in Figure 8.8a that a Cl—Cl bond is nonpolar covalent because the same atom on both ends of the bond means that ΔEN = 0. On the other end of the scale, electronegativity differences greater than about 2.0 are associated with the ionic compounds formed between metals and nonmetals. For example, calcium oxide is an ionic compound, and ΔEN = 2.5 for Ca (1.0) and O (3.5). Polar covalent bonds fall between nonpolar covalent and ionic bonds.

The scale in Figure 8.12 should be used judiciously because the correlation between ΔEN and bond character is far from perfect. Consider, for example, the bonds in LiCl and HF. The ΔEN values for both pairs of elements is 1.9, but the physical properties of LiCl (it is a crystalline solid that melts at 605°C) is typical of an ionic compound, whereas HF is a gas at room temperature as are nearly all molecular compounds with its small (20 g/mol) molar mass.

The presence of a double or triple bond, such as the carbon–oxygen double bond in the carbonyl group (C═O), does *not* change our assessment of the overall polarity of the bond. Oxygen is more electronegative than carbon, so the

CHEMTOUR
Bond Polarity and Polar Molecules

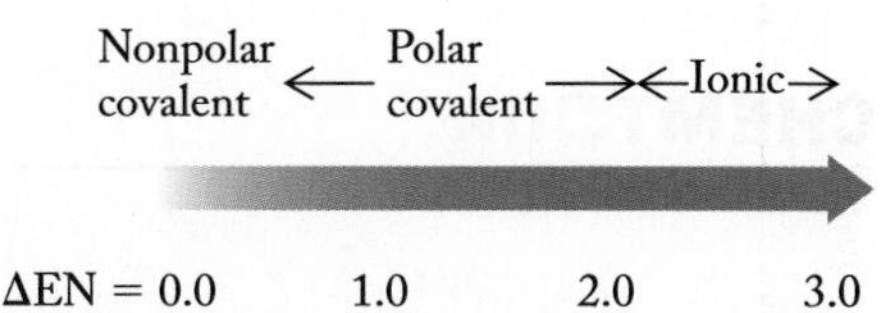

FIGURE 8.12 Electronegativity differences (ΔEN) near zero are associated with nonpolar covalent bonds. ΔEN values from near zero to about 2.0 are associated with polar covalent bonds. Bonds between atoms with ΔEN values greater than 2.0 are considered to be ionic.

carbonyl group is polar, with the carbon atom bearing a partial positive charge and the oxygen a partial negative charge, as shown here in formaldehyde:

$$H_2C{=}O \quad (\delta- \text{ on O}, \ \delta+ \text{ on C})$$

SAMPLE EXERCISE 8.6 Comparing the Polarity of Bonds **LO3**

Rank, in order of increasing polarity, the bonds formed between: O and C, Cl and Ca, N and S, and O and Si. Are any of these bonds ionic?

Collect and Organize We are given four pairs of atoms and asked to rank them according to the polarity of the bond each pair forms and to identify any ionic bonds in the set.

Analyze The polarity of a bond is related to the difference in electronegativities of the atoms in the bond. If the electronegativity difference is 2.0 or greater, the bond is considered ionic. We need to refer to the Pauling electronegativities (Figure 8.9) to judge the relative polarity.

Solve Calculate the electronegativity difference between the atoms:

O and C:	$\Delta EN = 3.5 - 2.5 = 1.0$
Cl and Ca:	$\Delta EN = 3.0 - 1.0 = 2.0$
N and S:	$\Delta EN = 3.0 - 2.5 = 0.5$
O and Si:	$\Delta EN = 3.5 - 1.8 = 1.7$

These electronegativity differences are proportional to the polarity of the bonds formed between the pairs of atoms. Therefore, ranking them in order of increasing polarity we have:

$$N{—}S < O{—}C < O{—}Si < Cl{—}Ca$$

The bond between Cl and Ca is considered ionic because $\Delta EN = 2.0$.

Think About It Calcium is a metal and chlorine is a nonmetal, so the result indicating that the bond between them is ionic is reasonable. Ionic bonds tend to be formed between metals and nonmetals. Two of the other bonds, N—S and O—C, connect pairs of nonmetals, and the O—Si bond connects a nonmetal with a metalloid. We expect these three bonds to be covalent.

Practice Exercise Which of the following pairs forms the most polar bond: O and S; Be and Cl; N and H; C and Br? Which, if any, of the pairs would form ionic bonds?

Vibrating Bonds and Greenhouse Gases

CHEMTOUR Vibrational Modes

Chemical bonds are not rigid (see the discussion of thermal energy in Chapter 6). They vibrate a little, stretching and bending like tiny atomic-sized springs (**Figure 8.13**). These vibrations have natural frequencies, which match the frequencies of infrared electromagnetic radiation. This match, coupled with the unequal sharing of bonding electrons, allows some atmospheric molecules containing polar covalent bonds—such as CO_2—to absorb infrared radiation emitted by Earth's surface. As a result, heat that might have dissipated into space is trapped in the atmosphere by these molecules, contributing to the atmospheric greenhouse effect (see Section 8.1). Understanding the energetics of bond vibrations goes beyond the scope of this book, but the basic ideas presented here

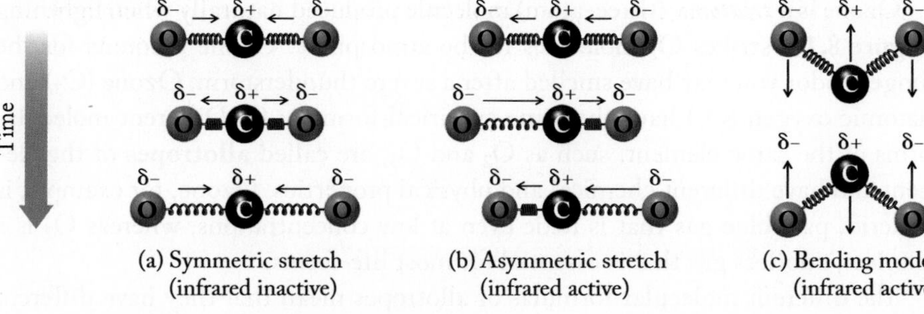

FIGURE 8.13 The arrows in these images show three of the ways in which the bonds in a molecule of CO_2 vibrate: (a) symmetric stretching of the C=O bonds, which produces no overall change in the polarity of the molecule; (b) asymmetric stretching, which produces side-to-side fluctuations in polarity that may result in absorption of infrared radiation; (c) bending, which produces up-and-down fluctuations that also may absorb infrared radiation.

provide a useful picture of the interaction between bond vibrations and infrared radiation.

Molecules with polar bonds may absorb photons, much like electrons in atoms absorb radiation and form excited states (see Section 7.4). The radiation absorbed by polar covalent bonds that result in vibrations is typically in the infrared region of the electromagnetic spectrum (see Figure 7.7). These vibrations result in tiny fluctuating electrical fields associated with the separation of partial charges in the polar covalent bonds in the molecule. These fluctuations can alter the strengths of the fields or even create new ones, depending on the nature of the vibration. When the frequencies of the polar covalent bond vibrations match the frequencies of photons of infrared radiation, the molecules may absorb those photons. Scientists say these vibrations are *infrared active*. This is the molecular mechanism behind the greenhouse effect. The presence of fluctuating electrical fields in molecules is further evidence that covalent bonds involve shared electrons.

CHEMTOUR
Greenhouse Effect

Not all polar bond vibrations result in absorption of infrared radiation. For example, two kinds of stretching vibrations can occur in a molecule of CO_2, which has two C=O bonds. One is a *symmetric* stretching vibration (Figure 8.13a) in which the two C=O bonds stretch and then compress at the same time. In this case the two fluctuating electrical fields produced by the two C=O bonds cancel each other out, and no infrared absorption or emission is possible. This vibration is said to be *infrared inactive*. However, when the bonds stretch such that one gets shorter as the other gets longer (Figure 8.13b), the changes in charge separation do *not* cancel. This *asymmetric* stretch produces a fluctuating electrical field that enables CO_2 to absorb infrared radiation, so this vibration is infrared active. Molecules can also bend (Figure 8.13c) producing fluctuating electrical fields. Because the frequencies of the asymmetric stretching and bending of the bonds in CO_2 are in the same range as the frequencies of infrared radiation emitted from Earth's surface, carbon dioxide is a potent greenhouse gas.

CONCEPT **TEST**

Nitrogen and oxygen make up about 99% of the gases in the atmosphere. Could the stretching of the N≡N and O=O bonds in these molecules result in the absorption of infrared radiation? Explain why or why not.

8.4 Resonance

The atmosphere contains two types of molecular oxygen. Most of it is O_2, but trace concentrations of ozone (O_3) are also present. Ozone in the lower atmosphere can damage crops, harm trees, and lead to human health problems. Ozone present in the upper atmosphere, on the other hand, shields life on Earth from potentially harmful ultraviolet radiation from the Sun.

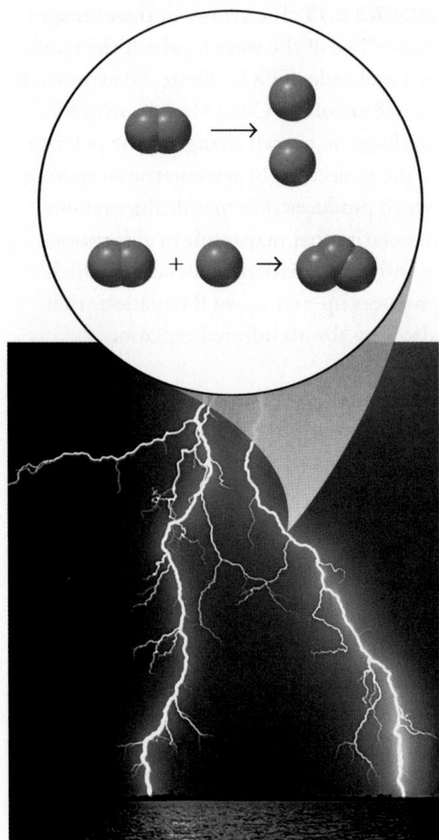

FIGURE 8.14 Lightning strikes contain sufficient energy to break O=O double bonds. The O atoms formed in this fashion collide with other O_2 molecules, forming ozone (O_3), an allotrope of oxygen.

Ozone is a *triatomic* (three-atom) molecule produced naturally when lightning (**Figure 8.14**) strikes O_2 molecules in the atmosphere. Ozone accounts for the pungent odor you may have smelled after a severe thunderstorm. Ozone (O_3) and diatomic oxygen (O_2) have the same empirical formula: O. Different molecular forms of the same element, such as O_2 and O_3, are called **allotropes** of the element and have different chemical and physical properties. Ozone, for example, is an acrid, pale blue gas that is toxic even at low concentrations, whereas O_2 is a colorless, odorless gas that is essential for most life-forms.

The different molecular formulas of allotropes mean that they have different molecular structures. Let's draw the Lewis structure for ozone (O_3) using our five-step approach. Oxygen is a group 16 element, so it has 6 valence electrons. The total number of valence electrons in an ozone molecule, then, is $3 \times 6 = 18$ (step 1). Connecting the three O atoms with single bonds (step 2), we have

O—O—O

Completing the octets of the two outer atoms (step 3) gives

:Ö—O—Ö:

This structure contains 16 electrons. We determined in step 1 that there are 18 valence electrons in the molecule, so we add a lone pair to the central oxygen atom (step 4):

:Ö—Ö—Ö:

This structure leaves the central atom two electrons short of an octet, so we convert one of the lone pairs on one of the outer O atoms (e.g., the one on the left end of the molecule) into a bonding pair (step 5):

Ö=Ö—Ö:

We could just as well have used a lone pair from the O atom on the right, which would have given us

:Ö—Ö=Ö

Which structure is correct? Interestingly, experimental evidence indicates that *neither* structure is accurate. Scientists have determined that both bonds in ozone have the same length. Because a double bond between two atoms is always shorter than a single bond between the same two atoms (see Section 8.7), the structure of ozone cannot consist of one single bond and one double bond. **Figure 8.15** shows that the length of the two bonds in O_3 (128 pm) is in between the length of an O—O single bond (148 pm) and an O=O double bond (121 pm). One way to explain this result is to assume that the bonding pattern of O_3 is an average of the two structures drawn previously:

O⋯O⋯O

In other words, each pair of oxygen atoms is held together by the equivalent of 1.5 bonds. However, this average does *not* mean that the molecule spends half its

allotropes different molecular forms of an element.

resonance a characteristic of electron distributions when two or more equivalent Lewis structures can be drawn for one compound.

resonance structure one of two or more Lewis structures with the same arrangement of atoms but different arrangements of bonding pairs of electrons.

time as the left-hand structure in Figure 8.15 and half as the right-hand structure. It always has three bonding pairs spread out evenly on the two sides of the central atom.

To better understand this bonding pattern, consider what happens when the bonding electrons and the lone pair electrons in structure (a) below are rearranged as shown by the red arrows:

(a) → (b)

This rearrangement produces structure (b). The process is completely reversible, so that the electron pairs in structure (b) could just as easily be rearranged into the pattern in structure (a):

(b) → (a)

The two structures are equivalent. We use a double-headed reaction arrow between the structures to indicate that they are equivalent and that the actual structure is a blend of the two:

(a) ⟷ (b)

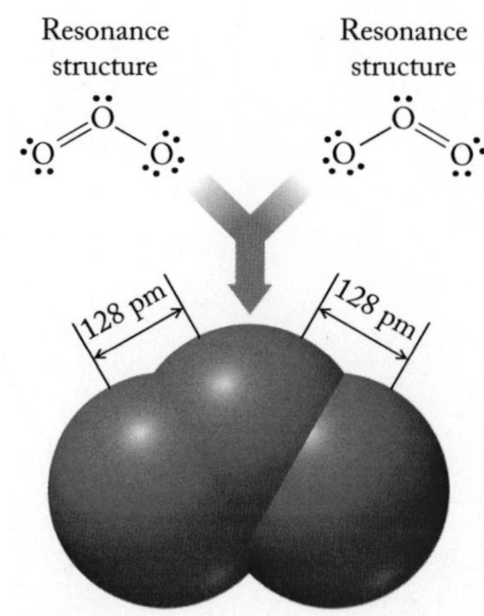

FIGURE 8.15 The molecular structure of ozone is an average of the two resonance structures shown at the top of the figure. Both bonds in ozone are 128 pm long, a value between the average length of an O—O single bond (148 pm) and the average length of an O═O double bond (121 pm). The intermediate value for the ozone bond length indicates that in an ozone molecule, the bonds are neither single bonds nor double bonds, but something in between.

CHEMTOUR
Resonance

The ability to draw two equivalent Lewis structures for the ozone molecule illustrates an important concept in Lewis's theory called **resonance**: the existence of multiple Lewis structures, called **resonance structures** (or *resonance hybrids*), for a given bonding arrangement of atoms. To determine whether resonance occurs in a molecule, we need to determine whether there can be alternative bond arrangements inside the molecule—that is, arrangements in which the positions of some electron pairs change but the positions of the atoms stay the same. One indicator of the possibility of resonance is the presence of both single and double bonds from a central atom to two or more atoms of the same element, as we just saw in ozone.

Because all resonance structures are Lewis structures by definition, the five-step guidelines still apply to drawing resonance structures. Because all the atoms in ozone are oxygen, an oxygen atom must be the central atom. Two additional issues, however, must be raised at this point, both of which relate to the concept of bonding capacity. First, oxygen atoms *typically* form two bonds, but Lewis theory allows both more and fewer bonds than are implied by the Lewis symbol of an atom. Remember that the resonance structures for ozone suggest that the oxygen atoms are held together by the equivalent of 1.5 bonds. Second, in previous examples we have used differences in typical bonding capacity as a criterion to select a central atom. As we deal with more molecules, the situation frequently arises that several atoms in one molecule may have the same bonding capacity. In that case, the selection of the central atom is based on electronegativity, with the least electronegative atom chosen as the central atom in the Lewis structure. Sample Exercise 8.7 explores the Lewis structure of just such a compound.

SAMPLE EXERCISE 8.7 Drawing Resonance Structures of a Molecule LO5

Sulfur trioxide (SO_3) is a pollutant produced in the atmosphere when the SO_2 from natural and industrial sources combines with O_2. Draw all the possible resonance structures of SO_3, assuming sulfur obeys the octet rule.

Collect and Organize Each SO_3 molecule contains one atom of sulfur and three atoms of oxygen. We need to draw all resonance structures, which means at least two different Lewis structures with the same atom placement but different bonding patterns.

Analyze Both sulfur and oxygen are in group 16, so each contributes 6 valence electrons, and each has a typical bonding capacity of 2. The electronegativity of sulfur (2.5) is less than that of oxygen (3.5).

Solve

1. The number of valence electrons in the SO_3 molecule is

Element:	S	+	3 O
Valence electrons:	6	+	$(3 \times 6) = 24$

2. Because sulfur is less electronegative, we select it as the central atom. Connecting it with single bonds to the three O atoms:

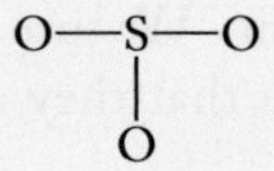

O—S—O
|
O

3. Each O atom gets three lone pairs of electrons to complete its octet:

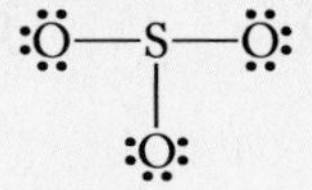

:Ö—S—Ö:
|
:Ö:

4. There are 24 valence electrons in the structure in step 3, which matches the number determined in step 1.
5. The central S atom has only 6 electrons in the structure of step 3. To complete its octet, we convert a lone pair on one of the oxygen atoms into a bonding pair:

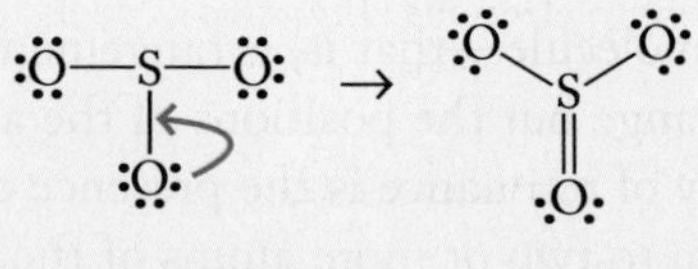

:Ö—S—Ö: → :Ö—S—Ö: (with S=O)

The sulfur atom now has a complete octet, and the Lewis structure is complete.

Because the three oxygen atoms are equivalent in the structure in step 3, we cannot arbitrarily choose one of them and ignore the others in forming a double bond. Therefore, three resonance structures describe the bonding in SO_3:

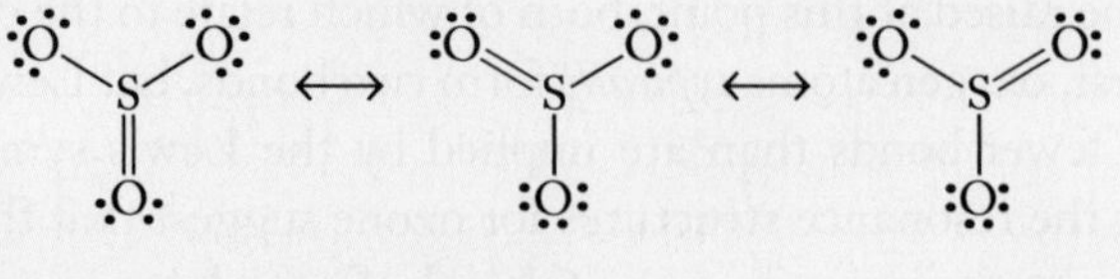

:Ö—S(=O)—Ö: ⟷ O=S(—Ö:)—Ö: ⟷ :Ö—S(=O)—Ö:

Think About It It makes sense that SO_3 has three resonance structures because three equivalent O atoms are bonded to the central S atom, and any one of the O atoms could be the one with the double bond.

Practice Exercise

Draw all possible resonance structures for sulfur dioxide.

We can also draw resonance structures for polyatomic ions. In doing so, we need to account for the charge on each ion, which means adding the appropriate number of valence electrons to a polyatomic anion and subtracting the appropriate number from a polyatomic cation.

SAMPLE EXERCISE 8.8 Drawing Resonance Structures of a Polyatomic Ion **LO5**

Draw all the resonance structures for the nitrate ion (NO_3^-).

Collect and Organize We are asked to draw the resonance structures for the NO_3^- ion, which contains one nitrogen atom and three oxygen atoms. The charge of the ion is 1−.

Analyze Nitrogen is a group 15 element, so it has 5 valence electrons per atom and a bonding capacity of 3. Oxygen is in group 16, so it has 6 valence electrons and a bonding capacity of 2. The 1− charge means the polyatomic ion has one additional valence electron.

Solve

1. The number of valence electrons is

Element:	N	+	3 O
Valence electrons:	5	+	(3 × 6) = 23
Additional electron from the 1− charge:			+ 1
Total valence electrons:			24

2. The nitrogen atom has the higher bonding capacity and the lower electronegativity, so N is the central atom in a NO_3^- ion. Connecting N with single bonds to the three O atoms gives:

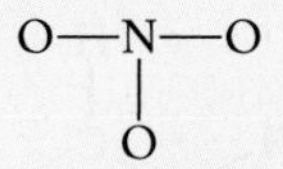

O—N—O (with a third O single-bonded below N)

3. Each O atom needs three lone pairs of electrons to complete its octet:

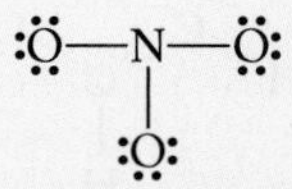

:Ö—N—Ö: (with :Ö: single-bonded below N)

4. There are 24 valence electrons in this structure, which matches the number determined in step 1.
5. The central N atom has only 6 electrons. To provide it with 2 more to complete its octet, we convert a lone pair on one of the oxygen atoms into a bonding pair:

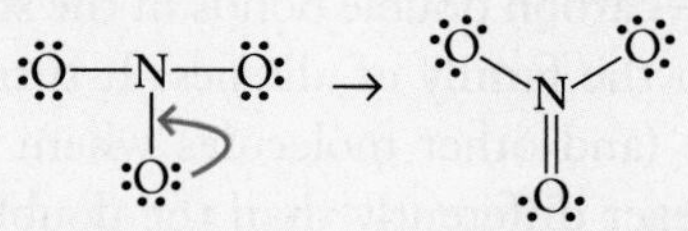

:Ö—N—Ö: (with :Ö: below N) → structure with N=O double bond to the lower O

The nitrogen atom now has a complete octet. Adding brackets and the ionic charge, we have a complete Lewis structure:

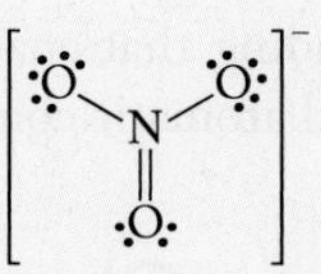

$[\ \ ]^-$ (N single-bonded to two O atoms with three lone pairs each, double-bonded to the lower O with two lone pairs)

Using the O atom to the left or right of the central N atom to form the double bond, instead of the O atom below N, creates two additional resonance structures (i.e., three in all):

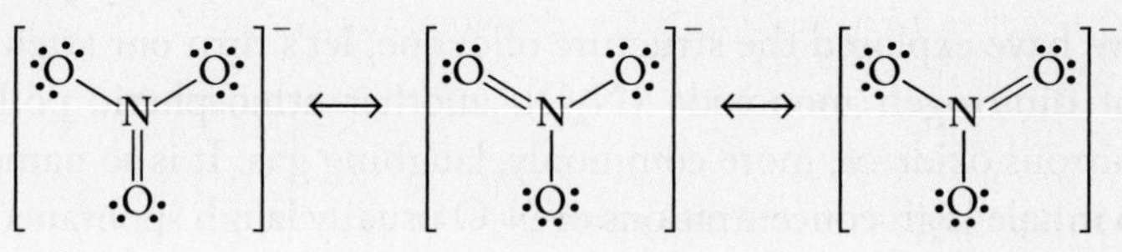

$[\ \ ]^- \longleftrightarrow [\ \ ]^- \longleftrightarrow [\ \ ]^-$

Think About It It makes sense that the NO_3^- ion has three resonance structures because three equivalent O atoms are bonded to the central N atom, and any one of the O atoms could be the one with the double bond. The additional electron from the negative charge on the ion means that NO_3^- has the same number of valence electrons as SO_3 (Sample Exercise 8.7), which also has the same number of resonance structures. Nitrogen in neutral molecules typically has a bonding capacity of 3, but in these resonance structures it forms four bonds.

Practice Exercise Draw all the resonance forms of the azide ion (N_3^-) and the nitronium ion (NO_2^+).

Resonance also occurs in larger organic molecules that have alternating single and double bonds. Molecules of benzene (C_6H_6), for example, contain six-membered rings of carbon atoms with alternating single and double carbon–carbon bonds (**Figure 8.16a**). When we fix the atoms in a molecule of benzene in place, we have two equivalent ways to draw the single and double bonds. To depict their equivalency, chemists frequently draw benzene molecules with circles in the centers (**Figure 8.16b**). This symbol emphasizes that the six carbon–carbon bonds in the ring are all identical and intermediate in character between single and double bonds. (Experimental evidence confirms that all six carbon–carbon bonds in benzene are equivalent.)

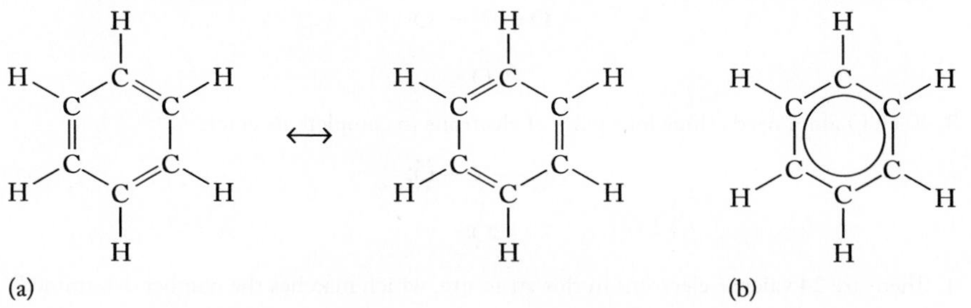

(a) (b)

FIGURE 8.16 (a) The molecular structure of benzene is an average of two equivalent resonance structures. (b) The average is frequently represented by a circle inside the hexagonal ring of single bonds, indicating completely uniform delocalization of the electrons in the bonds around the ring.

Because of the carbon–carbon double bonds in the structure, you might think that benzene would be in the family of alkenes. It turns out, however, that the double bonds in benzene (and other molecules where resonance influences the character of the bonds) react differently than the double bonds in alkenes. As a result, benzene and compounds with similar molecular structures are classified separately as *aromatic compounds*. We explore aromatic compounds in Chapter 9, where we develop models of bonding that enable us to envision the differences between alkenes like ethylene and aromatic compounds like benzene.

8.5 Formal Charge: Choosing among Lewis Structures

Now that we have explored the structure of ozone, let's turn our attention to the structure of dinitrogen monoxide (N_2O), another atmospheric pollutant also known as nitrous oxide or, more commonly, laughing gas. It is so named because people who inhale high concentrations of N_2O usually laugh spontaneously. Gaseous nitrous oxide has long been used as an anesthetic in dentistry because it has a narcotic effect.

In the lower atmosphere, nitrous oxide concentrations range between 0.1 and 1.0 ppm. It is produced naturally by bacterial action in soil and through human activities such as agriculture and sewage treatment. It is occasionally in the news because its concentration in the troposphere (the atmosphere at ground level) has been increasing. Along with CO_2 and other gases, N_2O may be contributing to global warming.

To draw the Lewis structure of nitrous oxide, first count the number of valence electrons: 5 each from the two nitrogen atoms and 6 from the oxygen atom for a total of $(2 \times 5) + 6 = 16$. The central atom is a nitrogen atom because N has a higher bonding capacity and is less electronegative than O. Connecting the atoms with single bonds, we have

$$\mathrm{N{-}N{-}O}$$

Completing the octets of the outer atoms gives a structure with 16 valence electrons, which is the number determined in step 1:

$$\mathrm{:\ddot{\underset{..}{N}}{-}N{-}\ddot{\underset{..}{O}}:}$$

However, there are only two bonds and thus only 4 valence electrons on the central N atom. To give it 4 more electrons, we need to convert lone pairs on the surrounding atoms to bonding pairs. Which lone pairs do we choose? We could use two lone pairs from the N atom on the left to form a N≡N triple bond:

$$\mathrm{:N{\equiv}N{-}\ddot{\underset{..}{O}}:}$$

Alternatively, we could use two lone pairs on the O atom to make a N≡O triple bond:

$$\mathrm{:\ddot{\underset{..}{N}}{-}N{\equiv}O:}$$

Or we could use one pair from each terminal atom to make two double bonds:

$$\mathrm{\ddot{N}{=}N{=}\ddot{O}}$$

Which of these resonance structures is best? We have seen that in some sets of resonance structures, such as those of O_3, SO_3, and NO_3^-, all the structures are equivalent, so no one of them is any more important than another in conveying the actual bonding in the molecule. This is not the case, however, with the resonance forms of N_2O. To decide which resonance form in a nonequivalent set is the most important and comes closest to representing the actual bonding pattern in a molecule, we use the concept of formal charge.

A **formal charge (FC)** is not a real charge but rather a measure of the number of electrons *formally assigned* to an atom in a molecular structure, compared with the number of electrons in the free (i.e., nonbonded) atom. To determine a formal charge, we follow a series of steps to calculate the number of electrons formally assigned to each atom in each resonance structure.

formal charge (FC) a value calculated for an atom in a molecule or polyatomic ion by determining the difference between the number of valence electrons in the free atom and the sum of lone-pair electrons plus half of the electrons in the atom's bonding pairs.

Calculating Formal Charge of an Atom in a Resonance Structure

For each atom:

1. Determine the number of valence electrons in the free atom.
2. Count the number of nonbonding electrons on the atom in the structure.
3. Count the number of electrons in bonds on the atom and divide that number by 2.
4. Sum the results of steps 2 and 3 and subtract that sum from the number determined in step 1.

Summarizing these steps in the form of an equation, we have

$$FC = \begin{pmatrix} \text{number of} \\ \text{valence e}^- \end{pmatrix} - \left[\begin{matrix} \text{number of} \\ \text{nonbonding e}^- \end{matrix} + \frac{1}{2} \begin{pmatrix} \text{number of e}^- \\ \text{in bonding pairs} \end{pmatrix} \right] \quad (8.1)$$

Once we have determined the formal charges, we then use the following three criteria to select the preferred structure:

1. The preferred structure is the one with formal charges of zero.
2. If no such structure can be drawn, or if the structure is that of a polyatomic ion, then the best structure is the one in which most atoms have formal charges equal to zero or closest to zero.
3. Any negative formal charges should be on the atom(s) of the most electronegative element(s).

The calculation of formal charge incorporates the assumption that each atom is formally assigned all its nonbonding electrons and shares its bonding electrons equally with the atoms at the other ends of the bonds. We can confirm that the three resonance forms of N_2O are not equivalent by calculating the formal charges in each structure.

In **Table 8.2** we have colored the lone pairs of electrons red and the shared pairs green to make it easier to track the quantities of these electrons in the formal charge calculation for N_2O. The numbers of valence electrons in the free atoms are shown in blue.

To illustrate one of the formal charge calculations in the table, consider the N atom at the left end of the left resonance structure. The Lewis dot symbol of nitrogen reminds us that free nitrogen atoms have 5 valence electrons. In this structure, the N atom has 2 electrons in a lone pair and 6 electrons in three shared (bonding) pairs. Using Equation 8.1 to calculate the formal charge on this N atom,

$$FC = 5 - [2 + \tfrac{1}{2}(6)] = 0$$

which is the first value in the bottom row of the table. The results of similar formal charge calculations for all the other atoms in the three resonance structures complete the row. The sum of the formal charges on the three atoms in each structure is zero, as it should be for a neutral molecule. When we analyze the formal charges of atoms in a polyatomic ion, the formal charges on its atoms must add up to the charge on the ion.

Now we must apply the three criteria for selecting the preferred resonance structure of N_2O. The first thing to note about these sets of formal charges is that in none of them are all three formal charges zero, meaning that criterion 1 is not met. The next step is to identify which structure has the most formal charge values that are the closest to zero, such as −1 or +1. On this count we have a tie between the structure on the left (0, +1, −1) and the one in the middle (−1, +1, 0).

TABLE 8.2 Formal Charge Calculations for the Resonance Structures of N_2O

Step	:N≡N—Ö:			N=N=O			:N—N≡O:		
1 Number of valence electrons	5	5	6	5	5	6	5	5	6
2 Number of electrons in lone pairs	2	0	6	4	0	4	6	0	2
3 Number of shared electrons	6	8	2	4	8	4	2	8	6
4 FC = [valence e⁻ − (lone pair e⁻ + ½ shared e⁻)]	0	+1	−1	−1	+1	0	−2	+1	+1

To break the tie, we invoke the third criterion and answer the question, "In which structure is the negative formal charge on the more electronegative atom?" The answer is the structure on the left, which has an oxygen atom with a formal charge of −1. This structure, then, is the best of the three in representing the actual bonding in a molecule of N_2O.

From experimental measurements, we know that although the N≡N—O structure contributes the most to the bonding in N_2O, the middle structure (N=N=O) also contributes to the bonding. We know this because the length of the bond between the two nitrogen atoms is between the length of a N=N bond and the length of a N≡N bond, and the nitrogen–oxygen bond is a little shorter than a typical N—O single bond.

CONCEPT TEST

What is the formal charge on a sulfur atom that has three lone pairs of electrons and one bonding pair?

SAMPLE EXERCISE 8.9 Selecting Resonance Structures on the Basis of Formal Charges LO5

Which of the following resonance forms best describes the actual bonding in a molecule of CO_2?

:Ö—C≡O: ⟷ Ö=C=Ö ⟷ :O≡C—Ö:

Collect, Organize, and Analyze The preferred structure is the one in which the formal charges are closest to zero and any negative formal charges are on the more electronegative atom. In this case, oxygen is a group 16 element, so it is more electronegative than carbon, a group 14 element. Each free carbon atom has 4 valence electrons, and free oxygen atoms have 6 valence electrons each.

Solve We use Equation 8.1 to find the formal charge on each atom, and we can illustrate the results in a table:

Formal Charge Calculations for the Resonance Structures of CO_2

Step	:Ö—	C	≡O:	Ö=	C	=Ö	:O≡	C	—Ö:
1 Number of valence electrons	6	4	6	6	4	6	6	4	6
2 Number of electrons in lone pairs	6	0	2	4	0	4	2	0	6
3 Number of shared electrons	2	8	6	4	8	4	6	8	2
4 FC = [valence e^- − (lone pair e^- + $\frac{1}{2}$ shared e^-)]	−1	0	+1	0	0	0	+1	0	−1

The formal charges are zero on all the atoms in the middle resonance structure with the two double bonds. Therefore, this structure best represents the actual bonding in a molecule of CO_2.

Think About It All three resonance structures are valid in that the atoms are in the same positions and the sum of the formal charges is zero, as they must be for a neutral molecule. The center structure best represents the actual bonding in CO_2 because its atoms all have formal charges of zero.

Practice Exercise Which resonance structure of the nitronium ion (NO_2^+) contributes the most to the actual bonding in this ion?

Using the resonance structures for N_2O once again, let's examine the link between the formal charge on an atom in a resonance structure and the bonding capacity of that atom. The Lewis symbol of nitrogen, which has three unpaired electrons, indicates that a nitrogen atom can complete its octet by forming three bonds. The two unpaired electrons in the Lewis symbol of oxygen indicate that the bonding capacity of an oxygen atom is 2. In the N_2O resonance structures in Table 8.2, the nitrogen atom with three bonds has a formal charge of zero (the left N in the first structure), and the oxygen atom with two bonds has a formal charge of zero (the O in the middle structure). As a rule—and assuming the octet rule is obeyed—atoms have formal charges of zero in resonance structures in which the numbers of bonds they form match their bonding capacities. If an atom forms one more bond than its bonding capacity, such as an oxygen atom with three bonds (O in the third structure in the table), the formal charge is +1. If the atom forms one fewer bond than the bonding capacity, such as an oxygen atom with one bond (O in the first structure in the table), the formal charge is −1.

8.6 Exceptions to the Octet Rule

FIGURE 8.17 The high temperatures of arc welding produce significant concentrations of nitrogen monoxide (NO) as a result of the highly endothermic reaction $N_2(g) + O_2(g) \rightarrow 2\,NO(g)$.

Nitric oxide (NO) is a simple and small molecule, and yet one with important roles in biology and medicine. It is crucial as a regulator of blood flow and is also a principal mediator of neurological signals for many physical and pathological processes. It is a known bioproduct in organisms ranging from bacteria to plants and higher animals, including humans. Pharmaceuticals such as nitroglycerine, used to relieve angina (chest pain caused by impaired blood flow to the heart), and Viagra, used to treat erectile dysfunction, act by producing NO *in vivo* to improve blood flow. Earth's atmosphere also contains trace concentrations of NO along with NO_2, both of which are generated in the environment by human activities such as welding (**Figure 8.17**) and driving gasoline-burning vehicles; these molecules contribute to photochemical smog formation in urban areas. NO and NO_2 illustrate the limitations of the octet rule. Each has an odd number of valence electrons per molecule, which means that at least one atom in each molecule cannot have a complete octet.

Odd-Electron Molecules

NO is produced in the environment when high temperatures in vehicle engines or other sources lead to the reaction

$$\underset{}{N_2(g)} + O_2(g) \rightarrow \underset{\text{Nitric oxide}}{2\,NO(g)} \qquad (8.2)$$

The nitric oxide then reacts with atmospheric oxygen to produce nitrogen dioxide:

$$\underset{\text{Nitric oxide}}{2\,NO(g)} + O_2(g) \rightarrow \underset{\text{Nitrogen dioxide}}{2\,NO_2(g)} \qquad (8.3)$$

Nitric oxide is highly reactive in living systems as well as in the atmosphere because it is an odd-electron molecule. To understand the implications of this, let's draw its Lewis structure. Nitric oxide has 11 valence electrons (5 from nitrogen and 6 from oxygen). There is no central atom, so we start with a single bond between N and O and then complete the octet around O, which is the more electronegative element:

N—Ö:

We then place the remaining 3 electrons around the N atom:

Ṅ—Ö:

This leaves the N atom short 3 valence electrons. We can increase the number of electrons around N from 5 to 7 by converting a lone pair on the O atom into a bonding pair:

.Ṅ=Ö:

This change has the added advantage of creating a double-bonded O atom, which gives it a formal charge of zero. The formal charge on the N atom is also zero. The only problem with the structure is that N does not have an octet: it has only 7 electrons. Because nitrogen is less electronegative than oxygen, it is reasonable that we "short-change" it when there are insufficient electrons to complete the octets of both atoms. The existence of NO is evidence that there are exceptions to the octet rule. When that happens, the most representative Lewis structures are those that come *as close as possible* to producing zero formal charges and complete octets.

free radical an odd-electron molecule with an unpaired electron in its Lewis structure.

Compounds that have odd numbers of valence electrons are called **free radicals**. They are typically very reactive species because it is often energetically favorable for them to acquire an electron from another molecule or ion. This characteristic makes them excellent oxidizing agents and is responsible for the damage they may cause to materials or living tissue with which they come in contact.

SAMPLE EXERCISE 8.10 Drawing Lewis Structures of Odd-Electron Molecules **LO5**

Draw the resonance structures of nitrogen dioxide (NO_2), assign formal charges to the atoms, and suggest which structure may best represent the actual bonding in the molecule.

Collect, Organize, and Analyze Each molecule contains one atom of nitrogen (bonding capacity 3) and two atoms of oxygen (bonding capacity 2). Nitrogen dioxide is an odd-electron molecule, so we anticipate that one of the atoms will have an incomplete octet. We analyze the resonance structures by assigning formal charges to select the one most representative of the actual bonding in the molecule.

Solve The number of valence electrons is

Element:	N	+	2 O
Valence electrons:	5	+	$(2 \times 6) = 17$

Nitrogen is less electronegative than oxygen and has the greater bonding capacity, so it is the central atom:

O—N—O

Completing the octets on the O atoms gives

:Ö—N—Ö:

There are 16 valence electrons in this structure, but we need 17 to match the number available in the molecule. We add 1 more electron to the N atom:

:Ö—N—Ö:

There are only 5 valence electrons around the N atom—3 fewer than the 8 we need. We can increase the number from 5 to 7 by converting a lone pair on one of the O atoms to a bonding pair, giving the formal charges shown in red:

[Lewis structure: O (0) double-bonded to N (+1), N single-bonded to O (−1)]

An equivalent resonance form can be drawn with the double bond on the right side:

[Lewis structure: O (−1) single-bonded to N (+1), N double-bonded to O (0)]

These structures are equivalent: each has a charge of 1+ on the nitrogen atom, one oxygen atom with a 1− charge, and one oxygen atom with a charge of 0 (zero).

Think About It The two Lewis structures are equivalent because the two O atoms in each structure are equivalent. The structures do not satisfy the octet rule, but the formal charges of the atoms are close to zero, and the negative formal charge is on the atom of the more electronegative element. Both O atoms have complete octets, leaving the less electronegative N atom one electron short in this odd-electron molecule.

Practice Exercise Nitrogen trioxide (NO_3) may form in polluted air when NO_2 reacts with O_3. Draw the resonance structures of NO_3 and use formal charges to select the preferred structure(s).

Molecules in Which Atoms Form More than Four Bonds

In some molecules, such as PCl_5 and SF_6, atoms of nonmetals in the third row and below in the periodic table appear to have more than an octet of valence electrons:

[Lewis structures of PCl_5 (P bonded to five Cl atoms) and SF_6 (S bonded to six F atoms)]

FIGURE 8.18 Electrical transformers use sulfur hexafluoride as an insulator because it is thermally stable and does not react with water. When SF_6 leaks into the atmosphere, however, it becomes a potent greenhouse gas. It absorbs much more infrared radiation than CO_2 and may remain in the atmosphere for thousands of years.

The five covalent bonds in PCl_5 and the six in SF_6 represent a total of 10 and 12 valence electrons around the central P and S atoms, respectively. All the atoms in both structures have formal charges of zero, so the two structures seem to be perfectly acceptable representations of the bonding in these molecules. How, though, can P and S atoms be surrounded by more than 8 valence electrons? Understanding the bonding in a compound such as SF_6 is important to understanding its potency as a greenhouse gas (**Figure 8.18**).

Over the years, chemists have debated several explanations for what many have called the *hypervalency* of phosphorus, sulfur, and other nonmetals in the third row and below in the periodic table. You might think that because elements in the third row of the periodic table and beyond have empty *d* orbitals that perhaps they use these to form additional bonds. However, recent studies employing high-speed computers and programs that allow chemists to predict the contributions that different atomic orbitals make in bond formation have shown that *d* orbitals contribute little to the bonding in molecules like PCl_5 and SF_6. We explore another theory of covalent bond formation in Chapter 9 that explains the bonding in these molecules without incorporating valence shell *d* orbitals.

Just because some atoms may have more than 8 valence electrons within certain compounds does not necessarily mean they always do. Rather, this tends to happen in these circumstances:

1. When they bond with strongly electronegative elements—particularly F, O, and Cl.
2. When doing so results in a structure in which the atoms have formal charges that are closer to zero.

Lewis Structures: Atoms with More than an Octet

Consider, for example, the S atom in a SO_4^{2-} ion. Following the usual steps to draw its Lewis structure and assign formal charges, we get

The sum of the formal charges on atoms in the ion is 2−, which is equal to the overall ionic charge, as it should be. Could this Lewis structure be redrawn so that at least some of the formal charges are zero? Converting one lone pair of electrons on each of two O atoms into bonding pairs gives the following:

Each O atom still has a complete octet, but the S atom accommodates 12 electrons. In this structure the formal charge on the S atom is 0, as are the formal charges on two of the four O atoms. There is still a formal charge of −1 on the other two oxygen atoms, which sums to the overall 2− charge of the ion. We could draw the two double bonds to any two of the O atoms, which means the structure is stabilized by resonance.

We can draw the Lewis structure of H_2SO_4 by combining two hydrogen ions (H^+) to the two oxygen atoms with 1− charges:

Each hydrogen atom has achieved its duet of electrons, each oxygen atom has an octet, and every atom has a formal charge of zero.

Based on the preceding formal charge analyses, we might conclude that the preferred bonding pattern in SO_4^{2-} ions and molecules of H_2SO_4 includes two S═O double bonds and zero formal charges all around. However, experimental evidence suggests that although the two structures with zero formal charges do contribute to the bonding of SO_4^{2-} and H_2SO_4, structures that obey the octet rule and have no S═O double bonds contribute as well. Thus, the actual bonding in these particles is an average of the structures with an octet and the structures with more than an octet of valence electrons.

SAMPLE EXERCISE 8.11 Drawing a Lewis Structure Containing an Atom with More than Four Bonds — LO5

Draw the Lewis structure for the phosphate ion (PO_4^{3-}) that minimizes the formal charges on its atoms.

Collect, Organize, and Analyze Each ion contains one atom of phosphorus and four atoms of oxygen and has an overall charge of 3−. Phosphorus and oxygen are in groups 15 and 16 and have bonding capacities of 3 and 2, respectively. Phosphorus has an atomic number greater than 12 ($Z = 15$), so it may exhibit hypervalency.

Solve The number of valence electrons is

Element:	P	+	4 O
Valence electrons:	5	+	(4 × 6) = 29
Additional electrons from the 3− charge:			+ 3
Total valence electrons:			32

Phosphorus has the greater bonding capacity (3), so it is the central atom:

```
      O
      |
  O — P — O
      |
      O
```

The addition of three lone pairs of electrons to each O atom completes their octets:

```
       ..
      :O:
       |
   ..      ..
  :O — P — O:
   ..      ..
       |
      :O:
       ..
```

There are 32 valence electrons in this structure, which matches the number determined for the ion. Therefore, it is a complete Lewis structure of a polyatomic ion once we add the brackets and electrical charge:

```
  [       ..      ]3−
  [      :O:      ]
  [       |       ]
  [   ..      ..  ]
  [  :O — P — O:  ]
  [   ..      ..  ]
  [       |       ]
  [      :O:      ]
  [       ..      ]
```

Each O has a single bond and a formal charge of −1; the four bonds around the P atom are one more than its bonding capacity, so its formal charge is +1. The sum of the formal charges, [+1 + 4(−1)], matches the 3− charge on the ion.

We can reduce the formal charge on P by increasing the number of bonds to it, and we can do that by converting a lone pair on one of the O atoms into a bonding pair:

$$\left[\text{:}\ddot{\text{O}}\text{—}\overset{+1}{\text{P}}(\text{—}\ddot{\text{O}}\text{:}^{-1})_3 \right]^{3-} \rightarrow \left[\text{O}^{-1}_3\overset{0}{\text{P}}\text{=}\ddot{\text{O}}^{0} \right]^{3-}$$

At the same time, we change a single-bonded O atom into a double-bonded O atom and thereby make its formal charge zero. Therefore, the structure on the right is the one that minimizes formal charges.

Think About It The structure on the right has the double bond drawn to the O atom at the bottom of the structure, but it could have been drawn to any one of the other O atoms. This means that the structure is stabilized by resonance, producing four equivalent P—O bonds in which each pair of atoms is held together by the equivalent of 1.25 bonds, much like the 1.5 bonds between each pair of O atoms in molecules of ozone (see Section 8.4).

Practice Exercise Draw the resonance structures of the selenite ion (SeO_3^{2-}) that minimize the formal charges on the atoms.

Lewis Structures: Atoms with Less than an Octet

When fluorine combines with boron, a compound with the formula BF_3 is isolated. The electronegativity difference between boron and fluorine is 2.0, which puts BF_3 on the borderline between ionic and covalent compounds. If BF_3 contains three fluoride anions and a B^{3+} cation, then the ions are isoelectronic with a noble gas. Boron trifluoride, however, does not have the physical properties associated with an ionic compound; it is a gas that condenses at −101°C. What does the Lewis structure for BF_3 look like?

Boron trifluoride has a total of 24 valence electrons (step 1):

Element:	B	+	3 F
Valence electrons:	3	+	(3 × 7) = 24

Boron has a greater bonding capacity (3) than fluorine (1), so B is the central atom (step 2). Connecting three fluorine atoms to the boron atom and completing the octets on fluorine (step 3) gives us a structure with the requisite number of valence electrons (step 4) on each F but not on the boron atom.

$$\overset{0}{\text{B}}(\text{—}\overset{0}{\ddot{\underset{..}{\text{F}}}}\text{:})_3$$

Formal charges are minimized in this structure; however, boron shares only 6 electrons and has no electrons in lone pairs. If we share an additional pair of electrons from one of the three F atoms, as shown in the resulting three resonance forms, we can complete the octets of all the atoms:

$$\overset{-1}{\text{B}}(\text{=}\overset{+1}{\text{F}})(\text{—}\overset{0}{\text{F}})_2 \longleftrightarrow \overset{-1}{\text{B}}(\text{=}\overset{+1}{\text{F}})(\text{—}\overset{0}{\text{F}})_2 \longleftrightarrow \overset{-1}{\text{B}}(\text{=}\overset{+1}{\text{F}})(\text{—}\overset{0}{\text{F}})_2$$

But forming a B═F double bond results in unfavorable formal charges, particularly for the F atom. We do not expect a positive formal charge on the most electronegative of all elements.

How do we resolve this conflict? The B—F bond lengths in BF_3 are all equal, but shorter than the B—F single bonds in the BF_4^- ion:

130.9 pm   137.9 pm

As we see in the next section, multiple bonds are generally shorter than single bonds. This suggests that all four of the following resonance forms contribute to the molecular structure of BF_3:

Dominant contributor

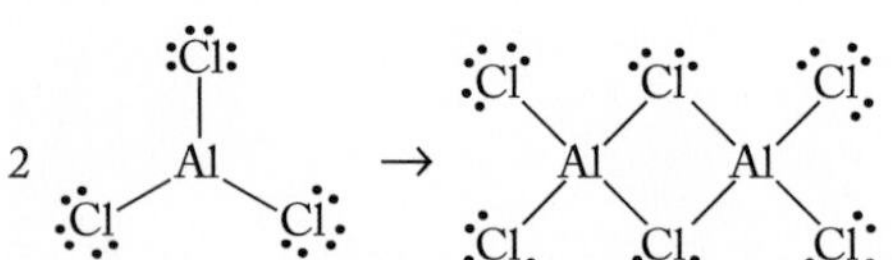

FIGURE 8.19 Aluminum chloride has a complete octet on aluminum by sharing a pair of electrons with one Cl atom from a second molecule of $AlCl_3$.

The atoms in the other group 13 halides complete their octets in different ways. For example, in aluminum chloride, the aluminum atom shares a lone pair of electrons with a chlorine atom in a second molecule, so that all the atoms in Al_2Cl_6 have a complete octet (**Figure 8.19**). The structure for Al_2Cl_6 is observed only in the liquid (or molten) phase. In the gas phase, aluminum chloride exists as discrete molecules of $AlCl_3$, like BF_3.

Do compounds containing elements other than group 13 ever have less than an octet around the central atom? The answer depends on which phase of the material we are considering. For example, as a solid, beryllium chloride consists of long chains of chlorine-bridged Be atoms (**Figure 8.20**) in which all atoms have completed octets. Above 800 K, $BeCl_2$ exists in the gas phase as discrete molecules in which the beryllium shares only two pairs of electrons with chlorine—much less than an octet. In general, H, Be, B, and Al are the elements most likely to have less than an octet of electrons.

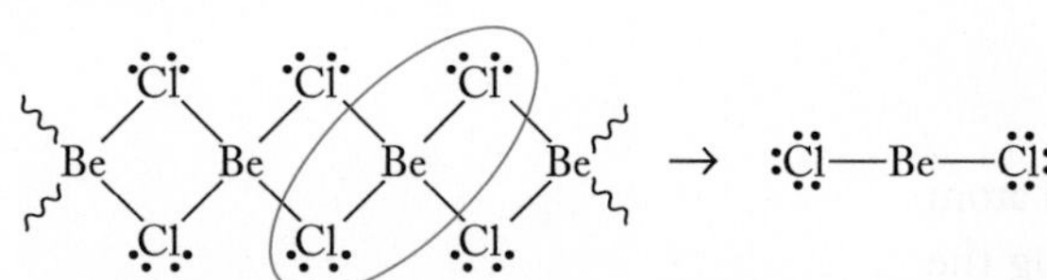

FIGURE 8.20 The chain structure of solid beryllium chloride, with bridging chlorine atoms, changes to discrete $BeCl_2$ molecules in the gas phase. Each beryllium atom in the solid has a completed octet. In the gas phase, each Be atom is surrounded by only four electrons.

SAMPLE EXERCISE 8.12 Drawing Lewis Structures of Compounds with Less than an Octet **LO5**

Two of the compounds used in the electronics industry to make thin films of aluminum metal are (a) $Li^+[AlH_4]^-$ and (b) AlH_3. Draw the Lewis structures of these ionic and molecular compounds.

Collect, Organize, and Analyze The formula of lithium aluminum hydride, $Li^+[AlH_4]^-$, indicates that it contains Li^+ cations and AlH_4^- anions. Lithium and hydrogen each have 1 valence electron and aluminum has 3 valence electrons. We use the steps outlined in Section 8.2 to draw Lewis structures for both compounds. Aluminum is the central atom in AlH_4^- and AlH_3 because each hydrogen atom can form only one bond.

Solve

a. The lithium atom has lost its valence electron to form a Li^+ ion, so its Lewis symbol is Li^+:

$$Li \xrightarrow{-e^-} Li^+$$

The total number of valence electrons in AlH_4^- is:

Element:	Al	+	4 H
Valence electrons:	3	+	$(4 \times 1) = 7$
Additional electron from 1− charge:			$+ 1$
Total valence electrons:			8

Connecting the four H atoms to the Al with single bonds results in a Lewis structure in which aluminum has a complete octet and each hydrogen atom has a complete duet.

$$Li^+ \left[\begin{array}{c} H \\ | \\ H{-}Al{-}H \\ | \\ H \end{array} \right]^-$$

b. The total number of valence electrons in AlH_3 is:

Element:	Al	+	3 H
Valence electrons:	3	+	$(3 \times 1) = 6$

In this case we can complete the duets of each H atom, but the aluminum atom is left with only 6 electrons, two fewer than a complete octet.

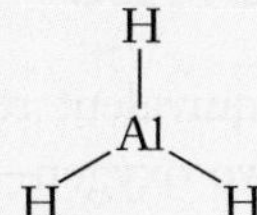

Think About It The aluminum atom in AlH_3 must have less than an octet because there are only 6 total valence electrons. As a result, AlH_3 reacts readily with the lone pair on nitrogen in $N(CH_3)_3$, for example, to form $AlH_3 \cdot N(CH_3)_3$. In AlH_4^-, the additional H atom and the overall negative charge on the anion each account for 1 electron to complete the octet on aluminum.

Practice Exercise The boiling point of sodium chloride is approximately 1700 K (≈ 1427°C). In the gas phase, NaCl particles with the formula Na_2Cl_2 have been identified. Draw the Lewis structure for this particle and identify which atom does not have a complete octet.

The Limits of Bonding Models

Given the structures of SF_6, BF_3, and other exceptions to the octet rule, does the octet rule have any validity at all? The point to remember is that the octet rule is just a *model*. The octet rule works remarkably well in many cases to predict the number of covalent bonds between main group elements and to predict the composition of ionic compounds. However, the true nature of the chemical bond is much more subtle and shows great variability among the myriad chemical compounds that have been discovered.

What about the validity of formal charges? How can F, the most electronegative atom, carry a +1 formal charge in BF_3? Like Lewis structures, formal charges represent a model for identifying resonance forms that contribute most to the bonding in a molecule. Sometimes our model does not fit the observed data as well as we would like. In the case of BF_3, the observed B—F distances suggest that resonance forms containing B=F double bonds are important.

If aluminum chloride is a covalently bonded molecule, why do aqueous solutions of $AlCl_3$ conduct electricity? Why doesn't BF_3, with an electronegativity difference of 2.0, behave similarly? Why does $AlCl_3$ display ionic properties when ΔEN for Al—Cl bonds is only 1.5? In part, the answer lies in the partial ionic character of polar bonds. The greater difference in electronegativity between B and F ($\Delta EN = 2.0$) suggests that B—F bonds have more ionic character than Al—Cl bonds. The fact that $AlCl_3$ does dissolve in water, forming a solution that conducts electricity, points to the limitations of the octet rule as a model. Other considerations must be taken into account—namely, it is not only the ionic character of the bond but also bond energy and the resultant chemical reactivity of each compound that matter.

The significance of molecules with atoms surrounded by less than or more than an octet is that they challenge our models for bonding and push us toward better explanations for what we observe. We explore questions about bonding in more detail in Chapter 9, but before we do, let's look at some experimental data about the bond lengths and bond energies of covalent bonds. As measurable quantities, these parameters allow us to test our model of the chemical bond as proposed by Gilbert Lewis.

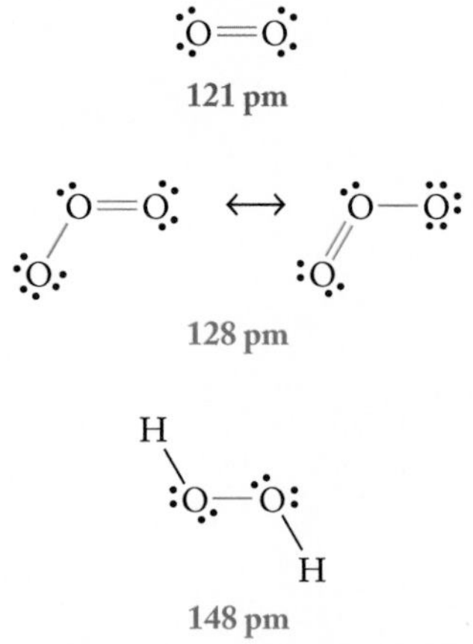

:Ö=Ö:
121 pm

Ö=Ö: ⟷ Ö—Ö:
128 pm

H—Ö—Ö—H
148 pm

FIGURE 8.21 Resonance influences bond length. Because of resonance, the lengths of the two bonds in ozone are identical and in between the lengths of the O=O double bond in O_2 and the O—O single bond in H_2O_2.

8.7 The Lengths and Strengths of Covalent Bonds

In Section 8.4 we discussed the equivalent resonance structures of ozone. We noted that the true nature of the two oxygen–oxygen bonds in O_3 is reflected in their equal bond length, 128 pm, which is between the length of a typical O=O double bond (121 pm) and an O—O single bond (148 pm), as shown in **Figure 8.21**. We used these results to conclude that there are effectively 1.5 bonds between the atoms in a molecule of O_3. In this section we explore this use of bond length and bond strength to rationalize and validate molecular structures. We also use bond strengths to estimate the enthalpy changes that occur in chemical reactions.

Bond Length

The length of the bond between any two atoms depends on the identity of the atoms and on whether the bond is single, double, or triple (**Table 8.3**). The number of bonds between two atoms is called the **bond order**. As bond order increases, bond length decreases, as we can see by comparing the lengths of the C—C, C=C, and C≡C bonds in Table 8.3. Similarly, for carbon–oxygen bonds, the C≡O triple bond in carbon monoxide is shorter than the C=O double bond in carbon dioxide (**Figure 8.22a**).

:C≡O:
113 pm
Carbon monoxide

:Ö=C=Ö:
123 pm
Carbon dioxide
(a)

110 pm
Methane

121 pm
111 pm
Formaldehyde
(b)

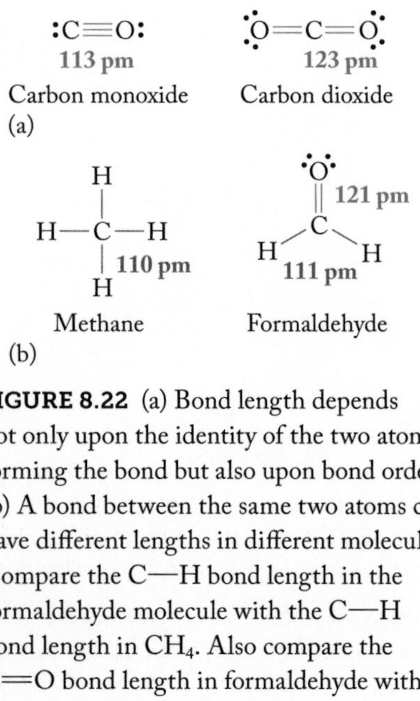

FIGURE 8.22 (a) Bond length depends not only upon the identity of the two atoms forming the bond but also upon bond order. (b) A bond between the same two atoms can have different lengths in different molecules. Compare the C—H bond length in the formaldehyde molecule with the C—H bond length in CH_4. Also compare the C=O bond length in formaldehyde with the C=O bond length in CO_2.

Measurements of the bond lengths in many molecules indicate that there are small differences in bond lengths for any given covalent bond. For example, the C—H and C=O bond lengths in formaldehyde (**Figure 8.22b**) are close to but not the same as the C—H bond length in CH_4 and the C=O bond length in CO_2.

CONCEPT **TEST**

Rank the following molecules in order of decreasing lengths of their nitrogen–oxygen bonds: NO, NO_2, N_2O.

TABLE 8.3 Selected Average Covalent Bond Lengths and Bond Energies

Bond	Bond Length (pm)	Bond Energy (kJ/mol)	Bond	Bond Length (pm)	Bond Energy (kJ/mol)
C—C	154	348	N≡O	106	678
C=C	134	614	O—O	148	146
C≡C	120	839	O=O	121	498
C—N	147	293	O—H	96	463
C=N	127	615	S—O	151	265
C≡N	116	891	S=O	143	523
C—O	143	358	S—S	204	266
C=O	123	743[a]	S—H	134	347
C≡O	113	1072	H—H	75	436
C—H	110	413	H—F	92	567
C—F	133	485	H—Cl	127	431
C—Cl	177	328	H—Br	141	366
N—H	104	388	H—I	161	299
N—N	147	163	F—F	143	155
N=N	124	418	Cl—Cl	200	243
N≡N	110	941	Br—Br	228	193
N—O	136	201	I—I	266	151
N=O	122	607			

[a]The bond energy of the C=O bond in CO_2 is 799 kJ/mol.

Bond Energies

The energy changes associated with chemical reactions depend on how much energy is required to break the bonds in the reactants and how much is released as the atoms recombine to form products. For example, in the methane combustion reaction

$$CH_4(g) + 2\,O_2(g) \rightarrow CO_2(g) + 2\,H_2O(g)$$

the C—H bonds in CH_4 and the O=O bonds in O_2 must be broken before the C=O bonds in CO_2 and the O—H bonds in H_2O can form. (In reality, some bond formation occurs simultaneously with bond breaking; it is not completely sequential.) Breaking bonds is endothermic (the blue "energy in" arrow in **Figure 8.23**), and forming bonds is exothermic (the red "energy out" arrow). If a chemical reaction is exothermic, as is methane combustion, more energy is released in forming the bonds in molecules of products than is consumed in breaking the bonds in molecules of reactants.

Bond energy, or *bond strength*, is usually expressed in terms of the enthalpy change (ΔH) that occurs when 1 mole of a bond in the gas phase is broken. Bond energies for some common covalent bonds are listed in Table 8.3. The values in Table 8.3 are all positive because bond breaking is always endothermic. As we first mentioned in Section 8.1, the quantity of energy needed to break a bond depends on the identity of the two bonded atoms and the bond order and is equal in magnitude but opposite in sign to the quantity of energy released when that same bond forms.

bond order the number of bonds between atoms: 1 for a single bond, 2 for a double bond, and 3 for a triple bond.

bond energy the energy needed to break 1 mole of a covalent bond in a molecule or in a polyatomic ion in the gas phase.

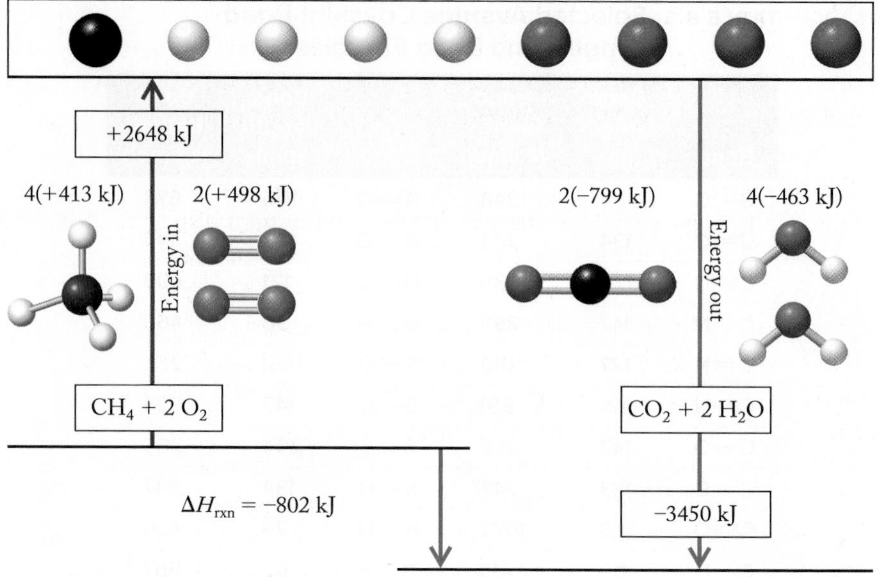

FIGURE 8.23 The combustion of 1 mole of methane requires that 4 moles of C—H bonds and 2 moles of O=O bonds be broken. These processes require an enthalpy change of about +2648 kJ. In the formation of 4 moles of O—H bonds and 2 moles of C=O bonds, there is an enthalpy change of about −3450 kJ. The overall reaction is exothermic: 2648 kJ − 3450 kJ = −802 kJ.

CHEMTOUR
Estimating Enthalpy Changes

Like bond lengths, the bond energies in Table 8.3 are average values because bond energies vary depending on the structure of the rest of the molecule (Figure 8.22). For example, the bond energy of a C=O bond in carbon dioxide is 799 kJ/mol, but the C=O bond energy in formaldehyde is only 743 kJ/mol.

Another view of the variability in bond energy comes from breaking the C—H bonds in CH_4 in a step-by-step fashion, as shown in **Figure 8.24**. The

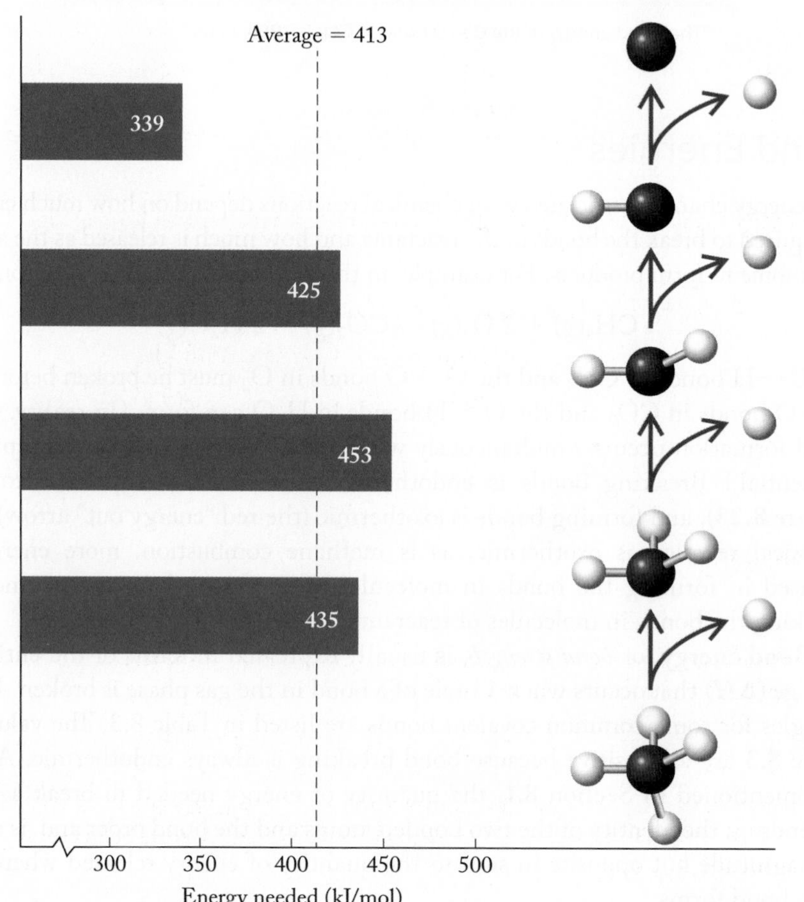

FIGURE 8.24 The energies needed to break the four C—H bonds in methane are not the same.

data show that the chemical environment of a bond affects the energy required to break it: breaking the first C—H bond in methane requires less energy than breaking the second but more than breaking the third or fourth. The total energy needed to break all four C—H bonds is 1652 kJ/mol, which is an average of 413 kJ/mol per bond.

The relationship between bond order and bond energy is also apparent in Table 8.3. The bond energy of the O═O double bond, at 498 kJ/mol, is more than three times that of the O—O single bond. In general, the bond energy for a pair of atoms increases as their bond order increases. For example, the bond energy of the N≡N triple bond (941 kJ/mol) is more than twice the bond energy of the N═N double bond and more than five times that of the N—N single bond. The large amount of energy required to break N≡N triple bonds is one reason why N_2 participates in so few chemical reactions.

In the combustion of 1 mole of CH_4 (Figure 8.23), 4 moles of C—H bonds and 2 moles of O═O bonds must be broken. The formation of 1 mole of CO_2 and 2 moles of H_2O requires the formation of 2 moles of C═O bonds and 4 moles of O—H bonds. The net change in energy resulting from breaking and forming these bonds can be estimated from average bond energies. We can take an inventory of the bond energies involved:

Bond Energies (ΔH) and Energy Changes in the Combustion of 1 Mol CH_4

Bond	Number of Bonds (mol)	Bond Energy (kJ/mol)	Bond Breaks or Forms	Energy Change
C—H	4	413	Breaks	+ 4 mol × 413 kJ/mol
O═O	2	498	Breaks	+ 2 mol × 498 kJ/mol
O—H	4	463	Forms	− 4 mol × 463 kJ/mol
C═O	2	799	Forms	− 2 mol × 799 kJ/mol

CONNECTION In Chapter 6 we calculated the difference between the sums of heats of formation of products and of reactants to estimate the heat of reaction.

Next, we estimate the enthalpy change of the overall reaction using. Equation 8.4:

$$\begin{aligned}\Delta H_{rxn} &= \sum\Delta H_{\text{bonds breaking}} - \sum\Delta H_{\text{bonds forming}} \quad (8.4)\\ &= [(4\ \cancel{\text{mol}} \times 413\ \text{kJ/}\cancel{\text{mol}}) + (2\ \cancel{\text{mol}} \times 498\ \text{kJ/}\cancel{\text{mol}})]\\ &\quad -[(4\ \cancel{\text{mol}} \times 463\ \text{kJ/}\cancel{\text{mol}}) + (2\ \cancel{\text{mol}} \times 799\ \text{kJ/}\cancel{\text{mol}})]\\ &= -802\ \text{kJ}\end{aligned}$$

SAMPLE EXERCISE 8.13 Estimating Heats of Reaction from Average Bond Energies **LO6**

Use the average bond energies in Table 8.3 to estimate ΔH_{rxn} for the reaction in which HCl(g) is formed from $H_2(g)$ and $Cl_2(g)$:

H—H + :C̤̈l—C̤̈l: → 2 H—C̤̈l:

Collect and Organize We are asked to estimate the value of ΔH_{rxn} of a gas-phase reaction from the bond energies of the reactants and products. Average bond energy values are listed in Table 8.3.

Analyze The Lewis structures of the reactants and products tell us that 1 mole of H—H bonds and 1 mole of Cl—Cl bonds are broken and 2 moles of H—Cl bonds form during the reaction. The average values of these bonds from Table 8.3 are:

H—H	436 kJ/mol
Cl—Cl	243 kJ/mol
H—Cl	431 kJ/mol

Energy is consumed when bonds break and is released when they form. Two moles of H—Cl bonds form during the reaction and twice the H—Cl bond energy value is more than the sum of the H—H and Cl—Cl bond energies, so there should be an overall release of energy, that is, $\Delta H_{rxn} < 0$.

Solve Using the above information in Equation 8.4:

$$\begin{aligned}\Delta H_{rxn} &= \sum \Delta H_{\text{bond breaking}} - \sum \Delta H_{\text{bond forming}} \\ &= [(1\ \cancel{\text{mol}} \times 436\ \text{kJ}/\cancel{\text{mol}}) + (1\ \cancel{\text{mol}} \times 243\ \text{kJ}/\cancel{\text{mol}})] \\ &\quad -[(2\ \cancel{\text{mol}} \times 431\ \text{kJ}/\cancel{\text{mol}}) \\ &= -183\ \text{kJ}\end{aligned}$$

Think About It In this exercise the Lewis structures provided information about the number and types of bonds that were broken and formed during the reaction. However, you may sometimes only know the identities of the reactants and products in a gas-phase reaction and will have to first write a balanced chemical reaction for the reaction and then draw Lewis structures of the reactants and products to develop an inventory of the bonds broken and formed.

As predicted, ΔH_{rxn} is less than zero, which means the reaction is exothermic. The reaction involves the formation of 2 moles of a gaseous compound from its component elements in their standard states. Therefore, ΔH_{rxn} should be close to two times the standard heat of formation (ΔH°_f) of HCl. That value (see Appendix 4) is −92.3 kJ/mol. Multiplying by 2 moles, we get −184.6 kJ, which is quite close to the calculated ΔH_{rxn} value.

Practice Exercise Use average bond energies to calculate ΔH_{rxn} for the reaction of H_2 and N_2 to form ammonia:

:N≡N: + 3 H—H → 2 H—N̈—H (with H bonded below N)

CONCEPT TEST

Suggest a reason, based on bond energies, why O_2 is much more reactive than N_2.

In this chapter we have explored the nature of the covalent bonds that hold together molecules and polyatomic ions. These bonds owe their strength to the presence of pairs of electrons shared between nuclei of atoms. Sharing does not necessarily mean equal sharing, and unequal sharing coupled with bond vibration accounts for the ability of some atmospheric gases to absorb and emit infrared radiation. As a result, these gases function as greenhouse gases.

Early in the chapter we noted that moderate concentrations of greenhouse gases are required for climate stability and to make our planet habitable. The escalating concern of many is that Earth's climate is being destabilized by too much of a good thing. Policies made by the world's governments will soon have a significant impact on the problem of global warming, one way or the other. As an informed member of the world community, you will have the opportunity to influence how those policy decisions are made.

SAMPLE EXERCISE 8.14 Integrating Concepts: Moth Balls

A compound often referred to by the abbreviation PDB is the active ingredient in most moth balls. It is also used to control mold and mildew, as a deodorant, and as a disinfectant. Tablets containing it are often stuck under the lids of garbage cans or placed in the urinals in public restrooms, producing a distinctive aroma. Molecules of PDB have the following skeletal structure:

```
        H     H
         \   /
          C—C
         /   \
  Cl—C       C—Cl
         \   /
          C—C
         /   \
        H     H
```

a. Draw the Lewis structure of PDB and note any nonzero formal charges.
b. Is the structure stabilized by resonance? If so, draw all resonance structures.
c. Which, if any, of the bonds in the structure you drew are polar?
d. Predict the average carbon–carbon bond length and bond strength in the structure you drew.

Collect and Organize We are given the skeletal structure of a molecule and are asked to draw its Lewis structure, including all resonance structures, and to perform a formal charge analysis. We are also asked to identify any nonpolar bonds in the structure and to predict the length and strength of the carbon–carbon bonds. Bond polarity depends on the difference in electronegativities of the bonded atoms, which are given in Figure 8.9. Table 8.3 lists average lengths and energies (strengths) of covalent bonds.

Analyze The five-step procedure used in Sample Exercises 8.2 through 8.5 to draw Lewis structures of other small molecules should be useful in drawing the Lewis structure of PDB. Resonance structures for PDB, like those for benzene (Figure 8.16), should be possible if there are alternating single and double carbon–carbon bonds in PDB's six-membered ring of carbon atoms.

Solve

a and b. The number of valence electrons is

Element:	6 C	+	2 Cl	+	4 H
Valence electrons:	(6×4)	+	(2×7)	+	$(4 \times 1) = 42$

Completing the octets on the Cl atoms:

```
         H     H
          \   /
           C—C
          /   \
  :Cl—C         C—Cl:
          \   /
           C—C
          /   \
         H     H
```

gives a structure with 36 valence electrons (12 bonding pairs and 6 lone pairs). We have 6 more electrons to place to reach 42, and none of the six carbons in the ring has an octet of electrons. Instead, each participates in three bonds, so each needs two more electrons. Placing a lone pair on every other C gives those three C atoms an octet of electrons but leaves the other three with only 6 electrons. If we convert the three lone pairs into three carbon–carbon double bonds, each C atom will have an octet, provided we distribute them evenly around the ring to avoid any C atoms with five bonds. Two equivalent resonance structures can be drawn to show the bonding pattern:

```
         H     H
          \   /
           C=C
          /   \
  :Cl—C         C—Cl:
          \\   //
           C—C
          /   \
         H     H
            ↕
         H     H
          \   /
           C—C
          //   \\
  :Cl—C         C—Cl:
          \   /
           C=C
          /   \
         H     H
```

Resonance stabilizes the structure of PDB. Each C atom has four bonds and each H and Cl atom has one bond, so every atom has the number of bonds that matches its bond capacity. This means that all formal charges are zero.

c. The differences in electronegativities for the bonded pairs of atoms are

C—C	$\Delta EN = 0$
Cl—C	$\Delta EN = 3.0 - 2.5 = 0.5$
C—H	$\Delta EN = 2.5 - 2.1 = 0.4$

Based on ΔEN values, we may conclude that the bonds between carbon and chlorine atoms are polar bonds and the bonds between pairs of carbon atoms are not polar. However, the bonds between pairs of elements with small ΔEN values (0.4 or less and especially C—H bonds) are considered *essentially* nonpolar.

d. The even distribution of a total of 9 bonding pairs of electrons among 6 C atoms means that, on average, each pair shares 1.5 pairs of bonding electrons. The corresponding bond length and bond strength should be about halfway between those of C—C single and C=C double bonds, given in Table 8.3:

Approximate bond length:

$$[(154 + 134)/2] \text{ pm} = 144 \text{ pm}$$

Approximate bond strength:

$$[(348 + 614)/2] \text{ kJ/mol} = 481 \text{ kJ/mol}$$

Think About It The resonance structures closely resemble those of benzene, which is reflected in the common name of PDB, *para*-dichlorobenzene. The two polar C—Cl bonds in PDB are oriented *in opposite directions*. Thus, the unequal sharing of the bonding pair of electrons in the Cl—C bond on the left side of the molecule is offset by the unequal sharing of the bonding pair of electrons in the C—Cl bond on the right side. In Chapter 9 we explore how offsetting bond polarities in symmetrical molecules like this one explain why substances such as PDB are nonpolar overall.

SUMMARY

LO1 A chemical bond results when two atoms share electrons (a covalent bond) or when two ions are attracted to each other (an **ionic bond**). The atoms in metallic solids pool their electrons to form **metallic bonds**. (Section 8.1)

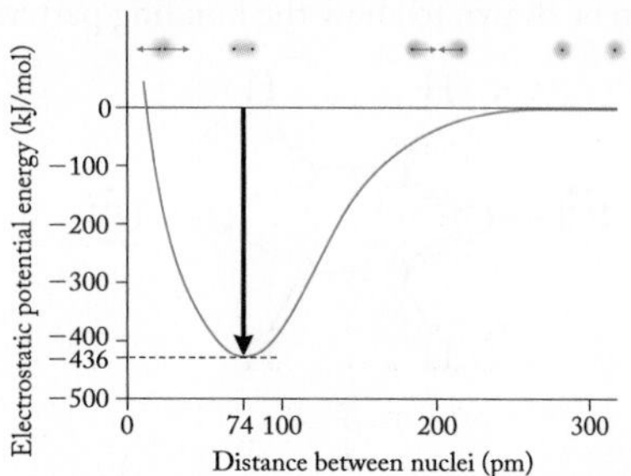

LO2 **Lewis symbols** use dots to represent paired and unpaired electrons in the ground states of atoms. The number of unpaired electrons indicates the number of bonds the element is likely to form—that is, its **bonding capacity**. Chemical stability is achieved when atoms have eight electrons in their valence *s* and *p* orbitals, following the **octet rule**. A **Lewis structure** shows the bonding pattern in molecules and polyatomic ions; pairs of dots represent **lone pairs** of electrons that do not contribute to bonding. A **single bond** consists of a single pair of electrons shared between two atoms; there are two shared pairs in a **double bond** and three shared pairs in a **triple bond**. (Section 8.2)

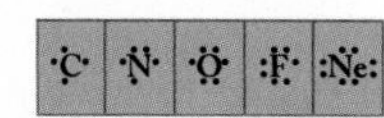

LO3 Unequal electron sharing between atoms of different elements results in **polar covalent bonds**. **Bond polarity** is a measure of how unequally the electrons in covalent bonds are shared. Greater polarity results from larger differences in the **electronegativities** of the bonded atoms. Electronegativity is a relative measure of the ability of an atom to attract electrons in a bond to itself. Polarity of bonds in molecules of atmospheric gases may lead to the absorption of infrared radiation, which contributes to the greenhouse effect. (Section 8.3)

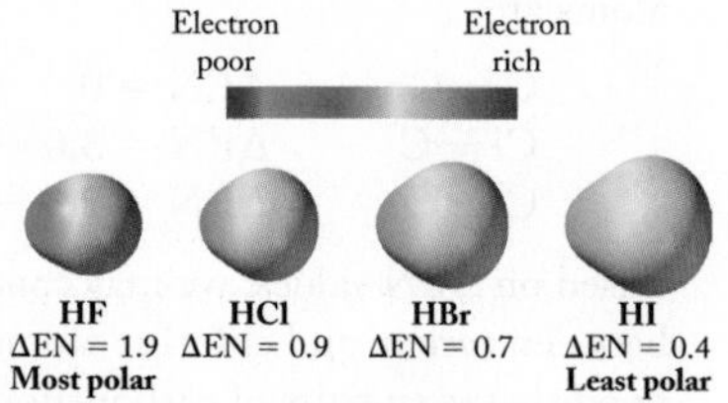

LO4 Covalent bonds behave more like flexible springs than rigid rods. They can undergo a variety of bond vibrations. The vibrations of polar bonds may create fluctuating electrical fields that allow molecules to absorb infrared (IR) electromagnetic radiation. When atmospheric gases absorb IR radiation, they contribute to the greenhouse effect. (Section 8.3)

LO5 Two or more Lewis structures—called **resonance structures**—can sometimes be drawn for one molecule or polyatomic ion. The actual bonding pattern in a molecule is a combination of resonance structures. The preferred resonance structure of a molecule is one in which the **formal charges (FC)** on its atoms are zero or as close to zero as possible, and any negative formal charges are on the more electronegative atoms. The formal charge on an atom in a Lewis structure is the difference between the number of valence electrons in the free atom and the sum of the number of electrons in lone pairs and half the number of electrons in bonding pairs on the bonded atom. **Free radicals** include reactive molecules that have an odd number of valence electrons and contain atoms with incomplete octets. Atoms of elements in the third row of the periodic table with $Z > 12$ and beyond may accommodate more than an octet of electrons. (Sections 8.4, 8.5, 8.6)

LO6 **Bond order** is the number of bonding pairs in a covalent bond. **Bond energy** is the enthalpy change, ΔH, required to break 1 mole of a covalent bond in the gas phase; energy changes in chemical reactions depend on the energy required to break bonds in the reactants and the energy released when new bonds are formed in the products. **Bond length** is the distance between the nuclear centers of two bonded atoms. As the bond order between two atoms increases, the bond length decreases and the bond energy increases. (Sections 8.1, 8.7)

PARTICULATE PREVIEW WRAP-UP

The space-filling models look identical and do not tell us which bonds are present in these molecules. O_2, F_2, and N_2 are the molecular formulas for the oxygen, fluorine, and nitrogen molecules, respectively, from left to right. The molecular formulas do not tell us which bonds are present in a molecule, either, but a ball-and-stick model would show a single bond as one stick, a double bond as two sticks, and a triple bond as three sticks.

PROBLEM-SOLVING SUMMARY

Type of Problem	Concepts and Equations	Sample Exercises
Describing differences in covalent, ionic, and metallic bonds	All bonds exist as a result of electrostatic attractions. Ionic bonds form between cations and anions, both of which can be polyatomic ions. Covalent bonds form between atoms of nonmetals or metalloids. Metallic bonds form from the delocalization of electrons across atoms of metallic elements.	**8.1**
Drawing Lewis structures for molecules, monatomic ions, and ionic compounds	Connect the atoms with single covalent bonds, distribute the valence electrons to give each outer atom (except H) eight valence electrons; use multiple bonds where necessary to complete the central atom's octet.	**8.2–8.5**

Type of Problem	Concepts and Equations	Sample Exercises
Comparing bond polarities	Calculate the difference in electronegativity (ΔEN) between the two bonded atoms; if $\Delta EN \geq 2.0$, the bond is considered ionic.	**8.6**
Drawing resonance structures of molecules and polyatomic ions	Include all possible arrangements of covalent bonds in the molecule if more than one equivalent structure can be drawn.	**8.7, 8.8**
Selecting resonance structures on the basis of formal charges	Calculate formal charge by using $$FC = \left(\begin{matrix}\text{number of}\\ \text{valence } e^-\end{matrix}\right) - \left[\begin{matrix}\text{number of}\\ \text{nonbonding } e^-\end{matrix} + \frac{1}{2}\left(\begin{matrix}\text{number of } e^-\\ \text{in bonding pairs}\end{matrix}\right)\right] \quad (8.1)$$ Select structures with formal charges closest to zero and with negative formal charges on the most electronegative atoms.	**8.9**
Drawing Lewis structures of odd-electron molecules	Distribute the valence electrons in the Lewis structure to leave the most electronegative atom(s) with eight valence electrons and the least electronegative atom with the odd number of electrons.	**8.10**
Drawing Lewis structures containing atoms with more than an octet of valence electrons	Distribute the valence electrons in the Lewis structure such that atoms of elements with $Z > 12$ have more than eight valence electrons if more than four bonds are needed or if doing so results in a structure with formal charges closer to zero.	**8.11**
Drawing Lewis structures with an incomplete octet	Distribute the valence electrons in the Lewis structure, allowing atoms of elements such as Be, B, and Al to have fewer than eight valence electrons if needed.	**8.12**
Estimating heats of reaction from average bond energies	Multiply bond energies by the number of bonds and calculate using $$\Delta H_{rxn} = \sum \Delta H_{\text{bonds breaking}} - \sum \Delta H_{\text{bonds forming}} \quad (8.4)$$	**8.13**

VISUAL PROBLEMS

(Answers to boldface end-of-chapter questions and problems are in the back of the book.)

8.1. Which of the elements highlighted in Figure P8.1 have Lewis symbols that contain (a) two dots, (b) four dots, (c) six dots, (d) eight dots?

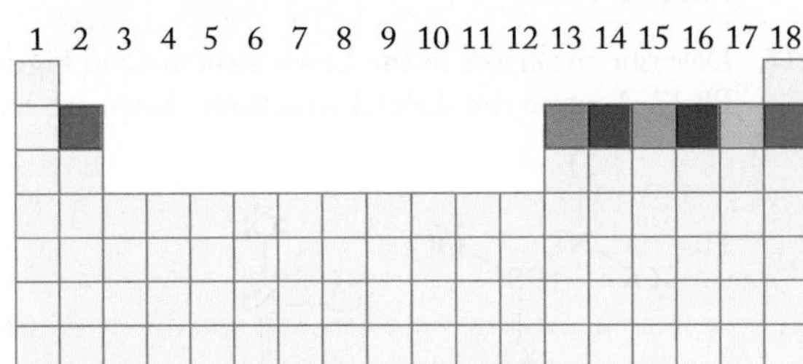

FIGURE P8.1

8.2. Which of the elements highlighted in Figure P8.1 form ions whose Lewis symbols have no dots?

8.3. Which of the elements highlighted in Figure P8.1 has the greatest bonding capacity?

8.4. Which of the elements highlighted in Figure P8.1 form monatomic ions whose Lewis symbols have four pairs of dots? (*Hint*: See Figure 2.20.)

8.5. Which of the elements highlighted in Figure P8.1 has the greatest electronegativity?

8.6. Which two of the highlighted elements in Figure P8.1 form the bonding pair with the most ionic character?

8.7. Which of the Lewis symbols in Figure P8.7 correctly portrays the most stable ion of magnesium?

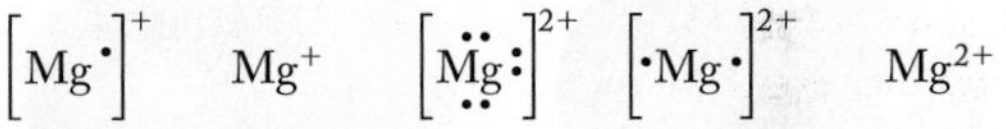

$[Mg^{\bullet}]^{+}$ $\quad Mg^{+}$ $\quad [\ddot{Mg}:]^{2+}$ $\quad [\cdot Mg \cdot]^{2+}$ $\quad Mg^{2+}$

FIGURE P8.7

8.8. Which Lewis symbol in Figure P8.8 correctly portrays a stable monatomic ion?

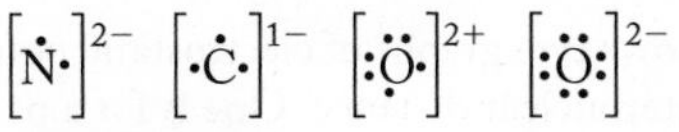

$[\dot{N}\cdot]^{2-}$ $\quad [\cdot\dot{C}\cdot]^{1-}$ $\quad [:\ddot{O}\cdot]^{2+}$ $\quad [:\ddot{O}:]^{2-}$

FIGURE P8.8

Note: The color scale used in Problems 8.9 through 8.11 is the same as in Figure 8.8, where violet means low electron density and red means high electron density.

***8.9.** Which of the drawings in Figure P8.9 best represents the distribution of electron density in LiF?

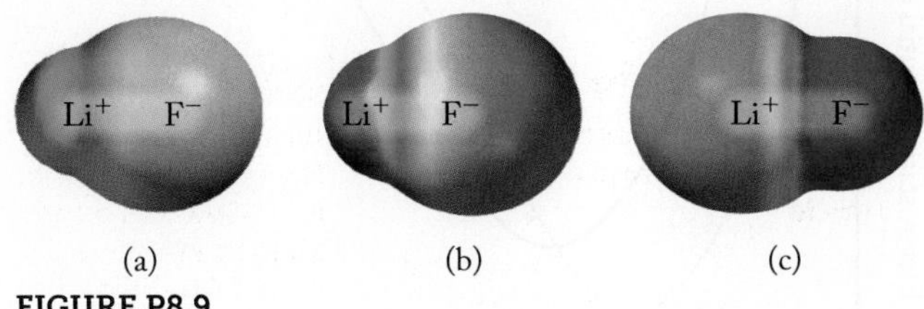

FIGURE P8.9

8.10. Which of the drawings in Figure P8.10 best represents the distribution of electron density in ClBr?

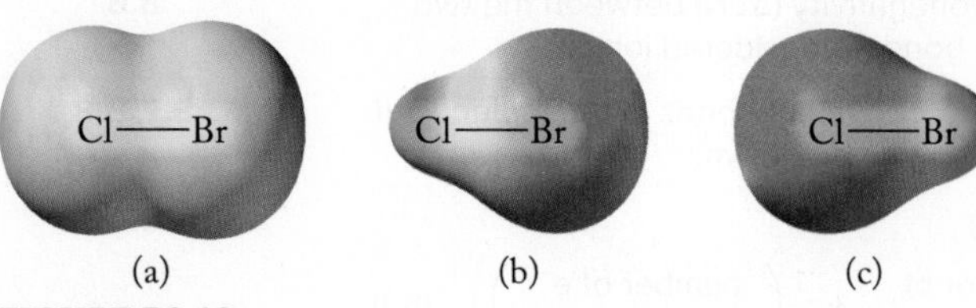

FIGURE P8.10

*8.11. Which of the drawings in Figure P8.11 most accurately represents the distribution of electron density in SO_2? Explain your answer.

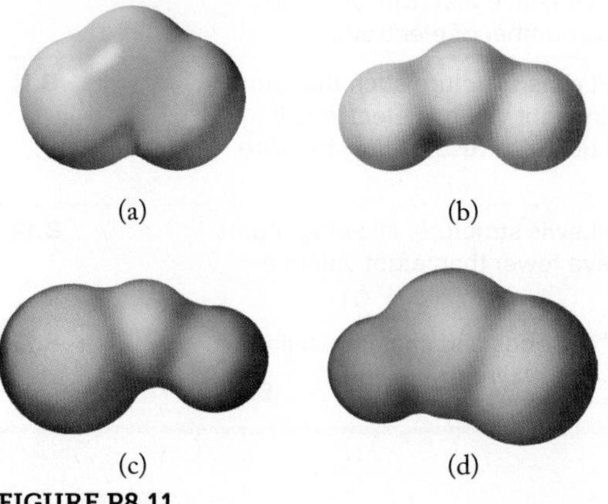

FIGURE P8.11

*8.12. Which of the drawings in Figure P8.12 most accurately represents the distribution of electrical charge in ozone? Explain your answer.

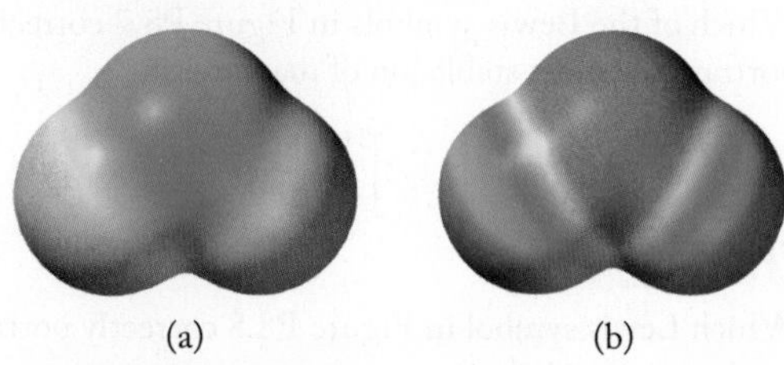

FIGURE P8.12

8.13. Figure P8.13 shows two graphs of electrostatic potential energy versus internuclear distance. One is for a pair of potassium and chloride ions, and the other is for a pair of potassium and fluoride ions. Which is which?

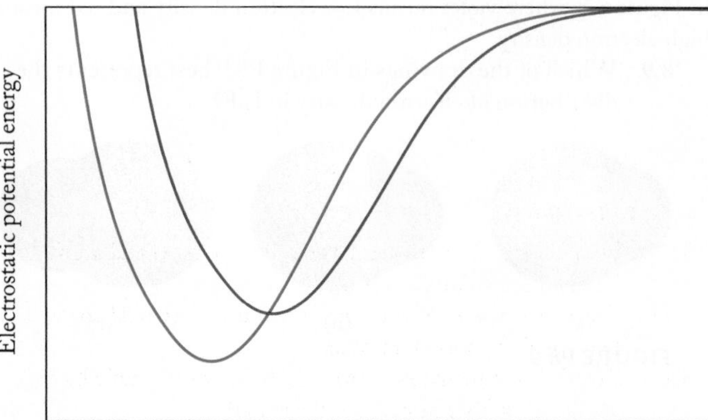

FIGURE P8.13

8.14. Water in the atmosphere is a greenhouse gas, which means its molecules are transparent to visible light but may absorb photons of infrared radiation. Which of the three modes of bond vibration shown in Figure P8.14 are infrared active? (The angle between the O—H bonds in H_2O is 104.5°. The reason why is discussed in Chapter 9.)

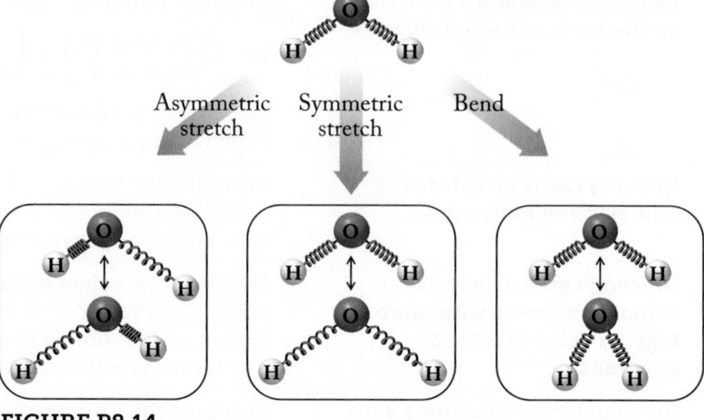

FIGURE P8.14

8.15. Explain why the structures in Figure P8.15 are not all resonance forms of the molecule S_2O.

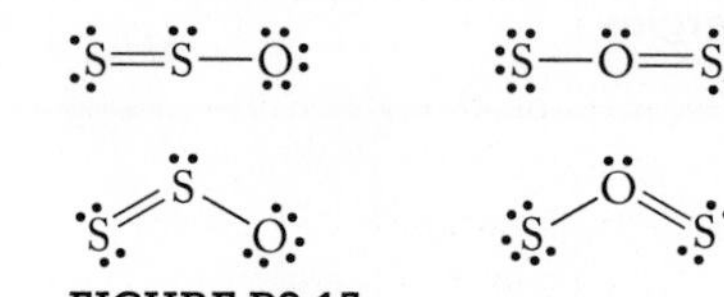

FIGURE P8.15

8.16. Are the three structures in Figure P8.16 resonance forms of the thiocyanate ion (SCN^-)? Explain why or why not.

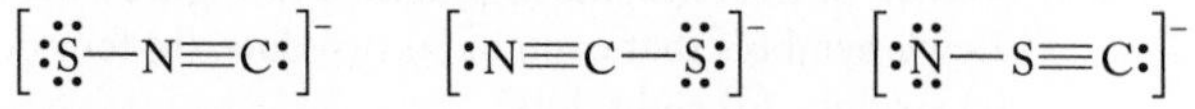

$[:\ddot{S}-N\equiv C:]^-$ $[:N\equiv C-\ddot{S}:]^-$ $[:\ddot{N}-S\equiv C:]^-$

FIGURE P8.16

8.17. Describe the errors in the Lewis structures in Figure P8.17. Assume the skeletal structures shown are correct.

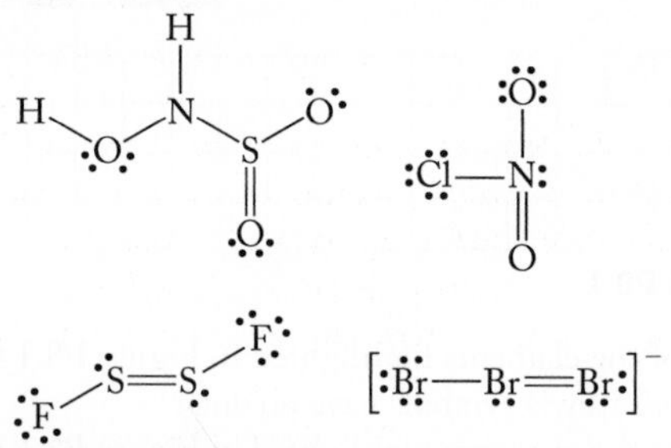

FIGURE P8.17

8.18. In each of the three pairs of resonance structures in Figure P8.18, which structure contributes more to the bonding in the molecule or molecular ion?

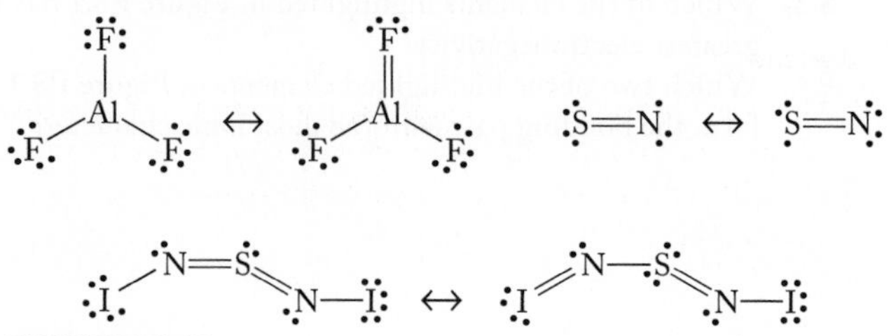

FIGURE P8.18

8.19. Which line in the graph in Figure P8.19 best represents the relationship between bond order and bond length? What would a graph of bond order versus bond energy look like?

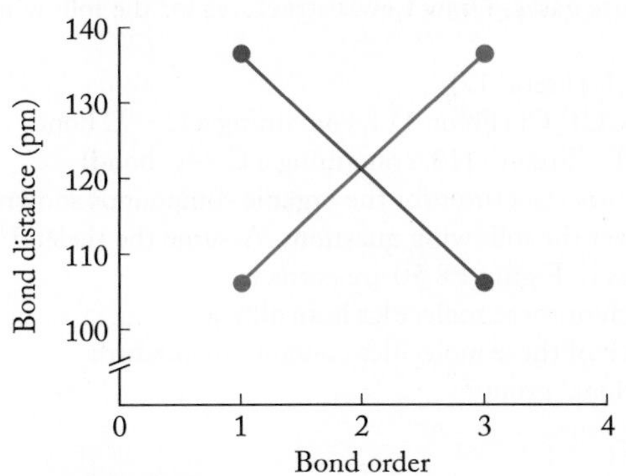

FIGURE P8.19

8.20. Use representations [A] through [I] in Figure P8.20 to answer questions (a)–(f).

a. Which hydrocarbon has the strongest carbon–carbon bond?
b. Which hydrocarbon has the weakest carbon–carbon bond?
c. Could [B] and/or [G] absorb an infrared photon through an asymmetric stretch and contribute to the greenhouse effect?
d. Label the most polar bond in [C] and in [D], using both the arrow (↔) and δ+ and δ− notations.
e. Which process is exothermic?
f. Which representation demonstrates a key limitation of Lewis structures?

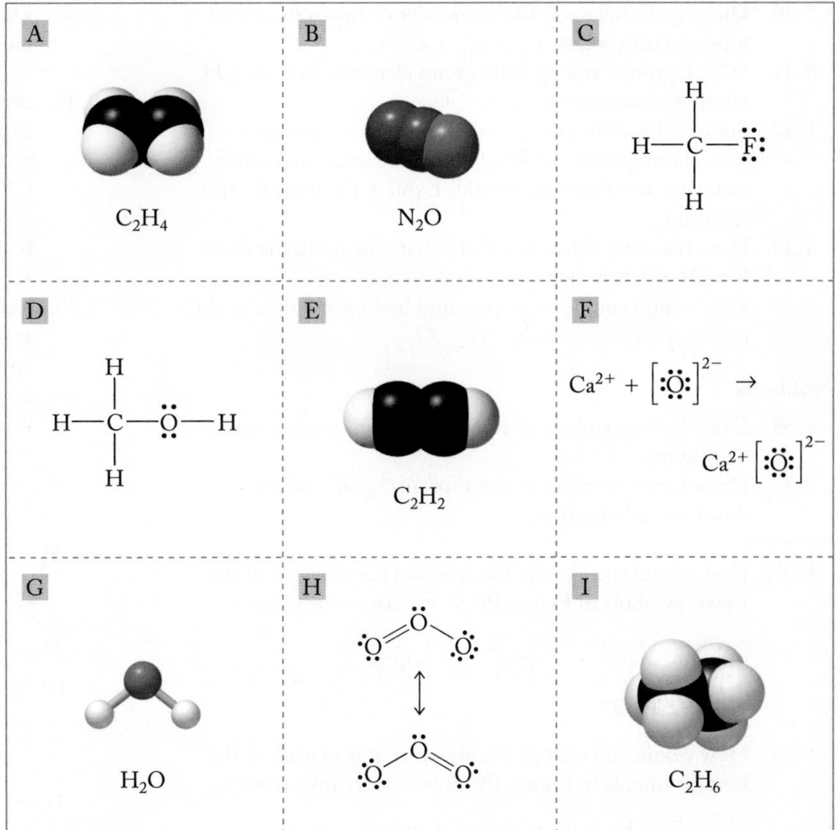

FIGURE P8.20

QUESTIONS AND PROBLEMS

Types of Chemical Bonds

Concept Review

8.21. In a hydrogen molecule, interactions between which subatomic particles cause electrostatic potential energy to decrease as two H atoms approach each other?

8.22. What name is given to the distance at which electrostatic potential energy between the atoms in a diatomic molecule is at a minimum?

8.23. In a halite (NaCl) crystal, which particles are electrostatically attracted to one another and which particles repel each other?

8.24. Describe two factors that explain why the magnesium and oxide ions in solid MgO are more strongly attracted to each other than the potassium and bromide ions in KBr?

Problems

8.25. Identify the type of bonding in each substance, and then for each pair of substances, describe the similarities and differences between the types of bonding.

a. Ni(*s*), S_8(*s*)
b. Ni(*s*), Na_2S(*s*)
c. SO_2(*g*), Na_2S(*s*)
d. Ni(*s*), Na_2SO_4(*s*)
e. S_8(*s*), Na_2SO_4(*s*)
f. Na_2S(*s*), Na_2SO_4(*s*)

8.26. Identify the type of bonding in each substance, and then for each pair of substances, describe the similarities and differences between the types of bonding.

a. Ag(*s*), Cl_2(*g*)
b. Ag(*s*), AgCl(*s*)
c. Cl_2(*g*), AgCl(*s*)
d. Ag(*s*), Ag_2CO_3(*s*)
e. Cl_2(*g*), Ag_2CO_3(*s*)
f. AgCl(*s*), Ag_2CO_3(*s*)

Lewis Structures

Concept Review

8.27. Does the number of valence electrons in a neutral atom ever equal the atomic number?

8.28. Does the number of valence electrons in a neutral atom ever equal the group number?

8.29. Do all the elements in a group in the periodic table have the same number of valence electrons?

8.30. Distinguish between an atom's valence electrons and its total electron count.

8.31. Which groups among main group elements have an odd number of valence electrons?

8.32. Some of his critics described G. N. Lewis's approach to explaining covalent bonding as an exercise in double counting and therefore invalid. Explain the basis for this criticism.

8.33. Does the octet rule mean that a diatomic molecule must have 16 valence electrons?

8.34. Why would you not expect to find hydrogen atoms in the bonding arrangement X—H—X?

Problems

8.35. Draw Lewis symbols of atoms of lithium, magnesium, and aluminum.

8.36. Draw Lewis symbols of atoms of nitrogen, oxygen, fluorine, and chlorine.

8.37. How would you change the arrays of dots in each of the Lewis symbols in Figure P8.37 to correct any errors?

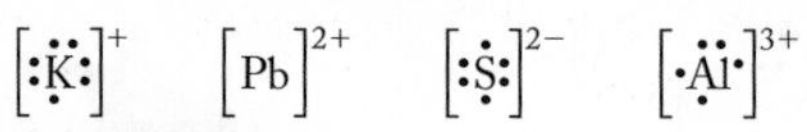

$[:\ddot{\underset{..}{K}}:]^{+}$ $[Pb]^{2+}$ $[:\dot{\underset{.}{S}}:]^{2-}$ $[\cdot\ddot{\underset{.}{Al}}\cdot]^{3+}$

FIGURE P8.37

8.38. How would you change the arrays of dots in each of the Lewis symbols in Figure P8.38 to correct any errors?

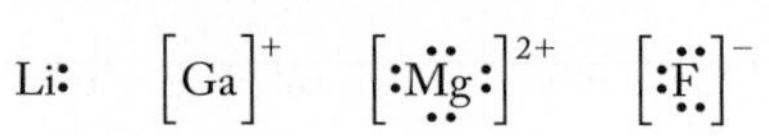

Li: $[Ga]^{+}$ $[:\ddot{\underset{..}{Mg}}:]^{2+}$ $[:\ddot{F}\,]^{-}$

FIGURE P8.38

8.39. Draw Lewis symbols for In^{+}, I^{-}, Ca^{2+}, and Sn^{2+}. Which ions have a complete valence-shell octet?

8.40. Draw Lewis symbols of Xe, Sr^{2+}, Cl, and Cl^{-}. How many valence electrons are in each atom or ion?

8.41. Sodium ions are isoelectronic with a noble gas.
a. Identify this noble gas.
b. Does this noble gas have the same Lewis symbol as a sodium ion?

8.42. In what ways are the Lewis symbols for the oxide anion and the argon atom similar?

8.43. Draw the Lewis symbol of an ion that has the following:
a. 1+ charge and 1 valence electron
b. 3+ charge and 0 valence electrons

8.44. Draw the Lewis symbol of an ion that has the following:
a. 1− charge and 8 valence electrons
b. 1+ charge and 5 valence electrons

8.45. How many valence electrons does each of the following species contain? (a) BN; (b) HF; (c) OH^{-}; (d) CN^{-}

8.46. How many valence electrons does each of the following species contain? (a) N_2^{+}; (b) CS^{+}; (c) CN; (d) CO

8.47. Draw Lewis structures for the following diatomic molecules and ions: (a) CO; (b) O_2; (c) ClO^{-}; (d) CN^{-}.

8.48. Draw Lewis structures for the following diatomic molecules and ions: (a) F_2; (b) NO^{+}; (c) SO; (d) HI.

8.49. **Greenhouse Gases** Chlorofluorocarbons (CFCs) are linked to the depletion of stratospheric ozone. They are also greenhouse gases. Draw Lewis structures for the following CFCs:
a. CF_2Cl_2 (Freon 12)
b. Cl_2FCCF_2Cl (Freon 113, containing a C—C bond)
c. C_2ClF_3 (Freon 1113, containing a C=C bond)

8.50. Draw Lewis structures for the organic compounds shown and answer the following questions. Assume the skeletal structures in Figure P8.50 are correct.
a. Which of these molecules is an alkyne?
b. Which of these molecules contains an aldehyde functional group?

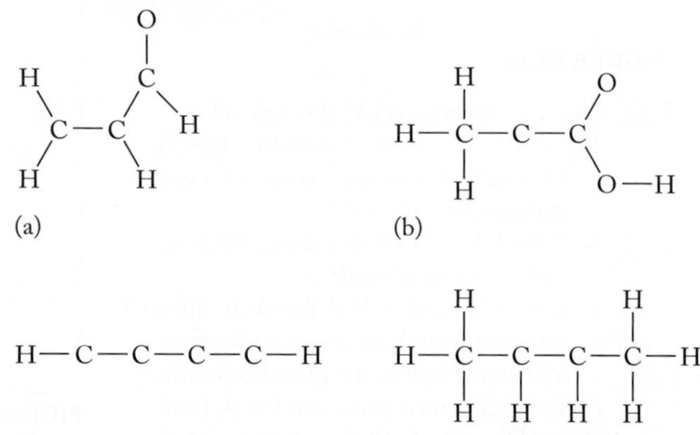

FIGURE P8.50

8.51. **Skunks and Rotten Eggs** Many sulfur-containing organic compounds have characteristically foul odors: butanethiol ($CH_3CH_2CH_2CH_2SH$) is responsible for the odor of skunks, and rotten eggs smell the way they do because they produce tiny amounts of pungent hydrogen sulfide (H_2S). Draw the Lewis structures for $CH_3CH_2CH_2CH_2SH$ and H_2S.

8.52. **Acid in Ants** Formic acid (HCOOH) is the smallest organic acid and was originally isolated by distilling red ants. Draw its Lewis structure if the atoms are connected as shown in Figure P8.52.

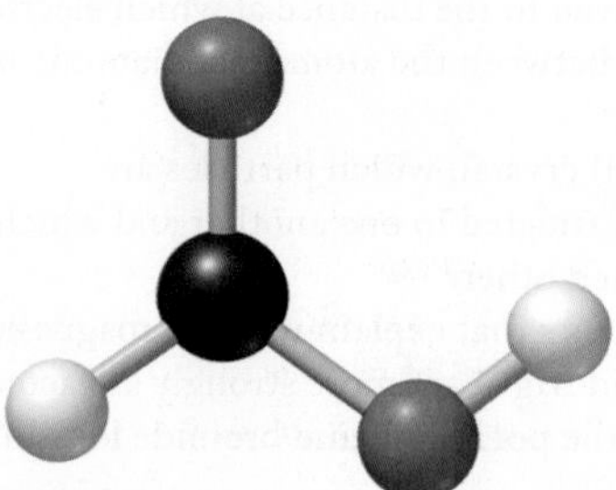

FIGURE P8.52

8.53. **Chlorine Bleach** Chlorine combines with oxygen in several proportions. Dichlorine monoxide (Cl_2O) is used in the manufacture of bleaching agents. Potassium chlorate ($KClO_3$) is used in oxygen generators aboard aircraft. Draw the Lewis structures for Cl_2O and ClO_3^- (Cl is the central atom).

8.54. **Dangers of Mixing Cleansers** Labels on household cleansers caution against mixing bleach with ammonia (Figure P8.54) because the reaction produces monochloramine (NH_2Cl) and hydrazine (N_2H_4), both of which are toxic:

$$NH_3(aq) + OCl^-(aq) \rightarrow NH_2Cl(aq) + OH^-(aq)$$

$$NH_2Cl(aq) + NH_3(aq) + OH^-(aq) \rightarrow N_2H_4(aq) + Cl^-(aq) + H_2O(\ell)$$

Draw the Lewis structures for monochloramine and hydrazine.

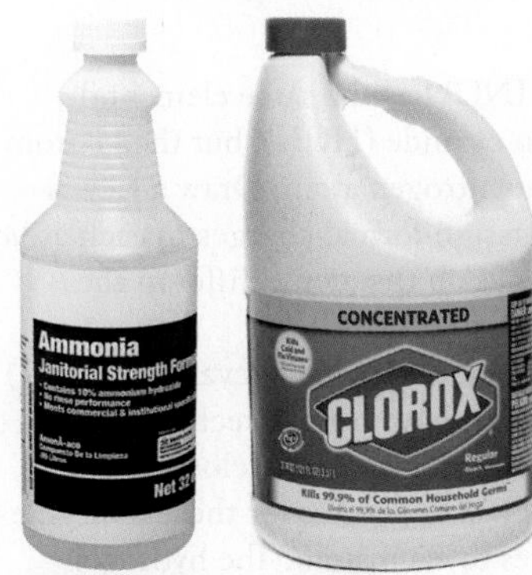

FIGURE P8.54

Polar Covalent Bonds

Concept Review

8.55. How can we use electronegativity to predict whether a bond between two atoms is likely to be covalent or ionic?

8.56. How do the electronegativities of the elements change across a period and down a group?

8.57. Explain, based on atomic structure, why trends in electronegativity are related to trends in atomic size.

8.58. Is the element with the most valence electrons in a period also the most electronegative? Explain.

8.59. What is meant by the term *polar covalent bond*?

8.60. What factor is responsible for the existence of polar covalent bonds?

8.61. Describe how atmospheric greenhouse gases act like the panes of glass in a greenhouse.

*8.62. **Understanding the Greenhouse Effect** Water vapor in the atmosphere contributes more to the greenhouse effect than carbon dioxide, yet water vapor is not considered an important factor in global warming. Propose a reason why.

8.63. Increasing concentrations of nitrous oxide in the atmosphere may be contributing to climate change. Is the ability of N_2O to absorb infrared radiation due to nitrogen–nitrogen bond stretching, nitrogen–oxygen bond stretching, or both? Explain your answer.

8.64. Atmospheric oxygen (O_2) is not a greenhouse gas, but atmospheric ozone (O_3) is. What types of bond vibration in molecules of O_3 are infrared active?

Problems

8.65. Which of the following bonds are polar: C—Se, C—O, Cl—Cl, O═O, N—H, C—H? In the bond or bonds that you selected, which atom has the greater electronegativity?

8.66. Which is the least polar bond: C—Se, C═O, Cl—Br, O═O, N—H, C—H?

8.67. In which of the following binary compounds is the bond expected to have the *least* ionic character? LiCl; CsI; KBr; NaF

8.68. In which of the following compounds is the bond between the atoms expected to have the most covalent character? $AlCl_3$; $AlBr_3$; AlI_3; GaF_3

8.69. Which bond in Figure P8.69 is polar and has a correctly drawn polarity arrow?

$\overset{+\!\!\longrightarrow}{\text{Br—C}}$ $\quad$ $\overset{+\!\!\longrightarrow}{\text{I—Cl}}$ $\quad$ $\overset{+\!\!\longrightarrow}{\text{O—S}}$

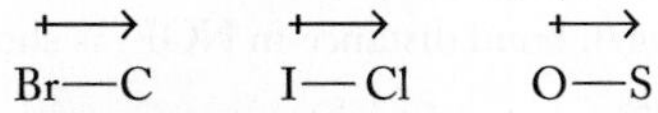

FIGURE P8.69

8.70. Which bond in Figure P8.70 is polar and has a correctly drawn polarity arrow?

$\overset{+\!\!\longrightarrow}{\text{F—C}}$ $\quad$ $\overset{+\!\!\longrightarrow}{\text{P—Cl}}$ $\quad$ $\overset{+\!\!\longrightarrow}{\text{O—O}}$

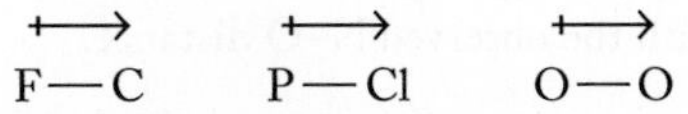

FIGURE P8.70

8.71. Which bond in the following is correctly labeled?

$\overset{\delta-\ \delta+}{\text{O—H}}$ $\quad$ $\overset{\delta-\ \delta+}{\text{C—Cl}}$ $\quad$ $\overset{\delta-\ \delta+}{\text{Br—Br}}$

8.72. Which bond in the following is correctly labeled?

$\overset{\delta+\ \delta-}{\text{Na—Cl}}$ $\quad$ $\overset{\delta+\ \delta-}{\text{Li—F}}$ $\quad$ $\overset{\delta+\ \delta-}{\text{Si—F}}$

8.73. Which substance has the most polar covalent bonds: PF_3, S_8, RbCl, or SF_2?

8.74. Atoms of which element are held together by nonpolar covalent bonds: lithium, phosphorus, or xenon?

Resonance

Concept Review

8.75. Explain the concept of resonance.

8.76. How does resonance influence the stability of a molecule or an ion?

8.77. What factors determine whether a molecule or ion exhibits resonance?

8.78. What structural features do all the resonance forms of a molecule or ion have in common?

8.79. Explain why NO_2 is more likely to exhibit resonance than CO_2.

8.80. Are these two skeletal structures resonance forms: X—X—O and X—O—X? Explain.

Problems

8.81. Draw Lewis structures for fulminic acid (HCNO), showing all resonance forms.

8.82. Draw Lewis structures for hydrazoic acid (HN_3), showing all resonance forms.

***8.83.** Oxygen and nitrogen combine to form a variety of nitrogen oxides, including the following two unstable compounds, each with two nitrogen atoms per molecule: N_2O_2 and N_2O_3. Draw Lewis structures for these molecules, showing all resonance forms.

*8.84. Oxygen and sulfur combine to form a variety of sulfur oxides. Some are stable molecules and some, including S_2O_2 and S_2O_3, decompose when they are heated. Draw Lewis structures for these two compounds, showing all resonance forms.

***8.85.** The O–O bond distance in F_2O_2 is about 20% shorter than in hydrogen peroxide.
a. Draw Lewis structures for both F_2O_2 and H_2O_2.
b. It was proposed that F_2O_2 has a resonance form of $[FO_2]^+F^-$. Draw a Lewis structure for this "ionic" form of F_2O_2 and explain how the structure is consistent with the observed O–O distance.

*8.86. The nitrogen–oxygen bond distance in NOF_3 is shorter than in $NO(CH_3)_3$.
a. Draw Lewis structures for NOF_3 and $NO(CH_3)_3$.
b. It was proposed that NOF_3 has a resonance form of $[NOF_2]^+F^-$. Draw a Lewis structure for this "ionic" form of NOF_3 and explain how the structure is consistent with the observed N–O distance.

8.87. Chemists can use the octet rule to predict the structures of new compounds to synthesize. Draw Lewis structures showing all resonance forms for the hypothetical compound ClSeNSO, where the atoms are connected in the order they are written.

8.88. Aromatic rings can connect to make larger molecules. The skeletal structures of two such molecules are shown in Figure P8.88. Draw Lewis structures for both molecules, showing all resonance forms.

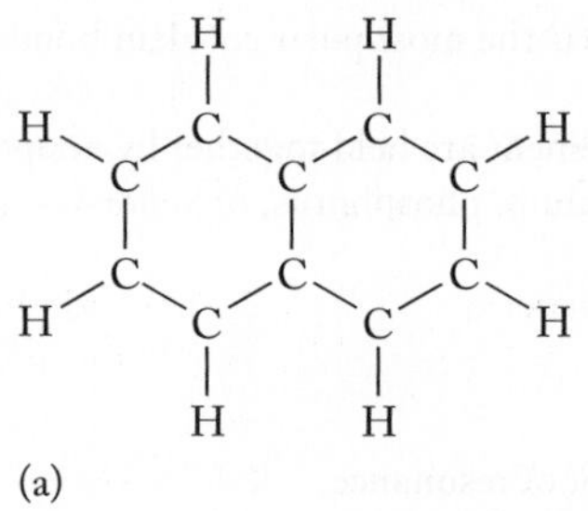

(a)

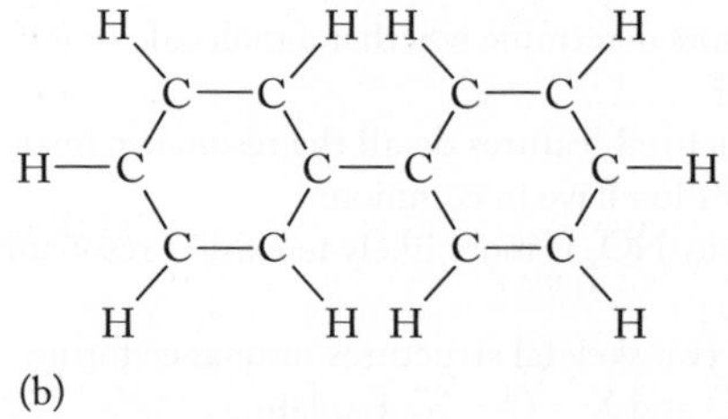

(b)

FIGURE P8.88

Formal Charge: Choosing among Lewis Structures

Concept Review

8.89. Describe how formal charges are used to choose between possible molecular structures.

8.90. How do the electronegativities of elements influence the selection of which Lewis structure is favored?

8.91. In a molecule containing S and O atoms, is a structure with a negative formal charge on sulfur more likely to contribute to bonding than an alternative structure with a negative formal charge on oxygen? Explain.

8.92. In a cation containing N and O, why do Lewis structures with a positive formal charge on nitrogen contribute more to the actual bonding in the molecule than do those structures with a positive formal charge on oxygen?

Problems

8.93. Hydrogen isocyanide (HNC) has the same elemental composition as hydrogen cyanide (HCN), but the H atom in HNC is bonded to the nitrogen atom. Draw a Lewis structure for HNC and assign formal charges to each atom. How do the formal charges on the atoms differ in the Lewis structures for HCN and HNC?

8.94. **Molecules in Interstellar Space** Hydrogen cyanide (HCN) and cyanoacetylene (HC_3N) have been detected in the interstellar regions of space and in comets close to Earth (Figure P8.94). Draw Lewis structures for these molecules and assign formal charges to each atom. The hydrogen atom is bonded to the carbon atom in both cases.

FIGURE P8.94

8.95. **Origins of Life** The discovery of polyatomic organic molecules such as cyanamide (H_2NCN) in interstellar space has led some scientists to believe that the molecules from which life began on Earth may have come from space. Draw Lewis structures for cyanamide and select the preferred structure based on formal charges.

8.96. Nitromethane (CH_3NO_2) reacts with hydrogen cyanide to produce $CNNO_2$ and CH_4:

$$HCN(g) + CH_3NO_2(g) \rightarrow CNNO_2(g) + CH_4(g)$$

a. Draw Lewis structures for CH_3NO_2, showing all resonance forms.
b. Draw Lewis structures for $CNNO_2$, showing all resonance forms, based on the two possible skeletal structures shown in Figure P8.96. Assign formal charges, and predict which structure is more likely to exist.

c. Are the two structures of $CNNO_2$ resonance forms of each other?

C—N—N(—O)(—O)   N—C—N(—O)(—O)

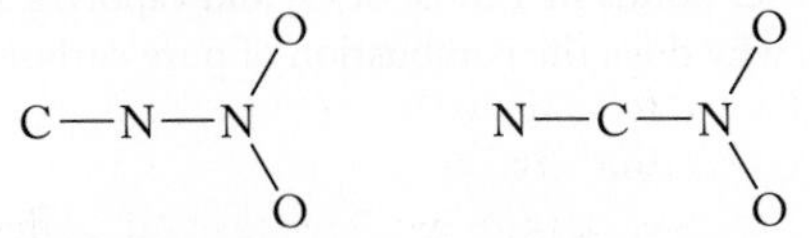

FIGURE P8.96

8.97. The formaldehyde molecule contains one carbon atom, one oxygen atom, and two hydrogen atoms. Lewis structures can be drawn that have either carbon or oxygen as the central atom.

a. Draw these structures and use formal charges to predict the more likely structure for formaldehyde.

b. Are the two structures you drew resonance forms of each other?

8.98. Draw all resonance forms of S_4N^- and assign formal charges. The atoms are arranged as SSNSS.

***8.99.** Nitrogen is the central atom in molecules of nitrous oxide (N_2O). Draw Lewis structures for another possible arrangement: N—O—N. Assign formal charges and suggest a reason that this structure is *un*likely to be stable.

8.100. Use formal charges to determine which resonance form of each of the following ions is preferred: (a) CNO^-; (b) NCO^-; (c) CON^-.

Exceptions to the Octet Rule

Concept Review

8.101. Are all odd-electron molecules exceptions to the octet rule?

8.102. Describe the factors that contribute to the stability of structures in which the central atoms have more than eight valence electrons.

8.103. Why do C, N, O, and F atoms in covalently bonded molecules and ions have no more than eight valence electrons?

8.104. Do atoms with $Z > 12$ *always* expand their valence shell? Explain your answer.

Problems

8.105. In which of the following molecules does the sulfur atom have more than eight valence electrons? (a) SF_6; (b) SF_5; (c) SF_4; (d) SF_2

8.106. In which of the following molecules does the phosphorus atom have more than eight valence electrons? (a) $POCl_3$; (b) PF_5; (c) PF_3; (d) P_2F_4 (which has a P—P bond)

8.107. How many electrons are there in the covalent bonds surrounding the central atom in the following species? (a) $(CH_3)_3Al$; (b) B_2Cl_4; (c) SO_3; (d) SF_5^-

8.108. How many electrons are there in the covalent bonds surrounding the central atom in the following species? (a) $POCl_3$; (b) $InCl_5^{2-}$; (c) FBO; (d) PF_4^-

***8.109.** Draw Lewis structures for NOF_3 and POF_3 in which the group 15 element is the central atom and the other atoms are bonded to it. What differences are there in the types of bonding in these molecules?

*8.110. The phosphate anion is common in minerals. The corresponding nitrogen-containing anion, NO_4^{3-}, is unstable but can be prepared by reacting sodium nitrate with sodium oxide at 300°C. Draw Lewis structures for each anion. What are the differences in bonding between these ions?

8.111. Dissolving NaF in selenium tetrafluoride (SeF_4) produces $NaSeF_5$. Draw Lewis structures for SeF_4 and SeF_5^-. In which structure does Se have more than eight valence electrons?

*8.112. Reaction between NF_3, F_2, and SbF_3 at 200°C and 100 atm pressure gives the ionic compound NF_4SbF_6:

$$NF_3(g) + 2\ F_2(g) + SbF_3(g) \rightarrow NF_4SbF_6(s)$$

Draw Lewis structures for the ions in this product.

8.113. **Ozone Depletion** The compound Cl_2O_2 plays a role in ozone depletion in the stratosphere, where it serves as a reservoir of Cl atoms that catalyze the destruction of O_3. Draw a Lewis structure for Cl_2O_2 based on the arrangement of atoms in Figure P8.113. Does either of the chlorine atoms in the structure have more than eight valence electrons?

FIGURE P8.113

*8.114. Trimethylaluminum reacts with dimethylamine, $(CH_3)_2NH$, forming methane and $Al_2(CH_3)_4[N(CH_3)_2]_2$ by the following balanced chemical equation:

$$2\ (CH_3)_3Al + 2\ HN(CH_3)_2 \rightarrow 2\ CH_4 + Al_2(CH_3)_4[N(CH_3)_2]_2$$

Draw the Lewis structures for the reactants and products. Must any of these structures contain atoms with fewer than eight valence electrons?

8.115. Which of the following chlorine oxides are odd-electron molecules? (a) Cl_2O_7; (b) Cl_2O_6; (c) ClO_4; (d) ClO_3; (e) ClO_2

8.116. Which of the following nitrogen oxides are odd-electron molecules? (a) NO; (b) NO_2; (c) NO_3; (d) N_2O_4; (e) N_2O_5

8.117. Which of the Lewis structures in Figure P8.117 contributes most to the bonding in CNO?

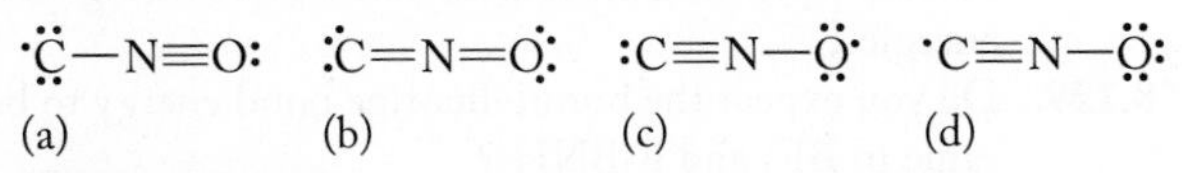

·C̈—N≡O: :C=N=Ö: :C≡N—Ö·   ·C≡N—Ö:

(a)   (b)   (c)   (d)

FIGURE P8.117

8.118. Which of the Lewis structures in Figure P8.118 are *un*likely to contribute to the bonding in NCO?

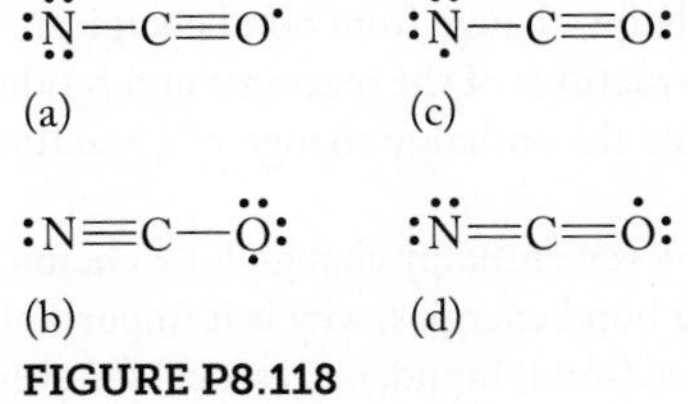

:N̈—C≡O·   :N̈—C≡O:

(a)   (c)

:N≡C—Ö: :N̈=C=Ö:

(b)   (d)

FIGURE P8.118

***8.119.** Using Lewis structures, explain why dimethylaluminum chloride is more likely to exist as $(CH_3)_4Al_2Cl_2$ than as $(CH_3)_2AlCl$.

*8.120. Some have argued that SF_6 has ionic resonance forms that do not require more than eight valence electrons for S. Draw a resonance structure consistent with this hypothesis and assign formal charges to each atom. Is this resonance form better than or the same as the one with more than eight valence electrons?

***8.121.** When left at room temperature, mixtures of BF_3 and BCl_3 are found to contain significant amounts of BF_2Cl and $BFCl_2$. Use Lewis structures to explain the origin of the latter two compounds.

*8.122. Reaction between $[XeF]^+[Sb_2F_{11}]^-$ and Xe in SbF_5 yields a product containing a bright blue $[Xe_4]^+$ cation with a linear arrangement of four Xe atoms (Figure P8.122).

Xe—Xe—Xe—Xe

353 pm | 319 pm | 353 pm

FIGURE P8.122

a. How many valence electrons are in XeF^+?
b. Draw Lewis structures for Xe_4^+, including any resonance forms.
c. How do the calculated bond distances help you choose between the resonance forms for Xe_4^+?

The Lengths and Strengths of Covalent Bonds

Concept Review

8.123. Do you expect the nitrogen–oxygen bond length in the nitrate ion to be the same as in the nitrite ion? Explain.

8.124. Why is the oxygen–oxygen bond length in O_3 different from the one in O_2?

8.125. Explain why the nitrogen–oxygen bond lengths in N_2O_4 (which has a nitrogen–nitrogen bond) and N_2O are nearly identical (118 and 119 pm, respectively).

8.126. Do you expect the sulfur–oxygen bond lengths in SO_3^{2-} and SO_4^{2-} ions to be about the same? Why?

8.127. Rank NO_2^-, NO^+, and NO_3^- in order of (a) increasing nitrogen–oxygen bond lengths and (b) increasing bond energies.

8.128. Rank CO, CO_2, and CO_3^{2-} in order of (a) increasing carbon–oxygen bond lengths and (b) increasing bond energies.

***8.129.** Do you expect the boron–fluorine bond energy to be the same in BF_3 and F_3BNH_3?

*8.130. The boron–oxygen distances in the BO_2^+ cation are equal. Does this mean the bond order of the B—O bond is two? Explain.

8.131. Why must the stoichiometry of a reaction be known to estimate the enthalpy change from bond energies?

8.132. Why must the structures of the reactants and products be known to estimate the enthalpy change of a reaction from bond energies?

***8.133.** When calculating the enthalpy change for a chemical reaction by using bond energies, why is it important to know the phase (solid, liquid, or gaseous) for every compound in the reaction?

*8.134. If the energy needed to break 2 moles of C═O bonds is greater than the sum of the energies needed to break the O═O bonds in 1 mole of O_2 and vaporize 1 mole of carbon, why does the combustion of pure carbon release heat?

Problems

Note: Use the average bond energies in Table 8.3 to answer Problems 8.135 through 8.146.

8.135. Use average bond energies to estimate the enthalpy changes of the following reactions:
a. $N_2(g) + 3\,H_2(g) \rightarrow 2\,NH_3(g)$
b. $N_2(g) + 2\,H_2(g) \rightarrow H_2NNH_2(g)$
c. $2\,N_2(g) + O_2(g) \rightarrow 2\,N_2O(g)$

8.136. Use average bond energies to estimate the enthalpy changes of the following reactions:
a. $CO_2(g) + H_2(g) \rightarrow H_2O(g) + CO(g)$
b. $N_2(g) + O_2(g) \rightarrow 2\,NO(g)$
*c. $C(s) + CO_2(g) \rightarrow 2\,CO(g)$
(*Hint*: The enthalpy of sublimation of graphite, C(*s*), is 719 kJ/mol.)

8.137. The combustion of CO to CO_2 releases 283 kJ/mol. What is the bond energy of the carbon–oxygen bond in carbon monoxide?

8.138. Use average bond energies to estimate the standard enthalpy of formation of HF gas.

8.139. Estimate how much less energy is released during the incomplete combustion of 1 mole of methane to carbon monoxide and water vapor than in the complete combustion to carbon dioxide and water vapor.

8.140. Estimate how much more energy is released by the reaction

$$C(s) + O_2(g) \rightarrow CO_2(g)$$

than by the reaction

$$C(s) + \tfrac{1}{2}\,O_2(g) \rightarrow CO(g)$$

***8.141.** Estimate ΔH_{rxn} for the following reaction:

$$4\,NH_3(g) + 7\,O_2(g) \rightarrow 4\,NO_2(g) + 6\,H_2O(g)$$

*8.142. The value of ΔH_{rxn} for the reaction

$$2\,H_2S(g) + 3\,O_2(g) \rightarrow 2\,SO_2(g) + 2\,H_2O(g)$$

is −1036 kJ. Estimate the energy of the bonds in SO_2.

8.143. A molecular view of the combustion of CS_2 is shown in Figure P8.143. If the standard enthalpy of combustion of CS_2 is −1102 kJ/mol, what is the average bond energy for the carbon–sulfur bonds in CS_2?

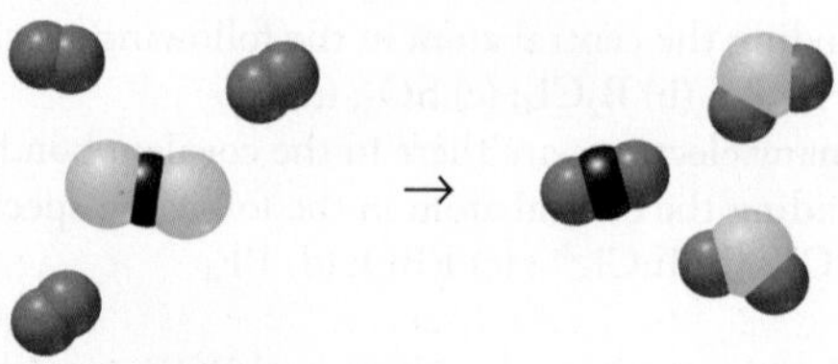

FIGURE P8.143

*8.144. The standard enthalpy of reaction for the decomposition of carbon oxysulfide (COS) to CO_2 and CS_2, as shown in Figure P8.144, is −1.9 kJ/mol. Are the apparent bond energies of the carbon–sulfur and carbon–oxygen bonds in COS stronger or weaker than in CS_2 and CO_2, respectively?

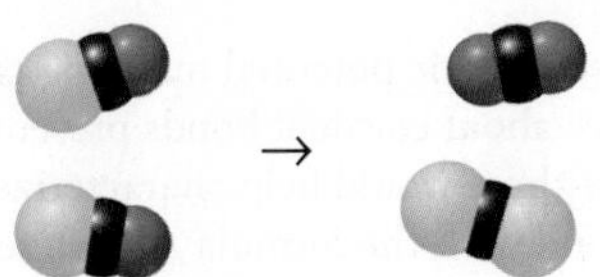

FIGURE P8.144

*8.145. Carbon and oxygen form three oxides: CO, CO_2, and carbon suboxide (C_3O_2).

a. Draw a Lewis structure for C_3O_2 in which the three carbon atoms have this bonding arrangement: C—C—C.

b. Predict whether the carbon–oxygen bond lengths in the carbon suboxide molecule are equal.

*8.146. Spectroscopic analysis of the linear molecule N_4O reveals that the nitrogen–oxygen bond length is 135 pm and that there are three nitrogen–nitrogen bond lengths: 148, 127, and 115 pm. Draw the Lewis structure for N_4O that is consistent with these observations.

Additional Problems

8.147. Why do we draw the Lewis symbol for the potassium ion without any "dots" to represent electrons?

8.148. Based on the Lewis symbols in Figure P8.148, predict to which group in the periodic table each element X belongs.

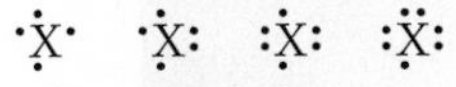

·X· ·X: :X: :X:

(a) (b) (c) (d)

FIGURE P8.148

8.149. Use formal charges to predict whether the atoms in carbon disulfide are arranged as CSS or SCS.

8.150. HClO, $HClO_2$, and $HClO_3$ represent a family of weak acids. Draw their Lewis structures, using formal charges to predict the best arrangement of atoms. Show any resonance forms of the molecules.

***8.151.** **Chemical Weapons** Phosgene is a poisonous gas first used in chemical warfare during World War I. It has the formula $COCl_2$ (C is the central atom).

a. Draw its Lewis structure.

b. Phosgene kills because it reacts with water in nasal passages, in the lungs, and on the skin to produce carbon dioxide and hydrogen chloride. Write a balanced chemical equation for this process, showing the Lewis structures for the reactants and products.

8.152. The dinitramide anion [$N(NO_2)_2^-$] was first isolated in 1996. The arrangement of atoms in $N(NO_2)_2^-$ is shown in Figure P8.152.

a. Complete the Lewis structure for $N(NO_2)_2^-$, including any resonance forms, and assign formal charges.

b. Explain why the nitrogen–oxygen bond lengths in $N(NO_2)_2^-$ and N_2O should (or should not) be similar.

c. $N(NO_2)_2^-$ was isolated as [NH_4^+][$N(NO_2)_2^-$]. Draw the Lewis structure for NH_4^+.

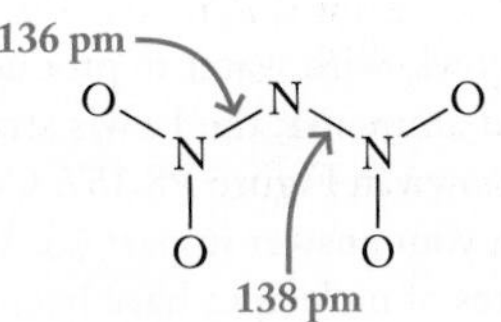

FIGURE P8.152

***8.153.** Sulfonylamine ($HNSO_2$) was first prepared in 2016. The two arrangements for the atoms are shown in Figure P8.153.

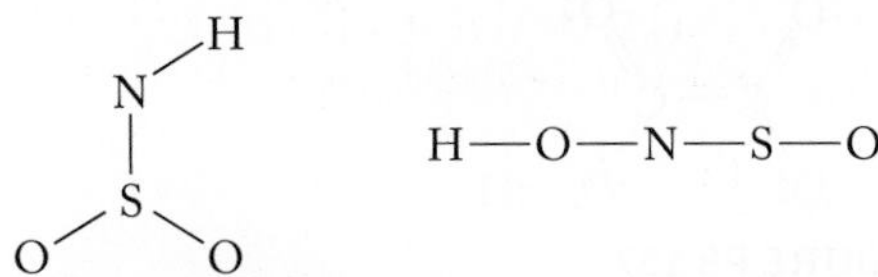

FIGURE P8.153

a. Complete the Lewis structures for each arrangement, including any resonance forms.

b. Do the two arrangements lead to the same S–N bond order?

c. Which arrangement might you expect based on formal charges?

d. Predict an approximate S–O bond length for the two arrangements.

*8.154. Sulfonylazide (HSO_3N_3) was first reported in 2016. Using the arrangement of the atoms shown in Figure P8.154, draw Lewis structures for HSO_3N_3 with

a. all atoms except H with a complete octet.

b. the lowest formal charges on sulfur.

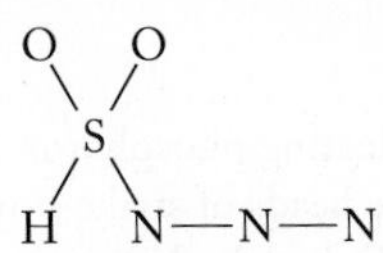

FIGURE P8.154

8.155. A compound with the formula Cl_2O_6 decomposes to a mixture of ClO_2 and ClO_4. Draw two Lewis structures for Cl_2O_6: one with a chlorine–chlorine bond and one with a Cl—O—Cl arrangement of atoms. Draw a Lewis structure for ClO_2.

$$Cl_2O_6 \rightarrow ClO_2 + ClO_4$$

*8.156. Dichlorine heptoxide (Cl_2O_7) decomposes by the following reaction:

$$Cl_2O_7 \rightarrow ClO_4 + ClO_3$$

a. Draw two Lewis structures for Cl_2O_7: one with a chlorine–chlorine bond and one with a Cl—O—Cl arrangement of atoms.

b. Draw a Lewis structure for ClO_3.

***8.157.** The odd-electron molecule CN dimerizes to give cyanogen (C_2N_2).

a. Draw a Lewis structure for CN and predict which arrangement for cyanogen is more likely: NCCN or CNNC.

b. Cyanogen reacts slowly with water to produce oxalic acid ($H_2C_2O_4$) and ammonia; the Lewis structure for oxalic acid is shown in Figure P8.157. Compare this structure with your answer in part (a). When the actual structures of molecules have been defined experimentally, the structures have been used to refine Lewis structures. Does this structure increase your confidence that the structure you selected in part (a) may be the better one?

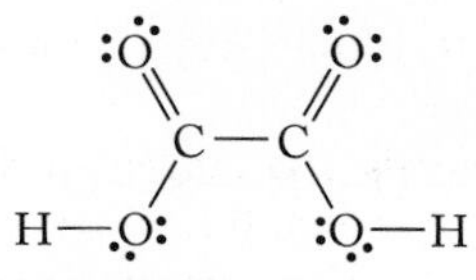

:Ö: :Ö:
C—C
H—Ö: :Ö—H

FIGURE P8.157

*8.158. Draw all resonance forms for the molecules NSF and HBS where S and B are the central atoms, respectively. Include possible ionic structures for NSF.

***8.159.** The molecular structure of sulfur cyanide trifluoride (SF_3CN) has been shown to have the arrangement of atoms with the bond lengths indicated in Figure P8.159. Complete the Lewis structure for SF_3CN and assign formal charges.

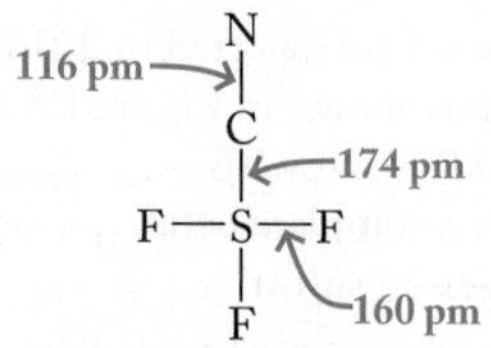

N
116 pm
C
174 pm
F—S—F
F
160 pm

FIGURE P8.159

8.160. **Strike-Anywhere Matches** Heating phosphorus with sulfur gives P_4S_3, a solid used in the heads of strike-anywhere matches (Figure P8.160). P_4S_3 has the Lewis structure framework shown. Complete the Lewis structure so that each atom has the optimal formal charge.

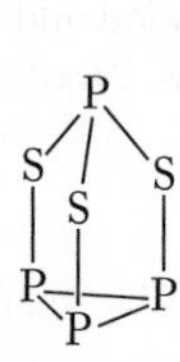

FIGURE P8.160

***8.161.** The heavier group 16 elements can expand their valence shell. The $TeOF_6^{2-}$ anion was first prepared in 1993. Draw the Lewis structure for $TeOF_6^{2-}$.

*8.162. **Sulfur in the Environment** Sulfur is cycled in the environment through compounds such as dimethyl sulfide (CH_3SCH_3), hydrogen sulfide (H_2S), and the sulfite and sulfate ions. Draw Lewis structures for these four species. Are more than eight valence electrons needed to minimize the formal charges for any of these species?

8.163. How many pairs of electrons does xenon share in the following molecules and ions? (a) XeF_2; (b) $XeOF_2$; (c) XeF^+; (d) XeF_5^+; (e) XeO_4

8.164. Consider a hypothetical structure of ozone that is cyclic (the atoms form a ring) such that its three O atoms are at the corners of a triangle. Draw the Lewis structure for this molecule.

***8.165.** Bond lengths and electrostatic potential mapping provide experimental evidence about covalent bonds present in a molecule. Which of these would help characterize the bonding in a compound with the formula A_2X in terms of the following?

a. distinguishing between these two bonding patterns: X—A—A and A—X—A

b. distinguishing between these resonance forms: A—X≡A, A=X=A, and A≡X—A

8.166. **Explosive Cation** The highly explosive linear N_5^+ cation was first isolated in 1999 by reaction of N_2F^+ with HN_3:

$$N_2F^+ + HN_3 \rightarrow N_5^+ + HF$$

Draw the Lewis structures for the reactants and products, including all resonance forms.

***8.167.** **Jupiter's Atmosphere** The ionic compound NH_4SH was detected in the atmosphere of Jupiter (Figure P8.167) by the *Galileo* space probe in 1995. Draw the Lewis structure for NH_4SH. Why couldn't there be a covalent bond between the nitrogen and sulfur atoms, making NH_4SH a molecular compound?

FIGURE P8.167

8.168. **Antacid Tablets** Antacids commonly contain calcium carbonate, magnesium hydroxide, or both. Draw the Lewis structures for calcium carbonate and magnesium hydroxide.

***8.169.** A *cyclic* polynitrogen anion, N_5^-, can be trapped as a salt with Co^{2+}.

a. Draw Lewis structures for N_5^-, including all resonance forms.

b. Predict the N=N bond order and estimate the N—N bond distance.

*8.170. Draw Lewis symbols for $B(CN)_3^-$, $B(CN)_4^-$, and $B_2(CN)_6^{2-}$. $B_2(CN)_6^{2-}$ contains a B—B bond. Do all three molecular ions follow the octet rule?

8.171. Draw a Lewis structure for $AlFCl_2$, the second product in the synthesis of Cl_2O_2 in the following reaction:

$$FClO_2(g) + AlCl_3(s) \rightarrow Cl_2O_2(g) + AlFCl_2(s)$$

*8.172. Draw Lewis structures for BF_3 and $(CH_3)_2BF$. The B—F distance in both molecules is the same (130 pm). Does this observation support the argument that all the boron–fluorine bonds in BF_3 are single bonds?

8.173. Which of the following molecules and ions contains an atom with more than eight valence electrons? (a) Cl_2; (b) ClF_3; (c) ClI_3; (d) ClO^-

8.174. Which of the following molecules contains an atom with more than eight valence electrons? (a) XeF_2; (b) $GaCl_3$; (c) ONF_3; (d) SeO_2F_2

***8.175.** A *linear* polynitrogen anion, N_5^-, was isolated for the first time in 1999.

a. Draw the Lewis structures for four resonance forms of linear N_5^-.
b. Assign formal charges to the atoms in the structures in part (a) and identify the structures that contribute the most to the bonding in N_5^-.
c. Compare the Lewis structures for N_5^- and N_3^-. In which ion do the nitrogen–nitrogen bonds have the higher average bond order?

*8.176. Carbon tetroxide (CO_4) was discovered in 2003.

a. Draw the Lewis structure for CO_4, assuming the atoms are arranged as shown in Figure P8.176.

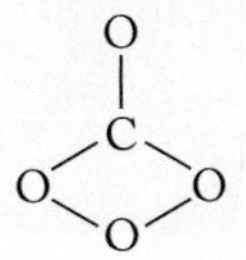

FIGURE P8.176

b. Are there any resonance forms for the structure you drew that have zero formal charges on all atoms?
c. Can you draw a structure in which all four oxygen atoms in CO_4 are bonded to carbon?

8.177. Plot the electronegativities of elements with $Z = 3$ to 9 (y-axis) versus their first ionization energies (x-axis). Is the plot linear? Use your graph to predict the electronegativity of neon, whose first ionization energy is 2080 kJ/mol.

8.178. Use the data in Figures 7.39 and 8.9 to plot electronegativity as a function of first ionization energy for the main group elements of the fifth row of the periodic table. From this plot estimate the electronegativity of xenon, whose first ionization energy is 1170 kJ/mol.

8.179. The cation N_2F^+ is isoelectronic with N_2O.

a. What does it mean to be isoelectronic?
b. Draw the Lewis structure for N_2F^+. (*Hint*: The molecule contains a nitrogen–nitrogen bond.)
c. Which atom has the +1 formal charge in the structure you drew in part (b)?
d. Does N_2F^+ have resonance forms?
e. Could the middle atom in the N_2F^+ ion be a fluorine atom? Explain your answer.

8.180. **Ozone Depletion** Methyl bromide (CH_3Br) is produced naturally by fungi. Methyl bromide has also been used in agriculture as a fumigant, but this use is being phased out because the compound has been linked to ozone depletion in the upper atmosphere.

a. Draw the Lewis structure for CH_3Br.
b. Which bond in CH_3Br is more polar, carbon–hydrogen or carbon–bromine?

***8.181.** All the carbon–fluorine bonds in tetrafluoroethylene (Figure P8.181) are polar, but experimental results show that the molecule is nonpolar. Draw arrows on each bond to show the direction of polarity and suggest a reason why the molecule is nonpolar overall.

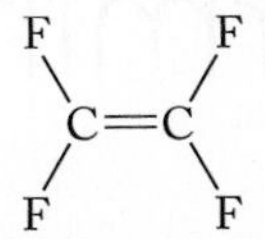

Tetrafluoroethylene

FIGURE P8.181

*8.182. Atoms of Xe have complete octets, yet the compound XeO_2 exists. How is this possible?

***8.183.** Free radicals increase in stability, and hence decrease in reactivity, if they have more than one atom in their structure that can carry the unpaired electron.

a. Draw Lewis structures for the three free radicals formed from CF_2Cl_2 in the stratosphere—ClO, Cl, and CF_2Cl—and use this principle to rank them in order of reactivity.
b. Free radicals are "neutralized" when they react with each other, forming an electron pair (a single covalent bond) between two atoms. This is called a *termination reaction* because it shuts down any reaction that was powered by the free radical. Predict the formula of the molecules that are produced when the chlorine free radical reacts with each of the free radicals in part (a).

9

Molecular Geometry

Shape Determines Function

CILANTRO IS A FRAGRANT HERB Its flavor appeals to many people, but to others it tastes like soap. These very different tastes can be attributed to genetic differences in how the shapes of molecules fit into biomolecular sites that send signals to the region of the brain that interprets taste.

PARTICULATE **REVIEW**

Smelly Molecules and Functional Groups

In Chapter 9 we explore the connection between molecular shape and properties such as taste and smell. Two molecules with very different smells are shown here. Malic acid is responsible for the smell and taste of apples, whereas putrescine is the smell associated with rotting flesh in a cadaver.

- What is the molecular formula for malic acid?
- What is the molecular formula for putrescine?
- Identify the functional groups present in both molecules.

(Review Section 2.8 if you need help.)

(Answers to Particulate Review questions are in the back of the book.)

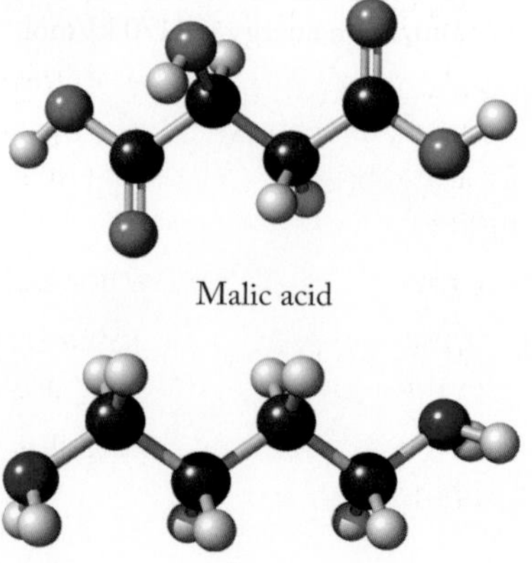

PARTICULATE **PREVIEW**

Smelly Molecules and Shapes

Your ability to detect a substance by smell depends on its elemental composition and on the shape of its molecules. As you read Chapter 9, look for ideas that will help you answer these questions about the shapes of the molecules shown here:

- Draw Lewis structures for both methyl mercaptan and acrolein.
- What are the approximate bond angles around the sulfur and carbon atoms in methyl mercaptan and around the central carbon atom in acrolein?
- Are either or both molecules planar; that is, do all their atoms lie in one plane?

Methyl mercaptan, present in human "bad breath"

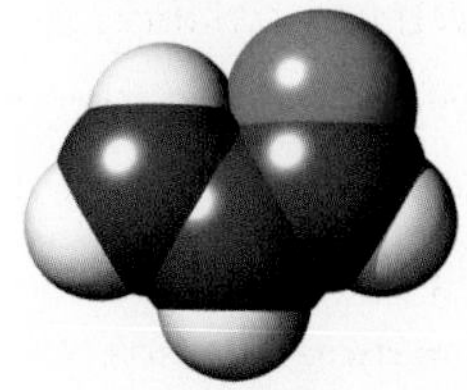

Acrolein, present in smoky barbeque

Learning Outcomes

LO1 Use VSEPR and the concept of steric number to predict the bond angles in molecules and the shapes of molecules with one central atom
Sample Exercises 9.1, 9.2, 9.3

LO2 Predict whether a substance is polar or nonpolar based on its molecular structure
Sample Exercise 9.4

LO3 Use atomic orbital hybridization and valence bond theory to explain orbital overlap, bond angles, and molecular shapes
Sample Exercises 9.5, 9.6, 9.8

LO4 Draw condensed and skeletal structures of organic compounds
Sample Exercise 9.7

LO5 Identify molecular structures that are stabilized by delocalized π electrons

LO6 Recognize chiral molecules

LO7 Draw molecular orbital (MO) diagrams of diatomic molecules and use MO theory to predict bond order and to explain magnetic properties and spectra
Sample Exercises 9.9, 9.10, 9.11

LO8 Use MO theory to describe metallic bonding and the properties of doped and undoped semiconductors
Sample Exercise 9.12

9.1 Biological Activity and Molecular Shape

Hold your hands out in front of you, palms up, fingers extended. Now rotate your wrists inward so that your thumbs point straight up. Your right hand looks the same as the image your left hand makes in a mirror. Does that mean your two hands have the same shape? If you have ever tried to put your right hand in a glove made for your left, you know that they do not have the same shape. Many other objects in our world have a "handedness" about them, from headphones to scissors to golf clubs to shoes.

This chapter focuses on the importance of shape at the molecular level. For example, the compound that produces the refreshing aroma of spearmint has the molecular formula $C_{10}H_{14}O$. The compound responsible for the musty aroma of caraway seeds has the same molecular formula *and* the same Lewis structure. To understand how two compounds can be so much alike and still have different properties, we must consider their structures in three dimensions.

We perceive a difference in aromas in part because each molecule has a unique site where it attaches to our nasal membranes. Just as a left hand only fits a left glove, the spearmint molecule fits only the spearmint-shaped site, and the caraway molecule fits only the caraway-shaped site. This behavior, called molecular recognition, enables biomolecular structures such as nasal membranes to recognize and react when a molecule with a particular shape fits into a part of the structure known as an *active site*. Many of the substances we ingest, from food to pharmaceuticals, exert physiological effects because their molecules are recognized by and become bound to active sites in biomolecules.

The shape of a compound's molecules can affect many of its properties, such as its physical state at room temperature, its solubility in water and other solvents, its aroma, its biological activity, and its reactivity with other compounds. In Chapter 8 we drew Lewis structures to describe bonding in molecules, but Lewis structures are only two-dimensional representations of how atoms and the electron pairs that surround them are arranged in molecules. Lewis structures show how atoms are connected in molecules, but they do not show how the atoms are arranged in three dimensions, nor do they necessarily show the overall shape of the molecules.

To illustrate this point, consider the Lewis structures and ball-and-stick models of carbon dioxide and methane (**Figure 9.1**). The linear arrangement of atoms and bonding electrons in the Lewis structure for CO_2 corresponds to the actual linear shape of the molecule, as represented by the ball-and-stick model. The angle between the two C=O bonds is 180°, just as in the Lewis structure. The Lewis structure of methane, on the other hand, does not convey the true three-dimensional orientation of the four C—H bonds in each molecule. The 90° angles between bonds in the planar Lewis structure are not close to the three-dimensional H—C—H **bond angles** of 109.5°.

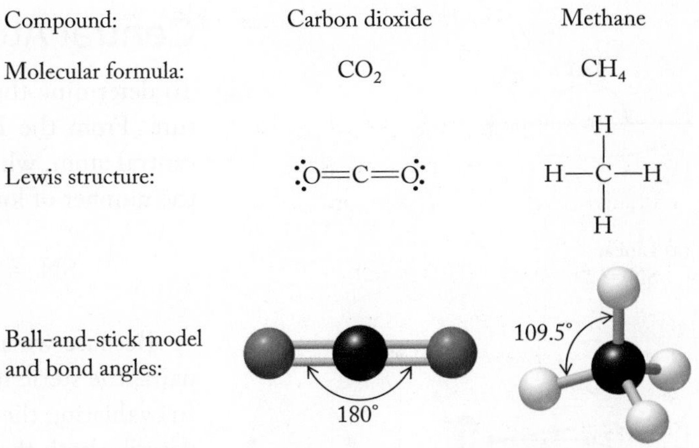

Compound:	Carbon dioxide	Methane
Molecular formula:	CO_2	CH_4
Lewis structure:	:Ö=C=Ö:	H—C(H)(H)—H
Ball-and-stick model and bond angles:	180°	109.5°

FIGURE 9.1 The Lewis structure of CO_2 matches its true molecular structure, but the Lewis structure of CH_4 does not because the C—H bonds in CH_4 extend in three dimensions.

Several theories of covalent bonding account for and predict the shapes of molecules. In this chapter we begin by examining small molecules with single central atoms, and we gradually apply these theories to larger molecules. We start with the shared pairs and lone pairs of electrons described by Lewis theory and then predict how those pairs should be oriented about a central atom to minimize their interactions and produce the most stable molecular structure. As we will see, each theory accurately predicts electron-pair orientations, molecular shapes, and the properties of molecular compounds.

9.2 Valence-Shell Electron-Pair Repulsion (VSEPR) Theory

Valence-shell electron-pair repulsion (VSEPR) theory is based on a fundamental chemical principle—namely, electrons have negative charges and repel each other. VSEPR theory applies this principle by incorporating the assumption that pairs of valence electrons are arranged about central atoms in ways that minimize repulsions between the pairs. To predict molecular shape by using VSEPR theory, we must consider two things:

- **Electron-pair geometry**, which defines the relative positions in three-dimensional space of all the bonding pairs and lone pairs of valence electrons on the central atom
- **Molecular geometry**, which defines the relative positions in three-dimensional space of the atoms in a molecule

The electron-pair geometry of a molecule may or may not be the same as its molecular geometry. The presence of lone pairs on the central atom makes the difference. To accurately predict molecular geometry, we first need to know the electron-pair geometry. If there are no lone pairs of electrons, then the process is simplified because the electron-pair geometry *is* the molecular geometry. Let's begin with this simple case and consider the shapes of molecules that have various numbers of bonds around a central atom and no lone pairs on the central atom.

bond angle the angle (in degrees) defined by lines joining the centers of two atoms to the center of a third atom to which they are chemically bonded.

valence-shell electron-pair repulsion (VSEPR) theory a model predicting that the arrangement of valence electron pairs around a central atom minimizes their mutual repulsion to produce the lowest-energy orientations.

electron-pair geometry the three-dimensional arrangement of bonding pairs and lone pairs of electrons about a central atom.

molecular geometry the three-dimensional arrangement of the atoms in a molecule.

CONCEPT TEST

In your own words, explain how electron-pair geometry and molecular geometry are related and how they are different.

(Answers to Concept Tests are in the back of the book.)

Central Atoms with No Lone Pairs

To determine the geometry of a molecule, we start by drawing its Lewis structure. From the Lewis structure, we determine the **steric number (SN)** of the central atom, which is the sum of the number of atoms bonded to that atom and the number of lone pairs in its valence shell:

$$\text{SN} = \begin{pmatrix}\text{number of atoms}\\ \text{bonded to central atom}\end{pmatrix} + \begin{pmatrix}\text{number of lone pairs}\\ \text{on central atom}\end{pmatrix} \quad (9.1)$$

Because we are focused on molecules in which the central atom has *no* lone pairs, the steric number equals the number of atoms bonded to the central atom. In evaluating the shapes of these molecules, we generate five common shapes that describe both the electron-pair geometries and the molecular geometries of many covalent compounds.

Let's start by thinking of the central atom as the center of a sphere, with all the other atoms in the molecule placed on the surface of the sphere and linked to the central atom by covalent bonds. If the central atom is bonded covalently to only two other atoms, then SN = 2. How do the electron pairs in the two bonds arrange themselves to minimize their mutual repulsion? They are as far from each other as possible—on opposite sides of the sphere (**Figure 9.2a**). This gives a **linear** electron-pair geometry and a linear molecular geometry. The three atoms in the molecule are arranged in a straight line, and the bond angle is 180°.

If three atoms are bonded to a central atom with no lone pairs, then SN = 3. The three bonding pairs are as far apart as possible when they are located at the three corners of an equilateral triangle. The angle between each pair of bonds is 120° (**Figure 9.2b**). The name of this electron-pair and molecular geometry is **trigonal planar**.

With four atoms around the central atom and no lone pairs (SN = 4), we have the first case in which the atoms are not all in the same plane. Instead, the atoms bonded to the central atom occupy the four vertices of a tetrahedron, which is a four-sided pyramid (*tetra* is Greek, meaning "four"). The bonding pairs form bond angles of 109.5° with each other, as shown in **Figure 9.2c**. The electron-pair geometry and molecular geometry are both **tetrahedral**.

When five atoms are bonded to a central atom with no lone pairs, SN = 5 and the atoms occupy the five corners of two triangular pyramids that share the same base. The central atom of the molecule is in the center of the sphere at the common center of the two bases, as shown in **Figure 9.2d**. One bond points to the tip of the top pyramid, one points to the tip of the bottom pyramid, and the other three point to the three vertices of the shared triangular base. These three vertices lie along the equator of the sphere, so the atoms that occupy these sites and the bonds that connect them to the central atom are called *equatorial* atoms and bonds, which are highlighted in orange in Figure 9.2d. The bond angles between the three equatorial bonds are 120° (just as in the triangle of the trigonal planar geometry in Figure 9.2b). The bond angle between an equatorial bond and either vertical, or *axial*, bond (highlighted in green) is 90°, and the angle between the two axial bonds is 180°. A molecule in which the atoms are arranged this way is said to have a **trigonal bipyramidal** electron-pair and molecular geometry.

For SN = 6, picture two pyramids that have a square base (**Figure 9.2e**). Put them together, base to base, and you form a shape in which all six positions around the sphere are equivalent.

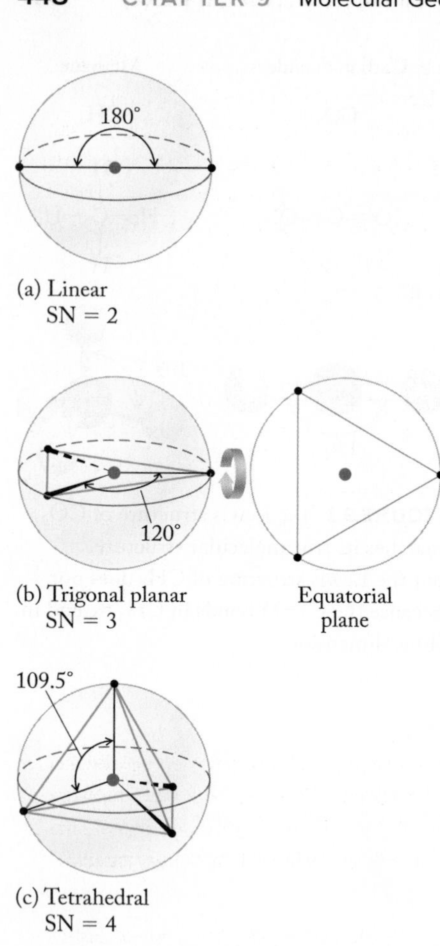

(a) Linear
SN = 2

(b) Trigonal planar
SN = 3

Equatorial plane

(c) Tetrahedral
SN = 4

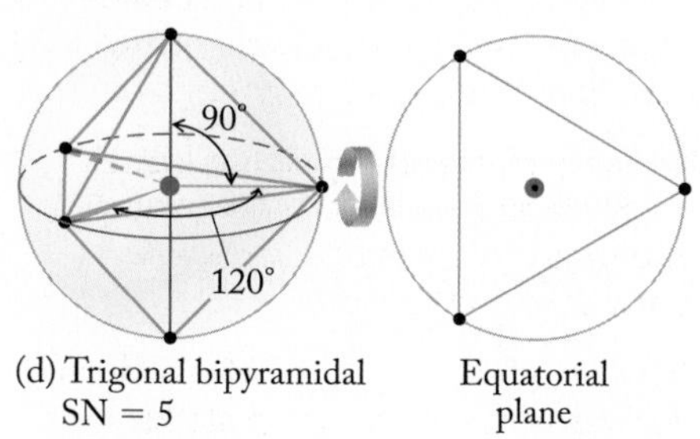

(d) Trigonal bipyramidal
SN = 5

Equatorial plane

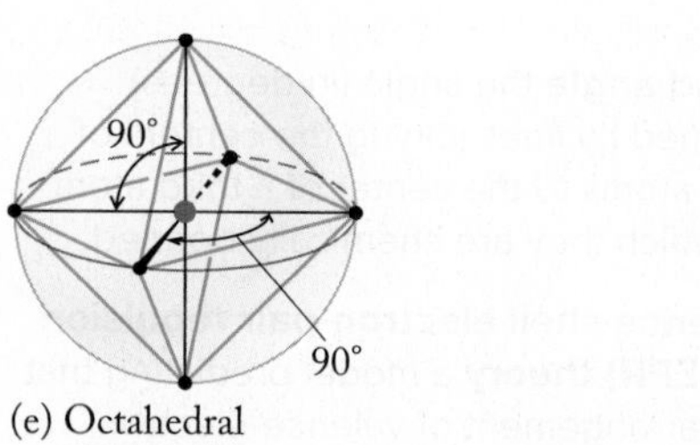

(e) Octahedral
SN = 6

FIGURE 9.2 Electron-pair geometries depend on the steric number (SN) of the central atom. In these images, there are no lone pairs of electrons on the central atoms (red dots), so the molecular geometries are the same as the electron-pair geometries. The images show bond angles for different numbers of atoms located on the surface of a sphere and bonded to an atom at the center of the sphere. The blue lines define the geometric forms that give the molecular shapes their names. Parts (b) and (d) show spheres rotated 90° so the equatorial plane is clearly visible.

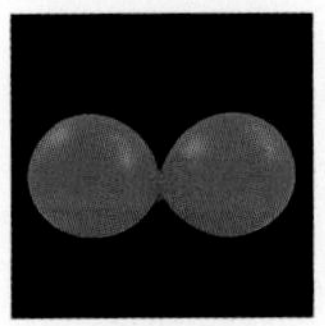
SN = 2

3

4

5

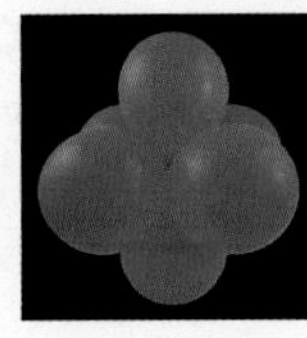
6

FIGURE 9.3 Balloons charged with static electricity repel each other and align themselves as far apart as possible when tied together. In doing so, they mimic the locations of electron pairs about a central atom. Each balloon represents one electron pair. Note the similarities between the balloons and the diagrams in Figure 9.2.

We can think of the six bonds as three sets of two bonds each. The two pairs in each set are oriented at 180° to each other and at 90° to the other two sets, just like the axes of an *x–y–z* coordinate system. Four equatorial atoms lie at the four vertices of the common square base and are 90° apart, another atom lies at the tip of the top pyramid, and a sixth lies at the tip of the bottom pyramid. This arrangement defines **octahedral** electron-pair and molecular geometries.

To demonstrate a real-world analogy to the bond orientations in Figure 9.2, we start with a batch of fully inflated yellow balloons. We tie them together in clusters of two, three, four, five, and six balloons (**Figure 9.3**). Assume that all the balloons have acquired a static electrical charge, so they repel each other. If the tie points of the clusters represent the central atom in our balloon models, then the opposite ends of the balloons represent the atoms that are bonded to the central atom. Note how our clusters of two, three, four, five, and six balloons produce the same orientations that resulted from selecting points on a sphere that were as far apart as possible. The long axes of the balloons in Figure 9.3 provide an accurate representation of the bond directions in Figure 9.2.

Now let's look at five simple molecules that have no lone pairs about the central atom and apply VSEPR theory and the concept of steric number to predict their electron-pair and molecular geometries. To do so, we follow these three steps:

1. Draw the Lewis structure, including any equivalent resonance structures.
2. Determine the steric number of the central atom.
3. Use the steric number to predict the electron-pair and molecular geometries by using the images in Figure 9.2.

Example I: Carbon Dioxide (CO_2)

1. Lewis structure:

$$:\ddot{O}=C=\ddot{O}:$$

2. The central carbon atom has two atoms bonded to it and no lone pairs, so the steric number is 2.
3. Because SN = 2, the O—C—O bond angle is 180° (see Figure 9.2a) and the electron-pair and molecular geometries are both linear.

Example II: Boron Trifluoride (BF_3)

1. Lewis structure:

:F̤: —B— :F̤: with :F̤: below B

2. Three fluorine atoms are bonded to the central boron atom, and there are no lone pairs on boron. Therefore, SN = 3.
3. Because SN = 3, all four atoms lie in the same plane, forming a triangle with F—B—F bond angles of 120°. The electron-pair and molecular geometries are trigonal planar (**Figure 9.4**).

CONNECTION We learned in Chapter 8 that some molecules have central atoms that do not have an octet of electrons. The boron atom in BF_3 shares only three pairs of electrons.

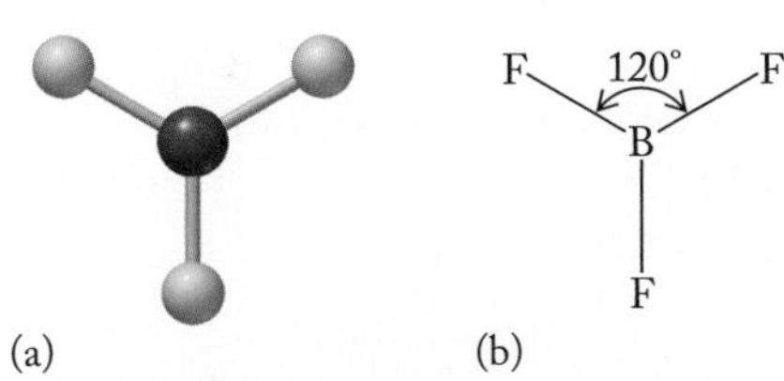

FIGURE 9.4 (a) The ball-and-stick model of BF_3 shows the orientation of the atoms. (b) All F—B—F bond angles are 120° in this trigonal planar molecular geometry.

steric number (SN) the sum of both the number of atoms bonded to a central atom and the number of lone pairs of electrons on the central atom.

linear the molecular geometry about a central atom with a steric number of 2 and no lone pairs of electrons; the bond angle is 180°.

trigonal planar the molecular geometry about a central atom with a steric number of 3 and no lone pairs of electrons; the bond angles are all 120°.

tetrahedral the molecular geometry about a central atom with a steric number of 4 and no lone pairs of electrons; the bond angles are all 109.5°.

trigonal bipyramidal the molecular geometry about a central atom with a steric number of 5 and no lone pairs of electrons; three atoms occupy equatorial sites (bond angles 120°) and two other atoms occupy axial sites (bond angle 180°) above and below the equatorial plane; the bond angle between the axial and equatorial bonds is 90°.

octahedral the molecular geometry about a central atom with a steric number of 6 and no lone pairs of electrons, in which all six sites are equivalent.

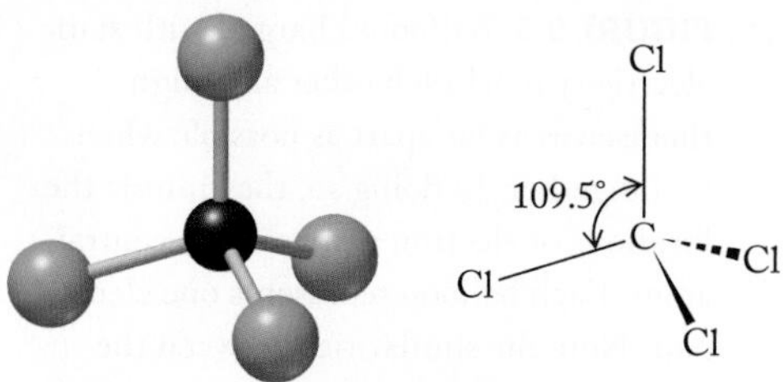

FIGURE 9.5 The ball-and-stick model shows the actual orientation of the atoms in carbon tetrachloride in three dimensions. All Cl—C—Cl bond angles are 109.5° in this tetrahedral molecular geometry.

Example III: Carbon Tetrachloride (CCl_4)

1. Lewis structure:

2. Four chlorine atoms are bonded to the central carbon atom, which gives carbon a full octet. Therefore, SN = 4.
3. Because SN = 4, the chlorine atoms are located at the vertices of a tetrahedron and all four Cl—C—Cl bond angles are 109.5°, producing tetrahedral electron-pair and molecular geometries (**Figure 9.5**).

Before we explore any more molecular geometries, you need to understand the conventions used in drawing three-dimensional structures in two dimensions. To convey the three-dimensional structure of a molecule, we use a solid wedge (◄) to indicate a bond that comes out of the page toward the viewer. The solid wedge in the CCl_4 structure in Figure 9.5, for example, means that the chlorine in that position projects out of the plane of the page and toward the viewer at a downward angle. A dashed wedge (┄◄) indicates a bond that goes behind the plane of the page and away from the viewer at a downward angle. Solid lines are used for bonds that lie in the plane of the page. We typically orient a structure so that the maximum number of bonds lie in the plane of the page.

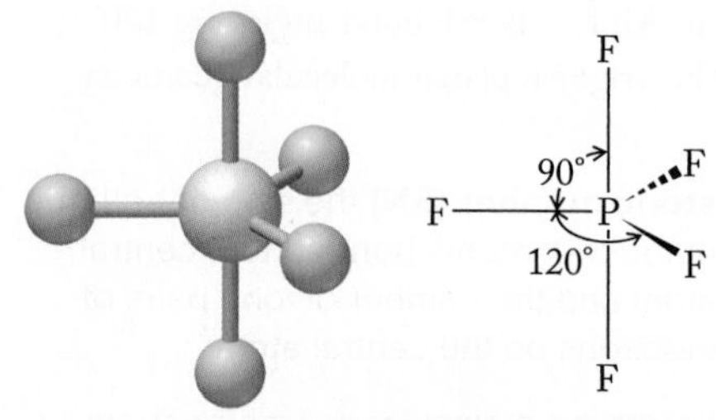

FIGURE 9.6 The ball-and-stick model shows the actual orientation of the atoms in phosphorus pentafluoride. The P—F bonds located around the equator of the imaginary sphere are 120° apart. Each of them is 90° from the two bonds connecting F atoms in the axial positions (at the north and south poles). The resulting molecular geometry is trigonal bipyramidal.

Example IV: Phosphorus Pentafluoride (PF_5)

1. Lewis structure:

2. Five fluorine atoms are bonded to the central phosphorus atom, which has no lone pairs. Therefore, SN = 5.
3. Because SN = 5, the fluorine atoms are located at the five vertices of a trigonal bipyramid, and the electron-pair geometry and molecular geometry are trigonal bipyramidal (**Figure 9.6**).

Example V: Sulfur Hexafluoride (SF_6)

1. Lewis structure:

2. Six fluorine atoms are bonded to the central sulfur atom, which has no lone pairs. Therefore, SN = 6.
3. Because SN = 6, the fluorine atoms are located at the vertices of an octahedron and the electron-pair and molecular geometries are octahedral (**Figure 9.7**).

SAMPLE EXERCISE 9.1 Using VSEPR Theory to Predict Geometry I LO1

Formaldehyde (CH_2O) is a gas at room temperature (boiling point −21°C). Aqueous solutions of formaldehyde are used to preserve biological samples. It is also a product of the incomplete combustion of hydrocarbons, and it was the first polyatomic molecule to be detected in interstellar space. Use VSEPR theory to predict the molecular geometry of formaldehyde.

Collect and Organize We are given the molecular formula of formaldehyde. The solution requires (1) drawing the Lewis structure, (2) determining the steric number, and (3) identifying the molecular geometry by using Figure 9.2.

Analyze Carbon is the likely central atom of the molecule because it has a bonding capacity of 4 and is less electronegative than oxygen, whose bonding capacity is 2. If there are no lone pairs of electrons, then SN = 3 and the three atoms bonded to C are as far from each other as possible.

Solve Using the five-step procedure developed in Section 8.2 for drawing Lewis structures, we obtain the following for formaldehyde:

O
C
H H

As predicted, SN = 3 and there are no lone pairs of electrons on the central atom. According to Figure 9.2, the molecular geometry is trigonal planar.

Think About It The key to predicting the correct molecular geometry of a molecule with no lone pairs on its central atom is to determine the steric number, which is simply a matter of counting the number of atoms bonded to the central atom.

Practice Exercise Use VSEPR to determine the molecular geometry of the chloroform molecule ($CHCl_3$) and draw the molecule by using the solid-wedge, dashed-wedge convention.

(Answers to Practice Exercises are in the back of the book.)

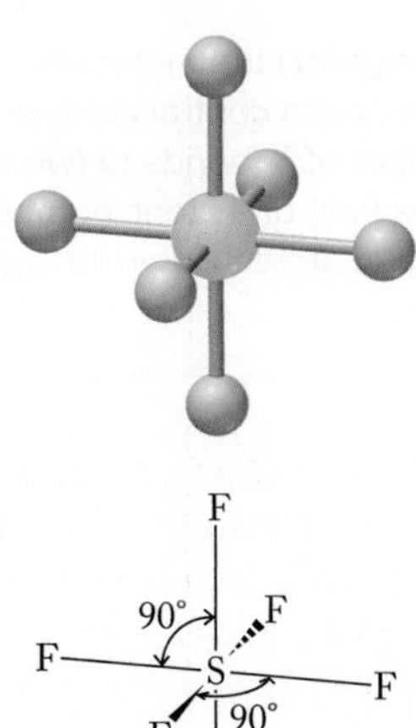

FIGURE 9.7 The ball-and-stick model of SF_6 shows how the six fluorine atoms are oriented in three dimensions about the sulfur atom. All the F—S—F bond angles are 90° in this octahedral molecular geometry.

Measurements of the bond angles in formaldehyde show that the H—C—H bond angle is slightly smaller than the 120° predicted for trigonal planar geometry (Figure 9.2b). The C=O double bond consists of two pairs of bonding electrons (four electrons) and exerts greater repulsion than a single bond would. This greater repulsion decreases the H—C—H bond angle. VSEPR theory does not enable us to predict the actual values of the bond angles in molecules containing double bonds to the central atom, but it does allow us to correctly predict how the presence of a double bond causes a bond angle to deviate from the ideal value.

CONCEPT TEST

Rank the following bond angles from largest to smallest:

a. The O—C—O bond angle in CO_2
b. The H—C—H bond angles in CH_4
c. The H—C—H bond angle in CH_2O

bent (or angular) the molecular geometry about a central atom with a steric number of 3 (bonds to two atoms + one lone pair) or a steric number of 4 (bonds to two atoms + two lone pairs).

Central Atoms with Lone Pairs

We now explore what happens when a central atom has one or more lone pairs. As we do, keep in mind that the impact of these lone pairs will be to create molecular geometries with names that are not the same as the molecules' electron pair geometries. For SN = 2, the only bonding pattern possible is two atoms bound to a central atom. If we replace one bonding pair with a lone pair, we have a molecule with only two atoms, which means there is no central atom and no bond angle. Because it takes three atoms to define a bond angle, we begin the discussion of molecules containing lone pairs with SN = 3.

Sulfur dioxide, one of the gases produced when high-sulfur coal is burned, has the following distribution of electrons in its two resonance structures:

$$:\ddot{\underset{..}{O}}-\ddot{S}=\ddot{O}: \longleftrightarrow :\ddot{O}=\ddot{S}-\ddot{\underset{..}{O}}:$$

To calculate the steric number of the central S atom we need to add up the number of atoms and lone pairs of electrons that surround it. In both resonance structures, the central S atom is bonded to two atoms and has one lone pair of electrons. Therefore, its SN value in either resonance structure is 2 + 1 = 3. When SN = 3, the arrangement of atoms and lone pairs about the sulfur atom is trigonal planar (Figure 9.2b). This means that the electron-pair geometry, which considers both the atoms and the nonbonding pairs of electrons around the central atom, is trigonal planar (as shown in **Figure 9.8a**). The molecular geometry, though, describes the relative position of only the atoms in the molecule. With SN = 3 and two atoms attached to the central S atom, the SO_2 molecule has the **bent** (or **angular**) molecular geometry shown in **Figure 9.8b**.

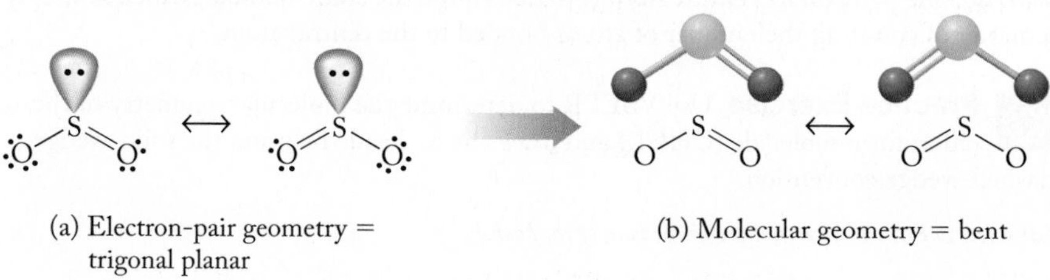

FIGURE 9.8 (a) The electron-pair geometry of SO_2 is trigonal planar because the steric number of the S atom is 3; it is bonded to two atoms and has one lone pair of electrons. (b) The molecular geometry is bent because the central sulfur atom in the structure has no bonded atom in the third position, only a lone pair of electrons.

Experimental measurements establish that SO_2 is indeed a bent molecule, but the O—S—O bond angle is a little smaller than 120°. We explain the smaller angle with VSEPR theory by comparing the space occupied by bonding electrons with the space occupied by electrons in a lone pair. Because the bonding electrons are attracted to two nuclei, they have a high probability of being located between the two atomic centers that share them. In contrast, the lone pair is *not* shared with a second atom and is spread out around the sulfur atom, as shown in **Figure 9.9**. This puts the lone pair of electrons closer to the bonding pairs and produces greater repulsion. As a result, the lone pair pushes the bonding pairs closer together, thereby reducing the bond angle. In general:

- Repulsion between lone pairs and bonding pairs is greater than the repulsion between bonding pairs.
- Repulsion caused by a lone pair is greater than that caused by a double bond.
- Repulsion caused by a double bond is greater than that caused by a single bond.
- Two lone pairs of electrons on a central atom exert a greater repulsive force on the atom's bonding pairs than does one lone pair.

FIGURE 9.9 The lone pair of electrons on the central sulfur atom in SO_2 occupies more space (larger purple region) than the electron pairs in the S—O bonds (smaller purple regions). The increased repulsion (represented by the double-headed arrows), resulting from the larger volume occupied by the lone pair, forces the oxygen atoms closer together, making the O—S—O bond angle slightly less than the ideal value of 120°.

SAMPLE EXERCISE 9.2 Predicting Relative Sizes of Bond Angles **LO1**

Rank NH_3, CH_4, and H_2O in order of decreasing bond angles in their molecular structures.

Collect and Organize We are asked to predict the relative sizes of the bond angles in three molecules, given their molecular formulas. The information in Figure 9.2 links steric numbers to electron-pair geometries and bond angles. The presence of double bonds and lone pairs of electrons on the central atom also influences bond angles.

Analyze We need to determine the steric numbers of the central atoms in these three molecules. To do that we first need to translate the molecular formulas into Lewis structures and determine how many lone pairs or bonds to other atoms surround the central atom.

Solve Using the method for drawing Lewis structures from Chapter 8, we obtain the following:

H—N̈—H with H below N; H—C—H with H above and below C; :Ö: bonded to two H

In each of these structures, the central atom has a steric number of 4: bonds to 3 atoms + 1 lone pair for NH_3, bonds to 4 atoms + 0 lone pairs for CH_4, and bonds to 2 atoms + 2 lone pairs for H_2O.

According to Figure 9.2, the electron-pair geometry for SN = 4 is tetrahedral. In a molecule such as CH_4, in which all four tetrahedral electron pairs are equivalent bonding pairs, all bond angles are 109.5°. In NH_3 and H_2O, however, repulsion from the lone pairs of electrons squeezes their bonds together, reducing the bond angles. The two lone pairs on the O atom in H_2O exert a greater repulsive force on its bonding pairs than the single lone pair on the N atom exerts on the bonding pairs in NH_3. Therefore, the bond angle in H_2O should be less than the bond angles in NH_3. Ranking the three molecules in order of decreasing bond angle, we have:

$$CH_4 > NH_3 > H_2O$$

Think About It The logic used in answering this problem is supported by experimental evidence: the bond angles in molecules of CH_4, NH_3, and H_2O are 109.5°, 107.0°, and 104.5°, respectively.

Practice Exercise
Are the O—S—O bond angles greater in SO_2 or SO_3?

As we saw in Sample Exercise 9.2, three combinations of atoms and lone pairs are possible about a central atom with a steric number of 4: four atoms and no lone pairs, three atoms and one lone pair, and two atoms and two lone pairs. The first case is illustrated by the molecular structure of methane (CH_4) in which a tetrahedral electron-pair geometry translates into a tetrahedral molecular geometry. A molecule of ammonia (NH_3) has only three bonding pairs and one lone pair. This means that the Lewis structure in **Figure 9.10a** translates into a tetrahedral electron-pair geometry in which one of the vertices is the lone pair on the N atom (**Figure 9.10b**). The resulting molecular geometry, which is essentially a flattened tetrahedron, is called **trigonal pyramidal** (**Figure 9.10c**). The strong repulsion produced by the diffuse lone pair of electrons on the N atom pushes the N—H bonds closer together in NH_3 and reduces the angles between them from the 109.5° of the H—C—H bond angles in CH_4 to 107.0°.

trigonal pyramidal the molecular geometry about a central atom with a steric number of 4 and one lone pair of electrons.

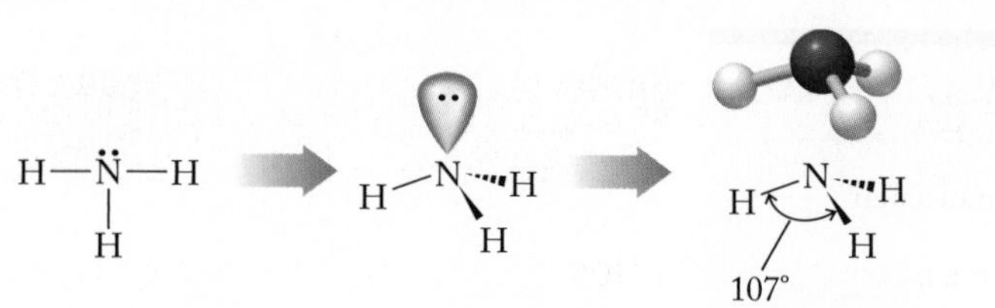

(a) Lewis structure (b) Tetrahedral electron-pair geometry (c) Trigonal pyramidal molecular geometry

FIGURE 9.10 (a) The steric number of the N atom in NH_3 is 4, (b) so its electron-pair geometry is tetrahedral. However, one of the vertices of the tetrahedron is occupied by a lone pair of electrons, not an atom, (c) so the molecular geometry is trigonal pyramidal.

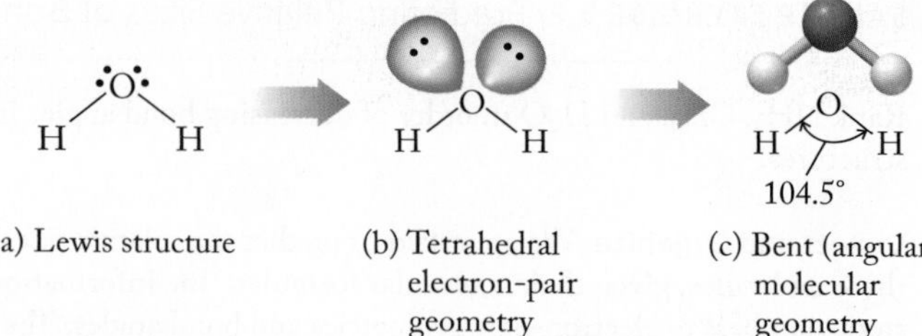

(a) Lewis structure (b) Tetrahedral electron-pair geometry (c) Bent (angular) molecular geometry

FIGURE 9.11 (a) The steric number of the O atom in H_2O is 4 (it is bonded to two atoms and has two lone pairs of electrons). (b) Therefore, its electron-pair geometry is tetrahedral. However, two of the vertices of the tetrahedron are occupied by lone pairs of electrons, meaning (c) only three atoms define the molecular geometry, which is bent.

The central O atom in a molecule of H_2O has a steric number of 4 because it is bonded to two atoms and has two lone pairs of electrons. The result is a tetrahedral electron-pair geometry (**Figure 9.11**). However, the presence of the two lone pairs means that two of the four tetrahedral vertices are not occupied by atoms, which leaves us with a bent (or angular) molecular geometry. Therefore, the H—O—H bond angle in water is reduced from 109.5° to 104.5° because of repulsion between the two lone pairs and each bonding pair.

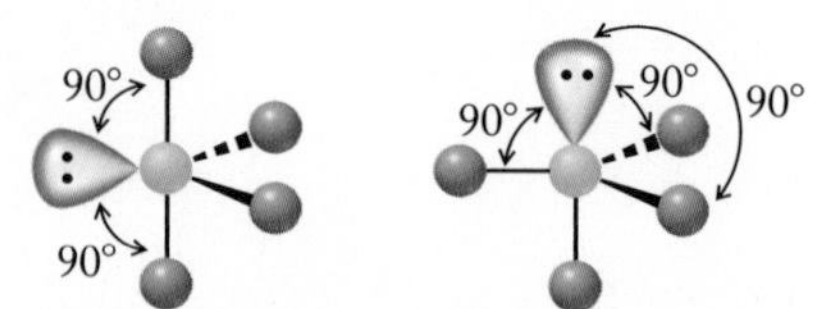

(a) Equatorial lone pair (b) Axial lone pair DOES NOT EXIST

FIGURE 9.12 (a) A lone pair of electrons in an equatorial position of a molecule with trigonal bipyramidal electron-pair geometry interacts through 90° with two other electron pairs. (b) A lone pair in an axial position interacts through 90° with *three* other electron pairs. Fewer 90° interactions reduce internal electron-pair repulsion and lead to greater stability.

Molecules with trigonal bipyramidal electron-pair geometry have four possible molecular geometries, depending on the number of lone pairs per molecule (**Table 9.1**). These options are possible because of the different axial and equatorial vertices in a trigonal bipyramid (Figure 9.2d). VSEPR theory enables us to predict which vertices are occupied by atoms and which are occupied by lone pairs. The key to these predictions is the fact that the repulsions between pairs of electrons decrease as the angle between them increases—that is, two electron pairs at 90° experience a greater mutual repulsion than two at 120°, which have a greater repulsion than two at 180°. To minimize repulsions involving lone pairs, VSEPR theory predicts that they preferentially occupy equatorial rather than axial vertices. Why? Because an equatorial lone pair has *two* 90° repulsions with the two axial electron pairs (**Figure 9.12a**), but an axial lone pair has *three* 90° repulsions with the three equatorial electron pairs (**Figure 9.12b**).

When we assign one, two, or three lone pairs of valence electrons to equatorial vertices, we get three of the molecular geometries for SN = 5 in Table 9.1. When a single lone pair occupies an equatorial site, we get a molecular geometry called **seesaw** (**Figure 9.13**) because its shape, when rotated 90° clockwise, resembles a playground seesaw. (The formal name for this shape is *disphenoidal*.)

seesaw the molecular geometry about a central atom with a steric number of 5 and one lone pair of electrons in an equatorial position.

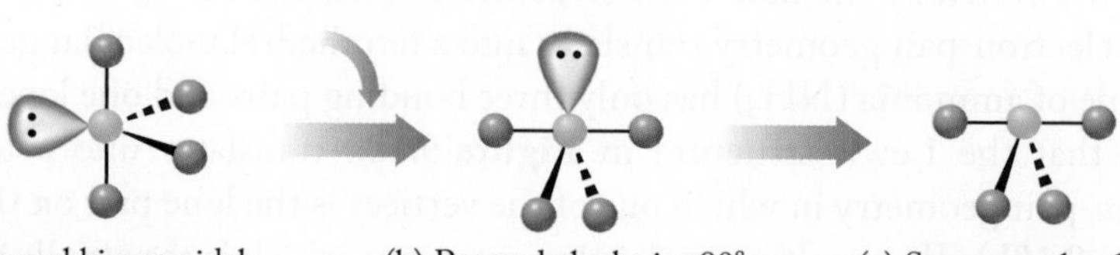

(a) Trigonal bipyramidal electron-pair geometry (b) Rotated clockwise 90° about horizontal axis (c) Seesaw molecular geometry

FIGURE 9.13 A single lone pair of electrons in an equatorial position of a trigonal bipyramidal electron-pair geometry produces a seesaw molecular geometry.

TABLE 9.1 Electron-Pair Geometries and Molecular Geometries

SN = 3	Electron-Pair Geometry	Number of Bonded Atoms	Number of Lone Pairs	Molecular Geometry	Theoretical Bond Angles	Example
	Trigonal planar	3	0	Trigonal planar	120°	CH_2O
	Trigonal planar	2	1	Bent (angular)	<120°	SO_2 ↔
SN = 4						
	Tetrahedral	4	0	Tetrahedral	109.5°	CCl_4
	Tetrahedral	3	1	Trigonal pyramidal	<109.5°	NH_3
	Tetrahedral	2	2	Bent (angular)	<109.5°	H_2O
SN = 5						
	Trigonal bipyramidal	5	0	Trigonal bipyramidal	90°, 120°, 180°	PF_3Cl_2
	Trigonal bipyramidal	4	1	Seesaw	<90°, <120°, <180°	SCl_4 =
	Trigonal bipyramidal	3	2	T-shaped	<90°, <180°	BrF_3 =
	Trigonal bipyramidal	2	3	Linear	180°	XeF_2 =
SN = 6						
	Octahedral	6	0	Octahedral	90°, 180°	SF_6
	Octahedral	5	1	Square pyramidal	<90°, <180°	IF_5
	Octahedral	4	2	Square planar	90°, 180°	XeF_4
	Octahedral	3	3	Although these geometries are possible, we will not encounter any molecules with them		
	Octahedral	2	4			

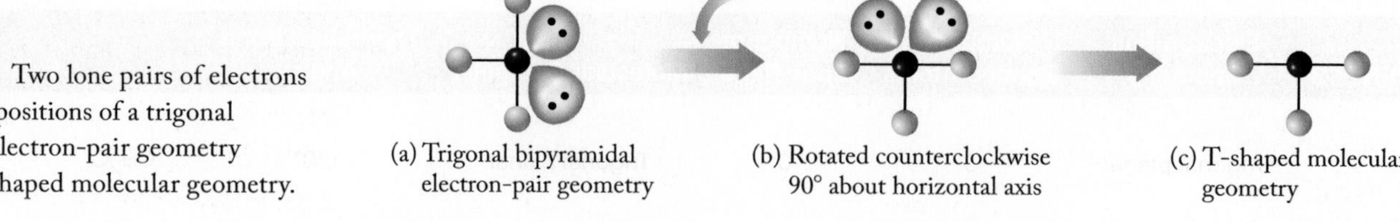

(a) Trigonal bipyramidal electron-pair geometry (b) Rotated counterclockwise 90° about horizontal axis (c) T-shaped molecular geometry

FIGURE 9.14 Two lone pairs of electrons in equatorial positions of a trigonal bipyramidal electron-pair geometry produce a T-shaped molecular geometry.

When two lone pairs occupy equatorial sites (**Figure 9.14a**), the molecular geometry that results is called **T-shaped**. This designation becomes more apparent in **Figure 9.14b**, where the structure in Figure 9.14a has been rotated 90° counterclockwise. Lone-pair repulsions result in bond angles in T-shaped molecules that are slightly less than the 90° and 180° angles we would expect from a perfectly shaped "T" geometry. Finally, a SN = 5 molecule with three lone pairs and two bonded atoms has a linear geometry because all three lone pairs occupy equatorial sites, and the bonding pairs are in the two axial positions (**Figure 9.15**).

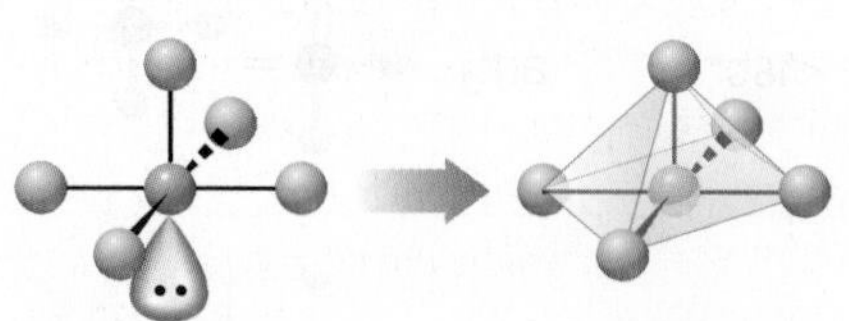

(a) Trigonal bipyramidal electron-pair geometry (b) Linear molecular geometry

FIGURE 9.15 Three lone pairs of electrons in equatorial positions of (a) a trigonal bipyramidal electron-pair geometry produce (b) a linear molecular geometry.

CONCEPT TEST

The bond angles in a trigonal bipyramidal molecular structure are 90°, 120°, and 180°. Are the corresponding bond angles in a seesaw structure likely to be larger than, the same as, or smaller than these values? Why?

Molecules that have a central atom with a steric number of 6 have an octahedral electron-pair geometry (Figure 9.2e) and the molecular geometries listed in Table 9.1. With one lone pair of electrons, there is only one possible molecular geometry because all the sites in an octahedron are equivalent (**Figure 9.16**). That one molecular geometry is called **square pyramidal**: a pyramid with a square base, four triangular sides, and the central atom "embedded" in the base.

Because of stronger repulsion between lone pairs and bonding pairs, we predict the bond angles in a square pyramidal molecule to be slightly less than the ideal angle of 90°. The molecule BrF_5, for example, has a square pyramidal molecular geometry, and the angles between its equatorial and axial bonds are 85°.

(a) Octahedral electron-pair geometry (b) Square pyramidal molecular geometry

FIGURE 9.16 (a) A lone pair of electrons in an octahedral electron-pair geometry produces (b) a square pyramidal molecular geometry.

When two lone pairs are present at the central atom in a molecule with octahedral electron-pair geometry, they occupy vertices on opposite sides of the octahedron to minimize the interactions between them. The resultant molecular geometry is called **square planar** because the molecule is shaped like a square and all five atoms reside in the same plane (**Figure 9.17**). Because the two lone pairs are on opposite sides of the bonding pairs, the presence of the lone pairs does not distort the bond angles: they are all 90°.

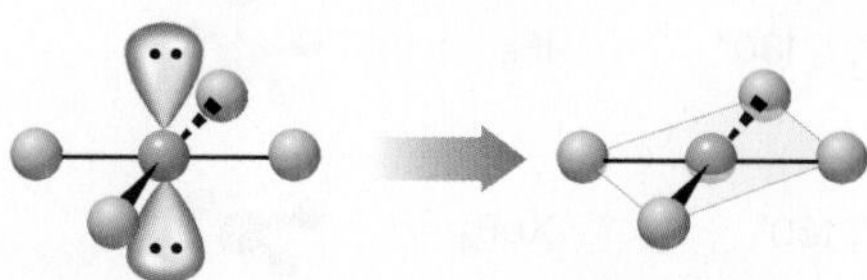

(a) Octahedral electron-pair geometry (b) Square planar molecular geometry

FIGURE 9.17 (a) Two lone pairs of electrons are on opposite sides in an octahedral electron-pair geometry. (b) The resulting molecular geometry is square planar.

SAMPLE EXERCISE 9.3 Using VSEPR Theory to Predict Geometry II LO1

The Lewis structure of sulfur tetrafluoride (SF_4) is

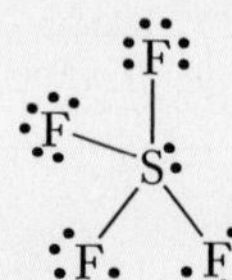

What is its molecular geometry and what are the angles between the S—F bonds?

Collect and Organize We are given the Lewis structure of SF_4. From it we can determine the steric number of the central atom in the molecule and its electron-pair geometry.

Analyze Four atoms are bonded to the central S atom, which also has one lone pair of electrons, so SN = 5.

Solve With a steric number of 5 for its central atom, the electron-pair geometry of SF_4 is trigonal bipyramidal. The presence of one lone pair of electrons on the S atom means that its molecular geometry is not the same as its electron-pair geometry. According to Table 9.1, a seesaw molecular geometry results when a lone pair occupies one of the three equatorial positions in a trigonal bipyramidal electron-pair geometry:

Electron-pair geometry trigonal bipyramidal = Molecular geometry seesaw = Frequently drawn from this perspective as well

The equatorial lone pair slightly reduces the bond angles from their normal values of 90° between the axial and equatorial bonds, 120° between the two equatorial bonds, and 180° between the two axial bonds.

Think About It Any molecule with SN = 5 and one lone pair should have the same geometry as SF_4.

Practice Exercise What are the molecular geometries and the bond angles of (a) SO_2Cl_2, (b) ClF_3, and (c) XeF_4?

9.3 Polar Bonds and Polar Molecules

In Chapter 8 we learned that differences in the electronegativities of the elements can cause the shared electrons in covalent bonds between their atoms to be distributed unevenly, with a higher density of the electrons toward the end of the bond that is occupied by the more electronegative atom. This unequal distribution of bonding electrons produces a partial negative charge in one region of the bond and a partial positive charge in the other. This charge separation creates a **bond dipole**.

In diatomic molecules with polar bonds, such as HF, unequal sharing of bonding electrons means the entire molecule is polar (see Figure 8.8). In other words, the H—F bond dipole gives a molecule of HF a permanent **dipole moment (μ)**, which is a measure of the uneven distribution of electron density over the entire molecule. The value of μ can be determined experimentally by measuring the degree to which molecules align with a strong electric field between parallel plates that have positive (+) and negative (−) charges (**Figure 9.18**). Polar molecules align with the field so that their negative regions are oriented toward the positive plate and their positive regions are oriented toward the negative plate. Molecules with larger dipole moments align more strongly with the field. Dipole moments are usually expressed in units of *debyes* (D), where 1 D = 3.34×10^{-30} coulomb-meter.

In molecules with multiple polar bonds around a center atom, the bond dipoles *may* combine to give the molecule an overall polarity. Such a molecule is said to have a *permanent dipole*. Whether a molecule has a permanent dipole depends on

T-shaped the molecular geometry about a central atom with a steric number of 5 and two lone pairs of electrons that occupy equatorial positions; the outer atoms occupy two axial sites and one equatorial site.

square pyramidal the molecular geometry about a central atom with a steric number of 6 and one lone pair of electrons; as typically drawn, the atoms occupy four equatorial and one axial site.

square planar the molecular geometry about a central atom with a steric number of 6 and two lone pairs of electrons that occupy axial sites; the atoms occupy four equatorial positions.

bond dipole the separation of electrical charge created when atoms with different electronegativities form a covalent bond.

dipole moment (μ) a measure of the degree to which a molecule aligns itself in a strong electric field; a quantitative expression of the polarity of a molecule.

CONNECTION See Figure 8.7 to review the use of arrows to indicate bond polarity and Figure 8.8 to review the use of electron-density surfaces to indicate bond polarity.

CONNECTION In Section 8.3 we introduced electronegativity and the unequal distribution of electrons in polar covalent bonds. See Figures 8.9 and 8.10 to review the Pauling electronegativity values and their periodic trends.

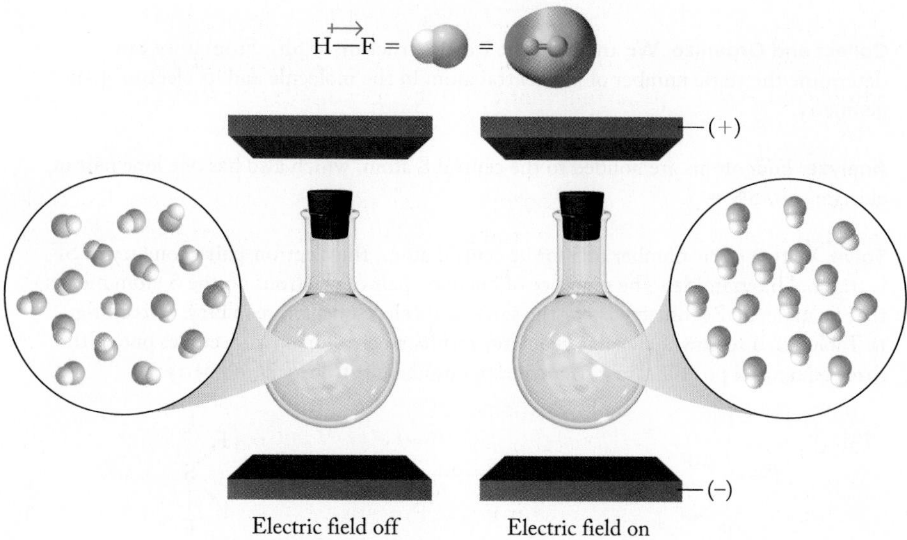

FIGURE 9.18 Gaseous HF molecules are randomly oriented in the absence of an electric field but align themselves when an electric field is applied to two metal plates. The negative (fluorine) end of each molecule is directed toward the positively charged plate, whereas the positive (hydrogen) end is directed toward the negative plate.

the strengths of its bond dipoles and how the dipoles are oriented with respect to one another. Determining how they are oriented starts with drawing the Lewis structure of the molecule, determining the steric number of the central atom, and then using this SN value and VSEPR to predict the geometry of the molecule and the orientation of its bonds and their dipoles.

Consider, for example, a molecule of CO_2. Its Lewis structure tells us that the steric number of the carbon atom is 2, which means it is a linear molecule with its two C═O bonds (and their bond dipoles) 180° from each other:

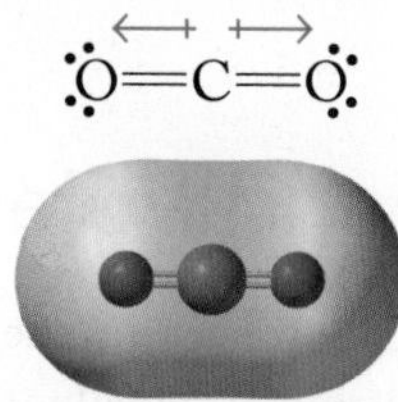

The color-coded distribution pattern of valence electrons in the molecular model tells us that the electron density is lower around the central C atom than around the two O atoms, but the electron densities around the two ends of the molecules are the same. This means the molecule has no overall polarity. The two bond dipoles, oriented in opposite directions, exactly offset each other. The result is no permanent dipole and a dipole moment of zero. In other words, carbon dioxide is a nonpolar substance.

Similarly, CF_4 is a nonpolar substance even though its molecules contain four polar C—F bonds. To understand why this is so, we again start with the Lewis structure of CF_4, which tells us that the carbon atom has a steric number of 4. According to VSEPR, the four C—F bonds (and their bond dipoles) are oriented toward the four corners of a tetrahedron. This means the angles between all the bonds (and bond dipoles) are exactly the same (109.5°), which makes the molecule and the orientation of its bond dipoles perfectly symmetric in three-dimensional space:

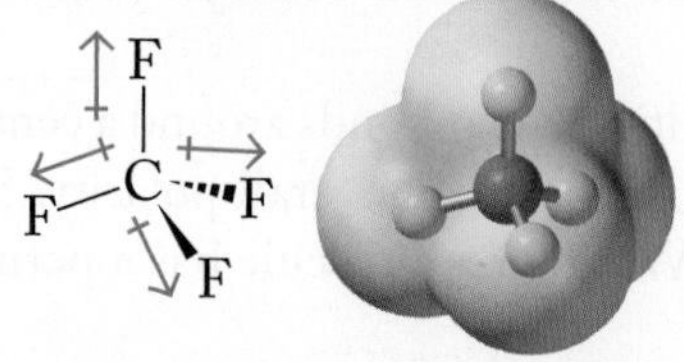

Molecular symmetry is a key characteristic of all nonpolar substances because symmetric orientation of their bond dipoles means the dipoles completely offset one another, producing dipole moments of zero.

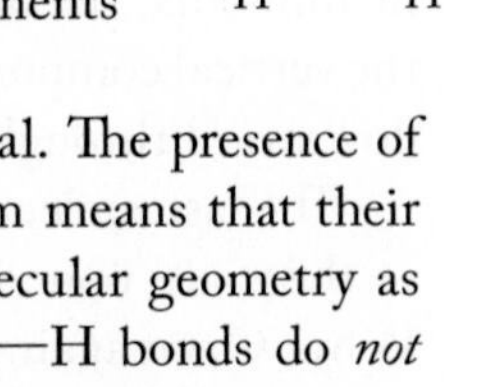

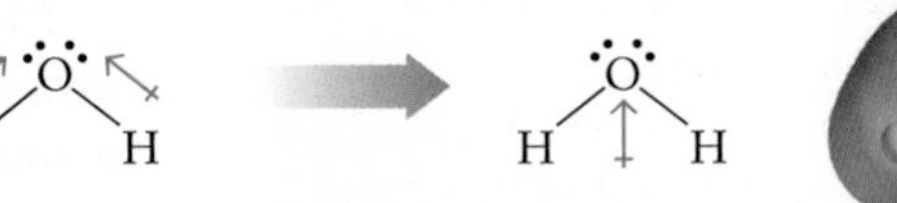

FIGURE 9.19 Water is a polar substance because the dipoles of the O—H bonds in its molecules are angled in such a way that they only partially offset one another with greater electron density around the O atom and lower electron density around the H atoms as shown in the electron density model on the right.

On the other hand, molecules of water are not symmetrical. The presence of two lone pairs of electrons in the valence shell of the O atom means that their tetrahedral electron-pair geometry translates into a bent molecular geometry as shown in **Figure 9.19**. Therefore, the dipoles of the two O—H bonds do *not completely* offset each other, giving the molecule a permanent dipole with a positive pole centered between the H atoms and negative pole in the O atom. The presence of this permanent dipole means that water is a polar substance. In Chapter 10 we discuss the considerable impact of the polarity of water on its physical properties.

The dipole moments of several polar substances are listed in **Table 9.2**. Let's examine their molecular structures to determine how their bond dipoles combine to give them permanent dipoles. It is not surprising that most (though not all) of these polar molecules contain polar bonds. Differences in electronegativity between the atoms in the bonds in these molecules range from $\Delta EN = 1.9$ for H—F, 1.4 for H—O, 0.9 for H—N and P—Cl, to 0.0 for the O—O bonds in ozone.

The polar bonds in HF and the other hydrogen halides make all these compounds polar. We have also discussed how the asymmetric shape of water molecules means the dipoles of their O—H bonds are oriented in such a way that they do not completely offset each other, giving H_2O molecules permanent dipoles. Similarly, the lone pair of electrons on the nitrogen atom in a molecule of ammonia gives the molecule the asymmetric shape of a trigonal pyramid. When viewed with the lone pair at the top, as in Table 9.2, the horizontal components of the N—H bond dipoles are arranged symmetrically around the N atom, which means they offset each other. However, all the vertical components are directed upward toward the N atom's side of the molecule, giving it a permanent dipole and a dipole moment only slightly smaller than that of water.

TABLE 9.2 Dipole Moments of Several Polar Molecules

Formula	Bond Dipole(s)	Permanent Dipole	Dipole Moment (debyes)
HF	H—F		1.91
H_2O	H—O—H		1.85
NH_3	H—N(—H)—H		1.47
PCl_3	Cl—P(—Cl)—Cl		0.97
O_3	O=O—O ↔ O—O=O		0.53

Molecules of PCl_3 are also trigonal pyramidal, but the P—Cl dipoles point toward the Cl atoms and away from the less electronegative (P) central atom. As in ammonia, the horizontal components of the three bond dipoles cancel out, but the vertical components do not. They combine to form a permanent dipole pointed downward through the center of the molecule as it is drawn in Table 9.2.

The last molecule in the table is ozone. There are no bond dipoles in O_3, just as there are none in O_2, but O_3 has a permanent dipole, whereas O_2 does not. Why? Once again, the key is symmetry: O_2 molecules are symmetrical because both O atoms are double bonded to the other O atom and have two lone pairs of electrons. However, we learned in Chapter 8 that ozone contains two equivalent O—O bonds that equally share a total of three bonding pairs of electrons. The steric number of the central O atom is 3 because it is bonded to the other two O atoms and it has one lone pair of valence-shell electrons. This SN value means that the molecule has a trigonal planar electron-pair geometry and a bent molecular geometry, which makes it asymmetric.

CONCEPT TEST

NCl_3 and $AlCl_3$ both contain polar bonds, yet NCl_3 is a polar molecule, whereas $AlCl_3$ is a nonpolar molecule. Explain how this is possible. (*Hint*: Begin by drawing their Lewis structures.)

SAMPLE EXERCISE 9.4 Predicting the Polarity of a Substance LO2

Does either of these molecules have a permanent dipole: (a) CH_2O (formaldehyde); (b) CH_2Cl_2 (dichloromethane, an organic solvent)?

Collect and Organize To predict whether a molecule has a permanent dipole, we need to determine whether there is asymmetry in its structure, particularly in the way polar bonds in the molecule are oriented relative to one another. Bond polarity depends on the difference in the electronegativity of the bonded pair of atoms. In Sample Exercise 9.1 we determined that formaldehyde has a trigonal planar structure:

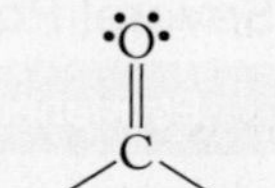

O
C
H H

Analyze The electronegativities of the atoms in formaldehyde are H = 2.1, C = 2.5, and O = 3.5, and the electronegativity of the chlorine atoms in dichloromethane is 3.0 (Figure 8.9). Therefore, bonding electrons will be drawn toward the O end of a C—O bond and the Cl end of a C—Cl bond. There is little polarity in C—H bonds ($\Delta EN = 0.4$); however, the presence of both C—H and C—O bonds in molecules of formaldehyde and C—H and C—Cl bonds in molecules of dichloromethane increases the probability that the bond dipoles in these molecules will not offset each other, at least not completely, and that the molecules will have permanent dipoles.

Solve

a. The hydrogen atoms are the least electronegative atoms in a molecule of formaldehyde and the oxygen atom is the most electronegative, so each of the bonds in the molecule has a bond dipole directed toward the oxygen atom:

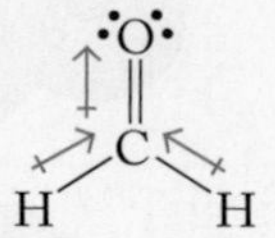

O
C
H H

This alignment of the bond dipoles means that the formaldehyde molecule has a permanent dipole moment.

b. To assess the symmetry of dichloromethane we need to draw its Lewis structure and determine the orientation of its bonds and their dipoles. Using the procedure for drawing Lewis structures that we learned in Chapter 8 yields this one for CH_2Cl_2:

```
      ..
     :Cl:
      |
  H—C—H
      |
     :Cl:
      ..
```

There are four atoms bonded to the central carbon atom and no lone pairs of electrons, so the SN value of the carbon atom is 4 and the molecular geometry is tetrahedral. One way to assess the symmetry of two C—H and two C—Cl bonds in three-dimensional space is to consult a ball-and-stick model of the molecule:

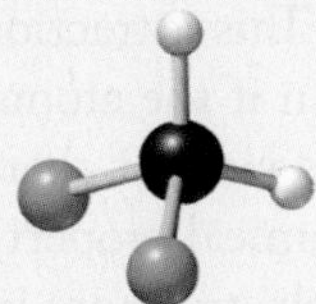

The model shows that the C—H and C—Cl bonds are arranged asymmetrically in three-dimensional space: the molecule has a hydrogen region (toward the right in this photo) that should have low electron density and a chlorine region (toward the left) that should have high electron density. Therefore, CH_2Cl_2 should have a permanent dipole oriented from lower right to upper left, which it does:

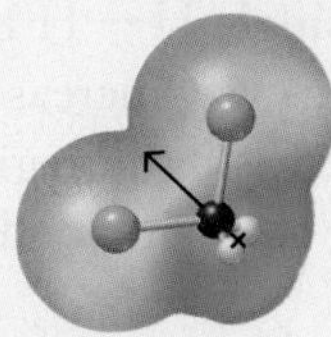

Think About It Molecular polarity cannot always be determined solely from a Lewis structure, but using a Lewis structure along with VSEPR helps us determine molecular shape and symmetry. Also, the presence of atoms of two or more elements bonded to the central atom of a molecule makes it likely that the molecule has a permanent dipole.

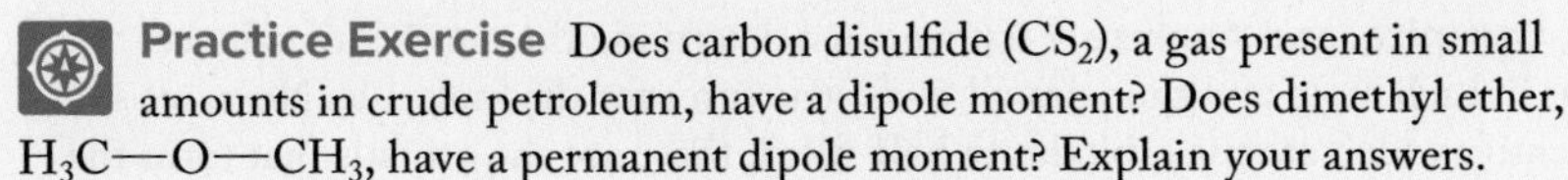

Practice Exercise Does carbon disulfide (CS_2), a gas present in small amounts in crude petroleum, have a dipole moment? Does dimethyl ether, $H_3C—O—CH_3$, have a permanent dipole moment? Explain your answers.

CONCEPT TEST

Water and hydrogen sulfide both have a bent molecular geometry with dipole moments of 1.85 D and 0.98 D, respectively. Why is the dipole moment of H_2S less than that of H_2O?

9.4 Valence Bond Theory

Drawing Lewis structures and applying VSEPR theory enable us to predict the geometry of many molecules reliably. However, we have yet to make a connection between molecular geometry and the arrangement of electrons in atoms in atomic orbitals as described in Chapter 7. The search for a connection between the

valence bond theory a quantum mechanics–based theory of bonding that incorporates the assumption that covalent bonds form when half-filled orbitals on different atoms overlap or occupy the same region in space.

overlap the degree to which atomic orbitals on adjoining atoms occupy the same region of space.

sigma (σ) bond a covalent bond in which the highest electron density lies between the two atoms along the bond axis.

electronic structure of atoms and the electronic structure of molecules led to the development of bonding theories that use atomic orbitals to account for the molecular geometries predicted by VSEPR theory. We explore one of these theories—called *valence bond theory*—next.

Bonds from Orbital Overlap

Valence bond theory arose in the late 1920s, largely because of the genius and efforts of Linus Pauling, who developed this theory of molecular bonding based on quantum mechanics. Valence bond theory incorporates the assumption that (a) a chemical bond between two atoms results from **overlap** of the atoms' atomic orbitals, and (b) the greater the overlap, the stronger and more stable the bond.

Shared electrons in a chemical bond are located between the nuclei of two atoms and are attracted to both. This attraction leads to lower potential energy and greater chemical stability than if the atoms were completely independent of one another (see Figure 8.2). This view of chemical bonding is especially useful for analyzing the physical and chemical properties of covalent substances because it provides a model of where the electrons are in a molecule. Just as the locations of electrons in atoms define atomic properties, the locations of electrons in molecules define the properties of those molecules.

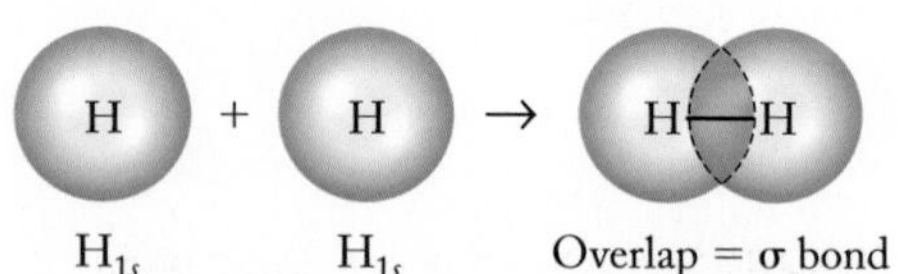

FIGURE 9.20 The overlap of the 1*s* orbitals on two hydrogen atoms produces a single σ bond that holds the two hydrogen atoms together in a H_2 molecule.

Let's begin by applying Pauling's valence bond theory to H_2, the simplest diatomic molecule. A single H atom has one electron in a 1*s* atomic orbital. According to valence bond theory, the overlap between two H atoms, each with a half-filled 1*s* orbital, produces a single H—H bond (**Figure 9.20**). Overlapping two orbitals from two atoms in this way increases electron density along the axis connecting the two nuclei. Whenever the region of high electron density lies along the bond axis, the resulting covalent bond is called a **sigma (σ) bond**.

SAMPLE EXERCISE 9.5 Identifying Overlapping Orbitals in a Molecule LO3

Identify the atomic orbitals responsible for the covalent bond in gaseous HCl.

Collect, Organize, and Analyze First we must determine the number of valence electrons in each atom and then draw Lewis structures for the molecule. Next, we must identify which atomic orbitals are half-filled; these are the orbitals that form the covalent bond between H and Cl.

Solve Atoms from group 17 (halogens) all have seven valence electrons, and hydrogen has one valence electron, so the Lewis structure for HCl is

$$\mathrm{H{-}\ddot{\underset{..}{Cl}}\!:}$$

The electron configurations of H and Cl atoms are

$$\mathrm{H{:}\ 1}s^1 \qquad \text{and} \qquad \mathrm{Cl{:}\ [Ne]3}s^2 3p^5$$

The covalent bond between H and Cl in HCl results from overlap between the half-filled hydrogen 1*s* orbital and a half-filled chlorine 3*p* atomic orbital.

Think About It The σ bond between hydrogen and chlorine in HCl that arises from overlap of two atomic orbitals is consistent with the prediction from its Lewis structure.

Practice Exercise The halogens (group 17 elements) form a series of interhalogen compounds such as IBr. Identify the atomic orbitals that overlap to form the σ bond in IBr.

Hybridization

In molecules other than simple ones such as H_2, an inconsistency arises between the atomic orbital shapes and orientations described in Chapter 7 and the molecular geometries we predict from VSEPR theory that are supported by experimental data. Methane, for example, has a carbon atom at its center. We established in Chapter 7 that carbon atoms have the electron configuration $[He]2s^22p^2$, that the 2*s* orbital is spherical, and the three 2*p* orbitals (i.e., p_x, p_y, and p_z) are oriented at 90° to one another (**Figure 9.21**). We also learned in Chapter 8 that carbon atoms form four covalent bonds oriented at 109.5° to one another in a tetrahedral arrangement. How can this be explained from the overlap of a filled 2*s* orbital and two partly filled 2*p* orbitals? Furthermore, how can there be two double bonds 180° apart in a molecule of CO_2, or one double bond and two single bonds 120° apart in a molecule of CH_2O?

To account for shapes of these and many other molecules, valence bond theory describes atomic orbitals of different shapes and energies as being mixed in a process called **hybridization** to form **hybrid atomic orbitals**. Covalent bonds then result either from overlap of a hybrid orbital on one atom with an unhybridized orbital on another atom, from overlap of two hybrid orbitals on two atoms, or from overlap of two unhybridized orbitals on two atoms. Let's examine the types of hybrid orbitals that form on carbon and other elements and how these orbitals account for observed molecular geometries.

hybridization in valence bond theory, the mixing of atomic orbitals to generate new sets of orbitals that are then available to form covalent bonds with other atoms.

hybrid atomic orbital in valence bond theory, one of a set of equivalent orbitals about an atom created when specific atomic orbitals are mixed.

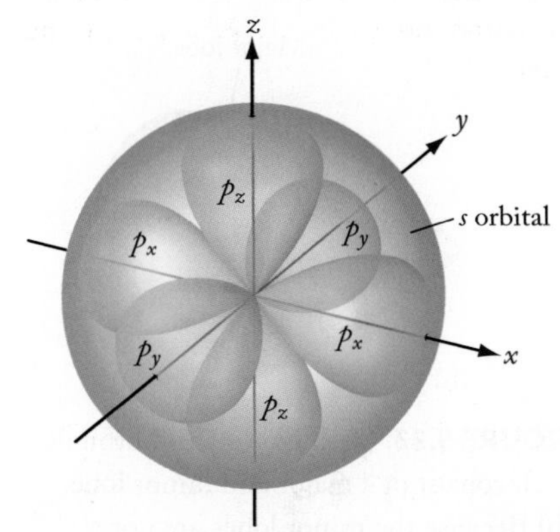

FIGURE 9.21 The relative orientation in space of one 2*s* orbital (spherical) and three 2*p* orbitals oriented at 90° to one another along the *x*-, *y*-, and *z*-axes of a coordinate system.

Tetrahedral Geometry: sp^3 Hybrid Orbitals

Methane has a tetrahedral molecular geometry with H—C—H bond angles of 109.5°. To account for this geometry by using orbital hybridization, we start with a single carbon atom with its $2s^22p^2$ valence-shell electron configuration (**Figure 9.22a**). We then promote one electron from the 2*s* orbital to the empty 2*p* orbital. Promotion raises the energy of the 2*s* electron; this gain in energy is balanced, however, by a slight decrease in the energies of the three *p* orbitals. We now have four orbitals of equal energy on the carbon atom, each containing one electron.

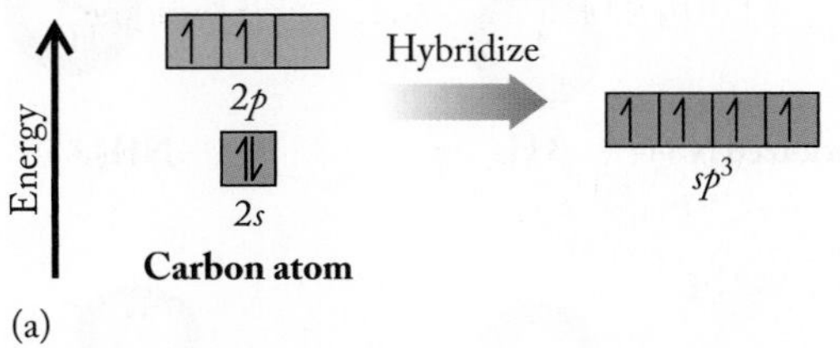

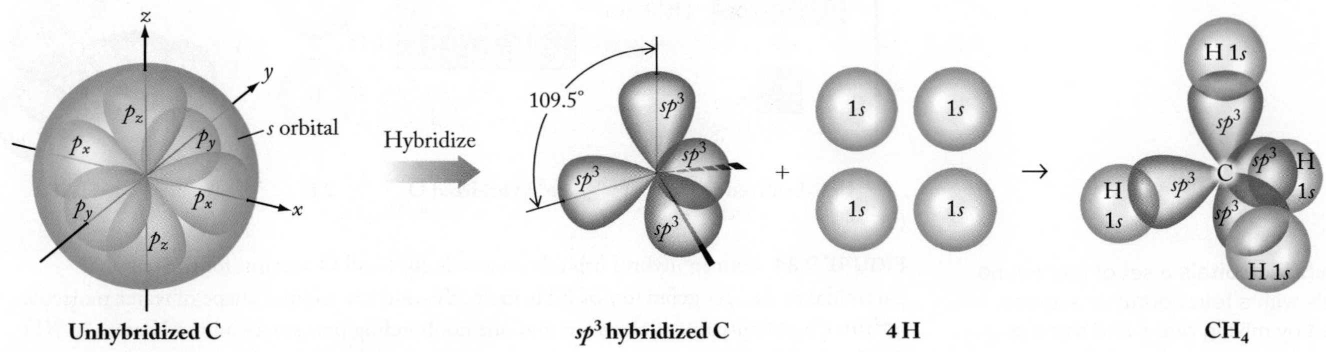

FIGURE 9.22 (a) When the 2*s* and 2*p* atomic orbitals of carbon are hybridized, one electron is promoted from the filled 2*s* orbital to an unoccupied 2*p* orbital. The four orbitals are then mixed to create four sp^3 hybrid orbitals. (b) The hybrid orbitals are oriented 109.5° from one another, exactly the orientation needed to form a tetrahedral methane molecule.

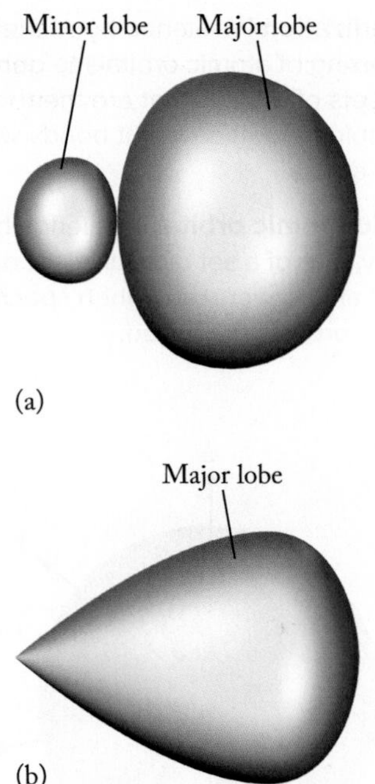

FIGURE 9.23 (a) The sp^3 hybrid orbitals each consist of a major and minor lobe. (b) Because the minor lobes are not involved in orbital overlap leading to bond formation, they have been omitted from the orbital models in this book. The major lobes have also been elongated in illustrations to better show bond angles and orientations.

These four orbitals have been *hybridized* (mixed and averaged) to make a set of four equivalent hybrid orbitals, each containing one electron. An atom's steric number always indicates how many of its orbitals must be mixed to generate the hybrid set, and the number of hybrid orbitals in a set is always equal to the number of atomic orbitals mixed. The carbon atom in methane has SN = 4, so four atomic orbitals on carbon are mixed to form four hybrid orbitals.

The four orbitals are called ***sp*³ hybrid orbitals** because they result from the mixing of one *s* and three *p* orbitals. Their energy level is the weighted average of the energies of the *s* and *p* orbitals that formed them. They are oriented 109.5° from one another and point toward the vertices of a tetrahedron (**Figure 9.22b**). Now four hydrogen atoms can form σ bonds with the sp^3 hybridized orbitals on carbon and form a methane molecule that has the observed tetrahedral geometry.

The sp^3 hybrid orbitals in Figure 9.22b are shown as having one lobe each, but in fact every sp^3 hybrid orbital consists of a major lobe and a minor lobe (**Figure 9.23a**). Because the minor lobe is not involved in bonding, we ignore it in this chapter. It becomes important in the discussion of chemical reactions that are beyond the scope of this book. Also, to better show the orientation of bonds formed by hybrid orbitals, we elongate the major lobe in illustrations (**Figure 9.23b**).

According to valence bond theory, any atom with a set of four equivalent sp^3 hybrid orbitals has a tetrahedral orientation of its valence electrons. This includes atoms in which one or more hybrid orbitals are filled before any bonding takes place. For example, sp^3 hybridization of the valence electrons on the nitrogen atom in ammonia produces three hybrid orbitals that are half-filled and one hybrid orbital that is filled (**Figure 9.24a**). The three half-filled orbitals form the three σ bonds in ammonia by overlapping with the 1*s* orbitals of three hydrogen atoms. The one filled hybrid orbital of N contains the lone pair of electrons. Similarly, the oxygen atom in water has four sp^3 hybrid orbitals in its valence shell,

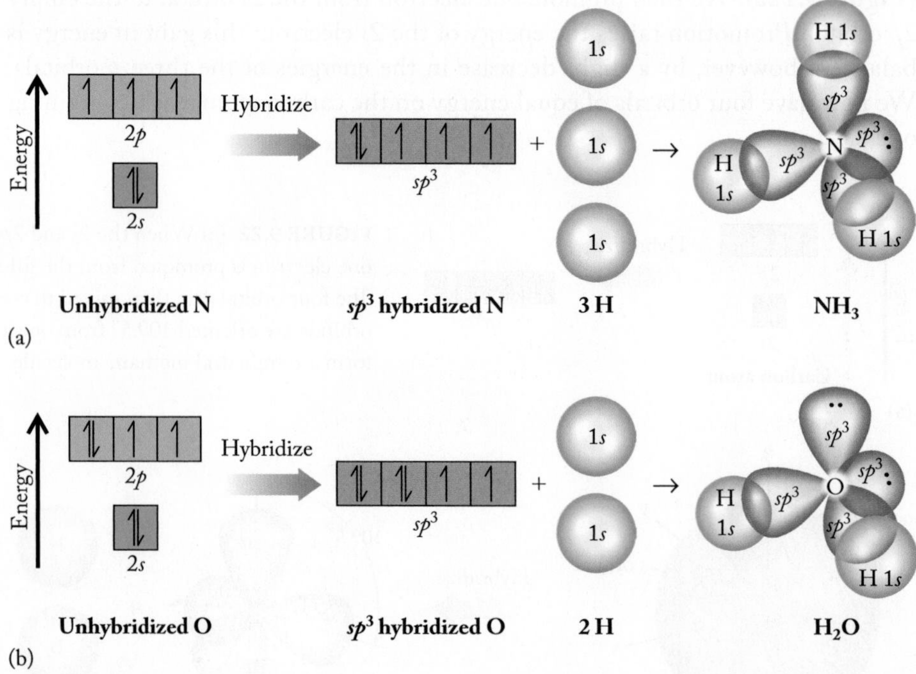

FIGURE 9.24 Four sp^3 hybrid orbitals on atoms of N and O account for the trigonal pyramidal molecular geometry of NH_3 molecules and the angular shape of water molecules. (a) Three bonding pairs of electrons and one nonbonding pair surround the N atom in NH_3. (b) Two bonding pairs and two lone pairs surround the O atom in H_2O.

***sp*³ hybrid orbitals** a set of four hybrid orbitals with a tetrahedral orientation, formed by mixing one *s* and three *p* atomic orbitals.

two filled before any bonding takes place and two half-filled and available for bond formation (**Figure 9.24b**). The two half-filled orbitals overlap with 1*s* orbitals from hydrogen atoms and form the two σ bonds in H_2O. Thus, a carbon atom forming four σ bonds, a nitrogen atom with three σ bonds and one lone pair of electrons, and an oxygen atom with two σ bonds and two lone pairs of electrons all have SN = 4 and all are sp^3 hybridized atoms. As we will see, the steric number of an atom and its hybridization are closely related in other electron-pair geometries as well.

***sp*² hybrid orbitals** three hybrid orbitals in a trigonal planar orientation, formed by mixing one *s* and two *p* atomic orbitals.

Trigonal Planar Geometry: *sp*² Hybrid Orbitals

Formaldehyde has a trigonal planar molecular geometry (**Figure 9.25**). A hybridization scheme other than sp^3 must be used to generate an orbital array with this molecular geometry because of the double bond. The VSEPR model defines SN = 3 for the carbon atom in a molecule of formaldehyde, so three atomic orbitals must be mixed to make three hybrid orbitals. To produce a trigonal planar geometry, the 2*s* orbital on carbon is mixed with two of the carbon 2*p* orbitals, and the third 2*p* orbital is left unhybridized (**Figure 9.26a**). This is the common hybridization scheme for all carbonyl (C═O) groups in organic molecules.

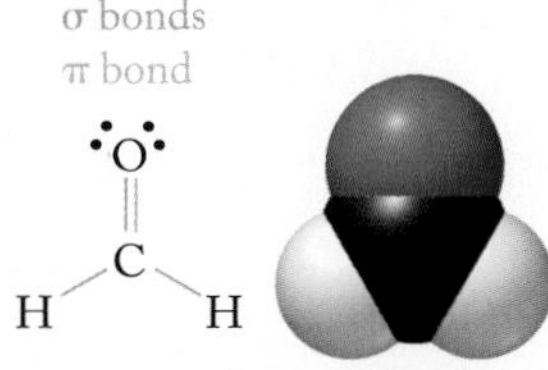

FIGURE 9.25 Formaldehyde has three σ bonds and one π bond.

Mixing and averaging one *s* and two *p* orbitals generates three hybrid orbitals called ***sp*² hybrid orbitals**. The energy level of sp^2 orbitals is slightly lower than that of sp^3 orbitals because only two *p* orbitals are mixed with an *s* orbital in a set of sp^2 orbitals. The orbitals in an sp^2 hybridized atom all lie in the same plane and are 120° apart. The two lobes of the unhybridized *p* orbital lie above and below the plane of the triangle defined by the sp^2 hybrid orbitals. An sp^2 hybridized carbon atom forms three σ bonds with its three hybridized orbitals.

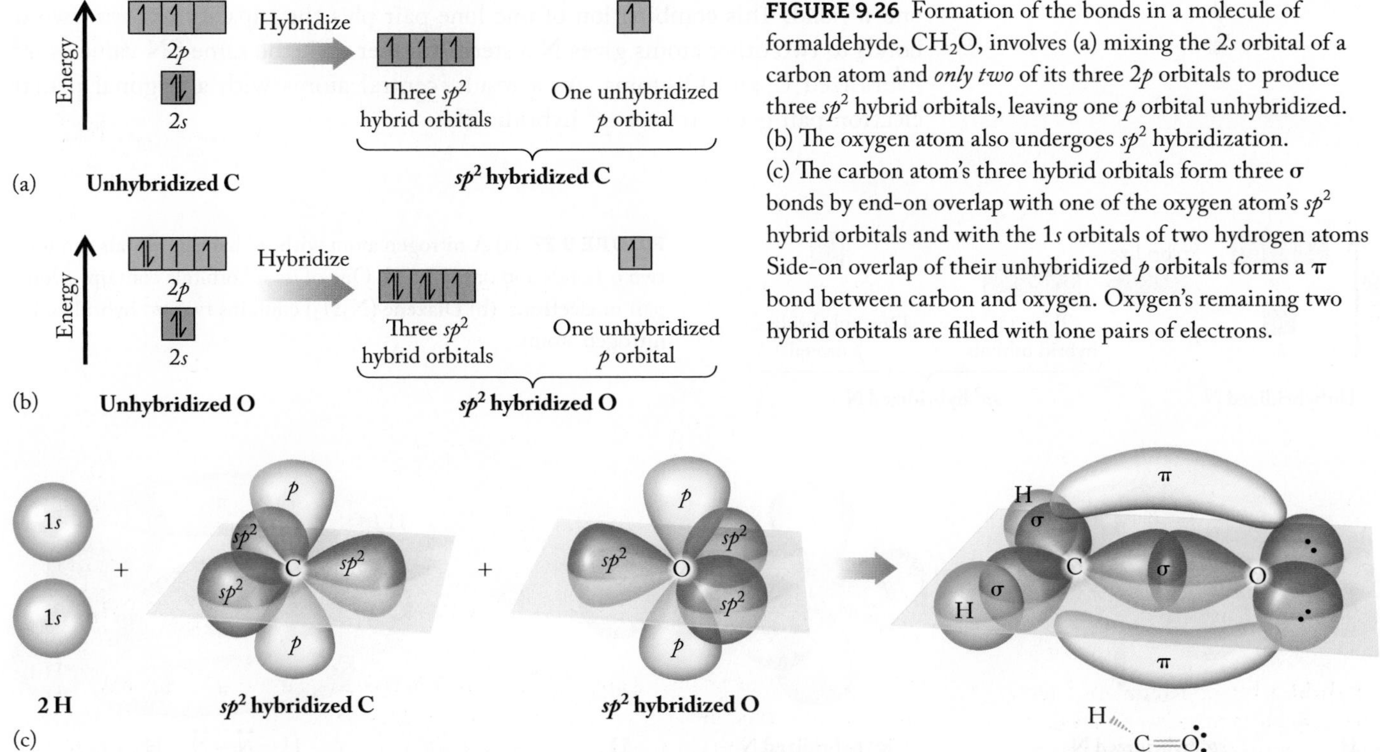

FIGURE 9.26 Formation of the bonds in a molecule of formaldehyde, CH_2O, involves (a) mixing the 2*s* orbital of a carbon atom and *only two* of its three 2*p* orbitals to produce three sp^2 hybrid orbitals, leaving one *p* orbital unhybridized. (b) The oxygen atom also undergoes sp^2 hybridization. (c) The carbon atom's three hybrid orbitals form three σ bonds by end-on overlap with one of the oxygen atom's sp^2 hybrid orbitals and with the 1*s* orbitals of two hydrogen atoms Side-on overlap of their unhybridized *p* orbitals forms a π bond between carbon and oxygen. Oxygen's remaining two hybrid orbitals are filled with lone pairs of electrons.

pi (π) bond a covalent bond in which electron density is greatest around—not along—the bonding axis; it forms from side-on overlap of two parallel *p* orbitals.

Valence electrons in *s* and *p* orbitals in other atoms can also be sp^2 hybridized. To complete the valence bond picture of formaldehyde or any carbonyl-containing compound, the oxygen atom must be sp^2 hybridized as well (**Figure 9.26b**). That both the carbon and the oxygen atoms in formaldehyde are hybridized points out another difference between VSEPR theory, which considers only the central atom in a molecule, and valence bond theory, which considers all the atoms in the molecule.

The valence bond view of the bonding in formaldehyde shows the 1*s* orbitals of two hydrogen atoms overlapping with two carbon sp^2 hybrid orbitals and the third carbon sp^2 hybrid orbital overlapping with one oxygen sp^2 hybrid orbital (**Figure 9.26c**). These overlapping orbitals constitute the σ-bonding framework of the molecule. The unhybridized carbon 2*p* orbital is parallel to the unhybridized oxygen 2*p* orbital, and the overlap of these two orbitals above and below the plane of the σ-bonding framework forms a **pi (π) bond**. The widths of the lobes on the *p* orbitals and the distances between atoms are not drawn to scale in this and similar figures. The lobes are much closer together than they appear, and they do overlap.

Pi bonds have their greatest electron density above and below the internuclear axis (or in front of and behind the internuclear axis). They can be formed only by the overlap of partially filled *p* orbitals that are parallel to each other and perpendicular to the σ bond joining the atoms. Pi bonds are inherently weaker than σ bonds because the side-on overlap of *p* orbitals is less efficient than the head-on overlap of σ bonds.

The lone pairs of electrons on the oxygen atom are in the two oxygen sp^2 hybrid orbitals not involved in the bonding with carbon. These two orbitals are oriented at an angle of about 120° in the plane of the molecule. A nitrogen atom can also form sp^2 hybrid orbitals (**Figure 9.27**). When it does, its lone pair occupies one of the three hybrid orbitals, the other two are half-filled and available to form two σ bonds, and the unhybridized half-filled *p* orbital is available to form one π bond. This combination of one lone pair plus the capacity to form two σ bonds to two other atoms gives N a steric number of 3, the same SN value as sp^2 hybridized C and O atoms. As a result, central atoms with a trigonal planar electron-pair geometry are sp^2 hybridized.

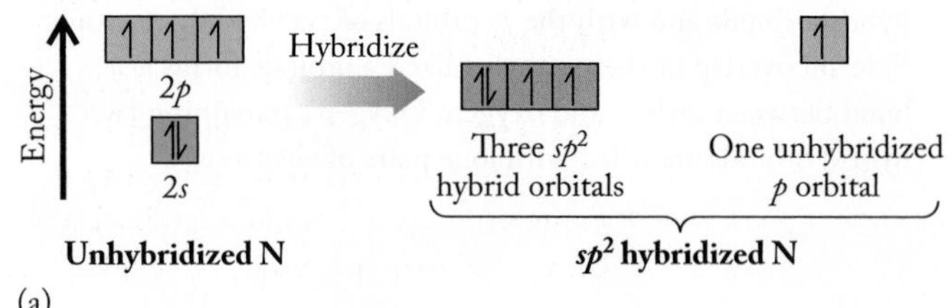

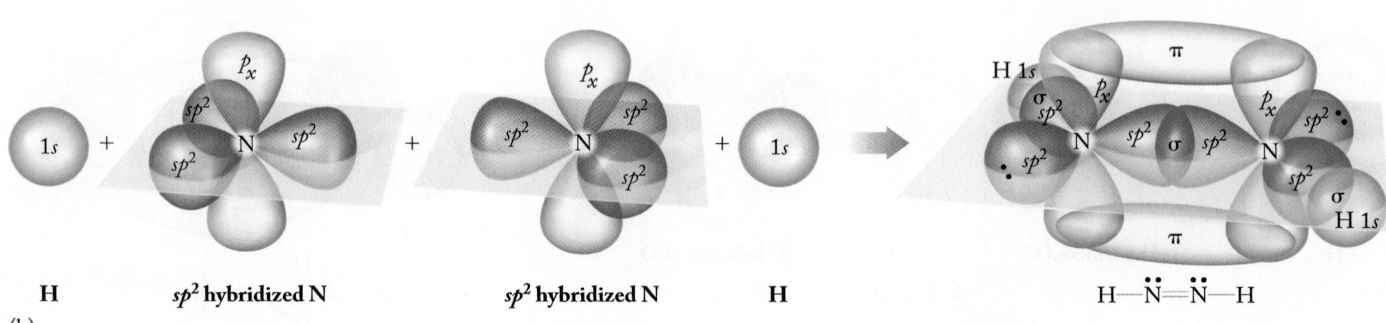

FIGURE 9.27 (a) A nitrogen atom with sp^2 hybrid orbitals can form two σ bonds and one π bond. One of its sp^2 orbitals contains a lone pair of electrons. (b) Diazene (N_2H_2) contains two sp^2 hybridized nitrogen atoms.

Linear Geometry: *sp* Hybrid Orbitals

sp hybrid orbitals two hybrid orbitals oriented 180° from one another, formed by mixing one *s* and one *p* orbital.

In the linear alkyne acetylene (C_2H_2), both carbon atoms have SN = 2 according to VSEPR theory. Because steric number indicates the number of hybrid orbitals that must be created, two atomic orbitals mix to make two hybrid orbitals on each carbon atom.

The results of mixing one 2*s* and one 2*p* orbital on carbon and leaving the other two 2*p* orbitals unhybridized are shown in **Figure 9.28a**. The two ***sp* hybrid orbitals** have major lobes that are on opposite sides of the carbon atom. One set of the unhybridized *p* orbitals has lobes above and below the axis of the *sp* hybrid orbitals; the second set is in front and behind the plane of the hybrid orbitals.

The valence bond view of C_2H_2 shows a σ-bonding framework in which a hydrogen 1*s* orbital overlaps with one *sp* hybrid orbital on each carbon atom. Each carbon atom's other *sp* hybrid orbital overlaps with one *sp* hybrid orbital on the other carbon atom (**Figure 9.28b**). This arrangement brings the two sets of unhybridized *p* orbitals on the two carbon atoms into parallel alignment with each other. They then overlap to form two π bonds between the two carbon atoms, so that the one σ bond and the two π bonds form the triple bond in HC≡CH. The steric number of the *sp* hybridized carbon atom in acetylene is 2 (bonds to two atoms and no lone pairs).

Figure 9.29 summarizes the differences in electron distributions for single, double, and triple bonds for the carbon atoms in alkanes, alkenes, and alkynes as described by valence bond theory. The structures of these molecules determine their physical and chemical properties, but "structure" in this sense refers to more than just the location of the atoms. The distribution of electrons in three-dimensional space also greatly influences these properties. For reactions to take place, particles must interact, and the location of the valence electrons in molecules influences which bonds will break in reactions and which new bonds will form. According to valence bond theory, the electron density in the carbon–carbon σ bonds in alkanes lies on the axis directly between the two atomic centers (Figure 9.29a). In contrast, the electron density in the π bonds in alkenes and alkynes is largest not between the two carbon atoms, but above and below or in front of and behind the bonding axis (Figure 9.29b, c). The electrons in these π bonds more easily interact with those on other molecules, which is one reason why alkenes and alkynes are more reactive than alkanes, which have only σ bonds.

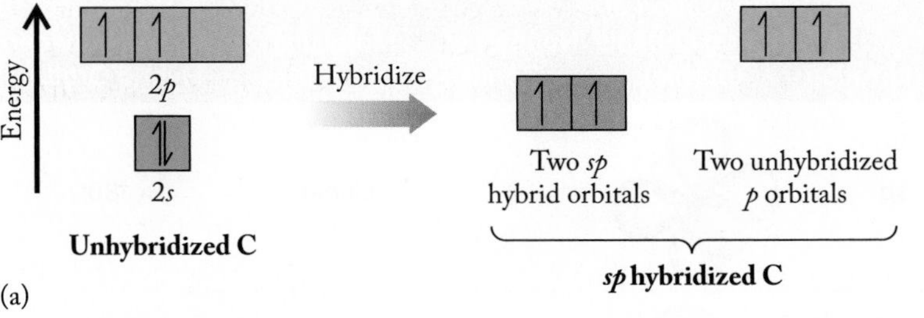

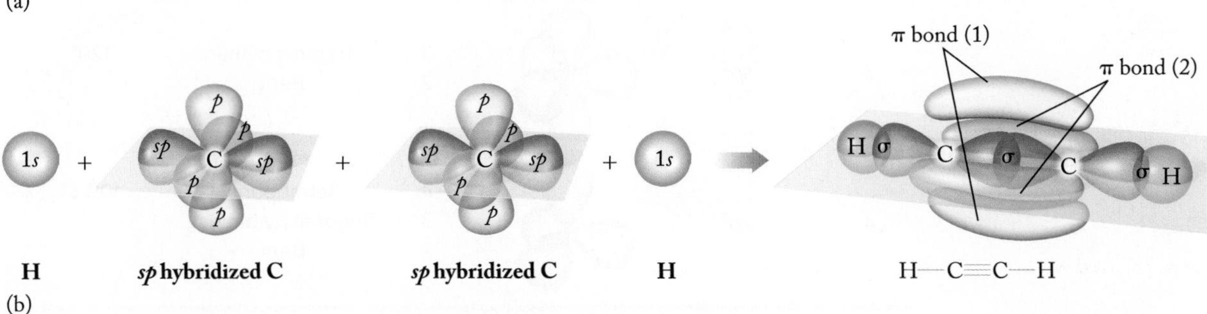

FIGURE 9.28 Formation of the bonds in a molecule of acetylene, C_2H_2, involves (a) mixing one 2*s* orbital and one 2*p* orbital on a carbon atom, creating two *sp* hybrid orbitals and leaving two of its 2*p* orbitals unhybridized. (b) Two *sp* hybridized carbon atoms each bond to one hydrogen atom and to the other carbon atom via σ bonds. Side-on overlap of the unhybridized *p* orbitals form two π bonds: one above and below, the other in front of and behind, the C—C σ bond.

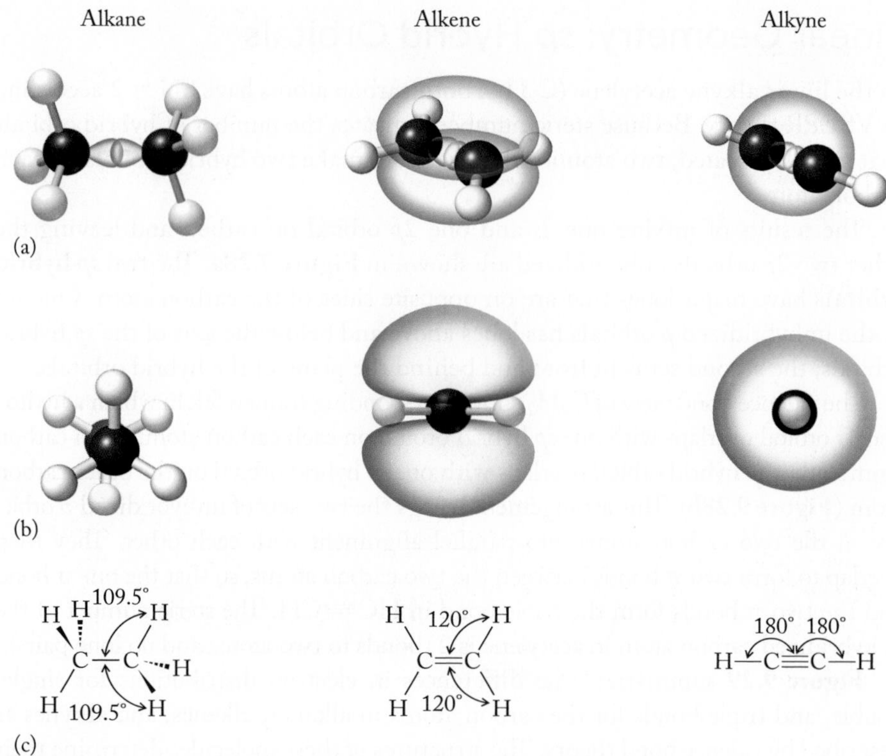

FIGURE 9.29 Carbon–carbon bonding in alkanes, alkenes, and alkynes. The electrons in the carbon–carbon bonds in alkanes are in σ bonds; the characteristic structural feature of alkenes and alkynes is the presence of additional electrons in π bonds. To simplify these drawings, orbitals are drawn only for the carbon–carbon bonds. (a) The electron distribution in a C—C single bond, a C═C double bond, and a C≡C triple bond. (b) View of these electron distributions looking down the carbon–carbon bond axis. (c) Idealized H–C–C bond angles for single, double, and triple bonds.

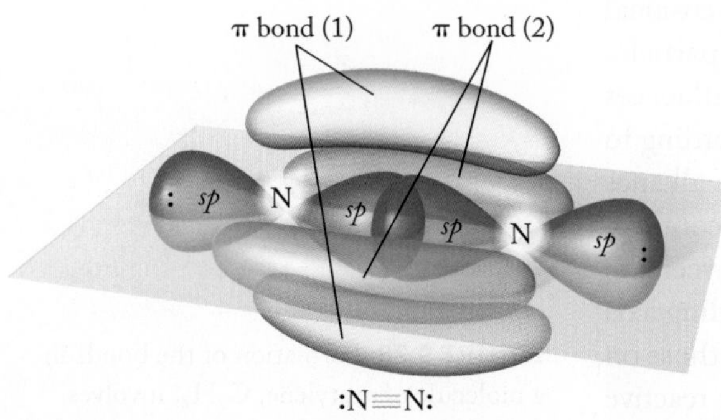

FIGURE 9.30 Valence bond representation of a N_2 molecule and its *sp* hybrid N atoms.

The link between SN = 2 and *sp* hybridization applies to atoms other than carbon—for example, the nitrogen atoms in N_2. A lone pair occupies one of the two *sp* orbitals on each nitrogen atom and the other *sp* hybrid orbital forms the σ bond (**Figure 9.30**).

A summary of the shapes associated with all the hybridization schemes we have discussed is provided in **Table 9.3**.

CHEMTOUR
Hybridization

TABLE 9.3 Hybrid Orbitals Based on Steric Number

Steric Number	Hybridization	Orientation of Hybrid Orbitals (purple)	Number of σ Bonds	Molecular Geometries	Theoretical Angles between Hybrid Orbitals
2	*sp*		2	Linear	180°
3	sp^2		3 2	Trigonal planar Bent	120°
4	sp^3		4 3 2	Tetrahedral Trigonal pyramidal Bent	109.5°

SAMPLE EXERCISE 9.6 Describing Bonding in a Molecule LO3

Use valence bond theory to account for the linear molecular geometry of CO_2, to determine the hybridization of carbon and oxygen in this molecule, and to describe the orbitals that overlap to form the bonds. Draw the molecule, showing the orbitals that overlap to form the bonds.

Collect and Organize We know that CO_2 is linear and that the Lewis structure of the molecule is

$$\ddot{O}=C=\ddot{O}$$

Analyze Each double bond is composed of one σ bond and one π bond. Therefore, the carbon atom forms two σ bonds and two π bonds. It has no lone pairs and so SN = 2. Each oxygen atom forms one σ bond and one π bond.

Solve The carbon atom must be *sp* hybridized because half-filled *sp* hybrid orbitals have the capacity to form two σ bonds, leaving two half-filled unhybridized *p* orbitals available to form two π bonds (**Figure 9.31**). The lobes of *sp* hybrid orbitals are at an angle of 180°, which is consistent with the linear molecular geometry of CO_2. The oxygen atoms must be sp^2 hybridized because that hybridization leaves them with one unhybridized *p* orbital to form a π bond. The σ bonds form when each *sp* orbital on the carbon atom overlaps with one sp^2 orbital on an oxygen atom.

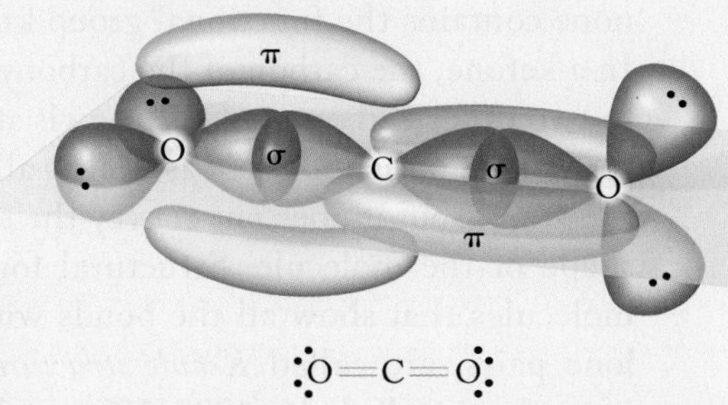

$$\ddot{O}=C=\ddot{O}$$

FIGURE 9.31 The bonding pattern in CO_2: the π bond to the left is drawn above and below the plane of the molecule; the π bond to the right is drawn in front of and behind the plane.

The two unhybridized *p* orbitals of carbon are oriented 90° to each other and the plane containing the three sp^2 orbitals of one oxygen atom is rotated 90° with respect to the plane containing the three sp^2 orbitals of the other oxygen atom. This orientation is necessary for the formation of the two π bonds because the two unhybridized *p* orbitals that overlap to form a π bond must be parallel. The unshared pairs of electrons on each oxygen atom and the σ bond to carbon lie in a trigonal plane 120° apart. The sp^2 hybridization of the oxygen accounts for all these features.

Think About It The *sp* hybridization of the carbon atom is consistent with its steric number (SN = 2), its capacity to form two σ and two π bonds, and the linear geometry of the molecule.

Practice Exercise In which of the following molecules does the central atom have sp^3 hybrid orbitals? (a) CCl_4; (b) HCN; (c) SO_2; (d) PH_3

CONCEPT TEST

Why can't a carbon atom with sp^3 hybrid orbitals form π bonds?

9.5 Shape, Large Molecules, and Molecular Recognition

Most of the molecules we have considered so far have small molar masses or a single central atom bonded to two or more atoms. Because we will see more examples of larger molecules (i.e., molecules with more than just one central atom) in this chapter and throughout the text, we need to learn some of the ways that chemists have developed to represent such structures in addition to Lewis structures. Then we can apply the concepts that we have discussed about valence bond theory to these larger molecules. It is important to learn how to represent larger molecules with more than one central atom because molecular shape is an important factor in determining the physical, chemical, and biological properties of all substances.

FIGURE 9.32 Different representations of the molecular structures of pentane and butanone: (a) Lewis structures; (b) Kekulé structures; (c) condensed structures; (d) skeletal structures.

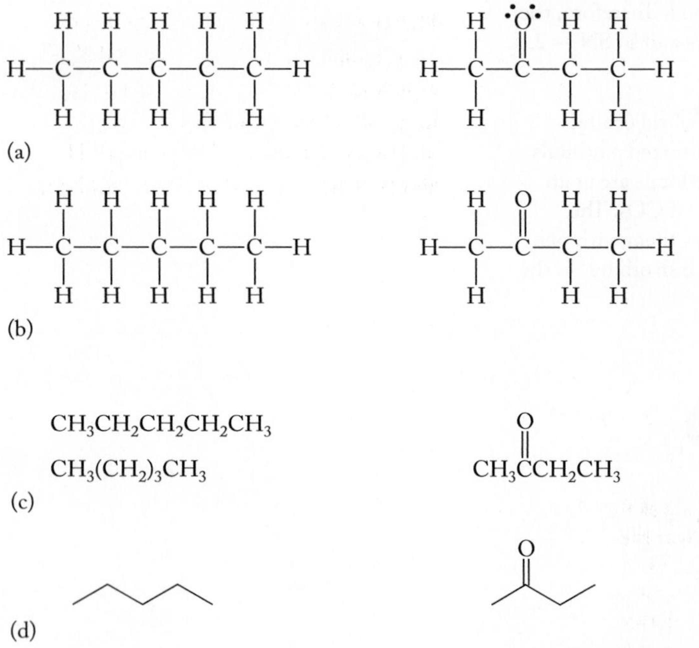

(a)

(b)

$CH_3CH_2CH_2CH_2CH_3$

$CH_3(CH_2)_3CH_3$

$CH_3\overset{O}{\overset{\|}{C}}CH_2CH_3$

(c)

(d)

Drawing Larger Molecules

Figure 9.32a shows the Lewis structures for both pentane and butanone. Pentane is a member of the alkane family that we discussed in Section 6.9, whereas butanone contains the functional group known as a **ketone**. In a ketone, the carbon of the carbonyl group is bonded to two other carbon atoms. A Lewis structure shows all the bonds in the molecule as well as all the lone pairs on the atoms, but it does not convey the three-dimensional shape of the molecule. Structural formulas of organic molecules that show all the bonds with lines, but omit lone pairs, are called *Kekulé structures* (**Figure 9.32b**) after August Kekulé (1829–1896), who first used this method for illustrating molecules. Alkanes do not have lone pairs on any atoms, so there is no difference between the Lewis structure and the Kekulé structure for pentane, but there is for butanone because the lone pairs on the oxygen atom are omitted in its Kekulé structure.

CONNECTION Kekulé structures are the same as the structural formulas we have been drawing throughout the text since Chapter 1.

Writing Lewis or Kekulé structures for larger and larger organic molecules gets tedious quickly. For this reason, chemists use even shorter notations called *condensed structures* (**Figure 9.32c**) to represent molecules. Condensed structures do not show most of the individual bonds between atoms as Lewis and Kekulé structures do. They use subscripts to indicate the number of times a subgroup is repeated. For example, the condensed structure of pentane can be written as $CH_3CH_2CH_2CH_2CH_3$ or as $CH_3(CH_2)_3CH_3$. The numerical subscript after the CH_2 group in parentheses means that three of these groups connect the terminal $-CH_3$ groups in this compound. Note that the subscript indicating the number of CH_2 groups comes *after* the closing parenthesis. The subscript 2 inside the parentheses is for the two H atoms bonded to each C atom of each CH_2 group.

CONNECTION We saw methyl groups ($-CH_3$) and methylene groups ($-CH_2-$) when we investigated alkanes in Section 6.9. We also classified alkanes as straight-chain or branched-chain hydrocarbons.

The most minimal notation, shown in **Figure 9.32d**, is the *skeletal structure* of a molecule, which has no element symbols for carbon and hydrogen atoms. Atoms other than C and H, such as the oxygen in butanone, are shown in skeletal structures. **Figure 9.33** shows how to create a skeletal structure for octane (C_8H_{18}). Short line segments are drawn at angles to one another in a "zigzag" manner, and each line segment symbolizes one carbon–carbon bond in the molecule. The

ketone organic molecule containing a carbonyl group in which the carbonyl carbon is bonded to two other carbon atoms.

angles represent the bond angle between the two carbon atoms (109.5° for sp^3 hybridized carbon atoms in alkanes). There is a $-CH_3$ group at each end of the zigzag line, and there is either a $-CH_2-$ group (as is the case in pentane) or a functional group (as is the case for butanone) at each vertex formed by the intersection of two line segments. Skeletal structures are assumed to contain the appropriate number of hydrogen atoms so that each carbon atom has a steric number of 4. Hydrogen atoms are not shown in a skeletal structure because all carbon atoms are known to make four bonds, and any bonds not shown are understood to be C—H bonds.

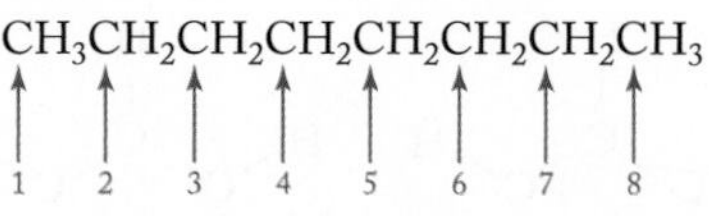

or $CH_3(CH_2)_6CH_3$

(a) Condensed structural formula

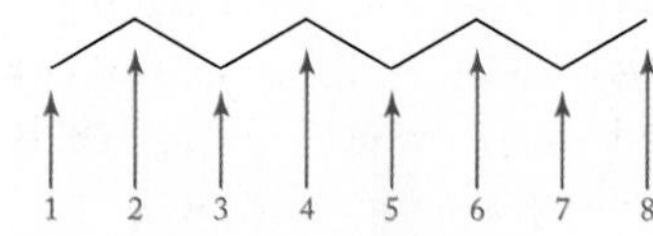

(b) Skeletal structure

FIGURE 9.33 When converting from (a) a condensed structure to (b) a skeletal structure, the carbon atoms are represented by junctions between lines, and each junction is assumed to have enough H atoms to give that carbon atom four bonds.

SAMPLE EXERCISE 9.7 Drawing Condensed and Skeletal Structures LO4

The Kekulé structure of a compound that contributes to the distinctive aroma of Gorgonzola cheese is shown in **Figure 9.34**. (a) Write a condensed structure and (b) draw the skeletal structure for this ketone.

FIGURE 9.34 Heptan-2-one, a compound found in blue cheese.

Collect, Organize, and Analyze Figure 9.32 summarizes the differences between Kekulé, condensed, and skeletal structures. Condensed structures show the symbol for each element in a molecule with subscripts indicating the numbers of atoms of each element, but they do not show C—H or C—C single bonds. We can group all $-CH_2-$ groups by using parentheses followed by a subscript to show the number of $-CH_2-$ groups. Skeletal structures use short line segments to represent the carbon–carbon bonds in molecules. Intersections of these segments represent carbon atoms.

Solve

a. The condensed structure comes from its Kekulé structure, where we write each carbon followed by H plus a subscript showing the number of H atoms bonded to that carbon atom. The ketone functional group is written as C=O, with the carbonyl carbon atom bonded to two other carbon atoms. There must be two CH_3- groups at the ends with a ketone functional group and four $-CH_2-$ groups between them:

$$CH_3\overset{O}{\overset{\|}{C}}CH_2CH_2CH_2CH_2CH_3 \quad \text{or} \quad CH_3\overset{O}{\overset{\|}{C}}(CH_2)_4CH_3$$

b. To draw the skeletal structure, first draw a short line slanted upward to the right; this represents the bond between the first carbon atom and the carbonyl carbon. Then without removing your pencil from the paper, draw a short line slanted downward to the right to represent the bond between the carbonyl carbon and the third carbon atom. Continue until you have drawn seven carbon atoms (including the carbonyl carbon).

Think About It Condensed structures and skeletal structures both represent the same number of atoms and the same number of bonds, even though the bonds are not shown in the condensed structure and not all the atoms are shown in the skeletal structure.

Practice Exercise Draw the skeletal structure of hexane, $CH_3(CH_2)_4CH_3$, and a condensed structure of heptane:

Heptane

(a) or

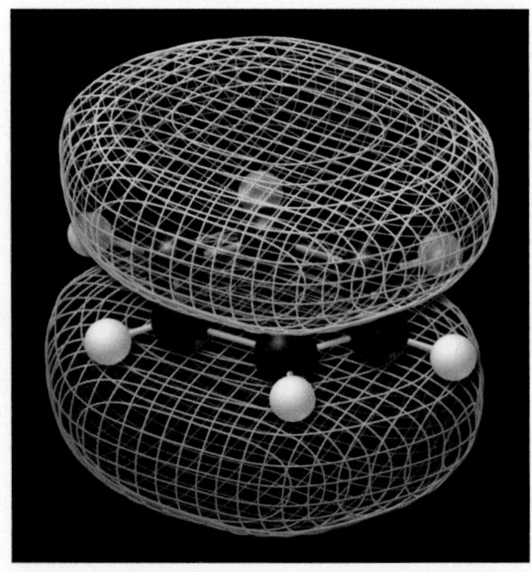

(b)

FIGURE 9.35 (a) Resonance leads to complete delocalization of the π bonds around the benzene ring. (b) A computer-generated view of benzene's σ bonds in the plane of the ring and its delocalized π bonds above and below the ring.

Molecules with More than One Functional Group

Now let's consider a biologically active molecule with three "central" atoms and more than one functional group. Acrolein is one of the components of barbeque smoke that contributes to the distinctive odor of a cookout. It is also a possible cancer-causing compound. The Lewis structure of acrolein is

Each molecule contains a C═O double bond, which is part of an aldehyde functional group, and a C═C double bond, which is an alkene functional group. Both double bonds are formed by trigonal planar, sp^2 hybridized carbon atoms.

Benzene (C_6H_6) is another molecule in barbeque smoke. As we saw in Section 8.4, each benzene molecule is a hexagon of six carbon atoms, each bonded to one hydrogen atom. The Lewis structure of benzene shows three C═C bonds. Resonance structures for benzene are shown in **Figure 9.35**, along with a view of the π bonds located above and below the plane of the ring. Each carbon in benzene has a trigonal planar geometry and forms sp^2 hybrid orbitals. The carbon ring is made of σ bonds formed by overlapping sp^2 hybrid orbitals on adjacent carbon atoms. The C—H bonds are formed by the overlap of carbon sp^2 hybrid orbitals with hydrogen 1*s* orbitals.

The two resonance forms of benzene shown in Figure 9.35a correspond to shifts in the locations of the π bonds formed by the overlap of carbon 2*p* orbitals. However, because all the carbon 2*p* orbitals are identical, they are all equally likely to overlap with their neighbors, so the electrons in the π bonds are *delocalized* over all six carbon atoms rather than fixed between alternating pairs of carbon atoms, as depicted in the computer-generated view in Figure 9.35b.

Delocalization reduces the potential energy of the electrons in π bonds and lowers the energy of the entire molecule, a phenomenon known as **resonance stabilization.**

As shown in Figure 8.16b, the presence of delocalized π electrons in benzene is often represented by a circle drawn in the middle of the hexagon of carbon atoms in the structural formula, corresponding to continuous rings of π electrons above and below the plane of the molecule. The skeletal structure of benzene is a simple hexagon with a circle in it.

CHEMTOUR
Structure of Benzene

CONNECTION We described delocalized electrons in Section 8.1 with respect to metallic bonds, but delocalization occurs in molecules and ions as well.

CONNECTION We mentioned aromatic compounds in Section 8.4 when we discussed resonance in Lewis structures.

CONCEPT TEST

In which of the molecules and polyatomic ions in **Figure 9.36** are the electrons in the π bonds delocalized?

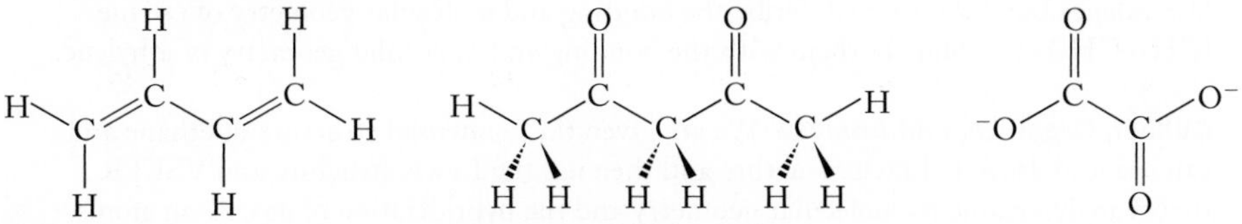

(a) (b) (c)

FIGURE 9.36 Organic compounds with double bonds: (a) butadiene, used to make polymers; (b) 2,4-pentanedione, a reactive organic compound used in synthesis; and (c) the oxalate ion, present in spinach and rhubarb.

The molecular structures of many other compounds in addition to benzene contain carbon rings with delocalized π electrons above and below the plane of the ring. They are called **aromatic compounds**. An important class of these compounds, known as *polycyclic aromatic hydrocarbons* (PAHs), consists of molecules containing several benzene rings joined together (**Figure 9.37a**). PAHs are formed any time coal, oil, gas, and most hydrocarbon fuels are burned, and they are also found in cigarette smoke. In 2004 they were discovered in interstellar space.

The shape of PAH molecules gives rise to a significant health hazard. After we inhale or ingest them, some PAHs may bind to the DNA in our cells in a process called *intercalation*. Because PAHs are flat, they can slide into the double helix that DNA forms (**Figure 9.37b**). Once there, they may alter or prevent DNA replication and thereby damage or kill cells. Intercalation in DNA is one step in the process by which PAHs induce cancer.

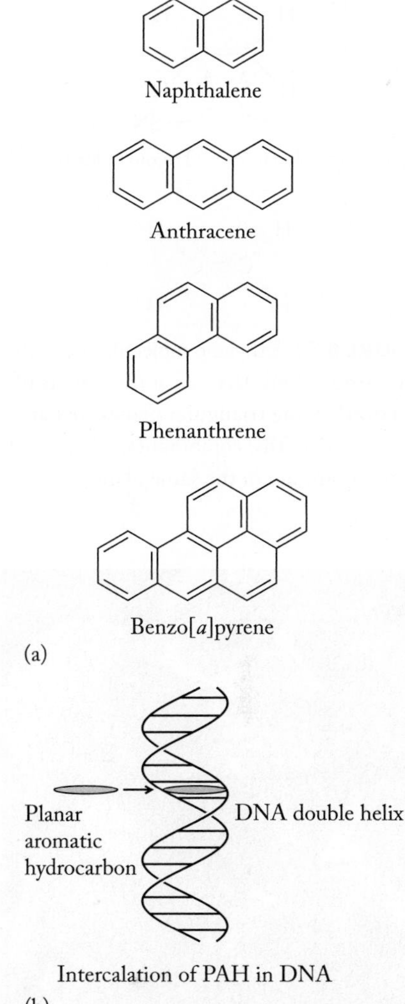

FIGURE 9.37 (a) The molecules of these polycyclic aromatic hydrocarbons consist of fused benzene rings whose π bonds are delocalized over all the rings in each molecule. (b) Molecules with this shape can slip between the strands of DNA and disrupt cell replication, which can lead to cell death or induce malignancy.

Chirality and Molecular Recognition

Living things respond to molecules that interact with regions in their tissues called *receptors* or *active sites*. The process by which these molecules and sites interact is known as **molecular recognition**. This recognition does not usually involve covalent bond formation. Instead, these noncovalent interactions require that the biologically active molecules and the receptors that respond to them fit tightly together, which means that they must have complementary three-dimensional shapes. Slight differences in these shapes can result in different biological responses, such as cilantro having the taste of a fresh herb to some people, but soap to other people.

Another example of molecular recognition is the role of the gaseous hydrocarbon ethylene, C_2H_4 (**Figure 9.38**), in speeding up the ripening of many fruits and vegetables, including the green tomatoes in **Figure 9.39**. Each of the carbon atoms in a molecule of ethylene is bonded to three other atoms and has no lone pairs of electrons. This means SN = 3, in which case the geometry around each carbon atom is trigonal planar and the atom is sp^2 hybridized (see Table 9.3). These three hybrid orbitals form three σ bonds with two hydrogen atoms and the other carbon atom. Side-on overlap of their unhybridized $2p$ orbitals forms a π bond between the carbon atoms. Taken together, the two trigonal planar carbon atoms produce an overall planar geometry for ethylene, which means that all six atoms lie in the same plane.

resonance stabilization the stability of a molecular structure due to delocalization of its electrons.

aromatic compound a cyclic, planar compound with delocalized π (pi) electrons above and below the plane of the molecule.

molecular recognition the process by which molecules interact with other molecules in living tissues to produce a biological effect.

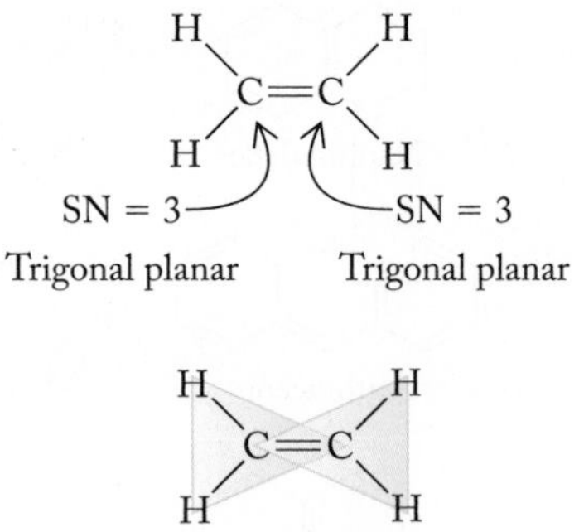

H H
C=C
H H
SN = 3 SN = 3
Trigonal planar Trigonal planar

H H
C=C
H H

FIGURE 9.38 Ethylene molecules contain two carbon atoms that are at the centers of two overlapping triangular planes that are also coplanar. This combination means that all the atoms are in the same plane.

FIGURE 9.39 Ripening tomatoes give off ethylene gas, which induces other green tomatoes nearby to ripen.

SAMPLE EXERCISE 9.8 Comparing Structures by Using Valence Bond Theory **LO3**

Use valence bond theory to describe the bonding and molecular geometry of ethane (CH_3–CH_3) and compare them with the bonding and molecular geometry of ethylene.

Collect, Organize, and Analyze We are given the condensed structure of ethane and can use it to draw its Lewis structure and then use the Lewis structure and VSEPR theory to determine its molecular geometry and the hybridization of its carbon atoms. These can be compared to the molecular structure of ethylene in Figure 9.38.

Solve Each carbon atom in ethane has four single bonds and no lone pairs, which means each carbon has SN = 4, tetrahedral geometry, and sp^3 hybridization. Comparing the overall molecular structures of ethane and ethylene (**Figure 9.40**), we see the distinctive three-dimensionality of tetrahedral environments: only two of the six C–H bonds in ethane are in the plane of the page, whereas all the bonds in ethylene are coplanar.

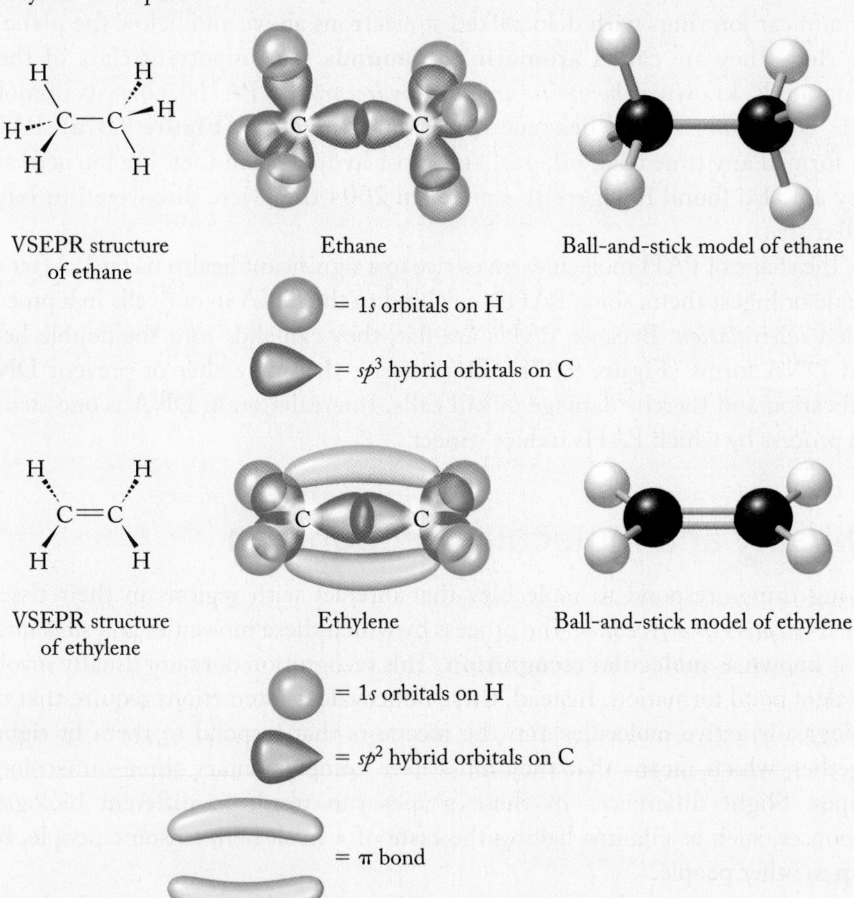

FIGURE 9.40 Molecular structures of ethane and ethylene.

Think About It Remember the analogy at the beginning of the chapter about how molecules fit receptors like hands fit gloves. Based on their structures, ethylene and ethane would fit into very different gloves.

Practice Exercise Diazene (N_2H_2) and hydrazine (NH_2NH_2) are reactive nitrogen compounds. Draw the Lewis structures of these compounds and use valence bond theory to describe the differences in their molecular structures.

As Sample Exercise 9.8 illustrates, ethane (an alkane) and ethylene (an alkene) have similar formulas and molar masses, but very different molecular geometries. Ethane does not trigger the ripening process because its shape does not allow it to fit the receptor in plant tissue that binds the planar molecule ethylene. This structural difference is important in larger molecules, where planar regions caused by the presence of sp^2 hybridized carbon atoms and double bonds produce overall molecular shapes that are very different from those associated with sp^3 hybridized carbon atoms.

Before ending our exploration of molecular shapes, we need to address the subject of handedness introduced at the beginning of the chapter. There we described how two molecules with the same molecular formula and Lewis structure can interact differently with receptors in our nasal membranes. As a result, one produces the smell of caraway seeds, and the other, spearmint leaves. You may find these different odors surprising when you consider the molecular structures of these two compounds (**Figure 9.41**). At first glance they may seem identical but look closely at the bonding pattern around the carbon atoms circled in red. The H atom bonded directly to the bottom carbon atom is on the front side of the ring in the spearmint compound (the C—H bond is shown as a solid wedge), but it is on the back side in the caraway compound (the C—H bond is shown as a dashed wedge). This minor difference in bond orientation creates a difference in molecular shape that is easily recognized by receptors in our noses.

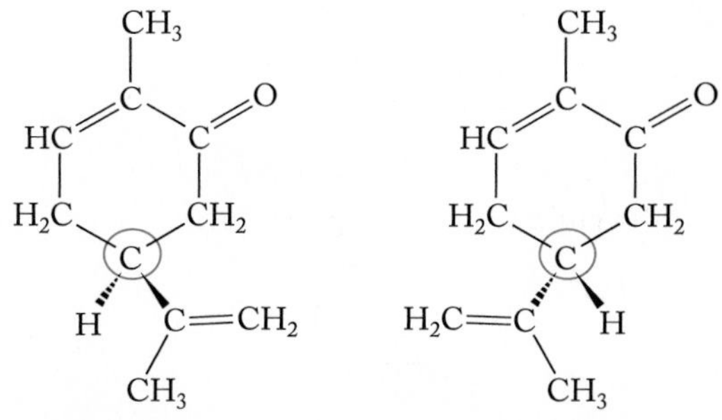

(a) (+)-Carvone (caraway)

(b) (−)-Carvone (spearmint)

FIGURE 9.41 The distinctive aromas of (a) caraway and (b) spearmint primarily result from two compounds with nearly identical molecular structures. The only difference between the two is the orientation of the two groups attached to the carbon atoms highlighted with the red circles. Note that the hydrogen atom is behind the plane of the paper and down in caraway (a) but is in front of the plane of the paper and down in spearmint (b).

The structures in Figure 9.41 are called *optical isomers*. The term "optical" refers to the ways these compounds interact with a special kind of light called *plane-polarized* light. We explore this topic in more detail in Chapter 20, but it is enough now for you to know that when plane-polarized light passes through a solution of the caraway compound, the light twists in one direction. When the light passes through a solution of the spearmint compound, it twists in the *opposite* direction.

Optical isomerism has another name: **chirality**. Many molecules of biological importance, including the proteins and carbohydrates that we consume each day, are composed of chiral compounds. Chirality comes from the Greek word *chier*, meaning "hand," and is quite correctly called "handedness." Although several features within molecules can lead to chirality, the most common is the presence of a carbon atom that has four different atoms or groups of atoms attached to it.

To see how chirality works, let's consider the molecular structure of bromochlorofluoromethane (CHBrClF; **Figure 9.42a**), which is used in fire extinguishers on airplanes. It contains a central carbon atom bonded to *four* different atoms. We will compare its molecular structure with that of dibromochloromethane ($CHBr_2Cl$), a compound that may form during the purification of municipal water supplies with chlorine and that has only *three* different atoms bonded to its central carbon atom (**Figure 9.42b**). First, we generate mirror images of both molecules, and then we rotate the mirror images 180° to superimpose each mirror image on its original image. If the reflected, rotated image is superimposable on the original image, then the substance is *not* chiral. Scientists call it *achiral*. The molecular images of $CHBr_2Cl$ can be superimposed in this way, so it is achiral. The images of CHBrClF, on the other hand, can*not* be superimposed. For example, when we superimpose the F, C, and H atoms of the two images in Figure 9.42a, the Br and Cl atoms are not aligned. This means that CHBrClF *is* a chiral compound.

Any structure like CHBrClF that has four different atoms or groups attached to an sp^3 hybridized carbon atom is chiral. It has two optical isomers that are

chirality a property of a molecule that is not superimposable on its mirror image.

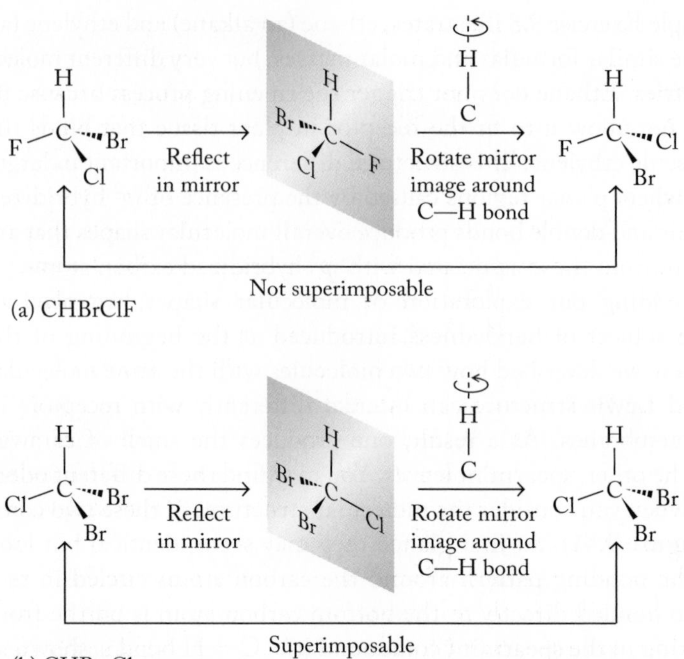

STEPWISE ANIMATION

Optical Activity

FIGURE 9.42 (a) A molecule of a chiral compound, such as CHBrClF, is *not* superimposable on its mirror image. (b) A molecule of a compound that is *not* chiral, such as $CHBr_2Cl$, *is* superimposable on its mirror image.

mirror images of each other and are distinctly different compounds. Where is the chiral carbon atom in the molecules of the caraway and spearmint compounds? If you guessed the circled carbon atoms, you were right. Two of the four different groups consist of a hydrogen atom and this group:

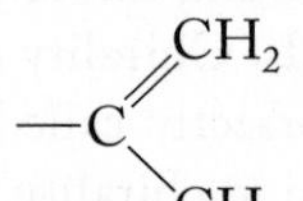

$$-C(=CH_2)CH_3$$

The other two "groups" on the bottom carbon atom are really the two sides of the ring. The left side contains a C═C double bond, whereas the right side contains a C═O double bond. These differences mean that the two sides are not equivalent, so the carbon atom is attached to four different groups. This makes it a chiral carbon atom, and the two compounds are optical isomers of each other.

CONCEPT TEST

Identify the molecules in **Figure 9.43** that are chiral.

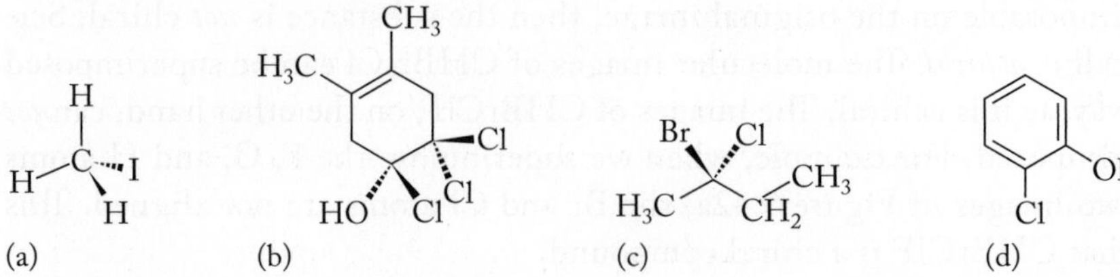

(a) (b) (c) (d)

FIGURE 9.43 Organic compounds.

9.6 Molecular Orbital Theory

Lewis structures and valence bond theory help us understand the bonding patterns in molecules and molecular ions; VSEPR theory and valence bond theory account for molecular shapes. Other phenomena, though, cannot be explained by any of these models. For example, none of these models explains why O_2 is attracted to a magnetic field whereas N_2 is slightly repelled by one, or how molecules and ions in Earth's upper atmosphere produce the shimmering, colorful displays known as *auroras* (**Figure 9.44**). To explain these phenomena, we need **molecular orbital (MO) theory**, another model that describes covalent bonding.

Like valence bond theory, molecular orbital theory invokes the mixing of atomic orbitals. In valence bond theory, the mixing results in hybrid atomic orbitals, whereas molecular orbital theory is based on the formation of **molecular orbitals**. A key difference between the two types of orbitals is that hybrid atomic orbitals are associated with a single atom in a molecule, but molecular orbitals are spread out over two or more atoms in a molecule.

Molecular orbitals represent discrete energy states in molecules, just as atomic orbitals represent allowed energy states in single atoms. As with atomic orbitals, electrons fill the lowest-energy MOs first; higher-energy MOs are filled as more electrons are added. Electrons in molecules can move to higher-energy MOs when molecules absorb quanta of electromagnetic radiation. When the electrons return to lower-energy MOs, distinctive wavelengths of UV and visible radiation are emitted, including some of the shimmering colors in an aurora.

The colors of an aurora are caused by collisions between atoms, molecules, and molecular ions in the atmosphere with electrons and positive particles in the solar wind that are attracted to Earth's magnetic poles. These collisions produce excited-state species. As they return to their ground states, these species emit characteristic colors of light. We discussed in Chapter 7 how excited-state atoms emit characteristic atomic spectra. In this section we use MO theory to explore how excited-state molecules and molecular ions, such as N_2 and N_2^+, do the same thing.

Unlike atomic orbitals, including hybrid atomic orbitals, MOs are not linked to single atoms but rather belong to all the atoms in a molecule. As a result, any valence electron in a molecule could be anywhere in the molecule. This delocalized view of covalent bonding is particularly effective at describing the bonding in molecules such as benzene, in which the electrons in the π bonds of a molecular orbital are spread out over all six carbon atoms.

In MO theory, molecular orbitals are formed by combining atomic orbitals from each of the atoms in the molecule. In this book, we limit our discussion of MO theory to simple molecules in which MOs are formed from the atomic orbitals of only a few atoms. Some MOs have lobes of high electron density that lie *between* bonded pairs of atoms; they are called **bonding orbitals**. The energies of bonding MOs are lower than the energies of the atomic orbitals that combined to form them, so populating them with electrons (each MO can hold two electrons, just like an atomic orbital) stabilizes the molecule and contributes to the strength of the bonds holding its atoms together.

There are other MOs with lobes of high electron density that are not located between the bonded atoms. They are **antibonding orbitals** and have energies that are higher than the atomic orbitals that combined to form them. When electrons are in antibonding orbitals, they destabilize the molecule. If a molecule were to have the same number of electrons in its bonding and antibonding orbitals, then there would be no net energy holding the molecule together, and it would never have formed.

FIGURE 9.44 Auroras are spectacular displays of color produced when the solar wind collides with Earth's upper atmosphere, producing excited-state atoms, ions, and molecules.

CONNECTION In Chapter 7 we discussed the role of atomic emission and absorption spectra in the development of quantum mechanics.

molecular orbital (MO) theory a bonding theory based on the mixing of atomic orbitals of similar shapes and energies to form molecular orbitals that extend across two or more atoms.

molecular orbital a region of characteristic shape and energy where electrons in a molecule are delocalized over two or more atoms in a molecule.

bonding orbital a molecular orbital in which electron density is highest in the region between atom centers.

antibonding orbital a molecular orbital in which electron density is highest outside the region between atom centers.

sigma (σ) molecular orbital a molecular orbital in which the greatest electron density is concentrated along an imaginary line drawn through the bonded atom centers.

molecular orbital diagram an energy-level diagram showing the relative energies and electron occupancy of the molecular orbitals for a molecule.

Another key point is that the *total number of molecular orbitals must match the number of atomic orbitals involved in forming them*. For example, combining two atomic orbitals, one from each of two atoms, produces one low-energy bonding orbital and one high-energy antibonding orbital. To further explore the distinction between bonding and antibonding molecular orbitals, let's apply MO theory to the simplest molecular compounds: hydrogen and helium.

CONCEPT TEST

In your own words, describe one way in which hybrid orbitals and molecular orbitals are similar and one way in which they are different.

Molecular Orbitals of Hydrogen and Helium

CHEMTOUR
Molecular Orbitals

According to MO theory, a hydrogen molecule is formed when the 1*s* atomic orbitals on two hydrogen atoms combine to form two molecular orbitals (**Figure 9.45a**). Molecular orbital theory also stipulates that these two orbitals represent two different energy states (**Figure 9.45b**).

The lower-energy bonding molecular orbital is oval and spans the two atomic centers. Its shape corresponds to enhanced electron density between the two atoms that donated their atomic orbitals. This enhanced electron density is a covalent bond. When two electrons occupy a bonding MO, a single bond is formed. When the region of highest density lies along the bond axis, as it does in the bonding MO in H_2, the MO is designated a **sigma (σ) molecular orbital** and the covalent bond is a σ bond. The bonding molecular orbital in H_2 is labeled σ_{1s} in Figure 9.45a because it is formed by mixing two 1*s* atomic orbitals.

The higher-energy (less stable) antibonding molecular orbital formed from two hydrogen atomic orbitals is designated σ^*_{1s} (pronounced "sigma star"). This antibonding orbital has two separate lobes of electron density and a region of zero electron density (a node) between the two hydrogen atoms (Figure 9.45a).

Figure 9.45b is a **molecular orbital diagram**, analogous to an energy-level diagram for atomic orbitals. It shows that the σ_{1s} MO is lower in energy and therefore more stable than the 1*s* atomic orbitals by nearly the same amount that the σ^*_{1s} MO is higher in energy than the 1*s* atomic orbitals. Therefore, the formation of the two MOs does not significantly change the total energy of the system. A hydrogen molecule has two valence electrons, one from each H atom, both residing in the lower-energy σ_{1s} orbital because that is the lowest-energy orbital available. As in atomic orbitals, these two σ_{1s} electrons must have opposite spins. The electron configuration that corresponds to the molecular orbital diagram in Figure 9.45b is written $(\sigma_{1s})^2$, where the superscript indicates that there are two

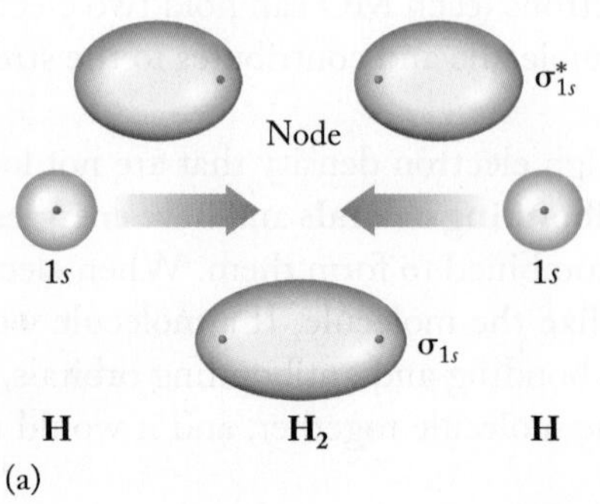

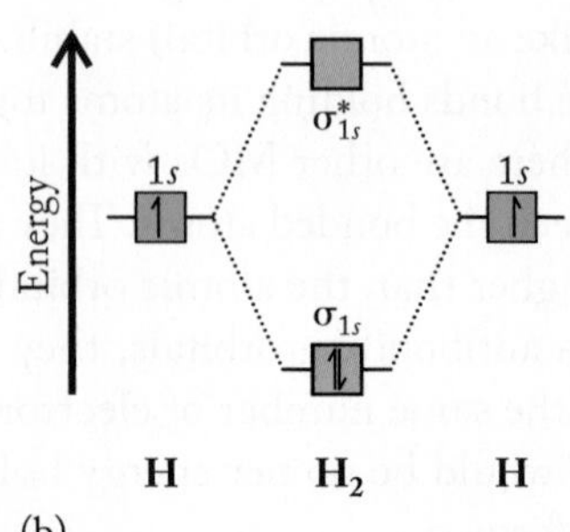

FIGURE 9.45 Mixing the 1*s* orbitals of two hydrogen atoms creates two molecular orbitals: a filled bonding σ_{1s} orbital containing two electrons and an empty antibonding σ^*_{1s} orbital. (a) The lower red oval is the bonding orbital. The two red ovals at the top together make up the antibonding orbital. The two top ovals represent only *one* molecular orbital with a node of zero electron density between. Dots show the locations of hydrogen nuclei. (b) A molecular orbital diagram shows the relative energies of bonding and antibonding molecular orbitals and of the atomic orbitals that formed them.

electrons in the σ_{1s} molecular orbital. Because the energy of the electrons in a $(\sigma_{1s})^2$ configuration is lower than the energy of the electrons in two isolated hydrogen atoms, MO theory explains why hydrogen is a diatomic gas—namely, H_2 molecules are lower in energy, so they are more stable than H atoms.

Hydrogen is a diatomic gas, but helium exists as free atoms and not as molecular He_2. MO theory explains why. One helium atom has two valence electrons in a 1*s* atomic orbital. Mixing two He 1*s* orbitals yields the same set of molecular orbitals we generated for H_2, as shown in **Figure 9.46**. Unlike H_2, the two helium atoms have a total of four valence electrons. Each molecular orbital in Figure 9.46—the bonding orbital and the antibonding orbital—has a maximum capacity of two electrons. Adding four valence electrons to the orbitals in Figure 9.46 means filling both orbitals and the two electrons in the σ^*_{1s} orbital cancel the stability gained from having two electrons in the σ_{1s} orbital. Because there is no net gain in stability, He_2 does not exist.

CONNECTION We used energy-level diagrams in Chapter 7 to show the electron transitions between energy levels as atoms absorb and emit electromagnetic radiation.

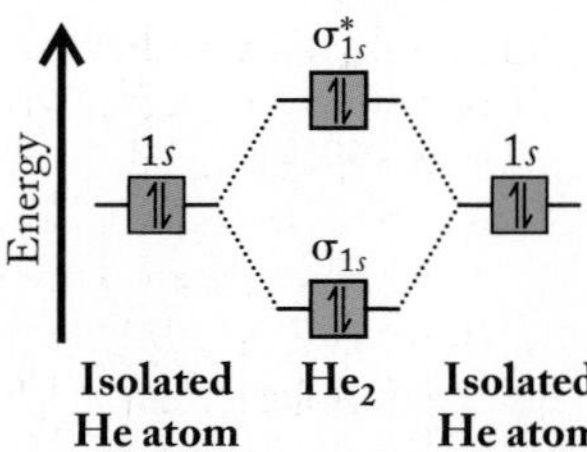

FIGURE 9.46 The molecular orbital diagram for the hypothetical molecule He_2 indicates that the same number of electrons occupy the antibonding orbital and the bonding orbital. Therefore, the bond order is 0, and the molecule is unstable.

Another way to compare the bonding in H_2 and He_2 is to look at the *bond order* in each. We previously defined bond order as the number of bonds between two atoms: a bond order of 1 for X—X, 2 for X=X, and 3 for X≡X. In MO theory, we define bond order as

$$\text{Bond order} = \frac{1}{2}\left(\begin{matrix}\text{number of}\\ \text{bonding electrons}\end{matrix} - \begin{matrix}\text{number of}\\ \text{antibonding electrons}\end{matrix}\right) \quad (9.2)$$

A molecule of H_2 has two electrons in the bonding MO and none in the antibonding MO, so

$$\text{Bond order in } H_2 = \tfrac{1}{2}(2 - 0) = 1$$

CONNECTION We first defined bond order in Section 8.7 when we related the length and strength of bonds to the number of electron pairs shared by two atoms.

For He_2, the bond order is 0 because an equal number of electrons reside in bonding and antibonding orbitals:

$$\text{Bond order in } He_2 = \tfrac{1}{2}(2 - 0) = 0$$

A bond order of 0 means that He_2 is not a stable molecule and does not exist. In general, the greater the bond order, the stronger the bond and the more stable the molecule.

SAMPLE EXERCISE 9.9 Using MO Diagrams to Predict Bond Order I **LO7**

Draw the MO diagram for the molecular ion H_2^-, determine the bond order of the ion, and predict whether the ion is stable.

Collect and Organize We are asked to draw the MO diagram for the molecular ion H_2^- and then determine bond order by using Equation 9.2. If the value of the bond order is greater than 0, the ion may be stable.

Analyze We should be able to base the MO diagram for H_2^- on the MO diagram for H_2 (Figure 9.45b) because the H_2^- ion has only one more electron than H_2 and the empty σ^*_{1s} orbital in H_2 can accommodate up to two more electrons.

Solve The σ_{1s} orbital is filled in H_2, so the third electron goes into the σ^*_{1s} orbital. The resulting MO diagram is shown in **Figure 9.47**. The notation for this electron configuration is $(\sigma_{1s})^2(\sigma^*_{1s})^1$ (listing the molecular orbitals in order of increasing energy). The bond order is

$$\text{Bond order} = \tfrac{1}{2}(2 - 1) = 0.5$$

We predict that the H_2^- ion is less stable than H_2 but more stable than He_2.

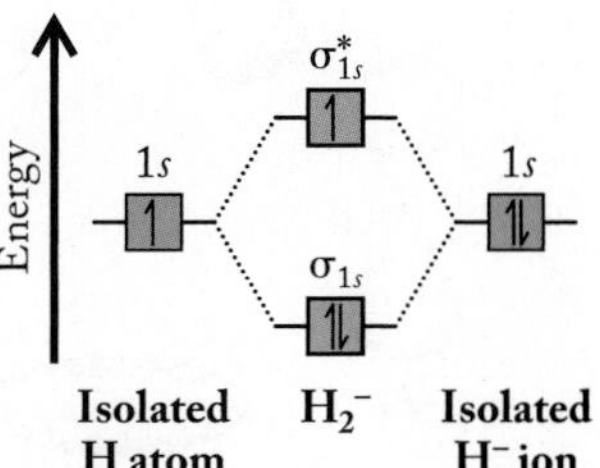

FIGURE 9.47 The molecular orbital diagram of H_2^-.

pi (π) molecular orbitals molecular orbitals formed by the mixing of atomic orbitals oriented above and below, or in front of and behind, the bonding axis

Think About It We encountered the idea of fractional bonds in Chapter 8 in the discussion of resonance, and we encounter it again here in the MO treatment of H_2^-. A bond order of 0.5 in MO theory means that the bond between the two atoms in H_2^- is weaker than the single bond in H_2, making H_2^- a less stable species.

Practice Exercise
Use MO theory to predict whether the H_2^+ ion can exist.

Molecular Orbitals of Homonuclear Diatomic Molecules

Molecular orbital diagrams for homonuclear (same atom) diatomic molecules such as N_2 and O_2 are more complex than that of H_2 because of the greater number and variety of atomic orbitals in N_2 and O_2. Not all combinations of atomic orbitals result in effective bonding, but there are some general guidelines for constructing the molecular orbital diagram for any molecule:

1. The number of molecular orbitals equals the number of atomic orbitals used to create them.

FIGURE 9.48 Mixing sets of three p atomic orbitals on two atoms forms six molecular orbitals. (a) The $2p_z$ atomic orbitals create a σ_{2p} bonding orbital and a σ^*_{2p} antibonding orbital. (b) The $2p_x$ and $2p_y$ atomic orbitals mix to form two π_{2p} bonding molecular orbitals and two π^*_{2p} antibonding molecular orbitals.

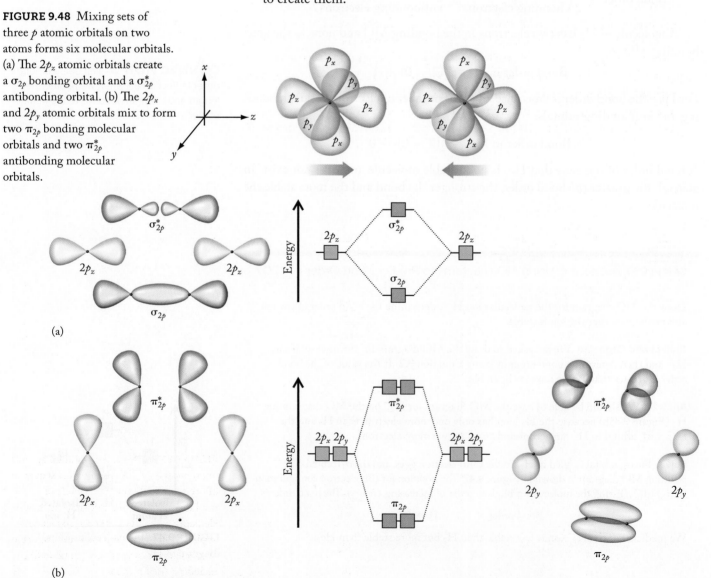

2. Atomic orbitals with similar energy and orientation mix more effectively than do those that have different energies and orientations. For example, an *s* atomic orbital mixes more effectively with another *s* atomic orbital than with a *p* orbital. The 1*s* and 2*s* orbitals have different sizes and energies, resulting in less effective mixing than two 1*s* or two 2*s* orbitals.
3. Better mixing leads to a larger energy difference between bonding and antibonding orbitals and thus greater stabilization of the bonding MOs.
4. A molecular orbital can accommodate a maximum of two electrons; two electrons in the same MO have opposite spins.
5. Electrons in ground-state molecules occupy the lowest-energy molecular orbitals available, following the aufbau principle and Hund's rule.

CONNECTION The aufbau principle (Section 7.8) states that the most stable atomic structures are those in which the electrons are in the lowest-energy orbitals available. Hund's rule (also Section 7.8) states that the lowest-energy electron configuration of a set of degenerate orbitals is the one with the maximum number of unpaired electrons, all having the same spin.

When mixing atomic orbitals to create molecular orbitals, we consider *only the valence electrons* on the atoms because core electrons do not participate in bonding. Focusing on N_2 and O_2 as examples, we first mix their 2*s* orbitals. The mixing process is analogous to the one we used for H_2, except that the resulting MOs are designated σ_{2s} and σ^*_{2s}.

Next, we mix the three pairs of 2*p* orbitals, producing a total of six MOs. The different spatial orientations of the $2p_x$, $2p_y$, and $2p_z$ atomic orbitals result in different kinds of MOs (**Figure 9.48**). The $2p_z$ atomic orbitals point toward each other. When they mix, two molecular orbitals form, a σ_{2p} bonding orbital and a σ^*_{2p} antibonding orbital. The lobes of the $2p_y$ and $2p_x$ atomic orbitals are oriented at 90° to the bonding axis and at 90° to each other. When the $2p_x$ orbitals mix, and when the $2p_y$ orbitals mix, they do so around the bonding axis instead of along it. This mixing produces two **pi (π) molecular orbitals** and two π* molecular orbitals. When electrons occupy a π orbital, they form a π bond.

The relative energies of σ and π molecular orbitals for N_2 and O_2 are shown in **Figure 9.49**. In each molecule, the energies of the MOs derived from two 2*s*

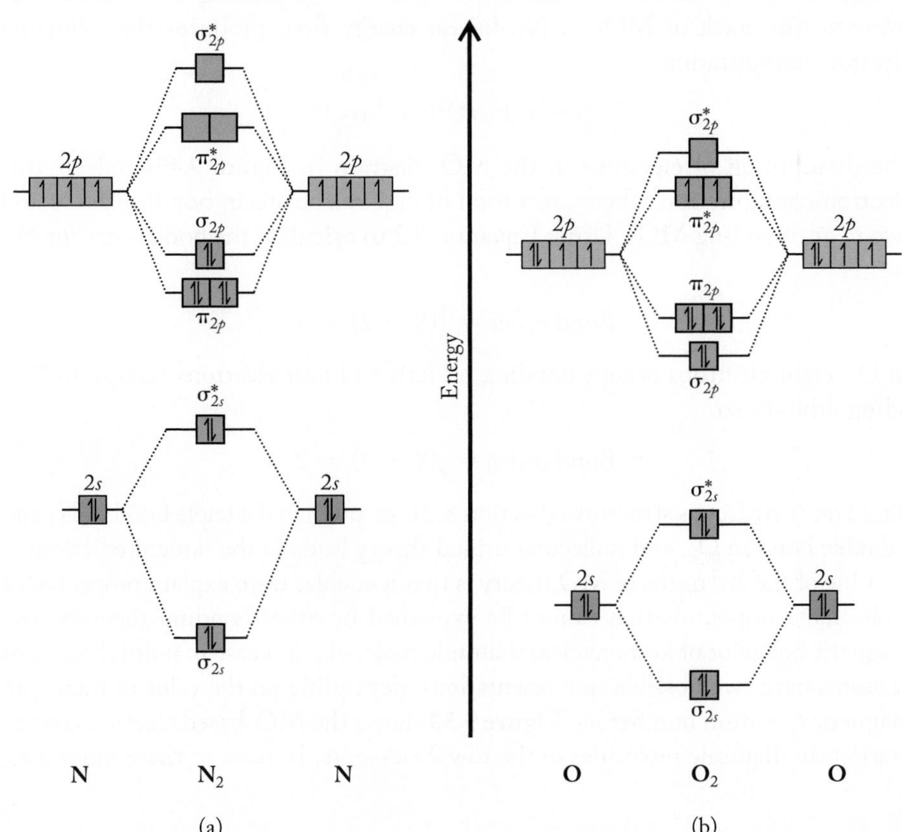

FIGURE 9.49 Molecular orbital diagrams of (a) N_2 and (b) O_2. The vertical sequence of orbitals for N_2 applies to MOs of the homonuclear diatomic molecules of elements with $Z \leq 7$. The O_2 sequence applies to homonuclear diatomic molecules of all elements beyond oxygen ($Z \geq 8$), including the halogens.

atomic orbitals (σ_{2s} and σ^*_{2s}) are lower than the energy of the σ_{2p} MO for the same reason that a 2*s* atomic orbital is lower in energy than a 2*p* atomic orbital.

Now let's consider the relative energies of the MOs formed by mixing the 2*p* orbitals in N_2 and O_2. We begin with O_2 (Figure 9.49b) because it is representative of most homonuclear diatomic molecules, including all the halogens. In order of increasing energy, the MOs are σ_{2p}, π_{2p}, π^*_{2p}, and σ^*_{2p}. Keep in mind that there are groups of two π_{2p} and two π^*_{2p} orbitals. This means that each group of two can hold four electrons. Adding 12 valence electrons (six from each O atom) in the O_2 molecule into these MOs, starting with the lowest-energy MO first and working our way up, we get the following valence electron configuration:

$$O_2\text{: } (\sigma_{2s})^2(\sigma^*_{2s})^2(\sigma_{2p})^2(\pi_{2p})^4(\pi^*_{2p})^2$$

The distribution of electrons in the MO diagram follows this sequence. The two π^*_{2p} orbitals are degenerate (equivalent in energy), so each contains a single electron, in accordance with Hund's rule.

According to the MO diagram, there are two unpaired electrons in a molecule of O_2. This finding contradicts the representation of bonding obtained from the Lewis structure of O_2, from VSEPR theory, and from valence bond theory, all of which predict that the valence electrons are paired. We will return to this point shortly, but for now let's consider the MO diagram for N_2.

The MO diagrams of N_2 and O_2 can be compared in Figure 9.49. There is a difference in the relative energies of two of their MOs—namely, the π_{2p} molecular orbital is lower in energy than the σ_{2p} orbital in N_2. These energy levels are switched from their relative positions in O_2 and many other diatomic molecules. As the atomic number increases from left to right, the energy difference between the 2*s* and 2*p* orbitals increases and there is no mixing of the 2*s* and 2*p*. In N_2, the 2*s* orbital does mix with the 2*p* orbital to a small extent. This proximity of energies has the effect of lowering the energy of the σ_{2s} molecular orbital and raising the energy of the σ_{2p} orbital—enough to put it above π_{2p}. Adding ten valence electrons to this stack of MOs in N_2, lowest energy first, produces the following electron configuration:

$$N_2\text{: } (\sigma_{2s})^2(\sigma^*_{2s})^2(\pi_{2p})^4(\sigma_{2p})^2$$

The distribution of electrons in the MO diagram in Figure 9.49a reflects this electron configuration. There are a total of eight electrons in bonding MOs and two in antibonding MOs. Using Equation 9.2 to calculate the bond order for N_2, we get

CONNECTION The spin magnetic quantum number (m_s; Section 7.6) has a value of either $+\frac{1}{2}$ or $-\frac{1}{2}$ and defines the orientation of an electron in a magnetic field.

$$\text{Bond order} = \tfrac{1}{2}(8 - 2) = 3$$

In O_2, eight electrons occupy bonding orbitals and four electrons occupy antibonding orbitals, so

$$\text{Bond order} = \tfrac{1}{2}(8 - 4) = 2$$

Based on their Lewis structures (Section 8.2), we predicted a triple bond in N_2 and a double bond in O_2, and molecular orbital theory leads to the same predictions.

diamagnetic a substance with no unpaired electrons that is weakly repelled by a magnetic field.

paramagnetic a substance with one or more unpaired electrons that is attracted to a magnetic field.

One of the strengths of MO theory is that it enables us to explain properties of molecular compounds that cannot be explained by other bonding theories. The magnetic behavior of homonuclear diatomic molecules is a case in point. Electrons in atoms have two possible spin orientations, depending on the value of their spin magnetic quantum number m_s. **Figure 9.50** shows the MO-based electron configurations in diatomic molecules of the row 2 elements. In most of these molecules,

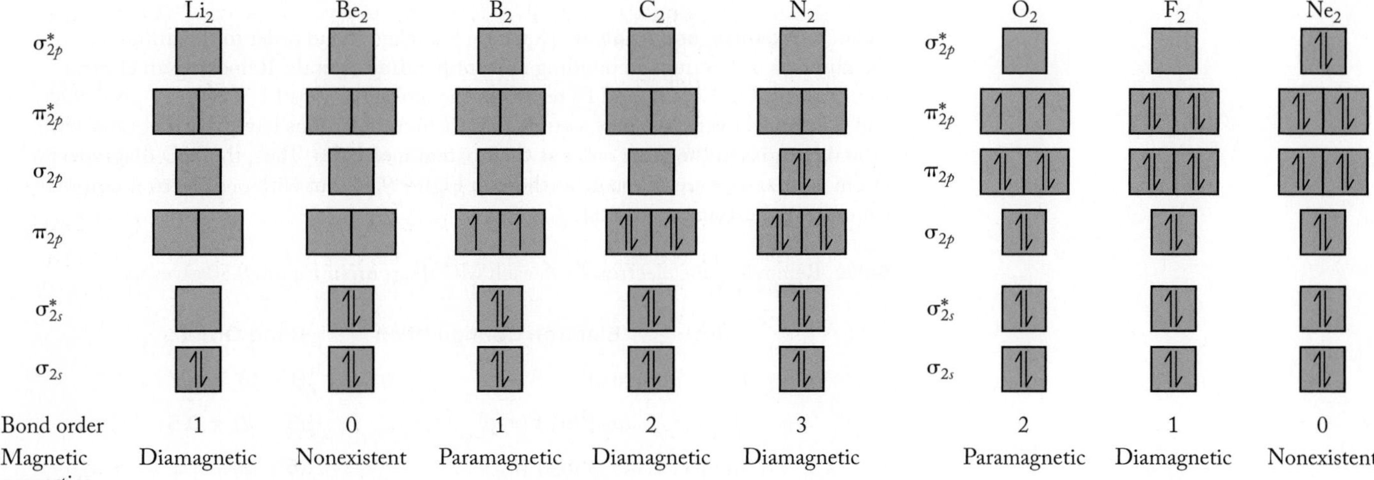

FIGURE 9.50 Valence-shell molecular orbital diagrams and magnetic properties of the homonuclear diatomic molecules of the second-row elements.

all the electrons are paired so that their spins cancel out. The molecules that make up most substances contain only paired electrons. This complete electron pairing means that these substances are repelled slightly by a magnetic field. These substances are said to be **diamagnetic**. If a substance's molecules contain unpaired electrons, as do those of B_2 and O_2, then it is attracted by a magnetic field, as shown in **Figure 9.51** for O_2, and is **paramagnetic**. The more unpaired electrons in a molecule, the greater its paramagnetism. Only MO theory accounts for the magnetic behavior of oxygen and of many other substances as well.

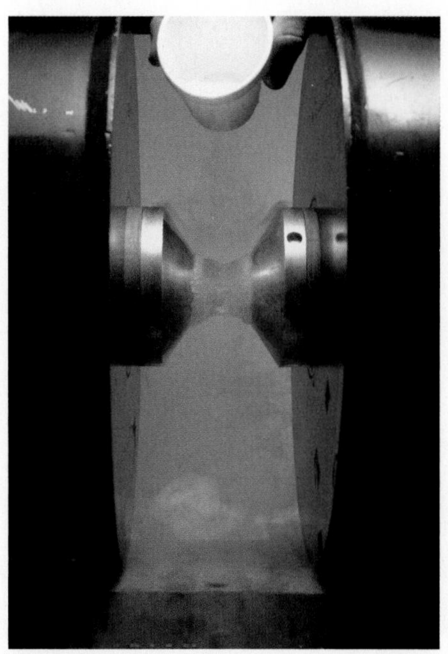

FIGURE 9.51 Liquid O_2 poured from a Styrofoam cup is suspended in the space between the poles of this magnet because the unpaired electrons in its molecules make O_2 paramagnetic. Paramagnetic substances are attracted to magnetic fields.

CONCEPT TEST

Can liquid N_2 and O_2 be separated from each other with a magnet? Explain your answer.

Figure 9.50 enables us to make several predictions about diatomic molecules of the second-row elements. First, Be_2 and Ne_2 do not exist for the same reason that He_2 does not exist: both Be_2 and Ne_2 have the same number of antibonding electrons as they have bonding electrons, so they have a net bond order of 0. Second, Li_2, B_2, and F_2 have a bond order of 1, whereas C_2 has a bond order of 2. Like O_2, B_2 is paramagnetic, whereas Li_2, C_2, N_2, and F_2 are diamagnetic. (We have not discussed C_2 molecules before, but they do exist in electric arcs and in comets, and they are responsible for the blue light of flames.)

SAMPLE EXERCISE 9.10 Using MO Diagrams to Predict Bond Order II LO7

In which of the molecules in Figure 9.50 does the bond order increase when one electron is removed from the molecule? You may exclude nonexistent Be_2 and Ne_2 from this analysis.

Collect, Organize, and Analyze Equation 9.2 relates bond order to the difference in the numbers of electrons in bonding and antibonding orbitals. Removing an electron from Li_2, B_2, C_2, N_2, O_2, and F_2 results in the molecular ions Li_2^+, B_2^+, C_2^+, N_2^+, O_2^+, and F_2^+, respectively. We may assume that the molecular ions have MO diagrams with orbital energies in the same order as their parent molecules. Thus, the MO diagrams for the molecular ions are the same as those in Figure 9.50, but with one electron removed from the highest-energy orbital.

Solve Removing one electron from each MO diagram in Figure 9.50 gives us

Ion	Electron Configuration	Bond Order
Li_2^+	$(\sigma_{2s})^1$	$\frac{1}{2}(1 - 0) = 0.5$
B_2^+	$(\sigma_{2s})^2(\sigma_{2s}^*)^2(\pi_{2p})^1$	$\frac{1}{2}(3 - 2) = 0.5$
C_2^+	$(\sigma_{2s})^2(\sigma_{2s}^*)^2(\pi_{2p})^3$	$\frac{1}{2}(5 - 2) = 1.5$
N_2^+	$(\sigma_{2s})^2(\sigma_{2s}^*)^2(\pi_{2p})^4(\sigma_{2p})^1$	$\frac{1}{2}(7 - 2) = 2.5$
O_2^+	$(\sigma_{2s})^2(\sigma_{2s}^*)^2(\sigma_{2p})^2(\pi_{2p})^4(\pi_{2p}^*)^1$	$\frac{1}{2}(8 - 3) = 2.5$
F_2^+	$(\sigma_{2s})^2(\sigma_{2s}^*)^2(\sigma_{2p})^2(\pi_{2p})^4(\pi_{2p}^*)^3$	$\frac{1}{2}(8 - 5) = 1.5$

Comparing these values with the bond orders listed in Figure 9.50, we see that bond order increases for only O_2^+ and F_2^+.

Think About It Removing an electron from O_2 or F_2 reduces the number of electrons in antibonding molecular orbitals while leaving the number of electrons in bonding molecular orbitals unchanged. The result is an increase in bond order. In the other four homonuclear diatomic molecules, removing an electron reduces the number of electrons in bonding molecular orbitals while leaving the number of electrons in antibonding orbitals unchanged. The result is a decrease in bond order for Li_2^+, B_2^+, C_2^+, and N_2^+.

Practice Exercise In which molecules in Figure 9.50 does the bond order increase when one electron is *added* to the molecule?

Molecular Orbitals of Heteronuclear Diatomic Molecules

Molecular orbital theory also enables us to account for the bonding in *heteronuclear* diatomic molecules, which are molecules containing two different atoms. The bonding in some of these molecules is difficult to explain using other bonding theories. For example, it is often difficult to draw a single Lewis structure for an odd-electron molecule such as nitrogen monoxide (NO). In Chapter 8 we considered two arrangements of its valence electrons:

$$:\dot{N}{=}\ddot{O}: \qquad :\dot{N}{=}\ddot{O}\cdot$$

We predicted that oxygen was more likely to have a complete octet of valence electrons because it is the more electronegative element. In addition, experimental evidence allows us to rule out structures with unpaired electrons on the oxygen atom. Our preferred structure was therefore the one shown in red. However, the bond length in NO (115 pm) is considerably shorter than the value in Table 8.3 for an average N═O double bond (122 pm). Molecular orbital theory is useful for explaining both the bonding in NO and the deviation from the expected bond length.

Let's consider the bonding first. The MO diagram for NO is different from the diagrams of homonuclear diatomic gases. Nitrogen and oxygen atoms have different numbers of protons and electrons, and the difference in effective nuclear charge in N and O atoms means that their atomic orbitals have different energies, as shown in **Figure 9.52** for their 2*s* and 2*p* orbitals.

For constructing the MO diagram for NO, the guidelines described previously still apply. The number of MOs formed must equal the number of atomic orbitals combined, and the energy and orientation of the atomic orbitals being mixed must be considered. One additional factor influences the energies of the MOs in heteronuclear diatomic molecules: *bonding* MOs tend to be closer in energy to the atomic orbitals of the *more* electronegative atom, whereas *antibonding* MOs tend to be closer in energy to the atomic orbitals of the *less* electronegative atom. In Figure 9.52, the energy of the bonding σ_{2s} orbital is closer to that of the 2*s* orbital of O (the more electronegative atom), whereas the energy of the antibonding σ^*_{2s} orbital is closer to that of the 2*s* orbital of N (the less electronegative atom). Similarly, the π_{2p} MOs in NO are closer in energy to the 2*p* orbitals of oxygen, whereas the π^*_{2p} MOs are closer in energy to the 2*p* orbitals of nitrogen. The proximity of the nitrogen 2*p* atomic orbitals to the π^*_{2p} MOs means that the single electron in the π^*_{2p} MO is more likely to be on nitrogen than on oxygen. This prediction is consistent with our Lewis structure in which the odd electron in NO is on the nitrogen atom.

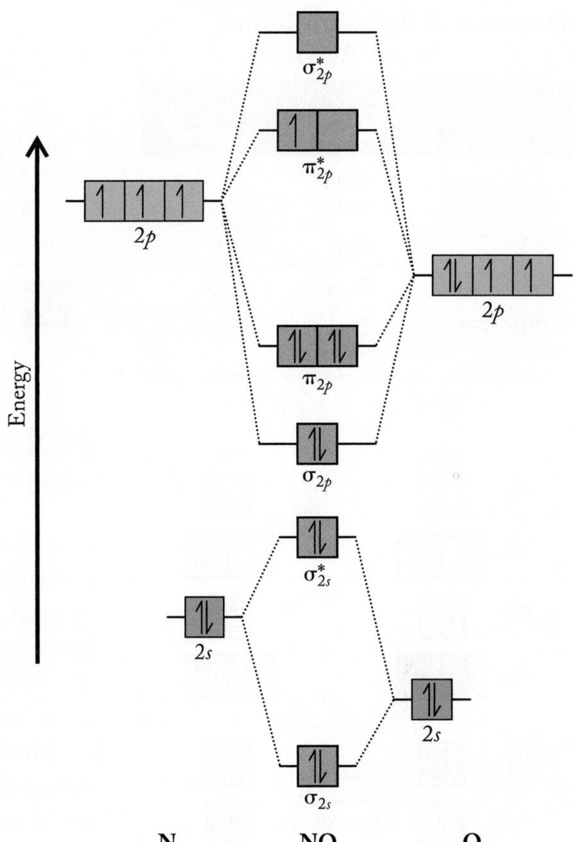

FIGURE 9.52 The molecular orbital diagram for NO shows that the unpaired electron occupies a π^*_{2p} antibonding orbital, which is closer in energy to the 2*p* atomic orbitals of nitrogen than to the 2*p* atomic orbitals of oxygen. Because of this proximity, the electron is more likely to be located on the nitrogen atom than on the oxygen atom.

Molecular orbital theory also enables us to rationalize the relatively short bond length in NO. Using Equation 9.2, we obtain a bond order of $\frac{1}{2}(8 - 3) = 2.5$. This is about halfway between the bond orders for N═O and N≡O, and it is consistent with a bond length of 115 pm, which is about halfway between the lengths of the N═O bond (122 pm) and the N≡O bond (106 pm).

SAMPLE EXERCISE 9.11 Using MO Diagrams for Heteronuclear Diatomic Molecules LO7

Nitrogen monoxide reacts with many transition metals, including the iron in our blood. In these compounds, NO is sometimes present as NO^+ and at other times as NO^-. Use Figure 9.52 to predict the bond order of NO^+ and NO^-.

Collect and Organize We are asked to predict the bond order of two diatomic ions based on the MO diagram of their parent molecule (Figure 9.52). Equation 9.2 relates bond order to the numbers of electrons in bonding and antibonding orbitals.

Analyze NO has 11 valence electrons. We can remove one electron from the MO diagram for NO to get the diagram for NO^+ and add one electron to get the diagram for NO^-.

Solve To generate NO^+, we remove the highest-energy electron in NO, which is the one in the π^*_{2p} molecular orbital. This gives NO^+ the electron configuration

$$NO^+: (\sigma_{2s})^2(\sigma^*_{2s})^2(\sigma_{2p})^2(\pi_{2p})^4$$

Adding an electron to the lowest-energy MO available in NO (also π^*_{2p}) yields NO^- with the valence electron configuration

$$NO^-: (\sigma_{2s})^2(\sigma^*_{2s})^2(\sigma_{2p})^2(\pi_{2p})^4(\pi^*_{2p})^2$$

The bond orders of the two ions are

TABLE 9.4 Origins of Colors in the Aurora

Wavelength (nm)	Color	Chemical Species
650–680	Deep red	N_2^*
630	Red	O*
558	Green	O*
391–470	Blue-violet	N_2^{+*}

$$NO^+\text{: bond order} = \tfrac{1}{2}(8 - 2) = 3$$

$$NO^-\text{: bond order} = \tfrac{1}{2}(8 - 4) = 2$$

Think About It The calculated bond orders make sense given the NO bond order of 2.5: NO^+ has one fewer electron in an antibonding orbital than NO, which increases its bond order by 0.5 to 3.0, whereas NO^- has one more electron in an antibonding orbital than NO, which decreases its bond order by 0.5 to 2.0.

Practice Exercise Using Figure 9.52 as a guide, draw the MO diagram for carbon monoxide, and determine the bond order of the carbon–oxygen bond.

Molecular Orbitals of N_2^+ and Spectra of Auroras

In addition to predicting the magnetic properties of molecules, MO theory is particularly useful for predicting their spectroscopic properties—and the colors of auroras. In Chapter 7 we learned that the light emitted by excited free atoms is quantized and can be related to the movement of electrons between atomic orbitals. Broadly speaking, the same is true in molecules—that is, electrons can move from one molecular orbital to another by absorbing or emitting light.

How, then, are the colors of the aurora produced? The principal chemical species involved are listed in **Table 9.4**. An asterisk indicates a molecule or molecular ion in an excited state. Excited N_2 molecules (i.e., N_2^*) produce deep crimson red (650–680 nm) light, and excited N_2^+ ions (i.e., N_2^{+*}) produce blue-violet (391–470 nm) light. The MO diagrams for these species are shown in **Figure 9.53**. Comparing the MO diagrams of N_2^* and N_2 in Figure 9.53a, we find that one of the two electrons originally in the σ_{2p} MO in N_2 has been raised to a π^*_{2p} orbital in N_2^*, leaving an unpaired σ_{2p} electron behind. Figure 9.53b shows us that N_2^{+*} also has one electron in a π^*_{2p} orbital, but its σ_{2p} orbital is empty because the other σ_{2p} electron originally in the N_2 molecule was lost when it was ionized to N_2^+. As π^*_{2p} electrons return from their antibonding, excited-state orbitals to the bonding σ_{2p} orbital in the ground state, the distinctive blue-violet and crimson emissions of N_2^+ and N_2 appear, as shown in the photograph in Figure 9.44.

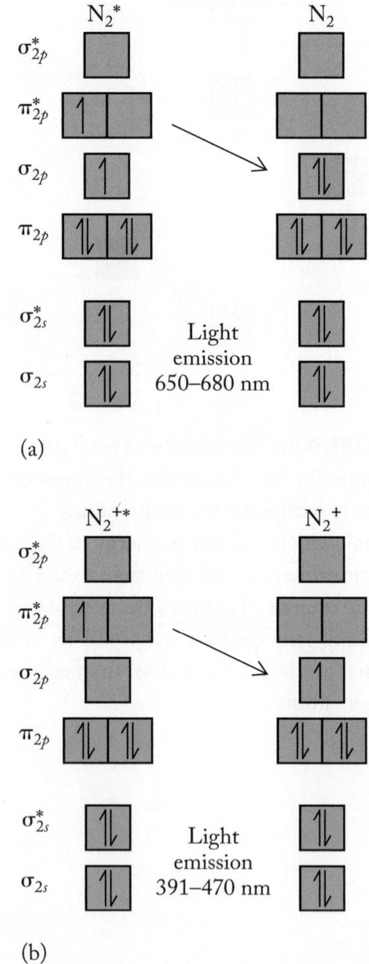

FIGURE 9.53 Molecular orbital diagrams for (a) N_2 and (b) N_2^+ show electronic transitions that result in the emission of visible light. Collisions with ions in the solar wind result in the promotion of electrons from the σ_{2p} orbitals in N_2 molecules to π^*_{2p} orbitals. Higher-energy collisions create N_2^{+*} molecular ions, which also have electrons in π^*_{2p} orbitals. When these electrons return to the ground state, red and blue-violet light are emitted.

CONCEPT TEST

Are the bond orders of the excited-state species in Figure 9.53 the same as the ground-state species?

Using MO Theory to Explain Fractional Bond Orders and Resonance

In Chapter 8 our attempts to draw the Lewis structure of ozone led to the conclusion that molecules of O_3 are held together by two O—O single bonds plus a shared pair of electrons distributed equally across both bonds. The concept of resonance was used to describe the delocalization of the second bonding pair and the equivalency of the bonds in O_3, which are shorter and stronger than O—O single bonds but not as short or strong as O=O double bonds. In this chapter we have learned how VSEPR theory explains the bent shape of O_3 molecules, and how valence bond theory and the formation of sp^2 hybrid orbitals help us visualize the trigonal planar orientation of bonding and lone pairs of electrons that account for ozone's molecular shape.

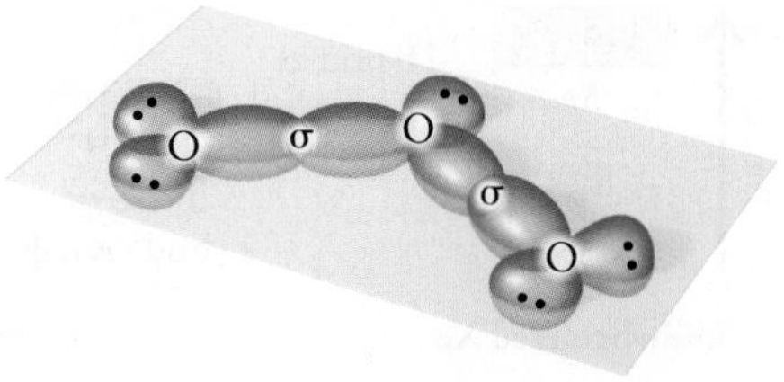

Now let's explore how MO theory provides another perspective on the delocalized bonding in O_3—one that does not require drawing resonance structures that do not accurately convey where all the bonding and lone pairs of electrons really are. MO theory is great at explaining bonding pair delocalization because MOs are spread across several atoms or even entire molecules, not isolated between pairs of atoms. We begin with three O atoms that are connected by two single (σ) bonds and that have a total of five lone pairs of valence electrons (**Figure 9.54**). This trigonal planar pattern of electron pairs is what we would expect if each of the O atoms had sp^2 hybridized atomic orbitals. There are 14 valence electrons in Figure 9.54 (5 lone pairs + 2 σ bonds = 10 + 4 = 14), leaving $(3 \times 6) - 14 = 4$ electrons unaccounted for. Let's distribute the four electrons among a set of molecular orbitals formed by mixing the three unhybridized *p* orbitals of the three O atoms.

FIGURE 9.54 Two sigma bonds and five lone pairs of electrons in a molecule of O_3.

In MO theory, mixing three *p* orbitals with lobes above and below the bonding plane produces three molecular orbitals. One of them is a low-energy π MO, and another is a high-energy π^* MO, but what about the third? It turns out that the third orbital is neither bonding nor antibonding. Instead, it is a *nonbonding* (*n*) molecular orbital. Electrons in nonbonding MOs neither lower the energy of a molecular structure, which would stabilize it, nor do they raise its energy and destabilize it. Rather, nonbonding MOs have the same energy as the atomic orbitals from which they formed, as shown in the partial MO diagram of ozone in **Figure 9.55**. Adding four electrons to these three molecular orbitals puts two electrons in the π orbital, two in the nonbonding orbital, and zero in π^*. Therefore, there is $\frac{1}{2}(2 - 0) = 1$ π bond distributed across the entire molecule. The overall bond order of 2 σ bonds + 1 π bond = 3 is divided equally between the two pairs of atoms, giving an average bond order for ozone of 1.5. The pair of nonbonding electrons is also delocalized. Each of the two terminal O atoms gets half ownership of this nonbonding pair for a total of 2.5 lone pairs of electrons each, which is the average of two lone pairs on one terminal O atom and three on the other in the resonance structures of O_3:

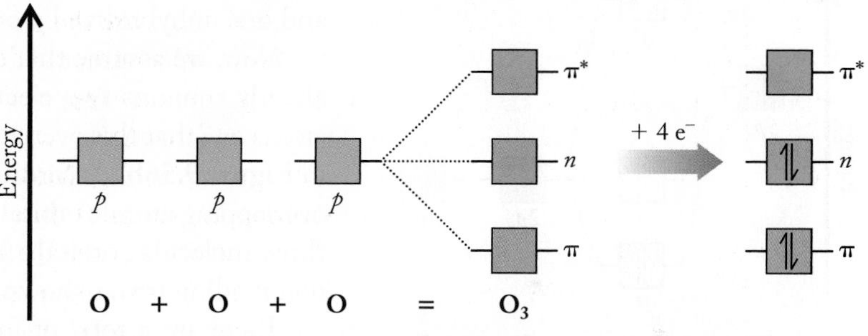

FIGURE 9.55 Forming molecular orbitals from unhybridized *p* orbitals in O_3.

A Bonding Theory for SN > 4

In Chapter 8 we noted that the central atoms in some molecules form more than four bonds, as shown in these Lewis structures of PCl_5 and SF_6:

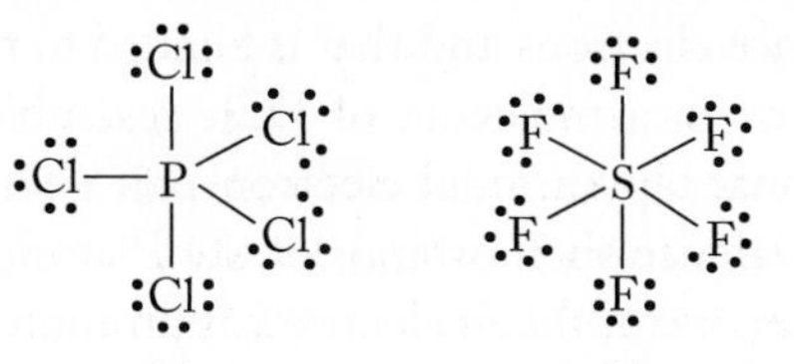

(a) (b)

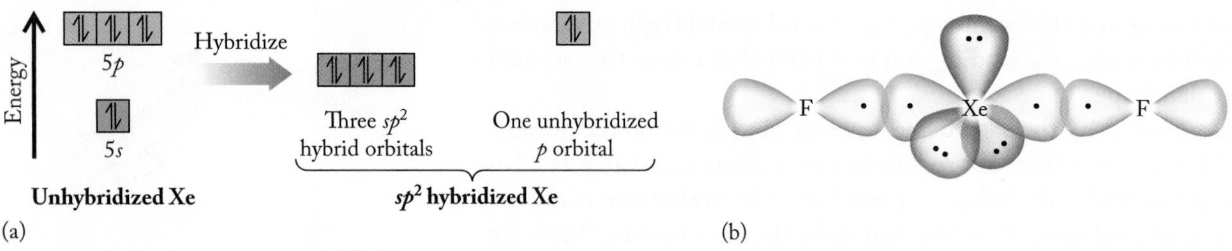

FIGURE 9.56 (a) Mixing one 5*s* orbital and two 5*p* orbitals on a xenon atom creates three sp^2 hybrid orbitals and leaves the xenon atom with one unhybridized *p* orbital. (b) The two lobes of the xenon atom's unhybridized 5*p* orbital overlap with the half-filled 2*p* orbital on each of the fluorine atoms.

Until recently, a popular explanation for how atoms such as phosphorus and sulfur can form more than four bonds was based on their using valence shell *d* orbitals to form hybrid sets of orbitals that expanded their capacities to form bonds and have more than an octet of valence electrons. Here we explore another way to explain the capacity of atoms to form more than four bonds, or, more generally, to have a steric number (consisting of bonds, lone pairs, or both) of more than four without expanding their octets.

Let's begin by examining the bonding in molecules of a noble gas compound, XeF_2, whose synthesis was first reported in 1962. The Lewis symbol of xenon is :Ẍe:, which shows that the atom has a full octet in its valence shell. It has no unpaired electrons, and, theoretically, no capacity to form covalent bonds. One way to explain the existence of XeF_2 begins by hybridizing the 5*s* and two of the 5*p* orbitals in xenon's valence shell, which yields a set of three sp^2 hybrid orbitals and one unhybridized *p* orbital as shown in **Figure 9.56a**.

Now, we assume that the two lobes of xenon's unhybridized 5*p* orbital (which already contains two electrons) overlap with the half-filled 2*p* orbitals of two F atoms *and* that this overlap results in the formation of two Xe—F bonds as shown in **Figure 9.56b**. According to MO theory, these assumptions are valid: the three overlapping atomic orbitals can mix together, and when they do they form a set of three molecular orbitals: one bonding (σ), one antibonding (σ^*), and one that is nonbonding (*n*), as shown in **Figure 9.57**.

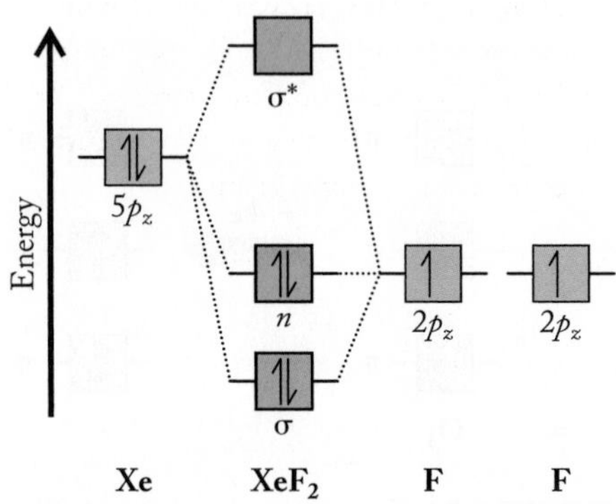

FIGURE 9.57 Partial molecular orbital diagram of XeF_2. These are the only MOs that contribute to bonding.

There are a total of four electrons in the three molecular orbitals, and they preferentially fill the lowest-energy bonding and nonbonding MOs, leaving the antibonding orbital empty. Electrons in the nonbonding orbital do not influence bond order, so the presence of two electrons in the bonding orbital together with the absence of any electrons in the antibonding orbital produces a bond order of $(2 - 0)/2 = 1$. This bond order of 1 is delocalized over both Xe—F bonds, leaving each one with an effective bond order of only $\frac{1}{2}$ (the two nonbonding electrons are located on the F atoms).

Thus, MO theory predicts that XeF_2 should exist, which it has since 1962. In addition, using xenon's unhybridized *p* orbital to make the two Xe—F bonds means that the molecule should be linear, which it is. This molecular shape is reinforced by offsetting interactions between the bonding electrons and the three lone pairs of electrons in xenon's sp^2 hybrid orbitals, which are evenly distributed in lobes that are all 120° from each other in the equatorial plane of the molecule and 90° from the axial bonding orbitals. This arrangement means that the electron-pair geometry of the molecule is trigonal bipyramidal, as is expected with a central atom that has three lone pairs of valence electrons and that is bonded to two other atoms.

The bonding pattern in a molecule of PCl_5 resembles the pattern in XeF_2 because PCl_5 has trigonal bipyramidal electron-pair geometry, too. As with the Xe atom in XeF_2, the valence shell orbitals of the P atom in PCl_5 are sp^2 hybridized, but before they are, one of the 3*s* electrons is promoted to a 3*p* orbital so that, after hybridization, the valence shell contains three half-filled sp^2 hybrid orbitals and one full *p* orbital (**Figure 9.58a**).

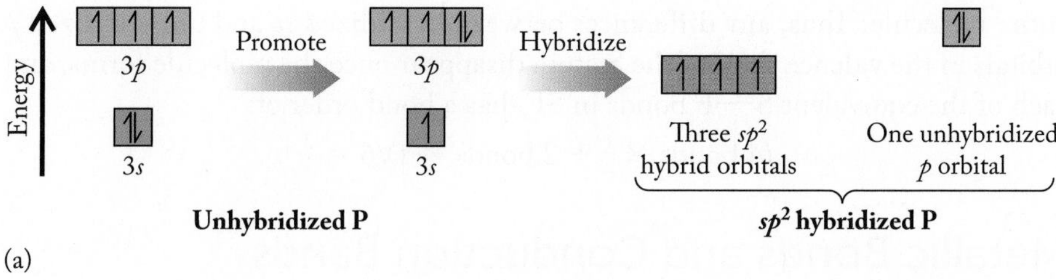

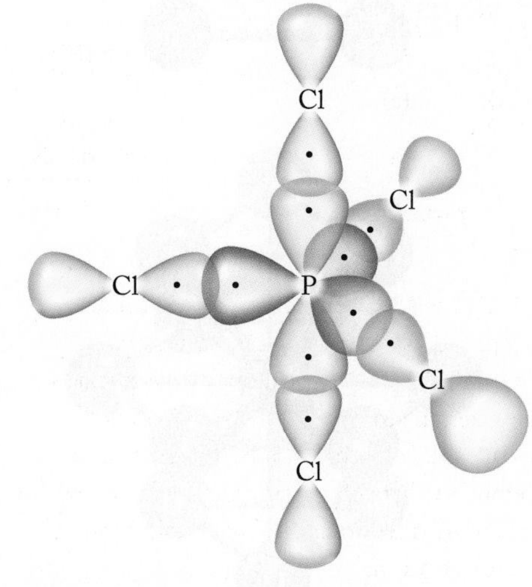

FIGURE 9.58 (a) The promotion of a 3*s* electron in a phosphorus atom, followed by the mixing of one 3*s* orbital and two 3*p* orbitals, creates three sp^2 hybrid orbitals and leaves the phosphorus atom with one unhybridized *p* orbital. (b) The two lobes of the phosphorus atom's unhybridized 3*p* orbital overlap with the half-filled 3*p* orbital on each of the chlorine atoms.

As in XeF_2, the two lobes of the P atom's full 3*p* orbital overlap with those of the half-filled 2*p* orbitals of two Cl atoms (**Figure 9.58b**), leading to the formation of two axial P—Cl bonds that each have a bond order of $\frac{1}{2}$. The half-filled sp^2 hybrid orbitals of the P atom overlap with the half-filled 2*p* orbitals of three Cl atoms to produce three equatorial P—Cl bonds that each have a bond order of 1. Because of their lower bond order, the axial bonds are weaker and longer than the equatorial single bonds: the equatorial bonds are 202 pm long and the axial bonds are 214 pm. There are no lone pairs of electrons in the valence shell of the P atom, so the molecular shape of PCl_5 matches its electron-pair geometry: trigonal bipyramidal, as predicted by VSEPR theory for SN = 5.

Finally, let's consider the bonding in a molecule of SF_6. How might we model an expansion of the bonding capacity of a sulfur atom from 2—as predicted by its Lewis symbol—to 6 using only the four orbitals in its 3*s* and 3*p* subshells? One such model uses the four lobes of two of sulfur's 3*p* orbitals to make a total of four S—F bonds.

The two lobes of each of the S atom's two full 3*p* orbitals (shown in blue in **Figure 9.59**) overlap with the half-filled 2*p* orbitals of the four F atoms to form four S—F bonds: two oriented above and below the S atom in the figure and two oriented behind and in front of the S atom. Each of these four bonds, according to MO theory, has a bond order of $\frac{1}{2}$. The two half-filled *sp* hybrid orbitals of the S atom (shown in purple) overlap with the half-filled 2*p* orbitals of two F atoms to produce the two S—F bonds to the left and right of the S atom. These bonds each have a bond order of 1. The six S—F bonds are all oriented 90° from each other and give the molecule an octahedral shape, as predicted for SN = 6.

Because of their different bond orders, you might expect two of the S—F bonds to be shorter than the other four. Actually, they are all the same length because all six bonds are directed toward equivalent fluorine atoms at the corners of an octahedron and all the bonding electrons are evenly distributed over the

Energy
3p
3s
Promote
3p
3s
Hybridize
Two *sp* hybrid orbitals
Two unhybridized *p* orbital
Unhybridized sulfur atom
sp hybridized S
(a)

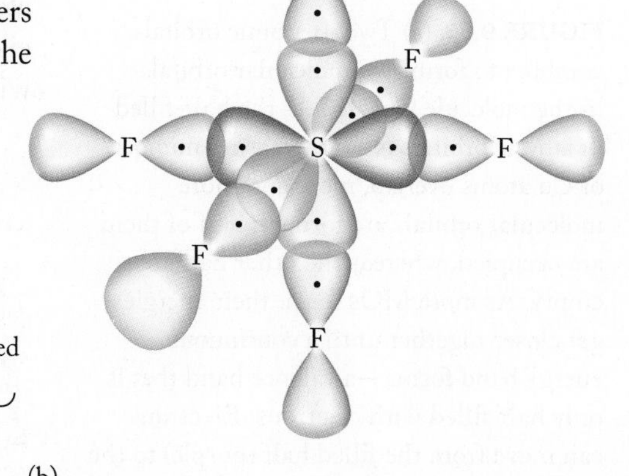

FIGURE 9.59 (a) The promotion of a 3*s* electron in a sulfur atom, followed by the mixing of the 3*s* orbital and one 3*p* orbital, creates two *sp* hybrid orbitals and leaves the sulfur atom with two unhybridized *p* orbitals. (b) The four lobes of the sulfur atom's unhybridized 3*p* orbitals overlap with the half-filled 2*p* orbital on each of the fluorine atoms.

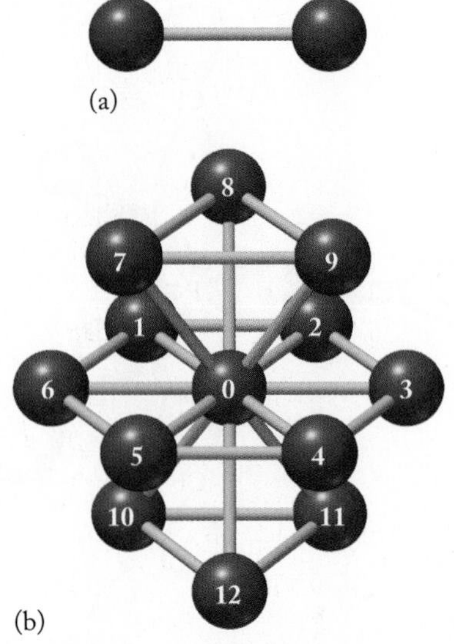

FIGURE 9.60 Covalent bonds differ from metallic bonds. (a) A ball-and-stick model of the molecule Cu_2 assumes that the two atoms share their 4*s* electrons to form a covalent bond. (b) In solid copper, the atom labeled 0 shares its 4*s* electron with 12 other atoms. As a result, the bonds in copper and other metals are much more diffuse than the covalent bonds in molecules.

entire molecule. Thus, any differences between hybridized *sp* and unhybridized *p* orbitals in the valence shell of the S atom disappear once the molecule forms, and each of the equivalent S—F bonds in SF_6 has a bond order of:

$$(4 \text{ bonds} \times \tfrac{1}{2} + 2 \text{ bonds} \times 1)/6 = \tfrac{2}{3}$$

Metallic Bonds and Conduction Bands

We introduced metallic bonding in Section 8.1 where we described the atoms in metals as "floating" in seas of mobile bonding electrons shared by all the atoms in a sample. In this section we use a more sophisticated bonding model to describe the forces that hold the densely packed atoms in metallic solids tightly in place. Dense packing means that the valence orbitals of these atoms overlap with the orbitals of many other nearby atoms. Many interactions between each atom and its nearest neighbors make metals strong but sharing a limited number of valence electrons per atom with many bonding partners makes the bond linking any two metal atoms relatively weak.

To understand this point, consider the bond that would result if two Cu atoms came together and formed a molecule of Cu_2. Copper atoms have the electron configuration $[Ar]3d^{10}4s^1$. When the partially filled 4*s* orbitals of the two atoms overlap, they form a diatomic molecule held together by a single covalent bond (**Figure 9.60a**). However, the Cu atoms in solid copper are each surrounded by a total of 12 other Cu atoms and each is bonded to all 12 of its neighbors (**Figure 9.60b**). This means that each Cu atom must share its 4*s* electron with 12 other atoms, not just one. Inevitably, this dispersion of bonding electrons weakens the Cu—Cu bond between each pair of Cu atoms.

CONNECTION The photoelectric effect is described in Chapter 7.

Metallic elements have low ionization energies, which reflects the diffuse nature of this bonding. Many metallic atoms emit photoelectrons when illuminated by photons of ultraviolet radiation or even lower-energy visible light.

We can also use an extension of MO theory called **band theory** to explain the bonding in metals and other solids. For example, when the 4*s* atomic orbitals on two Cu atoms overlap to form Cu_2, the atomic orbitals combine to form two molecular orbitals with different energies, equally spaced above and below the initial energy value (**Figure 9.61a**). This is analogous to the formation of low-energy bonding and high-energy antibonding molecular orbitals. If another two Cu atoms join the first two to form a molecule of Cu_4, the 4*s* atomic orbitals of four Cu atoms combine to form four molecular orbitals. If we add another four atoms to make Cu_8, a total of eight copper 4*s* atomic orbitals combine to form eight molecular orbitals. In all these molecules the lower-energy orbitals are filled with the available 4*s* electrons and the upper orbitals are empty (**Figure 9.61b**). If

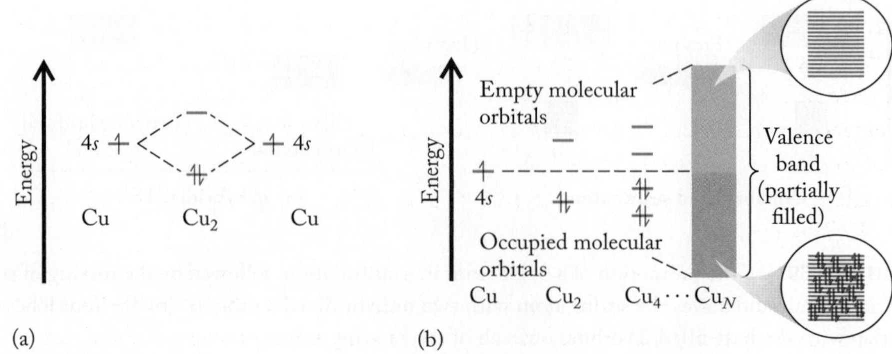

FIGURE 9.61 (a) Two 4*s* atomic orbitals combine to form two molecular orbitals in the molecule Cu_2. (b) As the half-filled 4*s* atomic orbitals of an increasing number of Cu atoms overlap, more and more molecular orbitals are formed; half of them are occupied, whereas the other half are empty. As more MOs form, their energies get closer together until a continuous energy band forms—a valence band that is only half-filled with electrons. Electrons can move from the filled half (purple) to the slightly higher-energy upper half (orange), where they are free to migrate through delocalized empty orbitals throughout the solid.

we apply this model to the enormous number of atoms in a piece of copper wire, an equally enormous number of molecular orbitals is created. The lower-energy half of them is occupied by electrons, whereas the higher-energy half is empty. There are so many of these orbitals that they form a continuous *band* of energies with no gap between the occupied lower half and the empty upper half. Because this band of MOs was formed by combining valence-shell orbitals, it is called a **valence band.**

Band theory explains the conductivity of copper and many other metals by incorporating the assumption that essentially no gap exists between the energy of the occupied lower portion of the valence band and the empty upper portion. As a result, valence electrons can move easily from the filled lower portion to the empty upper portion, where they are free to move from one empty orbital to the next and thus flow throughout the solid.

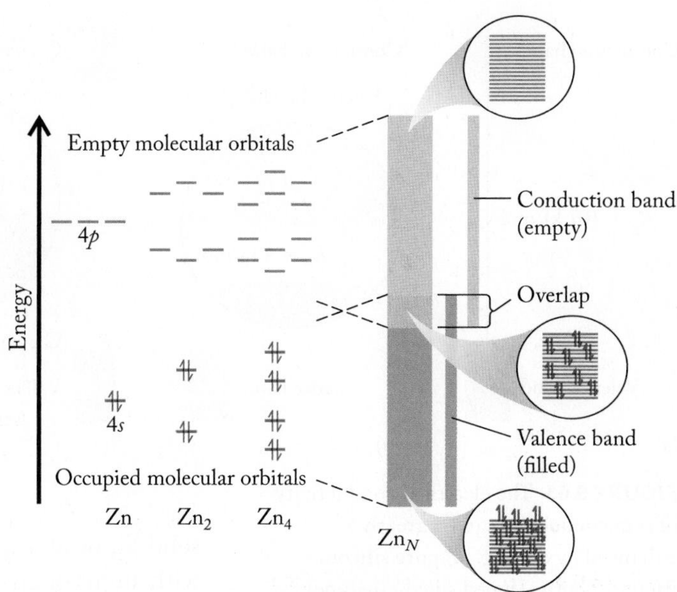

FIGURE 9.62 As the filled 4*s* atomic orbitals of an increasing number of Zn atoms overlap, they form a filled valence band (purple). An empty conduction band (gray) is produced by combining the empty 4*p* orbitals. The valence and conduction bands overlap each other, and electrons move easily from the filled valence band to the empty conduction band.

The model of a partially filled valence shell explains the conductivity of many metals, but not all. Zinc, for example, which is copper's neighbor in the periodic table, has an electron configuration of $[Ar]3d^{10}4s^2$. All its valence-shell electrons reside in filled orbitals, which means the valence band in solid zinc is filled (**Figure 9.62**). With no empty space in the valence band to accommodate additional electrons, it might seem that the valence-shell electrons in Zn would be immobile, making Zn a poor electrical conductor. However, electrons in the valence band of Zn *do* migrate through the solid, making zinc a good conductor of electricity. Band theory allows for *all* atomic orbitals of comparable shape and energy, including the empty 4*p* orbitals on zinc, to combine and form additional energy bands. The energy band produced by combining empty 4*p* orbitals, called a **conduction band,** is also empty and is broad enough to overlap the valence band. This overlap makes it possible for electrons from the valence band to move to the conduction band, where they are free to migrate from atom to atom in solid zinc, thereby conducting electricity.

CONCEPT **TEST**

Is the electrical conductivity of magnesium metal best explained in terms of overlapping conduction and valence bands or in terms of a partially filled valence band? Explain your answer.

Semiconductors

To the right of the metals in the periodic table is a "staircase" of elements that tend to have the physical properties of metals and the chemical properties of nonmetals. These metalloids are not as good at conducting electricity as metals, but they are much better at it than nonmetals. We can use band theory to explain this intermediate behavior. In metalloids, the conduction and valence bands do *not* overlap but instead are separated by an energy gap. In silicon, the most abundant metalloid, band theory predicts an energy gap, or **band gap (E_g),** of 107 kJ/mol at 298 K (**Figure 9.63a**).

Generally, only a few valence-band electrons in Si have sufficient energy to move to the conduction band, which limits silicon's ability to conduct electricity and makes it a **semiconductor**. However, we can enhance the conductivity of

band theory an extension of molecular orbital theory that describes bonding in solids.

valence band a band of orbitals that are filled or partially filled by valence electrons.

conduction band in metals, an unoccupied band higher in energy than a valence band, in which electrons are free to migrate.

band gap (E_g) the energy gap between the valence and conduction bands.

semiconductor a metalloid with electrical conductivity between that of metals and insulators that can be chemically altered to increase its electrical conductivity.

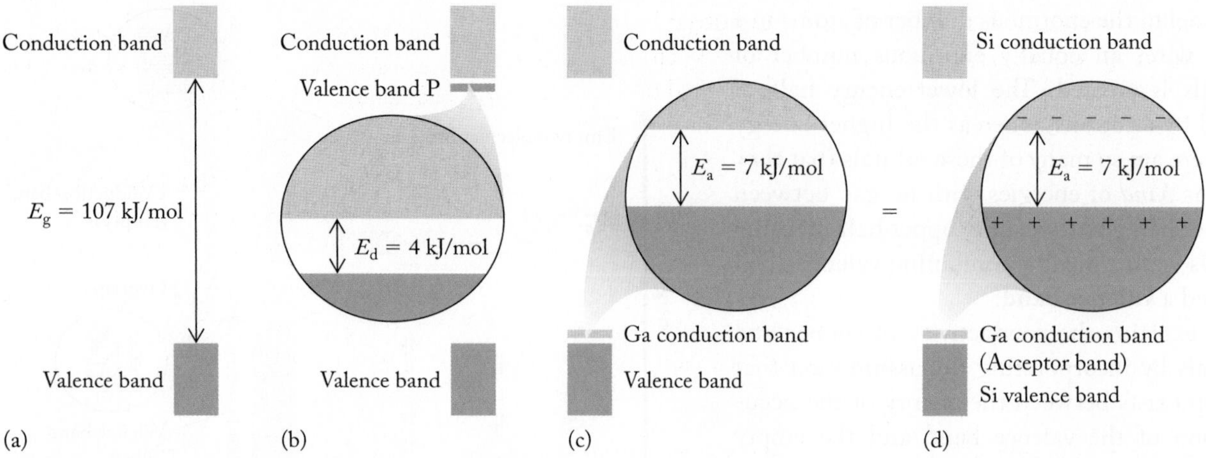

FIGURE 9.63 The electrical conductivity of semiconductors can be greatly enhanced by doping: (a) pure silicon, (b) phosphorus-doped silicon (n-type), (c) gallium-doped silicon (p-type). (d) In a p-type semiconductor, some electrons have enough energy to move from the valence band to the empty conduction band of the dopant, leaving behind positively charged vacancies, or "holes." The presence of these holes increases electron mobility and electrical conductivity.

solid Si, or of any other elemental metalloid, by replacing some of the Si atoms with atoms of an element of similar atomic radius but with a different number of valence electrons. The replacement process is called *doping*, and the added element is called a *dopant*. Suppose the dopant is a group 15 element such as phosphorus. Each P atom has one more electron than the atom of Si (group 14) that it replaced. The energy of these additional electrons is different from the energy of the silicon electrons. They populate a narrow band located in the silicon band gap (**Figure 9.63b**). This arrangement effectively reduces the size of the energy gap and increases electrical conductivity because electrons can move more easily across the remaining gaps between the valence and conduction bands. Phosphorus-doped silicon is an example of an **n-type semiconductor** because the dopant contributes extra negative charges (electrons) to the structure of the host element.

The conductivity of solid silicon can also be enhanced by replacing some Si atoms with atoms of a group 13 element such as gallium (**Figure 9.63c**). Because Ga atoms have one fewer valence electron than Si atoms, substituting them in the Si structure means fewer valence electrons in the solid. The result is the creation of a narrow Ga conduction band (*acceptor band*, **Figure 9.63d**) in the Si band gap. Because the gap between the Si valence band and the acceptor band is smaller than the band gap in pure Si, electrons from the valence band move more easily to the acceptor band, increasing electrical conductivity. This array of bands makes Si doped with Ga a **p-type semiconductor** because a reduction in the number of negatively charged electrons in the valence band is equivalent to the presence of positively charged "holes" as shown in Figure 9.63d. The semiconductors used in solid-state electronics are combinations of n-type and p-type.

Doping is not the only way to change the conductivity of semimetals. Compounds prepared from combinations of group 13 and group 15 elements also may behave as semiconductors. For example, gallium arsenide (GaAs) is a semiconductor that emits infrared radiation ($\lambda = 874$ nm) when connected to an electrical circuit. This emission is used in devices such as bar-code readers and DVD players. Like silicon, solid gallium arsenide has both a valence band and a conduction band separated by a characteristic band gap. The energy of each photon of light corresponds to the energy gap between the valence band and the conduction band. When electrical energy is applied to the material, electrons are raised to the conduction band. When they fall back to the valence band, they emit radiation. If some aluminum is substituted for gallium in

GaAs, the band gap increases, and the wavelength of emitted light decreases. For example, a material with the empirical formula $AlGaAs_2$ emits orange-red light ($\lambda = 620$ nm). Many of the multicolored indicator lights in electronic devices use $AlGaAs_2$ semiconductors.

CONNECTION We introduced the metalloids in Section 2.5 when we described the structure of the periodic table.

CONNECTION Emission spectra obtained from gas discharge tubes are discussed in Chapter 7.

SAMPLE EXERCISE 9.12 Distinguishing p- and n-Type Semiconductors **LO8**

Which kind of semiconductor—n-type or p-type—is created when germanium is doped with arsenic?

Collect and Organize We are asked to determine the type of semiconductor formed when germanium (Ge, group 14) is doped with arsenic (As, group 15). Figure 9.63 helps us distinguish between n- and p-type semiconductors.

Analyze When the dopant has *more* valence electrons than the host semimetal, the two form an n-type semiconductor (Figure 9.63b). If the dopant has *fewer* valence electrons than the host, the result is a p-type semiconductor (Figure 9.63c).

Solve Atoms of arsenic have one more valence electron than do the atoms of Ge, so As-doped Ge is an n-type semiconductor.

Think About It Other group 15 elements might also be used as dopants for making germanium-based, n-type semiconductors, but arsenic is a particularly good candidate because As atoms are nearly the same size as Ge atoms and fit easily into the structure of solid Ge.

Practice Exercise Gallium arsenide (GaAs) is a semiconductor used in optical scanners in retail stores. GaAs can be made an n-type or a p-type semiconductor by replacing some of the As atoms with another element. Which element—Se or Sn—would form an n-type semiconductor with GaAs?

In Chapter 8 and in this chapter we have presented several theories of chemical bonding. Each theory has its strengths and weaknesses. The best one to apply in a given situation depends on the questions being asked and on the level of sophistication required in the answer. If our focus is on how the atoms in the second row of elements in the periodic table bond together in molecules, Lewis structures usually suffice. If we are also interested in visualizing the shapes of molecules, then VSEPR theory and valence bond theory can be useful tools. However, if we want to explain the magnetic properties of molecular compounds and their ability to absorb and emit UV radiation and visible light, then molecular orbital theory is our most powerful model. Molecular orbital theory may provide the most complete picture of covalent bonding, but it is also the most difficult to apply to large molecules.

Although we have focused primarily on the small molecules found in the atmosphere—nitrogen, oxygen, water, carbon dioxide, methane, ozone—and somewhat larger molecules found in living systems, the principles described in this chapter apply to larger and more complex molecules and ions. We return to the importance of molecular shape, particularly in defining the biological activity of both large and small molecules, in later chapters of this book.

n-type semiconductor a semiconductor containing electron-rich dopant atoms that contribute excess electrons.

p-type semiconductor a semiconductor containing electron-poor dopant atoms that cause a reduction in the number of electrons, which is equivalent to the presence of positively charged holes.

SAMPLE EXERCISE 9.13 Integrating Concepts: Insect Pheromones

Social insects such as bees, wasps, and ants communicate with one another by secreting and detecting molecules called pheromones. Small differences in the structure of these compounds, including chiral carbon atoms, lead to differences in shape that signal for either finding a mate or avoiding a predator. This molecular recognition is the basis of traps used to capture destructive pests such as the Japanese beetle (**Figure 9.64**).

The sex pheromone for the Japanese beetle used in these traps is called japonilure. The condensed structure for this compound is

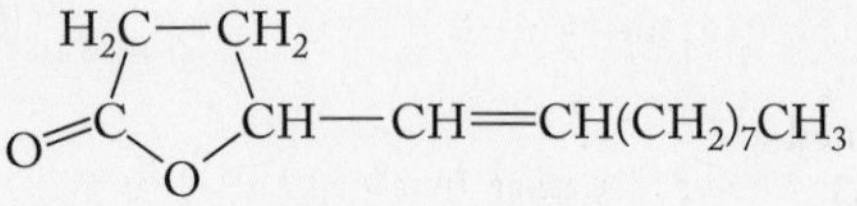

$H_2C—CH_2$

$O{=}C \quad CH—CH{=}CH(CH_2)_7CH_3$

O

a. Draw the skeletal structure of japonilure.
b. Determine the molecular geometry at the carbonyl carbon atom in the ring. What hybridization best explains this geometry? Identify two other atoms in japonilure that also have this hybridization.
c. How many π bonds are in this molecule? Do these contain delocalized π electrons?
d. Identify the chiral carbon atom in japonilure.

Collect, Organize, and Analyze We are given the condensed structure of japonilure. The molecule consists of a five-atom ring and a long hydrocarbon chain. Section 9.5 provides guidelines on how to draw skeletal structures and their relationship to condensed structures. The atoms at each vertex in the ring are carbon atoms. Sections 9.2 and 9.4 provide guidelines for determining the molecular geometry of an atom and for assessing the hybrid orbitals that are consistent with those geometries, including the formation of σ and π bonds. Section 9.5 also explains why carbon atoms bonded to four different groups are chiral.

(a)

(b)

FIGURE 9.64 (a) A Japanese beetle trap with pheromone-containing lure and (b) a Japanese beetle on a daisy.

Solve

a. The hydrocarbon chain attached to the ring can be drawn using the zigzag lines of a skeletal structure. There are 10 carbon atoms in the chain. There is a double bond between the first and second carbon atom in the chain, followed by seven methylene groups and ending with a methyl group:

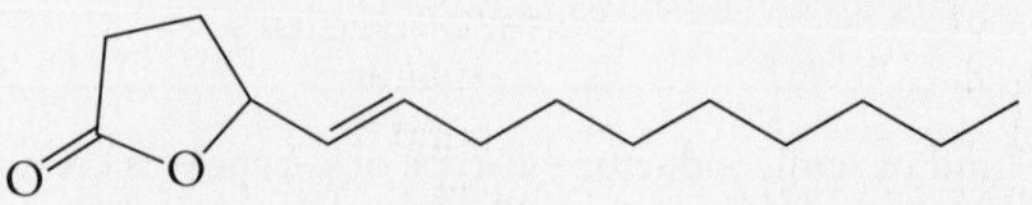

b. The carbon atom in the C=O carbonyl group has a steric number of 3 because it has the double bond to oxygen as well as single bonds to a carbon atom and a second oxygen atom. According to Table 9.1, the molecular geometry at the carbonyl carbon atom is trigonal planar. The carbonyl carbon is sp^2 hybridized. It forms three σ bonds: one to the adjacent carbon and one to each of the two oxygen atoms. The remaining p orbital is used to form a π bond to the oxygen atom that is part of the carbonyl group. The two carbon atoms in the hydrocarbon chain connected by a double bond are sp^2 hybridized, too.

c. As discussed in the solution to part (b), the carbonyl group contains a π bond. The double bond between the two carbon atoms in the hydrocarbon chain consists of one σ bond and one π bond. Therefore, there are two π bonds in the molecule, neither of which contains delocalized electrons because they are separated by more than one carbon atom.

d. A chiral carbon atom is bonded to four different groups. A carbon atom in a methylene ($–CH_2–$) or methyl ($–CH_3$) group cannot be chiral because it is bonded to multiple hydrogen atoms. Therefore, none of the carbon atoms in the hydrocarbon chain can be the chiral atom in japonilure. That leaves us with the four carbon atoms in the ring. The carbonyl carbon has SN = 3, so it is not bonded to four groups and therefore cannot be chiral. The carbon atom adjacent to the carbonyl atom (and the next carbon atom) are both ($–CH_2–$) groups. (In skeletal structures, hydrogen atoms are not explicitly drawn, and carbon atoms are assumed to be bonded to enough hydrogen atoms to have an octet.) That leaves us with the carbon atom in the ring to which the hydrocarbon chain is attached; it is marked with a *:

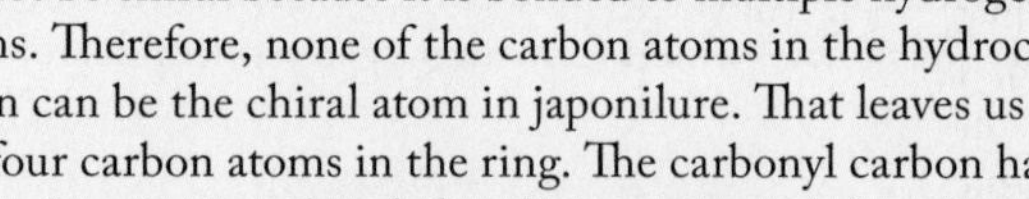

Think About It Japonilure contains one chiral carbon atom, which means that two optical isomers exist. One isomer is a sex attractant for a mate, whereas the other detects the sex attractant released by a predator. This important chiral carbon atom makes the difference between life and death for a Japanese beetle.

SUMMARY

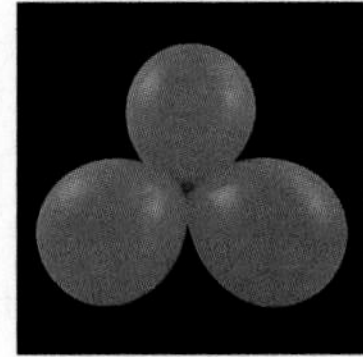

LO1 The shape of a molecule reflects the arrangement of the atoms in three-dimensional space and is determined largely by characteristic **bond angles**. Minimizing repulsion between pairs of valence electrons (the **VSEPR** model) results in the lowest-energy orientations of bonding and nonbonding electron pairs and accounts for the observed **molecular geometries** of molecules. The shape of a molecule can be determined by its **steric number** (the sum of the number of bonded atoms and lone pairs around a central atom) and the **electron-pair geometry** (the arrangement of its atoms and lone pairs). Molecules with SN = 2 and no lone pairs on the central atom have a **linear** electron-pair geometry and linear molecular geometry, while the electron-pair geometries of molecules with steric numbers 3 to 6 are **trigonal planar, tetrahedral, trigonal bipyramidal**, and **octahedral**, respectively. The presence of lone pairs of electrons in molecules with the preceding electron-pair geometries produce the following additional *molecular* geometries: **angular (bent), trigonal pyramidal, seesaw, T-shaped, square pyramidal**, and **square planar**. The observed bond angles in molecules deviate from the ideal values because of unequal repulsions between lone pairs and bonding pairs of electrons. (Sections 9.1 and 9.2)

LO2 Two covalently bonded atoms with different electronegativities have partial electrical charges of opposite sign, creating a **bond dipole**. If the individual bond dipoles in a molecule do not offset each other, the molecule is polar. If they do offset each other, the molecule is nonpolar. A polar molecule has a permanent **dipole moment (μ)**, which is a quantitative measure of the polarity of the molecule. (Section 9.3)

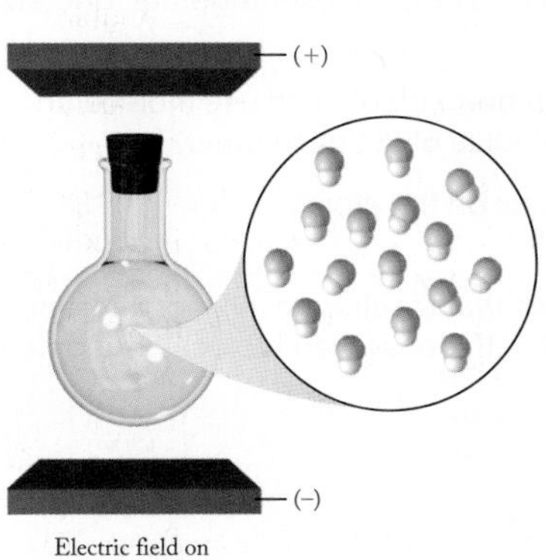

Electric field on

LO3 In **valence bond theory**, the **overlap** of half-filled atomic orbitals results in covalent bonds between pairs of atoms in molecules. Molecular geometry is explained by the mixing, or **hybridization**, of atomic orbitals to create **hybrid atomic orbitals**. Mixing one *s* and three *p* orbitals forms four $\boldsymbol{sp^3}$ **hybrid orbitals**. Overlap between sp^3 orbitals and other atomic or hybrid orbitals results in up to four **sigma (σ) bonds** and a tetrahedral orientation of valence electrons. Mixing one *s* and two *p* orbitals forms three $\boldsymbol{sp^2}$ **hybrid orbitals**. Overlap between sp^2 orbitals and other atomic or hybrid orbitals results in up to three σ bonds and a trigonal planar orientation of valence electrons. Mixing one *s* and one *p* orbital forms two ***sp*** **hybrid orbitals**. Overlap between two *sp* hybrid orbitals results in up to two σ bonds oriented linearly to one another. Covalent bonds in which the electron density is greatest either above and below or in front of and behind the bonding axis are **pi (π) bonds**. The shape of a molecule with more than one central atom is a result of overlapping geometries around the atoms. (Section 9.4)

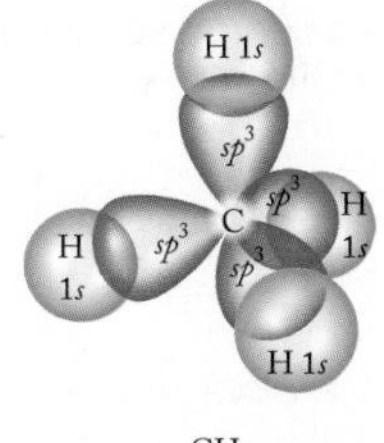

CH_4

LO4 Chemists use multiple representations to draw larger organic molecules, including condensed structures and skeletal structures. (Section 9.5)

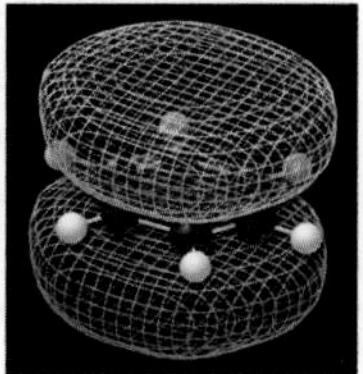

LO5 Molecules with only sp^2 hybridized central atoms have extended planar geometries. The molecules of **aromatic compounds** contain planar rings of six sp^2 hybridized carbon atoms with alternating π bonds whose electrons are delocalized over the entire ring system. Molecules with alternating single and double bonds are stabilized by electron delocalization over the system. (Section 9.5)

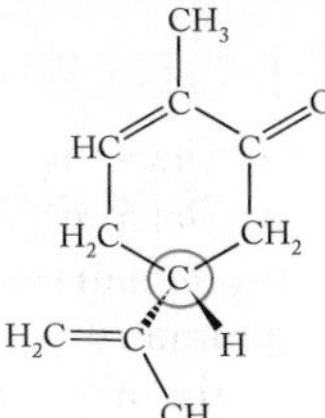

LO6 **Chiral** molecules exist in left- and right-handed forms that have different properties. Many contain an sp^3 hybridized carbon atom bonded to four different atoms or groups of atoms. (Section 9.5)

LO7 **Molecular orbital (MO) theory** is based on the formation of **molecular orbitals**, which are orbitals delocalized over two or more atoms in a molecule. MO theory does not explain shape, but it does explain the magnetic and spectroscopic properties of molecules in ways that valence bond theory cannot. Mixing two atomic orbitals creates one **bonding orbital** and one **antibonding orbital**. The region of highest electron density lies along the bond axis in a **sigma (σ) molecular orbital**. Electrons in σ bonding orbitals form σ bonds. The regions of highest electron density of **pi (π) molecular orbitals** are above and below or behind and in front of the bonding axis. Electrons occupying π bonding orbitals form π bonds. A **molecular orbital diagram** shows relative energies of the molecular orbitals in a molecule. MO electron configurations use the designations σ, σ*, π, and π* to describe the type of molecular orbitals occupied by electrons; subscripts to identify the atomic orbitals that combined to form the MOs; and superscripts to indicate the number of electrons in each MO. Atoms, ions, and molecules with no unpaired electrons are **diamagnetic** and are slightly repelled by an applied magnetic field. Atoms, ions, and molecules containing at least one unpaired electron are **paramagnetic** and are attracted by an external magnetic field. The bonding in molecules with more than an octet of valence electrons around a central atom is best explained using MO diagrams. (Section 9.6)

LO8 MO theory is useful to describe metallic bonding and semiconductors. The electrical conductivity of metals can be explained by **band theory** as the ease with which valence electrons can gain mobility by moving into empty energy levels in multiatom structures. Metalloids are **semiconductors**, intermediate in electrical conducting ability between metals and nonmetals. In semiconductors, the filled valence band and empty conduction band are separated by a **band gap** $(\boldsymbol{E_g})$. Substituting electron-rich atoms into a semiconductor results in **n-type semiconductors**. Substituting electron-poor atoms results in **p-type semiconductors**. Both types of substitution increase the conductivity of the semiconductor by decreasing its band gap. (Section 9.6)

PARTICULATE PREVIEW WRAP-UP

The Lewis structures of methyl mercaptan and acrolein are shown here. Lewis structures do not include shape information (bond angles), but they tell us, for example, that the steric numbers of the carbon atom and the sulfur atom in methyl mercaptan are both four, which means they have tetrahedral electron-pair geometries and bond angles near 109°, which means the molecule is *not* planar. The steric number of all three carbon atoms in acrolein is three, which gives all three atoms trigonal planar electron-pair and molecular geometries and bond angles near 120°. These linked planar molecular geometries make the entire molecule planar.

Methyl mercaptan

Acrolein

PROBLEM-SOLVING SUMMARY

Type of Problem	Concepts and Equations	Sample Exercises
Predicting molecular geometry	Draw a Lewis structure for the molecule. Determine the steric number (SN) of the central atom, where $$\text{SN} = \left(\begin{array}{c}\text{number of atoms}\\ \text{bonded to central atom}\end{array}\right) + \left(\begin{array}{c}\text{number of lone pairs}\\ \text{on central atom}\end{array}\right) \quad (9.1)$$ Choose a geometry that minimizes repulsion between electron pairs.	**9.1, 9.3**
Predicting relative sizes of bond angles	Lone pairs on a central atom push bonded atoms closer together, decreasing bond angles.	**9.2**
Predicting polarity of a substance	Assign the direction of polarity to each bond dipole and use molecular geometry to determine whether the dipoles offset each other.	**9.4**
Identifying overlapping orbitals in a molecule	Identify the partially filled atomic orbitals on the atoms.	**9.5**
Describing bonding in molecules and the shape of molecules by using hybrid orbitals	Identify the hybrid orbitals in molecules that result from mixing different numbers of *s* and *p* orbitals that result in the observed molecular geometry: $s + p$ = two sp hybrid orbitals s + two p = three sp^2 hybrid orbitals s + three p = four sp^3 hybrid orbitals	**9.6 and 9.8**
Drawing larger molecules	Write condensed structures and draw skeletal structures.	**9.7**
Using MO diagrams to predict bond order	$$\text{Bond order} = \frac{1}{2}\left(\begin{array}{c}\text{number of}\\ \text{bonding electrons}\end{array} - \begin{array}{c}\text{number of}\\ \text{antibonding electrons}\end{array}\right) \quad (9.2)$$	**9.9, 9.10, 9.11**
Describing n- and p-type semiconductors	Adding electron-rich dopants creates an n-type semiconductor. Adding electron-poor dopants creates a p-type semiconductor.	**9.12**

VISUAL PROBLEMS

(Answers to boldface end-of-chapter questions and problems are in the back of the book.)

9.1. The three compounds with the molecular structures shown in Figure P9.1 have the same molecular formula: $C_2H_2Cl_2$. Which, if any, of the molecules has a dipole moment of zero?

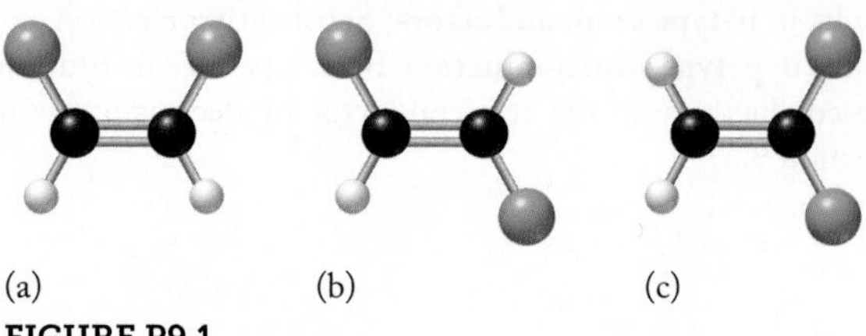

FIGURE P9.1

9.2. Figure P9.2 shows the molecular structures of (a) CCl_4, (b) SF_4, and (c) $XeCl_4$. What is the electron-pair geometry and the molecular geometry of each molecule?

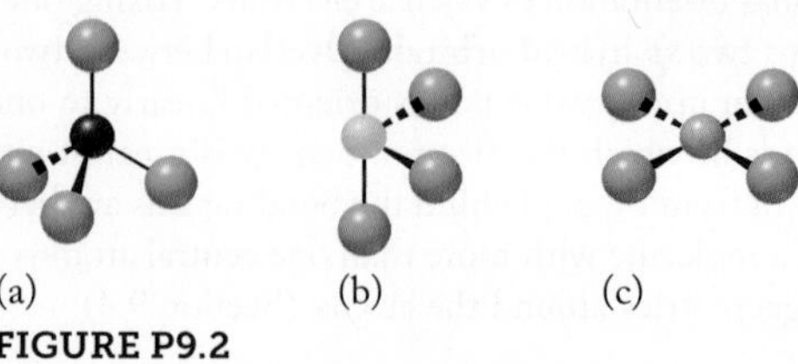

FIGURE P9.2

9.3. Figure P9.3 shows the molecular structures of (a) BF_3, (b) NF_3, and (c) ClF_3. What is the electron-pair geometry and the molecular geometry of each molecule?

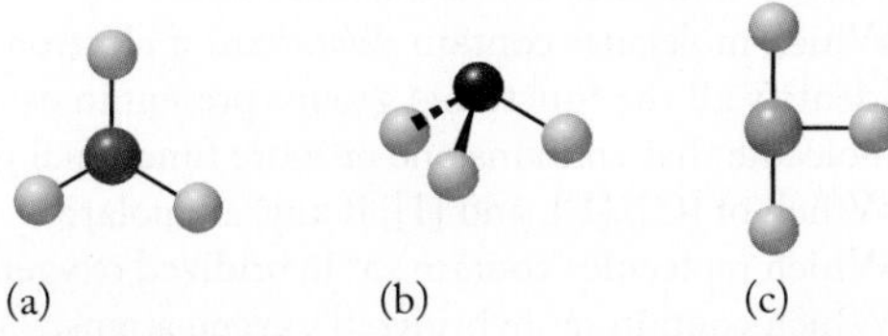

FIGURE P9.3

9.4. Which molecules in Figure P9.2 have polar bonds? Which molecules in Figure P9.2 have no dipole moment?

9.5. Which molecules in Figure P9.3 have polar bonds? Which molecules in Figure P9.3 have a permanent dipole?

9.6. Regarding the structures in Figure P9.6:

a. Which, if any, of the structures contains π electrons?

b. Which, if any, of the structures contains delocalized π electrons?

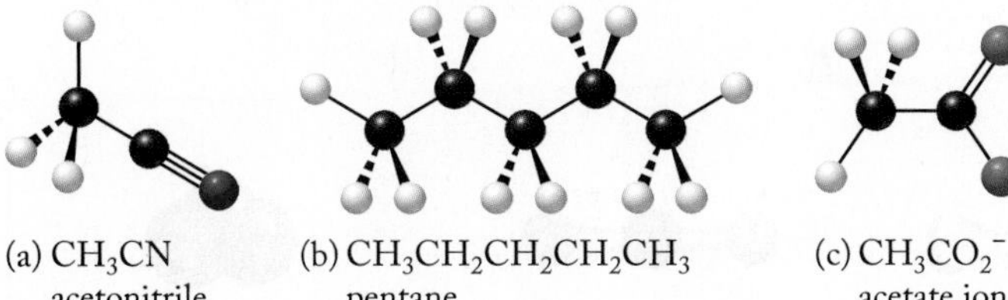

FIGURE P9.6

9.7. Regarding the molecular shape in Figure P9.7:

a. Which two electron-pair geometries could correspond to this molecular geometry?

b. For each of the two electron-pair geometries that could correspond to the shape in Figure P9.7, identify the hybridization of the central atom.

c. Explain why the hybridization of the central atom must be different for the two electron-pair geometries.

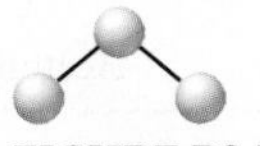

FIGURE P9.7

9.8. Could you distinguish between the two structures of N_2H_2 shown in Figure P9.8 by the magnitude of their dipole moments?

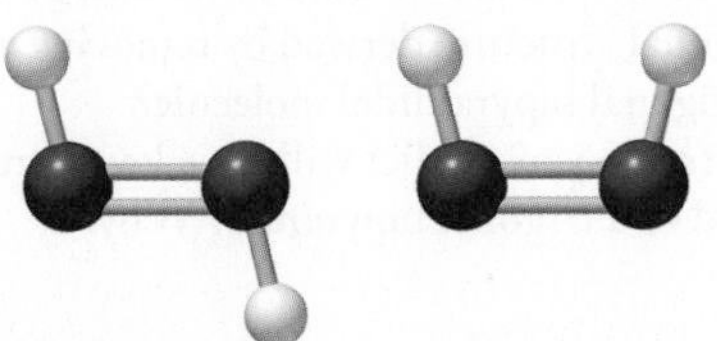

FIGURE P9.8

9.9. Which of the molecules shown in Figure P9.9 are planar—that is, all atoms are in a single plane? Are there delocalized π electrons in any of these molecules?

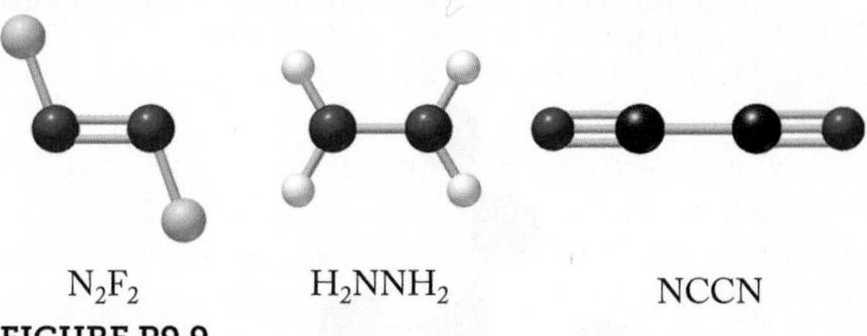

FIGURE P9.9

9.10. Which of the molecules shown in Figure P9.10 is *not* planar? Are there delocalized π electrons in any of these molecules?

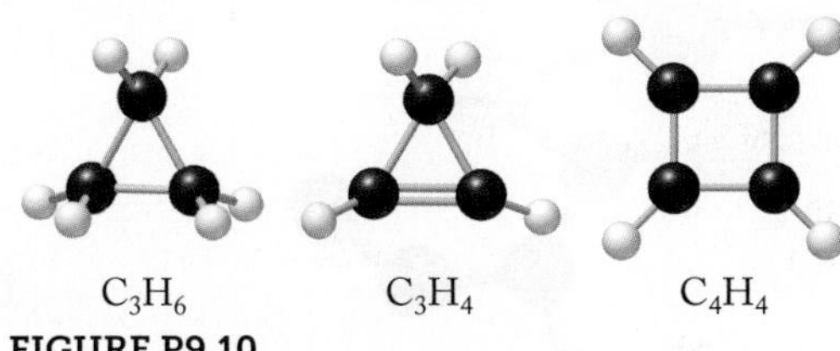

FIGURE P9.10

9.11. Use the MO diagram in Figure P9.11 to predict whether O_2^+ has more or fewer electrons in antibonding molecular orbitals than O_2^{2+}.

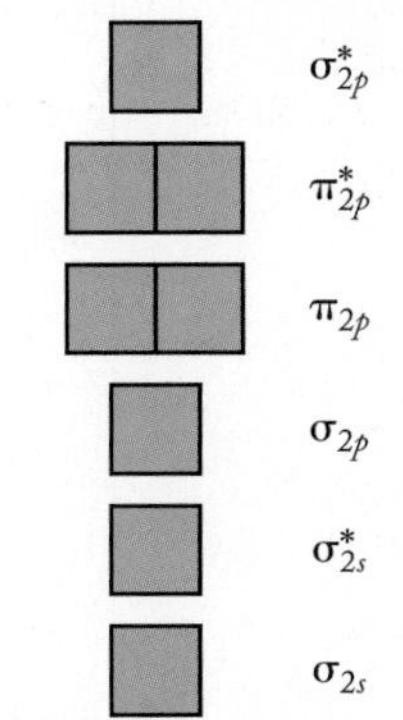

FIGURE P9.11

9.12. Under appropriate conditions, I_2 can be oxidized to I_2^+, which is bright blue. The corresponding anion, I_2^-, is not known. Use the molecular orbital diagram in Figure P9.12 to explain why I_2^+ is more stable than I_2^-.

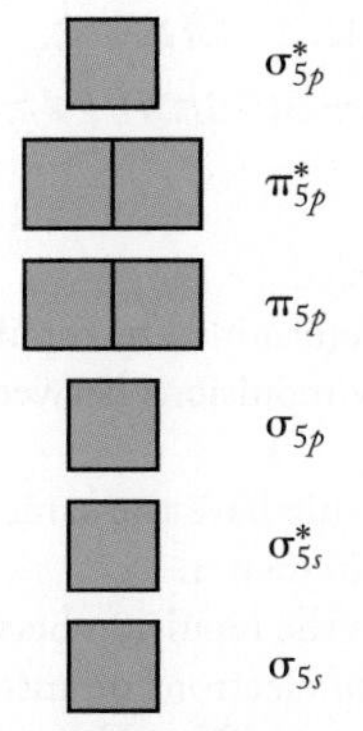

FIGURE P9.12

***9.13.** The structure of $CH_3Te(I)[S_2CN(Et_2)_2]$ is shown in Figure P9.13 (the hydrogen atoms are not shown). The geometry around the Te atom is called a pentagonal pyramid. What are the bond angles in a pentagonal pyramid?

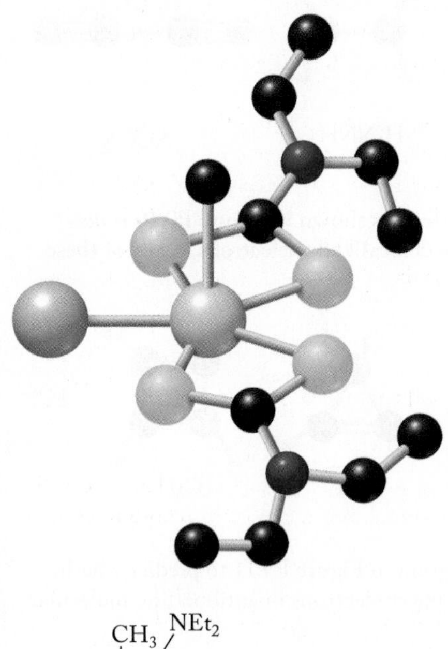

FIGURE P9.13

9.14. Use representations [A] through [I] in Figure P9.14 to answer questions (a)–(f).

a. Which representations depict orbital overlap?
b. Which molecules contain π electrons?
c. Which molecules contain *delocalized* π electrons?
d. Identify all the functional groups present in each molecule that contains one or more functional groups.
e. Which of [C], [F], and [I], if any, are polar?
f. Which molecules contain sp^2 hybridized oxygen atoms? Which contain sp^3 hybridized oxygen atoms?

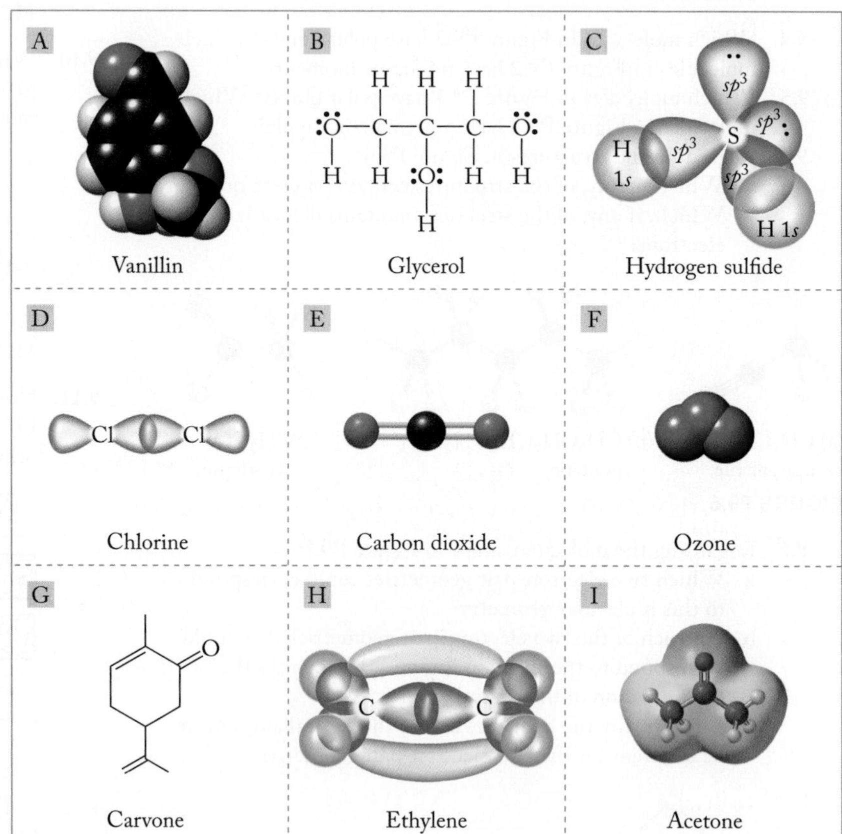

FIGURE P9.14

QUESTIONS AND PROBLEMS

Biological Activity and Molecular Shape; Valence-Shell Electron-Pair Repulsion (VSEPR) Theory

Concept Review

9.15. Why is the shape of a molecule determined by repulsions between electron pairs and not by repulsions between nuclei?

9.16. Do all resonance forms of a molecule have the same molecular geometry? Explain your answer.

9.17. In a molecule of ammonia, why is the repulsion between the lone pair and a bonding pair of electrons on nitrogen greater than the repulsion between two N—H bonding pairs?

9.18. Why is it important to draw a correct Lewis structure for a molecule before predicting its geometry?

9.19. Why does the seesaw structure have lower energy than a trigonal pyramidal structure derived by removing an axial atom from a trigonal bipyramidal molecule?

*9.20. Which geometry do you predict will have lower energy: a square pyramid or a trigonal bipyramid? Why?

Problems

9.21. Arrange the following molecular geometries in order of increasing bond angle: (a) trigonal planar; (b) octahedral; (c) tetrahedral.

9.22. Arrange the following molecular geometries in order of increasing bond angle: (a) square planar; (b) tetrahedral; (c) square pyramidal.

9.23. Which of the molecular geometries discussed in this chapter have more than one characteristic bond angle?

*9.24. Which molecular geometries for molecules of the general formula AB_x (x = 2 to 6) discussed in this chapter have the same bond angles when lone pairs replace one or more atoms?

9.25. Which of the following molecular geometries does *not* lead to linear triatomic molecules after the removal of one or more atoms? (a) tetrahedral; (b) octahedral; (c) T-shaped

9.26. Which of the following molecular geometries does *not* lead to linear triatomic molecules after the removal of one or more atoms? (a) trigonal bipyramidal; (b) seesaw; (c) trigonal planar

***9.27.** Describe the molecular geometries that result from replacing one atom with a lone pair of electrons in an AB_7 molecule with the pentagonal bipyramidal geometry shown in Figure P9.27.

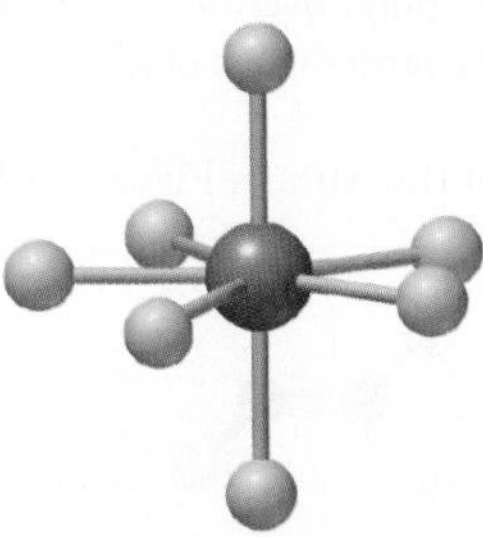

FIGURE P9.27

*9.28. Which numbered atoms would you have to remove from the cubic molecule shown in Figure P9.28 to create a geometry that approximates an octahedron?

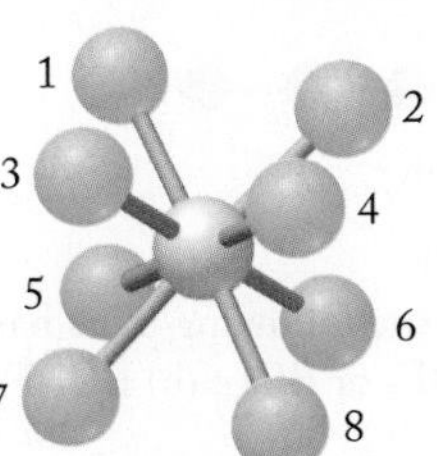

FIGURE P9.28

9.29. Determine the molecular geometries of the following molecules: (a) GeH_4; (b) PH_3; (c) H_2S; (d) $CHCl_3$.

9.30. Determine the molecular geometries of the following molecules and ions: (a) NO_3^-; (b) NO_4^{3-}; (c) S_2O; (d) NF_3.

9.31. Determine the bond angles in the following ions: (a) NH_4^+; (b) SO_3^{2-}; (c) NO_2^-; (d) XeF_5^+.

9.32. Determine the bond angles in the following ions: (a) SCN^-; (b) BF_2^+; (c) ICl_2^-; (d) PO_3^{3-}.

9.33. Determine the electron-pair and molecular geometries of the following ions: (a) $S_2O_3^{2-}$; (b) PO_4^{3-}; (c) NO_3^-; (d) NCO^-.

9.34. Determine the electron-pair and molecular geometries of the following molecules: (a) ClO_2; (b) ClO_3; (c) IF_3; (d) SF_4.

9.35. Which of the following triatomic molecules—O_3, SO_2, N_2O, S_2O, and CO_2—have the same molecular geometry?

9.36. Which of the following species—N_3^-, O_3, CO_2, SCN^-, CNO^-, and NO_2^-—have the same molecular geometry?

9.37. The anion $C(CN)_3^-$ has a trigonal planar geometry about the central carbon atom. Draw Lewis structures for $C(CN)_3^-$, including resonance forms, and determine which structure contributes the most to the bonding.

9.38. The anion $C(NO_2)_3^-$ has a trigonal planar geometry about the carbon atom. Draw Lewis structures for $C(NO_2)_3^-$, including resonance forms, and determine which structure contributes the most to the bonding.

***9.39.** The C—N—C bond angles in tri(methyl)amine, $N(CH_3)_3$, are approximately 109°, whereas the Si—N—Si and S—N—S bond angles in $N(SiH_3)_3$ and $N(SCF_3)_3$, respectively, are 120° as shown in Figure P9.39. Explain the difference in bond angle between $N(CH_3)_3$ and the other two compounds.

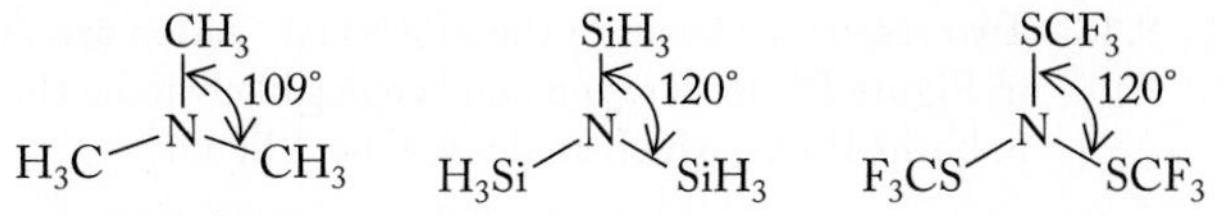

CH_3 109° H_3C—N—CH_3 SiH_3 120° H_3Si—N—SiH_3 SCF_3 120° F_3CS—N—SCF_3

FIGURE P9.39

*9.40. Complete the Lewis structures of $SCNCl_3$ in Figure P9.40. Is the geometry around nitrogen the same in both molecules?

Cl—C(Cl)—N—S—Cl Cl—S—C(Cl)—N—Cl

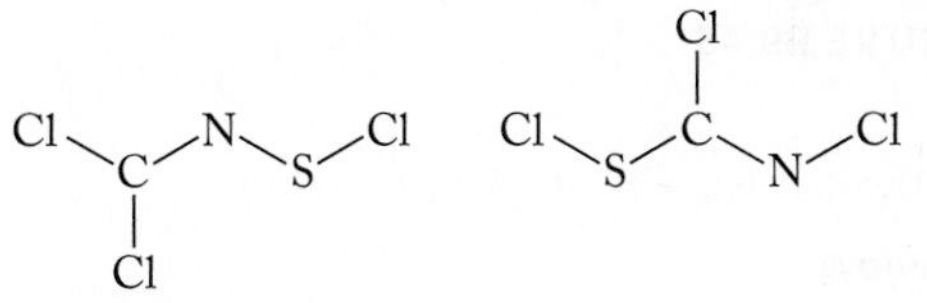

FIGURE P9.40

***9.41.** For many years, it was believed that the noble gases could not form covalently bonded compounds. However, xenon reacts with fluorine and oxygen. Reaction between xenon tetrafluoride and fluoride ions produces the pentafluoroxenate anion:

$$XeF_4 + F^- \rightarrow XeF_5^-$$

Draw Lewis structures for XeF_4 and XeF_5^-, and predict the geometry around xenon in XeF_4. The crystal structure of XeF_5^- compounds indicates a pentagonal bipyramidal orientation of valence pairs around Xe. Sketch the structure for XeF_5^-.

*9.42. The first compound containing a xenon–sulfur bond was isolated in 1998. Draw a Lewis structure for HXeSH and determine its molecular geometry at Xe.

***9.43.** The Cl—O distances in ClO_2^+, ClO_2, and ClO_2^- are 131 pm, 147 pm, and 156 pm, respectively. The corresponding O—Cl—O bond angles are 122°, 118°, and 110°. Draw Lewis structures consistent with these data.

*9.44. The skeletal structures of F_2PSN and F_2PNS are shown in Figure P9.44.

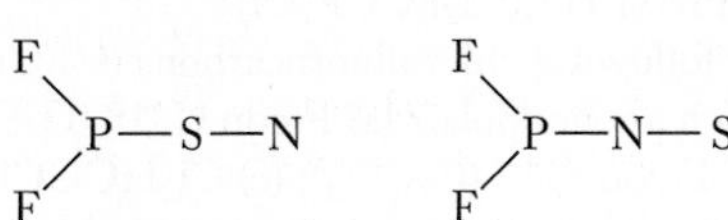

FIGURE P9.44

a. Complete the Lewis structures for both forms of this compound such that the P—S—N and P—N—S bond angles are 120°.
b. Is the N—S bond distance the same in both molecules?

9.45. The three group 16 tetrafluorides—SF_4, SeF_4, and TeF_4—all have the seesaw structure, but their bond angles in Figure P9.45 follow the order M = S > Se > Te. Explain this trend.

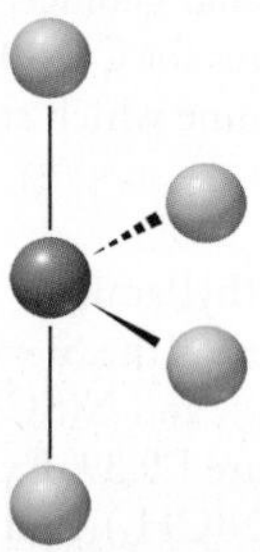

FIGURE P9.45

9.46. Two resonance forms of the $SF_4(NH_2)^+$ cation are shown in Figure P9.46. Are the bond angles likely to be the same in both? If not, which angles will be different?

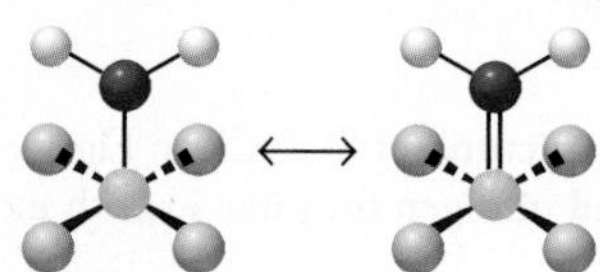

FIGURE P9.46

Polar Bonds and Polar Molecules

Concept Review

9.47. Describe the criteria for distinguishing between nonpolar covalent, polar covalent, and ionic bonds.

9.48. Describe how you would determine whether a molecule has a permanent dipole.

9.49. Can a nonpolar molecule contain polar covalent bonds?

9.50. Compare the dipole moments of CO_2 and OCS.

Problems

9.51. Consider the following molecules: (a) CCl_4; (b) $CHCl_3$; (c) CO_2; (d) H_2S; (e) SO_2.

a. Which of them contain polar bonds?
b. Which are polar molecules?
c. Which are nonpolar molecules?

9.52. Fluorine forms compounds with B, C, N, O, and itself: BF_3, CF_4, NF_3, OF_2, and F_2. Which of these molecules are polar?

9.53. **Freon Ban** Compounds containing carbon, chlorine, and fluorine are known as Freons or chlorofluorocarbons (CFCs). Widespread use of these substances was banned because they destroyed the ozone layer in the upper atmosphere. Which of the following CFCs are polar and which are nonpolar? (a) Freon 11 ($CFCl_3$); (b) Freon 12 (CF_2Cl_2); (c) Freon 113 (Cl_2FCCF_2Cl)

9.54. Which of the following chlorofluorocarbons (CFCs) are polar and which are nonpolar? (a) Freon C318 (C_4F_8, cyclic structure); (b) Freon 1113 (C_2ClF_3); (c) $Cl_2HCCClF_2$

9.55. Xenon gas reacts with ClCN and BrCN to form ClXeCN and BrXeCN, respectively (Figure P9.55).

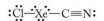

:C̤l̤—X̤e̤—C≡N: :B̤r̤—X̤e̤—C≡N:

FIGURE P9.55

a. Are the two compounds isoelectronic?
b. Which bond is the most polar in each molecule?
c. Which molecule has the larger value of μ?

9.56. The two compounds shown in Figure P9.56 are isoelectronic.

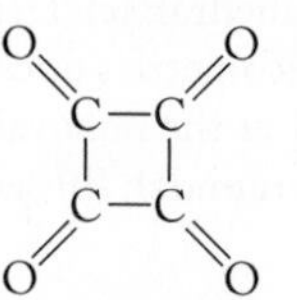

O=B—N=O, B—N ring: B—N / N—B, O=N—B=O

FIGURE P9.56

a. Why are the two molecules considered to be isoelectronic?
b. Which bond is the most polar in each molecule?
c. Which bond is the most polar overall?
d. Which molecule has the larger value of μ?

9.57. Which molecule in each of the pairs in Figure P9.57 has the larger dipole moment?

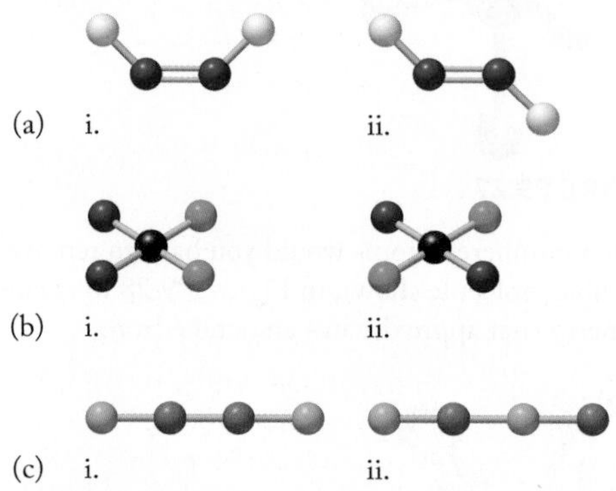

FIGURE P9.57

9.58. Which molecule in each of the following pairs has the larger dipole moment? (a) BF_3 or BCl_3; (b) BCl_2F or $BClF_2$

9.59. Aluminum chloride has the Lewis structure shown in Figure P9.59.

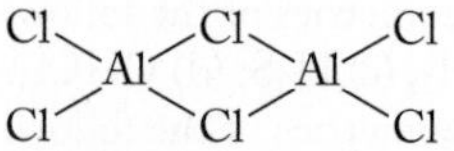

Cl, Cl, Cl / Al, Al / Cl, Cl, Cl

FIGURE P9.59

a. What is the geometry about the aluminum atom?
b. Is Al_2Cl_6 polar or nonpolar?

9.60. **Cleaning Silicon Chips** Nitrogen trifluoride (NF_3) is used in the electronics industry to clean surfaces. NF_3 is also a potent greenhouse gas.

a. Draw the Lewis structure of NF_3 and determine its molecular geometry.
b. BF_3 and NF_3 both have three covalently bonded fluorine atoms around a central atom. Do they have the same dipole moment?
c. Could BF_3 also behave as a greenhouse gas?

Valence Bond Theory

Concept Review

9.61. Why aren't the orbitals on isolated atoms hybridized?

*9.62. Can hybrid orbitals be associated with more than one atom? Explain your answer.

9.63. Describe in your own words the differences between sigma and pi bonds.

*9.64. What hybridization scheme of *s* and *p* orbitals are needed to form two σ and two π bonds to a carbon atom?

Problems

9.65. Draw a valence bond diagram like the ones in Figures 9.24 and 9.26 that describes the bonding in PCl_3.

9.66. Draw a valence bond diagram like the ones in Figures 9.24 and 9.26 that describes the bonding in NO_2^-.

9.67. **Airbags** Azides such as sodium azide (NaN_3) are used in automobile airbags as a source of nitrogen gas. Another compound with three nitrogen atoms bonded together is N_3F. What differences are there in the arrangement of the electrons around the nitrogen atoms in the azide ion (N_3^-) and N_3F? Is there a difference in the hybridization of the central nitrogen atom?

9.68. N_3F decomposes to nitrogen and N_2F_2 by the following reaction:

$$2\ N_3F \rightarrow 2\ N_2 + N_2F_2$$

N_2F_2 has two possible structures, as shown in Figure P9.68. Are the differences between these structures related to differences in the hybridization of nitrogen in N_2F_2? Identify the hybrid orbitals that account for the bonding in N_2F_2. Are they the same as those in acetylene (C_2H_2)?

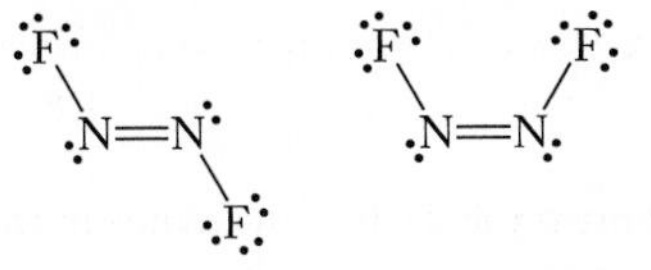

FIGURE P9.68

9.69. How does the hybridization of the sulfur atom change in the series SF_2, SF_4, and SF_6?

9.70. How does the hybridization of the central atom change in the series CO_2, NO_2, and O_3?

9.71. **Minoxidil** The drug minoxidil was originally developed for treating high blood pressure but is now used primarily for treating hair loss. The Lewis structure of minoxidil is shown in Figure P9.71. Complete the Lewis structure by adding lone pairs where needed. Assign formal charges to the nitrogen and oxygen highlighted in red. Describe the bonding around the nitrogen in the N–O group.

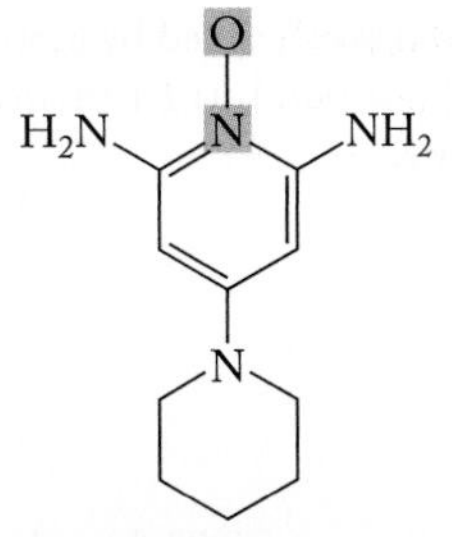

FIGURE P9.71

9.72. **Treating Diabetes** The drug metformin (Figure P9.72) has been used to treat type 2 diabetes for a half-century by suppressing glucose production. Metformin contains five nitrogen atoms. Determine the geometry around each nitrogen atom and describe the bonding according to valence bond theory.

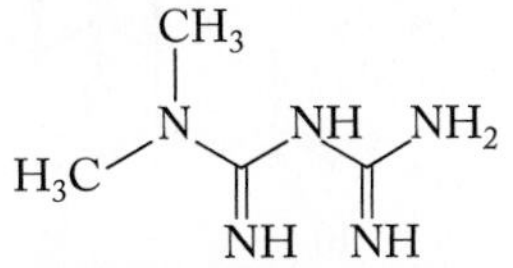

FIGURE P9.72

9.73. Draw a Lewis structure for CF_3PCF_2 where the fluorine atoms are all bonded to carbon atoms. Determine its molecular geometry at P and the hybridization of the phosphorus atom.

9.74. The Lewis structure of N_4O, with the skeletal structure O—N—N—N—N, contains one N—N single bond, one N=N double bond, and one N≡N triple bond. Is the hybridization of all the nitrogen atoms the same?

Shape, Large Molecules, and Molecular Recognition

Concept Review

9.75. What features do Lewis structures, Kekulé structures, condensed structures, and skeletal structures share? What features differentiate these four kinds of structures?

9.76. Explain why alkanes don't have optical isomers.

***9.77.** Are resonance structures examples of electron delocalization? Explain your answer.

9.78. Can sp^2 and *sp* hybridized carbon atoms be chiral centers? Explain your answer.

9.79. Which of the following objects are chiral? (a) a baseball bat with no lettering on it; (b) a pair of scissors; (c) a boot; (d) a fork

9.80. Why is it difficult to assign a single geometry to a molecule with more than one central atom?

Problems

9.81. Bombykol is the compound synthesized by female silkworm moths to attract mates. Convert the skeletal structure in Figure P9.81 to a condensed structure.

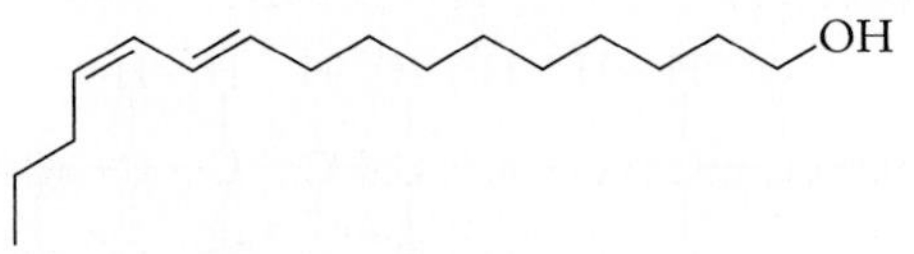

FIGURE P9.81

9.82. Fucoserratene is the compound synthesized by a brown alga to reproduce. Convert the skeletal structure in Figure P9.82 to a condensed structure.

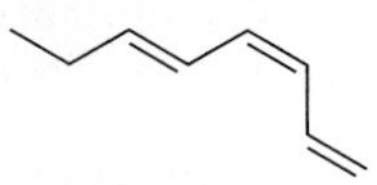

FIGURE P9.82

9.83. Sex Hormones Progesterone (Figure P9.83a) is a hormone involved in regulating menstrual cycles and pregnancy, whereas testosterone (Figure 9.83b) is the primary male sex hormone.

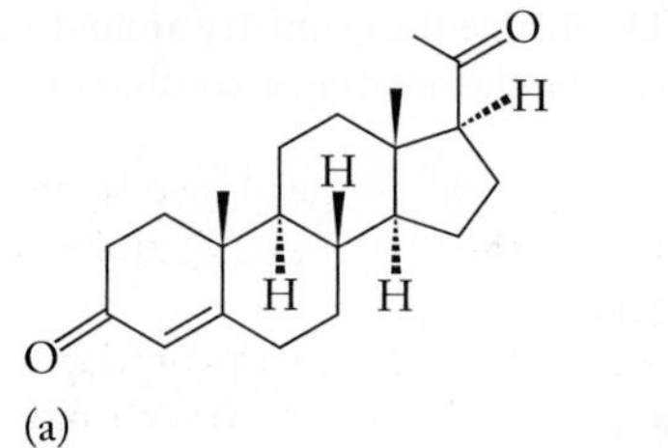

(a)

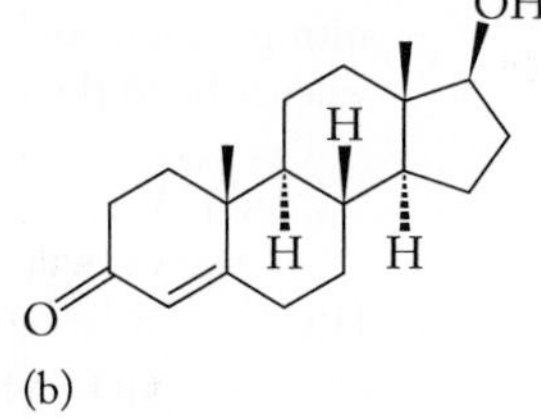

(b)

FIGURE P9.83

a. How many carbon atoms are in each of these molecules?
b. How many sp^2 hybridized carbon atoms are in each of these molecules?
c. What functional groups do these two hormones have in common?

9.84. **Steroid Hormones** Cortisone (Figure P9.84) is a steroid produced by the body in response to stress. It suppresses the immune system and can be administered as a drug to reduce inflammation, pain, and swelling.

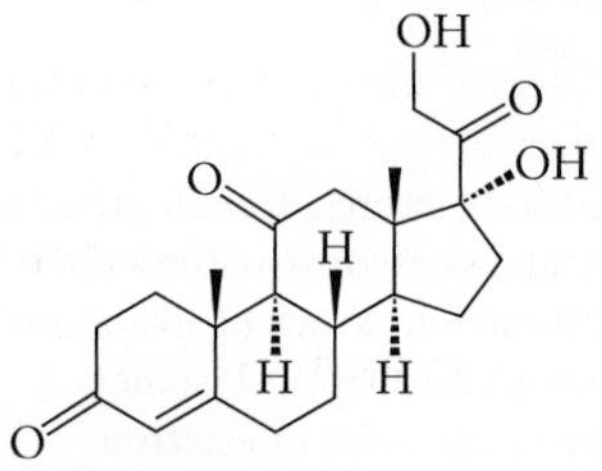

FIGURE P9.84

a. How many carbon atoms are in this molecule?
b. How many sp hybridized carbon atoms are in this molecule?
c. Identify all the functional groups in cortisone.

9.85. How many chiral carbon atoms are found in progesterone in Figure P9.83a? In testosterone in Figure P9.83b?

9.86. How many chiral carbon atoms are found in cortisone in Figure P9.84?

9.87. Oleic acid (Figure P9.87) is a fat derived from olive oil. Write a condensed structure for oleic acid.

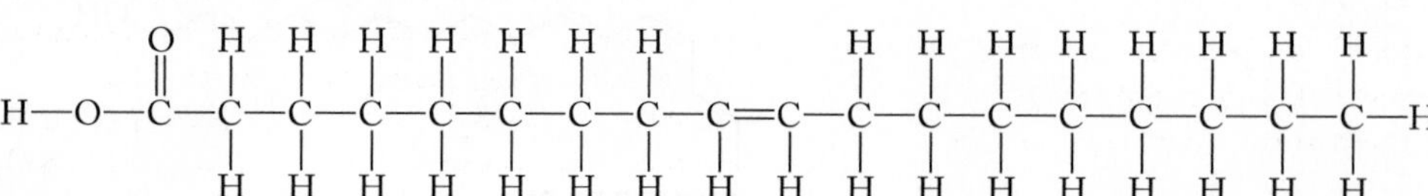

FIGURE P9.87

9.88. Draw a skeletal structure for oleic acid (Figure P9.87).

9.89. Prozac Fluoxetine (Figure P9.89) is an antidepressant medication sold commercially as Prozac. Identify the delocalized π electrons in fluoxetine.

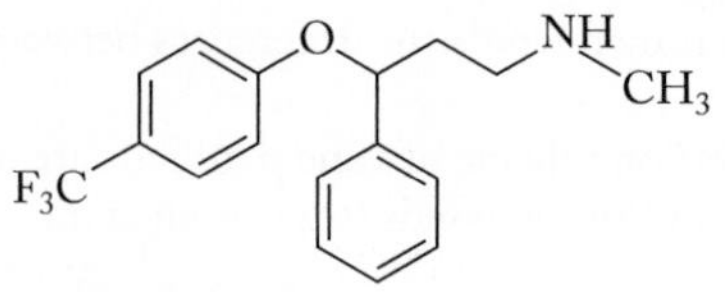

FIGURE P9.89

*9.90. Figure P9.90 shows the skeletal structure of the antifungal compound capillin. Are there delocalized electrons in capillin? If so, identify them.

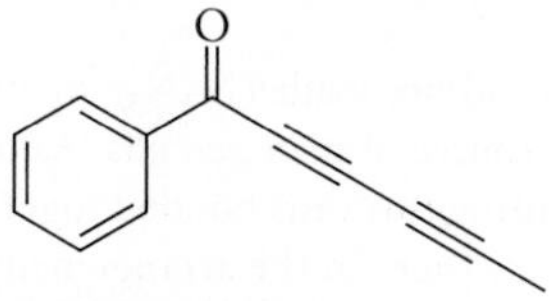

FIGURE P9.90

9.91. Identify any chiral carbon atoms in Prozac in Figure P9.89.

9.92. Is capillin (Figure P9.90) a chiral compound?

9.93. Artificial Sweeteners Acesulfame potassium is one of many artificial sweeteners used in food. It is 200 times sweeter than sucrose (table sugar) and has the structure shown in Figure P9.93.

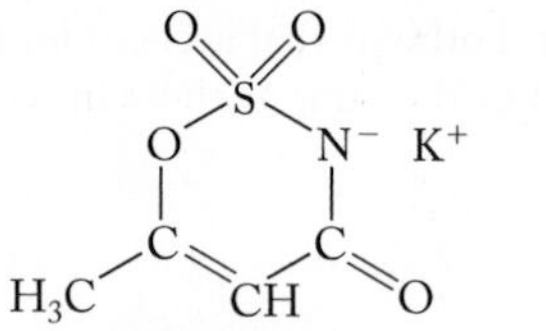

FIGURE P9.93

a. What is the geometry at each of the atoms in the six-membered ring?
b. Which atomic or hybrid orbitals overlap to form the C—O and C—N bonds?
c. In which atomic or hybrid orbital is the extra electron on N located?
d. Are there any delocalized electrons in acesulfame potassium?

*9.94. **First Artificial Sweetener** Saccharin (Figure P9.94) was the first artificial sweetener, discovered in 1879. Like acesulfame potassium, it contains a sulfur atom adjacent to a nitrogen atom.

a. Why don't all the atoms in the five-membered ring lie in the same plane?
b. Why is it difficult to explain the bonding between S and O by using the hybrid orbitals described in Section 9.4?

c. Are there any delocalized electrons in saccharin?

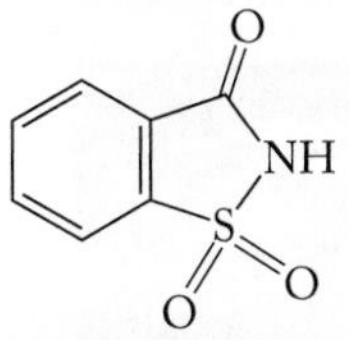

FIGURE P9.94

9.95. Which molecules in Figure P9.95 are chiral? Which ones contain delocalized electrons?

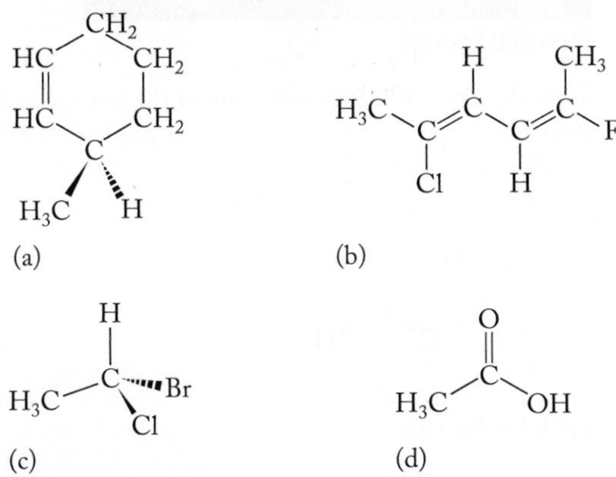

FIGURE P9.95

9.96. Which molecules in Figure P9.96 are chiral? Which ones contain delocalized electrons?

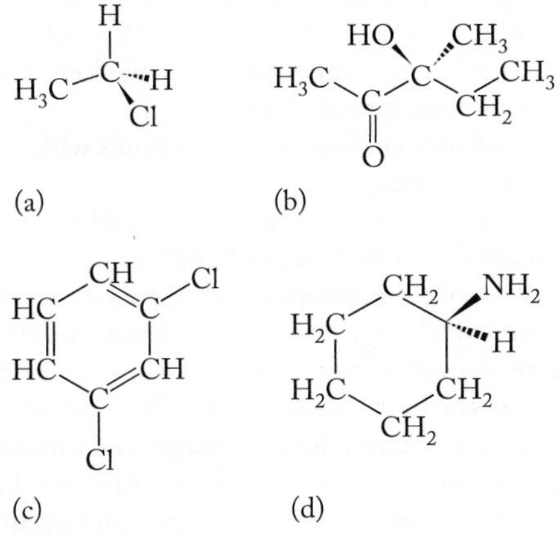

FIGURE P9.96

Molecular Orbital Theory

Concept Review

9.97. What is the difference between a bonding molecular orbital and an antibonding molecular orbital?

9.98. How many molecular orbitals can form from six atomic orbitals?

9.99. Are *s* atomic orbitals with different principal quantum numbers (n) as likely to overlap and form MOs as *s* atomic orbitals with the same value of n? Explain your answer.

9.100. Which atomic orbitals are more likely to mix to form a set of molecular orbitals—a 2*s* and a 3*p* orbital or a 4*s* and a 5*p* orbital?

9.101. How can molecules with even numbers of valence electrons be paramagnetic?

9.102. How does the molecular orbital diagram for a homonuclear diatomic species differ from that of a heteronuclear diatomic species?

9.103. How does the sea-of-electrons model (Chapter 8) explain the high electrical conductivity of gold? How does band theory explain this?

9.104. Some scientists believe that the solid hydrogen that forms at very low temperatures and high pressures may conduct electricity. Is this hypothesis supported by band theory?

9.105. Describe in general terms the differences in composition and conduction between n-type and p-type semiconductors.

9.106. How might doping of silicon with germanium affect the conductivity of silicon?

Problems

9.107. Make a sketch showing how two 1*s* orbitals overlap to form a σ_{1s} bonding molecular orbital and a σ^*_{1s} antibonding molecular orbital.

9.108. Make a sketch showing how two $2p_y$ orbitals overlap "side-on" to form a π_{2p} bonding molecular orbital and a π^*_{2p} antibonding molecular orbital.

9.109. Consider the following molecular ions: N_2^+, O_2^+, C_2^+, and Br_2^{2-}. Using MO theory, (a) write their orbital electron configurations; (b) predict their bond orders; (c) state whether you expect any of these species to exist.

9.110. Diatomic noble gas molecules, such as He_2 and Ne_2, do not exist.

a. Write their orbital electron configurations.
b. Does removing one electron from each of these molecules create molecular ions (He_2^+ and Ne_2^+) that are more stable than He_2 and Ne_2?

9.111. Which of the following molecular ions is expected to have one or more unpaired electrons? (a) N_2^+; (b) O_2^+; (c) C_2^{2+}; (d) Br_2^{2-}; (e) O_2^-; (f) O_2^{2-}; (g) N_2^{2-}; (h) F_2^+

9.112. Which of the following molecular ions have electrons in π antibonding orbitals? (a) O_2^-; (b) O_2^{2-}; (c) N_2^{2-}; (d) F_2^+; (e) N_2^+; (f) O_2^+; (g) C_2^{2+}; (h) Br_2^{2+}

9.113. The odd-electron molecule ClO affects the atmospheric chemistry of chlorofluorocarbons as illustrated by the reaction (where the * indicates an excited-state oxygen atom):

$$CF_2Cl_2 + O^* \rightarrow ClO + CF_2Cl$$

Draw a molecular orbital diagram for ClO. Is the odd electron in a bonding or antibonding orbital?

9.114. The elusive molecule boron monoxide (BO) can be stabilized by bonding to platinum. Draw a molecular orbital diagram for BO. Is the odd electron in a bonding or antibonding orbital?

9.115. For which of the following diatomic molecules does the bond order increase with the gain of two electrons, forming the corresponding anion with a 2− charge?

a. $B_2 + 2\,e^- \rightarrow B_2^{2-}$
b. $C_2 + 2\,e^- \rightarrow C_2^{2-}$
c. $N_2 + 2\,e^- \rightarrow N_2^{2-}$
d. $O_2 + 2\,e^- \rightarrow O_2^{2-}$

9.116. For which of the following diatomic molecules does the bond order increase with the loss of two electrons, forming the corresponding cation with a 2+ charge?
a. $B_2 \rightarrow B_2^{2+} + 2\,e^-$
b. $C_2 \rightarrow C_2^{2+} + 2\,e^-$
c. $N_2 \rightarrow N_2^{2+} + 2\,e^-$
d. $O_2 \rightarrow O_2^{2+} + 2\,e^-$

9.117. Do the 1+ cations of homonuclear diatomic molecules of the second-row elements always have shorter bond lengths than the corresponding neutral molecules?

9.118. Do any of the anions of the homonuclear diatomic molecules formed by B, C, N, O, and F have shorter bond lengths than those of the corresponding neutral molecules? Consider only the anions with 1− or 2− charge.

9.119. Thin films of doped diamond hold promise as semiconductor materials. Trace amounts of nitrogen impart a yellow color to otherwise colorless pure diamonds.
a. Are nitrogen-doped diamonds examples of semiconductors that are p-type or n-type?
b. Draw a picture of the band structure of diamond to indicate the difference between pure diamond and N-doped (nitrogen-doped) diamond.
*c. N-doped diamonds absorb violet light at about 425 nm. What is the magnitude of E_g that corresponds to this wavelength?

9.120. **Hope Diamond** Trace amounts of boron give diamonds (including the Smithsonian's Hope Diamond) a blue color (Figure P9.120).

FIGURE P9.120

a. Are boron-doped diamonds examples of semiconductors that are p-type or n-type?
b. Draw a picture of the band structure of diamond to indicate the difference between pure diamond and B-doped diamond.
*c. What is the band gap if a blue diamond absorbs red-orange light with a wavelength of 675 nm?

Additional Problems

9.121. In 1999 the ClO^+ ion, a potential contributor to stratospheric ozone depletion, was isolated in the laboratory.
a. Draw the Lewis structure for ClO^+.
b. Based on MO theory, what is the order of the Cl—O bond in ClO^+.

9.122. **Arsenic-Based DNA?** The waters of Mono Lake in the eastern Sierra Nevada of California (Figure P9.122) are rich in arsenate ion (AsO_4^{3-}). Some biochemists have proposed that microorganisms in this environment may incorporate arsenate into their DNA in place of the phosphate ion (PO_4^{3-}). What is the molecular geometry of the arsenate ion and the angle between its bonds?

FIGURE P9.122

9.123. Consider the molecular structure of the amino acid glycine shown in Figure P9.123. What is the angle formed by the N—C—C bonds in this structure? What are the O—C═O and C—O—H bond angles?

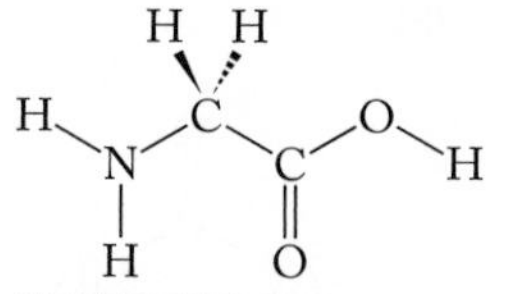

FIGURE P9.123

*9.124. Thermally unstable compounds can sometimes be synthesized using matrix isolation methods in which the compounds are isolated in a nonreactive medium such as frozen argon. The reaction of boron with carbon monoxide produces compounds with the skeletal structures B—B—C—O and O—C—B—B—C—O.
a. For each of these compounds, draw the Lewis structure that minimizes formal charges.
b. Do any of your structures contain atoms with incomplete octets?
c. Predict the molecular geometries around the central boron atoms in BBCO and OCBBCO.
d. Are either of these compounds nonpolar?

***9.125.** The products of the reaction between boron and NO can be trapped in solid argon matrices. Among the products is BNO. Draw the Lewis structure for BNO, including any resonance forms. Assign formal charges and predict which structure provides the best description of the bonding in this molecule. Do any of your structures contain atoms without complete octets? Predict the molecular geometry of BNO.

9.126. **Compounds May Help Prevent Cancer** Broccoli, cabbage, and kale contain compounds that break down in the human body to form isothiocyanates, whose presence may reduce the risk of certain types of cancer. The simplest isothiocyanate is methyl isothiocyanate (CH_3NCS). Draw the Lewis structure for CH_3NCS, including all resonance forms. Assign formal charges and determine which structure is likely to contribute the most to bonding. Predict the molecular geometry of the molecule at both carbon atoms.

9.127. **Toxic to Insects and People** Methyl thiocyanate (CH_3SCN) is used as an agricultural pesticide and fumigant. It is slightly water soluble and is readily absorbed through the skin; it is highly toxic if ingested. Its toxicity stems in part from its metabolism to cyanide ion. Draw three resonance structures for methyl thiocyanate. Assign formal charges and predict which structure would be the most stable. Predict the molecular geometry of the molecule at both carbon atoms.

9.128. **Skunks** The pungent smell of skunk spray is detected by receptors in the nose when the skunk secretes butanethiol, $CH_3(CH_2)_3SH$. Draw a skeletal structure for butanethiol.

9.129. **Grapefruit** Not all sulfur-containing compounds have unpleasant aromas. Figure P9.129 shows the structure of the compound primarily responsible for the aroma of grapefruit. Identify the chiral carbon in this structure.

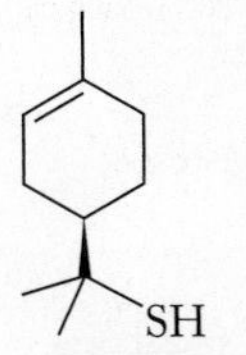

SH

FIGURE P9.129

9.130. Borazine ($B_3N_3H_6$), a cyclic compound with alternating B and N atoms in the ring, is isoelectronic with benzene (C_6H_6). Are there delocalized π electrons in borazine?

9.131. Unlike O_2, sulfur monoxide (SO) is highly unstable, decomposing to a mixture of S_2O and O_2 in less than 1 second. Using the O_{2s}, O_{2p}, S_{3s}, and S_{3p} atomic orbitals, construct an approximate molecular orbital diagram for SO. Is SO diamagnetic or paramagnetic?

9.132. Ozone (O_3) has a dipole moment of 0.53 D. How can a molecule made up of only one element have a dipole moment?

***9.133.** A group of Finnish scientists announced the discovery of the first argon compound, HArF, in 2000. Assuming one of argon's filled $3p$ orbitals is involved in bond formation (see Figure 9.56):

a. What is the shape of the HArF molecule?
b. What is the bond order of the H—Ar and Ar—F bonds?
c. Is HArF a polar substance?

9.134. Which of the unstable nitrogen oxides—N_2O_2, N_2O_5, and N_2O_3—are polar molecules? (N_2O_2 and N_2O_3 have N—N bonds; N_2O_5 does not.)

9.135. Using an appropriate molecular orbital diagram, show that the bond order in the disulfide anion (S_2^{2-}) is equal to 1. Is S_2^{2-} diamagnetic or paramagnetic?

9.136. **NO in Biology** Nitrogen monoxide (NO) acts as a signaling molecule in biological processes. The products from the reaction of NO with O_2 under biological conditions have not been confirmed, but N_2O_4 is believed to be one product. Two skeletal structures for N_2O_4 are shown in Figure P9.136. Complete the Lewis structures (including resonance forms) and determine the geometry at both nitrogen atoms in each case.

O—N—O—O—N—O $\qquad$ $O_2N—NO_2$

FIGURE P9.136

9.137. **Superoxide and Peroxide Ions** Some of the oxygen we breathe ends up as superoxide ion (O_2^-) and peroxide ion (O_2^{2-}). Cells use two classes of enzymes—superoxide dismutases and catalases—to degrade these reactive molecular ions. Use molecular orbital diagrams to determine the bond order of the O_2^{2-} and O_2^- ions. Are these bond order values consistent with those predicted from Lewis structures?

9.138. Elemental sulfur has several allotropic forms, including cyclic S_8 molecules. What is the orbital hybridization of sulfur atoms in this allotrope? The bond angles are about 108°.

***9.139.** The bond angle in H_2O is 104.5°, whereas the bond angles in H_2S, H_2Se, and H_2Te are very close to 90°. Which bonding theory—VSEPR, valence bond without invoking hybrid orbitals, or valence bond with hybrid orbitals—would you apply to describe the geometry in H_2S, H_2Se, and H_2Te? Why?

9.140. **Garlic** Garlic contains the molecule alliin (Figure P9.140). When garlic is crushed or chopped, a reaction occurs that converts alliin into the molecule allicin, which is primarily responsible for the aroma we associate with garlic.

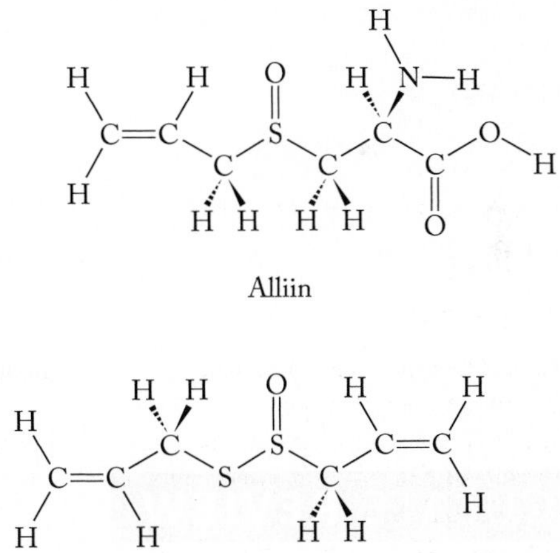

Alliin

Allicin

FIGURE P9.140

a. Describe the molecular geometry about the sulfur atoms in both compounds.
b. Do any of the sulfur atoms in allicin have the same geometry as the sulfur atoms in the volatile sulfur compounds that cause bad breath (i.e., H_2S, $CH_3—SH$, and $CH_3—S—CH_3$)?

10

Intermolecular Forces

The Uniqueness of Water

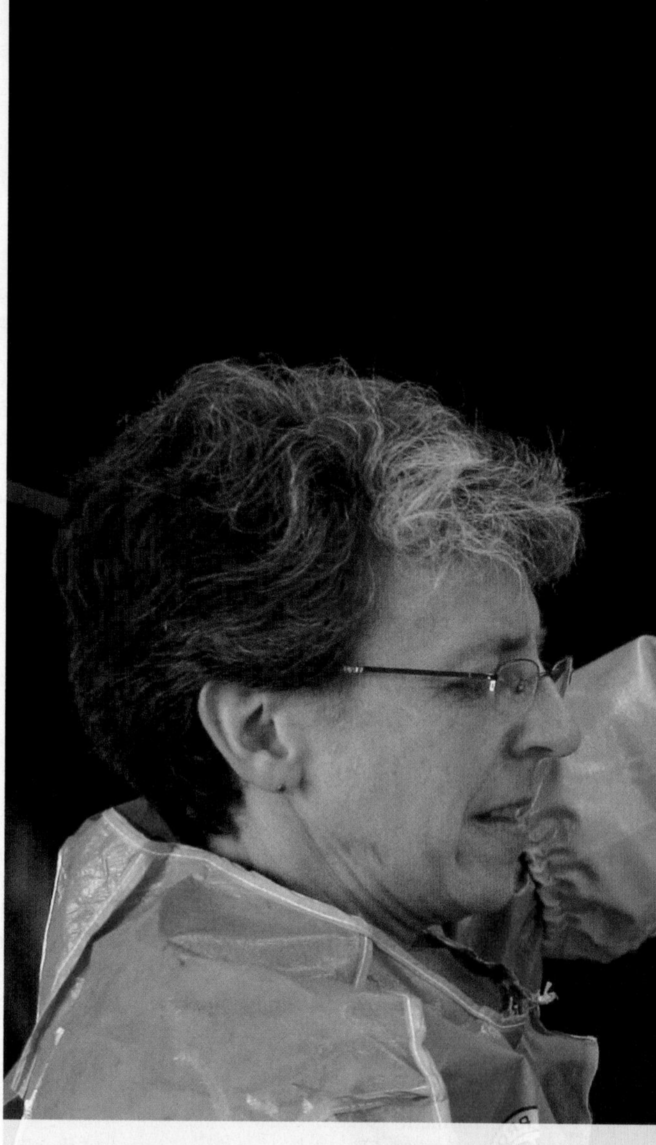

CLEANING UP AN OIL SPILL Volunteers use dishwashing detergent to clean the oiled feathers of this pelican.

PARTICULATE **REVIEW**

Polar Bonds versus Polar Molecules

In Chapter 10 we explore the connections between molecular polarity and attractive forces between molecules. Here are representations of four molecules: carbon dioxide, oxygen, water, and ozone.

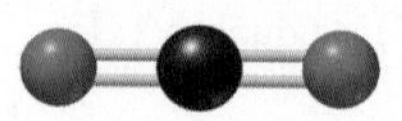   

- Label the polar covalent bonds in these molecules with partial charges ($\delta+$ and $\delta-$).
- Which molecule is nonpolar despite containing polar bonds?
- Which molecule is polar and contains polar bonds?
- Which molecule is polar even though it contains nonpolar bonds?

  (Review Section 9.3 if you need help.)

(Answers to Particulate Review questions are in the back of the book.)

PARTICULATE **PREVIEW**

One Formula, Two Structures

The physical and chemical properties of a compound depend on its structure. Here are ball-and-stick models of two compounds that have the same molecular formula. One compound is a liquid at room temperature, whereas the other is a gas. As you read this chapter, look for ideas that will help you answer these questions:

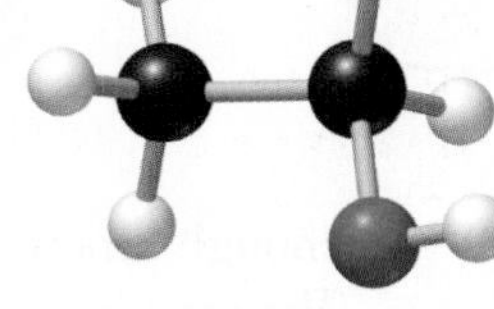

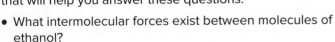

Ethanol

Dimethyl ether

- What intermolecular forces exist between molecules of ethanol?
- What intermolecular forces exist between molecules of dimethyl ether?

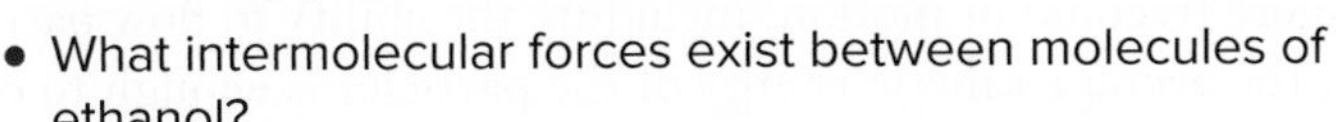

- Compare the relative strength of these intermolecular forces to determine which substance is a liquid at room temperature and which is a gas.

Learning Outcomes

LO1 Describe the types of intermolecular forces and explain how they and their strengths are related to the sizes of atoms and to the sizes, shapes, and dipole moments of molecules
Sample Exercise 10.1

LO2 Explain the effect of intermolecular forces on the physical properties of molecular substances, including their vapor pressures and boiling points
Sample Exercises 10.1, 10.2

LO3 Use the Clausius–Clapeyron equation to relate the vapor pressure of a liquid to its enthalpy of vaporization and temperature
Sample Exercise 10.3

LO4 Identify the regions of a phase diagram and explain the effect of temperature and pressure on phase stability and phase changes
Sample Exercise 10.4

LO5 Relate the unique physical properties of water to the hydrogen bonds between its molecules

LO6 Use intermolecular forces to explain and predict the relative solubilities of compounds in water and other solvents
Sample Exercises 10.5, 10.6

LO7 Explain how temperature, pressure, and intermolecular forces influence the solubilities of gases in liquids and use Henry's law to calculate those solubilities
Sample Exercise 10.7

CONNECTION The flow of energy accompanying the phase changes of water (solid ice $\rightleftharpoons$ liquid water $\rightleftharpoons$ water vapor) was observed in Chapter 6 in terms of the changes in kinetic and potential energies of the molecules.

10.1 Intramolecular Forces versus Intermolecular Forces

In Chapters 8 and 9 we examined the nature of the chemical bonds in molecules and showed how attractive and repulsive electrical forces between pairs of electrons determine molecular shape. We refer to these interactions *within* a molecule as *intramolecular forces*. In this chapter we begin the study of the forces that act *between* molecules and *between* molecules and ions. These *intermolecular forces* are also electrostatic in nature, but they are weaker than chemical bonds. Nevertheless, intermolecular forces strongly influence the physical properties of all substances, including their physical state (**Figure 10.1**) in the world around us.

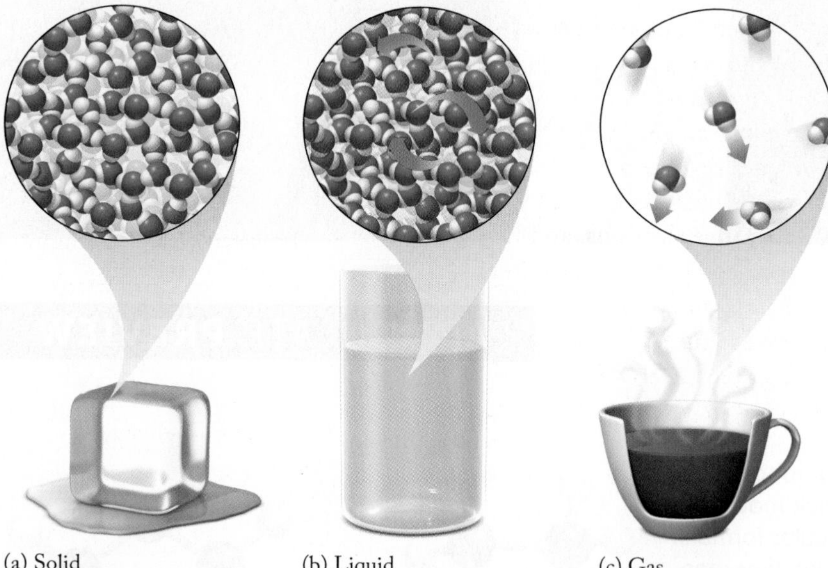

FIGURE 10.1 (a) The molecules of H_2O in solid ice are locked in place by the strength of intermolecular attractions. (b) In liquid water, molecules of H_2O have more energy and flow past one another. (c) In water vapor above a steaming cup of coffee, the molecules have enough energy to overcome nearly all intermolecular attractions and move freely throughout the space they occupy.

Substances that are solids, for example, consist of particles that are so strongly attracted to one another that they have limited ability to overcome those forces of attraction. Consequently, these particles have the same nearest neighbors over time, and their kinetic energy is principally vibrational energy. The particles in liquids have enough kinetic energy to overcome some of the attractive forces between them. They experience more freedom of motion, including the ability to flow past one another. In gases, the average kinetic energy of the particles is enough to overcome essentially all the attractive forces between them, imparting nearly complete freedom of motion to these widely separated particles.

The stronger the attractive forces among the particles in a solid, the greater the amount of energy needed to overcome those forces to cause melting or sublimation. Thus, a substance made of particles that interact relatively strongly has high melting and boiling points, which means the substance is likely to be a solid at room temperature and normal atmospheric pressure. Under the same conditions, a substance with somewhat weaker particle–particle interactions has a lower melting point and is more likely to be a liquid. A substance with very weak particle–particle interactions has even lower melting and boiling points and is more likely to be a gas. We begin our study of intermolecular forces by examining those that are experienced by all particles from individual atoms to small polar and nonpolar molecules to biological and synthetic molecules composed of thousands or even millions of atoms bonded together.

induced dipole the separation of charge produced in an atom or molecule by a momentary uneven distribution of electrons.

dispersion force (also called **London force**) an intermolecular force between molecules caused by the presence of temporary dipoles in the molecules.

10.2 Dispersion Forces

The macroscopic properties of substances such as their melting points and boiling points are related to microscopic interactions between the particles that make up those substances. Let's see how this happens starting with the simplest of particles: the single atoms that make up the noble gases. **Table 10.1** lists their atomic numbers (Z) and their boiling points. Notice that the boiling points increase as the atomic numbers increase. To understand why this correlation exists, think about what happens at the particle level when a liquid vaporizes. As we discussed above, the particles in liquids (and solids) are in direct contact with each other. When a liquid vaporizes, however, the contacts are broken: the gas-phase particles become essentially independent. Separating liquid-phase particles that are attracted to each other requires energy, and the stronger the particles' attractions for each other, the greater the amount of energy needed to separate them. In turn, the greater the energy required to separate these particles, the higher the boiling points.

TABLE 10.1 Boiling Points of Noble Gases

Noble Gas	Atomic View	Z	Boiling Point (K)
He		2	4
Ne		10	27
Ar		18	87
Kr		36	120
Xe		54	165
Rn		86	211

Why do single atoms interact more strongly as their atomic number increases? For that matter, *how* do atoms that are not bonded to each other interact? German-American physicist Fritz London (1900–1954) proposed one explanation for these interactions in 1930. It was based on the notion that when atoms approach each other (**Figure 10.2a**), they interact in ways that are like the electrostatic interactions involved in covalent bond formation. In other words, one atom's positive nucleus is attracted to the other atom's negative electrons, and vice versa, even as their electron clouds repel each other. These competing interactions can cause the electrons around each atom to be distributed unevenly, producing temporary **induced dipoles** of partial electrical charge (**Figure 10.2b**) that are attracted to regions of opposite partial charge on the adjacent atom. (This attraction is represented by the red dotted line in Figure 10.2b.) In Chapter 8 we used different colors to represent the partial electrical charges created by uneven sharing of bonding pairs of electrons. In this chapter we use those same colors (**Figure 10.2c**) to show partial electrical charges caused by uneven distributions of electrons in neutral atoms and over entire molecules.

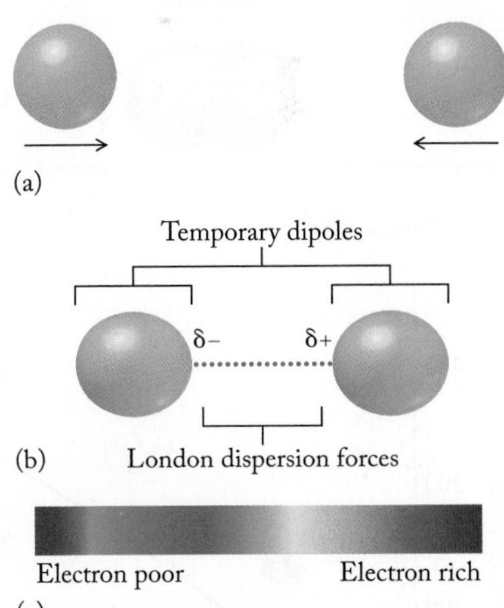

FIGURE 10.2 (a) Two atoms, each with a symmetrical distribution of electrons, approach each other and (b) create two temporary dipoles as their nuclei and electron clouds interact. (c) The strengths of temporary dipoles are shown using the same color scale as in Chapter 8 to represent the strengths of permanent dipoles in bonds.

The presence of temporary dipoles in atoms and molecules creates a way for them to interact with other atoms and molecules that are called **dispersion forces** (represented by ·····), also referred to as **London forces** in honor of Fritz London's pioneering work. The strengths of the interactions increase as the numbers of electrons in atoms and molecules increase. Because all atoms and molecules have electrons, all atoms and molecules experience London forces to some degree. The larger the cloud of electrons surrounding a nucleus in an atom or multiple nuclei in a molecule, the more likely those electrons are to be distributed unevenly or *polarized*. Electrons in larger atoms are held less tightly by the nucleus because of

CONNECTION The concept of screening of outer electrons by inner electrons was presented in Section 7.10, where we discussed trends in the sizes of atoms and ions.

their greater average distance from the nucleus as well as the screening of the nuclear charge by electrons in lower-energy orbitals. Consequently, they are polarized more easily than electrons in smaller atoms or molecules. Greater **polarizability** leads to stronger temporary dipoles and stronger intermolecular interactions, so dispersion forces become stronger as atoms and molecules become larger. This trend in polarizability accounts for the correlation between the boiling points and atomic numbers of the noble gases.

CONNECTION In Section 6.9 we considered the structure of straight-chain hydrocarbons.

CONCEPT TEST

Rank the following atoms in order of increasing polarizability: argon, hydrogen, krypton, and neon.

(Answers to Concept Tests are in the back of the book.)

TABLE 10.2 Boiling Points of the Halogens

Halogen	Molecular View	Molar Mass (g/mol)	Boiling Point (K)
F_2		38	85
Cl_2		71	239
Br_2		160	332
I_2		254	457
At_2		420	610

Polarizability also explains why the boiling points of the halogens increase as their molar masses increase (**Table 10.2**). These elements exist as diatomic molecules in which equal sharing of the bonding pairs of electrons by identical atoms means that the molecules have no permanent dipoles. In this case, the molar masses of the halogens represent a measure of particle size, so boiling point increases as particle size increases. A similar trend is also observed in the boiling points of a series of straight-chain hydrocarbons (**Figure 10.3**). London's explanation of these trends was also the same—namely, larger clouds of increasing numbers of electrons per molecule are more polarizable. Greater polarizability means they are more likely to form temporary dipoles that attract molecules to each other in the liquid phase and inhibit their vaporization.

CONCEPT TEST

Use intermolecular forces to explain why CF_4 is a gas at room temperature but CCl_4 is a liquid.

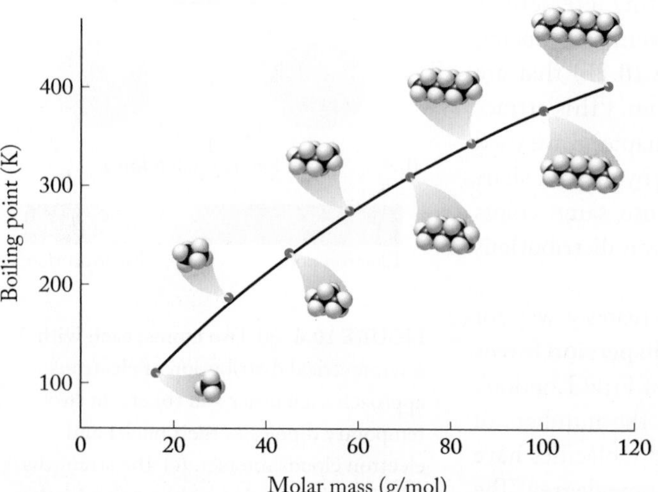

FIGURE 10.3 The boiling points of straight-chain alkanes increase as masses increase and the magnitude of the dispersion forces between molecules increases.

The Importance of Shape

Molecular shape, as well as size, plays a role in determining the strength of dispersion forces. Consider the molecular structures and boiling points of the three hydrocarbons in **Figure 10.4**. These compounds all have the same molecular formula (C_5H_{12}), so they all have the same molar mass (72.15 g/mol) and the same number of electrons. However, the bonds, shapes, and boiling points of the molecules are different. All three compounds are nonpolar, so none has a permanent dipole. Therefore, the only intermolecular interactions are dispersion forces due to temporary dipoles. Molecules of pentane are relatively long and straight—think of their shape as being like a piece of chalk. Pentane molecules can interact with one another over a relatively large surface area and therefore have more possibilities for dispersion forces to hold them together. In contrast, molecules of

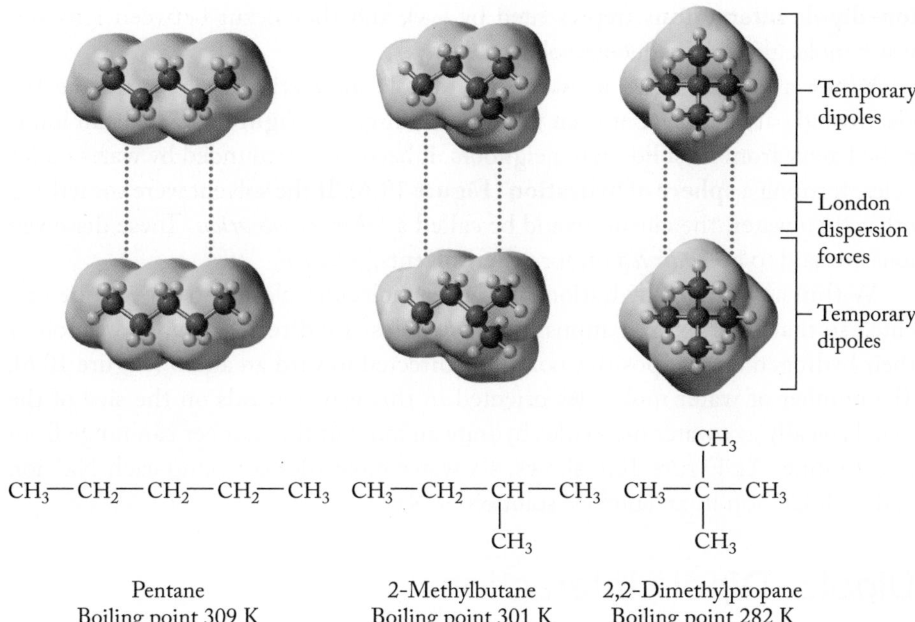

FIGURE 10.4 Three molecular shapes are possible for C_5H_{12}. The more spread out the atoms are in these molecules, the stronger the dispersion forces between them, and the higher the boiling point of the compound.

CHEMTOUR
Intermolecular Forces

2,2-dimethylpropane are almost spherical because of the considerable branching of the carbon atom chains. These molecules have less surface area to interact with adjacent molecules; less interaction means weaker dispersion forces, resulting in the lowest boiling point in this set of compounds. The remaining molecule, 2-methylbutane, is neither as straight as pentane nor as spherical as 2,2-dimethylpropane—think of it as shaped like a short, forked stick. 2-Methylbutane boils at a temperature lower than pentane but higher than 2,2-dimethylpropane. In general, molecules with more branching in their structures have lower boiling points.

Dispersion forces between small atoms or molecules are weak, but large molecules have many of these interactions that can sometimes dominate a much smaller number of strong interactions in a system, as we see in the discussion of polarity and solubility in Section 10.7.

10.3 Interactions Involving Polar Molecules

In Section 8.3 we saw how the unequal distributions of bonding pairs of electrons result in bond dipoles—that is, partial negative charges on some bonded atoms and partial positive charges on others. When bond dipoles are arranged asymmetrically within molecules, the molecules themselves have permanent dipoles. Molecules with permanent dipoles can interact with each other and with the ions in ionic compounds. The strengths of these interactions are much weaker than the strengths of ionic or covalent bonds, but they are strong enough to provide additional interactions beyond the dispersion forces that all molecules experience. For example, interactions between polar liquids and ionic solids play a key role in salts dissolving in water.

CONNECTION Bond energies of covalent bonds are discussed in Section 8.7.

CONNECTION In Chapter 4 we learned that some, but not all, ionic compounds are soluble in water.

Ion–Dipole Interactions

One reason why ionic compounds such as NaCl dissolve in water is that the sodium and chloride ions interact with the permanent dipoles of the polar water molecules, which carry both positive and negative partial charges. These attractions are called

polarizability the relative ease with which the electron clouds surrounding the nuclei of individual atoms or molecules can be distorted, thereby inducing temporary dipoles.

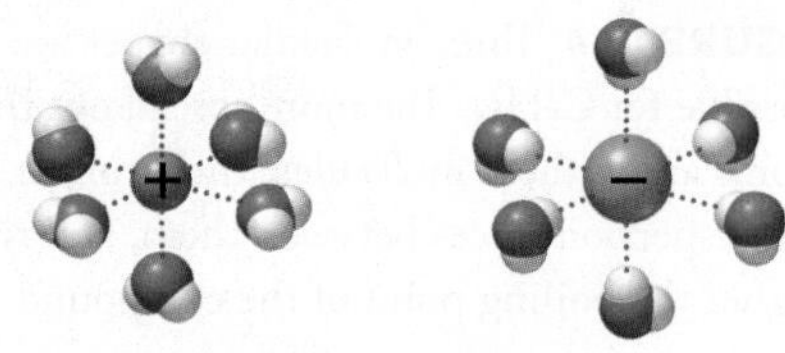

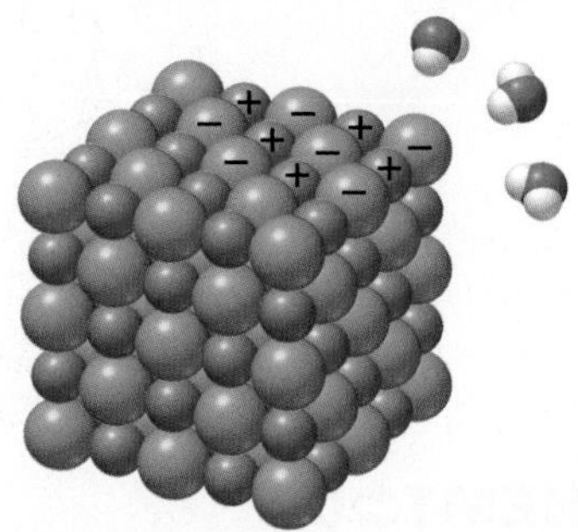

FIGURE 10.5 The hydrogen atoms (positive poles) of H_2O molecules are attracted to the Cl^- ions of NaCl; the O atoms (negative poles) are attracted to the Na^+ ions. Multiple ion–dipole interactions help overcome the attractive forces holding ions at the surface of the solid NaCl, causing it to dissolve.

CONNECTION In previous chapters we have indicated the presence of a hydration sphere around an ion in aqueous solution by writing (*aq*) after its symbol or formula.

ion–dipole interactions (represented by ·····), and they occur between ions and water molecules in all aqueous solutions.

When a salt dissolves in water, ion–dipole interactions help overcome the electrostatic attractions between the ions themselves (**Figure 10.5**). As an ion is pulled away from its solid-state neighbors, it becomes surrounded by water molecules, forming a **sphere of hydration** (**Figure 10.6**). If the solvent were something other than water, the cluster would be called a *sphere of solvation*. These dissolved ions are said to be *hydrated* or, for other solvents, *solvated*.

Within a sphere of hydration, the water molecules closest to the ion are oriented so that their oxygen atoms (negative poles) are directed toward a cation or their hydrogen atoms (positive poles) are directed toward an anion (Figure 10.6). The number of water molecules oriented in this way depends on the size of the ion. Typically, six water molecules hydrate an ion, but the number can range from four to nine. As Figure 10.6 shows, six water molecules surround each Na^+ ion and each Cl^- ion in an aqueous solution of NaCl.

Dipole–Dipole Interactions

The bulk water molecules further from the ions in Figure 10.6 are more randomly oriented than those in the inner hydration sphere. These molecules experience another type of intermolecular force, called a **dipole–dipole interaction** (represented by ·····), that is experienced by all polar molecules. Regions of partial positive charge and partial negative charge on neighboring polar molecules are attracted to each other. For example, the partial negative charge on the O atom of one water molecule is mutually attracted to the partial positive charges on the H atoms of up to two other water molecules. Dipole–dipole interactions are not as strong as ion–dipole interactions because dipole–dipole interactions involve only partial charges, caused by unequal sharing of the electrons in polar covalent bonds. In contrast, an ion involved in an ion–dipole interaction has lost or gained one or more electrons, so it has a full positive or negative charge (at a minimum).

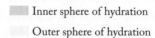

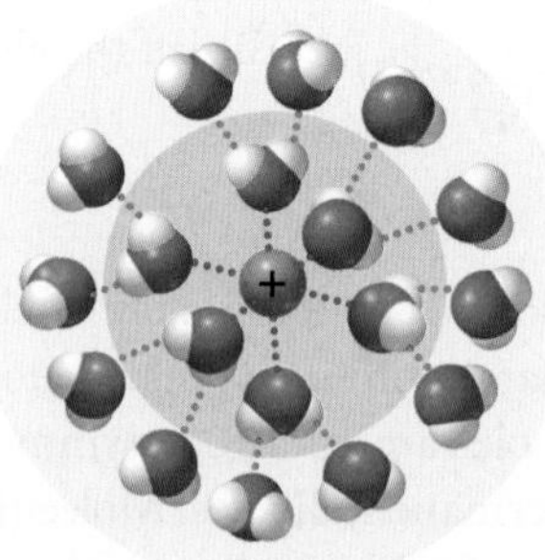

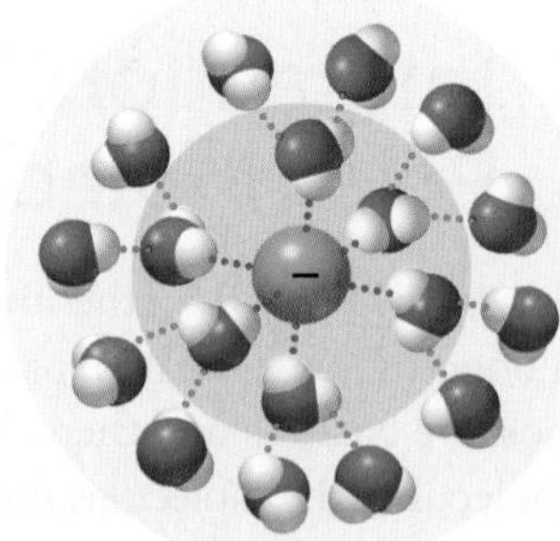

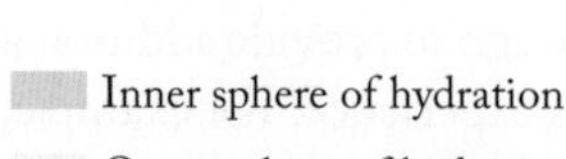

FIGURE 10.6 Each hydrated Na^+ ion and hydrated Cl^- ion is surrounded by six water molecules that create an inner hydration sphere. Water molecules in an outer hydration sphere surround the inner sphere. The outer sphere is the result of dipole–dipole interactions between the rest of the water (known as bulk water) and the water molecules of the inner sphere. Beyond the outer hydration sphere, dipole–dipole interactions also occur between outer-sphere water molecules and molecules in bulk water.

ion–dipole interaction an attractive force between an ion and a molecule that has a permanent dipole moment.

sphere of hydration the cluster of water molecules surrounding an ion in aqueous solution; the general term applied to such a cluster forming in any solvent is *sphere of solvation*.

dipole–dipole interaction an attractive force between polar molecules.

dipole–induced dipole interaction an attraction between a polar molecule and the oppositely charged pole it temporarily induces in another molecule.

To understand how dipole–dipole forces add to the dispersion interactions that exist among all molecules, let's consider the properties of 2-methylpropane and acetone (**Figure 10.7**). These two compounds have the same molar mass (58 g/mol) and have molecular shapes that are not all that different despite the different hybridizations of their central carbon atoms

We might expect these molecules to experience similar dispersion forces, yet the boiling points of 2-methylpropane and acetone are quite different: 261 K and 329 K, respectively. Why is the boiling point of acetone nearly 70 K higher? The answer lies in the ketone functional group in its molecular structure. The dipole created by the highly electronegative oxygen atom and the less electronegative carbon atom in the C═O bond gives acetone an overall dipole moment of 2.88 D and sets up dipole–dipole interactions between its molecules that nonpolar molecules of 2-methylpropane do not experience. Boiling a liquid requires that nearly all the intermolecular interactions between molecules in the liquid be overcome as these molecules are separated from one another entering the gas phase. Thus, polar substances with stronger interactions tend to have higher boiling points than substances with similar molar masses but weaker interactions.

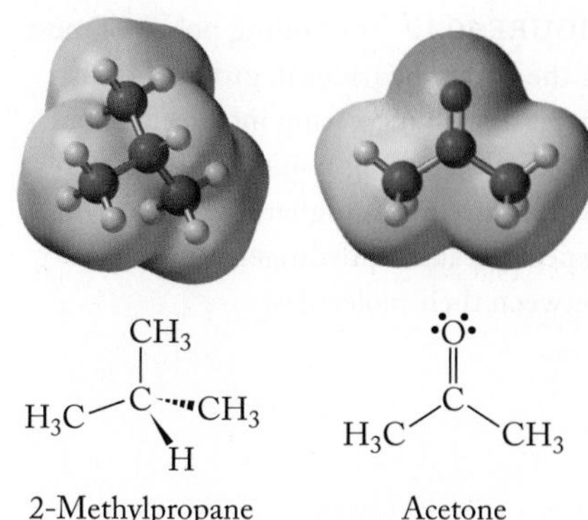

FIGURE 10.7 Molecular structures and boiling points of 2-methylpropane and acetone.

CONNECTION In Section 9.3 we learned that permanent dipole moments are experimentally measured values, expressed in units of debyes, that define the polarity of molecules.

CONCEPT TEST

Dimethyl ether (CH_3OCH_3) and acetone [$CH_3C(O)CH_3$] (**Figure 10.8**), have similar formulas and molar masses, but their dipole moments are quite different: 1.30 D for dimethyl ether and 2.88 D for acetone. Predict which compound has the higher boiling point and explain why.

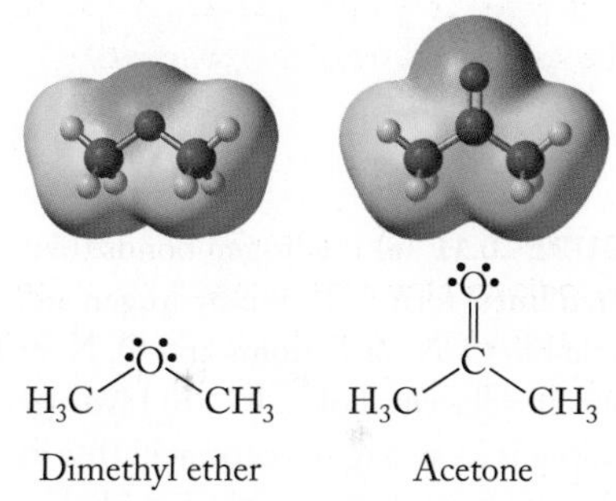

FIGURE 10.8 Molecular structures of dimethyl ether and acetone.

A molecule with a permanent dipole can create a **dipole–induced dipole interaction** (represented by ·····) when inducing a temporary dipole in a nonpolar molecule by perturbing the electron distribution in the nonpolar molecule (**Figure 10.9**). The intermolecular interaction in this case is weaker than the dipole–dipole force between two polar molecules and is of the same order of magnitude as the dispersion forces between temporary dipoles in nonpolar molecules. The magnitude of an induced dipole depends on the polarizability of the electrons in a molecule, ion, or atom.

Hydrogen Bonds

In the preceding section we introduced dipole–dipole interactions using the attraction between H and O atoms on adjoining water molecules. These dipole–dipole interactions are particularly strong, as are those among other molecules that contain O—H bonds and among molecules that contain N—H or F—H bonds. This unusual strength is illustrated in the boiling point data plotted in **Figure 10.10**. The compounds whose boiling points are plotted are composed of molecules that each have an atom of a group 14, 15, 16, or 17 element at its center that is bonded only to enough hydrogen atoms to complete its octet. Most of the data points follow a familiar trend: boiling points increase with increasing molar masses, as we saw with the halogens in Table 10.2 and the nonpolar hydrocarbons in Figure 10.3. All the data points for the group 14 compounds fit this trend. However, the boiling points of the other three groups' compounds with the lowest molecular mass—NH_3, H_2O, and HF—are unusually high compared with the others in their series. To understand why, we need to focus on the polar covalent bonds formed between

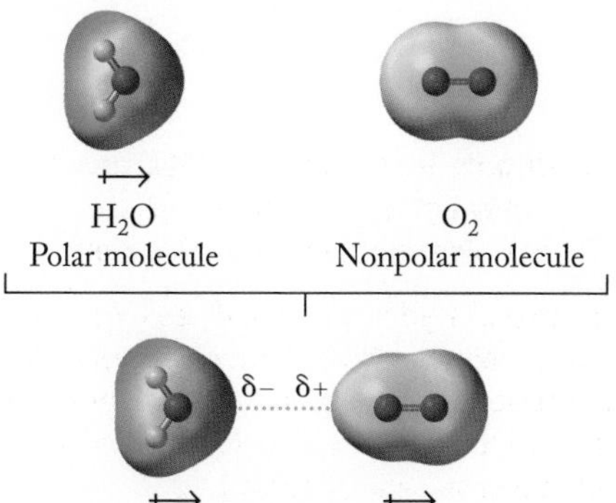

FIGURE 10.9 The approach of a polar water molecule induces a temporary dipole in an initially nonpolar oxygen molecule by distorting the distribution of the oxygen's electrons.

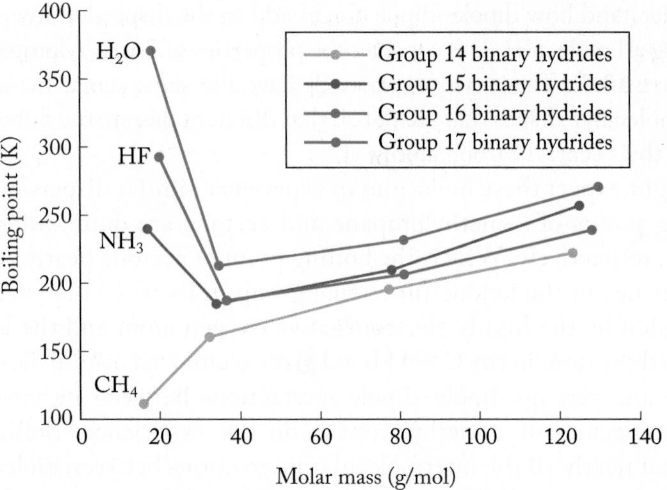

FIGURE 10.10 The boiling points of most of the binary hydrides in groups 14–17 increase with increasing molar mass, but not all. The boiling points of H_2O, NH_3, and HF are much higher than we might expect because of hydrogen bonding between their molecules.

H atoms and N, O, and F atoms. The H atoms share just two electrons, and when they are shared with a highly electronegative atom of N, O, or F, the electron density on the surface of the H atom is greatly reduced, leaving the nucleus of the hydrogen atom and its positive charge with very little electron density separating it from an approaching N, O, and F atom on a neighboring polar molecule. This allows the two atoms to get close together, which increases the strength of the interaction between their partial positive and negative charges.

These unusually strong dipole–dipole interaction have their own name: they are called **hydrogen bonds** (represented by ·····) (**Figure 10.11**). They are the

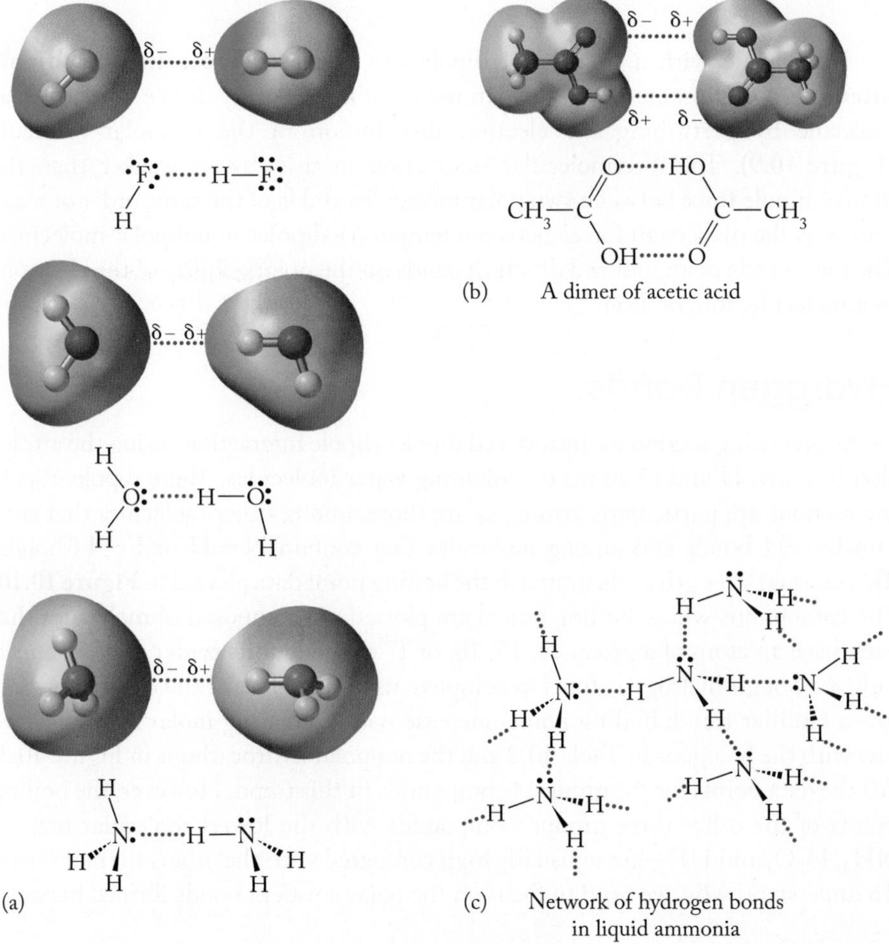

FIGURE 10.11 (a) Hydrogen bonds (blue dotted lines) form between hydrogen atoms bonded to O, N, or F atoms and O, N, or F atoms in adjacent molecules. (b) Hydrogen bonding is so strong in acetic acid that its molecules form dimers in the liquid phase. (c) Hydrogen bonds between nitrogen and hydrogen atoms on adjacent molecules form extensive three-dimensional networks in liquid ammonia.

strongest dipole–dipole interactions and can be nearly one-tenth the strength of some covalent bonds. The dipole–dipole interactions between water molecules in Figure 10.6 are hydrogen bonds, and they play a key role in defining the remarkable behavior of H_2O, which we explore in Section 10.6. **Table 10.3** summarizes the relative strengths of the intermolecular forces discussed in this chapter.

hydrogen bond the strongest dipole–dipole interaction. It occurs between a hydrogen atom bonded to a small, highly electronegative element (O, N, F) and an atom of oxygen or nitrogen in another polar molecule. Molecules of HF also form hydrogen bonds.

CONCEPT TEST

Use intermolecular forces to predict which compound has the higher boiling point in each of the following pairs: (a) CH_3Cl, CH_3Br; (b) CH_3CH_2OH, CH_3OH; (c) CH_3NH_2, $(CH_3)_2NH$.

TABLE 10.3 Relative Strengths of Intermolecular Forces[a] and Some Phenomena They Explain

Type of Force	Relative Strength	Phenomenon
Ion–dipole	↑	NaCl dissolves in water
Hydrogen bonding		Water expands when it freezes
Dipole–dipole		The boiling point of formaldehyde (dipole–dipole interactions; μ = 2.33 D) is 70°C higher than that of ethane (dispersion forces; μ = 0.00 D)
Dipole–induced dipole		O_2 dissolves in water
Dispersion		At 298 K: Cl_2 is a gas; Br_2 is a liquid; I_2 is a solid

[a]Between pairs of particles of similar combined masses.

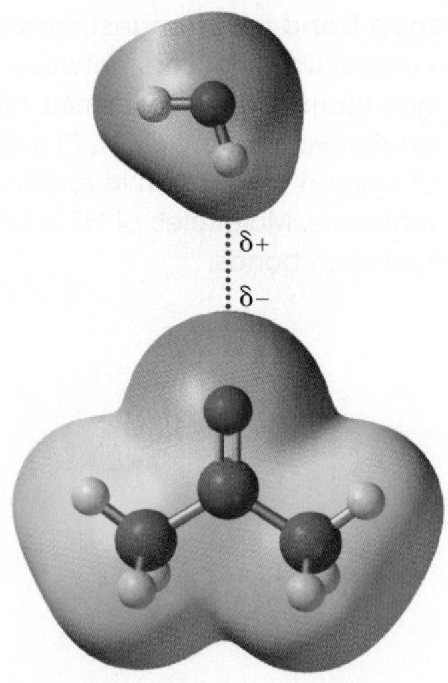

FIGURE 10.12 In solutions of acetone in water, hydrogen bonds form between the hydrogen atoms of water and the oxygen atoms of acetone.

CONCEPT TEST

Do you think attractive forces exist between ions and nonpolar molecules? How would you describe such interactions? Where would you place such intermolecular forces in Table 10.3?

Hydrogen bonds can also form between molecules of different substances, even when one of them has no H atoms bonded to N, O, or F atoms. For example, when acetone dissolves in water, hydrogen bonds form between the H atoms of water molecules and the O atoms of acetone molecules (**Figure 10.12**). These interactions happen even though the H atoms in acetone are bonded to C atoms and cannot form hydrogen bonds. The key to hydrogen bonding between acetone and water molecules is the negative partial charge of the O atoms in molecules of acetone, as indicated by the red color of the electrostatic potential map of acetone in Figure 10.12, and the blue-green color of the electron-deprived H atoms in molecules of H_2O.

Hydrogen bonds can also occur within the extended structures of many polymers and large biological molecules, such as proteins and DNA. As we have seen with small molecules, shape often determines behavior. Proteins are so large that their long chains of atoms tend to fold back and wrap around themselves, such that atoms may form hydrogen bonds with other atoms in an adjacent chain. The double strands of DNA (**Figure 10.13**) form a three-dimensional shape, called a double helix, in which pairs of the molecular building blocks of DNA, called nucleotides, form hydrogen bonds that keep the strands linked together. Pairs of two nucleotides named guanine (G) and cytosine (C) on adjacent DNA strands form three hydrogen bonds, whereas pairs of adenine (A) and thymine (T) form two hydrogen bonds. Because hydrogen bonds are weaker than covalent bonds, the DNA strands can be pulled apart during DNA replication (replication is discussed in greater detail in Section 20.7).

FIGURE 10.13 Hydrogen bonds (blue dotted lines) occur between hydrogen and nitrogen or oxygen in adjacent strands of DNA, contributing to its double-helix structure. The two detailed views of the DNA double helix contain the names of four of the building blocks of DNA: guanine, cytosine, adenine, and thymine. There are three hydrogen-bonding sites between pairs of guanine and cytosine but only two between adenine and thymine.

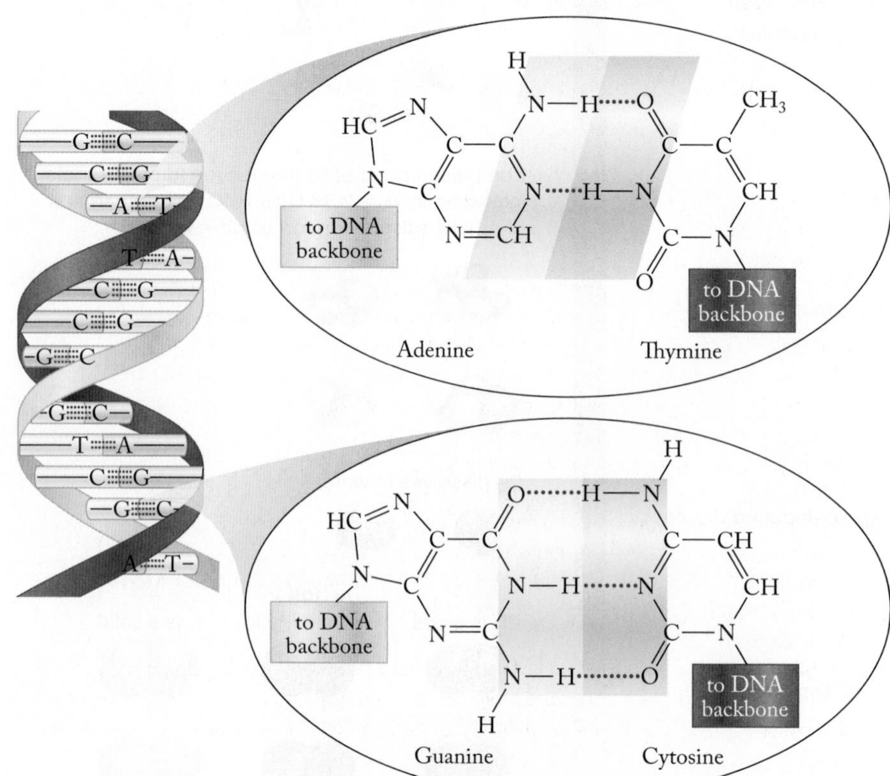

STEPWISE
ANIMATION
Interactions Involving Polar Molecules

SAMPLE EXERCISE 10.1 Identifying Intermolecular Forces LO1

Figure 10.14 shows the Lewis structures of five compounds: (a) sodium chloride, (b) methane, (c) dichloromethane, (d) methylamine, and (e) acetaldehyde. Identify the intermolecular forces that the ions or molecules of the compounds experience in a pure sample and when dissolved in water.

(a) $Na^+[:\ddot{Cl}:]^-$

(b) H—C—H with H above and below C

(c) :Cl—C—Cl: with H above and below C

(d) H—C—N—H with H on C above and below, H below N

(e) H—C—C—H with H above and below the first C, :O: double-bonded above the second C, H below the second C

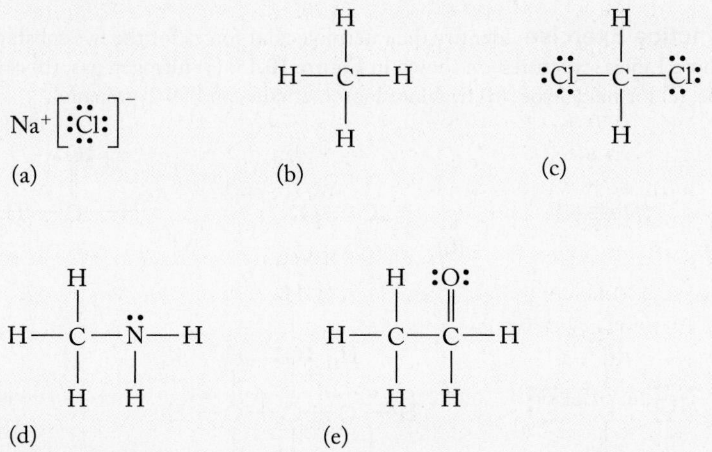

FIGURE 10.14 Lewis structures for Sample Exercise 10.1.

Collect, Organize, and Analyze We have the Lewis structures of five compounds and need to use them to identify the types of intermolecular forces (IMFs) the ions and molecules in those compounds experience as pure substances or when dissolved in water. The list of IMFs includes

- dispersion forces, which all individual atoms and molecules experience
- ion–dipole forces, which ions experience when they dissolve in water and other polar solvents
- dipole–dipole interactions, which molecules with permanent dipoles experience
- hydrogen bonds, which H atoms bonded to O, N, or F atoms form with O, N, or F atoms on other molecules
- dipole–induced dipole forces, which play a key role in the solubility of nonpolar molecules in polar solvents

Solve

a. The ions in solid NaCl are held together by ionic bonds, not IMFs, but when they dissolve in water they experience ion–dipole interactions in addition to ion–ion interactions.

b. Methane, like most hydrocarbons, is nonpolar, so its molecules experience only dispersion forces, but in an aqueous solution the permanent dipoles of water molecules induce temporary dipoles in methane molecules.

c. The Lewis structure of dichloromethane appears to be symmetrical, but we learned in Chapter 9 that its tetrahedral molecular shape means that the bond dipoles in the molecule are not arranged symmetrically in three-dimensional space; so, dichloromethane molecules have permanent dipoles. This means they experience dipole–dipole interactions among themselves and with polar water molecules.

d. Methylamine contains a methyl ($-CH_3$) group and a polar amine group due to the asymmetry produced by the lone pair of electrons on the nitrogen atom. The presence of N—H bonds also means that methylamine molecules form hydrogen bonds among themselves and with molecules of H_2O.

e. Acetaldehyde has a polar C=O group, but none of the H atoms in acetaldehyde are bonded to a N, O, or F atom. Therefore, it experiences dipole–dipole interaction but does not form hydrogen bonds *unless* it dissolves in water. When that happens the H atoms of water molecules form hydrogen bonds with O atoms in acetaldehyde molecules.

Think About It The key to recognizing the types of intermolecular forces that particles of a substance experience is to determine whether they are ions or polar or nonpolar molecules, and whether they are molecules that can form hydrogen bonds. In Chapter 2 we discussed how to distinguish between ionic and molecular compounds, and in Chapter 9 (and again in this exercise) we used the presence of bond dipoles and asymmetric molecular structures to identify polar molecules. The presence of H atoms bonded to N, O, or F atoms is the key to hydrogen bond formation.

Practice Exercise Identify the intermolecular forces for the five substances whose Lewis structures are shown in **Figure 10.15**: (a) nitrogen gas, (b) carbon monoxide, (c) formaldehyde, (d) trichlorofluoromethane, and (e) 2-propanol.

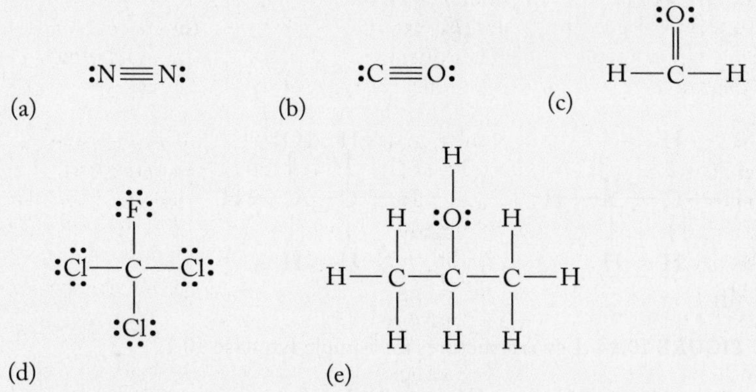

(a) :N≡N:

(b) :C≡O:

(c) H—C(=O:)—H

(d) CCl_3F: C bonded to :F:, :Cl:, :Cl:, :Cl:

(e) H—C(H)(H)—C(H)(:O:—H)—C(H)(H)—H

FIGURE 10.15 Lewis structures for Practice Exercise 10.1.

(Answers to Practice Exercises are in the back of the book.)

SAMPLE EXERCISE 10.2 Explaining Differences in Boiling Points **LO2**

Images of the molecular structures propane, dimethyl ether, and ethanol are shown in **Figure 10.16**. All three compounds have nearly the same molar mass and exactly the same number of electrons per molecule. However, the boiling points of propane, dimethyl ether, and ethanol are 231 K, 249 K, and 351 K, respectively. Explain these differences in boiling points.

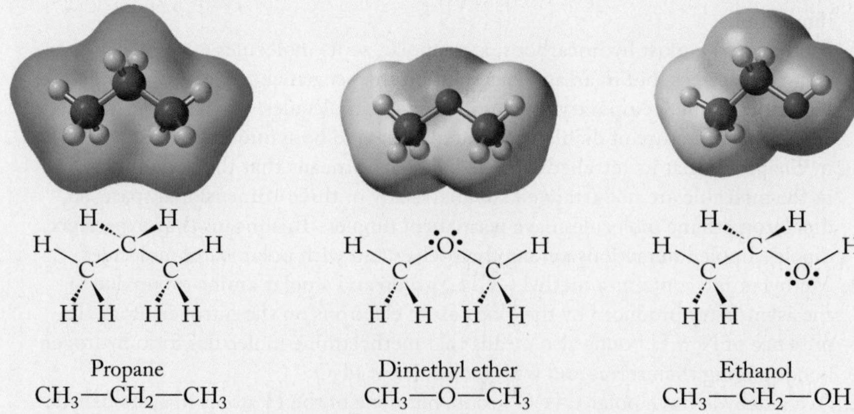

FIGURE 10.16 Molecular structures of propane, dimethyl ether, and ethanol.

Collect, Organize, and Analyze We are asked to explain the differences between the boiling points of three compounds with similar molar masses and the same number of electrons per molecule, which means they should experience similar dispersion forces. We are given their molecular structures and need to consider:

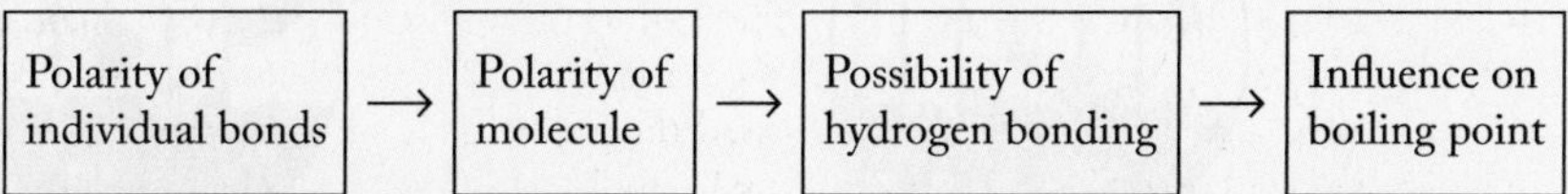

Propane is a hydrocarbon whose molecules contain only C—C bonds, which are not polar, and C—H bonds, which have little polarity. Therefore, the only intermolecular forces propane molecules are likely to experience are dispersion forces.

The other two molecules each contain an O atom bonded to atoms of less electronegative elements, so both molecules have bond dipoles. The shapes of the molecules show that the bond dipoles do not entirely offset each other, so both molecules have permanent dipoles (**Figure 10.17**). Their polarities mean that both experience dipole–dipole interactions. Ethanol molecules contain –OH groups, which means their principal dipole–dipole interactions are hydrogen bonds.

Solve Propane is essentially nonpolar, so its boiling point of 231 K is a measure of the strength of the dispersion forces that all three molecules experience because these forces must be overcome if molecules in the liquid state are to vaporize. Dipole–dipole interactions between molecules of dimethyl ether add a second attractive force between its molecules, which explains its slightly higher boiling point of 249 K. The much higher boiling point of ethanol (351 K) reflects the much stronger force of attraction produced by the hydrogen bonding between its molecules.

Think About It The only slightly higher boiling point of dimethyl ether over propane indicates that London dispersion is the primary intermolecular force for both compounds. The more than 100 K difference between the boiling points of dimethyl ether and ethanol attests to the strengths of the hydrogen bonds between ethanol molecules.

Practice Exercise 2-Propanol (molar mass 60.10 g/mol), the rubbing alcohol in your medicine cabinet, boils at 356 K. Ethylene glycol, used as automotive antifreeze, has nearly the same molar mass (62.07 g/mol) but boils at 470 K. Why do these substances (**Figure 10.18**) have such different boiling points?

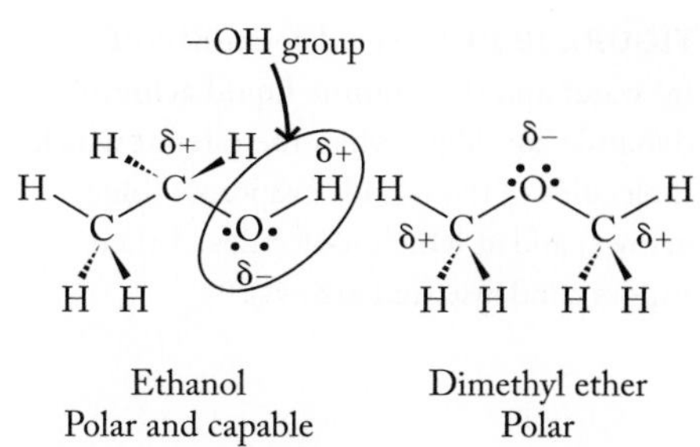

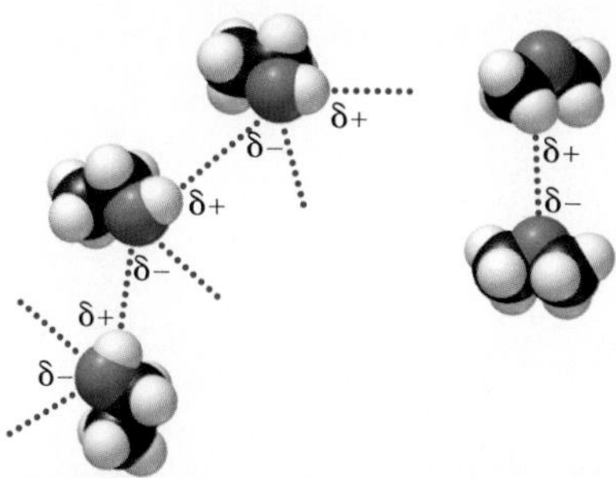

FIGURE 10.17 Hydrogen bonds between the –OH groups in ethanol molecules significantly add to the attraction between them. Molecules of dimethyl ether cannot form hydrogen bonds but do experience weaker dipole–dipole interactions.

10.4 Vapor Pressure of Pure Liquids

Water in a glass left on a kitchen countertop slowly disappears as molecules on the surface of the liquid vaporize (evaporate) over time. The rate at which the molecules make this transition from the liquid to the gas phase depends on the following factors:

1. Temperature: The higher the temperature, the greater the number of molecules with sufficient kinetic energy to break the attractive forces that hold them together in the liquid and to enter the gas phase.
2. Surface area: The larger the surface area of the liquid, the greater the number of molecules on the surface in a position to enter the gas phase.
3. Intermolecular forces: The stronger the intermolecular forces, the greater the kinetic energy needed for a molecule to escape the surface, and the smaller the number of molecules in the population that have this energy.

If a glass of water is covered (**Figure 10.19a**), molecules that evaporate are confined to the immediate space above the water. As the number of water molecules in the vapor increases at constant temperature (T), so does the partial pressure (P_{H_2O}).

Some of the molecules of water vapor collide with the surface and return to the liquid phase, that is, they condense. The rate of condensation increases as the

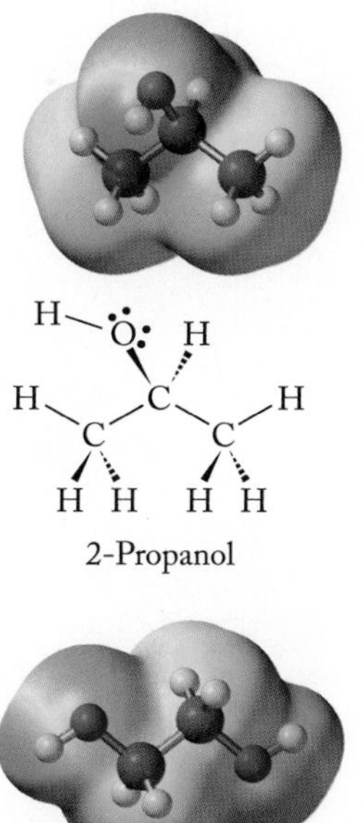

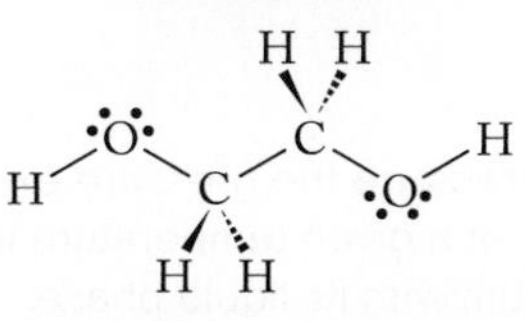

FIGURE 10.18 Structures of 2-propanol and ethylene glycol.

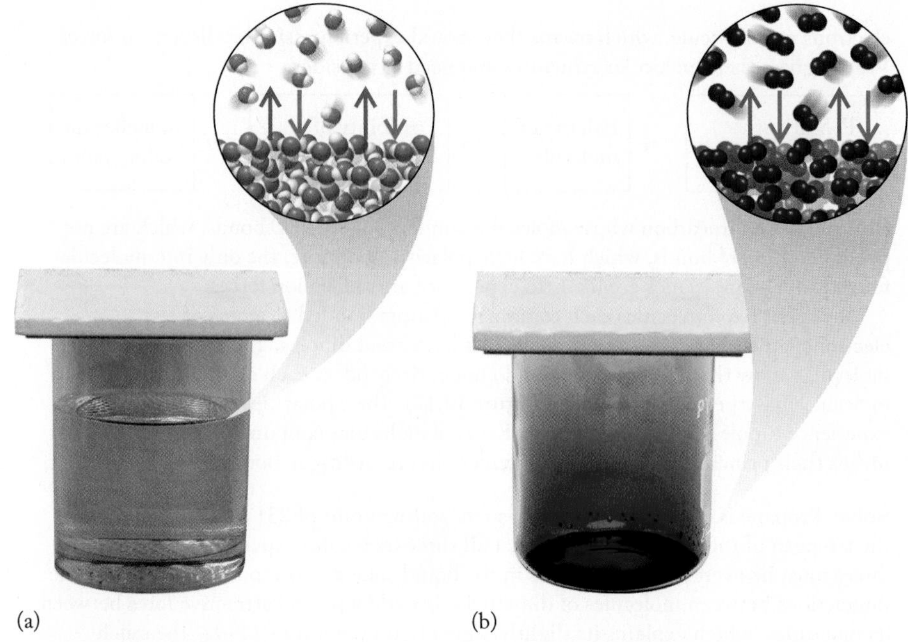

FIGURE 10.19 Covered containers of (a) water and (b) bromine liquid achieve dynamic equilibria when the rates at which molecules of the liquids evaporate (blue arrows) and at which molecules of their vapors condense (red arrows).

concentration of vapor increases because the frequency of these collisions increases. Eventually, the rates of evaporation and condensation equalize setting up a dynamic equilibrium of molecules leaving and returning to the liquid phase at the same rate with no net change in the level of the liquid in the glass over time. The resulting partial pressure of the water vapor in equilibrium with liquid water at a given temperature is called its **vapor pressure**. Other liquids have their own characteristic vapor pressures at room temperature. Some, such as bromine (**Figure 10.19b**), have higher pressures than that of water because, in this case, the strength of the intermolecular forces that keep Br_2 molecules together in the liquid state is weaker than the hydrogen bonds that keep molecules of liquid H_2O together. Others, such as ethylene glycol (Figure 10.18), have lower vapor pressures because the intermolecular forces between their molecules are stronger than those in water.

Vapor Pressure and Temperature

Figure 10.20 shows how the vapor pressures of several liquids increase with increasing temperature. For each of the vapor pressure curves, there is a temperature at which vapor pressure reaches 760 torr (1 atm). That temperature is called the **normal boiling point** of the substance. The vapor pressure of water (the dark blue curve in Figure 10.20) reaches 760 torr at a familiar temperature: 100.0°C. It is called the *normal* boiling point because most chemical reactions in nature, in our bodies, and in laboratories take place near an atmospheric pressure of 1 atm.

The sequence of the four curves and the normal boiling points of the liquids in Figure 10.20 depend on strengths of the interactions between their molecules in the liquid state. Molecules of diethyl ether experience the weakest interactions in this group of four compounds owing to their modest dipole moment (1.15 D) and because they cannot form hydrogen bonds with one another. They behave in this manner because their H atoms are bonded to C atoms, not O, N, or F atoms. On the other hand, molecules of the other three compounds *can* form hydrogen bonds. Ethanol is an alcohol with one –OH group per molecule and has a boiling point about 44°C higher than that of diethyl ether, even though ethanol molecules are smaller and experience small dispersion forces. The reason for its higher

vapor pressure the pressure exerted by a gas at a given temperature in equilibrium with its liquid phase.

normal boiling point the temperature at which the vapor pressure of a liquid equals 1 atm (760 torr).

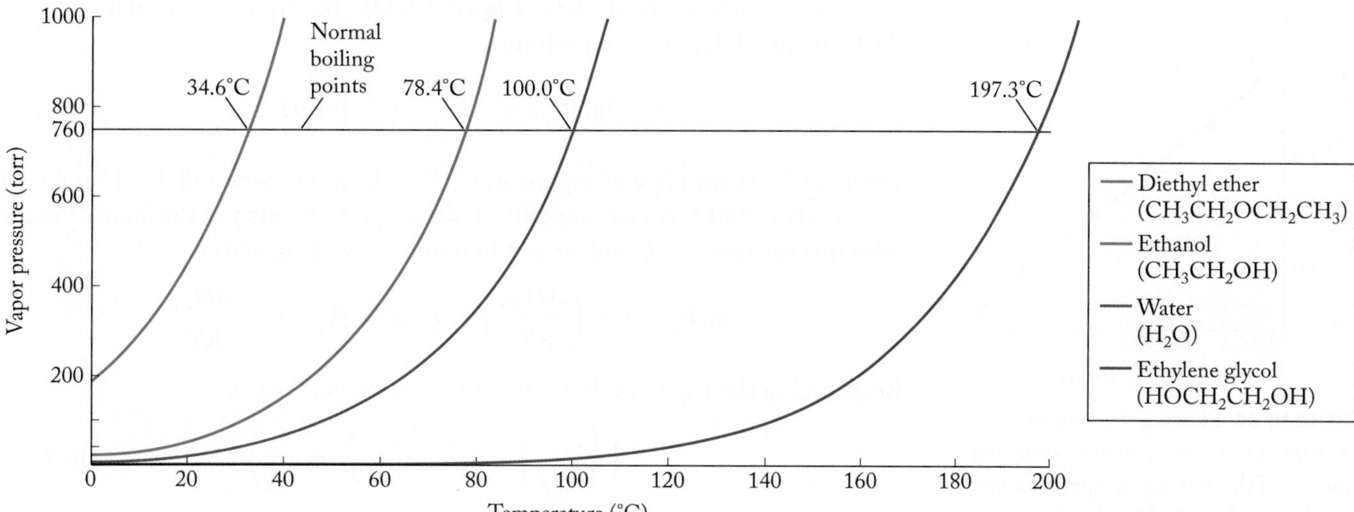

FIGURE 10.20 A graph of vapor pressure versus temperature for four liquids shows how vapor pressure increases with increasing temperature. The temperature at which each vapor pressure curve reaches 760 torr (1 atm) is the normal boiling point of the liquid.

boiling point is the strength of the hydrogen bonds between its molecules. The boiling point of water is nearly 22°C higher than that of ethanol even though water molecules are smaller. In this comparison, the higher boiling point of water occurs because each of its molecules contains *two* O—H bonds that can form up to *four* hydrogen bonds. This is so because the two lone pairs of electrons on one of its O atoms can form hydrogen bonds with two H atoms on two other water molecules as shown in **Figure 10.21**. The boiling point of ethylene glycol is nearly 100°C higher than that of water because its molecules are larger (and experience stronger dispersion forces) and because each of them contains *two* –OH groups that can form up to *six* hydrogen bonds.

The curves in Figure 10.20 also show that the relationship between vapor pressure and temperature is not linear. Even so, all these compounds have significant vapor pressures well below the temperatures at which they boil. For example, at room temperature (~22°C) the vapor pressures of water, ethanol, and diethyl ether are about 20, 50, and 475 torr, respectively. If left in open containers, all three compounds should evaporate at rates directly proportional to their vapor pressures.

Liquid substances with significant vapor pressures and evaporation rates at room temperature are called *volatile* liquids. Among the substances in Figure 10.20, diethyl ether is the most volatile followed by ethanol and water. Ethylene glycol is considered a nonvolatile liquid, which is one of the reasons why it is widely used in the cooling systems of automobile engines, which operate at temperatures approaching 100°C.

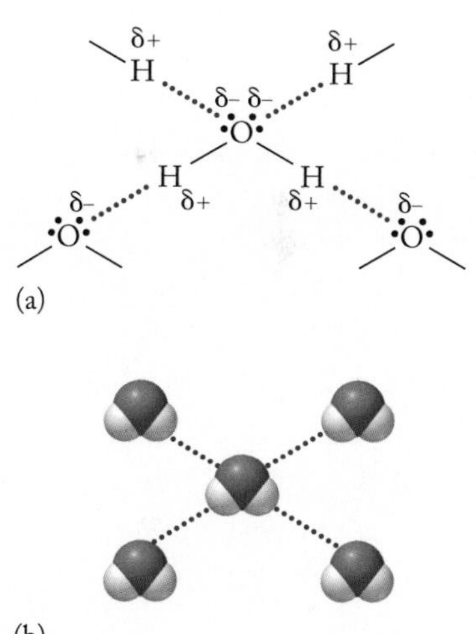

FIGURE 10.21 A single molecule of water can form up to four hydrogen bonds as shown using (a) Lewis structures and (b) molecular models.

CONCEPT **TEST**

Diesel fuel is made of hydrocarbons with an average of 13 carbon atoms per molecule, and gasoline is made of hydrocarbons with an average of 7 carbon atoms per molecule. Which fuel has the higher vapor pressure at room temperature?

CONNECTION Vapor pressure was introduced in Section 5.7 in the context of collecting gases over water.

Volatility and the Clausius–Clapeyron Equation

Although the vapor pressure of a pure substance is not a linear function of absolute temperature, plotting the natural logarithm of its vapor pressure (P_{vap}) versus

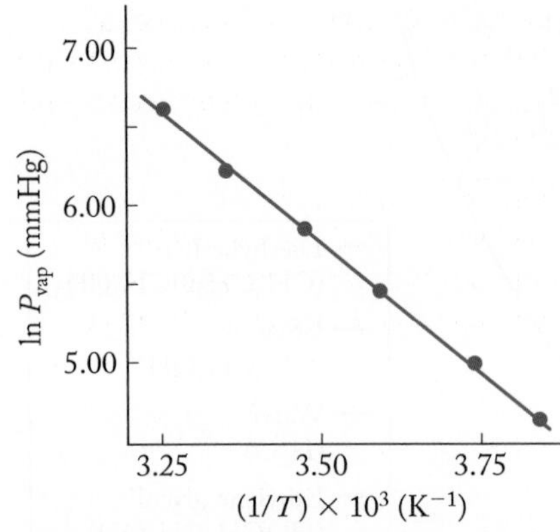

FIGURE 10.22 Plotting the natural logarithm of the vapor pressure versus the reciprocal of the absolute temperature gives a straight line described by the Clausius–Clapeyron equation. This graph shows the plot for *n*-pentane.

$1/T$ does produce a straight line (**Figure 10.22**). The equation of such a line fits the **Clausius–Clapeyron equation**:

$$\ln(P_{vap}) = -\frac{\Delta H_{vap}}{R}\left(\frac{1}{T}\right) + C \qquad (10.1)$$

where ΔH_{vap} is enthalpy of vaporization, R is the gas constant (8.314 J/(mol · K), T is in kelvin, and C is a constant that depends on the identity of the liquid. We can solve this equation for C and write it in terms of two temperatures:

$$\ln(P_{vap,T_1}) + \left(\frac{\Delta H_{vap}}{RT_1}\right) = C = \ln(P_{vap,T_2}) + \frac{\Delta H_{vap}}{RT_2}$$

Rearranging the right and left sides of this equation, we get

$$\ln\left(\frac{P_{vap,T_2}}{P_{vap,T_1}}\right) = -\frac{\Delta H_{vap}}{R}\left(\frac{1}{T_2} - \frac{1}{T_1}\right) \qquad (10.2)$$

This version of the Clausius–Clapeyron equation (Equation 10.2) can be used to calculate ΔH_{vap} if the vapor pressures at two temperatures are known or to calculate either the vapor pressure ($P_{vap,\,T_2}$) at any given temperature T_2, or the temperature at any given vapor pressure, if ΔH_{vap}, $P_{vap,\,T_1}$, and T_1 are known. Equation 10.2 is particularly useful when we know the normal boiling point (T_1) because P_{vap} = 760 torr = 1 atm.

SAMPLE EXERCISE 10.3 Calculating Vapor Pressure LO3

Dogs can't sweat, but they can still use evaporative cooling to regulate their body temperature by panting (taking rapid, shallow breaths). What is the vapor pressure of water in the air a panting dog exhales if the temperature of the air is 35°C? The enthalpy of vaporization of water is 40.7 kJ/mol.

Collect and Organize The vapor pressure of water at any temperature can be calculated using Equation 10.2 and inserting the value of water's heat of vaporization, which we are given, and its normal boiling point (100.0°C), which is the temperature at which P_{vap} is exactly equal to 1 atm = 760 torr.

Analyze To use Equation 10.2 we must convert 100.0°C to kelvin and convert the value of ΔH_{vap} from kJ/mol to J/mol because the units of R are J/(mol · K). As noted earlier, the vapor pressure of water at 20°C is 24 torr, so the value of P_{vap} at 35°C should be above this value, perhaps near 50 torr.

Solve The values of T_1 and T_2 on the Kelvin scale are

$$T_1 = 100.0°C + 273 = 373\ K \qquad \text{and} \qquad T_2 = 35°C + 273 = 308\ K$$

and the value of ΔH_{vap} in joules is

$$40.7\frac{\text{kJ}}{\text{mol}} \times \frac{1000\ \text{J}}{1\ \text{kJ}} = 40.7 \times 10^4 \frac{\text{J}}{\text{mol}}$$

Inserting these values into Equation 10.2:

$$\ln\left(\frac{P_{vap,T_2}}{P_{vap,T_1}}\right) = -\frac{\Delta H_{vap}}{R}\left(\frac{1}{T_2} - \frac{1}{T_1}\right)$$

$$\ln\left(\frac{P_{vap,T_2}}{760\ \text{torr}}\right) = -\frac{4.07 \times 10^4 \frac{\text{J}}{\text{mol}}}{8.314 \frac{\text{J}}{\text{mol} \cdot \text{K}}}\left(\frac{1}{308\ \text{K}} - \frac{1}{373\ \text{K}}\right) = -2.770$$

$$P_{vap,T_2} = 47.6\ \text{torr}$$

Think About It The calculated value is near our estimate and in agreement with the vapor pressure curve for water in Figure 10.20. Therefore, it is reasonable. We used three significant figures to report the result because normal boiling points correspond to vapor pressures of exactly 1 atm or exactly 760.

Practice Exercise The normal boiling point of pentane (C_5H_{12}) is 36°C. What is its molar heat of vaporization in kilojoules per mole if its vapor pressure at 25°C is 505 torr?

Clausius–Clapeyron equation a relationship between the vapor pressure of a substance at two temperatures and its heat of vaporization.

phase diagram a graphical representation of how the stabilities of the physical states of a substance depend on temperature and pressure.

10.5 Phase Diagrams: Intermolecular Forces at Work

In the previous section we discussed how the strengths of the attractive forces between particles determine the normal boiling points of liquids, noting that increasing the temperature of a liquid raises the fraction of its molecules that have enough kinetic energy to overcome these forces and to escape from the liquid phase into the gas phase. In this section we expand our examination of the effect of temperature on transitions between phases to include solid–liquid and solid–gas transitions, and we examine the effect of changing pressure on all phase changes, particularly those involving the gas phase.

Phases and Phase Transitions

Scientists use **phase diagrams** to show which phase of a substance is stable at given combinations of temperature (the *x*-coordinate) and pressure (the *y*-coordinate) values. Consider, for example, the phase diagram of water in **Figure 10.23**. There are three regions corresponding to the three common states of matter (solid, liquid, and gas) plus a fourth region called a *supercritical region*. The lines separating the regions represent *equilibrium lines* because the

CHEMTOUR Phase Diagrams

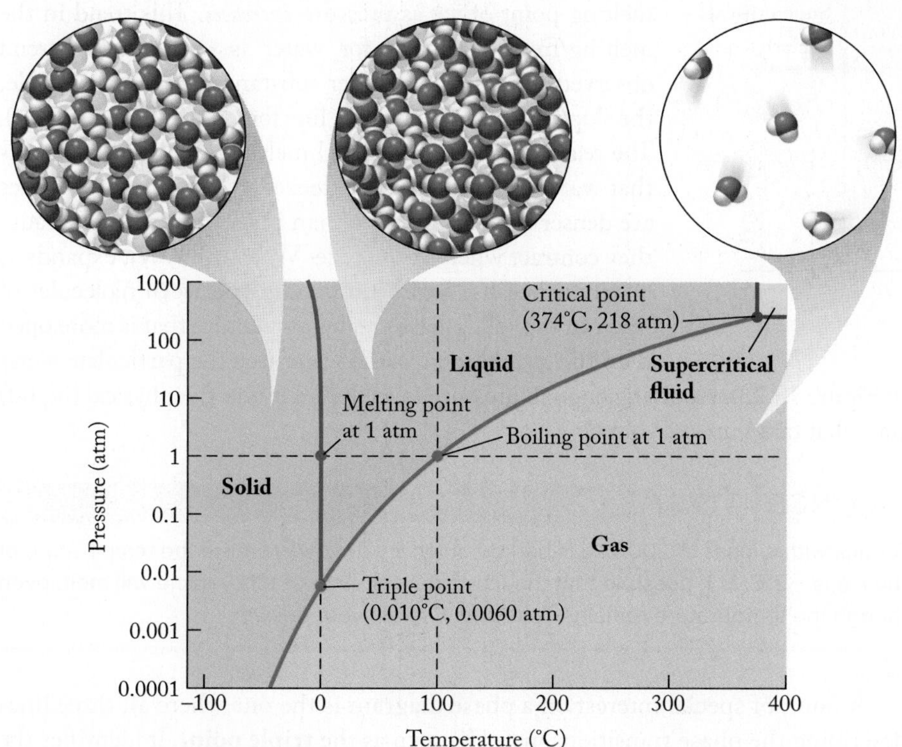

FIGURE 10.23 The phase diagram for water indicates in which phase water exists at various combinations of pressure and temperature.

two states bordering them are at equilibrium at the combinations of temperature and pressure on each line. The blue equilibrium line separating the solid and liquid regions represents a series of freezing (or melting) points; the points on this line are combinations of temperature and pressure at which the solid and liquid states coexist. The red line separating the liquid and gas regions represents a series of boiling points or condensation points; the points on this line are combinations of temperature and pressure at which the liquid and gaseous states coexist. The green line separating the solid and gaseous states represents a series of sublimation points (solid turning to gas) or deposition points (gas turning to solid); the points on this line are combinations of temperature and pressure at which the solid and gaseous states coexist.

Notice how the red line curves from the lower left to the upper right, separating the liquid and gas regions of the phase diagram. This line represents the changing boiling point of the liquid as a function of pressure. Its shape makes sense because when the pressure above a liquid increases, the temperature required for liquid molecules to overcome intermolecular attractive forces and enter the gas phase (i.e., the boiling point) increases. Conversely, applying more pressure to a gas forces its molecules closer together, which increases the frequency with which they collide with and interact with one another, and increases the likelihood that they form clusters of molecules that condense into the liquid phase. Moreover, when a liquid vaporizes, the volume it occupies increases enormously because the liquid (and solid) phases of a substance are many times denser than its gas phase. This expansion is made more difficult if there is an increase in the opposing pressure pushing against it.

The shape of the green solid–gas curve is much the same as the red curve and for much the same reason: higher pressures make it more difficult for ice to expand into an equal mass of water vapor that occupies hundreds of times more volume, and, on a molecular level, the greater interaction between gas-phase molecules under high pressure that promotes their condensation at higher temperatures also promotes their deposition at lower temperatures.

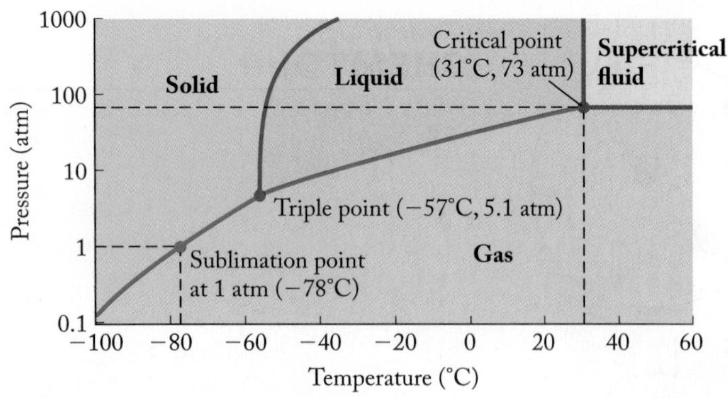

FIGURE 10.24 Phase diagram for carbon dioxide.

On the other hand, the trend in the blue melting/freezing equilibrium line for water shows a *decrease* in the melting point of ice as pressure *increases*. This trend in the melting/freezing points for water is opposite the trend observed for almost all other substances (see, for example, the slope direction of the blue line for CO_2 in **Figure 10.24**). The reason for water's unusual melting/freezing behavior is that water expands when it freezes. Most other substances are denser in the solid state than in the liquid state because they contract when they freeze. Water, however, expands as it freezes because hydrogen bonding between molecules of water in the solid phase creates a structure that is more open than the structure in liquid water (see the particulate views in Figure 10.23). Applying enough pressure to ice forces it into the physical (liquid) state that takes up less volume.

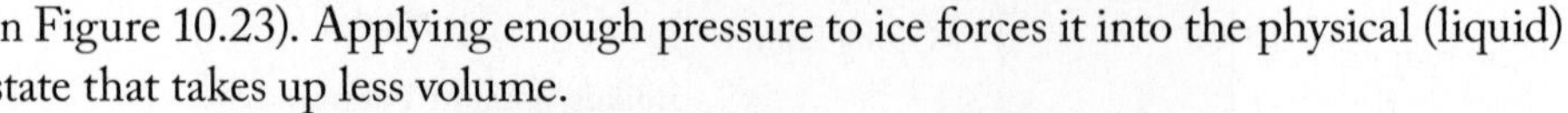

CONCEPT **TEST**

A truck with a mass of 2000 kg is parked on an icy driveway where the temperature of the ice is −2°C. Is it possible that the ice under the tires of this vehicle will melt, even though the temperature remains constant? Explain your answer.

A point of special interest on a phase diagram is the one where all three lines describing the phase transitions meet. Known as the **triple point**, it identifies the temperature and pressure at which all three states (liquid, solid, and gas) are in equilibrium with each other. The triple point of water is just above its normal melting temperature, at 0.010°C, but at a very low pressure of 0.0060 atm.

Another point of interest is the place where the boiling/condensation equilibrium line ends. Called the **critical point**, this is the temperature and pressure at which the liquid and gaseous states are indistinguishable from each other. This point is reached because thermal expansion at this high temperature causes the liquid to become less dense, while the high pressure compresses the gas into a small volume, increasing its density. At the critical point, the densities of the liquid and gaseous states are equal, so one cannot be distinguished from the other.

At temperature–pressure combinations above its critical point, a substance exists as a **supercritical fluid**. A supercritical fluid can penetrate materials like a gas but also dissolve substances in those materials like a liquid. Above the critical temperature, the kinetic energy of the molecules is sufficient to overcome the intermolecular forces holding the molecules together as a liquid. At the critical pressure, the molecules do not have the properties associated with ideal gases, so the properties of supercritical fluids represent a balance of forces acting on molecules. Supercritical carbon dioxide is used in the food-processing industry to decaffeinate coffee and remove fat from potato chips. Supercritical carbon dioxide and water are sometimes mixed to generate fluids that can selectively dissolve specified materials from mixtures while leaving other components untouched.

In the phase diagram of CO_2 (Figure 10.24), the dashed line at P = 1 atm defines the phases that exist at 1 atm pressure, but notice that the blue region representing liquid CO_2 does not extend below 5.1 atm. This means that solid CO_2 does not melt into a liquid at normal temperatures and pressures. Rather, it sublimes directly to CO_2 gas. This behavior explains why the common name for solid CO_2 is *dry ice*; it is a solid that keeps things cool, as water ice does, but it forms no "wet" liquid. The critical point of CO_2 is at 31°C and 73 atm, a pressure easily achieved with compressors in laboratories, factories, and food-processing plants, which means CO_2 is readily available for use as a supercritical fluid.

triple point the temperature and pressure at which all three phases of a substance coexist.

critical point the temperature and pressure at and above which the liquid and gas phases of a substance are indistinguishable from each other.

supercritical fluid a substance at conditions above its critical temperature and pressure, where the liquid and vapor phases are indistinguishable and have some characteristics of both a liquid and a gas.

SAMPLE EXERCISE 10.4 Interpreting a Phase Diagram **LO4**

Describe the phase changes that take place as the pressure on a sample of water at 0°C is increased from 0.0001 atm to 200 atm.

Collect and Organize We are asked to describe the phase changes water undergoes at a constant temperature as pressure is increased. We can use the phase diagram for water in **Figure 10.25a**. To read a phase diagram, remember that every point is characterized by a temperature and a pressure and that every time we cross an equilibrium line, the phase changes.

Analyze The changes take place along the vertical line (**Figure 10.25b**) that intersects the temperature axis at 0°C.

Solve Starting at the bottom of the phase diagram with the red line (Figure 10.25b) and following the dashed line at 0°C upward as pressure increases, we approximate the location of the starting pressure, 0.0001 atm, a little below and to the left of the triple point (T = 0.010°C and P = 0.0060 atm). Water at 0°C and 0.0001 atm is a gas, as indicated by the phase diagram. As we follow the 0°C dashed line up from the temperature axis, the point where this line intersects the green equilibrium line indicates that the gas solidifies to ice at a pressure below 1 atm. Increasing the pressure toward 1 atm, the dashed line intersects the blue equilibrium line at that pressure. At this pressure, then, the ice melts to liquid water. Extending the 0°C

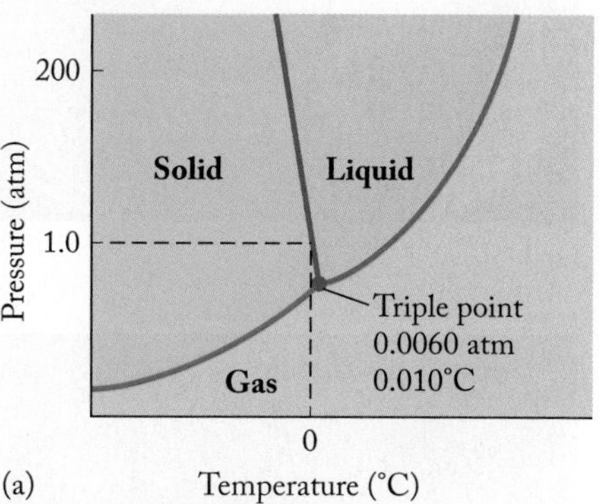

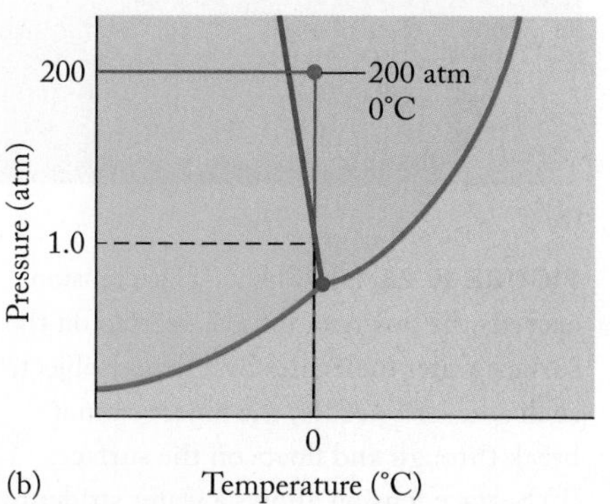

FIGURE 10.25 Phase diagrams for water.

STEPWISE
ANIMATION
Surface Tension

dashed line farther upward shows that the water remains a liquid at pressures up to and beyond 200 atm.

Think About It The phase changes in water, from gas to solid to liquid with increasing pressure at 0°C, make sense from a molecular point of view because higher pressures favor the densest phase. Liquid water is denser than ice, so the transitions from gas to solid to liquid are transitions from the least dense to the densest phase.

Practice Exercise Describe the phase changes that occur when the temperature of CO_2 (Figure 10.24) is increased from −100°C to 50°C at a pressure of 10 atm.

CONCEPT **TEST**

According to the phase diagram in Figure 10.24, does solid CO_2 float on liquid CO_2 under conditions described by any point along the blue line?

10.6 Some More Remarkable Properties of Water

Water has many remarkable properties, including melting and boiling points that are much higher than those of any other molecular substance with a similar molar mass. The boiling point of methane (CH_4), for example, is 112 K, whereas water boils at 373 K. Also, liquid water expands when it freezes; nearly all other substances found in nature contract when they solidify. These phenomena and others that are described in this section are directly related to the shape of water molecules and the strength of the hydrogen bonds they form with one another.

Surface Tension, Capillary Action, and Viscosity

Surface tension is the resistance of a liquid to any increase in its surface area. Surface tension represents the energy required to move molecules apart on the surface of a sample of liquid water so that an object denser than water can break through the surface and settle to the bottom.

Hydrogen bonding in water creates such a high surface tension, 7.29×10^{22} J/m^2 at 25°C, that a carefully placed steel needle floats on water and insects called water striders can walk on it (**Figure 10.26**). The same needle and insect would sink in oil or gasoline.

To understand how surface tension depends on intermolecular forces, consider the microscopic view of the steel needle on the surface of the water in Figure 10.26. Water molecules in the interior of the water are surrounded by other water molecules and form hydrogen bonds to them in all directions. However, there are no molecules of liquid water above those at the surface, so those water molecules form fewer hydrogen bonds per molecule than those in bulk liquid. Gravity pulls the steel needle downward, but to move downward the needle must penetrate the surface. This means pushing aside surface water molecules, increasing the area of the surface, and turning what were once interior water molecules into surface molecules. This transformation requires breaking some of the hydrogen bonds experienced by the interior molecules, which requires about 23 kJ of energy per mole of hydrogen

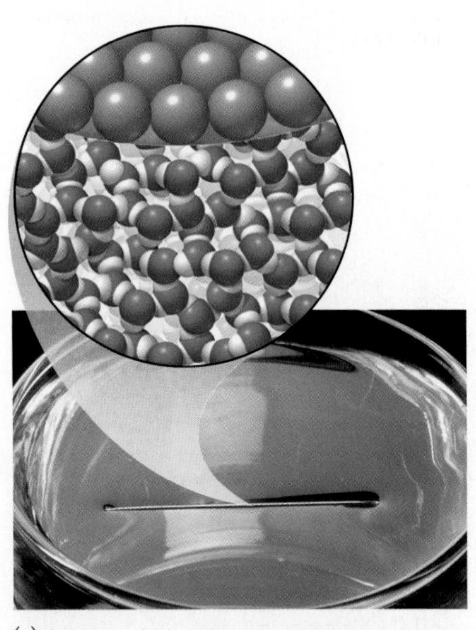

(a)

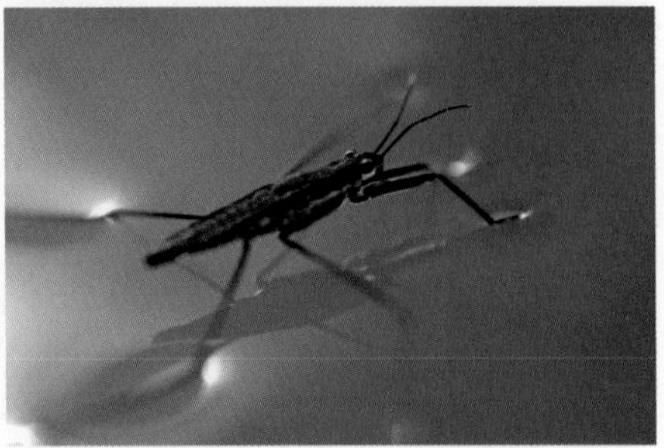

(b)

FIGURE 10.26 (a) When surface tension exceeds the downward force exerted on the surface water molecules by a denser object, such as a steel needle, the object cannot break through and floats on the surface. (b) Surface tension allows a water strider to rest on top of water without sinking.

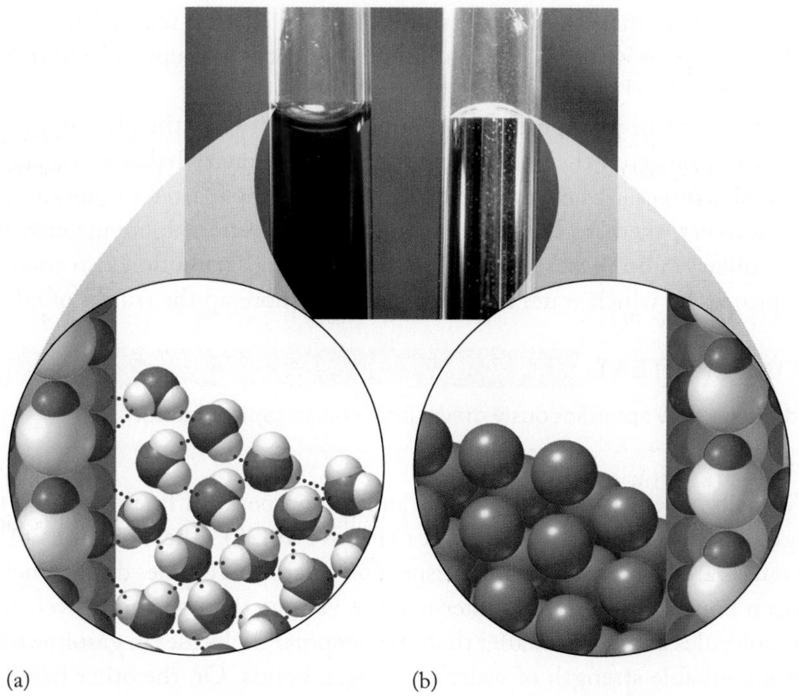

FIGURE 10.27 (a) Hydrogen bonds between the H atoms of water molecules and the O atoms of the silicon dioxide that makes up the glass are an adhesive force that causes water molecules to adhere to the glass surface and form a concave meniscus. (b) Because mercury atoms have no such attraction to the silicon dioxide in the glass, mercury forms a convex meniscus.

CHEMTOUR
Capillary Action

bonds. The downward pressure of the needle is insufficient to overcome this energy, so it floats on the surface.

Another illustration of the intermolecular forces acting in liquids is seen in the shape of the surface of the liquid when it is placed in a graduated cylinder or other small-diameter tube (**Figure 10.27**). The surface of water is concave in such a tube, but the surface of liquid mercury is convex. The curved surface, whether it is concave or convex, is called a **meniscus**. In both liquids, the meniscus is the result of two competing forces: *cohesive forces*, which are interactions between like particles, and *adhesive forces*, which are interactions between unlike particles. In the water sample in Figure 10.27a, the cohesive forces are hydrogen bonds between water molecules, whereas the adhesive forces are dipole–dipole interactions between water molecules and polar Si—O—Si groups on the surface of the glass. The adhesive forces are strong enough to cause the water to climb upward on the glass, creating the concave meniscus. The adhesive forces are greater than the cohesive forces in this case because surface water molecules next to the glass have less contact with other water molecules and therefore experience smaller cohesive forces.

FIGURE 10.28 Because of capillary action, colored water rises in (a) a capillary tube and (b) carnations.

In the mercury sample (Figure 10.27b), the cohesive forces are metallic bonds between mercury atoms, whereas the adhesive forces are interactions between induced dipoles in the mercury atoms and the polar Si—O—Si groups on the glass surface. In this case, the adhesive forces are much weaker than the cohesive forces, and mercury atoms are more attracted to one another than to the glass. As a result, they do not adhere well to the wall, forming a convex meniscus.

The fact that adhesive forces outweigh cohesive forces on the surface of water has consequences in other situations. In a narrow tube, adhesion to the tube wall draws the outer ring of water molecules upward. At the same time, cohesive forces between the outer ring molecules and those adjacent to them draw the molecules upward (**Figure 10.28**). If the tube is narrow enough (a capillary tube), this combination of adhesion and cohesion draws a column of water up the tube in a

surface tension the energy needed to separate the molecules at the surface of a liquid.

meniscus the concave or convex surface of a liquid in a small-diameter tube.

FIGURE 10.29 High viscosity of hot tar results from strong intermolecular forces between large polar molecules.

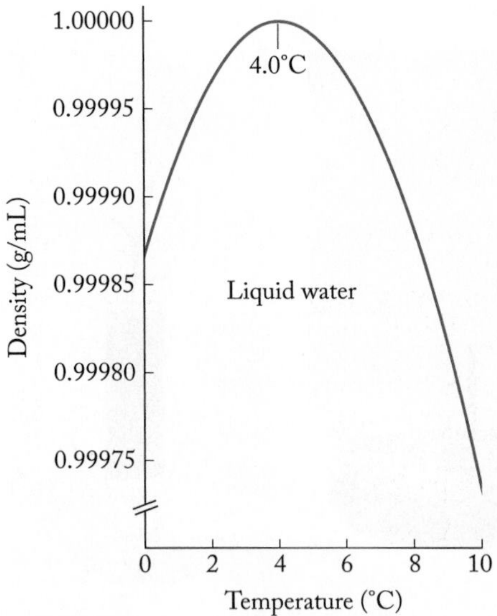

FIGURE 10.30 As water is cooled, its density increases until its temperature reaches 4°C. At this temperature the density has its maximum value. As the water cools from 4°C to its freezing point at 0°C, its density decreases.

capillary action the rise of a liquid in a narrow tube because of adhesive forces between the liquid and the tube and cohesive forces within the liquid.

viscosity the resistance to flow of a liquid.

phenomenon known as **capillary action**. The water column reaches its maximum height when the downward force of gravity balances the upward adhesive and cohesive forces.

In a test tube or pipet, water molecules in contact with the glass move only a very small distance up the glass before the force of gravity balances the adhesive and cohesive forces. When a nurse takes a blood sample after a pinprick in your finger, however, the blood (essentially an aqueous solution) spontaneously moves into a capillary tube because of capillary action. Capillary action also contributes to the process by which water rises 100 meters or more up the trunks of tall trees.

CONCEPT TEST

Would mercury be spontaneously drawn into a glass capillary tube?

Viscosity, the resistance to flow, is another property of liquids related to the strength of their intermolecular forces (**Figure 10.29**). The viscosities of polar compounds are influenced by both dispersion forces and dipole–dipole (including hydrogen bond) interactions. Water is more viscous than gasoline, even though water molecules are much smaller than the nonpolar molecules in gasoline because of the remarkable strength of water's hydrogen bonds. On the other hand, alcohols of different molar mass, such as 1-octanol [$CH_3(CH_2)_7OH$] and ethanol (CH_3CH_2OH), have different viscosities because in addition to the hydrogen bonds that both alcohols form, the much larger molar mass of 1-octanol means that its molecules experience much stronger dispersion forces, giving it viscosity over five times that of alcohol.

The Densities of Cold Water and Ice: Their Impact on Aquatic Life

The density of water increases as it is cooled to 4°C (**Figure 10.30**). This pattern is observed for most liquids and solids and for all gases, but water then expands as it is cooled from 4°C to 0°C as the pattern of its hydrogen bonds changes with temperature. As water freezes at 0°C, its density drops even more, to about 0.92 g/mL, which causes ice to float on liquid water (**Figure 10.31**). This unusual behavior is caused by the formation of a network of hydrogen bonds in ice. With each oxygen atom covalently bonded to two hydrogen atoms and hydrogen-bonded to two other hydrogen atoms, the molecules form an extensive and open hexagonal network in the solid phase that leaves, on average, more space between the molecules than in liquid water.

The expanded structure of ice plays a crucial ecological role in temperate and polar climates. The lower density of ice means that lakes, rivers, and polar oceans freeze from the top down, allowing fish and other aquatic life to survive in the liquid water below. Each fall, surface water cools first, and its density increases until its temperature reaches 4°C, the peak of the graph in Figure 10.30. This maximum-density water at 4°C sinks to the bottom, bringing warmer water to the surface, which in turn cools to 4°C and sinks, pushing the next layer of warm water to the surface in a continuing cycle (**Figure 10.32**). This autumnal turnover stirs up dissolved nutrients, making them available for life in the sunlit surface during the next growing season. When all the water has reached 4°C and the surface water cools further, it becomes less dense and ice may eventually form. The layer of ice insulates the 4°C water beneath it, allowing aquatic life to survive.

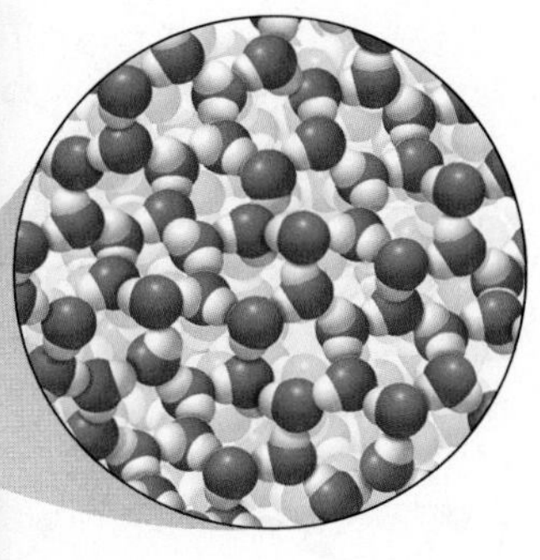

FIGURE 10.31 Because of the changes in the density of water near the freezing point and because ice has a lower density (0.92 g/mL) than water, ice floats on top of seawater.

In spring, the ice melts and the surface water warms to 4°C. At this temperature, the entire column of water has nearly the same temperature and density; dissolved nutrients for plant growth are evenly distributed, and the stage is set for a burst of photosynthesis and biological activity called the spring bloom.

Further warming of the surface water creates a warm upper layer separated from colder, denser water by a *thermocline*, which is a sharp change in temperature between the two layers. Biological activity depletes the pool of nutrients above the thermocline as decaying biomass settles to the bottom. Consequently, photosynthetic activity drops from its spring maximum during the summer, even though there is much more energy available from the Sun. The thermocline persists until the autumn turnover mixes the water column and the cycle begins anew.

10.7 Polarity and Solubility

In Section 10.3 we explained the importance of ion–dipole interaction in the process by which ionic compounds dissolve in water. These interactions allow ionic compounds to dissolve in other polar solvents. Similarly, polar molecular compounds, such as sugars and low-molar-mass alcohols and ethers, dissolve in water and other polar solvents because of dipole–dipole interactions (and sometimes hydrogen bond formation) between molecules of solute and solvent. On the other hand, nonpolar compounds have limited solubility in water and other polar solvents, because the dissolution process relies on dipole–induced dipole interactions, which are much weaker than dipole–dipole interactions (see Table 10.3). These solute–solvent interactions are too weak to overcome the stronger interactions that attract molecules of solvent to other molecules of solvent and that attract molecules of solute to one another. The key point is that the solubility of a particular solute in a particular solvent results from a competition between the strengths of solvent–solvent and solute–solute interactions that inhibit the dissolving process and the solute–solvent interactions that promote it.

Figure 10.33 illustrates this key point. Figure 10.33a shows that molecules of methanol form hydrogen bonds with one another as do molecules of water, which means that molecules of methanol and water can form hydrogen bonds with each other. It is reasonable to assume that the strengths of the water–methanol hydrogen bonds are like those that form in pure water and pure methanol. The validity

(a) Autumn

(b) Winter

FIGURE 10.32 (a) As the surface water of a pond cools to 4°C, its density increases, and it sinks to the bottom, bringing the warmer, less dense water to the surface. (b) Continued cooling of the surface water below 4°C produces a less dense layer that may eventually freeze while the denser water deeper in the pond remains at 4°C.

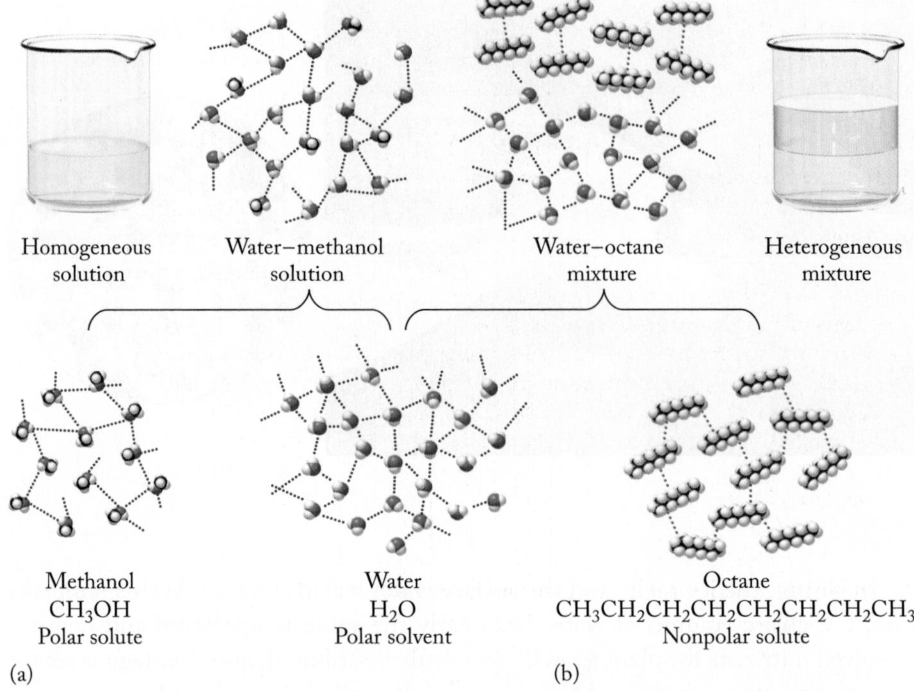

FIGURE 10.33 (a) A polar solvent such as water dissolves polar materials such as methanol because of favorable dipole–dipole interactions. Both molecules are also capable of forming hydrogen bonds among themselves and with each other, which is why methanol dissolves in water in all proportions. (b) Dispersion forces between long hydrocarbon chains hold octane molecules together in the liquid phase. The strong hydrogen bonds among water molecules keep them together, too. Solute–solvent interactions are too weak to compete with these strong solute–solute and solvent–solvent interactions. As a result, octane has little solubility in water.

CONNECTION *Miscible* was introduced in Section 5.1 to describe the homogeneous mixtures that gases form in all proportions.

of this assumption is supported by the observation that methanol dissolves in water in all proportions, that is, these two liquids are *miscible* in each other.

On the other hand, Figure 10.33b shows that the hydrocarbon octane, C_8H_{18}, does not dissolve in water. Apparently, solute–solvent interactions based on relatively weak dipole–induced dipole interactions (see Table 10.3) between polar molecules of water and nonpolar octane molecules are no match for the strong hydrogen bonds between water molecules that would have to be disrupted to accommodate molecules of octane in an aqueous solution, plus the dispersion forces between the long hydrocarbon chains of octane molecules that would need to be disrupted if they were to dissolve in water. Thus, strong solute–solute and solvent–solvent interactions and a lack of strong solute–solvent interactions limit the solubility of octane in water (it is only about 7 μg of octane per liter of water at 20°C).

SAMPLE EXERCISE 10.5 Predicting Solubility in Water — LO6

The compounds (a) carbon tetrachloride (CCl_4), (b) ammonia (NH_3), (c) hydrogen fluoride (HF), and (d) oxygen (O_2) are shown in **Figure 10.34**. Which ones are very soluble in water, and which ones have limited solubility in water?

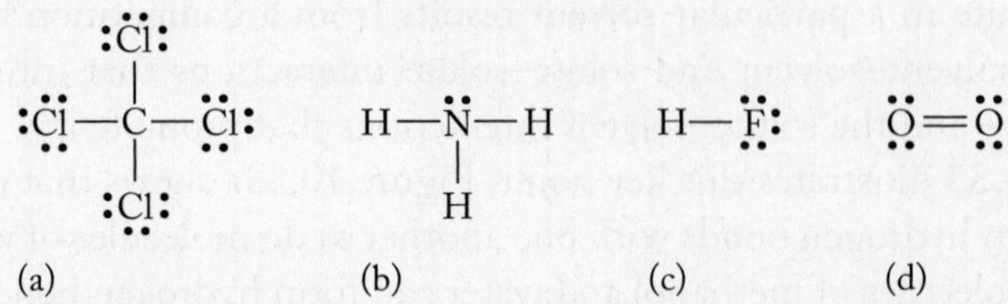

:C̈l:
:C̈l—C—C̈l: H—N̈—H H—F̈: Ö=Ö
:C̈l: H

(a) (b) (c) (d)

FIGURE 10.34 Lewis structures for Sample Exercise 10.5.

Collect and Organize We are asked to predict the relative solubilities of four substances in water. Water is polar, which means compounds that are polar are more likely to dissolve in it than nonpolar compounds.

Analyze To judge which of the four substances is polar, we need to draw their Lewis structures and determine (1) which of them have polar bonds and (2) whether the arrangement of these bonds gives these molecules permanent dipoles, as described in Chapter 9.

Solve The Lewis structures of the four compounds are shown in Figure 10.34. An analysis of their bond dipoles and the three-dimensional orientation of their bonds reveals that the tetrahedral molecular symmetry of CCl_4 makes it a nonpolar compound (as is O_2), but the asymmetry of the trigonal pyramidal molecular shape of NH_3 makes it a polar compound (as is HF). The polar compounds (NH_3 and HF) should be soluble in water, but the nonpolar ones (CCl_4 and O_2) should have only limited solubility in water. Also, NH_3 and HF form hydrogen bonds, another property they share with water, making them even more likely to be water soluble.

Think About It Our conclusion that the two nonpolar compounds should have limited solubility in water is supported by an analysis of the intermolecular forces involved, particularly the relatively high strength of the hydrogen bonds in water that must be disrupted to accommodate the molecules of these solutes and the relatively low strength of the dipole–induced dipole interactions between these solutes and water that promote the dissolution process.

Practice Exercise In Chapter 5 we mentioned that some deep-sea divers breathe a mixture of gases rich in helium because the solubility of helium gas in blood is lower than the solubility of nitrogen gas in blood. Assuming these gases dissolve in blood plasma as they do in water, why should helium be less soluble in water than nitrogen is?

The unlimited solubility of methanol in water and very limited solubility of octane in water illustrate an important solubility principle: polar solutes tend to dissolve in polar solvents, but nonpolar solutes tend not to dissolve in polar solvents. This principle may lead you to wonder whether nonpolar solutes tend to dissolve in *non*polar solvents. Indeed, they do. In fact, the ability of polar solvents to dissolve polar solutes and nonpolar solvents to dissolve nonpolar solutes leads to a simple, widely used description of solubility: *like dissolves like*. Among the common nonpolar solvents is the principal ingredient in some household cleaners (**Figure 10.35**) that remove grease, crayon wax, adhesives, and other nonpolar materials from surfaces and fabrics. It is often labeled *petroleum distillate*, which means a mixture of hydrocarbons derived from crude oil. These hydrocarbons are effective in dissolving nonpolar substances because they are also nonpolar.

Surfactants are organic compounds that are the principal cleaning agents in the detergents used to wash laundry and dirty dishes and, when needed, to clean the oiled feathers and fur of marine birds and mammals that encounter petroleum spills (see this chapter's opening photo). Surfactants are composed of long hydrocarbon chains bonded to ionic groups (**Figure 10.36**). The ionic group helps make these compounds soluble in water, but once they dissolve in water their molecules collect on the surface with their ionic *heads* in the water and the nonpolar hydrocarbon *tails* sticking above the surface. This orientation makes it possible for clusters of the tails to interact with and dissolve grease and oil spots on clothes, fatty food residues on dishes, and the oiled feathers and fur of marine birds and mammals.

FIGURE 10.35 Household cleaners such as this one are effective at removing grease stains and other nonpolar materials from surfaces and fabrics because the principal ingredient in them is a mixture of nonpolar hydrocarbons.

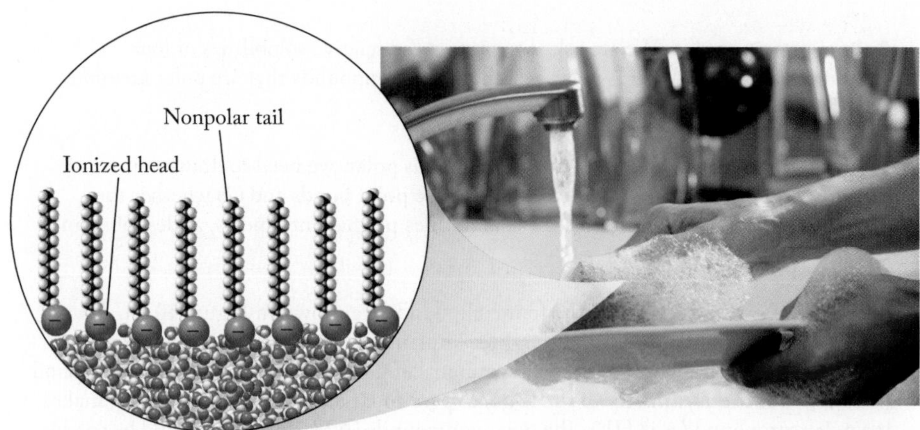

FIGURE 10.36 The nonpolar tails of surfactants create nonpolar surfaces on the bubbles they form in aqueous solutions. These surfaces attract and dissolve nonpolar substances such as those that coat dirty dishes and greasy pots and pans.

Let's consider a medical application of "like dissolves like." For more than 150 years diethyl ether, $CH_3CH_2OCH_2CH_3$, was widely used in medicine as an anesthetic. As we have seen in this chapter, **ethers** have the general formula R—O—R′, where R and R′ are straight-chain or branched hydrocarbons or aromatic rings (R and R′ may be the same or different organic groups). Because the C—O—C bond angle is close to the tetrahedral bond angle of 109.5°, the dipole moments of the two C—O bonds in an ether do not cancel, which means that ethers are polar molecules. This structural feature means that the solubilities of ethers in water are comparable to those of alcohols of similar molar mass. However, their boiling points are much lower than those of similar alcohols (see Figure 10.20) because molecules of ethers do not form hydrogen bonds with one another as molecules of alcohols do.

This combination of volatility and solubility helped make diethyl ether an effective anesthetic. Patients inhaled its vapor, which readily dissolved in blood passing through their lungs and was transported throughout the body. The presence of short hydrocarbon chains helped make diethyl ether soluble in nonpolar cell membranes where it blocked nerve stimuli and made the patients lose consciousness. Ether has the unfortunate effect of inducing nausea and headaches and it is flammable, which contributed to its being replaced by volatile, nonflammable halogenated hydrocarbons, such as CHClBrF, which are used for general anesthesia today.

Combinations of Intermolecular Forces

The surfactants used in detergents are an example of substances whose molecules are composed of functional groups (ionic or polar) that promote their solubility in water *and* nonpolar regions that inhibit aqueous solubility. Those that promote the molecule's solubility in water are called **hydrophilic** ("water-loving") interactions, and the nonpolar regions that inhibit it are called **hydrophobic** (literally, "water-fearing") regions. The solubility of a compound whose molecules contain both polar and nonpolar or ionizable groups is a balancing act between hydrophilic and hydrophobic interactions. As the hydrophobic portion of the molecule increases in size, the entire molecule becomes more hydrophobic and solubility in water decreases. At the same time, solubility of these molecules in nonpolar solvent increases.

One way that scientists determine the relative affinities of compounds for polar versus nonpolar solvents is to try dissolving them in a mixture of both types at the same time. A common combination of polar and nonpolar solvents that is used for such tests is water and 1-octanol, $CH_3(CH_2)_7OH$ (**Figure 10.37**). When a compound is added to a mixture of these two immiscible liquids and the mixture is vigorously shaken, some of the compound may dissolve in the water layer and some in the 1-octanol layer. After the compound has completely dissolved, its concentration is determined in each layer and the ratio of the concentrations is calculated. **Figure 10.38** shows the results of a set of these experiments in which the test compounds were alcohols with the generic formula $CH_3(CH_2)_nOH$, where n ranges from 0 to 9, that is, from methanol, CH_3OH, to 1-decanol, $CH_3(CH_2)_9OH$.

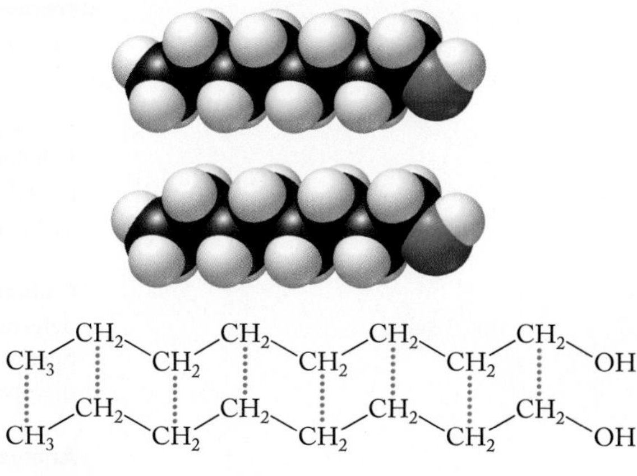

FIGURE 10.37 Strong dispersion forces between the lengthy hydrocarbon tails of 1-octanol molecules dominate the interactions between them.

The data plotted in Figure 10.38 show a steady increase in the ratio of 1-octanol to water as the length of the hydrocarbon chains of the alcohols increases from one carbon atom to ten. This trend makes sense because the longer hydrophobic chains increase their affinity for a nonpolar solvent and decrease their solubility in water. Also note that the y-axis is not linear. Each tick mark represents a factor of 10 change in the concentration ratio: the highest value (1-decanol, 3.7×10^4) is over 10^5 times that of the lowest (methanol, 0.17). The very small ratios for methanol and ethanol are consistent with the fact that both are miscible in water.

You may be wondering why scientists are interested in ratios such as those in Figure 10.38. One reason is the need to predict the impact of water pollution on aquatic organisms. Many of the cells in these organisms are surrounded by nonpolar membranes. Pollutants with high ratios of 1-octanol to water are likely to be composed of molecules that can penetrate these nonpolar cell membranes and disrupt cell function once inside.

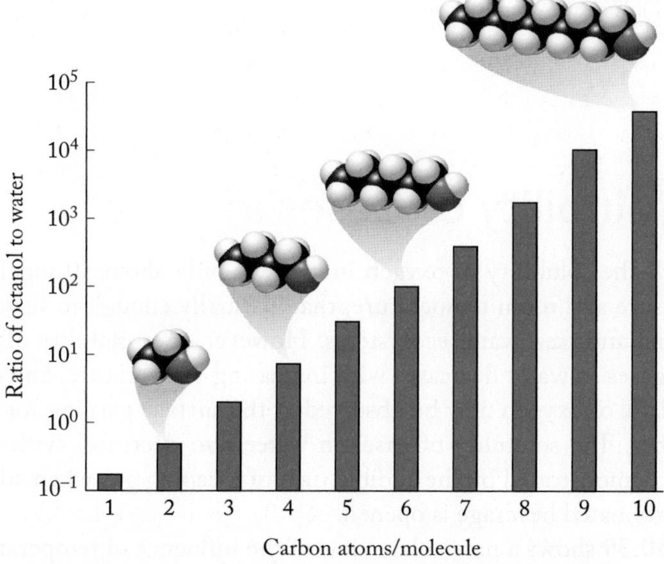

FIGURE 10.38 The ratio of the solubilities of these alcohols in nonpolar 1-octanol and in water increases by more than 10^5 from methanol, CH_3OH, to 1-decanol, $C_{10}H_{21}OH$.

ether an organic compound whose molecules have an R—O—R′ structure.

hydrophilic a "water-loving," or attractive, interaction between a solute and water that promotes water solubility.

hydrophobic a "water-fearing," or repulsive, interaction between a solute and water that diminishes water solubility.

SAMPLE EXERCISE 10.6 Predicting Miscibility of Liquids **LO6**

Predict which pairs of liquids are miscible. Explain your predictions. (a) Ethylene glycol ($HOCH_2CH_2OH$) and water; (b) benzene (C_6H_6) and pentane ($CH_3CH_2CH_2CH_2CH_3$); (c) octanol ($CH_3CH_2CH_2CH_2CH_2CH_2CH_2CH_2OH$) and octane ($CH_3CH_2CH_2CH_2CH_2CH_2CH_2CH_3$)

Collect and Organize We are given the formulas of different liquids and we want to determine their miscibility. We discussed the cyclic molecular structure of benzene in Section 8.4. We can apply a molecular-level understanding of the common phrase "like dissolves like" to predict miscibility.

Analyze To predict miscibility, we need to consider the composition of and molecular structure of each pair of compounds—particularly the presence of polar functional groups, and especially those capable of hydrogen bonding—and the length of nonpolar hydrocarbon chains.

Solve

a. Ethylene glycol is an organic liquid with two –OH groups per molecule that can form as many as six hydrogen bonds to nearby water molecules. Prediction: the two liquids are miscible.
b. Benzene and pentane are both nonpolar hydrocarbons. Given the "like dissolves like" mantra, the two should be miscible.
c. Octanol has a polar –OH group capable of forming hydrogen bonds, whereas octane does not and is completely nonpolar. However, octanol and octane both have hydrocarbon chains of the same length. The dispersion forces associated with these chains interacting should lead to their miscibility.

Think About It "Like dissolves like" was the guiding principle in all three predictions: in (a) both compounds form hydrogen bonds; in (b) and (c) the principal mode of interaction in the pure liquids and their miscible mixtures is dispersion forces. The fact that octanol molecules also form hydrogen bonds was overwhelmed by the length of their hydrocarbon chains, demonstrating the strength-in-numbers aspect of dispersion forces.

Practice Exercise Predict whether methanol (CH_3OH) or hexanol ($CH_3CH_2CH_2CH_2CH_2CH_2OH$) would be more soluble in acetone (Figure 10.7). Explain your answer.

10.8 Solubility of Gases in Water

CONNECTION The kinetic molecular theory of gases (Section 5.8) is a model that explains the behavior of ideal gases. Partial pressure (Section 5.7) is the contribution to the total pressure made by one gas in a mixture of gases.

Even though the solubility of oxygen in water is only about 10 mg/L at atmospheric pressure and room temperature, that is usually enough to sustain aquatic life in marine and freshwater ecosystems. However, the solubility of O_2 and of most other gases in water decreases with increasing temperature, and fish suffering from a lack of oxygen may be observed at the surface gasping for air in very warm weather. The solubility of gases in water also decreases with decreasing pressure, as demonstrated by the audible rush of escaping gas when a bottle containing a carbonated beverage is opened.

Figure 10.39 shows a particulate view of the influence of temperature on the solubility of O_2 (or any nonpolar gas). At low temperatures there is an equilibrium between O_2 in the air and dissolved O_2 in the water. At a higher temperature, O_2 molecules have more kinetic energy, which allows them to overcome the dipole–induced dipole interactions that keep them in solution. As a result, a greater fraction of the molecules escapes into the gas phase, decreasing their concentration in solution (Figure 10.39b).

The solubility of an atmospheric gas in a liquid such as water also depends on the partial pressure of the gas in the air above the surface of the liquid. For example, the partial pressure of O_2 at sea level is a steady 0.21 atm, but the lower partial pressure of oxygen at higher altitudes reduces O_2 concentrations in rivers and lakes and in other liquids, including the blood and tissues of humans and animals. When people travel to regions of low atmospheric pressure, such as the tops of mountains, they may need supplemental oxygen to function normally because of the decreased partial pressure of oxygen. Climbers of Earth's highest mountains may become weak and unable to think clearly from a lack of oxygen in the brain, a condition known as *hypoxia*.

To visualize this pressure dependence of gas solubility on a particulate level, consider the populations of gas molecules above the liquid surfaces in **Figure 10.40**. The gas in the container on the left is air in which $P_{O_2} = 0.21$ atm. The gas in the container on the right is pure O_2 at a pressure of 1.00 atm. The amount of O_2 that dissolves in either liquid depends on the frequency at which O_2 molecules collide with the surface. These collisions should occur nearly five times more frequently in the liquid on the right, which should result in five times the concentration of dissolved O_2 in that beaker. If the cover on the right is removed and the pure O_2 in it mixes with and is eventually replaced by air, decreasing P_{O_2} values produce decreasing concentrations of dissolved O_2, which eventually match that of the water in the container on the left. This is essentially what happens (though on a longer time scale) when opening carbonated beverages; as the lid opens, the pressure drops and the CO_2 gas escapes from the fluid, releasing bubbles and producing fizz.

The relationship between gas solubility in a liquid and the partial pressure of the gas in the environment surrounding the liquid applies to all sparingly soluble gases. The quantitative statement based on this observation is known as **Henry's law** in honor of William Henry (1775–1836), a British physician who first proposed the relationship:

$$C_{\text{gas}} = k_{\text{H}}P_{\text{gas}} \tag{10.3}$$

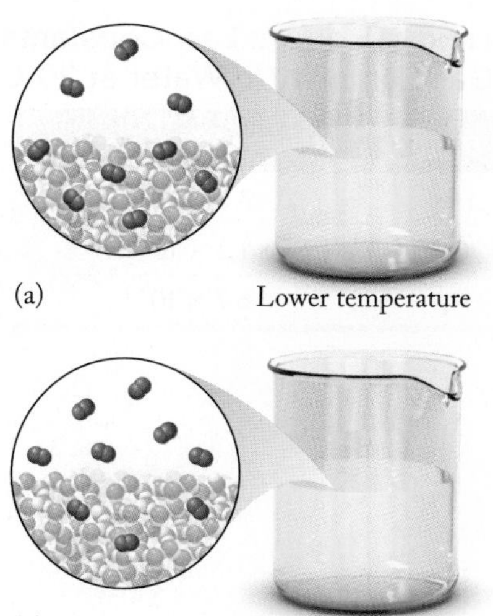

FIGURE 10.39 (a) At lower temperatures, molecules of gas dissolved in water have less kinetic energy, and dipole–induced dipole interactions between solute and solvent molecules keep more gas dissolved. (b) At higher temperatures, molecules have more kinetic energy, which overcomes solute–solvent interactions and reduces solubility.

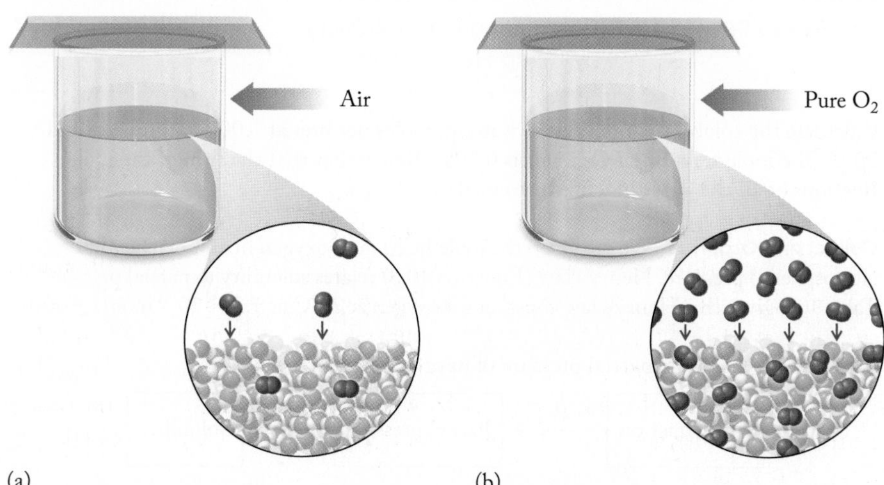

FIGURE 10.40 (a) Air contains a partial pressure of oxygen of 0.21 atm. (b) The partial pressure of pure oxygen is 1.0 atm. The partial pressure of oxygen (red particles) in the space above a liquid is five times as large in (b) as in (a). Consequently, the number of collisions the gas molecules have with the surface is greater in (b) than in (a), and the solubility of the gas in the liquid increases.

Henry's law the principle that the concentration of a sparingly soluble, chemically unreactive gas in a liquid is proportional to the partial pressure of the gas.

TABLE 10.4 Henry's Law Constants for Gas Solubility in Water at 20°C

Gas	k_H [mol/(L · atm)]
He	3.5×10^{-4}
O_2	1.3×10^{-3}
N_2	6.7×10^{-4}

where C_{gas} represents the concentration (solubility) of a gas in a particular solvent, k_H is the Henry's law constant for the gas in that solvent, and P_{gas} is the partial pressure of the gas in the environment surrounding the solvent. When C_{gas} is expressed in molarity and P_{gas} in atmospheres, the units of the Henry's law constant are moles per liter-atmosphere, mol/(L · atm). **Table 10.4** lists k_H values for several common gases in water.

Henry's law explains why the concentration of dissolved oxygen in blood is proportional to the partial pressure of oxygen in the air we inhale and thus is proportional to atmospheric pressure. Although this is an accurate statement of Henry's law, the residents of Denver, Colorado, or Kimberley, British Columbia, Canada (average atmospheric pressure 0.85 atm), do not live with less oxygen in their blood than the residents of New York City or Rome, Italy (average atmospheric pressure 1.00 atm). The oxygen transport system in our bodies must adjust to accommodate local atmospheric conditions.

The amount of oxygen in the blood is related to both the concentration of hemoglobin and the fraction of the hemoglobin sites that contain oxygen as the blood leaves the lungs. This saturation of binding sites depends on the partial pressure of oxygen, as well as on proper lung function. For most people, breathing air with $P_{O_2} > 0.11$ atm results in nearly 100% saturation of hemoglobin binding sites. If P_{O_2} decreases to about 0.066 atm (as it does on high mountains), the percent saturation decreases to about 80%. Over several weeks, the body responds to lower oxygen partial pressures by producing more red blood cells and more hemoglobin. This increase in the concentration of hemoglobin, and therefore in the number of O_2 binding sites, compensates for the lower partial pressure of O_2. Even though less than 100% of the oxygen binding sites are saturated, the actual number of oxygen binding sites has increased because of the production of more hemoglobin molecules, and therefore the same amount of O_2 is delivered to tissues. Some endurance athletes try to capitalize on these physiological effects by using a technique known as "live high, train low." They acclimate to higher altitudes, typically defined as any elevation above 1500 meters (5000 ft), to bring about the physiological changes, but they continue to train at lower elevations.

CHEMTOUR

Henry's Law

SAMPLE EXERCISE 10.7 Calculating Gas Solubility by Using Henry's Law **LO7**

Calculate the solubility of oxygen in water in moles per liter at 1.00 atm pressure and 20°C. The mole fraction of O_2 in air is 0.209. (Remember that the sum of the mole fractions of all the gases in a mixture equals 1.)

Collect and Organize We are given the mole fraction of oxygen in air and the total (atmospheric) pressure. Henry's law (Equation 10.3) relates solubility to partial pressure. Table 10.4 gives this Henry's law constant for oxygen at 20°C as 1.3×10^{-3} mol/(L · atm).

Analyze We need the partial pressure of oxygen for Henry's law:

$$\boxed{\text{Mole fraction}} \xrightarrow{\times P_{\text{total}}} \boxed{\text{Partial pressure}} \xrightarrow{\times k_H} \boxed{\text{Solubility}}$$

The product of the mole fraction of O_2 (X_{O_2}) times its total pressure gives us its partial pressure, which we use in Equation 10.3 to calculate the solubility of oxygen. We predict that oxygen is not very soluble in water because the interactions involved are weak dipole–induced dipole forces.

Solve We calculate the partial pressure of oxygen by using Equation 5.25:

$$P_{O_2} = X_{O_2}P_{total} = (0.209)(1.00\text{ atm}) = 0.209\text{ atm}$$

Substituting this value for P_{O_2} and k_H for O_2 in water in Equation 10.3 gives

$$C_{O_2} = k_H P_{O_2} = \left(\frac{1.3 \times 10^{-3}\text{ mol}}{\text{L} \cdot \cancel{\text{atm}}}\right)(0.209\ \cancel{\text{atm}}) = 2.7 \times 10^{-4}\text{ mol/L}$$

Think About It We predicted in Sample Exercise 10.5 that oxygen is not very soluble in water because O_2 is a nonpolar solute and water is a polar solvent, and the answer agrees with that prediction.

Practice Exercise Calculate the solubility of oxygen in water at the top of Mt. Everest, where atmospheric pressure is 0.35 atm. (Assume that the Henry's law constant for oxygen at 20°C is unchanged at the top of Mt. Everest.)

SAMPLE EXERCISE 10.8 Integrating Concepts: Testing Drug Candidates

In Section 10.7 we described how dissolving a substance in a mixture of two immiscible liquids, water and 1-octanol, allowed environmental scientists to evaluate the likelihood that the substance would be biologically available to aquatic organisms. The same test is also used by pharmaceutical scientists to determine whether a drug under development is hydrophilic enough to be transported in circulating blood plasma, but at the same time hydrophobic enough to pass through nonpolar cell membranes. The ratio of the concentration of the drug candidate in 1-octanol to its concentration in water provides a measure of how well-balanced are its hydrophobic versus hydrophilic properties.

a. Draw a rough sketch of a tube containing 10 mL of water (density 1.00 g/mL) and 10 mL of octanol (density 0.83 g/mL). Indicate each fluid clearly and suggest what the meniscus would look like.
b. Two anticancer drugs that are best known by the acronyms 5-FU and BCNU are shown in **Figure 10.41**. Which one should have the higher octanol–water concentration ratio. Explain your prediction.

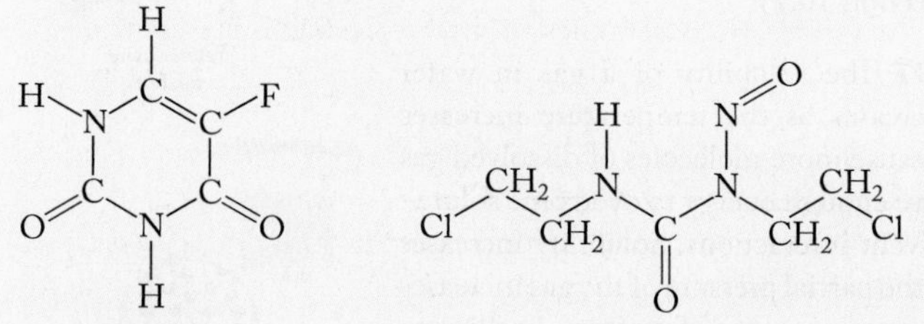

5-Fluorouracil
130.08 g/mol

1,3-Bis(2-chloroethyl)-1-nitrosourea (BCNU)
214.05 g/mol

FIGURE 10.41 Structures and molar masses of 5-FU and BCNU.

c. The octanol–water concentration ratio for 5-FU is 0.112. If 0.100 g 5-FU are mixed with and dissolved in equal volumes of 1-octanol and water, what mass of the drug should dissolve in the water and how much in the 1-octanol?

Collect and Organize To run the experiment described, we have octanol and water together in a test tube. We are given the structures of two anticancer drugs and the value of one of their octanol–water partition coefficients.

Analyze We know that 1-octanol and water are immiscible liquids and that 1-octanol is less dense than water. Therefore, it will be the top liquid in the test tube and the compound that forms a meniscus with the glass wall of the test tube. To predict solubility trends for the two drug candidates, we need to examine their structures for polar and nonpolar groups and for their capacity to form hydrogen bonds.

Solve

a. The nonpolar character of 1-octanol means it will experience little attraction to the polar glass surface of the test tube, resulting in a meniscus that is much less concave than the one that water would form (**Figure 10.42**).

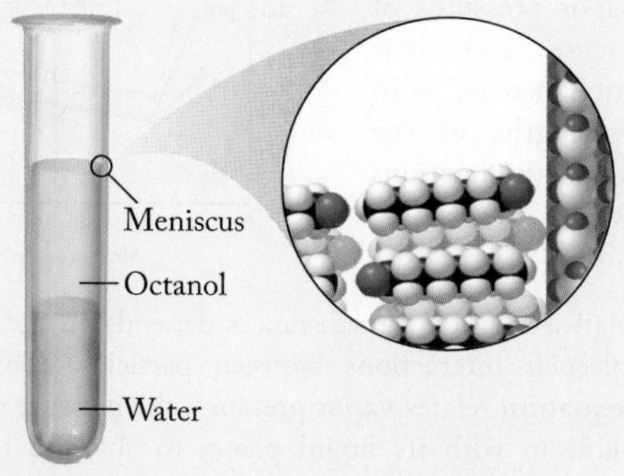

FIGURE 10.42 Test tube containing water and 1-octanol.

b. 5-FU has two –NH groups and two C═O groups capable of hydrogen bonding with water. BCNU has only one –NH group and one C═O group; it also has two other nitrogen atoms, but one is sp^2 hybridized and attached to a more electronegative oxygen atom, so it would not have electrons readily available to bond with a hydrogen atom in water. In addition, the molar mass of BCNU is almost twice that of 5-FU; contributing to this additional mass are four nonpolar CH_2 groups. All these factors suggest that 5-fluorouracil should be the more water-soluble of the two drugs and the one with the smaller 1-octanol–water partition coefficient.

c. Because equal volumes of the two solvents were used, the ratio of concentrations of 5-fluorouracil in them is the same as the ratio of the masses of 5-fluorouracil in them, so that:

$$0.112 = \frac{\text{g drug in octanol}}{\text{g drug in water}}$$

If the total amount of drug added to the test tube is 0.100 g, we can call the amount dissolved in water x and the amount dissolved in octanol $0.100 - x$. Substituting these expressions into the above ratio:

$$0.112 = \frac{0.100 - x}{x}$$

$$0.112x = 0.100 - x$$

$$x + 0.112x = 0.100$$

$$x = 0.0899 \text{ g}$$

Therefore, the amount of 5-FU dissolved in water is 0.0899 g and the amount dissolved in 1-octanol is 0.0101 g.

Think About It Our analysis of the molecular structure of 5-FU in part (b) disclosed a high density of polar groups, including four capable of forming hydrogen bonds, which should make this compound more hydrophilic than BCNU with a lower 1-octanol–water ratio. These conclusions were supported by the low value (0.112) of the concentration ratio we were given to use in calculating the distribution of 5-FU between the two solvents in part (c). Thus, the results calculated in part (c) served to validate the predictions in part (b).

SUMMARY

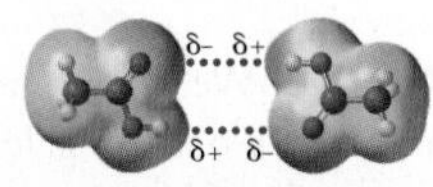

LO1 The solubility of ionic compounds in water is aided by strong **ion–dipole interactions**. Asymmetric molecules with permanent dipoles are attracted to each other via **dipole–dipole interactions**, the strengths of which are linked to their dipole moments. The strongest dipole–dipole interactions are **hydrogen bonds** that involve H atoms bonded to N, O, or F atoms. All molecules and individual atoms experience **dispersion forces** that grow in strength with increasing molar mass and particle surface area. These forces are the result of temporary **induced dipoles**, the strengths of which are linked to the **polarizability** of particles, which increases with increasing number of electrons surrounding their nuclei. (Sections 10.2 and 10.3)

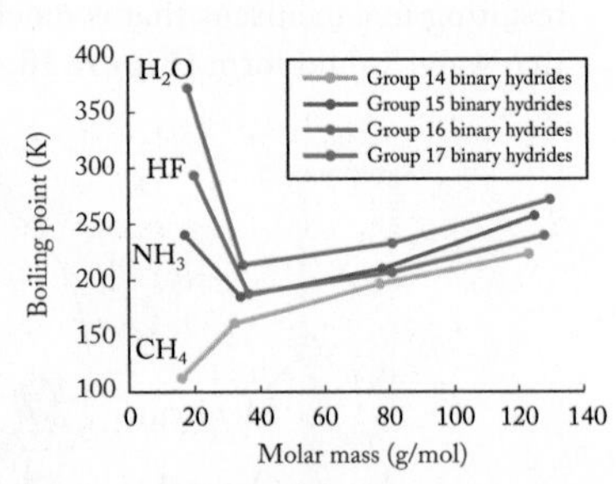

LO2 The vapor pressures of liquids decrease, and their boiling points increase with increasing strengths of the intermolecular interactions between their molecules. (Section 10.3)

LO3 The relative volatility of substances depends on the strength of the intermolecular interactions between particles. The **Clausius–Clapeyron equation** relates **vapor pressure**, the pressure exerted by a gas in equilibrium with its liquid phase, to absolute temperature. (Section 10.4)

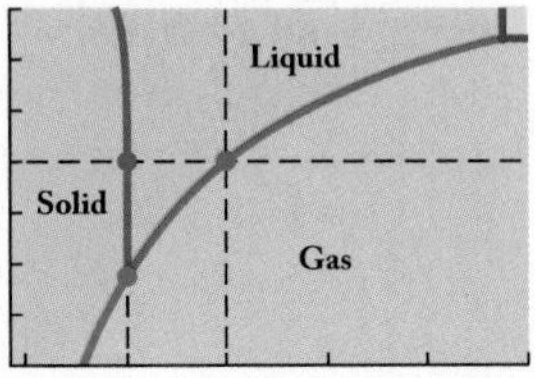

LO4 The **phase diagram** of a substance indicates whether the substance exists as a solid, liquid, gas, or **supercritical fluid** at particular combinations of pressure and temperature. For nearly all substances (except water), solid → liquid, solid → gas, and liquid → gas transitions occur at higher temperatures when pressure is increased. (Section 10.5)

LO5 The remarkable behavior of water, including its high melting and boiling points, its **surface tension**, its **capillary action**, and its **viscosity**, results from the strength of intermolecular hydrogen bonds. (Section 10.6)

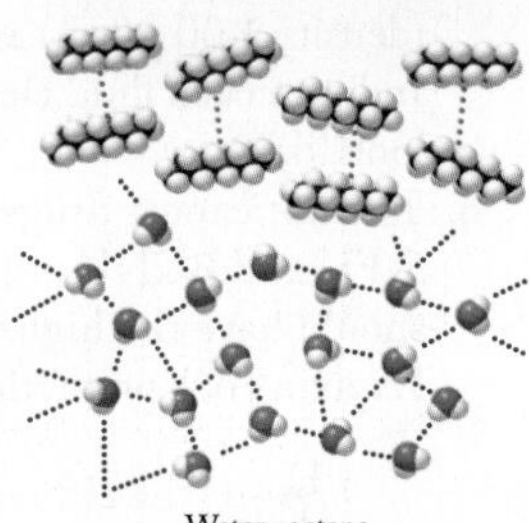
Water–octane mixture

LO6 Identifying the types of intermolecular forces between particles helps explain solubilities. The relationship is summarized by the phrase "like dissolves like." **Hydrophilic** (polar) substances are more soluble in water than are **hydrophobic** (nonpolar) substances. (Section 10.7)

LO7 The solubility of a gas in water decreases as the temperature increases because more molecules of dissolved gas have enough energy to overcome solute–solvent interactions. Solubility increases as the partial pressure of the gas increases because more frequent collisions between gas molecules and the water surface increase their concentration in solution. **Henry's law** gives the maximum concentration (the solubility) of a sparingly soluble gas in a liquid solvent. (Section 10.8)

PARTICULATE PREVIEW WRAP-UP

Molecules of ethanol and dimethyl ether experience similar dispersion forces because they have the same composition and molar mass. However, the hydrogen bonds between molecules of ethanol are stronger than the dipole–dipole interactions between molecules of dimethyl ether. The additional intermolecular attraction created by these hydrogen bonds is the reason that ethanol is a liquid at room temperature, whereas dimethyl ether is a gas.

PROBLEM-SOLVING SUMMARY

Type of Problem	Concepts and Equations	Sample Exercises
Explaining differences in boiling points of liquids and trends in boiling points of pure substances	Large molecules usually have higher boiling points than smaller molecules, with notable exceptions. The presence of polar –OH and –NH groups in molecules of a liquid leads to intermolecular hydrogen bonding that markedly increases the boiling point of the liquid.	**10.1, 10.2**
Calculating the vapor pressure, enthalpy of vaporization, or temperature of a pure liquid	Use the Clausius–Clapeyron equation: $$\ln\left(\frac{P_{\text{vap},T_2}}{P_{\text{vap},T_1}}\right) = -\frac{\Delta H_{\text{vap}}}{R}\left(\frac{1}{T_2} - \frac{1}{T_1}\right) \quad (10.2)$$ and solve for the unknown value.	**10.3**
Reading a phase diagram	Locate the point (T or P) specified by the question as the starting point for the exercise. Temperature is on the horizontal axis; pressure is on the vertical axis. When you draw a line either horizontally rightward from the pressure axis or vertically upward from the temperature axis, a phase change occurs wherever the line crosses an equilibrium line.	**10.4**
Predicting solubility in water and the miscibility of liquids	Polar molecules are more soluble in water than nonpolar molecules. Molecules that form hydrogen bonds are more soluble in water than molecules that cannot form hydrogen bonds. Like dissolves like.	**10.5, 10.6**
Calculating the solubility of a gas by using Henry's law	Use Henry's law: $$C_{\text{gas}} = k_{\text{H}}P_{\text{gas}} \quad (10.3)$$ where C_{gas} is the solubility of the gas; k_{H} is a constant that depends on the gas, the solvent, and the temperature; and P_{gas} is the pressure of the gas (or partial pressure if the gas is part of a mixture of gases).	**10.7**

VISUAL PROBLEMS

(Answers to boldface end-of-chapter questions and problems are in the back of the book.)

10.1. In Figure P10.1, identify the physical state (solid, liquid, or gas) of xenon and classify the attractive forces between the xenon atoms. How does the particulate representation in Figure P10.1 change as temperature increases?

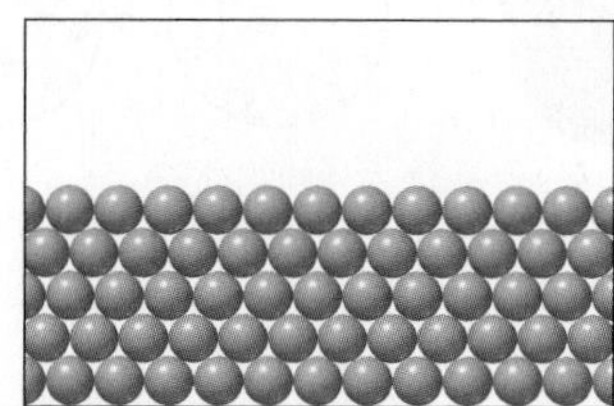

FIGURE P10.1

10.2. The molecules of three alkanes are depicted in Figure P10.2. Which of the three has the lowest boiling point? Explain your answer.

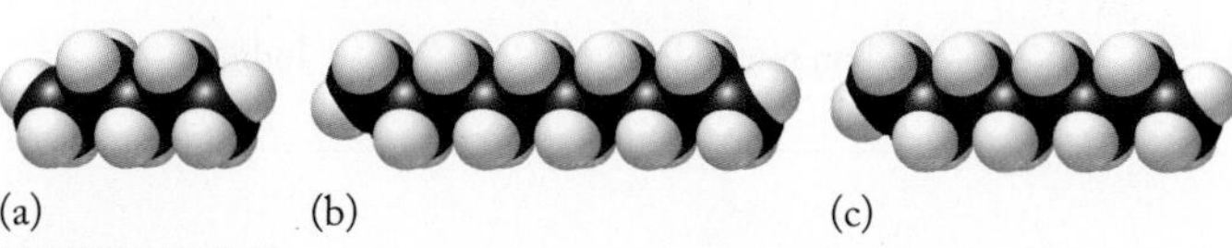

(a) (b) (c)

FIGURE P10.2

10.3. Which of the three alkanes in Figure P10.2 is the least soluble in water? Explain your answer.

10.4. Figure P10.4 shows molecules of XH_3 and ZH_3 (not shown to scale) and the boiling points of XH_3 and ZH_3 at 1 atm pressure. One substance is phosphine (PH_3) and the other substance is ammonia (NH_3). Which molecule is phosphine? Explain your answer.

XH_3	ZH_3
Boiling point −88°C	Boiling point −33°C

FIGURE P10.4

10.5. The graphs in Figure P10.5 have the same scales and describe the change in ln P_{vap} of two pure liquids as a function of temperature. Which liquid has the stronger intermolecular attractive forces? Explain your answer.

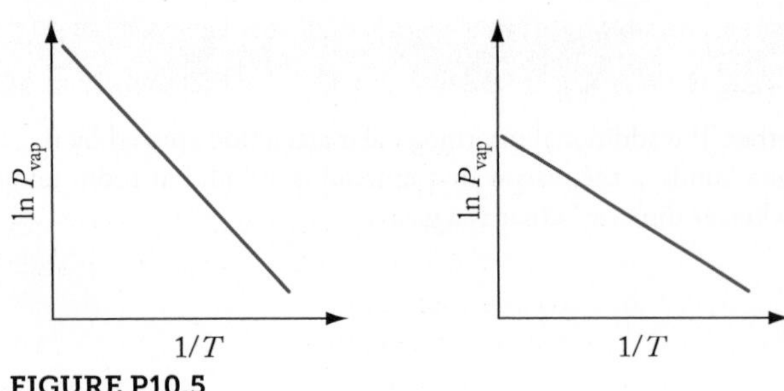

FIGURE P10.5

10.6. Which of the drawings in Figure P10.6, both of which are at constant temperature, most likely illustrates the pure liquid with the lower normal boiling point? Explain your choice.

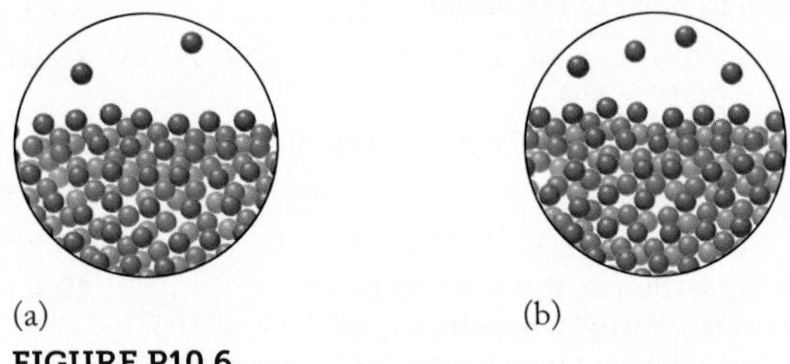

FIGURE P10.6

10.7. Examine the phase diagram of substance Z in Figure P10.7. (a) Does the freezing point of the substance increase or decrease with increasing pressure? (b) Do you predict the solid phase of substance Z would float on the liquid phase? Why?

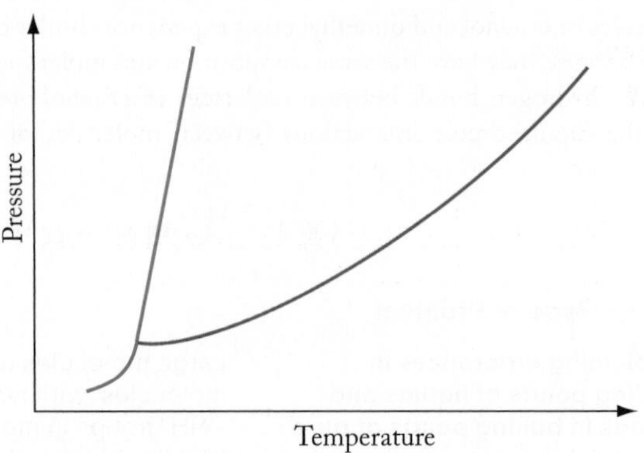

FIGURE P10.7

10.8. Use representations [A] through [I] in Figure P10.8 to answer questions (a)–(f).

a. What are the predominant intermolecular forces in each of the substances in the electrostatic potential maps?
b. Which substance requires the lowest temperature to condense?
c. Which representations are isomers of one another?
d. Consider ethylene glycol and iodine. Which will dissolve in ethanol? Which will dissolve in carbon tetrachloride? Name the intermolecular forces responsible.
e. Which process shows an increase in partial pressure?
f. Which process would occur due to an increase in temperature?

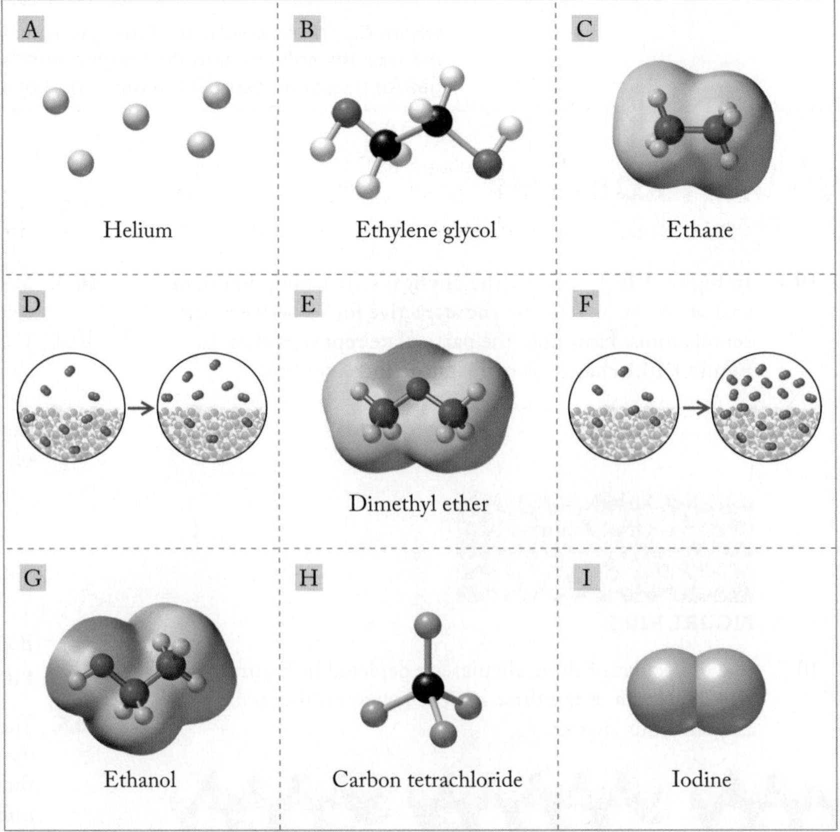

FIGURE P10.8

QUESTIONS AND PROBLEMS

Dispersion Forces

Concept Review

10.9. Which type of intermolecular force exists in all substances?

10.10. At room temperature, bromine (Br_2) is a corrosive red liquid, whereas iodine (I_2) is a volatile violet solid. The differences point to different strengths of intermolecular forces between these halogens, with those for I_2 being stronger. What kind of intermolecular force is responsible for these differences?

10.11. Why do gases behave nonideally at high pressures and low temperatures?

10.12. Why are normal boiling points generally lower for branched hydrocarbons than for straight-chain hydrocarbons of the same molecular mass?

Problems

10.13. In each of the following pairs of molecules, which one experiences the stronger dispersion forces? (a) CCl_4 or CF_4; (b) CH_4 or C_3H_8

10.14. What kinds of intermolecular forces must be overcome as solid CO_2 sublimes?

***10.15.** Consider the two molecules shown in Figure P10.15. One of these compounds is a liquid at room temperature; the other is a solid. Which is which? Explain why.

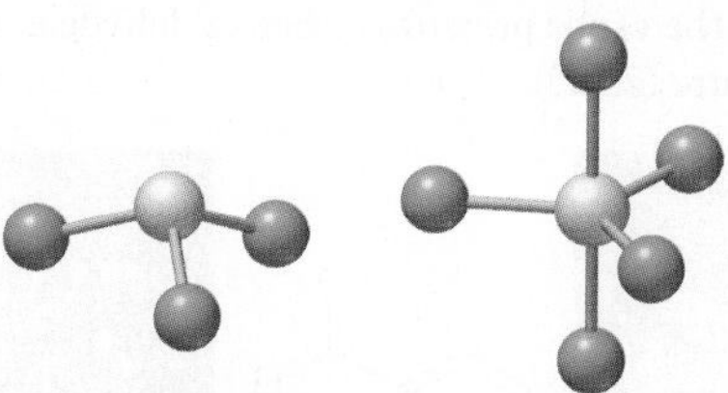

FIGURE P10.15

*10.16. Consider the two molecules shown in Figure P10.16. One has a boiling point of 57°C; the other, 84°C. Predict which has the higher boiling point and explain your choice.

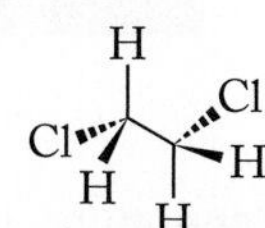

H, Cl, Cl, H, H, H

$ClCH_2CH_2Cl$ $CHCl_2CH_3$

FIGURE P10.16

Interactions Involving Polar Molecules

Concept Review

10.17. Why are dipole–dipole interactions generally weaker than ion–dipole interactions?

10.18. Two liquids—one polar, one nonpolar—have the same molar mass. Which one is likely to have the higher boiling point? Explain your answer.

10.19. Why are hydrogen bonds considered a special class of dipole–dipole interactions?

10.20. Can all polar hydrogen-containing molecules form hydrogen bonds? Why or why not?

Problems

10.21. The permanent dipole moment of CH_2F_2 (1.93 D) is larger than that of CH_2Cl_2 (1.60 D), yet the boiling point of CH_2Cl_2 (40°C) is much higher than that of CH_2F_2 (−52°C). Why?

10.22. How is it that the permanent dipole moment of HCl (1.08 D) is larger than the permanent dipole moment of HBr (0.82 D), yet HBr boils at a higher temperature?

10.23. **The Smell of the Sea** The distinct odor of the seashore at low tide results in part from the presence of dimethylsulfide (CH_3SCH_3), a molecule with the same structure as dimethyl ether (CH_3OCH_3). Their standard heats of vaporization ΔH°_{vap} are different: 28 kJ/mol versus 23 kJ/mol, respectively. Which substance has the stronger intermolecular forces? Can you suggest why?

10.24. **At the Fish Market** Fish emit increasing amounts of trimethylamine, $(CH_3)_3N$, as they decay, accounting for the distinctive odor one senses toward the end of the day at an open-air market. Opening a bottle of household ammonia soon leads to the smell of NH_3 in the immediate vicinity. Based on the thermochemical data for NH_3 (ΔH°_{vap} = 1.16 kJ/mol) and $(CH_3)_3N$ (ΔH°_{vap} = 27 kJ/mol), which substance has the higher boiling point?

10.25. **Detecting Natural Gas Leaks** Energy companies add small amounts of ethanethiol (CH_3CH_2SH) to the CH_4 in their pipelines because ethanethiol has a strong odor and will warn consumers of a leak in their equipment.

a. Ethanethiol and ethanol (CH_3CH_2OH) have the same structure and differ only in which group 16 element is present. Explain why the boiling point of ethanethiol (35°C) is lower than the boiling point of ethanol (78.5°C).

*b. To reach your nose, ethanethiol must diffuse through air when a leak occurs. How much faster will methane diffuse than ethanethiol?

10.26. Explain the differences in the physical properties of the following pairs of compounds.

a. Why is the melting point of methyl fluoride (CH_3F, −142°C) higher than the melting point of methane (CH_4, −182°C).

b. Why is the boiling point of Br_2 (59°C) lower than that of iodine monochloride (ICl, 97°C), even though they have nearly the same molar mass.

c. Will Br_2 and ICl have similar root-mean-square speeds at 100°C?

10.27. In which of the following compounds do the molecules experience the strongest dipole–dipole attractions? (a) CF_4; (b) CF_2Cl_2; (c) CCl_4

10.28. Which of the following compounds, CO_2, NO_2, SO_2, or H_2S, is expected to have the weakest interactions between its molecules?

10.29. Which of the following molecules can form hydrogen bonds among themselves in pure samples of bulk material? (a) methanol (CH_3OH); (b) ethane (CH_3CH_3); (c) dimethyl ether (CH_3OCH_3); (d) acetic acid (CH_3COOH)

10.30. Which of the following molecules can form hydrogen bonds with molecules of water? (a) methanol (CH_3OH); (b) ethane (CH_3CH_3); (c) dimethyl ether (CH_3OCH_3); (d) acetic acid (CH_3COOH)

Vapor Pressure of Pure Liquids

Concept Review

10.31. Why does the vapor pressure of a liquid increase as temperature increases?

10.32. Which of the following factors influences the vapor pressure of a pure liquid?
a. the volume of liquid present in a container
b. the temperature of the liquid
c. the surface area of the liquid

10.33. Is vapor pressure an intensive or extensive property of a liquid?

10.34. During the development of the U2 advanced reconnaissance aircraft a new jet fuel had to be formulated because existing fuels would boil away at the very high altitudes the aircraft could achieve. Why does boiling point decrease as a function of altitude?

Problems

10.35. Rank the following compounds in order of increasing vapor pressure at 298 K. (a) CH_3CH_2OH; (b) CH_3OCH_3; (c) $CH_3CH_2CH_3$

10.36. Rank the compounds in Figure P10.36 in order of increasing vapor pressure at 298 K.

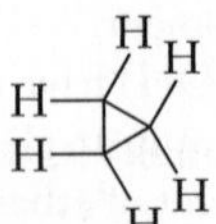
Cyclopropane

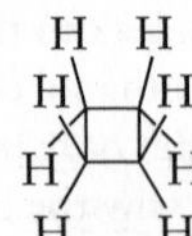
Cyclobutane

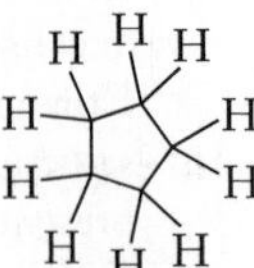
Cyclopentane

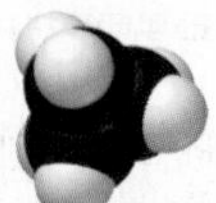

FIGURE P10.36

10.37. **Pine Oil** The smell of fresh cut pine arises in part from the cyclic alkene pinene, whose carbon-skeleton structure is shown in Figure P10.37.
a. Use the data in the table to calculate the heat of vaporization (ΔH_{vap}) of pinene.
b. Use the value of ΔH_{vap} determined in part (a) to calculate the vapor pressure of pinene at room temperature (23°C).

Pinene

FIGURE P10.37

Vapor Pressure (torr)	Temperature (K)
760	429
515	415
340	401
218	387
135	373

10.38. **Almonds and Cherries** Almonds and almond extracts are common ingredients in baked goods. Almonds contain the compound benzaldehyde (shown in Figure P10.38), which accounts for the odor of the nut. Benzaldehyde is also responsible for the aroma of cherries.
a. Use the data in the table to calculate the heat of vaporization (ΔH_{vap}) of benzaldehyde.
b. Use the value of ΔH_{vap} determined in part (a) to calculate the vapor pressure of benzaldehyde at room temperature (23°C).

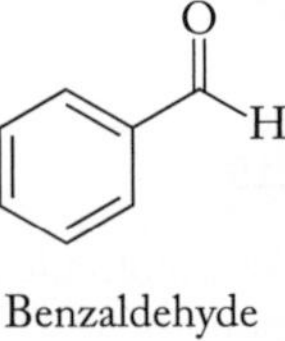

Benzaldehyde

FIGURE P10.38

Vapor Pressure (torr)	Temperature (K)
50	373
111	393
230	413
442	433
805	453

10.39. Skunks The distinctive odor of skunks results from the butanethiol, $CH_3(CH_2)_3SH$, that they spray when threatened. A plot of the natural logarithm of vapor pressure as a function of temperature for butanethiol is shown in Figure P10.39. Calculate the heat of vaporization for $CH_3(CH_2)_3SH$.

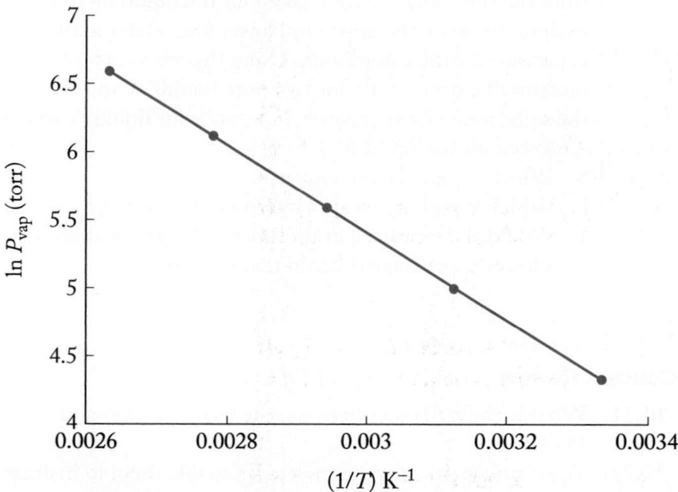

FIGURE P10.39

10.40. **Chemical Signaling** When some Saharan ants are injured, they emit a volatile sulfur compound called dimethyl disulfide (CH_3SSCH_3). A plot of the natural logarithm of vapor pressure as a function of temperature for dimethyl disulfide is shown in Figure P10.40. Calculate the heat of vaporization for CH_3SSCH_3.

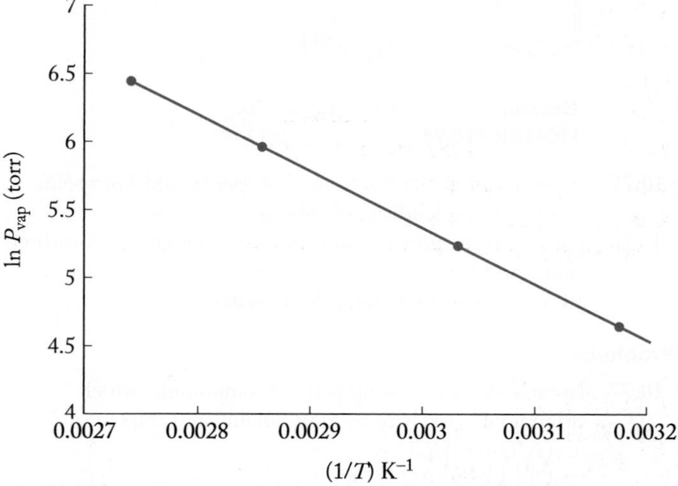

FIGURE P10.40

Phase Diagrams: Intermolecular Forces at Work

Concept Review

10.41. What phases of a substance are present at (a) its triple point and (b) its critical point?

10.42. Explain how the solid–liquid line in the phase diagram of water differs in character from the solid–liquid line in the phase diagrams of most other substances, such as CO_2.

10.43. Which phase of a substance (gas, liquid, or solid) is most likely to be the stable phase: (a) at low temperatures and high pressures; (b) at high temperatures and low pressures?

10.44. At what temperature and pressure does a substance behave as a supercritical fluid?

10.45. Preserving Food Freeze-drying is used to preserve food at low temperature with minimal loss of flavor. Freeze-drying works by freezing the food and then lowering the pressure with a vacuum pump to sublime the ice. Must the pressure be lower than the pressure at the triple point of H_2O? Why or why not?

10.46. Solid helium cannot be converted directly into the vapor phase. Does the phase diagram of helium have a triple point?

Problems

(For help in answering Problems 10.47 through 10.56, consult Figures 10.21, 10.23, and 10.24.)

10.47. What is the normal boiling point of bromine?

10.48. If water boils at 50°C, what is the pressure?

10.49. Which molecules have stronger intermolecular attractive forces: ethylene glycol or ethanol? Explain your choice.

10.50. What is the boiling point of diethyl ether at 300 mmHg of pressure?

10.51. List the steps you would take to convert a 10.0 g sample of water at 25°C and 1 atm of pressure to water at its triple point.

10.52. List the steps you would take to convert a 10.0 g sample of water at 25°C and 2 atm pressure to ice at 1 atm of pressure. At what temperature would the water freeze?

10.53. Predict which phase water exists in at 100°C and 5.0 atm. What phase changes, if any, does water undergo when the pressure is reduced to 0.5 atm at constant temperature?

10.54. Predict which phase CO_2 exists in at −80°C and 8.0 atm. What phase changes, if any, does CO_2 undergo when allowed to warm to −25°C at 5.0 atm?

10.55. Glaciers Melting glaciers in Greenland have the potential to significantly raise the level of the ocean. An article on the Jakobshavn Glacier contained the following passage: "Because of the extreme pressures involved, the freezing point of the ice itself changes, making it more susceptible to melting."

a. Is this statement valid?
b. Use the phase diagram of water to estimate the pressure required to reduce the melting point of ice by 0.5°C.

10.56. **Sequestration of Carbon Dioxide** One solution to combat the increasing concentration of CO_2 in Earth's atmosphere is to capture it and store it under high pressure.
a. What are the pressures required to convert CO_2 gas to solid CO_2 (dry ice) and to liquid CO_2 at −50°C?
b. What volume would 1 million grams of CO_2 vapor (d = 1.977 g/L at 0°C) occupy when cooled and condensed to liquid CO_2 (d = 1.101 g/mL at −37°C)?

Some More Remarkable Properties of Water

Concept Review

10.57. Explain why a needle floats on the surface of water but sinks in a container of methanol (CH_3OH).

10.58. Explain why different liquids do not reach the same height in capillary tubes of the same diameter.

10.59. Explain why pipes filled with water are in danger of bursting when the temperature drops below 0°C.

10.60. A hot needle sinks when put on the surface of cold water. Will a cold needle float in hot water? Explain your answer.

10.61. The meniscus of water in a glass tube is concave, but that of mercury (Figure P10.61) is convex. Explain why.

*10.62. The mercury level in a capillary tube inserted into a dish of mercury is below the level of the mercury in the dish. Explain why.

FIGURE P10.61

10.63. Describe the origin of surface tension at the molecular level.

10.64. The viscosities of alkane hydrocarbons increase as the number of carbon atoms per molecule increases. Explain why.

10.65. Describe how the surface tension and viscosity of a liquid are affected by increasing temperature.

10.66. Explain how strong intermolecular forces are expected to result in a relatively high surface tension and viscosity of a liquid.

Problems

10.67. One of two glass capillary tubes of the same diameter is placed in a dish of water and the other in a dish of ethanol (CH_3CH_2OH). Which liquid will rise higher in its tube? Explain your answer.

10.68. Would you expect water to rise to the same height in a tube made of a polyethylene plastic as it does in a glass capillary tube of the same diameter? Why or why not? The molecular structure of polyethylene is shown in Figure P10.68.

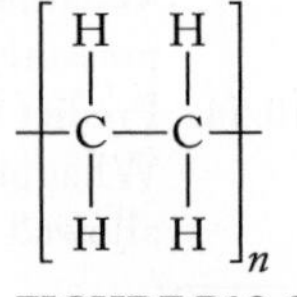

FIGURE P10.68

10.69. The normal boiling points of liquids A and B are 75.0°C and 151°C, respectively. Which of these liquids would you expect to have the higher surface tension and viscosity at 25°C? Explain your answer.

10.70. A simple viscometer consists of a thick-walled glass tube with a 0.5-mm bore. The tube has etched marks at one-quarter and three-quarters of its height. The tube is clamped with its lower end dipped in a container of the liquid to be tested. A pipet filler is used to draw liquid up the bore past the upper mark. The pipet filler is removed from the tube, and the time taken for the liquid meniscus to drain between the upper and lower viscometer marks is measured with a stopwatch. Using this viscometer to measure the drain times for two pure liquids A and B at the same temperature gives 3.45 seconds for liquid A and 4.64 seconds for liquid B.
a. Which liquid is more viscous?
b. Which liquid has weaker intermolecular forces?
c. Would the measured drain times be longer or shorter at a lower experimental liquid temperature?

Polarity and Solubility

Concept Review

10.71. What is the difference between the terms *miscible* and *insoluble*?

10.72. What properties of water molecules enable them to hydrate and separate cations and anions in aqueous solutions?

10.73. One of the compounds in Figure P10.16 is insoluble in water; the other has a water solubility of 0.87 g/100 mL at 20°C. Identify which is which and explain your reasoning.

10.74. Azaborine (C_4H_6BN; Figure P10.74) *is isoelectronic with benzene* (C_6H_6). Which of these two compounds is likely to have greater solubility in water?

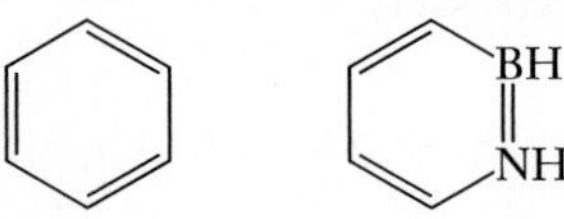

Benzene Azaborine
FIGURE P10.74

10.75. In what context do the terms *hydrophobic* and *hydrophilic* relate to the solubilities of substances in water?

10.76. How does the presence of increasingly longer hydrocarbon chains in the structure affect the solubility of a series of structurally related molecules in water?

Problems

10.77. In each of the following pairs of compounds, which compound is likely to be more soluble in water?
a. CCl_4 or $CHCl_3$
b. CH_3OH or $C_6H_{11}OH$
c. NaF or MgO
d. CaF_2 or BaF_2

10.78. In each of the following pairs of compounds, which compound is likely to be more soluble in CCl_4?
a. Br_2 or NaBr
b. CH_3CH_2OH or CH_3OCH_3
c. CS_2 or KOH
d. I_2 or CaF_2

10.79. Which of these pairs of substances is likely to be miscible?
a. Br_2 and C_6H_6 (benzene)
b. $CH_3CH_2OCH_2CH_3$ (diethyl ether) and CH_3COOH (acetic acid)
c. C_6H_{12} (cyclohexane) and hexane ($CH_3CH_2CH_2CH_2CH_2CH_3$)
d. CS_2 (carbon disulfide) and CCl_4 (carbon tetrachloride)

10.80. Which of these pairs of substances is likely to be miscible?
a. CH_3CH_2OH (ethanol) and $CH_3CH_2OCH_2CH_3$ (diethyl ether)
b. CH_3OH (methanol) and methyl amine (CH_3NH_2)
c. CH_3CN (acetonitrile) and acetone (CH_3COCH_3)
d. CF_3CHF_2 (a Freon replacement) and $CH_3CH_2CH_2CH_2CH_3$ (pentane)

10.81. Which of the following compounds is likely to be the most soluble in water? (a) NaCl; (b) KI; (c) $Ca(OH)_2$; (d) CaO

10.82. Based on the graph in Figure P10.82, which has a greater effect on the solubility of oxygen in water: (a) decreasing the temperature from 20°C to 10°C or (b) raising the pressure from 1.00 atm to 1.25 atm?

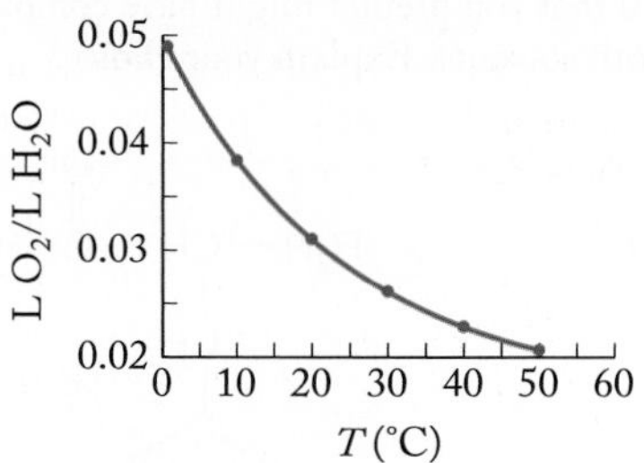

FIGURE P10.82

10.83. Improving the Solubility of Drugs Many common over-the-counter medications (Figure P10.83) are sold as salts to improve their water solubility. Explain why the salts of acetylsalicylic acid and pseudoephedrine are more soluble in water?

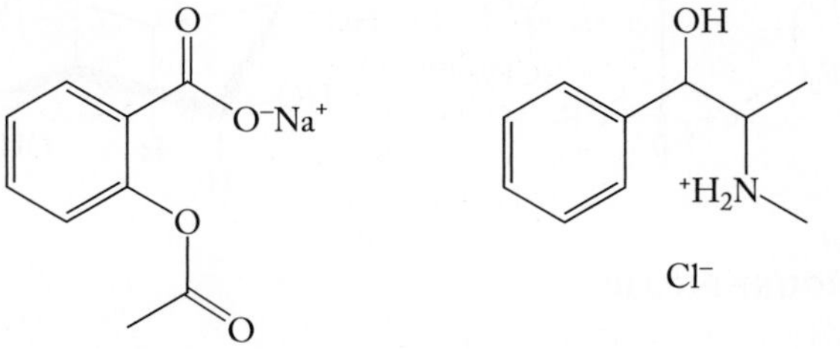

Sodium acetylsalicylate Pseudoephedrine hydrochloride

FIGURE P10.83

10.84. Predict the order of water solubility (from lowest to highest) for the following series of alcohols and explain your reasoning.
a. $CH_3(CH_2)_2CH_2OH$
b. $CH_3(CH_2)_4CH_2OH$
c. $CH_3(CH_2)_6CH_2OH$
d. $CH_3(CH_2)_8CH_2OH$

10.85. Which of the compounds in Figure P10.85 are alcohols and which ones are ethers? Place them in order of increasing boiling point.

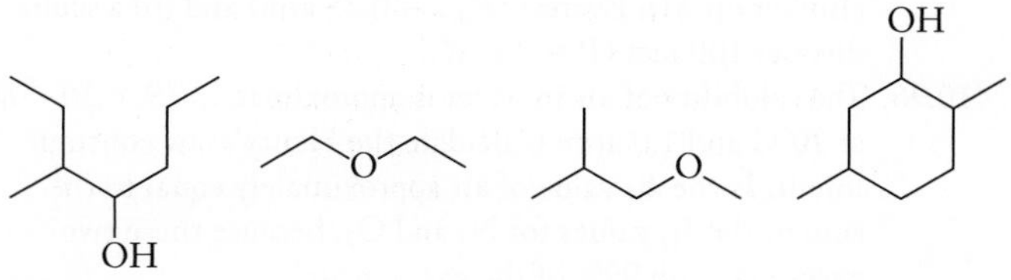

(a) (b) (c) (d)

FIGURE P10.85

10.86. Which of the compounds in Figure P10.86 are alcohols and which ones are ethers? Place them in order of increasing vapor pressure at 25°C.

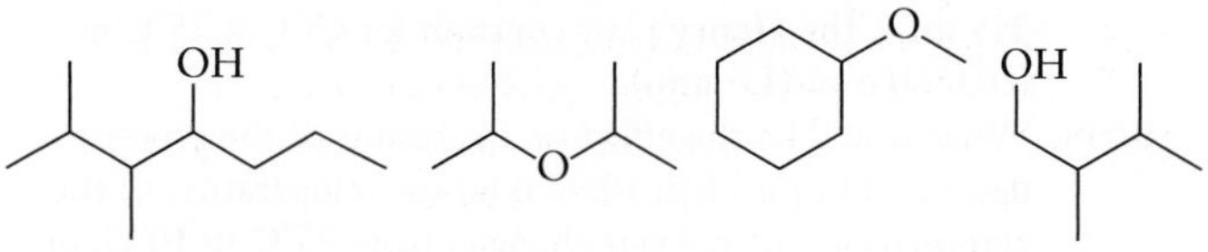

(a) (b) (c) (d)

FIGURE P10.86

Solubility of Gases in Water

Concept Review

10.87. Why does the solubility of most gases in most liquids increase with decreasing temperature?

10.88. Which term, k_H or P, in Henry's law is affected by temperature?

10.89. Air is primarily a mixture of nitrogen and oxygen. Is the Henry's law constant for the solubility of air in water the sum of k_H for N_2 and k_H for O_2? Explain why or why not.

10.90. Why is the Henry's law constant for CO_2 so much larger than those for N_2 and O_2 at the same temperature?

10.91. Which compound in each pair would you predict to be more soluble in nonpolar solvents: SO_2 or SO_3 and CO_2 or NO_2?

10.92. A chef observes bubbles while heating a pot of water to 60°C to poach vegetables. What are the gases in the bubbles and where did they come from?

Problems

10.93. Arterial Blood Arterial blood contains about 0.25 g of oxygen per liter at 37°C and standard atmospheric pressure. What is the Henry's law constant [in mol/(L · atm)] for O_2 dissolution in blood? The mole fraction of O_2 in air is 0.209.

10.94. The solubility of O_2 in water is 6.5 mg/L at an atmospheric pressure of 1 atm and temperature of 40°C. Calculate the Henry's law constant of O_2 at 40°C. The mole fraction of O_2 in air is 0.209.

*10.95. **Oxygen for Climbers and Divers** Use the Henry's law constant for O_2 dissolved in arterial blood from Problem 10.93 to calculate the solubility of O_2 in the blood of (a) a climber on Mt. Everest (P_{atm} = 0.35 atm) and (b) a scuba diver at 100 feet ($P \approx 3$ atm).

*10.96. The solubility of air in water is approximately 7.9×10^{-4} *M* at 20°C and 1.0 atm. Calculate the Henry's law constant for air. Is the k_H value of air approximately equal to the sum of the k_H values for N_2 and O_2, because these two gases make up 99% of the gases in air?

10.97. Carbonated Sodas Manufacturers of carbonated beverages dissolve $CO_2(g)$ in water under pressure to produce sodas. If a manufacturer uses a pressure of 4.16 atm at 25°C to carbonate the water, how many grams of CO_2 are in the average can of soda, the volume of which is 355 mL? The Henry's law constant for CO_2 at 25°C is 0.03360 mol/(L · atm).

10.98. What would be the effect on the results of the process described in Problem 10.97 if (a) the temperature of the carbonation process was changed from 25°C to 10°C, or (b) the temperature was held constant, but the pressure was doubled?

Additional Problems

10.99. Why do ethers typically boil at lower temperatures than alcohols with the same molecular formula?

*10.100. **Dry Gas** During the winter months in cold climates, water condensing in a vehicle's gas tank reduces engine performance. An auto mechanic recommends adding "dry gas" to the tank during your next fill-up. Dry gas is typically an alcohol that dissolves in gasoline and absorbs water. From the structures shown in Figure P10.100, which product do you predict would do a better job—methanol or 2-propanol? Why?

CH_3OH Methanol

OH 2-Propanol

FIGURE P10.100

10.101. Does the sublimation point of ice increase or decrease with increasing pressure? Explain why.

10.102. Does the sublimation point of CO_2 increase or decrease with increasing pressure? Explain why.

10.103. Liquid substances are often compared for their physical properties in different applications. Comparison of two liquids A and B at constant temperature and atmospheric pressure shows that liquid A has higher viscosity and surface tension, a higher boiling point, and lower vapor pressure than liquid B. Are these data all consistent with stronger intermolecular forces in liquid A than in liquid B?

10.104. Why is methanol (CH_3OH) miscible with water, but CH_4 is almost completely insoluble in water?

10.105. Sketch a phase diagram for element X, which has a triple point at 152 K and a pressure of 0.371 atm, a boiling point of 166 K at a pressure of 1.00 atm, and a normal melting point of 161 K.

*10.106. The melting point of hydrogen is 14.96 K at 1.00 atm of pressure. The temperature at its triple point is 13.81 K. Does H_2 expand or contract when it freezes?

10.107. Explain why water climbs higher in a capillary tube than in a test tube.

10.108. Explain why ice floats on water.

10.109. Fish Dying in Summer Heat Explain why fish in a pond die if water becomes too warm.

*10.110. **Evaluation of Pharmaceuticals** A test done on new pharmaceutical agents early in the development process required the observation of their relative solubilities in octanol and water. A drug had to be sufficiently soluble in water (hydrophilic) to be carried in the bloodstream but also sufficiently hydrophobic (octanol soluble) to move across cell membranes. Pick the molecule from Figure P10.110 that you predict might have comparable solubility in both solvents. Explain your choice.

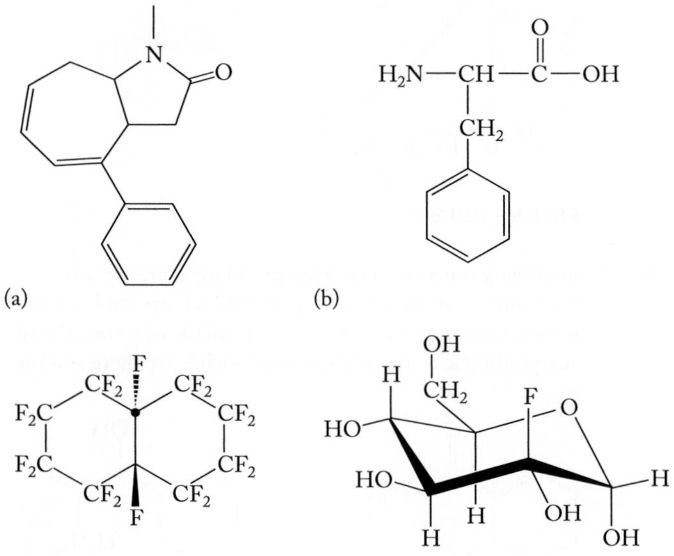

(a) N, O

(b) H_2N—CH—C(=O)—OH, CH_2

(c) F_2C, CF_2, F

(d) OH, H, CH_2, F, O, HO, HO, H, H, OH, H, OH

FIGURE P10.110

10.111. First Aid for Bruises Compounds with low boiling points may be sprayed on skin as a topical anesthetic—they chill it as they evaporate, providing short-term relief from injuries. Predict which compound among those in Figure P10.111 has the lowest boiling point.

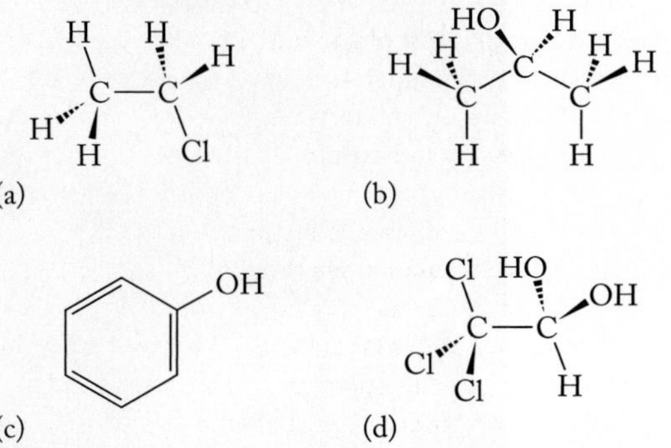

(a)

(b)

(c)

(d)

FIGURE P10.111

10.112. Refrigerators have a unit called a compressor that liquefies a gas. The refrigerator is cooled by a continuous cycle of compression of the gas to produce the liquid, followed by evaporation of the liquid to provide the cooling. Ammonia (NH_3) and sulfur dioxide (SO_2) were the gases used originally, and hexafluoroethane (C_2F_6) has been used since the 1990s. What are the intermolecular interactions that characterize these substances?

11

Solutions

Properties and Behavior

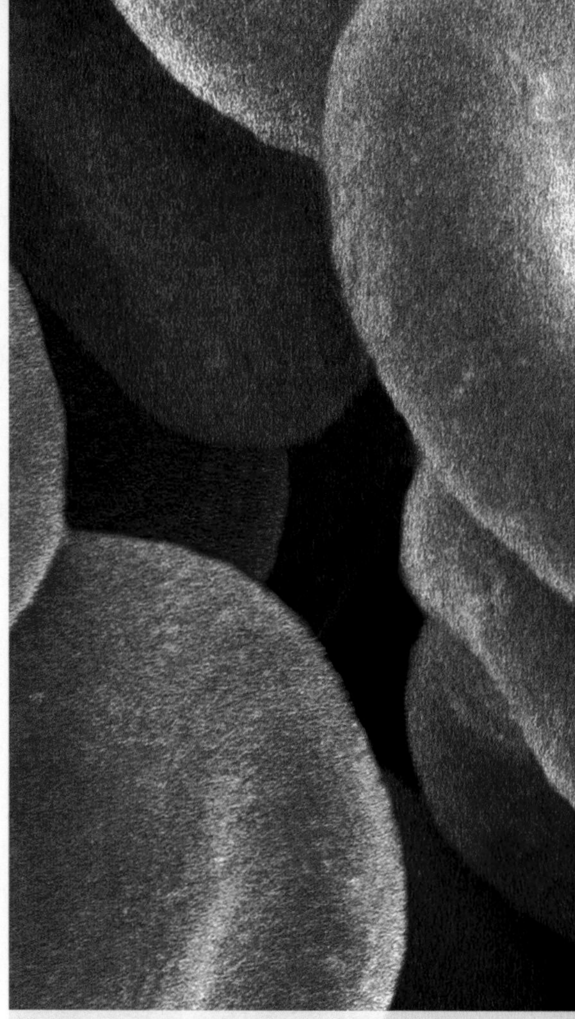

HEALTHY RED BLOOD CELLS
Blood is a complex aqueous mixture of dissolved and suspended components. The latter include red blood cells, which contain hemoglobin, a substance that binds oxygen and transports it from the lungs to tissues.

PARTICULATE **REVIEW**

Size Trends: Neutral Atoms versus Ions

In Chapter 11 we explore the properties of solutions containing dissolved solutes, both molecular and ionic. Colored spheres representing neutral atoms of potassium, sodium, sulfur, and fluorine, along with their most common ions, are shown here. In each case, the neutral atom is above the ion. Answer these questions based on the periodic trends for atomic size.

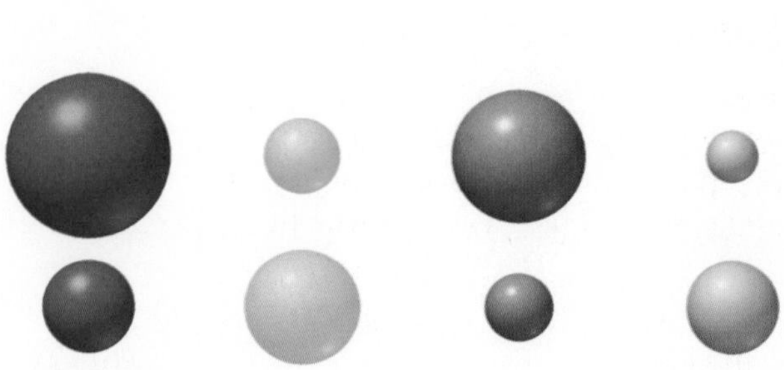

- Which pair is the potassium atom and its ion?
- Which pair is the fluorine atom and its ion?
- Which of the remaining pairs is sulfur and the sulfide ion? Which pair is sodium and its ion?

(Review Section 7.10 if you need help.)

(Answers to Particulate Review questions are in the back of the book.)

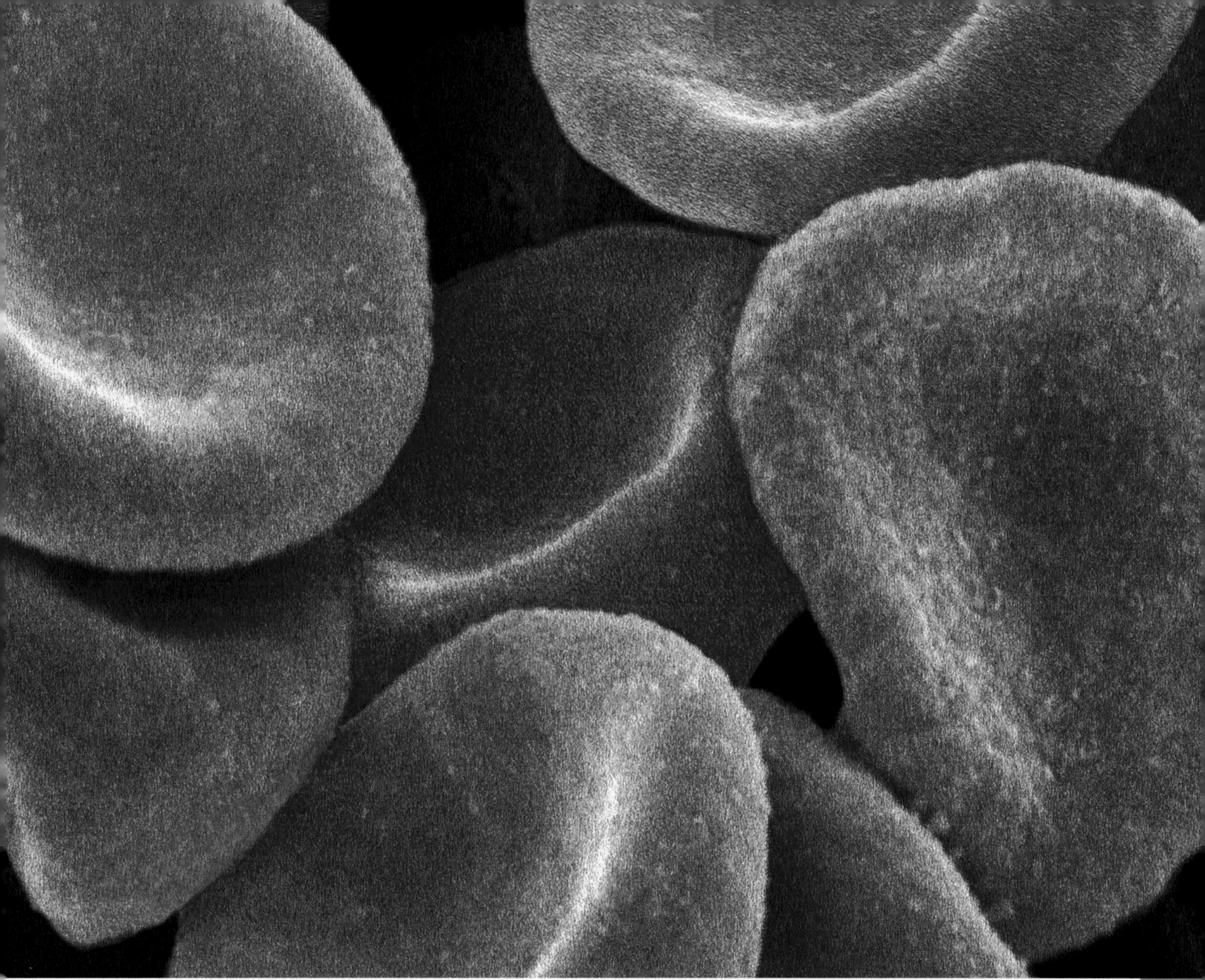

PARTICULATE **PREVIEW**

Solvent versus Solution

The covered containers shown here hold pure water (on the left) and an aqueous solution of NaCl (on the right). As you read Chapter 11, look for ideas that will help you answer these questions:

- When NaCl(*s*) is added to water, what intermolecular forces produce the solution on the right?
- What happens to the amount of water in the vapor phase when salt is added?
- How does the presence of a dissolved solute such as salt affect the boiling point of water?

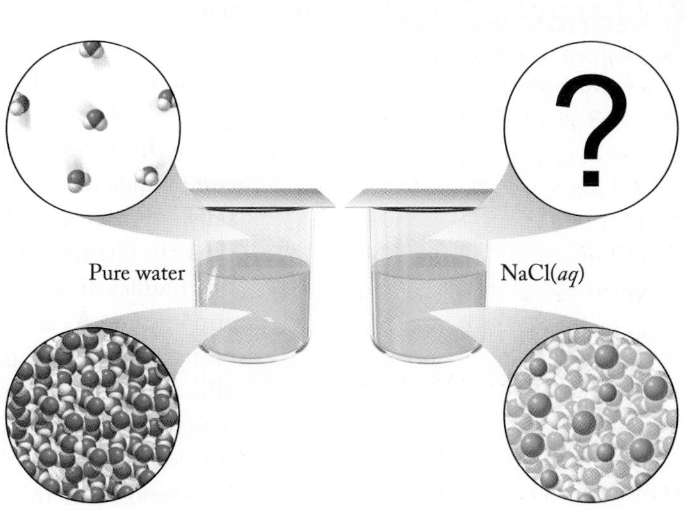

Learning Outcomes

LO1 Estimate the relative strengths of ion–ion interactions and explain how these interactions affect the physical properties of ionic compounds
Sample Exercise 11.1

LO2 Explain the energy changes that accompany the formation and dissolution of an ionic compound
Sample Exercises 11.2, 11.3, 11.4

LO3 Use Raoult's law to relate the vapor pressure of a solution to the concentration of solute particles
Sample Exercise 11.5, 11.7

LO4 Relate the fractional distillation of a mixture of volatile compounds to the vapor pressures of the compounds
Sample Exercise 11.6

LO5 Relate the molality of a solution to the quantity of solute and mass of solvent
Sample Exercise 11.8

LO6 Relate the freezing point or boiling point of a solution to the molal concentration of nonvolatile solute particles
Sample Exercises 11.9, 11.10

LO7 Explain the significance of the van 't Hoff factor and use it to predict the colligative properties of solutions
Sample Exercises 11.11, 11.12

LO8 Describe osmosis, predict the direction of solvent flow in osmosis, and calculate osmotic pressure
Sample Exercises 11.13, 11.14, 11.15

LO9 Use colligative properties such as osmotic pressure to determine solute molar mass
Sample Exercise 11.16

11.1 Interactions between Ions

We began Chapter 4 by noting that marine mammals can regulate the concentrations of electrolytes in their tissues to about one-third the concentrations of the same ions in seawater. Actually, electrolyte concentrations in the tissues and blood of most mammals, including us, must be maintained in narrow ranges for normal body function and good health. Therefore, the fluids dispensed intravenously in hospitals to deliver medications or to restore fluid levels in trauma victims must contain very nearly the same total number of dissolved particles (molecules and ions) in a given volume as there are in the same volume of blood plasma. Too high a concentration of solutes in the intravenous fluid causes dehydration; too low a concentration can cause edema, which is swelling that results from excess fluid in body tissues.

In this chapter we examine the properties of solutions in more detail, including some of the ways in which they differ from the properties of their pure solvents. For example, the solubilities of biopolymers, such as proteins, in aqueous solutions depends on the concentrations of dissolved salts such as NaCl that are also present in these solutions. If salt concentrations are high enough, water molecules that would normally cluster around the ionic groups on the surface of protein molecules, thereby stabilizing the protein in solution, cluster around Na^+ and Cl^- ions instead. The loss of ion–dipole interactions on the surface of the protein destabilizes it and causes it to precipitate. Scientists make use of this *salting out effect* to isolate proteins from samples of biological fluids.

CONNECTION We introduced solutions and several units of concentration in Section 4.2 when discussing how much solute is dissolved in the solvent.

In this chapter we also continue the discussion from Chapter 10 on why some compounds dissolve in water and others do not, but here we focus on why some salts are more soluble in water than others. To visualize this phenomenon, consider the changes of state that begin when an ocean wave crashes on a rocky coast. The plumes of sea spray the wave produces carry small drops of seawater into the atmosphere. As some of the water in these drops evaporates, their concentrations of dissolved ions (e.g., Cl^-, Na^+, Mg^{2+}, Br^-, Ca^{2+}, and SO_4^{2-}) increase. Eventually, the decreased volume of the drops produces saturated solutions of the salts, which then begin to precipitate. Among the first solids to form is $CaSO_4$. Among the last is NaCl, which does not precipitate until 90% of the seawater in a drop has evaporated. This sequence takes place even though the concentrations of Ca^{2+} and SO_4^{2-} ions are much lower than the concentrations of Na^+ and Cl^- ions in seawater. Why does $CaSO_4$ precipitate before NaCl? Put another way: why is $CaSO_4$ less soluble in water than NaCl?

There are several factors that influence the solubility of salts in water. One of the most important is the strength of the bonds that keep ions of opposite charge together in ionic solids and that must be overcome if the salt is to dissolve. These ionic bonds are among the strongest chemical bonds. We have seen (in Chapters 6 and 8) that the strengths of ionic bonds are defined by the electrostatic potential energies, E, between pairs of ions, which are functions of the product of their charges, $(Q_1 \times Q_2)$, divided by the distance between the centers of the ions (d):

$$E \propto \frac{(Q_1 \times Q_2)}{d} \qquad (6.3)$$

We can turn this expression into an equation by replacing its proportionality symbol with a constant of proportionality:

$$E = 2.31 \times 10^{-16}\,\text{J} \cdot \text{pm}\left(\frac{Q_1 \times Q_2}{d}\right) \qquad (11.1)$$

The value of this constant gives us energy in joules when we express the distance between ion centers in picometers. Calculating the value of d involves summing the appropriate ionic radii shown in **Figure 11.1** and listed in **Table 11.1**. The

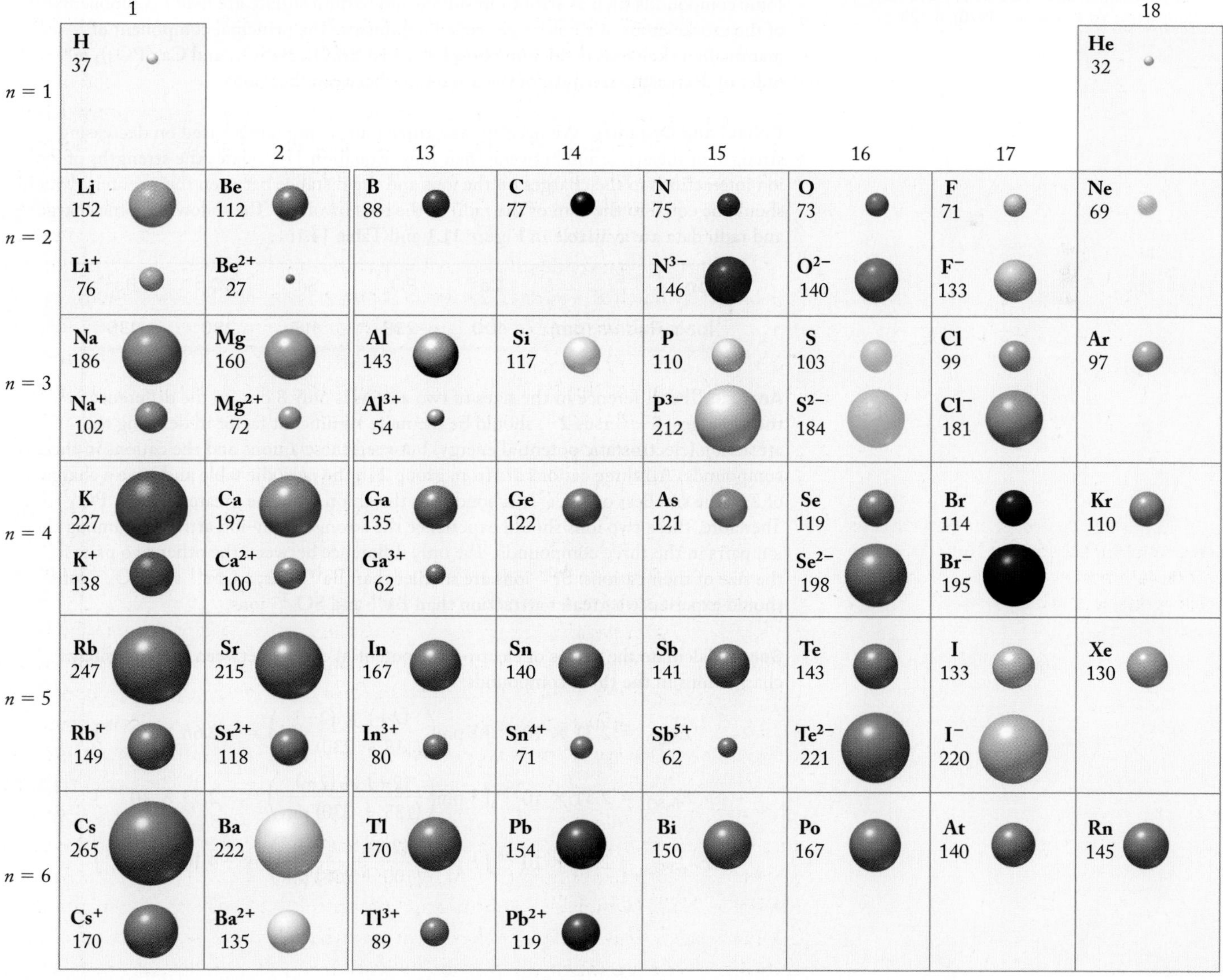

FIGURE 11.1 The radii of anions formed by the main group elements are larger than the radii of their parent atoms. The radii of cations are smaller than the radii of their parent atoms. All values are in picometers. Values from N. N. Greenwood and A. Earnshaw, *Chemistry of the Elements*, 2nd ed. (Boston: Butterworth-Heinemann, 1997); C. E. Housecroft and A. G. Sharpe, *Inorganic Chemistry*, 3rd ed. (Upper Saddle River, NJ: Pearson Prentice Hall, 2008).

CONNECTION We introduced ion–ion interactions in Section 6.2 and revisited them in Section 8.1 in discussing how they lead to the formation of ionic bonds.

CONNECTION In Section 7.10 we explained how ionic radii are used to calculate the distance between the nuclear centers in an ionic bond.

TABLE 11.1 Estimated Radii of Polyatomic Ions

Polyatomic Ion	Radius (pm)
CO_3^{2-}	185
NO_3^-	189
SO_4^{2-}	230
PO_4^{3-}	238

Values from R. B. Heslop and K. Jones. *Inorganic Chemistry: A Guide to Advanced Study* (New York: Elsevier, 1976), p. 123.

values of Q are the relative charges on the ions, such as 2+ for Ca^{2+} and 1− for Cl^-. Let's use Equation 11.1 to calculate the electrostatic potential energies between the pairs of oppositely charged ions in $CaSO_4$ and NaCl:

$$E_{CaSO_4} = 2.31 \times 10^{-16}\ J \cdot pm\left(\frac{(2+) \times (2-)}{(100 + 230)\ pm}\right) = -2.80 \times 10^{-18}\ J$$

$$E_{NaCl} = 2.31 \times 10^{-16}\ J \cdot pm\left(\frac{(1+) \times (1-)}{(102 + 181)\ pm}\right) = -8.16 \times 10^{-19}\ J$$

The ratio of these two E values: $(-2.80 \times 10^{-18}\ J)/(-8.16 \times 10^{-19}\ J) = 3.43$, tells us that the electrostatic potential energy between pairs of Ca^{2+} and SO_4^{2-} ions is nearly 3.5 greater than that between pairs of Na^+ and Cl^- ions. This greater energy of attraction between its ions plays a major role in making $CaSO_4$ less soluble in water than NaCl.

SAMPLE EXERCISE 11.1 Relate the Strengths of Ion–Ion Interactions to the Charges and Sizes of Ions **LO1**

Ionic compounds such as strontium sulfate and barium sulfate are major components of the exoskeletons of some single-celled organisms. The principal component of mammalian skeletons is calcium phosphate. List $SrSO_4$, $BaSO_4$, and $Ca_3(PO_4)_2$ in order of decreasing strength of the attractions between their ions.

Collect and Organize We need to rank three ionic compounds based on decreasing strength of the attraction between their ions. Equation 11.1 relates the strengths of ion–ion interactions to the charges on the ions and the distance between their centers, which should be equal to the sum of the radii of the ions involved. The following ionic charge and radii data are available in Figure 11.1 and Table 11.1:

Ion	Ca^{2+}	PO_4^{3-}	Sr^{2+}	SO_4^{2-}	Ba^{2+}
Ionic Radius (pm)	100	238	118	230	135

Analyze The difference in the sizes of two anions is only 8 pm, so the difference in their charges, 3− versus 2−, should be the more significant factor in defining the attraction (electrostatic potential energy) between these anions and the cations in their compounds. All three cations are from group 2 in the periodic table and have a charge of 2+. The smallest one, Ca^{2+}, is bonded to the anion with the greater charge, PO_4^{3-}. Therefore, these two ions should experience the strongest ion–ion attraction among the ion pairs in the three compounds. The only difference between the other two pairs is the size of their cations: Sr^{2+} ions are smaller than Ba^{2+} ions; so, Sr^{2+} and SO_4^{2-} ions should experience a greater attraction than Ba^{2+} and SO_4^{2-} ions.

Solve Calculate the values of electrostatic potential energy between the oppositely charges ions in the three compounds:

$$E_{SrSO_4} = 2.31 \times 10^{-16}\ J \cdot pm\left(\frac{(2+) \times (2-)}{(118 + 230)\ pm}\right) = -2.66 \times 10^{-18}\ J$$

$$E_{BaSO_4} = 2.31 \times 10^{-16}\ J \cdot pm\left(\frac{(2+) \times (2-)}{(135 + 230)\ pm}\right) = -2.53 \times 10^{-18}\ J$$

$$E_{Ca_3(PO_4)_2} = 2.31 \times 10^{-16}\ J \cdot pm\left(\frac{(2+) \times (3-)}{(100 + 238)\ pm}\right) = -4.10 \times 10^{-18}\ J$$

More negative E values represent greater attraction, so the three in decreasing order of ion–ion attraction are $Ca_3(PO_4)_2 > SrSO_4 > BaSO_4$.

Think About It The predicted order of decreasing attraction was confirmed by the calculated electrostatic potential energy values. The biggest gap in these values separates $Ca_3(PO_4)_2$ from the other two, which are sulfate compounds. Thus, the difference in the charges on the phosphate and sulfate ions is the primary factor influencing interaction strengths among these compounds. The much smaller difference in the E values of $SrSO_4$ and $BaSO_4$ arose from the different sizes of their cations, which resulted in a slightly smaller d value in the denominator of the E equation for $SrSO_4$ (348 pm) versus $BaSO_4$ (365 pm).

Practice Exercise Arrange the ionic compounds $CaCl_2$, BaO, and KCl in order of decreasing strength of the interaction between their ions.

(Answers to Practice Exercises are in the back of the book.)

The results of Sample Exercise 11.1 illustrate an important point about the strength of ion–ion interactions among the most common ions: differences in their radii tend to produce smaller differences in the strength of ion–ion interactions than do differences in their charges.

CONCEPT TEST

Rank the following sets of ions in order of increasing distance (d) between their nuclear centers: (a) KF, LiF, NaF; (b) $CaBr_2$, $CaCl_2$, CaF_2.

(Answers to Concept Tests are in the back of the book.)

11.2 Energy Changes during Formation and Dissolution of Ionic Compounds

In Section 10.3 we discussed the role of ion–dipole interactions in making salts soluble in water, and we described qualitatively how the combined effect of many ion–dipole interactions could overcome ion–ion interactions in a crystal of solute to produce hydrated ions in solution. Now we take the next step and *quantify* the actual changes in energy associated with these interactions. We then use these quantitative observations to understand further what happens at the particle level when solutes dissolve in solvents.

CONNECTION Figure 10.5 shows the dissociation of NaCl into Na^+ and Cl^- ions, each surrounded by a hydration sphere. Figure 10.6 shows the inner and outer spheres of hydration in more detail.

We can apply the qualitative picture of the dissolution process we developed in Chapter 10 to describe the factors that contribute to the overall enthalpy change that accompanies dissolution of an ionic compound in a polar solvent such as water. The ions must be separated from one another (**Figure 11.2a**), a process that requires energy to break the ion–ion interactions that hold the particles in the crystal lattice ($\Delta H_{\text{ion–ion}}$). Solvent molecules must also be separated from one another, so they can bind to the ions (**Figure 11.2b**); this process requires energy to break the dipole–dipole interactions—in water, these are hydrogen bonds—between solvent molecules ($\Delta H_{\text{dipole–dipole}}$). Last, energy is released when solvent

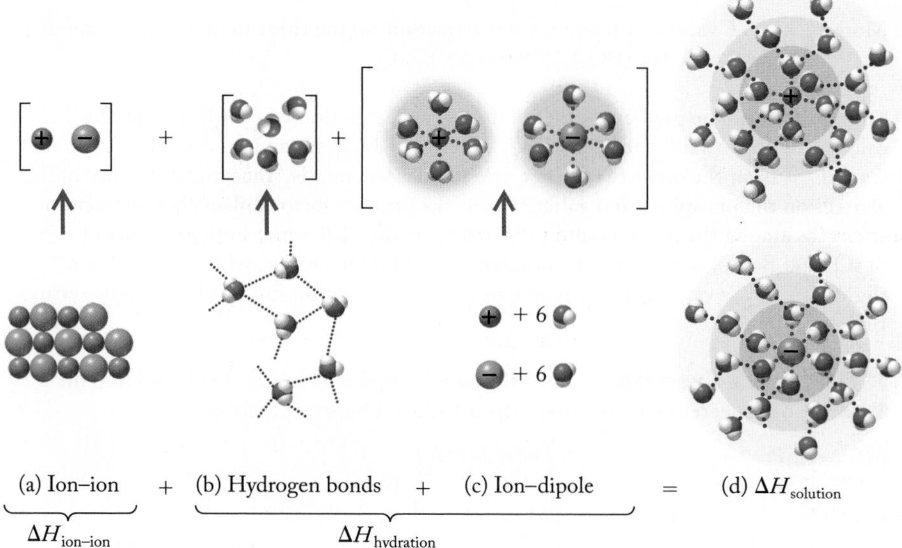

FIGURE 11.2 A homogeneous solution of an ionic compound in water consists of hydrated ions distributed throughout the solvent. We can account for the energy associated with establishing a solution by considering (a) the energy required to separate the ions from their lattice ($\Delta H_{\text{ion–ion}} = -U$), (b) the energy required to overcome hydrogen bonding in the solvent ($\Delta H_{\text{hydrogen bonding}}$; in water, these are hydrogen bonds), and (c) the energy released when ion–dipole bonds form between solvent and solute particles ($\Delta H_{\text{ion–dipole}}$). The sum of (b) and (c) is $\Delta H_{\text{hydration}}$, the enthalpy of hydration. (d) The sum of all three processes is $\Delta H_{\text{solution}}$, the enthalpy of solution.

molecules associate with the solute ions via ion–dipole interactions ($\Delta H_{\text{ion–dipole}}$) and form solvated (or *hydrated* if the solvent is water) ions (**Figure 11.2c**). The sum of the enthalpies of these interactions is the **enthalpy of solution ($\Delta H_{\text{solution}}$)**, which defines the overall change in enthalpy when an ionic solute dissolves in a polar solvent (**Figure 11.2d**).

Dissolution of Ammonium Nitrate

$$\Delta H_{\text{solution}} = \Delta H_{\text{ion–ion}} + \Delta H_{\text{hydrogen bonding}} + \Delta H_{\text{ion–dipole}} \quad (11.2)$$

We can combine the strengths of hydrogen bonding and ion–dipole interactions into a single term called the **enthalpy of hydration ($\Delta H_{\text{hydration}}$)** if the solvent is water, or the enthalpy of solvation ($\Delta H_{\text{solvation}}$) for any polar solvent:

$$\Delta H_{\text{hydration}} = \Delta H_{\text{hydrogen bonding}} + \Delta H_{\text{ion–dipole}} \quad (11.3)$$

Combining Equations 11.2 and 11.3 we have

$$\Delta H_{\text{solution}} = \Delta H_{\text{ion–ion}} + \Delta H_{\text{hydration}} \quad (11.4)$$

Let's examine each of these terms describing the enthalpy changes that accompany the dissolution of an ionic solid in water in more detail.

The strength of ion–ion interactions is contained within the **lattice energy (U)** of an ionic compound, which is the change in energy when free ions in the gas phase combine to form 1 mole of a solid ionic compound. The lattice energies of some common binary ionic compounds are listed in **Table 11.2**. The formula for lattice energy is

TABLE 11.2 Lattice Energies (U) of Common Binary Ionic Compounds

Compound	U (kJ/mol)
LiF	−1047
LiCl	−864
NaCl	−786
KCl	−720
KBr	−691
$MgCl_2$	−2540
MgO	−3791

$$U = \frac{k(Q_1 \times Q_2)}{d} \quad (11.5)$$

This equation resembles Equation 11.1 except that the value of its proportionality constant, k, depends on the structure of the ionic solid. We examine the structures of solids in Chapter 12, at which point we will learn about the multiple interactions that each ion experiences in an ionic solid. A key point here is that the same value of k may be used for all compounds that have the same or nearly the same arrangement of ions. We can then substitute the lattice energy, U, of an ionic compound in Equation 11.5 for $\Delta H_{\text{ion–ion}}$:

$$\Delta H_{\text{solution}} = -U + \Delta H_{\text{hydration}}$$

or

$$\Delta H_{\text{solution}} = \Delta H_{\text{hydration}} - U \qquad (11.6)$$

The lattice energy term in Equation 11.6 is negative because U is defined as the enthalpy change when gas-phase ions *combine* to form an ionic solid. In Equation 11.6 we are calculating an enthalpy change that includes *separating* the ions in an ionic compound. Recall from Section 6.4 that the enthalpy change for a process in one direction has the same magnitude but opposite sign of the process in the reverse direction.

Lattice energies affect not only the energetics of dissolving ionic compounds, but also their melting and boiling points. Melting ionic structures in which the ions are held together tightly should require more thermal energy (and higher temperatures) than melting structures in which the ions are held together less tightly. We can observe this trend experimentally in the properties of LiF ($U = -1047$ kJ/mol) and MgO ($U = -3791$ kJ/mol). The lattice energy of MgO is nearly four times more negative than that of LiF, which makes sense because the ($Q_1 \times Q_2$) term for MgO in Equation 11.5 is four times more negative than that of LiF. Also, the sums of the radii of their cations and anions are about the same: (76 + 133 = 209) pm for LiF and (72 + 140 = 212) pm for MgO. Thus, the difference in d values in Equation 11.5 has little to do with the big difference in U values. As expected, the melting point of MgO (2825°C) is also much higher than that of LiF (848°C). Likewise, the boiling point of MgO (3600°C) is much higher than that of LiF (1673°C).

enthalpy of solution ($\Delta H_{\text{solution}}$) the overall change in enthalpy that occurs when a solute dissolves in a solvent.

enthalpy of hydration ($\Delta H_{\text{hydration}}$) the energy change when gas-phase ions dissolve in water.

lattice energy (U) the enthalpy change that occurs when 1 mole of an ionic compound forms from its free ions in the gas phase.

SAMPLE EXERCISE 11.2 Ranking Lattice Energies and Melting Points LO2

Rank the ionic compounds NaF, KF, and RbF in order of (a) increasing (more negative) lattice energy and (b) increasing melting point. Assume these compounds have the same crystal structure and k value in Equation 11.5.

Collect, Organize, and Analyze We are asked to rank three group 1 fluorides in order of increasing lattice energies and increasing melting points. The ions in all three compounds have the same charges and their crystal structures are the same; so, significant differences in their lattice energies must be related to differences in the radii of the cations because the anions in all three compounds are fluoride ions. According to the images and numerical data in Figure 11.1, the trend in the ionic radii of the three cations is $Na^+ < K^+ < Rb^+$. Earlier in this section we described how increasingly negative lattice energies correlate with increasing melting points of ionic compounds.

Solve **Figure 11.3** shows the increasing sizes of the ion pairs in the three compounds: NaF < KF < RbF. Lattice energy is inversely proportional to the distance between the centers of paired cations and anions. Therefore, (a) the compounds in order of increasing lattice energy are RbF < KF < NaF and (b) the same trend should apply to the melting points of the compounds: RbF < KF < NaF.

Think About It The proposed order in part (a) is reasonable given the lattice energy trends in Table 11.1, and the link between increasingly negative lattice energy and increasing melting point makes sense because melting ionic solids in which the oppositely charged ions experience stronger attractions should require more energy and higher temperature. Moreover, the predicted order is confirmed by the compounds' observed melting points: 775°C for RbF, 846°C for KF, and 988°C for NaF.

Practice Exercise Predict which compound has the highest melting point: $CaCl_2$, $PbBr_2$, or TiO_2. All three compounds have nearly the same structure and therefore the same value of k in Equation 11.5. The radius of Ti^{4+} is 60.5 pm.

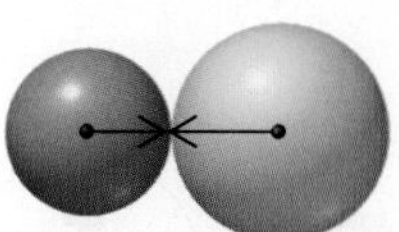

Na^+ = 102 pm
F^- = 133 pm

K^+ = 138 pm
F^- = 133 pm

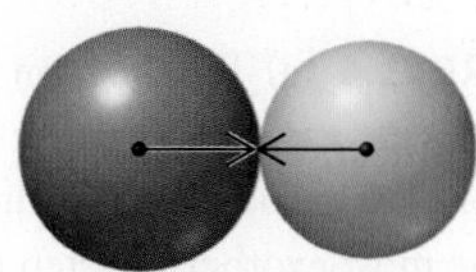

Rb^+ = 149 pm
F^- = 133 pm

FIGURE 11.3 The distance, d, between the nuclei of ions in NaF, KF, and RbF is in the order $d_{NaF} < d_{KF} < d_{RbF}$.

Calculating Lattice Energies by Using the Born–Haber Cycle

We predicted the relative order of lattice energies by considering atomic radii and bond distances, but we can also calculate values for lattice energies. Calculations are necessary because measuring lattice energies directly is difficult. We can calculate the lattice energy of a binary ionic compound from its standard enthalpy of formation (ΔH_f°). We use Hess's law to determine the enthalpies of reaction (ΔH_{rxn}) associated with a series of reactions that take the constituent elements from their standard states to ions in the gas phase and then to ions in the ionic solid.

Consider the formation of NaCl:

$$Na(s) + \tfrac{1}{2}\,Cl_2(g) \rightarrow NaCl(s) \qquad \Delta H_f^\circ = -411.2\ kJ$$

CONNECTION We introduced Hess's law in Section 6.7 to determine enthalpies of reactions that are difficult to measure experimentally. In Section 6.8 we defined a formation reaction as one in which 1 mole of a substance is produced from its constituent elements in their standard states.

The reaction is exothermic, and the enthalpy produced (411.2 kJ per mole of NaCl formed—see Table A4.3 in the Appendix) has been determined using calorimetry. In terms of Hess's law, this enthalpy of formation is the algebraic sum of all the enthalpy changes associated with five reactions that together form a **Born–Haber cycle** (**Figure 11.4**):

STEPWISE ANIMATION
Born–Haber Cycle

1. Sublimation of 1 mole of $Na(s)$ atoms into 1 mole of $Na(g)$ atoms: the molar enthalpy of sublimation of sodium, ΔH_{sub}.
2. Breaking covalent bonds in $\frac{1}{2}$ mole of $Cl_2(g)$ molecules to make 1 mole of $Cl(g)$ atoms: $\frac{1}{2}\Delta H_{BE}$, where ΔH_{BE} is the enthalpy change needed to break 1 mole of $Cl_2(g)$ bonds.
3. Ionization of 1 mole of $Na(g)$ atoms to 1 mole of $Na^+(g)$ ions and 1 mole of electrons: IE_1, the first ionization energy of sodium.
4. Combination of 1 mole of $Cl(g)$ atoms with 1 mole of electrons to form 1 mole of $Cl^-(g)$ ions: EA_1, the first electron affinity of chlorine.
5. Formation of 1 mole of $NaCl(s)$ from 1 mole of $Na^+(g)$ ions and 1 mole of $Cl^-(g)$ ions: $U = \Delta H_{lattice}$, the lattice energy of NaCl.

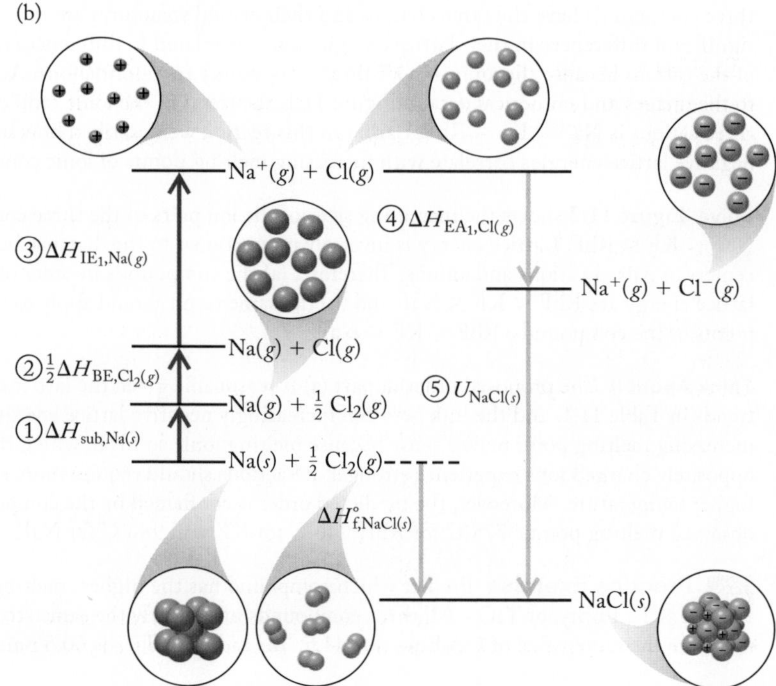

FIGURE 11.4 (a) The reaction between sodium metal and chlorine gas releases more than 400 kJ of energy per mole of NaCl produced. (b) The Born–Haber cycle shows that the most exothermic step is the combination of free sodium ions $Na^+(g)$ and free chloride ions $Cl^-(g)$ to form $NaCl(s)$. Although the particulate representations indicate whether each element or compound exists as atoms, molecules, or ions, they are not intended to represent the exact number of particles involved.

TABLE 11.3 Born–Haber Cycle for Formation of NaCl(*s*)

	PROCESS		
Step	Description	Chemical Equation	Enthalpy Change (kJ)
1	Sublime 1 mol Na(s)	$Na(s) \rightarrow Na(g)$	$\Delta H_{sub,Na} = +109$
2	Break $\frac{1}{2}$ mol Cl—Cl bonds	$\frac{1}{2}\,Cl_2(g) \rightarrow Cl(g)$	$\frac{1}{2}\Delta H_{BE,Cl_2} = \frac{1}{2}(240) = +120$
3	Ionize 1 mol Na(*g*) atoms, forming 1 mol $Na^+(g)$ ions	$Na(g) \rightarrow Na^+(g) + e^-$	$\Delta H_{IE_1,Na} = +495$
4	1 mol Cl(*g*) atoms acquires 1 mol electrons, forming 1 mol $Cl^-(g)$ ions	$Cl(g) + e^- \rightarrow Cl^-(g)$	$\Delta H_{EA_1,Cl} = -349$
5	1 mol $Na^+(g)$ ions combines with 1 mol $Cl^-(g)$ ions, forming 1 mol NaCl(s)	$Na^+(g) + Cl^-(g) \rightarrow NaCl(s)$	$\Delta H_{lattice} = U$

Born–Haber cycle a series of steps with corresponding enthalpy changes that describes the formation of an ionic solid from its constituent elements.

This reaction sequence is summarized in **Table 11.3**. To use the Born–Haber cycle to calculate the lattice energy of NaCl(*s*), we start with an equation relating the value of ΔH_f° for NaCl to the sum of the enthalpy changes of the five reactions in Table 11.3:

$$\begin{aligned}\Delta H_f^\circ &= \Delta H_{step\ 1} + \Delta H_{step\ 2} + \Delta H_{step\ 3} + \Delta H_{step\ 4} + \Delta H_{step\ 5}\\ &= \Delta H_{sub,Na} + \tfrac{1}{2}\Delta H_{BE,Cl_2} + \Delta H_{IE_1,Na} + \Delta H_{EA_1,Cl} + U_{NaCl}\end{aligned}$$

Inserting the values from Table 11.3 and solving for *U*:

$$-411.2\text{ kJ} = (+109\text{ kJ}) + \tfrac{1}{2}(+240\text{ kJ}) + (+495\text{ kJ}) + (-349\text{ kJ}) + U$$

$$\begin{aligned}U &= (-411.2\text{ kJ}) - (+109\text{ kJ} + 120\text{ kJ} + 495\text{ kJ} - 349\text{ kJ})\\ &= -786\text{ kJ}\end{aligned}$$

Besides lattice energies, the Born–Haber cycle can also be used to calculate other values, such as electron affinities, which can be difficult to measure, provided thermochemical values are known for all other steps in the cycle.

SAMPLE EXERCISE 11.3 Calculating Lattice Energy **LO2**

Based on the relative charges on the ions, the ion–ion attractions in CaF_2 should be greater than those in NaF. Confirm this prediction by calculating the lattice energies of (a) NaF and (b) CaF_2, given the following data:

$$\begin{aligned}&\Delta H_{sub}\ Na(s) = +109\text{ kJ/mol} &&\Delta H_{sub}\ Ca(s) = +154\text{ kJ/mol}\\ &\Delta H_{BE}\ F_2(g) = +154\text{ kJ/mol} &&\Delta H_{IE_1}\ Ca(g) = +590\text{ kJ/mol}\\ &\Delta H_{EA_1}\ F(g) = -328\text{ kJ/mol} &&\Delta H_{IE_2}\ Ca(g) = +1145\text{ kJ/mol}\\ &\Delta H_{IE_1}\ Na(g) = +495\text{ kJ/mol}\end{aligned}$$

Collect and Organize We are asked to calculate the lattice energies of two ionic compounds, using enthalpy values for processes that can be summed to describe an overall process that produces a salt from its constituent elements. We can use a Born–Haber cycle to calculate the unknown lattice energy. According to Table A4.3 in the Appendix, the standard enthalpy of formation values for NaF and CaF_2 are −569.0 kJ/mol and −1228.0 kJ/mol, respectively. These enthalpy changes apply to the formation of 1 mole of the two compounds by combining their component elements in their standard states.

Analyze **Figure 11.5** summarizes the Born–Haber cycles for calculating the lattice energies of NaF and CaF_2. The lattice energy U is the only unknown value in both cycles. We must include both the first and second ionization energies for calcium because Ca loses two electrons when it forms Ca^{2+} ions. Because two fluorine atoms are needed to react with a single calcium atom, we do not need the factor of $\frac{1}{2}$ in front of the term for energy to break a mole of F—F bonds. However, we need to multiply the electron affinity of F by 2 because 2 moles of fluorine atoms gain 2 moles of electrons to form 2 moles of fluoride ions. From the charges on the ions, their ionic radii, and Equation 11.5, we predict that the lattice energy of CaF_2 should be greater than the lattice energy of NaF.

Solve

a. The Born–Haber cycle for the formation of NaF(*s*) from Na(*s*) and $F_2(g)$ is illustrated in Figure 11.5a. The overall enthalpy change in the reaction producing NaF is

$$\Delta H_f^\circ = \Delta H_{\text{sub,Na}(s)} + \tfrac{1}{2}\Delta H_{\text{BE,F}_2(g)} + \Delta H_{\text{IE}_1,\text{Na}(g)} + \Delta H_{\text{EA}_1,\text{F}(g)} + U_{\text{NaF}(s)}$$

Substituting the values given:

$$-569.0\text{ kJ} = (+109\text{ kJ}) + \tfrac{1}{2}(+154\text{ kJ}) + (+495\text{ kJ}) + (-328\text{ kJ}) + U$$

Solving for U:

$$U = (-569\text{ kJ}) - (+109\text{ kJ} + 77\text{ kJ} + 495\text{ kJ} - 328\text{ kJ})$$

$$= -922\text{ kJ/mol NaF}(s)\text{ formed}$$

b. The Born–Haber cycle for the formation of $CaF_2(s)$ from Ca(*s*) and $F_2(g)$ is illustrated in Figure 11.5b. The overall enthalpy change in the reaction producing CaF_2 is

$$\Delta H_f^\circ = \Delta H_{\text{sub,Ca}(s)} + \Delta H_{\text{BE,F}_2(g)} + [\Delta H_{\text{IE}_1,\text{Ca}(g)} + \Delta H_{\text{IE}_2,\text{Ca}(g)}] + 2\Delta H_{\text{EA}_1,\text{F}(g)} + U_{\text{CaF}_2(s)}$$

Substituting the values given:

$$-1228.0\text{ kJ} = (+154\text{ kJ}) + (+154\text{ kJ}) + [(+590\text{ kJ}) + (+1145\text{ kJ})] + 2(-328\text{ kJ}) + U$$

Solving for U:

$$U = (-1228.0\text{ kJ}) - [+154\text{ kJ} + 154\text{ kJ} + (590\text{ kJ} + 1145\text{ kJ}) - 656\text{ kJ}]$$

$$= -2615\text{ kJ/mol CaF}_2(s)\text{ formed}$$

Think About It We predicted that the lattice energy of CaF_2 is stronger than that of NaF. The calculated values of U confirm this prediction.

Practice Exercise Burning magnesium metal in air produces MgO and a very bright white light, making the reaction popular in fireworks and signaling devices:

$$\text{Mg}(s) + \tfrac{1}{2}\text{O}_2(g) \rightarrow \text{MgO}(s) + \text{light}$$

The enthalpy change that accompanies this reaction is −602 kJ/mol MgO. Calculate the lattice energy of MgO from the following enthalpy changes:

Process	Enthalpy Change (kJ/mol)	Process	Enthalpy Change (kJ/mol)
$Mg(s) \rightarrow Mg(g)$	150	$Mg(g) \rightarrow Mg^{2+}(g) + 2\,e^-$	2188
$O_2(g) \rightarrow 2\,O(g)$	499	$O(g) + 2\,e^- \rightarrow O^{2-}(g)$	603

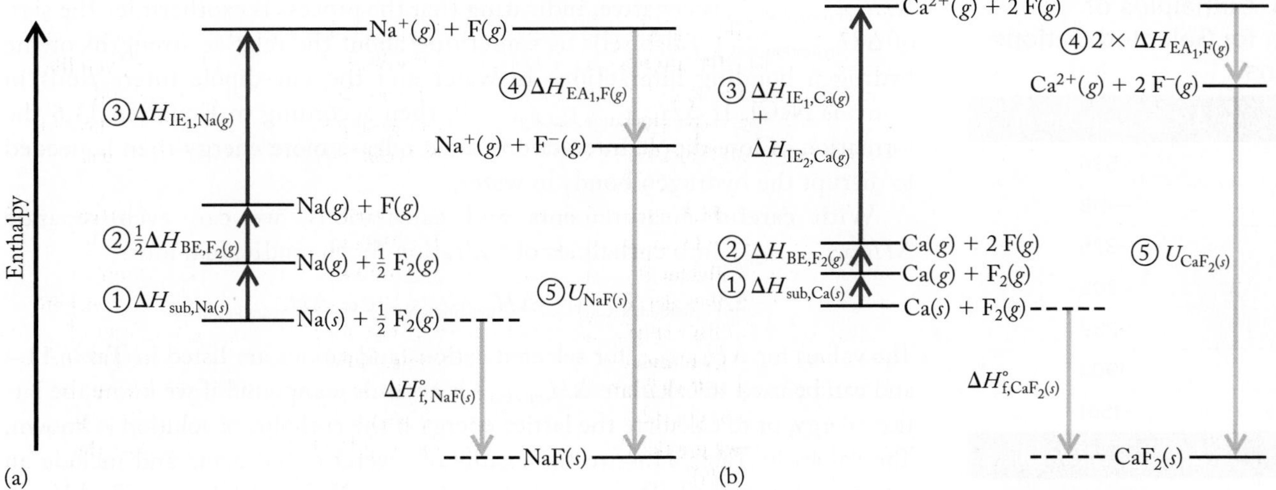

FIGURE 11.5 Born–Haber cycles for the formation of (a) NaF and (b) CaF_2.

Enthalpies of Hydration

Once we have calculated the lattice energy of an ionic compound, we can use an experimentally determined $\Delta H_{solution}$ value and Equation 11.6 to calculate a $\Delta H_{hydration}$ value as shown in **Figure 11.6**. Figure 11.6 is another example of a Born–Haber cycle. If the measured enthalpy of solution of sodium chloride is 4 kJ/mol and the lattice energy of NaCl is −786 kJ/mol, then:

$$\Delta H_{solution,NaCl(aq)} = \Delta H_{hydration,NaCl(g)} - U_{NaCl(s)}$$

$$\Delta H_{hydration,NaCl(aq)} = \Delta H_{solution,NaCl(g)} + U_{NaCl(s)}$$

$$= 4\text{ kJ/mol} + (-786\text{ kJ/mol}) = -782\text{ kJ/mol}$$

The quantity $\Delta H_{hydration,NaCl(g)}$ represents the enthalpy change that occurs when gas-phase Na^+ and Cl^- ions dissolve in water. In this case, the value for

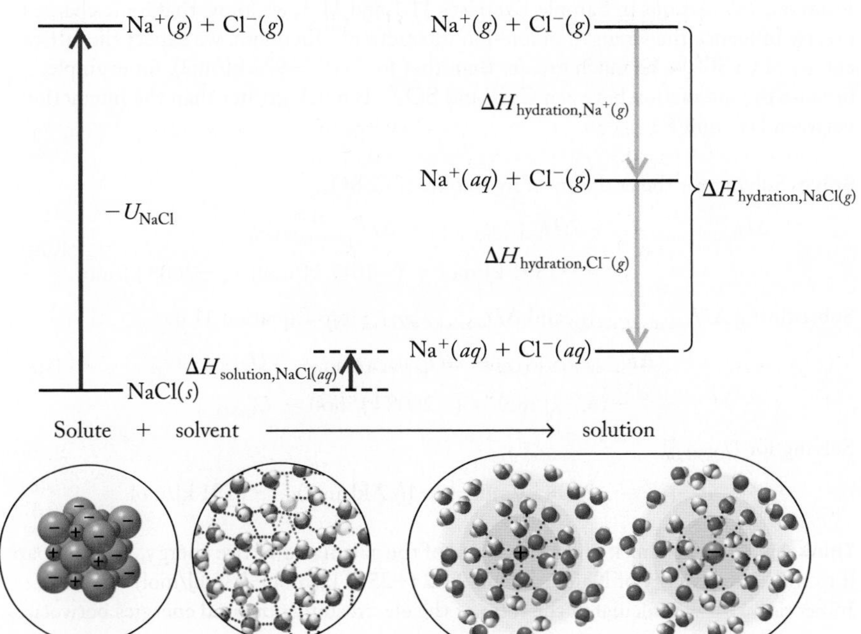

FIGURE 11.6 Born–Haber cycle for the dissolution of an ionic compound in water. The enthalpy of solution ($\Delta H_{solution}$) is the enthalpy of hydration ($\Delta H_{hydration}$) minus the lattice energy (U). The size of $\Delta H_{solution}$ is not drawn to scale; it is very small in relation to the other values.

TABLE 11.4 Enthalpies of Hydration for Selected Cations and Anions[a]

Cation	$\Delta H_{hydration}$ (kJ/mol)
Li^+	−536
Na^+	−418
K^+	−335
Rb^+	−305
Cs^+	−289
Mg^{2+}	−1903
Ca^{2+}	−1591
Anion	**$\Delta H_{hydration}$ (kJ/mol)**
F^-	−502
Cl^-	−368
Br^-	−335
I^-	−293
ClO_4^-	−238
NO_3^-	−301
SO_4^{2-}	−1017

[a]Based on enthalpy of hydration of H^+ equal to −1105 kJ/mol.

$\Delta H_{hydration,NaCl(g)}$ is negative, indicating that the process is exothermic. The sign of $\Delta H_{hydration,NaCl(g)}$ also tells us something about the relative strengths of the hydrogen bonding interactions in water and the ion–dipole interactions in aqueous NaCl. If $\Delta H_{hydration,NaCl(g)} < 0$, then according to Equation 11.6 the formation of ion–dipole interactions must release more energy than is needed to disrupt the hydrogen bonds in water.

With careful measurements and calculations, we can even separate $\Delta H_{hydration,NaCl(g)}$ into enthalpies of hydration for the individual ions:

$$\Delta H_{hydration,NaCl(g)} = \Delta H_{hydration,Na^+(g)} + \Delta H_{hydration,Cl^-(g)}$$

The values for $\Delta H_{hydration}$ for selected cations and anions are listed in **Table 11.4** and can be used to calculate $\Delta H_{solution}$ for an ionic compound if we know the lattice energy, or to calculate the lattice energy if the enthalpy of solution is known. The values in Table 11.4 are the results of several experiments and include an inherent uncertainty. The sum of the values for Na^+ and Cl^- is −786 kJ/mol, which is close to the −782 kJ/mol value calculated earlier.

SAMPLE EXERCISE 11.4 Calculating Lattice Energy by Using Enthalpies of Hydration — LO2

Calcium sulfate is used to prepare "plaster of Paris" for casting broken limbs. Use the appropriate enthalpy of hydration values in Table 11.4 and $\Delta H_{solution} = -16.7$ kJ/mol for $CaSO_4$ to calculate the lattice energy of $CaSO_4$.

Collect and Organize The enthalpies of hydration for Ca^{2+} and SO_4^{2-} are −1591 kJ/mol and −1017 kJ/mol, respectively. We can use the enthalpy of solution and Equation 11.6 to calculate the lattice energy.

Analyze To calculate U_{CaSO_4}, we must obtain a value for $\Delta H_{hydration}$ for $CaSO_4$:

$\Delta H_{hydration}$ of cation + $\Delta H_{hydration}$ of anion ⟶ $\Delta H_{hydration}$ of ionic compound

From our calculations in Sample Exercises 11.1 and 11.3, we know that ionic charges greatly influence the strength of ion–ion interactions. Therefore, we expect the lattice energy of $CaSO_4$ to be much greater than that for NaF (−922 kJ/mol), for example, because the interaction between Ca^{2+} and SO_4^{2-} is much greater than the interaction between Na^+ and F^-.

Solve Solving for the enthalpy of hydration of $CaSO_4$:

$$\Delta H_{hydration,CaSO_4(g)} = \Delta H_{hydration,Ca^{2+}(g)} + \Delta H_{hydration,SO_4^{2-}(g)}$$
$$= -1591\text{ kJ/mol} + (-1017\text{ kJ/mol}) = -2608\text{ kJ/mol}$$

Substituting $\Delta H_{hydration,CaSO_4(g)}$ and $\Delta H_{solution,CaSO_4(aq)}$ into Equation 11.6:

$$\Delta H_{solution,CaSO_4(aq)} = \Delta H_{hydration,CaSO_4(g)} - U_{CaSO_4(s)}$$
$$-16.7\text{ kJ/mol} = (-2608\text{ kJ/mol}) - U_{CaSO_4(s)}$$

Solving for $U_{CaSO_4(s)}$:

$$U_{CaSO_4(s)} = -2608\text{ kJ/mol} - (-16.7\text{ kJ/mol}) = -2591\text{ kJ/mol}$$

Think About It To check on the accuracy of the calculated lattice energy, let's compare it to the lattice energy of NaCl in Table 11.2 (−2591 kJ/mol/−786 kJ/mol) = 3.30. In Section 11.1 we calculated the ratio of the electrostatic potential energies between pairs of Ca^{2+} and SO_4^{2-} ions to pairs of Na^+ and Cl^- ions and found it to be 3.43. The similar values of these two ratios indicates that the calculated lattice energy of $CaSO_4$ is reasonable.

Practice Exercise Calculate the lattice energy for $NaClO_4$ by using the data in Table 11.4 and $\Delta H_{\text{solution,NaClO}_4(aq)} = 14$ kJ/mol.

11.3 Vapor Pressure of Solutions

In Section 10.4 we discussed how intermolecular interactions affect the vapor pressure and the normal boiling point of pure liquids. What happens to vapor pressure when nonvolatile solutes such as salts are dissolved in water?

When adjoining compartments of seawater (water that contains many dissolved nonvolatile solutes) and pure water are sealed in a chamber (**Figure 11.7**), the volume of fluid on the seawater side increases over time, while the volume on the side of pure water decreases at the same rate. Eventually, nearly all the water ends up in the seawater compartment. The transfer of the pure water is due to the dissolved solutes in the seawater.

CONNECTION For pure liquids, the vapor pressure is the pressure of a gas in equilibrium with its liquid, and the normal boiling point is the temperature at which the vapor pressure equals 1 atm (review Section 10.4).

As the water in both compartments evaporates, the concentration of water vapor in the air space of the sealed chamber increases. As the concentration of water vapor increases, the pressure the water vapor exerts on the two liquid surfaces increases. At constant temperature, this pressure eventually stabilizes at a value equal to the vapor pressure of water at that temperature. At this point, the rate of evaporation from the compartments is equal to the rate of condensation, which means a dynamic equilibrium has been established, as we discussed in Chapter 10.

STEPWISE ANIMATION
Vapor Pressure

If the evaporation and condensation rates for the pure water and the seawater were the same, the liquid levels in the compartments would not change over time. However, the liquid levels do change. The gas-phase water molecules in the sealed chamber are free to condense into either compartment, so the rate of condensation into both compartments is the same—but the rates of evaporation are not: the presence of dissolved solutes in the seawater affects its rate of evaporation.

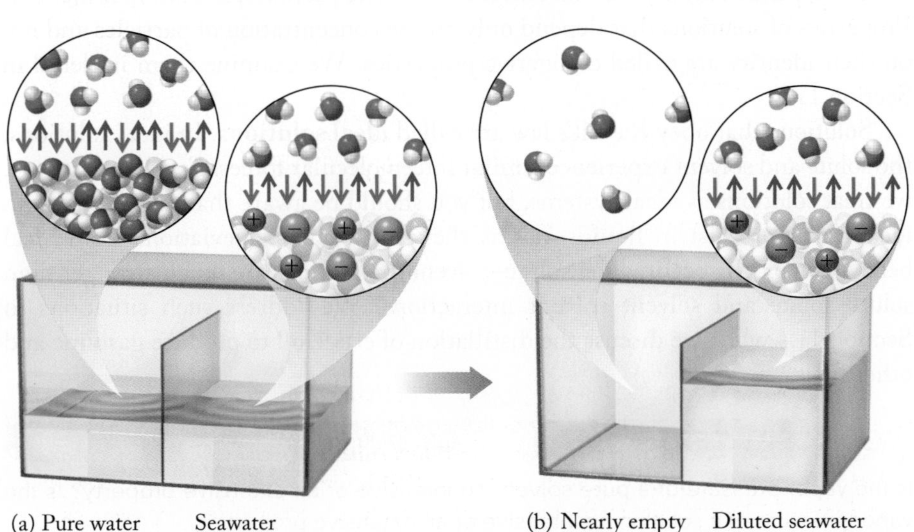

FIGURE 11.7 (a) Adjoining compartments are partially filled with pure water and seawater. The water in the left-hand compartment is in equilibrium with its vapor but the seawater is not. (b) The slightly higher vapor pressure of the pure water leads to a net transfer of water from the pure-water compartment to the seawater compartment. The water molecules in the diluted seawater are now in equilibrium with the water vapor above the surface.

Raoult's law the principle that the vapor pressure of the solvent in a solution is equal to the vapor pressure of the pure solvent multiplied by the mole fraction of the solvent in the solution.

ideal solution a solution that obeys Raoult's law.

Given that most of the pure water ends up in the seawater compartment, the pure water must have a higher rate of evaporation than the seawater. This conclusion leads to another—namely, if the seawater evaporates at a lower rate, it must have a lower vapor pressure than the pure water at the same temperature.

More water vapor enters the air in the chamber from the pure-water compartment than from the seawater compartment because the vapor pressure of the pure water is greater than the vapor pressure of the seawater. Because the condensation rates are the same, the pure water loses more water over time than is restored to it by condensation, while the seawater gains more water than it loses by evaporation. The process depicted in Figure 11.7 illustrates the following general observation: at a given temperature, the vapor pressure of the solvent in a solution containing nonvolatile solutes is *less* than the vapor pressure of the pure solvent.

Raoult's Law

CONNECTION Recall from the discussion of partial pressures of gases in Section 5.7 that the *mole fraction* is the ratio of the number of moles of a component in a mixture to the total number of moles in the mixture.

The connection between the vapor pressure of a solution and the concentration of nonvolatile solutes dissolved in the solvent was studied extensively by the French chemist François Marie Raoult (1830–1901). He discovered that the relation between the vapor pressure of a solution, P_{solution}, and that of the pure solvent, $P^{\circ}_{\text{solvent}}$, is

$$P_{\text{solution}} = X_{\text{solvent}}\, P^{\circ}_{\text{solvent}} \tag{11.7}$$

where X_{solvent} is the mole fraction of solvent. This relationship is now known as **Raoult's law**.

Let's take another look at the two compartments in Figure 11.7. When the temperature in the chamber is 20°C, the vapor pressure of pure water is 0.0231 atm. What is the vapor pressure produced by evaporation of water from seawater if the mole fraction of water in the sample is 0.980? We can use Raoult's law (Equation 11.7) to answer this question:

$$P_{\text{solution}} = (0.980)(0.0231 \text{ atm}) = 0.0226 \text{ atm}$$

Because the vapor pressure of seawater at 20°C is slightly *lower* than the vapor pressure of pure water at 20°C, seawater evaporates more slowly than pure water.

The lower vapor pressure of a solution in relation to the vapor pressure of pure solvent depends only on the concentration of solute particles, not on their identity. Properties of solutions that depend only on the concentration of particles and not on their identity are called colligative properties. We examine them in detail in Section 11.5.

CONNECTION As we first explained in Section 1.5, intensive properties of matter are independent of the amount of material present, whereas extensive properties depend on the quantity of substance present.

Solutions that obey Raoult's law are called **ideal solutions**. In ideal solutions, the solute and solvent experience similar intermolecular forces. For the most part, we treat solutions as ideal systems, but you should be aware that deviations from ideal behavior exist in liquids just as they do in gases. Deviations from ideal behavior typically occur when solute–solvent interactions are much stronger than solute–solute and solvent–solvent interactions. We address such situations in Section 11.4 when we discuss the distillation of crude oil to produce gasoline and other fuels.

CONCEPT TEST

Is the vapor pressure of a pure solvent an intensive or an extensive property? Is the vapor pressure of a solution an intensive or an extensive property?

SAMPLE EXERCISE 11.5 Calculating the Vapor Pressure of a Solution **LO3**

The liquid used in automobile cooling systems is prepared by dissolving ethylene glycol ($HOCH_2CH_2OH$, molar mass 62.07 g/mol) in water. What is the vapor pressure of a solution prepared by mixing 1.000 L of ethylene glycol (density 1.114 g/mL) with 1.000 L of water (density 1.000 g/mL) at 100.0°C? Assume that the mixture obeys Raoult's law.

Collect and Organize The vapor pressure of a solution is a colligative property that depends on the number of solute particles and hence the concentration. We have the volume and density of the components and can use them to determine the concentration of ethylene glycol. We may treat ethylene glycol as a nonvolatile solute whose aqueous solutions obey Raoult's law. According to Figure 10.20, the vapor pressure of pure water at 100.0°C (its normal boiling point) is 760 torr, whereas the vapor pressure of ethylene glycol at that temperature is less than 20 torr.

Analyze We have a mixture of equal volumes of two liquids. The solvent is the one present in the greater number of moles.

We can determine the numbers of moles of each by the following calculation:

$$\boxed{\text{Volume}} \xrightarrow{\text{density}} \boxed{\text{Mass}} \xrightarrow{\frac{1}{\text{molar mass}}} \boxed{\text{Moles}}$$

From these values we can decide which liquid is the solvent and calculate its mole fraction in the mixture. Water and ethylene glycol are both capable of hydrogen bonding, so their intermolecular interactions are similar, and we may treat the solution as ideal.

Solve Moles of ethylene glycol in 1.000 L:

$$1.000\ \cancel{\text{L}} \times \frac{1000\ \cancel{\text{mL}}}{1\ \cancel{\text{L}}} \times \frac{1.114\ \cancel{\text{g}}}{1\ \cancel{\text{mL}}} \times \frac{1\ \text{mol}}{62.07\ \cancel{\text{g}}} = 17.95\ \text{mol}$$

Moles of water in 1.000 L:

$$1.000\ \cancel{\text{L}} \times \frac{1000\ \cancel{\text{mL}}}{1\ \cancel{\text{L}}} \times \frac{1.000\ \cancel{\text{g}}}{1\ \cancel{\text{mL}}} \times \frac{1\ \text{mol}}{18.02\ \cancel{\text{g}}} = 55.49\ \text{mol}$$

Water, then, is the solvent, which allows us to use the model of a mixture that consists of a volatile solvent and nonvolatile solute. The mole fraction of water is

$$X_{\text{water}} = \frac{55.49\ \text{mol}}{55.49\ \text{mol} + 17.95\ \text{mol}} = 0.7556$$

Ethylene glycol is essentially nonvolatile, so the vapor pressure of the solution results only from the solvent, and $P_{\text{solvent}} = P^{\circ}_{H_2O} = 760$ torr. Using this value and the calculated mole fraction of water in the mixture yields

$$P_{\text{solution}} = X_{H_2O} \times P^{\circ}_{H_2O} = (0.7556)(760\ \text{torr}) = 574\ \text{torr}$$

Think About It The presence of a nonvolatile solute causes the vapor pressure of the solution to be less than the vapor pressure of a pure solvent, giving us confidence in our result.

Practice Exercise Glycerol [$HOCH_2CH(OH)CH_2OH$] can be treated as a nonvolatile, water-soluble liquid. Its density is 1.25 g/mL. Predict the vapor pressure of a solution of 275 mL of glycerol in 375 mL of water at the normal boiling point of water.

To boil the solution described in Sample Exercise 11.5, we must heat it above 100°C—to a temperature at which the vapor pressure of the solution is 1 atm. This is why antifreeze works in an automobile engine: it not only lowers the

fractional distillation a method of separating a mixture of compounds based on their different boiling points.

freezing point of water, protecting the radiator, but also raises the boiling point of the coolant above that of pure water, making it possible for the coolant to remove more energy from the engine while staying in the liquid state.

11.4 Mixtures of Volatile Solutes

CONNECTION We introduced simple distillation in Section 1.3 as a way to make drinking water from seawater.

Thus far we have considered only the effect of essentially *nonvolatile* solutes on the vapor pressure of a volatile solvent. In our everyday lives, however, we encounter many solutions containing *volatile* solutes. For example, the natural gas used for heating and the gasoline we use to power vehicles are mixtures of hydrocarbons. Gasoline comes from crude oil, a complex mixture of compounds composed mostly of carbon and hydrogen. Depending on its source, crude oil contains varying concentrations of hydrocarbons with five or more carbon atoms in their molecular structures. The hydrocarbons with one to four carbon atoms are usually found in deposits of natural gas, although they are also dissolved in crude oil. Most hydrocarbons in gasoline have from five to nine carbon atoms per molecule. Each of these compounds is volatile and has a measurable vapor pressure at 25°C. Because of this volatility, we can separate gasoline from crude oil by distillation. In this section we explore distillation on a molecular level and consider the effects of volatile solutes on the vapor pressure of a solution.

CHEMTOUR
Fractional Distillation

Vapor Pressures of Mixtures of Volatile Solutes

In Section 6.3 we discussed the process of distillation as a way to purify liquids (see Figure 6.15), and in Section 10.4 we discussed the vaporization of pure liquids and the role of intermolecular forces in determining the normal boiling point. In addition, we saw that the temperature at which a pure liquid boils remains constant throughout the vaporization process. Another type of distillation, called **fractional distillation** (**Figure 11.8**), is used to separate the volatile components of a mixture. This method is based on the observation that the boiling point of a mixture changes as the mixture is distilled.

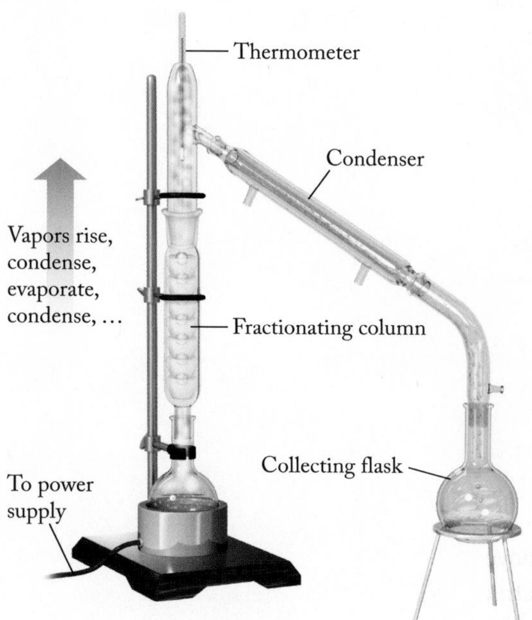

FIGURE 11.8 Fractional distillation separates mixtures of volatile components. Vapors rise through a fractionating column, where they repeatedly condense and revaporize. The most volatile component is the first one to pass through the fractionating column and reach the condenser and collecting flask. Increasingly less volatile, higher-boiling components are distilled in turn. The progress of the distillation process is monitored using the thermometer at the top of the fractionating column.

As a mixture of volatile liquids is heated, the vapor that rises and fills the space above the liquid has a different composition from the composition of the mixture: the concentration of the component with the lowest boiling point is higher in the vapor than in the liquid. If this enriched vapor is collected, condensed, and redistilled, the vapor this time is even richer in the component with the lowest boiling point. In a fractional distillation apparatus, repeated distillation steps allow components with only slightly different boiling points to be separated from one another.

To see how a mixture behaves when boiled and how fractional distillation uses that behavior to separate the components, let's compare the heating curves of a pure substance and a solution of two volatile liquids. **Figure 11.9a** is the heating curve for pure octane (C_8H_{18}, bp 126°C). As in the heating curves of Section 6.5, the phase change from liquid to vapor takes place at a constant temperature of 126°C. **Figure 11.9b** shows the heating curve for a mixture of heptane (C_7H_{16}, bp 98°C) and octane. Here the portion of the curve from point 1 to point 2 again represents a vapor–liquid phase change, but in this case the temperature is not constant during the phase change. The increasing temperature of the mixture as it vaporizes means that the vapors it produces as the mixture begins to boil

FIGURE 11.9 Heating curves describe the vaporization and distillation of volatile liquids. (a) When pure octane is distilled, the distillation takes place at a constant temperature, indicated by the horizontal line at $T = 126°C$. The liquid in the collecting flask is pure octane. (b) When a mixture of heptane (bp 98°C) and octane is distilled, the distillation takes place over a range of temperatures. (c) The blue line shows the temperatures where solutions of a given composition boil. The red line shows the composition of the vapor arising from those solutions. The stair-step (dashed) line illustrates what happens in a fractionating column. The shifting colors of the numbered circles track the changing composition of the two phases as fractional distillation proceeds toward the production of pure heptane.

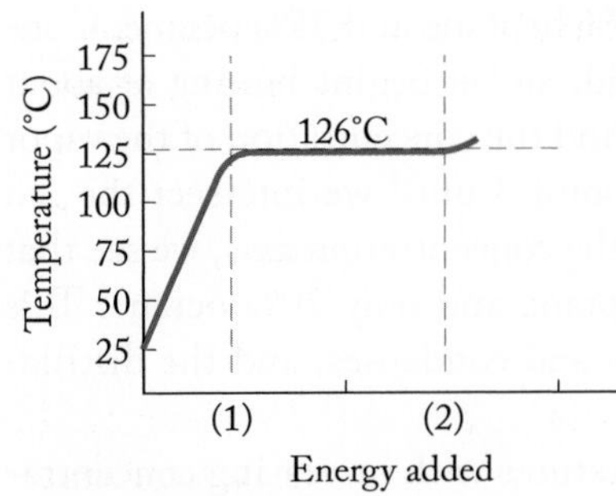

(a) Distillation of octane
(1) Octane begins to distill.
(2) Octane finishes distilling.

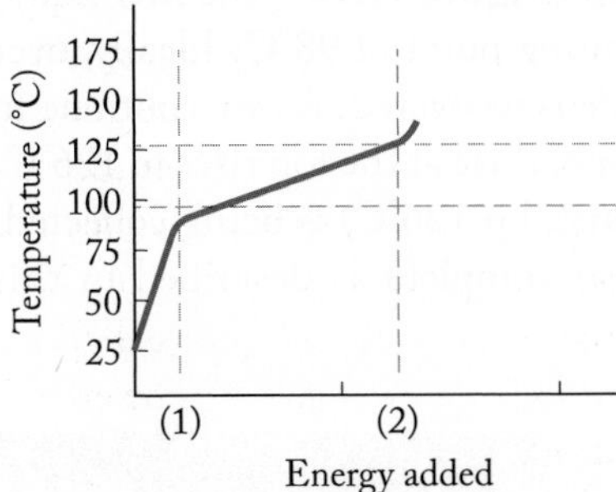

(b) Distillation of a mixture of heptane and octane
(1) Solution begins to distill.
(2) Solution finishes distilling.

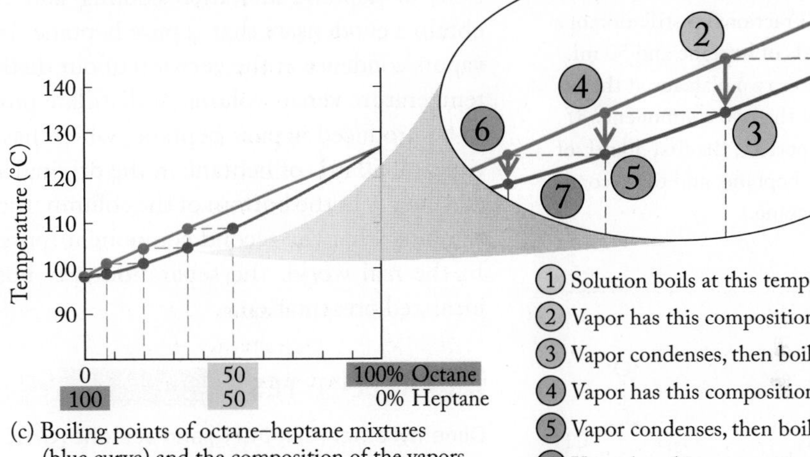

(c) Boiling points of octane–heptane mixtures (blue curve) and the composition of the vapors produced at those boiling points (red curve)

(1) Solution boils at this temperature.
(2) Vapor has this composition.
(3) Vapor condenses, then boils.
(4) Vapor has this composition.
(5) Vapor condenses, then boils.
(6) Vapor has this composition.
(7) Vapor condenses.

are richer in the more volatile (lower boiling point) component (heptane) leaving behind a boiling liquid that becomes richer in the higher-boiling component (octane). As the mixture continues to evaporate, the vapors it produces become richer in the higher-boiling component (octane). Thus, the two components co-distill over a range of temperatures, producing vapor that initially has a relatively high concentration of heptane, but that is never 100% heptane.

Now let's analyze the graphs describing how fractional distillation can achieve complete separation of two volatile liquids. **Figure 11.9c** shows how the composition of the vapor above a series of boiling solutions differs from the compositions of the solutions. Note that the concentration of the lower-boiling (more volatile) component is always greater in the vapor than in the solution that produced it. This is the key to how fractional distillation separates mixtures.

Suppose we use the distillation apparatus from Figure 11.8 to heat 100 mL of a solution that is 50% by volume heptane and 50% octane (50:50). The blue curve in Figure 11.9c gives the temperature of the boiling solution, and point 1 on the curve tells us that the boiling point of the 50:50 solution is about 108°C. The red line in the graph represents the composition of the vapor above the solution at a given temperature. To find the composition of the vapor at 108°C, we move horizontally to the left along the dashed line between points 1 and 2. This horizontal line corresponds to a temperature of 108°C on the y-axis. Point 2 corresponds to a different composition (on the x-axis) for the vapor than for the solution at point 1. The vapor at point 2 is enriched in the lower-boiling component: it is about 65% heptane and only 35% octane.

This 65:35 vapor rises in the distillation column, cools, and condenses in the next region of the column, a process represented by the red arrow from point 2 to

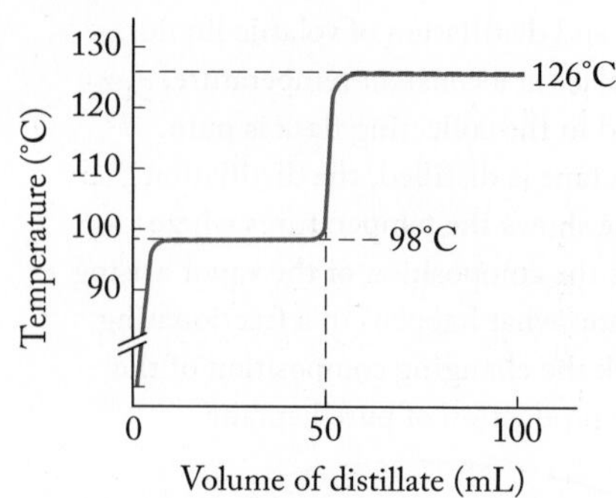

FIGURE 11.10 Fractional distillation of a mixture of 50 mL of heptane and 50 mL of octane produces two plateaus at the boiling points of the two components. If fractionation is perfect, the first 50 mL of distillate is pure heptane, and the second 50 mL is pure octane.

point 3. The condensed liquid in this flask is 65% heptane and 35% octane. Continued heating of the column warms this liquid, and it begins boiling at about 104°C, which is the temperature at point 3. To find the concentration of the vapor above this boiling liquid, we move left from point 3 until we intersect the red curve (point 4). Reading down from point 4 to the concentration axis, we see that the vapor concentration is now about 80% heptane and only 20% octane. This vapor with 80:20 composition rises, then cools and condenses, and the distillation cycle is repeated.

If we continue this process of redistilling mixtures with increasing concentrations of heptane and then cooling and condensing the vapors, we eventually obtain a condensate that is pure heptane. If we monitor the temperature at which vapors condense at the very top of our distillation column, we will see a profile of temperature versus volume of distillate produced (**Figure 11.10**). The first liquid to be produced is pure heptane, which has a boiling point of 98°C. Ideally, over time, all 50 mL of heptane in the original sample is recovered. As we continue to add energy at the bottom of the column, the temperature at the top rises to 126°C, signaling that the second component (pure octane, bp 126°C) is being collected. In the real world, the separation may not be as complete as described in this idealized presentation.

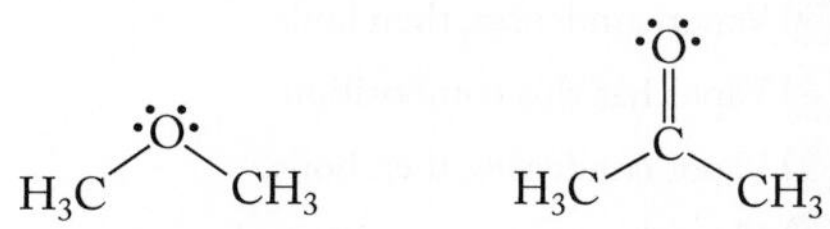

H_3C–Ö–CH_3 Dimethyl ether μ = 1.30 D

H_3C–C(=Ö)–CH_3 Acetone μ = 2.88 D

FIGURE 11.11 Condensed structures for dimethyl ether and acetone.

CONCEPT TEST

Dimethyl ether (CH_3OCH_3) and acetone [$CH_3C(O)CH_3$], shown in **Figure 11.11**, have similar molar masses but different dipole moments: 1.30 D and 2.88 D, respectively. Which compound would you expect to distill first from a mixture of acetone and dimethyl ether? Why?

SAMPLE EXERCISE 11.6 Predicting the Results of Fractional Distillation — LO4

How would the graph in Figure 11.10 be different if (a) we started with a mixture of 75 mL of octane (C_8H_{18}) and 25 mL of heptane (C_7H_{16}) and (b) we started with a mixture of 75 mL of octane (C_8H_{18}) and 25 mL of nonane (C_9H_{20})? The normal boiling points of heptane, octane, and nonane are 98°C, 126°C, and 151°C, respectively.

Collect and Organize The graph in Figure 11.10 is an idealized plot of the temperature of the solution as a function of the volume of distillate for a 50:50 mixture of C_7H_{16}:C_8H_{18}. We want to describe the changes in the graph if we change the ratio of the components and if we change the identity of one of the components. We are given the normal boiling points of all compounds.

Analyze Fractional distillation separates solutions of volatile liquids into pure substances based on their different boiling points. The components distill in the order of their boiling points, with the lowest-boiling component distilling first. Ideally, the total volume of distillate equals the total volume of the solution, and the volume of each fraction reflects the volume of the component in the original solution.

Solve

a. The first component that distills is the lower-boiling heptane. Ideally, 25 mL of heptane would distill at 98°C. Then, the only remaining component (75 mL of octane) would distill at 126°C. The idealized graph (**Figure 11.12**) shows the boiling points of the two substances along the *y*-axis and the volume of distillate along the *x*-axis. The lengths of the horizontal lines at 98°C and 126°C are in a 1:3 ratio, reflecting the composition of the mixture, 25 mL of C_7H_{16} and 75 mL of C_8H_{18}.

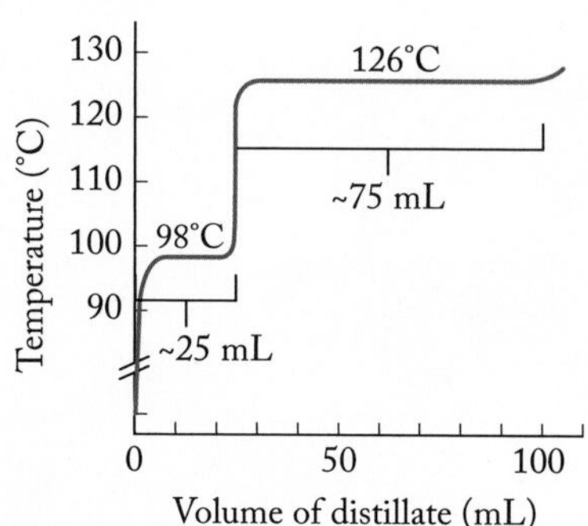

FIGURE 11.12 Temperature as a function of distillate volume for a mixture of 25 mL of C_7H_{16} and 75 mL of C_8H_{18}.

b. The first component that distills in the second mixture is the lower-boiling octane. Ideally, 75 mL of octane would distill at 126°C. Once the octane is removed from the solution, the only remaining component (25 mL of nonane) would distill at 151°C. The idealized graph (**Figure 11.13**) shows the boiling points of the two substances along the y-axis and the volume of distillate along the x-axis.

Think About It The constant temperature plateaus in the distillation graphs make sense because multiple vaporization–condensation cycles in the fractionating column mean that the vapor reaching the top of the column and flowing into the condenser is composed entirely of the most volatile component (remaining) in the distillation mixture. The constant temperature of the vapor matches the boiling point of that component.

Practice Exercise Draw an idealized graph of the temperature versus volume of distillate collected when a solution consisting of 30 mL of hexane (boiling point 69°C), 50 mL of heptane (boiling point 98°C), and 20 mL of nonane (boiling point 151°C) is fractionally distilled.

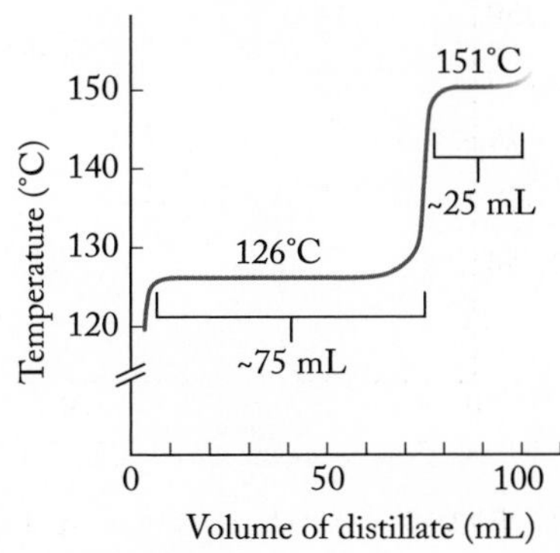

FIGURE 11.13 Temperature as a function of distillate volume for a mixture of 75 mL of C_8H_{18} and 25 mL of C_9H_{20}.

Now that we know how fractional distillation works, the next step is to understand *why* it works. We return to Raoult's law, which we used in Section 11.3 to describe the influence of nonvolatile solutes on the boiling point of a pure solvent. Raoult's law also applies to homogeneous mixtures of volatile compounds, such as crude oil. Because the solutes in a solution of crude oil are volatile, they contribute to the solution's overall vapor pressure. The total vapor pressure equals the sum of the vapor pressures of each component (P_x° is the equilibrium vapor pressure of the pure component at the temperature of interest) multiplied by the mole fraction of that component in the solution (X_x):

$$P_{\text{total}} = X_1P_1^\circ + X_2P_2^\circ + X_3P_3^\circ + \ldots \quad (11.8)$$

CHEMTOUR
Raoult's Law

SAMPLE EXERCISE 11.7 Calculating the Vapor Pressure of a Solution of Volatile Substances **LO3**

Ethanol (CH_3CH_2OH) is used to disinfect the skin prior to getting a shot, such as a flu vaccine. The solution quickly evaporates at 37°C, roughly the body temperature of a healthy human. Calculate the vapor pressure of a solution prepared by dissolving 5.0 g of water in 78.0 g of ethanol (CH_3CH_2OH) at 37°C. By what factor does the concentration of the more volatile component in the vapor exceed the concentration of this component in the liquid? The vapor pressures of ethanol and water at 37°C are 115 torr and 47 torr, respectively.

Collect and Organize We can use Equation 11.8 to determine the total vapor pressure of the solution from the vapor pressures of the components after we determine the composition of the solution in terms of mole fractions.

Analyze To calculate the mole fraction of each component, we need the molar masses of water and ethanol. The mole fraction is then equal to the number of moles of each component divided by the total number of moles of material in the solution.

$$\boxed{\text{Mass}} \xrightarrow{\frac{1}{\text{molar mass}}} \boxed{\text{Moles}} \xrightarrow{\frac{1}{\text{total moles}}} \boxed{\text{Mole fraction}}$$

The vapor pressure of the solution must lie between the vapor pressures of the pure compounds.

Solve The number of moles of each component is

$$78.0 \text{ g } C_2H_5OH \times \frac{1 \text{ mol } CH_3CH_2OH}{46.1 \text{ g } CH_3CH_2OH} = 1.69 \text{ mol } CH_3CH_2OH$$

$$5.0 \text{ g } H_2O \times \frac{1 \text{ mol } H_2O}{18.0 \text{ g } H_2O} = 0.28 \text{ mol } H_2O$$

The total number of moles of material is (1.69 mol + 0.28 mol) = 1.97 mol. Thus, the mole fraction of each component in the mixture is

$$X_{\text{ethanol}} = \frac{1.69 \text{ mol}}{1.97 \text{ mol}} = 0.858$$

$$X_{\text{water}} = 1 - X_{\text{ethanol}} = 0.142$$

Using these values and the vapor pressures of the two substances in Equation 11.8, we have

$$\begin{aligned} P_{\text{total}} &= X_{\text{water}}P°_{\text{water}} + X_{\text{ethanol}}P°_{\text{ethanol}} \\ &= 0.142(47 \text{ torr}) + 0.858(115 \text{ torr}) \\ &= 6.7 \text{ torr} + 98.7 \text{ torr} = 105.4 \text{ torr} \end{aligned}$$

To calculate how enriched the vapor phase is in the more volatile component (i.e., in ethanol, the component with the greater vapor pressure), we need to recall Dalton's law of partial pressures (Section 5.7) and the concept that the partial pressure of a gas in a mixture of gases is proportional to its mole fraction in the mixture. Therefore, the ratio of the mole fraction of ethanol to that of water is the ratio of their two vapor pressures:

$$\frac{98.7 \text{ torr}}{6.7 \text{ torr}} = 14.7$$

The mole ratio of ethanol to water in the liquid mixture is

$$\frac{1.69 \text{ mol}}{0.28 \text{ mol}} = 6.0$$

Therefore, the vapor phase is enriched in ethanol by a factor of

$$\frac{14.7}{6.0} = 2.5$$

Think About It The vapor pressure of the mixture (105.4 torr) is between the vapor pressures of the separate components (47 torr and 115 torr). This result illustrates how fractional distillation works. The vapor is enriched in the lower-boiling component.

Practice Exercise Benzene (C_6H_6) is a trace component of gasoline. What is the mole ratio of benzene to octane in the vapor above a solution of 10% benzene and 90% octane by mass at 25°C? The vapor pressures of octane and benzene at 25°C are 11 torr and 95 torr, respectively.

x_1
1.0 0.8 0.6 0.4 0.2 0.0
Vapor pressure
P_{total}
P_2
P_1
0.0 0.2 0.4 0.6 0.8 1.0
x_2

(a)

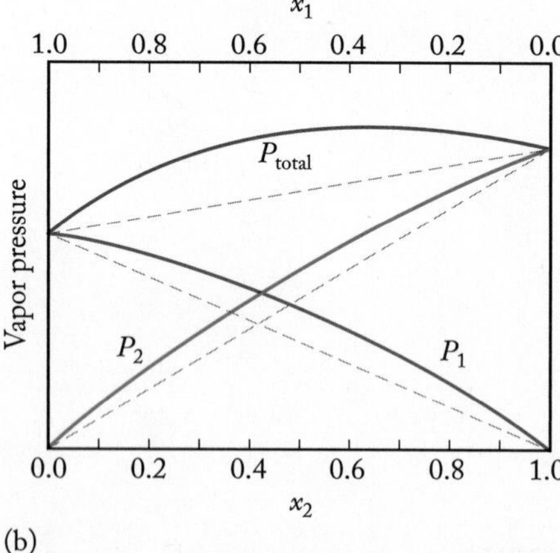

(b)

FIGURE 11.14 In a mixture of two volatile substances x_1 and x_2, the vapor pressures P_1 and P_2 may deviate from the ideal behavior predicted by Raoult's law and shown as dashed lines. (a) If solute–solvent interactions are *stronger* than solvent–solvent or solute–solute interactions, the deviations from Raoult's law are *negative*. (b) If solute–solvent interactions are *weaker* than solvent–solvent or solute–solute interactions, the deviations are *positive*.

Solutions such as the hydrocarbons in crude oil obey Raoult's law when the strengths of solvent–solvent, solute–solute, and solute–solvent interactions are similar. Under these conditions, a solution behaves ideally. A mixture of hydrocarbons is expected to behave like an ideal solution because intermolecular interactions between the components are all London forces acting on molecules of similar structure and size.

If the solute–solvent interactions are *stronger* than the solvent–solvent or solute–solute interactions, the solute inhibits the solvent from vaporizing and the solvent inhibits the solute from vaporizing. This situation produces negative deviations from the vapor pressures predicted by Raoult's law (**Figure 11.14a**). In such a solution, the rate of evaporation is slower because the vapor pressure of the mixture is lower than predicted. Because solvent and solute molecules are held at the surface by solute–solvent attractive interactions, more energy is required to separate them from the surface, and fewer vaporize at a given temperature.

If solute–solvent interactions are much *weaker* than solvent–solvent interactions, less energy is required to separate solute molecules from the surface, and more solute molecules vaporize. In this case the vapor pressure is greater than the value predicted by Raoult's law (**Figure 11.14b**).

colligative properties characteristics of solutions that depend on the concentration and not the identity of particles dissolved in the solvent.

CONCEPT TEST

Which of the following solutions is least likely to follow Raoult's law: (a) acetone–ethanol, (b) pentane–hexane, or (c) pentanol–water? The skeletal molecular structures of the compounds are given in **Figure 11.15**.

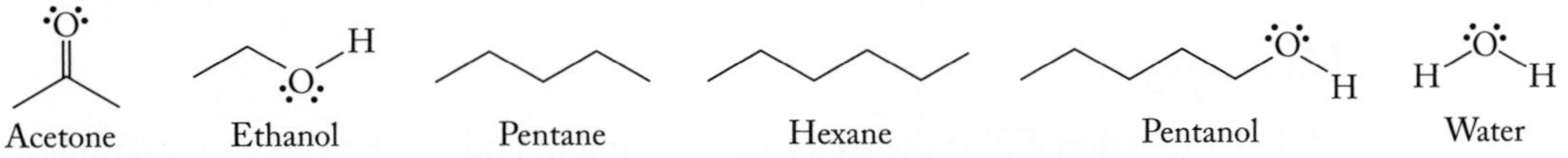

Acetone Ethanol Pentane Hexane Pentanol Water

FIGURE 11.15 Skeletal molecular structures of acetone, ethanol, pentane, hexane, pentanol, and water.

11.5 Colligative Properties of Solutions

We saw in Section 11.3 that the vapor pressure of a solution containing a nonvolatile solute is lower than the vapor pressure of the pure solvent. In this section we explain how many other physical properties of a solvent are changed when a nonvolatile solute is added to it. These properties of solutions that depend only on the concentration of solute particles, but not on the identity of the solute, are called **colligative properties**. Solutions generally have greater densities than the solvent alone, and an aqueous solution containing a nonvolatile solute has a higher boiling point and a lower freezing point than pure water. As we saw in Section 11.3, antifreeze, an aqueous solution of ethylene glycol used to cool automobile engines, boils above 100°C and freezes below 0°C. As a result, it has a higher boiling point and a lower freezing point than pure water.

Figure 11.16 shows the combined phase diagram of water and an aqueous solution of a nonvolatile solute. The red line representing boiling/condensation points of the solution lies below the orange line for pure water. The blue melting/

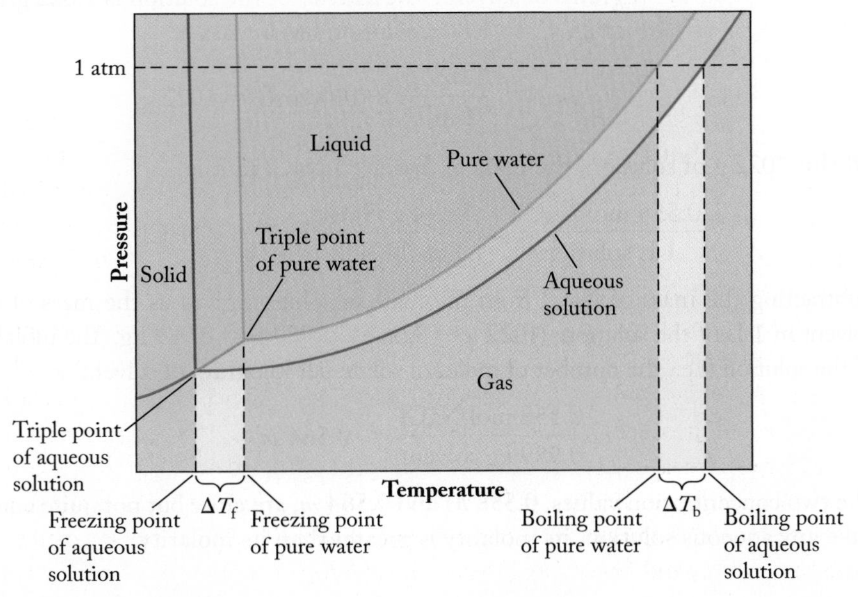

FIGURE 11.16 Combined phase diagram for pure water and a solution of a nonvolatile solute in water. Notice that the solution's boiling point is higher than the boiling point of water (the pure solvent) and that the solution's freezing point is lower than the freezing point of water. The line depicting the phase change from solid to liquid bends ever so slightly to the left. Compare this with Figure 10.23, where the negative slope of this line is clearer at pressures above 1 atm.

freezing line for the solution lies to the left of the light blue line for pure water. Following the dashed line from left to right at $P = 1$ atm, we see that the solution freezes at a lower temperature than pure water and boils at a higher temperature than pure water. Furthermore, the higher boiling point for the solution indicates that the solution has a lower vapor pressure than pure water.

CONNECTION Phase diagrams (Section 10.5) show how the physical states of substances change with changing temperatures and pressures.

Boiling point elevation and freezing point depression are both colligative properties. As noted in Section 11.3, only the number of particles in solution determines the impact of the solute on the colligative properties of the solvent. Before we can quantify these properties, though, we must introduce a new concentration unit.

Molality

Molarity (Section 4.2) is the concentration unit of choice when we run reactions in solutions, usually at constant temperature, so that we can measure volumes of solutions and know the number of moles of reactants we are dealing with. When we work with colligative properties, however, we frequently deal with the properties of systems as a function of temperature. Because volumes change with temperature, molarity changes with temperature. If we want to quantify certain colligative properties, therefore, the approach is to use a different concentration unit, called **molality (m)**, which is defined as the number of moles of solute (n_{solute}) per kilogram of solvent:

$$m = \frac{n_{\text{solute}}}{\text{kg solvent}} \tag{11.9}$$

The difference between these two similar-sounding concentration units is that molarity is the number of moles of solute *per liter of solution*, whereas molality is the number of moles of solute *per kilogram of solvent*. Because solvent mass does not change with temperature, a concentration expressed in molality does not change with temperature.

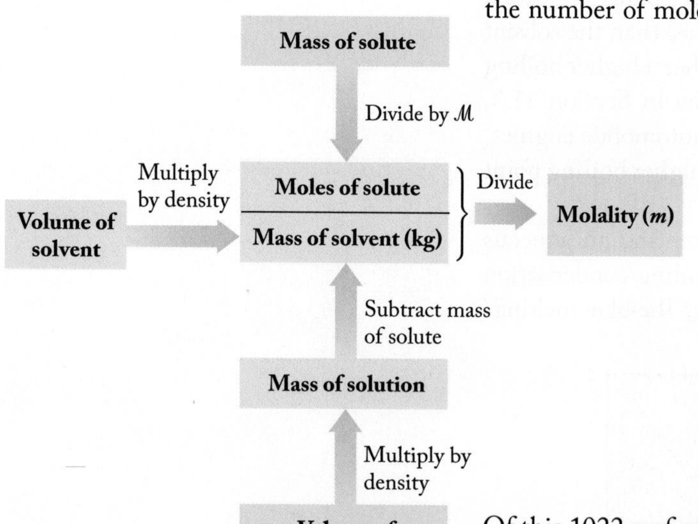

FIGURE 11.17 Flow diagram for calculating molality.

To illustrate the difference between the molarity and molality of a solution, let's use the procedure outlined in **Figure 11.17** to calculate the molality of 1 L of an aqueous solution of sodium chloride that is 0.558 M in NaCl (approximately the NaCl concentration in seawater) at 25°C. To calculate molality, we need the density of the solution so that we can convert liters of solution into kilograms of solvent. The density of the solution is 1.022 g/mL at 25°C, so 1 L of solution has a mass of

$$\frac{1.022 \text{ g}}{1 \text{ mL}} \times 1000 \text{ mL} = 1022 \text{ g}$$

Of this 1022 g of solution, the mass of dissolved NaCl in it is

$$\frac{0.558 \text{ mol NaCl}}{1 \text{ L solution}} \times \frac{58.44 \text{ g NaCl}}{1 \text{ mol NaCl}} = 32.6 \text{ g NaCl}$$

Subtracting the mass of NaCl from the mass of solution gives us the mass of the solvent in 1 L of the solution: (1022 g − 32.6 g) = 989 g = 0.989 kg. The molality of the solution (i.e., the number of moles of solute per kilogram of solvent) is

$$\frac{0.558 \text{ mol NaCl}}{0.989 \text{ kg solvent}} = 0.564\ m$$

The two concentration values, 0.558 M and 0.564 m, are close but not quite equal. Like any aqueous solution, its molality is greater than its molarity.

molality (m) concentration expressed as the number of moles of solute per kilogram of solvent.

CONCEPT TEST

The difference between the molarity and molality concentrations of a dilute aqueous solution is small. Why?

SAMPLE EXERCISE 11.8 Preparing a Solution of Known Molality LO5

How many grams of Na_2SO_4 should be added to 275 mL of water to prepare a 0.750 m solution of Na_2SO_4? Assume the density of water is 1.000 g/mL.

Collect and Organize We want to determine the mass of sodium sulfate (the solute) needed to prepare a 0.750 m solution. The volume of water (the solvent) is 275 mL.

Analyze Following the procedure shown in Figure 11.17:

$$\boxed{\text{Volume of water}} \xrightarrow{\text{density}} \boxed{\text{Mass of water}}$$

We can then work backward from molality to mass of solute:

$$\boxed{\text{Molality}} \xrightarrow{\text{mass of solvent}} \boxed{\text{Number of moles solute}} \xrightarrow{\text{molar mass}} \boxed{\text{Grams of solute}}$$

Our goal is to prepare a relatively small volume ($\sim\frac{1}{4}$ L) of a dilute solution (<1 m), so we predict that the mass of solute needed is probably less than $\frac{1}{4}$ mole of solute, which has a molar mass of about 142 g/mol, or about 35 g.

Solve Multiplying 275 mL of water by the density of water gives the mass of water we start with:

$$275\ \cancel{\text{mL water}} \times \frac{1.000\ \text{kg water}}{1000\ \cancel{\text{mL water}}} = 0.275\ \text{kg water}$$

We can rearrange Equation 11.9 to determine the number of moles of solute needed, using the fact that a molality of 0.750 is equivalent to 0.750 mol of solute in 1 kg of solvent:

$$n_{\text{solute}} = (m)(\text{kg solvent})$$

$$n_{Na_2SO_4} = \frac{0.750\ \text{mol}\ Na_2SO_4}{1\ \cancel{\text{kg water}}} \times 0.275\ \cancel{\text{kg water}} = 0.206\ \text{mol}\ Na_2SO_4$$

The molar mass of Na_2SO_4 is 142.04 g/mol, so the number of grams of Na_2SO_4 needed is

$$0.206\ \cancel{\text{mol}}\ Na_2SO_4 \times \frac{142.04\ \text{g}}{1\ \cancel{\text{mol}}} = 29.3\ \text{g}\ Na_2SO_4$$

Dissolving 29.3 g of Na_2SO_4 in 275 mL of water produces a 0.750 m solution.

Think About It The calculated value of 29.3 g of Na_2SO_4 is consistent with our prediction that about 35 g of solute would be required. Taking a different approach to this sample exercise, we could calculate the molality of a solution in which 29.3 g Na_2SO_4 is dissolved in 275 mL H_2O:

$$\frac{29.3\ \text{g}\ Na_2SO_4}{275\ \cancel{\text{mL water}}} \times \frac{1\ \text{mol}}{142.04\ \text{g}} \times \frac{1000\ \cancel{\text{mL water}}}{1.000\ \text{kg water}} = \frac{0.750\ \text{mol}\ Na_2SO_4}{1\ \text{kg water}} = 0.750\ m\ Na_2SO_4$$

Practice Exercise What is the molality of a solution prepared by dissolving 78.2 g of ethylene glycol ($HOCH_2CH_2OH$) in 1.50 L of water? Assume the density of water is 1.000 g/mL.

TABLE 11.5 Molal Freezing-Point-Depression and Boiling-Point-Elevation Constants for Selected Solvents

Solvent	Freezing Point (°C)	K_f (°C/m)	Boiling Point (°C)	K_b (°C/m)
Water (H_2O)	0.0	1.86	100.0	0.52
Benzene (C_6H_6)	5.5	4.90	80.1	2.53
Ethanol (CH_3CH_2OH)	−114.6	1.99	78.4	1.22
Carbon tetrachloride (CCl_4)	−22.3	29.8	76.8	5.02

Boiling Point Elevation

In Section 11.3 we discussed how the vapor pressure of a solution is reduced and the boiling point is elevated with respect to pure solvent, and at the beginning of this section we examined the phase diagrams of solutions. Now we can look quantitatively at how much the boiling points and freezing points of liquids change when solutes are present.

Boiling point elevation is a colligative property of the solvent. It is described in equation form as

$$\Delta T_b = K_b m \tag{11.10}$$

where ΔT_b is the increase in temperature above the boiling point of the pure solvent, K_b is the *boiling-point-elevation constant* of the solvent, and m is the molality of the solution. The values of K_b for several solvents are listed in **Table 11.5**. The units of K_b are °C/m, so the concentration of particles in solution must also have units of molality (m):

$$\Delta T_b = K_b m = \frac{°\text{C}}{m} \times m$$

Adding the increase in boiling point to the normal boiling point of pure solvent, T_b°, gives us an equation for calculating the boiling point (T_b) of any solution based on that solvent:

$$T_b = T_b^\circ + K_b m \tag{11.11}$$

The K_b of water is 0.52°C/m, which means that for every mole of particles that dissolves in 1 kg of water (1 m), the boiling point of the solution rises by 0.52°C:

$$\Delta T_b = K_b m = \frac{0.52°\text{C}}{m} \times 1\ m = 0.52°\text{C}$$

Therefore, the boiling point (T_b) of the solution is

$$T_b = T_b^\circ + \Delta T_b = 100.00°\text{C} + 0.52°\text{C} = 100.52°\text{C}$$

CHEMTOUR
Boiling and Freezing Points of Solutions

SAMPLE EXERCISE 11.9 Calculating the Boiling Point Elevation of an Aqueous Solution **LO6**

Maple syrup is produced by evaporating most of the water from sap collected from maple trees in early spring. Suppose a 100.0 g sample of maple syrup contains 61.2 g of sucrose ($\mathcal{M}$ = 342.3 g/mol), 6.8 g of fructose and glucose, which have the same molar mass (180.2 g/mol), and a variety of other trace organic and inorganic ingredients that do not contribute significantly to boiling point elevation. At what temperature does the syrup sample boil?

Collect and Organize We are asked to calculate the boiling point of maple syrup and are given the concentration of the three principal solutes in it. Equation 11.11 relates the boiling point of a solution to that of pure solvent and to the molal concentration of solute particles. The boiling-point-elevation constant of water is $K_b = 0.52°C/m$, and the normal boiling point of pure water is 100.0°C.

Analyze To use Equation 11.11, we must first convert concentration values in g solute/g solution to mol solute/kg solvent (H_2O). The following unit conversions should accomplish this task:

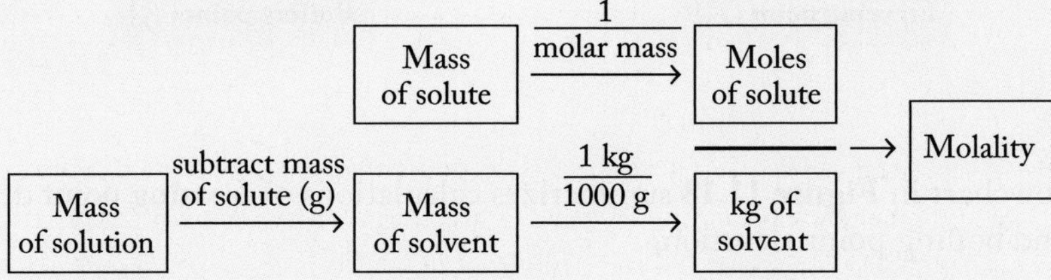

Sugars make up over $\frac{2}{3}$ of the mass of the sample, which means that less than $\frac{1}{3}$ of it is water. A kilogram of this syrup would have nearly 700 g, or about 2 moles, of the sugars in about $\frac{1}{3}$ of a kilogram of water for a molality around 6 m and a boiling point near (100 + 0.5 × 6) or 103°C.

Solve

1. Moles of sugars: sucrose $= 61.2 \text{ g} \times \frac{1 \text{ mol}}{342.3 \text{ g}} = 0.1788 \text{ mol}$;

   fructose + glucose $= 6.8 \text{ g} \times \frac{1 \text{ mol}}{180.2 \text{ g}} = 0.0377 \text{ mol}$

2. kg H_2O $= [100.0 \text{ g syrup} - (61.2 + 6.8) \text{ g solutes}] \times \frac{1 \text{ kg}}{1000 \text{ g}} = 0.0320 \text{ kg}$

3. Total concentration (m) of sugars $= \frac{(0.1788 + 0.0377) \text{ mol}}{0.0320 \text{ kg } H_2O} = 6.77\ m$

4. Calculating the boiling point:
   $T_b = T_b^\circ + K_b m = 100.0°C + 0.52°C/m \times 6.77\ m = 103.5°C$

Think About It The calculated boiling point is near the predicted value, so is reasonable.

Practice Exercise Crude oil pumped out of the ground may be accompanied by *formation water*, a solution that contains high concentrations of NaCl and other salts. If the boiling point of a sample of formation water is 2.3°C above the boiling point of pure water, what is the molality of dissolved particles in the sample?

Freezing Point Depression

As we mentioned when discussing Figure 11.16, *freezing point depression* is a colligative property that is put to good use in car radiators to ensure that their fluid does not freeze in cold weather. The magnitude of the freezing point depression is directly proportional to the molal concentration of dissolved solute:

$$\Delta T_f = K_f m \quad (11.12)$$

where ΔT_f is the change in the freezing temperature of the solvent, K_f is the *freezing-point-depression constant* of the solvent, and m is the molality of the solution. The values of K_f for several solvents are listed in Table 11.5. The freezing point (T_f) of a solution is

$$T_f = T_f^\circ - K_f m \quad (11.13)$$

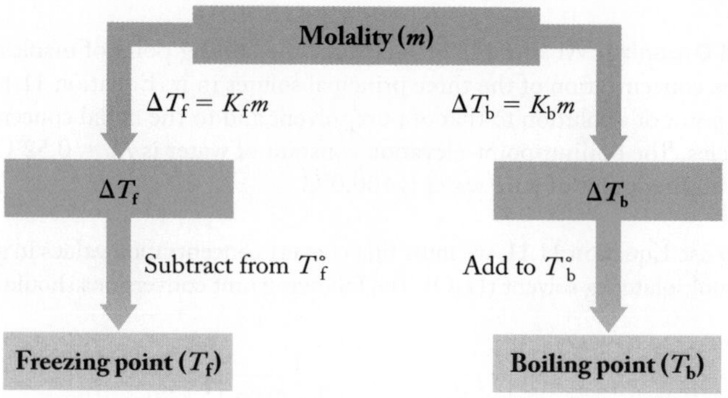

FIGURE 11.18 Flow diagram for calculating freezing points and boiling points of solutions of known molality (T_b = normal boiling point of pure solvent; T_f = normal freezing point of pure solvent).

The flowchart in **Figure 11.18** summarizes calculations of freezing point depression and boiling point elevation.

SAMPLE EXERCISE 11.10 Calculating the Freezing Point of a Solution **LO6**

What is the freezing point of radiator fluid prepared by mixing 1.00 L of ethylene glycol ($\mathcal{M}$ = 62.07 g/mol, density = 1.114 g/mL) with 1.00 L of water (density 1.000 g/mL)?

Collect and Organize We are asked to determine the freezing point of a solution of ethylene glycol in water from the volumes of the two liquids, their densities, and K_f for water. Equation 11.13 relates the freezing point of a solution to the normal freezing point of the solvent (0.0°C), the freezing-point-depression constant, K_f, which is 1.86°C/m for water, and the total molal concentration of all solute particles in solution.

Analyze To use Equation 11.13, we first need to calculate the molal concentration of the solute (ethylene glycol) in the solvent (water) using the following conversion steps:

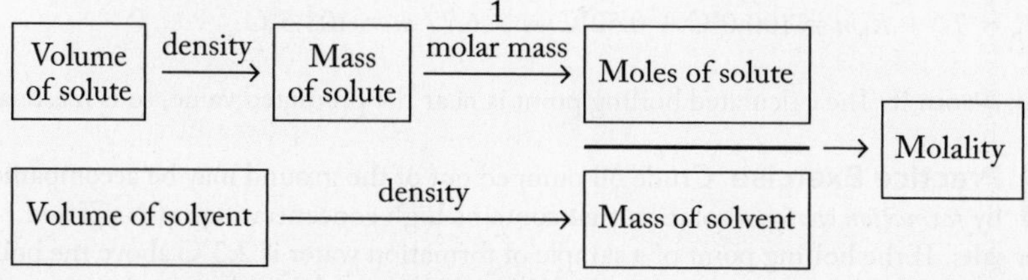

We have equal volumes of each substance, but it is logical for us to choose water as the solvent because ethylene glycol has a much higher molar mass (see Sample Exercise 11.5). The mass of 1 L of ethylene glycol is 1114 g. This mass divided by the molar mass (62 g/mol) is roughly 1000 g/60 g/mol or 17 mol, which forms a 17 m solution when dissolved in 1 kg of H_2O. Multiplying 17 m by 1.86°C/m produces a ΔT_f near −30°C.

Solve The solvent mass is

$$1.00\ \cancel{\text{L}} \times \frac{1000\ \cancel{\text{mL}}}{1\ \cancel{\text{L}}} \times \frac{1.000\ \cancel{\text{g}}}{1\ \cancel{\text{mL}}} \times \frac{0.001\ \text{kg}}{1\ \cancel{\text{g}}} = 1.00\ \text{kg solvent}$$

The moles of ethylene glycol in 1.00 L are

$$1.00\ \cancel{\text{L}}\ \text{solute} \times \frac{1000\ \cancel{\text{mL}}}{1\ \cancel{\text{L}}} \times \frac{1.114\ \cancel{\text{g}}}{1\ \cancel{\text{mL}}} \times \frac{1\ \text{mol}}{62.07\ \cancel{\text{g}}} = 17.9\ \text{mol solute}$$

Calculating the molal concentration of ethylene glycol:

$$m = \frac{17.9\ \text{mol solute}}{1.00\ \text{kg solvent}} = 17.9\ m$$

Using this concentration in Equation 11.13:

$$T_f = T_f^\circ - K_f m$$

$$= 0.0°\text{C} - \frac{1.86°\text{C}}{m} \times 17.9\ m = -33.3°\text{C}$$

Think About It The answer makes sense because it is close to the predicted value. Also, engine coolant with this freezing point would provide adequate protection against radiator fluid freezing during most winters in most of the United States, though not in much of Canada or Alaska where 60%–70% by volume solutions of ethylene glycol are often used to prevent engines from freezing up.

Practice Exercise Concentrated solutions of ionic compounds are used in biochemistry laboratories to disrupt the structure of proteins. What is the freezing point of a solution whose ion concentration is 8.15 *m*?

The van 't Hoff Factor

Recall from Section 4.4 that compounds called *electrolytes* dissociate into ions when in solution, whereas *nonelectrolytes* remain intact. If a solute is a nonelectrolyte, then every mole of solute yields 1 mole of particles. If the solute is an electrolyte, however, then the number of moles of particles depends on the total concentration of cations and anions in solution. For example, if we approximate seawater as 0.574 *m* sodium chloride, then each kilogram of water contains 2 × 0.574 = 1.148 moles of particles because NaCl forms Na^+ and Cl^- ions when it dissolves.

CONNECTION An electrolyte is a substance that dissociates into ions when it dissolves (Section 4.4).

Because freezing point depression and boiling point elevation are colligative properties, the dissolution of 1 mole of a strong electrolyte such as NaCl in a given quantity of water produces the same changes in freezing point and boiling point as 1 mole of the strong electrolyte KNO_3, even though the latter has a much higher molar mass. Each of these salts adds 2 moles of particles (1 mole of cations and 1 mole of anions) to the water for every 1 mole of salt that dissolves. When a nonelectrolyte such as ethylene glycol is dissolved in water, on the other hand, 1 mole of the solute produces only 1 mole of particles (1 mole of molecules) in the solution.

Dutch chemist Jacobus van 't Hoff (1852–1911) studied colligative properties and defined a term *i*, now called the **van 't Hoff factor** (or ***i* factor**), which is the ratio of the experimentally measured value of a colligative property to the value expected if the solute were a nonelectrolyte (i.e., no dissociation into ions). For example, suppose we make a solution that is 0.010 *m* in NaCl by dissolving 0.5844 g of NaCl in 1 kg of water. If NaCl was a nonelectrolyte, then the freezing point of the solution should be lower than that of pure water by

$$\Delta T_f = K_f m = \frac{1.86°\text{C}}{m} \times 0.0100\ m\ \text{NaCl} = 0.0186°\text{C}$$

However, the observed freezing point depression for 0.0100 *m* NaCl is not 0.0186°C, but instead is 0.0372°C: *twice* as much as that predicted for a nonelectrolyte. The factor of 2 difference in the observed change in freezing point compared to that of a nonelectrolyte is the van 't Hoff factor for NaCl. Its value makes sense because sodium chloride is a strong electrolyte and forms 2 moles of ions from each mole of NaCl. Like all colligative properties, the observed changes are proportional to *total concentration of dissolved particles* present in solution.

van 't Hoff factor (also called ***i* factor**) the ratio of the experimentally measured value of a colligative property to the theoretical value expected for that property if the solute were a nonelectrolyte.

Equations such as 11.10 and 11.12 can be modified to include the i factor:

$$\Delta T_b = iK_b m \quad (11.14)$$

$$\Delta T_f = iK_f m \quad (11.15)$$

If the solute is molecular (such as ethylene glycol) and therefore a nonelectrolyte, then $i = 1$ because each mole of solute produces 1 mole of dissolved particles. If the solute is a strong electrolyte, then $i =$ the number of ions in one formula unit. For NaCl, $i = 2$; for Na_2SO_4, $i = 3$ (two Na^+ ions and one SO_4^{2-} ion). We do *not* break a polyatomic ion such as SO_4^{2-} into its atoms when determining an i factor; polyatomic ions stay intact.

SAMPLE EXERCISE 11.11 Using the van 't Hoff Factor **LO7**

The salt lithium perchlorate ($LiClO_4$) is one of the most water-soluble salts known. At what temperature does a 0.130 m solution of $LiClO_4$ freeze? The K_f of water is 1.86°C/m; assume $i = 2$ for $LiClO_4$ and the freezing point of pure water is 0.00°C.

Collect and Organize We are asked to determine the freezing point of a salt solution. We know the formula of the solute, its molal concentration, and K_f of the solvent. We know that the freezing point of the pure solvent is 0.00°C. We are given the value of the van 't Hoff factor as $i = 2$.

Analyze We use Equation 11.15 to solve for the freezing point depression. The value of K_f is close to 2, the value of i is 2, and the concentration of solute is close to 0.1 m, so we predict that the freezing point of the solution will be about 0.4°C lower than that of pure water.

Solve

$$\Delta T_f = iK_f m = (2)\left(\frac{1.86°\text{C}}{m}\right)(0.130\ m) = 0.484°\text{C}$$

The freezing point of the solution is (0.00 − 0.484)°C = −0.48°C.

Think About It Lithium perchlorate dissolves in aqueous solution to form 2 moles of ions for every mole of solute that dissolves: 1 mole of Li^+ cations and 1 mole of ClO_4^- anions, consistent with $i = 2$. The calculated value is consistent with our prediction.

 **Practice Exercise** Determine the value of the van 't Hoff factor and calculate the boiling point of a 1.75 m aqueous solution of barium nitrate, $Ba(NO_3)_2$. $K_b = 0.52$°C/m for water.

CONCEPT TEST

Which aqueous solution has the lowest freezing point: (a) 3 m glucose ($C_6H_{12}O_6$), (b) 2 m potassium iodide (KI), or (c) 1 m sodium sulfate (Na_2SO_4)?

ion pair a cluster formed when a cation and an anion associate with each other in solution.

Using Equations 11.14 and 11.15 to calculate boiling point elevations and freezing point depressions for concentrated solutions of strong electrolytes often gives larger values than the experimentally measured values. The reason is that the cations and anions produced when strong electrolytes dissolve may not be totally independent of one another. As concentration increases, cations and anions may form ionic clusters. The simplest cluster, an **ion pair**, consists of a cation and an anion that associate in solution, acting as a single particle. Thus, the overall

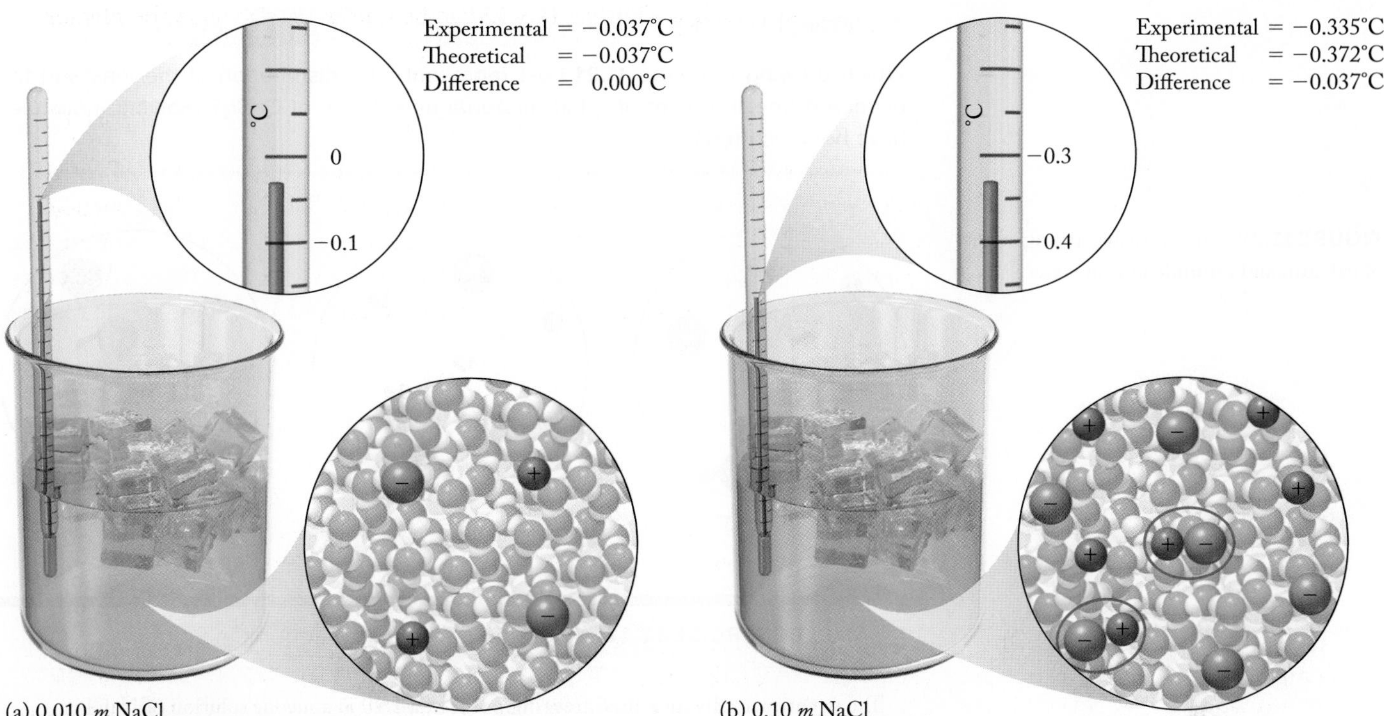

FIGURE 11.19 (a) The experimentally measured freezing point of a 0.010 *m* solution of NaCl is the same as the theoretical value obtained with Equation 11.15. This agreement means that the solution behaves ideally and little or no ion pairing takes place. (b) The experimentally measured freezing point of a 0.10 *m* solution is about 0.04°C higher than the theoretical value because some of the Na^+ and Cl^- ions form ion pairs, as shown inside the red ovals. The formation of ion pairs causes the concentration of solute particles to be less than the theoretical number—as a result, the van 't Hoff factor for the solution is less than the theoretical value of 2, and the decrease in the freezing point is less than expected.

concentration of particles is reduced when ion pairs form, and experimentally measured boiling point elevations and freezing point depressions are smaller than the theoretical values obtained with Equations 11.14 and 11.15 (**Figure 11.19**).

The extent to which free ions form when a strong electrolyte dissolves is expressed by the van 't Hoff factor. The van 't Hoff factor for NaCl in water is 2 if the solution behaves ideally, because ideally 2 moles of ions are produced for each mole of NaCl that dissolves. The value of i is 2 for 0.010 *m* NaCl, but a little less than 2 for 0.10 *m* NaCl. Whenever a calculation for i gives a noninteger value, solute particles are associating in solution, and the behavior is nonideal. **Figure 11.20** gives some theoretical and experimentally measured values of the van 't Hoff factor for several substances.

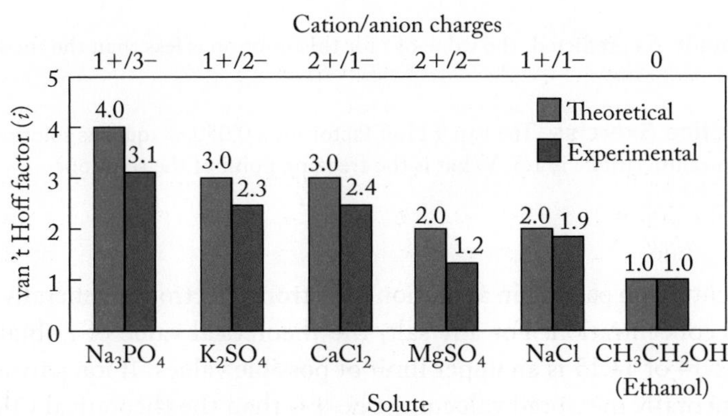

FIGURE 11.20 Theoretical and experimentally measured values for the van 't Hoff factors for 0.1 *m* solutions of several electrolytes and the nonelectrolyte ethanol. The higher the charge on the ions, the greater the difference between theoretical and experimentally measured values.

CONCEPT TEST

Which drawing in **Figure 11.21** best represents the distribution of the ions and the observed freezing point of a 1.26 *m* aqueous solution of $CaBr_2$? (Solvent molecules have been omitted.)

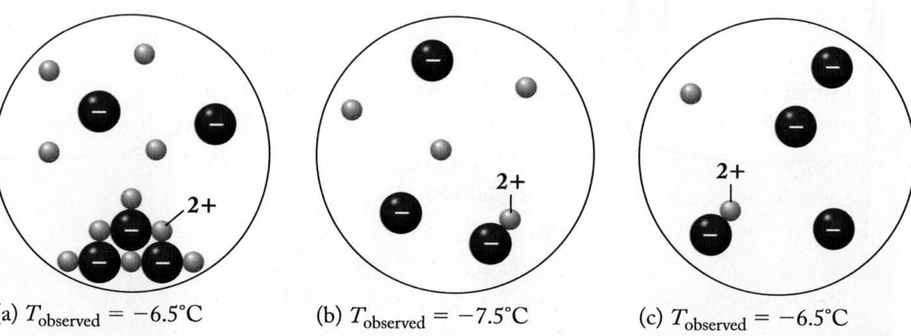

FIGURE 11.21 Three possible arrangements of calcium and bromide ions in water.

SAMPLE EXERCISE 11.12 Assessing Particle Interactions in Solution LO7

The experimentally measured freezing point of a 1.90 *m* aqueous solution of NaCl is −6.57°C. What is the value of the van 't Hoff factor for this solution? Is the solution behaving ideally, or is there evidence that solute particles are interacting with one another? The freezing-point-depression constant of water is $K_f = 1.86°C/m$, and the freezing point of pure water is 0.00°C.

Collect and Organize We are asked to calculate the *i* factor for a solution of known molality. We are given the measured freezing point and the K_f value. We are also asked whether the solution is behaving ideally.

Analyze We can rearrange Equation 11.15 ($\Delta T_f = iK_f m$) to solve for *i* before substituting the values for ΔT_f, K_f, and *m*. If the solution behaves ideally, we would expect $i = 2$ for a strong electrolyte like NaCl; however, given the high concentration of NaCl (1.90 *m*), we predict a value less than 2.

Solve Rearranging Equation 11.15 gives us

$$i = \frac{\Delta T_{f,measured}}{K_f m}$$

$$i = \frac{6.57\cancel{°C}}{\left(\frac{1.86\cancel{°C}}{\cancel{m}}\right)1.90\ \cancel{m}} = 1.86$$

The value of *i* is not an integer, so the solution is *not* behaving ideally. Ion pairs must be forming in solution.

Think About It As predicted, the value of *i* for this solution is less than the theoretical value of 2.

 **Practice Exercise** The van 't Hoff factor for a 0.050 *m* aqueous solution of magnesium sulfate is 1.3. What is the freezing point of the solution?

The extent of ion pairing in a solution of a strong electrolyte generally increases with solute concentration. For any salt, the theoretical value of *i* obtained with Equation 11.14 or 11.15 is an upper limit of possible values. If ion pairing occurs, the experimentally measured value must be less than the theoretical value.

Osmosis and Osmotic Pressure

The final colligative property we look at in this chapter is *osmotic pressure*, the result of the process called **osmosis**—the movement of a solvent through a semipermeable membrane from a region of lower solute concentration to a region of higher solute concentration. A *semipermeable membrane* allows particles of solvent to pass through it but not particles of solute.

To emphasize the importance of osmosis, we start with the observation that we all need water to survive. More than 97% of the water on Earth is seawater, however, and none of that is fit to drink. Why is that? The liquid inside each cell in the body is a complex solution of many solutes, with the average concentration of these solutes being about one-third the concentration of solutes in seawater. When cells are exposed to seawater, this substantial difference in solute concentration is the driving force for osmosis, with the cell membrane acting as the necessary semipermeable membrane (**Figure 11.22**). Because the solute concentration is higher outside the cell, water from inside the cell crosses the cell membrane and enters the seawater, moving from the low-solute-concentration side of the membrane to the high-solute-concentration side. As water leaves the cell, the cell shrivels and ultimately ceases to function.

Water molecules migrate through a cell membrane, or through any other semipermeable membrane, because a force makes it happen—a force caused by the different concentrations of solutes on the two sides of the membrane. When we divide the magnitude of this force, F, by the surface area, A, of the membrane, we get pressure: $P = F/A$. **Osmotic pressure (Π)** is the pressure required to halt the flow of solvent from a dilute solution through a semipermeable membrane into a more concentrated solution. Osmotic pressure exactly balances the pressure (F/A) driving solvent through the membrane so that no net flow of solvent takes place.

osmosis the flow of a fluid through a semipermeable membrane to balance the concentration of solutes in solutions on the two sides of the membrane. The solvent particles' flow proceeds from the more dilute solution into the more concentrated one.

osmotic pressure (Π) the pressure applied across a semipermeable membrane to stop the flow of solvent from the compartment containing pure solvent or a less concentrated solution to the compartment containing a more concentrated solution. The osmotic pressure of a solution increases with solute concentration, M, and with solution temperature, T.

CHEMTOUR
Osmotic Pressure

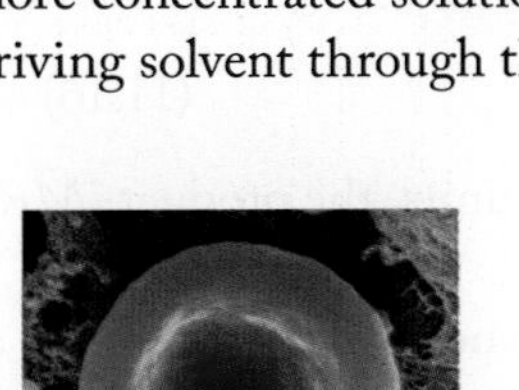

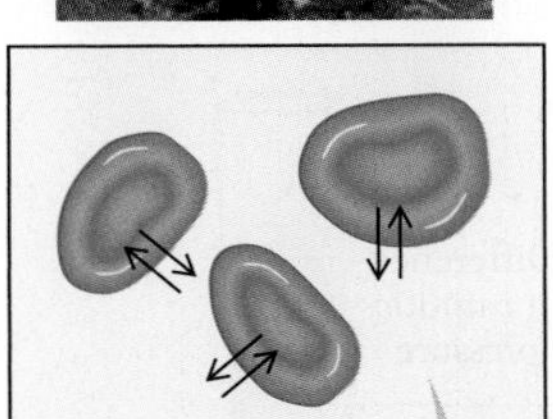

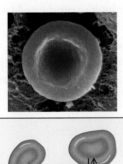

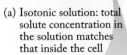

(a) Isotonic solution: total solute concentration in the solution matches that inside the cell

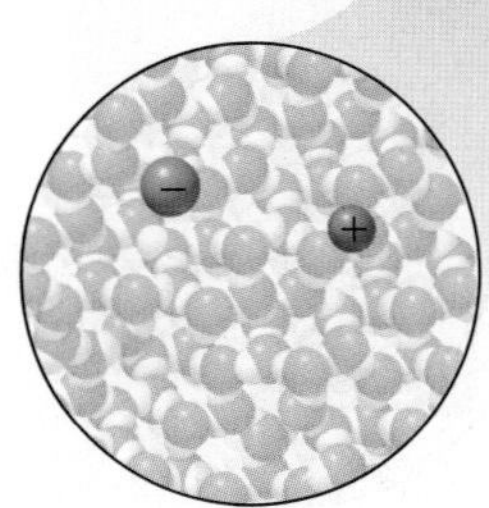

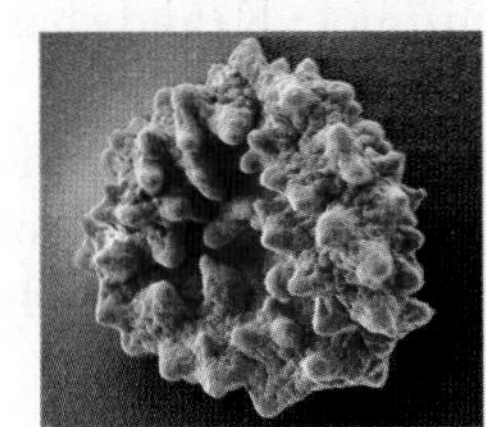

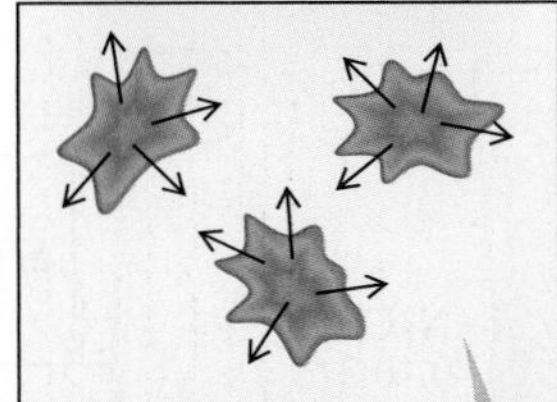

(b) Hypertonic solution: total solute concentration in the solution is greater than that inside the cell; water leaves cells, cells shrink

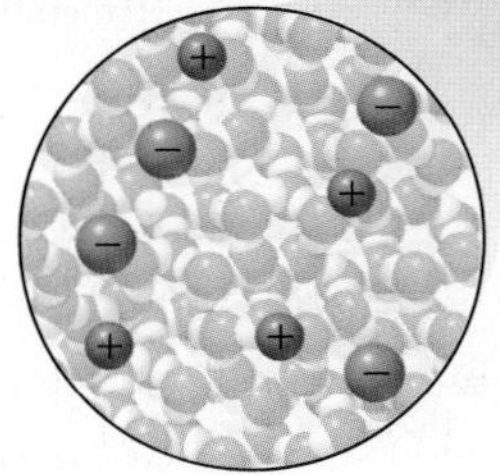

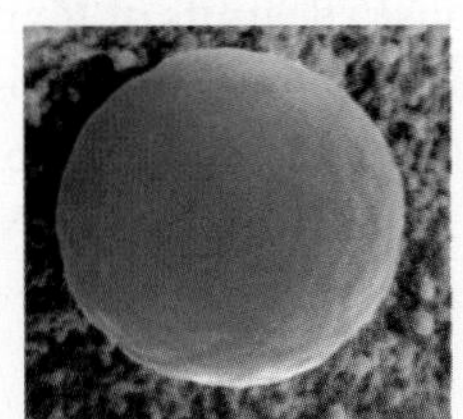

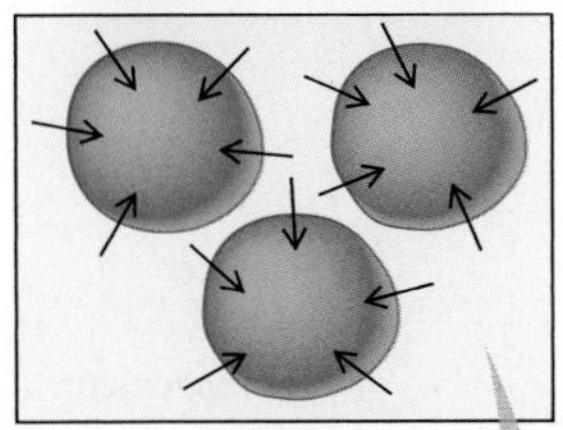

(c) Hypotonic solution: total solute concentration in the solution is less than that inside the cell; water enters cells, cells expand

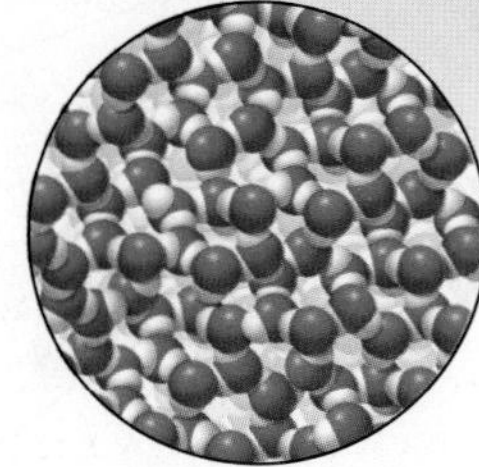

FIGURE 11.22 The membrane of a red blood cell is semipermeable, which means that water easily flows by osmosis into and out of the cell to equalize the solute concentrations on the two sides of the membrane. (a) When a cell is immersed in a solution in which the solute concentration equals the solute concentration inside the cell (isotonic conditions), the flow of water into the cell is exactly balanced by the flow of water out of the cell, and the cell size does not change. (b) When the cell is immersed in a solution in which the solute concentration is higher than the solute concentration inside the cell (hypertonic conditions), water flows by osmosis from the region of lower solute concentration to the region of higher solute concentration—in other words, out of the cell—and the cell shrinks. (c) When the cell is immersed in pure water, the solute concentration is higher inside the cell than outside (hypotonic conditions), and water flows by osmosis from the region of zero solute concentration to the region of high solute concentration—into the cell—and the cell expands.

The Greek letter pi (Π) is used to symbolize osmotic pressure to distinguish it from the pressure exerted by gases. In **Figure 11.23a**, the solution on the right has a lower concentration and lower osmotic pressure (Π_{NaCl}) than the solution on the left ($\Pi_{seawater}$): $\Pi_{NaCl} < \Pi_{seawater}$. The result is a net flow of solvent (water) from right to left until the pressure on both sides is equal. **Figure 11.23b** shows how the volumes of both solutions change: the difference in volume reflects the difference in osmotic pressures, $\Delta\Pi = \Pi_{seawater} - \Pi_{NaCl}$, in the original solutions. This process is profoundly important in living systems, where the solution inside each cell exerts an osmotic pressure on the cell membrane—a pressure that pushes toward the outside of the cell. At the same time, blood or other liquid outside the cell exerts an osmotic pressure on the cell membrane, and this pressure pushes toward the cell's interior.

The osmotic pressure Π depends on the solute concentration, the absolute temperature, and the constant R, 0.0821 L · atm/(mol · K):

$$\Pi = MRT$$

where M is the molarity of the solute. Molarity is used to express concentration in calculating Π because the expression for Π can be derived from an equation similar to the ideal gas equation:

$$\Pi = P = \left(\frac{n}{V}\right)RT = MRT$$

The term n/V has units of moles per liter, which matches the definition of molarity. Osmotic pressure is a colligative property because Π is proportional to the concentration of solute particles and does not depend on the identity of the solute. Therefore, molarity must be multiplied by the van 't Hoff factor i for the solute:

$$\Pi = iMRT \tag{11.16}$$

Because the units of R are L · atm/(mol · K) and i has no units, the product $iMRT$ gives Π in units of atmospheres.

A 1.0 M solution of NaCl should produce the same osmotic pressure as 1.0 M KCl or 1.0 M $NaNO_3$ because all three solutions are nearly 2.0 M in total ions ($i = 2$). These solutions should have nearly twice the osmotic pressure

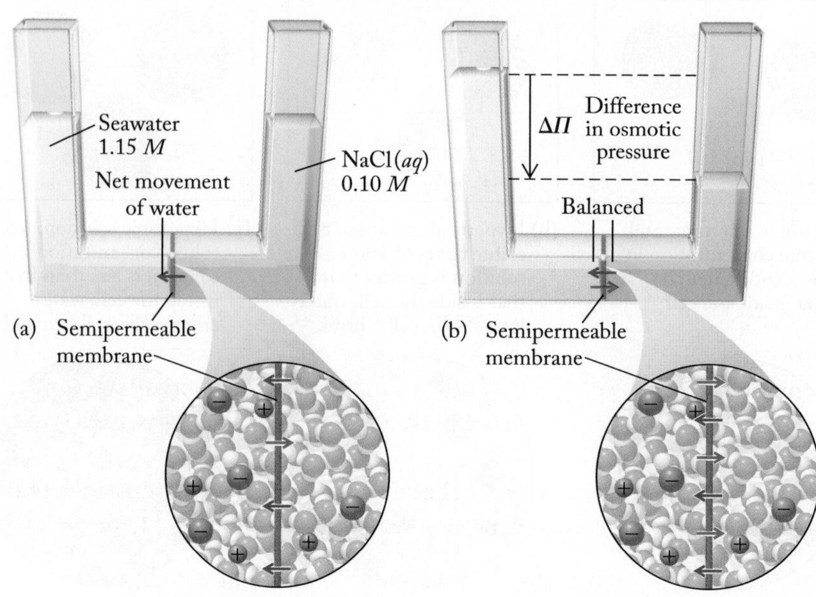

FIGURE 11.23 (a) The solute concentration of seawater is approximately 1.15 M. When equal volumes of seawater and a 0.10 M solution of NaCl are separated by a semipermeable membrane, water moves by osmosis from 0.10 M NaCl (low solute concentration) to the seawater (high solute concentration) side. (b) The volume on the NaCl side decreases until the osmotic pressure of the solution equals the osmotic pressure (and concentration) of the diluted seawater solution, resulting in a difference in the heights of the liquid levels. This difference in height is proportional to the original difference in osmotic pressures ($\Delta\Pi$) between the two solutions.

of a 1.0 M solution of glucose at a given temperature because glucose is a nonelectrolyte and produces a solution that is only 1.0 M in dissolved particles (glucose molecules).

CONCEPT TEST

If a 1.0 M glucose solution is on one side of a semipermeable membrane and a 1.0 M KCl solution is on the other side, in which direction does the water flow?

SAMPLE EXERCISE 11.13 Calculating Osmotic Pressure I — LO8

The concentration of solutes in a red blood cell is about a third of that of seawater—more precisely, about 0.30 M. If red blood cells are immersed in pure water, they swell, as shown in Figure 11.22c. Calculate the osmotic pressure at 25°C of red blood cells across the cell membrane from pure water.

Collect and Organize We are asked to calculate an osmotic pressure. We are given the total particle concentration (0.30 M) and temperature (25°C) of the solution. We must convert the temperature to kelvin.

Analyze Osmotic pressure is related to the total concentration of all particles in solution and the absolute temperature. The total particle concentration, 0.30 M, is the value of the $i \times M$ in Equation 11.16. Estimating the answer by assuming $i \times M = 0.3$, R is about 0.1, and T is about 300 K, and we get $0.3 \times 0.1 \times 300 = 9$ atm.

Solve First we need to convert the temperature from degrees Celsius to kelvin: $T(°\text{C}) + 273 = T(\text{K})$. Inserting the values of i, M, R, and T into Equation 11.16 gives

$$\Pi = iMRT = \frac{0.30\ \cancel{\text{mol}}}{\cancel{\text{L}}} \times \frac{0.0821\ \cancel{\text{L}} \cdot \text{atm}}{\cancel{\text{mol}} \cdot \cancel{\text{K}}} \times (25 + 273)\cancel{\text{K}} = 7.3\ \text{atm}$$

Think About It Our answer is in line with our estimation. The calculated pressure across the membranes of red blood cells in pure water is more than the gauge pressure in most bicycle tires and enough to rupture the membranes.

Practice Exercise Calculate the osmotic pressure across a semipermeable membrane separating pure water from seawater at 25°C. The total concentration of all the ions in seawater is 1.15 M. Assume $i = 0.92$.

CONCEPT TEST

Which has the greater effect on the osmotic pressure of a 1.0 M solution: increasing the temperature from 10°C to 20°C or adding enough solute to raise the concentration to 2.0 M?

In Sample Exercise 11.13, we calculated the osmotic pressure of a solution in relation to pure solvent. In Figure 11.23a, however, the two solutions have different osmotic pressures, Π_{NaCl} and Π_{seawater}. For these kinds of situations, we need to derive an equation for the difference between the osmotic pressures, $\Delta\Pi$, in terms of M, R, and T.

Earlier in this section we said that $\Delta\Pi = \Pi_{\text{seawater}} - \Pi_{\text{NaCl}}$, and that the osmotic pressures of both solutions are described by Equation 11.16:

$$\Pi_{\text{seawater}} = iM_{\text{seawater}}RT \qquad \text{and} \qquad \Pi_{\text{NaCl}} = iM_{\text{NaCl}}RT$$

Taking the difference in Π expressions:

$$\Delta\Pi = \Pi_{\text{seawater}} - \Pi_{\text{NaCl}} = iM_{\text{seawater}}RT - iM_{\text{NaCl}}RT$$

$$\Delta\Pi = (iM_{\text{seawater}} - iM_{\text{NaCl}})RT \qquad (11.17)$$

The total ion concentration in seawater is about 1.15 M, so $iM = 1.15\ M$. If a solution of that concentration is put on one side of a semipermeable membrane and a solution of 0.10 M NaCl on the other side, then $iM_{\text{NaCl}} = (2 \times 0.10\ M)$, and the osmotic pressure difference between the two solutions at 25°C is

$$\begin{aligned}\Delta\Pi &= (iM_{\text{seawater}} - iM_{\text{NaCl}})RT \\ &= \left(1.15\,\frac{\text{mol}}{\text{L}} - 2 \times 0.10\,\frac{\text{mol}}{\text{L}}\right)\left(0.0821\,\frac{\text{L}\cdot\text{atm}}{\text{mol}\cdot\text{K}}\right)(298\text{ K}) \\ &= 23\text{ atm}\end{aligned}$$

SAMPLE EXERCISE 11.14 Calculating Osmotic Pressure II **LO8**

Red blood cells placed in seawater shrivel, as shown in Figure 11.22(b). Calculate the pressure across the semipermeable cell membrane separating the solution inside a red blood cell from seawater at 25°C if the total concentration of all the particles inside the cell is 0.30 M and the total concentration of ions in seawater is 1.15 M. Compare the result with the answer from Sample Exercise 11.13.

Collect and Organize We can use Equation 11.17 to calculate the difference in osmotic pressure between the two solutions. As in Sample Exercise 11.13, we must convert the temperature to units of kelvin.

Analyze We know that the concentration of particles in seawater is greater than inside a red blood cell, so solvent will flow from the cell to the seawater. In addition, the osmotic pressure across the membrane is the difference between Π_{seawater} and Π_{cell}, or $\Delta\Pi$. The value of $\Delta\Pi$ should be greater than the Π calculated in Sample Exercise 11.13, where red blood cells were immersed in pure water, because the difference in solute concentrations is greater.

Solve First we need to convert the temperature from degrees Celsius to kelvin: T(°C) + 273 = T(K). Inserting the values of i, M, R, and T in Equation 11.17 for seawater and cells, we have

$$\begin{aligned}\Delta\Pi &= (iM_{\text{seawater}} - iM_{\text{cell}})RT \\ &= \left(1.15\,\frac{\text{mol}}{\text{L}} - 0.30\,\frac{\text{mol}}{\text{L}}\right)\left(0.0821\,\frac{\text{L}\cdot\text{atm}}{\text{mol}\cdot\text{K}}\right)(298\text{ K}) \\ &= 21\text{ atm}\end{aligned}$$

The pressure across a semipermeable membrane separating seawater from red blood cells, 21 atm, is greater than the pressure across the membrane separating red blood cells from pure water by almost a factor of three.

Think About It The values compare as we predicted, though we need to keep in mind that the osmotic pressure of 7.3 atm in Sample Exercise 11.13 would somehow have to be applied inside the red blood cell to keep pure water from entering it, whereas the 21 atm of osmotic pressure calculated in this sample exercise would have to be applied to the seawater outside the cell to keep water from inside the cell migrating through the cell wall and into the seawater.

Practice Exercise Calculate the osmotic pressure at 25°C across a semipermeable membrane separating seawater (1.15 M total particles) from a 0.50 M solution of aqueous NaCl.

The osmotic pressures across cell membranes in Sample Exercises 11.13 and 11.14 are very large. The pressure of more than 7 atm calculated in Sample Exercise 11.13 for red blood cells immersed in pure water is more than three times the air pressure in a typical automobile tire and about the same as the water pressure experienced by a diver at a depth of 80 m. A pressure of 21 atm across the walls of a steel-reinforced concrete building is enough to cause major structural damage or even collapse the building. The possibility of serious structural damage to cells because of such huge pressure differentials across membranes is one reason why solutions dispensed intravenously (IV) or intramuscularly (IM) must be carefully constituted.

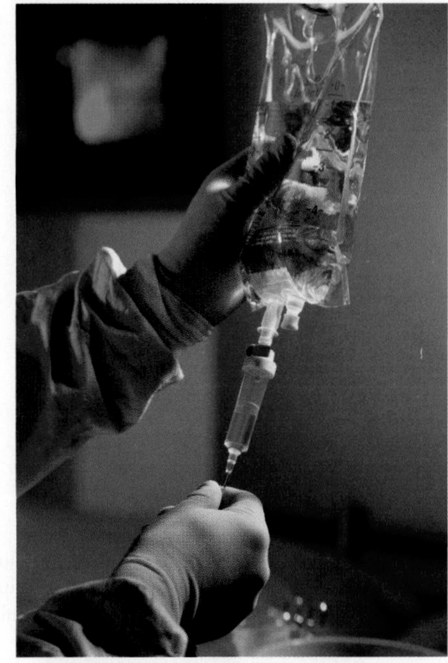

FIGURE 11.24 A solution of physiological saline has a concentration of 0.92 g of NaCl per 100 g of solution. The concentration of ions in this solution is equal to the concentration of ions in blood plasma. This solution is *isotonic* with blood plasma.

During a medical emergency, medication may need to be administered to a patient intravenously (**Figure 11.24**), and it is crucial that the osmotic pressure exerted by the intravenous solution on the body's cells be identical to the osmotic pressure exerted by the solution inside the cells. The solute concentrations in such solutions are said to be *isotonic* because they exert the same osmotic pressure as the blood exerts. As we saw in Figure 11.22, solutions with higher (*hypertonic*) or lower (*hypotonic*) solute concentrations cause the body's cells to either shrink as water leaves or swell as water enters the cell.

Physiological saline and D5W are two solutions that are widely used to administer intravenous medications, depending on the clinical situation. Physiological saline contains 0.92% NaCl by mass—that is, 0.92 g of NaCl for every 100 g of solution. The density of dilute aqueous solutions is close to 1.00 g/mL, so 100 g of this solution has a volume of 100 mL, which means a concentration of 0.92 g/100 g is nearly the same as 0.92 g/100 mL = 9.2 g/L.

D5W is a 5.5% solution by mass of dextrose (another name for glucose; molar mass 180.16 g/mol) in water. Because this solution must be isotonic with blood and therefore isotonic with physiological saline solution, it must contain about the same concentration of solute particles as saline solution. A 5.5% solution of D5W contains 5.5 g of dextrose in 100 g of water, 5.5 g of dextrose in 100 mL of water, or 55 g of dextrose/L.

To compare the solute levels of these two solutions, we compare molarities:

$$\text{Physiological saline: } \frac{9.2 \text{ g}}{1 \text{ L}} \times \frac{1 \text{ mol NaCl}}{58.44 \text{ g NaCl}} = 0.16\ M$$

$$\text{D5W: } \frac{55 \text{ g}}{1 \text{ L}} \times \frac{1 \text{ mol D}}{180.16 \text{ g D}} = 0.31\ M$$

The molar concentration of dextrose is about twice that of the saline solution, which makes sense because the theoretical i value for NaCl is 2, meaning that a 0.16 M NaCl solution contains about 0.32 M of Na^+ and Cl^- ions. Actually, the total concentration of ions should be slightly less than that because neutral NaCl ion pairs are present in the saline solution (see Figure 11.19). As a result, the solute particle concentrations in 0.92% NaCl and 5.5% D5W are essentially the same and so are their osmotic pressures.

The values used in these examples are for normal conditions in a healthy patient. Depending on the clinical condition presented, doctors may need to use solutions with different concentrations of solutes to respond most effectively to the patient's needs.

Reverse Osmosis

Many people in the world suffer from a lack of fresh water. A recent report from the United Nations (UN-Water Policy Brief on Water Quality, UN-Water 2011) suggested that by 2025, two out of three people in the world could be living in

reverse osmosis a process in which solvent is forced through semipermeable membranes, leaving a more concentrated solution behind.

water-stressed areas—that is, places without enough fresh water to drink or grow crops. The sea is already the source of drinking water in several desert countries bordering the Persian Gulf; to make the seawater drinkable, it must be *desalinated*, which means the salts must be removed. One way to desalinate saltwater is to distill it, but distillation requires a great deal of energy to heat seawater to its boiling point and convert it to steam.

The process of osmosis can also be applied to the desalination of seawater. As we have seen, osmosis is the movement of water from a region of low solute concentration to a region of high solute concentration. However, if a sufficiently high pressure is applied to the region of high solute concentration, the water can be forced to move from the region of high solute concentration to the region of low solute concentration. This technique, called **reverse osmosis**, is another way of desalinating seawater to make it drinkable.

The desalination apparatus shown in **Figure 11.25** consists of an outer metal tube containing many inner tubes made of a semipermeable membrane. Seawater is forced through the outer tube so that it washes over the exterior of all the inner tubes, which are initially filled with flowing pure water. Ordinarily, the direction of osmosis would be from the inner tubes (zero solute concentration) into the seawater (very high solute concentration), and indeed, the pure water does exert an osmotic pressure on the interior of the tube membranes. However, when an external pressure—a *reverse osmotic pressure*—greater than the osmotic pressure is exerted on the seawater side of the membranes, water molecules in the seawater move across the membranes into the inner tubes. The desalinated water entering the inner tubes flows into a collector and is ready for use.

Some municipal water-supply systems use reverse osmosis to make saline (brackish) water fit to drink. Some industries use this method to purify conventional tap water, and it is a common way for ships at sea to desalinate ocean water. However, very tough semipermeable membranes are necessary because reverse osmosis systems operate at very high pressures. The continued development of technologies using reverse osmosis has resulted in millions of people being supplied with sanitary drinking water. **Figure 11.26** shows the range of reverse osmosis facilities available, from one of the world's largest desalination plants capable of converting 625 million liters of seawater into drinkable water daily, to a portable unit, to a small device for a home drinking water supply.

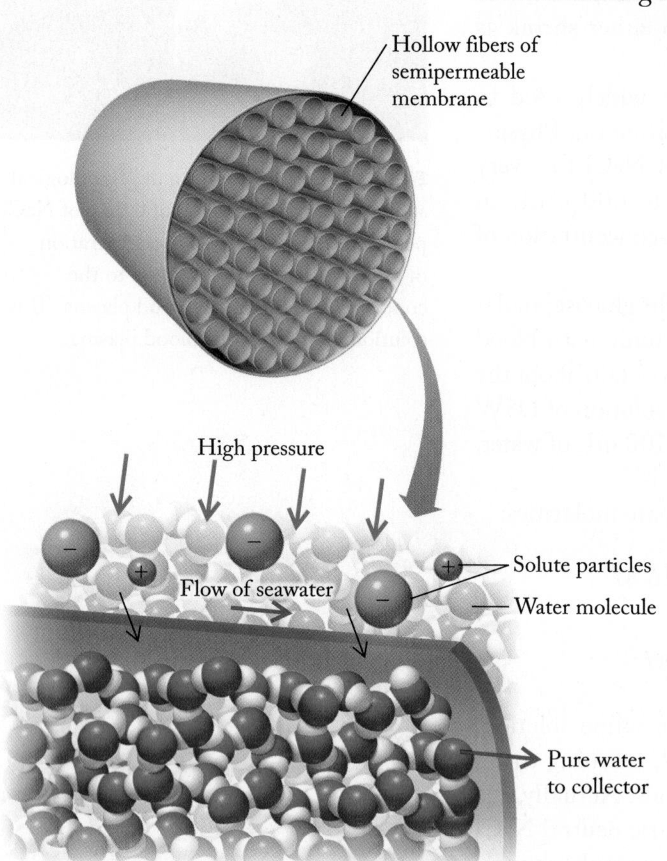

FIGURE 11.25 Seawater being desalinated by reverse osmosis flows at a pressure greater than its osmotic pressure around bundles of tubes with semipermeable walls. Water molecules pass from the seawater into the tubes and flow through the tubes to a collection vessel.

SAMPLE EXERCISE 11.15 Calculating Pressure for Reverse Osmosis LO8

What is the reverse osmotic pressure required at 20°C to purify brackish well water containing 0.355 *M* dissolved particles if the purified water is to contain no more than 87 mg of dissolved solids (measured as NaCl equivalents) per liter?

Collect and Organize We are asked to calculate the reverse osmotic pressure needed to purify water to a stated solute concentration. We are given the temperature, the solute concentration in the water to be purified, and the amount of solute (expressed as NaCl) tolerable in the product water. We can adapt Equation 11.17 to calculate the difference in osmotic pressure between the two solutions.

Analyze We need the molarity of the less concentrated solution (the drinkable water) to calculate the osmotic pressure exerted by the drinkable water, and we can calculate that from the given information. We can also calculate the osmotic pressure once we know the difference in the molarities of the solutions on the two sides of the semipermeable membrane.

Solve First we convert 87 mg of NaCl/L to molarity:

$$\frac{87\ \cancel{\text{mg NaCl}}}{1\ \text{L}} \times \frac{1\ \text{g}}{1000\ \cancel{\text{mg}}} \times \frac{1\ \text{mol NaCl}}{58.44\ \cancel{\text{g NaCl}}} = \frac{1.5 \times 10^{-3}\ \text{mol NaCl}}{1\ \text{L}}$$
$$= 1.5 \times 10^{-3}\ M\ \text{NaCl}$$

The total ion concentration for a 1.5×10^{-3} M NaCl solution is $2(1.5 \times 10^{-3}) = 0.0030\ M$ because $i = 2$ for NaCl. Using Equation 11.17:

$$\begin{aligned}\Delta\Pi &= \Pi_{\text{brackish water}} - \Pi_{\text{drinkable water}}\\ &= (iM_{\text{brackish water}} - iM_{\text{drinkable water}})RT\\ &= (0.355\ M - 0.0030\ \cancel{M})\left(0.0821\ \frac{\cancel{\text{L}} \cdot \text{atm}}{\cancel{\text{mol}} \cdot \cancel{\text{K}}}\right)(293\ \cancel{\text{K}})\\ &= 8.47\ \text{atm}\end{aligned}$$

If we maintain an osmotic pressure of exactly 8.47 atm on the side of the membrane with the brackish well water, no net flow of solvent will occur between the two solutions. Pressures greater than 8.47 atm will force water molecules from the well water through the membrane, producing water containing less than 87 mg of NaCl per liter. (We can never obtain pure water in this process; in practice, some Na^+ and Cl^- ions inevitably pass through the membrane from the well water to the drinkable water.)

Think About It The answer makes sense based on the results in Sample Exercise 11.13 in which we calculated the osmotic pressure (7.3 atm) of the liquid inside red blood cells (0.30 M in total ions). The osmotic pressure calculated here is (8.47/7.3) = 1.16 or 16% higher than the value calculated in Sample Exercise 11.13 and is based on a total ion concentration that is (0.355 − 0.0030)/0.30 = 1.17 or 17% higher.

Practice Exercise Calculate the minimum external pressure that must be applied in a reverse osmosis system to seawater with a total ion concentration of 1.15 M at 20°C if the maximum concentration allowed in the product water is 174 mg of NaCl per liter.

(a)

(b)

(c)

FIGURE 11.26 Reverse osmosis can supply water for personal, industrial, and commercial use. (a) The world's largest reverse osmosis desalination plant at Sorek, Israel, produces up to 625 million liters of water daily. (b) Portable reverse osmosis units can purify hundreds of liters of brackish water per day. (c) A small unit attached to a kitchen sink purifies a 16-oz glass of water in about 5 minutes.

CONCEPT **TEST**

If you were using the apparatus in Figure 11.26c to purify water by reverse osmosis, what advantage might there be in running the system at 50°C rather than 20°C?

Using Osmotic Pressure to Determine Molar Mass

In principle, we can determine the molar mass of any solute by dissolving a known quantity of the solute in a known quantity of solvent and then measuring the effect that the dissolved solute has on any colligative property of the solvent. In practice, methods based on colligative properties work only for nonelectrolytes that have a van 't Hoff factor of 1. There are also several compelling reasons for using osmotic pressure rather than boiling point elevation or freezing point

CONNECTION In Section 3.8 we used molar masses determined by mass spectrometry to convert empirical formulas derived from elemental analyses to molecular formulas. In Section 5.6 we combined density measurements with the ideal gas law to calculate molar masses of gases.

depression in these measurements: (1) very small osmotic pressures can be measured precisely, which means osmotic pressure can be used to accurately determine much smaller solute concentrations or larger molar masses than by the other methods; (2) pressure measurement equipment can be miniaturized so that only minute quantities of solute are needed; and (3) measurements can be made at room temperature and not near solvent freezing or boiling points at which solute stability may be compromised. The following sample exercise illustrates how osmotic pressure can be used to determine molar mass.

SAMPLE EXERCISE 11.16 Using Osmotic Pressure to Determine Molar Mass LO9

A molecular compound that is a nonelectrolyte was isolated from a South African tree. A 47 mg sample was dissolved in water to make 2.50 mL of solution at 25°C, and the osmotic pressure of the solution was 0.489 atm. Calculate the molar mass of the compound.

Collect and Organize We are given the mass of a substance, the volume of its aqueous solution, and the osmotic pressure and temperature of the solution. We can relate these parameters to the concentration of the solution using Equation 11.16, using the value $i = 1$ because the substance is a nonelectrolyte.

Analyze After calculating the molar concentration of the solution using Equation 11.16, we can then multiply that value by the volume of the sample to calculate the number of moles of solute in the solution. Dividing that value into the known mass of the solute in the sample will give us the solute's molar mass.

Solve Rearranging Equation 11.16 to isolate M and substituting the given values:

$$M = \frac{\Pi}{iRT} = \frac{0.489\ \cancel{\text{atm}}}{\{1[0.0821\ \text{L} \cdot \cancel{\text{atm}}/(\text{mol} \cdot \cancel{\text{K}})]\}(298\ \cancel{\text{K}})}$$

$$= \frac{0.0200\ \text{mol}}{\text{L}} = 0.0200\ M$$

Multiplying the solution's molarity by its volume:

$$n = MV = \frac{0.0200\ \text{mol}}{1\ \cancel{\text{L}}} \times 2.50\ \cancel{\text{mL}} \times \frac{1\ \cancel{\text{L}}}{1000\ \cancel{\text{mL}}} = 5.00 \times 10^{-5}\ \text{mol}$$

We know that this number of moles of solute has a mass of 47 mg. Therefore, the molar mass of the solute is

$$\mathcal{M} = \frac{\text{g solute}}{\text{mol solute}} = \frac{47 \times 10^{-3}\ \text{g}}{5.00 \times 10^{-5}\ \text{mol}} = 9.4 \times 10^{2}\ \text{g/mol}$$

Think About It The calculated concentration value (0.0200 M) used to determine molar mass makes sense because the results of Sample Exercise 11.13 showed that 7.3 atm of osmotic pressure is produced by a 0.30 M solution at 25°C. Therefore, an osmotic pressure of 0.489 atm corresponds to a concentration of 0.489 atm × (0.30 M)/7.3 atm = 0.0200 M. This exercise also demonstrates the capacity of an osmotic pressure measurement to determine the molar mass of a relatively small (47 mg) sample of a compound with relatively large molar mass (940 g/mol).

Practice Exercise A solution was made by dissolving 5.00 mg of a polysaccharide (a polymer made of sugar molecules) in water to give a final volume of 1.00 mL. The osmotic pressure of this solution was 1.91×10^{-3} atm at 25°C. Calculate the molar mass of the polysaccharide, which is a nonelectrolyte.

11.6 Ion Exchange

ion exchange a process by which one ion is displaced by another.

In Section 11.5 we examined how reverse osmosis is used to reduce the concentration of salts in water to make it potable (drinkable). In this section we discuss a technology that is used to *replace* some ions with others to make already drinkable water softer. What is *soft* water? To answer this question, let's turn it around and examine why the presence of certain metal ions—principally Ca^{2+} and Mg^{2+}—makes water *hard*. The presence of these ions causes several problems in industrial and residential water supplies: Ca^{2+} and Mg^{2+} ions combine with soap to precipitate a gray scum that can discolor clothes laundered in hard water. The limited solubility of the compounds these ions form with common anions, such as CO_3^{2-} and SO_4^{2-}, produces *scale* (an incrustation) in boilers, pipes, and fixtures, which diminishes their ability to conduct heat and carry water, and sometimes accumulation of these compounds produces unpleasant odors.

A common method of *water softening* (i.e., removing the ions responsible for water hardness) involves passing it through a system that replaces Ca^{2+} and Mg^{2+} ions with Na^+ ions (**Figure 11.27**) in a process called **ion exchange**. The equipment used consists of one or more cartridges packed with porous plastic resin (R)

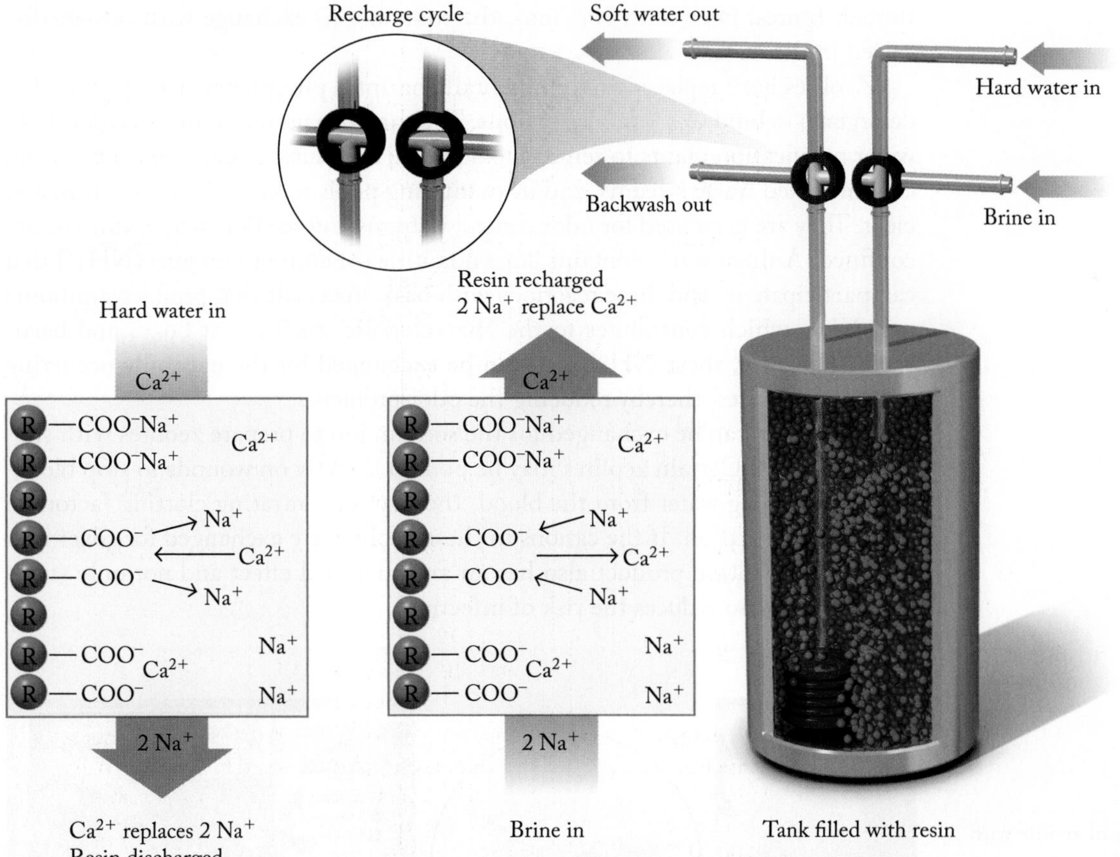

FIGURE 11.27 Residential water softeners use ion exchange to remove 2+ ions (such as Ca^{2+}) that make water hard. The ion-exchange resin contains cation-exchange sites that are initially occupied by Na^+ ions. These ions are replaced by 2+ "hardness" ions as water flows through the resin. Eventually most of the ion-exchange sites are occupied by 2+ ions, and the system loses its water-softening ability. The resin is then backwashed with a saturated solution of NaCl (*brine*), displacing the hardness ions (which wash down the drain) and restoring the resin to its Na^+ form.

zeolites natural crystalline minerals or synthetic materials consisting of three-dimensional networks of channels that contain sodium or other 1+ cations.

to which is bonded anions capable of binding with hard-water cations—mainly Ca^{2+}, Mg^{2+}, and Fe^{2+}. The carboxylate anion (COO^-) is often used, and water in the ion-exchange cartridge contains Na^+ to balance the negative charges on the anions, producing $(R–COO^-)Na^+$ ion-exchange sites. As hard water flows through the cartridge, 2+ ions in the water exchange places with sodium ions on the resin. This exchange takes place because cations with 2+ and 3+ charges bind more strongly to the $R–COO^-$ groups on the resin than do Na^+ ions. The ion-exchange reaction with calcium ions is

$$2\,(R–COO^-)Na^+(s) + Ca^{2+}(aq) \rightarrow (R–COO^-)_2Ca^{2+}(s) + 2\,Na^+(aq)$$

Hard water that has been softened in this way contains increased concentrations of sodium ions. Although this may not be a problem for healthy children and adults, people suffering from high blood pressure often must limit their intake of Na^+ and should not drink water softened by this kind of ion-exchange reaction.

Naturally occurring porous minerals called **zeolites** are also used as water softeners and purifiers and as livestock feed additives and odor suppressants. These zeolites are formed by chemical reactions between molten lava from volcanic eruptions and seawater. Synthetic zeolites have similar structures. All zeolites have a rigid three-dimensional structure like a honeycomb (**Figure 11.28**), consisting of a network of interconnecting tunnels and cages. As water flows through tunnels (pores) lined with Na^+ ions, the sodium ions exchange with cations dissolved in the water.

Zeolites have replaced environmentally harmful phosphates that are added to detergents to bind Ca^{2+} and Mg^{2+} ions. They are also used in municipal drinking water purification plants to remove toxic metal ions such as Cd^{2+} and Pb^{2+} from contaminated waste streams, and in swimming pools to keep the water clean and clear. They are even used for odor control in barns and feedlots where animals are confined. Animal waste contains large quantities of ammonium ions ($NH_4{}^+$) that can participate in acid–base reactions with basic materials that produce ammonia gas, NH_3, which contributes to the characteristic smell of cat boxes and barnyards. However, these $NH_4{}^+$ ions can be exchanged for the naturally occurring cations in zeolites, thereby reducing the odor problem.

Other ions can be exchanged for the sodium ion to prepare zeolites with special properties. Certain zeolites may be poured directly on wounds to stop bleeding by absorbing water from the blood, thereby concentrating clotting factors to promote coagulation. If the cations in these zeolites are exchanged for silver ions (Ag^+), the resulting product also has an antimicrobial effect and not only stems bleeding but also reduces the risk of infection.

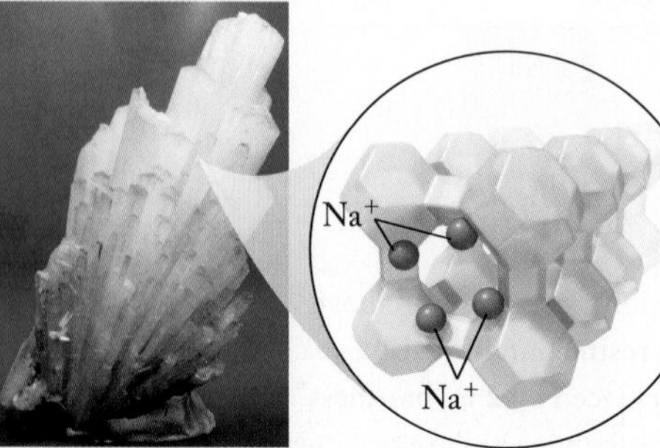

(a)

(b)

FIGURE 11.28 (a) A sample of zeolite with a drawing showing the regular pattern of pores containing sodium ions that exchange for other cations dissolved in water that flows through the material. (b) Commercial products that include zeolites to neutralize odors in and around cat litter boxes.

SAMPLE EXERCISE 11.17 Integrating Concepts: Fun with Eggs

The shell of a chicken's egg, which is mostly calcium carbonate [($\mathcal{M}$ = 100.1 g/mol)], is lined with a semipermeable membrane. Explain what happens during the following series of steps, all carried out at room temperature (21°C) and answer all the associated questions.

1. A chicken egg is weighed (its mass is 57 g) and then placed in 500 mL of vinegar (a 5.0 g/100 mL aqueous solution of acetic acid (CH_3COOH, $\mathcal{M}$ = 60.05 g/mol). Gas bubbles form on the outside of the shell (**Figure 11.29a**). (a) Write a balanced chemical equation describing the reaction. (b) How much egg shell ($CaCO_3$) could be dissolved by the vinegar? (c) Is the gaseous product more likely to dissolve in the vinegar or escape into the air? Explain your answer.
2. After the shell has dissolved, the egg is contained within a saclike membrane. It is gently removed from the vinegar bath, dried, and weighed. Its mass is 66 g, and the volume of the egg is greater than when it was in its shell (**Figure 11.29b**). (d) Explain what process caused the mass and volume of the egg to increase.
3. The egg is then covered in 250 mL of light corn syrup, which is a food syrup used in baking and to make jellies and jams. It is an aqueous solution that is about 75% sucrose ($C_{12}H_{22}O_{11}$) and other sugars. After 8 hours, the visibly shriveled egg (**Figure 11.29c**) is removed from the syrup, rinsed with water, dried, and weighed. Its mass is 46 g. (e) Why did its mass decrease?
4. Next, the shriveled egg is submerged in distilled water for 8 hours, during which time its volume increases (**Figure 11.29d**). The egg is then removed and reweighed; its mass is now 82 g. (f) Why did its mass and volume increase?

Collect and Organize We have the starting materials (calcium carbonate and dilute acetic acid) for a chemical reaction in which the egg's shell dissolves, and we are asked to identify the products and calculate the capacity of the vinegar bath to dissolve $CaCO_3$. Then we are asked to explain the mass and volume changes that the egg without its shell undergoes when it is immersed in corn syrup and in pure water.

Analyze The reaction between a carbonate and acetic acid is likely to be an acid–base neutralization reaction because, as discussed in Section 4.6, acetic acid is a H^+ donor (which makes it an acid) and CO_3^{2-} ions are H^+ acceptors (which makes them bases), forming HCO_3^- ions after accepting one H^+ and, after accepting a second, H_2CO_3, which decomposes into H_2O and CO_2 (see Sample Exercise 4.10). The observed changes in mass and volume probably involve osmosis across a semipermeable membrane (Section 11.5) because of different solute concentrations within and surrounding the egg.

Solve

a. A chemical equation describing the dissolution of the egg shell is

$$CaCO_3(s) + 2\,CH_3COOH(aq) \rightarrow Ca(CH_3COO)_2(aq) + H_2O(\ell) + CO_2(g)$$

b. The stoichiometry of this chemical equation allows us to calculate the mass of $CaCO_3$ that could dissolve in 500 mL of 5.0% acetic acid (which we abbreviate HAc in this calculation):

$$500\ \cancel{\text{mL HAc}} \times \frac{5.0\ \cancel{\text{g HAc}}}{100\ \cancel{\text{mL HAc}}} \times \frac{1\ \cancel{\text{mol HAc}}}{60.05\ \cancel{\text{g HAc}}} \times \frac{1\ \cancel{\text{mol CaCO}_3}}{2\ \cancel{\text{mol HAc}}} \times \frac{100.1\ \text{g CaCO}_3}{1\ \cancel{\text{mol CaCO}_3}} = 21\ \text{g CaCO}_3$$

c. The gas given off is CO_2, which forms bubbles on the egg shell, indicating that it has limited solubility in vinegar. Its limited solubility is also typical of other nonpolar gases (Table 10.4). Therefore, most of the gas escaped into the air.

d. One reason why the egg gained mass and volume in the vinegar bath as its shell dissolved is that solute particle concentrations were lower and the concentration of water in the bath was higher than in the egg. Therefore, water flowed from the bath into the egg by osmosis through the membrane.

e. The egg loses mass and volume when immersed in corn syrup because the solute concentration in the syrup is higher than the concentration of the solution within the egg sac. Osmosis causes water to leave the egg sac and enter the surrounding solution.

f. The egg gains mass and volume when immersed in distilled water because osmosis causes water molecules to flow from pure water (no solute) through the sac and into the egg.

(a)

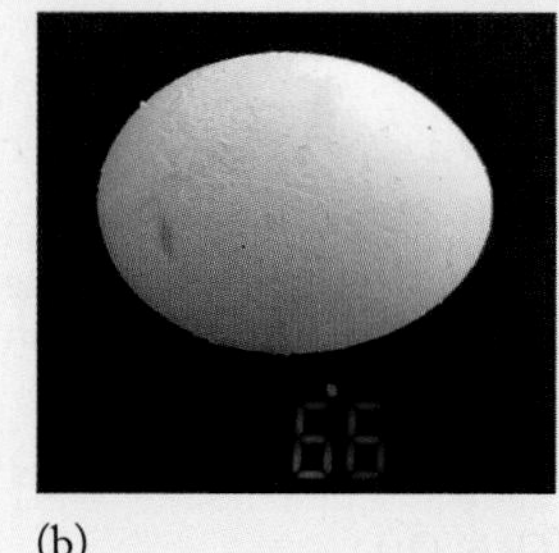

(b)

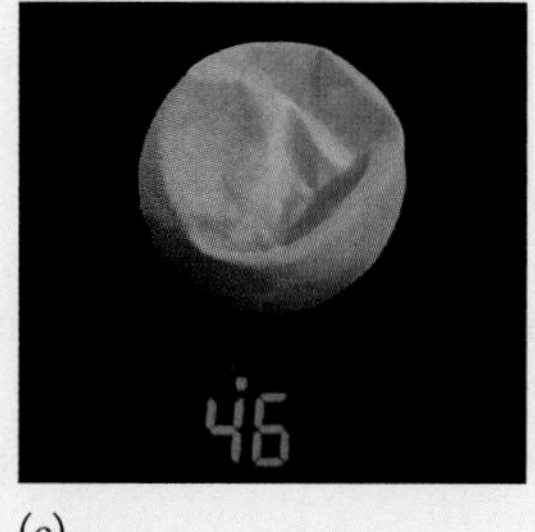

(c)

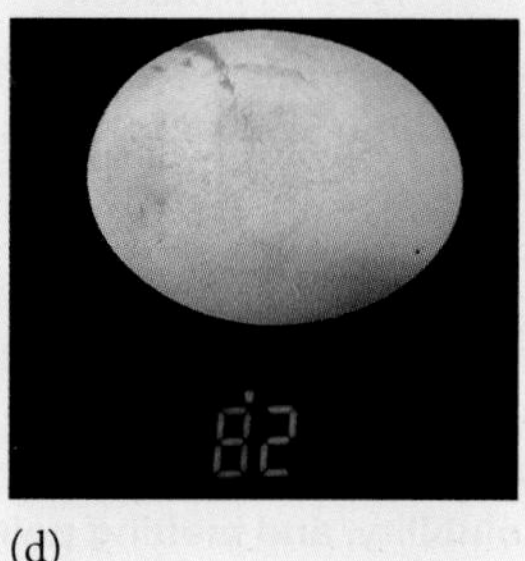

(d)

FIGURE 11.29 (a) The shell of a 57 g egg is dissolved in vinegar. (b) The shell-less egg is dried and weighed and then (c) immersed in corn syrup for 8 h, after which it is rinsed, dried, and weighed. (d) The egg is then immersed in distilled water for 8 h. All the weight values are in grams.

Think About It There were no calculations of osmotic pressure in this exercise because we did not know the total concentration of solute particles inside the egg or in the acetic acid solution after dissolving the shell, or in the corn syrup. However, the observed changes in egg volume and mass strongly suggest that osmosis added water to the egg when immersed in acetic acid, removed it from the egg when immersed in corn syrup, and added it to the egg when immersed in pure water. The results of the calculation in part (b) are expressed to 2 significant figures because that is the precision with which the vinegar volume could be measured (see Figure 11.29a).

SUMMARY

LO1 Ion–ion interactions hold ionic solids together. Their magnitude depends upon the charges of the ions and the distance between them. (Section 11.1)

LO2 The **enthalpy of solution ($\Delta H_{solution}$)** for an ionic compound is the sum of the **lattice energy (U)** and the **enthalpy of hydration ($\Delta H_{hydration}$)**. Lattice energies can be calculated with a **Born–Haber cycle**, an application of Hess's law. (Section 11.2)

LO3 The vapor pressure of a liquid is proportional to the fraction of its molecules that enter the gas phase. **Raoult's law** relates the vapor pressure of a solution to its composition and to the vapor pressure of the solvent. (Section 11.3)

LO4 The vapor pressure of an ideal solution of volatile compounds follows Raoult's law. **Fractional distillation** can be used to separate solutions of volatile compounds. (Section 11.4)

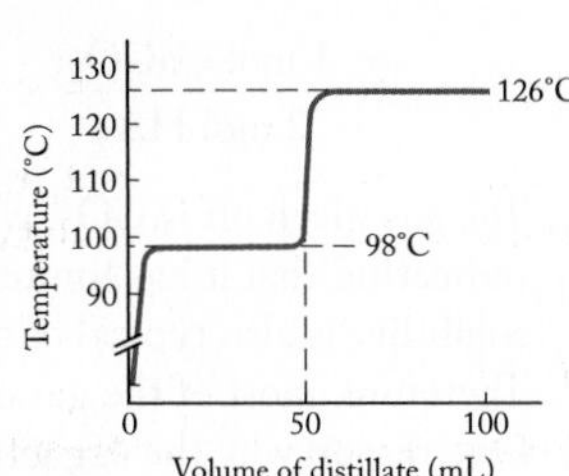

LO5 The concentration units used for **colligative property** measurements include molarity (M) and **molality (m)**. (Section 11.5)

LO6 Nonvolatile solutes in solution elevate the solvent's boiling point and depress its freezing point. (Section 11.5)

LO7 The **van 't Hoff factor** accounts for the colligative properties of electrolytes and the formation of solute **ion pairs** in concentrated solutions. (Section 11.5)

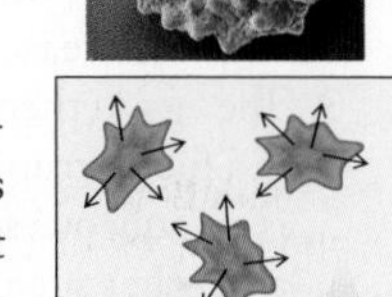

LO8 In **osmosis**, solvent flows through a semipermeable membrane. **Osmotic pressure (Π)** is the pressure required to halt the flow of solvent from the more dilute solution across the membrane. (Section 11.5)

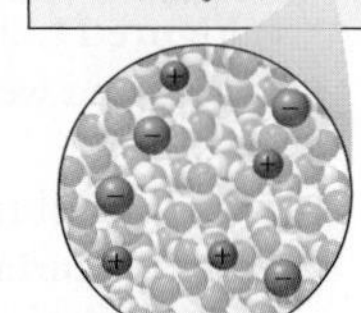

LO9 The molar mass of a compound can be determined by measuring the freezing point depression, boiling point elevation, or osmotic pressure of a solution of the compound. (Section 11.5)

PARTICULATE PREVIEW WRAP-UP

Ion–dipole forces form between the sodium cations and the $\delta-$ oxygen atoms and between the chloride anions and the $\delta+$ hydrogen atoms. The amount of vapor above a solution is reduced compared to that above pure solvent, and consequently the boiling point increases.

PROBLEM-SOLVING SUMMARY

Type of Problem	Concepts and Equations	Sample Exercises
Predicting relative strengths of ion–ion interactions	To predict relative interaction strengths, use $$E \propto \frac{(Q_1Q_2)}{d} \quad (6.3)$$ The value of E is negative for any interaction between oppositely charged particles. More negative values of E correspond to stronger ion–ion attractions.	**11.1**
Ranking lattice energies, solubility, and melting points	Predict relative lattice energies, solubilities, and melting points by using $$U = \frac{k(Q_1 \times Q_2)}{d} \quad (11.5)$$ More negative values of U correspond to stronger lattice energies, higher melting points, and generally lower solubilities.	**11.2**

Type of Problem	Concepts and Equations	Sample Exercises
Calculating lattice energy with a Born–Haber cycle	Use the standard enthalpy of formation of an ionic solid and equations such as this one for NaCl to calculate lattice energy (U). $$U_{NaCl} = \Delta H_f^\circ - (\Delta H_{sub,Na} + \tfrac{1}{2}\Delta H_{BE,Cl_2} + \Delta H_{IE_1,Na} + \Delta H_{EA_1,Cl})$$	**11.3**
Calculating lattice energy by using enthalpies of hydration	Use enthalpies of hydration ($\Delta H_{hydration}$) and enthalpies of solution ($\Delta H_{solution}$) to calculate lattice energy: $$\Delta H_{solution} = \Delta H_{hydration} - U \quad (11.6)$$ where $$\Delta H_{hydration} = \Delta H_{hydration,cation} + \Delta H_{hydration,anion}$$	**11.4**
Calculating vapor pressure of a solution	Use Raoult's law, $$P_{solution} = X_{solvent}P°_{solvent} \quad (11.7)$$ where $P_{solution}$ is the vapor pressure of the solution at a given temperature, $X_{solvent}$ is the mole fraction of the solvent in the solution, and $P°_{solvent}$ is the vapor pressure of the pure solvent at the same temperature. $$P_{total} = X_1P_1^\circ + X_2P_2^\circ + X_3P_3^\circ \ldots \quad (11.8)$$	**11.5, 11.6, 11.7**
Calculating molal concentrations	Molality is defined as $$m = \frac{n_{solute}}{\text{kg solvent}} \quad (11.9)$$ where n_{solute} is the number of moles of solute.	**11.8**
Calculating boiling point elevation or freezing point depression of a solution	For nonelectrolyte solutes, use $$\Delta T_b = K_b m \quad (11.10)$$ where ΔT_b is the elevation in the boiling point of the solvent, K_b is a constant that depends only on the solvent, and m is the molality of the solution. For the freezing point, use $$\Delta T_f = K_f m \quad (11.12)$$ where ΔT_f is the depression in the freezing point of the solvent, K_f is a constant that depends only on the solvent, and m is the molality of the solution.	**11.9, 11.10**
Assessing interactions among particles in solution by comparing the theoretical value of the van 't Hoff factor *i* with the experimentally measured value	For electrolytes, use $$\Delta T_b = iK_b m \quad (11.14)$$ and $$\Delta T_f = iK_f m \quad (11.15)$$ where the theoretical value of i is the number of particles created when an electrolyte dissociates completely: $i = 2$ for NaCl, 3 for $CaCl_2$, and 4 for Na_3PO_4. For real solutions of electrolytes, where ΔT_b, ΔT_f, K_b, K_f, and m are known, rearrange $\Delta T_b = iK_b m$ and $\Delta T_f = iK_f m$ to calculate the value of i and compare the result with the theoretical value.	**11.11, 11.12**
Calculating osmotic pressure and reverse osmotic pressure	Use $$\Pi = iMRT \quad (11.16)$$ where Π is the osmotic pressure, i is the van 't Hoff factor, M is the molar concentration of the solution, R is the ideal gas constant, and T is the absolute temperature of the solution.	**11.13, 11.14, 11.15**
Determining molar mass from osmotic pressure	Use $\Pi = MRT$ to calculate the molarity of a solution of a nonelectrolyte. Then use the molarity and volume of the solution and the mass of solute in it to calculate the molar mass of the solute.	**11.16**

VISUAL PROBLEMS

(Answers to boldface end-of-chapter questions and problems are in the back of the book.)

11.1. Figure P11.1 shows a particle-level view of a sealed container partially filled with a solution that has two components: X (blue spheres) and Y (red spheres). Which of the following statements about substances X and Y are true?

a. X is the solvent in this solution.
b. Pure Y is a volatile liquid.
c. If Y were *not* present, there would be fewer X particles in the gas above the liquid solution.
d. The presence of Y increases the vapor pressure of X.

FIGURE P11.1

11.2. Figure P11.2 shows a particle-level view of a sealed container partially filled with a solution of two miscible liquids: X (blue spheres) and Y (red spheres). Which of the following statements about substances X and Y are true?

FIGURE P11.2

a. Y is the solvent in this solution.
b. Pure Y has a higher vapor pressure than pure X.
c. The presence of Y in the solution lowers the vapor pressure of X.
d. If Y were *not* present, there would be fewer total particles in the gas above the liquid solution.

11.3. Figure P11.3 shows particle-level views of 0.001 *M* aqueous solutions of the following four solutes: $C_6H_{12}O_6$, NaCl, $MgCl_2$, and K_3PO_4. The blue spheres represent particles of solute.

a. Which compounds are represented in images (I)–(IV)?
b. Which of the four solutions in Figure P11.3 has the highest (i) vapor pressure; (ii) boiling point; (iii) freezing point; (iv) osmotic pressure?

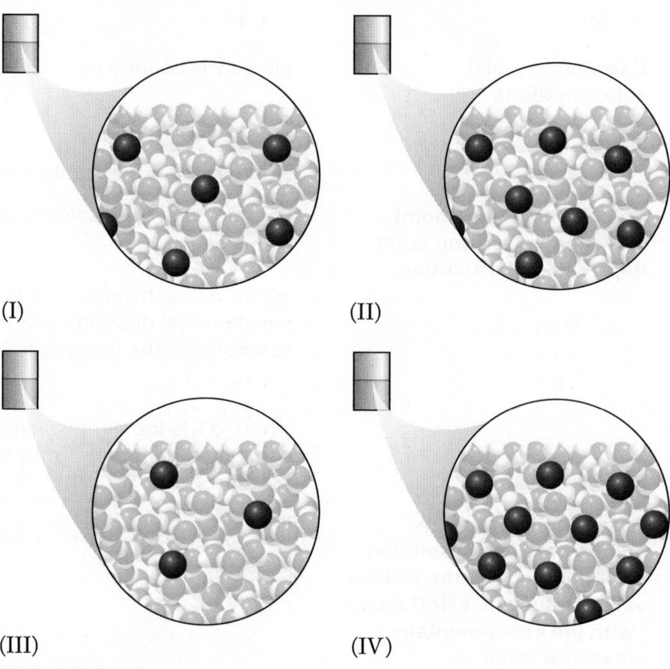

FIGURE P11.3

11.4. Which of the four solutions in Figure P11.4 has the highest (i) vapor pressure; (ii) boiling point; (iii) freezing point; (iv) osmotic pressure?

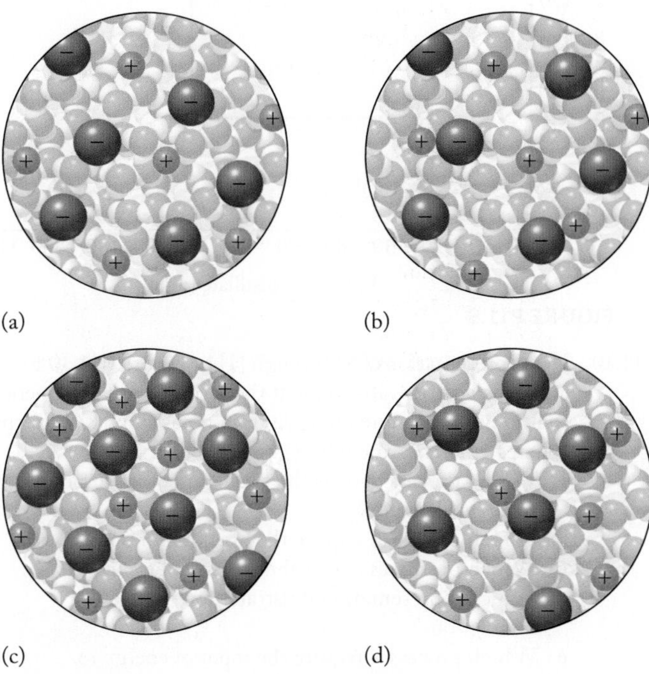

FIGURE P11.4

11.5. The graph in Figure P11.5 shows the values of ΔT_f for solutions of two different substances, A (triangles) and B (circles), in water. Explain how you can reasonably conclude that (a) A and B are nonelectrolytes and (b) the freezing-point-depression constant K_f of water is independent of the solute's identity.

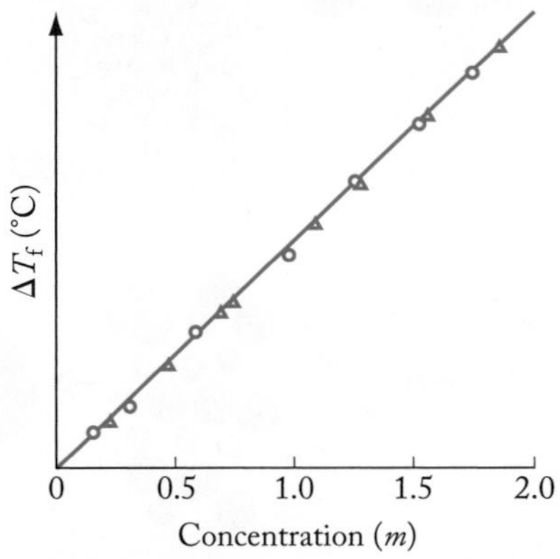

FIGURE P11.5

11.6. Figure P11.6 illustrates the relationship between the freezing points of three ideal aqueous solutions as a function of the molality of the solute. Which line represents a solution of glucose? Which line represents the solute with the largest value of i?

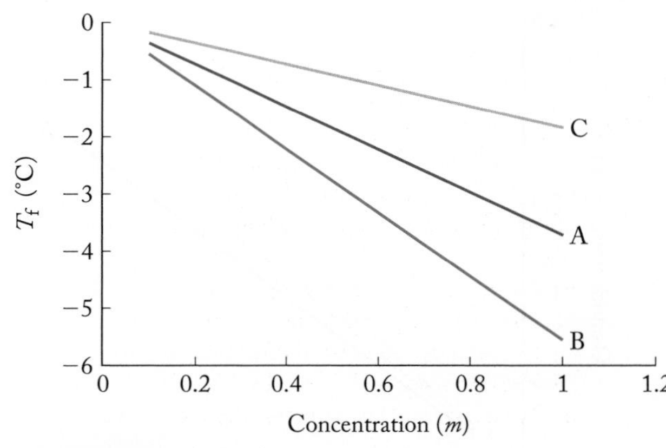

FIGURE P11.6

11.7. Kidney Dialysis Semipermeable membranes of the sort used in kidney dialysis do not allow large molecules and cells to pass but do allow small ions and water to pass. Figure P11.7 shows such a membrane separating fluids of various compositions.

a. In which direction does the water flow in each apparatus?
b. In which direction do sodium ions flow in each apparatus?
c. In which direction do the potassium ions flow in each apparatus?

(i)	(ii)
Pure H_2O	0.2 M KCl in H_2O
Membrane	Membrane
Proteins 0.5 M NaCl 0.1 M KCl in H_2O	Proteins 0.5 M NaCl 0.1 M KCl in H_2O

FIGURE P11.7

11.8. **Calibration Curves** A team of students measuring the molar masses of a series of proteins in water decides to make a graph of Π versus molar concentration to make their data analysis easier. Figure P11.8 shows the results for three different temperatures.

a. Which line represents the lowest temperature?
b. Which line represents $T = 37°C$?
*c. How would the graph change if the molar masses were measured in physiological saline rather than pure water?

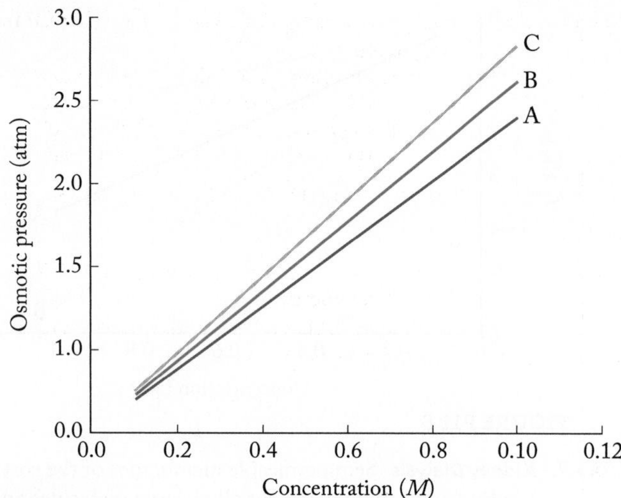

FIGURE P11.8

11.9. The graph in Figure P11.9 describes the volume of distillate collected during the fractional distillation of a liquid.

a. Is the sample a pure liquid or a mixture?
b. If it is a mixture: (i) How many components are in the mixture? (ii) What are the relative ratios of the volumes in the mixture? (iii) What are their approximate boiling points?

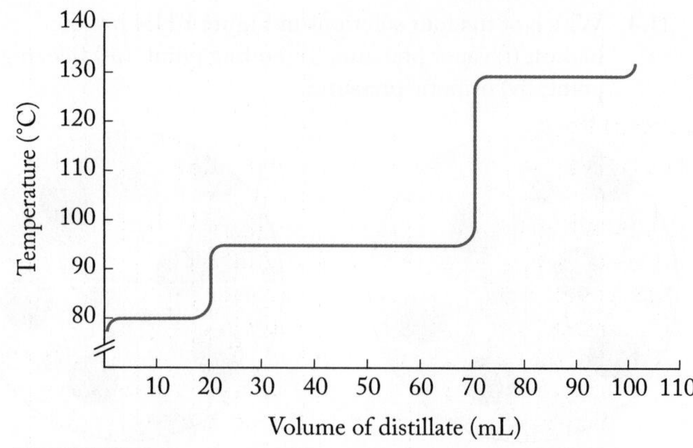

FIGURE P11.9

11.10. Use representations [A] through [I] in Figure P11.10 to answer questions (a)–(f) about the formation of an aqueous solution of potassium chloride from potassium [K(*s*), purple spheres] and chlorine [$Cl_2(g)$, green spheres].

a. Which process depicts the formation of a compound from its elements?
b. Which processes require the breaking of bonds?
c. Which processes depict the transfer of electrons?
d. Which representation illustrates ion–dipole interactions?
e. Which processes require the input of energy to overcome intermolecular forces?
f. Which processes are exothermic?

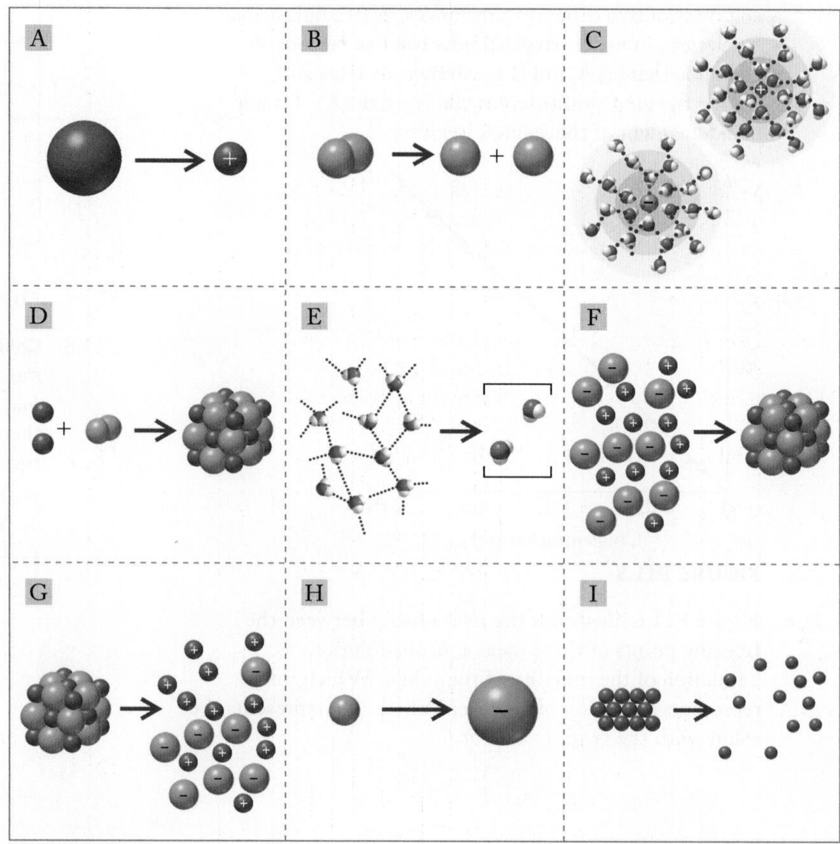

FIGURE P11.10

QUESTIONS AND PROBLEMS

Interactions between Ions

Concept Review

11.11. Which of these substances contains the largest anions? (a) NaCl; (b) CsF; (c) LiI; (d) KBr

11.12. Which of these substances contains the smallest cations? (a) $BaCl_2$; (b) Al_2O_3; (c) Mg_3N_2; (d) SrS

11.13. Describe how the strengths of ion–ion interactions help explain why $CaSO_4$ is less soluble in water than is NaCl?

11.14. For these three ionic compounds, KCl, $CaCl_2$, and $ScCl_3$, describe two reasons why the strengths of attraction between their cations and Cl^- ions increase with the number of Cl^- ions in their formulas?

Problems

11.15. Rank the following ionic compounds in order of increasing attraction between their ions: KBr, $SrBr_2$, CsBr.

11.16. Rank the following ionic compounds in order of increasing attraction between their ions: BaO, $BaCl_2$, CaO.

11.17. Which has the higher melting point, LiF or MgO?

11.18. Which has the higher boiling point, NaCl or $CaCl_2$?

Energy Changes during Formation and Dissolution of Ionic Compounds

Concept Review

11.19. Define the terms in the equation $\Delta H_{solution} = \Delta H_{hydration} - U$.

11.20. Explain why higher melting points of ionic solids are associated with more negative lattice energy values.

11.21. Explain why those ionic compounds with the most negative lattice energies tend to be the least soluble in water.

11.22. Which is the more important factor in explaining the difference in the lattice energies of LiCl and MgO: (1) the charges of their ions or (2) the radii of their ions?

Problems

11.23. How do the melting points of the series of sodium halides NaX (X = F, Cl, Br, I) relate to the atomic number of X?

11.24. Rank the following ionic compounds in order of (a) increasing melting point and (b) increasing water solubility: BaF_2, $CaCl_2$, $MgBr_2$, and SrI_2.

11.25. Which substance has the least negative lattice energy? (a) MgI_2; (b) $MgBr_2$; (c) $MgCl_2$; (d) MgF_2

11.26. Rank the following from lowest to highest lattice energy: NaBr, $MgBr_2$, $CaBr_2$, and KBr.

11.27. Use a Born–Haber cycle to calculate the lattice energy of potassium chloride (KCl) from the following data:

Ionization energy of K(g) = 425 kJ/mol
Electron affinity of Cl(g) = −349 kJ/mol
Energy to sublime K(s) = 89 kJ/mol
Bond energy of $Cl_2(g)$ = 240 kJ/mol

ΔH_f for $K(s) + \frac{1}{2} Cl_2(g) \rightarrow KCl(s) = -438$ kJ/mol

11.28. Calculate the lattice energy of sodium oxide (Na_2O) from the following data:

Ionization energy of Na(g) = 495 kJ/mol
Electron affinity of O(g) for 2 electrons = 603 kJ/mol
Energy to sublime Na(s) = 109 kJ/mol
Bond energy of $O_2(g)$ = 499 kJ/mol

ΔH_f for $2\ Na(s) + \frac{1}{2} O_2(g) \rightarrow Na_2O(s) = -416$ kJ/mol

---

11.29. Using the values in Table 11.4 and $\Delta H_{solution} = 19.9$ kJ/mol, calculate the lattice energy of KBr.

11.30. Using the values in Table 11.4 and $\Delta H_{solution} = -17.7$ kJ/mol, calculate the lattice energy of KF.

Vapor Pressure of Solutions

Concept Review

11.31. Explain the term *nonvolatile solute*.

11.32. Which has the higher vapor pressure at constant temperature, pure water or seawater? Explain your answer.

11.33. Why does the vapor pressure of a liquid increase with increasing temperature?

11.34. In the experiment shown in Figure 11.7, the vapor pressure of one of the solutions remains constant throughout the experiment, whereas the vapor pressure of the other solution changes. Which solution is which?

11.35. An experiment like that shown in Figure 11.7 is set up with the compartment containing pure ethanol full to the brim and the compartment containing a solution of sugar in ethanol half-full. Explain why the compartment that contains the ethanol–sugar solution will eventually overflow.

11.36. Will the vapor pressure of a solution containing 0.11 m NaCl and 0.11 m KCl be greater than, less than, or the same as a 0.22 m solution of just NaCl? Will $P_{solution}$ be greater than, less than, or the same as a 0.22 m solution of $CaCl_2$?

Problems

11.37. A solution contains 3.5 moles of water and 1.5 moles of nonvolatile glucose ($C_6H_{12}O_6$). What is the mole fraction of water in this solution? What is the vapor pressure of the solution at 25°C, given that the vapor pressure of pure water at 25°C is 23.8 torr?

11.38. A solution contains 4.5 moles of water, 0.3 moles of sucrose ($C_{12}H_{22}O_{11}$), and 0.2 moles of glucose. Sucrose and glucose are nonvolatile. What is the mole fraction of water in this solution? What is the vapor pressure of the solution at 35°C, given that the vapor pressure of pure water at 35°C is 42.2 torr?

11.39. Another way of stating Raoult's law is that the fractional lowering of the vapor pressure of the solvent in a solution $(P^\circ_{solvent} - P_{solvent})/P^\circ_{solvent}$ is equal to the mole fraction of the solute, X_{solute}. Use Equation 11.7 to show that this is true.

11.40. Use the statement of Raoult's law in Problem 11.39 to determine the mole fraction of glucose in Problem 11.37.

Mixtures of Volatile Solutes

Concept Review

11.41. In an equimolar mixture of C_5H_{12} and C_7H_{16}, which compound is present in higher concentration in the vapor above the solution?

11.42. Why does the boiling point of a mixture of volatile hydrocarbons increase over time during a simple distillation?

11.43. Would you expect a solution of cyclohexane (C_6H_{12}) in benzene (C_6H_6) to behave ideally? Explain your answer.

11.44. Explain the origin of negative deviations from Raoult's law for the predicted vapor pressure of a solution of two volatile liquids.

Problems

11.45. **Greener Fuel** Most gasoline sold in the United States contains ethanol (C_2H_5OH) in addition to a mixture of hydrocarbons such as octane (C_8H_{18}). At 20°C, the vapor pressure of ethanol is 45 torr and the vapor pressure of octane is 10 torr. What is the vapor pressure at 20°C of a solution prepared by mixing 8.16 g of ethanol and 63.3 g of octane?

11.46. **Alcohol Fuels** Using butanol (C_4H_9OH) rather than ethanol as a gasoline additive has an advantage in providing a higher fuel value when mixed with octane. At 20°C, the vapor pressure of butanol is 5 torr and the vapor pressure of octane is 10 torr. What is the vapor pressure at 20°C of a solution prepared by mixing 16.2 g of butanol and 127 g of octane?

***11.47.** At 90°C, the vapor pressure of styrene (C_8H_8) is 134 torr and that of ethylbenzene (C_8H_{10}) is 183 torr. A solution of 38% by weight styrene and 62% by weight ethylbenzene is separated by fractional distillation at reduced pressure so that it begins to boil when the solution in the distillation flask reaches 90°C.

a. What is the ratio of ethylbenzene to styrene in the vapor phase as the mixture first begins to boil?

b. What will be the temperature of the first distillate that comes off the top of the column? (i) lower than 90°C; (ii) 90°C; (iii) higher than 90°C

11.48. A bottle is half-filled with a 50:50 (mole-to-mole) mixture of heptane (C_7H_{16}) and octane (C_8H_{18}) at 25°C. What is the mole ratio of heptane vapor to octane vapor in the air space above the liquid in the bottle? The vapor pressures of heptane and octane at 25°C are 31 torr and 11 torr, respectively.

Colligative Properties of Solutions

Concept Review

11.49. Explain why freezing point depression, boiling point elevation, and the osmotic pressure of ionic solutes cannot be used to measure their formula masses.

11.50. What is the definition of the concentration scale called *molality* that is used to determine the molar mass of a nonvolatile solute by measuring the solute's effect on the freezing and boiling points of the solvent? Why must a mass-based concentration scale be used (i.e., *not* molarity)?

11.51. **Making Syrup** Evaporating maple tree sap converts the dilute sugar solution into the viscous amber syrup we pour on pancakes and waffles. How do the boiling points and vapor pressures of the sap change during the evaporation process?

*11.52. **Diet Soft Drinks** The thermostat in a refrigerator filled with cans of soft drinks malfunctions and the temperature of the refrigerator drops below 0°C. The contents of the cans of diet soft drinks freeze, rupturing many of the cans and causing an awful mess, whereas none of the cans containing regular, nondiet soft drinks rupture. Why?

11.53. Why is it important to know if a substance is a molecular compound or an ionic compound before predicting its effect on the boiling and freezing points of a solvent?

11.54. Refer to the phase diagram in Figure 11.16 and explain in your own words why the change in the vapor pressure of a solution caused by a nonvolatile solute results in a higher boiling point and a lower melting point.

11.55. Explain how the theoretical value of the van 't Hoff factor *i* for substances such as CH_3OH, NaBr, and K_2SO_4 can be predicted from their formulas.

11.56. Is it possible for an experimentally measured value of a van 't Hoff factor to be greater than the theoretical value? Explain your answer.

11.57. What is a semipermeable membrane?

11.58. A pure solvent is separated from a solution containing the same solvent by a semipermeable membrane. In which direction does the solvent flow across the membrane, and why?

11.59. A dilute solution is separated from a more concentrated solution containing the same solvent by a semipermeable membrane. In which direction does the solvent tend to flow across the membrane, and why?

11.60. How is the osmotic pressure of a solution related to its molar concentration and its temperature?

11.61. Explain the principle of reverse osmosis.

11.62. Explain how the minimum pressure for purification of seawater by reverse osmosis can be estimated from its composition.

11.63. Aqueous solutions of physiological saline (NaCl) and dextrose (glucose) are used to deliver intravenous medications. Why must their molar concentrations differ by a factor of two?

11.64. Why do red blood cells undergo hemolysis when they are placed in pure water?

Problems

11.65. Calculate the molality of each of the following solutions:

a. 0.433 mol of sucrose ($C_{12}H_{22}O_{11}$) in 2.1 kg of water

b. 71.5 mmol of acetic acid (CH_3COOH) in 125 g of water

c. 0.165 mol of baking soda ($NaHCO_3$) in 375.0 g of water

*11.66. Table 4.1 lists molarities of the major ions in seawater. Using a density of 1.022 g/mL for seawater, convert the concentrations into molalities.

11.67. What mass of the following solutions contains 0.100 mol of solute? (a) 0.135 *m* NH_4NO_3; (b) 3.92 *m* ethylene glycol, $HOCH_2CH_2OH$; (c) 1.07 *m* $CaCl_2$

11.68. How many moles of solute are there in the following solutions?
a. 0.750 m glucose solution made by dissolving the glucose in 10.0 kg of water
b. 0.183 m Na_2CrO_4 solution made by dissolving the Na_2CrO_4 in 900.0 g of water
c. 1.425 m urea solution made by dissolving the urea in 750.0 g of water

11.69. Fish Kills High concentrations of ammonia (NH_3), nitrite ions, and nitrate ions in water can kill fish. Lethal concentrations of these species for rainbow trout are 1.1 mg/L, 0.40 mg/L, and 1361 mg/L, respectively. Express these concentrations in molality units, assuming a solution density of 1.00 g/mL.

11.70. The concentrations of six important elements in a sample of river water are 0.050 mg/kg of Al^{3+}, 0.040 mg/kg of Fe^{3+}, 13.4 mg/kg of Ca^{2+}, 5.2 mg/kg of Na^+, 1.3 mg/kg of K^+, and 3.4 mg/kg of Mg^{2+}. Express each of these concentrations in molality units.

11.71. Cinnamon The distinctive aroma of cinnamon comes from cinnamaldehyde (C_9H_8O). Determine the boiling point elevation of a solution of 100 mg of cinnamaldehyde dissolved in 1.00 g of carbon tetrachloride (K_b = 5.02°C/m).

11.72. **Spearmint** Determine the boiling point elevation of a solution of 125 mg of carvone ($C_{10}H_{14}O$, oil of spearmint) dissolved in 1.50 g of carbon disulfide (K_b = 2.34°C/m).

11.73. What molality of a nonvolatile, nonelectrolyte solute is needed to lower the melting point of camphor by 1.000°C if K_f = 39.7°C/m?

11.74. What molality of a nonvolatile, nonelectrolyte solute is needed to raise the boiling point of water by 7.60°C if K_b = 0.52°C/m?

11.75. Saccharin Determine the melting point of an aqueous solution made by adding 186 mg of saccharin ($C_7H_5O_3NS$) to 1.00 mL of water (density = 1.00 g/mL, K_f = 1.86°C/m).

11.76. Determine the boiling point of an aqueous solution that is 2.50 m ethylene glycol ($HOCH_2CH_2OH$); K_b for water is 0.52°C/m. Assume that the boiling point of pure water is 100.00°C.

11.77. Arrange the following aqueous solutions in order of increasing boiling point:
a. 0.06 m $FeCl_3$ (i = 3.4)
b. 0.10 m $MgCl_2$ (i = 2.7)
c. 0.20 m KCl (i = 1.9)

11.78. Arrange the following solutions in order of increasing freezing point depression:
a. 0.10 m $MgCl_2$ in water, i = 2.7, K_f = 1.86°C/m
b. 0.20 m toluene in diethyl ether, i = 1.00, K_f = 1.79°C/m
c. 0.20 m ethylene glycol in ethanol, i = 1.00, K_f = 1.99°C/m

11.79. The following pairs of aqueous solutions are separated by a semipermeable membrane. In which direction will the solvent flow?
a. A = 1.25 M NaCl; B = 1.50 M KCl
b. A = 3.45 M $CaCl_2$; B = 3.45 M NaBr
c. A = 4.68 M glucose; B = 3.00 M NaCl

11.80. The following pairs of aqueous solutions are separated by a semipermeable membrane. In which direction will the solvent flow?
a. A = 1.00 L of 0.48 M NaCl; B = 55.85 g of NaCl dissolved in 1.00 L of solution
b. A = 1.00 mL of 0.982 M $CaCl_2$; B = 16 g of NaCl in 100 mL of solution
c. A = 1.00 mL of 6.56 mM $MgSO_4$; B = 5.24 g of $MgCl_2$ in 250 mL of solution

11.81. Calculate the osmotic pressure of each of the following aqueous solutions at 20°C:
a. 2.39 M methanol (CH_3OH)
b. 9.45 mM $MgCl_2$
c. 40.0 mL of glycerol ($C_3H_8O_3$) in 250.0 mL of aqueous solution (density of glycerol = 1.265 g/mL)
d. 25 g of $CaCl_2$ in 350 mL of solution

11.82. Calculate the osmotic pressure of each of the following aqueous solutions at 27°C:
a. 10.0 g of NaCl in 1.50 L of solution
b. 10.0 mg/L of $LiNO_3$
c. 0.222 M glucose
d. 0.00764 M K_2SO_4

11.83. Determine the molarity of each of the following solutions from its osmotic pressure at 25°C. Include the van 't Hoff factor for the solution when the factor is given.
a. Π = 0.674 atm for a solution of ethanol (CH_3CH_2OH)
b. Π = 0.0271 atm for a solution of aspirin ($C_9H_8O_4$)
c. Π = 0.605 atm for a solution of $CaCl_2$, i = 2.47

11.84. Determine the molarity of each of the following solutions from its osmotic pressure at 25°C. Include the van 't Hoff factor for the solution when the factor is given.
a. Π = 0.0259 atm for a solution of urea [$CO(NH_2)_2$]
b. Π = 1.56 atm for a solution of sucrose ($C_{12}H_{22}O_{11}$)
c. Π = 0.697 atm for a solution of KI, i = 1.90

11.85. How many milliliters of water must be added to 48.5 mL of a 1.00 M solution of glucose ($C_6H_{12}O_6$) to make it isotonic with a D5W solution whose osmotic pressure is 7.89 atm at 37°C?

11.86. How many milliliters of water must be added to 102.6 mL of a 1.00 M solution of sodium chloride to make it isotonic with a solution of physiological saline whose osmotic pressure is 7.89 atm at 37°C?

***11.87. Making Maple Syrup** Traditional methods for making maple syrup concentrate the sap from maple trees by boiling away much of the water. Some maple syrup producers have started to use reverse osmosis to reduce the concentration before heating. If the average concentration of sugars in the sap is 2% sucrose ($C_{12}H_{11}O_{12}$) by mass, what pressure must be applied to double the concentration by reverse osmosis?

11.88. Suppose you have 1.00 *M* aqueous solutions of each of the following solutes: glucose ($C_6H_{12}O_6$), NaCl, and acetic acid (CH_3COOH). Which solution has the highest pressure requirement for reverse osmosis?

Ion Exchange

Concept Review

11.89. Explain how a mixture of anion and cation exchangers can be used to deionize water.

11.90. Describe the process by which the ion exchanger in a home water-softening system is regenerated for further use.

11.91. (a) Use the solubility rules to write the balanced net ionic equation for each of the following reactions. If there is no net reaction, write "NR." (b) Which of these three reactions give clear visual evidence of the ion-exchange process?
1. $NaCl(aq) + AgNO_3(aq) \rightarrow AgCl(s) + NaNO_3(aq)$
2. $NaCl(aq) + KNO_3(aq) \rightarrow NaNO_3(aq) + KCl(aq)$
3. $MgCl_2(aq) + KOH(aq) \rightarrow Mg(OH)_2(s) + KCl(aq)$

11.92. (a) Use the solubility rules to write the balanced net ionic equation for each of the following reactions. If there is no net reaction, write "NR." (b) Which of these three reactions give clear visual evidence of the ion-exchange process?
1. $BaCl_2(aq) + Na_2CO_3(aq) \rightarrow BaCO_3(s) + NaCl(aq)$
2. $NaCl(aq) + KOH(aq) \rightarrow NaOH(aq) + KCl(aq)$
3. $Na_3PO_4(aq) + CaCl_2(aq) \rightarrow Ca_3(PO_4)_2(s) + NaCl(aq)$

Measuring the Molar Mass of a Solute by Using Colligative Properties

Concept Review

11.93. What effect does dissolving a solute have on the following properties of a solvent? (a) its osmotic pressure; (b) its freezing point; (c) its boiling point

11.94. How can measurements of osmotic pressure, freezing point depression, and boiling point elevation be used to find the molar mass of a solute? Why are such determinations usually carried out on molecular substances as opposed to ionic ones?

Problems

11.95. **Human Hemoglobin** The hemoglobin found in human red blood cells (see Figure 11.22) has a molar mass of about 6.2×10^4 g/mol. What concentration range of hemoglobin will result in an osmotic pressure between 0.35 and 0.55 atm at 37°C in aqueous solution?

11.96. **Antifreeze Proteins** Northern cod produce proteins that protect their cells from damage caused by subzero temperatures. Measurements of the osmotic pressure for two different "antifreeze" proteins at 18°C yielded the following data: 50.4 mg of protein A in 1.5 mL of water had an osmotic pressure of $\Pi = 0.299$ atm, whereas an osmotic pressure of 0.218 atm was measured for a solution of 52.6 mg of protein B in 1.75 mL of water. What is the apparent range of molar masses for these proteins?

***11.97.** **Cloves** Eugenol is one of the compounds responsible for the flavor of cloves. A 111 mg sample of eugenol was dissolved in 1.00 g of chloroform ($K_b = 3.63°C/m$), increasing the boiling point of the chloroform by 2.45°C. Calculate eugenol's molar mass. Eugenol is 73.17% C, 7.32% H, and 19.51% O by mass. What is the molecular formula of eugenol?

*11.98. **Caffeine** The freezing point of a solution prepared by dissolving 150 mg of caffeine in 10.0 g of camphor is lower than that of pure camphor ($K_f = 39.7°C/m$) by 3.07°C. What is the molar mass of caffeine? Elemental analysis of caffeine yields the following results: 49.49% C, 5.15% H, 28.87% N, and the remainder O. What is the molecular formula of caffeine?

Additional Problems

11.99. Consider the particulate views of the reaction between sodium metal and chlorine gas in Figure 11.4. Describe in your own words the chemical changes and the changes in energy that sodium atoms and chlorine molecules undergo as they form solid sodium chloride.

11.100. Calculate the lattice energy of magnesium chloride ($\Delta H_f^\circ = -641$ kJ/mol) by using (a) the Born-Haber cycle and (b) the enthalpy of solution of $MgCl_2$ ($\Delta H^\circ_{solution} = -158$ kJ/mol). Compare the values of *U* obtained from two methods. The following data should prove useful:

Ionization energy of Mg to Mg^{2+} = 2188 kJ/mol
Electron affinity of Cl = −349 kJ/mol
Energy to sublime Mg = 150 kJ/mol
Bond energy of Cl_2 = 240 kJ/mol
Enthalpy of hydration $Mg^{2+}(aq)$ = −1903 kJ/mol
Enthalpy of hydration of $Cl^-(aq)$ = −368 kJ/mol

***11.101.** **Melting Ice** $CaCl_2$ is often used to melt ice on sidewalks. Could $CaCl_2$ melt ice at −20°C? Assume that the solubility of $CaCl_2$ at this temperature is 70.1 g of $CaCl_2$/100.0 g of H_2O and that the van 't Hoff factor for a saturated solution of $CaCl_2$ is 2.5.

*11.102. **Making Ice Cream** A mixture of table salt and ice is used to chill the contents of hand-operated ice-cream makers. What is the melting point of a mixture of 2.00 lb of NaCl and 12.00 lb of ice if exactly half of the ice melts? Assume that all the NaCl dissolves in the melted ice and that the van 't Hoff factor for the resulting solution is 1.44.

11.103. The freezing points of 0.0935 *m* ammonium chloride and 0.0378 *m* ammonium sulfate in water were found to be −0.322°C and −0.173°C, respectively. What are the values of the van 't Hoff factors for these salts?

11.104. The following data were collected for three compounds in aqueous solution. Determine the value of the van 't Hoff factor for each salt (K_f for water = 1.86°C/*m*).

Compound	Concentration	Experimentally Measured ΔT_f
LiCl	5.0 g/kg	0.410°C
HCl	5.0 g/kg	0.486°C
NaCl	5.0 g/kg	0.299°C

11.105. **Physiological Saline** A solution of 100 mL of physiological saline (0.92% NaCl by mass) is diluted by the addition of 250.0 mL of water. What is the osmotic pressure of the final solution at 37°C? Assume that NaCl dissociates completely into $Na^+(aq)$ and $Cl^-(aq)$.

11.106. What is the osmotic pressure of 100.0 mL of 2.50 m*M* NaCl, 80.0 mL of 3.60 m*M* $MgCl_2$, and the solution that is produced when the first two are blended together at 20°C. Assume that the volumes are additive and that both salts dissociate completely into their component ions.

11.107. A solution of 7.50 mg of a small protein in 5.00 mL of aqueous solution has an osmotic pressure of 6.50 torr at 23.1°C. What is the molar mass of the protein?

11.108. **Kidney Dialysis** Hemodialysis, a method of removing waste products from the blood if the kidneys have failed, uses a tube made of a cellulose membrane that is immersed in a large volume of aqueous solution. Blood is pumped through the tube and is then returned to the patient's vein. The membrane does not allow passage of large protein molecules and cells but does allow small ions, urea, and water to pass through it. Assume that a physician wants to decrease the concentration of sodium ion and urea in a patient's blood while maintaining the concentration of potassium ion and chloride ion in the blood. What materials must be dissolved in the aqueous solution in which the dialysis tube is immersed? How must the concentrations of ions in the immersion fluid compare with those in blood?

11.109. **IV Solution** Another solution used clinically in the hospital setting for IV administration is Ringer's lactate, a solution of sodium, potassium, and calcium cations and chloride and lactate anions. This solution is isotonic with 0.9% saline and D5W described in Section 11.5. Write a mathematical statement that indicates the relationship between the concentrations of cations and anions in this solution compared with 0.9% saline.

11.110. **Injections** The injection of pharmaceutical solutions that are hypertonic in relation to human plasma can cause considerable pain at the site of injection. Why?

12

Solids

Crystals, Alloys, and Polymers

CRAB WALK Flexible electronics based on gold circuits printed onto synthetic polymers can be attached to crabs (*Portunis pelagicus*) to allow scientists to collect data on their movements.

PARTICULATE **REVIEW**

Ionic Compounds and Metals

In Chapter 12 we explore the structure and properties of solids. Most metals and ionic compounds are solids at room temperature, including potassium iodide and platinum, whose structures are depicted here.

- Which solid consists of bonds due to delocalized electrons?
- What type of particles experience electrostatic interactions in potassium iodide?
- What type of particles experience electrostatic interactions in platinum?

(Review Sections 8.1 and 11.1 if you need help.)

(Answers to Particulate Review questions are in the back of the book.)

PARTICULATE **PREVIEW**

Three-Dimensional Networks: Structure and Properties

Two allotropes of carbon are depicted, both as three-dimensional networks. As you read Chapter 12, look for ideas that will help you answer these questions:

- Which allotrope has a repeating pattern that is planar? Which allotrope has a repeating pattern that is not planar?
- Which allotrope consists of carbon atoms covalently bonded to one another? Which allotrope has *both* covalent bonds *and* weak interactions between layers?
- One allotrope is used as a lubricant because of its slipperiness. Is it A or B?

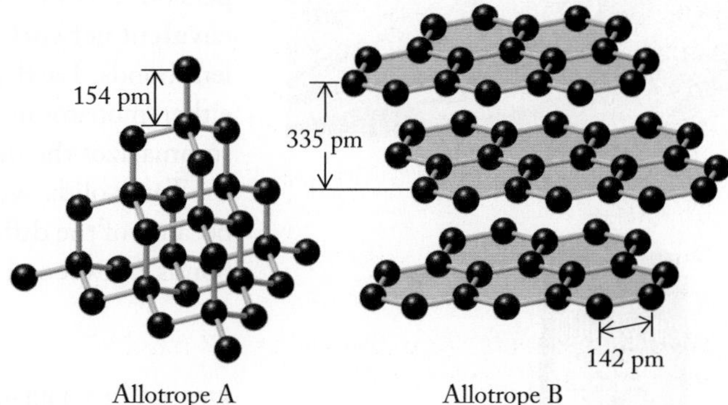

Learning Outcomes

LO1 Identify the differences in packing schemes for atoms in the solid state, including face-centered, body-centered, and simple cubic unit cells, and relate the unit cells to the density and atomic radii of elements
Sample Exercise 12.1

LO2 Describe the differences between substitutional and interstitial alloys
Sample Exercise 12.2

LO3 Relate the crystalline structures of molecular and ionic compounds to their formulas, particle radii, and densities
Sample Exercises 12.3, 12.4

LO4 Identify monomers given the structure of a polymer, and determine the structure of a polymer given the structure of a monomer
Sample Exercises 12.5, 12.7, 12.9

LO5 Analyze the properties of polymers
Sample Exercises 12.6, 12.8

12.1 The Solid State

People have placed a high value on gold for thousands of years and put gold to many different uses, including jewelry, coinage, and medicine (**Figure 12.1**). Today, research into solid materials is yielding applications that would have been unimaginable as recently as 50 years ago. **Nanoparticles** of gold (small particles with diameters less than 10^{-7} m) are used in medicine to treat rheumatoid arthritis and to bind drugs (using intermolecular forces) for delivery to specific target cells. Research on other solid elements has resulted in the breakthroughs responsible for miniaturized electronic devices and for the high-molar-mass organic compounds used as sutures and in knee or hip replacement surgery.

Gold is one of the few elements that exists in nature as a pure metal. Because metals are typically shiny, malleable (easily shaped), ductile (easily drawn out), and able to conduct electricity, they have many applications that take advantage of these properties. All metals except mercury exist in the solid state under standard conditions.

Many covalent and ionic compounds are solids at room temperature. Solids may exist as **crystalline solids**—that is, ordered arrays of atoms, ions, or molecules—or as **amorphous solids**, which have random or disordered arrangements of particles. There are several types of crystalline solids. Metallic elements generally crystallize as **metallic solids** consisting of ordered arrays of atoms held together by metallic bonds, whereas most nonmetals crystallize as **molecular solids** consisting of neutral, covalently bonded molecules held together by intermolecular forces. The noble gases crystallize as **atomic solids** with only weak dispersion forces between the atoms. A few elements and compounds crystallize as **covalent network solids**, which consist of extended arrays held together by covalent bonds. Lastly, ionic compounds form crystalline **ionic solids** in which ions, either monatomic or polyatomic, are held together by ionic bonds. **Table 12.1** summarizes the major classes of solids that we explore in this chapter.

Two solids with different compositions may have very different properties because of the different intermolecular forces between their molecules or different bonds between their atoms or ions. Molecular solids often have lower melting

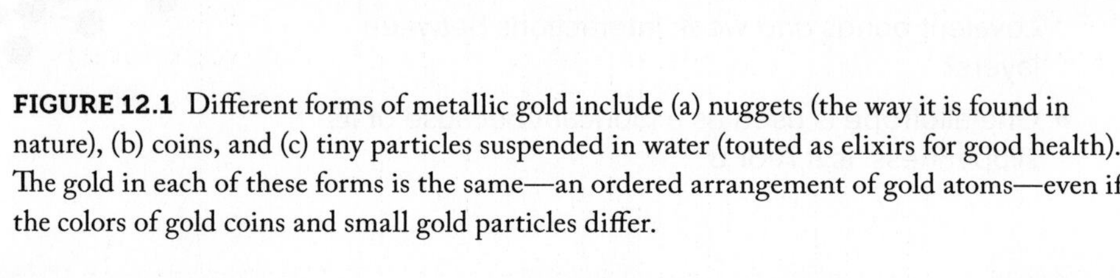

FIGURE 12.1 Different forms of metallic gold include (a) nuggets (the way it is found in nature), (b) coins, and (c) tiny particles suspended in water (touted as elixirs for good health). The gold in each of these forms is the same—an ordered arrangement of gold atoms—even if the colors of gold coins and small gold particles differ.

TABLE 12.1 Solid Characteristics and Classification by Particles and Electrostatic Attractions

Class	Particles	Electrostatic Attractions	Physical Properties	Examples
CRYSTALLINE SOLIDS				
Atomic	Atoms	London forces	Very low melting point, poor conductor, soft	Xe
Metallic	Atoms	Metallic bonds	Broad range of melting points, excellent conductor, malleable, ductile, soft to hard	Ag
Covalent network	Atoms	Covalent bonds	Very high melting point, poor conductor, very hard	C (diamond)
Ionic	Cations and anions	Ionic bonds	High melting point, good conductor (in molten state), hard and brittle	NaCl
Molecular	Molecules	London forces	Low melting point, poor conductor, soft	S_8, CO_2
		Dipole–dipole forces		$CHCl_3$
		Hydrogen bonds		H_2O
AMORPHOUS SOLIDS				
Covalent network	Atoms	Covalent bonds	Properties vary based on composition	SiO_2 (obsidian/volcano glass)

nanoparticle an approximately spherical sample of matter with dimensions smaller than 100 nm (1×10^{-7} m).

crystalline solid a solid made of an ordered array of atoms, ions, or molecules.

amorphous solid a solid that lacks long-range order for the atoms, ions, or molecules in its structure.

metallic solid a solid formed by metallic bonds among atoms of metallic elements.

molecular solid a solid formed by intermolecular attractive forces among neutral, covalently bonded molecules.

atomic solid a solid formed by weak attractions between noble gas atoms.

covalent network solid a solid formed by covalent bonds among nonmetal atoms in an extended array.

ionic solid a solid formed by ionic bonds among monatomic or polyatomic ions.

crystal lattice a three-dimensional repeating array of particles (atoms, ions, or molecules) in a crystalline solid.

crystal structure a particular arrangement in three-dimensional space that specifies the positions of the particles (atoms, ions, or molecules) in relation to one another in a crystalline solid.

hexagonal closest-packed (hcp) a crystal lattice in which the layers of atoms or ions in hexagonal unit cells have an *ababab* . . . stacking pattern.

CONNECTION Intermolecular forces are described in Chapter 10.

CONNECTION Some physical properties of metals were first described in Chapter 2.

points than ionic solids because the intermolecular forces that hold them together are relatively weak compared to the strength of ionic bonds. The melting points of metallic solids cover a broad range, from gallium at 29.76°C (just above room temperature but below human body temperature) to tungsten, which melts at 3422°C. (Mercury is the one metal that exists as a liquid at room temperature.) Metallic solids are malleable, so they can be bent into a variety of shapes, whereas ionic solids tend to be hard but brittle.

Mixtures of a host metal and one or more other elements are called *alloys*, and more than 500,000 alloys in the solid state have been made and characterized. Some are solid solutions, which means they are homogeneous at the atomic level. Other alloys are heterogeneous mixtures. The purpose of making alloys is to manipulate their properties by varying the proportions of their constituent metals. Alloys of nickel and titanium can be shaped into stents used to prop open arteries in heart patients. Brass, which is an alloy of copper and zinc, is used in hospitals and large kitchens because of its antibacterial properties. The transportation industry uses aluminum alloys that are ideal for making aircraft because they are strong and lighter than steel, resulting in less fuel consumption.

Polymers, which are discussed in greater detail in Section 12.6, are large solid organic compounds with molar masses up to and exceeding 1,000,000 g/mol. Like metals and alloys, different polymer compositions affect the physical properties of the material. Poly(vinyl chloride), for example, is used to make drainpipes and other building materials (Section 4.2), whereas Teflon is used in nonstick cookware. Abrasion-resistant, ultrahigh-molecular-weight polyethylene is used as a coating on some artificial joints used in hip replacement. A polymer formed by the reaction of glycolic acid and lactic acid is used in dissolving sutures and to support the growth of skin cells for burn victims.

In Chapter 12 we examine the major classes of solid materials and explore the links between their physical properties at the macroscopic level and their structures at the atomic level. Materials science is a fast-changing field, and we are able to discuss only a few of its many applications. The search for new materials and new applications is constant and constantly surprising. We start our exploration of the solid state by looking at the structures of some familiar metals.

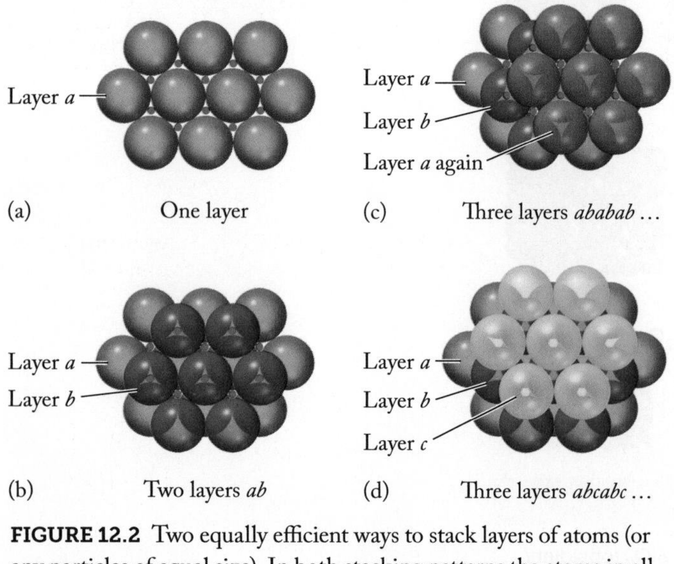

FIGURE 12.2 Two equally efficient ways to stack layers of atoms (or any particles of equal size). In both stacking patterns the atoms in all layers (shown in different colors to distinguish one layer from another) are packed as closely together as possible. The layers depicted in (a) and (b) are the same in both patterns. (c) In the *abababab* . . . pattern, atoms in the third layer are directly above the atoms in the first layer. (d) In the *abcabcabc* . . . pattern, atoms in the third layer are directly above spaces (marked by red dots) between the atoms in the first layer.

12.2 Structures of Metals

When a metal is heated above its melting point and then slowly allowed to cool, it solidifies into a crystalline solid in which the atoms are arranged in an ordered, repeating, three-dimensional array called a **crystal lattice**. In many crystal lattices, stacked layers (designated *a*, *b*, *c*, . . .) of particles are packed together as tightly as possible. Each atom in layer *a* touches six others in that layer (**Figure 12.2a**). The atoms in layer *b* nestle into some of the spaces created by the atoms of layer *a* (**Figure 12.2b**), just like oranges in a fruit-stand display or cannonballs at a 16th-century fort (**Figure 12.3**). Similarly, the atoms in a third layer, *c*, nestle among those in layer *b*. However, two different alignments are possible for the atoms in layer *c*. They can align directly above the atoms in layer *a* (**Figure 12.2c**), in which case we label them as layer *a* again, or they can nestle into the atoms of layer *b* in such a way that they are *not* aligned directly above the layer *a* atoms (**Figure 12.2d**). When a fourth layer is nestled into the spaces of layer *c*, the fourth layer atoms lie directly above the layer *a* atoms. In Figure 12.2c, we have an *ababab* . . . stacking pattern throughout the crystal, whereas in Figure 12.2d we have an *abcabc* . . . stacking pattern. As we will see in this and subsequent sections, the pattern in which metal atoms are packed and the available spaces between them affects their physical properties (e.g., their density) and their applications.

(a)

(b)

FIGURE 12.3 (a) Stacks of oranges in a grocery store and (b) cannonballs at a 16th-century fort illustrate closest-packed arrays of spherical objects. Both (a) and (b) exhibit an *abcabc* . . . stacking pattern.

Stacking Patterns and Unit Cells

Which of these two patterns do gold atoms adopt? Whether the atoms in a metal like gold are stacked in an *ababab* or an *abcabc* pattern determines the shape of its crystals. The **crystal structure** of an element or compound specifies the location of the atoms in three-dimensional space in relation to one another. To see how crystal structures are linked to stacking patterns, let's take a closer look at a cluster of atoms held together by the *ababab* stacking pattern (**Figure 12.4a**). This cluster forms a *hexagonal* (six-sided) prism of closely packed atoms. In fact, they are as tightly packed as they can be, so the crystal lattice is called **hexagonal closest-packed (hcp)**. Titanium metal used in surgical repair of fractures is one

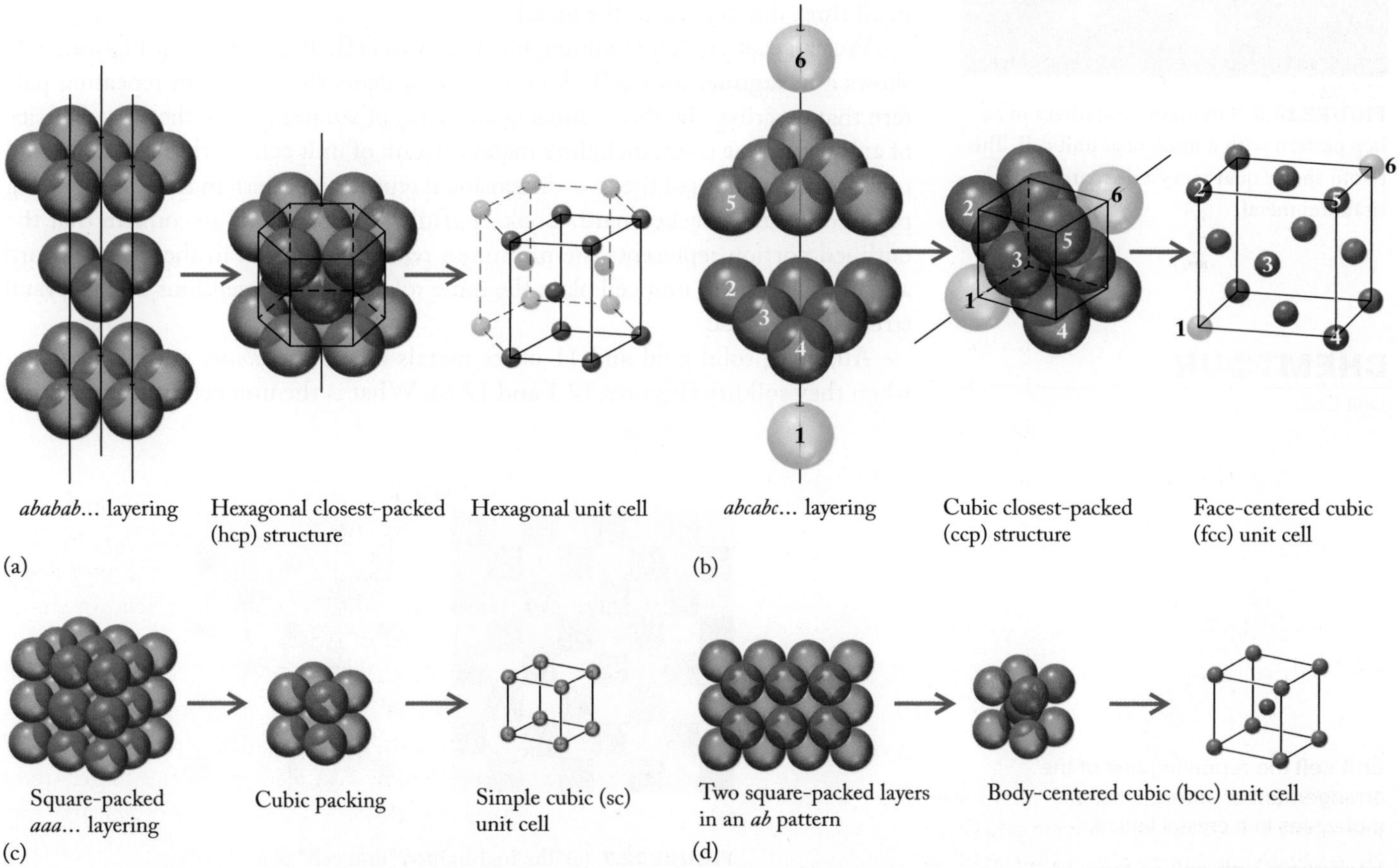

6
5
2
3
4
1
2
6
5
3
1
4
2
5
6
3
1
4

ababab... layering
Hexagonal closest-packed (hcp) structure
Hexagonal unit cell
(a)

abcabc... layering
Cubic closest-packed (ccp) structure
Face-centered cubic (fcc) unit cell
(b)

Square-packed *aaa*... layering
Cubic packing
Simple cubic (sc) unit cell
(c)

Two square-packed layers in an *ab* pattern
Body-centered cubic (bcc) unit cell
(d)

FIGURE 12.4 (a) A hexagonal closest-packed (hcp) crystal structure and its hexagonal unit cell. (b) The stacking pattern *abcabc* . . . produces a face-centered cubic (fcc) unit cell when tipped by 45° along the axis. (c) In *cubic packing*, the atoms in all layers are directly above those in the *a* layer. The repeating unit of this pattern is called a *simple cubic* (sc) unit cell. (d) The atoms (blue spheres) in the second *b* layer nestle into the spaces between the square-packed atoms (red spheres) in the *a* layer. Atoms in the third layer are directly above those in the first, producing an *ababab* . . . stacking pattern and a *body-centered cubic* (bcc) unit cell.

FIGURE 12.5 Titanium crystallizes in an hcp pattern with a hexagonal unit cell. This photo shows quartz crystals coated with titanium metal.

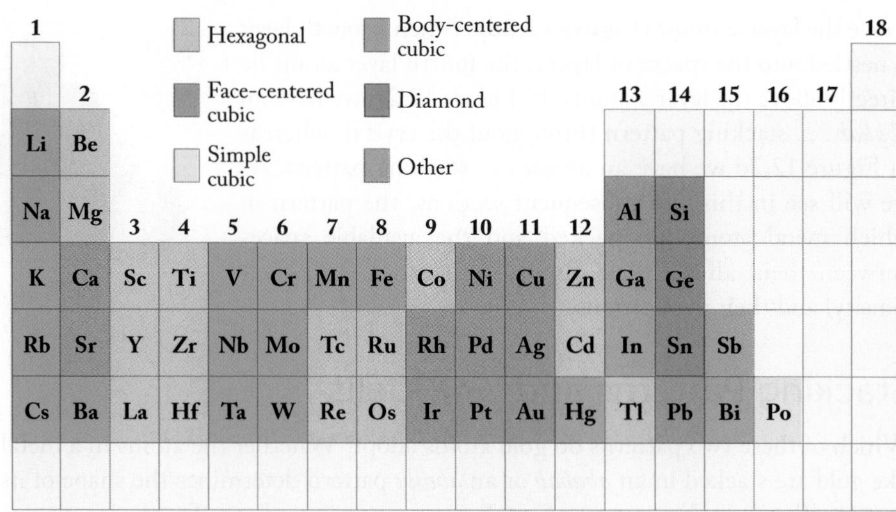

FIGURE 12.6 Unit cells of metals and metalloids in the periodic table. The five metals designated "Other" have unit cells more complicated than can easily be described in this book.

element that crystallizes in an hcp lattice, forming hexagonal crystals (**Figure 12.5**). As the periodic table in **Figure 12.6** shows, the atoms in 16 metallic elements have an hcp crystal lattice. In these metals, the cluster of atoms in Figure 12.4a serves as an atomic-scale building block—a pattern of atoms repeated over and over again in all three dimensions in the metal.

We call each of these building blocks a **unit cell**. The example in Figure 12.4a shows a **hexagonal unit cell**. A unit cell represents the minimum repeating pattern that describes the three-dimensional array of atoms forming the crystal lattice of any crystalline solid, including metals. Think of unit cells as three-dimensional microscopic analogs of the two-dimensional repeating pattern in fabrics, wrapping paper, or even a checkerboard. Look carefully at **Figure 12.7** to confirm that the outlined portion represents the minimum repeating pattern in the checkerboard and in the paper. A unit cell plays the same role in three dimensions in the crystal structure of a solid.

Atoms in solid gold and 11 other metals adopt the *abcabc* stacking pattern when they solidify (Figures 12.1 and 12.6). What is the unit cell in this stacking

CHEMTOUR
Unit Cell

unit cell the repeating unit of the arrangement of atoms, ions, or molecules in a crystal lattice.

hexagonal unit cell an array of closest-packed particles that includes parts of four particles on the top and four on the bottom faces of a hexagonal prism and one particle in a middle layer.

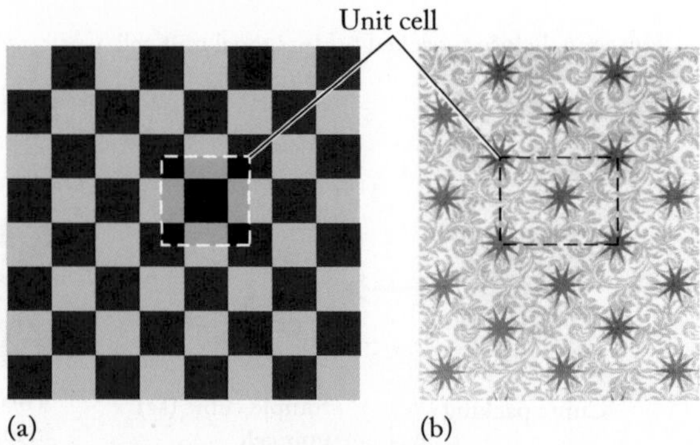

FIGURE 12.7 (a) The highlighted "unit cell" of a checkerboard is the smallest set of squares that defines the pattern repeated over the entire board. (b) This wrapping paper has a more complex pattern. One unit cell is highlighted. Can you outline another?

pattern? Consider what happens when we take the cluster of atoms on the left in **Figure 12.4b**, compact it, rotate the cluster, and tip it 45° to get the orientation shown in the center of Figure 12.4b. The black outline shows that the atoms form a cube: one atom at each of the eight corners of the cube and one at the center of each of the six faces. Atoms at adjacent corners (e.g., atoms 1 and 2, 1 and 4, 2 and 5, or 4 and 5) do *not* touch each other, but the three atoms along the diagonal of any face of the cube (e.g., atoms 2, 3, 4) do touch each other. Because the atoms are stacked together as closely as possible, this crystal lattice is called **cubic closest-packed (ccp)**, and the corresponding unit cell is called a **face-centered cubic (fcc) unit cell**. The unit cell edges are all of equal length, and the angle between any two edges is 90°.

cubic closest-packed (ccp) a crystal lattice in which the layers of atoms or ions in face-centered cubic unit cells have an *abcabc* . . . stacking pattern.

face-centered cubic (fcc) unit cell an array of closest-packed particles that has one particle at each of the eight corners of a cube and one more at the center of each face of the cube.

packing efficiency the percentage of the total volume of a unit cell occupied by the atoms, ions, or molecules.

simple cubic (sc) unit cell an array of atoms or molecules with one particle at the eight corners of a cube.

body-centered cubic (bcc) unit cell an array of atoms or molecules with one particle at each of the eight corners of a cube and one at the center of the cell.

The two *closest-packed* crystal lattices—hexagonal (hcp) and cubic (ccp)—represent the most efficient ways of arranging solid spheres of equal radius. We can express the **packing efficiency** as the percentage of the total volume of the unit cell occupied by the spheres:

$$\text{Packing efficiency (\%)} = \frac{\text{volume occupied by spheres}}{\text{volume of unit cell}} \times 100\% \qquad (12.1)$$

For both hcp and ccp crystal lattices, the packing efficiency is approximately 74%.

Other crystal lattices have stacking patterns in which the atoms are arranged close together, but not as efficiently as in hcp and ccp lattices. Two of these are shown in **Figure 12.4c and d**. When the atoms are arranged in an *a* layer so that each atom touches just four adjacent atoms in that layer, the arrangement is called *square packing* (Figure 12.4c). If we add a second layer of spheres directly above the first, we create the *aaa* . . . stacking pattern, which is called *cubic packing*. The three-dimensional repeating pattern of this arrangement is called a **simple cubic (sc) unit cell**. It is the least efficiently packed of the cubic unit cells and is quite rare among metals: only radioactive polonium (Po) forms a simple cubic unit cell.

For crystal lattices in which each atom in a second layer is nestled in the space created by four atoms in a square-packed *a* layer (Figure 12.4d), we have two layers in an *ab* stacking pattern. If the atoms in the third layer are directly above those in the first, then we have an *ababab* . . . stacking pattern based on layers of square-packed atoms. The simplest three-dimensional repeating unit of this pattern is called a **body-centered cubic (bcc) unit cell**. It consists of portions of nine atoms, one at each of the eight corners of a cube and one in the middle of the cube. All the group 1 metals and many transition metals have bcc unit cells (Figure 12.6). One of those transition metals is tantalum, which is used for wedding rings (**Figure 12.8**) and to coat artificial joints. **Table 12.2** summarizes the different stacking patterns, packing efficiencies, and unit cells described in this section.

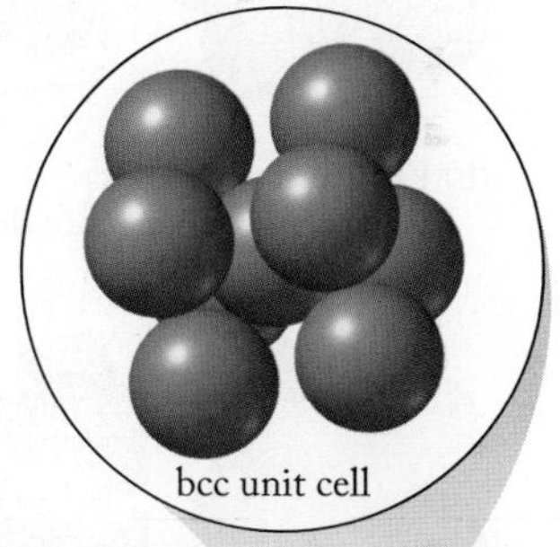

FIGURE 12.8 Tantalum metal is used for wedding rings because it is wear-resistant. The crystal lattice of tantalum is body-centered cubic.

CONCEPT **TEST**

What is the difference between a crystal lattice and a unit cell?

(Answers to Concept Tests are in the back of the book.)

Unit Cell Dimensions

Of all the metallic elements, only Li, Na, and K have densities less than 1 g/cm^3. The densities of the transition metals range from 3 g/cm^3 (Sc) to 22.5 g/cm^3 (Os). The number of atoms in the unit cell of a metallic element, its atomic radius, and its molar mass all contribute to its density. **Figure 12.9** shows whole-atom and

TABLE 12.2 Summary of Unit Cells, Stacking Patterns, and Packing Efficiencies for Solid Spheres

Lattice Name	Unit Cell	Type of Packing	Stacking Pattern	Number of Nearest Neighbors	Packing Efficiency
Hexagonal closest-packed (hcp)	Hexagonal	Close packing	*ababab*...	12	74%
Cubic closest-packed (ccp)	Face-centered cubic (fcc)	Close packing	*abcabc*...	12	74%
Body-centered cubic packing	Body-centered cubic (bcc)	Square packing	*ababab*...	8	68%
Cubic packing	Simple cubic (sc)	Square packing	*aaa*...	6	52%

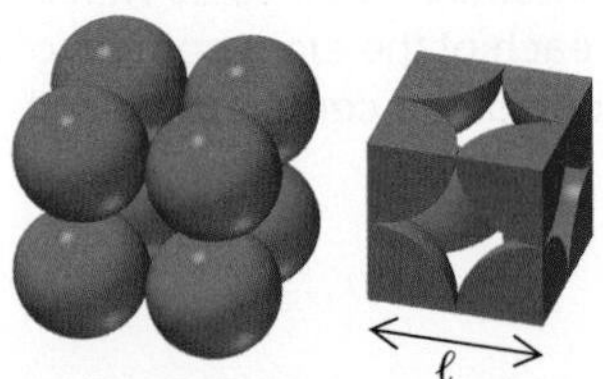

(a) Simple cubic:
Atoms touch along edge

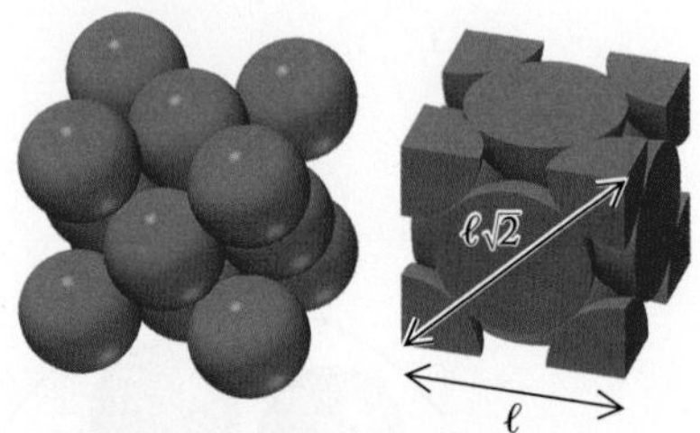

(b) Face-centered cubic:
Atoms touch along face diagonal

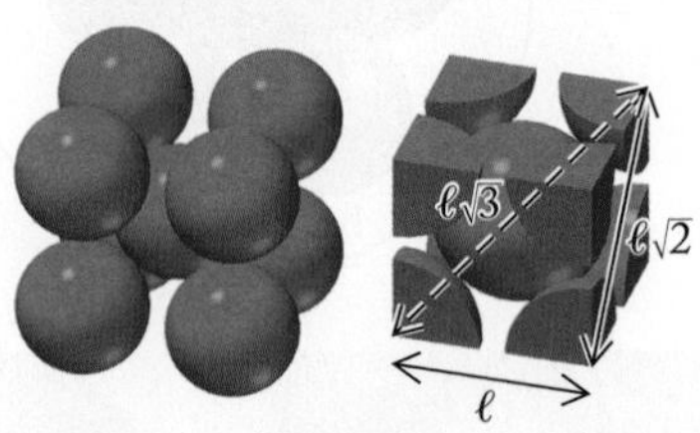

(c) Body-centered cubic:
Atoms touch along body diagonal

FIGURE 12.9 Whole-atom and cutaway views of cubic unit cells. (a) In a simple cubic unit cell, each corner atom of the unit cell is part of eight unit cells. Atoms along each edge touch with an edge length ℓ. (b) In a face-centered cubic unit cell, the face atoms are part of two unit cells. Atoms along the face diagonal touch. (c) In a body-centered cubic unit cell, one atom in the center lies entirely in one unit cell. The atoms along the body diagonal touch.

cutaway views of sc, fcc, and bcc unit cells. These views make it possible to determine how many atoms are in each type of cubic unit cell.

In the simple cubic unit cell (Figure 12.9a), only a fraction of each corner atom is inside the unit cell boundary. In a crystal lattice with this unit cell, each atom is a corner atom in eight unit cells (**Figure 12.10a**). Thus, each atom contributes the equivalent of one-eighth of an atom to the unit cell. A cube has eight corners, so there is a total of

$$\tfrac{1}{8}\ \text{corner atom}/\cancel{\text{corner}} \times 8\ \cancel{\text{corners}}/\text{unit cell} = 1\ \text{corner atom}/\text{unit cell}$$

This calculation applies to the corner atoms in any type of cubic unit cell. The two corner atoms along each edge in Figure 12.9a touch each other. Therefore, the edge length ℓ in the simple cubic unit cell is equal to twice the atomic radius:

$$\ell = 2r$$

If we can measure the length of the unit cell (ℓ), then we can calculate the atomic radius of an element with this unit cell. For example, an fcc unit cell (Figure 12.9b) has eight corner atoms (just like the sc unit cell) and one atom in the center of each of the six faces (**Figure 12.10b**). Each face atom is shared by

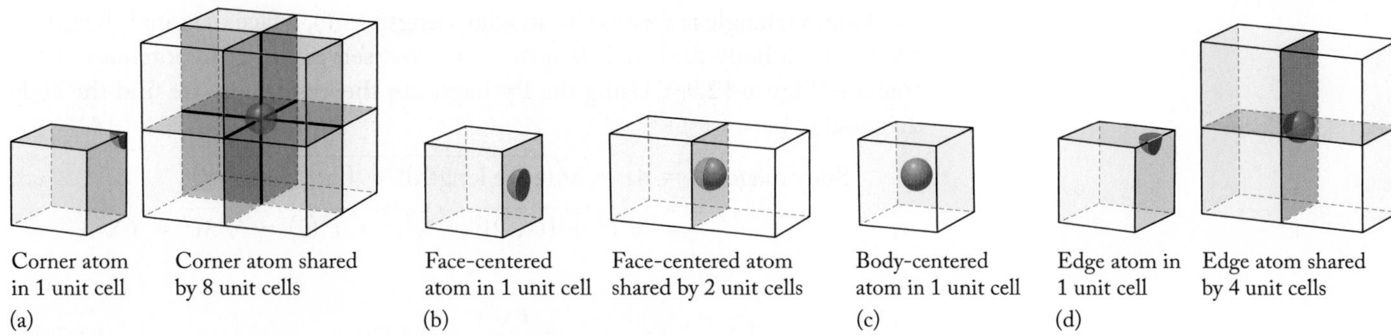

FIGURE 12.10 Crystal lattices illustrating (a) corner atoms shared by eight unit cells, (b) face atoms shared by two unit cells, (c) center atoms entirely in one unit cell, and (d) edge atoms shared by four unit cells.

the two unit cells that abut each other at that face. Therefore, each unit cell "owns" half of each face atom, making a total of

$$\frac{1}{2} \text{ face atom}/\cancel{\text{face}} \times 6 \ \cancel{\text{faces}}/\text{unit cell} = 3 \text{ face atoms/unit cell}$$

These three face atoms plus the equivalent of one corner atom means that an fcc unit cell consists of

$$1 \text{ corner atom} + 3 \text{ face atoms} = 4 \text{ atoms per fcc unit cell}$$

To relate the size of these atoms to the dimensions of the fcc unit cell, note in the cutaway view of Figure 12.9b that the corner atoms do not touch one another, but adjacent atoms along the face diagonal do touch each other. Therefore, a face diagonal spans the radius r of two corner atoms and the diameter (2 radii = $2r$) of a face atom. The length of a face diagonal, then, is $1 + 2 + 1 = 4$ atomic radii $= 4r$. A face diagonal connects the ends of two edges and forms a right triangle with those two edges, each of length ℓ (Figure 12.9b). According to the Pythagorean theorem, then, the length of a face diagonal is

$$\text{Face diagonal} = 4r = \sqrt{\ell^2 + \ell^2} = \sqrt{2\ell^2} = \ell\sqrt{2}$$

$$r = \frac{\ell\sqrt{2}}{4} = 0.3536\ell \qquad (12.2)$$

As in the case of the simple cubic unit cell, if we know the value of ℓ, we can calculate r by using Equation 12.2.

The bcc unit cell (Figure 12.9c) has one-eighth of an atom at each of the eight corners (just like the sc and fcc unit cells), as well as one atom in the center of the cell that is entirely within the cell (**Figure 12.10c**). This means a bcc unit cell consists of 1 corner atom + 1 center atom = 2 atoms per unit cell.

Relating unit cell edge length ℓ to atomic radius r in a bcc cell is complicated because, in addition to not touching along the edges, adjacent atoms along any face diagonal do not touch each other either. However, each corner atom does touch the atom in the center of the cell, which means that the atoms touch along a *body diagonal*, which runs between opposite corners through the center of the cube. In the cutaway view in Figure 12.9c, the body diagonal runs from the bottom left corner of the front face to the top right corner of the rear face. It spans (1) the radius of the front-face bottom left corner atom, (2) the diameter (2 radii) of the central atom, and (3) the radius of the rear-face top right atom, making the length of the body diagonal equivalent to $4r$.

A right triangle is formed by an edge (length = ℓ), a face diagonal (length = $\ell\sqrt{2}$), and a body diagonal (length = $4r$) that serves as the hypotenuse of the triangle (Figure 12.9c). Using the Pythagorean theorem again, we find the body diagonal to be:

$$\text{Body diagonal} = 4r = \sqrt{(\text{edge length})^2 + (\text{face diagonal})^2}$$
$$= \sqrt{\ell^2 + (\ell\sqrt{2})^2} = \sqrt{\ell^2 + \ell^2(2)} = \sqrt{3\ell^2} = \ell\sqrt{3}$$

so

$$r = \frac{\ell\sqrt{3}}{4} = 0.4330\ell \quad (12.3)$$

Table 12.3 summarizes how atoms in different locations in sc, bcc, and fcc unit cells contribute to the total number of atoms in each unit cell. **Table 12.4** summarizes the number of equivalent atoms and the relationship between r and ℓ for the three cubic unit cells.

TABLE 12.3 Contributions of Atoms to Cubic Unit Cells

Atom Position	Contribution to Unit Cell
Center	1 atom
Face	$\frac{1}{2}$ atom
Edge	$\frac{1}{4}$ atom
Corner	$\frac{1}{8}$ atom

TABLE 12.4 Summary of Unit Cells, Equivalent Atoms, and the Relationship between Radius of Atoms and Edge Length of Cubic Unit Cells

Unit Cell	Number of Equivalent Atoms per Unit Cell	Relationship between r and ℓ
Simple cubic (sc)	1 (8 corners)	$r = \frac{\ell}{2} = 0.5\ell$
Body-centered cubic (bcc)	2 (1 center + 8 corners)	$r = \frac{\ell\sqrt{3}}{4} = 0.4330\ell$
Face-centered cubic (fcc)	4 (6 faces + 8 corners)	$r = \frac{\ell\sqrt{2}}{4} = 0.3536\ell$

SAMPLE EXERCISE 12.1 Calculating Atomic Radius and Density from Unit Cell Dimensions **LO1**

Tantalum crystallizes with a bcc unit cell as shown in **Figure 12.11**. Its unit cell has an edge length of 330.3 pm. (a) Calculate the radius in picometers of the tantalum atoms. Check your answer against the data in Appendix 3. (b) Calculate the density of tantalum in grams per cubic centimeter at 25°C.

Collect and Organize We are given the unit cell (bcc—see Figure 12.9c) and edge length (ℓ = 330.3 pm) of tantalum. Together with the molar mass of tantalum (180.95 g/mol), we can calculate both the atomic radius of a Ta atom and the density of Ta metal.

Analyze The tantalum atoms do not touch along the unit cell edges or along any face diagonal, but they do touch along the body diagonals. Table 12.4 gives the relationship between r and ℓ for a bcc unit cell: $r = 0.4330\ell$. We recognize that the radius is a little less than half the edge length ($\ell/2$), so with ℓ =330 pm, we expect a value less than 165 pm for r.

We assume that the density of the unit cell is the same as the density of solid Ta. The density of the Ta bcc unit cell is the mass of two Ta atoms divided by the volume of the cell. We can calculate the mass of two Ta atoms from the molar mass, which is 180.95 g/mol. The conversion includes dividing by the Avogadro constant to calculate

FIGURE 12.11 The bcc unit cell of tantalum.

the mass of each Ta atom in grams. The formula for the volume of a cube of edge length ℓ is $V = \ell^3$.

Solve

a. Substituting the edge length into the formula for r from Table 12.4:

$$r = 0.4330 \times 330.3 \text{ pm} = 143.0 \text{ pm}$$

This is close to the reference value of 146 pm for Ta (Table A3.1).

b. We first calculate the mass m of two Ta atoms:

$$m = \frac{180.95 \text{ g Ta}}{1 \text{ mol Ta}} \times \frac{1 \text{ mol Ta}}{6.022 \times 10^{23} \text{ atoms Ta}} \times 2 \text{ atoms Ta} = 6.010 \times 10^{-22} \text{ g Ta}$$

The volume of the cell in cubic centimeters is

$$V = \ell^3 = (330.3 \text{ pm})^3 \times \frac{(10^{-10} \text{ cm})^3}{1 \text{ pm}^3} = 3.604 \times 10^{-23} \text{ cm}^3$$

The density is

$$d = \frac{m}{V} = \frac{6.010 \times 10^{-22} \text{ g}}{3.604 \times 10^{-23} \text{ cm}^3} = 16.68 \text{ g/cm}^3$$

Think About It The value of r is indeed less than half the value of the edge length, as predicted. If the calculated density is correct, then it should match the measured density in Table A3.2. The reported density of tantalum is 16.65 g/mL, so the result of our calculation (16.68 g/cm^3) is reasonable.

Practice Exercise Silver and gold both crystallize in face-centered cubic unit cells with edge lengths of 407.7 and 407.0 pm, respectively. Calculate the atomic radius and density of each metal, and compare your answers with the data listed in Appendix 3.

(Answers to Practice Exercises are in the back of the book.)

Earlier in this section we mentioned that ccp and hcp are the most efficient packing schemes for spheres. Now that we know more about the fcc unit cell, let's take another look at the calculation of packing efficiency by using Equation 12.1. The unit cell for a ccp arrangement of solid spheres of radius r is an fcc unit cell that contains four equivalent spheres. The volume occupied by these four spheres is

$$V_{\text{spheres}} = 4\left(\tfrac{4}{3}\pi r^3\right) = \tfrac{16}{3}\pi r^3$$

From Table 12.4, $r = (\ell\sqrt{2})/4$, in which case the unit cell edge (ℓ) is

$$\ell = \frac{4r}{\sqrt{2}} = 2.828r$$

This equation enables us to express the volume of the unit cell in terms of r:

$$V_{\text{unit cell}} = \ell^3 = (2.828r)^3 = 22.62r^3$$

Substituting V_{spheres} and $V_{\text{unit cell}}$ into Equation 12.1:

$$\text{Packing efficiency (\%)} = \frac{V_{\text{spheres}}}{V_{\text{unit cell}}} \times 100\% = \frac{\frac{16}{3}\pi r^3}{22.62r^3} \times 100\% = 74.1\%$$

This means that the most efficient packing of spheres results in about 26% of the total volume being empty.

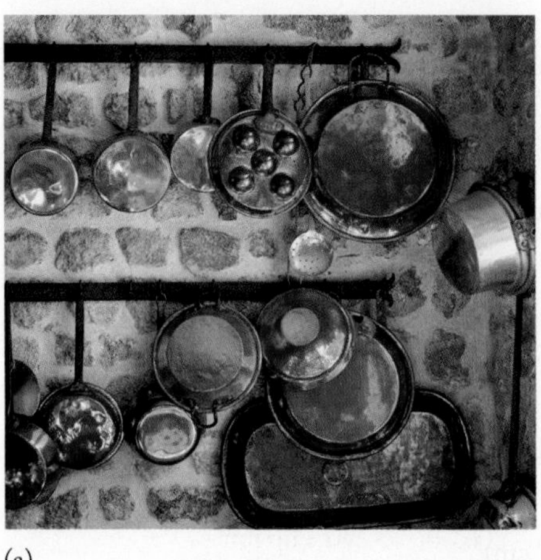

(a)

(b)

FIGURE 12.12 (a) Brass and (b) bronze are alloys of copper containing zinc and tin (up to 30% by mass), respectively. Both alloys have antibacterial properties, making them desirable coatings for fixtures in hospitals and restaurant kitchens.

STEPWISE ANIMATION
Alloys

12.3 Alloys and Medicine

The antibacterial properties of copper metal are attractive for coating surfaces in hospitals and in food service kitchens where an infection can prove deadly (**Figure 12.12**). However, pure copper has two disadvantages: it is both relatively soft and very malleable, which means that pure copper objects are easily bent and damaged. We can explain the malleability of Au, Cu, and other metals in terms of the relatively weak bonds between the atoms in their cubic closest-packed crystal structure. This arrangement gives the atoms in one layer the ability, under stress, to slip past atoms in an adjacent layer (**Figure 12.13**), but the overall crystal structure is still cubic closest-packed. The ease with which copper atoms slip past each other makes it easy to bend copper pipes used in plumbing, but it also makes them susceptible to damage. Additionally, copper reacts with air to produce blue-green copper hydroxides and carbonates.

CONNECTION The classification of homogeneous and heterogeneous mixtures and the difference between compounds and mixtures are described in Chapter 1.

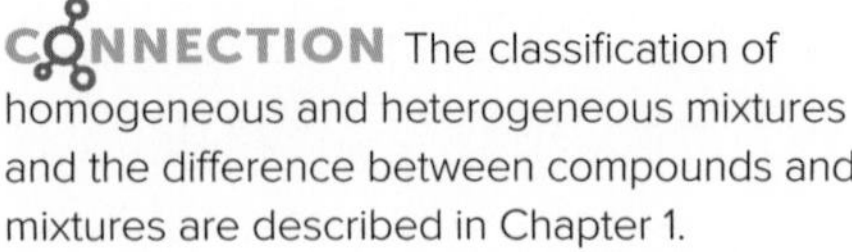

Metal is deformed

FIGURE 12.13 Copper and other metals are malleable because their atoms are stacked in layers that can slip past each other under stress. Slippage is possible because of the diffuse nature of metallic bonds and the relatively weak interactions between pairs of atoms in adjoining layers.

The physical and chemical properties of copper change when mixed with zinc or tin to produce mixtures known as brass and bronze, respectively. Brass and bronze are stronger than pure copper and have improved resistance to reactions with air or moisture while retaining the antibacterial properties of pure copper. A mixture of two metals represents an **alloy**: a metallic material made when a host metal is blended with one or more other elements, which may or may not be metals, thereby changing the properties of the host metal. Like the mixtures discussed in Chapter 1, alloys can be classified according to their composition as homogeneous or heterogeneous mixtures. Brass and bronze are both *homogeneous alloys*, solid solutions in which the atoms of the added elements (in this case tin or zinc) are randomly but uniformly distributed among the atoms of the host (copper). *Heterogeneous alloys* consist of matrices of atoms of host metals interspersed with small "islands" made up of individual atoms of other elements. For example, the different unit cells of lead and tin give solder its properties because it is a heterogeneous alloy. In both cases, the compositions may vary over a limited range.

Every year thousands of patients suffering from cardiovascular disease undergo a procedure called balloon angioplasty to open clogged arteries. This surgery involves the insertion of small metal supports called stents to prop open the artery. Increasingly, stents are manufactured from an alloy containing nickel and titanium. The alloy NiTi (nitinol) is an example of an *intermetallic compound*, a substance that has a reproducible stoichiometry and constant composition (just

alloy a blend of a host metal and one or more other elements, which may or may not be metals, that are added to change the properties of the host metal.

substitutional alloy an alloy in which atoms of the nonhost metal replace host atoms in the crystal lattice.

(a)

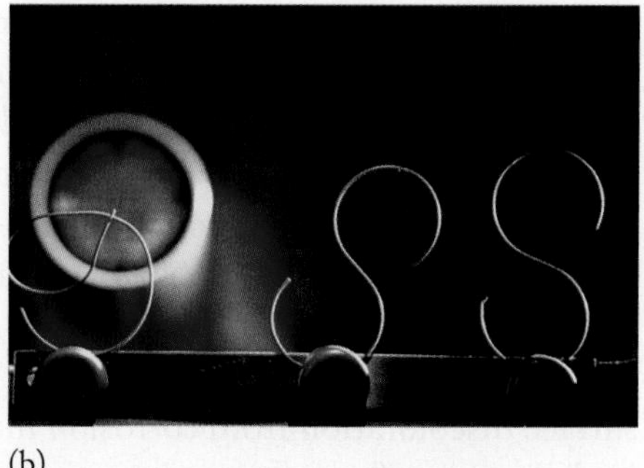
(b)

FIGURE 12.14 Shape memory alloys are used in stents for heart patients. This figure illustrates how shape memory alloys work. (a) The S shape on the right is the desired final shape of the wire. The left and middle wires have been coiled to take up less space. The white circle in the background is a hair dryer heating the wires. (b) When the wire in the middle is heated, it returns to its original shape, while the wire on the left is in the process of unraveling.

like chemical compounds). Intermetallic compounds are still commonly referred to as alloys and are considered to be a subgroup within homogeneous alloys. What makes nickel–titanium alloys particularly useful is that they demonstrate shape memory (**Figure 12.14**). A piece of NiTi wire is heated and formed into the desired shape for the stent. On cooling, the NiTi undergoes a phase change to a different structure. The structural change allows the stent to be easily straightened for insertion into an artery. After insertion, the coiled stent is heated, and it springs back into its original shape as it rapidly reverts to the high-temperature form.

Substitutional Alloys

If alloys are mixtures of metals, and pure metals have crystal lattices as described in Section 12.2, how are the metal atoms arranged in the crystal lattices and unit cells of alloys? How does the structure of NiTi memory metal change as a function of temperature? The answer to these questions gives rise to another classification system: a **substitutional alloy** is one in which atoms of the nonhost metal replace host atoms in the crystal lattice. Bronze and nitinol are both examples of *homogeneous, substitutional alloys.* Substitutional alloys may form between metals that have the same crystal lattice and atomic radii that are within about 15% of each other.

(a)

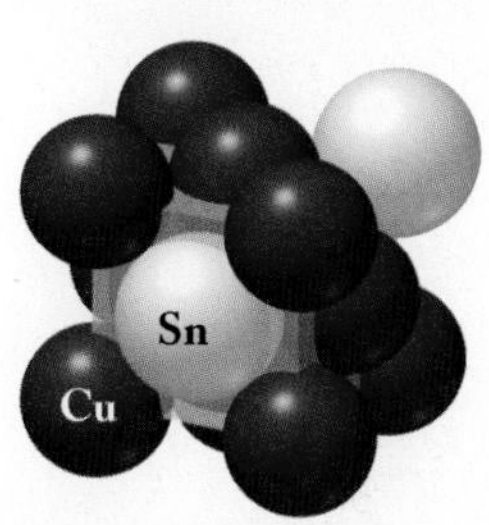

(b)

FIGURE 12.15 Two atomic-scale views of one type of bronze, a substitutional alloy. (a) A layer of close-packed copper (Cu) atoms interspersed with a few atoms of tin (Sn). (b) One possible unit cell for bronze. In this case, tin atoms have replaced one corner Cu atom and one face Cu atom.

CONCEPT TEST

Is it accurate to call bronze a *solution* of tin *dissolved* in copper? Explain why or why not.

Why is an alloy more difficult to bend than a pure metal? **Figure 12.15** illustrates one layer of the crystal lattice of bronze. The radii of copper and tin atoms are similar—128 pm and 140 pm, respectively. Inserting the slightly larger Sn atoms in the cubic closest-packed Cu crystal lattice disturbs the structure a little, making the planes of copper atoms "bumpy" instead of uniform (**Figure 12.16**), and making it more difficult for the copper atoms to slip past one another. Less slippage makes bronze less malleable than copper, but being less malleable also means that bronze is harder and stronger.

FIGURE 12.16 The larger Sn atoms in bronze disturb the Cu crystal lattice, producing atomic-scale bumps in the slip plane (wavy line) between layers of Cu atoms. These bumps make it more difficult for Cu atoms to slide past each other when an external force is applied.

interstitial alloy an alloy in which atoms of the added element occupy the spaces between atoms of the host.

Alloys with good strength and corrosion resistance are essential for the manufacture of surgical tools and biomedical implants. One family of such alloys is the stainless steels, and its high-grade medical version is considered "surgical steel." Steel is an alloy of iron, and stainless steel owes its useful properties to chromium; surgical steel contains 13%–16.5% chromium and 0.4%–0.6% carbon by mass. When atoms of Cr on the surface of a piece of stainless steel combine with oxygen, they form a layer of Cr_2O_3 that tightly bonds to the surface and protects the metallic material beneath from further oxidation. This resistance to surface discoloration from corrosion means that the surfaces of these alloys "stain less" than a surface of pure iron.

FIGURE 12.17 Carbon steel is an interstitial alloy of carbon in iron. The fcc form of iron (austenite) that forms at high temperatures can accommodate carbon atoms in its octahedral holes.

Interstitial Alloys

The major difference between iron and any kind of steel is the presence of carbon in steel. Molten iron produced in a blast furnace to which hot carbon has been added may contain up to 5% carbon. When the molten iron cools to its melting point of 1538°C, it crystallizes in a body-centered cubic structure before undergoing a phase transition at around 1390°C to austenite, a form of solid iron made up of face-centered cubic unit cells. The spaces, or *holes*, between iron atoms in austenite can accommodate carbon atoms, forming an **interstitial alloy** (**Figure 12.17**), so named because the carbon atoms occupy spaces, or *interstices*, between the iron atoms.

CONCEPT TEST

Is a substitutional alloy a homogeneous or a heterogeneous alloy, or could it be both? Is an interstitial alloy a homogeneous or a heterogeneous alloy, or could it be both?

Not all interstices in a crystal lattice are equivalent, however. Indeed, holes of two different sizes occur between the atoms in any closest-packed crystal lattice (**Figure 12.18**). The larger holes are surrounded by clusters of six host atoms in the shape of an octahedron and are called *octahedral holes*. The smaller holes are located between clusters of four host atoms and are called *tetrahedral holes*. The data in **Table 12.5** show which holes are more likely to be occupied based on the relative sizes of nonhost and host atoms. The atomic radii of C and Fe are 77 and 126 pm, respectively (see Appendix 3). According to Table 12.5, the ratio 77/126 = 0.61 means that C atoms should fit in the octahedral holes of austenite, but not in the tetrahedral holes.

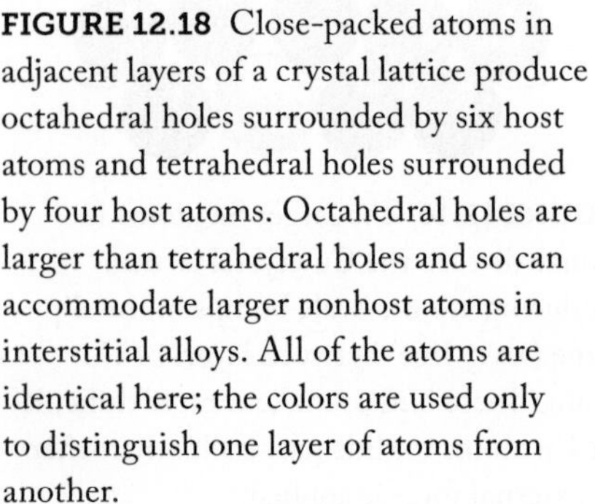

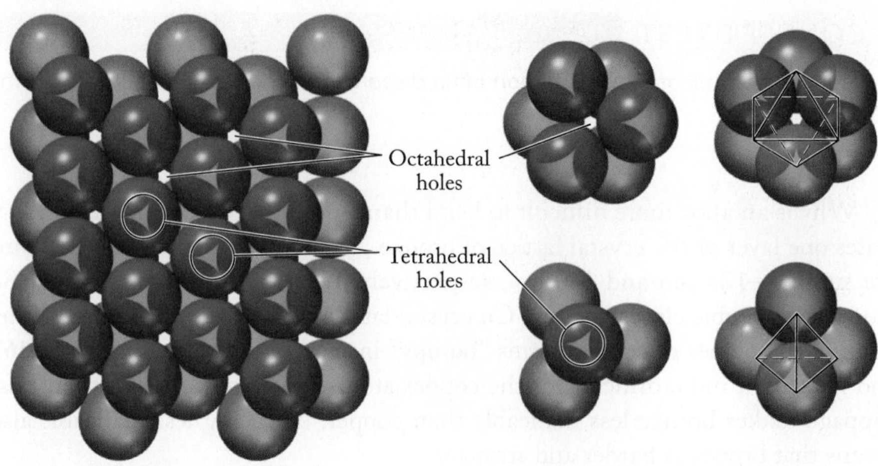

FIGURE 12.18 Close-packed atoms in adjacent layers of a crystal lattice produce octahedral holes surrounded by six host atoms and tetrahedral holes surrounded by four host atoms. Octahedral holes are larger than tetrahedral holes and so can accommodate larger nonhost atoms in interstitial alloys. All of the atoms are identical here; the colors are used only to distinguish one layer of atoms from another.

TABLE 12.5 Atomic Radius Ratios and Location of Nonhost Atoms in Unit Cells of Interstitial Alloys

Lattice Name	Hole Type	Atomic Radius Ratio $r_{nonhost}/r_{host}$ [a]
hcp or ccp	Tetrahedral	0.22–0.41
hcp or ccp	Octahedral	0.41–0.73
Cubic packing	Cubic	0.73–1.00

[a]Radius ratios as predictors of the crystal lattice in crystalline solids are of limited value because atoms are not truly solid spheres with a constant radius; given that the radius of an atom may differ in different compounds, these ranges are approximate.

As austenite (an fcc unit cell) cools to room temperature, it changes into the crystalline solid form of iron called ferrite, which is body-centered cubic. The octahedral holes in ferrite are smaller than those in austenite, and they are too small to accommodate carbon atoms. As a result, much of the carbon precipitates as clusters of carbon atoms, or it may react with iron to form iron carbide (Fe_3C). The clusters of carbon and Fe_3C disrupt ferrite's body-centered cubic lattice and inhibit the host iron atoms from slipping past each other when a stress is applied. This resistance to slippage, which is much like that experienced by the copper atoms in bronze (Figure 12.16), makes steel much harder and stronger than pure iron. In general, the greater the carbon concentration, the stronger the steel. However, there is a trade-off in this relationship; increased strength and hardness come at the cost of increased brittleness (**Table 12.6**).

TABLE 12.6 Effect of Carbon Content on the Properties of Steel

Carbon Content (%)	Designation	Properties	Use
0.05–0.19	Low carbon	Malleable, ductile	Nails, cables
0.20–0.49	Medium carbon	High strength	Construction girders
0.5–3.0	High carbon	Hard but brittle	Cutting tools

SAMPLE EXERCISE 12.2 Predicting the Crystal Structure of a Two-Element Alloy **LO2**

Gold nanoparticles containing 10%–50% platinum show enhanced antibiotic properties in comparison with pure gold or pure platinum nanoparticles. Both elements form cubic closest-packed crystal lattices with face-centered cubic unit cells. Are these gold–platinum alloys substitutional or interstitial alloys?

Collect, Organize, and Analyze Atoms of two or more elements form a substitutional alloy when all the atoms are of similar size (within 15%). The highest $r_{nonhost}/r_{host}$ ratio in Table 12.5 for a ccp lattice, 0.73, means that an interstitial alloy can form only when the radius of the nonhost atoms is *less* than 73% of the radius of the host atoms. The atomic radii given in Appendix 3 are 144 pm for Au and 139 pm for Pt.

Solve The ratio of the atomic radii of Pt to Au is

$$r_{nonhost}/r_{host} = \frac{139 \cancel{\text{pm}}}{144 \cancel{\text{pm}}} = 0.965$$

Based on this $r_{nonhost}/r_{host}$ ratio, the platinum atoms are too big to fit into interstices in the lattice of gold atoms, regardless of whether the holes are tetrahedral or octahedral, so an *interstitial* alloy is impossible. Instead, atoms of Pt can *substitute* for atoms of Au in the Au ccp lattice with some room to spare. Both elements form fcc unit cells (see Figure 12.6), so little disruption to the Au lattice should result from incorporating Pt atoms. Thus, gold and platinum form a *substitutional* alloy.

Think About It Substitution of one sphere for another in the fcc unit cell in Figure 12.9b is least disruptive to the packing of the spheres when the atoms are similar in size, that is, with atomic radii that are within 15% of each other. The radii of Au and Pt are nearly the same size, differing by only 3.5%, so we would expect gold and platinum to form a substitutional alloy.

Practice Exercise The potential for allergic reactions to nickel in nitinol shape memory alloys has spurred the exploration of titanium and niobium alloys. Would you expect niobium (atomic radius 146 pm) to form a substitutional alloy with titanium (atomic radius 147 pm)? With molybdenum (atomic radius 139 pm)?

12.4 Ionic Solids and Salt Crystals

Most of Earth's crust is composed of ionic solids, which consist of monatomic or polyatomic ions held together by ionic bonds. Most of these solids are crystalline. The simplest crystal structures are those of binary salts, such as NaCl (**Figure 12.19**). The cubic shape of large NaCl crystals results from the cubic shape of the NaCl unit cell. We can describe the unit cell of NaCl (Figure 12.19b, c) as a face-centered cubic arrangement of Cl^- ions at the corners and in the center of each face, with the smaller Na^+ ions occupying the 12 octahedral holes along the edges of the unit cell and the single octahedral hole in the middle of the cell. The ordered arrangement in Figure 12.19 maximizes the coulombic forces of attraction between oppositely charged particles and minimizes the repulsions between similarly charged ions. As a result, the potential energy of the crystal structure is minimized.

In Section 12.3 we used atomic radius ratios to determine the location of atoms in a crystal lattice. The same ratios that applied to interstitial alloys guide us in understanding the structures of binary ionic compounds. Rather than examine the ratio of nonhost radii to host radii, however, we consider the ratio of the

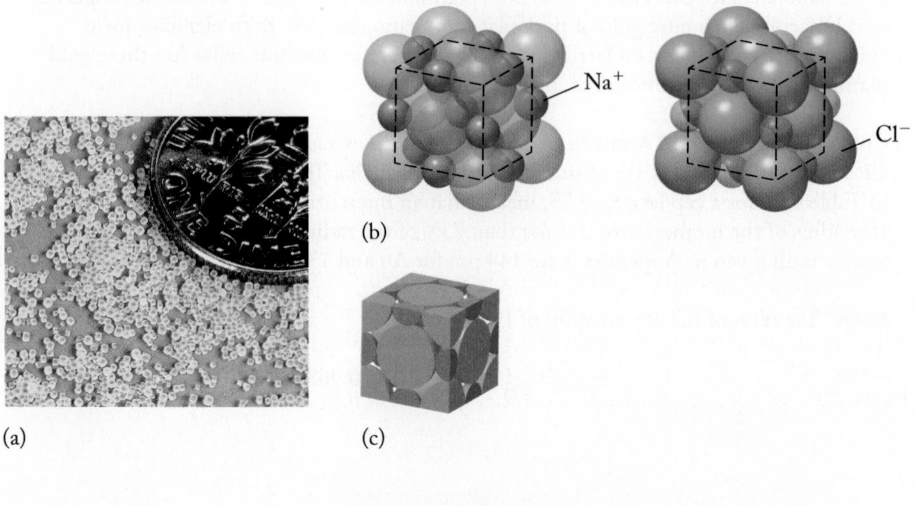

FIGURE 12.19 (a) Sodium chloride forms cubic crystals that grow to various sizes. (b) A NaCl crystal lattice is based on cubic closest packing: an fcc unit cell made of Cl^- ions, showing the Na^+ ions in the octahedral holes. (c) Cutaway view of the unit cell in part (b)

smaller ion to the larger ion. In the unit cell of NaCl (Figure 12.19b), for example, the smaller Na^+ cations occupy the interstices in the cubic closest-packed lattice of Cl^- anions. The Na^+ ions fit better into the octahedral holes because the radius ratio of Na^+ to Cl^- is

$$\frac{r_{\text{cation}}}{r_{\text{anion}}} = \frac{102 \text{ pm}}{181 \text{ pm}} = 0.564$$

The radius ratio 0.564 is too large for Na^+ to occupy a tetrahedral hole but well within the range for occupying an octahedral hole.

Let's take an inventory of the ions in a unit cell of NaCl. Like the metal atoms in the fcc unit cell in Figure 12.9b, an isolated unit cell contains portions of 14 Cl^- ions: one at each corner and one in each of the six faces of the cube. Therefore, accounting for partial ions gives us a total of four Cl^- ions in the unit cell in Figure 12.19c. To count the Na^+ ions, note that one Na^+ ion occupies the central octahedral hole, and one Na^+ ion fits into each of the 12 octahedral holes along the edges of the cell. Because each Na^+ ion along an edge is shared by four unit cells, only one-fourth of each Na^+ ion on an edge is in each cell (see Figure 12.10d). Only the Na^+ in the center belongs completely to the unit cell. As a result, the total number of Na^+ ions in the unit cell is

$$(12 \times \tfrac{1}{4}) + 1 = 4 \text{ Na}^+ \text{ ions}$$

The ratio of Na^+ to Cl^- ions in the unit cell is 4:4, consistent with the chemical formula NaCl. Because the four Na^+ ions occupy all the octahedral holes in the unit cell, the result of this calculation also means that each fcc unit cell contains the equivalent of four octahedral holes.

Note in Figure 12.19 that adjacent Cl^- ions along any face diagonal do *not* touch each other the way they do in Figure 12.9b because the Cl^- ions have to spread out a little to accommodate the Na^+ ions in the octahedral holes. Sodium ions and chloride ions touch along each edge of the unit cell, however, which means that each Na^+ ion touches six Cl^- ions and each Cl^- ion touches six Na^+ ions. This arrangement of positive and negative ions is common enough among binary ionic compounds to be assigned its own name: the *rock salt structure*.

CONNECTION Periodic trends in atomic radii are discussed in Section 7.10, and the sizes of common cations and anions are compared to their neutral parent atoms in Figure 11.1.

CONNECTION Electrostatic attractions, ion–ion forces, and potential energy in ionic compounds are discussed in Section 8.1.

SAMPLE EXERCISE 12.3 Determining the Formula of a Compound from Its Unit Cell **LO3**

Cesium chloride has the solid state structure shown in **Figure 12.20**, which is called the *cesium chloride* structure. The cation and anion have ionic radii of 170 and 181 pm, respectively. (a) Show that the structure in Figure 12.20 is consistent with the empirical formula of cesium chloride and (b) determine which type of hole the cation is found in.

Collect and Organize We are asked to determine the empirical formula of cesium chloride from its crystal structure using the information in Table 12.3. We are given the ionic radii of both ions and asked to assign the cation to one of the three types of holes identified in Table 12.5.

Analyze The empiricial formula of cesium chloride is CsCl. According to Figure 12.20, CsCl has a simple cubic unit cell that contains eight corner ions (chloride) arranged with the cesium ion in the center. The unit cell for CsCl resembles a bcc unit cell (e.g., tantalum in Figure 12.8) except that Cs^+ is found in the center of the unit cell rather than Cl^-. To determine the empirical formula of cesium chloride, we use Table 12.3 to

FIGURE 12.20 The unit cell of CsCl with Cs^+ in the cubic hole formed by eight Cl^- anions.

calculate the contribution each ion makes to the unit cell. The radius ratios applied in Section 12.3 to metal alloys can be used here to identify the type of hole in the simple cubic lattice of chloride ions suitable for the cation.

Solve

a. Table 12.3 indicates that particles (atoms or ions) on the corners of a cubic unit cell each contribute $\frac{1}{8}$ to the unit cell. The cesium ion lies in the middle of the unit cell and only touches eight corner atoms. Therefore, the unit cell consists of $\frac{1}{8}$(8 corner ions) = 1 Cl^- ion and 1 Cs^+ ion, which gives the empirical formula CsCl.

b. The cation:anion ratio for CsCl is

$$\frac{r_{cation}}{r_{anion}} = \frac{170 \text{ pm}}{181 \text{ pm}} = 0.939$$

Table 12.5 indicates that Cs^+ occupies a cubic hole in a simple cubic lattice of chloride ions.

Think About It From Section 2.6 we know that the empirical formula of cesium chloride is CsCl, so the formula of any unit cell for this salt must contain a 1:1 ratio of equivalent cations and anion. Cesium chloride contains the same ratio of cations to anions as sodium chloride (NaCl), but it adopts a different crystal structure because Cs^+ ions are 67% bigger than Na^+ ions and no longer fit into the octahedral holes in an fcc lattice of Cl^- ions. Only the cubic hole is big enough to accommodate the larger cation. The cesium chloride structure is not the same as a body-centered cubic unit cell because it does not contain identical particles.

Practice Exercise Thallium and bromine form two binary compounds, $TlBr_3$ and TlBr. Which compound crystallizes with the CsCl structure? Why doesn't this compound crystallize in the rock salt structure? The ionic radii of Tl^+ and Tl^{3+} are 159 and 89 pm, respectively.

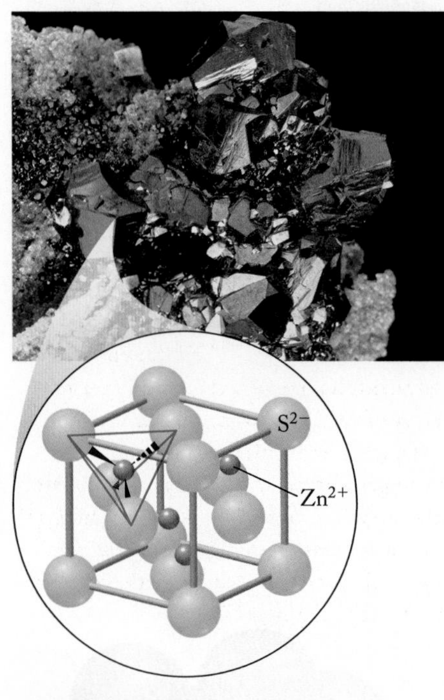

FIGURE 12.21 Many crystals of the mineral sphalerite (ZnS), like the largest ones in this photograph, have a tetrahedral shape. The crystal lattice of sphalerite is based on an fcc unit cell of S^{2-} ions with Zn^{2+} ions in four of the eight tetrahedral holes. In the expanded view of the sphalerite unit cell, each Zn^{2+} ion is in a tetrahedral hole formed by one corner S^{2-} ion and three face-centered S^{2-} ions.

In other binary ionic solids, the smaller ion is small enough to fit into the tetrahedral holes formed by the larger ions. In the unit cell of the mineral sphalerite (zinc sulfide), for example, the S^{2-} anions (ionic radius 184 pm) are arranged in an fcc lattice (**Figure 12.21**), and half of the eight tetrahedral holes inside the cell are occupied by Zn^{2+} cations (74 pm). Therefore, the unit cell contains four Zn^{2+} ions that balance the charges on the four S^{2-} ions. This pattern of half-filled tetrahedral holes in an fcc unit cell is sometimes called the *sphalerite structure.* Compounds containing zinc, cadmium, or mercury cations with a 2+ charge and heavier anions of the group 16 elements with a 2− charge (i.e., S^{2-}, Se^{2-}, and Te^{2-}) also have a sphalerite structure.

The crystal structure of the mineral fluorite (CaF_2), shown in **Figure 12.22**, is based on an fcc unit cell of smaller Ca^{2+} ions at the eight cube corners and six face centers, with all eight tetrahedral holes formed by neighboring Ca^{2+} ions filled by larger F^- ions. Because the unit cell has a total of four Ca^{2+} ions and the eight F^- ions are all completely inside the cell, this arrangement satisfies the 1:2 mole ratio of Ca^{2+} ions to F^- ions. This structure is so common that it too has its own name: the *fluorite structure.* Other compounds having this structure are SrF_2, $BaCl_2$, and PbF_2.

Some compounds in which the cation-to-anion mole ratio is 2:1 have an *antifluorite structure.* In the crystal lattices of these compounds, which include Li_2O and K_2S, the smaller cations occupy the tetrahedral holes in an fcc unit cell formed by cubic closest packing of the larger anions.

SAMPLE EXERCISE 12.4 Calculating an Ionic Radius and Density from a Unit Cell Dimension LO3

The unit cell of lithium chloride (LiCl) contains an fcc arrangement of Cl^- ions (**Figure 12.23**). In LiCl, the Li^+ cations (radius 76 pm) are small enough to allow adjacent Cl^- ions to touch along any face diagonal.

a. If the edge length (ℓ) of the LiCl fcc cell is 513 pm, what is the radius of the Cl^- ion?
b. Use that value to predict the type of hole the Li^+ ion occupies.
c. What is the density of LiCl at 25°C?

Collect and Organize We are given the edge length (513 pm), the radius of the Li^+ cation (76 pm), and the type of unit cell for LiCl (fcc). We are asked to calculate the ionic radius of the Cl^- ion, predict in which type of hole the Li^+ is found, and calculate the density of LiCl.

Analyze (a) The unit cell is an fcc array of Cl^- ions that touch along the face diagonal. Equation 12.2 relates the edge length (ℓ) of an fcc cell to the radius (r) of the atoms or ions that touch along its diagonals: $r = 0.3536\ell$. (b) Once we know the radius of the Cl^- ion, the radius ratio of Li^+ to Cl^- ions and Table 12.5 allow us to predict which type of hole Li^+ occupies. (c) LiCl has an fcc unit cell, which means that there are four Cl^- ions and four Li^+ ions in the cell. The molar masses of these ions are 35.45 and 6.941 g/mol, respectively. As in Sample Exercise 12.1, we need to convert from molar masses to the masses of individual particles by dividing by Avogadro's number. Density is the ratio of mass to volume, and the volume of a cubic cell is the cube of its edge length: $V = \ell^3$.

Solve

a. Substituting the edge length into Equation 12.2, we calculate the radius of Cl^-:

$$r = 0.3536 \times 513 \text{ pm} = 181 \text{ pm}$$

b. The ratio of the radius of a Li^+ ion to the radius of a Cl^- ion is

$$76 \text{ pm}/181 \text{ pm} = 0.42$$

According to Table 12.5, the Li^+ cations occupy octahedral holes in the lattice formed by the larger Cl^- anions.

c. Calculating the mass of four Cl^- ions:

$$m = \frac{35.45 \text{ g Cl}^-}{1 \cancel{\text{mol Cl}^-}} \times \frac{1 \cancel{\text{mol Cl}^-}}{6.022 \times 10^{23} \cancel{\text{ions Cl}^-}} \times 4 \cancel{\text{ions Cl}^-}$$
$$= 2.355 \times 10^{-22} \text{ g Cl}^-$$

The mass of four Li^+ ions is

$$m = \frac{6.941 \text{ g Li}^+}{1 \cancel{\text{mol Li}^+}} \times \frac{1 \cancel{\text{mol Li}^+}}{6.022 \times 10^{23} \cancel{\text{ions Li}^+}} \times 4 \text{ ions Li}^+$$
$$= 0.4610 \times 10^{-22} \text{ g Li}^+$$

Combining the masses of the two kinds of ions in the unit cell:

$$2.355 \times 10^{-22} \text{ g} + 0.4610 \times 10^{-22} \text{ g} = 2.816 \times 10^{-22} \text{ g}$$

The volume of the cell in cubic centimeters is

$$V = \ell^3 = (513 \cancel{\text{pm}})^3 \times \frac{(10^{-10} \text{ cm})^3}{(1 \cancel{\text{pm}})^3} = 1.35 \times 10^{-22} \text{ cm}^3$$

Taking the ratio of mass to volume, we have

$$d = \frac{m}{V} = \frac{2.816 \times 10^{-22} \text{ g}}{1.35 \times 10^{-22} \text{ cm}^3} = 2.09 \text{ g/cm}^3$$

FIGURE 12.22 The mineral fluorite (CaF_2) forms cubic crystals. The crystal lattice of CaF_2 is based on an fcc array of Ca^{2+} ions, with F^- ions occupying all eight tetrahedral holes. Because they are bigger than Ca^{2+} ions, the F^- ions do not fit in the tetrahedral holes of a cubic closest-packed array of Ca^{2+} ions. Instead, the Ca^{2+} ions, while maintaining the fcc unit cell arrangement, spread out to accommodate the larger F^- ions. Note how adjacent Ca^{2+} ions along any face diagonal do *not* touch each other the way they do in the ideal fcc unit cell in Figure 12.9b.

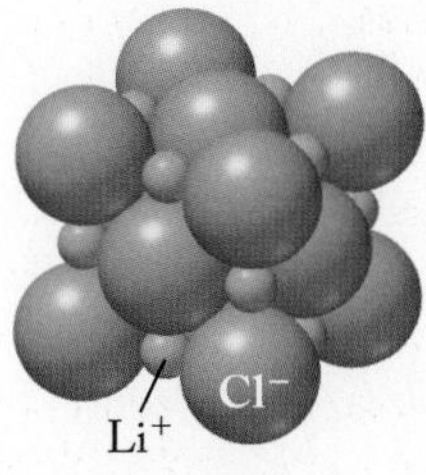

FIGURE 12.23 The face-centered cubic unit cell of LiCl.

Think About It The average ionic radius value in Figure 11.1 for Cl^- ions is 181 pm, so our calculation is correct. Figure 12.23 shows the structure of LiCl to be similar to the rock salt structure of NaCl in Figure 12.19b. The calculated density for LiCl is consistent with the measured density, $d = 2.068\ g/cm^3$, and the observation that most minerals are more dense than water but less dense than common metals.

Practice Exercise Assuming the Cl^- radius in NaCl is also 181 pm, what is the radius of the Na^+ ion in NaCl if the edge length of the NaCl unit cell is 564 pm but the anions do *not* touch along any face diagonal? What is the density of NaCl?

12.5 Allotropes of Carbon

CONNECTION Allotropes are structurally different forms of the same physical state of an element (Section 8.4).

CHEMTOUR

Allotropes of Carbon

CONNECTION Hybridization and σ bonds are described in Section 9.4.

According to the periodic table in Figure 12.6, the unit cells of silicon, germanium, and tin are "diamond." Carbon is a nonmetal, but the specific arrangement of carbon atoms in diamond (**Figure 12.24a**) is also found in some metallic materials. In addition, some diamonds may include metal impurities in the crystal lattice that give rise to distinctive colors.

Diamond is one of three allotropes of carbon, the other two being graphite (**Figure 12.24b**) and fullerenes (**Figure 12.25**). Diamond is classified as a crystalline covalent network solid because it consists of an extended three-dimensional network of atoms joined by covalent bonds. Each carbon atom in diamond bonds by overlapping one of its sp^3 hybrid orbitals with an sp^3 hybrid orbital in each of four neighboring carbon atoms, creating a network of carbon tetrahedra. The atoms in these tetrahedra are connected by localized σ bonds, making diamond a poor electrical conductor. The σ-bond network is extremely rigid, making diamond the hardest natural material known. The atoms of other group 14 elements,

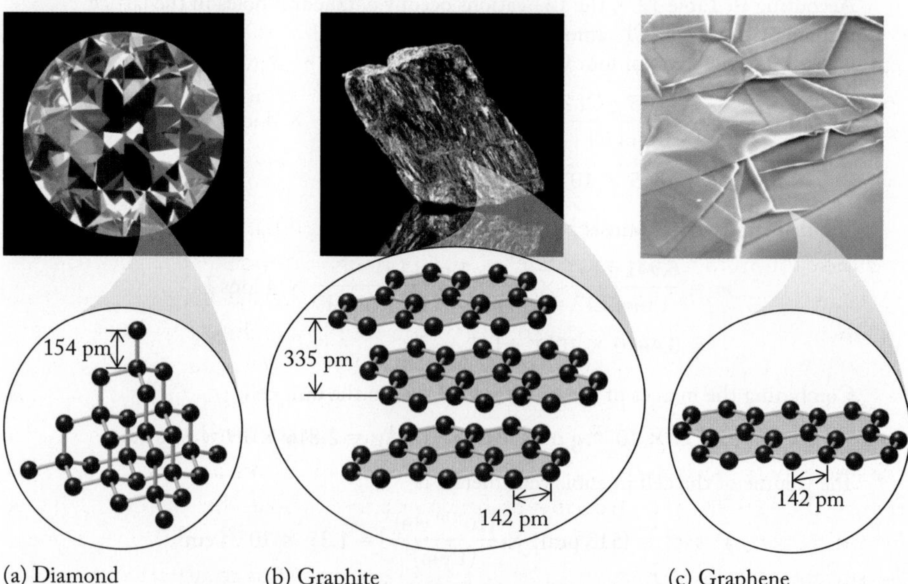

FIGURE 12.24 Two of carbon's three allotropes. (a) Diamond is a three-dimensional covalent network solid made of carbon atoms, each connected by σ bonds to four adjacent carbon atoms. (b) Graphite is a collection of layers of carbon atoms connected by σ bonds and delocalized π bonds. (c) Graphene is a single layer of C atoms arranged as the atoms are in a layer of graphite.

particularly silicon, germanium, and tin, also form covalent network solids based on the diamond crystal lattice.

Natural diamond forms from graphite under intense heat (>1700 K) and pressure (>50,000 atm) deep in Earth. Industrial diamonds are synthesized at high temperatures and pressures from graphite or any other source rich in carbon. Synthetic diamonds are used as abrasives and for coating the tips and edges of cutting tools. Diamond has the highest thermal conductivity of any natural substance (five times higher than copper and silver, the most thermally conductive metals), so tools made from diamond do not become overheated. Thin diamond films grown from methane (CH_4) and hydrogen have been used to reduce wear and extend the lifetime of prosthetic implants.

Graphite is by far the most abundant allotrope of carbon. It is a covalent network solid, too, which is frequently the principal ingredient in soot and smoke and is used to make pencils, lubricants, and gunpowder. Graphite contains sheets of carbon atoms in which each atom is connected by sp^2 orbitals to a like orbital in each of three neighboring carbon atoms, forming a two-dimensional covalent network of six-membered rings (Figure 12.24b). Each carbon–carbon bond is 142 pm, which is shorter than the C—C σ bond in diamond (154 pm). Overlapping unhybridized p orbitals on the carbon atoms form a network of π bonds that are delocalized across the plane defined by the rings. The mobility of these delocalized electrons makes graphite a conductor of electricity.

As shown in Figure 12.24b, the two-dimensional sheets in graphite are 335 pm apart. This distance is much too long to be a covalent bonding distance. Instead, the sheets are held together by dispersion forces only. These relatively weak interactions allow adjacent sheets to slide past each other, making graphite soft, flexible, and a good lubricant. In 2004, researchers in the United Kingdom and Russia found that they could isolate just a single layer of carbon atoms from graphite with a piece of adhesive tape. This material, called graphene (**Figure 12.24c**), behaves as a semiconductor.

CONCEPT TEST

The diamond form of carbon is a semiconductor with a much larger band gap than that of silicon. With what elements might you choose to dope diamonds to form an n-type semiconductor?

The third allotrope of carbon was discovered in the 1980s. Networks of five- and six-atom carbon rings form molecules of 60, 70, or more carbon atoms that look like miniature soccer balls (Figure 12.25a). They are called *fullerenes* because their shape resembles the geodesic domes designed by the American architect R. Buckminster Fuller (1895–1983). Many chemists call them *buckyballs* for the same reason. In terms of both size and properties, buckyballs are too small to be classified as covalent network solids but too large to be molecular solids (discussed below). They fall in an ambiguous zone between small molecules and large networks and are classified as *clusters*. The 70–100 pm diameter of fullerenes means that they can fit into the three-dimensional structure of enzymes like HIV protease, inhibiting its activity. The surface of fullerenes can also be chemically modified to transport drug molecules into cells. Some fullerenes stretch the meaning of the word cluster. These include structures known as nanotubes, so named because they are 1–2 nm in diameter (Figure 12.25b). Nanotubes have been fabricated that are 20 cm long. Like graphene, carbon nanotubes exhibit semiconductor properties, but they are also extremely strong and have been used to strengthen fibers in sports gear.

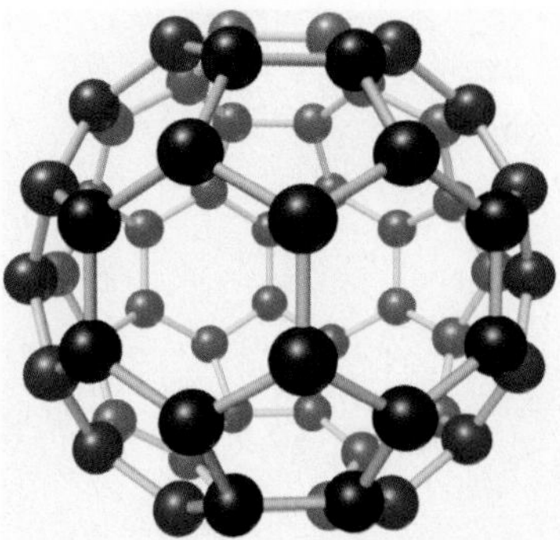

(a)

(b)

FIGURE 12.25 Some solids are described as clusters, a category between covalent network solids and molecular solids. Fullerenes, the third allotrope of carbon, are clusters. (a) One of them, buckminsterfullerene (C_{60}), is made up of 60 sp^2-hybridized carbon atoms. Both five- and six-membered rings are required to construct the nearly spherical molecule. (b) Nanotubes have diameters approximately equal to the diameter of a buckyball.

CONNECTION We introduced dispersion forces between molecules in Section 10.2. We discussed semiconductors in Section 9.6.

FIGURE 12.26 Items made of synthetic polymers are ubiquitous in modern society. They are also ubiquitous in our trash as shown in this photo of the Great Pacific Garbage Patch of waste plastic of all kinds that circulates with the currents in the Pacific Ocean.

12.6 Polymers

As part of our discussion of covalent bonding in Chapters 8 and 9, we studied small organic compounds containing carbon, hydrogen, nitrogen, and oxygen. These compounds had molar masses less than 1000 g/mol. Much larger organic compounds, with molar masses up to and exceeding 1,000,000 g/mol, represent a category of materials called **polymers**. The use of polymers in consumer products is underscored by the Great Pacific Garbage Patch (**Figure 12.26**), a vast field of largely nonbiodegradable debris that circulates in response to ocean currents. In between small molecules and polymers (also called **macromolecules**) are midsize molecules called *oligomers*. The root word *meros* is Greek for "part" or "unit," so *polymer* literally means "many units" and *oligomer* means "a few units." An example of a polymer from Chapter 4 is poly(vinyl chloride), used to make drain pipes and other building materials.

Polymers are composed of small structural units called **monomers**. Polymers may be formed from more than one type of monomer unit, they may have more than one functional group, and they may have shapes other than long chains. However, the single feature that distinguishes them as a class is their large size. In Section 12.6 we explore the preparation and properties of polymers, focusing on materials with biomedical applications.

CHEMTOUR
Polymers

CONNECTION We introduced many organic molecules in Chapters 4 through 11. These included acetic acid; methane, ethane, and other hydrocarbon fuels; acetylene used in welding; and ethanol and other alcohols.

CONNECTION The molecular structure and bonding in ethylene ($CH_2{=}CH_2$) is described in Section 9.5.

Polymers of Alkenes

Some widely used polymeric alkanes are produced industrially from small alkenes. The alkene polymer with the simplest structure is linear polyethylene (PE), produced from ethylene at high temperature and pressure:

$$n\ CH_2{=}CH_2 \rightarrow +\!\!\!\!-CH_2{-}CH_2\!-\!\!\!\!+_n \tag{12.4}$$

Polyethylene is a **homopolymer**, which means it is composed of only one type of monomer. Its condensed structure is $CH_3(CH_2)_nCH_3$, but there are so many more methylene groups than methyl groups that the structure is frequently written as $+\!\!\!\!-CH_2CH_2\!-\!\!\!\!+_n$ to highlight the structure and composition of the monomer. Polyethylene is also an example of an **addition polymer** because it is synthesized by adding many molecules together to form the polymer chain without the loss of any atoms or molecules. Correspondingly, the individual reactions that create an addition polymer are called **addition reactions**, in which two reactants combine to form one product.

In most products made of polyethylene, n is a very large number, ranging from 1000 to almost 1 million. Polyethylene has a wide range of properties that depend on the value of n and on whether the polymer chains are straight or branched. In low-density polyethylene (LDPE)—a stretchable, soft plastic used in films and wrappers—n ranges from 350 to 3500 and the chains are branched (**Figure 12.27**). When the bagger at the grocery store asks, "Paper or plastic?" the plastic in question is LDPE.

When the molar mass of a PE polymer is between 100,000 and 500,000 and the chains are straight, the polymer has physical properties that differ from those of LDPE grocery bags. This straight-chain polymer—a rigid, translucent solid called high-density polyethylene (HDPE)—is used in milk containers, electrical insulation, and toys.

Why are the properties of straight-chain HDPE (rigid, tough) so different from those of branched-chain LDPE (stretchable, soft)? Think of the branched polymer (Figure 12.27) as a tree branch with lots of smaller branches attached

Branched-chain LDPE

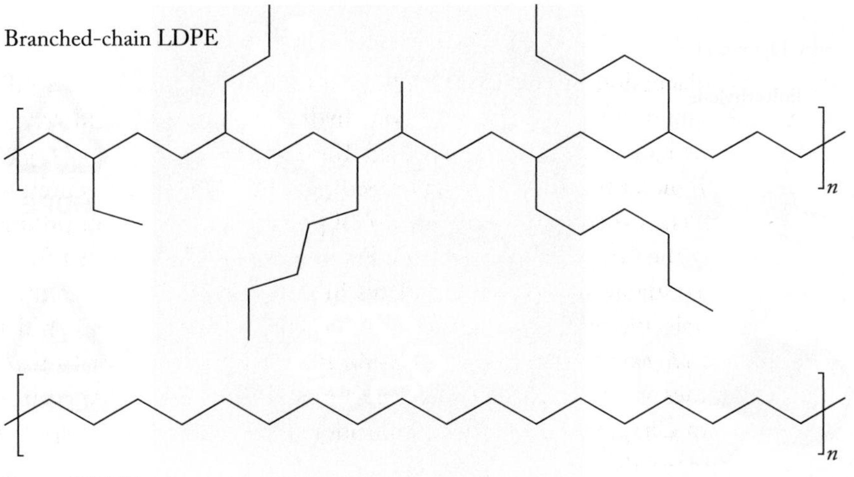

Linear HDPE

FIGURE 12.27 Low-density polyethylene (LDPE) consists of branched chains, but high-density polyethylene (HDPE) consists of straight chains. Efficient stacking—and thus high density—is possible in large polymer molecules of HDPE but not in large polymer molecules of LDPE.

CONNECTION The carbon-skeleton structures used to illustrate large molecules like those in Figure 12.27 were described in Section 9.5.

FIGURE 12.28 Devices used in knee-replacement surgery often incorporate a pad made of an ultrahigh molecular weight polyethylene (UHMWPE). UHMWPE is a very tough material. It is highly resistant to wear and it cushions the artificial joint. It consists of linear molecules with an average molar mass over 3 million.

to it. The polymer has three dimensions, so it can have branches that come out of the plane of the paper. In contrast, the straight-chain polymer is like a long, straight pole. Suppose you had a pile of 100 tree branches and a pile of 100 poles, and your task was to stack each pile into the smallest possible volume to fit into a truck. You can certainly pile the branches on top of one another, but they will not fit together neatly, and probably the best you can do is to make the pile a bit more compact. In contrast, you can stack the poles into a very compact pile.

The same situation arises with the branched and linear molecules of polyethylene. Branched-chain LDPE has a low density because the molecules do not line up neatly and so they stack less efficiently than do the molecules of HDPE. This makes LDPE more deformable and softer, whereas HDPE is more rigid and even has some regions that are crystalline because the packing is so uniform. Ultrahigh molecular weight polyethylene (UHMWPE; $n > 100{,}000$) is an even tougher material, because not only are the molecules straight, but they are also significantly *larger* than the molecules in HDPE. UHMWPE is used as a coating on some artificial ball-and-socket joints (**Figure 12.28**) because it is extremely resistant to abrasion and makes the joints last longer. The different forms of polyethylene illustrate how the size and shape of its molecules affect the physical properties of a material. The ethylene used to make polyethylene has traditionally come from petroleum but the availability of ethanol derived from biomass allows for a renewable source of this valuable monomer.

CONCEPT TEST

During recycling, articles made from LDPE are separated from those made from HDPE (**Figure 12.29**). Why?

Chemical composition also plays a role in determining physical properties. If the hydrogen atoms in PE are all replaced with fluorine atoms, the resultant polymer is chemically very unreactive, can withstand high temperatures, and has a

polymer (or **macromolecule**) a very large molecule with high molar mass formed by bonding many small molecules of low molecular mass.

monomer a small molecule that bonds with others like it to form polymers.

homopolymer a polymer composed of only one kind of monomer.

addition polymer a polymer formed by an addition reaction.

addition reaction a reaction in which two reactants couple to form one product without the loss of any atoms or molecules.

FIGURE 12.29 Products made of polyethylene bear a recycle symbol that identifies them as straight-chain molecules (high density) or branched-chain molecules (low density).

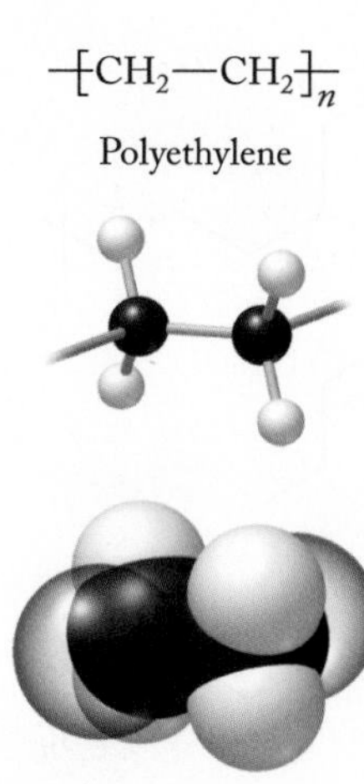

very low coefficient of friction, which means other things do not stick to it. This polymer is Teflon, $\text{+}CF_2CF_2\text{+}_n$ (**Figure 12.30**), most familiar for its use as a nonstick surface in cookware. However, Teflon tubing is also used in the grafts inserted into small-diameter blood vessels during vascular surgery on limbs.

(a)

(b)

FIGURE 12.30 (a) The monomer (C_2F_4) found in poly(tetrafluoroethylene) (Teflon). (b) Teflon-coated cookware provides a nonstick surface.

SAMPLE EXERCISE 12.5 Identifying Monomers — LO4

Polypropylene, $\text{+}CH_2CH(CH_3)\text{+}_n$, is an addition polymer used in the manufacture of fabrics, ropes, and other materials (**Figure 12.31**). Draw the condensed molecular structure of the monomer used to prepare polypropylene.

Collect and Organize We are given a condensed structure of a polymer and asked to identify the monomer used in its preparation. We know that polypropylene is an addition polymer, so the monomer must be an alkene.

Analyze To understand the relationship between the polymer and the monomer from which it is made, let's look at Equation 12.5, which is Equation 12.4 in the reverse direction:

$$\text{+}CH_2\text{—}CH_2)\text{+}_n \rightarrow n\, CH_2\text{=}CH_2 \qquad (12.5)$$

Breaking the blue bonds in Equation 12.5 and making the red carbon–carbon single bond a double bond illustrates the relationship between polyethylene and ethylene, the alkene monomer. We need to apply a similar analysis to polypropylene.

Solve The relationship between polypropylene and its monomer is illustrated by Equation 12.6:

$$\text{+}CH_2\text{—}\underset{\displaystyle CH_3}{\underset{|}{CH}}\text{+}_n \rightarrow n\, CH_2\text{=}CH(CH_3) \qquad (12.6)$$

Breaking the two blue bonds and making the red C—C bond a C═C double bond yields the alkene shown in **Figure 12.32**. This three-carbon alkene has the name propene.

Think About It The structural difference between polypropylene and polyethylene is the presence of a $-CH_3$ group bonded to one of the two sp^3 C atoms in polypropylene instead of an H atom in polyethylene. The $-CH_3$ group appears in a similar location in the propene monomer rather than in the polymer backbone making our answer reasonable.

Practice Exercise Draw the condensed molecular structure of the monomer used to make poly(methyl methacrylate) (PMMA), a polymer used in shatterproof transparent plastic that can be used in place of glass:

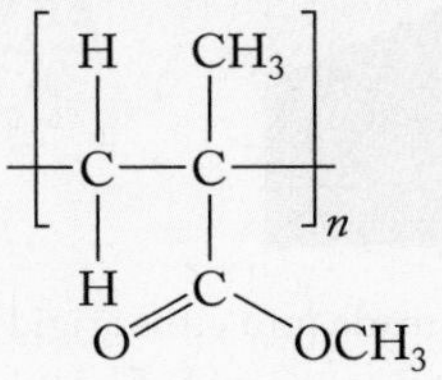

FIGURE 12.31 Polypropylene is a common material for furniture, containers, clothing, lighting fixtures, and even objects of art. In addition to being moldable into many shapes, polypropylene is a good thermal insulator and does not absorb water easily.

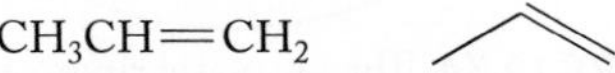

FIGURE 12.32 The monomer propene is polymerized to make polypropylene.

Polymers Containing Aromatic Rings

Benzene rings are flat molecules (see Figure 9.35 in Section 9.5). Replacing one hydrogen atom in benzene with a **vinyl group**, the $CH_2{=}CH-$ subunit, produces styrene. Polymerizing styrene produces polystyrene (PS) (**Figure 12.33**). Polystyrene molecules tend to stack neatly (**Figure 12.34**), which gives rise to useful properties in materials that incorporate benzene rings and other aromatic compounds.

Solid PS is a transparent, colorless, hard, inflexible plastic. In this form it is used for the handles of some disposable razors and plastic cutlery. A more common form of PS, however, is the *expanded solid* made by blowing carbon dioxide gas or pentane gas into molten polystyrene, which then expands and retains voids in its structure when it solidifies. One form of this expanded PS is the familiar material of coffee cups and takeout food containers (**Figure 12.35**).

The difference in properties between transparent, colorless, inflexible nonexpanded PS and the opaque, white, pliable material can be explained by considering the role of the aromatic ring in aligning the polymer chains. Branches in the chains have the same effect that we saw in polyethylene, but the aromatic rings and their tendency to stack (Figure 12.34) provide additional interactions that make chain alignment more favorable energetically. The aromatic rings along two neighboring chains can stack, and this stacking makes nonexpanded PS rigid. When the chains are blown apart by a gas, the stacking is disrupted and the

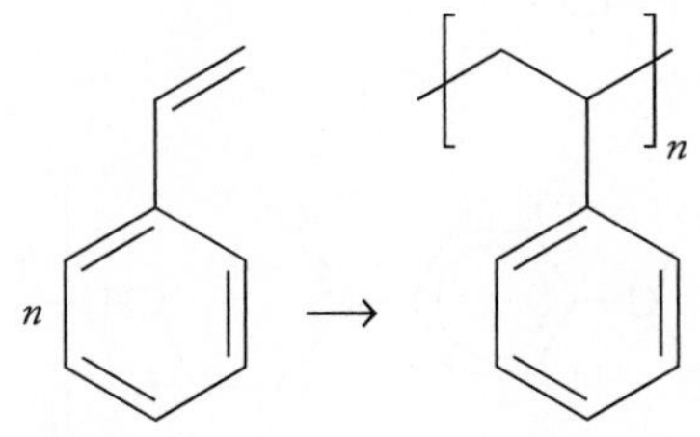

Styrene → Polystyrene

FIGURE 12.33 The monomer styrene is polymerized to make polystyrene.

CONNECTION The molecular structure and bonding in benzene (C_6H_6) was introduced in Section 8.4. We explored its shape and defined it as an aromatic compound in Section 9.5.

vinyl group the subgroup $CH_2{=}CH-$.

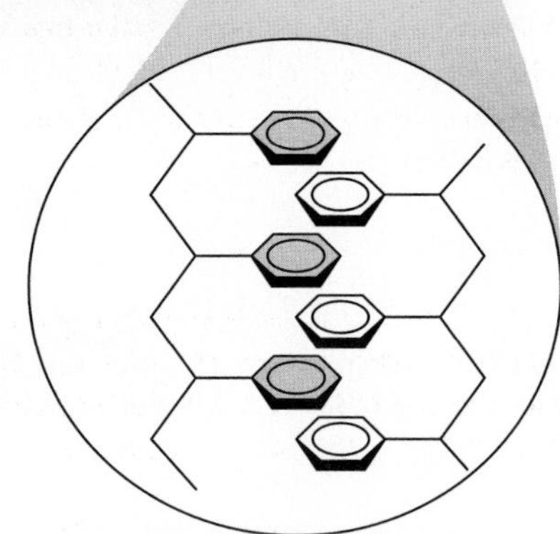

FIGURE 12.34 The aromatic rings on neighboring chains in polystyrene stack together and provide strength to the material.

FIGURE 12.35 Polystyrene can be made into either rigid or foamed products. In its nonexpanded form, it is rigid and strong, suitable for making such products as plastic knives, forks, and spoons. In its expanded form, it is used in takeout food containers, packing materials, and thermal insulation in buildings.

chains open to form cavities that fill with air, making expanded PS a good thermal insulator and packing material (Figure 12.35).

Polymers of Alcohols and Ethers

More than 400,000 tons of the addition polymer poly(vinyl alcohol) (PVAL) are produced annually in the United States. It is used in fibers, adhesives, and materials known as sizing, which changes the surface properties of textiles and paper to make them smooth, less porous, and less able to absorb liquids. Because the polymer chains in PVAL are studded with –OH groups (**Figure 12.36a**), the surface of the polymer chains in PVAL is very polar and very water-like, and hydrocarbon solvents that are not soluble in water do not penetrate PVAL barriers. PVAL is the material of choice for laboratory gloves that are resistant to organic solvents.

PVAL is also impenetrable to carbon dioxide, and this property has led to its use in soda bottles, in which it is blended with the polymer poly(ethylene terephthalate) (PETE; **Figure 12.36b**). The two polymers do not mix but rather separate into layers (**Figure 12.36c**). The PETE makes the bottle strong enough to bear pressure changes that result from temperature changes and survive the impact of falling off tables. Also, CO_2, the dissolved gas that makes soda fizz, passes readily through PETE but not through the PVAL layers, with the result that the soda does not go flat. Polymers with different properties are frequently combined in this way to create new materials with desired properties.

The blend of PVAL and PETE in early soda bottles was a physical mixture of the two polymers. New materials can also be made by reacting different monomer units in one polymer molecule. This type of molecule is called a **copolymer** when two different monomers are reacted and a **heteropolymer** when three or more different monomers are reacted. One example of an addition copolymer is a material called EVAL, made from ethylene and vinyl acetate (**Figure 12.37**). EVAL is used in food wrappings when the preservation of aroma and flavor is required. Food usually deteriorates in the presence of oxygen, and packages made of EVAL provide an excellent barrier to the entry of oxygen while retaining the flavor and fragrance of the packaged food.

Commercially important polymers made from ethers include poly(ethylene glycol) (PEG) and poly(ethylene oxide) (PEO), which contain the same subunit

$$\left[CH_2CH(OH) \right]_n$$

PVAL

(a)

$$\left[O{-}C(=O){-}C_6H_4{-}C(=O){-}O{-}CH_2{-}CH_2 \right]_n$$

Repeating unit in PETE

(b)

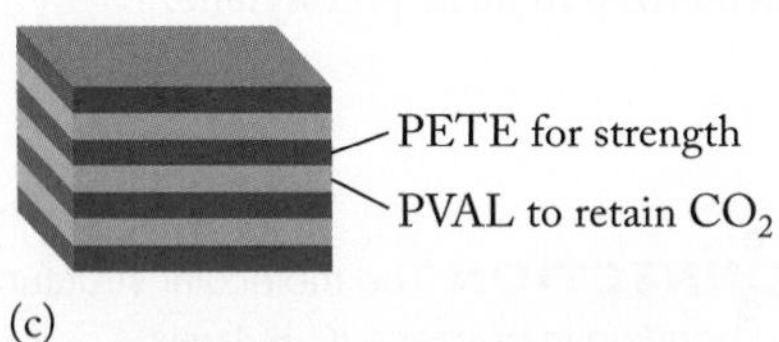

(c)

FIGURE 12.36 (a) Poly(vinyl alcohol) (PVAL). (b) The repeating unit in poly(ethylene terephthalate) (PETE). (c) Layers of the polymers PVAL and PETE are used to make soda bottles. PETE makes the bottle strong, and PVAL keeps the carbon dioxide from leaking out.

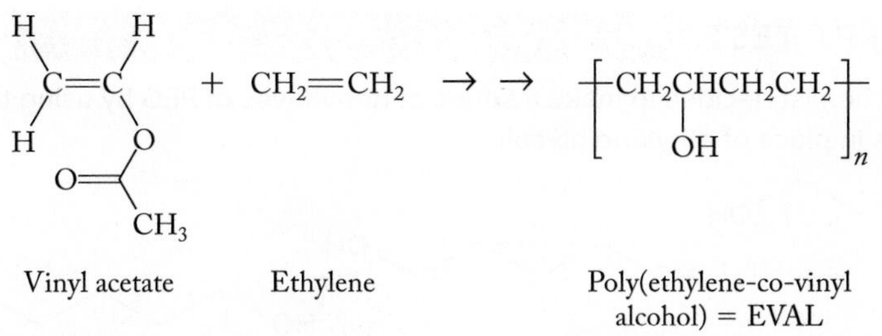

$$\mathrm{H_2C{=}CH(OCOCH_3)} + \mathrm{CH_2{=}CH_2} \rightarrow \rightarrow \left[\mathrm{CH_2CH(OH)CH_2CH_2}\right]_n$$

Vinyl acetate Ethylene Poly(ethylene-co-vinyl alcohol) = EVAL

FIGURE 12.37 EVAL is a copolymer of ethylene and vinyl acetate.

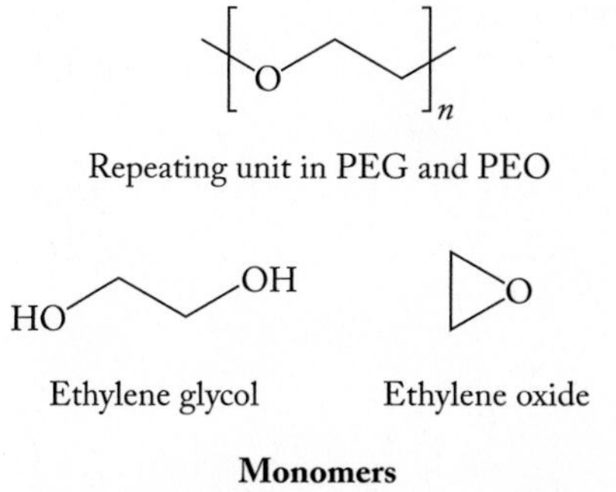

Repeating unit in PEG and PEO

Ethylene glycol Ethylene oxide

Monomers

FIGURE 12.38 Poly(ethylene glycol) (PEG) and poly(ethylene oxide) (PEO) have the same repeating unit. The two polymers differ only in their molar masses. PEG is typically made from ethylene glycol, whereas ethylene oxide is the monomer of choice for making PEO.

(**Figure 12.38**). PEG is a low-molar-mass liquid made from ethylene glycol, whereas PEO is a higher-molar-mass solid made from ethylene oxide. As a polyether, PEG has properties closely related to those of diethyl ether, in that it is soluble in both polar and nonpolar liquids. It is a common component in toothpaste because it interacts both with water and with the water-insoluble materials in the paste, so it keeps the toothpaste uniform both in the tube and during use. PEGs of many lengths are finding increasing use as attachments to pharmaceutical agents to improve their solubility and biodistribution: the enhanced circulation in the blood results from increased solubility. Like ethylene, the feedstock for PEG can now be prepared from ethanol obtained from plant-based sources relieving the demands on limited supplies of petroleum.

CONNECTION Alcohols are organic molecules with the general formula R—OH (Section 2.8). Some properties of alcohols are further investigated in Chapter 10.

SAMPLE EXERCISE 12.6 Analyzing Properties of Polymers **LO5**

PEG (Figure 12.38) is used to blend materials that are typically insoluble in one another. PEG is soluble in both water (a polar solvent) and benzene (a nonpolar solvent). Describe the structural features of PEG that make it soluble in these two liquids of very different polarities.

Collect, Organize, and Analyze The relationship between structure and the solubility of compounds, first discussed in Chapter 10, can be expressed as "like dissolves like." That is, polar substances will dissolve in polar solvents, because they both experience dipole–dipole interactions, and nonpolar substances will dissolve in nonpolar solvents as a result of dispersion forces. We need to describe the intermolecular forces in PEG and see whether one part is compatible with water and another part with benzene. Water is a polar molecule and benzene is a nonpolar molecule, so we predict that PEG must contain both polar and nonpolar regions in order to be soluble in both substances.

Solve The structure of PEG consists of $-CH_2CH_2-$ groups connected by oxygen atoms. The oxygen atoms can form hydrogen bonds with water molecules, so the hydrogen bonds provide the attractive force between PEG and water. Benzene is nonpolar and is attracted to the $-CH_2CH_2-$ groups in the polymer. Nonpolar groups interact via dispersion forces, so dispersion forces must be responsible for the solubility of PEG in nonpolar benzene.

Think About It As predicted, PEG contains both polar and nonpolar regions, allowing it to be solvated by both polar solvents such as water and nonpolar solvents such as benzene.

Practice Exercise When PEG is added to soft drinks it keeps CO_2, responsible for the fizz in soda, in solution longer when the soda is poured. What intermolecular attractive forces between PEG and CO_2 might make this application possible?

copolymer a polymer formed from the chemical reaction of two different monomers.

heteropolymer a polymer made of three or more different monomer units.

CONCEPT TEST

A polymer chemist decides to make a series of derivatives of PEG by using the following alcohols in place of ethylene glycol:

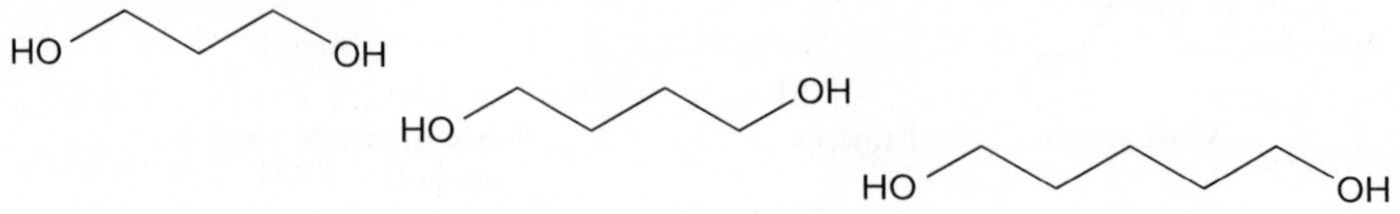

HO⁀⁀OH   HO⁀⁀⁀OH   HO⁀⁀⁀⁀OH

What effect will this have on the solubility of the resulting polymer in water?

Polyesters and Polyamides

Prior to this point, the polymers we have examined have been prepared from *monofunctional* monomers, in which the monomers have only one functional group. A wide range of polymers are derived from *difunctional* molecules, which contain two functional groups. Several classes of polymers are derived from difunctional carboxylic acids, but we consider only two of them here: polyesters and polyamides.

An **ester** is prepared by the *esterification* of a carboxylic acid with an alcohol. In esters, the –OH group of the acid is replaced by –OR, where R can be any organic group (**Figure 12.39**). Esterification is not an addition reaction but rather a **condensation reaction**—that is, two molecules react to create a larger molecule while a small molecule (typically water, hence "condensation reaction") is also formed. In Figure 12.39, the –COOH group of butyric acid reacts with the –OH group of ethanol to form a carbon–oxygen single bond in the resulting ester (ethyl butyrate) and releases a molecule of water.

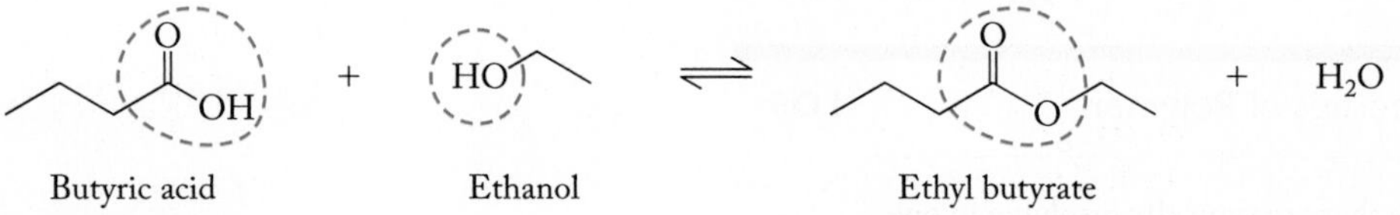

Butyric acid + Ethanol ⇌ Ethyl butyrate + H_2O

FIGURE 12.39 Condensation reactions between carboxylic acids and alcohols produce esters and water. Here butyric acid reacts with ethanol, forming ethyl butyrate. The functional groups (carboxylic acid, alcohol, and ester) are marked by the red, dashed circles.

What do you think happens at the molecular level when we have a single compound that contains a carboxylic acid functional group at one end and an alcohol functional group at the other (**Figure 12.40**)? The carboxylic acid group of one molecule can react with the alcohol group of another molecule in a condensation reaction to generate a molecule that has a carboxylic acid group

(a) Ester linkage

(b) Polyester

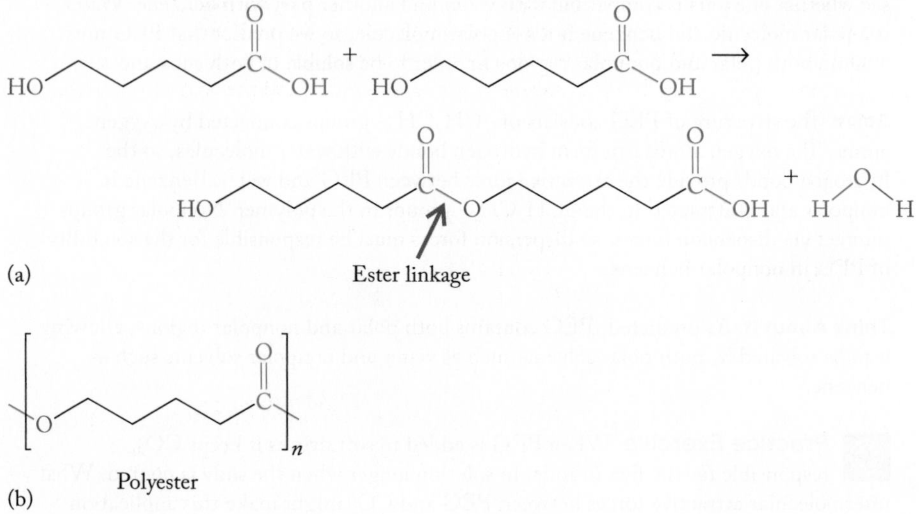

FIGURE 12.40 (a) Synthesis of an ester from a condensation reaction between two identical difunctional molecules, each containing an alcohol group and a carboxylic acid group. The diester can then react with additional difunctional molecules at its –OH and –COOH ends, forming a triester, and the reaction repeats over and over, forming (b) the polyester made up of the repeating monomer unit shown.

at one end, an alcohol at the other end, and an ester linkage between them. If this reaction takes place repeatedly, a monomer containing one carboxylic acid and one hydroxyl group (a hydroxy acid) polymerizes, as shown in Figure 12.40, to form a *polyester*, a **condensation polymer**. We have already encountered one condensation polymer, PETE. In general, condensation polymers are formed by the reaction of monomers yielding a polymer and water as a by-product of the reaction. In addition to its use in plastic soda bottles, PETE is used extensively in medicine. Artificial heart valves and grafts for arteries are also made from PETE.

CONNECTION The carboxylic acid functional group (–COOH) was first described in Section 4.5.

A copolymer formed by the reaction of glycolic acid and lactic acid (**Figure 12.41**) is used to support the growth of skin cells for burn victims. The polymer in Figure 12.41 is also used to make dissolving sutures. Esterification reactions used to make polyesters can be reversed by the addition of water, breaking their ester linkages and forming alcohol and acid functional groups.

x HO–(glycolic acid)–OH + y HO–(lactic acid)–OH → HO–[ ]$_x$–[ ]$_y$–H + H_2O

Glycolic acid Lactic acid A polyester

(a)

(b)

FIGURE 12.41 (a) The condensation polymer prepared from glycolic acid and lactic acid is used to make sutures that dissolve and as artificial skin that protects against infection while promoting the regrowth of skin cells. (b) Synthetic skin being applied to a burn patient.

SAMPLE EXERCISE 12.7 Making a Polyester **LO4**

Show how a polyester can be synthesized from the difunctional alcohol $HO(CH_2)_3OH$ and the difunctional carboxylic acid $HOOC(CH_2)_3COOH$.

Collect, Organize, and Analyze An ester is the product of a reaction between a carboxylic acid and an alcohol. A polyester is a polymer with a repeating unit containing an ester functional group. We are given an alcohol and a carboxylic acid to react to make the ester repeating monomer unit. Because the starting materials are difunctional, the alcohol can react with two molecules of carboxylic acid, and the acid can react with two molecules of alcohol.

Solve The reaction is

$$HOCH_2CH_2CH_2OH + HO{-}C({=}O)CH_2CH_2CH_2C({=}O){-}OH \rightarrow$$

$$HOCH_2CH_2CH_2O{-}C({=}O)CH_2CH_2CH_2C({=}O){-}OH + H_2O$$

ester an organic compound in which the –OH of a carboxylic acid group is replaced by –OR, where R can be any organic group.

condensation reaction a reaction in which two molecules combine to form a larger molecule and a small molecule, typically water.

condensation polymer a polymer formed by a condensation reaction.

The two –OH groups shown in blue react to form an ester at one end of the carboxylic acid. The product molecule has an alcohol group on one end (shown in red) that can react with another molecule of carboxylic acid and has a carboxylic acid group (green) on the other end that can react with another molecule of alcohol. Continuing these condensation reactions results in the formation of a polymer whose repeating unit is

$$\left[-CH_2CH_2CH_2O-\overset{O}{\overset{\|}{C}}CH_2CH_2CH_2\overset{O}{\overset{\|}{C}}-O- \right]_n$$

Think About It The repeating unit in the polyester is reasonable because it contains one section that came from the alcohol and a second section that came from the carboxylic acid, and esters are formed from alcohols and acids.

Practice Exercise A difunctional molecule may contain two different functional groups, such as this one with an alcohol group and a carboxylic acid group:

$$HO—CH_2CH_2CH_2\overset{O}{\overset{\|}{C}}—OH$$

Draw the repeating unit of the polyester made from this molecule.

SAMPLE EXERCISE 12.8 Comparing Properties of Polymers **LO5**

Clothes made from the polyester fabric known as Dacron can be less comfortable in hot weather than clothes made of cotton (also a polymer) because Dacron does not absorb perspiration as effectively as cotton. On the basis of the repeating units of these two polymers (**Figure 12.42**), suggest a structural reason why cotton absorbs perspiration (water) better than Dacron.

Collect, Organize, and Analyze Cotton and Dacron are polymers that differ in the functional groups in their repeating units. The absorption of water by a polymer depends on the intermolecular forces present. Water is a polar molecule and is attracted to polar groups. We need to compare the different functional groups in each monomer to see which has the greatest number of polar functional groups or atoms that can form hydrogen bonds. This will be the polymer more likely to absorb water.

Solve Each monomer unit in cotton has three –OH groups attached to it, all polar and capable of hydrogen bond formation, which means they are likely to interact with the water in perspiration and thereby draw it away from the body. The monomer unit in Dacron has oxygen atoms in it, but no –OH groups. Although regions in the Dacron monomer are polar, they are not nearly as polar as the –OH groups in the cotton monomer unit.

Think About It The principle of like interacting with like which works for small molecules in Chapter 11 also applies to polymers. The more polar groups on the polymer (cotton), the greater the interaction with polar molecules like water.

Practice Exercise Gloves made of a woven blend of cotton and polyester fibers protect the hands from exposure to oil and grease and are comfortable to wear because they "breathe"—they allow perspiration to evaporate and pass through them, thereby cooling the skin. Suggest how these gloves work at the molecular level.

FIGURE 12.42 Based on the molecular structures of (a) Dacron and (b) cotton, why is cotton the better material for making tee shirts worn during strenuous exercise?

Combining difunctional molecules in a condensation reaction can be used to make *polyamides*, another class of very useful synthetic polymers. The functional groups are a carboxylic acid and an amine. The difunctional monomer units can be identical (**Figure 12.43a**), each containing one carboxylic acid group and one amine group, or they can be different (**Figure 12.43b**), with one monomer containing two carboxylic acid groups (a dicarboxylic acid) and the other containing two amine groups (a diamine). Some of these monomers, such as adipic acid, are now available from renewable bio-based sources.

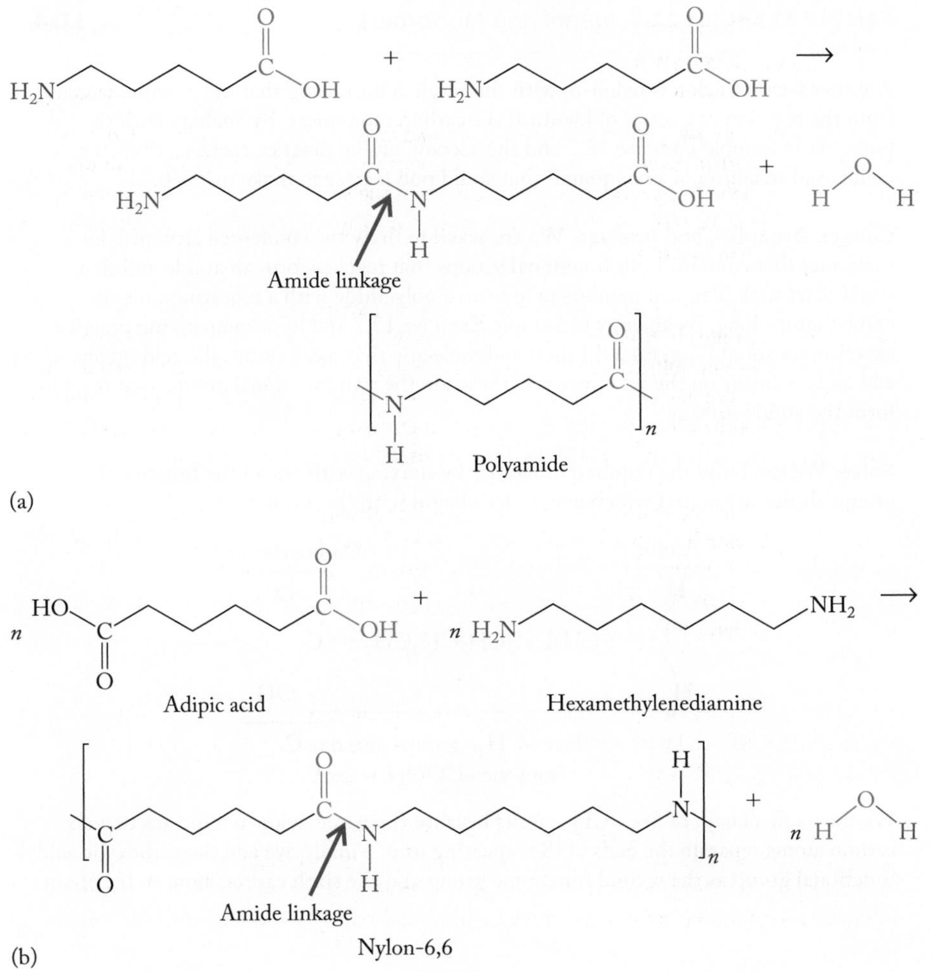

Amide linkage

Polyamide

(a)

Adipic acid

Hexamethylenediamine

Amide linkage

Nylon-6,6

(b)

FIGURE 12.43 (a) Synthesis of a polyamide from two identical monomers, each containing a carboxylic acid functional group and an amine functional group. (b) Synthesis of the polyamide nylon-6,6 from nonidentical monomers: adipic acid (a dicarboxylic acid) and hexamethylenediamine (a diamine).

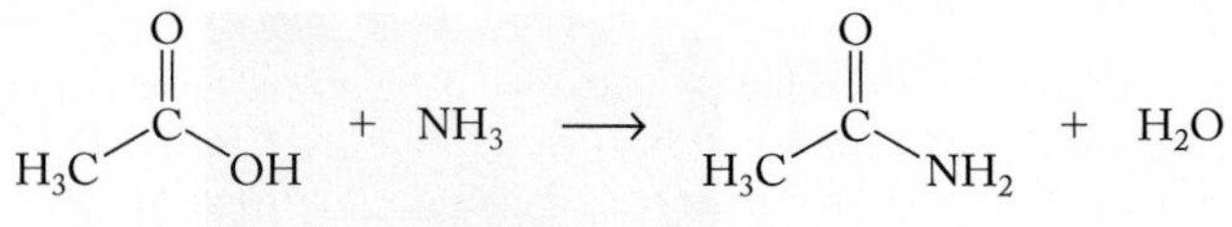

$$\underset{\text{Acetic acid}}{H_3C{-}C({=}O){-}OH} + NH_3 \longrightarrow \underset{\text{Acetamide}}{H_3C{-}C({=}O){-}NH_2} + H_2O$$

FIGURE 12.44 Condensation reactions between carboxylic acids and ammonia (or amines) produce amides. Here acetic acid reacts with ammonia, forming acetamide.

In **amides**, the –OH group of a carboxylic acid is replaced by an amine group, which can be $-NH_2$, –NHR, or $-NR_2$ (**Figure 12.44**). Amides are polar and can form intermolecular hydrogen bonds. The hydrogen atoms on the $-NH_2$ group can form hydrogen bonds with the oxygen atom in the carbonyl group of an adjacent molecule. This causes the boiling points of amides to be considerably higher than those of esters of comparable size. Although amines are bases, amides are actually weakly acidic. The carbonyl group pulls electron density away from the nitrogen, making it less favorable for the nitrogen to behave like a Brønsted–Lowry base and pick up a proton. It is actually more favorable for an amide to donate a proton and function as an acid.

Nylon-6,6 is a polyamide made from the monomers shown in Figure 12.43b. Each monomer contains six carbon atoms, which is what the numbers in the name represent.

amide an organic compound in which the same carbon atom is single bonded to a nitrogen atom and double bonded to an oxygen atom.

SAMPLE EXERCISE 12.9 Identifying Monomers LO4

Another form of nylon is nylon-6, with the single 6 indicating that the polymer is made from the reaction of a series of identical six-carbon monomers. By analogy with the polyester in Sample Exercise 12.7 and the accompanying practice exercise, draw the condensed structure of a compound that could polymerize to make nylon-6.

Collect, Organize, and Analyze We are asked to draw the condensed structure for a monomer that contains both functional groups that react to form an amide and that could react with identical monomers to form a polyamide with a repeating unit six carbon atoms long. By analogy to Sample Exercise 12.7 and its accompanying practice exercise, we should suggest a difunctional molecule that has a carboxylic acid on one end and an amine on the other because these are the two functional groups that react to form the amide linkage.

Solve We can build the required monomer by starting with one of the functional groups. It doesn't matter which one, so let's begin with the amine:

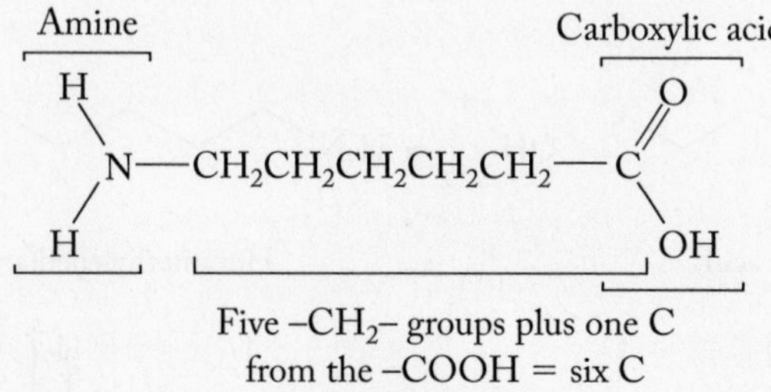

$$\underbrace{H_2N}_{\text{Amine}}{-}CH_2CH_2CH_2CH_2CH_2{-}\underbrace{C({=}O)OH}_{\text{Carboxylic acid}}$$

Five $-CH_2-$ groups plus one C from the –COOH = six C

We then add a chain of five $-CH_2-$ units because the name nylon-6 indicates that six carbon atoms separate the ends of the repeating unit. Finally, we add the carboxylic acid functional group as the second functional group and the sixth carbon atom in the chain.

Think About It Our monomer includes two different functional groups; an amine and a carboxylic acid, consistent with our strategy for solving this exercise. The five $-CH_2-$ units together with the carbon in the CO_2H group results in nylon-6, a polymer with a six-carbon repeating unit.

Practice Exercise Draw the carbon-skeleton structures of two monomers that could react with each other to make nylon-5,4. Draw the carbon-skeleton structure of the repeating unit in the polymer. (*Note*: The first number refers to the carboxylic acid monomer; the second refers to the amine monomer.)

FIGURE 12.45 Stephanie Kwolek (1923–2014) was the chemist at DuPont who created Kevlar, the first member of a family of exceptionally strong, stiff synthetic fibers.

Polymers of nylon make long, straight fibers that are quite strong and excellent for weaving into fabrics. Nylon is flexible and stretchable because the hydrocarbon chains can bend and curl, much like a telephone cord or a Slinky spring toy. To produce an even stronger nylon, researchers recognized they had to find some way to reduce the ability of the chains to form coils. They discovered this could be done by using monomer units containing functional groups that made it difficult for the chains to bend. One product of this work was a polyamide called Kevlar, invented by Stephanie Kwolek (**Figure 12.45**) at DuPont in 1965. Kevlar is formed from a dicarboxylic acid of benzene and a diamine of benzene (**Figure 12.46**). When these two monomers polymerize, the flat, rigid aromatic rings keep the chains straight.

Two additional intermolecular interactions orient the chains and hold them together very tightly (**Figure 12.47**). First, the –NH hydrogen atoms form hydrogen bonds with the oxygen atoms of carbonyl groups on adjacent chains. Second, the rings stack on top of one another (just as in polystyrene) and

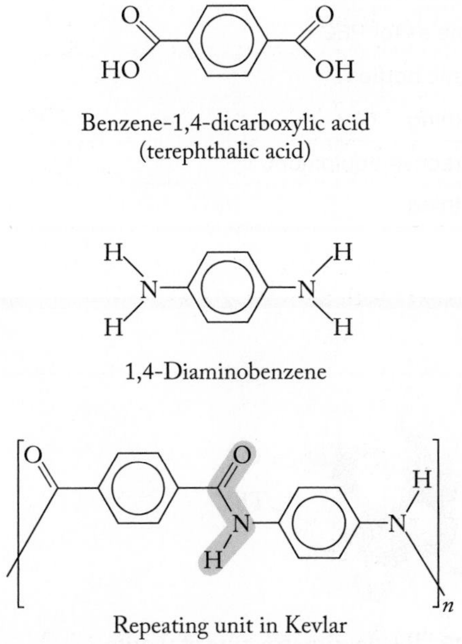

Benzene-1,4-dicarboxylic acid (terephthalic acid)

1,4-Diaminobenzene

Repeating unit in Kevlar

FIGURE 12.46 The monomers used to make Kevlar are a dicarboxylic acid and a diamine. The amide bond in the repeating unit is highlighted.

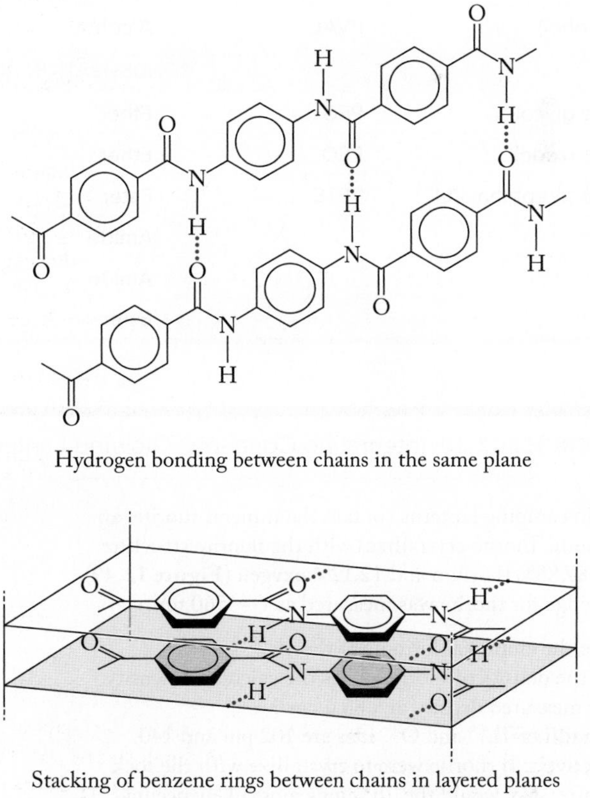

Hydrogen bonding between chains in the same plane

Stacking of benzene rings between chains in layered planes

FIGURE 12.47 Interactions between groups in Kevlar.

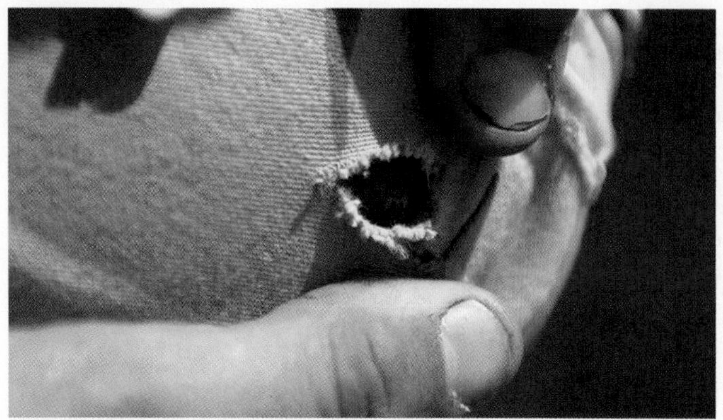

FIGURE 12.48 A bullet fired point-blank at a sheet of Kevlar wrapped in fabric can puncture the fabric, but the Kevlar is strong enough to stop bullets in many cases.

provide additional interactions that hold the chains together in parallel arrays. The result is a fiber that is very strong but still flexible. Fabrics and helmets made of Kevlar resist puncture, even when bullets (**Figure 12.48**) and hockey pucks are fired at them. They are resistant to flames and reactive chemicals, too.

In this section we have discussed many polymers and mentioned their uses in common products. The classification of these polymeric materials as addition or condensation polymers is summarized in **Table 12.7**. Table 12.7 also identifies the polymers by their distinctive functional groups and mentions some common uses in our lives.

TABLE 12.7 Summary of Common Polymers and Their Uses

Name	Abbreviation	Functional Group	Use
		ADDITION POLYMERS	
Polyethylene	PE	Alkane	Plastic bags and films
Poly(tetrafluoroethylene)	Teflon	Fluoroalkane	Nonstick coatings
Poly(1,1-dichloroethylene)	Saran	Chloroalkane	Plastic wrap
Poly(vinyl chloride)	PVC	Chloroalkane	Drain pipes
Poly(methyl methacrylate)	PMMA	Alkane and ester	Shatter-resistant glass (e.g., Plexiglas, Lucite)
Polystyrene	PS	Aromatic hydrocarbon	Cups, dishes, insulation
Poly(vinyl alcohol)	PVAL	Alcohol	Gloves, bottles
		CONDENSATION POLYMERS	
Poly(ethylene glycol)	PEG	Ether	Pharmaceuticals, consumer products
Poly(ethylene oxide)	PEO	Ether	Same as for PEG
Poly(ethylene terephthalate)	PETE	Ester	Plastic bottles
Nylon		Amide	Clothing
Kevlar		Amide	Protective equipment
Dacron		Ester	Clothing

SAMPLE EXERCISE 12.10 Integrating Concepts: Glowing Lantern Mantles

The mantles in camping lanterns contain the mineral thorite, an oxide of thorium. Thorite crystallizes with the fluorite structure and contains 87.88% thorium and 12.12% oxygen (**Figure 12.49**). The unit cell edge for thorite was measured as $\ell = 560$ nm.

a. Determine the empirical formula of thorite.
b. Calculate the density of thorite. Does the calculated density match the measured density of 9.86 g/cm^3?
c. The ionic radii of Th^{4+} and O^{2-} ions are 102 pm and 140 pm, respectively. If thorite were to crystallize with the rock salt structure, why would the Th^{4+} ions most likely occupy octahedral rather than tetrahedral holes?

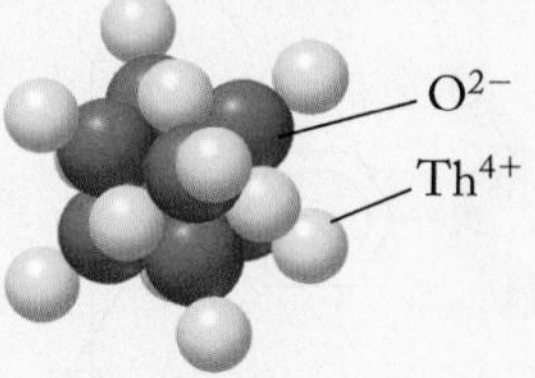

FIGURE 12.49 The fluorite structure of thorite.

d. Calculate the density of the rock salt polymorph of thorite by using the same unit cell edge length as in part (b). Compare the value with the observed and calculated values in part (b).

Collect and Organize We are given the percent composition and density of a thorium mineral. We are asked to determine the empirical formula of thorite and to calculate the density of thorite in two possible crystal structures—the fluorite (observed) and rock salt (hypothetical) structures. Methods for determining empirical formulas were described in Chapter 3. We will also make use of the information presented in Section 12.4.

Analyze Thorite contains only Th and O with the general formula Th_xO_y because the sum of the %Th and %O equals 100%, so we know the masses of each element in thorite. We can convert the masses to the equivalent number of moles and determine the simplest whole-number ratio of the moles of the elements. In the fluorite structure of thorite, the unit cell consists of a face-centered arrangement of Th ions with oxide ions in all of the tetrahedral holes. Following the procedure in Sample Exercise 12.4, we can calculate the density of thorite. If thorite were to crystallize in a ccp rock salt structure, the Th^{4+} ions are more likely to occupy a hole based on the cation:anion radius ratio.

Solve

a. A 100.00 g sample of thorite contains 87.88 g of Th and 12.12 g of O. Converting these masses to moles:

$$87.88 \text{ g Th} \times \frac{1 \text{ mol Th}}{232.04 \text{ g Th}} = 0.3787 \text{ mol Th}$$

$$12.12 \text{ g O} \times \frac{1 \text{ mol O}}{16.00 \text{ g O}} = 0.7575 \text{ mol O}$$

Dividing by the smallest number of moles gives us the empirical formula: ThO_2.

b. In the fluorite structure of ThO_2, the Th^{4+} ions adopt an fcc unit cell with O^{2-} ions in all the tetrahedral holes. The equivalent of four Th^{4+} and eight O^{2-} are in the unit cell. The mass of the ions in the unit cell is

$$m = \frac{232.04 \text{ g Th}^{4+}}{1 \text{ mol Th}^{4+}} \times \frac{1 \text{ mol Th}^{4+}}{6.022 \times 10^{23} \text{ ions Th}^{4+}} \times 4 \text{ ions Th}^{4+}$$
$$= 1.541 \times 10^{-21} \text{ g Th}^{4+}$$

$$m = \frac{16.00 \text{ g O}^{2-}}{1 \text{ mol O}^{2-}} \times \frac{1 \text{ mol O}^{2-}}{6.022 \times 10^{23} \text{ ions O}^{2-}} \times 8 \text{ ions O}^{2-}$$
$$= 0.2126 \times 10^{-21} \text{ g O}^{2-}$$

$$\text{Total mass of ions} = 1.541 \times 10^{-21} \text{ g Th}^{4+} + 0.2126 \times 10^{-21} \text{ g O}^{2-}$$
$$= 1.754 \times 10^{-21} \text{ g}$$

Using the unit cell edge length, $\ell = 560$ nm, the volume of the unit cell is

$$V = \ell^3 = (560 \text{ pm})^3 \times \frac{(10^{-10} \text{ cm})^3}{(1 \text{ pm})^3} = 1.756 \times 10^{-22} \text{ cm}^3$$

and the calculated density of ThO_2 is

$$d = \frac{m}{V} = \frac{1.754 \times 10^{-21} \text{ g}}{1.756 \times 10^{-22} \text{ cm}^3} = 9.99 \text{ g/cm}^3$$

The calculated density differs from the observed density of 9.86 g/cm^3 by less than 2%.

c. In a hypothetical rock salt structure of ThO_2, the lattice would be composed of fcc O^{2-} ions with Th^{4+} in holes. The radius ratio of Th^{4+} to O^{2-} would be

$$\frac{r_{\text{cation}}}{r_{\text{anion}}} = \frac{102 \text{ pm}}{140 \text{ pm}} = 0.729$$

From data in Table 12.5, the radius ratio indicates that Th^{4+} ions will fit in octahedral holes but not in tetrahedral holes. The four Th^{4+} ions would occupy one-half of the octahedral holes.

d. By analogy to part (b), the density of ThO_2 in a hypothetical rock salt structure would be

$$m = \frac{232.04 \text{ g Th}^{4+}}{1 \text{ mol Th}^{4+}} \times \frac{1 \text{ mol Th}^{4+}}{6.022 \times 10^{23} \text{ ions Th}^{4+}} \times 2 \text{ ions Th}^{4+}$$
$$= 7.706 \times 10^{-22} \text{ g Th}^{4+}$$

$$m = \frac{16.00 \text{ g O}^{2-}}{1 \text{ mol O}^{2-}} \times \frac{1 \text{ mol O}^{2-}}{6.022 \times 10^{23} \text{ ions O}^{2-}} \times 4 \text{ ions O}^{2-}$$
$$= 1.063 \times 10^{-22} \text{ g O}^{2-}$$

$$\text{Total mass of ions} = 7.706 \times 10^{-22} \text{ g Th}^{4+} + 1.063 \times 10^{-22} \text{ g O}^{2-}$$
$$= 8.769 \times 10^{-22} \text{ g ions}$$

The volume of the unit cell is the same as in part (b): $V = 1.756 \times 10^{-22}$ cm^3. Substituting into the equation for density:

$$d = \frac{m}{V} = \frac{8.769 \times 10^{-22} \text{ g}}{1.756 \times 10^{-22} \text{ cm}^3} = 4.99 \text{ g/cm}^3$$

The calculated density for ThO_2 in a rock salt structure would be about half of the observed value, and it is about half of the calculated value based on the fluorite structure.

Think About It By analogy to the fluorite structure shown in Figure 12.22, the larger oxide ions occupy the tetrahedral holes. The calculated value for the density of ThO_2 is very close to the observed value, supporting the assignment of the fluorite structure for ThO_2. The rock salt structure, where the smaller Th^{4+} ions occupy octahedral holes in the fcc lattice of the larger O^{2-} ions, can be ruled out because the calculated density is very different from the observed density.

SUMMARY

LO1 Many metallic crystals are based on **crystal lattices** of the **cubic closest-packed (ccp)** and **hexagonal closest-packed (hcp)** types, which are the two most efficient ways of packing atoms in a solid. **Crystalline solids** contain repeating **unit cells**, which can be **simple cubic (sc)**, **body-centered cubic (bcc)**, or **face-centered cubic (fcc)**. The dimensions of the unit cell in a crystalline lattice can be used to determine the radius of the atoms or ions and to predict density. (Sections 12.1 and 12.2)

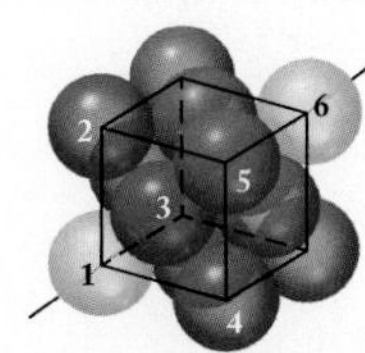

LO2 **Alloys** are blends of a host metal and one or more other elements (which may or may not be metals) added to enhance the properties of the host, including strength, hardness, and corrosion resistance. In **substitutional alloys,** atoms of the added elements replace atoms of the host metal in the crystal lattice. In **interstitial alloys,** atoms of added elements are located in the tetrahedral holes, in the octahedral holes of the host metal, or both. (Section 12.3)

LO3 Many **ionic solids** consist of crystals with some number of either cations or anions forming the unit cell with the opposite ion occupying cubic, octahedral, and tetrahedral holes in the unit cell. The unit cell edge lengths of ionic solids can be used to calculate ionic radii and to predict densities. Two allotropes of carbon are the **covalent network solids** graphite and diamond. (Sections 12.1, 12.4, and 12.5)

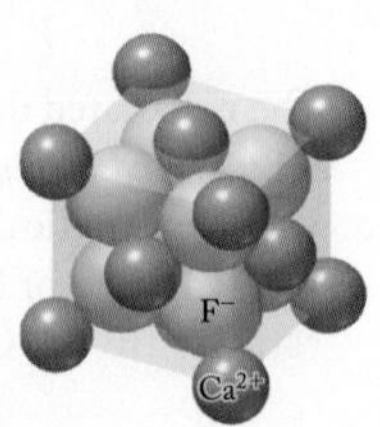

LO4 Alkenes undergo **addition reactions** to form **homopolymers** known as **addition polymers.** Monomer units derived from alcohols or ethers can be chemically reacted to make **heteropolymers** or **copolymers.** **Condensation reactions** of carboxylic acids and amines are used to prepare **condensation polymers** such as polyesters and polyamides. (Section 12.6)

LO5 Properties of polymers depend on the identity and arrangement of the monomer units in the polymer. (Section 12.6)

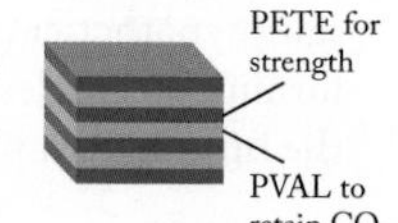

PARTICULATE PREVIEW WRAP-UP

Each carbon atom in allotrope A is covalently bonded to four other carbon atoms in a tetrahedral geometry, so the pattern is nonplanar. Each carbon atom in allotrope B is covalently bonded to three other carbon atoms in a trigonal planar geometry and is attracted to the planar layers above and below through dispersion forces. The pattern in allotope B is planar. Allotrope A (diamond) is very hard and resists compression, whereas allotrope B (graphite) can be used as a lubricant because the planar layers can slide past one another.

PROBLEM-SOLVING SUMMARY

Type of Problem	Concepts and Equations	Sample Exercises
Calculating atomic or ionic radii and density from unit cell dimensions	Determine the length of a unit cell edge, face diagonal, or body diagonal along which adjacent atoms touch. Use the relationship between edge length ℓ and the atomic radius r to calculate the value of r: $r = 0.5\ell$ for sc unit cells, $r = 0.3536\ell$ for fcc unit cells, and $r = 0.4330\ell$ for bcc unit cells. Determine the mass of the atoms in the unit cell from the molar mass and the volume of the unit cell from the unit cell edge length; then calculate the density: $d = \frac{m}{V}$	**12.1, 12.4**
Predicting the crystal structure of two-element alloys	Compare the radii of the alloying elements. Similarly sized radii (within 15%) suggest a substitutional alloy. When the radius of the smaller atom in a hcp or ccp lattice is <73% of the radius of the larger atom, an interstitial alloy forms.	**12.2**
Predicting the location of ions in ionic compounds	Determine the ratio of the radii of the guest and the host to predict whether the ion is more likely to reside in an octahedral or tetrahedral hole.	**12.3, 12.4**
Identifying monomers in polymers	Find the smallest portion of the polymer that is repeated.	**12.5, 12.9**
Analyzing properties of polymers	Evaluate the polarity of the functional groups in the polymer, and assess the relative importance of all types of intermolecular forces possible in the molecules.	**12.6, 12.8**
Making a polyester	Combine monomers with alcohol functional groups (ROH) and carboxylic acid functional groups (RCOOH) to form water (H_2O) and ester groups (RCOOR).	**12.7**

VISUAL PROBLEMS

(Answers to boldface end-of-chapter questions and problems are in the back of the book.)

12.1. Which drawings in Figure P12.1 are analogous to crystalline solids, and which are analogous to amorphous solids?

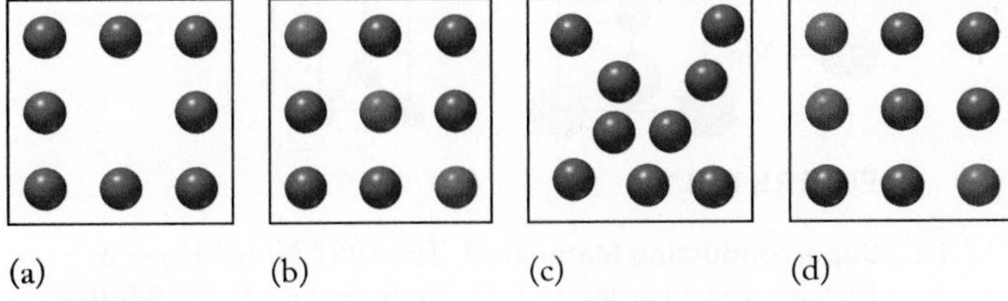

FIGURE P12.1

12.2. The unit cells in Figure P12.2 continue infinitely in two dimensions. Draw a box around a unit cell in each pattern. How many light squares and how many dark squares are in each unit cell?

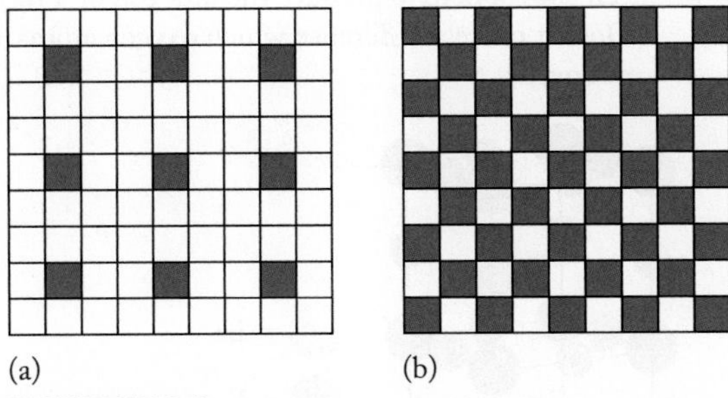

FIGURE P12.2

12.3. The pattern in Figure P12.3 continues indefinitely in three dimensions. Draw a box around a unit cell. If the red spheres represent element A and the blue spheres represent element B, what is the chemical formula of the compound?

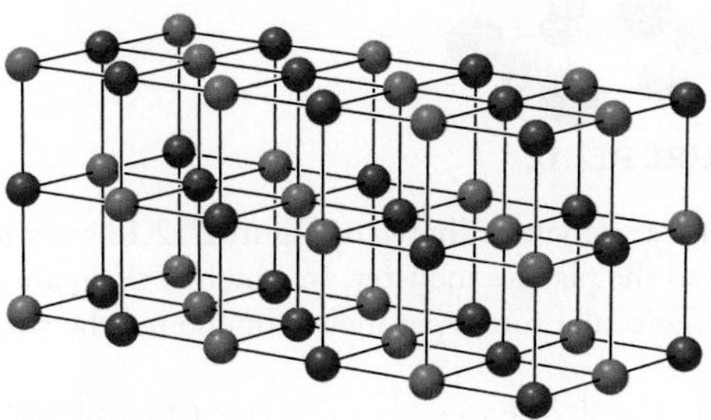

FIGURE P12.3

12.4. The unit cell of an ionic compound is shown in Figure P12.4. What is the chemical formula of the ionic compound, if A and X are cations and B is an anion?

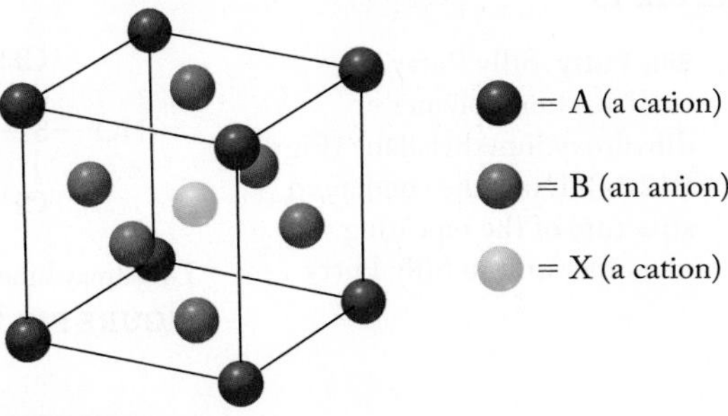

FIGURE P12.4

12.5. What is the formula of the compound that crystallizes in the cubic closest-packed arrangement of A atoms (silver) shown in Figure P12.5? The B atoms (red) occupy half of the octahedral holes and the C atoms (blue) occupy one-eighth of the tetrahedral holes.

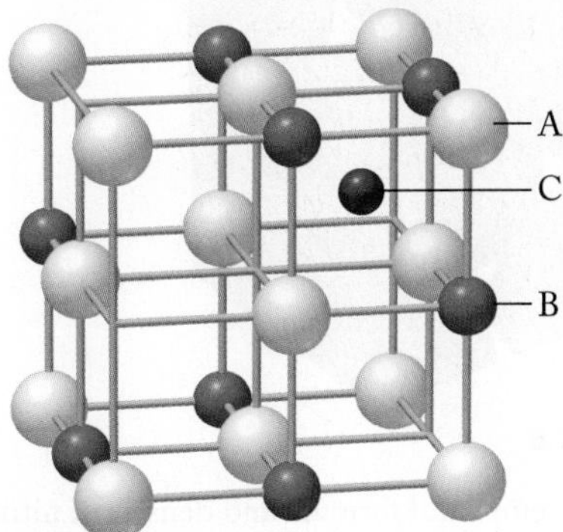

FIGURE P12.5

12.6. What is the formula of the compound that crystallizes with copper ions (the small spheres in Figure P12.6) occupying half of the tetrahedral holes in a cubic closest-packed arrangement of chloride ions?

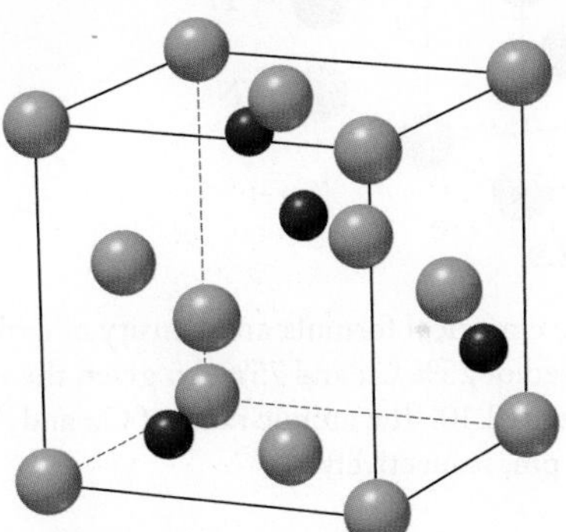

FIGURE P12.6

12.7. What is the formula of the compound that crystallizes with lithium ions occupying all of the tetrahedral holes in the cubic closest-packed arrangement of sulfide ions shown in Figure P12.7?

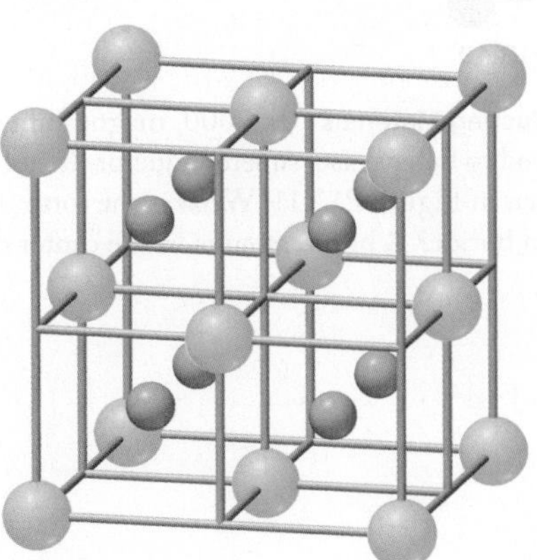

FIGURE P12.7

12.8. Figure P12.8 shows the unit cell of CsCl. From the information shown and given that the radius of the chloride (corner) ions is 181 pm, calculate the radius of Cs^+ ions.

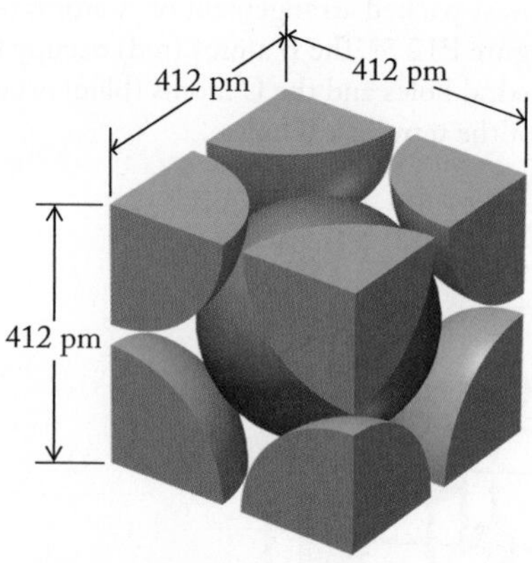

FIGURE P12.8

12.9. Calculate the empirical formula and density of nitinol, the memory alloy composed of nickel and titanium with the unit cell shown in Figure P12.9. The atomic radii of Ni and Ti are 124 and 147 pm, respectively.

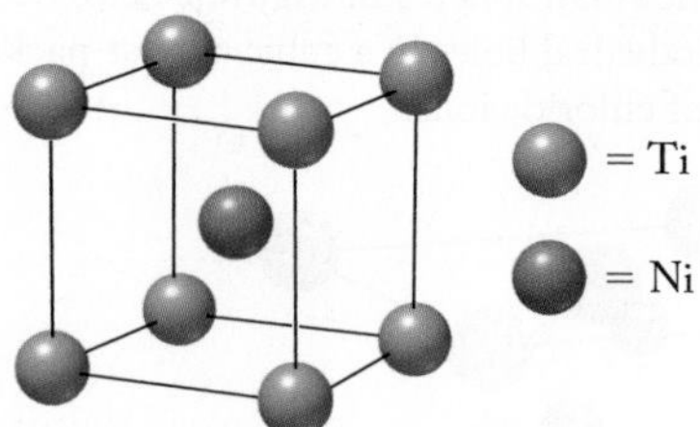

FIGURE P12.9

12.10. Calculate the empirical formula and density of a rose gold alloy composed of 25% Cu and 75% Au given the unit cell in Figure P12.10. The atomic radii of Cu and Au are 128 and 144 pm, respectively.

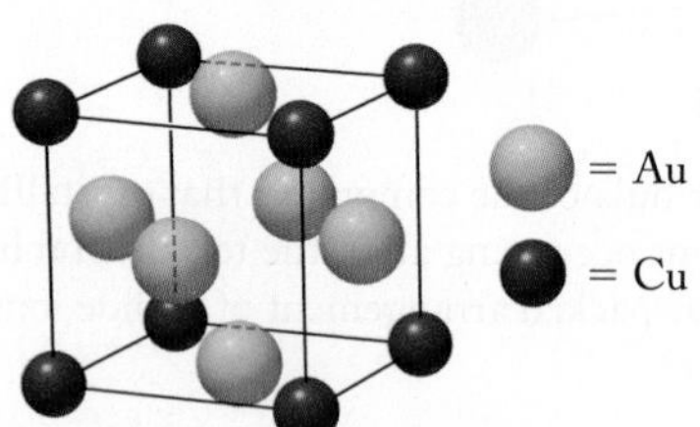

FIGURE P12.10

***12.11. Superconducting Materials I** In 2000, magnesium boride was observed to behave as a superconductor. Its unit cell is shown in Figure P12.11. What is the formula of magnesium boride? A boron atom is in the center of the unit cell (on the left), which is part of the hexagonal closest-packed crystal structure (right).

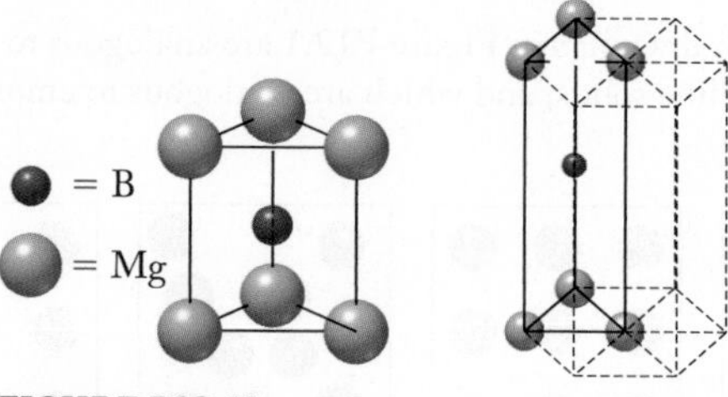

FIGURE P12.11

*12.12. **Superconducting Materials II** The 1987 Nobel Prize in Physics was awarded to J. G. Bednorz and K. A. Müller for their discovery of superconducting ceramic materials such as $YBa_2Cu_3O_7$. Figure P12.12 shows the unit cell of another yttrium–barium–copper oxide.

a. What is the chemical formula of this compound?
b. Eight oxygen atoms must be removed from the unit cell shown here to produce the unit cell of $YBa_2Cu_3O_7$. Does it make a difference which oxygen atoms are removed?

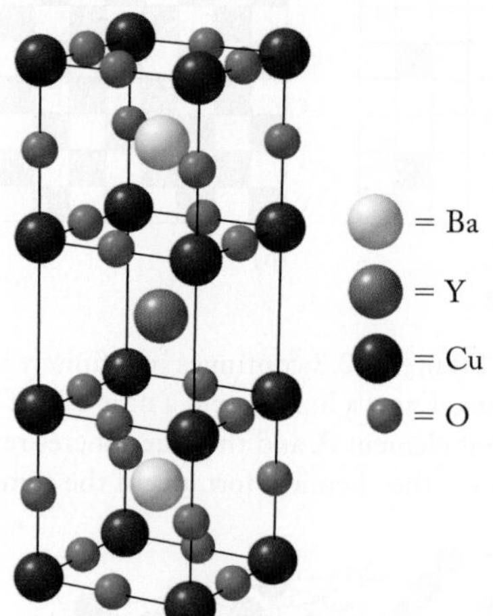

FIGURE P12.12

***12.13.** The three polymers shown in Figure P12.13 are widely used in the plastics industry. In which of them are the intermolecular forces per mole of monomer the strongest?

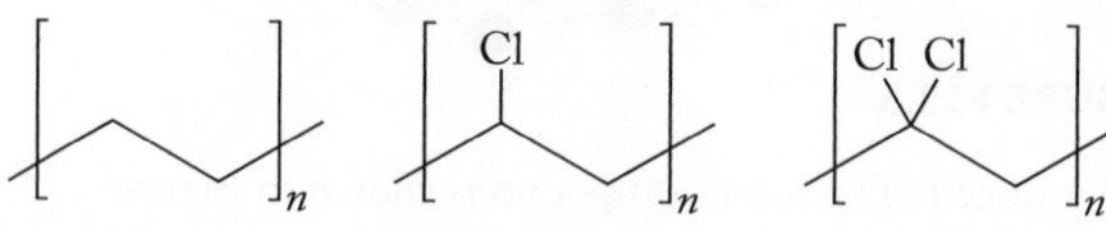

(a) Polyethylene (b) Poly(vinyl chloride) (c) Poly(1,1-dichloroethylene)

FIGURE P12.13

*12.14. **Silly Putty** Silly Putty is a condensation polymer of dihydroxydimethylsilane (Figure P12.14). Draw the condensed structure of the repeating monomer unit in Silly Putty.

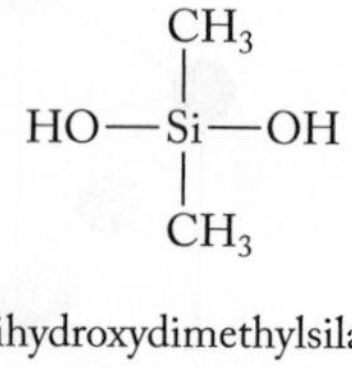

$$HO-\underset{CH_3}{\overset{CH_3}{|}}{Si}-OH$$

Dihydroxydimethylsilane

FIGURE P12.14

12.15. **Spider Silk** Polymer chemists have long sought to mimic the physical properties of spider silk, a remarkably strong but flexible material. The structure of spider silk is illustrated in Figure P12.15.

a. What are the intermolecular forces between the strands?
b. The strands are formed during a condensation reaction. What functional group is present along each strand? What two functional groups reacted to form this functional group?

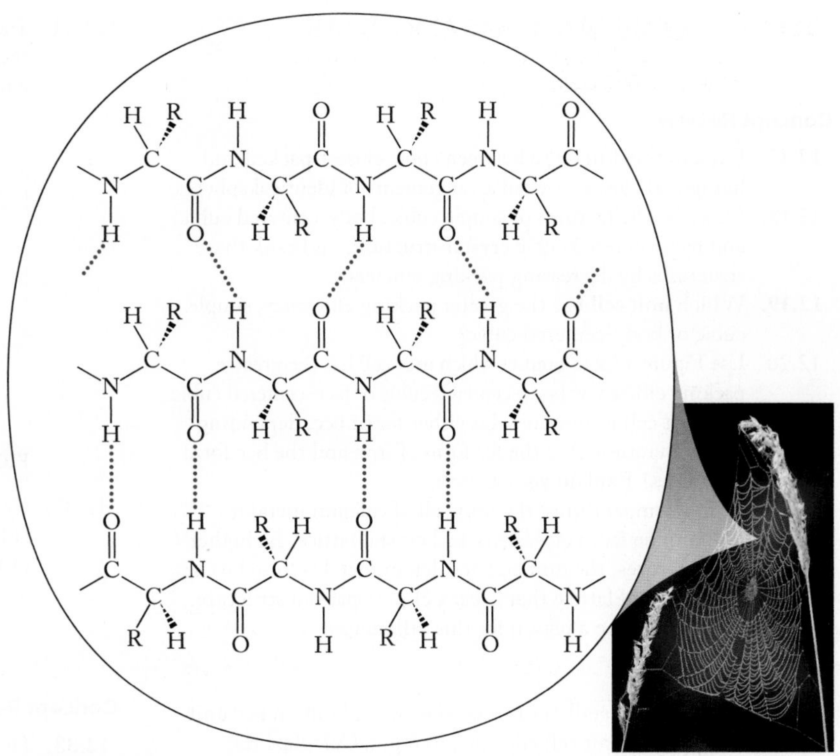

FIGURE P12.15

12.16. Use representations [A] through [I] in Figure P12.16 to answer questions (a)–(f).

a. Two forms of iron are ferrite and austenite. One has a body-centered cubic structure, whereas the other has a face-centered cubic structure. Which representations depict these common lattices for iron?
b. Which representation depicts an ionic compound?
c. Which representation depicts an interstitial alloy?
d. Which polymers were formed by an addition polymerization process? Which polymers were formed by a condensation polymerization process?
e. Which polymer structures are likely to result in rigid, hard materials?
f. Which polymer structures are likely to result in flexible, stretchable, soft materials?

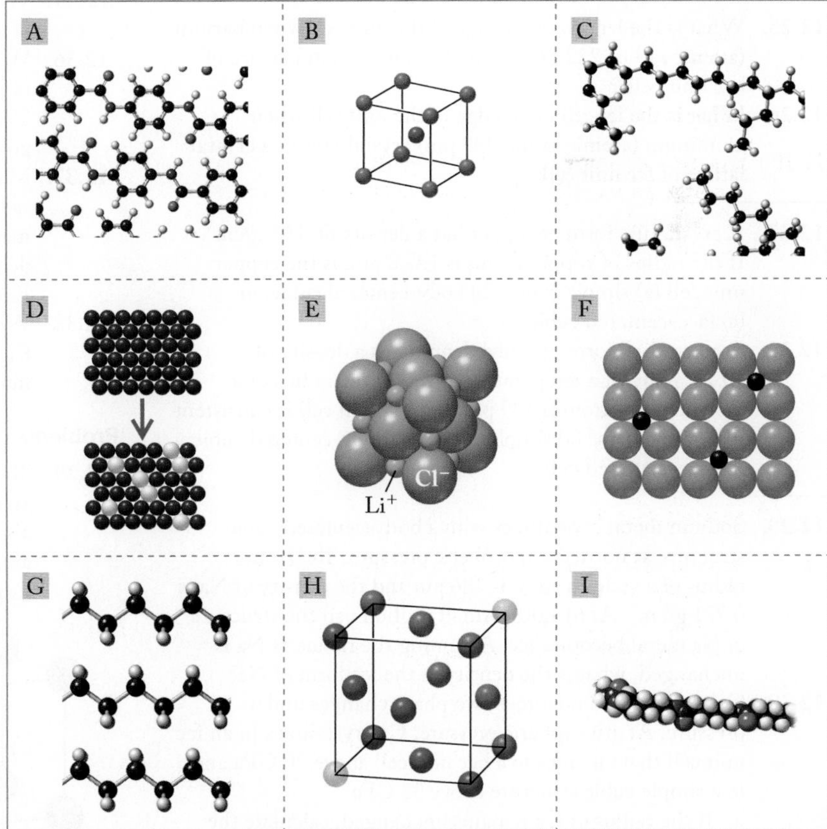

FIGURE P12.16

QUESTIONS AND PROBLEMS

Structures of Metals

Concept Review

12.17. Explain the difference between cubic closest-packed and hexagonal closest-packed arrangements of identical spheres.

*12.18. Describe the features of simple cubic, body-centered cubic, and face-centered cubic crystal structures and rank these structures by decreasing packing efficiency.

12.19. Which unit cell has the greater packing efficiency, simple cubic or body-centered cubic?

12.20. Use Figure 12.9 to predict which unit cell has the greater packing efficiency: body-centered cubic or face-centered cubic.

12.21. The unit cell in iron metal is either fcc or bcc, depending on temperature. Are the fcc form of iron and the bcc form allotropes? Explain your answer.

*12.22. At low temperatures, the unit cell of calcium metal is found to be fcc, a closest-packed crystal lattice. At higher temperatures, the unit cell of calcium metal is found to be bcc, a crystal lattice that is *not* a closest-packed structure. What might be a reason for this difference?

Problems

12.23. Europium crystallizes in a crystal lattice built on bcc unit cells, with a unit cell edge of 240.6 pm. Calculate the radius of a europium atom.

12.24. Nickel has an fcc unit cell with an edge length of 350.7 pm. Calculate the radius of a nickel atom.

12.25. What is the length of an edge of the unit cell when barium (atomic radius 222 pm) crystallizes in a crystal lattice of bcc unit cells?

12.26. What is the length of an edge of the unit cell when aluminum (atomic radius 143 pm) crystallizes in a crystal lattice of fcc unit cells?

12.27. A crystalline form of copper has a density of 8.95 g/cm^3. If the radius of copper atoms is 127.8 pm, is the copper unit cell (a) simple cubic, (b) body-centered cubic, or (c) face-centered cubic?

12.28. A crystalline form of molybdenum has a density of 10.28 g/cm^3 at a temperature at which the radius of a molybdenum atom is 139 pm. Which unit cell is consistent with these data? (a) simple cubic; (b) body-centered cubic; (c) face-centered cubic

---

12.29. Sodium metal crystallizes with a body-centered cubic structure at normal atmospheric pressure. The atomic radius of a sodium atom is 186 pm and the density of Na is 0.971 g/cm^3. At 613,000 atm (9 million psi) the structure of Na metal becomes fcc. Assuming the radius of Na is unchanged, what is the density of the fcc form of Na?

12.30. Calcium metal undergoes two phase changes under pressure. At atmospheric pressure, Ca crystallizes in an fcc unit cell that changes to a bcc unit cell above 20 GPa and to a simple cubic structure above 32 GPa.

a. If the radius of Ca remains unchanged, calculate the density of the simple cubic phase.

*b. Calculate the radius of a Ca atom in the simple cubic structure if the density of the sc and fcc phases is the same.

***12.31.** The unit cell for hcp Ti is shown in Figure P12.31. Given the unit cell dimensions, calculate the density of titanium (atomic radius 147 pm).

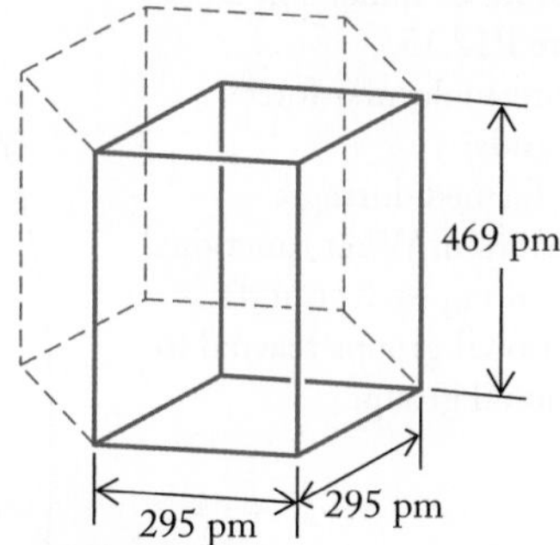

FIGURE P12.31

*12.32. Cobalt crystallizes with the same unit cell as titanium (Figure P12.31). The edge length (251 pm) and height (407 pm) are both shorter than for titanium. Is cobalt denser than titanium? Explain your answer.

Alloys and Medicine

Concept Review

12.33. Describe the structural differences between substitutional and interstitial alloys and give an example of each alloy.

12.34. Describe how homogeneous alloys and intermetallic compounds are similar and how they are different.

12.35. What effect does the substitution of Ni for Ti in the center of a bcc unit cell of Ti have on the unit cell edge length?

12.36. White gold was originally developed to give the appearance of platinum. One formulation of white gold contains 25% nickel and 75% gold. Which is more malleable, white gold or pure gold? Explain your answer.

12.37. Magnesium and hafnium have nearly the same atomic radius (within 1 pm). Does substitution of 25% of the magnesium by hafnium in an alloy increase or decrease the density of the alloy in comparison with pure magnesium? Explain your answer.

12.38. Why are the alloys that second-row nonmetals—such as B, C, and N—form with transition metals more likely to be interstitial than substitutional?

Problems

12.39. The unit cell of NiTi in Figure P12.39a shows a bcc arrangement of Ti atoms with Ni in the middle. Could we also draw the unit cell of NiTi as a bcc arrangement of Ni atoms with Ti in the middle as shown in Figure P12.39b? Explain why or why not.

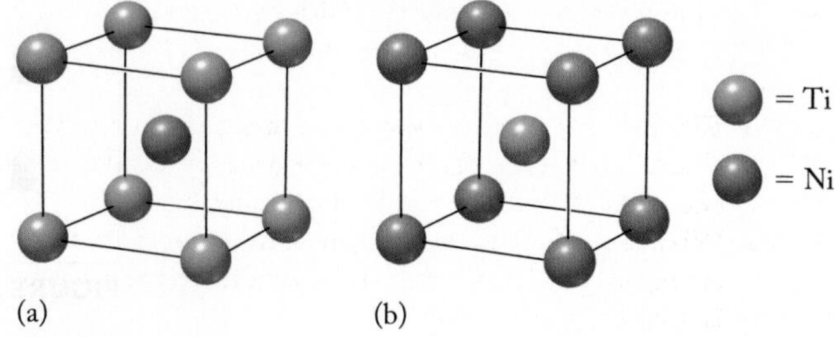

FIGURE P12.39

12.40. Can both of the unit cells in Figure P12.40 represent the same substitutional alloy? Explain why or why not.

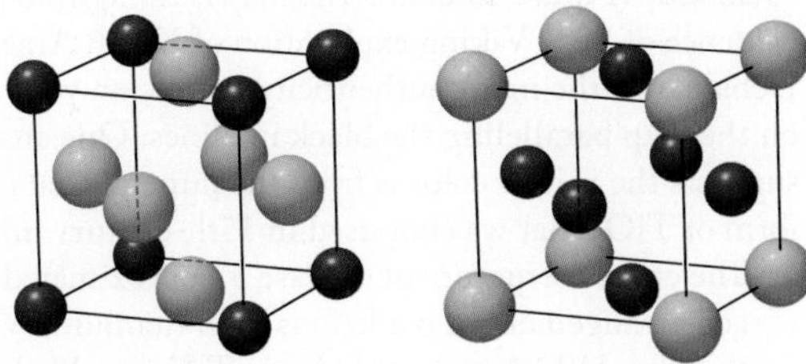

FIGURE P12.40

12.41. **Hydrogen Storage** Hydrogen is an attractive alternative fuel to hydrocarbons because it has a high fuel value and does not produce carbon dioxide when burned. One challenge to developing a hydrogen economy is storing hydrogen. Many transition metals absorb hydrogen by breaking the H—H bond and storing H atoms in the interstices between the metal atoms.

a. What is the minimum radius for the host metal for a H atom (radius 37 pm) to fit in a tetrahedral hole of an fcc unit cell?
b. Suppose the H is present as a hydride ion, H^- (ionic radius = 146 pm). Is the hydrogen then more likely to be found as an interstitial alloy in a tetrahedral or octahedral hole or as a substitutional alloy? *Hint*: Consult Appendix 3 for atomic radii of metals.

12.42. Metal borides MB_x exist for many transition metals.

a. What is the minimum radius for the host metal for a B atom (radius 88 pm) to fit in an octahedral hole of an fcc unit cell?
b. What is the value of x if an average of one octahedral hole is occupied per unit cell?

12.43. **Dental Fillings** Dental fillings are mixtures of several alloys, including one with the formula Ag_3Sn. Silver (radius 144 pm) and tin (140 pm) both crystallize in an fcc unit cell.

a. Is this alloy likely to be a substitutional alloy or an interstitial alloy?
*b. Draw an fcc unit cell for Ag_3Sn that corresponds to an *intermetallic* alloy of silver and tin.

12.44. **Joint Replacement** A titanium–gold alloy discovered in 2016 is four times harder than pure titanium, making the new material attractive for use in knee and hip replacements.

a. Is this titanium–gold alloy more likely to be a substitutional or an interstitial alloy?
b. The structure of the alloy can be described as a bcc arrangement of Au atoms with two Ti atoms on each face. What is the formula of this alloy?

12.45. What is the formula of vanadium carbide if the vanadium atoms have a cubic closest-packed structure and two of the octahedral holes are occupied by C atoms?

12.46. What is the formula of manganese nitride if the manganese atoms have an fcc unit cell and one of the tetrahedral holes is occupied by a N atom?

12.47. When a metallurgist allows a molten mixture of 96 mg molybdenum and 184 mg of tungsten to cool to 25°C, it forms a crystalline alloy.

a. The unit cell of the alloy contains eight Mo atoms on the corners and one tungsten atom in the center. What is the empirical formula of the alloy?
b. Is the unit cell consistent with the masses of the metals used in the reaction?
c. Calculate the density of the alloy given that the atomic radii of both metals are 139 pm.

12.48. Building on the success described in Problem 12.47, the metallurgist halved the amount of molybdenum and increased the amount of tungsten in the mixture 1.5 times to obtain a different alloy.

a. The unit cell of this alloy contains four W atoms on alternate corners and one tungsten atom in the center. Molybdenum atoms occupy the remaining four corners in the unit cell. What is the empirical formula of the alloy?
b. Calculate the density of the alloy given that the atomic radii of both metals are 139 pm.
*c. Would the density of the alloy change if the four tungsten and four molybdenum atoms in the unit cell were randomly distributed among all available sites in the lattice?

12.49. One of the many alloys of copper and zinc has the unit cell shown in Figure P12.49.

a. What type of crystal structure does this alloy have?
b. What are the proportions of copper and zinc in the alloy?

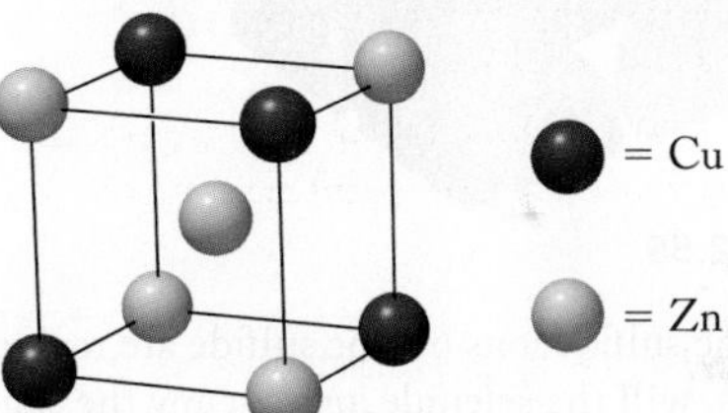

FIGURE P12.49

*12.50. The unit cell of an iron–aluminum alloy is shown in Figure P12.50. (*Hint*: Consult Figure 12.6 for help in identifying the atoms.)

a. Is this alloy substitutional or interstitial?
b. What is the composition of this alloy?

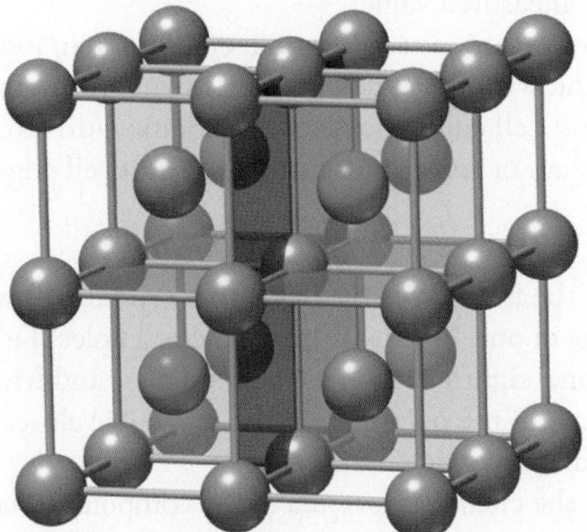

FIGURE P12.50

12.51. If the unit cell of a substitutional alloy of copper and tin has the same unit cell edge as the unit cell of copper, will the alloy have a greater density than copper? Explain your answer.

12.52. If the unit cell of an interstitial alloy of vanadium and carbon has the same unit cell edge as the unit cell of vanadium, will the alloy have a greater density than vanadium? Explain your answer.

Ionic Solids and Salt Crystals

Concept Review

12.53. Crystals of both LiCl and KCl have the rock salt structure. In the unit cell of LiCl, adjacent Cl^- ions touch each other. In KCl, they don't. Why?

12.54. Does the absolute size of an octahedral hole in an fcc lattice of halide ions change as we move down group 17 from fluoride to iodide?

***12.55.** In some books the unit cell of CsCl is described as being body-centered cubic (Figure P12.55); in others, as simple cubic (see Figure 12.9a). Explain how CsCl crystals might be described by either unit cell type.

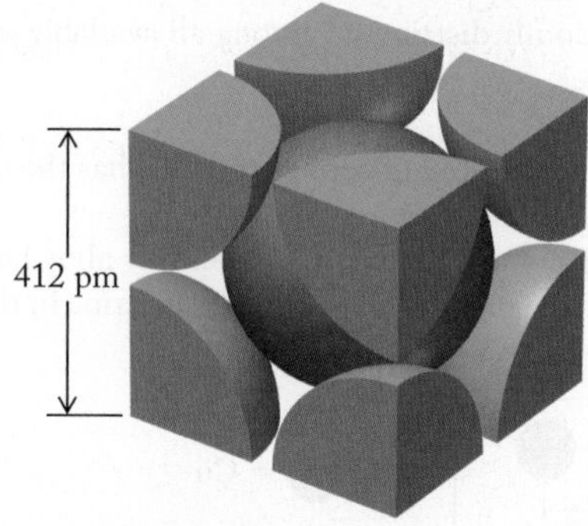

FIGURE P12.55

12.56. If some of the sulfide ions in zinc sulfide are replaced by selenide ions, will the selenide ions occupy the same sites as the sulfide ions?

12.57. Why can't calcium chloride have a rock salt structure?

12.58. Why can't sodium fluoride have exactly the same structure as calcium fluoride?

***12.59.** If the cation–anion radius ratio increases for an ionic compound with the rock salt crystal structure, is the calculated density more likely to be greater than or less than the measured value?

*12.60. If the cation–anion radius ratio increases for an ionic compound with the rock salt crystal structure, is the length of the unit cell edge calculated from ionic radii likely to be greater than or less than the observed unit cell edge length?

Problems

12.61. What is the formula of the oxide that crystallizes with Fe^{3+} ions in one-fourth of the octahedral holes, Fe^{3+} ions in one-eighth of the tetrahedral holes, and Mg^{2+} in one-fourth of the octahedral holes of a cubic closest-packed arrangement of oxide ions (O^{2-})?

12.62. What is the chemical formula of the compound that crystallizes in a simple cubic arrangement of fluoride ions with Ba^{2+} ions occupying half of the cubic holes?

12.63. **The Vinland Map** At Yale University there is a map, believed to date from the 1400s, of a landmass labeled "Vinland" (Figure P12.63). The map is thought to be evidence of early Viking exploration of North America. Debate over the map's authenticity centers on yellow stains on the map paralleling the black ink lines. One analysis suggests the yellow color is from the mineral anatase, a form of TiO_2 that was not used in 15th-century inks.

a. The crystal structure of anatase is approximated by a ccp arrangement of oxide ions with titanium(IV) ions in holes. Which type of hole are Ti^{4+} ions likely to occupy? The radius of Ti^{4+} is 60.5 pm.
b. What fraction of these holes is likely to be occupied?

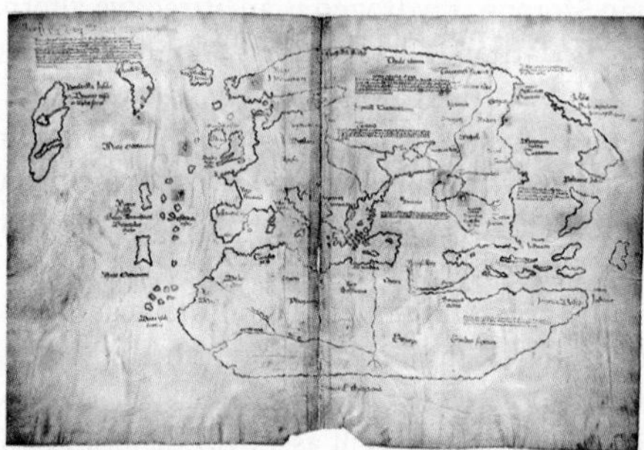

FIGURE P12.63

*12.64. The crystal structure of olivine—M_2SiO_4 (where M = Mg or Fe)—can be viewed as a ccp arrangement of oxide ions with silicon(IV) in tetrahedral holes and the metal ions in octahedral holes.

a. What fraction of each type of hole is occupied?
b. The unit cell volumes of Mg_2SiO_4 and Fe_2SiO_4 are 2.91×10^{-26} cm^3 and 3.08×10^{-26} cm^3, respectively. Why is the unit cell volume of Fe_2SiO_4 larger?

12.65. The rock salt structure (Figure 12.19) of magnesium selenide (MgSe) at atmospheric pressure changes to a cesium chloride structure (Figure P12.55) under pressure.

a. How do these structures differ?
b. Explain how both structures are consistent with the observed stoichiometry of MgSe.

12.66. The sphalerite structure of ZnS changes to a rock salt structure above 15 GPa.

a. Describe the differences between these two structures.
b. Explain how both structures are consistent with the observed stoichiometry of ZnS.

12.67. The unit cell of rhenium trioxide (ReO_3) consists of a cube with rhenium atoms at the corners and an oxygen atom on each of the 12 edges. The atoms touch along the edge of the unit cell. The radii of Re and O atoms in ReO_3 are 137 and 73 pm, respectively.

a. Draw the unit cell of ReO_3 and show that it is consistent with the empirical formula.
b. Calculate the density of ReO_3.

12.68. **New Forms of Salt** Sodium chloride is the prototypical example of the rock salt structure. In 2013, though, researchers discovered a new compound containing sodium and chlorine. Under high pressure, NaCl reacts with excess sodium metal to produce the compound shown in Figure P12.68.

a. What is the empirical formula of this compound?
b. Calculate the density of this compound if the atoms touch along the edges, using the atomic radii for the sodium and chlorine.

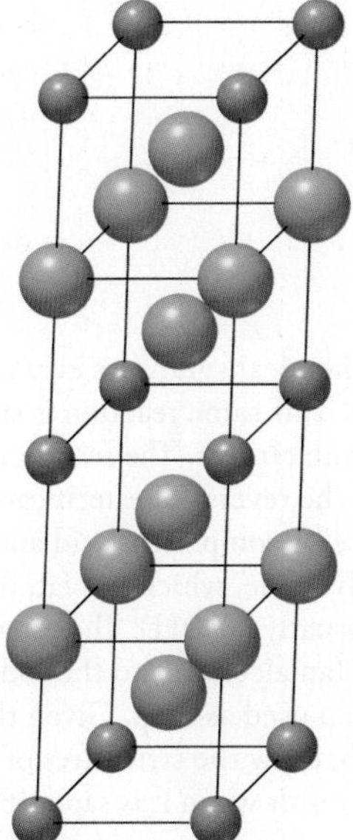

FIGURE P12.68

12.69. Magnesium oxide crystallizes in the rock salt structure. Its density is 3.60 g/cm^3. What is the edge length of the fcc unit cell of MgO?

12.70. Crystalline potassium bromide (KBr) has a rock salt structure and a density of 2.75 g/cm^3. Calculate its unit cell edge length.

Allotropes of Carbon

Concept Review

12.71. When amorphous red phosphorus is heated at high pressure, it is transformed into the allotrope black phosphorus, which can exist in one of several forms. One form consists of six-membered rings of phosphorus atoms (Figure P12.71a). Why are the six-atom rings in black phosphorus puckered, whereas the six-atom rings in graphene (Figure P12.71b) are planar?

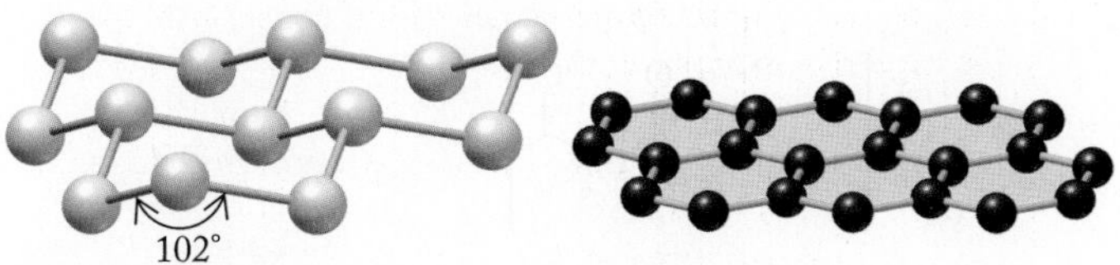

(a) Black phosphorus (b) Graphene

FIGURE P12.71

*12.72. If the carbon atoms in graphite are replaced by alternating B and N atoms, would the resulting structure contain puckered rings like black phosphorus or flat ones like graphite (see Figure P12.71)?

Problems

12.73. In the fullerene known as buckminsterfullerene (C_{60}), molecules of C_{60} form a cubic closest-packed array of spheres with a unit cell edge length of 1410 pm.

a. What is the density of crystalline C_{60}?
b. If we treat each C_{60} molecule as a sphere of 60 carbon atoms, what is the radius of the C_{60} molecule?

12.74. C_{60} reacts with alkali metals to form M_3C_{60} (where M = Na or K). The crystal structure of M_3C_{60} contains cubic closest-packed spheres of C_{60} with metal ions in holes. (Use the radius for C_{60} that you calculated in Problem 12.73.)

a. If the radius of a K^+ ion is 138 pm, which type of hole is a K^+ ion likely to occupy? What fraction of the holes will be occupied?
b. Under certain conditions, a different substance, K_6C_{60}, can be formed in which the C_{60} molecules have a bcc unit cell. Calculate the density of a crystal of K_6C_{60}.

12.75. Semiconducting Phosphorus A cubic form of phosphorus is drawing renewed interest as a semiconductor. The distance between atoms in this cubic form of phosphorus is 238 pm (Figure P12.75).

a. Calculate the density of this form of phosphorus.
b. Is the cubic hole in cubic P large enough to hold either a silicon or a boron atom?
*c. Would the material(s) from part (b) behave as n-type or p-type semiconductors?

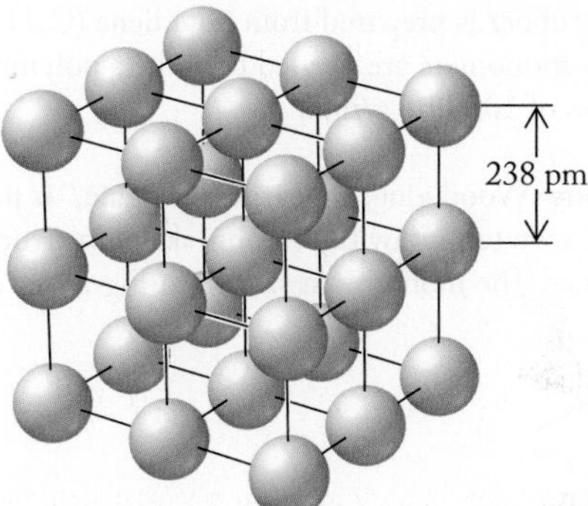

FIGURE P12.75

12.76. **Ice under Pressure** Kurt Vonnegut's novel *Cat's Cradle* describes an imaginary, high-pressure form of ice called "ice-nine." With the assumption that ice-nine has a cubic closest-packed arrangement of oxygen atoms with hydrogen atoms in the appropriate holes, what type of hole will accommodate the H atoms?

Polymers

Concept Review

12.77. Can a polymer be composed of more than one type of monomer? Explain why or why not.

12.78. **Body Wash** Personal care products such as body wash contain molecules like cetyl alcohol and glycol stearate (Figure P12.78). Are these two molecules considered to be polymers?

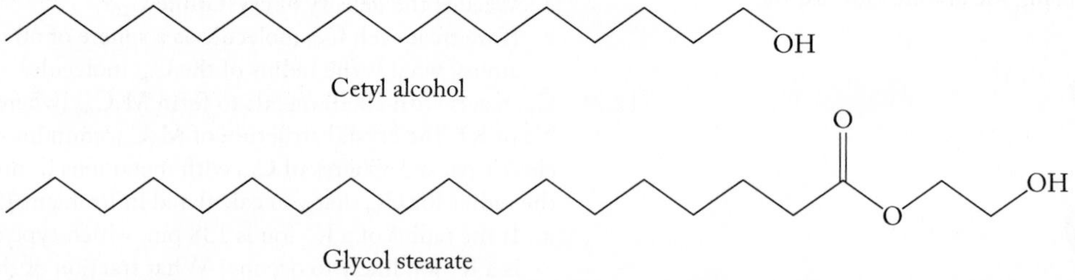

FIGURE P12.78

12.79. Compare the large hydrocarbon $C_{24}H_{50}$ with polyethylene. What structural feature(s) do they share in common? How do the structures of these compounds differ?

*12.80. The 2000 Nobel Prize in Chemistry was awarded for research on the electrically conductive polymer polyacetylene, an addition polymer that is prepared from the monomer acetylene (HC≡CH). How does polyacetylene differ from polyethylene?

Problems

12.81. Polyethylene is prepared from the monomer ethylene (C_2H_4). About how many monomers are needed to make a polymer with a molar mass of 100,000 g/mol?

12.82. Synthetic rubber is prepared from butadiene (C_4H_6). About how many monomers are needed to make a polymer with a molar mass of 100,000 g/mol?

12.83. Making Glue Wood glue, or "carpenter's glue," is made of poly(vinyl acetate). Draw the carbon-skeleton structure of this polymer. The monomer is shown in Figure P12.83.

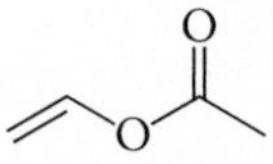

FIGURE P12.83

12.84. Cross-linked polymers based on boronic acids (Figure P12.84) show promise as a recyclable plastic. What is the other product of the reaction? Is this an addition or a condensation polymer?

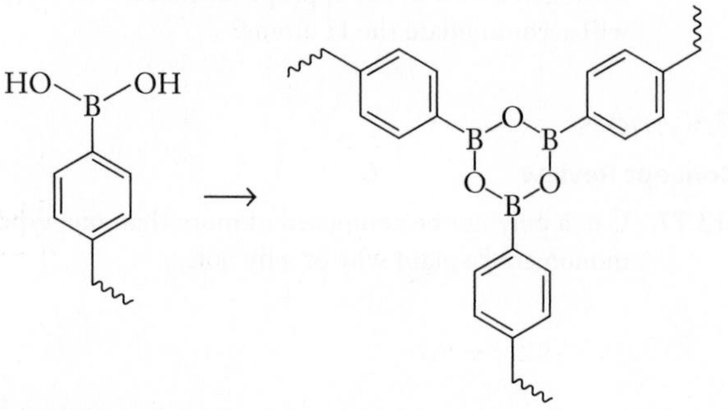

FIGURE P12.84

12.85. Soap One of the first chemical syntheses ever carried out was the making of soap. This same reaction is still carried out by modern-day manufacturers. The reaction is called saponification, and it is the reverse of esterification: an ester is taken apart to produce its component acid and alcohol. A compound called a glyceride, which is a fat, is reacted with an aqueous base (usually NaOH). The ester bonds are broken to yield glycerol (an alcohol) and the sodium salt of a fatty acid, which is then used as soap. Given the glyceride shown in Figure P12.85, draw the structures of the alcohol and the fatty acids that result when it is saponified.

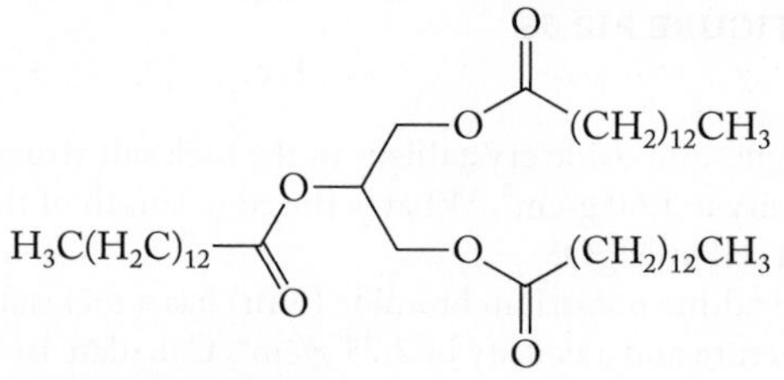

FIGURE P12.85

12.86. Reactions between 1,6-diaminohexane, $H_2N(CH_2)_6NH_2$, and different dicarboxylic acids, $HOOC(CH_2)_nCOOH$, are used to prepare polymers that have a structure similar to that of nylon. How many carbon atoms (n) were in the dicarboxylic acids used to prepare the polymers with the repeating units shown in Figure P12.86?

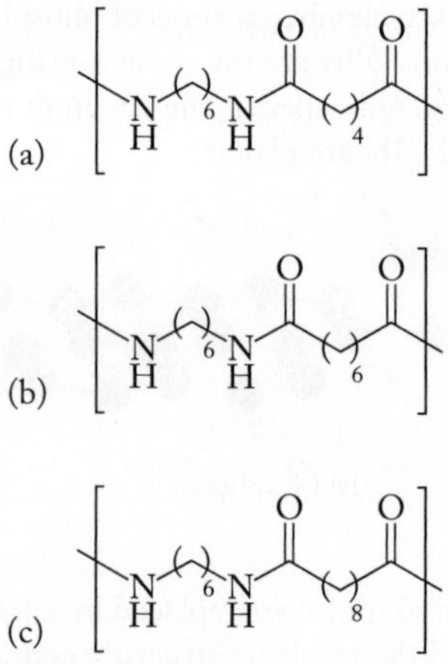

FIGURE P12.86

12.87. The two polymers in Figure P12.87 have the same empirical formula.
a. What pairs of monomers could be used to make each of them?
b. How might the physical properties of these two polymers differ?

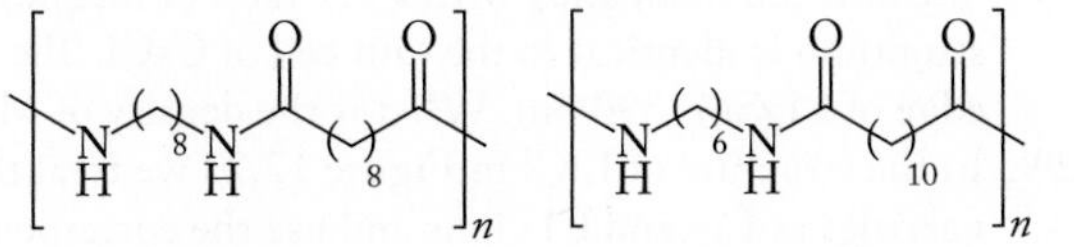

Polymer I Polymer II

FIGURE P12.87

12.88. The polyester called Kodel is made with polymeric strands prepared by the reaction of dimethyl terephthalate with 1,4-di(hydroxymethyl)cyclohexane (Figure P12.88).
a. Is Kodel a condensation polymer or an addition polymer? What is the other product of the reaction?
*b. Dacron (see Sample Exercise 12.8) is made from dimethyl terephthalate and ethylene glycol. What properties of Kodel fibers might make them better than Dacron as a clothing material?

Dimethyl terephthalate (dimethyl benzene-1,4-dicarboxylate) + 1,4-Di(hydroxymethyl)cyclohexane

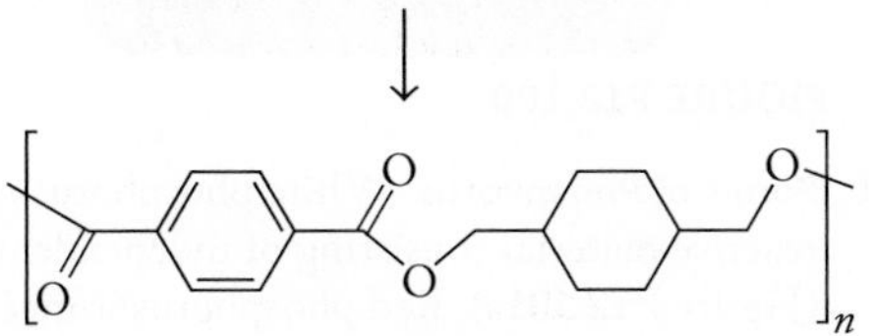

Kodel

FIGURE P12.88

12.89. Lexan is a polymer belonging to the class of materials called polycarbonates. Figure P12.89 shows the polymerization reaction for Lexan.
a. What other compound is formed in the polymerization reaction?
*b. Why is Lexan called a "polycarbonate"?

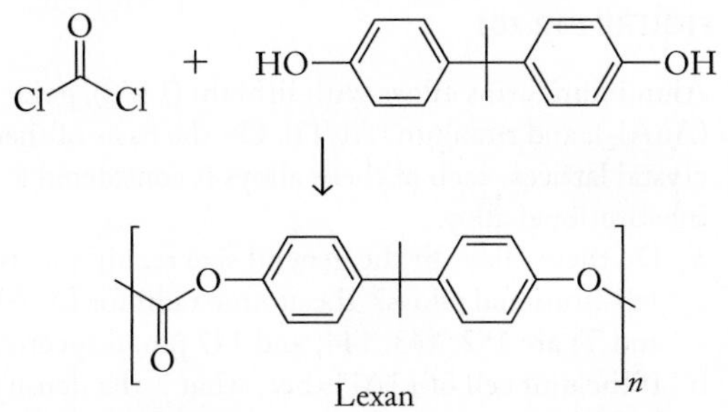

Lexan

FIGURE P12.89

12.90. **Nonbiodegradable Polymers** Biodegradable polymers are important in applications such as dissolving sutures, but biocompatible polymers that resist degradation can serve as drug delivery systems in implanted devices. Poly(caprolactone) (PCL; Figure P12.90) degrades slowly in the human body but is very permeable to the contraceptive levonorgestrel. This combination is used in implantable contraceptive devices that are effective for many years.
*a. Why is levonorgestrel soluble in PCL?
b. PCL does eventually degrade by reacting with water. What are the products of biodegradation?
c. Is PCL an addition or a condensation polymer?

PCL Levonorgestrel

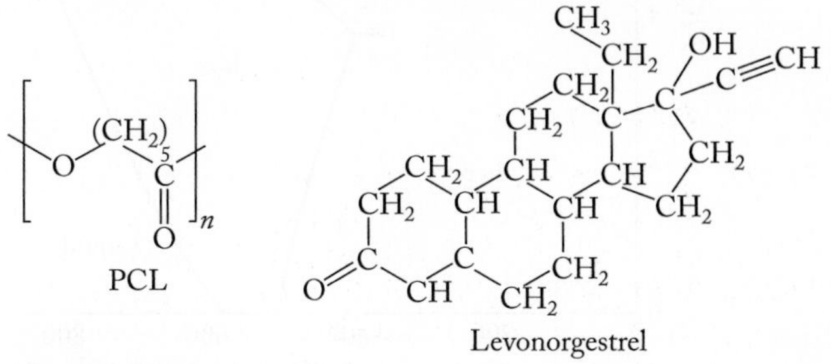

FIGURE P12.90

Additional Problems

12.91. A unit cell consists of a cube that has an ion of element X at each corner, an ion of element Y at the center of the cube, and an ion of element Z at the center of each face. What is the formula of the compound?

12.92. The unit cell of an oxide of uranium consists of cubic closest-packed uranium ions with oxide ions in all the tetrahedral holes. What is the formula of the oxide?

12.93. The phase diagram for titanium is shown in Figure P12.93.
a. Which structure does Ti metal have at 1500 K and 6 GPa of pressure?
b. How many phase changes does Ti metal undergo as pressure is increased at 725°C?

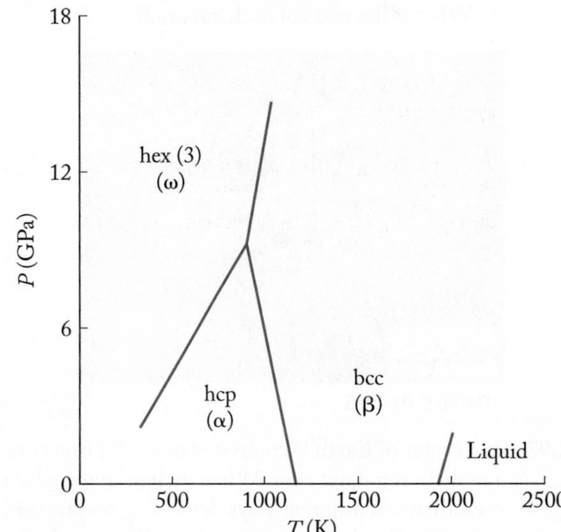

FIGURE P12.93

12.94. The phase diagram of thallium metal is shown in Figure P12.94.
a. At what temperature and pressure are all three solid phases of thallium metal present?

b. Does the melting point of thallium increase or decrease with increasing pressure?

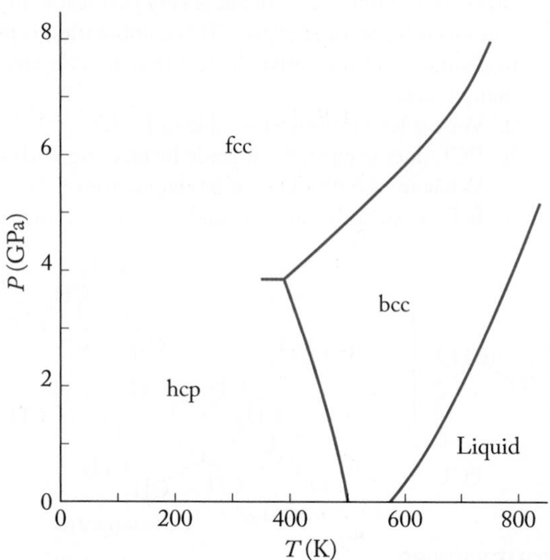

FIGURE P12.94

12.95. Silver nanoparticles are embedded in clothing fabric to kill odor-causing bacteria. Silver crystallizes in an fcc unit cell.

a. How many unit cells are present in a cubic silver particle with an edge length of 25 nm?

*b. How many silver atoms are there in this particle?

c. If there are 1360 μg of silver per sock, how many of these particles does this mass correspond to?

12.96. In 2011, researchers at Duke University reported on the use of cylindrical gold particles called "nanorods" to image brain tumors (Figure P12.96).

a. If the nanorods have a diameter of 32 nm and a height of 67 nm, approximately how many gold atoms are in each nanorod?

b. How many unit cells does this correspond to?

c. What is the mass of each nanorod?

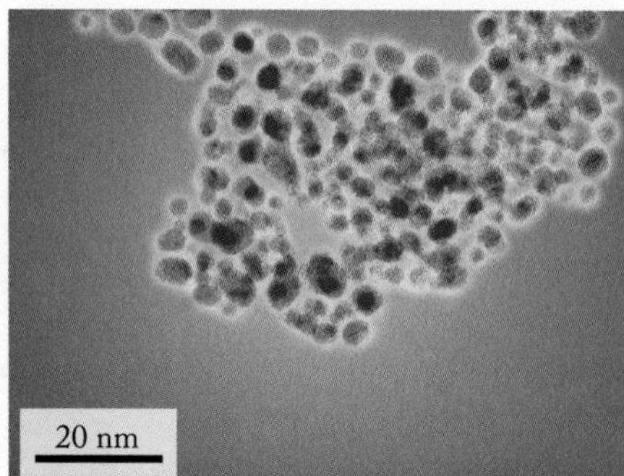

FIGURE P12.96

12.97. The center of Earth is composed of a solid iron core within a molten iron outer core. When molten iron cools, it crystallizes in different ways depending on pressure—in a bcc unit cell at low pressure and in a hexagonal unit cell at high pressures like those at Earth's center.

a. Calculate the density of bcc iron given that the radius of an iron atom is 126 pm.

b. Calculate the density of hexagonal iron given a unit cell volume of 5.414×10^{-23} cm^3.

*c. Seismic studies suggest that the density of Earth's solid core is only about 90% of that of hexagonal Fe. Laboratory studies have shown that up to 4% by mass of Si can be substituted for Fe without changing the hcp crystal structure built on hexagonal unit cells. Calculate the density of such a crystal.

12.98. The unit cell of an alloy with a 1:1 ratio of magnesium and strontium is identical to the unit cell of CsCl. The unit cell edge of MgSr is 390 pm. What is the density of MgSr?

***12.99.** In the structure of LiCl in Figure 12.23 we treat the particles as Li^+ and Cl^- ions and use the corresponding ionic radii to calculate the density of LiCl as 2.09 g/cm^3.

a. How would the structure change if we treated the particles as neutral atoms? Assume an fcc lattice composed of one type of atom with the other atoms in holes.

b. Calculate the density of LiCl if we treated the particles as atoms rather than ions.

12.100. Figure P12.100 shows four identical layers of atoms in an *aaaa* . . . pattern. Using the yellow triangle as a guide, how many nearest-neighbor atoms does any one hole between the layers have? Are there any other kinds of holes in this structure?

FIGURE P12.100

12.101. Forms of Phosphorus White phosphorus is a highly reactive material consisting of independent P_4 tetrahedra (Figure P12.101a). Red phosphorus is much less reactive with a structure consisting of linked P_4 tetrahedra (Figure P12.101b). What is the maximum-sized particle that could fit inside a P_4 tetrahedron? Is there any element that fits this criterion?

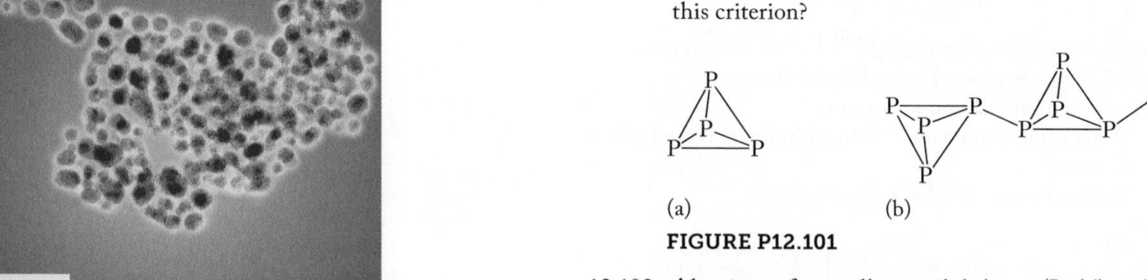

FIGURE P12.101

12.102. Aluminum forms alloys with lithium (LiAl), gold ($AuAl_2$), and titanium (Al_3Ti). On the basis of their crystal lattices, each of these alloys is considered a substitutional alloy.

a. Do these alloys fit the general size requirements for substitutional alloys? The atomic radii for Li, Al, Au, and Ti are 152, 143, 144, and 147 pm, respectively.

b. If the unit cell of LiAl is bcc, what is the density of LiAl?

***12.103.** The aluminum alloy Cu_3Al crystallizes in a bcc unit cell. Propose a way that the Cu and Al atoms could be allocated between bcc unit cells that is consistent with the formula of the alloy.

12.104. **Light-Emitting Diodes** The colored lights on many electronic devices are light-emitting diodes (LEDs). One of the compounds used to make them is aluminum phosphide (AlP), which crystallizes in a sphalerite crystal structure.

a. If AlP were an ionic compound, would the ionic radii of Al^{3+} and P^{3-} be consistent with the size requirements of the ions in a sphalerite crystal structure?

b. If AlP were a covalent compound, would the atomic radii of Al and P be consistent with the size requirements of atoms in a sphalerite crystal structure?

12.105. Identify the reactants in the polymerization reactions that produce the two polymers shown in Figure P12.105.

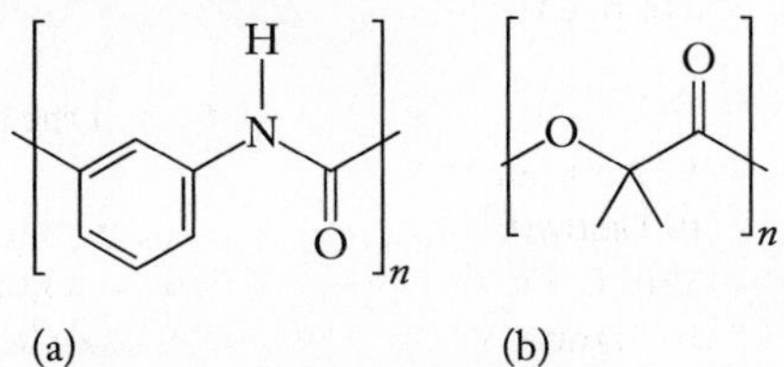

FIGURE P12.105

*12.106. **Raincoats** "Waterproof" nylon garments have a coating to prevent water from penetrating the hydrophilic fibers. Which functional groups in the nylon molecule make it hydrophilic?

12.107. Draw the carbon-skeleton structure of the condensation polymer of $H_2N(CH_2)_6COOH$. How does this polymer compare with nylon-6?

*12.108. Putrescine, $H_2N(CH_2)_4NH_2$, is one of the compounds that form in rotting meat.

a. Draw the carbon-skeleton structures of all the trimers (a molecule formed from three monomers) that can be formed from putrescine, adipic acid, and terephthalic acid (Figure P12.108). The three monomers forming the trimer do not have to be different from one another.

b. A chemist wants to make a putrescine polymer containing a 1:1 ratio of adipic acid to terephthalic acid. What should be the mole ratio of the three reactants?

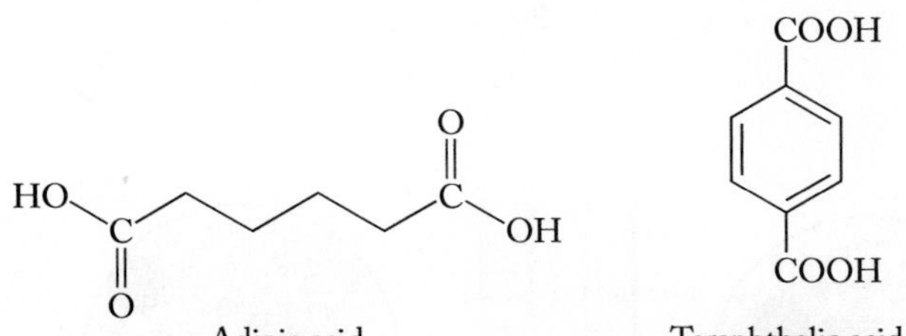

Adipic acid — Terephthalic acid (benzene-1,4-dicarboxylic acid)

FIGURE P12.108

***12.109.** Polymer chemists can modify the physical properties of polystyrene by copolymerizing divinylbenzene with styrene (Figure P12.109). The resulting polymer has strands of polystyrene cross-linked with divinylbenzene. Predict how the physical properties of the copolymer might differ from those of 100% polystyrene.

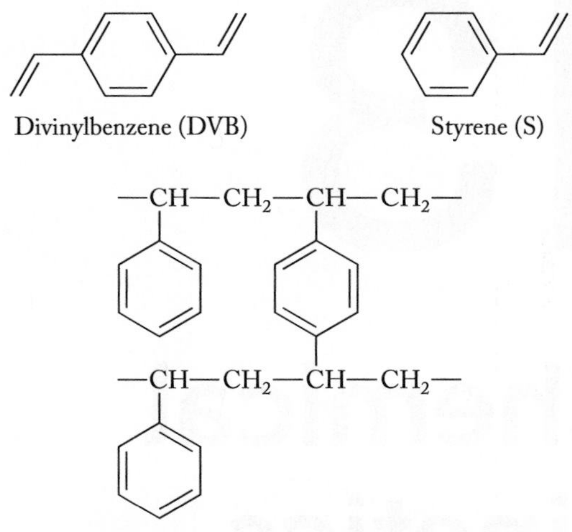

Divinylbenzene (DVB) — Styrene (S)

—CH—CH$_2$—CH—CH$_2$—

—CH—CH$_2$—CH—CH$_2$—

S cross-linked with DVB

FIGURE P12.109

12.110. Maleic anhydride and styrene (Figure P12.110) form a polymer with alternating units of each monomer.

a. Draw two repeating monomer units of the polymer.

b. From the structure of the copolymer, predict how its physical properties might differ from those of polystyrene.

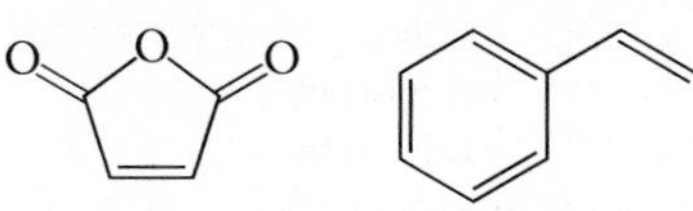

Maleic anhydride — Styrene

FIGURE P12.110

12.111. **Superglue** The active ingredient in superglue is methyl 2-cyanoacrylate (Figure P12.111). The liquid glue hardens rapidly when methyl 2-cyanoacrylate polymerizes. This happens when it contacts a surface containing traces of water or other compounds containing –OH or –NH– groups. Draw the carbon-skeleton structure of two repeating monomer units of poly(methyl 2-cyanoacrylate).

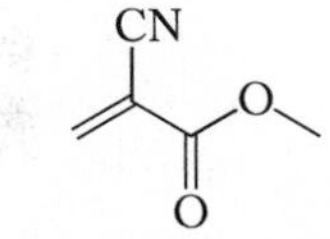

Methyl 2-cyanoacrylate

FIGURE P12.111

*12.112. Silicones are polymeric materials with the formula $[R_2SiO]_n$ (Figure P12.112). They are prepared by reaction of R_2SiCl_2 with water, yielding the polymer and aqueous HCl. Consider this reaction as taking place in two steps: (1) water reacts with 1 mole of R_2SiCl_2 to produce a new monomer and 1 mole of HCl(*aq*); (2) one new monomer molecule reacts with another new monomer molecule to eliminate one molecule of H_2O and make a dimer with a Si—O—Si bond.

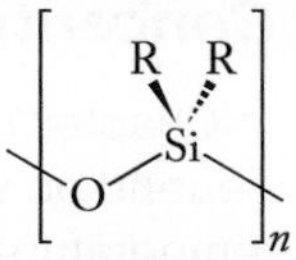

a. Suggest two balanced equations that describe these reactions that occur over and over again to produce a silicone polymer.

b. Why are silicones water repellent?

Silicone

FIGURE P12.112

13

Chemical Kinetics

Reactions in the Atmosphere

AIR QUALITY On a smoggy day, this iconic building in Tiananmen Square in Beijing is nearly invisible.

PARTICULATE **REVIEW**

Concentration and Solutions

In Chapter 13 we explore how the rates of reactions depend on variables such as concentration and temperature. The image shows two graduated cylinders that contain aqueous solutions of copper(II) nitrate. Cylinder 1 has a concentration of 0.01 *M*.

- Which cylinder contains the more concentrated solution?
- What is the concentration of the solution in cylinder 2?
- Which cylinder contains more dissolved solute?

(Review Section 4.2 if you need help.)

(Answers to Particulate Review questions are in the back of the book.)

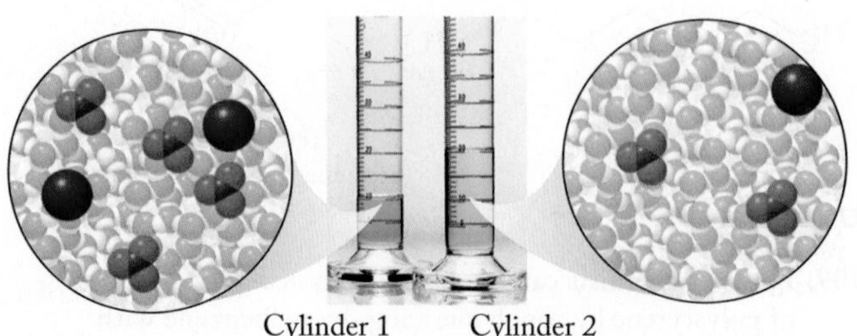

PARTICULATE **PREVIEW**

Collisions and Reactions

The images represent three samples of nitrogen dioxide, which is a component of the atmospheric pollution shown in the opening photograph of this chapter. At high temperatures, nitrogen dioxide decomposes to nitrogen monoxide and oxygen gas. As you read Chapter 13, look for ideas that will help you answer these questions:

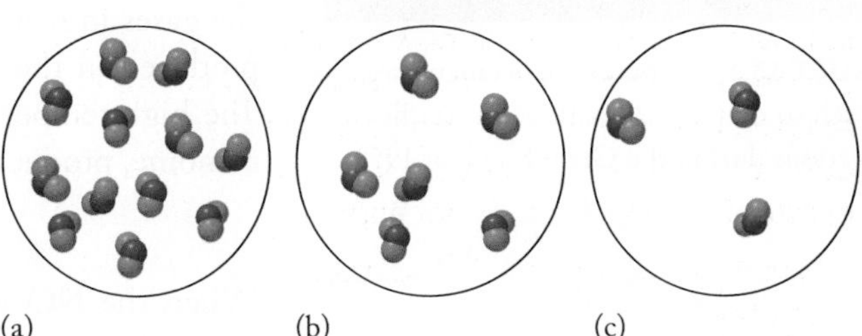

- Assuming the temperature of all three NO_2 samples is the same, in which one do the NO_2 molecules collide most frequently?
- In which sample do the NO_2 molecules collide four times more frequently than they do in sample (c)?
- In which of the three samples does the decomposition of NO_2 proceed the most rapidly?

Learning Outcomes

LO1 Relate the rates of change in the concentrations of reactants and products to each other and to reaction rate, reaction order, and rate constant units
Sample Exercises 13.1, 13.2

LO2 Determine average and instantaneous reaction rates from experimental data
Sample Exercise 13.3

LO3 Derive rate laws from initial reaction rate data
Sample Exercise 13.4

LO4 Use integrated rate laws to identify zero-, first-, and second-order reactions and to determine rate constants for such reactions
Sample Exercises 13.5, 13.8

LO5 Use rate laws to relate the half-lives of zero-, first-, and second-order reactions to reactant concentrations
Sample Exercises 13.6, 13.7

LO6 Relate the activation energy of a reaction and the effect of temperature on rate constants
Sample Exercises 13.9, 13.10

LO7 Relate a reaction mechanism (i.e., how a reaction occurs at the particle level) to its molecularity and overall rate law
Sample Exercises 13.11, 13.12

LO8 Describe catalysts and explain how they alter reaction mechanisms in ways that increase reaction rates
Sample Exercise 13.13

13.1 Cars, Trucks, and Air Quality

FIGURE 13.1 A London policeman uses a torch so that drivers will see his traffic directions during the Great Smog of 1952.

The common name for the haze hanging over the city in the chapter-opening photograph is *smog*, a name first used over a century ago to describe the air pollution that blanketed some cities during the winter when smoke from coal-fired stoves and furnaces combined with natural fog to produce air that had the appearance of pea soup. This greenish-yellow form of smog contained high concentrations of volatile sulfur compounds that combined with atmospheric moisture, forming highly acidic aerosols (suspended particles). Breathing in these pollutants can cause severe respiratory distress. During one particularly intense smog event in London, England, in December of 1952 (**Figure 13.1**), more than 100,000 people were sickened and about 12,000—mostly elderly or very young children—died from impaired lung function caused by the smog. In response to this environmental disaster, Parliament passed the Clean Air Act of 1956, which banned the use of high-sulfur coal for heating homes in cities and towns across the United Kingdom.

The smog in the chapter-opening photograph is different from the pea-soup variety that covered London. It is called **photochemical smog** because some of the gases in it form in the atmosphere when solar radiation interacts with gases produced in the internal combustion engines that power most cars and trucks. The high temperatures inside the engines in these vehicles allow N_2 and O_2 to combine, producing nitrogen monoxide:

$$N_2(g) + O_2(g) \rightarrow 2\,NO(g) \quad (13.1)$$

When the NO in engine exhaust enters the atmosphere, it slowly reacts with more oxygen, producing nitrogen dioxide:

$$2\,NO(g) + O_2(g) \rightarrow 2\,NO_2(g) \quad (13.2)$$

photochemical smog a mixture of gases formed in the lower atmosphere when sunlight interacts with compounds produced in internal combustion engines and other pollution sources.

which is the ingredient that often gives photochemical smog a brown color. Sunlight can break the bonds in NO_2, producing NO and very reactive oxygen atoms:

$$NO_2(g) \xrightarrow{\text{sunlight}} NO(g) + O(g) \quad (13.3)$$

These atoms of oxygen may react with molecules of water vapor, producing hydroxyl radicals,

$$O(g) + H_2O(g) \rightarrow 2\ OH(g) \qquad (13.4)$$

or they may combine with molecules of O_2, forming ozone,

$$O(g) + O_2(g) \rightarrow O_3(g) \qquad (13.5)$$

CONNECTION A free radical (Section 8.6), also called just a "radical," is a molecule with an odd number of electrons, which makes it chemically reactive.

Even trace concentrations of O_3 in the air can irritate your eyes and respiratory system. In addition, O_3 and OH radicals react with volatile organic compounds (VOCs) in the atmosphere, converting them to compounds containing C═O double bonds, such as acetaldehyde (**Figure 13.2**).

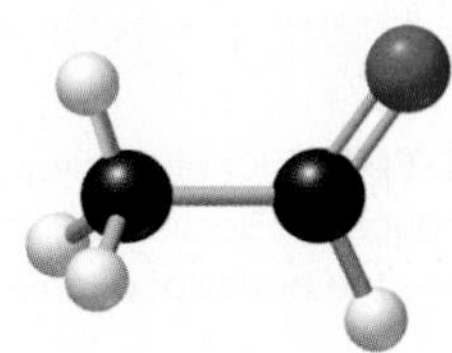

FIGURE 13.2 Acetaldehyde.

Acetaldehyde undergoes further oxidation and combines with nitrogen dioxide, forming a compound with the common name peroxyacetyl nitrate (PAN; **Figure 13.3**). PAN is an extremely potent eye and respiratory irritant and a principal contributor to the watery eyes, runny noses, sore throats, and difficulty breathing that many people experience during photochemical smog events.

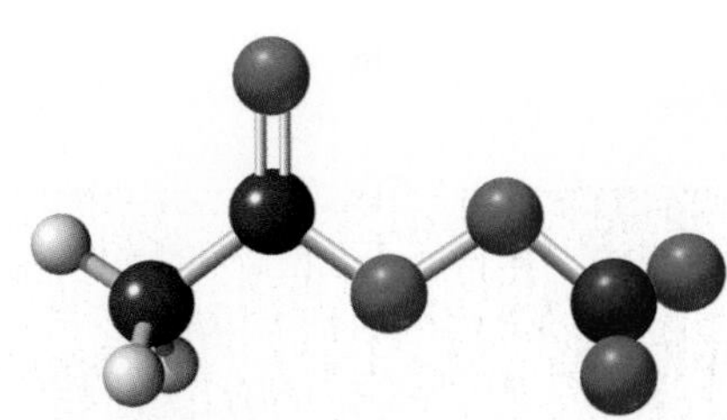

FIGURE 13.3 Peroxyacetyl nitrate (PAN).

The chemical reactions in Equations 13.1 to 13.5 are linked: the products of the early reactions in the sequence are reactants in later reactions. These relations are the reason why the atmospheric concentrations of different smog components reach their maximum values at different times during the day (**Figure 13.4**). Also, the *rates* of the reactions—that is, the rates at which reactants are consumed and products are formed—control the atmospheric concentrations of the components, including when they form and how long they last.

The formation of NO from N_2 and O_2 (Equation 13.1), for example, is an endothermic reaction ($\Delta H^\circ = 180.6$ kJ/mol) that proceeds only at extremely high temperatures. However, the gases in the combustion chamber of an automobile engine experience such temperatures for only a tiny fraction of a second before they flow into the car's exhaust system. Therefore, NO forms only if the rate of its formation reaction is very rapid. Unfortunately, the reaction is quite rapid.

To further explore the impact of reaction rates on smog formation, let's return to Figure 13.4 and see how NO emitted by vehicles during a morning rush hour affects air quality on a sunny day with little wind over a large urban area. Heavy traffic and exhaust emissions increase the concentration of NO, which reaches a maximum in the atmosphere early in the morning—near the height of traffic. By mid-morning, NO levels start to fall and NO_2 levels start to rise as NO combines with atmospheric O_2. By noon, the concentration of NO_2 is falling as it undergoes photodecomposition, releasing O atoms that combine with O_2, forming O_3. Ozone concentrations reach their maximum in mid-afternoon and then fall as this highly reactive allotrope of oxygen combines with VOCs and nitrogen oxides. A principal source of VOCs in urban air is unburnt hydrocarbons from car and truck fossil fuels. Their peak concentration also tracks with the morning rush hour. If we were to extend the graph in Figure 13.4, we would see a second maximum in VOC concentration after the evening rush hour.

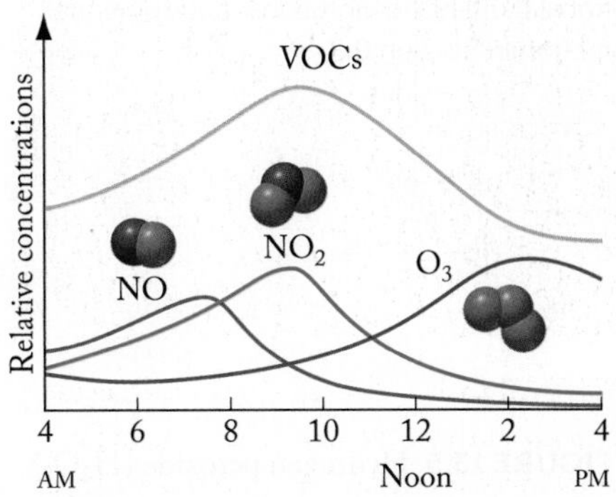

FIGURE 13.4 The concentrations of these components of photochemical smog reach their maxima in the atmosphere at different times as they are produced and then consumed in the many reactions that contribute to smog formation.

Smog conditions and ozone concentrations that posed a threat to human health were a worsening environmental crisis during the 1950s and 1960s in many American cities, particularly in metropolitan Los Angeles. In fact, the expression "brown L.A. haze" became so common it found its way into the lyrics of a popular song.[1] To combat smog pollution, exhaust systems of

[1]The chorus of "Come Monday" contains the line "I spent four lonely days in a brown L.A. haze." The song was written and recorded by Jimmy Buffett in 1974, the year that automobiles equipped with catalytic converters first went on sale.

chemical kinetics the study of the rates of change of concentrations of substances involved in chemical reactions.

automobiles sold in the United States since the 1975 model year have been equipped with devices called *catalytic converters*, which reduce the concentrations of NO in the exhaust gases.

In this chapter we explore how catalytic converters accelerate the rates of chemical reactions that turn NO and other pollutants into nonpolluting products. To understand this process, we need to investigate the factors that influence the rates of chemical reactions, but before we do that, we need to learn how to measure reaction rates and to express the results of those measurements. All of these concepts are part of **chemical kinetics**, the study of the rates of chemical reactions and the factors that influence those rates. Chemists use this knowledge to determine how reactions happen at the molecular level—that is, the *mechanisms* of reactions. With this knowledge, chemists can manipulate reaction conditions to increase reaction rates and, ultimately, product yields. This fundamental branch of chemistry is the focus of this chapter.

CONNECTION Kinetic molecular theory, introduced in Section 5.8, is a model that describes the behavior of gases.

13.2 Reaction Rates

CHEMTOUR
Reaction Rate

Previous chapters in this book provide a foundation on which we can build an understanding of chemical kinetics. In Chapter 5 we explored the kinetic molecular theory of gases and the relationship between temperature and the average speeds of molecules in the gas phase. Faster speeds mean more frequent and forceful collisions between molecules. As we see here in this chapter, more frequent collisions can mean more rapid reactions. Reaction rates vary over a wide range (**Figure 13.5**), from so slow that we barely perceive they are occurring to explosively fast. The following four factors influence reaction rates:

1. *Physical states of the reactants.* The more frequently that particles interact—that is, the more often they collide with one another—the faster they can react with each other. Particles in solids have the same nearest neighbors over time, whereas the particles in liquids move in relation to one another. The particles in gases are widely separated but move very rapidly. These differences in motion are one reason why reactions in the solid phase tend to be much slower than reactions in liquids and gases.
2. *Concentration of reactants.* The rates of most chemical reactions depend on the concentrations of one or more of the reactants. Greater concentrations often produce more rapid reaction rates.

CONNECTION We discussed translational (and vibrational and rotational) motion in Section 6.3.

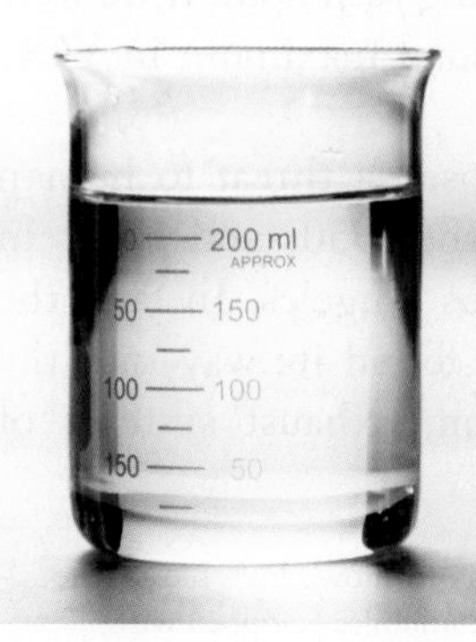

(a)

(b)

(c)

FIGURE 13.5 Hydrogen peroxide (H_2O_2) decomposes spontaneously to water and oxygen gas. (a) A 30% aqueous solution of H_2O_2 decomposes so slowly that no change is observable. (b) Pouring the solution on a wedge of potato, which contains the enzyme catalase, causes the reaction to proceed at a faster rate, as shown by the bubbles of O_2 gas given off. (c) The addition of a small amount of MnO_2 causes the reaction to proceed so rapidly that the energy liberated causes the solution to boil.

3. *Temperature.* As temperature increases, the rates of chemical reactions tend to increase. The average kinetic energy (KE) of particles increases as temperature rises, and as their KE increases, the number of collisions between the particles increases, as does the probability that those collisions will lead to reactions between them.
4. *Catalysts.* Catalytic converters contain materials that increase the rates of reactions that consume pollutants in automobile exhaust. In general, catalysts are materials or substances that accelerate reactions without themselves being consumed in the process. Catalysts also play a key role in biochemistry, where proteins called enzymes catalyze most reactions in living systems.

CONNECTION The connections between kinetic energy and temperature are discussed in Sections 5.8, 6.1, and 6.2.

With these four ideas in mind, let's begin our investigation of reaction rates by returning to the formation of NO in internal combustion engines:

$$N_2(g) + O_2(g) \rightarrow 2\ NO(g) \tag{13.1}$$

We can express the rate of this reaction in terms of the rate of change in the concentration of the product (NO) or the rate of change in the concentration of either reactant (N_2 or O_2). In equations describing concentration changes, we use square brackets around the formulas of substances to represent their concentrations expressed in moles per liter. If the concentration of NO increases from $[NO]_{initial}$ to $[NO]_{final}$ during the time between $t_{initial}$ and t_{final}, then the average rate of change in the concentration of NO during this time is

$$\frac{\Delta[NO]}{\Delta t} = \frac{[NO]_{final} - [NO]_{initial}}{t_{final} - t_{initial}} \tag{13.6}$$

Similarly, the rates of change in the concentrations of N_2 or O_2 over the same interval are

$$\frac{\Delta[N_2]}{\Delta t} = \frac{[N_2]_{final} - [N_2]_{initial}}{t_{final} - t_{initial}} \tag{13.7}$$

and

$$\frac{\Delta[O_2]}{\Delta t} = \frac{[O_2]_{final} - [O_2]_{initial}}{t_{final} - t_{initial}} \tag{13.8}$$

Now let's derive an equation that relates the three rate-of-change expressions in Equations 13.6, 13.7, and 13.8. The last two quantities, $\Delta[N_2]/\Delta t$ and $\Delta[O_2]/\Delta t$, are equal—that is, N_2 and O_2 are both reactants with the same stoichiometric coefficient (1) in Equation 13.1, so their concentrations decrease at the same rate as the reaction proceeds. The value of $\Delta[NO]/\Delta t$ during the same interval is different, however, because the coefficient of 2 for NO in Equation 13.1 indicates that 2 moles of NO form for every 1 mole of N_2 or O_2 that reacts. Therefore, the rate of increase in the concentration of NO is *twice* the rate of decrease in the concentrations of N_2 and O_2. We can express these relative rates of change by using an equation:

$$\frac{\Delta[NO]}{\Delta t} = -2\frac{\Delta[O_2]}{\Delta t} = -2\frac{\Delta[N_2]}{\Delta t} \tag{13.9}$$

Put another way, the rate at which either N_2 or O_2 is consumed is *half* the rate at which NO is produced, as we see if we divide Equation 13.9 through by 2:

$$-\frac{\Delta[N_2]}{\Delta t} = -\frac{\Delta[O_2]}{\Delta t} = \frac{1}{2}\frac{\Delta[NO]}{\Delta t} \tag{13.10}$$

Why are there minus signs on the terms describing the rates of consumption of N_2 and O_2? The rate of change in the concentration of a reactant ($\Delta[\text{reactant}]/\Delta t$)

has a negative value because reactant concentrations decrease as a reaction proceeds. In Equation 13.7, for example, $[N_2]_{initial}$ is greater than $[N_2]_{final}$, which means that the numerator of the equation has a negative value. On the other hand, the rate of change in the concentration of a product is always positive because its value increases as the reaction proceeds. The measured rate of any reaction is defined as a positive quantity, so a minus sign is used with Δ[reactant]/Δt values to obtain a positive value for the reaction rate.

Note that the number 2 in the denominator of the $\Delta[NO]/\Delta t$ term in Equation 13.10 matches the coefficient of NO in the balanced chemical equation. This pattern applies to all chemical reactions: the coefficient of each species in the balanced chemical equation appears in the *denominator* of its term in the relative rate expression if the numerator values are all 1.

SAMPLE EXERCISE 13.1 Predicting a Relative Reaction Rate **LO1**

The synthesis of ammonia:

$$N_2(g) + 3\,H_2(g) \rightarrow 2\,NH_3(g)$$

is an important reaction in the production of agricultural fertilizers. Derive an equation that relates the rates of change in the concentrations of N_2, H_2, and NH_3.

Collect and Organize We are given the balanced chemical equation describing the synthesis of NH_3 from N_2 and H_2, and we need to relate the rates of change in the concentrations of its reactants and product: $\Delta[N_2]/\Delta t$, $\Delta[H_2]/\Delta t$, and $\Delta[NH_3]/\Delta t$.

Analyze Nitrogen and hydrogen are consumed in the reaction, which means the values of $\Delta[N_2]/\Delta t$ and $\Delta[H_2]/\Delta t$ are negative. Therefore, we need to place minus signs in front of these terms in an equation that includes the positive $\Delta[NH_3]/\Delta t$ term. The coefficients of N_2, H_2, and NH_3 are 1, 3, and 2, respectively, so the coefficients in their rate of change terms are 1, 1/3, and 1/2.

Solve The relative rates are related as follows:

$$-\frac{\Delta[N_2]}{\Delta t} = -\frac{1}{3}\frac{\Delta[H_2]}{\Delta t} = \frac{1}{2}\frac{\Delta[NH_3]}{\Delta t}$$

Think About It The solution to the problem tells us that the rate of consumption of N_2 is one-third the rate of consumption of H_2, which makes sense given that three molecules of H_2 react for every one molecule of N_2. Likewise, the rate of consumption of N_2 is one-half the rate of production of NH_3 because two molecules of NH_3 are produced for every molecule of N_2 that reacts.

Practice Exercise How is the rate of change in the concentration of CO_2 related to the rate of change in the concentration of O_2 during the oxidation of carbon monoxide? The equation is

$$2\,CO(g) + O_2(g) \rightarrow 2\,CO_2(g)$$

(Answers to Practice Exercises are in the back of the book.)

Experimentally Determined Reaction Rates

Rates of product formation or reactant consumption are ratios of changes in concentration divided by changes in time, so they have units of concentration per unit time, such as molarity per second (M/s). Reaction rates can only be determined experimentally. Once the rate of a reaction is known, it can be used along with the coefficients in the balanced equation describing the reaction to calculate the rate of change in the concentration of any reactant or product.

For example, suppose we find that the rate of production of ammonia in the reaction in Sample Exercise 13.1 is 0.472 *M*/s, and we want to calculate the rate of consumption of N_2 during the reaction. To do so, we insert the known value of $\Delta[NH_3]/\Delta t$ into the equation derived in Sample Exercise 13.1 and solve for $\Delta[N_2]/\Delta t$:

$$\frac{\Delta[N_2]}{\Delta t} = -\frac{1}{2}\frac{\Delta[NH_3]}{\Delta t} = -\frac{1}{2}\left(\frac{0.472\ M}{s}\right) = -0.236\ M/s$$

This result makes sense because (1) it is negative, reflecting the decreasing concentration of a reactant, and (2) its absolute value is half the rate of production of ammonia, which fits the stoichiometry of the reaction equation.

Of the two values ($\Delta[NH_3]/\Delta t = 0.472\ M/s$ or $\Delta[N_2]/\Delta t = -0.236\ M/s$), which, if either, should we use to express the rate of the reaction? The answer is that rate of any reaction is based on the rate of change in the concentration of the reactant or product with a coefficient of 1 in the reaction equation. If none of the coefficients is 1, divide the rate of change in the concentration of any of the products by its coefficient to get the reaction rate. Furthermore, the reaction rate is always expressed as a positive value. Therefore, the rate of the ammonia reaction described above is 0.236 *M/*s.

We have just seen how a balanced chemical equation enables us to predict the *relative* rates at which reactants are consumed and products are formed in a particular reaction. It provides no information, however, on the numerical values of the rates. Although computational tools can be used to predict the rates of very simple reactions, in this chapter we limit our discussion of actual reaction rates to those based on experiments.

SAMPLE EXERCISE 13.2 Converting Reaction Rates **LO1**

Suppose that during the reaction between NO and O_2 to form NO_2,

$$2\ NO(g) + O_2(g) \rightarrow 2\ NO_2(g)$$

the rate of change in the concentration of O_2 is -0.033 *M*/s. What is the rate of formation of NO_2?

Collect and Organize We need to determine the rate of formation of a product in a reaction, knowing the balanced chemical equation and the rate of change in concentration of one of the reactants. The rate of formation of NO_2 is related to the rates of consumption of NO and O_2 by the balanced chemical equation. The coefficients of O_2 and NO_2 are 1 and 2, respectively.

Analyze We can write an equation that expresses the relative rates of change in $[O_2]$ and $[NO_2]$ from the coefficients in the balanced chemical equation:

$$-\frac{\Delta[O_2]}{\Delta t} = \frac{1}{2}\frac{\Delta[NO_2]}{\Delta t}$$

The negative sign is needed because the concentration of O_2 decreases as the concentration of NO_2 increases.

Solve Solving for the rate of change of $[NO_2]$, we get

$$\frac{\Delta[NO_2]}{\Delta t} = -2\frac{\Delta[O_2]}{\Delta t} = -2(-0.033\ M/s) = 0.066\ M/s$$

Think About It This result is twice the magnitude of $\Delta[O_2]/\Delta t$, which is consistent with the stoichiometry of the reaction: 2 moles of NO_2 are produced for every mole of O_2 that reacts.

 **Practice Exercise** The gas NO reacts with H_2, forming N_2 and H_2O:

$$2\ NO(g) + 2\ H_2(g) \rightarrow 2\ H_2O(g) + N_2(g)$$

If $\Delta[NO]/\Delta t = -21.5\ M/s$ under a given set of conditions, what are the rates of change of $[N_2]$ and $[H_2O]$?

TABLE 13.1 Changing Concentrations of Reactants and Product during the Reaction $N_2(g) + O_2(g) \rightarrow 2\ NO(g)$

Time (μs)	$[N_2]$, $[O_2]$ (μ*M*)	[NO] (μ*M*)
0.0	17.0	0.0
5.0	13.1	7.8
10.0	9.6	14.8
15.0	7.6	18.6
20.0	5.8	22.2
25.0	4.5	24.8
30.0	3.6	26.7

Average Reaction Rates

Suppose we run an experiment to determine the rate of formation of NO in an automobile engine. In the laboratory, we use a reaction vessel as hot as the combustion chambers in the engine, and we obtain the data listed in **Table 13.1** and plotted in **Figure 13.6**. We can use the data to calculate the reaction rate based on the change in the concentration of any participant in the reaction over a particular interval. Such calculations of reaction rates based on $-\Delta[\text{reactant}]/\Delta t$ or $\Delta[\text{product}]/\Delta t$ are *average* reaction rates. Suppose, for example, that we want to calculate the average rate of change in [NO] between 5.0 and 10.0 μs:

$$\frac{\Delta[NO]}{\Delta t} = \frac{[NO]_{10.0\mu s} - [NO]_{5.0\mu s}}{t_{10.0\mu s} - t_{5.0\mu s}} = \frac{(14.8 - 7.8)\mu M}{(10.0 - 5.0)\mu s} = 1.4\ M/s$$

During the same interval, the average rate of change in the concentration of N_2 (or O_2) is

$$\frac{\Delta[N_2]}{\Delta t} = \frac{[N_2]_{10.0\mu s} - [N_2]_{5.0\mu s}}{t_{10.0\mu s} - t_{5.0\mu s}} = \frac{(9.6 - 13.1)\mu M}{(10.0 - 5.0)\mu s} = -0.70\ M/s$$

These results give us two $\Delta[X]/\Delta t$ values for expressing the rate of the reaction. By convention we choose the value associated with the substance having a coefficient of 1 in the balanced equation, though we have to change the value's sign if that substance is a reactant. This means we can choose the rate of change of either $[N_2]$ or $[O_2]$. Therefore, the average rate of this reaction is

$$\text{Rate} = -\frac{\Delta[N_2]}{\Delta t} = 0.70\ M/s$$

The numerical value of this reaction rate applies only to the interval from $t = 5.0$ μs to $t = 10.0$ μs because the graph lines in Figure 13.6 are curved. Any other 5.0 μs interval would have a different average rate because the reaction slows as more time passes.

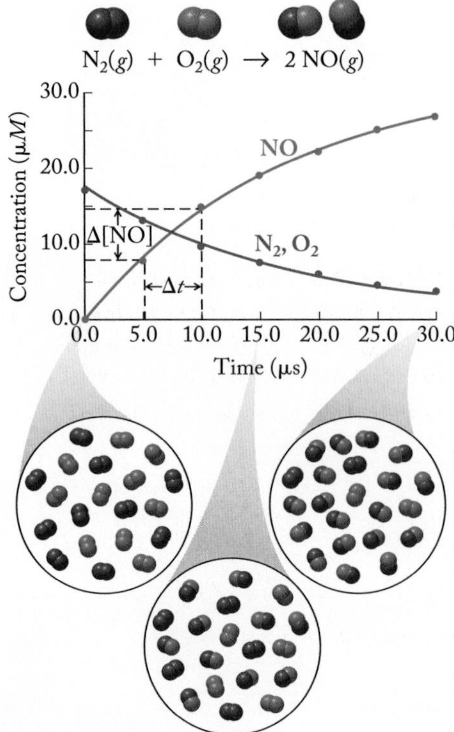

FIGURE 13.6 Concentrations of N_2, O_2, and NO over 30.0 μs for the reaction $N_2(g) + O_2(g) \rightarrow 2\ NO(g)$, plotted from data in Table 13.1.

Instantaneous Reaction Rates

We can also determine the *instantaneous* rate of a reaction—that is, its rate at a particular instant during the reaction. To see how, let's revisit the oxidation of NO to NO_2 in the atmosphere:

$$2\ NO(g) + O_2(g) \rightarrow 2\ NO_2(g) \qquad (13.2)$$

Suppose that chemical analyses of a reaction mixture of these three gases yield the concentration data listed in **Table 13.2**. Among the three gases, only O_2 has

TABLE 13.2 Changing Concentrations of Reactants and Product during the Reaction $2\ NO(g) + O_2(g) \rightarrow 2\ NO_2(g)$ at 25°C

Time (s)	[NO] (*M*)	$[O_2]$ (*M*)	$[NO_2]$ (*M*)
0.0	0.0100	0.0100	0.0000
285.0	0.0090	0.0095	0.0010
660.0	0.0080	0.0090	0.0020
1175.0	0.0070	0.0085	0.0030
1895.0	0.0060	0.0080	0.0040
2975.0	0.0050	0.0075	0.0050
4700.0	0.0040	0.0070	0.0060
7800.0	0.0030	0.0065	0.0070

a coefficient of 1 in the balanced chemical equation, so we will use the rate of change of $[O_2]$ to calculate the rate of the reaction.

What is the instantaneous rate of the reaction at t = 2000.0 s? We can plot the $[O_2]$ data from Table 13.2 versus time—this is the green curve in **Figure 13.7**—and then draw a tangent to the curve at t = 2000.0 s. Next, we select two convenient points along the tangent—for example, at t = 1000.0 s and t = 3000.0 s—to use in calculating the slope of the tangent. The slope is a measure of the instantaneous rate of change in $[O_2]$ at t = 2000.0 s, and the negative of the slope is the instantaneous rate of the reaction at that time.[2]

$$\text{Slope} = \frac{\Delta[O_2]}{\Delta t} = \frac{(0.0072 - 0.0084)\ M}{(3000.0 - 1000.0)\ \text{s}} = -6.0 \times 10^{-7}\ M/\text{s}$$

$$= -6.0 \times 10^{-7}\ M/\text{s}$$

$$\text{Rate} = -\frac{\Delta[O_2]}{\Delta t} = -(-6.0 \times 10^{-7}\ M/\text{s}) = 6.0 \times 10^{-7}\ M/\text{s}$$

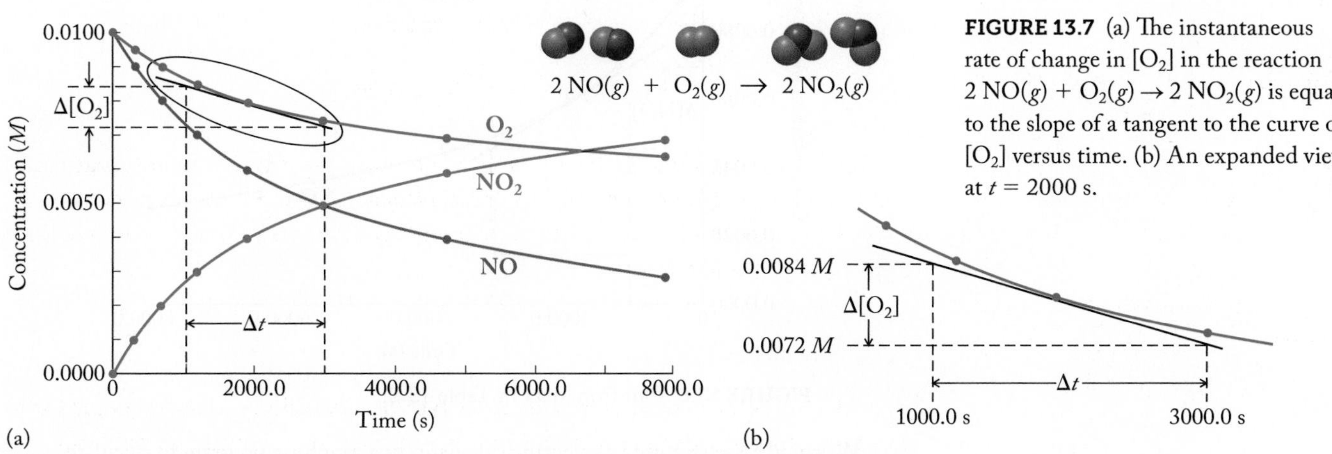

FIGURE 13.7 (a) The instantaneous rate of change in $[O_2]$ in the reaction $2\ NO(g) + O_2(g) \rightarrow 2\ NO_2(g)$ is equal to the slope of a tangent to the curve of $[O_2]$ versus time. (b) An expanded view at t = 2000 s.

[2]If you have studied calculus, you may know that the average rate of a reaction approaches the instantaneous rate as Δt approaches zero. The slope of the tangent to a curve at a given point is the derivative of the curve at that point and can be calculated using a scientific calculator. Appendix 1 discusses how to determine the slope and intercept of a line.

CONCEPT TEST

Which of the following statements is/are true about the instantaneous rate of the chemical reaction A → B as the reaction proceeds?

a. $-\Delta[A]/\Delta t$ increases, $\Delta[B]/\Delta t$ decreases
b. $-\Delta[A]/\Delta t$ decreases, $\Delta[B]/\Delta t$ increases
c. $-\Delta[A]/\Delta t$ and $\Delta[B]/\Delta t$ both increase
d. $-\Delta[A]/\Delta t$ and $\Delta[B]/\Delta t$ both decrease

(Answers to Concept Tests are in the back of the book.)

SAMPLE EXERCISE 13.3 Determining an Instantaneous Reaction Rate LO2

(a) What is the instantaneous rate of change of [NO] at t = 2000.0 s in the experiment that produced the data in Table 13.2? (b) What is the rate of the reaction based on your result in part (a)?

Collect and Organize We are asked to determine the instantaneous rate of change of [NO] and the corresponding reaction rate at t = 2000.0 s. The coefficient of NO is 2 in the balanced chemical equation:

$$2\,NO(g) + O_2(g) \rightarrow 2\,NO_2(g)$$

Analyze The instantaneous rate of change in [NO] at 2000.0 s can be determined from a graph of [NO] versus time and the slope of the curve at 2000.0 s. The corresponding reaction rate will be the slope value multiplied by −1/2.

Solve

a. First we plot [NO] versus time and draw a tangent to the curve at the point t = 2000.0 s (**Figure 13.8**). We then choose two points along the tangent, t = 1000.0 and 3000.0 s, and determine the concentrations corresponding to those times along the vertical axis. By using those values, we calculate the slope of the line:

$$\frac{\Delta[NO]}{\Delta t} = \frac{(0.0046 - 0.0070)\,M}{(3000.0 - 1000.0)\text{ s}} = -1.2 \times 10^{-6}\,M/\text{s}$$

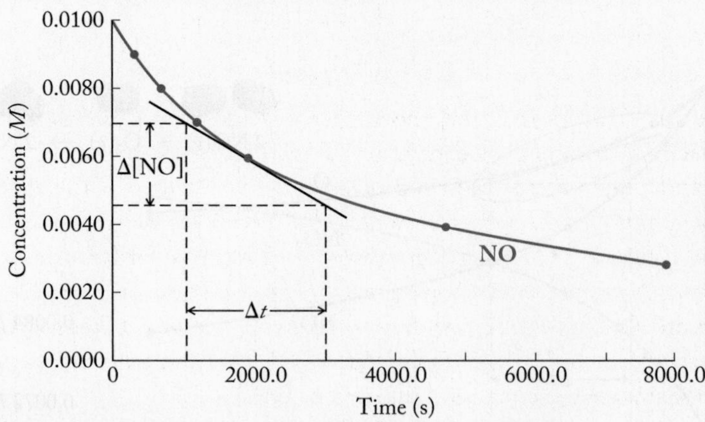

FIGURE 13.8 Plot from data in Table 13.2.

We could also have used a scientific calculator or a graphing program to calculate the slope of the tangent.

b. The corresponding reaction rate is

$$\text{Rate} = -\frac{1}{2}\frac{\Delta[NO]}{\Delta t} = -\frac{1}{2}(-1.2 \times 10^{-6}\,M/\text{s}) = 6.0 \times 10^{-7}\,M/\text{s}$$

Think About It The sign of $\Delta[NO]/\Delta t$ is negative because NO is a reactant whose concentration decreases with time. However, the rates of chemical reactions have positive values. Therefore, we needed a minus sign in front of the $\Delta[NO]/\Delta t$ term in the solution to part (b). Note that the instantaneous reaction rate calculated in this exercise is the same as the one we calculated previously based on the rate of change of $[O_2]$.

Practice Exercise What is the instantaneous rate of change in $[NO_2]$ at $t = 2000.0$ s in the experiment that produced the data in Table 13.2?

13.3 Effect of Concentration on Reaction Rate

Figure 13.9 shows a typical plot of reactant concentration as a function of time. Tangents have been drawn to the curve at three points: (a) at the instant the reaction begins ($t = 0$), (b) when the reaction is about halfway to completion, and (c) when the reaction is nearly over. The slope of the tangent at point (a) can be used to calculate the **initial rate** of the reaction.

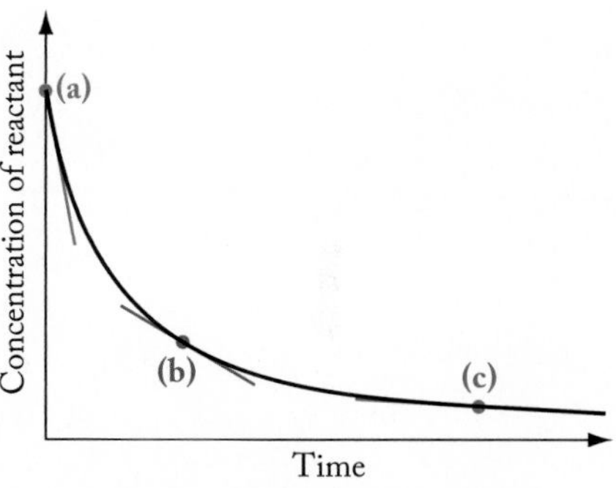

FIGURE 13.9 Typical plot of reactant concentration as a function of time with tangents drawn (a) at $t = 0$; (b) at an intermediate time; (c) when the reaction is nearly over.

As with nearly all plots of reactant or product concentration versus time, the most rapid changes in concentration take place at the beginning of the reaction. As the reaction proceeds, the slopes of tangents to the curve decrease, approaching zero. When the slope of the tangent to the curve reaches zero, there are no more changes in the concentrations of the product(s) or any remaining reactant(s).

Kinetic molecular theory explains why reaction rates change over time. If we assume that most reactions take place as a result of collisions between reactant molecules, then the more reactant molecules there are in a given volume, the more collisions per unit time there are, and there are more opportunities for reactants to turn into products. As reactants collide and change into products, fewer reactant molecules are present in the system (i.e., the concentrations of the reactants decrease), so the frequency of collisions decreases and the rate at which reactants change into products slows down.

Reaction Order and Rate Constants

CHEMTOUR
Reaction Order

Experimental observations and theoretical considerations tell us that reaction rates depend on reactant concentrations. However, they do not tell us *to what extent* rates depend on reactant concentrations. For example, if the concentration of a reactant doubles, does the reaction rate also double? The answer to this question comes from experimentally determining the **reaction order**, a parameter that tells us how reaction rate depends on reactant concentrations. Knowing the order of a reaction also provides insights into how the reaction takes place—that is, which molecules collide with which other molecules as bonds break, new bonds form, and reactants are converted into products.

Let's consider one way in which reaction order is determined by revisiting the reaction between nitrogen monoxide and oxygen:

$$2\,NO(g) + O_2(g) \rightarrow 2\,NO_2(g) \qquad (13.2)$$

To evaluate the kinetics of this reaction, multiple experiments can be conducted, each with different initial concentrations of NO and O_2 introduced into a reaction vessel at 25°C, and the initial reaction rate is determined in each case. **Table 13.3** shows the results of three such determinations.

initial rate the rate of a reaction at $t = 0$, immediately after the reactants are mixed.

reaction order an experimentally determined number defining the dependence of the reaction rate on the concentration of a reactant.

TABLE 13.3 Effect of Reactant Concentrations on Initial Reaction Rates at 25°C for the Reaction $2\ NO(g) + O_2(g) \rightarrow 2\ NO_2(g)$

Experiment	$[NO]_0$ (*M*)	$[O_2]_0$ (*M*)	Initial Reaction Rate (*M*/s)
1	0.0100	0.0100	1.0×10^{-6}
2	0.0100	0.0050	0.5×10^{-6}
3	0.0050	0.0100	2.5×10^{-7}

To interpret the data in Table 13.3, we select pairs of experiments in which the concentrations of one reactant differ but the concentrations of the other are the same. For example, $[NO]_0$ is the same in experiments 1 and 2, but $[O_2]_0$ in experiment 1 is twice $[O_2]_0$ in experiment 2. (The zero subscripts indicate that these are the concentrations of NO and O_2 at the start of the experiments, when $t = 0$.) The initial reaction rate in experiment 1 is twice that in experiment 2, allowing us to state that doubling the concentration of O_2 while holding the NO concentration constant doubles the initial reaction rate. We conclude that the initial reaction rate is directly proportional to the O_2 concentration:

$$\text{Rate} \propto [O_2]$$

In experiments 1 and 3, $[O_2]_0$ is the same, but $[NO]_0$ in experiment 1 is twice $[NO]_0$ in experiment 3. Comparing the initial reaction rates for these two experiments, we find that the rate in experiment 1 is four times the rate in experiment 3. Thus, the reaction rate is proportional to the square of the NO concentration:

$$\text{Rate} \propto [NO]^2$$

We can combine these two rate expressions to get an overall rate expression for the reaction by multiplying the right sides of the rate expressions for the two reactants:

CHEMTOUR
Collision Theory

$$\text{Rate} \propto [NO]^2[O_2]$$

To understand why we multiply the concentration terms together, let's consider another reaction between oxygen and nitrogen monoxide. This one involves the reaction of ozone (O_3) with NO that produces NO_2 and O_2:

$$NO(g) + O_3(g) \rightarrow NO_2(g) + O_2(g)$$

Figure 13.10 shows how different numbers of NO and O_3 molecules in a reaction vessel might collide with each other and how the probability of collisions between NO and O_3 molecules increases in proportion to the *product* of the number of NO molecules times the number of O_3 molecules. If the reaction occurs as a result of these collisions, then increasing the number of collisions should produce a

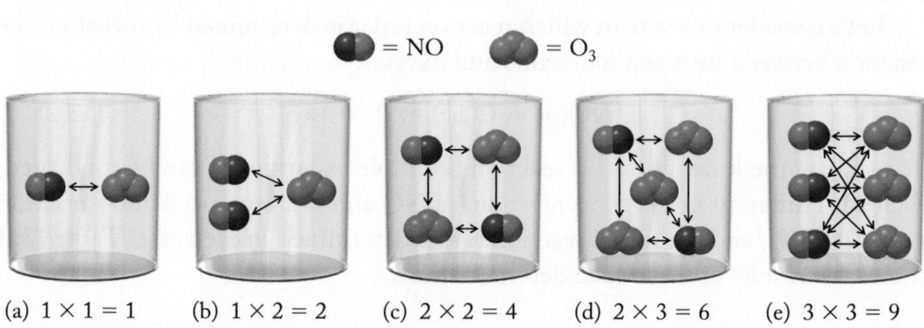

FIGURE 13.10 Increasing the concentration increases the number of possible collisions (double-headed arrows) and therefore the number of potential reaction events. Reaction rate depends on the number of collisions, which are shown for the reaction between NO and ozone (O_3) that produces NO_2 and O_2. With only one NO molecule and one O_3 molecule, as in (a), each molecule can collide only with the other, giving a relative reaction rate of $1 \times 1 = 1$. In (e), three molecules of NO can collide with three molecules of O_3 for a relative reaction rate of $3 \times 3 = 9$ times the rate in (a).

proportionate increase in the rate of the reaction. Therefore, the rate of the reaction should depend on the product of the concentrations of NO and O_3:

$$\text{Rate} \propto [\text{NO}][\text{O}_3] \tag{13.11}$$

Similar patterns exist for other chemical reactions whose rates depend on the frequency of molecular collisions. We can convert the expression in Equation 13.11 into the **rate law** for this reaction, which is an equation that relates reaction rate to reactant concentrations, by inserting a proportionality constant k, called the reaction's **rate constant**:

rate law an equation that defines the experimentally determined relation between the concentrations of reactants in a chemical reaction and the rate of that reaction.

rate constant the proportionality constant that relates the rate of a reaction to the concentrations of reactants at a particular temperature.

$$\text{Rate} = k[\text{NO}][\text{O}_3] \tag{13.12}$$

Now let's consider a generic chemical reaction with two reactants, A and B:

$$\text{A} + \text{B} \rightarrow \text{C}$$

The rate law expression for this reaction may be written

$$\text{Rate} = k[\text{A}]^m[\text{B}]^n \tag{13.13}$$

where m is the reaction order with respect to A, and n is the reaction order with respect to B. There is no [C] term in the rate law for this reaction because the rates of most chemical reactions depend on the concentrations of one or more reactants, and *never* on the concentrations of products.

We can determine reaction order values by comparing differences in reaction rates with differences in reactant concentrations. Sometimes these comparisons are easy to make, as with the data for the reaction between NO and O_2 in Table 13.3. Other times the comparisons are not so easy. On those occasions we can use a different mathematical approach. For example, suppose we conduct three experiments to solve for the values of m and n in Equation 13.13. In experiments 1 and 2 the value of [A] is the same, but the values of [B] are different. In experiments 2 and 3 we keep [B] constant and vary the concentrations of [A]. The ratio of the reaction rates in experiments 1 (Rate_1) and 2 (Rate_2) is related to the ratio of the concentrations of B:

$$\frac{\text{Rate}_1}{\text{Rate}_2} = \left(\frac{[\text{B}]_1}{[\text{B}]_2}\right)^n$$

We can solve this equation for n by taking the logarithm of both sides:

$$\log\left(\frac{\text{Rate}_1}{\text{Rate}_2}\right) = n \log\left(\frac{[\text{B}]_1}{[\text{B}]_2}\right)$$

Now rearrange the terms to solve for n:

$$n = \frac{\log\left(\frac{\text{Rate}_1}{\text{Rate}_2}\right)}{\log\left(\frac{[\text{B}]_1}{[\text{B}]_2}\right)}$$

An equation with this format can also be used to calculate the value of m from the results of experiments 2 and 3, or to calculate the order of any reaction with respect to a reactant (X) whose concentration differs in a pair of reaction rate experiments:

$$n = \frac{\log\left(\frac{\text{Rate}_1}{\text{Rate}_2}\right)}{\log\left(\frac{[\text{X}]_1}{[\text{X}]_2}\right)} \tag{13.14}$$

overall reaction order the sum of the exponents of the concentration terms in the rate law.

CONCEPT TEST

In the reaction A → B, the rate of the reaction triples when [A] is tripled.

a. What is the correct value of m in the rate law: rate = $k[A]^m$?

b. What is the value of m if the rate is unchanged when [A] is tripled?

The power to which a concentration term is raised (such as the m or n in Equation 13.13) is the order of the reaction in terms of that reactant. In our example from Table 13.3, the reaction of NO and O_2 is *second order* in NO and *first order* in O_2. The **overall reaction order** for a reaction is the sum of the powers in the rate equation, so the reaction of NO and O_2 is *third order* overall.

An exponent in a rate law may be a whole number, a fraction, zero, or, in rare cases, negative. It is important to remember that rate laws and reaction orders must be determined experimentally. They cannot be predicted from the coefficients in a balanced chemical equation, even though they sometimes do match the coefficients. The significance of reaction order in describing how a reaction takes place is addressed in Section 13.5.

The value of the rate constant k is unique to each particular reaction at a given temperature, and right now we will consider k simply as a proportionality constant. It does not change with concentration; in other words, *reaction rate depends on the concentration of the reactants, but the rate constant does not.* The rate constant changes only with changing temperature or in the presence of a catalyst.

We can calculate k for a reaction run at some specified temperature from initial reaction rate data. Let's do so for the reaction of O_2 and NO by selecting the results of one experiment in Table 13.3. Which experiment we use does not matter; as long as the temperature is the same, the value of k will be the same. The proportionality we derived for this reaction's overall rate expression was

$$\text{Rate} \propto [\text{NO}]^2[\text{O}_2]$$

When we convert this to an equation, we obtain

$$\text{Rate} = k[\text{NO}]^2[\text{O}_2]$$

We can insert the data from experiment 1 into this equation and solve for k:

$$1.0 \times 10^{-6}\ M/\text{s} = k(0.0100\ M)^2(0.0100\ M)$$

$$k = \frac{1.0 \times 10^{-6}\ M/\text{s}}{(0.0100\ M)^2(0.0100\ M)} = 1.0\ M^{-2}\,\text{s}^{-1} \tag{13.15}$$

This value, like any rate constant calculated from experimental data, is valid only at the temperature at which the experiments were carried out.

The units of k look a bit unusual in comparison with units we have seen before. Remember that reaction rates themselves have units of concentration per unit time, such as M/s. In a first-order reaction, then, in which concentration is expressed in molarity and time in seconds, the units of k must be per second (s^{-1}):

$$\text{Rate} = k[\text{X}]^1$$

$$k = \frac{\text{rate}}{[\text{X}]} = \frac{M/\text{s}}{M} = \frac{1}{\text{s}} = \text{s}^{-1}$$

The units of the rate constant for a reaction that is second order overall can be derived in a similar way. If the reaction is first order in reactants X and Y, we have

$$\text{Rate} = k[\text{X}][\text{Y}]$$

$$k = \frac{\text{rate}}{[\text{X}][\text{Y}]} = \frac{M/\text{s}}{M\,M} = M^{-1}\,\text{s}^{-1}$$

The units of k for a reaction that is third order overall are $M^{-2}\,\text{s}^{-1}$, as we determined in Equation 13.15. Therefore, the units of k for each reaction ultimately depend on its overall reaction order as summarized in **Table 13.4**.

TABLE 13.4 Units on Rate Constants for Different Reaction Orders

Overall Reaction Order	Units on k
1	s^{-1}
2	$M^{-1}\,\text{s}^{-1}$
3	$M^{-2}\,\text{s}^{-1}$
4	$M^{-3}\,\text{s}^{-1}$

CONCEPT TEST

The reaction A + 2 B → C is found to be third order overall. How many possible rate laws of the form

$$\text{Rate} = k[\text{A}]^m[\text{B}]^n$$

could we write, assuming that m and n are non-negative integers?

SAMPLE EXERCISE 13.4 Deriving a Rate Law from Initial Reaction Rate Data — LO3

Use the data in **Table 13.5** to write the rate law for the following reaction of N_2 with O_2:

$$N_2(g) + O_2(g) \rightarrow 2\,NO(g)$$

Determine the overall reaction order and the value of the rate constant.

TABLE 13.5 Initial Reaction Rates at Constant Temperature for the Reaction $N_2(g) + O_2(g) \rightarrow 2\,NO(g)$

Experiment	$[N_2]_0$ (M)	$[O_2]_0$ (M)	Initial Reaction Rate (M/s)
1	0.040	0.020	707
2	0.040	0.010	500
3	0.010	0.010	125

Collect and Organize We need to determine the rate law and rate constant for a reaction, given the initial reaction rate (note the subscript zero on the concentration terms in Table 13.5) for each of three sets of initial concentrations.

Analyze The general form of the rate law for any reaction between reactants A and B is

$$\text{Rate} = k[\text{A}]^m[\text{B}]^n$$

We can use the experimental data given to find the values of k, m, and n for the reaction in which $[\text{A}] = [N_2]$ and $[\text{B}] = [O_2]$. The overall order of the reaction is the sum of the reaction orders of the individual reactants. Once we have established the rate law, we can calculate the rate constant by using concentrations of reactants from any row in Table 13.5. The rate constant must have units that express the reaction rate in $M\,\text{s}^{-1}$.

Solve To determine the order of the reaction (m) with respect to N_2, we use the data from experiments 2 and 3 because in these two experiments the $[N_2]$ values are different, but the $[O_2]$ values are the same. When the concentration of N_2 is increased by a factor of 4, the rate increases by a factor of 4. Thus, the reaction rate is proportional to $[N_2]$, which means that $m = 1$. We can confirm the value of m by applying Equation 13.14:

$$m = \frac{\log\left(\frac{\text{Rate}_2}{\text{Rate}_3}\right)}{\log\left(\frac{[N_2]_2}{[N_2]_3}\right)} = \frac{\log\left(\frac{500\ M/\text{s}}{125\ M/\text{s}}\right)}{\log\left(\frac{0.040\ M}{0.010\ M}\right)} = 1$$

There are different values of $[O_2]$ in experiments 1 and 2, but $[N_2]$ is the same. Therefore, we can use these data to calculate the value of n. The ratio of the reaction rates (707/500) is not a whole number, so let's apply Equation 13.14:

$$n = \frac{\log\left(\frac{\text{Rate}_1}{\text{Rate}_2}\right)}{\log\left(\frac{[O_2]_1}{[O_2]_2}\right)} = \frac{\log\left(\frac{707\ M/\text{s}}{500\ M/\text{s}}\right)}{\log\left(\frac{0.020\ M}{0.010\ M}\right)} = 0.50$$

Thus, the reaction is $\frac{1}{2}$ order with respect to O_2, first order with respect to N_2, and $(\frac{1}{2} + 1) = \frac{3}{2}$ order overall:

$$\text{Rate} = k[N_2][O_2]^{1/2}$$

We can use the data from any experiment to obtain the value of k. Let's use experiment 1:

$$707\ M/\text{s} = k(0.040\ M)(0.020\ M)^{1/2}$$

$$k = 1.2 \times 10^5\ M^{-1/2}\ \text{s}^{-1}$$

Think About It The units of the rate constant, $M^{-1/2}\ \text{s}^{-1}$, are appropriate because when we substitute the units for each term into the rate law, we end up with the correct units for reaction rate: $(M^{-1/2}\ \text{s}^{-1})(M)(M^{1/2}) = M\ \text{s}^{-1}$. The reaction order of $\frac{1}{2}$ for oxygen is determined from experimental data, not from the balanced chemical equation.

Practice Exercise Nitrogen monoxide reacts rapidly with unstable nitrogen trioxide (NO_3) to form NO_2:

$$NO(g) + NO_3(g) \rightarrow 2\ NO_2(g)$$

Determine the rate law for the reaction and calculate the rate constant from the data in **Table 13.6**.

TABLE 13.6 Initial Reaction Rates at 25°C for the Reaction $NO(g) + NO_3(g) \rightarrow 2\ NO_2(g)$

Experiment	$[NO]_0$ (M)	$[NO_3]_0$ (M)	Initial Reaction Rate (M/s)
1	1.25×10^{-3}	1.25×10^{-3}	2.45×10^4
2	2.50×10^{-3}	1.25×10^{-3}	4.90×10^4
3	2.50×10^{-3}	2.50×10^{-3}	9.80×10^4

Integrated Rate Laws: First-Order Reactions

integrated rate law a mathematical expression that describes the change in concentration of a reactant in a chemical reaction with time.

Determining a rate law by using initial reaction rate data means that several experiments must be performed with different concentrations of reactants, and that these concentrations must be manipulated in a systematic fashion. We also must accurately determine the reaction rate at the instant the reaction begins. It would be much better if we could determine the rate law and calculate the rate constant of a reaction by using data from only a single experiment. In fact, we can do this for reactions in which the reaction rate depends on the concentration of

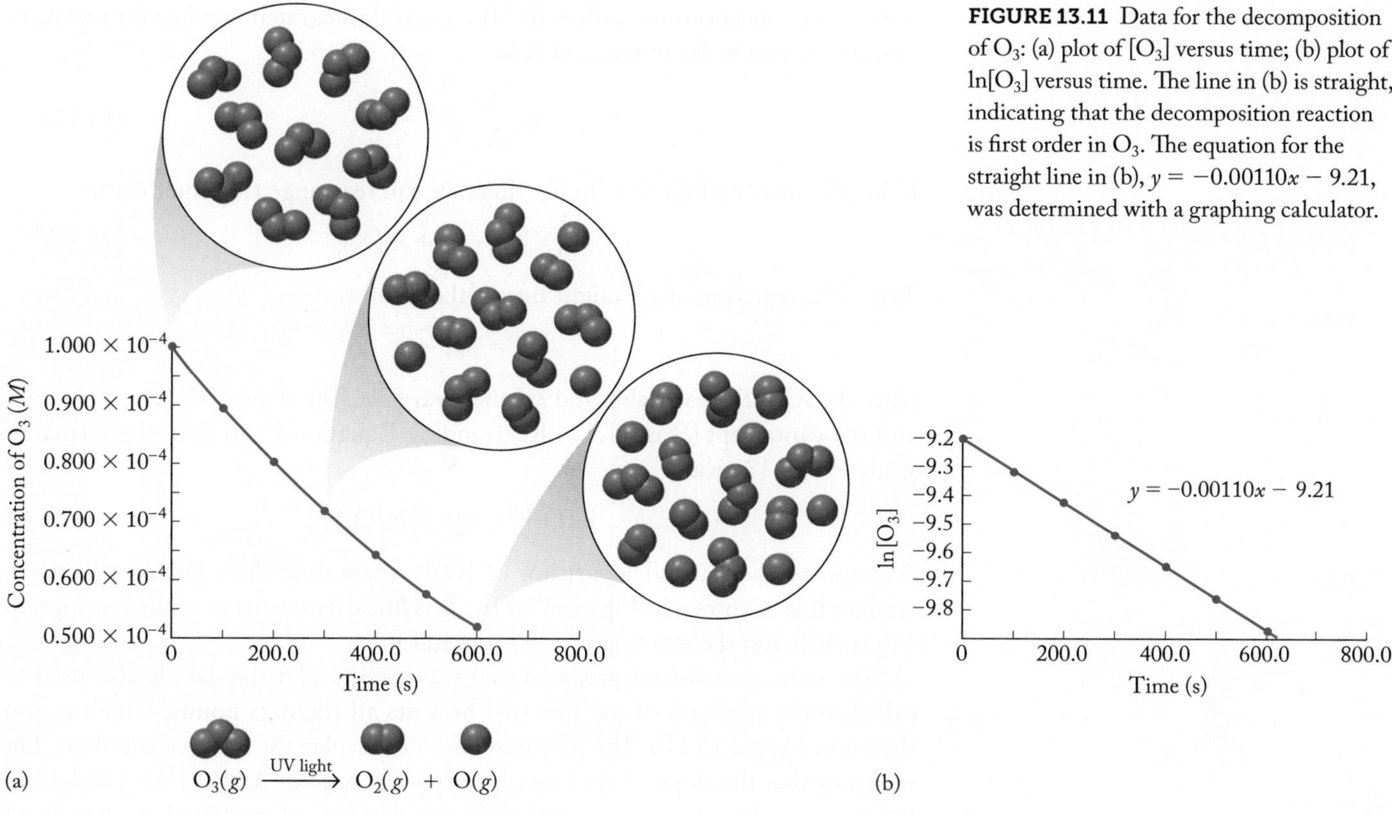

FIGURE 13.11 Data for the decomposition of O_3: (a) plot of $[O_3]$ versus time; (b) plot of $\ln[O_3]$ versus time. The line in (b) is straight, indicating that the decomposition reaction is first order in O_3. The equation for the straight line in (b), $y = -0.00110x - 9.21$, was determined with a graphing calculator.

only one substance. One such reaction is the photochemical decomposition of ozone in the stratosphere:

$$O_3(g) \xrightarrow{\text{sunlight}} O_2(g) + O(g)$$

This decomposition reaction can be studied in the laboratory by using high-intensity ultraviolet lamps to simulate solar radiation. One such study yielded the results listed in **Table 13.7**. The data were plotted (**Figure 13.11a**) using the graphing feature of a computer spreadsheet program (Excel). Because ozone is the only reactant, the rate law for the reaction should depend only on the ozone concentration. In our interpretation of the data in Table 13.7, we start with the assumption that the reaction is first order in O_3, which means that the rate law can be written

$$\text{Rate} = k[O_3]$$

The O_3 consumption rate $-\Delta[O_3]/\Delta t$ equals the reaction rate, and we can write

$$\text{Rate} = -\frac{\Delta[O_3]}{\Delta t} = k[O_3]$$

This rate law can be transformed into an expression that relates the concentration of ozone $[O_3]$ at any instant during the reaction to the initial concentration $[O_3]_0$:

$$\ln\frac{[O_3]}{[O_3]_0} = -kt \qquad (13.16)$$

Note that Equation 13.16 uses ln, the *natural* logarithm function, not log, the base 10 logarithm function. This version of the rate law is called an **integrated rate law** because integral calculus is used to derive it, and it describes the change

TABLE 13.7 Concentrations for Photochemical Decomposition of Ozone

Time (s)	$[O_3]$ (M)	$\ln[O_3]$
0.0	1.000×10^{-4}	−9.210
100.0	0.896×10^{-4}	−9.320
200.0	0.803×10^{-4}	−9.430
300.0	0.719×10^{-4}	−9.540
400.0	0.644×10^{-4}	−9.650
500.0	0.577×10^{-4}	−9.760
600.0	0.517×10^{-4}	−9.870

in reactant concentration with time. The general integrated rate law for any reaction that is first order in reactant X is

$$\ln\frac{[X]}{[X]_0} = -kt \tag{13.17}$$

Using the identity $\ln(a/b) = \ln a - \ln b$, we can rearrange this equation to

$$\ln[X] = -kt + \ln[X]_0 \tag{13.18}$$

This is the equation of a straight line of the form

$$y = mx + b$$

where ln[X] is the y variable and t is the x variable. The slope of the line (m) is $-k$, and the y intercept (b) is $\ln[X]_0$. Rearranging Equation 13.16 to fit the format of Equation 13.18 gives

$$\ln[O_3] = -kt + \ln[O_3]_0$$

A graph of the natural logarithm of $[O_3]$ versus time does indeed produce a straight line as shown in **Figure 13.11b**. This linearity means that our assumption was correct and the reaction is first order in O_3.

The same spreadsheet program that was used to plot the data is also used to calculate the equation of the line that best fits all the data points, which is also shown in Figure 13.11b. The program does not display the units of the slope, but we know that the slope of the line represents the ratio of $\Delta\ln[O_3]/\Delta t$. Logarithm terms are unitless and the time values in this data set are expressed in seconds, so the units on the slope value are actually 1/s or s^{-1}. Therefore, the value of m is $-0.00110\ s^{-1}$ and the value of k is $-(-0.00110\ s^{-1})$ or $1.10 \times 10^{-3}\ s^{-1}$.

TABLE 13.8 Concentration of N_2O_5 and $\ln[N_2O_5]$ as a Function of Time

Time (s)	$[N_2O_5]$ (*M*)	$\ln[N_2O_5]$
0.0	0.1000	−2.303
50.0	0.0649	−2.735
100.0	0.0561	−2.881
200.0	0.0250	−3.689
300.0	0.0136	−4.298
400.0	0.0059	−5.133

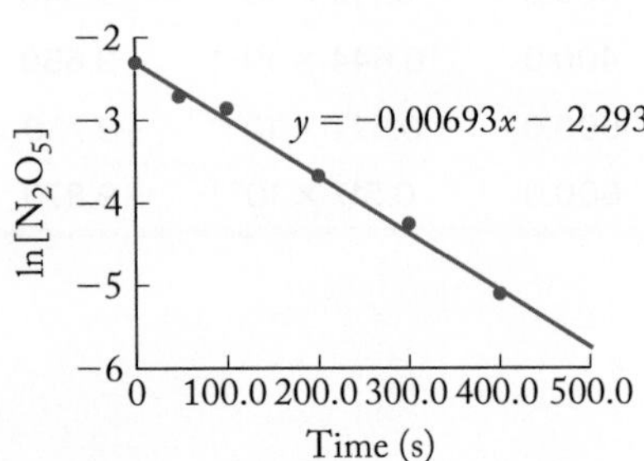

FIGURE 13.12 A graphing calculator was used to plot $\ln[N_2O_5]$ versus t and to determine the equation of the straight line (i.e., $y = -0.00693x - 2.293$) that best fits the data points.

SAMPLE EXERCISE 13.5 Using an Integrated Rate Law **LO4**

The concentration of dinitrogen pentoxide in the atmosphere is low in part because it rapidly decomposes to N_2O_4 and O_2:

$$2\ N_2O_5(g) \rightarrow 2\ N_2O_4(g) + O_2(g)$$

A kinetic study of the decomposition of N_2O_5 at a particular temperature yielded the data in the first two columns of **Table 13.8**. Assume that the decomposition of N_2O_5 is first order in N_2O_5. (a) Test the validity of your assumption, and (b) determine the value of the rate constant.

Collect and Organize We are given experimental data showing the concentration of a single reactant as a function of time and are told to assume a first-order reaction. We can verify the assumption by using the integrated rate law for a first-order reaction (Equation 13.18) and then calculate the rate constant.

Analyze Equation 13.18 has the form $y = mx + b$, which means that a plot of $\ln[N_2O_5](y)$ versus time (x) should be linear if the decomposition of N_2O_5 is first order. The slope (m) of the graph corresponds to $-k$, the negative of the rate constant.

Solve The first step is to convert the $[N_2O_5]$ values into $\ln[N_2O_5]$ values, which are listed in the third column of Table 13.8.

a. Plotting the time and $\ln[N_2O_5]$ values in Table 13.8 as x and y variables produces the graph shown in **Figure 13.12**. The same spreadsheet program that plots the data also calculates the equation of the straight line that best fits the data (Figure 13.12). All of the points lie near a straight line, confirming that the reaction is first order in N_2O_5.

b. The value of the rate constant (k) is the negative of the value of the slope of the line and has units of s^{-1}, which makes it $-(-0.00693\ s^{-1})$ or $6.93 \times 10^{-3}\ s^{-1}$.

Think About It Using a calculator or computer program to plot the data and determine the slope of the line that best fits it tends to generate the most accurate slope and rate constant values because these programs use all of the points in the data set. To check our answer, we can calculate the slope, $\Delta y/\Delta x$, using the coordinates of two of the plotted points. For example, had we estimated the slope using the coordinates for the points at $t = 100.0$ and 400.0 seconds, we would have obtained a value of k equal to $-[-5.133 - (-2.881)]/(400.0 - 100.0)\ s = 7.51 \times 10^{-3}\ s^{-1}$, which is 8.3% higher than the slope calculated using all of the data points. Calculating slope in this fashion is likely to be less accurate because there is inherent uncertainty in experimental results that can cause some of the plotted points to be slightly above or below the line that best fits them, as you can see in Figure 13.12.

 **Practice Exercise** Hydrogen peroxide (H_2O_2) decomposes into water and oxygen:

$$H_2O_2(\ell) \rightarrow H_2O(\ell) + \tfrac{1}{2}\,O_2(g)$$

Use the data in **Table 13.9** to determine whether the decomposition of H_2O_2 is first order in H_2O_2, and calculate the value of the rate constant at the temperature of the experiment that produced the data.

TABLE 13.9 Concentration of Hydrogen Peroxide as a Function of Time

Time (s)	$[H_2O_2]$ (M)
0.0	0.500
100.0	0.444
200.0	0.434
500.0	0.347
1000.0	0.205
1500.0	0.144

SAMPLE EXERCISE 13.6 Calculating the Concentration of a Reactant from an Integrated Rate Law LO5

In Sample Exercise 13.5 we confirmed that the decomposition of dinitrogen pentoxide

$$2\,N_2O_5(g) \rightarrow 2\,N_2O_4(g) + O_2(g)$$

is first order in N_2O_5 and has a rate constant $k = 0.00693\ s^{-1}$. If this reaction is run in the laboratory under the same conditions as in Sample Exercise 13.5 and the initial concentration of N_2O_5 in the reaction vessel is 0.375 M, what is the concentration of N_2O_5 after exactly 3 minutes?

Collect and Organize We are given the rate constant for a first-order reaction, the initial concentration of reactant, and the reaction time. We are asked to calculate the concentration of reactant remaining at that time. We also have the general rate equation for a first-order reaction (Equation 13.18).

Analyze We know all the terms in the rate equation

$$\ln[X] = -kt + \ln[X]_0$$

except [X], which is what we want to calculate. The rate constant k is given in terms of seconds, so we must convert $t = 3$ min into $t = 180$ s. In the data set for Sample Exercise 13.5, $[N_2O_5]$ dropped to about $\frac{1}{4}[N_2O_5]_0$ in 200 s, so we estimate that the concentration of N_2O_5 in this example will drop to about one-quarter of its initial value, or around 0.09 M.

Solve Substituting the given values into Equation 13.18:

$$\ln[X] = -kt + \ln[X]_0$$

$$\ln[N_2O_5] = -(0.00693\ s^{-1})(180\ s) + \ln(0.375)$$

$$\ln[N_2O_5] = -1.247 + (-0.981) = -2.228$$

$$[N_2O_5] = e^{-2.228} = 0.108\ M$$

half-life ($t_{1/2}$) the time in the course of a chemical reaction during which the concentration of a reactant decreases by half.

Think About It The answer is less than the starting concentration of N_2O_5, as we would expect, and close to our estimated value.

Practice Exercise Under a different set of conditions from those in Sample Exercise 13.5, dinitrogen pentoxide decomposes into nitrogen dioxide and O_2:

$$2\ N_2O_5(g) \rightarrow 4\ NO_2(g) + O_2(g)$$

This reaction is also first order in N_2O_5 and has a rate constant $k = 9.55 \times 10^{-4}\ s^{-1}$. What is the concentration of N_2O_5 in this system after exactly 10 min if $[N_2O_5]_0 = 0.763\ M$?

Reaction Half-Lives

A parameter frequently cited in kinetic studies is the **half-life ($t_{1/2}$)** of a reaction, which is the interval during which the concentration of a reactant decreases by half. Half-life is inversely related to the rate constant of a reaction: the higher the reaction rate, the shorter the half-life.

Let's consider reaction half-life in the context of another nitrogen oxide found in the atmosphere: dinitrogen monoxide, also called nitrous oxide or laughing gas, an anesthetic sometimes used by dentists. Atmospheric concentrations of this potent greenhouse gas have been increasing in recent years, although the principal source is not automotive emissions but rather bacterial degradation of nitrogen compounds in soil. Dinitrogen monoxide is not produced in internal combustion engines, because at high temperatures, N_2O rapidly decomposes (**Figure 13.13**) into nitrogen and oxygen:

$$N_2O(g) \rightarrow N_2(g) + \tfrac{1}{2}\ O_2(g)$$

We can derive a mathematical relation between half-life $t_{1/2}$ and rate constant k for this or any other first-order reaction by starting with Equation 13.17:

$$\ln \frac{[X]}{[X]_0} = -kt$$

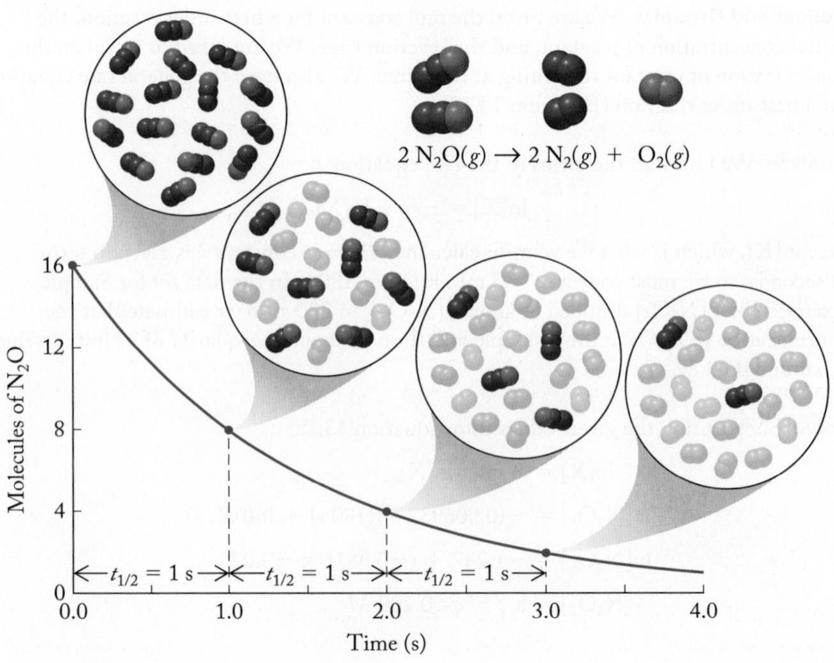

FIGURE 13.13 The decomposition of $N_2O(g)$ is first order in N_2O. At a particular temperature the half-life of the reaction is 1.0 s, which means that, on average, half of a population of 16 N_2O molecules decomposes in 1.0 s, half of the remaining 8 molecules decomposes in the next 1.0 s, and so on.

After one half-life has passed, $t = t_{1/2}$, and the concentration of X is half its original value: $[X]_0/2$. Inserting these values for [X] and t into the equation yields

$$\ln\frac{[X]_0/2}{[X]_0} = -kt_{1/2}$$

$$\ln(\tfrac{1}{2}) = -kt_{1/2}$$

The natural log of $\frac{1}{2}$ is −0.693, so

$$-0.693 = -kt_{1/2}$$

or

$$t_{1/2} = \frac{0.693}{k} \tag{13.19}$$

Thus, the half-life of a first-order reaction is inversely proportional to the rate constant, as noted at the beginning of this discussion. The absence of any concentration term in Equation 13.19 means that the half-life of a first-order reaction is constant throughout the reaction and independent of concentration: no matter the initial concentration of the reactant, half of it is consumed in one half-life.

SAMPLE EXERCISE 13.7 Calculating the Half-Life of a First-Order Reaction **LO5**

The rate constant for the decomposition of N_2O_5 at a particular temperature is $7.8 \times 10^{-3}\ s^{-1}$. What is the half-life of N_2O_5 at that temperature?

Collect, Organize, and Analyze We are asked to determine the half-life of N_2O_5 from the rate constant of its decomposition reaction. We know from Sample Exercise 13.5 that the decomposition of N_2O_5 is a first-order process, so the values of $t_{1/2}$ and k are related by Equation 13.19.

Solve

$$t_{1/2} = \frac{0.693}{k} = \frac{0.693}{7.8 \times 10^{-3}\ s^{-1}} = 89\ s$$

Think About It The calculated value makes sense because the k value is small—a little less than 0.01—so dividing 0.693 by it should produce a $t_{1/2}$ value a little larger than 69 s. Remember that Equation 13.19 is valid for first-order reactions only.

Practice Exercise Environmental scientists calculating half-lives of pollutants often define a *transport rate constant* that is analogous to a reaction rate constant and describes how rapidly a pollutant washes out of an ecosystem. In a study of the gasoline additive MTBE in Donner Lake, California, scientists from the University of California, Davis, found that in the summer the half-life of MTBE in the lake was 28 days. Assuming that the transport process is first order, what was the transport rate constant of MTBE out of Donner Lake during the study? Express your answer in reciprocal days.

CONCEPT **TEST**

Which has a shorter half-life, a fast reaction or a slow reaction?

Integrated Rate Laws: Second-Order Reactions

In Section 13.1 we described how NO_2 exposed to UV rays from the Sun decomposes to NO and atomic oxygen (O). Nitrogen dioxide may also undergo thermal decomposition, producing NO and molecular oxygen (O_2):

$$2\ NO_2(g) \rightarrow 2\ NO(g) + O_2(g) \qquad (13.20)$$

The data in **Table 13.10** describe the rate of the thermal decomposition reaction. In this case, the plot of $\ln[NO_2]$ versus time (**Figure 13.14a**) is *not* linear, which means the thermal decomposition of NO_2 is *not* first order.

TABLE 13.10 Decomposition of NO_2 as a Function of Time

Time (s)	$[NO_2]$ (*M*)	$\ln[NO_2]$	$1/[NO_2]$ (1/*M*)
0.0	1.00×10^{-2}	−4.605	100
1.00×10^2	6.48×10^{-3}	−5.039	154
2.00×10^2	4.79×10^{-3}	−5.341	209
3.00×10^2	3.80×10^{-3}	−5.573	263
4.00×10^2	3.15×10^{-3}	−5.760	317
5.00×10^2	2.69×10^{-3}	−5.918	372
6.00×10^2	2.35×10^{-3}	−6.057	426

What, then, is the order of the reaction? The answer is related to how the reaction takes place. If each NO_2 molecule simply fell apart, the reaction would be first order, much like the decomposition of N_2O_5. However, if the reaction happens as a result of collisions between pairs of NO_2 molecules, the reaction would be first order in each one of them and *second order* overall. In that case, the rate law expression would be

$$\text{Rate} = k[NO_2]^2 \qquad (13.21)$$

How can we determine whether this decomposition is really second order? One way is to assume that it is and then test that assumption. The test entails transforming the rate law in Equation 13.21 into the integrated rate law for a second-order reaction, again using calculus. The result of the transformation is

$$\frac{1}{[NO_2]} = kt + \frac{1}{[NO_2]_0} \qquad (13.22)$$

Like Equation 13.18, this has the form $y = mx + b$ and is the equation of a straight line, this time with $1/[NO_2]$ as the y variable and t as the x variable.

The graph obtained using data from columns 1 and 4 of Table 13.10 is shown in **Figure 13.14b**. The curve is linear, which means the decomposition of NO_2 is second order. The slope of the line provides a direct measure of k, which is $0.544\ M^{-1}\ s^{-1}$.

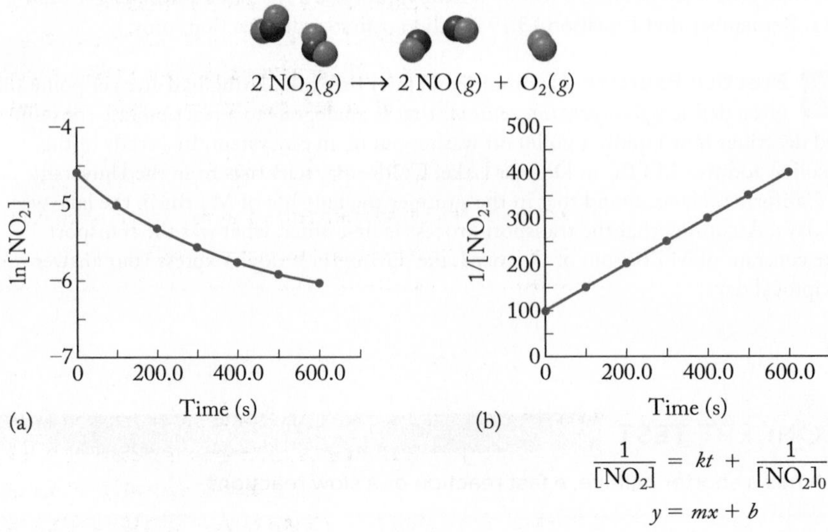

FIGURE 13.14 At high temperatures, NO_2 slowly decomposes into NO and O_2. (a) The plot of $\ln[NO_2]$ versus time is not linear, indicating that the reaction is not first order. (b) The plot of $1/[NO_2]$ versus time is linear, however, indicating that the reaction is second order in NO_2. The slope of the line in this graph equals the rate constant.

A general form of Equation 13.22 that applies to any reaction that is second order in a single reactant (X) is

$$\frac{1}{[X]} = kt + \frac{1}{[X]_0} \tag{13.23}$$

SAMPLE EXERCISE 13.8 Distinguishing between First- and Second-Order Reactions **LO4**

Chlorine monoxide accumulates in the stratosphere above Antarctica each winter and plays a key role in the formation of the ozone hole above the South Pole each spring. Eventually, ClO decomposes according to the equation

$$2\,ClO(g) \rightarrow Cl_2(g) + O_2(g)$$

A study of the kinetics of this reaction at 298 K yielded the data shown in the first two columns of **Table 13.11**. Determine the order of the reaction, the rate law, and the value of k at 298 K.

TABLE 13.11 Concentration of Chlorine Monoxide, ln[ClO], and 1/[ClO] as a Function of Time

Time (ms)	[ClO] (M)	ln[ClO]	1/[ClO] (1/M)
0.0	1.50×10^{-8}	−18.015	6.67×10^{7}
10.0	7.01×10^{-9}	−18.776	1.43×10^{8}
20.0	4.88×10^{-9}	−19.138	2.05×10^{8}
30.0	3.72×10^{-9}	−19.410	2.69×10^{8}
40.0	2.61×10^{-9}	−19.764	3.83×10^{8}
100.0	1.31×10^{-9}	−20.453	7.63×10^{8}
200.0	0.63×10^{-9}	−21.185	1.59×10^{9}

Collect and Organize We are given experimental data describing the variation in concentration of ClO with time at 298 K, and we are asked to determine the order (first or second) of the decomposition reaction of ClO.

Analyze To distinguish between first and second order for a reaction in which ClO is the single reactant, we need to calculate ln[ClO] and 1/[ClO] values and plot them versus time. If the ln[ClO] plot is linear, the reaction is first order; if the 1/[ClO] plot is linear, the reaction is second order. The rate law has the form

$$\text{Rate} = k[ClO]^m$$

where m equals 1 or 2 and, because there is only one reactant, m is also the overall order of the reaction. We determine the rate constant from the slope of whichever plot is linear.

Solve The last two columns of Table 13.11 contain the needed ln[ClO] and 1/[ClO] values. Plots of these values versus time, which were generated using Excel, are shown in **Figure 13.15**. The ln[ClO] plot is not linear, but the 1/[ClO] plot is, which means that the reaction is second order in ClO and second order overall. Thus, $m = 2$ and the rate law is

$$\text{Rate} = k[ClO]^2$$

Substituting [ClO] into the generic integrated rate law (Equation 13.23) gives

$$\frac{1}{[ClO]} = kt + \frac{1}{[ClO]_0}$$

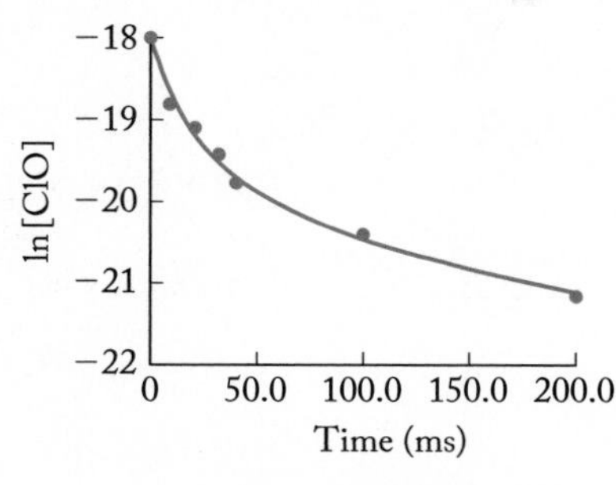

(a)

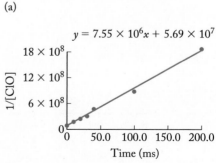

(b)

FIGURE 13.15 Plots using data from Table 13.11.

TABLE 13.12 Concentration of NO_2 as a Function of Time

Time (h)	$[NO_2]$ (*M*)
0.00	0.250
1.39	0.198
3.06	0.159
4.72	0.132
6.39	0.114
8.06	0.099
9.72	0.088
11.39	0.080

From the general equation $y = mx + b$, we know that k is the slope of the graph of 1/[ClO] versus time, which is $7.55 \times 10^6\ M^{-1}\ \text{ms}^{-1}$ or $7.55 \times 10^9\ M^{-1}\ \text{s}^{-1}$ to three significant figures.

Think About It We can distinguish between first- and second-order reactions involving a single reactant by transforming the concentration data and examining which gives a linear fit with the integrated rate law.

 **Practice Exercise** Experimental evidence shows that in the reaction

$$NO_2(g) + CO(g) \rightarrow NO(g) + CO_2(g)$$

the reaction rate depends only on the concentration of NO_2. Determine whether the reaction is first or second order in NO_2, and calculate the rate constant from the data in **Table 13.12**, which were obtained at 488 K.

The concept of half-life can also be applied to second-order reactions. The relationship between rate constant and half-life for the decomposition of NO_2 can be derived from Equation 13.23 if we first rearrange the terms to solve for kt:

$$kt = \frac{1}{[X]} - \frac{1}{[X]_0}$$

After one half-life has elapsed ($t = t_{1/2}$), [X] has decreased to half its initial concentration. Substituting this information into the preceding equation, we have

$$kt_{1/2} = \frac{1}{\frac{1}{2}[X]_0} - \frac{1}{[X]_0}$$

$$= \frac{2}{[X]_0} - \frac{1}{[X]_0} = \frac{1}{[X]_0}$$

or

$$t_{1/2} = \frac{1}{k[X]_0} \tag{13.24}$$

The value of $t_{1/2}$ for a second-order reaction, then, is inversely proportional to the initial concentration of X. This is unlike the $t_{1/2}$ values of first-order reactions, which are independent of concentration.

Zero-Order Reactions

In the practice exercise accompanying Sample Exercise 13.8, we introduced the reaction

$$NO_2(g) + CO(g) \rightarrow NO(g) + CO_2(g)$$

The rate law for the reaction is

$$\text{Rate} = k[NO_2]^2 \tag{13.25}$$

Because there is no concentration term for CO in the rate law, the rate of the reaction does not depend upon the concentration of CO. That is, the rate of the reaction does not change with changing [CO], even when the concentrations of CO and NO_2 are comparable.

One interpretation of Equation 13.25 is that it contains a [CO] term to the zeroth power, making the reaction *zero order* in that reactant. Because any value raised to the zeroth power equals 1, we have

$$\text{Rate} = k[\text{NO}_2]^2[\text{CO}]^0 = k[\text{NO}_2]^2(1) = k[\text{NO}_2]^2$$

Reactions with a true zero-order rate law are rare, but let's consider a hypothetical zero-order reaction involving a single reactant X that forms product Y:

$$\text{X} \rightarrow \text{Y}$$

If the reaction is zero order in X, then the rate law is

$$\text{Rate} = -\Delta[\text{X}]/\Delta t = k[\text{X}]^0 = k$$

and the integrated rate law is

$$[\text{X}] = -kt + [\text{X}]_0 \qquad (13.26)$$

The slope of a plot of reactant concentration versus time (**Figure 13.16**) is the negative of the zero-order rate constant k.

We can calculate the half-life of a zero-order reaction by substituting $t = t_{1/2}$ and $[\text{X}] = [\text{X}]_0/2$ into the integrated rate law:

$$[\text{X}]_0/2 = -kt_{1/2} + [\text{X}]_0$$

$$kt_{1/2} = [\text{X}]_0 - [\text{X}]_0/2 = [\text{X}]_0/2 \qquad (13.27)$$

$$t_{1/2} = [\text{X}]_0/2k$$

Figure 13.17 summarizes how to determine whether a reaction in which reactant X forms one or more products is zero, first, or second order. The figure also includes the rate laws and integrated rate laws for these reactions and the equations used to calculate the half-lives of X. The decision-making process shown

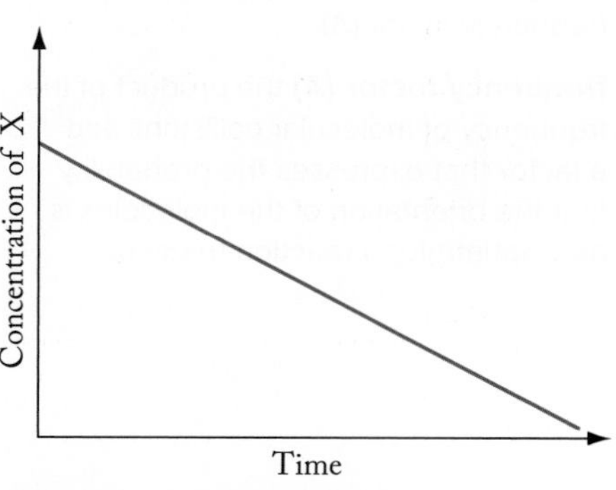

FIGURE 13.16 The change in concentration of the reactant X in the zero-order reaction X → Y is constant over time.

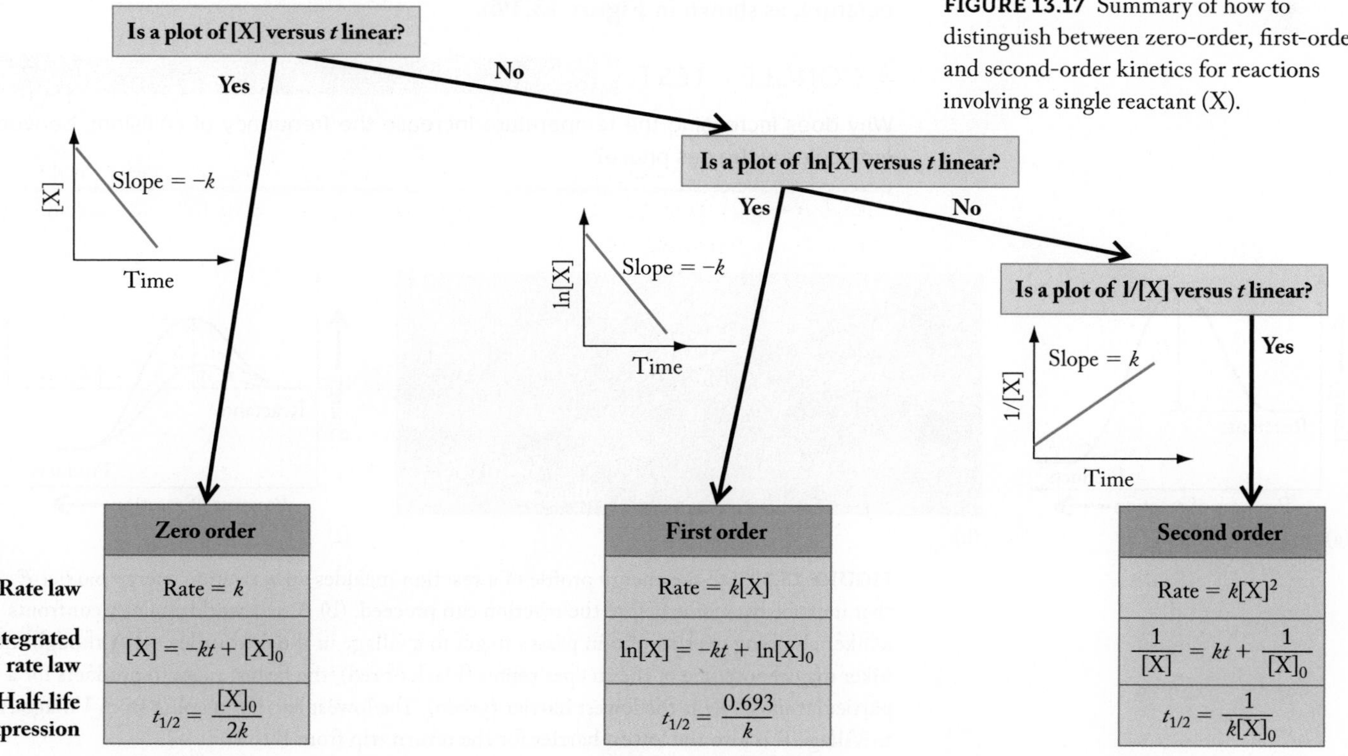

FIGURE 13.17 Summary of how to distinguish between zero-order, first-order, and second-order kinetics for reactions involving a single reactant (X).

activation energy (E_a) the minimum energy molecules need to react when they collide.

Arrhenius equation an equation relating the rate constant of a reaction to absolute temperature (T), the activation energy of the reaction (E_a), and the frequency factor (A).

frequency factor (A) the product of the frequency of molecular collisions and a factor that expresses the probability that the orientation of the molecules is appropriate for a reaction to occur.

in Figure 13.17 can also be used to determine the kinetics of reactions involving two reactants (X and Y) if the reaction is zero order in Y. In such cases, we focus on how the concentration of X changes with time. For now, we leave the discussion of zero-order reactions with this purely mathematical treatment. We return to these reactions and examine their meaning at the molecular level in Section 13.5.

13.4 Reaction Rates, Temperature, and the Arrhenius Equation

Chemical reactions take place when molecules collide with sufficient energy to break bonds in reactants and allow bonds to form in products. The minimum amount of energy that enables this to happen is called the **activation energy (E_a)**. Every chemical reaction has a characteristic activation energy, usually expressed in kilojoules per mole. Activation energy, which is always a positive value, is an energy barrier that must be overcome if a reaction is to proceed—like the mountain passes that must be climbed when hiking the trails connecting two valleys shown in **Figure 13.18**. Just as a hiker is likely to get from Village A to Village B more rapidly following the trail over the lowest pass, reactions tend to be more rapid when their activation energies are small.

CONNECTION Bond breaking is an endothermic process, whereas bond formation is an exothermic process (Section 6.3).

According to kinetic molecular theory, molecules have higher average kinetic energies at higher temperatures. Put another way, raising the temperature of a collection of molecules means that more of them have the minimum amount of energy needed to overcome the activation energy barrier of a reaction and react with each other. This principle is illustrated in **Figure 13.19a**, where the vertical dashed line represents activation energy, and the shaded areas to the right of the line represent the fraction of reactant molecules with enough energy to react with each other. Notice that the size of the shaded area is larger at higher temperature. This is why the rates of chemical reactions tend to increase with increasing temperature, as shown in **Figure 13.19b**.

CONCEPT TEST

Why does increasing the temperature increase the frequency of collisions between molecules in the gas phase?

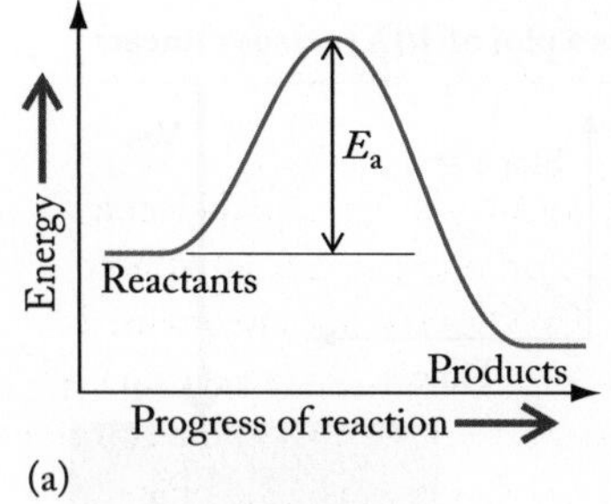

(a)

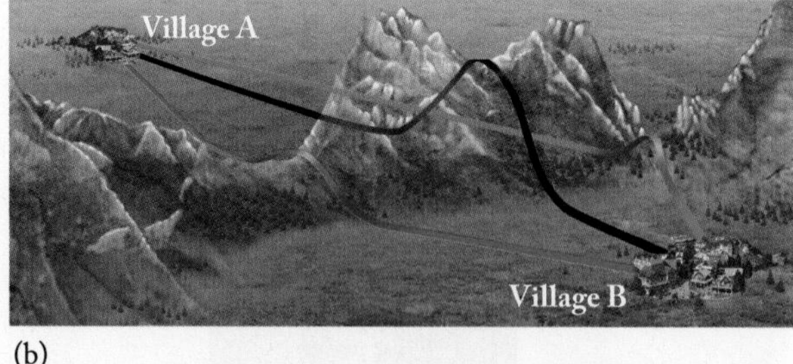

(b)

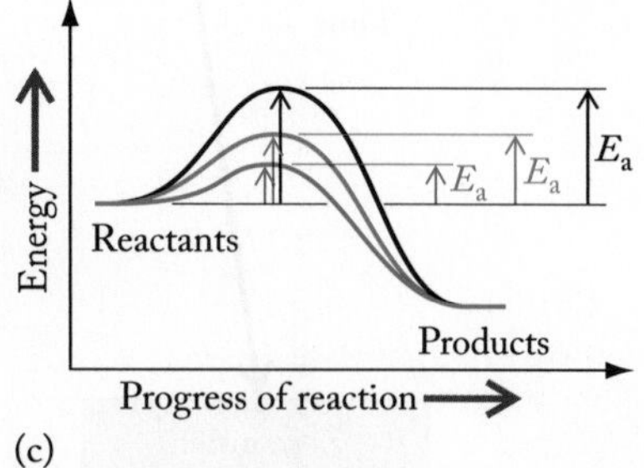

(c)

FIGURE 13.18 (a) The energy profile of a reaction includes an activation energy barrier E_a that must be overcome before the reaction can proceed. (b) A real-world analogy confronts a hiker climbing over mountain passes to get to a village in the next valley. (c) Although the hiker may choose one of the steeper routes (black or red), the fastest route to products for a particular molecule is the lowest barrier (green). The lowest barrier in going from Village A to Village B is also the lowest barrier for the return trip from B to A.

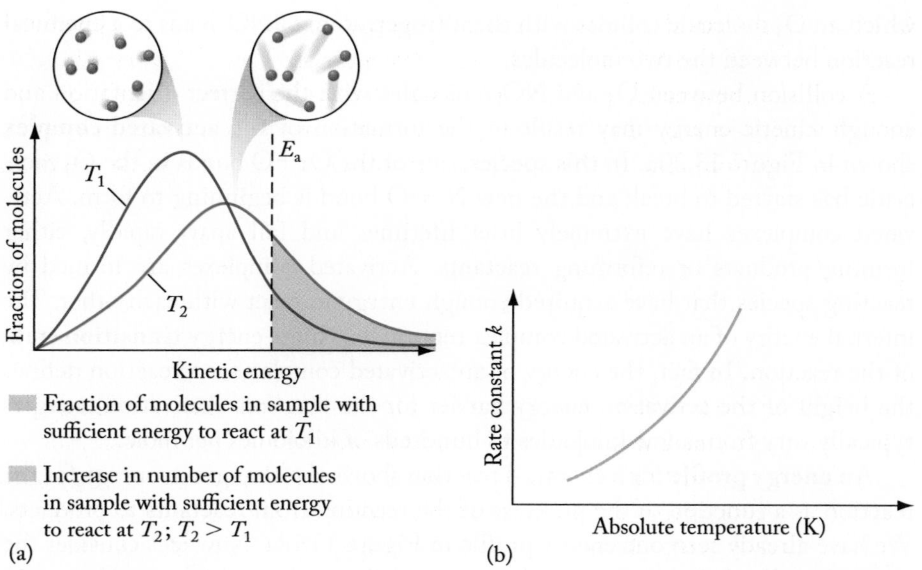

FIGURE 13.19 (a) According to kinetic molecular theory, a fraction of reactant molecules has kinetic energies equal to or greater than the activation energy (E_a) of the reaction. As temperature increases from T_1 to T_2, the number of molecules with energies exceeding E_a increases, leading to an increase in reaction rate. (b) The rate constant for any reaction increases with increasing temperature.

In the late 19th century, experiments carried out in the laboratories of Jacobus van 't Hoff and the Swedish chemist Svante Arrhenius (1859–1927) led to a fundamental advance in understanding how temperature affects the rates of chemical reactions. The mathematical connection between temperature, the rate constant k for a reaction, and its activation energy is given by the **Arrhenius equation:**

$$k = Ae^{-E_a/RT} \quad (13.28)$$

where R is the gas constant in J/(mol · K) and T is the reaction temperature in kelvin. The factor A, called the **frequency factor**, is the product of the collision frequency and a term that accounts for the fact that not every collision results in a chemical reaction.

CONNECTION The van 't Hoff factor introduced in Section 11.5 is named after the same Jacobus van 't Hoff who studied the temperature dependence of reaction rates.

CHEMTOUR Arrhenius Equation

Some collisions do not lead to products because the colliding molecules are not oriented with respect to each other in the right way. To examine the importance of molecular orientation during collisions, consider the reaction between O_3 and NO:

$$O_3(g) + NO(g) \rightarrow O_2(g) + NO_2(g)$$

Two ways in which ozone and nitric oxide molecules might approach each other are shown in **Figure 13.20**. Only the orientation in Figure 13.20a, the one in

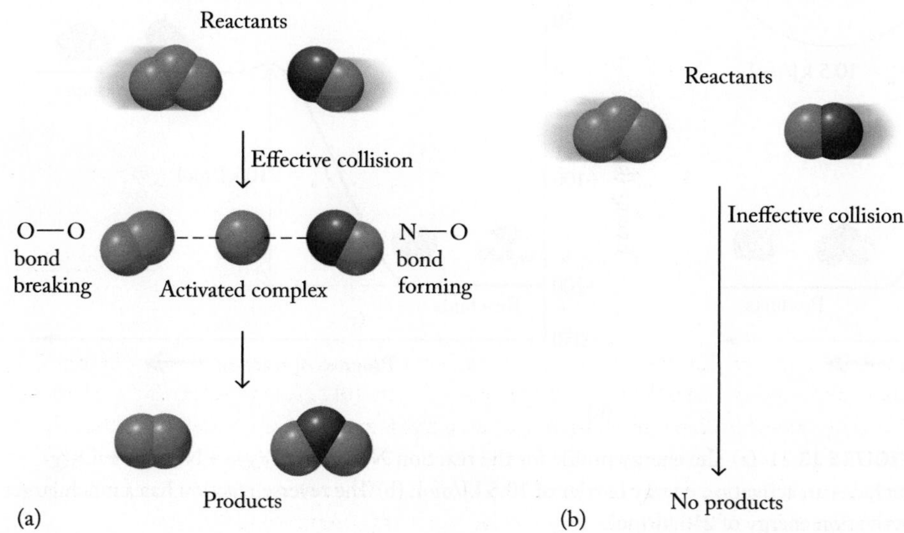

FIGURE 13.20 The effect of molecular orientation on reaction rate. (a) When an O_3 and a NO molecule are oriented such that the collision is between an O_3 oxygen and the NO nitrogen, the collision is effective and an activated complex forms, which then yields the two product molecules NO_2 and O_2. (b) When the reactant molecules are oriented such that the collision is between an O_3 oxygen and the NO oxygen, no activated complex forms and no reaction occurs.

activated complex a transient species in a chemical reaction in which molecules have both the proper orientation and enough energy to react with each other.

transition state a high-energy state between reactants and products in a chemical reaction.

energy profile a graph showing the changes in energy for a reaction as a function of the progress of the reaction from reactants to products.

which an O_3 molecule collides with the nitrogen atom of NO, leads to a chemical reaction between the two molecules.

A collision between O_3 and NO molecules with the correct orientation and enough kinetic energy may result in the formation of the **activated complex** shown in Figure 13.20a. In this species, one of the O—O bonds in the O_3 molecule has started to break and the new N—O bond is beginning to form. Activated complexes have extremely brief lifetimes and fall apart rapidly, either forming products or reforming reactants. Activated complexes are formed by reacting species that have acquired enough energy to react with each other. The internal energy of an activated complex represents a high-energy **transition state** of the reaction. In fact, the energy of an activated complex for a reaction defines the height of the activation energy barrier for the reaction. Activation energies typically vary from a few kilojoules to hundreds of kilojoules per mole.

An **energy profile** for a chemical reaction shows the changes in energy for the reaction as a function of the progress of the reaction from reactants to products. We have already seen one energy profile in Figure 13.18a. Now let's consider the energy profile for the reaction between nitric oxide and ozone, shown in **Figure 13.21a**. The x-axis represents the progress of the reaction, whereas the y-axis represents energy. The activation energy is equivalent to the difference in energy between the transition state and the reactants. The size of the activation energy barrier depends on the direction from which it is approached. In the forward direction ($NO + O_3 \rightarrow NO_2 + O_2$; Figure 13.21a), E_a is smaller than in the reverse direction ($NO_2 + O_2 \rightarrow NO + O_3$; **Figure 13.21b**). A smaller activation energy barrier means that the forward reaction proceeds at a higher rate than the reverse reaction if we have equal concentrations of reactants and products.

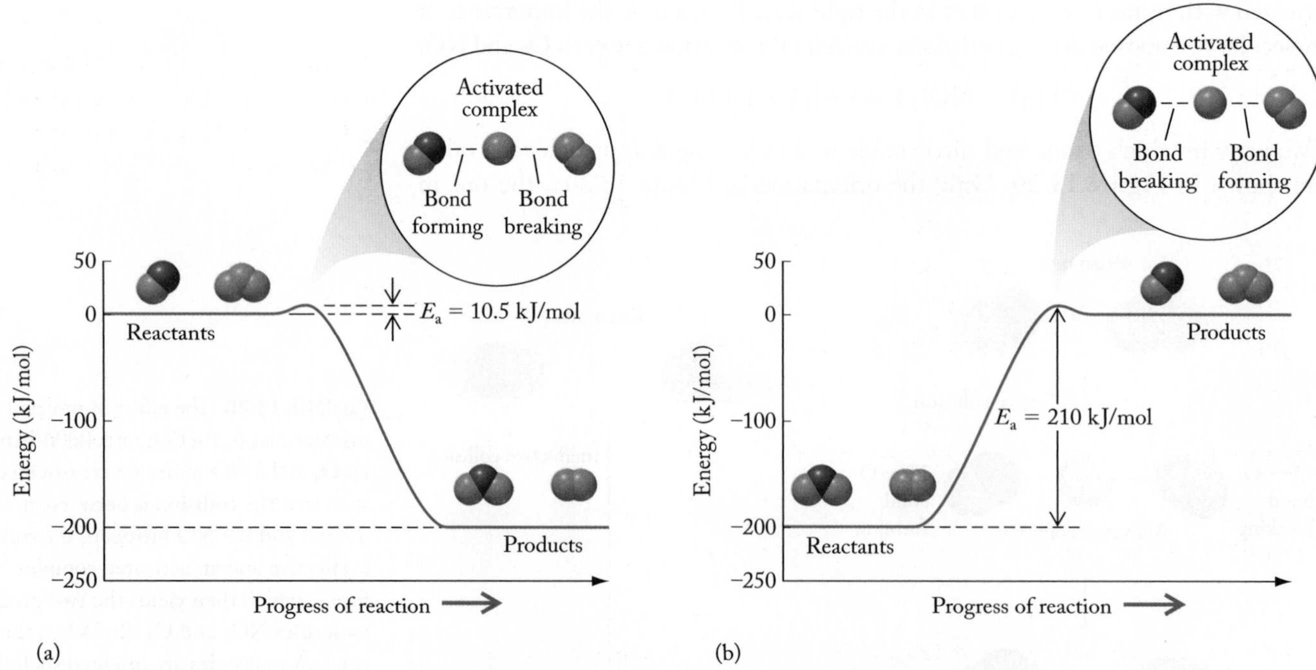

FIGURE 13.21 (a) The energy profile for the reaction $NO(g) + O_3(g) \rightarrow NO_2(g) + O_2(g)$ includes an activation energy barrier of 10.5 kJ/mol. (b) The reverse reaction has a much larger activation energy of 210 kJ/mol.

One of the many uses of the Arrhenius equation (Equation 13.28) is to calculate the value of E_a for a chemical reaction. When we take the natural logarithm of both sides of Equation 13.28,

$$\ln k = -\frac{E_a}{R}\left(\frac{1}{T}\right) + \ln A \qquad (13.29)$$

the result fits the general equation of a straight line ($y = mx + b$) if we make ln k the y variable and $1/T$ the x variable. We can calculate E_a by determining the rate constant k for the reaction at several temperatures. Plotting ln k versus $1/T$ should give a straight line, the slope of which is $-E_a/R$. **Table 13.13** and **Figure 13.22** show data for the reaction between NO and O_3 at six different temperatures.

The slope of the line in Figure 13.22 is equal to $-E_a/R$, so

$$E_a = -\text{slope} \times R$$
$$= -(-1256\ \cancel{\text{K}}) \times \left(\frac{8.314\ \text{J}}{\text{mol} \cdot \cancel{\text{K}}}\right) = 1.044 \times 10^4\ \text{J/mol}$$
$$= 10.44\ \text{kJ/mol}$$

The y intercept ($1/T = 0$) in Figure 13.22 is 27.39. From Equation 13.29, we know that this value represents ln A, which means

$$A = e^{27.39} = 7.86 \times 10^{11}$$

Now we can use the values of E_a and A to calculate k at any temperature. For example, at $T = 250$ K:

$$k = Ae^{-E_a/RT}$$
$$= (7.86 \times 10^{11})\, e^{-\left(\frac{1.044 \times 10^4\ \cancel{\text{J/mol}}}{8.314\ \frac{\cancel{\text{J}}}{\cancel{\text{mol}} \cdot \cancel{\text{K}}} \cdot 250\ \cancel{\text{K}}}\right)}$$
$$= 5.18 \times 10^9\ M^{-1}\,\text{s}^{-1}$$

TABLE 13.13 Temperature Dependence of the Rate of Reaction $NO(g) + O_3(g) \rightarrow NO_2(g) + O_2(g)$

T (K)	k (M^{-1} s^{-1})	$1/T$ (K^{-1})	ln k
300	1.20×10^{10}	3.33×10^{-3}	23.208
325	1.69×10^{10}	3.08×10^{-3}	23.551
350	2.01×10^{10}	2.86×10^{-3}	23.724
375	2.89×10^{10}	2.67×10^{-3}	24.087
400	3.32×10^{10}	2.50×10^{-3}	24.226
425	4.15×10^{10}	2.35×10^{-3}	24.449

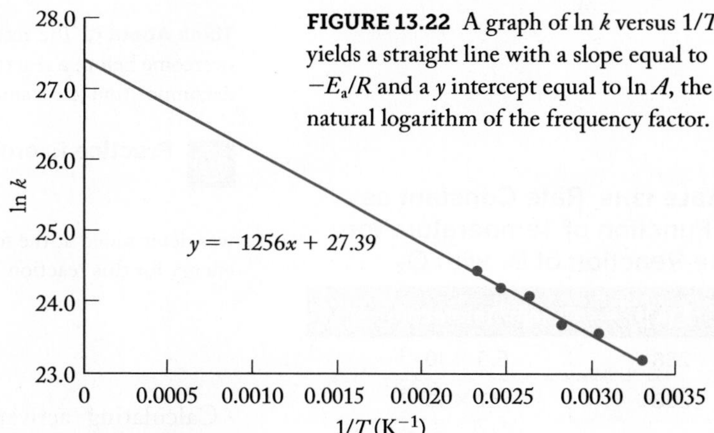

FIGURE 13.22 A graph of ln k versus $1/T$ yields a straight line with a slope equal to $-E_a/R$ and a y intercept equal to ln A, the natural logarithm of the frequency factor.

SAMPLE EXERCISE 13.9 Calculating an Activation Energy from Rate Constants **LO6**

The data in the first two columns of **Table 13.14** were collected in a study of the effect of temperature on the rate of the decomposition reaction

$$2\ ClO(g) \rightarrow Cl_2(g) + O_2(g)$$

Determine the activation energy for the reaction.

TABLE 13.14 Effect of Temperature on the Value of the Rate Constant for the Decomposition of ClO

T (K)	k (M^{-1} s^{-1})	$1/T$ (K^{-1})	ln k
238	1.90×10^9	4.20×10^{-3}	21.365
258	3.30×10^9	3.88×10^{-3}	21.855
278	4.60×10^9	3.60×10^{-3}	22.313
298	7.20×10^9	3.36×10^{-3}	22.697

Collect and Organize We can calculate activation energy by using the Arrhenius equation (Equation 13.28). We are given values of the rate constant k as a function of absolute temperature. According to Equation 13.29, the slope of a plot of ln k against $1/T$ is equal to $-E_a/R$, where E_a is the activation energy and R is the ideal gas constant with units of J/(mol · K).

Analyze First, we need to convert the T and k values in the first two columns in Table 13.14 to $1/T$ and ln k, respectively. We predict a positive value for E_a because activation energy represents a barrier that costs energy to overcome. The rate constant for the reaction is fairly large, about 10^9 M^{-1} s^{-1}, so we predict that the activation energy barrier will be relatively low.

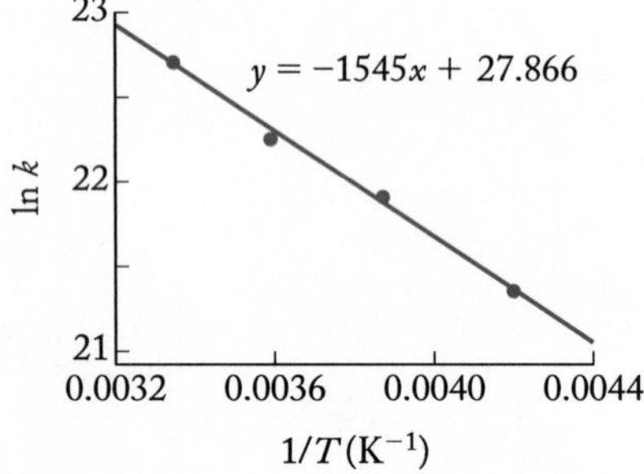

FIGURE 13.23 Plot of data in the last two columns of Table 13.14.

Solve The calculated values of $1/T$ and ln k (in the last two columns of Table 13.14) are used to plot ln k versus $1/T$ (**Figure 13.23**). The slope of the straight line that best fits the data is −1545 K.

We use the values of the slope and R [8.314 J/(mol · K)] to calculate the value of E_a:

$$E_a = -\text{slope} \times R$$

$$= -(-1545 \text{ K}) \times \left(\frac{8.314 \text{ J}}{\text{mol} \cdot \text{K}}\right) = 1.28 \times 10^4 \text{ J/mol}$$

$$= 12.8 \text{ kJ/mol}$$

Think About It The activation energy is the height of a barrier that has to be overcome before a reaction can proceed. As predicted, the activation energy for ClO decomposition has a small positive value.

Practice Exercise The rate constant for the reaction

$$Br(g) + O_3(g) \rightarrow BrO(g) + O_2(g)$$

was determined at the four temperatures shown in **Table 13.15**. Calculate the activation energy for this reaction.

TABLE 13.15 Rate Constant as a Function of Temperature for the Reaction of Br with O_3

T (K)	k [cm^3/(molecule · s)]
238	5.9×10^{-13}
258	7.7×10^{-13}
278	9.6×10^{-13}
298	1.2×10^{-12}

Calculating activation energies by using the graphical method generally requires measurement of the rate constant at a minimum of three different temperatures to verify that the plot of ln k versus $1/T$ is a straight line. Once we know the value of E_a, we can use it and the value of the rate constant (k_1) of a reaction at one temperature (T_1) to calculate the value of the rate constant (k_2) at another temperature (T_2). We start by substituting k_1, k_2, T_1, and T_2 into Equation 13.29:

$$\ln k_1 = -\frac{E_a}{R}\left(\frac{1}{T_1}\right) + \ln A \qquad \ln k_2 = -\frac{E_a}{R}\left(\frac{1}{T_2}\right) + \ln A$$

Subtracting these two expressions, $\ln k_2 - \ln k_1$, gives

$$\ln k_2 - \ln k_1 = \left[-\frac{E_a}{R}\left(\frac{1}{T_2}\right) + \ln A\right] - \left[-\frac{E_a}{R}\left(\frac{1}{T_1}\right) + \ln A\right]$$

Using the mathematical properties of logarithms, we can rearrange the terms to obtain Equation 13.30:

$$\ln\frac{k_2}{k_1} = -\frac{E_a}{R}\left(\frac{1}{T_2}\right) - \left[-\frac{E_a}{R}\left(\frac{1}{T_1}\right)\right]$$

$$\ln\frac{k_2}{k_1} = -\frac{E_a}{R}\left(\frac{1}{T_2} - \frac{1}{T_1}\right) \qquad (13.30)$$

SAMPLE EXERCISE 13.10 Determining the Effect of Temperature on Rate Constants **LO6**

The kinetics of decomposition of a potential drug have been tested in experiments at room temperature (25°C), where $k = 6.45 \times 10^{-6}\ M^{-1}\ s^{-1}$. If the activation energy for this reaction is 67.1 kJ/mol, what will happen to the reaction rate when the same kinetics experiments are conducted at body temperature (37°C)?

Collect and Organize We are asked to calculate what happens to the rate of the reaction when kinetics experiments are run at a higher temperature. We are given values of the activation energy E_a, one rate constant k, and two temperatures. We can use Equation 13.30 to solve for the second rate constant, where R is the ideal gas constant with units of J/(mol · K).

Analyze First, we need to convert the temperature values from Celsius degrees to kelvin. We have one (k, T) pairing—($6.45 \times 10^{-6}\ M^{-1}\ s^{-1}$, 298 K)—and we need to find the k value for the second temperature, 310 K. We predict that as the temperature increases from room temperature to body temperature, the reaction will go faster and k will increase.

Solve Substituting the given values into Equation 13.30 gives us:

$$\ln\left(\frac{k_2}{k_1}\right) = -\frac{E_a}{R}\left(\frac{1}{T_2} - \frac{1}{T_1}\right)$$

$$\ln\left(\frac{k_2}{6.45 \times 10^{-6}\ M^{-1}\ s^{-1}}\right) = -\frac{\left(\frac{67.1\ \text{kJ}}{\text{mol}}\right)\left(\frac{1000\ \text{J}}{1\ \text{kJ}}\right)}{\frac{8.314\ \text{J}}{\text{mol} \cdot \text{K}}}\left(\frac{1}{310\ \text{K}} - \frac{1}{298\ \text{K}}\right)$$

$$\ln\left(\frac{k_2}{6.45 \times 10^{-6}\ M^{-1}\ s^{-1}}\right) = 1.0484$$

$$\frac{k_2}{6.45 \times 10^{-6}\ M^{-1}\ s^{-1}} = e^{1.0484}$$

$$k_2 = 1.84 \times 10^{-5}\ M^{-1}\ s^{-1}$$

Think About It As predicted, k increases as T increases; in fact, it nearly triples. A higher temperature increases kinetic energy, which leads to more collisions of molecules with sufficient energy to exceed the threshold value of the activation energy and to a more rapid reaction rate.

Practice Exercise What temperature would be required to double the reaction rate when compared to the rate at room temperature for the potential drug tested in this sample exercise?

CONCEPT **TEST**

Which of the following statements is/are true about activation energies?

a. Exothermic reactions have negative activation energies.
b. Fast reactions have large rate constants *and* large activation energies.
c. The forward reaction sometimes has a lower activation energy than the reverse reaction.
d. Endothermic reactions always have large activation energies.

13.5 Reaction Mechanisms

CHEMTOUR
Reaction Mechanisms

Up to this point we have described reactions in terms of macroscopically observable quantities such as pressure, temperature, volume, and numbers of moles of substances. In this section we explore how reactions proceed at the molecular level. Being able to suggest what goes on at the molecular level by observing macroscopic properties is a remarkable achievement. We will connect the minimum amount of energy that is needed to break bonds in the reactants and form bonds in the products to an exact determination of which bonds break and which bonds form. The evaluation of the stepwise behavior of molecules—the mechanism of a reaction—provides valuable insights into reactions, ranging from the subtlest biochemical processes taking place in cells to the chemistry of industrial processes that yield tons of product.

Let's start by revisiting the thermal decomposition of NO_2:

$$2\ NO_2(g) \rightarrow 2\ NO(g) + O_2(g)$$

We noted in Section 13.3 that this reaction is second order in NO_2 because it takes place as a result of collisions between two NO_2 molecules at a time. How do the atoms in two colliding NO_2 molecules rearrange themselves to form two NO molecules and one O_2 molecule? The answer to this question is contained in the mechanism of the reaction. A **reaction mechanism** describes the stepwise manner in which the bonds in reactant molecules break and the bonds in product molecules form. Chemists use the results of reaction rate measurements to develop explanations of how reactions actually happen. Sometimes they can test for the presence of the products formed in preliminary steps, but, until recently, this was not possible for most reactions. Today, technological advances allow chemists to follow the transformation of reactants to products in the time that it takes for individual covalent bonds to break and new ones to form. These processes occur as rapidly as the bonds vibrate—that is, in femtoseconds (10^{-15} s), so the area of research based on monitoring these ultrafast processes is called *femtochemistry*.[3]

reaction mechanism a set of steps describing how a reaction occurs at the molecular level; the mechanism must be consistent with the experimentally determined rate law for the reaction.

intermediate a species produced in one step of a reaction and consumed in a subsequent step.

elementary step a molecular view of a collision taking place in a chemical reaction.

unimolecular step a step in a reaction mechanism involving only one molecule on the reactant side.

bimolecular step a step in a reaction mechanism involving a collision between two molecules.

termolecular step a step in a reaction mechanism involving a collision among three molecules.

molecularity the number of ions, atoms, or molecules involved in an elementary step in a reaction.

Elementary Steps

A reaction mechanism proposed for the decomposition of NO_2 is shown in **Figure 13.24**. In the first step of the mechanism, a collision between two NO_2 molecules produces a molecule of NO and a molecule of NO_3. In a second step,

[3]In 1999 Ahmed H. Zewail received the Nobel Prize in Chemistry for his pioneering research in the development of femtochemistry.

the NO_3 decomposes to NO and O_2. Both steps in the mechanism involve very short-lived activated complexes. In the activated complex of the first step, two molecules share an oxygen atom. The bonds in the activated complex of the second step rearrange so that two oxygen atoms bond together, forming a molecule of O_2 and leaving behind a molecule of NO. The NO_3 molecule is an **intermediate** in this mechanism because it is produced in one step and consumed in the next. Intermediates are not considered reactants or products and do not appear in the equation describing a reaction. In some cases, intermediates in chemical reactions are sufficiently long-lived to be isolated. In contrast, activated complexes have never been isolated, although they have been detected.

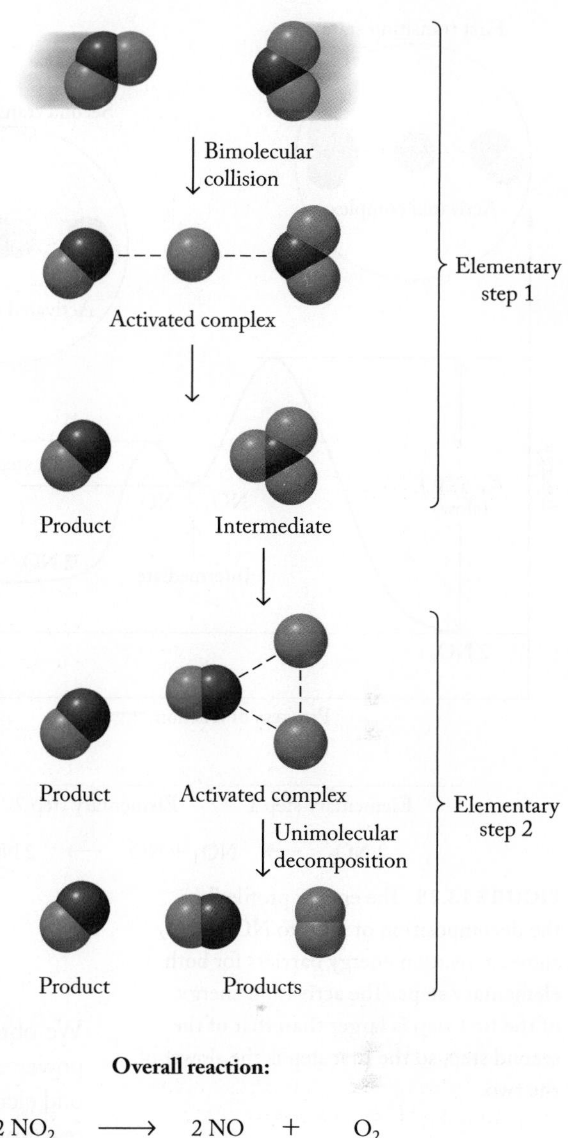

This reaction mechanism is a combination of two **elementary steps**. An elementary step that involves a single molecule is called **unimolecular**, and one that involves a collision between two molecules is **bimolecular**. Bimolecular elementary steps are much more common than **termolecular** (three-molecule) elementary steps because the chance of three molecules colliding at exactly the same time in the proper orientation is much smaller. The terms *uni-*, *bi-*, and *termolecular* are used by chemists to describe the **molecularity** of an elementary step, which refers to the number of atoms, ions, or molecules involved in that step.

A valid reaction mechanism must be consistent with the stoichiometry of the reaction. In other words, the sum of the elementary steps in Figure 13.24 must be consistent with the observed proportions of reactants and products as defined in the balanced chemical equation. In this case, the sum matches the overall stoichiometry:

Elementary step 1	$2\ NO_2(g) \rightarrow NO(g) + NO_3(g)$
Elementary step 2	$NO_3(g) \rightarrow NO(g) + O_2(g)$

Summing the two elementary steps and simplifying by canceling out the intermediate (NO_3) terms:

$$2\ NO_2(g) + \cancel{NO_3(g)} \rightarrow 2\ NO(g) + \cancel{NO_3(g)} + O_2(g)$$

we get the overall reaction

$$2\ NO_2(g) \rightarrow 2\ NO(g) + O_2(g)$$

Overall reaction:

$$2\ NO_2 \longrightarrow 2\ NO + O_2$$

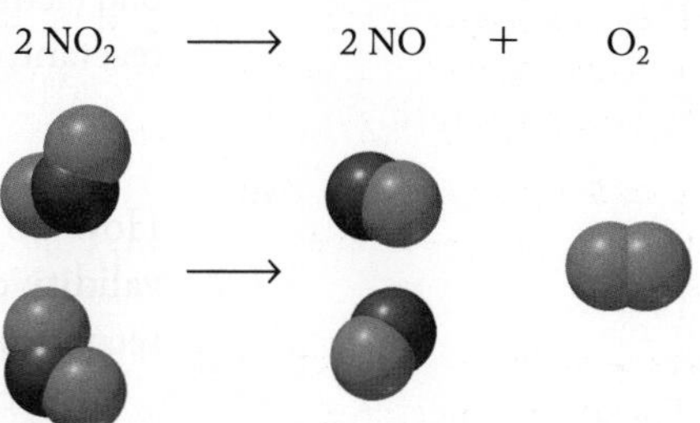

FIGURE 13.24 The decomposition of NO_2 begins when two NO_2 molecules collide, producing NO and NO_3 (elementary step 1). The NO_3 then rapidly decomposes into NO and O_2 (elementary step 2).

What does the energy profile of a two-step reaction such as this one look like? **Figure 13.25** shows that the two elementary steps produce an energy profile with two maxima, each with its own activation energy. In elementary step 1, collisions between pairs of NO_2 molecules result in the formation of an activated complex associated with the first transition state in Figure 13.25. As this activated complex transforms into NO and NO_3, the energy of the system drops to the bottom of the trough between the two maxima. In elementary step 2, NO_3 forms the activated complex associated with the second transition state. As this complex transforms into the final products NO and O_2, the energy of the system drops to its final level.

Figure 13.25 shows that the energy barrier for elementary step 1 is much greater than that for elementary step 2. This difference is consistent with the relative rates of the two steps: step 1 is slower than step 2. If the reaction were to proceed in the reverse direction (i.e., $2\ NO + O_2 \rightarrow 2\ NO_2$), the first energy

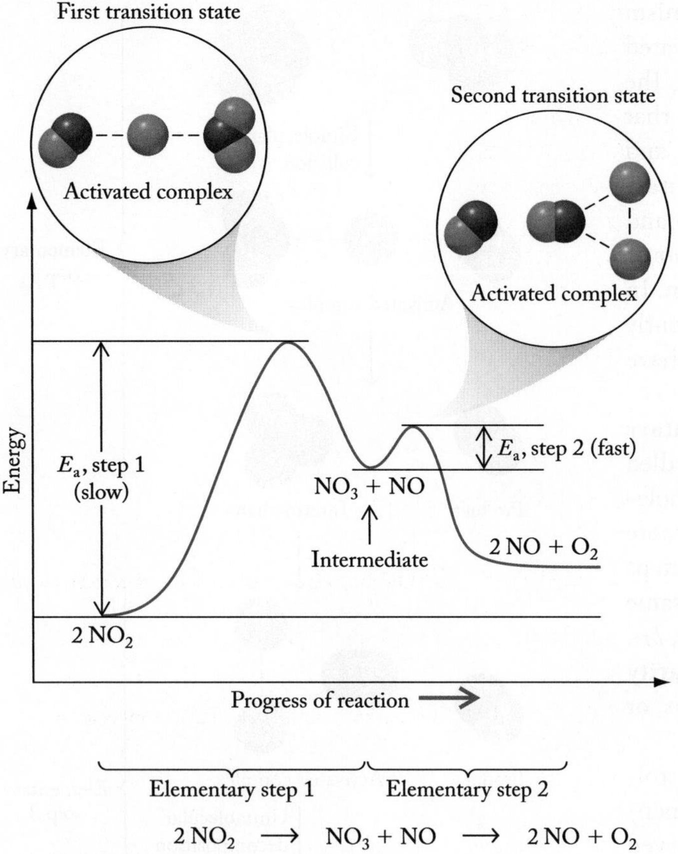

FIGURE 13.25 The energy profile for the decomposition of NO_2 to NO and O_2 shows activation energy barriers for both elementary steps. The activation energy of the first step is larger than that of the second step, so the first step is the slower of the two.

barrier would be the smaller of the two, and the first elementary step would be the more rapid one. Experimental evidence supports these expectations.

CONCEPT TEST

For a reaction mechanism with *n* elementary steps, how many activation energies does the overall reaction have?

Rate Laws and Reaction Mechanisms

Any mechanism proposed for a reaction must be consistent with the rate law derived from experimental data. We saw in Section 13.3 that the thermal decomposition of NO_2 is second order in NO_2:

$$\text{Rate} = k[NO_2]^2$$

We have also seen that the coefficients in the balanced chemical equation for any chemical reaction do not necessarily match the exponents in the rate law for the reaction. For any *elementary step* in a reaction mechanism, however, *we can* use the balanced chemical equation to write a rate law for that step. For the thermal decomposition of NO_2, the rate law for the first elementary step, $2 NO_2(g) \rightarrow NO_3(g) + NO(g)$, is

$$\text{Rate}_1 = k_1[NO_2]^2 \quad (13.31)$$

We obtain this rate law by writing the concentration for each reactant raised to a power equal to the coefficient of that reactant in the balanced equation. For the second elementary step in the NO_2 decomposition, $NO_3(g) \rightarrow NO(g) + O_2(g)$, the sole reactant, NO_3, has a coefficient of 1, so the rate law is

$$\text{Rate}_2 = k_2[NO_3] \quad (13.32)$$

How do we use the rate laws in Equations 13.31 and 13.32 to determine the validity of the mechanism—that is, to show that they conform to the observed second-order rate law for the decomposition of NO_2?

The two steps in the mechanism proceed at different rates, with different activation energies. One step is slower than the other, and this slower step is the **rate-determining step** in the reaction. The rate-determining step is the slowest elementary step in a chemical reaction, and the rate of this step controls the overall reaction rate. We can use the rate laws for the two steps to identify the rate-determining step. Because the rate law for step 1 matches the experimentally determined rate law, we may assume that step 1 defines how rapidly the reaction proceeds.

rate-determining step the slowest step in a multistep reaction mechanism.

One way of visualizing the concept of a rate-determining step is to analyze the flow of people through a busy airport. Many travelers arrive at the airport with their boarding passes in hand or print them from convenient kiosks that allow them to avoid lines at ticket counters and move quickly toward their departure gate. The next step in the process is passing through security. Typically, the number of people in line outside security points greatly exceeds the number of screening machines, so the time required to reach a departure gate depends mostly on the time needed to pass through security. Security screening is the rate-determining step on the way through an airport.

If the first step in the mechanism in Figure 13.24 is the rate-determining step, then the value of k for the overall reaction is equal to k_1 from Equation 13.31. In addition, the value of k_1 must be smaller than the value of k_2 from Equation 13.32. NO_3 is sufficiently stable that we can make small amounts of it and test whether $k_1 < k_2$. Experiments run at 300 K starting with NO_2 or NO_3 yield these values: $k_1 \approx 1 \times 10^{-10}\ M^{-1}\ s^{-1}$ and $k_2 \approx 6.3 \times 10^4\ s^{-1}$. Therefore, as soon as any NO_3 forms in step 1, it rapidly falls apart to NO and O_2 in step 2.

Now let's consider the reverse reaction of NO_2 decomposition—namely, the formation of NO_2 from NO and O_2:

$$\text{Overall reaction} \qquad 2\ NO(g) + O_2(g) \rightarrow 2\ NO_2(g)$$

We determined in Section 13.3 that this reaction is second order in NO and first order in O_2:

$$\text{Rate} = k[NO]^2[O_2] \qquad (13.33)$$

One proposed mechanism, shown in **Figure 13.26**, has two elementary steps:

$$\text{Step 1} \qquad NO(g) + O_2(g) \rightarrow NO_3(g) \qquad \text{Rate} = k_1[NO][O_2] \qquad (13.34)$$

$$\text{Step 2} \qquad NO_3(g) + NO(g) \rightarrow 2\ NO_2(g) \qquad \text{Rate} = k_2[NO_3][NO] \qquad (13.35)$$

We obtained the rate laws in Equations 13.34 and 13.35 by writing the concentration for each reactant raised to a power equal to the reactant's coefficient in the balanced elementary step. If step 1 were the rate-determining step, the reaction would be first order in NO and O_2, but that does not match the experimentally determined rate law. If step 2 were the rate-determining step, the reaction would be first order in NO and NO_3, but that does not match the rate law either. How, then, can we account for the experimental rate law?

If step 2 is slow while step 1 is fast *and reversible*, then NO_3 forms rapidly from NO and O_2 but decomposes just as rapidly back into NO and O_2. Expressing these rates in equation form:

$$\text{Rate of forward reaction} = k_f[NO][O_2] = \text{fast}$$

$$\text{Rate of reverse reaction} = k_r[NO_3] = \text{equally fast}$$

The subscripts "f" and "r" refer to the forward and reverse reactions, respectively.

Setting these two expressions equal to each other, we have

$$k_f[NO][O_2] = k_r[NO_3]$$

$$[NO_3] = \frac{k_f}{k_r}[NO][O_2] \qquad (13.36)$$

Now, if we replace the $[NO_3]$ term in the rate law of Equation 13.35 (which we select because step 2 is the rate-determining step) with the right side of Equation 13.36, we get

$$\text{Rate} = k_2 \frac{k_f}{k_r}[NO]^2[O_2]$$

The three rate constants can be combined,

$$k_{\text{overall}} = k_2 \frac{k_f}{k_r}$$

(1) Formation of intermediate: elementary step 1

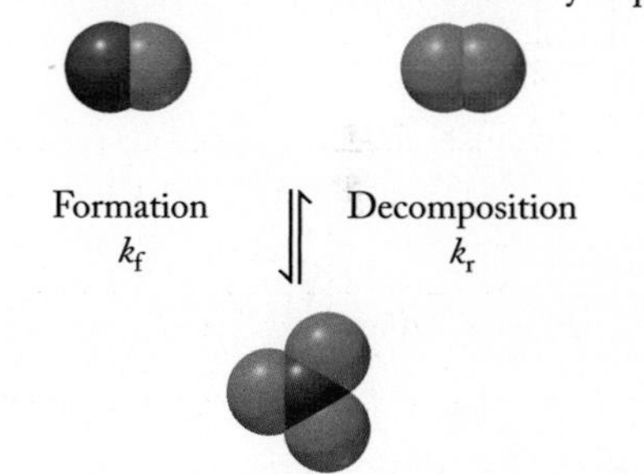

(2) Reaction of intermediate: elementary step 2

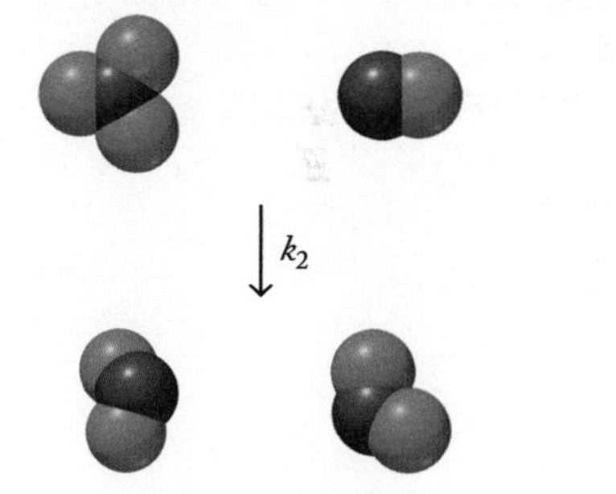

Overall reaction:

2 NO + O_2 ⟶ 2 NO_2

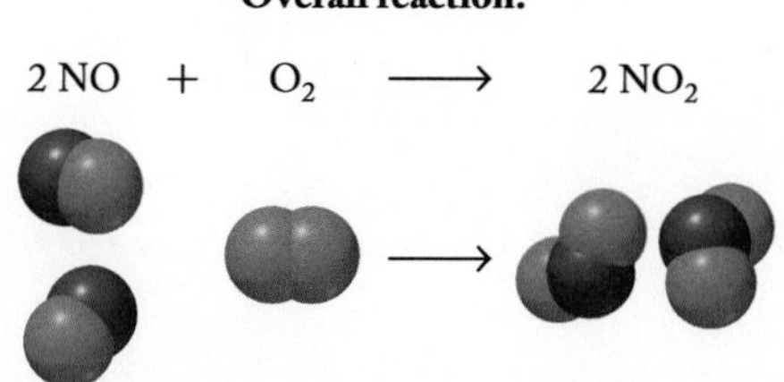

FIGURE 13.26 A mechanism for the formation of NO_2 from NO and O_2 has two elementary steps: (1) a fast, reversible bimolecular reaction in which NO and O_2 form NO_3 and (2) a slower, rate-determining bimolecular reaction in which NO_3 reacts with a molecule of NO to form two molecules of NO_2.

and the rate law for the overall reaction becomes

$$\text{Rate} = k_{\text{overall}}[\text{NO}]^2[\text{O}_2]$$

This expression matches the overall rate law in Equation 13.33, so the proposed mechanism—a fast and reversible step 1 followed by a slow step 2—may be valid.

Even though the proposed reaction mechanism is consistent with the overall stoichiometry of the reaction and with the experimentally derived rate law, these consistencies do not *prove* that the proposed mechanism is correct. Other mechanisms could be consistent with the rate law, too. On the other hand, *not* finding a reactive (and transient) intermediate would not necessarily disprove a reaction mechanism because we could be limited by our ability to detect such a short-lived species.

CONCEPT TEST

Could the products of an elementary step in a chemical reaction include activated complexes?

SAMPLE EXERCISE 13.11 Linking Reaction Mechanisms to Experimental Rate Laws LO7

The experimentally determined rate law for the reaction

$$2\,\text{NO}(g) + 2\,\text{H}_2(g) \rightarrow \text{N}_2(g) + 2\,\text{H}_2\text{O}(g)$$

is

$$\text{Rate} = k[\text{NO}]^2[\text{H}_2]$$

A proposed mechanism for the reaction is

$$\text{Elementary step 1} \quad 2\,\text{NO}(g) + \text{H}_2(g) \rightarrow \text{N}_2\text{O}(g) + \text{H}_2\text{O}(g)$$
$$\text{Elementary step 2} \quad \text{N}_2\text{O}(g) + \text{H}_2(g) \rightarrow \text{N}_2(g) + \text{H}_2\text{O}(g)$$

Is this reaction mechanism consistent with the stoichiometry of the reaction and with the rate law? If so, which is the rate-determining step?

Collect and Organize We need to determine whether the proposed two-step reaction mechanism is consistent with the overall reaction stoichiometry and the rate law of the overall reaction. We know that a rate law derived from the stoichiometry of the rate-determining step in a reaction mechanism should match the rate law of the overall reaction.

Analyze To answer the questions posed in this exercise, we need to

1. Determine whether the chemical equations of the elementary steps add up to the overall reaction equation.
2. Write rate laws for each elementary step.
3. Compare the rate laws of the elementary steps to the experimental rate law of the overall reaction to assess the validity of the mechanism.
4. Decide which step is rate determining by matching its rate law to the observed rate law.

Solve Adding up the elementary steps:

$$\begin{array}{ll} (1) & 2\,\text{NO}(g) + \text{H}_2(g) \rightarrow \cancel{\text{N}_2\text{O}(g)} + \text{H}_2\text{O}(g) \\ (2) & \underline{\cancel{\text{N}_2\text{O}(g)} + \text{H}_2(g) \rightarrow \text{N}_2(g) + \text{H}_2\text{O}(g)} \\ & 2\,\text{NO}(g) + 2\,\text{H}_2(g) \rightarrow \text{N}_2(g) + 2\,\text{H}_2\text{O}(g) \end{array}$$

This is indeed the equation of the overall reaction. The elementary steps are consistent with the stoichiometry of the overall reaction.

Next we need to focus on the reaction mechanism. Elementary step 1 involves two molecules of NO colliding with one molecule of H_2 in a termolecular reaction. Elementary step 2 is a bimolecular reaction between the N_2O produced in step 1 and another molecule of H_2. The rate law for any elementary step can be written directly from the balanced equations, using the equation coefficients as exponents in the rate law:

Step 1	$2\,NO(g) + H_2(g) \rightarrow N_2O(g) + H_2O(g)$	Rate $= k_1[H_2][NO]^2$
Step 2	$N_2O(g) + H_2(g) \rightarrow N_2(g) + H_2O(g)$	Rate $= k_2[H_2][N_2O]$

The rate law of step 1 matches the observed rate law of the overall reaction. Therefore, the proposed two-step mechanism is consistent with the experimental rate law, and step 1 is the rate-determining step.

Think About It Since the rate law we derived for our mechanism matches the observed rate law and data, the mechanism is reasonable, although we lack definitive proof that the proposed mechanism is the correct one.

Practice Exercise The following mechanism is proposed for a reaction between compounds A and B:

Step 1	$2\,A(g) + B(g) \rightleftharpoons C(g)$	fast and reversible
Step 2	$B(g) + C(g) \rightarrow D(g)$	slow
Overall	$2\,A(g) + 2\,B(g) \rightarrow D(g)$	

What is the rate law for the overall reaction based on the proposed mechanism?

SAMPLE EXERCISE 13.12 Testing a Proposed Reaction Mechanism LO7

A proposed mechanism for the decomposition of N_2O_5 to NO_2 and O_2 involves three elementary steps:

Step 1	$N_2O_5(g) \rightleftharpoons NO_2(g) + NO_3(g)$	slow
Step 2	$NO_2(g) + NO_3(g) \rightarrow NO(g) + NO_2(g) + O_2(g)$	fast
Step 3	$NO(g) + N_2O_5(g) \rightarrow 3\,NO_2(g)$	fast

Is the mechanism consistent with the stoichiometry of the overall reaction? What is the rate law of the overall reaction based on the proposed mechanism?

Collect and Organize We are given three elementary steps and their relative rates and are asked whether they constitute a reaction mechanism that is consistent with the overall stoichiometry of a chemical reaction, and we are asked to predict the rate law of the overall reaction based on the proposed mechanism. The rate law for the overall reaction is linked to the rate law for the rate-determining step.

Analyze The first step in the proposed mechanism is the slowest of the three and should be the rate-determining step. The rate law for an elementary step is linked directly to its molecularity. The reactant in step 1 is a single molecule of N_2O_5. To determine if the proposed mechanism is consistent with the overall reaction, we need to add the reactants consumed and the products formed in the three steps to see if they give a balanced chemical equation describing the overall reaction.

Solve Summing the reactants and products of the three elementary steps:

$$\begin{array}{l} N_2O_5(g) \rightleftharpoons NO_2(g) + NO_3(g) \\ + NO_2(g) + NO_3(g) \rightarrow NO(g) + NO_2(g) + O_2(g) \\ + NO(g) + N_2O_5(g) \rightarrow 3\,NO_2(g) \\ \hline N_2O_5(g) + \cancel{NO_2(g)} + \cancel{NO_3(g)} + \cancel{NO(g)} + N_2O_5(g) \rightarrow \\ \quad \cancel{NO_2(g)} + \cancel{NO_3(g)} + \cancel{NO(g)} + NO_2(g) + O_2(g) + 3\,NO_2(g) \end{array}$$

Combining the remaining terms:

$$2\,N_2O_5(g) \rightarrow 4\,NO_2(g) + O_2(g)$$

This balanced equation consists of just the reactant and products, and is consistent with the overall reaction.

The rate law for step 1 is

$$\text{Rate} = k[N_2O_5]$$

Step 1 is the rate-determining step, so its rate law is the rate law of the overall reaction.

Think About It The rate law we derived for our mechanism is consistent with a slow initial step that represents first-order, unimolecular decomposition of N_2O_5 followed by two fast steps to account for the observed stoichiometry of the reaction.

Practice Exercise The following is another proposed mechanism for the reaction of NO with H_2 in Sample Exercise 13.11:

Elementary step 1	$H_2(g) + NO(g) \rightarrow N(g) + H_2O(g)$
Elementary step 2	$N(g) + NO(g) \rightarrow N_2(g) + O(g)$
Elementary step 3	$H_2(g) + O(g) \rightarrow H_2O(g)$

Is this a valid mechanism? Explain why or why not.

Mechanisms and Zero-Order Reactions

Before we complete our discussion of reaction mechanisms, let's revisit the reaction between NO_2 and CO, which has an experimentally determined rate law that is zero order in CO, second order in NO_2, and second order overall:

$$NO_2(g) + CO(g) \rightarrow NO(g) + CO_2(g) \qquad \text{Rate} = k[NO_2]^2$$

What does this overall rate law tell us about how the reaction happens at the molecular level? Remember that the overall rate depends on the concentrations of the reactants in the rate-determining step, which means that CO is not a reactant in the rate-determining step. Carbon monoxide is involved in the reaction—it is converted into CO_2—but whatever step involves CO must occur *after* the rate-determining step. As a result, the reaction must have at least two elementary steps, one rate-determining and one not.

The proposed reaction mechanism is

(1)	$2\,NO_2(g) \rightarrow NO_3(g) + NO(g)$	$\text{Rate} = k_1[NO_2]^2$
(2)	$NO_3(g) + CO(g) \rightarrow NO_2(g) + CO_2(g)$	$\text{Rate} = k_2[NO_3][CO]$

The experimentally determined overall rate law matches the rate law for the first step, which must be the slower, rate-determining step. The overall reaction is zero order in CO because CO is not a reactant in that step.

CONCEPT **TEST**

The rate law for the reaction $XO_2(g) + M(g) \rightarrow XO(g) + MO(g)$, where X and M represent nonmetallic elements, is second order in $[XO_2]$ and zero order in [M] for a wide variety of compounds XO_2 and M. Why can we conclude that these reactions probably proceed by the same mechanism?

13.6 Catalysts

Slow reactions often have high activation energies (Section 13.4). How could we increase the rate of such a reaction? One way is to increase the temperature of the reaction mixture. In some chemical reactions, however, elevated temperatures can lead to undesired products or to lower yields. Another way is to add a **catalyst**, a substance that increases the rate of a reaction but is not consumed in the process.

Catalysts and the Ozone Layer

As we discussed in Chapter 8, the way we think about ozone depends on where the ozone is located. Ozone in the stratosphere between 10 and 40 km above Earth's surface is necessary to protect us from UV radiation, but ozone at ground level is hazardous to our health. In this section we discuss the role of catalysis in the loss of stratospheric ozone that has led to the annual formation of ozone holes over Antarctica.

The natural photodecomposition of ozone in the stratosphere occurs through the reaction

$$2\ O_3(g) \rightarrow 3\ O_2(g)$$

It begins with the absorption of UV radiation from the Sun and the generation of atomic oxygen:

$$O_3(g) \rightarrow O_2(g) + O(g)$$

The oxygen atom may react with another ozone molecule to form two more molecules of oxygen:

$$O_3(g) + O(g) \rightarrow 2\ O_2(g)$$

The second elementary step is slow because its activation energy is relatively high: 17.7 kJ/mol.

In 1974 two American scientists, F. Sherwood Rowland (1927–2012) and Mario J. Molina (1943–), predicted significant depletion of stratospheric ozone because of the release of a class of volatile compounds called chlorofluorocarbons (CFCs) into the atmosphere at ground level, which ultimately enter the stratosphere. This prediction was later supported by experimental evidence of a thinning of the ozone layer and the formation of annual ozone holes over Antarctica. By 2000 stratospheric ozone concentrations over Antarctica were less than half of what they were in 1980, and the ozone hole covered nearly all of Antarctica and the tip of South America (**Figure 13.27**). Less severe thinning of stratospheric ozone was observed in the Northern Hemisphere.

In 1989 an international agreement known as the Montreal Protocol, which called for an end to the production of ozone-depleting CFCs, went into effect. It has had a dramatic effect on CFC production and emission into the atmosphere, and as

catalyst a substance added to a reaction that increases the rate of the reaction but is not consumed in the process.

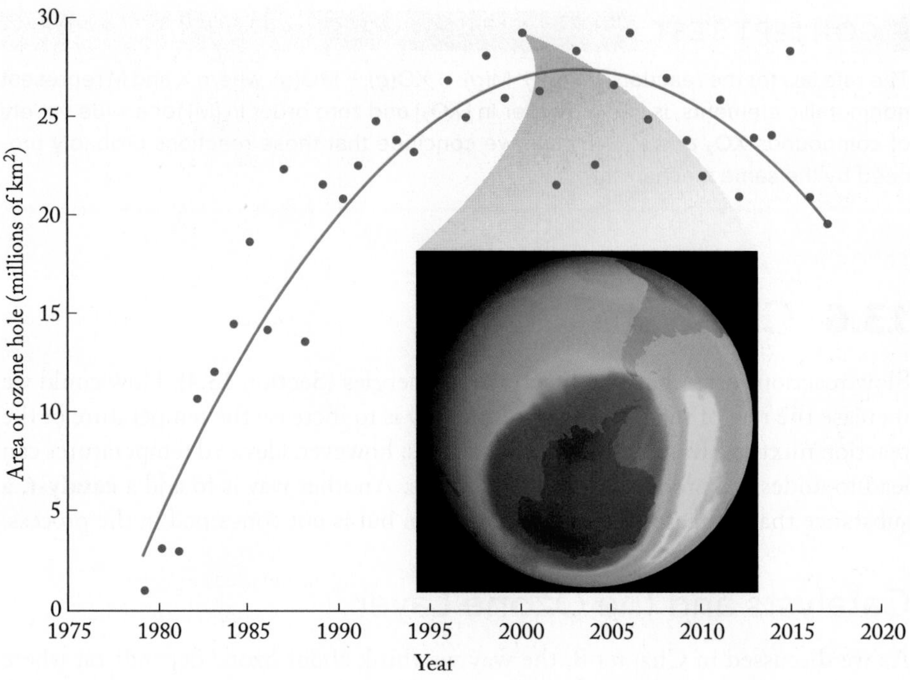

FIGURE 13.27 A 35-year trend in stratospheric ozone depletion over the South Pole. The data points represent the size of the Antarctic ozone hole observed each year. The deep blue and violet colors in the satellite image represent ozone concentrations that are less than half their normal values.

the trend line in Figure 13.27 indicates, the ozone layer over Antarctica appears to be slowly recovering. However, these compounds last for many years in the atmosphere, and full recovery of the ozone layer may take most of the 21st century.

How do CFCs contribute to the destruction of ozone? Three of the more widely used CFCs were CCl_2F_2, CCl_3F, and $CClF_3$. In the stratosphere, these molecules encounter UV radiation with enough energy to break C—Cl bonds, releasing chlorine atoms. For example,

$$CCl_3F(g) \xrightarrow{\text{sunlight}} CCl_2F(g) + Cl(g) \qquad (13.37)$$

Free chlorine atoms react with ozone, forming chlorine monoxide:

$$Cl(g) + O_3(g) \rightarrow ClO(g) + O_2(g) \qquad (13.38)$$

Chlorine monoxide then reacts with more ozone, producing oxygen and regenerating atomic chlorine:

$$ClO(g) + O_3(g) \rightarrow Cl(g) + 2\,O_2(g) \qquad (13.39)$$

If we add Equations 13.38 and 13.39 and cancel species as needed, we get

$$2\,O_3(g) \rightarrow 3\,O_2(g)$$

The overall reaction is exactly the same as the natural photodecomposition of ozone. The difference is the presence of chlorine atoms in Equations 13.38 and 13.39. Chlorine atoms act as a catalyst for the destruction of ozone because they speed up the reaction but are not consumed by it. Chlorine is a catalyst, not an intermediate, because it is consumed in an early elementary step of the mechanism and then regenerated in a later one, whereas an intermediate is produced and then consumed. A single chlorine atom can catalyze the destruction of hundreds to thousands of stratospheric O_3 molecules before it combines with other atoms and forms a less reactive molecule.

The activation energy for the O_3 decomposition following the Cl-catalyzed reaction is only 2.2 kJ/mol, whereas the activation energy for the uncatalyzed reaction pathway described in Equations 13.38 and 13.39 is 17.7 kJ/mol (**Figure 13.28**). Its smaller E_a value means that the catalyzed destruction of ozone is faster than the natural photodecomposition process. In the reaction describing the destruction of ozone, the catalyst, Cl(g), and reactant, O_3(g), exist in the same physical phase. When a catalyst and the reacting species are in the same phase, we call the catalyst a **homogeneous catalyst**.

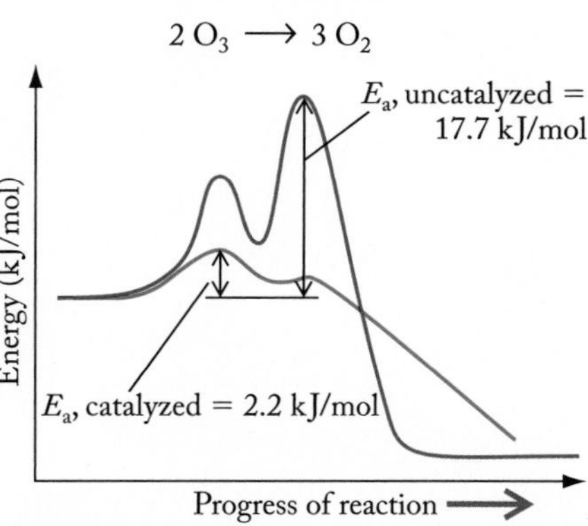

FIGURE 13.28 The decomposition of O_3 in the presence of chlorine atoms has a smaller activation energy (2.2 kJ/mol) than the naturally occurring photodecomposition of O_3 to O_2 (17.7 kJ/mol). The catalytic effect of chlorine is a key factor in the depletion of stratospheric ozone and the formation of an ozone hole over the South Pole.

CONCEPT **TEST**

Which of the following statements is/are true about the elementary step shown in Equation 13.38?

a. Its rate law is first order in [Cl].

b. Its rate law does not include [Cl] because Cl is a catalyst.

c. Its rate is zero order (i.e., independent of) [Cl].

The rates of the above reactions increase in the presence of drops of liquid, such as those in the clouds that cover much of Antarctica each spring. These clouds form from crystals of ice and tiny drops of nitric acid in the winter, when temperatures in the lower stratosphere dip to −80°C. The clouds become collection sites for HCl, ClO, and other compounds containing chlorine. In August (the end of the Antarctic winter), the ice in the clouds melts and ClO is free to react on the surface of the drops of liquid water. These drops catalyze reactions by ***adsorbing*** (binding to the surface of) the reactants. The drops are not reactants, but they provide a surface on which the reactants collect. The resulting proximity of the reactants increases the likelihood of reaction and, coupled with a decrease in activation energy, increases the reaction rate. When the catalyst is a liquid drop on which gas molecules adsorb, the drop is in a different phase than the reacting species and is called a **heterogeneous catalyst**. Both homogeneous and heterogeneous catalysts play a role in the reactions that diminish the amount of ozone in the stratosphere.

CONCEPT **TEST**

Is the ClO produced in Equation 13.38 a catalyst or an intermediate?

SAMPLE EXERCISE 13.13 Identifying Catalysts in Reaction Mechanisms **LO8**

A reaction mechanism proposed for the decomposition of ozone in the presence of NO at high temperatures consists of three elementary steps:

$$(1)\quad O_3(g) + NO(g) \rightarrow O_2(g) + NO_2(g)$$
$$(2)\quad NO_2(g) \rightarrow NO(g) + O(g)$$
$$(3)\quad O(g) + O_3(g) \rightarrow 2\,O_2(g)$$

If the rate of the overall reaction is higher than the rate of the uncatalyzed decomposition of ozone to oxygen, $2\,O_3(g) \rightarrow 3\,O_2(g)$, is NO a catalyst in the reaction?

Collect and Organize We are asked to determine whether NO is a catalyst in a reaction. A catalyst increases the rate of a reaction and is not consumed by the overall reaction. We are given the elementary steps of the reaction and are told that the reaction is more rapid in the presence of NO.

homogeneous catalyst a catalyst in the same phase as the reactants.

heterogeneous catalyst a catalyst in a different phase from the reactants.

Analyze We can sum the reactions to determine the overall reaction. If NO is consumed in an early step before it is regenerated in a later one, and it is not consumed in the overall process, then it is a catalyst.

Solve Summing the three elementary steps,

$$O_3(g) + \cancel{NO(g)} + \cancel{NO_2(g)} + \cancel{O(g)} + O_3(g) \rightarrow O_2(g) + \cancel{NO_2(g)} + \cancel{NO(g)} + \cancel{O(g)} + 2\,O_2(g)$$

gives the overall reaction,

$$2\,O_3(g) \rightarrow 3\,O_2(g)$$

This equation does not include NO, and the rate of the reaction is higher when NO is present. Thus, NO fulfills both requirements for being a catalyst.

Think About It NO behaves much like the Cl atoms in Equations 13.38 and 13.39. NO is *not* an intermediate because it is used in the reaction and then regenerated in a subsequent step. If it were an intermediate, it would have been produced and then consumed.

Practice Exercise The combustion of fossil fuels results in the release of SO_2 into the atmosphere, where it reacts with oxygen to form SO_3:

$$2\,SO_2(g) + O_2(g) \rightarrow 2\,SO_3(g)$$

In the atmosphere, SO_2 may react with NO_2, forming SO_3 and NO:

$$NO_2(g) + SO_2(g) \rightarrow NO(g) + SO_3(g)$$

The rate of reaction of SO_2 with NO_2 is faster than the rate of reaction of SO_2 with oxygen. If the NO produced in the reaction of NO_2 and SO_2 is then oxidized to NO_2,

$$2\,NO(g) + O_2(g) \rightarrow 2\,NO_2(g)$$

is NO_2 a catalyst in the reaction of SO_2 with O_2?

Catalysts and Catalytic Converters

We started this chapter discussing air pollution caused by vehicles and the technology that has been developed to clean the air. **Figure 13.29** shows a catalytic converter in a car's exhaust system and the reactions that take place in the converter to remove one representative pollutant, NO, from the engine exhaust. Hot exhaust gases flowing through the converter pass through a fine honeycomb mesh coated with one or more of the transition metals palladium, platinum, and rhodium. These metals are the catalysts, and they have two roles: (1) to speed up oxidation of carbon monoxide to CO_2 and of unburned hydrocarbons to CO_2 and water vapor; and (2) to convert NO and NO_2 into N_2 and O_2.

The metals used in making catalytic converters are dissolved as metal salts and dispersed on the mesh, and they are then reduced to clusters that are 2 to 10 nm in diameter. The large surface area of the metal clusters provides sites where the oxidation and reduction of the gases take place.

Catalysts not only speed up reactions but also allow them to take place at lower temperatures. For example, CO reacts rapidly with O_2 above 700°C, but the presence of a Pd or Pt/Rh catalyst enables this reaction to take place rapidly at the much lower temperature of automobile exhaust, about 250°C. The catalysts have similar effects on the reduction reactions taking place in the converters.

The catalysts are selective in terms of the molecules they interact with and specific in the reactions they promote. What makes the catalysts selective is that

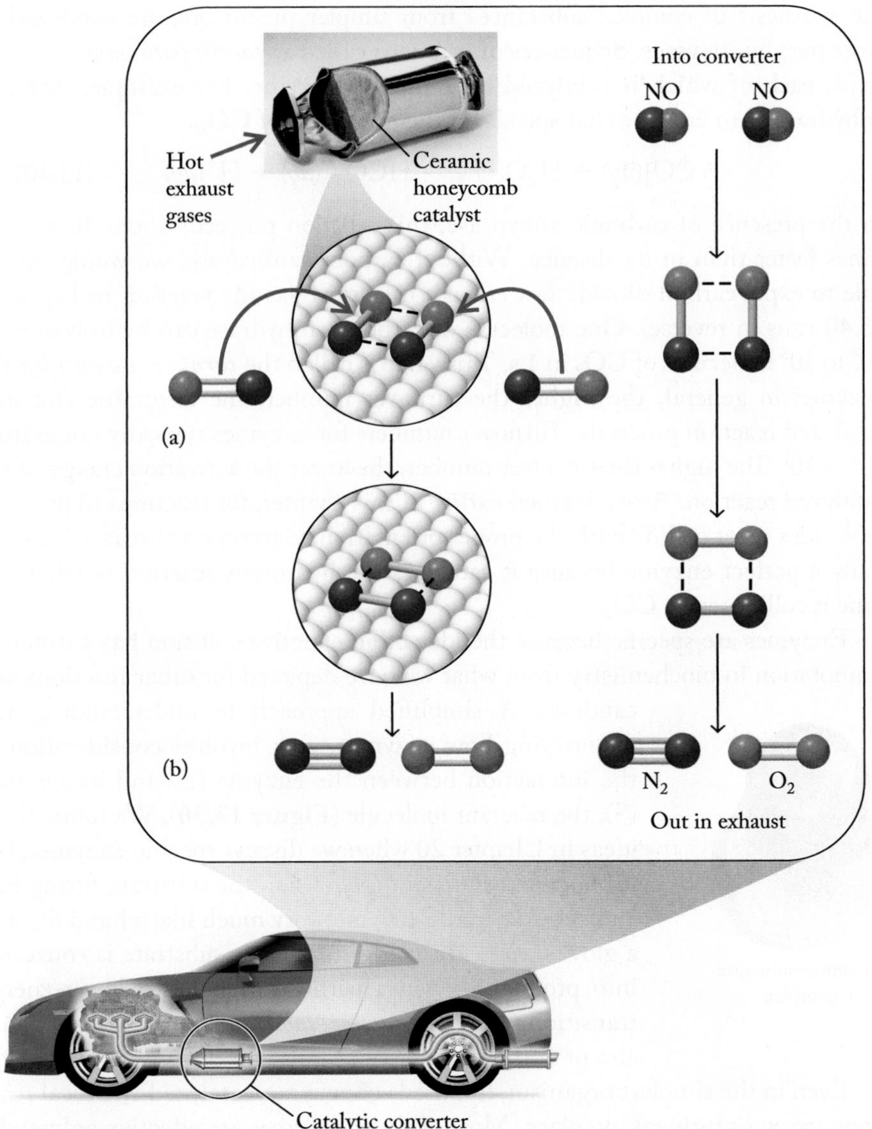

FIGURE 13.29 Catalytic converters in automobiles reduce emissions of NO by lowering the activation energy of its decomposition into N_2 and O_2. Metal catalysts are supported on a porous ceramic honeycomb. (a) NO molecules are adsorbed onto the surface of metal clusters where their NO bonds are broken, and (b) pairs of O atoms and N atoms form O_2 and N_2. The O_2 and N_2 desorb from the surface and are released to the atmosphere.

several reactions are possible for each pollutant, but one reaction proceeds more rapidly than the others. For example, the preferred reduction of the nitrogen in NO is to N_2, as shown in Figure 13.29, rather than to N_2O or to NH_3. However, carbon monoxide and hydrocarbons are also potential reducing agents for this reaction. Oxidizing CO and hydrocarbons is one of the two primary goals of a catalytic converter. If these compounds are oxidized, no additional NO reduction can take place via this process. Fortunately, the reduction of NO by CO and hydrocarbons is much faster than the reactions of CO and hydrocarbons with O_2. As a result, additional NO reduction by CO and hydrocarbons is essentially complete before any of the necessary reducing agents are consumed by reaction with oxygen.

Enzymes: Biological Catalysts

Large biomolecules called **enzymes** are highly selective catalysts; they mediate very specific reactions in biological systems. For example, the chemical reactions involved in metabolism, which includes both the breaking down of molecules and

enzyme a protein that catalyzes a reaction.

biocatalysis the use of enzymes to catalyze reactions on a large scale; it is becoming especially important in processes that involve chiral materials.

the synthesis of complex substances from simpler precursors, are catalyzed in large part by enzymes. Sequences of reactions called *metabolic pathways* consist of steps, each of which is catalyzed by a specific enzyme. For example, carbonic anhydrase is an enzyme that speeds up the hydrolysis of CO_2:

$$CO_2(aq) + H_2O(\ell) \rightleftharpoons HCO_3^-(aq) + H^+(aq) \qquad (13.40)$$

In the presence of carbonic anhydrase, this reaction proceeds about 10 million times faster than in its absence. Without carbonic anhydrase, we would not be able to expel carbon dioxide fast enough to survive (as the reaction in Equation 13.40 runs in reverse). One molecule of carbonic anhydrase can hydrolyze from 10^4 to 10^6 molecules of CO_2 in 1 s. This value is called the *turnover number* for the enzyme; in general, the higher the turnover number, the faster the enzyme-catalyzed reaction proceeds. Turnover numbers for enzymes typically range from 10^3 to 10^7. The higher the turnover number, the lower the activation energy of the catalyzed reaction. As we learned earlier in this chapter, for reactions to proceed, molecules must collide with the proper orientation. Carbonic anhydrase is essentially a perfect enzyme because it catalyzes the hydrolysis reaction nearly every time it collides with CO_2.

Enzymes are specific because the idea of an effective collision has a different connotation in biochemistry from what we have depicted for other reactions and catalysts. A simplified approach to understanding and quantifying how enzymes work involves consideration of the interaction between the enzyme (E) and its *substrate* (S), the reactant molecule (**Figure 13.30**). We refine these ideas in Chapter 20 when we discuss specific enzymes, but for now it is sufficient to envision the substrate fitting into an *active site* in the enzyme, very much like a hand fits into a glove. Once in the active site, the substrate is converted into product (P) via a pathway that has a lower-energy transition state (the *enzyme–substrate complex* ES), which is absent without the enzyme.

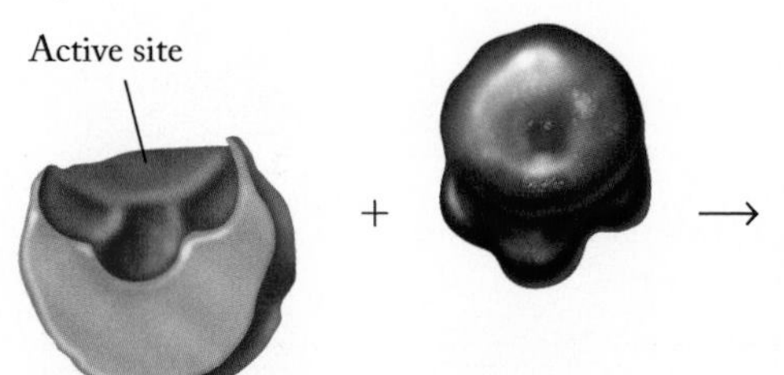

FIGURE 13.30 In this simplified view, the substrate fits into the active site of the enzyme that catalyzes the conversion of substrate into product.

Even in the simplest organism, hundreds of enzyme-catalyzed chemical reactions are constantly taking place. Most of these enzymes are effective only under limited reaction conditions: in aqueous media and at temperatures between 4°C and 37°C. These hundreds of catalyzed reactions require hundreds of different enzymes, many operating with exquisite efficiency and producing one pure product.

In the pharmaceutical industry, research is focused on using enzymes outside living systems to produce high yields of specific products. This research area, called **biocatalysis**, uses enzymes to catalyze chemical reactions run in industrial-sized reactors. Because it deals with both isolated enzymes and microorganisms, it is considered a special type of heterogeneous catalysis.

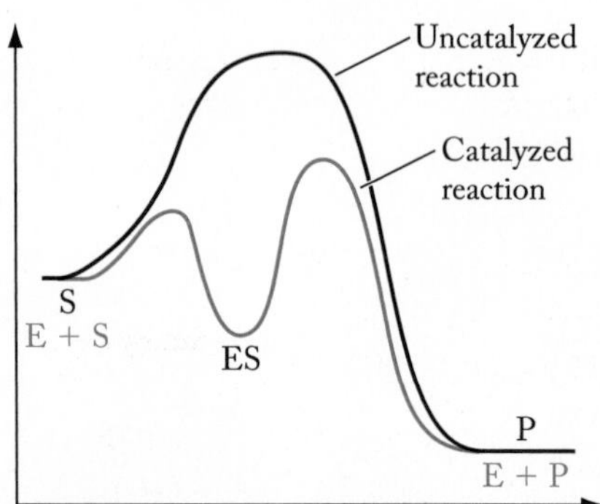

FIGURE 13.31 These superimposed reaction profiles show (black) the uncatalyzed reaction of S → P and (red) the enzyme-catalyzed reaction (E + S → E + P). The enzyme-catalyzed reaction takes place in two steps, in which the second step (ES → E + P) is rate determining.

The mathematical treatment of enzyme kinetics can be quite complex, but determining the rate of enzyme-catalyzed reactions is an important part of biochemical and medical research. We can think of the following elementary steps in the process:

$$\text{Step 1} \qquad E + S \underset{k_{-1}}{\overset{k_1}{\rightleftharpoons}} ES$$

$$\text{Step 2} \qquad ES \xrightarrow{k_2} E + P$$

Biochemists commonly assume that the formation of ES and its decomposition are both rapid and that step 2 is rate determining. A typical reaction profile for a system that meets these criteria is shown in **Figure 13.31**, and the rate of such a reaction is given by

$$\text{Rate} = \frac{\Delta[\text{P}]}{\Delta t} = k[\text{ES}]$$

Initially, the rate of reaction increases rapidly as the concentration of S increases (**Figure 13.32**). The rate of reaction is proportional to [S], so the reaction is first order and rate = k[S]. At some specific concentration of S, all of the active sites in available enzymes are occupied, and the rate of the reaction becomes constant. At this stage, the rate is zero order in S (rate = k); thus, adding more S has no effect because all the active sites are full.

In this chapter we have seen how studying the rates of chemical reactions can lead to an understanding of how reactions happen. Determinations of the rates of chemical reactions are crucial for us to understand processes on a molecular level. The mechanisms of the reactions of volatile oxides produced during combustion and by other natural events had to be thoroughly understood so that these substances in the environment could be effectively managed. The continued development of catalytic converters for vehicles to diminish the problems caused by burning fossil fuels and to clean our air requires a thorough knowledge of the kinetics and mechanisms of many of the reactions discussed in this chapter. Many biological catalysts are responsible for a myriad of reactions in living systems, and their behavior may be understood by applying the same methods of study we used for inorganic catalysts in reactions of small molecules.

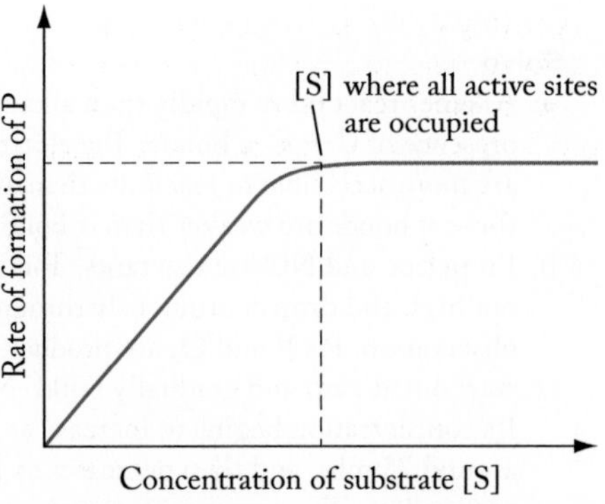

FIGURE 13.32 Plot of the rate of formation of product P versus concentration of substrate S in an enzyme-catalyzed reaction. The dotted line identifies the concentration of substrate at which all active sites are occupied.

SAMPLE EXERCISE 13.14 Integrating Concepts: Simulations of Smog

As we have seen throughout this chapter, reactions in the atmosphere are complex. **Figure 13.33** shows three graphs of data from programs that simulated atmospheric conditions arising from known reactions when the hydrocarbon propylene was introduced into air. Propylene enters the atmosphere from the evaporation of unburned petroleum fuels. The three graphs show simulated concentration versus time profiles on (a) a sunny day, (b) a cloudy day, and (c) a sunny day when the concentration of propylene is twice the amount in (a) .

a. Both alkanes and alkenes are present as dissolved gases in unburned fuels. In studies of both propylene ($CH_2{=}CHCH_3$) and butane ($CH_3CH_2CH_2CH_3$), the alkene was shown to react more rapidly than the alkane. Suggest a reason for this observation.

b. In Figure 13.33a, which substances are reactants and which are products? Are there any intermediates?

c. Compare Figures 13.33a and 13.33c. How does doubling the concentration of propylene affect the level of air pollutants such as ozone and peroxyacetyl nitrate (PAN)?

d. What are the major differences between the reactions on a sunny day (Figure 13.33a) and a cloudy day (Figure 13.33b)?

Collect and Organize We have three graphs for different conditions, and we can use them to get relative quantitative information.

Analyze The graphs show differences in the appearance and disappearance of reactants, intermediates, and products of reactions involving an alkene, NO_x, ozone, and PAN.

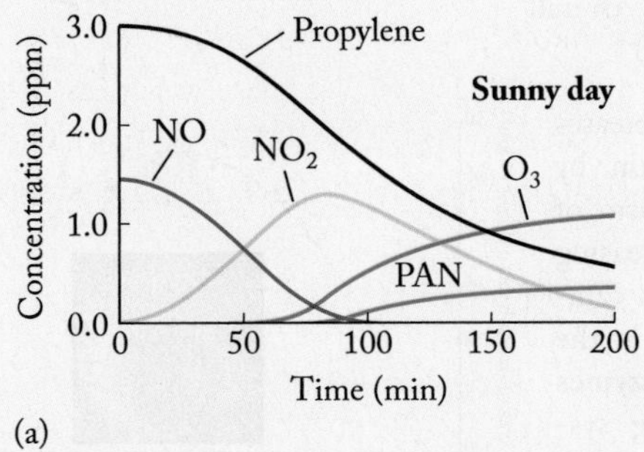

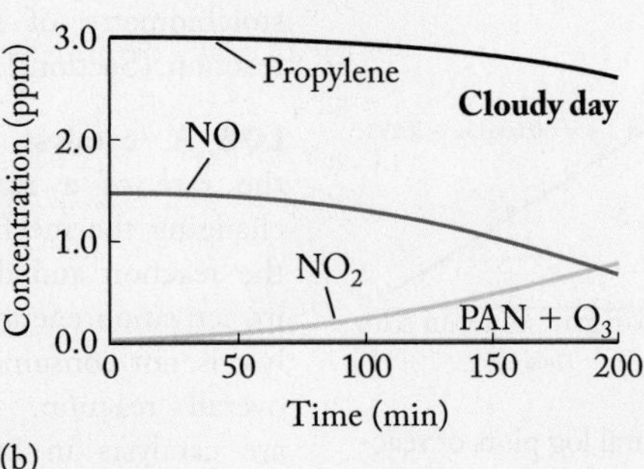

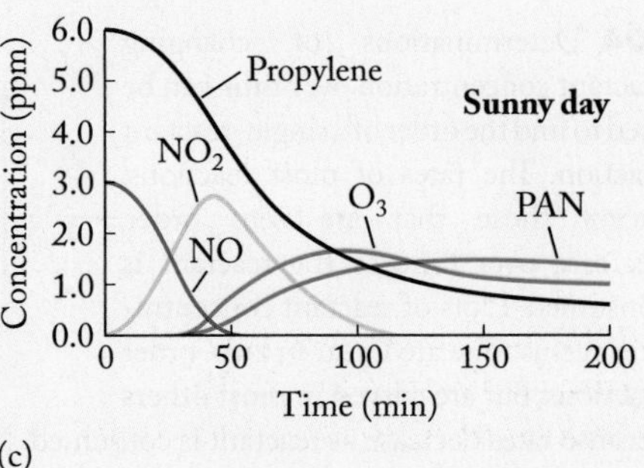

FIGURE 13.33 Graphs based on B. J. Hubert, *J. Chem. Educ.* 1974, *51*, 644–645; additional information from A. C. Baldwin, J. R. Barker, D. M. Golden, and D. G. Hendry, *J. Phys. Chem.* 1977, *81*(25), 2483–2492.

Solve

a. Alkenes react more rapidly than alkanes because of the presence of C—C π bonds. The electrons in these π bonds are more accessible to reactants than are those in σ bonds, and these π bonds are weaker than σ bonds.

b. Propylene and NO are reactants. Their concentrations start out high and drop continuously throughout the time of observation. PAN and O_3 are products. Their concentrations start out at zero and gradually build. NO_2 is an intermediate. Its concentration begins to increase at $t = 0$, reaches a peak around 75 min, and then decreases as PAN and O_3 form and accumulate. The decrease in propylene concentration begins immediately and accelerates as O_3 concentration builds. Some organic substances other than PAN must be forming to account for the drop; their concentrations are not shown on the graph.

c. Doubling the initial concentration of propylene caused it to be consumed at a faster rate and to accelerate the formation and consumption of other smog components. NO concentrations decreased more rapidly in (c); NO_2 concentrations reached a maximum in a shorter time and then decreased more rapidly than in (a). Ozone and PAN were detected earlier, reached much higher concentrations, and achieved concentration maxima in (c) that were not observed in (a).

d. PAN and O_3 are not evident at all in (b) on the cloudy day, and propylene and NO disappear more slowly than on a sunny day.

Think About It Our analysis of data presented in the three graphs is consistent with a key concept in this chapter: reaction rates depend on concentrations of reactants. Our answers are also consistent with recognized differences in reactivity between molecules such as butane and propylene as reflected in the kinetics of their reactions with ozone.

SUMMARY

LO1 The relative rates of the disappearance of reactants and the appearance of products are related by the stoichiometry of the reaction. **Chemical kinetics** is the study of these changing rates. (Sections 13.1 and 13.2)

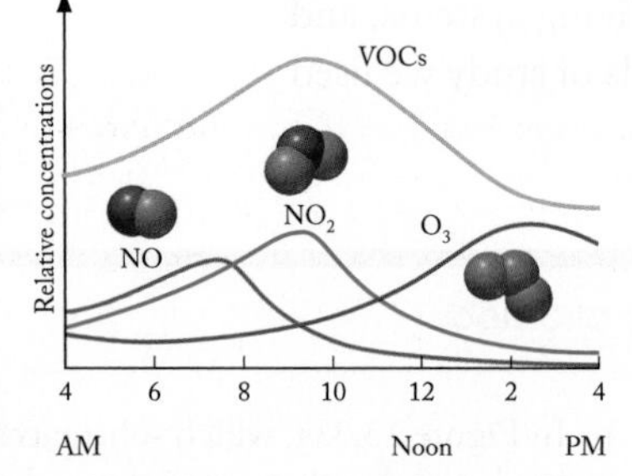

LO2 The rate of a reaction may be expressed as an average rate over a specified time interval based on the change in concentration of a reactant or product from the beginning to the end of that interval, or as an instantaneous rate at a particular time during the course of a reaction, based on the instantaneous rate of change in the concentration of a reactant or product. (Section 13.2)

LO3 The dependence of the rate of a reaction A + B → C on reactant concentrations is expressed in the **rate law** for the reaction:

$$\text{Rate} = k[\text{A}]^m[\text{B}]^n$$

where m and n are the **reaction order** with respect to reactants A and B, respectively, and k is the **rate constant.** The order of a reaction and the rate law for the reaction can be determined from differences in the **initial rates** of reaction. (Section 13.3)

LO4 Determinations of changing reactant concentration over time can be used to find the order of a single-reactant reaction. The rates of most reactions, except those that are zero order, decrease over time as the reactant is consumed. Plots of reactant concentration versus time are linear in zero-order reactions, but are curved in most others because rates decrease as reactant is consumed. Natural log plots of reactant concentration versus time are linear in first-order reactions, and plots of the reciprocal of reactant concentration are linear in second-order reactions. (Section 13.3)

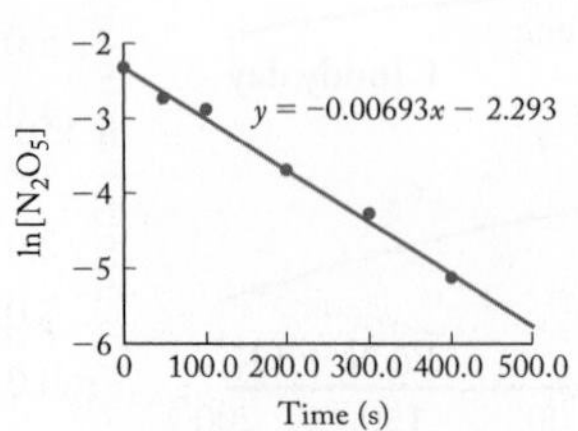

LO5 A first-order reaction has a characteristic **half-life** that is inversely proportional to the rate constant of the reaction. The half-life of a second-order reaction is inversely proportional to the rate constant and to the initial concentration of the reactant. The half-life of a zero order reaction is inversely proportional to the rate constant and directly proportional to the initial concentration. (Section 13.3)

LO6 Increasing the temperature of a chemical reaction increases its rate and rate constant because reactant particles collide more frequently and with enough energy to overcome the **activation energy** of the reaction, which is the minimum energy required for molecules to react when they collide. (Section 13.4)

LO7 Rate studies give insight into **reaction mechanisms,** which describe what is happening at the molecular level. A reaction mechanism consists of one or more **elementary steps** that describe how the reaction takes place on the molecular level. The proposed mechanism for any reaction must be consistent with the observed rate law and with the stoichiometry of the overall reaction. (Section 13.5)

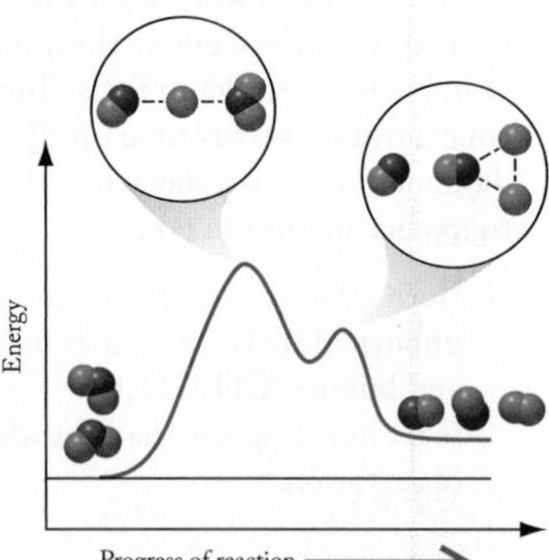

LO8 A **catalyst** increases the rate of a reaction by changing the mechanism of the reaction and decreasing its activation energy; a catalyst is not consumed in the overall reaction. **Enzymes** are catalysts in living systems. (Section 13.6)

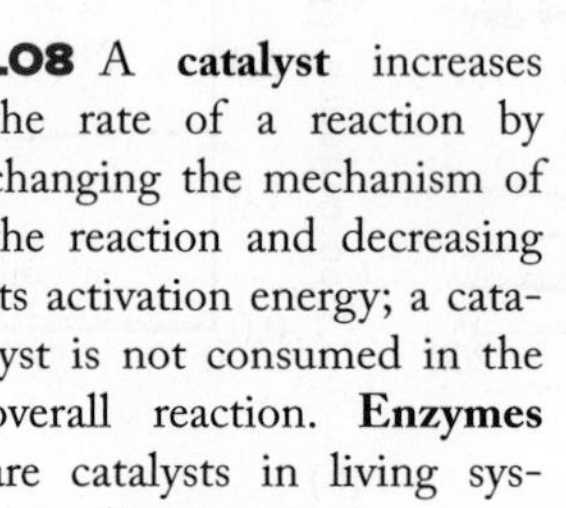

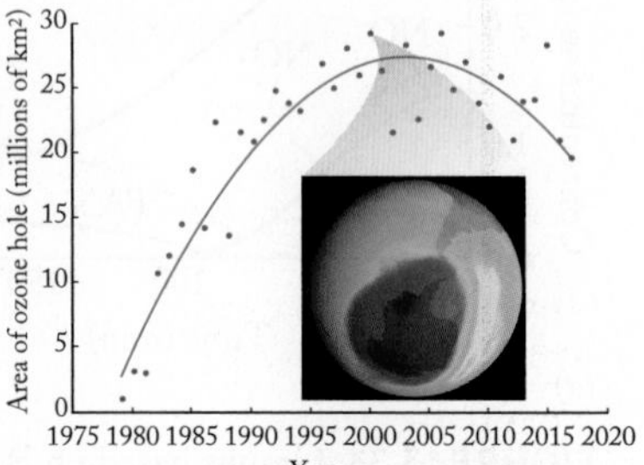

PARTICULATE PREVIEW WRAP-UP

The NO_2 molecules collide most frequently in sample (a) owing to the highest concentration of the three samples, so the reaction proceeds most rapidly for sample (a). The number of collisions in sample (b) is four times the number of collisions in sample (c) because the reaction is second order in NO_2.

PROBLEM-SOLVING SUMMARY

Type of Problem	Concepts and Equations	Sample Exercises
Relating rates of change in reactant and product concentrations	For the reaction $xX \rightarrow yY$, the rates of change of [X] and [Y] are related as follows: $$-\frac{1}{x}\frac{\Delta[X]}{\Delta t} = \frac{1}{y}\frac{\Delta[Y]}{\Delta t}$$	**13.1, 13.2**
Determining an instantaneous rate	Determine the slope of a line tangent to a point on the plot of concentration versus time.	**13.3**
Deriving a rate law from initial reaction rate data	Compare the change in rate when the concentration of one reactant is changed (while the concentrations of other reactants are kept constant) to determine the reaction order (usually whole numbers) with respect to that reactant: $$n = \frac{\log\left(\frac{\text{Rate}_1}{\text{Rate}_2}\right)}{\log\left(\frac{[X]_1}{[X]_2}\right)} \quad (13.14)$$	**13.4**
Using integrated rate laws to distinguish among zero-, first-, and second-order reactions and to calculate the value of *k*	A linear plot of concentration versus time indicates a zero-order reaction with a slope of $-k$, whereas a linear plot of the natural logarithm of concentration versus time indicates a first-order reaction with a slope of $-k$, and a linear plot of the reciprocal of reactant concentration 1/[X] versus time indicates a second-order reaction.	**13.5, 13.8**
Calculating remaining concentration of a reactant in a zero-, first-, or second-order reaction	Use the integrated rate law Zero order: $[X] = -kt + [X]_0$ (13.26) First order: $\ln[X] = -kt + \ln[X]_0$ (13.18) Second order: $\frac{1}{[X]} = kt + \frac{1}{[X]_0}$ (13.23) to calculate concentrations of reactant X at any given time *t*.	**13.6**
Calculating the half-life of a zero-, first-, or second-order reaction	Use the rate constant to calculate half-life: Zero order: $t_{1/2} = \frac{[X]_0}{2k}$ (13.27) First order: $t_{1/2} = \frac{0.693}{k}$ (13.19) Second order: $t_{1/2} = \frac{1}{k[X]_0}$ (13.24)	**13.7**
Calculating an activation energy from rate constants	Using the logarithmic form of the Arrhenius equation, $$\ln k = -\frac{E_a}{R}\left(\frac{1}{T}\right) + \ln A \quad (13.29)$$ plot ln *k* versus 1/*T*. The slope is $-E_a/R$. The rates of a reaction at two temperatures can be found using $$\ln\frac{k_2}{k_1} = -\frac{E_a}{R}\left(\frac{1}{T_2} - \frac{1}{T_1}\right) \quad (13.30)$$	**13.9, 13.10**
Linking reaction mechanisms to experimental rate laws and testing a proposed reaction mechanism	The order of each reactant in an elementary step equals its coefficient in that step. The rate law for the mechanism must be the same as the observed rate law and must not include intermediates.	**13.11, 13.12**
Identifying catalysts in reaction mechanisms	Determine whether a potential catalyst is present by summing the elementary-step reactions to get the overall reaction. If that procedure reveals a potential catalyst, determine whether it increases the rate of reaction and whether it is initially consumed and then regenerated in the process.	**13.13**

VISUAL PROBLEMS

(Answers to boldface end-of-chapter questions and problems are in the back of the book.)

13.1. Nitrous oxide decomposes to nitrogen and oxygen in the following reaction:

$$2\ N_2O(g) \rightarrow 2\ N_2(g) + O_2(g)$$

In Figure P13.1, which curve represents $[N_2O]$ and which curve represents $[O_2]$?

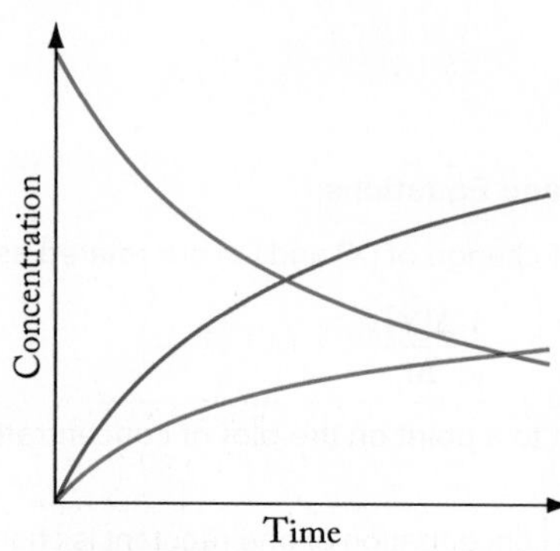

FIGURE P13.1

13.2. Sulfur trioxide is formed in the reaction

$$SO_2(g) + \tfrac{1}{2}O_2(g) \rightarrow SO_3(g)$$

In Figure P13.2, which curve represents $[SO_2]$ and which curve represents $[O_2]$? All three gases are present initially.

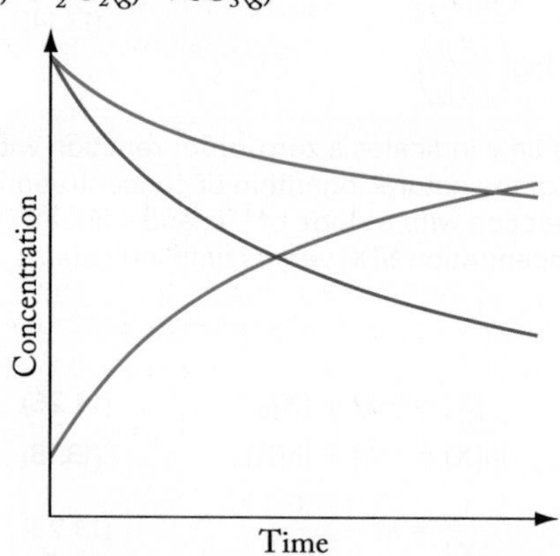

FIGURE P13.2

13.3. The rate law for the reaction $2\ A \rightarrow B$ is second order in A. Figure P13.3 represents samples with different concentrations of A (the red spheres). In which sample will the reaction $A \rightarrow B$ proceed most rapidly?

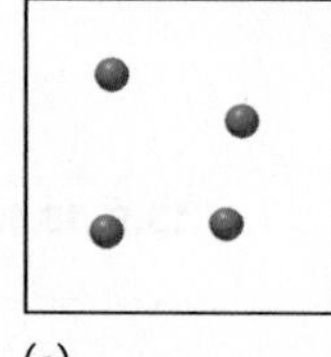
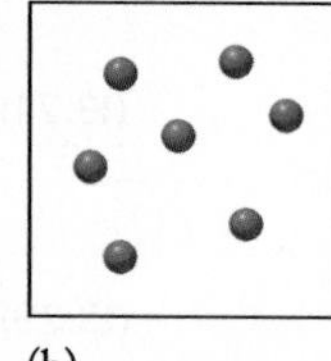
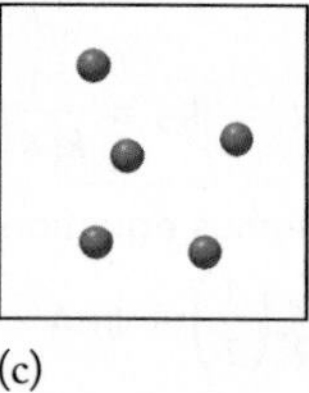

(a) (b) (c)

FIGURE P13.3

13.4. The rate law for the reaction $A + B \rightarrow C$ is first order in both A and B. Figure P13.4 represents samples with different concentrations of A (red spheres) and B (blue spheres). In which sample will the reaction $A + B \rightarrow C$ proceed most rapidly?

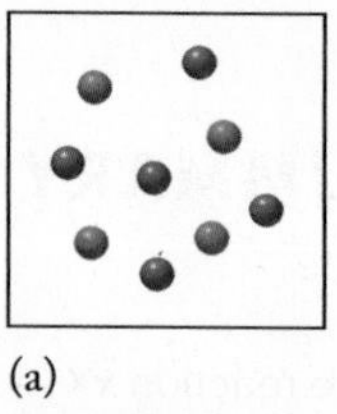
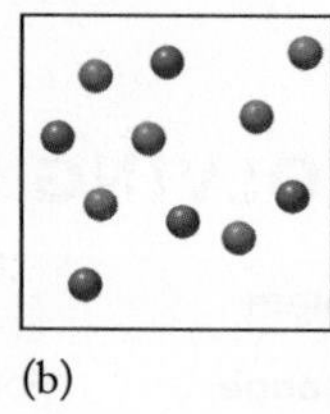
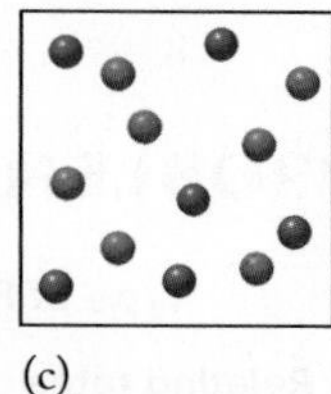

(a) (b) (c)

FIGURE P13.4

13.5. Figure P13.5 shows plots of reactant concentrations versus time for four reactions. Which one has the greatest initial rate?

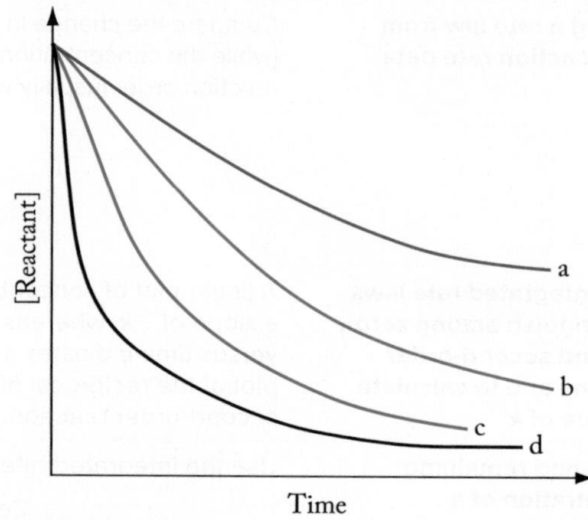

FIGURE P13.5

13.6. For a given temperature, which of the reaction profiles in Figure P13.6 represents (a) the slowest reaction? (b) the fastest reaction?

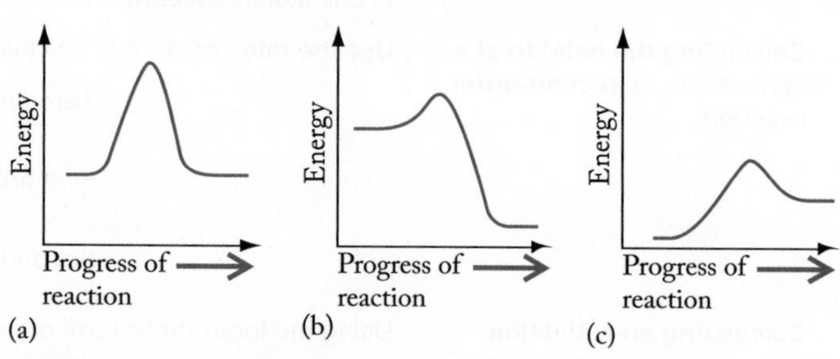

(a) (b) (c)

FIGURE P13.6

13.7. Which of the following mechanisms is consistent with the reaction profile shown in Figure P13.7?

a. $2\ A \xrightarrow{\text{slow}} B$
$B \xrightarrow{\text{fast}} C$

b. $A + B \rightarrow C$

c. $2\ A \underset{}{\overset{\text{fast}}{\rightleftharpoons}} B$
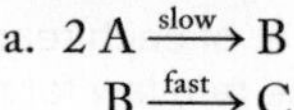
$B \xrightarrow{\text{slow}} C$

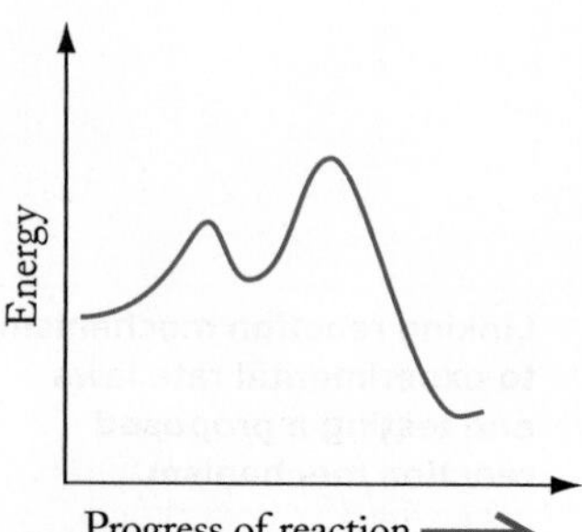

FIGURE P13.7

13.8. Which of the following mechanisms is consistent with the reaction profile shown in Figure P13.8?

a. $A + B \xrightarrow{\text{slow}} C$

$C \xrightarrow{\text{fast}} D$

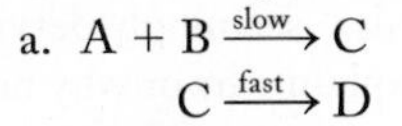

b. $A + B \rightarrow C$

c. $2\,A \xrightarrow{\text{fast}} B$

$B + C \xrightarrow{\text{slow}} D$

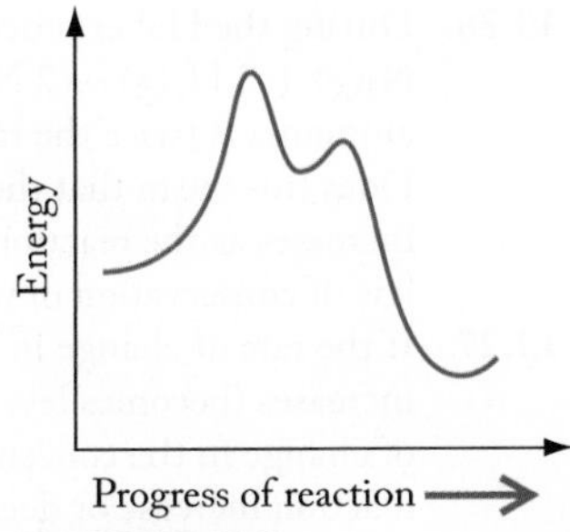

FIGURE P13.8

13.9. Use Figure P13.9 to answer the following questions:

a. Which asterisk identifies a transition state?

b. Which arrow identifies the activation energy of the reaction in the forward direction?

c. Which arrow identifies the activation energy of the reaction in the reverse direction?

d. Which arrow identifies the change in energy that accompanies the reaction—that is, the energy of the products less the energy of the reactants?

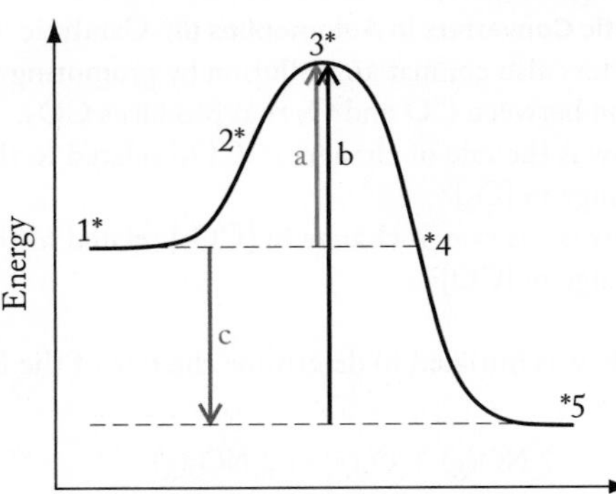

FIGURE P13.9

13.10. In the mechanism for decomposition of NO_2 to NO and O_2, through an NO_3 intermediate, why are the activated complexes in Figure P13.10 *unlikely* to lead to products?

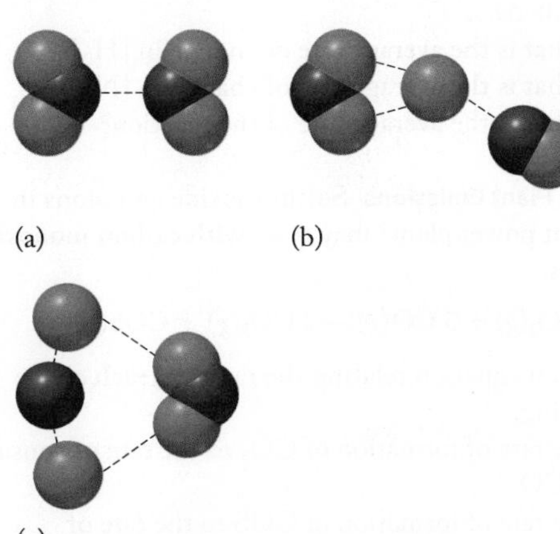

FIGURE P13.10

***13.11.** Figure P13.11 shows a plot of ln *k* versus $1/T$ for two reactions with different activation energies. Which reaction has a higher activation energy? The rate of which reaction is more sensitive to temperature changes?

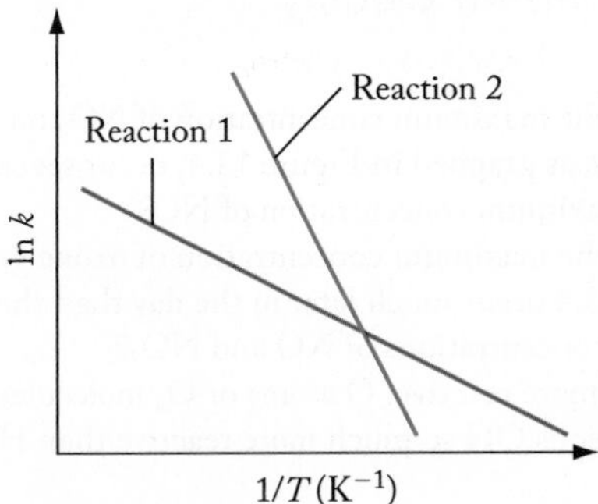

FIGURE P13.11

13.12. Use representations [A] through [I] in Figure P13.12 to answer questions (a)–(f).

a. Which two images describe elementary steps that, when combined, depict the chlorine-catalyzed destruction of ozone?

b. Which representation is the overall reaction for the chlorine-catalyzed destruction of ozone?

c. Which two images describe elementary steps that combine in an overall reaction in which NO_3 is an intermediate?

d. Write the chemical equation for the overall reaction described in question (c).

e. Which image shows the photodecomposition of a chlorofluorocarbon?

f. Rank images [C], [F], and [I] in decreasing order of number of collisions between Cl atoms and O_3 molecules in the stratosphere.

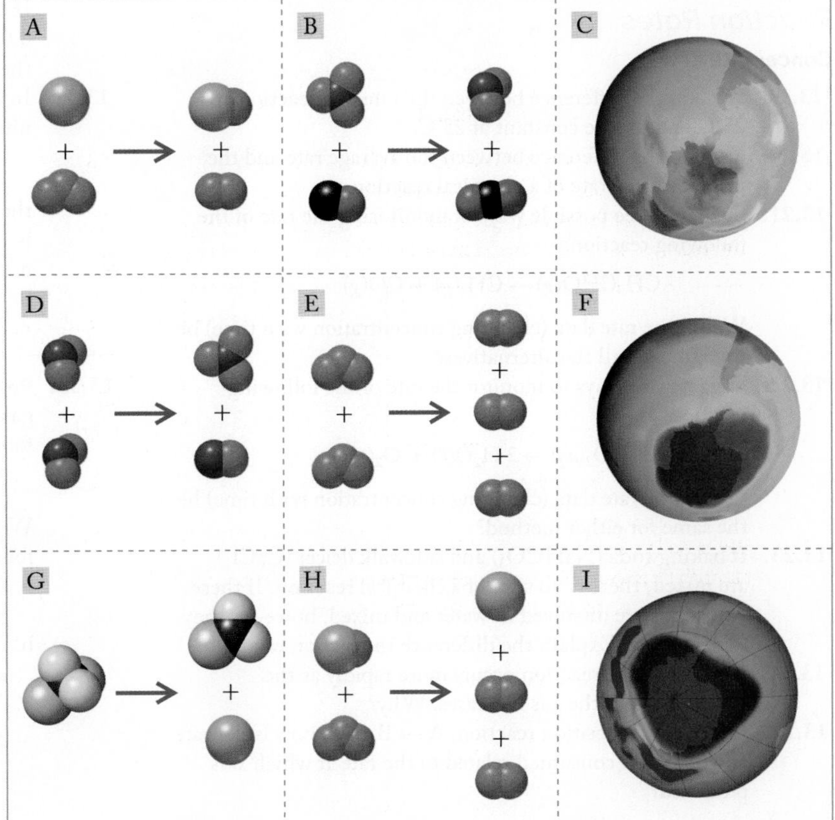

FIGURE P13.12

QUESTIONS AND PROBLEMS

Cars, Trucks, and Air Quality

Concept Review

13.13. Why does the maximum concentration of NO_2 on a smoggy day, as graphed in Figure 13.4, occur several hours after the maximum concentration of NO?

13.14. Why does the maximum concentration of ozone graphed in Figure 13.4 occur much later in the day than the maximum concentrations of NO and NO_2?

13.15. Which are more reactive: O atoms or O_2 molecules? Why?

13.16. Why is gaseous OH so much more reactive than H_2O vapor?

13.17. Which of the highlighted elements in Figure P13.17 forms volatile oxides associated with the formation of photochemical smog?

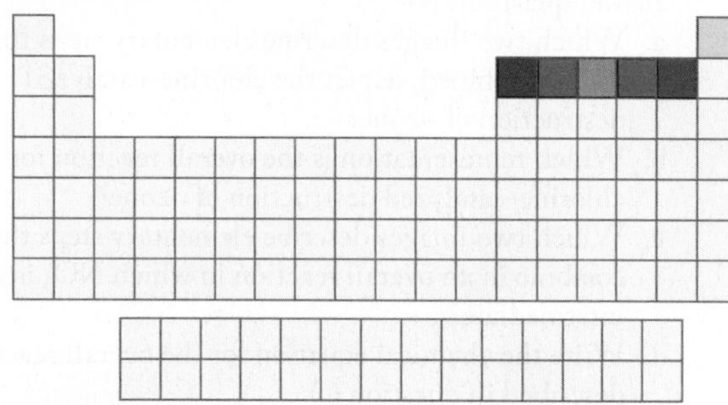

FIGURE P13.17

13.18. Which of the highlighted elements in Figure P13.17 has an allotrope associated with the formation of photochemical smog?

Reaction Rates

Concept Review

13.19. Explain the difference between the rate of a reaction at 25°C and its rate constant at 25°C.

13.20. Explain the difference between the average rate and the instantaneous rate of a chemical reaction.

13.21. Suggest three possible ways of monitoring the rate of the following reaction:

$$CH_3CHO(g) \rightarrow CH_4(g) + CO(g)$$

Would the rate data (changing concentration with time) be the same for all the alternatives?

13.22. Suggest two ways to monitor the rate of the following reaction:

$$2\ H_2O_2(aq) \rightarrow 2\ H_2O(\ell) + O_2(g)$$

Would the rate data (changing concentration with time) be the same for either method?

13.23. If baking soda ($NaHCO_3$) and sidewalk deicer ($CaCl_2$) are mixed, there is no sign of a chemical reaction. If these two solids are dissolved in water and mixed, however, they rapidly react. Explain the difference in reaction rate.

13.24. Any gas-phase reaction occurs more rapidly as the temperature of the gas increases. Why?

13.25. In the decomposition reaction, $A \rightarrow B + C$, how is the rate at which A is consumed related to the rate at which B is produced?

13.26. During the Haber process for synthesizing ammonia, $N_2(g) + 3\ H_2(g) \rightarrow 2\ NH_3(g)$, the rate of formation of ammonia is twice the rate at which nitrogen is consumed. Does this mean that the mass of the reaction mixture increases as the reaction proceeds—seemingly defying the law of conservation of mass? Explain why or why not.

13.27. If the rate of change in the concentration of a reactant increases (becomes less negative) with time, does the rate of change in the concentration of a product of the same reaction increase or decrease?

13.28. During a reaction, can there be a time when the instantaneous rate of the reaction does not change? If you think so, describe such a time.

Problems

13.29. **Catalytic Converters in Automobiles (I)** Catalytic converters combat air pollution by converting NO into N_2 and O_2.

a. How is the rate of formation of O_2 related to the rate of formation of N_2?
b. How is the rate of change in $[N_2]$ related to the rate of change in [NO]?

13.30. **Catalytic Converters in Automobiles (II)** Catalytic converters also combat air pollution by promoting the reaction between CO and O_2 that produces CO_2.

a. How is the rate of change in $[CO_2]$ related to the rate of change in $[O_2]$?
b. How is the rate of change in $[CO_2]$ related to the rate of change in [CO]?

13.31. A study was initiated to determine the rate of the following reaction:

$$2\ NO(g) + O_2(g) \rightarrow 2\ NO_2(g)$$

If the concentration of NO was 0.0300 *M* at $t = 5.0$ s and 0.0225 *M* at $t = 650.0$ s, then what was the average rate of the reaction during this period?

13.32. In a study of the thermal decomposition of ammonia into nitrogen and hydrogen:

$$2\ NH_3(g) \rightarrow N_2(g) + 3\ H_2(g)$$

the average rate of change in the concentration of ammonia is -0.38 *M*/s.

a. What is the average rate of change in $[H_2]$?
b. What is the average rate of change in $[N_2]$?
c. What is the average rate of the reaction?

13.33. **Power Plant Emissions** Sulfur dioxide emissions in stack gases at power plants may react with carbon monoxide as follows:

$$SO_2(g) + 3\ CO(g) \rightarrow 2\ CO_2(g) + COS(g)$$

Write an equation relating the rates for each of the following:

a. The rate of formation of CO_2 to the rate of consumption of CO
b. The rate of formation of COS to the rate of consumption of SO_2
c. The rate of consumption of CO to the rate of consumption of SO_2

13.34. **Reducing Power Plant Emissions** Nitrogen monoxide can be removed from gas-fired power plant emissions by reaction with methane as follows:

$$CH_4(g) + 4\,NO(g) \rightarrow 2\,N_2(g) + CO_2(g) + 2\,H_2O(g)$$

Write an equation relating the rates for each of the following:
a. The rate of formation of N_2 to the rate of formation of CO_2
b. The rate of formation of CO_2 to the rate of consumption of NO
c. The rate of consumption of CH_4 to the rate of formation of H_2O

13.35. Stratospheric Ozone Depletion Chlorine monoxide (ClO) plays a major role in the creation of the ozone holes in the stratosphere over Earth's polar regions.
a. If $\Delta[ClO]/\Delta t$ at 298 K is -2.3×10^7 M/s, what is the rate of change in $[Cl_2]$ and $[O_2]$ in the following reaction?

$$2\,ClO(g) \rightarrow Cl_2(g) + O_2(g)$$

b. If $\Delta[ClO]/\Delta t$ is -2.9×10^4 M/s, what is the rate of formation of oxygen and ClO_2 in the following reaction?

$$ClO(g) + O_3(g) \rightarrow O_2(g) + ClO_2(g)$$

13.36. The chemistry of smog formation includes NO_3 as an intermediate in several reactions.
a. If $\Delta[NO_3]/\Delta t$ is -2.2×10^5 mM/min in the following reaction, what is the rate of formation of NO_2?

$$NO_3(g) + NO(g) \rightarrow 2\,NO_2(g)$$

b. What is the rate of change of $[NO_2]$ in the following reaction if $\Delta[NO_3]/\Delta t$ is -2.3 mM/min?

$$2\,NO_3(g) \rightarrow 2\,NO_2(g) + O_2(g)$$

13.37. Nitrite ion reacts with ozone in aqueous solution, producing nitrate ion and oxygen:

$$NO_2^-(aq) + O_3(g) \rightarrow NO_3^-(aq) + O_2(g)$$

The following data were collected for this reaction at 298 K. Calculate the average reaction rate between 0.0 and 100.0 μs (microseconds) and between 200.0 and 300.0 μs.

Time (μs)	$[O_3]$ (M)
0.0	1.13×10^{-2}
100.0	9.93×10^{-3}
200.0	8.70×10^{-3}
300.0	8.15×10^{-3}

13.38. **Biomolecular Signaling** Hydrogen sulfide and nitrogen monoxide can act as signaling agents in living cells. NO is often bound in a molecule named 5-nitrosoglutathione (GSNO) and is released in a reaction with H_2S, producing glutathione (GSH) and thionitrous acid (HSNO):

$$GSNO(aq) + H_2S(aq) \rightarrow GSH(aq) + HSNO(aq)$$

The rate of the reaction at 25°C was determined by following the rate of consumption of H_2S:

Time (s)	$[H_2S]$ (μM)
0.0	250
25.0	158
50.0	98
75.0	71

a. Calculate the average rate of the reaction between $t = 0.0$ s and $t = 50.0$ s.
b. Estimate the instantaneous rate for the reaction at $t = 25.0$ s.

13.39. Biological Role of Nickel Nickel-containing compounds in living cells react with dissolved oxygen, forming a nickel peroxide (abbreviated as Ni_2O_2). An experiment to determine the rate decomposition of Ni_2O_2 at 298 K produced the following data:

Time (min)	$[Ni_2O_2]$ (mM)
5.0	0.59
10.0	0.40
15.0	0.33
20.0	0.21
25.0	0.15
30.0	0.10

Plot the data from this experiment and determine the instantaneous rates of change in the concentration of Ni_2O_2 at 10.0, 15.0, and 25.0 min.

13.40. **Tropospheric Ozone** Tropospheric (lower atmosphere) ozone is rapidly consumed in many reactions, including

$$O_3(g) + NO(g) \rightarrow NO_2(g) + O_2(g)$$

Use the following data to calculate the instantaneous rate of the preceding reaction at $t = 0.000$ s and $t = 0.052$ s.

Time (s)	[NO] (M)
0.000	2.0×10^{-8}
0.011	1.8×10^{-8}
0.027	1.6×10^{-8}
0.052	1.4×10^{-8}
0.102	1.2×10^{-8}

Effect of Concentration on Reaction Rate

Concept Review

13.41. Why do the rates of nearly all reactions decrease as reactants form products?

13.42. Why are the units of the rate constants different for reactions of different order?

13.43. A colleague has determined that the rate constant of the reaction you are studying is 2.5×10^2 M/s at 25.0°C. What is the order of the reaction?

13.44. The rate of a reaction does *not* decrease with time for a reaction with which order: zero, first, or second?

***13.45.** The reaction between A and B that produces C is first order in both A and B with a rate law of rate = k[A][B]. When the $[B]_0$ is 50 times the $[A]_0$, why does the rate law appear to be rate = k[A]?

13.46. When measuring the kinetics of gas-phase reactions, scientists often monitor changes in the partial pressures of reactants. How is this equivalent to measuring changes in concentration?

Problems

*13.47. For each of the following rate laws, determine the order with respect to each reactant and the overall reaction order.
a. Rate = $k[A][B]$
b. Rate = $k[A]^2[B]$
c. Rate = $k[A][B]^2$
*d. Rate = $k[A]/[B]$

*13.48. Determine the overall order of the following rate laws and the order with respect to each reactant.
a. Rate = $k[A]^2[B]^{1/2}$
b. Rate = $k[A]^2[B][C]$
c. Rate = $k[A][B]^2[C]^{0.5}$
*d. Rate = $k[A][B]/[C][D]^2$

13.49. Write rate laws and determine the units of the rate constant (using the units M for concentration and s for time) for the following reactions:
a. The reaction of oxygen atoms with NO_2 is first order in both reactants.
b. The reaction between NO and Cl_2 is second order in NO and first order in Cl_2.
c. The reaction between Cl_2 and chloroform ($CHCl_3$) is first order in $CHCl_3$ and one-half order in Cl_2.
*d. The decomposition of ozone (O_3) to O_2 is second order in O_3 and an order of −1 in O atoms.

13.50. Compounds A and B react to give a single product, C. Write the rate law for each of the following cases and determine the units of the rate constant by using the units M for concentration and s for time:
a. The reaction is first order in A and second order in B.
b. The reaction is first order in A and second order overall.
c. The reaction is zero order in A and second order overall.
d. The reaction is second order in both A and B.

13.51. Predict the rate law for the reaction $2\,BrO(g) \rightarrow Br_2(g) + O_2(g)$ if:
a. The rate doubles when [BrO] doubles.
b. The rate quadruples when [BrO] doubles.
c. The rate is halved when [BrO] is halved.
d. The rate is unchanged when [BrO] is doubled.

13.52. Predict the rate law for the reaction $NO(g) + Br_2(g) \rightarrow NOBr_2(g)$ if:
a. The rate doubles when [NO] is doubled and $[Br_2]$ remains constant.
b. The rate doubles when $[Br_2]$ is doubled and [NO] remains constant.
c. The rate increases by 1.56 times when [NO] is increased 1.25 times and $[Br_2]$ remains constant.
d. The rate is halved when [NO] is doubled and $[Br_2]$ remains constant.

13.53. The rate of the reaction:

$$NO(g) + O_3(g) \rightarrow NO_2(g) + O_2(g)$$

quadruples when the concentrations of NO and O_3 are doubled. Does this prove that the reaction is first order in both reactants? Why or why not?

13.54. Chlorine monoxide reacts with nitrogen dioxide to form chlorine nitrate ($ClONO_2$):

$$ClO(g) + NO_2(g) + M(g) \rightarrow ClONO_2(g) + M(g)$$

A third molecule (M) takes part in the reaction but is unchanged by it. The reaction is first order in both NO_2 and ClO, and it is second order overall.
a. Write the rate law for this reaction.
b. What is the reaction order with respect to M?

13.55. **Rate Laws for Destruction of Tropospheric Ozone** NO_2 reacts with ozone to produce NO_3 in a reaction that is second order overall:

$$NO_2(g) + O_3(g) \rightarrow NO_3(g) + O_2(g)$$

a. Write the rate law for the reaction if the reaction is first order in each reactant.
b. The rate constant for the reaction is $1.93 \times 10^4\ M^{-1}\,s^{-1}$ at 298 K. What is the rate of the reaction when $[NO_2] = 1.8 \times 10^{-8}\ M$ and $[O_3] = 1.4 \times 10^{-7}\ M$?
c. What is the rate of formation of NO_3 under these conditions?
d. What happens to the rate of the reaction if the concentration of $O_3(g)$ is doubled?

13.56. **Sources of Nitric Acid in the Atmosphere** The reaction between N_2O_5 and water,

$$N_2O_5(g) + H_2O(g) \rightarrow 2\,HNO_3(g)$$

is a source of nitric acid in the atmosphere.
a. The reaction is first order in each reactant. Write the rate law for the reaction.
b. When $[N_2O_5]$ is 0.132 mM and $[H_2O]$ is 230 mM, the rate of the reaction is 4.55×10^{-4} mM min^{-1}. What is the rate constant for the reaction?

13.57. Each of the following reactions is first order in the reactants and second order overall. Which reaction is fastest if the initial concentrations of the reactants are the same? All reactions are run at 298 K.
a. $ClO_2(g) + O_3(g) \rightarrow ClO_3(g) + O_2(g)$
$k = 3.0 \times 10^{-19}\ cm^3/(molecule \cdot s)$
b. $ClO_2(g) + NO(g) \rightarrow NO_2(g) + ClO(g)$
$k = 3.4 \times 10^{-13}\ cm^3/(molecule \cdot s)$
c. $ClO(g) + NO(g) \rightarrow Cl(g) + NO_2(g)$
$k = 1.7 \times 10^{-11}\ cm^3/(molecule \cdot s)$
d. $ClO(g) + O_3(g) \rightarrow ClO_2(g) + O_2(g)$
$k = 1.5 \times 10^{-17}\ cm^3/(molecule \cdot s)$

13.58. Two reactions in which there is a single reactant have nearly the same magnitude of rate constant. One is first order, whereas the other is second order.
a. If the initial concentrations of the reactants are both 1.0 mM, which reaction will proceed at the higher rate?
b. If the initial concentrations of the reactants are both 2.0 M, which reaction will proceed at the higher rate?

13.59. The rate constant for the decomposition of N_2O_5 to NO_2 and O_2,

$$2\,N_2O_5(g) \rightarrow 4\,NO_2(g) + O_2(g)$$

is $3.4 \times 10^{-5}\ s^{-1}$ at 298 K. What is the rate law expression for the reaction at 298 K?

13.60. **Hydroperoxyl Radicals in the Atmosphere** During a smog event, trace amounts of many highly reactive substances are present in the atmosphere. One of these is the hydroperoxyl radical (HO_2), which reacts with sulfur trioxide (SO_3). The rate constant for the reaction

$$2\ HO_2(g) + SO_3(g) \rightarrow H_2SO_3(g) + 2\ O_2(g)$$

is $2.6 \times 10^{11}\ M^{-1}\ s^{-1}$ at 298 K. The initial rate of the reaction doubles when the concentration of SO_3 or HO_2 is doubled. What is the rate law for the reaction?

13.61. **Disinfecting Municipal Water Supplies** Chlorine dioxide (ClO_2) is a disinfectant used in municipal water treatment plants (Figure P13.61). It dissolves in basic solution, producing ClO_3^- and ClO_2^-:

$$2\ ClO_2(g) + 2\ OH^-(aq) \rightarrow ClO_3^-(aq) + ClO_2^-(aq) + H_2O(\ell)$$

FIGURE P13.61

The following kinetic data were obtained at 298 K for the reaction:

Experiment	$[ClO_2]_0$ (*M*)	$[OH^-]_0$ (*M*)	Initial Rate (*M*/s)
1	0.060	0.030	0.0248
2	0.020	0.030	0.00827
3	0.020	0.090	0.0247

Determine the rate law and the rate constant for this reaction at 298 K.

13.62. **Oxygen Atom Transfer** A nickel compound containing peroxide (abbreviated as Ni_2O_2) can donate O atoms to other molecules. A model for such a reaction is the conversion of cyclooctene to 1,2-cyclooctene oxide shown in Figure P13.62.

2 [Cyclooctene] + Ni_2O_2 ⟶ 2 [1,2-Cyclooctene oxide] + 2 Ni

Cyclooctene 1,2-Cyclooctene oxide

FIGURE P13.62

The following rate data were collected for the reaction at 243 K:

Experiment	[Cyclooctene] (m*M*)	$[Ni_2O_2]$ (m*M*)	Initial Rate (m*M*/s)
1	4.80	4.00	3.46×10^{-3}
2	9.60	4.00	6.91×10^{-3}
3	9.60	2.00	3.46×10^{-3}

Determine the rate law and the rate constant for this reaction at 243 K.

13.63. Hydrogen gas reduces NO to N_2 in the following reaction:

$$2\ H_2(g) + 2\ NO(g) \rightarrow 2\ H_2O(g) + N_2(g)$$

The initial reaction rates of four mixtures of H_2 and NO were measured at 900°C with the following results:

Experiment	$[H_2]_0$ (*M*)	$[NO]_0$ (*M*)	Initial Rate (*M*/s)
1	0.212	0.136	0.0248
2	0.212	0.272	0.0991
3	0.424	0.544	0.793
4	0.848	0.544	1.59

Determine the rate law and the rate constant for the reaction at 900°C.

13.64. The rate of the reaction

$$NO_2(g) + CO(g) \rightarrow NO(g) + CO_2(g)$$

was determined in three experiments at 225°C. The results are listed in the following table:

Experiment	$[NO_2]_0$ (*M*)	$[CO]_0$ (*M*)	Initial Rate, $-\Delta[NO_2]/\Delta t$ (*M*/s)
1	0.263	0.826	1.44×10^{-5}
2	0.263	0.413	1.44×10^{-5}
3	0.526	0.413	5.76×10^{-5}

a. Determine the rate law for the reaction.
b. Calculate the value of the rate constant at 225°C.
c. Calculate the rate of formation of CO_2 when $[NO_2] = [CO] = 0.500\ M$.

13.65. Use the data for the decomposition of the nickel peroxide in Problem 13.39 at 298 K to answer the following questions:
a. What is the rate law for the reaction?
b. What is the rate constant for the reaction at 298 K?
c. What is the half-life of the reaction?

13.66. Two structural isomers of ClO_2 are shown in Figure P13.66.

FIGURE P13.66

The isomer with the Cl—O—O skeletal arrangement is unstable and rapidly decomposes according to the reaction $2\,ClOO(g) \rightarrow Cl_2(g) + 2\,O_2(g)$. The following data were collected for the decomposition of ClOO at 298 K:

Time (μs)	[ClOO] (*M*)
0.0	1.76×10^{-6}
0.7	2.36×10^{-7}
1.3	3.56×10^{-8}
2.1	3.23×10^{-9}
2.8	3.96×10^{-10}

Determine the rate law for the reaction and the value of the rate constant at 298 K.

13.67. Carbon Monoxide in Cancer Therapy Scientists are investigating a class of potential anticancer drugs that selectively releases low doses of carbon monoxide. One such compound reacts with myoglobin, forming a myoglobin–CO complex when the mixture is irradiated with UV light at 379 nm. The half-life for the first-order reaction at 293 K is 9.37 minutes.

a. What is the rate constant for the reaction?
b. What fraction of the drug remains after 30 minutes of irradiation?

13.68. **Yogurt Expiration Date** Labels of many food products have expiration dates, at which point they are typically removed from supermarket shelves. A particular natural yogurt degrades with a half-life of 45 days. The manufacturer of the yogurt wants unsold product pulled from the shelves when it degrades to no more than 80% of its original quality. Assume the degradation process is first order. What should be the "best if used before" date on the container with respect to the date the yogurt was packaged?

13.69. Acetoacetic acid (CH_3COCH_2COOH) decomposes in aqueous acidic solution to form acetone and carbon dioxide:

$$CH_3COCH_2COOH(aq) \rightarrow CH_3COCH_3(aq) + CO_2(g)$$

The reaction is first order. At room temperature, the half-life of the reactant is 139 min.

a. What is the rate constant of the decomposition reaction?
b. If the initial concentration of acetoacetic acid is 2.75 *M*, what is its concentration after 5 hours?

13.70. *p*-Toluenesulfinic acid undergoes a second-order redox reaction at room temperature (Figure P13.70).

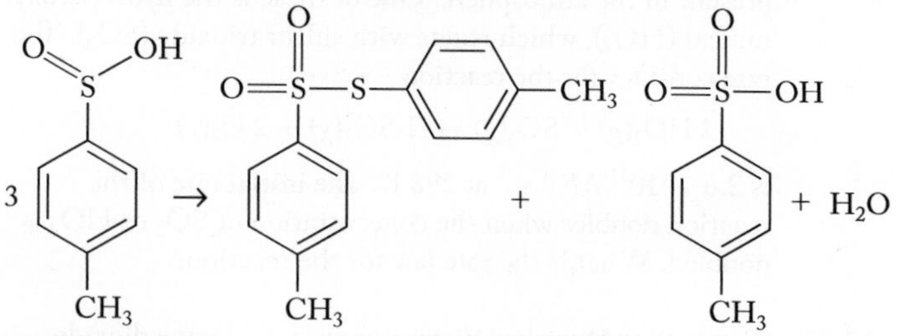

FIGURE P13.70

a. The value of the rate constant *k* for the reaction is 0.141 L mol^{-1} min^{-1}. What is the half-life of *p*-toluenesulfinic acid?
b. If the initial concentration of *p*-toluenesulfinic acid is 0.355 *M*, at what time will its concentration be 0.0355 *M*?

13.71. Laughing Gas Nitrous oxide (N_2O) is used as an anesthetic (laughing gas) and in aerosol cans to produce whipped cream. It is a potent greenhouse gas and decomposes slowly to N_2 and O_2:

$$2\,N_2O(g) \rightarrow 2\,N_2(g) + O_2(g)$$

a. If the plot of ln[N_2O] as a function of time is linear, what is the rate law for the reaction?
b. How many half-lives will it take for the concentration of the N_2O to reach 6.25% of its original concentration? [*Hint*: The amount of reactant remaining after time *t* (A_t) is related to the amount initially present (A_0) by the equation $A_t/A_0 = (0.5)^n$, where *n* is the number of half-lives in time *t*.]

13.72. The unsaturated hydrocarbon butadiene (C_4H_6) dimerizes to 4-vinylcyclohexene (C_8H_{12}). When data collected in studies of the kinetics of this reaction were plotted against reaction time, plots of [C_4H_6] or ln[C_4H_6] produced curved lines, but the plot of 1/[C_4H_6] was linear.

a. What is the rate law for the reaction?
b. How many half-lives will it take for the [C_4H_6] to decrease to 3.1% of its original concentration?

13.73. Tracing Phosphorus in Organisms Radioactive isotopes such as ^{32}P are used to follow biological processes. The following radioactivity data (in relative radioactivity values) were collected for a sample containing ^{32}P:

Time (days)	Radioactivity (relative radioactivity values)
0	10.0
1	9.53
2	9.08
5	7.85
10	6.16
20	3.79

a. Write the rate law for the decay of ^{32}P and determine the value of the first-order rate constant.
b. What is the half-life of ^{32}P?

13.74. Nitrous acid slowly decomposes to NO, NO_2, and water in the following second-order reaction:

$$2\ HNO_2(aq) \rightarrow NO(g) + NO_2(g) + H_2O(\ell)$$

a. Use the data in the table to determine the rate constant for this reaction at 298 K.

Time (min)	$[HNO_2]$ (μM)
0.0	0.1560
1000.0	0.1466
1500.0	0.1424
2000.0	0.1383
2500.0	0.1345
3000.0	0.1309

b. Determine the half-life for the decomposition of HNO_2.

13.75. The dimerization of ClO,

$$2\ ClO(g) \rightarrow Cl_2O_2(g)$$

is second order in ClO. Use the following data to determine the value of k at 298 K:

Time (s)	[ClO] (molecules/cm^3)
0.0	2.60×10^{11}
1.0	1.08×10^{11}
2.0	6.83×10^{10}
3.0	4.99×10^{10}
4.0	3.93×10^{10}

Determine the half-life for the dimerization of ClO.

13.76. Kinetic data for the reaction $Cl_2O_2(g) \rightarrow 2\ ClO(g)$ are summarized in the following table:

Time (μs)	$[Cl_2O_2]$ (M)
0.0	6.60×10^{-8}
172.0	5.68×10^{-8}
345.0	4.89×10^{-8}
517.0	4.21×10^{-8}
690.0	3.62×10^{-8}
862.0	3.12×10^{-8}

Determine the value of the first-order rate constant, then determine the half-life for the decomposition of Cl_2O_2.

Reaction Rates, Temperature, and the Arrhenius Equation

Concept Review

13.77. Suppose the high-energy reactants in a reaction form lower-energy products. Is the activation energy barrier higher in the forward or the reverse direction?

13.78. Explain why gas-phase reactions go faster at a higher temperature, yet rate laws such as rate = k[A] do not include T.

13.79. The order of a reaction is independent of temperature, but the value of the rate constant varies with temperature. Why?

13.80. Are exothermic reactions faster than endothermic reactions? Explain why or why not.

*13.81. Two first-order reactions have activation energies of 15 and 150 kJ/mol. Which reaction will show the larger increase in rate as temperature is increased?

*13.82. The activation energy for a particular reaction is nearly zero. Is its rate constant very sensitive to temperature changes? Explain why.

13.83. **Wine Chemistry** Some New Zealand wines have distinctive aromas that suggest guava and passion fruit. These aromas tend to disappear when the wines are stored because the esters that produce the aromas hydrolyze. Why does storing the wine at low temperature reduce the loss of the desired aromas?

13.84. **Ripening Bananas** Bananas ripen in the presence of ethylene (C_2H_4) gas. The source of the ethylene is the banana itself, although shippers will typically treat green bananas with ethylene to speed up the process. The same effect can be achieved by storing green bananas in a plastic bag. Why does storing them in plastic bags cause bananas to ripen faster?

Problems

13.85. The rate constant for the reaction of ozone with oxygen atoms was determined at four temperatures. Calculate the activation energy and frequency factor A for the reaction

$$O(g) + O_3(g) \rightarrow 2\ O_2(g)$$

given the following data:

T (K)	k ($M^{-1}\,s^{-1}$)
250	1.72×10^6
275	2.69×10^6
300	5.02×10^6
325	8.51×10^6

13.86. The rate constant for the reaction

$$NO_2(g) + O_3(g) \rightarrow NO_3(g) + O_2(g)$$

was determined over a temperature range of 40 K, with the following results:

T (K)	k ($M^{-1}\,s^{-1}$)
203	4.14×10^5
213	7.30×10^5
223	1.22×10^6
233	1.96×10^6
243	3.02×10^6

a. Determine the activation energy for the reaction.
b. Calculate the rate constant of the reaction at 300 K.

13.87. Activation Energy of a Smog-Forming Reaction The initial step in the formation of smog is the reaction between nitrogen and oxygen. The activation energy of the reaction can be determined from the temperature dependence of the rate constants. At the temperatures indicated, values of the rate constant of the reaction

$$N_2(g) + O_2(g) \rightarrow 2\ NO(g)$$

are as follows:

T (K)	k ($M^{-1/2}\ s^{-1}$)
2000	318
2100	782
2200	1770
2300	3733
2400	7396

a. Calculate the activation energy of the reaction.
b. Calculate the frequency factor for the reaction.
c. Calculate the value of the rate constant at ambient temperature, $T = 300$ K.

13.88. **Nitrogen Fixation** Nitrogen fixation is the biological process by which the N≡N bonds in molecules of atmospheric N_2 are broken and nitrogen compounds such an NH_3 are produced. For decades chemists have studied reactions that mimic biological nitrogen fixation. In one such investigation the rate constant for an N_2 dissociation reaction was measured at five different temperatures. The following results were obtained:

T (K)	k (s^{-1})
263	5.20×10^{-4}
268	9.91×10^{-4}
274	2.07×10^{-3}
278	3.29×10^{-3}
284	6.56×10^{-3}

a. Calculate the activation energy for the reaction.
b. What is the value of k at 298 K?

13.89. Activation Energy of Stratospheric Ozone Destruction The value of the rate constant for the reaction between chlorine dioxide and ozone was measured at four temperatures between 193 and 208 K. The following results were obtained:

T (K)	k ($M^{-1}\ s^{-1}$)
193	34.0
198	62.8
203	112.8
208	196.7

Calculate the values of the activation energy and the frequency factor for the reaction.

13.90. Chlorine atoms react with methane, forming HCl and CH_3. The rate constant for the reaction is $6.0 \times 10^7\ M^{-1}\ s^{-1}$ at 298 K. When the experiment was repeated at three other temperatures, the following data were collected:

T (K)	k ($M^{-1}\ s^{-1}$)
303	6.5×10^7
308	7.0×10^7
313	7.5×10^7

Calculate the values of the activation energy and the frequency factor for the reaction.

Reaction Mechanisms

Concept Review

13.91. The reaction between NO and Cl_2 is first order in each reactant. Does this mean that the reaction could occur in just one step? Explain your answer.

13.92. The reaction between NO and H_2 is second order in NO. Does this mean that the reaction could occur in just one step?

***13.93.** If a reaction is zero order in a reactant, does that mean the reactant is never involved in collisions with other reactants? Explain your answer.

*13.94. In question 13.45, is reactant B ever involved in collisions with the other reactant A?

Problems

13.95. Substance A decomposes slowly into substance B, which then rapidly decomposes into substances C and D. Sketch a reaction profile for the reaction A → C + D, adding labels in the appropriate locations for the four substances involved.

13.96. How would the reaction profile in Problem 13.95 change if the first step were fast and the second step were slow?

13.97. Write the rate laws for the following elementary steps and identify them as uni-, bi-, or termolecular steps:
a. $SO_2Cl_2(g) \rightarrow SO_2(g) + Cl_2(g)$
b. $NO_2(g) + CO(g) \rightarrow NO(g) + CO_2(g)$
c. $2\ NO_2(g) \rightarrow NO_3(g) + NO(g)$

13.98. Write the rate laws for the following elementary steps and identify them as uni-, bi-, or termolecular steps:
a. $Cl(g) + O_3(g) \rightarrow ClO(g) + O_2(g)$
b. $2\ NO_2(g) \rightarrow N_2O_4(g)$
*c. ${}^{14}_{6}C \rightarrow {}^{14}_{7}N + {}^{0}_{-1}\beta$

13.99. Write the overall reaction that consists of the following elementary steps:

(1) $N_2O_5(g) \rightarrow NO_3(g) + NO_2(g)$
(2) $NO_3(g) \rightarrow NO_2(g) + O(g)$
(3) $2\ O(g) \rightarrow O_2(g)$

13.100. What overall reaction consists of the following elementary steps?

(1) $ClO^-(aq) + H_2O(\ell) \rightarrow HClO(aq) + OH^-(aq)$
(2) $I^-(aq) + HClO(aq) \rightarrow HIO(aq) + Cl^-(aq)$
(3) $OH^-(aq) + HIO(aq) \rightarrow H_2O(\ell) + IO^-(aq)$

***13.101.** In the following mechanism for NO formation, oxygen atoms are produced by breaking O═O bonds at high temperature in a fast, reversible reaction. If $\Delta[\text{NO}]/\Delta t = k[\text{N}_2][\text{O}_2]^{1/2}$, which step in the mechanism is the rate-determining step?

(1) $O_2(g) \rightleftharpoons 2O(g)$
(2) $O(g) + N_2(g) \rightarrow NO(g) + N(g)$
(3) $N(g) + O(g) \rightarrow NO(g)$
Overall $N_2(g) + O_2(g) \rightarrow 2\,NO(g)$

13.102. A proposed mechanism for the decomposition of hydrogen peroxide consists of three elementary steps:

(1) $H_2O_2(g) \rightarrow 2\,OH(g)$
(2) $H_2O_2(g) + OH(g) \rightarrow H_2O(g) + HO_2(g)$
(3) $HO_2(g) + OH(g) \rightarrow H_2O(g) + O_2(g)$

If the rate law for the reaction is first order in H_2O_2, which step in the mechanism is the rate-determining step?

13.103. A complex silver ion whose formula is abbreviated $Ag(EBG)^+$ reacts with two moles of dithionate ion ($S_2O_8{}^{2-}$) in a two-step reaction to produce a compound with silver in the +3 oxidation state:

$$Ag(EBG)^+ + S_2O_8{}^{2-} \rightarrow Ag(EBG)^{2+} + S_2O_8{}^{3-}$$
$$Ag(EBG)^{2+} + S_2O_8{}^{2-} \rightarrow Ag(EBG)^{3+} + S_2O_8{}^{3-}$$

a. The activation energies for the two steps are 55 kJ/mol and 60 kJ/mol, respectively. Which step is the rate-determining step?
b. Sketch a graph of energy as a function of the reaction progress consistent with the two activation energies. Label the transition states and any intermediates in the reaction.

13.104. **Mechanism of Ozone Destruction** Ozone decomposes thermally to oxygen in the following reaction:

$$2\,O_3(g) \rightarrow 3\,O_2(g)$$

The following mechanism has been proposed:

$$O_3(g) \rightarrow O(g) + O_2(g)$$
$$O(g) + O_3(g) \rightarrow 2\,O_2(g)$$

The reaction is second order in ozone. What properties of the two elementary steps (specifically, relative rate and reversibility) are consistent with this mechanism?

13.105. Mechanism of NO_2 Destruction Which of the following mechanisms are possible for the thermal decomposition of NO_2, given that the rate $= k[NO_2]^2$?

a. $NO_2(g) \xrightarrow{\text{slow}} NO(g) + O(g)$
$O(g) + NO_2(g) \xrightarrow{\text{fast}} NO(g) + O_2(g)$

b. $NO_2(g) + NO_2(g) \xrightarrow{\text{fast}} N_2O_4(g)$
$N_2O_4(g) \xrightarrow{\text{slow}} NO(g) + NO_3(g)$
$NO_3(g) \xrightarrow{\text{fast}} NO(g) + O_2(g)$

c. $NO_2(g) + NO_2(g) \xrightarrow{\text{slow}} NO(g) + NO_3(g)$
$NO_3(g) \xrightarrow{\text{fast}} NO(g) + O_2(g)$

13.106. The rate laws for the thermal and photochemical decomposition of NO_2 are different. Which of the following mechanisms are possible for the photochemical decomposition of NO_2 given that the rate $= k[NO_2]$?

a. $NO_2(g) + NO_2(g) \xrightarrow{\text{slow}} N_2O_4(g)$
$N_2O_4(g) \xrightarrow{\text{fast}} N_2O_3(g) + O(g)$
$N_2O_3(g) + O(g) \xrightarrow{\text{fast}} N_2O_2(g) + O_2(g)$
$N_2O_2(g) \xrightarrow{\text{fast}} 2\,NO(g)$

b. $NO_2(g) + NO_2(g) \xrightarrow{\text{slow}} NO(g) + NO_3(g)$
$NO_3(g) \xrightarrow{\text{fast}} NO(g) + O_2(g)$

c. $NO_2(g) \xrightarrow{\text{slow}} N(g) + O_2(g)$
$N(g) + NO_2(g) \xrightarrow{\text{fast}} N_2O_2(g)$
$N_2O_2(g) \xrightarrow{\text{slow}} 2\,NO(g)$

Catalysts

Concept Review

***13.107.** Does a catalyst affect both the rate and the rate constant of a reaction? Explain your answer.

*13.108. Is the rate law for a catalyzed reaction the same as that for the uncatalyzed reaction?

13.109. Does a substance that increases the rate of a reaction also increase the rate of the reverse reaction?

13.110. The rate of the reaction between NO_2 and CO is zero order in [CO]. Does this mean that CO is a catalyst in the reaction?

***13.111.** Does the concentration of a homogeneous catalyst appear in the rate law for the reaction it catalyzes?

*13.112. The rate of a chemical reaction is too slow to measure at room temperature. We could either raise the temperature or add a catalyst. Which would be a better solution for making an accurate determination of the rate constant?

Problems

13.113. Is NO a catalyst for the decomposition of N_2O in the following two-step reaction mechanism, or is N_2O a catalyst for the conversion of NO to NO_2?

(1) $NO(g) + N_2O(g) \rightarrow N_2(g) + NO_2(g)$
(2) $2\,NO_2(g) \rightarrow 2\,NO(g) + O_2(g)$

13.114. **NO as a Catalyst for Ozone Destruction** Explain why NO is a catalyst in the following two-step process that results in the depletion of ozone in the stratosphere:

(1) $NO(g) + O_3(g) \rightarrow NO_2(g) + O_2(g)$
(2) $O(g) + NO_2(g) \rightarrow NO(g) + O_2(g)$
Overall $O(g) + O_3(g) \rightarrow 2\,O_2(g)$

13.115. On the basis of the frequency factors and activation energy values of the following two reactions, determine which one will have the larger rate constant at room temperature (298 K).

$O_3(g) + O(g) \rightarrow O_2(g) + O_2(g)$
$A = 8.0 \times 10^{-12}$ cm^3/(molecules · s) $\quad E_a = 17.1$ kJ/mol

$O_3(g) + Cl(g) \rightarrow ClO(g) + O_2(g)$
$A = 2.9 \times 10^{-11}$ cm^3/(molecules · s) $\quad E_a = 2.16$ kJ/mol

13.116. On the basis of the frequency factors and activation energy values of the following two reactions, determine which one will have the larger rate constant at room temperature (298 K).

$O_3(g) + Cl(g) \rightarrow ClO(g) + O_2(g)$
$A = 2.9 \times 10^{-11}$ cm^3/(molecules · s) $E_a = 2.16$ kJ/mol

$O_3(g) + NO(g) \rightarrow NO_2(g) + O_2(g)$
$A = 2.0 \times 10^{-12}$ cm^3/(molecules · s) $E_a = 11.6$ kJ/mol

Additional Problems

13.117. A student inserts a glowing wood splint into a test tube filled with O_2. The splint quickly catches fire (Figure P13.117). Why does the splint burn so much faster in pure O_2 than in air?

FIGURE P13.117

*13.118. A backyard chef turns on the propane gas to a barbecue grill. Even though the reaction between propane and oxygen is spontaneous, the gas does not begin to burn until the chef pushes an igniter button to produce a spark. Why is the spark needed?

***13.119.** On average, someone who falls through the ice covering a frozen lake is less likely to experience anoxia (lack of oxygen) than someone who falls into a warm pool and is underwater for the same length of time. Why?

13.120. **Ethanol Metabolism** The half-life of ethanol in the human body is $t_{1/2} = [\text{ethanol}]_0/2k$, where $[\text{ethanol}]_0$ is the initial concentration of ethanol and k is the zero-order rate constant. Write the rate law for the metabolism of ethanol and sketch the concentration–time curve.

***13.121.** If the rate of the reverse reaction is much slower than the rate of the forward reaction, does the method used to determine a rate law from initial concentrations and initial rates also work at some other time t? What concentrations would we use in the case where we use the rate when $t \neq 0$?

13.122. What is wrong with the following statement: The reaction rate and the rate constant for a reaction both depend on the number of collisions and on the concentrations of the reactants.

13.123. Why can't an elementary step in a mechanism have a rate law that is zero order in a reactant?

13.124. **Testing for a Banned Herbicide** Sodium chlorate was used in weed-control preparations, but its sale has been banned in EU countries since 2009. A simple colorimetric test for the presence of the chlorate ion in a solution of herbicide relies on the following reaction:

$$2\,MnO_4^-(aq) + 5\,ClO_3^-(aq) + 6\,H^+(aq) \rightarrow 2\,Mn^{2+}(aq) + 5\,ClO_4^-(aq) + 3\,H_2O(\ell)$$

The following table contains rate data for this reaction.

Experiment	$[MnO_4^-]_0$ (M)	$[ClO_3^-]_0$ (M)	$[H^+]_0$ (M)	Initial Rate (M/s)
1	0.10	0.10	0.10	5.2×10^{-3}
2	0.25	0.10	0.10	3.3×10^{-2}
3	0.10	0.30	0.10	1.6×10^{-2}
4	0.10	0.10	0.20	7.4×10^{-3}

Determine the rate law and the rate constant for this reaction.

13.125. The table contains reaction rate data for the reaction

$$2\,NO(g) + Cl_2(g) \rightarrow 2\,NOCl(g)$$

Experiment	$[NO]_0$ (M)	$[Cl_2]_0$ (M)	Initial Rate (M/s)
1	0.20	0.10	0.63
2	0.20	0.30	5.70
3	0.80	0.10	2.58
4	0.40	0.20	?

Predict the initial rate of reaction in experiment 4.

13.126. An important reaction in the formation of photochemical smog is the reaction between ozone and NO:

$$NO(g) + O_3(g) \rightarrow NO_2(g) + O_2(g)$$

The reaction is first order in NO and O_3. The rate constant of the reaction is 80 M^{-1} s^{-1} at 25°C and 3000 M^{-1} s^{-1} at 75°C.

a. If this reaction were to occur in a single step, would the rate law be consistent with the observed order of the reaction for NO and O_3?
b. What is the value of the activation energy of the reaction?
c. What is the rate of the reaction at 25°C when $[NO] = 3 \times 10^{-6}$ M and $[O_3] = 5 \times 10^{-9}$ M?
d. Predict the values of the rate constant at 10°C and 35°C.

13.127. Ammonia reacts with nitrous acid to form an intermediate, ammonium nitrite (NH_4NO_2), which decomposes to N_2 and H_2O:

$$NH_3(g) + HNO_2(aq) \rightarrow NH_4NO_2(aq) \rightarrow N_2(g) + 2\,H_2O(\ell)$$

a. The reaction is first order in ammonia and second order in nitrous acid. What is the rate law for the reaction? What are the units of the rate constant if concentrations are expressed in molarity and time in seconds?

*b. The rate law for the reaction has also been written as

$$\text{Rate} = k[NH_4^+][NO_2^-][HNO_2]$$

Why might this expression of the rate law be considered equivalent to the one you wrote in part (a)?

c. Use the data in Appendix 4 to calculate the value of ΔH°_{rxn} for the overall reaction (ΔH°_f for HNO_2 = −43.1 kJ/mol).

d. Draw an energy profile for the process with the assumption that E_a of the first step is lower than E_a of the second step.

13.128. **Lachrymators in Smog** The combination of ozone, volatile hydrocarbons, nitrogen oxide, and sunlight in urban environments produces peroxyacetyl nitrate (PAN), a potent lachrymator (eye irritant). PAN decomposes to peroxyacetyl radicals and nitrogen dioxide in a process that is second order in PAN, as shown in Figure P13.128:

a. The half-life of the reaction, at 23°C and $P_{CH_3CO_3NO_2}$ = 10.5 torr, is 100 h. Calculate the rate constant for the reaction.

b. Determine the rate of the reaction at 23°C and $P_{CH_3CO_3NO_2}$ = 10.5 torr.

c. Draw a graph showing P_{PAN} as a function of time from 0 to 200 h, starting with $P_{CH_3CO_3NO_2}$ = 10.5 torr.

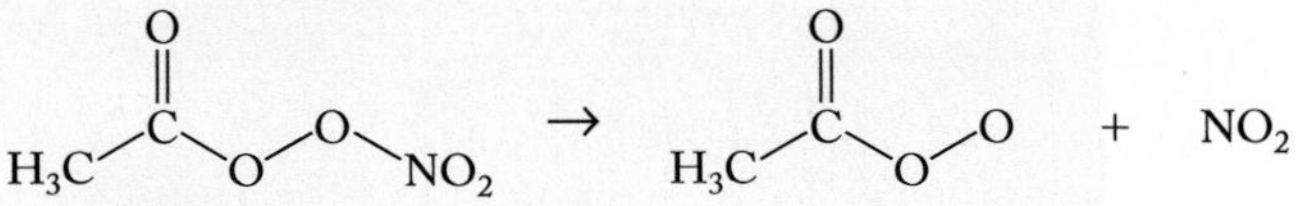

$$H_3C-C(=O)-O-O-NO_2 \rightarrow H_3C-C(=O)-O-O \cdot + NO_2$$

FIGURE P13.128

13.129. **Nitrogen Oxide in the Human Body** Nitrogen oxide is a free radical that plays many biological roles, including regulating neurotransmission and the human immune system. One of its many reactions involves the peroxynitrite ion ($ONOO^-$):

$$NO(aq) + ONOO^-(aq) \rightarrow NO_2(aq) + NO_2^-(aq)$$

a. Use the following data to determine the rate law and rate constant of the reaction at the experimental temperature at which these data were generated.

Experiment	$[NO]_0$ (*M*)	$[ONOO^-]_0$ (*M*)	Rate (*M*/s)
1	1.25×10^{-4}	1.25×10^{-4}	2.03×10^{-11}
2	1.25×10^{-4}	0.625×10^{-4}	1.02×10^{-11}
3	0.625×10^{-4}	2.50×10^{-4}	2.03×10^{-11}
4	0.625×10^{-4}	3.75×10^{-4}	3.05×10^{-11}

b. Draw the Lewis structure of peroxynitrite ion (including all resonance forms) and assign formal charges. Note which form is preferred.

c. Use the average bond energies in Table A4.1 to estimate the value of ΔH°_{rxn} by using the preferred structure from part (b).

13.130. **Reducing NO Emissions** Adding NH_3 to the stack gases at an electric-power-generating plant can reduce NO_x emissions. This selective noncatalytic reduction process depends on the reaction between NH_2 (an odd-electron molecule) and NO:

$$NH_2(g) + NO(g) \rightarrow N_2(g) + H_2O(g)$$

The following kinetic data were collected at 1200 K.

Experiment	$[NH_2]_0$ (*M*)	$[NO]_0$ (*M*)	Rate (*M*/s)
1	1.00×10^{-5}	1.00×10^{-5}	0.12
2	2.00×10^{-5}	1.00×10^{-5}	0.24
3	2.00×10^{-5}	1.50×10^{-5}	0.36
4	2.50×10^{-5}	1.50×10^{-5}	0.45

a. What is the rate law for the reaction?

b. What is the value of the rate constant at 1200 K?

14

Chemical Equilibrium

How Much Product Does a Reaction Really Make?

AMMONIA, FERTILIZERS, AND FOOD Nearly half of the world's food production relies on the use of fertilizers such as ammonia, which can be injected directly into the ground (as shown here) or used to make solid fertilizers such as urea and ammonium nitrate.

PARTICULATE **REVIEW**

Simultaneous Reactions

In Chapter 14 we explore reversible reactions, in which the reactants not only form products, but the products can also reform the reactants. Two such reactions take place simultaneously among the NO, O_2, and NO_2 molecules shown here.

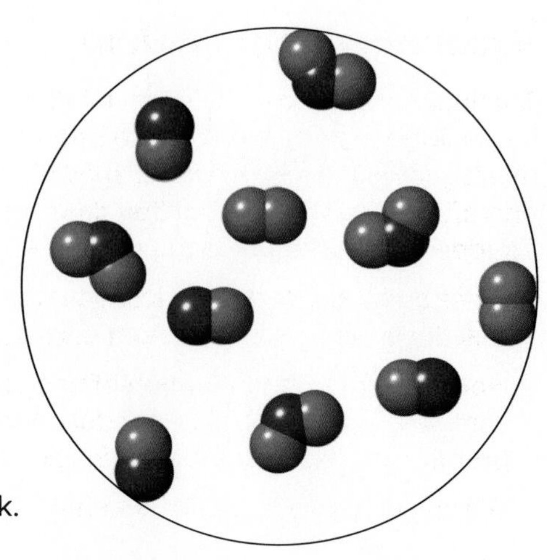

- Which of these molecules could combine, forming NO_2? Which of these molecules could be produced when molecules of NO_2 collide at high temperatures?
- Write one chemical equation showing that these two reactions could take place simultaneously.
- The first reaction proceeds slowly at room temperature. Describe how you would increase its rate and explain why your approach should work.

(Review Section 4.5 and Sections 13.2–13.4 if you need help.)

(Answers to Particulate Review questions are in the back of the book.)

PARTICULATE **PREVIEW**

Equal versus Equilibrium

The images show two different mixtures of steam and carbon monoxide reacting reversibly to produce carbon dioxide and hydrogen gas. Each sample is shown both in its initial state ($t = 0$) and after 5 seconds of reaction time. As you read Chapter 14, look for ideas that will help you answer these questions:

- Write one balanced chemical equation for the two simultaneous reactions taking place in each mixture.
- How does the concentration of each reactant change in sample A? In sample B? How does the concentration of each product change in sample A? In sample B?
- Which sample is more likely to be at equilibrium?

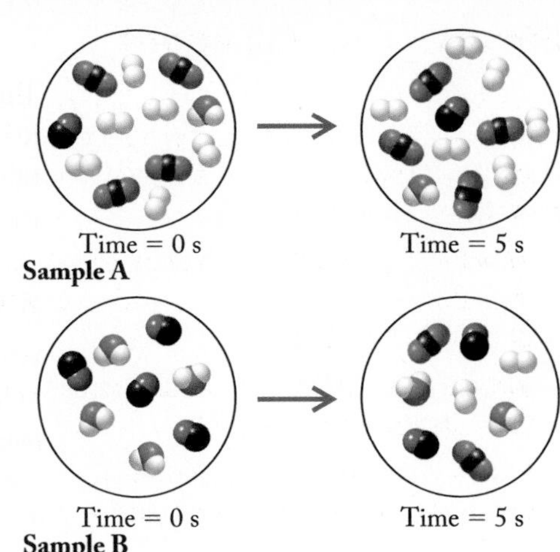

Learning Outcomes

LO1 Describe the dynamic nature of a chemical equilibrium and write mass action or equilibrium constant expressions for reversible reactions, including those involving heterogeneous equilibria
Sample Exercises 14.1, 14.9

LO2 Determine the value of an equilibrium constant from equilibrium concentrations or partial pressures of reactants and products
Sample Exercises 14.2, 14.3

LO3 Interconvert the K_c and K_p values of gas-phase reactions
Sample Exercise 14.4

LO4 Relate the values of K for a reaction, for the reverse reaction, for a reaction with different coefficients, and for combined reactions
Sample Exercises 14.5, 14.6, 14.7

LO5 Determine the value of a reaction quotient and use it to predict the direction of a reversible chemical reaction
Sample Exercise 14.8

LO6 Predict how a reaction at equilibrium responds to changes in reaction conditions, including adding or removing quantities of reactants and products or changing temperature
Sample Exercises 14.10, 14.11, 14.12

LO7 Relate the concentrations or partial pressures of reactants and products in a reaction mixture at equilibrium to their initial values and the value of K
Sample Exercises 14.13, 14.14, 14.15

14.1 The Dynamics of Chemical Equilibrium

Chemical fertilizers have become key components of modern agriculture and the global effort to feed the world's expanding population. Gaseous ammonia, NH_3, is used to produce solid nitrogen-based fertilizers such as urea and ammonium nitrate, and it is injected directly into soil as shown in the opening photo of this chapter. Over 150 million metric tons of ammonia are produced worldwide each year and over 80% of it is used in food production. Ammonia is produced in a reaction between N_2 from the atmosphere and H_2 gas. In most countries H_2 gas comes from high-temperature chemical reactions between methane (CH_4), which is the principal component in natural gas, and water vapor. We begin Chapter 14 by exploring these chemical reactions and return to the production of ammonia in Section 14.7.

CONNECTION Enthalpy and enthalpy changes were introduced in Section 6.4.

Most of the hydrogen gas needed to make ammonia is generated in a two-step process that begins with a reaction that is so widely used it has its own name—*steam–methane reforming*:

$$CH_4(g) + H_2O(g) \rightleftharpoons CO(g) + 3\,H_2(g) \qquad \Delta H^\circ = 206\text{ kJ/mol} \qquad (14.1)$$

The value of ΔH° indicates that the reaction is highly endothermic, which is why it is typically run at temperatures between 700°C and 1000°C. To increase the rate of the reaction, nickel or nickel-containing alloys are used as catalysts, and the reaction is run at pressures as high as 25 atm.

CONCEPT TEST

Why does raising the temperature or increasing the pressure of a gas-phase reaction help increase the rate of the reaction?

(Answers to Concept Tests are in the back of the book.)

In the second step, carbon monoxide produced in the first step reacts with more steam to produce more hydrogen gas:

$$CO(g) + H_2O(g) \rightleftharpoons CO_2(g) + H_2(g) \qquad \Delta H^\circ = -41\text{ kJ/mol} \qquad (14.2)$$

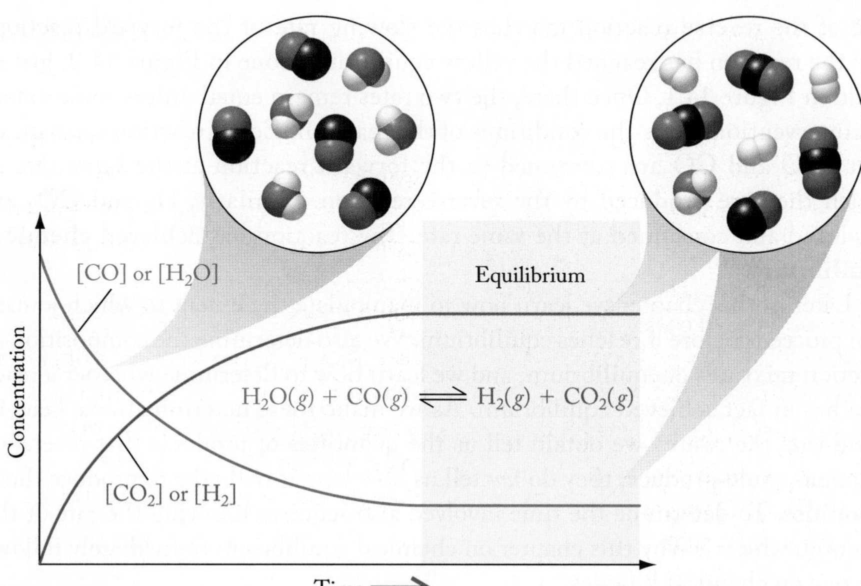

FIGURE 14.1 Concentrations of reactants (red curve) and products (blue curve) in the water–gas shift reaction change over time until equilibrium is reached (the yellow zone).

This reaction, called the water–gas shift reaction, also requires a catalyst and reaction temperatures greater than 350°C to achieve respectable reaction rates.

What is the purpose of the double arrow that connects the reactants and products in Equations 14.1 and 14.2? These arrows are meant to convey that steam–methane reforming and the water–gas shift reaction are reversible. Reversible reactions proceed in both the forward and the reverse directions at the same time.

CONNECTION We introduced the double arrow in Section 4.4 when discussing the dissociation of weak electrolytes.

To better understand how reactants become products and products become reactants, consider what happens if we run the water–gas shift reaction in a sealed vessel that initially contains equal numbers of moles of steam and carbon monoxide, but no products. As the reaction proceeds in the forward direction, the concentrations of H_2O and CO decrease, as shown by the red curve in **Figure 14.1**. The concentrations of H_2 and CO_2 start at zero and increase over time, as shown by the blue curve. Because H_2O and CO react in a 1:1 stoichiometric ratio, their concentrations decrease at the same rate. Similarly, H_2 and CO_2 are formed in a 1:1 ratio, so their concentrations increase at the same rate.

Eventually, both the red and blue curves level off (in the yellow equilibrium region of Figure 14.1), which means there is no net change in the concentrations of either the reactants or the products with time. Note that the unchanging (equilibrium) concentrations of H_2O and CO are significantly above zero. This means that the reaction mixture will contain some reactants no matter how long we let the reaction run.

To obtain another perspective on why the concentration curves in Figure 14.1 level off before all the reactants are consumed, recall from Chapter 13 that the rates of most chemical reactions depend on the concentrations of their reactant(s). Therefore, we expect the rate of the forward reaction to decrease as the concentrations of the reactants, H_2O and CO, decrease, which is shown by the downward trend of the red curve in **Figure 14.2**. As the reaction proceeds, the concentrations of H_2 and CO increase, which results in an increase in the rate of the reverse reaction as shown by the rising blue curve. Eventually, the increasing

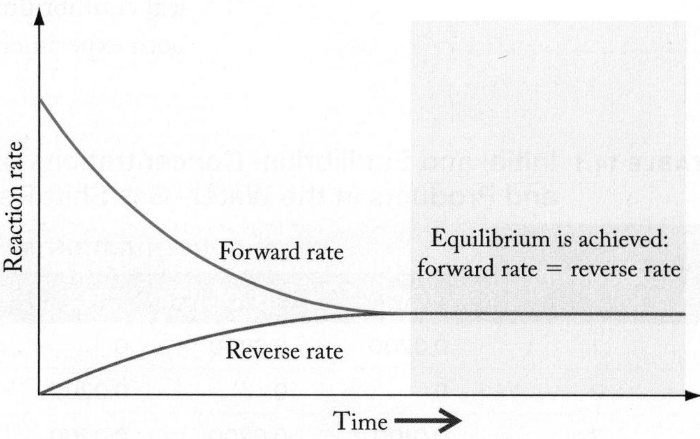

FIGURE 14.2 Equilibrium is achieved when the rates of a reversible reaction in the forward direction (red curve) and reverse direction (blue curve) become equal.

CHEMTOUR
Equilibrium

rate of the reverse reaction matches the slowing rate of the forward reaction, and the reaction has reached the yellow equilibrium zone in Figure 14.2, just as it did in Figure 14.1. Once there, the two rates remain equal unless some external intervention alters the conditions of the reaction. Equal reaction rates mean that H_2O and CO are consumed in the forward reaction at the same rate at which they are produced by the reverse reaction. Similarly, H_2 and CO_2 are produced and consumed at the same rate. The reaction has achieved **chemical equilibrium**.

Later in this chapter we learn how to manipulate the extent to which a reaction proceeds before it reaches equilibrium. We also determine the composition of reaction mixtures at equilibrium, and we learn how to determine whether a reaction has in fact achieved equilibrium. As we make these determinations, keep in mind that the results we obtain tell us the quantities of products that reversible reactions could produce; they do *not* tell us how long it will take to produce those quantities. To determine the time involved also requires knowing the rate of the reaction, which is why this chapter on chemical equilibrium immediately follows the one on chemical kinetics.

CONNECTION In Chapter 13 (Chemical Kinetics) we discussed how reactions occur when molecules collide; the higher the concentrations of reactants, the more frequently molecules collide, and the faster a reaction proceeds.

A final key point about chemical equilibria: they are *dynamic*. Just because the concentrations of reactants and products are not changing does not mean that a reaction has stopped. The reaction still proceeds, but it does so equally rapidly in both the forward and reverse directions simultaneously.

14.2 The Equilibrium Constant

In this section we take a quantitative look at the concentrations of products and reactants that are present when chemical reactions come to equilibrium. Let's return to the water–gas shift reaction:

$$CO(g) + H_2O(g) \rightleftharpoons CO_2(g) + H_2(g)$$

The data in **Table 14.1** describe the results of four experiments in which different quantities of H_2O gas, CO, H_2, and CO_2 are injected into a sealed reaction vessel heated to 350°C. The four gases are allowed to react and the final concentrations of all four are determined when the reaction has reached equilibrium.

In experiment 1 the reaction vessel initially contains equimolar concentrations of H_2O and CO but no H_2 or CO_2. In experiment 2 the vessel initially contains equimolar concentrations of H_2 and CO_2 but no H_2O or CO. Notice in Table 14.1 that when the reaction mixtures in experiments 1 and 2 achieve chemical equilibrium, the concentrations of H_2O and CO are the same (0.0036 *M*) in both experiments and so, too, are the concentrations of H_2 and CO_2 (0.0164 *M*).

TABLE 14.1 Initial and Equilibrium Concentrations of the Reactants and Products in the Water–Gas Shift Reaction at 350°C

Experiment	INITIAL CONCENTRATION (*M*)				EQUILIBRIUM CONCENTRATION (*M*)			
	[CO]	[H_2O]	[H_2]	[CO_2]	[CO]	[H_2O]	[H_2]	[CO_2]
1	0.0200	0.0200	0	0	0.0036	0.0036	0.0164	0.0164
2	0	0	0.0200	0.0200	0.0036	0.0036	0.0164	0.0164
3	0.0100	0.0200	0.0200	0.0100	0.0020	0.0120	0.0280	0.0180
4	0.0200	0.0100	0.0300	0.0400	0.0150	0.0050	0.0350	0.0450

The identical equilibrium concentrations of the reactants and products in experiments 1 and 2 confirm that the composition of a reaction mixture at equilibrium is independent of the direction in which a reaction proceeds to achieve equilibrium.

The data from experiment 3 show that the concentrations of products increase and the concentrations of reactants decrease when the total concentrations of products and reactants are initially the same. The data from experiment 4 show that the concentrations of products increase and the concentrations of reactants decrease even when the concentrations of the products are initially *greater than* the concentrations of the reactants.

The significance of the data from the four experiments in Table 14.1 can be appreciated if we multiply the equilibrium concentrations of the products (H_2 and CO_2) together and divide that product by the product of the equilibrium concentrations of the reactants (H_2O and CO):

$$\text{Experiments 1 and 2: } \frac{[H_2][CO_2]}{[H_2O][CO]} = \frac{(0.0164)(0.0164)}{(0.0036)(0.0036)} = 21$$

$$\text{Experiment 3: } \frac{[H_2][CO_2]}{[H_2O][CO]} = \frac{(0.0280)(0.0180)}{(0.0120)(0.0020)} = 21$$

$$\text{Experiment 4: } \frac{[H_2][CO_2]}{[H_2O][CO]} = \frac{(0.0350)(0.0450)}{(0.0050)(0.0150)} = 21$$

The ratio of the concentrations of products to reactants at equilibrium is the same in every experiment. It turns out that we would get the same ratio from *any* combination of initial concentrations of these four gases at 350°C. This concept, which applies to other reversible reactions, has been known since the mid-19th century when Norwegian chemists Cato Guldberg (1836–1902) and Peter Waage (1833–1900) discovered that any reversible reaction eventually reaches a state in which the ratio of the concentrations of products to reactants, with each value raised to a power corresponding to the coefficient for that substance in the balanced chemical equation for the reaction, has a characteristic value at a given temperature.

They called this phenomenon the **law of mass action** and the ratio of concentration terms the **mass action expression** for the reaction. This ratio is more commonly called the **equilibrium constant expression**, and its numerical value—which depends on temperature—is the **equilibrium constant (*K*)** of the reaction. Because *K* is a ratio of positive values, it is always positive. *K* values *much greater* than 1 are associated with reaction mixtures that consist of mostly products at equilibrium. *K* values *much less* than 1 are associated with reaction mixtures that contain few products at equilibrium.

For the water–gas shift reaction, the equilibrium constant expression and the value of *K* at 350°C are

$$K = \frac{[H_2][CO_2]}{[H_2O][CO]} = 21$$

None of the concentration terms in this expression has an exponent because all their coefficients are "1" in the balanced chemical equation, and a substance's coefficient *always* becomes its exponent in the equilibrium constant expression for a reaction. This contrasts with the rate law expressions we discussed in Chapter 13, where the exponents are frequently not the same as the coefficients in the balanced equation because the exponents reflect the stoichiometry of the rate-determining step, not necessarily the overall reaction.

chemical equilibrium a dynamic process in which the concentrations of reactants and products remain constant over time and the rate of the reaction in the forward direction matches its rate in the reverse direction.

law of mass action the principle relating the balanced chemical equation of a reversible reaction to its mass action expression (or equilibrium constant expression).

mass action expression an expression equivalent in form to the equilibrium constant expression but applied to reaction mixtures that may or may not be at equilibrium.

equilibrium constant expression the ratio of the concentrations or partial pressures of products to reactants at equilibrium, with each term raised to a power equal to the coefficient of that substance in the balanced chemical equation for the reaction.

equilibrium constant (*K*) the numerical value of the equilibrium constant expression of a reversible chemical reaction at a particular temperature.

CONCEPT TEST

Suppose equal numbers of moles of water vapor, carbon dioxide, carbon monoxide, and hydrogen gas are injected into a rigid, sealed reaction vessel and heated to 350°C. Which expression about the composition of the reaction mixture at equilibrium is true?

a. $[CO] = [H_2] = [CO_2] = [H_2O]$

b. $[CO_2] = [H_2] > [CO] = [H_2O]$

c. $[CO_2] = [H_2] < [CO] = [H_2O]$

d. $[H_2O] = [H_2] > [CO_2] = [CO]$

e. $[H_2O] = [H_2] < [CO_2] = [CO]$

For a generic reaction in which *a* moles of reactant A react with *b* moles of B to form *c* moles of substance C and *d* moles of D:

$$aA + bB \rightleftharpoons cC + dD$$

the equilibrium constant expression is

$$K_c = \frac{[C]^c[D]^d}{[A]^a[B]^b} \tag{14.3}$$

CHEMTOUR
Equilibrium in the Gas Phase

The subscript "c" on the equilibrium constant indicates that it is based on a ratio of concentration values. If substances A, B, C, and D are all gases, then the equilibrium constant (designated K_p in this case) may also be expressed in terms of their partial pressures:

$$K_p = \frac{(P_C)^c(P_D)^d}{(P_A)^a(P_B)^b} \tag{14.4}$$

We distinguish between K_c and K_p because their values for a given gas-phase reaction at a given temperature may or may not be the same (see Section 14.3). It depends on whether the number of moles of gaseous reactants is the same as the number of moles of gaseous products.

K values do not have units even when there are different numbers of moles of reactants and products. This may seem odd, but there is a reason why: the concentrations and partial pressures in K_c or K_p expressions are actually ratios of the ingredient concentrations or partial pressures in a reaction mixture to an ideal standard concentration (1.000 *M*) or partial pressure (1.000 bar). These ratios take into account the nonideal behavior of substances. Because the terms in equilibrium constant expressions are, strictly speaking, ratios of values having the same units, their units cancel out, producing unitless equilibrium constants. In practice, we use concentration and partial pressure values directly in equilibrium calculations. When we do, we are assuming that all the reactants and products are behaving ideally.

SAMPLE EXERCISE 14.1 Writing Equilibrium Constant Expressions LO1

A key reaction in the formation of photochemical smog involves the reversible combination of NO and O_2 in the atmosphere, producing NO_2:

$$2\,NO(g) + O_2(g) \rightleftharpoons 2\,NO_2(g)$$

Write the K_c and K_p expressions for this reaction.

Collect, Organize, and Analyze We are given the balanced chemical equation for a reaction and asked to write K_c and K_p expressions for it. Equilibrium constant expressions are ratios of the concentrations (K_c) or partial pressures (K_p) of products to reactants, with each term raised to the power equal to its coefficient in the balanced chemical equation of the reaction. Review Equations 14.3 and 14.4, if necessary.

Solve The coefficients of NO and NO_2 in the chemical equation are both 2, so the NO and NO_2 terms in the K_c and K_p expressions must be squared:

$$K_c = \frac{[NO_2]^2}{[NO]^2[O_2]}$$

$$K_p = \frac{(P_{NO_2})^2}{(P_{NO})^2(P_{O_2})}$$

Think About It The numerators and denominators of the K_c and K_p expressions contain terms for the same products and reactants, each raised to the same power. The difference between them is the nature of the terms: molar concentrations in the K_c expression and partial pressures in the K_p expression.

Practice Exercise Write K_c and K_p equilibrium constant expressions for the steam–methane reforming reaction:

$$CH_4(g) + H_2O(g) \rightleftharpoons CO(g) + 3\ H_2(g)$$

(Answers to Practice Exercises are in the back of the book.)

SAMPLE EXERCISE 14.2 Calculating the Value of K_c from Equilibrium Concentrations — LO2

Table 14.2 lists initial and equilibrium concentration data from four experiments on the dimerization of NO_2:

$$2\ NO_2(g) \rightleftharpoons N_2O_4(g)$$

The experiments were run at 100°C in a rigid, closed container. Use the data from each experiment to calculate a value of the equilibrium constant K_c for the reaction.

Collect and Organize We are given four sets of data that contain initial and equilibrium concentrations of a reactant and product. We are asked to determine the value of the equilibrium constant K_c in each experiment. The K_c expression for this reaction should have the same form as Equation 14.3—that is, the ratio of the equilibrium concentration of the product over that of the reactant, with each term raised to a power equal to its coefficient in the balanced chemical equation.

Analyze To calculate the value of K_c, we insert the pairs of equilibrium concentrations from each experiment into the K_c expression and calculate the ratio. The $[NO_2]$ values are greater than the $[N_2O_4]$ values in every experiment, so the equilibrium reaction

TABLE 14.2 Data for the Dimerization of NO_2 at 100°C

Experiment	INITIAL CONCENTRATION (*M*)		EQUILIBRIUM CONCENTRATION (*M*)	
	$[NO_2]$	$[N_2O_4]$	$[NO_2]$	$[N_2O_4]$
1	0.0200	0.0000	0.0172	0.00139
2	0.0300	0.0000	0.0244	0.00280
3	0.0400	0.0000	0.0310	0.00452
4	0.0000	0.0200	0.0310	0.00452

mixtures contain more reactant than product. Therefore, we might expect the value of K_c to be less than 1. However, all the gas concentrations are much less than 0.1 *M*, and the $[NO_2]$ values in the denominator are squared, which makes the value of the denominator even smaller. Therefore, the value of K_c is probably greater than 1.

Solve The K_c expression is

$$K_c = \frac{[N_2O_4]}{[NO_2]^2}$$

Inserting the data from the four experiments into this expression and doing the math:

$$\text{Experiment 1: } K_c = \frac{[N_2O_4]}{[NO_2]^2} = \frac{0.00139}{(0.0172)^2} = 4.70$$

$$\text{Experiment 2: } K_c = \frac{0.00280}{(0.0244)^2} = 4.70$$

$$\text{Experiment 3: } K_c = \frac{0.00452}{(0.0310)^2} = 4.70$$

$$\text{Experiment 4: } K_c = \frac{0.00452}{(0.0310)^2} = 4.70$$

Think About It The values calculated for K_c are the same, as they should be for the same reaction at the same temperature. Moreover, they are greater than 1, as we predicted.

Practice Exercise A mixture of gaseous CO and H_2, called *synthesis gas*, is used commercially to prepare methanol (CH_3OH), a compound considered an alternative fuel to gasoline. Under equilibrium conditions at 700 K, $[H_2] = 0.074$ *M*, $[CO] = 0.025$ *M*, and $[CH_3OH] = 0.040$ *M*. What is the value of K_c for this reaction at 700 K?

SAMPLE EXERCISE 14.3 Calculating the Value of K_p from Equilibrium Partial Pressures LO2

A sealed chamber contains an equilibrium mixture of NO_2 and N_2O_4 at 15°C. The partial pressures of NO_2 and N_2O_4 are 0.34 atm and 0.80 atm, respectively. What is the value of K_p at 15°C for the dimerization of NO_2: $2\ NO_2(g) \rightleftharpoons N_2O_4(g)$?

Collect, Organize, and Analyze We are asked to use the partial pressures of NO_2 and N_2O_4 at equilibrium to determine the value of K_p for the dimerization of NO_2. The K_p expression (shown for a general reaction in Equation 14.4) should have the same format as the K_c expression for this reaction that was used in Sample Exercise 14.2.

Solve

$$K_p = \frac{P_{N_2O_4}}{(P_{NO_2})^2} = \frac{0.80}{(0.34)^2} = 6.9$$

Think About It The value of K_p is not the same as the K_c value for this reaction that was calculated in Sample Exercise 14.2. This difference should not be surprising because the experiments in the two exercises were conducted at different temperatures, and *K* values change with changing temperature. Moreover, the K_p and K_c values of gas-phase reactions may differ, as noted previously in this section, depending on the number of moles of gaseous reactants and products.

Practice Exercise A reaction vessel contains an equilibrium mixture of SO_2, O_2, and SO_3. The partial pressures of the three gases are 0.0018 atm, 0.0032 atm, and 0.0166 atm, respectively. What is the value of K_p for the following reaction at the temperature inside the vessel?

$$2\ SO_2(g) + O_2(g) \rightleftharpoons 2\ SO_3(g)$$

As we end this section, let's more closely examine the connection between K values of reactions and the composition of their reaction mixtures at equilibrium. The values we have seen thus far—$K_c = 21$ for the water–gas shift reaction at 350°C, and $K_p = 6.9$ and $K_c = 4.70$ for the dimerization of NO_2 at 25°C and 100°C, respectively—are considered intermediate values. Because the range of K values is so large ($0 < K < \infty$), all three of these values are considered *close* to 1, which means that comparable concentrations of reactants and products are likely to be present at equilibrium.

On the other hand, many reactions reach equilibrium only after essentially all the reactants have formed products. We say that these equilibria *lie far to the right* (the direction of the forward reaction arrow). For example, the combustion of H_2 gas proceeds very rapidly (**Figure 14.3**) until H_2 is, for all practical purposes, completely consumed. This observation is consistent with an enormous value of K for the combustion reaction at 25°C:

FIGURE 14.3 The German passenger airship *Hindenburg* was filled with hydrogen gas to make it less dense than air. The flammable gas ignited as the airship docked in Lakehurst, New Jersey, on May 6, 1937. Of the 97 people aboard, 35 were killed.

$$2\,H_2(g) + O_2(g) \rightleftharpoons 2\,H_2O(g) \qquad K_c = 3 \times 10^{81}$$

In other reactions, little product is formed before equilibrium is reached. These equilibria are said to favor reactants and to *lie far to the left* (the direction of the reverse reaction arrow). For example, the decomposition of CO_2 to CO and O_2 at 25°C essentially doesn't happen, as predicted by our understanding of the stability of CO_2 and the minuscule K value for the decomposition reaction:

$$2\,CO_2(g) \rightleftharpoons 2\,CO(g) + O_2(g) \qquad K_c = 3 \times 10^{-92}$$

In the rest of this chapter, and in Chapters 15 and 16, we focus mainly on equilibria that do not lie far to the left or right but are more likely to be a little to the left or right. These reaction systems tend to have relatively small concentrations of either reactants or products at equilibrium, but not so small that they can be ignored.

14.3 Relationships between K_c and K_p Values

As we noted in Section 14.2, the values of K_c and K_p for a given reaction and temperature may or may not be the same, depending on the numbers of moles of gaseous reactants and products. To better understand this relationship, we begin with the ideal gas law:

$$PV = nRT$$

If we solve for P and express volume in liters, then n/V has units of moles per liter, which is the same as molarity (M):

$$P = \frac{n}{V}RT$$

$$P = M \cdot RT \qquad (14.5)$$

Let's apply Equation 14.5 to the gases in the NO_2–N_2O_4 equilibrium from Sample Exercise 14.3:

$$P_{NO_2} = \frac{n_{NO_2}}{V}RT = [NO_2]RT$$

$$P_{N_2O_4} = \frac{n_{N_2O_4}}{V}RT = [N_2O_4]RT$$

Substituting these values into the expression for K_p from Sample Exercise 14.3, we get

$$K_p = \frac{(P_{N_2O_4})}{(P_{NO_2})^2} = \frac{[N_2O_4]RT}{([NO_2]RT)^2} = \frac{[N_2O_4]RT}{[NO_2]^2(RT)^2}$$

The ratio of concentration terms in the expression on the right, $[N_2O_4]/[NO_2]^2$, is the same as the K_c expression for this reaction. Substituting K_c for those terms and simplifying the RT terms gives:

$$K_p = K_c \frac{1}{RT}$$

This last equation defines the relationship between K_c and K_p for this specific reaction. A more general expression can be derived for the generic reaction in which a moles of gas A reacts with b moles of gas B, producing c moles of gas C and d moles of gas D:

$$a\text{A} + b\text{B} \rightleftharpoons c\text{C} + d\text{D}$$

The K_p expression for this reaction is

$$K_p = \frac{(P_C)^c(P_D)^d}{(P_A)^a(P_B)^b} \quad (14.4)$$

Replacing each partial pressure term in Equation 14.4 with the corresponding [X]RT term from Equation 14.5 gives us the general expression

$$K_p = \frac{([C]RT)^c([D]RT)^d}{([A]RT)^a([B]RT)^b} \quad (14.6)$$

Combining the RT terms, we get

$$K_p = \frac{[C]^c[D]^d}{[A]^a[B]^b} \times (RT)^{[(c+d)-(a+b)]}$$

To simplify the right side of this equation we (1) replace the ratio of concentration terms with K_c (see Equation 14.3) and (2) replace the four values in the exponent of RT with Δn, which equals the difference between the number of moles of product gases ($c + d$) and the number of moles of reactant gases ($a + b$). With these substitutions, Equation 14.6 becomes

$$K_p = K_c(RT)^{\Delta n} \quad (14.7)$$

The message in Equation 14.7 is that the K_p and K_c values of a reaction are related by RT raised to the appropriate power. This relationship makes sense because K_p values are based on ratios of partial pressures, and K_c values are based on ratios of molar concentrations. Equation 14.5 reminds us that the partial pressure and molar concentration of an ideal gas are related by RT: $P = M \cdot RT$.

In reactions in which the number of moles of gas on both sides of the reaction arrow is the same, such as the water–gas shift reaction:

$$CO(g) + H_2O(g) \rightleftharpoons CO_2(g) + H_2(g)$$

$\Delta n = 0$ and $K_p = K_c$. However, in the steam–methane reforming reaction:

$$CH_4(g) + H_2O(g) \rightleftharpoons CO(g) + 3\,H_2(g)$$

Two moles of gaseous reactants form 4 moles of gaseous products. Therefore,

$$\Delta n = 4 \text{ mol} - 2 \text{ mol} = 2 \text{ mol}$$

Inserting this value for Δn in Equation 14.7 gives us the relationship between K_p and K_c for this reaction:

$$K_p = K_c(RT)^2$$

CONNECTION We determined the molar volume occupied by an ideal gas at standard temperature and pressure in Chapter 5.

SAMPLE EXERCISE 14.4 Calculating K_c from K_p **LO3**

In Sample Exercise 14.3 we determined that $K_p = 6.9$ at 15°C for the reaction

$$2\ NO_2(g) \rightleftharpoons N_2O_4(g)$$

What is the value of K_c for this reaction at 15°C?

Collect, Organize, and Analyze We are given a K_p value and asked to calculate the corresponding K_c value for the same reaction at the same temperature: 15°C or (15 + 273) = 288 K. Equation 14.7 relates the values of K_p and K_c:

$$K_p = K_c(RT)^{\Delta n}$$

where Δn represents the change in the number of moles of gas as the reaction proceeds.

Solve Two moles of gaseous reactants yield 1 mole of gaseous product. Therefore:

$$\Delta n = 1 \text{ mol gaseous product} - 2 \text{ mol gaseous reactants} = -1 \text{ mol}$$

Inserting this value of Δn, the given value of K_p, and the values of R and absolute temperature (T) into Equation 14.7:

$$K_p = K_c(RT)^{\Delta n}$$

$$6.9 = K_c(0.08206 \times 288)^{-1} = \frac{K_c}{0.08206 \times 288}$$

Cross multiplying and solving for K_c:

$$K_c = (6.9)(0.08206)(288) = 1.6 \times 10^2$$

Think About It The value of K_c is 170/6.9 or nearly 25 times larger than that of K_p, which makes sense because a Δn value of −1 means $K_c = K_p(RT)$ and the value of (RT) at 288 K is (0.08206 × 288) = 23.6.

Practice Exercise An important industrial process for synthesizing the ammonia used in agricultural fertilizers involves the combination of N_2 and H_2:

$$N_2(g) + 3\ H_2(g) \rightleftharpoons 2\ NH_3(g) \qquad K_p = 6.1 \times 10^5 \text{ at } 25°C$$

What is the value of K_c of this reaction at 25°C?

14.4 Manipulating Equilibrium Constant Expressions

In this section we learn how to write equilibrium constant expressions for the reverse direction of reversible reactions, for chemical equations that have been multiplied or divided by a value so that a key component has a coefficient of 1, and for overall reactions that are combinations of other reactions.

K for Reverse Reactions

The K values for the forward and reverse directions of a reversible reaction are related, just as the reactions are. In Sample Exercise 14.2, for example, we wrote the K_c expression for the dimerization of NO_2:

$$2\ NO_2(g) \rightleftharpoons N_2O_4(g) \qquad K_c = \frac{[N_2O_4]}{[NO_2]^2}$$

Now let's write the K_c expression for the reverse reaction—that is, for the decomposition of N_2O_4:

$$N_2O_4(g) \rightleftharpoons 2\,NO_2(g) \qquad K_c = \frac{[NO_2]^2}{[N_2O_4]}$$

Note how reversing the reaction turns the K_c expression upside down as reactants and products switch roles. Equation 14.8 generalizes this reciprocal relation between the forward ($K_{forward}$) and reverse ($K_{reverse}$) equilibrium constants to any reversible reaction:

$$K_{forward} = \frac{1}{K_{reverse}} \tag{14.8}$$

Equation 14.8 makes sense if we consider a generic reversible reaction in which the equilibrium favors the formation of product B over reactant A, which is reflected in its large K value:

$$A \rightleftharpoons B \qquad K > 1$$

If we write the reaction in reverse, B is still favored over A, but now B is the reactant and A is the product. Therefore, the value of K for the reverse reaction should be small because the reciprocal of a large (K) value is a small one:

$$B \rightleftharpoons A \qquad K < 1$$

SAMPLE EXERCISE 14.5 Calculating the Value of K for a Reverse Reaction **LO4**

Atmospheric NO combines with O_2 to form NO_2. The reverse reaction is the decomposition of NO_2 to NO and O_2. At 184°C, the value of K_c for the forward reaction is 1.48×10^4. Write the equilibrium constant expressions for both reactions and calculate the value of K_c for the decomposition of NO_2 at 184°C.

Collect and Organize We are asked to write equilibrium constant expressions for the formation and decomposition of NO_2 and to calculate the value of K_c of the decomposition reaction from the K_c value of the formation reaction. Each of these reactions is the reverse of the other, so their equilibrium constant expressions and K_c values are reciprocals of each other.

Analyze To write the K_c expressions for the two reactions, we first need to write a chemical equation describing one of them. That equation will provide the stoichiometric coefficients we need to use as exponents for the concentration terms in the K_c expressions for both reactions, with each expression being the reciprocal of the other.

Solve Writing a chemical equation for the formation of NO_2 from NO and O_2:

$$2\,NO(g) + O_2(g) \rightleftharpoons 2\,NO_2(g)$$

The K_c expression for this formation reaction is the same as K_c in Sample Exercise 14.1:

$$K_{formation} = \frac{[NO_2]^2}{[NO]^2[O_2]}$$

The K_c expression for the decomposition (reverse) reaction is

$$K_{decomposition} = \frac{[NO]^2[O_2]}{[NO_2]^2}$$

and its value is

$$K_{decomposition} = 1/K_{formation} = 1/1.48 \times 10^4 = 6.76 \times 10^{-5}$$

Think About It The value of $K_{formation}$ is large, so it makes sense that its reciprocal, $K_{decomposition}$, is small.

Practice Exercise At 300°C, the value of K_p for the combination reaction of N_2 and H_2 that produces NH_3 gas is 4.3×10^{-3}. What is the value of K_p at the same temperature for the decomposition reaction of NH_3 to produce N_2 and H_2?

K for an Equation Multiplied or Divided by a Number

As we saw in Chapter 6, sometimes we must represent a reaction by an equation containing fractional coefficients so that, for example, there is exactly 1 mole of a particular product. Suppose we want to rewrite the formation of NO_2 from NO and O_2:

$$2\,NO(g) + O_2(g) \rightleftharpoons 2\,NO_2(g) \qquad K_c = \frac{[NO_2]^2}{[NO]^2[O_2]}$$

to describe the formation of only 1 mole of NO_2. We do this by dividing all the coefficients by 2:

$$NO(g) + \tfrac{1}{2}\,O_2(g) \rightleftharpoons NO_2(g)$$

The corresponding K_c expression is

$$K_c = \frac{[NO_2]}{[NO][O_2]^{1/2}}$$

Comparing the original K_c expression with this one, we see that the concentration terms in the original expression correspond to those in the second expression *squared*. (The square root of a value is the same as raising the value to a power of $\tfrac{1}{2}$.) Put another way, dividing all the coefficients in the original reaction equation by 2 produces a K_c expression (and a *K* value) that is the *square root* of the original. We can extend this pattern to all chemical equilibria with the following general rule: if the balanced chemical equation of a reaction is multiplied by some factor *n*, then the value of *K* for the new reaction is the original *K* value raised to the *n*th power. Similarly, dividing all coefficients by *n* raises the value of *K* to the 1/*n* power.

SAMPLE EXERCISE 14.6 Calculating *K* for Different Coefficients **LO4**

An important reaction in the industrial production of sulfuric acid is the oxidation of SO_2 to SO_3. One way to write a chemical equation for the oxidation reaction is

$$SO_2(g) + \tfrac{1}{2}\,O_2(g) \rightleftharpoons SO_3(g)$$

If the value of K_c for this reaction at 298 K is 2.8×10^{12}, what is the value of K_c at 298 K for

$$2\,SO_2(g) + O_2(g) \rightleftharpoons 2\,SO_3(g)$$

Collect, Organize, and Analyze We are given the value of K_c for a reaction that describes the production of 1 mole of SO_3. We are asked to recalculate this K_c value for the same reaction but based on the production of 2 moles of SO_3. Multiplying the coefficients of a chemical equation by *n* raises the value of its equilibrium constant to the *n*th power.

Solve Multiplying the original chemical equation by 2 raises its K_c to the second power:

$$K_c = (2.8 \times 10^{12})^2 = 7.8 \times 10^{24}$$

Think About It Doubling the coefficients made the large original K_c value very large. However, if the original K_c value had been small (<1), doubling the coefficients and squaring the K_c would have made its value even smaller.

Practice Exercise The industrial production of ammonia for fertilizers involves the reaction of nitrogen and hydrogen. If $K_c = 2.4 \times 10^{-3}$ at 1000 K for the reaction

$$N_2(g) + 3\,H_2(g) \rightleftharpoons 2\,NH_3(g)$$

what is K_c at 1000 K for this reaction?

$$\tfrac{1}{3}\,N_2(g) + H_2(g) \rightleftharpoons \tfrac{2}{3}\,NH_3(g)$$

In Sample Exercise 14.6 we saw that two equivalent ways to balance a chemical equation for the same reaction produced two different K_c expressions and two different K_c values. Surely the same ingredients should be present in the same proportions at equilibrium, no matter how we choose to write a balanced equation describing their reaction. In fact, they are. The difference between the *K* values is not chemical; it is only mathematical. It does not affect the composition of a reaction mixture at equilibrium—only the arithmetic we use to predict it.

Combining *K* Values

In Section 6.7 we introduced Hess's law to calculate the enthalpies of combined reactions. We can also combine *K* values to obtain an overall *K* for a reaction that is the sum of two or more other reactions.

Consider two reactions, labeled (1) and (2) below, that were introduced in Chapter 13 and are involved in the formation of photochemical smog. As they proceed, NO produced in a car's engine at high temperature is oxidized to NO_2 in the atmosphere:

$$\begin{array}{lrcl} (1) & N_2(g) + O_2(g) & \rightleftharpoons & \cancel{2\,NO(g)} \\ (2) & \cancel{2\,NO(g)} + O_2(g) & \rightleftharpoons & 2\,NO_2(g) \\ \hline \text{Overall:} & N_2(g) + 2\,O_2(g) & \rightleftharpoons & 2\,NO_2(g) \end{array}$$

The K_c expressions for the reactions (1) and (2) are

$$K_1 = \frac{[NO]^2}{[N_2][O_2]} \quad \text{and} \quad K_2 = \frac{[NO_2]^2}{[NO]^2[O_2]}$$

How are these two individual equilibrium expressions related to the K_c expression for the overall reaction? We can derive the overall expression by multiplying K_1 and K_2 together:

$$K_1 \times K_2 = \frac{\cancel{[NO]^2}}{[N_2][O_2]} \times \frac{[NO_2]^2}{\cancel{[NO]^2}[O_2]} = \frac{[NO_2]^2}{[N_2][O_2]^2} = K_{overall}$$

This approach works for all series of reactions, and as a rule

$$K_{overall} = K_1 \times K_2 \times K_3 \times K_4 \times \ldots \times K_n \tag{14.9}$$

The overall equilibrium constant for a sum of two or more reactions is the product of the equilibrium constants of the individual reactions. Thus, the value of K_c for the overall reaction for the formation of NO_2 from N_2 and O_2 at 1000 K is the product of the equilibrium constants for reaction 1:

$$K_1 = \frac{[NO]^2}{[N_2][O_2]} = 7.2 \times 10^{-9}$$

and reaction 2:

$$K_2 = \frac{[NO_2]^2}{[NO]^2[O_2]} = 0.020$$

Using Equation 14.9:

$$K_{overall} = K_1 \times K_2 = 7.2 \times 10^{-9} \times 0.020 = 1.4 \times 10^{-10}$$

The equilibrium constant expression for the overall reaction must contain the appropriate terms for the products and reactants of that reaction. Just as with Hess's law in thermochemical calculations, we may need to reverse an equation or multiply an equation by a factor when we combine it with another to create the right overall equation. If we reverse a reaction, we must take the reciprocal of its K, and if we multiply or divide a reaction by n, we must raise its K to the n or $1/n$ power.

SAMPLE EXERCISE 14.7 Calculating an Overall *K* Value — LO4

At 1000 K, the K_c value of the following reaction is 1.5×10^6:

$$(1) \qquad N_2O_4(g) \rightleftharpoons 2\ NO_2(g)$$

The K_c value at 1000 K of the following reaction is 1.4×10^{-10}:

$$(2) \qquad N_2(g) + 2\ O_2(g) \rightleftharpoons 2\ NO_2(g)$$

What is the K_c value at 1000 K of the following reaction?

$$N_2(g) + 2\ O_2(g) \rightleftharpoons N_2O_4(g)$$

Collect and Organize We are given chemical equations describing two reactions and their K_c values, and we need to combine the equations in such a way that they describe a third reaction. When two equations are added, their K_c values are multiplied together to obtain $K_{overall}$.

Analyze The product of the overall reaction, N_2O_4, is a reactant in reaction (1), so we need to reverse it. N_2 and O_2 are reactants in reaction (2) and they are reactants with the same coefficients in the overall reaction, so we can combine reaction (2) as written with the reverse of reaction (1).

The K_c value of any reaction running in reverse is the reciprocal of the K_c value of the forward reaction. The K_c value of reaction (1) is large ($>10^6$), which means its reciprocal is small ($<10^{-6}$). The product of this small value and the smaller one for reaction (2), $\sim 10^{-10}$, should be even smaller still—about 10^{-16}.

Solve

Combining the two chemical equations:

Reaction (1) reversed	$\cancel{2\ NO_2(g)} \rightleftharpoons N_2O_4(g)$
+ Reaction (2)	$N_2(g) + 2\ O_2(g) \rightleftharpoons \cancel{2\ NO_2(g)}$
Summing:	$N_2(g) + 2\ O_2(g) + \cancel{2\ NO_2(g)} \rightleftharpoons N_2O_4(g) + \cancel{2\ NO_2(g)}$
Simplifying:	$N_2(g) + 2\ O_2(g) \rightleftharpoons N_2O_4(g)$

Calculating the value of K_{overall}:

$$K_{\text{overall}} = \frac{1}{K_1} \times K_2 = \frac{1}{1.5 \times 10^6} \times 1.4 \times 10^{-10} = 9.3 \times 10^{-17}$$

Think About It The calculated K_{overall} value is close to our estimate and indeed very small. As a result, the overall equilibrium lies far to the left, meaning very little N_2O_4 forms from N_2 and O_2 at 1000 K.

 **Practice Exercise** Calculate the value of K_c for the hypothetical reaction

$$W(g) + X(g) \rightleftharpoons M(g)$$

from the following information:

$$2\,M(g) \rightleftharpoons Z(g) \qquad K_c = 6.2 \times 10^{-4}$$
$$Z(g) \rightleftharpoons 2\,W(g) + 2\,X(g) \qquad K_c = 5.6 \times 10^{-2}$$

To summarize the key points for manipulating equilibrium constants:

- The K value of a reaction running in reverse is the reciprocal of the K of the forward reaction.
- If the original chemical equation describing a reversible reaction is multiplied by n, the value of K is the value of the original K raised to the nth power.
- If the original chemical equation describing a reversible reaction is divided by n, the value of K is the value of the original K raised to the $(1/n)$ power.
- If an overall chemical reaction is the sum of two or more other reactions, the overall value of K is the product of the K values of the other reactions.

14.5 Equilibrium Constants and Reaction Quotients

In Section 14.2 we introduced two key terms: *equilibrium constant expression* and *mass action expression*. Until now we have used the first term almost exclusively because we have been dealing with chemical reactions that have achieved equilibrium. Now we reintroduce the concept of a mass action expression because we can apply it not only to concentrations (or partial pressures) of products and reactants in reaction mixtures that have reached equilibrium, but also to reaction mixtures that are on their way to equilibrium but are not yet there.

Even if a reversible chemical reaction has not reached equilibrium, we can still insert reactant and product concentrations (or partial pressures) into its mass action expression. The mathematical result is not a K value because the reaction is not yet at equilibrium. Instead it is a Q value, where Q stands for **reaction quotient**. The value of Q provides us with a status report on the progress of a pair of reversible reactions.

reaction quotient (Q) the numerical value of the mass action expression based on the concentrations or partial pressures of the reactants and products present at any time during a reaction; at equilibrium, $Q = K$.

To see how this works, let's once again consider the water–gas shift reaction and the concentration data from experiment 4 in Table 14.1:

$$CO(g) + H_2O(g) \rightleftharpoons CO_2(g) + H_2(g) \qquad K_c = 21 \text{ at } 350°C$$

Inserting the initial concentration values into the mass action expression for the reaction yields a value for Q based on concentration; that is, Q_c:

$$Q_c = \frac{[H_2][CO_2]}{[H_2O][CO]} = \frac{(0.0300)(0.0400)}{(0.0100)(0.0200)} = 6.00$$

In this case, $Q_c = 6.00$ and $K_c = 21$. These Q_c and K_c values mean there are lower concentrations of products and higher concentrations of reactants in the initial reaction mixture than there will be at equilibrium. To achieve equilibrium, some of the reactants must form products. Mathematically, that will increase the numerator and decrease the denominator in Q_c, thereby increasing the value of Q_c until it matches the value of K_c and equilibrium has been achieved.

To put the results from the data in Table 14.1 in context, let's consider the curves in **Figure 14.4**. Starting at the left side of the graph in zone (a), we have the initial conditions of experiment 1 from Table 14.1: equal concentrations of reactants are present, but no products. Over time, reactant concentrations (the red curve) decrease as product concentrations (the blue curve) increase. At the right side of the graph in zone (c), we have the initial conditions of experiment 2: products are present, but no reactants. Over time, reactant concentrations increase as product concentrations decrease. In zone (b) in the middle of the graph, no net change occurs in the composition because the reaction is at equilibrium.

We can also characterize the three zones in Figure 14.4 based on the value of Q compared with K. In zone (a), $Q < K$ and there is a net conversion of reactants into products as the forward reaction dominates. In zone (c), $Q > K$ and a net conversion of products into reactants takes place as the reverse reaction dominates. In the middle zone (b), $Q = K$ and no net change in the composition of the

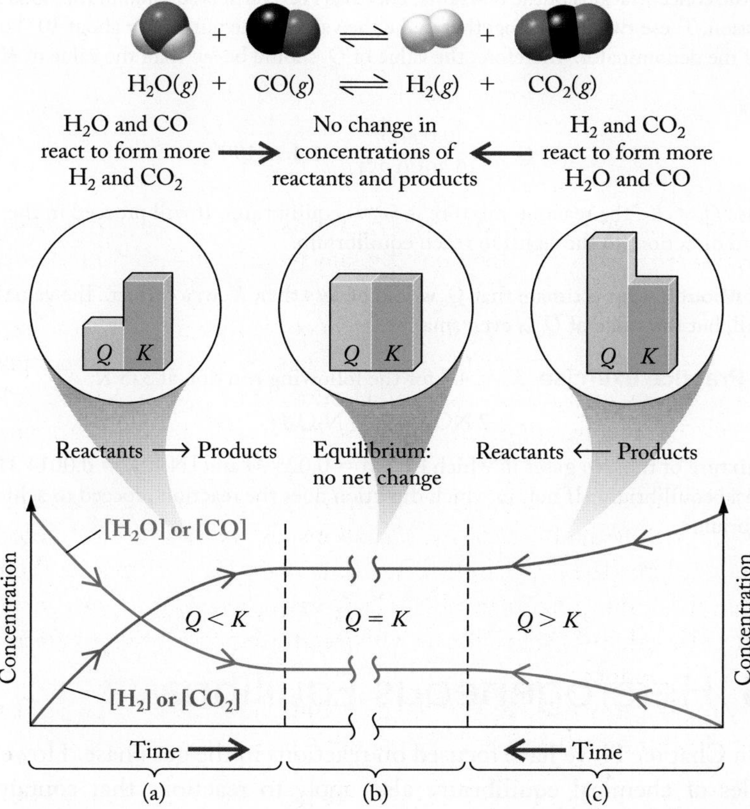

FIGURE 14.4 The value of the reaction quotient Q in relation to the equilibrium constant K for the water–gas shift reaction. (a) Reactant concentrations (red) are higher than they are once equilibrium is reached, and product concentrations (blue) are lower than at equilibrium; $Q < K$, and the reactants form more products. (b) Equilibrium concentrations are achieved; $Q = K$, and no net change in concentrations takes place. (c) Product concentrations are higher than what they are at equilibrium, and reactant concentrations are lower than at equilibrium; $Q > K$, and products form more reactants as the reaction runs in reverse.

TABLE 14.3 Comparison of Q and K Values

Value of Q	What It Means
$Q < K$	Reaction as written proceeds in forward direction ($\rightarrow$)
$Q = K$	Reaction is at equilibrium ($\rightleftharpoons$)
$Q > K$	Reaction as written proceeds in reverse direction ($\leftarrow$)

reaction mixture occurs over time. The relative values of Q and K and their consequences are summarized in **Table 14.3**.

CONCEPT **TEST**

In which of the three zones in Figure 14.4 do the initial reaction conditions of experiments 3 and 4 from Table 14.1 fall?

SAMPLE EXERCISE 14.8 Using Q and K Values to Predict the Direction of a Reaction **LO5**

At 2300 K, $K_c = 1.5 \times 10^{-3}$ for the following reaction:

$$N_2(g) + O_2(g) \rightleftharpoons 2\,NO(g)$$

At the instant when a reaction vessel at 2300 K contains 0.50 M N_2, 0.25 M O_2, and 0.0042 M NO, is the reaction mixture at equilibrium? If not, in which direction will the reaction proceed to reach equilibrium?

Collect and Organize We are asked whether a reaction mixture is at equilibrium. We are given the value of K and the concentrations of reactants and product.

Analyze The mass action expression for this reaction based on concentrations is

$$Q_c = \frac{[NO]^2}{[N_2][O_2]}$$

Inserting the given concentration values into this expression and calculating the value of Q_c will allow us to determine whether the reaction is at equilibrium ($Q_c = K_c$), will proceed in the forward direction ($Q_c < K_c$), or will proceed in the reverse direction ($Q_c > K_c$). To estimate our answer, we note that the given value of [NO] is about 10^{-2} times the concentrations of the reactants, and the [NO] term is squared in the mass action expression. These two factors together make the value of the numerator about 10^{-4} times that of the denominator. Therefore, the value of Q_c should be *less* than the value of K_c.

Solve

$$Q_c = \frac{(0.0042)^2}{(0.50)(0.25)} = 1.4 \times 10^{-4}$$

Because $Q_c < K_c$, the reaction mixture is *not* at equilibrium. It will proceed in the forward direction (to the right) to reach equilibrium.

Think About It Our estimate that Q_c would be less than K_c was correct. The value of K_c is small, but the value of Q_c is even smaller.

 **Practice Exercise** $K_c = 4.7$ for the following reaction at 373 K:

$$2\,NO_2(g) \rightleftharpoons N_2O_4(g)$$

Is a mixture of the two gases in which $[NO_2] = 0.025$ M and $[N_2O_4] = 0.0014$ M in chemical equilibrium? If not, in which direction does the reaction proceed to achieve equilibrium?

homogeneous equilibria equilibria that involve reactants and products in the same phase.

heterogeneous equilibria equilibria that involve reactants and products in more than one phase.

14.6 Heterogeneous Equilibria

So far in Chapter 14 we have focused on reactions in the gas phase. However, the principles of chemical equilibrium also apply to reactions that contain solids and liquids, particularly reactions in solution. Equilibria in which products and reactants are all in the same phase are called **homogeneous equilibria**. Equilibria in which reactants and products are in different phases are **heterogeneous equilibria**.

Let's consider a heterogeneous reaction that plays a key role in preventing SO_2 produced by coal-burning power plants from escaping into the atmosphere. The reaction occurs when tiny particles of solid calcium oxide are sprayed into exhaust gases that contain SO_2 gas:

$$CaO(s) + SO_2(g) \rightleftharpoons CaSO_3(s)$$

The large quantities of lime (CaO) needed for this reaction and for many other industrial and agricultural uses come from heating pulverized limestone, which is mostly $CaCO_3$, in kilns (**Figure 14.5**) operated at temperatures near 900°C. At these temperatures, $CaCO_3$ decomposes into solid lime and CO_2 gas in another reversible two-phase reaction:

$$CaCO_3(s) \rightleftharpoons CaO(s) + CO_2(g)$$

We might be tempted to write the concentration-based equilibrium constant expression for this reaction as follows:

$$K_c = \frac{[CaO][CO_2]}{[CaCO_3]}$$

This expression contains concentration terms for two solids: CaO and $CaCO_3$. However, the terms on the right side of an equilibrium concentration expression are reserved for reactants and products whose values change during a chemical reaction. The concentrations of the solids do not change: as long as these solids are present in the reaction mixture, their concentrations are defined by their densities, which do not change during the reaction. The constant concentration values of the solids are built into the value of K_c, so there is no need of terms for them in the right side of the equilibrium constant expression. This leaves us with

$$K_c = [CO_2]$$

This expression means that, if some CaO and $CaCO_3$ are present, the concentration of CO_2 gas does not vary at a given temperature, as shown in **Figure 14.6**. Instead, the concentration of CO_2 is the same as the value of K_c at that temperature.

The concept of constant concentration also applies to pure liquids involved in reversible chemical reactions. Assuming the liquid is present, its "concentration" is considered constant during the reaction and does not appear in the equilibrium

(a)

(b)

FIGURE 14.5 (a) For centuries, primitive kilns were used to decompose limestone ($CaCO_3$) into lime (CaO) and CO_2. These kilns were charged with layers of fuel (originally wood, later coal) and crushed limestone, and the fuel was ignited. Farmers use lime to "sweeten" (neutralize the acidity of) soil. It is also used to make mortar and cement. (b) A modern plant for converting limestone into lime.

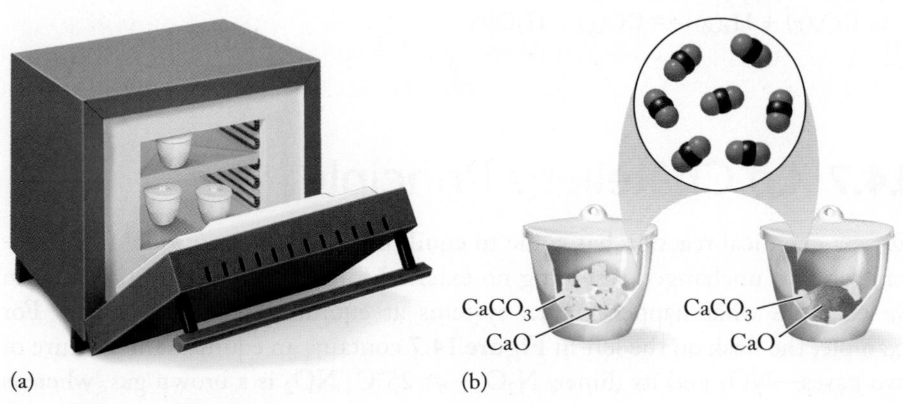

FIGURE 14.6 Two views of the thermal decomposition of $CaCO_3$ to CaO and CO_2. (a) Muffle furnace for heating crucibles containing $CaCO_3$ samples. (b) After heating and decomposition, the two crucibles now contain both $CaCO_3$ and CaO. Different crucibles with different amounts of $CaCO_3$ and CaO still have the same equilibrium concentration of CO_2 gas.

constant expression. Similarly, most K expressions for reactions in aqueous solutions do not include a term for $[H_2O]$, even when water is a reactant or product, because its concentration is assumed to not change significantly. In summary, we follow the rules we learned earlier when writing equilibrium constant expressions for heterogeneous equilibria, with the additional rule that pure solids and liquids do not appear in the expression.

SAMPLE EXERCISE 14.9 Writing Equilibrium Constant Expressions for Heterogeneous Equilibria **LO1**

Write K_c expressions for the following reactions:

a. $CaO(s) + SO_2(g) \rightleftharpoons CaSO_3(s)$
b. $CO_2(g) + H_2O(\ell) \rightleftharpoons H_2CO_3(aq)$

Collect, Organize, and Analyze We are asked to write equilibrium constant expressions for equilibria involving reactants and products in more than one phase. We do not include terms for pure liquids and pure solids, nor do we include a concentration term for H_2O in reactions occurring in aqueous solutions because the value of $[H_2O]$ is assumed not to change significantly during the reaction. The first equilibrium involves two solids: CaO and $CaSO_3$. The second involves liquid H_2O.

Solve

a. An expression with terms for all reactants and products is

$$K_c = \frac{[CaSO_3]}{[CaO][SO_2]}$$

After removing the terms representing solids, we have

$$K_c = \frac{1}{[SO_2]}$$

b. In the equilibrium constant expression for reaction b, $[H_2O]$ is nearly constant and is not included, leaving

$$K_c = \frac{[H_2CO_3]}{[CO_2]}$$

Think About It Writing these equilibrium constant expressions is like writing those for homogeneous reactions, except we leave out terms for pure liquids and solids. When CO_2 dissolves in water, the water is, technically, no longer a *pure* liquid. However, its decrease in "concentration" is much too small to be considered a variable, so there is no $[H_2O]$ term in the K_c expression in part (b).

Practice Exercise Write K_p expressions for the following reactions:

a. $C(s) + CO_2(g) \rightleftharpoons 2\ CO(g)$
b. $CO_2(g) + H_2(g) \rightleftharpoons CO(g) + H_2O(\ell)$

14.7 Le Châtelier's Principle

Once a chemical reaction has come to equilibrium, the composition of the system remains unchanged assuming no external forces perturb it. In this section we examine what happens when systems at equilibrium are perturbed. For example, the flask on the left in **Figure 14.7** contains an equilibrium mixture of two gases—NO_2 and its dimer, N_2O_4—at 25°C. NO_2 is a brown gas, whereas

(a) (b)

FIGURE 14.7 A flask contains an equilibrium reaction mixture of brown $NO_2(g)$ and colorless $N_2O_4(g)$ at (a) 25°C and (b) 0°C. A liquid forms in the bottom of the flask in the ice bath and the pressure inside drops as do the concentrations of both gases. We explore why these changes happen later in this section.

N_2O_4 is colorless. The corresponding dimerization reaction and its equilibrium constant are

$$2\ NO_2(g) \rightleftharpoons N_2O_4(g) \qquad K_p = 3.1$$

The flask on the right in Figure 14.7 initially contained the same mixture of NO_2 and N_2O_4 as the one on the left, but was then placed in an ice bath at 0°C. Why did the temperature change cause a color change in the flask on the right? What other factors can perturb an equilibrium, and how do we explain any changes in the system?

CHEMTOUR
Le Châtelier's Principle

One of the first scientists to study and then successfully predict how chemical equilibria respond to such perturbations was French chemist Henri Louis Le Châtelier (1850–1936). He articulated **Le Châtelier's principle**, which states that, if a system at equilibrium is perturbed (or subjected to an external *stress*), the position of the equilibrium shifts in either the forward or reverse direction as a response to reduce that stress. We now consider the effects of several stresses, including changes in concentration, pressure, volume, and temperature.

Effects of Adding or Removing Reactants or Products

When a reactant or product is added or removed, a system at chemical equilibrium is perturbed. Following Le Châtelier's principle, the system responds in such a way as to restore equilibrium. Through the years, chemists have used Le Châtelier's principle to increase the yields of chemical reactions that would otherwise have provided very little of a desired product.

To explore how industrial chemists exploit Le Châtelier's principle, let's look again at the water–gas shift reaction for making hydrogen:

$$CO(g) + H_2O(g) \rightleftharpoons CO_2(g) + H_2(g) \qquad (14.2)$$

To perturb the equilibrium and shift the system toward the production of more H_2 (**Figure 14.8**), chemists pass the reaction mixture through a gas scrubber

Le Châtelier's principle the principle that a system at equilibrium responds to a stress in such a way that it relieves that stress.

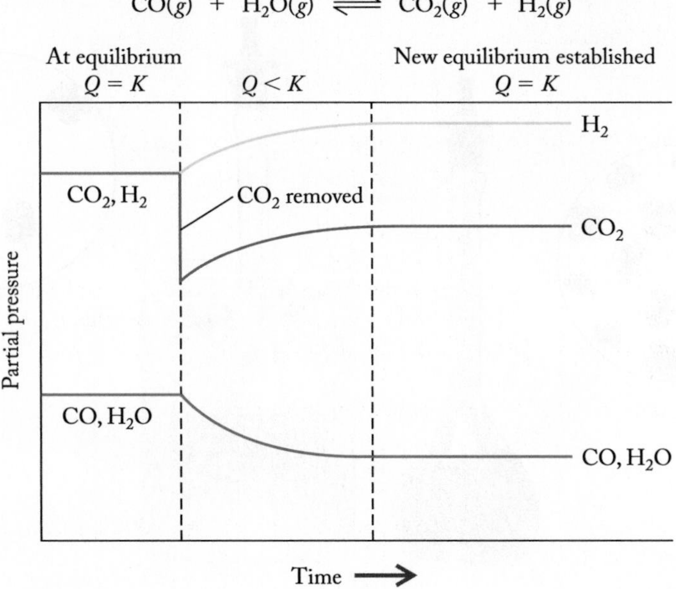

FIGURE 14.8 Removing CO_2 from an equilibrium mixture of the water–gas shift reaction results in fewer collisions between CO_2 and H_2 (slows the rate of the reverse reaction), while the forward reaction continues at the same rate and generates additional CO_2 and H_2 until the mixture establishes a new equilibrium.

containing a concentrated aqueous solution of K_2CO_3. Doing this removes CO_2 from the gaseous mixture because of the following reaction:

$$CO_2(g) + H_2O(\ell) + K_2CO_3(aq) \rightleftharpoons 2\ KHCO_3(s)$$

Removing CO_2 means fewer molecules of it are available to collide with molecules of H_2 in the reverse reaction in Equation 14.2. As a result, the rate of the reverse reaction becomes slower than the rate of the forward reaction, and the system is no longer in equilibrium. The same conclusion is reached by calculating the value of the reaction quotient, Q_p:

$$Q_p = \frac{(P_{H_2})(P_{CO_2})}{(P_{CO})(P_{H_2O})}$$

after much of the CO_2 has been removed. The depleted term in the numerator results in a Q_p value that is less than K_p, which means that the reaction proceeds in the forward direction to restore equilibrium. Chemists say the reaction *shifts to the right*, making more CO_2 to replace some of what was removed in the reaction with K_2CO_3, and in the process more H_2 gas is made.

Another way to shift a reaction mixture at equilibrium is to add more reactant or product. If the goal is to form more products, then adding more reactants increases their concentration in the reaction mixture, which increases the rate of the forward reaction. Some of the added reactants are converted into additional products. This approach works even when only one of multiple reactants is added, as long as there are sufficient quantities of all the other reactants. This makes sense mathematically because adding one reactant increases the value of one of the terms in the denominator of the mass action expression. Therefore, Q is reduced to a value less than K and the reaction proceeds in the forward reaction, forming more products until equilibrium is restored.

SAMPLE EXERCISE 14.10 Adding or Removing Reactants or Products to Stress an Equilibrium **LO6**

Suggest three ways the production of ammonia via the reaction

$$N_2(g) + 3\ H_2(g) \rightleftharpoons 2\ NH_3(g)$$

can be increased without changing the reaction temperature.

Collect and Organize We are given the balanced chemical equation of a reversible reaction. We are asked to suggest three ways to increase its yield—that is, to shift its equilibrium to the right to form more product.

Analyze The mass action expression for this reaction is

$$Q_p = \frac{(P_{NH_3})^2}{(P_{H_2})^3(P_{N_2})}$$

Changes in the reaction system that reduce the value of Q_p, so that $Q_p < K_p$, will result in the formation of more NH_3.

Solve Interventions that (1) increase the partial pressure of N_2, (2) increase the partial pressure of H_2, or (3) remove NH_3 from the system will all shift the equilibrium to the right and increase the production of ammonia.

Think About It The industrial synthesis of ammonia relies on shifting the equilibrium to the right by (1) running the reaction at high partial pressures of the reactants and (2) removing the product NH_3 by passing the reaction mixture through chilled condensers. Chilling the mixture removes ammonia because it condenses at a higher temperature than N_2 or H_2, reducing the value of Q_p and shifting the equilibrium toward producing more ammonia.

Practice Exercise Describe the changes that occur in a gas-phase equilibrium according to the reaction

$$2\ H_2S(g) + 3\ O_2(g) \rightleftharpoons 2\ SO_2(g) + 2\ H_2O(g)$$

if (a) the reaction mixture is cooled, and water vapor condenses; (b) SO_2 gas dissolves in liquid water as it condenses; or (c) more O_2 is added.

Effects of Pressure and Volume Changes

A reaction involving gaseous reactants or products may be perturbed by altering the volume of the system, thereby changing the partial pressures of the reactants and products in accordance with Boyle's law. To see how, let's revisit the equilibrium between NO_2 and its dimer:

$$2\ NO_2(g) \rightleftharpoons N_2O_4(g)$$

Suppose a syringe is filled with an equilibrium mixture of the two gases, as shown in **Figure 14.9a**. The plunger on the syringe is then depressed until the volume of the reaction mixture is half what it was initially, as shown in **Figure 14.9b**. At first the concentrations and partial pressures of the gases both double because the same number of molecules of each gas have been compressed into half their original volume. However, the composition of the mixture rapidly changes and some of the increase in brown color intensity fades, as shown in **Figure 14.9c**. This change happens because doubling the partial pressures of both gases perturbed

CONNECTION Boyle's law (Section 5.3) states that the volume of a given amount of a gas is inversely proportional to its pressure when kept at a constant temperature.

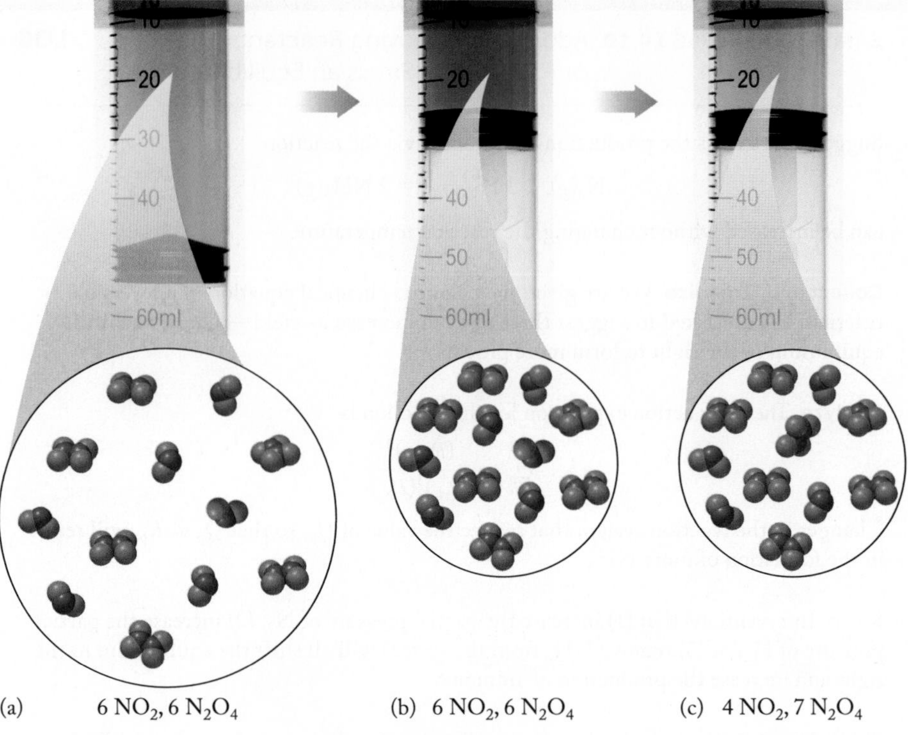

FIGURE 14.9 Changing pressure affects equilibrium in a gas-phase reaction. (a) A gastight syringe contains an equilibrium reaction mixture of brown NO_2 and colorless N_2O_4. (b) The plunger is rapidly pushed in, decreasing the sample volume by half. The color is a darker brown than in (a) because the concentration of NO_2 has doubled. (c) Some of the darker brown color in (b) quickly fades because brown NO_2 rapidly forms colorless N_2O_4, restoring chemical equilibrium.

the equilibrium of the reaction. To understand why, let's assume the initial equilibrium partial pressures of NO_2 and N_2O_4 are x and y, respectively. The value of K_p, therefore, is

$$K_p = \frac{(P_{N_2O_4})}{(P_{NO_2})^2} = \frac{y}{x^2}$$

When the volume of the gases is halved, then their partial pressures double to $P_{NO_2} = 2x$ and $P_{N_2O_4} = 2y$. Inserting these pressures into the mass action expression gives us a reaction quotient that is half the value of K_p:

$$Q_p = \frac{2y}{(2x)^2} = \frac{2y}{4x^2} = \tfrac{1}{2}K_p$$

Because $Q_p < K_p$, the reaction proceeds in the forward direction, consuming reactants and forming products. The deeper brown color of the gas mixture in Figure 14.9b fades a little in Figure 14.9c, confirming that some of the brown NO_2 has dimerized, forming colorless N_2O_4.

Converting NO_2 into N_2O_4 reduces the total number of moles of gas in the reaction mixture because 2 moles of NO_2 are consumed for every 1 mole of N_2O_4 produced. Fewer moles of gas results in fewer collisions with the inner walls of the syringe, which means reduced pressure. Whenever a reaction mixture at equilibrium is compressed—and the partial pressures of its gas-phase reactants and products increase—the equilibrium shifts toward the side of the reaction equation with fewer moles of gases. In doing so, the reaction mixture relieves some of the stress induced by the increase in pressure resulting from compression. Conversely, increasing the volume of a system at equilibrium at constant temperature results in fewer collisions, lowers the partial pressures of all the gaseous reactants and products, and shifts the equilibrium toward the side of the reaction equation with more moles of gases (**Figure 14.10**).

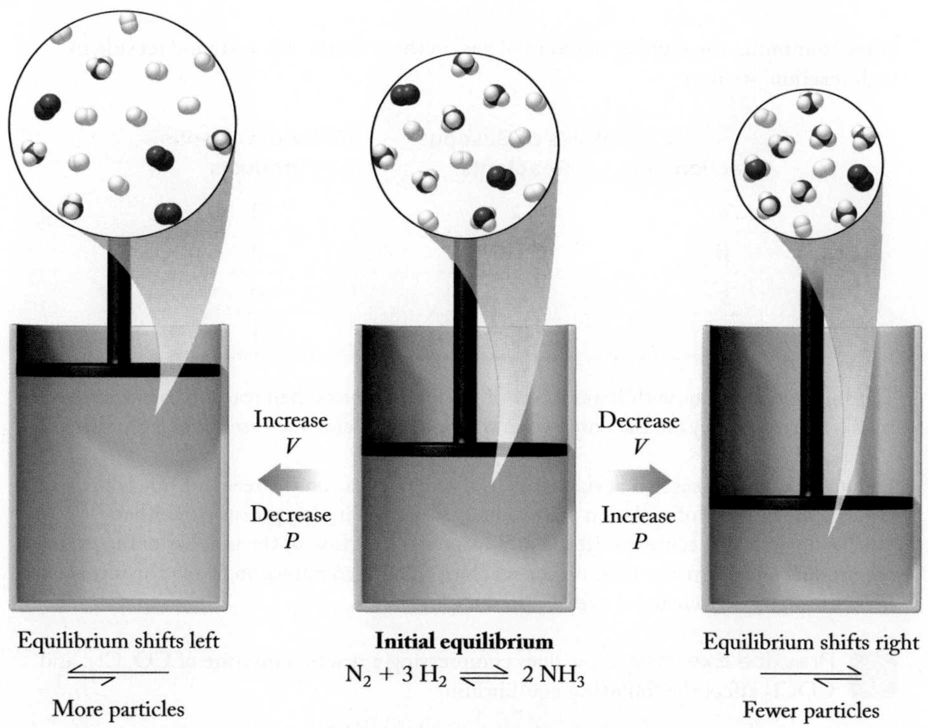

FIGURE 14.10 The effect of compressing and expanding an equilibrium mixture of the gases in the reaction $N_2(g) + 3\,H_2(g) \rightleftharpoons 2\,NH_3(g)$. An increase in volume (a decrease in pressure) causes the reaction to shift to the left: the side with the greater number of particles. A decrease in volume (an increase in pressure) causes the reaction to shift to the right: the side with the smaller number of particles.

CONCEPT TEST

Changing the overall volume of the reaction of the water–gas shift reaction at equilibrium:

$$CO(g) + H_2O(g) \rightleftharpoons CO_2(g) + H_2(g)$$

does not shift the equilibrium. Why?

SAMPLE EXERCISE 14.11 Assessing the Impact of Compressing or Expanding the Volume of a Gas-Phase Reaction Mixture at Equilibrium LO6

In which of the following equilibria would compressing an equilibrium reaction mixture promote the formation of more product(s)?

a. $N_2(g) + O_2(g) \rightleftharpoons 2\,NO(g)$
b. $2\,NO(g) + O_2(g) \rightleftharpoons 2\,NO_2(g)$
c. $H_2O(\ell) + CO_2(g) \rightleftharpoons H_2CO_3(aq)$
d. $CaCO_3(s) \rightleftharpoons CaCO(s) + CO_2(g)$

Collect, Organize, and Analyze We are asked to identify the reactions for which higher pressure causes an increase in product formation. We know that compression shifts a chemical equilibrium involving gases toward the side of the reaction with fewer moles of gas.

Solve Summing the number of moles of gas on the reactant side and product side in each reaction, we have:

Reaction	Moles of Gaseous Reactants	Moles of Gaseous Products
a	2	2
b	3	2
c	1	0
d	0	1

The only two reactions with fewer moles of gaseous products than reactants are reactions b and c. Therefore, they are the only two in which compression increases product formation.

Think About It In reaction a, the number of moles of gas on the reactant side is the same as the number of moles on the product side, so changing pressure in either direction would not cause a shift in equilibrium. In reaction d, the number of moles on the product side is greater than on the reactant side, so compression favors the reverse of the reaction as written and decreases product formation.

Practice Exercise How does compressing a reaction mixture of CO, Cl_2, and $COCl_2$ affect the following equilibrium?

$$CO(g) + Cl_2(g) \rightleftharpoons COCl_2(g)$$

Effect of Temperature Changes

Let's return now to Figure 14.7 and reconsider what happens when a volumetric flask containing an equilibrium mixture of NO_2 and N_2O_4 at 25°C is placed into an ice bath. The color of the gas mixture in Figure 14.7b becomes much lighter, indicating that the concentration of NO_2 has decreased. Why? For one thing, the boiling point of N_2O_4 is 21°C. Therefore, lowering the temperature from 25°C to 0°C removed most of the dimer from the gas phase, creating a pool of colorless N_2O_4 liquid in the bottom of the flask. The resulting decrease in the concentration and partial pressure of N_2O_4 makes $Q_p < K_p$, so the reaction shifts toward the formation of additional N_2O_4. This shift drives down the concentration of NO_2 and makes the gas phase in Figure 14.7b a lighter color.

CONNECTION According to Amontons's law (Section 5.3), the pressure of a quantity of an ideal gas at constant volume is proportional to its absolute temperature.

In Section 14.1 we noted how important the production of ammonia is for the world's food supply. What is the effect of decreasing the temperature upon the synthesis of ammonia, which is an exothermic reaction in the forward direction?

$$N_2(g) + 3\,H_2(g) \rightleftharpoons 2\,NH_3(g) \qquad \Delta H^\circ_{rxn} = -92.2 \text{ kJ/mol}$$

If we think of energy as a product of an exothermic reaction, then lowering the temperature of the reaction mixture favors the forward reaction (because a product is effectively removed), whereas increasing the temperature favors the reverse reaction (because the amount of a product is increased). The situation is reversed for endothermic reactions, in which energy is treated as a reactant, so raising the temperature would favor the forward reaction, whereas lowering the temperature would favor the reverse reaction.

There is an important difference between applying Le Châtelier's principle to explain shifts in equilibrium position when changing concentrations or partial pressures of reactants or products versus applying Le Châtelier's principle to explain temperature changes: *changes in temperature change the value of K.*

Increasing temperature reduces the yield of ammonia synthesis because increasing temperature reduces the value of *K*.

In general, the value of *K* decreases as temperature increases for exothermic reactions and increases as temperature increases for endothermic reactions. We revisit the influence of temperature on *K* values in Chapter 17, but for now this general analysis enables us to predict the direction of shifts in equilibria with changing temperature.

SAMPLE EXERCISE 14.12 Predicting Changes in Equilibrium with Changing Temperature — LO6

The color of cobalt(II) chloride dissolved in dilute hydrochloric acid depends on temperature, as shown in **Figure 14.11**. The solution is pink at 0°C and royal blue at 75°C. Is the reaction that produces the pink-to-blue color change exothermic or endothermic?

Collect, Organize, and Analyze We are asked to determine whether the reaction in Figure 14.11 is exothermic or endothermic. If it is exothermic, then increasing its temperature should shift the reaction toward the formation of the pink reactant. If the reaction is endothermic, then increasing temperature should shift the reaction toward the formation of the blue product.

Solve The blue product is favored at higher temperatures, so the reaction as written must be endothermic.

Think About It The reaction mixture in this exercise is magenta at room temperature, indicating that both the pink and blue forms are present. As a result, the value of *K* at room temperature must be close to 1, but the concentration of $Cl^-(aq)$ is also an important factor because it is raised to the fourth power in the equilibrium constant expression.

 **Practice Exercise** Does the value of the equilibrium constant of the reaction

$$N_2(g) + O_2(g) \rightleftharpoons 2\,NO(g) \qquad \Delta H^\circ_{rxn} = 180.6 \text{ kJ/mol}$$

increase or decrease with increasing temperature?

Temperature = 0°C Temperature = 75°C

$$\underbrace{Co(H_2O)_6^{\,2+}(aq)}_{\text{Pink}} + 4\,Cl^-(aq) \rightleftharpoons \underbrace{CoCl_4^{\,2-}(aq)}_{\text{Royal blue}} + 6\,H_2O(\ell)$$

FIGURE 14.11 Two cobalt(II) species, one pink and one blue, are in equilibrium in aqueous hydrochloric acid solution. The equilibrium shifts in favor of the blue species as the temperature increases.

TABLE 14.4 Response of the Reaction $2\ A(g) \rightleftharpoons B(g)$ to Different Kinds of Stress

Kind of Stress	How System Responds	Direction of Shift
Add A	Consume A	To the right
Remove A	Produce A	To the left
Remove B	Produce B	To the right
Add B	Consume B	To the left
Compress the reaction mixture	Reduce moles of gas	To the right
Increase the volume of the reaction mixture	Increase moles of gas	To the left

CONNECTION In Section 13.6 we learned that a catalyst simultaneously increases the rate of a reaction in both the forward and reverse directions, when we discussed the effect of temperature on the rates of reactions occurring in a catalytic converter.

Table 14.4 summarizes how a generic gas-phase reversible reaction in which 2 moles of reactants form 1 mole of product responds to various stresses.

Catalysts and Equilibrium

The industrial production of ammonia was developed by the German chemists Fritz Haber (1868–1934) and Carl Bosch (1874–1940) in the early 20th century and is still widely referred to as the Haber–Bosch process. This process is commercially feasible because of the use of iron-based catalysts. As discussed in Section 13.6, a catalyst increases the rate of a chemical reaction by lowering its activation energy. The synthesis of ammonia involves both an equilibrium and a catalyst, so if a catalyst increases the rate of a reaction, does that catalyst affect the equilibrium constant of the reaction?

To answer this question, look at the energy profiles of the catalyzed and uncatalyzed reaction in **Figure 14.12**. The catalyst increases the rate of the reaction by decreasing the height of the energy barrier. However, the barrier height is reduced by the same amount whether the reaction proceeds in the forward direction or in reverse. As a result, the increase in reaction rate produced by the catalyst is the same in both directions. Therefore, a catalyst has *no* effect on the equilibrium constant of a reaction or on the composition of an equilibrium reaction mixture. A catalyst does, however, shorten the time needed for a system to reach equilibrium.

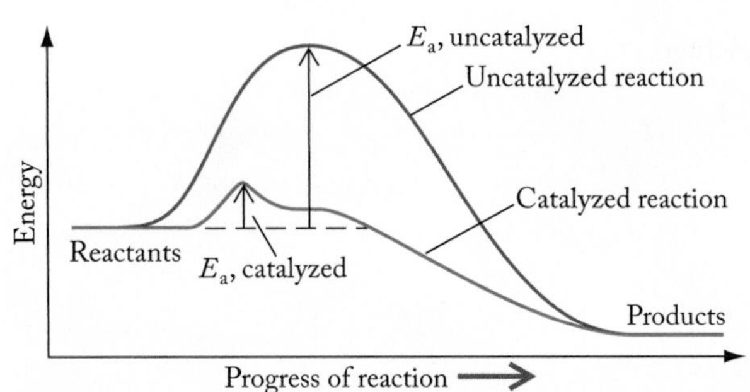

FIGURE 14.12 The effect of a catalyst on a reaction. A catalyst lowers the activation energy barrier, and as a result the rate of the reaction increases. Both the forward reaction and the reverse reaction occur more rapidly, however, so the position of equilibrium (i.e., the value of K) does not change. The system comes to equilibrium more rapidly, but the relative amounts of product and reactant present at equilibrium do not change.

14.8 Calculations Based on K

Reference books and the tables in Appendix 5 of this book contain lists of equilibrium constants for chemical reactions. We can use these values in several kinds of calculations, including those in which:

1. We want to determine whether a reaction mixture has reached equilibrium (Sample Exercise 14.8).
2. We know the value of K and the starting concentrations or partial pressures of reactants or products, and we want to calculate their equilibrium concentrations or pressures.

In this section we focus on the second type of calculation and introduce a useful way of handling such problems: a table of reactant and product concentrations (or partial pressures) called a *RICE table*. The acronym RICE means that the table starts with the balanced chemical equation describing the **R**eaction followed by rows that contain **I**nitial concentration values, **C**hanges in those initial values as the reaction proceeds toward equilibrium, and **E**quilibrium values.

In our first example, we calculate how much hydrogen iodide forms from the reaction of hydrogen gas and iodine gas at 445°C. At this temperature, the K_p value of the reaction is 50.2:

$$H_2(g) + I_2(g) \rightleftharpoons 2\,HI(g)$$

The initial partial pressures are $P_{H_2} = 1.00$ atm and $P_{I_2} = 1.03$ atm, and no HI is present.

We start by writing the given (initial) information in our RICE table:

Reaction (R)	$H_2(g)$	+	$I_2(g)$	$\rightleftharpoons$	$2\,HI(g)$
	P_{H_2} (atm)		P_{I_2} (atm)		P_{HI} (atm)
Initial (I)	1.00		1.03		0
Change (C)					
Equilibrium (E)					

We know the reaction will proceed in the forward direction because there is no product initially present, so $Q_p = 0 < K_p$.

We need to use algebra to fill in rows C and E. We don't know how much H_2 or I_2 will be consumed or how much HI will be made. We can define the change in partial pressure of H_2 as $-x$ because H_2 is consumed during the reaction. Because the mole ratio of H_2 to I_2 in the reaction is 1:1, the change in P_{I_2} is also $-x$. Two moles of HI are produced from each mole of H_2 and I_2, so the change in P_{HI} is $+2x$. Inserting these values in the C row, we have

CHEMTOUR
Solving Equilibrium Problems

Reaction (R)	$H_2(g)$	+	$I_2(g)$	$\rightleftharpoons$	$2\,HI(g)$
	P_{H_2} (atm)		P_{I_2} (atm)		P_{HI} (atm)
Initial (I)	1.00		1.03		0
Change (C)	$-x$		$-x$		$+2x$
Equilibrium (E)					

Combining the values in the I and C rows, we obtain the three partial pressures at equilibrium:

Reaction (R)	$H_2(g)$	+	$I_2(g)$	$\rightleftharpoons$	$2\,HI(g)$
	P_{H_2} (atm)		P_{I_2} (atm)		P_{HI} (atm)
Initial (I)	1.00		1.03		0
Change (C)	$-x$		$-x$		$+2x$
Equilibrium (E)	$1.00 - x$		$1.03 - x$		$2x$

The next step is to substitute the terms from the E row into the K_p expression for the reaction:

$$K_p = \frac{(P_{HI})^2}{(P_{H_2})(P_{I_2})} = \frac{(2x)^2}{(1.00 - x)(1.03 - x)}$$

Expanding the terms in the numerator and denominator gives

$$K_p = \frac{4x^2}{1.03 - 2.03\,x + x^2} = 50.2$$

Cross multiplying, we get

$$1.03 \times 50.2 - (2.03 \times 50.2)x + 50.2x^2 = 4x^2$$

Combining the x^2 terms and rearranging, we have

$$46.2x^2 - 101.9x + 51.7 = 0$$

This equation fits the general form of a quadratic equation:

$$ax^2 + bx + c = 0$$

We can solve for x with a scientific calculator or by using the quadratic formula as described in detail in Appendix 1 and illustrated here in the following steps:

1. Insert the three coefficients $a = 46.2$, $b = -101.9$, and $c = 51.7$ into the quadratic formula:

$$x = \frac{-b \pm \sqrt{b^2 - 4ac}}{2a} = \frac{-(-101.9) \pm \sqrt{(-101.9)^2 - 4 \times 46.2 \times 51.7}}{2 \times 46.2}$$

2. Do the math within each of the terms:

$$x = \frac{101.9 \pm \sqrt{10383.61 - 9554.16}}{92.4} = \frac{101.9 \pm \sqrt{829.45}}{92.4}$$

3. Take the square root, combine the $\pm$ terms, and simplify:

$$x = \frac{101.9 \pm 28.8}{92.4} = \frac{130.7}{92.4} = 1.415 \quad \text{and} \quad \frac{73.1}{92.4} = 0.791$$

The $\pm$ sign yields two solutions for x: 0.791 atm and 1.415 atm. We focus on 0.791 atm because it is the only one that is physically possible. Recall that the algebraic terms for the equilibrium concentrations of H_2 and I_2 gas are $(1.00 - x)$ and $(1.03 - x)$. Inserting $x = 1.415$ atm into either of these produces an equilibrium partial pressure with a negative value, which is impossible. Therefore, we use $x = 0.791$ atm to calculate the equilibrium partial pressures:

$$P_{H_2} = 1.00 - x = 1.00 - 0.791 = 0.21 \text{ atm}$$

$$P_{I_2} = 1.03 - x = 1.03 - 0.791 = 0.24 \text{ atm}$$

$$P_{HI} = 2x = 2(0.791) = 1.582 = 1.58 \text{ atm}$$

The value of x makes sense because it produces pressures of H_2 and I_2 that are less than their initial values but still positive. The sum of the partial pressures of the components in the system is $(0.21 + 0.24 + 1.58) = 2.03$ atm, which is the same as the starting pressure. This result is expected, because we have 2 moles of gas on the reactant side and 2 moles of gas on the product side, so we do not expect the pressure to change as the reaction proceeds. As a final check, we can use the calculated partial pressures in the expression for K_p and calculate its value. When we do so, we get $K_p = (1.58)^2/[(0.21)(0.24)] = 50$, which is not significantly different from the value given of 50.2.

SAMPLE EXERCISE 14.13 Calculating an Equilibrium Partial Pressure I **LO7**

Most of the H_2 used in the Haber–Bosch process for making ammonia is produced by the steam–gas reforming reaction followed by the water–gas shift reaction:

$$CO(g) + H_2O(g) \rightleftharpoons CO_2(g) + H_2(g)$$

If a reaction vessel at 675 K is filled with an equimolar mixture of CO and steam, such that the partial pressure of each gas is 2.00 atm, what is the partial pressure of H_2 at equilibrium if $K_p = 11.8$ at 675 K?

Collect, Organize, and Analyze We are asked to find the partial pressure of a gaseous product in an equilibrium mixture, given the initial partial pressures of reactants and the value of K_p. The system initially contains no product, which means that the reaction quotient Q_p is equal to zero. A K_p value of 11.8 means that reactants should be converted into products, but there should be significant partial pressures of both reactants and products at equilibrium. The coefficients of the species in the chemical equation describing the reaction are all one, so in our RICE table we let $+x$ be the increase in the partial pressures of H_2 and CO_2 and $-x$ be the decreases in both reactants.

Solve

Reaction (R)	$CO(g)$	+	$H_2O(g)$	$\rightleftharpoons$	$CO_2(g)$	+	$H_2(g)$
	P_{CO} (atm)		P_{H_2O} (atm)		P_{CO_2} (atm)		P_{H_2} (atm)
Initial (I)	2.00		2.00		0.00		0.00
Change (C)	$-x$		$-x$		$+x$		$+x$
Equilibrium (E)	$2.00 - x$		$2.00 - x$		x		x

Inserting these equilibrium terms into the K_p expression:

$$K_p = \frac{(P_{CO_2})(P_{H_2})}{(P_{CO})(P_{H_2O})} = \frac{(x)(x)}{(2.00 - x)(2.00 - x)} = 11.8$$

Note that the algebraic terms for both products are equal to x, and the terms for both reactants are equal to $(2.00 - x)$. These equalities do not happen often, but solving for x is much easier when they do because the two terms can be written this way:

$$\left(\frac{x}{2.00 - x}\right)^2 = 11.8$$

Taking the square root of both sides removes the exponent on the left:

$$\left(\frac{x}{2.00 - x}\right) = \sqrt{11.8} = 3.435$$

Cross multiplying:

$$x = 3.435\,(2.00 - x) = 6.870 - 3.435x$$

Solving for x:

$$(1 + 3.435)x = 6.870$$

$$x = \frac{6.870}{4.435} = 1.55 \text{ atm}$$

This is the equilibrium partial pressure of both products: H_2 and CO_2.

Think About It Our prediction that significant quantities of both reactants and products would be present at equilibrium is confirmed. Let's check the accuracy of our solution by inserting the equilibrium partial pressures of H_2 and CO_2 (1.55 atm) and of steam and CO (2.00 − 1.55 = 0.45 atm) into the equilibrium constant expression:

$$K_p = \frac{(P_{CO_2})(P_{H_2})}{(P_{CO})(P_{H_2O})} = \frac{(1.55\text{ atm})(1.55\text{ atm})}{(0.45\text{ atm})(0.45\text{ atm})} = 11.8$$

This result matches the K_p value we started with, confirming that our calculation is correct.

Practice Exercise The chemical equation for the reaction of chlorine and bromine to produce BrCl is

$$Cl_2(g) + Br_2(g) \rightleftharpoons 2\ BrCl(g)$$

If $K_p = 4.7 \times 10^{-2}$ for the reaction at a given temperature, what is the partial pressure of each reactant and product in a sealed reaction vessel at equilibrium? The initial partial pressures of Cl_2 and Br_2 are both 0.100 atm and no BrCl is present.

SAMPLE EXERCISE 14.14 Calculating an Equilibrium Partial Pressure II — LO7

The K_p value for the reaction $PCl_3(g) + Cl_2(g) \rightarrow PCl_5(g)$ is 24.2 at 250°C. If the initial partial pressures of PCl_3 and Cl_2 are 0.43 and 0.87 atm, respectively, and no PCl_5 is initially present, what are the partial pressures of the three gases when equilibrium is achieved?

Collect and Organize We are asked to determine the equilibrium partial pressures of the three gases from the initial partial pressures of the reactants and the equilibrium constant. The chemical equation describing the reaction is

$$PCl_3(g) + Cl_2(g) \rightleftharpoons PCl_5(g) \qquad K_p = 24.2$$

Analyze The system initially contains no product, so $Q < K_p$ and the reaction will proceed in the forward direction as written. We can construct a RICE table from the information given. If $+x$ is the increase in the partial pressure of the product formed, then, given the 1:1:1 stoichiometry of the reactants and product, the decreases in both reactants will be $-x$.

Solve Setting up the RICE table in which the equilibrium (E) terms are the sum of the (I) + (C) terms:

Reaction (R)	$PCl_3(g)$	+	$Cl_2(g)$	$\rightleftharpoons$	$PCl_5(g)$
	P_{PCl_3} (atm)		P_{Cl_2} (atm)		P_{PCl_5} (atm)
Initial (I)	0.43		0.87		0
Change (C)	$-x$		$-x$		$+x$
Equilibrium (E)	$0.43 - x$		$0.87 - x$		x

Inserting the terms from row (E) into the K_p expression for the reaction:

$$K_p = \frac{(P_{PCl_5})}{(P_{PCl_3})(P_{Cl_2})} = \frac{x}{(0.43 - x)(0.87 - x)} = 24.2$$

There is only one partial pressure term in the numerator and the two terms in the denominator do not match, so solving for x will not be as simple as in Sample Exercise 14.13. Here we need to expand the terms in the denominator:

$$K_p = \frac{x}{0.3741 - 1.30x + x^2} = 24.2$$

and cross multiply:

$$0.3741 \times 24.2 - (1.30 \times 24.2)x + 24.2x^2 = x$$

Combining the x terms and rearranging, we have

$$24.2x^2 - 32.46x + 9.05322 = 0$$

Using a scientific calculator or the quadratic formula to solve for x yields two values: 0.3955 and 0.9458. Only 0.3955 gives positive partial pressure values for all three gases, which are

$$P_{PCl_3} = 0.43 - 0.3955 = 0.03 \text{ atm}$$
$$P_{Cl_2} = 0.87 - 0.3955 = 0.47 \text{ atm}$$
$$P_{PCl_5} = 0.3955 = 0.40 \text{ atm}$$

Think About It The values make sense because the partial pressures of both reactants have decreased and that of the product has increased. When we check our answers by substituting them into the equation for K_p:

$$K_p = \frac{(P_{PCl_5})}{(P_{PCl_3})(P_{Cl_2})} = \frac{0.40}{(0.03)(0.47)} = 28$$

we get a value that is reasonably close to the K_p value used in the calculation (24.2), given that the partial pressure of PCl_3 was rounded off to only one significant figure. To obtain a more useful accuracy check, let's *not* round off the calculated partial pressures to calculate K_p:

$$K_p = \frac{0.3955}{(0.0345)(0.4745)} = 24.2$$

Practice Exercise Consider the reverse reaction in Sample Exercise 14.14 at the same temperature. If the vessel initially contains 1.35 atm PCl_5, what are the partial pressures of the three gases when equilibrium is achieved?

The equilibrium calculations we have done so far have been based on K_p values in the 10^{-2} to 10^2 range. As we noted in Section 14.2, reactions with these K_p values produce reaction mixtures with significant quantities of both reactants and products at equilibrium. When equilibrium constants are very small, however, we can frequently make an approximation at the outset that enables us to simplify our mathematical operations by considering the effect that the size of an equilibrium constant has on the values in our equation. For example, the K_p value for the following reaction at 1500 K is small:

$$N_2(g) + O_2(g) \rightleftharpoons 2\,NO(g) \qquad K_p = 1.0 \times 10^{-5}$$

If the initial partial pressures of N_2 and O_2 are 0.79 and 0.21 atm, respectively, and no NO is present, we can set up a RICE table and calculate the partial pressure of NO at equilibrium as we have done previously:

Reaction (R)	$N_2(g)$	+	$O_2(g)$	$\rightleftharpoons$	$2\,NO(g)$
	P_{N_2} (atm)		P_{O_2} (atm)		P_{NO} (atm)
Initial (I)	0.79		0.21		0
Change (C)	$-x$		$-x$		$+2x$
Equilibrium (E)	$0.79 - x$		$0.21 - x$		$2x$

Inserting the equilibrium partial pressure terms in the K_p expression:

$$K_p = 1.0 \times 10^{-5} = \frac{(P_{NO})^2}{(P_{N_2})(P_{O_2})}$$

$$= \frac{(2x)^2}{(0.79 - x)(0.21 - x)} = \frac{4x^2}{(0.79 - x)(0.21 - x)}$$

Before expanding the equation and solving for x, stop and think about the size of x. The equilibrium constant is very small, which means very little product will form. As a result, the value of x will probably be much smaller than the values of the initial partial pressures, so that $(0.79 - x)$ and $(0.21 - x)$ will be approximately the same as 0.79 and 0.21. If we assume that the $-x$ terms in the denominator are negligible and simply use the initial values of P_{N_2} and P_{O_2} instead, we obtain an equation that is much easier to solve:

$$\frac{4x^2}{(0.79)(0.21)} = 1.0 \times 10^{-5}$$

$$4x^2 = (0.79)(0.21)(1.0 \times 10^{-5}) = 1.659 \times 10^{-6}$$

$$x^2 = 4.148 \times 10^{-7}$$

$$x = 6.4 \times 10^{-4} \text{ atm}$$

This value is in fact the same as that obtained by solving for x without neglecting the $-x$ term, at least to two significant figures, which are all we are allowed in this problem.

In equilibrium calculations, we can ignore the $-x$ or $+x$ component of an equilibrium concentration or partial pressure term if the value of x is less than 5% of the initial value. In this problem the smaller initial partial pressure was 0.21 atm, and the ratio of x/0.21 atm = 6.4×10^{-4} atm/0.21 atm = 0.0030 or 0.3%, which is well less than the 5% guideline.

SAMPLE EXERCISE 14.15 Calculating an Equilibrium Partial Pressure III **LO7**

The value of K_p for the decomposition of phosgene ($COCl_2$) gas:

$$COCl_2(g) \rightleftharpoons CO(g) + Cl_2(g)$$

is 2.2×10^{-10} at 100°C. If a sealed vessel contains only phosgene at a partial pressure of 2.75 atm, what are the partial pressures of the reactant and its decomposition products when the system comes to equilibrium?

Collect and Organize We are asked to find the partial pressures of $COCl_2$ and its decomposition products in a gas-phase equilibrium mixture. We know the initial value of P_{COCl_2} and the value of K_p.

Analyze The value of K_p is very small, so we may be able to assume that the $-x$ component of the P_{COCl_2} term at equilibrium is negligible.

Solve Setting up a RICE table for this reaction system:

Reaction (R)	$COCl_2(g)$	$\rightleftharpoons$	$CO(g)$	+	$Cl_2(g)$
	P_{COCl_2} (atm)		P_{CO} (atm)		P_{Cl_2} (atm)
Initial (I)	2.75		0		0
Change (C)	$-x$		$+x$		$+x$
Equilibrium (E)	$2.75 - x$		x		x

Using the equilibrium values in the mass action expression and then simplifying and solving for x:

$$K_p = \frac{(P_{CO})(P_{Cl_2})}{(P_{COCl_2})} = \frac{(x)(x)}{(2.75 - x)} = 2.2 \times 10^{-10}$$

Making the simplifying assumption that the $-x$ part of the P_{COCl_2} term is negligible:

$$\frac{x^2}{(2.75)} = 2.2 \times 10^{-10}$$

Cross multiplying and taking the square root to solve for x:

$$x^2 = 2.75 \times (2.2 \times 10^{-10}) = 6.05 \times 10^{-10}$$

$$x = 2.5 \times 10^{-5}$$

Therefore, at equilibrium P_{COCl_2} is not significantly less than its initial value (2.75 atm) and $P_{CO} = P_{Cl_2} = 2.5 \times 10^{-5}$ atm.

Think About It The ratio of x to the initial partial pressure of $COCl_2$ is 2.5×10^{-5} atm/2.75 atm $= 9.1 \times 10^{-6}$, which is much less than 5%, so our simplifying assumption was valid. To check on the accuracy of our arithmetic, let's insert the calculated partial pressures into the K_p expression:

$$K_p = \frac{(P_{CO})(P_{Cl_2})}{(P_{COCl_2})} = \frac{(2.5 \times 10^{-5})(2.5 \times 10^{-5})}{(2.75)} = 2.3 \times 10^{-10}$$

This value is very close to the K_p value (2.2×10^{-10}) used in the calculation and supports the accuracy of the results.

 **Practice Exercise** What effect will doubling the partial pressure of phosgene in the sample exercise have on the partial pressures of CO and Cl_2 at equilibrium?

SAMPLE EXERCISE 14.16 Integrating Concepts: Making Nitric Acid

About 15% of the ammonia produced industrially by the Haber–Bosch process is converted into nitric acid by the Ostwald process, named after Wilhelm Ostwald (1853–1932), the German chemist who developed it. In the Ostwald process, ammonia is burned in air at 900°C in the presence of a platinum–rhodium catalyst:

Step 1: $4\,NH_3(g) + 5\,O_2(g) \rightleftharpoons 4\,NO(g) + 6\,H_2O(g)$

Under these conditions, about 90% of the starting ammonia is converted into NO.

a. Air is 21% O_2 by volume. Typical instructions for step 1 specify that the volume of air should be 10 times the volume of ammonia reacted. (i) Using this volume ratio, which is the limiting reactant? (ii) Write the K_p expression for step 1 and explain how the partial pressure of oxygen influences the position of the equilibrium.

b. In the second step of the process, the temperature is lowered, and more air is mixed with the products of step 1.

Step 2: $2\,NO(g) + O_2(g) \rightleftharpoons 2\,NO_2(g)$

$\Delta H^\circ_{rxn} = -114.0$ kJ/mol

Predict what would happen to the yield in step 2 if the temperature were increased instead of lowered.

c. In the third and final step, NO_2 is passed through liquid water to form a solution of nitric acid (HNO_3); nitrogen monoxide gas is also produced. Write the K_c expression for this reaction.

Collect and Organize In part (a), we are provided with the balanced chemical equation describing a reversible gas-phase reaction and the volume ratio of the two reactants. We are asked to identify the limiting reactant, to write the K_p expression for the reaction, and to describe how altering the partial pressure of one of the reactants influences the equilibrium's position. In part (b), we are asked to predict the effect of raising reaction temperature on the yield of an exothermic reaction. In part (c), we need to write the K_c expression for a reaction happening in solution. We know at least some of the reactants and products, but we are not provided with a balanced chemical equation for this final step.

Analyze If we assume that both reactants in part (a) behave as ideal gases, then their volume ratio is the same as their mole ratio. Increasing the partial pressure of a reactant (O_2) shifts the equilibrium to the right (formation of more product). The reaction in part (b) is exothermic ($\Delta H^\circ < 0$), so the value of K_p and the yield of the reaction should decrease as temperature increases. The reaction in step 3 is a redox reaction because the oxidation state of nitrogen is initially +4 in NO_2, but it is +5 in HNO_3 and +2 in NO. Thus, nitrogen is both oxidized and reduced. To write a K_c expression we first need to write a balanced chemical equation, and that requires balancing the loss and gain of electrons by the N atoms in molecules of NO_2 as they form molecules of HNO_3 and NO. To do so, we need to use the procedure developed in Section 4.8.

Solve

a. (i) A volume ratio of 10:1 for air:ammonia translates into a mole ratio of $(10 \times 0.21) = 2.1$ mol O_2/mol NH_3. The stoichiometric ratio is 5 moles of O_2 for every 4 moles of NH_3, or 1.25 mol O_2/mol NH_3. Therefore, the reaction mixture contains excess O_2, making NH_3 the limiting reactant. (ii) The

K_p expression (partial pressures of products over reactants each raised to a power equal to its stoichiometric coefficient) is

$$K_p = \frac{(P_{NO})^4(P_{H_2O})^6}{(P_{NH_3})^4(P_{O_2})^5}$$

The presence of excess O_2 at equilibrium shifts the position of the equilibrium toward the production of more NO.

b. The reaction is exothermic, so energy is a product of the reaction. If the temperature were increased, the reaction would shift in the direction of more reactants, and less NO_2 would be present at equilibrium.

c. The known reactants and products of the final reaction step are

$$NO_2(g) + H_2O(\ell) \rightleftharpoons NO(g) + HNO_3(aq)$$

During the reaction, the oxidation state of nitrogen simultaneously changes by +1 (from +4 in NO_2 to +5 in HNO_3) and by −2 (from +4 in NO_2 to +2 in NO). To balance the redox equation, we will use the guideline developed in Section 4.8. First, we divide the overall reaction into two half-reactions:

$$\text{Oxidation: } H_2O(\ell) + NO_2(g) \rightarrow HNO_3(aq)$$
$$\text{Reduction: } NO_2(g) \rightarrow NO(g) + H_2O(\ell)$$

All of the elements are balanced except for hydrogen and oxygen. We balance the number of oxygen atoms in each half-reaction by adding water, and we balance the hydrogen atoms by adding protons, H^+, because the reaction takes place in acidic solution. To balance the charges, we add an appropriate number of electrons. Before adding the half-reactions, we need to multiply the oxidation half-reaction by 2 so that the number of electrons lost equals the number of electrons gained.

$$\text{Oxidation: } 2[H_2O(\ell) + NO_2(g) \rightarrow HNO_3(aq) + H^+(aq) + e^-]$$
$$\text{Reduction: } 2\,H^+(aq) + 2\,e^- + NO_2(g) \rightarrow NO(g) + H_2O(\ell)$$

$$\cancel{2\,H^+(aq)} + \cancel{2\,e^-} + \cancel{NO_2(g)} + 2\,H_2O(\ell) + 2\,NO_2(g) \rightarrow NO(g) + \cancel{H_2O(\ell)} + 2\,HNO_3(aq) + \cancel{2\,H^+(aq)} + \cancel{2\,e^-}$$

Cancelling out the protons, electrons, and water molecules that appear on both sides of the equation, we arrive at a balanced, overall equation:

$$H_2O(\ell) + 3\,NO_2(g) \rightarrow NO(g) + 2\,HNO_3(aq)$$

A check of the numbers of H and O atoms on both sides of the equation reveals that it is balanced. Because the concentration of H_2O is unlikely to decrease significantly during the reaction, there is no $[H_2O]$ term in the K_c expression:

$$K_c = \frac{[NO][HNO_3]^2}{[NO_2]^3}$$

Think About It Writing a chemical equation describing the conversion of NO_2 into HNO_3 and NO was simplified by recognizing that the reaction is a redox reaction that takes place in acidic solution.

SUMMARY

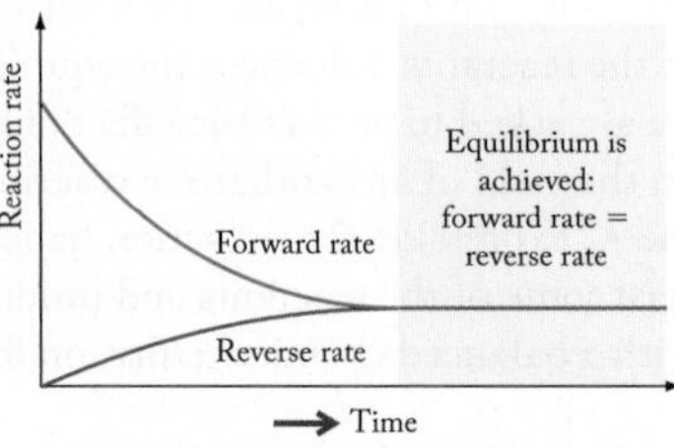

LO1 According to the **law of mass action**, the ratio of the concentrations of the products of a reversible reaction divided by the concentrations of the reactants, with each raised to its stoichiometric coefficient from the balanced equation—a quantity called the **mass action expression**, or the **equilibrium constant expression**—has a characteristic value at a given temperature. The concentrations of pure liquids and solids that do not significantly change during a reaction are omitted from equilibrium constant expressions. (Sections 14.1, 14.2, 14.6)

LO2 The value of the **equilibrium constant, K_c**, of a reversible reaction can be calculated from the composition of a reaction mixture at equilibrium by inserting the molar concentrations of each reactant and product into the equilibrium constant expression, raising each value to the appropriate power and calculating the ratio of the product to reactant terms. Similarly, the equilibrium partial pressures of the reactants and products of a reversible gas–phase reaction can be used to calculate the K_p value of the reaction at that temperature. (Section 14.2)

LO3 The relationship between K_c and K_p for a reversible gas-phase reaction depends on the number of moles of gaseous reactants and products in the balanced chemical equation. (Section 14.3)

LO4 The reverse of a reaction has an equilibrium constant that is the reciprocal of K for the forward reaction. If the balanced equation for a reaction is multiplied by some factor n, the value of K for that reaction is raised to the nth power. If reactions are summed to give an overall reaction, their equilibrium constants are multiplied together to obtain an overall equilibrium constant. (Section 14.4)

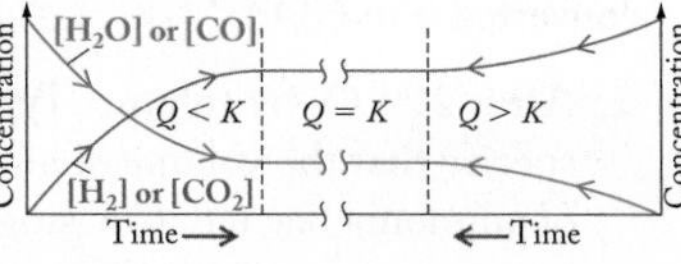

LO5 The **reaction quotient (Q)** is the value of the mass action expression at any instant during a reaction. At equilibrium, $Q = K$, but for nonequilibrium conditions, Q indicates how a reaction is proceeding. (Sections 14.2, 14.5)

LO6 According to **Le Châtelier's principle**, chemical reactions at equilibrium respond to stress by shifting the position of the equilibrium to relieve the stress. A catalyst decreases the time it takes a system to achieve equilibrium but does not change the value of the equilibrium constant. (Section 14.7)

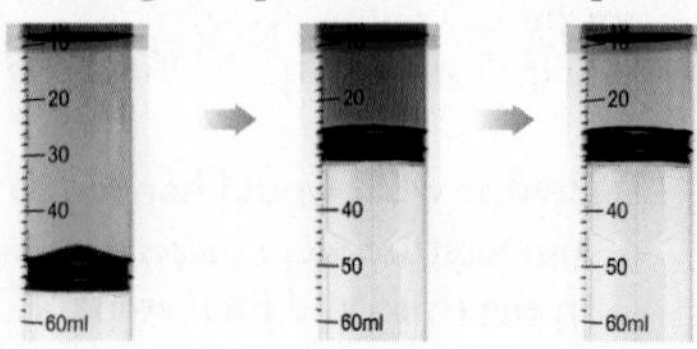

LO7 Equilibrium concentrations or partial pressures of reactants and products can be calculated from initial concentrations or pressures, the reaction stoichiometry, and the value of the equilibrium constant. (Section 14.8)

PARTICULATE PREVIEW WRAP-UP

The chemical equation for both samples is

$$CO(g) + H_2O(g) \rightleftharpoons CO_2(g) + H_2(g)$$

In sample A, the concentrations of the reactants are constant; in sample B, the concentrations of the reactants decrease. In sample A, the concentrations of the products are constant; in sample B, the concentrations of the products increase. Sample A is probably at equilibrium.

PROBLEM-SOLVING SUMMARY

Type of Problem	Concepts and Equations	Sample Exercises
Writing equilibrium constant expressions	For the reaction $aA + bB \rightleftharpoons cC + dD$ $K_c = \frac{[C]^c[D]^d}{[A]^a[B]^b}$ (14.3) and $K_p = \frac{(P_C)^c(P_D)^d}{(P_A)^a(P_B)^b}$ (14.4)	**14.1**
Calculating and interconverting K_c and K_p	Insert equilibrium molar concentrations or partial pressures into the equilibrium constant expression and use the relation $K_p = K_c(RT)^{\Delta n}$ (14.7) where $R = 0.08206$ L · atm/(mol · K), T is the absolute temperature, and Δn is the number of moles of product gas minus the number of moles of reactant gas in the balanced chemical equation.	**14.2, 14.3, 14.4**
Calculating K values of related reactions	$K_{reverse} = 1/K_{forward}$ for the forward and reverse reactions in an equilibrium system. If all the coefficients in a chemical equation are multiplied by n, the value of K increases by the power of n. If reactions are summed to give an overall reaction, their equilibrium constants are multiplied together to obtain an overall K.	**14.5, 14.6, 14.7**
Using Q and K values to predict the direction of a reaction	If $Q < K$, the reaction proceeds in the forward direction to make more products; if $Q = K$, the reaction is at equilibrium; if $Q > K$, the reaction proceeds in the reverse direction to make more reactants.	**14.8**
Writing equilibrium constant expressions for heterogeneous equilibria	Molar concentrations of pure liquids and pure solids are omitted from equilibrium constant expressions because their concentrations are constant.	**14.9**
Adding or removing reactants or products to stress an equilibrium	Decreasing the concentration of a substance involved in an equilibrium shifts the equilibrium toward the production of more of that substance. Increasing the concentration of a substance shifts the equilibrium to react some of that substance.	**14.10**
Predicting the effect of changing volume on gas-phase equilibria	Equilibria involving different numbers of moles of gaseous reactants and products shift in response to an increase (or decrease) in volume caused by a decrease (or increase) in pressure toward the side with more (or fewer) moles of gases.	**14.11**
Predicting changes in equilibrium with temperature	The value of K for an endothermic reaction increases with increasing temperature; the value of K for an exothermic reaction decreases with increasing temperature.	**14.12**
Calculating concentrations or partial pressures of reactants and products at equilibrium	Use a RICE table to develop algebraic terms for each reactant's and product's partial pressure or concentration at equilibrium. Substitute these terms into the expression for K and solve.	**14.13, 14.14, 14.15**

VISUAL PROBLEMS

(Answers to boldface end-of-chapter questions and problems are in the back of the book.)

14.1. Consider the graph of concentration versus time in Figure P14.1.

a. What is the mass action expression for the reaction?

b. What is the value of K_c?

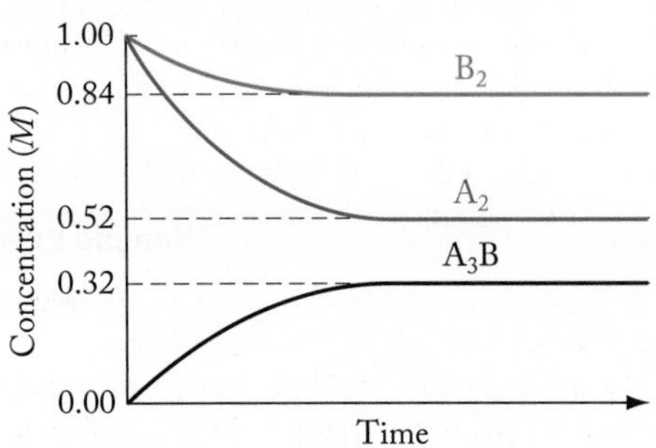

FIGURE P14.1

14.2. In Figure P14.2, the red spheres represent reactant A and the blue spheres represent product B in equilibrium with A.

a. Write a chemical equation that describes the equilibrium.

b. What is the value of the equilibrium constant K_c?

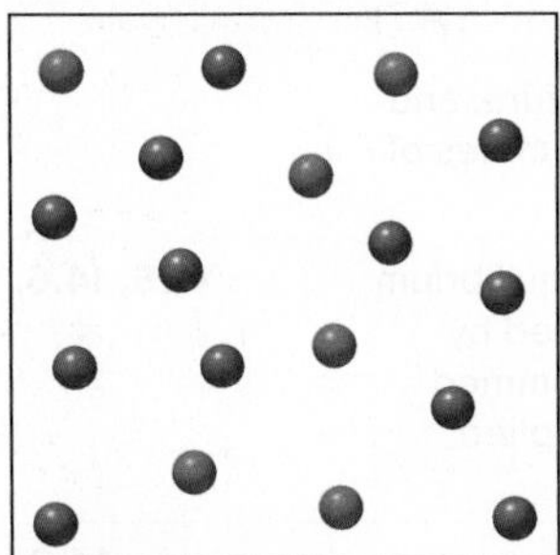

FIGURE P14.2

14.3. The equilibrium constant K_c for the reaction

$$\text{A (red spheres)} + \text{B (blue spheres)} \rightleftharpoons \text{AB}$$

is 3.0 at 300.0 K. Does the situation depicted in Figure P14.3 correspond to equilibrium? If not, in what direction (to the left or to the right) will the system shift to attain equilibrium?

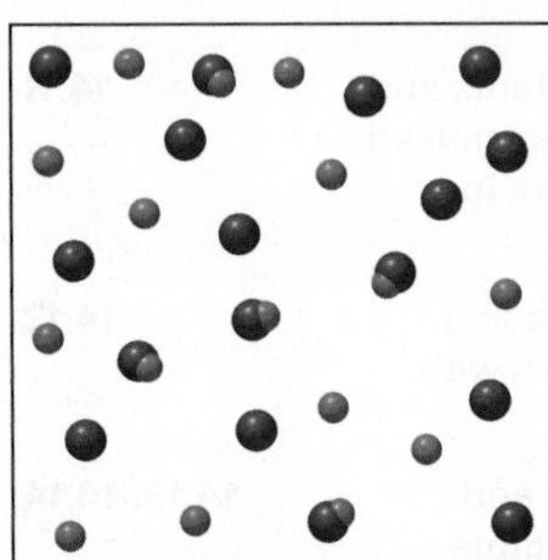

FIGURE P14.3

14.4. The diagrams in Figure P14.4 represent equilibrium states of the reaction

$$\text{A (red spheres)} + \text{B (blue spheres)} \rightleftharpoons \text{AB}$$

at 300 K and 400 K, respectively. Is this reaction endothermic or exothermic? Explain.

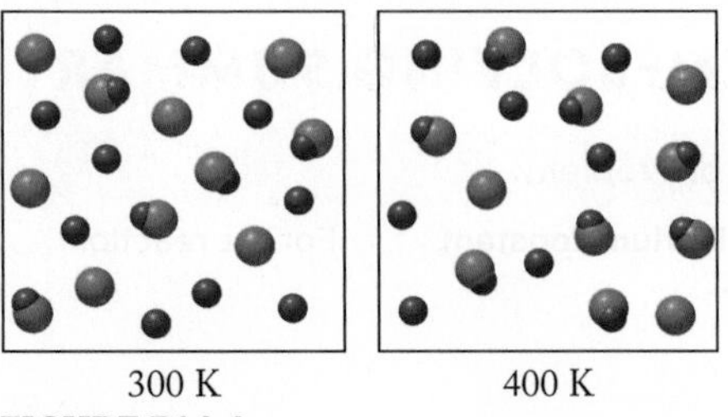

FIGURE P14.4

14.5. Does the reaction $A \rightarrow 2\,B$ represented in Figure P14.5 reach equilibrium in 20 μs? Explain your answer.

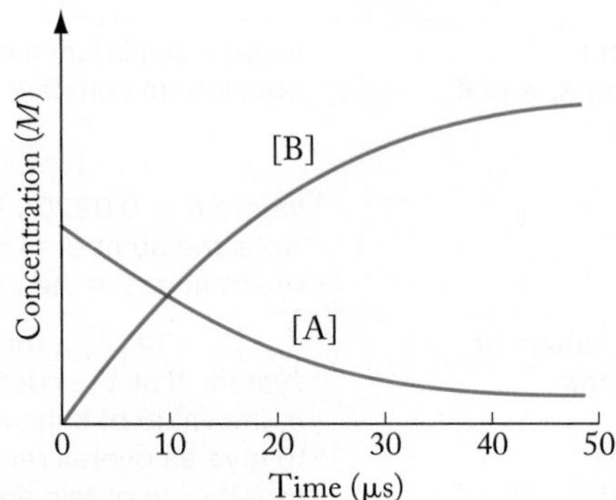

FIGURE P14.5

14.6. How does the cartoon in Figure P14.6 illustrate the concept of dynamic equilibrium?

FIGURE P14.6

14.7. There is a temperature at which the equilibrium constant $K_p = 4$ for the dimerization of NO_2 to N_2O_4. Does the particle view (Figure P14.7) of a reaction mixture of these two gases at this temperature represent (a) $Q_p < K_p$, (b) $Q_p = K_p$, or (c) $Q_p > K_p$?

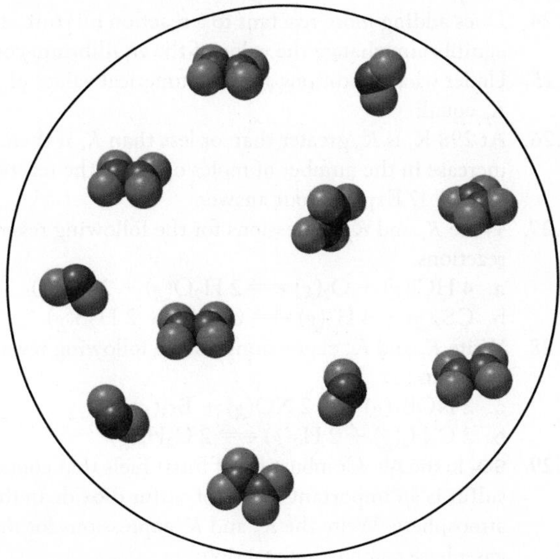

FIGURE P14.7

14.8. The graph in Figure P14.8 shows the results of an experiment in which the progress of the decomposition of NO_2 into NO and O_2 was monitored over time. (a) Which line corresponds to which gas? (b) Use the results shown in the graph to estimate the value of K_c for the decomposition reaction at the temperature used in the experiment.

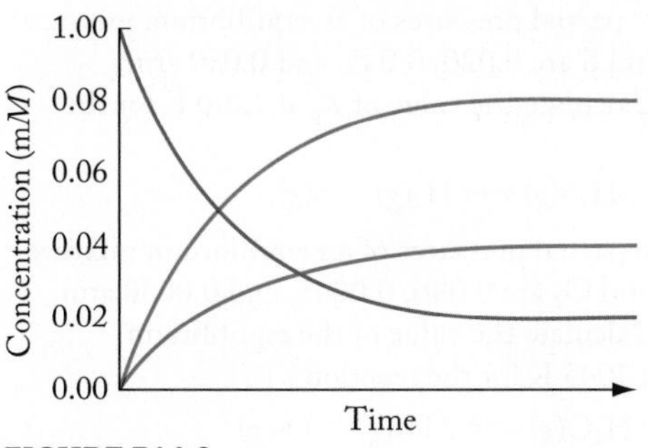

FIGURE P14.8

14.9. Sketch a graph similar to the one in Figure P14.8, but depicting the progress of the reaction $CaO(s) + SO_2(g) \rightleftharpoons CaSO_3(s)$ in which the initial partial pressure of SO_2 is 0.1 atm. Assume that the reaction mixture contains more than enough CaO to react with all of the SO_2 present and that the reaction occurs at a temperature at which its K_p value is exactly 50.

14.10. Use representations [A] through [I] in Figure P14.10 to answer questions (a)–(f). The center cell [E] represents an equilibrium mixture of $NO_2(g)$ and $N_2O_4(g)$. Assume that the temperature remains constant.

a. Write a balanced chemical equation describing the reversible reaction (NO_2 is the reactant) taking place in [E].
b. Identify one stress to the system that would transform [E] to [C].
c. Although [E] is an equilibrium mixture, the reaction mixture in [C] is not yet at equilibrium. Which representation depicts how [C] transforms to reestablish equilibrium?
d. Are [G] and [I] at equilibrium or not? How can you tell?
e. Given that N_2O_4 is colorless and NO_2 is brown, match the pictures of the syringes to their respective particulate images.
f. Does the syringe in [H] contain any N_2O_4?

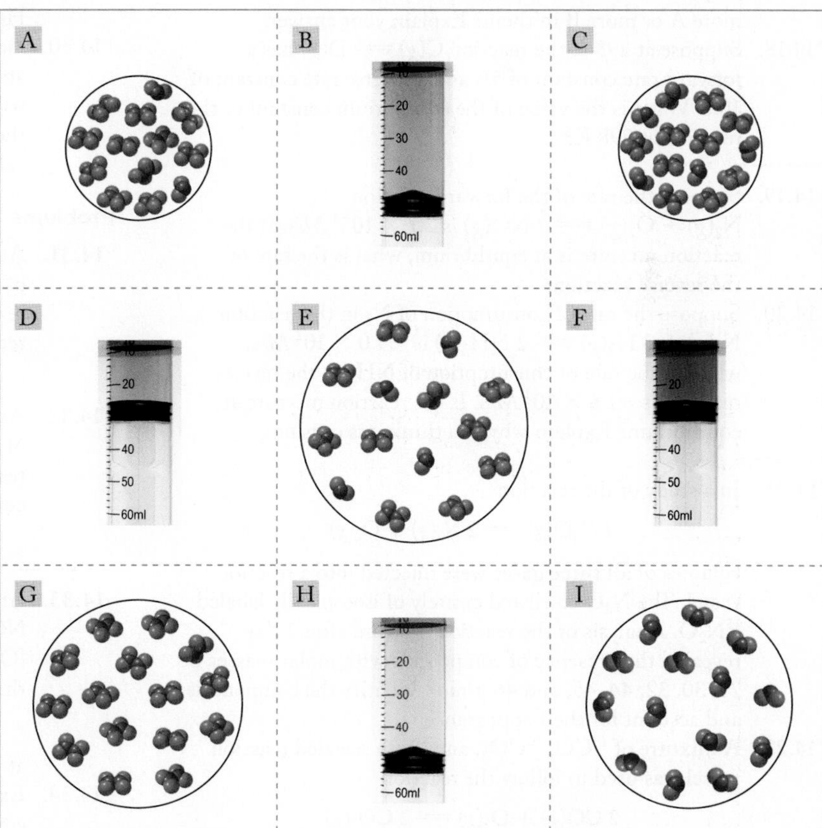

FIGURE P14.10

QUESTIONS AND PROBLEMS

The Dynamics of Chemical Equilibrium

Concept Review

14.11. Describe one way in which all the graphs used in Chapter 13 to illustrate the progress of chemical reactions differ from a graph of the progress of a reaction that has come to equilibrium.

14.12. What does the word *dynamic* mean when it is used to describe chemical equilibrium?

14.13. Do rapid reversible reactions always have greater yields of product than slow reversible reactions? Explain why or why not.

14.14. At equilibrium, is the sum of the concentrations of all the reactants always equal to the sum of the concentrations of the products? Explain why or why not.

14.15. How are forward and reverse reaction rates related in a system at chemical equilibrium?

14.16. Are ice cubes floating in 0°C water in an insulated container an example of a system in dynamic equilibrium? Explain why or why not.

Problems

***14.17.** Suppose the rate constant of the forward reaction $A(g) \rightleftharpoons B(g)$ is smaller than the rate constant of the reverse reaction. Do equilibrium mixtures of A and B have more A or more B in them? Explain your answer.

*14.18. Suppose at 298 K the reaction $C(g) \rightleftharpoons D(g)$ has a forward rate constant of 5/s and a reverse rate constant of 10/s. What is the value of the equilibrium constant of the reaction at 298 K?

14.19. Suppose the rate of the forward reaction $N_2(g) + O_2(g) \rightleftharpoons 2\,NO(g)$ is 2.7×10^{-3} *M*/s. If the reaction mixture is at equilibrium, what is the rate of the reverse reaction?

14.20. Suppose the rate of consumption of N_2 in the reaction $N_2(g) + 3\,H_2(g) \rightleftharpoons 2\,NH_3(g)$ is -4.0×10^2 *M*/s, whereas the rate of consumption of NH_3 in the reverse reaction is -1.6×10^3 *M*/s. Is the reaction mixture at equilibrium? Explain why you think it is or is not.

***14.21.** In a study of the reaction

$$2\,N_2O(g) \rightleftharpoons 2\,N_2(g) + O_2(g)$$

volumes of all three gases were injected into a reaction vessel. The N_2O consisted entirely of isotopically labeled $^{15}N_2O$. Analysis of the reaction mixture after 1 day revealed the presence of compounds with molar masses 28, 29, 30, 32, 44, 45, and 46 g/mol. Identify the compounds and account for their appearance.

*14.22. A mixture of ^{13}CO, $^{12}CO_2$, and O_2 in a sealed reaction vessel was used to follow the reaction

$$2\,CO(g) + O_2(g) \rightleftharpoons 2\,CO_2(g)$$

Analysis of the reaction mixture after 1 day revealed the presence of compounds with molar masses 28, 29, 32, 44, and 45 g/mol. Identify the compounds and account for their appearance.

The Equilibrium Constant; Relationships between K_c and K_p Values

Concept Review

14.23. What is the difference between an *equilibrium constant* and an *equilibrium constant expression*?

14.24. Does adding more reactant to a reaction mixture at equilibrium change the value of the equilibrium constant?

14.25. Under what conditions are the numerical values of K_c and K_p equal?

14.26. At 298 K, is K_p greater than or less than K_c if there is a net increase in the number of moles of gas in the reaction and if $K_c > 1$? Explain your answer.

14.27. Write K_c and K_p expressions for the following reversible reactions.

a. $4\,HCl(g) + O_2(g) \rightleftharpoons 2\,H_2O(g) + 2\,Cl_2(g)$
b. $CS_2(g) + 4\,H_2(g) \rightleftharpoons CH_4(g) + 2\,H_2S(g)$

14.28. Write K_c and K_p expressions for the following reversible reactions.

a. $2\,NOBr(g) \rightleftharpoons 2\,NO(g) + Br_2(g)$
b. $2\,C_2H_4(g) + 2\,H_2(g) \rightleftharpoons 2\,C_2H_6(g)$

14.29. **SO_2 in the Air** Combustion of fossil fuels that contain sulfur is an important source of sulfur dioxide in the atmosphere. Write the K_p and K_c expressions for the gas-phase combustion reaction:

$$S(g) + O_2(g) \rightleftharpoons SO_2(g)$$

How are the values of K_p and K_c related?

14.30. **More SO_2 in the Air** Sulfur dioxide released into the atmosphere during the combustion of fossil fuels may react with atmospheric oxygen, forming sulfur trioxide. Write the K_p and K_c expressions for this reaction. How are the values of K_p and K_c related?

Problems

14.31. At 1200 K the partial pressures of an equilibrium mixture of H_2S, H_2, and S are 0.020, 0.045, and 0.030 atm, respectively. Calculate the value of K_p at 1200 K for the reaction

$$H_2S(g) \rightleftharpoons H_2(g) + S(g)$$

14.32. At 1045 K the partial pressures of an equilibrium mixture of H_2O, H_2, and O_2 are 0.040, 0.0045, and 0.0030 atm, respectively. Calculate the value of the equilibrium constant K_p at 1045 K for the reaction

$$2\,H_2O(g) \rightleftharpoons 2\,H_2(g) + O_2(g)$$

14.33. At equilibrium, the concentrations of gaseous N_2, O_2, and NO in a sealed reaction vessel are $[N_2] = 3.3 \times 10^{-3}$ *M*, $[O_2] = 5.8 \times 10^{-3}$ *M*, and $[NO] = 3.1 \times 10^{-3}$ *M*. What is the value of K_c for the reaction

$$N_2(g) + O_2(g) \rightleftharpoons 2\,NO(g)$$

at the temperature of the reaction mixture?

14.34. Exactly 2 moles of ammonia are heated in a sealed 1.00 L container to 650°C. At this temperature, ammonia decomposes to nitrogen and hydrogen gas:

$$2\,NH_3(g) \rightleftharpoons N_2(g) + 3\,H_2(g)$$

At equilibrium, the concentration of ammonia in the container is 1.00 *M*. What is the value of K_c for the decomposition reaction at 650°C?

14.35. **Synthesis of Hydrogen (Step 1)** Hydrogen gas production often begins with the steam–methane reforming reaction

$$CH_4(g) + H_2O(g) \rightleftharpoons CO(g) + 3\,H_2(g)$$

At 1000 K the partial pressures of the gases in an equilibrium mixture are 0.71 atm CH_4, 1.41 atm H_2O, 1.00 atm CO, and 3.00 atm H_2. What is the value of K_p for the reaction at 1000 K?

14.36. **Synthesis of Hydrogen (Step 2)** When the CO produced in the steam–methane reforming reaction is reacted with more steam at 450 K, the water–gas shift reaction:

$$CO(g) + H_2O(g) \rightleftharpoons CO_2(g) + H_2(g)$$

produces more hydrogen. If the equilibrium partial pressures of the gases in the reactor are 0.35 atm H_2O, 0.24 atm CO, 4.47 atm H_2, and 4.36 atm CO_2, what is the value of K_p?

14.37. For which of the following reactions are the values of K_p and K_c equal?
a. $2\,NOCl(g) \rightleftharpoons 2\,NO(g) + Cl_2(g)$
b. $SO_2Cl_2(\ell) \rightleftharpoons SO_2(g) + Cl_2(g)$
c. $2\,CaSO_3(s) + O_2(g) \rightleftharpoons 2\,CaSO_4(s)$

14.38. For which of the following reactions are the values of K_c and K_p equal?
a. $2\,SO_2(g) + O_2(g) \rightleftharpoons 2\,SO_2(g)$
b. $O_3(g) + SO_2(g) \rightleftharpoons SO_3(g) + O_2(g)$
c. $4\,SO(g) \rightleftharpoons 2\,S_2O(g) + O_2(g)$

14.39. How is the K_c value of the reaction

$$3\,NO(g) \rightleftharpoons NO_2(g) + N_2O(g)$$

related to the K_p value for the reverse reaction?

$$NO_2(g) + N_2O(g) \rightleftharpoons 3\,NO(g)$$

14.40. If the value of the equilibrium constant K_c for the following reaction is 5×10^5 at 298 K, what is the value of K_p at 298 K?

$$2\,CO(g) + O_2(g) \rightleftharpoons 2\,CO_2(g)$$

14.41. **Bulletproof Glass** Phosgene ($COCl_2$) is used in the manufacture of foam rubber and bulletproof glass. It is formed from carbon monoxide and chlorine in the following reaction:

$$Cl_2(g) + CO(g) \rightleftharpoons COCl_2(g)$$

If $K_c = 5.0$ for this reaction at 327°C, then what is the value of K_p at 327°C?

14.42. If $K_p = 3.45$ at 298 K for the following reaction,

$$SO_2(g) + NO_2(g) \rightleftharpoons NO(g) + SO_3(g)$$

what is the value of K_c for the reverse reaction?

14.43. The K_c value for the reaction

$$2\,NOBr(g) \rightleftharpoons 2\,NO(g) + Br_2(g)$$

is 3.0×10^{-4} at 298 K. What is the value of K_p at 298 K for the following reaction?

$$NOBr(g) \rightleftharpoons NO(g) + \tfrac{1}{2}\,Br_2(g)$$

14.44. How is the value of the equilibrium constant K_c for the reaction

$$2\,H_2O(g) + N_2(g) \rightleftharpoons 2\,H_2(g) + 2\,NO(g)$$

related to the value of K_p for the following reaction at the same temperature?

$$H_2O(g) + \tfrac{1}{2}\,N_2(g) \rightleftharpoons H_2(g) + NO(g)$$

Manipulating Equilibrium Constant Expressions

Concept Review

14.45. Explain why representing the same reaction with different chemical equations, like this:

(1) $N_2(g) + 2\,O_2(g) \rightleftharpoons 2\,NO_2(g)$ K_1

(2) $\tfrac{1}{2}\,N_2(g) + O_2(g) \rightleftharpoons NO_2(g)$ K_2

results in different equilibrium constant values [i.e., $K_1 \neq K_2$].

14.46. The reversible reaction $X(g) + Y(g) \rightleftharpoons Z(g)$ occurs in two reversible steps:

(1) $X(g) + Y(g) \rightleftharpoons Q(g) + R(g)$

(2) $Q(g) + R(g) \rightleftharpoons Z(g)$

How is the equilibrium constant for the overall reaction related to K_1 and K_2?

14.47. The thermal decomposition of NO_2 can be written as

(1) $2\,NO_2(g) \rightleftharpoons 2\,NO(g) + O_2(g)$

or it can be written as

(2) $NO_2(g) \rightleftharpoons NO(g) + \tfrac{1}{2}\,O_2(g)$

How is the value of K_2 related to the value of K_1?

14.48. How is the value of the equilibrium constant K_p for the reaction

$$2\,H_2O(g) + N_2(g) \rightleftharpoons 2\,H_2(g) + 2\,NO(g)$$

related to the value of K_p for the following reaction at the same temperature?

$$H_2O(g) + \tfrac{1}{2}\,N_2(g) \rightleftharpoons H_2(g) + NO(g)$$

Problems

14.49. If $K_c = 3.0 \times 10^{-4}$ at 298 K for

$$2\,NOBr(g) \rightleftharpoons 2\,NO(g) + Br_2(g)$$

what is the value of K_c at 298 K for the following reaction?

$$NOBr(g) \rightleftharpoons NO(g) + \tfrac{1}{2}\,Br_2(g)$$

14.50. **Producing Ammonia** At 450°C, $K_p = 4.5 \times 10^{-5}$ for the formation of ammonia:

$$N_2(g) + 3\,H_2(g) \rightleftharpoons 2\,NH_3(g)$$

What is the value of K_p at 450°C for the following reaction?

$$2\,NH_3(g) \rightleftharpoons N_2(g) + 3\,H_2(g)$$

14.51. At a particular temperature, $K_c = 2.4 \times 10^{-3}$ for the reaction

$$2\ SO_2(g) + O_2(g) \rightleftharpoons 2\ SO_3(g)$$

What is the value of the equilibrium constant K_c for each of the following reactions at the same temperature?

a. $SO_2(g) + \frac{1}{2}O_2(g) \rightleftharpoons SO_3(g)$
b. $2\ SO_3(g) \rightleftharpoons 2\ SO_2(g) + O_2(g)$
c. $SO_3(g) \rightleftharpoons SO_2(g) + \frac{1}{2}O_2(g)$

14.52. If the equilibrium constant K_c for the reaction

$$2\ NO(g) + O_2(g) \rightleftharpoons 2\ NO_2(g)$$

is 5×10^{12} at a given temperature, what is the value of the equilibrium constant K_c for each of the following reactions at the same temperature?

a. $NO(g) + \frac{1}{2}O_2(g) \rightleftharpoons NO_2(g)$
b. $2\ NO_2(g) \rightleftharpoons 2\ NO(g) + O_2(g)$
c. $NO_2(g) \rightleftharpoons NO(g) + \frac{1}{2}O_2(g)$

14.53. Calculate the value of the equilibrium constant K_p at 298 K for the reaction

$$N_2(g) + 2\ O_2(g) \rightleftharpoons 2\ NO_2(g)$$

from the following K_p values at 298 K:

$$N_2(g) + O_2(g) \rightleftharpoons 2\ NO(g) \qquad K_p = 4.4 \times 40^{-31}$$
$$2\ NO(g) + O_2(g) \rightleftharpoons 2\ NO_2(g) \qquad K_p = 2.4 \times 40^{12}$$

14.54. Calculate the value of the equilibrium constant K_p at 298 K for the reaction

$$\tfrac{1}{4}\ S_8(s) + 3\ O_2(g) \rightleftharpoons 2\ SO_3(g)$$

from the following K_p values at 298 K:

$$\tfrac{1}{8}\ S_8(g) + O_2(g) \rightleftharpoons SO_2(g) \qquad K_p = 4.0 \times 10^{52}$$
$$2\ SO_2(g) + O_2(g) \rightleftharpoons 2\ SO_3(g) \qquad K_p = 7.8 \times 10^{24}$$

Equilibrium Constants and Reaction Quotients

Concept Review

14.55. How is an equilibrium constant different from a reaction quotient?

14.56. Explain how comparing the values of reaction quotient Q and equilibrium constant K for a given reaction and temperature enables chemists to predict whether a reversible reaction will proceed in the forward direction, in the reverse direction, or in neither direction.

Problems

14.57. If $K_c = 22$ for the hypothetical reaction $A(g) \rightleftharpoons B(g)$ at a given temperature, and if $[A] = 0.10\ M$ and $[B] = 2.0\ M$ in a reaction mixture at that temperature, is the reaction at chemical equilibrium? If not, in which direction will the reaction proceed to reach equilibrium?

14.58. The equilibrium constant K_c for the hypothetical reaction

$$2\ C(g) \rightleftharpoons D(g) + E(g)$$

is 3×10^{-3}. If the composition of the reaction mixture is $[C] = [D] = [E] = 5 \times 10^{-4}\ M$, in which direction will the reaction proceed to reach equilibrium?

14.59. If $K_c = 1.5 \times 10^{-3}$ for the reaction

$$N_2(g) + O_2(g) \rightleftharpoons 2\ NO(g)$$

in which direction will the reaction proceed if the partial pressures of the three gases are all 1.00×10^{-3} atm?

14.60. At 650 K, $K_p = 4.3 \times 10^{-4}$ for the ammonia synthesis reaction:

$$N_2(g) + 3\ H_2(g) \rightleftharpoons 2\ NH_3(g)$$

If a vessel at 650 K contains a reaction mixture in which $[N_2] = 0.010\ M$, $[H_2] = 0.030\ M$, and $[NH_3] = 0.00020\ M$, will more ammonia form?

14.61. Use the information below to determine whether a reaction mixture in which the partial pressures of PCl_3, Cl_2, and PCl_5 are 0.20, 0.40, and 0.60 atm, respectively, is at equilibrium at 450 K.

$$PCl_3(g) + Cl_2(g) \rightleftharpoons PCl_5(g) \qquad K_p = 3.8 \text{ at } 450\ K$$

If the reaction mixture is *not* at equilibrium, in which direction does the reaction proceed to achieve equilibrium?

14.62. If enough PCl_3 were injected into the reaction mixture in Problem 14.61 to double its partial pressure, would the mixture be closer to equilibrium? Explain why or why not.

Heterogeneous Equilibria

Concept Review

14.63. Write the K_c expression for the oxidation of calcium sulfite to gypsum (calcium sulfate):

$$2\ CaSO_3(s) + O_2(g) \rightleftharpoons 2\ CaSO_4(s)$$

14.64. **Plaster of Paris** The mineral gypsum, a hydrated form of calcium sulfate, is used to make casts for broken limbs. Gypsum can be converted to anhydrite ($CaSO_4$) by heating. Write the K_c and K_p expressions for the conversion reaction:

$$CaSO_4 \cdot 2\ H_2O(s) \xrightleftharpoons{\text{energy}} CaSO_4(s) + 2\ H_2O(g)$$

14.65. The brown residues that form on the surfaces of plumbing fixtures, such as inside toilet tanks, are the result of the oxidation of more soluble iron(II) compounds with dissolved oxygen, forming less soluble iron(III) compounds. Write the K_c expression for one such reaction:

$$4\ Fe(OH)_2(aq) + O_2(aq) + 2\ H_2O(\ell) \rightleftharpoons 4\ Fe(OH)_3(s)$$

*14.66. **Testing Minerals** Write the K_c expression for this reaction, which is used to test whether a white, shiny mineral is marble (calcium carbonate):

$$2\ HCl(aq) + CaCO_3(s) \rightleftharpoons CaCl_2(aq) + H_2O(\ell) + CO_2(g)$$

14.67. **Carbon Capture** A process for removing CO_2 from the atmosphere is based on converting CO_2 to carbonate and then precipitating it as calcium carbonate. The following equilibria are relevant to this process:

$$CO_2(g) + 2\ OH^-(aq) \rightleftharpoons CO_3^{2-}(aq) + H_2O(\ell)$$
$$Ca(OH)_2(aq) + CO_3^{2-}(aq) \rightleftharpoons CaCO_3(s) + 2\ OH^-(aq)$$

Write K_c expressions for these equilibria and for the overall reaction of CO_2 with calcium hydroxide.

14.68. The calcium carbonate produced in the process described in Problem 14.67 is heated to a temperature at which it thermally decomposed:

$$CaCO_3(s) \rightleftharpoons CaO(s) + CO_2(g)$$

and the products are collected and stored. The CaO that is produced is mixed with room-temperature water to regenerate calcium hydroxide:

$$CaO(s) + H_2O(\ell) \rightleftharpoons Ca(OH)_2(aq)$$

The equilibrium constant for the second equilibrium is very large. Does this mean that the equilibrium constant for the overall reaction:

$$CaCO_3(s) + H_2O(\ell) \rightleftharpoons Ca(OH)_2(aq) + CO_2(g)$$

is also very large at room temperature? Explain why or why not.

Le Châtelier's Principle

Concept Review

14.69. Does adding reactants to a system at equilibrium increase the value of the equilibrium constant? Why or why not?

14.70. Increasing the concentration of a reactant shifts the position of chemical equilibrium toward formation of more products. What effect does adding a reactant have on the rates of the forward and reverse reactions?

14.71. **Carbon Monoxide Poisoning** Patients suffering from carbon monoxide poisoning are treated with pure oxygen to remove CO from the hemoglobin (Hb) in their blood. The two relevant equilibria are

$$Hb(aq) + 4\,CO(g) \rightleftharpoons Hb(CO)_4(aq)$$

$$Hb(aq) + 4\,O_2(g) \rightleftharpoons Hb(O_2)_4(aq)$$

The value of the equilibrium constant for CO binding to Hb is greater than that for O_2. How, then, does this treatment work?

14.72. Is the equilibrium constant K_p for the reaction:

$$2\,NO_2(g) \rightleftharpoons N_2O_4(g)$$

in air the same in Los Angeles as in Denver if the atmospheric pressure in Denver is lower but the temperature is the same? Explain your answer.

14.73. Henry's law predicts that the solubility of a gas in a liquid increases with its partial pressure. Explain Henry's law in relation to Le Châtelier's principle.

*14.74. Why does adding an inert gas such as argon to an equilibrium mixture of CO, O_2, and CO_2 in a sealed vessel increase the total pressure of the system but not shift the following equilibrium?

$$2\,CO(g) + O_2(g) \rightleftharpoons 2\,CO_2(g)$$

Problems

14.75. Which of the following equilibria will shift toward formation of more products if an equilibrium mixture is compressed into half its volume?

a. $2\,N_2O(g) \rightleftharpoons 2\,N_2(g) + O_2(g)$
b. $2\,CO(g) + O_2(g) \rightleftharpoons 2\,CO_2(g)$
c. $N_2(g) + O_2(g) \rightleftharpoons 2\,NO(g)$
d. $2\,NO(g) + O_2(g) \rightleftharpoons 2\,NO_2(g)$

14.76. Which of the following equilibria will shift toward formation of more products if the volume of a reaction mixture at equilibrium increases by a factor of 2?

a. $2\,SO_2(g) + O_2(g) \rightleftharpoons 2\,SO_3(g)$
b. $NO(g) + O_3(g) \rightleftharpoons NO_2(g) + O_2(g)$
c. $2\,N_2O_5(g) \rightleftharpoons 4\,NO_2(g) + O_2(g)$
d. $N_2O_4(g) \rightleftharpoons 2\,NO_2(g)$

14.77. How will the changes listed affect the equilibrium concentrations of reactants and products in the following reaction?

$$2\,O_3(g) \rightleftharpoons 3\,O_2(g)$$

a. O_3 is added to the system.
b. O_2 is added to the system.
c. The mixture is compressed to one-tenth its initial volume.

14.78. How will the changes listed affect the position of the following equilibrium?

$$2\,NO_2(g) \rightleftharpoons NO(g) + NO_3(g)$$

a. The concentration of NO is increased.
b. The concentration of NO_2 is increased.
c. The volume of the system expands to 5 times its initial value.

14.79. How will reducing the partial pressure of O_2 affect the position of the equilibrium in the following reaction?

$$2\,SO_2(g) + O_2(g) \rightleftharpoons 2\,SO_3(g)$$

*14.80. Ammonia is added to a gaseous reaction mixture containing H_2, Cl_2, and HCl that is at chemical equilibrium. How will the addition of ammonia affect the relative concentrations of H_2, Cl_2, and HCl if the equilibrium constant of reaction 2 is much greater than the equilibrium constant of reaction 1?

(1) $H_2(g) + Cl_2(g) \rightleftharpoons 2\,HCl(g)$
(2) $HCl(g) + NH_3(g) \rightleftharpoons NH_4Cl(s)$

14.81. In which of the following equilibria does an increase in temperature produce a shift toward the formation of more product?

a. $PCl_3(g) + Cl_2(g) \rightleftharpoons PCl_5(g)$ $\quad \Delta H° < 0$
b. $CH_4(g) + H_2O(g) \rightleftharpoons CO(g) + 3\,H_2(g)$ $\quad \Delta H° > 0$
c. $CO(g) + H_2O(g) \rightleftharpoons CO_2(g) + H_2(g)$ $\quad \Delta H° < 0$

14.82. In which of the following equilibria does an increase in temperature produce a shift toward the formation of more product?

a. $N_2(g) + 3\,H_2(g) \rightleftharpoons 2\,NH_3(g)$ $\quad \Delta H° < 0$
b. $2\,NO_2(g) \rightleftharpoons 2\,NO(g) + O_2(g)$ $\quad \Delta H° > 0$
c. $C_2H_4(g) + H_2(g) \rightleftharpoons C_2H_6(g)$ $\quad \Delta H° < 0$

Calculations Based on K

Concept Review

14.83. Why are calculations based on K often simpler when the value of K is very small?

14.84. The following reaction is carried out in a sealed, rigid vessel at constant temperature.

$$2\,NO(g) + O_2(g) \rightarrow 2\,NO_2(g)$$

a. If the change in the partial pressure of O_2 is $-x$, what are the changes in the partial pressures of NO and NO_2?
b. As the reaction proceeds, what happens to the total pressure in the reaction vessel?

Problems

14.85. For the reaction

$$PCl_5(g) \rightleftharpoons PCl_3(g) + Cl_2(g) \qquad K_p = 23.6 \text{ at } 500 \text{ K}$$

a. Calculate the equilibrium partial pressures of the reactants and products at 500 K if the initial pressures are $P_{PCl_5} = 0.560$ atm and $P_{PCl_3} = 0.500$ atm.
b. If more chlorine is added after equilibrium is reached, how will the concentrations of PCl_5 and PCl_3 change?

14.86. Enough NO_2 gas is injected into a cylindrical vessel to produce a partial pressure, P_{NO_2}, of 0.900 atm at 298 K. Calculate the equilibrium partial pressures of NO_2 and N_2O_4, given

$$2\ NO_2(g) \rightleftharpoons N_2O_4(g) \qquad K_p = 4 \text{ at } 298 \text{ K}$$

14.87. At 25°C, $K_c = 0.0900$ for the reaction between water vapor and dichlorine monoxide:

$$H_2O(g) + Cl_2O(g) \rightleftharpoons 2\ HOCl(g)$$

Determine the equilibrium concentrations of all three compounds at 25°C if the starting concentrations of both reactants are 0.00432 *M* and no HOCl is present.

14.88. At 648 K, $K_p = 4.3 \times 10^{-4}$ for the reaction

$$3\ H_2(g) + N_2(g) \rightleftharpoons 2\ NH_3(g)$$

Determine the equilibrium partial pressure of NH_3 in a reaction vessel that initially contained 0.900 atm N_2 and 0.500 atm H_2 at 648 K.

14.89. At 25°C, $K_p = 1.5 \times 10^6$ for the reaction

$$NO(g) + \tfrac{1}{2}\ O_2(g) \rightleftharpoons NO_2(g)$$

At equilibrium, what is the ratio of P_{NO_2} to P_{NO} in air at 25°C? Assume that $P_{O_2} = 0.21$ and does not change.

*14.90. **The Water–Gas Reaction** Passing steam over hot carbon produces a mixture of carbon monoxide and hydrogen known as water gas:

$$H_2O(g) + C(s) \rightleftharpoons CO(g) + H_2(g)$$

The value of K_c for the reaction at 1000°C is 3.0×10^{-2}.
a. Calculate the equilibrium partial pressures of the products and reactants at 1000°C if $P_{H_2O} = 0.442$ and $P_{CO} = 5.0$ atm at the start of the reaction. Assume that the carbon is in excess.
b. Determine the equilibrium partial pressures of the reactants and products after enough CO and H_2 are added to the equilibrium mixture in part (a) to initially increase the partial pressures of both gases by 0.075 atm.

14.91. At 700°C, $K_p = 1.5$ for the reaction

$$CO_2(g) + C(s) \rightleftharpoons 2\ CO(g)$$

Calculate the equilibrium partial pressures of CO and CO_2 at 700°C if initially $P_{CO_2} = 5.0$ atm and CO is not present. Pure graphite is present initially and when equilibrium is achieved.

14.92. **Composition of Jupiter's Atmosphere** Ammonium hydrogen sulfide (NH_4SH) has been detected in the atmosphere of Jupiter. The equilibrium between ammonia, hydrogen sulfide, and NH_4SH is described by the following equation:

$$NH_4SH(s) \rightleftharpoons NH_3(g) + H_2S(g)$$

The value of K_p for the reaction at 24°C is 0.126. Suppose a sealed flask contains an equilibrium mixture of NH_4SH, NH_3, and H_2S at 24°C. At equilibrium, the partial pressure of H_2S is 0.355 atm. What is the partial pressure of NH_3?

***14.93.** A flask containing pure NO_2 was heated to 1000 K, a temperature at which $K_p = 158$ for the decomposition of NO_2:

$$2\ NO_2(g) \rightleftharpoons 2\ NO(g) + O_2(g)$$

The partial pressure of O_2 at equilibrium is 0.136 atm.
a. Calculate the partial pressures of NO and NO_2.
b. Calculate the total pressure in the flask at equilibrium.

14.94. At 830°C, $K_p = 7.69$ for the reaction

$$2\ SO_3(g) \rightleftharpoons 2\ SO_2(g) + O_2(g)$$

If a vessel at this temperature initially contains pure SO_3 and if the partial pressure of SO_3 at equilibrium is 0.100 atm, what is the partial pressure of O_2 in the flask at equilibrium?

***14.95.** **NO_x Pollution** In a study of the formation of NO_x in air pollution, a chamber heated to 2200°C was filled with air (0.79 atm N_2, 0.21 atm O_2). What are the equilibrium partial pressures of N_2, O_2, and NO if $K_p = 0.050$ for the following reaction at 2200°C?

$$N_2(g) + O_2(g) \rightleftharpoons 2\ NO(g)$$

*14.96. At 450°C, $K_p = 6.5 \times 10^{-6}$ for the thermal decomposition of NO_2:

$$2\ NO_2(g) \rightleftharpoons 2\ NO(g) + O_2(g)$$

If a reaction vessel at this temperature initially contains only 0.500 atm NO_2, what will be the partial pressures of NO_2, NO, and O_2 in the vessel when equilibrium has been attained?

14.97. At 1400 K, $K_c = 2.2 \times 10^{-4}$ for the thermal decomposition of hydrogen sulfide:

$$2\ H_2S(g) \rightleftharpoons 2\ H_2(g) + S_2(g)$$

A sample of gas in which $[H_2S] = 6.00\ M$ is heated to 1400 K in a sealed high-pressure vessel. After chemical equilibrium has been achieved, what is the value of $[H_2S]$? Assume that no H_2 or S_2 was present in the original sample.

14.98. **Urban Air** On a very smoggy day, the equilibrium concentration of NO_2 in the air over an urban area reaches 2.2×10^{-7} *M*. If the temperature of the air is 25°C, what is the concentration of the dimer N_2O_4 in the air?

$$N_2O_4(g) \rightleftharpoons 2\ NO_2(g) \qquad K_c = 6.1 \times 10^{-3}$$

***14.99.** **Chemical Weapon** Phosgene ($COCl_2$) gained notoriety as a chemical weapon in World War I. Phosgene is produced by the reaction of carbon monoxide with chlorine:

$$CO(g) + Cl_2(g) \rightleftharpoons COCl_2(g)$$

If $K_c = 5.0$ at 600 K for this reaction, what are the equilibrium partial pressures of the three gases if a reaction vessel initially contains a mixture of the reactants in which $P_{CO} = P_{Cl_2} = 0.265$ atm and there is no $COCl_2$?

*14.100. At 2000°C, $K_c = 1.0$ for the reaction

$$2\,CO(g) + O_2(g) \rightleftharpoons 2\,CO_2(g)$$

What is the ratio of [CO] to [CO_2] at 2000°C in an atmosphere in which [O_2] = 0.0045 *M* at equilibrium?

***14.101.** The water–gas shift reaction is an important source of hydrogen:

$$CO(g) + H_2O(g) \rightleftharpoons CO_2(g) + H_2(g)$$

If $K_c = 5.1$ for this reaction at 700 K, calculate the equilibrium concentrations of the four gases at 700 K if the initial concentration of each of them is 0.050 *M*.

*14.102. Sulfur dioxide reacts with NO_2, forming SO_3 and NO:

$$SO_2(g) + NO_2(g) \rightleftharpoons SO_3(g) + NO(g)$$

If $K_c = 2.50$ for the reaction at a given temperature, what are the equilibrium concentrations of the products when the reaction mixture was initially 0.50 *M* SO_2, 0.50 *M* NO_2, 0.0050 *M* SO_3, and 0.0050 *M* NO?

Additional Problems

***14.103. CO as a Fuel** Is carbon dioxide a viable source of the fuel CO? Pure carbon dioxide (P_{CO_2} = 1 atm) decomposes at high temperatures. For the system

$$2\,CO_2(g) \rightleftharpoons 2\,CO(g) + O_2(g)$$

the percentage of decomposition of $CO_2(g)$ changes with temperature as follows:

Temperature (K)	Decomposition (%)
1500	0.048
2500	17.6
3000	54.8

Is the reaction endothermic? Calculate the value of K_p at each temperature and discuss the results. Is the decomposition of CO_2 an antidote for global warming?

14.104. Ammonia decomposes at high temperatures. In an experiment to explore this behavior, 2.00 moles of gaseous NH_3 is sealed in a rigid 1 liter vessel. The vessel is heated at 800 K and some of the NH_3 decomposes according to the following reaction:

$$2\,NH_3(g) \rightleftharpoons N_2(g) + 3\,H_2(g)$$

The system eventually reaches equilibrium and is found to contain 1.74 moles of NH_3. What are the values of K_p and K_c for the decomposition reaction at 800 K?

***14.105.** Elements of group 16 form hydrides with the generic formula H_2X. At a certain temperature, when gaseous H_2X is bubbled through a solution containing 0.3 *M* hydrochloric acid, the solution becomes saturated and [H_2X] = 0.1 *M*. The following equilibria exist in this solution:

$$H_2X(aq) + H_2O(\ell) \rightleftharpoons HX^-(aq) + H_3O^+(aq) \qquad K_1 = 8.3 \times 10^{-8}$$

$$HX^-(aq) + H_2O(\ell) \rightleftharpoons X^{2-}(aq) + H_3O^+(aq) \qquad K_2 = 1 \times 10^{-14}$$

Calculate the concentration of X^{2-} in the solution.

*14.106. Nitrogen dioxide reacts with SO_2 to form NO and SO_3:

$$NO_2(g) + SO_2(g) \rightleftharpoons NO(g) + SO_3(g)$$

An equilibrium mixture is analyzed at a certain temperature and found to contain [NO_2] = 0.100 *M*, [SO_2] = 0.300 *M*, [NO] = 2.00 *M*, and [SO_3] = 0.600 *M*. At the same temperature, extra $SO_2(g)$ is added to make [SO_2] = 0.800 *M*. Calculate the composition of the mixture when equilibrium has been reestablished.

***14.107. Controlling Air Pollution** Calcium oxide is used to remove the pollutant SO_2 from smokestack gases. The overall reaction is

$$CaO(s) + SO_2(g) + \tfrac{1}{2}\,O_2(g) \rightleftharpoons CaSO_4(s)$$

and $K_p = 2.38 \times 10^{73}$.

a. What is P_{SO_2} in equilibrium with air and solid CaO if P_{O_2} in air is 0.21 atm?
b. Consider a sample of gas that contains 100.0 moles of gas. How many molecules of SO_2 are in that sample?

*14.108. A 100 mL reaction vessel initially contains 2.60×10^{-2} mol of NO and 1.30×10^{-2} mol of H_2. At equilibrium, the concentration of NO in the vessel is 0.161 *M*. The vessel also contains N_2, H_2O, and H_2 at equilibrium. What is the value of the equilibrium constant K_c for the following reaction?

$$2\,H_2(g) + 2\,NO(g) \rightleftharpoons 2\,H_2O(g) + N_2(g)$$

14.109. Thermal decomposition of the mineral dolomite (a mixture of calcium and magnesium carbonates) is described by the following equations:

$$CaCO_3(s) \rightleftharpoons CaO(s) + CO_2(g)$$

$$MgCO_3(s) \rightleftharpoons MgO(s) + CO_2(g)$$

a. Write K_p expressions for both reactions.
*b. Write a chemical equation and K_p expression describing the decomposition of a sample of dolomite that contains 75% $CaCO_3$ and 25% $MgCO_3$ by mass.
c. Why is essentially all the $MgCO_3$ and $CaCO_3$ in a dolomite sample converted to the corresponding oxides when heated in an open container but not necessarily in a closed container?
d. Would the K_p values for these reactions depend on the volume of a closed reaction vessel?

*14.110. **Scrubbing Sulfur Dioxide from Power Plants** Sulfur dioxide in the smokestack gases of power plants can be reduced by passing the gases through a spray of pulverized limestone ($CaCO_3$) suspended in water:

$$CaCO_3(s) + SO_2(g) \rightleftharpoons CaSO_3(s) + CO_2(g)$$

a. Are the values of K_p and K_c the same for this reaction?
b. The efficiency of scrubbing (the rate of SO_2 removal) increases with decreasing $CaCO_3$ particle size. Does this affect the value of K_p for the reaction?
c. Write a balanced chemical equation for the SO_2 sequestration reaction when a slurry of solid $Ca(OH)_2$ is used instead of $CaCO_3$.
d. Calcium hydroxide also has the capacity to remove CO_2 from stack gases, producing $CaCO_3$. Write a chemical equation describing the overall reaction for the removal of both SO_2 and CO_2.
e. How is the K_p value for the overall reaction related to the K_p values of the single-gas reactions?

15

Acid–Base Equilibria

Proton Transfer in Biological Systems

SHADES OF PINK AND BLUE AND IN BETWEEN The color of hydrangea blossoms depends on the acidity of the soil in which they grow.

PARTICULATE **REVIEW**

Donating H^+ Ions

In this chapter we use the principles of chemical equilibrium from Chapter 14 to explore the structure and properties of acids and bases. Consider these models of four molecules that each contain at least one hydrogen atom.

- Which molecules produce H^+ ions when dissolved in water? Identify the acidic hydrogen atoms.
- Which molecule dissociates completely in water, meaning that each molecule in a sample produces a H^+ ion?
- Which molecule partially dissociates in water, meaning that only a small fraction of molecules in solution releases H^+ ions, whereas most do not?

(Review Section 4.5 if you need help.)

(Answers to Particulate Review questions are in the back of the book.)

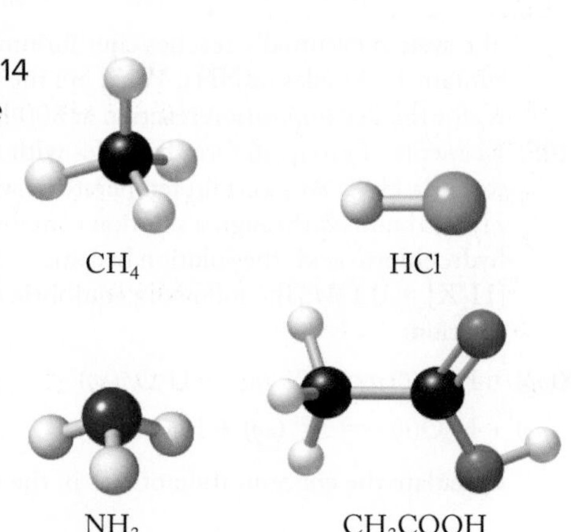

PARTICULATE **PREVIEW**

Accepting H⁺ Ions

Bases form covalent bonds as they accept H^+ ions donated by acids. As you read Chapter 15, look for ideas that will help you answer these questions:

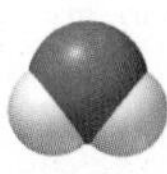

- Which of the three species depicted form covalent bonds as they accept H^+ ions?
- When each species accepts a proton, does it produce a cation? An anion? A neutral molecule?
- What feature do these particles have in common to form bonds to H^+ ions?

Learning Outcomes

LO1 Relate the strengths of acids and bases to their concentrations and to their K_a and K_b values
Sample Exercise 15.1

LO2 Relate the strengths of acids and bases to their molecular structures
Sample Exercise 15.2

LO3 Recognize conjugate acid–base pairs, and predict their formulas and complementary acidic and basic strengths
Sample Exercises 15.3, 15.4

LO4 Interconvert $[H_3O^+]$, $[OH^-]$, pH, and pOH
Sample Exercises 15.5, 15.6, 15.7

LO5 Relate the pH of solutions of weak acids and bases to their percent ionization and to their K_a and K_b values
Sample Exercises 15.8, 15.9

LO6 Calculate the pH of solutions of strong and weak acids and bases
Sample Exercises 15.10, 15.11, 15.12, 15.13

LO7 Calculate the pH of solutions of polyprotic acids
Sample Exercises 15.14, 15.15

LO8 Predict whether a salt is acidic, basic, or neutral, and calculate the pH of a salt solution
Sample Exercises 15.16, 15.17, 15.18

15.1 Acids and Bases: A Balancing Act

The pink and blue colors of the hydrangea blossoms in the chapter-opening photograph are produced by the same species of plant grown in the same garden, but in soils having slightly different composition. The colors of hydrangea blossoms are controlled by the availability of aluminum ions (Al^{3+}) in the soil in which they grow. Aluminum ions are soluble in acidic soils, and hydrangeas grown in acidic soils form blue blossoms. In soils that are neutral or slightly basic, however, aluminum ions precipitate as aluminum hydroxide [$Al(OH)_3$] and are unavailable. As a result, hydrangea blossoms grown in those soils lack blue pigment and are pink.

For many biological systems, including the human body, acid–base balance is vital to normal function and good health. Our blood, for example, is normally slightly basic, regulated by the proportions of CO_2 gas and bicarbonate (HCO_3^-) ions dissolved in it. This regulation happens because of the following reversible chemical reactions:

CONNECTION When describing aqueous solutions, the terms *hydrogen ion*, *proton*, and *hydronium ion* all refer to the same species (Section 4.5). Thus, the symbols H^+ and $H^+(aq)$ are simply abbreviated forms of H_3O^+.

$$CO_2(g) + H_2O(\ell) \rightleftharpoons H_2CO_3(aq) \quad (15.1)$$

$$H_2CO_3(aq) + H_2O(\ell) \rightleftharpoons HCO_3^-(aq) + H_3O^+(aq) \quad (15.2)$$

Note how molecules of CO_2 and H_2O combine, forming molecules of carbonic acid (H_2CO_3), which react with more water molecules, donating H^+ ions to them and forming bicarbonate and hydronium (H_3O^+) ions.

Compounds that produce hydronium ions when dissolved in water are acids, whereas compounds that produce hydroxide ions (OH^-) in aqueous solution are bases. More precisely, such compounds are referred to respectively as **Arrhenius acids** and **Arrhenius bases** in honor of Svante Arrhenius, whose research on the behavior of electrolytic solutions was recognized with the 1903 Nobel Prize in Chemistry.

Arrhenius acid a compound that produces H_3O^+ ions in aqueous solution

Arrhenius base a compound that produces OH^- ions in aqueous solution

In Section 4.5 we introduced the *Brønsted–Lowry* model of acids and bases, in which acids are defined as substances that *donate* H^+ ions, and bases are defined as substances that *accept* H^+ ions. We also learned in Chapter 4 that acids and bases are classified as strong or weak, depending on the degree to which their molecules either donate or accept electrons. Here in Section 15.1 we use the concept of chemical equilibrium to further explore what it means to be a weak or strong acid or base.

If the concentration of H_2CO_3 in Equation 15.2 increases, then Le Châtelier's principle predicts that the concentration of H_3O^+ ions in a solution that contains carbonic acid and bicarbonate ions will increase (as will the acidity of the solution). The equilibrium described in Equation 15.1 indicates, moreover, that the concentration of H_2CO_3 increases when the concentration of dissolved CO_2 increases. Therefore, an increase in the concentration of dissolved CO_2 shifts the equilibria in *both* Equations 15.1 and 15.2 to the right, increasing the concentration of H_3O^+ ions and making the solution more acidic. On the other hand, Le Châtelier's principle also predicts that an increase in the concentration of HCO_3^- ions shifts both equilibria to the left, decreasing the concentration of H_3O^+ ions and reducing the acidity of the solution.

CONNECTION In Section 4.5 we learned that the greater the concentration of H_3O^+ ions in a solution, the more acidic the solution.

The equilibria in Equations 15.1 and 15.2 play a key role in human health. Normally, the concentration of HCO_3^- ions in our blood is about 20 times the dissolved concentration of CO_2 because the dissociation of carbonic acid is not our blood's only source of HCO_3^-. The abundance of HCO_3^- ions drives down the concentration of H_3O^+ ions in our blood below 10^{-7} *M*. As we discuss later in this chapter, a $[H_3O^+]$ value this small means that our blood is slightly basic.

Unfortunately, some people suffer from medical problems that alter the normal balance of CO_2 and HCO_3^- in their blood. For example, chronic lung disease impairs not only a patient's ability to inhale O_2 but also the ability to exhale CO_2. As a result, CO_2 builds up in the blood and in the tissues where it is produced. The result is a condition called *respiratory acidosis*. Its symptoms include fatigue, lethargy, and shortness of breath. In severe cases, the condition can be fatal.

CONCEPT TEST

When people hyperventilate, they breathe too rapidly. Hyperventilation results in carbon dioxide being removed from the bloodstream more quickly than the body produces it through metabolism. What effect does this have on the equilibria in Equations 15.1 and 15.2?

(Answers to Concept Tests are in the back of the book.)

In this chapter we examine acid–base balance, both in environmental samples and in our bodies. We study why changes in this balance occur, some of the consequences of these changes, and what we can do to control them.

15.2 The Molecular Structures and Strengths of Acids and Bases

In this section we explore how and why molecules of acids dissociate when they dissolve in water, producing hydrogen ions and anions with 1− charges. We also examine why the molecules of a few of these compounds completely ionize in this way, making them strong acids, whereas those of many others only partially ionize, making them weak acids. Later in the section we examine the capacity of molecules of basic compounds to accept hydrogen ions, and we compare their strengths as bases to their molecular structures. We start by taking a molecular view of the process of acid ionization in aqueous solutions.

Strong and Weak Acids

To understand why molecules of acidic substances ionize when they dissolve in water, we need to consider the structures of their molecules and the nature and strength of the intermolecular forces they experience. The molecular structures of the common strong acids are shown in **Table 15.1**. The H atoms that can be released as H^+ ions are highlighted in red. In each of these strong acids one or more hydrogen atoms is bonded to the atom of an electronegative element: O, Cl, Br, or I. The bonds that these pairs of atoms form are polar covalent, which results in strong dipole–dipole interactions between the H atoms in these bonds and the O atoms of H_2O molecules. In aqueous solutions of HCl, for example, the strength of these dipole–dipole interactions results in every H—Cl bond breaking in such a way that the bonding pair of electrons moves to the Cl atom, forming a Cl^- ion. The H^+ ion that is simultaneously produced is drawn to the water molecule's O atom, where it uses one of the lone pairs of electrons on O to form a covalent bond. The product of this union is an H_3O^+ ion as shown in **Figure 15.1**.

CONNECTION Electronegativity was introduced in Section 8.3 and intermolecular forces were described in Chapter 10.

As with gas-phase equilibria (Chapter 14), the extent to which an acid ionization reaction takes place is reflected in the value of its equilibrium constant, K: the greater the value of K, the higher the concentrations of products and the lower the concentrations of reactants at equilibrium. The equilibrium constants that describe acid ionization reactions are given the symbol K_a, where the subscript "a" indicates that the equilibrium involves ionization of an *acid*.

Table 15.2 shows the molecular structures of a few of the many compounds that are weak acids and lists their K_a values at 25°C. All these values are much

TABLE 15.1 Strong Acids and Their Ionization Reactions in Water

Strong Acid	Molecular Structure	Reaction in Water
Hydrobromic	H—Br	$HBr(aq) + H_2O(\ell) \rightarrow Br^-(aq) + H_3O^+(aq)$
Hydrochloric	H—Cl	$HCl(aq) + H_2O(\ell) \rightarrow Cl^-(aq) + H_3O^+(aq)$
Hydroiodic	H—I	$HI(aq) + H_2O(\ell) \rightarrow I^-(aq) + H_3O^+(aq)$
Nitric	[structure: O, N, O, O, H]	$HNO_3(aq) + H_2O(\ell) \rightarrow NO_3^-(aq) + H_3O^+(aq)$
Perchloric	[structure: O, Cl, O, O, O, H]	$HClO_4(aq) + H_2O(\ell) \rightarrow ClO_4^-(aq) + H_3O^+(aq)$
Sulfuric	[structure: O, S, O, O, O, H, H]	$H_2SO_4(aq) + H_2O(\ell) \rightarrow HSO_4^-(aq) + H_3O^+(aq)$ $HSO_4^-(aq) + H_2O(\ell) \rightleftharpoons SO_4^{2-}(aq) + H_3O^+(aq)$

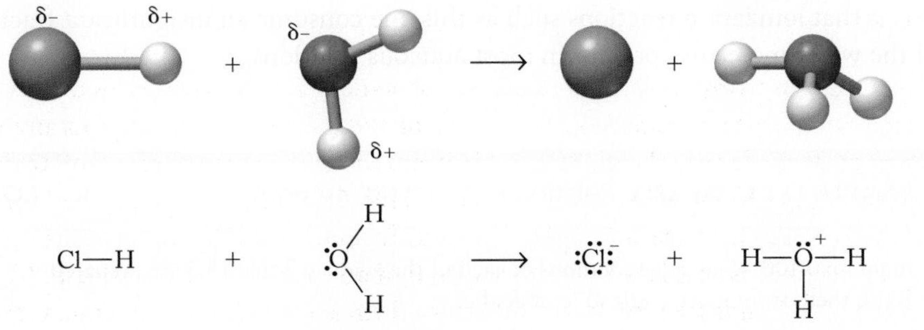

FIGURE 15.1 When hydrogen chloride dissolves in water, each molecule of HCl ionizes by reacting with a molecule of H_2O, forming Cl^- and H_3O^+ ions.

smaller than 1, which means that only a fraction of their molecules donate H^+ ions in most aqueous solutions. Let's write the equilibrium constant expressions for the first acid in the table—acetic acid—following the same process used in Chapter 14 for gas-phase reactions. The numerator contains concentration terms for the products each raised to a power equal to its coefficient in the acid ionization reaction (both are one), and the denominator contains the concentration value of the reactant: un-ionized acetic acid. Its coefficient is also one.

CHEMTOUR
Acid–Base Ionization

$$K_a = \frac{[H_3O^+][CH_3COO^-]}{[CH_3COOH]}$$

Note that the K_a expression does not contain a term for the concentration of water, even though molecules of H_2O reacted to form H_3O^+ ions. The reason for

TABLE 15.2 Some Common Weak Acids and Their Ionization Reactions in Water

Weak Acid	Molecular Structure	Reaction in Water	K_a
Acetic	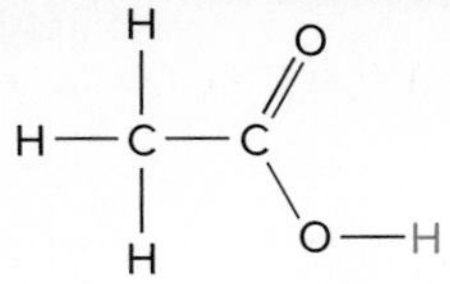 H_3C—C(=O)—O—H	$CH_3COOH(aq) + H_2O(\ell) \rightleftharpoons CH_3COO^-(aq) + H_3O^+(aq)$	1.76×10^{-5}
Formic	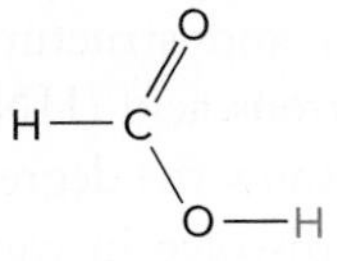 H—C(=O)—O—H	$HCOOH(aq) + H_2O(\ell) \rightleftharpoons HCOO^-(aq) + H_3O^+(aq)$	1.77×10^{-4}
Hydrofluoric	F—H	$HF(aq) + H_2O(\ell) \rightleftharpoons F^-(aq) + H_3O^+(aq)$	6.8×10^{-4}
Hypochlorous	Cl—O—H	$HClO(aq) + H_2O(\ell) \rightleftharpoons ClO^-(aq) + H_3O^+(aq)$	2.9×10^{-8}
Nitrous	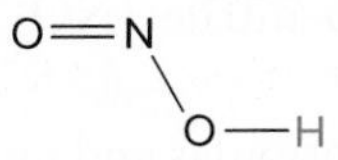 O=N—O—H	$HNO_2(aq) + H_2O(\ell) \rightleftharpoons NO_2^-(aq) + H_3O^+(aq)$	4.0×10^{-4}

this is that ionization reactions such as this one consume an insignificant fraction of the water molecules present in most aqueous solutions.

SAMPLE EXERCISE 15.1 Relating K_a and Acid Strength **LO1**

Suppose 0.100 *M* aqueous solutions of each of the acids in Table 15.2 are prepared. Rank them from most acidic to least acidic.

Collect, Organize, and Analyze The acidity of an aqueous solution is proportional to the concentration of H_3O^+ ions in it. The larger the value of K_a, the greater the concentration of H_3O^+ ions produced by a given concentration of acid.

Solve Ranking the acids in order of decreasing K_a values, we have: hydrofluoric acid > nitrous acid > formic acid > acetic acid > hypochlorous acid.

Think About It It makes sense that concentrations of H_3O^+ ions in the samples should be proportional to the K_a values of the acids because larger K_a values mean larger numerator values of the acids' K_a expressions (assuming the denominators of these 0.100 *M* solutions were essentially the same), and the numerators contain the $[H_3O^+]$ terms.

Practice Exercise Rank the weak acids in the table shown here in order of decreasing acid strength.

Weak Acid	K_a
HN_3	1.9×10^{-5}
$CH_2BrCOOH$	2.0×10^{-3}
$HClO_2$	1.1×10^{-2}
Lactic	1.4×10^{-4}
C_6H_5COOH	6.5×10^{-5}

(Answers to Practice Exercises are in the back of the book.)

Now let's compare the molecular formulas and structures of nitric acid (HNO_3), a strong acid from Table 15.1, and nitrous acid (HNO_2), a weak acid from Table 15.2. The bar graphs in **Figure 15.2** show the degree to which 0.1 *M* solutions of these two acids ionize when they dissolve in water. Molecules of HNO_3 ionize completely, each one donates a H^+ ion to a molecule of H_2O as all the HNO_3 molecules become NO_3^- ions:

$$HNO_3(aq) + H_2O(\ell) \rightarrow NO_3^-(aq) + H_3O^+(aq) \qquad (15.3)$$

On the other hand, only a small fraction of the molecules of HNO_2 ionize (Figure 15.2b) as described by the following chemical equation:

$$HNO_2(aq) + H_2O(\ell) \rightleftharpoons NO_2^-(aq) + H_3O^+(aq) \qquad (15.4)$$

At equilibrium, molecules of HNO_2 donate H^+ ions to molecules of H_2O at the same rate that H_3O^+ ions donate H^+ ions to nitrite (NO_2^-) ions, reforming HNO_2 and H_2O.

Why do these two acids, with such similar formulas and molecular structures, have such different acid strengths? The key is in the number of O atoms bonded

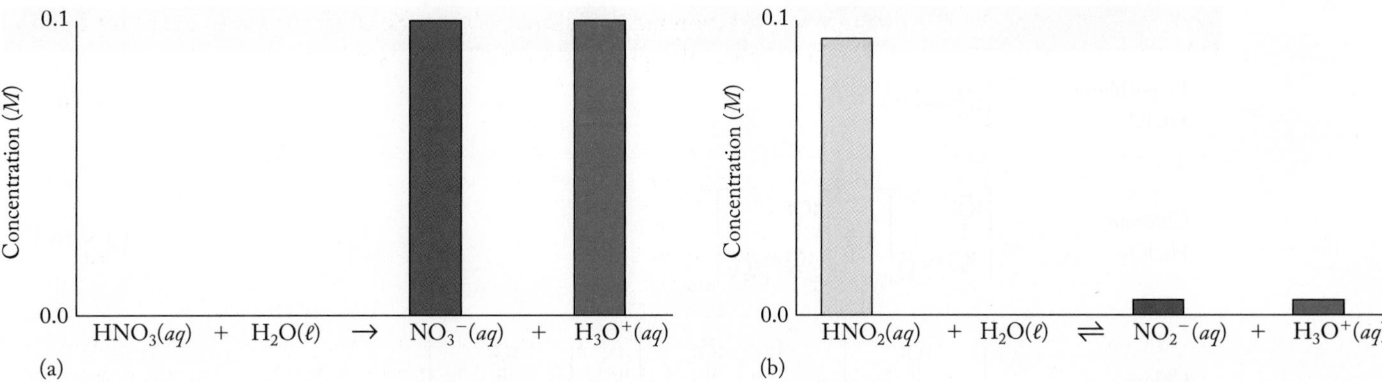

FIGURE 15.2 Degrees of ionization of a strong acid versus a weak acid. (a) The strong acid HNO_3 is ionized completely, existing only as H_3O^+ and NO_3^- ions. (b) The weak acid HNO_2 ionizes very little. [*Note*: Although $H_2O(\ell)$ is part of the balanced chemical equation in both (a) and (b), the amount is quite large relative to the other species and has been omitted from these graphs.]

to their central N atoms. Recall from Section 8.3 (see Figure 8.9) that oxygen is the second-most-electronegative element (after fluorine). This means that oxygen atoms bonded to the central atoms in molecules of oxoacids attract electron density toward themselves and away from the H atoms in the molecules' O—H groups. This allows these H atoms to interact more strongly with the O atoms in molecules of H_2O.

In addition, the presence of more O atoms in an oxoacid allows the negative charge of the oxoanion it forms when it ionizes to be spread out (or *delocalized*) over a greater number of O atoms, as shown by the resonance structures in **Figure 15.3**. Single-bonded O atoms (highlighted in red) have formal charges of −1. The three resonance structures of the nitrate ions (Figure 15.3a) each have two single-bonded O atoms and a central N atom (highlighted in blue) with a formal charge of +1, for a net ionic charge of 1−. Thus, the nitrate ion has *three* ways to distribute its charge. The nitrite ion, on the other hand, has only *two* (Figure 15.3b). Greater delocalization of the charge on a nitrate ion helps make nitric acid stronger than nitrous acid.

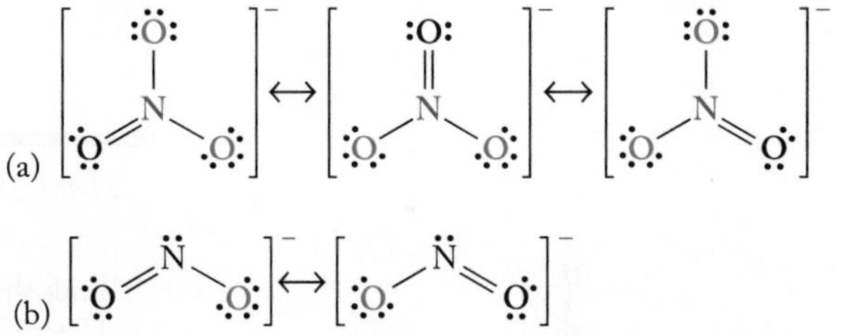

FIGURE 15.3 Nitric acid (HNO_3) is a stronger acid than nitrous acid (HNO_2) in part because more resonance structures more effectively delocalize the negative charge on a nitrate ion shown in (a) compared to a nitrite ion shown in (b). This delocalization leads to greater chemical stability and promotes the acid ionization reaction in which molecules of HNO_3 donate H^+ ions and form NO_3^- ions.

In a similar fashion, SO_4^{2-} ions are more stable than SO_3^{2-} ions, helping to make H_2SO_4 a stronger acid than H_2SO_3. This trend of increasing acid strength with increasing numbers of oxygen atoms bonded to the central atom (i.e., with increasing oxidation number of the central atom) is true for all groups of oxoacids with the same central atom, as illustrated by the strengths of the oxoacids of chlorine (**Figure 15.4**).

The strength of an oxoacid is also related to the electron-withdrawing power of the central atom. Consider, for example, the relative strengths of the three hypohalous acids in **Figure 15.5**. The most electronegative of the three halogen atoms (Cl) has the greatest attraction for the pair of electrons it shares with oxygen. This attraction helps draw electron density toward chlorine and away from the hydrogen end of the already polar O—H bond. These greater shifts in electron density also mean that the negative charge on a hypochlorite (ClO^-) ion is more delocalized than on other hypohalite ions. Thus, hypochlorous acid [$HClO(aq)$] is the strongest of the three acids, followed by hypobromous acid [$HBrO(aq)$] and then hypoiodous acid [$HIO(aq)$].

CONNECTION Introduced in Section 4.8, oxidation numbers are positive or negative numbers assigned to an atom based on the number of electrons that it gains or loses when it forms an ion or that it shares when it forms covalent bonds with the atoms of other elements.

CONNECTION Figure 8.9 depicts the electronegativities of the elements.

Acid	Resonance Structures of Oxoanion	K_a
Hypochlorous $HClO$	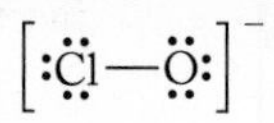 $[:\ddot{Cl}—\ddot{O}:]^-$	2.9×10^{-8}
Chlorous $HClO_2$	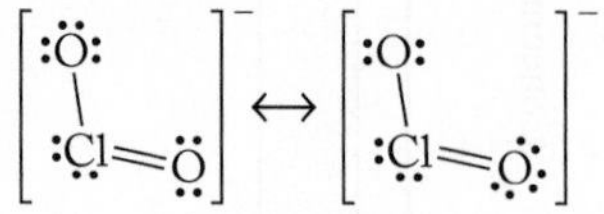 $[:\ddot{O}:—Cl=\ddot{O}]^- \leftrightarrow [:O:—Cl=\ddot{O}:]^-$	1.1×10^{-2}
Chloric $HClO_3$	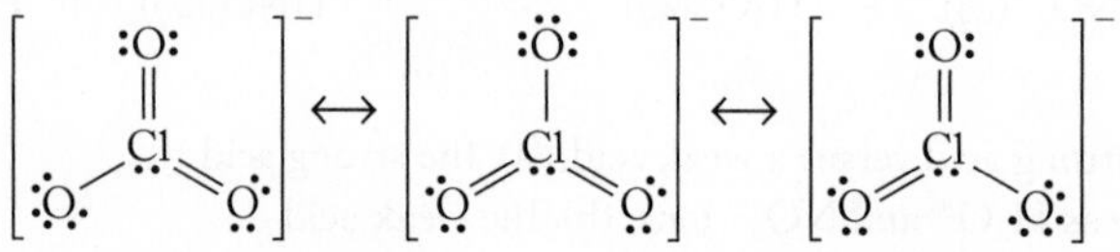 $[O=Cl(=O)—\ddot{O}:]^- \leftrightarrow [:\ddot{O}—Cl(=O)=O]^- \leftrightarrow [O=Cl(=O)—\ddot{O}:]^-$	· 1
Perchloric $HClO_4$	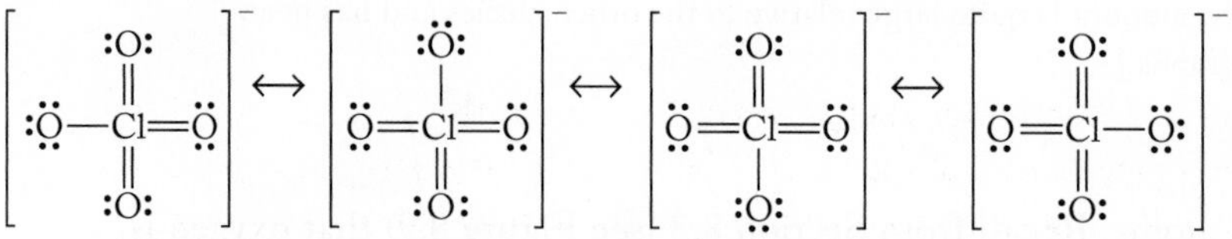 $[:\ddot{O}—Cl(=O)_2=O]^- \leftrightarrow [O=Cl(—\ddot{O}:)(=O)=O]^- \leftrightarrow [O=Cl(=O)(—\ddot{O}:)=O]^- \leftrightarrow [O=Cl(=O)_2—\ddot{O}:]^-$	Strong acid

FIGURE 15.4 In the oxoacids of chlorine, acid strength increases with increasing number of O atoms bonded to the Cl atom and the greater ability of the oxoanions they form to delocalize their negative charge.

CHEMTOUR
Acid Strength and Molecular Structure

SAMPLE EXERCISE 15.2 Ranking Oxoacid Strength — LO2

Rank the following compounds in order of decreasing acid strength: phosphoric acid (H_3PO_4), arsenic acid (H_3AsO_4), and antimonic acid (H_3SbO_4).

Collect and Organize Each of these oxoacids contains a different group 15 element. In each acid, this central atom is bonded to the same number of O atoms and has the same oxidation number (+5). The electronegativities of these elements decrease with increasing atomic number and row number.

Analyze In oxoacids, more electronegative central atoms draw electron density away from the H ends of the O—H bonds and help disperse the negative charges that form when oxoacids release H^+ ions to form oxoanions.

Solve Ranking the oxoacids in order of decreasing acid strength is a matter of ranking them in order of decreasing electronegativity of the central atom—that is, in order of increasing atomic number (and row number) in the periodic table: $H_3PO_4 > H_3AsO_4 > H_3SbO_4$.

Think About It The trend of decreasing oxoacid strength with increasing atomic number (decreasing electronegativity) of their central atoms is confirmed by the K_a values of these acids as each of their molecules loses one H^+ ion, that is, $K_{a,H_3PO_4} = 6.9 \times 10^{-3}$; $K_{a,H_3AsO_4} = 5.5 \times 10^{-3}$; $K_{a,H_3SbO_4} = 1.9 \times 10^{-3}$.

Practice Exercise Of the following acids, which is the strongest and which is the weakest? Sulfuric acid (H_2SO_4), sulfurous acid (H_2SO_3), and selenous acid (H_2SeO_3)

Acid	Structure	Electronegativity of Halogen Atom	K_a
Hypochlorous HClO	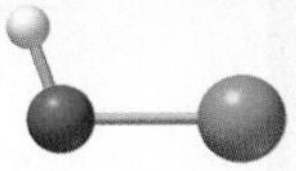	3.0	2.9×10^{-8}
Hypobromous HBrO		2.8	2.3×10^{-9}
Hypoiodous HIO	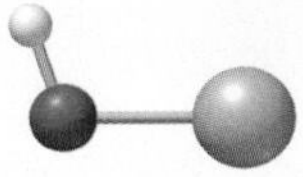	2.5	2.3×10^{-11}

FIGURE 15.5 The strengths of these three hypohalous acids increase with the increasing electronegativities of their halogen atoms. Thus, the relative acidities are HClO > HBrO > HIO.

Strong and Weak Bases

Most strong bases are not molecular compounds, but rather are the soluble hydroxides of the group 1 and 2 elements. **Table 15.3** lists the ions present when these ionic compounds completely dissociate as they dissolve in water. Note that a mole of a group 1 hydroxide such as NaOH produces 1 mole of OH^- ions, but a mole of a group 2 hydroxide such as $Ba(OH)_2$ produces 2 moles of OH^- ions. Their complete dissociation makes these hydroxides strong bases because OH^- ions are the strongest H^+ ion acceptors that can exist in aqueous solutions. We explore why this is true in Section 15.4.

TABLE 15.3 Strong Bases and Their Dissociation in Water

Strong Base	Formula	Ions in Aqueous Solutions
Lithium hydroxide	LiOH	$Li^+(aq) + OH^-(aq)$
Sodium hydroxide	NaOH	$Na^+(aq) + OH^-(aq)$
Potassium hydroxide	KOH	$K^+(aq) + OH^-(aq)$
Calcium hydroxide	$Ca(OH)_2$	$Ca^{2+}(aq) + 2\ OH^-(aq)$
Barium hydroxide	$Ba(OH)_2$	$Ba^{2+}(aq) + 2\ OH^-(aq)$
Strontium hydroxide	$Sr(OH)_2$	$Sr^{2+}(aq) + 2\ OH^-(aq)$

Molecular bases such as those listed in **Table 15.4** are weakly basic. To understand how they function as bases, let's focus on what happens when a 0.1 *M* solution of ammonia is prepared. Collisions between water and ammonia molecules sometimes result in the transfer of a proton from a water molecule (functioning as

TABLE 15.4 Some Common Weak Bases and Their Ionization Reactions in Water

Weak Base	Reaction in Water	K_b
Ammonia	$NH_3(aq) + H_2O(\ell) \rightleftharpoons NH_4^+(aq) + OH^-(aq)$	1.8×10^{-5}
Aniline	$C_6H_5NH_2(aq) + H_2O(\ell) \rightleftharpoons C_6H_5NH_3^+(aq) + OH^-(aq)$	4.0×10^{-10}
Dimethylamine	$(CH_3)_2NH(aq) + H_2O(\ell) \rightleftharpoons (CH_3)_2NH_2^+(aq) + OH^-(aq)$	5.9×10^{-4}
Methylamine	$CH_3NH_2(aq) + H_2O(\ell) \rightleftharpoons CH_3NH_2^+(aq) + OH^-(aq)$	4.4×10^{-4}
Pyridine	$C_5H_5N(aq) + H_2O(\ell) \rightleftharpoons C_5H_5NH^+(aq) + OH^-(aq)$	1.7×10^{-9}

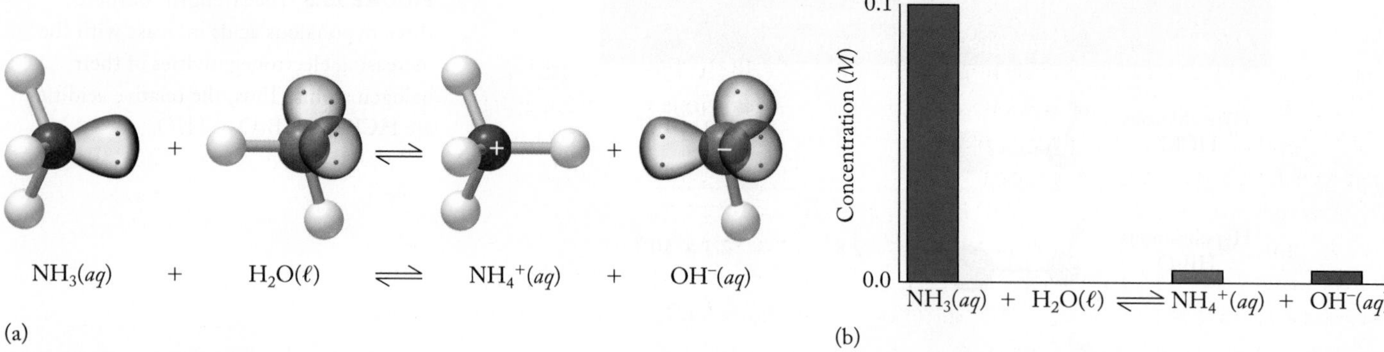

FIGURE 15.6 (a) Molecules of ammonia can react with molecules of water, producing NH_4^+ and OH^- ions. (b) In a 0.1 *M* solution, ammonia exists mostly as molecules of NH_3 in equilibrium with much smaller concentrations of NH_4^+ and OH^- ions.

a Brønsted–Lowry acid) to an ammonia molecule (a Brønsted–Lowry base). The transferred proton bonds to the lone pair on the nitrogen atom, producing NH_4^+ and OH^- ions as shown in **Figure 15.6a**. Many other weak bases also contain nitrogen atoms with lone pairs of electrons that can use these electrons to form covalent bonds with H^+ ions (see Table A5.3 in Appendix 5).

Only a small fraction of the ammonia molecules in a 0.1 *M* solution react with molecules of water, as shown by the bar graph in **Figure 15.6b**. This tells us that the equilibrium constant for the reaction between NH_3 and H_2O, which has the symbol K_b to indicate that it is an ionization reaction involving a basic compound and water, must be very small. In fact, it is only 1.76×10^{-5} at 25°C.

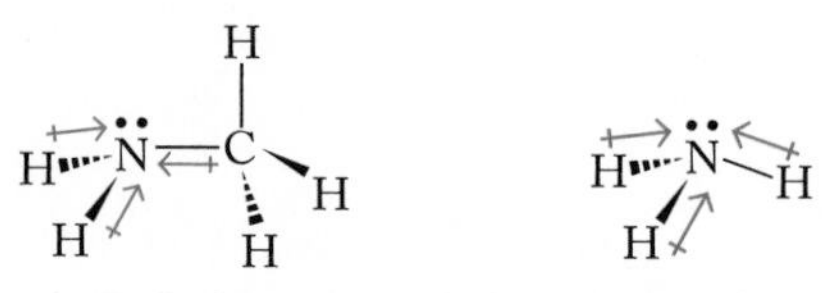

(a) Methylamine (b) Ammonia

(c) Aniline Pyridine

FIGURE 15.7 (a) The greater number of valence electrons in the methyl group of a molecule of CH_3NH_2 produces a greater density of electrons that are drawn toward its N atom as compared to the electron density in a molecule of NH_3 (b), which makes methylamine a stronger base than ammonia. The electron density surrounding the N atoms in the aromatic compounds (c) aniline and pyridine is much less than that around the N atom in ammonia, because the electrons are delocalized around the aromatic rings.

The other nitrogen-containing bases in Table 15.4 have K_b values that are either an order of magnitude or more greater than that of ammonia (methylamine and dimethylamine) or several orders of magnitude less (aniline and pyridine). As with the strengths of acids, differences in base strength are related to molecular structures. Molecules of methylamine and dimethylamine resemble molecules of ammonia in which methyl groups have replaced one or two of the H atoms in NH_3. These replacements mean there are more valence electrons in the two amines than in ammonia, and these electrons are in polar covalent bonds whose electrons are drawn to the electronegative N atoms of the amines (**Figure 15.7**). This shift means there is more electron density surrounding the N atoms, which enhances the strength of their interactions with the H atoms of water molecules, which makes them better H^+ ion acceptors and, therefore, stronger bases.

There are many more valence electrons in aniline and pyridine, compared to ammonia, but these electrons are *not* drawn to the N atoms in these molecules. Why? One reason is that both aniline and pyridine are aromatic compounds (Figure 15.7c), which means they have six-atom rings of highly delocalized π electrons. The enhanced stability of these delocalized electrons draws electron density away from the N atoms and into the rings.

15.3 Conjugate Pairs and Their Complementary Strengths as Acids and Bases

In this section we explore the relationship between acidic compounds and the anions they form when they undergo acid ionization. If these ionization reactions are reversible, then the anions must have the capacity to accept H^+ ions as they reform the parent acid. In other words, the anions are bases!

Recognizing Conjugate Pairs

Let's revisit the nitrous acid equilibrium in Equation 15.4. In the forward direction, HNO_2 functions as a Brønsted–Lowry acid by donating H^+ ions to molecules of H_2O (**Figure 15.8a**). In the reverse reaction, NO_2^- ions accept H^+ ions and function as a Brønsted–Lowry base. Structurally, the difference between HNO_2 and NO_2^- is the H^+ ion that a molecule of HNO_2 donates, resulting in the formation of a NO_2^- ion, and that a NO_2^- ion accepts in the reverse reaction to reform a molecule of HNO_2. An acid and a base that are related in this way are called a **conjugate acid–base pair**, or *conjugate pair.* An acid forms its **conjugate base** when it donates a H^+ ion, and a base forms its **conjugate acid** when it accepts a H^+ ion.

Figure 15.8b depicts the base ammonia and its conjugate acid. As we saw in Figure 15.6, when ammonia is added to water, some molecules of NH_3 accept H^+ ions from molecules of H_2O. In this way NH_3 functions as a Brønsted–Lowry base. In the reverse reaction, NH_4^+ ions donate H^+ ions and function as Brønsted–Lowry acids. Because the structure of NH_3 molecules and the structure of NH_4^+ ions differ only by the H^+ ion that a molecule of NH_3 accepts and that an NH_4^+ ion donates, NH_3 and NH_4^+ are also a conjugate acid–base pair.

Now let's consider the acid–base behavior of the water molecules in the reactions in Figure 15.8. H_2O functions as a *base* in the nitrous acid reaction, accepting a H^+ ion and forming its conjugate acid, H_3O^+. This behavior makes H_2O and H_3O^+ a conjugate pair. In the ammonia reaction, however, H_2O functions as an *acid* by donating a H^+ ion to form its conjugate base, the OH^- ion. These two reactions illustrate that water can act as an acid or a base, depending on the acid–base properties of the substance dissolved in it. We revisit this acid–base duality of water in Section 15.4.

The following generic chemical equations summarize the relationships between conjugate pairs in aqueous solutions.

$$\text{Acid}(aq) + H_2O(\ell) \rightleftharpoons \text{conjugate base}(aq) + H_3O^+(aq)$$

$$\text{Base}(aq) + H_2O(\ell) \rightleftharpoons \text{conjugate acid}(aq) + OH^-(aq)$$

The difference within each conjugate acid–base pair is the H^+ ion that the acid donates to form its conjugate base and that the base accepts to form its conjugate acid.

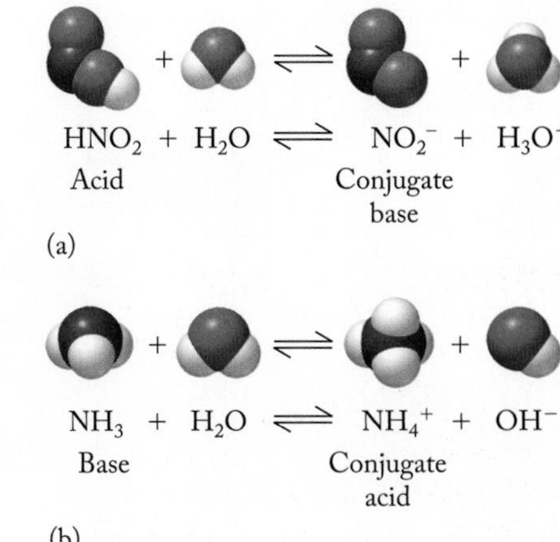

FIGURE 15.8 Two acid–base conjugate pairs.

STEPWISE ANIMATION
Conjugate Acids and Bases

SAMPLE EXERCISE 15.3 Identifying a Conjugate Acid–Base Pair **LO3**

Identify the conjugate acid–base pair in the ionization reactions that occur when (a) perchloric acid ($HClO_4$) and (b) formic acid (HCOOH) dissolve in water.

Collect, Organize, and Analyze Brønsted–Lowry acids form their conjugate bases when they donate H^+ ions to molecules of H_2O. Therefore, the formulas of their conjugate bases are the formulas of the original acids, minus a H^+ ion.

Solve (a) Perchloric acid ($HClO_4$) has only one H atom per molecule, which must be the one that it loses as a H^+ ion:

$$\underset{\text{Acid}}{HClO_4(aq)} + H_2O(\ell) \rightarrow \underset{\text{Conjugate base}}{ClO_4^-(aq)} + H_3O^+(aq)$$

(b) Formic acid (HCOOH) has two H atoms per molecule, but only the one bonded to an O atom in the carboxylic acid group (see Table 15.2) is ionizable in water:

$$\underset{\text{Acid}}{HCOOH(aq)} + H_2O(\ell) \rightleftharpoons \underset{\text{Conjugate base}}{HCOO^-(aq)} + H_3O^+(aq)$$

conjugate acid–base pair a Brønsted–Lowry acid and base that differ from each other only by a H^+ ion: acid ⇌ conjugate base + H^+.

conjugate base the base formed when a Brønsted–Lowry acid donates a H^+ ion.

conjugate acid the acid formed when a Brønsted–Lowry base accepts a H^+ ion.

In both reactions, H_3O^+ and H_2O are also a conjugate acid–base pair: H_2O is the base and H_3O^+ is its conjugate acid.

Think About It The formulas of the two conjugate bases make sense because each has one fewer H atom than the parent acid molecule and each is an anion with a 1− charge. These properties are consistent with an acid molecule losing one H^+ ion.

Practice Exercise Identify the conjugate acid–base pairs in the reaction that occurs when acetic acid (CH_3COOH) dissolves in water.

Relative Strengths of Conjugate Acids and Bases

HCl is a strong acid (see Table 15.1), and like $HClO_4$ in Sample Exercise 15.3, its acid ionization reaction goes to completion:

$$HCl(aq) + H_2O(\ell) \rightarrow Cl^-(aq) + H_3O^+(aq) \quad (15.5)$$

As a result, the reverse reaction essentially does not happen at all, which means that the Cl^- ion (the conjugate base of HCl) must be a very weak base. This contrast in relative strengths applies to all conjugate pairs: strong acids have very weak conjugate bases and strong bases have very weak conjugate acids, as shown in **Figure 15.9**.

Between these extremes exist many weak acids with weak conjugate bases. For example, HNO_2 is a weak acid ($K_a = 4.0 \times 10^{-4}$), which means that its conjugate base, NO_2^-, is a weak base. This pairing of weak acids and weak bases applies to most of the conjugate acids and bases that occur in nature and in ourselves.

All the strong acids in Figure 15.9 ionize completely in water; that is, every molecule transfers one H^+ ion to a molecule of H_2O. In this context, water is said to *level* the strengths of these acids; they all are equally strong in water because they cannot be more than 100% ionized. This **leveling effect** means that H_3O^+, the conjugate acid of H_2O, is the strongest acid that can exist in water. An acid that is stronger than H_3O^+ simply donates all its ionizable H atoms to water molecules, forming H_3O^+ ions, when it dissolves in water. On the other hand, weak acids are differentiated by the fact that only a small fraction of the molecules donate their ionizable H atoms to water molecules. The weak acids higher on the list in Figure 15.9 form more acidic aqueous solutions than do acids lower on the list because the larger the K_a, the greater the fraction of molecules that dissociate at a given concentration.

A similar leveling effect exists for strong bases. The strongest base that can exist in water is the OH^- ion, the conjugate base of H_2O. Any base that is stronger than OH^- hydrolyzes in water to produce OH^- ions. The oxide ion (O^{2-}), for example, is a very strong base, and it reacts with water to produce two OH^- ions:

$$O^{2-}(aq) + H_2O(\ell) \rightarrow 2\ OH^-(aq)$$

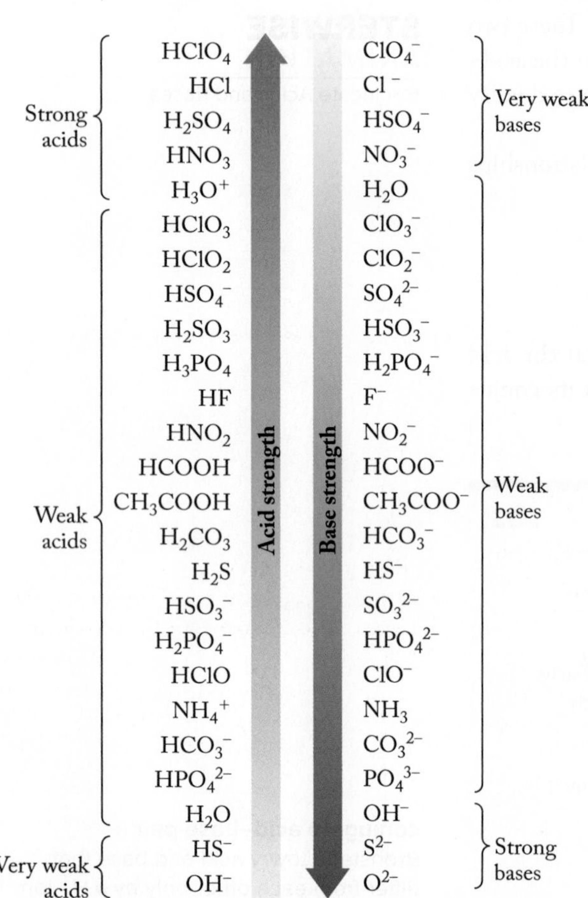

FIGURE 15.9 Opposing trends characterize the relative strengths of acids and their conjugate bases: the stronger the acid, the weaker its conjugate base. The same is true for bases: the stronger the base, the weaker its conjugate acid.

The strengths of bases weaker than OH^- ions can be differentiated by the fraction of their molecules that accept H^+ ions from water molecules in aqueous solutions. Weaker bases are higher on the list in Figure 15.9; stronger bases are lower on the list.

leveling effect the observation that all strong acids have the same strength in water and are completely converted into solutions of H_3O^+ ions; strong bases are likewise leveled in water and are completely converted into solutions of OH^- ions.

autoionization the process that produces equal and very small concentrations of H_3O^+ and OH^- ions in pure water.

SAMPLE EXERCISE 15.4 Relating the Strengths of a Conjugate Pair **LO3**

List the following anions in order of decreasing strength as Brønsted–Lowry bases: F^-, Cl^-, OH^-, $HCOO^-$, and NO_2^-.

Collect, Organize, and Analyze All five anions are listed among the bases in Figure 15.9. Therefore, all we must do is rank them according to their location in the figure: the strongest one will be the closest to the bottom and the weakest will be the closest to the top.

Solve The anions in decreasing order of strength as bases (H^+ ion acceptors) are OH^-, $HCOO^-$, NO_2^-, F^-, and Cl^-.

Think About It According to Figure 15.9, we have also sorted the anions in *increasing* order for the strength of their conjugate acids. The complementary sorting makes sense because the stronger the acid, the weaker its conjugate base, and vice versa.

Practice Exercise List the following anions in order of decreasing strength as Brønsted–Lowry bases: Br^-, S^{2-}, ClO^-, CH_3COO^-, and HSO_3^-.

15.4 pH and the Autoionization of Water

The acidity of an aqueous solution is directly related to its concentration of H_3O^+ ions. In Section 15.4 we examine another way to quantify acidity. To understand this alternative approach, we begin with the **autoionization** of water, which is the process that can happen when two water molecules collide to produce equal and very small concentrations of H_3O^+ and OH^- ions in pure water:

$$H_2O(\ell) + H_2O(\ell) \rightleftharpoons H_3O^+(aq) + OH^-(aq) \qquad (15.6)$$

One water molecule, acting as an acid, donates a H^+ ion to another water molecule that functions as a base (**Figure 15.10**). The donor H_2O molecule forms its conjugate base (OH^-), and the acceptor H_2O molecule forms its conjugate acid (H_3O^+). We have already encountered this dual nature of water: molecules of H_2O act as H^+ ion acceptors in solutions of acidic solutes and as H^+ ion donors in solutions of basic solutes. The autoionization of water is an example of amphiprotic behavior.

CONNECTION *Amphiprotic* compounds (Section 4.5) can behave as either a proton acceptor (a Brønsted–Lowry base) or a proton donor (a Brønsted–Lowry acid).

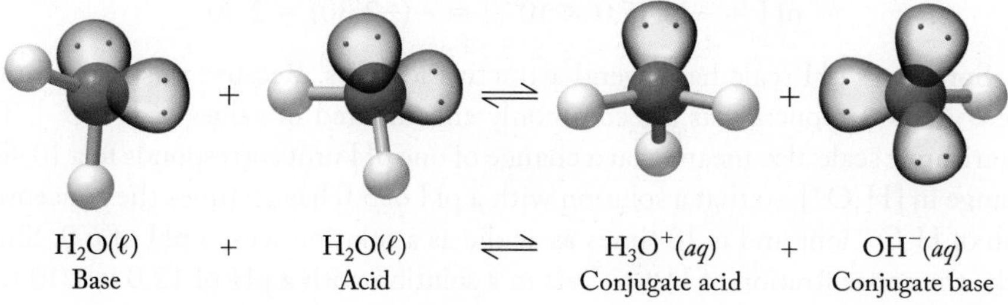

FIGURE 15.10 The autoionization of water takes place when a proton (H^+) is transferred from one water molecule to another. Both molecules are converted to ions.

The equilibrium constant expression for the autoionization of water might be written

$$K_c = \frac{[H_3O^+][OH^-]}{[H_2O][H_2O]} \quad (15.7)$$

Water is a pure liquid, however, so we do not include its concentration in equilibrium constant expressions. This reduces Equation 15.7 to an equilibrium constant expression that is given the symbol K_w:

CHEMTOUR
Autoionization of Water

$$K_w = [H_3O^+][OH^-] \quad (15.8)$$

In pure water at 25°C $[H_3O^+] = [OH^-] = 1.0 \times 10^{-7}$ *M*. Inserting these values into Equation 15.8 gives us

$$K_w = [H_3O^+][OH^-] = (1.0 \times 10^{-7})(1.0 \times 10^{-7}) = 1.0 \times 10^{-14} \quad (15.9)$$

This tiny K_w value confirms that only a very small fraction of water molecules undergoes autoionization at room temperature. The reverse of autoionization—the reaction between $[H_3O^+]$ and $[OH^-]$ to produce H_2O—has an equilibrium constant of $1/K_w = 1.0 \times 10^{14}$ at 25°C and essentially goes to completion:

$$H_3O^+(aq) + OH^-(aq) \rightleftharpoons H_2O(\ell) + H_2O(\ell) \qquad K = 1/K_w = 1.0 \times 10^{14}$$

Equation 15.8 means that an inverse relationship exists between $[H_3O^+]$ and $[OH^-]$ in any aqueous sample—namely, as the concentration of one increases, the concentration of the other must decrease so that the product of the two is always 1.0×10^{-14}. A solution in which $[H_3O^+] > [OH^-]$ is acidic, a solution in which $[H_3O^+] < [OH^-]$ is basic, and a solution in which $[H_3O^+] = [OH^-] = 1.0 \times 10^{-7}$ *M* is neutral (neither acidic nor basic).

The tiny value of K_w means that autoionization of water does not contribute significantly to $[H_3O^+]$ in solutions of most acids or to $[OH^-]$ in solutions of most bases, so we can ignore the contribution of autoionization in most calculations of acid or base strength. If acids or bases are extremely weak, however, or if their concentrations are extremely low, then H_2O autoionization may need to be considered.

The pH Scale

In the early 1900s scientists developed a device called the *hydrogen electrode* to determine the concentrations of hydronium ions in solutions. The electrical voltage, or *potential*, produced by the hydrogen electrode is a linear function of the logarithm of $[H_3O^+]$. This relationship led Danish biochemist Søren Sørensen (1868–1939) to propose a scale for expressing acidity based on what he termed "the potential of the hydrogen ion," abbreviated **pH**. Mathematically, we define pH as the base-10 negative logarithm of $[H_3O^+]$:

CHEMTOUR
pH Scale

$$pH = -\log[H_3O^+] \quad (15.10)$$

For example, the pH of a solution in which $[H_3O^+] = 5.0 \times 10^{-3}$ *M* is

$$pH = -\log(5.0 \times 10^{-3}) = -(-2.30) = 2.30$$

Sørensen's pH scale has several attractive features. Because it is logarithmic, there are no exponents, as are commonly encountered in values of $[H_3O^+]$. The logarithmic scale also means that a change of one pH unit corresponds to a 10-fold change in $[H_3O^+]$, so that a solution with a pH of 5.0 has 10 times the concentration of H_3O^+ ions and is 10 times as acidic as a solution with a pH of 6.0. Similarly, the concentration of H_3O^+ ions in a solution with a pH of 12.0 is 1/10 that

pH the negative logarithm of the hydronium ion concentration in an aqueous solution.

of a solution with a pH of 11.0. Conversely, the concentration of OH^- ions in a solution with a pH of 12.0 is 10 times that of a solution with a pH of 11.0.

The negative sign in front of the logarithmic term in Equation 15.10 means that pH values of most aqueous solutions, except for very concentrated solutions of strong acids or bases, are positive numbers between 0 and 14. It also means that *large* pH values correspond to *small* values of $[H_3O^+]$. Acidic solutions have pH values less than 7.00 ($[H_3O^+] > 1.0 \times 10^{-7}$ *M*), and basic solutions have pH values greater than 7.00 ($[H_3O^+] < 1.0 \times 10^{-7}$ *M*). A solution with a pH of exactly 7.00 is neutral. The pH values for some common aqueous solutions are shown in **Figure 15.11**.

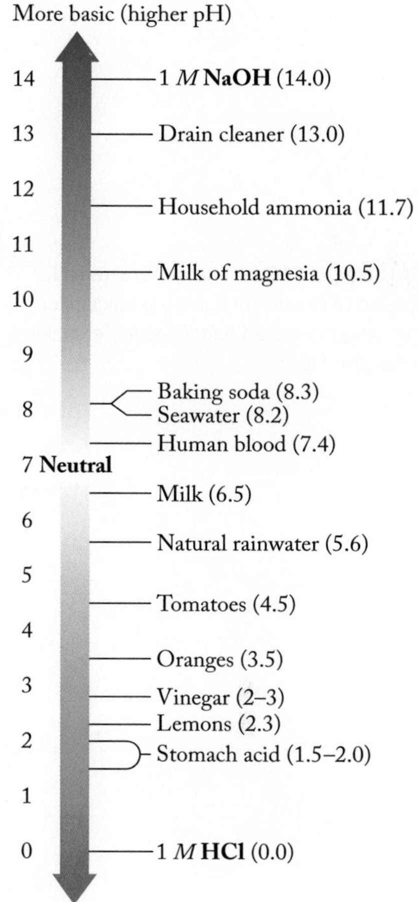

FIGURE 15.11 The pH scale is a convenient way to express the range of acidic or basic properties of some common materials.

CONCEPT TEST

Match the pH values on the left with the descriptors on the right.

13.77	strongly acidic
10.03	weakly acidic
7.00	weakly basic
4.37	strongly basic
0.22	neutral

SAMPLE EXERCISE 15.5 Relating pH and $[H_3O^+]$ LO4

Suppose the pH of solution A is 8.0 and the pH of solution B is 4.0. Which of the following statements about the two solutions are true and which are false?

a. Solution A is 2 times as acidic as solution B.
b. Solution B is 10,000 times as acidic as solution A.
c. The concentration of OH^- ions in solution A is half their concentration in solution B.
d. The values of $[OH^-]$ and $[H_3O^+]$ are closer to each other in solution A than in solution B.

Collect, Organize, and Analyze The pH scale is logarithmic, so a decrease (or increase) of one pH unit corresponds to a 10-fold increase (or a 90% decrease) in $[H_3O^+]$. The negative sign in the pH formula ($pH = -\log[H_3O^+]$) means that a higher pH value corresponds to a (much) lower $[H_3O^+]$.

Solve

a. False, because pH is a log scale, and because higher pH values mean lower $[H_3O^+]$.
b. True; a decrease of 4 pH units corresponds to a 10^4 increase in $[H_3O^+]$ and acidity.
c. False, because pH is a log scale. Actually, $[OH^-]$ in solution A is 10^4 times that of solution B.
d. True, because a pH of 8.0 is only one unit away from the pH (7.0) where $[H_3O^+] = [OH^-]$, so $[OH^-] = 10 \times [H_3O^+]$ in solution A. A pH of 4.0 is 3 units below 7.0, so $[H_3O^+] = 1000 \times [OH^-]$ in solution B.

Think About It There are two key aspects of the equation $pH = -\log[H_3O^+]$ that affect which of the four statements are true: (1) the negative sign means that the higher the pH of a solution, the lower its $[H_3O^+]$; (2) the log scale means that a decrease in pH of one unit corresponds to a 10-fold increase in $[H_3O^+]$.

Practice Exercise Which of the following changes in pH corresponds to the greatest percent increase in $[H_3O^+]$? Which change corresponds to the greatest percent decrease in $[H_3O^+]$? (a) pH 1 → pH 3; (b) pH 7 → pH 4; (c) pH 12 → pH 14; (d) pH 5 → pH 9; (e) pH 2 → pH 0

When we convert $[H_3O^+]$ to pH values and pH values to $[H_3O^+]$, we need to express the resulting values with the appropriate number of significant digits. Because a pH value is the negative logarithm of a hydronium ion concentration, the digit or digits before the decimal point in the pH value define the location of the decimal point in the $[H_3O^+]$ value. They do *not* define how precisely we know the concentration value, so they are not considered significant digits. For example, $[H_3O^+] = 2.7 \times 10^{-4}$ has two significant digits: the 2 and 7. The corresponding pH value should also have two significant digits, but those two should come after the decimal point, which means the pH value is 3.57, not 3.6. The 3 in pH 3.57 is not a significant digit because it serves only to indicate that the corresponding $[H_3O^+]$ value falls between 10^{-3} and 10^{-4} *M*.

CONNECTION The general rules for the use of significant figures in calculations involving measured quantities are explained in Section 1.10.

SAMPLE EXERCISE 15.6 Converting $[H_3O^+]$ to pH and pH to $[H_3O^+]$ **LO4**

Cells located in the upper region of the human stomach called the *fundus* secrete gastric acid (about 0.16 *M* HCl) during digestion. After this acid mixes with the food being digested, the pH of the contents of the stomach is often between 3 and 5.

a. How much more acidic is 0.16 *M* HCl than stomach contents with a pH of 3.20?
b. Which of the two samples in part (a) has the lower $[H_3O^+]$?

Collect and Organize We are asked to compare the acidities (the ratio of the $[H_3O^+]$ values) of two solutions that both contain hydrochloric acid. We know the pH of one and the HCl concentration of the other. We must also determine which sample has the lower $[H_3O^+]$.

Analyze Hydrochloric acid is a strong acid that is 100% ionized in aqueous solutions. Therefore, the $[H_3O^+]$ value of a solution of HCl is the same as the concentration of the acid. We can use Equation 15.10 to calculate the pH of 0.16 *M* HCl and to convert pH 3.20 into its corresponding $[H_3O^+]$ value. The result of the second calculation, when divided into 0.16 *M*, will give us the desired ratio of acidities.

Solve

a. To convert pH to $[H_3O^+]$, we need to solve Equation 15.10 for $[H_3O^+]$:

$$\text{pH} = -\log[H_3O^+]$$

or,

$$\log[H_3O^+] = -\text{pH}$$

Taking the antilog (10^x) of both sides:

$$[H_3O^+] = 10^{-\text{pH}}$$

Inserting pH = 3.20:

$$[H_3O^+] = 10^{-3.20} = 6.3 \times 10^{-4}\ M$$

Calculating the ratio of acidities ($[H_3O^+]$ values):

$$\frac{[H_3O^+](\text{gastric acid})}{[H_3O^+](\text{stomach contents})} = \frac{0.16\ M}{6.3 \times 10^{-4}\ M} = 2.5 \times 10^2$$

Thus, secreted gastric acid is 2.5×10^2 times more acidic than the digestion mixture in the stomach.

b. Using the values calculated in part (a), the stomach contents have a lower $[H_3O^+]$.

Think About It The results of the two parts to this sample exercise are consistent: 0.16 *M* HCl is 2.5×10^2 times more acidic than the stomach content sample, and its pH is 3.20 − 0.8 = 2.4 units lower. Note how this seemingly small change in pH corresponds to a more than 200-fold difference in acidity.

pOH the negative logarithm of the hydroxide ion concentration in an aqueous solution.

Practice Exercise The results of several oceanographic studies indicate that the average pH of Earth's oceans is now 8.07. The hydrogen ion concentration of the oceans at the start of the industrial revolution was 6.61×10^{-9} *M*. How much more acidic is the ocean now than then?

CONCEPT **TEST**

Is the pH of a 1.00 *M* solution of a weak acid higher or lower than the pH of a 1.00 *M* solution of a strong acid?

pOH, pK_a, and pK_b Values

The letter *p* as used in pH is also used with other symbols to mean *the negative logarithm* of the variable that follows it. For example, just as every aqueous solution has a pH value, it also has a **pOH** value, defined as

$$\text{pOH} = -\log[\text{OH}^-] \tag{15.11}$$

We can use the K_w expression (Equation 15.9) to relate pOH to pH. We start by taking the negative logarithm of both sides of the equation and then rearrange the terms:

$$K_w = [\text{H}_3\text{O}^+][\text{OH}^-] = 1.0 \times 10^{-14}$$
$$-\log K_w = -\log([\text{H}_3\text{O}^+][\text{OH}^-]) = -\log(1.0 \times 10^{-14})$$
$$\text{p}K_w = -\log[\text{H}_3\text{O}^+] + -\log[\text{OH}^-] = -(-14.00)$$
$$\text{p}K_w = \text{pH} + \text{pOH} = 14.00 \tag{15.12}$$

Many tables of equilibrium constants list p*K* values rather than *K* values because doing so does not require the use of exponential notation and saves space. The tables in Appendix 5 of this book contain both sets of values. Use them whenever you need a *K* or p*K* value that is not provided in a problem. For example, we can define both pK_a and pK_b values for acetic acid and methylamine, respectively, and for the other acids and bases in Tables 15.2 and 15.4:

$$\text{p}K_a = -\log K_a = -\log(1.8 \times 10^{-5}) = 4.74 \qquad \text{p}K_b = -\log K_b = -\log(4.4 \times 10^{-4}) = 3.36$$

SAMPLE EXERCISE 15.7 Interconvert $[\text{H}_3\text{O}^+]$, $[\text{OH}^-]$, pH, and pOH **LO4**

Digested food travels from your stomach, where the pH is 4.50, and moves to your small intestine, where the pH increases to 7.50. Convert these pH values to pOH, $[\text{H}_3\text{O}^+]$, and $[\text{OH}^-]$ values.

Collect, Organize, and Analyze We are given two pH values, and we want to determine the corresponding pOH, $[\text{H}_3\text{O}^+]$, and $[\text{OH}^-]$ values.

- We can use Equation 15.10 to convert from pH to $[\text{H}_3\text{O}^+]$, as we did in Sample Exercise 15.6.
- We can use Equation 15.12 to convert from pH to pOH.
- We can use Equation 15.11 to convert from pOH to $[\text{OH}^-]$.

The $[H_3O^+]$ value we calculate for pH 4.50 should be 10^3, or 1000, times the $[H_3O^+]$ value we calculate for pH 7.50 (because 7.50 − 4.50 = 3.00). Correspondingly, the $[OH^-]$ value we calculate for pH 7.50 should be 10^3, or 1000, times the $[OH^-]$ value we calculate for pH 4.50. The inverse relationship between pH and pOH means that the pOH value we calculate for pH 4.50 should be 3 units larger than the one we get for pH 7.50.

Solve

1. Converting pH to $[H_3O^+]$:

$$\text{pH} = -\log[H_3O^+] \quad (15.10)$$

or

$$[H_3O^+] = 10^{-\text{pH}}$$

Stomach: $[H_3O^+] = 10^{-4.50} = 3.2 \times 10^{-5}\ M$

Small intestine: $[H_3O^+] = 10^{-7.50} = 3.2 \times 10^{-8}\ M$

2. Converting pH to pOH:

$$\text{pH} + \text{pOH} = 14.00 \quad (15.12)$$

$$\text{pOH} = 14.00 - \text{pH}$$

Stomach: $\text{pOH} = 14.00 - 4.50 = 9.50$

Small intestine: $\text{pOH} = 14.00 - 7.50 = 6.50$

3. Converting pOH to $[OH^-]$:

$$\text{pOH} = -\log[OH^-] \quad (15.11)$$

or

$$[OH^-] = 10^{-\text{pOH}}$$

Stomach: $[OH^-] = 10^{-9.50} = 3.2 \times 10^{-10}\ M$

Small intestine: $[OH^-] = 10^{-6.50} = 3.2 \times 10^{-7}\ M$

Think About It As predicted, the $[H_3O^+]$ in the stomach is 1000 times that of the small intestine. The differences in pOH values are also as predicted, as are the differences in $[OH^-]$. All calculated concentration values were rounded off to two significant digits because each starting pH value had only two digits after the decimal point.

Practice Exercise What are the values of $[H_3O^+]$ and $[OH^-]$ in household ammonia, an aqueous solution of NH_3 that has a pH of 11.70?

15.5 K_a, K_b, and the Ionization of Weak Acids and Bases

Most of the acids and bases in nature are weak acids and bases. In Sections 15.5 and 15.6 we explore how to quantify the degree to which these molecules react with molecules of water to form either H_3O^+ or OH^- ions.

Weak Acids

Nitrous acid (HNO_2) is a weak acid (see Section 15.2), which means it is only partially ionized when it dissolves in water:

$$HNO_2(aq) + H_2O(\ell) \rightleftharpoons NO_2^-(aq) + H_3O^+(aq) \quad (15.4)$$

To what degree does this reaction proceed before it reaches equilibrium? We begin by translating Equation 15.4 into an equilibrium constant expression, with products in the numerator and reactants in the denominator:

$$K_a = \frac{[NO_2^-][H_3O^+]}{[HNO_2]}$$

As noted in Section 15.2, there is no concentration term for water in the denominator because it is the solvent in this reaction and a pure liquid.

We can generalize this K_a expression for the ionization of any weak acid, HA:

$$HA(aq) + H_2O(\ell) \rightleftharpoons A^-(aq) + H_3O^+(aq) \quad (15.13)$$

The corresponding K_a expression is

$$K_a = \frac{[A^-][H_3O^+]}{[HA]} \quad (15.14)$$

Equation 15.14 can be used to calculate the K_a value of an unknown weak acid if we know the concentration of the acid, [HA], in a solution and the pH of the solution. Suppose, for example, we determine that a 0.100 *M* solution of HA has a pH of 2.20. To calculate K_a, we first convert the pH value to $[H_3O^+]$:

$$[H_3O^+] = 10^{-2.20} = 6.3 \times 10^{-3}\ M$$

Next, we assume that the only source of H_3O^+ ions is the ionization of HA. If so, then $[A^-]$ must also be 6.3×10^{-3} *M* because each mole of HA that ionizes produces 1 mole of H_3O^+ and 1 mole of A^-. Moreover, if $[H_3O^+]$ increases by 6.3×10^{-3} *M*, then [HA] must have decreased by 6.3×10^{-3} *M*. Therefore, the concentration of HA once equilibrium has been achieved is

$$[HA] = (0.100 - 6.3 \times 10^{-3})\ M = 0.094\ M$$

Using the three calculated equilibrium concentrations in the K_a expression gives us:

$$K_a = \frac{[A^-][H_3O^+]}{[HA]} = \frac{(6.3 \times 10^{-3})(6.3 \times 10^{-3})}{(0.094)} = 4.2 \times 10^{-4}$$

The small value of K_a confirms that HA is a weak acid.

The ratio of the equilibrium $[A^-]$ value to the initial [HA] value represents the **degree of ionization** of HA, which is usually expressed as a percentage of the initial acid concentration. For this reason it is also called *percent ionization*. In equation form, this relationship is

$$\text{Percent ionization} = \frac{[A^-]_{\text{equilibrium}}}{[HA]_{\text{initial}}} \times 100\% \quad (15.15)$$

Inserting the data from the previous calculation into Equation 15.15 yields the percent ionization of 0.100 *M* HA:

$$\text{Percent ionization} = \frac{6.3 \times 10^{-3}\ M}{0.100\ M} \times 100\% = 6.3\%$$

CONCEPT **TEST**

Describe how the percent ionization of a weak acid is related to its K_a value.

Now let's examine how the degree of ionization of a weak acid depends on its concentration in solution. A plot of percent ionization as a function of the initial

degree of ionization the ratio of the quantity of a substance that is ionized to the concentration of the substance before ionization; when expressed as a percentage, it is often called *percent ionization*.

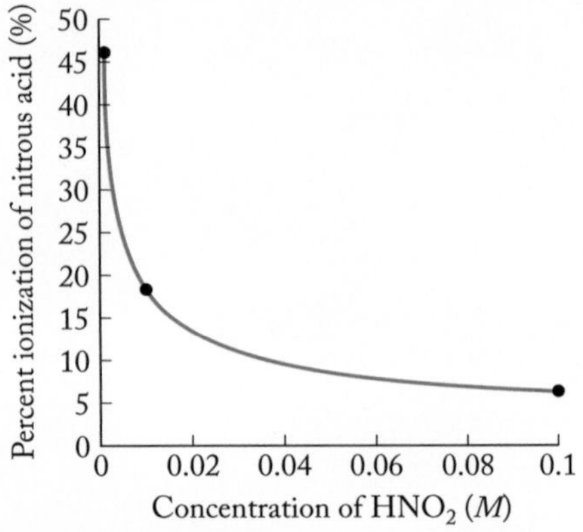

FIGURE 15.12 The degree of ionization of a weak acid increases with decreasing acid concentration. Here the degree of ionization of nitrous acid increases from about 6% in a 0.100 *M* solution to 18% in a 0.010 *M* solution to 46% in a 0.001 *M* solution.

concentration of nitrous acid at 25°C is shown in **Figure 15.12**. Note that the degree of ionization *increases* as the concentration of the acid *decreases*. This same pattern is observed for all weak acids.

Suppose, for example, we have a 1.0 *M* solution of a generic weak acid, HA, in which the concentrations of A^- and H_3O^+ ions at equilibrium are both 0.0010 *M*. Using these concentrations in Equation 15.14 to solve for K_a, we have:

$$K_a = \frac{[A^-][H_3O^+]}{[HA]} = \frac{(0.0010)(0.0010)}{(1.0 - 0.0010)} = 1.0 \times 10^{-6}$$

Now suppose that more water is added to the HA solution, increasing its volume by a factor of 10. If no change in the degree of ionization occurred, then [HA] would be 0.10 *M* and the concentrations of A^- and H_3O^+ ions would both be 1.0×10^{-4} *M*. However, if we insert these values into the equilibrium constant expression, we get a reaction quotient (Q) value of:

$$Q = \frac{(1.0 \times 10^{-4})(1.0 \times 10^{-4})}{(0.10 - 1.0 \times 10^{-4})} = 1.0 \times 10^{-7}$$

This Q value is only 1/10 the value of K_a, so the acid ionization reaction is no longer at equilibrium. Because $Q < K_a$, the reaction proceeds in the forward direction. When this happens, the concentrations of A^- and H_3O^+ ions increase, and so does the percentage of HA molecules that ionize. Further dilutions would produce even greater percent ionization values, following the trend we see for nitrous acid in Figure 15.12.

SAMPLE EXERCISE 15.8 Relating pH, K_a, and Percent Ionization of a Weak Acid **LO5**

The pH of a 1.00 *M* solution of formic acid (HCOOH) is 1.88.

a. What is the percent ionization of 1.00 *M* HCOOH?
b. What is the K_a value of formic acid?

Collect, Organize, and Analyze We are asked to determine the K_a value of formic acid and its percent ionization in a solution of known concentration and pH. There are two H atoms in a molecule of formic acid, but only one is in a carboxylic acid group and can ionize in an aqueous solution. As a result, the acid ionization reaction is

$$HCOOH(aq) + H_2O(\ell) \rightleftharpoons HCOO^-(aq) + H_3O^+(aq)$$

The 1:1:1 stoichiometry of the ionization reaction indicates that $[HCOO^-] = [H_3O^+]$ in a solution of HCOOH at equilibrium, so using Equation 15.10 (i.e., pH = $-\log[H_3O^+]$) will give us the values of $[H_3O^+]$ and $[HCOO^-]$ at equilibrium and will allow us to calculate the degree of ionization by using Equation 15.15. At equilibrium, [HCOOH] is equal to its initial concentration minus the portion that ionized, or

$$[HCOOH]_{equilibrium} = [HCOOH]_{initial} - [H_3O^+]_{equilibrium}$$

We can then calculate the value of K_a by using the following equilibrium constant expression:

$$K_a = \frac{[HCOO^-][H_3O^+]}{[HCOOH]}$$

The pH of the 1.00 *M* solution is close to 2, which corresponds to $[H_3O^+] = 10^{-2}$ *M*. Therefore, the percent ionization of formic acid in this solution should be about $[(10^{-2}\ M)/(1\ M)](100\%) = 1\%$.

Solve

a. The pH of the 1.00 *M* solution is 1.88. The corresponding $[H_3O^+] = [HCOO^-]$ is

$$10^{-1.88} = 1.32 \times 10^{-2}\ M$$

Inserting this value and the initial concentration of HCOOH in Equation 15.15:

$$\text{Percent ionization} = \frac{[HCOO^-]_{\text{equilibrium}}}{[HCOOH]_{\text{initial}}} \times 100\%$$

$$= \frac{1.32 \times 10^{-2}\ M}{1.00\ M} \times 100\% = 1.3\%$$

b. A RICE table can be used to determine the value of K_a, though not quite in the way these tables were used to calculate equilibrium concentrations in Chapter 14. In this exercise we are given $[HCOOH]_{\text{initial}}$ and, before any acid ionization occurs: $[HCOO^-]_{\text{initial}} = [H_3O^+]_{\text{initial}} = 0$. We've already determined the value of $[H_3O^+]$ at equilibrium: 1.32×10^{-2} *M*. In other words, we have already solved for *x*. The ionization reaction produces equal numbers of moles of H_3O^+ and $HCOO^-$, so $[HCOO^-]$ must also be 1.32×10^{-2} *M*. Also, *the change* in the concentrations of $HCOO^-$ and H_3O^+ must be $+x$, or $+1.32 \times 10^{-2}$ *M*, and the change in [HCOOH] must be $-x$, or -1.32×10^{-2} *M*. Filling in the RICE table with these concentration values and calculating $[HCOOH]_{\text{equilibrium}}$:

Reaction (R)	$HCOOH(aq) + H_2O(\ell)$ $\rightleftharpoons$	$HCOO^-(aq)$ +	$H_3O^+(aq)$
	[HCOOH] (*M*)	$[HCOO^-]$ (*M*)	$[H_3O^+]$ (*M*)
Initial (I)	1.00	0	0
Change (C)	-1.32×10^{-2}	$+1.32 \times 10^{-2}$	$+1.32 \times 10^{-2}$
Equilibrium (E)	0.987	1.32×10^{-2}	1.32×10^{-2}

Inserting the equilibrium concentration values in the K_a expression for HCOOH and doing the math:

$$K_a = \frac{[HCOO^-][H_3O^+]}{[HCOOH]} = \frac{(1.32 \times 10^{-2})(1.32 \times 10^{-2})}{(0.987)} = 1.8 \times 10^{-4}$$

Think About It The result of the percent ionization calculation is close to our estimate of about 1%, and the calculated K_a value matches the K_a value for formic acid listed in Table 15.2. In calculating the value of K_a, we used equilibrium concentrations of the products to calculate their change in concentration during the ionization reaction, then reversed the sign on that value to calculate the change in [HCOOH], which was subtracted from $[HCOOH]_{\text{initial}}$ to calculate $[HCOOH]_{\text{equilibrium}}$.

Practice Exercise The value of $[H_3O^+]$ in a 0.050 *M* solution of an organic acid is 5.9×10^{-3} *M*. What is the pH of the solution, the percent ionization of the acid, and its K_a value?

CONCEPT TEST

Three weak acids have the following K_a values:

Acid	K_a
A	3.6×10^{-5}
B	4.9×10^{-4}
C	9.2×10^{-4}

Which acid is the most extensively ionized in a 0.10 *M* solution of the acid? Which acid has the lowest percent ionization in a 1.00 *M* solution of the acid?

Weak Bases

Now let's examine the degree of ionization of weakly basic compounds when they dissolve in water. For example, ammonia bonds to hydrogen ions transferred from water:

$$NH_3(aq) + H_2O(\ell) \rightleftharpoons NH_4^+(aq) + OH^-(aq)$$

The limited strength of ammonia as a base is reflected in its small K_b value at 25°C:

$$K_b = \frac{[NH_4^+][OH^-]}{[NH_3]} = 1.76 \times 10^{-5}$$

As with K_a expressions, there is no $[H_2O]$ term in this K_b expression because the concentration of water does not change significantly during the reaction.

More generally, the transfer of a proton to any Brønsted–Lowry weak base in aqueous solution can be written as:

$$B(aq) + H_2O(\ell) \rightleftharpoons HB^+(aq) + OH^-(aq)$$

with a corresponding equilibrium constant expression:

$$K_b = \frac{[HB^+][OH^-]}{[B]} \tag{15.16}$$

Just as with weak acids, we can calculate the degree of ionization of a weak base and its K_b value if we know the pH and the initial concentration of a solution of the base, $[B]_{initial}$. We first need to convert pH into pOH using Equation 15.12, then pOH into $[OH^-]$ using Equation 15.11, and then, assuming $[OH^-] = [HB^+]$ at equilibrium, use $[HB^+]_{equilibrium}$ to calculate percent ionization:

$$\text{Percent ionization} = \frac{[HB^+]_{equilibrium}}{[B]_{initial}} \times 100\% \tag{15.17}$$

SAMPLE EXERCISE 15.9 Relating pH, K_b, and Percent Ionization of a Weak Base **LO5**

Trimethylamine, $(CH_3)_3N$, is a particularly foul-smelling volatile organic compound that forms during the decay of plant and animal matter. Its base ionization reaction in water is

$$(CH_3)_3N(aq) + H_2O(\ell) \rightleftharpoons (CH_3)_3NH^+(aq) + OH^-(aq)$$

The pH of a 0.500 *M* solution of trimethylamine is 11.75 at 25°C.

a. What is the percent ionization of trimethylamine in a 0.500 *M* solution?
b. What is the K_b value of trimethylamine?

Collect and Organize We are asked to determine the K_b value of trimethylamine and its percent ionization in a solution of known concentration and pH. We can use Equation 15.12 to convert pH into pOH and Equation 15.11 to convert pOH into $[OH^-]$. Equation 15.17 relates percent ionization to the equilibrium concentration of $(CH_3)_3NH^+$ ions and the initial concentration of $(CH_3)_3N$.

Analyze The stoichiometry of the ionization reaction indicates that $[(CH_3)_3NH^+] = [OH^-]$ at equilibrium assuming the initial concentrations of both were essentially zero. After using Equations 15.12 and 15.11 to calculate these concentrations we can use them in a RICE table (see Sample Exercise 15.8) to calculate $[(CH_3)_3N]$ at equilibrium. The equilibrium concentrations of the three reaction components can be used in the K_b expression

$$K_b = \frac{[(CH_3)_3NH^+][OH^-]}{[(CH_3)_3N]}$$

to calculate the value of K_b. The pH of the 0.500 M solution is a little less than 12, which corresponds to a pOH that is a little more than $(14 - 12) = 2$ and a value for $[OH^-]$ that is a little less than 10^{-2} M. Therefore, the percent ionization of trimethylamine in this solution should be 1%–2%.

Solve

a. The pH of the 0.500 M solution is 11.75. Calculating the corresponding pOH:

$$\text{pH} + \text{pOH} = 14.00 \quad \text{or} \quad \text{pOH} = 14.00 - \text{pH} = 14.00 - 11.75 = 2.25$$

Calculating $[OH^-]$:

$$[OH^-] = 10^{-\text{pOH}} = 10^{-2.25} = 5.6 \times 10^{-3}\ M$$

Assuming $[(CH_3)_3NH^+] = [OH^-]$, then the percent ionization of $(CH_3)_3N$ is

$$\text{Percent ionization} = \frac{[(CH_3)_3NH^+]_{\text{equilibrium}}}{[(CH_3)_3N]_{\text{initial}}} \times 100\%$$

$$= \frac{5.6 \times 10^{-3}\ M}{0.500\ M} \times 100\% = 1.1\%$$

b. Setting up a RICE table based on the base ionization reaction and inserting $[(CH_3)_3NH^+] = [OH^-] = 5.6 \times 10^{-3}$ M at equilibrium, we use this value to calculate the changes in concentration of the products and reactant and the equilibrium $[(CH_3)_3N]$ value:

Reaction (R)	$(CH_3)_3N(aq) + H_2O(\ell)$ $\rightleftharpoons$	$(CH_3)_3NH^+(aq)$ +	$OH^-(aq)$
	$[(CH_3)_3N]$ (M)	$[(CH_3)_3NH^+]$ (M)	$[OH^-]$ (M)
Initial (I)	0.500	0	0
Change (C)	-5.6×10^{-3}	$+5.6 \times 10^{-3}$	$+5.6 \times 10^{-3}$
Equilibrium (E)	0.494	5.6×10^{-3}	5.6×10^{-3}

Inserting these equilibrium concentration values in the K_b expression:

$$K_b = \frac{[(CH_3)_3NH^+][OH^-]}{[(CH_3)_3N]} = \frac{(5.6 \times 10^{-3})(5.6 \times 10^{-3})}{0.494}$$

$$= 6.3 \times 10^{-5}$$

Think About It The result of the percent ionization calculation is in our estimated range of 1%–2%, and the calculated K_b is close to the value listed in Appendix 5: 6.46×10^{-5}. The calculated K_b value was rounded to only 2 significant figures (as was the calculated percent ionization value) because there were only 2 digits to the right of the decimal point in the initial pH value.

Practice Exercise The pH of a 0.100 M aqueous solution of ethylamine ($CH_3CH_2NH_2$) is 11.86 at 25°C. What percentage of the $CH_3CH_2NH_2$ molecules in this solution are ionized as described in the chemical equation below, and what is the K_b value of ethylamine?

$$CH_3CH_2NH_2(aq) + H_2O(\ell) \rightleftharpoons CH_3CH_2NH_3^+(aq) + OH^-(aq)$$

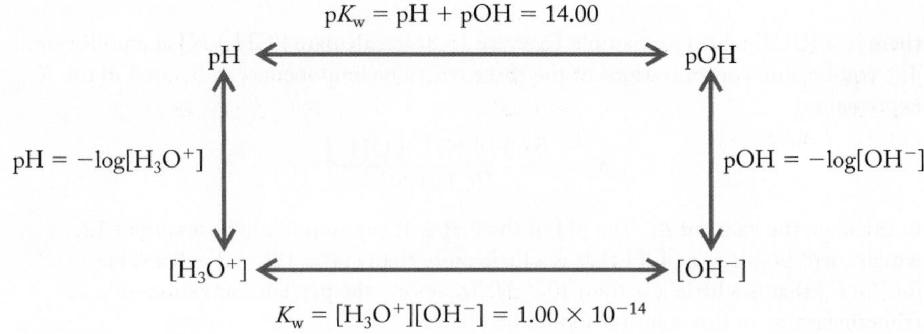

FIGURE 15.13 Pathways for interconverting $[H_3O^+]$, $[OH^-]$, pH, and pOH values.

Figure 15.13 summarizes the ways we can interconvert $[H_3O^+]$, $[OH^-]$, pH, and pOH. The diagram may leave you wondering which path to follow to make cross-corner (two-step) conversions, such as $[OH^-]$ to pH. Many students find the math simpler if they travel the "northern" route—that is, via a pOH-to-pH conversion.

15.6 Calculating the pH of Acidic and Basic Solutions

In this section we calculate the pH values of solutions of acids and bases. We begin with solutions of strong acids and strong bases.

Strong Acids and Strong Bases

Strong acids have large K_a values and ionize essentially completely in aqueous solutions. In nearly all their solutions, then, the concentration of H_3O^+ ions is the same as the initial concentration of the strong acid. For example, the pH of muriatic acid (7.4 *M* HCl), which is sold in hardware and building supply stores to clean concrete surfaces, is

$$pH = -\log[H_3O^+] = -\log(7.4) = -0.87$$

Strong bases are ionic compounds that dissociate completely in aqueous solutions. For a group 1 hydroxide such as NaOH, then, $[OH^-]$ will be the same as the strong base's initial concentration. For a group 2 hydroxide such as $Ca(OH)_2$, $[OH^-]$ will be twice the strong base's initial concentration. To calculate the pH of a solution of a strong base, therefore, we need to calculate the concentration of OH^- ions given the initial concentration of the base and then convert $[OH^-]$ to pOH, and then to pH. Sample Exercise 15.10 illustrates the steps involved.

FIGURE 15.14 Cleaners used to remove grease and hair clogs from water pipes typically contain large amounts of sodium hydroxide.

SAMPLE EXERCISE 15.10 Calculating the pH of a Solution of a Strong Base **LO6**

Liquids such as the one in the bottle in **Figure 15.14** are used to remove clogs from bathroom and kitchen drains. They contain sodium hydroxide at concentrations as high as 1.2 *M*. What is the pH of 1.2 *M* NaOH?

Collect, Organize, and Analyze NaOH is a strong base that produces 1 mole of hydroxide ions per mole of NaOH in solution, so $[OH^-]$ is the same as the initial concentration of the base. Therefore, $[OH^-] = 1.2\ M$. We can convert this value to pOH and then to pH following the steps shown in Figure 15.13.

Solve

Calculating pOH: $\text{pOH} = -\log[OH^-] = -\log(1.2) = -0.079$

Calculating pH: $\text{pH} = 14.00 - \text{pOH} = 14.00 - (-0.079) = 14.08$

Think About It The calculated pH is slightly above 14 because $[OH^-]$ is slightly greater than 1 *M*.

Practice Exercise *Kalkwasser* is the German word for "lime water," which is the common name for saturated solutions of $Ca(OH)_2$. It has many uses, including being added to water in aquarium tanks to adjust their pH (and to provide Ca^{2+} ions for the plants and animals living in the tank). What is the pH of a Kalkwasser solution whose concentration is 0.0225 *M* $Ca(OH)_2$?

Weak Acids and Weak Bases

In Section 15.5 we learned how pH and concentration can be used to calculate the percent ionization and the K_a or K_b of a weak acid or weak base, respectively. If we know the concentration of a weak acid solution and its K_a value (or a weak base and its K_b value), can we determine the pH of that solution? Sample Exercises 15.11 and 15.12 take us through these calculations.

CONNECTION RICE tables were introduced in Section 14.8 to help us calculate the changes (C) from initial (I) to equilibrium (E) concentrations for a given reaction (R).

SAMPLE EXERCISE 15.11 Calculating the pH of a Solution of a Weak Acid **LO6**

Most of the acids on our planet and in our bodies are weak acids, including all carboxylic acids. Formic acid (HCOOH) is one of the more common carboxylic acids and the one with the simplest molecular structure. Secretions of the carabid beetle shown in **Figure 15.15** contain 0.040 *M* HCOOH. What is the pH of 0.040 *M* formic acid? The K_a value for formic acid is 1.77×10^{-4}.

Collect and Organize We are asked to determine the pH of a known concentration of a weak acid. We also know its K_a value. The molecular structure of formic acid in Table 15.2 indicates that only one H atom per molecule is ionizable in aqueous solution.

Analyze The acid ionization reaction is

$$HCOOH(aq) + H_2O(\ell) \rightleftharpoons HCOO^-(aq) + H_3O^+(aq)$$

and the corresponding equilibrium constant expression is

$$K_a = \frac{[HCOO^-][H_3O^+]}{[HCOOH]} = 1.77 \times 10^{-4}$$

Calculating pH involves first solving for $[H_3O^+]$, which we can do by setting up a RICE table based on the above reaction, as we did for several equilibrium calculations in Chapter 14.

FIGURE 15.15 The carabid beetle secretes formic acid (HCOOH) as a defense against predators.

Solve First we set up a RICE table based on the acid ionization reaction. The object of the table is to solve for $[H_3O^+]$ at equilibrium, so we give it the symbol x. Filling in the other cells in the table according to the 1:1:1 stoichiometry of the reaction gives us

Reaction (R)	$HCOOH(aq) + H_2O(\ell)$	$\rightleftharpoons$	$HCOO^-(aq)$	+	$H_3O^+(aq)$
	$[HCOOH]$ (M)		$[HCOO^-]$ (M)		$[H_3O^+]$ (M)
Initial (I)	0.040		0		0
Change (C)	$-x$		$+x$		$+x$
Equilibrium (E)	$0.040 - x$		x		x

Inserting the equilibrium terms into the K_a expression,

$$K_a = \frac{[HCOO^-][H_3O^+]}{[HCOOH]} = \frac{(x)(x)}{(0.040 - x)} = 1.77 \times 10^{-4}$$

Solving for x by first cross multiplying:

$$x^2 = 7.08 \times 10^{-6} - (1.77 \times 10^{-4})x$$

and rearranging the terms:

$$x^2 + (1.77 \times 10^{-4})x - 7.08 \times 10^{-6} = 0$$

gives us a quadratic equation with two solutions:

$$x = -0.00275 \quad \text{and} \quad 0.00257$$

The negative x value has no physical meaning because it results in negative concentration values, so $[H_3O^+] = 0.00257$ M.

Solving for pH:

$$\text{pH} = -\log[H_3O^+] = -\log(0.00257) = 2.59$$

CONNECTION Directions for using the quadratic formula to solve quadratic equations are given in Appendix 1.

Think About It We did not attempt to simplify the calculation of $[H_3O^+]$ by eliminating "$-x$" from the denominator of the equilibrium constant expression. This was a sensible decision because the value of x is 0.00257/0.040, or 6.4%, of the initial concentration of formic acid. Put another way, the formic acid in the sample was 6.4% ionized. This value of x should be less than 5% of the initial value of [HA] to use the simplification.

Practice Exercise Acetic acid, the main ingredient in vinegar, is a carboxylic acid. What is the pH of a 0.035 M solution of acetic acid, given that its $K_a = 1.76 \times 10^{-5}$?

SAMPLE EXERCISE 15.12 Calculating the pH of a Solution of a Weak Base — LO6

The concentration of NH_3 in the household ammonia used to clean windows ranges between 50 and 100 g/L, or from about 3 M to almost 6 M. What is the pH of 3.0 M NH_3? The K_b value for ammonia is 1.76×10^{-5}.

Collect and Organize We are asked to determine the pH of a 3.0 M solution of ammonia. Figure 15.6 described the basic behavior of ammonia in aqueous solutions.

Analyze The hydrolysis of ammonia produces OH^- ions. We can calculate the equilibrium concentration of OH^- ions by using the K_b value and expression and then convert $[OH^-]$ to pOH and finally to pH. Given the 1:1:1 stoichiometry of NH_3, NH_4^+, and OH^-, we know that $[NH_4^+] = [OH^-]$ at equilibrium. If that equilibrium value is x, then the change in $[NH_3]$ during the reaction is $-x$. The pH value of a concentrated solution of a base with a K_b value near 10^{-5} should be well above 7 but below 14.

Solve We begin by setting up a RICE table in which we let $[NH_4^+] = [OH^-] = x$ at equilibrium:

Reaction (R)	$NH_3(aq) + H_2O(\ell)$	$\rightleftharpoons$	$NH_4^+(aq)$	+	$OH^-(aq)$
	$[NH_3]$ (M)		$[NH_4^+]$ (M)		$[OH^-]$ (M)
Initial (I)	3.0		0		0
Change (C)	$-x$		$+x$		$+x$
Equilibrium (E)	$3.0 - x$		x		x

Because K_b is small (1.76×10^{-5}) in comparison with the initial concentration of base (3.0 M), we can make the simplifying assumption that x will be less than 5% of 3.0 M, so the concentration term in the denominator can be simplified from $(3.0 - x)$ to 3.0. With this assumption, our equilibrium constant expression is

$$K_b = \frac{[NH_4^+][OH^-]}{[NH_3]} = \frac{(x)(x)}{3.0} = \frac{x^2}{3.0} = 1.76 \times 10^{-5}$$

so

$$x^2 = 5.28 \times 10^{-5}$$

Solving for x gives us

$$x = [OH^-] = \sqrt{5.28 \times 10^{-5}} = 7.3 \times 10^{-3}\ M$$

Taking the negative logarithm of $[OH^-]$ to calculate pOH:

$$\text{pOH} = -\log[OH^-] = -\log(7.3 \times 10^{-3}\ M) = 2.14$$

Then we subtract this value from 14.00 to obtain the pH:

$$\text{pH} = 14.00 - \text{pOH} = 14.00 - 2.14 = 11.86$$

Think About It The calculated pH value falls in the range we predicted given the small K_b value but relatively high initial concentration of ammonia. To check our simplifying assumption, let's compare the value of x with the initial $[NH_3]$ value of 3.0 M:

$$\frac{7.3 \times 10^{-3}\ M}{3.0\ M} \times 100\% = 0.24\%$$

This percentage is much less than 5%, which means our simplifying assumption was justified. Moreover, substituting $x = 7.3 \times 10^{-3}$ into the expression for K_b yields 1.78×10^{-5}, which is very close to the value given.

 **Practice Exercise** What is the pH of a 0.200 M solution of methylamine (CH_3NH_2, $K_b = 4.4 \times 10^{-4}$)?

CONNECTION As noted in Section 14.8, the $-x$ or $+x$ component of an equilibrium concentration or partial pressure term can be ignored if x is less than 5% of the initial concentration or partial pressure.

pH of Very Dilute Solutions of Strong Acids

Despite the autoionization of water (Section 15.4), water typically contributes very little to equilibrium concentrations of H_3O^+ or OH^- ions because K_w is only 1.00×10^{-14} at 25°C. Now that we have developed the concept of pH and explored how it helps us quantify the acidity of a solution, let's consider a very dilute solution of acid, a situation where the concentration of hydronium ions produced by the autoionization of water is greater than the concentration of hydronium ions produced by the acid.

SAMPLE EXERCISE 15.13 pH Calculations Involving the Autoionization of Water **LO6**

What is the pH of 1.0×10^{-8} *M* HCl?

Collect and Organize We are asked to calculate the pH of a very dilute solution of HCl, which is a strong acid and ionizes completely. The autoionization of pure water at 25°C produces an equilibrium $[H_3O^+] = 1.0 \times 10^{-7}$ *M*.

Analyze In a 1.0×10^{-8} *M* HCl solution $[H_3O^+] = 1.0 \times 10^{-8}$ *M*, and the pH should be

$$pH = -\log[H_3O^+] = -\log(1.0 \times 10^{-8}) = 8.00$$

This answer is not reasonable because a solution of a strong acid, no matter how dilute it is, cannot be basic (pH > 7). To calculate pH, we must also consider the autoionization of water. Let's set up a RICE table based on the autoionization equilibrium in which *x* represents the increase in $[H_3O^+]$ resulting from autoionization and in which the initial value of $[H_3O^+]$ is 1.0×10^{-8} *M*.

Solve Set up the RICE table and fill in the rows as described above:

Reaction (R)	$H_2O(\ell) + H_2O(\ell)$	$\rightleftharpoons$	$H_3O^+(aq)$	+	$OH^-(aq)$
			$[H_3O^+]$ (*M*)		$[OH^-]$ (*M*)
Initial (I)			1.0×10^{-8}		0
Change (C)			$+x$		$+x$
Equilibrium (E)			$(1.0 \times 10^{-8}) + x$		x

Substituting equilibrium values into Equation 15.8, and solving for *x*:

$$K_w = [H_3O^+][OH^-]$$
$$1.0 \times 10^{-14} = (1.0 \times 10^{-8} + x)(x)$$

Rearranging this equation to solve for *x* gives:

$$x^2 + (1.0 \times 10^{-8})x - 1.0 \times 10^{-14} = 0$$

There are two solutions to this quadratic equation:

$$x = 9.5 \times 10^{-8}\ M \quad \text{or} \quad x = -1.1 \times 10^{-7}\ M$$

Only the first of these numbers makes physical sense, because a negative value for *x* would mean a negative concentration of hydroxide ions at equilibrium. Thus, the total concentration of hydrogen ions in the solution is

$$[H_3O^+] = (9.5 \times 10^{-8}\ M) + (1.0 \times 10^{-8}\ M) = 10.5 \times 10^{-8}\ M = 1.05 \times 10^{-7}\ M$$

and pH is

$$pH = -\log(1.05 \times 10^{-7}\ M) = 6.98$$

Think About It This value agrees with our prediction that the solution should be (very) slightly acidic.

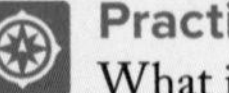

Practice Exercise What is the pH of 1.5×10^{-7} *M* $Ca(OH)_2$?

CONCEPT TEST

In the pH calculations in this chapter, we routinely ignore the concentrations of H_3O^+ and OH^- ions produced by the autoionization of water. Suppose calculations of the pH of six different solutions produced the results shown in the following table. Which, if any, of the calculations should have taken into account the autoionization of water to obtain an accurate result?

Solution	pH
A	2.66
B	4.12
C	6.39
D	7.27
E	9.10
F	12.88

15.7 Polyprotic Acids

Up to this point we have dealt with **monoprotic acids**, which have only one ionizable hydrogen atom per molecule. Acids that contain more than one ionizable hydrogen atom—such as sulfuric acid (H_2SO_4) and phosphoric acid (H_3PO_4)—are called **polyprotic acids**. For molecules with two and three ionizable hydrogen atoms, we use the more specific terms *diprotic acids* and *triprotic acids*, respectively. Let's consider the acidic properties of a strong diprotic acid, sulfuric acid.

Acid Rain

Coal naturally contains sulfur impurities, which combustion releases into the atmosphere as SO_2. For this reason, burning coal for energy production has significant environmental consequences: in the atmosphere, some SO_2 is oxidized to SO_3, which then combines with water vapor to form particles of sulfuric acid (H_2SO_4), a principal component of acid rain in many parts of the world. Sulfuric acid is a strong acid (see Table 15.1) because it is completely ionized in water:

$$H_2SO_4(aq) + H_2O(\ell) \rightarrow HSO_4^-(aq) + H_3O^+(aq)$$

However, the second ionization step may not be complete:

$$HSO_4^-(aq) + H_2O(\ell) \rightleftharpoons SO_4^{2-}(aq) + H_3O^+(aq) \qquad K_{a_2} = 0.012$$

The equilibrium constant symbol K_{a_2} corresponds to the donation of the second H^+ ion.

The combination of one complete and one incomplete ionization reaction means that many solutions of H_2SO_4 contain more than 1 mole but less than 2 moles of H_3O^+ ions for every mole of H_2SO_4 dissolved. One such solution is the subject of Sample Exercise 15.14.

CHEMTOUR
Acid Rain

SAMPLE EXERCISE 15.14 Calculating the pH of a Solution of a Strong Diprotic Acid **LO7**

What is the pH of a 0.100 M solution of H_2SO_4?

Collect and Organize We are given the concentration of a solution of sulfuric acid and asked to calculate its pH. Sulfuric acid is a strong diprotic acid in that one hydrogen atom ionizes completely but a second hydrogen atom may not be ionized for every molecule ($K_{a_2} = 0.012$).

monoprotic acid an acid that has one ionizable hydrogen atom per molecule.

polyprotic acid an acid that has two or more ionizable hydrogen atoms per molecule.

Analyze We start with the assumption that the ionization of H_2SO_4 is complete:

$$H_2SO_4(aq) + H_2O(\ell) \rightarrow HSO_4^-(aq) + H_3O^+(aq)$$

As a result, the second ionization step begins with $[HSO_4^-] = [H_3O^+] = 0.100\ M$. Ionization of HSO_4^- then produces additional H_3O^+ ions:

$$HSO_4^-(aq) + H_2O(\ell) \rightleftharpoons SO_4^{2-}(aq) + H_3O^+(aq)$$

We can use a RICE table to calculate the increase in $[H_3O^+]$ (represented by x). The second ionization adds to the total $[H_3O^+]$ value. This increase in $[H_3O^+]$ will have a value between 0 and 0.1 M, which means the total $[H_3O^+]$ value will be between 0.1 and 0.2 M and the pH of the solution at equilibrium should be a little less than 1.

Solve We begin by setting up a RICE table in which initially $[HSO_4^-] = [H_3O^+] = 0.100\ M$. We let the change in $[H_3O^+]$ during the second ionization step be $+x$. Filling in the other cells of the table according to the stoichiometry of the second ionization step:

Reaction (R)	$HSO_4^-(aq) + H_2O(\ell)$	$\rightleftharpoons$	$SO_4^{2-}(aq)$	+	$H_3O^+(aq)$
	$[HSO_4^-]$ (M)		$[SO_4^{2-}]$ (M)		$[H_3O^+]$ (M)
Initial (I)	0.100		0		0.100
Change (C)	$-x$		$+x$		$+x$
Equilibrium (E)	$0.100 - x$		x		$0.100 + x$

Inserting the equilibrium concentrations in the equilibrium constant expression for K_{a_2}:

$$K_{a_2} = \frac{[H_3O^+][SO_4^{2-}]}{[HSO_4^-]} = \frac{(0.100 + x)(x)}{(0.100 - x)} = 1.2 \times 10^{-2}$$

Cross multiplying and rearranging the terms:

$$x^2 + 0.112x - 1.2 \times 10^{-3} = 0$$

Solving this quadratic equation for x yields these two values:

$$x = 0.00985 \quad \text{and} \quad -0.122$$

The negative value of x represents a negative $[SO_4^{2-}]$ value and has no physical meaning. Therefore, we use only the positive x value to calculate the total $[H_3O^+]$ at equilibrium:

$$[H_3O^+] = (0.100 + x) = (0.100 + 0.00985) = 0.10985 = 0.110\ M$$

The corresponding pH is

$$\text{pH} = -\log[H_3O^+] = -\log(0.110) = 0.96$$

Think About It As predicted, the value of $[H_3O^+]$ is between 0.1 and 0.2 M and the pH of the solution is a little less than one. The degree of ionization of HSO_4^- is

$$\frac{[SO_4^{2-}]_{\text{equilibrium}}}{[HSO_4^-]_{\text{initial}}} = \frac{0.00985\ M}{0.100\ M} = 0.0985 = 9.8\%$$

This value exceeds 5%, so ignoring the decrease in HSO_4^- to avoid solving a quadratic equation would have been a bad idea.

Practice Exercise What is the pH of a 0.200 M solution of H_2SO_4? How did doubling the concentration of H_2SO_4 affect the pH when compared with the 0.100 M solution of H_2SO_4 in Sample Exercise 15.14?

CONCEPT TEST

List all the phosphorus-containing species present in an aqueous solution of phosphoric acid (H_3PO_4) in decreasing order of concentration.

Normal Rain

Rainwater falling from the sky is naturally acidic. The fourth-most-abundant gas in the atmosphere is CO_2, which dissolves in water to form a small amount of carbonic acid, as we discussed in Section 15.1:

$$CO_2(g) + H_2O(\ell) \rightleftharpoons H_2CO_3(aq) \qquad (15.1)$$

The equilibrium constant for this reaction is only about 10^{-3}, so most of the dissolved carbon dioxide remains in the form of hydrated molecules of CO_2. However, some molecules of H_2CO_3 then go on to release H^+ ions:

$$H_2CO_3(aq) + H_2O(\ell) \rightleftharpoons HCO_3^-(aq) + H_3O^+(aq) \qquad K_{a_1} = 4.3 \times 10^{-7}$$

When using this K_{a_1} value, keep in mind that its value is based on the total concentration of dissolved CO_2 present in a sample. To simplify these calculations, $[H_2CO_3]$ is used to represent the total concentration of CO_2 dissolved in the sample. The second ionization constant, K_{a_2}, has an even smaller value:

$$HCO_3^-(aq) + H_2O(\ell) \rightleftharpoons CO_3^{2-}(aq) + H_3O^+(aq) \qquad K_{a_2} = 4.7 \times 10^{-11}$$

It turns out that calculating the pH of a solution of carbonic acid is simpler than the pH calculation for sulfuric acid. The reason is that K_{a_1}, although small, is still much larger than K_{a_2}. We can explain this difference based on electrostatic attractions between oppositely charged ions. The first ionization step produces a negatively charged oxoanion (HCO_3^-). The second requires a H^+ ion to dissociate from HCO_3^- to produce an even more negative oxoanion (CO_3^{2-}). Separating oppositely charged ions that are naturally attracted to each other is not a process that we would expect to be favored, and the much smaller value of K_{a_2} confirms our expectation.

In general, the K_{a_2} value of any diprotic acid is smaller—often much smaller—than its K_{a_1} value. A consequence of these large differences is that essentially all the limited strength of many weak diprotic acids is derived from the first ionization reaction. For those acids whose K_{a_1} value is at least a thousand times greater than its K_{a_2} value, we can ignore the second ionization reaction.

SAMPLE EXERCISE 15.15 Calculating the pH of a Solution of a Weak Diprotic Acid **LO7**

What is the pH of rainwater at 25°C in equilibrium with atmospheric CO_2, producing a total dissolved CO_2 concentration of 1.4×10^{-5} *M*?

Collect and Organize We are asked to determine the pH of a dilute solution of dissolved CO_2. There are two ionizable H atoms in H_2CO_3. The K_{a_1} and K_{a_2} values are 4.3×10^{-7} and 4.7×10^{-11}, respectively. Any H_2CO_3 that dissociates to produce bicarbonate and hydronium ions will be replaced by more CO_2 dissolving from the atmosphere, so the value of $[H_2CO_3]$ in the denominator in the K_{a_1} expression will be constant at 1.4×10^{-5} *M*.

Analyze The value of K_{a_1} is more than 1000 times the K_{a_2} value, so the second ionization step can be ignored in calculating the pH of a solution of dissolved CO_2. Therefore, we focus on the first ionization equilibrium:

$$H_2CO_3(aq) + H_2O(\ell) \rightleftharpoons HCO_3^-(aq) + H_3O^+(aq) \qquad K_{a_1} = 4.3 \times 10^{-7}$$

Because of the small value of K_{a_1} and the small concentration of dissolved CO_2, we should obtain a pH value that is less than 7, but a lot closer to 7 than 0.

Solve First we set up a RICE table in which $x = [H_3O^+] = [HCO_3^-]$ at equilibrium and the value of $[H_2CO_3]$ is a constant 1.4×10^{-5} *M*.

Reaction (R)	$H_2CO_3(aq) + H_2O(\ell)$	$\rightleftharpoons$	$HCO_3^-(aq)$	+	$H_3O^+(aq)$
	$[H_2CO_3]$ (*M*)		$[HCO_3^-]$ (*M*)		$[H_3O^+]$ (*M*)
Initial (I)	1.4×10^{-5}		0		0
Change (C)	0		$+x$		$+x$
Equilibrium (E)	1.4×10^{-5}		x		x

$$K_{a_1} = \frac{[HCO_3^-][H_3O^+]}{[H_2CO_3]} = \frac{(x)(x)}{1.4 \times 10^{-5}} = 4.3 \times 10^{-7}$$

so

$$x^2 = 6.0 \times 10^{-12}$$

Therefore,

$$x = [H_3O^+] = 2.4 \times 10^{-6}\ M$$

Taking the negative logarithm of $[H_3O^+]$ to calculate pH:

$$\text{pH} = -\log[H_3O^+] = -\log(2.4 \times 10^{-6}\ M) = 5.62$$

Think About It Carbonic acid is a weak acid, and its concentration here is small, so obtaining a pH value that is only about 1.4 units below neutral pH is reasonable.

Practice Exercise The proximity of the calculated pH value to 7.00 raises the question of whether the autoionization of water contributes significantly to $[H_3O^+]$ in the rainwater sample. Recalculate the pH of the rainwater sample in Sample Exercise 15.15, assuming the initial concentration of $H_3O^+ = 1.00 \times 10^{-7}$ *M*.

Some acids have three ionizable H atoms per molecule. Two important triprotic acids are phosphoric acid (H_3PO_4) and citric acid, the acid responsible for the tart flavor of citrus fruits. According to **Table 15.5**, $K_{a_1} > K_{a_2} > K_{a_3}$ for both acids. This pattern is much like that for the $K_{a_1} > K_{a_2}$ values of diprotic acids. The reason is the same, too—namely, it is more difficult to remove a second H^+ ion from the negatively charged ion formed after the first H^+ ion is removed, and it is even more difficult to remove a third H^+ ion from an ion with a 2− charge.

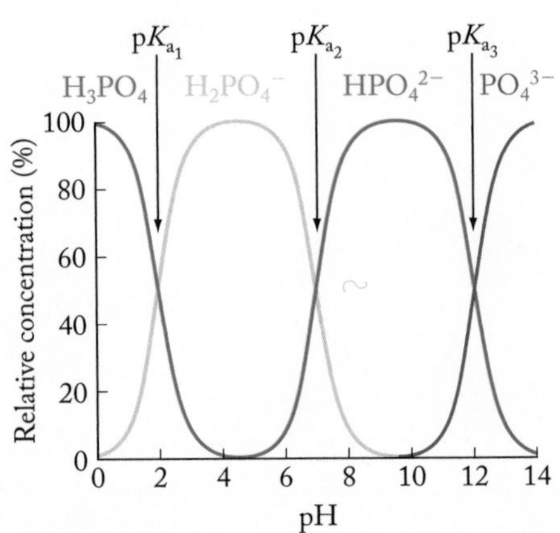

FIGURE 15.16 The relative concentrations of H_3PO_4 and the oxoanions it forms at different pH values.

Figure 15.16 illustrates the separations between the pK_a values of phosphoric acid and their impact on the relative concentrations of H_3PO_4 and the oxoanions it forms at different pH values. At pH values below about 4.5, the principal species present in solution are H_3PO_4 and $H_2PO_4^-$. The concentrations of HPO_4^{2-} and PO_4^{3-} that are produced in the second and third acid ionization reactions (see Table 15.5) are insignificant, which is why we can ignore them in calculating the pH of a solution of phosphoric acid.

The pH values at the three cross-over points in the concentration curves in Figure 15.16 correspond to the three p*K* values of phosphoric acid. This makes sense when we consider, for example, phosphoric acid's second ionization step:

$$H_2PO_4^-(aq) + H_2O(\ell) \rightleftharpoons HPO_4^{2-}(aq) + H_3O^+(aq)$$

TABLE 15.5 Ionization Equilibria of Two Triprotic Acids

Reaction	K_a	pK_a
Phosphoric Acid		
(1) $HO{-}P(=O)(OH){-}OH + H_2O \rightleftharpoons HO{-}P(=O)(OH){-}O^- + H_3O^+$	$K_{a_1} = 7.1 \times 10^{-3}$	$pK_{a_1} = 2.15$
(2) $HO{-}P(=O)(OH){-}O^- + H_2O \rightleftharpoons {}^-O{-}P(=O)(OH){-}O^- + H_3O^+$	$K_{a_2} = 6.3 \times 10^{-8}$	$pK_{a_2} = 7.20$
(3) ${}^-O{-}P(=O)(OH){-}O^- + H_2O \rightleftharpoons {}^-O{-}P(=O)(O^-){-}O^- + H_3O^+$	$K_{a_3} = 4.5 \times 10^{-13}$	$pK_{a_3} = 12.35$
Citric Acid		
(1) $HO{-}C(CH_2COOH)(CH_2COOH){-}COOH + H_2O \rightleftharpoons HO{-}C(CH_2COO^-)(CH_2COOH){-}COOH + H_3O^+$	$K_{a_1} = 7.4 \times 10^{-4}$	$pK_{a_1} = 3.13$
(2) $HO{-}C(CH_2COO^-)(CH_2COOH){-}COOH + H_2O \rightleftharpoons HO{-}C(CH_2COO^-)(CH_2COOH){-}COO^- + H_3O^+$	$K_{a_2} = 1.7 \times 10^{-5}$	$pK_{a_2} = 4.76$
(3) $HO{-}C(CH_2COO^-)(CH_2COOH){-}COO^- + H_2O \rightleftharpoons HO{-}C(CH_2COO^-)(CH_2COO^-){-}COO^- + H_3O^+$	$K_{a_3} = 4.1 \times 10^{-7}$	$pK_{a_3} = 6.39$

The equilibrium constant expression for this reaction is

$$K_{a_2} = \frac{[HPO_4^{2-}][H_3O^+]}{[H_2PO_4^-]}$$

At the point where the yellow and green curves cross over, $[H_2PO_4^-] = [HPO_4^{2-}]$, which means these two terms cancel out of the K_{a_2} expression and we are left with $K_{a_2} = [H_3O^+]$. Taking the negative log values of this equality gives us $pK_{a_2} = -\log[H_3O^+]$. In other words, the pH at which the yellow and green curves intersect in Figure 15.16 should equal the value of pK_{a_2} (7.20), and indeed it does. To test your understanding of these interesting curves, do a similar analysis of the pH at the intersection of the green and blue curves in Figure 15.16.

CONCEPT TEST

Do you expect the second or third acid ionization steps in phosphoric acid and citric acid to influence the pH of 0.100 *M* solutions of either acid?

15.8 Acidic and Basic Salts

Seawater and the freshwater in many rivers and lakes have pH values that range from weakly basic to weakly acidic. How can these waters be more basic than the acidic rainwater (pH $\leq$ 5.6) that serves, directly or indirectly, as their water supply? When rain soaks into the ground, its pH changes as it flows through soils that contain basic components. To understand the chemical processes that produce neutral or slightly basic groundwater, we first need to examine the acid–base properties of some common ionic compounds present in these waters.

As we discussed in Chapter 4, soluble ionic compounds separate into their component ions when they dissolve in water. For example, a 0.01 *M* solution of NaCl contains 0.01 *M* Na^+ ions and 0.01 *M* Cl^- ions. It is also a neutral solution. Neither Na^+ ions nor Cl^- ions hydrolyze (react with water) to form H_3O^+ or OH^- ions when they dissolve in water. The Cl^- ion is the conjugate base of a strong acid (HCl); therefore, the strength of the Cl^- ion as a base can be considered negligible.

When NaF, another sodium salt, dissolves in water, it produces F^- ions, which are the conjugate base of the weak acid, HF. Therefore F^- ions are weakly basic. When they dissolve in water some of them react with molecules of water forming molecules of HF and OH^- ions:

$$:\ddot{\underset{..}{F}}:^- + H_2\ddot{\underset{..}{O}} \rightleftharpoons H\text{—}F + [:\ddot{\underset{..}{O}}\text{—}H]^-$$

or,

$$F^-(aq) + H_2O(\ell) \rightleftharpoons HF(aq) + OH^-(aq)$$

The small concentrations of OH^- ions produced by this reaction make NaF solutions weakly basic.

If the salts that contain the conjugate bases of weak acids are basic, it is only reasonable that salts that contain the conjugate acids of weak bases are acidic. An example of such a salt is NH_4Cl. The Cl^- ions that are produced when NH_4Cl dissolves have negligible strengths as Brønsted–Lowry bases, but NH_4^+ ions are the conjugate acid of NH_3, a weak base. As a result, NH_4^+ ions are weakly acidic, producing at least some H_3O^+ ions as ammonium ions dissociate: donating one of their H atoms as an H^+ ion to a molecule of water, forming a hydronium ion and leaving behind a neutral molecule of ammonia:

$$[NH_4]^+ + H_2\ddot{\underset{..}{O}} \rightleftharpoons H_3\ddot{N} + [H\text{—}\ddot{O}(H)\text{—}H]^+$$

or,

$$NH_4^+(aq) + H_2O(\ell) \rightleftharpoons NH_3(aq) + H_3O^+(aq)$$

Consequently, solutions of NH_4Cl are weakly acidic.

Table 15.6 summarizes how salts can be acidic, basic, or neutral depending on whether they include cations that are the conjugate acids of weak bases, anions that are the conjugate bases of weak acids, or both. Salts that contain both the conjugate base of a weak acid *and* the conjugate acid of a weak base may be acidic, basic, or neutral, depending on the relative strengths of the acid and base.

TABLE 15.6 Acid–Base Properties of Salts

Source of Cation	Source of Anion	pH of Solutions	Example
Strong base	Strong acid	7	NaCl
Weak base	Strong acid	<7	NH_4Cl
Strong base	Weak acid	>7	NaF
Weak base	Weak acid	Neutral,[a] acidic,[b] or basic[c]	NH_4CH_3COO, NH_4F, NH_4HCO_3

[a]If K_a (of weak acid) $= K_b$ (of weak base)
[b]If K_a (of weak acid) $> K_b$ (of weak base)
[c]If K_a (of weak acid) $< K_b$ (of weak base)

Ammonium acetate represents the rare example of a salt in which the strengths of the conjugate acid (acetic acid) and the base (ammonia) happen to be *exactly the same*: the K_a of acetic acid $= K_b$ of ammonia 1.76×10^{-5}. As a result, ammonium acetate is a neutral salt.

SAMPLE EXERCISE 15.16 Predicting Whether a Salt Is Acidic, Basic, or Neutral **LO8**

NaClO is the active ingredient in chlorine bleach such as Clorox (**Figure 15.17**). Is an aqueous solution of NaClO acidic, basic, or neutral?

Collect, Organize, and Analyze Sodium ions do not hydrolyze and do not affect the pH of aqueous solutions. However, ClO^- ions are the conjugate base of HClO, which is a weak acid (Table 15.2). Therefore, ClO^- ions are weak bases that partially hydrolyze, forming OH^- ions:

$$ClO^-(aq) + H_2O(\ell) \rightleftharpoons HClO(aq) + OH^-(aq)$$

Solve Because hydrolysis of ClO^- ions produces OH^- ions, solutions of NaClO are weakly basic.

Think About It Any sodium salt that contains an anion that is the conjugate base of a weak acid produces weakly basic aqueous solutions.

Practice Exercise Write a chemical equation for the hydrolysis reaction that explains why an aqueous solution of K_2SO_4 is slightly basic.

FIGURE 15.17 Clorox is a common hypochlorite-based bleach used for cleaning and disinfection.

Having established that salts can be acidic, basic, or neutral, let's now address how to calculate the pH of their aqueous solutions. Our approach is much like that used to calculate the pH of solutions of weak acids and bases, but there is an extra step. For example, if we want to calculate the pH of a solution of ammonium chloride, we need to know the equilibrium constant of the reaction in which NH_4^+ ion functions as a Brønsted–Lowry acid:

$$NH_4^+(aq) + H_2O(\ell) \rightleftharpoons NH_3(aq) + H_3O^+(aq)$$

K_a values are typically *not* listed for the conjugate acids of weak bases, but K_b values for the corresponding weak bases are (see Appendix 5). For ammonia,

$$NH_3(aq) + H_2O(\ell) \rightleftharpoons NH_4^+(aq) + OH^-(aq) \qquad K_b = 1.76 \times 10^{-5}$$

The strengths of conjugate acids and bases are complementary—that is, the stronger one is, the weaker the other. Therefore, we can derive the K_a value for ammonium ions from the K_b value for ammonia. To see how, we first write the K_a and K_b equilibrium constant expressions for NH_4^+ and NH_3:

$$K_a = \frac{[NH_3][H_3O^+]}{[NH_4^+]} \qquad K_b = \frac{[NH_4^+][OH^-]}{[NH_3]}$$

These expressions are similar, although any shared terms that appear in the numerator of one expression are found in the denominator of the other. We can take advantage of this situation by multiplying the two expressions together:

$$K_a \times K_b = \frac{\cancel{[NH_3]}[H_3O^+]}{\cancel{[NH_4^+]}} \times \frac{\cancel{[NH_4^+]}[OH^-]}{\cancel{[NH_3]}} = [H_3O^+][OH^-]$$

The simplified product is the K_w expression for the autoionization of water. Thus,

$$K_a \times K_b = K_w \tag{15.18}$$

This is a very handy equation because (1) it works for all conjugate acid–base pairs, and (2) it allows us to calculate either the K_b of the anion in a basic salt from the K_a of its conjugate acid, or the K_a of the cation in an acidic salt from the K_b of its conjugate base.

SAMPLE EXERCISE 15.17 Calculating the pH of a Solution of a Basic Salt **LO8**

The chlorine bleach shown in Figure 15.17 contains 82.5 g/L of NaClO. What is the pH of this solution?

Collect and Organize We know the concentration of an aqueous solution of hypochlorite (NaClO) and we are asked to calculate its pH. We determined in Sample Exercise 15.16 that NaClO is a basic salt because the ClO^- ion is the conjugate base of HClO, a weak acid. Sodium ions do not hydrolyze and play no role in the acid–base properties of sodium salts. The K_a value for HClO is 2.9×10^{-8} (see Figure 15.5 and Appendix 5).

Analyze We need to first convert the K_a value for HClO into the K_b value for the ClO^- ion by using Equation 15.18: $K_a \times K_b = K_w$. We also need to convert the concentration of NaClO from g/L to molarity. That value will be the initial value of the reactant in a RICE table based on its hydrolysis reaction:

$$ClO^-(aq) + H_2O(\ell) \rightleftharpoons HClO(aq) + OH^-(aq)$$

We will solve for $[OH^-]$, then pOH, and then pH. The concentration of the solution should be about 1 *M*, and K_b should be a little more than 10^{-7}. Therefore, $[OH^-]$ of a ~1 *M* solution should be near the square root of that value, or ~10^{-3}, making pOH ≈ 3 and pH ≈ 11.

Solve Calculating K_b:

$$K_b = \frac{K_w}{K_a} = \frac{1.0 \times 10^{-14}}{2.9 \times 10^{-8}} = 3.4 \times 10^{-7}$$

and then calculating the molar concentration of NaClO to use in the following RICE table:

$$\frac{82.5\ \text{g}}{\text{L}} \times \frac{1\ \text{mol}}{74.44\ \text{g}} = 1.11\ \text{mol/L}$$

Reaction (R)	$ClO^-(aq) + H_2O(\ell)$	$\rightleftharpoons$	$HClO(aq)$	+	$OH^-(aq)$
	$[ClO^-]$ (*M*)		$[HClO]$ (*M*)		$[OH^-]$ (*M*)
Initial (I)	1.11		0		0
Change (C)	$-x$		$+x$		$+x$
Equilibrium (E)	$1.11 - x$		x		x

To solve for x, we make the simplifying assumption that x will be small compared with 1.11 M because the value of K_b is so small ($<<10^{-5}$):

$$K_b = \frac{[HClO][OH^-]}{[ClO^-]} = \frac{(x)(x)}{1.11 - x} \approx \frac{x^2}{1.11} = 3.4 \times 10^{-7}$$

$$x = [OH^-] = 6.1 \times 10^{-4}$$

Calculating pOH:

$$\text{pOH} = -\log[OH^-] = -\log(6.1 \times 10^{-4}) = 3.21$$

and pH:

$$\text{pH} = 14.00 - \text{pOH} = 14.00 - 3.21 = 10.79$$

Think About It The calculated pH (10.79) matches our estimate of 11. Also, the calculated $[OH^-]$ value is much less than 5% of the initial $[ClO^-]$ value:

$$\frac{6.1 \times 10^{-4}\ M}{1.11\ M} \times 100\% = 0.055\%$$

Our assumption that we could ignore the $-x$ term in the denominator of the K_b expression was justified.

Practice Exercise The pH of swimming pools is made slightly basic by spreading solid Na_2CO_3 across the surface. Another approach involves preparing concentrated solutions of Na_2CO_3 and slowly adding them to the water circulating through the pool pump and filter. What is the pH of an aqueous solution of 0.100 M Na_2CO_3?

SAMPLE EXERCISE 15.18 Calculating the pH of a Solution of an Acidic Salt — LO8

What is the pH of 0.25 M NH_4Cl?

Collect and Organize We are asked to calculate the pH of a solution of NH_4Cl. When NH_4Cl dissolves in water, NH_4^+ and Cl^- ions are released into solution. The NH_4^+ ion is the conjugate acid of NH_3, which is a weak base. The Cl^- ion is the conjugate base of HCl, which is a strong acid.

Analyze The Cl^- ion has negligible strength as a Brønsted–Lowry base, so it does not contribute to the acid–base properties of NH_4Cl. Ammonia is a weak base ($K_b = 1.76 \times 10^{-5}$), which means that its conjugate acid, NH_4^+, is a weak acid, and the pH of the solution will be controlled by its hydrolysis:

$$NH_4^+(aq) + H_2O(\ell) \rightleftharpoons NH_3(aq) + H_3O^+(aq)$$

The K_a value of NH_4^+ can be calculated by dividing K_w by the K_b of ammonia (i.e., by using Equation 5.18). The K_b value of ammonia is close to 10^{-5}, so the K_a value of the ammonium ion will be close to 10^{-9}. Therefore, $[H_3O^+]$ of a 0.25 M solution should be near the square root of about 10^{-10}, or about 10^{-5}, giving a pH $\approx$ 5.

Solve The simplified K_a expression for the NH_4^+ ion is

$$K_a = \frac{[NH_3][H_3O^+]}{[NH_4^+]}$$

Rearranging Equation 15.18 to solve for K_a:

$$K_a = \frac{K_w}{K_b} = \frac{1.0 \times 10^{-14}}{1.76 \times 10^{-5}} = 5.68 \times 10^{-10} = \frac{[NH_3][H_3O^+]}{[NH_4^+]}$$

We set up a RICE table in which we make the usual assumptions that the reaction is the only significant source of H^+ and that $x = [H_3O^+] = [NH_3]$ at equilibrium:

Reaction (R)	$NH_4^+(aq) + H_2O(\ell)$	$\rightleftharpoons$	$NH_3(aq)$	+	$H_3O^+(aq)$
	$[NH_4^+]$ (M)		$[NH_3]$ (M)		$[H_3O^+]$ (M)
Initial (I)	0.25		0		0
Change (C)	$-x$		$+x$		$+x$
Equilibrium (E)	$0.25 - x$		x		x

$$K_a = 5.68 \times 10^{-10} = \frac{[NH_3][H_3O^+]}{[NH_4^+]} = \frac{(x)(x)}{0.25 - x}$$

Given the very small value of K_a, we can make the simplifying assumption that $(0.25\ M - x) \approx 0.25\ M$, which gives us

$$\frac{x^2}{0.25} = 5.68 \times 10^{-10}$$

$$x^2 = 1.42 \times 10^{-10}$$

$$x = 1.19 \times 10^{-5} = [H_3O^+]$$

$$pH = -\log[H_3O^+] = -\log(1.19 \times 10^{-5}) = 4.92$$

Think About It This result matches our prediction quite closely. The calculated $[H_3O^+]$ is much less than 5% of the initial concentration of NH_4^+,

$$\frac{1.19 \times 10^{-5}\ M}{0.25\ M} \times 100\% = 0.0048\%$$

so our simplifying assumption was valid.

Practice Exercise What is the pH of a 0.25 M solution of methylamine hydrochloride (CH_3NH_3Cl)?

SAMPLE EXERCISE 15.19 Integrating Concepts: The pH of Human Blood

We began this chapter by noting how essential it is that our circulation and respiration systems efficiently remove from our bodies the CO_2 produced in our cells. An enzyme, carbonic anhydrase, plays a key role in this process by speeding up the rate at which CO_2 hydrolyzes:

$$CO_2(g) + H_2O(\ell) \rightleftharpoons H_2CO_3(aq)$$

and the rate at which H_2CO_3 ionizes:

$$H_2CO_3(aq) + H_2O(\ell) \rightleftharpoons HCO_3^-(aq) + H_3O^+(aq)$$

a. Does carbonic anhydrase increase the acid strength of dissolved CO_2—that is, the K_{a_1} of carbonic acid based on the total concentration of CO_2 in solution?

b. Suppose that, during strenuous exercise, the total concentration of dissolved CO_2 in the blood flowing through muscle tissues is 2.7×10^{-3} M. If the equilibrium constant of the hydrolysis reaction is 3.4×10^{-2}, what is the equilibrium concentration of carbonic acid in the blood?

c. Suppose the concentration of HCO_3^- ions in the blood in part (b) is 0.028 M. What is the pH of the blood?

Collect and Organize We are asked whether the presence of an enzyme that increases the rates at which CO_2 is hydrolyzed and undergoes acid ionization makes H_2CO_3 a stronger acid. We then need to calculate the equilibrium value of $[H_2CO_3]$ from an initial value of $[CO_2]$; we are also given the K value for the hydrolysis reaction and the pH of blood in which $[HCO_3^-]$ and $[CO_2]$ are known. Carbonic acid is a weak diprotic acid; its K_{a_1}, based on the total concentration of CO_2 in solution, is 4.3×10^{-7} (see Appendix 5).

Analyze (a) We learned in Section 14.7 that catalysts speed up reactions but do not alter their equilibrium constants. (b) Calculating $[H_2CO_3]$ will involve solving for the numerator of the equilibrium constant expression for the hydrolysis of CO_2, knowing the $[CO_2]$ value in the denominator. If continued exercise maintains $[CO_2]$ at a constant value, there will be no $-x$ term in the denominator. With a K value of 0.034 and $[CO_2] \approx 2 \times 10^{-3}\ M$, $[H_2CO_3]$ should be about $7 \times 10^{-5}\ M$. (c) Calculating the pH of a carbonic acid solution will require a RICE table based on K_{a_1} in which $[H_3O^+] = x$ at equilibrium but in which the initial concentration of HCO_3^- is 0.028 M, not zero. The normal pH of human blood is close to 7.4, so the calculated pH should be close to that.

Solve (a) Carbonic anhydrase should not affect the value of K_{a_1}, though it should produce increases in both $[H_2CO_3]$ and $[HCO_3^-]$. These increases in the numerator and denominator of the K_{a_1} expression offset each other and do not affect pH.

(b) Inserting the known values of K and $[CO_2]$ in the hydrolysis equilibrium constant expression:

$$K = \frac{[H_2CO_3]}{[CO_2]} = \frac{x}{2.7 \times 10^{-3}} = 3.4 \times 10^{-2}$$

and solving for x:

$$x = [H_2CO_3] = 9.2 \times 10^{-5}\ M$$

(c) Setting up a RICE table based on the K_{a_1} reaction, where the values in the $[H_2CO_3]$ column represent a constant total concentration of $CO_2(aq) + H_2CO_3(aq)$:

Reaction (R)	$H_2CO_3(aq)$ + $H_2O(\ell)$ ⇌	$HCO_3^-(aq)$ +	$H_3O^+(aq)$
	$[H_2CO_3]$ (M)	$[HCO_3^-]$ (M)	$[H_3O^+]$ (M)
Initial (I)	2.7×10^{-3}	0.028	0
Change (C)	0	$+x$	$+x$
Equilibrium (E)	2.7×10^{-3}	$0.028 + x$	x

In solving for x, we make the simplifying assumption that its value will be much less than 0.028 M (as it must be, if the calculated pH is close to 7.4, which means $[H_3O^+] \approx 10^{-7}$), and we ignore the very small x^2 term in the numerator of the following calculation:

$$K_{a_1} = \frac{[HCO_3^-][H_3O^+]}{[H_2CO_3]} = \frac{(0.028 + x)(x)}{2.7 \times 10^{-3}} \approx \frac{0.028x}{2.7 \times 10^{-3}}$$

$$= 4.3 \times 10^{-7}$$

$$x = 4.15 \times 10^{-8}$$

Therefore, the pH of the blood in the muscle tissue is

$$\text{pH} = -\log[H_3O^+] = -\log(4.15 \times 10^{-8}) = 7.38$$

Think About It The pH value calculated in (c) is reasonable because it matches our estimate based on the normal pH of human blood.

SUMMARY

LO1 The strengths of acids and bases are related to the values of their acid and base ionization constants, K_a and K_b. Most acids are weak, which means their K_a is much less than 1 and they ionize only partially in water. (Section 15.2)

LO2 The strengths of acids are related to both the electronegativity of the atom adjacent to the hydrogen atom and the stability of the anions they form when they release H^+ ions. The stability of oxoanions is enhanced by multiple oxygen atoms bonded to the central atom, which disperse the negative charge(s) over the anion. (Section 15.2)

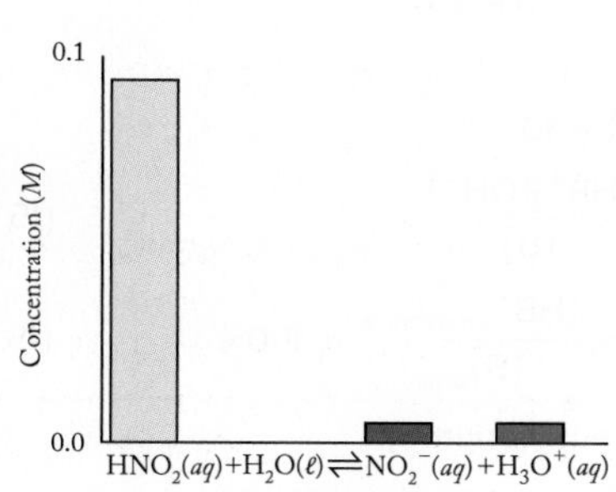

LO3 When an acid (HA) ionizes, it forms its **conjugate base**, A^-. When base B bonds to a H^+ ion, it forms its **conjugate acid**, HB^+. (Section 15.3)

LO4 In a neutral solution at 25°C, $[H_3O^+] = [OH^-] = 1.0 \times 10^{-7}$. The **pH** scale is a logarithmic scale for expressing the acidic or basic strength of solutions. Acidic solutions have pH values less than 7, whereas basic solutions have pH values greater than 7. Because pH is the negative logarithm of H_3O^+ concentration, the higher the pH, the lower the H_3O^+ concentration. An increase in one pH unit represents a decrease in $[H_3O^+]$ to 1/10 of its initial value. Likewise, **pOH** is the negative logarithm of OH^- concentration. The sum of pH and pOH equals 14 in an aqueous solution at 25°C. (Section 15.4)

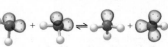

LO5 The K_a and K_b values of acids and bases can be used to calculate the extent to which their solutions are ionized—that is, their **degree of ionization** or *percent ionization*—and vice versa. (Section 15.5)

LO6 To calculate the pH of a weak acid or base, use a RICE table based on the acid or base ionization reaction to determine the equilibrium value of $[H_3O^+]$ or $[OH^-]$ in a solution of a weak acid or base. (Section 15.6)

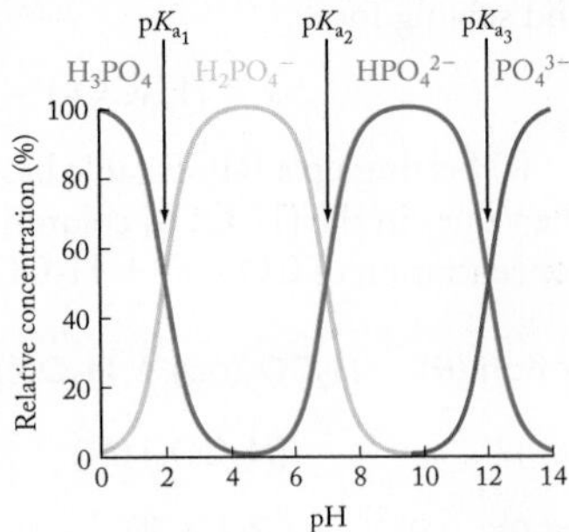

LO7 Polyprotic acids can undergo more than one acid ionization reaction, but for most, the first ionization reaction is the one that controls pH. (Section 15.7)

LO8 A salt solution is acidic if the cation in the salt is the conjugate acid of a weak base and the anion is the conjugate base of a strong acid. A salt solution is basic if the anion in the salt is the conjugate base of a weak acid and the cation is the conjugate acid of a strong base. A salt solution is neutral if neither the cation nor the anion hydrolyzes (reacts with water) to form H_3O^+ or OH^- ions when it dissolves in water. (Section 15.8)

PARTICULATE **PREVIEW WRAP-UP**

All three species form covalent bonds when they accept donated protons. ClO^- becomes a neutral molecule (HClO), whereas NH_3 and H_2O become cations (NH_4^+ and H_3O^+, respectively). All three reactants contain atoms with at least one lone pair of electrons (i.e., the O atoms in ClO^- and H_2O, and the N atom in NH_3) that can bond to H^+ ions.

PROBLEM-SOLVING SUMMARY

Type of Problem	Concepts and Equations	Sample Exercises
Rank acids and bases from strongest to weakest	The strengths of acids and bases are proportional to their concentrations and K_a and K_b values, which are related to their molecular structures.	**15.1, 15.2**
Identifying acid–base conjugate pairs and the complementary nature of their strengths	The formula of the base in a conjugate pair is the formula of the acid less one H^+ ion. The stronger the acid, the weaker its conjugate base; the stronger the base, the weaker its conjugate acid.	**15.3, 15.4**
Interconverting $[H_3O^+]$, $[OH^-]$, pH, and pOH	Use the following: $\mathrm{pH} = -\log[H_3O^+]$ (15.10) $\mathrm{pOH} = -\log[OH^-]$ (15.11) $pK_w = \mathrm{pH} + \mathrm{pOH} = 14.00$ (15.12)	**15.5, 15.6, 15.7**
Determining percent ionization and K_a given the pH of a weak acid	Use the following: $[H_3O^+] = 10^{-\mathrm{pH}}$ $K_a = \frac{[A^-][H_3O^+]}{[HA]}$ (15.14) $\text{Percent ionization} = \frac{[A^-]_{\text{equilibrium}}}{[HA]_{\text{initial}}} \times 100\%$ (15.15)	**15.8**
Determining percent ionization and K_b given the pH of a weak base	Use the following: $[OH^-] = 10^{-\mathrm{pOH}}$ $K_b = \frac{[HB^+][OH^-]}{[B]}$ (15.16) $\text{Percent ionization} = \frac{[HB^+]_{\text{equilibrium}}}{[B]_{\text{initial}}} \times 100\%$ (15.17)	**15.9**
Calculating the pH of a solution of weak acid HA	Set up a RICE table based on the K_a equilibrium $HA(aq) + H_2O(\ell) \rightleftharpoons H_3O^+(aq) + A^-(aq)$ Let $x = [H_3O^+] = [A^-]$ at equilibrium. Calculate x by using $K_a = \frac{[A^-][H_3O^+]}{[HA]}$ Then calculate $\mathrm{pH} = -\log[H_3O^+]$.	**15.11**

<table>
<tr><th>Type of Problem</th><th>Concepts and Equations</th><th>Sample Exercises</th></tr>
<tr><td>Calculating the pH of a solution of weak base B</td><td>Set up a RICE table based on the equilibrium
$$B(aq) + H_2O(\ell) \rightleftharpoons HB^+(aq) + OH^-(aq)$$
Let $x = [OH^-] = [HB^+]$ at equilibrium. Calculate x by using
$$K_b = \frac{[HB^+][OH^-]}{[B]}$$
Then use
$$pOH = -\log[OH^-]$$
$$pH = 14.00 - pOH$$</td><td>15.12</td></tr>
<tr><td>Calculating the pH of a solution of very dilute acidic (or very dilute basic) solution while considering the autoionization of water</td><td>Set up a RICE table based on the equilibrium
$$H_2O(\ell) + H_2O(\ell) \rightleftharpoons H_3O^+(aq) + OH^-(aq)$$
Let $x = [H_3O^+] = [OH^-]$ because of autoionization. Add x to the $[H_3O^+]$ because of the dilute acid (or to the $[OH^-]$ because of the weak base). Solve for x by using
$$K_w = [H_3O^+][OH^-] \quad (15.8)$$
Then calculate $pH = -\log[H_3O^+]$.</td><td>15.13</td></tr>
<tr><td>Calculating the pH of a solution of a strong diprotic acid</td><td>Assume
$$H_2A(aq) + H_2O(\ell) \rightarrow H_3O^+(aq) + HA^-(aq)$$
is complete.
Set up a RICE table based on the K_{a_2} equilibrium
$$HA^-(aq) + H_2O(\ell) \rightleftharpoons H_3O^+(aq) + A^{2-}(aq)$$
Let x = additional $[H^+]$ from second ionization step; $[HA^-]_{initial\ (2nd\ step)} = [H_2A]_{initial}$. Calculate x by using
$$K_{a_2} = \frac{[H_3O^+][A^{2-}]}{[HA^-]}$$
Then calculate $pH = -\log[H_3O^+]$.</td><td>15.14</td></tr>
<tr><td>Calculating the pH of a solution of a weak diprotic acid</td><td>Set up a RICE table based on the K_{a_1} equilibrium
$$H_2A(aq) + H_2O(\ell) \rightleftharpoons H_3O^+(aq) + HA^-(aq)$$
Let $x = [H_3O^+] = [HA^-]$ at equilibrium. Calculate x by using
$$K_{a_1} = \frac{x^2}{[H_2A] - x}$$
Then calculate $pH = -\log[H_3O^+]$.</td><td>15.15</td></tr>
<tr><td>Predicting whether a salt is acidic, basic, or neutral</td><td>The cations in acidic salts are the conjugate acids of weak bases. The anions in basic salts are the conjugate bases of weak acids. The cations and anions in most neutral salts do not hydrolyze.</td><td>15.16</td></tr>
<tr><td>Calculating the pH of a solution of a basic salt</td><td>Assume the salt (MX) completely dissociates into M^+ and X^-. Set up a RICE table for the equilibrium
$$X^-(aq) + H_2O(\ell) \rightleftharpoons HX(aq) + OH^-(aq)$$
Let $x = [HX] = [OH^-] =$ at equilibrium.
Calculate x by using
$$K_w = K_a \times K_b$$
$$K_b = \frac{K_w}{K_{a\ (of\ conjugate\ acid,\ HA)}}$$
$$K_b = \frac{K_w}{K_a} = \frac{x^2}{[X^-] - x}$$
Convert $[OH^-]$ to pOH, and then pH.</td><td>15.17</td></tr>
</table>

Type of Problem	Concepts and Equations	Sample Exercises
Calculating the pH of a solution of an acidic salt	Assume the salt (HBX) completely dissociates into HB^+ and X^-. Set up a RICE table for the equilibrium $HB^+(aq) + H_2O(\ell) \rightleftharpoons B(aq) + H_3O^+(aq)$ Let $x = [H_3O^+] = [B]$ at equilibrium. Calculate x by using $K_a = \frac{K_w}{K_{b\,(\text{of conjugate base, B})}}$ $K_a = \frac{K_w}{K_b} = \frac{x^2}{[HB^+] - x}$ Then calculate pH.	**15.18**

VISUAL PROBLEMS

(Answers to boldface end-of-chapter questions and problems are in the back of the book.)

15.1. Which of the lines in Figure P15.1 best represents the dependence of the degree of ionization of acetic acid on its concentration in aqueous solution?

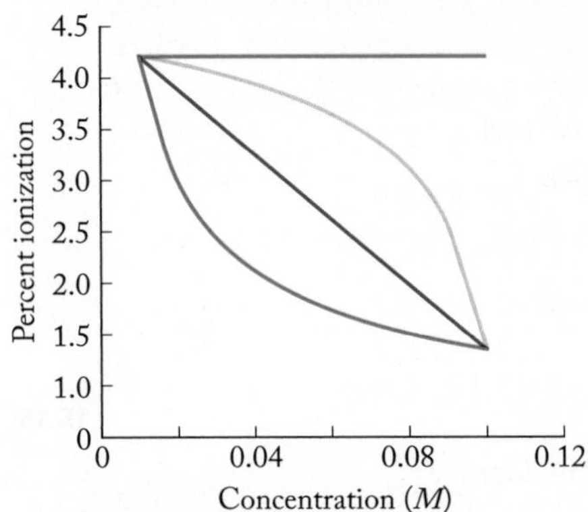

FIGURE P15.1

15.2. The graph in Figure P15.2 shows the percent ionization of two acids as a function of concentration in water. Which line describes the behavior of HNO_3, and which line describes the behavior of acetic acid (CH_3COOH)?

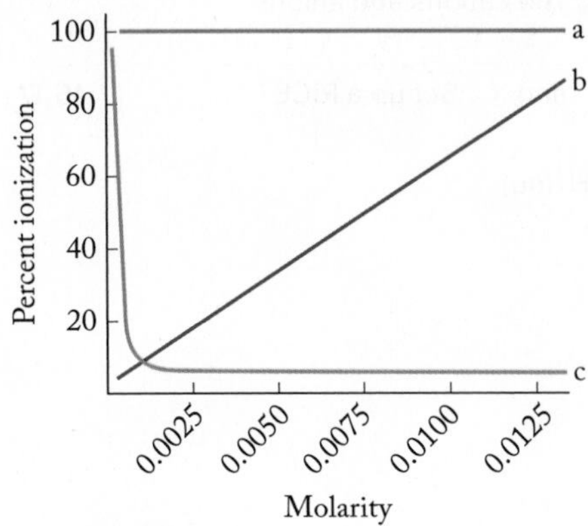

FIGURE P15.2

15.3. The bar graph in Figure P15.3 shows the degree of ionization of 1×10^{-3} *M* solutions of HClO, HBrO, and HIO. Which bar corresponds to HIO?

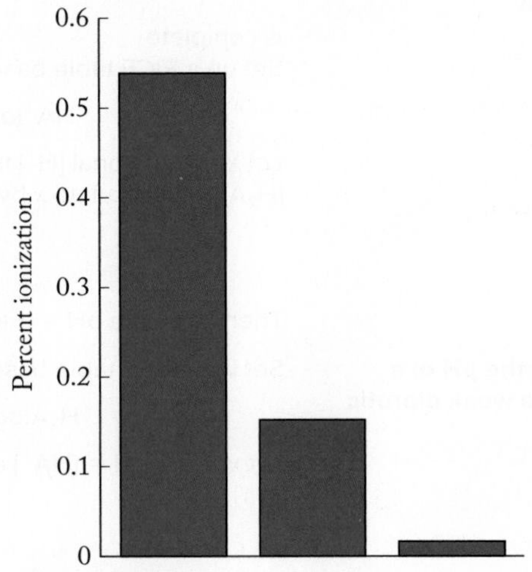

FIGURE P15.3

15.4. The bar graph in Figure P15.4 shows the degree of ionization of 1 *M* solutions of HClO, $HClO_2$, and $HClO_3$. Which bar corresponds to HClO?

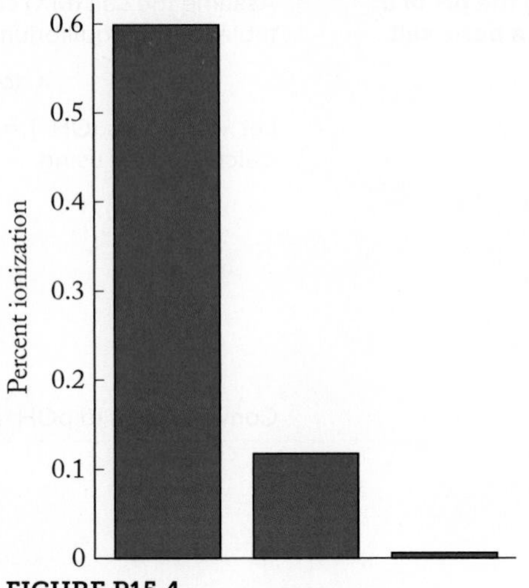

FIGURE P15.4

15.5. Figure P15.5 shows a condensed molecular structure of pseudoephedrine, a widely used decongestant and stimulant.

a. Is pseudoephedrine an acidic, basic, or neutral compound?
b. Which functional group in its structure gives it the property you selected in part (a)?

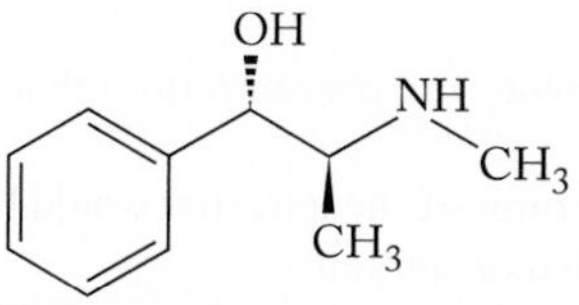

OH
NH
CH_3
CH_3

FIGURE P15.5

15.6. Figure P15.6 shows the molecular structure of alanine. Is an aqueous solution of alanine acidic, basic, or neutral? Explain your selection, or explain what additional information you need to make a more informed selection.

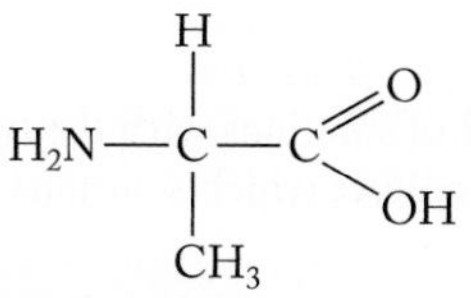

H
$H_2N—C—C$ (=O) (OH)
CH_3

FIGURE P15.6

***15.7.** Figure P15.7 shows the skeletal structures of piperidine and morpholine. Which is the stronger base? Explain your selection.

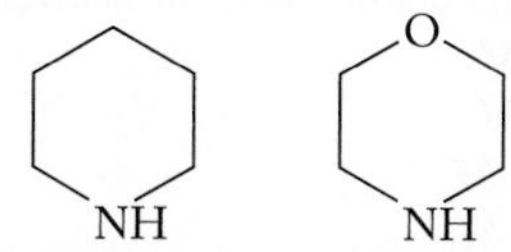

FIGURE P15.7

15.8. Based on the skeletal structures shown in Figure P15.8, is ethanolamine or ethylamine the stronger base? Explain why you think so.

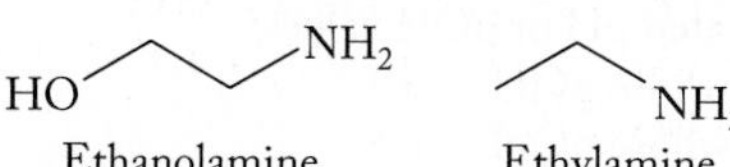

FIGURE P15.8

15.9. Figure P15.9 shows molecular models of acetic acid and trichloroacetic acid. Use these two models to explain why the K_a of trichloroacetic acid is about 10^4 times that of acetic acid.

Acetic acid Trichloroacetic acid

FIGURE P15.9

15.10. Use representations [A] through [I] in Figure P15.10 to answer questions (a)–(f). Note that water has been omitted from the solutions in [D], [E], [F], and [G] for ease of viewing.

a. Which molecules are weak acids? Which are weak bases?
b. Which pairs of molecules represent conjugate acid–base pairs? (Give the letter of the acid first.)
c. Write net ionic equations describing the equilibria involving the conjugate pairs identified in part (b).
d. Which images represent acidic solutions?
e. Which images represent basic solutions?
f. Which image represents a neutral solution?

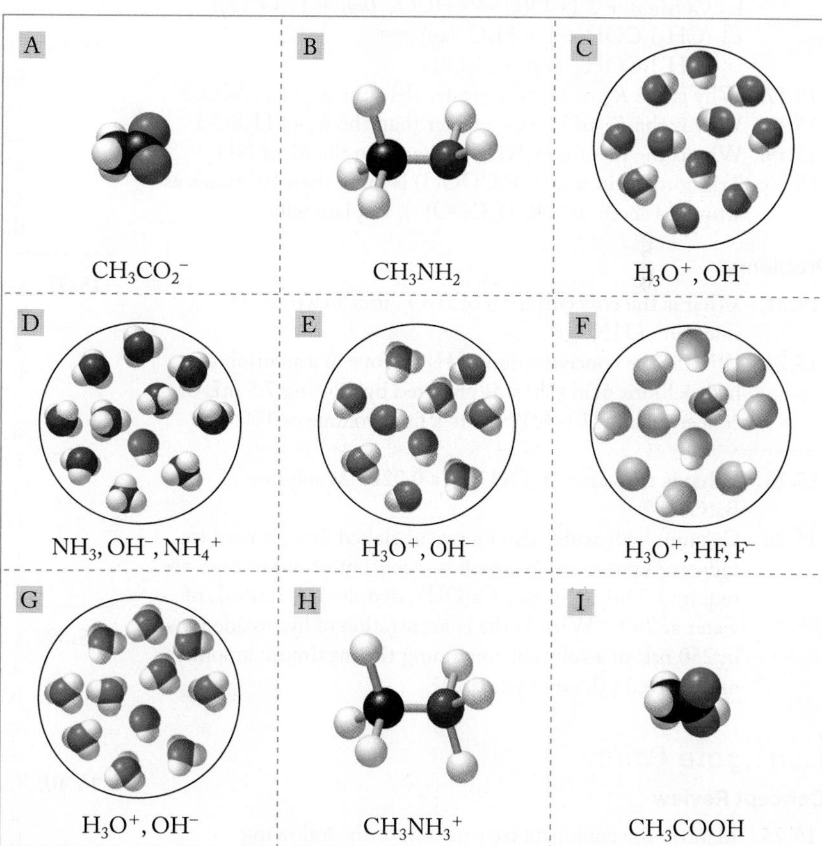

FIGURE P15.10

QUESTIONS AND PROBLEMS

The Molecular Structures and Strengths of Acids and Bases

Concept Review

15.11. In an aqueous solution of LiOH, which compound acts as a Brønsted–Lowry acid and which is the Brønsted–Lowry base?

15.12. In an aqueous solution of H_2SO_3, which compound acts as a Brønsted–Lowry acid and which is the Brønsted–Lowry base?

15.13. In an aqueous solution of methylamine, which species acts as a Brønsted–Lowry acid and which is the Brønsted–Lowry base?

15.14. Both NaOH and $Ba(OH)_2$ are strong bases. Does this mean that solutions of the two compounds with the same molarity have the same capacity to accept hydrogen ions? Why or why not?

15.15. Identify the acids and bases in the following reactions:
a. $HCl(aq) + NaOH(aq) \rightarrow NaCl(aq) + H_2O(\ell)$
b. $MgCO_3(s) + 2\,HCl(aq) \rightarrow MgCl_2(aq) + CO_2(g) + H_2O(\ell)$
c. $2\,NH_3(aq) + H_2SO_4(aq) \rightarrow (NH_4)_2SO_4(aq)$

15.16. Identify the acids and bases in the following reactions:
a. $(CH_3)_3N(aq) + H_2O(\ell) \rightleftharpoons (CH_3)_3NH^+(aq) + OH^-(aq)$
b. $CO_2(aq) + 2\,H_2O(\ell) \rightleftharpoons HCO_3^-(aq) + H_3O^+(aq)$
c. $(CH_3)_3COH(aq) + H_3O^+(aq) \rightleftharpoons (CH_3)_3COH_2^+(aq) + H_2O(\ell)$

15.17. Why is the K_{a_2} of H_2SO_4 greater than the K_{a_1} of H_2SeO_4?

15.18. Why is the K_{a_2} of H_2SO_4 greater than the K_{a_1} of H_2SO_3?

15.19. Why is the K_b of CH_3NH_2 greater than the K_b of NH_3?

15.20. Trifluoroacetic acid (CF_3COOH) is more than 10^4 times as strong as acetic acid (CH_3COOH). Explain why.

Problems

15.21. What is the concentration of H_3O^+ ions in a 0.65 M solution of HNO_3?

15.22. What is the concentration of H_3O^+ ions in a solution of hydrochloric acid that was prepared by diluting 7.5 mL of concentrated (11.6 M) HCl to a final volume of 100.0 L?

15.23. What is the value of $[OH^-]$ in a 0.0205 M solution of $Ba(OH)_2$?

15.24. Calcium hydroxide, also known as slaked lime, is used in industrial processes in which low concentrations of base are required. Only 0.16 g of $Ca(OH)_2$ dissolves in 100 mL of water at 25°C. What is the concentration of hydroxide ions in 250 mL of a solution containing the maximum amount of dissolved calcium hydroxide?

Conjugate Pairs

Concept Review

15.25. Identify the conjugate base of each of the following compounds: HF, $HBrO_2$, H_2SO_4, and $HClO_4$.

15.26. Identify the conjugate acid of each of the following species: $(CH_3)_2NH$, CH_3COO^-, ClO^-, and OH^-.

15.27. What is the conjugate acid of the dihydrogen phosphate ion ($H_2PO_4^-$), and what is its conjugate base?

15.28. Why is it unnecessary to pubiish tables of K_b values of the conjugate bases of weak acids whose K_a values are known?

pH and the Autoionization of Water

Concept Review

15.29. Explain why high pH values correspond to low acidities.

15.30. Solution A is 1000 times more acidic than solution B. What is the difference in the pH values of solution A and solution B?

15.31. Describe a solution (solute and concentration) that has a negative pH value.

15.32. Describe a solution (solute and concentration) that has a negative pOH value.

***15.33.** Draw the Lewis structures of the ions that would be produced if pure methanol autoionizes.

*15.34. Liquid ammonia autoionizes at a temperature of 223 K. The value of the equilibrium constant for the autoionization of ammonia is considerably less than that of water. Write an equation for the autoionization of ammonia and suggest a reason why the value of K for the process is less than that of water.

Problems

15.35. Calculate the pH and pOH of solutions with the following $[H_3O^+]$ or $[OH^-]$ values. Indicate which solutions are acidic, basic, or neutral.
a. $[H_3O^+] = 2.4 \times 10^{-4}\,M$
b. $[H_3O^+] = 5.5 \times 10^{-8}\,M$
c. $[OH^-] = 2.6 \times 10^{-2}\,M$
d. $[OH^-] = 1.0 \times 10^{-14}\,M$

15.36. Calculate the pH and pOH of the solutions with the following hydrogen ion or hydroxide ion concentrations. Indicate which solutions are acidic, basic, or neutral.
a. $[OH^-] = 1.0 \times 10^{-7}\,M$
b. $[OH^-] = 3.9 \times 10^{-2}\,M$
c. $[H_3O^+] = 8.8 \times 10^{-4}\,M$
d. $[H_3O^+] = 5.9 \times 10^{-8}\,M$

15.37. Calculate the concentration of the following ions in the solution described:
a. $[OH^-]$ in $7.7 \times 10^{-4}\,M$ NaOH
b. $[OH^-]$ in $6.2 \times 10^{-3}\,M$ $Ca(OH)_2$
c. $[H_3O^+]$ in $1.9 \times 10^{-2}\,M$ HCl
d. $[H_3O^+]$ in $4.4 \times 10^{-4}\,M$ HCl

15.38. Determine the indicated pH or pOH values:
a. pH of a solution whose pOH = 5.5
b. pH of a solution whose pOH = 6.8
c. pOH of a solution whose pH = 9.7
d. pOH of a solution whose pH = 4.4

15.39. Calculate the pH and pOH of the following solutions:
a. stomach acid in which [HCl] = 0.155 M
b. 0.00500 M HNO_3
c. a 2:1 mixture of 0.0125 M HCl and 0.0125 M NaOH
d. a 3:1 mixture of 0.0125 M H_2SO_4 and 0.0125 M KOH

15.40. Calculate the pH and pOH of the following solutions:
a. 0.0450 M NaOH
b. 0.160 M $Ca(OH)_2$
c. a 1:1 mixture of 0.0125 M HCl and 0.0125 M $Ca(OH)_2$
d. a 2:3 mixture of 0.0125 M HNO_3 and 0.0125 M KOH

15.41. Calculate the pH of a $1.33 \times 10^{-9}\,M$ solution of LiOH.

15.42. Calculate the pH of a $6.9 \times 10^{-8}\,M$ solution of HBr.

Calculations Involving pH, K_a, and K_b

Concept Review

15.43. One-molar solutions of the following acids are prepared: CH_3COOH, HNO_2, $HClO$, and HCl.
a. Rank them in order of decreasing $[H_3O^+]$.
b. Rank them in order of increasing strength as acids (weakest to strongest).

15.44. On the basis of the following degree-of-ionization data for 0.100 *M* solutions, select which acid has the largest K_a.

Acid	Degree of Ionization (%)
C_6H_5COOH	2.5
HF	8.5
HN_3	1.4
CH_3COOH	1.3

15.45. A 1.0 *M* aqueous solution of HCl conducts electricity better than a 1.0 *M* solution of HF. Explain why.

15.46. Hydrogen chloride and water are molecular compounds, yet a solution of HCl dissolved in H_2O is an excellent conductor of electricity. Explain why.

***15.47.** When 1,2-diaminoethane ($H_2NCH_2CH_2NH_2$) dissolves in water, the resulting solution is basic. Write the formula of the ionic compound that is formed when hydrochloric acid is added to a solution of 1,2-diaminoethane.

15.48. **Early Antiseptic** The use of phenol, also known as carbolic acid, was pioneered in the 19th century by Sir Joseph Lister (after whom Listerine was named) as an antiseptic in surgery. Its formula is C_6H_5OH; the red hydrogen atom is ionizable. Write the mass action expression for the acid ionization equilibrium of phenol.

***15.49.** The K_a values of weak acids depend on the solvent in which they dissolve. For example, the K_a of alanine in aqueous ethanol is less than its K_a in water.
a. In which solvent does alanine ionize more?
b. Which is the stronger Brønsted–Lowry base: water or ethanol?

*15.50. The K_a of proline is 2.5×10^{-11} in water, 2.8×10^{-11} in an aqueous solution that is 28% ethanol, and 1.66×10^{-8} in aqueous formaldehyde at 25°C.
a. In which solvent is proline the strongest acid?
b. Rank water, ethanol, and formaldehyde based on their strengths as Brønsted–Lowry bases.

Problems

15.51. **Muscle Physiology** During strenuous exercise, lactic acid builds up in muscle tissues. In a 1.00 *M* aqueous solution, 2.94% of lactic acid is ionized. What is the value of its K_a?

15.52. **Rancid Butter** The odor of spoiled butter is caused in part by butanoic acid, which results from the chemical breakdown of butterfat. A 0.100 *M* solution of butanoic acid is 1.23% ionized. Calculate the value of K_a for butanoic acid.

15.53. At equilibrium, the value of $[H_3O^+]$ in a 0.250 *M* solution of an unknown acid is 3.36×10^{-3} *M*. Determine the degree of ionization of the acid and its K_a value.

15.54. **Poisonous Plant** Gifblaar is a small South African shrub and one of the most poisonous plants known because it contains fluoroacetic acid. If a 0.480 *M* solution of fluoroacetic acid has a pH of 1.44, what is the K_a of the acid?

15.55. **Acid Rain I** A weather system moving through the American Midwest produced rain with an average pH of 5.02. By the time the system reached New England, the rain it produced had an average pH of 4.66. How much more acidic was the rain falling in New England?

15.56. **Acid Rain II** A newspaper reported that the "level of acidity" in a sample taken from an extensively studied watershed in New Hampshire in 1998 was "an astounding 200 times lower than the worst measurement" taken in the preceding 23 years. What is this difference expressed in units of pH?

15.57. The K_b of aminoethanol ($HOCH_2CH_2NH_2$) is 3.1×10^{-5}.
a. Is aminoethanol a stronger or weaker base than ethylamine, $pK_b = 3.36$?
b. Calculate the pH of a 1.67×10^{-2} *M* solution of aminoethanol.
c. Calculate the $[OH^-]$ concentration of a 4.25×10^{-4} *M* solution of aminoethanol.

15.58. **Food Dye** Quinoline is a weakly basic liquid used in the manufacture of quinolone yellow, a greenish-yellow dye for foods, and in the production of niacin. Its pK_b is 9.15.
a. What is the pH of a 0.0752 *M* solution of quinolone?
b. What is the hydroxide ion concentration of the solution in part (a)?

15.59. **Painkillers** Morphine is an effective painkiller but is also highly addictive. Codeine is a popular prescription painkiller because it is much less addictive than morphine. Codeine contains a basic nitrogen atom that can be protonated to give the conjugate acid of codeine.
a. Calculate the pH of a 1.8×10^{-3} *M* solution of morphine if its $pK_b = 5.79$.
b. Calculate the pH of a 2.7×10^{-4} *M* solution of codeine if the pK_a of the conjugate acid is 8.21.

15.60. The awful odor of dead fish is caused mostly by trimethylamine $[(CH_3)_3N]$, one of three compounds related to ammonia in which methyl groups replace one, two, or all three of the H atoms in ammonia.
a. The K_b of trimethylamine $[(CH_3)_3N]$ is 6.5×10^{-5} at 25°C. Calculate the pH of a 3.00×10^{-4} *M* solution of trimethylamine.
b. The K_b of methylamine $[(CH_3)NH_2]$ is 4.4×10^{-4} at 25°C. Calculate the pH of a 2.88×10^{-3} *M* solution of methylamine.
*c. The K_b of dimethylamine $[(CH_3)_2NH]$ is 5.9×10^{-4} at 25°C. What concentration of dimethylamine is needed for the solution to have the same pH as the solution in part (b)?

15.61. **Nicotine Addiction** Nicotine is responsible for the addictive properties of tobacco. What is the pH of a 1.00×10^{-3} *M* solution of nicotine?

15.62. Pseudoephedrine hydrochloride (Figure P15.62) is a common ingredient in cough syrups and decongestants. Its $pK_a = 9.22$. What is the pH of a solution that is 0.0295 *M* in pseudoephedrine hydrochloride?

[C_6H_5–CH(OH)–CH(CH_3)–NH_2^+–CH_3] Cl^- (structure: H H H above C—C—N—CH_3; OH CH_3 H below; bracket charge +)

FIGURE P15.62

Polyprotic Acids

Concept Review

15.63. Why is the K_{a_2} value of phosphoric acid less than its K_{a_1} value but greater than its K_{a_3} value?

15.64. In calculating the pH of a 1.0 M solution of H_2SO_3, we can ignore the H^+ ions produced by the ionization of the bisulfate (HSO_3^-) ion; however, in calculating the pH of a 1.0 M solution of sulfuric acid, we cannot ignore the H^+ ions produced by the ionization of the bisulfate ion. Why?

15.65. Carbonic acid (H_2CO_3) is a very weak diprotic acid ($K_{a_1} = 4.3 \times 10^{-7}$), but germanic acid ($H_2GeO_3$) is even weaker ($K_{a_1} = 9.8 \times 10^{-10}$). Suggest a reason why.

15.66. Figure P15.66 contains skeletal structures of the dicarboxylic acids malonic acid and oxalic acid. The K_{a_1} of malonic acid is about 10^4 times larger than its K_{a_2}, whereas the K_{a_1} of oxalic acid is about 10^3 times larger than its K_{a_2}. Suggest a reason why the separation in K_a values is greater for malonic acid.

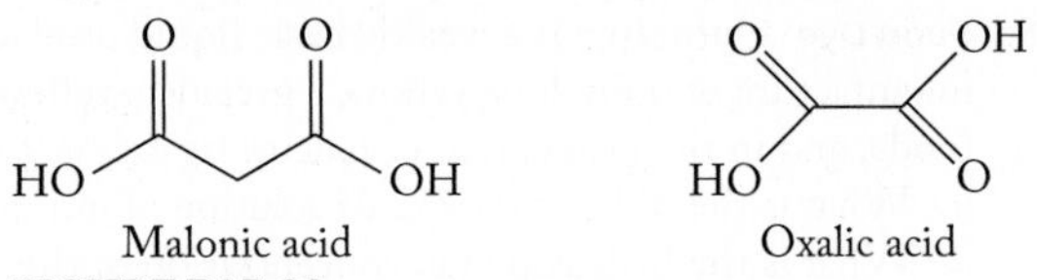

FIGURE P15.66

Problems

15.67. What is the pH of a 2.00 M solution of H_2SO_4?

15.68. What is the pH of a 2.00 M solution of H_2SO_3?

15.69. Ascorbic acid (vitamin C) is a weak diprotic acid. What is the pH of a 0.250 M solution of ascorbic acid?

15.70. **Rhubarb Pie** The leaves of the rhubarb plant contain high concentrations of diprotic oxalic acid (HOOCCOOH) and must be removed before the stems are used to make rhubarb pie. What is the pH of a 0.0288 M solution of oxalic acid?

15.71. **Malaria Treatment** Quinine occurs naturally in the bark of the cinchona tree. For centuries it was the only treatment for malaria. Calculate the pH of a 0.01050 M solution of quinine in water.

15.72. Dozens of pharmaceuticals, ranging from cyclizine for motion sickness to Viagra for impotence, are derived from the organic compound piperazine (Figure P15.72).

a. Solutions of piperazine are basic ($K_{b_1} = 5.28 \times 10^{-5}$; $K_{b_2} = 2.15 \times 10^{-9}$). What is the pH of a 0.0125 M solution of piperazine?

*b. Draw the structure of the ionic form of piperazine that would be present in stomach acid (about 0.16 M HCl).

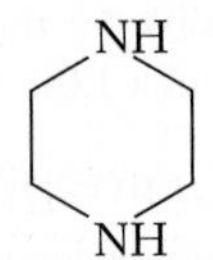

FIGURE P15.72

Acidic and Basic Salts

Concept Review

15.73. Which of the following salts produces an acidic solution in water: ammonium acetate, ammonium nitrate, or sodium formate?

15.74. Which of the following salts produces a basic solution in water: NaF, KCl, NH_4Cl?

15.75. **Neutralizing the Smell of Fish** Trimethylamine [$(CH_3)_3N$], $K_b = 6.5 \times 10^{-5}$ at 25°C, is a contributor to the "fishy" odor of not-so-fresh seafood. Some people squeeze fresh lemon juice (which contains a high concentration of citric acid) on cooked fish to reduce the fishy odor. Why is this practice effective?

*15.76. **Nutritional Value of Beets** Beets contain high concentrations of the calcium salt of malonic acid (see Figure P15.66). Could the presence of the calcium salt of malonic acid affect the pH balance of beets? If so, in which direction? Explain.

Problems

15.77. The K_a of the conjugate acid of the artificial sweetener saccharin is 5.9×10^{-11}. What is the pK_b for saccharin?

15.78. The K_{a_1} value for oxalic acid (HOOCCOOH) is 5.9×10^{-2}, and the K_{a_2} value is 6.4×10^{-5}. What are the values of K_{b_1} and K_{b_2} of the oxalate anion ($^-OOCCOO^-$)?

15.79. **Dental Health** Sodium fluoride is added to many municipal water supplies to reduce tooth decay. Calculate the pH of a 0.00339 M solution of NaF at 25°C.

15.80. Calculate the pH of a 1.25×10^{-2} M solution of the decongestant ephedrine hydrochloride if the pK_b of ephedrine (its conjugate base) is 3.86.

Additional Problems

15.81. Consider the following compounds: CH_3NH_2, CH_3COOH, $Ca(OH)_2$, and $HClO_4$.

a. Identify the Arrhenius acid(s).
b. Identify the Arrhenius base(s).
c. Identify the Brønsted–Lowry acid(s).
d. Identify the Brønsted–Lowry base(s).

15.82. Are all Arrhenius acids also Brønsted–Lowry acids? Are all Brønsted–Lowry acids also Arrhenius acids? If yes, explain why. If not, give a specific example to demonstrate the difference.

15.83. Are all Arrhenius bases also Brønsted–Lowry bases? Are all Brønsted–Lowry bases also Arrhenius bases? If yes, explain why. If not, give a specific example to demonstrate the difference.

15.84. Describe the intermolecular forces and changes in bonding that lead to the formation of a basic solution when methylamine (CH_3NH_2) dissolves in water.

***15.85.** Describe the chemical reactions of sulfur that begin with the burning of high-sulfur fossil fuel and that end with the reaction between acid rain and building exteriors made of marble ($CaCO_3$).

15.86. The K_{a_1} of phosphorous acid (H_3PO_3) is nearly the same as the K_{a_1} of phosphoric acid (H_3PO_4).

a. Draw the Lewis structure of phosphorous acid.
b. Identify the ionizable hydrogen atoms in the structure.
c. Explain why the K_{a_1} values of phosphoric and phosphorous acid are similar.

***15.87.** **pH of Natural Waters** In a 1985 study of Little Rock Lake in Wisconsin, 400 gallons of 18 M sulfuric acid were added to the lake over six years. The initial pH of the lake was 6.1 and the final pH was 4.7. If none of the acid was consumed in chemical reactions, estimate the volume of the lake.

15.88. **Acid–Base Properties of Pharmaceuticals I** Sertraline (also known by the trade name Zoloft) is a prescription drug for the treatment of depression. It is sold as its hydrochloride salt, which has the structure shown in Figure P15.88. When the hydrochloride dissolves in water, will the resulting solution be acidic or basic?

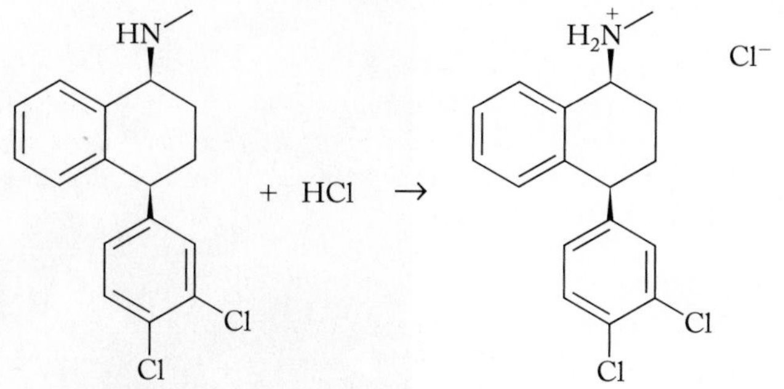

FIGURE P15.88

15.89. **Acid–Base Properties of Pharmaceuticals II** Fluoxetine (also known by the trade name Prozac; Figure P15.89) is a popular antidepressant drug.

a. Is a solution of fluoxetine in water likely to be acidic or basic? Explain your answer.
b. Fluoxetine is also sold as a hydrochloride salt. Which functional group is most likely to react with HCl?
c. Fluoxetine is sold as its hydrochloride because the solubility of the salt in water is higher than that of unreacted fluoxetine. Why is the salt more soluble?

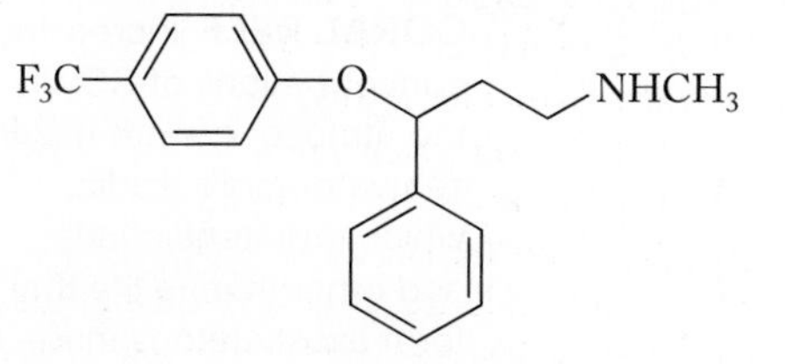

FIGURE P15.89

15.90. Naproxen (also known as Aleve; Figure P15.90) is an anti-inflammatory drug used to reduce pain, fever, inflammation, and stiffness caused by conditions such as osteoarthritis and rheumatoid arthritis. Naproxen is an organic acid that has limited solubility in water, so it is sold as its sodium salt.

a. Draw the molecular structure of the sodium salt.
b. Is an aqueous solution of the salt acidic or basic? Explain why.
c. Explain why the salt is more soluble in water than naproxen itself.

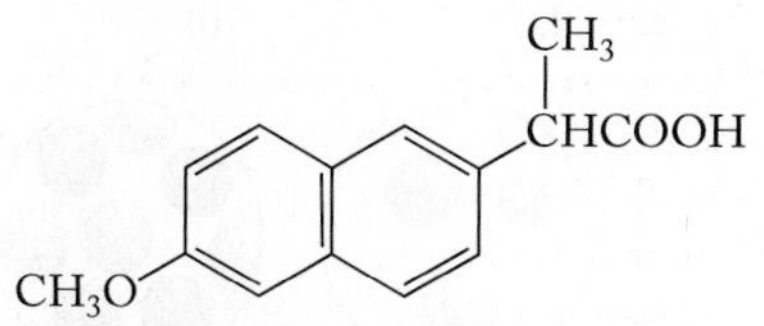

FIGURE P15.90

***15.91.** Pentafluorocyclopentadiene (Figure P15.91) is a strong acid.

a. Draw the conjugate base of C_5F_5H.
b. Pentafluorocyclopentadiene is very acidic, whereas most organic acids are weak. Why?

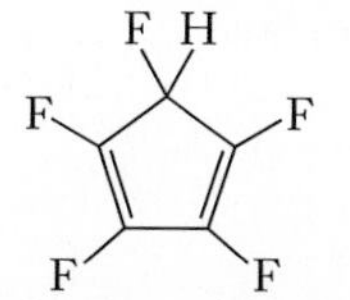

FIGURE P15.91

15.92. **Ocean Acidification** Some climate models predict a decrease in the pH of the oceans of 0.3 to 0.5 pH units by 2100 because of increases in atmospheric carbon dioxide.

a. Explain, by using the appropriate chemical reactions and equilibria, how an increase in atmospheric CO_2 could produce a decrease in oceanic pH.
b. How much more acidic would the oceans be if their pH dropped this much?
c. Oceanographers are concerned about how a drop in oceanic pH could affect the survival of coral reefs. Why?

***15.93.** Sulfuric acid reacts with nitric acid as follows:

$$HNO_3(aq) + 2\,H_2SO_4(aq) \rightarrow NO_2^+(aq) + H_3O^+(aq) + 2\,HSO_4^-(aq)$$

a. Is the reaction a redox process?
b. Identify the acid, base, conjugate acid, and conjugate base in the reaction. (*Hint*: Draw the Lewis structures for each.)

15.94. Thiosulfuric acid ($H_2S_2O_3$) can be prepared by the reaction of H_2S with HSO_3Cl:

$$HSO_3Cl(\ell) + H_2S(g) \rightarrow HCl(g) + H_2S_2O_3(\ell)$$

a. Draw a Lewis structure for $H_2S_2O_3$, given that it is isostructural with H_2SO_4.
b. Do you expect $H_2S_2O_3$ to be a stronger or weaker acid than H_2SO_4? Explain your answer.

***15.95.** The pK_a values of the conjugate acids of pyridine derivatives shown in Figure P15.95 increase as more methyl groups are added. Do more methyl groups increase or decrease the strength of the parent pyridine bases?

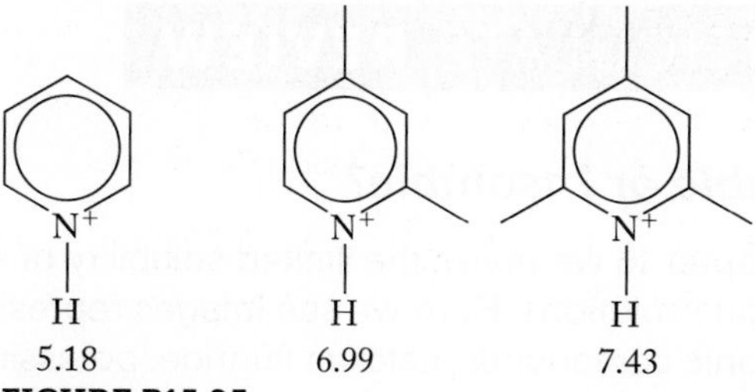

FIGURE P15.95

*15.96. Compounds that do not ionize in water have been known to ionize in nonaqueous solvents. In such a solvent, what would be the conjugate acid and conjugate base of ethanol (CH_3CH_2OH)?

16

Additional Aqueous Equilibria

Chemistry and the Oceans

CORAL REEF Increasing concentrations of CO_2 in the atmosphere are making seawater more acidic, which threatens corals and other marine life that form exoskeletons made of calcium carbonate.

PARTICULATE **REVIEW**

Soluble or Insoluble?

In Chapter 16 we revisit the limited solubility of some ionic compounds in aqueous solutions. Here we see images representing aqueous solutions of four ionic compounds: calcium fluoride, potassium fluoride, lithium sulfide, and lead(II) sulfide. (To simplify the images, solvent water molecules are not shown.)

- What is the chemical formula of each compound?
- Which two of these compounds are much less soluble in water than the other two?
- Match each particulate representation to one of the four compounds.

(Review Section 4.7 if you need help.)

(Answers to Particulate Review questions are in the back of the book.)

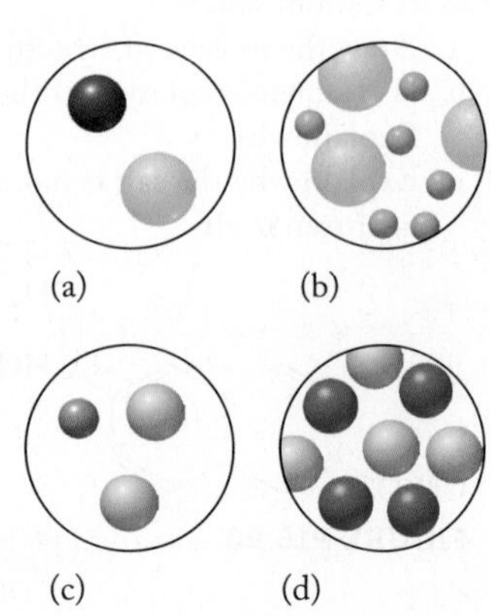

PARTICULATE **PREVIEW**

pH: To Change or Not to Change

Here are two representations of solutions that contain (a) hydrofluoric acid and (b) equal concentrations of both hydrofluoric acid and sodium fluoride. As you read Chapter 16, look for ideas that will help you answer these questions:

- Write a chemical equation describing the proton transfer equilibrium that occurs in each solution.
- Which solution(s) would resist a sharp rise in pH if some strong base were added?
- Which solution(s) would resist a sharp drop in pH if some strong acid were added?

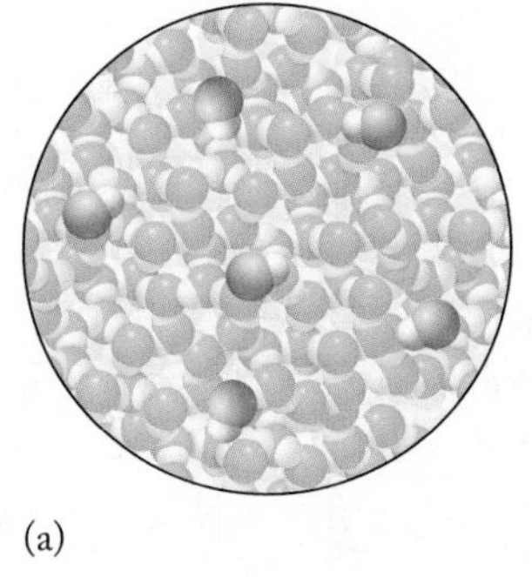

(a)

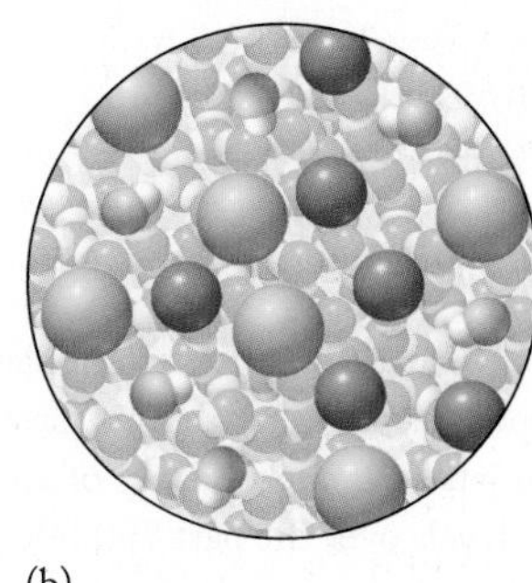

(b)

Learning Outcomes

LO1 Use Le Châtelier's principle to explain the common-ion effect, and calculate the pH of a solution containing a weak acid or base and its conjugate base or acid
Sample Exercises 16.1, 16.2

LO2 Describe pH buffering and prepare a buffer with a desired pH and buffer capacity
Sample Exercises 16.3, 16.4, 16.5

LO3 Evaluate the capacity of a buffer to resist changes in its pH
Sample Exercises 16.6, 16.7, 16.8

LO4 Predict and interpret the results of an acid–base titration and describe how color indicators can be used to detect the equivalence points in these titrations
Sample Exercises 16.9, 16.10, 16.11, 16.12

LO5 Identify a compound as a Lewis acid or Lewis base in a reaction
Sample Exercise 16.13

LO6 Describe the bonding in complex ions and use their formation constants to calculate the concentrations of free and complexed metal ions in solution
Sample Exercise 16.14

LO7 Relate the acid strength of hydrated metal ions and their limited solubility in basic solutions to the charges on their ions

LO8 Relate the solubility of an ionic compound to its solubility product constant and to solution pH
Sample Exercises 16.15, 16.16, 16.17, 16.18

LO9 Determine which mixtures of ionic compounds can be separated by selective precipitation reactions
Sample Exercise 16.19

16.1 Ocean Acidification: Equilibrium under Stress

Fossil fuel combustion is increasing the concentration of carbon dioxide in the atmosphere. We know this from historical records of atmospheric CO_2 concentrations. Part of this record is based on analyzing bubbles of air trapped in layers of glacial ice, which show that atmospheric CO_2 concentrations cycled between about 175 and 300 parts per million (ppm) by volume for at least 800,000 years (**Figure 16.1**). About 250 years ago, however, when CO_2 levels were already at the top of one of these cycles, they began to rise even more. Since the late 1950s, CO_2 concentrations have increased from 330 to over 410 ppm with no sign of slowing down (**Figure 16.2**).

In Chapter 8 we discussed the impact of increasing concentrations of CO_2 in the atmosphere on climate. It turns out that elevated atmospheric CO_2 concentration poses another environmental problem: ocean acidification. As we saw in Chapter 15, CO_2 gas dissolves in water, where it can form carbonic acid:

$$CO_2(g) + H_2O(\ell) \rightleftharpoons H_2CO_3(aq) \tag{16.1}$$

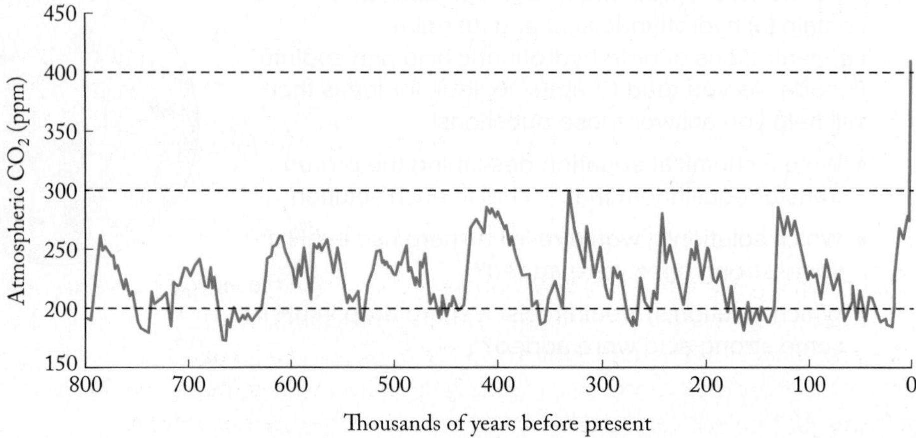

FIGURE 16.1 Analyses of bubbles of air trapped in Antarctic glaciers have enabled scientists to create a history of atmospheric CO_2 levels over the past 800,000 years, showing that atmospheric concentrations of CO_2 have generally oscillated between about 175 and 300 ppm. In recent years, however, CO_2 concentrations have sharply increased.

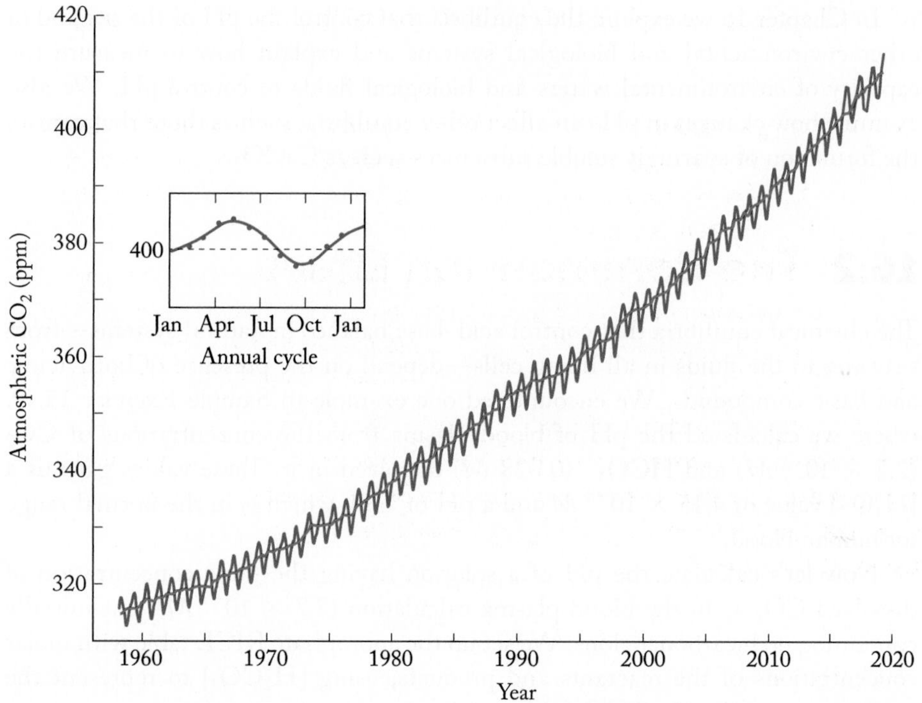

FIGURE 16.2 This graph of CO_2 concentrations in the atmosphere is based on analyses done since 1958 at the Mauna Loa Observatory in Hawaii—work begun by Charles David Keeling (1928–2005). The annual cycle in the data is caused by seasonally changing rates of photosynthesis by trees and other vegetation in the Northern Hemisphere.

Although H_2CO_3 is unstable in aqueous solutions, we write $[H_2CO_3]$ in this chapter to represent the total concentration of $CO_2(g)$ dissolved in water and to remind us that when $CO_2(g)$ dissolves in water, it forms an acidic solution:

$$H_2CO_3(aq) + H_2O(\ell) \rightleftharpoons HCO_3^-(aq) + H_3O^+(aq) \quad K_{a_1} = 4.3 \times 10^{-7} \quad (16.2)$$

We also saw in Chapter 15 that bicarbonate ions can donate a second H^+ ion in aqueous solution:

$$HCO_3^-(aq) + H_2O(\ell) \rightleftharpoons CO_3^{2-}(aq) + H_3O^+(aq) \quad K_{a_2} = 4.7 \times 10^{-11} \quad (16.3)$$

The value of K_{a_2} is much smaller than K_{a_1}, so the concentration of CO_3^{2-} ions in slightly basic seawater is much smaller than the concentration of HCO_3^- ions.

The solubility of atmospheric CO_2 increases as its partial pressure increases, so higher concentrations of CO_2 in the atmosphere mean higher concentrations of dissolved CO_2 in the sea. Scientists estimate that about 30% of the "fossil" CO_2 added to the atmosphere dissolves in the sea. Increasing concentrations of dissolved CO_2 have shifted the equilibrium in Equation 16.1 to the right. This, in turn, has shifted the equilibrium in Equation 16.2 to the right resulting in an increase in the acidity of the sea. As a result, the average pH of seawater has dropped by about 0.11 pH units since the 18th century.

Are oceans with lower pH a problem? They are—especially if the pH continues to drop—because many marine organisms such as plankton, shellfish, and corals have exoskeletons made of $CaCO_3$. To make $CaCO_3$, these organisms need a supply of dissolved Ca^{2+} and CO_3^{2-} ions so that the following precipitation reaction can take place:

$$Ca^{2+}(aq) + CO_3^{2-}(aq) \rightleftharpoons CaCO_3(s)$$

Unfortunately, increasing concentrations of H_3O^+ ions in the sea drive down already small concentrations of CO_3^{2-} ions by shifting the equilibrium in Equation 16.3 to the left. There is a growing concern that the oceanic $[CO_3^{2-}]$ may become too low for marine organisms to form their $CaCO_3$ skeletons and that they—and the aquatic systems that rely on them, such as tropical coral reefs—will not survive.

CONNECTION As discussed in Section 8.3, atmospheric CO_2 is a greenhouse gas because asymmetric stretching of the bonds in its molecules absorbs infrared radiation emitted by Earth's surface.

CONNECTION Henry's law (Section 10.8) states that the solubility of a gas (C_{gas}) is proportional to its partial pressure: $C_{gas} = k_H P_{gas}$.

CONNECTION Le Châtelier's principle (Section 14.7) predicts how a system at equilibrium responds to stress.

In Chapter 16 we explore the equilibria that control the pH of the sea and of other environmental and biological systems and explain how to measure the capacity of environmental waters and biological fluids to control pH. We also examine how changes in pH can affect other equilibria, such as those that control the formation of sparingly soluble substances such as $CaCO_3$.

16.2 The Common-Ion Effect

The chemical equilibria that control acid–base balance in natural systems—from seawater to the fluids in all living cells—depend on the presence of both acidic and basic compounds. We encountered one example in Sample Exercise 15.19, where we calculated the pH of blood plasma from the concentrations of CO_2 (2.7×10^{-3} *M*) and HCO_3^- (0.028 *M*) dissolved in it. These values gave us a $[H_3O^+]$ value of 4.15×10^{-8} *M* and a pH of 7.38, which is in the normal range for human blood.

Now let's calculate the pH of a solution having the same concentration of dissolved CO_2 as in the blood plasma calculation (2.7×10^{-3} *M*), but initially containing no bicarbonate ions. We set up the appropriate RICE table with molar concentrations of the reactants and products, using $[H_2CO_3]$ to represent the total amount of dissolved $CO_2(g)$ in solution:

Reaction (R)	$H_2CO_3(aq) + H_2O(\ell)$	$\rightleftharpoons$	$HCO_3^-(aq)$	+	$H_3O^+(aq)$
	$[H_2CO_3]$ (*M*)		$[HCO_3^-]$ (*M*)		$[H_3O^+]$ (*M*)
Initial (I)	2.7×10^{-3}		0		0
Change (C)	$-x$		$+x$		$+x$
Equilibrium (E)	$(2.7 \times 10^{-3}) - x$		x		x

Taking the K_{a_1} value for carbonic acid (4.3×10^{-7}) from Equation 16.2, and solving for x:

$$K_{a_1} = \frac{[HCO_3^-][H_3O^+]}{[H_2CO_3]} = \frac{x \cdot x}{(2.7 \times 10^{-3}) - x} \approx \frac{x^2}{2.7 \times 10^{-3}} = 4.3 \times 10^{-7}$$

$$x = [H_3O^+] = 3.4 \times 10^{-5}\ M$$

Calculating pH:

$$pH = -\log[H_3O^+] = -\log(3.4 \times 10^{-5}) = 4.47$$

Thus, the new solution has a pH value nearly 3 units lower than the blood plasma sample (7.38), corresponding to a $[H_3O^+]$ value *nearly 10^3 larger.* This difference makes sense because the presence of bicarbonate ions in the plasma sample inhibited the forward reaction in Equation 16.2, as predicted by Le Châtelier's principle. In this new solution, which lacks a second source of bicarbonate ions, the result is higher $[H_3O^+]$ and lower pH values.

The difference in pH between these two solutions illustrates a principle known as the **common-ion effect**: in any ionic equilibrium, a reaction produces less of an ion when that ion is already in the system or if more of it is added to the system. One result of this suppression of the ionization reaction is that the initial concentrations of the conjugate acid–base pair change very little before equilibrium is achieved. Therefore, we can use their initial concentrations to calculate the pH of

common-ion effect a shift in the position of an equilibrium caused by an ion taking part in the reaction.

Henderson–Hasselbalch equation an equation used to calculate the pH of solutions containing comparable concentrations of acid and conjugate base.

their solution, as illustrated for this generic acid ionization equilibrium involving a solution of acid, HA, and its conjugate base, A^-:

$$K_a = \frac{[H_3O^+][A^-]_{initial}}{[HA]_{initial}}$$

If we take the negative logarithm of both sides of the K_a expression, we transform $[H_3O^+]$ into pH and K_a into pK_a:

$$pK_a = pH - \log\frac{[A^-]}{[HA]} \quad \text{or} \quad pH = pK_a + \log\frac{[A^-]}{[HA]}$$

Generalizing this equation by replacing [HA] and $[A^-]$ with [acid] and [base], respectively:

$$pH = pK_a + \log\frac{[\text{base}]}{[\text{acid}]} \tag{16.4}$$

Equation 16.4 is used to calculate the pH of solutions in which there are comparable concentrations of both an acid and its conjugate base (or both a base and its conjugate acid). It is called the **Henderson–Hasselbalch equation**. The presence of the conjugate base (or conjugate acid) suppresses the ionization reaction because of the common-ion effect. As a result, the equilibrium concentrations of the conjugate pairs are not significantly different from their initial concentrations.

What happens to the logarithm term in the Henderson–Hasselbalch equation when the concentrations of acid and base are equal? The numerator and denominator are the same, so the value of the fraction in Equation 16.4 is 1. The log of 1 is 0, so pH = pK_a. This equality serves as a handy reference point in working with solutions of conjugate acid–base pairs. If the concentration of the basic component is greater than that of the acid, then the logarithmic term is positive and pH > pK_a. If [base] < [acid], on the other hand, then the logarithmic term is negative, and pH < pK_a.

Suppose the concentration of the base is 10 times the concentration of the acid; that is, [base] = 10[acid]. Substituting this equality into Equation 16.4, we have

$$pH = pK_a + \log\frac{10[\text{acid}]}{[\text{acid}]}$$

$$= pK_a + \log 10 = pK_a + 1$$

Thus, a 10-fold higher concentration of base produces a pH that is 1 unit above the pK_a value. Similarly, if the concentration of the acid component is 10 times that of the base, then pH = pK_a − 1. Sample Exercises 16.1 and 16.2 further illustrate how the Henderson–Hasselbalch equation simplifies pH calculations when we know the concentrations of both components of conjugate pairs.

STEPWISE ANIMATION
Common-Ion Effect

CONNECTION The pH of a solution is the negative logarithm of its hydronium ion concentration (see Equation 15.10 in Section 15.4).

CONNECTION The pK_a notation (Section 15.4) is an equivalent expression of K_a values but without using scientific notation and negative exponents.

SAMPLE EXERCISE 16.1 Calculating the pH of a Solution of a Weak Acid and Its Conjugate Base **LO1**

What is the pH of a sample of river water in which $[HCO_3^-] = 1.0 \times 10^{-4}$ *M* and the concentration of dissolved CO_2 in equilibrium with atmospheric CO_2 is 1.4×10^{-5} *M*?

Collect and Organize We are given the concentration of $CO_2(aq)$, which we will represent as $[H_2CO_3]$. We also know the concentration of the conjugate base of

H_2CO_3—namely, HCO_3^-. The pH of a solution of a weak acid and its conjugate base can be calculated using the Henderson–Hasselbalch equation:

$$\text{pH} = \text{p}K_\text{a} + \log\frac{[\text{base}]}{[\text{acid}]}$$

Analyze The conjugate acid–base pair relationship in this reaction is described by Equation 16.2:

$$H_2CO_3(aq) + H_2O(\ell) \rightleftharpoons HCO_3^-(aq) + H_3O^+(aq)$$

This reaction has a pK_{a_1} value (Table A5.1) of 6.37. The concentration of the base is nearly 10 times the concentration of the acid, so the log term in the Henderson–Hasselbalch equation should be almost 1, and the calculated pH should be about 7.

Solve Inserting the concentration and pK_{a_1} values into the Henderson–Hasselbalch equation:

$$\text{pH} = \text{p}K_{\text{a}_1} + \log\frac{[\text{base}]}{[\text{acid}]} = 6.37 + \log\frac{1.0 \times 10^{-4}}{1.4 \times 10^{-5}}$$
$$= 6.37 + 0.85 = 7.22$$

Think About It This result is near the pH value we expected. Remember that a pH value of 7.22 contains only two significant figures (the 2 and the 2) because there were two significant figures in all three of the values used to calculate it. The 7 in pH 7.22 is not significant because its function is to tell us that the corresponding $[H_3O^+]$ value is between 10^{-7} and 10^{-8} *M*.

 **Practice Exercise** What is the pH of a solution in which [HCOOH] = 2.5×10^{-2} *M* and $[HCOO^-] = 7.8 \times 10^{-3}$ *M*?

(Answers to Practice Exercises are in the back of the book.)

We can also use the Henderson–Hasselbalch equation to calculate the pH of a solution of a weak base and its conjugate acid. An extra step is usually involved because we may know the pK_b value of the base but not the pK_a value of its conjugate acid, and only pK_a values are used in the Henderson–Hasselbalch equation. However, we learned in Section 15.8 that the equilibrium constants for conjugate acid–base pairs are related by Equation 15.18 ($K_\text{a} \times K_\text{b} = K_\text{w}$). To recast this as an equation with p*K* terms, we insert the numerical value of K_w at 25°C (1.0×10^{-14}):

$$K_\text{a} \times K_\text{b} = K_\text{w} = 1.0 \times 10^{-14}$$

and take the –log values of all three terms:

$$\text{p}K_\text{a} + \text{p}K_\text{b} = 14.00 \qquad (16.5)$$

Thus, converting the pK_b of a base into the pK_a of its conjugate acid is simply a matter of subtracting the pK_b value from 14.00.

SAMPLE EXERCISE 16.2 Calculating the pH of a Solution of a Weak Base and Its Conjugate Acid **LO1**

What is the pH of a solution that is 0.200 *M* in NH_3 and 0.300 *M* in NH_4Cl?

Collect and Organize We are asked to calculate the pH of a solution containing known concentrations of a weak base (NH_3) and a salt of its conjugate acid (NH_4^+).

We can use the Henderson–Hasselbalch equation to calculate the pH of such a solution from the concentrations of the two components and the pK_a of the acid. Table A5.3 contains K_b and pK_b values of common bases. The pK_a and pK_b values of a conjugate acid–base pair are related by Equation 16.5:

$$pK_a + pK_b = 14.00$$

Analyze Our approach involves converting the pK_b value for NH_3 from Table A5.3 (4.75) into a pK_a value. Addition of ammonium ion to a solution of ammonia should produce a solution that is still basic, but not as basic as a solution that contains only ammonia.

Solve Inserting the pK_b value into Equation 16.5, and solving for pK_a:

$$pK_a = 14.00 - pK_b = 14.00 - 4.75 = 9.25$$

We can then use this value and the given concentrations of NH_3 and NH_4^+ in Equation 16.4:

$$\text{pH} = pK_a + \log\frac{[\text{base}]}{[\text{acid}]}$$

$$= 9.25 + \log\frac{0.200}{0.300} = 9.07$$

Think About It We predicted the solution would be basic, but not as basic as a solution of ammonia alone. To check whether this prediction was true, we can calculate the pH of 0.200 *M* NH_3 by using the approach we followed in Sample Exercise 15.12. The result is a pH of 11.27—more than 2 pH units higher (more basic) than the solution of ammonia and ammonium chloride in this exercise.

 **Practice Exercise** Calculate the pH of a solution that is 0.100 *M* in methylamine and 0.150 *M* in methylammonium chloride.

16.3 pH Buffers

The common-ion effect plays a key role in controlling the pH of solutions that contain relatively high concentrations of both a weak acid and its conjugate base. These solutions have the capacity to withstand additions of acidic or basic substances with little or no measurable change in their pH, and they are known as **pH buffers**. A buffer solution maintains a constant pH because the weak acid component of the buffer gives it the capacity to neutralize additions of basic substances, whereas the conjugate base component gives it the capacity to neutralize acids. Ideally, a buffer has similar concentrations of both components of its conjugate pairs so that it can neutralize additions of acids or bases equally well. When the ratio of the conjugate pair is close to 1, the log term in the Henderson–Hasselbalch equation is close to 0, and the pH of the buffer is close to the pK_a of the weak acid in it. Actually, a buffer with different concentrations of its conjugate acid–base pair can still be effective at controlling pH over a range of pH values up to about 1 unit above or below its acid's pK_a value.

pH buffer a solution that resists changes in pH when acids or bases are added to it; typically a solution of a weak acid and its conjugate base or a weak base and its conjugate acid.

SAMPLE EXERCISE 16.3 Building an Acidic Buffer **LO2**

Select a weak acid in Table A5.1 of Appendix 5 that, when mixed with the sodium salt of its conjugate base in approximately equimolar proportions, produces a buffer with a pH of 2.80. Will the buffer contain *the same* concentrations of acid and conjugate base, or slightly more acid or base?

CHEMTOUR
Buffers

CONNECTION The limited solubility of hydrophobic compounds in water is explained in Section 10.7.

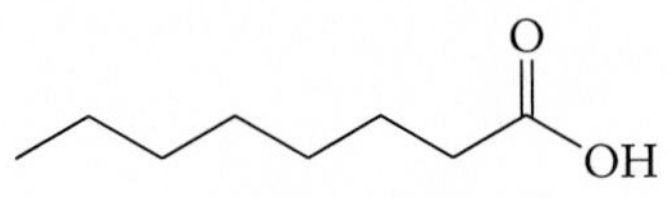

Collect, Organize, and Analyze The weak acid we seek is one whose pK_a is close to the target pH, 2.80.

Solve Among the acids in Table A5.1 with pK_a values near 2.80 are bromoacetic acid ($pK_a = 2.70$) and chloroacetic acid ($pK_a = 2.85$). Either could be used to prepare a buffer at pH 2.80, though neither would contain the same concentration of the acid and its conjugate base: the bromoacetic acid buffer would require a slightly higher concentration of the conjugate base, and the chloroacetic acid buffer would contain a little more of the acid.

Think About It A key factor in preparing an effective buffer is to select an acid with a pK_a that is close (within 1 pH unit) to the target pH of the buffer. In this range the concentrations of the acid and its conjugate base, though perhaps not equal, are at least comparable, which means the buffer has the capacity to withstand additions of acidic or basic substances with only minor changes in pH. Another criterion for selecting weak acids for aqueous buffers is that they are soluble in water. Both acids selected in this sample exercise are carboxylic acids with relatively small molar masses and are quite soluble in water. However, organic acids with large hydrocarbon (hydrophobic) regions in their molecular structures, such as octanoic acid and benzoic acid (**Figure 16.3**), are only slightly soluble in water.

O
OH
Octanoic acid

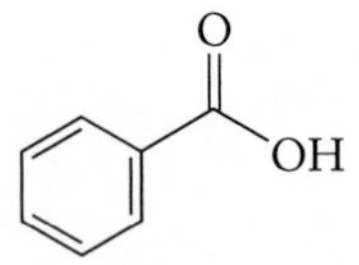

O
OH
Benzoic acid

FIGURE 16.3 Octanoic acid and benzoic acid are organic acids that are only slightly water soluble.

Practice Exercise Select a weak acid in Table A5.1 that, when mixed with the sodium salt of its conjugate base in approximately equimolar proportions, produces a buffer with a pH of 1.77.

SAMPLE EXERCISE 16.4 Building a Basic Buffer **LO2**

Select a weak base in Table A5.3 that, when mixed with the chloride salt of its conjugate acid in approximately equimolar proportions, produces a buffer with a pH of 9.25. Indicate whether the buffer will contain *the same* concentrations of base and conjugate acid, or slightly more base or acid.

Collect, Organize, and Analyze We can follow a strategy like that used in Sample Exercise 16.3, but we need to search the table for a weak base *that has a conjugate acid whose* pK_a is close to the target pH of 9.25. This means we are looking for a base with a pK_b value of $14.00 - 9.25 = 4.75$.

Solve According to Table A5.3, NH_3 has a pK_b of 4.75, which exactly matches the target pH. This match means that the buffer should contain equimolar proportions of aqueous ammonia (NH_3) and ammonium chloride (NH_4Cl).

Think About It A buffer with equal concentrations of an acid–base conjugate pair can withstand additions of acid or base with equally small changes in pH.

Practice Exercise Select a weak base in Table A5.3 that, when mixed with the chloride salt of its conjugate acid in approximately equimolar proportions, produces a buffer with a pH of 10.93.

Having gained experience in selecting the components that are needed to prepare pH buffers, let's now calculate the quantities of these components that are needed to prepare a volume of a buffer with a desired pH.

SAMPLE EXERCISE 16.5 Preparing a Buffer Solution with a Desired pH **LO2**

STEPWISE ANIMATION
The Buffer System

A buffer system containing dihydrogen phosphate ($H_2PO_4^-$) and hydrogen phosphate (HPO_4^{2-}) helps regulate the pH of cytoplasm in living cells.

a. What is the mole ratio of HPO_4^{2-} ions to $H_2PO_4^-$ ions in a buffer with a pH of 6.75?
b. If the combined concentration of $H_2PO_4^-$ and HPO_4^{2-} ions in the buffer is to be 0.200 *M*, how many grams of NaH_2PO_4 ($\mathcal{M}$ = 120.0 g/mol) and of Na_2HPO_4 ($\mathcal{M}$ = 142.0 g/mol) are needed to prepare 20.0 L of the buffer?

Collect and Organize We need to determine the mole ratio of HPO_4^{2-} ions to $H_2PO_4^-$ ions in a pH 6.75 buffer and to calculate the masses of the sodium salts of these ions that are needed to make 20.0 L of 0.200 *M* buffer. According to Table A5.1, the pK_a of $H_2PO_4^-$ (the pK_{a_2} of H_3PO_4) is 7.19. The Henderson–Hasselbalch equation relates the pH of a buffer to the pK_a of its acid component and to the concentrations of that acid and its conjugate base.

Analyze The pH of this buffer is controlled by the acid ionization equilibrium of dihydrogen phosphate ions:

$$H_2PO_4^-(aq) + H_2O(\ell) \rightleftharpoons HPO_4^{2-}(aq) + H_3O^+(aq)$$

The target pH (6.75) is less than the pK_a (7.19), so the buffer must contain a higher concentration of $H_2PO_4^-$ ions (the acid) than HPO_4^{2-} ions (the base).

Solve

a. Let's rearrange the Henderson–Hasselbalch equation to solve for the ratio of base to acid.

$$\log\frac{[\text{base}]}{[\text{acid}]} = \text{pH} - pK_a$$

Substituting the values for pH and pK_a gives us:

$$\log\frac{[HPO_4^{2-}]}{[H_2PO_4^-]} = \text{pH} - pK_a = 6.75 - 7.19 = -0.44$$

Then the ratio of $[HPO_4^{2-}]$ to $[H_2PO_4^-]$ is

$$\frac{[HPO_4^{2-}]}{[H_2PO_4^-]} = 10^{-0.44} = 0.36$$

b. The sum of $[H_2PO_4^-]$ and $[HPO_4^{2-}]$ is 0.200 *M*, so if we let $x = [HPO_4^{2-}]$, then $[H_2PO_4^-] = (0.200 - x)$ *M*. Calculating the dissolved concentrations of HPO_4^{2-} and $H_2PO_4^-$:

$$\frac{x}{0.200 - x} = 0.36$$

$$x = 0.36(0.200 - x)$$

$$1.36x = 0.072$$

$$x = [HPO_4^{2-}] = 0.053\ M$$

$$(0.200 - x) = [H_2PO_4^-] = 0.147\ M$$

The masses of their sodium salts in 20.0 L of buffer are

$$Na_2HPO_4\text{:}\quad 20.0\ \cancel{\text{L}} \times 0.053\ \frac{\cancel{\text{mol}}}{\cancel{\text{L}}} \times 142.0\ \frac{\text{g}}{\cancel{\text{mol}}} = 150\ \text{g}$$

$$NaH_2PO_4\text{:}\quad 20.0\ \cancel{\text{L}} \times 0.147\ \frac{\cancel{\text{mol}}}{\cancel{\text{L}}} \times 120.0\ \frac{\text{g}}{\cancel{\text{mol}}} = 353\ \text{g}$$

Think About It The buffer contains a higher concentration of $H_2PO_4^-$ ions than HPO_4^{2-} ions, which is consistent with our prediction and with the ratio of acid to conjugate base of any buffer that has a pH below the pK_a of the acid used to make it.

Practice Exercise How many grams of sodium ascorbate and ascorbic acid are needed to make 10.0 L of pH 5.25 buffer if the total concentration of the two buffer components is 0.500 *M*?

Buffer Capacity

In addition to selecting the appropriate conjugate acid–base pair to prepare a buffer, chemists also need to decide how concentrated the buffer should be. The greater the concentrations of the conjugate pair components, the greater is its **buffer capacity**—the ability of the buffer to withstand additions of acid or base without a significant change in pH (**Figure 16.4**).

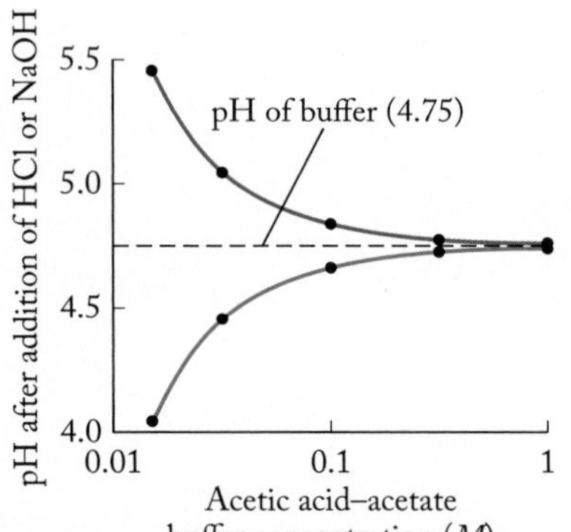

FIGURE 16.4 When strong acid (red line) or strong base (blue line) is added to a buffer solution, the extent to which the pH changes is inversely proportional to buffer concentration: the higher the concentrations of the buffer components, the smaller the change in pH. In this illustration, 100 mL samples of five solutions that are 0.015, 0.030, 0.100, 0.300, and 1.000 *M* acetic acid and sodium acetate all have an initial pH of 4.75 (dashed line). The graph shows the pH values of these solutions after 1.00 mL of 1.00 *M* HCl or 1.00 *M* NaOH has been added.

buffer capacity the quantity of acid or base that a pH buffer can neutralize while keeping its pH within a desired range.

SAMPLE EXERCISE 16.6 Calculating Buffer Response to Additions of Acid or Base — LO3

a. What is the change in pH of the buffer prepared in Sample Exercise 16.5 if 1.00 mL of 0.200 *M* HNO_3 is added to 100.0 mL of the buffer?
b. Compare the pH change in part (a) to the pH change when the same quantity of acid is added to 100.0 mL pure water (pH = 7.00).

Collect and Organize We know the volume and composition of a pH buffer, and we are asked to determine how much the pH of the buffer changes when a known volume and concentration of acid is added. The buffer in Sample Exercise 16.5 contains 0.147 *M* $H_2PO_4^-$ ions and 0.053 *M* HPO_4^{2-} ions. Its pH is controlled by the acid ionization of $H_2PO_4^-$ ions

$$H_2PO_4^-(aq) + H_2O(\ell) \rightleftharpoons HPO_4^{2-}(aq) + H_3O^+(aq)$$

and can be calculated using the Henderson–Hasselbalch equation in which $H_2PO_4^-$ ions are the acid and HPO_4^{2-} ions are the base:

$$\text{pH} = \text{p}K_a + \log\frac{[\text{base}]}{[\text{acid}]} = 7.19 + \log\frac{[HPO_4^{2-}]}{[H_2PO_4^-]}$$

Analyze

a. When strong acid is added to a pH buffer, the acid is neutralized by the base component of the buffer. In this exercise, HPO_4^{2-} ions neutralize added H_3O^+ ions as the acid ionization reaction runs in reverse:

$$HPO_4^{2-}(aq) + H_3O^+(aq) \rightarrow H_2PO_4^-(aq) + H_2O(\ell)$$

The single reaction arrow is used to indicate that the added acid is completely neutralized. Therefore, x mol H_3O^+ ions are completely consumed by x mol HPO_4^{2-} ions, producing x mol $H_2PO_4^-$ ions. To calculate the effect of these changes on buffer pH, we need to calculate the initial numbers of moles of $H_2PO_4^-$ and HPO_4^{2-} ions in 100.0 mL of buffer and then add or subtract the moles of H_3O^+ ions from these initial values as described above. The resulting mole ratio of HPO_4^{2-} ions to $H_2PO_4^-$ ions can be used in the above Henderson–Hasselbalch equation. The ratio will be less after some of the HPO_4^{2-} ions are consumed and $H_2PO_4^-$ ions are produced. Therefore, the final pH of the buffer should be less than its initial value of 6.75.

b. Adding a strong acid to pure water is an exercise in dilution, a concept introduced in Section 4.3 and for which we can use Equation 4.4:

$$V_{\text{initial}} \times M_{\text{initial}} = V_{\text{diluted}} \times M_{\text{diluted}}$$

Solve

a. A table that resembles a RICE table but that is based on neutralization of the added acid provides guidance in solving buffer capacity problems such as this one. It is sometimes called a ICA table because its rows contain the **I**nitial quantities of the reactants and product, the **C**hanges in those quantities during the reaction, and the quantities in solution **A**fter the reaction is complete. Using the given volume and concentration values and the strategy described in the Analyze section above:

	$HPO_4^{2-}(aq)$	+ $H_3O^+(aq)$	→ $H_2PO_4^-(aq)$	+ $H_2O(\ell)$
Initial	100.0 mL × 0.053 mol/L = 5.3 mmol	1.00 mL × 0.200 mol/L = 0.200 mmol	100.0 mL × 0.147 mol/L = 14.7 mmol	
Change	−0.200 mmol	−0.200 mmol	+0.200 mmol	
After	5.1 mmol	0 mmol	14.9 mmol	

The mole ratio of HPO_4^{2-} and HPO_4^{2-} in the bottom row, 5.1 mmol/14.9 mmol, can be inserted directly in the log term of the Henderson–Hasselbalch equation to calculate pH because these two mole (n) quantities are dissolved in the same volume of solution. Therefore, the mole ratio has the same value as the concentration ratio (because the identical volume terms in the numerator and denominator cancel out).

$$\text{pH} = 7.19 + \log\frac{[HPO_4^{2-}]}{[H_2PO_4^-]} = 7.19 + \log\frac{n_{HPO_4^{2-}}}{n_{H_2PO_4^-}}$$

$$= 7.19 + \log\frac{5.1\ \text{mol}}{14.9\ \text{mol}} = 7.19 - 0.47 = 6.72$$

Thus, the addition of acid changes the pH of the buffer by (6.72 − 6.75) = −0.03 pH units.

b. Solving for M_{diluted} in Equation 4.4:

$$M_{\text{diluted}} = [H_3O^+] = \frac{V_{\text{initial}} \times M_{\text{initial}}}{V_{\text{diluted}}}$$

$$= \frac{1.00\ \text{mL} \times (0.200\ M)}{(100.0\ \text{mL} + 1.00\ \text{mL})} = 0.00198\ M$$

$$\text{pH} = -\log[H_3O^+] = -\log 0.00198 = 2.70$$

Thus, the change in pH of pure water is (2.70 − 7.00) = −4.30 pH units.

Think About It Addition of a strong acid lowered the pH of the buffer by only 0.03 pH units because it consumed only 0.2 mmol/5.3 mmol = 0.038 or 3.8% of the basic component of the buffer system. This decrease in pH is a tiny fraction of the change in pH that happened when the same quantity of acid was added to unbuffered pure water.

Practice Exercise What is the pH of a buffer that contains 0.225 M acetic acid and 0.375 M sodium acetate? What is the pH of 100.0 mL of the buffer after 1.00 mL of 0.318 M NaOH is added to it?

SAMPLE EXERCISE 16.7 Effect of Concentration on Buffer Capacity LO3

Calculate the buffer pH after 0.0100 moles of H_3O^+ ions are added to (a) 0.100 L of buffer A, containing 1.00 M acetic acid and 1.00 M sodium acetate, and (b) 0.100 L of buffer B, containing 0.150 M acetic acid and 0.150 M sodium acetate.

Collect and Organize We are asked to calculate the pH of equal volumes of two buffer solutions after the same quantity of acid (0.0100 mol H_3O^+) is added to both. According to Table A5.1, the pK_a of CH_3COOH is 4.75. The pH of a buffer with known concentrations of a conjugate acid–base pair can be calculated using the Henderson–Hasselbalch equation.

Analyze When strong acid is added to a pH buffer, the acid is neutralized by the base component of the buffer. In this example, the conjugate base of acetic acid, acetate (CH_3COO^-) ions, neutralizes the added H_3O^+ ions as the acid ionization reaction

$$CH_3COOH(aq) + H_2O(\ell) \rightleftharpoons CH_3COO^-(aq) + H_3O^+(aq)$$

runs in reverse

$$CH_3COO^-(aq) + H_3O^+(aq) \rightarrow CH_3COOH(aq) + H_2O(\ell)$$

The stoichiometry of the second equation tells us that adding (and reacting) 0.0100 mol H_3O^+ to either of two buffers consumes 0.0100 mol CH_3COO^- and produces 0.0100 mol CH_3COOH. To calculate the effect of these changes on buffer pH, we need to calculate the initial numbers of moles of CH_3COOH and CH_3COO^- ions in both buffers and then add or subtract 0.100 mol from these values as described above. As we saw in Sample Exercise 16.6, the mole ratio of CH_3COO^- to CH_3COOH after the neutralization reaction is complete can be used in the Henderson–Hasselbalch equation to calculate the final pH. The trends shown in Figure 16.4 suggest that the more concentrated buffer will experience a smaller change in pH upon the addition of acid.

Solve Set up ICA tables for buffers A and B like that used in Sample Exercise 16.6.

Buffer A

	$CH_3COO^-(aq)$ +	$H_3O^+(aq)$	→ $CH_3COOH(aq)$ +	$H_2O(\ell)$
Initial	0.100 L × 1.00 mol/L = 0.100 mol	0.0100 mol	0.100 L × 1.00 mol/L = 0.100 mol	
Change	−0.0100 mol	−0.0100 mol	+0.0100 mol	
After	0.090 mol	0 mol	0.110 mol	

Buffer B

	$CH_3COO^-(aq)$ +	$H_3O^+(aq)$	→ $CH_3COOH(aq)$ +	$H_2O(\ell)$
Initial	0.100 L × 0.150 mol/L = 0.0150 mol	0.0100 mol	0.100 L × 0.150 mol/L = 0.0150 mol	
Change	−0.0100 mol	−0.0100 mol	+0.0100 mol	
After	0.0050 mol	0 mol	0.0250 mol	

The mole ratio of CH_3COO^- to CH_3COOH in each buffer can be substituted for the $[CH_3COO^-]/[CH_3COOH]$ ratio in the Henderson–Hasselbalch equation because both components are dissolved in the same volume of buffer. Using these substitutions and solving for pH:

$$\text{Buffer A:} \quad pH = pK_a + \log\frac{n_{CH_3COO^-}}{n_{CH_3COOH}} = 4.75 + \log\frac{0.090\ \cancel{mol}}{0.110\ \cancel{mol}} = 4.66$$

$$\text{Buffer B:} \quad pH = pK_a + \log\frac{n_{CH_3COO^-}}{n_{CH_3COOH}} = 4.75 + \log\frac{0.0050\ \cancel{mol}}{0.0250\ \cancel{mol}} = 4.05$$

Think About It Adding the same quantity of acid produced pH changes of 0.09 units in buffer A and 0.70 units in an equal volume of buffer B. As predicted, the buffer with a higher concentration was better able to resist pH change because the relative changes in the concentrations of its components were smaller.

Practice Exercise Calculate the change in pH when 0.145 mol OH^- is added to 1.00 L of two buffers: (a) a solution containing 1.16 M NaH_2PO_4 and 1.16 M Na_2HPO_4 and (b) a solution containing 0.58 M NaH_2PO_4 and 0.58 M Na_2HPO_4.

Does the mole ratio of acid to conjugate base make a difference in how well a buffer controls pH? Suppose, for example, you want to prepare an acidic buffer and you know that contamination by basic substances is more likely than contamination by acids. You might use a 1:1 mixture if you could find a weak acid with a pK_a that exactly matched the target pH—but what if there was another conjugate acid–base pair available that required the acid component to be three times as concentrated as the base to achieve the desired pH? Would the second buffer, with the greater proportion of acid in it, do a better job of controlling pH against additions of base than the 1:1 buffer? Sample Exercise 16.8 helps answer this question.

SAMPLE EXERCISE 16.8 Effect of Base/Acid Ratio on Buffer Capacity LO3

Equal volumes of two pH 3.75 buffers are prepared. Buffer A contains 1.00 mol formic acid ($pK_a = 3.75$) and 1.00 mol sodium formate. Buffer B is prepared using the same overall concentration of conjugate pair components, but the pK_a of the weak acid in this buffer is 4.23, so it contains 1.5 mol acid and 0.5 mol conjugate base to achieve the target pH of 3.75. What are the changes in pH of the two buffers if 0.25 mol, 0.50 mol, and 0.75 mol NaOH are added to both? Assume the NaOH additions do not significantly increase the volumes of the buffers.

Collect, Organize, and Analyze The composition of the buffer A changes due to this neutralization reaction:

$$HCOOH(aq) + OH^-(aq) \rightarrow HCOO^-(aq) + H_2O(\ell)$$

A similar equation could be written for buffer B based on its acid and conjugate base. The three additions of NaOH will reduce the moles of the acid components in both buffers by 0.25, 0.50, and 0.75 moles and increase the moles of the base components by the same amounts. The Henderson–Hasselbalch equation can be used to calculate the pH values of the two buffers, but the [base]/[acid] term in the equation will be replaced with the mole ratio of base/acid because the volumes of the buffers are unknown. We know, however, that they are the same. Therefore, the ratio of the concentration values of base to acid in each of the solutions is equal to the mole ratio of base to acid.

Solve Setting up Henderson–Hasselbalch equations with the initial moles of acid and base values in the two buffers and the changes in these initial values from the additions of NaOH:

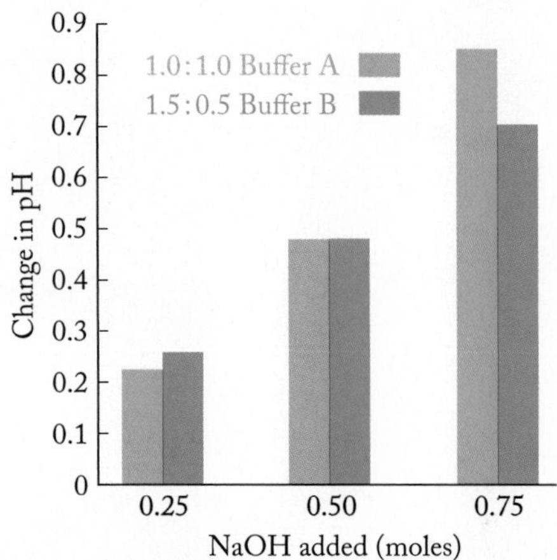

FIGURE 16.5 Changes in pH produced by additions of NaOH to two buffers. Buffer A contains equal molar concentration of a weak acid and its conjugate base; buffer B contains the same total concentration of acid and conjugate base as in A, but the mole ratio of acid to conjugate base is 3 to 1.

$$\text{Buffer A, 0.25 mol NaOH added:}\quad \text{pH} = \text{p}K_a + \log\frac{n_{\text{base}}}{n_{\text{acid}}} = 3.75 + \log\frac{(1.00 + 0.25)}{(1.00 - 0.25)} = 3.97$$

$$\text{0.50 mol NaOH added:}\quad \text{pH} = \text{p}K_a + \log\frac{n_{\text{base}}}{n_{\text{acid}}} = 3.75 + \log\frac{(1.00 + 0.50)}{(1.00 - 0.50)} = 4.23$$

$$\text{0.75 mol NaOH added:}\quad \text{pH} = \text{p}K_a + \log\frac{n_{\text{base}}}{n_{\text{acid}}} = 3.75 + \log\frac{(1.00 + 0.75)}{(1.00 - 0.75)} = 4.60$$

$$\text{Buffer B, 0.25 mol NaOH added:}\quad \text{pH} = \text{p}K_a + \log\frac{n_{\text{base}}}{n_{\text{acid}}} = 4.23 + \log\frac{(0.50 + 0.25)}{(1.50 - 0.25)} = 4.01$$

$$\text{0.50 mol NaOH added:}\quad \text{pH} = \text{p}K_a + \log\frac{n_{\text{base}}}{n_{\text{acid}}} = 4.23 + \log\frac{(0.50 + 0.50)}{(1.50 - 0.50)} = 4.23$$

$$\text{0.75 mol NaOH added:}\quad \text{pH} = \text{p}K_a + \log\frac{n_{\text{base}}}{n_{\text{acid}}} = 4.23 + \log\frac{(0.50 + 0.75)}{(1.50 - 0.75)} = 4.45$$

The changes in buffer pH from the target pH of 3.75 are shown in **Figure 16.5**.

Think About It The greater capacity of buffer B to neutralize additions of base is evident when 0.75 moles of NaOH are added to both: neutralizing the base consumes 75% of the formic acid in buffer A, but only 50% of the acid in buffer B. Therefore, the pH change experienced by buffer B is less than that of buffer A. However, Figure 16.5 also shows that buffer A does a better job of controlling pH when small quantities of base are added. From this exercise we can conclude that the most effective buffers for most applications are those with nearly equal concentrations of their acidic and basic components.

Practice Exercise Small volumes of strong acid, each containing 0.025 mol H_3O^+ ions, are added to 1.00 L samples of two basic buffers. Buffer A contains 0.150 mol diethylamine and 0.150 mol diethylammonium chloride. Buffer B contains 0.200 mol methylamine and 0.100 mol methylammonium chloride. Calculate the changes in pH in the two buffers because of adding the strong acid.

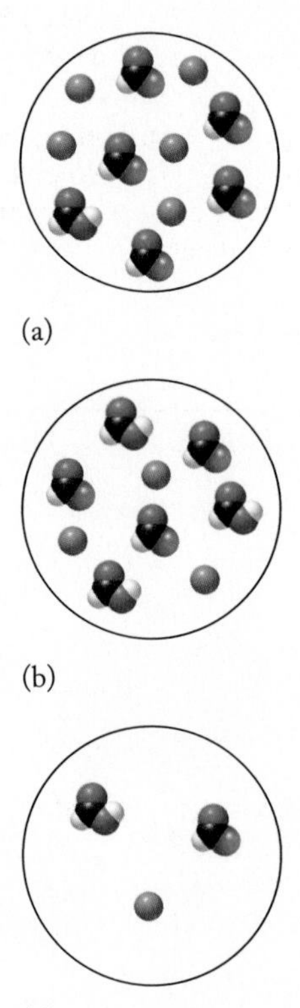

FIGURE 16.6 Possible representations of buffers.

CONCEPT TEST

Which representation in **Figure 16.6** depicts the formic acid–sodium formate buffer that will have the smallest change in pH upon the addition of a strong acid or strong base? Solvent molecules have been omitted for clarity.

(Answers to Concept Tests are in the back of the book.)

16.4 Indicators and Acid–Base Titrations

Swimming pool operators routinely check the pH of pool water to make sure it is close to 7.3, which is the average pH of our eyes. They often add sodium carbonate to raise the water's pH after acidic summertime precipitation sends it below 7.3. To determine how much sodium carbonate to add, they test the pH of the water by using a kit that includes a **pH indicator** (**Figure 16.7**), which is a substance that changes color as its pH changes.

pH indicator a water-soluble weak organic acid that changes color as it ionizes.

Phenol red is a pH indicator. It is a weak acid ($\text{p}K_a = 7.6$) that is yellow in its un-ionized form (which, for convenience, we assign the generic formula HIn) and violet in its ionized (In^-) form. At a pH that is 1 unit above the $\text{p}K_a$ (i.e., at pH 8.6), the ratio $[In^-]/[HIn]$ is 10:1 and a phenol red solution is violet. At a pH less

(a)

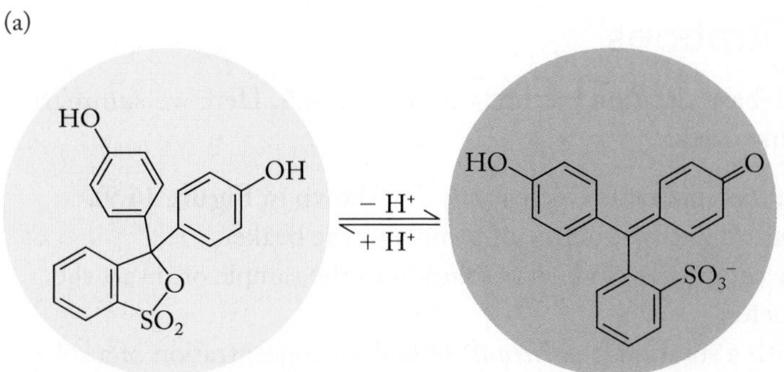

(b)

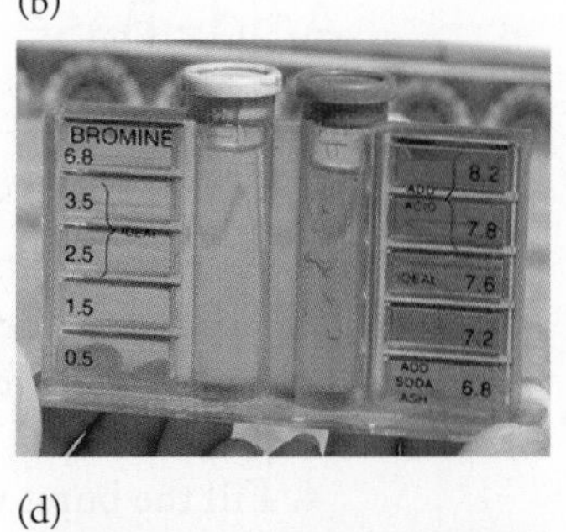

(d)

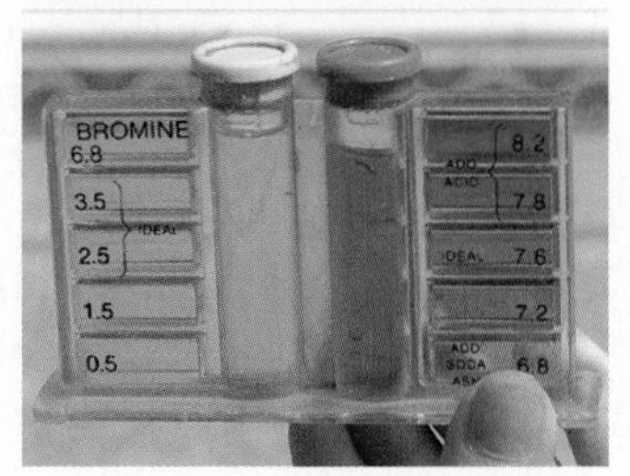

(c)

FIGURE 16.7 Many pool test kits include the pH indicator phenol red. (a) The molecular structures of the acid and base forms of phenol red. A few drops of phenol red indicator are added to a sample of pool water collected in the tube with the red cap. (b) After a rainstorm, the pH of the pool water is 6.8 (or less), as indicated by the yellow color of the sample. (c) Sodium carbonate is added to the pool to raise the pH. (d) A follow-up test produces a red-orange color, indicating the pH of the pool has been properly adjusted.

than 6.6, phenol red is largely un-ionized, and a solution of it is yellow. In the pH range from about 6.8 to 8.6, the color changes from yellow to orange to red to violet with increasing pH (**Figure 16.8**). These color changes allow pH to be determined to within about ±0.2 unit.

As with buffers, a pH indicator is useful over a pH range from about 1 pH unit below to 1 pH unit above the indicator's pK_a value. In addition to their role in determining pH values, indicators are also used to detect the large changes in pH that occur at the equivalence points in acid–base titrations. The ideal indicator for a titration has a pK_a value that is near the pH of the titration reaction mixture at the equivalence point.

CONNECTION In Section 4.6 we learned that when the number of moles of titrant is stoichiometrically equal to the number of moles of the analyte, the titration is at its equivalence point. When the indicator begins to change color, the titration is at its (observable) end point.

STEPWISE ANIMATION

Indicators

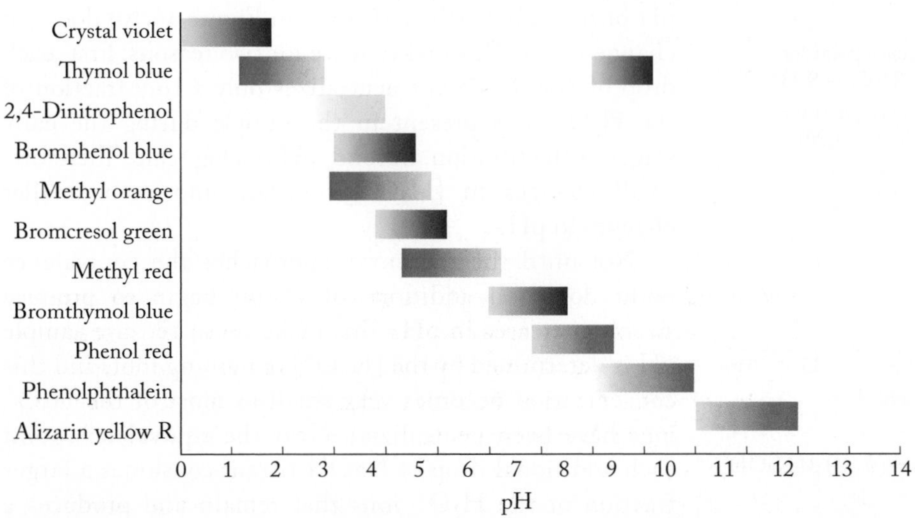

FIGURE 16.8 A pH indicator is useful within a range of 1 pH unit above and below the pK_a value of the indicator. This array of indicators could be used to determine pH values from 0 to 12.

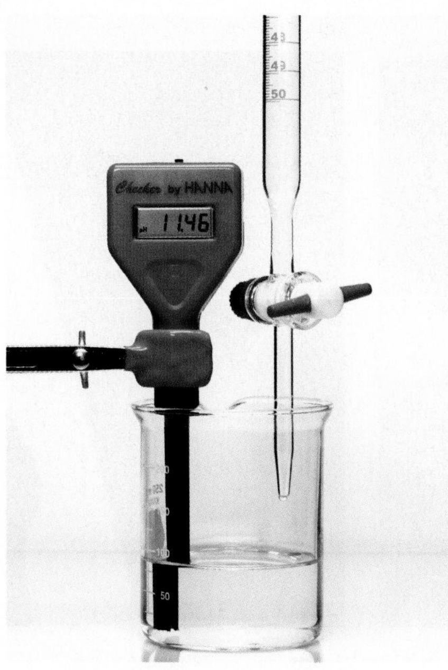

FIGURE 16.9 A digital pH meter is used to measure pH during a titration.

Acid–Base Titrations

We introduced acid–base titration methods in Section 4.6. Here we summarize the principal steps involved:

1. Assemble a titration apparatus (such as the one shown in **Figure 16.9**).
2. Accurately transfer a known volume of sample to the beaker.
3. Either add a few drops of an indicator solution to the sample or insert the probe of a pH meter.
4. Fill the buret with a solution (the *titrant*) of known concentration of a substance that reacts with a solute (the *analyte*) in the sample.
5. Slowly add titrant to the sample and monitor the change in pH. When the volume of titrant needed to completely consume the analyte has been added, the equivalence point has been reached, as indicated by either a change in indicator color or a large change in pH as sensed by a pH electrode. This volume of titrant is a measure of the concentration of analyte in the sample.

The neutralization titrations in Chapter 4 involved titrating strong acids with strong bases and vice versa. In this section we begin with titrations of aqueous samples containing weak as well as strong monoprotic acids or *monobasic* bases. (A monobasic base accepts one hydrogen ion per molecule.) In the examples that follow, we monitor changes in pH during the titration by using a pH electrode, and we plot the pH of the titration reaction mixture against the volume of titrant added.

CHEMTOUR
Acid–Base Titrations

First, let's compare the titration curves of two 20.0 mL samples: one contains 0.100 M HCl and the other contains 0.100 M acetic acid (CH_3COOH). Both are titrated with 0.100 M NaOH. The two graphs of pH versus titrant volume are shown in **Figure 16.10**. The initial pH of the HCl solution is lower than that of the acetic acid solution because HCl is a stronger acid, and the HCl titration curve stays below the acetic acid curve until they reach their equivalence points, where both acids have been completely neutralized.

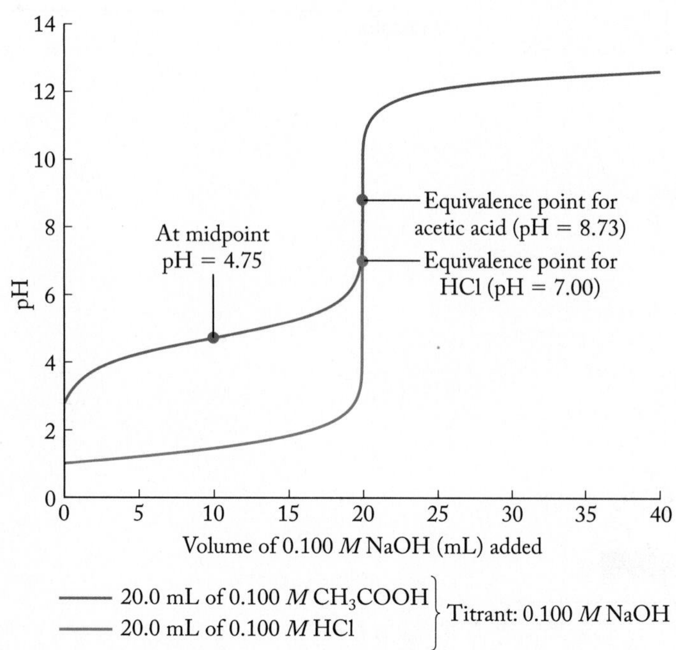

FIGURE 16.10 The titration curves of solutions of HCl and acetic acid with a standard solution of NaOH as the titrant.

As NaOH titrant is first added to the HCl sample, the pH of the sample (the red curve in Figure 16.10) does not change much. This makes sense for two reasons: first, each drop of NaOH titrant neutralizes only a tiny fraction of the H_3O^+ ions present in the sample during the early stages of the titration. Second, pH is a log scale. Therefore, small changes in $[H_3O^+]$ translate into even smaller changes in pH.

Not until the red curve approaches the equivalence point do small additions of titrant begin to produce sizable increases in pH. This make sense because sample pH is determined by the $[H_3O^+]$ remaining in it, and this concentration becomes very small as most of the H_3O^+ ions have been neutralized. Near the equivalence point each additional drop of NaOH titrant consumes a larger fraction of the H_3O^+ ions that remain and produces a larger increase in pH until the sample is completely neutralized.

At the equivalence point, the sample consists of Na^+ and Cl^- ions dissolved in water. All the H_3O^+ ions in the original sample and all the OH^- ions in the added titrant have reacted to form molecules of H_2O. Sodium and chloride ions do not hydrolyze and do not influence pH; therefore, the pH of the sample at the equivalence point is 7.00.

SAMPLE EXERCISE 16.9 Calculating pH during Titration of a Strong Acid with a Strong Base **LO4**

What is the pH of the solution being titrated in Figure 16.10 just before the equivalence point when 19.0 mL of 0.100 *M* NaOH has been added to 20.0 mL of 0.100 *M* HCl?

Collect and Organize We are asked to calculate the pH in the titration of a sample of a strong acid, HCl, with a strong base, NaOH, at a point before the equivalence point, which means that some of the acid remains unreacted. HCl is completely ionized in water

$$HCl(aq) + H_2O(\ell) \rightarrow H_3O^+(aq) + Cl^-(aq)$$

and NaOH completely dissociates into its ions in water

$$NaOH(aq) \rightarrow Na^+(aq) + OH^-(aq)$$

Combining these equations and eliminating spectator ions yields this net ionic equation for the titration reaction:

$$H_3O^+(aq) + OH^-(aq) \rightarrow 2\,H_2O(\ell)$$

Analyze To solve the problem, we will need to calculate the concentration of H_3O^+ ions remaining in the reaction mixture, which means calculating how many moles of H_3O^+ ions were in the original 20.0 mL sample and how many of those moles were neutralized by the OH^- ions in 19.0 mL of titrant. Taking the difference between these values will determine the remaining moles of H_3O^+ ions and dividing by the total volume of the reaction mixture will gives us $[H_3O^+]$. Taking the −log of this value will give us pH.

Solve The amount of acid initially in the sample is

$$\text{moles } H_3O^+ = 20.0\ \cancel{\text{mL}} \times \frac{1\ \cancel{\text{L}}}{1000\ \cancel{\text{mL}}} \times \frac{0.100\ \text{mol HCl}}{\cancel{\text{L}}\ \text{solution}} = 2.00 \times 10^{-3}\ \text{mol HCl}$$

The amount of base added is

$$\text{moles } OH^- = 19.0\ \cancel{\text{mL}} \times \frac{1\ \cancel{\text{L}}}{1000\ \cancel{\text{mL}}} \times \frac{0.100\ \text{mol NaOH}}{\cancel{\text{L}}\ \text{solution}} = 1.90 \times 10^{-3}\ \text{mol NaOH}$$

Each H_3O^+ ion reacts with one OH^- ion, so there is an excess of H_3O^+ ions:

$$\text{excess } H_3O^+ \text{ ions} = 2.00 \times 10^{-3}\ \text{mol } H_3O^+ \text{ ions} - 1.90 \times 10^{-3}\ \text{mol } H_3O^+ \text{ ions reacted}$$
$$= 1.0 \times 10^{-4}\ \text{mol } H_3O^+ \text{ remain in solution}$$

These H_3O^+ ions are dissolved in (20.0 mL + 19.0 mL) = 39.0 mL solution. Therefore, the molarity of the unreacted ions is

$$\frac{1.0 \times 10^{-4}\ \text{moles } H_3O^+}{0.0390\ \text{L solution}} = 2.6 \times 10^{-3}\ M$$

and

$$\text{pH} = -\log(2.6 \times 10^{-3}) = 2.59$$

Think About It The pH of the solution before the equivalence point in a strong acid–strong base titration is determined by the concentration of H_3O^+ ions not yet neutralized. In this exercise, $[H_3O^+]$ remaining in the sample was enough to produce a pH of 2.59: well below the equivalence point pH of 7.00. This difference reflects the rapid pH change that occurs near the equivalence points in titrations of strongly acidic solutions.

Practice Exercise What is the pH of the solution in Figure 16.10 just after the equivalence point when 21.0 mL of 0.100 *M* NaOH has been added to 20.0 mL of 0.100 *M* HCl?

CONCEPT TEST

The representations in **Figure 16.11** depict the particles present during the titration of LiOH with HI. Which representation depicts the starting solution? The equivalence point? Put the representations in order from the beginning to the end of the titration. The solvent has been omitted for clarity.

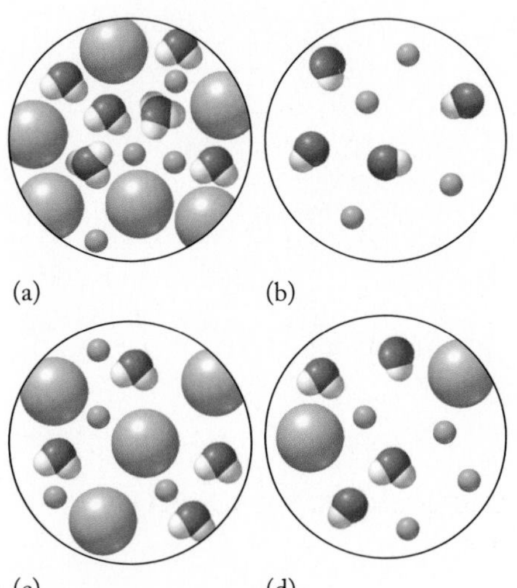

FIGURE 16.11 Particles present during titration.

CHEMTOUR
Titrations of Weak Acids

The pH of the acetic acid sample (the blue curve in Figure 16.10) changes abruptly with the first few drops of added base, but then the changes become smaller and the titration curve levels out. In the nearly flat region, the sample acts like a pH buffer as additions of NaOH titrant neutralize acetic acid, forming the sodium salt of its conjugate base:

$$CH_3COOH(aq) + NaOH(aq) \rightarrow CH_3COONa(aq) + H_2O(\ell)$$

As long as there are significant concentrations of both acetic acid and acetate ions in the sample, the changes in pH with added titrant are small.

When enough titrant has been added to consume nearly all the acid in the CH_3COOH sample, its pH rises sharply, much like the red curve and for the same reason: small additions of titrant consume proportionately more of the remaining acid in the sample as the blue curve approaches the equivalence point. At their equivalence points, the same volume of NaOH titrant has been added to both acid solutions because the initial concentrations of the two acids and their sample volumes were the same. Therefore, the number of moles of acid in each sample was the same, and that is the quantity that defines the number of moles of base needed to reach the equivalence point. It does not matter whether the acids were initially completely ionized, because even a weak acid such as acetic acid is completely neutralized at its equivalence point. Why is that? Adding NaOH to a solution of acetic acid (or any weak acid) consumes H_3O^+ ions, which, in accordance with Le Châtelier's principle, shifts the position of the acid ionization equilibrium to the right:

$$CH_3COOH(aq) + H_2O(\ell) \rightleftharpoons CH_3COO^-(aq) + H_3O^+(aq)$$

Even though the volumes of titrant needed to reach the equivalence points of the red and blue curves in Figure 16.10 are the same, the pH values at the two equivalence points are not. The pH values of the equivalence points are 7.00 (neutral) in the HCl titration, but 8.73 (weakly basic) in the acetic acid titration, because the product of this titration reaction is a solution of sodium acetate (CH_3COONa). Acetate ions react with water, forming molecules of acetic acid and OH^- ions:

$$CH_3COO^-(aq) + H_2O(\ell) \rightleftharpoons CH_3COOH(aq) + OH^-(aq)$$

In general, the pH at the equivalence point in the titration of any weak acid (HA) with a strong base is greater than 7.00 because hydrolysis of the anion (A^-) of the salt formed by the neutralization reaction produces OH^- ions.

Beyond their equivalence points, the red and blue curves overlap because the pH values of the solutions in this region are controlled only by the increasing concentration of OH^- ions as excess NaOH beyond that needed to neutralize the acids is added to the reaction mixtures.

An important reference point in the acetic acid titration in Figure 16.10 is the *midpoint*, which is located at the point where half the volume needed to reach the equivalence point has been added. At the midpoint, half of the acetic acid initially in the sample has been converted into its acetate ions. As a result, the concentration of acetate ions produced by the neutralization reaction and the concentration of acetic acid remaining from the original sample *are the same*. If we insert this equality ([base] = [acid]) into the log term of the Henderson–Hasselbalch equation, the value of the term is zero, and the equation reduces to pH = pK_a. In an acetic acid titration, this pH value is 4.75.

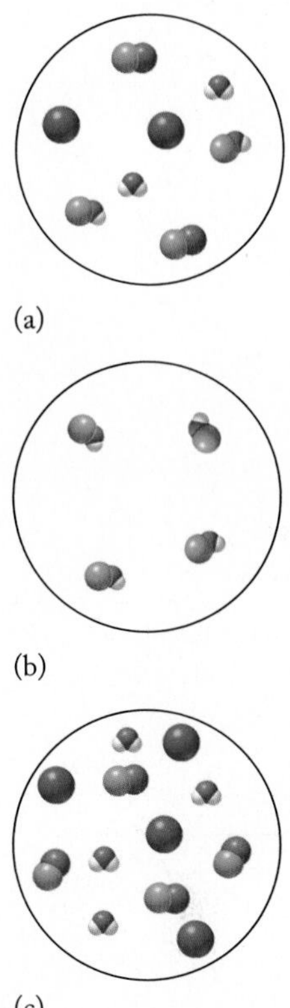

FIGURE 16.12 Particles present during titration.

CONCEPT TEST

The representations in **Figure 16.12** depict the particles present during a titration of HClO with KOH. Which represents the weak acid at the start of the titration? At the midpoint? At the equivalence point? The solvent has been omitted for clarity.

SAMPLE EXERCISE 16.10 Calculating pH during the Titration of a Weak Acid with a Strong Base LO4

What is the pH of the acetic acid sample in Figure 16.10 after (a) 5.0 mL and (b) 15.0 mL of 0.100 *M* NaOH titrant have been added to a 20.0 mL sample in which $[CH_3COOH]$ is initially 0.100 *M*?

Collect, Organize, and Analyze We are asked to calculate the pH at the points along the blue titration curve in Figure 16.10 at which 5.0 mL and 15.0 mL of titrant have been added. These volumes are less than that needed to reach the equivalence point, meaning that all the OH^- ions added to the sample will react with CH_3COOH and convert the acid into CH_3COO^- ions as described by this net ionic equation:

$$CH_3COOH(aq) + OH^-(aq) \rightarrow CH_3COO^-(aq) + H_2O(\ell)$$

We can use the 1:1 stoichiometry of the reaction along with the volume and concentration of the original sample and the volumes and concentration of the titrant added to calculate how many moles of CH_3COOH were in the original sample, how many moles of OH^- ions have been added (which is equal to the moles of CH_3COO^- ions produced), and how many moles of CH_3COOH remain in the sample. These last two quantities can be used with the Henderson–Hasselbalch equation to determine the two pH values. The two volumes are 5.0 mL less than and 5.0 mL greater than the midpoint titration volume of 10.0 mL. It is likely that the calculated pH values will also be equally less than and greater than the midpoint pH (4.75 = the pK_a of acetic acid).

Solve The following ICA tables (introduced in Sample Exercise 16.6) provide a guide in determining how many moles of CH_3COOH remain after each addition of titrant, how many moles of CH_3COO^- are produced, and sample pH.

a. 5.0 mL of titrant:

	$CH_3COOH(aq)$ +	$OH^-(aq)$	→ $CH_3COO^-(aq)$ + $H_2O(\ell)$
Initial	20.0 mL × 0.100 mol/L = 2.00 mmol	5.0 mL × 0.100 mol/L = 0.50 mmol	0
Change	−0.50 mmol	−0.50 mmol	+0.50 mmol
After	1.50 mmol	0	0.50 mmol

The total sample volume is (20.0 mL + 5.0 mL) = 25.0 mL. We could divide the moles of CH_3COOH and CH_3COO^- ions in the After row by 25.0 mL to calculate $[CH_3COOH]$ and $[CH_3COO^-]$ values to use in the Henderson–Hasselbalch equation. However, we can skip this step and use the moles of CH_3COO^- ions and moles of CH_3COOH in the After row, as we did in Sample Exercise 16.7 and for the same reason: the two mole quantities are dissolved in the same volume of solution, so their mole ratio has the same value as their concentration ratio:

$$\frac{[CH_3COO^-]}{[CH_3COOH]} = \frac{n_{CH_3COO^-}}{n_{CH_3COOH}}$$

Therefore:

$$pH = pK_a + \log\frac{n_{CH_3COO^-}}{n_{CH_3COOH}} = 4.75 + \log\frac{0.50\ \text{mmol}}{1.50\ \text{mmol}} = 4.27$$

b. 15.0 mL of titrant:

	$CH_3COOH(aq)$ +	$OH^-(aq)$	→ $CH_3COO^-(aq)$ + $H_2O(\ell)$
Initial	20.0 mL × 0.100 mol/L = 2.00 mmol	15.0 mL × 0.100 mol/L = 1.50 mmol	0
Change	−1.50 mmol	−1.50 mmol	+1.50 mmol
After	0.50 mmol	0	1.50 mmol

Using the mole values in the After row in the Henderson–Hasselbalch equation:

$$pH = 4.75 + \log\frac{n_{CH_3COO^-}}{n_{CH_3COOH}}$$

$$= 4.75 + \log\frac{1.50\ \text{mmol}}{0.50\ \text{mmol}} = 4.75 + \log(3.00) = 5.23$$

Think About It As predicted, the pH values are spaced equally below and above the midpoint pH: 4.75 − 4.27 = 0.48 and 5.23 − 4.75 = 0.48. The procedure used in this sample exercise worked because the reaction mixture acted like a pH buffer. Letting the initial quantity of acetate ions be zero meant ignoring the small concentration of them present in a 0.100 *M* solution of acetic acid due to ionization of the acid. This is reasonable because the initial pH of the solution is about 3, which means the initial concentrations of $[H_3O^+]$ and $[CH_3COO^-]$ are about 1×10^{-3} *M* and the initial quantity of CH_3COO^- ions in 20.0 mL is about 0.02 mmol. This value is less than 5% of the quantities of the ion we used to calculate pH.

Practice Exercise If 30.0 mL of titrant is required to reach the equivalence point in the titration with a strong base of an aqueous sample of formic acid, what is the pH of the titration mixture after (a) 10.0 mL, (b) 15.0 mL, and (c) 20.0 mL of titrant have been added?

CONCEPT **TEST**

Why is it better to titrate a weak acid with a strong base instead of titrating a weak acid with a weak base?

Titration curves of weak or strong base analytes with strong acid titrants resemble the curves in Figure 16.10 but are inverted, starting with high initial pH values and ending with low ones. **Figure 16.13** illustrates two such titrations. One sample contains a strong base, 0.100 *M* NaOH; the other contains a weak base, 0.100 *M* NH_3. Both are titrated with 0.100 *M* HCl. The initial pH of the NaOH titration curve is higher than that of the NH_3 sample because NaOH is a stronger base. The pH of the NaOH sample does not change much as acid is added until near the equivalence point. As the equivalence point is approached, the principal ions present in the reaction mixture are Na^+, Cl^-, and decreasing concentrations of OH^-. The sample pH, which is determined only by the $[OH^-]$ still present, starts to fall steeply. When enough acid has been added to consume all the OH^- ions in the sample, the equivalence point is reached. The solution now consists of water and NaCl, which is neutral, so pH = 7.00 at the equivalence point.

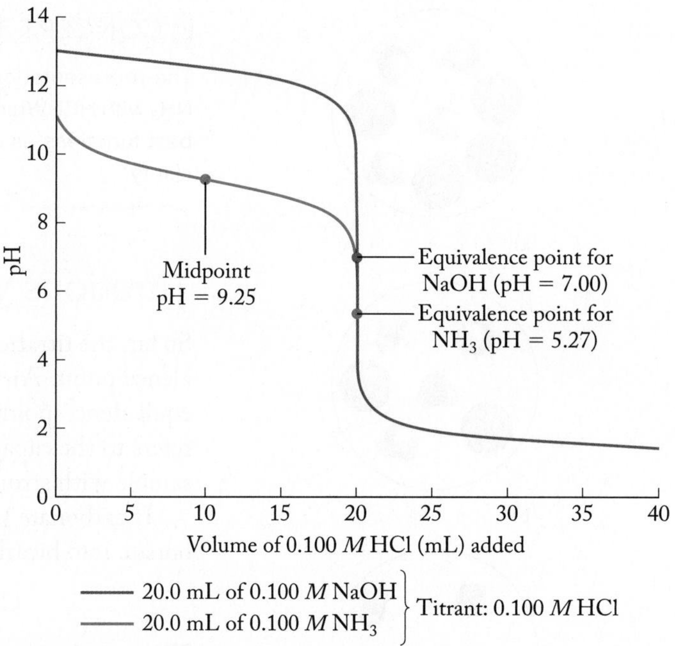

FIGURE 16.13 The titration curves of solutions of NaOH and NH_3 with a standard solution of HCl as the titrant.

The pH of the NH_3 solution changes abruptly with the first few drops of added acid, but then the changes become smaller and the titration curve levels out. In this nearly flat region, additions of acidic titrant are consumed as NH_3 reacts with H_3O^+ ions to form $NH_4{}^+$ ions. In this region of the titration curve, the sample acts like a pH buffer, just as the mixture of acetic acid and acetate ions did in the titration in Figure 16.10. The pH value at the equivalence point in the ammonia titration is 5.27 (slightly acidic) because the product of the titration reaction is NH_4Cl, an acidic salt, which reacts with water, forming ammonia molecules and H_3O^+ ions. Beyond their equivalence points, the NaOH and NH_3 curves are identical because the pH is determined only by the moles of HCl added after the equivalence point and the total volume of the titration reaction mixtures.

As in the acetic acid titration, the midpoint in the NH_3 titration curve is significant because it is the point at which half of the NH_3 initially in the sample has been converted into $NH_4{}^+$ ions. As a result, $[NH_3] = [NH_4{}^+]$. This concentration equality can be used in the Henderson–Hasselbalch equation based on the hydrolysis of ammonium ions:

$$NH_4{}^+(aq) + H_2O(\ell) \rightleftharpoons NH_3(aq) + H_3O^+(aq)$$

$$pH = pK_a + \log \frac{[NH_3]}{[NH_4{}^+]}$$

This pK_a value can be calculated from the pK_b value for ammonia, which is 4.75 (see Table A5.3), using Equation 16.5: $pK_a = 14.00 - 4.75 = 9.25$. Because the value of the log term in the Henderson–Hasselbalch equation is zero, $pH = pK_a = 9.25$ at the midpoint of this titration.

CONCEPT **TEST**

What is the pH at the midpoint of the titration of an aqueous sample of methylamine with a standard solution of hydrochloric acid?

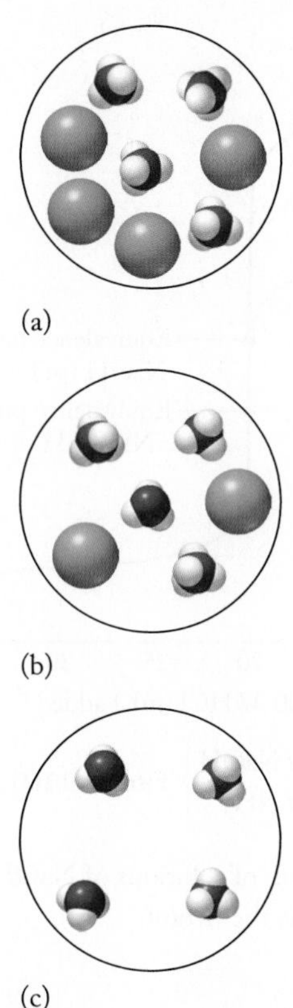

FIGURE 16.14 Particles present during titration.

CONCEPT TEST

The representations in **Figure 16.14** depict the particles present during a titration of NH_3 with HCl. Which represents the start of the titration? The region where the solution best functions as a buffer? The equivalence point? The solvent has been omitted for clarity.

Titrations with Multiple Equivalence Points

So far, the titration curves discussed in this chapter have each had only one equivalence point. An alkalinity titration (**Figure 16.15**), on the other hand, has two equivalence points. In environmental science, the alkalinity of a water sample refers to the capacity of the water to neutralize additions of acid. Titration of the sample with strong acid provides a way to determine that capacity.

If carbonate is present in the sample, the first additions of titrant convert carbonate into bicarbonate:

$$CO_3^{2-}(aq) + H_3O^+(aq) \rightarrow HCO_3^-(aq) + H_2O(\ell)$$

This reaction continues until all the carbonate initially in the sample has been converted into bicarbonate. This represents the first equivalence point in the titration and is marked by a sharp drop in pH. In the second stage of the titration, the bicarbonate ions formed in the first stage plus any bicarbonate present in the original sample react with additional acidic titrant, forming carbonic acid, most of which decomposes into carbon dioxide and water:

$$HCO_3^-(aq) + H_3O^+(aq) \rightarrow CO_2(aq) + 2\,H_2O(\ell)$$

The solubility of CO_2 in water is limited, so its production in the second step of an alkalinity titration often results in formation of bubbles of carbon dioxide as it comes out of solution:

$$CO_2(aq) \rightarrow CO_2(g)$$

In the first stage of the alkalinity titration, the titration curve has a region in which added acid has little effect on pH. This is a buffering region where HCO_3^- and CO_3^{2-} function as a weak acid–conjugate base pair. At the first equivalence point, the dominant carbon-containing species is HCO_3^- and the sample is still slightly basic (pH $>$ 8). This basic pH tells us that the bicarbonate is more effective as a base (Equation 16.6) than as an acid (Equation 16.7):

$$HCO_3^-(aq) + H_2O(\ell) \rightleftharpoons H_2CO_3(aq) + OH^-(aq) \qquad (16.6)$$

$$HCO_3^-(aq) + H_2O(\ell) \rightleftharpoons CO_3^{2-}(aq) + H_3O^+(aq) \qquad (16.7)$$

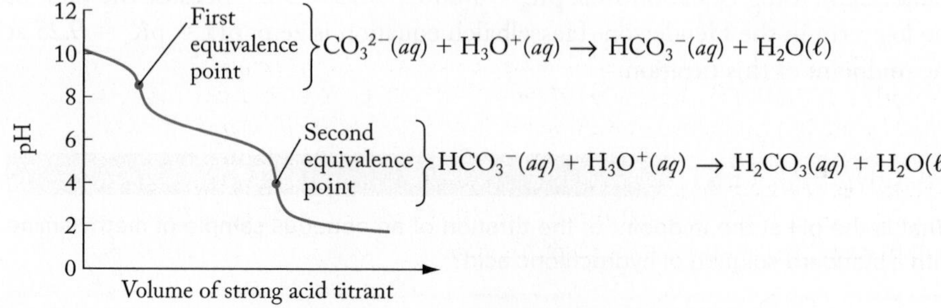

FIGURE 16.15 An alkalinity titration curve can have two equivalence points. The first marks the complete conversion of any carbonate in the sample into bicarbonate, and the second marks the conversion of bicarbonate into carbonic acid.

We can confirm the dominance of the reaction in Equation 16.6 by considering the two-step acid ionization equilibria of carbonic acid. Equation 16.7 represents the second step and its equilibrium constant is $K_{a_2} = 4.7 \times 10^{-11}$. The K_b of the base ionization described in Equation 16.6 is linked to the first ionization of carbonic acid:

$$H_2CO_3(aq) + H_2O(\ell) \rightleftharpoons HCO_3^-(aq) + H_3O^+(aq) \qquad K_{a_1} = 4.3 \times 10^{-7}$$

Convert K_{a_1} into K_b using Equation 15.18:

$$K_a \times K_b = K_w$$

$$K_b = \frac{K_w}{K_{a_1}} = \frac{1.0 \times 10^{-14}}{4.3 \times 10^{-7}} = 2.3 \times 10^{-8}$$

This K_b value is much greater than the value of K_{a_2}:

$$\frac{K_b}{K_{a_2}} = \frac{2.3 \times 10^{-8}}{4.7 \times 10^{-11}} = 4.9 \times 10^2$$

Thus, bicarbonate is nearly 500 times stronger a base than it is an acid.

During the second stage of the titration, conversion of HCO_3^- ions to H_2CO_3 produces a second pH buffer and another plateau of slowly changing pH. When the HCO_3^- ions are completely consumed, pH drops sharply for a second time at the second equivalence point. Its pH is below 7, reflecting the acidic character and pK_{a_1} value of carbonic acid.

The initial pH of the sample in Figure 16.15 is slightly above 10, which is quite basic and above the pH range tolerated by many species of aquatic life. Such highly basic water may be found in arid regions such as the U.S. Southwest, where rocks containing $CaCO_3$ and other basic compounds are in contact with water. The pH of most freshwater is lower than 8, which means that the dominant carbonate species is bicarbonate. The alkalinity titration curves of these waters have only one equivalence point, coinciding with the pH of the second equivalence point in Figure 16.15.

SAMPLE EXERCISE 16.11 Interpreting the Results of a Titration with Multiple Equivalence Points I — LO4

If the volume of titrant needed to reach the first equivalence point in the titration shown in Figure 16.15 is 9.00 mL, and *a total of* 27.00 mL is required to reach the second equivalence point, what was the ratio of carbonate to bicarbonate in the original sample?

Collect, Organize, and Analyze Carbonate (CO_3^{2-}) ions in the sample combine with H_3O^+ ions from the titrant and are converted to bicarbonate (HCO_3^-) ions in the first stage of the titration. Bicarbonate ions—including any in the original sample plus all those produced in the first-stage reaction—are converted to carbonic acid (H_2CO_3) in the second stage.

Solve The volume of titrant needed to titrate the CO_3^{2-} in the sample (9.00 mL) is exactly half the additional volume (27.00 − 9.00 = 18.00 mL) needed to titrate the HCO_3^- in the reaction mixture. However, 9.00 mL of the titrant consumed in the second stage was needed just to titrate the HCO_3^- produced in the first stage. This means the volume of titrant needed to titrate the HCO_3^- that was in the original sample was only (18.00 − 9.00) = 9.00 mL. Therefore, the original sample contained equal concentrations of CO_3^{2-} and HCO_3^- ions.

Think About It It would be tempting to interpret the volumes of titrant consumed in the two stages to mean that there was twice as much bicarbonate as carbonate in the original sample. It's important to remember that the HCO_3^- ions titrated in the second stage come from the original sample *and* from HCO_3^- ions produced from CO_3^{2-} ions during the first stage of the titration.

 **Practice Exercise** Is the midpoint pH in the first stage of an alkalinity titration always the same as the pK_{a_2} of carbonic acid? Explain why or why not.

We can detect both equivalence points in an alkalinity titration by using a pH electrode, or we can use appropriate indicators. Phenol red would *not* be a good choice for the titration in Figure 16.15 because it changes color between pH 6.8 and 8.4. This range is just below the pH of the first equivalence point and well above the pH of the second equivalence point. To detect the first equivalence point, we need an indicator with a pK_a near the pH of the solution at the first equivalence point, which is about 8.5. One candidate in Figure 16.8 is phenolphthalein ($pK_a = 9.7$), which is pink in its basic form and colorless at low pH.

To detect the second equivalence point, we could add bromcresol green ($pK_a = 4.6$) after the first equivalence point has been reached. We would not add it earlier because its blue-green color in basic solutions would obscure the pink-to-colorless transition of phenolphthalein. We do not need to be concerned about the phenolphthalein obscuring the bromcresol green color change because phenolphthalein is colorless in acidic solutions.

FIGURE 16.16 The water in Sapphire Pool in Yellowstone National Park is slightly alkaline mostly as a result of carbonate and bicarbonate ions. It is also crystal clear and hot.

SAMPLE EXERCISE 16.12 Interpreting the Results of a Titration with Multiple Equivalence Points II — LO4

A 100.0 mL sample of water from the Sapphire Pool in Yellowstone National Park (**Figure 16.16**) is titrated with 0.0300 *M* HCl. A few drops of phenolphthalein are added at the beginning of the titration, and the solution turns pink. It takes 5.91 mL of titrant to reach the pink-to-clear equivalence point. Then a few drops of bromcresol green are added, and it takes an additional 26.02 mL of titrant before the blue-green color changes to yellow. What were the initial concentrations of carbonate and bicarbonate in the sample?

Collect and Organize We are asked to determine the concentrations of CO_3^{2-} and HCO_3^- ions in a water sample from the results of a titration. These determinations are based on the volumes of titrant needed to reach two equivalence points. In the first stage, CO_3^{2-} ions in the sample are converted to HCO_3^- ions:

$$(1) \quad H_3O^+(aq) + CO_3^{2-}(aq) \rightarrow HCO_3^-(aq) + H_2O(\ell)$$

In the second stage, HCO_3^- ions are converted to H_2CO_3:

$$(2) \quad H_3O^+(aq) + HCO_3^-(aq) \rightarrow H_2CO_3(aq) + H_2O(\ell)$$

Analyze According to the stoichiometries of the reactions, it takes 1 mole of HCl to titrate 1 mole of carbonate to bicarbonate in the first stage, and it takes 1 mole of HCl to titrate 1 mole of bicarbonate to carbonic acid in the second stage. The HCO_3^- ions titrated in stage 2 include any in the original sample plus all the HCO_3^- ions produced by reaction 1. The difference between the titrant volumes needed to reach the two equivalence points, (26.02 − 5.91) = 20.11 mL, is the volume of acid required to react with the HCO_3^- that was present in the original sample. This difference is between 3 and 4 times the volume of titrant consumed in stage 1, so we can predict that there was 3 to 4 times as much HCO_3^- in the original sample as CO_3^{2-}.

Solve Calculating the concentrations of CO_3^{2-} and HCO_3^- in the original sample from the volumes and molarity of the HCl titrant consumed in stages 1 and 2:

$$[CO_3^{2-}] = \frac{5.91\ \cancel{\text{mL titrant}} \times \dfrac{0.0300\ \cancel{\text{mmol HCl}}}{\cancel{\text{mL titrant}}} \times \dfrac{1\ \text{mmol } CO_3^{2-}}{1\ \cancel{\text{mmol HCl}}}}{100.0\ \text{mL solution}} = \frac{0.1773\ \text{mmol } CO_3^{2-}}{100.0\ \text{mL solution}}$$

$$= 1.77 \times 10^{-3}\ M$$

$$[HCO_3^-] = \frac{20.11\ \cancel{\text{mL titrant}} \times \dfrac{0.0300\ \cancel{\text{mmol HCl}}}{\cancel{\text{mL titrant}}} \times \dfrac{1\ \text{mmol } HCO_3^-}{1\ \cancel{\text{mmol HCl}}}}{100.0\ \text{mL solution}} = \frac{0.6033\ \text{mmol } HCO_3^-}{100.0\ \text{mL solution}}$$

$$= 6.03 \times 10^{-3}\ M$$

Think About It The titration results confirm that the bicarbonate concentration in the original sample was just over 3 times the carbonate concentration.

Practice Exercise Suppose you titrate a 100.0 mL sample of water from another pool in Yellowstone National Park with a ratio of carbonate to bicarbonate of 3:1. If the total volume of titrant used was 24.04 mL, approximately what volume of titrant was used to reach the first equivalence point? What additional volume of titrant was used to reach the second equivalence point?

CONCEPT TEST

In a titration that initially contains both CO_3^{2-} and HCO_3^-, the volume of titrant required to reach the first equivalence point is less than that required to titrate from the first equivalence point to the second. Why?

16.5 Lewis Acids and Bases

Until now we have used the Brønsted–Lowry definitions of acids (H^+ ion donors) and bases (H^+ ion acceptors). However, the time has come to expand our concept of acids and bases to include acid–base interactions that may or may not involve the transfer of H^+ ions. Let's begin by revisiting what happens when ammonia gas dissolves in water:

$$NH_3(g) + H_2O(\ell) \rightleftharpoons NH_4^+(aq) + OH^-(aq)$$

Figure 16.17a shows a Brønsted–Lowry interpretation of this reaction: in donating H^+ ions to ammonia, H_2O acts as a Brønsted–Lowry acid. In accepting H^+ ions, NH_3 acts as a Brønsted–Lowry base.

Another way to view this reaction is illustrated in **Figure 16.17b**. Instead of focusing on the transfer of hydrogen ions, consider the two reactants as a donor and an acceptor of a *pair of electrons*. In this view, the N atom in NH_3 donates its lone pair of electrons to one of the H atoms in H_2O. In the process, one of the H—O bonds in H_2O is broken in such a way that the bonding pair of electrons remains with the O atom. The donated lone pair from the N atom forms a fourth N—H covalent bond. The result is the same as in the Brønsted–Lowry model of

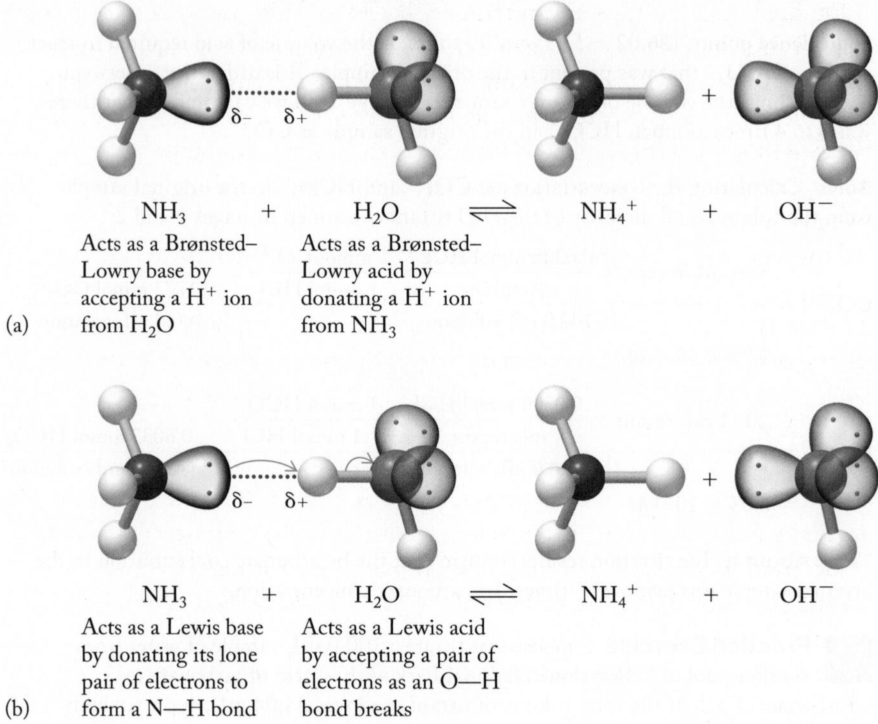
δ−
δ+
NH_3 + H_2O ⇌ NH_4^+ + OH^-
Acts as a Brønsted–Lowry base by accepting a H^+ ion from H_2O
Acts as a Brønsted–Lowry acid by donating a H^+ ion from NH_3
(a)
δ−
δ+
NH_3 + H_2O ⇌ NH_4^+ + OH^-
Acts as a Lewis base by donating its lone pair of electrons to form a N—H bond
Acts as a Lewis acid by accepting a pair of electrons as an O—H bond breaks
(b)

FIGURE 16.17 (a) Brønsted–Lowry view of the reaction between H_2O (proton donor) and NH_3 (proton acceptor). (b) Lewis view of the reaction: H_2O acts as a Lewis acid (electron-pair acceptor) and NH_3 acts as a Lewis base (electron-pair donor).

CONNECTION Lewis's pioneering theories of the nature of covalent bonding were described in Section 8.2.

acid–base behavior: a molecule of NH_3 bonds to a H^+ ion, forming an NH_4^+ ion, and a molecule of H_2O loses a H^+ ion, becoming a OH^- ion.

Viewing this process as the donation and acceptance of an electron pair provides the following basis for defining acids and bases:

- A **Lewis base** is a substance that *donates* a lone pair of electrons in a chemical reaction.
- A **Lewis acid** is a substance that *accepts* a lone pair of electrons in a chemical reaction.

These definitions are named after their developer, Gilbert N. Lewis, who also pioneered research into the nature of chemical bonds. The Lewis definition of a base is consistent with the Brønsted–Lowry model we have used because a substance must be able to donate a pair of electrons if it is to bond with a H^+ ion. However, the same parallelism is not true for acids. The Brønsted–Lowry model defines an acid as a hydrogen-ion donor, but the Lewis definition includes species that have no hydrogen ions to donate, but that can still accept electrons.

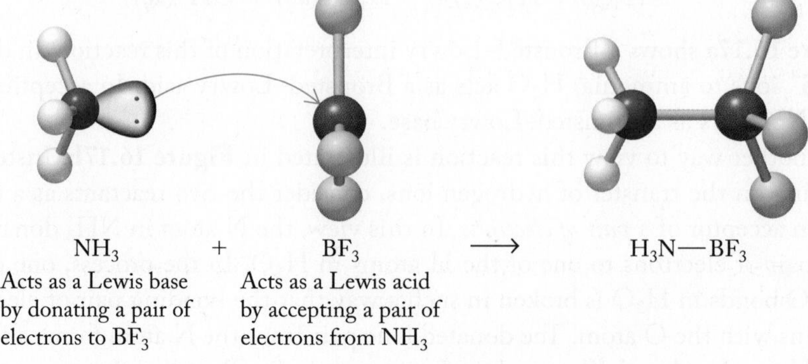
NH_3 + BF_3 ⟶ H_3N—BF_3
Acts as a Lewis base by donating a pair of electrons to BF_3
Acts as a Lewis acid by accepting a pair of electrons from NH_3

FIGURE 16.18 In the reaction between NH_3 and BF_3, NH_3 acts as a Lewis base and BF_3 acts as a Lewis acid. This is an acid–base reaction in the Lewis sense because we focus on the acceptance and donation of an electron pair, not the transfer of a proton, as in the Brønsted–Lowry system.

One such compound is boron trifluoride (BF_3). With only six valence electrons, the boron atom in BF_3 can accept another pair to complete its octet. NH_3 is a suitable electron-pair donor, as shown in **Figure 16.18**. There is no transfer of H^+ ions in this reaction, so it is not an acid–base reaction according to the Brønsted–Lowry model. However, NH_3 donates a lone pair of electrons and BF_3 accepts them, so it is an acid–base reaction according to the broader Lewis model.

Lewis base a substance that *donates* a lone pair of electrons in a chemical reaction.

Lewis acid a substance that *accepts* a lone pair of electrons in a chemical reaction.

Many important Lewis bases are anions, including the halide ions (i.e., F^-, Cl^-, Br^-, and I^-), OH^-, and O^{2-}. To see how O^{2-} functions as a Lewis base, let's revisit the reaction described in Section 14.6 between SO_2 and CaO that is used to reduce SO_2 emissions from coal-burning power stations:

$$CaO(s) + SO_2(g) \rightleftharpoons CaSO_3(s)$$

The oxide ion in CaO is the electron-pair donor, so O^{2-} is a Lewis base. Sulfur dioxide is the electron-pair acceptor and is therefore a Lewis acid. As an SO_2 molecule is adsorbed onto the surface of solid CaO, the oxide ion donates an electron pair to the sulfur atom, resulting in an additional S—O covalent bond and the formation of a sulfite anion, $SO_3^{\ 2-}$. The formal charges on the S atom and the double-bonded O atom in $SO_3^{\ 2-}$ are both zero; the formal charges on the two single-bonded O atoms are −1, giving an ion with an overall charge of 2−.

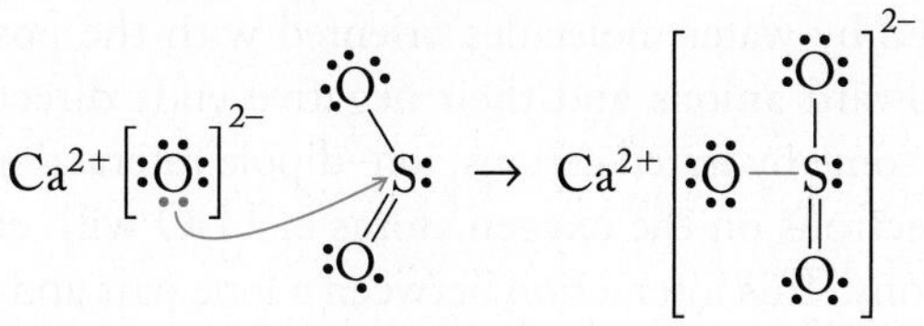

CONNECTION Formal charge and how to calculate it are described in Section 8.5.

SAMPLE EXERCISE 16.13 Identifying Lewis Acids and Bases LO5

In the following reaction, which species is a Lewis acid and which is a Lewis base?

$$AlCl_3 + Cl^- \rightarrow AlCl_4^-$$

Collect and Organize We are given a chemical reaction and asked to identify the Lewis acid (i.e., the reactant that accepts a pair of electrons) and the Lewis base (i.e., the reactant that donates that pair of electrons).

Analyze A Cl^- ion has four lone pairs of electrons in its valence shell, so it has the capacity to *donate* one of them to form a covalent bond to aluminum. To analyze the capacity of $AlCl_3$ to act as a Lewis acid, we need to draw its Lewis structure, which includes a total of (3 × 7) electrons from 3 Cl atoms, plus 3 from 1 Al atom, or 24 electrons in all. Using 6 electrons to draw three Al—Cl bonds leaves 18 with which to complete the octets around each of the three Cl atoms:

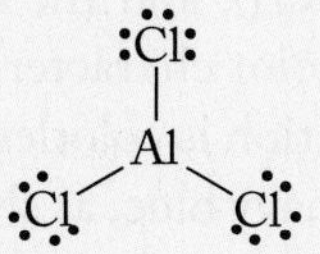

This structure accounts for all the valence electrons, but it leaves Al with only 6 valence electrons and the capacity to accept one more pair—that is, to act as a Lewis acid.

Solve In this reaction, $AlCl_3$ is a Lewis acid and the Cl^- ion is a Lewis base as shown in these Lewis structures:

$$\text{AlCl}_3 \leftarrow [\text{:}\ddot{\text{Cl}}\text{:}]^- \rightarrow [\text{AlCl}_4]^-$$

Think About It Drawing the Lewis structure of $AlCl_3$ and determining that the central Al atom has an incomplete octet is the key to identifying its capacity to accept another pair of electrons and to act as a Lewis acid.

Practice Exercise In the following reaction, which reactant is the Lewis acid, and which is the Lewis base?

$$CO_2(g) + CaO(s) \rightarrow CaCO_3(s)$$

16.6 Formation of Complex Ions

In Chapter 10 we described how ions dissolved in water are *hydrated*; that is, they are surrounded by water molecules oriented with the positive ends of their dipoles directed toward anions and their negative ends directed toward cations (**Figure 16.19**). In some hydrated cations, ion–dipole interactions lead to the sharing of lone-pair electrons on the oxygen atoms of H_2O with empty valence-shell orbitals on the cations. This interaction between a lone pair and an empty orbital is yet another example of Lewis acid–base behavior. These shared electron pairs meet our definition of covalent bonds, but these bonds are called *coordinate covalent bonds*, or simply **coordinate bonds**. They are also called *dative covalent bonds.*

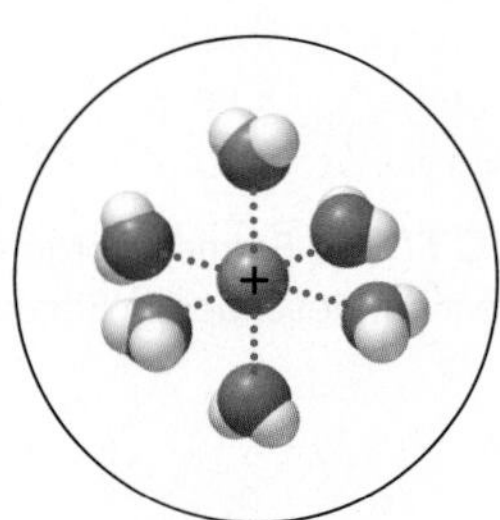

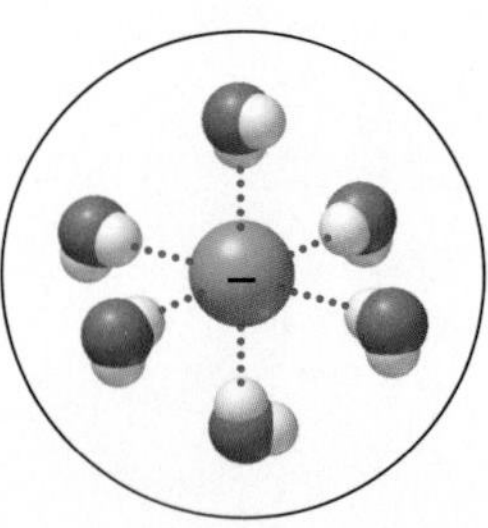

FIGURE 16.19 Ion–dipole interactions in a hydrated cation and a hydrated anion.

Coordinate bonds form when either a molecule or an anion donates a lone pair of electrons to an empty valence-shell orbital of an atom, cation, or molecule. Once formed, a coordinate bond is indistinguishable from any other kind of covalent bond. When a cation forms coordinate bonds with one or more molecular or ionic electron-pair donors, the resulting structure is called a **complex ion**. The electron-pair donors in complex ions are called **ligands**. For example, when six molecules of water form coordinate bonds to a Ni^{2+} cation in an aqueous solution, they form a complex ion with the formula $Ni(H_2O)_6^{2+}$ to show that there are six ligands (six water molecules bonded to the metal) in this complex ion.

We can investigate the formation of complex ions with the mathematical tools we have used in examining acid–base equilibria. These tools are appropriate because complex formation processes are reversible, and many of them reach chemical equilibrium rapidly. **Figure 16.20** shows two aqueous solutions: one contains copper(II) sulfate ($CuSO_4$), and the other contains NH_3. The $CuSO_4$ solution is robin's-egg blue, the color characteristic of $Cu^{2+}(aq)$ ions dissolved in water, whereas the ammonia solution is colorless. When the solutions are mixed, the robin's-egg blue turns a dark navy blue, as shown on the right in Figure 16.20. This is the color of $Cu(NH_3)_4^{2+}$ complex ions. The change in color provides visual evidence that the following equilibrium lies far to the right, favoring complex ion formation:

CONNECTION Spheres of hydration around cations and anions were introduced in Section 10.3.

$$Cu^{2+}(aq) + 4\,NH_3(aq) \rightleftharpoons Cu(NH_3)_4^{2+}(aq)$$

This conclusion is supported by the large equilibrium constant for the reaction:

$$K_f = \frac{[Cu(NH_3)_4^{2+}]}{[Cu^{2+}][NH_3]^4} = 5.0 \times 10^{13}$$

The equilibrium constant K_f is called a **formation constant** because it describes the formation of a complex ion. For the general case in which 1 mole of metal ions (M^{m+}) combines with n moles of ligand (X^{x-}) to form the complex ion $MX_n^{(m-nx)+}$, the formation constant expression is

$$K_f = \frac{[MX_n^{(m-nx)+}]}{[M^{m+}][X^{x-}]^n}$$

Formation constants can be used to calculate the concentration of free (uncomplexed) metal ions, $M^{m+}(aq)$, in equilibrium with a given (usually larger) concentration of a ligand. Because K_f values are usually very large (see Table A5.5), equilibrium concentrations of uncomplexed metal ions are usually very small. One approach to calculating the concentration of an uncomplexed metal ion is to consider the reverse of the formation reaction and calculate how much of the complex ion, in this case $Cu(NH_3)_4^{2+}(aq)$, dissociates. In Section 14.4 we learned that the equilibrium constant for the reverse reaction is the reciprocal of the original equilibrium constant.

$$Cu(NH_3)_4^{2+}(aq) \rightleftharpoons Cu^{2+}(aq) + 4\ NH_3(aq)$$

$$K_c = \frac{1}{K_f} = \frac{[Cu^{2+}][NH_3]^4}{[Cu(NH_3)_4^{2+}]}$$

The calculation of concentration of uncomplexed copper ion, $[Cu^{2+}]$, is illustrated in Sample Exercise 16.14.

FIGURE 16.20 The beaker on the left contains a solution of $Cu^{2+}(aq)$, which is a characteristic robin's-egg blue. As a colorless solution of ammonia is added (from the beaker in the middle), the mixture of the two solutions turns dark navy blue (beaker on the right), which is the color of the $Cu(NH_3)_4^{2+}$ complex ion.

SAMPLE EXERCISE 16.14 Calculating the Concentration of Free Metal Ions in Equilibrium with a Complex Ion **LO6**

Ammonia gas is dissolved in a $1.00 \times 10^{-4}\ M$ solution of $CuSO_4$ to give an equilibrium concentration of $[NH_3] = 1.60 \times 10^{-3}\ M$. Calculate the concentration of $Cu^{2+}(aq)$ ions in the solution.

Collect and Organize The concentration of $CuSO_4$ means that $[Cu^{2+}]$ before complex formation is $1.00 \times 10^{-4}\ M$. The equilibrium concentration of the ligand (NH_3) is $1.60 \times 10^{-3}\ M$. The values of $[NH_3]$, $[Cu^{2+}]$, and $[Cu(NH_3)_4^{2+}]$ at equilibrium are related by the formation constant expression:

$$K_f = \frac{[Cu(NH_3)_4^{2+}]}{[Cu^{2+}][NH_3]^4} = 5.0 \times 10^{13}$$

Analyze Because K_f is large, we can assume that essentially all the Cu^{2+} ions are converted to complex ions and $[Cu(NH_3)_4^{2+}] = 1.00 \times 10^{-4}\ M$. Only a tiny

coordinate bond a covalent bond formed when one anion or molecule donates a pair of electrons to another ion or molecule.

complex ion an ionic species consisting of a metal ion bonded to one or more Lewis bases.

ligand a Lewis base bonded to the central metal ion of a complex ion.

formation constant (K_f) an equilibrium constant describing the formation of a metal complex from a free metal ion and its ligands.

concentration of free Cu^{2+} ions, x, remains at equilibrium. We can calculate the $[Cu^{2+}]$ by considering how much $Cu(NH_3)_4^{2+}$ dissociates:

$$Cu(NH_3)_4^{2+}(aq) \rightleftharpoons Cu^{2+}(aq) + 4\,NH_3(aq)$$

The equilibrium constant, K_c, for this equilibrium is the reciprocal of K_f:

$$K_c = \frac{1}{K_f} = \frac{[Cu^{2+}][NH_3]^4}{[Cu(NH_3)_4^{2+}]} = \frac{1}{5.0 \times 10^{13}} = 2.0 \times 10^{-14}$$

We can construct a RICE table incorporating the concentrations and concentration changes of the products and reactants. Given the small value of K_c, we should obtain a $[Cu^{2+}]$ value at equilibrium that is much less than $1.00 \times 10^{-4}\,M$.

Solve First, we complete the row of equilibrium concentrations of Cu^{2+}, NH_3, and $Cu(NH_3)_4^{2+}$ in the following RICE table:

Reaction (R)	$Cu(NH_3)_4^{2+}(aq)$	$\rightleftharpoons$	$Cu^{2+}(aq)$	+	$4\,NH_3(aq)$
	$[Cu(NH_3)_4^{2+}]$ (*M*)		$[Cu^{2+}]$ (*M*)		$[NH_3]$ (*M*)
Initial (I)	1.00×10^{-4}		0		
Change (C)	$-x$		$+x$		
Equilibrium (E)	$1.00 \times 10^{-4} - x$		x		1.60×10^{-3}

Next, we make the simplifying assumption that x is much smaller than $1.00 \times 10^{-4}\,M$. Therefore, we can ignore the x terms in the equilibrium value of $[Cu(NH_3)_4^{2+}]$ and use the simplified values in the K_c expression:

$$K_c = \frac{[Cu^{2+}][NH_3]^4}{[Cu(NH_3)_4^{2+}]} = \frac{(x)(1.60 \times 10^{-3})^4}{(1.0 \times 10^{-4})} = 2.0 \times 10^{-14}$$

$$x = 3.1 \times 10^{-7} = [Cu^{2+}]$$

STEPWISE
ANIMATION
Hydrated Metal Ions

Think About It This result validates our simplifying assumption and confirms our prediction that $[Cu^{2+}]$ at equilibrium is much less than $[Cu^{2+}]$ initially. In fact, the ratio of the free to complexed copper(II) ions is only $(3.1 \times 10^{-7})/(1.00 \times 10^{-4}) = 0.0031$ or 0.31%.

Practice Exercise Calculate the equilibrium concentration of $Ag^+(aq)$ in a solution that is initially 0.100 *M* $AgNO_3$ and 0.800 *M* NH_3 after the following reaction takes place:

$$Ag^+(aq) + 2\,NH_3(aq) \rightleftharpoons Ag(NH_3)_2^+(aq) \qquad K_f = 1.7 \times 10^7$$

16.7 Hydrated Metal Ions as Acids

In Chapter 15 we saw that the strength of an oxoacid depends on the electronegativity of its central atom. For example, the relative strengths of the three hypohalous acids, HOCl > HOBr > HOI, align with the relative electronegativities of their halogen atoms: Cl > Br > I. This order makes sense because the more electronegative the halogen, the more it draws electron density away from the oxygen in the polar –OH bond (see Figure 15.5). These shifts make the negative ion formed by dissociation better able to bear a negative charge because of increased delocalization.

CONNECTION Electronegativity (Section 8.3) increases as you go left-to-right across a row of the periodic table, and it decreases as you go down a column (see Figure 8.9).

A similar shift in electron density occurs in hydrated metal ions having the generic formula $M(H_2O)_6^{n+}$ when $n \geq 2$. The electrons in the O—H bonds of

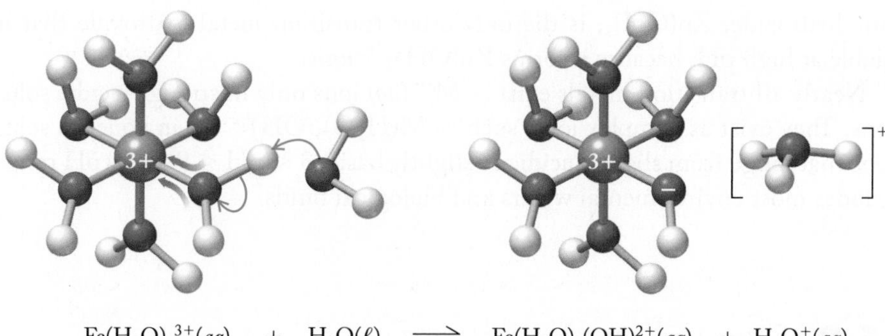

$$Fe(H_2O)_6^{3+}(aq) + H_2O(\ell) \rightleftharpoons Fe(H_2O)_5(OH)^{2+}(aq) + H_3O^+(aq)$$

FIGURE 16.21 A hydrated Fe^{3+} cation draws electron density away from the water molecules of its inner coordination sphere, which makes it possible for one or more of these molecules to donate a H^+ ion to a water molecule outside the sphere. The hydrated Fe^{3+} ion has one less H_2O ligand with an electrostatic attraction between the complex ion and the OH^- ion.

the water molecules surrounding the metal ions are attracted to the positively charged ions. The resulting distortion in electron density increases the likelihood of one of these O—H bonds ionizing and donating a H^+ ion to a neighboring molecule of water:

$$M(H_2O)_6^{n+}(aq) + H_2O(\ell) \rightleftharpoons M(H_2O)_5(OH)^{(n-1)+}(aq) + H_3O^+(aq)$$

Figure 16.21 provides a molecular view of this reaction, using $Fe^{3+}(aq)$ as the central ion. Similar reactions allow other hydrated metal ions, particularly those with charges of 3+, to function as Brønsted–Lowry acids. The K_a values of several metal ions are listed in **Table 16.1**. Note how much stronger the 3+ ions are than the 2+ ions, as we would expect given the greater electron-withdrawing power of the more highly charged central ions.

Figure 16.21 shows how one of the six water molecules of hydration surrounding a Fe^{3+} ion is converted into a hydroxide ion because of the acid ionization reaction. This reduces the charge of the complex ion from 3+ to 2+. If the pH of a solution of $Fe(H_2O)_5(OH)^{2+}$ ions is raised by adding a small quantity of a strong base such as NaOH, the ions undergo additional acid ionization reactions to form $Fe(H_2O)_4(OH)_2^+$:

$$Fe(H_2O)_5(OH)^{2+}(aq) + H_2O^-(aq) \rightleftharpoons Fe(H_2O)_4(OH)_2^+(aq) + H_2O(\ell)$$

and, at still higher pH, solid iron(III) hydroxide:

$$Fe(H_2O)_4(OH)_2^+(aq) + H_2O^-(aq) \rightleftharpoons Fe(H_2O)_3(OH)_3(s) + H_2O(\ell)$$

For simplicity, we usually write the formula of iron(III) hydroxide as $Fe(OH)_3(s)$, even though each formula unit also contains three water molecules of hydration.

TABLE 16.1 K_a Values of Hydrated Metal Ions

Ion	K_a
$Fe^{3+}(aq)$	3×10^{-3}
$Cr^{3+}(aq)$	1×10^{-4}
$Al^{3+}(aq)$	1×10^{-5}
$Cu^{2+}(aq)$	3×10^{-5}
$Pb^{2+}(aq)$	3×10^{-5}
$Zn^{2+}(aq)$	1×10^{-9}
$Co^{2+}(aq)$	2×10^{-10}
$Ni^{2+}(aq)$	1×10^{-10}

Acid strength (increases upward)

CONCEPT **TEST**

Would $Fe^{2+}(aq)$ be a stronger or weaker Lewis acid than $Fe^{3+}(aq)$? Where should the K_a of $Fe^{2+}(aq)$ be listed in Table 16.1 relative to that of $Fe^{3+}(aq)$?

Two other 3+ cations, Cr^{3+} and Al^{3+}, display similar behavior, but they differ from Fe^{3+} in that they are more soluble in strongly basic solutions than in weakly basic solutions. Why? Because solid $Cr(OH)_3$ and $Al(OH)_3$ may accept additional OH^- ions at high pH, forming soluble anionic complex ions:

$$Cr(OH)_3(s) + OH^-(aq) \rightleftharpoons Cr(OH)_4^-(aq) = Cr(H_2O)_2(OH)_4^-(aq)$$

$$Al(OH)_3(s) + OH^-(aq) \rightleftharpoons Al(OH)_4^-(aq) = Al(H_2O)_2(OH)_4^-(aq)$$

Zinc hydroxide, $Zn(OH)_2$, is the only other transition metal hydroxide that is soluble at high pH, because it forms $Zn(OH)_4^{2-}$ ions.

Nearly all transition metals exist as $M^{n+}(aq)$ ions only in strongly acidic solutions. They exist as complex ions, such as $M(H_2O)_5(OH)^{(n-1)+}$, in aqueous solutions that range from slightly acidic to slightly basic ($3 < pH < 9$). This pH range includes most environmental waters and biological fluids.

16.8 Solubility Equilibria

In Section 15.2 we noted that the common strong bases include the hydroxides of alkaline earth elements. One exception is $Mg(OH)_2$, which is a weak base because it has limited solubility in water. Magnesium hydroxide is the active ingredient in a product found in many medicine cabinets: the antacid called *milk of magnesia*. This liquid appears "milky" because it is an aqueous *suspension* (not solution) of solid, white $Mg(OH)_2$. We can express the limited solubility of solid $Mg(OH)_2$ with the following equation:

$$Mg(OH)_2(s) \rightleftharpoons Mg^{2+}(aq) + 2\,OH^-(aq)$$

Because $Mg(OH)_2$ is a solid, its effective concentration does not change as long as some of it is present in the system. Therefore, the equilibrium constant for the dissolution of $Mg(OH)_2$ is

$$K_{sp} = [Mg^{2+}][OH^-]^2$$

where $\mathbf{K_{sp}}$ represents an equilibrium constant called the *solubility-product constant* or simply the **solubility product.**

The K_{sp} values of $Mg(OH)_2$ and other slightly soluble compounds are listed in Table A5.4. We can use these values to calculate the concentrations of these compounds in aqueous solutions. Two terms are widely used to describe how much of a solid dissolves in a solvent: *solubility*, which is often expressed in grams of solute per 100 mL of solution, and *molar solubility*, which is expressed in moles of solute per liter of solution. In Sample Exercise 16.15 we use the K_{sp} of $Mg(OH)_2$ to calculate its solubility and molar solubility.

solubility product (K_{sp}) (also called *solubility-product constant*) an equilibrium constant that describes the formation of a saturated solution of a slightly soluble salt.

SAMPLE EXERCISE 16.15 Calculating the Solubility of an Ionic Compound from Its K_{sp} — LO8

What is the molar solubility of $Mg(OH)_2$ at 25°C, and what is its solubility expressed in grams $Mg(OH)_2$ per 100 mL of solution? The K_{sp} of $Mg(OH)_2$ from Table A5.4 is 5.6×10^{-12}.

Collect, Organize, and Analyze The RICE table below is based on the dissolution of $Mg(OH)_2$ in water:

$$Mg(OH)_2(s) \rightleftharpoons Mg^{2+}(aq) + 2\,OH^-(aq)$$

Converting molar solubility (x mol/L) into (g/100 mL) requires multiplying x by the molar mass of $Mg(OH)_2$, 58.32 g/mol, and factoring in the smaller volume (100 mL versus 1 L).

Solve

Reaction (R)	$Mg(OH)_2(s)$	$\rightleftharpoons$	$Mg^{2+}(aq)$	+	$2\ OH^-(aq)$
			$[Mg^{2+}]$ (*M*)		$[OH^+]$ (*M*)
Initial (I)			0		0
Change (C)			$+x$		$+2x$
Equilibrium (E)			x		$2x$

Inserting the equilibrium terms in the K_{sp} expression and solving for x:

$$K_{sp} = [Mg^{2+}][OH^-]^2 = (x)(2x)^2 = (x)(4x^2) = 4x^3 = 5.6 \times 10^{-12}$$

$$x = 1.1 \times 10^{-4}\ M$$

Converting this molar solubility into g/100 mL:

$$\frac{1.1 \times 10^{-4}\ \cancel{mol}}{\cancel{L}} \times \frac{58.32\ g}{1\ \cancel{mol}} \times \frac{1\ \cancel{L}}{1000\ \cancel{mL}} \times 100\ \cancel{mL} = 6.4 \times 10^{-4}\ g$$

Think About It Note the lack of concentration terms for $Mg(OH)_2(s)$ in the RICE table and its absence from the K_{sp} expression. Like all pure solids, its concentration has no relevance in equilibrium calculations as long as it is present in the reaction mixture. Also, the entire algebraic expression for $[OH^-]$, $2x$, is squared in this calculation: $(2x)^2 = 4x^2$. Forgetting to square the coefficient is a common mistake. Finally, the calculated molar solubility is much larger than the K_{sp} value. This difference is true for all sparingly soluble ionic compounds because K_{sp} values are the products of small concentration values multiplied together, producing even smaller K_{sp} values.

Practice Exercise What is the solubility of $Ca_3(PO_4)_2$ at 25°C expressed in mol/L and in g/100 mL? See Table A5.4 for the appropriate K_{sp} value.

CONCEPT TEST

In Sample Exercise 16.15 we calculated molar solubility from K_{sp}, but the reverse calculation is also possible. If the molar solubility of BaF_2 is $1.10 \times 10^{-3}\ M$, what is the K_{sp} for barium fluoride?

SAMPLE EXERCISE 16.16 Evaluating the Common-Ion Effect on Solubility — LO8

The mineral barite is mostly barium sulfate ($BaSO_4$) and is widely used in industry and in medical imaging. Calculate the molar solubility at 25°C of $BaSO_4$ in (a) pure water and (b) seawater in which the concentration of sulfate ions is 2.8 g/L.

Collect and Organize We want to calculate the molar solubility of $BaSO_4$ in both pure water and in seawater that already contains sulfate ions. The K_{sp} value of $BaSO_4$ given in Table A5.4 is 1.08×10^{-10}.

Analyze As each mole of $BaSO_4$ dissolves, 1 mole of Ba^{2+} ions and 1 mole of SO_4^{2-} ions go into solution:

$$BaSO_4(s) \rightleftharpoons Ba^{2+}(aq) + SO_4^{2-}(aq)$$

The presence of a background concentration of sulfate ions means that the common-ion effect will shift the above equilibrium to the left, which means the solubility of $BaSO_4$ we calculate for seawater should be less than its solubility in pure water.

Solve

a. Setting up a RICE table based on the above dissolution equilibrium in pure water:

Reaction (R)	$BaSO_4(s)$	$\rightleftharpoons$	$Ba^{2+}(aq)$	+	$SO_4^{2-}(aq)$
			$[Ba^{2+}]$ (M)		$[SO_4^{2-}]$ (M)
Initial (I)			0		0
Change (C)			$+x$		$+x$
Equilibrium (E)			x		x

$$K_{sp} = [Ba^{2+}][SO_4^{2-}] = (x)(x) = 1.08 \times 10^{-10}$$
$$x = 1.04 \times 10^{-5}\ M$$

b. The RICE table for seawater must include an initial concentration value for SO_4^{2-} ions, which requires converting the given concentration in g/L into mol/L:

$$[SO_4^{2-}]_{\text{initial}} = \frac{2.8\ \text{g}}{\text{L}} \times \frac{1\ \text{mol}}{96.06\ \text{g}} = \frac{0.029\ \text{mol}}{\text{L}}$$

Reaction (R)	$BaSO_4(s)$	$\rightleftharpoons$	$Ba^{2+}(aq)$	+	$SO_4^{2-}(aq)$
			$[Ba^{2+}]$ (M)		$[SO_4^{2-}]$ (M)
Initial (I)			0		0.0290
Change (C)			$+x$		$+x$
Equilibrium (E)			x		$(0.0290 + x)$

Incorporating the equilibrium terms into the K_{sp} expression:

$$K_{sp} = [Ba^{2+}][SO_4^{2-}] = (x)(0.029 + x) = 1.08 \times 10^{-10}$$

Solving for x is simplified if we assume that the K_{sp} of $BaSO_4$ is so small that we can ignore its contribution to the total SO_4^{2-} concentration. Therefore,

$$(x)(0.029 + x) \approx (x)(0.029) = 1.08 \times 10^{-10}$$
$$x = 3.7 \times 10^{-9}\ M$$

Think About It The calculated value of x is much less than 5% of the initial $[SO_4^{2-}]$ value: $(3.7 \times 10^{-9}/0.029 = 1.9 \times 10^{-7})$, so our simplifying assumption was justified. The solubility of $BaSO_4$ is much lower in seawater, as predicted, and is another illustration of the common-ion effect: the dissolution of $BaSO_4$ is suppressed by the SO_4^{2-} ions already present in seawater.

Practice Exercise What is the molar solubility of $MgCO_3$ in alkaline spring water at 25°C in which $[CO_3^{2-}] = 0.0075\ M$? See Table A5.4 for the appropriate K_{sp} value.

CONNECTION As noted in Section 14.8, the $-x$ or $+x$ component of an equilibrium concentration or partial pressure term can be ignored if x is less than 5% of the initial concentration of partial pressure.

Sample Exercise 16.16 shows how the common-ion effect can suppress the solubility of an ionic compound. Other perturbations to solubility equilibria can promote solubility, as occurs when the anion of the compound is the conjugate base of a weak acid. The molar solubilities of such compounds increase in acidic solutions, as shown in Sample Exercise 16.17.

SAMPLE EXERCISE 16.17 Calculating the Effect of pH on Solubility **LO8**

What is the molar solubility of CaF_2 at 25°C in (a) pure water and (b) an acidic buffer in which $[H_3O^+]$ is a constant 0.050 *M*?

Collect and Organize We are asked to calculate the solubility of CaF_2 in both pure water and in an acidic buffer. The dissolution process is described by the following equilibrium:

$$(1) \qquad CaF_2(s) \rightleftharpoons Ca^{2+}(aq) + 2\,F^-(aq) \qquad K_{sp} = 5.3 \times 10^{-9}$$

Analyze To account for the effect of acid on the solubility of a fluoride salt, we need to consider the chemical equilibrium in which the fluoride ion acts as a Brønsted–Lowry base (H^+ ion acceptor):

$$(2) \qquad H_3O^+(aq) + F^-(aq) \rightleftharpoons HF(aq) + H_2O(\ell)$$

This reaction is the reverse of the acid ionization reaction:

$$HF(aq) + H_2O(\ell) \rightleftharpoons H_3O^+(aq) + F^-(aq)$$

As a result, the equilibrium constant for reaction 2 is the reciprocal of the K_a (see Table A5.1) of HF:

$$K_2 = \frac{[HF]}{[H_3O^+][F^-]} = \frac{1}{6.8 \times 10^{-4}} = 1.47 \times 10^3$$

As reaction 2 proceeds, F^- ions are consumed, which shifts the equilibrium in reaction 1 to the right, increasing CaF_2 solubility. Therefore, the solubility of CaF_2 should increase when acid is present.

Solve

a. Let *x* be the molar solubility of CaF_2 in pure water. According to the stoichiometry of reaction 1, $[Ca^{2+}] = x$ mol/L and $[F^-] = 2x$ mol/L. Inserting these symbols in the K_{sp} expression:

$$K_{sp} = [Ca^{2+}][F^-]^2 = (x)(2x)^2 = 5.3 \times 10^{-9}$$

$$4x^3 = 5.3 \times 10^{-9}$$

$$x = 1.1 \times 10^{-3}\ M$$

b. In the acidic buffer, $[H_3O^+] = 0.050\ M$. The F^- and H_3O^+ ions combine as shown in reaction 2. Assuming $[H_3O^+]$ is a constant 0.050 *M*, then:

$$K_2 = \frac{[HF]}{[0.050][F^-]} = 1.47 \times 10^3$$

$$\frac{[HF]}{[F^-]} = 73.5$$

$$(3) \qquad [HF] = 73.5[F^-]$$

This calculation tells us that most of the F^- produced when calcium fluoride dissolves is converted into HF. The tiny fraction that remains free F^- ions is defined by this ratio:

$$(4) \qquad \frac{[F^-]}{[F^-] + [HF]}$$

Here the numerator is the concentration of free F^- ions at equilibrium and the denominator is the concentration of all the F^- ions produced when CaF_2 dissolved. Combining expressions 3 and 4 to calculate the fraction of dissolved fluoride ions that are free F^- ions and not molecules of HF:

$$\frac{[F^-]}{[F^-] + [HF]} = \frac{[F^-]}{[F^-] + 73.5[F^-]} = \frac{[F^-]}{74.5[F^-]} = 0.0134$$

If x is the molar solubility of CaF_2 in the acid, x mol/L Ca^{2+} and $2x$ mol/L F^- are produced. However, most of the fluoride ions are converted into HF, and the free F^- ion concentration is only $(0.0134 \times 2x) = 0.0268x$. Inserting this value in the K_{sp} expression:

$$K_{sp} = [Ca^{2+}][F^-]^2 = (x)(0.0268\,x)^2 = 7.18 \times 10^{-4}\,x^3 = 5.3 \times 10^{-9}$$

$$x = 1.9 \times 10^{-2}\,M$$

Think About It A comparison of the results from parts (a) and (b) reveals that the molar solubility of CaF_2 is $(1.9 \times 10^{-2}\,M/1.1 \times 10^{-3}\,M)$ or 17 times higher in the acidic buffer, as we predicted, because most of the F^- ions produced when CaF_2 dissolves in the buffer are converted into molecules of HF. This conversion removes a product from the K_{sp} equilibrium mixture. According to Le Châtelier's principle, the result will be a shift in the position of the equilibrium to the right, in favor of forming product and increasing solubility.

Practice Exercise What is the molar solubility of $ZnCO_3$ at 25°C in pure water and in a pH 7.00 buffer solution? Assume the pH of the buffer is unaffected by the presence of $ZnCO_3$. See Table A5.4 for the appropriate K_{sp} value.

CONCEPT TEST

Would adding a solution of sodium fluoride or nitric acid increase the solubility of PbF_2? Would either solution decrease the solubility of PbF_2? Which would have little effect?

According to Table A5.4, the hydroxides of many metals, including all transition metals, have very small K_{sp} values. However, these tiny K_{sp} values apply to equilibrium concentrations of *free metal ions*. As we have seen in this chapter, many metals form stable complex ions in aqueous solution, reducing their free metal ion concentration. What, for example, is the solubility of Al^{3+} ions in pH 7.00 water? The K_{sp} of $Al(OH)_3$ is

$$K_{sp} = [Al^{3+}][OH^-]^3 = 1.3 \times 10^{-33}$$

Solving the K_{sp} expression for $[Al^{3+}]$ and inserting $[OH^-] = 1.0 \times 10^{-7}$, we get

$$[Al^{3+}] = \frac{K_{sp}}{[OH^-]^3} = \frac{1.3 \times 10^{-33}}{(1.0 \times 10^{-7})^3} = 1.3 \times 10^{-12}\,M \qquad (16.8)$$

This very small value seems to imply that no aluminum salt is soluble in water because the $Al^{3+}(aq)$ ions that it releases as it dissolves would immediately precipitate as $Al(OH)_3$. However, that conclusion is incorrect. For example, $Al(NO_3)_3$, like all nitrate salts, is quite soluble in water. This solubility can be explained by the acidic properties of hydrated Al^{3+} ions, $Al(H_2O)_6{}^{3+}$, which make $Al(NO_3)_3$ an acidic salt. The pH of 0.1 M $Al(NO_3)_3$ is about 3.0. At this pH, $[OH^-] = 1 \times 10^{-11}$. Inserting this value in Equation 16.8 and solving for $[Al^{3+}]$, we get:

$$[Al^{3+}] = \frac{K_{sp}}{[OH^-]^3} = \frac{1.3 \times 10^{-33}}{(1.0 \times 10^{-11})^3} = 1.3\,M$$

The maximum value is more than 10 times the concentration of Al^{3+} ions in a 0.1 M solution of $Al(OH)_3$.

K_{sp} and Q

We can also use K_{sp} values to predict whether a concentration of an ionic compound is possible or whether a precipitate will form when the solutions of two salts are mixed. In making these predictions, it is convenient to use the concept of the reaction quotient Q that we developed in Chapter 14. When applied to the equilibrium governing a slightly soluble salt, Q is sometimes called the *ion product*, because it is the product of the concentrations of the ions in solution after each is raised to a power equal to its subscript in the formula of the compound. If the calculated Q value is greater than the K_{sp} of the compound ($Q > K_{sp}$), the reaction will favor reactant formation, and the compound will precipitate (or never dissolve in the first place). If $Q < K_{sp}$, the reaction will favor product formation, and the compound will be soluble and will not precipitate.

SAMPLE EXERCISE 16.18 Predicting Whether a Precipitate Forms When Two Solutions Are Mixed **LO8**

Lead(II) chloride is a white pigment used in 15th-century European painting. Will $PbCl_2$ precipitate when 275 mL of a 0.134 M solution of $Pb(NO_3)_2$ is added to 125 mL of a 0.0339 M solution of NaCl?

Collect and Organize We are asked whether $PbCl_2$ will precipitate when two solutions containing Pb^{2+} ions and Cl^- ions are mixed. The solubility product of $PbCl_2$ is

$$K_{sp} = [Pb^{2+}][Cl^-]^2 = 1.70 \times 10^{-5}$$

Analyze To determine whether a precipitate forms, we need to calculate Q and compare its value to K_{sp}. If $Q > K_{sp}$, $PbCl_2$ will precipitate; if $Q < K_{sp}$, it will not precipitate. Q has the same form as the equilibrium constant, $[Pb^{2+}][Cl^-]^2$, but the concentration values used in it are the values for a given system that may or may not be at equilibrium.

Solve First we calculate the concentrations of the lead ions and chloride ions in the two solutions immediately after they are mixed. Mixing the two solutions dilutes both, so the volumes of the solutions (275 mL and 125 mL) must be added to get the final solution volume (400 mL = 0.400 L).

$$Pb^{2+}(aq): \quad 0.134\ \frac{\text{mol}}{\cancel{\text{L}}} \times 0.275\ \cancel{\text{L}} = 0.03685\ \text{mol}$$

$$[Pb^{2+}] = \frac{0.03685}{0.400\ \text{L}} = 0.0921\ M$$

$$Cl^-(aq): \quad 0.0339\ \frac{\text{mol}}{\cancel{\text{L}}} \times 0.125\ \cancel{\text{L}} = 0.004238\ \text{mol}$$

$$[Cl^-] = \frac{0.004238}{0.400\ \text{L}} = 0.0106\ M$$

The value of Q is

$$Q = [Pb^{2+}][Cl^-]^2 = (0.0921)(0.0106)^2 = 1.03 \times 10^{-5}$$

Because $Q < K_{sp}$, no precipitate forms.

Think About It Lead(II) chloride was categorized as a *slightly soluble* compound in Table 4.5. In this scenario, $PbCl_2$ does not precipitate because the solutions of Pb^{2+} and Cl^- ions are too dilute to provide the concentrations required for the precipitate to form.

Practice Exercise Will calcium fluoride precipitate when 175 mL of a 4.78×10^{-3} M solution of $Ca(NO_3)_2$ is added to 135 mL of a 7.35×10^{-3} M solution of KF? See Table A5.4 for the appropriate K_{sp} value.

We can use differences in the solubilities of ionic compounds to selectively separate ions, particularly cations, in solution. For example, suppose an aqueous solution contains 0.10 *M* Ca^{2+} ion and 0.020 *M* Mg^{2+} ion. Is it possible to selectively remove the Mg^{2+} ions from solution by precipitating them as $Mg(OH)_2$ while leaving the Ca^{2+} ions in solution? This approach might work because $Mg(OH)_2$ is much less soluble than $Ca(OH)_2$, as indicated by their K_{sp} values:

$$K_{sp} = [Mg^{2+}][OH^-]^2 = 5.6 \times 10^{-12}$$

$$K_{sp} = [Ca^{2+}][OH^-]^2 = 5.5 \times 10^{-6}$$

One way to answer this ion separation question involves calculating the maximum concentration of OH^- ions that will *not* cause the 0.10 *M* Ca^{2+} ion to precipitate and then determining whether that concentration is high enough to precipitate all the Mg^{2+} ions. We can calculate the target $[OH^-]$ value from the K_{sp} of calcium hydroxide:

$$K_{sp} = 5.5 \times 10^{-6} = [Ca^{2+}][OH^-]^2 = (0.10)(x)^2$$

$$x = \sqrt{\frac{5.5 \times 10^{-6}}{0.10}} = 7.4 \times 10^{-3}\ M$$

Now we need to determine whether all the Mg^{2+} ions in solution would have precipitated as $Mg(OH)_2$ if $[OH^-]$ reached 7.4×10^{-3} *M*. Inserting this value in the K_{sp} expression for $Mg(OH)_2$ and solving for the $[Mg^{2+}]$ that would be in equilibrium with it:

$$K_{sp} = 5.6 \times 10^{-12} = [Mg^{2+}][OH^-]^2 = (x)(7.4 \times 10^{-3})^2$$

$$x = \frac{5.6 \times 10^{-12}}{(7.4 \times 10^{-3})^2} = 1.0 \times 10^{-7}\ M$$

The concentration of Mg^{2+} ions in the original solution is 0.020 *M*. If we assume that quantitative removal of the Mg^{2+} ions is achieved if their concentration drops to below 0.1% of the original value, or 2.0×10^{-5} *M*, then quantitative removal was achieved before $Ca(OH)_2$ precipitation began. This approach of precipitation of $Mg(OH)_2$ has been used to selectively separate these ions from seawater, where $[Ca^{2+}] = 0.0106$ *M* and $[Mg^{2+}] = 0.054$ *M*.

STEPWISE ANIMATION

Selective Precipitation

CONCEPT TEST

In evaluating the use of a precipitation reaction to separate magnesium ions from calcium ions in solution, we did not specify the concentration of the hydroxide ion solution used to form the precipitates. Why did we not need this value?

SAMPLE EXERCISE 16.19 Separating Anions in Solution — LO9

Both lead(II) chloride and lead(II) fluoride are slightly soluble salts. A solution of lead(II) nitrate is added to a solution that is 0.275 *M* in both $Cl^-(aq)$ and $F^-(aq)$. Can we use this method to separate the two halide ions? If "complete precipitation" is defined as there being less than 0.1% of a particular ion left in solution, is the precipitation of the first salt complete before the second salt begins to precipitate?

Collect and Organize We are given a solution that contains two ions that form slightly soluble lead(II) salts, and we are asked whether one ion can be completely removed before the second one starts to precipitate when lead(II) ion is added to the solution. We have the K_{sp} values for both salts and the initial concentrations of both ions.

Analyze The equilibrium constant expressions for both ions are

$$K_{sp} = [Pb^{2+}][Cl^-]^2 = 1.7 \times 10^{-5}$$

$$K_{sp} = [Pb^{2+}][F^-]^2 = 3.3 \times 10^{-8}$$

The K_{sp} of $PbCl_2$ is $(1.7 \times 10^{-5}/3.3 \times 10^{-8}) = 500$ times the K_{sp} of PbF_2. Therefore, PbF_2 should precipitate first when Pb^{2+} ions are added to a solution containing the same concentrations of F^- and Cl^- ions. We need to determine the maximum $[Pb^{2+}]$ that could be added to precipitate PbF_2 and not cause $PbCl_2$ to precipitate. When we determine that value, we can calculate $[F^-]$ remaining in solution to determine whether the precipitation of F^- as PbF_2 was complete.

Solve The maximum concentration of Pb^{2+} in the solution that will not cause the chloride ion to precipitate is

$$K_{sp} = 1.7 \times 10^{-5} = [Pb^{2+}][Cl^-]^2 = (x)(0.275)^2$$

$$x = \frac{1.7 \times 10^{-5}}{(0.275)^2} = 2.25 \times 10^{-4}\ M$$

The concentration of $F^-(aq)$ in the solution at this concentration of lead(II) ion is

$$K_{sp} = 3.3 \times 10^{-8} = (2.25 \times 10^{-4})(x)^2$$

$$x = \sqrt{\frac{3.3 \times 10^{-8}}{2.25 \times 10^{-4}}} = 0.0121\ M$$

The original solution was 0.275 M in F^- ions. A residual concentration of 0.0121 M F^- represents $(0.0121/0.275) \times 100\% = 4.4\%$ of the original $[F^-]$, which is greater than our 0.1% residual value that represents complete removal. Therefore, precipitation of $PbF_2(s)$ is *not* complete and we cannot use this method to separate the two ions.

Think About It This attempt at selective precipitation did not work because the ratio of the K_{sp} values was only 500 *and* because the halide concentration terms in the K_{sp} expressions were both squared. These squared concentration terms had the effect of producing a F^-/Cl^- ion ratio at equilibrium that was only the square root of their K_{sp} ratio: $(1/500)^{0.5} = 0.045$, or 4.5%.

Practice Exercise A water sample contains barium ions (0.0375 M) and calcium ions (0.0667 M). Can they be completely separated by selective precipitation of CaF_2? See Table A5.4 for the appropriate K_{sp} values.

We end Chapter 16 with Sample Exercise 16.20, which integrates acid–base and solubility equilibria by revisiting ocean acidification from Section 16.1. In this exercise we use the ionization equilibria of carbonic acid, as we have done in several other exercises, but this time we use different K_{a_1} and K_{a_2} values. Why? Because the K values we have used until now have all been *theoretical* values, which means they apply to ideal solutions in which each solute ion behaves as freely and independently as if it were the only ion present. Theoretical K values work best for very dilute solutions, but they don't perform as well for solutions as concentrated (salty) as seawater. They do not, for example, account for ion pair formation. Therefore, we use *apparent* K'_{a_1} and K'_{a_2} values at 25°C in Sample Exercise 16.20 that apply to chemical equilibria in typical seawater, which contains 35 g of dissolved sea salts per kilogram of seawater. Their symbols contain a prime (′) after the K to indicate that their values apply only to that particular sample.

CONNECTION Ion pair formation in 1.0 M solutions of ionic compounds significantly reduces the number of free ions in these solutions (see Section 11.5).

SAMPLE EXERCISE 16.20 Integrating Concepts: Evaluating the Impact of Ocean Acidification

As we discussed at the beginning of this chapter, increasing concentrations of atmospheric CO_2 have increased the concentration of CO_2 dissolved in the sea. As we saw in Equations 16.1–16.3, an increase in $[CO_2]$ will then shift the following equilibria to the right:

(1) $$H_2CO_3(aq) + H_2O(\ell) \rightleftharpoons HCO_3^-(aq) + H_3O^+(aq) \qquad pK'_{a_1} = 5.85$$

(2) $$HCO_3^-(aq) + H_2O(\ell) \rightleftharpoons CO_3^{2-}(aq) + H_3O^+(aq) \qquad pK'_{a_2} = 9.00$$

a. The average pH of the Pacific Ocean near Hawaii dropped from 8.12 to 8.07 between 1989 and 2014. By how much did the average acidity ($[H_3O^+]$) of the ocean increase? Express your answer as a percentage of the 1989 acidity.
b. Ocean pH is expected to drop by at least another 0.30 units by the end of this century. Assuming the concentrations of most other dissolved ions remain the same as today—that is, $[Ca^{2+}] = 0.0106\ M$ and $[HCO_3^-] + [CO_3^{2-}] = 0.00211\ M$—will the skeletal structures of marine organisms that are composed of $CaCO_3$ be soluble or insoluble in seawater in 2100? The K'_{sp} value for aragonite (the crystalline form of $CaCO_3$ in corals and seashells) is 6.46×10^{-7}.

Collect and Organize We are asked to convert a decrease in pH into an increase in acidity and to evaluate the impact of that increase on the solubility of $CaCO_3$. We know the concentration of Ca^{2+} ions and the total ($[CO_3^{2-}] + [HCO_3^-]$) value, as well as the appropriate K'_{sp} value. The last two ions are a conjugate acid–base pair whose concentrations are linked by reaction (2) and the Henderson–Hasselbalch equation based on it:

$$pH = pK_a + \log\frac{[\text{base}]}{[\text{acid}]} = 9.00 + \log\frac{[CO_3^{2-}]}{[HCO_3^-]}$$

Analyze The ratio of the acidity of the seawater near Hawaii in 2014 compared with that in 1989 can be calculated by applying the definition $[H_3O^+] = 10^{-pH}$. To determine whether $CaCO_3$ dissolves in seawater in 2100, we need to calculate the $[CO_3^{2-}]/[HCO_3^-]$ ratio in equilibrium with seawater at pH (8.07 − 0.30 = 7.77) and then use that ratio and the total carbonate and bicarbonate value to calculate $[CO_3^{2-}]$. The product of that value and $[Ca^{2+}]$ will give us a Q value that can be compared with the K'_{sp} of $CaCO_3$ to determine whether seawater will be saturated with $CaCO_3$.

Solve

a. Converting the decrease in pH values to an increase in $[H_3O^+]$, expressed as a ratio of the 2014 value to the 1989 value:

$$\frac{[H^+]_{2014}}{[H^+]_{1989}} = \frac{10^{-8.07}\ M}{10^{-8.12}\ M} = 10^{0.05} = 1.12$$

Therefore, the acidity of the 2014 sample is 1.12 times the acidity of the 1989 sample, which means acidity increased 12% during the 25 years before 2014.

b. The $[CO_3^{2-}]/[HCO_3^-]$ ratio in 2100 will be

$$7.77 = 9.00 + \log\frac{[CO_3^{2-}]}{[HCO_3^-]}$$

$$\frac{[CO_3^{2-}]}{[HCO_3^-]} = 0.0589$$

If we let x be $[CO_3^{2-}]$, then

$$\frac{x}{(0.00211 - x)} = 0.0589$$

$$x = 1.17 \times 10^{-4}\ M$$

Calculating Q for $CaCO_3$:

$$Q = [Ca^{2+}][CO_3^{2-}] = 0.0106 \times 1.17 \times 10^{-4} = 1.24 \times 10^{-6}$$

This value of Q is $(1.24 \times 10^{-6})/(6.46 \times 10^{-7}) = 1.9$ times the value of K'_{sp}, which means seawater will still be supersaturated with respect to $CaCO_3$.

Think About It The results of the calculations suggest that coral exoskeletons and seashells should still be able to form in pH 7.77 seawater because it will be supersaturated with respect to $CaCO_3$. However, the predicted degree of supersaturation (essentially 100%) in 2100 will have decreased from 400% in 1989 [you may repeat the part (b) calculation, using pH 8.12 to confirm this value]. Moreover, some climate change models predict even greater decreases in oceanic pH by 2100. There appears to be reason for concern.

SUMMARY

LO1 As predicted by Le Châtelier's principle, adding conjugate base to a solution of a weak acid inhibits ionization of the acid, causing the pH of the solution to be greater than in the acid alone. Adding conjugate acid to a solution of a weak base lowers the pH of the solution. These shifts are examples of the **common-ion effect**. (Section 16.2)

LO2 A **pH buffer** is a solution that contains either a weak acid and a salt of its conjugate base or a weak base and a salt of its conjugate acid. The acidic component should have a pK_a close to the desired pH of the buffer. (Section 16.3)

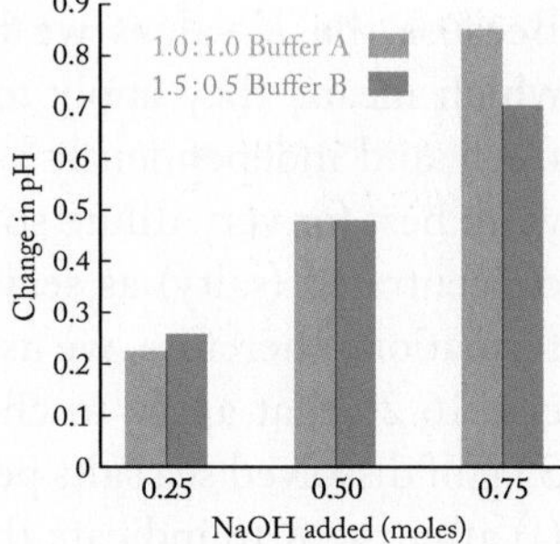

LO3 pH buffers have the capacity to resist pH change because their acid components neutralize additions of bases and their base components neutralize additions of acids. The pK_a of the buffer's acid component should be within 1 pH unit of the buffer's target pH. (Section 16.3)

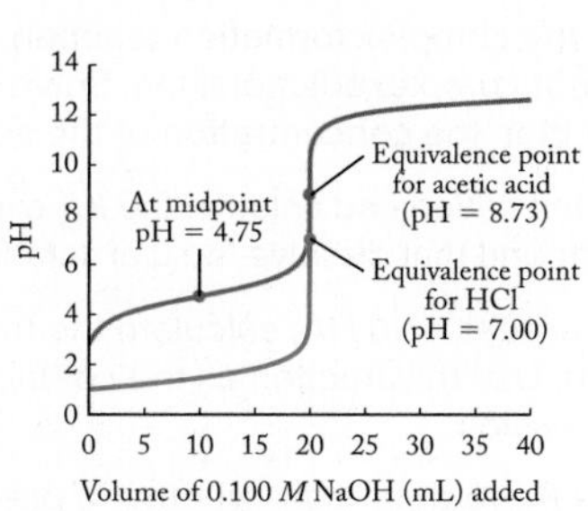

LO4 Color **pH indicators** or pH electrodes are used to detect the equivalence points in pH titrations, which are used to determine the concentrations of acids or bases in aqueous samples. (Section 16.4)

LO5 A **Lewis base** is a substance that donates pairs of electrons to a **Lewis acid**, defined as an electron-pair acceptor. The donated electron pair forms a **coordinate bond**. In some Lewis acid–Lewis base reactions, other bonds must break to accommodate the new one. (Sections 16.5 and 16.6)

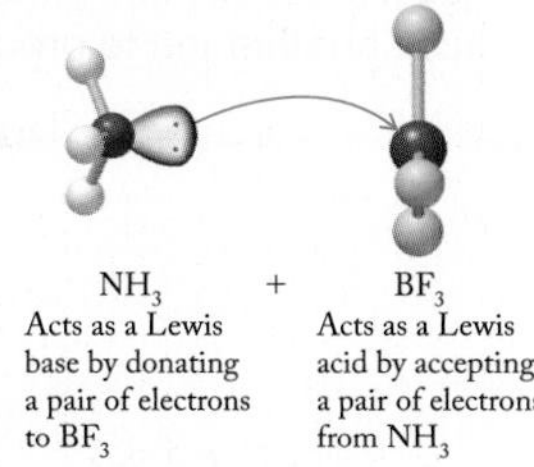

LO6 Coordinate covalent bonds are formed by ligands sharing lone pairs of electrons with empty orbitals of metal ions. The stability of the **complex ions** formed in this way is expressed mathematically by their **formation constant (K_f)**, which can be used to calculate the equilibrium concentration of free metal ions in a solution of complex ions. (Section 16.6)

LO7 Highly charged (e.g., 3+) hydrated metal ions are weak acids because of the ionization of water molecules covalently bonded to them. (Section 16.7)

LO8 The solubility of slightly soluble ionic compounds is described by their K_{sp}, or **solubility product**. Their solubility can be influenced by the common-ion effect, complex ion formation, and pH, especially if the anion is the conjugate base of a weak acid. (Section 16.8)

LO9 The relative solubilities of two slightly soluble ionic compounds can be used to selectively precipitate one from solution while the other remains soluble. (Section 16.8)

PARTICULATE PREVIEW WRAP-UP

The proton transfer reactions are described in the following reaction equation:

$$HF(aq) + H_2O(\ell) \rightleftharpoons F^-(aq) + H_3O^+(aq)$$

The only source of fluoride ions in (a) is HF molecules that ionize as the reaction proceeds more rapidly in the forward direction until equilibrium is achieved. In (b) there is a second source of fluoride ions in the reaction mixture, which gives it the capacity to act as a pH buffer: more fluoride ions give it the capacity to neutralize additions of acid, as the above reaction runs in reverse, and the HF component neutralizes additions of base, forming F^- ions and H_2O:

$$HF(aq) + OH^-(aq) \rightleftharpoons F^-(aq) + H_2O(\ell)$$

PROBLEM-SOLVING SUMMARY

Type of Problem	Concepts and Equations	Sample Exercises
Calculating pH of a solution of a weak base and its conjugate acid (or a weak acid and its conjugate base)	Insert the concentrations of the base and acid components and the acid's pK_a in the Henderson–Hasselbalch equation: $pH = pK_a + \log \frac{[base]}{[acid]}$ (16.4)	**16.1, 16.2**
Preparing a buffer of given pH	Select a weak acid with a pK_a within 1 pH unit of the target pH. Add a Na^+ salt of its conjugate base in a proportion calculated using the Henderson–Hasselbalch equation.	**16.3, 16.4, 16.5**
Evaluating the effect of concentration and [base]:[acid] ratio on buffer capacity	Assume additions of strong acid or base react completely with the base or acid components of the buffer. Calculate pH from the concentrations of the components that remain by using the Henderson–Hasselbalch equation.	**16.6, 16.7, 16.8**
Interpreting results of acid–base titrations	Use the volume and molarity of the titrant needed to reach the equivalence point to calculate the number of moles of it consumed by the analyte, and, from the stoichiometry of the titration reaction, the number of moles of analyte in the sample.	**16.9, 16.10, 16.11, 16.12**

Type of Problem	Concepts and Equations	Sample Exercises
Identifying Lewis acids and bases	Lewis bases donate pairs of electrons, whereas Lewis acids accept these pairs of electrons.	**16.13**
Using formation constants to calculate the concentration of free or complexed ion	Set up a RICE table based on the complex formation reaction. Let *x* be the concentration of free (not complexed) metal ion. Solve for *x*, which is usually much smaller than the concentration of the complex.	**16.14**
Calculating the solubility of an ionic compound from its K_{sp}	Express the concentrations of the cation and anion in the K_{sp} expression in terms of *x* moles of the compound that dissolve in 1 L of solution.	**16.15, 16.16**
Calculating the effect of pH on solubility	If A^- is the conjugate base of a weak acid HA, calculate the fraction of A^- that remains as the free ion. Use this fraction as the coefficient for molar solubility in the K_{sp} expression.	**16.17**
Determining whether a precipitate forms when solutions are mixed, and which precipitate forms first if more than one is possible	Compare the ion product *Q* to K_{sp} to determine whether a precipitate will form; use K_{sp} expressions to calculate maximum concentrations of one ion in solution that will not cause another ion to precipitate.	**16.18, 16.19**

VISUAL PROBLEMS

(Answers to boldface end-of-chapter questions and problems are in the back of the book.)

16.1. The graph in Figure P16.1 shows the titration curves of a 1 *M* solution of a weak acid with a strong base and a 1 *M* solution of a strong acid with the same base. Which curve is which?

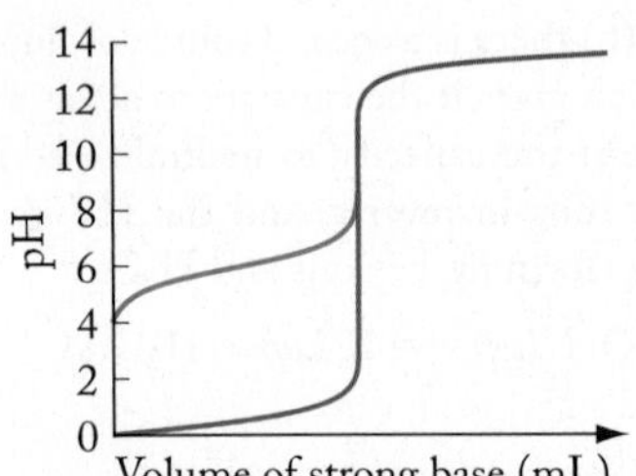

FIGURE P16.1

16.2. Estimate (to two significant figures) the pK_a of the weak acid in Problem 16.1.

16.3. Suppose you have four color indicators to choose from to detect the equivalence point of the titration reaction represented by the red curve in Figure P16.1. The pK_a values of the four indicators are 3.3, 5.0, 7.0, and 9.0. Which indicator would be the best one to choose?

16.4. Explain why the slope of the red titration curve in Figure P16.1 is nearly flat in the region extending about halfway from the start of the titration to its equivalence point.

16.5. One of the titration curves in Figure P16.5 represents the titration of an aqueous sample of Na_2CO_3 with strong acid; the other represents the titration of an aqueous sample of $NaHCO_3$ with the same acid. Which curve is which?

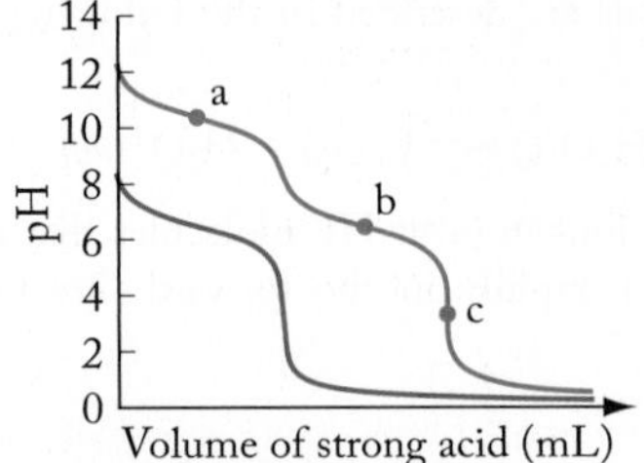

FIGURE P16.5

16.6. Identify the principal carbon-containing species in solution at points a, b, and c on the red titration curve in Figure P16.5.

16.7. Each of the three beakers in Figure P16.7 contains a few drops of the color indicator bromthymol blue, which is yellow in acidic solutions and blue in basic solutions. One beaker contains a solution of ammonium chloride, one contains ammonium acetate, and the third contains sodium acetate. Which beaker contains which salt?

FIGURE P16.7

*16.8. The graphs in Figure P16.8 show the conductivity of a solution as a function of the volume of titrant added. Which of the graphs best represents the titration of (a) a strong acid with a strong base and (b) a weak acid with a strong base?

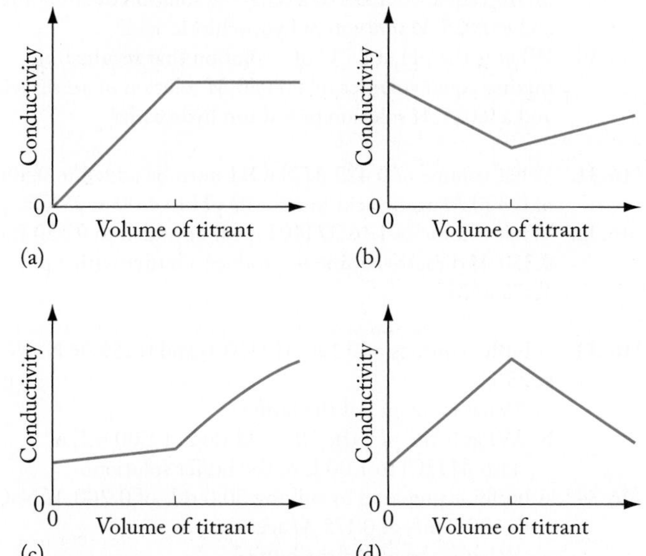

FIGURE P16.8

16.9. (a) What kind of aqueous solution is represented in Figure P16.9: a weak acid, a weak base, or a buffer? The solvent molecules and cations have been omitted for clarity. (b) What particles depicted in Figure P16.9 will change, and how will they change, upon addition of a strong base such as NaOH?

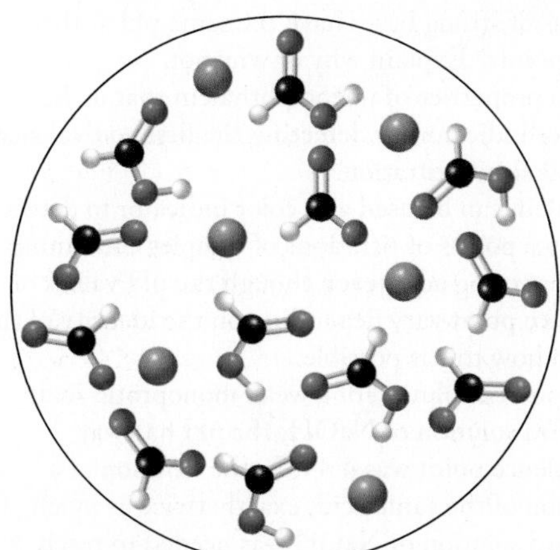

FIGURE P16.9

16.10. Use representations [A] through [I] in Figure P16.10 to answer questions (a)–(f).

a. Images [C] and [G] depict the different molecular structures of the indicator methyl red at high and low pH. Which is which?
b. If methyl red is red at low pH and yellow at higher pH, match the flasks in images [A] and [I] to the species in [C] and [G].
c. Which image represents a buffer with equal concentrations of a weak acid and its conjugate base?
d. Which image represents the buffer from your answer to part (c) after the addition of strong acid, and which represents that buffer after the addition of strong base?
e. Saturated solutions of two sparingly soluble salts, calcium sulfide and calcium fluoride, are shown in [D] and [F]. Which is which?
f. Will acidifying the solution in [D] to pH 2.00 by adding strong acid increase the solubility of the solid? Will adding the same amount of acid to [F] increase the solubility of the solid?

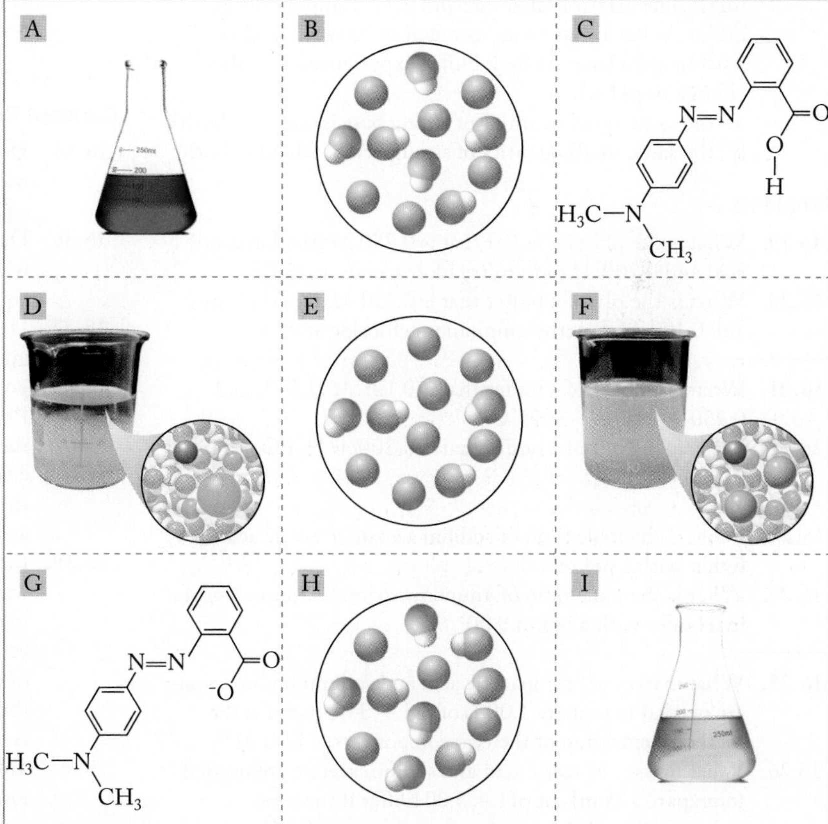

FIGURE P16.10

QUESTIONS AND PROBLEMS

Note: Tables A5.1 and A5.3 in Appendix 5 contain K_a, K_b, pK_a, pK_b, and K_f values that may be useful in answering the questions and solving the problems in Chapter 16.

The Common-Ion Effect; pH Buffers

Concept Review

16.11. Why does a solution of a weak acid and its conjugate base control pH better than a solution of the weak acid alone?

16.12. Why does a solution of a weak base and its conjugate acid control pH better than a solution of the weak base alone?

16.13. Identify a suitable buffer system to maintain a pH of 3.00 in an aqueous solution.

16.14. Identify a suitable buffer system to maintain a pH of 9.0 in an aqueous solution.

16.15. What does "buffer capacity" mean?

16.16. Three buffers are prepared using equal concentrations of formic acid and sodium formate, hydrofluoric acid and sodium fluoride, and acetic acid and sodium acetate. Rank the three buffers from highest to lowest pH.

16.17. How does diluting a pH 9.00 buffer with an equal volume of pure water affect its pH?

*16.18. Buffer A contains nearly equal concentrations of its conjugate acid–base pair. Buffer B contains the same total concentration of acidic and basic components as buffer A, but B has twice as much of its weak acid as its conjugate base. Which buffer experiences a smaller change in pH when:

a. the same small quantity of strong base is added to both?
b. the same small quantity of strong acid is added to both?

Problems

16.19. What is the pH of a buffer that is 0.250 *M* trichloroacetic acid and 0.200 *M* sodium trichloroacetate at 25°C?

16.20. What is the pH of a buffer that is 0.150 *M* dimethylamine and 0.100 *M* dimethylammonium chloride at 25°C?

16.21. What is the pH of a buffer that is 0.125 *M* H_3PO_4 and 0.250 *M* $H_2PO_4^-$ at 25°C?

16.22. What is the pH of a buffer that is 0.200 *M* $H_2PO_4^-$ and 0.250 *M* HPO_4^{2-} at 25°C?

16.23. What is the mole ratio of sodium acetate to acetic acid in a buffer with a pH of 4.00?

16.24. What is the mole ratio of ammonia to ammonium chloride in a buffer with a pH of 9.00?

16.25. What masses of bromoacetic acid and sodium bromoacetate are needed to prepare 1.00 L of pH = 3.00 buffer if the total concentration of the two components is 1.00 *M*?

16.26. What masses of acetic acid and sodium acetate are needed to prepare 125 mL of pH = 5.00 buffer if the total concentration of the two components is 0.500 *M*?

16.27. What masses of dimethylamine and dimethylammonium chloride do you need to prepare 1.000 L of pH = 12.00 buffer if the total concentration of the two components is 0.500 *M*?

16.28. What masses of ethylamine and ethylammonium chloride do you need to prepare 1.000 L of pH = 11.00 buffer if the total concentration of the two components is 0.500 *M*?

16.29. What is the pH at 25°C of a solution that results from mixing equal volumes of a 0.050 *M* solution of ammonia and a 0.025 *M* solution of hydrochloric acid?

16.30. What is the pH at 25°C of a solution that results from mixing equal volumes of a 0.050 *M* solution of acetic acid and a 0.025 *M* solution of sodium hydroxide?

***16.31.** What volume of 0.422 *M* NaOH must be added to 0.500 L of 0.300 *M* acetic acid to raise its pH to 4.00 at 25°C?

*16.32. What volume of 1.16 *M* HCl must be added to 0.250 L of 0.350 *M* dimethylamine to produce a buffer with a pH of 10.75 at 25°C?

***16.33.** A buffer consists of 0.120 *M* HNO_2 and 0.150 *M* $NaNO_2$ at 25°C.

a. What is the pH of the buffer?
b. What is the pH after the addition of 1.00 mL of 11.6 *M* HCl to 1.00 L of the buffer solution?

*16.34. A buffer is prepared by mixing 50.0 mL of 0.200 *M* NaOH with 100.0 mL of 0.175 *M* acetic acid.

a. What is the pH of the buffer?
b. What is the pH of the buffer after 1.00 g of NaOH is dissolved in it?

Indicators and Acid–Base Titrations

Concept Review

16.35. Do all titrations of samples of strong monoprotic acids with solutions of strong bases have the same pH at their equivalence points? Explain why or why not.

16.36. Do all titrations of samples of weak monoprotic acids with solutions of strong bases have the same pH at their equivalence points? Explain why or why not.

16.37. Describe two properties of phenolphthalein that make it a good choice of indicator for detecting the first equivalence point in an alkalinity titration.

*16.38. Phenolphthalein can be used as a color indicator to detect the equivalence points of titrations of samples containing either weak or strong acids even though the pH values of the equivalence point vary depending on the identity of the acid. Explain how this is possible.

16.39. In the titration of a solution of a weak monoprotic acid with a standard solution of NaOH, the pH halfway to the equivalence point was 4.44. In the titration of a second solution of the same acid, exactly twice as much of the standard solution of NaOH was needed to reach the equivalence point. What was the pH halfway to the equivalence point in this titration?

16.40. The pH of a solution of a strong monoprotic acid is lower than the pH of an equal concentration of a weak monoprotic acid, yet equal volumes of both require the same volume of basic titrant to reach the equivalence point. Explain why.

Problems

16.41. A 25.0 mL sample of 0.100 *M* acetic acid is titrated with 0.125 *M* NaOH at 25°C. What is the pH of the solution after 10.0, 15.0, 20.0, 25.0, and 35.0 mL of the base have been added?

16.42. A 25.0 mL sample of a 0.100 *M* solution of aqueous trimethylamine is titrated with a 0.125 *M* solution of HCl. What is the pH of the solution after 10.0, 15.0, 20.0, 25.0, and 30.0 mL of acid have been added?

***16.43.** **Window Cleaner** (a) What is the concentration of ammonia in a popular window cleaner if 25.34 mL of 1.162 *M* HCl is needed to titrate a 10.00 mL sample of the cleaner? (b) Suppose that the sample was diluted to about 50 mL with deionized water prior to the titration to make it easier to mount a pH electrode in it. What effect did this dilution have on the volume of titrant needed?

*16.44. In an alkalinity titration of a 100.0 mL sample of water from a hot spring, 2.56 mL of a 0.0355 *M* solution of HCl is needed to reach the first equivalence point (pH = 8.3) and another 10.42 mL is needed to reach the second equivalence point (pH = 4.0). If the alkalinity of the spring water results only from the presence of carbonate and bicarbonate, what are the concentrations of each?

16.45. Sketch a titration curve for the titration of 50.0 mL of 0.250 *M* HNO_2 with 1.00 *M* NaOH. What is the pH at the equivalence point?

16.46. Sketch a titration curve for the titration of 40.0 mL of a 0.100 *M* solution of oxalic acid with a 0.100 *M* solution of NaOH. What is the pH of the titration reaction mixture at the last equivalence point?

16.47. For each titration, predict whether the equivalence point is less than, equal to, or greater than pH = 7.

a. Quinine titrated with nitric acid
b. Pyruvic acid titrated with calcium hydroxide
c. Hydrobromic acid titrated with strontium hydroxide

16.48. For each titration, predict whether the equivalence point is less than, equal to, or greater than pH = 7.

a. HCN titrated with $Ca(OH)_2$
b. LiOH titrated with HI
c. C_5H_5N titrated with KOH

16.49. When 100 mL of 0.0125 *M* ascorbic acid is titrated with 0.010 *M* NaOH, how many equivalence points will the titration curve have, and what pH indicator(s) could be used? Refer to Figure 16.8 for colors of indicators.

16.50. Red cabbage juice is a sensitive acid–base indicator; its colors range from red at acidic pH to yellow in alkaline solutions. What color would red cabbage juice have at the equivalence point when 25 mL of a 0.10 *M* solution of acetic acid is titrated with 0.10 *M* NaOH?

Lewis Acids and Bases

Concept Review

16.51. Are all Lewis bases also Brønsted–Lowry bases? Explain why or why not.

16.52. Are all Brønsted–Lowry bases also Lewis bases? Explain why or why not.

16.53. Are all Brønsted–Lowry acids also Lewis acids? Explain why or why not.

16.54. Why is BF_3 a Lewis acid but not a Brønsted–Lowry acid?

Problems

16.55. Draw Lewis structures that show how electron pairs move and bonds form and break during the autoionization of water. Label the appropriate H_2O molecules as the Lewis acid and Lewis base.

16.56. Draw Lewis structures that show how electron pairs move and bonds form and break in the following reaction and identify the Lewis acid and Lewis base.

$$MgO(s) + CO_2(g) \rightarrow MgCO_3(s)$$

16.57. Draw Lewis structures that show how electron pairs move and bonds form and break in the following reaction and identify the Lewis acid and Lewis base.

$$SO_2(g) + H_2O(\ell) \rightarrow H_2SO_3(aq)$$

16.58. Draw Lewis structures that show how electron pairs move and bonds form and break in the following reaction and identify the Lewis acid and Lewis base.

$$SeO_3(g) + H_2O(\ell) \rightarrow H_2SeO_4(aq)$$

16.59. Draw Lewis structures that show how electron pairs move and bonds form and break in the following reaction and identify the Lewis acid and Lewis base.

$$B(OH)_3(aq) + H_2O(\ell) \rightleftharpoons B(OH)_4^-(aq) + H^+(aq)$$

*16.60. Draw Lewis structures that show how electron pairs move and bonds form and break in the following reaction and identify the Lewis acid and Lewis base. (*Note*: $HSbF_6$ is an ionic compound and one of the strongest Brønsted–Lowry acids known.)

$$SbF_5(s) + HF(g) \rightarrow HSbF_6(s)$$

Formation of Complex Ions

Concept Review

16.61. When $CaCl_2$ dissolves in water, which molecules or ions occupy the inner coordination sphere around the Ca^{2+} ions?

16.62. When $AgNO_3$ dissolves in water, which molecules or ions occupy the inner coordination sphere around the Ag^+ ions?

***16.63.** A lab technician cleaning glassware that contains residues of AgCl washes the glassware with an aqueous solution of ammonia. The AgCl, which is insoluble in water, rapidly dissolves in the ammonia solution. Why?

*16.64. The procedure used in Problem 16.63 dissolves AgCl but not AgI. Why?

Problems

16.65. A solution is prepared in which 0.00100 mol of $Ni(NO_3)_2$ and 0.500 mol of NH_3 are dissolved in a total volume of 1.00 L. What is the concentration of $Ni(H_2O)_6^{2+}$ ions in the solution at equilibrium?

16.66. A 1.00 L solution contains 5.00×10^{-5} *M* $Cu(NO_3)_2$ and 1.00×10^{-3} *M* ethylenediamine, $H_2NCH_2CH_2NH_2$. What is the concentration of $Cu(H_2O)_6^{2+}$ ions in the solution at equilibrium?

***16.67.** Suppose a solution contains 1.00 mmol of $Co(NO_3)_2$, 0.100 mol of NH_3, and 0.100 mol of ethylenediamine in a total volume of 0.250 L. What is the concentration of $Co(H_2O)_6{}^{2+}$ ions in the solution?

*16.68. If 1.00 mL of 0.0100 *M* $AgNO_3$, 1.00 mL of 0.100 *M* NaBr, and 1.00 mL of 0.100 *M* NaCN are diluted to 250 mL with deionized water in a volumetric flask and shaken vigorously, will the contents of the flask be cloudy or clear? Support your answer with the appropriate calculations. (*Hint*: The K_{sp} of AgBr is 5.4×10^{-13}.)

Hydrated Metal Ions as Acids

Concept Review

16.69. Which, if any, aqueous solutions of the following chloride compounds are acidic? (a) $CaCl_2$; (b) $CrCl_3$; (c) NaCl; (d) $FeCl_3$

16.70. If 0.100 *M* aqueous solutions of each of these compounds were prepared, which one would have the lowest pH? (a) $BaCl_2$; (b) LiCl; (c) KCl; (d) $TiCl_4$

16.71. When ozone is bubbled through an aqueous solution of Fe^{2+} ions, the ions are oxidized to Fe^{3+} ions. How does the oxidation process affect the pH of the solution?

16.72. As an aqueous solution of KOH is slowly added to a stirred solution of $AlCl_3$, the mixture becomes cloudy but then clears when more KOH is added.
  a. Explain the chemical changes responsible for the changes in the appearance of the mixture.
  b. Would you expect to observe the same changes if KOH were added to a solution of $FeCl_3$? Explain why or why not.

16.73. Chromium(III) hydroxide is amphiprotic. Write chemical equations showing how an aqueous suspension of this compound reacts to the addition of a strong acid and a strong base.

16.74. Zinc hydroxide is amphiprotic. Write chemical equations showing how an aqueous suspension of this compound reacts to the addition of a strong acid and a strong base.

16.75. **Refining Aluminum** To remove impurities such as calcium and magnesium carbonates and iron(III) oxides from aluminum ore (which is mostly Al_2O_3), the ore is treated with a strongly basic solution. In this treatment, Al^{3+} dissolves but the other metal ions do not. Why?

*16.76. Exactly 1.00 g of $FeCl_3$ is dissolved in each of four 0.500 L samples: 1 *M* HNO_3, 1 *M* HNO_2, 1 *M* CH_3COOH, and pure water. Is the concentration of $Fe(H_2O)_6{}^{3+}$ ions the same in all four solutions? Explain why or why not.

Problems

16.77. What is the pH of 0.25 *M* $Al(NO_3)_3$?

16.78. What is the pH of 0.50 *M* $CrCl_3$?

16.79. What is the pH of 0.100 *M* $Fe(NO_3)_3$?

16.80. What is the pH of 1.00 *M* $Cu(NO_3)_2$?

16.81. Sketch the titration curve (pH versus volume of 0.50 *M* NaOH) for a 25 mL sample of 0.25 *M* $FeCl_3$.

16.82. Sketch the titration curve that results from the addition of 0.50 *M* NaOH to a sample containing 0.25 *M* $KFe(SO_4)_2$.

Solubility Equilibria

Concept Review

16.83. What is the difference between *molar solubility* and *solubility product*?

16.84. Give an example of how the common-ion effect limits the dissolution of a sparingly soluble ionic compound.

16.85. Which cation will precipitate first as a carbonate mineral from an equimolar solution of Mg^{2+}, Ca^{2+}, and Sr^{2+}?

16.86. If the solubility of a compound increases with increasing temperature, does K_{sp} increase or decrease?

16.87. The K_{sp} of strontium sulfate increases from 2.8×10^{-7} at 37°C to 3.8×10^{-7} at 77°C. Is the dissolution of strontium sulfate endothermic or exothermic?

16.88. Identify any of the following solids that are more soluble in acidic solution than in neutral water: $CaCl_2$, $Ba(HCO_3)_2$, $PbSO_4$, $Cu(OH)_2$. Explain your choices.

16.89. **Chemistry of Tooth Decay** Tooth enamel is composed of a mineral known as hydroxyapatite, which has the formula $Ca_5(PO_4)_3(OH)$. Explain why tooth enamel can be eroded by acidic substances released by bacteria growing in the mouth.

16.90. **Fluoride and Dental Hygiene** Fluoride ions in drinking water and toothpaste convert hydroxyapatite in tooth enamel into fluorapatite:

$$Ca_5(PO_4)_3(OH)(s) + F^-(aq) \rightleftharpoons Ca_5(PO_4)_3F(s) + OH^-(aq)$$

Why is fluorapatite less susceptible than hydroxyapatite to erosion by acids?

Problems

16.91. At a particular temperature the $[Ba^{2+}]$ in a saturated solution of barium sulfate is 1.04×10^{-5} *M*. Starting with this information, calculate the K_{sp} value of barium sulfate at this temperature.

16.92. If only 0.160 g of $Ca(OH)_2$ dissolves in 0.100 L of water, what is the K_{sp} value for calcium hydroxide at that temperature?

16.93. What are the equilibrium concentrations of Cu^+ and Cl^- in a saturated solution of copper(I) chloride at 25°C?

16.94. What are the equilibrium concentrations of Pb^{2+} and F^- in a saturated solution of lead fluoride at 25°C?

16.95. What is the pH at 25°C of a saturated solution of silver hydroxide?

16.96. **pH of Milk of Magnesia** What is the pH at 25°C of a saturated solution of magnesium hydroxide (the active ingredient in the antacid milk of magnesia)?

16.97. Suppose you have 100 mL of each of the following solutions. In which will the most $CaCO_3$ dissolve? (a) 0.1 *M* NaCl; (b) 0.1 *M* Na_2CO_3; (c) 0.1 *M* NaOH; (d) 0.1 *M* HCl

16.98. In which of the following solutions will CaF_2 be most soluble? (a) 0.010 *M* $Ca(NO_3)_2$; (b) 0.01 *M* NaF; (c) 0.001 *M* NaF; (d) 0.10 *M* $Ca(NO_3)_2$

16.99. Composition of Seawater The average concentration of sulfate in surface seawater is about 0.028 *M*. The average concentration of Sr^{2+} is 9×10^{-5} *M*. Is the concentration of strontium in the sea significantly controlled by the insolubility of its sulfate salt?

16.100. **Fertilizing the Sea to Combat Climate Change** Some scientists have proposed adding iron(III) compounds to large expanses of the open ocean to promote the growth of phytoplankton that would in turn remove CO_2 from the atmosphere through photosynthesis. Assuming the average pH of open ocean water is 8.13, what is the maximum value of $[Fe^{3+}]$ in seawater if the K_{sp} value of $Fe(OH)_3$ is 1.1×10^{-36}?

16.101. Will calcium fluoride precipitate when 125 mL of 0.375 *M* $Ca(NO_3)_2$ is added to 245 mL of 0.255 *M* NaF at 25°C?

16.102. Will lead(II) chloride precipitate if 185 mL of 0.025 *M* sodium chloride is added to 235 mL of 0.165 *M* lead(II) perchlorate at 25°C?

16.103. A solution is 0.010 *M* in both Br^- and SO_4^{2-}. A 0.250 *M* solution of lead(II) nitrate is slowly added to it with a buret.

a. Which anion will precipitate first?
b. What is the concentration in the solution of the first ion when the second one starts to precipitate at 25°C?

*16.104. Solution A is 0.0200 *M* in Ag^+ ions and Pb^{2+} ions. You have access to two other solutions: (B) 0.250 *M* NaCl and (C) 0.250 *M* NaBr.

a. Which solution, B or C, would be the better one to add to solution A to separate Ag^+ ions from Pb^{2+} by selective precipitation?
b. Using the solution you selected in part (a), is the separation of the two ions complete?

Additional Problems

16.105. Fluoride in Drinking Water Hydrogen fluoride (HF) behaves as a weak acid in aqueous solution. Two equilibria influence which fluorine-containing species are present in solution.

$$HF(aq) + H_2O(\ell) \rightleftharpoons H_3O^+(aq) + F^-(aq) \qquad K_a = 6.8 \times 10^{-4}$$
$$F^-(aq) + HF(aq) \rightleftharpoons HF_2^-(aq) \qquad K_c = 2.6 \times 10^{-1}$$

a. Is fluoride in pH 7.00 drinking water more likely to be present as F^- or HF_2^-?
b. What is the equilibrium constant for the following equilibrium?

$$2\,HF(aq) + H_2O(\ell) \rightleftharpoons H_3O^+(aq) + HF_2^-(aq)$$

c. What are the pH and equilibrium concentration of HF_2^- in a 0.150 *M* solution of HF?

16.106. A 125.0 mg sample of an unknown monoprotic acid was dissolved in 100.0 mL of distilled water and titrated with a 0.050 *M* solution of NaOH. The pH of the solution was monitored throughout the titration, and the following data were collected.

a. What is the K_a value for the acid?
b. What is the molar mass of the acid?

Volume of OH^- Added (mL)	pH	Volume of OH^- Added (mL)	pH
0	3.09	22	5.93
5	3.65	22.2	6.24
10	4.10	22.6	9.91
15	4.50	22.8	10.2
17	4.55	23	10.4
18	4.71	24	10.8
19	4.94	25	11.0
20	5.11	30	11.5
21	5.37	40	11.8

***16.107.** When silver oxide dissolves in water, the following reaction occurs:

$$Ag_2O(s) + H_2O(\ell) \rightarrow 2\,Ag^+(aq) + 2\,OH^-(aq)$$

If the pH of a saturated aqueous solution of silver oxide is 10.20, what is the K_{sp} of silver oxide?

*16.108. **Greenhouse Gases and Ocean pH** Some climate models predict the pH of the oceans will decrease by as much as 0.77 pH units as a result of increases in atmospheric carbon dioxide.

a. Explain, by using the appropriate chemical reactions and equilibria, how an increase in atmospheric CO_2 could produce a decrease in oceanic pH.
b. How much more acidic (in terms of $[H_3O^+]$) would the oceans be if their pH dropped this much?
c. Oceanographers are concerned about how a drop in oceanic pH would affect the survival of oysters. Why?

17

Thermodynamics

Spontaneous and Nonspontaneous Reactions and Processes

CORROSION Most metal objects in the sea, including the remains of this shipwreck, corrode as a result of spontaneous chemical reactions. Iron metal is converted into iron(III) oxide.

PARTICULATE **REVIEW**

Endothermic or Exothermic?

In Chapter 17 we examine fundamental ideas about why some reactions happen spontaneously but others do not. In doing so, we revisit the concepts of thermochemistry from Chapter 6. Propane is a common fuel used for cooking on outdoor grills.

Propane

- Is the combustion of propane an endothermic or exothermic process?
- What bonds are broken in the combustion of propane? Is the breaking of bonds an endothermic or exothermic process?
- What bonds are formed in the combustion of propane? Is the forming of bonds an endothermic or exothermic process?

(Review Sections 6.3 and 8.7 if you need help.)

(Answers to Particulate Review questions are in the back of the book.)

PARTICULATE PREVIEW

Motion and Dispersion of Energy

We learned in Chapter 6 that the total energy of atoms and molecules is the sum of their kinetic energy (from random motion of particles) and potential energy (from their arrangement). Consider the elements copper, bromine, and helium at room temperature. As you read Chapter 17, look for ideas that will help you answer the following questions:

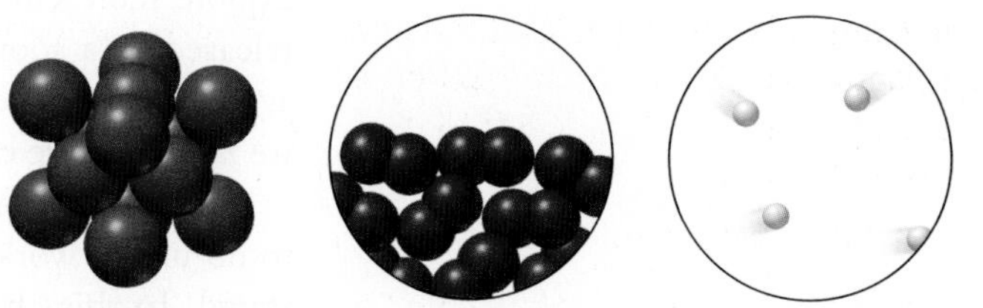

Copper

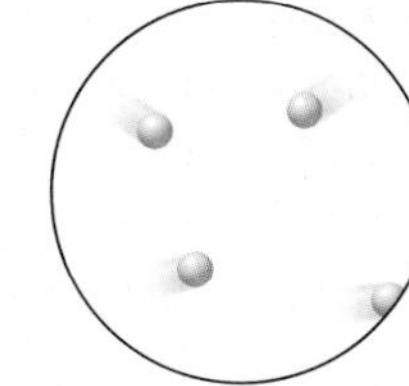

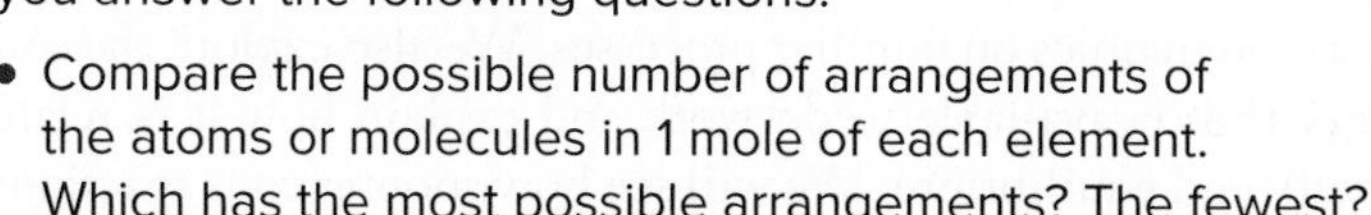

- Compare the possible number of arrangements of the atoms or molecules in 1 mole of each element. Which has the most possible arrangements? The fewest?

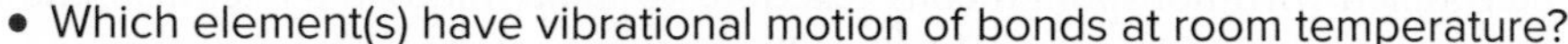

- Which element(s) have vibrational motion of bonds at room temperature?
- Which element can disperse its energy in the most ways?

Learning Outcomes

LO1 Predict the signs of entropy changes for spontaneous and nonspontaneous chemical reactions and physical processes
Sample Exercise 17.1

LO2 Predict the relative entropies of substances on the basis of their molecular structures
Sample Exercise 17.2

LO3 Calculate entropy changes in chemical reactions using standard molar entropies
Sample Exercise 17.3

LO4 Calculate free-energy changes and standard free-energy changes in chemical reactions and physical processes
Sample Exercises 17.4, 17.5

LO5 Predict the spontaneity of a chemical reaction or physical process as a function of temperature and in terms of changes in its enthalpy and entropy
Sample Exercises 17.6, 17.7

LO6 Relate the value of the equilibrium constant of a reaction to its change in free energy under standard conditions
Sample Exercise 17.8

LO7 Determine the value of the equilibrium constant K at any temperature from thermodynamic data
Sample Exercise 17.9

LO8 Calculate the net change in free energy of coupled spontaneous and nonspontaneous reactions
Sample Exercise 17.10

LO9 Use microstates to explain why a perfect crystalline solid has zero entropy

17.1 Spontaneous Processes

Some processes are so familiar that we rarely consider why they happen. If a car tire is punctured, the air inside rushes out and the tire goes flat; air does not rush back into a punctured tire and reinflate it. Objects made of iron left on the ground or underwater for months or years become clumps of rust; they do not turn back into shiny metal after even more time. A tray of ice cubes left on a kitchen counter melts into a tray of liquid water. As long as the kitchen remains at the same temperature, there is no way the water in the tray will resolidify into ice cubes.

STEPWISE ANIMATION
Spontaneous Processes

All three of these processes are **spontaneous**, which means that they all happen without any ongoing intervention and without work being done on the system. In each case, the reverse process is **nonspontaneous** because it cannot happen on its own. Why are some processes spontaneous and others not?

The first law of thermodynamics (Section 6.3) states that energy cannot be created or destroyed. This means that when we play the game of energy conversion, we can't win. Furthermore, the second law of thermodynamics—which we explore more closely in Section 17.2—says, in effect, that not all of the energy released by a spontaneous reaction, such as burning gasoline in a car engine, is available to do useful work. In the game of energy conversion, then, not only can we *not* win—we can't even break even.

If energy cannot be destroyed, what happens to the energy that is unavailable to do useful work? It turns out this energy spreads out, becoming less concentrated. In this chapter we explore the meaning and some of the impacts of the second law of thermodynamics on familiar processes. We also explore the component of the energy that is available to do work and explain how it is related to reaction spontaneity and equilibrium. We will see how spontaneous reactions can be coupled with nonspontaneous reactions to make multistep chemical and biochemical processes happen. This coupling is important to us because it powers the molecular processes that sustain life.

spontaneous process a process that occurs without outside intervention.

nonspontaneous process a process that occurs only as long as energy is continually added to the system.

The word "spontaneous" can be a bit misleading because many spontaneous reactions do not start all by themselves; they need a little energy boost, such as a spark or external flame. For example, before Thomas Edison invented the lightbulb in 1879, city streets were often illuminated at night by gas-fueled lamps. Once lit, they burned through the night until their fuel was cut off as dawn approached. Chemical reactions such as the combustion of gas-lamp fuel are examples of spontaneous reactions: once started, they proceed without outside intervention, as long as reactants are available. It is the self-sustaining nature of these reactions that earns them the label *spontaneous*. The reverse reaction—in this case, converting carbon dioxide and water into fuel and oxygen—is nonspontaneous: it cannot happen on its own without the continuous addition of energy from an external source.

In addition, spontaneous does not necessarily mean rapid. Though combustion reactions are fast, other spontaneous reactions, such as the formation of a layer of rust on an object made of iron, can take a very long time, depending on temperature and the rate at which O_2 reaches the iron surface:

$$4\ Fe(s) + 3\ O_2(g) \rightarrow 2\ Fe_2O_3(s) \qquad \Delta H^\circ = -1648\ kJ$$

Spontaneity has nothing to do with kinetics. Still, rust formation does proceed without intervention and meets the definition of thermodynamic spontaneity.

Spontaneous is also *not* a synonym for exothermic, although many scientists once thought so. In the mid-19th century, many chemists thought that exothermic reactions, in which high-enthalpy reactants formed low-enthalpy (more stable) products, should always be spontaneous. Many exothermic reactions, including combustion reactions and metal corrosion, *are* spontaneous. However, some exothermic processes may not be spontaneous, and some endothermic processes may be spontaneous, depending on reaction conditions.

Familiar examples of spontaneous endothermic processes include the phase changes that occur when ice melts and water boils. Both are endothermic processes, but both can be spontaneous, depending on the temperature and pressure. Another spontaneous endothermic reaction occurs when room-temperature vinegar (dilute acetic acid) is added to baking soda (sodium bicarbonate), as shown in **Figure 17.1**. The foaming mixture indicates that a reaction is taking place, and the decrease in temperature of the mixture indicates that the reaction is endothermic ($\Delta H > 0$).

CONNECTION Enthalpy change (ΔH) is a thermodynamic quantity that describes energy flow into or out of a system (Section 6.4).

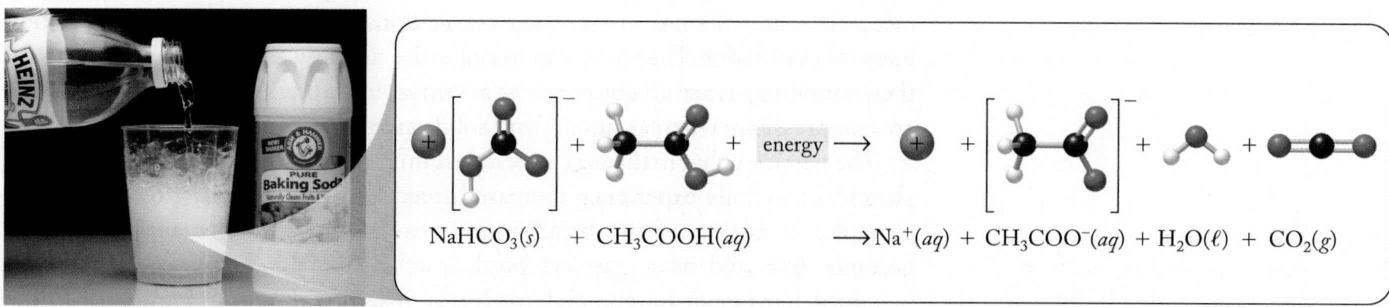

FIGURE 17.1 Adding vinegar to baking soda produces sodium acetate, water, and carbon dioxide in an endothermic ($\Delta H^\circ = +48.5$ kJ), yet spontaneous, reaction.

FIGURE 17.2 Instant cold packs get cold when the water-filled pouch inside is ruptured and the water-soluble compound (e.g., ammonium nitrate) within the pack dissolves. The NH_4^+ and NO_3^- ions in solid NH_4NO_3 experience increased freedom of motion as they form hydrated NH_4^+ and NO_3^- ions in solution. The temperature of the solution decreases because the dissolution process is endothermic.

CONNECTION The ion–dipole interactions that promote the solubility of ionic compounds in water are described in Section 10.3.

Yet another spontaneous endothermic process makes instant cold packs cold (**Figure 17.2**). An instant cold pack has two compartments. One is filled with water, whereas the other contains a water-soluble compound such as ammonium nitrate that has a positive enthalpy of solution ($\Delta H_{soln} > 0$). When the membrane separating the two compartments is ruptured, the solid compound (NH_4NO_3) mixes with and dissolves in the water. The resulting solution gets very cold, becoming an effective anti-inflammation treatment for bruises and muscle sprains.

Why are endothermic processes such as these spontaneous? The answer is that the particles that make up their products are more spread out and have more freedom of motion than the particles in their starting materials—something that all these processes have in common. Consider the molecular changes that accompany ice melting and water boiling (**Figure 17.3**). The molecules of H_2O in ice occupy fixed positions. Their motion (Figure 17.3a), like that of particles in all crystalline solids, is limited to vibrating in place, not going anywhere. When ice melts, its H_2O molecules become more mobile, acquiring translational motion and rotational motion (Figure 17.3b and c). When liquid water evaporates and the molecules enter the gas phase, their translational motion and rotational motion increase even more. The water vapor molecules are now also able to expand to fill their container, as are all gas particles at atmospheric pressure, which is why gases are compressible, whereas liquids and solids are not.

The particles that make up the reaction mixtures in spontaneous endothermic chemical reactions experience increased freedom of motion. In Figure 17.1, for example, a portion of the bicarbonate ions in the solid reactant ($NaHCO_3$) becomes liberated as a gaseous product of the reaction (CO_2), resulting in increased movement for the system. In the instant cold pack (Figure 17.2), the particles of the solid solute acquire more freedom of motion and can move throughout the resulting solution. These particles experience increases in motion similar to those of a melting solid.

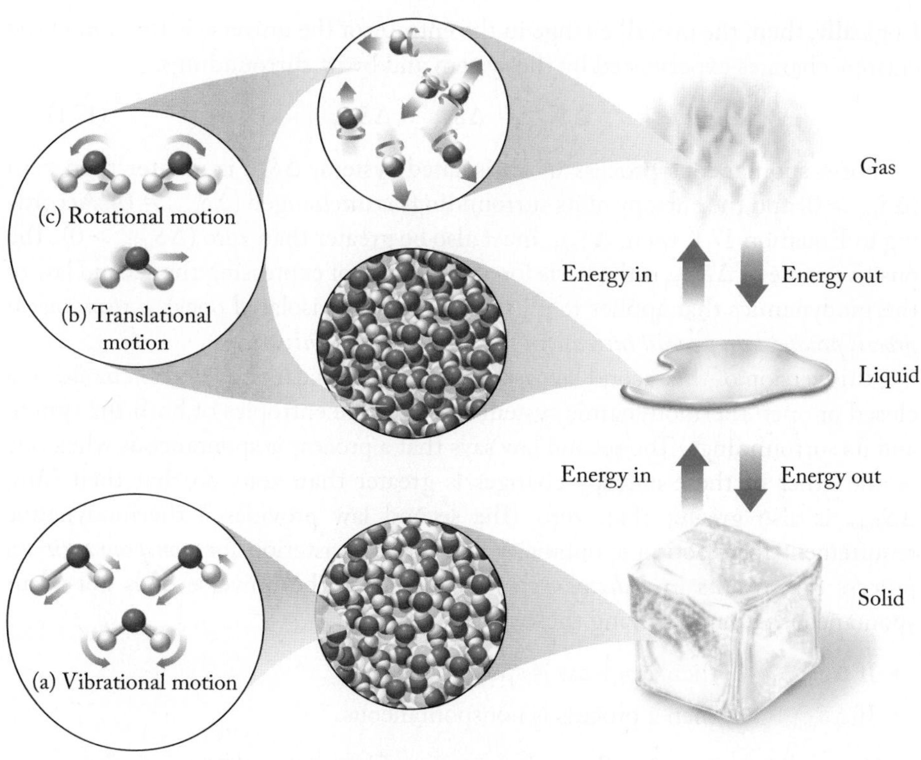

FIGURE 17.3 Molecules of water have three types of motion: (a) vibrational—which is their only motion in solids; (b) translational; and (c) rotational. Molecules in all three phases experience vibrations; translational and rotational motion is also seen in the liquid phase, with much more of both in the gas phase.

CONCEPT TEST

In which of the following processes do particles experience an increase in freedom of motion?

a. A glass of water evaporates.
b. Dew forms overnight on grass and other surfaces.
c. Table salt is added to water for cooking spaghetti.
d. A log of wood burns in a fireplace.

(Answers to Concept Tests are in the back of the book.)

17.2 Entropy and the Second Law of Thermodynamics

When particles spread out and gain freedom of motion, the kinetic energy associated with their motion also spreads out. This spreading out, or dispersion, of energy turns out to be a key characteristic of all spontaneous processes. It even has a name: **entropy (S)**. Entropy is a thermodynamic property that provides a measure of the dispersal of energy in a system at a specific temperature. The **second law of thermodynamics** states that entropy of an *isolated* thermodynamic system *always increases* during a spontaneous process.

The second law also covers thermodynamic systems that are not isolated (most systems are not isolated; they are either open or closed, as discussed in Section 6.3). Recall from Chapter 6 that in thermodynamics we divide the universe into two parts: the part we are interested in (the system) and everything else (the system's surroundings). Mathematically:

$$\text{Universe} = \text{System} + \text{Surroundings}$$

CHEMTOUR
Entropy

entropy (S) a measure of the dispersion of energy in a system at a specific temperature.

second law of thermodynamics the principle that the total entropy of the universe increases in any spontaneous process.

Logically, then, the overall change in the entropy of the universe is the sum of the entropy changes experienced by the system and by its surroundings:

$$\Delta S_{univ} = \Delta S_{sys} + \Delta S_{surr} \tag{17.1}$$

For a spontaneous process in an isolated system, ΔS_{sys} is greater than zero ($\Delta S_{sys} > 0$) and the entropy of its surroundings is unchanged ($\Delta S_{surr} = 0$). According to Equation 17.1, then, ΔS_{univ} must also be greater than zero ($\Delta S_{univ} > 0$). The positive value of ΔS_{univ} is the basis for another way of expressing the second law of thermodynamics that applies to all systems (not just isolated ones): *a spontaneous process produces an overall increase in the entropy of the universe.*

This version of the second law assumes that a physical or chemical change in a closed or open thermodynamic system can alter the entropies of both the system and its surroundings. The second law says that a process is spontaneous when one or the other of these entropy changes is greater than zero, so that their sum, ΔS_{univ}, is also greater than zero. The second law provides a thermodynamic requirement for reaction spontaneity as well as a criterion for *nonspontaneity*: a process that results in a *decrease* in the entropy of the universe does not occur spontaneously. Summarizing these relationships:

CONNECTION Recall from Section 6.3 that an isolated thermodynamic system exchanges neither energy nor matter with its surroundings, a closed system exchanges energy but not matter, and an open system exchanges both.

- If $\Delta S_{univ} > 0$, then a process is spontaneous.
- If $\Delta S_{univ} < 0$, then a process is nonspontaneous.

To see how a process affects the entropy of its surroundings, let's focus on a familiar exothermic reaction, the combustion of natural gas (methane):

$$CH_4(g) + 2\,O_2(g) \rightarrow CO_2(g) + 2\,H_2O(\ell) \qquad \Delta H^\circ = -890 \text{ kJ}$$

As written, the reaction consumes 3 moles of gases and produces 2 moles of a liquid product and 1 mole of gas. As a result, there are fewer moles of gas on the product side of the reaction equation than on the reactant side. Given the much greater freedom of motion of particles in the gas phase, we can accurately predict that there will be a decrease in entropy of the reaction mixture, which is our thermodynamic system:

$$\Delta S_{sys} < 0$$

The reaction is spontaneous, however, so

$$\Delta S_{univ} > 0$$

How do we reconcile these opposing inequalities? Equation 17.1 supplies an explanation. If $\Delta S_{univ} > 0$, then $\Delta S_{sys} + \Delta S_{surr}$ must also be greater than zero. Because $\Delta S_{sys} < 0$, then ΔS_{surr} is not only greater than zero, but it must also have a large enough positive value to more than offset the negative value of ΔS_{sys}. Expressing this relationship in terms of the absolute values of ΔS_{surr} and ΔS_{sys}:

$$|\Delta S_{surr}| > |\Delta S_{sys}|$$

Is the combustion of methane likely to produce a large, positive ΔS_{surr}? Absolutely, because the reaction is highly exothermic: combustion of only 1 mole (16 g) of methane generates 890 kJ of thermal energy. As energy flows from the system into its surroundings, a dispersion of energy occurs, producing a positive ΔS_{surr} that more than compensates for the unfavorable (negative) value of ΔS_{sys}.

All exothermic reactions have the capacity to increase the entropy of their surroundings. The more energy that flows into the surroundings in the form of heat (q), or into any collection of particles, the greater the dispersion of energy among the particles and the greater the increase in their entropy (ΔS). However, adding energy to particles that are at a higher temperature produces a smaller change in entropy than does adding the same quantity of energy to the same particles at a lower temperature. Equation 17.2 represents this inverse relationship between entropy change and temperature:

$$\Delta S = \frac{q_{rev}}{T} \quad (17.2)$$

The subscript "rev" means that the heating process is reversible. Theoretically, a **reversible process** happens so slowly that equilibrium is constantly maintained. For example, after an incremental change in the system has occurred in the forward direction, the process can be reversed by an incremental change in reaction conditions that restores the original state of the system with no net flow of energy into or out of the system. In other words, everything about the system goes back to exactly as it was before the process began.

reversible process a process that can be run in the reverse direction in such a way that, once the system has been restored to its original state, no net energy has flowed either to the system or to its surroundings.

STEPWISE
ANIMATION
Reversible Processes

A reversible process is an idealization; it describes a theoretical limit. All real chemical reactions and physical processes are irreversible, but many of them approximate reversibility closely enough that we can adapt Equation 17.2 to calculate ΔS_{sys}:

$$\Delta S_{sys} = \frac{q_{sys}}{T} \quad (17.3)$$

For example, it takes 6.01 kJ (or 6.01×10^3 J) of energy to melt 1.00 mol of ice at 0°C, a physical process. Assuming the process occurs reversibly, then

$$\Delta S_{sys} = \frac{q_{sys}}{T} = \frac{(1.00\ \cancel{\text{mol}})(6.01 \times 10^3\ \text{J}/\cancel{\text{mol}})}{273\ \text{K}} = 22.0\ \text{J/K}$$

Because energy flows into the ice, q_{sys} is positive, which also makes ΔS_{sys} positive, as we would expect given the greater freedom of motion of particles in the liquid phase. The units of entropy in this calculation are joules per kelvin. We use these units in all entropy calculations in Chapter 17.

Suppose 1.00 mol of ice melts as 6.01×10^3 J of energy flows into it from room-temperature (22°C or 295 K) surroundings. We can calculate the change in entropy of the surroundings by using another adaptation of Equation 17.2:

$$\Delta S_{surr} = \frac{q_{surr}}{T} = \frac{(1.00\ \cancel{\text{mol}})(-6.01 \times 10^3\ \text{J}/\cancel{\text{mol}})}{295\ \text{K}} = -20.4\ \text{J/K}$$

The sign of q_{surr} is negative because energy flows from the surroundings into the system (ice cube). Moreover, the value of ΔS_{surr} is less than zero, and the magnitude of the decrease in ΔS_{surr} is less than the increase in ΔS_{sys} because the same absolute value of q was divided by a higher temperature to calculate ΔS_{surr}. Therefore, when we sum ΔS_{sys} and ΔS_{surr} (**Figure 17.4a**), we get a positive value of ΔS_{univ}:

$$\Delta S_{univ} = \Delta S_{sys} + \Delta S_{surr} = (22.0 - 20.4)\ \text{J/K} = 1.6\ \text{J/K}$$

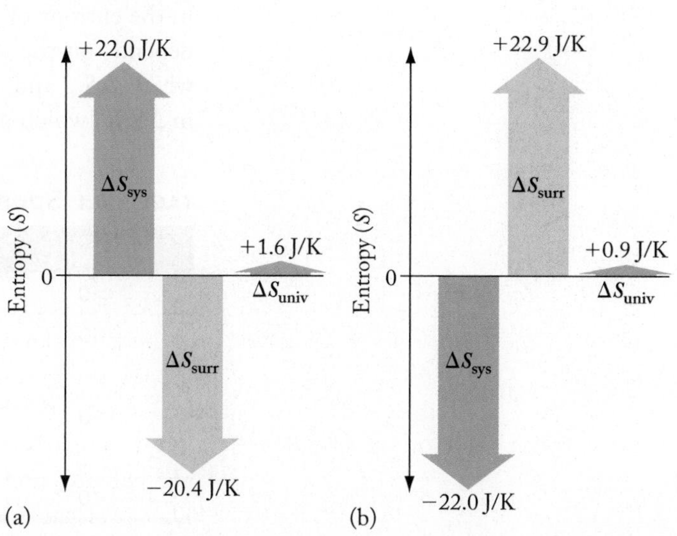

FIGURE 17.4 (a) If the surroundings experience a decrease in entropy ($\Delta S_{surr} = -20.4$ J/K), and the system (ice) experiences an increase in entropy ($\Delta S_{sys} = +22.0$ J/K) that more than offsets the decrease, then the process (melting) is spontaneous ($\Delta S_{univ} = \Delta S_{sys} + \Delta S_{surr} = 22.0\ \text{J/K} - 20.4\ \text{J/K} = +1.6$ J/K). (b) $\Delta S_{univ} = \Delta S_{sys} + \Delta S_{surr} = +0.9$ J/K because the decrease in entropy for the system ($\Delta S_{sys} = -22.0$ J/K) is more than offset by the increase in entropy for the surroundings ($\Delta S_{surr} = +22.9$ J/K).

The result of this calculation is one example of a general truth about energy flow—namely, energy flows spontaneously into a system that is cooler than its surroundings. Even more generally, energy flows spontaneously from a warm object to an adjacent cooler object. Entropy and the second law of thermodynamics simply provide a mathematical explanation of why.

Why didn't we consider the change in temperature of the surroundings as energy flowed from it into the melting ice? The answer lies in the sheer size of the surroundings (the universe minus a small cube of ice). The temperature of such an enormous mass is not likely to change significantly.

Now let's consider how entropy changes when the temperature of the surroundings is *lower* than the temperature of the system. Suppose a tray containing 1.00 mol of liquid water at 0°C (273 K) is placed in a freezer at −10°C (263 K). Because the temperature of the surroundings (the freezer) is lower than the temperature of the liquid water, energy spontaneously flows from the water into its surroundings. The water freezes and the net entropy change for this process is

$$\begin{aligned}\Delta S_{univ} &= \Delta S_{sys} + \Delta S_{surr} \\ &= \frac{(1.00\ \cancel{mol})(-6.01 \times 10^3\ \text{J}/\cancel{mol})}{273\ \text{K}} + \frac{(1.00\ \cancel{mol})(+6.01 \times 10^3\ \text{J}/\cancel{mol})}{263\ \text{K}} \\ &= (-22.0\ \text{J/K}) + (+22.9\ \text{J/K}) \\ &= 0.9\ \text{J/K}\end{aligned}$$

Once again the entropy of the universe increases (**Figure 17.4b**) as energy flows spontaneously from the warmer object (liquid water at 0°C) into the colder surroundings (the freezer at −10°C).

FIGURE 17.5 Water vapor exhaled by this Inuit hunter in the Northwest Territories of Canada was spontaneously deposited on his facial hair as frost—evidence of how cold it was when this photo was taken.

CONCEPT TEST

Is ΔS_{univ} greater than, less than, or equal to zero when water vapor exhaled by the Inuit hunter in **Figure 17.5** is deposited as crystals of ice on his beard?

Entropy-change calculations based on Equation 17.2 assume process reversibility, which, as we have discussed, is an idealized, theoretical concept. In reality, the ΔS values calculated in this way are *minimum* ΔS values. When processes take place in the real world, the accompanying changes in entropy are inevitably greater than those calculated using Equation 17.2.

According to the second law, spontaneous processes *always* produce an increase in the entropy of the universe. **Table 17.1** shows how the spontaneity of a process depends on the sign and magnitude of ΔS_{sys} and ΔS_{surr}. Note how processes in which ΔS_{sys} and ΔS_{surr} are both greater than zero inevitably result in an increase in ΔS_{univ}, which means they are always spontaneous. On the other hand, processes

TABLE 17.1 Spontaneity of Process as a Function of ΔS_{sys} and ΔS_{surr}

ΔS_{sys}	ΔS_{surr}	Spontaneity of Process
>0	>0	Always spontaneous
<0	>0	Spontaneous if $\lvert\Delta S_{sys}\rvert < \lvert\Delta S_{surr}\rvert$ Nonspontaneous if $\lvert\Delta S_{sys}\rvert > \lvert\Delta S_{surr}\rvert$
>0	<0	Spontaneous if $\lvert\Delta S_{sys}\rvert > \lvert\Delta S_{surr}\rvert$ Nonspontaneous if $\lvert\Delta S_{sys}\rvert < \lvert\Delta S_{surr}\rvert$
<0	<0	Always nonspontaneous

in which ΔS_{sys} and ΔS_{surr} are both less than zero always produce a decrease in ΔS_{univ}, which means they are always nonspontaneous. Between these extremes are four possible combinations of ΔS_{sys} and ΔS_{surr} in which these changes have opposite signs. These combinations may or may not produce positive ΔS_{univ} values, depending on the absolute values of ΔS_{sys} and ΔS_{surr}. If the larger of the two is the one with the positive value, then $\Delta S_{univ} > 0$ and the process is spontaneous.

SAMPLE EXERCISE 17.1 Predicting the Sign of Entropy Change **LO1**

Predict whether ΔS_{sys} is greater or less than zero when each of these processes occurs at constant temperature:

a. $H_2O(\ell) \rightarrow H_2O(g)$
b. $NH_3(g) + HCl(g) \rightarrow NH_4Cl(s)$
c. $C_{12}H_{22}O_{11}(s) \rightarrow C_{12}H_{22}O_{11}(aq)$

Collect, Organize, and Analyze To predict the signs of the accompanying entropy changes, we can compare the freedom of motion of the reactant particles to the freedom of motion of the product particles:

a. Molecules of liquid water become molecules of water vapor, which means the molecules have increased freedom of motion.
b. Two gaseous compounds form a single solid compound, which means the particles of product occupy much less space and have much less freedom of motion.
c. A solid dissolves in water, forming molecules dispersed in an aqueous solution that have more freedom of motion.

Solve

a. $\Delta S_{sys} > 0$, because the kinetic energies of water molecules are more dispersed in the gas state than they are in the liquid state.
b. $\Delta S_{sys} < 0$, because formation of a solid causes the loss of the translational and rotational motion the gas-phase particles had.
c. $\Delta S_{sys} > 0$, because particles in a solution have translational and rotational motion that particles in crystalline solids do not have.

Think About It Entropy increases when solids melt and liquids vaporize because of the increased freedom of motion of their particles and the increased dispersion of their particles' kinetic energies. When gases combine to form a solid, they lose freedom of motion and entropy decreases. When solids dissolve in liquids, however, they gain freedom of motion and experience an increase in entropy.

Practice Exercise Predict whether these chemical reactions result in an increase or decrease in the entropy of the system. Assume the reactants and products are at the same temperature and pressure.

a. $CaCO_3(s) + 2\,HCl(aq) \rightarrow CaCl_2(aq) + CO_2(g) + H_2O(\ell)$
b. $NH_3(g) + BF_3(g) \rightarrow NH_3BF_3(s)$

(Answers to Practice Exercises are in the back of the book.)

17.3 Absolute Entropy and the Third Law of Thermodynamics

The entropy of a system depends on temperature because higher temperatures mean higher particle kinetic energies, which mean the particles have more vibrational, rotational, and translational motion—and more entropy. Conversely, lower temperatures mean that all these quantities are smaller. If we lower the temperature of a substance to absolute zero, in principle all translational motion

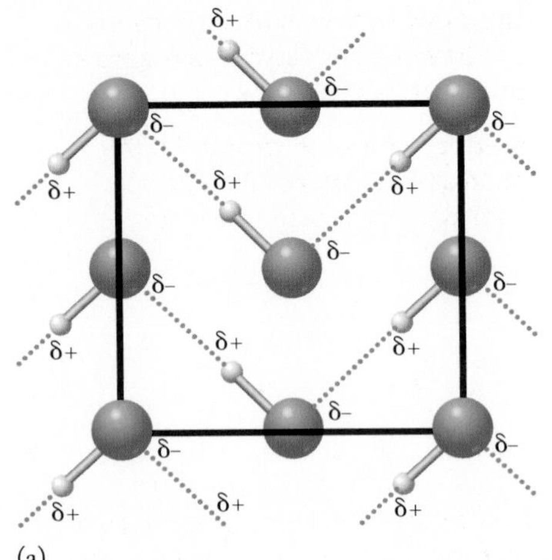

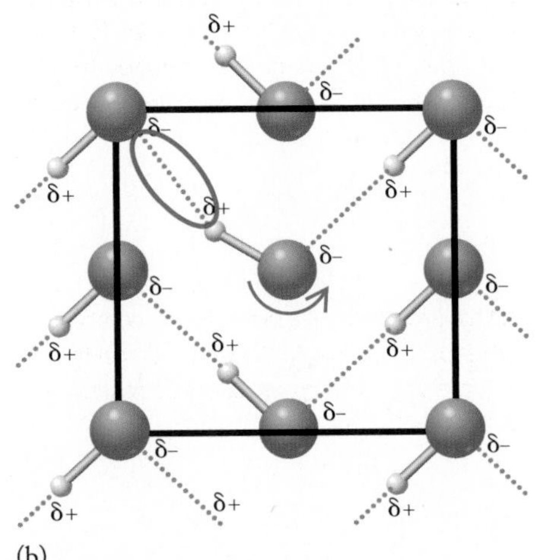

FIGURE 17.6 Two views of the layers of molecules in solid HCl. (a) A perfect crystal of HCl at $T = 0$ K has zero entropy because all the molecules have zero translational kinetic energy and are arranged uniformly, which means they all experience equivalent dipole–dipole interactions. (b) In this imperfect crystal at $T > 0$ K the center HCl molecule is rotated a little, creating a slightly different dipole–dipole interaction and a disorderd structure. The entropy of this crystal is greater than 0.

should cease. If we assume that the substance forms a perfect crystalline solid (**Figure 17.6**), in which each particle is locked in one and only one site within the crystal, then each particle has zero freedom of motion. There is no dispersion of translational kinetic energy because there is none to disperse. As a result, the *absolute* entropy of a perfect crystalline solid is zero at absolute zero (0 K). This conclusion is known as the **third law of thermodynamics.**

Setting a zero point on the entropy scale allows scientists to establish absolute entropy values for pure substances at any temperature. The absolute entropy of a substance is often expressed as its **standard molar entropy ($S°$)**—that is, the entropy of 1 mole of the substance at 298 K and 1 bar (~1 atm) of pressure in its standard state (**Table 17.2**). For example, the standard molar entropy of solid NaCl is 72.1 J/(mol · K) at 298 K. Absolute entropies are determined from careful determinations of the molar heat capacity (or specific heat) of substances as a function of temperature. Our ability to determine absolute entropy values of substances and systems contrasts with the observation in Chapter 6 that we cannot measure absolute enthalpy (H), and the best we can do is calculate *changes* in enthalpy, ΔH.

TABLE 17.2 Standard States of Pure Substances and Solutions[a]

Physical State	Standard State	Pressure[b]
Solid	Pure solid, most stable allotrope of an element	1 bar
Liquid	Pure liquid	1 bar
Gas	Pure gas	1 bar
Solution	1 *M*	1 bar

[a]The thermodynamic data in Appendices 4 to 6 and used elsewhere in this book are based on a temperature of 298.15 K (25°C). Note that this temperature is not the STP temperature we use for gases (see Chapter 5), which is 273 K.

[b]Since 1982, 1 bar has been the standard pressure for tabulating all thermodynamic data. Before 1982, standard pressure was 1 atmosphere (atm) = 1.01325 bar.

The $S°$ values for liquid water and water vapor in **Table 17.3** illustrate an important difference between the entropies of liquids and gases that we have discussed before in this chapter: the molecules in a gas under standard conditions are

TABLE 17.3 Selected Standard Molar Entropy Values[a]

Substance	$S°$, J/(mol · K)	Substance	Name	$S°$, J/(mol · K)
$Br_2(g)$	245.5	$CH_4(g)$	Methane	186.2
$Br_2(\ell)$	152.2	$CH_3CH_3(g)$	Ethane	229.5
$C_{diamond}(s)$	2.4	$CH_3OH(g)$	Methanol	239.9
$C_{graphite}(s)$	5.7	$CH_3OH(\ell)$		126.8
$CO(g)$	197.7	$CH_3CH_2OH(g)$	Ethanol	282.6
$CO_2(g)$	213.8	$CH_3CH_2OH(\ell)$		160.7
$H_2(g)$	130.6	$CH_3CH_2CH_3(g)$	Propane	269.9
$N_2(g)$	191.5	$CH_3(CH_2)_2CH_3(g)$	Butane	310.0
$O_2(g)$	205.0	$CH_3(CH_2)_2CH_3(\ell)$		231.0
$H_2O(g)$	188.8	$C_6H_6(g)$	Benzene	269.2
$H_2O(\ell)$	69.9	$C_6H_6(\ell)$		172.9
$NH_3(g)$	192.5	$C_{12}H_{22}O_{11}(s)$	Sucrose	360.2

[a]Values for additional substances are given in Appendix 4.

much more widely dispersed than the molecules in a liquid, and the entropies of the different phases of a given substance at a given temperature follow the order $S_{solid} < S_{liquid} < S_{gas}$.

third law of thermodynamics the entropy of a perfect crystal is zero at absolute zero.

standard molar entropy ($S°$) the absolute entropy of 1 mole of a substance in its standard state.

The entropy changes that occur as 1 mole of ice at 0 K is heated are shown in **Figure 17.7**. Note the jump in entropy as the ice melts and the even bigger jump as the liquid water vaporizes. Note, too, that the lines between the phase changes are curved. The change in entropy (ΔS) with temperature is not linear because, as described by Equation 17.2, heating a substance at a higher temperature produces a smaller entropy increase than does transferring the same quantity of heat to the same substance at a lower temperature.

Let's summarize the factors that affect entropy change:

1. Entropy increases when temperature increases.
2. Entropy increases when volume increases.[1]
3. Entropy increases when the number of independent particles increases.

CONNECTION As defined in Section 6.5, molar heat capacity (c_P) is the quantity of energy required to raise the temperature of 1 *mole* of a substance by 1°C at constant pressure, whereas specific heat (c_s) is the quantity of energy required to raise the temperature of 1 *gram* of a substance by 1°C at constant pressure.

In all three cases, entropy increases because each change increases the dispersion of the kinetic energy of a system's particles. We can often make qualitative predictions about entropy changes that accompany chemical reactions on the basis of these three factors, even if we have no thermodynamic data about the reactants and products. For example, when propane burns in air, 6 moles of gaseous reactants react to form 7 moles of gaseous products:

$$CH_3CH_2CH_3(g) + 5\,O_2(g) \rightarrow 3\,CO_2(g) + 4\,H_2O(g)$$

CONNECTION According to Avogadro's law (Section 5.3), the number of moles of a gas is directly proportional to the volume occupied by the gas at constant temperature and pressure.

The number of moles of gaseous particles increases as the reaction proceeds, so entropy also increases ($\Delta S_{sys} > 0$).

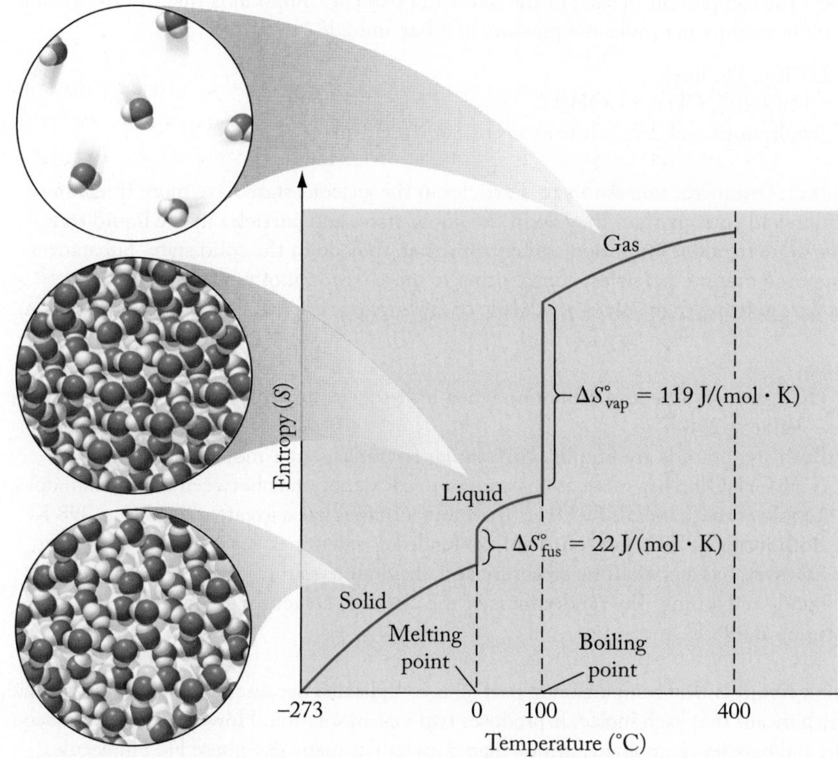

FIGURE 17.7 Abrupt increases in entropy accompany changes of state, with the greater increase occurring during the transition from liquid to gas.

[1]Water is a notable exception to this observation; the molar volume of ice is larger than for liquid water even though the absolute entropy of ice is less than the absolute entropy of liquid water.

Entropy and Structure

The data in Table 17.3 demonstrate that the standard molar entropies of substances are strongly linked to molecular structure. Note, for example, how the standard molar entropies of the C_1 to C_4 alkanes in natural gas (**Figure 17.8**) increase as the number of atoms per molecule increases. This trend can be explained by the freedom of motion of the atoms inside their molecules. The more bonds within a molecule, the more opportunities for internal (vibrational) motion, and the greater the standard molar entropy.

Another structural feature that influences entropy is rigidity. The two most common forms of carbon—diamond and graphite—are both polymeric network solids. However, the rigid three-dimensional structure of diamonds results in much less entropy [$S° = 2.4$ J/(mol · K)] than that of the less rigid, layered structure of graphite [$S° = 5.7$ J/(mol · K)].

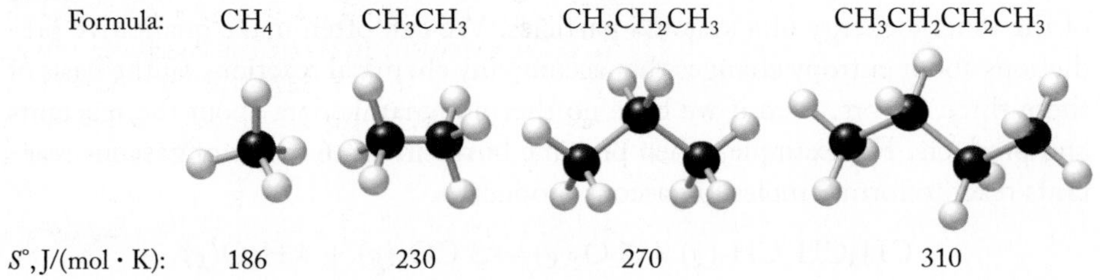

Formula:	CH_4	CH_3CH_3	$CH_3CH_2CH_3$	$CH_3CH_2CH_2CH_3$
$S°$, J/(mol · K):	186	230	270	310

FIGURE 17.8 Among methane, ethane, propane, and butane, standard molar entropies increase as the numbers of atoms and chemical bonds in the molecules increase.

SAMPLE EXERCISE 17.2 Comparing Absolute Entropy Values **LO2**

Select the component in each of the following pairs of compounds that has the greater absolute entropy per mole at a pressure of 1 bar and 298 K.

a. HCl(*g*), HCl(*aq*)
b. $CH_3OH(\ell)$, $CH_3CH_2OH(\ell)$
c. Amorphous and crystalline forms of $SiO_2(s)$ (**Figure 17.9**)

Collect, Organize, and Analyze Particles in the gaseous state have more freedom of motion and entropy than they do in the liquid state, and particles in the liquid state have more freedom of motion and entropy than they do in the solid state. Substances composed of more particles, contributing to more freedom of motion, have more absolute entropy than substances made of fewer particles with less freedom of motion.

Solve

a. HCl gas loses freedom of motion when it dissolves in water, so HCl gas has a greater $S°$ value at 298 K.
b. Both compounds are liquids with similar formulas, but 1 mole of ethanol (CH_3CH_2OH) has more atoms and more covalent bonds between atoms than does 1 mole of methanol (CH_3OH). Therefore, ethanol has a greater $S°$ value at 298 K.
c. Both forms of SiO_2 are solids with extended covalent networks of atoms. Quartz, however, has a crystalline structure and obsidian has an irregular arrangement of bonds and atoms. The randomness of the obsidian structure gives it a greater $S°$ value at 298 K.

Think About It The comparison in part (a) is complicated because HCl(*aq*) is a strong acid, which means that each molecule produces two ions in solution. However, they are aqueous ions and have less combined entropy than does half as many gas-phase HCl molecules. (Compare the $S°$ values of HCl(*g*), $H^+(aq)$, and $Cl^-(aq)$ in Appendix 4 to see for yourself.)

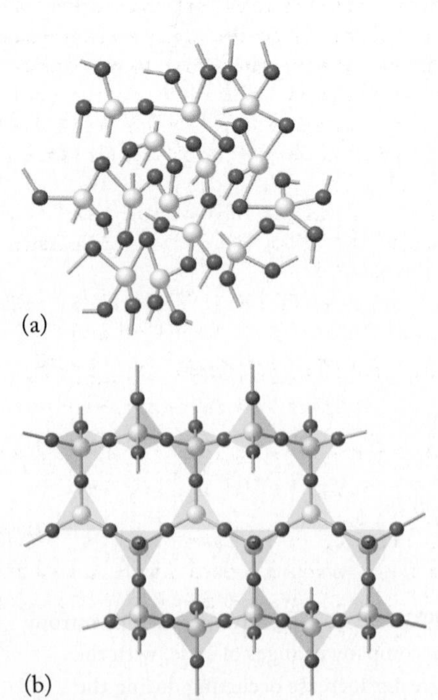

FIGURE 17.9 (a) Obsidian (amorphous) and (b) quartz (crystalline) are two forms of SiO_2.

Practice Exercise Ball-and-stick models of four hydrocarbons that each contain six carbon atoms per molecule are shown in **Figure 17.10**. Rank these compounds in order of decreasing $S°$ values.

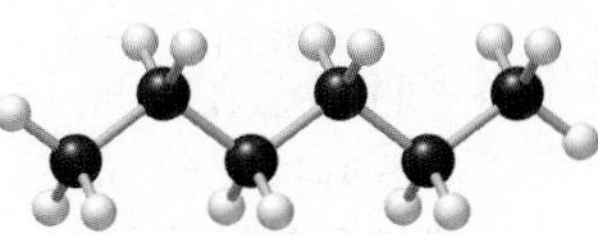

(a) Hexane

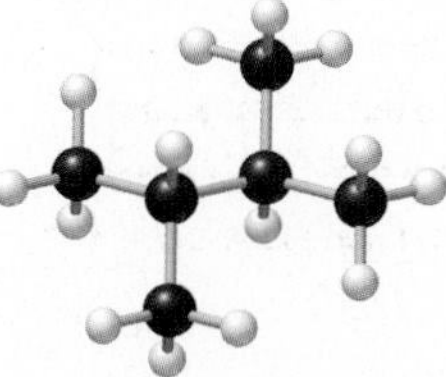

(b) 2,3-Dimethylbutane

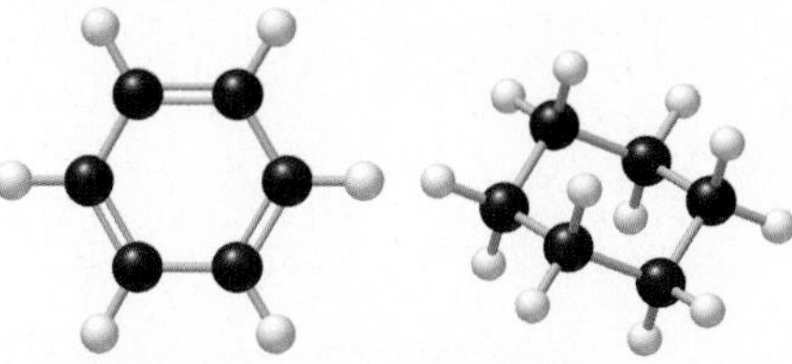

(c) Benzene (d) Cyclohexane

FIGURE 17.10 Four six-carbon hydrocarbons.

17.4 Calculating Entropy Changes

The entropy of a system (like its enthalpy and internal energy) is a state function, which means that the change in entropy that accompanies a process depends only on the initial and final states of the system, not on the pathway of the process. As a result, the change in entropy experienced by a system is simply the difference between its initial and final absolute entropy levels:

$$\Delta S_{sys} = S_{final} - S_{initial} \qquad (17.4)$$

We can adapt Equation 17.4 to calculate the change in entropy that accompanies a chemical reaction under standard conditions, $\Delta S°_{rxn}$, from the difference in the standard molar entropies of moles of reactants, $n_{reactants}$ (the equivalent of $S_{initial}$ in Equation 17.4), and the molar entropies of moles of products, $n_{products}$ (i.e., S_{final}):

$$\Delta S°_{rxn} = \sum n_{products} S°_{products} - \sum n_{reactants} S°_{reactants} \qquad (17.5)$$

Each $S°$ value for a product or reactant is multiplied by the appropriate number of moles from the balanced chemical equation. In other words, just as we saw for $\Delta H°$ in Section 6.8 (see Equation 6.19), entropy is an extensive thermodynamic property that depends on the quantities of substances consumed or produced in a reaction. Standard molar entropies of selected substances are listed in Appendix 4.

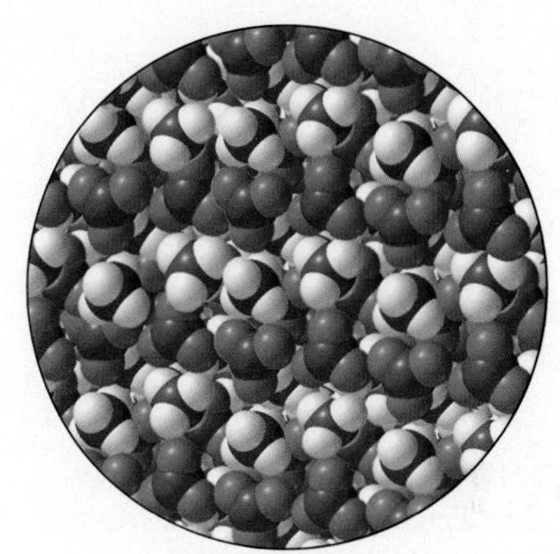

$NH_4NO_3(s)$

$\downarrow H_2O$

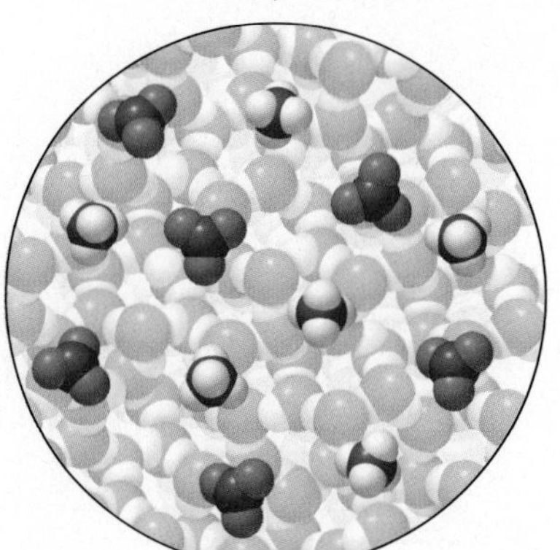

$NH_4NO_3(aq)$

FIGURE 17.11 Dissolution of solid ammonium nitrate into NH_4^+ and NO_3^- ions is endothermic but spontaneous under standard conditions because $\Delta S°$ is greater than zero, which contributes to $\Delta S_{univ} > 0$.

SAMPLE EXERCISE 17.3 Calculating Entropy Changes **LO3**

What is $\Delta S°$ when ammonium nitrate dissolves (**Figure 17.11**, $\Delta H° > 0$) under standard conditions, given the following standard molar entropy values:

	$NH_4NO_3(s) \rightarrow$	$NH_4^+(aq)$ +	$NO_3^-(aq)$
$S°$ [J/(mol · K)]	151.1	113.4	146.4

Collect, Organize, and Analyze We are given the standard molar entropy values of a solid ionic compound and its ions in aqueous solution, and we want to calculate the change in entropy that occurs when the compound dissolves under standard conditions. Entropy changes associated with chemical reactions depend on the entropies of the reactants and products, as described in Equation 17.5. Dissolving an ionic solid in water increases the freedom of motion of the solute ions, so $\Delta S°$ should be greater than zero.

Solve

$$\begin{aligned}\Delta S° &= \sum n_{products} S°_{products} - \sum n_{reactants} S°_{reactants} \\ &= \left[1\ \cancel{mol} \times \left(\frac{113.4\ J}{\cancel{mol} \cdot K}\right) + 1\ \cancel{mol} \times \left(\frac{146.4\ J}{\cancel{mol} \cdot K}\right)\right] - 1\ \cancel{mol} \times \left(\frac{151.1\ J}{\cancel{mol} \cdot K}\right) \\ &= 108.7\ J/K\end{aligned}$$

CONNECTION State functions are defined in Section 6.2.

Think About It As predicted, $\Delta S°$ is greater than zero because the freedom of motion and the dispersion of the kinetic energy of solute particles increase when the solid solute dissolves. The dissolution of ammonium nitrate is an example of an endothermic reaction that results in an increase in entropy.

Practice Exercise Calculate the standard molar entropy change for the combustion of methane gas by using $S°$ values from Appendix 4. Before carrying out the calculation, predict whether the entropy of the system increases or decreases.

$$CH_4(g) + 2\,O_2(g) \rightarrow CO_2(g) + 2\,H_2O(\ell)$$

17.5 Free Energy

In Sample Exercise 17.3 we used standard molar entropy values to calculate $\Delta S°$. The solute, solvent, and resulting solution together constituted a closed thermodynamic system, which meant that energy could flow into it as its temperature dropped, but matter was not exchanged with its surroundings. Therefore, the $\Delta S°$ value that we calculated applied only to the system. There was no evaluation of the change in the entropy of the system's surroundings accompanying the dissolution process, though we might predict that energy flowing from the surroundings into the system would produce a decrease in S_{surr}. (Had the dissolution process been exothermic, energy would have flowed from the system into its surroundings, and ΔS_{surr} would have been greater than zero.)

Thus, the entropy change experienced by the surroundings of any chemical thermodynamic system depends on whether the process occurring in the system is exothermic or endothermic. When energy in the form of heat flows from an exothermic process occurring at constant pressure into a system's surroundings, the quantity of heat is equal in magnitude but opposite in sign to the enthalpy change of the system:

CONNECTION In Section 6.4 we defined a change in enthalpy (ΔH) as the heat gained or lost in a reaction carried out at constant pressure.

$$q_{surr} = -\Delta H_{sys} \tag{17.6}$$

When an endothermic process occurs in the system, as in Sample Exercise 17.3, the direction of energy flow is reversed, but Equation 17.6 still applies. Assuming these transfers of energy occur reversibly, we can calculate the value of ΔS_{surr} by using the following modification of Equation 17.2:

$$\Delta S_{surr} = \frac{q_{surr}}{T} \tag{17.7}$$

Combining Equations 17.6 and 17.7 yields:

$$\Delta S_{surr} = -\frac{\Delta H_{sys}}{T}$$

We can then substitute this expression into Equation 17.1 ($\Delta S_{univ} = \Delta S_{sys} + \Delta S_{surr}$):

$$\Delta S_{univ} = \Delta S_{sys} - \frac{\Delta H_{sys}}{T} \tag{17.8}$$

The beauty of Equation 17.8 is that it allows us to predict whether a process is spontaneous at a particular temperature once we calculate the enthalpy and entropy changes accompanying the process. The downside of Equation 17.8 is that spontaneity relies on the value of a parameter (ΔS_{univ}) that is impossible to determine directly and that has little physical meaning. It would be better if we could substitute a thermodynamic parameter for ΔS_{univ} that is based only on the system and not the entire universe. Such a parameter exists and is called the change in the system's *free energy.*

In chemistry we focus on a particular kind of free energy called **Gibbs free energy (*G*)**—named in honor of American scientist J. Willard Gibbs (1839–1903). Gibbs free energy is the energy released by processes happening at constant temperature and pressure that is available to do useful work.

Gibbs free energy (*G*) the maximum energy released by a process occurring at constant temperature and pressure that is available to do useful work.

CHEMTOUR

Gibbs Free Energy

Like many of the thermodynamic properties we have examined, absolute free energy values of substances are often of less interest than the *changes* in free energy that accompany chemical reactions and other processes. Gibbs defined the change in free energy (ΔG_{sys}) of a process occurring at constant temperature and pressure in terms of T and ΔS_{univ} as:

$$\Delta G_{sys} = -T\Delta S_{univ}$$

Because of the $-T$ multiplier, *negative* values of ΔG_{sys} correspond to *positive* values of ΔS_{univ}. Therefore,

- If $\Delta G_{sys} < 0$, then $\Delta S_{univ} > 0$, and the reaction is spontaneous.
- If $\Delta G_{sys} > 0$, then $\Delta S_{univ} < 0$, and the reaction is nonspontaneous. Instead, the reaction running in reverse is spontaneous.
- If $\Delta G_{sys} = 0$, then $\Delta S_{univ} = 0$, and the composition of the reaction mixture does not change. In other words, the reaction has reached chemical equilibrium.

We can combine Gibbs' equation with Equation 17.8 by multiplying all of the terms in Equation 17.8 by $-T$:

$$-T\Delta S_{univ} = -T\Delta S_{sys} + \Delta H_{sys} \quad (17.9)$$

The left side of Equation 17.9 is equal to ΔG_{sys}. Making that substitution and rearranging the terms on the right side gives us

$$\Delta G_{sys} = \Delta H_{sys} - T\Delta S_{sys}$$

Because all of the parameters in this equation apply to the system, we typically simplify the equation by eliminating the "sys" subscripts:

$$\Delta G = \Delta H - T\Delta S \quad (17.10)$$

CONCEPT TEST

(a) Given the thermodynamic data in Table A4.3 in the Appendix, is the conversion of diamond to graphite spontaneous? Explain your answer. (b) If yes, does knowing that the conversion is spontaneous tell you how rapid the conversion is?

Equation 17.10 highlights the two thermodynamic driving forces that contribute to a decrease in free energy and to making a process spontaneous at constant temperature and pressure:

1. The system experiences an increase in entropy ($\Delta S > 0$).
2. The process is exothermic ($\Delta H < 0$).

One or both of these conditions must be true for a reaction to be spontaneous.

Using Equation 17.10, we can calculate the change in Gibbs free energy of a process if we first calculate the values of ΔH and ΔS. For a chemical reaction occurring under standard conditions, we can calculate the *standard* change in Gibbs free energy ΔG°_{rxn} by using the following modified version of Equation 17.10:

$$\Delta G^\circ_{rxn} = \Delta H^\circ_{rxn} - T\Delta S^\circ_{rxn} \quad (17.11)$$

In Section 6.8 we calculated ΔH°_{rxn} values from the differences in standard enthalpies of formation ΔH°_f of products and reactants:

$$\Delta H^\circ_{rxn} = \sum n_{products}\Delta H^\circ_{f,products} - \sum n_{reactants}\Delta H^\circ_{f,reactants} \quad (6.19)$$

The value of ΔS°_{rxn} can be calculated using Equation 17.5:

$$\Delta S^\circ_{rxn} = \sum n_{products}S^\circ_{products} - \sum n_{reactants}S^\circ_{reactants} \quad (17.5)$$

We can combine the results of these two calculations in Equation 17.11 to calculate ΔG°_{rxn}, as illustrated in Sample Exercise 17.4.

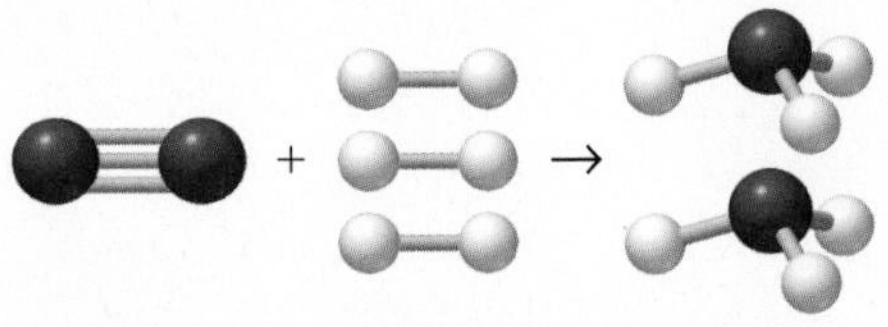

N$_2$ + 3 H$_2$ → 2 NH$_3$

FIGURE 17.12 In the synthesis of ammonia, one molecule of nitrogen reacts with three molecules of hydrogen to yield two molecules of ammonia. All substances are gases.

SAMPLE EXERCISE 17.4 Predicting Reaction Spontaneity under Standard Conditions LO4

Consider the reaction of nitrogen gas and hydrogen gas (**Figure 17.12**) at 298 K to make ammonia at the same temperature:

$$N_2(g) + 3\,H_2(g) \rightarrow 2\,NH_3(g)$$

a. Before doing any calculations, predict the sign of ΔS°_{rxn}.
b. What is the actual value of ΔS°_{rxn}?
c. What is the value of ΔH°_{rxn}?
d. What is the value of ΔG°_{rxn} at 298 K?
e. Is the reaction spontaneous under standard conditions?

Collect and Organize For a given reaction, we want to predict the sign of the standard entropy change and then calculate the changes in entropy, enthalpy, and Gibbs free energy. We can then determine the spontaneity of the reaction under standard conditions. Standard molar entropies and standard enthalpies of formation of the reactants and product are found in Table 17.3 and Appendix 4, respectively. Figure 17.12 reinforces the point that there are more molecules of gaseous reactants than products in the reaction.

Analyze We can use Equation 17.5 to calculate entropy changes under standard conditions, Equation 6.19 to calculate ΔH°_{rxn} from standard enthalpy of formation values (in Appendix 4), and then use these values in Equation 17.11 to calculate the free energy change in the reaction. The sign of ΔG°_{rxn} will indicate whether the reaction is spontaneous under standard conditions ($T = 298$ K).

Solve

a. The number of gas-phase molecules decreases as the reaction proceeds, so ΔS°_{rxn} will be negative.
b. We use data from Table 17.3 in Equation 17.5 to calculate ΔS°_{rxn}:

$$\Delta S^\circ_{rxn} = \sum n_{products}S^\circ_{products} - \sum n_{reactants}S^\circ_{reactants} = \Delta S^\circ_{sys}$$

$$= \left\{\left[2\ \cancel{mol} \times \left(\frac{192.5\ J}{\cancel{mol} \cdot K}\right)\right] - \left[1\ \cancel{mol} \times \left(\frac{191.5\ J}{\cancel{mol} \cdot K}\right) + 3\ \cancel{mol} \times \left(\frac{130.6\ J}{\cancel{mol} \cdot K}\right)\right]\right\}$$

$$= -198.3\ J/K$$

The entropy change is negative, as predicted in part (a).

c. The change in enthalpy that accompanies the reaction under standard conditions is

$$\Delta H^\circ_{rxn} = \sum n_{products}\Delta H^\circ_{f,products} - \sum n_{reactants}\Delta H^\circ_{f,reactants}$$

$$= \left\{\left[2\ \text{mol} \times \left(\frac{-46.1\ \text{kJ}}{\text{mol}}\right)\right] - \left[1\ \text{mol} \times \left(\frac{0.0\ \text{kJ}}{\text{mol}}\right) + 3\ \text{mol} \times \left(\frac{0.0\ \text{kJ}}{\text{mol}}\right)\right]\right\}$$

$$= -92.2\ \text{kJ}$$

d. We insert the values of ΔS°_{rxn} and ΔH°_{rxn} calculated in parts (a) and (b) into Equation 17.11:

$$\Delta G^\circ_{rxn} = \Delta H^\circ_{rxn} - T\Delta S^\circ_{rxn}$$

$$= -92.2\ \text{kJ} - \left[(298\ \text{K}) \times \left(-198.3\ \frac{\text{J}}{\text{K}} \times \frac{1\ \text{kJ}}{1000\ \text{J}}\right)\right]$$

$$= -33.1\ \text{kJ}$$

Note that the units on ΔH°_{rxn} and ΔS°_{rxn} are different: kJ and J/K, respectively. We needed to multiply the ΔS°_{rxn} value by the temperature to convert to units of kJ.

e. The decrease in Gibbs free energy tells us that the reaction is spontaneous under standard conditions.

Think About It We predicted that the sign of ΔG°_{rxn} would be diagnostic of the spontaneity of a reaction. After substituting for ΔH°_{rxn} and ΔS°_{rxn} in Equation 17.11, $\Delta G^\circ_{rxn} < 0$ indicating that the reaction is spontaneous at 298°C and $P = 1$ bar. In this reaction, a decrease in entropy ($\Delta S^\circ_{rxn} < 0$) is more than offset by a favorable enthalpy change ($\Delta H^\circ_{rxn} < 0$) so that overall, $\Delta G^\circ_{rxn} < 0$.

Practice Exercise

For the reaction $2\ H_2(g) + O_2(g) \rightarrow 2\ H_2O(\ell)$

a. Predict the sign of the entropy change for the reaction.
b. What is the value of ΔS°_{rxn}?
c. What is the value of ΔH°_{rxn}?
d. Is the reaction spontaneous under standard conditions?

CONCEPT **TEST**

The preparation of ammonia from nitrogen and hydrogen in Sample Exercise 17.4 is determined to be spontaneous under standard conditions, yet if we mix the two gases at 298 K, no reaction is observed. Suggest a reason why.

Another way to calculate the change in Gibbs free energy of a reaction under standard conditions is based on another thermodynamic property of substances listed in Appendix 4: **standard free energy of formation (ΔG°_f)**. A compound's ΔG°_f value is the change in free energy associated with the formation of 1 mole of it in its standard state from its elements in their standard states.

In Equation 6.19 we calculated standard enthalpies of reaction (ΔH°_{rxn}) from the difference in the standard enthalpies of formation (ΔH°_f) of their products and reactants. We can also calculate the change in standard free energy of a reaction under standard conditions from the difference in the standard free energies of formation of its products and reactants. As with standard enthalpies of formation, standard free energies of formation of the most stable forms of elements in their standard states are defined to be zero. The similarities between the two calculations can be seen from the similar formats of the equations used to calculate ΔG°_{rxn},

$$\Delta G^\circ_{rxn} = \sum n_{products}\Delta G^\circ_{f,products} - \sum n_{reactants}\Delta G^\circ_{f,reactants} \quad (17.12)$$

and ΔH°_{rxn},

$$\Delta H^\circ_{rxn} = \sum n_{products}\Delta H^\circ_{f,products} - \sum n_{reactants}\Delta H^\circ_{f,reactants} \quad (6.19)$$

standard free energy of formation (ΔG°_f) the change in free energy associated with the formation of 1 mole of a compound in its standard state from its component elements.

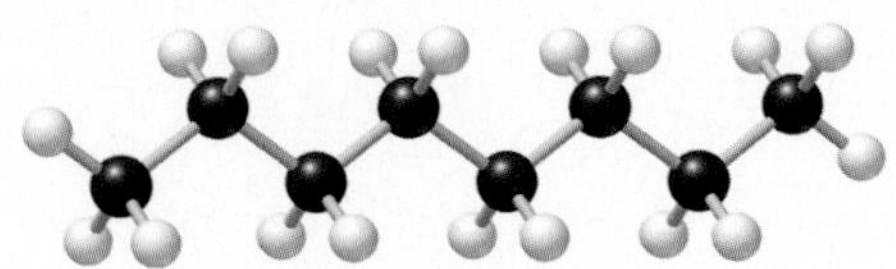
Octane
$\Delta G^\circ_f = 16.3$ kJ/mol

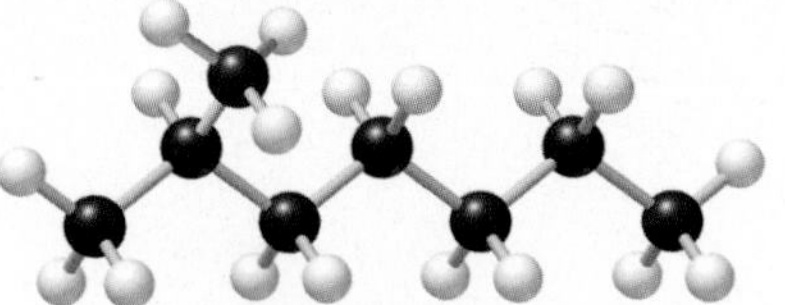
2-Methylheptane
$\Delta G^\circ_f = 11.7$ kJ/mol

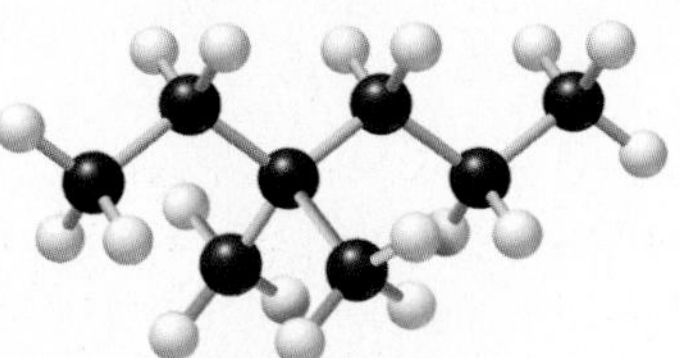
3,3-Dimethylhexane
$\Delta G^\circ_f = 12.6$ kJ/mol

FIGURE 17.13 Molecular structures and ΔG°_f values of three C_8H_{18} isomers.

Sample Exercise 17.5 illustrates just how similar the two calculations are.

CONCEPT TEST

Molecular models and standard free energies of formation for three structural isomers with the molecular formula C_8H_{18} are shown in **Figure 17.13**. All three isomers burn in air, as described by the same chemical equation:

$$2\ C_8H_{18}(\ell) + 25\ O_2(g) \rightarrow 16\ CO_2(g) + 18\ H_2O(g)$$

Are the ΔG°_{rxn} values for the three combustion reactions also the same? Why or why not?

SAMPLE EXERCISE 17.5 Calculating ΔG°_{rxn} by Using Appropriate ΔG°_f Values — LO4

Use the appropriate standard free energy of formation values from Appendix 4 to calculate the change in Gibbs free energy as ethanol burns under standard conditions. Assume the reaction proceeds as described by the following chemical equation:

$$CH_3CH_2OH(\ell) + 3\ O_2(g) \rightarrow 2\ CO_2(g) + 3\ H_2O(\ell)$$

Collect and Organize We can calculate the value of ΔG°_{rxn} for the combustion of ethanol according to Equation 17.12, using the ΔG°_f values of the reactants and products in the combustion reaction:

Substance	$CH_3CH_2OH(\ell)$	$O_2(g)$	$CO_2(g)$	$H_2O(\ell)$
ΔG°_f (kJ/mol)	−174.9	0	−394.4	−237.2

Analyze The reaction consumes 1 mole of liquid ethanol and 3 moles of oxygen, and it produces 2 moles of CO_2 gas and 3 moles of *liquid* H_2O. Substituting the above ΔG°_f values and the appropriate numbers of moles into Equation 17.12 yields the value of ΔG°_{rxn}. The combustion of ethanol, a common additive in gasoline in the United States and Canada, is spontaneous, so ΔG°_{rxn} should be less than zero.

Solve We insert the appropriate numbers of moles and ΔG°_f values into Equation 17.12:

$$\begin{aligned}\Delta G^\circ_{rxn} &= \sum n_{products}\Delta G^\circ_{products} - \sum n_{reactants}\Delta G^\circ_{reactants}\\ &= [2\ \cancel{mol\ CO_2} \times (-394.4\ kJ/\cancel{mol\ CO_2}) + 3\ \cancel{mol\ H_2O} \times (-237.2\ kJ/\cancel{mol\ H_2O})]\\ &\quad -1\ \cancel{mol\ CH_3CH_2OH} \times (-174.9\ kJ/\cancel{mol\ CH_3CH_2OH}) + 3\ \cancel{mol\ O_2}\\ &\quad \times (0.0\ kJ/\cancel{mol\ O_2})\\ &= -1325.5\ kJ\end{aligned}$$

Think About It The calculated value represents that part of the total energy released by the combustion of 1 mole of ethanol under standard conditions that is available to do useful work.

Practice Exercise Use the appropriate standard free energy of formation values in Appendix 4 to calculate the value of ΔG°_{rxn} for the steam-reforming reaction used to produce H_2 gas:

$$CH_4(g) + H_2O(g) \rightarrow CO(g) + 3\ H_2(g)$$

What exactly does "energy available to do useful work" mean? Let's attempt to answer this question, using as our model the internal combustion (gasoline) engines that power most automobiles.

A combustion reaction is a thermodynamic system that experiences a decrease in internal energy (ΔE) as energy in the form of heat (q) flows from it into its surroundings and as it does work (w) on its surroundings. These three variables are related by Equation 6.5:

$$\Delta E = q + w \tag{6.5}$$

From the perspective of the system, all three quantities are less than zero. Internal combustion engines have cooling systems to manage the dissipation of heat, which is wasted energy that does nothing to power the car. The energy that moves the car is derived from the rapid expansion of the gaseous products of combustion in the cylinders of the engine. As **Figure 17.14** shows, this expansion pushes down on the piston of a cylinder, increasing the volume of the reaction mixture. The pressure exerted by the reacting gases and their products on the pistons in a car's engine, multiplied by the volumes they displace, constitutes the P–V work (see Equation 6.6), which propels the car:

$$w = -P\Delta V$$

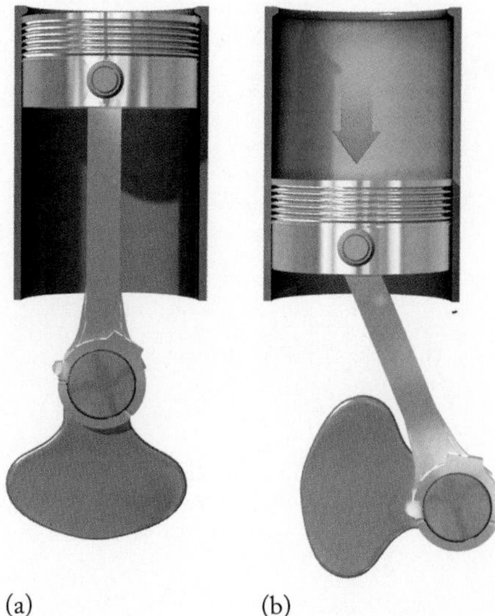

(a) (b)

FIGURE 17.14 (a) In a car engine, thermal expansion causes the gases in a cylinder to (b) push down on a piston with a pressure (P) represented by the blue arrow. The product of P and the change in volume of the gases (ΔV) is the work done by the expanding gases that propels the car.

Gibbs free energy is a measure of the *maximum* amount of work that can be done by the energy released during combustion. To see how this theoretical quantity of work compares with the total energy released, let's rearrange Equation 17.10 by isolating the ΔH term:

$$\Delta H = \Delta G + T\Delta S \tag{17.13}$$

Equation 17.13 tells us that the enthalpy change that accompanies making and breaking chemical bonds during a chemical reaction may be divided into two parts. The ΔG part is the energy that can theoretically be converted into motion and other useful work (such as propelling a car or generating electricity for its electrical system). The $T\Delta S$ part, on the other hand, is not usable: it is the portion of energy that disperses when, for example, hot gases flow out of an automobile exhaust pipe. This part of ΔH is wasted.

Consequently, the conversion of chemical energy into useful mechanical energy (ΔG) is never 100% efficient. A portion of ΔG is also wasted because combustion and the energy conversion happen quickly and therefore irreversibly. Maximum efficiency comes with very slow and reversible reactions (see Section 17.2), but that is not how automobile engines operate. It turns out that gasoline engines convert only about 30% of the energy produced during combustion into useful work.

17.6 Temperature and Spontaneity

Let's revisit the process of ice melting, this time focusing on how the values of the three terms ΔH, $T\Delta S$, and ΔG in Equation 17.10 change as the temperature of a mixture of ice and water increases from −10°C to +10°C (**Figure 17.15**). As the temperature rises over this range, there is little impact on the enthalpy of fusion, ΔH, as shown by the nearly flat green line in Figure 17.15. However, increasing the value of T does increase the value $T\Delta S$ (as shown by the upward slope of the purple line). After all, the ΔS of melting ice (or any melting solid)[2] has a positive value, so $T\Delta S$ must increase as T increases.

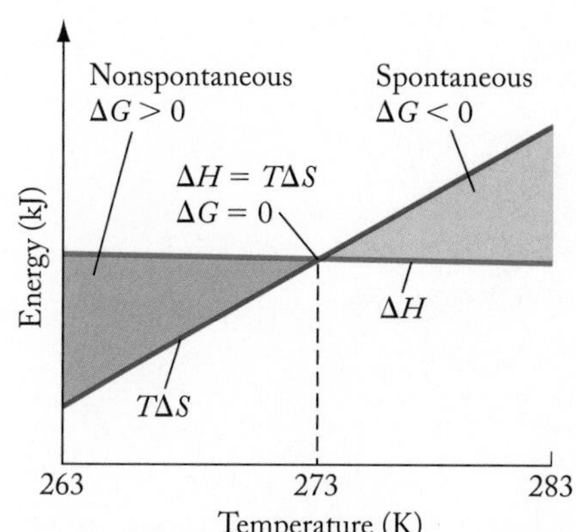

FIGURE 17.15 Changes in the values of ΔH, the quantity $T\Delta S$, and ΔG for ice melting as temperature increases from −10°C (263 K) to +10°C (283 K).

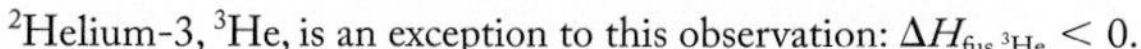

[2]Helium-3, ^3He, is an exception to this observation: $\Delta H_{fus,^3He} < 0$.

The green and purple lines intersect at 0°C, which means $\Delta H = T\Delta S$ at that temperature. Put another way, the difference between ΔH and $T\Delta S$ at 0°C is zero, and so is ΔG (because $\Delta G = \Delta H - T\Delta S$). The temperature at the point in Figure 17.15 where $\Delta G = 0$ defines the melting (or freezing) point, which by definition is the temperature at which the solid melts at the same rate as the liquid freezes. The two phases are in *equilibrium*. At 0°C, no net change takes place, and both phases coexist.

We also know that ice melts spontaneously above 0°C (273 K), which means that ΔG for the melting process must be less than zero. The graph in Figure 17.15 shows that indeed it is. Above 0°C, the value of $T\Delta S$ is greater than ΔH. Therefore, the difference between them ($\Delta H - T\Delta S$) is less than zero and becomes more negative with increasing temperature, as illustrated by the increasing distance from the purple line to the green line in that region of the graph.

Below 0°C, ice does not melt spontaneously, which means ΔG is greater than zero. The graph shows why this is true: at $T < 273$ K (0°C), $T\Delta S$ values (purple line) are less than ΔH values (green line), which means that $\Delta H - T\Delta S$ (and ΔG) is greater than zero. The positive ΔG values mean the process is nonspontaneous below 273 K, and ice does not melt below its freezing point. However, the opposite process—liquid water freezing—*is* spontaneous because reversing a process keeps the absolute values but switches the signs of ΔH, ΔS, and ΔG. Therefore, if the melting process is nonspontaneous, then the exothermic freezing process *is* spontaneous at low temperatures.

CONNECTION In Chapter 10 we discussed the equilibrium between phases of a substance and used phase diagrams to illustrate which physical states are stable at various combinations of temperature and pressure.

CONCEPT TEST

A given process is spontaneous at lower temperature and nonspontaneous at higher temperature. What does a graph like the one in Figure 17.15 look like for such a process?

Equilibrium also exists for water at 100°C and 1 atm pressure, which are the temperature and pressure at which $\Delta G_{\text{vaporization}}$ and $\Delta G_{\text{condensation}}$ both equal zero. At 100°C, liquid water vaporizes and water vapor condenses at the same rate; the two phases coexist.

The temperature at which $\Delta G = 0$ for a process can be calculated from Equation 17.10 if we know the values of ΔH and ΔS. For example, the values for the fusion of water are

$$H_2O(s) \rightarrow H_2O(\ell) \qquad \Delta H^\circ_{\text{fus}} = 6.01 \times 10^3 \text{ J/mol};\ \Delta S^\circ_{\text{fus}} = 22.0 \text{ J/(mol} \cdot \text{K)}$$

In doing this calculation, we assume that the values of ΔH° and ΔS° do not change significantly with small changes in temperature. For this process, as well as for most other physical and chemical processes, this assumption is acceptable. We can assume, therefore, that $\Delta G = \Delta H - T\Delta S \approx \Delta H^\circ - T\Delta S^\circ$. Inserting the values of ΔH° and ΔS° and using $\Delta G = 0$ for a process at equilibrium, we get

$$\Delta G = (6.01 \times 10^3 \text{ J/mol}) - T[22.0 \text{ J/(mol} \cdot \text{K)}] = 0$$

$$T = \frac{6.01 \times 10^3 \text{ J/}\cancel{\text{mol}}}{22.0 \text{ J/(}\cancel{\text{mol}} \cdot \text{K)}} = 273 \text{ K} = 0°\text{C}$$

This is the familiar value for the melting point of ice.

Like physical processes, chemical reactions under standard conditions may be spontaneous below or above a characteristic temperature (T) at which $\Delta G^\circ = 0$, and

$$T = \frac{\Delta H^\circ}{\Delta S^\circ}$$

TABLE 17.4 Effects of $\Delta H°$ and $\Delta S°$ on $\Delta G°$ and Spontaneity

$\Delta H°$	$\Delta S°$	$\Delta G°$	Spontaneity
$-$	$+$	Always <0	Always spontaneous
$-$	$-$	<0 at lower temperature	Spontaneous at lower temperature
$+$	$+$	<0 at higher temperature	Spontaneous at higher temperature
$+$	$-$	Always >0	Never spontaneous

As summarized in **Table 17.4**, an exothermic process ($\Delta H° < 0$) for which $\Delta S°$ is also less than zero is spontaneous only below a characteristic temperature, such as the freezing of liquid water. An endothermic process ($\Delta H° > 0$) for which $\Delta S°$ is also greater than zero is spontaneous only above a characteristic temperature, such as the melting of ice into liquid water. This pattern makes sense because lower temperatures make the value of the $T\Delta S°$ term in Equation 17.11 less negative if $\Delta S° < 0$. This helps make $\Delta G°$ more negative because of the minus sign in front of the $T\Delta S°$ term. Higher temperatures make the value of $T\Delta S°$ more positive if $\Delta S° > 0$, which again helps make $\Delta G°$ more negative.

SAMPLE EXERCISE 17.6 Relating Reaction Spontaneity to ΔH_{rxn} and ΔS_{rxn} **LO5**

The gas-phase reaction between methanol and water that produces carbon dioxide and 3 moles of hydrogen is accompanied by an increase in entropy, $\Delta S°_{rxn} = 176.9$ J/K, under standard conditions.

$$CH_3OH(g) + H_2O(g) \rightarrow CO_2(g) + 3\ H_2(g)$$

At what temperatures is this reaction spontaneous?

Collect and Organize We are asked to determine the temperatures at which a reaction with a positive entropy change is spontaneous. Equation 17.11 relates spontaneity ($\Delta G°_{rxn} < 0$) to $\Delta H°_{rxn}$, $\Delta S°_{rxn}$, and T.

$$\Delta G°_{rxn} \approx \Delta H°_{rxn} - T\Delta S°_{rxn} < 0$$

Analyze According to Table 17.4, a reaction in which entropy increases ($\Delta S°_{rxn} > 0$) will be spontaneous at all temperatures if it is also exothermic ($\Delta H°_{rxn} < 0$). However, if the reaction is accompanied by a positive enthalpy change, (endothermic), then the reaction will only be spontaneous above a certain temperature. We need to calculate using standard enthalpies of formation and then solve the relationship:

$$T > \frac{\Delta H°_{rxn}}{\Delta S°_{rxn}}$$

Solve The enthalpy change for the reaction is calculated as follows:

$$\begin{aligned}\Delta H°_{rxn} &= [\Delta H°_{CO_2,g} + 3\ \Delta H°_{H_2,g}] - [\Delta H°_{CH_3OH,g} + \Delta H°_{H_2O,g}]\\ &= [(1\ \cancel{\text{mol CO}_2})(-393.5\ \text{kJ}/\cancel{\text{mol}}) + (3\ \cancel{\text{mol H}_2})(0.0\ \text{kJ}/\cancel{\text{mol}})]\\ &\quad -[(1\ \cancel{\text{mol CH}_3\text{OH}})(-200.7\ \text{kJ}/\cancel{\text{mol}}) + (1\ \cancel{\text{mol H}_2\text{O}})(-241.8\ \text{kJ}/\cancel{\text{mol}})]\\ &= 49.0\ \text{kJ}\end{aligned}$$

Substituting into the relationship between T, $\Delta H°_{rxn}$, and $\Delta S°_{rxn}$:

$$T > \frac{\Delta H°_{rxn}}{\Delta S°_{rxn}} > \frac{49.0\ \cancel{\text{kJ}}}{0.1769\ \cancel{\text{kJ}}/\text{K}} = 277\ \text{K}$$

Thus, the conversion of methanol vapor and steam to hydrogen and carbon dioxide should be spontaneous above 277 K or 4°C.

Think About It The enthalpy change for the reaction is a relatively small value so we do not require an enormously high temperature for the reaction to be spontaneous.

Practice Exercise Hydrazine, N_2H_4, is used as a rocket fuel reacting with oxygen to produce nitrogen gas and liquid water.

$$N_2H_4(g) + O_2(g) \rightarrow N_2(g) + 2\ H_2O(\ell)$$

At what temperatures is this reaction spontaneous?

SAMPLE EXERCISE 17.7 Relating Reaction Spontaneity to ΔH and ΔS **LO5**

A certain chemical reaction is spontaneous at low temperatures but not at high temperatures. Use Equation 17.10 ($\Delta G = \Delta H - T\Delta S$) to determine the signs of the enthalpy and entropy changes for this reaction.

Collect, Organize, and Analyze We are asked to determine the signs of ΔH (enthalpy change) and ΔS (entropy change) on the basis of the change in spontaneity of a reaction as temperature changes. This means we need to think about not only the signs of ΔH and ΔS but also how the relative magnitude of ΔH and $T\Delta S$ varies with temperature.

Solve The importance of ΔS increases with increasing temperature because the product $T\Delta S$ appears in Equation 17.10. The given reaction is known to be spontaneous ($\Delta G < 0$) at low temperatures, meaning that the magnitude of the enthalpy term is more likely to be larger than the magnitude of $T\Delta S$, in which case ΔH must be negative for the reaction to be spontaneous. The given reaction is known to be nonspontaneous ($\Delta G > 0$) at higher temperatures, where the magnitude of $T\Delta S$ is more likely to be larger than the magnitude of ΔH, in which case ΔS must also be negative in order for $\Delta G > 0$.

Think About It In this reaction, the impact of a negative ΔS value is more than offset by a decrease in enthalpy—a change that favors the reaction. Table 17.4 confirms that a process that is spontaneous only at low temperatures is one in which there is a decrease in both entropy and enthalpy. Recall that the freezing of water below 0°C is an exothermic reaction ($\Delta H < 0$) that is accompanied by a decrease in entropy ($\Delta S < 0$).

Practice Exercise At high temperatures, ammonia decomposes to nitrogen and hydrogen gases:

$$2\ NH_3(g) \rightarrow N_2(g) + 3\ H_2(g)$$

ΔH for the reaction is positive and ΔS is positive. Predict whether the reaction is spontaneous at all temperatures, or only at high temperatures.

17.7 Free Energy and Chemical Equilibrium

We have just seen that equilibrium exists when the free energy change that accompanies a process is zero. For chemical reactions *not* at equilibrium, the magnitude of ΔG—how far it is from zero in either a negative or positive direction—indicates how far a system is from its equilibrium position.

In Chapter 14 we showed how the reaction quotient (Q) can be used to determine whether a reaction is at chemical equilibrium. When the value of Q is much larger or smaller than the value of the reaction's equilibrium constant (K), we know that the reaction is far from chemical equilibrium. When $Q < K$, for example, the ratio of product concentrations to reactant concentrations (raised to the appropriate powers) is less than it would be at equilibrium. To reach equilibrium, the reaction proceeds (spontaneously!) in the forward direction until the ratio of product to reactant concentrations matches the value of K. In other words, when $Q < K$, ΔG_{rxn} must be less than zero.

On the other hand, when $Q > K$, the ratio of product to reactant concentrations is greater than it would be at equilibrium. To reach equilibrium, the reaction must proceed in reverse, so that products are consumed and reactants form, until the ratio of product to reactant concentrations matches the value of K. When $Q > K$, then, the forward reaction is *not* spontaneous because the *reverse* reaction is spontaneous, which means ΔG_{rxn} for the forward reaction must be greater than zero.

To summarize these points for any reaction:

- When $Q = K$, $\Delta G_{rxn} = 0$ and the reaction is at equilibrium.
- When $Q < K$, $\Delta G_{rxn} < 0$ and the reaction is spontaneous.
- When $Q > K$, $\Delta G_{rxn} > 0$ and the reaction is nonspontaneous.

CONNECTION In Section 14.5 the reaction quotient (Q) is defined as the ratio of the concentrations (or partial pressures) of products to reactants in a chemical reaction, each term raised to a power equal to the coefficient of that substance in the balanced chemical equation describing the reaction.

CHEMTOUR
Equilibrium and Thermodynamics

The key concept here is that the value of ΔG_{rxn} for any chemical reaction depends not only on the standard free energy of formation values of its reactants and products—that is, on the value of ΔG°_{rxn}—but also on the concentrations of the reactants and products in the reaction mixture. In other words, the value of ΔG_{rxn} depends on the value of Q. This dependency is expressed in Equation 17.14:

$$\Delta G_{rxn} = \Delta G^\circ_{rxn} + RT \ln Q \qquad (17.14)$$

To illustrate the dependency of ΔG_{rxn} on Q, let's use as our chemical model the dimerization of NO_2:

$$2\,NO_2(g) \rightleftharpoons N_2O_4(g)$$

We can calculate the change in standard free energy for the reaction (ΔG°_{rxn}) using standard free energy of formation (ΔG°_f) values from Table A4.3 in Appendix 4 and Equation 17.12:

$$\begin{aligned}\Delta G^\circ_{rxn} &= \sum n_{products}\Delta G^\circ_{f,products} - \sum n_{reactants}\Delta G^\circ_{f,reactants}\\ &= 1\,\cancel{mol}\,(99.8\text{ kJ}/\cancel{mol}) - 2\,\cancel{mol}\,(51.3\text{ kJ}/\cancel{mol}) = -2.8\text{ kJ}\end{aligned}$$

The negative value of ΔG°_{rxn} indicates that the reaction as written should, under standard conditions, proceed spontaneously in the forward direction at 298 K. However, this value of ΔG°_{rxn} applies only when $P_{NO_2} = P_{N_2O_4} = 1$ bar, which rarely happens.

Let's consider what happens in a reaction vessel at 298 K if it initially contains pure NO_2 at a partial pressure of 2.0 bar with no (or hardly any) N_2O_4 present. Under these conditions, $Q_p = 0$:

$$Q_p = \frac{P_{N_2O_4}}{(P_{NO_2})^2} = \frac{0}{4.0} = 0$$

The natural log (ln) of zero is undefined, so we can't use Equation 17.14 to calculate the value of ΔG_{rxn}. However, if the reaction mixture contained a trace of N_2O_4 so that, for example, its partial pressure was 1.0 mbar, then Q_p would be $0.0010/4.0 = 0.00025$, and ΔG_{rxn} would be

$$\begin{aligned}\Delta G_{rxn} &= \Delta G^\circ_{rxn} + RT\ln Q\\ &= -2.8\text{ kJ/mol} + \left[8.314\,\cancel{J}/(\cancel{mol\cdot K}) \times 298\,\cancel{K} \times \ln 0.00025 \times \frac{1\text{ kJ}}{1000\,\cancel{J}}\right]\\ &= -23.3\text{ kJ}\end{aligned}$$

This value of ΔG_{rxn} is plotted as point ① on the graph in **Figure 17.16**. Its negative value tells us that the reaction will proceed spontaneously, consuming molecules of NO_2 at twice the rate at which it produces molecules of N_2O_4. As the reaction proceeds, P_{NO_2} decreases and $P_{N_2O_4}$ increases as does the value of Q, which makes the $RT\ln Q$ term in Equation 17.14 greater and the value ΔG_{rxn} less negative. Eventually, P_{NO_2} falls to 0.49 bar; $P_{N_2O_4}$ increases to 0.75 bar, and the value of $Q_p = 0.75/(0.49)^2 = 3.1$. At point ② on the graph, the value of $RT\ln Q = 8.314 \times 298 \times \ln 3.1 = +2.8$ kJ, which means $\Delta G_{rxn} = (-2.8 + 2.8) = 0$. The reaction has come to chemical equilibrium, which means $Q_p = 3.1 = K_p$.

At equilibrium Equation 17.14 becomes

$$\Delta G_{rxn} = \Delta G^\circ_{rxn} + RT\ln K = 0$$

or

$$\Delta G^\circ_{rxn} = -RT\ln K \tag{17.15}$$

We can use Equation 17.15 to confirm that the above K_p value is consistent with the value of ΔG°_{rxn} for the dimerization reaction:

$$\begin{aligned}\Delta G^\circ_{rxn} &= -RT\ln K\\ &= -8.314\ \text{J/(mol}\cdot\text{K)} \times 298\ \text{K} \times \ln(3.1) \times \frac{1\ \text{kJ}}{1000\ \text{J}} = -2.8\ \text{kJ/mol}\end{aligned}$$

Starting with a reaction mixture that is essentially pure reactant is a common way to run a reaction, but Figure 17.16 shows the results of another approach: starting with (nearly) pure product. Consider what would happen if a reaction vessel at 298 K contained a reaction mixture in which the partial pressure of N_2O_4 was 1.0 bar with a trace amount (2.0 mbar) of NO_2. Using these values in Equation 17.14 yields a ΔG_{rxn} value of +28.0 kJ for the reaction. This positive value, which is plotted at point ③ in Figure 17.16, indicates that the forward reaction is not spontaneous. However, the reverse reaction is: the dimer spontaneously dissociates into molecules of NO_2, causing $P_{N_2O_4}$ to decrease, P_{NO_2} to increase, and the values of Q_p (of the dimerization reaction) and $RT\ln Q_p$ to decrease. When ($RT\ln Q_p$) falls to +2.8 kJ/mol, the value of ΔG_{rxn} has again reached $(-2.8 + 2.8) = 0$ at point ② on the graph, but this time as N_2O_4 decomposed into NO_2:

$$N_2O_4(g) \rightleftharpoons 2\ NO_2(g)$$

Rearranging Equation 17.15 allows us to calculate the K value for a reaction from its change in standard free energy and absolute temperature. First, we rearrange the terms:

$$\ln K = \frac{-\Delta G^\circ_{rxn}}{RT} \tag{17.16}$$

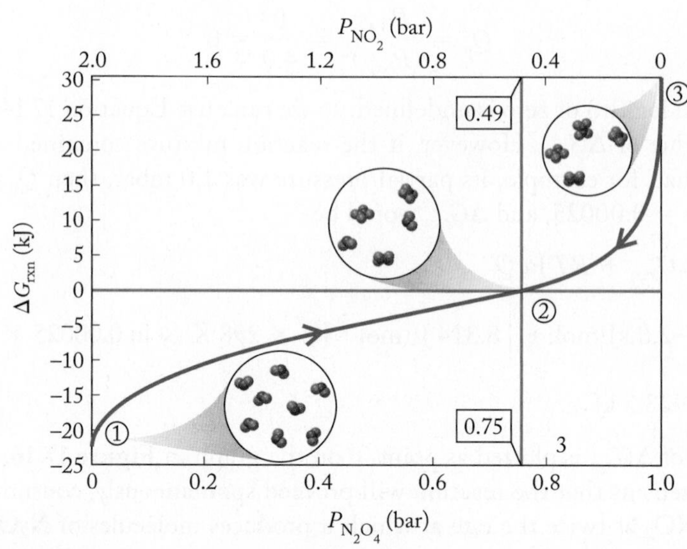

FIGURE 17.16 Changes in the value of ΔG_{rxn} during the dimerization reaction $2\ NO_2(g) \rightleftharpoons N_2O_4(g)$. At point ① the system consists of nearly pure reactant, NO_2 gas; ΔG_{rxn} has a large negative value, and the dimerization reaction proceeds spontaneously. As it does, P_{NO_2} decreases and $P_{N_2O_4}$ increases, which means Q_p and the value ΔG_{rxn} also increase until ΔG_{rxn} reaches zero at point ②. Here $Q_p = K_p$; the reaction has reached equilibrium, and no further change in the composition of the reaction mixture occurs. At point ③ the system consists of nearly pure N_2O_4, and ΔG_{rxn} has a large positive value, which means the dimerization reaction is nonspontaneous, but the reverse reaction, the decomposition of N_2O_4, does proceed spontaneously. As it does, $P_{N_2O_4}$ decreases and P_{NO_2} increases, which means Q_p and the value ΔG_{rxn} both decrease until ΔG_{rxn} reaches zero at point ②, and the system is again at equilibrium.

Then we take the antilogarithm of both sides:

$$K = e^{-\Delta G^\circ_{\text{rxn}}/RT} \tag{17.17}$$

Equation 17.17 provides the following interpretation of reaction spontaneity under standard conditions: whenever ΔG° is negative, the exponent $-\Delta G^\circ/RT$ in Equation 17.17 is positive, and $e^{-\Delta G^\circ/RT} > 1$, making $K > 1$. Therefore, any reversible reaction with an equilibrium constant greater than 1 is spontaneous under standard conditions, as shown in **Figure 17.17a**. This spontaneity has its

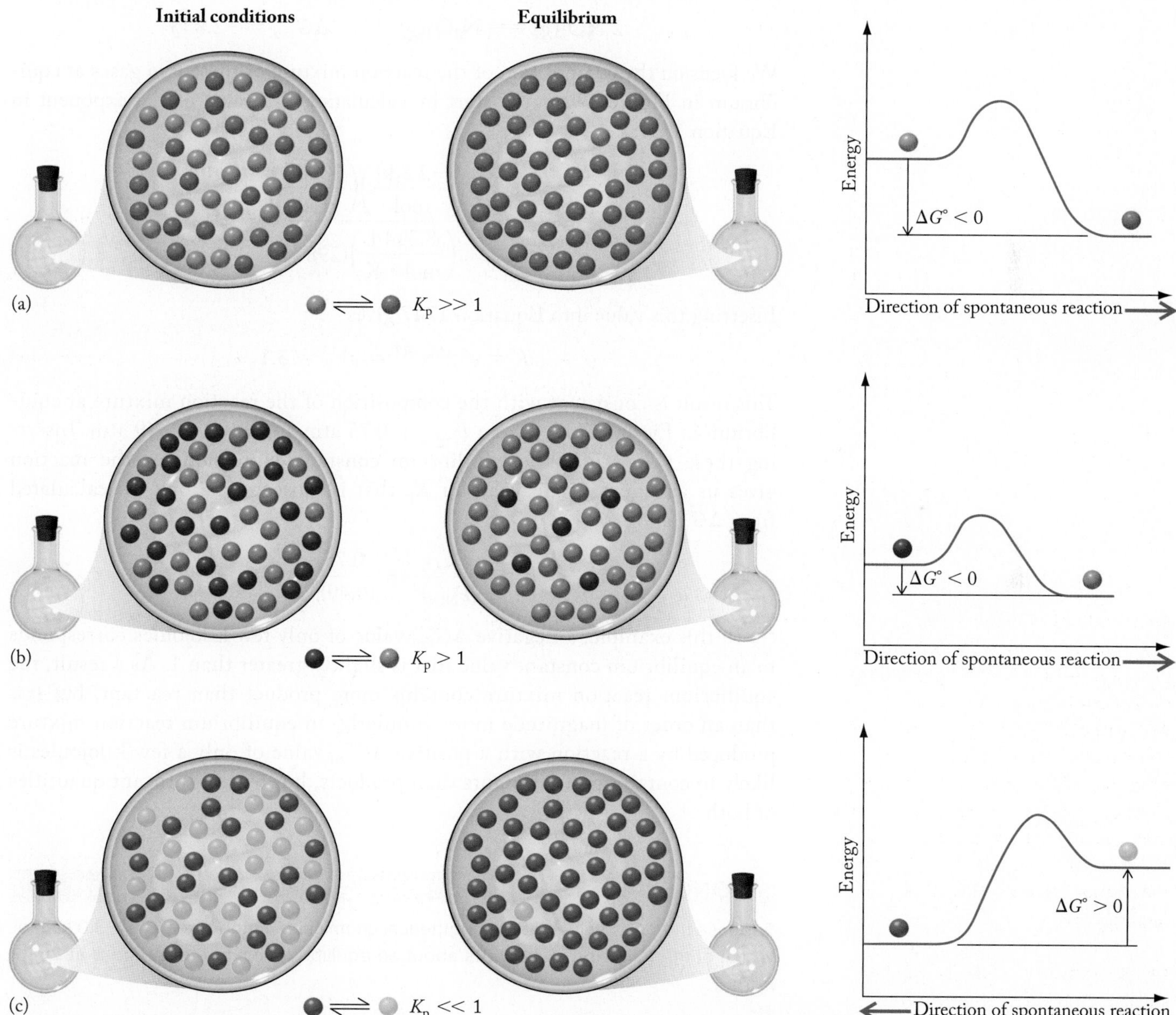

FIGURE 17.17 The equilibrium constant of a chemical reaction is linked to its ΔG° value. Initially the three flasks on the left each contain equimolar mixtures of different pairs of gases. The initial partial pressure of each gas is 1 atm. The three images on the right represent the mixtures in these same flasks at equilibrium. (a) The value of ΔG° for the formation of "red" gas from "green" gas has a large negative value, which makes K_p much greater than 1. The reaction proceeds in the forward direction, leaving only a little green gas left over. (b) If the value of ΔG° for the formation of "orange" gas from "blue" gas has a smaller negative value than in part (a), the value of K_p is smaller, though still greater than 1, and there is more orange gas present at equilibrium than blue gas. (c) If the value of ΔG° for the formation of "yellow" gas from "purple" gas is positive, K_p is less than 1 and the reaction runs in the reverse direction, forming purple gas from yellow gas.

limits. As reactants are consumed and products are formed, the value of the reaction quotient increases, making the value of ΔG less negative. When it reaches zero, there is no further change in the composition of the reaction mixture because chemical equilibrium has been achieved.

It follows that a reversible reaction with a less negative value of $\Delta G°$ (**Figure 17.17b**) is still spontaneous but has a smaller equilibrium constant, so less reactant is consumed, and less product is formed before the value of ΔG reaches 0. Finally, a reaction that has a positive value of $\Delta G°$ (**Figure 17.17c**) has $K < 1$, and the reaction is not spontaneous under standard conditions. Instead, the reverse of the reaction is spontaneous.

Let's apply this concept to the equilibrium between NO_2 and N_2O_4:

$$2\ NO_2(g) \rightleftharpoons N_2O_4(g) \qquad \Delta G°_{rxn} = -2.8\ kJ$$

We focus on the composition of the reaction mixture of these two gases at equilibrium in Figure 17.16. We start by calculating the value of the exponent in Equation 17.17:

$$-\frac{\Delta G°_{rxn}}{RT} = -\frac{\left(\frac{-2.8\ \cancel{kJ}}{\cancel{mol}}\right)\left(\frac{1000\ \cancel{J}}{1\ \cancel{kJ}}\right)}{\left(\frac{8.314\ \cancel{J}}{\cancel{mol}\cdot\cancel{K}}\right)(298\ \cancel{K})} = 1.13$$

Inserting this value into Equation 17.17 gives

$$K = e^{-\Delta G°_{rxn}/RT} = e^{1.13} = 3.1$$

This result is consistent with the composition of the reaction mixture at equilibrium in Figure 17.16, where $P_{N_2O_4} \approx 0.75$ atm and $P_{NO_2} \approx 0.49$ atm. Inserting these values into the equilibrium constant expression for the reaction gives us an approximate value of K_p that is close to the one we calculated from $\Delta G°_{rxn}$:

$$K_p = \frac{P_{N_2O_4}}{(P_{NO_2})^2} \approx \frac{0.75}{(0.49)^2} = 3.1$$

In this example, a negative $\Delta G°_{rxn}$ value of only few kilojoules corresponds to an equilibrium constant value that is slightly greater than 1. As a result, the equilibrium reaction mixture contains more product than reactant, but less than an order of magnitude more. Similarly, an equilibrium reaction mixture produced by a reaction with a positive $\Delta G°_{rxn}$ value of only a few kilojoules is likely to contain more reactants than products, but with significant quantities of both.

CONCEPT TEST

Suppose the $\Delta G°_{rxn}$ value of the hypothetical chemical reaction $A \rightleftharpoons B$ is −3.0 kJ/mol. Which of the following statements about an equilibrium mixture of A and B at 298 K is true?

a. There is only A present.
b. There is only B present.
c. There is an equimolar mixture of A and B present.
d. There is more A than B present.
e. There is more B than A present.

SAMPLE EXERCISE 17.8 Calculating the Value of K from ΔG_f° Values **LO6**

Use ΔG_f° values from Table A4.3 in the Appendix to calculate ΔG_{rxn}° and the value of K_p for the formation of NO_2 from NO and O_2 at 298 K:

$$NO(g) + \tfrac{1}{2} O_2(g) \rightleftharpoons NO_2(g)$$

Collect, Organize, and Analyze The ΔG_f° values we need from Table A4.3 are 51.3 kJ/mol for NO_2 and 86.6 kJ/mol for NO. Because O_2 gas is the most stable form of the element, its ΔG_f° value is 0.0 kJ/mol. We can use Equation 17.12 to calculate ΔG_{rxn}° and then use Equation 17.17 to calculate the value of K. We predict that a large, negative value for ΔG_{rxn}° will lead to a large equilibrium constant, K. Similarly, if $\Delta G_{rxn}^\circ > 0$, then K will be small.

Solve

$$\begin{aligned}\Delta G_{rxn}^\circ &= [\Delta G_{f,NO_2}^\circ] - [\Delta G_{f,NO}^\circ + \tfrac{1}{2}\Delta G_{f,O_2}^\circ] \\ &= [1\ \cancel{mol}\ (51.3\ kJ/\cancel{mol})] - [1\ \cancel{mol}\ (86.6\ kJ/\cancel{mol}) + \tfrac{1}{2}\ \cancel{mol}\ (0.0\ kJ/\cancel{mol})] \\ &= (51.3 - 86.6)\ kJ \\ &= -35.3\ kJ, \text{ or } -35{,}300\ J \text{ per mol of } NO_2 \text{ produced}\end{aligned}$$

The exponent in Equation 17.17 is

$$-\frac{\Delta G_{rxn}^\circ}{RT} = -\frac{\left(\dfrac{-35{,}300\ \cancel{J}}{\cancel{mol}}\right)}{\left(\dfrac{8.314\ \cancel{J}}{\cancel{mol} \cdot \cancel{K}}\right)(298\ \cancel{K})} = 14.2$$

The corresponding value of K_p is

$$K_p = e^{\frac{-\Delta G_{rxn}^\circ}{RT}} = e^{14.2} = 1.47 \times 10^6$$

Think About It The value for K, 10^6, is much greater than 1, which is consistent with $\Delta G_{rxn}^\circ = -35.3$ kJ, a large, negative value.

 **Practice Exercise** The standard free energy of formation of ammonia at 298 K is −16.5 kJ/mol. What is the value of K_p for the following reaction at 298 K?

$$N_2(g) + 3\ H_2(g) \rightleftharpoons 2\ NH_3(g)$$

We have yet to explain which kind of equilibrium constant, K_c or K_p, is related to ΔG° by Equation 17.17. The symbol ΔG° represents a change in free energy under standard conditions. The standard state of a gaseous reactant or product is one in which its *partial pressure* is 1 bar. Thus, the ΔG° of a reaction *in the gas phase* is linked by Equation 17.17 to its K_p value. However, standard conditions for reactions in solution (the focus of Chapters 15 and 16) mean that all dissolved reactants and products are present at a concentration of 1.00 *M*. Thus, the ΔG° of a reaction *in solution* is related by Equation 17.17 to its K_c value.

CONNECTION The relationship between K_c and K_p, derived in Section 14.3, is $K_p = K_c(RT)^{\Delta n}$ (Equation 14.7), where Δn is the change in the number of moles of gases between the products and reactants. $K_p = K_c$ when $\Delta n = 0$.

17.8 Influence of Temperature on Equilibrium Constants

We noted on many occasions in Chapter 14 that the value of K changes with temperature. In this section we use the thermodynamics of chemical reactions to explain how and why.

Let's begin with a simplified version of Equation 17.11 (leaving out all the "rxn" subscripts):

$$\Delta G^\circ = \Delta H^\circ - T\Delta S^\circ$$

We can combine it with a similarly simplified version of Equation 17.16:

$$\ln K = \frac{-\Delta G^\circ}{RT}$$

The result is an equation that relates K to ΔH° and ΔS°:

$$\ln K = \frac{-\Delta G^\circ}{RT} = -\frac{\Delta H^\circ}{RT} + \frac{T\Delta S^\circ}{RT}$$

$$= -\frac{\Delta H^\circ}{RT} + \frac{\Delta S^\circ}{R} \quad (17.18)$$

Note how a negative value of ΔH° or a positive value of ΔS° contributes to a large value of K. These dependencies are consistent with negative values of ΔH° and positive values of ΔS° because these two factors contribute to making reactions spontaneous.

Because we are discussing the influence of temperature on K, let's identify the factors affected by changes in T in Equation 17.18. The ΔS° term is unaffected, but the influence of ΔH° does depend on temperature: the higher the temperature, the larger the denominator of the ΔH° term, and the smaller the influence of a favorable ΔH°—that is, a ΔH° value that is negative. This temperature dependence is consistent with Le Châtelier's principle and the notion that energy is produced in exothermic reactions and required for endothermic reactions to proceed. Increasing temperature inhibits exothermic reactions, but promotes endothermic reactions.

CONNECTION Recall from Section 6.3 that exothermic reactions give off energy—they have a negative enthalpy change ($\Delta H < 0$)—whereas endothermic reactions absorb energy ($\Delta H > 0$).

If ΔH° and ΔS° do not vary much with temperature, then Equation 17.18 predicts that $\ln K$ will be a linear function of $1/T$. Furthermore, we expect a graph of $\ln K$ versus $1/T$ to have a positive slope for an exothermic process and a negative slope for an endothermic process. We can determine ΔH° from the slope of the line and ΔS° from its y intercept. Thus, we can calculate fundamental thermodynamic values of a reaction at equilibrium by determining its equilibrium constant at different temperatures.

What, for example, are the values of ΔH° and ΔS° for the following reaction?

$$2\,CO_2(g) \rightleftharpoons 2\,CO(g) + O_2(g)$$

We start with the K_p values of 2.75×10^{-11} at 1500 K, 1.42×10^{-3} at 2500 K, and 0.112 at 3000 K. K_p increases as temperature increases, indicating that energy is a reactant and the reaction is endothermic. We can use these data to determine the values of ΔH° and ΔS° by plotting $\ln K_p$ values versus $1/T$. To see how this graphical method works, let's rewrite Equation 17.18 so that it fits the form of the equation for a straight line ($y = mx + b$):

$$\ln K = -\frac{\Delta H^\circ_{rxn}}{R}\left(\frac{1}{T}\right) + \frac{\Delta S^\circ_{rxn}}{R} \quad (17.19)$$

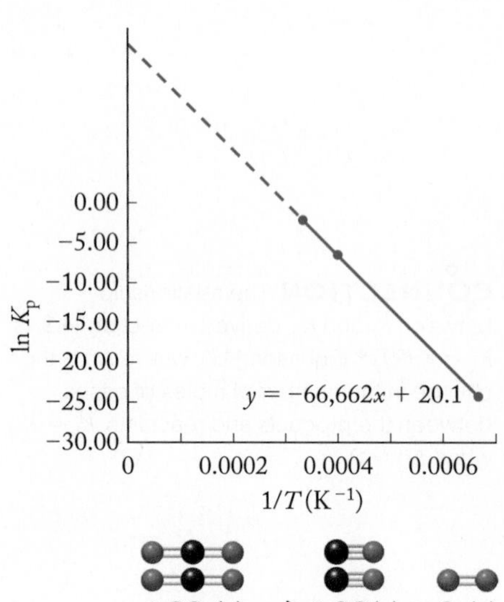

FIGURE 17.18 The graph of $\ln K_p$ versus $1/T$ is a straight line. We can calculate ΔH°_{rxn} from the slope and extrapolate the line to the y intercept, from which we can calculate ΔS°_{rxn}.

The graph of $\ln K_p$ versus $1/T$ (**Figure 17.18**) is indeed a straight line. The slope of this line is −66,662 K, which is equal to $-\Delta H^\circ_{rxn}/R$. The corresponding value of ΔH°_{rxn} is

$$\Delta H^\circ_{rxn} = -\text{slope} \times R$$

$$= -(-66{,}662\ \cancel{K})\left(\frac{8.314\ J}{mol \cdot \cancel{K}}\right)$$

$$= 554{,}228\ J/mol = 554\ kJ/mol$$

The y intercept ($\Delta S^\circ_{rxn}/R$) of the graph is 20.1. The corresponding value of ΔS°_{rxn} is

$$\Delta S^\circ_{rxn} = (20.1)\left(\frac{8.314\ \text{J}}{\text{mol} \cdot \text{K}}\right) = 167\ \text{J/(mol} \cdot \text{K)}$$

This positive entropy change is logical because the forward reaction converts 2 moles of gaseous CO_2 into 3 moles of gaseous products.

We can use a modified version of Equation 17.19 to relate the values of K at two different temperatures:

$$\ln\left(\frac{K_2}{K_1}\right) = -\frac{\Delta H^\circ_{rxn}}{R}\left(\frac{1}{T_2} - \frac{1}{T_1}\right) \tag{17.20}$$

Equation 17.20 is called the *van 't Hoff equation* because it was first derived by Jacobus van 't Hoff. It is particularly useful for calculating the value of K at a very high or low temperature if we know what it is at a standard reference temperature (e.g., 298 K). We perform one such calculation in Sample Exercise 17.9.

SAMPLE EXERCISE 17.9 Calculating the Value of K at a Specific Temperature **LO7**

Use data from Appendix 4 to calculate the equilibrium constant K_p for the exothermic reaction

$$N_2(g) + 3\ H_2(g) \rightleftharpoons 2\ NH_3(g)$$

at 298 K and at 773 K, a typical temperature used in the Haber–Bosch process for synthesizing ammonia.

Collect and Organize We are asked to calculate the K_p value for a reaction at two temperatures. One of the temperatures is 298 K, which is the reference temperature for the standard thermodynamic data in Appendix 4. The ΔG°_f value of NH_3 is −16.5 kJ/mol.

Analyze We can use Equation 17.17 to calculate the value of K_p at 298 K from the value of ΔG°_{rxn}. Then we can use Equation 17.20 to calculate the value of K_p at 773 K from the value of K_p at 298 K and the value of ΔH°_{rxn}. The value of ΔH°_{rxn} can be calculated from the enthalpy of formation of NH_3 (−46.1 kJ/mol). The negative value of ΔH°_f means that ΔH°_{rxn} is also negative, indicating that the reaction is exothermic. Because energy is a product of the reaction, we predict that the value of K_p is smaller at 773 K than at 298 K.

Solve The value of ΔG°_{rxn} for the reaction is

$$\frac{2\ \cancel{\text{mol NH}_3}}{\text{mol N}_2} \times \frac{-16.5\ \text{kJ}}{\cancel{\text{mol NH}_3}} \times \frac{1000\ \text{J}}{\text{kJ}} = -33{,}000\ \frac{\text{J}}{\text{mol N}_2} = -3.30 \times 10^4\ \frac{\text{J}}{\text{mol N}_2}$$

Using this value in Equation 17.17 to calculate the value of K_p gives us

$$K_p = e^{\frac{-\Delta G^\circ_{rxn}}{RT}}$$

$$= e^{-\left(\frac{-3.30\times10^4 \frac{\text{J}}{\text{mol}}}{8.314\frac{\text{J}}{\text{mol}\cdot\text{K}}\times 298\ \text{K}}\right)}$$

$$= 6.09 \times 10^5$$

Once we know the value of K_p at 298 K, we can calculate K_p at 773 K by using Equation 17.20. To do this, we must first calculate the value of ΔH°_{rxn}, which is twice the ΔH°_f of NH_3 (use Equation 6.19 to confirm this for yourself):

$$\Delta H^\circ_{rxn} = \frac{2\ \cancel{\text{mol NH}_3}}{\text{mol N}_2} \times \frac{-46.1\ \text{kJ}}{\cancel{\text{mol NH}_3}} \times \frac{1000\ \text{J}}{\text{kJ}}$$

$$= -92{,}200\ \frac{\text{J}}{\text{mol N}_2},\ \text{or}\ -92.2\ \frac{\text{kJ}}{\text{mol N}_2}$$

After substituting $K_1 = 6.09 \times 10^5$, $T_1 = 298$ K, and $T_2 = 773$ K into Equation 17.20, we solve for K_2:

$$\ln\left(\frac{K_2}{6.09 \times 10^5}\right) = \frac{\left(-92{,}200 \frac{\text{J}}{\text{mol}}\right)}{8.314 \frac{\text{J}}{\text{mol} \cdot \text{K}}}\left(\frac{1}{773\ \text{K}} - \frac{1}{298\ \text{K}}\right)$$

$$\ln\left(\frac{K_2}{6.09 \times 10^5}\right) = -22.87$$

$$\frac{K_2}{6.09 \times 10^5} = e^{-22.87} = 1.17 \times 10^{-10}$$

$$K_2 = 7.12 \times 10^{-5} \text{ at } 773\ \text{K}$$

Think About It The equilibrium constant decreases 10 orders of magnitude when the temperature of the reaction is raised from 298 K to 773 K. This huge decrease fits our prediction for this exothermic reaction. Note that the standard thermodynamic data for the reaction (ΔH°_{rxn} and ΔG°_{rxn}) are expressed per mole of the reactant with a coefficient of 1 in the balanced chemical equation for N_2.

Practice Exercise Use data from Appendix 4 to calculate the value of K_p for the reaction

$$2\ N_2(g) + O_2(g) \rightleftharpoons 2\ N_2O(g)$$

at 298 K and 2000 K.

17.9 Driving the Human Engine: Coupled Reactions

The laws of thermodynamics that govern chemical reactions in the laboratory also govern all the chemical reactions that take place in living systems (**Figure 17.19**). Organisms carry out reactions that release energy by oxidizing compounds derived from the nutrients they consume and then use that energy to do work and sustain an array of other essential biological functions. Just like mechanical engines, however, humans and other life-forms are far from 100% efficient, which means that life requires a continual input of energy in terms of the caloric content of the food we eat. Thus, we must constantly absorb energy in the form of food and release energy and waste products into our surroundings. Young women have an average daily nutritional need of 2100 Cal; for young men the need is 2900 Cal. This level of caloric intake provides the energy we need in order to function at all levels, from thinking to getting out of bed in the morning.

CONNECTION In nutrition, 1 Calorie, the "calorie" used in discussing food, is equivalent to 1 kcal (see Section 6.3): 1 Cal = 1 kcal = 1000 cal.

In living systems, spontaneous reactions ($\Delta G < 0$) typically involve metabolizing food, whereas nonspontaneous reactions ($\Delta G > 0$) involve building molecules needed by the body. Living systems use the energy from spontaneous reactions to run nonspontaneous reactions; we say that the spontaneous reactions are *coupled* to the nonspontaneous reactions. Part of the study of biochemistry involves deciphering the molecular mechanisms that enable reaction coupling. This topic is covered in Chapter 20, but for now it is sufficient to know that elegant molecular processes have evolved to enable living systems to couple chemical reactions so that the energy obtained from spontaneous reactions can be used to drive the nonspontaneous reactions that maintain life. The metabolic chemical

glycolysis a series of reactions that converts glucose into pyruvate; a major anaerobic (no oxygen required) pathway for the metabolism of glucose in the cells of almost all living organisms.

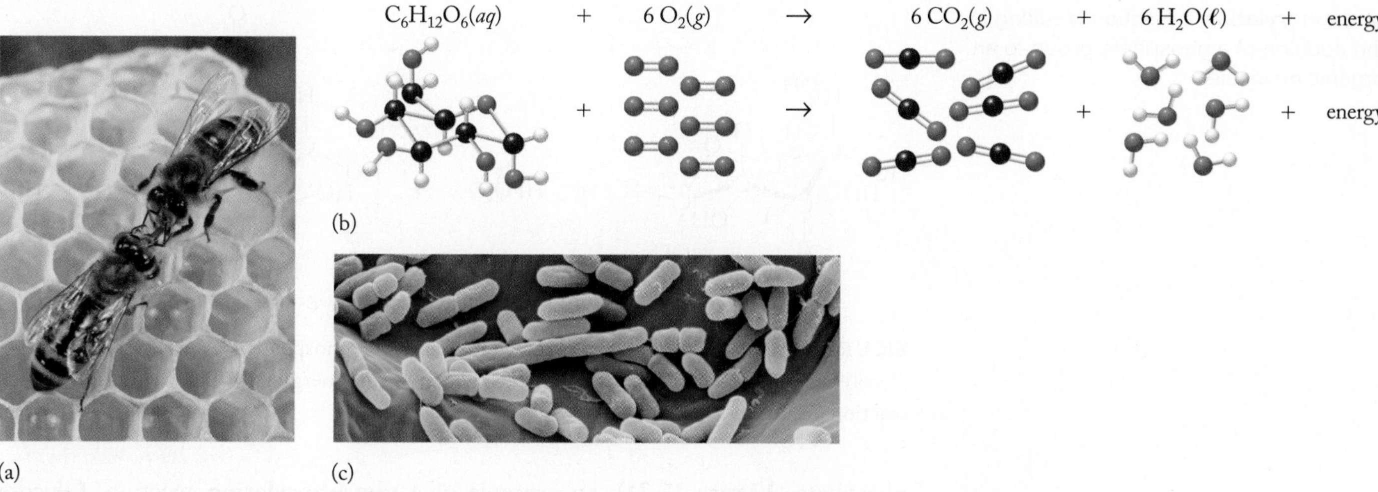

FIGURE 17.19 The rules of thermodynamics apply to all living systems. (a) Honeybees extract energy from nutrients to support life. The bees store this energy as honey (a mixture containing levulose and dextrose), which is then consumed by the bees, by humans, or by other animals to generate energy. (b) When honey is consumed, energy, water, and carbon dioxide are released, increasing the entropy of the universe. (c) Microscopic organisms—such as the *Escherichia coli* bacteria that live in our gastrointestinal tracts—consume nutrients to live and generate energy in the process. Their expenditure of energy also increases the entropy of the universe.

reactions we look at here are *not* presented for you to memorize. Rather, the intent is to aid your understanding of how changes in free energy allow spontaneous reactions to drive nonspontaneous reactions and to illustrate operationally what the phrase "coupled reactions" actually means.

As noted at the beginning of Chapter 6, all the energy contained in the food we eat has sunlight as its ultimate source. Green plants store energy from sunlight in their tissues as molecules such as glucose ($C_6H_{12}O_6$), which they produce from CO_2 and H_2O during photosynthesis.

Production of glucose by green plants is a nonspontaneous process, which is why the plants require the energy of sunlight to carry out this reaction. Animals that consume plants use the energy stored in the chemical bonds of glucose and other molecules and release CO_2 and H_2O back into the environment. This reaction,

$$C_6H_{12}O_6(s) + 6\ O_2(g) \rightarrow 6\ CO_2(g) + 6\ H_2O(\ell)$$

also releases energy, an event that increases the entropy of the universe, and is therefore spontaneous. Thus, the processes of life increase the entropy of the universe by converting chemical energy into energy (heat) that flows into the surroundings.

As a result, the free-energy change for the breakdown of glucose must be less than zero—it is actually $\Delta G° = -880$ kJ/mol. This spontaneous process is highly controlled in living systems, however, so that the energy it produces can be directed into the nonspontaneous processes essential to organisms that do not carry out photosynthesis. **Figure 17.20** summarizes one portion of glucose metabolism—**glycolysis**—that involves coupled reactions.

In glycolysis, each mole of glucose is converted into 2 moles of pyruvate ion (CH_3COCOO^-). An early step is the conversion of glucose into glucose-6-

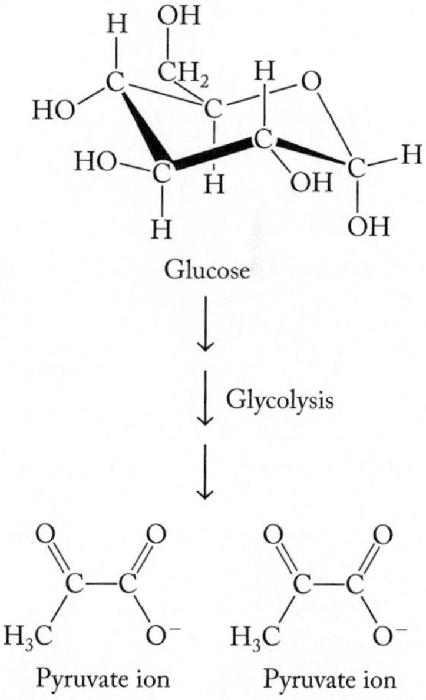

Glucose

Glycolysis

Pyruvate ion   Pyruvate ion

FIGURE 17.20 In glycolysis, molecules of glucose are converted into twice their number of pyruvate ions.

CONNECTION Photosynthesis and the carbon cycle were introduced in Section 3.5 (see Figure 3.21).

phosphorylation a reaction resulting in the addition of a phosphate group to an organic molecule.

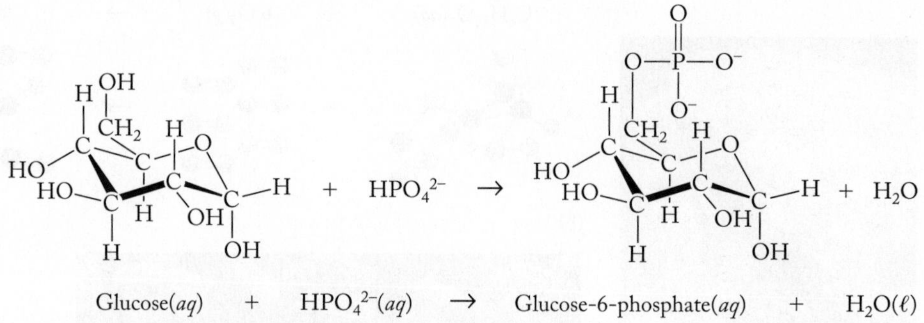

FIGURE 17.21 The conversion of glucose into glucose-6-phosphate is an early step in glycolysis. This reaction is nonspontaneous, which means energy must be added to make the reaction go: $\Delta G^\circ_{rxn} = 13.8$ kJ/mol.

phosphate (**Figure 17.21**), an example of a **phosphorylation** reaction. Glucose reacts with the hydrogen phosphate ion (HPO_4^{2-}), producing glucose-6-phosphate and water. This reaction is not spontaneous ($\Delta G^\circ = +13.8$ kJ/mol), and the energy needed to make it happen comes from adenosine triphosphate (ATP), which functions in our cells both as a storehouse of energy and as an energy-transfer agent. ATP hydrolyzes to adenosine diphosphate (ADP) in a reaction (**Figure 17.22**) that produces a hydrogen phosphate ion and energy: $\Delta G^\circ_{rxn} = -30.5$ kJ/mol. As can be seen in Figure 17.22, both ATP and ADP are ions, and as such their formulas are written as ATP^{4-} and ADP^{3-}, respectively, when used in chemical equations.

In a living system, spontaneous hydrolysis of ATP consumes water and produces HPO_4^{2-} and H^+, whereas the nonspontaneous phosphorylation of glucose consumes HPO_4^{2-} and produces water. The two reactions are coupled so that the spontaneous ATP → ADP reaction supplies the energy that drives the nonspontaneous formation of glucose-6-phosphate (**Figure 17.23**).

This example demonstrates that the ΔG° values for coupled reactions (and for sequential reactions) are additive. This is true for any set of reactions, not just those occurring in living systems.

The first two steps in glycolysis are

(1) $$ATP^{4-}(aq) + H_2O(\ell) \rightarrow ATP^{3-}(aq) + HPO_4^{2-}(aq) + H^+(aq) \qquad \Delta G^\circ = -30.5 \text{ kJ}$$

(2) $$\underset{\text{Glucose}}{C_6H_{12}O_6(aq)} + \underset{\text{Hydrogen phosphate}}{HPO_4^{2-}(aq)} \rightarrow \underset{\text{Glucose-6-phosphate}}{C_6H_{11}O_6PO_3^{2-}(aq)} + H_2O(\ell) \qquad \Delta G^\circ = -13.8 \text{ kJ}$$

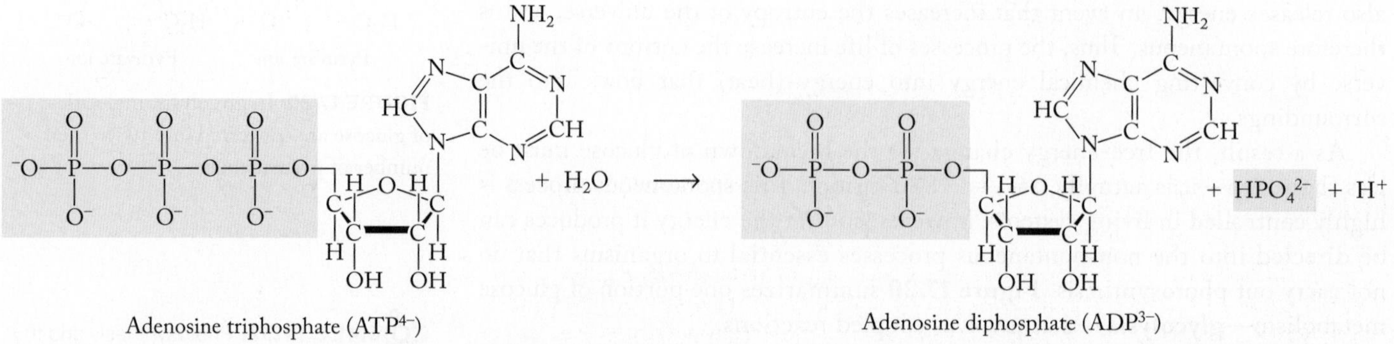

FIGURE 17.22 The hydrolysis of ATP to ADP is a spontaneous reaction: $\Delta G^\circ = -30.5$ kJ/mol. The body couples this reaction to nonspontaneous reactions so that the energy released can drive them (see Figure 17.23).

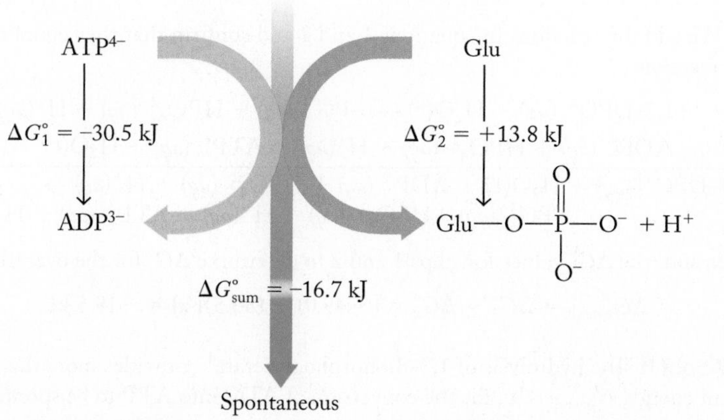

ATP^{4-} Glu

$\Delta G_1^\circ = -30.5$ kJ $\Delta G_2^\circ = +13.8$ kJ

ADP^{3-} Glu—O—PO_3^{2-} + H^+

$\Delta G_{sum}^\circ = -16.7$ kJ

Spontaneous

FIGURE 17.23 The spontaneous hydrolysis of ATP is coupled to the nonspontaneous phosphorylation of glucose. The overall reaction—the sum of the two individual reactions—is spontaneous.

CONNECTION Hess's law (i.e., the enthalpy change of a reaction that is the sum of two or more reactions equals the sum of the enthalpy changes of the constituent reactions) was introduced in Section 6.7 to calculate ΔH. We apply a similar principle here when adding $\Delta G°$ values of coupled reactions.

If we add these reactions and their free energies, we get

$$ATP^{4-}(aq) + \cancel{H_2O(\ell)} + C_6H_{12}O_6(aq) + \cancel{HPO_4^{2-}(aq)} \rightarrow$$
$$ATP^{3-}(aq) + \cancel{HPO_4^{2-}(aq)} + H^+(aq) + C_6H_{11}O_6PO_3^{2-}(aq) + \cancel{H_2O(\ell)}$$
$$\Delta G° = (-30.5 + 13.8)\text{ kJ} = -16.7\text{ kJ}$$

Because equal quantities of H_2O and HPO_4^{2-} appear on both sides of the combined equation, they cancel out, leaving the following net reaction:

$$C_6H_{12}O_6(aq) + ATP^{4-}(aq) \rightarrow ADP^{3-}(aq) + C_6H_{11}O_6PO_3^{2-}(aq) + H^+(aq)$$
$$\Delta G° = -16.7\text{ kJ}$$

Because $\Delta G°$ for the net reaction is negative, the reaction is spontaneous.

SAMPLE EXERCISE 17.10 Calculating $\Delta G°$ of Coupled Reactions LO8

The body would rapidly run out of ATP if there were not some process for regenerating ATP from ADP. That process is the hydrolysis of 1,3-diphosphoglycerate^{4-} (1,3-DPG^{4-}) to 3-phosphoglycerate^{3-} (3-PG^{3-}; **Figure 17.24**):

$$ADP^{3-} + \text{1,3-diphosphoglycerate}^{4-} \rightarrow \text{3-phosphoglycerate}^{3-} + ATP^{4-}$$

This hydrolysis is spontaneous. Calculate its $\Delta G°$ value from the following values:

(1) $\text{1,3-DPG}^{4-}(aq) + H_2O(\ell) \rightarrow \text{3-PG}^{3-}(aq) + HPO_4^{2-}(aq) + H^+(aq)$ $\quad \Delta G_1^\circ = -49.0$ kJ

(2) $ADP^{3-}(aq) + HPO_4^{2-}(aq) + H^+(aq) \rightarrow ATP^{4-}(aq) + H_2O(\ell)$ $\quad \Delta G_2^\circ = 30.5$ kJ

Collect and Organize We can calculate $\Delta G°$ for a reaction that is the sum of two reactions. If the reactions in equations 1 and 2 add up to the overall reaction, then the overall $\Delta G°$ is the sum of the $\Delta G°$ values for the individual reactions.

Analyze First we add the reactions described by equations 1 and 2. Assuming that the overall reaction between ADP and 1,3-diphosphoglycerate^{4-} is the sum of the reactions describing the hydrolysis of 1,3-diphosphoglycerate^{4-} and the phosphorylation of ADP, we know that the sum of ΔG_1° and ΔG_2° will be less than zero because the overall reaction is spontaneous.

1,3-Diphosphoglycerate^{4-} (1,3-DPG^{4-})

3-Phosphoglycerate^{3-} (3-PG^{3-})

Glucose

Lactic acid

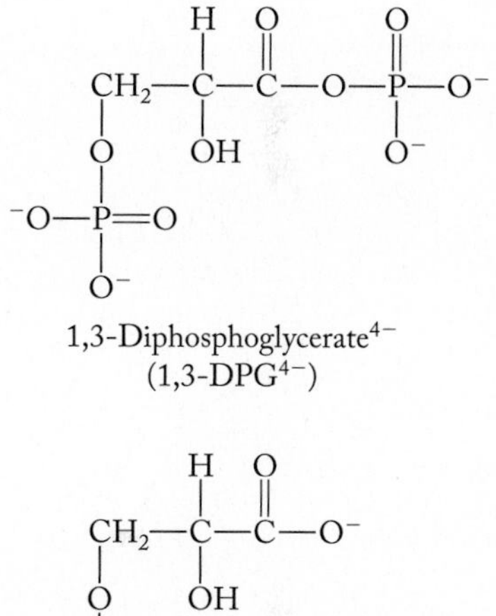

FIGURE 17.24 Structures of 1,3-diphosphoglycerate, 3-phosphoglycerate, glucose, and lactic acid.

Solve We add the reactions in equations 1 and 2 and confirm that they equal the overall reaction:

$$\begin{array}{ll}(1) & \text{1,3-DPG}^{4-}(aq) + H_2O(\ell) \rightarrow \text{3-PG}^{3-}(aq) + HPO_4^{\,2-}(aq) + H^+(aq) \\ (2) & ADP^{3-}(aq) + HPO_4^{\,2-}(aq) + H^+(aq) \rightarrow ATP^{4-}(aq) + H_2O(\ell) \\ \hline \end{array}$$

$$\text{1,3-DPG}^{4-}(aq) + \cancel{H_2O(\ell)} + ADP^{3-}(aq) + \cancel{HPO_4^{\,2-}(aq)} + \cancel{H^+(aq)} \rightarrow \text{3-PG}^{3-}(aq) + \cancel{HPO_4^{\,2-}(aq)} + \cancel{H^+(aq)} + ATP^{4-}(aq) + \cancel{H_2O(\ell)}$$

We then sum the $\Delta G°$ values for steps 1 and 2 to determine $\Delta G°$ for the overall reaction:

$$\Delta G°_{\text{overall}} = \Delta G°_1 + \Delta G°_2 = [(-49.0) + (30.5)]\text{ kJ} = -18.5\text{ kJ}$$

Think About It The hydrolysis of 1,3-diphosphoglycerate^{4-} provides more than sufficient energy, $G°_{\text{overall}} < 0$, for the conversion of ADP into ATP to be spontaneous.

 **Practice Exercise** The conversion of glucose into lactic acid drives the phosphorylation of 2 moles of ADP to ATP:

$$\underset{\text{Glucose}}{C_6H_{12}O_6(aq)} + 2\ HPO_4^{\,2-}(aq) + 2\ ADP^{3-}(aq) + 2\ H^+(aq) \rightarrow$$

$$2\ \underset{\text{Lactic acid}}{CH_3CH(OH)COOH(aq)} + 2\ ATP^{4-}(aq) + 2\ H_2O(\ell) \qquad \Delta G° = -135\text{ kJ}$$

What is $\Delta G°$ for the conversion of glucose into lactic acid?

$$C_6H_{12}O_6(aq) \rightarrow 2\ CH_3CH(OH)COOH(aq)$$

The ATP produced from the breakdown of glucose (such as that formed via the first reaction in the preceding practice exercise) is used to drive nonspontaneous reactions in cells. The metabolism of fats and proteins relies on a series of cycles, all of which involve coupled reactions. The ATP–ADP system is a carrier of chemical energy because ADP requires energy to accept a phosphate group and thus is coupled to reactions that yield energy, whereas ATP donates a phosphate group, releases energy, and is coupled to reactions that require energy. All energy changes in living systems are governed by the first and second laws of thermodynamics, as are all the energy changes in the inanimate world.

17.10 Microstates: A Quantized View of Entropy

In Chapter 17 we have described entropy as a measure of how energy is distributed throughout a system: the more spread out the energy is, the greater the system's entropy. This thermodynamic view evolved in the middle of the 19th century, largely through the work of the German physicist Rudolf Clausius (1822–1888), who was a leader in establishing the field of thermodynamics and who introduced the concept of entropy in 1865.

Clausius's thermodynamic approach views entropy from a macroscopic, systemwide perspective. In much the same way as we use such parameters as pressure, volume, and temperature to describe a quantity of a gaseous substance, the change in a system's entropy is defined in terms of the quantity of energy flowing out from or into the system and its temperature (Equation 17.2). However, just as the kinetic molecular theory of gases provides a particle-based explanation of why gases exhibit the macroscopic properties they do, we can also understand entropy changes from the viewpoint of the dispersion of particles and their energies.

In 1877 an Austrian physicist named Ludwig Boltzmann (1844–1906) developed a microscopic view of Clausius's concept of entropy. His theory would eventually incorporate quantum mechanics to explain how energy was dispersed not only throughout a system but also within its molecules. Let's examine Boltzmann's microscopic definition of entropy by using two models: a single molecule of O_2 and a room full of them.

As shown in **Figure 17.25**, a molecule of a gas such as O_2 is free to undergo three types of motion: (1) *translational motion* as it zips around the space it occupies; (2) *rotational motion* as it spins about imaginary axes perpendicular to the O═O bond; and (3) *vibrational motion* as its bonded atoms move toward and away from each other, like balls on the ends of a spring. Each of these modes of motion increases as the thermal energy of the molecule increases.

In our discussion of quantum mechanics in Chapter 7, we saw that energy is not continuous on the atomic scale. Instead, the energies and motion of atoms and molecules are quantized. At temperatures near room temperature, the differences between the translational energy levels of atoms and molecules in the gas phase are so small that we may consider them a continuum of energy. However, the gaps in vibrational and rotational energy levels are large enough that we must take quantization into account.

To investigate these gaps, let's revisit the change in electrostatic potential energy that occurs when two oxygen atoms approach each other and a covalent bond forms between them (see Figure 8.2). Energy reaches a minimum (**Figure 17.26a**) when the nuclei of the two atoms are 121 pm apart—a distance that corresponds to the length of the O═O bond. Figure 17.26a also contains additional energy levels represented by horizontal red lines. These are vibrational energy levels. Notice that the length of the red lines increases as the vibrational energy increases. Greater length corresponds to greater variation in intermolecular distance; that is, more energetic oscillations of the O═O bond occur with increasing vibrational energy.

Superimposed on each vibrational energy level is a set of rotational energy levels, shown for the first vibrational energy level in **Figure 17.26b**. Similar sets of rotational energy levels are associated with all the other vibrational energy levels, creating a multitude of energy states that are accessible to every O_2 molecule at room temperature.

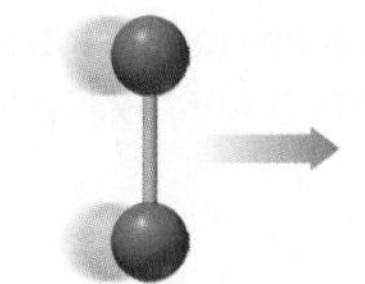

(a) Translational motion

(b) Rotational motion

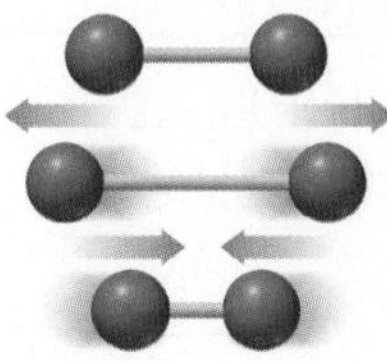

(c) Vibrational motion

FIGURE 17.25 A diatomic molecule has three fundamental types of motion: (a) translational motion, (b) rotational motion, and (c) vibrational motion.

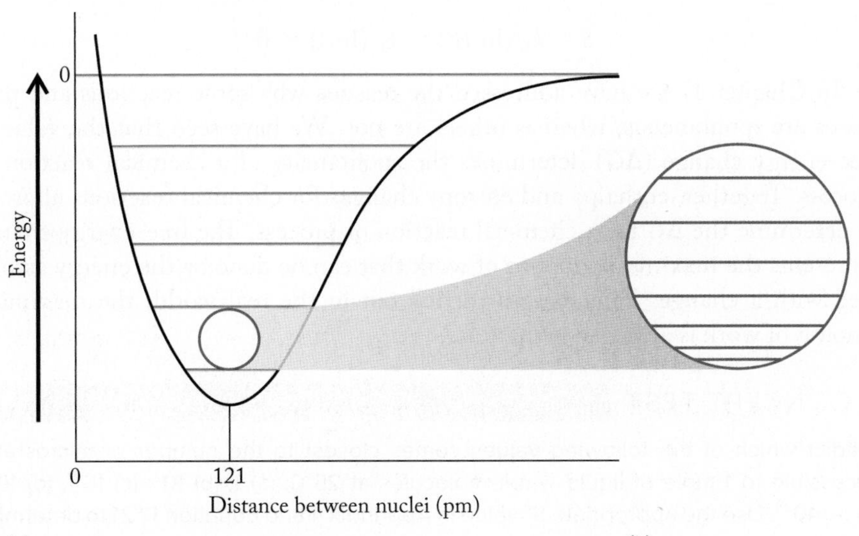

FIGURE 17.26 Vibrational and rotational energy levels in a molecule of O_2 are quantized. (a) The electrostatic potential energy between two oxygen atoms reaches a minimum when their nuclei are 121 pm apart—the length of an O═O bond. (b) Quantized rotational energy levels (represented by the blue horizontal lines) are superimposed on each vibrational energy level, represented by a red horizontal line in part (a). (Not all the accessible rotational and vibrational states are shown, and the gaps between them are not to scale.)

microstate a unique distribution of particles among energy levels.

Now let's focus on one oxygen molecule as it collides with others, sometimes gaining and other times losing translational kinetic energy. At room temperature and pressure, a single O_2 molecule experiences billions of collisions per second, each collision altering its speed and changing its vibrational and rotational energies. At any instant, the overall quantized energy of the molecule is a particular combination of translational, vibrational, and rotational energy states. The number of possible combinations of energy states accessible to an O_2 molecule is enormous at room temperature, and the number increases as temperature increases.

In a room full of O_2 molecules (our second model), the number of different quantized total-molecule energy states is so large (*e* raised to a power that is many times Avogadro's number) that it is beyond the capacity of your calculator to compute. The entirety of all the quantized energy states occupied by all the atoms in the room at a particular instant is called a **microstate**. Each microstate represents one discrete way for the system to disperse all the energy in the system. The vast number of microstates to which a system has access defines its entropy: the more microstates, the higher its entropy. This correlation between the number of accessible microstates (W) and entropy (S) is defined by a simple mathematical equation developed by Boltzmann:

STEPWISE ANIMATION

Microstates

$$S = k_B \ln W \qquad (17.21)$$

Here k_B is the Boltzmann constant, which is equal to the gas constant divided by Avogadro's number:

$$k_B = \frac{R}{N_A} = \frac{8.314 \dfrac{\text{J}}{\cancel{\text{mol}} \cdot \text{K}}}{6.022 \times 10^{23}/\cancel{\text{mol}}} = 1.381 \times 10^{-23} \text{ J/K}$$

Boltzmann's definition of entropy, based on access to energy microstates, is conceptually compatible with the thermodynamic, macroscopic view based on energy dispersion because each microstate represents one of the multitude of ways that energy can be dispersed in a thermodynamic system. In fact, some facets of entropy are easier to understand using the microstate model. One of them is the concept of absolute entropy, the notion embedded in the third law of thermodynamics that a perfect crystalline solid has zero entropy at a temperature of absolute zero. If the particles of a crystalline solid are perfectly aligned at 0 K, only one distribution of the particles is possible in the space occupied by the crystal. Therefore, the crystal has only one microstate, and, according to Equation 17.21, its entropy is zero:

$$S = k_B (\ln W) = k_B (\ln 1) = 0$$

In Chapter 17 we have addressed the reasons why some reactions and processes are spontaneous, whereas others are not. We have seen that the value of free-energy change (ΔG) determines the spontaneity of a chemical reaction or process. Together, enthalpy and entropy changes for chemical reactions allow us to determine the ΔG for a chemical reaction or process. The free-energy change represents the maximum amount of work that can be done by the energy associated with a change. For changes carried out in the real world, the maximum amount of work is always less than ΔG.

CONCEPT TEST

Predict which of the following values comes closest to the number of microstates accessible to 1 mole of liquid water molecules at 25°C: (a) 1; (b) 10^2; (c) 10^{10}; (d) 10^{23}; (e) $>>10^{23}$. Use the appropriate $S°$ value in Appendix 4 and Equation 17.21 to determine whether your selection was the best one.

SAMPLE EXERCISE 17.11 Integrating Concepts: Ötzi's Axe and the Chemistry of Copper Refining

In 1991 two hikers discovered the remains of a prehistoric man frozen in a glacier high in the Ötztal Alps, near the Italy–Austria border. Ötzi the Iceman, as he came to be known, lived about 5300 years ago. Among his possessions was an axe with a head made of nearly pure (99.7%) copper (**Figure 17.27**). The purity of the copper, as well as other clues, told scientists that the copper had been skillfully extracted and refined. Though copper does occur in its native elemental state in Earth's crust, ancient artisans usually had to extract the metal from copper-containing minerals such as chalcocite (Cu_2S). The process required that they find a way to add a lot of energy to a sulfide mineral like Cu_2S because its decomposition reaction has a large, positive $\Delta G°$ value:

$$Cu_2S(s) \rightarrow 2\ Cu(s) + \tfrac{1}{8}\ S_8(s) \qquad \Delta G° = 86.2\ \text{kJ}$$

However, when Cu_2S is roasted in a furnace that provides an ample supply of very hot air, any sulfur produced by the decomposition reaction is then oxidized to SO_2 gas in a decidedly spontaneous reaction:

$$\tfrac{1}{8}\ S_8(s) + O_2(g) \rightarrow SO_2(g) \qquad \Delta G° = -300.1\ \text{kJ}$$

a. Can these two reactions be coupled so that the spontaneous second reaction drives the nonspontaneous first reaction?
b. Is there likely to be an increase in entropy during the overall reaction?
c. What is the equilibrium constant of the first reaction under standard conditions?
d. Does the value of the equilibrium constant, K_P, for the overall reaction change if we carry out the reaction in two sealed vessels, one with twice the volume of the other?

Collect and Organize We are given two chemical reactions and asked to determine whether the free energy released by the spontaneous one can be coupled to and drive the nonspontaneous reaction. We also need to calculate the equilibrium constant of the nonspontaneous reaction and to decide how reaction conditions shift the equilibrium state. Equation 17.17 relates the equilibrium constant of a chemical reaction to its $\Delta G°_{rxn}$ value.

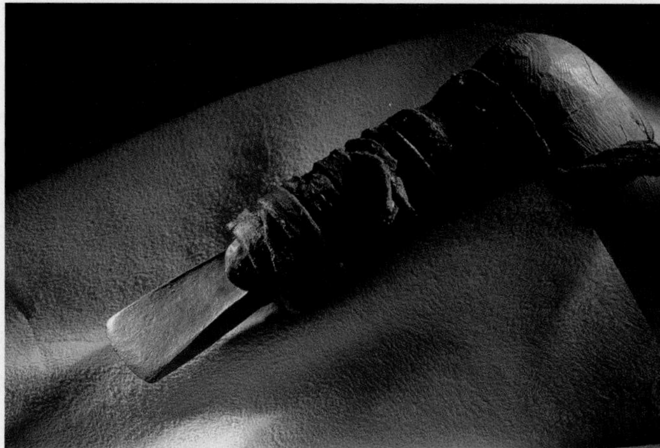

FIGURE 17.27 Copper axe found alongside Ötzi the Iceman, who lived about 5300 years ago.

Analyze Sulfur is a product of the first reaction and a reactant in the second one, so it should be possible to couple the two reactions. Given the large positive value of $\Delta G°_{rxn}$ for the first reaction, its K value should be much less than 1. An effect of volume on an equilibrium constant, K_P, is observed if there are different numbers of moles of gas on either side of the chemical equation.

Solve

a. One mole of sulfur is produced in the first reaction and consumed in the second, so the two reactions can be coupled. The result is an overall reaction with a $\Delta G°_{rxn}$ value that is the sum of the first two:

$$\begin{array}{lr} Cu_2S(s) \rightarrow 2\ Cu(s) + \cancel{\tfrac{1}{8}S_8(s)} & \Delta G°_{rxn} = 86.2\ \text{kJ} \\ + \cancel{\tfrac{1}{8}S_8(s)} + O_2(g) \rightarrow SO_2(g) & \Delta G°_{rxn} = -300.1\ \text{kJ} \\ \hline Cu_2S(s) + O_2(g) \rightarrow 2\ Cu(s) + SO_2(g) & \Delta G°_{rxn} = -213.9\ \text{kJ} \end{array}$$

The calculated negative $\Delta G°_{rxn}$ value means that the overall reaction is spontaneous.

b. The reactants and products both contain 1 mole of gas, but SO_2 is triatomic and has a larger standard molar entropy [248.2 J/(mol · K)] than that of diatomic O_2 [205.0 J/(mol · K)]. Also, the other reactant is 1 mole of a binary ionic solid, but the products include 2 moles of solid metal. Therefore, from the $S°$ values in Appendix 4, we can safely predict that the system entropy increases in the overall reaction.

c. Using Equation 17.17 to calculate the equilibrium constant for the decomposition reaction under standard conditions (and letting $T = 298$ K) gives us:

$$K = e^{-\Delta G°_{rxn}/RT}$$

First calculate the exponent:

$$-\frac{\Delta G°_{rxn}}{RT} = -\frac{86.2\ \frac{\cancel{\text{kJ}}}{\cancel{\text{mol}}} \times 1000\ \frac{\text{J}}{\cancel{\text{kJ}}}}{\left(8.314\ \frac{\text{J}}{\cancel{\text{mol}} \cdot \cancel{\text{K}}}\right)(298\ \cancel{\text{K}})} = -34.79$$

Then the value of K is

$$K = e^{-34.79} = 7.8 \times 10^{-16}$$

d. The chemical equation for the overall reaction has equal numbers of moles of gas on the reactant and product sides, which means that there will be no effect of volume changes on K_P.

Think About It As predicted, the nonspontaneous first reaction has a small K value; however, the reaction becomes spontaneous when coupled to the spontaneous second reaction. Le Châtelier's principle predicts that there will be no change in K_P if we change the volume, but here 1 mole of gaseous reactant is converted to 1 mole of a gas-phase product.

SUMMARY

LO1 Spontaneous processes happen on their own without continual outside intervention. **Nonspontaneous processes**, which are spontaneous processes in reverse, do *not* happen on their own. Spontaneous processes may be exothermic or endothermic and are often accompanied by an increase in the freedom of motion of the particles involved in the process. Spontaneous reactions are not necessarily rapid. (Section 17.1)

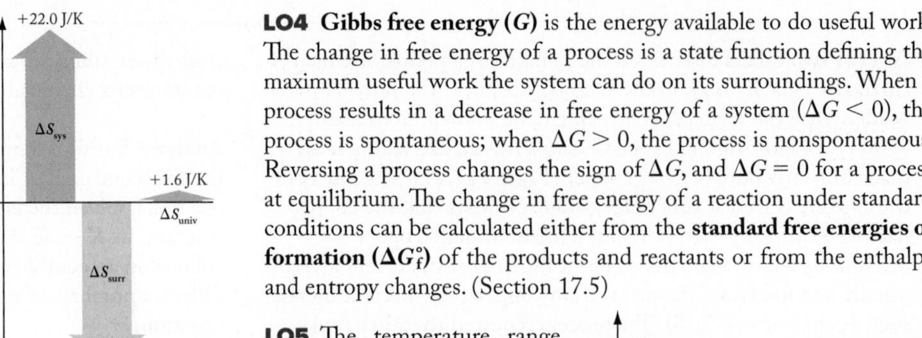

LO2 According to the **third law of thermodynamics**, a perfect crystal of a pure substance has zero entropy at absolute zero. All substances have positive entropies at temperatures above absolute zero. **Standard molar entropies ($S°$)** are entropy values for substances in their standard states. The entropy of a system increases with increasing molecular complexity and with increasing temperature. (Section 17.3)

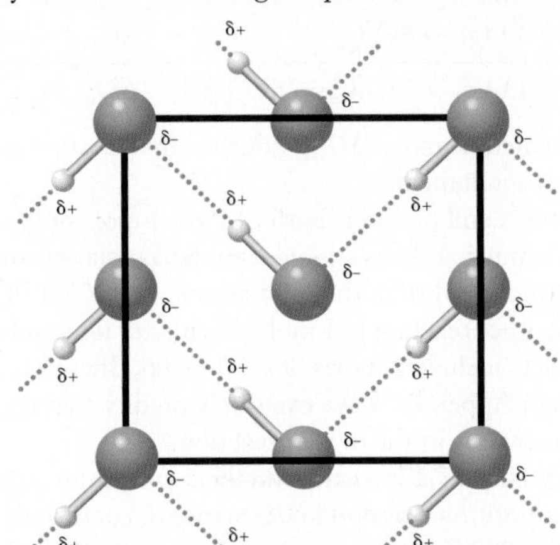

LO3 Entropy (S) is a thermodynamic property that provides a measure of how dispersed the energy is in a system at a given temperature. According to the **second law of thermodynamics**, a spontaneous process is accompanied by an increase in the entropy of an isolated system, or an increase in entropy of the universe for a process occurring in any system. A **reversible process** takes place in very small steps and very slowly so that the system can be restored to its initial state with no net flow of energy to or from its surroundings. The entropy change in a reaction under standard conditions can be calculated from the standard entropies of the products and reactants and their coefficients in the balanced chemical equation. (Sections 17.2 and 17.4)

LO4 Gibbs free energy (G) is the energy available to do useful work. The change in free energy of a process is a state function defining the maximum useful work the system can do on its surroundings. When a process results in a decrease in free energy of a system ($\Delta G < 0$), the process is spontaneous; when $\Delta G > 0$, the process is nonspontaneous. Reversing a process changes the sign of ΔG, and $\Delta G = 0$ for a process at equilibrium. The change in free energy of a reaction under standard conditions can be calculated either from the **standard free energies of formation (ΔG_f°)** of the products and reactants or from the enthalpy and entropy changes. (Section 17.5)

LO5 The temperature range over which a process is spontaneous depends on the relative magnitudes of ΔH and ΔS. (Section 17.6)

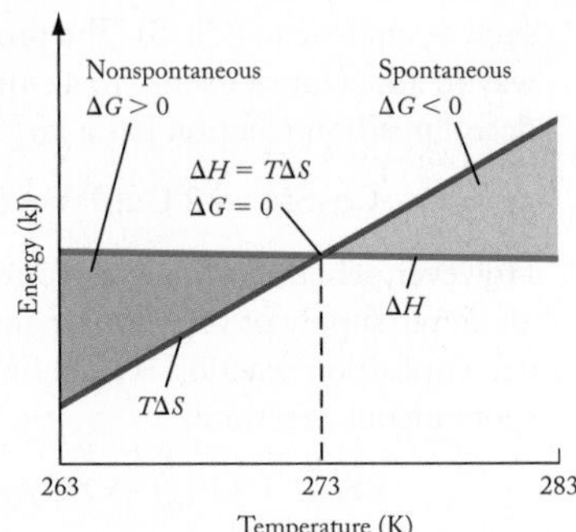

LO6 Negative values of $\Delta G°$ correspond to $K > 1$ and equilibrium reaction mixtures composed mostly of products. Positive values of $\Delta G°$ correspond to $K < 1$. As spontaneous reactions proceed, the free energy of the reaction mixture increases from an initial negative value and reaches zero when the reaction comes to chemical equilibrium. (Section 17.7)

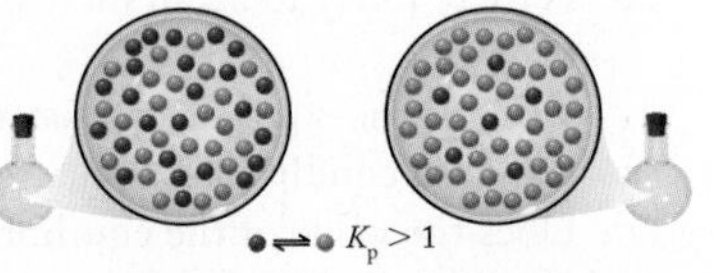

LO7 Higher reaction temperatures increase the equilibrium constant of an endothermic reaction but decrease the equilibrium constant of an exothermic reaction. The slope of a plot of ln K versus $1/T$ for an equilibrium system is used to determine the standard enthalpy for the equilibrium, and the y intercept of the plot is used to determine the standard entropy change. (Section 17.8)

LO8 Many important biochemical processes, including **glycolysis** and **phosphorylation**, are made possible by coupled spontaneous and nonspontaneous reactions. The free energy released in the spontaneous processes going on in the body is used to drive nonspontaneous processes. (Section 17.9)

LO9 As temperature increases, the translational motion, rotational motion, and vibrational motion of molecules increase. A **microstate** is defined as one discrete way for a system to disperse all its energy. As the number of microstates increases, so does the entropy of a system. (Section 17.10)

PARTICULATE PREVIEW WRAP-UP

Cu is a solid, Br_2 is a liquid, and He is a gas. The number of possible arrangements increases from the copper atoms (with the fewest) to the bromine molecules to the helium atoms (with the most). Br_2 molecules in the liquid phase and Cu atoms in the solid phase have vibrational motion, whereas atoms of He(g) do not. Br_2 molecules also have rotational and translational motion (Cu atoms do not), so Br_2 molecules have more ways to disperse their energy than the atoms of the other two elements.

PROBLEM-SOLVING SUMMARY

Type of Problem	Concepts and Equations	Sample Exercises
Predicting the sign of entropy change	Look for the net removal of gas molecules or precipitation of a solute from solution ($\Delta S < 0$ for both). For the reverse processes, $\Delta S > 0$.	**17.1**
Comparing absolute entropy values	Substances composed of larger, less rigid molecules have more entropy. Among substances with similar molar masses, gases have more entropy than liquids, which have more entropy than solids.	**17.2**
Calculating entropy changes	Use $$\Delta S^\circ_{rxn} = \sum n_{products}S^\circ_{products} - \sum n_{reactants}S^\circ_{reactants} \quad (17.5)$$ where $S^\circ_{products}$ and $S^\circ_{reactants}$ are the standard molar entropies and $n_{products}$ and $n_{reactants}$ are the stoichiometric coefficients for the process.	**17.3**
Predicting reaction spontaneity under standard conditions	A reaction is spontaneous if $$\Delta S_{univ} = (\Delta S_{sys} + \Delta S_{surr}) > 0$$	**17.4**
Calculating ΔG°_{rxn} by using appropriate ΔG°_f values	Use $$\Delta G^\circ_{rxn} = \sum n_{products}\Delta G^\circ_{f,products} - \sum n_{reactants}\Delta G^\circ_{f,reactants} \quad (17.12)$$ where $\Delta G^\circ_{f,products}$ and $\Delta G^\circ_{f,reactants}$ are the standard molar free energies of formation, and $n_{products}$ and $n_{reactants}$ are the stoichiometric coefficients for the process.	**17.5**
Relating reaction spontaneity to ΔH and ΔS	Use $$\Delta G^\circ_{rxn} = \Delta H^\circ_{rxn} - T\Delta S^\circ_{rxn} \quad (17.11)$$	**17.6, 17.7**
Calculating the value of K from ΔG°_f values	$$K = e^{-\Delta G^\circ_{rxn}/RT} \quad (17.17)$$	**17.8**
Calculating the value of K at a specific temperature	Use $$\ln\left(\frac{K_2}{K_1}\right) = -\frac{\Delta H^\circ_{rxn}}{R}\left(\frac{1}{T_2} - \frac{1}{T_1}\right) \quad (17.20)$$ Convert ΔH°_{rxn} from kilojoules per mole to joules per mole to match the units of R.	**17.9**
Calculating ΔG° of coupled reactions	Free-energy changes are additive. If adding two reactions gives the desired overall reaction, add the free-energy changes of the two reactions to obtain the free-energy change of the overall reaction.	**17.10**

VISUAL PROBLEMS

(Answers to boldface end-of-chapter questions and problems are in the back of the book.)

17.1. Two balloons are inflated at the same temperature to the same volume (Figure P17.1), though it takes more gas to inflate the balloon on the right. In which balloon is the gas under greater internal pressure? In which balloon does the gas have greater total entropy?

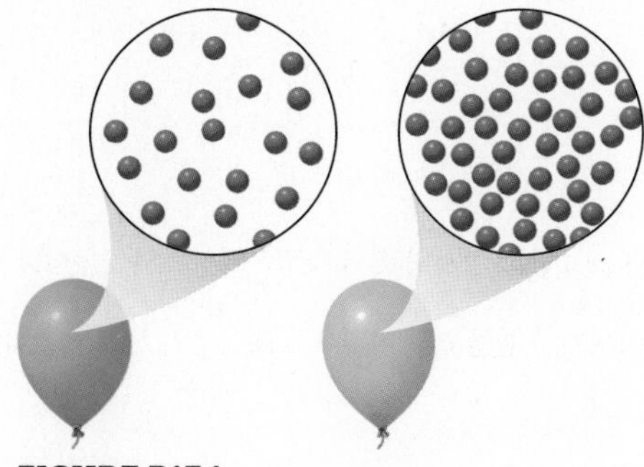

FIGURE P17.1

17.2. Two cubic containers (Figure P17.2) contain the same number of moles of gas at the same temperature. Which cube contains gas with more entropy? If the sample in container (b) is left unchanged but the sample in container (a) is cooled so that it condenses, which sample has the higher entropy?

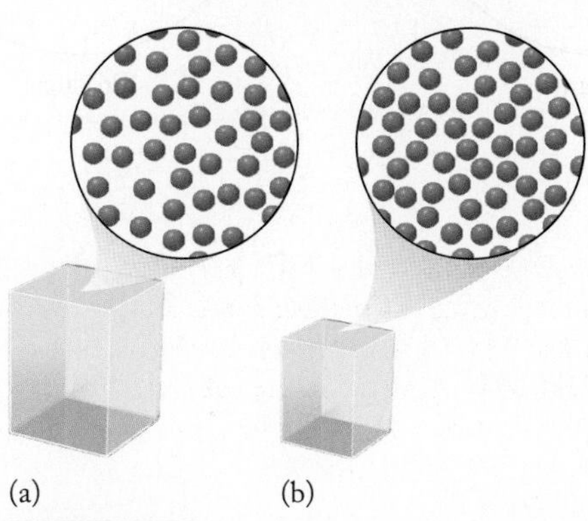

FIGURE P17.2

17.3. Figure P17.3 shows a tank that has just been filled with a mixture of two ideal gases: A (red spheres) and B (blue spheres). If the molar mass of A is twice that of B, will the atoms of A eventually fill the bottom of the tank and the atoms of B fill the top of the tank? Why or why not? Does the tank represent an isolated system?

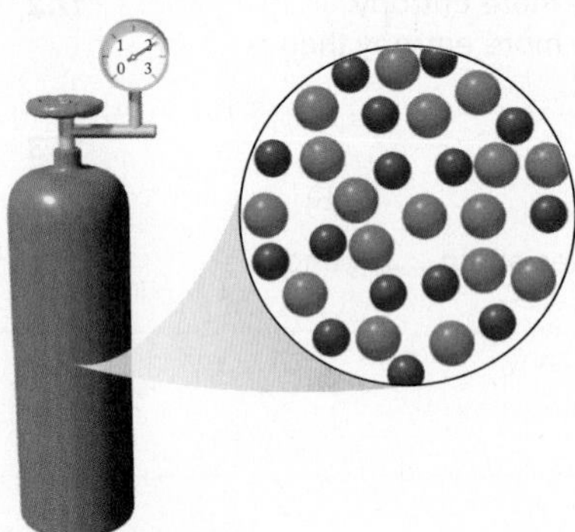

FIGURE P17.3

17.4. Predict whether an increase or decrease in entropy accompanies the physical change in Figure P17.4 when it occurs at constant temperature.

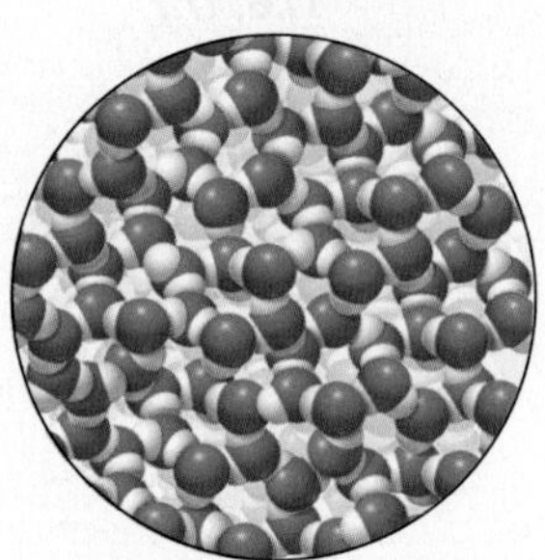

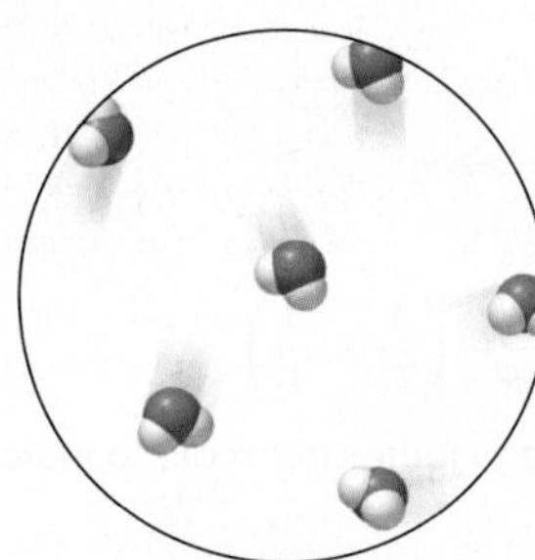

Reactants Products

FIGURE P17.4

17.5. Predict whether an increase or decrease in entropy accompanies the chemical change in Figure P17.5 when it occurs at constant temperature.

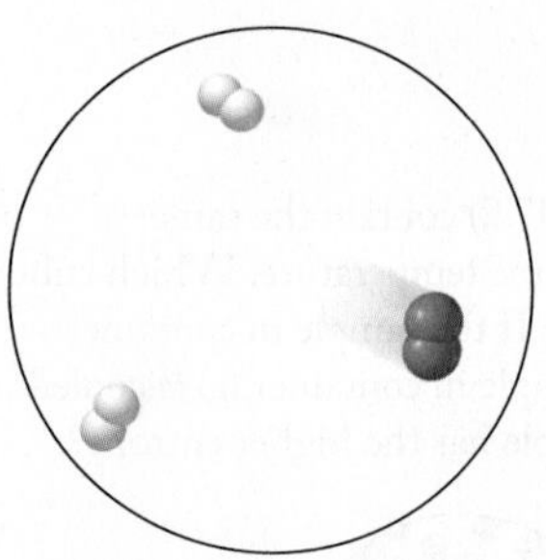

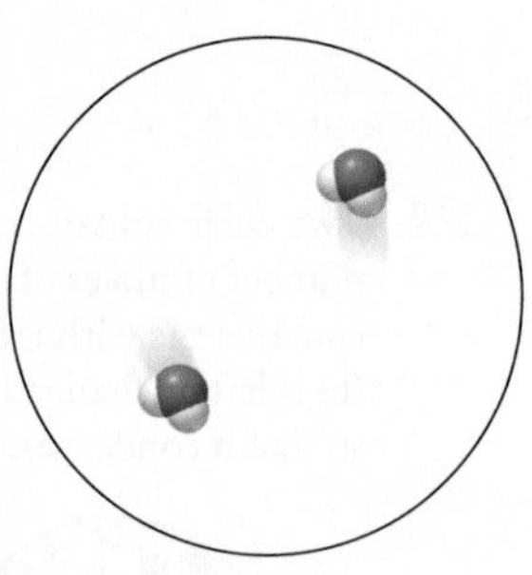

Reactants Products

FIGURE P17.5

17.6. Does the chemical reaction in Figure P17.6 result in an increase or decrease in the entropy of the system? Assume the reactants and products are at the same temperature.

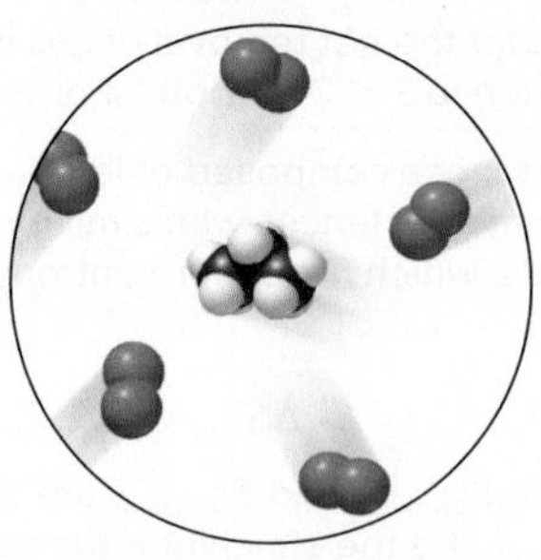

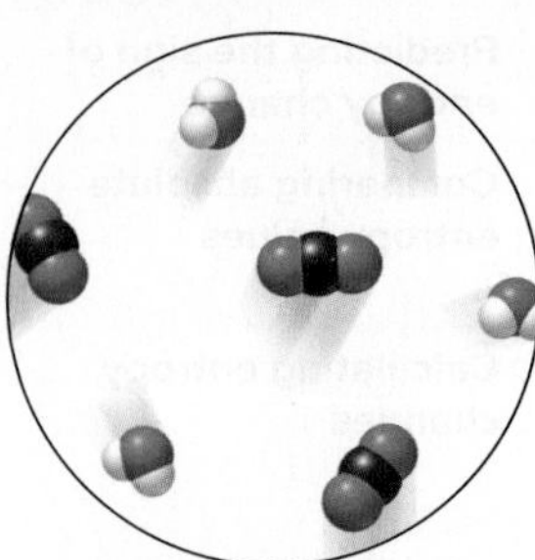

Reactants Products

FIGURE P17.6

17.7. Does the entropy increase or decrease in the reaction in Figure P17.7 between hydrazine (N_2H_4) and dinitrogen tetroxide (N_2O_4) to yield a mixture of nitrogen and water?

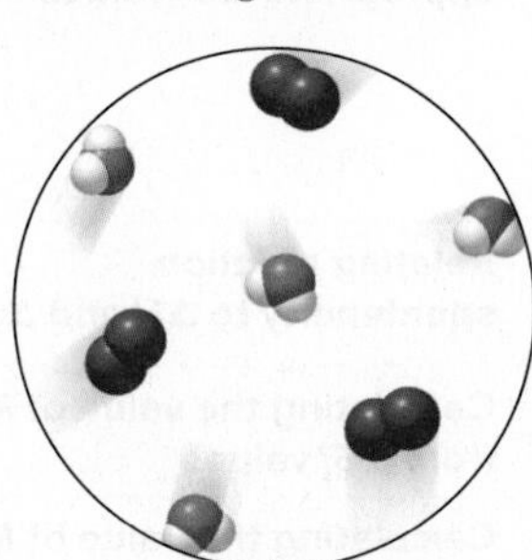

Reactants Products

FIGURE P17.7

17.8. The box on the left of Figure P17.8 represents a mixture of two diatomic gases: A_2 (red spheres) and B_2 (blue spheres). As a result of the process depicted by the arrow, how do the entropies of A_2 and B_2 change?

FIGURE P17.8

17.9. Is the process in Figure P17.8 more likely to be spontaneous at high temperature or low temperature, or is it unaffected by changing temperature?

17.10. Figure P17.10 shows the plots of ΔH and $T\Delta S$ for a phase change as a function of temperature.

a. What is the status of the process at the point where the two lines intersect?
b. Over what temperature range is the process spontaneous?

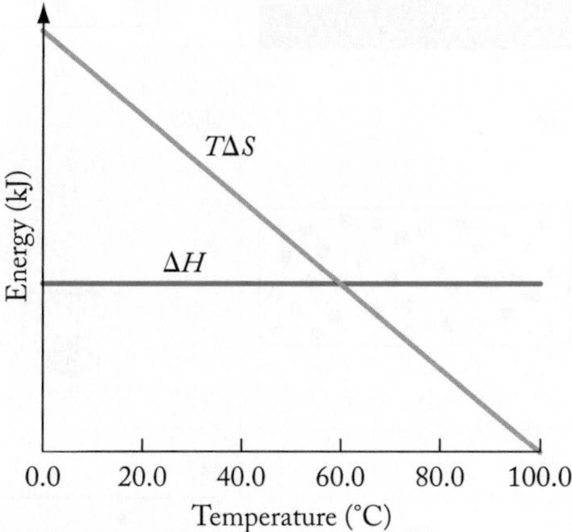

FIGURE P17.10

17.11. Figure P17.11 presents the ΔG°_f values of several elements and compounds selected from Appendix 4. Which of the following conversions are spontaneous? (a) $C_6H_6(\ell)$ to $CO_2(g)$ and $H_2O(\ell)$; (b) $CO_2(g)$ to $C_2H_2(g)$; (c) $H_2(g)$ and $O_2(g)$ to $H_2O(\ell)$. Explain your reasoning. *Hint*: Begin by writing a balanced equation describing the conversion. Add other substances as needed to complete the equation.

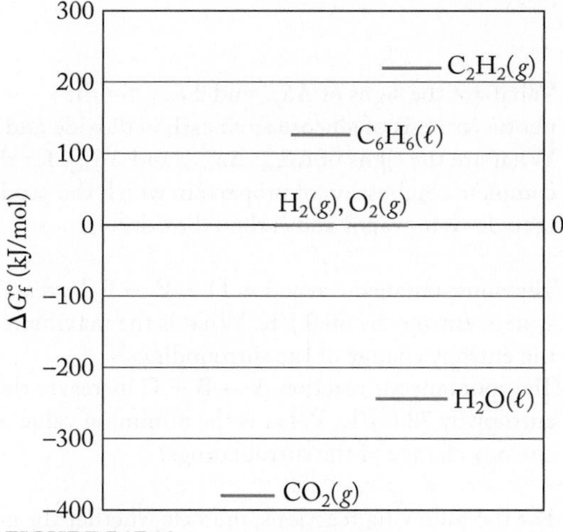

FIGURE P17.11

17.12. The three balloons in Figure P17.12 contain the same number of particles at the same temperature. Rank the balloons in order of increasing number of microstates accessible to the particles inside them.

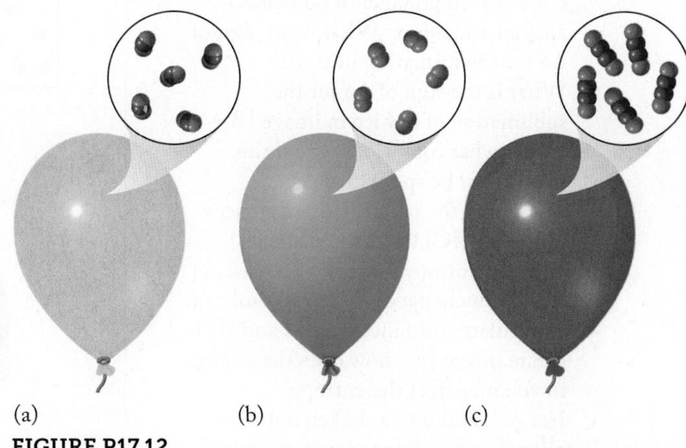

FIGURE P17.12

17.13. Figure P17.13a shows a cylinder within a cylinder that contains a population of gaseous molecules. The volume occupied by the molecules can be increased by pulling the inside cylinder out (Figure P17.13b), much like a telescope. The number of molecules within the cylinder remains the same during this operation. Compare the number of microstates available to the molecules in Figure P17.13a and b.

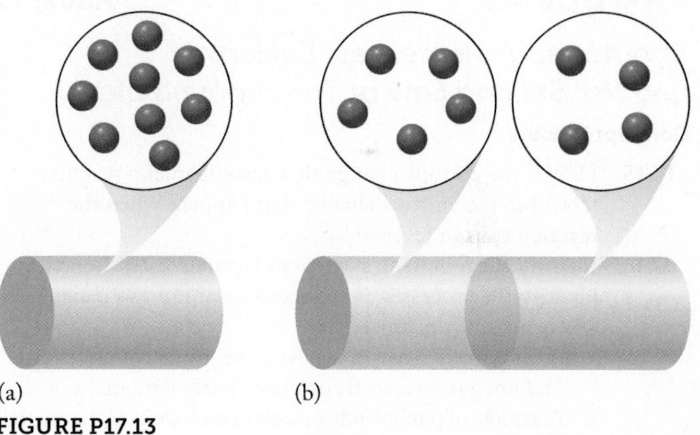

FIGURE P17.13

17.14. Use representations [A] through [I] in Figure P17.14 to answer questions (a)–(f).
  a. What is the sign of ΔS for image [F]?
  b. The slime in image [H] is a cross-linked polymer of polyvinyl alcohol and borax. What is the sign of ΔS for the formation of slime?
  c. What is the sign of ΔS for the sublimation of dry ice in image [B]? Under what conditions would this process *not* be spontaneous?
  d. If you mix the particulate substances in [A] and [C] to create image [E], does the entropy increase, decrease, or remain unchanged? When you mix the particulate substances in [G] and [I] to create image [E], how does the change in volume affect the entropy?
  e. Image [D] shows a shaken bottle of oil and water. After some time passes, what will happen to the oil and water? Will this process be spontaneous? Will the entropy increase or decrease?
  f. Does all mixing lead to an increase in entropy of the system?

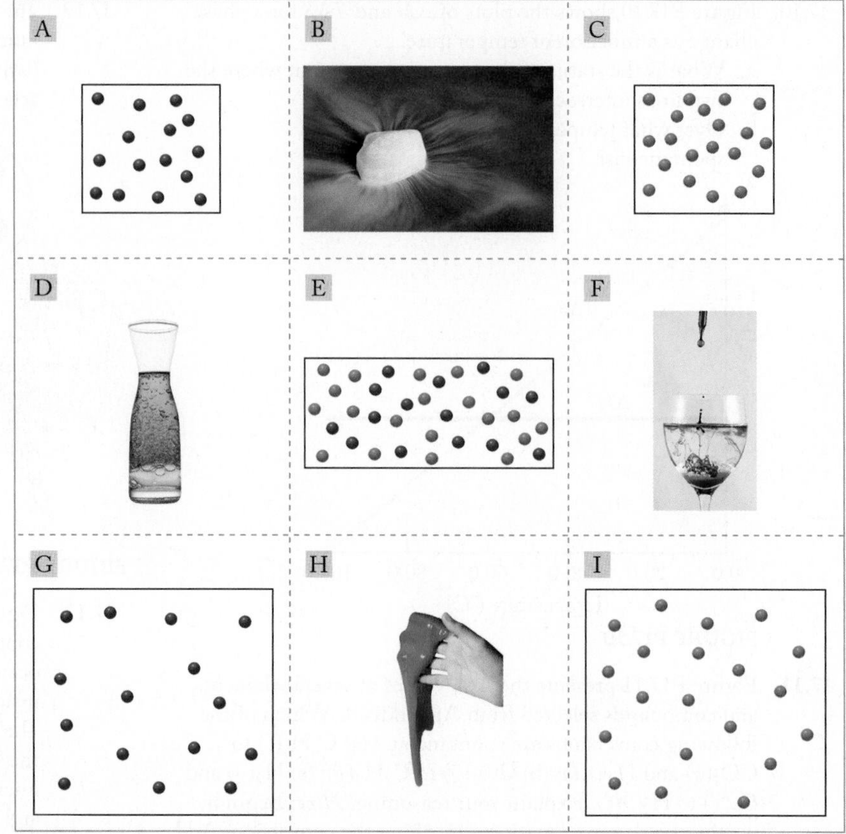

FIGURE P17.14

QUESTIONS AND PROBLEMS

Spontaneous Processes; Entropy and the Second Law of Thermodynamics

Concept Review

17.15. How is the entropy change that accompanies a reaction related to the entropy change that happens when the reaction runs in reverse?

17.16. Identify the following systems as isolated or not isolated, identify the processes as spontaneous or nonspontaneous, and explain your choice.
  a. A photovoltaic cell in a solar panel produces electricity.
  b. Helium gas escapes from a latex party balloon.
  c. A sample of pitchblende (uranium ore) emits alpha particles.

17.17. Ice cubes melt in a glass of lemonade, cooling the lemonade from 10.0°C to 0.0°C. If the ice cubes are the system, what are the signs of ΔS_{sys} and ΔS_{surr}?

17.18. Adding sidewalk deicer (calcium chloride) to water lowers the freezing point of the water. If solid $CaCl_2$ is the system, what are the signs of ΔS_{sys} and ΔS_{surr}?

Problems

17.19. Which of the following combinations of entropy changes for a process are mathematically possible?
  a. $\Delta S_{sys} > 0, \Delta S_{surr} > 0, \Delta S_{univ} > 0$
  b. $\Delta S_{sys} > 0, \Delta S_{surr} < 0, \Delta S_{univ} > 0$
  c. $\Delta S_{sys} > 0, \Delta S_{surr} > 0, \Delta S_{univ} < 0$

17.20. Which of the following combinations of entropy changes for a process are mathematically possible?
  a. $\Delta S_{sys} < 0, \Delta S_{surr} > 0, \Delta S_{univ} > 0$
  b. $\Delta S_{sys} < 0, \Delta S_{surr} < 0, \Delta S_{univ} > 0$
  c. $\Delta S_{sys} < 0, \Delta S_{surr} > 0, \Delta S_{univ} < 0$

17.21. What are the signs of ΔS_{sys} and ΔS_{univ} for the photosynthesis of glucose from carbon dioxide and water?

17.22. What are the signs of ΔS_{sys}, ΔS_{surr}, and ΔS_{univ} for the complete combustion of propane in which the products include water vapor and carbon dioxide?

17.23. The nonspontaneous reaction $D + E \rightarrow F$ decreases the system entropy by 66.0 J/K. What is the maximum value of the entropy change of the surroundings?

17.24. The spontaneous reaction $A \rightarrow B + C$ increases the system entropy by 72.0 J/K. What is the minimum value of the entropy change of the surroundings?

17.25. For the following reactions, indicate whether the entropy of the system increases, decreases, or remains nearly the same.
  a. $Al^{3+}(aq) + 3\,OH^-(aq) \rightarrow Al(OH)_3(s)$
  b. $CaCO_3(s) \rightarrow CaO(s) + CO_2(g)$
  c. $Mg(s) + Cu^{2+}(aq) \rightarrow Mg^{2+}(aq) + Cu(s)$

17.26. Which of the following processes would *not* result in an entropy increase for the indicated system?
  a. Melting of an ice cube
  b. Evaporation of a sample of an alcohol
  c. Sublimation of a mothball
  d. Cooling of hot water to room temperature

Absolute Entropy and the Third Law of Thermodynamics

Concept Review

17.27. Which component in each of the following pairs has the greater entropy?
a. 1 mole of $S_2(g)$ or 1 mole of $S_8(g)$
b. 1 mole of $S_2(g)$ or 1 mole of $S_8(s)$
c. 1 mole of $O_2(g)$ or 1 mole of $O_3(g)$
d. 1 gram of $O_2(g)$ or 1 gram of $O_3(g)$

17.28. **Digestion** During digestion, complex carbohydrates decompose into simple sugars. Do the carbohydrates experience an increase or decrease in entropy?

***17.29.** Diamond and the fullerenes are two allotropes of carbon. On the basis of their different structures and properties, predict which has the higher standard molar entropy.

17.30. **Superfluids** The 1996 Nobel Prize in Physics was awarded to Douglas Osheroff, Robert Richardson, and David Lee for discovering *superfluidity* (apparently frictionless flow) in ^{3}He. When ^{3}He is cooled to 2.7 mK, the liquid settles into an *ordered* superfluid state. What is the predicted sign of the entropy change for the conversion of liquid ^{3}He into its superfluid state?

17.31. Rank the compounds in each of the following groups in order of increasing standard molar entropy ($S°$):
a. $CH_4(g)$, $CF_4(g)$, and $CCl_4(g)$
b. $CH_2O(g)$, $CH_3CHO(g)$, and $CH_3CH_2CHO(g)$
c. $HF(g)$, $H_2O(g)$, and $NH_3(g)$

17.32. The hydrocarbon C_5H_{12} has three isomers, shown in Figure P17.32. Which isomer has the largest molar entropy, $S°$?

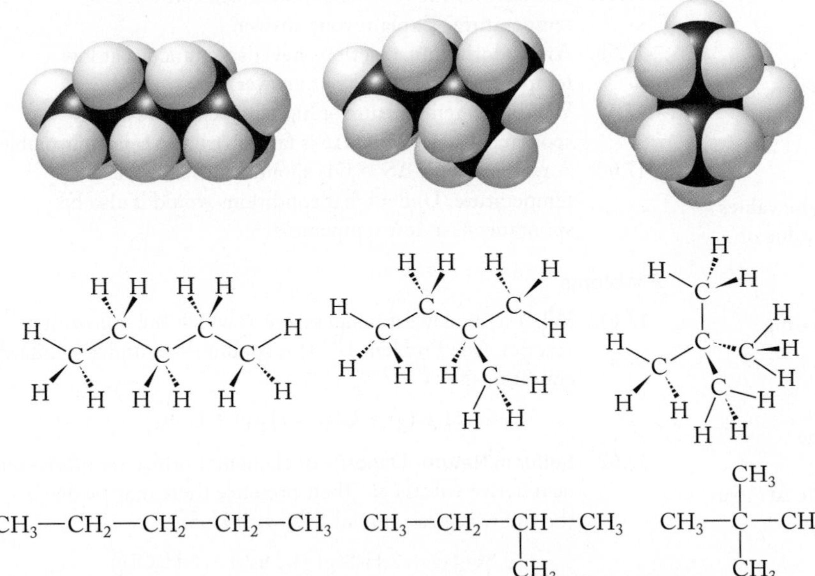

FIGURE P17.32

Calculating Entropy Changes

Concept Review

17.33. Under standard conditions, the products of a reaction have, overall, greater entropy than the reactants. What is the sign of $\Delta S°_{rxn}$?

17.34. Do polymerization reactions tend to have $\Delta S°_{rxn}$ values that are greater than zero or less than zero? Why?

17.35. Do precipitation reactions tend to have $\Delta S°_{rxn}$ values that are greater than zero or less than zero? Why?

*17.36. Suppose compound A(s) decomposes to substances B(ℓ) and C(ℓ) at moderately high temperatures. How would running the decomposition reaction at even higher temperatures (above the melting point of A, but still below the boiling points of B and C) affect the value of $\Delta S°_{rxn}$?

Problems

17.37. **Smog** Use the standard molar entropies in Appendix 4 to calculate $\Delta S°$ values for each of the following atmospheric reactions that contribute to the formation of photochemical smog.
a. $N_2(g) + O_2(g) \rightarrow 2\,NO(g)$
b. $2\,NO(g) + O_2(g) \rightarrow 2\,NO_2(g)$
c. $NO(g) + \frac{1}{2}O_2(g) \rightarrow NO_2(g)$
d. $2\,NO_2(g) \rightarrow N_2O_4(g)$

17.38. Use the standard molar entropies in Appendix 4 to calculate the $\Delta S°$ value for each of the following reactions.
a. $CH_4(g) + N_2(g) \rightarrow HCN(g) + NH_3(g)$
b. $Cu_2S(s) + O_2(g) \rightarrow 2\,Cu(s) + SO_2(g)$
c. $SO_3(g) + H_2O(\ell) \rightarrow H_2SO_4(aq)$
d. $S(g) + O_2(g) \rightarrow SO_2(g)$

17.39. What is the entropy change to the surroundings when a small decorative ice sculpture at a temperature of 0°C and weighing 456 g melts on a granite tabletop, if the temperature of the granite is 12°C and the process occurs reversibly? Assume a final temperature for the water of 0°C. The enthalpy of fusion of ice is 6.01 kJ/mol.

17.40. Another decorative "ice" sculpture is carved from dry ice (solid CO_2) and held at its sublimation point of −78.5°C. What is the entropy change to the universe when the CO_2 sculpture, weighing 389 g, sublimes on a granite tabletop if the temperature of the granite is 12°C and the process occurs reversibly? Assume a final temperature for the CO_2 vapor of −78.5°C. The enthalpy of sublimation of CO_2 is 26.1 kJ/mol.

Free Energy

Concept Review

17.41. What single criterion allows us to determine whether a process is spontaneous at constant temperature and pressure?
a. The sign of the equilibrium constant, K
b. The sign of the enthalpy change, ΔH
c. The sign of the free-energy change, ΔG
d. The sign of the entropy change, ΔS

17.42. If ΔG for a reaction is negative, then which of the following is true?
a. The reaction is spontaneous in the forward direction as written.
b. The reaction system is at equilibrium.
c. The reverse of the given reaction is spontaneous.
d. No reaction would ever have a negative ΔG.

***17.43.** Many 19th-century scientists believed that all exothermic reactions were spontaneous. Why did so many scientists share this belief?

17.44. In which direction does a reaction proceed when its ΔG value is (a) less than zero; (b) equal to zero; (c) greater than zero?

17.45. What are the signs of ΔS, ΔH, and ΔG for the sublimation of dry ice (solid CO_2) at 25°C?

17.46. What are the signs of ΔS, ΔH, and ΔG for the formation of dew on a cool night?

17.47. Which of the following processes is/are spontaneous? (a) a hurricane forms; (b) a corpse decomposes; (c) you get an A in this course; (d) ice cream melts on a hot summer day

17.48. Which of the following processes is/are spontaneous? (a) wood burns in air; (b) water vapor condenses on the sides of a glass of iced tea; (c) salt dissolves in water; (d) photosynthesis

Problems

17.49. Calculate the free-energy change for the dissolution in water of 1 mole of NaBr and 1 mole of NaI at 298 K from the values in the following table.

	ΔH°_{soln} (kJ/mol)	ΔS°_{soln} [J/(mol · K)]
NaBr	−0.86	57
NaI	−7.5	74

17.50. The values of ΔH°_{rxn} and ΔS°_{rxn} for the reaction

$$2\ NO(g) + O_2(g) \rightarrow 2\ NO_2(g)$$

are −12 kJ and −146 J/K, respectively. (a) Use these values to calculate ΔG°_{rxn} at 298 K. (b) Explain why the value of ΔG°_{rxn} is negative.

17.51. A mixture of $CO(g)$ and $H_2(g)$ is produced by passing steam over hot charcoal:

$$H_2O(g) + C(s) \rightarrow H_2(g) + CO(g)$$

Calculate the ΔG°_{rxn} value for the reaction from the appropriate ΔG°_f data in Appendix 4.

17.52. Using appropriate data from Appendix 4, calculate ΔG° for the following reaction:

$$N_2O_3(g) \rightarrow NO(g) + NO_2(g)$$

17.53. Determine the value of ΔG° for the reduction of iron ore with hydrogen gas:

$$Fe_2O_3(s) + 3\ H_2(g) \rightarrow 2\ Fe(s) + 3\ H_2O(g)$$

given the following thermodynamic properties at 25°C:

	$Fe_2O_3(s)$	$H_2(g)$	$Fe(s)$	$H_2O(g)$
ΔH°_f (kJ/mol)	−824.2	0	0	−241.8
S° [J/(mol · K)]	87.4	130.6	27.3	188.8

17.54. Determine the value of ΔG° for the reaction of the fuel gas acetylene with hydrogen gas:

$$C_2H_2(g) + 2\ H_2(g) \rightarrow C_2H_6(g)$$

given the following thermodynamic properties at 25°C:

	$C_2H_2(g)$	$H_2(g)$	$C_2H_6(g)$
ΔH°_f (kJ/mol)	226.7	0	−84.7
S° [J/(mol · K)]	200.8	130.6	229.5

17.55. Acid Precipitation Aerosols (fine droplets) of sulfuric acid form in the atmosphere as a result of the following reaction.

$$SO_3(g) + H_2O(g) \rightarrow H_2SO_4(\ell)$$

Use the appropriate data from Appendix 4 to calculate ΔG° for the combination reaction.

17.56. One source of sulfuric acid aerosols in the atmosphere is combustion of high-sulfur fuels, which releases SO_2 gas that then is further oxidized to SO_3:

$$2\ SO_2(g) + O_2(g) \rightarrow 2\ SO_3(g)$$

Use the appropriate ΔG°_f data from Appendix 4 to calculate ΔG°_{rxn} for this combination reaction at 25°C. Is it spontaneous under standard conditions?

Temperature and Spontaneity

Concept Review

17.57. Are exothermic reactions spontaneous only at low temperature? Explain your answer.

17.58. Are endothermic reactions never spontaneous at low temperature? Explain your answer.

17.59. Would low temperature or high temperature cause a spontaneous reaction if ΔS is favorable but ΔH is unfavorable?

17.60. A reaction with $\Delta S > 0$ is spontaneous at high temperature. Under what conditions would it also be spontaneous at low temperature?

Problems

17.61. What is the lowest temperature at which the following reaction (see Problem 17.51) is spontaneous under standard conditions?

$$H_2O(g) + C(s) \rightarrow H_2(g) + CO(g)$$

17.62. **Sulfur in Nature** Deposits of elemental sulfur are often seen near active volcanoes. Their presence there may be due to the following reaction of SO_2 with H_2S:

$$SO_2(g) + 2\ H_2S(g) \rightarrow \tfrac{3}{8}\ S_8(s) + 2\ H_2O(g)$$

Assuming the values of ΔH°_{rxn} and ΔS°_{rxn} do not change appreciably with temperature, over what temperature range is the reaction spontaneous under standard conditions?

17.63. Use the data in Appendix 4 to calculate ΔH° and ΔS° for the vaporization of hydrogen peroxide:

$$H_2O_2(\ell) \rightarrow H_2O_2(g)$$

Assuming that the calculated values are independent of temperature, what is the boiling point of hydrogen peroxide at $P = 1.00$ atm?

17.64. Determine the normal melting point of carbon tetrachloride in degrees Celsius given the following data:

$$\Delta H^\circ_{fus} = 2.67 \text{ kJ/mol} \qquad \Delta S^\circ_{fus} = 10.86 \text{ J/(mol} \cdot \text{K)}$$

17.65. Which of the following reactions is spontaneous (i) only at low temperatures; (ii) only at high temperatures; (iii) at all temperatures?
a. $2\,NO(g) + O_2(g) \rightarrow 2\,NO_2(g)$
b. $2\,NH_3(g) + 2\,O_2(g) \rightarrow N_2O(g) + 3\,H_2O(g)$
c. $NH_4NO_3(s) \rightarrow 2\,H_2O(g) + N_2O(g)$

17.66. Which of the following reactions is spontaneous (i) only at low temperatures; (ii) only at high temperatures; (iii) at all temperatures?
a. $2\,H_2S(g) + 3\,O_2(g) \rightarrow 2\,H_2O(g) + 2\,SO_2(g)$
b. $SO_2(g) + H_2O_2(\ell) \rightarrow H_2SO_4(\ell)$
c. $S(g) + O_2(g) \rightarrow SO_2(g)$

Free Energy and Chemical Equilibrium

Concept Review

17.67. If the value of K for a reaction is less than 1, what is the sign of ΔG°_{rxn}?

*17.68. The equation $\Delta G^\circ = -RT \ln K$ relates the value of K_p, not K_c, to ΔG° for gas-phase reactions. Explain why.

17.69. If a reaction mixture contains only reactants and no products, will the reaction proceed in the forward direction even if $\Delta G^\circ > 0$? Explain why or why not.

*17.70. If a gas-phase reaction mixture contains 1 mole of each reactant and product and has a 1:1 stoichiometry, is the ΔG°_{rxn} of the reaction mixture the same as ΔH°_{rxn}? Explain why or why not.

Problems

17.71. Which of the following reactions has the largest K_p value at 25°C?
a. $Cl_2(g) + F_2(g) \rightleftharpoons 2\,ClF(g)$ $\quad \Delta G^\circ_{rxn} = 115.4$ kJ
b. $Cl_2(g) + Br_2(g) \rightleftharpoons 2\,ClBr(g)$ $\quad \Delta G^\circ_{rxn} = -2.0$ kJ
c. $Cl_2(g) + I_2(g) \rightleftharpoons 2\,ICl(g)$ $\quad \Delta G^\circ_{rxn} = -27.9$ kJ

17.72. Use the appropriate ΔG°_f value from Appendix 4 to calculate the value of K_p at 298 K for the reaction

$$N_2(g) + 2\,O_2(g) \rightleftharpoons 2\,NO_2(g)$$

17.73. Use the appropriate equilibrium constant from Appendix 5 to calculate the value of ΔG°_{rxn} for the reaction

$$NH_3(g) + H_2O(\ell) \rightleftharpoons NH_4^+(aq) + OH^-(aq)$$

17.74. Use the appropriate equilibrium constant from Appendix 5 to calculate the value of ΔG°_{rxn} for the reaction

$$HClO(aq) + H_2O(\ell) \rightleftharpoons ClO^-(aq) + H_3O^+(aq)$$

Influence of Temperature on Equilibrium Constants

Concept Review

17.75. The value of the equilibrium constant of a reaction decreases with increasing temperature. Is the reaction endothermic or exothermic?

17.76. The reaction

$$2\,CO(g) + O_2(g) \rightleftharpoons 2\,CO_2(g)$$

is exothermic. Does the value of K_p increase or decrease with increasing temperature?

17.77. The value of K_p for the water–gas shift reaction

$$CO(g) + H_2O(g) \rightleftharpoons H_2(g) + CO_2(g)$$

increases as the temperature decreases. Is the reaction exothermic or endothermic?

17.78. Does the value of K_p for the reaction

$$CH_4(g) + H_2O(g) \rightleftharpoons 3\,H_2(g) + CO(g) \qquad \Delta H^\circ = 206 \text{ kJ}$$

increase, decrease, or remain unchanged as temperature increases?

Problems

17.79. **Air Pollution** Automobiles and trucks pollute the air with NO. At 2000°C, the value of K_c for the reaction

$$N_2(g) + O_2(g) \rightleftharpoons 2\,NO(g)$$

is 4.10×10^{-4} and $\Delta H^\circ = 180.6$ kJ. What is the value of K_c at 25°C?

17.80. At 400 K the value of K_p for the reaction

$$N_2(g) + 3\,H_2(g) \rightleftharpoons 2\,NH_3(g)$$

is 41 and $\Delta H^\circ = -92.2$ kJ. What is the value of K_p at 700 K?

17.81. The equilibrium constant for the reaction

$$NO(g) + O_2(g) \rightleftharpoons 2\,NO_2(g)$$

decreases from 1.5×10^5 at 430°C to 23 at 1000°C. From these data, calculate the value of ΔH° for the reaction.

17.82. The value of K_c for the reaction $A \rightleftharpoons B$ is 0.455 at 50°C and 0.655 at 100°C. Calculate ΔH° for the reaction.

Driving the Human Engine: Coupled Reactions

Concept Review

17.83. Describe the ways in which two chemical reactions must complement each other so that the decrease in free energy of the spontaneous reaction can drive the nonspontaneous reaction.

17.84. Why is it important that at least some of the spontaneous steps in glycolysis convert ADP to ATP?

17.85. The second step in glycolysis converts glucose-6-phosphate into fructose-6-phosphate (Figure P17.85). Suggest a reason why ΔG° for this reaction is close to zero.

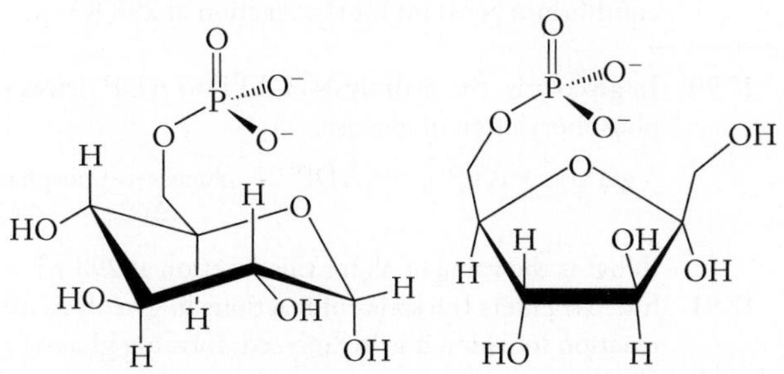

FIGURE P17.85

Problems

17.86. **Uses of Methane** The methane in natural gas is an important starting material, or feedstock, for producing industrial chemicals, including H_2 gas.

a. Use the appropriate ΔG°_f value(s) from Appendix 4 to calculate ΔG°_{rxn} for the reaction known as *steam–methane reforming*:

$$CH_4(g) + H_2O(g) \rightarrow CO(g) + 3\,H_2(g)$$

b. To drive this nonspontaneous reaction, the CO that is produced can be oxidized to CO_2 by using more steam:

$$CO(g) + H_2O(g) \rightarrow CO_2(g) + H_2(g)$$

Use the appropriate ΔG°_f value(s) from Appendix 4 to calculate ΔG°_{rxn} for this reaction, which is known as the *water–gas shift reaction*.

c. Combine these two reactions and write the chemical equation of the overall reaction in which methane and steam combine to produce hydrogen gas and carbon dioxide.

d. Calculate the ΔG°_{rxn} value of the overall reaction. Is it spontaneous under standard conditions?

17.87. In addition to its role in the reactions described in Problem 17.86, methane can, in theory, be used to produce hydrogen gas by a process in which it decomposes into elemental carbon and hydrogen:

$$CH_4(g) \rightarrow C(s) + 2\,H_2(g)$$

The carbon produced in the first step is then oxidized to CO_2:

$$C(s) + O_2(g) \rightarrow CO_2(g)$$

a. Calculate the ΔG°_{rxn} values of these two reactions.

b. Write a balanced chemical equation describing the overall reaction obtained by coupling the two reactions, and calculate its ΔG°_{rxn} value. Is the coupled reaction spontaneous under standard conditions?

17.88. Which of the following steps in glycolysis has the largest equilibrium constant?

a. Fructose-1,6-diphosphate $\rightleftharpoons$ 2 glyceraldehyde-3-phosphate $\Delta G^\circ_{rxn} = 24$ kJ

b. 3-Phosphoglycerate $\rightleftharpoons$ 2-phosphoglycerate $\Delta G^\circ_{rxn} = 4.4$ kJ

c. 2-Phosphoglycerate $\rightleftharpoons$ phosphoenolpyruvate $\Delta G^\circ_{rxn} = 1.8$ kJ

17.89. The value of ΔG° for the phosphorylation of glucose in glycolysis is 13.8 kJ/mol. What is the value of the equilibrium constant for the reaction at 298 K?

17.90. In glycolysis, the hydrolysis of ATP to ADP drives the phosphorylation of glucose:

$$\text{glucose} + ATP^{4-} \rightleftharpoons ADP^{3-} + \text{glucose-6-phosphate} \quad \Delta G^\circ_{rxn} = -17.7 \text{ kJ}$$

What is the value of K_c for this reaction at 298 K?

17.91. Sucrose enters the series of reactions in glycolysis after the reaction in which it is hydrolyzed, forming glucose and fructose:

$$\text{sucrose} + H_2O \rightleftharpoons \text{glucose} + \text{fructose} \qquad K_c = 5.3 \times 10^{12} \text{ at } 298 \text{ K}$$

What is the value of ΔG°_{rxn}?

Microstates: A Quantized View of Entropy

Concept Review

17.92. You flip three coins, assigning the values +1 for heads and −1 for tails. Each outcome of the three flips constitutes a microstate. How many different microstates are possible from flipping the three coins? Which value or values for the sums in the microstates are most likely? (*Hint*: The sequence HHT [+1 +1 −1] is one possible outcome, or microstate. Note, however, that this outcome differs from THH [−1 +1 +1], even though the two sequences sum to the same value.)

17.93. Imagine you have four identical chairs to arrange on four steps leading up to a stage, one chair on each step. The chairs have numbers on their backs: 1, 2, 3, and 4. How many different microstates for the chairs are possible? (Notice that when viewed from the front, all the microstates look the same. Viewed from the back, you can identify the different microstates because you can distinguish the chairs by their numbers.)

Problems

17.94. Use the appropriate standard molar entropy value from Appendix 4 to calculate how many microstates are accessible to a single molecule of N_2 at 298 K.

17.95. Use the appropriate standard molar entropy value from Appendix 4 to calculate how many microstates are accessible to a single molecule of liquid H_2O at 298 K.

Additional Problems

17.96. Methanogenic bacteria convert aqueous acetic acid (CH_3COOH) into $CO_2(g)$ and $CH_4(g)$. (a) Is this process endothermic or exothermic under standard conditions? (b) Is the reaction spontaneous under standard conditions?

17.97. Chlorofluorocarbons (CFCs) are no longer used as refrigerants because they catalyze the decomposition of stratospheric ozone. Trichlorofluoromethane (CCl_3F) boils at 23.8°C, and its molar enthalpy of vaporization is 24.8 kJ/mol. What is the molar entropy of vaporization of $CCl_3F(\ell)$?

17.98. Consider the precipitation reactions described by the following net ionic equations:

$$Mg^{2+}(aq) + 2\,OH^-(aq) \rightarrow Mg(OH)_2(s)$$
$$Ag^+(aq) + Cl^-(aq) \rightarrow AgCl(s)$$

a. Predict the sign of ΔS°_{rxn} for the reactions.

b. Using the appropriate values for S° from Appendix 4, calculate ΔS° for these reactions.

c. Do your calculations support your predictions?

17.99. (a) Using the appropriate values for S° from Appendix 4, calculate ΔS° for the reaction of Na(s) and $Cl_2(g)$ to form NaCl(s). (b) Explain why this reaction takes place given your answer to part (a).

*17.100. At what temperature is the standard free-energy change for the following reaction equal to zero?

$$NH_4Cl(s) \rightarrow NH_3(g) + HCl(g)$$

17.101. Which of these processes result in an entropy decrease of the system?

a. Diluting hydrochloric acid with water

b. Boiling water

c. $2\,NO(g) + O_2(g) \rightarrow 2\,NO_2(g)$

d. Making ice cubes in the freezer

*17.102. Calculate the standard free-energy change of the following reaction. Is it spontaneous?

$$2\,NO(g) + 2\,H_2(g) \rightleftharpoons N_2(g) + 2\,H_2O(g)$$

***17.103.** Estimate the standard free-energy change of the following reaction at 225°C:

$$C_2H_4(g) + 3\ O_2(g) \rightarrow 2\ CO_2(g) + 2\ H_2O(g)$$

*17.104. Show that hydrogen cyanide (HCN) is a gas at 25°C by estimating its normal boiling point from the following data:

	ΔH_f° (kJ/mol)	S° [J/(mol · K)]
HCN(ℓ)	108.9	113
HCN(g)	135.1	202

17.105. Making Methanol The element hydrogen (H_2) is not abundant on Earth, but it is a useful reagent in, for example, the potential synthesis of the liquid fuel methanol from gaseous carbon monoxide:

$$2\ H_2(g) + CO(g) \rightarrow CH_3OH(\ell)$$

Under what temperature conditions is this reaction spontaneous?

17.106. **Lightbulb Filaments** Tungsten (W) was the favored metal for incandescent lightbulb filaments, partly because of its high melting point of 3422°C. The enthalpy of fusion of tungsten is 35.4 kJ/mol. What is its entropy of fusion?

***17.107.** Over what temperature range is the reduction of tungsten(VI) oxide by hydrogen to give metallic tungsten and water spontaneous? The standard enthalpy of formation of $WO_3(s)$ is −843 kJ/mol, and its standard molar entropy is 76 J/(mol · K).

17.108. Two allotropes (A and B) of sulfur interconvert at 369 K and 1 atm of pressure:

$$S_8(s, A) \rightarrow S_8(s, B)$$

The enthalpy change in this transition is 297 J/mol. What is the entropy change?

***17.109.** Copper forms two oxides, Cu_2O and CuO.

a. Name these oxides.

b. Predict over what temperature range this reaction,

$$Cu_2O(s) \rightarrow CuO(s) + Cu(s)$$

is spontaneous by using the following thermodynamic data:

	ΔH_f° (kJ/mol)	S° [J/(mol · K)]
$Cu_2O(s)$	−170.7	92.4
CuO(s)	−156.1	42.6

c. Why is the standard molar entropy of $Cu_2O(s)$ larger than that of CuO(s)?

*17.110. **Lime** Enormous amounts of lime (CaO) are used in steel industry blast furnaces to remove impurities from iron. Lime is made by heating limestone and other solid forms of $CaCO_3(s)$. Why is the standard molar entropy of $CaCO_3(s)$ higher than that of CaO(s)? At what temperature is the pressure of $CO_2(g)$ over $CaCO_3(s)$ equal to 1.0 atm?

	ΔH_f° (kJ/mol)	S° [J/(mol · K)]
$CaCO_3(s)$	−1207	93
CaO(s)	−636	40
$CO_2(g)$	−394	214

***17.111.** According to *Trouton's rule*, the ratio $\Delta H^\circ_{vap}/T_b$ for a liquid is approximately 88 J/(mol · K). Here, ΔH°_{vap} is the molar enthalpy of vaporization of a liquid, and T_b is its normal boiling point. (a) What idea suggests that $\Delta H^\circ_{vap}/T_b$ for a range of liquids should be approximately constant? (b) Check Trouton's rule against the data in Figure P17.111. Which liquids deviate from Trouton's rule, and why?

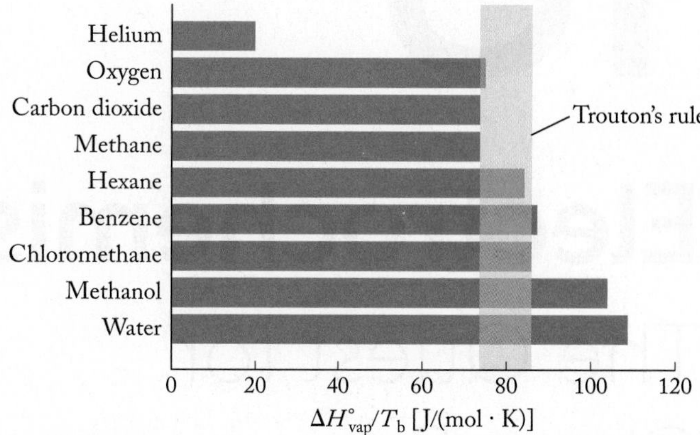

FIGURE P17.111

*17.112. **Melting DNA** When a solution of DNA in water is heated, the DNA double helix separates into two single strands:

$$1\ \text{DNA double helix} \rightleftharpoons 2\ \text{single strands}$$

a. What is the sign of ΔS for the forward process as written?

b. The DNA double helix reforms as the system cools. What is the sign of ΔS for the process by which two single strands reform the double helix?

c. The melting point of DNA is defined as the temperature at which $\Delta G = 0$. At that temperature, the forward reaction produces two single strands as fast as two single strands recombine to form the double helix. Write an equation that defines the melting temperature (T) of DNA in terms of ΔH and ΔS.

***17.113. Melting Organic Compounds** When dicarboxylic acids (compounds with two –COOH groups in their structures) melt, they frequently decompose to produce 2 moles of CO_2 gas for every 1 mole of dicarboxylic acid melted (shown in Figure P17.113). (a) What are the signs of ΔH and ΔS for the process as written? (b) Problem 17.112 describes the DNA double helix reforming when the system cools after melting. Do you think the dicarboxylic acid will reform when the melted material cools? Why or why not?

$$HOOC{-}(CH_2)_n{-}COOH(s) \rightarrow H{-}(CH_2)_n{-}H(\ell) + 2\ CO_2(g)$$

FIGURE P17.113

17.114. **Computer Chips** The absolute entropy (S) of a perfect, defect-free solid equals zero and has one accessible microstate. It is impossible to make such a material, but silicon chip manufacturers strive for as few defects as possible in their products. Calculate the absolute molar entropy for a piece of silicon with a number of probable arrangements of (a) $W = 16$, (b) $W = 625$, and (c) $W = 2500$ per atom of Si.

18

Electrochemistry

The Quest for Clean Energy

AN ELECTRIFYING SEDAN The 2019 Chevrolet Bolt has a 60-kWh lithium-ion battery and driving range of about 383 km (238 mi).

PARTICULATE **REVIEW**

Redox: Metal versus Nonmetal

In Chapter 18 we investigate transformations between chemical energy and electrical energy via redox reactions. Rust (mostly Fe_2O_3) forms on old cars such as the one in this photo through a series of reactions between iron in the car and oxygen in the atmosphere.

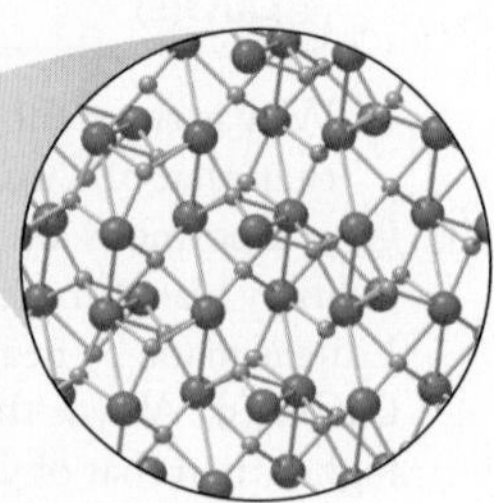

- Write a balanced chemical equation for the formation of Fe_2O_3 from its elements.
- Which element is oxidized and which is reduced during this reaction?
- Describe the direction of the electron transfer in the formation of Fe_2O_3.

(Review Section 4.8 if you need help.)

(Answers to Particulate Review questions are in the back of the book.)

PARTICULATE **PREVIEW**

Redox: Electricity and Clean Fuel

The redox chemistry used to create fuel cells involves hydrogen gas and oxygen gas as pictured here. As you read Chapter 18, look for ideas that will help you answer these questions:

- Determine which half-reaction represents oxidation and which represents reduction. Write an overall equation for the chemistry of fuel cells.
- How do the electrons transferred in this reaction generate electricity to power cars?
- Cars powered by fuel cells are considered "zero-emission vehicles." Explain this designation based on the chemistry of fuel cells.

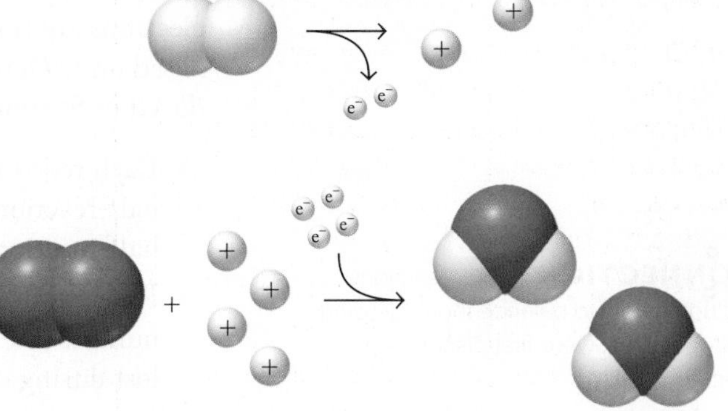

Learning Outcomes

LO1 Combine the appropriate half-reactions to write net ionic equations of spontaneous redox reactions
Sample Exercises 18.1, 18.2, 18.3, 18.4

LO2 Use cell diagrams to describe the components of electrochemical cells and their roles in interconverting chemical and electrical energy
Sample Exercise 18.2

LO3 Use the standard reduction potentials of two half-reactions to decide which occurs at the cathode and which at the anode and to calculate the standard potential of the electrochemical cell based on the half-reactions
Sample Exercise 18.3

LO4 Interconvert a cell's potential and the change in free energy of the cell reaction
Sample Exercise 18.4

LO5 Use the Nernst equation to relate the concentrations of cell reaction components to cell potentials
Sample Exercise 18.5

LO6 Relate standard cell potentials to the equilibrium constants of the cell reactions
Sample Exercise 18.6

LO7 Use the Faraday constant to relate quantity of charge to changes in the quantities of reactants and products in cell reactions
Sample Exercises 18.7, 18.8

LO8 Explain the applicability of the term zero-emission to electric vehicles

18.1 Running on Electrons: Redox Chemistry Revisited

In previous chapters we explored some of the environmental impacts of producing energy by burning fossil fuels. Concerns about these impacts have spurred the development of innovative propulsion systems for cars and trucks that reduce or eliminate the use of fossil fuels. Some vehicles are hybrids, which means that they are propelled by a combination of gasoline engines and electric motors powered by rechargeable batteries. Plug-in hybrids (such as the one in this chapter's opening photograph) are powered exclusively by electric motors, but they have a small gasoline engine to generate electricity and extend their driving range. Completely electric vehicles are powered by banks of high-performance batteries or by fuel cells. If these all-electric cars are to replace most of the vehicles powered by fossil fuels, scientists and engineers will need to develop lighter-weight, higher-capacity, more powerful, more reliable, and less expensive batteries and fuel cells. It will not be easy, but progress is being made.

Batteries and fuel cells are devices based on **electrochemistry**, the branch of chemistry that links chemical reactions to the production or consumption of electrical energy. At the heart of electrochemistry are chemical reactions in which electrons are transferred between substances. In other words, electrochemistry is based on *red*uction and *ox*idation reactions, or *redox* reactions, which were introduced in Section 4.8. There we noted the following:

- Each redox reaction is the sum of two half-reactions: a reduction half-reaction, in which a reactant gains electrons, and an oxidation half-reaction, in which a reactant loses electrons.
- Reduction and oxidation half-reactions happen simultaneously, and the number of electrons gained during reduction must exactly match the number lost during oxidation.

CONNECTION Redox reactions, including how to balance them by using half-reactions, were first discussed in Section 4.8.

We begin our exploration of electrochemistry by reviewing the oxidation and reduction half-reactions involved in a redox reaction between copper and zinc.

FIGURE 18.1 A strip of zinc is immersed in an aqueous solution of blue copper(II) sulfate. Over time it becomes encrusted with a dark layer of copper as Cu^{2+} ions are reduced to Cu atoms and Zn atoms are oxidized to colorless Zn^{2+} ions.

When a strip of Zn metal is placed in a solution of $CuSO_4$ (**Figure 18.1**), electrons spontaneously transfer from Zn atoms to Cu^{2+} ions, forming Zn^{2+} ions and Cu atoms. The shiny zinc surface turns dark brown as a textured layer of copper metal accumulates on it, and the distinctive blue color of $Cu^{2+}(aq)$ ions fades as these ions gain electrons and become atoms of copper metal.

The electron transfer in this spontaneous reaction can be represented through two half-reactions that occur simultaneously—the oxidation half-reaction of zinc atoms and the reduction half-reaction of copper ions:

$$Zn(s) \rightarrow Zn^{2+}(aq) + 2\ e^-$$

$$Cu^{2+}(aq) + 2\ e^- \rightarrow Cu(s)$$

For 1 mole of zinc atoms, 2 moles of electrons are "lost," or transferred to 1 mole of copper ions. Because the number of moles of electrons lost and gained are the same in the two half-reactions, writing a net ionic equation to describe the overall redox reaction is simply a matter of adding the two half-reactions together:

$$\begin{array}{r} Zn(s) \rightarrow Zn^{2+}(aq) + 2\ e^- \\ Cu^{2+}(aq) + 2\ e^- \rightarrow Cu(s) \\ \hline Zn(s) + Cu^{2+}(aq) + \cancel{2\ e^-} \rightarrow Cu(s) + Zn^{2+}(aq) + \cancel{2\ e^-} \end{array}$$

Canceling out the equal numbers of electrons gained and lost, we get

$$Zn(s) + Cu^{2+}(aq) \rightarrow Cu(s) + Zn^{2+}(aq) \tag{18.1}$$

Combining half-reactions is a convenient way to write net ionic equations for redox reactions. In Chapter 18 we use a valuable resource in this equation-writing process: a table of common half-reactions in Appendix 6. All these half-reactions are written as reduction half-reactions. There is no need for a separate table of *oxidation* half-reactions because any reduction half-reaction can always be reversed to obtain the corresponding oxidation half-reaction.

Having established that the redox reaction between zinc metal and $Cu^{2+}(aq)$ ions is the result of two distinct half-reactions, let's examine how chemists can *physically separate* the two half-reactions to produce electricity. To do this we use a device called an **electrochemical cell** (**Figure 18.2a**). Like nearly all electrochemical cells, it consists of two compartments. One compartment contains a strip of zinc metal immersed in 1.00 *M* $ZnSO_4$; the other compartment has a strip of copper metal immersed in 1.00 *M* $CuSO_4$. Sulfate ions are spectator ions in this reaction and are not included in the net ionic equation, although we will see that they do play an important role in cell function. The

electrochemistry the branch of chemistry that examines the transformations between chemical and electrical energy.

electrochemical cell an apparatus that can either generate electrical energy from a chemical reaction or use electrical energy to produce a chemical reaction.

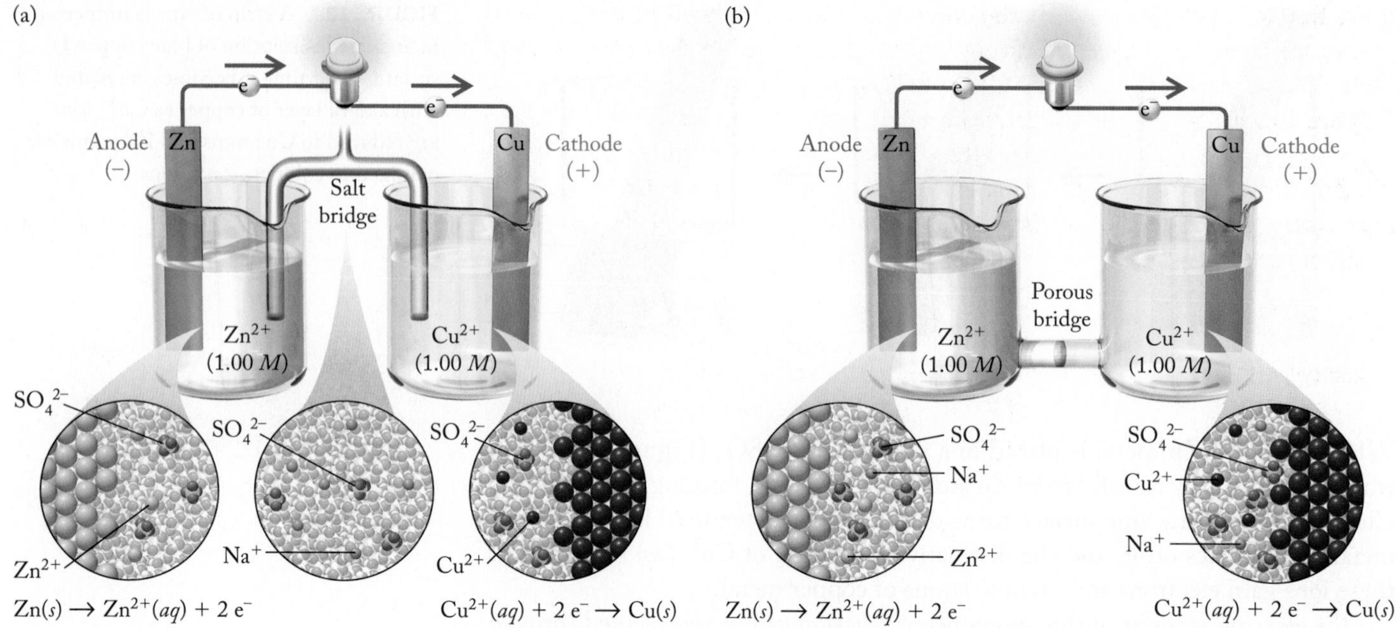

FIGURE 18.2 (a) This electrochemical cell consists of two compartments: the one on the left contains a zinc metal anode immersed in 1.00 *M* $ZnSO_4$; the one on the right contains a copper metal cathode immersed in 1.00 *M* $CuSO_4$. A salt bridge containing an aqueous solution of Na_2SO_4 provides an electrical connection through which ions (and their charges) can move from one compartment to the other. (b) In this apparatus, a porous bridge substitutes for the salt bridge, thus allowing ions from the background electrolyte (Na_2SO_4) to move between the two cells.

CHEMTOUR
Zinc–Copper Cell

STEPWISE ANIMATION
Electricity and Water Analogy

two metal strips function as the *electrodes* of the cell, providing pathways through which the electrons produced and consumed in the two half-reactions flow to and from an external circuit, converting chemical energy into electrical energy.

As the cell reaction proceeds, oxidation of Zn atoms produces electrons, which travel from the Zn electrode through the external circuit to the surface of the Cu electrode, where they combine with Cu^{2+} ions, forming atoms of Cu metal. In an electrochemical cell, the electrode at which the oxidation half-reaction takes place (the zinc electrode in this case) is called the **anode**, and the electrode at which the reduction half-reaction takes place (the copper electrode in this case) is called the **cathode**.

You might think that production of Zn^{2+} ions in the left compartment in Figure 18.2a would result in a buildup of positive charge on that side of the cell, and that conversion of Cu^{2+} ions to Cu metal would result in an excess of SO_4^{2-} ions and thus a negative charge in the Cu compartment. Creation of a net charge in each compartment would cause electron transfer to stop. However, no buildup of charge occurs because the two compartments in Figure 18.2a are connected by a salt bridge, a bent glass tube filled with a strong electrolyte (Na_2SO_4 in this example) not involved in either half-reaction. Porous plugs on both ends of the tube allow ions to migrate through the bridge, which allows electrical charge to

anode an electrode at which an oxidation half-reaction (loss of electrons) takes place.

cathode an electrode at which a reduction half-reaction (gain of electrons) takes place.

voltaic cell an electrochemical cell in which chemical energy is transformed into electrical work by a spontaneous cell reaction.

electrolysis a process in which electrical energy is used to drive a nonspontaneous chemical reaction.

electrolytic cell a device in which an external source of electrical energy does work on a chemical system, turning reactant(s) into higher-energy product(s).

cell diagram symbols that show how the components of an electrochemical cell are connected.

move from one side of the cell to the other. Often a salt bridge is inconvenient to use, so scientists add a background electrolyte not involved in the cell reaction to each compartment and connect the compartments with a porous bridge (**Figure 18.2b**) that allows electrolyte ions to move from one side of the cell to the other. Migration of Na^+ ions toward the Cu compartment and SO_4^{2-} ions toward the Zn compartment through the salt bridge or porous bridge balances the flow of electrons in the external circuit and eliminates any accumulation of ionic charge in either compartment.

CONCEPT TEST

As the cell reaction in Figure 18.2 proceeds, is the increase in mass of the copper strip the same as the decrease in mass of the zinc strip? Explain your answer.

(Answers to Concept Tests are in the back of the book.)

(a)

(b)

FIGURE 18.3 (a) Alessandro Volta and (b) Luigi Galvani, two pioneers in electrochemistry, were contemporary Italian scientists. Galvani's discovery of bioelectricity by making dead frogs' legs twitch with electricity was the basis for Volta's belief that bioelectricity was a special case of a much broader phenomenon of conducting electricity by the movement of ions dissolved in electrolytes. Volta's own research led to the development of the first battery: a stack of alternating layers of zinc, blotter paper soaked in salt water, and silver.

18.2 Voltaic and Electrolytic Cells

Cells like the Zn/Cu^{2+} cell, in which the chemical reactions inside the cell pump electrons from the anode, through an external circuit, and into the cathode, are called **voltaic cells** in honor of the Italian physicist Alessandro Volta (1745–1827; **Figure 18.3a**), who is credited with building the first battery. His battery and all the modern batteries that power familiar electric devices, from cell phones to flashlights to laptop computers, are examples of voltaic cells. Voltaic cells are also known as *galvanic cells* after Luigi Galvani (1737–1798; **Figure 18.3b**) whose discovery of *bio*electricity contributed to Volta's development of the battery shown in Figure 18.3a.

In this chapter we also investigate cells in which an external electrical power supply drives a nonspontaneous chemical reaction inside the cell. A reaction that is driven by the consumption of electrical energy is called **electrolysis**, and a cell in which electrolysis occurs is called an **electrolytic cell** (**Figure 18.4**). In electrolytic cells, electrons are pumped into the cathodes (making them the negative electrodes) and pulled away from the anodes (making them the positive electrodes). As we explain in Section 18.9, many of the batteries used to power familiar electronic devices are rechargeable, which means that not only do their cell reactions produce electricity, but these same reactions can also be forced to run in reverse when connected to an external power supply. When this happens, these voltaic cells become electrolytic cells as the products of the voltaic reaction become the reactants in the electrolytic reaction, regenerating the original reactants and recharging the battery.

Cell Diagrams

Chemists have developed a notation to efficiently convey what components are present in an electrochemical cell such as the Zn/Cu^{2+} one depicted in Figure 18.2: A **cell diagram** consists of both chemical formulas and symbols to show how the components of the cell are connected. A cell diagram does not convey stoichiometry, so any coefficients in the balanced equation for the cell reaction do not appear in the cell diagram. A cell diagram has the following components:

anode | anode solution || cathode solution | cathode

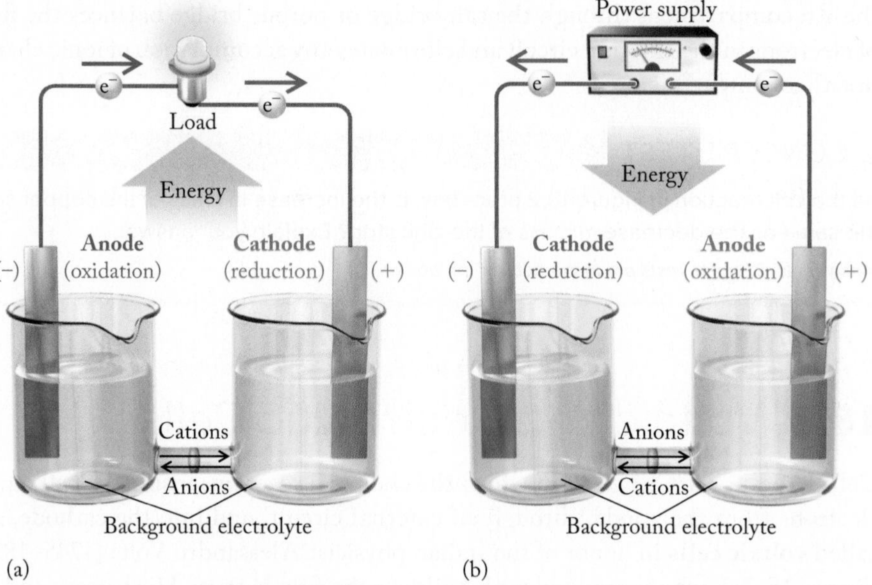

FIGURE 18.4 Voltaic versus electrolytic cells. (a) In a voltaic cell, a spontaneous chemical reaction generates electrical energy and does electrical work on its surroundings, such as lighting an LED lightbulb. (b) In an electrolytic cell, an external supply of electrical energy does work on the chemical system in the cell, driving a nonspontaneous cell reaction.

STEPWISE ANIMATION
Voltaic vs. Electrolytic Cells

There are three steps to drawing a cell diagram. (The symbols used to represent the cell diagram for the Zn/Cu^{2+} electrochemical cell in Figure 18.2 are included as an example with each step.)

1. Write the chemical symbol of the anode at the far left of the diagram, the symbol of the cathode at the far right, and double vertical lines for the connecting bridge halfway between them:

$$Zn(s) \ldots\ldots\ldots \| \ldots\ldots\ldots Cu(s)$$

2. Work inward from the electrodes toward the connecting bridge, using vertical lines to indicate phase changes (such as that between a solid metal electrode and an aqueous solution). Represent the electrolytes surrounding the electrode by using the symbols of the ions or compounds that are changed by the cell reaction. Use commas to separate species in the same phase:

$$Zn(s) \mid Zn^{2+}(aq) \| Cu^{2+}(aq) \mid Cu(s)$$

3. If known, use the concentrations of the dissolved species in place of (*aq*) phase symbols, and add the partial pressures of any gases within their (*g*) phase symbols:

$$Zn(s) \mid Zn^{2+}(1.00\ M) \| Cu^{2+}(1.00\ M) \mid Cu(s)$$

CONCEPT TEST

Describe in your own words the meaning of the cell diagram for Figure 18.2. Start your description with: "The cell consists of a zinc anode, which is oxidized to . . .".

To draw a cell diagram, we need to know the identities of both the reactants and products in both the anode and cathode half-reactions. As noted in Section 18.1, the half-reactions listed in Table A6.1 are a very handy reference in identifying the atoms, ions, and molecules involved not only in electrochemical reactions but also in redox reactions in general. Many of the half-reactions in Table A6.1 include H^+ ions because they occur in acidic solutions, and H^+ ions are often used to balance the number of H atoms in half-reactions in which water is a reactant or product. Although H^+ ions in aqueous solution are better represented as hydronium ions, $H_3O^+(aq)$, it is customary to write redox half-reactions by using the more simplistic $H^+(aq)$. For example, the reduction of O_2 to H_2O is written as

$$O_2(g) + 4\,H^+(aq) + 4\,e^- \rightarrow 2\,H_2O(\ell)$$

However, a few half-reactions in Table A6.1 contain OH^- ions, which tells us that these reactions occur in basic solutions. One such reaction involves another reduction of O_2:

$$O_2(g) + 2\,H_2O(\ell) + 4\,e^- \rightarrow 4\,OH^-(aq)$$

In Sample Exercise 18.1 and its accompanying practice exercise, we need to select half-reactions that match the pH conditions of the reaction.

SAMPLE EXERCISE 18.1 Writing Net Ionic Equations of Cell Reactions by Combining Half-Reactions **LO1**

Identify two half-reactions listed in Table A6.1 that could be combined to produce a net ionic equation describing the electrolysis of water into hydrogen gas and oxygen gas. Assume the pH of the solution is 12.0.

Collect and Organize The products of the electrolysis of water are $H_2(g)$ and $O_2(g)$, so our target net ionic equation is

$$2\,H_2O(\ell) \rightarrow 2\,H_2(g) + O_2(g)$$

Table A6.1 contains four half-reactions in which H_2O, H_2, and O_2 are either reactants or products:

$$2\,H^+(aq) + 2\,e^- \rightarrow H_2(g)$$

$$2\,H_2O(\ell) + 2\,e^- \rightarrow H_2(g) + 2\,OH^-(aq)$$

$$O_2(g) + 2\,H_2O(\ell) + 4\,e^- \rightarrow 4\,OH^-(aq)$$

$$O_2(g) + 4\,H^+(aq) + 4\,e^- \rightarrow 2\,H_2O(\ell)$$

Analyze The solution has a basic pH, so the appropriate half-reactions from Table A6.1 are those that contain OH^-, not H^+, ions. This condition eliminates the first and last half-reactions just listed, leaving the middle two to use in writing the net ionic equation:

$$2\,H_2O(\ell) + 2\,e^- \rightarrow H_2(g) + 2\,OH^-(aq)$$

$$O_2(g) + 2\,H_2O(\ell) + 4\,e^- \rightarrow 4\,OH^-(aq)$$

Water is a reactant and H_2 is a product in the first of these half-reactions, as in the target electrolysis reaction. However, O_2 is a reactant in the second half-reaction but a product in the target electrolysis reaction. Therefore, the second half-reaction must be reversed and written as an oxidation half-reaction:

$$4\,OH^-(aq) \rightarrow O_2(g) + 2\,H_2O(\ell) + 4\,e^-$$

The number of electrons gained and lost in the two half-reactions must balance, which means we must multiply the reduction half-reaction by 2, so that both involve 4 moles of electrons.

Solve Multiplying the reduction half-reaction by 2 and adding it to the oxidation half-reaction gives:

$$4\,H_2O(\ell) + 4\,e^- \rightarrow 2\,H_2(g) + 4\,OH^-(aq)$$

$$+ \quad 4\,OH^-(aq) \rightarrow O_2(g) + 2\,H_2O(\ell) + 4\,e^-$$

$$4\,H_2O(\ell) + 4\,OH^-(aq) + 4\,e^- \rightarrow 2\,H_2(g) + 4\,OH^-(aq) + O_2(g) + 2\,H_2O(\ell) + 4\,e^-$$

Simplifying by eliminating the terms common to both sides of the reaction arrow:

$$\overset{2}{\cancel{4}}\,H_2O(\ell) + \cancel{4\,OH^-(aq)} + \cancel{4\,e^-} \rightarrow 2\,H_2(g) + \cancel{4\,OH^-(aq)} + O_2(g) + \cancel{2\,H_2O(\ell)} + \cancel{4\,e^-}$$

or,

$$2\,H_2O(\ell) \rightarrow 2\,H_2(g) + O_2(g)$$

Think About It Had the hydrolysis reaction been run under acidic conditions, the other two half-reactions would have been combined but the overall cell reaction would have been the same: multiplying the first half-reaction by 2 before combining it with the reverse of the last half-reaction gives us

$$4\,H^+(aq) + \cancel{4\,e^-} \rightarrow 2\,H_2(g)$$

$$-2\,H_2O(\ell) \rightarrow + O_2(g) + \cancel{4\,H^+(aq)} + \cancel{4\,e^-}$$

$$2\,H_2O(\ell) \rightarrow 2\,H_2(g) + O_2(g)$$

The same result is convincing evidence that both are correct.

Practice Exercise Use the appropriate half-reactions from Table A6.1 to write a net ionic equation describing the oxidation of HNO_2 to HNO_3 by O_2 in an acidic solution.

(Answers to Practice Exercises are in the back of the book.)

SAMPLE EXERCISE 18.2 Diagramming an Electrochemical Cell LO1, LO2

Figure 18.5 depicts an electrochemical cell in which a copper electrode immersed in a 1.00 *M* solution of Cu^{2+} ions is connected to a silver electrode immersed in a 1.00 *M* solution of Ag^+ ions. Write a balanced chemical equation for this cell reaction and draw a cell diagram for this cell.

Collect and Organize We have a cell reaction in which electrons spontaneously flow from a copper electrode in contact with a solution of Cu^{2+} ions through an external circuit to a silver electrode in contact with a solution of Ag^+ ions. The half-reactions in Table A6.1 involving these metals and ions are

$$Cu^{2+}(aq) + 2\,e^- \rightarrow Cu(s)$$

$$Ag^+(aq) + e^- \rightarrow Ag(s)$$

In a cell diagram, the anode and the species involved in the oxidation half-reaction are written on the left, whereas the cathode and the species involved in the reduction half-reaction are written on the right. We use single lines to separate phases and a double line to represent the porous bridge separating the two compartments of the cell.

Analyze In an electrochemical cell, electrons flow from the anode through an external circuit to the cathode. In the cell in Figure 18.5, then, copper is the anode and silver is the cathode. This means that the Cu reduction half-reaction must run in reverse as an oxidation half-reaction:

$$Cu(s) \rightarrow Cu^{2+}(aq) + 2\,e^-$$

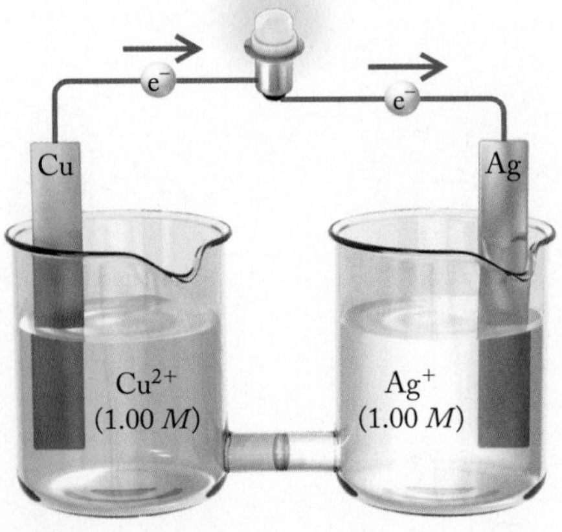

FIGURE 18.5 A Cu/Ag electrochemical cell.

Two moles of electrons are produced in the anode half-reaction, but only 1 mole of electrons is consumed in the cathode half-reaction at the Ag electrode. We therefore need to multiply the silver half-reaction by 2 before combining the two equations so that all the electrons are accounted for.

Solve Multiplying the Ag^+ half-reaction by 2 and adding it to the Cu half-reaction, we get

$$2\ Ag^+(aq) + 2\ e^- \rightarrow 2\ Ag(s)$$
$$Cu(s) \rightarrow Cu^{2+}(aq) + 2\ e^-$$
$$2\ Ag^+(aq) + \cancel{2\ e^-} + Cu(s) \rightarrow 2\ Ag(s) + Cu^{2+}(aq) + \cancel{2\ e^-}$$
$$2\ Ag^+(aq) + Cu(s) \rightarrow 2\ Ag(s) + Cu^{2+}(aq)$$

The equation is now balanced.

Applying the rules for drawing a cell diagram:

1. Anode on the left, cathode on the right, bridge in the middle:

$$Cu(s) \qquad || \qquad Ag(s)$$

2. Adding electrode–solution boundaries and the formulas of the ions produced and consumed in the cell reaction:

$$Cu(s)\ |\ Cu^{2+}(aq)\ ||\ Ag^+(aq)\ |\ Ag(s)$$

3. Adding concentration terms:

$$Cu(s)\ |\ Cu^{2+}(1.00\ M)\ ||\ Ag^+(1.00\ M)\ |\ Ag(s)$$

Think About It To test the validity of the cell diagram, translate it into a sentence. *A copper anode is oxidized to aqueous Cu^{2+} ions, which are separated by a porous bridge from an aqueous solution of Ag^+ ions that are reduced to Ag atoms at a silver cathode.* This description matches the net ionic equation for the reaction and is consistent with the cell layout and flow of electrons in Figure 18.5.

Practice Exercise Write a balanced chemical equation and draw the cell diagram for an electrochemical cell that has a copper cathode immersed in a solution of Cu^{2+} ions and an aluminum anode immersed in a solution of Al^{3+} ions.

18.3 Standard Potentials

Table A6.1 lists half-reactions in order of their **standard reduction potentials ($E°$)**. The superscript (°) has its usual thermodynamic meaning—namely, all reactants and products are in their standard states, which means the concentrations of all dissolved substances are 1 M and the partial pressures of all gases are 1 bar (≈ 1 atm). The more positive the value of $E°$, the greater the probability that the reduction half-reaction will couple with an oxidation half-reaction to produce a spontaneous redox reaction. The most positive $E°$ value in Table A6.1 is for the reduction of fluorine:

$$F_2(g) + 2\ e^- \rightarrow 2\ F^-(aq) \qquad E° = +2.866\ V$$

This means fluorine is the most easily reduced substance in Table A6.1. It also means that F_2 is the strongest oxidizing agent in the table. It can oxidize any of the substances on the product side of the half-reactions lower in the table. Likewise, F^- is the weakest reducing agent.

CONNECTION In a redox reaction, the substance that contains the element being reduced is the oxidizing agent and the substance that contains the element being oxidized is the reducing agent (see Section 4.8).

standard reduction potential ($E°$) the potential of a reduction half-reaction in which all reactants and products are in their standard states at 25°C.

The half-reaction at the very bottom of the table with the most negative $E°$ value

$$Li^+(aq) + e^- \rightarrow Li(s) \qquad E° = -3.05 \text{ V}$$

is least likely to proceed as written because Li^+ is the weakest oxidizing agent in the table. However, that also means that Li is the strongest reducing agent in this table, so the reverse reaction:

$$Li(s) \rightarrow Li^+(aq) + e^-$$

occurs more readily than the reverse of any other half-reaction in Table A6.1. Lithium metal is a very powerful reducing agent that can reduce any of the substances on the reactant side of the half-reactions in Table A6.1.

Substances with very negative $E°$ values at the bottom of Table A6.1 include the major cations in biological systems and environmental waters: Na^+, K^+, Mg^{2+}, and Ca^{2+}. Their negative $E°$ values indicate that these ions are not easily reduced to their free metals in aqueous solutions. They are chemically very stable (which explains the presence of these cations in nature).

CONCEPT **TEST**

Compare the order of the metals listed in the activity series in Table 4.6 to the order of the metals listed in Table A6.1. Describe the relationship between these two tables.

CONCEPT **TEST**

Use the order of the half-reactions listed in Table A6.1 to predict which of the following reactions is/are spontaneous under standard conditions:

a. $Cu(s) + 2\ Fe^{3+}(aq) \rightarrow Cu^{2+}(aq) + 2\ Fe^{2+}(aq)$

b. $2\ Ag(s) + Zn^{2+}(aq) \rightarrow 2\ Ag^+(aq) + Zn(s)$

c. $Hg(\ell) + 2\ H^+(aq) \rightarrow Hg^{2+}(aq) + H_2(g)$

Standard reduction potentials in Table A6.1 can be used to calculate the **standard cell potential ($E°_{cell}$)** of an electrochemical cell. Standard cell potential, as the name suggests, is related to the chemical (potential) energy stored in a cell and specifically to how forcefully the cell can pump electrons out from their anodes, through external circuits, and into their cathodes.

Consider the hypothetical case in which two half-reactions both have the same standard reduction potential. That is, there would be no difference between the abilities of their reactants to function as oxidizing agents and no electrons would spontaneously flow. For two half-reactions with different standard reduction potentials, the larger the $E°_{cell}$ difference between those potentials, the larger the $E°_{cell}$ of the electrochemical cell that could be built from them. Therefore, $E°_{cell}$ can be determined by calculating the difference between the standard reduction potentials of a voltaic cell's cathode and anode:

standard cell potential ($E°_{cell}$) a measure of how forcefully an electrochemical cell in which all reactants and products in their standard states can pump electrons through an external circuit.

$$E°_{cell} = E°_{cathode} - E°_{anode} \qquad (18.2)$$

Let's use Equation 18.2 to calculate $E°_{cell}$ for the Zn/Cu^{2+} voltaic cell in Figure 18.2. The standard reduction potential of the cathode half-reaction is

$$Cu^{2+}(aq) + 2\ e^- \rightarrow Cu(s) \qquad E° = 0.342 \text{ V}$$

To obtain the standard potential for the oxidation half-reaction at the zinc anode, we find the standard reduction potential of Zn^{2+} ions in Table A6.1:

$$Zn^{2+}(aq) + 2\,e^- \rightarrow Zn(s) \qquad E^\circ = -0.762\ V$$

Now we use Equation 18.2 to calculate E°_{cell}:

$$E^\circ_{cell} = E^\circ_{cathode} - E^\circ_{anode}$$
$$= 0.342 - (-0.762) = 1.104\ V$$

This is the cell potential we would measure if we connected a device called a voltmeter across the two electrodes (**Figure 18.6**) at 25°C under standard conditions.

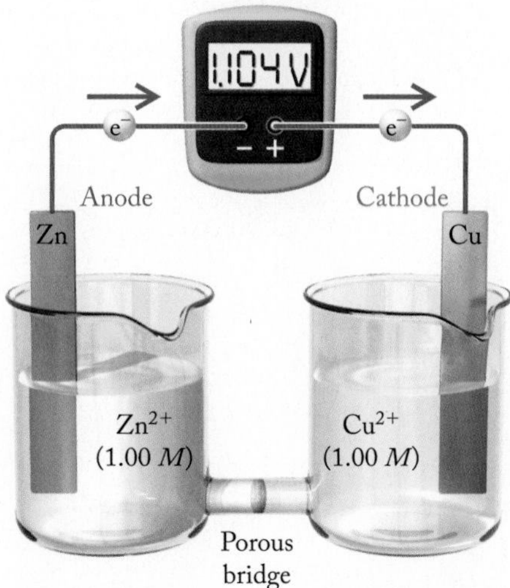

FIGURE 18.6 A voltmeter displays a cell potential of 1.104 V between a Zn electrode immersed in a 1.00 *M* solution of Zn^{2+} ions and a Cu electrode immersed in a 1.00 *M* solution of Cu^{2+} ions.

To use Equation 18.2, we need to know which half-reaction occurs at the cathode (reduction) and which occurs at the anode (oxidation). In other words, we need to know which component of the spontaneous cell reaction is more likely to be oxidized and which is more likely to be reduced. This decision can be made based on the data in Table A6.1. In our Zn/Cu^{2+} cell, for instance, the value of E° for the reduction of Cu^{2+} ions to Cu metal is 0.342 V, which is greater than the value of E° for reducing Zn^{2+} ions to Zn metal (−0.762 V). In the Zn/Cu^{2+} voltaic cell, therefore, Cu^{2+} ions are reduced, and Zn metal is oxidized. We can generalize this observation to the cell reaction of any voltaic cell: the half-reaction with the more positive value of E° runs as a reduction and the other half-reaction occurs as an oxidation. As a result, E°_{cell} will always be a positive value for a spontaneous cell reaction.

SAMPLE EXERCISE 18.3 Identifying Anode and Cathode Half-Reactions and Calculating the Value of E°_{cell} **LO1, LO3**

The half-reactions and the standard reduction potentials of typical AA or AAA single-use alkaline batteries are

$$Zn(OH)_2(s) + 2\,e^- \rightarrow Zn(s) + 2\,OH^-(aq) \qquad E^\circ = -1.249\ V$$
$$2\,MnO_2(s) + H_2O(\ell) + 2\,e^- \rightarrow Mn_2O_3(s) + 2\,OH^-(aq) \qquad E^\circ = 0.15\ V$$

What is the net ionic equation for the cell reaction and the value of E°_{cell}?

Collect and Organize We can calculate E°_{cell} by using Equation 18.2:

$$E^\circ_{cell} = E^\circ_{cathode} - E^\circ_{anode}$$

However, first we need to decide which half-reaction occurs at the cathode and which at the anode. The equation for a cell reaction is written by combining half-reactions once the loss or gain of electrons in the two half-reactions is balanced.

Analyze The MnO_2 half-reaction has the more positive E° value, making it our reduction half-reaction. We must reverse the $Zn(OH)_2$ half-reaction, turning it into an oxidation half-reaction. The two half-reactions both involve the transfer of 2 moles of electrons, so we may combine them by simply adding them together.

Solve The oxidation half-reaction at the anode is

$$Zn(s) + 2\,OH^-(aq) \rightarrow Zn(OH)_2(s) + 2\,e^-$$

The reduction half-reaction at the cathode is

$$2\,MnO_2(s) + H_2O(\ell) + 2\,e^- \rightarrow Mn_2O_3(s) + 2\,OH^-(aq)$$

CHEMTOUR
Alkaline Battery

Combining these half-reactions to obtain the overall cell reaction, we get

$$2\,MnO_2(s) + H_2O(\ell) + Zn(s) + \cancel{2\,OH^-(aq)} + \cancel{2\,e^-} \rightarrow Mn_2O_3(s) + \cancel{2\,OH^-(aq)} + Zn(OH)_2(s) + \cancel{2\,e^-}$$

Simplifying gives us the net ionic equation for the cell reaction:

$$2\,MnO_2(s) + H_2O(\ell) + Zn(s) \rightarrow Mn_2O_3(s) + Zn(OH)_2(s)$$

The overall E°_{cell} for this reaction is obtained by using Equation 18.2:

$$E^\circ_{cell} = E^\circ_{cathode} - E^\circ_{anode}$$
$$= 0.15\ V - (-1.249\ V) = 1.40\ V$$

Think About It The E°_{cell} value is reasonable because in our everyday experience the potential of most alkaline batteries is nominally 1.5 V. In this cell reaction, the net ionic equation is also the complete molecular equation.

Practice Exercise
The half-reactions in nicad (nickel–cadmium) batteries are

$$Cd(OH)_2(s) + 2\,e^- \rightarrow Cd(s) + 2\,OH^-(aq) \qquad E^\circ = -0.81\ V$$
$$2\,NiO(OH)(s) + 2\,H_2O(\ell) + 2\,e^- \rightarrow 2\,Ni(OH)_2(s) + 2\,OH^-(aq) \qquad E^\circ = 0.52\ V$$

Write the net ionic equation for the cell reaction and calculate the value of E°_{cell}.

Let's now examine what happens when two half-reactions in which different numbers of electrons are gained and lost are combined into an electrochemical cell. This combination occurs in the zinc–air battery (**Figure 18.7**), a type of battery that has a limitless supply of one of its reactants. This battery powers devices in which small battery size and low mass are high priorities, such as hearing aids. Most of the internal volume of one of these batteries is occupied by an anode consisting of a paste of zinc particles packed in an aqueous solution of KOH. As in alkaline batteries (Sample Exercise 18.3), the anode half-reaction is

$$Zn(s) + 2\,OH^-(aq) \rightarrow Zn(OH)_2(s) + 2\,e^-$$

which is the reverse of the reaction in Table A6.1:

$$Zn(OH)_2(s) + 2\,e^- \rightarrow Zn(s) + 2\,OH^-(aq) \qquad E^\circ = -1.249\ V$$

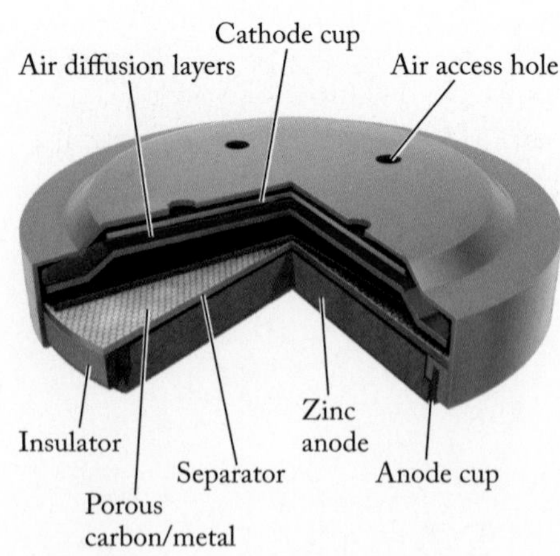

FIGURE 18.7 Most of the internal volume of a zinc–air battery is occupied by the anode: a paste of Zn particles in an aqueous solution of KOH, surrounded by a metal cup that serves as the negative terminal of the battery. Oxygen from the air is reduced at the cathode.

The cathode consists of porous carbon supported by a metal screen. Air diffuses through small holes in the battery and across a layer of Teflon that lets gases pass through but keeps electrolyte from leaking out. As air passes through the cathode, oxygen is reduced to hydroxide ions:

$$O_2(g) + 2\,H_2O(\ell) + 4\,e^- \rightarrow 4\,OH^-(aq) \qquad E^\circ_{cathode} = 0.401\ V$$

To write the overall cell reaction, we need to multiply the oxidation half-reaction by 2 before combining it with the reduction half-reaction:

$$2[Zn(s) + 2\,OH^-(aq) \rightarrow Zn(OH_2)(s) + 2\,e^-]$$
$$O_2(g) + 2\,H_2O(\ell) + 4\,e^- \rightarrow 4\,OH^-(aq)$$
$$2\,Zn(s) + \cancel{4\,OH^-(aq)} + O_2(g) + 2\,H_2O(\ell) + \cancel{4\,e^-} \rightarrow 2\,Zn(OH)_2(s) + \cancel{4\,OH^-(aq)} + \cancel{4\,e^-}$$

This simplifies to

$$2\,Zn(s) + O_2(g) + 2\,H_2O(\ell) \rightarrow 2\,Zn(OH)_2(s)$$
$$E^\circ_{cell} = E^\circ_{cathode} - E^\circ_{anode} = 0.401\ V - (-1.25\ V) = 1.65\ V$$

Note that when we multiply the anode half-reaction by 2 and add it to the cathode half-reaction, *we do not multiply* the $E°$ of the anode half-reaction by 2. $E°$ is an *intensive* property of a half-reaction or a complete cell reaction, so it does *not* change when the quantities of reactants and products change. Thus, a zinc–air battery the size of a pea has the same $E°$ as one the size of a book (such as those being developed for electric vehicles). On the other hand, the amount of electrical work a zinc–air battery can do *does* depend on how much zinc is inside it because, as we explain in Section 18.4, the electrical work that a voltaic cell can do depends on both cell potential *and* the quantity of charge it can deliver at that potential.

18.4 Chemical Energy and Electrical Work

When we connect the Zn and Cu electrodes in Figure 18.6 to a digital voltmeter—the Zn electrode to the negative terminal of the meter and the Cu electrode to the positive terminal—the meter reads 1.104 V. These connections indicate that under standard conditions the battery can pump electrons from the Zn electrode through an external circuit to the Cu electrode with a potential of 1.104 V.

Where does the energy come from to pump electrons this forcefully through an external circuit? A hint at the answer comes from the fact that these moving electrons can do electrical work (w_{elec}), such as lighting a lightbulb or turning an electric motor. We saw in Chapter 17 that the ability of a chemical reaction to do work (different from expansion or compression) is expressed by the change in free energy (ΔG_{cell}) that accompanies the cell reaction. When a thermodynamic system does work (w) on its surroundings, w is less than zero. Similarly, the change in free energy of the reaction (system) that did this work is also less than zero. Thus, the two quantities would be identical if this energy conversion process were 100% efficient:

$$\Delta G_{cell} = w_{elec} \qquad (18.3)$$

The work done by a voltaic cell (in this case the Zn/Cu^{2+} cell reaction) on its surroundings is defined as the product of the quantity of electrical charge (C) that the cell pumps through an external circuit times the cell potential:

$$w_{elec} = -CE_{cell} \qquad (18.4)$$

CONNECTION The sign conventions used for work done *on* a thermodynamic system (+) and the work done *by* the system (−) were explained in Section 6.3 (see Figure 6.20).

The negative sign indicates that work done *by* a voltaic cell on its surroundings (the external circuit) corresponds to free energy lost by the cell.

Quantities of electrical charge are typically expressed in coulombs (C), not in moles. As noted in Section 2.2, the magnitude of the charge on a single electron is 1.602×10^{-19} coulombs (C). (Note, too, the distinction between italic C, the symbol for the variable "charge," and nonitalic C, the abbreviation for the unit "coulomb.") The magnitude of electrical charge on 1 mole of electrons is

$$\frac{1.602 \times 10^{-19}\ \text{C}}{\cancel{e^-}} \times \frac{6.022 \times 10^{23}\ \cancel{e^-}}{\text{mol e}^-} = \frac{9.65 \times 10^{4}\ \text{C}}{\text{mol e}^-}$$

This quantity of charge, 9.65×10^4 C/mol e^-, is called the **Faraday constant (F)** after Michael Faraday (1791–1867), the English chemist and physicist who discovered that redox reactions take place when electrons are transferred from one species to another. The quantity of charge (C) flowing through an electrical circuit is the product of the number of moles (n) of electrons times the Faraday constant:

$$C = nF \qquad (18.5)$$

Faraday constant (F) the magnitude of electrical charge in 1 mole of electrons; its value to three significant figures is 9.65×10^4 C/mol e^-.

Combining Equations 18.4 and 18.5 gives us an equation relating w_{elec} and E_{cell}:

$$w_{cell} = -nFE_{cell} \tag{18.6}$$

If we combine Equations 18.3 and 18.6, we connect the quantity of electrical work a voltaic cell can do on its surroundings with the change in free energy in the cell:

$$\Delta G_{cell} = -nFE_{cell} \tag{18.7}$$

How is the product on the right side of Equation 18.7 the equivalent of energy? The units on the right side are

$$\cancel{\text{mole e}^-} \times \frac{\text{coulomb}}{\cancel{\text{mole e}^-}} \times \text{volt} = \text{coulomb-volt}$$

So, the change in free energy term on the left side of Equation 18.7, which is usually expressed in joules (or kilojoules), is equivalent to an electrical energy term on the right expressed in coulomb-volts. This equivalence makes sense because of the following important unit equality:

$$1 \text{ joule} = 1 \text{ coulomb-volt} \quad \text{or} \quad 1\text{ J} = 1\text{ C}\cdot\text{V}$$

The negative sign on the right side of Equation 18.7 indicates that the E_{cell} of any voltaic cell must have a positive value because the sign of ΔG_{cell} for the spontaneous chemical reaction inside the cell must be negative.

Let's calculate the change in standard free energy of the Zn/Cu^{2+} cell reaction. We start with the standard cell potential calculated in Section 18.3:

$$E^\circ_{cell(Zn/Cu^{2+})} = 1.104 \text{ V}$$

We can convert this standard cell potential into a change in standard free energy (ΔG°_{cell}) by using Equation 18.7 under standard conditions, so that $\Delta G = \Delta G^\circ$ and $E_{cell} = E^\circ_{cell}$:

$$\begin{aligned}\Delta G^\circ_{cell} &= -nFE^\circ_{cell}\\ &= -\left(2\ \cancel{\text{mol e}^-} \times \frac{9.65 \times 10^4\text{ C}}{\cancel{\text{mol e}^-}} \times 1.104\text{ V}\right) = -2.13 \times 10^5\text{ C}\cdot\text{V}\\ &= -2.13 \times 10^5\text{ J} = -213\text{ kJ}\end{aligned}$$

To put this value in perspective, the Zn/Cu^{2+} reaction produces nearly as much useful energy as the combustion of 1 mole of hydrogen gas:

$$H_2(g) + \tfrac{1}{2}O_2(g) \rightarrow H_2O(g) \qquad \Delta G^\circ = -228.6\text{ kJ}$$

CONCEPT TEST

When a rechargeable battery, such as the one used to start a car's engine, is recharged, an external source of electrical power forces the voltaic cell reaction to run in reverse. What are the signs of E°_{cell} and ΔG°_{cell} during the recharging process?

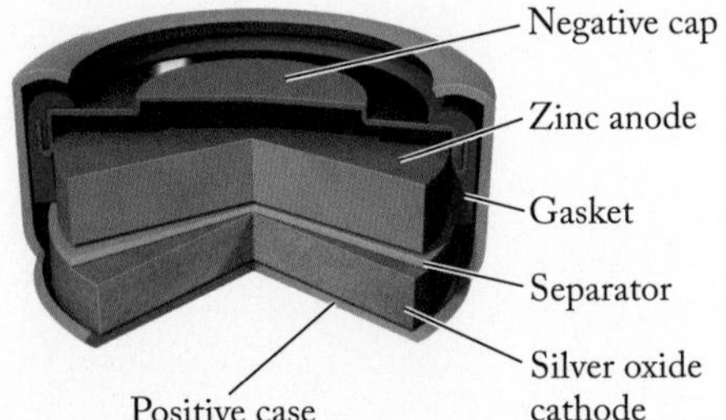

FIGURE 18.8 Many of the button batteries that power small electronic devices incorporate a Zn anode and Ag_2O cathode separated by a membrane containing KOH electrolyte.

SAMPLE EXERCISE 18.4 Relating ΔG°_{cell} and E°_{cell} **LO1, LO4**

Many of the "button" batteries used in electric watches consist of a Zn anode and a Ag_2O cathode, separated by a membrane soaked in a concentrated solution of KOH (**Figure 18.8**). At the cathode, Ag_2O is reduced to Ag metal; at the anode, Zn is oxidized to solid $Zn(OH)_2$. Write the net ionic equation for the reaction and use the appropriate standard reduction potentials from Table A6.1 to calculate the values of E°_{cell} and ΔG°_{cell}.

Collect and Organize We know the reactants and products of the anode and cathode reactions and that the reaction occurs in a basic solution. Equations 18.2 and 18.7 should be useful in calculating E°_{cell} and ΔG°_{cell} from the appropriate standard potentials:

$$E^\circ_{cell} = E^\circ_{cathode} - E^\circ_{anode}$$

$$\Delta G^\circ_{cell} = -nFE^\circ_{cell}$$

Analyze The half-reaction at the cathode is based on the reduction of Ag_2O to Ag. The appropriate half-reaction in Table A6.1 is

$$Ag_2O(s) + H_2O(\ell) + 2\,e^- \rightarrow Ag(s) + 2\,OH^-(aq) \qquad E^\circ_{cathode} = 0.342\text{ V}$$

The following entry in Table A6.1 has $Zn(OH)_2$ as the reactant and Zn as the product:

$$Zn(OH)_2(s) + 2\,e^- \rightarrow Zn(s) + 2\,OH^-(aq) \qquad E^\circ_{anode} = -1.249\text{ V}$$

We must reverse this half-reaction to represent the oxidation occurring at the anode before combining it with the cathode half-reaction. The two half-reactions involve the transfer of the same number of electrons ($n = 2$), so combining them simply means adding them together. The value of E°_{cell} is about $0.35 - (-1.25) = 1.60$ V. This value is about 50% larger than the E°_{cell} of the Zn/Cu^{2+} cell (which is 1.10 V). Therefore, compared to the Zn/Cu cell in Figure 18.6, the magnitude of its ΔG°_{cell} value should be 50% larger than −212 kJ/mol, or about −300 kJ/mol.

Solve Reversing the $Zn(OH)_2$ half-reaction and adding it to the Ag_2O half-reaction, we get

$$\begin{aligned} Zn(s) + 2\,OH^-(aq) &\rightarrow Zn(OH)_2(s) + 2\,e^- \\ Ag_2O(s) + H_2O(\ell) + 2\,e^- &\rightarrow 2\,Ag(s) + 2\,OH^-(aq) \\ \hline Ag_2O(s) + H_2O(\ell) + Zn(s) + \cancel{2\,OH^-(aq)} + \cancel{2\,e^-} &\rightarrow \\ 2\,Ag(s) + \cancel{2\,OH^-(aq)} &+ Zn(OH)_2(s) + \cancel{2\,e^-} \end{aligned}$$

This simplifies to

$$Ag_2O(s) + H_2O(\ell) + Zn(s) \rightarrow 2\,Ag(s) + Zn(OH)_2(s)$$

Calculating E°_{cell}:

$$\begin{aligned} E^\circ_{cell} &= E^\circ_{cathode} - E^\circ_{anode} \\ &= 0.342\text{ V} - (-1.249\text{ V}) \\ &= 1.591\text{ V} \end{aligned}$$

Converting E°_{cell} to ΔG°_{cell}:

$$\begin{aligned} \Delta G^\circ_{cell} &= -nFE^\circ_{cell} \\ &= -(2\text{ }\cancel{\text{mol e}^-} \times 9.65 \times 10^4\text{ C/}\cancel{\text{mol e}^-} \times 1.591\text{ V}) \\ &= -3.07 \times 10^5\text{ C} \cdot \text{V} = -3.07 \times 10^5\text{ J} = -307\text{ kJ} \end{aligned}$$

Think About It The positive value of E°_{cell} and negative value of ΔG°_{cell} are expected because voltaic cell reactions are spontaneous. The calculated values are close to those we estimated.

 **Practice Exercise** If a single-use alkaline battery (see Sample Exercise 18.3) produces a cell potential of 1.50 V, what is the value of ΔG°_{cell}?

The ΔG°_{cell} value of −307 kJ for the silver oxide battery reaction in Sample Exercise 18.4 is based on the reaction of 1 mole of Ag_2O and 1 mole of Zn, which correspond to 232 g of Ag_2O and 65 g of Zn. The energy stored in a button battery (Figure 18.8), which has a mass of only 1 or 2 g, would be a tiny fraction of this calculated value. Moreover, no ions appear in the net ionic equation. Because all the reactants and products are solids, the net ionic equation and molecular equation are identical.

standard hydrogen electrode (SHE) a reference electrode based on the half-reaction $2\ H^+(aq) + 2\ e^- \rightarrow H_2(g)$ that produces a standard electrode potential of 0.000 V.

18.5 A Reference Point: The Standard Hydrogen Electrode

We can measure the value of E_{cell} by using a voltmeter, but can we measure the individual electrode potentials of the cathode and anode? It turns out that we can assign potentials to an individual electrode by arbitrarily assigning a value of zero volts to a half-reaction that serves as a reference point—namely, the standard potential for the reduction of hydrogen ions to hydrogen gas:

$$2\ H^+(aq) + 2\ e^- \rightarrow H_2(g) \qquad E° = 0.000\ V \qquad (18.8)$$

An electrode that generates this reference potential, called the **standard hydrogen electrode (SHE)**, consists of a platinum electrode in contact with a solution of a strong acid ($[H^+] = 1.00\ M$) and hydrogen gas at a pressure of 1.00 bar (≈ 1 atm; **Figure 18.9**). The platinum is not changed by the electrode reaction. Rather, it serves as a chemically inert conveyor of electrons. Electrons are conveyed to the electrode surface when H^+ ions are reduced to hydrogen gas; electrons are conveyed away from the electrode surface when hydrogen gas is oxidized to H^+ ions. The potential of the SHE is the same for both half-reactions: 0.000 V.

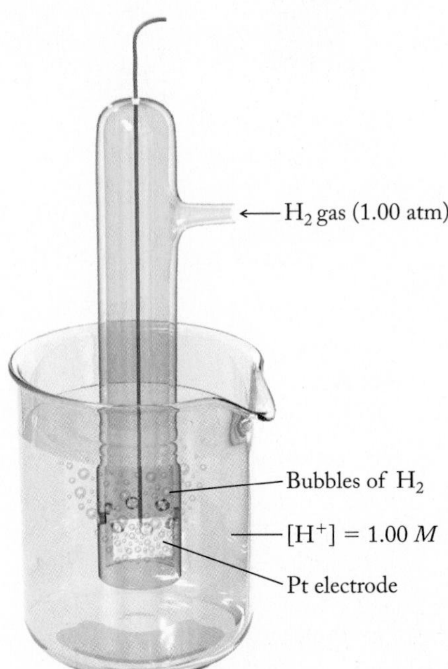

FIGURE 18.9 The standard hydrogen electrode (SHE) consists of a platinum electrode immersed in a 1.00 *M* solution of $H^+(aq)$ and bathed in a stream of pure H_2 gas at a pressure of 1.00 bar (≈ 1 atm). Its potential is the same (0.000 V) whether $H^+(aq)$ ions are reduced or H_2 gas is oxidized.

STEPWISE ANIMATION
Standard Hydrogen Electrode (SHE)

To draw the cell diagram for a cell in which the SHE serves as the anode, we represent the SHE half of the cell as follows:

$$Pt(s) \mid H_2(g,\ 1.00\ atm) \mid H^+(1.00\ M) \parallel$$

This indicates that the anode half-reaction involves the oxidation of H_2 gas to H^+ ions. If the SHE is the cathode, then we diagram its half of the cell this way:

$$\parallel H^+\ (1.00\ M) \mid H_2(g,\ 1.00\ atm) \mid Pt(s)$$

This indicates that the cathode half-reaction involves the reduction of H^+ ions to H_2 gas.

Because the standard reduction (or oxidation) potential of the SHE is assigned a reference potential of 0.000 V, the measured $E°_{cell}$ of any voltaic cell in which a SHE is one of the two electrodes—either cathode or anode—can be considered the potential produced by the other electrode. This means that if we attach a voltmeter to the cell, the meter reading is the electrode potential of the other electrode. Suppose, for example, that a voltaic cell consists of a strip of zinc metal immersed in a 1.00 *M* solution of Zn^{2+} ions in one compartment and a SHE in the other (**Figure 18.10a**). Also suppose that a voltmeter is connected to the cell so that it measures the potential at which the cell pumps electrons from the zinc

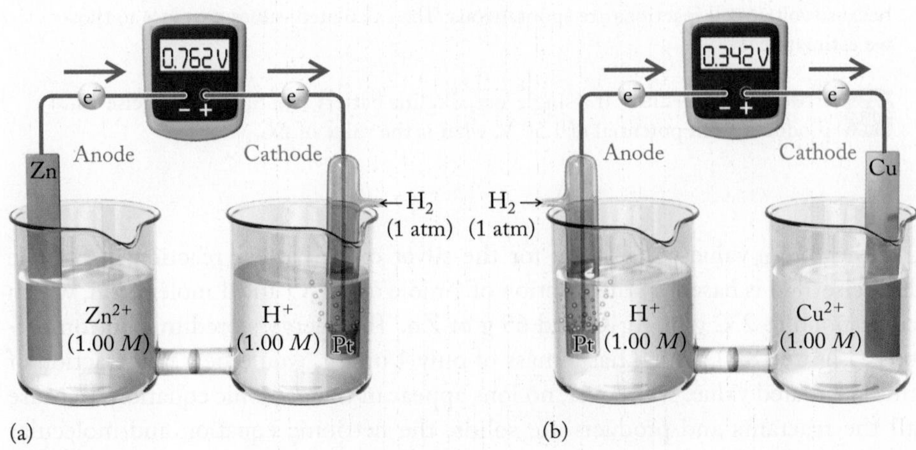

FIGURE 18.10 (a) When a standard hydrogen electrode is coupled to a Zn electrode under standard conditions, the SHE is the cathode (H^+ is reduced) and the Zn electrode is the anode. When the SHE is connected to the positive terminal of a voltmeter and the Zn electrode to the negative terminal, the meter measures a cell potential of 0.762 V. (b) When coupled to a Cu electrode under standard conditions, the SHE is the anode (H_2 is oxidized), the Cu electrode is the cathode, and the meter measures a cell potential of 0.342 V.

electrode to the SHE. This direction of electron flow means that the zinc electrode is the cell's anode and the SHE is the cathode of the cell. At 25°C the meter reads 0.762 V. We know that the value of E°_{cathode} is that of the SHE (0.000 V) and that E°_{anode} is E°_{Zn}.

Inserting these values and symbols into Equation 18.2,

$$E^\circ_{\text{cell}} = E^\circ_{\text{cathode}} - E^\circ_{\text{anode}}$$
$$E^\circ_{\text{cell}} = E^\circ_{\text{SHE}} - E^\circ_{\text{Zn}}$$
$$0.762\ \text{V} = 0.000\ \text{V} - E^\circ_{\text{Zn}}$$
$$E^\circ_{\text{Zn}} = -0.762\ \text{V}$$

This value is equal to the standard reduction potential of Zn^{2+} in Table A6.1:

$$Zn^{2+}(aq) + 2\ e^- \rightarrow Zn(s) \qquad E^\circ = -0.762\ \text{V}$$

In **Figure 18.10b**, the SHE is coupled to a copper electrode immersed in a 1.00 *M* solution of Cu^{2+} ions. In this cell, electrons flow from the SHE through an external circuit to the copper electrode at a cell potential of 0.342 V at 25°C. The direction of current flow means that the electrons are consumed at the copper electrode, making it the cathode. The value of E° for the copper half-reaction is calculated as follows:

$$E^\circ_{\text{cell}} = E^\circ_{\text{cathode}} - E^\circ_{\text{anode}}$$
$$= E^\circ_{\text{Cu}} - E^\circ_{\text{SHE}}$$
$$0.342\ \text{V} = E^\circ_{\text{Cu}} - 0.000\ \text{V}$$
$$E^\circ_{\text{Cu}} = 0.342\ \text{V}$$

This half-reaction potential matches the value of E° for the reduction of Cu^{2+} to Cu metal in Table A6.1.

Even though the SHE serves as the reference for all E° values, it is cumbersome to use and maintain. Therefore, scientists have developed other reliable and more convenient reference electrodes for use in electrochemistry. One common choice is the silver–silver chloride electrode (**Figure 18.11**), which is based on the following half-reaction:

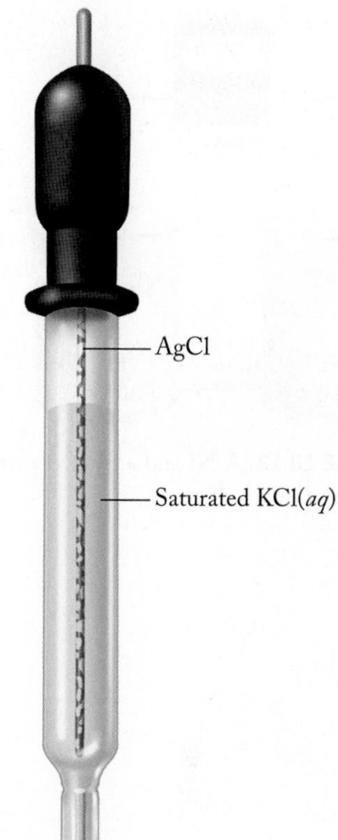

FIGURE 18.11 A Ag/AgCl reference electrode.

$$AgCl(s) + e^- \rightarrow Ag(s) + Cl^-(aq) \qquad E^\circ = 0.222\ \text{V versus SHE}$$

The Ag/AgCl electrode consists of a piece of silver wire coated with a thin layer of AgCl and immersed in a solution containing a high concentration (often a saturated solution) of KCl. Typically, the electrode connects with the solution in the cell through an ion-permeable porous ceramic plug like those that connect the two halves of the electrochemical cells in Figure 18.10.

CONNECTION A saturated solution contains the maximum concentration of a solute possible at a given temperature, as discussed in Section 4.7.

What would happen if we replaced the SHE in Figure 18.10 with a Ag/AgCl electrode in which [KCl] = 1.00 *M*? The voltmeter in Figure 18.10a would now read 0.984 V:

Anode: $Zn(s) \rightarrow Zn^{2+}(aq) + 2\ e^- \qquad E^\circ_{\text{anode}} = -0.762\ \text{V}$

Cathode: $AgCl(s) + e^- \rightarrow Ag(s) + Cl^-(aq) \qquad E^\circ_{\text{cathode}} = 0.222\ \text{V}$

$$E^\circ_{\text{cell}} = E^\circ_{\text{cathode}} - E^\circ_{\text{anode}} = 0.222\ \text{V} - (-0.762\ \text{V}) = 0.984\ \text{V}$$

The voltmeter in Figure 18.10b would read:

$$E^\circ_{\text{cell}} = E^\circ_{\text{cathode}} - E^\circ_{\text{anode}} = 0.342\ \text{V} - 0.222 = 0.120\ \text{V}$$

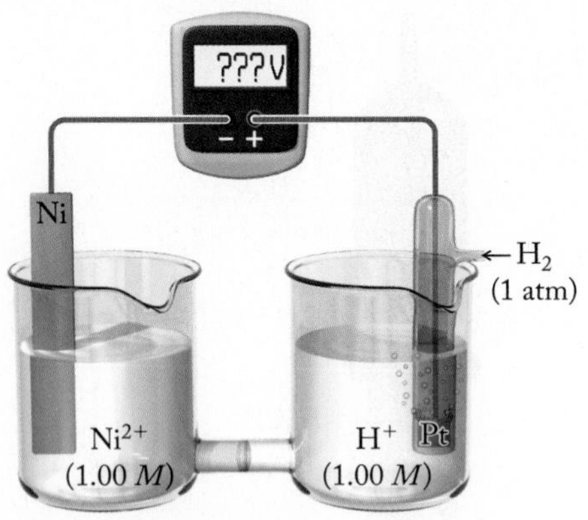

FIGURE 18.12 A Ni and a SHE electrode connected by a voltmeter.

The Ag/AgCl reference electrode and others of similar design are so widely used in electrochemistry that the electrode potentials used in many electrochemical experiments are not reported relative to a SHE, but rather to one of these other reference electrodes.

CONCEPT TEST

A cell consists of a SHE in one compartment and a Ni electrode immersed in a 1.00 *M* solution of Ni^{2+} ions in the other. If a voltmeter is connected to the electrodes as shown in **Figure 18.12**, what will be the value on the voltmeter's display? What would the voltmeter read if we replaced the SHE with an Ag/AgCl reference electrode?

18.6 The Effect of Concentration on E_{cell}

Reactions stop when one of the reactants is completely consumed. This concept was the basis for our discussion of limiting reactants in Chapter 3. A commercial battery, on the other hand, usually stops operating at its rated cell potential—1.5 V for a flashlight battery—before its reactants are completely consumed. In other words, the battery "goes dead" before all the reactants are consumed. This happens because the cell potential of a voltaic cell depends on the concentrations of the reactants and products.

The Nernst Equation

In 1889, the German chemist Walther Nernst (1864–1941) derived an expression, now called the **Nernst equation**, that describes how cell potentials depend on reactant and product concentrations. We can reconstruct his derivation starting with Equation 17.14, which relates the change in free energy ΔG of any reaction to its change in free energy under standard conditions $\Delta G°$:

CONNECTION In Section 17.7 we introduced the relationship between change in free energy and the reaction quotient *Q*. Writing expressions for reaction quotients was first explained in Section 14.5.

$$\Delta G = \Delta G° + RT \ln Q \qquad (17.14)$$

Q represents the reaction quotient, which does not contain terms for pure solids and pure liquids.

As a spontaneous reaction proceeds, concentrations of products increase, and concentrations of reactants decrease until the positive value of $RT \ln Q$ offsets the negative value of $\Delta G°$. At that point, $\Delta G = 0$ and the reaction has reached chemical equilibrium.

CHEMTOUR
Cell Potential

Now let's write an expression analogous to Equation 17.14 that relates the **cell potential** under nonstandard conditions (i.e., $\boldsymbol{E_{cell}}$) to $E°_{cell}$. We start by substituting $-nFE_{cell}$ for ΔG_{cell} and $-nFE°_{cell}$ for $\Delta G°_{cell}$:

$$-nFE_{cell} = -nFE°_{cell} + RT \ln Q$$

Dividing all terms by $-nF$ gives

$$E_{cell} = E°_{cell} - \frac{RT \ln Q}{nF} \qquad (18.9)$$

Nernst equation an equation relating the potential of a cell (or half-cell) reaction to its standard potential ($E°$) and to the concentrations of its reactants and products.

cell potential (E_{cell}) the force with which an electrochemical cell can pump electrons through an external circuit.

This is the equation Walther Nernst developed in 1889. We can obtain a very useful form of Equation 18.9 if we insert values for *R* [8.314 J/(mol · K)] and

F (9.65×10^4 C/mol), assume $T = 298$ K, and convert the natural logarithm to a base-10 logarithm: $\ln Q = 2.303 \log Q$. With these changes, the Nernst equation becomes

$$E_{cell} = E^\circ_{cell} - \frac{0.0592\ \text{V}}{n} \log Q \qquad (18.10)$$

Equation 18.10 allows us to predict how the potential (E_{cell} in V) of a voltaic cell at 298 K changes as the concentrations of products inside the cell increase and the concentrations of reactants decrease. As they do, Q increases and so does the log term in Equation 18.10. The negative sign in front of this term means that the value of E_{cell} decreases as reactants are converted into products. Eventually, E_{cell} approaches zero. When it reaches zero, the cell reaction has achieved chemical equilibrium. The cell can no longer pump electrons through an external circuit, because it's dead.

CONCEPT TEST

We can also use Equation 18.10 to calculate the potential of a single electrode. Consider the half-reaction at the Ag/Ag$^+$ electrode:

$$Ag^+(aq) + e^- \rightarrow Ag(s) \qquad E^\circ = 0.800\ \text{V}$$

Is the potential of this half-reaction at 25°C greater than or less than 0.800 V when the concentration of Ag$^+$ is 0.100 *M*?

STEPWISE ANIMATION

Lead–Acid Battery

Batteries are voltaic cells, so their cell potential should drop with usage. Let's consider how much a cell potential can drop by focusing on the *lead–acid* battery used to start most car engines. These batteries each contain six electrochemical cells. Their anodes are made of Pb and their cathodes are made of PbO_2. Both electrodes are immersed in 4.5 *M* H_2SO_4 (**Figure 18.13**). The E_{cell} value of a fully charged battery is about 2.0 V. The six cells are connected in series so that the operating potential of the battery is the sum of the six cell potentials, or about 12 V.

As the battery discharges, $PbO_2(s)$ is reduced to $PbSO_4(s)$ at the cathodes:

$$PbO_2(s) + 3\ H^+(aq) + HSO_4^-(aq) + 2\ e^- \rightarrow PbSO_4(s) + 2\ H_2O(\ell) \qquad E^\circ = 1.685\ \text{V}$$

Also, Pb(s) is oxidized to $PbSO_4(s)$ at the anodes:

$$Pb(s) + HSO_4^-(aq) \rightarrow PbSO_4(s) + H^+(aq) + 2\ e^-$$

The reduction half-reaction consumes 2 moles of electrons, and the oxidation half-reaction involves the loss of 2 moles of electrons for each mole of lead that is oxidized.

The net ionic equation for the overall cell reaction is the sum of the two half-reactions:

$$PbO_2(s) + Pb(s) + 2\ H^+(aq) + 2\ HSO_4^-(aq) \rightarrow 2\ PbSO_4(s) + 2\ H_2O(\ell)$$

To calculate the value of E°_{cell}, we find the reduction potential for the half-reaction that corresponds to the reverse reaction at the anode in Table A6.1 (Appendix 6):

$$PbSO_4(s) + H^+(aq) + 2\ e^- \rightarrow Pb(s) + HSO_4^-(aq) \qquad E^\circ = -0.356\ \text{V}$$

Therefore,

$$E^\circ_{cell} = E^\circ_{cathode} - E^\circ_{anode} = 1.685\ \text{V} - (-0.356\ \text{V}) = 2.041\ \text{V}$$

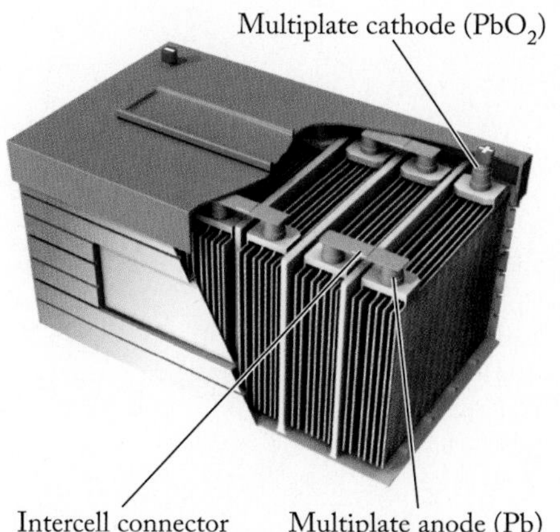

FIGURE 18.13 The lead–acid battery that provides power to start most motor vehicles contains six cells. Each has an anode made of lead and a cathode made of PbO_2 immersed in a solution of 4.5 *M* H_2SO_4, which serves as both a reactant and background electrolyte. The electrodes are formed into plates and held in place by grids made of a lead alloy. The grids connect the cells in series so that the operating potential of the battery (12.0 V) is the sum of six E_{cell} values (each 2.0 V).

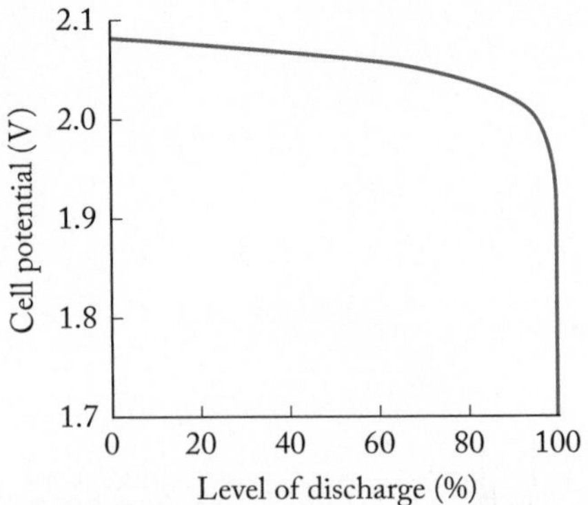

FIGURE 18.14 The potential of a cell in a lead–acid battery decreases as reactants are converted into products, but the change in potential is small until the battery is nearly completely discharged. Although the value of E°_{cell} is 2.041 V, a fully charged commercial battery has a slightly higher potential (shown here as 2.08 V) because the concentration of H_2SO_4 in its cells' electrolytes is greater than 1 *M*.

As the battery discharges, the concentration of sulfuric acid decreases, and so does the value of E_{cell} calculated from the Nernst equation:

$$E_{cell} = 2.041 \text{ V} - \frac{0.0592 \text{ V}}{2} \log \frac{1}{[H^+]^2 [HSO_4^-]^2}$$

Note that there is no term in the numerator representing the quantity of $PbSO_4$ in the cell, nor are there terms in the denominator for the quantities of PbO_2 and Pb because all three battery components are solids. Assuming all three are present, the voltage produced by the battery depends only on the concentration of sulfuric acid in the solution in which the three solid components are immersed.

The decrease in E_{cell} is very gradual, however, not falling below 2.0 V until the battery is about 97% discharged, as shown in **Figure 18.14**. It decreases gradually because of the logarithmic relationship between Q and E_{cell}. If, for example, the concentration of sulfuric acid decreased by an order of magnitude, say, from 1.00 *M* to 0.100 *M*, then the value of E_{cell} would decrease by less than 6%—from 2.041 V to

$$E_{cell} = 2.041 \text{ V} - \frac{0.0592 \text{ V}}{2} \log \frac{1}{0.100^2 \times 0.100^2} = 1.923 \text{ V}$$

As a result, most batteries can deliver current at a cell potential close to their "design" potential until they are almost completely discharged.

SAMPLE EXERCISE 18.5 Calculating E_{cell} from E°_{cell} and the Concentrations of Reactants and Products **LO5**

The standard potential (E°_{cell}) of a voltaic cell based on the Zn/Cu^{2+} ion reaction:

$$Zn(s) + Cu^{2+}(aq) \rightarrow Zn^{2+}(aq) + Cu(s)$$

is 1.104 V. What is the value of E_{cell} at 25°C when the concentration of Cu^{2+} is 0.100 *M* and the concentration of Zn^{2+} is 1.90 *M*?

Collect and Organize We are given the standard cell potential and the concentrations of Cu^{2+} and Zn^{2+}, and we are asked to determine the value of E_{cell}. The Nernst equation enables us to calculate E_{cell} values for different concentrations of reactants and products. This equation requires us to work with the reaction quotient Q, which we know from Section 14.5 to be the mass action expression for the reaction. Solid copper and zinc are also part of the reaction system, but no terms for pure solids appear in reaction quotients.

Analyze The only term in the numerator of the Q expression for this cell reaction is $[Zn^{2+}]$, and the only one in the denominator is $[Cu^{2+}]$. The two solids, metallic Cu and Zn, are not included. Each Cu^{2+} ion gains two electrons, and each Zn atom donates two electrons, so the value of n in the Nernst equation is 2. The value of $[Zn^{2+}]$ is greater than $[Cu^{2+}]$, which makes $Q > 1$. The negative sign in front of the $0.0592/n \times \log Q$ term in Equation 18.10 means that the calculated value of E_{cell} should be less than the value of E°_{cell}.

Solve Substituting the values of $[Zn^{2+}]$ and $[Cu^{2+}]$ in the Nernst equation gives

$$E_{cell} = E^\circ_{cell} - \frac{0.0592 \text{ V}}{n} \log Q = 1.104 \text{ V} - \frac{0.0592 \text{ V}}{2} \log \frac{1.90}{0.100}$$

$$E_{cell} = 1.104 \text{ V} - \frac{0.0592 \text{ V}}{2}(1.279) = 1.066 \text{ V}$$

Think About It The calculated E_{cell} value is less than E°_{cell} (as predicted), but it is only 0.038 V less because the logarithmic dependence of cell potential on reactant and product concentrations minimizes the impact of changing concentrations.

Practice Exercise The standard cell potential of the zinc–air battery (Figure 18.7) is 1.65 V. If the partial pressure of oxygen in the air at 25°C diffusing through its cathode is 0.21 atm, what is the cell potential? Assume the cell reaction is

$$2\ Zn(s) + O_2(g) + 2\ H_2O(\ell) \rightarrow 2\ Zn(OH)_2(s)$$

$E°$ and K

CHEMTOUR
Cell Potential, Equilibrium, and Free Energy

STEPWISE ANIMATION
Concentration Cell

When the cell reaction of a voltaic cell reaches chemical equilibrium, $\Delta G_{cell} = E_{cell} = 0$ and $Q = K$. Equation 18.10 then becomes

$$0 = E°_{cell} - \frac{0.0592\ V}{n} \log K$$

Rearranging this equation gives

$$\log K = \frac{nE°_{cell}}{0.0592\ V} \quad (18.11)$$

We can use Equation 18.11 to calculate the equilibrium constant for any redox reaction at 25°C, not just those in electrochemical cells. For the more general case, we substitute $E°_{rxn}$ for $E°_{cell}$:

$$\log K = \frac{nE°_{rxn}}{0.0592\ V} \quad (18.12)$$

SAMPLE EXERCISE 18.6 Calculating K for a Redox Reaction from the Standard Potentials of Its Half-Reactions **LO6**

Many procedures for determining mercury levels in environmental samples involve reducing Hg^{2+} ions to elemental Hg by using Sn^{2+} ions. Use the appropriate $E°$ values from Table A6.1 to calculate the equilibrium constant at 25°C for the reaction

$$Sn^{2+}(aq) + Hg^{2+}(aq) \rightarrow Sn^{4+}(aq) + Hg(\ell)$$

Collect and Organize Equation 18.12 relates the equilibrium constant for any redox reaction to the standard potential $E°_{rxn}$. To calculate $E°_{rxn}$, we need to find the difference between the appropriate standard reduction potentials. Table A6.1 lists two half-reactions involving our reactants and products:

$$Hg^{2+}(aq) + 2\ e^- \rightarrow Hg(\ell) \qquad E° = 0.851\ V$$

$$Sn^{4+}(aq) + 2\ e^- \rightarrow Sn^{2+}(aq) \qquad E° = 0.154\ V$$

Analyze The problem states that Hg^{2+} ions are reduced by Sn^{2+} ions, so Sn^{2+} is the reducing agent in the reaction, which means that it must be oxidized. As a result, the mercury half-reaction occurs at the cathode and the second reaction occurs at the anode, so we must subtract its standard potential from that of the mercury half-reaction. The difference between the two half-reaction potentials is about +0.7 V, so the right side of Equation 18.12 will be about $(2 \times 0.7)/0.06 \approx 23$, and the value of K should be about 10^{23}.

Solve We obtain the standard potential for the reaction from a modified version of Equation 18.2:

$$E°_{rxn} = E°_{cathode} - E°_{anode} = 0.851\ V - 0.154\ V = 0.697\ V$$

Using this value for E°_{rxn} in Equation 18.12 and a value of 2 for n (because there are two moles of electrons), we have

$$\log K = \frac{nE^\circ_{rxn}}{0.0592\ \text{V}} = \frac{2(0.697\ \cancel{\text{V}})}{0.0592\ \cancel{\text{V}}} = 23.55$$

$$K = 10^{23.55} = 4 \times 10^{23}$$

Think About It The calculated value is quite close to what we estimated. Note how a E°_{rxn} value of less than 1 V corresponds to a huge equilibrium constant, indicating that the reaction essentially goes to completion. The calculated value of K has only one significant figure, which may seem like too few because the uncertain starting values both had three. However, the value of K was calculated from $10^{23.55}$, and the two digits to the left of the decimal point became the exponent in the final K value. Only the first digit after the decimal was significant (the second "5" was carried through the calculation to avoid rounding error), which left the final K value with only one digit before the exponent.

Practice Exercise Use the appropriate standard reduction potentials from Table A6.1 to calculate the value of K at 25°C for the reaction

$$5\ Fe^{2+}(aq) + MnO_4^-(aq) + 8\ H^+(aq) \rightarrow 5\ Fe^{3+}(aq) + Mn^{2+}(aq) + 4\ H_2O(\ell)$$

TABLE 18.1 Relationships between K, E°_{cell}, and ΔG°_{cell} Values of Electrochemical Reactions

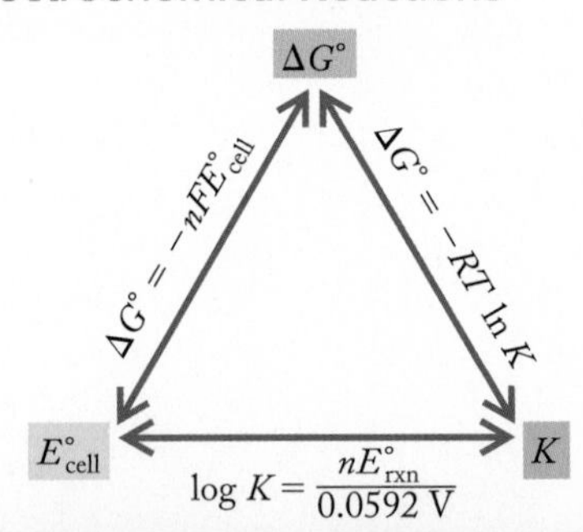

K	E°_{cell}	ΔG°_{cell}	Favors Formation of
<1	<0	>0	Reactants
>1	>0	<0	Products
1	0	0	Neither

Before ending our discussion of how to derive equilibrium constant values at 25°C from E°_{cell} values, we should note that measuring the potential of an electrochemical reaction allows us to calculate equilibrium constant values that may be too large or too small to determine from the equilibrium concentrations of reactants and products. A value of K as large as that calculated in Sample Exercise 18.6 could not be obtained by analyzing the composition of an equilibrium reaction mixture because the concentrations of the reactants would be too small to be determined accurately. Similarly, a cell potential of about −1 V would correspond to a tiny K value and concentrations of products that are too small to be determined quantitatively.

Table 18.1 summarizes how the values of K and E°_{cell} are related to each other and to the change in free energy (ΔG°_{cell}) of a cell reaction under standard conditions. If we know any one of the quantities in Table 18.1, we can calculate the other two. Spontaneous electrochemical reactions have $E^\circ_{cell} > 0$ and $K > 1$. The connection between positive cell potential (E_{cell}) and reaction spontaneity applies even under nonstandard conditions. Keep in mind, too, that even small positive values of E°_{cell} (only a fraction of a volt, for example) correspond to very large K values and to cell reactions that go nearly to completion.

18.7 Relating Battery Capacity to Quantities of Reactants

An important performance characteristic of a battery is its capacity to do electrical work—that is, to deliver electrical charge at the designed cell potential. This capacity—the amount of electrical work done—is defined by Equation 18.4, $w_{elec} = -CE_{cell}$, where w_{elec} is the quantity of electrical charge (C) expressed in coulombs (C) delivered at a cell potential (E_{cell}) expressed in volts.

Another important unit in electricity is the *ampere* (A), which is the SI base unit (see Table 1.2) of electrical current and equal to a flow of one coulomb of charge per second. In equation form:

$$1\ \text{A} = 1\ \text{C/s} \tag{18.13}$$

Rearranging the terms in Equation 18.13 gives us

$$1\ \text{C} = 1\ \text{A} \cdot \text{s} \qquad (18.14)$$

Combining this expression with the link between free energy and electrical energy, $1\ \text{J} = 1\ \text{C} \cdot \text{V}$:

$$1\ \text{J} = 1\ \text{C} \cdot \text{V} = 1\ (\text{A} \cdot \text{s})\ \text{V} \qquad (18.15)$$

A coulomb (ampere-second) is a small quantity of charge, so the capacities of batteries even as small as those used to power LED flashlights are often expressed in units such as ampere-hours. For example, the rechargeable AA batteries in **Figure 18.15** are rated at about 2.5 Ah or 2500 mAh (milliampere-hours).

FIGURE 18.15 The electrical energy rating of these rechargeable AA batteries is 2500 milliampere-hours at 1.2 V.

The units used to express the energy content of large, fully charged batteries, such as batteries used to power all-electric or hybrid vehicles, incorporate the quantity of electrical charge they can store and the voltage with which that charge is delivered to the vehicle's electric motor. These ratings are based on a unit of electrical *power*, that is, the *rate* at which energy is delivered. The SI unit of power is the *watt* (W), which is equal to one joule of energy produced (or consumed) per second. In other words:

$$1\ \text{W} = 1\ \text{J/s}$$

Combining this definition with Equation 18.15:

$$1\ \text{W} = 1\ \text{J/s} = 1\ \frac{\text{A} \cdot \text{s}}{\text{s}}\ \text{V} = 1\ \text{A} \cdot \text{V}$$

Like a joule of energy, a watt of power is a small quantity, so the power of large electric devices, such as the electric motors in all-electric vehicles, are usually expressed in kilowatts (kW), and the energy capacities of the batteries that power those motors are usually expressed in kilowatt-hours (kWh). To appreciate how much energy this unit represents, let's calculate how many joules there are in one kilowatt-hour:

$$1\ \cancel{\text{kWh}} \times \frac{1000\ \cancel{\text{W}}}{1\ \cancel{\text{kW}}} \times \frac{1\ \text{J/s}}{1\ \cancel{\text{W}}} \times \frac{60\ \text{s}}{1\ \cancel{\text{min}}} \times \frac{60\ \cancel{\text{min}}}{1\ \cancel{\text{h}}} = 3.6 \times 10^6\ \text{J}$$

Nickel–Metal Hydride Batteries

Hybrid vehicles such as the Toyota Prius are powered by combinations of small gasoline engines and electric motors. Electricity for the motors comes from battery packs (**Figure 18.16**) made of dozens of nickel–metal hydride (NiMH) cells. At the cathodes in these cells, NiO(OH) is reduced to $Ni(OH)_2$, and at the anodes, made of one or more transition metals, hydrogen atoms are oxidized to H^+ ions. The electrodes are separated by aqueous KOH.

The cathode half-reaction is

$$\text{NiO(OH)}(s) + \text{H}_2\text{O}(\ell) + \text{e}^- \rightarrow \text{Ni(OH)}_2(s) + \text{OH}^-(aq)$$

At the anode, hydrogen is present as a metal hydride. To write the anode half-reaction, we use the generic formula MH, where M stands for a transition metal or metallic compound that forms a hydride. In a basic background electrolyte, the anode oxidation half-reaction is

$$\text{MH}(s) + \text{OH}^-(aq) \rightarrow \text{M}(s) + \text{H}_2\text{O}(\ell) + \text{e}^-$$

The standard potential of this half-reaction depends on the chemical properties of MH, but generally the value is near that of the SHE, or about 0.0 V.

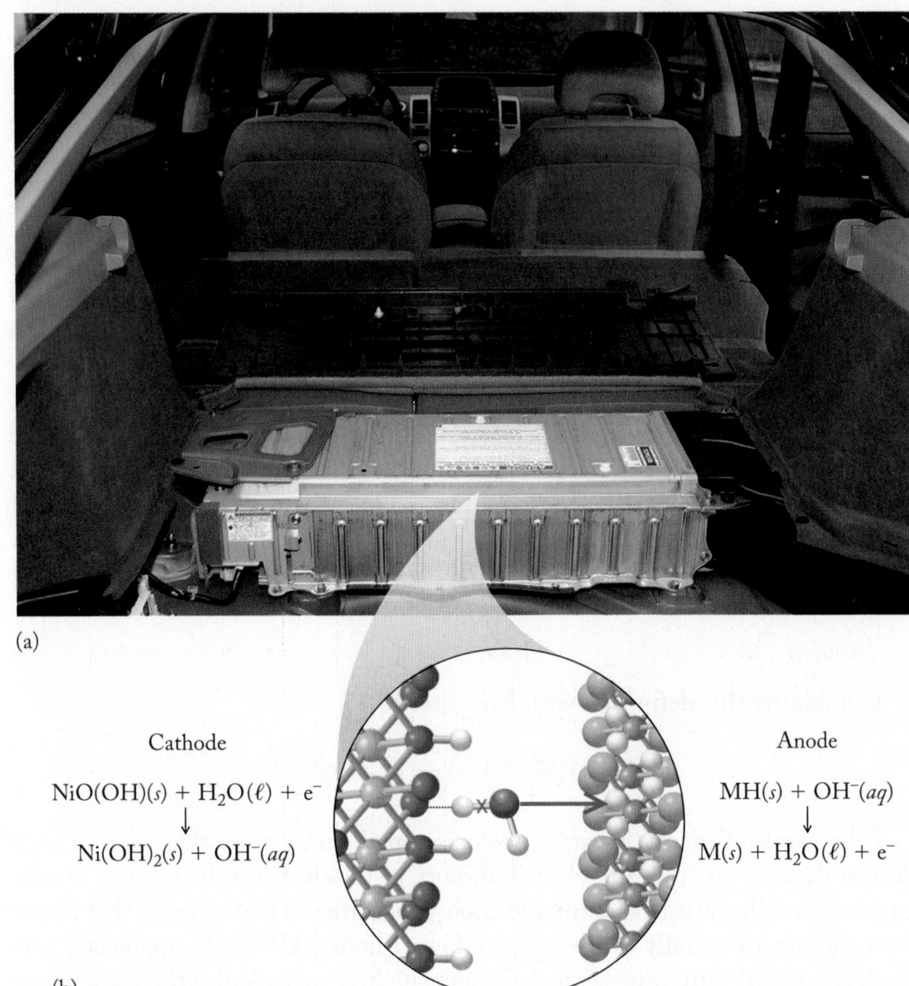

FIGURE 18.16 The Toyota Prius and other hybrid vehicles are powered by combinations of gasoline engines and electric motors. (a) Electricity for the motor of this Prius is stored in a nickel–metal hydride battery pack under the back seat. (b) The cell of a nickel–metal hydride battery contains a cathode made of NiO(OH), which is reduced to $Ni(OH)_2$ as electrons are consumed and OH^- ions are produced (as indicated by the red "X" to show a bond breaking within a water molecule). These ions migrate through a porous membrane soaked with KOH(*aq*) to the anode, where they react with H atoms held in the crystal structure of a transition metal or a compound composed of two or more metals, such as $LaNi_5$. During this reaction the H atoms are oxidized to H^+ ions that combine with the incoming OH^- ions, forming molecules of H_2O. The anode's compounds are called metal hydrides.

The overall cell reaction from these two half-reactions is

$$MH(s) + NiO(OH)(s) \rightarrow M(s) + Ni(OH)_2(s)$$

The value of E°_{cell} for the NiMH battery cannot be calculated precisely because we have only an approximate value of E°_{anode}. Most NiMH cells are rated at about 1.2 V.

Now let's relate the electrical energy stored in a battery (in other words, its capacity) to the quantities of reactants needed to produce that energy. Suppose a rechargeable AA NiMH battery is rated to deliver 2.5 ampere-hours of electrical charge at 1.2 V. How much NiO(OH) must be converted to $Ni(OH)_2$ to deliver this much charge? We need to relate the quantity of charge to a number of moles of electrons, then convert that to an equivalent number of moles of reactant, and finally to a mass of reactant. An ampere is defined as a coulomb per second, so the quantity of electrical charge delivered is

$$2.5\ \cancel{A} \cdot \cancel{h} \times \frac{1\ C}{\cancel{A} \cdot \cancel{s}} \times \frac{60\ \cancel{min}}{1\ \cancel{h}} \times \frac{60\ \cancel{s}}{1\ \cancel{min}} = 9.0 \times 10^3\ C$$

The Faraday constant (9.65×10^4 C/mol e^-) can then be used to convert quantity of electrical charge (in C) to the number of moles of charge, which is equal to the number of moles of electrons that flow from the battery:

$$9.0 \times 10^3\ \cancel{C}\left(\frac{1\ mol\ e^-}{9.65 \times 10^4\ \cancel{C}}\right) = 0.0933\ mol\ e^-$$

According to the stoichiometry of the cathode half-reaction, the mole ratio of NiO(OH) to electrons is 1:1. Therefore, the mass of NiO(OH) consumed is

$$0.0933 \text{ mol e}^- \left(\frac{1 \text{ mol NiO(OH)}}{1 \text{ mol e}^-}\right)\left(\frac{91.70 \text{ g NiO(OH)}}{1 \text{ mol NiO(OH)}}\right) = 8.6 \text{ g NiO(OH)}$$

The mass of an AA battery is about 30 g, so this mass of NiO(OH) is reasonable if we allow for the mass of the anode, background electrolyte, and exterior shell.

Lithium-Ion Batteries

The NiMH batteries used in hybrid vehicles do not have the capacity to power them at highway speeds or for extended distances. Nor do these batteries have the energy capacity to power all-electric vehicles such as the Chevrolet Bolt shown on the opening page of this chapter. The electrical power demands of these vehicles require batteries with much greater ratios of energy capacity to battery size. The current technology of choice in these applications is the lithium-ion battery (**Figure 18.17**), the same kind of battery that powers laptop computers, cell phones, and digital cameras.

In a lithium-ion battery, Li^+ ions are stored in a graphite or silicon anode. During discharge, these ions move through a nonaqueous electrolyte to a porous cathode. These cathodes are made of transition metal oxides or phosphates that can form stable complexes with Li^+ ions. One popular cathode material is cobalt(IV) oxide. Lithium-ion batteries with these cathodes have cell potentials of about 3.6 V (three times that of a NiMH battery). The cell reaction for a lithium-ion battery with a cobalt oxide cathode is

CONNECTION Crystal structures and octahedral holes were discussed in Section 12.3.

$$Li_{1-x}CoO_2(s) + Li_xC_6(s) \rightarrow 6\ C(s) + LiCoO_2(s) \qquad (18.16)$$

In a fully charged cell, $x = 1$, which makes the cathode lithium-free CoO_2. As the cell discharges and Li^+ ions move from the carbon anode to the cobalt oxide cathode, the value of x falls toward zero. To balance this flow of positive charges inside the cell, electrons flow from the anode to the cathode through an external circuit. When fully discharged, the cathode is $LiCoO_2$, and the oxidation number of Co is reduced to +3. The electrodes in a lithium-ion battery may react with oxygen and water, so the background electrolytes (e.g., $LiPF_6$) are dissolved in

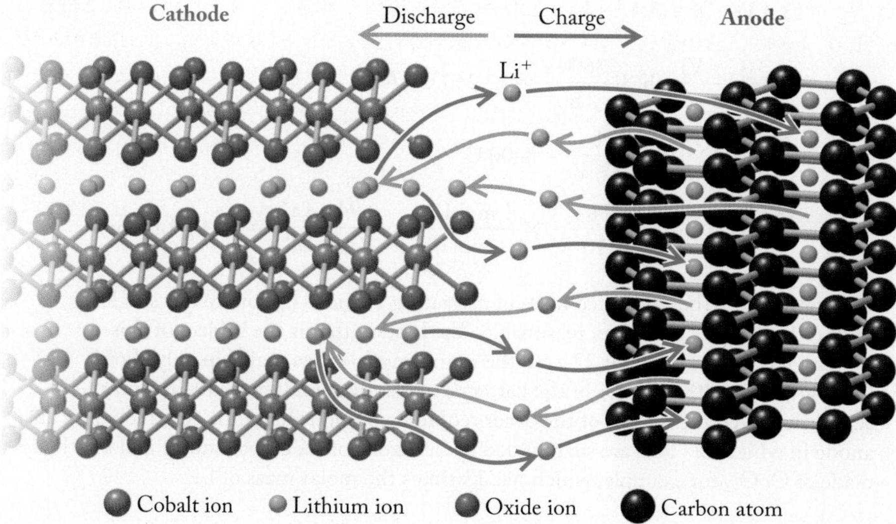

FIGURE 18.17 Most lithium-ion batteries, including the one used to power the Chevrolet Bolt shown in the opening photograph of this chapter, contain Li^+ ions stored in graphite layers of the anode. During discharge these ions travel to the cathode, which contains a transition metal compound such as CoO_2. During recharging, the direction of ion migration reverses. The crystal structure of this cathode is a cubic closest-packed array of oxide ions in which the Co^{4+} ions occupy half the octahedral holes. Li^+ ions move in and out of the remaining holes.

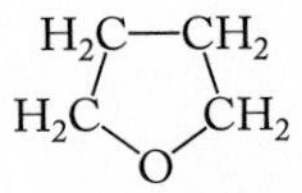

$$\text{Tetrahydrofuran}$$

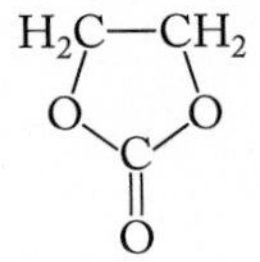

Tetrahydrofuran

Ethylene carbonate

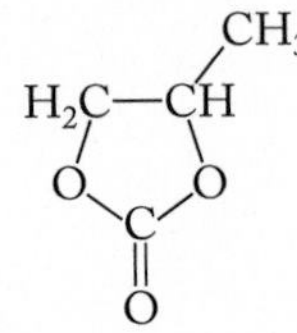

Propylene carbonate

FIGURE 18.18 Polar organic compounds, such as tetrahydrofuran, ethylene carbonate, and propylene carbonate, are among the solvents used for the background electrolytes in lithium-ion batteries.

polar organic solvents, such as tetrahydrofuran, ethylene carbonate, or propylene carbonate (**Figure 18.18**).

CONCEPT TEST

Which element is oxidized, and which is reduced in the Li^+ ion cell reaction (Equation 18.16)?

SAMPLE EXERCISE 18.7 Relating Mass of Reactant in an Electrochemical Reaction to Quantity of Electrical Charge — LO7

The capacity of the lithium-ion battery in a digital camera is 3.4 W · h at 3.6 V. How many grams of Li^+ ions must move from anode to cathode to produce this much electrical energy?

Collect and Organize We are asked to relate the electrical energy generated by an electrochemical cell to the mass of the ions involved in generating that energy. We know the cell potential and its capacity in the energy unit of watt-hours. Given these starting points and the eventual need to calculate moles and then grams of Li^+ ions, the following equivalencies may be useful:

1 watt = 1 joule/second (1 W = 1 J/s) and 1 joule = 1 coulomb-volt (1 J = 1 C · V)

We may also need to use the Faraday constant, 9.65×10^4 C/mol e^-.

Analyze Combining the two equations above gives us one that relates watts to coulombs, volts, and seconds: 1 W = 1 (C · V)/s. Multiplying a value with the units on the right side of this equation by a time (in seconds) cancels out the denominator and gives us a quantity in (C · V). Dividing this quantity by the voltage (V) of the battery yields coulombs (C) of charge. Faraday's constant relates coulombs of charge to moles of electrons, which are related 1:1 to moles of Li^+ ions, which can be converted into grams of Li^+ ions by multiplying by the molar mass of Li.

The initial value, 3.4 W · h, when multiplied by 3600 s/h and then divided by 3.6 V produces roughly 3000 C of charge. Divided by the Faraday constant ($\sim 10^5$) gives us about 0.03 mol of electrons = 0.03 mol Li^+ ions, which, when multiplied by a molar mass of about 7 g/mol, should produce a mass of about 0.2 g Li^+ ions.

Solve Using the given energy and cell-potential values in the above unit conversion series, we get

$$3.4\ \text{W}\cdot\text{h} = 3.4\frac{\text{J}}{\text{s}}\cdot\text{h} = 3.4\frac{(\text{C}\cdot\text{V})}{\text{s}}\cdot\text{h}$$

$$3.4\frac{(\text{C}\cdot\text{V})}{\cancel{\text{s}}}\cdot\cancel{\text{h}}\times\frac{3600\ \cancel{\text{s}}}{\cancel{\text{h}}} = 12{,}240\ \text{C}\cdot\text{V}$$

$$12{,}240\ \text{C}\cdot\cancel{\text{V}}\times\frac{1}{3.6\ \cancel{\text{V}}} = 3400\ \text{C}$$

$$3400\ \cancel{\text{C}}\times\frac{1\ \cancel{\text{mol e}^-}}{9.65\times10^4\ \cancel{\text{C}}}\times\frac{1\ \cancel{\text{mol Li}^+}}{1\ \cancel{\text{mol e}^-}}\times\frac{6.94\ \text{g Li}^+}{1\ \cancel{\text{mol Li}^+}} = 0.24\ \text{g Li}^+\ \text{ions}$$

Think About It The calculated mass of migrating Li^+ ions is close to our single-digit estimate and, therefore, reasonable. The battery that is the subject of this exercise has a mass of about 22 g, so the migrating Li^+ ions make up only about (0.24 g/22 g) × 100% = 1% of the battery's mass. This small percentage is not surprising given the masses of the other required components of the cell, including an anode in which Li^+ ions are surrounded by hexagons of six carbon atoms and a cathode made of CoO_2, for example, which has 13 times the molar mass of Li.

Practice Exercise Magnesium metal is produced by passing an electrical current through molten $MgCl_2$. The reaction at the cathode is

$$Mg^{2+}(\ell) + 2\ e^- \rightarrow Mg(s)$$

How many grams of magnesium metal are produced if an average current of 63.7 A flows for 4.50 h? Assume all the current is consumed by the half-reaction shown.

18.8 Corrosion: Unwanted Electrochemical Reactions

We began this chapter with a Particulate Review question based on the corrosion of the metal surface of a car. We now examine this process and some of the half-reactions that contribute to it in more detail. We have seen how half-reactions are physically separated in electrochemical cells. It turns out they are often separated in corrosion reactions as well; in this respect, the chemistry of corrosion is much like the electrochemical reactions in voltaic cells. **Corrosion** is the process in which metals are oxidized by substances in the environment, leading to deterioration from spontaneous electrochemical reactions. This definition is reflected in several of the factors that promote corrosion:

1. *The presence of water.* Metals that are left out in the rain and snow rust or corrode more rapidly than those that are under cover. For example, objects made from iron spontaneously rust in moist air, as iron reacts with O_2 from the air in a series of reactions that includes

$$4\ Fe(s) + 3\ O_2(g) + 2\ H_2O(\ell) \rightarrow 4\ FeO(OH)(s)$$

2. *The presence of electrolytes.* Just as electrolytes carry electrical current between anodes and cathodes and facilitate cell reactions, corrosion is much more rapid in seawater, for example, than in freshwater.
3. *Contact between dissimilar metals.* Metals corrode more rapidly when in contact with other metals that are less likely to be oxidized; that is, other metals that have higher reduction potentials.

Let's explore the impact of these factors using the Statue of Liberty as a model. As the statue was built, its exterior copper sheets were attached to and supported by an interior network of iron beams (**Figure 18.19**). The French designers of the statue knew that these two metals in contact with each other might someday pose a corrosion problem because the two have very different electrochemical properties. The difference is reflected in the standard reduction potentials of these elements when they oxidize under neutral to slightly basic conditions, as occur in marine environments:

$$Cu(OH)_2(s) + 2\ e^- \rightarrow Cu(s) + 2\ OH^-(aq) \qquad E° = -0.230\ V$$

$$FeO(OH)(s) + H_2O(\ell) + 3\ e^- \rightarrow Fe(s) + 3\ OH^-(aq) \qquad E° = -0.87\ V$$

As we discussed in Section 18.3, the greater (less negative) $E°$ of $Cu(OH)_2$ means that it is more easily reduced under standard conditions than FeO(OH), and Fe is more easily oxidized than Cu. We can confirm this by reversing the iron half-reaction and combining it with the copper half-reaction. We also need to

corrosion a process in which a metal is oxidized by substances in its environment.

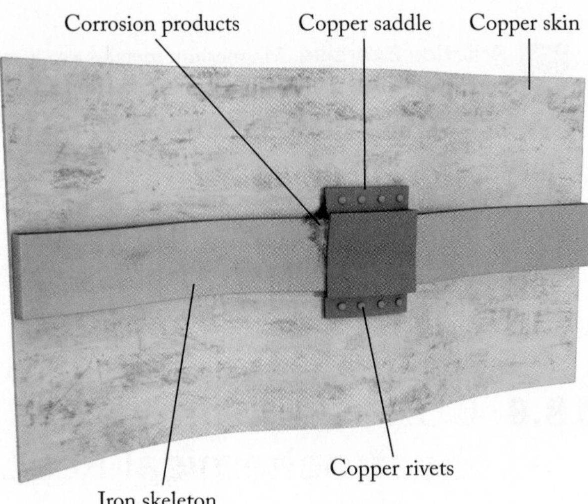

FIGURE 18.19 The light green patina of the Statue of Liberty is caused by the accumulation of copper(II) compounds on the surface of the copper sheets that make up its exterior. When the statue was built, these sheets were supported by an iron skeleton that corroded near the points of contact with the copper saddles that held the sheets and skeleton together.

multiply the iron half-reaction by 2 and the copper half-reaction by 3 to balance the loss and gain of electrons:

$$2[Fe(s) + 3\,OH^-(aq) \rightarrow FeO(OH)(s) + H_2O(\ell) + 3\,e^-]$$
$$3[Cu(OH)_2(s) + 2\,e^- \rightarrow Cu(s) + 2\,OH^-(aq)]$$

$$2\,Fe(s) + 3\,Cu(OH)_2(s) + \cancel{6\,OH^-(aq)} + \cancel{6\,e^-} \rightarrow 2\,FeO(OH)(s) + 3\,Cu(s) + 2\,H_2O(\ell) + \cancel{6\,OH^-(aq)} + \cancel{6\,e^-}$$

or

$$3\,Cu(OH)_2(s) + 2\,Fe(s) \rightarrow 3\,Cu(s) + 2\,FeO(OH)(s) + 2\,H_2O(\ell) \quad (18.17)$$

The resulting equation could be that of an electrochemical cell that has a copper cathode coated in $Cu(OH)_2$ and an iron anode. We obtain the standard potential for the reaction by substituting the reduction potentials for each half-reaction (Appendix A6.1) into Equation 18.2:

$$E^\circ_{rxn} = E^\circ_{cathode} - E^\circ_{anode} = -0.230\text{ V} - (-0.87\text{ V}) = 0.64\text{ V}$$

The reaction has a positive standard cell potential, which means that under standard conditions, iron in contact with copper will spontaneously oxidize to $FeO(OH)$ as $Cu(OH)_2$ is reduced.

To suppress the reaction in Equation 18.17, insulators made of asbestos mats soaked in shellac were used to separate the statue's iron skeleton from its copper exterior. Unfortunately, these insulators did not stand up to the humid, marine environment of New York Harbor. They absorbed water vapor and seawater spray, eventually turning into electrolyte-soaked sponges that promoted rather than retarded iron oxidation.

What are the sources of the $Cu(OH)_2$ that are the oxidizing agents in Equation 18.17? The light green patina of the Statue of Liberty (see Figure 18.19) results from the presence of copper(II) compounds such as $Cu(OH)_2$, formed by the oxidation of Cu metal by atmospheric oxygen:

$$O_2(g) + 2\,H_2O(\ell) + 2\,Cu(s) \rightarrow 2\,Cu(OH)_2(s) \quad (18.18)$$

This reaction occurs on an expansive surface of copper metal, which is also an excellent conductor of electricity. Unfortunately, this copper surface was in contact with the statue's iron skeleton through the water-logged, ion-rich asbestos mats. This connection meant that the reactions in Equations 18.17 and 18.18 were linked together. Equation 18.19 describes the resulting interior and exterior reactions:

$$2[3\ Cu(OH)_2(s) + 2\ Fe(s) \rightarrow 3\ Cu(s) + 2\ FeO(OH)(s) + 2\ H_2O(\ell)] \qquad (18.17)$$

$$3[O_2(g) + 2\ H_2O(\ell) + 2\ Cu(s) \rightarrow 2\ Cu(OH)_2(s)] \qquad (18.18)$$

$$\cancel{6\ Cu(OH)_2(s)} + 4\ Fe(s) + 3\ O_2(g) + \overset{2}{\cancel{6}}\ H_2O(\ell) + \cancel{6\ Cu(s)} \rightarrow \cancel{6\ Cu(s)} + 4\ FeO(OH)(s) + \cancel{4\ H_2O(\ell)} + \cancel{6\ Cu(OH)_2(s)}$$

or

$$4\ Fe(s) + 3\ O_2(g) + 2\ H_2O(\ell) \rightarrow 4\ FeO(OH)(s) \qquad (18.19)$$

Note how the copper half-reaction has disappeared from the overall reaction. Thus, the statue's copper exterior functions as a giant electron delivery system, allowing electrons to flow from iron atoms—as they oxidize to FeO(OH) inside the statue—to molecules of atmospheric O_2 on the exterior surface. The overall effect was a dramatic increase in the rate of oxidation of the original iron skeleton. The reaction in Equation 18.19 resulted in severe deterioration of the skeletal network that held up the Statue of Liberty, so much so that in the 1980s the iron skeleton had to be replaced with one made of corrosion-resistant stainless steel.

Deterioration of the Statue of Liberty is not an isolated incident. According to one industrial estimate, the direct and indirect cost of corrosion to the U.S. economy exceeds \$1 trillion per year. Worldwide, the cost is over \$2 trillion. These figures include the costs to repair or replace corroded equipment and structures and to protect them against corrosion. The latter category includes money spent on protective coatings, including paint and chemical or electrochemical modification of metal surfaces to make them less reactive.

Another widely used method for inhibiting corrosion involves chemically bonding metal oxide coatings to metal surfaces. For example, a method called *bluing* is widely used to protect steel tools, gun barrels, wood-burning stoves, and other steel materials. The name comes from the distinctive very dark blue color of a surface layer of magnetite (Fe_3O_4), which can be reaction-bonded to steel. Formation of a protective oxide layer is also the mechanism that makes stainless steel resistant to corrosion. Although the main ingredient in all forms of stainless steel is iron, chromium is also present. Oxidation of surface Cr atoms forms a durable protective layer of Cr_2O_3. Similarly, objects made of aluminum are protected by a surface layer of Al_2O_3 that strongly adheres to the underlying metal and inhibits further oxidation.

Another way to protect metal structures, especially those in contact with seawater (**Figure 18.20**), involves attaching to them objects made of even more reactive metals. These objects are called *sacrificial* anodes. As their name implies, their role is to form a voltaic cell with the protected structure in which the object is the anode and the structure is the cathode, so that the sacrificial anodes oxidize and the structure does not. This preservation technique is called *cathodic protection*. Many of the sacrificial anodes used in the marine industry are made of zinc or aluminum/magnesium alloys.

$$Zn(s) + 2\ OH^-(aq) \rightarrow ZnO(s) + H_2O(\ell) + 2\ e^-$$

$$Mg(s) + 2\ OH^-(aq) \rightarrow MgO(s) + H_2O(\ell) + 2\ e^-$$

$$2\ Al(s) + 6\ OH^-(aq) \rightarrow Al_2O_3(s) + 3\ H_2O(\ell) + 6\ e^-$$

These materials have the benefit of forming oxide coatings as they oxidize, which partially protect them and slow the rate at which they oxidize further, thereby extending their life span.

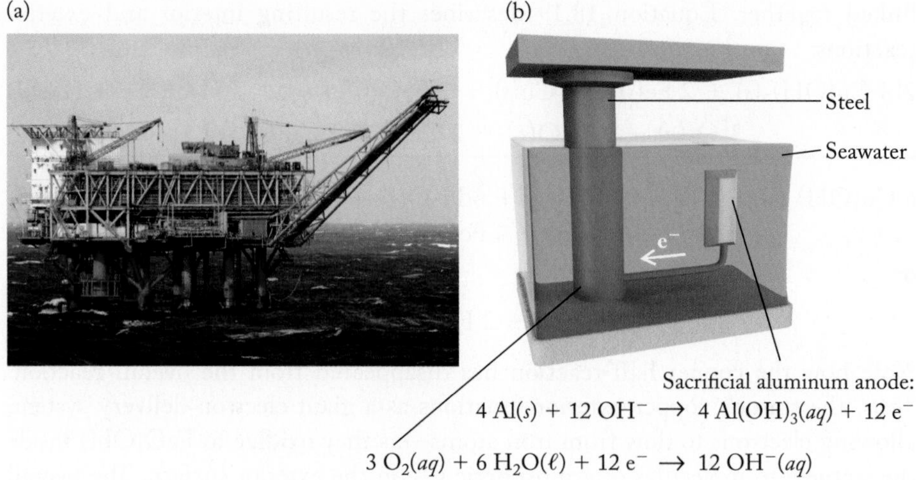

FIGURE 18.20 (a) Oceangoing metal structures are likely to have serious corrosion problems unless (b) "sacrificial" anodes are attached to them that corrode (oxidize) before the structure does.

18.9 Electrolytic Cells and Rechargeable Batteries

Lead–acid, NiMH, and lithium-ion batteries are rechargeable, which means that their spontaneous ($\Delta G < 0$) cell reactions that convert chemical energy into electrical work can be forced to run in reverse. Recharging happens when external sources of electrical energy are applied to the batteries. This electrical energy is converted into chemical energy as it drives nonspontaneous ($\Delta G > 0$) reverse cell reactions, reforming reactants from products. To make this possible, the products of the original cell reactions must be substances that either adhere to, or are embedded in, the electrodes and are available to react with the electrons supplied to the cathodes and drawn away from the anodes by the external power supply.

When a car's lead–acid battery (**Figure 18.21**) discharges, electrons flow out of the (−) terminal, which is connected to the battery's Pb electrodes, as atoms of Pb are oxidized to Pb^{2+} ions, forming $PbSO_4$ and releasing electrons that flow toward the (−) terminal. At the same time, electrons flow into the positive terminal, which is connected to the battery's PbO_2 electrodes. The incoming electrons are consumed as Pb^{4+} ions in PbO_2 are reduced to Pb^{2+} ions in $PbSO_4$. Thus, the Pb electrodes serve as anodes during discharge and the PbO_2 electrodes serve as cathodes.

To recharge the battery, the flow of electrons is reversed as they are pumped from the (+) electrode and into the (−) electrode by an alternator driven by the car's engine. The alternator creates an opposing electrical potential that is greater than that of the battery, forcing the two half-reactions that occur during the discharge cycle to run in reverse, as shown in Figure 18.22. Any $PbSO_4$ that formed on the Pb electrodes during discharge is reduced back to Pb metal:

$$PbSO_4(s) + H^+(aq) + 2\ e^- \rightarrow Pb(s) + HSO_4^-(aq)$$

and any $PbSO_4$ that formed on the PbO_2 electrodes is oxidized back to PbO_2:

$$PbSO_4(s) + 2\ H_2O(\ell) \rightarrow PbO_2(s) + 3\ H^+(aq) + HSO_4^-(aq) + 2\ e^-$$

In other words, the roles of the two sets of electrodes have been reversed when the battery is recharged: the Pb electrodes connected to the (−) terminal are the cathodes because they are the electrodes where reduction takes place, and the PbO_2 electrodes connected to the (+) terminal are the anodes because they are the electrodes where oxidation takes place.

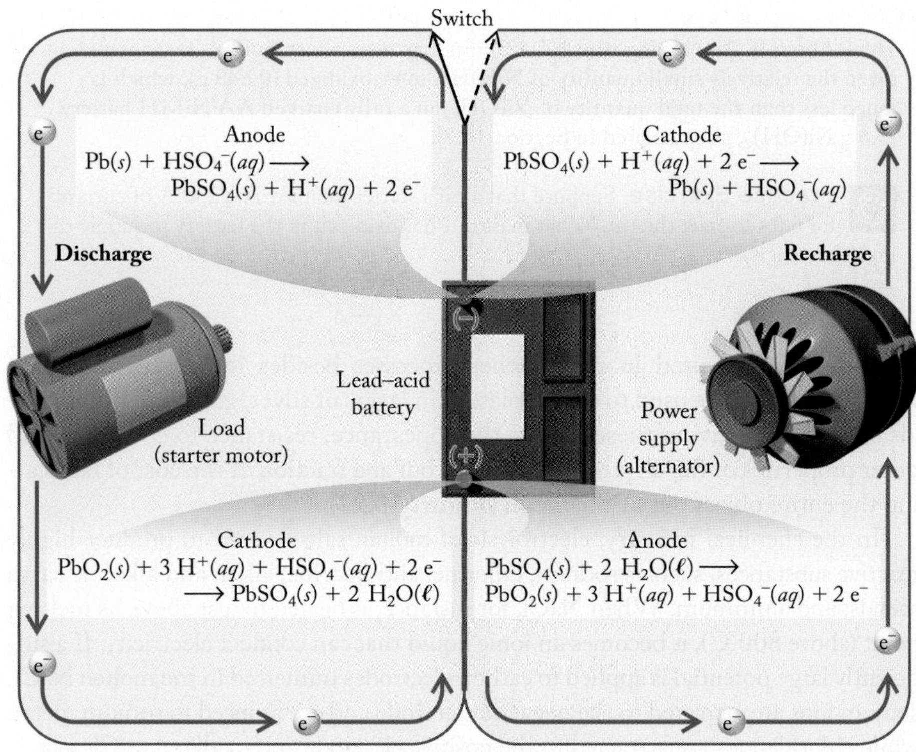

FIGURE 18.21 The lead–acid battery used in many vehicles is based on oxidation of Pb and reduction of PbO_2. As the battery discharges (circuit on left), Pb is oxidized to $PbSO_4$ and PbO_2 is reduced to $PbSO_4$. When the engine is running, a device called an alternator generates electricity that flows into the battery, as shown in the circuit on the right, recharging the battery as both electrode reactions are reversed: $PbSO_4$ is oxidized to PbO_2, and $PbSO_4$ is reduced to Pb.

SAMPLE EXERCISE 18.8 Calculating the Time Required to Oxidize a Quantity of Reactant **LO7**

If a battery charger for AA NiMH batteries supplies a charging current of 1.00 A, how many minutes does it take to oxidize 0.649 g of $Ni(OH)_2$ to $NiO(OH)$?

Collect, Organize, and Analyze During the discharge of a NiMH battery, the spontaneous cathode half-reaction is

$$NiO(OH)(s) + H_2O(\ell) + e^- \rightarrow Ni(OH)_2(s) + OH^-(aq)$$

During recharging, the reverse of this half-reaction takes place:

$$Ni(OH)_2(s) + OH^-(aq) \rightarrow NiO(OH)(s) + H_2O(\ell) + e^-$$

One mole of electrons is produced for each mole of $Ni(OH)_2$ consumed. Our first steps are to convert 0.649 g of $Ni(OH)_2$ into moles of $Ni(OH)_2$ and then into moles of electrons. We can use the Faraday constant to convert moles of electrons to coulombs of charge. A coulomb is the same as an ampere-second (Equation 18.14), so dividing by the charging current gives us seconds of charging current. The unit conversions to minutes of charging time are summarized as follows:

$$\boxed{\text{g Ni(OH)}_2} \xrightarrow{\frac{1}{\text{molar mass Ni(OH)}_2}} \xrightarrow{\frac{1 \text{ mol e}^-}{1 \text{ mol Ni(OH)}_2}} \xrightarrow{\frac{9.65 \times 10^4 \text{ C}}{\text{mol e}^-}} \xrightarrow{\frac{\text{A} \cdot \text{s}}{\text{C}}} \xrightarrow{\frac{1}{\text{A}}} \xrightarrow{\frac{1 \text{ min}}{60 \text{ s}}}$$

$$= \text{charging time (min)}$$

Solve

$$0.649 \text{ g Ni(OH)}_2 \times \frac{1 \text{ mol Ni(OH)}_2}{92.71 \text{ g Ni(OH)}_2} \times \frac{1 \text{ mol e}^-}{1 \text{ mol Ni(OH)}_2} \times \frac{9.65 \times 10^4 \text{ C}}{1 \text{ mol e}^-}$$

$$\times \frac{1 \text{ A} \cdot \text{s}}{1 \text{ C}} \times \frac{1}{1.00 \text{ A}} \times \frac{1 \text{ min}}{60 \text{ s}} = 11.3 \text{ min}$$

Think About It A charging time of 11.3 min may seem short, but it is reasonable given the relatively small quantity of $Ni(OH)_2$ to be oxidized (0.649 g), which is much less than the total quantity of $Ni(OH)_2$ in a fully charged AA NiMH battery [8.6 g $Ni(OH)_2$, as calculated in Section 18.7].

Practice Exercise Suppose that a car's starter motor draws 230 A of current for 6.0 s to start the car. What mass of Pb is oxidized in the battery to supply this much electricity?

FIGURE 18.22 The Oscar statuettes given out at Academy Awards ceremonies are made of an alloy of tin, antimony, and copper that is electroplated with three different materials: copper, nickel silver (a silvery alloy of copper, nickel, and zinc), and a final layer of 24-karat gold.

Electrolysis is used in many other processes besides recharging batteries. Electrolytic cells are used to electroplate thin layers of silver, gold, and other metals onto objects, giving these objects the appearance, resistance to corrosion, and other properties of the electroplated metal, but at a fraction of the cost of fabricating the entire object out of the metal (**Figure 18.22**).

In the chemical industry, electrolysis of molten salts is used to produce highly reactive substances, such as sodium, chlorine, and fluorine; alkali and alkaline earth metals; and aluminum. When NaCl, for instance, is heated to just above its melting point (above 800°C), it becomes an ionic liquid that can conduct electricity. If a sufficiently large potential is applied to carbon electrodes immersed in the molten NaCl, sodium ions are attracted to the negative electrode and are reduced to sodium metal, while chloride ions are attracted to the positive electrode and oxidized to Cl_2 gas:

$$2\ Na^+(\ell) + 2\ Cl^-(\ell) \rightarrow 2\ Na(\ell) + Cl_2(g)$$

A final note about how the anode and cathode vary in voltaic and electrolytic cells is in order. As we saw in Section 18.2, the reactions in voltaic cells are spontaneous. These cells pump electrical current through external circuits and electric devices with a force equal to their cell potentials. The anode in a voltaic cell is negative because an oxidation half-reaction supplies negatively charged electrons to the device powered by the cell. Electrons flow from the device into the positive battery terminal that is connected to the cathode, where these electrons are consumed in a reduction half-reaction.

The reactions in electrolytic cells are nonspontaneous. They require electrical energy from an external power supply. When the negative terminal of such a power supply is connected to the cathode of the battery, the power supply pumps electrons into the cathode, where they are consumed in reduction half-reactions. Electrons are pumped away from the anode, where they were generated in an oxidation half-reaction, toward the positive terminal of the power supply. As a result, the cathode of an electrolytic cell is the negative electrode, but the cathode of a voltaic cell is the positive electrode. Similarly, the anode of an electrolytic cell is the positive electrode, but the anode of a voltaic cell is the negative electrode. These electrode differences between voltaic and electrolytic cells make sense if we keep in mind the fundamental definitions:

- Anodes are electrodes where oxidation takes place.
- Cathodes are electrodes where reduction takes place.

CONCEPT **TEST**

The electrolysis of molten NaCl produces liquid Na metal at the cathode and Cl_2 gas at the anode. However, the electrolysis of an aqueous solution of NaCl also produces a gas at the cathode. Based on the standard potentials in Table A6.1, predict which gas forms at the cathode.

18.10 Fuel Cells and Flow Batteries

fuel cell a voltaic cell based on the oxidation of a continuously supplied fuel; the reaction is the equivalent of combustion, but chemical energy is converted directly into electrical energy.

Fuel cells are promising energy conversion devices for many applications, from powering office buildings to cruise ships to electric vehicles. Fuel cells are voltaic cells, but they differ from batteries in that their supplies of reactants are constantly renewed. Fuel cells are energy conversion devices that continue to supply electricity as long as fuel and oxygen are delivered to the anode and cathode, respectively. Therefore, they do not "discharge": they never run down, and they don't die unless their fuel supply is cut off. From a thermodynamic perspective, most batteries are closed systems, whereas fuel cells are open systems.

CHEMTOUR
Fuel Cell

In a typical fuel cell, electrons are supplied to an external circuit by the oxidation of H_2 at the anode. Electrons are consumed by the reaction of O_2 with protons migrating through the electrolyte to the cathode. Chemical energy from the reaction of hydrogen with oxygen is converted directly into electrical energy, and the net fuel cell reaction is

$$2\,H_2(g) + O_2(g) \rightarrow 2\,H_2O(\ell)$$

The fuel cells used to power electric vehicles consist of metallic or graphite electrodes separated by a hydrated polymeric material called a *proton-exchange membrane* (PEM). The PEM serves as both an electrolyte and a barrier that prevents crossover and mixing of the fuel and oxidant (**Figure 18.23**). The electrodes are carbon-supported transition metal catalysts dispersed in a layer upon the PEM. The catalytic layers speed up the electrode half-reactions. Platinum catalysts promote the breaking of H—H bonds during the oxidation of H_2 gas to H^+ ions at the anode:

$$H_2(g) \rightarrow 2\,H^+(aq) + 2\,e^- \qquad E° = 0.000\ V$$

At the cathode, a platinum–nickel alloy with the nominal composition Pt_3Ni is particularly effective in catalyzing the formation of free O atoms from O_2 molecules that are part of the reduction half-reaction:

$$O_2(g) + 4\,H^+(aq) + 4\,e^- \rightarrow 2\,H_2O(\ell) \qquad E° = 1.229\ V$$

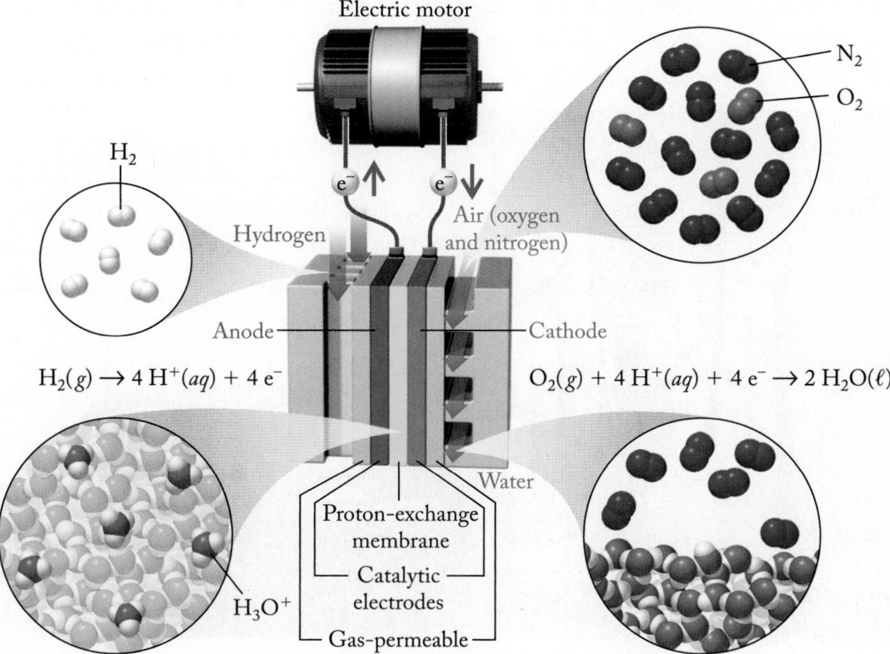

FIGURE 18.23 Fuel cells used in vehicles have a proton-exchange membrane between the two halves of the cell. Hydrogen gas diffuses to the anode, while oxygen gas diffuses to the cathode. These electrodes are made of porous material, such as carbon nanofibers, that has a high surface area for a given mass of material. Catalysts on the electrode surfaces also increase the rates of anode and cathode half-reactions.

Hydrogen ions that form at the anode move through the PEM to the cathode, where they combine with the free O atoms formed on the Pt_3Ni catalyst surface and electrons from the external circuit. This migration of positive charges across the PEM electrolyte is concurrent with the flow of electrons through the device in the external circuit.

A single PEM fuel cell typically delivers useful currents at cell potentials of about 0.8 V. The operating cell voltage is less than the expected E°_{cell} value of 1.23 V because some of the energy of combustion is wasted as heat. When these cells are assembled into fuel cell *stacks*, they can produce 100 kW of electrical power. That is enough to give a midsize car, such as the one in **Figure 18.24**, a top speed of 160 km/h (100 mi/h).

PEM fuel cells are well suited for use in vehicles because they are compact and lightweight, and they operate at moderate temperatures (60°C–80°C). It turns out that the performance of nearly all fuel cells is better at above-ambient temperatures because the rates of the half-reactions increase with temperature, as predicted by the Arrhenius equation. The increase in rate results in higher efficiency and more power.

Some fuel cells use basic electrolytes such as concentrated KOH. Pure O_2 is supplied to a cathode made of porous graphite containing a nickel catalyst, and H_2 gas is supplied to a graphite anode containing nickel(II) oxide. Hydroxide ions formed during O_2 reduction at the cathode,

$$O_2(g) + 2\,H_2O(\ell) + 4\,e^- \rightarrow 4\,OH^-(aq) \qquad E^\circ = 0.401\text{ V}$$

move through the cell to the anode, where they combine with H_2 as it is oxidized to water:

$$2\,H_2(g) + 4\,OH^-(aq) \rightarrow 4\,H_2O(\ell) + 4\,e^-$$

This oxidation half-reaction is the reverse of the following reduction half-reaction in Table A6.1:

$$H_2O(\ell) + 4\,e^- \rightarrow 2\,H_2(g) + 4\,OH^-(aq) \qquad E^\circ = -0.828\text{ V}$$

This basic pair of standard electrode potentials yields the same E°_{cell} value,

$$E^\circ_{cell} = 0.401\text{ V} - (-0.828\text{ V}) = 1.229\text{ V}$$

FIGURE 18.24 The fuel tank of this Honda Clarity FCV (fuel cell vehicle) holds 5.5 kg of compressed H_2 gas, which gives the car a driving range of about 588 km (366 mi).

as the acidic pair described previously:

$$E^\circ_{cell} = 1.229 \text{ V} - 0.000 \text{ V} = 1.229 \text{ V}$$

This equality is logical because the energy (ΔG°) released under standard conditions by the oxidation of hydrogen gas to form liquid water should have only one value, which means that E°_{cell} should have only one value, independent of electrolyte pH. What, then, is the advantage of the alkaline fuel cell? Although the *thermodynamics* of the cell are independent of pH, the *kinetics* of oxygen reduction are much better in an alkaline environment.

CONNECTION The temperature dependence of reaction rates (the Arrhenius equation, $k = Ae^{-E_a/RT}$) was first discussed in Section 13.4.

The same chemical energy that is released in fuel cells could also be obtained by burning hydrogen gas in an internal combustion engine. However, typically only about 20%–25% of the chemical energy in the fuel burned in such an engine is converted into mechanical energy; most is lost to the surroundings as heat instead. In contrast, fuel-cell technologies can convert up to about 80% of the energy released in a fuel-cell redox reaction into electrical energy. The electric motors they power are also about 80% efficient at converting electrical energy into mechanical energy. Thus, the overall conversion efficiency of a fuel-cell–powered car is theoretically as high as (80% × 80%), or 64%. Actually, the measured efficiency of the propulsion system of the car in Figure 18.24 is about 60%, which is still more than twice that of an internal combustion engine. In addition, H_2-fueled vehicles emit only water vapor—they produce no oxides of nitrogen, no carbon monoxide, and no CO_2, hence the name "zero-emission vehicles."

As fuel cells become even more efficient and less expensive, the principal limit on their use in passenger cars will be the availability and cost of hydrogen fuel. Some people worry about the safety of storing hydrogen in a high-pressure tank in a car, although H_2 has a higher ignition temperature than gasoline and spreads through the air more quickly, reducing the risk of fire. Still, hydrogen in air burns over a much wider range of concentrations than does gasoline, and its flame is almost invisible.

Because of the lack of an extensive hydrogen distribution network, many fuel-cell–powered vehicles are buses and fleet vehicles operating from a central location where hydrogen gas is available. With their very large fuel tanks, buses powered by fuel cells have an operating range of 400 km (≈250 mi). Longer ranges would be possible if better methods for storing hydrogen gas were available.

Hydrogen gas is produced from methane in a two-step process introduced in Chapter 14:

$$\text{(1)} \quad CH_4(g) + H_2O(g) \rightleftharpoons CO(g) + 3\,H_2(g)$$
$$\text{(2)} \quad CO(g) + H_2O(g) \rightleftharpoons CO_2(g) + H_2(g)$$
$$\text{Overall:} \quad CH_4(g) + 2\,H_2O(g) \rightleftharpoons CO_2(g) + 4\,H_2(g)$$

In addition to 4 moles of H_2, the process produces 1 mole of CO_2 per mole of CH_4—the same amount of carbon dioxide as the complete combustion of methane produces. From the perspective of climate change, it is the greater efficiency of fuel cells powering electric motors coupled with the higher fuel value of CH_4 compared to the hydrocarbons that make up gasoline and diesel oil that reduces the impact of fuel-cell–powered cars compared to those that rely on the combustion of liquid fossil fuels.

CONNECTION Processes for producing hydrogen gas by reacting methane and other fuels with high-temperature steam are described in Section 14.1.

Switching to the all-electric vehicles described in Section 18.7 avoids the problems of CO_2 production if alternative sources of electricity can replace power plants that burn fossil fuels. In principle, solar and wind power can meet all our energy needs but these too suffer from certain disadvantages. Among the issues most relevant to the current chapter is that the Sun does not shine all day and the wind

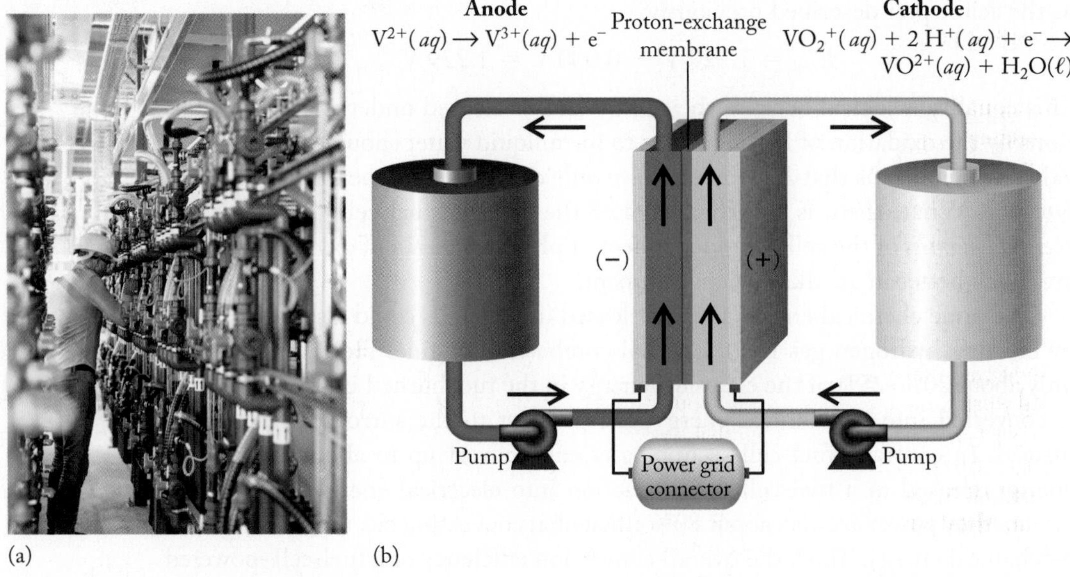

FIGURE 18.25 (a) A bank of redox flow batteries. (b) Circulating pumps continually resupply the reactants for the half-reactions that together generate electrical energy in a redox flow battery. Large arrays of these batteries and huge holding tanks filled with their reacting electrolytes can store in chemical form the equivalent of 10^7 to 10^8 Wh of electrical energy.

does not blow continually. We need ways to store the excess electricity produced on sunny and windy days for use when demand peaks. The lithium-ion and other batteries described in Section 18.7 lack the necessary energy storage capacity, so scientists are exploring **redox flow batteries** such as the one shown in **Figure 18.25**.

A redox flow battery consists of two large storage tanks containing solutions of electrolytes. The two solutions are pumped through an electrochemical cell resembling the fuel cell, except that a cation-exchange membrane separates the two solutions instead of two gases. The electrons flow through an external circuit and into a device or into a power grid. Pumping in fresh electrolyte keeps the battery charged. Eventually, decreasing reactant concentrations in the holding tanks require that other energy sources, such as solar- or wind-generated electricity, be used to reverse the voltaic reaction and regenerate the reactants. In this way, charging the redox flow battery allows a utility to store excess solar and wind energy by recharging redox flow batteries and then relying on battery power at night or when winds are calm. Commercial installations of redox flow batteries are being built that have energy storage capacities in the 10s to 100s of megawatt hours (MWh); enough to meet peak-hour demands for electricity of a small city.

A popular electrolyte system is based on the different oxidation states of vanadium. At the anode, V^{2+} ions are oxidized to V^{3+}:

$$V^{2+}(aq) \rightarrow V^{3+}(aq) + e^- \qquad E^\circ_{anode} = -0.255\ V$$

while $VO_2^{\ +}$ ions are reduced to VO^{2+} ions at the cathode:

$$VO_2^{\ +}(aq) + 2\,H^+(aq) + e^- \rightarrow VO^{2+}(aq) + H_2O(\ell) \qquad E^\circ_{cathode} = +0.991\ V$$

One electron is produced in the first half-reaction and one is consumed in the second, so we simply combine the two half-reactions to obtain an overall cell reaction:

$$V^{2+}(aq) + VO_2^{\ +}(aq) + 2\,H^+(aq) \rightarrow V^{3+}(aq) + VO^{2+}(aq) + H_2O(\ell)$$

Using Equation 18.2 to calculate E°_{cell}:

$$E^\circ_{cell} = E^\circ_{cathode} - E^\circ_{anode} = 0.991\ V - (-0.255\ V) = 1.246\ V$$

redox flow battery an electrochemical cell consisting of two solutions separated by an ion-exchange membrane.

CONCEPT **TEST**

Why must redox flow batteries be large to be practical as energy storage devices?

SAMPLE EXERCISE 18.9 Integrating Concepts: Corrosion at Sea

In the 18th century, the British Royal Navy began the practice of adding copper sheathing to the bottoms of its sailing vessels to inhibit the growth of barnacles and seaweed that slowed them down. Unfortunately, the first copper sheets they used were fastened with iron nails. (a) Explain why using iron nails was not a good idea. (b) Which metal would corrode?

Modern navies continue to encounter shipboard corrosion problems, sometimes with disastrous results. In 1963, the USS *Thresher*, the lead boat in a new class of nuclear submarines, was conducting deep diving tests about 200 miles off the coast of Cape Cod when it suddenly sank. All 129 sailors and civilians aboard perished. It was later determined that loss of the *Thresher* may have been caused by failure of the silver solder used to connect metal pipes that transported seawater to cool the *Thresher*'s nuclear power plant. Metal pipes that carry seawater are made of copper/nickel alloys. (c) Why might silver solder have been considered a logical choice for connecting them? (d) How is the oxidation of silver in seawater influenced by the presence of chloride ions? (e) Use the appropriate half-reactions and standard reduction potentials from Table A6.1 (see Appendix 6) to calculate the cell potential for the air-oxidation of Ag in pH 8.00 seawater in which $[Cl^-] = 0.56\ M$. Assume $P_{O_2} = 0.21$ atm. Is the reaction spontaneous?

Collect and Organize First we are asked to explain why nailing one metal to another is a bad idea when they will be immersed in seawater and to identify which one will corrode. Then we have another corrosion problem involving Ag and Cu/Ni alloys, where we need to address their electrochemical properties and to explain how the oxidation of Ag in seawater depends on the presence of Cl^- ions. Finally, we are asked to determine whether Ag spontaneously oxidizes in aerated seawater, using half-reactions from Table A6.1 to calculate E_{cell}.

Analyze The problem with nailing copper sheets with iron nails is like that encountered in the Statue of Liberty (see Figure 18.19), in which the iron skeleton of the statue rusted so badly that it had to be replaced. Copper and iron metal in contact with each other and surrounded by a strong electrolyte (seawater) form an electrochemical cell in which copper is the cathode and iron is the anode as shown in Equation 18.17.

The logic behind using silver solder to connect pipes made of another metal (or metal alloy) must have been based on their similar electrochemical properties. To determine similarity, we need to use the appropriate half-reaction for silver. Table A6.1 lists two half-reactions: one that is simply $Ag^+(aq) + e^- \rightarrow Ag(s)$ and the other in which the half-reaction occurs in a solution of chloride ions: $AgCl(s) + e^- \rightarrow Ag(s) + Cl^-(aq)$. Given the high $[Cl^-]$ in seawater and the limited solubility of AgCl, we should use the latter to compare electrochemical properties and to determine whether the oxidation of Ag is spontaneous.

Solve

a. Iron and copper should *not* be allowed to contact each other in seawater because they are dissimilar metals, as shown by the standard reduction potentials of the following half-reactions:

$$Fe^{2+}(aq) + 2\,e^- \rightarrow Fe(s) \qquad E^\circ = -0.447\ V$$

$$Cu^{2+}(aq) + 2\,e^- \rightarrow Cu(s) \qquad E^\circ = 0.342\ V$$

Cu^{2+} ions are produced by air oxidation of the Cu metal to $Cu(OH)_2$. When Cu and Fe are in contact with each other in an electrolytic solution, the two become the cathode and anode, respectively, of a voltaic cell with the cell reaction:

$$Cu^{2+}(aq) + Fe(s) \rightarrow Cu(s) + Fe^{2+}(aq)$$

This reaction has an $E^\circ_{cell} = E^\circ_{cathode} - E^\circ_{anode} = 0.342\ V - (-0.447\ V) = 0.789\ V$.

b. The spontaneous cell reaction indicates that Fe is the metal that is oxidized (corrodes).

c. Table A6.1 contains the following half-reactions involving Ag, Cu, and Ni:

$$AgCl(s) + e^- \rightarrow Ag(s) + Cl^-(aq) \qquad E^\circ = 0.222\ V$$

$$Cu^{2+}(aq) + 2\,e^- \rightarrow Cu(s) \qquad E^\circ = 0.342\ V$$

$$Ni^{2+}(aq) + 2\,e^- \rightarrow Ni(s) \qquad E^\circ = -0.257\ V$$

The E° value of the Ag reaction is in between the other two values. If the actual electrochemical properties of the alloy are closer to those of Cu than Ni (which, it turns out, they are), then Ag would have been a logical choice to use in connecting pipes made from the alloy.

d. The difference in the potentials of the two silver half-reactions in Table A6.1,

$$AgCl(s) + e^- \rightarrow Ag(s) + Cl^-(aq) \qquad E^\circ = 0.222\ V$$

$$Ag^+(aq) + e^- \rightarrow Ag(s) \qquad E^\circ = 0.800\ V$$

indicates that the presence of Cl^- ions has a significant impact on the electrochemical properties of Ag. As Ag atoms are oxidized, they precipitate as AgCl, which keeps the concentration of $Ag^+(aq)$ low. The Nernst equation for the Ag^+/Ag half-reaction is

$$E = E^\circ - 0.0592\ V \log(1/[Ag^+])$$

A small $[Ag^+]$ value in the denominator means a large value for the log term and an E value for the Ag^+/Ag half-reaction that is much less than its E°. A smaller potential for the reduction half-reaction means a greater potential for the oxidation of Ag. Therefore, Ag is more likely to be oxidized in the presence of Cl^- ions.

e. The appropriate half-reactions for the corrosion of silver in seawater are

$$O_2(g) + 4\,H^+(aq) + 4\,e^- \rightarrow 2\,H_2O(\ell) \qquad E^\circ = 1.229\ V$$

$$AgCl(s) + e^- \rightarrow Ag(s) + Cl^-(aq) \qquad E^\circ = 0.222\ V$$

To assess the spontaneity of the air-oxidation of Ag, we need to reverse the second half-reaction and multiply it by 4 to balance the loss and gain of electrons:

$$O_2(g) + 4\,H^+(aq) + 4\,e^- \rightarrow 2\,H_2O(\ell)$$

$$4[Ag(s) + Cl^-(aq) \rightarrow AgCl(s) + e^-]$$

$$\overline{O_2(g) + 4\,H^+(aq) + 4\,Ag(s) + 4\,Cl^-(aq) \rightarrow 4\,AgCl(s) + 2\,H_2O(\ell)}$$

The silver reaction takes place at the anode and the standard cell potential is

$$E^\circ_{cell} = E^\circ_{cathode} - E^\circ_{anode} = 1.229\ V - 0.222\ V = 1.007\ V$$

Using this E°_{cell} value and setting up the Nernst equation for the reaction gives us

$$E_{\text{cell}} = E^\circ_{\text{cell}} - \left(\frac{0.0592\text{ V}}{4}\right)\log\left[\frac{1}{(P_{O_2})[\text{H}^+]^4[\text{Cl}^-]^4}\right]$$

We take the antilog of pH = 8.00 and use the P_{O_2} and $[\text{Cl}^-]$ values for aerated seawater:

$$E_{\text{cell}} = 1.007\text{ V} - \left(\frac{0.0592\text{ V}}{4}\right)\log\left[\frac{1}{(0.21)(1.00\times10^{-8})^4(0.56)^4}\right]$$

$$= 0.51\text{ V}$$

The positive cell potential means that the oxidation of silver in seawater should be spontaneous.

Think About It The above analysis makes it clear why using iron nails to fasten copper sheathing to the hulls of wooden sailing ships was a bad idea. However, we were unable to definitively establish the role of the silver solder joints in the sinking of the *Thresher*. According to modern naval engineering references, silver solder and Cu/Ni alloys should have similar electrochemical properties when immersed in seawater.

SUMMARY

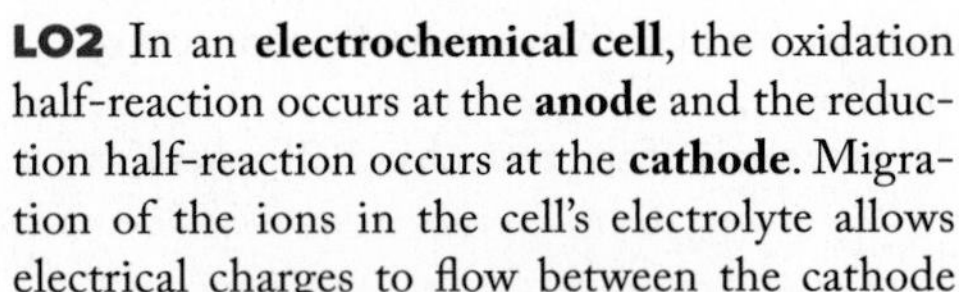

LO1 Electrochemistry is the branch of chemistry that links redox reactions to the production or consumption of electrical energy. Any redox reaction can be broken down into oxidation and reduction half-reactions. (Section 18.1)

LO2 In an **electrochemical cell**, the oxidation half-reaction occurs at the **anode** and the reduction half-reaction occurs at the **cathode**. Migration of the ions in the cell's electrolyte allows electrical charges to flow between the cathode and anode compartments as electrons flow through an external electrical circuit. A **cell diagram** shows how the components of the cathodic and anodic compartments of the cell are connected. (Sections 18.1 and 18.2)

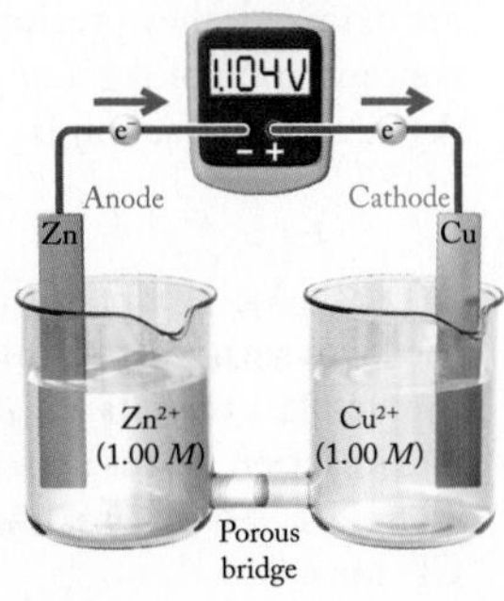

LO3 The difference between the **standard reduction potentials ($E°$)** of a cell's cathode and anode half-reactions is equal to the **standard cell potential (E°_{cell})**, with the more positive $E°$ half-reaction occurring at the cathode. The value of E°_{cell} is a measure of how forcefully a voltaic cell, in which all reactants and products are in their standard states, can pump electrons through an external circuit. All standard cell potentials are referenced to the cell potential of the **standard hydrogen electrode** ($E^\circ_{\text{SHE}} = 0.000$ V). (Sections 18.3, 18.4, 18.5)

LO4 A voltaic cell has a positive cell potential ($E_{\text{cell}} > 0$) and its cell reaction has a negative change in free energy ($\Delta G_{\text{cell}} < 0$). This decrease in free energy in a voltaic cell is available to do work in an external electrical circuit. The **Faraday constant** relates the quantity of electrical charge to the number of moles of electrons and indirectly to the number of moles of reactants. (Section 18.4)

LO5 The potential of a voltaic cell decreases as reactants turn into products. The **Nernst equation** describes how cell potential changes with concentration changes. (Section 18.6)

LO6 The potential of a voltaic cell approaches zero as the cell reaction approaches chemical equilibrium, at which point $E_{\text{cell}} = 0$ and $Q = K$. (Sections 18.6, 18.7, 18.10)

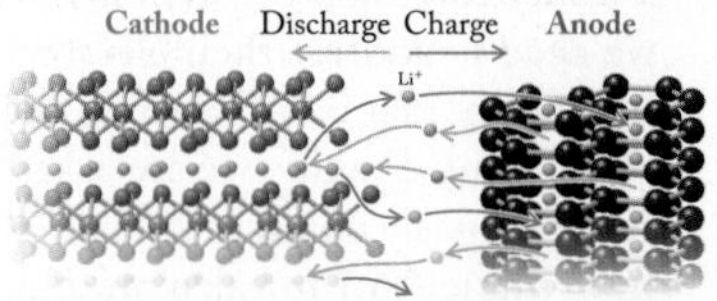

LO7 The quantities of reactants consumed in a voltaic cell reaction are related by the Faraday constant to the coulombs of electrical charge delivered by the cell. Nickel–metal hydride batteries supply electricity when H atoms are oxidized to H^+ ions at the anodes and NiO(OH) is reduced to $Ni(OH)_2$ at the cathodes. In lithium-ion batteries, electricity is produced when Li^+ ions stored in graphite anodes move toward and are incorporated into transition metal oxide or phosphate cathodes. (Sections 18.7, 18.8, 18.9)

LO8 Cars and trucks that are propelled by electric motors are considered "zero-emission" vehicles because they efficiently convert the chemical energy in their batteries or fuel cells into mechanical work with no release of pollutants. However, these vehicles still depend on fossil fuels that release fossil CO_2 for some of the electricity needed to charge their batteries and to produce the hydrogen consumed in their fuel cells.

PARTICULATE **PREVIEW WRAP-UP**

The two half-reactions depict the oxidation of hydrogen and the reduction of oxygen:

$$H_2(g) \rightarrow 2\,H^+(aq) + 2\,e^-$$

$$O_2(g) + 4\,H^+(aq) + 4\,e^- \rightarrow 2\,H_2O(\ell)$$

which sum to a balanced, overall reaction of

$$2\,H_2(g) + O_2(g) \rightarrow 2\,H_2O(\ell)$$

When chemists physically separate two half-reactions, electrons can be forced to flow through an external circuit from one half-cell to the other. This external flow of electrons can be harnessed to do work as electricity. Chemists call hydrogen a "clean" fuel because it emits only water vapor, unlike traditional fuels, which when combusted, produce oxides of nitrogen and carbon that pollute the atmosphere.

PROBLEM-SOLVING SUMMARY

Type of Problem	Concepts and Equations	Sample Exercises
Writing net ionic equations of cell reactions by combining half-reactions	Combine the reduction and oxidation half-reactions after balancing the gain and loss of electrons.	**18.1–18.4**
Diagramming an electrochemical cell	Use the format: anode | anode solution || cathode solution | cathode. Insert solution concentrations if they are known.	**18.2**
Identifying anode and cathode half-reactions and calculating the value of E°_{cell}	The half-reaction with the more positive standard reduction potential is the cathode half-reaction in a voltaic cell. $E^\circ_{cell} = E^\circ_{cathode} - E^\circ_{anode}$ (18.2)	**18.3**
Relating ΔG_{cell} and E_{cell}	$\Delta G_{cell} = -nFE_{cell}$ (18.7) where n is the number of moles of electrons transferred in the cell reaction and F is the Faraday constant, 9.65×10^4 C/mol e^-.	**18.4**
Calculating E_{cell} from E°_{cell} and the concentrations of reactants and products	E_{cell} at 25°C is related to E°_{cell} and the cell reaction quotient by the Nernst equation: $E_{cell} = E^\circ_{cell} - \frac{0.0592\text{ V}}{n}\log Q$ (18.10)	**18.5**
Calculating K for a redox reaction from the standard potentials of its half-reactions	E°_{rxn} is related to K at 298 K by $\log K = \frac{nE^\circ_{rxn}}{0.0592\text{ V}}$ (18.12)	**18.6**
Relating mass of reactant in an electrochemical reaction to quantity of electrical charge	Determine the ratio of moles of reactants to moles of electrons transferred; use the Faraday constant to relate coulombs of charge to moles of electrons.	**18.7**
Calculating the time required to oxidize a quantity of reactant	Use the Faraday constant and the relation between moles of electrons and moles of reactants to describe an electrolytic process.	**18.8**

VISUAL PROBLEMS

(Answers to boldface end-of-chapter questions and problems are in the back of the book.)

18.1. In the voltaic cell shown in Figure P18.1, the greater density of a concentrated solution of $CuSO_4$ allows a less concentrated solution of $ZnSO_4$ solution to be (carefully) layered on top of it. Why is a porous bridge not needed in this cell?

FIGURE P18.1

18.2. In the voltaic cell shown in Figure P18.2, the concentrations of Cu^{2+} and Cd^{2+} are 1.00 *M*. Based on the standard potentials in Appendix 6, identify which electrode is the anode and which is the cathode. Indicate the direction of electron flow.

FIGURE P18.2

18.3. In the voltaic cell shown in Figure P18.3, $[Ag^+] = [H^+] = 1.00\ M$ and $P_{H_2} = 1.00$ atm. From the standard potentials in Table A6.1 (Appendix 6), identify which electrode is the anode and which is the cathode. Indicate the direction of electron flow.

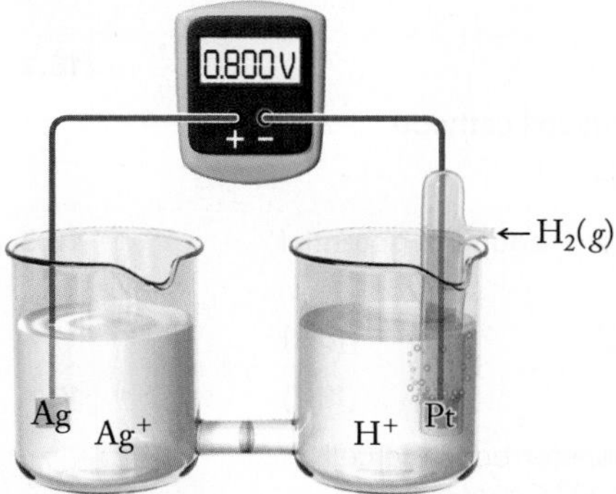

FIGURE P18.3

18.4. In many electrochemical cells the electrodes are metals that carry electrons to and from the cell but are not chemically changed by the cell reaction. Each of the highlighted clusters in the periodic table in Figure P18.4 consists of three metals. Which of the highlighted clusters is best suited to form inert electrodes?

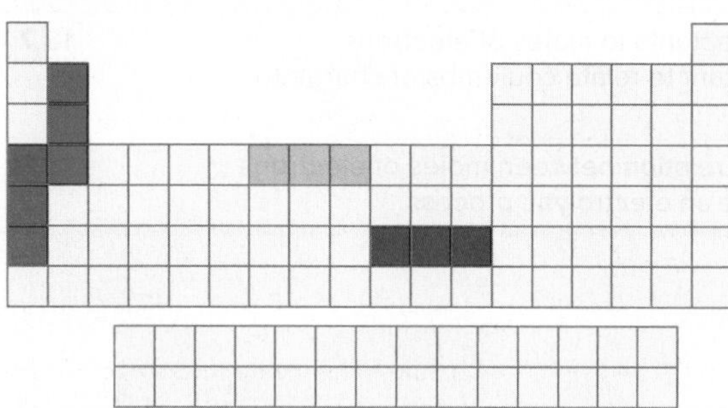

FIGURE P18.4

18.5. Which of the four curves in Figure P18.5 best represents the dependence of the potential of a lead–acid battery on the concentration of sulfuric acid? The scale of the x-axis is logarithmic.

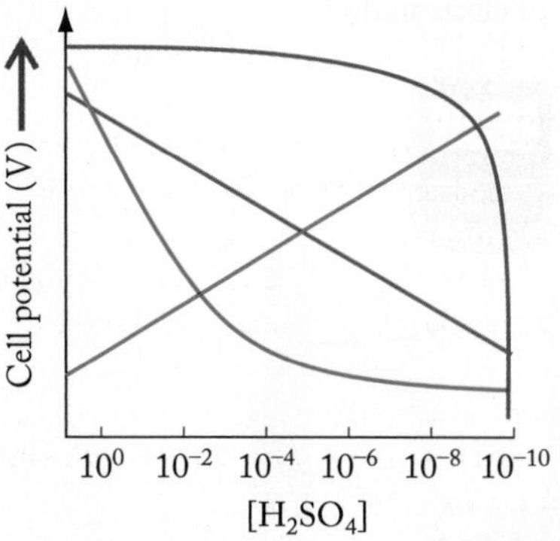

FIGURE P18.5

18.6. From top to bottom in Figure P18.6, the sizes of the batteries are D, C, AA, and AAA. The performance of batteries such as these is often expressed in units such as (a) volts, (b) watt-hours, or (c) milliampere-hours. Which of the values differ significantly between the four batteries?

FIGURE P18.6

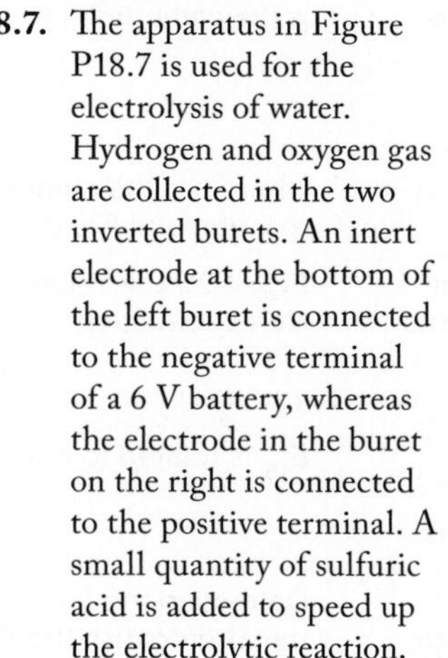

18.7. The apparatus in Figure P18.7 is used for the electrolysis of water. Hydrogen and oxygen gas are collected in the two inverted burets. An inert electrode at the bottom of the left buret is connected to the negative terminal of a 6 V battery, whereas the electrode in the buret on the right is connected to the positive terminal. A small quantity of sulfuric acid is added to speed up the electrolytic reaction.

a. What are the half-reactions at the left and right electrodes and their standard potentials?
b. Why does sulfuric acid make the electrolysis reaction go more rapidly?

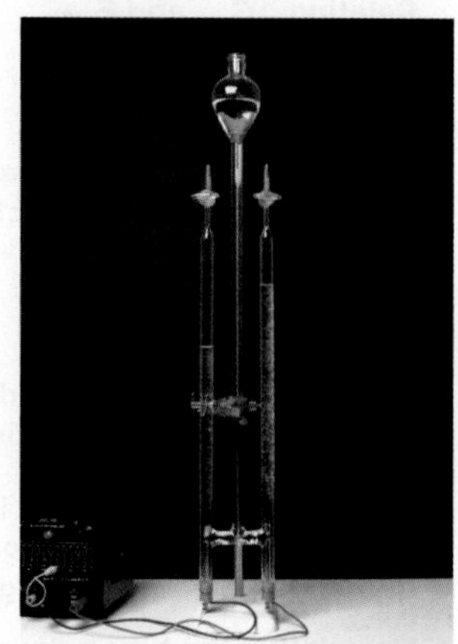

Overall cell reaction
$H_2O(\ell) \rightarrow H_2(g) + \frac{1}{2}\,O_2(g)$

FIGURE P18.7

18.8. An electrolytic apparatus identical to the one shown in Problem 18.7 is used to electrolyze water, but the reaction is speeded up by the addition of sodium carbonate instead of sulfuric acid.

a. What are the half-reactions and the standard potentials for the electrodes on the left and right?
b. Why does sodium carbonate make the electrolysis reaction go more rapidly?

18.9. Most classic cars, such as the one shown in Figure P18.9, have chromium-electroplated, or *chrome*, bumpers.

a. Is the bumper the anode or the cathode in the electroplating process?
b. How does the presence of a layer of chromium protect the bumper from corroding?

FIGURE P18.9

18.10. Use representations [A] through [I] in Figure P18.10 to answer questions (a)–(f). The photo in image [E] shows silver deposited onto copper. Suppose the materials in [A], [C], [G], and [I] were combined, along with a porous bridge and external circuit, to generate an electrochemical cell that produced [E].

a. Which metal would be the cathode? Which solution would surround it?
b. Which metal would be the anode? Which solution would surround it?
c. When the reaction in [E] is finished, what will happen to the light blue color of the solution?
d. How could [E] be produced without a porous bridge or an external circuit?
e. Of the four particulate images [B], [D], [F], and [H] in Figure P18.10, which two correspond to the solutions [G] and [I]?
f. Which particulate image represents the copper wire with silver deposited onto it? What does the fourth particulate image depict?

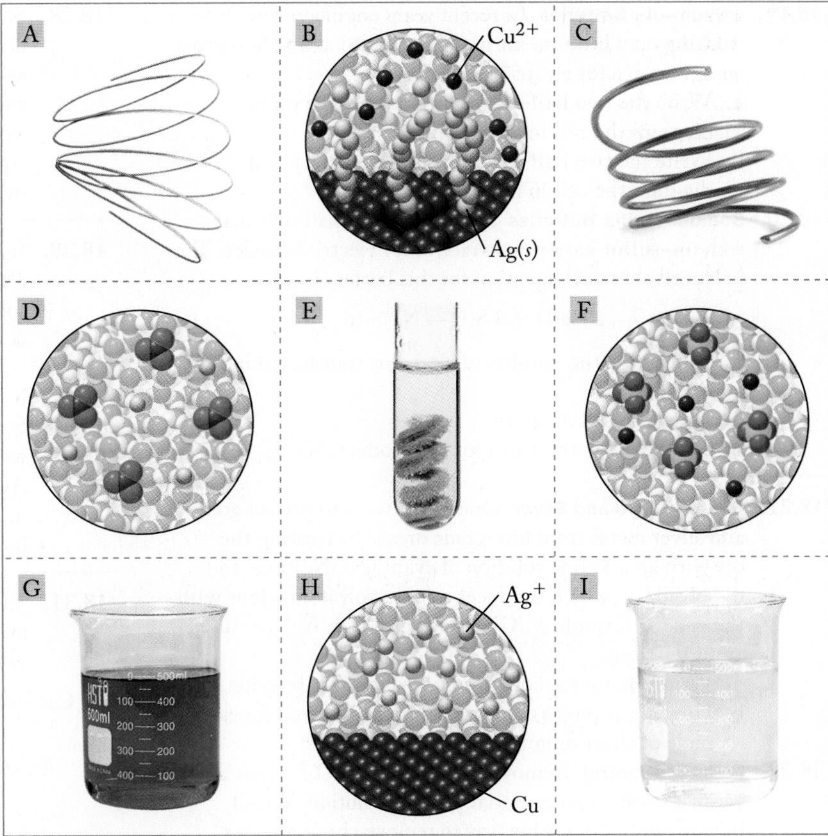

FIGURE P18.10

QUESTIONS AND PROBLEMS

Redox Chemistry Revisited; Electrochemical Cells

Concept Review

18.11. An element with a strong tendency to gain electrons is also ________.

a. easily oxidized
b. a good oxidizing agent
c. a good reducing agent
d. a reactive metal

18.12. An element that is a good reducing agent is also ________.

a. easily oxidized
b. a good oxidizing agent
c. easily reduced
d. a noble gas

18.13. Regarding the porous separator between the two halves of an electrochemical cell:

a. Describe how it allows electrical charge to flow between the two half-cells.
b. Explain why a piece of wire could not perform the same function.

18.14. The Zn/Cu^{2+} reactions in Figures 18.1 and 18.2 are the same; however, the reaction in the cell in Figure 18.2 generates electricity, whereas the reaction in the beaker in Figure 18.1 does not. Why?

Problems

18.15. Complete and balance the following partial chemical equation, using the appropriate half-reactions in Table A6.1 for acidic solutions.

$$Cr_2O_7^{2-}(aq) + Fe^{2+}(aq) \rightarrow Cr^{3+}(aq) + Fe^{3+}(aq)$$

18.16. Complete and balance the following partial net ionic equation, using the appropriate half-reactions in Table A6.1 for acidic solutions.

$$MnO_4^-(aq) + H_2O_2(aq) \rightarrow MnO_2(s) + O_2(g)$$

18.17. An electrochemical cell with an aqueous electrolyte is based on the reaction between $Ni^{2+}(aq)$ and $Cd(s)$, producing $Ni(s)$ and $Cd^{2+}(aq)$.

a. Write half-reactions for the anode and cathode.
b. Write a balanced net ionic equation describing the cell reaction.
c. Draw the cell diagram.

18.18. A voltaic cell is based on the reaction between $Cu^{2+}(aq)$ and $Ni(s)$, producing $Cu(s)$ and $Ni^{2+}(aq)$.

a. Write the anode and cathode half-reactions.
b. Write a balanced cell reaction.
c. Draw the cell diagram.

18.19. **Lithium–Air Batteries** In recent years engineers have been working on a lithium–air battery (Li–O_2) as an alternative energy source for electric vehicles.
a. Write the two half-reactions for the battery and diagram the cell in acid solution.
b. Write the two half-reactions for the battery and diagram the cell in basic solution.

18.20. **Sodium–Sulfur Batteries** Less expensive materials make sodium-sulfur batteries attractive for electric vehicles. The balanced chemical equation for this battery is

$$2\ Na(\ell) + 3\ S(\ell) \rightarrow Na_2S_3(s)$$

a. Determine the number of electrons transferred in the cell reaction.
b. Draw the cell diagram.
c. Draw Lewis structures for the product, Na_2S_3.

***18.21.** **Refining Gold and Silver** One of the ways to extract gold and silver metal from low-grade ores is by treating the ore with an alkaline solution of cyanide (CN^-) ions and dissolved O_2, which produces soluble polyatomic ions with the generic formula $M(CN)_2^-$ (where M = Ag^+ or Au^+) and OH^- ions.
a. Write a net ionic equation describing this reaction.
b. How many moles of electrons are transferred for each mole of silver or gold metal that dissolves?

18.22. **Pollution Control** Removing chromate, CrO_4^{2-}, ions from waste streams is an essential part of pollution control in many industries. One way to remove chromate ions involves reacting them with SO_3^{2-} ions, producing solid $Cr(OH)_3$ and SO_4^{2-} ions under alkaline conditions.
a. Write a net ionic equation describing this reaction.
b. How many moles of electrons are transferred for each mole of chromate that reacts?

Standard Potentials

Concept Review

18.23. Is oxygen or chlorine a stronger oxidizing agent under standard conditions? Use the standard reduction potentials in Table A6.1 to support your answer.

*18.24. Of the group 1 elements Li, K, and Na, which is the strongest reducing agent?

Problems

18.25. If a piece of silver is placed in a solution in which $[Ag^+] = [Cu^{2+}] = 1.00\ M$, will the following reaction proceed spontaneously?

$$2\ Ag(g) + Cu^{2+}(aq) \rightarrow 2\ Ag^+(aq) + Cu(s)$$

18.26. A piece of cadmium is placed in a solution in which $[Cd^{2+}] = [Sn^{2+}] = 1.00\ M$. Will the following reaction proceed spontaneously?

$$Cd(s) + Sn^{2+}(aq) \rightarrow Cd^{2+}(aq) + Sn(s)$$

18.27. Sometimes the anode half-reaction in the zinc–air battery (Figure 18.7) is written with the zincate ion, $Zn(OH)_4^{2-}$, as the product. Write a balanced equation for the cell reaction based on this product.

*18.28. Sometimes the cell reaction of nickel–cadmium batteries is written with Cd metal as the anode and solid NiO_2 as the cathode. If the products of the reactions are a solid hydroxide of cadmium(II) at the anode and a solid hydroxide of nickel(II) at the cathode, write balanced equations for the cathode and anode half-reactions and the overall cell reaction.

18.29. In a voltaic cell like the Cu–Zn cell in Figure 18.2, the Cu electrode is replaced with one made of Ni immersed in a solution of $NiSO_4$.
a. Will the standard potential of this Ni/Zn cell be greater than, the same as, or less than 1.10 V?
b. Which electrode would be the anode?

18.30. Suppose the copper half of the Cu/Zn cell in Figure 18.2 were replaced with a silver wire in contact with 1 M $Ag^+(aq)$.
a. What would be the value of E°_{cell}?
b. Which electrode would be the anode?

18.31. The half-reactions and standard potentials for a nickel–metal hydride battery with a titanium–zirconium anode are as follows:

$$\text{Cathode:}\quad NiO(OH)(s) + H_2O(\ell) + e^- \rightarrow Ni(OH)_2(s) + OH^-(aq) \qquad E^\circ = 0.52\ V$$

$$\text{Anode:}\quad TiZr_2H(s) + OH^-(aq) \rightarrow TiZr_2(s) + H_2O(\ell) + e^- \qquad E^\circ = -0.80\ V$$

a. Write the overall cell reaction for this battery.
b. Calculate the standard cell potential.

*18.32. **Lithium-Ion Batteries** There are lithium-ion batteries that have cathodes composed of $FePO_4$ when fully charged.
a. What is the formula of the cathode when the battery is fully discharged?
b. Is Fe oxidized or reduced as the battery discharges?
c. Is the cell potential of a lithium-ion battery with an iron phosphate cathode likely to differ from one with a cobalt oxide cathode? Explain your answer.

Chemical Energy and Electrical Work

Concept Review

18.33. Equation 18.4 tells us that the electrical work done by a voltaic cell on its surroundings (w_{elec}) is equal to the quantity of electrical charge (C) the cell pumps through an external circuit times the cell potential (E_{cell}): $w_{elec} = -CE_{cell}$. Why is there a negative sign in the equation?

18.34. Some people see an analogy between a battery pumping electrons through a circuit and an actual pump pumping water through a pipe. In this analogy, what electrical quantities correspond to each of the following? (*Hint*: See Table 1.2.)
a. the volume of water flowing through the pipe
b. the pressure forcing the water through the pipe
c. the rate at which a certain volume of water flows through the pipe each second

Problems

18.35. The value of E°_{cell} for the following reaction is 0.500 V. What is the value of ΔG°_{cell}?

$$2\,Mn^{3+} + 2\,H_2O \rightarrow Mn^{2+} + MnO_2 + 4\,H^+$$

18.36. What is the value of ΔG°_{cell} for an electrochemical cell based on a cell reaction described by the following net ionic equation?

$$Mg + 2\,Cu^+ \rightarrow Mg^{2+} + 2\,Cu$$

18.37. For many years the 1.50 V batteries used to power flashlights were based on the following cell reaction:

$$Zn(s) + 2\,NH_4Cl(s) + 2\,MnO_2(s) \rightarrow Zn(NH_3)_2Cl_2(s) + Mn_2O_3(s) + H_2O(\ell)$$

What is the value of ΔG°_{cell}?

18.38. The first generation of laptop computers was powered by nickel–cadmium (nicad) batteries, which generated 1.20 V in accordance with the following cell reaction:

$$Cd(s) + 2\,NiO(OH)(s) + 2\,H_2O(\ell) \rightarrow Cd(OH)_2(s) + 2\,Ni(OH)_2(s)$$

What is the value of ΔG°_{cell}?

18.39. The cells in the nickel–metal hydride battery packs used in many hybrid vehicles produce 1.20 V in accordance with the following cell reaction:

$$MH(s) + NiO(OH)(s) \rightarrow M(s) + Ni(OH)_2(s)$$

What is the value of ΔG°_{cell}?

18.40. A cell in a lead–acid battery delivers exactly 2.00 V of cell potential in accordance with the following cell reaction:

$$Pb(s) + PbO_2(s) + 2\,H_2SO_4(aq) \rightarrow 2\,PbSO_4(s) + 2\,H_2O(\ell)$$

What is the value of ΔG°_{cell}?

A Reference Point: The Standard Hydrogen Electrode

Concept Review

18.41. What is the function of platinum in the standard hydrogen electrode?

18.42. Platinum is very expensive, so why is it used in standard hydrogen electrodes where the half-reaction is $2\,H^+(aq) + 2\,e^- \rightarrow H_2(g)$?

18.43. The potential of the standard hydrogen electrode (SHE) is the reference against which other half-reaction potentials are expressed. Why, then, is the SHE not widely used as a reference electrode in electrochemical cells?

*18.44. Suggest a replacement metal for platinum in the standard hydrogen electrode. Explain why you selected the metal you did.

Problems

18.45. An electrochemical cell consists of a standard hydrogen electrode and a second half-cell in which a magnesium electrode is immersed in a 1.00 *M* solution of Mg^{2+} ions.

a. What is the value of E_{cell}?
b. Which electrode is the anode?
c. Which is a product of the cell reaction: H^+ ions or H_2 gas?

18.46. An electrochemical cell consists of a standard hydrogen electrode and a second half-cell in which a cadmium electrode is immersed in a 1.00 *M* solution of Cd^{2+} ions.

a. What is the value of E_{cell}?
b. Which electrode is the anode?
c. Which is a product of the cell reaction: Cd^{2+} ions or Cd metal?

The Effect of Concentration on E_{cell}

Concept Review

18.47. Why does the operating cell potential of most batteries change little until the battery is nearly discharged?

18.48. Does the potential of the standard hydrogen electrode increase, decrease, or stay the same if it contains 1.0 *M* HBr instead of 1.0 *M* HCl?

Problems

18.49. **Glucose Metabolism** Muscle cells convert glucose to pyruvate through a multistep process called *glycolysis*. During periods of strenuous exercise, when cellular oxygen levels are low, pyruvate is converted to lactate using the reduced form of nicotinamide adenine dinucleotide (NAD^H) in a redox reaction (Figure P18.49). The standard potentials for the reduction of the oxidized form of nicotinamide adenine dinucleotide (NAD^+) and pyruvate are as follows:

$$NAD^+(aq) + 2\,H^+(aq) + 2\,e^- \rightarrow NADH(aq) \qquad E^\circ = -0.320\text{ V}$$

$$\text{pyruvate}(aq) + 2\,H^+(aq) + 2\,e^- \rightarrow \text{lactate}(aq) \qquad E^\circ = -0.190\text{ V}$$

a. Calculate the standard potential for the reaction in Figure P18.49.
b. Calculate the equilibrium constant for the reaction at 25°C.

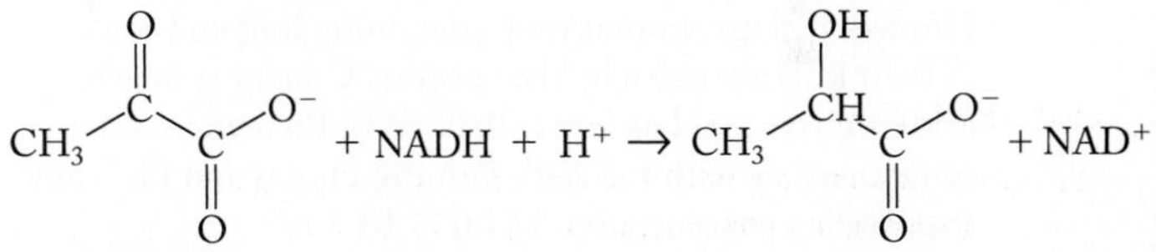

$$CH_3C(=O)COO^- + NADH + H^+ \rightarrow CH_3CH(OH)COO^- + NAD^+$$

Pyruvate — Lactate

FIGURE P18.49

18.50. **Citric Acid Cycle** Pyruvate can also enter the multistep citric acid cycle (also called the *Krebs cycle*), which is a pathway for oxidizing pyruvate to CO_2 and H_2O. One of the first steps in the citric acid cycle is the redox reaction shown in Figure P18.50 that converts pyruvate to acetyl-CoA using the biological reducing agent nicotinamide adenine dinucleotide (NADH) and the coenzyme CoA–SH.

a. Calculate the standard potential for the reaction in Figure P18.50 given that $\Delta G^\circ = -33.4$ kJ/mol for the reaction.
b. Given the reduction potential for NAD^+,

$$NAD^+(aq) + 2\,H^+(aq) + 2\,e^- \rightarrow NADH(aq) \qquad E^\circ = -0.320\text{ V}$$

calculate the reduction potential for acetyl-CoA:

$$\text{acetyl-CoA} + CO_2 + 2\,e^- \rightarrow \text{pyruvate}(aq) \qquad E^\circ = ?\text{ V}$$

c. Calculate the equilibrium constant for the reaction at 25°C.

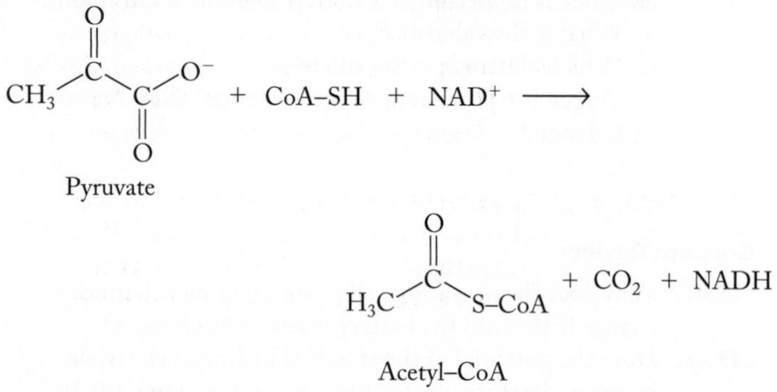

$$CH_3C(=O)C(=O)O^- + CoA{-}SH + NAD^+ \longrightarrow H_3C{-}C(=O){-}S{-}CoA + CO_2 + NADH$$

Pyruvate

Acetyl–CoA

FIGURE P18.50

18.51. Permanganate ion oxidizes sulfite to sulfate in basic solution as follows:

$$2\ MnO_4^-(aq) + 3\ SO_3^{2-}(aq) + H_2O(\ell) \rightarrow 2\ MnO_2(s) + 3\ SO_4^{2-}(aq) + 2\ OH^-(aq)$$

Determine the potential for the reaction (E_{rxn}) at 25°C when the concentrations of the reactants and products are as follows: $[MnO_4^-] = 0.250\ M$, $[SO_3^{2-}] = 0.425\ M$, $[SO_4^{2-}] = 0.075\ M$, and $[OH^-] = 0.0200\ M$. Will the value of E_{rxn} increase or decrease as the reaction proceeds?

*18.52. A *concentration cell* can be constructed by using the same half-reaction for both the cathode and anode. What is the value of E_{cell} for a concentration cell that combines copper electrodes in contact with 0.35 *M* copper(II) nitrate and 0.00075 *M* copper(II) nitrate solutions?

18.53. California Condors The continued recovery of endangered California condors is threatened by lead poisoning. Hunters still use ammunition containing lead and some of their kills are eaten by the condors. Condor stomach acid can have a pH as low as 0.80. Will Pb metal react spontaneously with the acid, forming H_2 gas and Pb^{2+} ions that reach a concentration of $1.0 \times 10^{-6}\ M$?

18.54. Chlorine dioxide (ClO_2) is produced by the following reaction of chlorate (ClO_3^-) with Cl^- in acid solution:

$$2\ ClO_3^-(aq) + 2\ Cl^-(aq) + 4\ H^+(aq) \rightarrow 2\ ClO_2(g) + Cl_2(g) + 2\ H_2O(\ell)$$

a. Determine $E°$ for the reaction.
b. The reaction produces a mixture of gases in the reaction vessel in which $P_{ClO_2} = 2.0$ atm; $P_{Cl_2} = 1.00$ atm. Calculate $[ClO_3^-]$ if, at equilibrium ($T = 25°C$), $[H^+] = [Cl^-] = 10.0\ M$.

Relating Battery Capacity to Quantities of Reactants

Concept Review

18.55. One 12 V lead–acid battery has a higher ampere-hour rating than another. Which of the following parameters are likely to be different for the two batteries?

a. individual cell potentials
b. anode half-reactions
c. total masses of electrode materials
d. number of cells
e. electrolyte composition
f. combined surface areas of their electrodes

18.56. Suppose a voltaic cell based on the Cu/Zn cell reaction

$$Zn(s) + Cu^{2+}(aq) \rightarrow Cu(s) + Zn^{2+}(aq)$$

contains exactly 1 mole of each reactant and product and a second cell based on the Cd/Cu cell reaction

$$Cd(s) + Cu^{2+}(aq) \rightarrow Cu(s) + Cd^{2+}(aq)$$

also has exactly 1 mole of each reactant and product. Which of the following statements about these two cells is true?

a. Their cell potentials are the same.
b. The masses of their electrodes are the same.
c. The quantities of electrical charge that they can produce are the same.
d. The quantities of electrical energy that they can produce are the same.

Problems

18.57. Which of the following voltaic cells will produce the greater quantity of electrical charge per gram of anode material?

$$Cd(s) + 2\ NiO(OH)(s) + 2\ H_2O(\ell) \rightarrow 2\ Ni(OH)_2(s) + Cd(OH)_2(s)$$

or

$$4\ Al(s) + 3\ O_2(g) + 6\ H_2O(\ell) + 4\ OH^-(aq) \rightarrow 4\ Al(OH)_4^-(aq)$$

18.58. Which of the following voltaic cells will produce the greater quantity of electrical charge per gram of anode material?

$$Zn(s) + MnO_2(s) + 2\ H_2O(\ell) \rightarrow Zn(OH)_2(s) + Mn(OH)_2(s)$$

or

$$Li(s) + MnO_2(s) \rightarrow LiMnO_2(s)$$

***18.59.** Which of the following voltaic cells delivers more electrical energy per gram of anode material at 25°C?

$$Zn(s) + 2\ NiO(OH)(s) + 2\ H_2O(\ell) \rightarrow 2\ Ni(OH)_2(s) + Zn(OH)_2(s) \qquad E°_{cell} = 1.20\ V$$

or

$$Li(s) + MnO_2(s) \rightarrow LiMnO_2(s) \qquad E°_{cell} = 3.15\ V$$

*18.60. Which of the following voltaic cell reactions delivers more electrical energy per gram of anode material at 25°C?

$$Zn(s) + Ni(OH)_2(s) \rightarrow Zn(OH)_2(s) + Ni(s) \qquad E°_{cell} = 1.50\ V$$

or

$$2\ Zn(s) + O_2(g) + H_2O(\ell) \rightarrow 2\ Zn(OH)_2(s) \qquad E°_{cell} = 2.08\ V$$

Corrosion: Unwanted Electrochemical Reactions

Concept Review

18.61. When the iron skeleton of the Statue of Liberty was replaced with stainless steel, the asbestos mats that had separated the skeleton from the copper exterior were replaced with Teflon spacers. Why was Teflon a good choice?

18.62. What does a sacrificial anode do to protect a metal structure, and why is the process called *cathodic* protection?

18.63. **Aquarium Windows** The windows of the giant ocean tank at the New England Aquarium (Figure P18.63) are held in place with aluminum frames. What would be a good material to use to make sacrificial anodes for the frames?

FIGURE P18.63

18.64. **Corrosion of Copper Pipes** The copper pipes frequently used in household plumbing may corrode and eventually leak. The corrosion reaction is believed to involve the formation of copper(I) chloride:

$$2\ Cu(s) + Cl_2(aq) \rightarrow 2\ CuCl(s)$$

a. Write balanced equations for the half-reactions in this redox reaction.

b. Calculate E°_{rxn} and ΔG°_{rxn} for the reaction.

Electrolytic Cells and Rechargeable Batteries

Concept Review

18.65. The positive terminal of a voltaic cell is the cathode. However, the cathode of an electrolytic cell is connected to the negative terminal of a power supply. Explain this difference in polarity.

18.66. The anode in an electrochemical cell is defined as the electrode where oxidation takes place. Why is the anode in an electrolytic cell connected to the positive (+) terminal of an external supply, whereas the anode in a voltaic cell battery is connected to the negative (−) terminal?

18.67. The salts obtained from the evaporation of seawater can be a source of halogens, principally Cl_2 and Br_2, through the electrolysis of the molten alkali metal halides. As the potential of the anode in an electrolytic cell is increased, which of these two halogens forms first?

18.68. **Quantitative Analysis** Electrolysis can be used to determine the concentration of Cu^{2+} in a given volume of solution by electrolyzing the solution in a cell equipped with a platinum cathode. If all the Cu^{2+} is reduced to Cu metal at the cathode, the increase in mass of the electrode provides a measure of the concentration of Cu^{2+} in the original solution. To ensure the complete (99.99%) removal of the Cu^{2+} from a solution in which $[Cu^{2+}]$ is initially about 1.0 *M*, will the potential of the cathode (vs. SHE) have to be more or less negative than 0.34 V (the standard potential for $Cu^{2+} + 2\ e^- \rightarrow Cu$)?

Problems

18.69. A high school chemistry student wants to demonstrate how water can be separated into hydrogen and oxygen by electrolysis. She knows that the reaction will proceed more rapidly if an electrolyte is added to the water. She has access to 2.00 *M* solutions of H_2SO_4, HBr, NaI, Na_2SO_4, and Na_2CO_3. Which one(s) should she use? Explain your selection(s).

18.70. A battery charger used to recharge the NiMH batteries used in a digital camera can deliver as much as 0.75 A of current to each battery. If it takes 100 min to recharge one battery, how many grams of $Ni(OH)_2$ are oxidized to NiO(OH)?

18.71. A NiMH battery containing 4.10 g of NiO(OH) was 75% discharged when it was connected to a charger with an output of 2.00 A at 1.3 V. How long does it take to recharge the battery?

*18.72. How long does it take to deposit a coating of gold 1.00 μm thick on a disk-shaped medallion 2.0 cm in diameter and 3.0 mm thick at a constant current of 45 A? The density of gold is 18.3 g/cm³. The gold solution contains gold(III).

***18.73.** **Oxygen Supply in Submarines** Nuclear submarines can stay under water nearly indefinitely because they can produce their own oxygen by the electrolysis of water.

a. How many liters of O_2 at 25°C and 1.00 bar are produced in 1 hour in an electrolytic cell operating at a current of 0.025 A?

b. Could seawater be used as the source of oxygen in this electrolysis? Explain why or why not.

18.74. In the electrolysis of water, how long will it take to produce 1.00×10^2 L of H_2 at STP (273 K and 1.00 atm) by using an electrolytic cell through which the current is 52 mA?

18.75. Calculate the minimum (least negative) cathode potential (versus SHE) needed to begin electroplating nickel from 0.35 *M* Ni^{2+} onto a piece of iron.

*18.76. What is the minimum (least negative) cathode potential (versus SHE) needed to electroplate silver onto cutlery in a solution of Ag^+ and NH_3 in which most of the silver ions are present as the complex, $Ag(NH_3)_2^+$, and the concentration of $Ag^+(aq)$ is only 2.50×10^{-4} *M*?

Fuel Cells and Flow Batteries

Concept Review

18.77. Describe two advantages of hybrid (gasoline engine–electric motor) power systems over all-electric systems based on fuel cells. Describe two disadvantages.

18.78. Describe three factors limiting widespread use of cars powered by fuel cells.

18.79. Methane can serve as the fuel for electric cars powered by fuel cells. Carbon dioxide is a product of the fuel cell reaction. All cars powered by internal combustion engines burning natural gas (mostly methane) produce CO_2. Why are electric vehicles powered by fuel cells likely to produce less CO_2 per mile?

18.80. To make the refueling of fuel cells easier, several manufacturers offer converters that turn readily available fuels—such as natural gas, propane, and methanol—into H_2 for the fuel cells and CO_2. Although vehicles with such power systems are not truly "zero emission," they still offer significant environmental benefits over vehicles powered by internal combustion engines. Describe a few of those benefits.

Problems

18.81. Fuel cells with molten alkali metal carbonates as electrolytes can use methane as a fuel. The methane is first converted into hydrogen in a two-step process:

$$CH_4(g) + H_2O(g) \rightarrow CO(g) + 3\ H_2(g)$$

$$CO(g) + H_2O(g) \rightarrow H_2(g) + CO_2(g)$$

a. Assign oxidation numbers to carbon and hydrogen in the reactants and products.
b. Using the standard free energy of formation values in Table A4.3 (Appendix 4), calculate the standard free-energy changes in the two reactions and the overall $\Delta G°$ for the formation of $H_2 + CO_2$ from methane and steam.

*18.82. A direct methanol fuel cell uses the oxidation of methanol by oxygen to generate electrical energy. The overall reaction,

$$CH_3OH(\ell) + \tfrac{3}{2}O_2(g) \rightarrow CO_2(g) + 2\ H_2O(\ell)$$

has a $\Delta G°$ value of −702.4 kJ/mol of methanol oxidized. What is the standard cell potential for this fuel cell?

18.83. There is interest in developing redox flow batteries based on the reduction of VO_2^{+} to VO^{2+}:

$$VO_2^{+}(aq) + 2\ H^{+}(aq) + e^{-} \rightarrow VO^{2+}(aq) + H_2O(l) \qquad E° = 0.991\ V$$

Which combination of this half-reaction with another involving the oxidation of Cr^{3+}, Sn^{2+}, or V^{2+} ions will produce the largest electrochemical potential, $E°_{cell}$?

18.84. Which combination of anode reactions from Problem 18.83 and the following cathode reactions in a redox flow battery would lead to the largest electrochemical potential, $E°_{cell}$?
a. reduction of oxygen in alkaline solution
b. reduction of bromine
c. reduction of Ce^{4+} to Ce^{3+}

Additional Problems

18.85. Calculate the E_{cell} value at 298 K for the cell based on the reaction,

$$Fe^{3+}(aq) + Cu^{+}(aq) \rightarrow Fe^{2+}(aq) + Cu^{2+}(aq)$$

when $[Fe^{3+}] = [Cu^{+}] = 1.50 \times 10^{-3}\ M$ and $[Fe^{2+}] = [Cu^{2+}] = 2.5 \times 10^{-4}\ M$.

18.86. Calculate the E_{cell} value at 298 K for the cell based on the reaction,

$$Cu(s) + 2\ Ag^{+}(aq) \rightarrow Cu^{2+}(aq) + 2\ Ag(s)$$

when $[Ag^{+}] = 2.56 \times 10^{-3}\ M$ and $[Cu^{2+}] = 8.25 \times 10^{-4}\ M$.

18.87. Using the appropriate standard potentials in Appendix 6, determine the equilibrium constant for the following reaction at 298 K:

$$Fe^{3+}(aq) + Cr^{2+}(aq) \rightarrow Fe^{2+}(aq) + Cr^{3+}(aq)$$

18.88. Using the appropriate standard potentials in Appendix 6, determine the equilibrium constant at 298 K for the following reaction between MnO_2 and Fe^{2+} in acid solution:

$$4\ H^{+}(aq) + MnO_2(s) + 2\ Fe^{2+}(aq) \rightarrow Mn^{2+}(aq) + 2\ Fe^{3+}(aq) + 2\ H_2O(\ell)$$

***18.89.** **Electrolysis of Seawater** Magnesium metal is obtained by the electrolysis of molten Mg^{2+} salts from evaporated seawater.
a. Would elemental Mg form at the cathode or anode?
b. Do you think the principal ingredient in sea salt (NaCl) would need to be separated from the Mg^{2+} salts before electrolysis? Explain your answer.
c. Would electrolysis of an aqueous solution of $MgCl_2$ also produce elemental Mg?
d. If your answer to part (c) was no, what would be the products of electrolysis?

*18.90. **Silverware Tarnish** Low concentrations of hydrogen sulfide in air react with silver to form Ag_2S, more familiar to us as tarnish. Silver polish contains aluminum metal powder in a basic suspension.
a. Write a balanced net ionic equation for the redox reaction between Ag_2S and Al metal that produces Ag metal and $Al(OH)_3$.
b. Calculate $E°$ for the reaction.

18.91. A magnesium battery can be constructed from an anode of magnesium metal and a cathode of molybdenum sulfide, Mo_3S_4. The standard reduction potentials of the electrode half-reactions are

$$Mg^{2+}(aq) + 2\ e^{-} \rightarrow Mg(s) \qquad E° = -2.37\ V$$

$$Mg^{2+}(aq) + Mo_3S_4(s) + 2\ e^{-} \rightarrow MgMo_3S_4(s) \qquad E° = ?$$

a. If the standard cell potential for the battery is 1.50 V, what is the value of $E°$ for the reduction of Mo_3S_4?
b. What are the apparent oxidation states of Mo in Mo_3S_4 and in $MgMo_3S_4$?
*c. The electrolyte in the battery contains a complex magnesium salt, $Mg(AlCl_3CH_3)_2$. Why is it necessary to include Mg^{2+} ions in the electrolyte?

18.92. Suppose there were a scale for expressing electrode potentials in which the standard potential for the reduction of water in base,

$$2\ H_2O(\ell) + 2\ e^{-} \rightarrow H_2(g) + 2\ OH^{-}(aq)$$

is assigned an $E°$ value of 0.000 V. How would the standard potential values on this new scale differ from those in Appendix 6?

***18.93.** **Clinical Chemistry** The concentrations of Na^{+} ions in red blood cells (11 mM) and in the surrounding plasma (140 mM) are quite different. Calculate the potential difference across the cell membrane because of this concentration gradient at 37°C.

*18.94. **Waterline Corrosion** The photo in Figure P18.94 shows a phenomenon known as waterline corrosion. Assuming the oxidizing agent in the corrosion process is O_2, propose a reason why metal pilings such as these tend to corrode the most at the waterline and corrode less at heights above and below the waterline.

FIGURE P18.94

18.95. The Particulate Review exercise at the beginning of this chapter includes a photograph of a rusting automobile. Our daily experience with rust tells us that wet iron surfaces are more likely to rust. The following equations describe the rusting of iron:

$$\text{(i) } 4\,Fe(s) + 3\,O_2(g) + 6\,H_2O(\ell) \rightarrow 4\,Fe(OH)_3(s)$$

$$\text{(ii) } 2\,Fe(OH)_3(s) \rightarrow Fe_2O_3(s) + 3\,H_2O(\ell)$$

a. Identify the half-reactions for the overall process in equation (i).
b. Calculate $E°$ for reaction (i) given $E° = -0.775$ V for the reduction of $Fe(OH)_3(s)$ to iron metal.
c. Calculate ΔG and $E°$ for the process in the Particulate Review exercise:

$$4\,Fe(s) + 3\,O_2(g) \rightarrow 2\,Fe_2O_3(s)$$

18.96. **Rusting Shipwreck** The *Titanic* sank in the waters of the North Atlantic on its maiden voyage in 1912. When the wreck was finally located in 1985, much of the vessel was covered by *rusticles*, which have the formula FeO(OH) (Figure P18.96). Reactions responsible for the formation of FeO(OH) include:

$$\text{(i) } 2\,Fe(s) + O_2(g) + 2\,H_2O(\ell) \rightarrow 2\,Fe(OH)_2(s)$$

$$\text{(ii) } 4\,Fe(OH)_2(s) + O_2(g) + 2\,H_2O(\ell) \rightarrow 4\,Fe(OH)_3(s)$$

$$\text{(iii) } Fe(OH)_3(s) \rightarrow FeO(OH)(s) + H_2O(\ell)$$

FIGURE P18.96

a. Which of these are redox reactions?
b. Use the following half-reaction and $E°$ value to calculate $E°_{cell}$ for reaction (ii):

$$Fe(OH)_3(s) + e^- \rightarrow Fe(OH)_2(s) + OH^-(aq) \qquad E° = -0.56\text{ V}$$

*18.97. **Olympic Games** The diving competition at the 2016 Olympic Games in Rio de Janeiro was plagued by green, turbid, and odorous water in the diving pool (Figure P18.97). Chlorine was used as a disinfectant in the pool water. One hypothesis for the green color is that 160 L of a 30% hydrogen peroxide solution (9.8 M H_2O_2) was added to the pool by mistake, deactivating the chlorine disinfectant and allowing green algae to grow.

a. When chlorine dissolves in slightly alkaline pool water, it hydrolyzes, forming hypochlorite and chloride ions:

$$Cl_2(g) + 2\,OH^-(aq) \rightarrow OCl^-(aq) + Cl^-(aq) + H_2O(\ell)$$

Use the appropriate half-reactions in Appendix 6 to calculate $E°_{rxn}$ for the reaction between OCl^- ions and H_2O_2.

b. Calculate the value of K for the reaction between OCl^- ions and H_2O_2.
c. If the swimming pool had dimensions of 25.0 m × 25.0 m × 10.0 m and contained 2.0 mg/L of OCl^- ions before H_2O_2 was added, what was the concentration of OCl^- ions after the reaction with H_2O_2?

FIGURE P18.97

18.98. **Flint, Michigan, Water Supply** Flint, Michigan, made headlines in 2017 when the municipal water supply was found to contain dangerous levels of dissolved lead(II) ion. The problem arose when the city began using water from the Flint River in its treatment plant and made changes in the purification process. The source of the lead turned out to be the lead and iron in the pipes.

a. One source of $Pb^{2+}(aq)$ is the reaction between dissolved oxygen and Pb in pipes. Write a balanced redox equation for this reaction in acidic solution.
b. Calculate $E°_{cell}$ for the reaction in part (a) and determine whether it is spontaneous at 298 K.
c. Is the reaction more or less favorable in basic solution?
d. The pH of the water in Flint decreased from 8.0 in December 2014 to 7.3 in August 2015. How does this affect the cell potential if the $[Pb^{2+}(aq)] = 2000$ ppb and $P_{O_2} = 0.21$ atm at 298 K? (*Hint*: Use Henry's law to calculate $[O_2]$.)
e. In principle, adding Cl^- to the water would precipitate any Pb^{2+} as $PbCl_2$ ($K_{sp} = 1.7 \times 10^{-5}$). What is the concentration of chloride in a saturated solution of $PbCl_2$? At what chloride concentration would $[Pb^{2+}]$ drop below 10 ppb?

19

Nuclear Chemistry

Applications in Science and Medicine

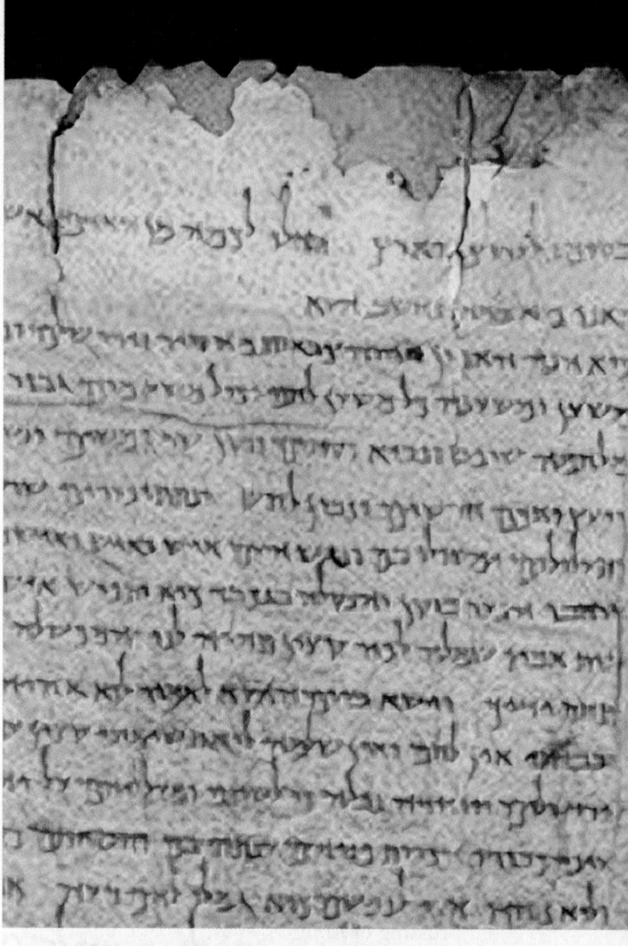

THE DEAD SEA SCROLLS These artifacts were discovered in the late 1940s and early 1950s in caves near the northwest coast of the Dead Sea. Made of parchment and papyrus, many contained passages from the Hebrew Bible, including one with the entire Book of Isaiah, a portion of which is shown. By measuring the natural radioactivity of the scrolls due to the presence of carbon-14, scientists were able to determine that the scrolls were written between 200 BCE and 100 CE.

PARTICULATE **REVIEW**

Isotopes Revisited

In Chapter 19 we investigate the stability and properties of radioactive nuclei. Three atomic nuclei are depicted here; each is drawn so that all the subatomic particles in the nucleus are visible.

- How many protons and how many neutrons does each nuclide contain?
- What is the mass number of each nuclide?
- Which two are isotopes of each other?

(Review Sections 2.2 and 2.3 if you need help.)

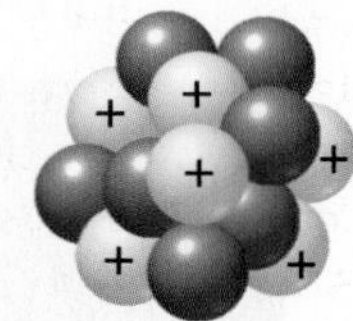

(Answers to Particulate Review questions are in the back of the book.)

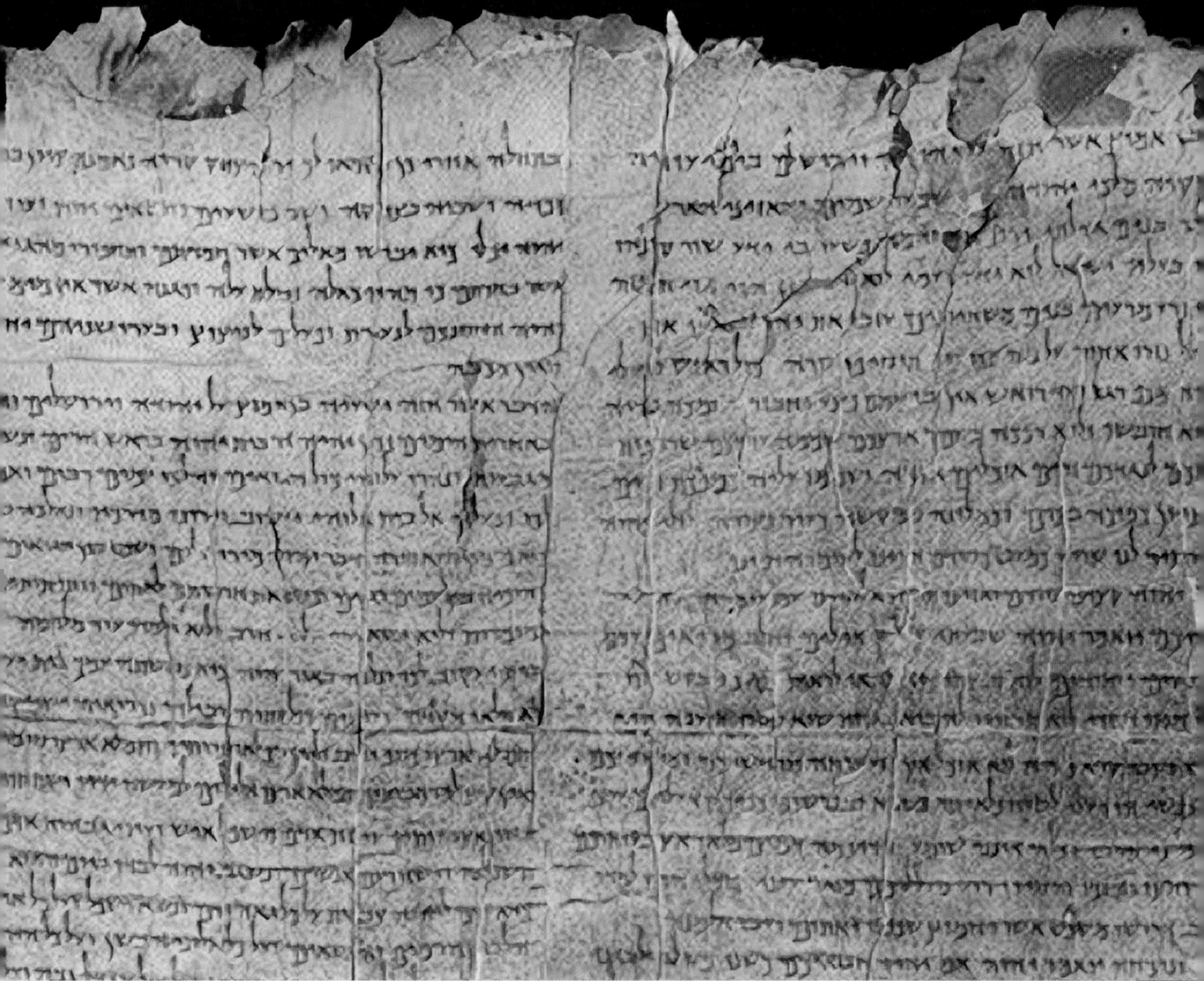

PARTICULATE **PREVIEW**

Unstable and Stable Nuclides

Radiocarbon dating measures the amount of carbon-14, the nuclide depicted here, in carbon-containing artifacts such as the Dead Sea Scrolls to determine the age of artifacts. Accurate measurements depend upon knowing how this unstable nuclide decays. As you read Chapter 19, look for ideas that will help you answer these questions.

- Does carbon-14 have more neutrons or protons?
- How might the ratio of neutrons to protons affect the decay of an unstable nuclide?
- What nuclide is produced when carbon-14 undergoes decay? What is the neutron-to-proton ratio of this nuclide?

Learning Outcomes

LO1 Relate the binding energy of atomic nuclei to their composition and stability

LO2 Predict the decay modes of radionuclides and the products they form
Sample Exercise 19.1

LO3 Compare the radioactivity of a sample of a radionuclide to the quantity of the nuclide in the sample
Sample Exercise 19.2

LO4 Compare the quantity of a radionuclide remaining after a defined decay time to the initial quantity and the half-life of the nuclide
Sample Exercise 19.3

LO5 Determine the age of a sample by radiometric dating
Sample Exercise 19.4

LO6 Describe the dangers of exposure to nuclear radiation and calculate effective radiation doses
Sample Exercise 19.5

LO7 Relate the energy change in a nuclear reaction to the masses of the products and reactants
Sample Exercise 19.6

19.1 Energy and Nuclear Stability

In 1898 Marie Curie separated and purified two new radioactive elements from uranium—polonium and radium—helping launch a new branch of chemistry now known as **nuclear chemistry**. Interest in the applications of radium grew rapidly. Paint containing radium was used to create luminous dials for watches and gauges. Ointments containing radium were prescribed as treatments for skin lesions. People were encouraged to visit health spas such as those in Saratoga Springs, New York, where the waters contained low concentrations of dissolved radium salts.

The proliferation of radium-containing products led to the discovery that nuclear radiation is dangerous. In one tragic example, young women hired to paint the dials of watches during World War I were instructed to "point" the tips of their brushes by licking them, inadvertently introducing radium into their bodies, where it concentrated in their teeth and bones. Within a few years, many of these young women developed bone cancer and died. Marie Curie herself succumbed to aplastic anemia caused by years of research with radioactive materials.

CONNECTION In Section 2.2 we described Henri Becquerel's discovery of radioactive uranium, and we saw how Ernest Rutherford's experiment, using α particles from the radioactive decay of uranium, led to the model of a dense atomic nucleus composed of protons and neutrons.

In recent years a better understanding of the biological impacts of radiation has led to the development of nuclear methods for diagnosing and treating disease that minimize the hazards of radiation to patients and those who treat them. Nuclear reactors provide doctors and scientists with radioactive atoms that decay over minutes to hours by predictable pathways. Selective uptake of these nuclides by different organs in the body allows doctors to evaluate organ function and prescribe treatment when function is impaired. Radiation from other nuclides that concentrate in cancerous tissues can be used, often in conjunction with other therapies, to destroy malignant tumors.

CONNECTION In Section 1.7 we described the rapid transformation of energy into matter that followed the Big Bang. The relationship between the energy and matter involved is described by Einstein's equation, $E = mc^2$.

Nuclear radiation results from reactions that take place in the unstable nuclei of some atoms. The energy released by these reactions can be used to help meet the world's need for electrical power and to diagnose and treat disease. However, these radioactive substances also pose a threat to human health, so shielding ourselves from them is essential. We begin by looking at the energy associated with nuclear transformations.

CONNECTION In Section 2.9 we explored the nucleosynthesis responsible for the formation of our early universe, including how early elements such as hydrogen and helium were transformed into more massive nuclei in the cores of the giant stars that populated the first generation of galaxies.

We first explored nuclear transformations in Section 2.9, focusing on the reactions that take place in stars. Beginning with the lightest elements, hydrogen and helium, stars ultimately produce nuclei as heavy as iron atoms, which contain 26 protons. Why does stellar nucleosynthesis end with the formation of iron nuclei? The answer has to do with the energy that binds the neutrons and protons together in atomic nuclei. This form of energy was discovered in the 1930s when scientists determined that the masses of stable nuclei are always less than the sum of the free masses of their nucleons. (The masses and symbols of subatomic particles are listed in **Table 19.1**.) For example, the total mass of the free nucleons in one ^{4}He nucleus—that is, 2 neutrons and 2 protons—is

TABLE 19.1 Symbols and Masses of Subatomic Particles and Small Nuclei

Particle	Symbol	Mass (kg)
Neutron	$^1_0$n	1.67493×10^{-27}
Proton	$^1_1$p or $^1_1$H	1.67262×10^{-27}
Electron (β particle)	$^{\ 0}_{-1}\beta$ or $^{\ 0}_{-1}$e	9.10939×10^{-31}
Deuteron	$^2_1$D or $^2_1$H	3.34358×10^{-27}
α Particle	$^4_2\alpha$ or $^4_2$He	6.64466×10^{-27}
Positron	$^0_1\beta$	9.10938×10^{-31}

$$\begin{aligned} \text{mass of 2 neutrons} &= 2(1.67493 \times 10^{-27}\ \text{kg}) \\ \text{mass of 2 protons} &= 2(1.67262 \times 10^{-27}\ \text{kg}) \\ \hline \text{total mass} &= 6.69510 \times 10^{-27}\ \text{kg} \end{aligned}$$

The difference between this value and the mass of a ^{4}He nucleus (from Table 19.1) is

$$\begin{aligned} &6.69510 \times 10^{-27}\ \text{kg} \\ -\ &6.64465 \times 10^{-27}\ \text{kg} \\ \hline &0.05045 \times 10^{-27}\ \text{kg} = 5.045 \times 10^{-29}\ \text{kg} \end{aligned}$$

This difference is called the **mass defect (Δ*m*)** of the nucleus. The energy equivalent to the mass defect of a nucleus is called its **binding energy (BE)** and can be calculated using Einstein's equation:

$$\begin{aligned} \text{BE} &= (\Delta m)c^2 \\ &= 5.045 \times 10^{-29}\ \text{kg} \times (2.998 \times 10^8\ \text{m/s})^2 \\ &= 4.534 \times 10^{-12}\ \text{kg} \cdot (\text{m/s})^2 = 4.534 \times 10^{-12}\ \text{J} \end{aligned}$$

This value may seem very small, but it is the binding energy of a single nucleus. Its value per mole of ^{4}He is equivalent to billions of kilojoules:

$$\frac{4.534 \times 10^{-12}\ \text{J}}{\cancel{\text{atom}}} \times \frac{6.022 \times 10^{23}\ \cancel{\text{atoms}}}{\text{mol}} \times \frac{1\ \text{kJ}}{1000\ \cancel{\text{J}}} = 2.730 \times 10^9\ \text{kJ/mol}$$

For comparison, the bond energy of the H—H bond in H_2 is only 436 kJ/mol, which means the binding energy in a mole of He atoms is more than 6,000,000 times greater!

Most of the stable nuclei with more nucleons than helium-4 have even larger binding energies. This makes sense because a nucleus with many protons in proximity must be held together by an enormous energy to overcome the coulombic repulsion these positively charged particles exert on one another. When close together, nucleons come under the influence of a fundamental force of nature known as the **strong nuclear force**. It operates only over very small distances, such as the diameters of atomic nuclei, but it is many times stronger than the repulsions the protons experience.

Despite the enormous difference in the magnitudes of the forces involved, binding energies and bond energies are similar in that their zero reference points are free nucleons and free atoms, respectively. When these particles come together, forming nuclei or molecules, they are attracted to each other by the strong nuclear force in nuclei or by the electrostatic attractions between negatively charged electrons and the positively charged nuclei of other atoms that are the source of covalent bond energies. To separate nuclei or molecules into their component nucleons or atoms requires adding energy to overcome these forces

nuclear chemistry a subdiscipline of chemistry that explores the properties and reactivity of atomic nuclei.

mass defect (Δ*m*) the difference between the mass of a nucleus and the masses of the individual nucleons that make it up.

binding energy (BE) the energy that would be released when free nucleons combine to form the nucleus of an atom; it is also the energy needed to split the nucleus into free nucleons.

strong nuclear force the fundamental force of nature that keeps nucleons together in atomic nuclei.

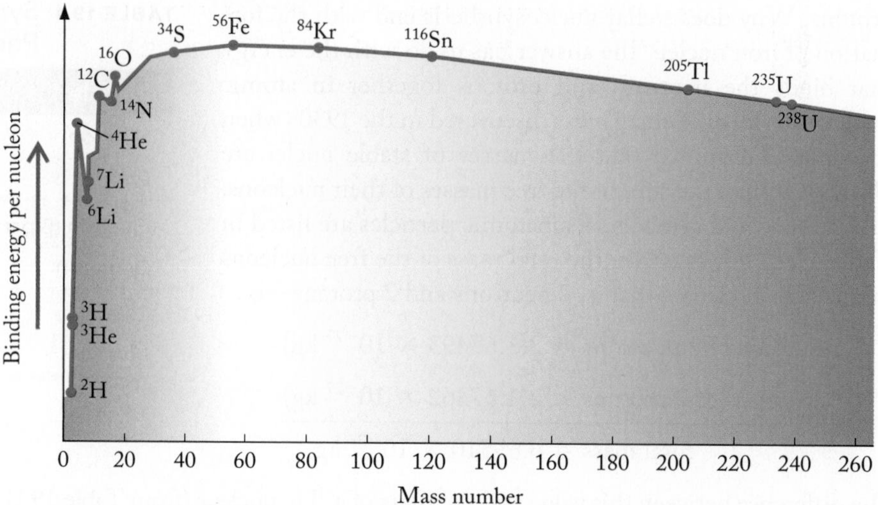

FIGURE 19.1 Trends in the binding energy per nucleon of atomic nuclei. For all nuclides up to ^{56}Fe, fusion reactions lead to products that have greater binding energies per nucleon than do the reactants. Therefore, these fusion reactions release energy. However, fusion of nuclei with atomic numbers greater than 26 produces nuclei that have less binding energy per nucleon than do the reactants, and thus consumes energy.

and return the particles to their free, zero-energy states. Thus, both binding energies and bond energies have positive values.

To identify trends in the stabilities of different nuclei, their binding energies are usually divided by the number of nucleons in each nucleus. A graph of binding energy per nucleon plotted against mass number is shown in **Figure 19.1**. These energy values reach a maximum with ^{56}Fe. This pattern means that ^{56}Fe is the most stable nuclide in the universe, which fits what we know about stellar nucleosynthesis. The nuclear furnaces of giant stars are fueled by the energy released when lighter nuclei fuse to form heavier nuclei, but only when the heavier ones that form are more stable than the lighter ones. Energy is not released if one of the reacting particles is ^{56}Fe or a heavier nuclide because, according to Figure 19.1, these heavier fusion products are *less stable*, not more stable. Their formation consumes energy instead of producing it. Thus, an aged giant star whose core has turned into iron as a result of multiple nuclear fusion processes has essentially run out of fuel. Without heating from its nuclear furnace, the star's fusion reactions cease and its enormous gravity forces it to collapse into itself. The result is rapid heating as the star's mass is compressed, which causes the star to explode, releasing elements from $Z = 1$ to $Z = 26$. In rare cases stars explode with so much energy that they synthesize and release elements with atomic numbers greater than 26. Such an event is called a supernova, such as "Tycho's supernova" shown in Figure 2.26.

CONCEPT **TEST**

Why does a negative change in mass when nucleons combine to form a nucleus produce a positive binding energy?

(Answers to Concept Tests are in the back of the book.)

CHEMTOUR
Modes of Radioactive Decay

19.2 Unstable Nuclei and Radioactive Decay

The values of the atomic masses and mass numbers of the elements in the periodic table indicate the ratios of neutrons to protons in the nuclei of their stable isotopes. The lighter elements have atomic masses that are about twice their atomic numbers

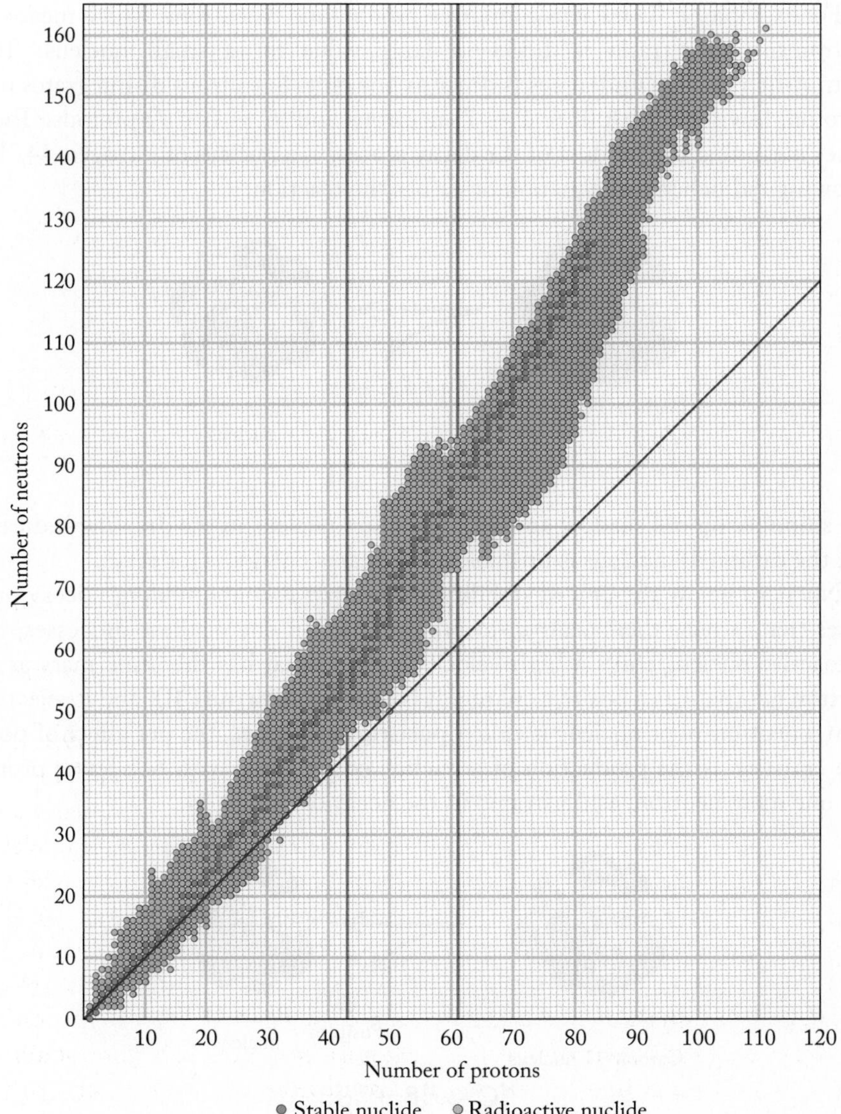

FIGURE 19.2 The belt of stability. Green dots represent stable combinations of protons and neutrons. Orange dots represent known radioactive (unstable) nuclides. Nuclides that fall along the purple line have equal numbers of neutrons and protons. There are no stable nuclides (no green dots) for $Z = 43$ (technetium) and $Z = 61$ (promethium), as indicated by the two vertical red lines.

STEPWISE ANIMATION
Belt of Stability

and have neutron-to-proton ratios close to unity. For example, ^{12}C has 6 neutrons and 6 protons, and most oxygen atoms have 8 neutrons and 8 protons.

With increasing values of Z, however, the ratio of neutrons to protons increases. This trend is illustrated in **Figure 19.2**, where the green dots represent combinations of neutrons and protons that form stable nuclides. The band of green dots runs diagonally through the graph, defining the **belt of stability**. Notice how the belt curves upward away from the purple straight line, which represents a neutron-to-proton ratio of 1. Based on this curvature, the neutron:proton ratio increases from about 1:1 for the lightest stable nuclides to about 1.5:1 for the most massive ones.

CONNECTION We encountered beta (β) decay, the spontaneous ejection of a β particle by a neutron-rich nucleus, in Section 2.9.

The nuclides represented by orange dots in Figure 19.2 are **radionuclides**. They are not stable but instead undergo **radioactive decay**, which is the spontaneous disintegration of radioactive nuclei, accompanied by the release of nuclear radiation. Modes of radioactive decay depend on whether a nuclide is above or below the belt of stability. Those above, such as carbon-14, are *neutron rich* and tend to undergo *beta (β) decay*: a **nuclear reaction** in which a neutron disintegrates, producing a proton that remains in the nucleus and a high-speed, high-energy electron, called a β particle, that is emitted from the atom.

belt of stability the region on a graph of number of neutrons versus number of protons that includes all stable nuclei.

radionuclide an unstable nuclide that undergoes radioactive decay.

radioactive decay the spontaneous disintegration of unstable particles accompanied by the release of radiation.

nuclear reaction a process that changes the number of neutrons or protons in the nucleus of an atom.

For example, a ^{14}C nucleus contains 6 protons and 8 neutrons, which means its neutron-to-proton ratio is 1.333—a large value for a small nucleus. This neutron-rich nucleus undergoes β decay as one of its 8 neutrons disintegrates into a proton, leaving it with (8 − 1 = 7) neutrons and (6 + 1 = 7) protons. These values mean that the product of the decay process is a nucleus of nitrogen-14. The following radiochemical equation describes the reaction:

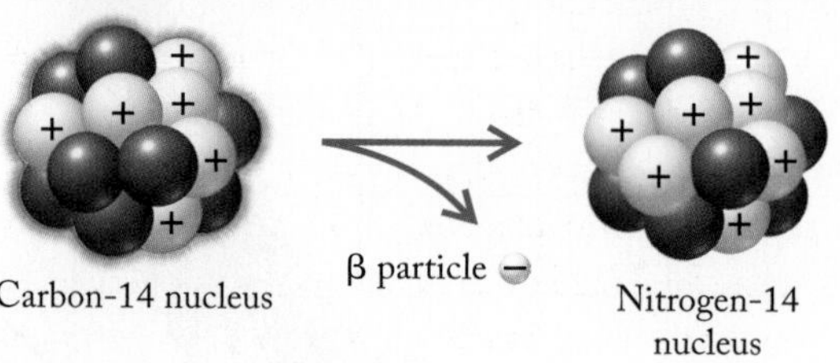

$$^{14}_{6}C \rightarrow {}^{14}_{7}N + {}^{\;0}_{-1}\beta$$

CONNECTION Section 2.9 explains how to write and balance nuclear equations.

The shimmering red shadow around the carbon nucleus in the drawing indicates that the carbon-14 nucleus is unstable and releases energy as it decays.

Nuclides below the belt of stability are *neutron poor* and undergo decay processes that *increase* their neutron-to-proton ratio. In one of these processes, the radioactive nucleus emits a high-velocity particle that has the same mass as an electron but that has a positive charge. It is called a **positron ($^{0}_{1}\beta$)**, and its ejection from a neutron-poor nucleus is called **positron emission**. The net effect of positron emission is the production of a nuclide of a nucleus with one fewer proton and one more neutron, as illustrated in the decay of carbon-11:

CHEMTOUR
Balancing Nuclear Equations

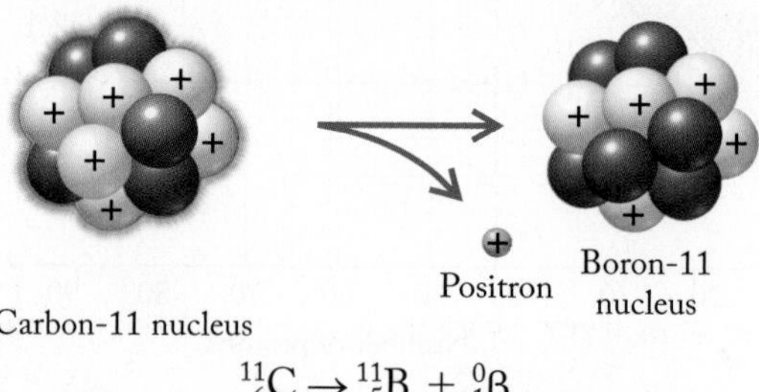

$$^{11}_{6}C \rightarrow {}^{11}_{5}B + {}^{0}_{1}\beta$$

Positrons belong to a group of subatomic particles that have the opposite charge but the same mass as particles typically found in atoms. In addition to positrons there are also protons with negative charges, called *antiprotons*. These charge opposites are particles of **antimatter**.

Particles of matter and their antimatter opposites are like mortal enemies. If they collide, they instantly annihilate each other. In their mutual destruction, they cease to exist as matter, and all of their mass is converted to energy in the form of two or more gamma (γ) rays:

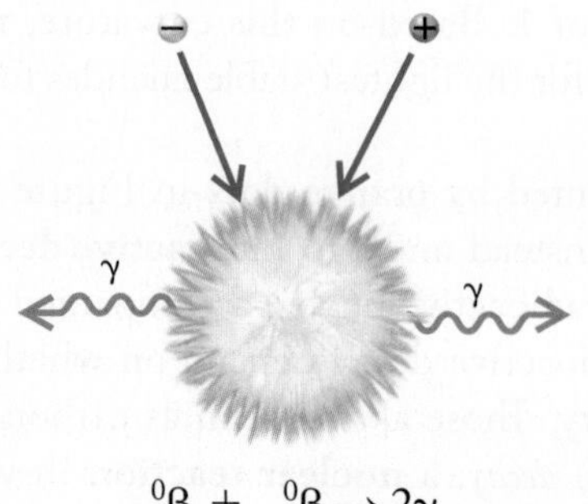

$$^{0}_{1}\beta + {}^{\;0}_{-1}\beta \rightarrow 2\gamma$$

The yellow "sunburst" here symbolizes the energy released in the process depicted.

positron ($^{0}_{1}\beta$) a particle with the mass of an electron but with a positive charge.

positron emission the spontaneous emission of a positron from a neutron-poor nucleus.

antimatter particles that are the charge opposites of normal subatomic particles.

electron capture a nuclear reaction in which a neutron-poor nucleus draws in one of its surrounding electrons, which transforms a proton in the nucleus into a neutron.

Emission of γ rays accompanies all nuclear reactions, not just positron emission. The γ rays represent quantities of energy that are equivalent to the loss in mass that occurs when reactants form products in nuclear reactions. They are generated by the nuclear furnaces of stars and permeate outer space. Those that reach Earth are absorbed by the gases in our atmosphere. In the process, molecular gases are broken up into their component atoms, and atomic nuclei may be broken up into subatomic particles.

CONNECTION Figure 7.7 reveals that γ rays are the highest energy (shortest wavelength) form of electromagnetic radiation.

There is another way to increase the neutron-to-proton ratio of a neutron-poor nucleus: it can capture one of the inner-shell electrons of its atom. When it does, the negatively charged electron combines with a positively charged proton. The product of this combination reaction is a neutron. The effect of this **electron capture** process on the nucleus is the same as positron emission: the number of protons *decreases* by one and the number of neutrons *increases* by one. When a nucleus of carbon-11 undergoes electron capture, the product is boron-11, the same nuclide formed in positron emission:

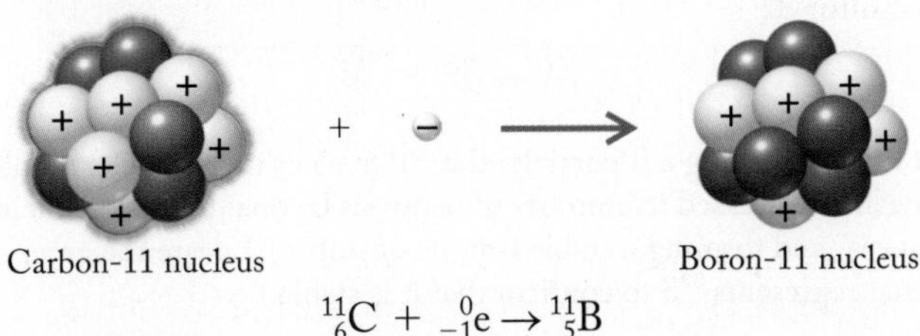

$$^{11}_{6}C + {}^{0}_{-1}e \rightarrow {}^{11}_{5}B$$

Table 19.2 summarizes the effects of being neutron rich, neutron poor, or neither on isotopes of carbon. Carbon has two stable isotopes: ^{12}C and ^{13}C. The isotopes with mass numbers *greater* than 13 are neutron rich and undergo beta decay, whereas those with mass numbers *less* than 12 are neutron poor and undergo either positron emission or electron capture.

TABLE 19.2 Isotopes of Carbon and Their Radioactive Decay Products

Name	Symbol	Mass (u)	Mode(s) of Decay	Half-Life	Natural Abundance (%)
Carbon-10	$^{10}_{6}C$		Positron emission	19.45 s	
Carbon-11	$^{11}_{6}C$		Positron emission, electron capture	20.3 min	
Carbon-12	$^{12}_{6}C$	12.00000		(Stable)	98.89
Carbon-13	$^{13}_{6}C$	13.00335		(Stable)	1.11
Carbon-14	$^{14}_{6}C$		β Decay	5730 yr	Trace
Carbon-15	$^{15}_{6}C$		β Decay	2.4 s	
Carbon-16	$^{16}_{6}C$		β Decay	0.74 s	

SAMPLE EXERCISE 19.1 Predicting the Modes and Products of Radioactive Decay **LO2**

Predict the mode of radioactive decay of ^{32}P, which is one of the most widely used radionuclides in biomedical research and treatment. Identify the nuclide that is produced in the decay process.

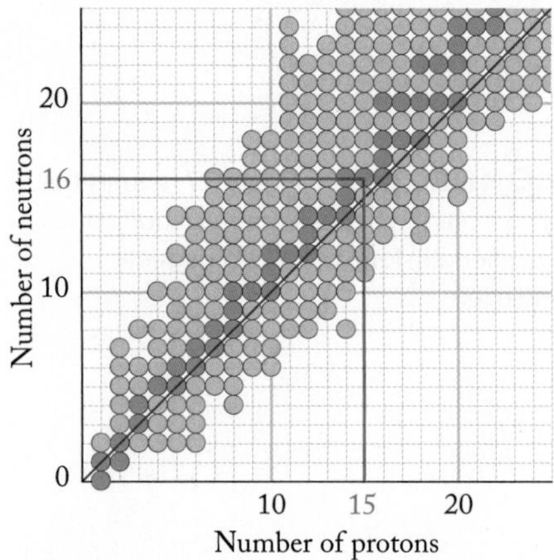

FIGURE 19.3 Nuclides in the belt of stability. The red lines at 16 neutrons and 15 protons intersect at the green dot that represents ^{31}P, the only stable isotope of phosphorus. The orange dot directly above ^{31}P is ^{32}P.

Collect, Organize, and Analyze We are asked to predict the mode of decay of a radionuclide, which depends on whether it is neutron rich or neutron poor. The mass number of ^{32}P is greater than the average atomic mass of all phosphorus atoms (30.974 u), which indicates that ^{32}P is neutron rich. Neutron-rich radioisotopes of lighter elements tend to undergo β decay in their nuclei to decrease the number of neutrons and increase the number of protons. Alternatively, we can consult the graph in **Figure 19.3**, which is a close up of the relevant portion of the belt of stability first shown in Figure 19.2. In Figure 19.3, ^{32}P (17 neutrons + 15 protons) is represented by the orange dot directly above the green dot for ^{31}P (the one and only stable phosphorus nuclide), which means that ^{32}P is radioactive and neutron rich.

Solve A β particle must be one product of the decay reaction, giving the incomplete nuclear equation:

$$^{32}_{15}P \rightarrow ? + ^{0}_{-1}\beta$$

The missing product must have an atomic number of 16 (so that the subscripts on the right side add up to 15), which makes it an isotope of S. Its mass number must be 32, so the product is sulfur-32:

$$^{32}_{15}P \rightarrow ^{32}_{16}S + ^{0}_{-1}\beta$$

Think About It By emitting a β particle, the ^{32}P nucleus increased its number of protons by one and decreased its number of neutrons by one, thereby reducing its neutron "richness" and forming a stable isotope of sulfur. (Locate the green dot in Figure 19.3 that represents ^{32}S to confirm that it is stable.)

Practice Exercise What is the mode of radioactive decay of ^{28}Al? Identify the nuclide produced by the decay process.

(Answers to Practice Exercises are in the back of the book.)

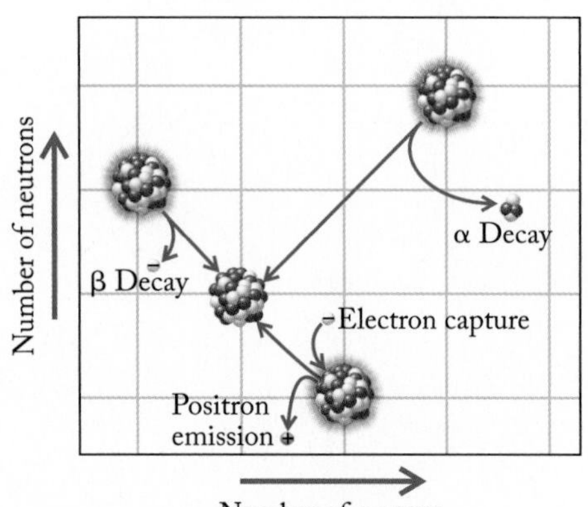

FIGURE 19.4 Radioactive decay results in predictable changes in the number of protons and neutrons in a nucleus. In α decay, the nucleus loses 2 neutrons and 2 protons, resulting in a decrease of 2 in atomic number and 4 in mass number. Beta decay leads to an increase of 1 proton at the expense of 1 neutron, so the atomic number increases by 1, but the mass number is unchanged. In positron emission and electron capture, the number of protons decreases by 1 and the number of neutrons increases by 1, so the atomic number decreases by 1, but the mass number remains the same.

All known nuclides with more than 83 protons are radioactive. Because there is no stable reference point in the pattern of green dots in Figure 19.2, it is hard to say whether any given nuclide with $Z > 83$ is neutron rich or neutron poor. We can make one general statement though: these most massive nuclides tend to undergo either β decay or **alpha (α) decay**. In α decay they produce a nuclide with two fewer protons and two fewer neutrons, as shown here for uranium-238:

$$^{238}_{92}U \rightarrow ^{234}_{90}Th + ^{4}_{2}\alpha$$

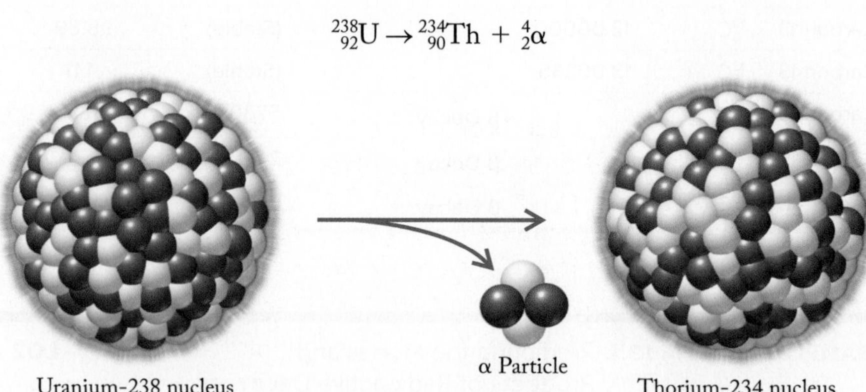

Figure 19.4 illustrates the changes in atomic number and mass number caused by the various modes of decay.

Among the most massive radionuclides, one radioactive decay process often leads to another in what is referred to as a *radioactive decay series*. Consider, for

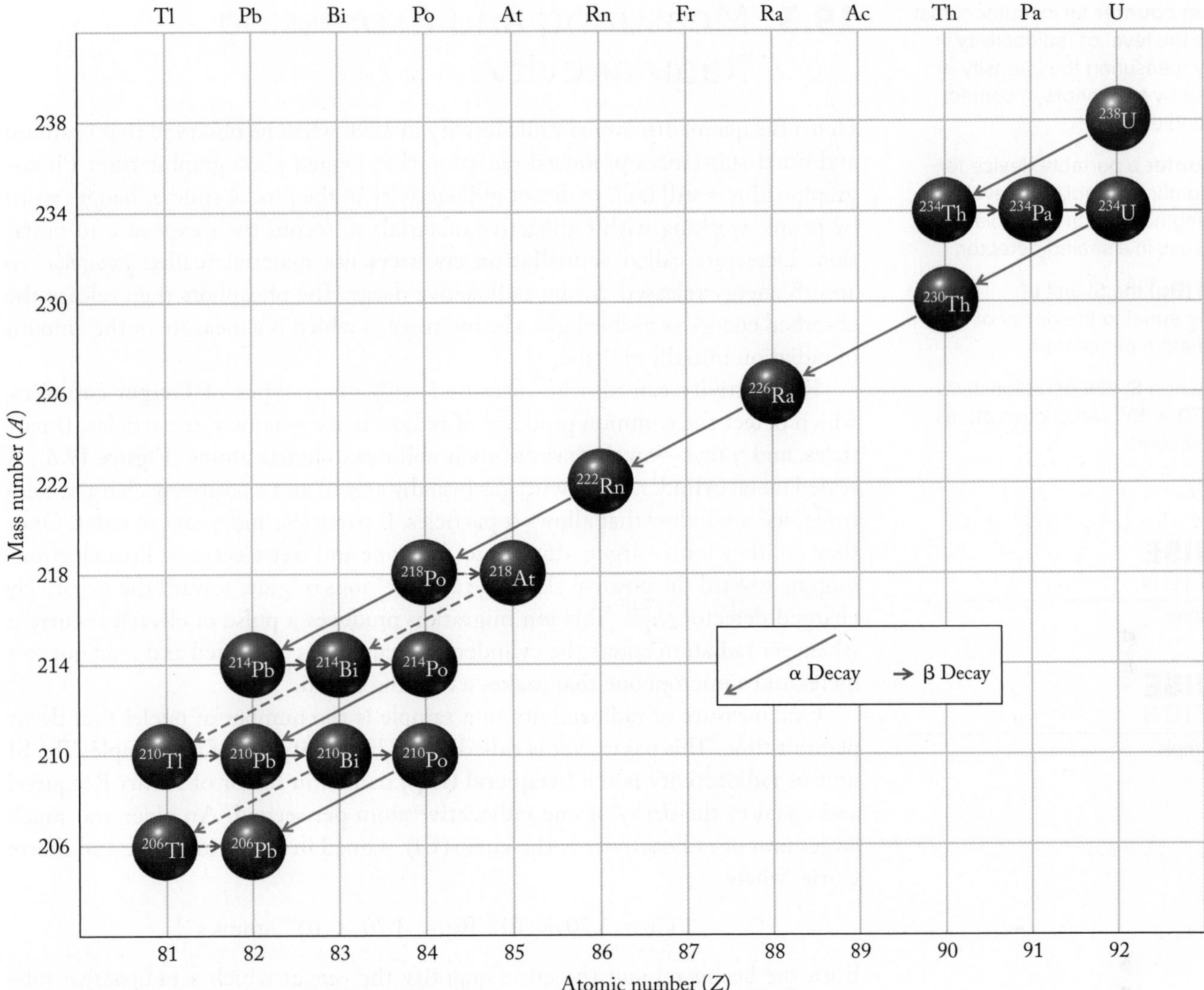

FIGURE 19.5 Uranium-238 radioactive decay series. The red diagonal arrows represent α decay events, whereas the blue horizontal ones represent β decay events. The dashed arrows are alternative pathways representing less than 1% of the decay events in this series. Whether decay proceeds by the solid-line pathways or the dashed-line pathways, the end product is always stable lead-206.

example, the decay series that begins with the α decay of ^{238}U to ^{234}Th (**Figure 19.5**). Thorium has no stable isotopes and undergoes two β decay steps to produce ^{234}U. In a series of subsequent α decay steps, ^{234}U turns into thorium-230, radium-226, radon-222, polonium-218, and finally lead-214. Although some isotopes of lead ($Z = 82$) are stable, ^{214}Pb is not one of them. Therefore, the radioactive decay series continues, as shown at the bottom left of Figure 19.5, and does not end until the stable nuclide ^{206}Pb is produced.

CONCEPT TEST

In the ^{238}U radioactive decay series, five *α* decay steps in a row transform ^{234}U into ^{214}Pb. Given the shape of the belt of stability, why does it make sense that the product of these *α* decay steps would be a neutron-rich nuclide that undergoes β decay?

alpha (α) decay a nuclear reaction in which an unstable nuclide spontaneously emits an alpha particle.

scintillation counter an instrument that determines the level of radioactivity in samples by measuring the intensity of light emitted by phosphors in contact with the samples.

Geiger counter a portable device for determining nuclear radiation levels by measuring how much the radiation ionizes the gas in a sealed detector.

becquerel (Bq) the SI unit of radioactivity equal to the decay of one radioactive atom per second.

curie (Ci) a non-SI unit of radioactivity in which 3.70×10^{10} radioactive atoms decay per second.

STEPWISE ANIMATION
Geiger Counter

STEPWISE ANIMATION
Activity Example

19.3 Measuring and Expressing Radioactivity

Henri Becquerel discovered radioactivity in 1896 when he observed that uranium and other substances produced radiation that fogged photographic film. Photographic film is still used to detect radioactivity in the film dosimeter badges worn by people working with radioactive materials to record their exposure to radiation. Detectors called **scintillation counters** use materials called *phosphors* to absorb energy released during radioactive decay. The phosphors then release the absorbed energy as visible light, the intensity of which is a measure of the amount of radiation initially emitted.

Radioactivity can also be measured with many types of **Geiger counters**, which detect the common products of radioactivity—namely, α particles, β particles, and γ rays—on the basis of their abilities to ionize atoms (**Figure 19.6**). A sealed metal cylinder, filled with gas (usually argon) and a positively charged electrode, has a window that allows α particles, β particles, and γ rays to enter. Once they do, they ionize argon atoms into Ar^+ ions and free electrons. Free electrons migrate toward the positive electrode and Ar^+ ions migrate toward the negatively charged detector shell. This ion migration produces a pulse of electrical current whenever radiation enters the cylinder. The current is amplified and read out to a meter and a microphone that makes a clicking sound.

One measure of radioactivity in a sample is the number of nuclei that decay per unit time. This parameter is called the radioactivity (A) of the sample. The SI unit of radioactivity is the **becquerel (Bq)**, named in honor of Henri Becquerel and equal to the decay of one radioactive atom per second. An older and much larger unit of radioactivity is the **curie (Ci)**, named in honor of Marie and Pierre Curie, where

$$1\ \text{Ci} = 3.70 \times 10^{10}\ \text{Bq} = 3.70 \times 10^{10}\ \text{atoms s}^{-1}$$

Both the becquerel and the curie quantify the *rate* at which a radioactive substance decays, which provides a measure of how radioactive a sample is. All radioactive decay processes follow first-order kinetics, in which the reaction rate is equal to the rate constant (k) times the concentration of a reactant (R):

$$\text{Rate} = k[\text{R}]$$

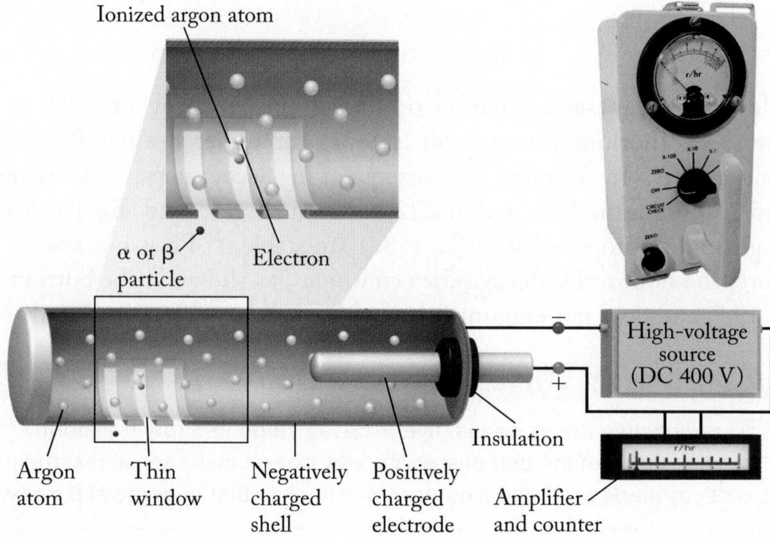

FIGURE 19.6 In a Geiger counter, a particle produced by radioactive decay passes through a thin window, usually made of beryllium or a plastic film. Inside the tube, the particle ionizes atoms of argon gas. The resulting argon cations migrate toward the negatively charged tube housing, and the electrons migrate toward a positive electrode, creating a pulse of current through the tube. The current pulses are amplified and recorded via a meter and a speaker that produces an audible "click" for each pulse.

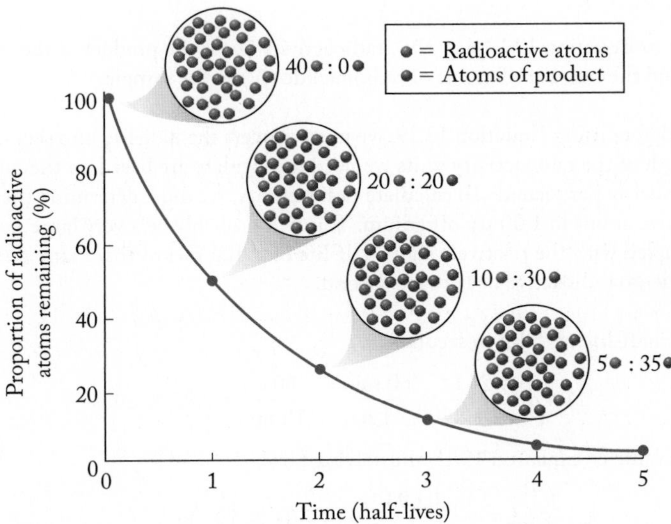

FIGURE 19.7 Radioactive decay follows first-order kinetics, which means, for example, that if a sample initially contains 40 radioactive atoms, it will contain only one-half that number after an interval equal to one half-life. Half of the remaining half, or 10 radioactive atoms, remain after two half-lives, and so on.

We refer to radioactivity (A) instead of reaction rate when describing radioactive decay, and to the number of atoms (N) of a radionuclide in a sample instead of their concentration:

$$A = kN \tag{19.1}$$

Because radioactivity (A) represents the number of radioactive atoms that decay per second, the units of the *decay rate constant* (k) are per second, which means radioactivity itself has units of atoms (that decay) per second, or atoms s^{-1}.

Because radioactive decay is a first-order reaction, the decay rate constant is related to the radionuclide's half-life (**Figure 19.7**) by Equation 13.19:

$$t_{1/2} = \frac{0.693}{k} \tag{13.19}$$

CONNECTION The *half-life* ($t_{1/2}$) is the time in the course of a chemical reaction during which the concentration of a reactant decreases by half (see Section 13.3). For a first-order reaction, $t_{1/2}$ is inversely proportional to the rate constant (k).

19.4 Calculations Involving Half-Lives of Radionuclides

CHEMTOUR Half-Life

In Section 19.4 we use first-order reaction kinetics to *calculate* the level of radioactivity in a sample initially and after it has decayed for a specified time and to calculate how long it takes for a sample to decay to a target percentage of its initial radioactivity. We start by calculating the radioactivity of a sample based on the quantity and half-life of a radionuclide in the sample.

SAMPLE EXERCISE 19.2 Calculating the Radioactivity of a Sample **LO3**

Radium-223 undergoes decay with a half-life of 11.4 days. What is the radioactivity of a sample that contains 1.00 μg of ^{223}Ra? Express your answer in becquerels and in curies.

Collect and Organize We are given the half-life and quantity of a radioactive substance and are asked to determine its radioactivity—that is, its rate of decay. The decay rate constant (k) is related to half-life by Equation 13.19:

$$t_{1/2} = \frac{0.693}{k}$$

According to Equation 19.1 ($A = kN$), radioactivity (A) is the product of the rate constant and the number of atoms of radionuclide (N) in the sample.

Analyze Before using Equation 13.19, we must convert the half-life into seconds because both of the radioactivity units we need to calculate are based on the number of atoms that decay per second. To calculate radioactivity, we must determine the number of radioactive atoms in 1.00 μg of radium. This will probably be a very large number, which, coupled with the relatively short half-life of ^{223}Ra, means that many nuclei should undergo radioactive decay each second.

Solve The half-life of ^{223}Ra in seconds is

$$11.4\ \text{d} \times \frac{24\ \text{h}}{1\ \text{d}} \times \frac{60\ \text{min}}{1\ \text{h}} \times \frac{60\ \text{s}}{1\ \text{min}} = 9.85 \times 10^5\ \text{s}$$

Using this value in Equation 13.19 and solving for k:

$$k = \frac{0.693}{9.85 \times 10^5\ \text{s}} = 7.04 \times 10^{-7}\ \text{s}^{-1}$$

The number of atoms (N) of ^{223}Ra is

$$N = 1.00\ \mu\text{g} \times \frac{1\ \text{g}}{10^6\ \mu\text{g}} \times \frac{1\ \text{mol}}{223\ \text{g}} \times \frac{6.022 \times 10^{23}\ \text{atoms}}{1\ \text{mol}} = 2.70 \times 10^{15}\ \text{atoms}$$

Inserting these values of k and N into Equation 19.1 gives us

$$A = kN = 7.04 \times 10^{-7}\ \text{s}^{-1} \times 2.70 \times 10^{15}\ \text{atoms} = 1.90 \times 10^9\ \text{atoms/s}$$

Because 1 Bq = 1 atom/s, the radioactivity of the sample is 1.90×10^9 Bq. Expressing radioactivity in curies:

$$\frac{1.90 \times 10^9\ \text{atoms}}{\text{s}} \times \frac{1\ \text{Ci}}{3.70 \times 10^{10}\ \text{atoms/s}} = 0.0514\ \text{Ci}$$

Think About It The large number of atoms that decay each second meets our expectation of a high level of radioactivity.

Practice Exercise In March 2011 an earthquake and tsunami off the coast of northeast Japan crippled nuclear reactors at a power station in Fukushima, Japan. The resulting explosions and fires released ^{133}Xe into the atmosphere. Determine the radioactivity of 1.00 μg of this radionuclide ($t_{1/2}$ = 5.25 days) in becquerels and in millicuries.

In Section 13.3 we used the integrated rate law for first-order reactions to develop Equation 13.18, which relates the concentration of reactant X, $[X]$, at any time t during the course of a first-order reaction to its initial concentration, $[X]_0$, and the rate constant (k) of the reaction:

$$\ln[\text{X}] = -kt + \ln[\text{X}]_0 \tag{13.18}$$

To apply this relationship to radioactive decay reactions, we replace the concentration terms with numbers of radioactive atoms at time 0 (N_0) and at any later time t (N_t):

$$\ln N_t = -kt + \ln N_0 \tag{19.2}$$

Frequently the kinetics of radioactive decay processes are expressed using their half-lives rather than rate constants, so we use Equation 13.19 to substitute $t_{1/2}$ for k in Equation 19.2:

$$\ln N_t = -0.693 \frac{t}{t_{1/2}} + \ln N_0 \tag{19.3}$$

The fraction $t/t_{1/2}$ represents the number of half-lives (including fractions of half-lives) that have passed since the sample contained N_0 atoms of the radionuclide. Sample Exercise 19.3 illustrates how useful Equation 19.3 can be.

SAMPLE EXERCISE 19.3 Calculating the Quantity of a Radionuclide Remaining in a Sample **LO4**

Free neutrons are radioactive, undergoing β decay with a half-life of 10.25 min. Starting with a population of 7.25×10^5 free neutrons, how many are left after 6.67 min?

Collect, Organize, and Analyze Equation 19.3 relates quantities of radioactive particles to decay times. In this problem the initial number of neutrons (N_0) is 7.25×10^5, $t = 6.67$ min, and $t_{1/2} = 10.25$ min. We need to solve for N_t. The value of t is less than $t_{1/2}$, so fewer than half of the initial number of neutrons will have decayed after 6.67 min. In other words, more than half of the neutrons will still be present.

Solve Substituting the data into Equation 19.3 and solving for N_t:

$$\ln N_t = -0.693\frac{t}{t_{1/2}} + \ln N_0 = -0.693\frac{6.67\ \cancel{\text{min}}}{10.25\ \cancel{\text{min}}} + \ln(7.25 \times 10^5) = 13.043$$

$$N_t = e^{13.043} = 4.62 \times 10^5 \text{ neutrons}$$

Think About It The value of N_t is reasonable because, as predicted, less than half of the initial quantity of free neutrons decayed in a time (6.67 min) that was less than the half-life.

Practice Exercise Cesium-131 is a short-lived radionuclide ($t_{1/2} = 9.7$ d) used to treat prostate cancer. If the therapeutic strength of the radionuclide is directly proportional to the number of nuclei present, how much therapeutic strength does a cesium-131 source lose over exactly 60 days? Express your answer as a percentage of the strength the source had at the beginning of the first day.

19.5 Radiometric Dating

Radiometric dating refers to methods for determining the age of objects on the basis of the tiny concentrations of radionuclides that occur naturally in them and the rates of radioactive decay of these nuclides. The concept originated in the early 1900s with the work of Ernest Rutherford, who recognized that radioactive decay processes have characteristic half-lives. Rutherford proposed to use this concept to determine the age of rocks and even the age of Earth itself. The basis for his initial attempt was the emission of α particles from uranium ore. He correctly suspected that α particles were part of helium atoms, and he proposed to determine the age of uranium ore samples by determining the concentration of helium gas trapped inside them.

Rutherford's helium method did not yield accurate results, but it did inspire a young American chemist, Bertram Boltwood (1870–1927), who had determined that the decay of radioactive uranium involves a series of decay events ending with the formation of stable lead (see Figure 19.5). In 1907 Boltwood published the results of dating 43 samples of uranium-containing minerals on the basis of the ratio of lead to uranium in them. The ages he reported spanned hundreds of millions to over a billion years and probably represented the first successful attempt at radiometric dating.

radiometric dating a method for determining the age of an object on the basis of the quantity of a radioactive nuclide, the products of its decay that the object contains, or both.

The development of the mass spectrometer for accurately determining the abundances of individual isotopes of elements, coupled with more accurate half-life values for decay events such as those in Figure 19.5, has allowed scientists to use the ratio of ^{206}Pb to ^{238}U in geological samples to determine their ages with a precision of about ±1%. Other methods, including one based on the decay of ^{235}U to ^{207}Pb ($t_{1/2} = 7.0 \times 10^6$ yr), may be used to analyze the same samples, providing independent determinations that mutually ensure more accurate results. In 1956 American geochemist Claire Patterson (1922-1995) showed that the oldest rocks on Earth are more than 4.0 billion years old and that meteorites formed as the solar system formed are 4.5 billion years old.

CONNECTION In Section 2.3 we saw that mass spectrometry (i.e., Francis W. Aston's positive-ray analyzer, Figure 2.10) can be used to determine the abundances of the isotopes of elements in a sample. Mass spectrometry is discussed in greater detail in Section 3.8.

These radiometric methods for dating geological samples yield reliable results only when the sample is a closed system, which means that the only loss of the radionuclide is through radioactive decay and that all of the nuclides produced by the decay processes remain in the sample. In addition, those decay processes must be the only source of the product nuclides. For these reasons, scientists must exercise care in selecting the types of samples they subject to radiometric dating analysis. For example, the presence of the mineral zircon ($ZrSiO_4$) in a geological sample is good news for scientists interested in using radiometric dating because U^{4+} ions readily substitute for Zr^{4+} ions as crystals of $ZrSiO_4$ solidify from the molten state, but Pb^{2+} ions do not. Therefore, the only source of ^{206}Pb and ^{207}Pb in a zircon sample should be the decay of ^{238}U and ^{235}U, respectively.

In 1947 American chemist Willard Libby (1908–1980) developed a radiometric dating technique, called **radiocarbon dating**, for determining the age of artifacts from prehistory and early civilizations. The method is based on determining the carbon-14 content of samples derived from plants or the animals that consumed them. Carbon-14 originates in the upper atmosphere, where cosmic rays break apart the nuclei of atoms, forming free protons and neutrons. When one of these neutrons collides with a nitrogen-14 atom, it forms an atom of radioactive carbon-14 and a proton:

$$^{14}_{7}N + ^{1}_{0}n \rightarrow ^{14}_{6}C + ^{1}_{1}p$$

Atmospheric carbon-14 combines with oxygen, forming $^{14}CO_2$. The atmospheric concentration of $^{14}CO_2$ amounts to only about 10^{-12} of all the molecules of CO_2 in the air. These traces of radioactive CO_2, along with the stable forms, $^{12}CO_2$ and $^{13}CO_2$, are incorporated into the structures of green plants during photosynthesis. The tiny fraction of the plant's mass that is ^{14}C gets even tinier after a plant dies, or after a part of it stops growing and photosynthesizing, because ^{14}C undergoes β decay, as we described in Section 19.2:

$$^{14}_{6}C \rightarrow ^{14}_{7}N + ^{0}_{-1}\beta$$

The half-life of this decay process is 5730 years.

The dating process starts by determining the ^{14}C content (N_t) of an object of historical interest, such as a piece of wood from an ancient building, charcoal from a prehistoric campfire, or the parchment used in the Dead Sea Scroll shown in the opening photo of this chapter. The analyst also needs to know (or be able to predict) the object's ^{14}C content when the material in it was alive (N_0). This information can be used in an equation based on Equation 19.3 in which the N terms are combined in the ln term:

radiocarbon dating a method for establishing the age of a carbon-containing object by measuring the amount of radioactive carbon-14 remaining in the object.

$$\ln \frac{N_t}{N_0} = -0.693 \frac{t}{t_{1/2}}$$

Next, the terms are rearranged to solve for t:

$$t = -\frac{t_{1/2}}{0.693} \ln \frac{N_t}{N_0} \tag{19.4}$$

SAMPLE EXERCISE 19.4 Radiocarbon Dating **LO5**

The ^{14}C content of a wooden harpoon handle found in the remains of an Inuit encampment in western Alaska is 61.9% of the ^{14}C content of the same type of wood from a recently cut tree. How old is the harpoon?

Collect, Organize, and Analyze The half-life of carbon-14 is 5730 years. Equation 19.4 provides the age t of the artifact if we know the ratio of the ^{14}C in it today to its initial ^{14}C content. The ^{14}C content of the modern sample can be used as a surrogate for the initial ^{14}C content of the artifact. Therefore, 61.9% (or 0.619) represents the ratio N_t/N_0. This value is greater than 0.5, which means that the age of the sample is less than one half-life (5730 yr).

Solve Substituting into Equation 19.4:

$$\begin{aligned} t &= -\frac{t_{1/2}}{0.693} \ln \frac{N_t}{N_0} \\ &= -\frac{5730 \text{ yr}}{0.693} \ln(0.619) \\ &= 3966 \text{ yr} = 3.97 \times 10^3 \text{ yr} \end{aligned}$$

Think About It The resulting age is less than one half-life, which is reasonable because it contained more than half the original carbon-14 content. The result is expressed with three significant figures to match that of the starting composition (61.9%).

Practice Exercise The carbon-14:carbon-12 ratio in papyrus growing along the Nile River today is 1.8 times greater than in a papyrus scroll found near the Great Pyramid at Giza. How old is the scroll?

The accuracy of radiocarbon dating can be checked by determining the ^{14}C content of the annual growth rings of very old trees, such as the bristlecone pines that grow in the American Southwest (**Figure 19.8**). When scientists plot the radiocarbon ages of these rings against their actual ages obtained by counting rings starting from the outer growth layer of the tree (representing $t = 0$), they find that the two sets of ages do not agree exactly, as shown in **Figure 19.9**. There are several reasons for this lack of agreement, including variability in the rate of ^{14}C production because of changing intensity of the cosmic rays striking Earth's upper atmosphere. To ensure accurate ^{14}C results, scientists must correct for these and other variations, and they must be careful to avoid contaminating ancient samples with modern carbonaceous material. With proper analytical technique, modern radiocarbon dating results are generally accurate to within 40 years for samples that are 500–50,000 years old.

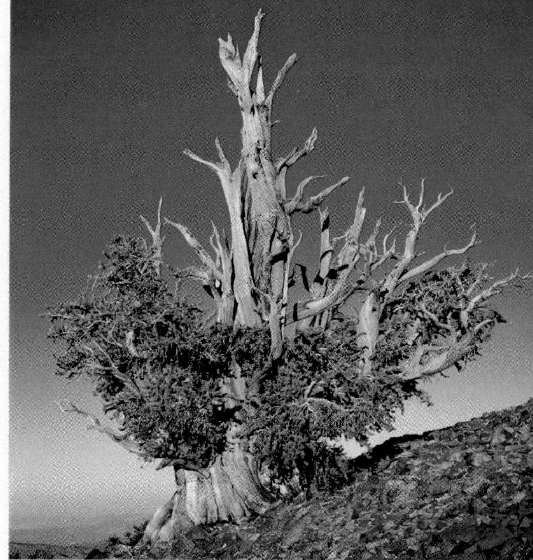

FIGURE 19.8 Radiocarbon dating relies on knowing the atmospheric concentration of carbon-14 over time. Annual growth rings in the trunks of ancient living trees, such as the bristlecone pines in the American Southwest, act as a check of the atmospheric carbon-14 levels over thousands of years. The ages of the rings can be determined by counting them, and their carbon-14 content can be determined by mass spectrometry.

CONCEPT **TEST**

How might the increased consumption of fossil fuels over the last century affect the ^{14}C content of growing plant tissues?

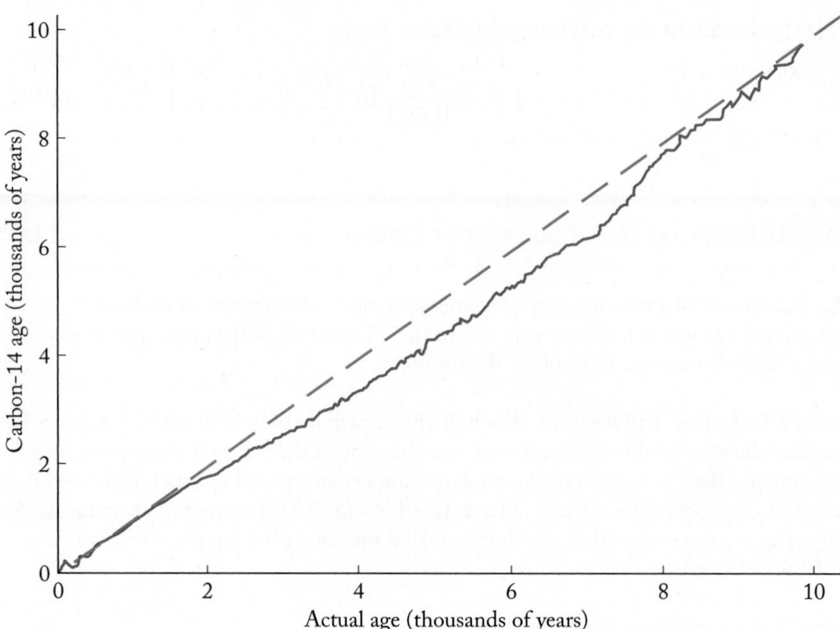

FIGURE 19.9 Calibration curves for radiocarbon dating allow scientists to accurately calculate the ages of archeological objects. If the rate of ^{14}C production in the upper atmosphere were constant, then the age of objects on the basis of their ^{14}C content would match their actual age—a condition represented by the red dashed line. However, analyses of tree rings, corals, and lake sediments indicate that the rate of ^{14}C production in the upper atmosphere is variable, so a real plot of ^{14}C age versus actual age produces the jagged blue line. This plot allows scientists to convert ^{14}C ages into actual ages.

19.6 Biological Effects of Radioactivity

The γ rays and most of the α and β particles produced by nuclear reactions have more than enough energy to tear chemical bonds apart, producing odd-electron radicals or free electrons and cations. Consequently, these rays and particles are classified as **ionizing radiation**. Other examples include X-rays and short-wavelength ultraviolet rays. The ionization of atoms and molecules in living tissue can lead to radiation sickness, cancer, birth defects, and death. The scientists who first worked with radioactive materials were unaware of these hazards, and some of them suffered for it. Marie Curie died from radiation exposure, and so did her daughter Irène Joliot-Curie, who continued the research started by her parents.

In medicine, the term *ionizing radiation* is limited to photons and particles that have sufficient energy to remove an electron from water[1]:

$$H_2O(\ell) \xrightarrow{1216\ \text{kJ/mol}} H_2O^+(aq) + e^-$$

The logic behind this definition is that the human body is composed largely of water. Therefore, water molecules are the most abundant ionizable targets when we are exposed to nuclear radiation. The cation H_2O^+ reacts with another water molecule in the body to form a hydronium ion and a hydroxyl free radical:

$$H_2O^+(aq) + H_2O(\ell) \rightarrow H_3O^+(aq) + OH$$

The rapid reactions of free radicals with biomolecules can disrupt cell function, sometimes with life-threatening consequences.

Radiation-induced alterations to the biochemical machinery that controls cell growth are most likely to occur in tissues in which cells grow and divide rapidly. One such tissue is bone marrow, where billions of white blood cells are produced each day to fortify the body's immune system. Molecular damage to bone marrow

ionizing radiation high-energy products of radioactive decay that can ionize molecules.

gray (Gy) the SI unit of absorbed radiation; 1 Gy = 1 J/kg of tissue.

relative biological effectiveness (RBE) a factor that accounts for the differences in physical damage caused by different types of radiation.

sievert (Sv) the SI unit used to express the amount of biological damage caused by ionizing radiation.

[1]Even relatively low-energy γ rays have 10^4 times the energy needed to ionize water.

can lead to leukemia—the uncontrolled production of nonfunctioning white blood cells that spread throughout the body, crowding out healthy cells. Ionizing radiation can also cause molecular alterations in the genes and chromosomes of sperm and egg cells, increasing the chances of birth defects.

Radiation Dosage

The biological impact of ionizing radiation depends on how much of it is absorbed by an organism. If the radiation is coming from one radioactive source, then the amount absorbed depends on the radioactivity of the source and the energy of the radiation that is produced per decay event. Tables of radioactive isotopes often include information about their modes of decay and the energies of the particles and γ rays they emit.

Absorbed dose is the quantity of ionizing radiation absorbed by a unit mass of living tissue. The SI unit of absorbed dose is the **gray (Gy)**. One gray is equal to the absorption of 1 J of radiation energy per kilogram of body mass:

$$1\ \text{Gy} = 1\ \text{J/kg}$$

Grays express dosage, but they do not indicate the amount of *tissue damage* caused by that dosage. Different products of nuclear reactions affect living tissue differently. Exposure to 1 Gy of γ rays produces about the same amount of tissue damage as exposure to 1 Gy of β particles. However, 1 Gy of α particles, which move about 100 times slower than β particles but have nearly 10^4 times the mass, causes up to 20 times as much damage as 1 Gy of γ rays. Neutrons cause 3 to 5 times as much damage. To account for these differences, values of **relative biological effectiveness (RBE)** have been established for the various forms of ionizing radiation (**Table 19.3**). When an absorbed dose in grays is multiplied by an RBE factor, the product is called the *effective* dose, a measure of tissue damage. The SI unit of effective dose is the **sievert (Sv)**.

TABLE 19.3 RBE Values of Nuclear Radiation

Radiation	RBE
γ Rays	1.0
β Particles	1.0–1.5
Neutrons	3–5
Protons	10
α Particles	20

Table 19.4 summarizes the various units used to express quantities of radiation and their biological impact. Two non-SI units are listed that predate their SI counterparts but are still often used. They are *radiation absorbed dose* (*rad*), which is equivalent to 0.01 Gy, and *roentgen equivalent man* (*rem*) for tissue damage. One rem is the product of one rad of ionization times the appropriate RBE factor. There are 100 rems in 1 Sv.

The RBE of 20 for α particles may lead you to believe that these particles pose the greatest health threat from radioactivity. Not exactly. Because they are so big, α particles have little penetrating power; they are stopped by a sheet of paper, clothing, or even a layer of dead skin (**Figure 19.10**). If you ingest or inhale an α emitter, however, then the relatively massive α particles cause severe damage to the small number of cells each of them can penetrate. In contrast, γ rays emanating from sources outside the body are dangerous because they have the greatest penetrating power.

STEPWISE ANIMATION
Radiation Penetration

TABLE 19.4 Units for Expressing Quantities of Ionizing Radiation

Parameter	SI Unit	Description	Alternative Common Unit	Description
Radioactivity	Becquerel (Bq)	1 decay event/s	Curie (Ci)	3.70×10^{10} decay events/s
Ionizing energy absorbed	Gray (Gy)	1 J/kg of tissue	Rad	0.01 J/kg of tissue
Amount of tissue damage	Sievert (Sv)	1 Gy × RBE[a]	Rem	1 rad × RBE[a]

[a]RBE = relative biological effectiveness.

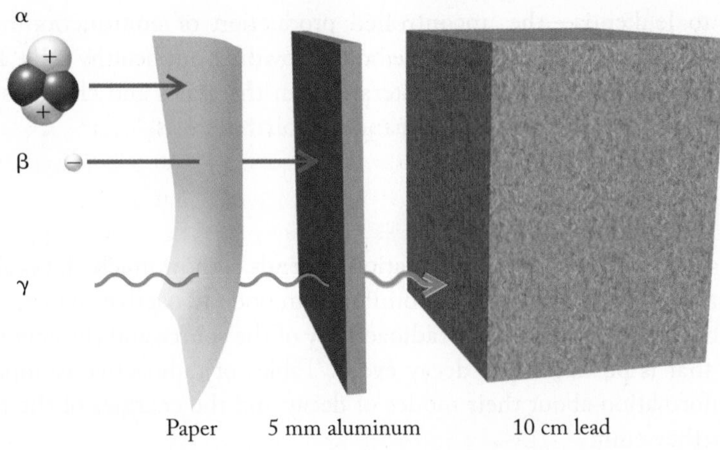

FIGURE 19.10 The tissue damage caused by α particles, β particles, and γ rays depends on their ability to penetrate materials that shield the tissues. Alpha particles are stopped by paper or clothing but are extremely dangerous if formed inside the body because their low penetrating power traps them inside, where they do not have to travel far to cause cell damage. Stopping γ rays requires a thick layer of lead or several meters of concrete or soil.

FIGURE 19.11 The ruins of the nuclear reactor at Chernobyl, Ukraine, which exploded in 1986.

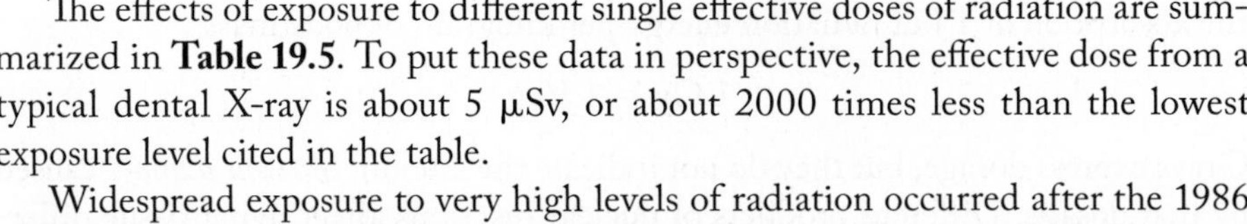

The effects of exposure to different single effective doses of radiation are summarized in **Table 19.5**. To put these data in perspective, the effective dose from a typical dental X-ray is about 5 μSv, or about 2000 times less than the lowest exposure level cited in the table.

Widespread exposure to very high levels of radiation occurred after the 1986 explosion at the Chernobyl nuclear reactor in what is now Ukraine (**Figure 19.11**). Many plant workers and first responders were exposed to more than 1.0 Sv of radiation. At least 30 of them died in the weeks after the accident. Many of the more than 300,000 workers who cleaned up the area around the reactor exhibited symptoms of radiation sickness, and at least 5 million people in Ukraine, Belarus, and Russia were exposed to fallout in the days following the accident. Studies conducted in the early 1990s uncovered high incidences of thyroid cancer in children in southern Belarus as a result of ^{131}I released in the Chernobyl accident, and children born in the region nearly a decade after the accident had unusually high rates of mutations in their DNA because of their parents' exposure to ionizing radiation. Genetic damage was also widespread among plants and animals living in the region.

Radiation exposure was not confined to Ukraine and Belarus. After the accident, a cloud of radioactive material spread across northern Europe. (Its detection came as a surprise because news of the incident had been covered up by the leaders of what was then the Soviet Union.) Within two weeks increased levels of radioactivity were detected throughout the Northern Hemisphere (**Figure 19.12**). The

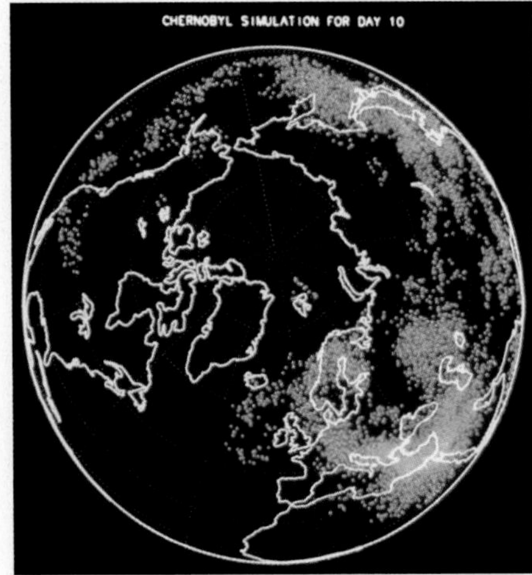

FIGURE 19.12 Radioactive fallout (shown in pink) from the Chernobyl accident in 1986 was detected throughout the Northern Hemisphere.

TABLE 19.5 Acute Effects of Single Whole-Body Effective Doses of Ionizing Radiation

Effective Dose (Sv)	Toxic Effect
0.05–0.25	No acute effect, possible carcinogenic or mutagenic damage to DNA
0.25–1.0	Temporary reduction in white blood cell count
1.0–2.0	Radiation sickness: fatigue, vomiting, diarrhea, impaired immune system
2.0–4.0	Severe radiation sickness: intestinal bleeding, bone marrow destruction
4.0–10.0	Death, usually through infection, within weeks
>10.0	Death within hours

accident produced a global increase in human exposure to ionizing radiation estimated to be equivalent to 50 μSv per year, which is comparable to ten dental X-rays.

As a result of an earthquake and tsunami in March 2011, nuclear reactors at a power station in Fukushima, Japan, released between 340 and 780 PBq (i.e., 340×10^{15} to 780×10^{15} Bq) of radiation into the environment (**Figure 19.13**), with an estimated 80% landing in the Pacific Ocean. This represents a slightly smaller release of radiation than the Chernobyl disaster, estimated at 5200 PBq.

FIGURE 19.13 The cleanup after the 2011 accident at the Fukushima nuclear power plant included washing away any radioactive substances that accumulated on sidewalks. The worker in the background is monitoring the level of radioactivity.

Evaluating the Risks of Radiation

To put global radiation exposure from Chernobyl in perspective, we need to consider typical annual exposure levels. For many people, the principal source of radiation is radon gas in indoor air and in well water (**Figure 19.14**). Like all noble gases, radon ($Z = 86$) is chemically inert. Unlike the others, all of its isotopes are radioactive (as are all known nuclides with $Z > 83$—see Section 19.2). The most common isotope, radon-222, is produced when uranium-238 in rocks and soil decays to lead-206 (see Figure 19.5). The radon gas formed in this decay series percolates upward and can enter a building through cracks and pores in its foundation.

If you breathe air contaminated with radon and then exhale before it decays, no harm is done. If radon-222 decays inside the lungs, however, then it emits an α particle that can attack lung tissue. The nuclide produced by the α decay of ^{222}Rn is radioactive polonium-218, which may become attached to tissue in the respiratory system and undergo a second α decay, forming lead-214:

$$^{222}_{86}\text{Rn} \rightarrow {}^{218}_{84}\text{Po} + {}^{4}_{2}\alpha \qquad t_{1/2} = 3.8 \text{ d}$$

$$^{218}_{84}\text{Po} \rightarrow {}^{214}_{82}\text{Pb} + {}^{4}_{2}\alpha \qquad t_{1/2} = 3.1 \text{ min}$$

As we have seen, α particles are the most damaging product of nuclear decay when formed *inside the body*. How big of a threat does radon pose to human health? Concentrations of indoor radon depend on local geology (**Figure 19.15**) and on how gastight building foundations are. The air in many buildings contains concentrations of radon with radioactivity levels in the range of 1 pCi per liter of air. How hazardous are such tiny concentrations? There appears to be no simple answer. The U.S. Environmental Protection Agency has established 4 pCi/L as

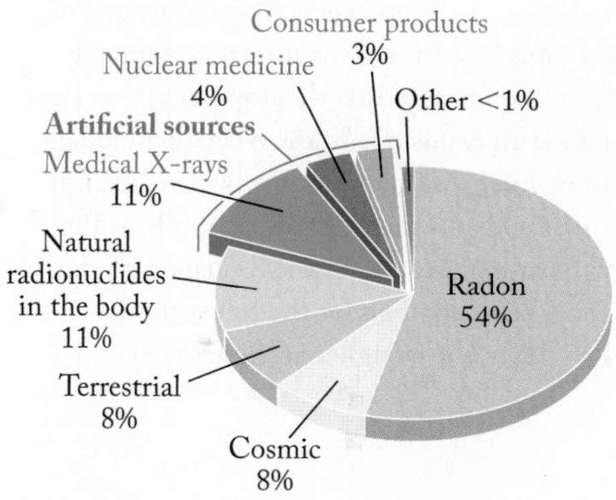

FIGURE 19.14 Sources of radiation exposure of the U.S. population. On average, a person living in the United States is exposed to 0.0036 Sv of radiation each year. More than 80% of this exposure comes from natural sources—mainly radon in the air and water. Artificial sources account for about 18% of the total exposure.

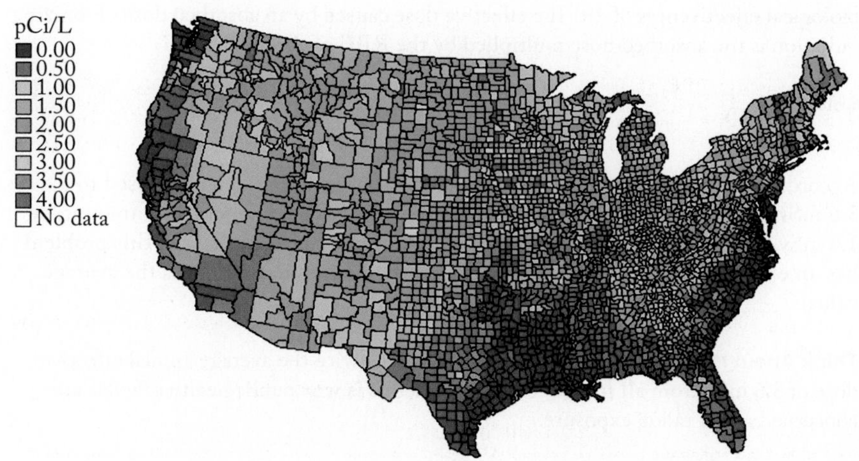

FIGURE 19.15 Levels of radon gas in soils and rocks across the United States.

an "action level," meaning that people occupying houses with higher concentrations should take measures to minimize their exposure.

CONCEPT TEST

The Environment Canada guideline for radon in indoor air is 200 Bq/m^3. Is this value more or less than the American action level?

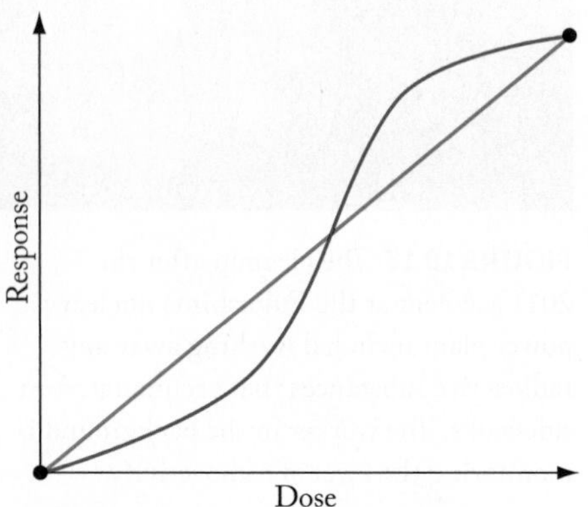

FIGURE 19.16 The risk of death from radiation-induced cancer may follow one of two models. In the linear response model (red line), risk is directly proportional to the radiation exposure. In the S-shaped model (blue line), risk remains low below a critical threshold and then increases rapidly as the exposure increases. In the S-shaped model, the risk is less than for the linear model at low doses but is higher at higher doses.

This action level is based on studies of the incidence of lung cancer in workers in uranium mines. These workers are exposed to radon concentrations (and concentrations of other radionuclides) that are much higher than the concentrations in homes and other buildings. However, many scientists believe that people exposed to very low levels of radon for many years are as much at risk as miners exposed to high levels of radiation for shorter periods. Some researchers use a model that assumes a linear relation between radon exposure and incidence of lung cancer. This model is represented by the red line in **Figure 19.16**, which graphs the risk of cancer deaths as a function of radiation absorbed. On the basis of this dose–response model, an estimated 15,000 Americans die of lung cancer each year because of exposure to indoor radon. This number comprises 10% of all lung-cancer fatalities and 30% of those among nonsmokers.

Is this linear model valid? Perhaps—but some scientists believe that there may be a threshold of exposure below which radon poses no significant threat to public health. They advocate an S-shaped dose–response curve, shown by the blue line in Figure 19.16. The risk of death from cancer in the S-shaped curve is much lower than in the linear response model at low radiation exposure but rises rapidly above a critical value.

SAMPLE EXERCISE 19.5 Calculating Effective Dose — LO6

It has been estimated that a person living in a home where the air radon concentration is 4.0 pCi/L receives an annual absorbed dose of ionizing radiation equivalent to 0.40 mGy increasing the risk of dying from lung cancer by 1%. What is the person's annual effective dose, in millisieverts (mSv), from this radon? Use information from Figure 19.14 to compare this annual effective dose from radon with the average annual effective dose from radon estimated for persons living in the United States.

Collect, Organize, and Analyze We are given an absorbed radiation dose of 0.40 mGy. Radon isotopes emit α particles, which we know from Table 19.3 have a relative biological effectiveness of 20. The effective dose caused by an absorbed dose of ionizing radiation is the absorbed dose multiplied by the RBE of the radiation.

Solve

$$0.40 \text{ mGy} \times 20 = 8.0 \text{ mSv}$$

According to the caption to Figure 19.14, the average American is exposed to 3.6 mSv of radiation per year, and the graph indicates that 54% of that amount, or 1.9 mSv, comes from radon. A person living in the home described in this problem has an effective dose from radon that is slightly more than four times the average value.

Think About It The calculated value is more than twice the average annual effective dose of 3.6 mSv from all sources of radiation. This is why public health officials are concerned about radon exposure.

Practice Exercise Imaging the lower gastrointestinal tract involves the administration of a barium salt followed by radiation with X-rays. A typical effective dose in these studies is 8 μSv. If an X-ray machine emits X-rays with an energy of 6.0×10^{-17} J each, how many of these X-rays must be absorbed per kilogram of tissue to produce an effective dose of 8 μSv? Assume the RBE of these X-rays is 1.2.

19.7 Medical Applications of Radionuclides

Radionuclides are used in both the detection and the treatment of diseases: they are key agents in the respective medical fields of *diagnostic radiology* and *therapeutic radiology*. In diagnostic radiology, radionuclides are used alongside magnetic resonance imaging (MRI) and other imaging systems that involve only nonionizing radiation. Therapeutic radiology, however, is based almost entirely on the ionizing radiation that comes from nuclear processes.

Therapeutic Radiology

Because ionizing radiation causes the most damage to cells that grow and divide rapidly, it is a powerful tool in the fight *against* cancer. Radiation therapy consists of exposing cancerous tissue to γ radiation.

Surgically inaccessible tumors can be treated with beams of γ rays from a radiation source outside the body. Unfortunately, γ radiation destroys both cancer cells and healthy ones. Thus, patients receiving radiation therapy frequently suffer symptoms of radiation sickness, including nausea, vomiting (the tissues that make up intestinal walls are especially susceptible to radiation-induced damage), and hair loss. To reduce the severity of these side effects, radiologists must carefully control the dosage a patient receives.

Often the radiation source is external to the patient, but sometimes it is encased in a platinum capsule and surgically implanted in a cancerous tumor. The platinum provides a chemically inert outer layer and acts as a filter, absorbing α and β particles emitted by the radionuclide but allowing γ rays to pass into the tumor.

A nuclide's chemical properties can sometimes be exploited to direct it to a tumor site. For example, most iodine in the body is concentrated in the thyroid gland, so an effective therapy against thyroid cancer starts with ingestion of potassium iodide containing radioactive iodine-131. Some radionuclides used in cancer therapy are listed in **Table 19.6**.

TABLE 19.6 Some Radionuclides Used in Radiation Therapy

Nuclide	Radiation	Half-Life	Treatment
^{32}P	β	14.3 d	Leukemia therapy
^{60}Co	β, γ	5.27 yr	Cancer therapy
^{131}I	β	8.1 d	Thyroid therapy
^{131}Cs	γ	9.7 d	Prostate cancer therapy
^{192}Ir	β, γ	74 d	Coronary disease

TABLE 19.7 Selected Radionuclides Used for Medical Imaging

Nuclide	Radiation	Half-Life (h)	Use
^{99m}Tc	γ	6.0	Bones, circulatory system, various organs
^{123}I	γ	13.2	Thyroxine production in thyroid gland
^{201}Tl	γ	73	Coronary arteries, heart muscle
^{67}Ga	γ	78	Tumors in the brain and other organs

Diagnostic Radiology

The transport of radionuclides in the body and their accumulation in certain organs also provide ways to assess organ function. In some applications, a tiny quantity of a radioactive isotope is used, together with a much larger amount of a stable isotope of the same element. The radioactive isotope is called a *tracer*, and the stable isotope is the *carrier*. For example, the circulatory system can be imaged by injecting into the blood a solution of sodium chloride containing a trace amount of $^{24}NaCl$. Circulation is monitored by measuring the γ rays emitted by ^{24}Na as it decays.

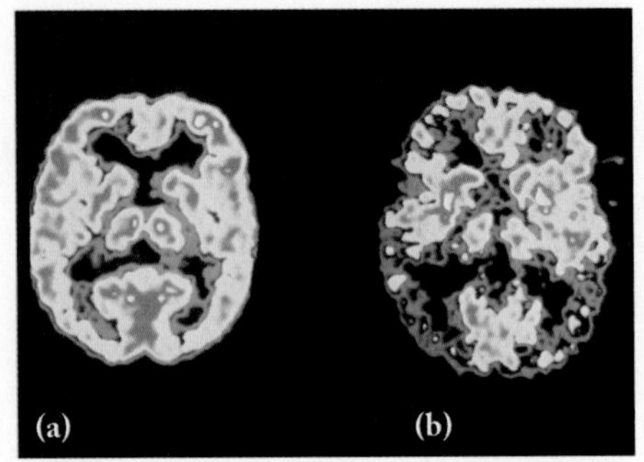

FIGURE 19.17 Positron emission tomography (PET) is used to monitor cell activity in organs such as the brain. (a) Brain function in a healthy person. The red and yellow regions indicate high brain activity; blue and black indicate low activity. (b) Brain function in a patient with Alzheimer's disease.

The ideal isotope for medical imaging is one that has a half-life about equal to the length of time required to perform the imaging measurements. It should emit moderate-energy γ rays or β particles but no α particles that might cause tissue damage. Sodium-24 (a β emitter with a half-life of 15 h) meets both these criteria. **Table 19.7** lists several other radionuclides that are used in medical imaging.

Positron emission tomography (PET) is a powerful tool for diagnosing organ and cell function. PET uses short-lived, neutron-poor, positron-emitting radionuclides such as carbon-11, oxygen-15, and fluorine-18. A patient might be administered a solution of glucose in which some of the sugar molecules contain atoms of ^{11}C, ^{15}O, or an atom of ^{18}F in place of a hydrogen atom. The rate at which glucose is metabolized in various regions of the brain is monitored by detecting the γ rays produced by positron–electron annihilations. Unusual patterns in PET images of brains (**Figure 19.17**) can indicate schizophrenia, bipolar disorder, Alzheimer's disease, damage from strokes, and even nicotine addiction in tobacco smokers.

19.8 Nuclear Fission

STEPWISE ANIMATION
Transmutation

The energy stored in atomic nuclei can also be put to practical use for electrical power generation. When an atom of uranium-235 captures a neutron, the nucleus of the unstable product, uranium-236, splits into two lighter nuclei in a process called **nuclear fission**. Several uranium-235 fission reactions can occur, including these three:

$$^{235}_{92}U + ^{1}_{0}n \rightarrow ^{141}_{56}Ba + ^{92}_{36}Kr + 3\,^{1}_{0}n$$

$$^{235}_{92}U + ^{1}_{0}n \rightarrow ^{137}_{52}Te + ^{97}_{40}Zr + 2\,^{1}_{0}n$$

$$^{235}_{92}U + ^{1}_{0}n \rightarrow ^{138}_{55}Cs + ^{96}_{37}Rb + 2\,^{1}_{0}n$$

In all of these reactions, the sums of the masses of the products are slightly less than the sums of the masses of the reactants. As we observed for nucleosynthesis in Section 19.1, this difference in mass is converted to energy in accordance with Einstein's equation ($E = mc^2$).

These reactions also produce additional neutrons, which can be absorbed by other uranium-235 nuclei and initiate more fission events in a **chain reaction** (**Figure 19.18**). The reaction proceeds as long as there are enough uranium-235 nuclei present to absorb the neutrons being produced. On average, at least one neutron from each fission event must cause another nucleus to split apart for the chain reaction to be self-sustaining. The quantity of fissionable material needed to ensure that every fission event produces another is called the **critical mass.** For uranium-235, the critical mass is about 52 kg for a sphere of the pure isotope. The critical mass decreases to 15 kg if surrounded by a substance that reflects neutrons back toward the ^{235}U.

Uranium-235 is the most abundant fissionable isotope, but it makes up only 0.72% of the uranium in the principal uranium ore, pitchblende (**Figure 19.19a**). The uranium in nuclear reactors must be at least 3% to 4% uranium-235, and enrichment to about 85% is needed for nuclear weapons. The most common method for enriching uranium ore involves extracting the uranium in a process that yields a material called yellowcake, which is mostly U_3O_8 (**Figure 19.19b**). This oxide is then converted to UF_6, which, despite a molar mass of more than 300 g per mole, is a volatile solid that sublimes at 56°C. The volatility of this nonpolar molecular compound can be explained by the relatively weak London dispersion forces experienced by its compact, symmetrical molecules (**Figure 19.20**). Fissionable $^{235}UF_6$ is separated from $^{238}UF_6$ on the basis of their slightly different densities. Elaborate centrifuge systems are used to exploit this difference (**Figure 19.19c**).

nuclear fission a nuclear reaction in which the nucleus of an element splits into two lighter nuclei; the process is usually accompanied by the release of one or more neutrons and energy.

chain reaction a self-sustaining series of fission reactions in which the neutrons released when nuclei split apart initiate additional fission events and sustain the reaction.

critical mass the minimum quantity of fissionable material needed to sustain a chain reaction.

STEPWISE ANIMATION

Induced Fission and Chain Reactions

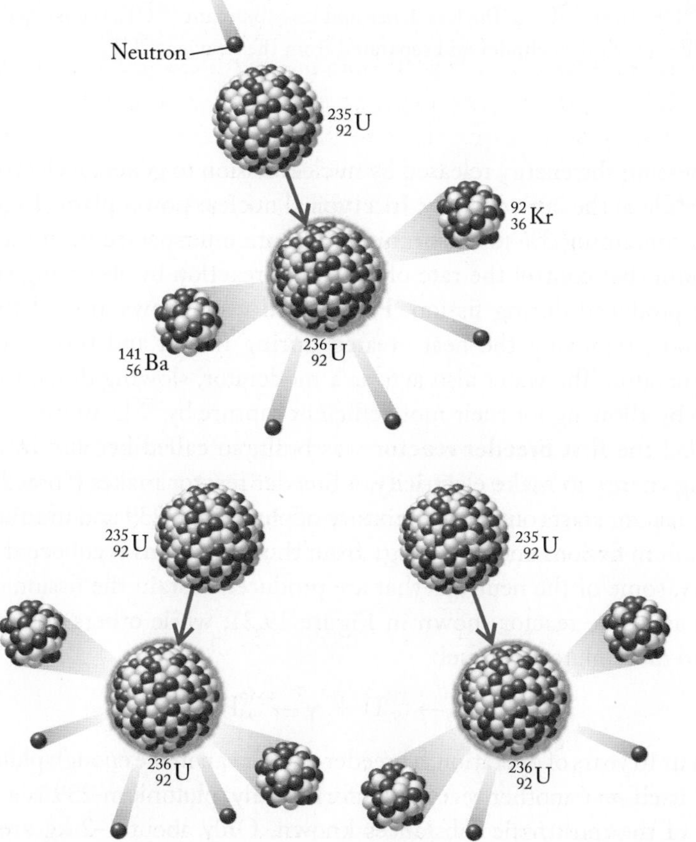

FIGURE 19.18 Each fission event in the chain reaction of a uranium-235 nucleus begins when the nucleus captures a neutron, forming an unstable uranium-236 nucleus that then splits apart (fissions) in one of several ways. In the first process shown here, the uranium-236 nucleus splits into krypton-92, barium-141, and three neutrons. If, on average, at least one of the three neutrons from each fission event causes the fission of another uranium-235 nucleus, then the process is sustained in a chain reaction.

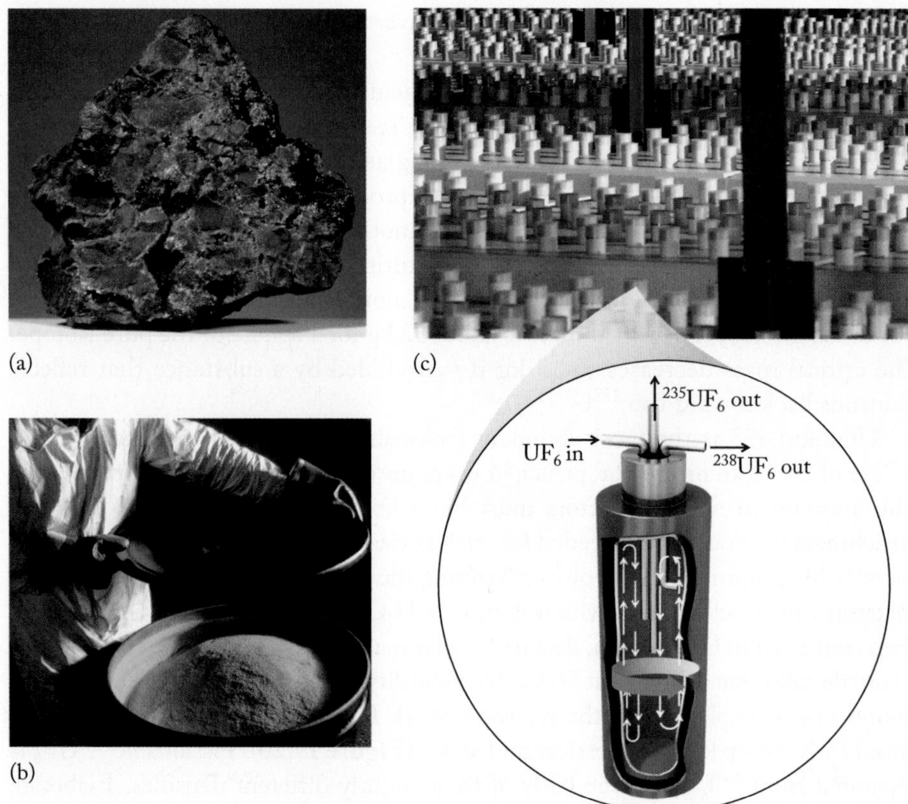

FIGURE 19.19 Preparing uranium fuel. (a) A piece of pitchblende, source of the uranium fuel for nuclear reactors. Pitchblende ore is ground up and extracted with strong acid. (b) The uranium compounds (mostly U_3O_8) obtained from the extract are called yellowcake. (c) Uranium oxides are converted to volatile UF_6, which is centrifuged at very high speed to separate $^{235}UF_6$ from $^{238}UF_6$. The less dense and less abundant $^{235}UF_6$ is enriched near the center of the centrifuge cylinder and separated from the heavier $^{238}UF_6$.

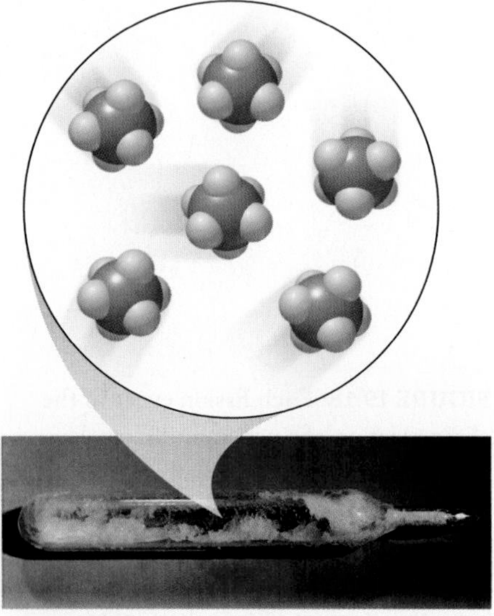

FIGURE 19.20 Uranium hexafluoride is a volatile solid that sublimes at only 56°C because it is composed of compact, symmetrical molecules that experience relatively weak London dispersion forces despite their considerable mass.

Harnessing the energy released by nuclear fission to generate electricity began in the middle of the 20th century. In a typical nuclear power plant (**Figure 19.21**), fuel rods containing 3% to 4% uranium-235 are interspersed with rods of boron or cadmium that control the rate of the chain reaction by absorbing some of the neutrons produced during fission. Pressurized water flows around the fuel and control rods, removing the heat created during fission and transferring it to a steam generator. The water also acts as a moderator, slowing down the neutrons and thereby allowing for their more efficient capture by ^{235}U atoms.

In 1952 the first **breeder reactor** was built, so called because in addition to producing energy to make electricity, a breeder reactor makes ("breeds") its own fuel. The reactor starts out with a mixture of plutonium-239 and uranium-238. As the plutonium fissions and the energy from those reactions is collected to produce electricity, some of the neutrons that are produced sustain the fission chain reaction just as in the reactor shown in Figure 19.21, while others convert the uranium into more plutonium fuel:

$$^{238}_{92}U + ^{1}_{0}n \rightarrow ^{239}_{92}U + \gamma \rightarrow ^{239}_{94}Pu + 2\ ^{0}_{-1}\beta$$

In less than 10 years of operation, a breeder reactor can make enough plutonium-239 to refuel itself *and* another reactor. Unfortunately, plutonium-239 is a carcinogen and one of the most toxic substances known. Only about 1–2 kg are needed to

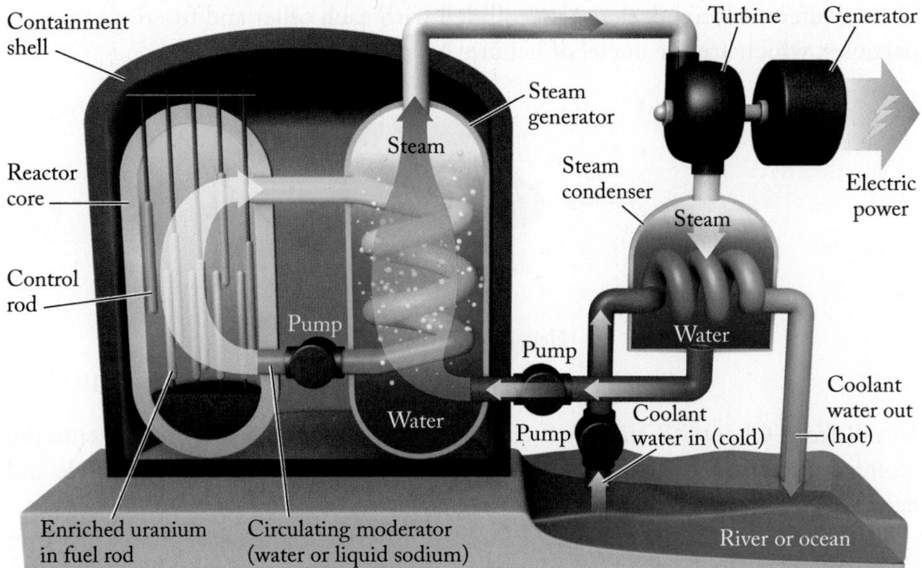

FIGURE 19.21 A pressurized, water-cooled nuclear power plant uses fuel rods containing uranium enriched to about 4% uranium-235. The fission chain reaction is regulated with control rods and a moderator that is either water or liquid sodium. The moderator slows down the neutrons released by fission so that they can be captured more efficiently by other uranium-235 nuclei. It also transfers the heat produced by the fission reaction to a steam generator. The steam generated by this heat drives a turbine that generates electricity. Nuclear power plants are usually situated near coasts or large rivers, where large amounts of coolant water are readily available.

make an atomic bomb, and it has a long half-life: 2.4×10^4 years. Understandably, extreme caution and tight security surround the handling of plutonium fuel and the transportation and storage of nuclear wastes containing even small amounts of plutonium. Health and safety matters related to reactor operation and spent-fuel disposal are the principal reasons there are no breeder power stations in the United States, although they have been built in at least seven other countries.

19.9 Nuclear Fusion and the Quest for Clean Energy

The energy of the Sun is derived from the high-speed collision and reaction of hydrogen nuclei to form helium. This **nuclear fusion** process involves more steps than the process that probably took place during primordial nucleosynthesis (see Chapter 2), when protons (hydrogen nuclei) and neutrons fused to form deuterons (nuclei of the hydrogen isotope deuterium, D):

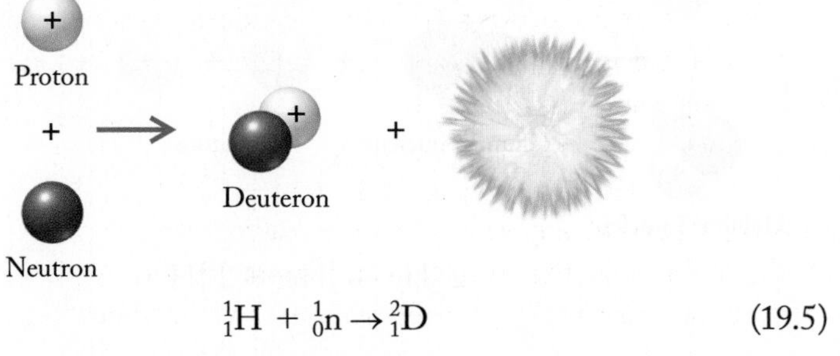

$$^{1}_{1}\text{H} + ^{1}_{0}\text{n} \rightarrow ^{2}_{1}\text{D} \tag{19.5}$$

breeder reactor a nuclear reactor in which fissionable material is produced during normal reactor operation.

nuclear fusion a nuclear reaction in which subatomic particles or atomic nuclei collide with each other at very high speeds and fuse, forming more massive nuclei and releasing energy.

Once deuterons formed, they also collided with each other and fused, forming α particles, which are the nuclei of helium-4 atoms:

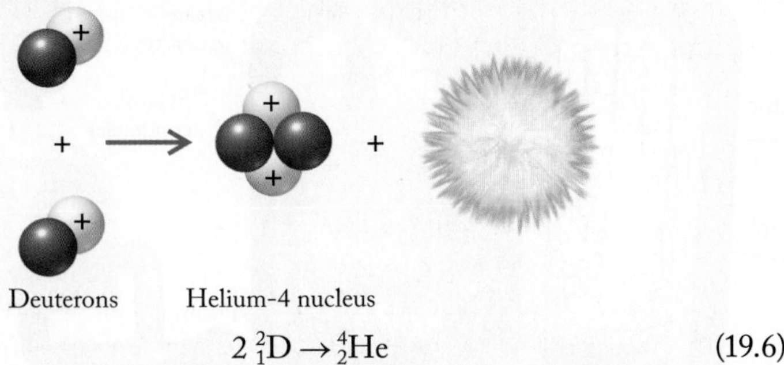

$$2\,{}^{2}_{1}D \rightarrow {}^{4}_{2}He \quad (19.6)$$

Recall from Chapter 2 that each subscript in a nuclear equation represents the atomic number (Z) of an atom or the electrical charge of a subatomic particle and each superscript represents the atom or particle's mass number (A).

CHEMTOUR
Fusion of Hydrogen

Hydrogen fusion in our Sun follows a different path because free neutron concentrations are far lower there than they were in the primordial universe. In the Sun, colliding protons may fuse to form a deuteron and a positron:

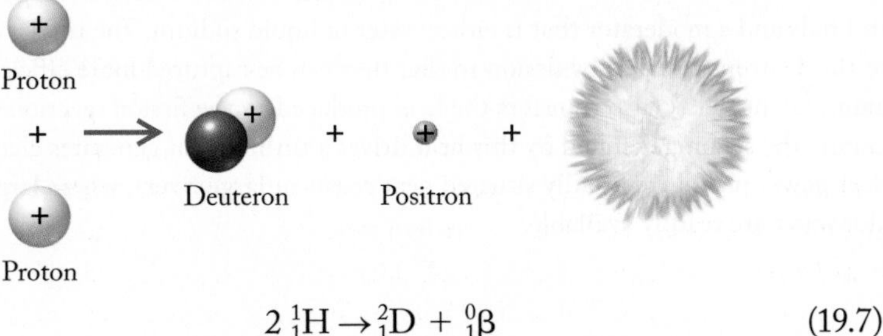

$$2\,{}^{1}_{1}H \rightarrow {}^{2}_{1}D + {}^{0}_{1}\beta \quad (19.7)$$

In the second stage of solar fusion, protons fuse with deuterons to form helium-3 nuclei:

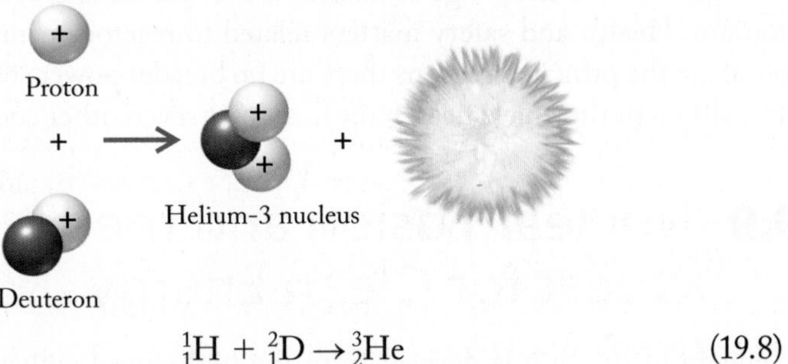

$${}^{1}_{1}H + {}^{2}_{1}D \rightarrow {}^{3}_{2}He \quad (19.8)$$

Finally, fusion of two helium-3 nuclei produces a helium-4 nucleus and two protons:

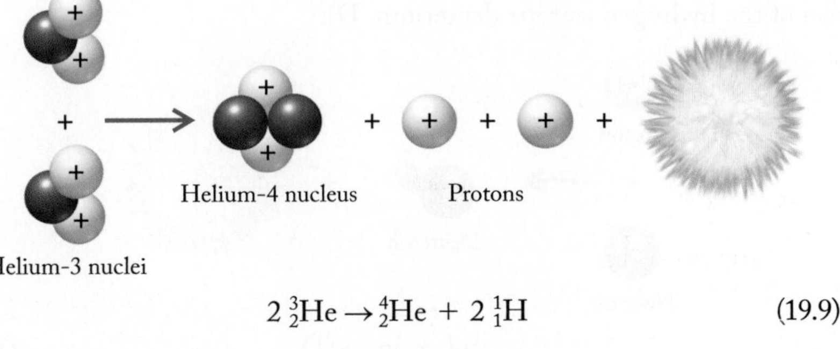

$$2\,{}^{3}_{2}He \rightarrow {}^{4}_{2}He + 2\,{}^{1}_{1}H \quad (19.9)$$

Deuterium and ^{3}He nuclei are intermediates in the hydrogen-fusion process because they are made in one step but then consumed in another. To write an overall equation for solar fusion, we combine Equations 19.7, 19.8, and 19.9, multiplying Equations 19.7 and 19.8 by 2 to balance the production and consumption of the intermediate particles:

$$\begin{array}{l} 2[2\,{}^1_1\text{H} \rightarrow {}^2_1\text{D} + {}^0_1\beta] \\ +\ 2[{}^1_1\text{H} + {}^2_1\text{D} \rightarrow {}^3_2\text{He}] \\ +\qquad 2\,{}^3_2\text{He} \rightarrow {}^4_2\text{He} + 2\,{}^1_1\text{H} \\ \hline 4\,\cancel{6}\,{}^1_1\text{H} + \cancel{2\,{}^2_1\text{D}} + \cancel{2\,{}^3_2\text{He}} \rightarrow \cancel{2\,{}^2_1\text{D}} + \cancel{2\,{}^3_2\text{He}} + {}^4_2\text{He} + 2\,{}^0_1\beta + \cancel{2\,{}^1_1\text{H}} \end{array}$$

This equation reduces to:

$$4\,{}^1_1\text{H} \rightarrow {}^4_2\text{He} + 2\,{}^0_1\beta \qquad (19.10)$$

Protons Helium-4 nucleus Positrons

Annihilation reactions between the positrons produced in Equation 19.10 and electrons in the matter surrounding the reactants release considerable energy, but most of the energy from hydrogen fusion comes from the loss in mass as four protons are transformed into an α particle (i.e., a helium-4 nucleus) and two positrons (see Table 19.1). In Sample Exercise 19.6 we use Einstein's equation, $E = mc^2$, to calculate how much energy this is.

For decades scientists and engineers have sought to harness the enormous energy released during hydrogen fusion for peaceful purposes. In 2010 construction began on ITER (originally an acronym for International Thermonuclear Experimental Reactor), a project to build the world's largest nuclear fusion reactor. Located at the Cadarache facility in southern France, the project is funded and run by the European Union, India, Japan, China, Russia, South Korea, and the United States. When it is operational, ITER will use a device called a *tokamak* (**Figure 19.22**) to heat a mixture of deuterium ($^2_1$H) and tritium ($^3_1$H) to temperatures near 1.5×10^8 K. At such temperatures, all these atoms are ionized, forming an incandescent plasma that is confined by the tokamak's powerful magnets to the center of a donut-shaped tunnel. High-speed collisions between deuterium and tritium nuclei in the plasma produce nuclei of helium-4:

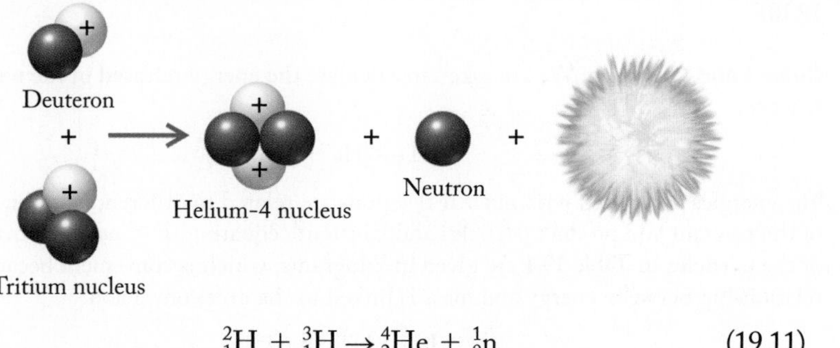

$${}^2_1\text{H} + {}^3_1\text{H} \rightarrow {}^4_2\text{He} + {}^1_0\text{n} \qquad (19.11)$$

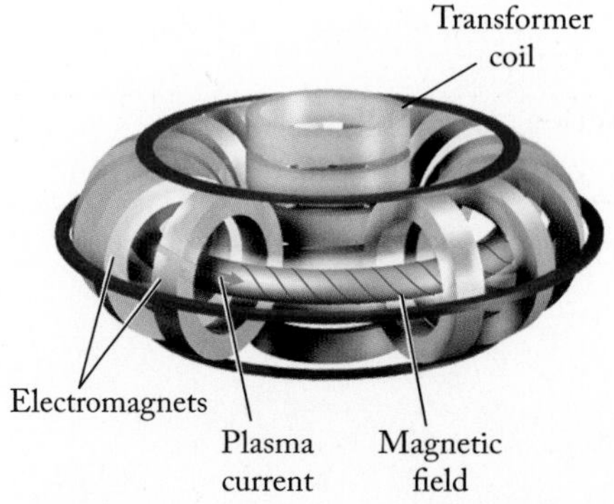

(a)

(b)

FIGURE 19.22 (a) A tokamak transmits electrical energy into a toroidal (donut-shaped) chamber containing deuterium and tritium, causing these isotopes of hydrogen to ionize and form a plasma of nuclei and free electrons with a temperature above 10^8 K. Combinations of electromagnets confine the plasma to the interior of the torus, where collisions between ^{2}H and ^{3}H nuclei result in fusion, forming ^{4}He nuclei and free neutrons. The neutrons then collide with Li atoms in the walls of the chamber, initiating additional nuclear reactions that produce more tritium fuel. (b) The tokamak at Princeton University fills a small warehouse.

The neutrons produced in the reaction collide with the nuclei of Li atoms in "breeder" blankets surrounding the hydrogen plasma. Two nuclear reactions are initiated by these collisions, depending on which isotope of Li is involved:

Neutron + Lithium-6 nucleus → Helium-4 nucleus + Tritium nucleus

$$^{1}_{0}n + ^{6}_{3}Li \rightarrow ^{4}_{2}He + ^{3}_{1}H \quad (19.12)$$

Neutron + Lithium-7 nucleus → Helium-4 nucleus + Tritium nucleus + Neutron

$$^{1}_{0}n + ^{7}_{3}Li \rightarrow ^{4}_{2}He + ^{3}_{1}H + ^{1}_{0}n \quad (19.13)$$

The reactions in Equations 19.12 and 19.13 produce tritium nuclei. In this way the reactions supply more fuel for the primary fusion reaction. The world's supply of deuterium, the other fuel, is enormous (seawater contains about 15 mg of deuterium per kilogram). On the other hand, tritium is not abundant in nature because it is radioactive, with a half-life of only 12.3 years. One disadvantage of these reactions is their reliance on lithium during a time when expanding production of lithium-ion batteries (see Section 18.7) is increasing global demand for the element. The National Ignition Facility located at the Lawrence Livermore National Laboratory in California has taken a different approach to initiating fusion reactions. Powerful lasers are focused on a small sample containing deuterium and tritium to start the fusion reaction.

CONCEPT TEST

Nuclear reactors powered by the energy released by the fission of uranium-235 have been operating since the 1950s, but a reactor powered by the energy released by the fusion of hydrogen has yet to be built. Why is it taking so long to build a fusion reactor?

SAMPLE EXERCISE 19.6 Calculating the Energy Released in a Nuclear Reaction — LO7

How much energy in joules is released by the overall fusion process in which four protons undergo nuclear fusion, producing an α particle and two positrons (Equation 19.10)?

Collect and Organize We are asked to calculate the energy released in the nuclear reaction:

$$4\,^{1}_{1}H \rightarrow ^{4}_{2}He + 2\,^{0}_{1}\beta \quad (19.10)$$

The energies associated with nuclear reactions are related to differences in the masses of the reactant and product particles and Einstein's equation, $E = mc^2$. The masses of the particles in Table 19.1 are given in kilograms, which is convenient because the relationship between energy and mass is linked to the unit conversion

$$1\ J = 1\ kg \cdot (m/s)^2$$

Analyze Given the value of the masses in Table 19.1, the difference in mass will probably be less than 10^{-27} kg. When multiplied by the square of the speed of light, $(2.998 \times 10^8 \text{ m/s})^2 \approx 10^{17}$, the calculated value of E should be less than 10^{-10} J.

Solve First we calculate the change in mass:

$$\Delta m = (m_{\alpha\text{ particle}} + 2\,m_{\text{positron}}) - 4\,m_{\text{proton}}$$
$$= [6.64465 \times 10^{-27} + (2 \times 9.10939 \times 10^{-31})]\text{ kg} - (4 \times 1.67262 \times 10^{-27})\text{ kg}$$
$$= -4.40081 \times 10^{-29}\text{ kg}$$

The energy corresponding to this loss in mass is calculated using Einstein's equation where $m = -4.40081 \times 10^{-29}$ kg:

$$E = mc^2$$
$$= -4.40081 \times 10^{-29}\text{ kg} \times (2.998 \times 10^8\text{ m/s})^2$$
$$= -3.955 \times 10^{-12}\text{ kg} \cdot (\text{m/s})^2 = -3.955 \times 10^{-12}\text{ J}$$

Think About It The decrease in mass translates into energy lost by the reaction system to its surroundings. As we predicted, the absolute value of this energy is less (actually much less) than 10^{-10} J, which seems like an awfully small value compared with the world's energy needs. However, this value applies to the formation of a single α particle. If we multiply it by Avogadro's number and convert it into a value in kilojoules per mole, a unit we typically use in thermochemistry, we get

$$\frac{-3.955 \times 10^{-12}\text{ J}}{\cancel{\alpha\text{ particle}}} \times \frac{6.022 \times 10^{23}\ \cancel{\alpha\text{ particles}}}{\text{mol}} \times \frac{1\text{ kJ}}{1000\text{ J}} = -2.382 \times 10^9\text{ kJ/mol}$$

To put this value in perspective, it is about 10^7 times the change in free energy from the combustion of 1 mole of hydrogen gas.

Practice Exercise How much energy is released in the nuclear reaction described by Equation 19.11? Express your answer in kilojoules per mole. (*Note*: The mass of a tritium nucleus is 5.00736×10^{-27} kg.)

SAMPLE EXERCISE 19.7 Integrating Concepts: Radium Girls and Safety in the Workplace

Radium was discovered by Pierre and Marie Curie in 1898. By 1902, the new element had its first practical use: radium compounds were mixed with zinc sulfide (ZnS) to make paint that glowed in the dark as α particles emitted by the decay of ^{226}Ra ($t_{1/2} = 1.6 \times 10^3$ yr) caused ZnS crystals to emit a greenish fluorescence (**Figure 19.23**). The paint was used to make dials for watches, clocks, and instruments used on naval vessels and, a few years later, in military and civilian airplanes.

By 1914, U.S. companies were making radium-painted dials and employing young women in their late teens and early 20s as dial painters. Soon after their employment, many of the women became very sick, suffering from anemia and other symptoms we now associate with exposure to nuclear radiation. Some developed bone cancer and more than 100 of them, who became known around the world as the Radium Girls, died. Their deaths were linked to the practice of "pointing" the fine paint brushes they used, which meant using their lips to make fine points on the brushes to help them paint the tiny numerals and hands on watch faces (**Figure 19.24**). In the process, they ingested some of the radioactive paint. Tests later determined that about 20% of the radium ingested was

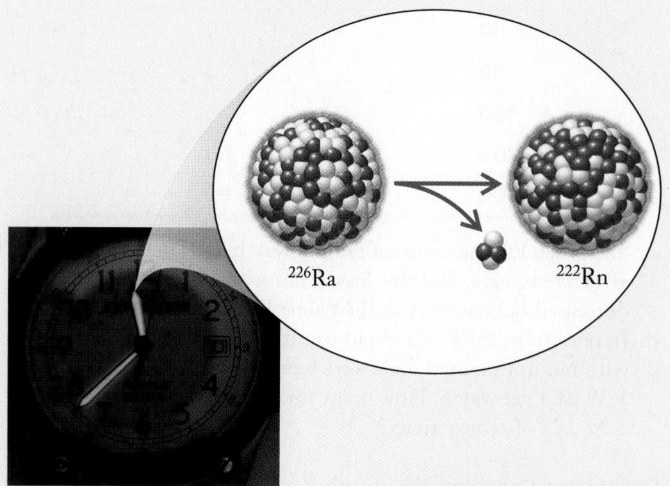

FIGURE 19.23 During the 20th century, many millions of watches and clocks had dials that glowed in the dark as high-energy α particles emitted by ^{226}Ra caused crystals of ZnS to fluoresce.

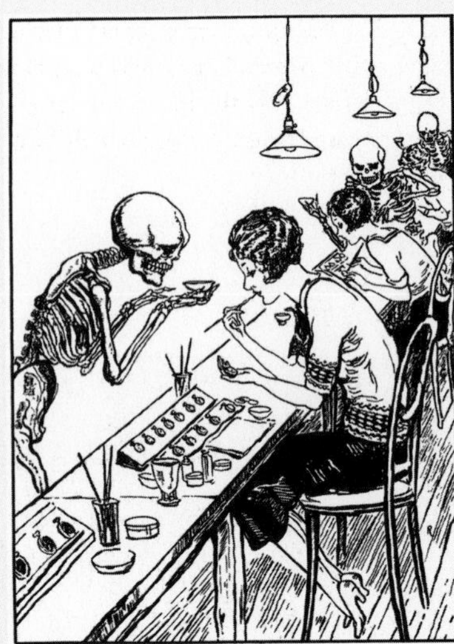

FIGURE 19.24 This editorial cartoon appeared in Sunday newspapers on February 28, 1926. It portrayed the deadly consequences of young women "pointing" their brushes with their lips as they painted the dials of watches with paint that contained radioactive radium.

incorporated into their bones, where it attacked bone marrow and caused malignancies known as osteosarcomas.

a. Suggest a reason why radium was concentrated in the victims' bones.
b. Studies of radiation levels and incidence of cancer in more than 1000 female dial painters yielded the results in the table presented here. Which of the two dose–response curves in Figure 19.16 best fits these results?

Radium Exposure (mg ingested)	Occurrence of Malignancy (% of workers exposed)
1	0
3	0
10	0
30	0
100	5
300	53
1000	85

c. The green luminescence of radium watch dials began to fade after a few years. Did this loss in luminosity result from decreased radioactivity in the paint? Explain why or why not.
d. In one study, the levels of radioactivity in pocket watches with radium-painted dials were found to be between 0.6 and 1.39 μCi per watch. How many micrograms of ^{226}Ra produce 1.39 μCi of radioactivity?

Collect and Organize We are asked (a) why ingested ^{226}Ra concentrates in bones; (b) whether malignancy in dial painters was proportional to their exposure to ^{226}Ra radiation or followed an S-shaped dose–response curve; (c) why the luminosity of radium-activated paint fades after a few years; and (d) how many micrograms of ^{226}Ra are needed to produce 1.39 mCi of radiation. One curie (Ci) is equal to 3.70×10^{10} decay events/s. The half-life ($t_{1/2}$) of ^{226}Ra is 1600 years and is related to the first-order rate constant (k) of the decay reaction by the equation $t_{1/2} = 0.693/k$. The level of radioactivity (A) in a sample of radium is equal to the product of the rate constant and the number (N) of ^{226}Ra atoms: $A = kN$.

Analyze Radium is a group 2 element and should have chemical and biochemical properties that are similar to those of the other elements in that group, including its association with biological tissues. High concentrations of another group 2 element, calcium, occur in teeth and bones. Worker exposure levels in the data table cover a wide range, but exposure up to nearly 100 mg ^{226}Ra caused few malignancies, whereas concentrations above 100 mg caused many. The half-life of ^{226}Ra is 1600 years, so the radioactivity of a sample decreases little over a few years or even over many decades. Relating a half-life expressed in years to a level of radioactivity expressed in a multiple of decay events per second requires converting units of time and then quantities of radioactive atoms to moles and then micrograms.

Solve

a. Radium probably accumulates in bones because its chemistry is similar to that of calcium, which means that $^{226}Ra^{2+}$ ions are likely to take the place of Ca^{2+} ions in bone.
b. Malignancies did not occur among the dial painters who ingested less than 100 mg of ^{226}Ra; however, the percentage of the women who suffered malignancies increased sharply with exposure between 100 and 1000 mg. This pattern is described by the S-shaped (blue) curve in Figure 19.16.
c. Given the 1600-year half-life of ^{226}Ra, the loss of watch dial luminescence was not the result of depleted radioactivity. Rather, it must have resulted from less efficient conversion of the energy of radioactive decay into visible light by ZnS crystals.
d. Let's first convert the half-life of ^{226}Ra into a decay rate constant in units of s^{-1}:

$$k = \frac{0.693}{t_{1/2}} = \frac{0.693}{1.6 \times 10^3 \cancel{\text{yr}}} \times \frac{1 \cancel{\text{yr}}}{365.25 \cancel{\text{d}}} \times \frac{1 \cancel{\text{d}}}{24 \cancel{\text{h}}} \times \frac{1 \cancel{\text{h}}}{3600 \text{ s}}$$
$$= 1.372 \times 10^{-11} \text{ s}^{-1}$$

Next, we solve the equation $A = kN$ for N, and we use the rate constant just calculated and the radioactivity of the watch to calculate the number of ^{226}Ra atoms in the dial:

$$N = \frac{A}{k} = \frac{1.39 \cancel{\mu\text{Ci}}}{1.372 \times 10^{-11} \text{ s}^{-1}} \times \frac{1 \cancel{\text{Ci}}}{10^6 \cancel{\mu\text{Ci}}} \times \frac{3.70 \times 10^{10} \text{ atoms Ra s}^{-1}}{\cancel{\text{Ci}}} = 3.75 \times 10^{15} \text{ atoms of Ra}$$

The corresponding mass in micrograms is

$$3.75 \times 10^{15} \cancel{\text{atoms Ra}} \times \frac{1 \cancel{\text{mol Ra}}}{6.022 \times 10^{23} \cancel{\text{atoms Ra}}} \times \frac{226 \text{ g Ra}}{1 \cancel{\text{mol Ra}}} \times \frac{10^6\ \mu\text{g}}{1 \text{ g}} = 1.4\ \mu\text{g of Ra}$$

Think About It The similarity in the level of radioactivity (1.39 μCi) in the watch dial and the mass of radium (1.41 μg) producing it is not a coincidence. When the curie was adopted as the standard unit of radioactivity in the early 20th century, it was chosen to honor the pioneering work of Marie and Pierre Curie, and it was based on what was then believed to be the level of radioactivity in 1 g of radium. Newspaper articles published in the 1920s made clear that Marie Curie was deeply troubled by the tragedy of the Radium Girls. Sadly, in 1934 she herself died from aplastic anemia—a disease caused by the inability of bone marrow to produce red blood cells.

SUMMARY

LO1 **Nuclear chemistry** is the study and application of reactions that involve changes in atomic nuclei. The **mass defect (Δ*m*)** of a nucleus is the difference between its mass and the sum of the masses of its nucleons. **Binding energy (BE)** is the energy needed to separate a nucleus into its component nucleons. Binding energy per nucleon is a measure of the relative stability of a nucleus compared to others of similar atomic number. When a particle of matter encounters a particle of antimatter, both are converted into energy (they annihilate one another), yielding γ rays. (Section 19.1)

LO2 Stable nuclei have neutron-to-proton ratios that fall within a range of values called the **belt of stability**. Unstable nuclides undergo **radioactive decay**. Neutron-rich nuclides (mass number greater than the average atomic mass) undergo β decay, whereas neutron-poor nuclides undergo **positron emission** or **electron capture**. Very heavy nuclides ($Z > 83$) may undergo β decay or **α decay**. (Section 19.2)

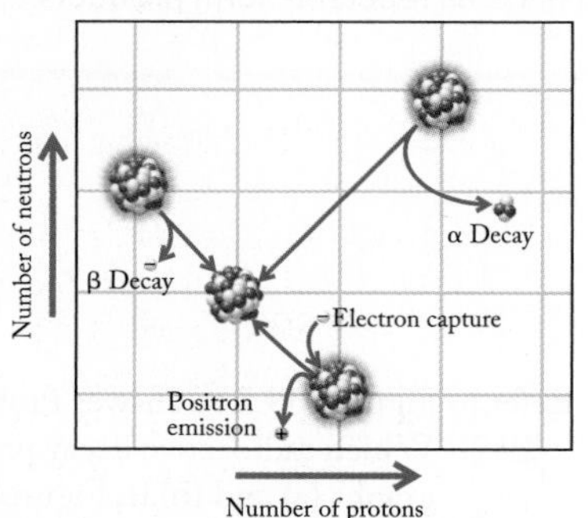

LO3 **Scintillation counters** and **Geiger counters** are used to measure levels of nuclear radiation. Radioactivity is the number of decay events per unit time and is related to the radionuclide's half-life. Common units are the **becquerel** (**Bq**; 1 decay event/s) and the **curie** (**Ci**; 1 Ci = 3.70×10^{10} Bq). (Section 19.3)

LO4 Radioactive decay follows first-order kinetics, so the half-life ($t_{1/2}$) of a radionuclide is a characteristic value of the decay process. (Section 19.4)

LO5 **Radiometric dating** is used to determine the age of an object from its content of a radionuclide or its decay product, or both. **Radiocarbon dating** involves measuring the amount of radioactive carbon-14 that remains in an object derived from plant or animal tissue to calculate the age of the object. The accuracy of the technique relies on calibration of the data with results of radiometric analyses of samples of known age, such as the trunks of trees that have lived for thousands of years and whose age can be confirmed by their growth rings. (Section 19.5)

LO6 The products of radioactive decay, α particles, β particles, and γ rays, have enough energy to break up molecules into electrons and cations and are examples of **ionizing radiation**, which can damage body tissue and DNA. The quantity of ionizing radiation energy absorbed per kilogram of body mass is called the *absorbed dose* and is expressed in **grays** (**Gy**; 1 Gy = 1.00 J/kg). The effective dose of any type of ionizing radiation is the product of the absorbed dose in grays and the **relative biological effectiveness (RBE)** of the radiation; the unit of effective dose is the **sievert (Sv)**. Alpha particles have a larger RBE than do β particles and γ rays but have the least penetrating power of these three types of ionizing radiation. Selected radioactive isotopes are useful as tracers in the human body to map biological activity and diagnose diseases. Other radioactive isotopes are used to treat cancers. (Sections 19.6, 19.7)

LO7 Neutron absorption by uranium-235 and a few other massive isotopes may lead to **nuclear fission**, creating lighter nuclei, accompanied by the release of energy that can be harnessed to generate electricity. A **chain reaction** happens when the neutrons released during fission collide with other fissionable nuclei. They require a **critical mass** of a fissionable isotope. A **breeder reactor** is used to make plutonium-239 from uranium-238, while also producing energy to make electricity. **Nuclear fusion** occurs when subatomic particles or atomic nuclei collide and fuse. Facilities that harness the enormous energy of hydrogen fusion must operate at temperatures greater than 10^8 K. (Sections 19.8, 19.9)

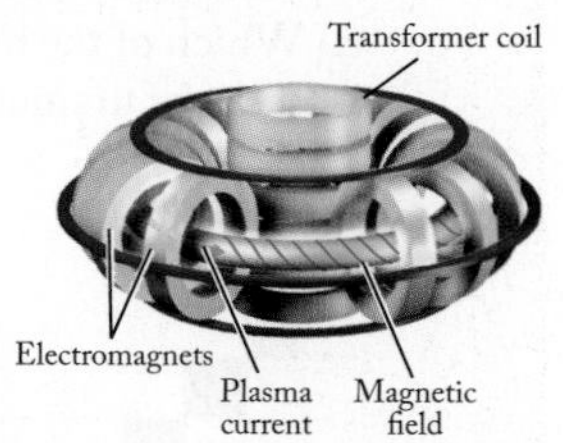

PARTICULATE PREVIEW WRAP-UP

Carbon-14 has 6 protons and 8 neutrons; therefore, with a neutron-to-proton ratio greater than 1, it will decay to reduce that ratio. When ^{14}C radioactively decays by emitting a β particle, it produces a nitrogen nuclide with a 1:1 neutron:proton ratio (7 protons and 7 neutrons).

PROBLEM-SOLVING SUMMARY

Type of Problem	Concepts and Equations	Sample Exercises
Predicting the modes and products of radioactive decay	Neutron-rich nuclides tend to undergo β decay, whereas neutron-poor nuclides undergo positron emission or electron capture.	**19.1**
Calculating the radioactivity (*A*) of a sample	$A = kN$ (19.1) where N is the number of atoms of a radionuclide and k, the rate constant of the radioactive decay process, is inversely proportional to the half-life of the process: $k = \frac{0.693}{t_{1/2}}$	**19.2**
Calculations involving half-lives and radiocarbon dating	$\ln N_t = -0.693 \frac{t}{t_{1/2}} + \ln N_0$ (19.3) $t = -\frac{t_{1/2}}{0.693} \ln \frac{N_t}{N_0}$ (19.4) where N_t/N_0 is the ratio of the quantity of radionuclide present in a sample at time t (N_t) to the quantity at $t = 0$ (N_0) and $t_{1/2}$ is the half-life of the radionuclide.	**19.3, 19.4**
Calculating effective dose	Effective dose = absorbed dose × RBE	**19.5**
Calculating the energy released in a nuclear reaction	$E = mc^2$, where m is the loss in mass as reactants form products.	**19.6**

VISUAL PROBLEMS

(Answers to boldface end-of-chapter questions and problems are in the back of the book.)

19.1. Figure P19.1 highlights six elements whose isotopes play important roles in nuclear chemistry.
a. Which of the highlighted elements currently plays a key role in the controlled fusion of hydrogen?
b. Exposure to which of the highlighted elements could cause anemia and bone disease?
c. Which of the highlighted elements are part of the decay series of uranium-238?

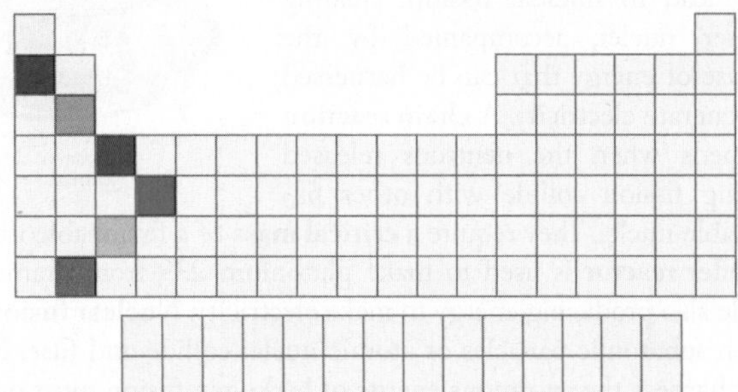

FIGURE P19.1

Refer to Figure P19.2 to answer Problems 19.2–19.5.

19.2. Which radioactive decay processes are represented by graphs (a) and (b) in Figure P19.2?

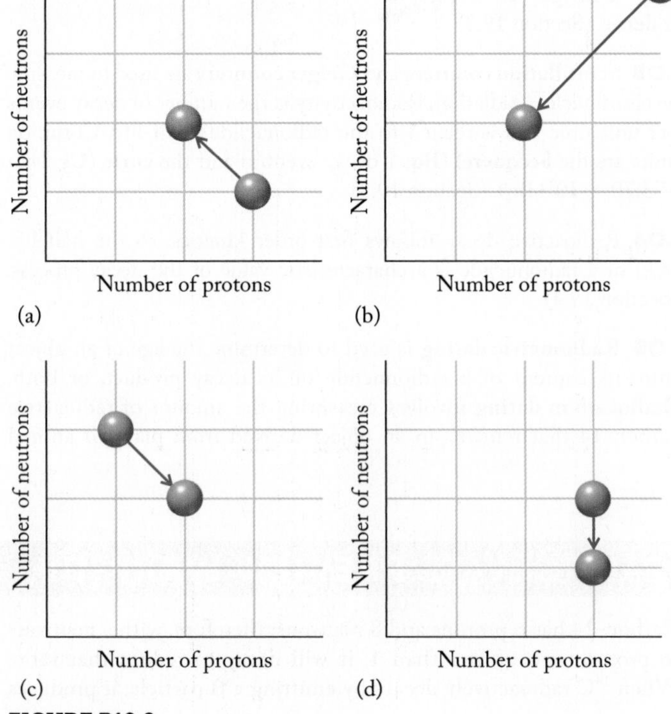

FIGURE P19.2

19.3. Which of the graphs in Figure P19.2 illustrates β decay?

19.4. Which of the graphs in Figure P19.2 illustrates the likely decay pathway for
a. ^{131}I
b. carbon-11
c. ^{222}Rn

19.5. In 1932, James Chadwick discovered neutrons by bombarding ^{9}Be with α particles to form ^{13}C, which then decayed to ^{12}C and a neutron. Which of the graphs in Figure P19.2 describes this decay process?

19.6. Which of the curves in Figure P19.6 represents the decay of an isotope that has a half-life of 2.0 days?

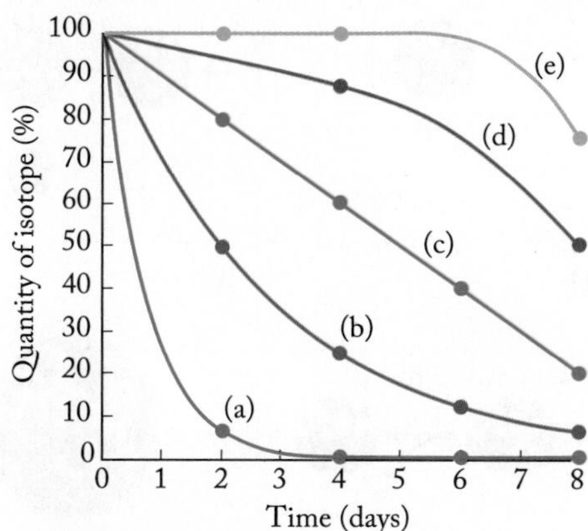

FIGURE P19.6

19.7. Which of the curves in Figure P19.6 do(es) *not* represent a radioactive decay curve?

19.8. Which of the models in Figure P19.8 represents fission and which represents fusion?

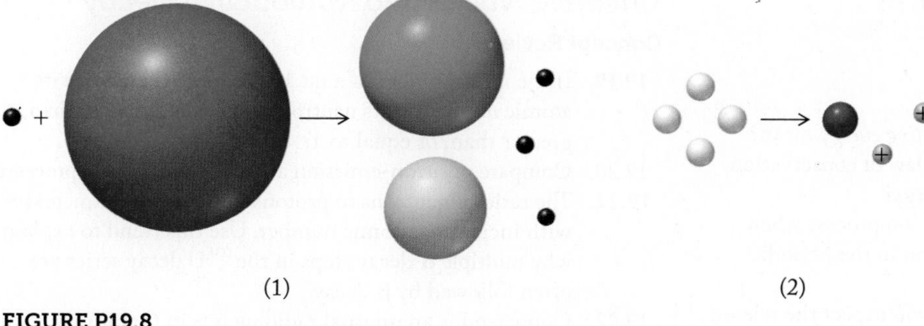

FIGURE P19.8

19.9. Isotopes in a nuclear decay series emit particles with a positive charge and particles with a negative charge. The two kinds of particles penetrate a column of water as shown in Figure P19.9. Is the "X" particle the positive or the negative one?

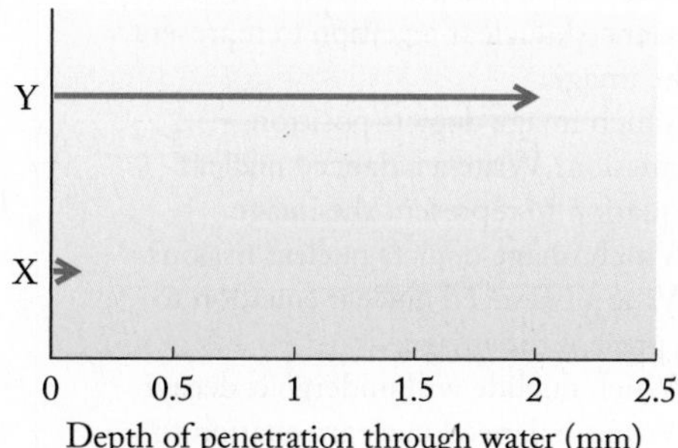

FIGURE P19.9

19.10. Use representations [A] through [I] in Figure P19.10 to answer questions (a)–(f).

a. Which image depicts β decay? Write a balanced nuclear equation to represent the image.
b. Which image depicts fusion? Write a balanced nuclear equation to represent the image.
c. Which image depicts positron emission? Write a balanced nuclear equation to represent the image.
d. Which image depicts nuclear fission? Write a balanced nuclear equation to represent the image.
e. Which nuclide will undergo α decay? Write a balanced nuclear equation to depict these decay processes.
f. Which nuclides will undergo β decay? Write a balanced nuclear equation to depict these decay processes.

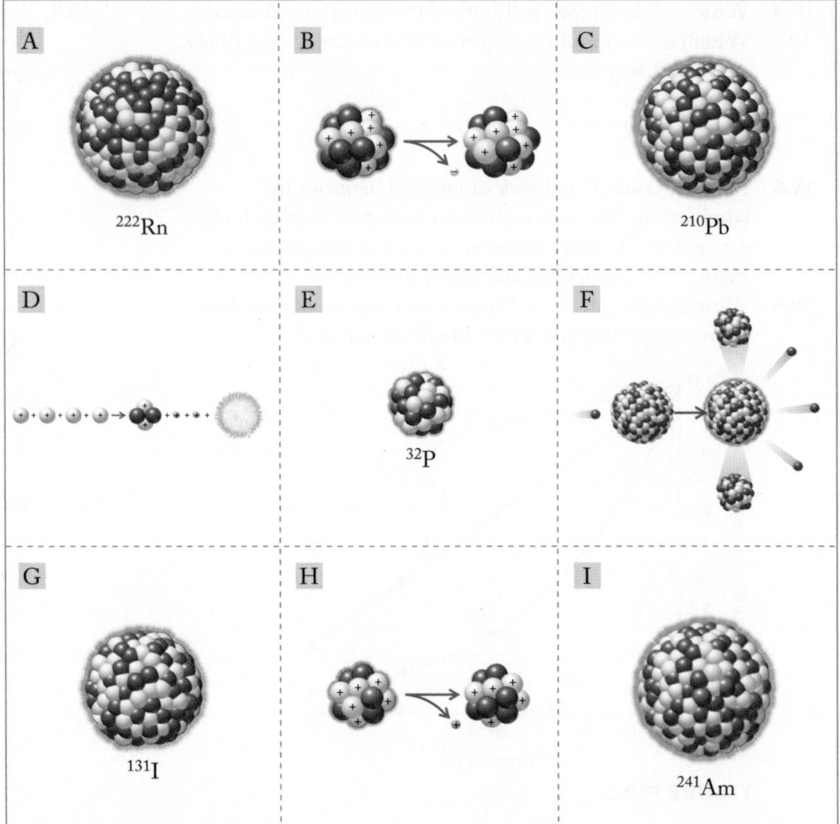

FIGURE P19.10

QUESTIONS AND PROBLEMS

Energy and Nuclear Stability

Concept Review

19.11. What do the terms *mass defect* and *binding energy* mean?

19.12. Do nuclear decay processes follow the law of conservation of mass and law of conservation of energy?

19.13. Why is energy released in a nuclear fusion process when the product is an element preceding iron in the periodic table?

*19.14. Titanium has five stable isotopes. Do you expect the release of energy in nuclear fusion processes that produce any of these isotopes?

Problems

19.15. What is the binding energy of a deuteron?

19.16. What is the binding energy of ^{6}Li, which has a nuclear mass of 9.98561×10^{-27} kg?

19.17. Our Sun is a fairly small star that has barely enough mass to fuse hydrogen into helium. Along the way, an isotope of helium, helium-3, is formed. Calculate the binding energy per nucleon of helium-3 on the basis of the following masses: $^{3}_{2}$He (3.01693 u), $^{1}_{1}$p (1.00728 u), and $^{1}_{0}$n (1.00866 u).

19.18. What is the binding energy per nucleon of ^{12}C, which has an atomic mass of 12.00000 u? (*Note*: Atomic mass includes the mass of 6 electrons.)

Unstable Nuclei and Radioactive Decay

Concept Review

19.19. If the mass number of a nuclide is more than twice its atomic number, is its neutron-to-proton ratio less than, greater than, or equal to 1?

19.20. Compare positron-emission and electron-capture processes.

19.21. The ratio of neutrons to protons in stable nuclei increases with increasing atomic number. Use this trend to explain why multiple α decay steps in the ^{238}U decay series are often followed by β decay.

19.22. Copper-64 is an unusual radionuclide in that it may undergo β decay, positron emission, or electron capture. Using Figure 19.2 as a guide, explain why ^{64}Cu decays in these multiple ways?

Problems

19.23. Iodine-137 decays to give xenon-137, which decays to give cesium-137. What are the modes of decay in these two reactions?

19.24. Write a balanced nuclear equation describing (a) α decay of curium-242; (b) β decay of magnesium-28; (c) positron emission by xenon-118; (d) electron capture by cadmium-104.

19.25. Arrange the following nuclides in order of increasing neutron:proton ratio. Which has the greatest number of neutrons? (a) ^{25}Al; (b) ^{24}Na; (c) ^{11}B, (d) ^{14}C

19.26. In each of the following pairs of nuclides, select the one that has more protons and the one that has more neutrons. Also indicate which, if any, of the pairs have the same number of neutrons or protons. (a) ^{63}Cu and ^{65}Cu; (b) ^{71}Ga and ^{71}Ge; (c) ^{39}K and ^{40}Ar; (d) ^{94}Sr and ^{88}Y.

19.27. Medical Applications of ^{64}Cu and ^{89}Zr Copper-64 and zirconium-89 are two of the radionuclides used in positron emission tomography. Both are produced in nuclear reactors by bombarding targets of other metals with protons.
a. Which transition metals would you choose as the targets to make ^{64}Cu and ^{89}Zr by proton bombardment?
b. Write radiochemical equations describing the production of ^{64}Cu and ^{89}Zr. Assume one proton was captured by each target nucleus and one free neutron was produced in addition to the desired radionuclide.
c. Do the target nuclides you used to answer part (b) have to be stable isotopes of the target metals? Explain why or why not?

19.28. **Manufacturing Isotopes** The isotopes used in positron emission tomography are all prepared in nuclear reactors. One route to these isotopes is through proton bombardment followed by loss of a neutron in a cyclotron.
a. Which metal isotopes would yield ^{67}Ga and ^{111}In by this process?
b. Would bombarding target metals with neutrons instead of protons be another way to produce radionuclides for PET? Explain why you do or don't think so.

19.29. Predict the mode(s) of decay for the following radioactive isotopes: (a) ^{10}C; (b) ^{19}Ne; (c) ^{51}Ti.

19.30. Predict the mode(s) of decay of the following radionuclides: (a) ^{24}Ne; (b) ^{38}K; (c) ^{45}Ti; (d) ^{237}Np.

19.31. Elements in a Supernova The isotopes ^{56}Co and ^{44}Ti were detected in supernova SN 1987A. Predict the decay pathway for these radioactive isotopes.

19.32. There are isotopes of nitrogen that have as few as 5 or as many as 11 neutrons in each of their nuclei. Write a balanced nuclear equation describing the decay of the isotope with 11 neutrons.

***19.33.** Chlorine has isotopes with mass numbers from 32 through 39. Two of them, ^{35}Cl and ^{37}Cl, are stable.
a. Which three of the other isotopes emit positrons?
b. Which of the other isotopes emit β particles?
c. Which one of the other isotopes can emit *either* positrons or β particles?

*19.34. Bromine has isotopes with mass numbers from 74 through 90. Two of them, ^{79}Br and ^{81}Br, are stable.
a. How many of the others emit positrons or undergo electron capture?
b. How many of the others emit β particles?
c. Which one of the other isotopes can emit *either* positrons or β particles?

Measuring and Expressing Radioactivity; Calculations Involving Half-Lives of Radionuclides

Concept Review

19.35. Why does nuclear decay follow first-order reaction kinetics?

19.36. Barium-140 undergoes β decay to lanthanum-140 with a half-life of 12.8 d. Lanthanum-140 also undergoes β decay to cerium-140 with a half-life of 40 h. Which step is rate-determining for the overall decay of ^{140}Ba to ^{140}Ce?

19.37. What property of radioactivity allows it to be detected by a Geiger counter?

19.38. Which of the following quantities of radioactivity is most likely to be found in a household appliance such as a smoke detector: (a) 1 KCi; (b) 1 Ci; (c) 1 mCi; (d) 1 μCi?

Problems

19.39. The half-life of radon-222, a radioactive gas found in some basements, is 3.82 d. Calculate the decay rate constant of radon-222.

19.40. The decay rate constant of sodium-24, a tracer used in blood studies, is $4.6 \times 10^{-2}\ h^{-1}$. What is the value of its half-life?

19.41. Explosions at a disabled nuclear power station in Fukushima, Japan, in 2011 may have released more cesium-137 ($t_{1/2} = 30.2$ yr) into the ocean than any other single event. How long will it take the radioactivity of this radionuclide to decay to 5.0% of the level released in 2011?

19.42. Spent fuel removed from nuclear power stations contains plutonium-239 ($t_{1/2} = 2.41 \times 10^4$ yr). How long will it take a sample of this radionuclide to reach a level of radioactivity that is 2.5% of the level it had when it was removed from a reactor?

Radiometric Dating

Concept Review

19.43. Explain why radiocarbon dating is reliable only for artifacts less than about 50,000 years old.

19.44. Which of the following statements about ^{14}C dating are true?
a. The amount of ^{14}C in all sources of carbon is the same.
b. Carbon-14 is unstable and decays rapidly in the atmosphere.
c. The ratio of ^{14}C to ^{12}C in the atmosphere is a constant.
d. Green plants incorporate ^{12}C, ^{13}C, and ^{14}C during photosynthesis.

19.45. Why is ^{40}K dating ($t_{1/2} = 1.28 \times 10^9$ yr) useful only for rocks older than 300,000 years?

19.46. Where does the ^{14}C in biological samples come from?

Problems

19.47. First Humans in South America Archeologists continue to debate the arrival of the first humans in the Western Hemisphere. Radiocarbon dating of charcoal from a cave in Chile was used to establish the earliest date of human habitation in South America as 8700 years ago. What fraction of the ^{14}C initially present remained in the charcoal after 8700 years?

19.48. **Early Financial Records** For thousands of years Native Americans living along the north coast of Peru used knotted cotton strands called *quipu* (Figure P19.48) to record financial transactions and governmental actions. A particular quipu sample is 4800 years old. Compared with the fibers of cotton plants growing today, what is the ratio of carbon-14 to carbon-12 in the sample?

FIGURE P19.48

*19.49. **Rings in Sequoia Trees** Figure P19.49 shows a close-up of the center of a giant sequoia tree cut down in 1891 in what is now Kings Canyon National Park. It contained 1342 annual growth rings. If samples of the tree were removed for radiocarbon dating today, what would be the difference in $^{14}C/^{12}C$ ratio in the innermost (oldest) ring compared with that ratio in the youngest ring?

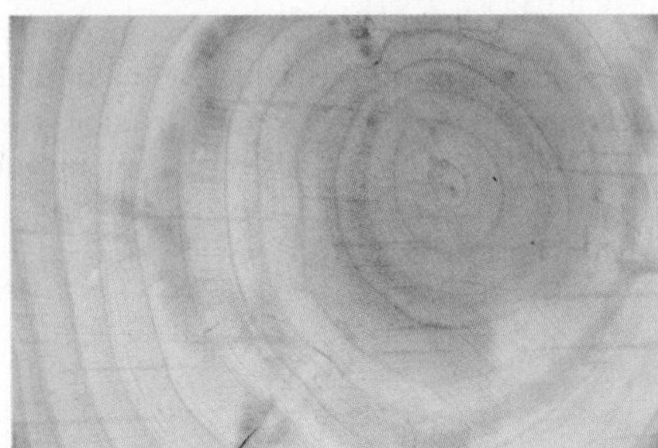

FIGURE P19.49

*19.50. **Dating Volcanic Eruptions** Geologists who study volcanoes can develop historical profiles of previous eruptions by determining the $^{14}C/^{12}C$ ratios of charred plant remains entrapped in old magma and ash flows. If the uncertainty in determining these ratios is 0.1%, could radiocarbon dating distinguish between debris from the eruptions of Mt. Vesuvius that occurred in the years 472 and 512? (*Hint*: Calculate the $^{14}C/^{12}C$ ratios for samples from the two dates.)

19.51. **Age of Mammoth Tusk** Figure P19.51 shows a carved mammoth tusk that was uncovered at an ancient campsite in the Ural Mountains in 2001. The $^{14}C/^{12}C$ ratio in the tusk was only 1.19% of that in modern elephant tusks. How old is the mammoth tusk?

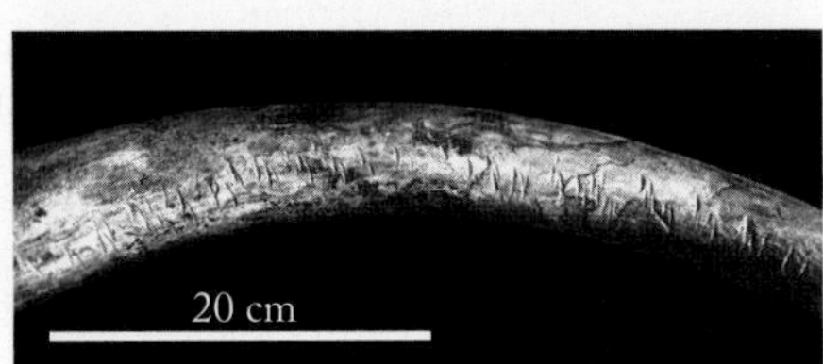

FIGURE P19.51

19.52. **Destruction of Jericho** The Bible describes the Exodus as a period of 40 years that began with plagues in Egypt and ended with the destruction of Jericho. Archeologists seeking to establish the exact dates of these events have proposed that the plagues coincided with a huge eruption of the volcano Thera in the Aegean Sea.

a. Radiocarbon dating suggests that the eruption occurred around 1360 BCE, although other records place the eruption of Thera in the year 1628 BCE. What is the percent difference in the ^{14}C decay rate in biological samples from these two dates?
b. Radiocarbon dating of blackened grains from the site of ancient Jericho provides a date of 1315 BCE $\pm$ 13 years for the fall of the city. What is the $^{14}C/^{12}C$ ratio in the blackened grains compared with that of grain harvested last year?

Biological Effects of Radioactivity

Concept Review

19.53. What is the difference between a *level* of radioactivity and a *dose* of radioactivity?

19.54. What are some of the molecular effects of exposure to radioactivity?

19.55. Describe the dangers of exposure to radon-222.

19.56. **Food Safety** Periodic outbreaks of food poisoning from meat contaminated by *E. coli* have renewed the debate about irradiation as an effective treatment of food. In one newspaper article on the subject, the following statement appeared: "Irradiating food destroys bacteria by breaking apart their molecular structure." How would you improve or expand on this explanation?

Problems

19.57. **Radiation Exposure from Dental X-rays** Dental X-rays expose patients to about 5 μSv of radiation. Given an RBE of 1 for X-rays, how many grays of radiation does 5 μSv represent? For a 50 kg person, how much energy does 5 μSv correspond to?

*19.58. **Radiation Exposure at Chernobyl** Some workers responding to the explosion at the Chernobyl nuclear power plant were exposed to 5 Sv of radiation, killing many of them. If the exposure was primarily in the form of γ rays with an energy of 3.3×10^{-14} J and an RBE of 1, how many γ rays did an 80-kg person absorb?

*19.59. **Strontium-90 in Milk** In the years immediately following the explosion at the Chernobyl nuclear power plant, the concentration of ^{90}Sr in cow's milk in southern Europe was slightly elevated. Some samples contained as much as 1.25 Bq/L of ^{90}Sr radioactivity. The half-life of strontium-90 is 28.8 yr.

a. Write a balanced nuclear equation describing the decay of ^{90}Sr.
b. How many atoms of ^{90}Sr are in a 200 mL glass of milk with 1.25 Bq/L of ^{90}Sr radioactivity?
c. Why would strontium-90 be more concentrated in milk than other foods, such as grains, fruits, or vegetables?

*19.60. **Radium Watch Dials** If exactly 1.00 μg of ^{226}Ra was used to paint the glow-in-the-dark dial of a wristwatch made in 1914, how radioactive is the watch today? Express your answer in microcuries and becquerels. The half-life of ^{226}Ra is 1.60×10^3 years.

19.61. In 1999 the U.S. Environmental Protection Agency set a maximum radon level for drinking water at 4.0 pCi per milliliter.

a. How many decay events occur per second in a milliliter of water for this level of radon radioactivity?
b. If the 4.0 pCi/mL level were due to decay of ^{222}Rn ($t_{1/2}$ = 3.8 days), how many ^{222}Rn atoms would there be in 1.0 mL of water?

19.62. **Death of a Spy** A former Russian spy died from radiation sickness in 2006 after dining at a London restaurant where he apparently ingested polonium-210. The other people at his table did not suffer from radiation sickness, even though they were very near the radioactive tea the victim drank. Why were they unaffected?

Medical Applications of Radionuclides

Concept Review

19.63. How does the selection of an isotope for radiotherapy relate to (a) its half-life, (b) its mode of decay, and (c) the properties of the products of decay?

19.64. For imaging, why is an isotope with a shorter half-life preferable to one with a longer $t_{1/2}$?

19.65. Are the same radioactive isotopes likely to be used for both imaging and cancer treatment? Why or why not?

19.66. Iridum-192 is used in some cancer treatments. ^{192}Ir decays by three pathways: β decay, positron emission, and electron capture. Could all three decay pathways be used in imaging?

Problems

19.67. Predict the most likely mode of decay for the following isotopes used as imaging agents in nuclear medicine: (a) ^{197}Hg (kidney); (b) ^{75}Se (parathyroid gland); (c) ^{18}F (bone).

19.68. Predict the most likely mode of decay for the following isotopes used as imaging agents in nuclear medicine: (a) ^{133}Xe (cerebral blood flow); (b) ^{57}Co (tumor detection); (c) ^{51}Cr (red blood cell mass); (d) ^{67}Ga (tumor detection).

19.69. A 1.00 mg sample of ^{192}Ir was inserted into the artery of a heart patient. After 30 days, 0.756 mg remained. What is the half-life of ^{192}Ir?

19.70. In a treatment that decreases pain and reduces inflammation of the lining of the knee joint, a sample of dysprosium-165 with a radioactivity of 1100 counts per second was injected into the knee of a patient suffering from rheumatoid arthritis. After 24 h, the radioactivity had dropped to 1.14 counts per second. Calculate the half-life of ^{165}Dy.

19.71. **Study of Tourette's Syndrome** Tourette's syndrome is a condition whose symptoms include sudden movements and vocalizations. Iodine isotopes are used in brain imaging of people with Tourette's syndrome. To study the uptake and distribution of iodine in cells, researchers treated mammalian brain cells in culture with a solution containing ^{131}I with an initial radioactivity of 108 counts per minute. The cells were removed after 30 days, and the remaining solution was found to have a radioactivity of 4.1 counts per minute. What percentage of the ^{131}I ($t_{1/2}$ = 8.1 days) is absorbed by the brain?

19.72. **Mercury Test of Kidney Function** A patient is administered mercury-197 to evaluate kidney function. Mercury-197 has a half-life of 65 h. What fraction of an initial dose of mercury-197 remains after 6 days?

19.73. Carbon-11 is an isotope used in positron emission tomography and has a half-life of 20.4 min. How long will it take for 99% of the ^{11}C injected into a patient to decay?

19.74. **Leukemia Treatment Using Sodium** Sodium-24 is used to treat leukemia and has a half-life of 15 h. In a patient injected with a salt solution containing sodium-24, what percentage of the ^{24}Na remains after 48 h?

***19.75.** **Boron Neutron-Capture Therapy** In boron neutron-capture therapy (BNCT), a patient is given a compound containing ^{10}B that accumulates inside cancer tumors. Then the tumors are irradiated with neutrons, which are absorbed by ^{10}B nuclei. The product of neutron capture is an unstable form of ^{11}B that undergoes α decay to ^{7}Li.

a. Write a balanced nuclear equation for the neutron absorption and the α decay process.
b. Calculate the energy released by each nucleus of boron-10 that captures a neutron and undergoes α decay, given the following masses of the particles in the process: ^{10}B (10.0129 u), ^{7}Li (7.01600 u), ^{4}He (4.00260 u), and ^{1}n (1.00866 u).
c. Why is the formation of a nuclide that undergoes α decay a particularly effective cancer therapy?

19.76. **Balloon Angioplasty and Arteriosclerosis** Balloon angioplasty is a common procedure for unclogging arteries in patients suffering from arteriosclerosis. Iridium-192 therapy is being tested as a treatment to prevent reclogging of the arteries. In the procedure, a thin ribbon containing pellets of ^{192}Ir is threaded into the artery. The half-life of ^{192}Ir is 74 days. How long will it take for 99% of the radioactivity from 1.00 mg of ^{192}Ir to disappear?

Nuclear Fission

Concept Review

19.77. How is the rate of fission controlled in a nuclear reactor used to generate electrical power?

19.78. How does a breeder reactor create fuel and energy at the same time?

***19.79.** Why are neutrons always by-products of the fission of the most massive nuclides? (*Hint*: Look closely at the neutron-to-proton ratios shown in Figure 19.2.)

19.80. Seaborgium (Sg, element 106) is prepared by the bombardment of curium-248 with neon-22, which produces two isotopes, ^{265}Sg and ^{266}Sg. Write balanced nuclear reactions for the formation of both isotopes. Are these reactions better described as fusion or fission processes?

Problems

19.81. The fission of uranium produces dozens of isotopes. For each of the following fission reactions, determine the identity of the unknown nuclide:

a. $^{235}U + ^1_0n \rightarrow ^{96}Zr + ? + 2\,^1_0n$
b. $^{235}U + ^1_0n \rightarrow ^{99}Nb + ? + 4\,^1_0n$
c. $^{235}U + ^1_0n \rightarrow ^{90}Rb + ? + 3\,^1_0n$

19.82. For each of the following fission reactions, determine the identity of the unknown nuclide:

a. $^{235}U + ^1_0n \rightarrow ^{137}I + ? + 2\,^1_0n$
b. $^{235}U + ^1_0n \rightarrow ^{137}Cs + ? + 3\,^1_0n$
c. $^{235}U + ^1_0n \rightarrow ^{141}Ce + ? + 2\,^1_0n$

19.83. Thorium-Based Reactors Thorium-232 has been proposed as an alternative to uranium as the fuel for nuclear reactors. The relevant nuclear reactions are

$$^{232}_{90}Th \xrightarrow{^1_0n} {}^{233}_{90}Th \xrightarrow[t_{1/2}\,=\,22.2\text{ min}]{^0_{-1}\beta} {}^{233}_{91}Pa \xrightarrow[t_{1/2}\,=\,27\text{ d}]{^0_{-1}\beta} {}^{233}_{92}U$$

a. Do any of these reactions represent fission?
b. Why is there no danger of an uncontrolled chain reaction in a ^{232}Th-powered reactor?
c. The exact masses of ^{232}Th and ^{233}U are 232.03805 and 233.0395 u, respectively. How much energy is released per atom of ^{232}Th?

19.84. **The Discovery of Fission** The fission of uranium was suspected by Lise Meitner, Fritz Strassman, and Otto Hahn in 1938 after they observed that bombardment of uranium with neutrons produced barium.

a. How is the observation of Ba from this reaction evidence for fission?
b. If the ^{140}Ba is the result of fission of ^{235}U, what is the other product if three neutrons are also produced?
c. Explain how this process could lead to a chain reaction?

Nuclear Fusion and the Quest for Clean Energy

Concept Review

19.85. In what ways are the fusion reactions that formed α particles during primordial nucleosynthesis different from those that fuel our Sun today?

19.86. How are the fusion reactions that are the basis for power production in the tokamak described in Section 19.9 different from those that power our Sun?

Problems

19.87. All of the following fusion reactions produce ^{28}Si. Calculate the energy released in each reaction from the masses of the isotopes: ^2H (2.0146 u), ^4He (4.00260 u), ^{10}B (10.0129 u), ^{12}C (12.00000 u), ^{14}N (14.00307 u), ^{16}O (15.99491 u), ^{24}Mg (23.98504 u), ^{28}Si (27.97693 u).

a. $^{14}N + ^{14}N \rightarrow ^{28}Si$
b. $^{10}B + ^{16}O + ^2H \rightarrow ^{28}Si$
c. $^{16}O + ^{12}C \rightarrow ^{28}Si$
d. $^{24}Mg + ^4He \rightarrow ^{28}Si$

19.88. All of the following fusion reactions produce ^{32}S. Calculate the energy released in each reaction from the masses of the isotopes: ^4He (4.00260 u), ^6Li (6.01512 u), ^{12}C (12.00000 u), ^{14}N (14.00307 u), ^{16}O (15.99491 u), ^{24}Mg (23.98504 u), ^{28}Si (27.97693 u), ^{32}S (31.97207 u).

a. $^{16}O + ^{16}O \rightarrow ^{32}S$
b. $^{28}Si + ^4He \rightarrow ^{32}S$
c. $^{14}N + ^{12}C + ^6Li \rightarrow ^{32}S$
d. $^{24}Mg + 2\,^4He \rightarrow ^{32}S$

19.89. Tokamak Radiochemistry What is the change in energy per nucleus of tritium during the following reactions?

a. $^1_0n + ^6_3Li \rightarrow ^4_2He + ^3_1H$
b. $^1_0n + ^7_3Li \rightarrow ^4_2He + ^3_1H + ^1_0n$

19.90. It has been proposed that electrical power production in the future might be based on the fusion of deuterium to helium-4.

a. Write a radiochemical equation describing the reaction (assume that ^4He is the only product).
b. Calculate how much energy is released during the formation of 1 mole of ^4He.

Additional Problems

19.91. Powering a Starship Thirty years before the creation of antihydrogen, television producer and *Star Trek* creator Gene Roddenberry (1921–1991) proposed to use this form of antimatter to fuel the powerful "warp" engines of the fictional starship *Enterprise*.

a. Why would antihydrogen have been a particularly suitable fuel?
b. Describe the challenges of storing such a fuel on a starship.

19.92. **Accuracy in Labeling** Tiny concentrations of radioactive tritium (^3_1H) occur naturally in rain and groundwater. The half-life of ^3_1H is 12 years. Assuming that tiny concentrations of tritium can be determined accurately, could the isotope be used to determine whether a bottle of wine with the year 1969 on its label actually contained wine made from grapes that were grown in 1969? Explain your answer.

19.93. The energy released during the fission of ^{235}U is about 3.2×10^{-11} J per atom of the isotope. Compare this quantity of energy with that released by the fusion of four hydrogen atoms to make an atom of helium-4:

$$4\,^1_1H \rightarrow ^4_2He + 2\,^0_1\beta$$

Assume that the positrons are annihilated in collisions with electrons so that the masses of the positrons are converted into energy. In your comparison, express the energies released by the fission and fusion processes in joules per nucleon for ^{235}U and ^4He, respectively.

19.94. How much energy is required to remove a neutron from the nucleus of an atom of carbon-13 (mass = 13.00335 u)? (*Hint*: The mass of an atom of carbon-12 is exactly 12.00000 u.)

19.95. Smoke Detectors Americium-241 ($t_{1/2}$ = 433 yr) is used in smoke detectors. The α particles from this isotope ionize nitrogen and oxygen in the air, creating an electric current. When smoke is present, the current decreases, setting off the alarm.

a. Does a smoke detector bear a closer resemblance to a Geiger counter or to a scintillation counter?
b. How long will it take for the radioactivity of a sample of ^{241}Am to drop to 1% of its original radioactivity?
c. Why are smoke detectors containing ^{241}Am safe to handle without protective equipment?

*19.96. **Colorectal Cancer Treatment** Cancer therapy with radioactive rhenium-188 shows promise in patients with colorectal cancer.

a. Write the symbol for rhenium-188 and determine the number of neutrons, protons, and electrons.

b. Are most rhenium isotopes likely to have fewer neutrons than rhenium-188?

c. The half-life of rhenium-188 is 17 h. If it takes 30 min to bind the isotope to an antibody that delivers the rhenium to the tumor, what percentage of the rhenium remains after binding to the antibody?

d. The effectiveness of rhenium-188 is thought to result from penetration of β particles as deep as 8 mm into the tumor. Why wouldn't an α emitter be more effective?

e. Using an appropriate reference text, such as the *CRC Handbook of Chemistry and Physics*, pick out the two most abundant isotopes of rhenium. List their natural abundances and explain why the one that is radioactive decays by the pathway that it does.

19.97. Synthesis of a New Element In 2006 an international team of scientists confirmed the synthesis of a total of three atoms of $^{294}_{118}Og$ in experiments run in 2002 and 2005. They had bombarded a ^{249}Cf target with ^{48}Ca nuclei.

a. Write a balanced nuclear equation describing the synthesis of oganesson, Og.

b. The synthesized isotope of Og undergoes α decay ($t_{1/2}$ = 0.9 ms). What nuclide is produced by the decay process?

c. The nuclide produced in part (b) also undergoes α decay ($t_{1/2}$ = 10 ms). What nuclide is produced by this decay process?

d. The nuclide produced in part (c) also undergoes α decay ($t_{1/2}$ = 0.16 s). What nuclide is produced by this decay process?

e. If you had to select an element that occurs in nature and that has physical and chemical properties similar to Og, which element would it be?

*19.98. **Imaging with Thulium** Thulium-165 emits positrons, producing an isotope of erbium that decays, in turn, by electron capture to a stable isotope of holmium. These properties could be exploited in PET. The half-lives for the two decay processes are 30.1 h and 10.3 h, respectively. Which isotope will be present in the greatest amount after 24 h?

19.99. Which element in the following series will be present in the greatest amount after one year?

$$^{214}_{83}Bi \xrightarrow{\alpha} {}^{210}_{81}Tl \xrightarrow{\beta} {}^{210}_{82}Pb \xrightarrow{\beta} {}^{210}_{83}Bi \longrightarrow$$

$t_{1/2}$ = 20 min   1.3 min   20 yr   5 d

*19.100. **Dating Cave Paintings** Cave paintings in Gua Saleh Cave in Borneo have been dated by measuring the amount of ^{14}C in calcium carbonate that formed over the pigments used in the paint. The source of the carbonate ion was atmospheric CO_2.

a. What is the ratio of the ^{14}C radioactivity in calcium carbonate that formed 9900 years ago to that in calcium carbonate formed today?

b. The archeologists also used a second method, uranium–thorium dating, to confirm the age of the paintings by measuring trace quantities of these elements present as contaminants in the calcium carbonate. Shown below are two candidates for the U–Th dating method. Which isotope of uranium do you suppose was chosen? Explain your answer.

$$^{235}_{92}U \longrightarrow {}^{231}_{90}Th \longrightarrow {}^{231}_{91}Pa$$

$$^{234}_{92}U \longrightarrow {}^{230}_{90}Th \longrightarrow {}^{226}_{91}Pa$$

Isotope	$t_{1/2}$	Isotope	$t_{1/2}$
^{234}U	2.24×10^5 yr	^{231}Th	25.6 h
^{235}U	7.04×10^8 yr	^{226}Pa	1600 yr
^{230}Th	7.7×10^4 yr	^{231}Pa	3.26×10^4 yr

19.101. The synthesis of new elements and specific isotopes of known elements in linear accelerators involves the fusion of smaller nuclei.

a. An isotope of platinum can be prepared from nickel-64 and tin-124. Write a balanced equation for this nuclear reaction. (You may assume that no neutrons are ejected in the fusion reaction.)

b. Substituting tin-132 for tin-124 increases the rate of the fusion reaction 10 times. Which isotope of Pt is formed in this reaction?

19.102. **Radon in Drinking Water** A sample of drinking water collected from a suburban Boston municipal water system in 2002 contained 0.5 pCi/L of radon. Assume that this level of radioactivity resulted from the decay of ^{222}Rn ($t_{1/2}$ = 3.8 days).

a. What was the level of radioactivity (Bq/L) of this nuclide in the sample?

b. How many decay events per hour would occur in 2.5 L of the water?

19.103. Stone Age Skeletons The discovery of six skeletons in an Italian cave at the beginning of the 20th century was considered a significant find in Stone Age archeology. The age of these bones has been debated. The first attempt at radiocarbon dating indicated an age of 15,000 years. Redetermination of the age in 2004 indicated that two bones were between 23,300 and 26,400 years old. What is the ratio of ^{14}C in a sample 15,000 years old to one 25,000 years old?

*19.104. Atmospheric testing of nuclear weapons in the 1950s and 1960s produced an increase in the concentration of carbon-14 in the atmosphere. Use one or more balanced nuclear equations to explain how this could have happened.

19.105. Dating Prehistoric Bones In 1997 anthropologists uncovered three partial skulls of prehistoric humans in the Ethiopian village of Herto. From the amount of ^{40}Ar in the volcanic ash in which the remains were buried, their age was estimated at between 154,000 and 160,000 years.

a. ^{40}Ar is produced by the decay of ^{40}K ($t_{1/2}$ = 1.28×10^9 yr). Propose a decay mechanism for ^{40}K to ^{40}Ar.

b. Why did the researchers choose ^{40}Ar rather than ^{14}C as the isotope for dating these remains?

19.106. The rates of chemical reactions increase with increasing temperature, but the rates of nuclear decay do not. Why not?

20

Organic and Biological Molecules

The Compounds of Life

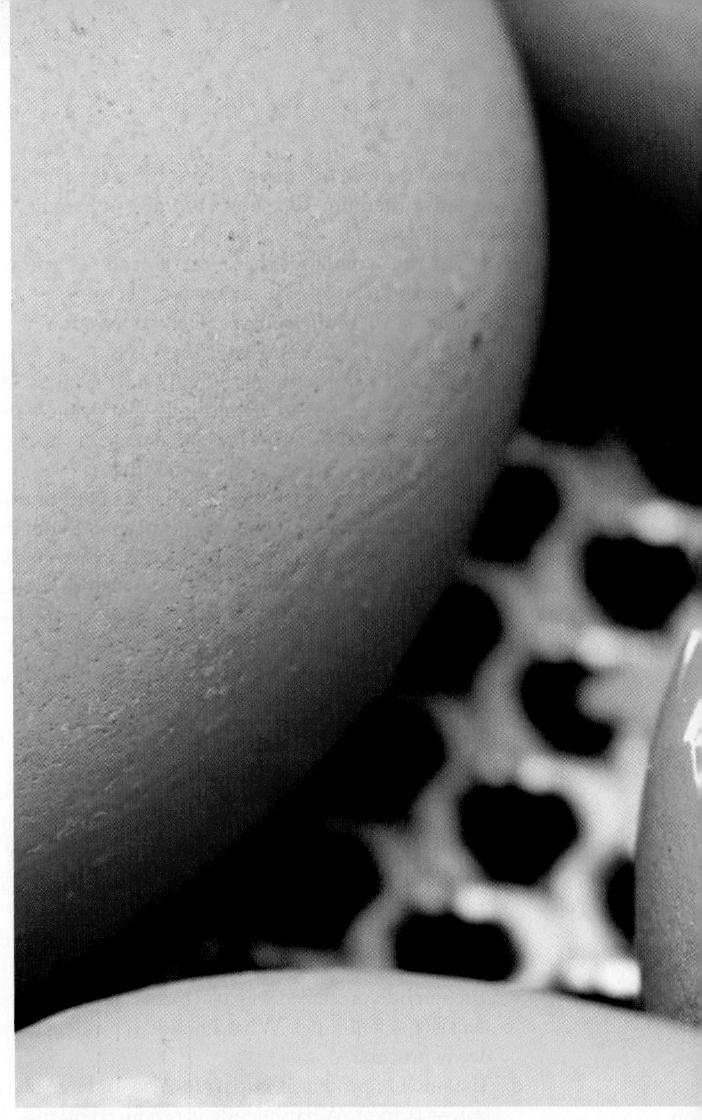

SIMILARITY, DIVERSITY, INTERDEPENDENCE The abundant diversity of life on Earth arises from different combinations of 40 or 50 small molecules. These basic building blocks of life are responsible for the similarity, diversity, and interdependence of all forms of life.

PARTICULATE **REVIEW**

Finding Functional Groups

In Chapter 20 we explore the relationship between structure and function for carbon-containing compounds.

- Name the functional group present in each molecule shown here.
- Which compound functions as a Brønsted–Lowry acid in water? Which compound functions as a Brønsted–Lowry base in water?
- Which functional group is formed when the Brønsted–Lowry acid and base react?

(Review Sections 2.8, 4.5, and 15.2 if you need help.)

(Answers to Particulate Review questions are in the back of the book.)

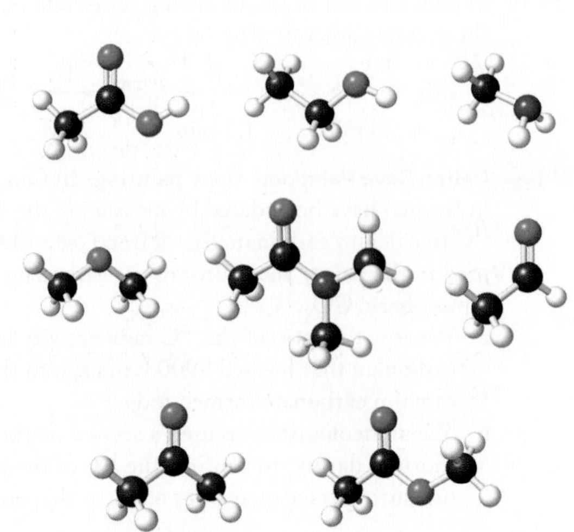

PARTICULATE **PREVIEW**

Protonated or Deprotonated?

Glutamic acid is used in the biosynthesis of proteins and contains three functional groups, as shown here. As you read Chapter 20, look for ideas that will help you answer these questions:

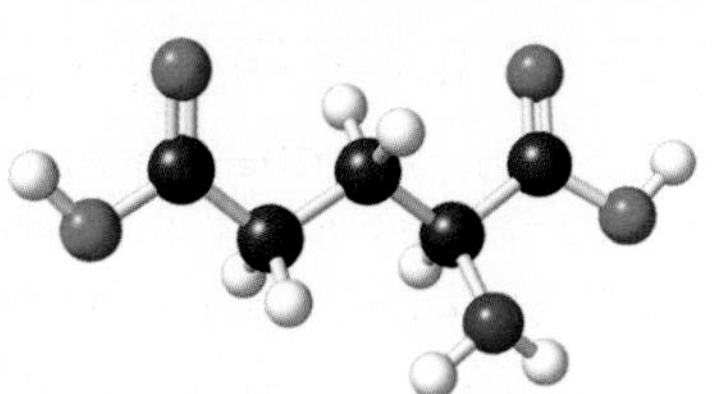

- What are the three functional groups in glutamic acid?
- Which functional groups are deprotonated at physiological pH? Which ones are protonated at physiological pH?
- Does glutamic acid exist as a negatively charged, neutral, or positively charged molecule at physiological pH?

Learning Outcomes

LO1 Recognize constitutional isomers, draw their molecular structures, and identify chiral molecules
Sample Exercises 20.1, 20.2, 20.3

LO2 Interpret titration curves for amino acids and use them to define pK_a and pI values
Sample Exercise 20.4

LO3 Relate the names and structures of small peptides
Sample Exercises 20.5, 20.6

LO4 Describe the four levels of protein structure and explain how intermolecular forces and covalent bonds stabilize these structures

LO5 Describe the molecular structures of simple sugars and the types of bonds linking them in polysaccharides, and explain how these compounds are used as energy sources and for energy storage

LO6 Relate the molecular structures of lipids to their physical and chemical properties
Sample Exercise 20.7

LO7 Describe the structures of DNA and RNA and how they function together to translate genetic information
Sample Exercise 20.8

20.1 Molecular Structure and Functional Groups

For all the stunning diversity of the biosphere, from single-cell organisms to elephants, whales, and giant redwoods, all life forms consist of substances made from only about 40 or 50 different small molecules. This small number of starting materials can link together to form large molecules that display astonishing variations in structure and function.

Most of these compounds are organic molecules containing C—C and C—H bonds. The designation *organic* for carbon-containing compounds was first applied to substances produced by living organisms, but that definition has been broadened for two reasons. First, scientists have learned how to synthesize many materials previously thought to be the products only of living systems. Second, chemists also synthesize many carbon-based materials that have *never* been produced by living systems. Today the study of *organic chemistry* encompasses the chemistry of all compounds containing carbon–carbon and carbon–hydrogen bonds, regardless of their origin.

In contrast to organic molecules, **biomolecules** retain a link to life: a biomolecule is any molecule produced by a living organism. This definition encompasses the four classes of large molecules that we describe in this chapter—proteins, carbohydrates, lipids, and nucleic acids—as well as small molecules that result from biochemical processes such as metabolism. **Biochemistry**, or biological chemistry, is the study of the chemical processes taking place in living organisms, with a focus on understanding how biomolecules participate in the reactions occurring in living cells. In Chapter 20 we concentrate on the composition and structure of organic molecules and biomolecules that compose our bodies and the other organisms with which we share this planet, as well as the foods we eat and the medicines we take. Many of these molecules are large and complex, but our knowledge of the behavior of small molecules can serve as a framework for understanding the functions and reactions of these compounds in living systems.

biomolecule any molecule produced by a living organism.

biochemistry the study of chemical processes taking place in living organisms.

An important aspect of studying the biochemical processes associated with life is finding out what happens when something goes wrong. It has been estimated that 70% of inherited diseases in people are caused by the absence of particular proteins or some other defect that causes proteins to function improperly. For example, the malformation or the total absence of the protein dystrophin causes several debilitating diseases collectively referred to as muscular dystrophy (MD). In mild forms of MD, misshapen molecules of dystrophin are unable to build muscle fibers sufficient to function normally. As a result, muscle tissue wastes away and the patient becomes physically disabled. In Duchenne MD, a severe form of the disease, functional dystrophin is absent, and the disability often results in early death. Knowledge of the mechanisms that produce dystrophin has led to promising therapies. Similar approaches based on what we know about disease mechanisms have resulted in the development of new drugs and cell therapies for other inherited and contagious diseases. Scientists frequently use information about the way biomolecules behave in healthy individuals to develop drugs and treatments for all manner of medical problems.

Families Based on Functional Groups

To understand biomolecules and their reactions, we need to review some fundamental concepts and develop additional ideas about the composition, structure, and reactivity of organic molecules. Much of the variety in the chemistry of carbon arises from the carbon atom's ability to form covalent bonds to other carbon atoms. These carbon atoms, in turn, can bond to more carbon atoms or atoms of other elements, yielding molecules that consist of only a few atoms to molecules that consist of many thousands of atoms. Additional variety is introduced because these molecules may contain a multitude of linear and branched structures as well as rings. In addition to carbon and hydrogen, organic compounds can also contain nonmetallic heteroatoms such as nitrogen, oxygen, sulfur, phosphorus, or the halogens.

CONNECTION We first encountered heteroatoms and functional groups in Section 2.8.

Learning to understand the structures and reactivity of millions of organic compounds requires some organizing concepts. One organizing principle is based on the chemical composition and structure of subunits within molecules. Chemists group organic compounds into families on the basis of *functional groups*—that is, subunits in a molecule that confer particular chemical and physical properties.

We have encountered many of the most important functional groups in organic chemistry in earlier chapters. These functional groups are reviewed in **Table 20.1**. When discussing functional groups, the convention is to use R to represent the entire molecule except the functional group. All the functional groups in Table 20.1 appear in biologically important molecules. A key learning outcome in this chapter is being able to connect the properties of biomolecules to their structures, including the functional groups within them.

A second organizing principle for organic molecules and biomolecules is based on molecular size. In Section 12.6 we discussed the structure of polymers and the influence of molecular size on the physical properties of large molecules. All of the principles introduced with respect to polymeric materials in Chapter 12 are pertinent to the discussion of large *biopolymers* in this chapter.

TABLE 20.1 Functional Groups of Organic Compounds

Name	Structural Formula of Group[a]	Example	Name
Alkane (Section 2.8)	R—H	$CH_3CH_2CH_3$	Propane
Alkene (Section 2.8)	>C=C<	$H_2C{=}CH_2$	Ethylene (ethene)
Alkyne (Section 2.8)	—C≡C—	H—C≡C—H	Acetylene (ethyne)
Alcohol (Section 2.8)	R—OH	H_3C—OH	Methanol
Carboxylic acid (Section 4.5)	R—C(=O)—OH	H_3C—C(=O)—OH	Acetic acid
Amine (Section 4.5)	$R—NH_2$ R—NHR′ R—NRR′	$H_3C—NH_2$	Methylamine
Aldehyde (Section 8.2)	R—C(=O)—H	H_3C—C(=O)—H	Acetaldehyde
Ketone (Section 9.5)	R—C(=O)—R′	H_3C—C(=O)—CH_3	Acetone
Aromatic (Section 9.5)	e.g., (benzene ring)	(benzene ring)	Benzene
Ether (Section 10.7)	R—O—R′	H_3C—O—CH_3	Dimethyl ether
Ester (Section 12.6)	R—C(=O)—OR′	H_3C—C(=O)—OCH_3	Methyl acetate
Amide (Section 12.6)	R—C(=O)—NH_2 [b]	H_3C—C(=O)—NH_2 [b]	Acetamide

[a]In compounds with two R groups, the two may be the same (R = R′) or they may be different (R ≠ R′).
[b]One or both H atoms may be replaced by R groups: –NHR or –NRR′.
Note: Functional groups are highlighted in yellow.

20.2 Organic Molecules, Isomers, and Chirality

In Section 3.1 we explained that Friedrich Wöhler's discovery of urea (H_2NCONH_2; **Figure 20.1a**) occurred during his attempt to synthesize ammonium cyanate, NH_4NCO (**Figure 20.1b**). The chemical formula of both structures can be written CH_4N_2O, but because their structures are not the same, they represent different compounds with different physical and chemical properties. Compounds that have the same number and same kinds of atoms but differ in the arrangement of those atoms are called *isomers*. The existence of isomers is partly responsible for the enormous number and wide variety of organic compounds in the world.

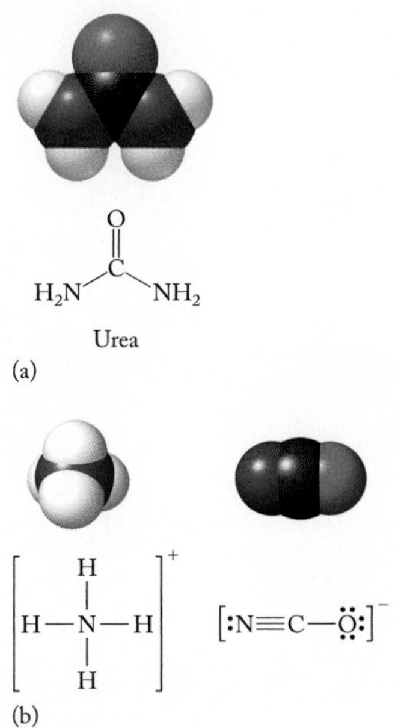

FIGURE 20.1 (a) Urea and (b) ammonium cyanate can be written with the same chemical formula, CH_4N_2O.

CONNECTION We first mentioned isomers in Section 3.1 in the context of Wöhler's discovery of two compounds with the same composition.

To explore isomers further, let's revisit hydrocarbons, the organic compounds composed exclusively of carbon atoms and hydrogen atoms. We have seen that hydrocarbons in which each carbon atom forms four single bonds to four other atoms are called alkanes. In Table 6.5 we compared the properties of two four-carbon alkanes: butane ($CH_3CH_2CH_2CH_3$) and 2-methylpropane [$CH_3CH(CH_3)CH_3$]. The molecular formula of both structures is C_4H_{10}, but their atoms are connected to each other in different ways, making them **constitutional isomers** (also called *structural isomers*).

CONNECTION In Section 2.8 we defined hydrocarbons, alkanes, alkenes, and alkynes. In Section 6.9 we explored straight-chain and branched-chain hydrocarbons. Additional information regarding the rules for naming these compounds (i.e., nomenclature) can be found in Appendix 7.

When presented with the condensed structures or carbon-skeleton structures of two alkanes, how do we determine whether they represent two compounds with different molecular formulas, a single compound, or two constitutional isomers? We begin by translating each structure into a molecular formula. If the molecular formulas are different, then the structures represent two compounds. If the molecular formulas are the same, then we must compare the bonds in the two structures; that is, we must compare the way the atoms are connected to one another. There are two ways we can do this:

STEPWISE ANIMATION
Naming Branched Alkanes

1. Simply inspecting the two structures may reveal that they are identical; however, we may have to reverse or rotate one of the structures. For example, structures (1) and (2) below may seem to be constitutional isomers, but rotating structure (2) by 180° shows that it is the same as structure (1). The two represent the same compound because they have the same composition and the same structure (bonds): a four-carbon *main chain* with a one-carbon *side chain* connected to the carbon atom next to a terminal carbon.

(1) (2) 180° rotation = After rotation

CONNECTION In Section 9.5 we introduced how to draw Kekulé structures that use lines to represent bonds and omit lone pairs.

2. A second method to determine whether two structures are identical involves drawing the structures being compared with the longest chains positioned horizontally. The longest chains in structures (3) and (4) below are highlighted in yellow. Number the carbons in the longest chain so that the branches have the lowest possible numbers, and then check to see whether the same side chains are attached at the same numbered positions along the longest chain. In this example, the longest chain in both structures is five carbon atoms long, and both chains have methyl ($-CH_3$)

constitutional isomers molecules with the same chemical formula but different connections between the atoms.

groups on carbon 2 and carbon 3, so they are identical. It is permissible to number some chains right-to-left, as in structure (3), or left-to-right, as in structure (4).

4 3 2 5 1

(3)

4 3 5 2 1

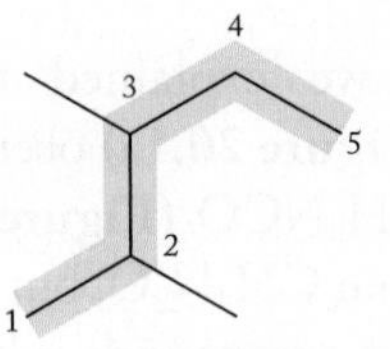

(4)

SAMPLE EXERCISE 20.1 Recognizing Constitutional Isomers LO1

Do the two structures in each set describe the same compound, constitutional isomers, or compounds with different molecular formulas?

a. $(CH_3)_2CHCH_2CH(CH_3)_2$

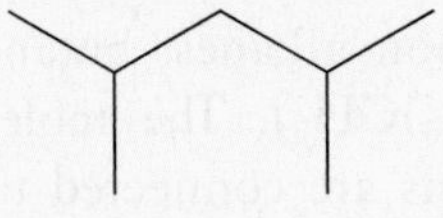

b.

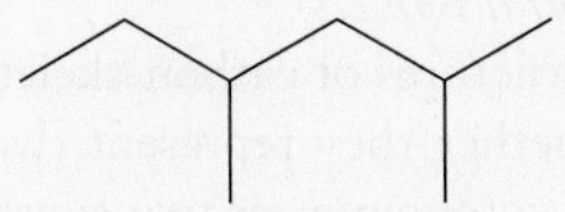

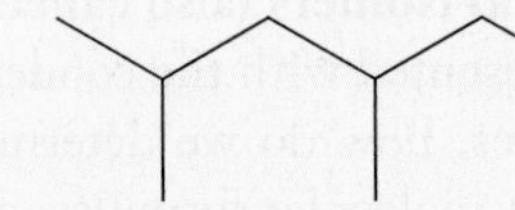

c.

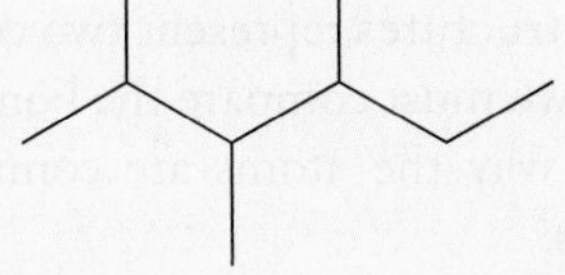

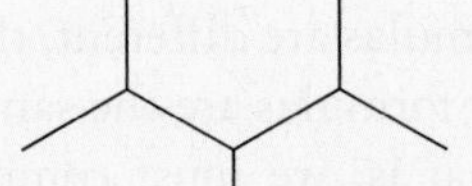

Collect and Organize We have structures showing the connectivity of the atoms and from which we can determine molecular formulas. Identical compounds have the same molecular formula and the same connectivity of the atoms. Constitutional isomers have the same molecular formula but different connectivity.

Analyze In each pair, we check first to see whether the molecular formulas are the same. That means we count the number of each kind of atom present (C and H atoms, in these cases). If the molecular formulas are the same, we may have either one compound drawn two ways or a pair of constitutional isomers. If the carbon skeletons are the same in any pair, the drawings represent the same hydrocarbon.

Solve

a. The molecular formulas are the same: C_7H_{16}. Converting the condensed structure to a carbon-skeleton structure gives us

$$(CH_3)_2CHCH_2CH(CH_3)_2 \quad \rightarrow$$

This structure is identical to the carbon-skeleton structure in this set. Therefore, the condensed structure and carbon-skeleton structure represent the same molecule.

b. Both molecules in this set have the molecular formula C_8H_{18}. If we rotate the second structure 180° about a vertical axis, we get the first structure, so the two compounds are identical.

(1) (2) 180° rotation After rotation

c. The left structure contains nine carbon atoms, but the right structure contains only eight, so the two structures represent different compounds.

Think About It Two compounds may have the same molecular formula but they are not necessarily identical; they may be constitutional isomers.

Practice Exercise Do the two structures in each set describe the same compound, constitutional isomers, or compounds with different molecular formulas?

a.

b.

c.

(Answers to Practice Exercises are in the back of the book.)

Hydrocarbons containing one or more carbon–carbon double bonds (alkenes) or triple bonds (alkynes) may also have branches, and their structures are treated just like alkanes to determine their identity. However, another situation arises with alkenes that involves different orientations of groups about the double bond. For example, consider the straight-chain isomers of the alkene that contains five carbon atoms and one double bond (**Figure 20.2**).

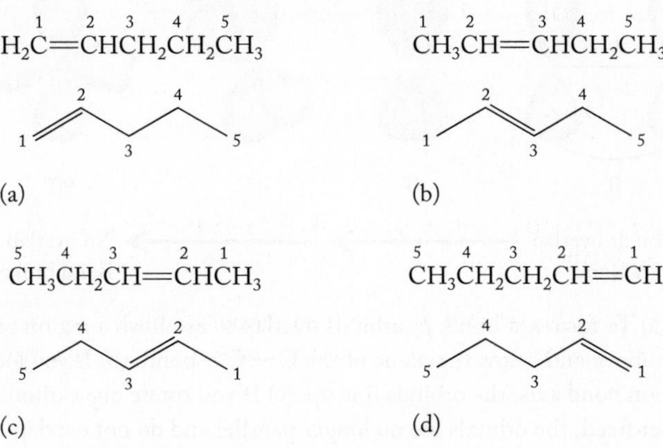

FIGURE 20.2 Four possible constitutional isomers of pentene (C_5H_{10}). Structures (a) and (d) are identical, as are (b) and (c). Structures (a) and (d) are constitutional isomers of (b) and (c).

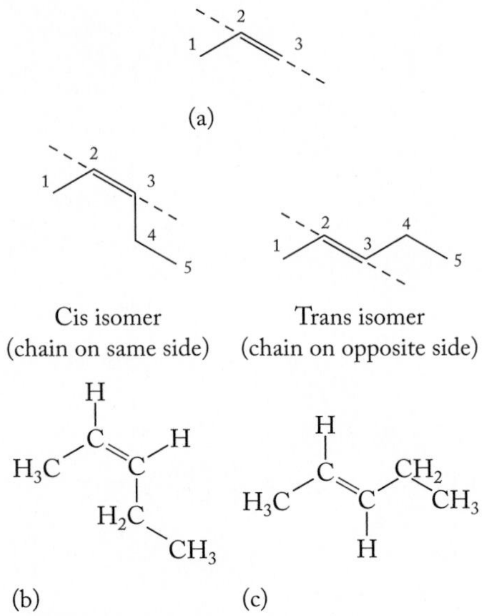

(a)

Cis isomer
(chain on same side)

Trans isomer
(chain on opposite side)

(b)

(c)

FIGURE 20.3 (a) The first three atoms of a carbon chain with a double bond between C2 and C3. (b) The chain continues on the same side of the double bond as C1 in the cis isomer. (c) The chain continues on the opposite side of the double bond from C1 in the trans isomer.

Structures (a) and (d) in Figure 20.2 are the same. This is easier to see if we look at the carbon-skeleton structures in the figure. In both cases the double bond is between C1 and C2. As with branched-chain alkanes, we number the carbons from whichever end gives the carbon attached to the branch the lowest number. The same holds for functional groups, as shown here, where the carbons in structure (d) must be numbered from right to left. Structures (b) and (c) are different from (a) and (d), but equivalent to each other. Moreover, they are constitutional isomers of (a) and (d) because they have the same chemical formula, but their double bond is in a different location.

Drawing the carbon-skeleton structures of (b) and (c) in Figure 20.2, however, presents us with a new situation. After we draw the first three atoms of structure (b), we have two options for how we orient the rest of the molecule with respect to the double bond. If we draw a straight dashed line through the double bond, as in **Figure 20.3a**, we can place the bond between C3 and C4 on the same side of the dashed line as the methyl group at C1 (**Figure 20.3b**) or on the opposite side (**Figure 20.3c**).

These two molecules are isomers of each other because they have the same molecular formula but different structures. The isomer in Figure 20.3b is called either the ***Z* isomer** (*Z* stands for the German word *zusammen*, or "together") or the **cis isomer** (*cis* is Latin for "on this side"), which in this case translates to "the methyl group and the chain after the double bond are both *together* or *on this side* of the structure." The isomer in Figure 20.3c is called either the ***E* isomer** (*E* for *entgegen*, or "opposite") or the **trans isomer** (*trans* is Latin for "across"). The system of naming using cis and trans is in wide use and is sufficient for the simple molecules discussed in this text. The *E*/*Z* system is routinely used for more complex molecules in which more than two different substituents are attached to a double bond.

Cis–trans isomers are *not* constitutional isomers: their atoms are actually connected to each other in the same way, but they differ in their three-dimensional structure. Cis–trans isomers are a type of **stereoisomer**. What makes stereoisomers special is that the arrangement of their atoms in space differs. That is why the prefix "stereo-" is used in the name: stereoisomers are molecules with bonds that connect the same sequence of atoms, but those bonds are oriented in different directions.

The stereoisomers in Figure 20.3 exist because there is no free rotation about the double bond. In Chapter 9 we learned that a double bond consists of a σ bond and a π bond and that the π bond forms from the overlap of two unhybridized *p* orbitals on adjacent carbon atoms. As **Figure 20.4** shows, if the carbon atoms

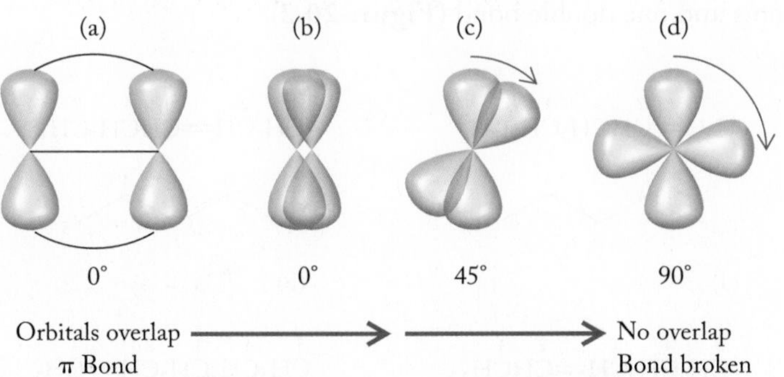

FIGURE 20.4 (a) To form a π bond, p_z orbitals overlap to establish a region of shared electron density above and below the plane of the C—C σ bond. (b) If you look down the carbon–carbon bond axis, the orbitals line up. (c) If you rotate one carbon atom while keeping the other fixed, the orbitals are no longer parallel and do not overlap. (d) If you rotate one of the two bonded C atoms far enough, the π bond breaks.

cis isomer (also called ***Z* isomer**) a molecule with two like groups (such as two R groups or two hydrogen atoms) on the same side of the molecule.

trans isomer (also called ***E* isomer**) a molecule with two like groups (such as two R groups or two hydrogen atoms) on opposite sides of the molecule.

stereoisomers isomers created by differences in the orientations of the bonds in molecules.

optical activity a property of a compound that refers to the rotation of a beam of plane-polarized light when it passes through a solution of the compound.

optical isomers molecules that are not superimposable on their mirror images.

joined in a double bond were to rotate freely, they would have to twist about the bond axis, eliminating the orbital overlap and breaking the π bond. Breaking the π bonds in 1 mole of an alkene requires about 290 kJ of energy, and that much energy is not available to the molecules at room temperature. As a result, rotation about a carbon–carbon double bond is restricted, which leads to the existence of stereoisomers.

CONCEPT TEST

Which of the following alkenes has cis and trans isomers?

a. $CH_2{=}CHCH_2CH_3$

b. CH_2 bonded to both carbons of $CH{=}CH$ (ring)

c. $(CH_3)_2C{=}C(CH_3)_2$

d. $(CH_3)_2C{=}CH_2$

e. $CH_3CH{=}CHCH_3$

(Answers to Concept Tests are in the back of the book.)

Chirality and Optical Activity

Before we discuss the major classes of compounds in living systems, we need to review *chirality*, a property that characterizes many biomolecules. Chirality was introduced in Section 9.5 when we learned about the three-dimensional shapes of molecules. A tangible measure of chirality is called **optical activity**. As a review, let's look at some familiar items that illustrate the basic principles of chirality and then turn to some important molecules in living systems that have this characteristic. A key point to understand is what is meant by an object being superimposable on its mirror image.

Figure 20.5 illustrates the concept of chirality by starting with the reflection of a plain coffee mug that is *achiral* (not chiral) because the reflected image can be superimposed on the original object. Your two hands, however, are mirror images that cannot be superimposed. Molecules that *can* be superimposed on their mirror images are achiral, whereas molecules that *cannot* be superimposed on their mirror images are chiral.

Chiral molecules exist as **optical isomers**; a molecule that has one chiral center exists as two optical isomers. The term "optical" refers to the ways these compounds interact with *plane-polarized* light (**Figure 20.6**). In plane-polarized light,

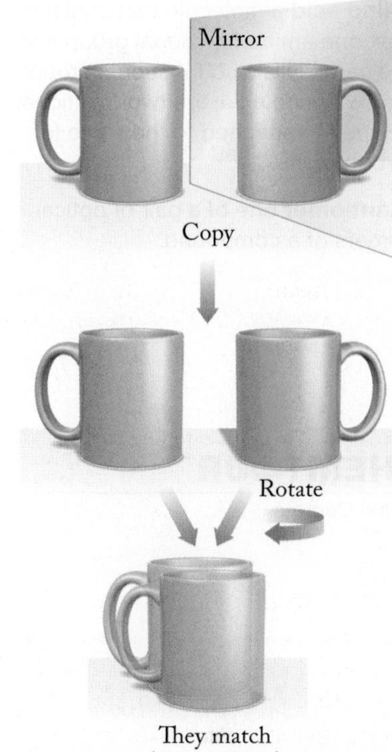

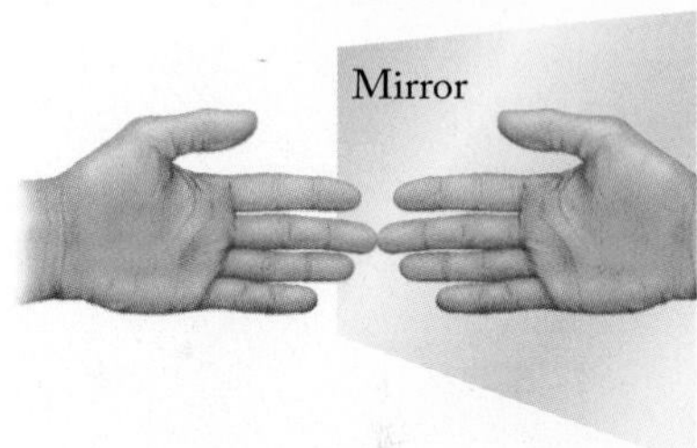

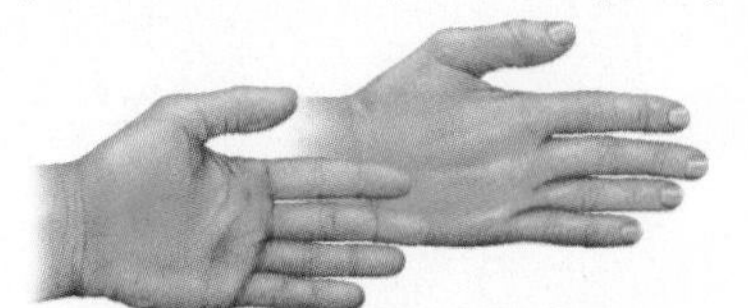

FIGURE 20.5 A plain coffee mug is superimposable on its mirror image, so it is achiral. The mirror image of your left hand is your right hand, so the two are *not* superimposable.

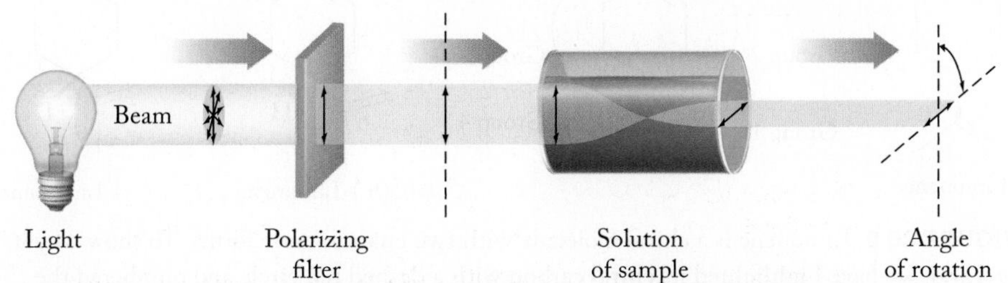

FIGURE 20.6 A beam of plane-polarized light contains an electric field that oscillates in only one direction. The plane of oscillation rotates if the beam passes through a solution of one enantiomer of an optically active compound. The (+)-enantiomer causes the beam to rotate clockwise; the (−)-enantiomer causes the beam to rotate the same amount but counterclockwise.

CONNECTION In Section 9.5 we defined chiral molecules as being not superimposable on their mirror images.

CONNECTION In Section 9.2 we introduced the use of solid and dashed wedges to indicate bonds that should be viewed as coming out of the plane of the paper and projecting behind the plane of the paper, respectively.

amino acid a molecule that contains at least one amine functional group and one carboxylic acid functional group; in an α-amino acid, the two functional groups are attached to the same (α) carbon atom.

enantiomer one of a pair of optical isomers of a compound.

CHEMTOUR
Chiral Centers

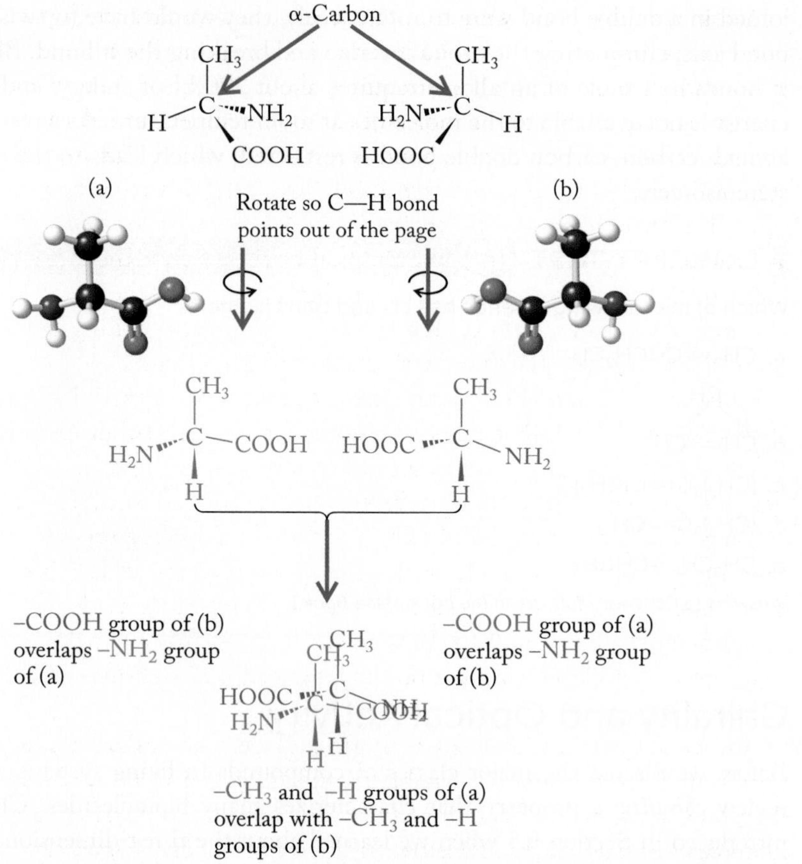

α-Carbon

(a) (b)

Rotate so C—H bond points out of the page

–COOH group of (b) overlaps $-NH_2$ group of (a)

–COOH group of (a) overlaps $-NH_2$ group of (b)

$-CH_3$ and –H groups of (a) overlap with $-CH_3$ and –H groups of (b)

FIGURE 20.7 Alanine is an α-amino acid. Like most α-amino acids, alanine is also chiral, which means that the molecular structure of one form (a) is not superimposable on (b) its mirror image. Rotating the mirror image so that the $-CH_3$ and –H groups overlap when the structures are superimposed produces an image in which the $-NH_2$ and –COOH do not overlap.

the electric fields that compose the beam oscillate in only one plane. When plane-polarized light passes through a solution of one enantiomer of a chiral pair, the light twists in one direction. However, when the light passes through a solution of the other member of the pair, it twists the same amount but in the opposite direction.

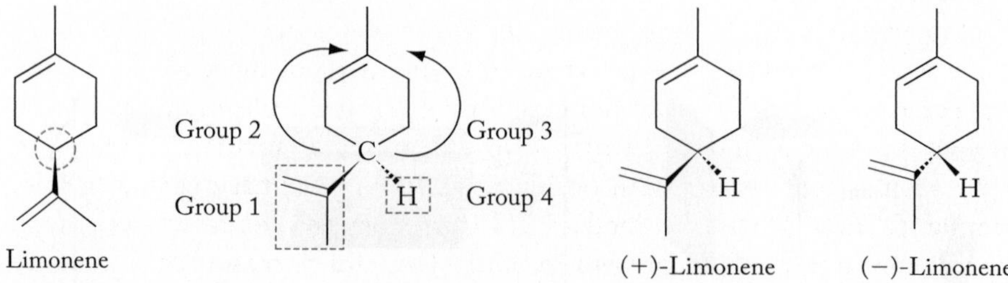

FIGURE 20.8 Limonene is a chiral molecule with two enantiomeric forms. To show why it is chiral, we have highlighted its chiral carbon with a dashed red circle and numbered the four groups bonded to it. Groups 1 and 4 are different. Groups 2 and 3 are two halves of the same ring but they are different because group 2 has a C=C bond and group 3 does not. As a result, the circled carbon atom is bonded to four different groups and is chiral.

Look carefully at the structure of the molecule alanine in **Figure 20.7**. Alanine is an **amino acid**, so named because it contains at least one amine ($-NH_2$) group and at least one carboxylic acid (–COOH) group. Alanine is called an *α-amino acid* because one carbon atom in its structure, called the α-carbon, is bonded to both an $-NH_2$ group and a –COOH group. The α-carbon atom in alanine is bonded to four *different* groups, making this α-carbon atom a *chiral center*. The two molecules shown in Figure 20.7a and b are both alanine. They have the same formula and the same bonds; they differ only in the three-dimensional orientation of the groups attached to the chiral center. For the most part they have the same physical properties, such as melting point and boiling point; however, they differ in how they interact with plane-polarized light.

Limonene (**Figure 20.8**) is an alkene found in oranges and turpentine. The dashed red circle identifies its chiral center. This chiral carbon atom is part of a six-membered ring, and if we examine the structure of limonene closely, we see that this carbon atom has four different groups bonded to it. The three-carbon alkene group (group 1 in Figure 20.8) and the hydrogen atom represent two of those groups. If we then follow the carbon atoms around the ring, we find a C=C bond three bonds from the circled carbon on the left side, but C—C single bonds on the right side. The presence of the C=C bond makes groups 2 and 3 in Figure 20.8 different. As a result, limonene is chiral, in which case optical isomers of limonene exist.

The two optical isomers of a single molecule are called **enantiomers**, so enantiomers are nonsuperimposable mirror images. In no orientation do all of the groups on the chiral center of superimposed enantiomers coincide, because the molecules have different shapes in three dimensions. Figure 20.7 shows the two enantiomers of alanine, whereas Figure 20.8 shows the two enantiomers of limonene. If we try to superimpose the two enantiomers of limonene (**Figure 20.9**), then the substituents bonded to the chiral carbon do not coincide. Because enantiomers differ in their three-dimensional structures, they are a type of stereoisomer.

To distinguish between a pair of enantiomers, (+) and (−) signs are added to their names to refer to the specific effect each isomer has on plane-polarized light. When a beam of plane-polarized light passes through a solution containing one member of an enantiomeric pair, the beam rotates as in Figure 20.6. If the beam rotates to the left (counterclockwise), the enantiomer is the *levorotatory* form of the molecule and a (−) sign precedes its name. If the beam rotates to the right (clockwise), the enantiomer is the *dextrorotatory* form and a (+) sign precedes its name. The magnitude of the rotation of plane-polarized light by the (+) isomer will be the same for the (−) isomer but in the opposite direction.

The molecules in an enantiomeric pair share the same physical and chemical properties except for those that relate to a few specialized types of behavior. As we saw in Chapter 9, molecular recognition is one of those special behaviors. For example, the (+)-enantiomer of limonene smells like oranges, whereas the (−)-enantiomer smells like turpentine. Part of the process of sensing different aromas involves recognition by receptors in your nasal passages. One receptor is shaped to accommodate (+)-limonene; the other, (−)-limonene. Each enantiomer binds to its own receptor. For similar reasons, the enantiomers of certain chiral drugs have different biological effects. The enantiomer of naproxen (**Figure 20.10**) that rotates light to the right relieves pain and swelling, whereas its mirror image, which rotates light to the left, is toxic to the liver and has no analgesic effects. **Figure 20.11** summarizes the classification of isomers.

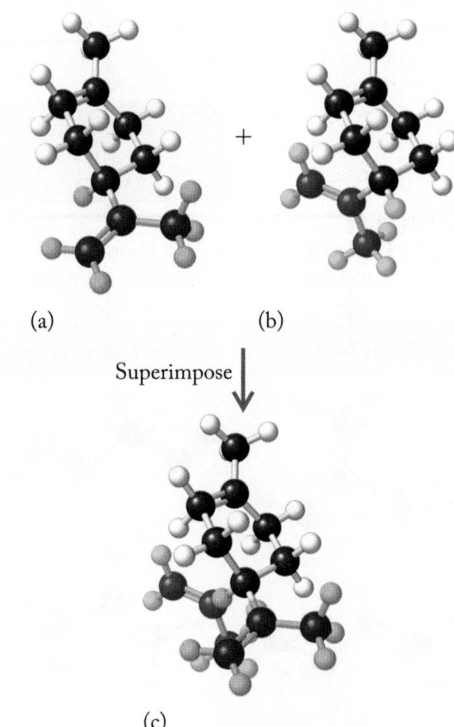

FIGURE 20.9 The two enantiomers of limonene, (a) and (b), are mirror images of each other but cannot be superimposed upon each other (c) because the substituents bonded to the chiral carbon do not coincide.

CONNECTION In Section 9.5 we saw how molecular recognition of ethylene gas triggers the ripening process for tomatoes.

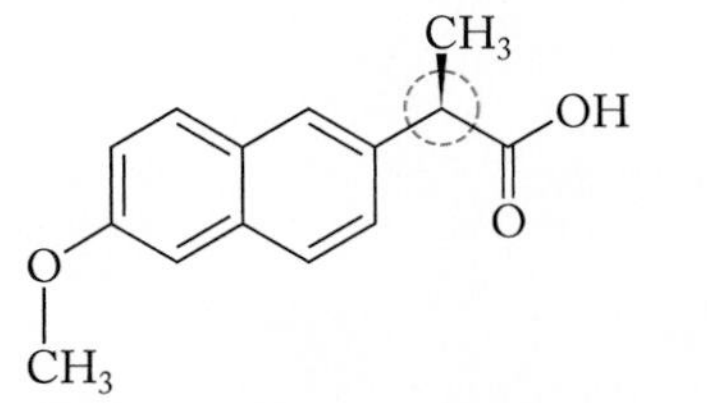

FIGURE 20.10 The chiral carbon in naproxen is indicated by the dashed red circle. All carbon atoms have four bonds and, in this type of structural representation, any bond not shown is understood to be a hydrogen atom. The chiral carbon in the circle is bonded to four different groups: the ring, the $-CH_3$ group, the –COOH group, and a hydrogen atom that is not shown.

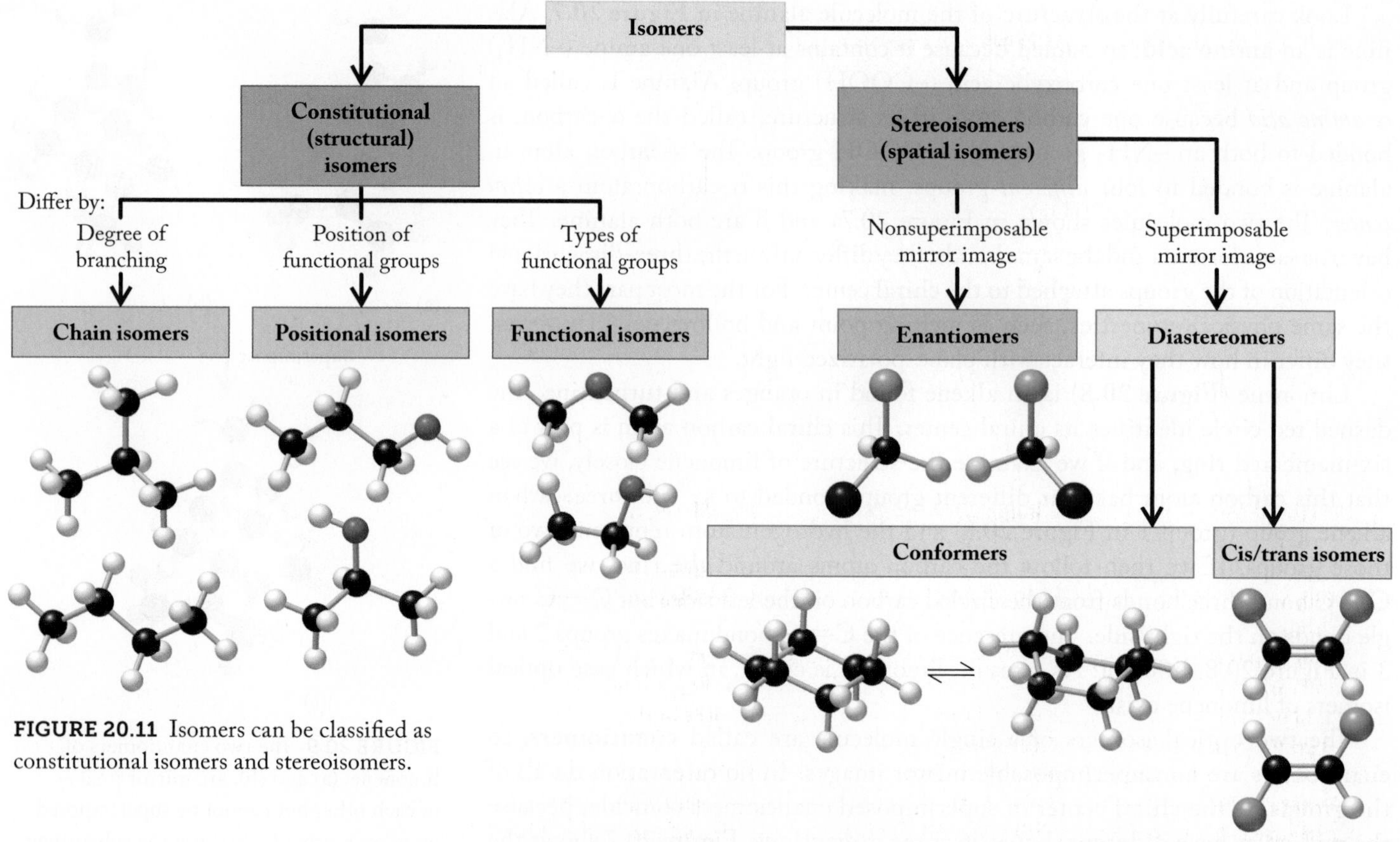

FIGURE 20.11 Isomers can be classified as constitutional isomers and stereoisomers.

SAMPLE EXERCISE 20.2 Recognizing Chiral Molecules **LO1**

Identify which of the molecules shown are chiral and circle the chiral centers. Structures may have more than one chiral center.

a. H; H_2N—C—CH_2CH_3; COOH

b.

c.

d. O; OH

e. O

Collect and Organize Chiral molecules are nonsuperimposable on their mirror images. If molecules have carbon atoms with four different groups attached, those carbon atoms are chiral centers and the molecules are chiral.

Analzye In the line structures shown in compounds (b), (c), (d), and (e), some carbon atoms have four bonds drawn, but some have fewer. Although hydrogen atoms are implied in these structures, be sure to consider an implicit hydrogen when counting the number of different groups attached to any one carbon.

Solve The chiral carbon atoms in each structure are circled.

a. b. c.

d. e.

The circled carbon atom in (a) is bonded to four different groups: $-H$, $-NH_2$, $-COOH$, and $-CH_2CH_3$, so this compound is chiral.

The circled carbon atom in compound (b) is bonded to a $-CH_3$ group, a $-CH_2CH_3$ group, a $-CH(CH_3)_2$ group, and an H atom, so the compound, 2,3-dimethylpentane, is chiral.

The chiral center in compound (c) is similar to the chiral center in limonene. The chiral carbon is bonded to a methyl group and an H. Tracing your finger around the ring in both directions from the circled carbon reveals the differences in the remaining two groups bonded to the chiral carbon.

All of the carbon atoms in compound (d) are sp^2 hybridized, giving compound (d) a planar molecular geometry. Planar molecules cannot be chiral because they have a superimposable mirror image. Think of a mirror plane that contains all nine carbons in compound (d): the mirror image is exactly the same.

Compound (e) has three chiral centers. Working from right to left, the first chiral center is similar to the chiral center in compound (a). The other two chiral centers are in the cyclohexane ring. In each case, the carbon atom is bonded to four different groups, so compound (e) is chiral.

Think About It The presence of a chiral center in a molecule means that the molecule has two enantiomeric forms. The molecule is not superimposable on its mirror image. Although structure (d) has no chiral center, it can have cis–trans isomers about the double bond in the chain. The isomer shown has the trans configuration.

Practice Exercise Identify which of the following molecules are chiral. Circle the chiral centers in each structure.

a. b. c.

d. e.

racemic mixture a sample containing equal amounts of both enantiomers of a compound.

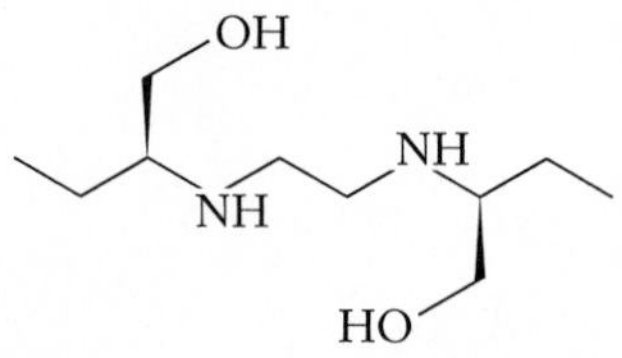

Chiral Mixtures

Chirality is ubiquitous in the organic compounds formed by living systems, and usually only one enantiomer occurs naturally in a particular organism. A molecule might even have more than one chiral center, which means it could exist as two or more optical isomers. Whatever the origin of these compounds, processes requiring molecular recognition in living systems often depend on the selectivity conveyed by chirality. The human body is a chiral environment, so handedness of molecules matters. As many as half of the drugs made by large pharmaceutical companies are chiral and owe their function to recognition by a receptor that favors one enantiomer over the other; in 2014 each of the ten most-prescribed drugs in the United States was chiral. Typically only one of the enantiomers of a chiral drug is active. A classic example of two enantiomers with different activity is the drug ethambutol (**Figure 20.12**). The (+)-enantiomer is effective against tuberculosis, whereas the (−)-enantiomer causes blindness.

OH NH NH HO

FIGURE 20.12 The (+)-enantiomer of ethambutol shown here is active against tuberculosis infections; the (−)-enantiomer causes blindness.

When living systems produce chiral compounds, typically only one enantiomer is produced. However, when a molecule with a chiral center is made in a laboratory, both isomers are produced unless special methods are used. When both enantiomers are present in equal amounts in a sample, the material is known as a **racemic mixture**. Because one isomer rotates the plane of polarized light in one direction, and the other to the same extent in the opposite direction, a racemic mixture does not rotate the plane of polarized light at all.

Because any pair of enantiomers may interact differently with receptors, the pharmaceutical industry routinely faces two choices: devise a special synthetic procedure that yields only the isomer of interest, or separate the two isomers at the end of the manufacturing process. Both approaches are widely used. For the chiral drug sotalol (**Figure 20.13**), one isomer is physiologically active for hypertension and cardiac arrhythmias, whereas the other is inactive. Sotalol is routinely prepared as a racemic mixture that is then separated to isolate the active enantiomer. For a drug such as ethambutol (Figure 20.12), where the biological consequences of administering the undesired isomer are tragic, it is essential to exclude the incorrect isomer from the preparations administered to a patient. This is accomplished by a synthesis that produces a single pure enantiomer.

O O S H_3C NH OH NH CH_3 CH_3

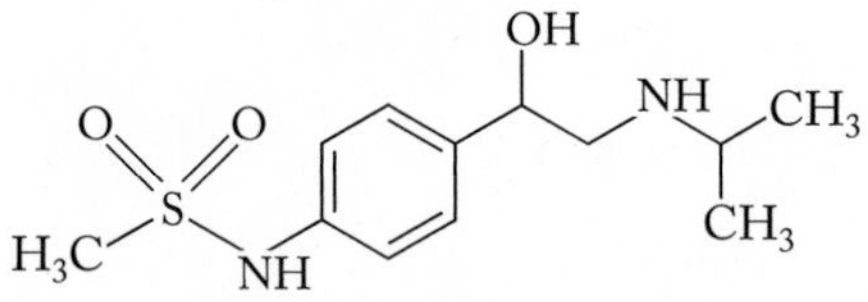

Sotalol

FIGURE 20.13 The chiral drug sotalol is administered as a racemic mixture but only one isomer is active.

SAMPLE EXERCISE 20.3 Recognizing Optical Properties of Chiral Molecules **LO1**

The structure of the antidepressant drug bupropion is shown in **Figure 20.14a** along with two other compounds. Explain which of the following statements is/are true:

1. A mixture of the three compounds in any proportion has the same percent composition.
2. All three compounds rotate a beam of polarized light.
3. A 1:1 mixture of bupropion and the molecule in **Figure 20.14b** rotates a beam of polarized light.
4. The mass spectra of the three compounds are identical.

Collect and Organize We are given the structure of three compounds and asked which of the four statements about their composition and their response to plane-polarized light are true. Mass spectrometry is discussed in Section 3.8.

Analyze Isomers have the same molecular formula but differ in their structures. Compounds with a chiral center can exist as optical isomers called enantiomers. The enantiomers of a compound rotate plane-polarized light equally in opposite directions. The mass spectrum of a compound reflects the connectivity of the atoms.

Solve

1. All three compounds in Figure 20.14 have the same chemical formula, $C_{13}H_{18}NOCl$, so they will have the same percent composition.
2. The chiral centers are circled in Figure 20.14a and b. Chiral molecules are optically active and will rotate plane-polarized light. The compound in **Figure 20.14c** is not chiral and is not optically active.
3. The compounds in Figure 20.14a and b represent a pair of enantiomers. A 1:1 mixture will show no net rotation of plane-polarized light because each enantiomer will rotate plane-polarized light equally but in opposite directions.
4. The compound in Figure 20.14c is a constitutional isomer of the other molecules because the atoms are connected differently. This means that its mass spectrum will also differ from that of the enantiomeric pair. The latter will have identical mass spectra.

Think About It Enantiomers have the same chemical composition and the same molar mass. Their physical properties are identical except for the direction in which their solutions cause plane-polarized light to rotate when it passes through them.

Practice Exercise Identify the chiral carbon atoms in ethambutol (Figure 20.12) and sotalol (Figure 20.13).

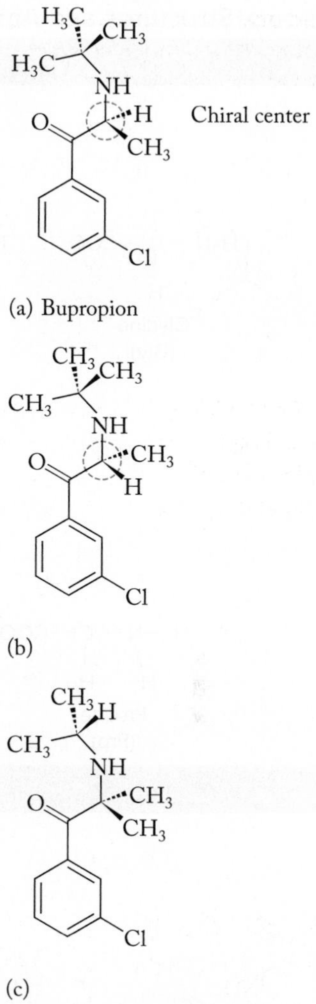

(a) Bupropion

(b)

(c)

FIGURE 20.14 The antidepressant drug bupropion and two similar molecules. Chiral centers are indicated by the red circles.

20.3 The Composition of Proteins

We opened our discussion of chirality with the α-amino acid alanine. We chose that molecule because amino acids are the small-molecule building blocks of **proteins**, the most abundant class of biomolecules in all animals, including us. Proteins account for about half the mass of the human body that is not water. They are the major component in skin, muscles, cartilage, hair, and nails. Most of the enzymes that catalyze biochemical reactions are proteins, as are the molecules that transport oxygen to our cells and many of the hormones that regulate cell function and growth. Most proteins are large, with molar masses of 10^5 g/mol or more.

protein a biological polymer made of amino acids.

Amino Acids

The molecular structures and biological functions of big biomolecules depend on the identities of their amino acids and the sequence in which those small molecules are connected. About 500 amino acids are known, and about 240 of these occur free in nature, but only 20 common amino acids form proteins in humans (**Table 20.2**).

In addition to its single bonds to the $-NH_2$ and $-COOH$ groups, the α-carbon atom in each of the amino acids that make up human proteins is bonded to a hydrogen atom and to one of 20 R groups. The structures of these R groups, often called *side-chain groups*, are highlighted in pink in the amino acid structures in Table 20.2. The amino acids are arranged in four categories based on their R groups. In the first category, containing 9 amino acids, the R groups contain mostly carbon and hydrogen atoms and are nonpolar. The R groups of the remaining 11 amino acids contain at least one heteroatom (O, N, or S) bonded to an H atom and are polar. Of these amino acids, two (aspartic acid and glutamic acid)

TABLE 20.2 Structures and Abbreviations of the 20 Common Amino Acids[a]

Nonpolar R Groups

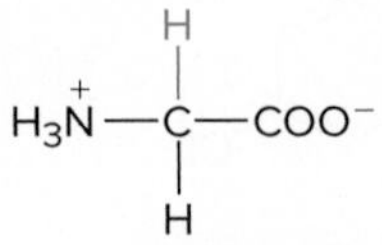

Glycine (Gly)

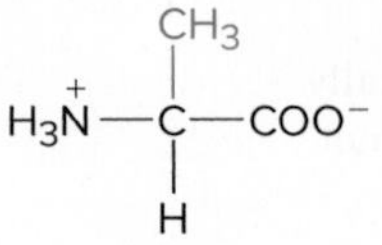

Alanine (Ala)

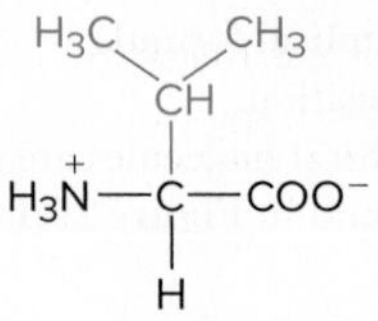

Valine[b] (Val)

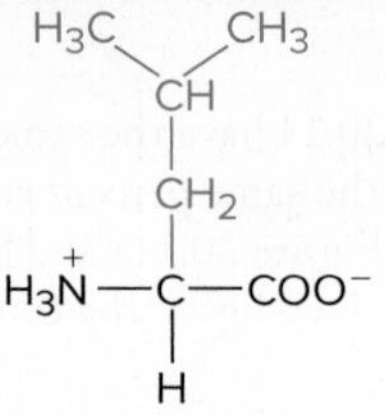

Leucine[b] (Leu)

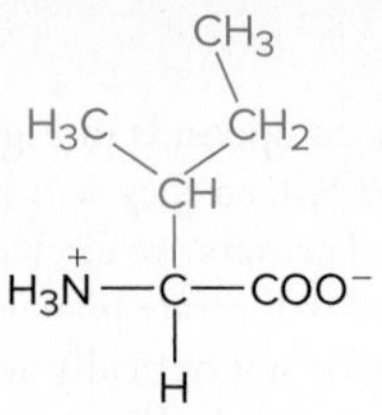

Isoleucine[b] (Ile)

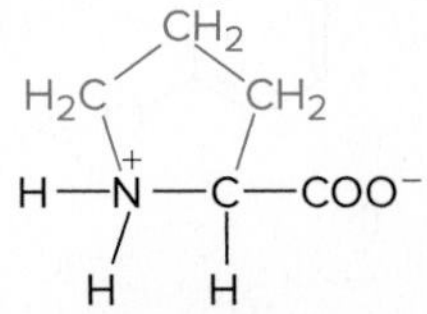

Proline (Pro)

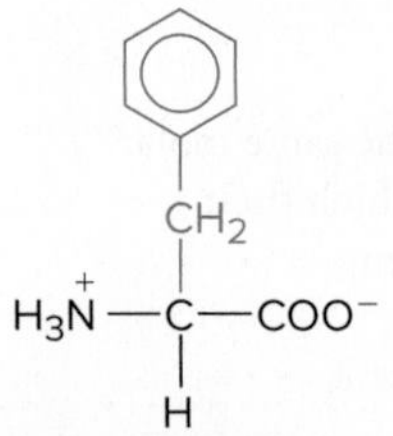

Phenylalanine[b] (Phe)

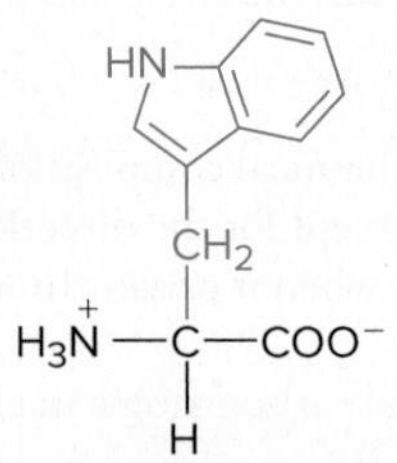

Tryptophan[b] (Trp)

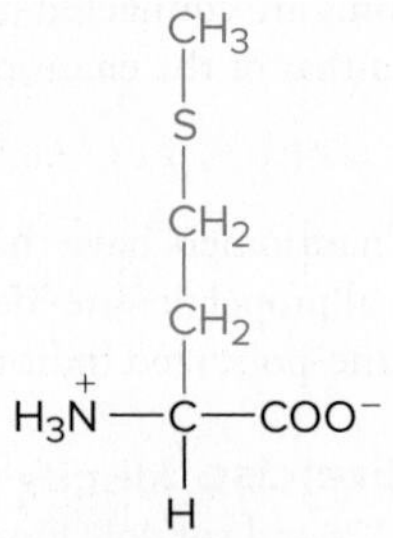

Methionine[c] (Met)

Polar R Groups

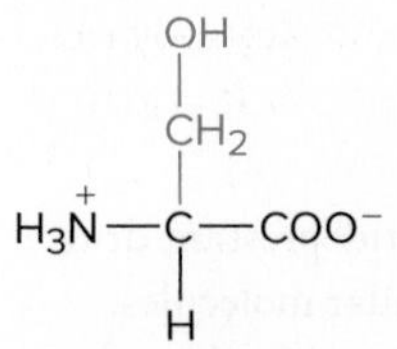

Serine (Ser)

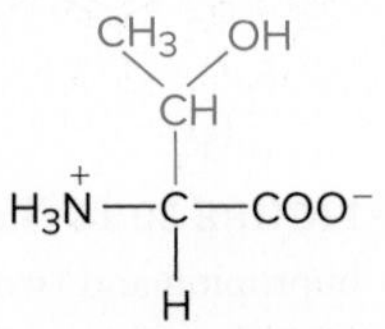

Threonine[b] (Thr)

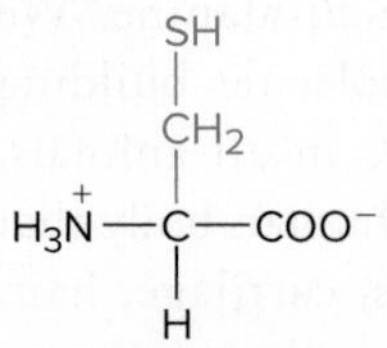

Cysteine (Cys)

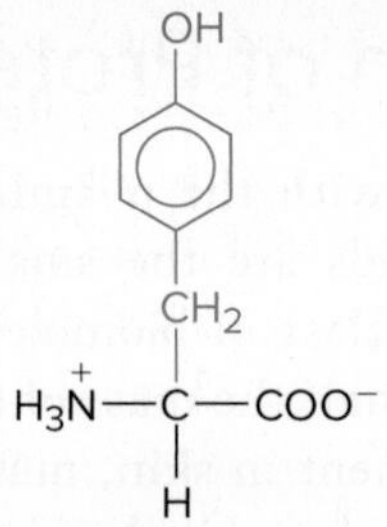

Tyrosine[c] (Tyr)

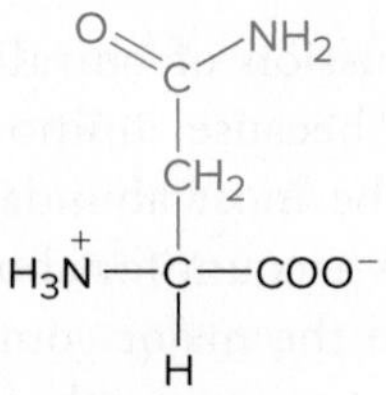

Asparagine (Asn)

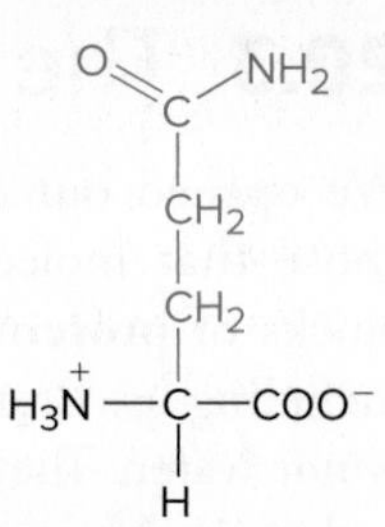

Glutamine (Gln)

Basic R Groups

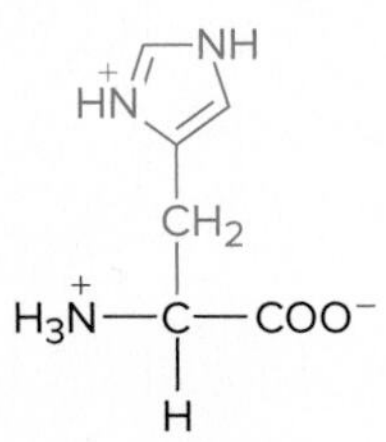

Histidine[b] (His)

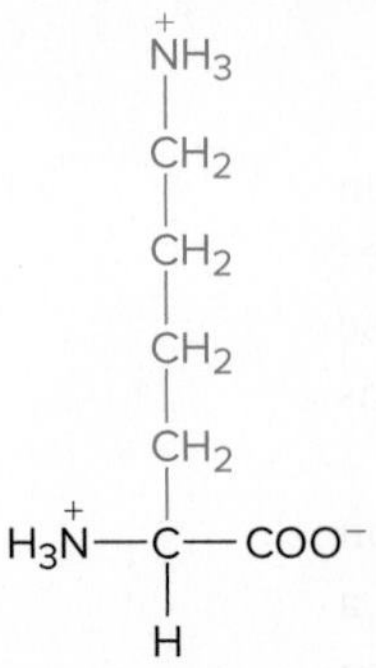

Lysine[b] (Lys)

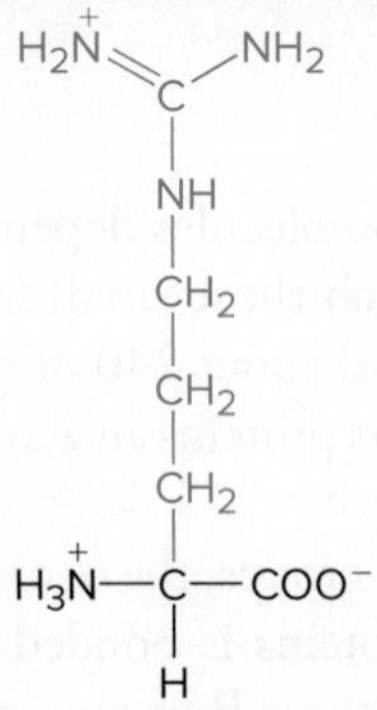

Arginine[b, d] (Arg)

Acid R Groups

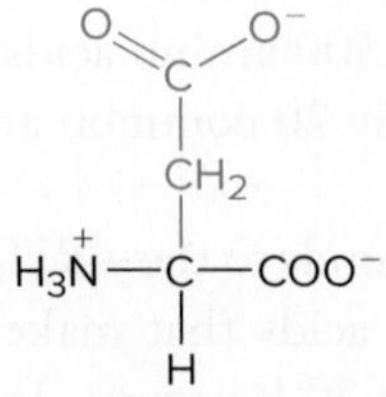

Aspartic acid (Asp)

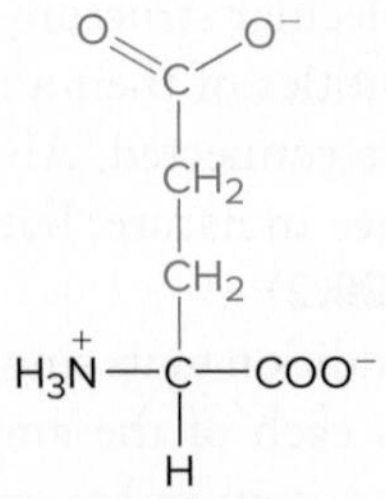

Glutamic acid (Glu)

[a]R groups in pink; ionized forms at pH near 7 shown.
[b]The ten essential amino acids for adults.
[c]In animals, tyrosine is synthesized from phenylalanine.
[d]Arginine is synthesized in animals but is mostly consumed before it can be used in proteins so it is an essential amino acid.

have R groups that contain carboxylic acid functional groups, and three (histidine, lysine, and arginine) have R groups with nitrogen atoms that are weakly basic.

essential amino acid any of the ten amino acids that make up peptides and proteins but are not synthesized in the human body and must be obtained through the food we eat.

CONCEPT TEST

Identify the organic functional groups in the R groups of the following amino acids in Table 20.2: (a) leucine; (b) phenylalanine; (c) serine; (d) glutamine; (e) aspartic acid; (f) lysine.

Our bodies can synthesize 10 of the 20 amino acids in Table 20.2, but the other 10 must be present in the food we eat. These 10 are marked with a superscript *b* and are referred to as **essential amino acids** in adults. Arginine is also an essential amino acid because most arginine produced in our bodies is unavailable for protein synthesis. Most proteins from animal sources, including those in meats, eggs, and dairy products, contain all the essential amino acids needed by the human body, and in close to the correct proportions. These foods are sometimes referred to as *perfect foods* or, more precisely, *complete proteins*. In contrast, most plant proteins from foods such as legumes and vegetables do *not* contain all the essential amino acids, so vegetarians must be careful to eat a combination of foods that provide all the essential amino acids. A good example is red beans and rice (**Figure 20.15**), a traditional dish in Latin American cuisine that provides a balance of essential amino acids: rice has all of them but lysine, and beans lack only methionine.

FIGURE 20.15 A meal of rice and beans provides all the essential amino acids.

For historical reasons, amino acid enantiomers are designated by the prefixes D- (for *dextro-*, right) and L- (*levo-*, left). These labels refer to how the four groups bonded to each chiral carbon atom are oriented in three-dimensional space. They *do not* refer to the direction in which plane-polarized light rotates as it passes through a solution of the amino acid. In other words, there is no connection between the D- and L- prefixes in the names of amino acids and the *dextrorotatory* and *levorotatory* enantiomers that are designated with (+) and (−) signs based on their optical properties. All the chiral amino acids in the proteins in our bodies are L-enantiomers, even though 9 of the 19 are actually dextrorotatory.

CONCEPT TEST

One of the 20 amino acids in Table 20.2 is not chiral. Identify it and explain why it is not. If any of the amino acids have more than one chiral center, identify them.

Zwitterions

If we dissolve an amino acid in a solution that already contains a strong acid and has a low pH, the carboxylic acid group on each molecule does not ionize. In addition, each amine group, being a weak base, accepts a H^+ ion, forming a *protonated* $-NH_3^+$ group. The result is a molecular ion with an overall positive charge, as illustrated for alanine in **Figure 20.16a**.

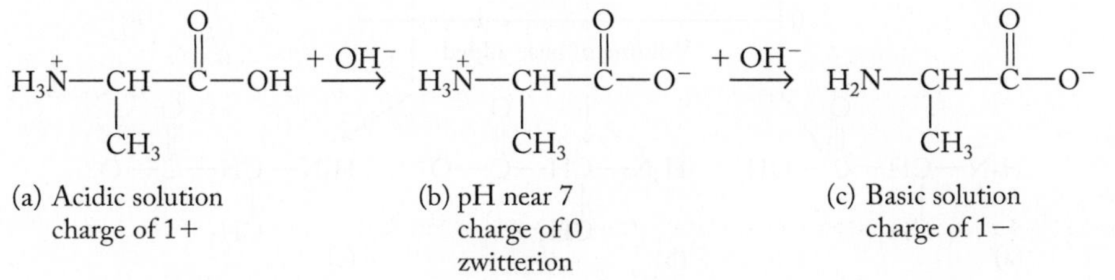

$$H_3\overset{+}{N}{-}CH(CH_3){-}C(=O){-}OH \xrightarrow{+\,OH^-} H_3\overset{+}{N}{-}CH(CH_3){-}C(=O){-}O^- \xrightarrow{+\,OH^-} H_2N{-}CH(CH_3){-}C(=O){-}O^-$$

(a) Acidic solution charge of 1+

(b) pH near 7 charge of 0 zwitterion

(c) Basic solution charge of 1−

FIGURE 20.16 (a) At low pH, alanine (like many other amino acids) exists as a 1+ ion. (b) At physiological pH (7.4), the $-COOH$ group is ionized and the molecule becomes a zwitterion with an overall charge of zero. (c) In basic solutions, the charge decreases to 1− as the $-NH_3^+$ group deprotonates.

zwitterion a molecule that has both negatively and positively charged groups in its structure.

p*I* (also called **isoelectric point**) the pH at which molecules of the amino acid have, on average, zero charge.

Now suppose we add a strong base to this solution, neutralizing the strong acid and raising the pH to about 7.4. We choose this value because it is close to the pH of human blood and of the fluids in most of our tissues. At this *physiological pH*, the $-COOH$ groups in amino acids are mostly ionized, but nearly all the amine groups are still in the protonated $-NH_3^+$ form because $-NH_3^+$ is the conjugate acid of a weak base and is itself a very weak acid. As a result, most of the protonated amine groups still have positive charges and most of the ionized carboxylic acid groups have negative charges. This combination of negatively charged $-COO^-$ groups and positively charged $-NH_3^+$ groups means that these amino acid molecules are **zwitterions** (literally, *hybrid ions*). This term is used to describe molecules that contain both negatively and positively charged functional groups. For example, if an amino acid contains one $-COO^-$ group and one $-NH_3^+$ group (**Figure 20.16b**), it would have no net electrical charge.

If we add more base, alanine loses a H^+ from its NH_3^+ group (we say that it *deprotonates*), and we have a molecular ion with an overall charge of 1− (**Figure 20.16c**). The titration curve for alanine in **Figure 20.17** shows this two-step neutralization process: first the carboxylic acid ionizes, and then the protonated amine deprotonates.

Amino acids with acidic or basic R groups have three-step neutralization processes. For example, an amino acid like glutamic acid or aspartic acid, each of which has a second carboxylic acid group in R, will experience the neutralization of both acidic groups and then the deprotonation of the amine. Which acidic group is the stronger; which deprotonates at lowest pH? Think about the structure of the molecule to answer this. The $-COOH$ group bonded to the α-carbon atom is only one carbon atom away from a positively charged NH_3^+ group. The side-chain $-COOH$ group is at least two carbon atoms away, and the carboxylate ion that it forms when it ionizes is less stabilized by delocalization of its negative charge toward the cationic group. Therefore, the $-COOH$ group bonded to the α-carbon atom is the stronger acid and ionizes first.

For an amino acid with a basic R group, such as asparagine or glutamine, the order of ionization in a titration also depends on the relative strengths of the groups as proton donors. The $-COOH$ group always ionizes first as base is added; it is the strongest acid in the molecule. The protonated nitrogen groups deprotonate in order of their relative strengths as proton donors.

Each amino acid has a characteristic **p*I*** value, called the **isoelectric point**. It represents the pH at which molecules of the amino acid have, on average, zero charge. The pH at the equivalence point between the two pK_a values is the p*I*. Its

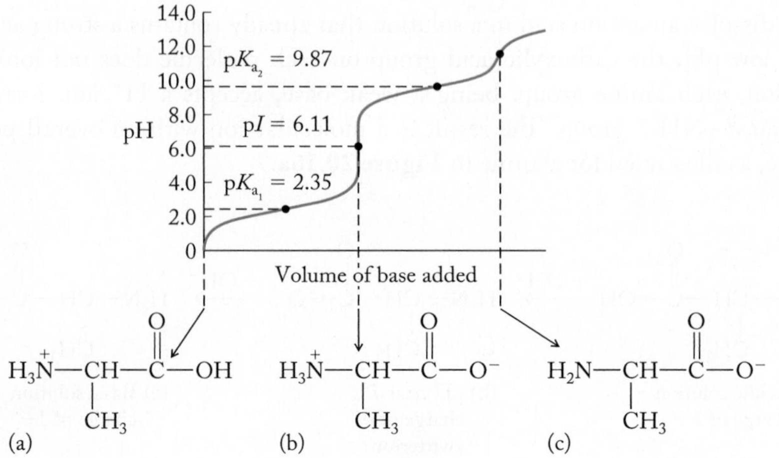

FIGURE 20.17 The titration curve of alanine resembles that of a weak diprotic acid in Figure 16.15. The carboxylic acid group ($pK_{a_1} = 2.35$) is neutralized first. The protonated amine group ($pK_{a_2} = 9.87$) is neutralized in the second step of the titration. The dominant forms of alanine present in solution are shown at (a) the starting point, and at (b) and (c), the two equivalence points.

value is the average of the neighboring pK_a values. At this pH, the amino acid does not migrate in an electric field. If the pH is more acidic than the p*I*, the amino acid is positively charged and migrates to the cathode (negative electrode). At pH values more basic than the p*I*, the amino acid has a net negative charge and migrates toward the anode (positive electrode). This behavior is the basis of several analytical techniques used to characterize and identify amino acids. The p*I* values of neutral amino acids are around pH 7; acidic amino acids have p*I* values much lower than 7, and basic amino acids, much higher.

CONNECTION The inverse relationship between the strengths of acid–base conjugate pairs was introduced in Section 15.3 (see Figure 15.9).

SAMPLE EXERCISE 20.4 Interpreting Acid–Base Titration Curves of Amino Acids **LO2**

Figure 20.18 shows the titration curve for aspartic acid and its pK_{a_1}, pK_{a_2}, and pK_{a_3} values. Draw the molecular structures of the principal form of aspartic acid that is in solution at the start of the titration and at each equivalence point. Estimate the p*I* of aspartic acid.

Collect, Organize, and Analyze Aspartic acid (Table 20.2) has two carboxylic acid groups and one amine group. At low pH the carboxylic acid groups are not ionized and the amine group is protonated. As pH is raised, the –COOH groups ionize first and then, under basic conditions, the protonated amine group releases its proton. The –COOH group bonded to the α-carbon atom should be the stronger acid and ionize first.

Solve At (a), the beginning of the titration, the fully protonated 1+ ion is present:

O＝C—OH
|
CH_2
|
$H_3\overset{+}{N}$—C—C(＝O)—OH
|
H

At (b), the first equivalence point, the –COOH group bonded to the α-carbon atom is ionized; the charge on the ion is 0.

O＝C—OH
|
CH_2
|
$H_3\overset{+}{N}$—C—C(＝O)—O^-
|
H

At (c), the second equivalence point, the side-chain –COOH group is ionized; the charge on the ion is 1−:

O＝C—O^-
|
CH_2
|
$H_3\overset{+}{N}$—C—C(＝O)—O^-
|
H

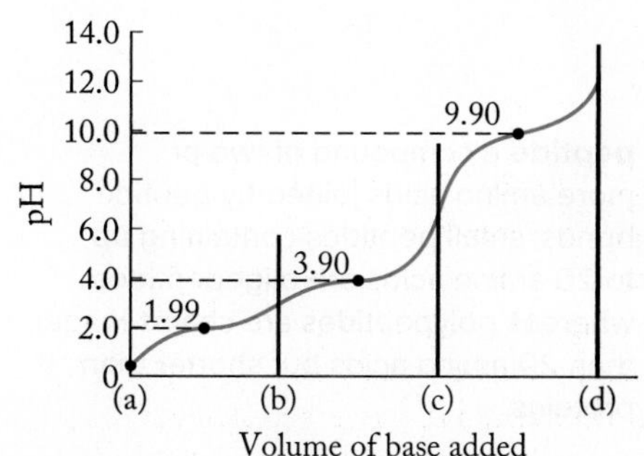

FIGURE 20.18 Titration curve of aspartic acid. The vertical black lines identify the equivalence points.

At (d), the third equivalence point in the titration, the amine group is deprotonated, resulting in an ion with a 2− charge:

$$\mathrm{H_2N{-}CH(CH_2COO^-){-}COO^-}$$

Although there are three equivalence points, the pI is calculated from the first two pK_a values because the pI indicates the pH at which molecules of the amino acid, on average, have a net neutral charge. The pI of aspartic acid can be estimated by taking the value midway between pK_{a_1} and pK_{a_2} in Figure 20.18: (1.99 + 3.90)/2 = 2.94.

Think About It We predicted that the α-COOH group would ionize first, consistent with the observation of two values of pK_{a_1} and pK_{a_2}.

Practice Exercise Sketch the titration curve for lysine starting at pH = 1.0, and draw the molecular structure of the principal form of lysine that is in solution at the beginning of the titration and at each equivalence point.

The acid–base characteristics of amino acids are important for their function in enzymes, where protonation/deprotonation controls their activity as catalysts. In addition, the structures of proteins depend in part on hydrogen bonding, which can be disrupted by changes in pH. This process is used in food preparation, when acids such as lemon juice are used to "cook" raw fish in dishes such as ceviche.

Peptides

When amino acids bond they form chainlike molecules with a wide range of sizes. Amino acid *residues* (the name we give to amino acids that are part of a larger molecule) make up each link in the chain. The longest chains, some with molar masses as high as 3 million g/mol, are called proteins, but the shortest chains, called **peptides**, are only a few links long. The smallest peptides contain only two or three amino acid residues. These molecules are called *dipeptides* and *tripeptides*, respectively. Peptides up to 20 residues long are called *oligopeptides.* Those made of more than 20 are called *polypeptides.* The size at which a polypeptide becomes a protein, or biopolymer, is arbitrary but is typically set around 50–75 amino acid residues.

CONNECTION The formation of amide bonds in reactions between carboxylic acids and amines was described in Section 12.6 (see Figures 12.43 and 12.46).

The type of bond linking the amino acids in peptides and proteins is called a **peptide bond** (highlighted in blue in **Figure 20.19**) or *peptide linkage.* It forms when the α-carboxylic acid group of one amino acid condenses with the α-amine group of another, with the loss of a molecule of water. The peptide bonds have the same structure as the amide bonds that hold together the monomeric units of synthetic polyamides such as nylon and Kevlar (see Section 12.6). This process is not spontaneous and is driven by being coupled to ATP hydrolysis (see Section 17.9).

peptide a compound of two or more amino acids joined by peptide bonds; small peptides containing up to 20 amino acids are oligopeptides, whereas polypeptides are chains longer than 20 amino acids but shorter than proteins.

peptide bond the result of a condensation reaction between the carboxylic acid group of one amino acid and the amine group of another.

The convention for drawing the structures of peptides begins by placing the amino acid that has a free α-amine group at the left end of the peptide chain and the amino acid with a free α-carboxylic acid at the right end. The left end is called the *amine* (or *N-*) *terminus* of the peptide, and the right end is called the

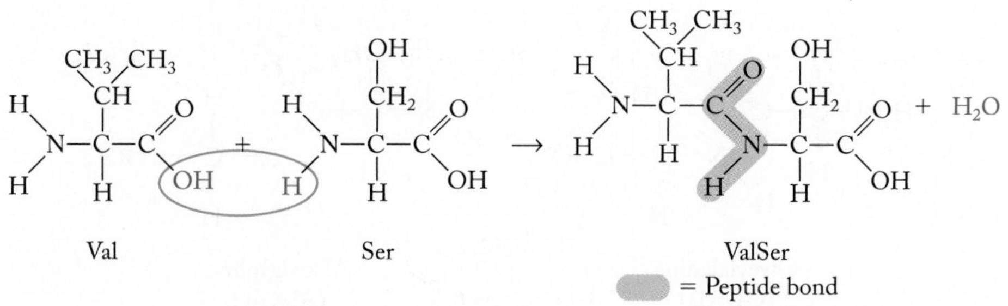

FIGURE 20.19 When the carboxylic acid group of valine reacts with the amine group of serine (green oval) to form a peptide bond (blue highlight), the products are the dipeptide valylserine and water.

carboxylic acid (or *C-*) *terminus*. The name of a peptide is based on the names of its amino acids, starting with the one at the N-terminus. The names of all the amino acids except the one at the C-terminus are changed to end in *-yl*. For example, if a peptide bond forms between the α-COOH group of valine (Val) and the α-NH_2 group of serine (Ser), as in Figure 20.19, the resulting dipeptide is called valylserine, or more typically, using the three-letter abbreviation, ValSer.

CHEMTOUR
Condensation of Biological Polymers

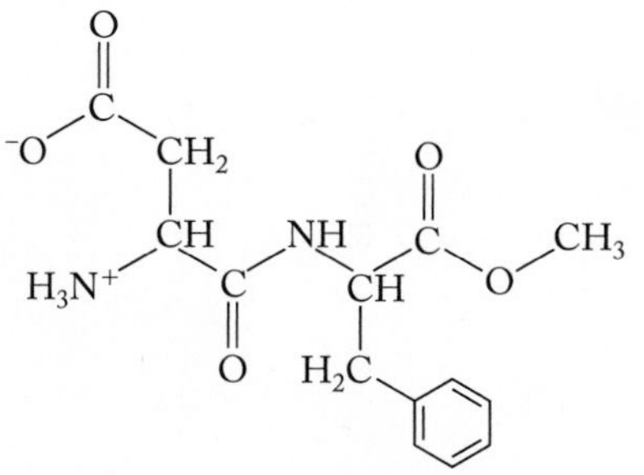

The artificial sweetener aspartame is the methyl ester of the dipeptide aspartylphenylalanine (**Figure 20.20**). At pH 7.4, aspartame exists as a zwitterion because the aspartic acid amine group is protonated and the carboxylic acid in its R group is ionized. In contrast, its parent dipeptide has a net charge of 1– because both –COOH groups are ionized at that pH. In an amino acid that has an amine group in its R group, such as lysine or arginine, that amine group is probably protonated at physiological pH, giving the amino acid a net charge of 1+. The overall charge on the peptide at physiological pH is the sum of the positive charges on protonated amine groups and the negative charges of ionized carboxylic acid groups.

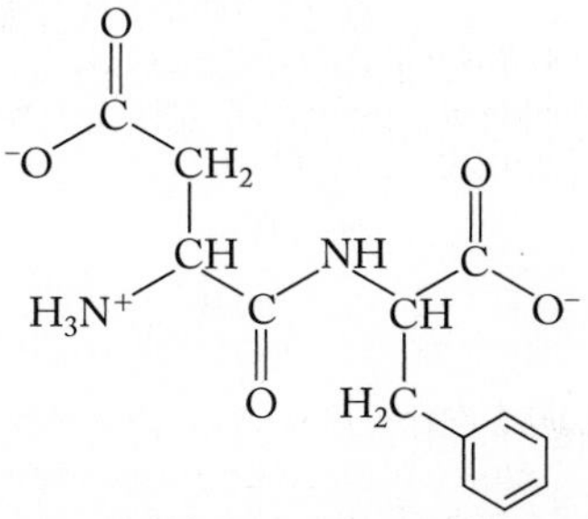

FIGURE 20.20 Aspartame is the methyl ester of the dipeptide aspartylphenylalanine, so it has one fewer –COOH group than its parent dipeptide. Therefore, the overall charge of a molecule of aspartame at pH 7.4 is zero, whereas the charge on aspartylphenylalanine is 1–.

SAMPLE EXERCISE 20.5 Drawing and Naming Peptides **LO3**

(a) Name all the dipeptides that can be made by reacting an amino acid mixture containing alanine and glycine, and (b) draw their molecular structures in solution at pH 7.4.

Collect, Organize, and Analyze The structures of alanine and glycine are shown in Table 20.2. Four different peptides can be made from two amino acids: AlaAla, GlyGly, AlaGly, and GlyAla. The R groups in alanine (–CH_3) and glycine (–H) have no acidic or basic properties. At pH 7.4, the carboxylic acid terminus –COOH groups should be ionized, and the amine terminus –NH_2 groups should be protonated.

Solve

a. The names of the four peptides with the amino acid sequence AlaAla, GlyGly, GlyAla, and AlaGly are alanylalanine, glycylglycine, glycylalanine and alanylglycine, respectively.
b. At pH 7.4, the –NH_2 groups of amino acids are protonated and the –COOH groups are ionized, so the principal forms of these dipeptides are

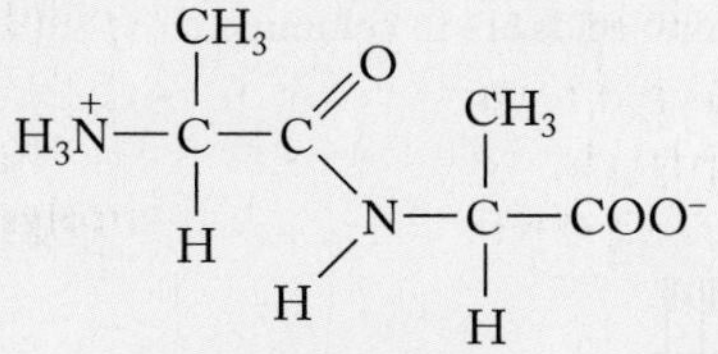

Alanylalanine (AlaAla)

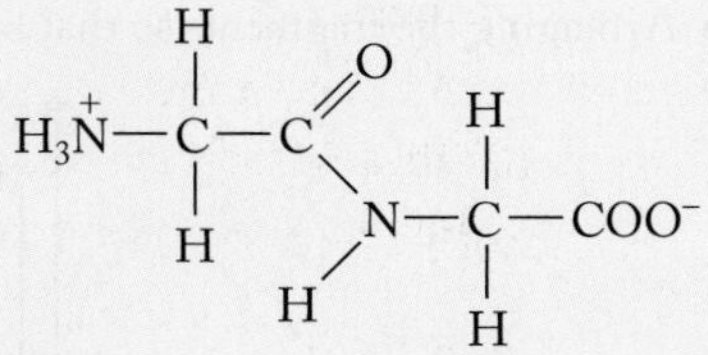

Glycylglycine (GlyGly)

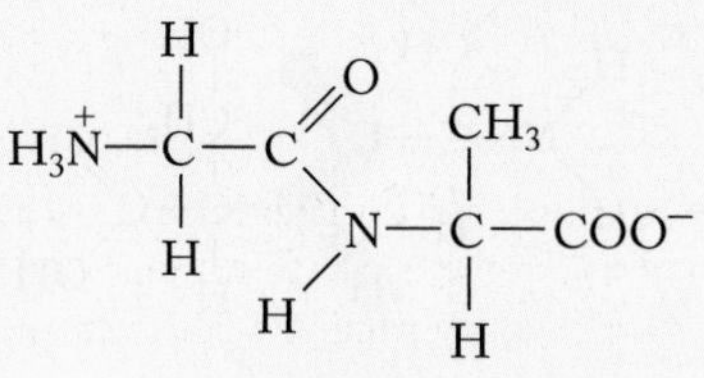

$$H_3\overset{+}{N}-CH_2-C(=O)-N(H)-CH(CH_3)-COO^-$$

Glycylalanine
(GlyAla)

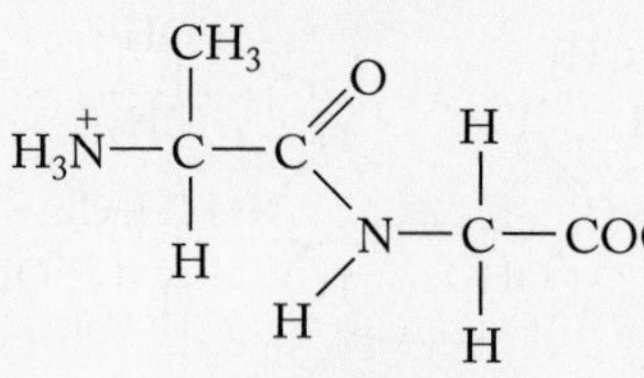

$$H_3\overset{+}{N}-CH(CH_3)-C(=O)-N(H)-CH_2-COO^-$$

Alanylglycine
(AlaGly)

Think About It Only the N-terminus $-NH_2$ and C-terminus $-COOH$ groups can be protonated or ionized in these dipeptides. Thus, all four of the dipeptides are zwitterionic with a net charge of $(1+) + (1-) = 0$ at pH 7.4.

Practice Exercise How many different tripeptides can be synthesized from one molecule of each of three different amino acids, Gly, Tyr, and Ser?

Only two sequences are possible for a peptide containing two different amino acids. Amino acid A and amino acid B can make dipeptide AB or BA. For a peptide made from three different amino acids—A, B, and C—six sequences are possible. For longer peptides, the number of possibilities becomes enormous. The sequence matters in determining the properties of the peptides, and figuring out the amino acid sequence of peptides and proteins is a significant challenge in biochemistry. One of the fundamental experimental methods used to determine sequence involves finding the identities of individual component amino acids, and then identifying fragments of the structure containing several amino acids and piecing them together like a puzzle. For example, a tripeptide could be broken down into amino acids A, B, and C. Dipeptide fragments BA and AC could be identified, which means the overall sequence of the tripeptide must be BAC.

SAMPLE EXERCISE 20.6 Determining the Amino Acid Sequence of a Small Peptide **LO3**

A small peptide was hydrolyzed completely to its component amino acids: A, B, C, D, E, F, and G. Alternative ways to treat the peptide yielded the larger fragments BA, CDF, EC, AG, and GEC. What is the sequence of the amino acids in the peptide?

Collect, Organize, and Analyze We know the identity of seven amino acids that make up a peptide, and we have several dipeptide and tripeptide fragments. We need to arrange them to reveal the complete sequence of the peptide.

Solve Arranging the fragments so that like amino acids are in columns:

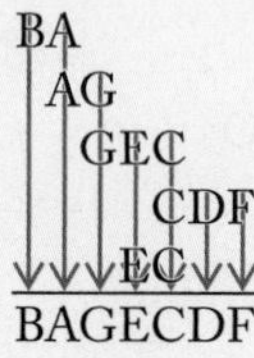

The amino acid sequence of the peptide is BAGECDF.

Think About It Complete hydrolysis of the peptide provides less information about the amino acid sequence than longer fragments do.

Practice Exercise The hydrolysis of a peptide yielded the following amino acids: Gly, Ala, Val, Thr, and 2 Leu; and the following larger fragments: LeuThr, AlaLeuLeu, ThrGly, and ValAla. What is the sequence of the amino acids in the peptide?

20.4 Protein Structure and Function

CHEMTOUR Fiber Strength and Elasticity

The structure of a protein is crucial to its function. This fact is easiest to appreciate for structural proteins such as the collagens, which impart flexibility, strength, and elasticity to skin and tendons. The functions of nonstructural proteins, such as the ability of enzymes to catalyze biochemical reactions, are also closely linked to their structures. These large biomolecules must assume particular three-dimensional conformations to interact with other molecules and to function properly (**Figure 20.21**). **Table 20.3** lists some of the important functional classifications of proteins and gives examples of each.

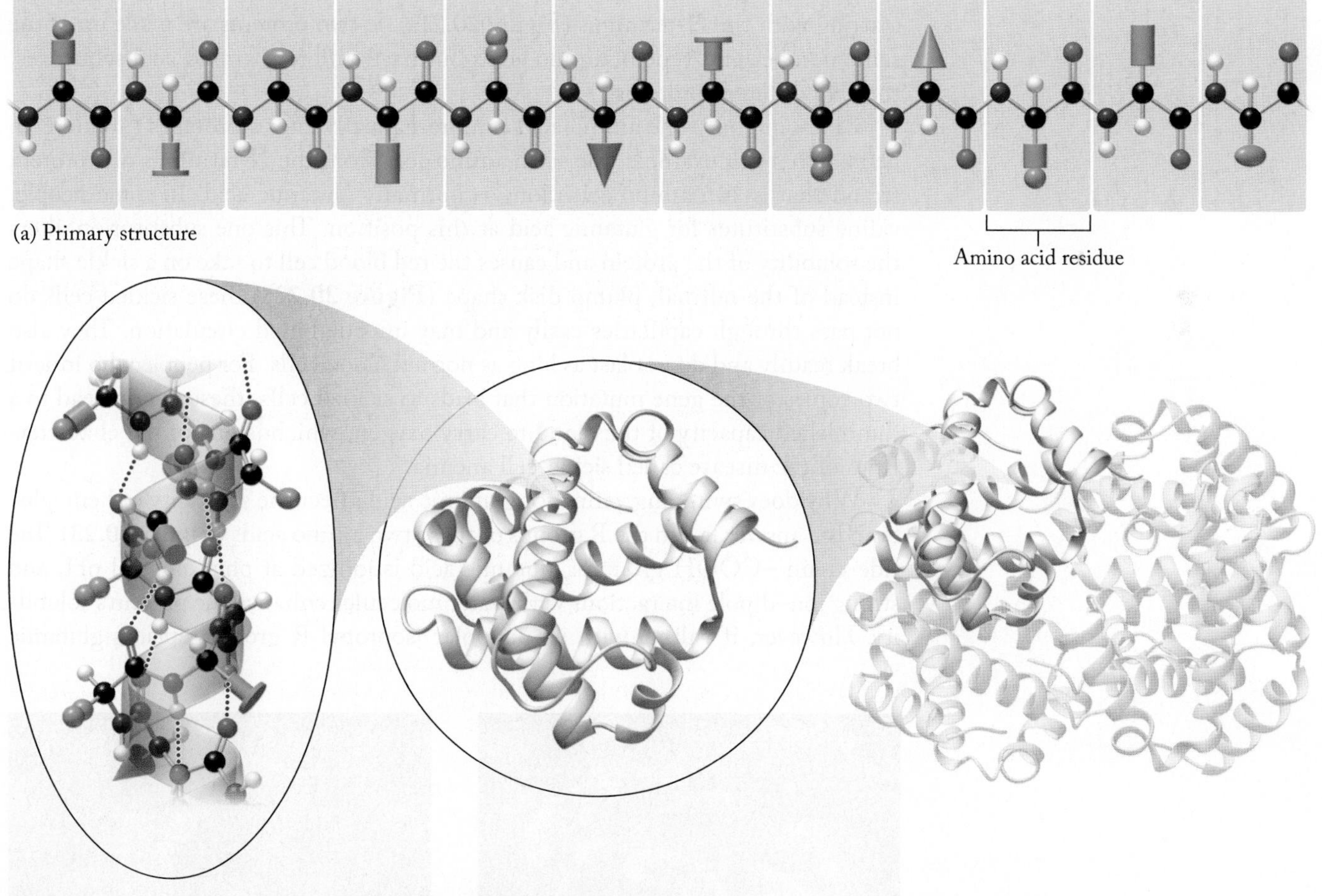

FIGURE 20.21 The four levels of protein structure. (a) A protein's primary structure is its amino acid sequence. The green shapes represent different R groups. (b) Secondary structure (here, an α helix) describes the three-dimensional pattern adopted by segments of the protein strand. (c) Tertiary structure is the overall shape of the molecule as segments of it bend and fold. (d) Quaternary structure refers to the overall shape adopted by multiple protein strands that assemble into a single unit.

primary (1°) structure the sequence in which amino acid monomers occur in a protein chain.

secondary (2°) structure the pattern of arrangement of segments of a protein chain.

α helix a coil in a protein chain's secondary structure held together by hydrogen bonds.

β-pleated sheet a puckered two-dimensional array of protein strands held together by hydrogen bonds.

TABLE 20.3 Functional Classes of Proteins

Class of Protein	Function	Example
Structural	Gives strength and flexibility to tissues	Collagen: a component of connective tissue
Enzymes	Catalyze reactions in cells	Sucrase: promotes conversion of sucrose into simpler sugars
Hormones	Regulate processes in organisms	Insulin: helps cells regulate absorption of glucose
Signaling	Coordinates the activity of cells	GTPase: regulates intracellular dynamics
Immune response	Defense	Antibodies: attack bacteria
Storage	Releases energy during metabolism	Ovalbumin: present in eggs
Transport	Transports molecules from place to place	Hemoglobin: transports oxygen from lungs to cells

Primary Structure

The **primary (1°) structure** of a protein is the sequence of the amino acids in it, starting with the N-terminus (Figure 20.21a). If two proteins are made up of the same number and type of amino acids but have different amino acid sequences, they are different proteins.

Changing only one amino acid can dramatically alter a protein's function. In hemoglobin, for example, the sixth amino acid from the N-terminus of a protein strand that is 146 amino acids long is normally glutamic acid. In some people, valine substitutes for glutamic acid at this position. This one substitution alters the solubility of the protein and causes the red blood cell to take on a sickle shape instead of the normal, plump disk shape (**Figure 20.22**). These sickled cells do not pass through capillaries easily and may impede blood circulation. They also break readily and do not last as long as normal blood cells. For people who inherit two copies of the gene mutation that leads to sickled cells, these factors lead to a diminished capacity of the blood to carry oxygen, which is one of the characteristics of the disease called sickle-cell anemia.

Why does switching valine for glutamic acid affect the solubility of hemoglobin? The answer lies in the R groups of these two amino acids (**Figure 20.23**). The side-chain –COOH group of glutamic acid is ionized at physiological pH, and strong ion–dipole interactions with water molecules enhance the protein's solubility. However, if valine, with its nonpolar isopropyl R group, replaces glutamic

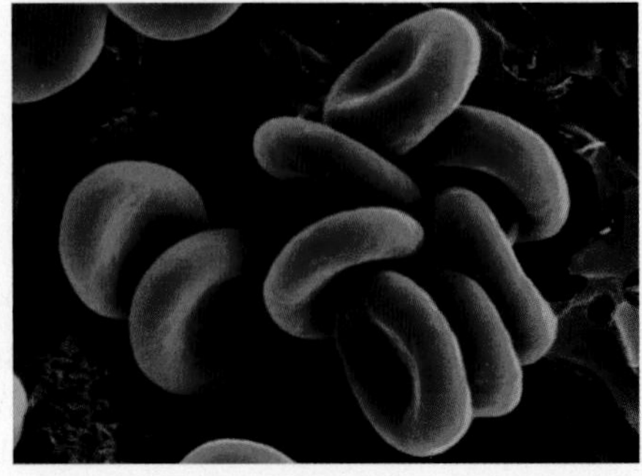

(a)

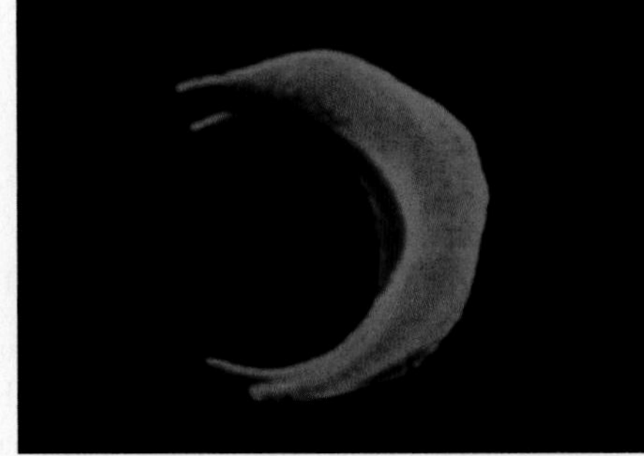

(b)

FIGURE 20.22 (a) Normal red blood cells are plump disks, whereas (b) those in patients with sickle-cell anemia are distorted and incomplete.

acid, that strong intermolecular interaction with water is lost. The valine creates a hydrophobic patch on the surface of the protein that results in the deoxygenated forms of hemoglobin sticking to each other, producing stiff fibers inside red blood cells. This, in turn, deforms the red blood cell from its usual smooth disk shape (Figure 20.22a) into a sickle shape (Figure 20.22b).

Sickle-cell anemia is a debilitating disease, but it provides a survival advantage in regions where malaria is endemic. Having sickle-cell anemia does not protect people from contracting malaria or make them invulnerable to the parasite that causes it. However, children infected with malaria are more likely to survive the illness if they have sickle-cell anemia. Exactly why sickle-cell anemia has this effect is not completely understood, but recent work points toward sickle-hemoglobin increasing the production of an enzyme that catalyzes the production of carbon monoxide; the carbon monoxide protects the host from succumbing to malaria without actually interfering with the parasite.

Normal protein:

Val - His - Leu - Thr - Pro - Glu - Lys - . . .

Abnormal protein:

Val - His - Leu - Thr - Pro - Val - Lys - . . .

(a) Primary structure

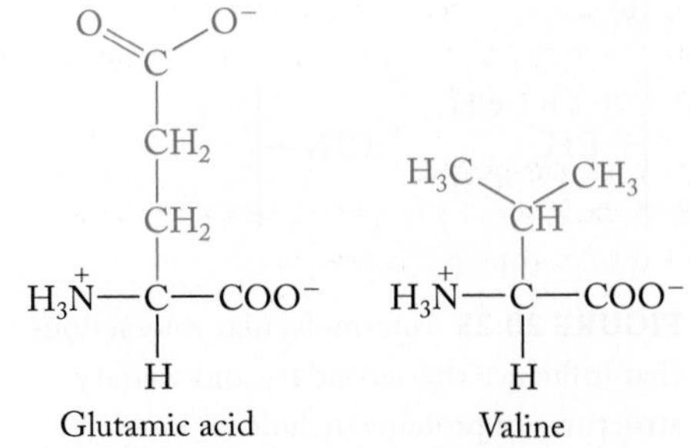

$O=C(O^-)-CH_2-CH_2-C(H)(COO^-)-\overset{+}{N}H_3$ Glutamic acid

$(H_3C)(CH_3)CH-C(H)(COO^-)-\overset{+}{N}H_3$ Valine

(b)

FIGURE 20.23 (a) The primary structure of the amine end of a protein in normal hemoglobin and in the abnormal hemoglobin responsible for sickle-cell anemia. (b) The replacement of glutamic acid (with its hydrophilic R group) by valine (with its hydrophobic group) is responsible for the disease.

CONCEPT TEST

Which other amino acids besides valine might result in the sickling of the cell when substituted for glutamic acid?

Secondary Structure

The next level of protein structure, the **secondary (2°) structure**, describes the geometric patterns made by segments of amino acid chains. One common pattern is the **α helix**, a coiled arrangement with the R groups pointing outward (Figure 20.21b). The helical structure is maintained by hydrogen bonds between –NH groups on one part of the chain and C═O groups on amino acids four residues away from them in the sequence. The α helix looks very much like a spring. One group of proteins that are mostly α-helical is the keratins, the proteins in hair and fingernails.

CONNECTION We discussed in Chapter 10 how ion–dipole and dipole–dipole interactions are the key to the solubility of solutes in water.

Another common pattern of 2° structure is called the **β-pleated sheet**. These sheets are assemblies of multiple amino acid chains aligned side by side. The pleats are caused by the tetrahedral molecular geometries of the atoms along the chains (**Figure 20.24**). Adjacent chains are linked by hydrogen bonds, and the collection of side-by-side zigzag chains forms a continuous β-pleated sheet. R groups extend above and below the pleats. Sheets may stack on top of one another like two pieces of corrugated roofing. Stacked sheets are held together by the same interactions

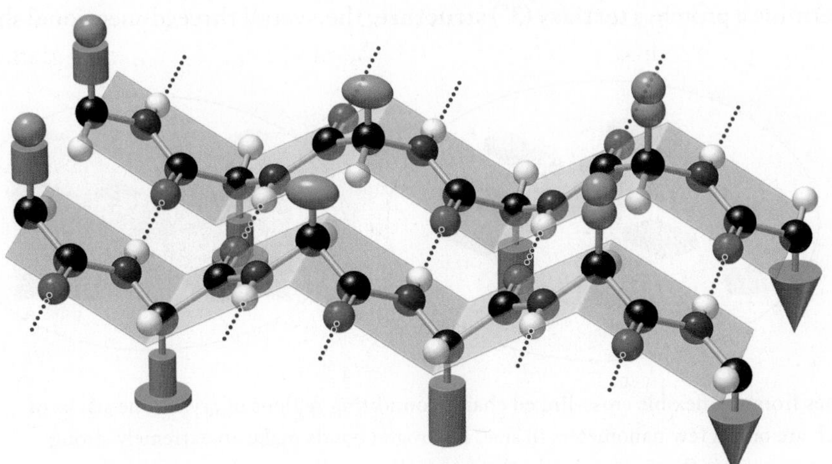

FIGURE 20.24 Each amino acid chain in a β-pleated sheet is folded in a zigzag pattern. Adjacent chains in a sheet are held together by hydrogen bonds (blue dotted lines). The R groups (green shapes) extend above and below the sheet, linking it to adjacent sheets via noncovalent intermolecular interactions.

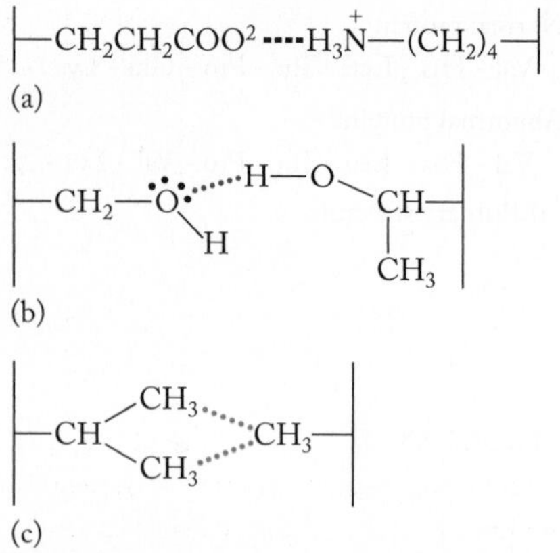

$-CH_2CH_2COO^{2}\cdots H_3\overset{+}{N}-(CH_2)_4-$

(a)

$-CH_2-\ddot{O}(H)\cdots H-O-CH(CH_3)-$

(b)

$-CH(CH_3)(CH_3)\cdots CH_3-$

(c)

FIGURE 20.25 Intermolecular interactions that influence the secondary and tertiary structures of proteins include (a) ion–ion interactions between acidic and basic R groups, (b) hydrogen bonding, and (c) dispersion forces between nonpolar side chains.

that hold all proteins together, including ion–ion forces, hydrogen bonds, and dispersion forces, depending on the pairs of R groups involved (**Figure 20.25**).

The proteins that make up strands of silk form thin, planar crystals of β-pleated sheets that are only a few nanometers on a side. Enormous numbers of these crystals form long arrays of sheets stacked together like nanoscale pancakes (**Figure 20.26**). Hydrogen bonds hold the stacks together and reinforce adjacent sheets. As a result of these interactions, strands of silk are stronger than strands of steel with the same mass; indeed, silk is one of the toughest known materials—natural or synthetic.

Some single-stranded proteins exist as α helices on their own but form β-pleated sheets when they clump together in multistrand aggregates. One consequence of this clumping is the formation of insoluble protein deposits called plaques. Abnormal accumulation of plaque can be a serious health risk: plaque formed by a protein called amyloid β has been linked to the onset of Alzheimer's disease.

Large protein molecules may contain both types of 2° structure as well as other loop and turn structures. In describing a protein, scientists may indicate the percentage of amino acids involved in each type—for example, 50% α-helical, 30% β-pleated sheet, and 20% other structures. **Figure 20.27** shows a model of a protein called carbonic anhydrase, which contains several different types of 2° structure.

CONCEPT TEST

The aqueous solution of proteins in egg whites turns into a solid mass when eggs are cooked or are dropped into an organic solvent such as acetone. Does the solidification process occur as a result of a change in the primary structure of the proteins? Explain your answer.

Tertiary and Quaternary Structure

Large proteins have structure beyond the 1° and 2° levels. Their molecules can fold back on themselves as a result of ion–ion, ion–dipole, or dispersion forces between R groups on amino acids that are considerable distances apart along the protein chain. They may even involve the formation of covalent bonds. For example, the –SH groups on two cysteine residues may combine to form a disulfide linkage that holds two parts of the protein strand together via an intrastrand –S—S– covalent bond (**Figure 20.28**). Interactions and reactions such as these determine a protein's **tertiary (3°) structure**, the overall three-dimensional shape

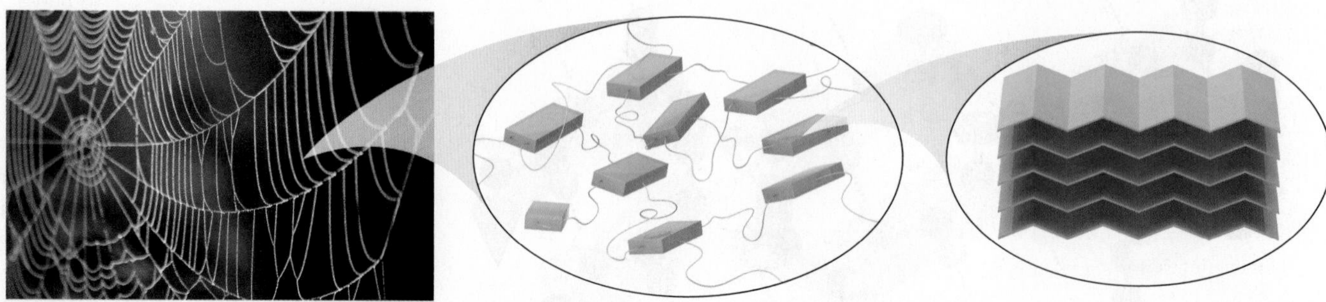

FIGURE 20.26 The strength of spider silk comes from the flexible cross-linked chains connecting regions of crystalline stacks of β-sheets, shown here as blue-green boxes, which are only a few nanometers in size. Hydrogen bonds make an extremely strong network, and if a hydrogen bond breaks, many more are left that can maintain the material's overall strength.

of the protein that is key to its biological activity (Figure 20.21c). Hydrophobic R groups tend to be positioned in the interiors of proteins in the aqueous environment of living cells, whereas hydrophilic groups (as we saw in normal hemoglobin) are oriented toward the outside, where they interact with nearby molecules of water. Hydrophobic interactions are the primary force that causes protein folding and compaction, but all the other modes of interaction help a large protein form its unique 3° structure.

Hemoglobin and some other proteins exhibit an even higher order of structure. One hemoglobin unit (**Figure 20.29a**) contains four protein strands, each of which enfolds a porphyrin ring containing one Fe^{2+} ion. The combination of four protein strands to make one hemoglobin assembly is an example of **quaternary (4°) structure** (Figure 20.21d). The strands in these structures are held together by many of the same intermolecular forces and covalent bonds that determine tertiary structures, such as disulfide linkages that help protein strands with α-helical 2° structures coil around each other to make even larger coils (**Figure 20.29b**). When the protein strands in keratin structures are held together mostly by dispersion forces, as they are in the keratin in skin tissue, the structures are flexible and elastic. If they are also restrained by many covalent bonds, as in Figure 20.29b, they produce tissues that are hard and less flexible, like fingernails and the beaks of birds.

Enzymes: Proteins as Catalysts

We introduced *enzymes* as biological catalysts in Section 13.6. As described there, the chemical reactions involved with metabolism—both *catabolism* (breaking down of molecules) and *anabolism* (synthesis of complex materials from simple feedstocks)—are mediated in large part by enzymes. Both catabolism and anabolism are organized in sequences of reactions called *metabolic pathways*, and each step in a metabolic pathway is catalyzed by a specific enzyme.

Enzymes are highly selective: each catalyzes a particular reaction involving particular reactants. For example, an enzyme called lactase catalyzes only the reaction by which lactose (the sugar in milk) is broken down during digestion. People who lack this enzyme cannot metabolize this sugar; they are said to be

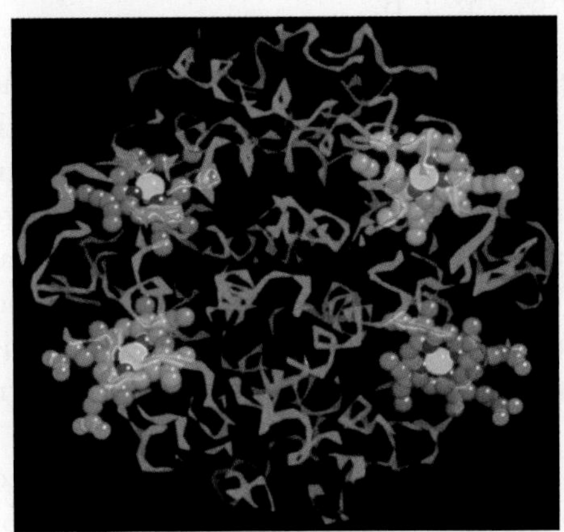

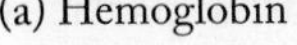

(a) Hemoglobin

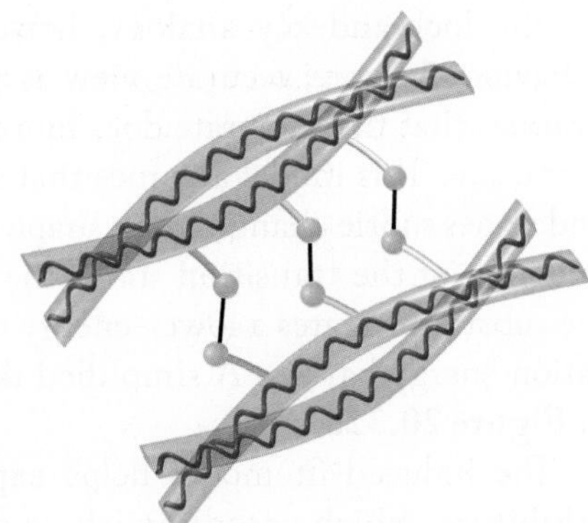

(b) Keratin

FIGURE 20.29 Quaternary structure of proteins. (a) Four protein chains form a single unit in the quaternary structure of hemoglobin. The iron-containing porphyrins are bright green. (b) Pairs of α-helical chains (blue) wound together and linked by interstrand –S—S– bonds (yellow) stabilize the quaternary structure of the keratin in hair and fingernails.

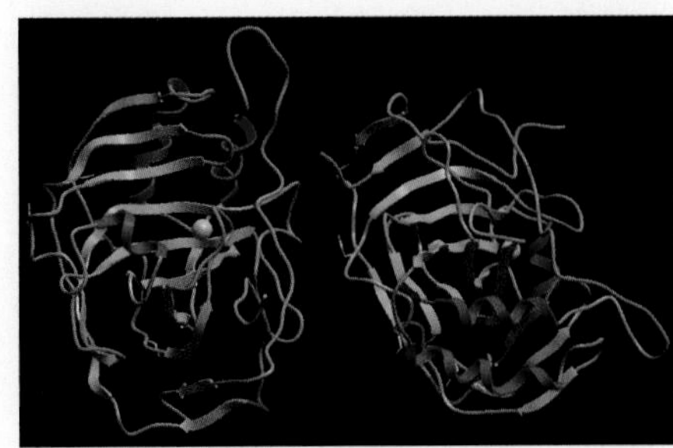

FIGURE 20.27 The structure of the protein carbonic anhydrase contains both α-helical regions (red) and β-pleated sheet regions (green). The structure of the blue region is more complicated. The light blue sphere in the center is a zinc ion.

$$\vdash CH_2—SH \quad HS—CH_2 \dashv$$

$$\downarrow$$

$$\vdash CH_2—S—S—CH_2 \dashv$$

$$+ 2\,H^+(aq) + 2\,e^-$$

FIGURE 20.28 The tertiary structure of some proteins is stabilized by intrastrand –S—S– bonds that form between the –SH groups on the side chains of cysteine residues.

CONNECTION All the types of intermolecular forces described in Chapter 10 are involved in the *intra*molecular interactions that give proteins their unique 2° and 3° structures.

CONNECTION In Section 13.6 we described enzymes as homogeneous catalysts that selectively speed up biochemical reactions.

tertiary (3°) structure the three-dimensional, biologically active structure of a protein that arises because of interactions between the R groups on its amino acids.

quaternary (4°) structure the larger structure functioning as a single unit that results when two or more proteins associate.

lactose intolerant. If they consume dairy products, unmetabolized lactose passes into their large intestines where bacteria ferment it, and unpleasant and painful abdominal disturbances result.

Synthetic reaction pathways catalyzed by enzymes are usually more rapid and involve fewer steps than uncatalyzed pathways to make the same product. The products also tend to be optically pure materials. These advantages can significantly reduce the cost of production of, for example, biological pharmaceuticals. However, biocatalytic reactions often run best in very dilute solutions, which limits production. A key issue driving interest in such processes is that an enantiomerically pure pharmaceutical is likely to be more potent and produce fewer side effects than a racemic mixture of the same product, as we saw with the drug ethambutol in Section 20.2.

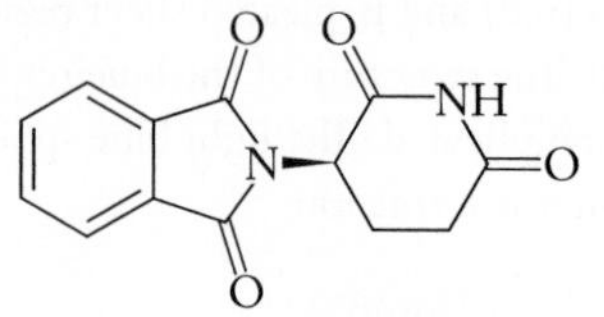

The drug thalidomide (**Figure 20.30**) is a racemic mixture of two enantiomeric forms: the (+)-enantiomer is a sedative and is effective in treating morning sickness, but the (−)-enantiomer is a teratogen (an agent that disturbs the development of a fetus). More than 10,000 children were born in the late 1950s and early 1960s with serious, frequently fatal, deformities as a result of their mothers taking the racemic drug. The individual isomers can be interconverted *in vivo*, so administering only the (+)-enantiomer does not avoid the serious effects. Today, thalidomide is prescribed to treat multiple myeloma (a type of cancer) and Hansen's disease (leprosy).

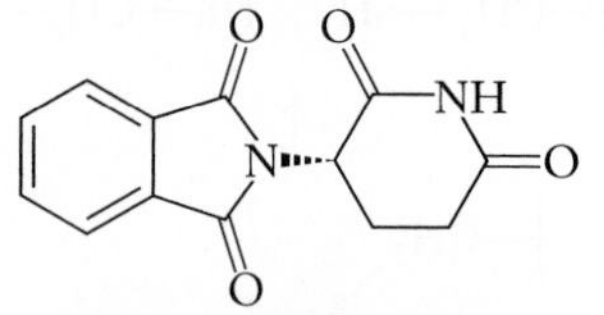

(+)-Thalidomide

(−)-Thalidomide

FIGURE 20.30 Enantiomers of thalidomide.

The phosphorylation of glucose in Section 17.9 (Figure 17.21) is the first step in glycolysis, one of the pathways by which we metabolize glucose. The reaction requires both ATP so that $\Delta G < 0$ and an enzyme to catalyze the reaction. The molecular structure of enzymes contains a region called an **active site** that binds the reactant molecule, called the **substrate**. The action of enzymes was originally explained by a lock-and-key analogy in which the substrate is the key and the active site is the lock (**Figure 20.31**). The substrate is held in the active site by the same kinds of intermolecular interactions that hold any biomolecules together. Some enzymes become covalently bonded to intermediates in the catalytic process. Once in the active site, the substrate is converted into product via a reaction having a lower-energy transition state than it would without the enzyme. The reaction of a substrate (S) with an enzyme (E) produces an *enzyme–substrate* (ES) *complex* that decomposes, forming a product P and regenerating the enzyme:

$$E + S \rightleftharpoons ES \rightarrow E + P$$

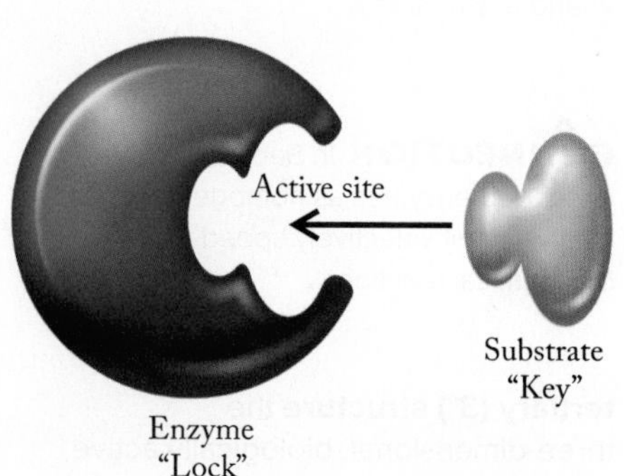

FIGURE 20.31 In the lock-and-key model of enzyme activity, the substrate (yellow) fits exactly into the active site of the enzyme (purple) that catalyzes a chemical reaction involving the substrate.

The lock-and-key analogy, however, does not fully account for enzyme behavior. A more accurate view is provided by the *induced-fit model*, which assumes that the substrate does more than just fit into the existing shape of an active site. This model assumes that as the ES complex forms, the binding site undergoes subtle changes in its shape to more precisely fit the three-dimensional structure of the transition state. The binding energy between the enzyme and the substrate creates a lower-energy transition state, thereby lowering the activation energy barrier. A simplified illustration of such an interaction is shown in **Figure 20.32a**.

The induced-fit model helps explain the behavior of compounds called **inhibitors**, which can diminish or destroy the effectiveness of enzymes. For example, as the concentration of glucose-6-phosphate exceeds a particular threshold, the compound starts to inhibit the enzyme. An inhibitor may bind to an active site and block it from interacting with the substrate (**Figure 20.32b**) and, from that position, prevent the enzyme from achieving its active shape (i.e., the shape necessary to bind the substrate). Alternatively, an inhibitor may disable the enzyme by preventing it from assuming its active shape. That is, the inhibitor may bind to the enzyme at a site other than the active site (**Figure 20.32c**).

Natural enzyme inhibitors play important roles in regulating the rates of reactions that are catalyzed by enzymes. In a multistep reaction pathway, for example, the product of a later step may inhibit an enzyme that catalyzes an earlier reaction. This kind of negative feedback keeps the sequence of reactions from running too quickly and perhaps jeopardizing the health of the organism as the result of an accumulation of undesirable products or intermediates. Enzyme inhibitors may also be used to fight disease. Powerful drugs have been developed that inhibit enzymes called proteases involved in virus maturation. Several such drugs have been particularly effective in treating HIV.

active site the location on an enzyme where a reactive substance binds.

substrate the reactant that binds to the active site in an enzyme-catalyzed reaction.

inhibitor a compound that diminishes or destroys the ability of an enzyme to catalyze a reaction.

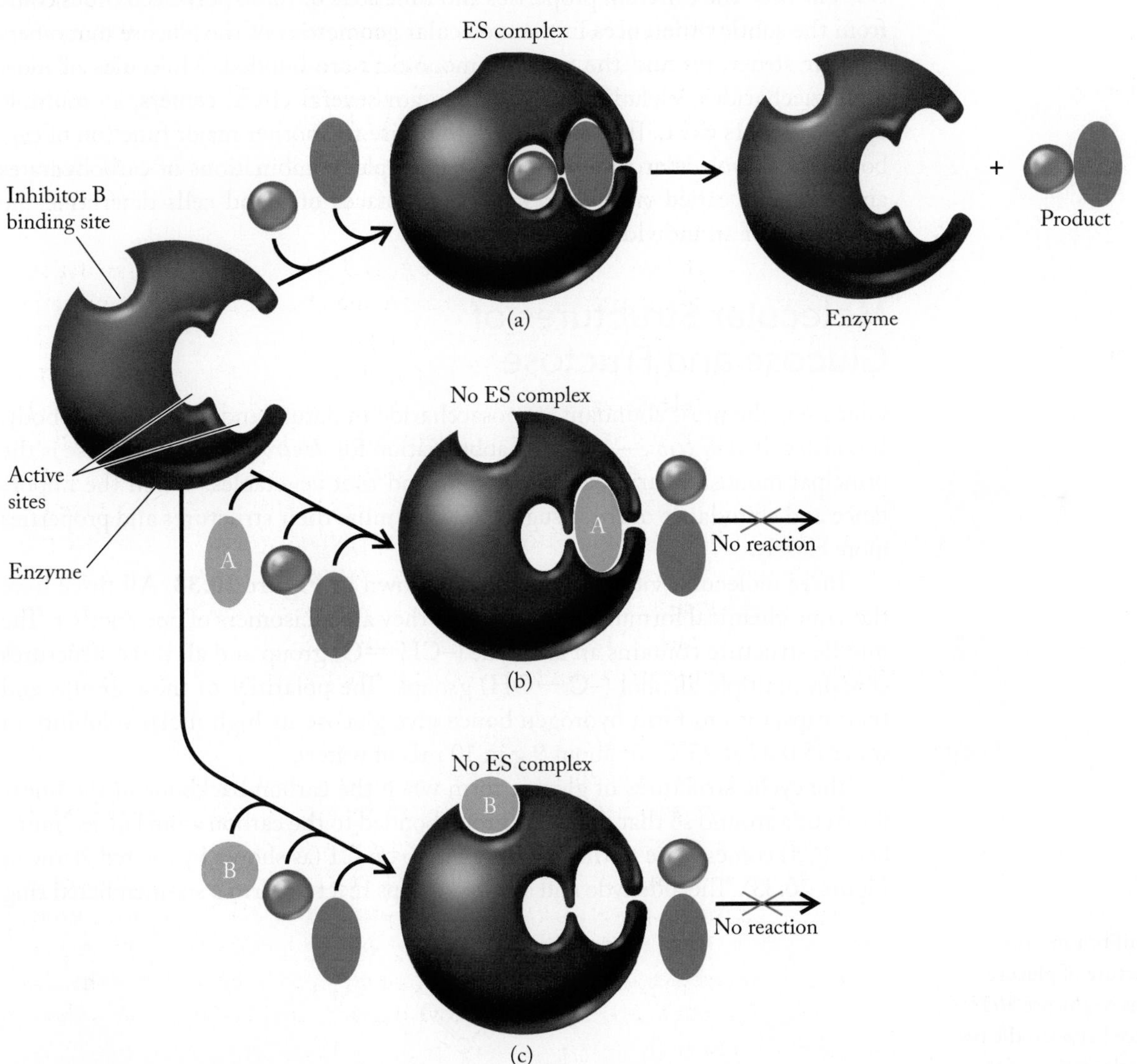

FIGURE 20.32 (a) The induced-fit model assumes that the shape of the enzyme (purple) changes to allow the enzyme and two substrates (red and green) to form an enzyme-substrate (ES) complex that promotes the formation of the product. In this example the change in shape of the active site distorts the inhibitor B binding site. (b) Inhibitor A blocks the enzyme's active sites. No ES complex can form. (c) Inhibitor B occupies a binding site, preventing the enzyme from attaining the shape needed to form the ES complex.

carbohydrate an organic molecule with the generic formula $C_x(H_2O)_y$.

monosaccharide a single-sugar unit and the simplest carbohydrate.

polysaccharide a polymer of monosaccharides.

20.5 Carbohydrates

Carbohydrates have the generic formula $C_x(H_2O)_y$. This formula gives us a clue about where the name *carbohydrate*, or *hydrate of carbon*, comes from. The smallest carbohydrates are **monosaccharides** (the name means "one sugar"), which bond to form more complex carbohydrates called **polysaccharides**. Many organisms use monosaccharides as their main energy source but convert them to polysaccharides for the purpose of energy storage. Starch is the most abundant energy-storage polysaccharide in plants. Polysaccharides in the form of cellulose also provide structural support in plants. Plants produce over 100 billion tons of cellulose each year—the woody parts of trees are more than 50% cellulose and cotton is 99% cellulose.

The principal building block of both starch and cellulose is the monosaccharide glucose. The different properties and functions of these polysaccharides come from the subtle differences in the molecular geometries of the glucose monomers in their structures and the way the monomers are bonded. Molecules of most monosaccharides, including glucose, contain several chiral centers, so multiple optical isomers exist. This complexity gives rise to another major function of carbohydrates: molecular recognition. For example, combinations of carbohydrates and proteins, called *glycoproteins*, on the surfaces of blood cells determine the blood type of an individual.

Molecular Structures of Glucose and Fructose

Glucose is the most abundant monosaccharide in nature and in the human body. It is also called *dextrose*—a kind of abbreviation for *dextro*-glucose. Fructose is the principal monosaccharide in many fruits and root vegetables. Given the importance and abundance of these sugars, let's examine their structures and properties more closely.

Three molecular views of glucose are shown in **Figure 20.33**. All three have the same chemical formula ($C_6H_{12}O_6$), so they are all isomers of one another. The middle structure contains an aldehyde (–CH=O) group and all three structures contain multiple alcohol (–C—OH) groups. The polarities of these groups and their capacities to form hydrogen bonds give glucose its high molar solubility in water (5.0 *M* at 25°C, or about 9 g in 10 mL of water).

The cyclic structures of glucose form when the carbon backbone of the linear form curls around so that the –OH group bonded to the carbon atom labeled number 5 (C5) comes close to the aldehyde group on C1 (as shown by the red arrow in Figure 20.33). The aldehyde and alcohol groups react to form a six-membered ring

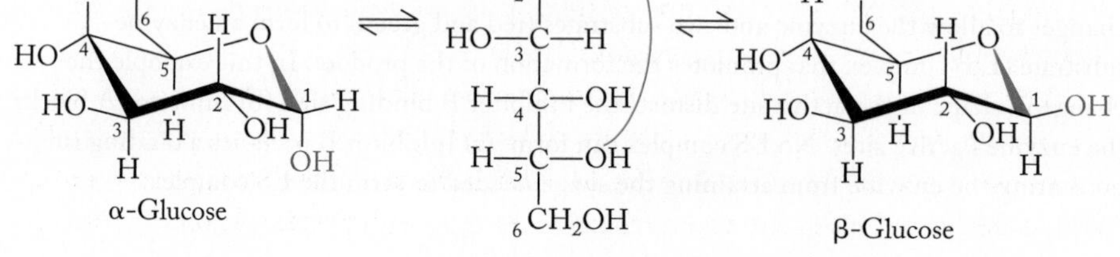

FIGURE 20.33 An equilibrium exists between the linear structure of glucose and the two cyclic forms α-glucose and β-glucose. The difference between the two cyclic structures is the orientation of the –OH group on C1 (highlighted in blue), which is formed by the O atom on C1 as indicated by the blue arrow. The new bond between oxygen and carbon atoms is shown in red in the cyclic structures and is formed by the reaction of the –OH on C5 and C1 as indicated by the red arrow.

α-Glucose

β-Glucose

made up of five carbon atoms (C1 to C5) and the oxygen atom of the C5 alcohol. Notice a small but significant difference in the two cyclic structures. In the molecule on the left, called α-glucose, the –OH group on C1 points down. In the molecule on the right, called β-glucose, the C1 –OH group points up. These two orientations are possible because there are two ways that the carbon chain in the middle structure can form a cyclic structure: the –OH group on C5 can approach C1 from either of the two sides of the plane defined by the C1 carbon atom and the O and H atoms bonded to it. When the C5 –OH group approaches C1 as shown in **Figure 20.34**, the α isomer is produced. Approaching C1 from the other side yields the β isomer. In the linear form, C1 is an aldehyde and is not a chiral center, but cyclization creates a new chiral center in the cyclic molecule.

The β form is slightly more stable than the α form and accounts for 64% of glucose molecules in aqueous solution; the α form accounts for the remaining 36%. Both cyclic forms are more stable than the straight-chain form, which exists only as an intermediate between the two cyclic forms. The energy differences are small, however, so glucose molecules in solution are constantly opening and closing in a dynamic structural equilibrium.

Figure 20.35 shows the structures of the linear and cyclic forms of fructose, commonly known as fruit sugar and contained in honey and many fruits. The C=O group in the linear form is not on the terminal carbon atom because fructose is a ketone rather than an aldehyde. It forms a five-membered ring, not a six-membered ring, when the –OH group at C5 reacts with the carbonyl carbon atom at the C2 position. The product is a ring with two $-CH_2OH$ groups that are either on the same side of the ring (α-fructose) or on opposite sides (β-fructose).

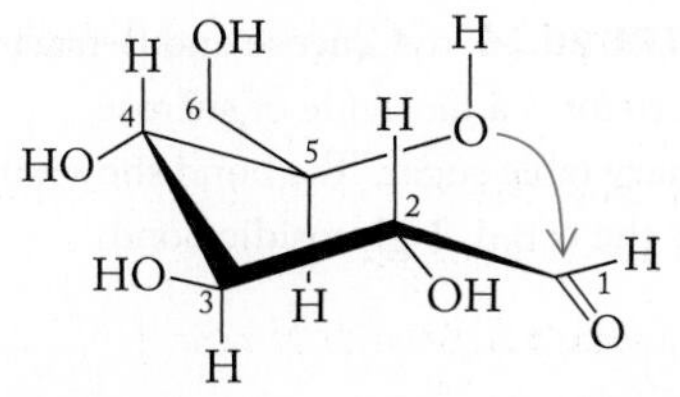

H OH H
4 6 H O
HO 5
2
HO 3 H OH 1 H
H O

FIGURE 20.34 The –OH group on C5 may approach the aldehyde at C1 from either side of the plane defined by the C, H, and O atoms in the aldehyde group. In the approach shown here the product is the α isomer.

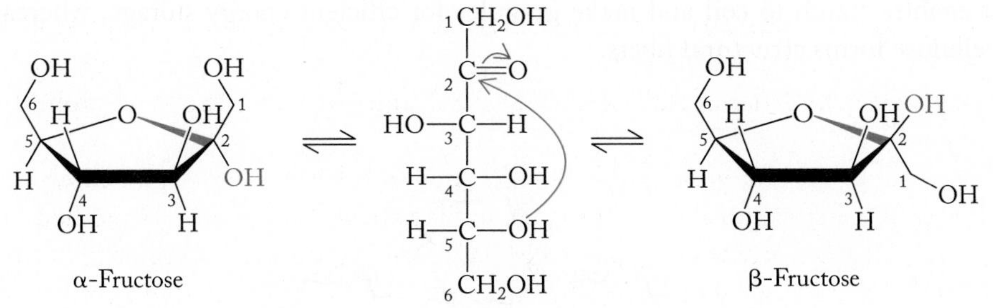

OH OH
6 H O OH 1
5 2
H 4 3 OH
OH H
α-Fructose

$_1CH_2OH$
2 C=O
HO—$_3$C—H
H—$_4$C—OH
H—$_5$C—OH
$_6CH_2OH$

OH
6 H O OH OH
5 2
H 4 3 1 OH
OH H
β-Fructose

FIGURE 20.35 The cyclization of fructose proceeds via the –OH group on C5 and the ketone on C2 as indicated by the red arrow. The new bond is shown in red in the cyclic structures. The oxygen atom in the –OH group on C2 comes from the carbonyl as indicated by the blue arrow.

Disaccharides and Polysaccharides

Sucrose, or ordinary table sugar, is a *disaccharide* ("two sugars") that consists of one molecule of α-glucose bonded to one molecule of β-fructose (**Figure 20.36**). The bond between them forms when the –OH group on the C1 carbon atom of α-glucose reacts with the C2 carbon atom of β-fructose, producing a C—O—C **glycosidic bond** or *glycosidic linkage*. Water is also produced, making this reaction another example of a condensation reaction, as is peptide bond formation. The glycosidic bond in sucrose is called an α,β-1,2 linkage because of the orientations (α and β) of the two –OH groups involved and their positions (C1 and C2) in the cyclic structures of the two monosaccharides.

CHEMTOUR
Formation of Sucrose

Another important glycosidic bond involves the –OH groups on the C1 and C4 carbon atoms of glucose molecules. When glucose molecules link at these positions they can form long-chain polysaccharides. If the starting monomer is α-glucose, the bonds between them are α-1,4-glycosidic linkages, and the

glycosidic bond a C—O—C bond between sugar molecules.

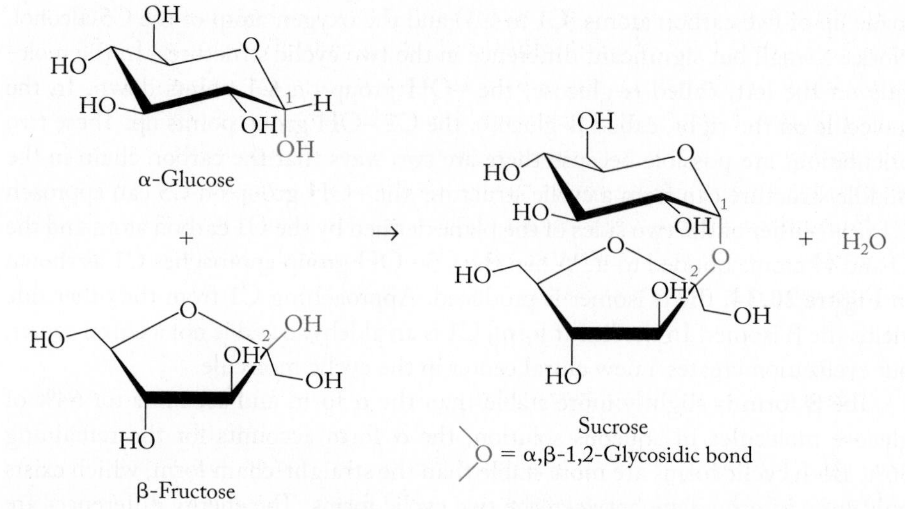

FIGURE 20.36 α-Glucose and β-fructose react to form a molecule of sucrose, ordinary table sugar. The bond shown in red is the α,β-1,2-glycosidic bond.

product of bond formation is starch (**Figure 20.37a**). The conversion of α-glucose into starch is an effective way for plants to store energy because formation of the α-1,4 linkage is reversible. With the aid of digestive enzymes, α-1,4 bonds can be hydrolyzed and starch converted back into glucose by plants that make the starch or by animals that eat the plants.

The cellulose that plants synthesize to build stems and other organs has a structure (**Figure 20.37b**) slightly different from that of starch because the building blocks of cellulose are β-glucose instead of α-glucose, so the monomers are linked by β-1,4-glycosidic bonds. This structural difference is important because it enables starch to coil and make granules for efficient energy storage, whereas cellulose forms structural fibers.

(a) Starch

Granules

(b) Cellulose

Fibers

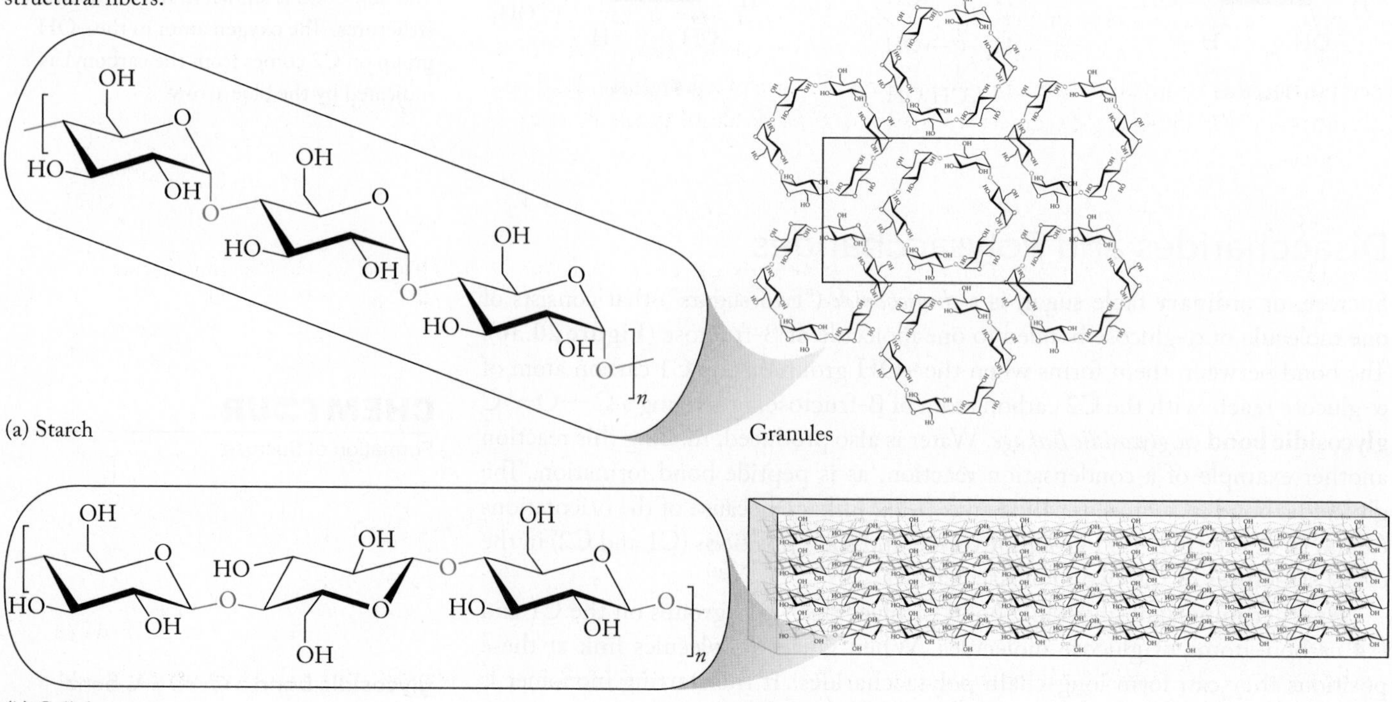

FIGURE 20.37 (a) Starch, found in potatoes, is a polysaccharide of α-glucose molecules joined by α-1,4-glycosidic bonds, shown in red. Starch molecules form spirals, much like coiled springs, that pack together to form granules. (b) Cellulose, found in trees, is a polysaccharide of β-glucose molecules joined by β-1,4-glycosidic bonds (in red). Cellulose chains pack together much more tightly to form structural fibers.

Carbohydrates in plants are a major part of the total organic matter in any given ecological system; that is, they make up much of the system's **biomass**. People have used various forms of biomass, including wood and animal dung, as fuel for thousands of years. More recently, we have begun to convert biomass into a liquid *biofuel*, ethanol, for use as a gasoline additive and substitute. Ethanol contains oxygen, which improves the combustion of a hydrocarbon fuel such as gasoline and reduces carbon monoxide emissions. An important industrial application of starch hydrolysis is the conversion of cornstarch into glucose and then, through fermentation, into ethanol:

biomass the total mass of organic matter in any given ecological system.

$$C_6H_{12}O_6(aq) \rightarrow 2\ CH_3CH_2OH(aq) + 2\ CO_2(g)$$

This exothermic reaction provides energy to the yeast cells whose biological processes drive fermentation.

Unlike grazing animals, humans cannot digest cellulose because we do not have microorganisms in our digestive tracts that have enzymes called cellulases, which catalyze hydrolysis of β-glycosidic bonds. The challenge of reproducing what the cellulose-eating bacteria do in a laboratory or on an industrial scale is the focus of an enormous research effort as scientists try to develop efficient procedures for converting cellulose to glucose and then to ethanol. This research has focused on more efficient, less energy-intensive ways to break apart cellulose fibers. In addition, scientists are genetically engineering microorganisms like those in cattle stomachs to increase the supply of cellulases.

If this research is successful, it will address several major problems associated with ethanol as a gasoline additive or alternative fuel, including the cost of production. Today it costs more to produce ethanol from cornstarch than to produce gasoline from crude oil. One reason for this is that most of the mass of a corn plant, or any plant, is cellulose, not starch. Ethanol production from cornstarch is also energy intensive: more than 70% of the energy contained in ethanol is expended in producing it; therefore, the net energy value of ethanol from corn is less than 30%. If ethanol could be produced from cellulose instead of starch, its net energy value could be as high as 80%. Finally, the use of edible cornstarch in fuel production has driven up the price of foods derived from corn, including livestock feed. The impacts of this inflation have been felt worldwide and have been particularly painful in developing countries. The use of agricultural land and consumption of increasingly scarce water resources for ethanol production raise further concerns.

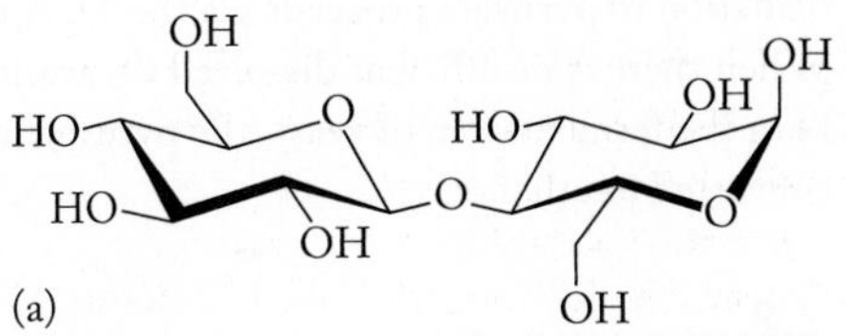

(a)

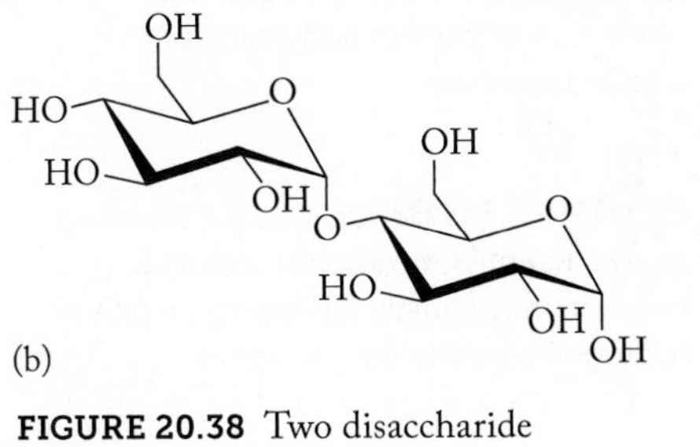

(b)

FIGURE 20.38 Two disaccharide molecules.

CONCEPT TEST

Cellobiose is a disaccharide made from the degradation of cellulose. We cannot digest cellobiose. Which of the two structures in **Figure 20.38** represents a molecule of cellobiose?

Energy from Glucose

Glycolysis is the series of reactions by which most animal cells, including those in our bodies, metabolize glucose. The product of these reactions is pyruvate ions (**Figure 20.39**), the conjugate base of pyruvic acid. Pyruvate sits at a metabolic crossroads and can be converted into different products, depending on the type of cell in which it is generated, the enzymes present, and the availability of oxygen (**Figure 20.40**). In yeast

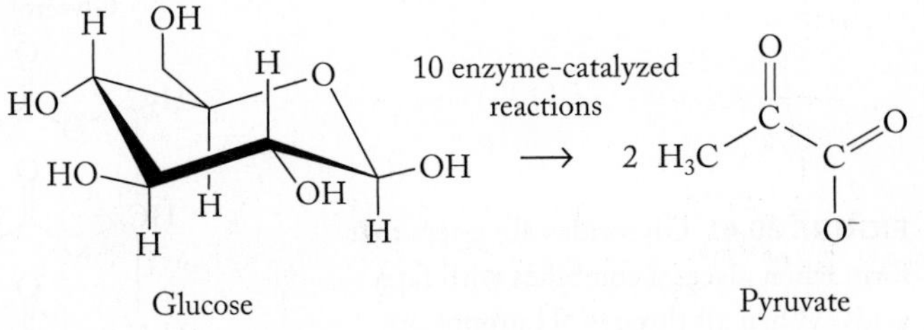

10 enzyme-catalyzed reactions

Glucose

Pyruvate

FIGURE 20.39 In glycolysis, molecules of glucose are broken down to pyruvate ions.

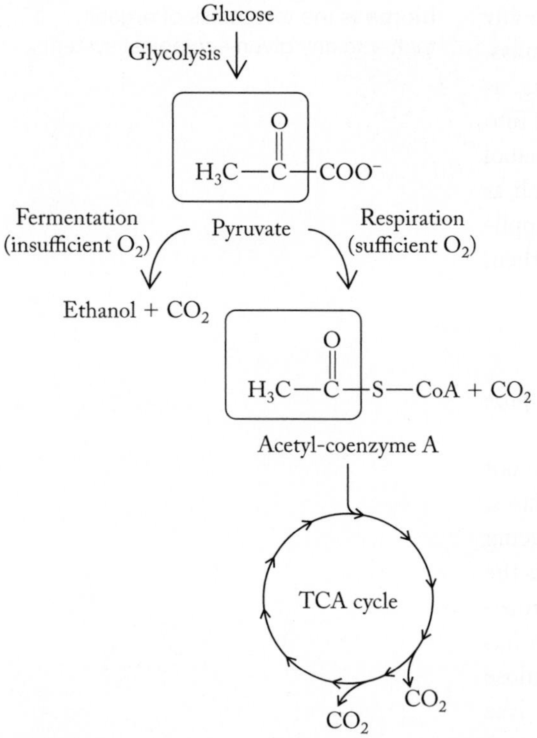

cells growing under low-oxygen conditions, pyruvate is converted into ethanol and CO_2.

Another series of reactions, called the **tricarboxylic acid (TCA) cycle** or the *citrate cycle*, occurs in the presence of sufficient dissolved O_2 and is fundamental to the conversion of glucose to energy in humans and other animals. A key step prior to the TCA cycle occurs when pyruvate loses CO_2 and forms an acetyl group, which then combines with coenzyme A. The resulting product, acetyl-coenzyme A, is a reactant in many biosynthetic pathways, including the production of fats.

Glucose
Glycolysis
$H_3C-C(=O)-COO^-$
Pyruvate
Fermentation (insufficient O_2)
Respiration (sufficient O_2)
Ethanol + CO_2
$H_3C-C(=O)-S-CoA + CO_2$
Acetyl-coenzyme A
TCA cycle
CO_2
CO_2

FIGURE 20.40 The fate of the pyruvate ions formed during glycolysis depends on the partial pressure of O_2 in the system. In the presence of sufficient O_2, the oxidation of pyruvate proceeds via the TCA cycle. When there is insufficient dissolved O_2 available, as in the fermentation of yeast, the pyruvate may be converted to ethanol.

CONNECTION We introduced glycolysis in Section 17.9 as part of a discussion of the thermodynamics of coupled reactions.

CONNECTION Alkanes are classified as saturated hydrocarbons (Section 6.9) because they contain the maximum ratio of hydrogen atoms to carbon atoms.

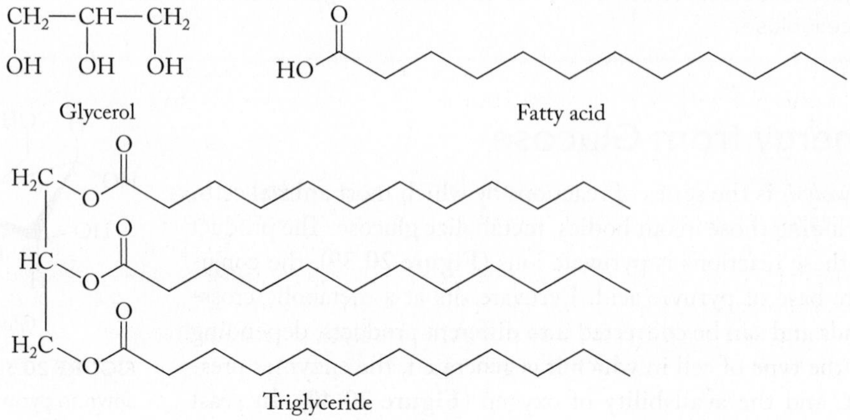

20.6 Lipids

Lipids, unlike carbohydrates and proteins, are *not* biopolymers. Lipids are best described by their physical properties, such as their hydrophobicity, rather than by any common structural subunit. Because they are insoluble in water, they are ideal components of cell membranes, which separate the aqueous solutions within cells from the aqueous environments outside them. An important class of lipids called **glycerides** are esters formed between glycerol and long-chain **fatty acids** (**Figure 20.41**). The three –OH groups on glycerol allow for mono-, di-, and triglycerides, with the last group being the most abundant. Glycerides account for over 98% of the lipids in the fatty tissues of mammals. When we look at lipids, glycerides, and fatty acids, we will see examples of both kinds of stereoisomers we discussed in Section 20.2 (enantiomers and cis–trans isomers), frequently all in one molecule.

Table 20.4 lists some common fatty acids. The most abundant ones have an even number of carbon atoms because the fatty acids are built *in vivo* from two-carbon subunits. Their biosynthesis begins with the conversion of pyruvate to acetyl-coenzyme A. Most fatty acids contain between 14 and 22 carbon atoms. Some have no carbon–carbon double or triple bonds and are called *saturated* because they have as much hydrogen in their structures as the carbon atoms can hold. Other fatty acids have carbon–carbon double or triple bonds, or both, and are called *unsaturated* because they do not have all the hydrogen their carbon atoms could possibly hold; they have the capacity to add hydrogen to those multiple bonds.

A type of fatty acid much in the news because of its alleged health benefits is the family of omega-3 fatty acids. These fatty acids are *polyunsaturated*, meaning

$CH_2-CH-CH_2$ (OH OH OH)
Glycerol
HO–C(=O)–
Fatty acid
H_2C–O–C(=O)–
HC–O–C(=O)–
H_2C–O–C(=O)–
Triglyceride

FIGURE 20.41 Glycerides are esters that form when glycerol combines with fatty acids. When all three –OH groups on glycerol react to form ester bonds, the product is a *tri*glyceride.

TABLE 20.4 Names and Structural Formulas of Common Fatty Acids

Common Name (chemical name) (source)	Formula
SATURATED FATTY ACIDS	
Lauric acid (dodecanoic acid) (coconut oil)	$CH_3(CH_2)_{10}COOH$
Myristic acid (tetradecanoic acid) (nutmeg butter)	$CH_3(CH_2)_{12}COOH$
Palmitic acid (hexadecanoic acid) (animal and vegetable fats)	$CH_3(CH_2)_{14}COOH$
Stearic acid (octadecanoic acid) (animal and vegetable fats)	$CH_3(CH_2)_{16}COOH$
UNSATURATED FATTY ACIDS	
Oleic acid (*cis*-9-octadecenoic acid) (animal and vegetable fats)	$CH_3(CH_2)_7CH{=}CH(CH_2)_7COOH$
Linoleic acid (*cis,cis*-9,12-octadecadienoic acid) (linseed oil, cottonseed oil)	$CH_3(CH_2)_4CH{=}CHCH_2CH{=}CH(CH_2)_7COOH$
α-Linolenic acid (*cis,cis,cis*-9,12,15-octadecatrienoic acid) (linseed oil)	$CH_3CH_2CH{=}CHCH_2CH{=}CHCH_2CH{=}CH(CH_2)_7COOH$

that they have more than one carbon–carbon multiple bond. The name "omega-3" comes from the location of the first double bond counted from the methyl end of the chain, which is known as the omega (ω) end of the molecule. Omega-3 fatty acids are found in fish oils and some plant oils. α-Linolenic acid (see Table 20.4) is an omega-3 fatty acid.

SAMPLE EXERCISE 20.7 Identifying Triglycerides **LO6**

How many different triglycerides (including constitutional isomers and stereoisomers) can be made from glycerol combining with two different fatty acids (symbolized by the letters X and Y to simplify the structures) if each molecule of triglyceride contains at least one molecule of each fatty acid?

Collect and Organize We are asked to determine how many different triglycerides can be made from glycerol and two different fatty acids. A triglyceride contains three fatty acid units.

Analyze Each fatty acid may bond to one of three –OH groups in glycerol. Each triglyceride has at least one X and one Y residue, and two formulas are possible: X_2Y and Y_2X. Each of these formulas has two constitutional isomers, depending on whether the single fatty acid in the formula is bonded to the middle carbon or to one of the end carbon atoms. Finally, if a structure has an X on one end carbon atom and a Y on the other, then the middle carbon atom is a chiral center, which means that there are two enantiomeric forms of that compound.

tricarboxylic acid (TCA) cycle a series of reactions that continue the oxidation of pyruvate formed in glycolysis.

lipid a class of water-insoluble, oily organic compounds used as common structural materials in cells.

glyceride a lipid consisting of esters formed between fatty acids and the alcohol glycerol.

fatty acid a carboxylic acid with a long hydrocarbon chain that may be either saturated or unsaturated.

Solve We can generate four different molecular structures by attaching X and Y in four different sequences to the glycerol –OH groups:

(1)	(2)	(3)	(4)
$H_2C—O—X$	$H_2C—O—X$	$H_2C—O—Y$	$H_2C—O—Y$
$HC—O—X$	$HC—O—Y$	$HC—O—Y$	$HC—O—X$
$H_2C—O—Y$	$H_2C—O—X$	$H_2C—O—X$	$H_2C—O—Y$

Structures (1) and (3) have chiral centers, so each of these isomers has two enantiomeric forms. As a result, there are a total of six different triglycerides possible.

Think About It Reacting three –OH groups in glycerol with two different fatty acids should allow for more than two products because there are two different sites in glycerol for reaction with the fatty acids even if we restrict each product to contain at least one of each fatty acid.

Practice Exercise How many different triglycerides can be made from glycerol and one molecule each of three different fatty acids? To simplify the drawings, use the letters A, B, and C to symbolize the different fatty acids.

Function and Metabolism of Lipids

Lipids are an important energy source in our diets, providing more energy per gram than carbohydrates or proteins. **Fats** are glycerides composed primarily of saturated fatty acids. They are solids at room temperature because the molecules can pack very tightly together. **Oils** are glycerides composed predominantly of unsaturated fatty acids and are liquids at room temperature because their chains have kinks in them caused by the double bonds, which prevent the chains from packing together as closely as the saturated chains can. Oils can be converted into solid, saturated glycerides by hydrogenation. In the hydrogenation of corn oil, for example, which is a mixture of mostly two unsaturated fatty acids (oleic and linoleic acids), hydrogen is added to convert some or all of the –CH═CH– subunits into $–CH_2—CH_2–$ subunits. Hydrogenation converts the oil into a solid at room temperature that is whipped with skim milk, coloring agents, and vitamins to produce the food spread we know as margarine.

The consumption of certain fats is associated with increased risk of coronary heart disease. The Mediterranean diet, which reduces the amount of saturated fats, promotes the inclusion of greater amounts of vegetables and wild-caught fish and the use of olive oil in cooking. Olive oil is a liquid composed of glycerides containing more than 80% oleic acid, an unsaturated fatty acid. In contrast, animal fats such as butter and lard contain more saturated fats. Fish in particular are rich in omega-3 fatty acids such as linolenic acid. The origins of the cardiovascular benefits of omega-3 fatty acids aren't fully understood, but one hypothesis suggests that they may be more readily metabolized than other glycerides. Sunflower oil also contains unsaturated fats but fewer omega-3 fatty acids and more omega-6 (ω-6) fatty acids such as linoleic acid. The latter do not appear to carry the same health benefits and may actually contribute to poor heart health. The difference between these two fatty acids is the absence of the double bond located between the third and fourth carbon in linoleic acid and other ω-6 fatty acids.

fat a solid triglyceride containing primarily saturated fatty acids.

oil a liquid triglyceride containing primarily unsaturated fatty acids.

Another problem with hydrogenating vegetable oil arises when the oils are only partially hydrogenated. Partial hydrogenation alters the molecular structure around their remaining C═C double bonds, changing them from their natural cis isomers into trans isomers (**Figure 20.42**). Unsaturated trans fats such as elaidic acid tend to be solids at room temperature because their molecules pack together more uniformly than do molecules of cis unsaturated fatty acids. Consumption of trans fatty acids is associated with increased levels of cholesterol in the blood and other health risks.

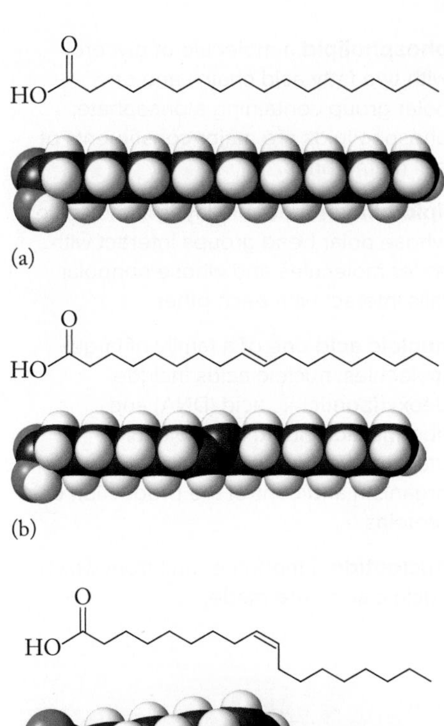

FIGURE 20.42 Types of fatty acids. (a) Stearic acid, a saturated C_{18} fatty acid. (b) Elaidic acid, an unsaturated C_{18} fatty acid (trans isomer). (c) Oleic acid, an unsaturated C_{18} fatty acid (cis isomer).

CONCEPT TEST

The mixture of lipids in the legs of reindeer living close to the Arctic Circle changes as a function of distance from the hoof: the closer to the hoof, the higher the percentage of unsaturated fatty acids. Why would this be an advantage for these reindeer?

The lipids in some prepared foods have been modified to reduce their caloric content but still provide the taste, aroma, and "mouth feel" we associate with lipid-rich foods. The active sites of enzymes that break down natural lipids accommodate triglycerides formed from glycerol and fatty acids. However, chemically modified esters made from the same fatty acids, but attached to an alcohol other than glycerol, cannot be metabolized by these enzymes. Such molecules, if they have the appropriate physical properties and are nontoxic, can be incorporated into foods without adding any calories because they are not metabolized.

Olestra is one such product (**Figure 20.43**). It is an ester made from long-chain fatty acids and the carbohydrate sucrose. (Remember, sugars have –OH groups and are technically alcohols.) Each of the eight –OH groups in a sucrose molecule reacts with a molecule of fatty acid to make the ester in olestra. The resultant material is used to deep-fry potato chips. Any olestra that remains on the chip does not add calories because it cannot be processed by enzymes that recognize only fatty acid esters on a glycerol scaffold.

CONCEPT TEST

Olestra may be "calorie-free" as a food subject to metabolism in the living system, but how would it compare with a common triglyceride in terms of kilojoules of heat released per mole in a calorimeter experiment?

FIGURE 20.43 Olestra has a very different shape from that of the triglycerides typically metabolized by our bodies. Consequently, it cannot be processed by the enzymes that digest triglycerides, leading to unpleasant side effects such as abdominal cramping and diarrhea.

phospholipid a molecule of glycerol with two fatty acid chains and one polar group containing a phosphate; phospholipids are major constituents of cell membranes.

lipid bilayer a double layer of molecules whose polar head groups interact with water molecules and whose nonpolar tails interact with each other.

nucleic acid one of a family of large molecules; nucleic acids include deoxyribonucleic acid (DNA) and ribonucleic acid (RNA), molecules that store the genetic blueprint of an organism and control the production of proteins.

nucleotide a monomer unit from which nucleic acids are made.

The enzymes that metabolize triglycerides hydrolyze the esters and release glycerol and the fatty acids that were bonded to them. Glycerol enters the metabolic pathway for glucose. The fatty acids are oxidized in a series of reactions known as β-oxidation—a process that removes two carbon atoms at a time. For example, stearic acid (the saturated C_{18} fatty acid) is transformed into a C_2 fragment and the C_{16} acid, palmitic acid. Palmitic acid yields another C_2 fragment and myristic acid, and so forth, until the fatty acid is completely degraded. Electrons released from this oxidative process eventually are donated to O_2. The energy released by this process powers metabolism.

Other Types of Lipids

Cells contain other types of lipids in addition to triglycerides. One type, **phospholipids** (**Figure 20.44a**), plays a key role in cell structure. A phospholipid molecule consists of a glycerol molecule bonded to two fatty acid chains and to one phosphate group that is also bonded to polar substituents. The presence of nonpolar fatty acid chains and a polar region in the same molecule

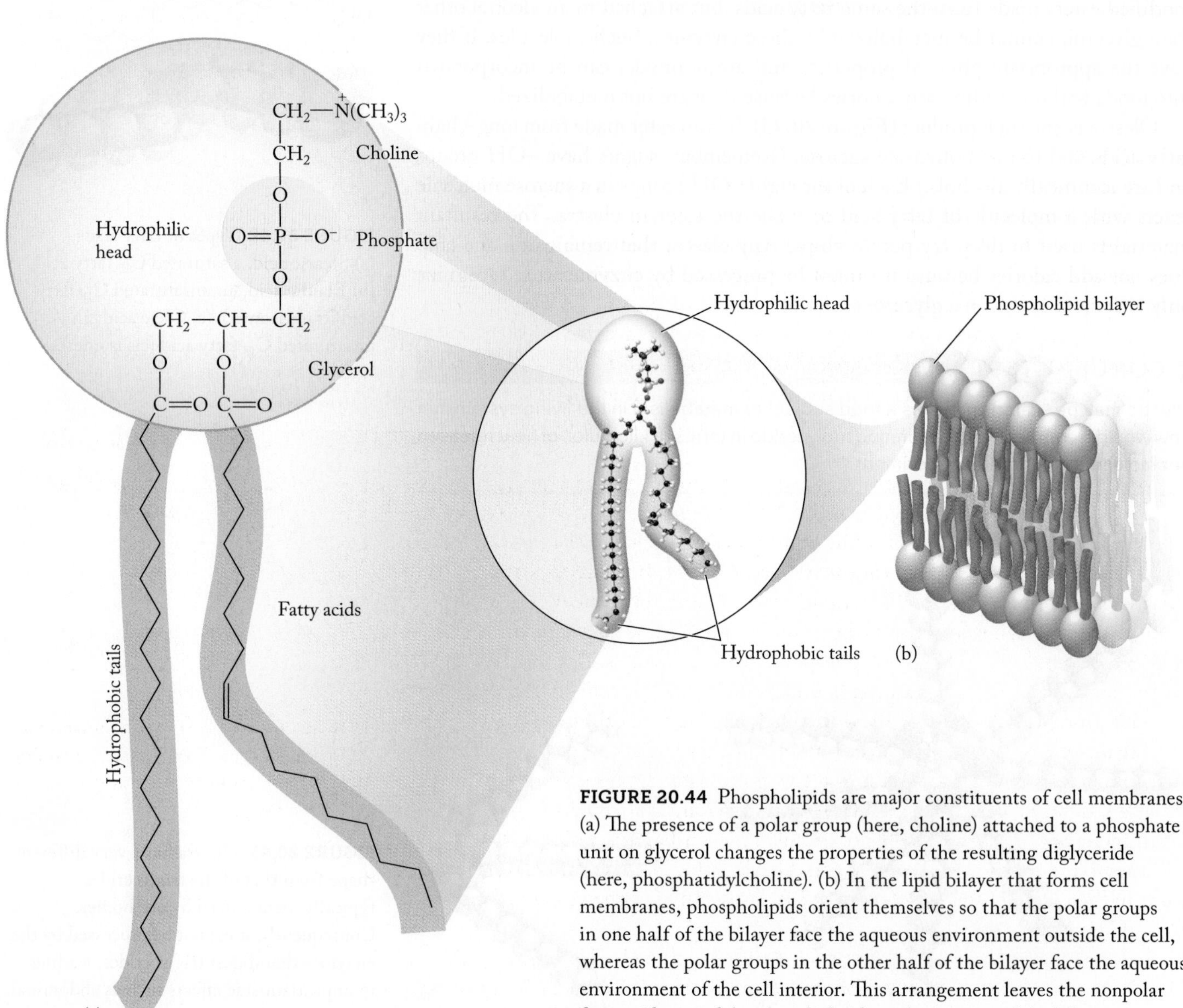

FIGURE 20.44 Phospholipids are major constituents of cell membranes. (a) The presence of a polar group (here, choline) attached to a phosphate unit on glycerol changes the properties of the resulting diglyceride (here, phosphatidylcholine). (b) In the lipid bilayer that forms cell membranes, phospholipids orient themselves so that the polar groups in one half of the bilayer face the aqueous environment outside the cell, whereas the polar groups in the other half of the bilayer face the aqueous environment of the cell interior. This arrangement leaves the nonpolar fatty acid part of the phospholipids in the interior of the bilayer.

makes phospholipids ideal for forming cell membranes. In an aqueous medium, phospholipids form a **lipid bilayer**, a double layer enclosing each cell and isolating its interior from the outside environment. The phospholipid molecules of the bilayer align so that the nonpolar groups interact with each other inside the membrane, whereas the polar groups interact with water molecules outside it (**Figure 20.44b**). Membranes exist both to isolate the contents of cells and to serve as the locus of communication between processes that occur within and outside the cells.

Cholesterol is a lipid and a key component in the structure of cell membranes. It is also a precursor of the bile acids that aid in digestion and of steroid hormones, which regulate the development of the sex organs and secondary sexual traits, stimulate the biosynthesis of proteins, and regulate the balance of electrolytes in the kidneys. Biosynthesis of cholesterol, like the biosynthesis of fatty acids, begins with the conversion of pyruvate to acetyl-coenzyme A (Figure 20.40). In a healthy person, synthesis and use of cholesterol are tightly regulated to prevent overaccumulation and consequent deposition of cholesterol in coronary arteries (**Figure 20.45**). We clearly need cholesterol, but deposition in the arteries can lead to serious coronary disease.

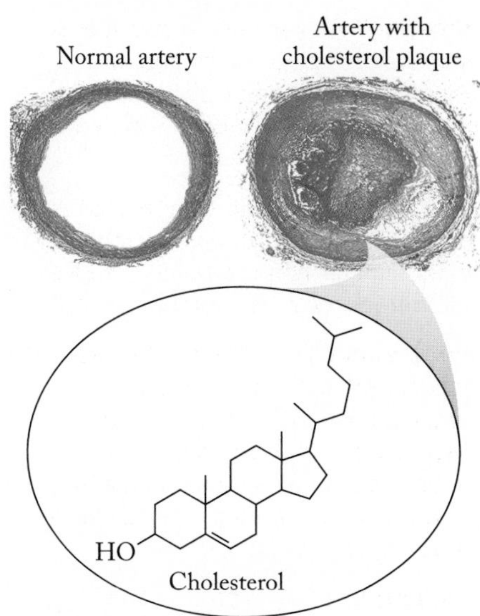

FIGURE 20.45 Cholesterol deposits called plaques are responsible for restricted blood flow, which results in a variety of sometimes catastrophic cardiovascular problems.

20.7 Nucleotides and Nucleic Acids

Nucleic acids are our fourth class of biomolecules and third class of biopolymers. We focus on two types: deoxyribonucleic acid (DNA) and ribonucleic acid (RNA). Even though nucleic acids make up only about 1% of a higher organism's mass, they control the metabolic activity of all its cells. The fraction is much higher in yeast and bacteria because those organisms are packed with ribosomes, which are themselves half RNA. DNA carries the genetic blueprint of an organism, and a variety of RNAs use that DNA blueprint to guide the production of proteins.

A nucleic acid is a polymer composed of monomeric units called **nucleotides**. Each nucleotide unit consists of three subunits: a five-carbon sugar, a phosphate group, and a nitrogen-containing base (**Figure 20.46**). The phosphate group in each nucleotide is attached to a carbon in the side chain of the sugar called the 5′ carbon atom (the prime number refers to the position of the carbon atom in the sugar molecule). The nitrogen-containing base is attached to the 1′ carbon atom in each sugar molecule. The sugar in Figure 20.46 is called *ribose*, which makes this a nucleotide in a strand of *ribo*nucleic acid (RNA). If the sugar were *deoxyribose* instead, there would be a H atom in place of the –OH group on the 2′ carbon atom, and the nucleotide would be a building block of *deoxyribo*nucleic acid (DNA). Because of the ionized phosphate groups, both DNA and RNA are polyanions. Their anionic character is important because it enables them to interact with proteins but does not allow them to pass through cell membranes.

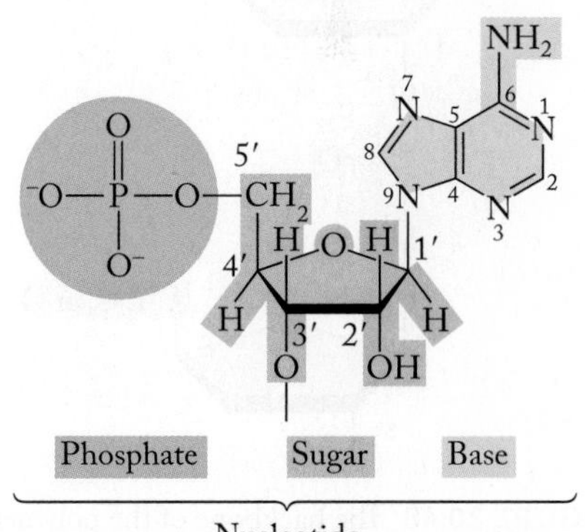

FIGURE 20.46 A nucleotide consists of a phosphate group and a nitrogen-containing base, both of which are bonded to a five-carbon sugar.

The nitrogen-containing base in Figure 20.46 is called adenine (A). Structural formulas of adenine and the other four bases in nucleic acids—cytosine (C), guanine (G), thymine (T), and uracil (U)—are shown in **Figure 20.47**. The point of attachment of the sugar–phosphate groups on each base is indicated by –R. In addition to the difference in their sugars, RNA and DNA also differ in one of the bases in their nucleotides: RNA contains A, C, G, and U, whereas DNA contains A, C, G, and T.

In a polymeric strand of nucleic acid, each phosphate is also linked to the 3′ carbon atom in the sugar of the monomer that precedes it in the chain, as shown for a strand of DNA in **Figure 20.48**. Both DNA and RNA strands are

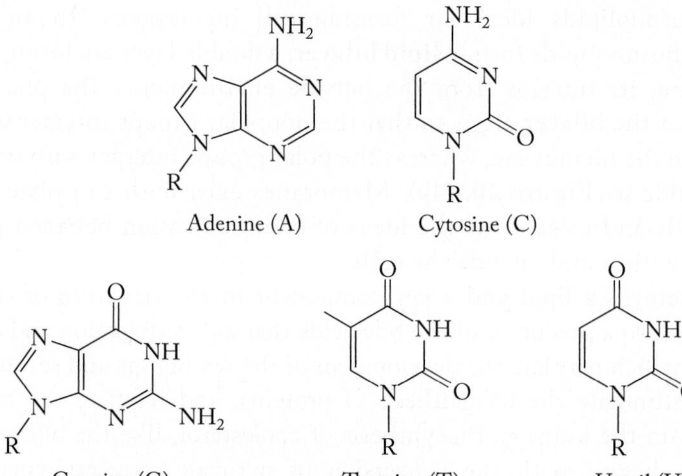

Adenine (A) Cytosine (C)

Guanine (G) Thymine (T) Uracil (U)

FIGURE 20.47 Structural formulas of the five nitrogen-containing bases in nucleotides. The nucleotides in DNA contain A, C, G, and T; those in RNA contain A, C, G, and U. The R group identifies the point of attachment of the sugar residue.

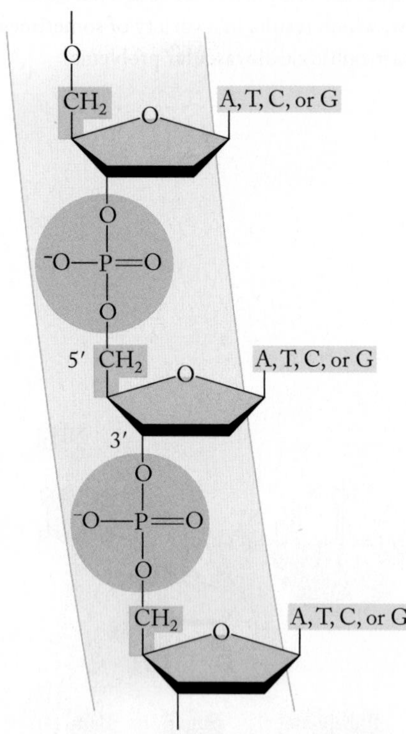

FIGURE 20.48 The backbone of the polymer chain in one of the two strands of DNA consists of alternating sugar units (yellow) and phosphate units (pink). The bases (blue) are attached to the backbone through the 1′ carbon atom of the sugar unit.

replication the process by which one double-stranded DNA forms two new DNA molecules, each one containing one strand from the original molecule and one new strand.

synthesized in the cell from the 5′ to the 3′ direction (downward in Figure 20.48). The structures of DNA and RNA are frequently written using only the single-letter labels of their bases, beginning with the free phosphate group on the 5′ end of the chain and reading toward the free 3′ hydroxyl group at the other terminus.

When scientists first isolated DNA and began to analyze its composition, they made a pivotal observation about the abundance of the nitrogen-containing bases. A typical molecule of DNA consists of thousands of nucleotides, and the percentages of the four bases in different samples of DNA can vary over a wide range. However, the percentage of A in a sample always matches the percentage of T. Likewise, the percentage of C always matches that of G. This result makes sense if the bases are paired because a molecule of A can form two hydrogen bonds to a molecule of T, whereas a molecule of C can form three hydrogen bonds with a molecule of G. Therefore, A–T and G–C pairings maximize the number of hydrogen bonds possible (**Figure 20.49a**). More important, the base pairs assembled this way all have the same width and fit together in a regular structure.

The normal structure of DNA has two strands of nucleotides wrapped around each other in a form that is called a *double helix* (**Figure 20.49b**). The nucleotide backbone is on the outside of the spiraling strands, with hydrogen bonds between the complementary bases keeping the two strands together. Notice also that the base pairs are parallel to each other and perpendicular to the helical axis. The fidelity of this base-pairing—A always with T, and C always with G—gives DNA the ability to copy itself. If a pair of complementary strands is unzipped into two single strands, each strand provides a template on which a new complementary strand can be synthesized via the process called **replication** (**Figure 20.50**).

During replication, the two strands are separated by enzymes called helicases that break the hydrogen bonds between the two strands. This process forms a structure called the *replication fork*. The fork advances through the DNA as replication proceeds. One of the two resulting strands, the leading strand, is unzipped in the 3′–5′ direction, which allows the new complementary strand to be continuously synthesized in the 5′–3′ direction. The replication of the second strand, called the lagging strand, is more complicated. It proceeds in short fragments, which are then assembled into a continuous strand later in the process.

(a)

Adenine Thymine Guanine Cytosine

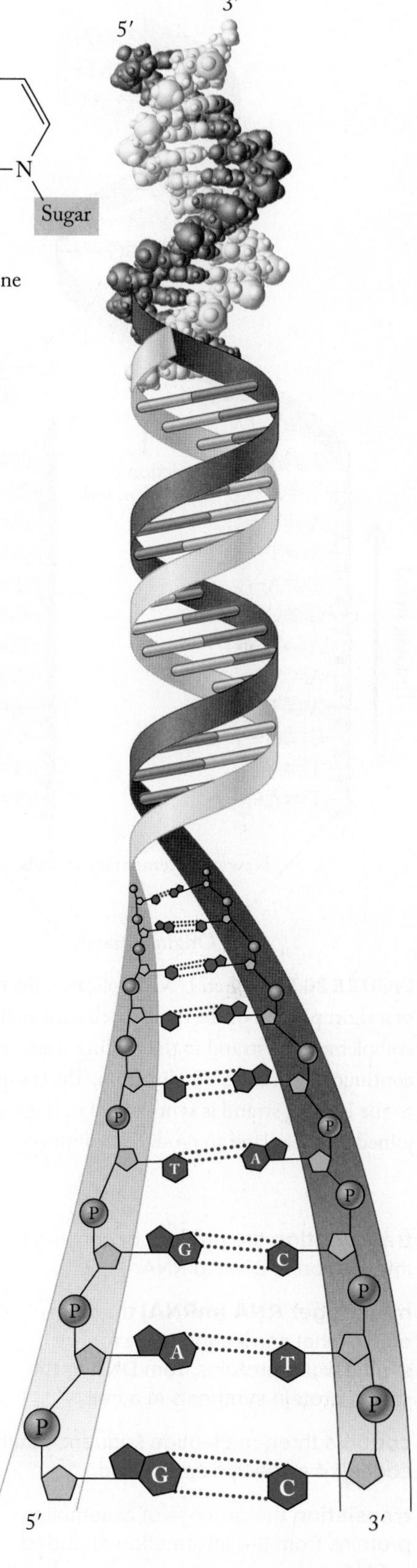

FIGURE 20.49 The nitrogen-containing bases on one strand of DNA pair with the bases on a second strand by hydrogen bonding. (a) Adenine and thymine pair via two hydrogen bonds, whereas guanine and cytosine pair via three hydrogen bonds. (b) DNA as a double helix with the sugar–phosphate backbone on the outside and the base pairs on the inside.

SAMPLE EXERCISE 20.8 Using Base Complementarity in DNA **LO7**

If 31.6% of the nucleotides in a sample of DNA are adenine, what are the percentages of cytosine, guanine, and thymine?

Collect, Organize, and Analyze We know how much adenine is in a DNA sample and need to calculate the rest of the nucleotide composition. We also know that nucleotides are paired so that A always pairs with T, and C always pairs with G. So, the percentage of T must equal the percentage of A in the sample, and the percentage of C must equal the percentage of G.

Solve If A = 31.6%, then T = 31.6%. This leaves (100 − 2 × 31.6) = 36.8% left to be equally distributed between G and C. Therefore, G = C = 36.8/2 = 18.4%.

Think About It The percentages should total 100%, and they do.

Practice Exercise Indicate the sequence of the complementary strand on the double helix formed by each of the following sequences of nucleotides:

a. CGGTATCCGAT
b. TTAAGCCGCTAG

DNA's double-stranded structure is also the key to its ability to preserve genetic information. The two strands carry the same information, much like an old-fashioned photograph and its negative. Genetic information is duplicated every time a DNA molecule is replicated, a process that is essential whenever a cell divides into two new cells.

From DNA to New Proteins

Proteins are formed from amino acids in accordance with the *genetic code* contained in the base sequences of DNA strands. The bases A, T, G, and C are the alphabet in this code, and the "words" in the code are three-letter combinations of these four letters, with each word representing a particular amino acid or a signal to begin or end protein synthesis. Using four letters to write three-letter words means there are 4^3 = 64 combinations possible, more than enough to

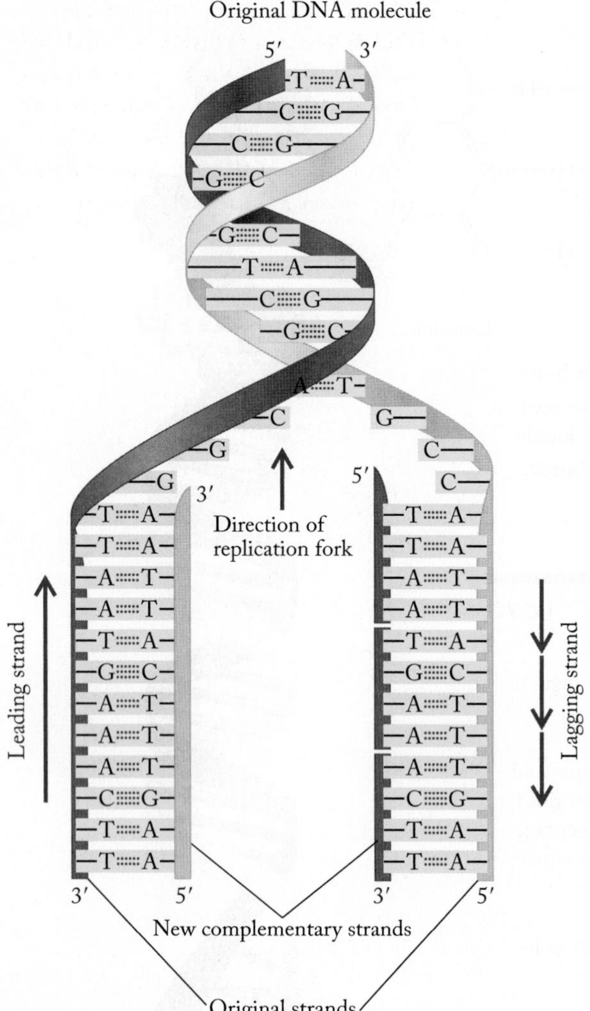

FIGURE 20.50 When DNA replicates, the two strands of a short portion of the double helix are unzipped. The complementary strand to the leading strand is synthesized continuously in the 5′–3′ direction. The complementary strand to the lagging strand is synthesized in fragments, which are joined together later to produce a continuous strand.

encode the 20 amino acids found in cells. The function of the genetic code is to specify the protein's primary structure—the sequence of amino acids in proteins. The flow of genetic information goes from DNA to RNA to proteins, a sequence sometimes called the *central dogma of molecular biology*.

Protein synthesis begins with a process called **transcription** (**Figure 20.51a**), in which double-stranded DNA unwinds and its genetic information guides the synthesis of a single strand of a molecule called **messenger RNA (mRNA)**. This strand of mRNA has the complementary base sequence of the original DNA. It carries the 3-letter words of that DNA, in the form of three-base sequences called **codons** (**Table 20.5**), from the nucleus of the cell into the cytoplasm, where the mRNA binds with a cellular structure called a ribosome. A sequence of A, C, G, and T in the original DNA is transcribed into the following sequence in mRNA:

DNA: . . . ACGT . . .
mRNA: . . . UGCA . . .

At the ribosome, the genetic information in the mRNA directs the synthesis of particular proteins in a process called **translation**. Another type of RNA, called **transfer RNA (tRNA)**, plays a key role in translation. There are 20 different forms of tRNA in the cell, one for each amino acid. To see how tRNA works, look at **Figure 20.51b**. The first codon in this piece of an mRNA strand is AUG, which codes for the amino acid methionine (see Table 20.5). In the cytoplasm surrounding the ribosome, molecules of tRNA are reversibly bonded to molecules of every amino acid. The particular tRNA molecules that are bonded to methionine also contain the sequence UAC, the complement of AUG, at a site that allows the tRNA to interact with mRNA. As Figure 20.51b shows, the segment of mRNA with the AUG codon links with the complementary strand on the tRNA molecule bonded to methionine. In doing so, the methionine is put into a position to unlink from the tRNA and to be the first amino acid residue in the protein being synthesized.

transcription the process of copying the information in DNA to RNA.

messenger RNA (mRNA) the form of RNA that carries the code for synthesizing proteins from DNA to the site of protein synthesis in a cell.

codon a three-nucleotide sequence that codes for a specific amino acid.

translation the process of assembling proteins from the information encoded in RNA.

transfer RNA (tRNA) the form of the nucleic acid RNA that delivers amino acids, one at a time, to polypeptide chains being assembled by the ribosome–mRNA complex.

TABLE 20.5 mRNA Codons

Amino Acid	Codons	Amino Acid	Codons
Ala	GCU, GCC, GCA, GCG	Leu	UUA, UUG, CUU, CUC, CUA, CUG
Arg	CGU, CGC, CGA, CGG, AGA, AGG	Lys	AAA, AAG
Asn	AAU, AAC	Met	AUG
Asp	GAU, GAC	Phe	UUU, UUC
Cys	UGU, UGC	Pro	CCU, CCC, CCA, CCG
Gln	CAA, CAG	Ser	UCU, UCC, UCA, UCG, AGU, AGC
Glu	GAA, GAG	Thr	ACU, ACC, ACA, ACG
Gly	GGU, GGC, GGA, GGG	Trp	UGG
His	CAU, CAC	Tyr	UAU, UAC
Ile	AUU, AUC, AUA	Val	GUU, GUC, GUA, GUG
Start	AUG	Stop	UAG, UGA, UAA

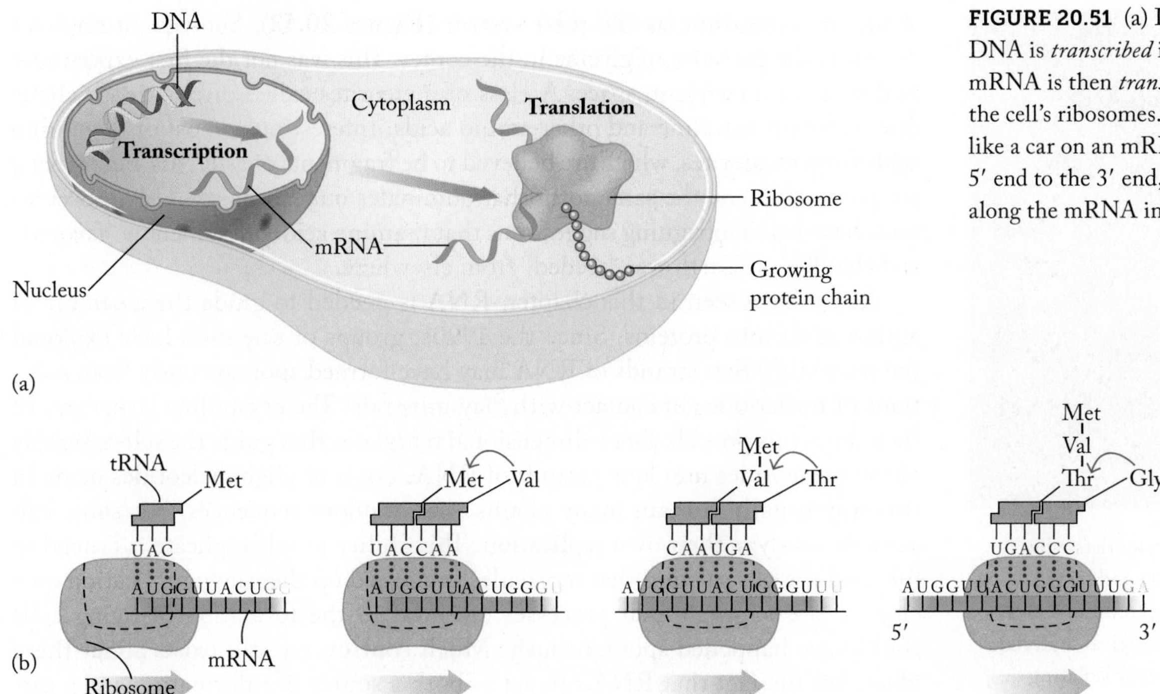

FIGURE 20.51 (a) In protein synthesis, DNA is *transcribed* into mRNA, and (b) the mRNA is then *translated* into proteins on the cell's ribosomes. The ribosome moves like a car on an mRNA track from the 5′ end to the 3′ end, or from left to right along the mRNA in the illustration.

This sequence of events is repeated many times. In Figure 20.51b, the second codon, GUU, links up with a molecule of tRNA that has a CAA binding site and a molecule of valine in tow. In this way valine moves into position to become the next amino acid residue in the protein and to form a peptide bond with the N-terminal methionine. Valine is followed by threonine, which is followed by glycine, and so on, until a Stop codon finally signals the end of the translation process.

CONCEPT TEST

If a GUU codon attracts valine to the translation site, does a UUG codon do the same thing? Explain your answer.

20.8 From Biomolecules to Living Cells

We end Chapter 20 by addressing two fundamental questions about the major classes of biomolecules and their roles in sustaining life: (1) how were they first formed on prebiotic Earth, and (2) how did they assemble into living cells? Experiments conducted in the 1950s at the University of Chicago by chemistry professor Harold Urey (1893–1981) and his student Stanley Miller (1930–2007) showed that amino acids could form from H_2O, CH_4, NH_3, and H_2. Although the reactants the two scientists chose are now thought to be different from those present on early Earth, the result still stands: inorganic molecules can react to produce the organic molecules found in living systems.

CONNECTION The results of the Miller–Urey experiment were discussed in Section 3.1.

Evidence also exists that some biomolecules may have reached Earth from extraterrestrial origins. In 2006 the NASA spacecraft *Stardust* returned to Earth with samples collected from the tail of a comet that is believed to have formed at

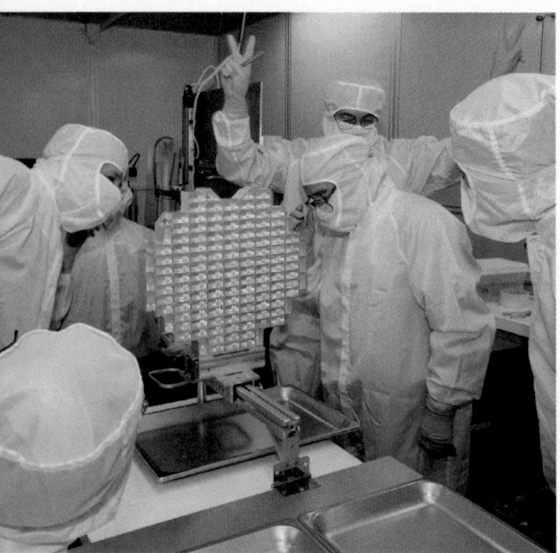

FIGURE 20.52 NASA scientists who analyzed the samples collected by the *Stardust* spacecraft were careful to avoid contaminating it with biological material from Earth. Doing so meant isolating themselves from the sample as they prepared it for analysis.

about the same time as the solar system (**Figure 20.52**). Subsequent analyses disclosed the presence of glycine in the comet. This was not the first experiment to detect amino acids in space. A class of meteorites called carbonaceous chondrites contain isovaline and other amino acids. Interestingly, most of the amino acids from meteorites, which are believed to be fragments of asteroids and comets, are L-enantiomers, the same form that dominates our biosphere. These observations have led to intriguing suggestions that L-amino acids are somehow "favored" and that life on Earth was "seeded" from elsewhere.

As we have seen in this chapter, RNA is needed to guide the assembly of amino acids into proteins. Since the 1990s, groups of scientists have explored the possibility that strands of RNA may have formed spontaneously from solutions of nucleotides in contact with clay minerals. The crystalline structures of these minerals provide three-dimensional templates that guide the self-assembly of the nucleotides into long strands of RNA. Pools of oligonucleotides made in this way usually contain many chains with random sequences, but some can actually catalyze their own replication. This ability to self-replicate is crucial to life, and the observation that molecules can speed up their own replication on a clay surface suggests that processes essential to the formation of living cells could have happened spontaneously. Much controversy still exists about these ideas, but the fact that RNA can act as both a source of information and a catalyst is part of the *RNA world hypothesis*. This hypothesis proposes that a world filled with life based on RNA predates the current world of life based on DNA and proteins. The capacity of RNA to both store information like DNA *and* act as a catalyst like an enzyme suggests that RNA alone could have supported cellular or precellular life-forms.

Current research is also testing the hypothesis that life on Earth may have evolved near deep-ocean hydrothermal vents. Entire ecosystems have been discovered at these locations since they were first explored in the 1970s. They are sustained by geothermal and chemical energy rather than energy from the Sun. It may be that life actually began in such environments, with hydrothermal energy driving reactions in which inorganic compounds such as carbon dioxide and hydrogen sulfide formed small organic compounds. As with the reactions on the surfaces of clay minerals, the synthesis reactions at hydrothermal vents may have been catalyzed and guided by reactants adsorbed on solid compounds such as FeS and MnO_2, which pour into the sea in dense black clouds near some vents (**Figure 20.53**). Among the known products of these reactions are acetate ions (CH_3COO^-). Acetate is a key intermediate in many biosynthetic pathways in living organisms. In modern bacteria, the systems that make acetate depend on a catalyst made of iron, nickel, and sulfur that has a structure much like that of particles produced by "black smokers" on the ocean floor.

FIGURE 20.53 Black clouds of transition metal oxides and sulfides flow into the sea through chimneys like this one at a deep-ocean hydrothermal vent. Some scientists believe that these particles may have guided and catalyzed the formation of the first self-replicating molecules on Earth.

To take the next step toward forming living cells, large biomolecules must have assembled themselves into even larger structures, such as membranes, that allow cells and structures within them to collect materials and retain them at concentrations different from those in the surrounding medium. Molecules in these assemblies are not necessarily connected by covalent bonds, but rather are held together by the intermolecular interactions we have discussed in this chapter.

SAMPLE EXERCISE 20.9 Integrating Concepts: Liquid Oil to Solid Fat

Imagine we have a research project to turn soybean oil into a solid for use in a butter substitute. We can accomplish this by hydrogenating the oil. Suppose we are working at the level of a small-scale industrial facility to test the feasibility of this process. A particular sample of soybean oil contains the following percentages by mass:

Composition of Soybean Oil

Fatty Acid	Percentage
Linoleic $CH_3(CH_2)_4CH{=}CHCH_2CH{=}CH(CH_2)_7COOH$	54
Oleic $CH_3(CH_2)_7CH{=}CH(CH_2)_7COOH$	24
Palmitic $CH_3(CH_2)_{14}COOH$	18

We need to hydrogenate 10.0 kg of soybean oil. Our source of hydrogen is the steam-reforming of methane:

$$H_2O(g) + CH_4(g) \rightarrow CO(g) + 3\,H_2(g)$$

a. What volume of methane at 20°C and 1.00 atm of pressure do we need to produce the amount of hydrogen required to completely hydrogenate 10.0 kg of soybean oil?
b. How many different compounds could result from the hydrogenation of a triglyceride that contained one of each of the three fatty acids in this study: linoleic acid, oleic acid, and palmitic acid?

Collect and Organize We need to determine how much methane is needed in the steam-reforming reaction to supply enough hydrogen to react completely with the unsaturated fats present in a 10.0-kg sample of soybean oil. We know the oil contains three fatty acids: linoleic, oleic, and palmitic acid.

Analyze According to the structures in Table 20.4, palmitic acid is a saturated fatty acid, which means it does not react with hydrogen. Oleic acid has one C=C double bond per molecule and linoleic acid has two. Therefore, 1 mole of oleic acid combines with 1 mole of H_2, and 1 mole of linoleic acid combines with 2 moles of H_2. We can use the given weight percentages (24% oleic acid and 54% linoleic acid) to calculate that a 10.0-kg sample contains 2.4 kg of oleic acid and 5.4 kg of linoleic acid. The structures of these two fatty acids can be used to write their molecular formulas, and from those formulas we can calculate their molar masses. Converting these masses into grams and dividing these masses by the molar masses of the two fatty acids yields the moles of each in the sample. We then use the 1:1 and 1:2 hydrogenation reaction stoichiometries to calculate the number of moles of H_2 we need, and we use the stoichiometry of the steam-reforming reaction (1 mol CH_4:3 mol H_2) to convert moles of H_2 into moles of CH_4. The number of moles of CH_4 needed, along with the given temperature and pressure of the gas, can then be used in the ideal gas law equation to calculate the volume of CH_4 needed.

Solve

a. Using the condensed structure of the two unsaturated fatty acids to determine their molecular formulas and calculate their molar masses:

Fatty Acid	Molecular Formula	Molar Mass (g/mol)
Oleic acid	$C_{18}H_{34}O_2$	282.46
Linoleic acid	$C_{18}H_{32}O_2$	280.45

Converting the masses of these fatty acids in the sample into moles:

$$2.4\ \cancel{\text{kg}}\ \text{oleic acid} \times \frac{1000\ \cancel{\text{g}}}{1\ \cancel{\text{kg}}} \times \frac{1\ \text{mol}}{282.46\ \cancel{\text{g}}} = 8.5\ \text{mol oleic acid}$$

$$5.4\ \cancel{\text{kg}}\ \text{linoleic acid} \times \frac{1000\ \cancel{\text{g}}}{1\ \cancel{\text{kg}}} \times \frac{1\ \text{mol}}{280.45\ \cancel{\text{g}}} = 19\ \text{mol linoleic acid}$$

Calculating the total moles of CH_4 needed:

$$8.5\ \cancel{\text{mol oleic acid}} \times \frac{1\ \cancel{\text{mol } H_2}}{1\ \cancel{\text{mol oleic acid}}} \times \frac{1\ \text{mol } CH_4}{3\ \cancel{\text{mol } H_2}} = 2.8\ \text{mol } CH_4$$

$$19\ \cancel{\text{mol linoleic acid}} \times \frac{2\ \cancel{\text{mol } H_2}}{1\ \cancel{\text{mol linoleic acid}}} \times \frac{1\ \text{mol } CH_4}{3\ \cancel{\text{mol } H_2}} = 13\ \text{mol } CH_4$$

Total moles of CH_4 = 2.8 + 13 = 15.8 mol CH_4.

$$PV = nRT$$

$$(1.00\ \text{atm})V = 15.8\ \text{mol} \times \frac{0.08206\ \text{L} \cdot \text{atm}}{\text{K} \cdot \text{mol}} \times 293\ \text{K}$$

$$V = \frac{15.8\ \cancel{\text{mol}} \times \dfrac{0.08206\ \text{L} \cdot \cancel{\text{atm}}}{\cancel{\text{K}} \cdot \cancel{\text{mol}}} \times 293\ \cancel{\text{K}}}{1.00\ \cancel{\text{atm}}} = 3.80 \times 10^2\ \text{L}$$

Reported to the correct number of significant figures, the reaction requires 3.8×10^2 L of methane.

b. If a single triglyceride contained one of each of the saturated fatty acids that result from this process, it would contain one unit of palmitic acid, which was saturated in the original material and unchanged by the hydrogenation, and two units of stearic acid, the C_{18} saturated fatty acid that is the product of hydrogenation of oleic acid and linoleic acid. We can represent this symbolically, letting A = palmitic acid and B = stearic acid. Two constitutional isomers of triglycerides result:

```
CH2—CH—CH2     CH2—CH—CH2
 |    |   |     |    |   |
 A    B   B     B    A   B
```

The structure on the left also has two enantiomeric forms because the carbon in the –CH– unit is a chiral center. Therefore, the products are two constitutional isomers, one of which also has two enantiomers (stereoisomers), for a total of three different triglycerides.

Think About It The number of moles of methane required seems reasonable according to our estimate. Actual soybean oil probably has several different triglycerides in it that are a combination of saturated and unsaturated fatty acid components, resulting in the observed composition by weight. Natural oils can be quite complex mixtures.

SUMMARY

LO1 Isomers may be **constitutional isomers**, which are distinguished by the connectivity of their atoms, or they may be **stereoisomers**, which are distinguished by their arrangement in three-dimensional space. Many biologically important molecules are chiral, resulting in **optical isomers**, a type of stereoisomer. (Section 20.2)

LO2 The acid–base properties of **amino acids** (including their pK_a and pI values) are important for structural and functional reasons. We can define these properties by using data derived from titrations. (Sections 20.2 and 20.3)

LO3 The **proteins** and **peptides** in the human body are composed of 20 α-amino acids covalently linked by **peptide bonds**. The sequence of amino acids in peptide and protein chains matters in determining their properties. (Section 20.3)

LO4 The structure of a protein is crucial to its function. It is defined by the sequence of amino acids in the chain (its **primary structure**), the geometric pattern that segments of a chain adopt (its **secondary structure**), the overall three-dimensional shape of the protein (its **tertiary structure**), and any larger structure formed when two or more proteins interact and function as a single unit (its **quaternary structure**). (Section 20.4)

LO5 **Carbohydrates** are produced from CO_2 and H_2O, and organisms derive energy from glycolysis and the **tricarboxylic acid (TCA) cycle**, the reaction pathways by which glucose is oxidized back to CO_2 and H_2O. **Monosaccharides** are joined through **glycosidic bonds** into **polysaccharides** such as starch for energy storage and cellulose for structural support in plants. (Section 20.5)

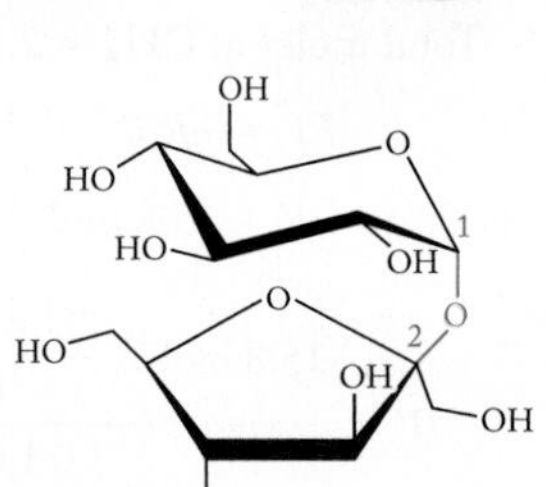

LO6 **Lipids** include important families of molecules, including **glycerides, fats,** and **oils**. Some lipids are a major source of energy in our diet, and others play key roles in cell structure. The physical properties of oils (liquids) and fats (solids) reflect differences in the amounts of unsaturated versus saturated fatty acids. (Section 20.6)

Olestra

LO7 The **nucleic acids** DNA and RNA contain an organism's genetic information and control protein synthesis through **transcription** and **translation**. Living cells may have formed as a result of chemical reactions in which inorganic molecules combined to form small organic molecules such as amino acids and **nucleotides**. (Section 20.7)

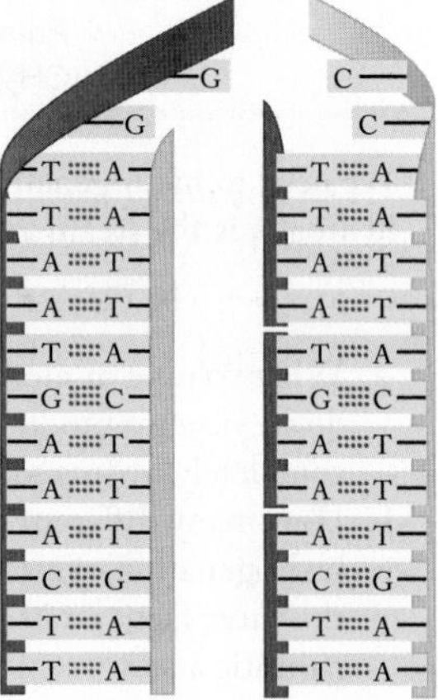

PARTICULATE PREVIEW WRAP-UP

Glutamic acid contains an amine (R–NH_2) functional group and two carboxylic acid (R–COOH) functional groups. At physiological pH (~7.4), the carboxylic acid groups are deprotonated (R–COO^-), but the amine is still protonated (R–NH_3^+), meaning that the glutamic acid molecule has an overall negative charge at physiological pH.

PROBLEM-SOLVING SUMMARY

Type of Problem	Concepts and Equations	Sample Exercises
Recognizing constitutional isomers	Establish that the compounds have the same molecular formula, and if they do, look for different arrangements of bonds.	**20.1**
Identifying chiral molecules	Identify carbon atoms with four different groups attached.	**20.2, 20.3**
Interpreting acid–base titration curves of amino acids	At low pH all amino acids have at least two ionizable H atoms, one each from –COOH and –NH_3^+. Side-chain carboxylic acid and amine groups may also impart acidic and basic strength to amino acids.	**20.4**

Type of Problem	Concepts and Equations	Sample Exercises
Drawing and naming peptides	Connect the α-amine of one amino acid to the α-carboxylic acid of another with a peptide bond. Starting with the free amine (N-) terminus on the left, name each amino acid residue by changing the ending of the name of the parent amino acid to *-yl* in all but the last (C-terminal) amino acid.	**20.5**
Determining the amino acid sequence of a small peptide	Arrange fragments of the peptide in a way that reveals the sequence of the peptide.	**20.6**
Identifying triglycerides	The –COOH groups of fatty acids react with the –OH groups in glycerol to form triglycerides and water.	**20.7**
Using base complementarity in DNA	Identify the base pairs: A pairs with T; G pairs with C. The percentage of T should equal the percentage of A, and the percentage of C should equal the percentage of G.	**20.8**

VISUAL PROBLEMS

(Answers to boldface end-of-chapter questions and problems are in the back of the book.)

20.1. The nucleotides in DNA contain the bases with the structures shown in Figure P20.1. Identify the basic functional groups in the structures.

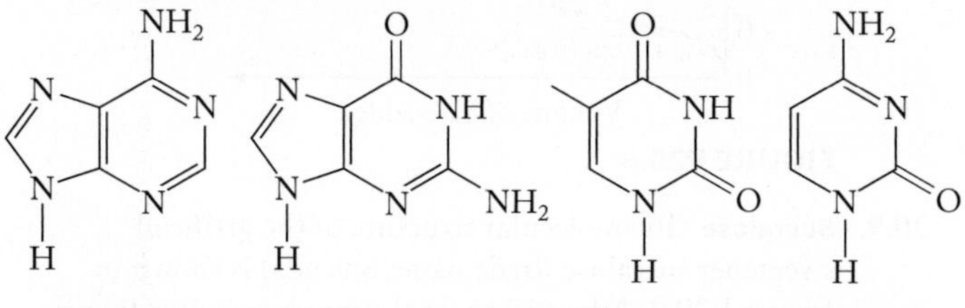

Adenine Guanine Thymine Cytosine

FIGURE P20.1

20.2. **Experimental Agent** Bentiromide (Figure P20.2) is a peptide that was once evaluated as an agent to monitor the function of the pancreas during therapy. Draw the structures of the amino acids that form bentiromide and indicate whether they are α-amino acids.

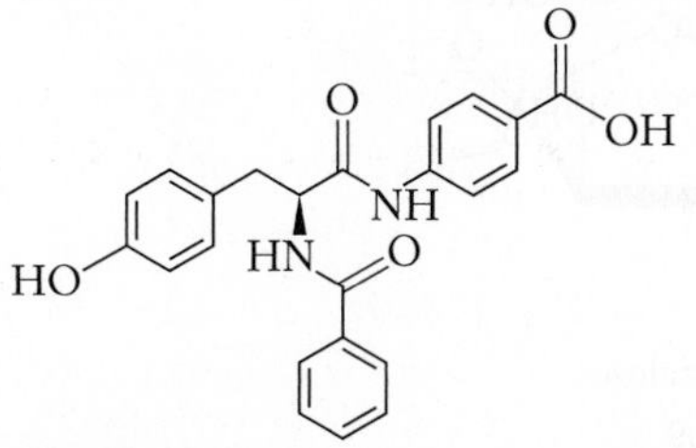

FIGURE P20.2

20.3. Olive Oil Olive oil contains triglycerides such as those shown in Figure P20.3. Which of the fatty acids in these triglycerides is/are saturated?

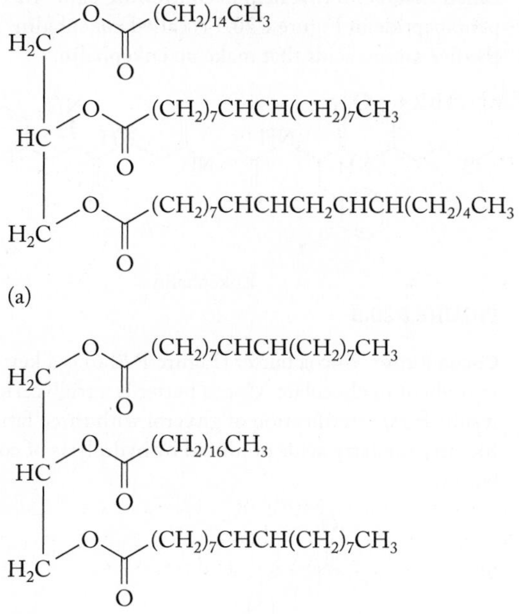

(a)

(b)

FIGURE P20.3

20.4. **Treating Infections** The major component of the antibiotic ointment bacitracin is a cyclic polypeptide called bacitracin A (Figure P20.4). It was first isolated in 1943 from a knee scrape from a girl named Margaret Tracy, after whom it is named. Bacitracin is effective topically and is used to treat skin, eye, and wound infections. Identify the amino acids found in human proteins that are also part of the structure of bacitracin A.

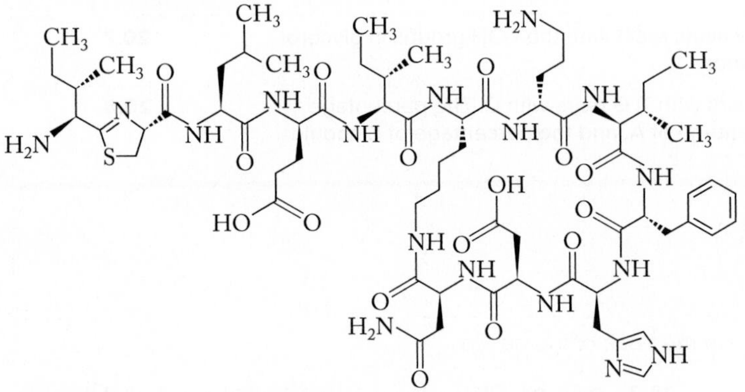

FIGURE P20.4

20.5. **Natural Painkillers** The human brain produces polypeptides called *endorphins* that help in controlling pain. The pentapeptide in Figure P20.5 is called enkephalin. Identify the five amino acids that make up enkephalin.

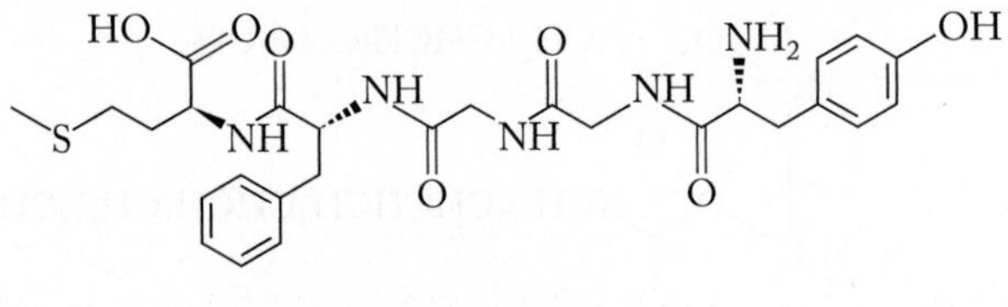

Enkephalin

FIGURE P20.5

20.6. **Cocoa Butter** Cocoa butter (Figure P20.6) is a key ingredient in chocolate. Cocoa butter is a triglyceride that results from esterification of glycerol with three fatty acids. Identify the fatty acids produced by hydrolysis of cocoa butter.

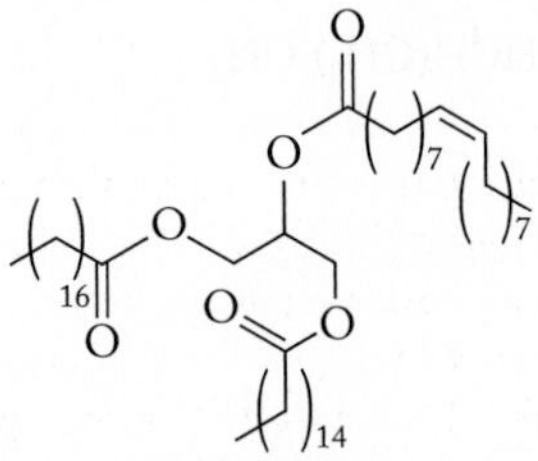

Cocoa butter

FIGURE P20.6

20.7. **Trans Fats** The role of "trans fats" in human health has been extensively debated both in the scientific community and in the popular press. What type of isomerism does the word "trans fat" refer to? Which of the molecules in Figure P20.7 are considered trans fats?

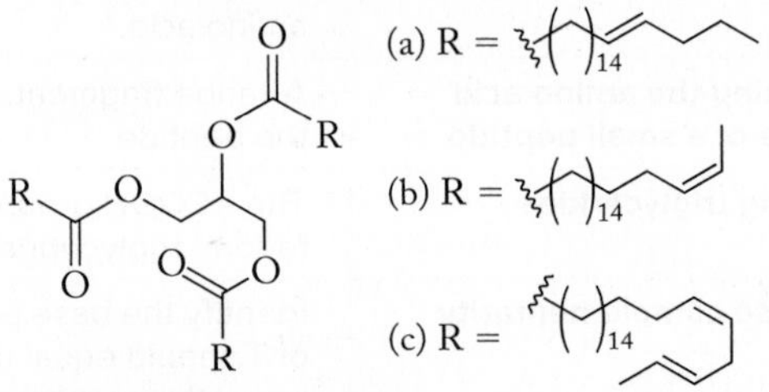

(a) R =

(b) R =

(c) R =

FIGURE P20.7

20.8. Figure P20.8 shows the titration curve of which of these amino acids: leucine, histidine, or lysine?

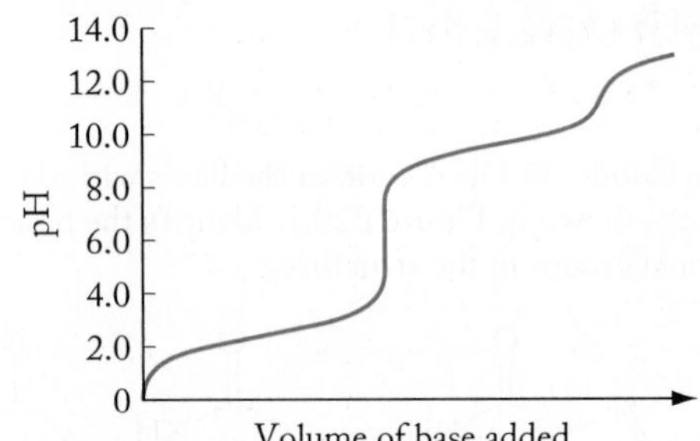

FIGURE P20.8

20.9. **Sucralose** The molecular structure of the artificial sweetener sucralose (trade name Splenda) is shown in Figure P20.9. Advertising for this product claims that it is made from sugar, implying that it is a natural product. What sugar might it be made from? Comment on the implication that it is a "natural" product.

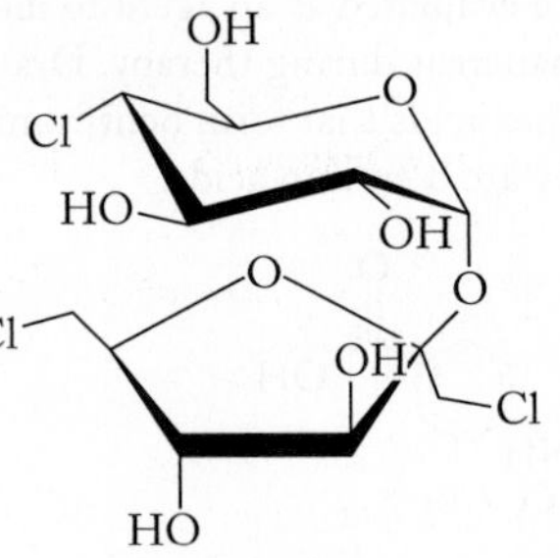

Sucralose

FIGURE P20.9

20.10. Covalent bonding leads to primary structure and intermolecular forces, both of which ultimately lead to secondary, tertiary, and quaternary structure. Use representations [A] through [I] in Figure P20.10 to answer questions (a)–(f).

a. [A] depicts two strands of DNA in a double helix. Are the two strands held together by covalent bonds, intermolecular forces, or both? [B] depicts two alpha helices linked together. Are the two helices held together by covalent bonds, intermolecular forces, or both?

b. [C] depicts a phospholipid bilayer. Is the bilayer held together by covalent bonds, intermolecular forces, or both?

c. [D] depicts a disaccharide. Are the two sugars each held together by covalent bonds, intermolecular forces, or both?

d. Which of the single molecules depicted are an important structural component of [C]?

e. Which of the single molecules depicted react to form [E]?

f. What intermolecular forces are typically involved for the amino acid depicted in [F] when it is incorporated into a protein?

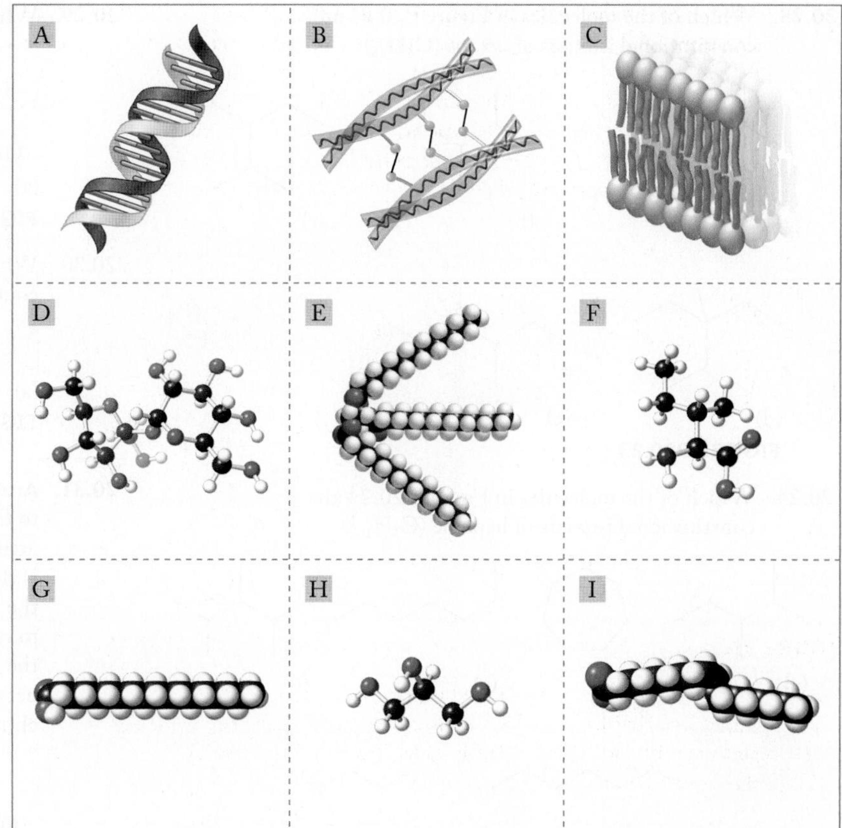

FIGURE P20.10

QUESTIONS AND PROBLEMS

Organic Molecules, Isomers, and Chirality

Concept Review

20.11. Figure P20.11 shows the carbon-skeleton structures of hexane and cyclohexane. Are hexane and cyclohexane constitutional isomers?

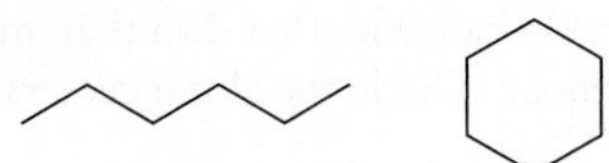

FIGURE P20.11

20.12. Consider the properties of constitutional isomers:

a. Do constitutional isomers always have the same molecular formula?

b. Do constitutional isomers always have the same chemical properties?

20.13. Can all of the terms *enantiomer*, *achiral*, and *optically active* be used to describe a single compound? Explain.

20.14. Two compounds have the same structure and the same physical properties but also have the same optical activity. Are they enantiomers or the same molecule?

20.15. How do constitutional isomers differ from stereoisomers?

*20.16. Could a racemic mixture be distinguished from an achiral compound on the basis of optical activity? Explain your answer.

20.17. Why is the amino acid glycine (Figure P20.17) achiral?

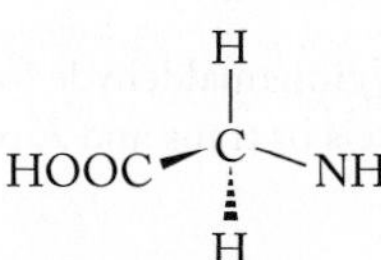

FIGURE P20.17

20.18. Can stereoisomers of molecules such as cis and trans RHC=CHR also have optical isomers? (R may be any of the functional groups we have encountered in this textbook.) Explain your answer.

20.19. Which type of hybrid orbitals on a carbon atom—*sp*, sp^2, or sp^3—can give rise to enantiomers?

*20.20. Could an oxygen atom in an alcohol, ketone, or ether ever be a chiral center in the molecule?

Problems

20.21. Draw and name all the constitutional isomers of C_5H_{12}.

20.22. Draw and name all the constitutional isomers of C_6H_{14}.

20.23. Which of the molecules in Figure P20.23 are constitutional isomers of octane (C_8H_{18})?

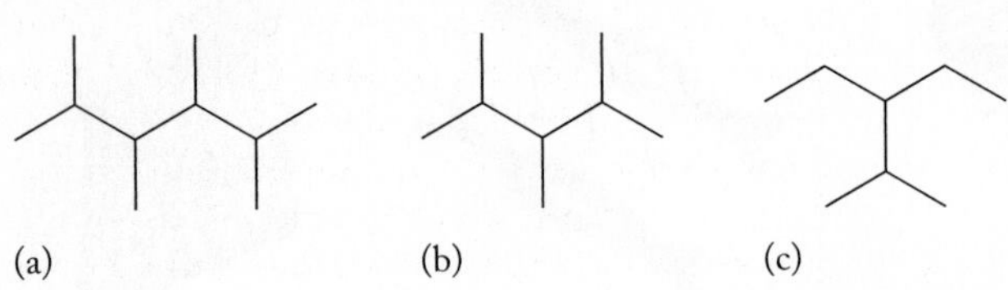

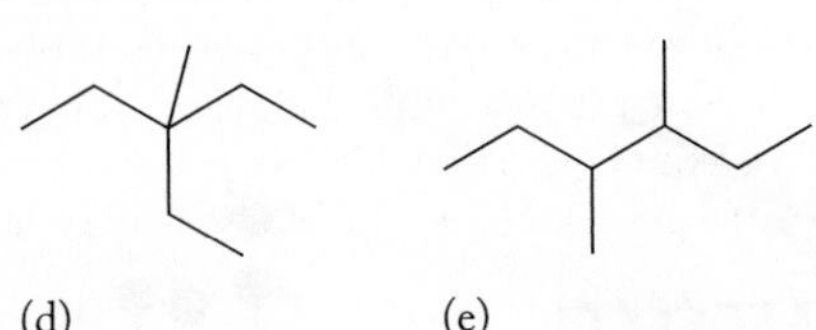

FIGURE P20.23

20.24. Which of the molecules in Figure P20.24 are constitutional isomers of heptane (C_7H_{16})?

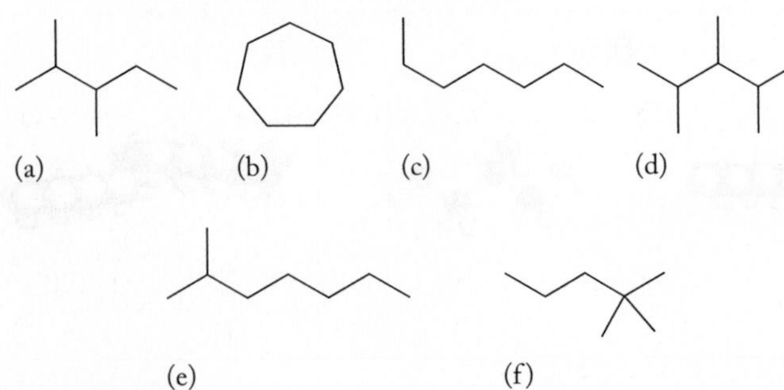

FIGURE P20.24

20.25. Why don't the alkenes in Figure P20.25 have cis and trans isomers?

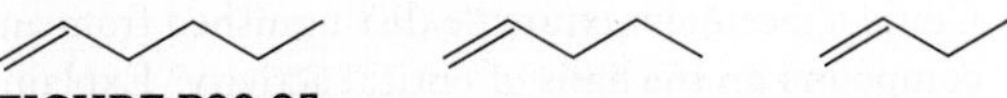

FIGURE P20.25

20.26. Why don't alkynes have cis and trans isomers?

20.27. Cinnamon Label the isomers of cinnamaldehyde (oil of cinnamon) in Figure P20.27 as cis or trans and *E* or *Z*.

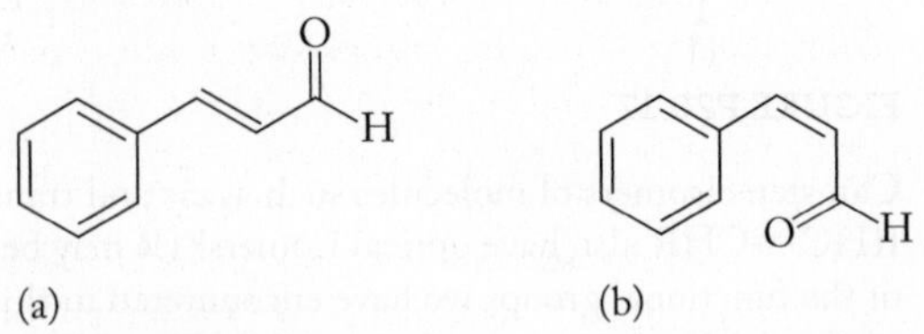

FIGURE P20.27

*20.28. Figure P20.28 shows the carbon-skeleton structure of carvone, which is found in oil of spearmint. Why doesn't the molecule carvone have cis and trans isomers?

Carvone (oil of spearmint)

FIGURE P20.28

20.29. Which of the molecules in Figure P20.29 are chiral?

(a) (b) (c)

FIGURE P20.29

20.30. Which, if any, of the molecules shown in Figure P20.30 contains a chiral center?

(a) (b) (c)

FIGURE P20.30

20.31. Artificial Sweeteners Artificial sweeteners are fundamental to the diet food industry. Figure P20.31 shows three artificial sweeteners that have been used in food. Saccharin is the oldest, dating to 1879. Cyclamates were banned in the United States in 1969 after research suggested they led to tumors. Aspartame may be more familiar to you under the name NutraSweet. Each of these sweeteners contains between zero and two chiral carbon atoms. Circle the chiral center(s) in each compound.

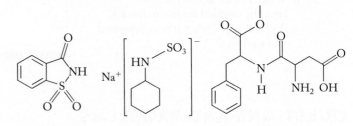

Saccharin — Sodium cyclamate — Aspartame

FIGURE P20.31

20.32. **Smoother Skin** Researchers at Tufts University have found that a fungal toxin called cytochalasin B (Figure P20.32) can restore elasticity and shrink skin cells in mice without causing any ill effects. Mice treated with a cream containing this molecule had smoother skin than mice treated with a plain cream. Circle the chiral centers in the molecule.

FIGURE P20.32

20.33. Lowering Cholesterol Statins such as rosuvastatin calcium (Figure P20.33) are the most widely prescribed class of drugs for lowering concentrations of cholesterol and other lipids in the blood.

a. Is the anion in the figure chiral? If so, how many chiral centers does it have?

b. Are any other kinds of stereoisomerism possible for the anion?

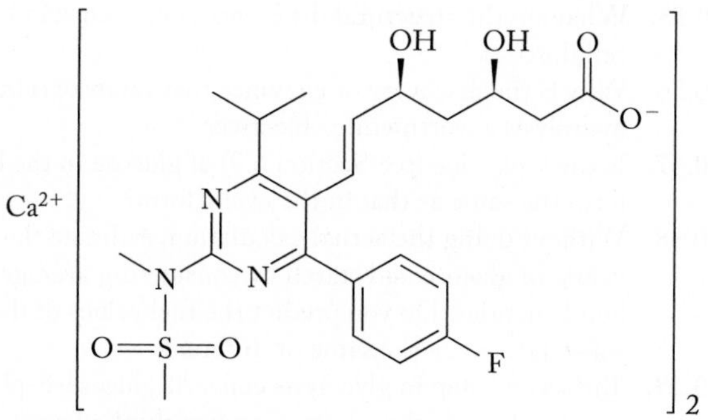

FIGURE P20.33

*20.34. **Preventing Blood Clots** In 2009 Plavix (Figure P20.34) was the second-highest-selling pharmaceutical in the world. It is prescribed to prevent blood clots in heart attack and stroke patients.

a. Is Plavix optically active? If so, identify the chiral center(s) in the structure.
b. Plavix contains an ester group. Identify it. Draw the carboxylic acid and the alcohol you could react together to make Plavix.

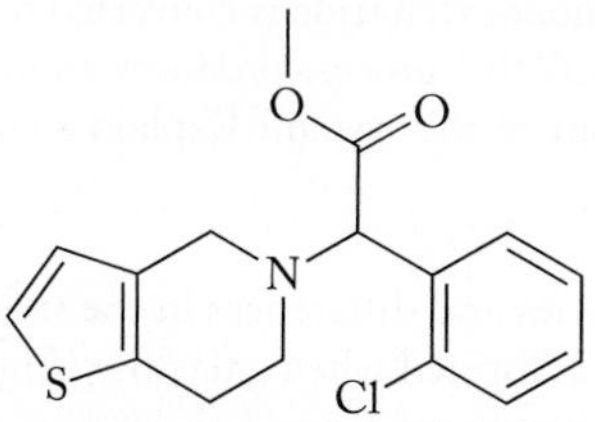

FIGURE P20.34

The Composition of Proteins

Concept Review

20.35. In living cells, amino acids combine to make peptides and proteins. Are these processes accompanied by increases or decreases in entropy of the reaction system?

20.36. What is the difference between a peptide bond and an amide bond?

20.37. **Into the Wild** Jon Krakauer, the author of *Into the Wild*, postulated that the main character in the book died in the Alaskan wilderness because he accidentally ate seeds containing a toxin called β-ODAP (Figure P20.37). This substance is a neurotoxin; its mechanism of action is not fully understood, but it is referred to as a "glutamic acid mimic." Compare the structure of β-ODAP to that of glutamic acid and describe the similarities between the two molecules.

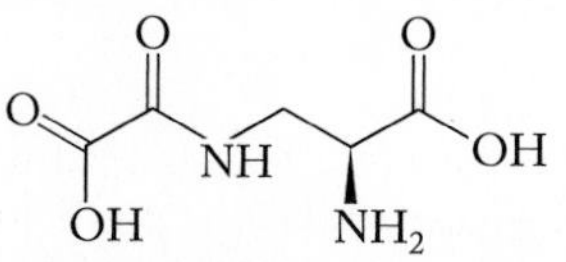

FIGURE P20.37

20.38. **Metabolic Disease in Dogs** Cystinuria is a metabolic disease that occurs in some breeds of dogs. It is characterized by the presence of kidney stones made of cystine, a dimer formed when two cysteine residues covalently link through an –S—S– bond (see Figure 20.28).

a. Draw the structure of the dipeptide Cys-Cys.
b. Is cystine a dipeptide? Why or why not?

20.39. Meteorites contain more L-amino acids, which are the forms that make up the proteins in our bodies, than D-amino acids. What do the prefixes L- and D- mean?

20.40. Do any of the amino acids in Table 20.2 have more than one chiral carbon atom per molecule?

20.41. Which of the compounds in Figure P20.41 is/are *not* an α-amino acid?

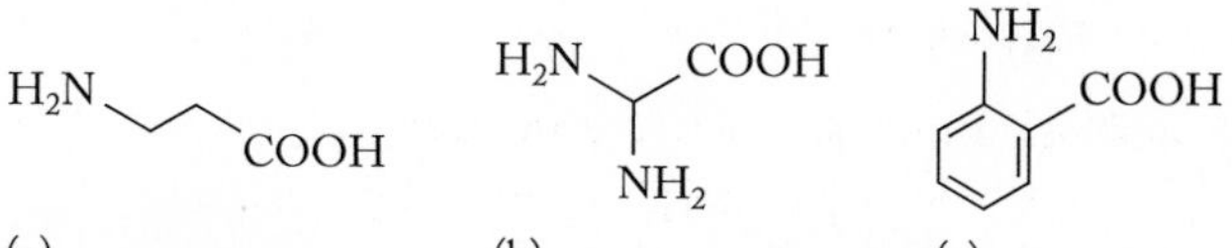

FIGURE P20.41

20.42. Which of the compounds in Figure P20.42 is/are α-amino acids?

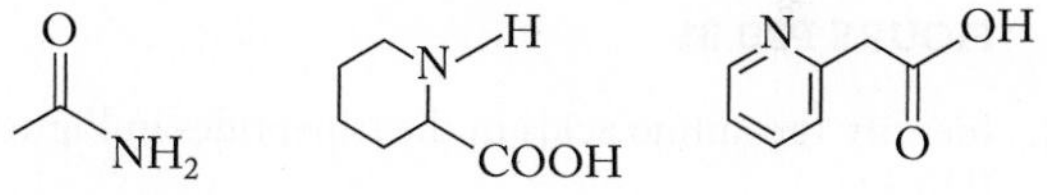

FIGURE P20.42

20.43. Why do most amino acids exist in the zwitterionic form at physiological pH (~7.4)?

20.44. Draw the condensed structural formulas of the amino acid tyrosine that you would expect to predominate in:

a. strongly acidic aqueous solution
b. strongly basic aqueous solution
c. an aqueous solution in which pH = p*I*

Protein Structure and Function

Concept Review

20.45. Which type of intermolecular interaction plays the dominant role in holding strands of proteins together in β-pleated sheets and stabilizing α helices?

20.46. Which level of protein structure is associated with ion–ion interactions and disulfide bond formation?

20.47. When protein strands fold back on themselves in forming stable tertiary structures, lysine residues are often paired up with glutamic acid residues. Why?

20.48. Ion–ion interactions are particularly effective at stabilizing tertiary structures of proteins. Suggest a pair of amino acid residues that would be attracted to each other via ion–ion interactions at pH = 7.4.

Problems

20.49. Draw structures and name all possible dipeptides produced from condensation reactions of the following L-amino acids:

a. alanine + isoleucine
b. serine + tyrosine
c. valine + phenylalanine

20.50. Draw structures of the peptides produced from condensation reactions of the following L-amino acids. Assume the amino acids bond in the order given.
a. glycine + tyrosine + aspartic acid
b. serine + threonine + leucine
c. asparagine + lysine + histidine

20.51. Identify the amino acids in the dipeptides shown in Figure P20.51.

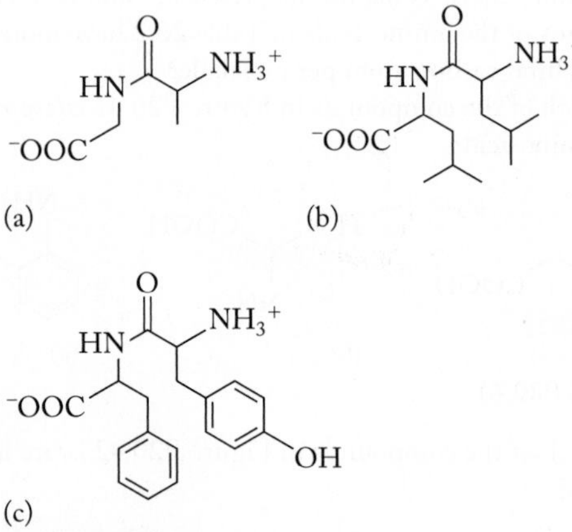

(a) (b) (c)

FIGURE P20.51

20.52. Identify the amino acids in the tripeptides in Figure P20.52.

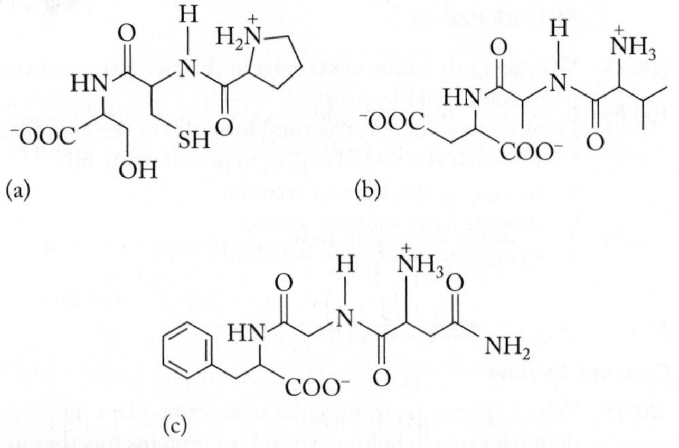

(a) (b) (c)

FIGURE P20.52

20.53. Identify the missing product in the metabolic reaction shown in Figure P20.53.

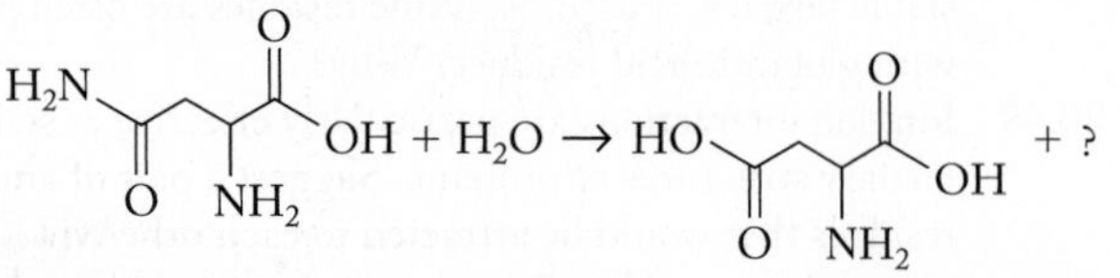

FIGURE P20.53

20.54. The molecular formula for glycine is $C_2H_5NO_2$. What is the molecular formula of the linear peptide formed when ten glycine molecules are linked together in peptide bonds?

Carbohydrates

Concept Review

20.55. What are the structural differences between starch and cellulose?

20.56. Why is the discovery of enzymes that catalyze cellulose hydrolysis a worthwhile objective?

20.57. Is the fuel value (see Section 6.9) of glucose in the linear form the same as that in the cyclic form?

*20.58. Without doing the actual calculation, estimate the fuel values of glucose and starch by considering average bond energies. Do you predict the fuel values of the two substances to be the same or different?

20.59. The second step in glycolysis converts glucose-6-phosphate into fructose-6-phosphate. Can you think of a reason why ΔG° for this reaction is close to zero?

*20.60. Which of the following statements are correct about glycosidic bonds in carbohydrates?
a. The glycosidic bond in maltose is hydrolyzed by people who are lactose intolerant.
b. A glycosidic bond links glucose and fructose together to form sucrose.
c. A glycosidic bond is an ether linkage, but all ether linkages are not glycosidic bonds.

20.61. How do we calculate the overall free-energy change of a process consisting of two steps?

20.62. During glycolysis a monosaccharide is converted to pyruvate. Do you think this process produces an increase or decrease in the entropy of the system? Explain your answer.

Problems

20.63. Describe the similarities and differences in the structures of the α and β isomers formed when galactose (Figure P20.63) forms a six-membered ring.

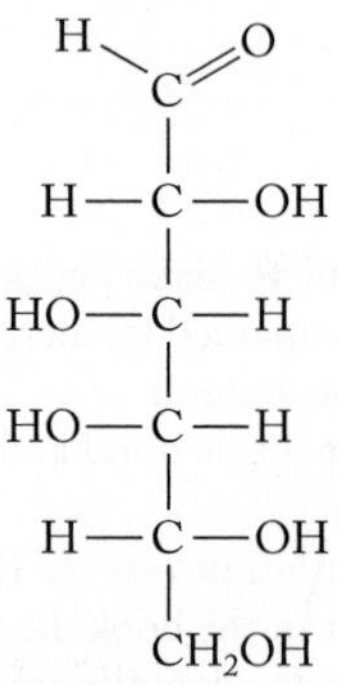

Galactose

FIGURE P20.63

20.64. Describe the similarities and differences in the structures of the α and β isomers formed when ribose (Figure P20.64) forms a five-membered ring.

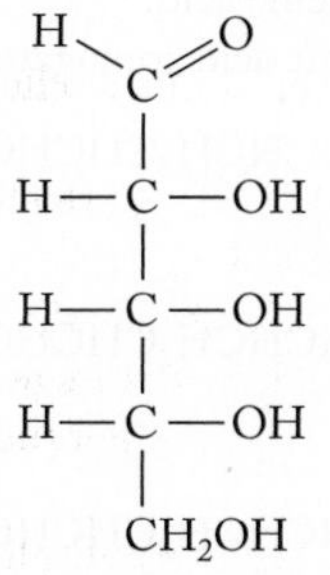

H–C=O
H–C–OH
H–C–OH
H–C–OH
CH_2OH

Ribose

FIGURE P20.64

20.65. Which, if any, of the structures in Figure P20.65 are β isomers of a monosaccharide?

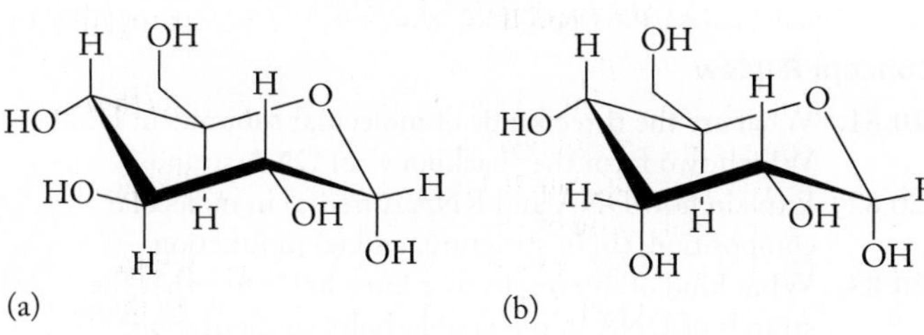

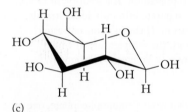

(c)

FIGURE P20.65

20.66. Which, if any, of the structures in Figure P20.66 are α isomers?

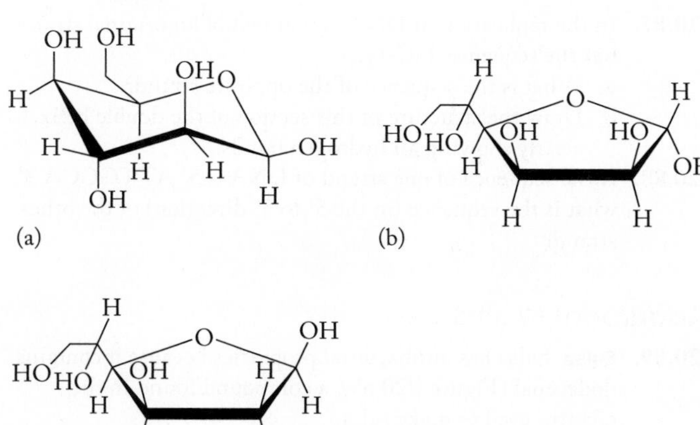

FIGURE P20.66

20.67. Which of the saccharides in Figure P20.67 is digestible by humans?

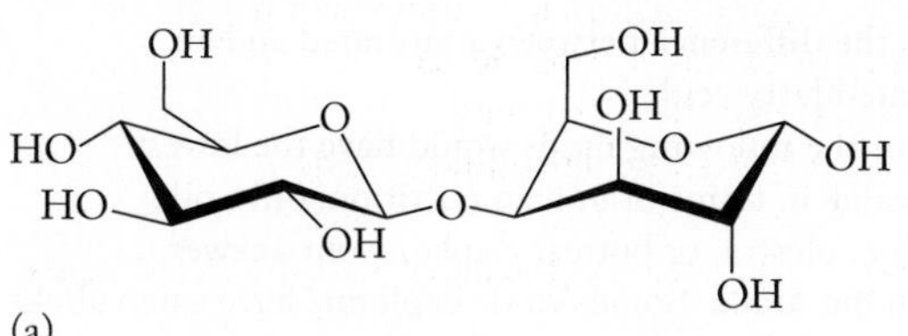

(b)

(c)

FIGURE P20.67

*20.68. For any of the disaccharides in Problem 20.67 that are not digestible by humans, draw an isomer that would be digested.

20.69. The structure of the disaccharide maltose appears in Figure P20.69. Hydrolysis of 1 mole of maltose ($\Delta G_f^\circ = -2246.6$ kJ/mol) produces 2 moles of glucose ($\Delta G_f^\circ = -1274.4$ kJ/mol):

$$\text{Maltose} + H_2O \rightarrow 2\ \text{glucose}$$

If the value of ΔG_f° for water is −285.8 kJ/mol, what is the change in free energy of the hydrolysis reaction?

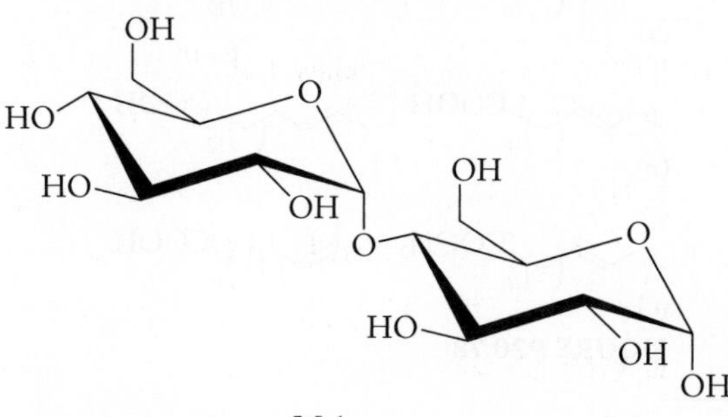

Maltose

FIGURE P20.69

20.70. If the maltose in Problem 20.69 were replaced by another disaccharide, would you expect the free-energy change for the hydrolysis to be exactly the same or just similar in value? Explain your answer.

Lipids

Concept Review

20.71. What is the difference between a saturated and an unsaturated fatty acid?

20.72. Which of the following lipids would have the lowest energy value in terms of human nutrition: olive oil, margarine, olestra, or butter? Explain your answer.

20.73. **Dining in the Arctic** Some Arctic explorers have eaten sticks of butter on their explorations. Give a nutritional reason for this unusual cuisine.

*20.74. If you agitate a mixture of fatty acids in water, an emulsion forms, in which spherical structures called micelles are dispersed throughout the water. Micelles form when the carboxylic acid groups of the fatty acids face the solvent and their hydrocarbon tails are directed toward the inside of the sphere.
a. Explain why these structures form with this orientation.
b. It is sometimes possible to "break" an emulsion, destroying the micelles by adding a strong acid to the mixture. Why would this destroy the micelles?
c. One can also sometimes break an emulsion by adding salt (NaCl) to the mixture. Why would this destroy the micelles?

20.75. Do triglycerides have a chiral center? Explain your answer.

*20.76. Using your knowledge of molecular geometry and intermolecular forces, why might polyunsaturated triglycerides be more likely to be liquid than saturated triglycerides?

Problems

20.77. Oleic acid and α-linolenic acid (Table 20.4) are both unsaturated fats and are liquids at room temperature. They can be converted into solid saturated fats by hydrogenation.
a. Which would consume more hydrogen: 1.0 kg of oleic acid or 0.50 kg of α-linolenic acid?
b. Could you distinguish between the two fatty acids by determining the identity of their hydrogenation products? Explain your answer.

20.78. For each of the pairs of fatty acids in Figure P20.78, indicate whether they are constitutional isomers, stereoisomers, or unrelated compounds.

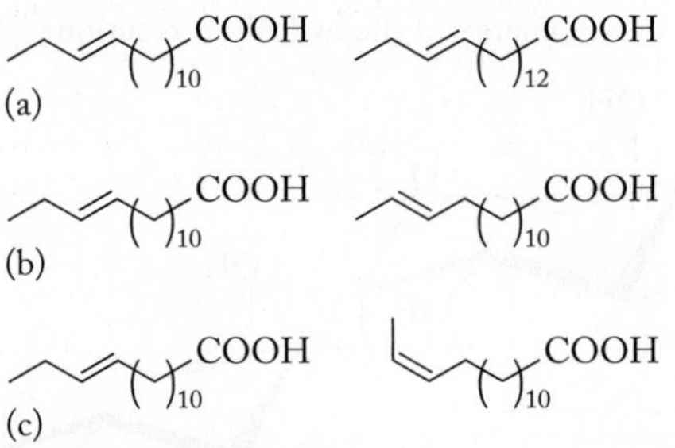

(a) COOH, 10 — COOH, 12

(b) COOH, 10 — COOH, 10

(c) COOH, 10 — COOH, 10

FIGURE P20.78

20.79. Draw the structures of the three fats formed by reaction of glycerol with (a) octanoic acid ($C_7H_{15}COOH$), (b) decanoic acid ($C_9H_{19}COOH$), and (c) dodecanoic acid ($C_{11}H_{23}COOH$).

20.80. **Oil-Based Paints** Oil-based paints contain linseed oil, a triglyceride formed by esterification of glycerol with linolenic acid (Figure P20.80).
a. Draw the line structure of linolenic acid.
*b. Are the double bonds in linolenic acid conjugated?

H_2C–O–C(=O)–$(CH_2)_7CHCHCH_2CHCHCH_2CHCHCH_2CH_3$
HC–O–C(=O)–$(CH_2)_7CHCHCH_2CHCHCH_2CHCHCH_2CH_3$
H_2C–O–C(=O)–$(CH_2)_7CHCHCH_2CHCHCH_2CHCHCH_2CH_3$

Linseed oil

FIGURE P20.80

Nucleotides and Nucleic Acids

Concept Review

20.81. What are the three kinds of molecular subunits in DNA? Which two form the "backbone" of DNA strands?

20.82. Explain how DNA and RNA differ (a) in molecular composition, (b) in structure, and (c) in function.

20.83. What kind of intermolecular force holds together the strands of DNA in the double-helix configuration?

20.84. DNA is a highly charged polyanion. If a solution of DNA is heated, the DNA will separate into individual strands, a process called denaturation. If the salt concentration of the solution is increased, the temperature at which denaturation occurs increases. Suggest a reason why.

Problems

20.85. Draw the structure of adenosine 5′-monophosphate, one of the four ribonucleotides in a strand of RNA.

20.86. Draw the structure of deoxythymidine 5′-monophosphate, one of the four nucleotides in a strand of DNA.

20.87. In the replication of DNA, a segment of an original strand has the sequence T-C-G.
a. What is the sequence of the opposite strand?
b. Draw the structure of this section of the double helix, clearly showing all hydrogen bonds.

20.88. If the sequence of one strand of DNA is 5′ ATTGCCA 3′, what is the sequence (in the 5′ to 3′ direction) of the other strand?

Additional Problems

20.89. **Salsa** Salsa has antibacterial properties because it contains dodecenal (Figure P20.89), a compound found in the cilantro used to make salsa.

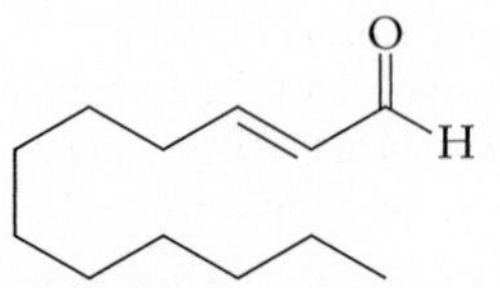

O, H — Dodecenal

FIGURE P20.89

a. How many carbon atoms are in dodecenal?
b. What functional groups are present in dodecenal?
c. What types of isomerism are possible in dodecenal?

20.90. **Turmeric** Turmeric is commonly used as a spice in Indian and Southeast Asian dishes. Turmeric contains a high concentration of curcumin (Figure P20.90), a potential anticancer drug and a possible treatment for cystic fibrosis.
a. Are the substituents on the C=C double bonds highlighted in red in cis or trans configurations?
b. Draw two other stereoisomers of this compound.

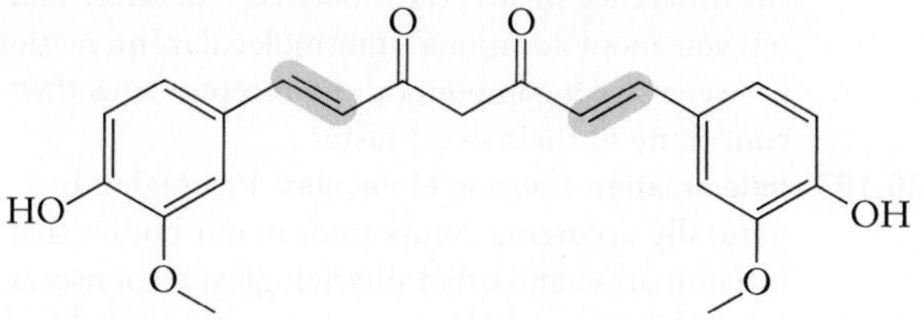

Curcumin

FIGURE P20.90

20.91. **Fat Substitutes** Olestra is a calorie-free fat substitute. The core of the olestra molecule (Figure P20.91) is a disaccharide that has reacted with a carboxylic acid; this results in the conversion of hydroxyl groups on the disaccharide into the depicted structure.
a. What is the name of the disaccharide core of the olestra molecule?
b. What functional group has replaced the hydroxyl groups on the disaccharide?
c. What is the formula of the carboxylic acid used to make olestra?

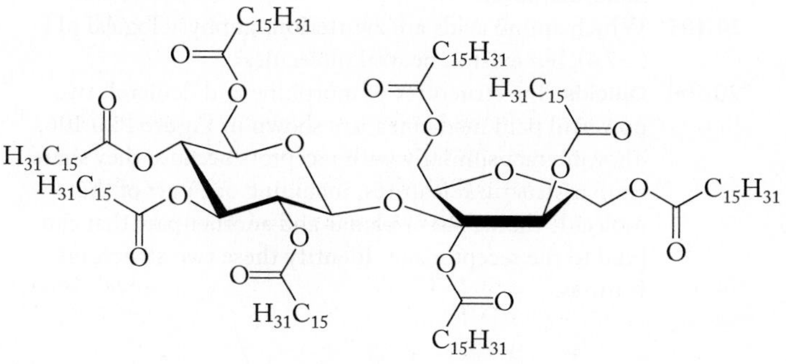

Olestra

FIGURE P20.91

20.92. When scientists at the University of California, Santa Cruz, directed UV radiation at an ice crystal containing methanol, ammonia, and hydrogen cyanide, three amino acids (glycine, alanine, and serine) were detected among the products of photochemical reactions. The formation of these amino acids suggests that they may also be synthesized in comets approaching the Sun (and Earth). Determine the standard free-energy change of the hypothetical formation of glycine in comets, using standard free energies of formation of the reactants and products in this reaction [ΔG°_f for HCN(g) is +125 kJ/mol and for solid glycine is −368.4 kJ/mol; other ΔG°_f values are in Appendix 4].

$$CH_3OH(\ell) + HCN(g) + H_2O(\ell) \rightarrow H_2NCH_2COOH(s) + H_2(g)$$

20.93. Homocysteine (Figure P20.93) is formed during the metabolism of amino acids. A mutation in some people's genes leads to high concentrations of homocysteine in the blood and a consequent increase in their risk of heart disease and incidence of bone fractures in old age.
a. What is the structural difference between homocysteine and cysteine?
b. Cysteine is a chiral compound. Is homocysteine chiral?

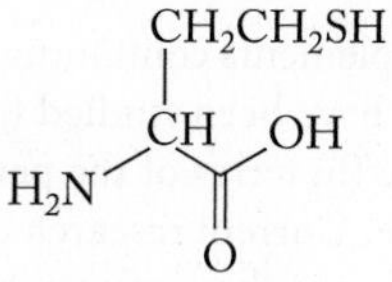

Homocysteine

FIGURE P20.93

20.94. **Amino Acids in Comets** Some scientists believe life on Earth can be traced to amino acids and other molecules brought to Earth by comets and meteorites. In 2004, a new class of amino acids called diamino acids (Figure P20.94) were found in the Murchison meteorite.
a. Which of these diamino acids is/are *not* an α-amino acid?
b. Which of these diamino acids is/are chiral?

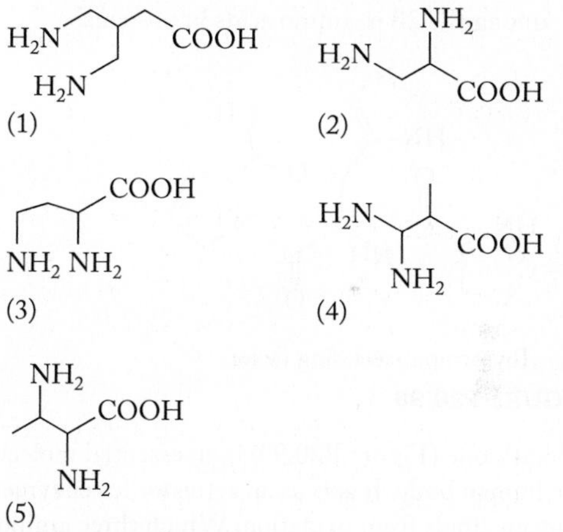

(1) (2) (3) (4) (5)

FIGURE P20.94

20.95. **Jamaican Fruit** Ackee, the national fruit of Jamaica, is a staple in many Jamaican diets. Unfortunately, a potentially fatal sickness known as Jamaican vomiting disease is caused by the consumption of unripe ackee fruit, which contains the amino acid hypoglycin (Figure P20.95). Is hypoglycin an α-amino acid?

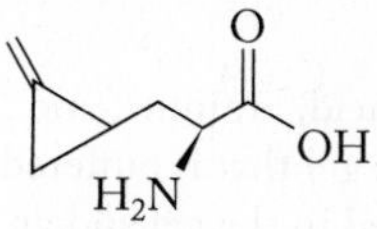

Hypoglycin

FIGURE P20.95

*20.96. In Section 20.2 we discussed the rotation of a beam of plane-polarized light when it passes through a solution containing an optically active molecule. Equimolar solutions of α-glucose and β-glucose rotate plane-polarized light by +112° and +18.7°, respectively. If these two solutions are then mixed and allowed to reach equilibrium, the solution then rotates the polarized light by +53.4°. Calculate the percent glucose in the α and β forms in this solution.

20.97. Amino Acid Supplements Supplements containing BCAAs (branched-chain amino acids) have been studied for their effects on athletic performance (in terms of the perception of exertion) and mental fatigue. Current research does not support that they have either of these effects, but they are used medically to slow muscle wasting in bedridden patients. BCAAs are amino acids having hydrocarbon side chains with a branch, a carbon atom bound to more than two other carbon atoms. Consult Table 20.2 and draw the structures of the BCAAs among the common amino acids.

20.98. In response to specific neural messages, the human hypothalamus may secrete several polypeptides, including the tripeptide shown in Figure P20.98.

a. Sketch the structures of the three amino acids that combine to make this tripeptide.

b. Which, if any, of the constituent amino acids are among the 20 α-amino acids in proteins?

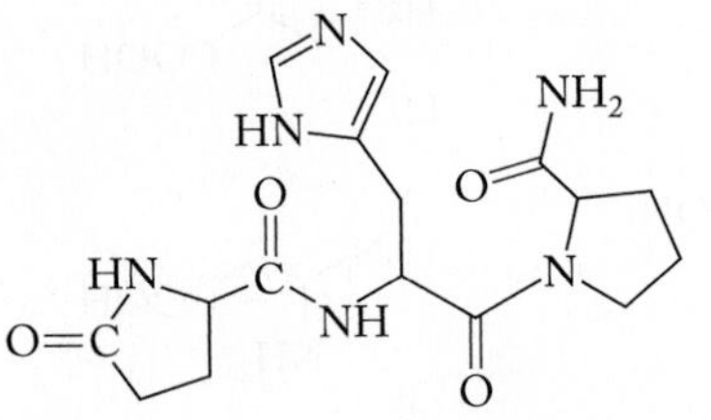

Thyrotropin-releasing factor

FIGURE P20.98

20.99. Glutathione (Figure P20.99) is an essential molecule in the human body. It acts as an activator for enzymes and protects lipids from oxidation. Which three amino acids combine to make glutathione?

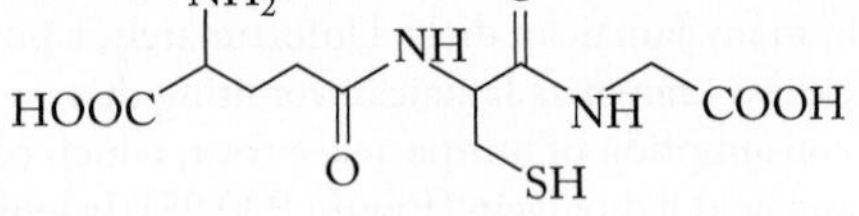

Glutathione

FIGURE P20.99

*20.100. Three amino acids—glutamic acid, arginine, and tryptophan—are dissolved in a gel that is buffered at a pH of 5.9. Two electrodes are placed in the gel and an electric current is applied.

a. Toward which electrode does each amino acid migrate?

b. Draw the forms of each amino acid present in the gel at a pH of 5.9.

***20.101.** Without doing the actual calculation, estimate the fuel values of leucine and isoleucine by considering average bond energies. Should the fuel values of the two amino acids be the same? Actual calorimetric measurements show that isoleucine has a lower fuel value than leucine. Explain why.

20.102. Sucralose (see Figure P20.9) is about 600 times sweeter than sucrose (see Figure 20.36). All substances that taste sweet have functional groups that form hydrogen bonds with "sweetness" receptor sites on the tongue. What does the difference in sweetness between sucralose and sucrose tell you about additional intermolecular interactions between sweet compounds and receptor sites that contribute to their sweet taste?

***20.103. Inflammation-Causing Molecules** Prostaglandins, naturally occurring compounds in our bodies that cause inflammation and other physiological responses, are formed from arachidonic acid, an unsaturated hydrocarbon containing four C=C double bonds and a carboxylic acid functional group. The stereoisomer containing all cis double bonds is shown in Figure P20.103. How many stereoisomers other than this one are possible? Draw the isomer containing all trans double bonds.

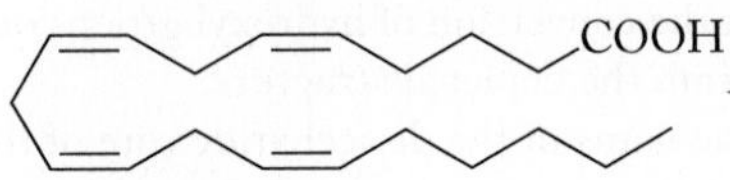

Arachidonic acid

FIGURE P20.103

20.104. A newspaper article contains the wording, "Made primarily by the liver, cholesterol begins with tiny pieces of sugar . . ." What does this statement mean at the molecular level?

20.105. Which amino acids are zwitterions at physiological pH (~7.4), but exist as neutral molecules?

*20.106. **Opioids** The structures of morphine and demerol, two powerful pain medicines, are shown in Figure P20.106. They interact similarly with receptors because they share similar structural features, including one part of the molecule that is flat or planar and another part that can bind to the receptor site. Identify these two structural features.

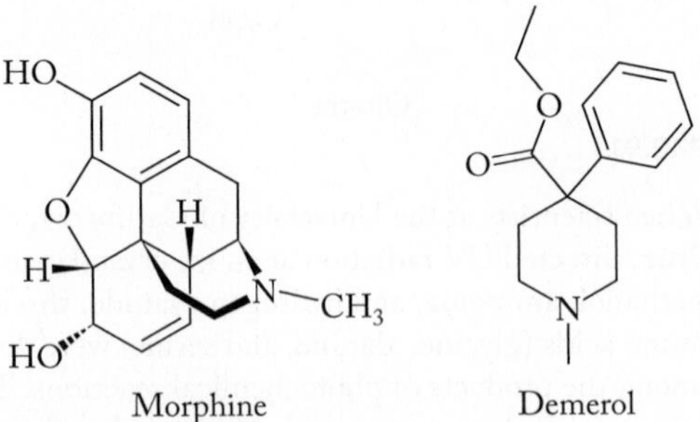

FIGURE P20.106

***20.107. Antihistamines** Figure P20.107 shows the structure of the histamine molecule that causes sneezing and itching in allergy sufferers. Antihistamines are a class of molecules that reduce allergy symptoms by preferentially binding to the same receptor sites as histamines. What structural features would permit these molecules to bind to the same receptor?

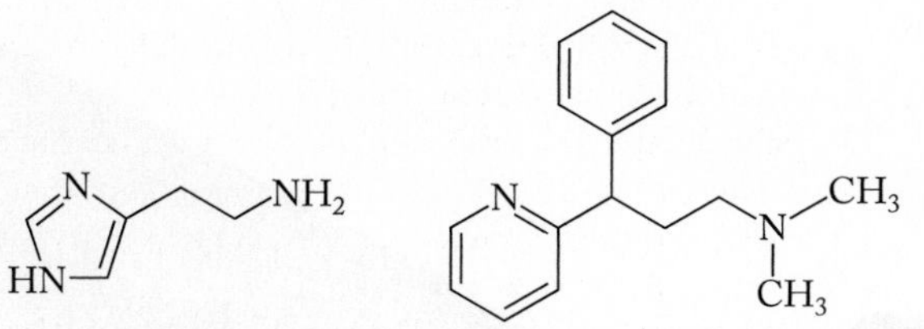

Histamine Antihistamine

FIGURE P20.107

***20.108. Pain Relievers** The structures of aspirin, acetaminophen, and ibuprofen are shown in Figure P20.108. What structural features do these molecules share?

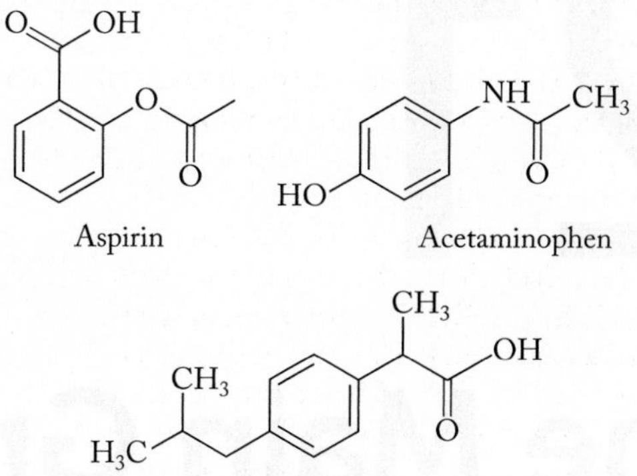

Aspirin Acetaminophen

Ibuprofen

FIGURE P20.108

21

The Main Group Elements

Life and the Periodic Table

DIETARY SUPPLEMENTS Calcium is included in many vitamins and dietary supplements to promote healthy bones and teeth.

PARTICULATE **REVIEW**

Nitrogen, Sulfur, and Phosphorus: One, Two, or All Three?

In Chapter 21 we survey the roles of main group elements in living organisms. In addition to carbon, hydrogen, and oxygen, atoms of nitrogen, sulfur, and phosphorus are fundamental building blocks for important biological molecules in plants and animals.

- Which of these structures contains sulfur atoms? Where are the sulfur atoms?
- Which two structures contain phosphorus atoms? Where are the phosphorus atoms?
- All three structures contain nitrogen atoms. List the functional group(s) that contains nitrogen atoms in each structure.

(Review Chapter 20 if you need help.)

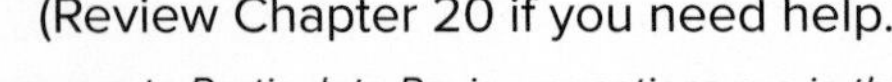

(Answers to Particulate Review questions are in the back of the book.)

PARTICULATE **PREVIEW**

Charge versus Size: Selective Transport

Shown here are representations of four essential cations involved in selective ion transport through channels in cell membranes: Ca^{2+}, H_3O^+, K^+, and Na^+. The size of each cation is listed. As you read Chapter 21, look for ideas that will help you answer these questions:

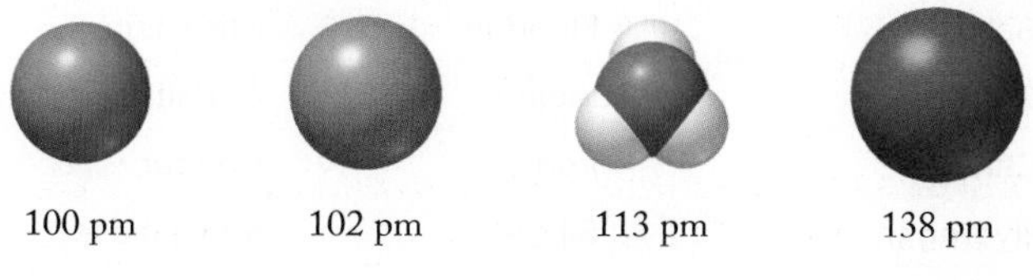

100 pm

102 pm

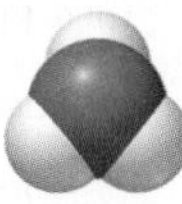
113 pm

138 pm

- Identify each of the four ions pictured.
- The Na^+ channel is selective, allowing only Na^+ cations and one of the other three cations to pass through it. Which other ion—Ca^{2+}, H_3O^+, or K^+—is most likely to undergo selective transport through the sodium channel?
- Which of the other three ions, if any, is likely to pass through the K^+ channel?

Learning Outcomes

LO1 Distinguish between essential and nonessential elements and between major, trace, and ultratrace elements

LO2 Describe the pathways and calculate the free energy and electrochemical potential driving ion transport across cell membranes
Sample Exercise 21.1

LO3 Balance equations, draw structures, and carry out calculations relevant to the behavior of major, trace, and ultratrace elements and their compounds *in vivo*
Sample Exercises 21.2, 21.3, 21.4

LO4 Describe the function of the essential and nonessential group 14–16 elements in the human body

LO5 Describe the health hazards posed by some radioactive isotopes of main group elements and the use of others in medical diagnosis and therapy
Sample Exercise 21.5

LO6 Describe how compounds containing main group elements are used in therapy and in other biomedical applications

21.1 Main Group Elements and Human Health

How many of the elements in the periodic table are in the human body? Moreover, which of them are important to the health of humans?

Roughly one-third of the 90 naturally occurring elements have an identifiable role in human health and in organisms in general. **Essential elements** are defined as those that have a beneficial physiological effect, including those whose absence impairs functioning of the organism (**Table 21.1**). Some—such as carbon, hydrogen, oxygen, nitrogen, sulfur, and phosphorus—are the principal constituents of all plants and animals. The alkali metal cations Na^+ and K^+ act as charge carriers, maintain osmotic pressure, and transmit nerve impulses. The alkaline earth cation Mg^{2+} is important in photosynthesis, and Ca^{2+} ions are components of bones and teeth. Chloride ions balance the charge of Na^+ and K^+ ions to maintain electrical neutrality in living cells. Other main group elements, such as iodine and selenium, are required in tiny amounts in our bodies to regulate metabolism and as components of the enzymes that catalyze biochemical reactions.

TABLE 21.1 Essential Elements Found in the Human Body

Major (>1 mg/g of body mass)	Trace (1–1000 μg/g of body mass)	Ultratrace (<1 μg/g of body mass)
Calcium	Fluorine	Chromium
Carbon	Iodine	Cobalt
Chlorine	Iron	Copper
Hydrogen	Silicon	Manganese
Magnesium	Zinc	Molybdenum
Nitrogen		Nickel
Oxygen		Selenium
Phosphorus		Vanadium
Potassium		
Sodium		
Sulfur		

The essential elements are further classified as **major, trace**, or **ultratrace essential elements**. Major essential elements are present in gram quantities in the human body and are required in large amounts in our diet. Almost all foods are rich in compounds containing carbon, hydrogen, oxygen, nitrogen, sulfur, and phosphorus. Salt is perhaps the most familiar dietary source of sodium and chloride ions, although both are ubiquitous in food. Vegetables such as broccoli and Brussels sprouts and fruits such as bananas are rich in potassium. Calcium is found in dairy products and is often added to orange juice as a dietary supplement.

Many **nonessential elements** are also present in the body but have no known function (**Table 21.2**). Some are useful in medicine as either diagnostic tools or therapeutic agents. Some radioactive isotopes may be used as imaging agents for organs and tumors, for example, whereas others are used to treat disease. Drugs containing lithium ions are used to treat depression. Compounds of bismuth act as mild antibacterial agents for treatment of diarrhea, and antimony compounds represent one of the few options for patients suffering from the tropical disease leishmaniasis, caused by protozoan parasites.

In some cases, the presence of a nonessential element has a **stimulatory effect**, which means that the consumption of small amounts of the element causes increased activity or growth in an organism. The effect may be beneficial or not, and often the mechanism of the effect is not understood. For example, small amounts of the nonessential element antimony promote growth in some mammals when added to their diets. Nonessential elements, and even toxic elements, are often incorporated into our bodies because their chemical properties are similar to those of an essential element. For example, Rb^+ ions are retained by the human body because they are similar to K^+ ions in size, charge, and chemistry, making rubidium the most abundant nonessential element in humans.

Oxygen, in the form of O_2 gas, occurs in the body in elemental form; oxygen is also incorporated into many compounds, such as H_2O, and many ions, such as HCO_3^-. When we speak of any element in the body other than oxygen, however, we usually refer to an ion or compound containing that element rather than the pure element. For example, when we describe calcium as an essential element, we are referring to calcium ions, Ca^{2+}, not calcium metal.

Table 21.3 compares the elemental compositions of the human body, the universe, Earth's crust, and seawater. The composition of our bodies most closely

TABLE 21.2 Nonessential Elements Found in the Human Body

Stimulatory	Unknown Role	No Role
Boron	Antimony	Barium
Titanium	Arsenic	Bromine
		Cesium
		Germanium
		Rubidium
		Strontium

TABLE 21.3 Comparative Composition[a] of the Universe, Earth's Crust, Seawater, and the Human Body

Element	Universe (%)	Crust (%)	Seawater (%)	Human Body (%)
Hydrogen	91	0.22	66	63
Oxygen	0.57	47	33	25.5
Carbon	0.021	0.019	0.0014	9.5
Nitrogen	0.042			1.4
Calcium		3.5	0.006	0.31
Phosphorus				0.22
Chlorine			0.33	0.03
Potassium		2.5	0.006	0.06
Sulfur	0.001	0.034	0.017	0.05
Sodium		2.5	0.28	0.01
Magnesium	0.002	2.2	0.033	0.01
Helium	9.1			
Silicon	0.003	28		
Aluminum		7.9		
Neon	0.003			
Iron	0.002	6.2		
Bromine			0.0005	
Titanium		0.46		
All other elements	<0.1	<0.1	<0.1	<0.1

[a]Compositions are expressed as the percentage of the total number of atoms. Because of rounding, the totals do not equal exactly 100%.

essential element an element present in tissue, blood, or other body fluids that has a physiological function.

major essential element an essential element present in the body in average concentrations greater than 1 mg of element per gram of body mass, resulting in gram quantities in the human body.

trace essential element an essential element present in the body in average concentrations between 1 and 1000 μg of element per gram of body mass.

ultratrace essential element an essential element present in the body in average concentrations less than 1 μg of element per gram of body mass.

nonessential element an element present in humans that has no known function.

stimulatory effect increased activity, growth, or other biological response to the presence of a nonessential element.

TABLE 21.4 Dietary Reference Intake (DRI) and Recommended Dietary Allowance (RDA) for Selected Essential Elements[a]

Element	DRI	RDA
Calcium	1000 mg	1200 mg
Chlorine	2300 mg	2300 mg
Chromium	25–35 μg	35 μg
Copper	900 μg	900 μg
Fluorine	3–4 mg	4 mg
Iodine	150 μg	150 μg
Iron	8–18 mg	18 mg
Magnesium	420 mg	320–400 mg
Manganese	1.8–2.3 mg	2–5 mg
Molybdenum	45 μg	45 μg
Phosphorus	700 mg	700 mg
Potassium	4700 mg	4700 mg
Selenium	55 μg	55 μg
Sodium	1500 mg	1500 mg
Zinc	8–11 mg	11 mg

[a]DRI and RDA values in milligrams or micrograms per day from the U.S. Department of Agriculture (2009) and from the Council on Responsible Nutrition (CRN) for 19- to 30-year-olds.

resembles the composition of seawater. The match would be even closer if it were not for the biological processes in the sea that remove essential elements such as nitrogen and phosphorus and that store others in solid structures like the $CaCO_3$ that makes up corals and mollusk shells.

Our diet should supply us with sufficient quantities of all essential elements. In the United States and Canada, these quantities are called *dietary reference intake* (DRI) values. They are based on the recommendations of the Food and Nutrition Board of the National Academy of Sciences and are frequently updated in response to research. For many essential elements, DRI values have replaced the *recommended dietary allowance* (RDA) values you may be familiar with from labels on food and vitamin packages (**Figure 21.1**). **Table 21.4** compares the DRI and RDA values for several major, trace, and ultratrace essential elements.

Some elements, such as radon, beryllium, and lead, are toxic. As described in Section 19.6, inhaled radon gas poses serious health hazards when it undergoes α decay inside the body. Beryllium toxicity is most often encountered in industrial settings where beryllium-contaminated dust is inhaled; the Be^{2+} ion replaces Mg^{2+} in the body, where it inhibits Mg^{2+}-catalyzed RNA and DNA synthesis in cells. Lead ions, Pb^{2+}, are incorporated into teeth and bones because they are similar to Ca^{2+} ions in size and charge. Lead also interferes with the functioning of enzymes that require calcium ions and thereby causes chronic neurological problems and blood-based disorders, especially in children. This issue became a major news story early in 2016 when it was discovered that a budget-cutting decision to switch drinking-water sources in Flint, Michigan, had exposed as many as 8000 children under age 6 to unsafe levels of lead. In June 2014, the city's source of water was changed from Lake Huron to the Flint River. The river water proved more corrosive than the lake water and leached lead from old water pipes, causing lead levels to rise to more than 100 ppb; the U.S. Environmental Protection Agency's action level for lead in drinking water is 15 ppb. Unsafe levels of lead have been identified in other cities with aging water systems.

In this chapter we review the periodic properties of the main group elements and survey the roles of selected elements in the human body, as well as their importance to good health. At the same time, we call on the knowledge and skills

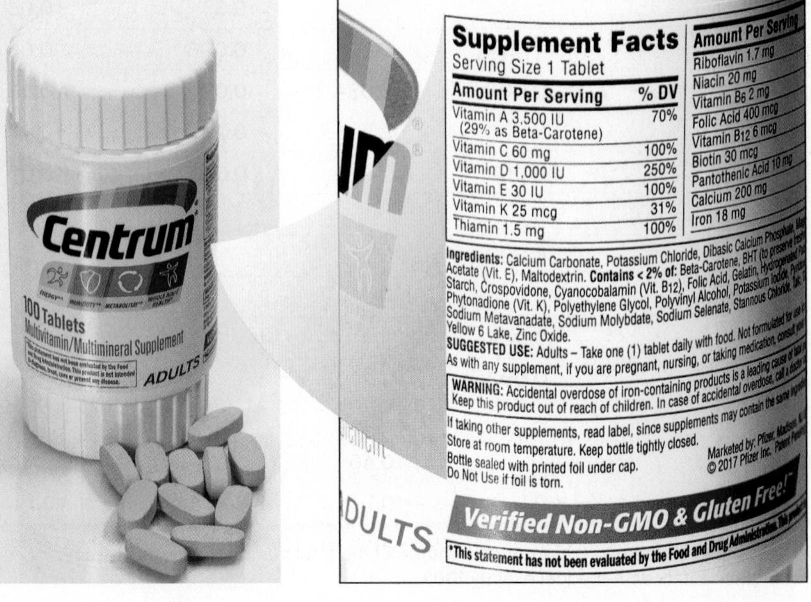

FIGURE 21.1 The labels on multivitamin supplements may not list DRI or RDA values but rather *% daily values* (DVs). Daily values are based on RDA or DRI values, but there can be inconsistencies, particularly among the ultratrace essential elements.

you have acquired in your study of general chemistry to solve problems that link concepts from prior chapters to the central question of this chapter: What are the roles of the main group elements in the chemistry of life? The roles of transition metal ions such as Fe^{2+}, Fe^{3+}, and Zn^{2+} in biology are addressed in Chapter 22.

21.2 Periodic Properties of Main Group Elements

The main group or representative elements are found in groups 1, 2, and 13–18 in the periodic table. These eight groups, comprising 44 naturally occurring elements, include five elements for which no stable isotopes are known: Fr, Ra, Po, At, and Rn. Relatively little is known about the chemistry of francium or astatine, both of which are exceedingly rare. Estimates indicate that the outermost kilometer of Earth's crust contains at most 44 mg of At and only 15 g of Fr. They have no proven uses, so we will largely ignore them in our discussions. Before turning to the role of the main group elements in life, it is useful to review what we have already learned about the physical and chemical properties of these elements and to look for periodic trends in these properties.

CONNECTION The origin of some periodic trends is discussed in Sections 7.10 to 7.12.

The atomic radii of the main group elements increase as we descend a group and increase right to left across a period (**Figure 21.2a**). In other words, the largest elements are found in the lower left-hand corner of the periodic table. The first ionization energies and electronegativities of the main group elements increase as we ascend a group and increase left to right across a period (**Figure 21.2b**). An overall trend toward more negative electron affinities is seen left to right across a period, although several anomalies are also observed. Each of these periodic trends reflects the changes in the effective nuclear charge and shielding of the valence electrons by core electrons as atomic numbers increase and inner shells fill with electrons.

A survey of the main group elements reveals a variety of physical properties. The alkali metal and alkaline earth elements are all solids, whereas all the group 18 elements (the noble gases) are gases at standard temperature and pressure. All group 17 elements (the halogens) exist as diatomic molecules. Group 17 is also the only group that contains elements in all three phases of matter at room temperature: bromine is one of two liquid elements in the periodic table, iodine is a volatile solid, and the remaining elements are gases. Groups 13–16 include seven elements classified as semimetals (B, Si, Ge, As, Sb, Te, and At), as well as elements with metallic properties (Al, Ga, In, Tl, Sn, Pb, Bi, and Po). The remaining elements in these groups behave as nonmetals with the properties described in Chapter 2: they are gases (N_2, O_2) or brittle solids that are poor conductors of electricity (C, P, S, and Se). For groups 13–16, metallic properties increase down a group.

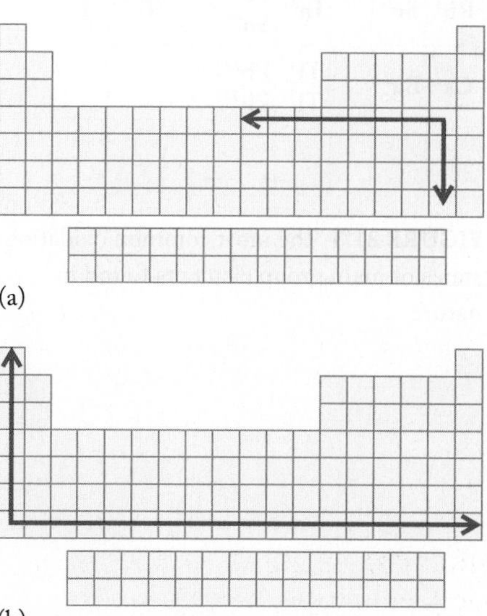

FIGURE 21.2 The blue arrows point in the direction of (a) increasing atomic radii and (b) increasing first ionization energies for the main group elements.

The melting and boiling points of the metallic and semimetallic main group elements decrease down groups 1 and 2, but this trend is reversed for the nonmetals in groups 17 and 18 (**Table 21.5**). These trends can be understood in terms of the different types of forces holding the atoms together. The increase in the size of the metallic elements when descending groups 1 and 2 leads to weaker metallic bonds and lower boiling points. In groups 17 and 18, however, increasing size leads to greater London dispersion forces as the polarizability of the atoms increases. The trends in melting points for groups 13–16 do not fit a clear pattern, in part because the properties of the elements change from nonmetallic to metallic down these groups.

TABLE 21.5 Summary of Periodic Trends for the Main Group Elements

Property	Group 1	Group 2	Group 13	Group 14	Group 15	Group 16	Group 17	Group 18
Melting point	Decreases down the group	Decreases down the group	No single trend	No single trend	No single trend	No single trend	Increases down the group	Increases down the group
Boiling point	Decreases down the group	Decreases down the group	No single trend	Decreases down the group	No single trend	No single trend	Increases down the group	Increases down the group

CONNECTION Ionic compounds were defined in Section 2.6, and organic compounds were introduced in Section 2.8. Covalent bonding was discussed in detail in Chapters 8 and 9.

CONNECTION The organization of the periodic table was introduced in Chapter 2. Periodic properties were discussed in Chapter 7, and the stability of nuclei was addressed in Chapter 19.

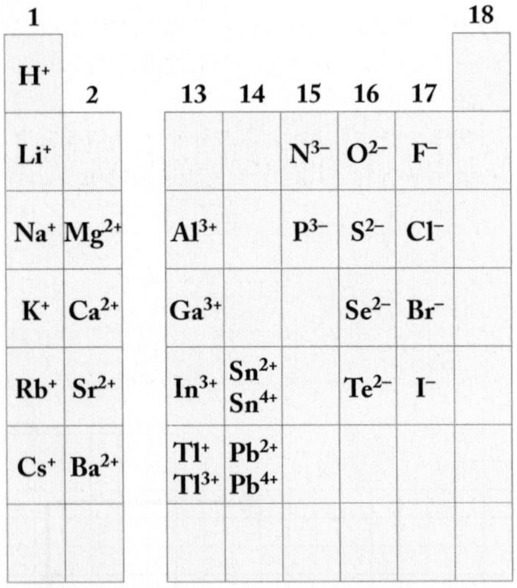

1	2	13	14	15	16	17	18
H^+							
Li^+				N^{3-}	O^{2-}	F^-	
Na^+	Mg^{2+}	Al^{3+}		P^{3-}	S^{2-}	Cl^-	
K^+	Ca^{2+}	Ga^{3+}			Se^{2-}	Br^-	
Rb^+	Sr^{2+}	In^{3+}	Sn^{2+} Sn^{4+}		Te^{2-}	I^-	
Cs^+	Ba^{2+}	Tl^+ Tl^{3+}	Pb^{2+} Pb^{4+}				

FIGURE 21.3 The most common oxidation states of main group elements found in nature.

Of the main group elements, only hydrogen, carbon, nitrogen, oxygen, sulfur, and the six noble gases exist in nature in elemental form. The remaining elements are found exclusively in ionic or covalent compounds. The group 1 and 2 elements readily lose their valence electrons, forming 1+ and 2+ cations, respectively, which can react with group 17 anions to produce familiar ionic compounds such as NaCl and KI (**Figure 21.3**). Halogen compounds of groups 13–17 generally contain covalent bonds; however, the heavier group 13 and 14 elements form insoluble ionic salts such as $PbCl_2$ and TlCl. The metallic elements in groups 1, 2, and 13 react with oxygen to yield oxides, including K_2O, CaO (quicklime, used in the manufacture of steel), and Al_2O_3 (alumina, used in orthodontics).

The lighter elements of groups 13–17 tend to form covalent bonds rather than ionic bonds. The result is a library of more than 150 million organic compounds, covalently bonded substances composed primarily of carbon, nitrogen, hydrogen, and oxygen. More than 95% of the new compounds registered in a given year are classified as organic compounds.

Hydrogen, the smallest and lightest element, is difficult to classify. Most periodic tables include hydrogen in group 1 on the basis of its electron configuration of a half-filled *s* orbital and on the dissociation of acids to protons and anions, analogous to the dissolution of alkali metal salts to 1+ cations and anions. Hydrogen also has a complete valence shell after gaining an electron to form a hydride ion, H^-, which is isoelectronic with He, a noble gas. Compounds called metal hydrides form between hydrogen and group 1, 2, or 13 metals, and they behave as salts containing a metal cation and a hydride anion. However, hydrogen is most commonly covalently bonded in compounds to oxygen (as water, H_2O) or to another group 14–16 element.

21.3 Major Essential Elements

The 11 elements shown in red in **Figure 21.4** and listed in the first column of Table 21.1 are the major essential elements. Together they account for more than 99% of the mass of the human body. Oxygen is the most abundant element by mass, followed by carbon and hydrogen. Although life depends on the presence of elemental oxygen in the form of O_2 gas, much of the oxygen in our bodies is combined with hydrogen in water molecules.

The most abundant elements in the human body include seven nonmetals: C, H, O, Cl, S, P, and N. They are the building blocks for most of the body's molecular compounds and its principal polyatomic ions, HCO_3^-, SO_4^{2-}, and $H_2PO_4^-$, which are dissolved in body fluids. The average concentrations of the four major metals in the human body—Ca^{2+}, K^+, Na^+, and Mg^{2+}—are listed in **Table 21.6**. In this section we explore some of the roles that sodium, potassium, magnesium, calcium, chlorine, nitrogen, phosphorus, and sulfur play in the biochemistry of the human body. As we do, we revisit several of the chemical principles discussed in earlier chapters.

ion channel a group of helical proteins that penetrate cell membranes and allow selective transport of ions.

Sodium and Potassium

Regulated concentrations of sodium and potassium ions are crucial to cell function. For some people, consuming too much Na^+ has been linked to hypertension (high blood pressure). To maintain a constant concentration of these two alkali metal ions in body fluids, the ions must be able to move into and out of cells. As noted in Section 20.6, the membrane surrounding a typical cell is a lipid bilayer, with polar groups containing phosphate groups on the two surfaces of the cell membrane and nonpolar fatty acids oriented toward the interior of the membrane. Direct diffusion of Na^+ and K^+ through the lipid bilayer is difficult because these polar cations do not dissolve in the nonpolar interior.

As **Figure 21.5** shows, the cell membrane is pierced by **ion channels**, which are groups of protein complexes that allow selective transport of ions. The ion channels control which ions pass through the membrane on the basis of the size and charge of the ion as well as the shape of the protein. For example, the protein of the potassium ion channel has amino acids oriented in such a fashion that favorable ion–dipole interactions occur only for ions with the radius of a K^+ ion (138 pm) but not for Na^+ (102 pm) or any other cation. The sodium ion channel is also selective, excluding K^+ and Ca^{2+} even though the radii of Na^+ and Ca^{2+} (100 pm) differ by only 2 pm. Another difference between the Na^+ and K^+ channels is the ability of H_3O^+ (hydronium ion, radius 113 pm) to pass through Na^+ channels but not K^+ channels.

Living organisms also contain oxygen-rich molecules such as the Lewis base nonactin (**Figure 21.6**) that bond to Ca^{2+}, K^+, Na^+, and Mg^{2+} ions (Lewis acids) through strong ion–dipole forces. The resulting complex ions consist of a polar, charged alkali metal ion encapsulated in a nonpolar exterior. Because the complex ion has both a polar portion and a nonpolar portion, it does not require a channel for passage through a cell membrane. Instead, the complex ion carries its alkali metal cation through both the polar and nonpolar regions of the bilayer, providing an alternative to ion channels for the transport of these metal ions. We revisit the bonding between Lewis acids and bases in more detail in Chapter 22.

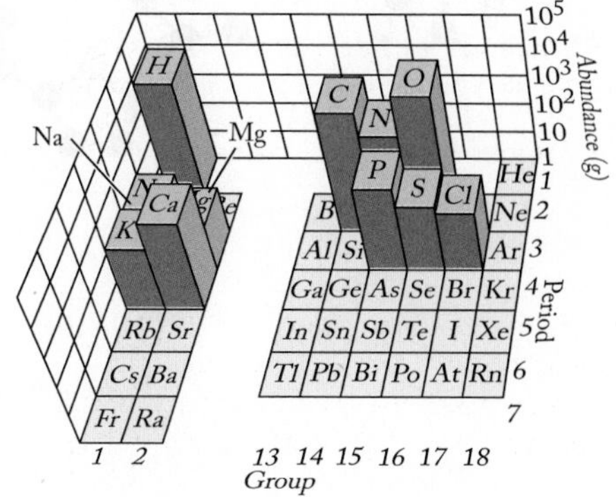

FIGURE 21.4 The 11 elements shown in red are the major essential elements. Their abundances range from 35 g of magnesium to 46 kg of oxygen in a 70-kg adult human.

TABLE 21.6 Average Concentration of Four Metallic Elements in the Human Body

Element	mg/g of Body Mass
Calcium	15.0
Potassium	2.0
Sodium	1.5
Magnesium	0.5

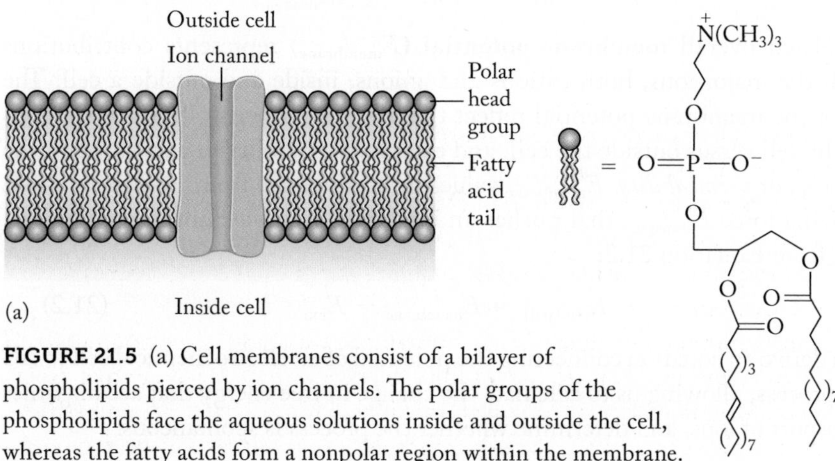

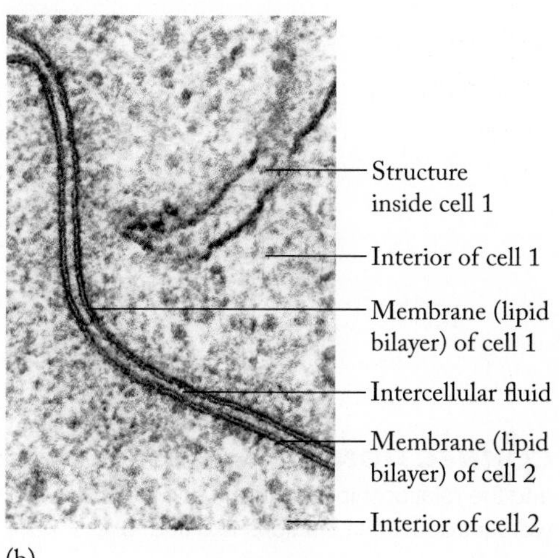

FIGURE 21.5 (a) Cell membranes consist of a bilayer of phospholipids pierced by ion channels. The polar groups of the phospholipids face the aqueous solutions inside and outside the cell, whereas the fatty acids form a nonpolar region within the membrane. (b) An electron micrograph of the membranes separating two adjacent cells.

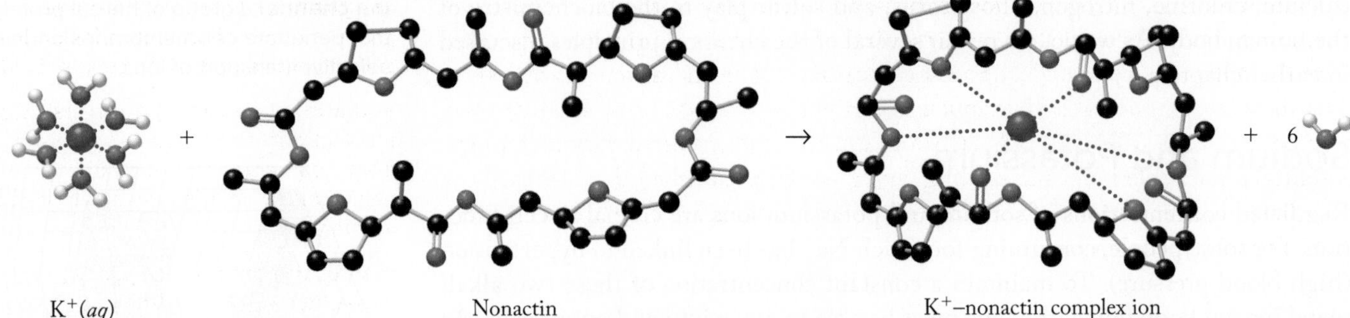

FIGURE 21.6 In living organisms, ligands such as nonactin can form a complex ion with any one of the four alkali metal or alkaline earth metal major essential ions and carry the ion directly through a cell membrane. An ion channel is *not* required in this transport pathway.

In addition to ion channels and diffusion of complex ions, alkali metal cations can be transported by a third mechanism, one involving Na^+–K^+ ion pumps. An **ion pump** is a system of membrane proteins that exchange ions inside the cell (e.g., Na^+) with those in the intercellular fluid (e.g., K^+). Unlike diffusion or transport through ion channels, transport via the Na^+–K^+ pump requires energy, which is provided by the hydrolysis of ATP to ADP. An example of how the Na^+–K^+ ion pump works is the response of a nerve cell to touch. Stimulation of the nerve cell causes Na^+ to flow into the cell and K^+ to flow out via ion channels; this two-way flow of ions produces the nerve impulse. The ion pump then "recharges" the system by pumping Na^+ out of the cell and K^+ into the cell so that another impulse can immediately be transmitted along the nerve.

CONNECTION In Section 10.3 we described how alkali metal cations dissolved in water are surrounded by six water molecules (a sphere of hydration). Each water molecule is oriented so that the oxygen atoms point toward the cation (see Figure 10.6). In Sections 16.5 and 16.6 we described this interaction as an example of a Lewis acid (cation, electron-pair acceptor) interacting with a Lewis base or ligand (water, electron-pair donor).

The unequal concentrations of Na^+, K^+, and other ions on opposite sides of a cell membrane result in an electrochemical **equilibrium potential** or **reversal potential (E_{ion})** for each ion that can be calculated using the following variation of the Nernst equation (introduced in Section 18.6), in which n now represents the charge on an ion and Q is now the ratio of the concentrations of the ion outside the cell to inside the cell:

$$E_{\text{ion}} = \frac{RT \ln Q}{nF} = \frac{RT}{nF} \ln\left(\frac{[\text{M}^{x+}_{\text{outside}}]}{[\text{M}^{x+}_{\text{inside}}]}\right) \tag{21.1}$$

A related overall **membrane potential ($E_{membrane}$)** represents contributions from all the major ions, both cations and anions, inside and outside a cell. The values of the membrane potential reflect the different concentrations of the ions inside the cell versus outside the cell, and their relative ability to pass through the membrane, or *permeability*. E_{membrane} values typically range from −50 to −70 mV. The driving force $E_{\text{transport}}$ that pushes an ion across a cell membrane can be calculated from Equation 21.2:

$$E_{\text{transport}} = E_{\text{membrane}} - E_{\text{ion}} \tag{21.2}$$

We can derive an equation connecting the potential of an electrochemical cell to ΔG for the process, allowing us to calculate the change in free energy that accompanies the transport of ions, and determine whether the process is spontaneous:

CONNECTION The Nernst equation and the relationship between E and ΔG were introduced in Sections 18.6 and 18.4, respectively.

$$\Delta G_{\text{transport}} = -nFE_{\text{transport}} \tag{21.3}$$

Ion channels, transporters, and pumps, and the ion concentrations they control, are a finely tuned system that regulates the membrane potential of cells. Rapidly multiplying normal cells such as fertilized eggs and differentiating stem cells have membrane potentials much less negative than the typical values and may be as low as 0 to −10 mV. It has recently been suggested that membrane potential plays a role in wound healing, and it is well established that some cancer cells have depolarized cell membranes and that their membrane potentials are also in the 0 to −10 mV range.

ion pump a system of membrane proteins that exchange ions inside the cell with those in the intercellular fluid.

equilibrium (reversal) potential, E_{ion} an electrochemical potential that results from a concentration gradient of a particular ion on opposite sides of a cell membrane.

membrane potential, $E_{membrane}$ a weighted average of the equilibrium (reversal) potentials of the major ions based on concentration and permeability of the individual ions.

SAMPLE EXERCISE 21.1 Calculating E_{ion} for Ion Transport **LO2**

Nerve cells contain structures called axons whose function relies on concentration differences on opposite sides of a membrane. The concentration of Na^+ inside a squid axon is 0.050 *M* (50 m*M*), in comparison with $[Na^+]$ = 0.440 *M* (440 m*M*) in the fluid surrounding the cell.

a. Calculate the equilibrium potential E_{ion} for the Na^+ ion at 280 K.

b. If the membrane potential $E_{membrane}$ = −0.050 V, is the transport of Na^+ from inside to outside the cell membrane spontaneous?

Collect, Organize, and Analyze We are given the concentrations of Na^+ on opposite sides of a cell membrane, and we are asked to calculate E_{ion} for Na^+ and to determine whether the transport across the membrane is spontaneous. Equation 21.1 allows us to calculate the equilibrium potential that arises from an unequal concentration of an ion on either side of a permeable membrane. If $[Na^+_{outside}] > [Na^+_{inside}]$, then the natural logarithm (ln) term in Equation 21.1 will be greater than 1 and E_{ion} will be positive. Using Equations 21.2 and 21.3, we can calculate the useful work done during transport: if ΔG is negative, the process is spontaneous; if positive, it is nonspontaneous.

Solve

a. Substitution into Equation 21.1 gives

$$E_{Na^+} = \frac{RT}{nF}\ln\frac{[Na^+_{outside}]}{[Na^+_{inside}]} = \frac{[8.314\ J/(\cancel{mol}\cdot\cancel{K})](280\ \cancel{K})}{(1/\cancel{mol})(96{,}500\ J/V)}\ln\left(\frac{440\ \cancel{mM}}{50\ \cancel{mM}}\right)$$
$$= 0.052\ V$$

b. Using Equation 21.2, we can calculate $E_{transport}$, which is the driving force available to push the ions across the membrane:

$$E_{transport} = E_{membrane} - E_{ion} = -0.050\ V - (0.052\ V) = -0.102\ V$$

Substituting −0.102 V into Equation 21.3:

$$\Delta G_{transport} = -nFE_{transport} = -(1/mol)(96{,}500\ J/\cancel{V})(-0.102\ \cancel{V}) = 9888\ J/mol$$
$$= 9.88\ kJ/mol$$

Because $\Delta G > 0$, the transport is not spontaneous.

Think About It The equilibrium potential for Na^+ under these conditions is positive, consistent with our prediction. The transport of Na^+ from the inside to the outside of the cell membrane is not spontaneous because it goes against the concentration gradient and the negative membrane potential.

Practice Exercise The concentration of K^+ inside a frog muscle cell is 124 m*M*, in comparison with $[K^+]$ = 2.3 m*M* in the fluid surrounding the cell. Calculate the equilibrium potential for the K^+ ion at 310 K. If the membrane potential $E_{membrane}$ is −73 mV, how much useful work must be done to transport K^+ into the cell at 310 K?

(Answers to Practice Exercises are in the back of the book.)

CONCEPT TEST

Do ion pumps represent spontaneous or nonspontaneous processes?

(Answers to Concept Tests are in the back of the book.)

Magnesium and Calcium

The biological roles of Mg^{2+} and Ca^{2+} are more varied than those of Na^+ and K^+. Calcium is a major component of teeth and bones. A prolonged deficiency of calcium can lead to osteoporosis (a disease characterized by low bone density), whereas high concentrations of calcium in muscle cells contribute to cramps. Most kidney stones are made of calcium oxalate or calcium phosphate. Magnesium deficiencies can reduce physical and mental capacity because of the role of Mg^{2+} in the transfer of phosphate groups to and from ATP (Mg^{2+} binds to ATP); slowing this transfer diminishes the amount of energy available to cells. The cellular concentrations of Mg^{2+} and Ca^{2+} are maintained by ion pumps.

Magnesium is a component of chlorophyll (**Figure 21.7**), which is one of several molecules used by plants to collect and capture light energy across the visible portion (400 to 700 nm) of the electromagnetic spectrum. Chlorophylls from different plants vary slightly in composition, but all of them contain magnesium coordinated to four nitrogen atoms. The presence of magnesium in chlorophyll does not account for the green color of the molecule, nor does it play a direct role in absorption of sunlight. The function of the Mg^{2+} ion is to orient the molecules in positions that allow energy to be transferred to the reaction centers where H_2O is consumed and O_2 is produced during photosynthesis. Carotene and related compounds are responsible for the orange colors (but not yellow or red) of autumn leaves on deciduous trees when chlorophyll production ceases. Mg^{2+} ions play important roles in ATP hydrolysis and ADP phosphorylation. The many Mg^{2+}-mediated ATP $\rightarrow$ ADP processes include transferring phosphate to glucose in the conversion of glucose to pyruvate and driving Na^+–K^+ ion pumps.

CONNECTION The catabolism of glucose was described in Sections 17.9 and 20.5.

To some extent, calcium ions can also mediate ATP hydrolysis, but these ions play other roles in the cell. They are necessary to trigger muscle contractions, for example—the calcium ions used for this purpose are stored in proteins. Recall that the action of Na^+–K^+ pumps is responsible for the generation of nerve impulses. One effect of nerve impulses is to trigger the release of Ca^{2+} ions from their storage proteins into the intracellular fluid. In a multistep process, muscle cells contract and relax as Ca^{2+} ions are released. Once the muscle action is complete, the ions are returned to their storage proteins in a process coupled to Mg^{2+}-mediated ATP hydrolysis.

CONNECTION The role of ATP and ADP in metabolism was described in Section 17.9.

Of Na, K, Mg, and Ca—the alkali metal and alkaline earth metal major essential elements (see Figure 21.4)—only calcium plays a major role in the formation of teeth and bones. Mammalian bones are a *composite material*, defined as a material containing a mixture of different substances. About 30% of dry bone mass is elastic protein fibers. The rest of the mass consists of calcium compounds, including the mineral hydroxyapatite, $Ca_5(PO_4)_3(OH)$, which is also a principal component of teeth. Hydroxyapatite crystals are bound to the protein fibers in bone through phosphate groups.

The shells of marine organisms are mostly calcium carbonate ($CaCO_3$) in a matrix of proteins and polysaccharides. Some magnesium is incorporated into the calcium carbonate outer shell of marine organisms that are capable of photosynthesis, such as algae and phytoplankton.

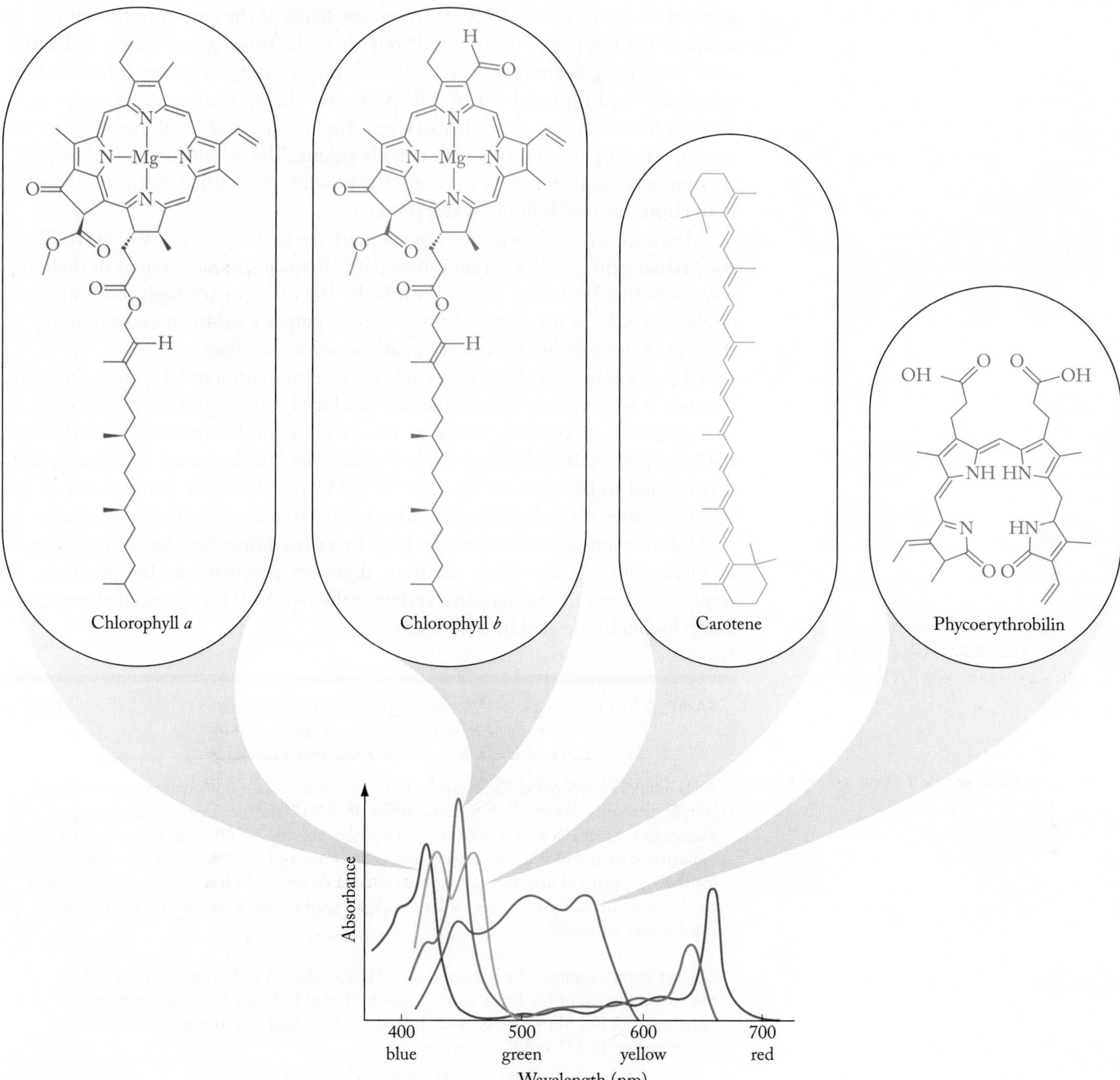

FIGURE 21.7 Photosynthetic bacteria, green plants, and algae use a variety of molecules to absorb all the visible wavelengths in sunlight. Among them, only chlorophylls contain magnesium and absorb blue-green and red-orange light. Carotene also absorbs in the blue-green region, whereas phycoerythrobilin absorbs a broad range of wavelengths from 400 to 600 nm.

Chlorine

Of all the halogens, only chlorine (as chloride ions) is present at high enough cocentrations to be considered a major essential element in humans. Chloride ions are the most abundant anions in the human body and are involved in many processes. The concentration of chloride ions in the human body (1.5 mg per gram of body mass) is slightly less than one-tenth of the concentration of Cl^- in seawater (19 mg per gram of water) but about 12 times greater than in Earth's crust (0.13 mg per gram of crust). Like the major essential cations, chloride ions are transported into and out of cells primarily via ion channels and ion pumps. To maintain electrical neutrality in a cell, the transport of alkali metal cations is accompanied by the transport of chloride anions. The *cotransport* of Na^+ and Cl^- is essential in kidney function, where the ions are reabsorbed by the body rather than eliminated with liquid waste products.

Malfunctioning chloride ion channels are the underlying cause of cystic fibrosis, a lethal genetic disease that causes patients to accumulate mucus in their airways such that breathing becomes difficult. The discovery of high concentrations of Na^+ and Cl^- in the sweat of cystic fibrosis patients led to an understanding of the role of chloride ion transport in patients with this disease.

Chloride ions also play a major role in the elimination of CO_2 from the body. Because it is nonpolar, carbon dioxide produced during glucose catabolism can pass from muscle cells (for example) into red blood cells, moving easily through the largely nonpolar cell membranes of these cells. Inside the red blood cells, CO_2 is converted to bicarbonate ion (HCO_3^-). When HCO_3^- is pumped out of the cell, Cl^- enters the cell through an ion channel to maintain charge balance.

Chloride ion concentrations are high in gastric juices because of the presence of hydrochloric acid, which catalyzes digestive processes in the stomach. In response to food in the digestive system, cells tap ATP for the needed energy to pump hydrochloric acid into the stomach.

SAMPLE EXERCISE 21.2 Calculating the Concentration of HCl in Stomach Acid — LO3

Acid reflux (sometimes called heartburn, though the heart is not involved) affects many people. It results from acid in the stomach leaking into the esophagus and causing discomfort. Stomach acid is primarily an aqueous solution of HCl. (a) Calculate the molarity of hydrochloric acid in gastric juice that has a pH of 0.80. (b) One treatment for the symptoms of acid reflux is to take an antacid tablet. What volume of gastric juice can be neutralized by a 750 mg tablet of calcium carbonate (a typical size for an over-the-counter antacid)?

Collect and Organize We are given the pH of a solution and are asked to calculate the concentration of HCl that corresponds to that pH. As we explained in Section 15.4 (Equation 15.10), pH = $-\log[H_3O^+]$. Hydrochloric acid is a strong acid and ionizes completely to H_3O^+ and Cl^- in water:

$$HCl(aq) + H_2O(\ell) \rightarrow H_3O^+(aq) + Cl^-(aq)$$

We are also asked to calculate the volume of HCl solution that can be neutralized by a 750 mg tablet of calcium carbonate ($CaCO_3$). We need to write a balanced chemical equation for the neutralization reaction.

Analyze The equation describing the ionization of hydrochloric acid indicates that 1 mole of H_3O^+ ions is formed for every mole of HCl present. The pH of gastric juice falls between 1 and 0, so $[H_3O^+]$ will be between 10^{-1} (= 0.1) M and 10^0 (= 1) M.

The net ionic equation for the neutralization reaction is

$$CaCO_3(s) + 2\ H_3O^+(aq) \rightarrow Ca^{2+}(aq) + CO_2(g) + 3\ H_2O(\ell)$$

This equation indicates that 2 moles of H_3O^+ are consumed for every mole of $CaCO_3$. We are told that the tablet size is typical of an antacid tablet, so common sense suggests that the volume of acid this tablet can neutralize will not be excessively large (greater than 1 L) or small (less than 10 mL): too large a tablet would be a waste of antacid, and too small a tablet would not relieve the symptoms.

Solve

a. Substitution into Equation 15.10 gives

$$pH = -\log[H_3O^+] = 0.80$$

We take the antilog of both sides to solve for $[H_3O^+]$:

$$[H_3O^+] = 10^{-0.80} = 0.16\ M\ H_3O^+$$

Therefore, the concentration of HCl is

$$0.16\ M\ H_3O^+ \times \frac{1\ \text{mol HCl}}{1\ \text{mol } H_3O^+} = 0.16\ M\ \text{HCl}$$

b. First, we calculate the number of moles of $CaCO_3$ present in 750 mg:

$$0.750\ \cancel{\text{g } CaCO_3} \times \frac{1\ \text{mol } CaCO_3}{100.09\ \cancel{\text{g } CaCO_3}} = 7.49 \times 10^{-3}\ \text{mol } CaCO_3$$

Next, we use the stoichiometry of the neutralization reaction to calculate the volume of 0.16 M HCl this quantity of $CaCO_3$ can neutralize:

$$7.49 \times 10^{-3}\ \cancel{\text{mol } CaCO_3} \times \frac{2\ \cancel{\text{mol } H_3O^+}}{1\ \cancel{\text{mol } CaCO_3}} \times \frac{1\ \text{L}}{0.16\ \cancel{\text{mol } H_3O^+}} = 9.36 \times 10^{-2}\ \text{L}$$

$$= 94\ \text{mL of } 0.16\ M\ \text{HCl}$$

Think About It A concentration of 0.16 M seems reasonable because it is indeed within the range of values predicted for a solution with pH <1. The volume of 0.16 M acid that a 750 mg tablet of $CaCO_3$ can neutralize is also reasonable; 94 mL represents about 3 ounces of gastric juice.

Practice Exercise Calculate the pH of a solution prepared by mixing 10.0 mL of 0.160 M HCl with 15.0 mL of water. How much antacid containing 4.00×10^2 mg of $Mg(OH)_2$ in 5.00 mL of water is needed to neutralize this volume of acid?

CONCEPT TEST

Taking an antacid tablet is often sufficient to treat an occasional case of mild acid reflux. Another remedy is a drug such as Prilosec, which inhibits a cell's proton pumps by binding to the site of the pump and disabling it for more than 24 hours. Is the equilibrium constant for the binding of a proton pump inhibitor likely to be less than or greater than 1?

Nitrogen

Nitrogen is a major essential element found primarily in proteins but also in DNA and RNA. Nitrogen is available in the atmosphere as N_2, and soil and water contain nitrate ions, but neither of these forms of nitrogen can be directly incorporated into amino acids, the building blocks of proteins. The biosynthesis of amino acids requires ammonia or ammonium ions. For example, glycine (NH_2CH_2COOH), the simplest amino acid, is formed by reaction of CO_2 and NH_3 in the presence of the appropriate enzyme. Interconversion of nitrogen-containing compounds in the environment is described by the nitrogen cycle (**Figure 21.8**). Certain bacteria use enzymes called *nitrogenases* to convert N_2 to NH_3. Plants convert NO_3^- ions to NO_2^- and then to NH_3 by using enzymes called *reductases*. Ultimately, the chemical reactions in these organisms begin a food chain that supplies the essential amino acids for human diets.

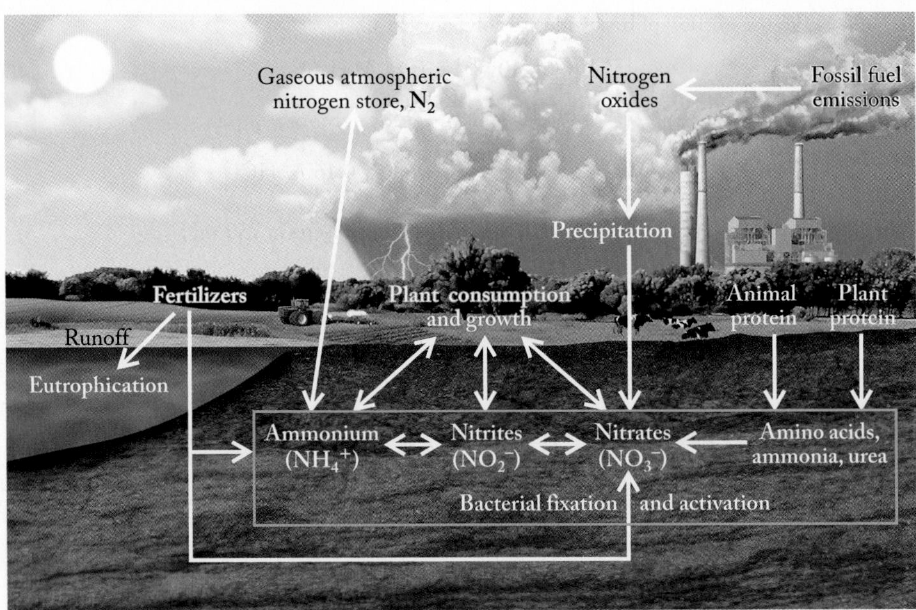

FIGURE 21.8 The nitrogen cycle. Enzymes interconvert the nitrogen-containing molecules and ions found in nature. Bacteria convert atmospheric nitrogen to ammonium ion, which is oxidized to nitrite (NO_2^-) and nitrate (NO_3^-) ions before being reduced back to N_2.

CONNECTION The Haber–Bosch process for making ammonia was discussed in Chapter 14.

The reactions in **Figure 21.9** are all redox reactions that illustrate the wide range of oxidation numbers found among nitrogen compounds in the nitrogen cycle. By definition, each atom in nitrogen gas (N_2) is assigned an oxidation number of zero. When N_2 is converted to NH_3, the oxidation number of N decreases to −3, a reduction. The oxidation numbers of N in nitrite (NO_2^-) and nitrate (NO_3^-) are +3 and +5, respectively, as a result of oxidation.

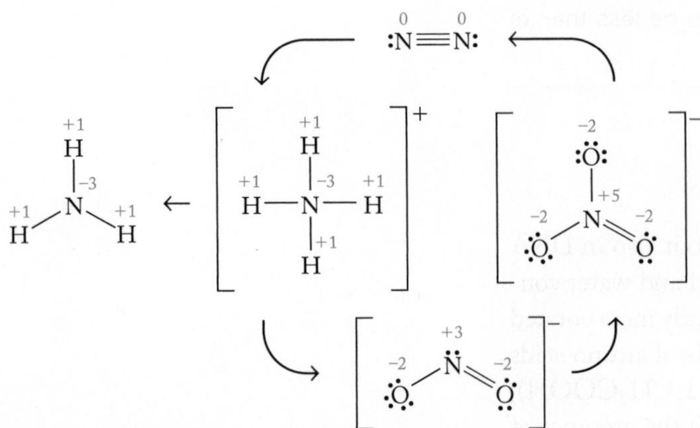

FIGURE 21.9 Nitrogen-containing molecules and ions in the nitrogen cycle exhibit oxidation numbers ranging from −3 to +5.

Modern agriculture depends heavily on the use of nitrogen-based fertilizers, which include ammonia and salts containing ammonium salts and nitrate ions. Most of the fertilizer used in the 21st century is derived from the high-temperature, high-pressure Haber–Bosch process. There is considerable research in progress to emulate the nitrogen cycle in Figure 21.8 and produce fertilizer more efficiently. In Chapter 13 we encountered a different cycle for nitrogen in the environment: the conversion of N_2 and O_2 to NO and NO_2 in the engines of automobiles. Dinitrogen monoxide (N_2O) is a greenhouse gas. The nitrogen atoms in these volatile nitrogen oxides are assigned oxidation numbers of +1, +2, and +4 for N_2O, NO, and NO_2, respectively.

SAMPLE EXERCISE 21.3 Writing a Balanced Chemical Equation Describing the Reaction of Nitrate Reductases **LO3**

Nitrate ion can be reduced to ammonia by enzymes called nitrate reductases. The first step is conversion of nitrate ion to nitrite ion. Assign oxidation numbers to the elements in these ions, and write a balanced net ionic equation for the conversion of nitrate to nitrite in a basic solution:

$$NO_3^-(aq) \rightarrow NO_2^-(aq)$$

Collect, Organize, and Analyze We need to assign oxidation numbers according to the guidelines in Section 4.8. We know that the oxidation numbers of nitrogen and oxygen in each polyatomic ion must add up to the charge on the ion. Oxygen in compounds usually has an oxidation number of −2.

Solve The oxidation number of nitrogen is unknown, so we call it x. The oxidation number of nitrogen in NO_3^- is

$$x + 3(-2) = -1$$
$$x = +5$$

The oxidation number of nitrogen in NO_2^- is

$$x + 2(-2) = -1$$
$$x = +3$$

In the preliminary expression:

$$NO_3^-(aq) \rightarrow NO_2^-(aq)$$

The number of nitrogen atoms is balanced; we balance the number of oxygen atoms by adding water:

$$NO_3^-(aq) \rightarrow NO_2^-(aq) + H_2O(\ell)$$

Then we balance hydrogen by adding hydrogen ions:

$$2\,H^+(aq) + NO_3^-(aq) \rightarrow NO_2^-(aq) + H_2O(\ell)$$

We balance charge by adding electrons:

$$2\,e^- + 2\,H^+(aq) + NO_3^-(aq) \rightarrow NO_2^-(aq) + H_2O(\ell)$$

We switch to a basic solution by adding just enough OH^- ions to neutralize any hydrogen ions; we must add the same number of OH^- ions to both sides of the equation:

$$2\,OH^-(aq) + 2\,e^- + 2\,H^+(aq) + NO_3^-(aq) \rightarrow NO_2^-(aq) + H_2O(\ell) + 2\,OH^-(aq)$$

The hydrogen ions combine with the hydroxide ions to form water, and we cancel the species that are the same on both sides:

$$2\,e^- + 2\,H_2O(\ell) + NO_3^-(aq) \rightarrow NO_2^-(aq) + \cancel{H_2O(\ell)} + 2\,OH^-(aq)$$

This gives us a final net ionic equation for the reduction half-reaction:

$$2\,e^- + H_2O(\ell) + NO_3^-(aq) \rightarrow NO_2^-(aq) + 2\,OH^-(aq)$$

Think About It Assigning oxidation numbers is a convenient way of identifying which element is reduced or oxidized in a half-reaction and of determining how many electrons are gained or lost. The balanced half-reaction confirms that electrons are added to nitrate to reduce it to nitrite ion.

Practice Exercise The reduction half-reaction catalyzed by one type of nitrogenase produces 1 mole of $H_2(g)$ for every 2 moles of $NH_4^+(aq)$ from molecular nitrogen under acidic conditions. Write a balanced equation for this half-reaction.

In humans and other mammals, excess nitrogen is converted to urea in the liver and excreted via the kidneys. Plants use urea as a source of ammonia by the action of *ureases* via the reaction:

$$H_2N{-}C({=}O){-}NH_2 + H_2O \rightarrow 2\ NH_3 + CO_2$$

Unlike reactions catalyzed by nitrogenases and nitrate reductases, the conversion of urea to ammonia and carbon dioxide is *not* a redox reaction. It is a hydrolysis reaction, similar to the reaction of nonmetal oxides with water described in Chapter 4. Acidic solutions are observed when NH_4^+, NO_2, and N_2O_5 dissolve in water:

$$NH_4^+(aq) + H_2O(\ell) \rightleftharpoons NH_3(aq) + H_3O^+(aq)$$

$$2\ NO_2(g) + H_2O(\ell) \rightarrow HNO_2(aq) + HNO_3(aq)$$

$$N_2O_5(g) + H_2O(\ell) \rightarrow 2\ HNO_3(aq)$$

Hydrolysis of nitrite ion, however, leads to weakly basic solutions:

$$NO_2^-(aq) + H_2O(\ell) \rightleftharpoons HNO_2(aq) + OH^-(aq)$$

Phosphorus and Sulfur

Phosphorus and sulfur are major essential elements found primarily in proteins and DNA, but they are also present in polyatomic anions prevalent in the environment. In comparison with the nitrogen cycle, the biological phosphorus cycle contains only a single major species: the phosphate ion (PO_4^{3-}) and its conjugate acids HPO_4^{2-} and $H_2PO_4^-$, along with phosphate esters. Reduction of phosphorus(V) in PO_4^{3-} to phosphine (PH_3) does occur in swamps. An oxygen-free environment is needed because PH_3 spontaneously ignites in humid air, yielding phosphoric acid:

$$PH_3(g) + 2\ O_2(g) \rightarrow H_3PO_4(\ell)$$

Gases analogous to those in the nitrogen cycle, such as NO and NO_2, are absent from the phosphorus cycle. Slow weathering of insoluble phosphate minerals by weak acids in soil introduces phosphate ion into the environment, where it is eventually taken up by plants. Some of this phosphate is incorporated into biominerals such as hydroxyapatite [$Ca_5(PO_4)_3(OH)$], as seen previously in our discussion of calcium.

CONCEPT TEST

Why are phosphates more likely to precipitate with cations from aqueous solution than are nitrates?

In the discussion of pH buffers (Section 16.3), we explained how phosphate ion is in equilibrium with its conjugate acid, HPO_4^{2-}, which hydrolyzes, in turn, to $H_2PO_4^-$:

$$PO_4^{3-}(aq) + H_2O(\ell) \rightleftharpoons HPO_4^{2-}(aq) + OH^-(aq)$$

$$HPO_4^{2-}(aq) + H_2O(\ell) \rightleftharpoons H_2PO_4^-(aq) + OH^-(aq)$$

Aqueous phosphate ion can be transported into cells, where it is incorporated into organic molecules such as ATP, ADP, glucose-6-phosphate, and nucleic acids (**Figure 21.10**). Notice the carbon–oxygen bond between the monosaccharide and the phosphate group in all four of the molecules in Figure 21.10. In glucose-6-phosphate and nucleic acids, this new bond forms through a condensation reaction of an –OH group on the sugar molecule and HPO_4^{2-} that produces water as a product. Adenosine diphosphate (ADP) is converted to ATP through a condensation reaction between HPO_4^{2-} and a phosphate group on ADP.

The processes in Figure 21.10 are all reversible; the P–O bond can be hydrolyzed, releasing HPO_4^{2-}. For ATP, this reaction is exothermic and provides the energy for many cellular processes. Given the importance of phosphorus to life, significant amounts of phosphates are used as fertilizer in agriculture. Agricultural runoff may stimulate rapid growth of algae in freshwater ponds and lakes. The explosive growth of algae can use up all the dissolved oxygen, killing other higher aquatic organisms.

The sulfur cycle in **Figure 21.11** illustrates the array of sulfur compounds found in the environment. We have already encountered volatile sulfur oxides, SO_2 and SO_3, in the context of acid rain on early Earth in Section 3.1. Like the nonmetal oxides of groups 14 and 15, SO_2 and SO_3 dissolve in water to produce

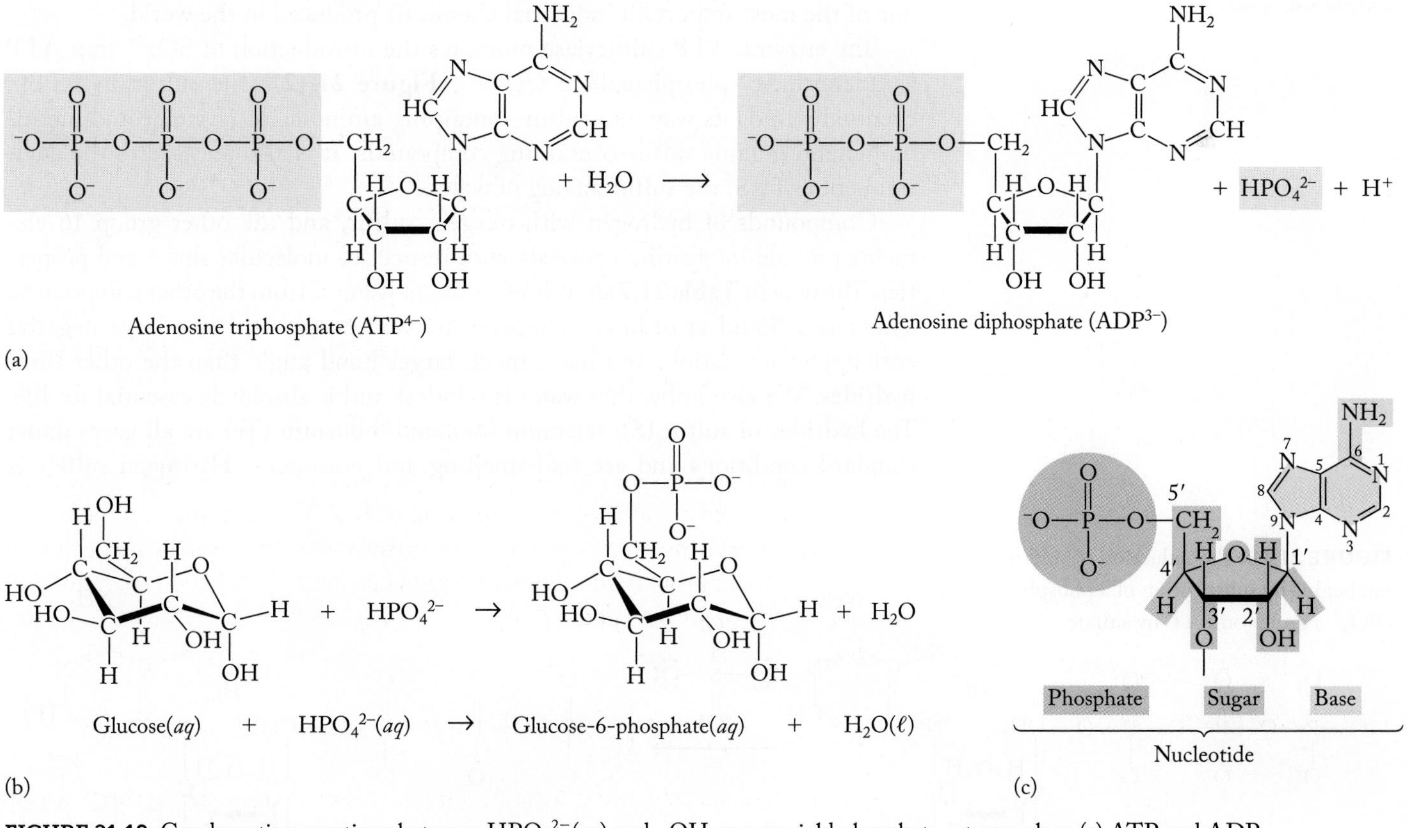

(a) Adenosine triphosphate (ATP^{4-}) + H_2O → Adenosine diphosphate (ADP^{3-}) + HPO_4^{2-} + H^+

(b) Glucose(aq) + $HPO_4^{2-}(aq)$ → Glucose-6-phosphate(aq) + $H_2O(\ell)$

(c) Phosphate Sugar Base — Nucleotide

FIGURE 21.10 Condensation reactions between $HPO_4^{2-}(aq)$ and –OH groups yield phosphate esters such as (a) ATP and ADP, (b) glucose-6-phosphate, and (c) nucleic acids.

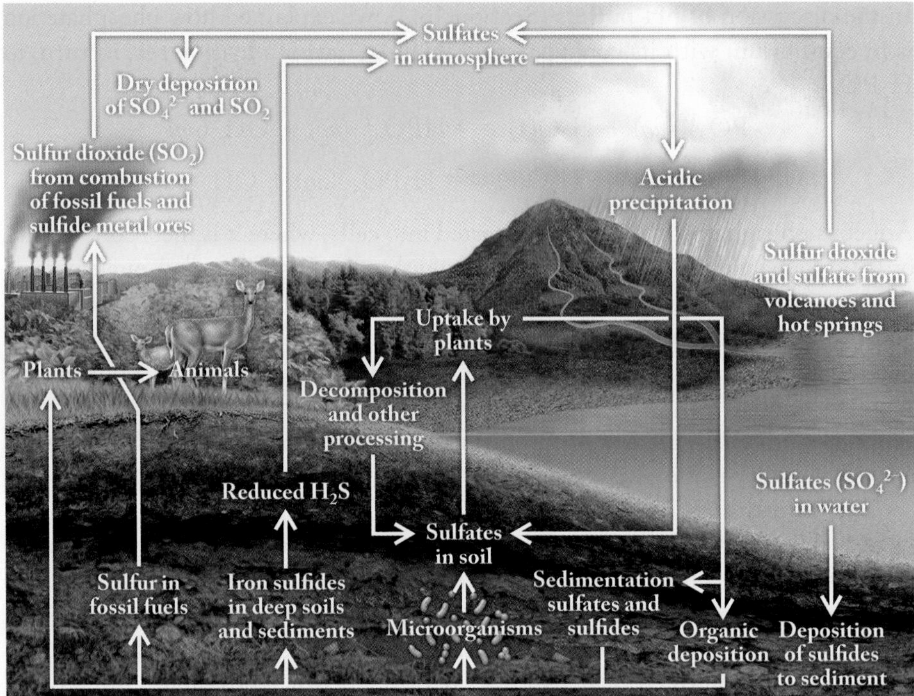

FIGURE 21.11 The key components in the sulfur cycle are sulfide ion (S^{2-}) in sediments and metal sulfides, sulfur oxides (SO_2 and SO_3), hydrogen sulfide (H_2S), and sulfate (SO_4^{2-}).

CONNECTION The role of nitrogen and sulfur oxides in acid rain is discussed in Chapters 4, 8, and 15.

the weak acid H_2SO_3 and the strong acid H_2SO_4, respectively. The sulfate ion produced by the dissociation of H_2SO_4 and H_2SO_3, the sulfate ions derived from minerals in soil, and the sulfate ions produced by the oxidation of H_2S are all absorbed by plants. Sulfuric acid, made by the reaction between SO_3 and water, is one of the most important industrial chemicals produced in the world.

The enzyme ATP sulfurylase promotes the introduction of SO_4^{2-} into ATP as adenosine-5′-phosphosulfate (APS^{2-}; **Figure 21.12**). The sulfur in APS^{2-} eventually finds its way into sulfur-containing amino acids (cysteine and methionine) and organic sulfur-containing compounds. It is also released to the environment as H_2S, the sulfur analog of water.

Compounds of hydrogen with oxygen, sulfur, and the other group 16 elements provide interesting contrasts with respect to molecular shape and properties. The data in **Table 21.7** show how different water is from the other compounds. Water is a liquid at ordinary temperatures and pressures, has a large negative enthalpy of formation, and has a much larger bond angle than the other three hydrides. We also know that water is odorless and is absolutely essential for life. The hydrides of sulfur (S), selenium (Se), and tellurium (Te) are all gases under standard conditions and are foul-smelling and poisonous. Hydrogen sulfide is

FIGURE 21.12 The reduction of sulfate ion begins by substitution of a phosphate (PO_4^{3-}) group on ATP by sulfate.

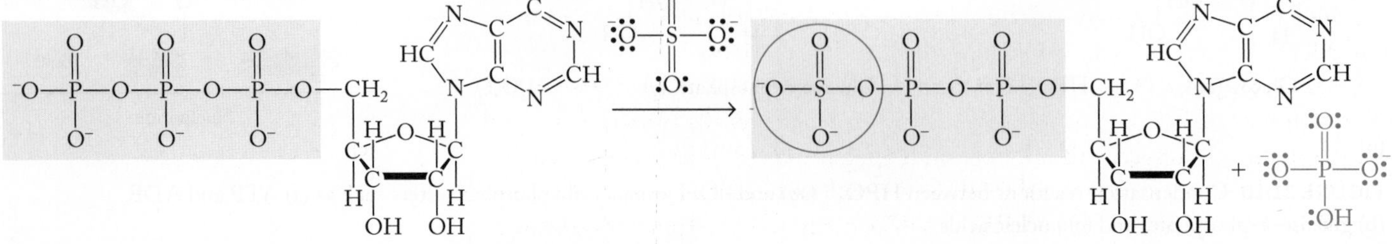

TABLE 21.7 Comparison of the Properties of Group 16 Dihydrides

Hydride[a]	Melting Point (°C)	Boiling Point (°C)	Bond Length (pm)	Bond Angle (degrees)	Enthalpy of Formation (kJ/mol)
H_2O	0	100	96	104.5	−285.8
H_2S	−86	−60	134	92	−20.17
H_2Se	−66	−41	146	91	73.0
H_2Te	−51	−4	169	90	99.6

[a]H_2Po is excluded; too little is known of its chemistry. Polonium has no stable isotopes and is present on Earth only in very small quantities.

responsible for the smell of rotten eggs. It is especially dangerous because it tends to very quickly fatigue the nasal sensory sites responsible for detecting it. This means that the intensity of the odor is a very poor indicator of the concentration of H_2S in the air. Headache and nausea begin at air concentrations of H_2S as low as 5 ppm, and exposure to 100 ppm leads to paralysis and may result in death.

Similarly, organic compounds containing sulfur have properties that differ from those of their oxygen-containing counterparts. Many of them also have characteristic odors. Methanol is an alcohol with the formula CH_3OH. It is a liquid at room temperature and has an odor usually described as slightly alcoholic. Methanethiol ($CH_3—SH$) is a gas at room temperature and has the pungent odor of rotten cabbage. It is produced in the intestinal tract of animals by the action of bacteria on proteins and is one of the sulfur compounds responsible for the characteristic aroma of a feedlot or a barnyard.

If we consider compounds with two carbon atoms, ethanol (CH_3CH_2OH; beverage-grade alcohol) is a liquid at room temperature. The corresponding sulfur compound is a very low-boiling liquid called ethanethiol, which has a penetrating and unpleasant odor, like very powerful green onions. The human nose can detect the presence of ethanethiol at levels as low as 1 ppb (part per billion) in the air. This gives rise to its use as an odorant in natural gas. Natural gas has no odor, and natural gas leaks are such enormous fire hazards that ethanethiol is added to natural gas streams to make even small leaks detectable.

If we rearrange the atoms in ethanol and ethanethiol, we produce two new compounds. For ethanol we get dimethyl ether ($CH_3—O—CH_3$), a colorless gas used in refrigeration systems. Its counterpart, dimethyl sulfide ($CH_3—S—CH_3$), is one of the compounds responsible for the "low-tide" smell of ocean shorelines.

Three of the sulfur compounds described—hydrogen sulfide, methanethiol, and dimethyl sulfide—are referred to as volatile sulfur compounds (VSCs) by dentists. They are produced by bacteria in the mouth and are the principal compounds responsible for bad breath. One of the reasons the odors of these compounds differ from those of their oxygen counterparts is that their molecular sizes and shapes are slightly different. Also, their polarities differ because of the electronegativity difference between oxygen (EN = 3.5) and sulfur (EN = 2.5). Recall that we discussed the importance of molecular shape in determining the extent of interaction of a compound with receptors in nasal membranes in Section 9.5. In part, the vast differences in odor and sensory detectability of these compounds are due to their shapes and electron distributions.

Many sulfur compounds have an odor. The characteristic and unpleasant smell of urine produced by some people after eating asparagus results from the inability of their bodies to convert sulfur compounds (**Figure 21.13**) into odor-free sulfate ions. Not all people can convert the sulfur compounds in asparagus to sulfate, and not all people can smell the odiferous sulfur compounds. Apparently, genetic

FIGURE 21.13 Structures of some of the volatile sulfur compounds responsible for the smell of "asparagus" urine. Compounds shown here have stronger (and more unpleasant) odors.

$H_3C—SH$ Methanethiol

$H_3C—S—S—CH_3$ Dimethyl disulfide

$H_3C—S—CH_2—S—CH_3$ Bis(methylthio)methane

FIGURE 21.14 The pungent smell of skunk spray is caused by butanethiol, $CH_3(CH_2)_3SH$.

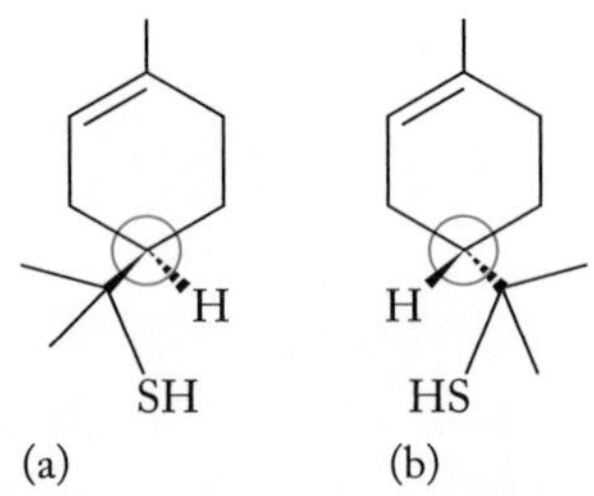

FIGURE 21.15 (a) This compound is responsible for the aroma of grapefruit. (b) If the orientation of the substituents at one of the carbon atoms in the molecule is switched, the resulting molecule has the same Lewis structure but no aroma at all. The carbon atoms circled in red identify the chiral carbon in each molecule.

differences determine how we metabolize these compounds and how well we can sense their odors.

The odor of skunk is caused mostly by butanethiol (**Figure 21.14**), and the odor of well-used athletic shoes is caused primarily by the presence of sulfur compounds produced by bacteria. Not all sulfur compounds have aromas as unpleasant as these, however. A compound with the formula $C_{10}H_{18}S$ is responsible for the aroma of grapefruit. If the orientation of the substituents at the carbon atom circled in red in the molecule in **Figure 21.15a** is switched, the resulting molecule has the same Lewis structure but no aroma at all (**Figure 21.15b**). The carbon atom circled in red in Figure 21.15 is a chiral center, and the two compounds represent an enantiomeric pair (see Section 20.2).

SAMPLE EXERCISE 21.4 Drawing Lewis Structures for Molecules in the Sulfur Cycle — LO3

Dimethyl sulfide (CH_3SCH_3) is one of the products of the sulfur cycle in Figure 21.11. In marine environments, dimethyl sulfide can be oxidized by bacteria to dimethyl sulfoxide [$(CH_3)_2SO$].

a. Draw Lewis structures for CH_3SCH_3 and $(CH_3)_2SO$, and determine the molecular geometry about the S atom in each.
b. Methanethiol (CH_3SH) is the simplest of the thiols and has a boiling point of 6°C. Dimethyl sulfide has a boiling point of 38°C. Explain the difference in boiling point between CH_3SH and CH_3SCH_3.
c. Which hybrid orbitals does sulfur use in bonding to carbon in CH_3SCH_3 and CH_3SH?

Collect and Organize We need to draw Lewis structures for two sulfur compounds, determine their geometry, and compare their boiling points. Guidelines for drawing Lewis structures were introduced in Section 8.2. We are also asked to describe the hybrid orbitals of S in these compounds. Hybrid orbitals were described in Section 9.4. The effect of structure on boiling points was discussed in Section 10.2.

Analyze The Lewis structure for a molecule depends on the total number of valence electrons available, distributed over the atoms so that each atom has a complete octet (except H, which has a duet). Atoms with $Z \geq 13$ may have an expanded octet if such an arrangement leads to lower formal charges on the atoms. The arrangement of the bonding and lone pairs on the central atom allow us to predict intermolecular forces such as dipole–dipole interactions, hydrogen bonds, and dispersion forces, all of which can affect the boiling point of a substance. We can account for observed molecular geometries by combining *s* and *p* orbitals to form hybrid atomic orbitals such as *sp*, sp^2, and sp^3.

Solve (a) CH_3SCH_3 has a total of 20 valence electrons: 6 from the S atom, 6 from the H atoms, and 8 from the C atoms. These electrons can be distributed in six C—H single bonds and two S—C single bonds, with 4 electrons remaining as 2 nonbonding pairs on S. $(CH_3)_2SO$ has an oxygen atom bonded to sulfur, which brings an additional 6 valence electrons to the molecule, giving $(CH_3)_2SO$ a total of 26 valence electrons. Sharing one of the S lone pairs with O will complete the octets of both S and O but will leave O with a formal charge of −1 and sulfur with a formal charge of +1. Forming a S═O double bond makes the formal charges on both S and O equal to zero but requires that sulfur have an expanded octet. The two structures for $(CH_3)_2SO$ represent resonance forms:

(b) CH_3SH has 14 valence electrons: 6 from the S atom, 4 from the H atoms, and 4 from the C atom. The electrons are distributed in three C—H single bonds, one C—S single bond, and one S—H single bond. As in CH_3SCH_3, there are 2 nonbonding pairs left on S. Both CH_3SCH_3 and CH_3SH contain a sulfur atom surrounded by 2 bonding pairs and 2 nonbonding pairs of electrons for a total of 4 electron pairs. The electron-pair geometry is tetrahedral, and the molecular geometry is bent:

Both CH_3SH and CH_3SCH_3 are polar as a result of their molecular geometry and experience dipole–dipole interactions. Both molecules interact through dispersion forces as well. One might expect that the stronger dipole–dipole forces in CH_3SH would lead to a higher boiling point than for CH_3SCH_3, but we observe the opposite. We conclude that the dispersion forces in CH_3SCH_3 have a greater effect than the dipole–dipole interactions in CH_3SH, leading to CH_3SCH_3 boiling at a temperature about 32°C higher than CH_3SH.

(c) The similar geometry for CH_3SCH_3 and CH_3SH, with each molecule's central sulfur atom having 2 bonded atoms and 2 lone pairs, is consistent with sp^3 hybrid orbitals on the sulfur atom in both cases.

Think About It Methanethiol and dimethyl sulfide differ only in that CH_3SCH_3 has a CH_3 group in place of a hydrogen atom. (This relationship between thiols and sulfides corresponds to the relationship between alcohols and ethers.) Sulfur does not expand its octet in these compounds because using the lone pairs on S to form multiple bonds is not needed. The boiling points of CH_3SCH_3 and CH_3SH reveal an important observation: many weaker bonds (dispersion forces) in CH_3SCH_3 can outweigh a few stronger forces (dipole–dipole forces) in CH_3SH.

Practice Exercise The nitrogen cycle in Figure 21.8 involves both neutral compounds such as NO_2 and polyatomic ions such as NO_2^-. Draw Lewis structures for both species and determine whether they have the same molecular geometry about nitrogen. Identify which hybrid orbitals contain nitrogen lone pairs in NO_2 and NO_2^-.

21.4 Trace and Ultratrace Essential Elements

Figure 21.16 shows the DRI values of four main group essential elements: silicon, selenium, fluorine, and iodine. Silicon, fluorine, and iodine are present in the body in average concentrations between 1 and 1000 μg of element per gram of body mass and are considered trace elements. Selenium is considered ultratrace, meaning that it is present in an average concentration of less than 1 μg of element per gram of body mass.

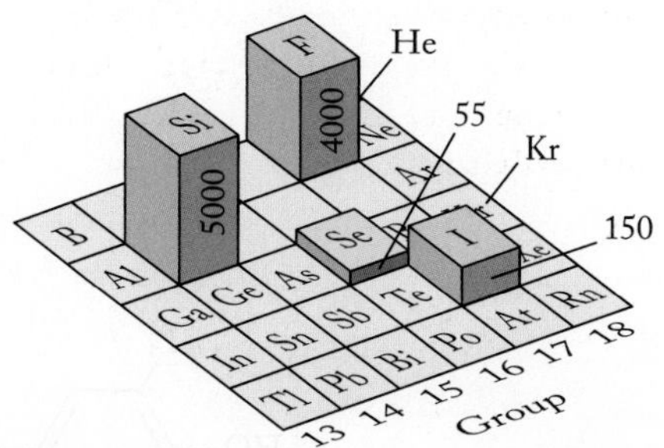

FIGURE 21.16 Silicon, fluorine, and iodine are trace essential elements, and selenium is an ultratrace essential element. The remaining labeled elements are nonessential. The vertical bars show DRI values for these elements in micrograms per day.

Selenium

The volatile selenium analog to water, H_2Se, is toxic; however, selenium is considered an ultratrace essential element. The average concentration of Se in the human body is 0.3 μg per gram of body mass. Mounting scientific evidence points to a need for a minimum daily dose of approximately 55 μg of selenium. The effects of selenium toxicity, however, are apparent in people who ingest more than 500 μg per day. Most of the selenium we need is obtained from selenium-rich produce such as garlic, mushrooms, and asparagus, or from fish. Selenium occurs in the body as the amino acid selenocysteine (**Figure 21.17**) and is incorporated into enzymes.

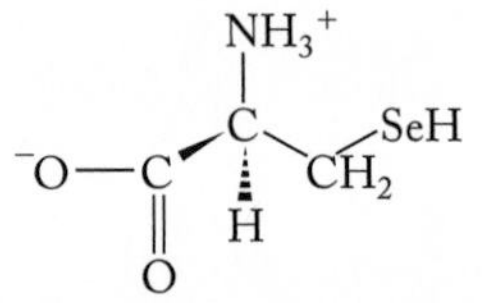

FIGURE 21.17 Selenocysteine is the selenium-containing analog of the amino acid cysteine. Much of the selenium in the human body is found in proteins containing selenocysteine.

Selenocysteine is an antioxidant. Our bodies need oxygen to survive, yet living in an oxygen-rich atmosphere can lead to the formation of potentially dangerous oxidizing agents in cells. For example, metabolism of fatty acids forms oxidizing agents called alkyl hydroperoxides, which can attack the lipid bilayer of cell membranes. It is believed that aging is related to the inability of the body to inhibit oxidative degradation of tissue. Selenocysteine participates in a series of reactions that result in the decomposition of these alkyl hydroperoxides.

Fluorine and Iodine

Fluorine is a trace essential element, and fluoride ions have significant benefits for dental health. Tooth enamel is composed of the mineral hydroxyapatite, $Ca_5(PO_4)_3(OH)$, which is essentially insoluble in water:

$$Ca_3(PO_4)_3(OH)(s) \rightleftharpoons Ca_5(PO_4)_3^{+}(aq) + OH^-(aq) \qquad K_{sp} \approx 2.4 \times 10^{-59}$$

When hydroxyapatite comes into contact with weak acids in your mouth, this equilibrium shifts to the right as the acid reacts with the hydroxide ions. This shift effectively increases the solubility of hydroxyapatite, so that your tooth enamel becomes pitted, and dental caries form. Fluoride ions reduce the likelihood of caries by displacing the OH^- ions in hydroxyapatite to form fluorapatite:

$$Ca_3(PO_4)_3(OH)(s) + F^-(aq) \rightleftharpoons Ca_5(PO_4)_3F(s) + OH^-(aq) \qquad K = 8.48$$

CONNECTION The solubility product (K_{sp}) was introduced in Section 16.8. Le Châtelier's principle was discussed in Section 14.7.

The solubility of fluorapatite depends less on pH than the solubility of hydroxyapatite, so changing tooth enamel to fluorapatite makes your teeth more resistant to decay. This is why toothpaste contains fluoride compounds and why fluoride is added to drinking water in many communities in North America and Europe.

Of all the trace essential elements, iodine may have the best-defined role in human health. The body concentrates iodide ions in the thyroid gland, where they are incorporated into two hormones—thyroxine and 3,5,3′-triiodothyronine (**Figure 21.18**)—whose role is to regulate energy production and use. The con-

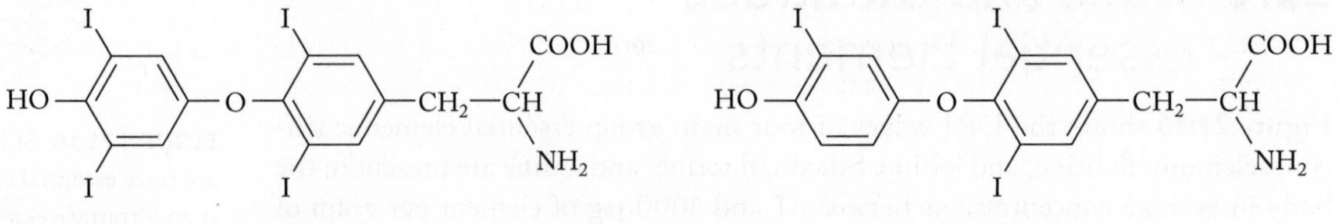

Thyroxine

3,5,3′-Triiodothyronine

FIGURE 21.18 Thyroxine and 3,5,3′-triiodothyronine, two iodine-containing hormones found in the thyroid gland, regulate metabolism.

version of thyroxine to 3,5,3′-triiodothyronine is catalyzed by selenocysteine-containing proteins. A deficiency of iodine or of either hormone can cause fatigue or feeling cold and can ultimately lead to an enlarged thyroid gland, a condition known as goiter. To help prevent iodine deficiency, table salt sold in the United States and many other countries is "iodized" with a small amount of sodium iodide. An excess of either hormone can cause a person to feel hot and is linked to Graves' disease, an autoimmune disease. The immune system in a patient with Graves' disease attacks the thyroid gland and causes it to overproduce the two hormones.

Silicon

The role of silicon in biological systems is less clear than for selenium and the halides. In mammals, a lack of the trace essential element silicon stunts growth. The presence of silicon as silicic acid, $Si(OH)_4$, is believed to reduce the toxicity of Al^{3+} ions in organisms by precipitating the aluminum as aluminosilicate minerals. Amorphous silica, SiO_2, is found in the exoskeletons of diatoms and in the cell membranes of some plants, such as the tips of stinging nettles.

21.5 Nonessential Elements

The ten elements listed in Table 21.2 are found in the human body but are classified as nonessential. In Section 21.5 we discuss how some of these elements may end up in our bodies, working our way from left to right across the periodic table.

Rubidium and Cesium

Rubidium is generally regarded as nonessential in humans, yet it is the 15th most abundant element in the body. It is believed that Rb^+ is retained by the body because of the similarity of its size and chemistry to that of K^+. Like the other cations of group 1, cesium ions (Cs^+) are also readily absorbed by the body. Cesium cations have no known function, although they can substitute for K^+ and interfere with potassium-dependent functions. Usually, the concentration of cesium in the environment is low, so exposure to Cs^+ is not a health concern. The nuclear accident at Chernobyl in 1986, however, released significant quantities of radioactive ^{137}Cs into the environment. The ability of Cs^+ to substitute for K^+ led to the incorporation of $^{137}Cs^+$ into plants, which rendered crops grown in the immediate area unfit for human consumption because of the radiation hazard posed by this long-lived ($t_{1/2} \approx 30$ yr) β emitter.

Strontium and Barium

Some single-celled organisms build exoskeletons made with $SrSO_4$ and $BaSO_4$, but the human body appears to have no use for Sr^{2+} and Ba^{2+} ions. These ions do find their way into human bones, where they replace Ca^{2+} ions. At the low concentrations of Sr^{2+} and Ba^{2+} that are typically present in the human body, these elements appear to be benign. As with radioactive ^{137}Cs, however, incorporation of ^{90}Sr ($t_{1/2} = 29$ yr) in bones can lead to leukemia. Atmospheric testing of nuclear weapons over the Pacific Ocean and in sparsely populated regions of the American West in the 1950s released ^{90}Sr into the environment. The full extent of the toxic effects of the fallout from these tests did not become apparent for several decades.

CONNECTION The biological effects of different types of nuclear radiation were described in Section 19.6.

CONNECTION We discussed the St. Louis Baby Tooth Survey and the effect of nuclear testing on children's teeth in Section 2.1.

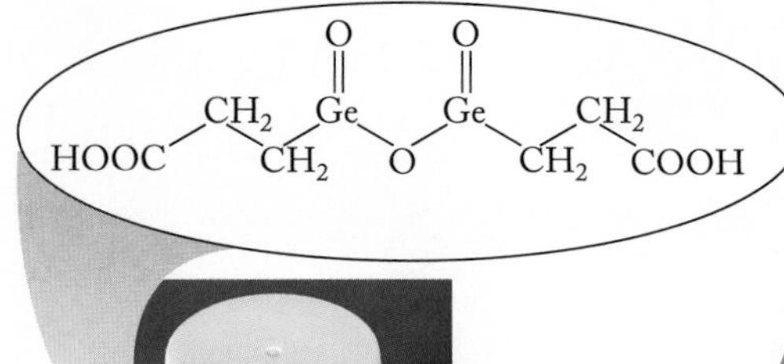

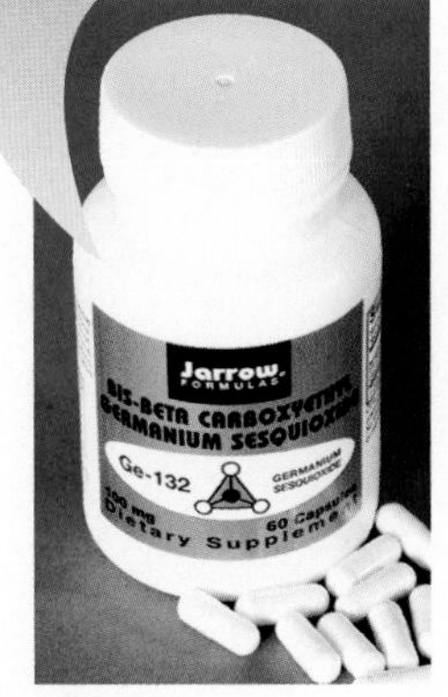

FIGURE 21.19 Bis(carboxyethyl) germanium sesquioxide has been sold as a nutrition supplement, but its benefits are not well established. The claims made for many dietary supplements remain controversial and have not been fully explored by scientists.

Germanium

Germanium is a nonessential element and is barely detectable in the human body. Bis(carboxyethyl)germanium sesquioxide (**Figure 21.19**) has been touted as a nutritional supplement, but its efficacy remains in doubt.

Antimony

The role of antimony is also poorly understood. Most antimony compounds are toxic because they cause liver damage. However, ultratrace amounts of antimony may have a stimulatory effect, and selected antimony compounds have been used medically as antiparasitic agents, as discussed in Section 21.6.

Bromine

Bromine has no known function in the human body but is consumed in foods such as grains, nuts, and fish in amounts ranging from 2 to 8 mg per day, leading to average concentrations of Br^- in blood of about 6 mg/L. Bromide ion has sedative and anticonvulsive properties but becomes toxic at concentrations around 100 mg/L, limiting its use to veterinary medicine. Bromide ion concentrations in seawater typically range from 65 to 80 mg/L. A select group of aquatic species can metabolize Br^- into bromomethane (CH_3Br) and other brominated organic compounds.

CONCEPT TEST

Looking at groups 1, 2, 14, 15, and 17, what periodic trend do you see in the location of the nonessential elements compared with the essential elements in the same group?

21.6 Elements for Diagnosis and Therapy

So far we have talked about the biological roles of several essential and nonessential main group elements found in our bodies. Some of these elements are also useful in diagnosing or treating diseases, as are some of the other elements in groups 1, 2, and 13–18 that we have not mentioned (**Figure 21.20**). In this

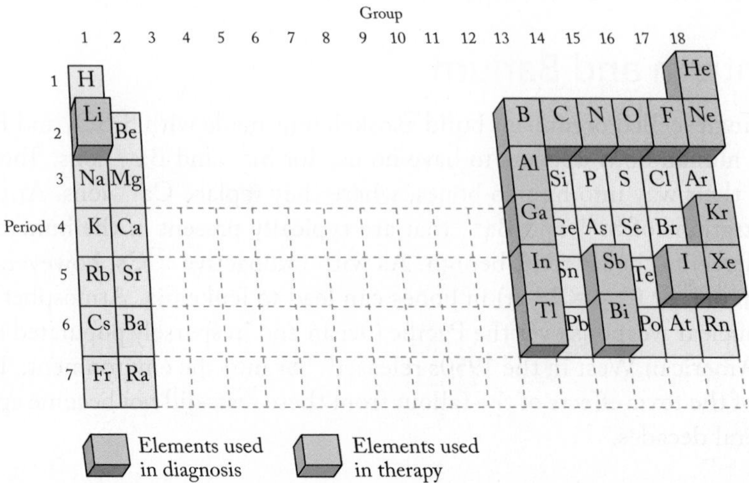

FIGURE 21.20 The elements shown in red are used in diagnostic imaging and those shown in green are used in therapy. Gallium is used in both diagnostic imaging and therapy.

section we describe some of the applications of radioactive isotopes in the diagnosis of diseases. We also explore how compounds of essential and nonessential elements have found application in the treatment of a wide variety of illnesses.

Any diagnostic or therapeutic compound that is injected intravenously must be sufficiently soluble in blood to be delivered to the target. While in transit, the compound must be stable enough not to undergo chemical reactions that result in its precipitation or rapid elimination from the body. A medicinal chemist can also take advantage of substances that occur naturally in the body, such as antibodies, to carry a diagnostic or therapeutic metal ion to its target. We focus here on the main group elements and their compounds (groups 1, 2, and 13–18). Medical applications of compounds containing the transition elements (groups 3–12) are described in Chapter 22.

Diagnostic Applications

Physicians in the 21st century have an array of imaging agents to help in diagnosing disease. Some methods use radionuclides with short half-lives that emit easily detectable gamma rays. Examples include the use of iodine-131 to image the thyroid gland (**Figure 21.21**) and of neutron-poor isotopes such as carbon-11 and fluorine-18 for positron emission tomography (PET). Not all imaging depends on radionuclides, however. In magnetic resonance imaging (MRI), for instance, which can diagnose soft-tissue injuries, stable isotopes of gadolinium (a lanthanide) are heavily used as contrast agents to enhance images despite their toxicity to kidneys.

Imaging with Radionuclides The radionuclides used in medicine have short half-lives to limit the patient's exposure to ionizing radiation. If the half-life is too short, however, the nuclide may either decay before it can be administered or not reach the target organ rapidly enough to provide an image. Emission of relatively low-energy γ rays is essential to preventing collateral tissue damage.

Nuclide selection is also governed by the toxicity of both the parent element and the daughter nuclide. The speed at which the imaging agent is eliminated from the body can help mitigate toxic effects. The cost and availability of a particular nuclide also factor into its usefulness in a clinical setting.

CONNECTION Chapter 19 gives a more detailed discussion of nuclear chemistry and nuclear medicine, including an assessment of the effects of different types of radiation on living tissue.

CONCEPT TEST

Why is it important to consider the nature of the decay products—α, β, or γ particles, or positrons—when choosing a radionuclide for medical imaging?

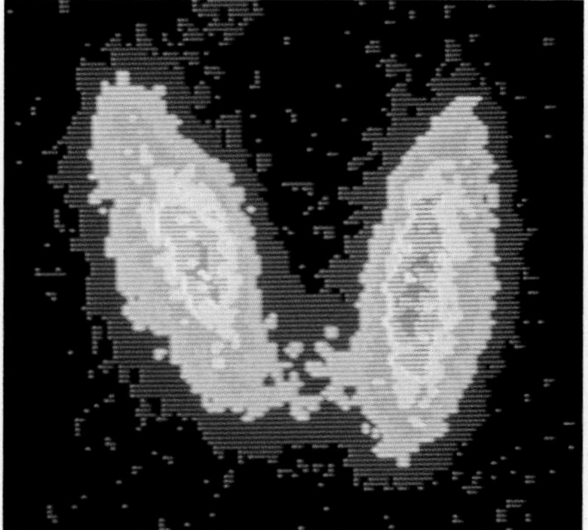

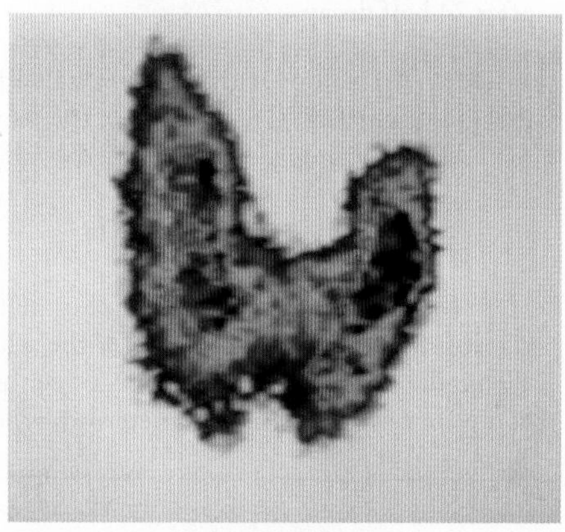

FIGURE 21.21 The γ radiation that accompanies the decay of iodine-131 can be used to image the two butterfly-shaped lobes of the thyroid gland.

Gallium, Indium, and Thallium Gallium-66, gallium-67, gallium-68, and indium-111 are used as imaging agents for tumors and leukemia. All four nuclides decay by electron capture, and the γ radiation emitted in this nuclear reaction produces the images. All three gallium isotopes also decay by positron emission, which makes compounds containing these isotopes attractive for positron-emission tomography. Their half-lives range from just over 1 h for gallium-68 to 78 h for gallium-67. The indium-111–containing compound Zevalin is currently used to treat some forms of non-Hodgkin's lymphoma. It is formed by reacting indium(III) chloride with the molecule in **Figure 21.22**. The indium 3+ ion acts as a Lewis acid and forms coordinate bonds of the type described in Section 16.6. Ibritumomab in Figure 21.22 is an antibody used to transport the radioactive indium to its target.

The use of the γ emitter thallium-201 ($t_{1/2}$ = 73 h) in diagnosing heart disease presents an interesting case for balancing the risks and benefits of using a particular isotope in medicine. Although thallium compounds are among the most toxic metal-containing compounds known, the nanogram quantities required for diagnosis pose few, if any, health hazards, meaning that the benefits outweigh the risks.

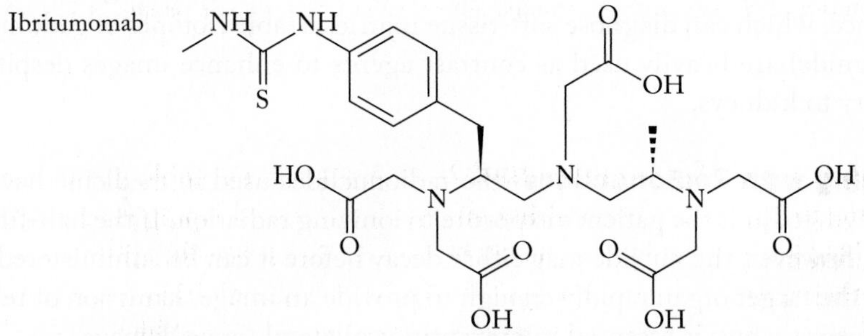

FIGURE 21.22 Zevalin contains the large organic molecule tiuxetan covalently bonded to an antibody called ibritumomab. Addition of indium-111 results in a useful imaging agent.

SAMPLE EXERCISE 21.5 Calculating Quantities of Radioactive Isotopes **LO5**

Indium-111 ($t_{1/2}$ = 2.805 d) and gallium-67 ($t_{1/2}$ = 3.26 d) are both used in radioimaging to diagnose chronic infections. Which isotope decays faster? If we start with 10.0 mg of each isotope, how much of each remains after 24 h (1 d)?

Collect and Organize We are given the half-lives of two radionuclides and asked to predict which one will decay faster and to calculate how much of each isotope remains after 24 h of decay.

Analyze An isotope with a shorter half-life decays faster. Radioactive decay follows first-order kinetics. Quantitatively, the relationship between half-life and the amount of material remaining is described by the following equation from Section 19.5:

$$\ln \frac{N_t}{N_0} = -0.693 \frac{t}{t_{1/2}}$$

Here N_0 and N_t refer to the amount of material present initially and the amount at time t, respectively. If our prediction for the relative decay rates of the two isotopes is correct, then more of the isotope with the longer half-life should remain after 24 h. We need to complete two calculations to determine the amount of each sample present after 24 h.

Solve Indium-111 has the shorter half-life, so it should decay faster.

For the amount of indium-111 remaining after 24 h, we have

$$\ln \frac{N_t}{10.0\text{ mg}} = \frac{(-0.693)(1\text{ d})}{(2.805\text{ d})} = -0.247$$

Taking the antilog of both sides, we get

$$\frac{N_t}{10.0\text{ mg}} = 0.781$$

$$N_t = (0.781)(10.0\text{ mg}) = 7.81\text{ mg}$$

For gallium-67:

$$\ln \frac{N_t}{10.0\text{ mg}} = \frac{(-0.693)(1\text{ d})}{(3.26\text{ d})} = -0.213$$

$$\frac{N_t}{10.0\text{ mg}} = 0.808$$

$$N_t = (0.808)(10.0\text{ mg}) = 8.08\text{ mg}$$

Think About It We predicted that indium-111 would decay faster, which means that after 24 h the quantity of this isotope should be less than the quantity of gallium-67, and it is.

Practice Exercise Two radioactive isotopes of bismuth are used to treat cancer. The half-lives are 61 min for bismuth-212 and 46 min for bismuth-213. If we start with 25.0 mg of each isotope, how much of each sample remains after 24 h?

CONNECTION Tritium is produced in nuclear reactors by bombarding ^{6}Li and ^{7}Li with high-energy neutrons. The reactions involved (Equations 19.13 and 19.14) are discussed in Section 19.9.

Imaging with Noble Gases and MRI None of the noble gas elements are essential to the human body, although the World Anti-Doping Agency (WADA) has added both Xe and Ar to the list of banned substances for athletes at Olympic and other sporting events since 2014. Apparently, in addition to behaving as an anesthetic, xenon increases the oxygen-carrying capacity of blood, providing an advantage in aerobically demanding sports. Similar effects are believed to occur with argon.

The lack of chemical reactivity of the group 18 elements and their ease of introduction into the body by inhalation, however, make selected isotopes of the noble gases—including helium-3, krypton-83, and xenon-129—attractive as agents for enhancing MRI images, particularly of the lungs. Krypton-83 provides greater sensitivity than xenon-129, allowing for better resolution in the images and, in principle, requiring the use of less gas.

Helium-3 has no known side effects and is preferable to xenon-129 for MRI, but it is present in only trace natural abundance. This isotope is obtained from β decay of tritium (^{3}H):

$${}^{3}_{1}\text{H} \rightarrow {}^{3}_{2}\text{He} + {}^{\ 0}_{-1}\beta$$

Neon-19 has been used in PET despite its short half-life (17.5 s). A patient positioned in a PET scanner breathes air containing a small amount of this isotope. Positron emission from the neon is recorded, and an image is created.

Therapeutic Applications

We next examine therapeutic agents that contain metallic and heavier main group elements in addition to carbon, hydrogen, nitrogen, oxygen, and sulfur.

Lithium, Boron, Aluminum, and Gallium The similar size of Li^+ (76 pm) and Mg^{2+} (72 pm) means that lithium ions can compete with magnesium ions in biological systems. The substitution of lithium for magnesium may account for its toxicity at high concentrations. Nevertheless, lithium carbonate is used to treat bipolar disorder, and other lithium compounds have been used to treat hyperactivity. In all cases, however, the use of lithium-containing drugs must be carefully monitored.

Of the elements of group 13, only boron and aluminum have been detected in humans. The role of boron in our bodies is not fully understood, but it appears to play a role in nucleic acid synthesis and carbohydrate metabolism. Selected boron compounds appear to concentrate in human brain tumors. This property has opened the door to a treatment known as boron neutron-capture therapy (BNCT). Once a suitable boron compound has been injected and has made its way to a tumor, irradiation of the tumor with low-energy neutrons leads to the nuclear reaction

$$^{10}_{5}B + ^{1}_{0}n \rightarrow ^{7}_{3}Li + ^{4}_{2}He$$

CONNECTION The relative biological effectiveness (RBE) of radioactive particles is a factor that accounts for the differences in physical damage caused by different types of radiation. It was introduced in Section 19.6.

The α particles generated in the reaction have a short penetration depth but high relative biological effectiveness (RBE), so they can kill the tumor cells without harming surrounding tissue. The identification of compounds suitable for BNCT remains an area of active research.

Aluminum is found in some antacids as aluminum hydroxide, $Al(OH)_3$, or aluminum carbonate, $Al_2(CO_3)_3$. Some brands of baking powder contain sodium aluminum sulfate, $NaAl(SO_4)_2 \cdot 12\ H_2O$. Most of the aluminum in the human body can be traced to these sources. Aluminum is not considered essential to humans, but low-aluminum diets have been observed to harm goats and chickens. High concentrations of aluminum are clearly toxic; the effects are most noticeable in patients with impaired kidney function. The role of aluminum in Alzheimer's disease has been extensively debated but remains unresolved.

Simple gallium compounds such as gallium(III) nitrate and gallium(III) chloride, either alone or in combination with other drugs, have shown activity on bladder and ovarian cancers. The similar ionic radii of Ga^{3+} (62 pm) and Fe^{3+} (64.5 pm) allow gallium to block DNA synthesis by replacing iron in a protein called transferrin and in other enzymes. Because gallium compounds accumulate in tumors at a higher rate than in healthy tissue, the disruption of DNA synthesis in the tumor cells inhibits tumor growth.

Antimony and Bismuth Antimony compounds are generally considered toxic. It has been reported, for instance, that exposure of infants to antimony compounds used as fire retardants in mattresses may contribute to sudden infant death syndrome (SIDS). However, this element does appear to have a medical use. Leishmaniasis, an insect-borne disease characterized by the formation of boils or skin lesions, is resistant to most treatments, but some patients have been successfully treated with sodium stibogluconate, one of the few applications of antimony compounds in human health.

Popular over-the-counter remedies for indigestion, diarrhea, and other gastrointestinal disorders contain bismuth subsalicylate (**Figure 21.23**). The bismuth in these compounds acts as a mild antibacterial agent that reduces the number of bacteria that cause diarrhea.

The human body requires about 30 elements to function properly. To manage the problems of disease and injury, scientists, physicians, engineers, and scores of other people have turned to the properties of these and many other elements in

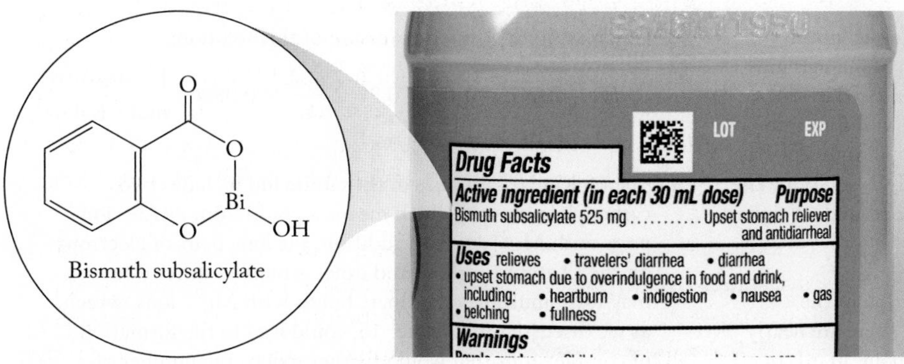

FIGURE 21.23 Bismuth subsalicylate is found in some antacids.

the periodic table to develop treatments and to enhance quality of life. In this chapter we have briefly introduced the roles of the main group elements in establishing and maintaining living systems.

SAMPLE EXERCISE 21.6 Integrating Concepts: Colonoscopy

Many patients undergoing gastrointestinal medical procedures, such as a colonoscopy to screen for colon cancer, must drink about 230 mL of a magnesium citrate solution ($MgC_6H_6O_7$; $\mathcal{M}$ = 214.41 g/mol), which serves as a laxative that clears the large intestine. Magnesium citrate (**Figure 21.24**) is an electrolyte composed of magnesium(II) ions and citrate ions.

a. As indicated on the label in Figure 21.24, the solution contains 1.745 g magnesium citrate per fluid ounce (fl oz). What is the molarity of this solution? (There are 29.57 mL in 1.000 fl oz.)
b. What is the osmotic pressure (Π) of the solution at 37°C? Its van 't Hoff factor (i) is 1.09.
c. Most electrolytes composed of 2+ cations and 2− anions have van 't Hoff factors closer to 2 than to 1 at the concentration calculated in part (a). Propose a reason why the i value for sodium citrate is so low. In developing your answer, consider the structure of citrate ions and how they might interact with Mg^{2+} ions in aqueous solutions.
d. The walls of the small intestine act as a semipermeable membrane, allowing water to pass from the cells of the tissues surrounding the small intestine into the fluid within it or in the reverse direction from inside the small intestine into the cells of the surrounding tissues. If the osmotic pressure in these cells is about 3.5 atm, in which direction will osmosis occur if the fluid passing through the small intestine has an osmotic pressure near that of the solution in the bottle?

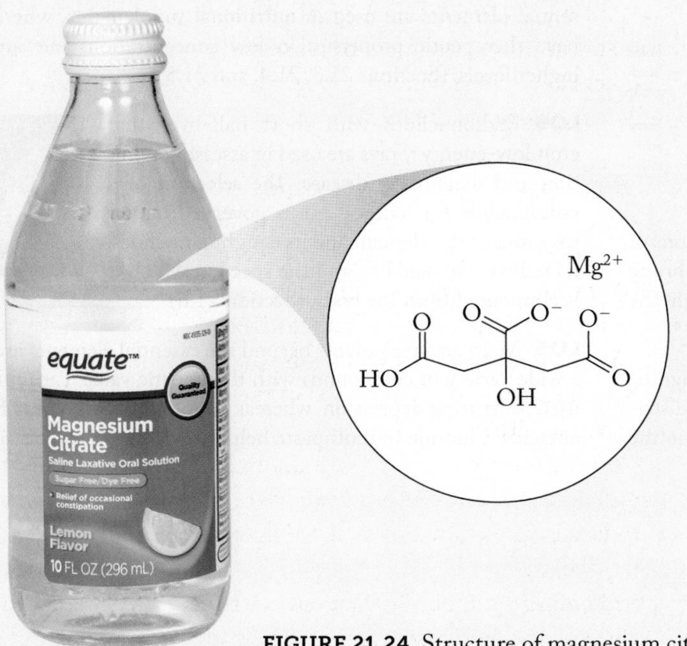

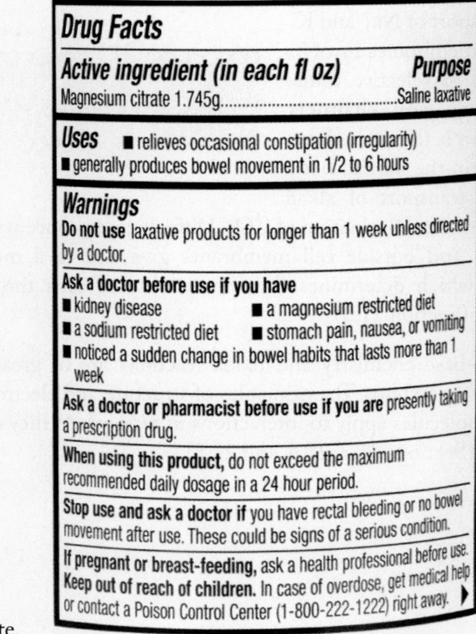

FIGURE 21.24 Structure of magnesium citrate.

Collect and Organize We are asked to calculate the molar concentration and osmotic pressure of a solution of magnesium citrate. These two properties of the solution are related by the equation $\Pi = iMRT$. We are also asked if water will flow into or out from the small intestine if the fluid in the small intestine has nearly the same osmotic pressure as the $MgC_6H_6O_7$ solution. The osmotic pressure of the fluid in the cells of the surrounding tissues is about 3.5 atm.

Analyze We first need to convert the concentration of the $MgC_6H_6O_7$ solution from g/fl oz into molarity. There are nearly 2 g $MgC_6H_6O_7$/fl oz in the solution; there are about 30 mL in 1 fl oz, and $MgC_6H_6O_7$ has a molar mass near 200 g/mol. Therefore, the molarity of the solution should be about 2 g/fl oz × 1 fl oz/30 mL × 1000 mL/L × 1 mol/200 g = 0.3 mol/L. The van 't Hoff factor i is about 1, and the temperature is 37 + 273 = 310 K, so the value of Π should be about 0.3 mol/L × 0.08 L · atm/(mol · K) × 300 K = 7 atm. This value is about twice that of the fluids surrounding the small intestine.

Solve

a. Calculating the molarity of the $MgC_6H_6O_7$ solution:

$$\frac{1.745\ \text{g}}{1\ \cancel{\text{fl oz}}} \times \frac{1.000\ \cancel{\text{fl oz}}}{29.57\ \cancel{\text{mL}}} \times \frac{1000\ \cancel{\text{mL}}}{1\ \text{L}} \times \frac{1\ \text{mol}}{214.41\ \text{g}} = 0.2752\ \text{mol/L}$$

b. Calculating the osmotic pressure of this solution:

$$\Pi = iMRT = 1.09 \times 0.2752\ \frac{\text{mol}}{\cancel{\text{L}}} \times 0.08206\ \frac{\cancel{\text{L}} \cdot \text{atm}}{\cancel{\text{mol}} \cdot \cancel{\text{K}}} \times 310\ \cancel{\text{K}} = 7.63\ \text{atm}$$

c. The molecular structure of the citrate ion includes two carboxylate groups, which means it could form 2+/2− ion pairs with Mg^{2+} ions. In addition, the lone pairs of electrons on the O atoms in these and other groups in the citrate ions may form multiple coordinate bonds with Mg^{2+} ions, which, as we discussed in Chapter 16, could lead to the formation of complex ions that would further reduce the number of independent solute particles in solution.

d. The osmotic pressure of the $MgC_6H_6O_7$ solution is more than twice that in the cells of the tissues surrounding the walls of the small intestine. This means water will spontaneously flow from those cells into the small intestine if the fluid inside the small intestine has nearly the same osmotic pressure as the $MgC_6H_6O_7$ solution.

Think About It Osmotic laxatives such as magnesium citrate work by drawing water into the intestines. This softens the waste in the large intestine, which makes bowel movements easier, and it also stimulates muscle action that helps speed up the clearance process.

SUMMARY

LO1 **Essential elements** have a physiological function in the body. **Nonessential elements** are present in the body but have no known functions. Some may have **stimulatory effects**. Essential elements are categorized as **major, trace,** or **ultratrace essential elements** depending on their concentrations in the body. (Sections 21.1 and 21.2)

LO2 Transport of Na^+ and K^+ across cell membranes involves **ion pumps** or selective transport through **ion channels.** Chloride ion is the most abundant anion in the human body, facilitating transport of alkali metal cations and elimination of CO_2. Differences in concentrations of ions inside and outside cell membranes give rise to a **membrane potential,** which determines the permeability of ions through the membrane. (Section 21.3)

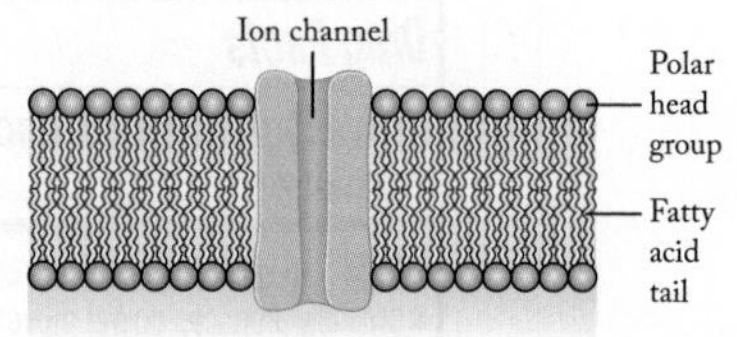

LO3 Acid–base chemistry and redox reactions are of great significance in living systems. The principles of structure and electron distribution in molecules apply to interactions *in vivo* just as they do in the laboratory. (Sections 21.3, 21.4, and 21.5)

LO4 The most abundant elements in the human body include seven nonmetals (C, H, O, Cl, S, P, and N) and four metals (Ca^{2+}, K^+, Na^+, and Mg^{2+}). Nonessential elements from groups 1 and 2 may be absorbed into cells or substituted into bone or other tissues because they are similar in size and charge to essential elements. Some nonessential elements are used in nutritional supplements, whereas others have therapeutic properties in low concentrations but are toxic at higher levels. (Sections 21.3, 21.4, and 21.5)

LO5 Radionuclides with short half-lives that emit low-energy γ rays are used in assessing function and diagnosing disease. The selection of a radionuclide for medical use is governed by the toxicities of the element and its daughter nuclides, its radioactive half-life, and the speed at which it is eliminated from the body. (Section 21.6)

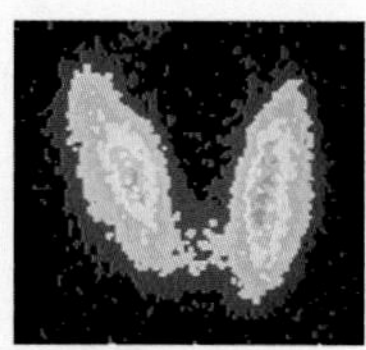

LO6 Main group elements beyond the essential elements are useful in a wide variety of compounds with therapeutic value. Lithium salts are used in treating depression, whereas aluminum compounds find use as antacids. Fluoride in toothpaste helps prevent cavities. (Section 21.6)

PARTICULATE PREVIEW WRAP-UP

From left to right, the ions are Ca^{2+}, Na^+, H_3O^+, and K^+. The H_3O^+ ion is most likely to undergo selective transport through the sodium ion channel as a result of its similar charge and size. None of the other three ions is likely to pass through the K^+ ion channel because those ions differ in size from the K^+ ion.

PROBLEM-SOLVING SUMMARY

Type of Problem	Concepts and Equations	Sample Exercises
Calculating equilibrium potential (E_{ion}) for ions and ΔG for ion transport	Calculate E_{ion} by using a modified form of the Nernst equation: $$E_{ion} = \frac{RT}{nF}\ln\left(\frac{[M^{x+}_{outside}]}{[M^{x+}_{inside}]}\right) \quad (21.1)$$ Calculate ΔG by using the relationships between ΔG and E: $$E_{transport} = E_{membrane} - E_{ion} \quad (21.2)$$ $$\Delta G_{transport} = -nFE_{transport} \quad (21.3)$$	**21.1**
Calculating an acid concentration from its pH	Relate the pH of a solution to the $[H_3O^+]$ by the equation $$pH = -\log[H_3O^+] \quad (15.10)$$	**21.2**
Assigning oxidation numbers and writing half-reactions	Use the guidelines in Section 4.8.	**21.3**
Drawing Lewis structures for molecules	Use the guidelines in Sections 8.2 and 9.4.	**21.4**
Calculating quantities of radioactive isotopes	Use the equation $$\ln\frac{N_t}{N_0} = -0.693\frac{t}{t_{1/2}}$$ where N_0 and N_t are the amounts of material present initially and at time t, respectively.	**21.5**

VISUAL PROBLEMS

(Answers to boldface end-of-chapter questions and problems are in the back of the book.)

21.1. Which part of Figure P21.1 best describes the periodic trend in monatomic cation radii moving up or down a group or across a period in the periodic table? (Arrows point in the direction of *increasing* radii.)

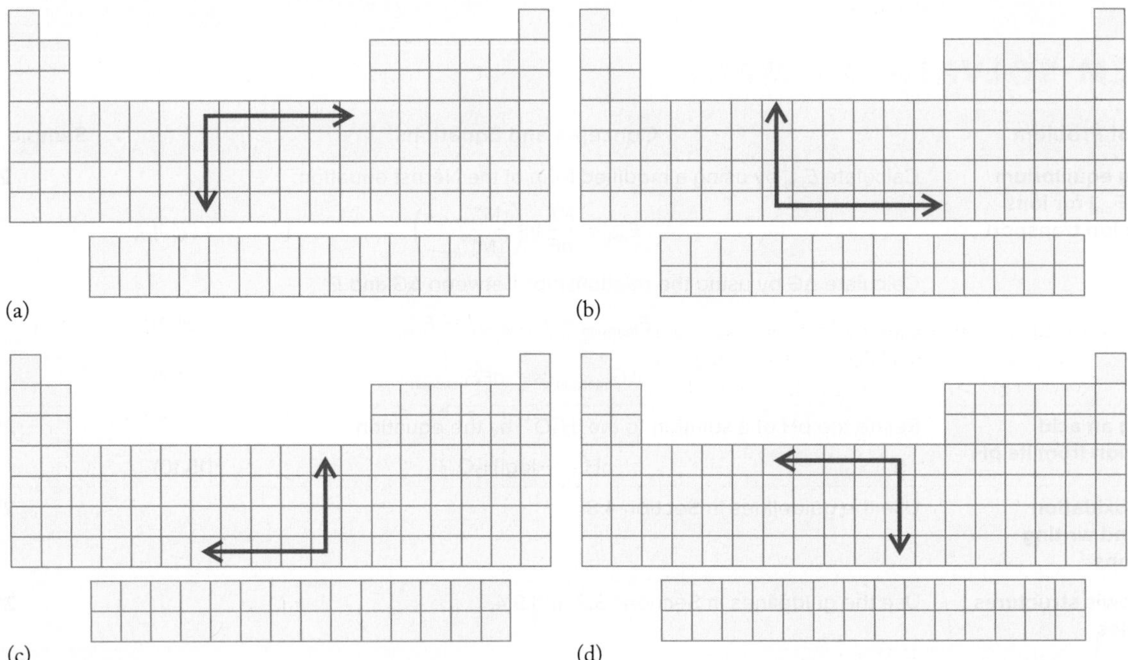

FIGURE P21.1

21.2. Which part of Figure P21.1 best describes the periodic trend in monatomic anion radii moving up or down a group or across a period in the periodic table? (Arrows point in the direction of *increasing* radii.)

21.3. Which of the two groups highlighted in the periodic table in Figure P21.3 typically forms ions that have *larger* radii than those of the corresponding neutral atoms?

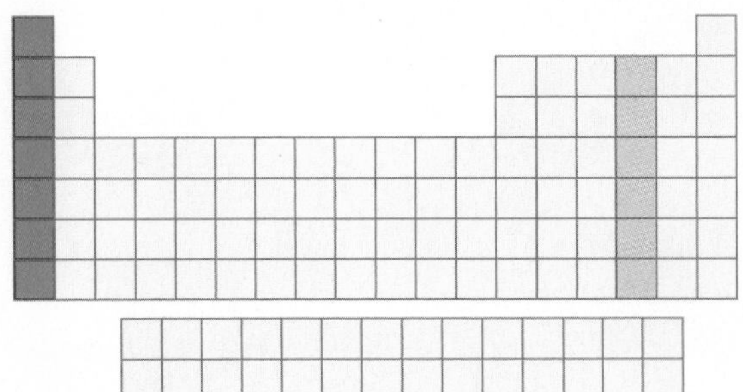

FIGURE P21.3

21.4. Which of the two groups highlighted in the periodic table in Figure P21.4 typically forms ions that have *smaller* radii than those of the corresponding neutral atoms?

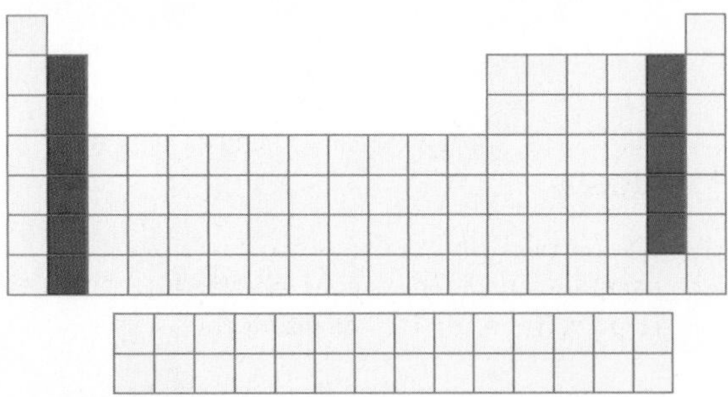

FIGURE P21.4

***21.5.** As we saw in Chapter 18, the free energy (ΔG) of a reaction is related to the cell potential by the equation $\Delta G = -nFE$. In Figure P21.5, two solutions of Na^+ of different concentrations are separated by a semipermeable membrane. Calculate ΔG for the transport of Na^+ from the side with higher concentration to the side with lower concentration.

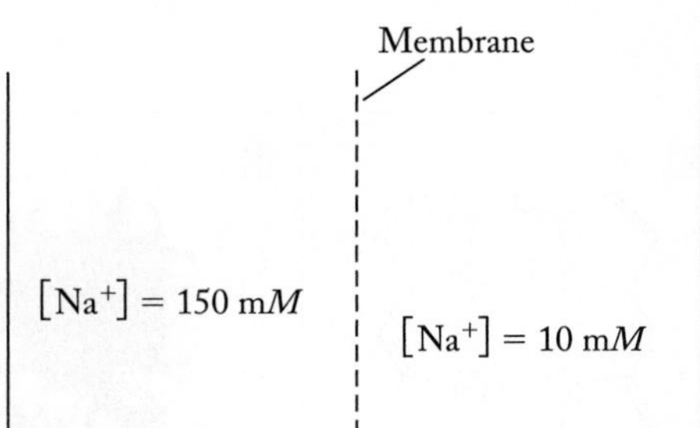

FIGURE P21.5

21.6. Two solutions of K^+ are separated by a semipermeable membrane in Figure P21.6. Calculate ΔG for the transport of K^+ from the side with lower concentration to the side with higher concentration.

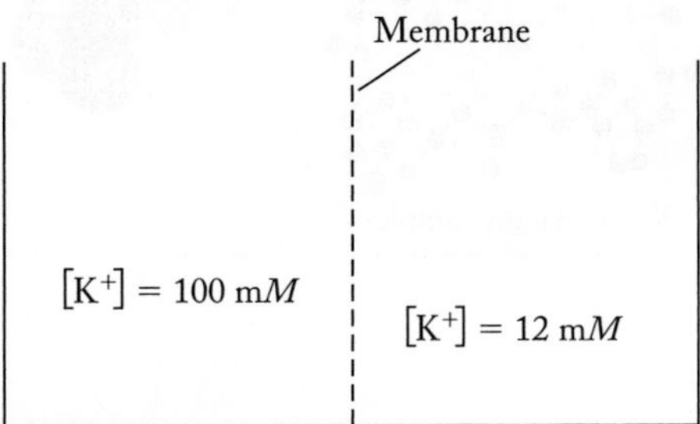

FIGURE P21.6

21.7. Describe the molecular geometry around each germanium atom in the compound shown in Figure P21.7.

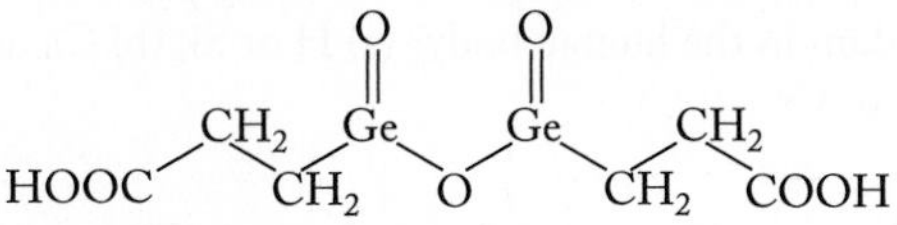

HOOC–CH_2–CH_2–Ge(=O)–O–Ge(=O)–CH_2–CH_2–COOH

FIGURE P21.7

21.8. Selenocysteine can exist as two enantiomers (stereoisomers). Identify the atom in Figure P21.8 responsible for the two enantiomers.

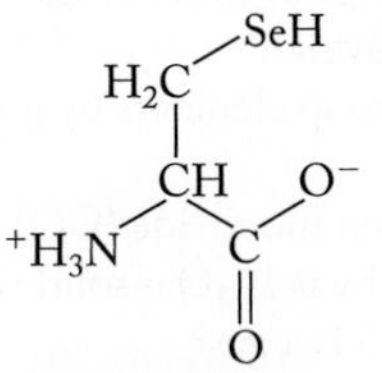

$^{+}H_3N$–CH(CH_2–SeH)–C(=O)–O^-

FIGURE P21.8

***21.9.** The sizes in pm of the main group atoms and ions are shown in Figure P21.9. Using this figure as a guide, which of the following polyatomic ions is likely to be the largest: sulfate, phosphate, or perchlorate?

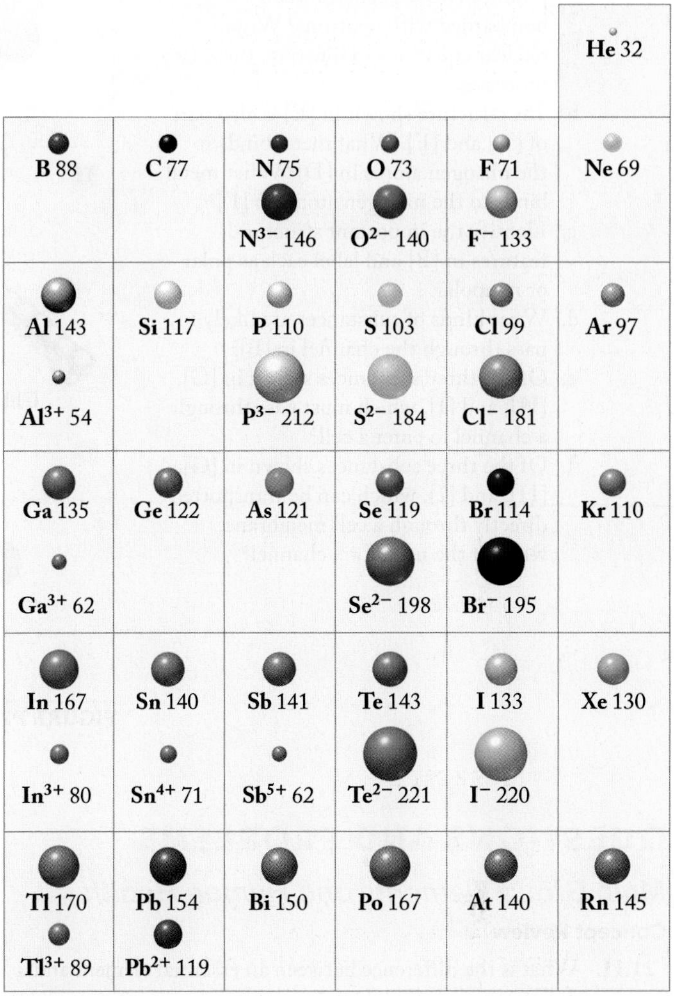

					He 32
B 88	C 77	N 75 N^{3-} 146	O 73 O^{2-} 140	F 71 F^- 133	Ne 69
Al 143 Al^{3+} 54	Si 117	P 110 P^{3-} 212	S 103 S^{2-} 184	Cl 99 Cl^- 181	Ar 97
Ga 135 Ga^{3+} 62	Ge 122	As 121	Se 119 Se^{2-} 198	Br 114 Br^- 195	Kr 110
In 167 In^{3+} 80	Sn 140 Sn^{4+} 71	Sb 141 Sb^{5+} 62	Te 143 Te^{2-} 221	I 133 I^- 220	Xe 130
Tl 170 Tl^{3+} 89	Pb 154 Pb^{2+} 119	Bi 150	Po 167	At 140	Rn 145

FIGURE P21.9

21.10. Use representations [A] through [I] in Figure P21.10 to answer questions (a)–(f).

a. Which nuclide is the product of β decay from tritium? Which will produce two α particles when bombarded with neutrons? Write nuclear equations to illustrate these two processes.
b. The structure shown in [E] is also part of [D] and [F]. What metal binds to the nitrogen atoms in [D]? What metal binds to the nitrogen atoms in [F]?
c. Identify the important structural features in [B] and label each as polar or nonpolar.
d. What kinds of substances are likely to pass through the channel in [B]?
e. Of the three substances shown in [G], [H], and [I], which must pass through a channel to enter a cell?
f. Of the three substances shown in [G], [H], and [I], which can be transported directly through a cell membrane, without the need for a channel?

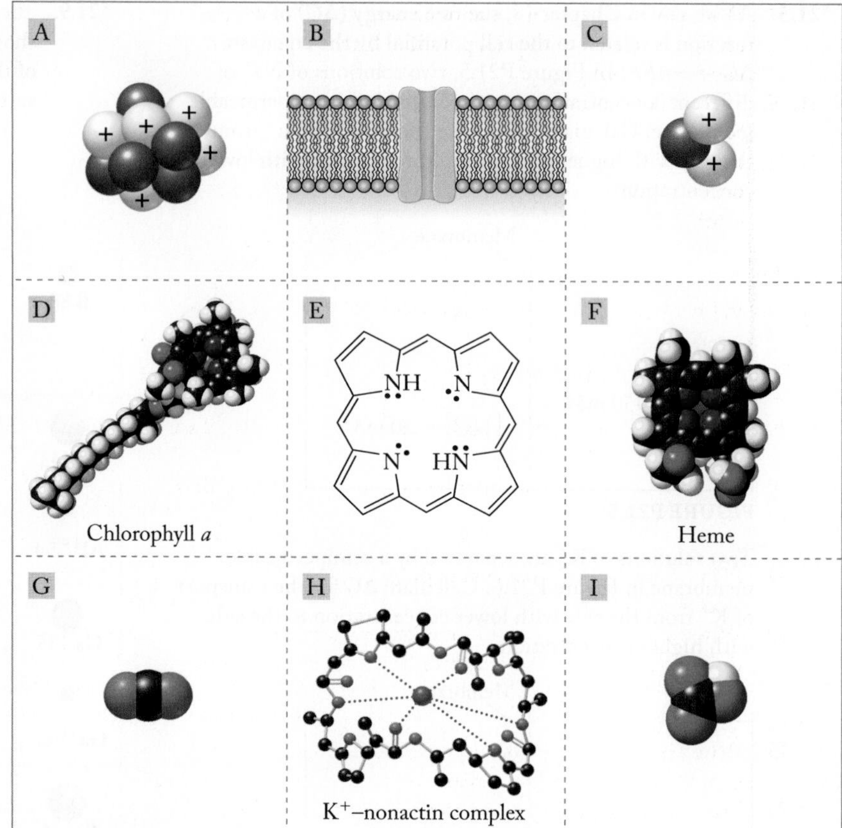

FIGURE P21.10

QUESTIONS AND PROBLEMS

Main Group Elements and Human Health

Concept Review

21.11. What is the difference between an essential element and a nonessential element?

21.12. Are all essential elements major essential elements?

21.13. What is the main criterion that distinguishes major, trace, and ultratrace essential elements from one another?

21.14. Should trace essential elements also be considered to be stimulatory?

Problems

21.15. The concentrations of very dilute solutions are sometimes expressed as parts per million. Express the concentration of each of the following trace and ultratrace essential elements in parts per million:

a. Fluorine, 110 mg in 70 kg
b. Silicon, 525 mg/kg
c. Iodine, 0.043 g in 100 kg

21.16. In the human body, the concentrations of ultratrace essential elements are even lower than those of trace essential elements and therefore are sometimes expressed in parts per billion. Express the concentrations of each of the following elements in parts per billion:

a. Bromine, 6 mg/L
b. Boron, 0.014 g/100 kg
c. Selenium, 5.0 mg/70 kg

21.17. Which element in each of the following pairs is more abundant in the human body? (a) silicon or oxygen; (b) iron or oxygen; (c) carbon or aluminum

21.18. Which element in each of the following pairs is more abundant in the human body? (a) H or Si; (b) Ca or Fe; (c) N or Cr

Periodic Properties of Main Group Elements

Concept Review

21.19. In Chapter 2 we defined main group elements as those elements found in groups 1, 2, and 13–18 in the periodic table. Why do some chemists refer to these as the "*s*-block" and "*p*-block" elements, respectively?

21.20. Why do we classify the main group elements by group rather than by period?

21.21. Lithium oxide (Li_2O) and carbon monoxide (CO) have nearly the same molar mass. Why is Li_2O a solid with a high melting point, whereas CO is a gas?

21.22. The nonradioactive group 17 elements are found as diatomic molecules, X_2 (X = F, Cl, Br, I). Why is Br_2 a liquid at room temperature, whereas Cl_2 is a gas?

21.23. Which of the following properties can be used to distinguish a metallic element from a semimetallic element: atomic radius, electrical conductivity, and/or molar mass?

21.24. Which of the following cannot be measured: ionization energy, electron affinity, ionic radius, atomic radius, or electronegativity?

21.25. Why is Be^{2+} more likely than Ca^{2+} to displace Mg^{2+} in biomolecules?

21.26. PbS, $PbCO_3$, and PbCl(OH) have limited solubility in water. Which of them is/are more likely to dissolve in acidic solutions?

Problems

21.27. Which ion channel—K^+ or Na^+—must accommodate the larger cation?

21.28. Which ion is larger: Cl^- or I^-?

21.29. Place the following ions in order of increasing ionic radius: Mg^{2+}, Li^+, Al^{3+}, and Cl^-.

21.30. Place the following ions in order of increasing ionic radius: Br^-, O^{2-}, K^+, and Ca^{2+}.

21.31. Place the following elements in order of increasing electronegativity: K, S, F, and Mg.

21.32. When we compare any two main group elements, is the element with the smaller atomic radius always more electronegative?

21.33. Why can we estimate the electron affinity of Cl atoms by measuring the ionization energy of a Cl^- anion?

21.34. Place the following ions in order of increasing ionization energy: Na^+, S^{2-}, F^-, and Mg^+.

Major Essential Elements

Concept Review

21.35. **Ion Transport in Cells** Describe three ways in which ions of major essential elements (such as Na^+ and K^+) enter and exit cells.

21.36. Which transport mechanism for ions requires ATP: diffusion, ion channels, or ion pumps?

21.37. Why is it difficult for ions to diffuse across cell membranes?

21.38. Why does Sr^{2+} substitute for Ca^{2+} in bones?

21.39. Which alkali metal ion is Rb^+ most likely to substitute for?

21.40. Why don't alkaline earth metal cations substitute for alkali metal cations in cases where the ionic radii are similar?

***21.41.** Why might nature have selected calcium carbonate over calcium sulfate as the major exoskeleton material in shells?

21.42. Bromide ion and fluoride ion are nonessential elements in the body. Do you expect their concentrations to be more similar to the concentrations of major essential elements or to the concentrations of ultratrace essential elements?

Problems

21.43. **Osmotic Pressure of Red Blood Cells** One of the functions of the alkali metal cations Na^+ and K^+ in cells is to maintain the cells' osmotic pressure. The concentration of NaCl in red blood cells is approximately 11 m*M*. Calculate the osmotic pressure of this solution at body temperature (37°C).

21.44. Calculate the osmotic pressure exerted by a 92 m*M* solution of KCl in a red blood cell at body temperature (37°C).

***21.45.** **Electrochemical Potentials across Cell Membranes** Very different concentrations of Na^+ ions exist in red blood cells (11 m*M*) and the blood plasma (160 m*M*) surrounding those cells. Solutions with two different concentrations separated by a membrane constitute a concentration cell. Calculate the electrochemical potential created by the unequal concentrations of Na^+ at 310 K.

21.46. The concentration of K^+ in red blood cells is 92 m*M*, and the concentration of K^+ in plasma is 10 m*M*. Calculate the electrochemical potential created by the two concentrations of K^+ at 310 K.

21.47. **Calcium Ion Pumps** Transport of Ca^{2+} across cell membranes is critical for cardiac muscle cells. The $[Ca^{2+}]$ in these cells increases by $\approx 10^2$ during contraction of the heart at 310 K. What electrochemical potential (E°_{ion}) does this difference in concentration correspond to?

*21.48. Removing excess Na^+ from a cell by an ion pump requires energy. How many moles of ATP must be hydrolyzed to overcome a cell potential of −0.07 V? The hydrolysis of 1 mole of ATP provides 34.5 kJ of energy.

21.49. **Genetic Engineering** Plants require phosphate (PO_4^{3-}) to grow, and application of fertilizers that contain phosphate promotes the growth of the desired crops but also weeds, requiring the application of herbicides to kill the latter. In 2018 geneticists announced the development of a strain of cotton that could grow on *phosphite* (HPO_3^{2-}) rather than phosphate, while common weeds could not.

a. Draw Lewis structures for both anions and assign oxidation states to phosphorus in each. (The phosphite ion contains a P–H bond).

*b. Why might phosphite-based fertilizers make it possible to use less herbicide?

21.50. **Naming Molecular Ions** The name "phosphite" for HPO_3^{2-} is misleading. By analogy to the nomenclature for polyoxoanions in Chapter 2, what molecular formula and charge does the term "phosphite" suggest? Draw a Lewis structure for this molecular ion.

Trace and Ultratrace Essential Elements; Nonessential Elements

Concept Review

21.51. What danger to human health is posed by ^{137}Cs ($t_{1/2} \approx 30$ yr)?

21.52. Why is ^{137}Cs ($t_{1/2} \approx 30$ yr) considered dangerous to human health, whereas naturally occurring ^{40}K ($t_{1/2} = 1.28 \times 10^6$ yr) is benign?

21.53. What are the likely signs of ΔS and ΔH for the dissolution of tooth enamel?

21.54. Why does fluorapatite resist acid better than hydroxyapatite if both are insoluble in water?

***21.55.** Why do superoxide ions (O_2^-) act as strong oxidizing agents?

*21.56. Why are thyroxine and 3,5,3′-triiodothyronine (Figure 21.18) considered amino acids? Why aren't they *essential* amino acids?

Problems

21.57. The Aroma of Coffee More than 1000 compounds have been identified in roasted coffee beans, many of which contribute to the aroma of brewed coffee when the beans are ground and extracted with water. Four sulfur compounds are included among the known aroma compounds in coffee (Figure P21.57).

a. Which compound do you expect to have the highest vapor pressure?
b. If we replaced the S with O, would you expect the vapor pressure of the resulting compounds to increase or decrease? Explain.

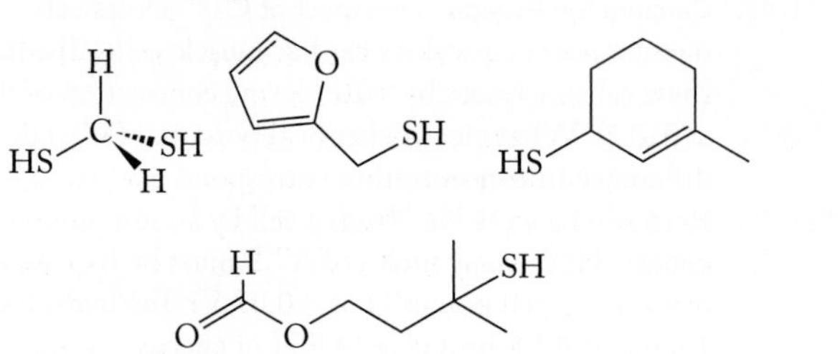

FIGURE P21.57

21.58. Identify the functional group that contains the sulfur in Figure P21.57 and determine whether any of these compounds are chiral. Are these compounds likely to be polar or nonpolar?

21.59. Calculate the pH of a 1.00×10^{-3} *M* solution of selenocysteine ($pK_{a_1} = 2.21$, $pK_{a_2} = 5.43$).

21.60. Calculate the pH of a 1.00×10^{-3} *M* solution of cysteine ($pK_{a_1} = 1.7$, $pK_{a_2} = 8.3$). Is selenocysteine a stronger acid than cysteine?

21.61. Composition of Tooth Enamel Tooth enamel contains the mineral hydroxyapatite. Hydroxyapatite reacts with fluoride ion in toothpaste to form fluorapatite. The equilibrium constant for the reaction between hydroxyapatite and fluoride ion is $K = 8.48$. Write the equilibrium constant expression for the following reaction. In which direction does the equilibrium lie?

$$Ca_3(PO_4)_3(OH)(s) + F^-(aq) \rightleftharpoons Ca_5(PO_4)_3(F)(s) + OH^-(aq)$$

*21.62. **Effects of Excess Fluoridation on Teeth** Too much fluoride might lead to the formation of calcium fluoride according to the reaction

$$Ca_5(PO_4)_3(OH)(s) + 10\ F^-(aq) \rightleftharpoons 5\ CaF_2(s) + 3\ PO_4^{3-}(aq) + OH^-(aq)$$

Write the equilibrium constant expression for the reaction. Given the K_{sp} values for the following two reactions, calculate K for the reaction between $Ca_5(PO_4)_3(OH)$ and fluoride ion that forms CaF_2.

$$Ca_5(PO_4)_3(OH)(s) \rightleftharpoons 5\ Ca^{2+}(aq) + 3\ PO_4^{3-}(aq) + OH^-(aq) \qquad K_{sp} = 2.3 \times 10^{-59}$$

$$CaF_2(s) \rightleftharpoons Ca^{2+}(aq) + 2\ F^-(aq) \qquad K_{sp} = 3.9 \times 10^{-11}$$

21.63. Tooth enamel is actually a composite material containing both hydroxyapatite and a calcium phosphate, $Ca_8(HPO_4)_2(PO_4)_4 \cdot 6\ H_2O$ ($K_{sp} = 1.1 \times 10^{-47}$).

a. Is this calcium mineral more or less soluble than hydroxyapatite ($K_{sp} = 2.3 \times 10^{-59}$)?
b. Calculate the solubility in moles per liter of hydroxyapatite $[Ca_5(PO_4)_3(OH)]$, where $K_{sp} = 2.3 \times 10^{-59}$ in water at 25°C and pH = 7.00.
c. What is the solubility of hydroxyapatite at pH = 5.00?

21.64. The K_{sp} of actual tooth enamel is reported to be 1×10^{-58}.

a. Does this mean that tooth enamel is more soluble than pure hydroxyapatite ($K_{sp} = 2.3 \times 10^{-59}$)?
b. Does the measured value of K_{sp} for tooth enamel support the idea that tooth enamel is a mixture of hydroxyapatite $[Ca_5(PO_4)_3(OH)]$ and a calcium phosphate, $Ca_8(HPO_4)_2(PO_4)_4 \cdot 6\ H_2O$ ($K_{sp} = 1.1 \times 10^{-47}$)?
c. Calculate the solubility in moles per liter of $Ca_8(HPO_4)_2(PO_4)_4 \cdot 6\ H_2O$ ($K_{sp} = 1.1 \times 10^{-47}$) in water at 25°C and pH = 7.00.

21.65. Lead in Drinking Water The municipal water systems of many older cities, including Flint, Michigan, still rely on water pipes made of lead or lead alloys. In principle, this is not as dangerous as it seems because lead corrodes, forming a protective coating of lead(II) carbonate ($PbCO_3$; Figure P21.65).

a. Calculate the molar solubility of $PbCO_3$.
b. The EPA limit for dissolved Pb^{2+} is 15 ppb. Does your answer to part (a) meet this standard?
c. How does the solubility of $PbCO_3$ depend on pH?

FIGURE P21.65

21.66. **Removing Lead from Water** To reduce the concentration of Pb^{2+} in drinking water, many treatment plants add phosphate ion (PO_4^{3-}) to the water to precipitate lead(II) phosphate.

a. Is $Pb_3(PO_4)_2$ ($K_{sp} = 3.0 \times 10^{-44}$) more or less soluble than $PbCO_3$?
b. Why does the addition of phosphate also regulate the pH of the solution?
c. Calculate the pH of a 0.125 *M* solution of Na_3PO_4.

21.67. All the group 16 elements form compounds with the generic formula H_2E (E = O, S, Se, or Te). Which compound is the most polar? Which compound is the least polar?

*21.68. All the group 15 elements form compounds with the generic formula H_3E (E = N, P, As, Sb, and Bi). Which compound is the most polar? Which compound do you predict to have the smallest H–E–H bond angle?

Elements for Diagnosis and Therapy

Concept Review

21.69. When choosing an isotope for imaging, why is it important to consider the decay mode of the isotope as well as the half-life?

21.70. Why might an α emitter be a good choice for radiation therapy?

***21.71.** What advantage does a β emitter have over an α emitter for imaging?

21.72. Why do we sometimes use radioisotopes of toxic elements, such as thallium, for imaging?

21.73. Why does ^{213}Bi undergo β decay but ^{111}In decays by electron capture?

21.74. Several isotopes of arsenic are used in medical imaging. Which isotope, ^{72}As or ^{77}As, is more likely to be useful for PET imaging?

21.75. The World Anti-Doping Agency (WADA) added xenon and argon to the list of banned substances in 2014. Which intermolecular forces account for the solubility of Xe and Ar in blood?

21.76. Helium is used in SCUBA gear to prevent nitrogen narcosis. Do you expect the solubility of He in blood to be greater than or less than the solubility of Xe and Ar in blood?

Problems

21.77. PET Imaging with Gallium A patient is injected with a 5 μM solution of gallium citrate containing ^{68}Ga ($t_{1/2}$ = 9.4 h) for a PET study. How long is it before the activity of the ^{68}Ga drops to 5% of its initial value?

21.78. Iodine-123 ($t_{1/2}$ = 13.3 h) has replaced iodine-131 ($t_{1/2}$ = 8.1 d) for diagnosis of thyroid conditions. How long is it before the activity of ^{123}I drops to 5% of its initial value?

21.79. Like aqueous solutions of Fe^{3+} in Chapter 16, bismuth ions (Bi^{3+}) form acidic solutions with a $pK_a = 1.5$.

a. Write a balanced net ionic equation for the reaction of Bi^{3+} with water to form a dication, $Bi(H_2O)_5OH^{2+}$.
b. What is the pH of a 0.00364 M solution of Bi^{3+}?
c. Is this solution more or less acidic than an equimolar solution of vinegar? (*Note*: No calculation is necessary to answer this question.)

21.80. Some medicines used in treating depression contain lithium carbonate.

a. Draw the Lewis structure for Li_2CO_3.
b. Explain whether or not Li_2CO_3 hydrolyzes in water.
c. Calculate the solubility of Li_2CO_3 in water in moles/L.

21.81. Aluminum hydroxide is used in some antacids. Write a balanced net ionic equation for the reaction of aluminum hydroxide with HCl.

21.82. Aluminum carbonate is used in some antacids. Write a balanced net ionic equation for the reaction of aluminum carbonate with hydrochloric acid.

21.83. Maalox The antacid known as Maalox contains a mixture of magnesium and aluminum hydroxides.

a. Which substance will neutralize more acid on a per-mole basis?
b. Does the same substance also neutralize more acid on a per-gram basis?
c. How many grams of each are required to neutralize 115 mL of 0.75 M stomach acid?

21.84. **Tums** Sodium bicarbonate and calcium carbonate both act as antacids and are found in common stomach remedies such as Tums.

a. Which substance will neutralize more acid on a per-mole basis?
b. Does the same substance also neutralize more acid on a per-gram basis?
c. How many grams of each are required to neutralize 115 mL of 0.75 M stomach acid?

21.85. Il Campanello The 1836 opera *Il Campanello* by Gaetano Donizetti features the farcical troubles of a newly wed wealthy pharmacist and his much younger wife and her meddlesome old boyfriend. The pharmacist turns to a mixture of antimony(III) chloride and mercury(II) sulfide to solve his problems.

a. In which group of the periodic table is antimony found?
b. Draw a Lewis structure for antimony(III) chloride.
c. Is antimony(III) chloride more likely to behave as a Lewis acid or a Lewis base?

21.86. The chemistry of antimony(III) chloride is similar to that of the other elements in its group. Antimony(III) chloride can be converted to Sb_2O_3, which has been used as a flame retardant.

a. Write a balanced chemical equation for the reaction of antimony(III) oxide with water to form H_3SbO_3.
b. Write a balanced chemical equation for the reaction of antimony(III) oxide with sodium hydroxide to form $NaSbO_2$ and water.
c. Identify the role of Sb_2O_3 in parts (a) and (b) as either an acid or a base.

22

Transition Metals

Biological and Medical Applications

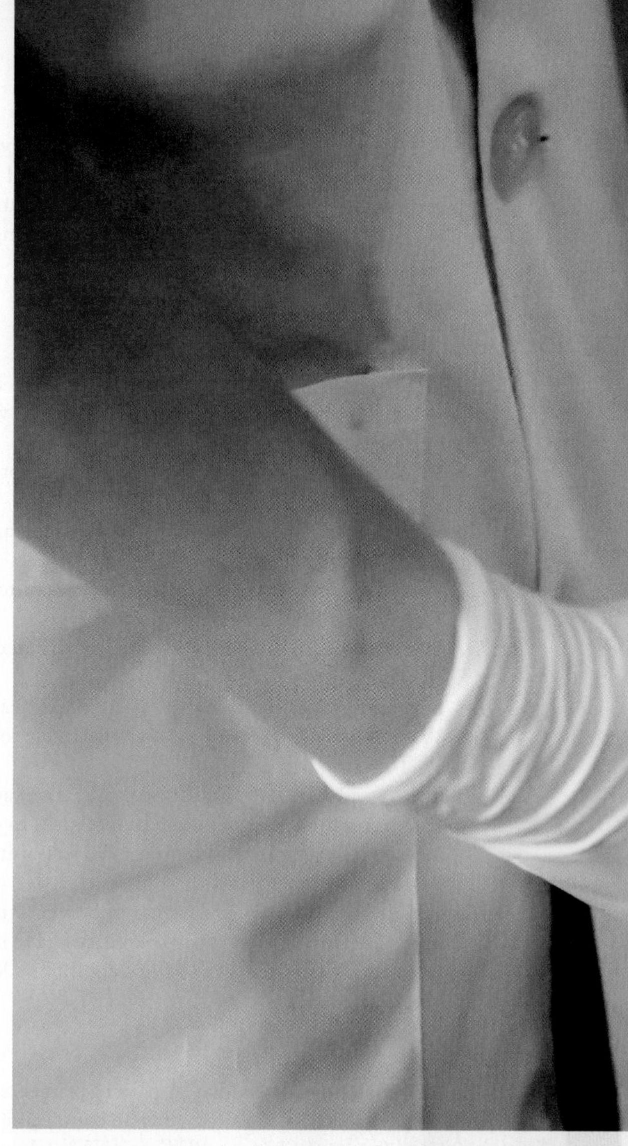

RED COLOR OF BLOOD The red color of blood comes from the heme group, a molecule containing a ring with four nitrogen atoms that bind to a central Fe^{2+} ion.

PARTICULATE **REVIEW**

Lewis Acid or Lewis Base?

In Chapter 22 we discuss the bonding and structure of compounds and complex ions formed by transition metals and learn about their presence in biological systems and their applications to medicine. The compounds shown here are ammonia, borane, and water.

- Draw the Lewis structure for each compound.
- Which compound(s) has/have at least one lone pair of electrons on the central atom?
- Which compound(s) can function as a Lewis acid? Which can function as a Lewis base?

(Review Chapters 4, 8, and 16 if you need help.)

(Answers to Particulate Review questions are in the back of the book.)

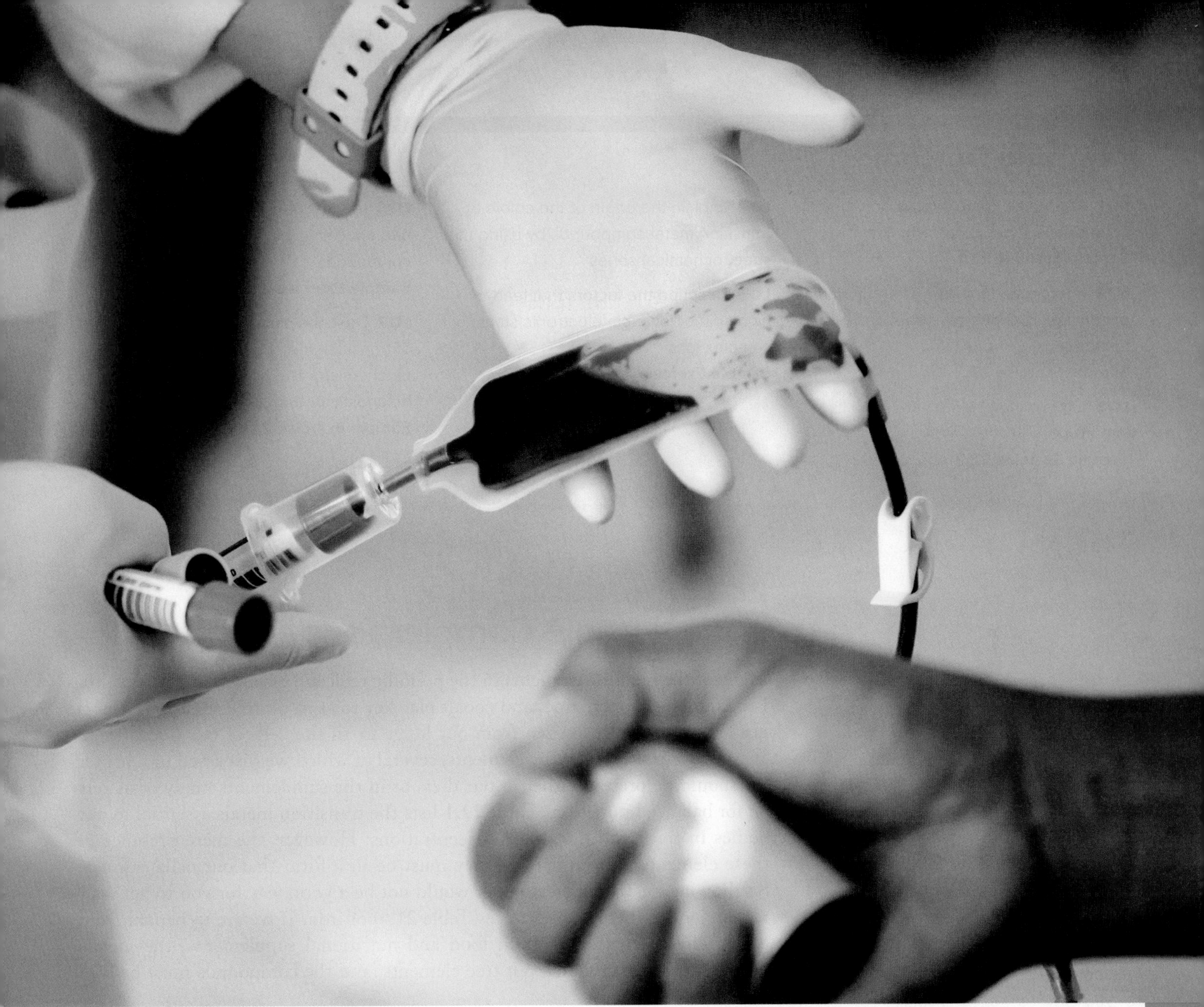

PARTICULATE **PREVIEW**

One Molecule, One Bond versus One Molecule, Two Bonds

Here are two complex ions that contain metal cations bonded to different molecules. As you read Chapter 22, look for ideas that will help you answer these questions:

- Two different molecules form coordinate bonds to Cu^{2+} and Ni^{2+} in the figure to the right. Which element coordinates to the metal?
- What molecule is bonded to the copper cation? How many of these molecules form coordinate bonds with the copper cation?
- What molecule is bonded to the nickel cation? How many of these molecules form coordinate bonds with the nickel cation?

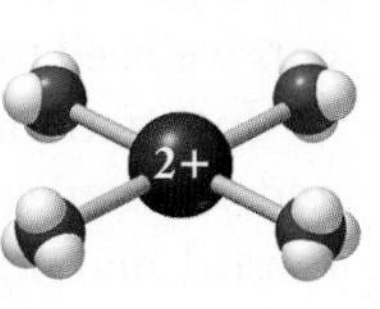

Cu^{2+} complex

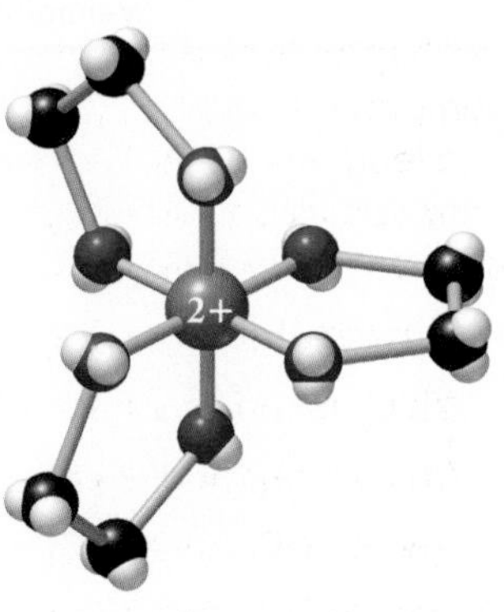

Ni^{2+} complex

Learning Outcomes

LO1 Recognize complex ions and their counterions in chemical formulas
Sample Exercise 22.1

LO2 Interconvert the names and formulas of complex ions and coordination compounds
Sample Exercise 22.2

LO3 Explain the chelate effect and its importance
Sample Exercise 22.3

LO4 Explain the origin of the colors of transition metal compounds by using the spectrochemical series

LO5 Describe the factors that lead to high-spin or low-spin electronic states of complex ions
Sample Exercise 22.4

LO6 Identify geometric, linkage, and optical isomers of coordination compounds
Sample Exercise 22.5

LO7 Describe several of the roles of transition metal complexes in biochemistry and how they are used as diagnostic or therapeutic compounds
Sample Exercise 22.6

22.1 Transition Metals in Biology: Complex Ions

Many of the metallic elements in the periodic table are essential to good health. For example, copper, zinc, and cobalt play key roles in protein function. Iron is needed to transport oxygen from our lungs to all the cells of our body. These and other essential metallic elements, several of which we discussed in Chapter 21, should be present either in our diets or in the supplements many of us rely on for balanced nutrition. **Table 22.1** lists the transition metals essential to our bodies in trace and ultratrace concentrations. However, the mere presence of these elements is insufficient—they must be in a form that our cells can use. Swallowing an 18-mg steel pellet would not be a good way for you to get your recommended dietary allowance (Table 21.4) of iron. If we are to benefit from consuming essential metals in food and nutritional supplements, the metals need to be in compounds, not free elements, and the compounds must be bioavailable to the body.

TABLE 22.1 Essential Transition Elements Found in the Human Body

Trace (1–1000 μg/g Body Mass)	Ultratrace (<1 μg/g Body Mass)
Iron	Chromium
Zinc	Cobalt
	Copper
	Manganese
	Molybdenum
	Nickel
	Vanadium

All the metallic transition elements essential to human health occur in nature in ionic compounds, but not all ionic forms are absorbed equally well. For example, most of the iron in fish, poultry, and red meat is readily absorbed because it is present in a form called *heme iron*. The iron in plants, however, is mostly nonheme and is not as readily absorbed. Eating a meal that includes both meat and vegetables improves the absorption of the nonheme iron in the vegetables. Fruits, vegetables and other foods high in vitamin C increase the availability of iron, possibly by forming complex ions of the type described in Section 16.6. All these dietary factors work together at the molecular level to provide us with the nutrients we need to survive.

Interactions between transition metals and accompanying nonmetal ions and molecules influence the solubility of the metals, which is a key factor toward making them chemically reactive and biologically available to plants and animals. These interactions also influence other properties, including the wavelengths of visible light that the metals absorb and therefore the colors of their compounds and solutions. In this chapter we explore how the chemical environment of transition metal ions in solids and in solutions affects their physical, chemical, and biological properties. We answer questions such as why many, but not all, metal compounds have distinctive colors, and how, through the formation of complex ions with biomolecules, transition metals play key roles in many biological processes.

inner coordination sphere the ligands that are bound directly to a metal via coordinate bonds.

We begin by examining the interactions between transition metal ions and the other ions and molecules that surround them in solids and solutions. To understand these interactions, we need to review the definitions of Lewis acids and Lewis bases we first used in Section 16.5:

- A *Lewis base* is a substance that *donates* a lone pair of electrons in a chemical reaction.
- A *Lewis acid* is a substance that *accepts* a lone pair of electrons in a chemical reaction.

CONNECTION Lewis's pioneering theories of the nature of covalent bonding were described in Chapter 8, and the formation of complex ions between metal ions (Lewis acids) and ligands (Lewis bases) was discussed in Section 16.6.

In Section 10.3, we described how ions dissolved in water are *hydrated*, that is, surrounded by water molecules oriented with their positive dipoles directed toward anions and their negative dipoles directed toward cations (see Figure 10.6). When these ion–dipole interactions lead to the sharing of lone-pair electrons with empty valence-shell orbitals on the cations, they meet our definition of covalent bonds, and in this case are called *coordinate covalent bonds*, or simply *coordinate bonds*. Much of the chemistry of the transition metals is associated with their ability to form coordinate bonds with molecules or anions. Review Chapter 10 if you need help with these concepts.

As we saw in Section 16.6, molecules or anions that function as Lewis bases and form coordinate bonds with metal cations are called *ligands*. The resulting species, which are composed of central metal ions and the surrounding ligands, are called *complex ions* or simply *complexes*. Direct bonding to a central cation means that the ligands in a complex occupy the **inner coordination sphere** of the cation. Take another look at Sample Exercise 14.12, in which we discussed the equilibrium between two forms of cobalt(II), one pink and one blue (see Figure 14.11), in an aqueous solution of HCl (**Figure 22.1**). Both the pink and blue forms are complexes; the pink cobalt(II) species has six water ligands in its inner coordination sphere, whereas the blue species has four chloride ions. The charge on each complex ion is the sum of the charges of the metal ion and the ligands: 2+ for $Co(H_2O)_6^{2+}$ because the charge of the cobalt ion is 2+ and water molecules are neutral, and 2− for $CoCl_4^{2-}$ because the sum of the 2+ charge on the cobalt ion and four 1− charges on the chloride ions is 2−.

The foundation of our understanding of the bonding and structure of complex ions comes from the pioneering research of Swiss chemist Alfred Werner (1866–1919), for which he was awarded the Nobel Prize in Chemistry in 1913. Some of Werner's research addressed the unusual behavior of different compounds

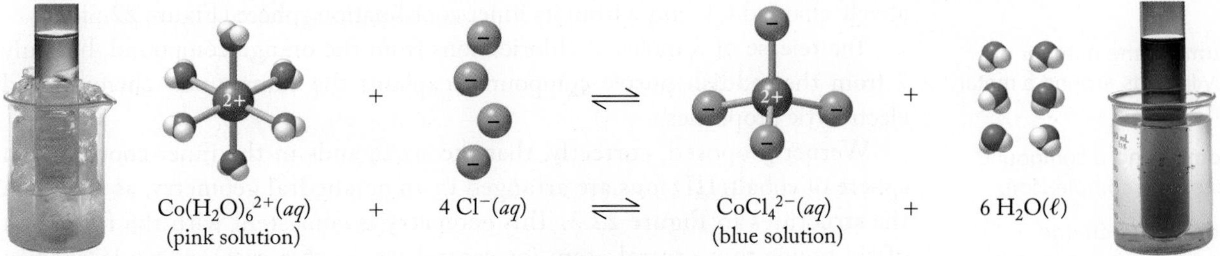

FIGURE 22.1 Equilibrium between two forms of cobalt in aqueous HCl solution.

CONNECTION We discussed in Chapter 4 how the electrical conductivity of aqueous solutions depends on the concentrations of dissolved ions in the solutions.

formed by dissolving cobalt(II) chloride in aqueous ammonia and oxidizing it to cobalt(III) by bubbling air through the solution. One of the redox reactions produces an orange compound (**Figure 22.2a**) that contains 3 moles of Cl^- ions and 6 moles of ammonia for every 1 mole of Co^{3+} ions. Another reaction produces a reddish purple compound (**Figure 22.2b**) that has the same proportions of Cl^- and Co^{3+} ions, but with only 5 moles of ammonia per mole of cobalt(III). Both compounds are water-soluble solids that react with aqueous solutions of $AgNO_3$, forming solid AgCl. However, 1 mole of the orange compound produces 3 moles of solid AgCl, whereas 1 mole of the reddish purple one produces only 2 moles of AgCl. Results like these inspired Werner to study the electrical conductivity of aqueous solutions of the two compounds. He found that the orange compound was the better conductor, indicating that it produced more ions in solution than the reddish purple one.

Werner concluded that the differences in the two compounds' composition, in their capacities to react with Ag^+ ions, and in their electrolytic properties are all caused by the presence of two types of Co–Cl bonds inside them. He proposed that these bonds were all ionic in the orange compound, but that only two-thirds of them were ionic in the reddish purple compound. The remaining one-third are coordinate covalent bonds.

(a) $[Co(NH_3)_6]Cl_3$

(b) $[Co(NH_3)_5Cl]Cl_2$

FIGURE 22.2 Two compounds of cobalt(III) chloride and ammonia.

CONCEPT TEST

How would the electrical conductivity and freezing points of 0.010 *m* aqueous solutions of the two compounds in Figure 22.2 differ?

(Answers to Concept Tests are in the back of the book.)

This capacity to form two kinds of bonds means that Co^{3+} and other transition metal ions have two kinds of bonding capacity, or *valence*. The first kind involves ionic bonds and is based on the number of electrons that a metal atom loses when it forms an ion. This valence is equivalent to its oxidation number, which is +3 for cobalt in both the orange and reddish purple compounds. The second kind of valence is based on the capacity of metal ions to form coordinate bonds. This property corresponds to an ion's **coordination number**, the number of sites around a central metal ion where bonds to ligands form.

When the two compounds in Figure 22.2a and b dissolve in water, they form an orange and a reddish purple solution. The respective solutions contain the ions shown in **Figure 22.3**. The orange solution contains $[Co(NH_3)_6]^{3+}$ and three chloride ions, whereas the reddish purple solution contains $[Co(NH_3)_5Cl]^{2+}$ and two chloride ions. Notice how brackets in the formulas of these **coordination compounds** set off the complex ions from the ionically bonded chloride **counterions**, which balance the charges on the complex ions: 3+ for the orange one (Figure 22.3a), but only 2+ for the reddish purple one because it contains a negatively charged Cl^- ion within its inner coordination sphere (Figure 22.3b).

The release of 3 moles of chloride ions from the orange compound, but only 2 from the reddish purple compound, explains the compounds' chemical and electrolytic properties.

Werner proposed, correctly, that the six ligands in the inner coordination sphere of cobalt(III) ions are arranged in an octahedral geometry, as shown in the structures in Figure 22.3. This geometry is consistent with the formation of six bonds to a central atom (or central ion in this case), as we learned in Chapter 9 (see Table 9.1). Many other transition metal ions form six coordinate

coordination number the number of sites occupied by ligands around a metal ion in a complex.

coordination compound a compound made up of at least one complex ion.

counterion an ion whose charge balances the charge of a complex ion in a coordination compound.

(a) $[Co(NH_3)_6]Cl_3(aq)$; orange solution (b) $[Co(NH_3)_5Cl]Cl_2(aq)$; reddish purple solution

FIGURE 22.3 Structures of cobalt(III) complex ions and counterions in aqueous solution.

bonds and octahedral complex ions. Some are listed in **Table 22.2**, as are complex ions that have only two or four ligands. Those with two have linear structures, whereas those with four may be tetrahedral or square planar. Both geometries occur in cobalt(II) ions: $CoCl_4^{2-}$ is tetrahedral, whereas $Co(CN)_4^{2-}$ is square planar.

CONCEPT **TEST**

In the coordination compound $Na_3[Fe(CN)_6]$, which ions occupy the inner coordination sphere of the Fe^{3+} ion and which ions are counterions? Of the four cobalt complexes described in this section, which would have conductivity in aqueous solution similar to that of $Na_3[Fe(CN)_6]$?

TABLE 22.2 Common Coordination Numbers and Shapes of Complex Ions

Coordination Number	Steric Number	Shape	Structure	Examples
6	6	Octahedral		$Fe(H_2O)_6^{3+}$ $Ni(H_2O)_6^{2+}$ $Co(H_2O)_6^{3+}$
4	6	Square planar		$Pt(NH_3)_4^{2+}$
4	4	Tetrahedral		$Zn(H_2O)_4^{2+}$
2	2	Linear		$Ag(NH_3)_2^{+}$

FIGURE 22.4 The source of the blue color in *The Great Wave off Kanagawa*, a woodblock print by the Japanese artist Hokusai, is the pigment Prussian blue.

SAMPLE EXERCISE 22.1 Assigning Oxidation States in Coordination Compounds **LO1**

The first modern synthetic color, Prussian blue, was made in 1704 in Berlin, Germany. It was heavily used in the 19th-century Japanese woodblock print *The Great Wave off Kanagawa* (**Figure 22.4**), and it continues to be popular with artists worldwide. The formula for the insoluble pigment is $Fe_4[Fe(CN)_6]_3$.

a. What is the formula of the complex ion in this compound and what is its charge if the counterion is Fe^{3+}?
b. What is the oxidation state of iron in the complex ion?

Collect, Organize, and Analyze We have the formula of a coordination compound and are asked to identify the complex ion and the counterion. The complex ion is contained in brackets, the four Fe^{3+} counterions are outside the brackets, and the compound must be electrically neutral.

Solve

a. The combined charge of four Fe^{3+} is 12+. To achieve electrical neutrality, the combined charge of the three complex anions must be 12−. Therefore, the charge of each anion is (12−)/3 = 4−, and the formula for the anion is $[Fe(CN)_6]^{4-}$.
b. Six cyanide ligands, CN^-, occupy the inner coordination sphere for a total charge of 6−. Therefore, the oxidation state of Fe must be 2+ for the charge on the complex ion to equal 4−.

Think About It Complex ions may have positive or negative charges. Correspondingly, counterions will have charges opposite to those of the complex ions to ensure electrical neutrality.

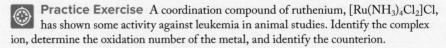

Practice Exercise A coordination compound of ruthenium, $[Ru(NH_3)_4Cl_2]Cl$, has shown some activity against leukemia in animal studies. Identify the complex ion, determine the oxidation number of the metal, and identify the counterion.

(Answers to Practice Exercises are in the back of the book.)

STEPWISE ANIMATION
Naming Coordination Compounds

22.2 Naming Complex Ions and Coordination Compounds

The names of complex ions and coordination compounds tell us the identity and oxidation state of the central ion, the names and numbers of ligands, the charge in the case of complex ions, and the identity of counterions. To convey all this information, we need to follow some naming rules.

Complex Ions with a Positive Charge

1. Start with the identities of the ligand(s). Names of common ligands appear in **Table 22.3**. If there is more than one kind of ligand, list the names alphabetically.
2. Use the usual prefix(es) in front of the name(s) written in step 1 to indicate the number of each type of ligand (**Table 22.4**).
3. Write the name of the metal ion with a Roman numeral indicating its oxidation state.

TABLE 22.3 Names and Structures of Common Ligands

Ligand	Name within Complex Ion	Structure	Charge	Number of Donor Groups
Iodide	Iodo	I^-	1–	1
Bromide	Bromo	Br^-	1–	1
Chloride	Chloro	Cl^-	1–	1
Fluoride	Fluoro	F^-	1–	1
Nitrite	Nitro	$[O{\cdots}N{\cdots}O]^-$	1–	1
Hydroxide	Hydroxo	$[O—H]^-$	1–	1
Water	Aqua	H–O–H	0	1
Pyridine (py)	Pyridyl	(pyridine ring, N)	0	1
Ammonia	Ammine	NH_3	0	1
Ethylenediamine (en)	(same)[a]	H_2N–CH_2–CH_2–NH_2	0	2
2,2'-Bipyridine (bipy)	Bipyridyl	(two linked pyridine rings, N N)	0	2
1,10-Phenanthroline (phen)	(same)[a]	(fused ring structure, N N)	0	2
Cyanide[b]	Cyano	$[C≡N]^-$	1–	1
Carbon monoxide[b]	Carbonyl	C≡O	0	1

[a]The names of some electrically neutral ligands in complexes are the same as the names of the molecules.

[b]Carbon atoms are the lone pair donors in these ligands.

TABLE 22.4 Common Prefixes Used in the Names of Complex Ions

Number of Ligands	Prefix
2	di-
3	tri-
4	tetra-
5	penta-
6	hexa-

Examples:

Formula	Name	Structure
$Ni(H_2O)_6^{2+}$	Hexaaquanickel(II)	
$Co(NH_3)_6^{3+}$	Hexaamminecobalt(III)	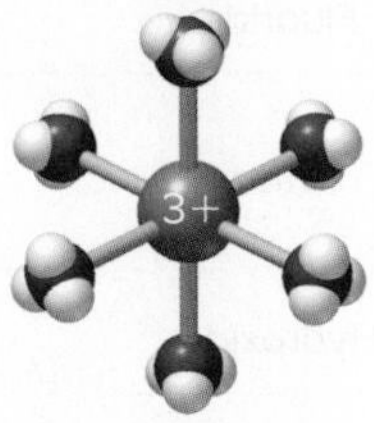
$Cu(NH_3)_4(H_2O)_2^{2+}$	Tetraamminediaquacopper(II)	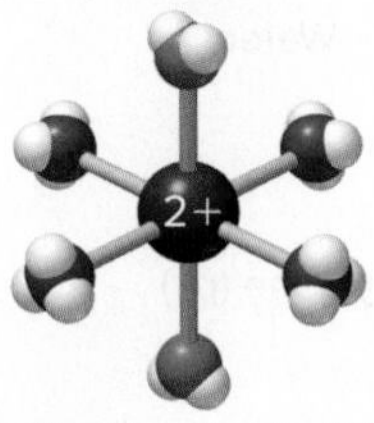

It may seem strange to have two *a*'s together in these names, but it is consistent with current naming rules. Prefixes are ignored in determining alphabetical order, which is why *ammine* comes before *aqua* rather than *di* before *tetra* in tetraamminediaquacopper(II).

The ligands are all electrically neutral in these three examples. This makes determining the oxidation state of the central metal ion a simple task because the charge on the complex ion is the same as the charge on the metal ion, which is the oxidation state of the metal. When the ligands are anions, determining the oxidation state of the central metal ion requires us to account for these charges.

Complex Ions with a Negative Charge

1. Follow the steps for naming positively charged complexes.
2. Add *-ate* to the name of the central metal ion to indicate that the complex ion carries a negative charge (just as we use *-ate* to end the names of oxoanions). For some metals, the base name changes, too. The two most common examples are iron, which becomes *ferrate*, and copper, which becomes *cuprate*.

Examples:

Formula	Name	Structure
$Fe(CN)_6^{3-}$	Hexacyanoferrate(III)	
$[Fe(H_2O)(CN)_5]^{3-}$	Aquapentacyanoferrate(II)	
$[VCl_4(NH_3)_2]^-$	Diamminetetrachlorovanadate(III)	

In the first two examples we must determine the oxidation state of Fe. We start with the charge on the complex ion and then take into account the charges on the ligand anions to calculate the charge on the metal ion. For example, the overall charge of the aquapentacyanoferrate(II) ion is 3−. It contains five CN^- ions. To reduce the combined charge of 5− from these cyanide ions to an overall charge of 3−, the charge on Fe must be 2+. The most common ionic charges for some transition metals are shown in **Figure 22.5**.

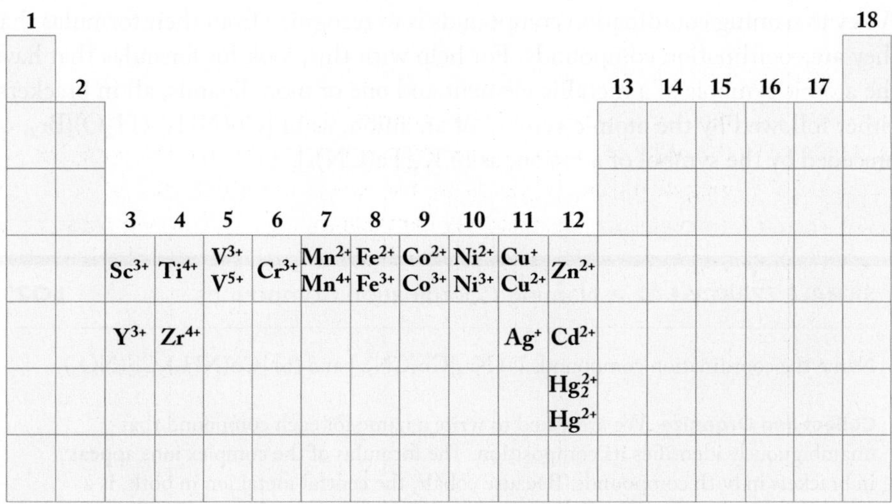

FIGURE 22.5 Common ionic charges of some transition metals.

CONCEPT TEST

What is the name of the complex anion with the formula $PtCl_4^{2-}$?

Coordination Compounds

1. If the counterion of the complex ion is a cation, the cation's name goes first, followed by the name of the anionic complex ion.
2. If the counterion of the complex ion is an anion, the name of the cationic complex ion goes first, followed by the name of the anion.

Examples:

Formula	Name	Structure
$[Ni(NH_3)_6]Cl_2$	Hexaamminenickel(II) chloride	$[$Ni with NH_3, H_3N, NH_3, H_3N, NH_3, $NH_3$$]Cl_2$
$K_3[Fe(CN)_6]$	Potassium hexacyanoferrate(III)	$K_3[$Fe with CN, NC, CN, NC, CN, CN$]$
$[Co(NH_3)_5(H_2O)]Br_2$	Pentaammineaquacobalt(II) bromide	$[$Co with NH_3, H_3N, NH_3, H_3N, NH_3, $H_2O$$]Br_2$

A key to naming coordination compounds is to recognize from their formulas that they are coordination compounds. For help with this, look for formulas that have the atomic symbols of a metallic element and one or more ligands, all in brackets, either followed by the atomic symbol of an anion, as in $[Co(NH_3)_5(H_2O)]Br_2$, or preceded by the symbol of a cation, as in $K_3[Fe(CN)_6]$.

SAMPLE EXERCISE 22.2 Naming Coordination Compounds **LO2**

Name the coordination compounds (a) $Na_4[Co(CN)_6]$ and (b) $[Co(NH_3)_5Cl](NO_3)_2$.

Collect and Organize We are asked to write a name for each compound that unambiguously identifies its composition. The formulas of the complex ions appear in brackets in both compounds. Because cobalt, the central metal ion in both, is a transition metal, we express its oxidation state by using Roman numerals. The names of common ligands are given in Table 22.3.

Analyze It is useful to take an inventory of the ligands and counterions:

		LIGAND			
Compound	**Counterion**	**Formula**	**Name**	**Number**	**Prefix**
$Na_4[Co(CN)_6]$	Na^+	CN^-	Cyano	6	Hexa-
$[Co(NH_3)_5Cl](NO_3)_2$	NO_3^-	NH_3	Ammine	5	Penta-
		Cl^-	Chloro	1	—

The oxidation state of each cobalt ion can be calculated by setting the sum of the charges on all the ions in both compounds equal to zero:

a. Ions: (4 Na^+ ions) + (1 Co^x ion) + (6 CN^- ions)
   Charges: $4+ \quad + \quad x \quad + \quad 6- \quad = 0$
   $x = 2+$

b. Ions: (1 Co^x ion) + (1 Cl^- ion) + (2 NO_3^- ions)
   Charges: $x \quad + \quad 1- \quad + \quad 2- \quad = 0$
   $x = 3+$

Solve

a. Because the counterion, sodium, is a cation, its name comes first. The complex ion is an anion. To name it, we begin with the ligand cyano, to which we add the prefix *hexa-* and write *hexacyano*. This is followed by the name of the transition metal ion: hexacyano*cobalt*. We add *-ate* to the ending of the name of the complex ion because it is an anion: hexacyanocobalt*ate*. We then add a Roman numeral to indicate the oxidation state of the cobalt: hexacyanocobaltate(*II*). Putting it all together, we get sodium hexacyanocobaltate(II).
b. The complex ion is the cation in this compound, and we begin by naming the ligands directly attached to the metal ion in alphabetical order: ammine and chloro. We indicate the number (5) of NH_3 ligands with the appropriate prefix: *penta*amminechloro. We name the metal next and indicate its oxidation state with a Roman numeral: pentaamminechloro*cobalt(III)*. Finally, we name the anionic counterion: *nitrate*. Putting it all together, we obtain the name: pentaamminechlorocobalt(III) nitrate.

The structures of the complex ions in the named coordination compounds are shown in **Figure 22.6**.

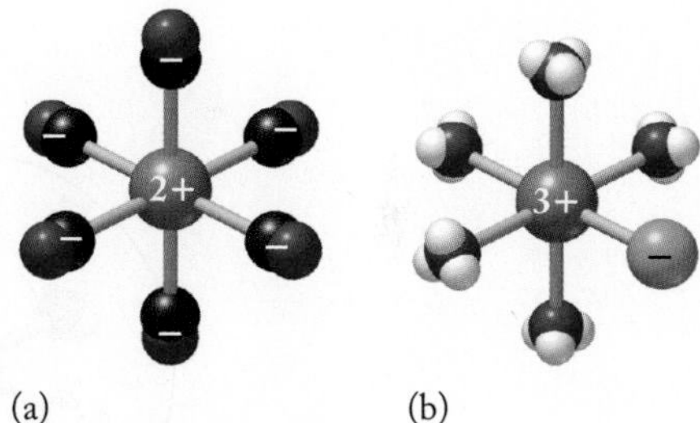

FIGURE 22.6 The structures of the complex ions in (a) sodium hexacyanocobaltate(II) and (b) pentaamminechlorocobalt(III) nitrate.

Think About It Naming coordination compounds requires us to write a name for each compound that unambiguously identifies its composition. The names sodium hexacyanocobaltate(II) and pentaamminechlorocobalt(III) nitrate are unique to the compounds $Na_4[Co(CN)_6]$ and $[Co(NH_3)_5Cl](NO_3)_2$, respectively.

Practice Exercise Identify the ligands and counterions in (a) $[Zn(NH_3)_4]Cl_2$ and (b) $[Co(NH_3)_4(H_2O)_2](NO_2)_2$, and name each compound.

22.3 Polydentate Ligands and Chelation

Ligands are electron-pair donors—that is, Lewis bases. Let's explore the strengths of several ligands as Lewis bases by considering their affinity for $Ni^{2+}(aq)$ ions. Suppose we dissolve crystals of nickel(II) chloride hexahydrate, $NiCl_2 \cdot 6\ H_2O$ (**Figure 22.7a**), in water. The dot connecting the two halves of the formula and the prefix *hexa* indicate that each Ni^{2+} ion in crystals of nickel(II) chloride is surrounded by six water molecules. Lime green crystals of $NiCl_2 \cdot 6\ H_2O$ form green aqueous solutions (**Figure 22.7b**). The solid and its solution have the same

monodentate ligand a species that forms only a single coordinate bond to a metal ion in a complex.

polydentate ligand a species that can form more than one coordinate bond per molecule.

chelation the interaction of a metal with a polydentate ligand (chelating agent); pairs of electrons on one molecule of the ligand occupy two or more coordination sites on the central metal.

color, which tells us that the same arrangement of water molecules around Ni^{2+} ions occurs in both solid $NiCl_2 \cdot 6\ H_2O$ and aqueous solutions of Ni^{2+} ions. Thus, the Ni^{2+} ion in a solution of nickel(II) chloride is most likely $Ni(H_2O)_6{}^{2+}$. The hydrated ion $Ni(H_2O)_6{}^{2+}$ can also be expressed as $Ni^{2+}(aq)$.

Now let's bubble colorless ammonia gas through a green solution of $Ni(H_2O)_6{}^{2+}$ ions. As shown in the middle beaker in Figure 22.7b, the green solution turns blue. The color change means that different ligands are bonded to the Ni^{2+} ions. We may conclude that NH_3 molecules have displaced at least some H_2O molecules around the Ni^{2+} ions. If all the molecules of H_2O are displaced, the complex $Ni(NH_3)_6{}^{2+}$ is formed. The following chemical equation describes this change:

$$Ni(H_2O)_6{}^{2+}(aq) + 6\ NH_3(g) \rightleftharpoons Ni(NH_3)_6{}^{2+}(aq) + 6\ H_2O(\ell) \qquad K_f = 5 \times 10^8$$

This *ligand displacement* reaction illustrates that Ni^{2+} ions have a greater affinity for molecules of NH_3 than for molecules of H_2O. Many other transition metal ions also have a greater affinity for ammonia than for water. We may conclude that ammonia is inherently a better electron-pair donor and hence a stronger Lewis base than water. This conclusion is reasonable because we saw in Chapter 15 that ammonia was also a stronger Brønsted–Lowry base than H_2O.

Next we add the compound ethylenediamine (see Table 22.3) to the blue solution of $Ni(NH_3)_6{}^{2+}$ ions. The solution changes color again, from blue to purple (Figure 22.7b), indicating yet another change in the ligands surrounding the Ni^{2+} ions. Molecules of ethylenediamine displace ammonia molecules from the inner coordination sphere of Ni^{2+} ions. This affinity of Ni^{2+} ions for ethylenediamine molecules is reflected in the large formation constant for $Ni(en)_3{}^{2+}$ (where "en" represents ethylenediamine):

$$Ni(H_2O)_6{}^{2+}(aq) + 3\ en(aq) \rightleftharpoons Ni(en)_3{}^{2+}(aq) + 6\ H_2O(\ell) \qquad K_f = 1.1 \times 10^{18}$$

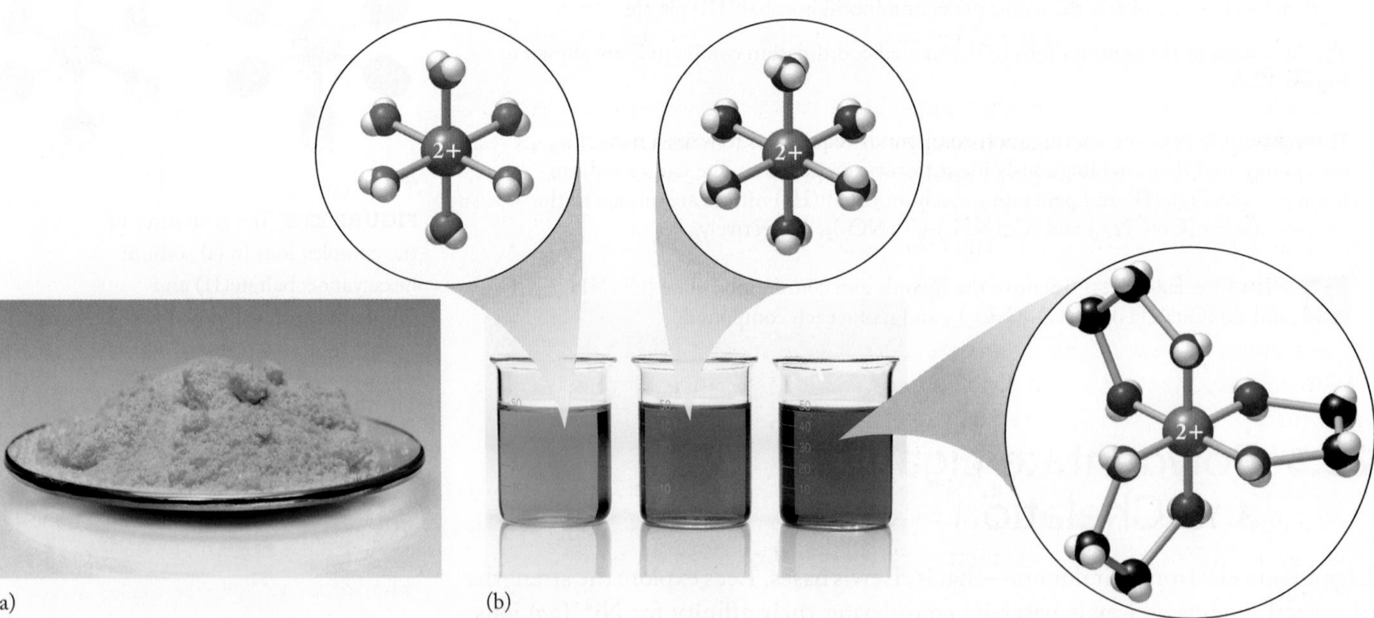

FIGURE 22.7 Structures of the complex ions in solid and dissolved nickel(II) chloride. (a) Solid nickel(II) chloride hexahydrate. (b) When it dissolves in water, the resulting solution has the same green color, indicating that each Ni^{2+} ion (gold sphere) must be surrounded by H_2O molecules both in the solid and in the solution. When ammonia gas is bubbled through a solution of $Ni(H_2O)_6{}^{2+}$, the color changes to blue as NH_3 replaces H_2O in the Ni^{2+} ion's inner coordination sphere. When ethylenediamine (en) is added to a solution of $Ni(NH_3)_6{}^{2+}$, the color turns from blue to purple as the ethylenediamine displaces the ammonia ligands and the $Ni(en)_3{}^{2+}$ complex forms.

This value is more than 10^9 times the K_f value for $Ni(NH_3)_6^{2+}$. Why should the affinity of Ni^{2+} ions for ethylenediamine be so much greater than their affinity for ammonia? After all, in both ligands the coordinate bonds are formed by lone pairs of electrons on N atoms. To answer this question, we need to look at the differences between ligands such as ammonia, which occupy only one site on a metal ion, and ligands such as ethylenediamine, which occupy two or more sites.

Many ligands in Table 22.3 can donate only one pair of electrons to a single metal ion. Even atoms with more than one lone pair usually donate only one pair at a time to a given metal ion because the other lone pair or pairs are oriented away from the metal ion. Because these ligands have effectively only one donor group, they are called **monodentate ligands**, which literally means "single-toothed."

Certain molecules larger than ammonia and water may be able to donate more than one lone pair of electrons and therefore form more than one coordinate bond to a central metal ion. Ligands in this category are called **polydentate ligands**, or more specifically *bidentate*, *tridentate*, and so on. One group of polydentate ligands is the polyamines, which include ethylenediamine, a bidentate ligand that has the structure:

$$\begin{array}{ccc} H_2\ddot{N} & & \ddot{N}H_2 \\ \backslash & & / \\ H_2C & — & CH_2 \end{array}$$

The lone pairs on the two $-NH_2$ groups are separated from each other by two $-CH_2-$ groups. This combination means that a molecule of ethylenediamine can partially encircle a metal ion so that both lone pairs can bond to the same metal ion.

The structure of an ethylenediamine complex of $Ni^{2+}(aq)$ is shown in **Figure 22.8a**. The two orbitals that share the lone pairs of electrons from a molecule of ethylenediamine must be on the same side of the Ni^{2+} ion. However, two more ethylenediamine molecules can bond to other pairs of bonding sites, displacing additional pairs of water molecules and forming a complex in which the Ni^{2+} ion is surrounded by three bidentate ethylenediamine molecules, as shown in **Figure 22.8b**.

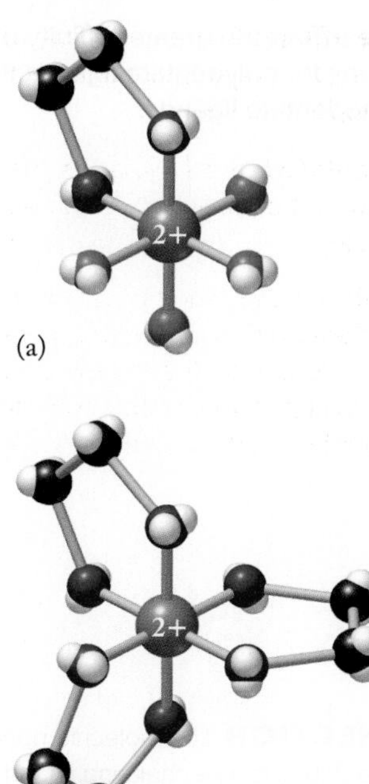

FIGURE 22.8 (a) The bidentate ligand ethylenediamine has two N atoms that can each donate a pair of electrons to empty orbitals of adjacent octahedral bonding sites on the same $Ni^{2+}(aq)$ ion (gold sphere), displacing two molecules of water. (b) Three ethylenediamine molecules occupy all six octahedral coordination sites of a Ni^{2+} ion.

Each ethylenediamine molecule forms a five-atom ring with the metal ion. If the ring were a regular pentagon (meaning all bond lengths and bond angles were exactly the same), each of its bond angles would be 108°. These pentagons are not perfect, but each ring's preferred octahedral bond angles of 90° for the N–Ni–N bond, and of 107° to 109° for all the other bonds, are accommodated with only a little strain on the ideal bond angles. The compound in Figure 22.8b is also chiral, a topic we address in Section 22.6.

An even larger ligand, diethylenetriamine ($H_2NCH_2CH_2NHCH_2CH_2NH_2$), is shown in **Figure 22.9a**. The lone pairs of electrons on its three nitrogen atoms give diethylenetriamine the capacity to form three coordinate bonds to a metal ion, meaning this is a tridentate ligand (**Figure 22.9b**).

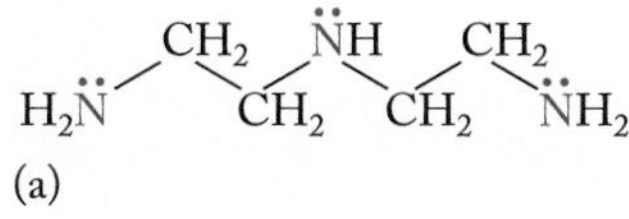

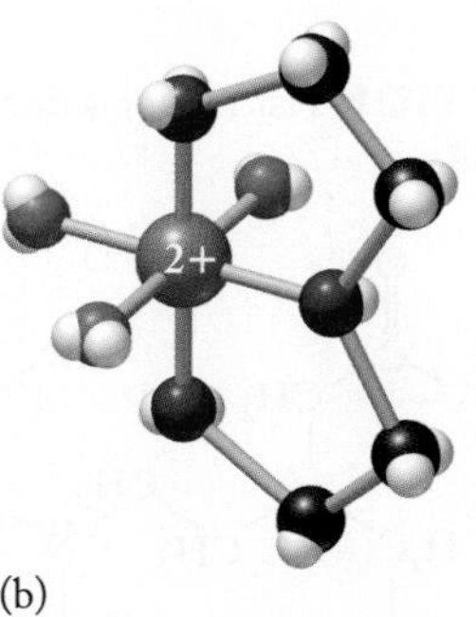

FIGURE 22.9 Tridentate chelation. (a) The three amine groups in the tridentate ligand diethylenetriamine are all potential electron-pair donor groups. (b) When these groups donate their lone pairs of electrons to a $Ni^{2+}(aq)$ ion (gold sphere), they occupy three of the six coordination sites on the ion.

The interaction of a metal ion with a ligand having multiple donor atoms is called **chelation** (pronounced *key-LAY-shun*). The word comes from the Greek *chele*, meaning "claw." The polydentate ligands that take part in these interactions are called *chelating agents*.

When we added ethylenediamine to the blue solution of $Ni(NH_3)_6^{2+}$ ions in Figure 22.7b, molecules of ethylenediamine (en) displaced ammonia molecules from the inner coordination sphere of Ni^{2+} ions as described in the following chemical equation:

$$Ni(NH_3)_6^{2+}(aq) + 3\ en(aq) \rightleftharpoons Ni(en)_3^{2+}(aq) + 6\ NH_3(aq) \qquad (22.1)$$

chelate effect the greater affinity of metal ions for polydentate ligands than for monodentate ligands.

The color change tells us that this reaction as written is spontaneous. As we discussed in Chapter 17, spontaneous reactions are those in which free energy decreases (i.e., $\Delta G < 0$). Furthermore, under standard conditions the change in free energy ΔG° is related to the changes in enthalpy and entropy that accompany the reaction:

$$\Delta G^\circ = \Delta H^\circ - T\Delta S^\circ$$

The displacement of NH_3 by ethylenediamine is exothermic, but only slightly ($\Delta H^\circ = -12$ kJ/mol). More important, $\Delta S^\circ = +185$ J/mol · K. At 25°C, then,

$$T\Delta S^\circ = 298\ \text{K} \times \frac{185\ \text{J}}{\text{mol} \cdot \text{K}} \times \frac{1\ \text{kJ}}{1000\ \text{J}} = 55.1\ \text{kJ/mol}$$

There is such a large increase in entropy because there are 4 moles of reactants but 7 moles of products in Equation 22.1. Nearly doubling the number of moles of aqueous products over reactants translates into a large gain in entropy. It is this positive ΔS°, more than the negative ΔH° value, that drives the reaction and makes it spontaneous. Entropy gains drive many complexation reactions that involve polydentate ligands. The entropy-driven affinity of metal ions for polydentate ligands is called the **chelate effect**.

CONNECTION The molecule nonactin in Figure 21.6 acts as a chelating ligand when it forms a complex ion with K^+.

Many chelating agents have more than one kind of electron-pair–donating group. *Aminocarboxylic acids* represent one family of such compounds. The most important of them is ethylenediaminetetraacetic acid (EDTA), which is shown in **Figure 22.10a**. One molecule of EDTA contains two amine (nitrogen-containing) groups and four carboxylic acid (–COOH) groups. When the acid groups release their H^+ ions, they form four carboxylate anions, $-COO^-$, in which either of the O atoms can donate a pair of electrons to a central metal ion. When O atoms on all four groups do so and the two amine groups do as well, six octahedral bonding sites around the metal ion can be occupied, as shown in **Figure 22.10b**.

EDTA forms very stable complex ions and is used as a metal ion *sequestering agent*—that is, as a chelating agent that binds metal ions so tightly that they are "sequestered" and prevented from reacting with other substances. For example, EDTA is used as a preservative in many beverages and prepared foods because it sequesters iron, copper, zinc, manganese, and other transition metal ions often present in these foods that can catalyze the degradation of ingredients in the foods. Many foods are fortified with ascorbic acid (vitamin C), which is particularly vulnerable to metal-catalyzed degradation because it is also a polydentate ligand and is more likely to be oxidized when chelated to one of the above metal ions. EDTA effectively shields vitamin C from these ions. EDTA is also a common additive to shampoos because it can remove metal cations from hard water.

CONNECTION Hard water is discussed in Section 11.5.

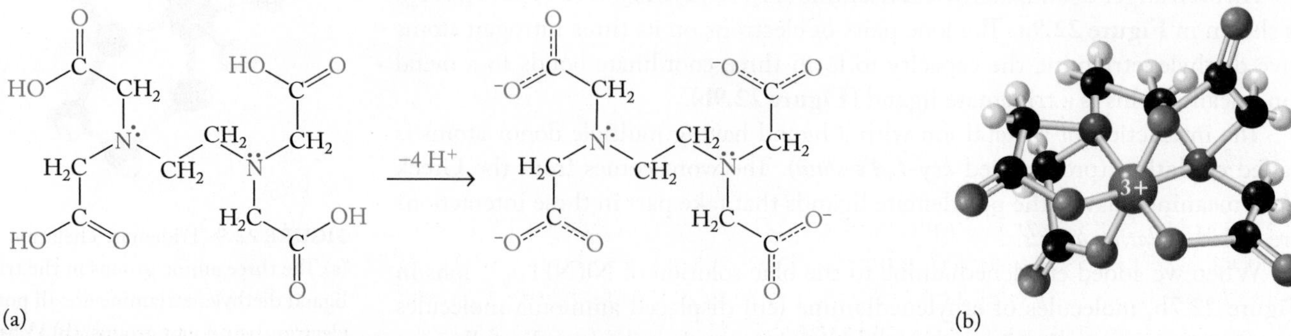

FIGURE 22.10 (a) In the hexadentate ligand EDTA, the six donor groups are the two amine groups and the four carboxylic acid groups. The acid groups ionize to form carboxylate anions. (b) All six Lewis base groups in ionized EDTA can form a coordinate bond with the same metal ion, such as Co^{3+} (the gold sphere) shown here. In the process they form four 5-membered rings.

SAMPLE EXERCISE 22.3 Identifying the Potential Electron-Pair–Donor Groups in a Molecule **LO3**

How many donor groups does nitrilotriacetic acid (NTA) have?

```
            O
             \\
              C—OH
             /
          CH2
             \
   HO         N—CH2
     \       /     \
      C—CH2         C=O
     //            /
    O            HO
```

Collect, Organize, and Analyze We need to examine the molecular structure of this polydentate ligand to find electron pairs that can be donated. Because there are three single bonds around the N atom, the atom's fourth sp^3 orbital must contain a lone pair of electrons. When all three carboxylic acid groups are ionized, there are three carboxylate groups in the molecule, and each carboxylate group can donate one nonbonding pair of electrons from one of its oxygen atoms to a metal atom.

Solve The central N atom and an O atom from each of the three carboxylate groups form a total of four coordinate bonds. Therefore, NTA is potentially a tetradentate ligand with four donor groups. If we also consider the lone pairs on the oxygen atoms of the C=O group, NTA could coordinate a total of six oxygens to a transition metal cation, making NTA a hexadentate ligand. If we coordinate the nitrogen in NTA through its lone pair, one can imagine a seven-coordinate, heptadentate NTA ligand. The three carboxylate groups, though, are all bonded to a single N through a CH_2 group, so the bond distances and bond angles in NTA may prevent coordination numbers greater than four.

Think About It The tetradentate capacity of NTA is reasonable because, like EDTA, it is an aminocarboxylic acid. It has one fewer amino group and one fewer carboxylic acid group than the hexadentate EDTA.

 **Practice Exercise** How many potential donor groups are there in citric acid, a component of citrus fruits and a widely used preservative in the food industry?

```
              O    OH
        O      \\ /      O
        ||      C        ||
        C       |        C
       / \     C        / \
     HO   CH2 / \  CH2     OH
               OH
```

22.4 Crystal Field Theory

Why does the formation of complex ions change the color of solutions of transition metals? The colors of transition metal compounds and ions in solution result from transitions of *d*-orbital electrons. Let's explore these transitions by using Cr^{3+} as our model transition metal ion (**Figure 22.11**).

A Cr^{3+} ion has the electron configuration $[Ar]3d^3$ in the gas phase. When a Cr^{3+} ion (or any atom or ion) is in the gas phase, all the orbitals in a given subshell have the same energy (**Figure 22.12a**). However, when a Cr^{3+} ion is in an aqueous

CONNECTION In Section 7.3 we first used the equation $E = h\nu = hc/\lambda$ when discussing the energy of light in the electromagnetic spectrum (Figure 7.7).

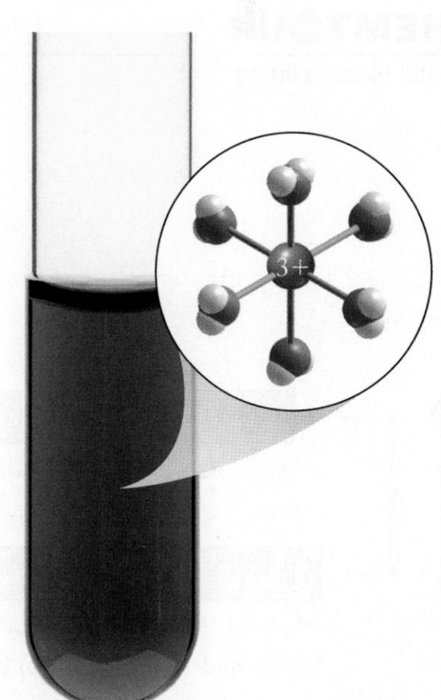

FIGURE 22.11 When chromium(III) nitrate dissolves in water, the resulting solution has a distinctive violet color caused by the presence of $Cr(H_2O)_6{}^{3+}$ ions.

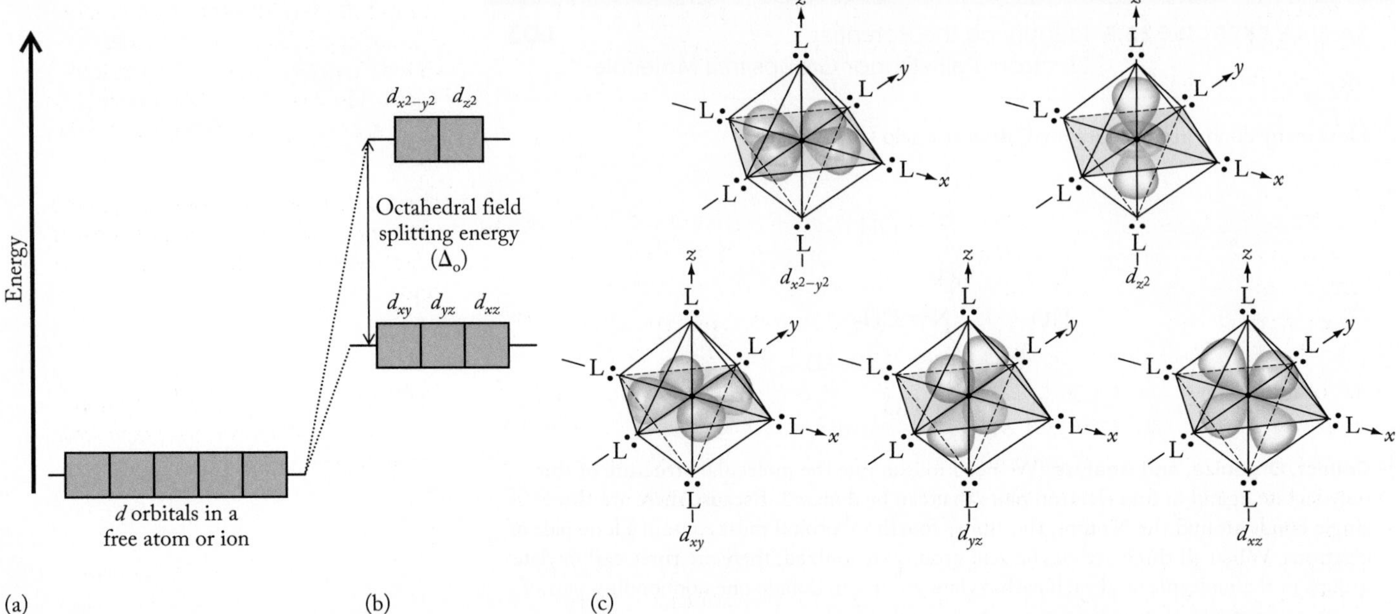

FIGURE 22.12 Octahedral crystal field splitting. (a) In an atom or ion in the gas phase, all orbitals in a subshell are degenerate, as shown here for the five 3*d* orbitals. (b) When an ion is part of a complex ion in a compound or solution, repulsions between electrons in the ion's *d* orbitals and ligand electrons raise the energy of the orbitals, as shown here for an octahedral field. (c) The greatest repulsion is experienced by electrons in the $d_{x^2-y^2}$ and d_{z^2} orbitals because the lobes of these orbitals are directed toward the corners of the octahedron and so are closest to the lone pairs on the ligands (L). The lobes of the lower-energy d_{xy}, d_{yz}, and d_{xz} orbitals are directed toward points that lie between the corners of the octahedron, so electrons in them experience less repulsion.

solution and surrounded by an octahedral array of water molecules in $Cr(H_2O)_6^{3+}$, the energies of its 3*d* orbitals are no longer all the same. The $3d_{xy}$, $3d_{yz}$, and $3d_{xz}$ orbitals experience some increase in energy, but the energies of the $3d_{x^2-y^2}$ and $3d_{z^2}$ orbitals increase even more (**Figure 22.12b**) because the lobes of the $3d_{x^2-y^2}$ and $3d_{z^2}$ orbitals point directly toward the H_2O molecules' oxygen atoms at the corners of the octahedron formed by the ligands and are repelled by the electrons on those O atoms (**Figure 22.12c**). The energies of the $3d_{xy}$, $3d_{yz}$, and $3d_{xz}$ orbitals are not raised as much because the lobes of these three orbitals do not point directly toward the corners of the octahedron, so the electron repulsion they experience is weaker.

CHEMTOUR
Crystal Field Splitting

This process of changing degenerate (equal-energy) *d* orbitals into orbitals with different energies is known as **crystal field splitting**, and the difference in energy created by crystal field splitting is called **crystal field splitting energy (Δ)**. The name was originally used to describe splitting of *d*-orbital energies in minerals, but the theory is also routinely applied to species in aqueous solutions.

In a $Cr(H_2O)_6^{3+}$ ion, three electrons are distributed among five 3*d* orbitals. According to Hund's rule, each of the three electrons should occupy one of the three lower-energy orbitals, leaving the two higher-energy orbitals unoccupied, as shown in **Figure 22.13a**. The energy difference between the two subsets of orbitals is symbolized by Δ_o, where the subscript "o" indicates that the energy split was caused by an *o*ctahedral array of electron repulsions.

What if an aqueous Cr^{3+} ion absorbs a photon whose energy is exactly equal to Δ_o? As the photon is absorbed, a 3*d* electron moves from a lower-energy orbital to a higher-energy orbital (**Figure 22.13b**).

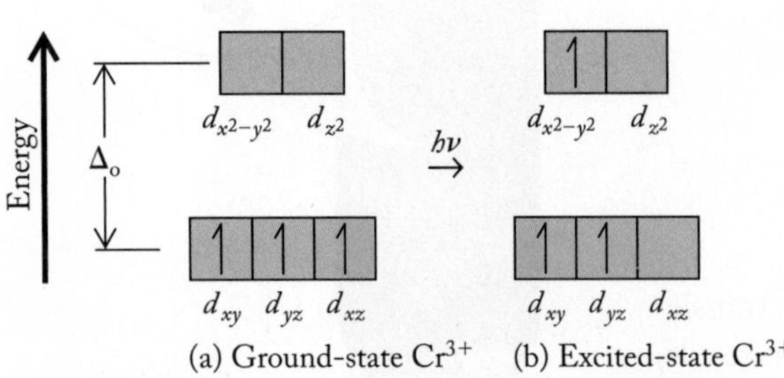

FIGURE 22.13 A Cr^{3+} ion, $[Ar]3d^3$, in an octahedral field can absorb a photon of light that has energy ($h\nu$) equal to Δ_o. This energy raises a 3*d* electron from (a) one of the lower-energy *d* orbitals to (b) one of the higher-energy *d* orbitals.

The wavelength λ of the absorbed photon is related to the energy difference between the two groups of orbitals—in other words, to the crystal field splitting energy—as follows:

$$E = \frac{hc}{\lambda} = \Delta_o \tag{22.2}$$

The energy and wavelength of a photon are inversely proportional to each other (see Equation 7.3 in Section 7.3). Therefore, the larger the crystal field splitting in a complex ion, the shorter the wavelength of the photons the ion absorbs.

The size of the energy gap between split *d* orbitals often corresponds to radiation in the visible region of the electromagnetic spectrum. This means that the colors of solutions of metal complexes depend on the strengths of metal–ligand interactions that affect Δ_o. When white light (which contains all colors of visible light) passes through a solution containing complex ions, the ions may absorb energy corresponding to one or more colors. So, the light leaving the solution and reaching our eyes is missing some colors.

The color we perceive for any transparent object is not the color it absorbs but rather the color(s) that it transmits. To relate the color of a solution to the wavelengths of light it absorbs, we need to consider complementary colors as defined by a simple color wheel (**Figure 22.14**). For example, red and green are complementary colors, so a solution that absorbs green light appears red to us.

Aqueous solutions of Cu^{2+} (Figure 18.1) are blue because $Cu(H_2O)_6{}^{2+}$ absorbs orange light, the color that is complementary to blue. A solution of $Cu(NH_3)_4{}^{2+}$ ions also has a distinctive deep blue color (**Figure 22.15**). The spectrum of a solution containing $Cu(NH_3)_4{}^{2+}$ features a rather broad absorption band between 580 and 590 nm that spans yellow, orange, and red wavelengths. Once again, the complementary color is in the blue-violet region. If more than one color is absorbed, then our brain processes the bands of transmitted colors and signals to us as the average of these colors. The violet color of the solution containing $Cr^{3+}(aq)$ in Figure 22.11 is the result of color averaging, as are the colors of the aqueous Ni^{2+}, $Ni(NH_3)_6{}^{2+}$, and $Ni(en)_3{}^{2+}$ solutions in Figure 22.7.

The nickel(II) and chromium(III) complexes we have examined up to this point were octahedral, and their central ions had a coordination number of 6. In the solution of $Cu(NH_3)_4{}^{2+}$, however, the copper(II) ion has a coordination number of 4. This means that the deep blue color of this complex ion is caused by a different crystal field.

Four ligands around a central metal have either a tetrahedral arrangement or a square planar arrangement (Table 22.2). Square planar geometries tend to be limited to the transition metal ions with nearly filled valence-shell *d* orbitals, particularly those with d^8 or d^9 electron configurations. Cu^{2+} has the electron

crystal field splitting the separation of a set of *d* orbitals into subsets with different energies as a result of interactions between electrons in those orbitals and lone pairs of electrons in ligands.

crystal field splitting energy (Δ) the difference in energy between subsets of *d* orbitals split by interactions in a crystal field.

CONNECTION Crystal field theory is an example of molecular orbital theory, which was introduced in Section 9.6.

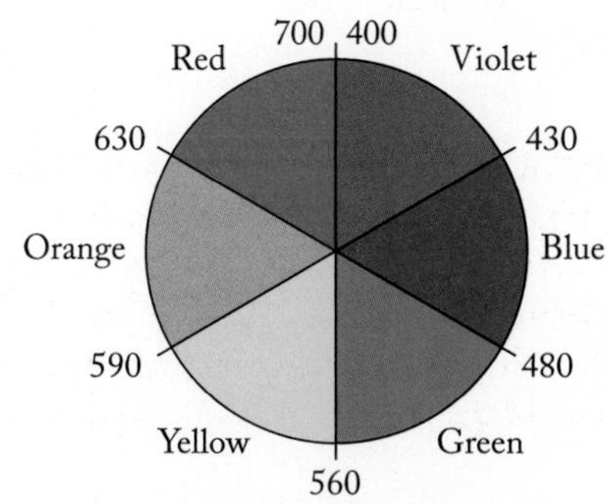

FIGURE 22.14 A color wheel. Colors on opposite sides of the wheel are complementary to each other. When we look at a solution or object that absorbs light corresponding to a given color, we see the complementary color. Wavelengths are in nanometers.

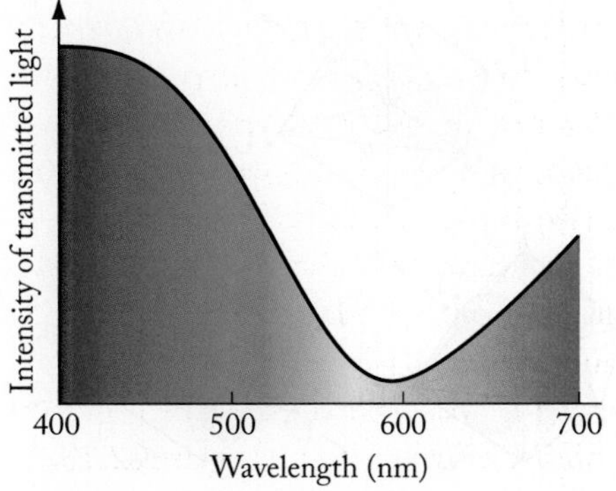

FIGURE 22.15 The visible light transmitted by a solution of $Cu(NH_3)_4{}^{2+}$ ions is missing much of the yellow, orange, and red portions of the visible spectrum because of a broad absorption band centered at 590 nm. Our eyes and brain perceive the transmitted colors as navy blue.

spectrochemical series a list of ligands rank-ordered by their ability to split the energies of the *d* orbitals of transition metal ions.

configuration [Ar]$3d^9$, and the $Cu(NH_3)_4^{2+}$ complex is square planar—which means that the strongest interactions occur between the 3*d* orbitals on the central ion and the nitrogen atom lone pairs at the four corners of the equatorial plane of the octahedron, as shown in **Figure 22.16**. The $d_{x^2-y^2}$ orbital has the strongest interactions and the highest energy because its lobes are oriented directly at the four corners of the plane. The d_{xy} orbital has slightly less energy because its lobes, although in the *xy* plane, are directed 45° away from the corners. Electrons in the three *d* orbitals with most of their electron density out of the *xy* plane (i.e., d_{z^2}, d_{xz}, and d_{yz}) interact even less with the lone pairs of the ligand and thus have even lower energies.

Finally, let's consider the *d* orbital crystal field splitting that occurs in a tetrahedral complex (**Figure 22.17a**). In this geometry, the greatest electron–electron repulsions are experienced in the d_{xy}, d_{yz}, and d_{xz} orbitals because the lobes of these orbitals are oriented most directly to the corners of the tetrahedron, which are occupied by ligand electron pairs (**Figure 22.17b**). The two other *d* orbitals ($d_{x^2-y^2}$ and d_{z^2}) are less affected because their lobes do not point toward the corners. The difference in energy between the two subsets of *d* orbitals in a tetrahedral geometry is labeled Δ_t.

Before ending this discussion on the colors of transition metal ions, let's revisit the color changes we saw in Figure 22.7, when first ammonia and then ethylenediamine were added to a solution of $Ni^{2+}(aq)$ ions. A green solution of $Ni^{2+}(aq)$ ions absorbs at red light (725 nm) as shown in **Table 22.5**. Similarly, a blue solution of $Ni(NH_3)_6^{2+}$ or a violet solution of $Ni(en)_3^{2+}$ absorbs yellow light (570 nm) or green light (545 nm), respectively. Notice how the colors these three solutions absorb run from longest to shortest wavelength. Radiant energy is inversely proportional to wavelength (Equation 22.2); therefore, the crystal field splitting of the *d* orbitals of Ni^{2+} ions is en > NH_3 > H_2O. Table 22.5 summarizes the

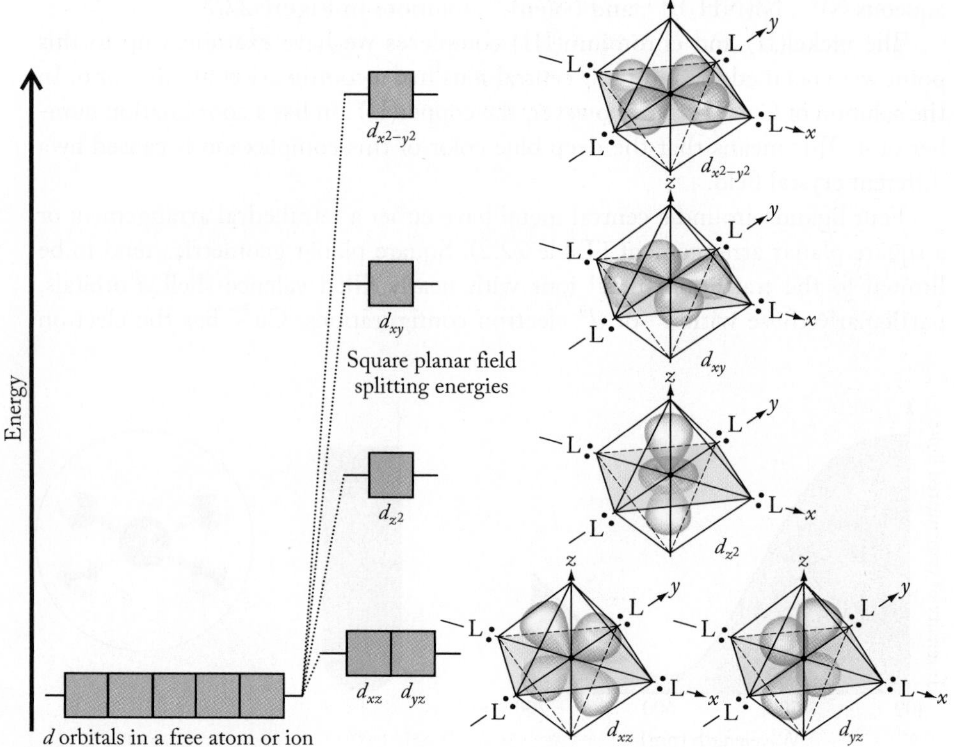

FIGURE 22.16 Square planar crystal field splitting. The *d* orbitals of a transition metal ion in a square planar field are split into several energy levels depending on the relative orientations of the metal orbitals and ligand electrons at the four corners of the square. The $d_{x^2-y^2}$ orbital has the highest energy because its lobes are directed right at the four corners of the square plane.

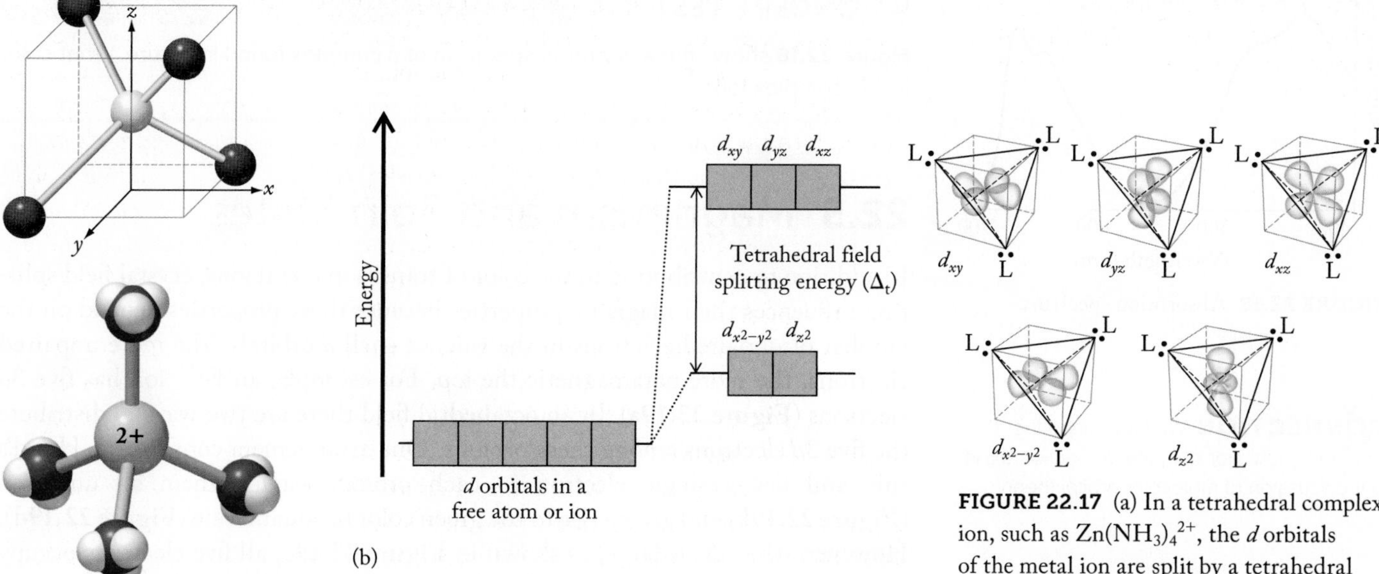

FIGURE 22.17 (a) In a tetrahedral complex ion, such as $Zn(NH_3)_4^{2+}$, the *d* orbitals of the metal ion are split by a tetrahedral crystal field. The tetrahedral ion fits inside a cube so that NH_3 molecules occupy opposite corners on the top and bottom face of the cube. The NH_3 ligands lie between the x, y, and z axes. (b) The lobes of the higher-energy orbitals (d_{xy}, d_{yz}, and d_{xz}) are closer to the ligands at the four corners of the tetrahedron than the lobes of the lower-energy orbitals ($d_{x^2-y^2}$ and d_{z^2}) are. (One of the four corners of the tetrahedron is hidden in these drawings.)

observed colors of these Ni^{2+} ion complexes and the meaning we can infer from their colors.

Chemists use the parameter *field strength* to describe the relative magnitude of the split in the energies of the *d* orbitals in metal ions, ranking ligands in what is called a **spectrochemical series. Table 22.6** contains one such series. As the field strength of the ligand increases from the bottom to the top of Table 22.6, the crystal field splitting energy (Δ) increases. Consequently, high-field-strength ligands form complexes that absorb short-wavelength, high-energy light, whereas complexes of low-field-strength ligands absorb long-wavelength, low-energy light. The refinements of crystal field theory that have been made to better explain the spectrochemical series are beyond the scope of this book.

TABLE 22.5 Light Transmitted and Absorbed by Three Ni^{2+} Complexes

Complex	$Ni(H_2O)_6^{2+}$	$\xrightarrow{NH_3}$	$Ni(NH_3)_6^{2+}$	$\xrightarrow{en}$	$Ni(en)_3^{2+}$
Appearance	Green		Blue		Violet
Absorbs	Red		Yellow		Green
Absorbed λ (nm)	725	>	570	>	545
$E = hc/\lambda$ (J × 10^{19})	2.7	<	3.5	<	3.6

TABLE 22.6 Spectrochemical Series of Some Common Ligands

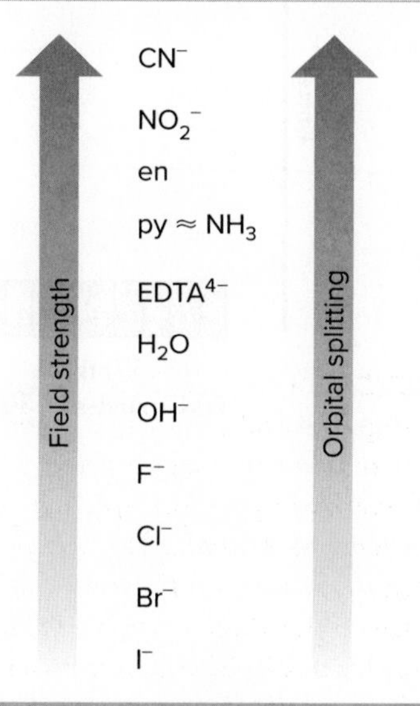

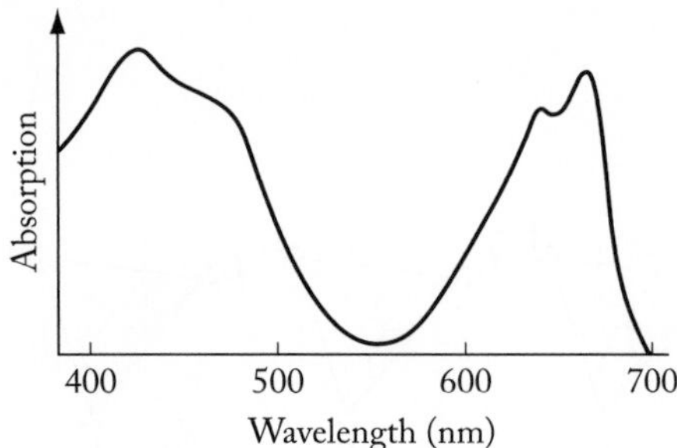

FIGURE 22.18 Absorption spectrum.

CONNECTION We introduced the magnetic behavior of matter in Section 9.6 in our discussion of molecular orbital theory.

CONCEPT TEST

Figure 22.18 shows the absorption spectrum of a complex found in nature. What color is this complex ion?

22.5 Magnetism and Spin States

In addition to contributing to the color of transition metal ions, crystal field splitting influences their magnetic properties because these properties depend on the number of unpaired electrons in the valence shell *d* orbitals. The more unpaired electrons, the more paramagnetic the ion. For example, an Fe^{3+} ion has five 3*d* electrons (**Figure 22.19a**). In an octahedral field there are two ways to distribute the five 3*d* electrons among these orbitals. One arrangement conforms to Hund's rule and has a single electron in each orbital, leaving them all unpaired (**Figure 22.19b**) and giving rise to the green color in aquamarine (**Figure 22.19d**). However, when Δ_o is large, as shown in **Figure 22.19c**, all five electrons occupy the three lower-energy orbitals. This pattern of electron distribution occurs when the energy of repulsion between two electrons in the same orbital is less than the energy needed to promote an electron to a higher-energy orbital. In this configuration, only one electron is unpaired.

The configuration with all five electrons unpaired is called the *high-spin state* because the spin on all five electrons is in the same direction, resulting in the maximum magnetic field produced by the spins. The configuration with only one electron unpaired is called the *low-spin state*. Both configurations are paramagnetic because both have at least one unpaired electron, but material containing high-spin iron(III) ions would be more strongly attracted to an external magnet than a material containing low-spin iron(III) ions.

Not all transition metal ions can have both high-spin and low-spin states. A Cr^{3+} ion in an octahedral field, for instance, has only three 3*d* electrons (Figure 22.13), so each electron is unpaired whether the orbital energies are split a

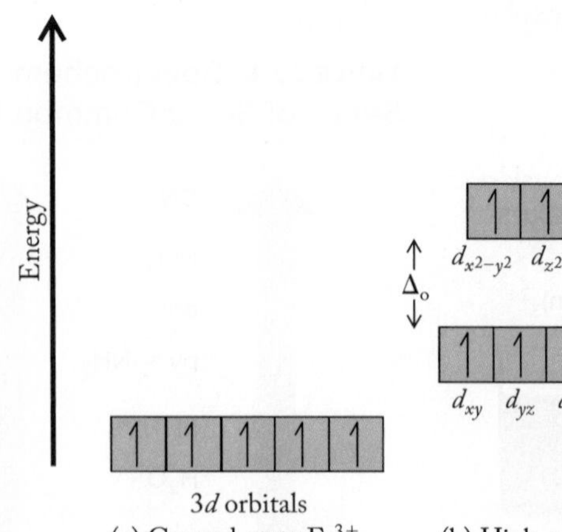

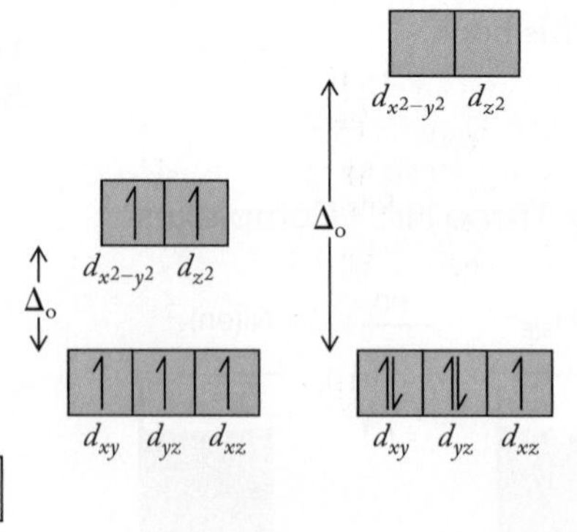

FIGURE 22.19 Low-spin and high-spin complexes. (a) The ground state of a free Fe^{3+} ion has a degenerate, half-filled set of 3*d* orbitals. (b) A weak octahedral field (Δ_o < electron-pairing energy) produces the high-spin state: five unpaired electrons, each in its own orbital. (c) In a strong octahedral field (Δ_o > electron-pairing energy), the energies of the 3*d* orbitals are split enough to produce the low-spin state: two sets of paired electrons, one unpaired electron, and two empty, higher-energy orbitals. (d) The Fe^{3+} ions in crystals of aquamarine are high spin.

lot or only a little. Therefore, Cr^{3+} ions and any ion with less than four *d* electrons will have only one spin state. Metal ions with eight or more electrons in *d* orbitals also have only one spin state because there cannot be more than one or two unpaired electrons in these orbitals no matter how the electrons are distributed.

SAMPLE EXERCISE 22.4 Predicting Spin States **LO5**

Determine which of the following ions can have high-spin and low-spin configurations when surrounded by six Lewis bases in an octahedral complex: (a) Mn^{4+}; (b) Mn^{2+}; (c) Cu^{2+}.

Collect and Organize Mn and Cu are in groups 7 and 11 of the periodic table, so their atoms have 7 and 11 valence electrons, respectively. In an octahedral field, a set of five *d* orbitals splits into a low-energy subset of three orbitals and a high-energy subset of two orbitals.

Analyze To determine whether high-spin and low-spin states are possible, we need to determine the number of *d* electrons in each ion. Then we need to distribute them among sets of *d* orbitals split by an octahedral field to see whether it is possible for the ions to have different spin states. The electron configurations for manganese and copper atoms are $[Ar]3d^54s^2$ for Mn and $[Ar]3d^{10}4s^1$, respectively. When the transition metals Mn and Cu form cations, their atoms lose their 4*s* electrons first and then their 3*d* electrons. Therefore, the numbers of *d* electrons in the ions are 3 in Mn^{4+}, 5 in Mn^{2+}, and 9 in Cu^{2+}. It is likely that the ion with the fewest *d* electrons (Mn^{4+}) and the one with nearly the most *d* electrons possible (Cu^{2+}) will each have only one spin state.

Solve

a. Mn^{4+}: Following Hund's rule and putting three electrons into the lowest-energy 3*d* orbitals available and keeping them as unpaired as possible gives this orbital distribution of electrons (**Figure 22.20a**). There is only one spin state for Mn^{4+}.
b. Mn^{2+}: There are two options for distributing 5 electrons among the five 3*d* orbitals (**Figure 22.20b**). Thus, Mn^{2+} can have a high-spin (on the left) or a low-spin (on the right) configuration in an octahedral field.
c. Cu^{2+}: The 9 3*d* electrons fill the lower-energy orbitals and nearly fill the higher-energy ones. Only one arrangement is possible, so Cu^{2+} has only one spin state (**Figure 22.20c**).

Think About It In an octahedral field, metal ions with 4, 5, 6, or 7 *d* electrons can exist in high-spin and low-spin states. Those ions with 3 or fewer *d* electrons have only one spin state, in which all the electrons are unpaired and in the lower-energy set of orbitals. Ions with 8 or more *d* electrons have only one spin state because their lower-energy set of orbitals is filled. The magnitude of the crystal field splitting energy, Δ_o, determines which spin state an ion with 4, 5, 6, or 7 *d* electrons occupies.

Practice Exercise Which of the following ions can have high-spin and low-spin configurations when part of an octahedral complex: (a) V^{4+}; (b) Cr^{3+}; (c) Ni^{3+}? Are any of the possible spin configurations diamagnetic?

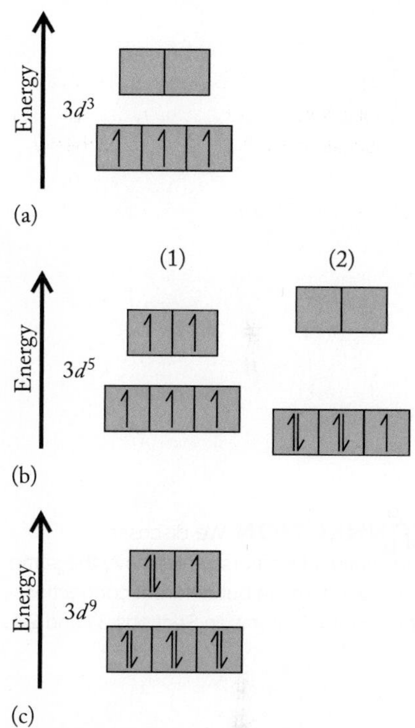

FIGURE 22.20 Options for electron distribution.

If the energy needed to promote an electron to a higher-energy orbital (Δ_o) is greater than the energy from the repulsion experienced by two electrons sharing the same lower-energy orbital, a low spin state is observed. If Δ_o is less than the energy of repulsion, a high spin state is observed. Several factors affect the size of Δ_o. We have already discussed a major one in the context of the spectrochemical series shown in Table 22.6: the different field strengths of different ligands. Because molecules containing nitrogen atoms with lone pairs of electrons are stronger field-splitting ligands than H_2O molecules, hydrated metal ions (small Δ_o) are more likely to be in high-spin states, and metals surrounded by nitrogen-containing ligands (large Δ_o) are more likely to be in low-spin states. Another factor affecting spin state is the oxidation state of the metal ion. The higher the oxidation number (and ionic charge), the stronger the attraction of the electron pairs on the ligands for the ion. Greater attraction leads to more ligand–*d* orbital interaction and therefore to a larger Δ_o.

Complexes of transition metals in the fifth and sixth rows tend to be low spin because their 4*d* and 5*d* orbitals extend farther from the nucleus than do 3*d* orbitals. These larger *d* orbitals overlap more and interact more strongly with the lone pairs of electrons on the ligands, leading to greater crystal field splitting.

Our discussion of high-spin and low-spin states has focused entirely on *d* orbitals split by octahedral fields. What about spin states in tetrahedral fields? Almost all tetrahedral complexes are high spin because tetrahedral fields are weaker than octahedral fields. Weaker field strength means less *d*-orbital splitting—not enough to offset the energies associated with pairing two electrons in the same orbitals. Therefore, Hund's rule is obeyed.

CONCEPT TEST

Explain the following:

a. $Mn(py)_6^{2+}$ is a high-spin complex ion, but $Mn(CN)_6^{4-}$ is low spin.

b. $Fe(NH_3)_6^{2+}$ is high spin, but $Ru(NH_3)_6^{2+}$ is low spin.

CONNECTION We discussed constitutional isomers, which have the same molecular formula but different connections between their atoms, in Sections 3.1 and 20.2.

22.6 Isomerism in Coordination Compounds

The coordination compound $Pt(NH_3)_2Cl_2$ was first described in 1849 and is now widely used as a chemotherapeutic agent to treat several types of cancer. Both molecules of ammonia and the two chloride ions are coordinately bonded to the Pt^{2+} ion, so the name of the compound is diamminedichloroplatinum(II). No counterions are present because the sum of the charges on the ligands and the platinum is zero and the complex is neutral. The compound is square planar, and when we draw it, there are two ways to arrange the ligands about the central metal ion: the two chloride ions and the two ammonia ligands can be at adjacent corners of the square (**Figure 22.21a**) or at opposite corners (**Figure 22.21b**). Coordination compounds like these two that have the same composition and the same connections between parts but differ in the three-dimensional arrangement of those parts are stereoisomers.

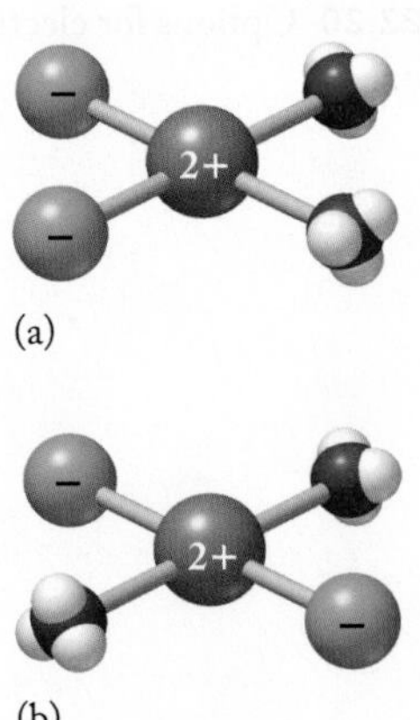

FIGURE 22.21 Two ways to orient the Cl^- ions and NH_3 molecules around a Pt^{2+} ion (gold sphere) in the square planar coordination compound $Pt(NH_3)_2Cl_2$. (a) The two members of each pair of ligands are on the same side of the square in *cis*-diamminedichloroplatinum(II). (b) The two members of each pair are at opposite corners in *trans*-diamminedichloroplatinum(II).

The two molecules in Figure 22.21 have different physical and chemical properties, and we must find a way to distinguish between them when we name them. Isomer (a), with two members of each pair of ligands at adjacent corners, is called *cis*-diamminedichloroplatinum(II). Isomer (b), with pairs of ligands at opposite corners, is *trans*-diamminedichloroplatinum(II). By analogy to the stereoisomers of alkenes we saw in Chapter 20, these types of stereoisomers are called cis–trans isomers.

To illustrate the importance of stereoisomerism in coordination compounds, consider this: *cis*-diamminedichloroplatinum(II) is a widely used anticancer drug with the common name *cisplatin*, but the trans isomer is much less effective in fighting cancer. The therapeutic power of cisplatin comes from its structurally specific reactions with DNA. During these reactions the two Cl^- ions of

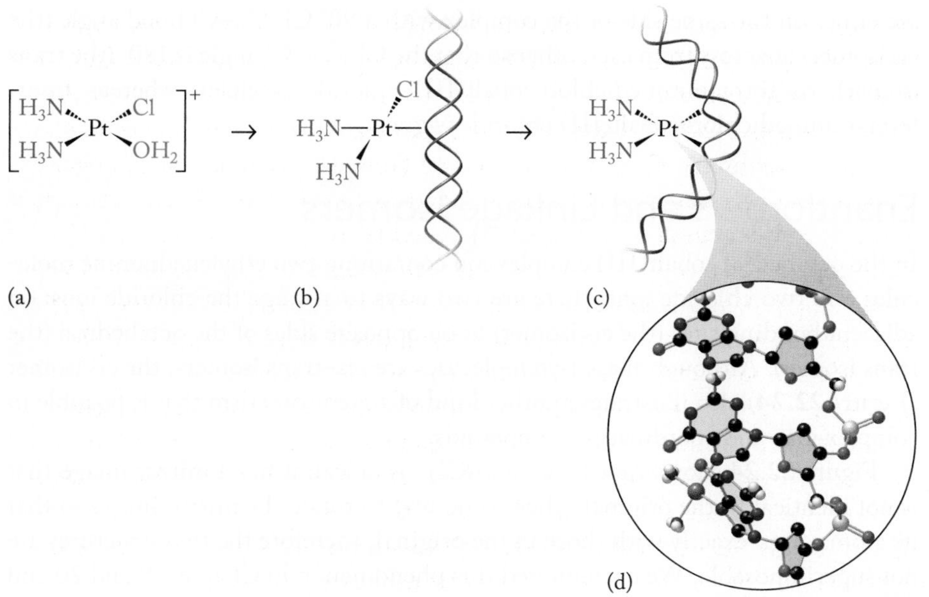

FIGURE 22.22 The reaction of *cis*-diamminedichloroplatinum(II) with DNA. (a) After the drug is administered, one of its chloride ions is displaced by a molecule of water, turning each molecule into a 1+ ion. (b) The ion is close to a nitrogen-containing base on a strand of DNA, which displaces the water molecule and forms a coordinate bond to Pt. (c) A nearby base forms a bond to Pt by displacing the chloride ion. Forming bonds to the two DNA bases distorts the DNA molecule so much that it cannot function properly. (d) A magnified view of the bonds that form between platinum(II) and nitrogen atoms in DNA. We explored the structure and function of DNA in Section 20.7.

$Pt(NH_3)_2Cl_2$ are replaced by two nitrogen-containing bases on a strand of DNA in the nucleus of a cell, as shown in **Figure 22.22**. The ability of cisplatin to cross-link these bases distorts the molecular shape of the DNA (Figure 22.22d) and disrupts its normal function. Most important, this kind of cell DNA damage inhibits the ability of cancerous cells to grow and replicate. The trans isomer forms complexes with other intracellular compounds more readily than with DNA so it is much less effective as an anticancer agent.

Cis–trans isomers are also possible in octahedral complexes containing more than one type of ligand. For example, there are two possible stereoisomers of $[Co(NH_3)_4Cl_2]Cl$ (**Figure 22.23**). The two chloro ligands in $[Co(NH_3)_4Cl_2]^+$

CONNECTION We introduced the concept of isomers in Section 3.1 and again in Section 9.5 when we discussed chirality. Enantiomers, also called optical isomers, are one type of stereoisomer; they are molecules that are not superimposable on their mirror image and that rotate the plane of polarized light, as we saw in Section 20.2.

(a) *cis*-Tetraamminedichlorocobalt(III) chloride (b) *trans*-Tetraamminedichlorocobalt(III) chloride

FIGURE 22.23 Structures of the complex ions in the two stereoisomers of the coordination compound with the formula $[Co(NH_3)_4Cl_2]Cl$: (a) *cis*-tetraamminedichlorocobalt(III) chloride and (b) *trans*-tetraamminedichlorocobalt(III) chloride.

are either on the same side of the complex with a 90° Cl–Co–Cl bond angle (the cis isomer) or across from each other so that the Cl–Co–Cl angle is 180° (the trans isomer). *cis*-Tetraamminedichlorocobalt(III) chloride is violet, whereas *trans*-tetraamminedichlorocobalt(III) chloride is green.

Enantiomers and Linkage Isomers

In the octahedral cobalt(III) complex ion containing two ethylenediamine molecules and two chloride ions, there are two ways to arrange the chloride ions: on adjacent bonding sites (the cis isomer) or on opposite sides of the octahedron (the trans isomer). Although these two molecules are cis–trans isomers, the cis isomer (**Figure 22.24**) also illustrates another kind of stereoisomerism that is possible in complex ions and coordination compounds.

Figure 22.24 shows that *cis*-$Co(en)_2Cl_2^+$ is chiral: it has a mirror image that is not identical to the original. There is no way to rotate the mirror image so that its atoms align exactly with those in the original, therefore the two structures are not superimposable. We encountered this phenomenon in Chapters 9 and 20 and noted that such nonsuperimposable stereoisomers are called *enantiomers*. If you look closely at Figure 22.8b, you will recognize a similar situation. The arrangement of the three ethylenediamine ligands around Ni^{2+} makes this compound chiral, too.

Naming *cis*-$Co(en)_2Cl_2^+$ also requires an additional rule. The complex ion is named *cis*-dichlorobis(ethylenediamine)cobalt(III). Note the prefix *bis-* just before (ethylenediamine). In naming complex ions containing a polydentate ligand, we use *bis-* instead of *di-* to indicate that two molecules of the ligand are present and to avoid the use of two *di-* prefixes in the same ligand name. Other prefixes used in this fashion are *tris-* for three polydentate ligands and *tetrakis-* for four.

CONCEPT TEST

Would four different ligands arranged in a square planar geometry produce a chiral complex ion? What about four different ligands in a tetrahedral geometry?

A third kind of isomerism called *linkage isomerism* occurs in coordination compounds when a ligand can bind to a central metal ion by using either of two possible electron-pair–donating atoms. The two complexes of Co^{3+} shown in

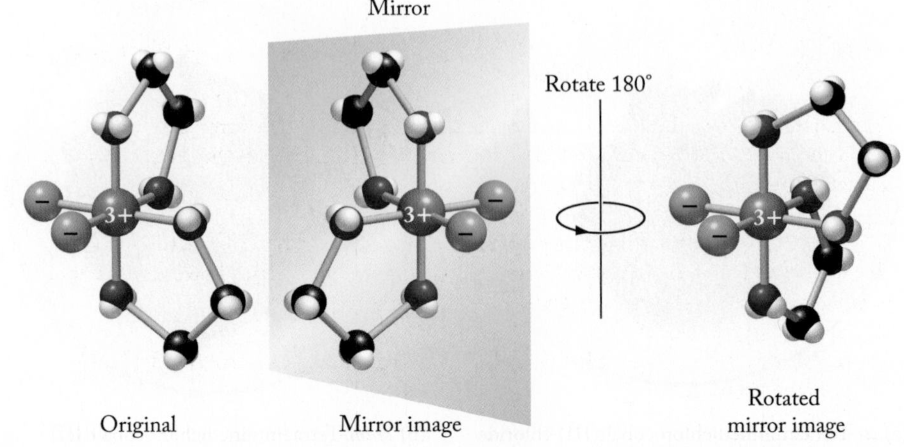

FIGURE 22.24 The complex ion *cis*-dichlorobis(ethylenediamine)cobalt(III) is chiral, which means that its mirror image is *not* superimposable on the original complex. To illustrate this point, we rotate the mirror image 180° about its vertical axis so that it looks as much like the original as possible. However, note that the top ethylenediamine ligand is located behind the plane of the page in the original but in front of the plane of the page in the rotated mirror image. The mirror images, therefore, are *not* superimposable.

Figure 22.25 are a pair of linkage isomers. One complex contains the thiocyanate (SCN^-) ion as a ligand in which the S atom covalently bonds to the metal (Figure 22.25a). The other contains the isothiocyanate (NCS^-) ion as a ligand in which the N atom forms the bond (Figure 22.25b). Another ligand that forms linkage isomers with metals is NO_2^-, called nitro when the N atom donates an electron pair and nitrito when an O atom is the donor. Because these pairs of molecules have different connectivities, linkage isomers are *not* stereoisomers. Instead, linkage isomers are a type of constitutional isomer.

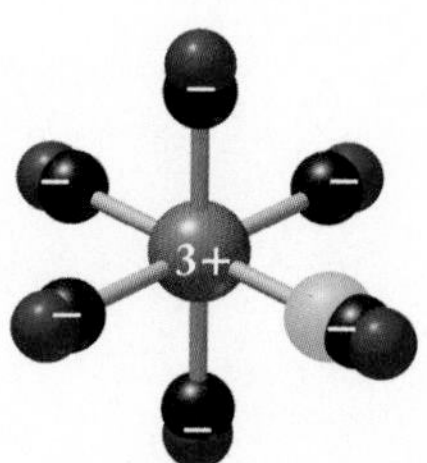

(a) $[Co(CN)_5SCN]^{3-}$

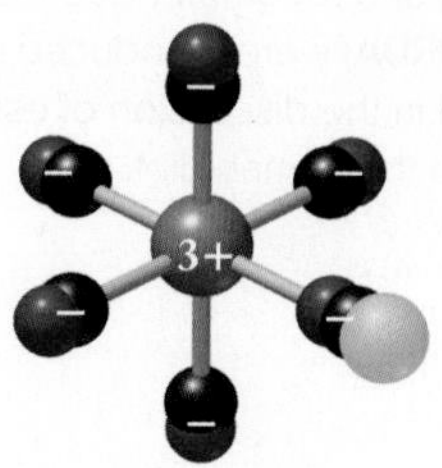

(b) $[Co(CN)_5NCS]^{3-}$

FIGURE 22.25 The SCN^- ligand is bound to the Co^{3+} ion through the S atom in (a) the pentacyanothiocyanatocobaltate(III) ion and through the N atom in (b) the pentacyanoisothiocyanatocobaltate(III) ion.

SAMPLE EXERCISE 22.5 Identifying Isomers of Coordination Compounds LO6

Sketch the structures and name the isomers of $Ni(NH_3)_4(NO_2)_2$.

Collect and Organize We are given the formula and asked to name and draw the structures of the isomers of a coordination compound. The Ni^{2+} complexes we have seen so far in the chapter have all been octahedral. Ammonia molecules and NO_2^- ions are both monodentate ligands, but NO_2^- can bond to Lewis bases through either the lone pair on N or through one of the lone pairs on O.

Analyze The formula contains no brackets, so the NO_2^- ions are not counterions; they must be covalently bonded to the Ni ion. There are two NO_2^- ions, so the charge on Ni must be 2+. There are a total of six ligands, which confirms that the compound is octahedral.

Solve There are two ways to orient the nitrite ions relative to the four ammonia ligands: opposite each other with a $O_2N–Ni–NO_2$ bond angle of 180° or on the same side of the octahedron with a $O_2N–Ni–NO_2$ bond angle of 90°:

NO_2 (top); H_3N, NH_3, H_3N, NH_3 around Ni; NO_2 (bottom)	NH_3 (top); H_3N, NO_2, H_3N, NO_2 around Ni; NH_3 (bottom)

The first case represents a *trans* arrangement of the NO_2^- ligands, whereas the second is a *cis* arrangement. The first isomer is *trans*-tetraamminedinitronickel(II); the second is *cis*-tetraamminedinitronickel(II). Both names include the word *nitro* to signify coordination of the NO_2^- group through N, but the nitrite ligand can also bond through the O, which changes the ligand name to *nitrito* and generates two additional linkage isomers for both the trans and cis isomers.

ONO (top); H_3N, NH_3, H_3N, NH_3 around Ni; NO_2 (bottom)	ONO (top); H_3N, NH_3, H_3N, NH_3 around Ni; ONO (bottom)	NH_3 (top); H_3N, ONO, H_3N, NO_2 around Ni; NH_3 (bottom)	NH_3 (top); H_3N, ONO, H_3N, ONO around Ni; NH_3 (bottom)

The names for these four isomers are *trans*-tetraammine(nitrito)(nitro)nickel(II), *trans*-tetraamminedinitritonickel(II); *cis*-tetraammine(nitrito)(nitro)nickel(II), and *cis*-tetraamminedinitritonickel(II), respectively.

Think About It We can draw other tetraamminedichloronickel(II) structures that do not look exactly like these two structures. If we flip or rotate them, however, they will match one of the two structures shown.

Practice Exercise Sketch the stereoisomers of $[CoBr_2(en)(NH_3)_2]^+$ and name them.

22.7 Coordination Compounds in Biochemistry

CONNECTION Dietary reference intake (DRI) and recommended dietary allowance (RDA) were introduced in Section 21.1 in the discussion of essential elements in the human diet.

In Section 22.1, we noted that metals essential to human health must be present in foods in forms the body can absorb. In this section we explore some biological polydentate ligands that help metal ions participate in processes that are essential to nutrition and good health. DRI/RDA values for trace essential elements, including iron and zinc, are in the range of 10–20 mg/d. There are DRI/RDA values for some, but not all, of the ultratrace essential elements, ranging from 45 μg/d for molybdenum up to 5 mg/d for manganese. Compounds of iron and zinc are present in the body in average concentrations between 1 and 1000 μg of element per gram of body mass and are considered to be trace elements. Other transition metals (chromium, cobalt, copper, manganese, molybdenum, nickel, and vanadium) are considered ultratrace, meaning essential elements present in the body in average concentrations less than 1 μg of element per gram of body mass. Here we survey the roles of some of the other ultratrace transition metals in biology.

Manganese and Photosynthesis

Let's begin with photosynthesis, a chemical process at the foundation of our food chain. Green plants can harness solar energy because they contain large biomolecules we collectively call *chlorophyll.* All molecules of chlorophyll contain ring-shaped tetradentate ligands called *chlorins* (**Figure 22.26a**). The structures of chlorins are similar to those of **porphyrins**, another class of tetradentate ligands found in biological systems (**Figure 22.26b**). We saw porphyrins in Section 20.4 as the Fe^{2+}-binding molecules in hemoglobin (Figure 20.29a). Chlorins and porphyrins are members of a larger category of polydentate compounds known as **macrocyclic ligands**. (*Macrocycle* means, literally, "big ring.")

Two of the four nitrogen atoms in porphyrins and chlorins are sp^3 hybridized and bound to hydrogen atoms, whereas the other two are sp^2 hybridized with no hydrogen atoms. When either compound forms coordinate bonds with a metal ion M^{n+} (**Figure 22.26c**), the two hydrogen atoms ionize, giving the ring a charge of 2− and the complex ion an overall charge of $(n - 2)$. The lone pairs of electrons on the N atoms in the ionized structure are oriented toward the ring center. These lone pairs can occupy either the four equatorial coordination sites in an octahedral complex ion or all four coordination sites in a square planar complex ion. In octahedral complex ions, each central metal ion still has two axial sites available for bonding to other ligands. Depending on the charge of the central ion, the coordination compound may be either ionic (a complex ion) or electrically neutral.

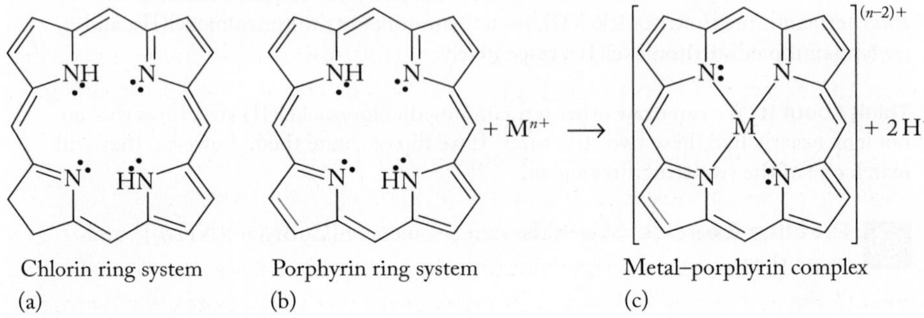

FIGURE 22.26 Skeletal structures of (a) chlorin and (b) porphyrin. The principal difference between the structures is a C═C in the porphyrin structure (shown in red) that is a single bond in chlorin rings. The innermost atoms in each ring are four nitrogen atoms with lone pairs of electrons. All four N atoms form coordinate bonds with a metal ion, as shown with the porphyrin ring in (c).

Porphyrin and chlorin rings are widespread in nature and play many biochemical roles. Their chemical and physical properties depend on

1. the identity of the central metal ion
2. the species that occupy the axial coordination sites of octahedral complexes
3. the number and identity of organic groups attached to the outside of the ring

porphyrin a type of tetradentate macrocyclic ligand.

macrocyclic ligand a ring containing multiple electron-pair donors that bind to a metal ion.

The role of the major essential element Mg^{2+} in photosynthesis was described in Section 21.3. Another metal, the ultratrace essential element manganese, also plays a role in the production of oxygen during photosynthesis. To examine that role, let's write an equation for photosynthesis that is slightly different from the equation we are used to seeing. Instead of writing the formula $C_6H_{12}O_6$ for glucose, we use the generic carbohydrate formula $(CH_2O)_n$ so that the coefficient is 1 for all other species in the reaction (rather than 6):

$$H_2O(\ell) + CO_2(g) \rightarrow \tfrac{1}{n}\,(CH_2O)_n(aq) + O_2(g)$$

Writing the equation in this form makes it easier to see how the overall reaction between water and carbon dioxide involves electron transfer: photosynthesis is a redox reaction in which oxygen is oxidized and carbon is reduced.

This redox process requires manganese-containing enzymes in which Mn ions in the +3 and +4 oxidation states mediate the transfer of electrons from water in photosynthesis. Although the exact structure of these manganese compounds remains undetermined, it is believed that two manganese(III) ions and two manganese(IV) ions are present at the site of O_2 production. Copper-containing enzymes are also involved in the series of reactions that make up photosynthesis.

CONCEPT TEST

Manganese ions involved in photosynthesis are often surrounded by six ligands in an octahedral geometry. Is reduction of a manganese(IV) ion to manganese(III) in an octahedral complex (see Section 22.5) accompanied by a change in spin state of the Mn ion?

Transition Metals in Enzymes

Transition metal cations can form complexes by bonding to the nitrogen atoms of the amino acids in proteins and other Lewis bases in biological systems. Many of the enzymes in our bodies contain zinc or iron ions; these enzymes, along with others containing transition metal ions, are called *metalloenzymes.*

Zinc The enzyme carbonic anhydrase catalyzes the reaction between water and carbon dioxide to form bicarbonate ions:

$$H_2O(\ell) + CO_2(g) \rightleftharpoons HCO_3^-(aq) + H^+(aq)$$

The α form of carbonic anhydrase contains 260 amino acid residues and a zinc ion at the active site. Notice in **Figure 22.27** that the zinc ion is coordinately bonded to three nitrogen atoms on histidine side chains and to one molecule of water. The presence of these ligands and a fourth histidine nearby facilitates ionization of the water molecule. Ionization leaves a OH^- ion attached to the Zn^{2+} ion and a H^+ ion bonded to the side-chain nitrogen atom of the fourth histidine. In addition, a space just the right size and shape to accommodate a CO_2 molecule is next to the active site. When a CO_2 molecule in this space bonds to a hydroxide

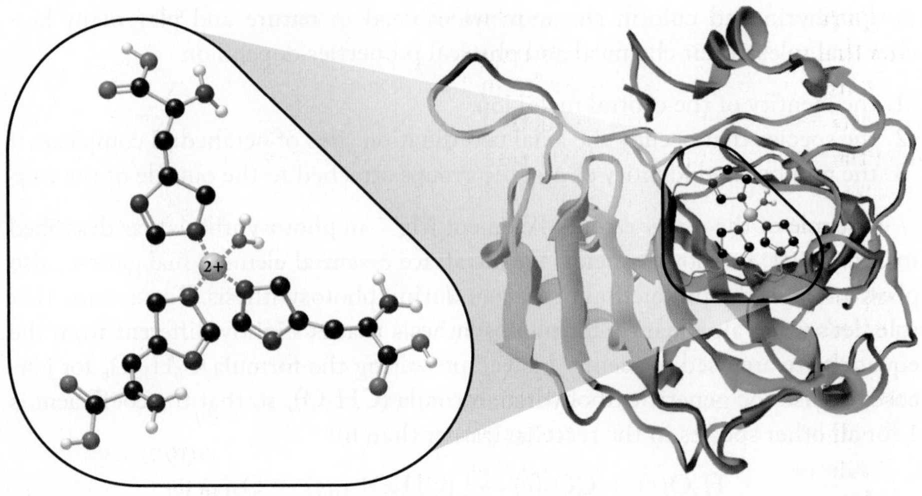

FIGURE 22.27 The active site of one form of carbonic anhydrase consists of a zinc ion (gray sphere) bonded to three histidine molecules and one H_2O molecule. The OH^- ion produced when this H_2O ionizes combines with a molecule of CO_2, forming an HCO_3^- ion.

ion, a HCO_3^- ion forms. As the bicarbonate ion pulls away, another water molecule occupies the fourth coordination site on the Zn^{2+} ion, another CO_2 molecule enters, the histidine is protonated, and the catalytic cycle repeats. This reaction is important because it helps eliminate CO_2 from cells during respiration and mediates the uptake of CO_2 during photosynthesis in some plants.

Iron Iron-containing peroxidases and catalases are integral to the transfer of oxygen to biomolecules. For example, plants use a fatty acid peroxidase to catalyze the stepwise degradation of fatty acids, where R represents a long hydrocarbon chain. One $-CH_2-$ group at a time is removed from the fatty acid by using hydrogen peroxide:

$$\underset{\text{Fatty acid}}{R{-}CH_2{-}COOH(aq)} + \underset{\substack{\text{Hydrogen}\\\text{peroxide}}}{2\,H_2O_2(aq)} \xrightarrow{\substack{\text{fatty acid}\\\text{peroxidase}}} 3\,H_2O(\ell) + \underset{\text{Aldehyde}}{R{-}CHO(aq)} + CO_2(aq)$$

Subsequent oxidation of the aldehyde back to a carboxylic acid yields a new fatty acid with one fewer $-CH_2-$ group:

$$\underset{\text{Aldehyde}}{R{-}CHO(aq)} + \tfrac{1}{2}\,O_2(aq) \rightarrow \underset{\text{Fatty acid}}{RCOOH(aq)}$$

> **CONNECTION** Fatty acids were described in Section 20.6.

Iron-containing enzymes also catalyze the reduction of nitrite (NO_2^-) and sulfite (SO_3^{2-}) ions:

$$NO_2^-(aq) + 6\,e^- + 8\,H^+(aq) \xrightarrow{\text{nitrite reductase}} NH_4^+(aq) + 2\,H_2O(\ell)$$

$$SO_3^{2-}(aq) + 6\,e^- + 7\,H^+(aq) \xrightarrow{\text{sulfite reductase}} HS^-(aq) + 3\,H_2O(\ell)$$

In each case, the iron in the enzyme is oxidized, providing the electrons needed for reduction.

Proteins called *cytochromes* also contain one or more heme groups (**Figure 22.28a**). Cytochromes mediate oxidation and reduction processes connected with energy production in cells. The heme group conveys electrons as the half-reaction

$$Fe^{3+} + e^- \rightleftharpoons Fe^{2+}$$

rapidly and reversibly consumes or releases electrons needed in the biochemical reactions that sustain life. Cytochromes catalyze electron transport in photosynthesis and the metabolism of glucose to CO_2 and water. Their role in electron transport makes them key participants in *in vivo* reactions that produce ATP, which is required for intracellular energy production and transfer in living things. As catalysts for oxidation and reduction, cytochromes are essential for the removal of toxic substances and for the process of programmed cell death, both of which are required for the health of multicellular organisms. Cytochrome *c* (**Figure 22.28b**) is a component of the electron transport chain in mitochondria that is ubiquitous in life forms ranging in complexity from microbes to human beings. Its amino acid sequence is very similar or the same across a spectrum of plants, animals, and many unicellular organisms. Consequently, studies of the cytochrome *c* molecule are seminal in evolutionary biology.

Many other kinds of cytochrome proteins have different substituents on the porphyrin rings and different axial ligands, each of which influences the function of the complex. This last point has been repeated several times in this chapter: the chemical properties and biological functions of transition metals that are essential to living organisms are linked to their molecular environments and to the formation of stable complex ions with ligands that are strong electron donors—that is, strong Lewis bases.

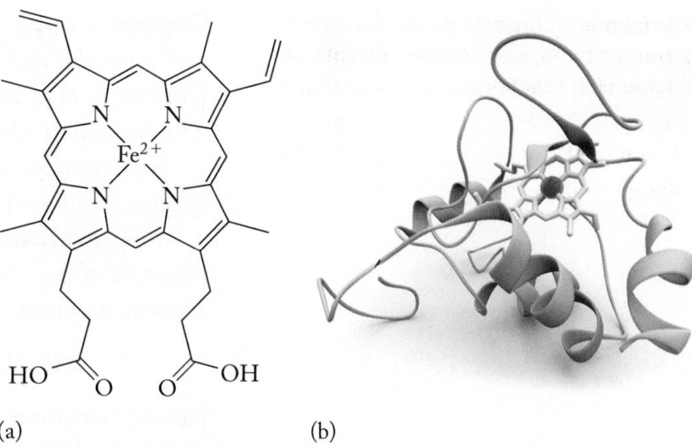

FIGURE 22.28 (a) Heme is the specific porphyrin that binds iron in the oxygen-transport protein hemoglobin and also in many enzymes of great significance in all life on Earth. (b) The structure of cytochrome proteins, such as cytochrome *c* shown here, includes one or more heme complexes (shown in gray) that mediate energy production and redox reactions in living cells. Different cytochromes have different axial ligands occupying the fifth and sixth octahedral coordination sites, and different groups in the protein may be attached to the porphyrin ring.

CONCEPT TEST

Which of the following small molecules could *not* function as a Lewis base and hence would not be expected to bond to iron in a heme protein? CO; H_2; NO_2; H_2S

CONNECTION The role of adenosine triphosphate (ATP) as an energy source in biology is discussed in Section 17.9. The binding of oxygen (O_2) to hemoglobin is described in Section 20.4.

Molybdenum and Vanadium Many metalloenzymes are involved in transformations of nitrogen. Molybdenum-containing reductases are responsible for converting NO_3^- ions to NO_2^- ions and then to NH_3 in the nitrogen cycle (see Figure 21.8). The active site of sulfite oxidase, which converts SO_3^{2-} ions to SO_4^{2-} ions in the sulfur cycle (see Figure 21.11), also contains molybdenum. Sometimes more than one type of transition metal is found in a metalloenzyme. For example, both molybdenum and iron are required by xanthine oxidase, an important enzyme along the pathway for degradation of excess nucleic acids (adenine and guanine) to xanthine and then to uric acid for elimination through the kidneys.

$$\text{Xanthine} + H_2O + O_2 \xrightarrow{\text{xanthine oxidase}} \text{Uric acid} + H_2O_2$$

Xanthine — Uric acid

The combination of iron and vanadium is essential to the function of haloperoxidases, a class of enzymes found in some algae, lichens, and fungi that replace C—H bonds with carbon–halogen bonds. The reaction products are thought to function in the defense systems of these organisms.

coenzyme an organic molecule that, like an enzyme, accelerates the rate of biochemical reactions.

Copper Copper-containing proteins perform several functions in both plants and animals, including oxygen transport in mollusks such as clams and oysters. Copper is also an essential element in the enzymes azurin and plastocyanin, which mediate electron transfer during photosynthesis.

Reactions catalyzed by xanthine oxidase may produce other reactive oxygen species besides H_2O_2. One of them is the superoxide ion, O_2^-. Superoxide ion is a strong oxidizing agent and must be eliminated to prevent cell damage. The removal of superoxide begins with the action of superoxide dismutases, which convert superoxide to hydrogen peroxide:

$$2\,O_2^-(aq) + 2\,H^+(aq) \xrightarrow{\text{superoxide dismutase}} H_2O_2(aq) + O_2(aq)$$

Researchers have isolated superoxide dismutases containing a variety of transition metals, including a copper–zinc enzyme. The hydrogen peroxide produced in this reaction is decomposed to water and oxygen by iron-containing catalase.

Nickel Ureases, the enzymes responsible for the conversion of urea to ammonia in plants, contain nickel(II). Nickel is also found with iron in enzymes called *hydrogenases*. Hydrogenases oxidize hydrogen gas to protons:

$$H_2(g) \xrightarrow{\text{hydrogenase}} 2\,H^+(aq) + 2\,e^-$$

Nickel and iron also combine in CO dehydrogenase, an enzyme that catalyzes the formation of acetyl-CoA, which is a key component in the tricarboxylic acid cycle discussed in Section 20.5. Finally, nickel-containing enzymes are found among the enzymes responsible for methane generation by bacteria.

Cobalt and Coenzymes Many enzymes require the presence of a **coenzyme,** which is an organic compound that cocatalyzes a biochemical reaction. The coenzyme B_{12} (**Figure 22.29**) contains the ultratrace essential element cobalt(III) and is a derivative of vitamin B_{12}. The cobalt(III) in coenzyme B_{12} is easily reduced to cobalt(II) and even cobalt(I) in the course of enzyme-catalyzed redox reactions. The change in oxidation state of Co allows for facile transfer of methyl groups, as in the conversion of methionine to homocysteine:

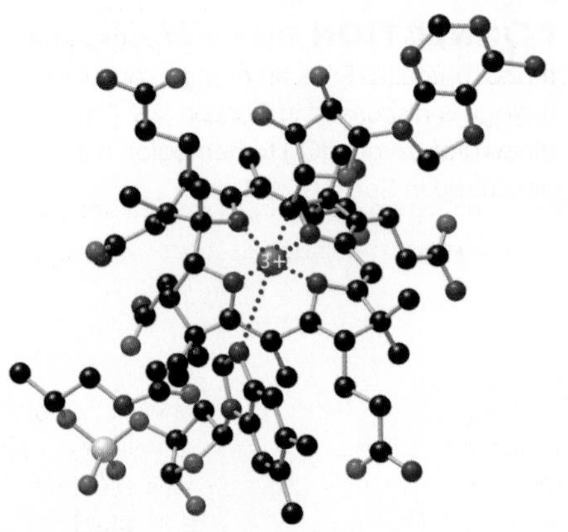

FIGURE 22.29 Coenzyme B_{12} contains cobalt(III) (gold sphere), an ultratrace essential metal.

$$\text{Methionine } ({}^-OOC\text{–}CH(\overset{+}{N}H_3)\text{–}CH_2CH_2\text{–}S\text{–}CH_3) \longrightarrow \text{Homocysteine } ({}^-OOC\text{–}CH(\overset{+}{N}H_3)\text{–}CH_2CH_2\text{–}SH)$$

Coenzyme B_{12} is also critical to the function of *mutases,* which are enzymes that catalyze the rearrangement of the skeleton of a molecule, as in the interconversion of glutamate and methylaspartate:

$$\text{Glutamate } ({}^-OOC\text{–}CH(\overset{+}{N}H_3)\text{–}CH_2CH_2\text{–}COO^-) \underset{\text{coenzyme } B_{12}}{\overset{\text{glutamate mutase}}{\rightleftharpoons}} \text{Methylaspartate } ({}^-OOC\text{–}CH(\overset{+}{N}H_3)\text{–}CH(CH_3)\text{–}COO^-)$$

Chromium Chromium in the +3 oxidation state is an ultratrace essential element in our diets. It is involved in regulating glucose levels in the blood through a molecule called chromodulin. Chromodulin is a polypeptide containing only 4 of the 20 naturally occurring amino acids: glycine, cysteine, glutamic acid, and aspartic acid. Four Cr^{3+} ions are bound to the peptide chain. Cereals and grains contain enough chromium for our daily needs, but certain plants (such as shepherd's purse) concentrate chromium and have been used as herbal remedies in diabetes treatment.

Chromium in the +6 oxidation state, as found in chromate ions (CrO_4^{2-}), is acutely toxic and also carcinogenic. Chromate ion enters cells through ion channels that transport SO_4^{2-} ions. Once inside, CrO_4^{2-} is reduced to Cr^{3+}, which binds to the phosphate backbone of DNA.

22.8 Coordination Compounds in Medicine

Transition metal coordination compounds are becoming important in both the diagnosis and treatment of diseases (**Figure 22.30**). Any diagnostic or therapeutic compound that is injected intravenously must be sufficiently soluble in blood to be delivered to the target. While in transit, the compound must be stable enough not to undergo chemical reactions that result in its precipitation or rapid elimination from the body. Occasionally, the compound can be in the form of a simple salt, but more often a metal ion is introduced as a coordination complex or coordination compound. Ligands used in forming biologically active coordination complexes include amino acids and simple anions such as the citrate ion. Chelating ligands such as diethylenetriaminepentaacetate ($DTPA^{5-}$; **Figure 22.31**) are often used in biological applications. A medicinal chemist can also take advantage of substances that occur naturally in the body, such as antibodies, to transport a diagnostic or therapeutic metal ion to its target.

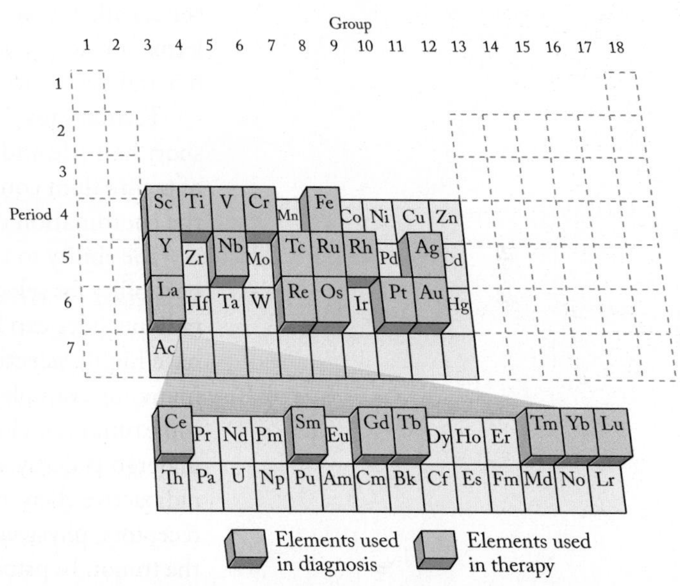

FIGURE 22.30 Radioisotopes of the elements shown in red are used in diagnostic imaging and those shown in green are used in therapy.

Transition Metals in Diagnosis

As discussed in Section 19.7, radionuclides of main group elements with short half-lives that emit easily detectable γ rays are useful in positron emission tomography (PET). An equally wide array of transition metal isotopes is available for imaging. Magnetic resonance imaging (MRI), for instance, can be

CONNECTION Chapter 19 gave a more detailed discussion of nuclear chemistry and nuclear medicine, including an assessment of how different types of radiation affect living tissue.

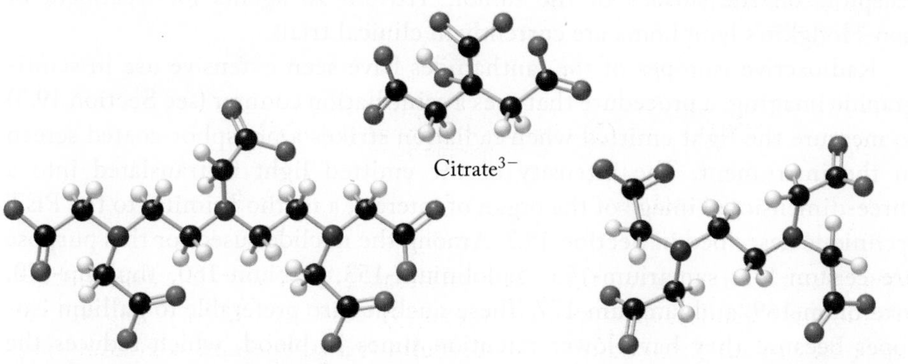

FIGURE 22.31 Diethylenetriaminepentaacetate ($DTPA^{5-}$), citrate^{3-}, and ethylenediaminetetraacetate ($EDTA^{4-}$) are often used as chelating ligands for diagnostic and therapeutic agents based on transition metals. These ions form stable complex ions with 2+ and 3+ metal cations. The solubilities of the complex ions are typically much greater than those of the hydrated ions at physiological pH (7.4).

used to diagnose soft-tissue injuries and uses stable isotopes of gadolinium to enhance images.

Technetium and Rhenium Technetium, just below manganese in group 7 of the periodic table, is the most widely used radioactive isotope in imaging. Technetium is unusual in that it does not exist naturally on Earth in easily measurable amounts because it has no stable isotopes; in other words, all technetium isotopes are radioactive. These isotopes can be produced in nuclear reactors for use in medicine. The ^{99m}Tc used in hospitals for imaging is prepared in technetium generators in which a stable isotope of molybdenum, ^{98}Mo (23.78% natural abundance), is bombarded with neutrons. Technetium-99 has a half-life greater than 20,000 years; when it is produced in a nuclear reactor, however, its nucleus is in an excited state, called a *metastable* nucleus. The metastable state, designated by adding the letter "m" to the mass number, as in technetium-99m or ^{99m}Tc, has a half-life of 6 h and decays to the more stable technetium-99 nucleus.

Technetium-99m has been widely used as an imaging agent because it has a short half-life and emits low-energy γ rays. Patients can be injected with a variety of technetium compounds, depending on the target organ. For imaging the heart, the coordination compounds shown in **Figure 22.32** are used.

The ability to target the delivery of radionuclides to a particular organ opens the possibility for selective irradiation of a tumor located in that organ. Therefore, some radionuclides can be used to not only image but also treat cancers. Certain tumors have highly selective receptor sites for particular molecules on their surfaces. Rhenium, for example, is being studied as both an imaging and a therapeutic agent for some tumors, including breast, liver, and skin cancer. Rhenium-186 and rhenium-188 undergo β decay with half-lives of 3.72 d and 17.0 h, respectively. By including a radioactive rhenium ion in a molecule that binds strongly and specifically to these receptors, physicians can deliver both an imaging agent and a therapeutic agent to the tumor. In principle, the β particles destroy the tumor. Several patents have been issued for the use of compounds containing rhenium isotopes, but therapies based on these compounds remain in the experimental stage. One example of a rhenium compound used in these applications is shown in **Figure 22.33**.

Scandium, Yttrium, and Lanthanide Elements Scandium, yttrium, and lanthanum are in group 3 in the periodic table, the first group in the transition metal series. Scandium-46 has been used to image the spleen, and scandium-47 shows promise in diagnosing breast cancer. Among the available radioactive isotopes of yttrium, yttrium-90 has been applied to imaging a variety of tumors, including intestinal, breast, and thyroid cancers. Chelating ligands are used to complex the $^{90}Y^{3+}$ cation and deliver it to the tumor, where it binds strongly to receptors on the surface of the tumor. Yttrium-90 agents for treatment of non-Hodgkin's lymphoma are currently in clinical trials.

Radioactive isotopes of the lanthanides have seen extensive use in scintigraphic imaging, a procedure that uses a scintillation counter (see Section 19.3) to measure the light emitted when radiation strikes a phosphor-coated screen in the instrument. The intensity of the emitted light is translated into a three-dimensional image of the organ of interest, a method similar to the PET technique described in Section 19.7. Among the nuclides used for this purpose are cerium-141, samarium-153, gadolinium-153, terbium-160, thulium-170, ytterbium-169, and lutetium-177. These nuclides are preferable to gallium isotopes because they have lower retention times in blood, which reduces the

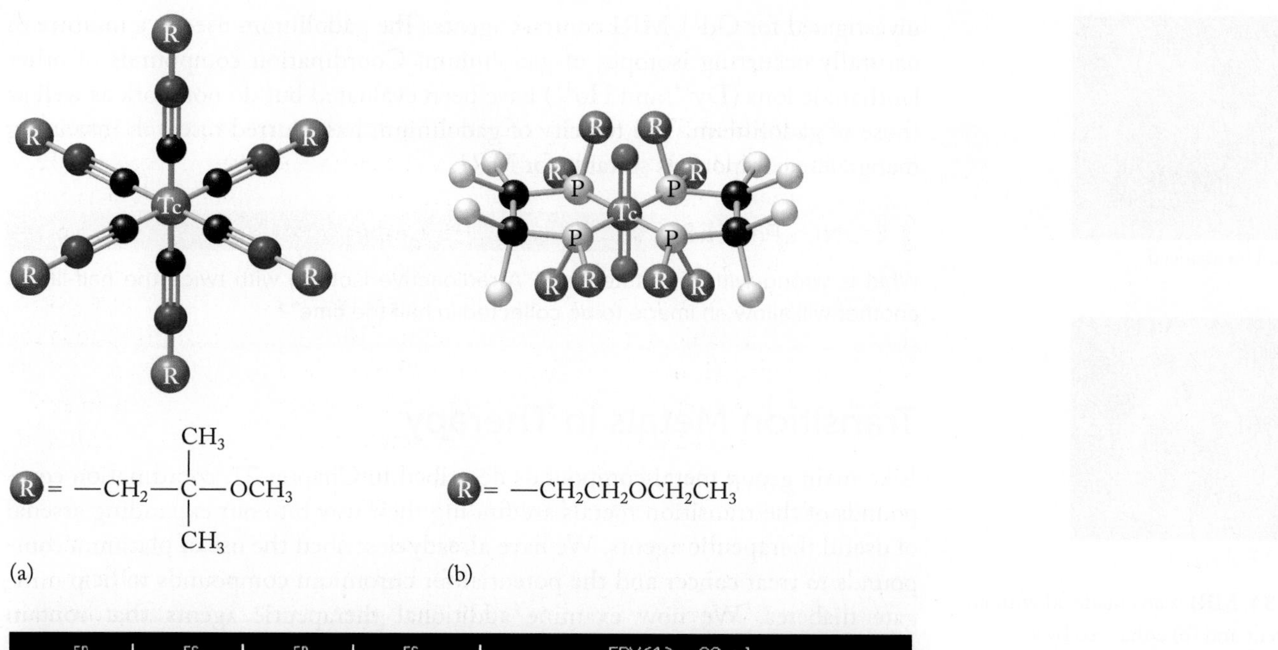

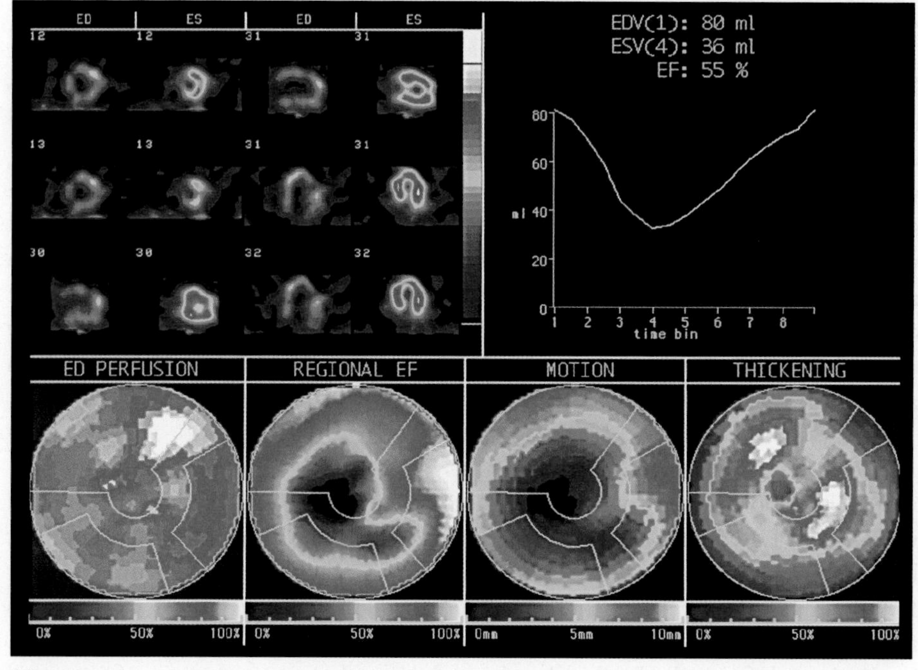

FIGURE 22.32 (a) Cardiolite $[Tc(CNR)_6]$ and (b) Myoview $[TcO_2(RPCH_2CH_2PR)_2]$ are used for imaging the heart. (c) Images of a patient's heart after intravenous injection with a technetium-containing drug. The images along the bottom show four measurements of heart function. Radioactive technetium compounds emit γ rays that allow the blood to be tracked as it is pumped through the heart.

possibility of potentially harmful biological side effects. In some cases, lutetium-177 derivatives have replaced yttrium-90 compounds under investigation as radioactive antitumor agents.

As noted previously, the quality of an MRI scan (**Figure 22.34**) can be improved when gadolinium is used as a contrast agent. Before the procedure, the patient is injected with a gadolinium compound. Many ligands have been

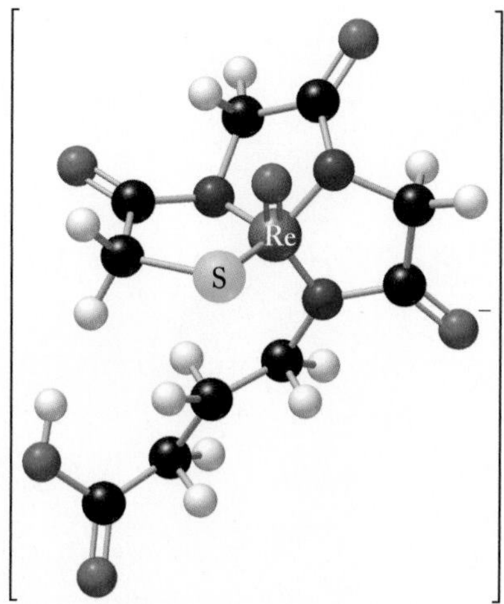

FIGURE 22.33 This rhenium–mercaptoacetylglycylglycyl-γ-amino acid complex is used to attach ^{186}Re or ^{188}Re to monoclonal antibodies or peptides.

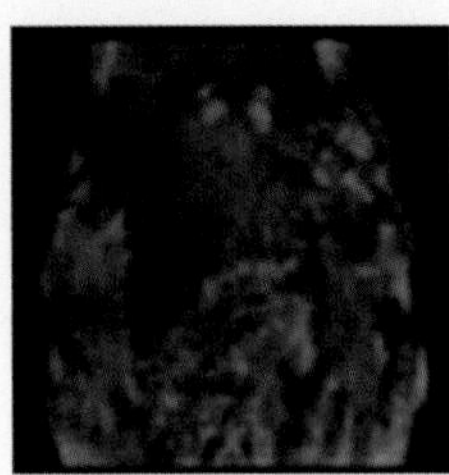

(a) Unenhanced

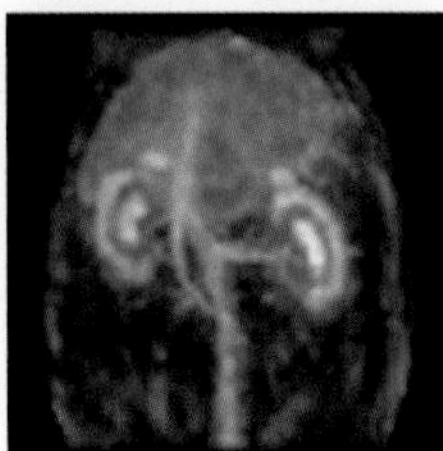

(b) Enhanced

FIGURE 22.34 MRI scans made (a) without a contrast agent and (b) enhanced by a gadolinium contrast agent.

investigated for Gd^{3+} MRI contrast agents. The gadolinium used is a mixture of naturally occurring isotopes of gadolinium. Coordination compounds of other lanthanide ions (Dy^{3+} and Ho^{3+}) have been evaluated but do not work as well as those of gadolinium. The toxicity of gadolinium has spurred research into using manganese compounds suitable for MRI.

CONCEPT **TEST**

What is wrong with the statement, "A radioactive isotope with twice the half-life of another will allow an image to be collected in half the time"?

Transition Metals in Therapy

Like main group metal compounds described in Chapter 21, coordination compounds of the transition metals are finding their way into our expanding arsenal of useful therapeutic agents. We have already described the use of platinum compounds to treat cancer and the potential for chromium compounds to help mitigate diabetes. We now examine additional therapeutic agents that contain transition metal elements.

Iron The body must regulate the amount of iron in cells in order to produce enough hemoglobin to maintain good health. Deficiencies in iron lead to several diseases broadly classified as anemia. Mild forms of anemia are common among women of child-bearing age and are treated with oral iron supplements. A more serious genetic form of anemia, however, known as thalassemia, must be treated with blood transfusions that leave patients with too much iron in their blood. To remove the excess iron, these patients are treated with chelating ligands that complex some of the iron and transport it out of the red blood cells and eventually out of the body in what is known as chelation therapy. Chelation therapy relies on a large equilibrium constant for the reaction

$$\mathrm{FeL}(aq) + \mathrm{L'}(aq) \rightleftharpoons \mathrm{FeL'}(aq) + \mathrm{L}(aq)$$

where iron coordinated to ligand L (such as hemoglobin) in the blood is treated with ligand L′, which also binds iron. If the formation constant for the complex between iron and L′ is greater than for iron and L, then the equilibrium constant for the equation is greater than one, and formation of the iron–L′ complex is favored. The ligand L′ is chosen so that the complex formed between iron and L′ is eliminated from the body, thereby reducing the concentration of iron and relieving the symptoms caused by thalassemia treatments. Often the chelating ligand leads to chiral metal complexes of the kind described in Section 22.6. Chelation therapy is also the treatment of choice for heavy metal poisoning from toxic metals, including lead and mercury.

SAMPLE EXERCISE 22.6 Calculating the Equilibrium Constant of a Ligand Exchange Reaction **LO7**

The protein transferrin is involved in the transport of iron into cells. Iron accumulation in the human body has been implicated in diseases such as Parkinson's, Alzheimer's, and thalassemia. The chelating ligand deferoxamine (DFO) is used to treat thalassemia; the complex between iron and DFO is eliminated from the body. Given the formation constants for the complexation of iron(III) by transferrin (Equation 1) and the reaction of DFO with iron (Equation 2), calculate the equilibrium constant for the ligand exchange reaction between DFO and transferrin (Equation 3).

$$Fe^{3+}(aq) + \text{transferrin}(aq) \rightleftharpoons \text{Fe(transferrin)}^{3+}(aq) \qquad K_{f,1} = 4.7 \times 10^{20} \quad (1)$$

$$Fe^{3+}(aq) + \text{DFO}(aq) \rightleftharpoons \text{Fe(DFO)}^{3+}(aq) \qquad K_{f,2} = 4.0 \times 10^{30} \quad (2)$$

$$\text{Fe(transferrin)}^{3+}(aq) + \text{DFO}(aq) \rightleftharpoons \text{Fe(DFO)}^{3+}(aq) + \text{transferrin}(aq) \qquad K_3 = ? \quad (3)$$

Collect and Organize We are given the formation constants of two reactions involving iron and different ligands and are asked to calculate the equilibrium constant for the ligand exchange reaction. You may want to review Section 16.6 for calculations involving formation constants of complexes, as well as Chapters 14 and 15, where we manipulated equilibrium constants.

Analyze Transferrin is a reactant in Equation 1 and a product in the exchange reaction (3). Therefore, we need to reverse Equation 1 before adding it to Equation 2 to obtain Equation 3.

$$\text{Fe(transferrin)}^{3+}(aq) \rightleftharpoons \cancel{Fe^{3+}(aq)} + \text{transferrin}(aq)$$

$$\cancel{Fe^{3+}(aq)} + \text{DFO}(aq) \rightleftharpoons \text{Fe(DFO)}^{3+}(aq)$$

$$\text{Fe(transferrin)}^{3+}(aq) + \text{DFO}(aq) \rightleftharpoons \text{Fe(DFO)}^{3+}(aq) + \text{transferrin}(aq)$$

Reversing a reaction means taking the reciprocal of its equilibrium constant. Combining the reverse of Equation 1 with Equation 2 to obtain Equation 3 means multiplying $(1/K_{f,1})$ and $K_{f,2}$ together to obtain K_3. The reciprocal of $K_{f,1}$ has a value of about 10^{-20}. Therefore, the value of K_3 should be about $10^{-20} \times 10^{30}$, or about 10^{10}.

Solve Multiplying $(1/K_{f,1})$ by $K_{f,2}$ to obtain K_3:

$$K_3 = \frac{1}{4.7 \times 10^{20}} \times (4.0 \times 10^{30}) = 8.5 \times 10^{9}$$

Think About It The equilibrium constant for the ligand exchange between the $\text{Fe(transferrin)}^{3+}$ complex and DFO is indeed $\sim 10^{10}$ and illustrates why DFO is an effective treatment for thalassemia—the large value for the equilibrium constant means that the equilibrium in Equation 3 lies toward the products (to the right), removing iron from the blood.

Practice Exercise The equilibrium constants for the reactions of penicillamine and methionine with methyl mercury are

$$CH_3Hg^+ + \text{penicillamine} \rightleftharpoons CH_3Hg(\text{penicillamine})^+ \qquad K_f = 6.3 \times 10^{13}$$

$$CH_3Hg^+ + \text{methionine} \rightleftharpoons CH_3Hg(\text{methionine})^+ \qquad K_f = 2.5 \times 10^{7}$$

Calculate the equilibrium constant for the exchange reaction:

$$CH_3Hg(\text{methionine})^+ + \text{penicillamine} \rightleftharpoons CH_3Hg(\text{penicillamine})^+ + \text{methionine}$$

Titanium, Vanadium, and Niobium Cancer Drugs Compounds of a surprisingly large number of transition metals demonstrate antitumor activity. Titanium(IV) complexes such as budotitane (**Figure 22.35**) show promise against colon cancer. Encouraging results against breast, lung, and colon cancers were observed with a series of compounds with formulas $(C_5H_5)_2TiCl_2$, $(C_5H_5)_2VCl$, and $(C_5H_5)_2NbCl_2$. The delivery of titanium and many other

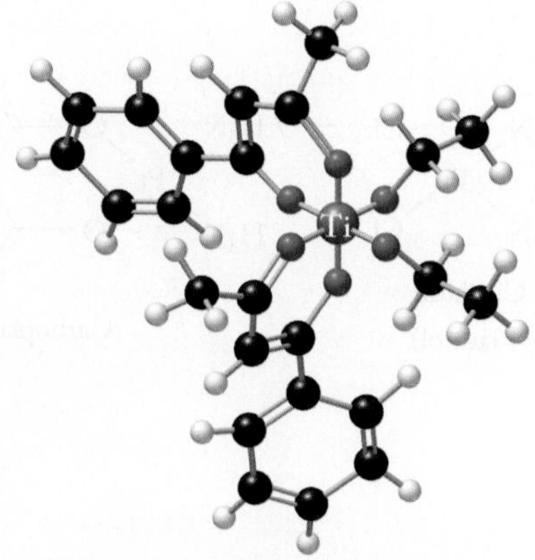

Budotitane

FIGURE 22.35 Budotitane is a titanium(IV) complex that shows activity against colon cancer.

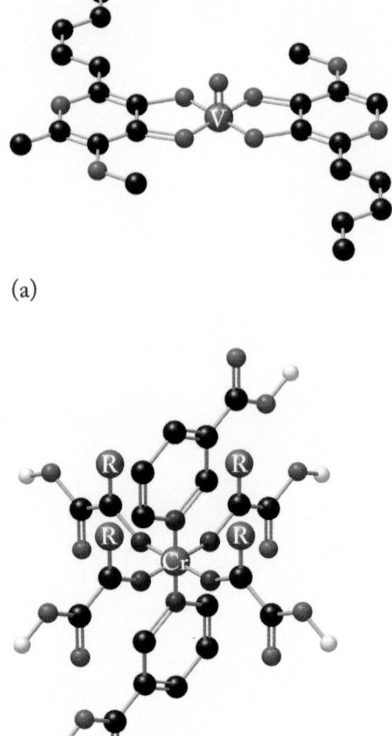

FIGURE 22.36 (a) Vanadium(IV) compounds such as bis(allixinato)-oxovanadium(IV) can act as insulin mimics. (b) The chromium(III) complex in a glucose tolerance factor contributes to our bodies' ability to regulate insulin levels. The light brown spheres are the side-chain R groups of the amino acids in the structure. (To simplify the structures, the hydrogen atoms bonded to carbon atoms are not shown.)

transition metal complexes to cells is believed to be mediated by the protein transferrin described in Sample Exercise 22.6. The precise way in which these compounds inhibit the growth of tumor cells is not yet understood. One working hypothesis proposes a mechanism similar to that for platinum compounds discussed in Section 22.6. Unfortunately, at therapeutically useful doses, the potential for liver damage outweighs the benefits of these compounds. These compounds contain carbon–metal bonds and belong to a class of substances called **organometallic compounds**.

Chromium, Vanadium, and Controlling Blood Sugar Insulin is a hormone needed for proper glucose metabolism and protein synthesis. People with diabetes either cannot produce insulin or produce the hormone but cannot use it effectively. The suggestion that vanadium plays a role in insulin production has prompted investigation of vanadium compounds as oral diabetes drugs. Redox activity and selective inhibition of enzymes at the insulin receptor site are possible modes of action for vanadium coordination compounds. Encouraging results from animal studies using bis(allixinato)oxovanadium(IV) (**Figure 22.36a**) have been reported, but the compound is not currently approved for use in humans. Chromium(III) compounds have been shown to lower fasting blood sugar levels and reduce the amount of insulin required by some diabetics (**Figure 22.36b**).

Platinum Group and Coinage Metals The period 5 and period 6 elements in groups 8, 9, and 10 (Ru, Rh, Pd, Os, Ir, and Pt) are often referred to as the *platinum group metals*. The group 11 elements of these two periods (Ag and Au) along with Cu are called the *coinage metals*. We now explore examples of soluble compounds of these metals used in medications for arthritis, cancer, and other diseases.

The serendipitous discovery in the 1960s that *cis*-diamminedichloroplatinum(II) (cisplatin in **Figure 22.37**) is effective in treating testicular, ovarian, and other cancers spurred the development of a host of cancer drugs based on both platinum group metals and coinage metals. The results of this research include a compound known as carboplatin that shows the same activity as cisplatin, whose mechanism of activity is shown in Figure 22.22, but has fewer side effects. In addition to platinum compounds, the antitumor activities of

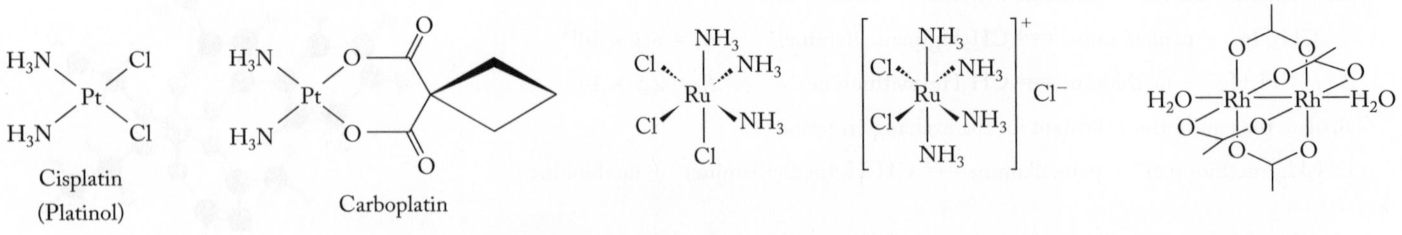

Cisplatin (Platinol)

Carboplatin

$[Ru(NH_3)_4Cl_2]^+Cl^-$

FIGURE 22.37 Cisplatin and carboplatin are effective antitumor agents. The ruthenium and rhodium compounds show activity against leukemia but have not yet seen widespread use.

complexes of gold, rhodium, ruthenium, and silver have been explored. The effectiveness of all of these drugs lies in their ability to bind to the nitrogen-containing bases in DNA and inhibit cell replication. If their cells cannot divide, tumors cannot grow. The greater toxicity of rhodium compounds compared to compounds of platinum and ruthenium has limited their clinical application.

organometallic compound a molecule containing direct carbon–metal covalent bonds.

Selected osmium compounds reduce inflammation in joints resulting from arthritis, although the use of such compounds has diminished with the development of other anti-inflammatory agents. The therapeutic effects of aqueous solutions of osmium tetroxide (OsO_4) were first investigated in the 1950s. The use of osmium tetroxide was superseded by the use of glucose polymers containing osmium, known as osmarins. The use of osmarins reduces the toxic effects of osmium and illustrates how even toxic metals can be adapted to therapy.

Although the historic use of gold for medicinal purposes dates back millennia, the effective use of gold-containing pharmaceuticals originated with the discovery in the 1920s and 1930s that a gold thiosulfate compound, sanochrysin (**Figure 22.38a**), alleviates the symptoms of rheumatoid arthritis. The most commonly used gold drugs for the treatment of arthritis today are sold under the names myochrysine and auranofin (**Figure 22.38b, c**). Myochrysine is injected; auranofin can be taken orally.

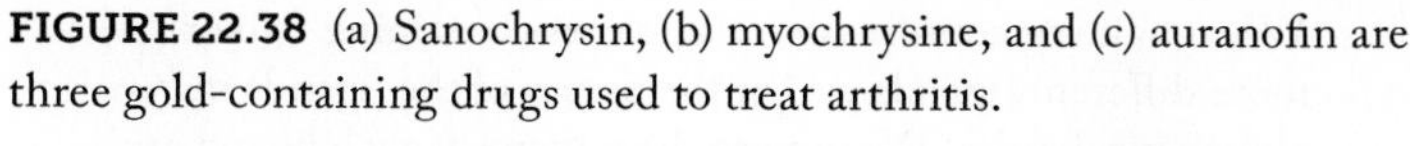
FIGURE 22.38 (a) Sanochrysin, (b) myochrysine, and (c) auranofin are three gold-containing drugs used to treat arthritis.

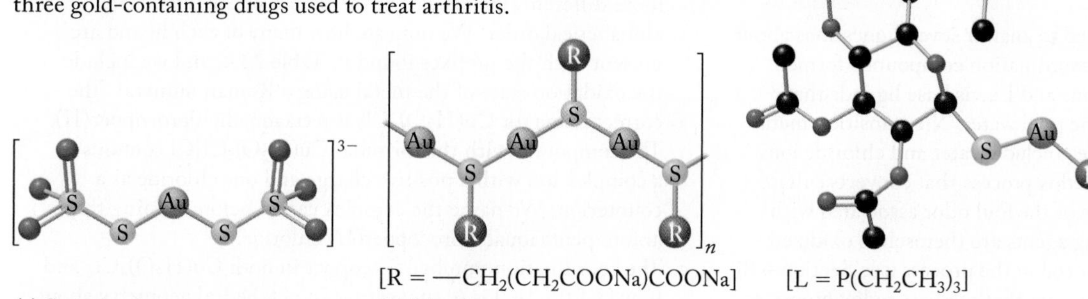

Eye drops containing silver salts are used to treat eye infections. Silver sulfadiazine (**Figure 22.39**) is a broad-spectrum "sulfa" drug used to prevent and treat bacterial and fungal infections. It is also an active ingredient in creams used to treat thermal and chemical burns.

Although present in smaller amounts in the human body than the main group elements, the transition metals play crucial roles in our physiological system and in those of many organisms. Their rich and diverse chemistry has led to many applications of both essential and nonessential transition metal compounds to maintaining health and diagnosing and treating disease. Without a doubt, the smorgasbord of the periodic table will continue to provide us with useful substances and materials.

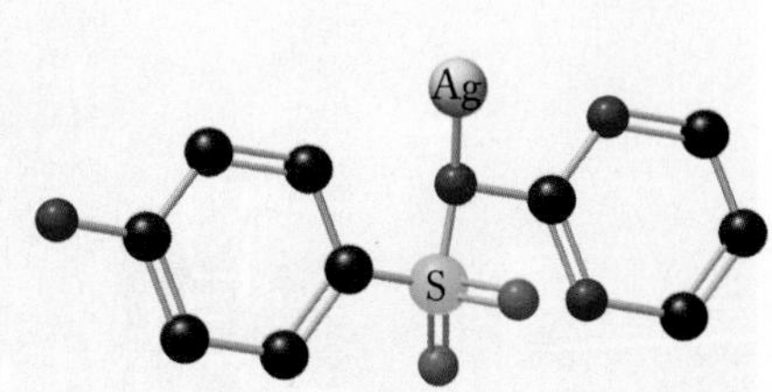

FIGURE 22.39 Silver sulfadiazine is an effective antibiotic when applied to burns. (H atoms are not shown.)

SAMPLE EXERCISE 22.7 Integrating Concepts: Water Quality in Swimming Pools

The diving competition at the 2016 Olympic Games in Rio de Janeiro was plagued by smelly green pool water (**Figure 22.40**). Several explanations were proposed including one based on the addition of copper(II) sulfate to kill the algae growing under the tropical conditions. The color in the water was attributed to the formation of green copper(II) chloride complexes and the odor from the conversion of sulfate to hydrogen sulfide by cytochromes. (The chloride ions were derived from ClO^- ions added to the water as a disinfectant.)

a. Aqueous solutions of copper(II) sulfate contain blue $[Cu(H_2O)_6]^{2+}$ ions. How do the neutral water molecules bond to the Cu^{2+} ion?
b. Two possible copper(II) chloride complexes are $Cu(H_2O)_4Cl_2$ and $[Cu(H_2O)_5Cl]Cl$. How would we name these two coordination complexes?
c. How many isomers can be drawn for the copper(II) complexes in part (c)?
d. Are any of these complexes paramagnetic?
e. What is the origin of the blue color in solutions of $[Cu(H_2O)_6]^{2+}$?
f. Why does the color of the soluble copper complexes change from blue to green in the presence of chloride ion?
g. Does the conversion of SO_4^{2-} to H_2S require a reducing agent or an oxidizing agent—perhaps cytochromes?

Collect and Organize We need to answer several questions about the structure and bonding in coordination compounds formed between transition metal cations and Lewis base ligands that may account for the color of the pool water. The transition metal ion is Cu^{2+} and the Lewis bases include water and chloride ions. We also need to consider the redox process that converts sulfate to hydrogen sulfide as a source of the foul odor associated with the green pool water. Reducing agents are themselves oxidized while oxidizing agents are reduced in the process (see Section 4.8). Cytochromes are a class of enzymes that mediate redox processes.

Analyze Transition metal cations act as Lewis acids, accepting electron pairs from Lewis bases. The Lewis bases have lone pairs available to share with Cu^{2+} in a coordinate covalent bond. The coordination number of copper equals 6 in each of the copper complexes. The number and identity of the ligands determines whether isomers are possible and provides clues to the color of the complexes. We will likely need to consult the *d*-orbital splitting in Figure 22.12 and the color wheel in Figure 22.14 when addressing questions about magnetism and color in copper complexes.

FIGURE 22.40 The aquatic events at the 2016 Olympic Games were plagued by green, odorous water.

Solve

a. Based on the Lewis structure for water, each oxygen atom can share a lone pair of electrons with the Cu^{2+} ion to form a coordinate covalent bond.

H–Ö–H $\quad [Cu(H_2O)_6]^{2+}$ (octahedral: Cu bonded to six OH_2/H_2O ligands)

b. The two complexes, $Cu(H_2O)_4Cl_2$ and $[Cu(H_2O)_5Cl]Cl$, have different numbers of water and chloride ligands in their inner coordination sphere, which leads to different names. When naming coordination complexes containing two or more different ligands we list the names of the ligands in alphabetical order. We indicate how many of each ligand are present with the prefixes found in Table 22.4, and we include the oxidation state of the metal using a Roman numeral. The correct name for $Cu(H_2O)_4Cl_2$ is tetraaquadichlorocopper(II). The compound with the formula $[Cu(H_2O)_5Cl]Cl$ contains a complex ion with a positive charge and one chloride as a counterion. We name the complex cation before naming the anion: pentaaquachlorocopper(II) chloride.

c. The coordination number for copper in both $Cu(H_2O)_4Cl_2$ and $[Cu(H_2O)_5Cl]Cl$ is 6, suggesting an octahedral geometry about the copper cation. There are two ways to arrange four water molecules and two chloride ions around one Cu^{2+}: a cis and a trans arrangement. We can modify the name for $Cu(H_2O)_4Cl_2$ by adding the prefix *cis*- or *trans*- to indicate which isomer is present. Only one isomer is possible for $[Cu(H_2O)_5Cl]^+$.

$[Cu(H_2O)_5Cl]^+$ — Pentaquachloro-copper(II)

$Cu(H_2O)_4Cl_2$ (Cl ligands adjacent) — *cis*-Tetraquadichloro-copper(II)

$Cu(H_2O)_4Cl_2$ (Cl ligands opposite) — *trans*-Tetraquadichloro-copper(II)

d. The magnetic properties of six-coordinate complexes can be determined from the energy levels of the *d* orbitals of their central atoms (see Figure 22.12). The 9 valence electrons of Cu^{2+} fill the lower-energy d_{xy}, d_{xz}, and d_{yz} orbitals, but one of the two higher-energy orbitals, the d_{z^2} or $d_{x^2-y^2}$, contains a single, unpaired electron, which means that both Cu(II) complexes are paramagnetic.

e. The colors of six-coordinate complexes are also linked to their crystal field splitting energy Δ. The absorption of visible light of the proper wavelength by $[Cu(H_2O)_6]^+$ leads to promotion of an electron from one of the fully occupied lower-energy orbitals to the partially filled higher-energy orbital. The color we see is the complementary color to the wavelength absorbed by $[Cu(H_2O)_6]^+$ (see Figure 22.14). In this case, absorption of orange light accounts for the blue color.

f. The magnitude of Δ depends on the nature of the ligands that surround Cu^{2+}, as shown in the spectrochemical series (Table 22.6). Chloride is found below water in Table 22.6, which means that Δ for $Cu(H_2O)_4Cl_2$ and $[Cu(H_2O)_5Cl]^+$ will be less than for $[Cu(H_2O)_6]^+$. The result is that the energy of light required to move an electron from the d_{xy}, d_{xz}, and d_{yz} set of orbitals to the d_{z^2} or $d_{x^2-y^2}$ orbitals will be *less* for copper chloride complexes than for $[Cu(H_2O)_6]^+$. Lower-energy light corresponds to a *longer* wavelength, so copper chloride complexes will absorb the red region of the electromagnetic spectrum shown in Figure 22.14. The complementary color of red is green.

g. The conversion of sulfate (SO_4^{2-}) to hydrogen sulfide (H_2S) is a redox process because the oxidation number of sulfur changes from +6 in SO_4^{2-} to −2 in H_2S, a net gain of 8 electrons. The gain of electrons is defined as reduction, so one or more reducing agents must have been present in the pool. Cytochromes could provide the necessary electrons for the reduction of sulfate to hydrogen sulfide.

Think About It If copper sulfate was added to the diving pool water at the 2016 Olympic Games, the observed green color is consistent with the formation of copper(II) chloride complexes. The redox activity of cytochromes could account for the formation of H_2S, consistent with the hypothesis presented in this exercise. The data do not prove, however, that this was the origin of the green color.

SUMMARY

LO1 Molecules or anions that function as Lewis bases and form coordinate bonds with metal cations are called *ligands*. The resulting species, which are composed of central metal ions and the surrounding ligands, are called *complex ions* or simply complexes. Direct bonding to a central cation means that the ligands in a complex occupy the **inner coordination sphere** of the cation. (Section 22.1)

LO2 The names of complex ions and coordination compounds provide information about the identities and numbers of ligands, the identity and oxidation state of the central metal ion, and the identity and number of counterions. (Section 22.2)

LO3 A **monodentate ligand** donates one pair of electrons in a complex ion; a **polydentate ligand** donates more than one pair in a process called **chelation**. Polydentate ligands are particularly effective at forming complex ions. This phenomenon is called the **chelate effect**. EDTA is a particularly effective sequestering agent, which is a chelating agent that prevents metal ions in solution from reacting with other substances. (Section 22.3)

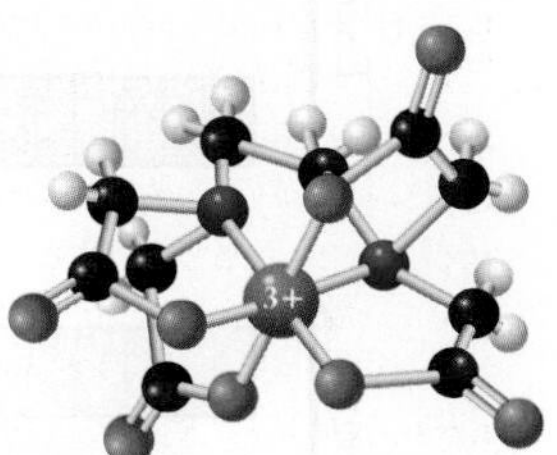

LO4 The colors of transition metal compounds can be explained by the interactions between electrons in different *d* orbitals and the lone pairs of electrons on surrounding ligands. These interactions create **crystal field splitting** of the energies of the *d* orbitals. A **spectrochemical series** ranks ligands on the basis of their field strength and the wavelengths of electromagnetic radiation absorbed by their complex ions. (Section 22.4)

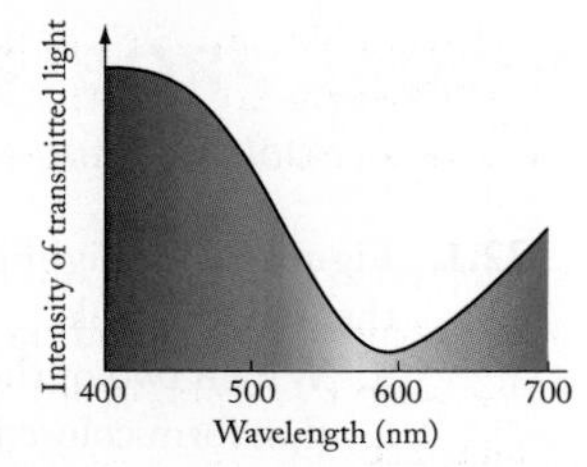

LO5 Strong repulsions and large values of crystal field splitting energy can lead to electron pairing in lower-energy orbitals and an electron configuration called a low-spin state. Metals and their ions with their *d* electrons evenly distributed across all the *d* orbitals in the valence shell represent a high-spin state. (Section 22.5)

LO6 Complex metal ions containing more than one type of ligand may form different kinds of isomers. When one type occupies two adjacent corners of a square planar complex, the complex is a *cis* isomer; when the same ligand occupies opposite corners, it is a *trans* isomer. When ligands contain two different Lewis base atoms, linkage isomers are possible. Octahedral complexes may form as a pair of enantiomers. (Sections 22.6 and 22.7)

LO7 Many enzymes contain transition metal ions. Soluble coordination compounds and **organometallic compounds** of the transition metals appear in enzymes. Such compounds are used for imaging and in medications for arthritis, cancer, and other diseases. (Section 22.8)

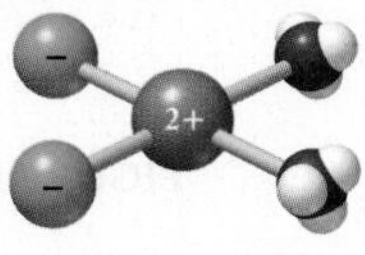

PARTICULATE PREVIEW WRAP-UP

Nitrogen; ammonia, 4; ethylenediamine, 3.

PROBLEM-SOLVING SUMMARY

Type of Problem	Concepts and Equations	Sample Exercises
Interpreting and writing formulas of coordination compounds	Identify complex ions, counterions, and their charges. Determine the oxidation states of metals in complex ions. Use square brackets to represent a complex ion.	**22.1**
Naming coordination compounds	Follow the naming rules in Section 22.2.	**22.2**
Identifying the potential electron-pair–donor groups in a molecule	Examine the molecular structure of a compound and find lone pairs that can be donated to a metal.	**22.3**
Predicting spin states	Sketch a *d*-orbital diagram based on crystal field splitting. Fill the lowest-energy orbitals with one valence electron each. If there are two ways of adding the remaining valence electrons to the diagram, then multiple spin states are possible.	**22.4**
Identifying isomers of coordination compounds	If ligands of one type are all on the same side of the complex ion, it is a cis isomer. If ligands of one type are on opposite sides, it is a trans isomer.	**22.5**
Calculating the equilibrium constant for a ligand exchange reaction	Use the rules for manipulating equilibrium constant expressions to calculate the equilibrium constant for a ligand exchange reaction.	**22.6**

VISUAL PROBLEMS

(Answers to boldface end-of-chapter questions and problems are in the back of the book.)

22.1. Figure P22.1 highlights six elements in the fourth row of the periodic table.

a. Which two of the highlighted elements have cations that form colored compounds with Cl^-?

b. Which of the highlighted transition metals form M^{2+} cations that cannot have high-spin and low-spin states?

c. Which of the highlighted transition metals have M^{2+} cations that form colorless tetrahedral complex ions?

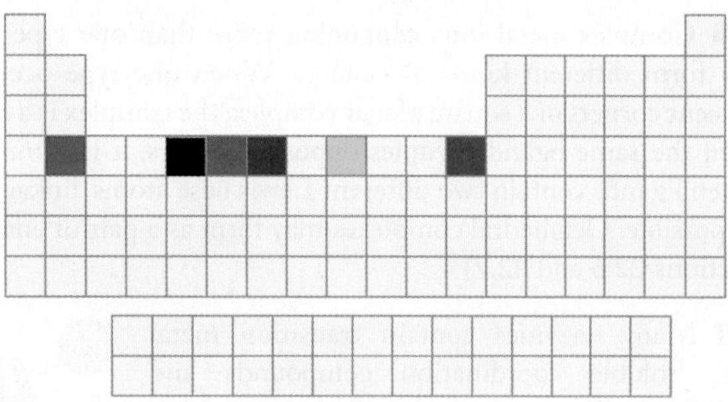

FIGURE P22.1

22.2. Predict the number of unpaired electrons for each of the transition metal complexes in Figure P22.2.

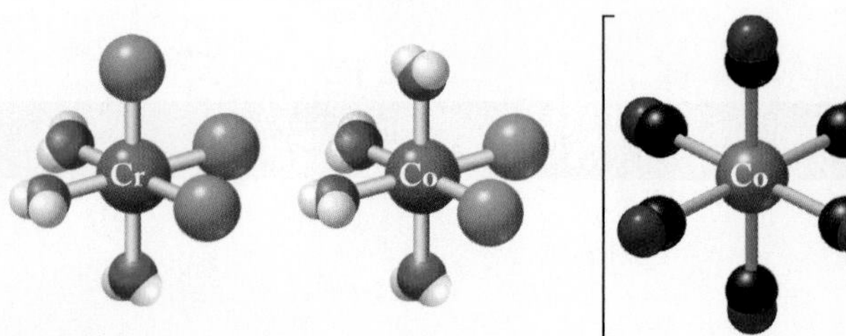

FIGURE P22.2

22.3. The presence of manganese impurities in crystals of silicon dioxide gives smoky quartz its distinctive lavender and purple colors. Which of the orbital diagrams in Figure P22.3 best describes the Mn^{2+} ion in a tetrahedral field?

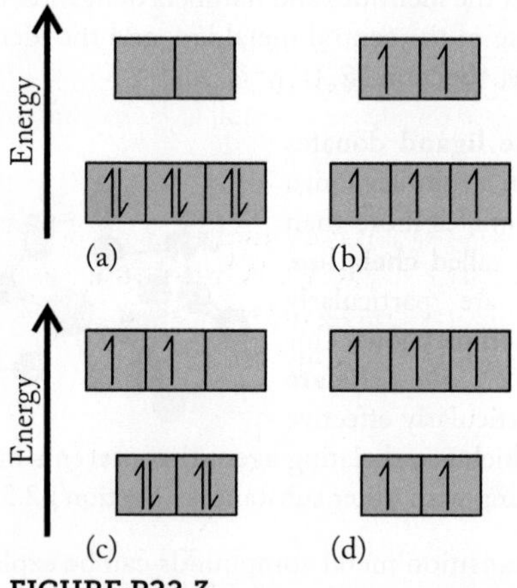

FIGURE P22.3

22.4. The nickel(II) complex $(PMePh_2)_2NiX_2$, where X = Cl, has no unpaired electrons, but it has two when X = Br. Which of the orbital diagrams in Figure P22.4 best explains the difference between the two complexes?

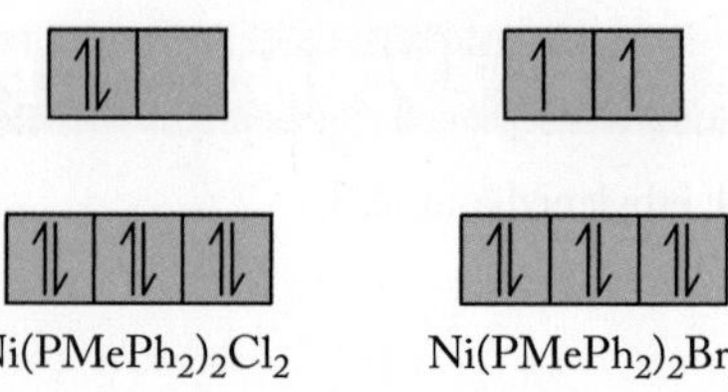

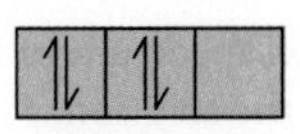
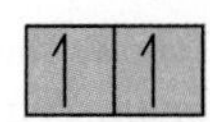
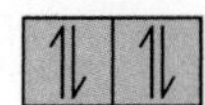

$Ni(PMePh_2)_2Cl_2$ $Ni(PMePh_2)_2Br_2$

(b)

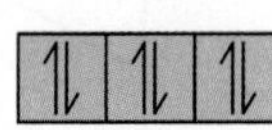

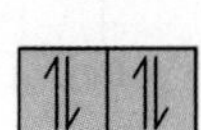
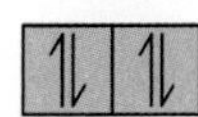
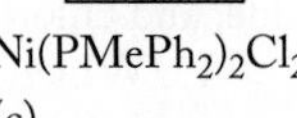

$Ni(PMePh_2)_2Cl_2$ $Ni(PMePh_2)_2Br_2$

(c)

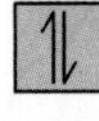

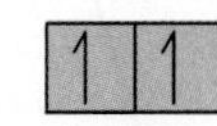
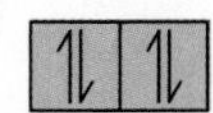
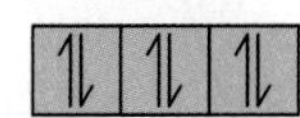

$Ni(PMePh_2)_2Cl_2$ $Ni(PMePh_2)_2Br_2$

(d)

FIGURE P22.4

22.5. **Platinum Drugs** There is great interest in developing new cancer drugs based on Pt(IV). Which of the compounds in Figure P22.5 contain ligands that are in a trans arrangement? In a cis arrangement? Both?

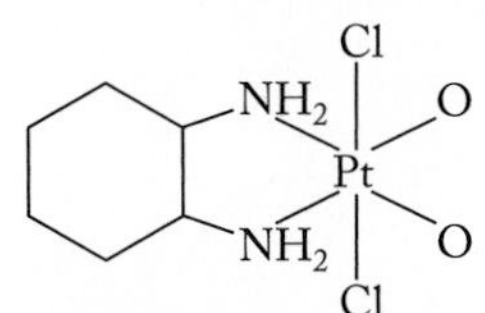

Ormaplatin 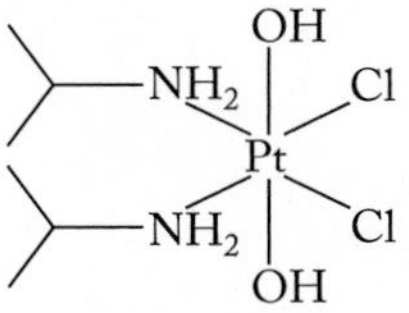 Iproplatin

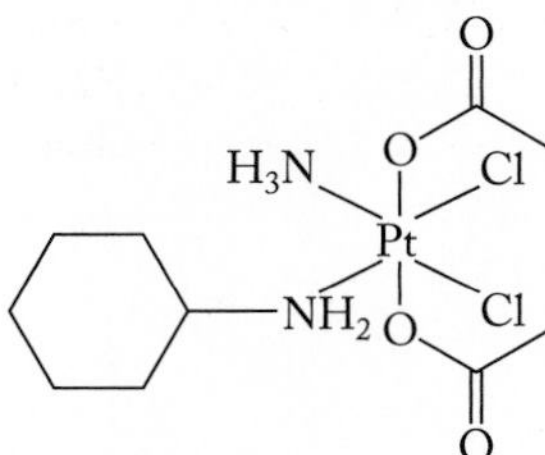

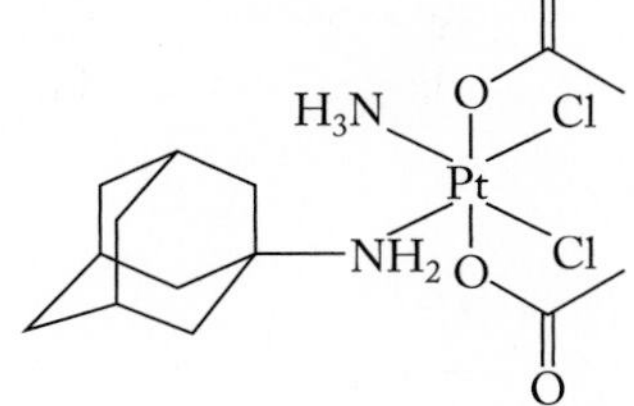

Satraplatin LA-12

FIGURE P22.5

22.6. For each pair of complexes in Figure P22.6, indicate whether the complexes are (i) identical, (ii) isomers, or (iii) neither.

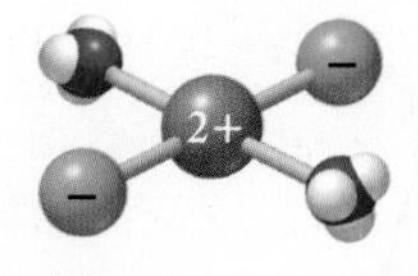

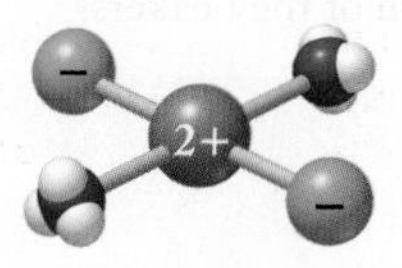

(a)

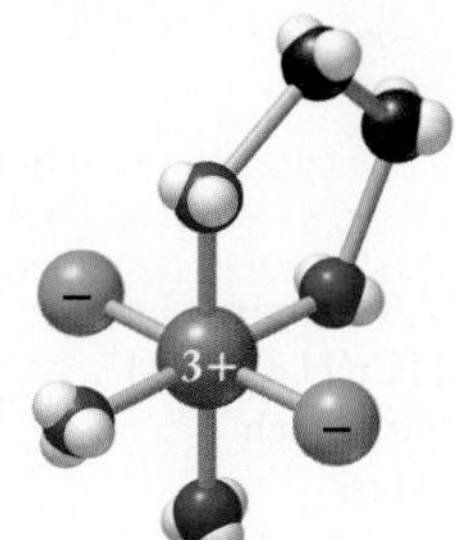

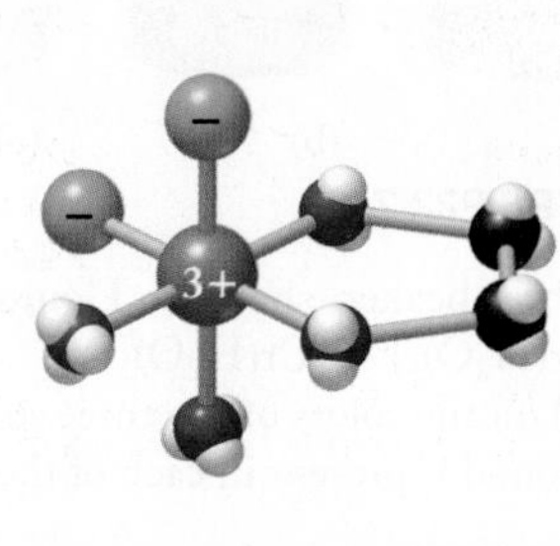

(b)

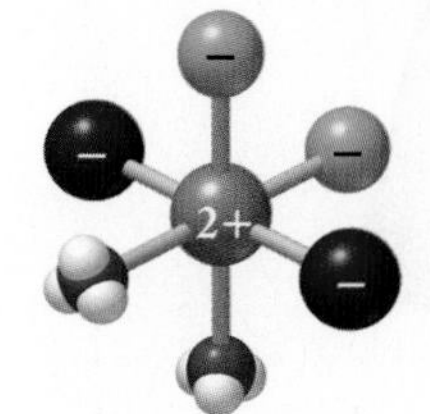

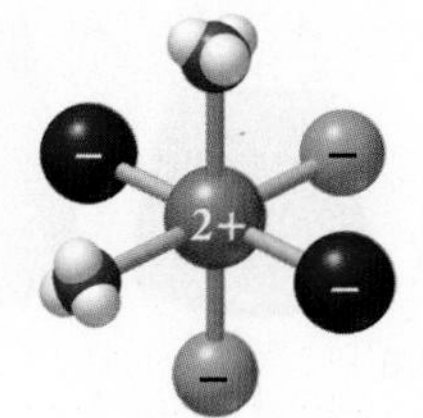

(c)

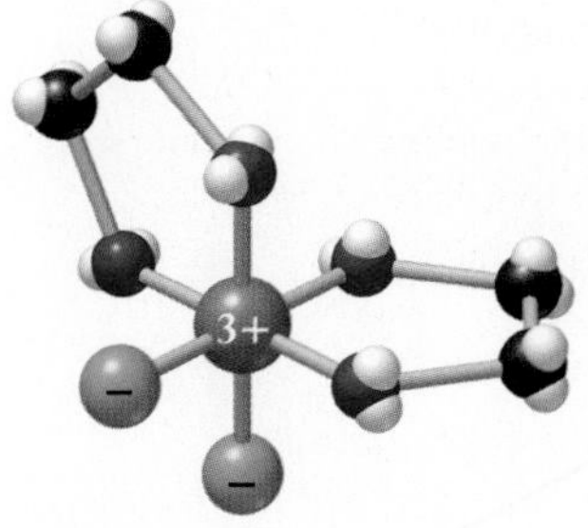

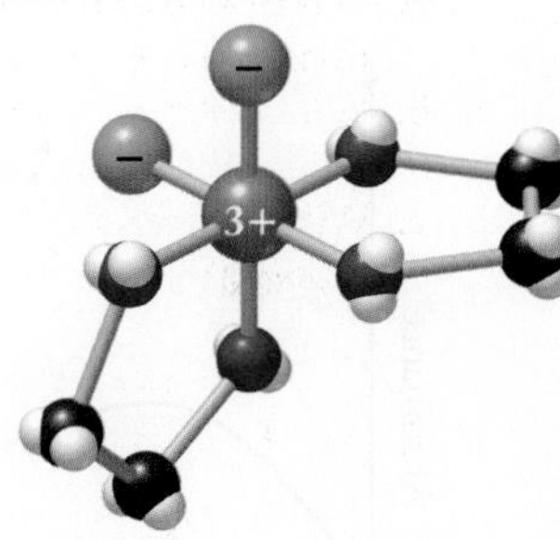

(d)

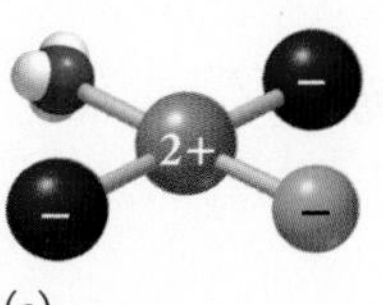

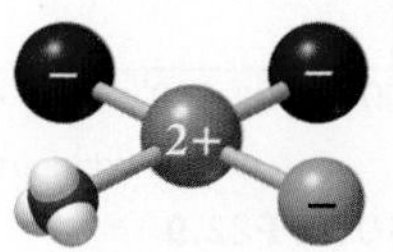

(e)

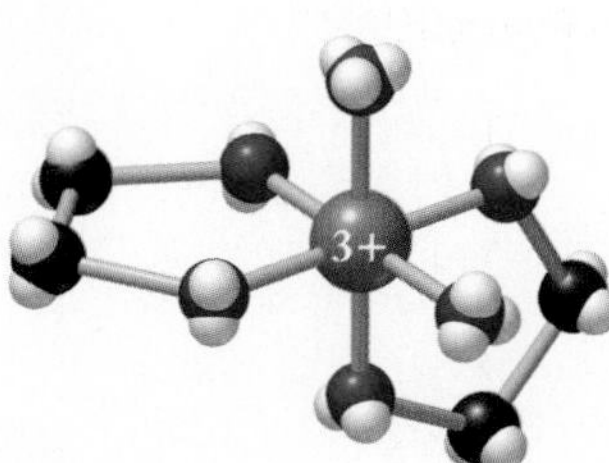

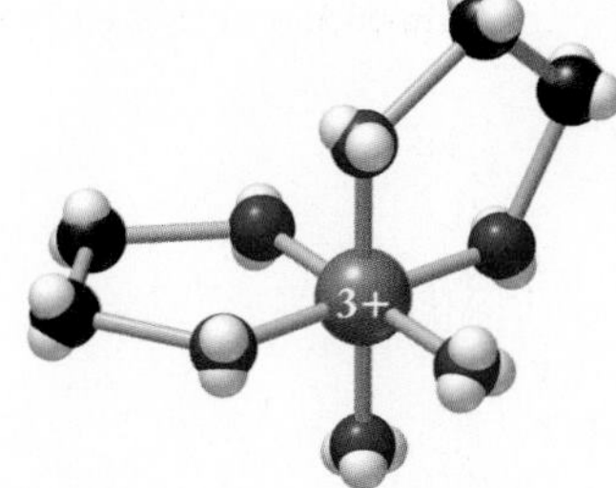

(f)

FIGURE P22.6

22.7. The three beakers shown in Figure P22.7 contain solutions of $[Co(H_2O)_6]^{3+}$, $[Co(NH_3)_6]^{3+}$, and $[Co(CN)_6]^{3-}$. Based on the colors of the three solutions, which compound is present in each of the beakers?

FIGURE P22.7

22.8. The three beakers shown in Figure P22.8 contain solutions of $[Cr(H_2O)_6]^{3+}$, $[Cr(H_2O)_5Cl]^{2+}$, and $[Cr(H_2O)_4Cl_2]^{+}$. Based on the colors of the three solutions, which compound is present in each of the beakers?

FIGURE P22.8

22.9. Figure P22.9 shows the absorption spectrum of a solution of $Ti(H_2O)_6^{3+}$. What color is the solution?

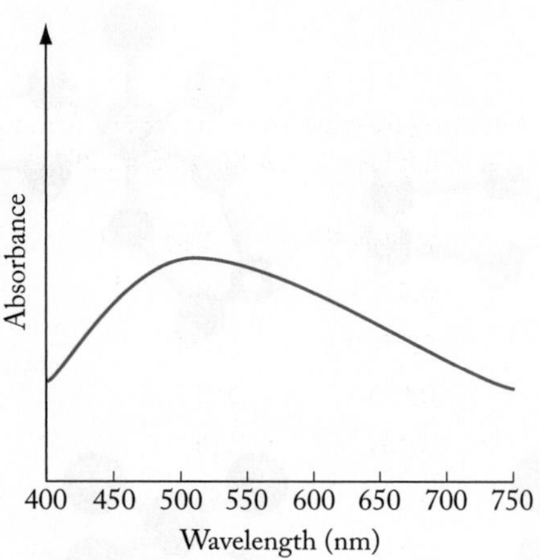

FIGURE P22.9

22.10. Figure P22.10 shows the absorption spectrum of a solution of $TiCl_3(pyridine)_3$. What color is the solution?

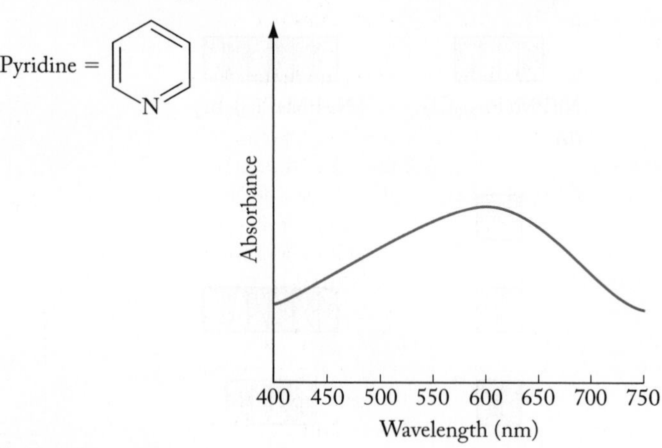

FIGURE P22.10

***22.11.** The periodic trend for iodide, bromide, and chloride in the spectrochemical series is $Cl^- > Br^- > I^-$. Which curve in Figure P22.11 represents the spectrum for $CoCl_4^{2-}$? Which curve represents CoI_4^{2-}?

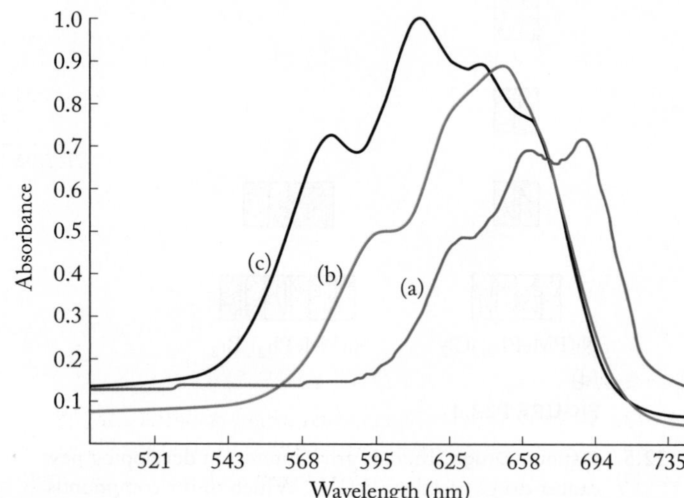

FIGURE P22.11

22.12. Use representations [A] through [I] in Figure P22.12 to answer questions (a)–(f).

a. What are the donor atoms in each ligand depicted?
b. Which ligands are monodentate?
c. Which ligands are bidentate?
d. What is the coordination number for nickel in $Ni(CN)_4^{2-}$? In $Ni(dmg)_2$?
e. If the field strength of cyanide is greater than that of dmg, which complex forms the yellow solution in [E]: $Ni(CN)_4^{2-}$ or $Ni(dmg)_2$? Which forms the red solution?
f. Given that both $Ni(CN)_4^{2-}$ and $Ni(dmg)_2$ are diamagnetic complexes, which orbital diagram depicts the *d*-orbital energies for $Ni(CN)_4^{2-}$? For $Ni(dmg)_2$?

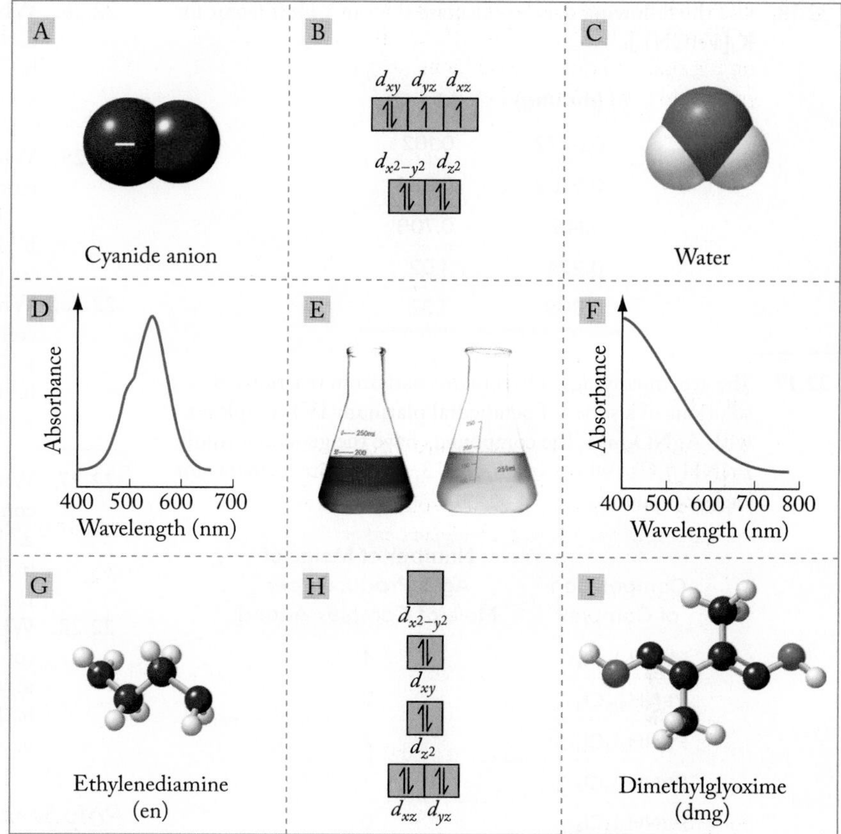

FIGURE P22.12

QUESTIONS AND PROBLEMS

Complex Ions

Concept Review

22.13. When NaCl dissolves in water, which molecules or ions occupy the inner coordination sphere around the Na^+ ions?

22.14. When $Cr(NO_3)_3$ dissolves in water, which of the following species are nearest the Cr^{3+} ions?

a. Other Cr^{3+} ions
b. NO_3^- ions
c. H_2O molecules with the O atoms closest to the Cr^{3+}
d. H_2O molecules with the H atoms closest to the Cr^{3+}

22.15. The coordination compound $Co_3[Cr(CN)_6]_2$ contains a Co^{2+} cation and a complex anion. What is the likely oxidation state for Cr in the anion?

22.16. The coordination compound $Fe_3[Fe(CN)_6]_3$ contains a Fe^{2+} cation and a complex anion. What is the likely oxidation state for Fe in the anion?

Problems

22.17. Use the following freezing point depression data to calculate the van 't Hoff factor for $K_3[Fe(CN)_6]$.

m (mol/kg)	Δ*T* (°C)
0.0301	0.180
0.0780	0.334
0.160	0.807
0.247	1.19
0.338	1.58
0.414	1.90
0.535	2.41
0.664	3.00
0.754	3.43

22.18. Use the following data to calculate the van 't Hoff factor for $K_4[Fe(CN)_6]$.

m (mol/kg)	Δ*T* (°C)
0.0272	0.162
0.0702	0.385
0.145	0.709
0.224	1.02
0.309	1.32

22.19. The accompanying table contains data from reactions of solutions of a series of octahedral platinum(IV) complexes with $AgNO_3(aq)$. The compounds have the general formula $Pt(NH_3)_xCl_4$, where $x = 6, 5, 4, 3$, or 2. Write formulas for each compound.

Composition of Complex	Number of Moles of AgCl Produced per Mole of Complex Added
$Pt(NH_3)_6Cl_4$	4
$Pt(NH_3)_5Cl_4$	3
$Pt(NH_3)_4Cl_4$	2
$Pt(NH_3)_3Cl_4$	1
$Pt(NH_3)_2Cl_4$	0

22.20. The compositions of three compounds of chromium(III) and chloride ion are known: $Cr(H_2O)_6Cl_3$, $Cr(H_2O)_5Cl_3$, and $Cr(H_2O)_4Cl_3$. The following table summarizes their properties.

Compound	Color of Aqueous Solution	Number of Moles of AgCl Produced per Mole of Complex Added
A	Dark green	1
B	Blue-green	3
C	Green	2

Write the correct formulas for compounds A, B, and C.

Naming Complex Ions and Coordination Compounds

Problems

22.21. Name the complex ions of platinum(IV) in Problem 22.19.

22.22. Name the complex ions of chromium(III) in Problem 22.20.

22.23. What are the names of the following complex ions?
a. $Cr(NH_3)_6^{3+}$
b. $Co(H_2O)_6^{3+}$
c. $Ni(CN)_5^{3-}$

22.24. What are the names of the following complex ions?
a. $Cu(NH_3)_2^{+}$
b. $Ti(H_2O)_4(OH)_2^{2+}$
c. $Cr(en)(OH)_4^{-}$

22.25. What are the names of the following coordination compounds?
a. K_2CoBr_4
b. $NH_4[Zn(H_2O)(OH)_3]$
c. $[Fe(NH_3)_5Cl]Cl_2$

22.26. What are the names of the following coordination compounds?
a. Na_2CoI_4
b. $BaCuCl_4$
c. $[Ni(NH_3)_4(H_2O)_2]SO_4$

22.27. What are the names of the following coordination compounds?
a. $[Zn(en)_2]SO_4$
b. $[Ni(NH_3)_5(H_2O)]Cl_2$
c. $K_4[Fe(CN)_6]$

22.28. What are the names of the following coordination compounds?
a. $(NH_4)_3[Co(CN)_6]$
b. $[Co(en)_2Cl](NO_3)_2$
c. $[Fe(H_2O)_4(OH)_2]Cl$

Polydentate Ligands and Chelation

Concept Review

22.29. What is meant by the term *sequestering agent*? What properties make a substance an effective sequestering agent?

22.30. The structures of two compounds that each contain two $-NH_2$ groups are shown in Figure P22.30. The one on the left is ethylenediamine, a bidentate ligand. Does the molecule on the right have the same ability to donate two pairs of electrons to a metal ion? Explain why you think it does or does not.

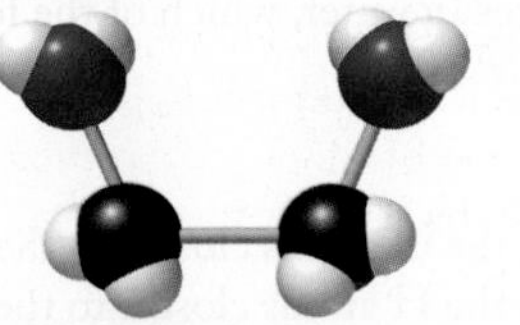
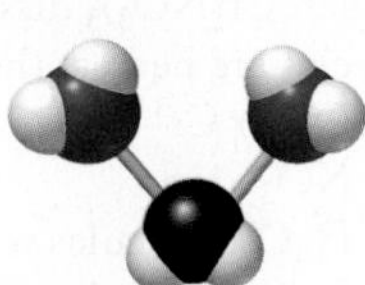

FIGURE P22.30

22.31. How does the chelating ability of an aminocarboxylic acid vary with changing pH?

*22.32. **Food Preservative** The EDTA that is widely used as a food preservative is added to food, not as the undissociated acid, but rather as the calcium disodium salt: $Na_2[CaEDTA]$. This salt is actually a coordination compound with a Ca^{2+} ion at the center of a complex ion. Draw the structure of this compound.

Crystal Field Theory

Concept Review

22.33. Explain why the compounds of most of the first-row transition metals are colored.

22.34. Unlike the compounds of most transition metal ions, those of Ti^{4+} are colorless. Why?

22.35. Why is the d_{xy} orbital higher in energy than the d_{xz} and d_{yz} orbitals in a square planar crystal field?

22.36. On average, the *d* orbitals of a transition metal ion in an octahedral field are higher in energy than they are when the ion is in the gas phase. Why?

Problems

22.37. Aqueous solutions of one of the following complex ions of chromium(III) are violet; solutions of the other are yellow. Which is which? (a) $Cr(H_2O)_6{}^{3+}$; (b) $Cr(NH_3)_6{}^{3+}$

22.38. Which of the following complex ions should absorb the shortest wavelengths of electromagnetic radiation? (a) $CuCl_4{}^{2-}$; (b) $CuF_4{}^{2-}$; (c) $CuI_4{}^{2-}$; (d) $CuBr_4{}^{2-}$

22.39. The octahedral crystal field splitting energy Δ_o of $Co(phen)_3{}^{3+}$ is 5.21×10^{-19} J/ion. What is the color of a solution of this complex ion?

22.40. The octahedral crystal field splitting energy Δ_o of $Co(CN)_6{}^{3-}$ is 6.74×10^{-19} J/ion. What is the color of a solution of this complex ion?

22.41. Solutions of $NiCl_4{}^{2-}$ and $NiBr_4{}^{2-}$ absorb light at 702 nm and 756 nm, respectively. In which ion is the split of *d*-orbital energies greater?

22.42. Chromium(III) chloride forms six-coordinate complexes with bipyridine, including *cis*-$[Cr(bipy)_2Cl_2]^+$, which reacts slowly with water to produce *cis*-$[Cr(bipy)_2(H_2O)Cl]^{2+}$ and *cis*-$[Cr(bipy)_2(H_2O)_2]^{3+}$. In which of these complexes should Δ_o be the largest?

Magnetism and Spin States

Concept Review

22.43. What determines whether a transition metal ion is in a *high-spin* configuration or a *low-spin* configuration?

22.44. Would you expect a solution of a high-spin complex of a transition metal ion to be the same color as a solution of a low-spin complex? Why?

Problems

22.45. How many unpaired electrons are there in the following transition metal ions in an octahedral field? High-spin Fe^{2+}, Cu^{2+}, Co^{2+}, and Mn^{3+}

22.46. Which of the following cations can have either a high-spin or a low-spin electron configuration in an octahedral field? Fe^{2+}, Co^{3+}, Mn^{2+}, and Cr^{3+}

***22.47.** A solid compound containing iron(II) in an octahedral crystal field has four unpaired electrons at 298 K. When the compound is cooled to 80 K, the same sample appears to have no unpaired electrons. How do you explain this change in the compound's properties?

22.48. The iron(II) compound $Fe(bipy)_2(SCN)_2$ is paramagnetic, but the corresponding cyanide compound $Fe(bipy)_2(CN)_2$ is diamagnetic. Why do these two compounds have different magnetic properties?

22.49. The manganese minerals pyrolusite (MnO_2) and hausmannite (Mn_3O_4) contain manganese ions surrounded by oxide ions.

a. What are the charges of the Mn ions in each mineral?
b. In which of these compounds could there be high-spin and low-spin Mn ions?

22.50. **Dietary Supplement** Chromium picolinate is an over-the-counter diet aid sold in many pharmacies. The Cr^{3+} ions in this coordination compound are in an octahedral field. Is the compound paramagnetic or diamagnetic?

***22.51.** **Refining Cobalt** One method for refining cobalt involves the formation of the complex ion $CoCl_4{}^{2-}$. This anion is tetrahedral. Is this complex paramagnetic or diamagnetic?

*22.52. Why is it that $Ni(CN)_4{}^{2-}$ is diamagnetic, but $NiCl_4{}^{2-}$ is paramagnetic?

Isomerism in Coordination Compounds

Concept Review

22.53. What do the prefixes *cis-* and *trans-* mean in the context of an octahedral complex ion?

22.54. What do the prefixes *cis-* and *trans-* mean in the context of a square planar complex?

22.55. What is the minimum number of donor groups required in order to have stereoisomers of a square planar complex?

22.56. With respect to your answer to Problem 22.55, do all square planar complexes with this number of donor groups have stereoisomers?

Problems

22.57. Does the complex $Co(en)(H_2O)_2Cl_2$ have stereoisomers?

22.58. Does the complex ion $Fe(en)_3{}^{3+}$ have stereoisomers?

***22.59.** Sketch the stereoisomers of the square planar complex ion $CuCl_2Br_2{}^{2-}$. Are any of these isomers chiral?

*22.60. Sketch the stereoisomers of the complex ion $Ni(en)Cl_2(CN)_2{}^{2-}$. Are any of these isomers chiral?

Coordination Compounds in Biochemistry

Concept Review

22.61. Enzymes are large proteins.

a. What is the function of enzymes?
b. Are all proteins enzymes?

*22.62. Why is Cd^{2+} more likely than Cr^{2+} to replace Zn^{2+} in an enzyme such as carbonic anhydrase?

***22.63.** What effect does an enzyme have on the activation energy of a biochemical reaction?

*22.64. Why might reductases also be described as reducing agents?

***22.65.** When a transition metal ion such as Cu^{2+} is incorporated into a metalloenzyme, is the formation constant likely to be much greater than one ($K >> 1$) or much less than one ($K << 1$)?

$$Cu^{2+} + \text{protein} \rightleftharpoons \text{metalloenzyme} \qquad K = \frac{[\text{metalloenzyme}]}{[Cu^{2+}][\text{protein}]}$$

*22.66. Carbon monoxide poisoning derives from competitive binding of CO versus O_2 to the Fe^{2+} in the coordination sphere of hemoglobin. If CO can be displaced by breathing pure O_2, which of the following is likely to be correct under these conditions: $K_{O_2}/K_{CO} < 1$ or $K_{O_2}/K_{CO} > 1$?

22.67. What is the likely sign of ΔS for the reaction in Figure P22.67, which is catalyzed by carboxypeptidase?

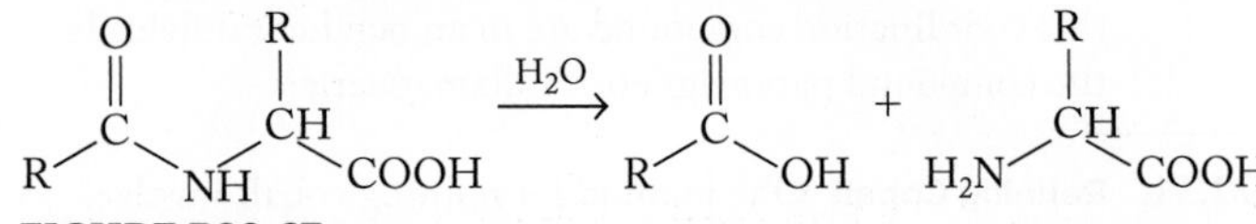

FIGURE P22.67

22.68. What is the likely sign of ΔG for the reaction in Problem 22.67?

Problems

***22.69.** The activation energy for the uncatalyzed decomposition of hydrogen peroxide at 20°C is 75.3 kJ/mol. In the presence of the enzyme catalase, the activation energy is reduced to 29.3 kJ/mol. By using the following form of the Arrhenius equation, $RT\ln(k_1/k_2) = E_{a_1} - E_{a_2}$, how much faster is the catalyzed reaction?

*22.70. **Enzymatic Activity of Urease** Urease catalyzes the decomposition of urea to ammonia and carbon dioxide (Figure P22.70). The rate constant for the uncatalyzed reaction at 20°C and pH = 8 is $k = 3 \times 10^{-10}\ s^{-1}$. A urease isolated from the jack bean increases the rate constant to $k = 3 \times 10^4\ s^{-1}$. By using the $RT\ln(k_1/k_2) = E_{a_1} - E_{a_2}$ form of the Arrhenius equation, calculate the difference between the activation energies.

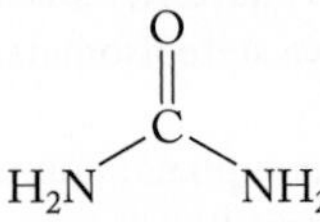

$$H_2N{-}C(=O){-}NH_2(aq) + H_2O(\ell) \rightarrow 2\,NH_3(aq) + CO_2(g)$$

FIGURE P22.70

***22.71.** The initial rate of an enzyme-catalyzed reaction depends on the concentration of substrate as shown in Figure P22.71. What is the apparent reaction order at the far-right side of the graph?

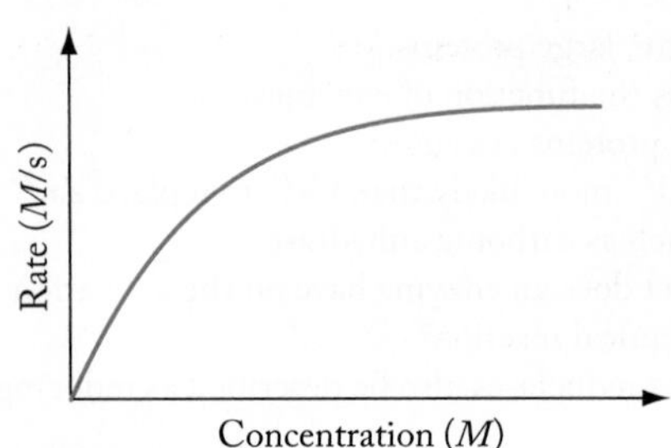

FIGURE P22.71

22.72. The rate of an enzyme-catalyzed reaction depends on the concentration of substrate. Which line in Figure P22.72 has the highest reaction rate?

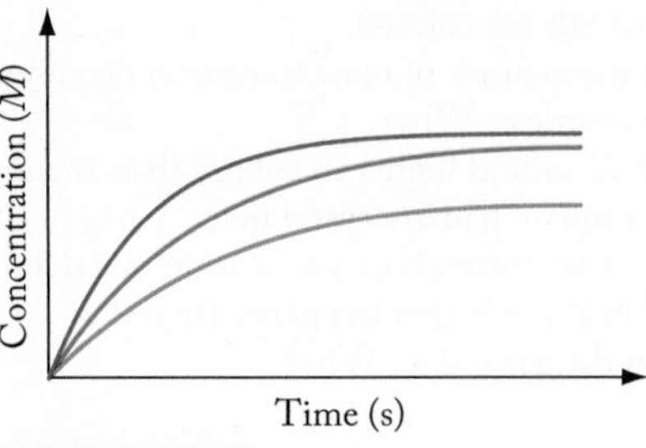

FIGURE P22.72

Coordination Compounds in Medicine

Concept Review

22.73. Under what circumstances might a coordination complex of a toxic metal be useful in medicine?

22.74. **Cancer Treatment** Brachytherapy is a cancer treatment that involves surgical implantation of a small capsule of ^{192}Ir into a tumor. Iridium-192 lies between two stable isotopes of iridium, ^{193}Ir and ^{191}Ir. How does this account for the fact that ^{192}Ir decays by β decay, electron capture, and positron emission?

22.75. Gadolinium-153 decays by electron capture. What type of radiation does ^{153}Gd produce that makes it useful for imaging?

22.76. Gadolinium-153 and samarium-153 both have the same mass number. Why might ^{153}Gd decay by electron capture, whereas ^{153}Sm decays by emitting β particles?

22.77. How do platinum- and ruthenium-containing drugs fight cancer?

*22.78. Many transition metal complexes are brightly colored. Why might the titanium(IV) compound budotitane be colorless?

***22.79.** Is the glucose tolerance factor that contains chromium(III) paramagnetic or diamagnetic?

22.80. Coordination complexes containing paramagnetic transition metal ions make the best MRI contrast agents.

a. Which of the first-row transition metal cations with a 2+ charge have the most unpaired valence electrons in the gas phase?
b. Which of the first-row transition metal cations with a 2+ charge have the most unpaired valence electrons in octahedral coordination complexes?

Problems

22.81. The complexation of mercury(II) ion with methionine,

$$Hg^{2+} + \text{methionine} \rightleftharpoons \text{Hg(methionine)}^{2+}$$

has a formation constant of $\log K_f = 14.2$, whereas the formation constant for the Hg^{2+} complex with penicillamine,

$$Hg^{2+} + \text{penicillamine} \rightleftharpoons \text{Hg(penicillamine)}^{2+}$$

is $\log K_f = 16.2$. Calculate the equilibrium constant for the reaction,

$$\text{Hg(methionine)}^{2+} + \text{penicillamine} \rightleftharpoons \text{Hg(penicillamine)}^{2+} + \text{methionine}$$

22.82. The complexation of mercury(II) ion with cysteine in aqueous solution,

$$Hg^{2+} + \text{cysteine} \rightleftharpoons Hg(\text{cysteine})^{2+}$$

has a formation constant of log K_f =14.2, whereas the formation constant for the Hg^{2+} complex with glycine,

$$Hg^{2+} + \text{glycine} \rightleftharpoons Hg(\text{glycine})^{2+}$$

is log K_f = 10.3. Calculate the equilibrium constant for the reaction,

$$Hg(\text{cysteine})^{2+} + \text{glycine} \rightleftharpoons Hg(\text{glycine})^{2+} + \text{cysteine}$$

22.83. The equilibrium constant of the reaction,

$$CH_3Hg(\text{penicillamine})^+(aq) + \text{cysteine}(aq) \rightleftharpoons CH_3Hg(\text{cysteine})^+(aq) + \text{penicillamine}(aq)$$

is K = 0.633. Calculate the equilibrium concentrations of cysteine and penicillamine if we start with a 1.00 M solution of cysteine and a 1.00 mM solution of $CH_3Hg(\text{penicillamine})^+$.

22.84. The equilibrium constant of the reaction,

$$CH_3Hg(\text{glutathione})^+(aq) + \text{cysteine}(aq) \rightleftharpoons CH_3Hg(\text{cysteine})^+(aq) + \text{glutathione}(aq)$$

is K = 5.0. Calculate the equilibrium concentrations of cysteine and glutathione if we start with a 1.20 mM solution of cysteine and a 1.20 mM solution of $CH_3Hg(\text{glutathione})^+$.

22.85. The two platinum compounds shown in Figure P22.85 have been studied in cancer therapy.
a. What is the difference between the two compounds?
b. Name the two complex ions.
c. Do both compounds have the same orbital diagram?
d. Sketch the d orbital diagram for the platinum compound(s), including the appropriate number of electrons.

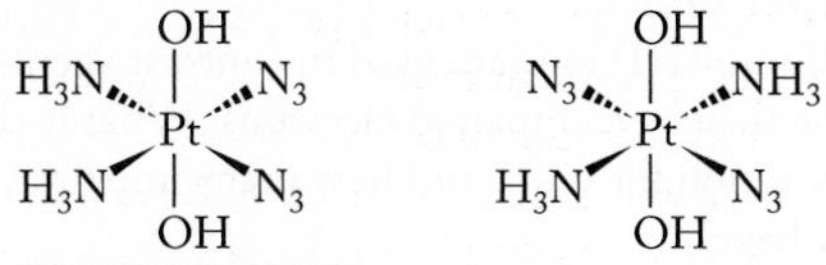

FIGURE P22.85

22.86. Anticancer treatments based on coordination complexes of ruthenium may have less severe side effects than cisplatin and other platinum-based medications. The complex shown in Figure P22.86 was tested against a colon cancer cell line with promising results.
a. Draw the other stereoisomer of $[Ru(py)_2Cl_4]^-$.
b. Name both isomers.
c. Sketch the orbital diagram for the ruthenium compound, including the appropriate number of electrons.

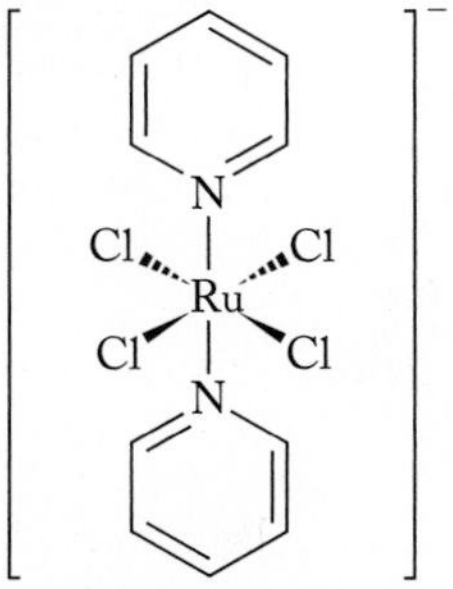

FIGURE P22.86

***22.87.** Two square planar gold and platinum complexes with medicinal properties are shown in Figure P22.87.
a. Do these two complexes have the same number of d electrons in their orbital diagram?
b. Are these compounds diamagnetic or paramagnetic?

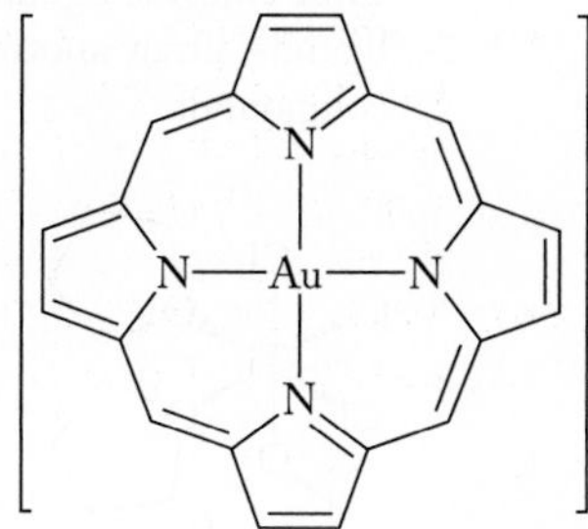

FIGURE P22.87

*22.88. In Section 9.5 we saw that polycyclic aromatic compounds can intercalate into DNA. Compounds such as the rhodium and ruthenium complexes shown in Figure P22.88 behave in a similar fashion toward DNA.
a. What is the hybridization at the nitrogen and the carbon atoms in these ligands?
b. Why are these ligands planar?
c. Are these compounds diamagnetic or paramagnetic?

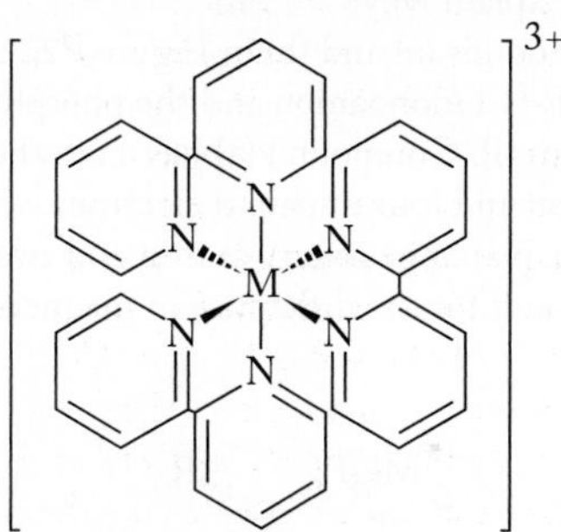

M = Ru, Rh

FIGURE P22.88

Additional Problems

22.89. Dissolving cobalt(II) nitrate in water gives a beautiful red-purple solution. There are three unpaired electrons in this cobalt(II) complex. When cobalt(II) nitrate is dissolved in aqueous ammonia and oxidized with air, the resulting yellow complex has no unpaired electrons.
a. Which cobalt complex has the larger crystal field splitting energy Δ_o?
b. Draw orbital diagrams for the d orbitals of both complexes.

22.90. Adding pyridine to a solution of $TiCl_3(THF)_3$ (Figure P22.90) leads to a decrease in the wavelength at which the solution absorbs.
a. Is this consistent with the formation of $TiCl_3(py)_3$?
b. Figure P22.90 shows one possible arrangement of three chloride ligands and three THF or three pyridine ligands. Draw another geometric isomer of these compounds.

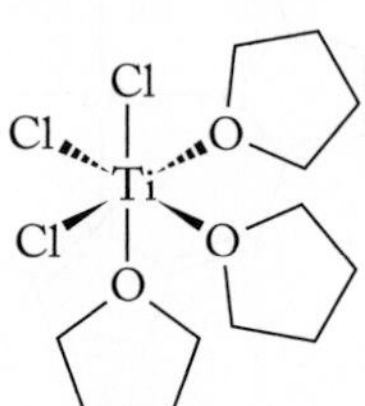

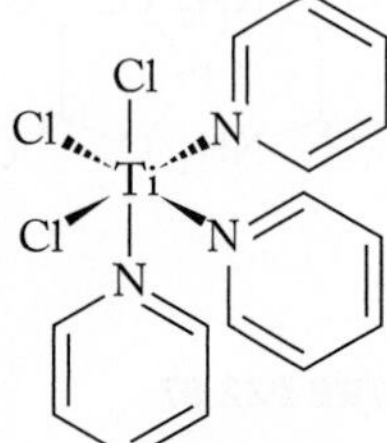

FIGURE P22.90

22.91. When Ag_2O reacts with peroxodisulfate ($S_2O_8^{2-}$) ion (a powerful oxidizing agent), AgO is produced. Crystallographic and magnetic analyses of AgO suggest that it is not simply silver(II) oxide, but rather a blend of silver(I) and silver(III) in a square planar environment. The Ag^{2+} ion is paramagnetic but, like AgO, Ag^+ and Ag^{3+} are diamagnetic. Explain why?

*22.92. Two iron compounds (a) and (b) in Figure P22.92 both contain Fe^{2+} (R is a monoanion and the phosphorus ligands are neutral). Compound (a) has a tetrahedral geometry at iron and four unpaired electrons. Compound (b) has a square planar geometry at iron and two unpaired electrons. Account for the difference in magnetic properties.

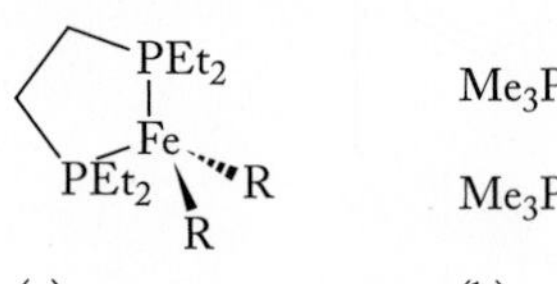

FIGURE P22.92

22.93. **Ruthenium Drugs** One of the earliest ruthenium compounds to be investigated for medicinal use was *trans*-$RuCl_2(DMSO)_4$, where DMSO represents dimethyl sulfoxide, $(CH_3)_2S{=}O$. When dissolved in water, some of the ligands are replaced by water to form $[RuCl(DMSO)_2(H_2O)_3]^+$.

$$trans\text{-}RuCl_2(DMSO)_4 + H_2O \rightarrow [RuCl(DMSO)_2(H_2O)_3]Cl + 2\ DMSO$$

a. The product has an octahedral geometry (coordination number = 6) and can exist as two geometric isomers. Draw these isomers.
*b. Why might the starting material and product have linkage isomers?

22.94. When the compound in Figure P22.94 is exposed to light at 400 nm, it undergoes linkage isomerism.
a. Draw the linkage isomer.
b. Are either or both of the isomers chiral?

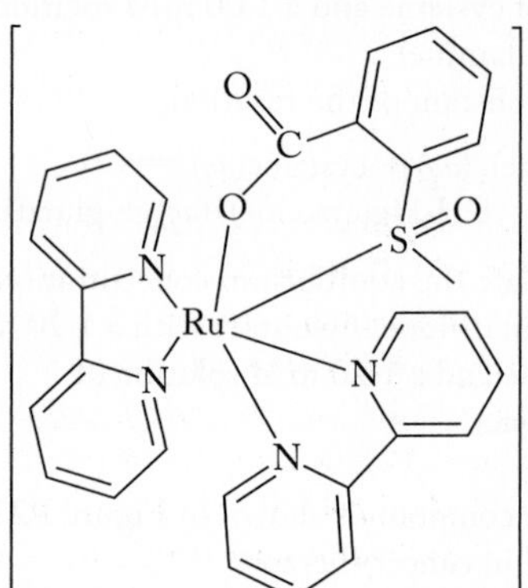

FIGURE P22.94

22.95. Manganese(II) chloride was one of the first compounds to be investigated as an MRI contrast agent. How many unpaired electrons does the complex have when $MnCl_2$ dissolves in water to form the coordination compound $MnCl_2(H_2O)_4$?

22.96. Gadolinium(III) ions are used in contrast agents for MRI because they have unpaired electrons. What is the electron configuration for Gd^{3+} and how many unpaired electrons does it have?

Appendix 1

Mathematical Procedures

Working with Scientific Notation

Quantities that scientists work with are sometimes very large, such as Earth's mass, and other times very small, such as the mass of an electron. It is easier to work with these values when they are expressed in scientific notation.

The general form of standard scientific notation is a value between 1 and 10 multipled by 10 raised to an integral power. According to this definition, 598×10^{22} kg (Earth's mass) is not in standard scientific notation, but 5.98×10^{24} kg is. It is good practice to use and report values in standard scientific notation.

1. **To convert an "ordinary" number to standard scientific notation,** move the decimal point to the left for a large number, or to the right for a small one, so that the decimal point is located after the first nonzero digit.
   - **A.** For example, to express Earth's average density (5517 kg/m^3) in scientific notation requires moving the decimal point three places to the left. Doing so is the same as dividing the number by 1000, or 10^3. To keep the value the same we multiply it by 10^3. So, Earth's density in standard scientific notation is 5.517×10^3 kg/m^3.
   - **B.** If we move the decimal point of a value less than 1 to the right to express it in scientific notation, then the exponent is a negative integer equal to the number of places we moved the decimal point to the right. For example, the value of R used in solving ideal gas law problems is 0.08206 L · atm/(mol · K). Moving the decimal point two places to the right converts the value of R to scientific notation: 8.206×10^{-2} L · atm/(mol · K).
   - **C.** Another value of R, 8.314 J/(mol · K), does not need an exponent, though it could be written 8.314×10^0 J/(mol · K) because any value raised to the zero power is equal to 1.
2. **For calculations with numbers in scientific notation,** most calculators have a function key for entering the exponents of values expressed in scientific notation. In many calculators it is labeled "E" or "EE" or "Exp." To enter, for example, the speed of light in meters per second, 2.998×10^8, we enter the value before the exponent, 2.998, followed by the exponent key and then the value of the exponent (8). To enter a value with a negative exponent, use the sign-change key. Sometimes it is labeled "(−)" or "+/−" or "±."

Working with Logarithms

A logarithm to the base 10 has the following form:

$$\log_{10} x = \log x = a, \text{ where } x = 10^a$$

We usually abbreviate the logarithm function "log" if the logarithm is to the base 10, which means the scale in which log 10 = 1.

A logarithm to the base e, called a *natural logarithm*, has the following form:

$$\log_e x = \ln x = b, \text{ where } x = e^b$$

Scientific calculators have (LOG) and (LN) keys, so it is easy to convert a number into its log or ln form. The directions below apply to most calculators.

Sample Exercise 1 Find the logarithm to the base 10 of 2.247 (log 2.247).

Solution In some calculators you enter 2.247 first and then press the (LOG) key. In others, such as the TI 84/89 series, you press the (LOG) key first followed by 2.247 and then the (ENTER) key. Either way, the answer should be 0.3516 (to four significant figures).

Sample Exercise 2 Find the natural logarithm of 2.247 (ln 2.247).

Solution Follow the same procedure as in Sample Exercise 1 except use the (LN) key. The answer should be 0.8096. This answer is (0.8096/0.3516) = 2.303 times larger than log value. That is,

$$\ln x = 2.303 \log x$$

Sample Exercise 3 Find the log of 6.0221×10^{23}.

Solution Following the procedures described above for entering a value with an exponent into your calculator and then taking its log to the base 10, you should obtain the value 23.77974796. We know the original value to five significant figures; the log value should have the same precision, but how do we express it? You might think the log value should be 23.780; however, the 23 to the left of the decimal point reflects the value of the exponent in the original value, and exponents don't count in determining the number of significant figures in a value. Therefore, the value of the logarithm to five significant figures is 23.77975. To understand better why this is so, calculate the log of 6.0221. You should get 0.77975 to five significant figures. Note how the difference between the two log values (23.77975 and 0.77975) is simply the value of the exponent in the first value.

Combining Logs The following equations summarize how logarithms of the products or quotients of two or more values are related to the individual logs of those values:

$$\text{logarithm } (ab) = \text{logarithm } a + \text{logarithm } b$$

and

$$\text{logarithm } (a/b) = \text{logarithm } a - \text{logarithm } b$$

Converting Logarithms into Numbers

If we know the value of log x, what is the value of x? This question is frequently asked when working with pH, which is the negative log of the concentration of hydrogen ions, $[H_3O^+]$, in solution:

$$\text{pH} = -\log[H_3O^+]$$

Suppose the pH of a solution of a weak acid is 2.50. The concentration of H_3O^+ is related to this pH value as follows:

$$2.50 = -\log[H_3O^+]$$

or

$$-2.50 = \log[H_3O^+]$$

To find the value of $[H_3O^+]$, we enter 2.5 into the calculator and press the appropriate key to change its sign to −2.5. The next step depends on the type of calculator. If yours has a (10^x) key (often accessible using a second function key), use it to find the value of $10^{-2.5}$, which is the value we are looking for. The corresponding keystrokes with many graphing calculators are (10^x), (−), 2.5, (ENTER). Some calculators, including the virtual one in many Windows operating systems, have an (x^y) key. We can use it for this problem by entering 10 and pushing the (x^y) key, then entering 2.5 and pushing the (+/−) key followed by the (=) key. All of these approaches do the same calculation, taking 10 to the −2.50 power, and give the same answer, $[H_3O^+] = 3.2 \times 10^{-3}$ to two significant figures. (Remember, pH is a log value, so the digit before the decimal point is not significant.)

Solving Quadratic Equations

If the terms in an equation can be rearranged so that they take the form

$$ax^2 + bx + c = 0$$

they have the form of a quadratic equation. The value(s) of x can be determined from the values of the coefficients a, b, and c by using the equation

$$x = \frac{-b \pm \sqrt{b^2 - 4ac}}{2a}$$

For example, if the solution to a problem yields the following expression where x is the concentration of a solute:

$$x^2 + 0.112x - 1.2 \times 10^{-3} = 0$$

Then the value of x can be determined as follows:

$$x = \frac{-b \pm \sqrt{b^2 - 4ac}}{2a}$$
$$= \frac{-0.112 \pm \sqrt{(0.112)^2 - 4(1)(-1.2 \times 10^{-3})}}{2(1)}$$
$$= \frac{-0.112 \pm \sqrt{0.01254 + 0.0048}}{2}$$
$$= \frac{-0.112 \pm 0.132}{2} = +0.010 \text{ or } -0.122$$

In this example, the negative value for x satisfies the equation, but it has no meaning because we cannot have negative concentration values; therefore we use only the +0.010 value.

Expressing Data in Graphical Form

Fitting curves to plots of experimental data is a powerful tool in determining the relationships between variables. Many natural phenomena obey exponential functions. For example, the rate constant (k) of a chemical reaction increases exponentially with increasing absolute temperature (T). This relationship is described by the Arrhenius equation:

$$k = A\,e^{-E_a/RT}$$

where A is a constant for a particular reaction (called the frequency factor), E_a is the activation energy of the reaction, and R is the ideal gas constant. Taking the natural logarithms of both sides of the Arrhenius equation gives

$$\ln k = \ln A - \left(\frac{E_a}{RT}\right)$$

This equation fits the general equation of a straight line ($y = mx + b$) if ($\ln k$) is the y-variable and ($1/T$) is the x-variable. Plotting ($\ln k$) versus ($1/T$) should give a straight line with a slope equal to $-E_a/R$. The slopes of these plots are negative because the activation energies, E_a, of chemical reactions are positive. The data for a reaction given in columns 2 and 4 of Table A1.1 are plotted in Figure A1.1. The program that generated the graph also gives us the equation of the straight line that best fits the data. The slope (−1281 K) of this line is used to calculate the value of E_a:

$$-1281 \text{ K} = -\frac{E_a}{R}$$
$$E_a = -(-1281 \text{ K})[8.314 \text{ J/(mol} \cdot \text{K)}]$$
$$= 10{,}650 \text{ J/mol} = 10.65 \text{ kJ/mol}$$

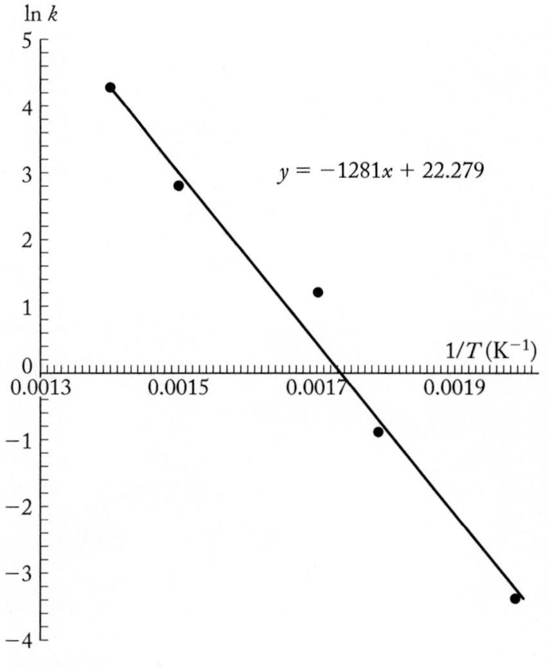

FIGURE A1.1

Expressing Precision and Accuracy

The precision in the results of a set of replicate measurements or analyses is determined by calculating the arithmetic mean ($\bar{x}$) of the data set and its standard deviation (s), which is a measure of the variation between the mean and the results of each measurement or analysis (x_i). The equation for calculating s is

$$s = \sqrt{\frac{\sum_i (x_i - \bar{x})^2}{n - 1}}$$

TABLE A1.1 Rate Constant k as a Function of Temperature T

Temperature T (K)	$1/T$ (K^{-1})	Rate Constant k	ln k
500	0.0020	0.030	−3.51
550	0.0018	0.38	−0.97
600	0.0017	2.9	1.06
650	0.0015	17	2.83
700	0.0014	75	4.32

where n is the number of data points. To calculate the confidence interval, which is the range that is predicted to contain the true mean value (μ) of the data set, we use this equation:

$$\mu = \bar{x} \pm \frac{t\,s}{\sqrt{n}}$$

where t is a statistic that depends on the value of n and the degree of certainty, or *confidence level*, in the prediction. Table A1.2 contains values of t for confidence levels of 90.0, 95.0, 99.0, and 99.9%.

TABLE A1.2 Values of t

	CONFIDENCE LEVEL			
$n-1$	90.0%	95.0%	99.0%	99.9%
1	6.314	12.71	63.66	636.62
2	2.920	4.303	9.925	31.599
3	2.353	3.182	5.841	12.924
4	2.132	2.776	4.604	8.610
5	2.015	2.571	4.032	6.869
6	1.943	2.447	3.707	5.959
7	1.895	2.365	3.499	5.408
8	1.860	2.306	3.355	5.041
9	1.833	2.262	3.250	4.781
10	1.812	2.228	3.169	4.587
11	1.796	2.201	3.106	4.437
12	1.782	2.179	3.055	4.318
13	1.771	2.160	3.012	4.221
14	1.761	2.145	2.977	4.140
15	1.753	2.131	2.947	4.073
16	1.746	2.120	2.921	4.015
17	1.740	2.110	2.898	3.965
18	1.734	2.101	2.878	3.922
19	1.729	2.093	2.861	3.883
20	1.725	2.086	2.845	3.850
25	1.708	2.060	2.787	3.725
30	1.697	2.042	2.750	3.646
40	1.684	2.021	2.704	3.551
60	1.671	2.000	2.660	3.460
80	1.664	1.990	2.639	3.416
100	1.660	1.984	2.626	3.390
∞	1.645	1.960	2.576	3.291

Combining Vector Quantities

Unequal sharing of the electrons in covalent bonds between atoms of different elements produces partial negative and positive charges on the ends of those bonds. This separation of partial charge creates a *bond dipole*. The quantity called a *dipole moment* assigns a value to the magnitude of the negative and positive charges that are separated and to the distance between the charges. Polar bonds have dipole moments and so, too, do polar molecules.

Dipole moments are *vector* quantities. A vector has two components: *magnitude*, which is the product of the partial charges times the distance between them, and *direction*, which is usually indicated by an arrow, ↦, pointing toward the negative end of a polar bond, or toward the negative side of a polar molecule. Adding the dipole moments of all the polar bonds in a molecule allows us to predict whether the molecule itself is polar—that is, whether it has a permanent dipole.

For example, a molecule of CO_2 has two polar C=O bonds with dipole moments directed toward the oxygen atoms at each end of the molecule. The two dipole moments are equivalent because they involve the same kinds of atoms separated by the same kind of bond, as shown in the Lewis structure below. However, the linear shape of the molecule means that the direction of one C=O dipole is opposite the direction of the other, so the vectors corresponding to the two dipole moments offset each other, as shown in the graph in the middle image. As a result, a molecule of CO_2 has no permanent dipole, which means that carbon dioxide is a nonpolar compound.

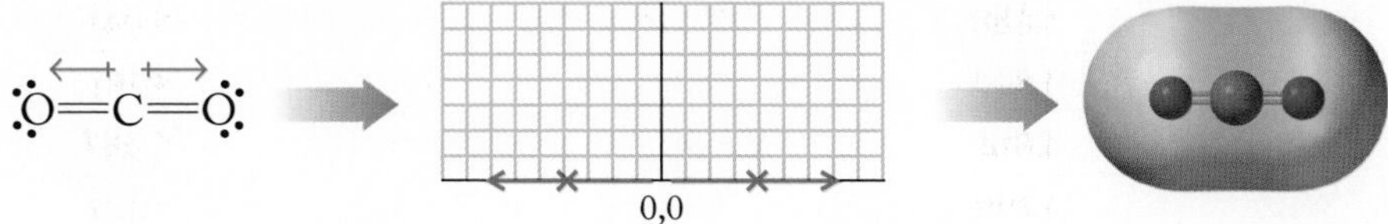

On the other hand, a molecule of H_2O has a bent molecular geometry, as shown in the Lewis structure below. Therefore, the dipole moments of the two identical O—H bonds *do not completely* offset each other: their *x* components (shown in blue in the graph below) point directly toward each other, and they do offset each other. However, their *y* components (also shown in blue) are pointed in the same direction, and they add together as shown, giving the molecule a permanent dipole with a positive end centered between the H atoms and a negative pole in the O atom:

The above method of adding the *x* and *y* components of the vectors corresponding to dipole moments can be extended to three-dimensional molecular structures such as the trigonal pyramidal structure of NH_3. In the figure below, three: N—H bond dipoles are shown for a molecule of NH_3 drawn with its N atom at the top. Its three H atoms form a trigonal base in a two-dimensional plane as shown in the vector diagram. One component of each bond dipole vector is directed from the H atom toward the center of the base, directly below the N atom. All these vectors are spaced equally about the center and have the same magnitude. Therefore, they offset each other. However, the vertical components of all three dipole moments point in the same direction (upward), which means they are additive, giving NH_3 a permanent dipole centered between the three H atoms and pointing toward the N atom.

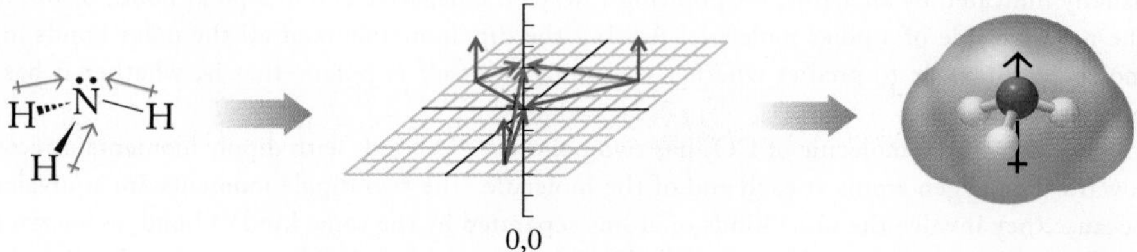

Appendix 2

SI Units and Conversion Factors

TABLE A2.1 Six SI Base Units

SI Base Quantity	Unit	Symbol
length	meter	m
mass	kilogram	kg
time	second	s
amount of substance	mole	mol
temperature	kelvin	K
electric current	ampere	A

TABLE A2.2 Some SI-Derived Units

SI-Derived Quantity	Unit	Symbol	Dimensions
electric charge	coulomb	C	A · s
electric potential	volt	V	J/C
force	newton	N	kg · m/s^2
frequency	hertz	Hz	s^{-1}
momentum	newton-second	N · s	kg · m/s
power	watt	W	J/s
pressure	pascal	Pa	N/m^2
radioactivity	becquerel	Bq	s^{-1}
speed or velocity	meter per second	m/s	m/s
energy	joule (newton-meter)	J (N · m)	kg · m^2/s^2

TABLE A2.3 SI Prefixes

Prefix	Symbol	Multiplier	Prefix	Symbol	Multiplier
deci	d	10^{-1}	deka	da	10^{1}
centi	c	10^{-2}	hecto	h	10^{2}
milli	m	10^{-3}	kilo	k	10^{3}
micro	μ	10^{-6}	mega	M	10^{6}
nano	n	10^{-9}	giga	G	10^{9}
pico	p	10^{-12}	tera	T	10^{12}
femto	f	10^{-15}	peta	P	10^{15}
atto	a	10^{-18}	exa	E	10^{18}
zepto	z	10^{-21}	zetta	Z	10^{21}

TABLE A2.4 Special Units and Conversion Factors

Quantity	Unit	Symbol	Conversion[a]
energy[a]	electron-volt	eV	1 eV = 1.6022×10^{-19} J
energy[a]	kilowatt-hour	kWh	1 kWh = 3600 kJ
energy	calorie	cal	1 cal = 4.184 J
mass	pound	lb	1 lb = 453.59 g
mass[a]	unified atomic mass unit	u	1 u = 1.66054×10^{-27} kg
length	angstrom	Å	1 Å = 10^{-8} cm = 10^{-10} m
length	inch	in	1 in = 2.54 cm
length	mile	mi	1 mi = 5280 ft = 1.6093 km
pressure	atmosphere	atm	1 atm = 1.01325×10^{5} Pa
pressure	torr	torr	1 torr = 1/760 atm
temperature	Celsius scale	°C	T(°C) = T(K) − 273.15
temperature	Fahrenheit scale	°F	T(°F) = $\frac{9}{5}$ T(°C) + 32
volume	liter	L	1 L = 1 dm^3 = 10^{-3} m^3
volume	cubic centimeter	cm^3, cc	1 cm^3 = 1 mL = 10^{-3} L
volume	cubic foot	ft^3	1 ft^3 = 7.4805 gal
volume	gallon (U.S.)	gal	1 gal = 3.785 L

[a]From http://physics.nist.gov/cuu/constants/.

TABLE A2.5 Physical Constants[a]

Quantity	Symbol	Value
acceleration due to gravity (Earth)	g	9.807 m/s^2
Avogadro constant	N_A	6.0221×10^{23} mol^{-1}
Bohr radius	a_0	5.29×10^{-11} m
Boltzmann constant	k_B	1.3806×10^{-23} J/K
electron charge-to-mass ratio	$-e/m_e$	1.7588×10^{11} C/kg
elementary charge	e	1.602×10^{-19} C
Faraday constant	F	9.65×10^{4} C/mol
mass of an electron	m_e	9.10938×10^{-31} kg
mass of a neutron	m_n	1.67493×10^{-27} kg
mass of a proton	m_p	1.67262×10^{-27} kg
molar volume of ideal gas at 0°C and 1 atm	V_m	22.4 L/mol
Planck constant	h	6.626×10^{-34} J · s
speed of light in vacuum	c	2.998×10^{8} m/s
universal gas constant	R	8.314 J/(mol · K) 0.08206 L · atm/(mol · K)

[a]From http://physics.nist.gov/cuu/constants/.

Appendix 3

The Elements and Their Properties

TABLE A3.1 Ground-State Electron Configurations, Atomic Radii, and First Ionization Energies of the Elements

Element	Symbol	Atomic Number *Z*	Ground-State Configuration	Atomic Radius (pm)	First Ionization Energy (kJ/mol)
hydrogen	H	1	$1s^1$	37	1312.0
helium	He	2	$1s^2$	32	2372.3
lithium	Li	3	$[He]2s^1$	152	520.2
beryllium	Be	4	$[He]2s^2$	112	899.5
boron	B	5	$[He]2s^22p^1$	88	800.6
carbon	C	6	$[He]2s^22p^2$	77	1086.5
nitrogen	N	7	$[He]2s^22p^3$	75	1402.3
oxygen	O	8	$[He]2s^22p^4$	73	1313.9
fluorine	F	9	$[He]2s^22p^5$	71	1681.0
neon	Ne	10	$[He]2s^22p^6$	69	2080.7
sodium	Na	11	$[Ne]3s^1$	186	495.3
magnesium	Mg	12	$[Ne]3s^2$	160	737.7
aluminum	Al	13	$[Ne]3s^23p^1$	143	577.5
silicon	Si	14	$[Ne]3s^23p^2$	117	786.5
phosphorus	P	15	$[Ne]3s^23p^3$	110	1011.8
sulfur	S	16	$[Ne]3s^23p^4$	103	999.6
chlorine	Cl	17	$[Ne]3s^23p^5$	99	1251.2
argon	Ar	18	$[Ne]3s^23p^6$	97	1520.6
potassium	K	19	$[Ar]4s^1$	227	418.8
calcium	Ca	20	$[Ar]4s^2$	197	589.8
scandium	Sc	21	$[Ar]3d^14s^2$	162	633.1
titanium	Ti	22	$[Ar]3d^24s^2$	147	658.8
vanadium	V	23	$[Ar]3d^34s^2$	135	650.9
chromium	Cr	24	$[Ar]3d^54s^1$	128	652.9
manganese	Mn	25	$[Ar]3d^54s^2$	127	717.3
iron	Fe	26	$[Ar]3d^64s^2$	126	762.5
cobalt	Co	27	$[Ar]3d^74s^2$	125	760.4
nickel	Ni	28	$[Ar]3d^84s^2$	124	737.1
copper	Cu	29	$[Ar]3d^{10}4s^1$	128	745.5
zinc	Zn	30	$[Ar]3d^{10}4s^2$	134	906.4
gallium	Ga	31	$[Ar]3d^{10}4s^24p^1$	135	578.8
germanium	Ge	32	$[Ar]3d^{10}4s^24p^2$	122	762.2
arsenic	As	33	$[Ar]3d^{10}4s^24p^3$	121	947.0
selenium	Se	34	$[Ar]3d^{10}4s^24p^4$	119	941.0

Continued on next page

TABLE A3.1 Ground-State Electron Configurations, Atomic Radii, and First Ionization Energies of the Elements *(Continued)*

Element	Symbol	Atomic Number Z	Ground-State Configuration	Atomic Radius (pm)	First Ionization Energy (kJ/mol)
bromine	Br	35	$[Ar]3d^{10}4s^24p^5$	114	1139.9
krypton	Kr	36	$[Ar]3d^{10}4s^24p^6$	110	1350.8
rubidium	Rb	37	$[Kr]5s^1$	247	403.0
strontium	Sr	38	$[Kr]5s^2$	215	549.5
yttrium	Y	39	$[Kr]4d^15s^2$	180	599.8
zirconium	Zr	40	$[Kr]4d^25s^2$	160	640.1
niobium	Nb	41	$[Kr]4d^45s^1$	146	652.1
molybdenum	Mo	42	$[Kr]4d^55s^1$	139	684.3
technetium	Tc	43	$[Kr]4d^55s^2$	136	702.4
ruthenium	Ru	44	$[Kr]4d^75s^1$	134	710.2
rhodium	Rh	45	$[Kr]4d^85s^1$	134	719.7
palladium	Pd	46	$[Kr]4d^{10}$	137	804.4
silver	Ag	47	$[Kr]4d^{10}5s^1$	144	731.0
cadmium	Cd	48	$[Kr]4d^{10}5s^2$	151	867.8
indium	In	49	$[Kr]4d^{10}5s^25p^1$	167	558.3
tin	Sn	50	$[Kr]4d^{10}5s^25p^2$	140	708.6
antimony	Sb	51	$[Kr]4d^{10}5s^25p^3$	141	833.6
tellurium	Te	52	$[Kr]4d^{10}5s^25p^4$	143	869.3
iodine	I	53	$[Kr]4d^{10}5s^25p^5$	133	1008.4
xenon	Xe	54	$[Kr]4d^{10}5s^25p^6$	130	1170.4
cesium	Cs	55	$[Xe]6s^1$	265	375.7
barium	Ba	56	$[Xe]6s^2$	222	502.9
lanthanum	La	57	$[Xe]5d^16s^2$	187	538.1
cerium	Ce	58	$[Xe]4f^15d^16s^2$	182	534.4
praseodymium	Pr	59	$[Xe]4f^36s^2$	182	527.2
neodymium	Nd	60	$[Xe]4f^46s^2$	181	533.1
promethium	Pm	61	$[Xe]4f^56s^2$	183	535.5
samarium	Sm	62	$[Xe]4f^66s^2$	180	544.5
europium	Eu	63	$[Xe]4f^76s^2$	208	547.1
gadolinium	Gd	64	$[Xe]4f^75d^16s^2$	180	593.4
terbium	Tb	65	$[Xe]4f^96s^2$	177	565.8
dysprosium	Dy	66	$[Xe]4f^{10}6s^2$	178	573.0
holmium	Ho	67	$[Xe]4f^{11}6s^2$	176	581.0
erbium	Er	68	$[Xe]4f^{12}6s^2$	176	589.3
thulium	Tm	69	$[Xe]4f^{13}6s^2$	176	596.7
ytterbium	Yb	70	$[Xe]4f^{14}6s^2$	193	603.4
lutetium	Lu	71	$[Xe]4f^{14}5d^16s^2$	174	523.5
hafnium	Hf	72	$[Xe]4f^{14}5d^26s^2$	159	658.5
tantalum	Ta	73	$[Xe]4f^{14}5d^36s^2$	146	761.3
tungsten	W	74	$[Xe]4f^{14}5d^46s^2$	139	770.0
rhenium	Re	75	$[Xe]4f^{14}5d^56s^2$	137	760.3
osmium	Os	76	$[Xe]4f^{14}5d^66s^2$	135	839.4

TABLE A3.1 Ground-State Electron Configurations, Atomic Radii, and First Ionization Energies of the Elements *(Continued)*

Element	Symbol	Atomic Number Z	Ground-State Configuration	Atomic Radius (pm)	First Ionization Energy (kJ/mol)
iridium	Ir	77	$[Xe]4f^{14}5d^{7}6s^{2}$	136	878.0
platinum	Pt	78	$[Xe]4f^{14}5d^{9}6s^{1}$	139	868.4
gold	Au	79	$[Xe]4f^{14}5d^{10}6s^{1}$	144	890.1
mercury	Hg	80	$[Xe]4f^{14}5d^{10}6s^{2}$	151	1007.1
thallium	Tl	81	$[Xe]4f^{14}5d^{10}6s^{2}6p^{1}$	170	589.4
lead	Pb	82	$[Xe]4f^{14}5d^{10}6s^{2}6p^{2}$	154	715.6
bismuth	Bi	83	$[Xe]4f^{14}5d^{10}6s^{2}6p^{3}$	150	703.3
polonium	Po	84	$[Xe]4f^{14}5d^{10}6s^{2}6p^{4}$	167	812.1
astatine	At	85	$[Xe]4f^{14}5d^{10}6s^{2}6p^{5}$	140	924.6
radon	Rn	86	$[Xe]4f^{14}5d^{10}6s^{2}6p^{6}$	145	1037.1
francium	Fr	87	$[Rn]7s^{1}$	242	380
radium	Ra	88	$[Rn]7s^{2}$	211	509.3
actinium	Ac	89	$[Rn]6d^{1}7s^{2}$	188	499
thorium	Th	90	$[Rn]6d^{2}7s^{2}$	179	587
protactinium	Pa	91	$[Rn]5f^{2}6d^{1}7s^{2}$	163	568
uranium	U	92	$[Rn]5f^{3}6d^{1}7s^{2}$	156	587
neptunium	Np	93	$[Rn]5f^{4}6d^{1}7s^{2}$	155	597
plutonium	Pu	94	$[Rn]5f^{6}7s^{2}$	159	585
americium	Am	95	$[Rn]5f^{7}7s^{2}$	173	578
curium	Cm	96	$[Rn]5f^{7}6d^{1}7s^{2}$	174	581
berkelium	Bk	97	$[Rn]5f^{9}7s^{2}$	170	601
californium	Cf	98	$[Rn]5f^{10}7s^{2}$	186	608
einsteinium	Es	99	$[Rn]5f^{11}7s^{2}$	186	619
fermium	Fm	100	$[Rn]5f^{12}7s^{2}$	167	627
mendelevium	Md	101	$[Rn]5f^{13}7s^{2}$	173	635
nobelium	No	102	$[Rn]5f^{14}7s^{2}$	176	642
lawrencium	Lr	103	$[Rn]5f^{14}6s^{1}7s^{2}$	161	—
rutherfordium	Rf	104	$[Rn]5f^{14}6d^{2}7s^{2}$	157	—
dubnium	Db	105	$[Rn]5f^{14}6d^{3}7s^{2}$	149	—
seaborgium	Sg	106	$[Rn]5f^{14}6d^{4}7s^{2}$	143	—
bohrium	Bh	107	$[Rn]5f^{14}6d^{5}7s^{2}$	141	—
hassium	Hs	108	$[Rn]5f^{14}6d^{6}7s^{2}$	134	—
meitnerium	Mt	109	$[Rn]5f^{14}6d^{7}7s^{2}$	129	—
darmstadtium	Ds	110	$[Rn]5f^{14}6d^{8}7s^{2}$	128	—
roentgenium	Rg	111	$[Rn]5f^{14}6d^{9}7s^{2}$	121	—
copernicium	Cn	112	$[Rn]5f^{14}6d^{10}7s^{2}$	122	—
nihonium	Nh	113	$[Rn]5f^{14}6d^{10}7s^{2}7p^{1}$	136	—
flerovium	Fl	114	$[Rn]5f^{14}6d^{10}7s^{2}7p^{2}$	143	—
moscovium	Mc	115	$[Rn]5f^{14}6d^{10}7s^{2}7p^{3}$	162	—
livermorium	Lv	116	$[Rn]5f^{14}6d^{10}7s^{2}7p^{4}$	175	—
tennessine	Ts	117	$[Rn]5f^{14}6d^{10}7s^{2}7p^{5}$	165	—
oganesson	Og	118	$[Rn]5f^{14}6d^{10}7s^{2}7p^{6}$	157	—

TABLE A3.2 Miscellaneous Physical Properties of the Naturally Occurring Elements[a]

Element	Symbol	Atomic Number	Physical State[b,c]	Density[d] (g/cm^3)	Melting Point (°C)	Boiling Point (°C)
hydrogen	H	1	gas	0.000090	−259.14	−252.87
helium	He	2	gas	0.000179	<−272.2	−268.93
lithium	Li	3	solid	0.534	180.5	1347
beryllium	Be	4	solid	1.848	1283	2484
boron	B	5	solid	2.34	2300	3650
carbon	C	6	solid (gr)	1.9–2.3	~3350	sublimes
nitrogen	N	7	gas	0.00125	−210.00	−195.8
oxygen	O	8	gas	0.00143	−218.8	−182.95
fluorine	F	9	gas	0.00170	−219.62	−188.12
neon	Ne	10	gas	0.00090	−248.59	−246.08
sodium	Na	11	solid	0.971	97.72	883
magnesium	Mg	12	solid	1.738	650	1090
aluminum	Al	13	solid	2.6989	660.32	2467
silicon	Si	14	solid	2.33	1414	2355
phosphorus	P	15	solid (wh)	1.82	44.15	280
sulfur	S	16	solid	2.07	115.21	444.60
chlorine	Cl	17	gas	0.00321	−101.5	−34.04
argon	Ar	18	gas	0.00178	−189.3	−185.9
potassium	K	19	solid	0.862	63.28	759
calcium	Ca	20	solid	1.55	842	1484
scandium	Sc	21	solid	2.989	1541	2380
titanium	Ti	22	solid	4.54	1668	3287
vanadium	V	23	solid	6.11	1910	3407
chromium	Cr	24	solid	7.19	1857	2671
manganese	Mn	25	solid	7.3	1246	1962
iron	Fe	26	solid	7.874	1538	2750
cobalt	Co	27	solid	8.9	1495	2870
nickel	Ni	28	solid	8.902	1455	2730
copper	Cu	29	solid	8.96	1084.6	2562
zinc	Zn	30	solid	7.133	419.53	907
gallium	Ga	31	solid	5.904	29.76	2403
germanium	Ge	32	solid	5.323	938.25	2833
arsenic	As	33	solid (gy)	5.727	614	sublimes
selenium	Se	34	solid (gy)	4.79	221	685
bromine	Br	35	liquid	3.12	−7.2	58.78
krypton	Kr	36	gas	0.00373	−157.36	−153.22
rubidium	Rb	37	solid	1.532	39.31	688
strontium	Sr	38	solid	2.64	777	1382
yttrium	Y	39	solid	4.469	1526	3336
zirconium	Zr	40	solid	6.506	1855	4409
niobium	Nb	41	solid	8.57	2477	4744
molybdenum	Mo	42	solid	10.22	2623	4639

TABLE A3.2 Miscellaneous Physical Properties of the Naturally Occurring Elements[a] *(Continued)*

Element	Symbol	Atomic Number	Physical State[b,c]	Density[d] (g/cm³)	Melting Point (°C)	Boiling Point (°C)
technetium	Tc	43	solid	11.50	2157	4538
ruthenium	Ru	44	solid	12.41	2334	3900
rhodium	Rh	45	solid	12.41	1964	3695
palladium	Pd	46	solid	12.02	1555	2963
silver	Ag	47	solid	10.50	961.78	2212
cadmium	Cd	48	solid	8.65	321.07	767
indium	In	49	solid	7.31	156.60	2072
tin	Sn	50	solid (wh)	7.31	231.9	2270
antimony	Sb	51	solid	6.691	630.63	1750
tellurium	Te	52	solid	6.24	449.5	998
iodine	I	53	solid	4.93	113.7	184.4
xenon	Xe	54	gas	0.00589	−111.75	−108.0
cesium	Cs	55	solid	1.873	28.44	671
barium	Ba	56	solid	3.5	727	1640
lanthanum	La	57	solid	6.145	920	3455
cerium	Ce	58	solid	6.770	799	3424
praseodymium	Pr	59	solid	6.773	931	3510
neodymium	Nd	60	solid	7.008	1016	3066
promethium	Pm	61	solid	7.264	1042	~3000
samarium	Sm	62	solid	7.520	1072	1790
europium	Eu	63	solid	5.244	822	1596
gadolinium	Gd	64	solid	7.901	1314	3264
terbium	Tb	65	solid	8.230	1359	3221
dysprosium	Dy	66	solid	8.551	1411	2561
holmium	Ho	67	solid	8.795	1472	2694
erbium	Er	68	solid	9.066	1529	2862
thulium	Tm	69	solid	9.321	1545	1946
ytterbium	Yb	70	solid	6.966	824	1194
lutetium	Lu	71	solid	9.841	1663	3393
hafnium	Hf	72	solid	13.31	2233	4603
tantalum	Ta	73	solid	16.654	3017	5458
tungsten	W	74	solid	19.3	3422	5660
rhenium	Re	75	solid	21.02	3186	5596
osmium	Os	76	solid	22.57	3033	5012
iridium	Ir	77	solid	22.42	2446	4130
platinum	Pt	78	solid	21.45	1768.4	3825
gold	Au	79	solid	19.3	1064.18	2856
mercury	Hg	80	liquid	13.546	−38.83	356.73
thallium	Tl	81	solid	11.85	304	1473
lead	Pb	82	solid	11.35	327.46	1749
bismuth	Bi	83	solid	9.747	271.4	1564

Continued on next page

TABLE A3.2 Miscellaneous Physical Properties of the Naturally Occurring Elements[a] *(Continued)*

Element	Symbol	Atomic Number	Physical State[b,c]	Density[d] (g/cm^3)	Melting Point (°C)	Boiling Point (°C)
polonium	Po	84	solid	9.32	254	962
astatine	At	85	solid	unknown	302	337
radon	Rn	86	gas	0.00973	−71	−61.7
francium	Fr	87	solid	unknown	27	677
radium	Ra	88	solid	5	700	1737
actinium	Ac	89	solid	10.07	1051	~3200
thorium	Th	90	solid	11.72	1750	4788
protactinium	Pa	91	solid	15.37	1572	unknown
uranium	U	92	solid	19.05	1132	3818

[a]For relative atomic masses and alphabetical listing of the elements, see the flyleaf at the front of this volume.
[b]Normal state at 25°C and 1 atm.
[c]Allotropes: gr = graphite, gy = gray, wh = white.
[d]Liquids and solids at 25°C and 1 atm; gases at 0°C and 1 atm (STP).

TABLE A3.3 A Selection of Stable Isotopes[a]

Isotope ^{A}X	Natural Abundance (%)	Atomic Number Z	Neutron Number N	Mass Number A	Atomic Mass (u)	Binding Energy per Nucleon (MeV)[b]
^{1}H	99.985	1	0	1	1.007825	—
^{2}H	0.015	1	1	2	2.014000	1.160
^{3}He	0.000137	2	1	3	3.016030	2.572
^{4}He	99.999863	2	2	4	4.002603	7.075
^{6}Li	7.5	3	3	6	6.015121	5.333
^{7}Li	92.5	3	4	7	7.016003	5.606
^{9}Be	100.0	4	5	9	9.012182	6.463
^{10}B	19.9	5	5	10	10.012937	6.475
^{11}B	80.1	5	6	11	11.009305	6.928
^{12}C	98.90	6	6	12	12.000000	7.680
^{13}C	1.10	6	7	13	13.003355	7.470
^{14}N	99.634	7	7	14	14.003074	7.476
^{15}N	0.366	7	8	15	15.000108	7.699
^{16}O	99.762	8	8	16	15.994915	7.976
^{17}O	0.038	8	9	17	16.999131	7.751
^{18}O	0.200	8	10	18	17.999160	7.767
^{19}F	100.0	9	10	19	18.998403	7.779
^{20}Ne	90.48	10	10	20	19.992435	8.032
^{21}Ne	0.27	10	11	21	20.993843	7.972
^{22}Ne	9.25	10	12	22	21.991383	8.081
^{23}Na	100.0	11	12	23	22.989770	8.112
^{24}Mg	78.99	12	12	24	23.985042	8.261
^{25}Mg	10.00	12	13	25	24.985837	8.223
^{26}Mg	11.01	12	14	26	25.982593	8.334

TABLE A3.3 A Selection of Stable Isotopes[a] *(Continued)*

Isotope ^{A}X	Natural Abundance (%)	Atomic Number Z	Neutron Number N	Mass Number A	Atomic Mass (u)	Binding Energy per Nucleon (MeV)[b]
^{27}Al	100.0	13	14	27	26.981538	8.331
^{28}Si	92.23	14	14	28	27.976927	8.448
^{29}Si	4.67	14	15	29	28.976495	8.449
^{30}Si	3.10	14	16	30	29.973770	8.521
^{31}P	100.0	15	16	31	30.973761	8.481
^{32}S	95.02	16	16	32	31.972070	8.493
^{33}S	0.75	16	17	33	32.971456	8.498
^{34}S	4.21	16	18	34	33.967866	8.584
^{36}S	0.02	16	20	36	35.967080	8.575
^{35}Cl	75.77	17	18	35	34.968852	8.520
^{37}Cl	24.23	17	20	37	36.965903	8.570
^{36}Ar	0.337	18	18	36	35.967545	8.520
^{38}Ar	0.063	18	20	38	37.962732	8.614
^{40}Ar	99.600	18	22	40	39.962384	8.595
^{39}K	93.258	19	20	39	38.963707	8.557
^{41}K	6.730	19	22	41	40.961825	8.576
^{40}Ca	96.941	20	20	40	39.962591	8.551
^{42}Ca	0.647	20	22	42	41.958618	8.617
^{43}Ca	0.135	20	23	43	42.958766	8.601
^{44}Ca	2.086	20	24	44	43.955480	8.658
^{46}Ca	0.004	20	26	46	45.953689	8.669
^{48}Ca	0.187	20	28	48	47.952533	8.666
^{45}Sc	100.0	21	24	45	44.955910	8.619
^{46}Ti	8.0	22	24	46	45.952629	8.656
^{47}Ti	7.3	22	25	47	46.951764	8.661
^{48}Ti	73.8	22	26	48	47.947947	8.723
^{49}Ti	5.5	22	27	49	48.947871	8.711
^{50}Ti	5.4	22	28	50	49.944792	8.756
^{51}V	99.750	23	28	51	50.943962	8.742
^{50}Cr	4.345	24	26	50	49.946046	8.701
^{52}Cr	83.789	24	28	52	51.940509	8.776
^{53}Cr	9.501	24	29	53	52.940651	8.760
^{54}Cr	2.365	24	30	54	53.938882	8.778
^{55}Mn	100.0	25	30	55	54.938049	8.765
^{54}Fe	5.9	26	28	54	53.939612	8.736
^{56}Fe	91.72	26	30	56	55.934939	8.790
^{57}Fe	2.1	26	31	57	56.935396	8.770
^{58}Fe	0.28	26	32	58	57.933277	8.792
^{59}Co	100.0	27	32	59	58.933200	8.768
^{204}Pb	1.4	82	122	204	203.973020	7.880

Continued on next page

TABLE A3.3 A Selection of Stable Isotopes[a] *(Continued)*

Isotope ^{A}X	Natural Abundance (%)	Atomic Number Z	Neutron Number N	Mass Number A	Atomic Mass (u)	Binding Energy per Nucleon (MeV)[b]
^{206}Pb	24.1	82	124	206	205.974440	7.875
^{207}Pb	22.1	82	125	207	206.975872	7.870
^{208}Pb	52.4	82	126	208	207.976627	7.868
^{209}Bi	100.0	83	126	209	208.980380	7.848

[a]Selection is complete through cobalt-59. Where natural abundances do not add to 100%, the differences are made up by radioactive isotopes with exceedingly long half-lives: potassium-40 (0.0117%, $t_{1/2} = 1.3 \times 10^9$ yr); vanadium-50 (0.250%, $t_{1/2} > 1.4 \times 10^{17}$ yr).
[b]1 MeV (mega electron-volt) = 1.6022×10^{-13} J.

TABLE A3.4 A Selection of Radioactive Isotopes

Isotope ^{A}X	Decay Mode[a]	Half-Life $t_{1/2}$	Atomic Number Z	Neutron Number N	Mass Number A	Atomic Mass (u)	Binding Energy per Nucleon (MeV)[b]
^{3}H	β^-	12.3 yr	1	2	3	3.01605	2.827
^{8}Be	α	$\sim 7 \times 10^{-17}$ s	4	4	8	8.005305	7.062
^{14}C	β^-	5.7×10^3 yr	6	8	14	14.003241	7.520
^{22}Na	β^+	2.6 yr	11	11	22	21.994434	7.916
^{24}Na	β^-	15.0 hr	11	13	24	23.990961	8.064
^{32}P	β^-	14.3 d	15	17	32	31.973907	8.464
^{35}S	β^-	87.2 d	16	19	35	34.969031	8.538
^{59}Fe	β^-	44.5 d	26	33	59	58.934877	8.755
^{60}Co	β^-	5.3 yr	27	33	60	59.933819	8.747
^{90}Sr	β^-	29.1 yr	38	52	90	89.907738	8.696
^{99}Tc	β^-	2.1×10^5 yr	43	56	99	98.906524	8.611
^{109}Cd	EC	462 d	48	61	109	108.904953	8.539
^{125}I	EC	59.4 d	53	72	125	124.904620	8.450
^{131}I	β^-	8.04 d	53	78	131	130.906114	8.422
^{137}Cs	β^-	30.3 yr	55	82	137	136.907073	8.389
^{222}Rn	α	3.82 d	86	136	222	222.017570	7.695
^{226}Ra	α	1600 yr	88	138	226	226.025402	7.662
^{232}Th	α	1.4×10^{10} yr	90	142	232	232.038054	7.615
^{235}U	α	7.0×10^8 yr	92	143	235	235.043924	7.591
^{238}U	α	4.5×10^9 yr	92	146	238	238.050784	7.570
^{239}Pu	α	2.4×10^4 yr	94	145	239	239.052157	7.560

[a]Modes of decay include alpha emission (α), beta emission (β^-), positron emission (β^+), and electron capture (EC).
[b]1 MeV (mega electron-volt) = 1.6022×10^{-13} J.

Appendix 4

Chemical Bonds and Thermodynamic Data

TABLE A4.1 Average Lengths and Energies of Covalent Bonds

Atom	Bond	Bond Length (pm)	Bond Energy (kJ/mol)
H	H—H	75	436
	H—F	92	567
	H—Cl	127	431
	H—Br	141	366
	H—I	161	299
C	C—C	154	348
	C═C	134	614
	C≡C	120	839
	C—H	110	413
	C—N	147	293
	C═N	127	615
	C≡N	116	891
	C—O	143	358
	C═O	123	743[a]
	C≡O	113	1072
	C—F	133	485
	C—Cl	177	328
	C—Br	179	276
	C—I	215	238
N	N—N	147	163
	N═N	124	418
	N≡N	110	945
	N—H	104	391
	N—O	136	201
	N═O	122	607
	N≡O	106	678
O	O—O	148	146
	O═O	121	498
	O—H	96	463
S	S—O	151	265
	S═O	143	523
	S—S	204	266
	S—H	134	347
F	F—F	143	155
Cl	Cl—Cl	200	243
Br	Br—Br	228	193
I	I—I	266	151

[a]The bond energy of C═O in CO_2 is 799 kJ/mol.

TABLE A4.2 Critical Temperatures (T_c) and van der Waals Parameters (a, b) of Real Gases

Gas[a]	Molar Mass (g/mol)	T_c (K)	a ($L^2 \cdot atm/mol^2$)	b (L/mol)
H_2O	18.015	647.14	5.46	0.0305
Br_2	159.808	588	9.75	0.0591
CCl_3F	137.367	471.2	14.68	0.1111
Cl_2	70.906	416.9	6.343	0.0542
CO_2	44.010	304.14	3.59	0.0427
Kr	83.798	209.41	2.325	0.0396
CH_4	16.043	190.53	2.25	0.0428
O_2	31.999	154.59	1.36	0.0318
Ar	39.948	150.87	1.34	0.0322
F_2	37.997	144.13	1.171	0.0290
CO	28.010	132.91	1.45	0.0395
N_2	28.013	126.21	1.39	0.0391
H_2	2.016	32.97	0.244	0.0266
He	4.003	5.19	0.0341	0.0237

[a]Listed in descending order of critical temperature.

TABLE A4.3 Thermodynamic Properties at 25°C

Substance[a,b]	Molar Mass (g/mol)	ΔH_f° (kJ/mol)	S° [J/(mol · K)]	ΔG_f° (kJ/mol)
	ELEMENTS AND MONATOMIC IONS			
$Ag^+(aq)$	107.87	105.6	72.7	77.1
Ag(*g*)	107.87	284.9	173.0	246.0
Ag(*s*)	107.87	0.0	42.6	0.0
$Al^{3+}(aq)$	26.982	−531	−321.7	−485
Al(*g*)	26.982	330.0	164.6	289.4
Al(*s*)	26.982	0.0	28.3	0.0
Al(*ℓ*)	26.982	10.6	39.6	−1.2
Ar(*g*)	39.948	0.0	154.8	0.0
Au(*g*)	196.97	366.1	180.5	326.3
Au(*s*)	196.97	0.0	47.4	0.0
B(*g*)	10.811	565.0	153.4	521.0
B(*s*)	10.811	0.0	5.9	0.0
$Ba^{2+}(aq)$	137.33	−537.6	9.6	−560.8
Ba(*g*)	137.33	180.0	170.2	146.0
Ba(*s*)	137.33	0.0	62.8	0.0
Be(*g*)	9.0122	324.0	136.3	286.6
Be(*s*)	9.0122	0.0	9.5	0.0
$Br^-(aq)$	79.904	−121.6	82.4	−104.0
Br(*g*)	79.904	111.9	175.0	82.4
$Br_2(g)$	159.808	30.9	245.5	3.1
$Br_2(\ell)$	159.808	0.0	152.2	0.0
C(*g*)	12.011	716.7	158.1	671.3
C(*s*, diamond)	12.011	1.9	2.4	2.9

TABLE A4.3 Thermodynamic Properties at 25°C *(Continued)*

Substance[a,b]	Molar Mass (g/mol)	ΔH_f° (kJ/mol)	S° [J/(mol · K)]	ΔG_f° (kJ/mol)
C(*s*, graphite)	12.011	0.0	5.7	0.0
Ca^{2+}(*aq*)	40.078	−542.8	−55.3	−553.6
Ca(*g*)	40.078	177.8	154.9	144.0
Ca(*s*)	40.078	0.0	41.6	0.0
Cl^-(*aq*)	35.453	−167.2	56.5	−131.2
Cl(*g*)	35.453	121.3	165.2	105.3
Cl_2(*g*)	70.906	0.0	223.0	0.0
Co^{2+}(*aq*)	58.933	−58.2	−113	−54.4
Co^{3+}(*aq*)	58.933	92	−305	134
Co(*g*)	58.933	424.7	179.5	380.3
Co(*s*)	58.933	0.0	30.0	0.0
Cr(*g*)	51.996	396.6	174.5	351.8
Cr(*s*)	51.996	0.0	23.8	0.0
Cs^+(*aq*)	132.91	−258.3	133.1	−292.0
Cs(*g*)	132.91	76.5	175.6	49.6
Cs(*s*)	132.91	0.0	85.2	0.0
Cu^+(*aq*)	63.546	71.7	40.6	50.0
Cu^{2+}(*aq*)	63.546	64.8	−99.6	65.5
Cu(*g*)	63.546	337.4	166.4	297.7
Cu(*s*)	63.546	0.0	33.2	0.0
F^-(*aq*)	18.998	−332.6	−13.8	−278.8
F(*g*)	18.998	79.4	158.8	62.3
F_2(*g*)	37.997	0.0	202.8	0.0
Fe^{2+}(*aq*)	55.845	−89.1	−137.7	−78.9
Fe^{3+}(*aq*)	55.845	−48.5	−315.9	−4.7
Fe(*g*)	55.845	416.3	180.5	370.7
Fe(*s*)	55.845	0.0	27.3	0.0
H^+(*aq*)	1.0079	0.0	0.0	0.0
H(*g*)	1.0079	218.0	114.7	203.3
H_2(*g*)	2.0158	0.0	130.6	0.0
He(*g*)	4.0026	0.0	126.2	0.0
Hg_2^{2+}(*aq*)	401.18	172.4	84.5	153.5
Hg^{2+}(*aq*)	200.59	171.1	−32.2	164.4
Hg(*g*)	200.59	61.4	175.0	31.8
Hg(ℓ)	200.59	0.0	75.9	0.0
I^-(*aq*)	126.90	−55.2	111.3	−51.6
I(*g*)	126.90	106.8	180.8	70.2
I_2(*g*)	253.81	62.4	260.7	19.3
I_2(*s*)	253.81	0.0	116.1	0.0
K^+(*aq*)	39.098	−252.4	102.5	−283.3
K(*g*)	39.098	89.0	160.3	60.5
K(*s*)	39.098	0.0	64.7	0.0
Li^+(*aq*)	6.941	−278.5	13.4	−293.3

Continued on next page

TABLE A4.3 Thermodynamic Properties at 25°C *(Continued)*

Substance[a,b]	Molar Mass (g/mol)	ΔH_f° (kJ/mol)	S° [J/(mol · K)]	ΔG_f° (kJ/mol)
Li(*g*)	6.941	159.3	138.8	126.6
Li^+(*g*)	6.941	685.7	133.0	648.5
Li(*s*)	6.941	0.0	29.1	0.0
Mg^{2+}(*aq*)	24.305	−466.9	−138.1	−454.8
Mg(*g*)	24.305	147.1	148.6	112.5
Mg(*s*)	24.305	0.0	32.7	0.0
Mn^{2+}(*aq*)	54.938	−220.8	−73.6	−228.1
Mn(*g*)	54.938	280.7	173.7	238.5
Mn(*s*)	54.938	0.0	32.0	0.0
N(*g*)	14.007	472.7	153.3	455.5
N_2(*g*)	28.013	0.0	191.5	0.0
Na^+(*aq*)	22.990	−240.1	59.0	−261.9
Na(*g*)	22.990	107.5	153.7	77.0
Na^+(*g*)	22.990	609.3	148.0	574.3
Na(*s*)	22.990	0.0	51.3	0.0
Ne(*g*)	20.180	0.0	146.3	0.0
Ni^{2+}(*aq*)	58.693	−54.0	−128.9	−45.6
Ni(*g*)	58.693	429.7	182.2	384.5
Ni(*s*)	58.693	0.0	29.9	0.0
O(*g*)	15.999	249.2	161.1	231.7
O_2(*g*)	31.999	0.0	205.0	0.0
O_3(*g*)	47.998	142.7	238.8	163.2
P(*g*)	30.974	314.6	163.1	278.3
P_4(*s*, red)	123.895	−17.6	22.8	−12.1
P_4(*s*, white)	123.895	0.0	41.1	0.0
Pb^{2+}(*aq*)	207.2	−1.7	10.5	−24.4
Pb(*g*)	207.2	195.2	162.2	175.4
Pb(*s*)	207.2	0.0	64.8	0.0
Rb^+(*aq*)	85.468	−251.2	121.5	−284.0
Rb(*g*)	85.468	80.9	170.1	53.1
Rb(*s*)	85.468	0.0	76.8	0.0
S(*g*)	32.065	277.2	167.8	236.7
S_8(*g*)	256.520	102.3	430.2	49.1
S_8(*s*)	256.520	0.0	32.1	0.0
Sc(*g*)	44.956	377.8	174.8	336.0
Sc(*s*)	44.956	0.0	34.6	0.0
Si(*g*)	28.086	450.0	168.0	405.5
Si(*s*)	28.086	0.0	18.8	0.0
Sn(*g*)	118.71	301.2	168.5	266.2
Sn(*s*, gray)	118.71	−2.1	44.1	0.1
Sn(*s*, white)	118.71	0.0	51.2	0.0
Sr^{2+}(*aq*)	87.62	−545.8	−32.6	−559.5
Sr(*g*)	87.62	164.4	164.6	130.9
Sr(*s*)	87.62	0.0	52.3	0.0

TABLE A4.3 Thermodynamic Properties at 25°C *(Continued)*

Substance[a,b]	Molar Mass (g/mol)	ΔH_f° (kJ/mol)	S° [J/(mol · K)]	ΔG_f° (kJ/mol)
Ti(*g*)	47.867	473.0	180.3	428.4
Ti(*s*)	47.867	0.0	30.7	0.0
V(*g*)	50.942	514.2	182.2	468.5
V(*s*)	50.942	0.0	28.9	0.0
W(*s*)	183.84	0.0	32.6	0.0
Zn^{2+}(*aq*)	65.38	−153.9	−112.1	−147.1
Zn(*g*)	65.38	130.4	161.0	94.8
Zn(*s*)	65.38	0.0	41.6	0.0
POLYATOMIC IONS				
CH_3COO^-(*aq*)	59.045	−486.0	86.6	−369.3
CO_3^{2-}(*aq*)	60.009	−677.1	−56.9	−527.8
$C_2O_4^{2-}$(*aq*)	88.020	−825.1	45.6	−673.9
CrO_4^{2-}(*aq*)	115.994	−881.2	50.2	−727.8
$Cr_2O_7^{2-}$(*aq*)	215.988	−1490.3	261.9	−1301.1
$HCOO^-$(*aq*)	45.018	−425.6	92	−351.0
HCO_3^-(*aq*)	61.017	−692.0	91.2	−586.8
HSO_4^-(*aq*)	97.072	−887.3	131.8	−755.9
MnO_4^-(*aq*)	118.936	−541.4	191.2	−447.2
NH_4^+(*aq*)	18.038	−132.5	113.4	−79.3
NO_3^-(*aq*)	62.005	−205.0	146.4	−108.7
OH^-(*aq*)	17.007	−230.0	−10.8	−157.2
PO_4^{3-}(*aq*)	94.971	−1277.4	−222	−1018.7
SO_4^{2-}(*aq*)	96.064	−909.3	20.1	−744.5
INORGANIC COMPOUNDS				
AgCl(*s*)	143.32	−127.1	96.2	−109.8
AgI(*s*)	234.77	−61.8	115.5	−66.2
$AgNO_3$(*s*)	169.87	−124.4	140.9	−33.4
Al_2O_3(*s*)	101.961	−1675.7	50.9	−1582.3
B_2H_6(*g*)	27.669	35.0	232.0	86.6
B_2O_3(*s*)	69.622	−1263.6	54.0	−1184.1
$BaCO_3$(*s*)	197.34	−1216.3	112.1	−1137.6
$BaSO_4$(*s*)	233.39	−1473.2	132.2	−1362.2
$CaCO_3$(*s*)	100.087	−1206.9	92.9	−1128.8
$CaCl_2$(*s*)	110.984	−795.4	108.4	−748.8
CaF_2(*s*)	78.075	−1228.0	68.5	−1175.6
CaO(*s*)	56.077	−634.9	38.1	−603.3
$Ca(OH)_2$(*s*)	74.093	−985.2	83.4	−897.5
$CaSO_4$(*s*)	136.142	−1434.5	106.5	−1322.0
CO(*g*)	28.010	−110.5	197.7	−137.2
CO_2(*g*)	44.010	−393.5	213.8	−394.4
CO_2(*aq*)	44.010	−412.9	121.3	−386.2
CS_2(*g*)	76.143	115.3	237.8	65.1
CS_2(*ℓ*)	76.143	87.9	151.0	63.6

Continued on next page

TABLE A4.3 Thermodynamic Properties at 25°C *(Continued)*

Substance[a,b]	Molar Mass (g/mol)	ΔH_f° (kJ/mol)	S° [J/(mol · K)]	ΔG_f° (kJ/mol)
$CsCl(s)$	168.358	−443.0	101.2	−414.6
$CuSO_4(s)$	159.610	−771.4	109.2	−662.2
$Cu_2S(s)$	159.16	−79.5	120.9	−86.2
$FeCl_2(s)$	126.750	−341.8	118.0	−302.3
$FeCl_3(s)$	162.203	−399.5	142.3	−334.0
$FeO(s)$	71.844	−271.9	60.8	−255.2
$Fe_2O_3(s)$	159.688	−824.2	87.4	−742.2
$HBr(g)$	80.912	−36.3	198.7	−53.4
$HCl(g)$	36.461	−92.3	186.9	−95.3
$HCN(g)$	27.02	135.1	201.81	124.7
$HF(g)$	20.006	−273.3	173.8	−275.4
$HI(g)$	127.912	26.5	206.6	1.7
$HNO_2(g)$	47.014	−79.5	254.1	−46.0
$HNO_3(g)$	63.013	−135.1	266.4	−74.7
$HNO_3(\ell)$	63.013	−174.1	155.6	−80.7
$HNO_3(aq)$	63.013	−206.6	146.0	−110.5
$HgCl_2(s)$	271.50	−224.3	146.0	−178.6
$Hg_2Cl_2(s)$	472.09	−265.4	191.6	−210.7
$H_2O(g)$	18.015	−241.8	188.8	−228.6
$H_2O(\ell)$	18.015	−285.8	69.9	−237.2
$H_2S(g)$	34.082	−20.17	205.6	−33.01
$H_2O_2(g)$	34.015	−136.3	232.7	−105.6
$H_2O_2(\ell)$	34.015	−187.8	109.6	−120.4
$H_2SO_4(\ell)$	98.079	−814.0	156.9	−690.0
$H_2SO_4(aq)$	98.079	−909.2	20.1	−744.5
$KBr(s)$	119.002	−393.8	95.9	−380.7
$KCl(s)$	74.551	−436.5	82.6	−408.5
$KHCO_3(s)$	100.115	−963.2	115.5	−863.6
$K_2CO_3(s)$	138.205	−1151.0	155.5	−1063.5
$LiBr(s)$	86.845	−351.2	74.3	−342.0
$LiCl(s)$	42.394	−408.6	59.3	−384.4
$Li_2CO_3(s)$	73.891	−1215.9	90.4	−1132.1
$MgCl_2(s)$	95.211	−641.3	89.6	−591.8
$Mg(OH)_2(s)$	58.320	−924.5	63.2	−833.5
MgO	40.30	−601.1	27.0	−630.9
$MgSO_4(s)$	120.369	−1284.9	91.6	−1170.6
$MnO_2(s)$	86.937	−520.0	53.1	−465.1
$CH_3COONa(s)$	82.034	−708.8	123.0	−607.2
$NaBr(s)$	102.894	−361.1	86.82	−349.0
$NaCl(s)$	58.443	−411.2	72.1	−384.2
$NaCl(g)$	58.443	−181.4	229.8	−201.3
$Na_2CO_3(s)$	105.989	−1130.7	135.0	−1044.4
$NaHCO_3(s)$	84.007	−950.8	101.7	−851.0
$NaNO_3(s)$	84.995	−467.9	116.5	−367.0

TABLE A4.3 Thermodynamic Properties at 25°C *(Continued)*

Substance[a,b]	Molar Mass (g/mol)	ΔH_f° (kJ/mol)	S° [J/(mol · K)]	ΔG_f° (kJ/mol)
$NaOH(s)$	39.997	−425.6	64.5	−379.5
$Na_2SO_4(s)$	142.043	−1387.1	149.6	−1270.2
$NF_3(g)$	71.002	−132.1	260.8	−90.6
$NH_3(aq)$	17.031	−80.3	111.3	−26.50
$NH_3(g)$	17.031	−46.1	192.5	−16.5
$NH_4Cl(s)$	53.491	−314.4	94.6	−203.0
$NH_4NO_3(s)$	80.043	−365.6	151.1	−183.9
$N_2H_4(g)$	32.045	95.35	238.5	159.4
$N_2H_4(\ell)$	32.045	50.63	121.52	149.3
$NiCl_2(s)$	129.60	−305.3	97.7	−259.0
$NiO(s)$	74.60	−239.7	38.0	−211.7
$NO(g)$	30.006	90.3	210.7	86.6
$NO_2(g)$	46.006	33.2	240.0	51.3
$N_2O(g)$	44.013	82.1	219.9	104.2
$N_2O_3(g)$	76.01	86.6	314.7	142.4
$N_2O_4(g)$	92.011	11.1	304.2	99.8
$NOCl(g)$	65.459	51.7	261.7	66.1
$PCl_3(g)$	137.33	−288.07	311.7	−269.6
$PCl_3(\ell)$	137.33	−319.6	217	−272.4
$PF_5(g)$	125.96	−1594.4	300.8	−1520.7
$PH_3(g)$	33.998	5.4	210.2	13.4
$PbCl_2(s)$	278.1	−359.4	136.0	−314.1
$PbSO_4(s)$	303.3	−920.0	148.5	−813.0
$SO_2(g)$	64.065	−296.8	248.2	−300.1
$SO_3(g)$	80.064	−395.7	256.8	−371.1
$ZnCl_2(s)$	136.30	−415.1	111.5	−369.4
$ZnO(s)$	81.37	−348.0	43.9	−318.2
$ZnSO_4(s)$	161.45	−982.8	110.5	−871.5
ORGANIC COMPOUNDS				
$CCl_4(g)$	153.823	−102.9	309.7	−60.6
$CCl_4(\ell)$	153.823	−135.4	216.4	−65.3
$CH_4(g)$	16.043	−74.8	186.2	−50.8
$CH_3COOH(g)$	60.053	−432.8	282.5	−374.5
$CH_3COOH(\ell)$	60.053	−484.5	159.8	−389.9
$CH_3OH(g)$	32.042	−200.7	239.9	−162.0
$CH_3OH(\ell)$	32.042	−238.7	126.8	−166.4
$C_2H_2(g)$	26.038	226.7	200.8	209.2
$C_2H_4(g)$	28.054	52.4	219.5	68.1
$C_2H_6(g)$	30.070	−84.67	229.5	−32.9
$CH_3CH_2OH(g)$	46.069	−235.1	282.6	−168.6
$CH_3CH_2OH(\ell)$	46.069	−277.7	160.7	−174.9
$CH_3CHO(g)$	44.053	−166	266	−133.7
$C_3H_8(g)$	44.097	−103.8	269.9	−23.5

Continued on next page

TABLE A4.3 Thermodynamic Properties at 25°C *(Continued)*

Substance[a,b]	Molar Mass (g/mol)	ΔH_f° (kJ/mol)	S° [J/(mol · K)]	ΔG_f° (kJ/mol)
$CH_3(CH_2)_2CH_3(g)$	58.123	−125.6	310.0	−15.7
$CH_3(CH_2)_2CH_3(\ell)$	58.123	−147.6	231.0	−15.0
$CH_3COCH_3(\ell)$	58.079	−248.4	199.8	−155.6
$CH_3COCH_3(g)$	58.079	−217.1	295.3	−152.7
$CH_3(CH_2)_2CH_2OH(\ell)$	74.122	−327.3	225.8	
$(CH_3CH_2)_2O(\ell)$	74.122	−279.6	172.4	
$(CH_3CH_2)_2O(g)$	74.122	−252.1	342.7	
$(CH_3)_2C{=}C(CH_3)_2(\ell)$	84.161	66.6	362.6	−69.2
$(CH_3)_2NH(\ell)$	45.084	−43.9	182.3	
$(CH_3)_2NH(g)$	45.084	−18.5	273.1	
$(CH_3CH_2)_2NH(\ell)$	73.138	−103.3		
$(CH_3CH_2)_2NH(g)$	73.138	−71.4		
$(CH_3)_3N(\ell)$	59.111	−46.0	208.5	
$(CH_3)_3N(g)$	59.111	−23.6	287.1	
$(CH_3CH_2)_3N(\ell)$	101.191	−134.3		
$(CH_3CH_2)_3N(g)$	101.191	−95.8		
$C_6H_6(g)$	78.114	82.9	269.2	129.7
$C_6H_6(\ell)$	78.114	49.0	172.9	124.5
$C_6H_{12}O_6(s)$	180.158	−1274.4	212.1	−910.1
$CH_3(CH_2)_6CH_3(\ell)$	114.231	−249.9	361.1	6.4
$CH_3(CH_2)_6CH_3(g)$	114.231	−208.6	466.7	16.4
$C_{12}H_{22}O_{11}(s)$	342.300	−2221.7	360.2	−1543.8
$HCOOH(\ell)$	46.026	−424.7	129.0	−361.4

[a]Substances are arranged alphabetically by chemical formula within each class: (1) elements and monatomic ions; (2) polyatomic ions; (3) inorganic compounds (including CO and CO_2); (4) organic compounds (hydrocarbon-based).
[b]Symbols denote standard enthalpy of formation (ΔH_f°), standard third-law entropy (S°), and standard Gibbs free energy of formation (ΔG_f°). Entropies in aqueous solution are referred to $S^\circ[H^+(aq)] = 0$, not to absolute zero.

TABLE A4.4 Vapor Pressure of Water as a Function of Temperature

T (°C)	P (torr)
0.0	4.579
10.0	9.209
20.0	17.535
25.0	23.756
30.0	31.824
40.0	55.324
60.0	149.4
70.0	233.7
90.0	525.8
100	760.0
105	906.0

Appendix 5

Equilibrium Constants

TABLE A5.1 Ionization Constants of Selected Acids at 25°C

Acid	Step	Aqueous Equilibrium[a]	K_a	pK_a
acetic	1	$CH_3COOH(aq) + H_2O(\ell) \rightleftharpoons H_3O^+(aq) + CH_3COO^-(aq)$	1.76×10^{-5}	4.75
arsenic	1	$H_3AsO_4(aq) + H_2O(\ell) \rightleftharpoons H_3O^+(aq) + H_2AsO_4^-(aq)$	5.5×10^{-3}	2.26
	2	$H_2AsO_4^-(aq) + H_2O(\ell) \rightleftharpoons H_3O^+(aq) + HAsO_4^{2-}(aq)$	1.7×10^{-7}	6.77
	3	$HAsO_4^{2-}(aq) + H_2O(\ell) \rightleftharpoons H_3O^+(aq) + AsO_4^{3-}(aq)$	5.1×10^{-12}	11.29
ascorbic	1	$H_2C_6H_6O_6(aq) + H_2O(\ell) \rightleftharpoons H_3O^+(aq) + HC_6H_6O_6^-(aq)$	9.1×10^{-5}	4.04
	2	$HC_6H_6O_6^-(aq) + H_2O(\ell) \rightleftharpoons H_3O^+(aq) + C_6H_6O_6^{2-}(aq)$	5×10^{-12}	11.3
benzoic	1	$C_6H_5COOH(aq) + H_2O(\ell) \rightleftharpoons H_3O^+(aq) + C_6H_5COO^-(aq)$	6.25×10^{-5}	4.20
boric	1	$H_3BO_3(aq) + H_2O(\ell) \rightleftharpoons H_3O^+(aq) + H_2BO_3^-(aq)$	5.4×10^{-10}	9.27
	2	$H_2BO_3^-(aq) + H_2O(\ell) \rightleftharpoons H_3O^+(aq) + HBO_3^{2-}(aq)$	$<10^{-14}$	>14
bromoacetic	1	$CH_2BrCOOH(aq) + H_2O(\ell) \rightleftharpoons H_3O^+(aq) + CH_2BrCOO^-(aq)$	2.0×10^{-3}	2.70
butanoic	1	$CH_3CH_2CH_2COOH(aq) + H_2O(\ell) \rightleftharpoons H_3O^+(aq) + CH_3CH_2CH_2COO^-(aq)$	1.5×10^{-5}	4.82
carbonic	1	$H_2CO_3(aq) + H_2O(\ell) \rightleftharpoons H_3O^+(aq) + HCO_3^-(aq)$	4.3×10^{-7}	6.37
	2	$HCO_3^-(aq) + H_2O(\ell) \rightleftharpoons H_3O^+(aq) + CO_3^{2-}(aq)$	4.7×10^{-11}	10.33
chloric	1	$HClO_3(aq) + H_2O(\ell) \rightleftharpoons H_3O^+(aq) + ClO_3^-(aq)$	~1	~0
chloroacetic	1	$CH_2ClCOOH(aq) + H_2O(\ell) \rightleftharpoons H_3O^+(aq) + CH_2ClCOO^-(aq)$	1.4×10^{-3}	2.85
chlorous	1	$HClO_2(aq) + H_2O(\ell) \rightleftharpoons H_3O^+(aq) + ClO_2^-(aq)$	1.1×10^{-2}	1.96
citric	1	$CH_2(COOH)C(OH)(COOH)CH_2COOH(aq) + H_2O(\ell) \rightleftharpoons H_3O^+(aq) + CH_2(COOH)C(OH)(COO^-)CH_2COOH(aq)$	7.4×10^{-4}	3.13
	2	$CH_2(COOH)C(OH)(COO^-)CH_2COOH(aq) + H_2O(\ell) \rightleftharpoons H_3O^+(aq) + CH_2(COO^-)C(OH)(COO^-)CH_2COOH(aq)$	1.7×10^{-5}	4.77
	3	$CH_2(COO^-)C(OH)(COO^-)CH_2COOH(aq) + H_2O(\ell) \rightleftharpoons H_3O^+(aq) + CH_2(COO^-)C(OH)(COO^-)CH_2COO^-(aq)$	4.0×10^{-7}	6.40
dichloroacetic	1	$CHCl_2COOH(aq) + H_2O(\ell) \rightleftharpoons H_3O^+(aq) + CHCl_2COO^-(aq)$	5.5×10^{-2}	1.26
ethanol	1	$CH_3CH_2OH(aq) + H_2O(\ell) \rightleftharpoons H_3O^+(aq) + CH_3CH_2O^-(aq)$	1.3×10^{-16}	15.9
fluoroacetic	1	$CH_2FCOOH(aq) + H_2O(\ell) \rightleftharpoons H_3O^+(aq) + CH_2FCOO^-(aq)$	2.6×10^{-3}	2.59
formic	1	$HCOOH(aq) + H_2O(\ell) \rightleftharpoons H_3O^+(aq) + HCOO^-(aq)$	1.77×10^{-4}	3.75
germanic	1	$H_2GeO_3(aq) + H_2O(\ell) \rightleftharpoons H_3O^+(aq) + HGeO_3^-(aq)$	9.8×10^{-10}	9.01
	2	$HGeO_3^-(aq) + H_2O(\ell) \rightleftharpoons H_3O^+(aq) + GeO_3^{2-}(aq)$	5×10^{-13}	12.3
hydr(o)azoic	1	$HN_3(aq) + H_2O(\ell) \rightleftharpoons H_3O^+(aq) + N_3^-(aq)$	1.9×10^{-5}	4.72
hydrobromic	1	$HBr(aq) + H_2O(\ell) \rightarrow H_3O^+(aq) + Br^-(aq)$	≫1 (strong)	<0
hydrochloric	1	$HCl(aq) + H_2O(\ell) \rightarrow H_3O^+(aq) + Cl^-(aq)$	≫1 (strong)	<0
hydrocyanic	1	$HCN(aq) + H_2O(\ell) \rightleftharpoons H_3O^+(aq) + CN^-(aq)$	6.2×10^{-10}	9.21
hydrofluoric	1	$HF(aq) + H_2O(\ell) \rightleftharpoons H_3O^+(aq) + F^-(aq)$	6.8×10^{-4}	3.17
hydr(o)iodic	1	$HI(aq) + H_2O(\ell) \rightarrow H_3O^+(aq) + I^-(aq)$	≫1 (strong)	<0

Continued on next page

TABLE A5.1 Ionization Constants of Selected Acids at 25°C *(Continued)*

Acid	Step	Aqueous Equilibrium[a]	K_a	pK_a
hydrosulfuric	1	$H_2S(aq) + H_2O(\ell) \rightleftharpoons H_3O^+(aq) + HS^-(aq)$	8.9×10^{-8}	7.05
	2	$HS^-(aq) + H_2O(\ell) \rightleftharpoons H_3O^+(aq) + S^{2-}(aq)$	$\sim 10^{-19}$	~19
hypobromous	1	$HBrO(aq) + H_2O(\ell) \rightleftharpoons H_3O^+(aq) + BrO^-(aq)$	2.3×10^{-9}	8.64
hypochlorous	1	$HClO(aq) + H_2O(\ell) \rightleftharpoons H_3O^+(aq) + ClO^-(aq)$	2.9×10^{-8}	7.54
hypoiodous	1	$HIO(aq) + H_2O(\ell) \rightleftharpoons H_3O^+(aq) + IO^-(aq)$	2.3×10^{-11}	10.64
iodic	1	$HIO_3(aq) + H_2O(\ell) \rightleftharpoons H_3O^+(aq) + IO_3^-(aq)$	1.7×10^{-1}	0.77
iodoacetic	1	$CH_2ICOOH(aq) + H_2O(\ell) \rightleftharpoons H_3O^+(aq) + CH_2ICOO^-(aq)$	7.6×10^{-4}	3.12
lactic	1	$CH_3CHOHCOOH(aq) + H_2O(\ell) \rightleftharpoons H_3O^+(aq) + CH_3CHOHCOO^-(aq)$	1.4×10^{-4}	3.85
maleic (*cis*-butenedioic)	1	$HOOCCH{=}CHCOOH(aq) + H_2O(\ell) \rightleftharpoons H_3O^+(aq) + HOOCCH{=}CHCOO^-(aq)$	1.2×10^{-2}	1.92
	2	$HOOCCH{=}CHCOO^-(aq) + H_2O(\ell) \rightleftharpoons H_3O^+(aq) + {}^-OOCCH{=}CHCOO^-(aq)$	4.7×10^{-7}	6.33
malonic	1	$HOOCCH_2COOH(aq) + H_2O(\ell) \rightleftharpoons H_3O^+(aq) + HOOCCH_2COO^-(aq)$	1.5×10^{-3}	2.82
	2	$HOOCCH_2COO^-(aq) + H_2O(\ell) \rightleftharpoons H_3O^+(aq) + {}^-OOCCH_2COO^-(aq)$	2.0×10^{-6}	5.70
nitric	1	$HNO_3(aq) + H_2O(\ell) \rightarrow H_3O^+(aq) + NO_3^-(aq)$	$\gg 1$ (strong)	<0
nitrous	1	$HNO_2(aq) + H_2O(\ell) \rightleftharpoons H_3O^+(aq) + NO_2^-(aq)$	4.0×10^{-4}	3.40
oxalic	1	$HOOCCOOH(aq) + H_2O(\ell) \rightleftharpoons H_3O^+(aq) + HOOCCOO^-(aq)$	5.9×10^{-2}	1.23
	2	$HOOCCOO^-(aq) + H_2O(\ell) \rightleftharpoons H_3O^+(aq) + {}^-OOCCOO^-(aq)$	6.4×10^{-5}	4.19
perchloric	1	$HClO_4(aq) + H_2O(\ell) \rightarrow H_3O^+(aq) + ClO_4^-(aq)$	$\gg 1$ (strong)	<0
periodic	1	$HIO_4(aq) + H_2O(\ell) \rightleftharpoons H_3O^+(aq) + IO_4^-(aq)$	2.3×10^{-2}	1.64
phenol	1	$C_6H_5OH(aq) + H_2O(\ell) \rightleftharpoons H_3O^+(aq) + C_6H_5O^-(aq)$	1.3×10^{-10}	9.89
phosphoric	1	$H_3PO_4(aq) + H_2O(\ell) \rightleftharpoons H_3O^+(aq) + H_2PO_4^-(aq)$	6.9×10^{-3}	2.16
	2	$H_2PO_4^-(aq) + H_2O(\ell) \rightleftharpoons H_3O^+(aq) + HPO_4^{2-}(aq)$	6.4×10^{-8}	7.19
	3	$HPO_4^{2-}(aq) + H_2O(\ell) \rightleftharpoons H_3O^+(aq) + PO_4^{3-}(aq)$	4.8×10^{-13}	12.32
propanoic	1	$CH_3CH_2COOH(aq) + H_2O(\ell) \rightleftharpoons H_3O^+(aq) + CH_3CH_2COO^-(aq)$	1.4×10^{-5}	4.85
pyruvic	1	$CH_3C(O)COOH(aq) + H_2O(\ell) \rightleftharpoons H_3O^+(aq) + CH_3C(O)COO^-(aq)$	2.8×10^{-3}	2.55
sulfuric	1	$H_2SO_4(aq) + H_2O(\ell) \rightarrow H_3O^+(aq) + HSO_4^-(aq)$	$\gg 1$ (strong)	<0
	2	$HSO_4^-(aq) + H_2O(\ell) \rightleftharpoons H_3O^+(aq) + SO_4^{2-}(aq)$	1.2×10^{-2}	1.92
sulfurous	1	$H_2SO_3(aq) + H_2O(\ell) \rightleftharpoons H_3O^+(aq) + HSO_3^-(aq)$	1.7×10^{-2}	1.77
	2	$HSO_3^-(aq) + H_2O(\ell) \rightarrow H_3O^+(aq) + SO_3^{2-}(aq)$	6.2×10^{-8}	7.21
thiocyanic	1	$HSCN(aq) + H_2O(\ell) \rightleftharpoons H_3O^+(aq) + SCN^-(aq)$	$\gg 1$ (strong)	<0
trichloroacetic	1	$CCl_3COOH(aq) + H_2O(\ell) \rightleftharpoons H_3O^+(aq) + CCl_3COO^-(aq)$	2.3×10^{-1}	0.64
trifluoroacetic	1	$CF_3COOH(aq) + H_2O(\ell) \rightleftharpoons H_3O^+(aq) + CF_3COO^-(aq)$	5.9×10^{-1}	0.23
water	1	$H_2O(aq) + H_2O(\ell) \rightleftharpoons H_3O^+(aq) + OH^-(aq)$	1.0×10^{-14}	14.00

[a]The formulas of most carboxylic acids are written in an RCOOH format to highlight their molecular structures.

TABLE A5.2 Acid Ionization Constants of Hydrated Metal Ions at 25°C

Free Ion	Hydrated Ion	K_a
Fe^{3+}	$Fe(H_2O)_6{}^{3+}$	3×10^{-3}
Sn^{2+}	$Sn(H_2O)_6{}^{2+}$	4×10^{-4}
Cr^{3+}	$Cr(H_2O)_6{}^{3+}$	1×10^{-4}
Al^{3+}	$Al(H_2O)_6{}^{3+}$	1×10^{-5}
Cu^{2+}	$Cu(H_2O)_6{}^{2+}$	3×10^{-8}
Pb^{2+}	$Pb(H_2O)_6{}^{2+}$	3×10^{-8}
Zn^{2+}	$Zn(H_2O)_6{}^{2+}$	1×10^{-9}
Co^{2+}	$Co(H_2O)_6{}^{2+}$	2×10^{-10}
Ni^{2+}	$Ni(H_2O)_6{}^{2+}$	1×10^{-10}

TABLE A5.3 Ionization Constants of Selected Bases at 25°C

Base	Aqueous Equilibrium	K_b	pK_b
ammonia	$NH_3(aq) + H_2O(\ell) \rightleftharpoons NH_4{}^+(aq) + OH^-(aq)$	1.76×10^{-5}	4.75
aniline	$C_6H_5NH_2(aq) + H_2O(\ell) \rightleftharpoons C_6H_5NH_3{}^+(aq) + OH^-(aq)$	4.0×10^{-10}	9.40
diethylamine	$(CH_3CH_2)_2NH(aq) + H_2O(\ell) \rightleftharpoons (CH_3CH_2)_2NH_2{}^+(aq) + OH^-(aq)$	8.6×10^{-4}	3.07
dimethylamine	$(CH_3)_2NH(aq) + H_2O(\ell) \rightleftharpoons (CH_3)_2NH_2{}^+(aq) + OH^-(aq)$	5.9×10^{-4}	3.23
ethylamine	$CH_3CH_2NH_2(aq) + H_2O(\ell) \rightleftharpoons CH_3CH_2NH_3{}^+(aq) + OH^-(aq)$	5.6×10^{-4}	3.25
methylamine	$CH_3NH_2(aq) + H_2O(\ell) \rightleftharpoons CH_3NH_3{}^+(aq) + OH^-(aq)$	4.4×10^{-4}	3.36
nicotine (1)	N, N, H_3C + H_2O ⇌ N, N^+, H_3C, H + OH^-	1.0×10^{-6}	6.0
(2)	N, N^+, H_3C, H + H_2O ⇌ N^+, H, N^+, H_3C, H + OH^-	1.3×10^{-11}	10.9
pyridine	$C_5H_5N(aq) + H_2O(\ell) \rightleftharpoons C_5H_5NH^+(aq) + OH^-(aq)$	1.7×10^{-9}	8.77
quinine			
(1)	HO, N, H_3CO, N + H_2O ⇌ HO, N^+, H, H_3CO, N + OH^-	3.3×10^{-6}	5.48
(2)	HO, N^+, H, H_3CO, N + H_2O ⇌ HO, N^+, H, H_3CO, N^+, H + OH^-	1.4×10^{-10}	9.9
trimethylamine	$(CH_3)_3N(aq) + H_2O(\ell) \rightleftharpoons (CH_3)_3NH^+(aq) + OH^-(aq)$	6.46×10^{-5}	4.19
urea	$H_2NCONH_2(aq) + H_2O(\ell) \rightleftharpoons H_2NCONH_3{}^+(aq) + OH^-(aq)$	1.3×10^{-14}	13.9

TABLE A5.4 Solubility-Product Constants at 25°C

Cation	Anion	Heterogeneous Equilibrium[a]	K_{sp}[b]
aluminum	hydroxide	$Al(OH)_3(s) \rightleftharpoons Al^{3+}(aq) + 3\ OH^-(aq)$	1.3×10^{-33}
	phosphate	$AlPO_4(s) \rightleftharpoons Al^{3+}(aq) + PO_4^{3-}(aq)$	9.84×10^{-21}
barium	carbonate	$BaCO_3(s) \rightleftharpoons Ba^{2+}(aq) + CO_3^{2-}(aq)$	2.58×10^{-9}
	fluoride	$BaF_2(s) \rightleftharpoons Ba^{2+}(aq) + 2\ F^-(aq)$	1.84×10^{-7}
	sulfate	$BaSO_4(s) \rightleftharpoons Ba^{2+}(aq) + SO_4^{2-}(aq)$	1.08×10^{-10}
calcium	carbonate	$CaCO_3(s) \rightleftharpoons Ca^{2+}(aq) + CO_3^{2-}(aq)$	2.8×10^{-9}
	fluoride	$CaF_2(s) \rightleftharpoons Ca^{2+}(aq) + 2\ F^-(aq)$	5.3×10^{-9}
	hydroxide	$Ca(OH)_2(s) \rightleftharpoons Ca^{2+}(aq) + 2\ OH^-(aq)$	5.5×10^{-6}
	phosphate	$Ca_3(PO_4)_2(s) \rightleftharpoons 3\ Ca^{2+}(aq) + 2\ PO_4^{3-}(aq)$	2.07×10^{-29}
	sulfate	$CaSO_4(s) \rightleftharpoons Ca^{2+}(aq) + SO_4^{2-}(aq)$	4.93×10^{-5}
cobalt(II)	carbonate	$CoCO_3(s) \rightleftharpoons Co^{2+}(aq) + CO_3^{2-}(aq)$	1.4×10^{-3}
	phosphate	$Co_3(PO_4)_2(s) \rightleftharpoons 3\ Co^{2+}(aq) + 2\ PO_4^{3-}(aq)$	2.05×10^{-7}
	sulfide	$CoS(s) \rightleftharpoons Co^{2+}(aq) + S^{2-}(aq)$	2.0×10^{-25}
copper(I)	bromide	$CuBr(s) \rightleftharpoons Cu^+(aq) + Br^-(aq)$	6.27×10^{-9}
	chloride	$CuCl(s) \rightleftharpoons Cu^+(aq) + Cl^-(aq)$	1.72×10^{-7}
	iodide	$CuI(s) \rightleftharpoons Cu^+(aq) + I^-(aq)$	1.27×10^{-12}
copper(II)	phosphate	$Cu_3(PO_4)_2(s) \rightleftharpoons 3\ Cu^{2+}(aq) + 2\ PO_4^{3-}(aq)$	1.4×10^{-37}
	hydroxide	$Cu(OH)_2(s) \rightleftharpoons Cu^{2+}(aq) + 2\ OH^-(aq)$	2.2×10^{-20}
iron(II)	carbonate	$FeCO_3(s) \rightleftharpoons Fe^{2+}(aq) + CO_3^{2-}(aq)$	3.13×10^{-11}
	fluoride	$FeF_2(s) \rightleftharpoons Fe^{2+}(aq) + 2\ F^-(aq)$	2.36×10^{-6}
	hydroxide	$Fe(OH)_2(s) \rightleftharpoons Fe^{2+}(aq) + 2\ OH^-(aq)$	4.87×10^{-17}
	sulfide	$FeS(s) \rightleftharpoons Fe^{2+}(aq) + S^{2-}(aq)$	6.3×10^{-18}
lead	bromide	$PbBr_2(s) \rightleftharpoons Pb^{2+}(aq) + 2\ Br^-(aq)$	6.60×10^{-6}
	carbonate	$PbCO_3(s) \rightleftharpoons Pb^{2+}(aq) + CO_3^{2-}(aq)$	7.4×10^{-14}
	chloride	$PbCl_2(s) \rightleftharpoons Pb^{2+}(aq) + 2\ Cl^-(aq)$	1.7×10^{-5}
	fluoride	$PbF_2(s) \rightleftharpoons Pb^{2+}(aq) + 2\ F^-(aq)$	3.3×10^{-8}
	iodide	$PbI_2(s) \rightleftharpoons Pb^{2+}(aq) + 2\ I^-(aq)$	9.8×10^{-9}
	sulfate	$PbSO_4(s) \rightleftharpoons Pb^{2+}(aq) + SO_4^{2-}(aq)$	2.53×10^{-8}
lithium	carbonate	$Li_2CO_3(s) \rightleftharpoons 2\ Li^+(aq) + CO_3^{2-}(aq)$	2.5×10^{-2}
magnesium	carbonate	$MgCO_3(s) \rightleftharpoons Mg^{2+}(aq) + CO_3^{2-}(aq)$	6.82×10^{-6}
	fluoride	$MgF_2(s) \rightleftharpoons Mg^{2+}(aq) + 2\ F^-(aq)$	5.16×10^{-11}
	hydroxide	$Mg(OH)_2(s) \rightleftharpoons Mg^{2+}(aq) + 2\ OH^-(aq)$	5.61×10^{-12}
manganese(II)	carbonate	$MnCO_3(s) \rightleftharpoons Mn^{2+}(aq) + CO_3^{2-}(aq)$	2.34×10^{-11}
	hydroxide	$Mn(OH)_2(s) \rightleftharpoons Mn^{2+}(aq) + 2\ OH^-(aq)$	1.9×10^{-13}
mercury(I)	bromide	$Hg_2Br_2(s) \rightleftharpoons Hg_2^{2+}(aq) + 2\ Br^-(aq)$	6.40×10^{-23}
	carbonate	$Hg_2CO_3(s) \rightleftharpoons Hg_2^{2+}(aq) + CO_3^{2-}(aq)$	3.6×10^{-17}
	chloride	$Hg_2Cl_2(s) \rightleftharpoons Hg_2^{2+}(aq) + 2\ Cl^-(aq)$	1.43×10^{-18}
	iodide	$Hg_2I_2(s) \rightleftharpoons Hg_2^{2+}(aq) + 2\ I^-(aq)$	5.2×10^{-29}
	sulfate	$Hg_2SO_4(s) \rightleftharpoons Hg_2^{2+}(aq) + SO_4^{2-}(aq)$	6.5×10^{-7}
mercury(II)	hydroxide	$Hg(OH)_2(s) \rightleftharpoons Hg^{2+}(aq) + 2\ OH^-(aq)$	3.2×10^{-26}
	iodide	$HgI_2(s) \rightleftharpoons Hg^{2+}(aq) + 2\ I^-(aq)$	2.9×10^{-29}
nickel(II)	carbonate	$NiCO_3(s) \rightleftharpoons Ni^{2+}(aq) + CO_3^{2-}(aq)$	1.42×10^{-7}
	phosphate	$Ni_3(PO_4)_2(s) \rightleftharpoons 3\ Ni^{2+}(aq) + 2\ PO_4^{3-}(aq)$	4.74×10^{-32}
	sulfide	$NiS(s) \rightleftharpoons Ni^{2+}(aq) + S^{2-}(aq)$	1×10^{-24}
silver	bromide	$AgBr(s) \rightleftharpoons Ag^+(aq) + Br^-(aq)$	5.35×10^{-13}
	carbonate	$Ag_2CO_3(s) \rightleftharpoons 2\ Ag^+(aq) + CO_3^{2-}(aq)$	8.46×10^{-12}
	chloride	$AgCl(s) \rightleftharpoons Ag^+(aq) + Cl^-(aq)$	1.77×10^{-10}
	chromate	$Ag_2CrO_4(s) \rightleftharpoons 2\ Ag^+(aq) + CrO_4^{2-}(aq)$	1.12×10^{-12}
	hydroxide	$AgOH(s) \rightleftharpoons Ag^+(aq) + OH^-(aq)$	2.0×10^{-8}
	iodide	$AgI(s) \rightleftharpoons Ag^+(aq) + I^-(aq)$	8.52×10^{-17}
	phosphate	$Ag_3PO_4(s) \rightleftharpoons 3\ Ag^+(aq) + PO_4^{3-}(aq)$	8.89×10^{-17}
	sulfate	$Ag_2SO_4(s) \rightleftharpoons 2\ Ag^+(aq) + SO_4^{2-}(aq)$	1.20×10^{-5}
	sulfide	$Ag_2S(s) \rightleftharpoons 2\ Ag^+(aq) + S^{2-}(aq)$	6.3×10^{-50}
strontium	carbonate	$SrCO_3(s) \rightleftharpoons Sr^{2+}(aq) + CO_3^{2-}(aq)$	5.60×10^{-10}
	fluoride	$SrF_2(s) \rightleftharpoons Sr^{2+}(aq) + 2\ F^-(aq)$	4.33×10^{-9}
	sulfate	$SrSO_4(s) \rightleftharpoons Sr^{2+}(aq) + SO_4^{2-}(aq)$	3.44×10^{-7}
zinc	carbonate	$ZnCO_3(s) \rightleftharpoons Zn^{2+}(aq) + CO_3^{2-}(aq)$	1.46×10^{-10}
	hydroxide	$Zn(OH)_2(s) \rightleftharpoons Zn^{2+}(aq) + 2\ OH^-(aq)$	3.0×10^{-17}

[a]Equilibrium is between solid phase and aqueous solution.

[b]From Dean, J. *Lange's Handbook of Chemistry* (The McGraw-Hill Companies, 1998).

TABLE A5.5 Formation Constants of Complex Ions at 25°C

Complex Ion	Aqueous Equilibrium	K_f
$[Ag(NH_3)_2]^+$	$Ag^+(aq) + 2\ NH_3(aq) \rightleftharpoons Ag(NH_3)_2^+(aq)$	1.7×10^7
$[AgCl_2]^-$	$Ag^+(aq) + 2\ Cl^-(aq) \rightleftharpoons AgCl_2^-(aq)$	2.5×10^5
$[Ag(CN)_2]^-$	$Ag^+(aq) + 2\ CN^-(aq) \rightleftharpoons Ag(CN)_2^-(aq)$	1.0×10^{21}
$[Ag(S_2O_3)_2]^{3-}$	$Ag^+(aq) + 2\ S_2O_3^{2-}(aq) \rightleftharpoons Ag(S_2O_3)_2^{3-}(aq)$	4.7×10^{13}
$[AlF_6]^{3-}$	$Al^{3+}(aq) + 6\ F^-(aq) \rightleftharpoons AlF_6^{3-}(aq)$	4.0×10^{19}
$[Al(OH)_4]^-$	$Al^{3+}(aq) + 4\ OH^-(aq) \rightleftharpoons Al(OH)_4^-(aq)$	7.7×10^{33}
$[Au(CN)_2]^-$	$Au^+(aq) + 2\ CN^-(aq) \rightleftharpoons Au(CN)_2^-(aq)$	2.0×10^{38}
$[Co(NH_3)_6]^{2+}$	$Co^{2+}(aq) + 6\ NH_3(aq) \rightleftharpoons Co(NH_3)_6^{2+}(aq)$	7.7×10^4
$[Co(NH_3)_6]^{3+}$	$Co^{3+}(aq) + 6\ NH_3(aq) \rightleftharpoons Co(NH_3)_6^{3+}(aq)$	5.0×10^{31}
$[Co(en)_3]^{2+}$	$Co^{2+}(aq) + 3\ en(aq) \rightleftharpoons Co(en)_3^{2+}(aq)$	8.7×10^{13}
$[Co(C_2O_4)_3]^{4-}$	$Co^{2+}(aq) + 3\ C_2O_4^{2-}(aq) \rightleftharpoons Co(C_2O_4)_3^{4-}(aq)$	4.5×10^6
$[Cu(NH_3)_4]^{2+}$	$Cu^{2+}(aq) + 4\ NH_3(aq) \rightleftharpoons Cu(NH_3)_4^{2+}(aq)$	5.0×10^{13}
$[Cu(en)_2]^{2+}$	$Cu^{2+}(aq) + 2\ en(aq) \rightleftharpoons Cu(en)_2^{2+}(aq)$	3.2×10^{19}
$[Cu(CN)_4]^{2-}$	$Cu^{2+}(aq) + 4\ CN^-(aq) \rightleftharpoons Cu(CN)_4^{2-}(aq)$	1.0×10^{25}
$[Cu(C_2O_4)_2]^{2-}$	$Cu^{2+}(aq) + 2\ C_2O_4^{2-}(aq) \rightleftharpoons Cu(C_2O_4)_2^{2-}(aq)$	1.7×10^{10}
$[Fe(C_2O_4)_3]^{4-}$	$Fe^{2+}(aq) + 3\ C_2O_4^{2-}(aq) \rightleftharpoons Fe(C_2O_4)_3^{4-}(aq)$	6×10^6
$[Fe(C_2O_4)_3]^{3-}$	$Fe^{3+}(aq) + 3\ C_2O_4^{2-}(aq) \rightleftharpoons Fe(C_2O_4)_3^{3-}(aq)$	3.3×10^{20}
$[HgCl_4]^{2-}$	$Hg^{2+}(aq) + 4\ Cl^-(aq) \rightleftharpoons HgCl_4^{2-}(aq)$	1.2×10^{15}
$[Ni(NH_3)_6]^{2+}$	$Ni^{2+}(aq) + 6\ NH_3(aq) \rightleftharpoons Ni(NH_3)_6^{2+}(aq)$	5.5×10^8
$[PbCl_4]^{2-}$	$Pb^{2+}(aq) + 4\ Cl^-(aq) \rightleftharpoons PbCl_4^{2-}(aq)$	2.5×10^1
$[Zn(NH_3)_4]^{2+}$	$Zn^{2+}(aq) + 4\ NH_3(aq) \rightleftharpoons Zn(NH_3)_4^{2+}(aq)$	2.9×10^9
$[Zn(OH)_4]^{2-}$	$Zn^{2+}(aq) + 4\ OH^-(aq) \rightleftharpoons Zn(OH)_4^{2-}(aq)$	2.8×10^{15}

Appendix 6

Standard Reduction Potentials

TABLE A6.1 Standard Reduction Potentials at 25°C

Half-Reaction	n	$E°$ (V)
$F_2(g) + 2\ e^- \rightarrow 2\ F^-(aq)$	2	2.866
$H_2N_2O_2(s) + 2\ H^+(aq) + 2\ e^- \rightarrow N_2(g) + 2\ H_2O(\ell)$	2	2.65
$O(g) + 2\ H^+(aq) + 2\ e^- \rightarrow H_2O(\ell)$	2	2.421
$Cu^{3+}(aq) + e^- \rightarrow Cu^{2+}(aq)$	1	2.4
$XeO_3(s) + 6\ H^+(aq) + 6\ e^- \rightarrow Xe(g) + 3\ H_2O(\ell)$	6	2.10
$O_3(g) + 2\ H^+(aq) + 2\ e^- \rightarrow O_2(g) + H_2O(\ell)$	2	2.076
$OH(g) + e^- \rightarrow OH^-(aq)$	1	2.02
$Co^{3+}(aq) + e^- \rightarrow Co^{2+}(aq)$	1	1.92
$H_2O_2(\ell) + 2\ H^+(aq) + 2\ e^- \rightarrow 2\ H_2O(\ell)$	2	1.776
$N_2O(g) + 2\ H^+(aq) + 2\ e^- \rightarrow N_2(g) + H_2O(\ell)$	2	1.766
$Ce(OH)^{3+}(aq) + H^+(aq) + e^- \rightarrow Ce^{3+}(aq) + H_2O(\ell)$	1	1.70
$Au^+(aq) + e^- \rightarrow Au(s)$	1	1.692
$PbO_2(s) + SO_4^{2-}(aq) + 4\ H^+(aq) + 2\ e^- \rightarrow PbSO_4(s) + 2\ H_2O(\ell)$	2	1.691
$PbO_2(s) + HSO_4^-(aq) + 3\ H^+(aq) + 2\ e^- \rightarrow PbSO_4(s) + 2\ H_2O(\ell)$	2	1.685
$MnO_4^-(aq) + 4\ H^+(aq) + 3\ e^- \rightarrow MnO_2(s) + 2\ H_2O(\ell)$	3	1.673
$NiO_2(s) + 4\ H^+(aq) + 2\ e^- \rightarrow Ni^{2+}(aq) + 2\ H_2O(\ell)$	2	1.678
$HClO(\ell) + H^+(aq) + e^- \rightarrow \frac{1}{2}\ Cl_2(g) + H_2O(aq)$	1	1.63
$Ce^{4+}(aq) + e^- \rightarrow Ce^{3+}(aq)$	1	1.61
$Mn^{3+}(aq) + e^- \rightarrow Mn^{2+}(aq)$	1	1.542
$MnO_4^-(aq) + 8\ H^+(aq) + 5\ e^- \rightarrow Mn^{2+}(aq) + 4\ H_2O(\ell)$	5	1.507
$BrO_3^-(aq) + 6\ H^+(aq) + 5\ e^- \rightarrow \frac{1}{2}\ Br_2(\ell) + 3\ H_2O(\ell)$	5	1.52
$ClO_3^-(aq) + 6\ H^+(aq) + 5\ e^- \rightarrow \frac{1}{2}\ Cl_2(g) + 3\ H_2O(\ell)$	5	1.47
$PbO_2(s) + 4\ H^+(aq) + 2\ e^- \rightarrow Pb^{2+}(aq) + 2\ H_2O(\ell)$	2	1.455
$Au^{3+}(aq) + 3\ e^- \rightarrow Au(s)$	3	1.40
$Cl_2(g) + 2\ e^- \rightarrow 2\ Cl^-(aq)$	2	1.358
$Cr_2O_7^{2-}(aq) + 14\ H^+(aq) + 6\ e^- \rightarrow 2\ Cr^{3+}(aq) + 7\ H_2O(\ell)$	6	1.33
$MnO_2(s) + 4\ H^+(aq) + 2\ e^- \rightarrow Mn^{2+}(aq) + 2\ H_2O(\ell)$	2	1.23
$O_2(g) + 4\ H^+(aq) + 4\ e^- \rightarrow 2\ H_2O(\ell)$	4	1.229
$IO_3^-(aq) + 6\ H^+(aq) + 5\ e^- \rightarrow \frac{1}{2}\ I_2(s) + 3\ H_2O(\ell)$	5	1.195
$IO_3^-(aq) + 6\ H^+(aq) + 6\ e^- \rightarrow I^-(aq) + 3\ H_2O(\ell)$	6	1.085
$Br_2(\ell) + 2\ e^- \rightarrow 2\ Br^-(aq)$	2	1.066
$HNO_2(\ell) + H^+(aq) + e^- \rightarrow NO(g) + H_2O(\ell)$	1	1.00
$VO_2^+(aq) + 2\ H^+(aq) + e^- \rightarrow VO^{2+}(aq) + H_2O(\ell)$	1	0.991
$NO_3^-(aq) + 4\ H^+(aq) + 3\ e^- \rightarrow NO(g) + 2\ H_2O(\ell)$	3	0.96
$2\ Hg^{2+}(aq) + 2\ e^- \rightarrow Hg_2^{2+}(aq)$	2	0.92

TABLE A6.1 Standard Reduction Potentials at 25°C *(Continued)*

Half-Reaction	*n*	*E*° (V)
$ClO^-(aq) + H_2O(\ell) + 2\ e^- \rightarrow Cl^-(aq) + 2\ OH^-(aq)$	2	0.89
$HO_2^-(aq) + H_2O(\ell) + 2\ e^- \rightarrow 3\ OH^-(aq)$	2	0.88
$Hg^{2+}(aq) + 2\ e^- \rightarrow Hg(\ell)$	2	0.851
$Ag^+(aq) + e^- \rightarrow Ag(s)$	1	0.800
$Hg_2^{2+}(aq) + 2\ e^- \rightarrow 2\ Hg(\ell)$	2	0.797
$Fe^{3+}(aq) + e^- \rightarrow Fe^{2+}(aq)$	1	0.770
$PtCl_4^{2-}(aq) + 2\ e^- \rightarrow Pt(s) + 4\ Cl^-(aq)$	2	0.73
$O_2(g) + 2\ H^+(aq) + 2\ e^- \rightarrow H_2O_2(\ell)$	2	0.68
$MnO_4^-(aq) + 2\ H_2O(\ell) + 3\ e^- \rightarrow MnO_2(s) + 4\ OH^-(aq)$	3	0.59
$H_3AsO_4(s) + 2\ H^+(aq) + 2\ e^- \rightarrow H_3AsO_3(aq) + H_2O(\ell)$	2	0.559
$I_2(s) + 2\ e^- \rightarrow 2\ I^-(aq)$	2	0.536
$Cu^+(aq) + e^- \rightarrow Cu(s)$	1	0.521
$2\ NiO(OH)(s) + 2\ H_2O(\ell) + 2\ e^- \rightarrow 2\ Ni(OH)_2(s) + 2\ OH^-(aq)$	2	0.52
$H_2SO_3(\ell) + 4\ H^+(aq) + 4\ e^- \rightarrow S(s) + 3\ H_2O(\ell)$	4	0.449
$Ag_2CrO_4(s) + 2\ e^- \rightarrow 2\ Ag(s) + CrO_4^{2-}(aq)$	2	0.447
$O_2(g) + 2\ H_2O(\ell) + 4\ e^- \rightarrow 4\ OH^-(aq)$	4	0.401
$Fe(CN)_6^{3-}(aq) + e^- \rightarrow Fe(CN)_6^{4-}(aq)$	1	0.36
$Ag_2O(s) + H_2O(\ell) + 2\ e^- \rightarrow 2\ Ag(s) + 2\ OH^-(aq)$	2	0.342
$Cu^{2+}(aq) + 2\ e^- \rightarrow Cu(s)$	2	0.342
$BiO^+(aq) + 2\ H^+(aq) + 3\ e^- \rightarrow Bi(s) + H_2O(\ell)$	3	0.32
$AgCl(s) + e^- \rightarrow Ag(s) + Cl^-(aq)$	1	0.222
$HSO_4^-(aq) + 3\ H^+(aq) + 2\ e^- \rightarrow H_2SO_3(\ell) + H_2O(\ell)$	2	0.17
$Sn^{4+}(aq) + 2\ e^- \rightarrow Sn^{2+}(aq)$	2	0.154
$Cu^{2+}(aq) + e^- \rightarrow Cu^+(aq)$	1	0.153
$2\ MnO_2(s) + H_2O(\ell) + 2\ e^- \rightarrow Mn_2O_3(s) + 2\ OH^-(aq)$	2	0.15
$S(s) + 2\ H^+(aq) + 2\ e^- \rightarrow H_2S(g)$	2	0.141
$HgO(s) + H_2O(\ell) + 2\ e^- \rightarrow Hg(\ell) + 2\ OH^-(aq)$	2	0.0977
$AgBr(s) + e^- \rightarrow Ag(s) + Br^-(aq)$	1	0.095
$Ag(S_2O_3)_2^{3-}(aq) + e^- \rightarrow Ag(s) + 2\ S_2O_3^{2-}(aq)$	1	0.01
$NO_3^-(aq) + H_2O(\ell) + 2\ e^- \rightarrow NO_2^-(aq) + 2\ OH^-(aq)$	2	0.01
$2\ H^+(aq) + 2\ e^- \rightarrow H_2(g)$	2	0.000
$Pb^{2+}(aq) + 2\ e^- \rightarrow Pb(s)$	2	−0.126
$CrO_4^{2-}(aq) + 4\ H_2O(\ell) + 3\ e^- \rightarrow Cr(OH)_3(s) + 5\ OH^-(aq)$	3	−0.13
$Sn^{2+}(aq) + 2\ e^- \rightarrow Sn(s)$	2	−0.136
$AgI(s) + e^- \rightarrow Ag(s) + I^-(aq)$	1	−0.152
$CuI(s) + e^- \rightarrow Cu(s) + I^-(aq)$	1	−0.185
$N_2(g) + 5\ H^+(aq) + 4\ e^- \rightarrow N_2H_5^+(aq)$	4	−0.23
$Ni^{2+}(aq) + 2\ e^- \rightarrow Ni(s)$	2	−0.257
$PbSO_4(s) + H^+(aq) + 2\ e^- \rightarrow Pb(s) + HSO_4^-(aq)$	2	−0.356
$Co^{2+}(aq) + 2\ e^- \rightarrow Co(s)$	2	−0.277
$Ag(CN)_2^-(aq) + e^- \rightarrow Ag(s) + 2\ CN^-(aq)$	1	−0.31
$Cd^{2+}(aq) + 2\ e^- \rightarrow Cd(s)$	2	−0.403
$Cr^{3+}(aq) + e^- \rightarrow Cr^{2+}(aq)$	1	−0.41

Continued on next page

TABLE A6.1 Standard Reduction Potentials at 25°C *(Continued)*

Half-Reaction	*n*	*E*° (V)
$Fe^{2+}(aq) + 2\ e^- \rightarrow Fe(s)$	2	−0.447
$2\ CO_2(g) + 2\ H^+(aq) + 2\ e^- \rightarrow H_2C_2O_4(s)$	2	−0.49
$Ni(OH)_2(s) + 2\ e^- \rightarrow Ni(s) + 2\ OH^-(aq)$	2	−0.72
$Cr^{3+}(aq) + 3\ e^- \rightarrow Cr(s)$	3	−0.74
$Zn^{2+}(aq) + 2\ e^- \rightarrow Zn(s)$	2	−0.762
$Cd(OH)_2(s) + 2\ e^- \rightarrow Cd(s) + 2\ OH^-(aq)$	2	−0.81
$2\ H_2O(\ell) + 2\ e^- \rightarrow H_2(g) + 2\ OH^-(aq)$	2	−0.828
$SO_4^{2-}(aq) + H_2O(\ell) + 2\ e^- \rightarrow SO_3^{2-}(aq) + 2\ OH^-(aq)$	2	−0.92
$N_2(g) + 4\ H_2O(\ell) + 4\ e^- \rightarrow 4\ OH^-(aq) + N_2H_4(\ell)$	4	−1.16
$Mn^{2+}(aq) + 2\ e^- \rightarrow Mn(s)$	2	−1.185
$Zn(OH)_2(s) + 2\ e^- \rightarrow Zn(s) + 2\ OH^-(aq)$	2	−1.249
$Al^{3+}(aq) + 3\ e^- \rightarrow Al(s)$	3	−1.662
$Mg^{2+}(aq) + 2\ e^- \rightarrow Mg(s)$	2	−2.37
$Na^+(aq) + e^- \rightarrow Na(s)$	1	−2.71
$Ca^{2+}(aq) + 2\ e^- \rightarrow Ca(s)$	2	−2.868
$Ba^{2+}(aq) + 2\ e^- \rightarrow Ba(s)$	2	−2.912
$K^+(aq) + e^- \rightarrow K(s)$	1	−2.95
$Li^+(aq) + e^- \rightarrow Li(s)$	1	−3.05

Appendix 7

Naming Organic Compounds

While organic chemistry was becoming established as a discipline within chemistry, many compounds were given trivial names that are still commonly used and recognized. We refer to many of these compounds by their nonsystematic names throughout this book, and their names and structures are listed in Table A7.1.

TABLE A7.1 Organic Compounds and Their Commonly Used Nonsystematic Names

Name	Formula	Structure
ethylene	C_2H_4	$H_2C{=}CH_2$
acetylene	C_2H_2	$HC{\equiv}CH$
benzene	C_6H_6	
toluene	$C_6H_5CH_3$	CH_3
ethyl alcohol	CH_3CH_2OH	$H_3C{-}CH_2{-}OH$
acetone	CH_3COCH_3	$H_3C{-}C({=}O){-}CH_3$
acetic acid	CH_3COOH	$H_3C{-}C({=}O){-}OH$
formaldehyde	CH_2O	$H{-}C({=}O){-}H$

The International Union of Pure and Applied Chemistry (IUPAC) has proposed a set of rules for the systematic naming of organic compounds. When naming compounds or drawing structures based on names, we need to keep in mind that the IUPAC system of nomenclature is based on two fundamental ideas: (1) the name of a compound must indicate how the carbon atoms in the skeleton are bonded together, and (2) the name must identify the location of any functional groups in the molecule.

Alkanes

Table A7.2 contains the prefixes used for carbon chains ranging in size from C_1 to C_{20} and gives the names for compounds consisting of unbranched chains. The name of a compound consists of a prefix identifying the number of carbons in the chain and a suffix defining the type of hydrocarbon. The suffix *-ane* indicates that the compounds are alkanes and that all carbon–carbon bonds are single bonds.

TABLE A7.2 Prefixes for Naming Carbon Chains

Prefix	Example	Name	Prefix	Example	Name
meth	CH_4	methane	undec	$C_{11}H_{24}$	undecane
eth	C_2H_6	ethane	dodec	$C_{12}H_{26}$	dodecane
pro	C_3H_8	propane	tridec	$C_{13}H_{28}$	tridecane
but	C_4H_{10}	butane	tetradec	$C_{14}H_{30}$	tetradecane
pent	C_5H_{12}	pentane	pentadec	$C_{15}H_{32}$	pentadecane
hex	C_6H_{14}	hexane	hexadec	$C_{16}H_{34}$	hexadecane
hept	C_7H_{16}	heptane	heptadec	$C_{17}H_{36}$	heptadecane
oct	C_8H_{18}	octane	octadec	$C_{18}H_{38}$	octadecane
non	C_9H_{20}	nonane	nonadec	$C_{19}H_{40}$	nonadecane
dec	$C_{10}H_{22}$	decane	eicos	$C_{20}H_{42}$	eicosane

Branched-Chain Alkanes

The alkane drawn here is used to illustrate each step in the naming rules:

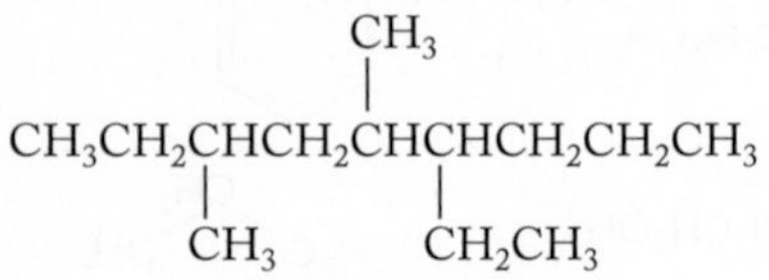

$$CH_3$$
$$CH_3CH_2CHCH_2CHCHCH_2CH_2CH_3$$
$$CH_3 \qquad CH_2CH_3$$

1. **Identify and name the longest continuous carbon chain.**

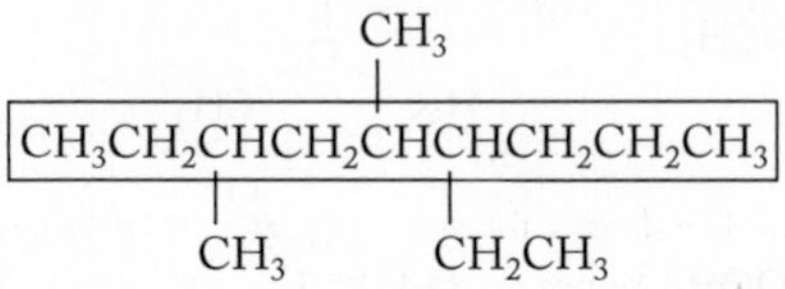

$$CH_3$$
$$\boxed{CH_3CH_2CHCH_2CHCHCH_2CH_2CH_3}\ \text{Nonane}$$
$$CH_3 \qquad CH_2CH_3$$

2. **Identify the groups attached to this chain and name them.** Names of substituent groups consist of the prefix from Table A7.2 that identifies the length of the group and the suffix *-yl* that identifies it as an alkyl group.

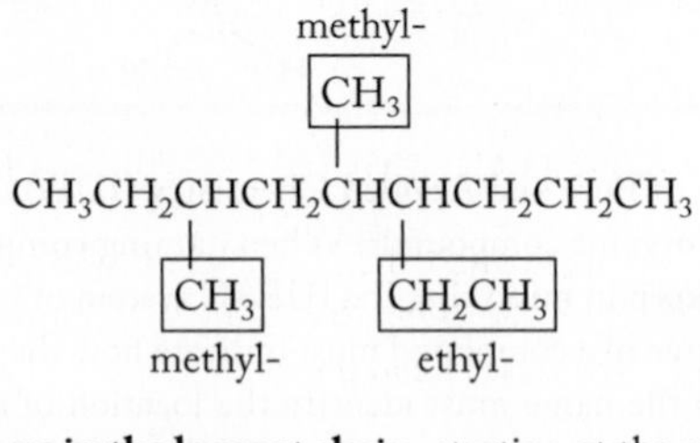

methyl-
$$\boxed{CH_3}$$
$$CH_3CH_2CHCH_2CHCHCH_2CH_2CH_3$$
$$\boxed{CH_3} \qquad \boxed{CH_2CH_3}$$
methyl- ethyl-

3. **Number the carbon atoms in the longest chain,** starting at the end nearest a substituent group. Doing this identifies the points of attachment of the alkyl groups with the lowest possible numbers.

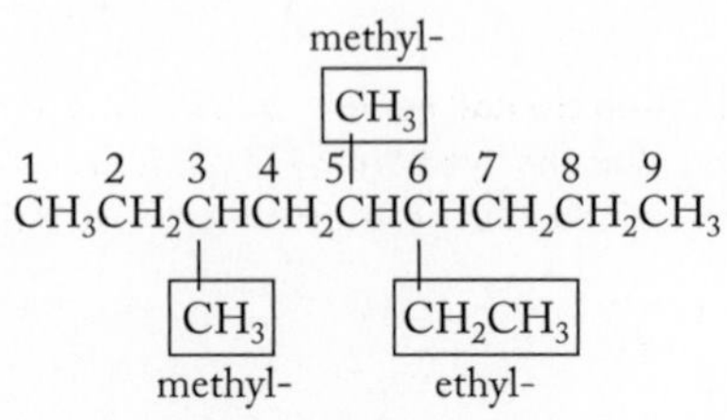

4. **Designate the location and identity of each substituent group with a number,** followed by a hyphen, and its name.

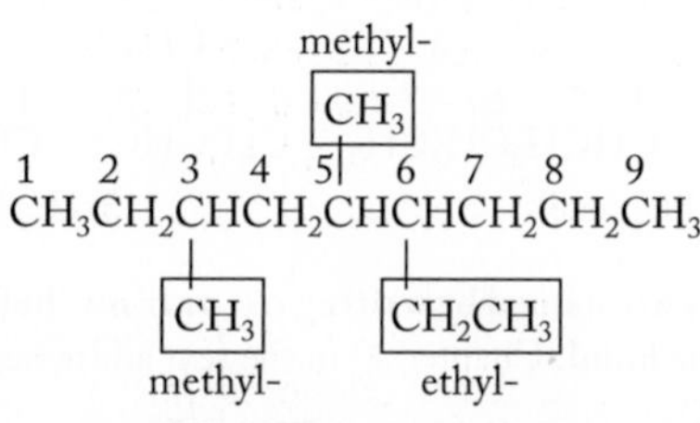

3-methyl-, 5-methyl-, 6-ethyl-

5. **Put together the complete name by listing the substituent groups in alphabetical order.** If more than one of a given type of substituent group is present, prefixes *di-*, *tri-*, *tetra-*, and so forth, are appended to the names, but these numerical prefixes are not considered when determining the alphabetical order. The name of the last substituent group is written together with the name identifying the longest carbon chain.

$$\begin{array}{c} \quad\quad CH_3 \\ \quad\quad | \\ CH_3CH_2CHCH_2CHCHCH_2CH_2CH_3 \\ |\quad\quad\quad | \\ CH_3\quad\quad CH_2CH_3 \end{array}$$

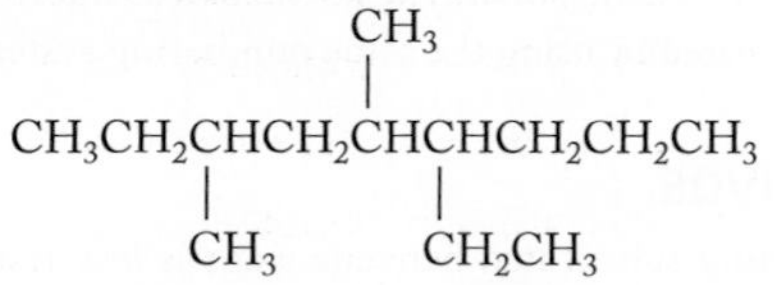

6-Ethyl-3,5-dimethylnonane

Cycloalkanes

The simplest examples of this class of compounds consist of one unsubstituted ring of carbon atoms. The IUPAC names of these compounds consist of the prefix *cyclo-* followed by the parent name from Table A7.2 to indicate the number of carbon atoms in the ring. As an illustration, the names, formulas, and line structures of the first three cycloalkanes in the homologous series are

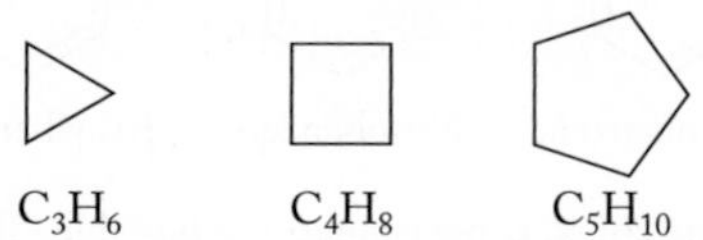

C_3H_6	C_4H_8	C_5H_{10}
Cyclopropane	Cyclobutane	Cyclopentane

Alkenes and Alkynes

Alkenes have carbon–carbon double bonds and alkynes have carbon–carbon triple bonds as functional groups. The names of these types of compounds consist of (1) a parent name that identifies the longest carbon chain that includes the double or triple bond, (2) a suffix that identifies the class of compound, and (3) names of any substituent groups attached to the longest carbon chain. The suffix *-ene* identifies an alkene; *-yne* identifies an alkyne.

The alkene and alkyne drawn here are used to illustrate each step in the naming rules:

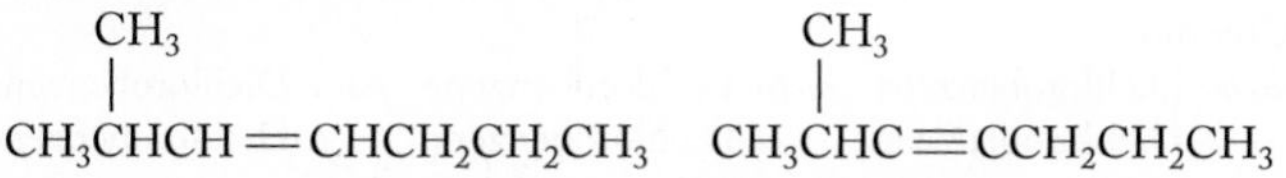

$$\begin{array}{c} CH_3 \\ | \\ CH_3CHCH{=}CHCH_2CH_2CH_3 \end{array} \qquad \begin{array}{c} CH_3 \\ | \\ CH_3CHC{\equiv}CCH_2CH_2CH_3 \end{array}$$

1. **To determine the parent name,** identify the longest chain that contains the unsaturation. Name the parent compound with the prefix that defines the number of carbons in that chain and the suffix that identifies the class of compound.

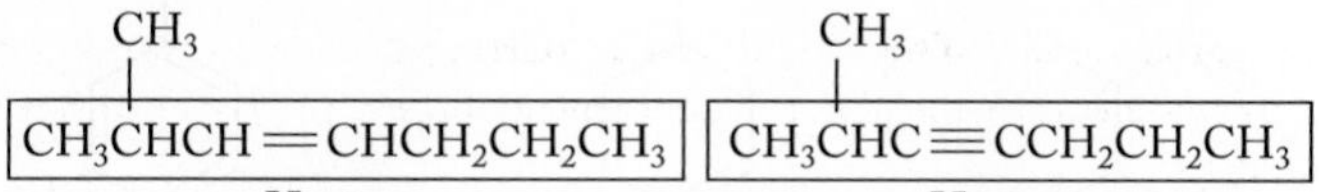

$$\begin{array}{c} CH_3 \\ | \\ \boxed{CH_3CHCH{=}CHCH_2CH_2CH_3} \\ \text{Heptene} \end{array} \qquad \begin{array}{c} CH_3 \\ | \\ \boxed{CH_3CHC{\equiv}CCH_2CH_2CH_3} \\ \text{Heptyne} \end{array}$$

2. **Number the parent chain from the end nearest the unsaturation so that the first carbon in the double or triple bond has the lowest number possible.** (If the unsaturation is in the middle of a chain, the location of any substituent group is used to determine where the

numbering starts.) The smaller of the two numbers identifying the carbon atoms involved in the unsaturation is used as the locator of the multiple bond.

$$\overset{}{\overset{1}{CH_3}}\overset{CH_3}{\overset{|}{\overset{2}{C}}}H\overset{3}{C}H{=}\overset{4}{C}H\overset{5}{C}H_2\overset{6}{C}H_2\overset{7}{C}H_3 \qquad \overset{1}{CH_3}\overset{CH_3}{\overset{|}{\overset{2}{C}}}H\overset{3}{C}{\equiv}\overset{4}{C}\overset{5}{C}H_2\overset{6}{C}H_2\overset{7}{C}H_3$$

3-Heptene 3-Heptyne

3. **Stereoisomers of alkenes are named by writing *cis-* or *trans-* before the number identifying the location of the double bond.** Chapter 20 in the text addresses naming stereoisomers.
4. **The rules for naming substituted alkanes are followed to name and locate any other groups on the chain.**

$$CH_3\overset{CH_3}{\overset{|}{C}}HCH{=}CHCH_2CH_2CH_3 \qquad CH_3\overset{CH_3}{\overset{|}{C}}HC{\equiv}CCH_2CH_2CH_3$$

2-Methyl-3-heptene 2-Methyl-3-heptyne

Halogens attached to an alkane, alkene, or alkyne are named as fluoro- (F–), chloro- (Cl–), bromo- (Br–), or iodo- (I–) and are located by using the same numbering system described for alkyl groups.

Benzene Derivatives

Naming compounds containing substituted benzene rings is less systematic than naming hydrocarbons. Many compounds have common names that are incorporated into accepted names, but for simple substituted benzene rings, the following rules may be applied.

1. **For monosubstituted benzene rings,** a prefix identifying the group is appended to the parent name benzene:

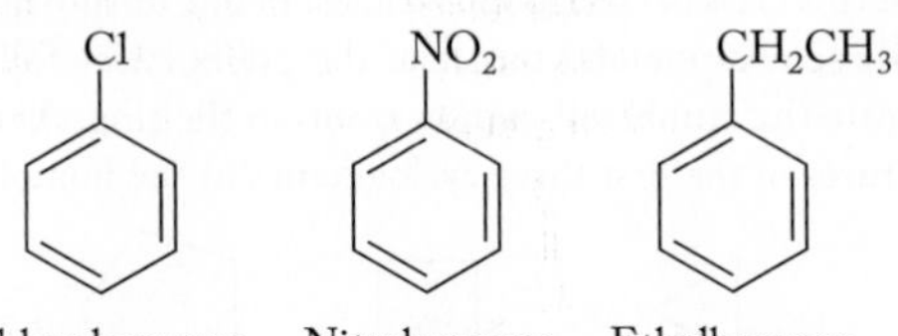

Chlorobenzene Nitrobenzene Ethylbenzene

2. **For disubstituted benzene rings,** three isomers are possible. The relative position of the substituent groups is indicated by numbers in IUPAC nomenclature, but the set of prefixes shown are very commonly used as well:

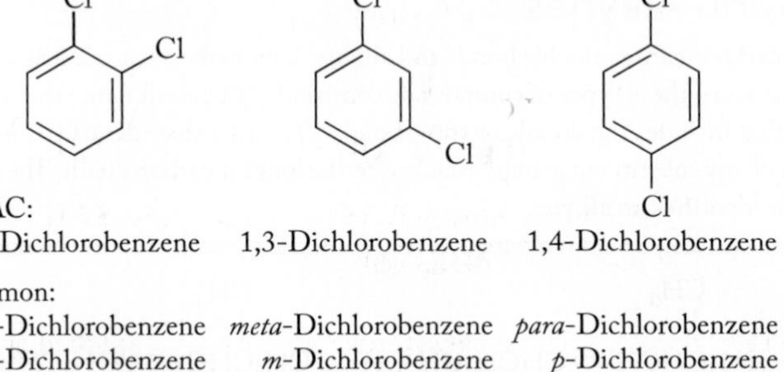

IUPAC:
1,2-Dichlorobenzene 1,3-Dichlorobenzene 1,4-Dichlorobenzene

Common:
ortho-Dichlorobenzene *meta*-Dichlorobenzene *para*-Dichlorobenzene
o-Dichlorobenzene *m*-Dichlorobenzene *p*-Dichlorobenzene

3. **When three or more groups are attached to a benzene ring,** the lowest possible numbers are assigned to locate the groups with respect to each other.

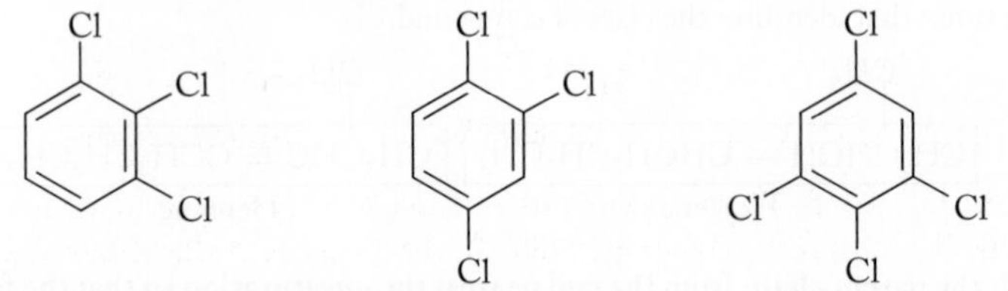

1,2,3-Trichlorobenzene 1,2,4-Trichlorobenzene 1,2,3,5-Tetrachlorobenzene

(*Note*: Not 1,3,4-trichlorobenzene; and not 1,3,4,5-tetrachlorobenzene.)

Hydrocarbons Containing Other Functional Groups

The same basic principles developed for naming alkanes apply to naming hydrocarbons with functional groups other than alkyl groups. The name must identify the carbon skeleton, locate the functional group, and contain a suffix that defines the class of compound. The following examples give the suffixes for some common functional groups; when suffixes are used, they replace the final *-e* in the name of the parent alkane. Other functional groups may be identified by including the name of the class of compounds in the name of the molecule.

Alcohols: Suffix ***-ol***

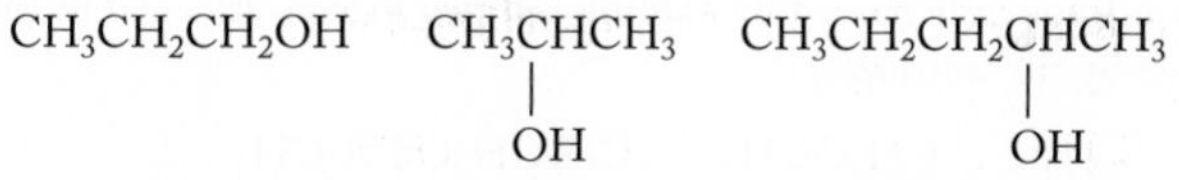

$CH_3CH_2CH_2OH$ — 1-Propanol

$CH_3CH(OH)CH_3$ — 2-Propanol

$CH_3CH_2CH_2CH(OH)CH_3$ — 2-Pentanol

Aldehydes: Suffix ***-al***

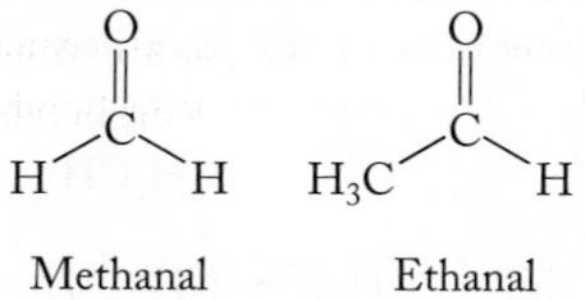

	HCHO	H_3C–CHO
IUPAC:	Methanal	Ethanal
Common:	Formaldehyde	Acetaldehyde

Because the aldehyde group can only be on a terminal carbon, no number is necessary to locate it on the carbon chain.

Ketones: Suffix ***-one*** The location of the carbonyl is given by a number, and the chain is numbered so that the carbonyl carbon has the lowest possible value. Many ketones also have common names generated by identifying the hydrocarbon groups on both sides of the carbonyl group.

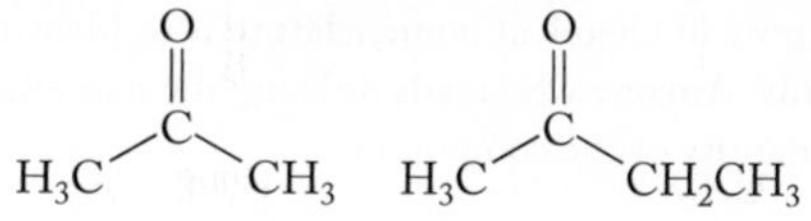

	H_3C–CO–CH_3	H_3C–CO–CH_2CH_3
IUPAC:	Propan-2-one	Butan-2-one
Common:	Acetone	Methyl ethyl ketone

Carboxylic Acids: Suffix ***-oic acid*** The carboxylic acid group is by definition carbon 1, so no number identifying its location is included in the name.

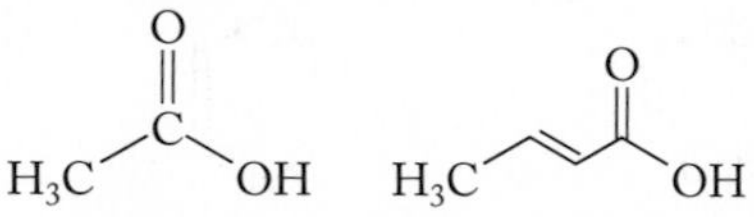

	H_3C–CO–OH	H_3C–CH=CH–CO–OH
IUPAC:	Ethanoic acid	*trans*-2-Butenoic acid
Common:	Acetic acid	

Salts of Carboxylic Acids Salts are named with the cation first, followed by the anion name of the acid from which *-ic acid* is dropped and the suffix *-ate* is added. The sodium salt of acetic acid is sodium acetate.

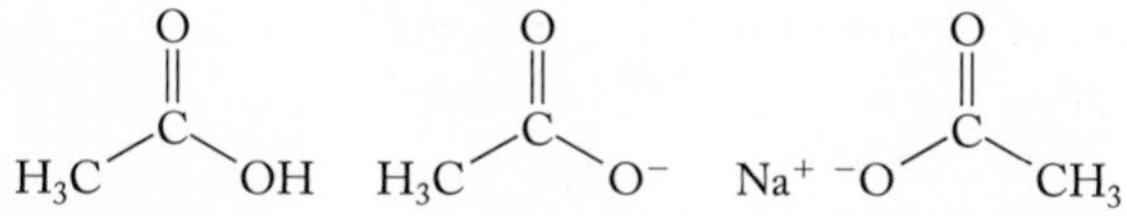

H_3C–CO–OH — Acetic acid

H_3C–CO–O^- — Acetate ion

Na^+ ^-O–CO–CH_3 — Sodium acetate

Esters Esters are viewed as derivatives of carboxylic acids. They are named in a manner analogous to that of salts. The alkyl group comes first, followed by the name of the carboxylate anion.

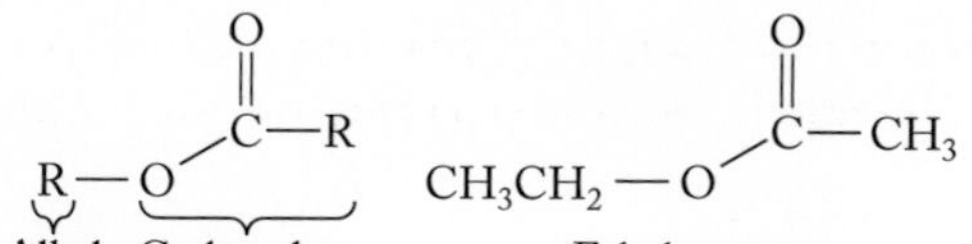

R—O–C(=O)—R (R = Alkyl; O–C(=O)—R = Carboxylate)

CH_3CH_2—O–C(=O)—CH_3 — Ethyl acetate

Amides Amides are also derivatives of carboxylic acids. They are named by replacing *-ic acid* (of the common names) or *-oic acid* (of the IUPAC names) with *-amide.*

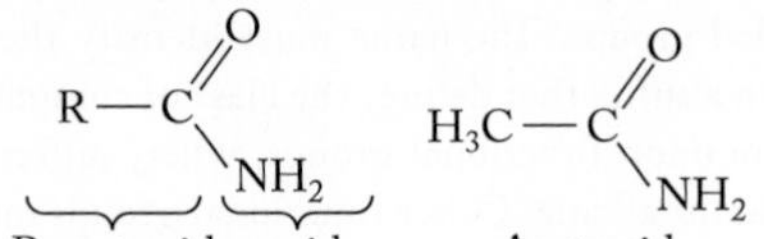

$$\underbrace{R-C}_{\text{Parent acid}}(=O)\underbrace{NH_2}_{\text{-amide}} \qquad H_3C-C(=O)NH_2$$

Parent acid -amide

Acetamide (ethanamide)

Ethers Ethers are frequently named by naming the two groups attached to the oxygen and following those names by the word *ether.*

CH_3OCH_3	$CH_3CH_2OCH_2CH_3$
Dimethyl ether	Diethyl ether

Amines Aliphatic amines are usually named by listing the group or groups attached to the nitrogen and then appending *-amine* as a suffix. They may also be named by prefixing *amino-* to the name of the parent chain.

H_3C-NH_2	$H_3C-NH(CH_2CH_3)$	$H_2NCH_2CH_2OH$
Methylamine	Ethylmethylamine	2-Aminoethanol

This brief summary will enable you to understand the names of organic compounds used in this book. IUPAC rules are much more extensive than this and can be applied to all varieties of carbon compounds, including those with multiple functional groups. It is important to recognize that the rules of systematic nomenclature do not necessarily lead to a unique name for each compound, but they do always lead to an unambiguous one. Furthermore, common names are still used frequently in organic chemistry because the systematic alternatives do not improve communication. Remember that the main purpose of chemical nomenclature is to identify a chemical species by means of written or spoken words. Anyone who reads or hears the name should be able to deduce the structure and thereby the identity of the compound.

Glossary

A

absolute temperature Temperature expressed in kelvins on the absolute (Kelvin) temperature scale, on which 0 K is the lowest possible temperature. (Ch. 5)

absolute zero (0 K) The zero point on the Kelvin temperature scale; theoretically the lowest temperature possible. (Ch. 1)

absorbance (*A*) A measure of the quantity of light absorbed by a sample. (Ch. 4)

accuracy Agreement between one or more experimental values and the true (or accepted) value. (Ch. 1)

acid (Brønsted–Lowry acid) A proton donor. (Ch. 4)

activated complex A transient species in a chemical reaction in which molecules have both the proper orientation and enough energy to react with each other. (Ch. 13)

activation energy (E_a) The minimum energy molecules need to react when they collide. (Ch. 13)

active site The location on an enzyme where a reactive substance binds. (Ch. 20)

activity series A high-to-low ranking of metals based on their strengths as reducing agents. (Ch. 4)

actual yield The amount of product obtained from a chemical reaction, which may be less than the theoretical yield. (Ch. 3)

addition polymer A polymer formed by an addition reaction. (Ch. 12)

addition reaction A reaction in which two reactants couple to form one product without the loss of any atoms or molecules. (Ch. 12)

alcohol An organic compound containing the OH functional group. (Ch. 2)

aldehyde An organic compound having a carbonyl group with a single bond to a hydrogen atom and a single bond to another atom or group of atoms designated as R– in the general formula RCHO. (Ch. 8)

alkali metals The elements in group 1 of the periodic table. (Ch. 2)

alkaline earth metals The elements in group 2 of the periodic table. (Ch. 2)

alkane A hydrocarbon in which all the bonds are single bonds. (Ch. 2)

alkene A hydrocarbon containing one or more carbon–carbon double bonds. (Ch. 2)

alkyne A hydrocarbon containing one or more carbon–carbon triple bonds. (Ch. 2)

allotropes Different molecular forms of the same element. (Ch. 8)

alloy A blend of a host metal and one or more other elements, which may or may not be metals, that are added to change the properties of the host metal. (Ch. 12)

alpha (α) decay A nuclear reaction in which an unstable nuclide spontaneously emits an alpha particle. (Ch. 19)

α helix A coil in a protein chain's secondary structure held together by hydrogen bonds. (Ch. 20)

alpha (α) particle A radioactive emission with a charge of 2+ and a mass equivalent to that of a helium nucleus. (Ch. 2)

amide An organic compound in which the same carbon atom is single bonded to a nitrogen atom and double bonded to an oxygen atom. (Ch. 12)

amine An organic compound that functions as a base and has the general formula RNH_2, R_2NH, or R_3N, where R is any organic group. (Ch. 4)

amino acid A molecule that contains at least one amine functional group and one carboxylic acid functional group; in an α-amino acid, the two functional groups are attached to the same (α) carbon atom. (Ch. 20)

Amontons's law The principle that the pressure of a quantity of gas at constant volume is directly proportional to its absolute temperature. (Ch. 5)

amorphous solid A solid that lacks long-range order for the atoms, ions, or molecules in its structure. (Ch. 12)

amphiprotic Describes a substance that can behave as either a proton acceptor or a proton donor. (Ch. 4)

amplitude The height of the crest or depth of the trough of a wave with respect to the center line of the wave. (Ch. 7)

angular The molecular geometry about a central atom with a steric number of 3 (bonds to two atoms + one lone pair) or a steric number of 4 (bonds to two atoms + two one pairs). (Ch. 9)

angular momentum quantum number (ℓ) An integer having any value from 0 to $n - 1$ that defines the shape of an orbital. (Ch. 7)

anion An ion with a negative charge. (Ch. 1)

anode An electrode at which an oxidation half-reaction (loss of electrons) takes place. (Ch. 18)

antibonding orbital A molecular orbital in which electron density is highest outside the region between atom centers. (Ch. 9)

antimatter Particles that are the charge opposites of normal subatomic particles. (Ch. 19)

aromatic compound A cyclic, planar compound with delocalized π (pi) electrons above and below the plane of the molecule. (Ch. 9)

Arrhenius acid A compound that produces H_3O^+ ions in aqueous solution. (Ch. 15)

Arrhenius base A compound that produces OH^- ions in aqueous solution. (Ch. 15)

Arrhenius equation An equation relating the rate constant of a reaction to absolute temperature (*T*), the activation energy of the reaction (E_a), and the frequency factor (*A*). (Ch. 13)

atmosphere (atm) A unit of pressure based on Earth's mean atmospheric pressure at sea level. (Ch. 5)

atmospheric pressure (P_{atm}) The force exerted by the gases surrounding Earth on Earth's surface and on all surfaces of all objects. (Ch. 5)

atom The smallest particle of an element that retains the element's properties. (Ch. 1)

atomic absorption spectrum A characteristic series of dark lines produced when free, gaseous atoms are illuminated by a continuous source of radiation. (Ch. 7)

atomic emission spectrum A characteristic series of bright lines produced by high-temperature atoms. (Ch. 7)

atomic number (*Z*) The number of protons in the nucleus of an atom. (Ch. 2)

atomic solid A solid formed by weak attractions between noble gas atoms. (Ch. 12)

aufbau principle The method of building electron configurations of atoms by adding one electron at a time as atomic number increases across the rows of the periodic table; each electron goes into the lowest-energy orbital available. (Ch. 7)

autoionization The process that produces equal and very small concentrations of H_3O^+ and OH^- ions in pure water. (Ch. 15)

average atomic mass A weighted average of the masses of all the isotopes of an element, calculated by multiplying the natural abundance of each isotope by its mass in atomic mass units and then summing these products. (Ch. 2)

Avogadro constant (N_A) The fundamental constant representing the number of characteristic particles in 1 mole of a substance: $6.02214076 \times 10^{23}$ particles/mol. (Ch. 3)

Avogadro's law The principle that the volume of a gas at a given temperature and pressure is proportional to the quantity of the gas. (Ch. 5)

B

band gap (E_g) The energy gap between the valence and conduction bands. (Ch. 9)

band theory An extension of molecular orbital theory that describes bonding in solids. (Ch. 9)

barometer An instrument that measures atmospheric pressure. (Ch. 5)

base (Brønsted–Lowry base) A proton acceptor. (Ch. 4)

becquerel (Bq) The SI unit of radioactivity equal to the decay of one radioactive atom per second. (Ch. 19)

Beer's Law The relation of the absorbance of a solution (A) to concentration (c), the light's path length (b), and the solute's molar absorptivity (ε) by the equation $A = \varepsilon bc$. (Ch. 4)

belt of stability The region on a graph of number of neutrons versus number of protons that includes all stable nuclei. (Ch. 19)

bent See *angular.*

beta (β) decay A spontaneous process by which a neutron in a radioactive nuclide is transformed into a proton and emits a high-energy electron (β-particle). (Ch. 2, Ch. 19)

beta (β) particle A radioactive emission equivalent to a high-energy electron. (Ch. 2)

β-pleated sheet A puckered two-dimensional array of protein strands held together by hydrogen bonds. (Ch. 20)

bimolecular step A step in a reaction mechanism involving a collision between two molecules. (Ch. 13)

binding energy (BE) The energy that would be released when free nucleons combine to form the nucleus of an atom; it is also the energy needed to split the nucleus into free nucleons. (Ch. 19)

biocatalysis The use of enzymes to catalyze reactions on a large scale; it is becoming especially important in processes that involve chiral materials. (Ch. 13)

biochemistry The study of chemical processes taking place in living organisms. (Ch. 20)

biomass The total mass of organic matter in any given ecological system. (Ch. 20)

biomolecule Any molecule produced by a living organism. (Ch. 20)

body-centered cubic (bcc) unit cell An array of atoms or molecules with one particle at each of the eight corners of a cube and one at the center of the cell. (Ch. 12)

bomb calorimeter A constant-volume device used to measure the energy released during a combustion reaction. (Ch. 6)

bond angle The angle (in degrees) defined by lines joining the centers of two atoms to the center of a third atom to which they are chemically bonded. (Ch. 9)

bond dipole The separation of electrical charge created when atoms with different electronegativities form a covalent bond. (Ch. 9)

bond energy The energy needed to break 1 mole of a covalent bond in a molecule or in a polyatomic ion in the gas phase. (Ch. 8)

bond length The distance between the nuclear centers of two atoms joined in a bond. (Ch. 8)

bond order The number of bonds between atoms: 1 for a single bond, 2 for a double bond, and 3 for a triple bond. (Ch. 8)

bond polarity A measure of the extent to which bonding electrons are unequally shared. (Ch. 8)

bonding capacity The number of covalent bonds an atom forms to have an octet of electrons in its valence shell. (Ch. 8)

bonding orbital A molecular orbital in which electron density is highest in the region between the atom centers. (Ch. 9)

bonding pair A pair of electrons shared between two atoms. (Ch. 8)

Born–Haber cycle A series of steps with corresponding enthalpy changes that describes the formation of an ionic solid from its constituent elements. (Ch. 11)

Boyle's law The principle that the volume of a given amount of gas at constant temperature is inversely proportional to its pressure. (Ch. 5)

branched-chain hydrocarbon A hydrocarbon in which the chain of carbon atoms is not linear. (Ch. 6)

breeder reactor A nuclear reactor in which fissionable material is produced during normal reactor operation. (Ch. 19)

buffer capacity The quantity of acid or base that a pH buffer can neutralize while keeping its pH within a desired range. (Ch. 16)

C

calibration curve A graph showing how a measurable property, such as absorbance, varies for a set of standard samples of known concentration that can later be used to determine the unknown concentration of a sample from its absorbance. (Ch. 4)

calorie (cal) The amount of energy necessary to raise the temperature of 1 g of water by 1°C; for example, from 14.5°C to 15.5°C. (Ch. 6)

calorimeter A device used to measure the absorption or release of energy by a physical change or chemical process. (Ch. 6)

calorimeter constant ($C_{calorimeter}$) The heat capacity of a calorimeter. (Ch. 6)

calorimetry The measurement of the quantity of energy transferred during a physical change or chemical process. (Ch. 6)

capillary action The rise of a liquid in a narrow tube because of adhesive forces between the liquid and the tube and cohesive forces within the liquid. (Ch. 10)

carbohydrate An organic molecule with the generic formula $C_x(H_2O)_y$. (Ch. 20)

carbonyl group A carbon atom with a double bond to an oxygen atom. (Ch. 8)

carboxylic acid An organic compound containing the –COOH group. (Ch. 4)

catalyst A substance added to a reaction that increases the rate of the reaction but is not consumed in the process. (Ch. 13)

cathode An electrode at which a reduction half-reaction (gain of electrons) takes place. (Ch. 18)

cathode rays Streams of electrons emitted by the cathode in a partially evacuated tube. (Ch. 2)

cation An ion with a positive charge. (Ch. 1)

cell diagram Symbols that show how the components of an electrochemical cell are connected. (Ch. 18)

cell potential (E_{cell}) The force with which an electrochemical cell can pump electrons through an external circuit. (Ch. 18)

chain reaction A self-sustaining series of fission reactions in which the neutrons released when nuclei split apart initiate additional fission events and sustain the reaction. (Ch. 19)

Charles's law The principle that the volume of a fixed quantity of gas at constant pressure is directly proportional to its absolute temperature. (Ch. 5)

chelate effect The greater affinity of metal ions for polydentate ligands than for monodentate ligands. (Ch. 22)

chelation The interaction of a metal with a polydentate ligand (chelating agent); pairs of electrons on one molecule of the ligand occupy two or more coordination sites on the central metal. (Ch. 22)

chemical bond A force that holds two atoms or ions in a compound together. (Ch. 1)

chemical equation Notation in which chemical formulas express the identities and their coefficients express the quantities of substances involved in a chemical reaction. (Ch. 1)

chemical equilibrium A dynamic process in which the concentrations of reactants and products remain constant over time and the rate of the reaction in the forward direction matches its rate in the reverse direction. (Ch. 14)

chemical formula Notation for representing elements and compounds; consists of the symbols of the constituent elements, each followed by subscripts that identify the number of atoms of each element present. (Ch. 1)

chemical kinetics The study of the rates of change of concentrations of substances involved in chemical reactions. (Ch. 13)

chemical nomenclature The rules that are followed in naming substances. (Ch. 2)

chemical property A property of a substance that can be observed only by reacting it to form another substance. (Ch. 1)

chemical reaction The transformation of one or more substances into different substances. (Ch. 1)

chemistry The study of the composition, structure, and properties of matter, and of the energy consumed or given off during electrochemical reactions. (Ch. 1)

chirality A property of a molecule that is not superimposable on its mirror image. (Ch. 9)

cis isomer (also called *Z isomer*) A molecule with two like groups (such as two R groups or two hydrogen atoms) on the same side of the molecule. (Ch. 20)

Clausius–Clapeyron equation A relationship between the vapor pressure of a substance at two temperatures and its heat of vaporization. (Ch. 10)

closed system A system that exchanges energy but not matter with the surroundings. (Ch. 6)

codon A three-nucleotide sequence that codes for a specific amino acid. (Ch. 20)

coenzyme An organic molecule that, like an enzyme, accelerates the rate of biochemical reactions. (Ch. 22)

colligative properties Characteristics of solutions that depend on the concentration and not the identity of particles dissolved in the solvent. (Ch. 11)

combination reaction A reaction in which two (or more) substances combine to form one product. (Ch. 3)

combined gas law (also called *general gas equation*) The principle relating the pressure, volume, and temperature of a quantity of an ideal gas:

$$\frac{P_1V_1}{T_1} = \frac{P_2V_2}{T_2}$$

(Ch. 5)

combustion analysis A laboratory procedure for determining the composition of a substance by burning it completely in excess oxygen to produce known compounds whose masses are used to determine the composition of the original material. (Ch. 3)

combustion reaction A heat-producing reaction between oxygen and another element or compound. (Ch. 3)

common-ion effect A shift in the position of an equilibrium caused by an ion taking part in the reaction. (Ch. 16)

complex ion An ionic species consisting of a metal ion bonded to one or more Lewis bases. (Ch. 16)

compound A pure substance that is composed of two or more elements combined in fixed proportions and that can be broken down into those elements by a chemical reaction. (Ch. 1)

concentration The amount of a solute in a given mass or volume of solvent or solution. (Ch. 4)

condensation polymer A polymer formed by a condensation reaction. (Ch. 12)

condensation reaction A reaction in which two molecules combine to form a larger molecule and a small molecule, typically water. (Ch. 12)

conduction band In metals, an unoccupied band higher in energy than a valence band, in which electrons are free to migrate. (Ch. 9)

confidence interval A range of values that has a specified probability of containing the true value of a measurement. (Ch. 1)

conjugate acid The acid formed when a Brønsted–Lowry base accepts a H^+ ion. (Ch. 15)

conjugate acid–base pair A Brønsted–Lowry acid and base that differ from each other only by a H^+ ion: acid $\rightleftharpoons$ conjugate base + H^+. (Ch. 15)

conjugate base The base formed when a Brønsted–Lowry acid donates a H^+ ion. (Ch. 15)

constitutional isomers Molecules with the same chemical formula but different connections between the atoms. (Ch. 20)

conversion factor A fraction in which the numerator is equivalent to the denominator, even though they are expressed in different units, making the value of the fraction one. (Ch. 1)

coordinate bond A covalent bond formed when one anion or molecule donates a pair of electrons to another ion or molecule. (Ch. 16)

coordination compound A compound made up of at least one complex ion. (Ch. 22)

coordination number The number of sites occupied by ligands around a metal ion in a complex. (Ch. 22)

copolymer A polymer formed from the chemical reaction of two different monomers. (Ch. 12)

core electrons Electrons in the filled, inner shells of an atom or ion that are not involved in chemical reactions. (Ch. 7)

corrosion A process in which a metal is oxidized by substances in its environment. (Ch. 18)

counterion An ion whose charge balances the charge of a complex ion in a coordination compound. (Ch. 22)

covalent bond A bond between two atoms created by sharing one or more pairs of electrons. (Ch. 2)

covalent network solid A solid formed by covalent bonds among nonmetal atoms in an extended array. (Ch. 12)

critical mass The minimum quantity of fissionable material needed to sustain a chain reaction. (Ch. 19)

critical point The temperature and pressure at and above which the liquid and gas phases of a substance are indistinguishable from each other. (Ch. 10)

crystal field splitting The separation of a set of *d* orbitals into subsets with different energies as a result of interactions between electrons in those orbitals and lone pairs of electrons in ligands. (Ch. 22)

crystal field splitting energy (Δ) The difference in energy between subsets of *d* orbitals split by interactions in a crystal field. (Ch. 22)

crystal lattice A three-dimensional repeating array of particles (atoms, ions, or molecules) in a crystalline solid. (Ch. 12)

crystal structure A particular arrangement in three-dimensional space that specifies the positions of the particles (atoms, ions, or molecules) in relation to one another in a crystalline solid. (Ch. 12)

crystalline solid A solid made of an ordered array of atoms, ions, or molecules. (Ch. 12)

cubic closest-packed (ccp) A crystal lattice in which the layers of atoms or ions in face-centered cubic unit cells have an *abcabc . . .* stacking pattern. (Ch. 12)

curie (Ci) A non-SI unit of radioactivity in which 3.70×10^{10} radioactive atoms decay per second. (Ch. 19)

D

dalton (Da) A unit of mass identical to 1 unified atomic mass unit (u); thus 1 Da = 1 u. (Ch. 2)

Dalton's law of partial pressures The principle that the total pressure of any mixture of gases equals the sum of the partial pressure of each gas in the mixture. (Ch. 5)

degenerate orbitals Describes orbitals of the same energy. (Ch. 7)

degree of ionization The ratio of the quantity of a substance that is ionized to the concentration of the substance before ionization; when expressed as a percentage, called *percent ionization.* (Ch. 15)

delocalized electrons Electrons that are shared among more than two atoms. (Ch. 8)

density (d) The ratio of the mass (m) of an object to its volume (V). (Ch. 1)

deposition Transformation of a vapor (gas) directly into a solid. (Ch. 1)

diamagnetic A substance with no unpaired electrons that is weakly repelled by a magnetic field. (Ch. 9)

diffusion The spread of one substance (usually a gas or liquid) through another. (Ch. 5)

dilution The process of lowering the concentration of a solution by adding more solvent. (Ch. 4)

dipole moment (μ) A measure of the degree to which a molecule aligns itself in a strong electric field; a quantitative expression of the polarity of a molecule. (Ch. 9)

dipole–dipole interaction An attractive force between polar molecules. (Ch. 10)

dipole–induced dipole interaction An attraction between a polar molecule and the oppositely charged pole it temporarily induces in another molecule. (Ch. 10)

dispersion force (also called *London force*) An intermolecular force between molecules caused by the presence of temporary dipoles in the molecules. (Ch. 10)

double bond A chemical bond in which two atoms share two pairs of electrons. (Ch. 8)

E

***E* isomer** See *trans isomer.*

effective nuclear charge (Z_{eff}) The attraction toward the nucleus experienced by a valence-shell electron, which is approximately equal to the sum of the positive charge on the nucleus and the total negative charge of all inner-shell electrons. (Ch. 7)

effusion The process by which a gas escapes from its container through a tiny hole into a region of lower pressure. (Ch. 5)

electrochemical cell An apparatus that can either generate electrical energy from a chemical reaction or use electrical energy to produce a chemical reaction. (Ch. 18)

electrochemistry The branch of chemistry that examines the transformations between chemical and electrical energy. (Ch. 18)

electrolysis A process in which electrical energy is used to drive a nonspontaneous chemical reaction. (Ch. 18)

electrolyte A substance that dissociates into ions when it dissolves in water, enhancing the conductivity of the solvent. (Ch. 4)

electrolytic cell A device in which an external source of electrical energy does work on a chemical system, turning reactant(s) into higher-energy product(s). (Ch. 18)

electromagnetic radiation Any form of radiant energy in the electromagnetic spectrum. (Ch. 7)

electromagnetic spectrum A continuous range of radiant energy that includes radio waves, microwaves, infrared radiation, visible light, ultraviolet radiation, X-rays, and gamma rays. (Ch. 7)

electron A subatomic particle that has a negative charge and little mass. (Ch. 2)

electron affinity (EA) The energy change that occurs when 1 mole of electrons combines with 1 mole of atoms or monatomic cations in the gas phase. (Ch. 7)

electron capture A nuclear reaction in which a neutron-poor nucleus draws in one of its surrounding electrons, which transforms a proton in the nucleus into a neutron. (Ch. 19)

electron configuration The distribution of electrons among the orbitals of an atom or ion. (Ch. 7)

electron transition The movement of an electron between energy levels. (Ch. 7)

electronegativity A relative measure of the ability of an atom to attract electrons in a bond to itself. (Ch. 8)

electron-pair geometry The three-dimensional arrangement of bonding pairs and lone pairs of electrons about a central atom. (Ch. 9)

electrostatic potential energy (E_{el}) The energy a particle has because of its electrostatic charge and its position with respect to another particle; it is directly proportional to the product of the charges of the particles and inversely proportional to the distance between them. (Ch. 6)

element A pure substance that cannot be separated into simpler substances. (Ch. 1)

elementary step A molecular view of a collision taking place in a chemical reaction. (Ch. 13)

empirical formula A formula showing the smallest whole-number ratio of the elements in a compound. (Ch. 2)

enantiomer One of a pair of optical isomers of a compound. (Ch. 20)

end point The point in a titration that is reached when just enough standard solution has been added to cause the indicator to change color. (Ch. 4)

endothermic process A thermochemical or physical process in which energy flows from the surroundings into the system. (Ch. 6)

energy The capacity to do work. (Ch. 1)

energy profile A graph showing the changes in energy for a reaction as a function of the progress of the reaction from reactants to products. (Ch. 13)

enthalpy (H) The sum of the internal energy and the pressure–volume product of a system; $H = E + PV$. (Ch. 6)

enthalpy change (ΔH) The heat absorbed by an endothermic process or given off by an exothermic process occurring at constant pressure. (Ch. 6)

enthalpy of hydration ($\Delta H_{hydration}$) The energy change when gas-phase ions dissolve in water. (Ch. 11)

enthalpy of reaction (ΔH_{rxn}) The energy absorbed or given off by a chemical reaction under conditions of constant pressure; also called *heat of reaction.* (Ch. 6)

enthalpy of solution ($\Delta H_{\text{solution}}$) The overall change in enthalpy that occurs when a solute dissolves in a solvent. (Ch. 11)

entropy (S) A measure of the dispersion of energy in a system at a specific temperature. (Ch. 17)

enzyme A protein that catalyzes a reaction. (Ch. 13)

equilibrium constant (K) The numerical value of the equilibrium constant expression of a reversible chemical reaction at a particular temperature. (Ch. 14)

equilibrium constant expression The ratio of the concentrations or partial pressures of products to reactants at equilibrium, with each term raised to a power equal to the coefficient of that substance in the balanced chemical equation for the reaction. (Ch. 14)

equilibrium (reversal) potential (E_{ion}) An electrochemical potential that results from a concentration gradient of a particular ion on opposite sides of a cell membrane. (Ch. 21)

equivalence point The point in a titration at which just enough titrant has been added to completely react the substance being analyzed. (Ch. 4)

essential amino acid Any of the ten amino acids that make up peptides and proteins but are not synthesized in the human body and must be obtained through the food we eat. (Ch. 20)

essential element An element present in tissue, blood, or other body fluids that has a physiological function. (Ch. 21)

ester An organic compound in which the –OH of a carboxylic acid group is replaced by –OR, where R can be any organic group. (Ch. 12)

ether An organic compound whose moecules have an R—O—R′ structure. (Ch. 10)

excited state Any energy state above the ground state in an atom, ion, or molecule. (Ch. 7)

exothermic process A thermochemical or physical process in which energy flows from a system into its surroundings. (Ch. 6)

extensive property A property that varies with the quantity of the substance present. (Ch. 1)

F

face-centered cubic (fcc) unit cell An array of closest-packed particles that has one particle at each of the eight corners of a cube and one more at the center of each face of the cube. (Ch. 12)

Faraday constant (F) The magnitude of electrical charge in 1 mole of electrons; its value to three significant figures is 9.65×10^4 C/mol e^-. (Ch. 18)

fat A solid triglyceride containing primarily saturated fatty acids. (Ch. 20)

fatty acid A carboxylic acid with a long hydrocarbon chain that may be either saturated or unsaturated. (Ch. 20)

first law of thermodynamics The principle that the energy gained or lost by a system must equal the energy lost or gained by its surroundings. (Ch. 6)

food value The quantity of energy produced when a material consumed by an organism for sustenance is burned completely; it is typically reported in Calories (kilocalories) per gram of food. (Ch. 6)

formal charge (FC) A value calculated for an atom in a molecule or polyatomic ion by determining the difference between the number of valence electrons in the free atom and the sum of lone-pair electrons plus half of the electrons in the atom's bonding pairs. (Ch. 8)

formation constant (K_f) An equilibrium constant describing the formation of a metal complex from a free metal ion and its ligands. (Ch. 16)

formation reaction A reaction in which 1 mole of a substance is formed from its component elements in their standard states. (Ch. 6)

formula mass The mass of one formula unit of an ionic compound. (Ch. 3)

formula unit The smallest electrically neutral unit of an ionic compound. (Ch. 2)

fractional distillation A method of separating a mixture of compounds based on their different boiling points. (Ch. 11)

Fraunhofer lines A set of dark lines in the otherwise continuous solar spectrum. (Ch. 7)

free radical An odd-electron molecule with an unpaired electron in its Lewis structure. (Ch. 8)

frequency (ν) The number of crests of a wave that pass a stationary point of reference per second. (Ch. 7)

frequency factor (A) The product of the frequency of molecular collisions and a factor that expresses the probability that the orientation of the molecules is appropriate for a reaction to occur. (Ch. 13)

fuel cell A voltaic cell based on the oxidation of a continuously supplied fuel; the reaction is the equivalent of combustion, but chemical energy is converted directly into electrical energy. (Ch. 18)

fuel density The quantity of energy released during the complete combustion of 1 L of a liquid fuel. (Ch. 6)

fuel value The quantity of energy released during the complete combustion of 1 g of a substance. (Ch. 6)

functional group A group of atoms in the molecular structure of an organic compound that imparts characteristic chemical and physical properties. (Ch. 2)

G

gas A form of matter that has neither definite volume nor shape and that expands to fill its container; also called *vapor.* (Ch. 1)

gauge pressure The pressure of a gas that is above atmospheric pressure. (Ch. 5)

Geiger counter A portable device for determining nuclear radiation levels by measuring how much the radiation ionizes the gas in a sealed detector. (Ch. 19)

general gas equation See *combined gas law.*

Gibbs free energy (G) The maximum energy released by a process occurring at constant temperature and pressure that is available to do useful work. (Ch. 17)

glyceride A lipid consisting of esters formed between fatty acids and the alcohol glycerol. (Ch. 20)

glycolysis A series of reactions that converts glucose into pyruvate; a major anaerobic (no oxygen required) pathway for the metabolism of glucose in the cells of almost all living organisms. (Ch. 17)

glycosidic bond A C—O—C bond between sugar molecules. (Ch. 20)

Graham's law of effusion The principle that the rate of effusion of a gas is inversely proportional to the square root of its molar mass. (Ch. 5)

gray (Gy) The SI unit of absorbed radiation; 1 Gy = 1 J/kg of tissue. (Ch. 19)

ground state The most stable, lowest-energy state available to an atom, ion, or molecule. (Ch. 7)

group All the elements in the same column of the periodic table; also called *family.* (Ch. 2)

Grubbs' test A statistical test used to detect an outlier in a set of data. (Ch. 1)

H

half-life ($t_{1/2}$) The time in the course of a chemical reaction during which the concentration of a reactant decreases by half. (Ch. 13)

half-reaction One of the two halves of an oxidation–reduction reaction; one half-reaction is the oxidation component, and the other is the reduction component. (Ch. 4)

halogens The elements in group 17 of the periodic table. (Ch. 2)

heat The energy transferred between objects because of a difference in their temperatures. (Ch. 5)

heat capacity (C_P) The quantity of energy needed to raise the temperature of an object by 1°C at constant pressure. (Ch. 6)

Heisenberg uncertainty principle The principle that we cannot determine both the position and the momentum of a particle in an atom at the same time. (Ch. 7)

Henderson–Hasselbalch equation An equation used to calculate the pH of solutions containing comparable concentrations of acid and conjugate base. (Ch. 16)

Henry's law The principle that the concentration of a sparingly soluble, chemically unreactive gas in a liquid is proportional to the partial pressure of the gas. (Ch. 10)

hertz (Hz) The SI unit of frequency, equivalent to 1 cycle per second, or simply 1/s. (Ch. 7)

Hess's law The principle that the enthalpy of reaction ΔH_{rxn} for a reaction that is the sum of two or more reactions is equal to the sum of the ΔH_{rxn} values of the constituent reactions; also called *Hess's law of constant heat of summation*. (Ch. 6)

heteroatom An atom of an element other than carbon and hydrogen within a molecule of an organic compound. (Ch. 2)

heterogeneous catalyst A catalyst in a different phase from the reactants. (Ch. 13)

heterogeneous equilibria Equilibria that involve reactants and products in more than one phase. (Ch. 14)

heterogeneous mixture A mixture in which the components are not distributed uniformly, so that the mixture contains distinct regions of different compositions. (Ch. 1)

heteropolymer A polymer made of three or more different monomer units. (Ch. 12)

hexagonal closest-packed (hcp) A crystal lattice in which the layers of atoms or ions in hexagonal unit cells have an *ababab* . . . stacking pattern. (Ch. 12)

hexagonal unit cell An array of closest-packed particles that includes parts of four particles on the top and four on the bottom faces of a hexagonal prism and one particle in a middle layer. (Ch. 12)

homogeneous catalyst A catalyst in the same phase as the reactants. (Ch. 13)

homogeneous equilibria Equilibria that involve reactants and products in the same phase. (Ch. 14)

homogeneous mixture A mixture in which the components are distributed uniformly throughout and have no visible boundaries or regions. (Ch. 1)

homopolymer A polymer composed of only one kind of monomer. (Ch. 12)

Hund's rule The lowest-energy electron configuration of an atom has the maximum number of unpaired electrons, all of which have the same spin, in degenerate orbitals. (Ch. 7)

hybrid atomic orbital In valence bond theory, one of a set of equivalent orbitals about an atom created when specific atomic orbitals are mixed. (Ch. 9)

hybridization In valence bond theory, the mixing of atomic orbitals to generate new sets of orbitals that are then available to form covalent bonds with other atoms. (Ch. 9)

hydrocarbon An organic compound whose molecules are composed only of carbon and hydrogen atoms. (Ch. 2)

hydrogen bond The strongest dipole–dipole interaction. It occurs between a hydrogen atom bonded to a small, highly electronegative element (O, N, F) and an atom of oxygen or nitrogen in another polar molecule. Molecules of HF also form hydrogen bonds. (Ch. 10)

hydrolysis The reaction of water with another material. The hydrolysis of nonmetal oxides produces acids. (Ch. 4)

hydronium ion (H_3O^+) A H^+ ion bonded to a molecule of water, H_2O; the chemical form of a hydrogen ion in an aqueous solution. (Ch. 4)

hydrophilic A "water-loving," or attractive, interaction between a solute and water that promotes water solubility. (Ch. 10)

hydrophobic A "water-fearing," or repulsive, interaction between a solute and water that diminishes water solubility. (Ch. 10)

hypothesis A tentative and testable explanation for an observation or a series of observations. (Ch. 1)

I

***i* factor** See *van 't Hoff factor.*

ideal gas A gas whose behavior is predicted by the linear relationships defined by Boyle's, Charles's, Avogadro's, and Amontons's laws. (Ch. 5)

ideal gas equation (also called *ideal gas law*) The principle relating the pressure, volume, number of moles, and temperature of an ideal gas; expressed as $PV = nRT$, where R is the universal gas constant. (Ch. 5)

ideal gas law See *ideal gas equation*.

ideal solution A solution that obeys Raoult's law. (Ch. 11)

induced dipole The separation of charge produced in an atom or molecule by a momentary uneven distribution of electrons. (Ch. 10)

inhibitor A compound that diminishes or destroys the ability of an enzyme to catalyze a reaction. (Ch. 20)

initial rate The rate of a reaction at $t = 0$, immediately after the reactants are mixed. (Ch. 13)

inner coordination sphere The ligands that are bound directly to a metal via coordinate bonds. (Ch. 22)

integrated rate law A mathematical expression that describes the change in concentration of a reactant in a chemical reaction with time. (Ch. 13)

intensive property A property that is independent of the amount of substance present. (Ch. 1)

intermediate A species produced in one step of a reaction and consumed in a subsequent step. (Ch. 13)

internal energy (E) The sum of all the kinetic and potential energies of all the components of a system. (Ch. 6)

interstitial alloy An alloy in which atoms of the added element occupy the spaces between atoms of the host. (Ch. 12)

ion A particle consisting of one or more atoms that has a net positive or negative electrical charge. (Ch. 1)

ion channel A group of helical proteins that penetrate cell membranes and allow selective transport of ions. (Ch. 21)

ion exchange A process by which one ion is displaced by another. (Ch. 11)

ion pair A cluster formed when a cation and an anion associate with each other in solution. (Ch. 11)

ion pump A system of membrane proteins that exchange ions inside the cell with those in the intercellular fluid. (Ch. 21)

ion–dipole interaction An attractive force between an ion and a molecule that has a permanent dipole moment. (Ch. 10)

ionic bond A chemical bond that results from the electrostatic attraction between a cation and an anion. (Ch. 8)

ionic character An estimate of the magnitude of charge separation in a covalent bond. (Ch. 8)

ionic compound A compound composed of positively and negatively charged ions held together by electrostatic attraction. (Ch. 2)

ionic solid A solid formed by ionic bonds among monatomic or polyatomic ions. (Ch. 12)

ionization energy (IE) The quantity of energy needed to remove 1 mole of electrons from 1 mole of ground-state atoms or ions in the gas phase. (Ch. 7)

ionizing radiation High-energy products of radioactive decay that can ionize molecules. (Ch. 19)

isoelectric point See *pI.*

isoelectronic Atoms or ions having identical electron configurations. (Ch. 7)

isolated system A system that exchanges neither energy nor matter with the surroundings. (Ch. 6)

isomers Compounds with the same molecular formula but different arrangements of the atoms in their molecules. (Ch. 3)

isotopes Atoms of the same element that contain different numbers of neutrons. (Ch. 2)

J

joule (J) The SI unit of energy; 4.184 J = 1 cal. (Ch. 6)

K

kelvin (K) The SI unit of temperature. (Ch. 1)

ketone Organic molecule containing a carbonyl group in which the carbonyl carbon is bonded to two other carbon atoms. (Ch. 9)

kinetic energy The energy of an object in motion due to its mass (m) and its speed (u): $KE = \frac{1}{2}mu^2$. (Ch. 5)

kinetic molecular theory A model that describes the behavior of ideal gases; all equations defining relationships between pressure, volume, temperature, and number of moles of gases can be derived from the theory. (Ch. 5)

L

lattice energy (U) The enthalpy change that occurs when 1 mole of an ionic compound forms from its free ions in the gas phase. (Ch. 11)

law of conservation of energy The principle that energy cannot be created or destroyed but can be converted from one form into another. (Ch. 6)

law of conservation of mass The principle that the sum of the masses of the reactants in a chemical reaction is equal to the sum of the masses of the products. (Ch. 3)

law of constant composition The principle that all samples of a compound contain the same elements combined in the same proportions. (Ch. 1)

law of mass action The principle relating the balanced chemical equation of a reversible reaction to its mass action expression (or equilibrium constant expression). (Ch. 14)

law of multiple proportions The principle that, when two masses of one element react with a given mass of another element to form two different compounds, the two masses of the first element have a ratio of two small whole numbers. (Ch. 2)

Le Châtelier's principle The principle that a system at equilibrium responds to a stress in such a way that it relieves that stress. (Ch. 14)

leveling effect The observation that all strong acids have the same strength in water and are completely converted into solutions of H_3O^+ ions; strong bases are likewise leveled in water and are completely converted into solutions of OH^- ions. (Ch. 15)

Lewis acid A substance that *accepts* a lone pair of electrons in a chemical reaction. (Ch. 16)

Lewis base A substance that *donates* a lone pair of electrons in a chemical reaction. (Ch. 16)

Lewis dot symbol See *Lewis symbol.*

Lewis structure A two-dimensional representation of the bonds and lone pairs of valence electrons in a molecule or polyatomic ion. (Ch. 8)

Lewis symbol (also called *Lewis dot symbol*) A system in which the chemical symbol for an element is surrounded by one or more dots representing each of its atom's valence electrons. (Ch. 8)

ligand A Lewis base bonded to the central metal ion of a complex ion. (Ch. 16)

limiting reactant A reactant that is consumed completely in a chemical reaction. The amount of product formed depends on the amount of the limiting reactant available. (Ch. 3)

linear The molecular geometry about a central atom with a steric number of 2 and no lone pairs of electrons; the bond angle is 180°. (Ch. 9)

lipid A class of water-insoluble, oily organic compounds that are common structural materials in cells. (Ch. 20)

lipid bilayer A double layer of molecules whose polar head groups interact with water molecules and whose nonpolar tails interact with each other. (Ch. 20)

liquid A form of matter that occupies a definite volume but flows to assume the shape of its container. (Ch. 1)

London force See *dispersion force.*

lone pair A pair of electrons that is not shared. (Ch. 8)

M

macrocyclic ligand A ring containing multiple electron-pair donors that bind to a metal ion. (Ch. 22)

macromolecule See *polymer.*

magnetic quantum number (m_ℓ) An integer that may have any value from $-\ell$ to $+\ell$, where ℓ is the angular momentum quantum number; it defines the orientation of an orbital in space. (Ch. 7)

main group elements (also called *representative elements*) The elements in groups 1, 2, and 13 through 18 of the periodic table. (Ch. 2)

major essential element An essential element present in the body in average concentrations greater than 1 mg of element per gram of body mass, resulting in gram quantities in the human body. (Ch. 21)

manometer An instrument for measuring the pressure exerted by a gas. (Ch. 5)

mass The property that defines the quantity of matter in an object. (Ch. 1)

mass action expression An expression equivalent in form to the equilibrium constant expression but applied to reaction mixtures that may or may not be at equilibrium. (Ch. 14)

mass defect (Δm) The difference between the mass of a nucleus and the masses of the individual nucleons that make it up. (Ch. 19)

mass number (A) The number of nucleons in an atom. (Ch. 2)

mass spectrometer An instrument that separates and counts ions according to their mass-to-charge ratios. (Ch. 3)
mass spectrum A graph of the data from a mass spectrometer. (Ch. 3)
matter Anything that has mass and occupies space. (Ch. 1)
matter wave The wave associated with any particle. (Ch. 7)
mean (arithmetic mean, $\bar{x}$) An average calculated by summing a set of related values and dividing the sum by the number of values in the set. (Ch. 1)
membrane potential ($E_{membrane}$) A weighted average of the equilibrium (reversal) potentials of the major ions based on concentration and permeability of the individual ions. (Ch. 21)
meniscus The concave or convex surface of a liquid in a small-diameter tube. (Ch. 10)
messenger RNA (mRNA) The form of RNA that carries the code for synthesizing proteins from DNA to the site of protein synthesis in a cell. (Ch. 20)
metallic bond A chemical bond consisting of metal atoms sharing a "sea" of electrons. (Ch. 8)
metallic solid A solid formed by metallic bonds among atoms of metallic elements. (Ch. 12)
metalloids (also called *semimetals*) The elements along the border between the metals and nonmetals in the periodic table; they have some metallic and some nonmetallic properties. (Ch. 2)
metals The elements on the left side of the periodic table that are typically shiny solids that conduct heat and electricity well and are malleable and ductile. (Ch. 2)
meter The standard unit of length, equivalent to 39.37 inches. (Ch. 1)
methyl group ($—CH_3$) A structural unit that can make only one bond. (Ch. 6)
methylene group ($—CH_2—$) A structural unit that can make two bonds. (Ch. 6)
microstate A unique distribution of particles among energy levels. (Ch. 17)
miscible Capable of being mixed in any proportion. (Ch. 5)
mixture A combination of pure substances in variable proportions in which the individual substances retain their chemical identities and can be separated from one another by a physical process. (Ch. 1)
molality (m) Concentration expressed as the number of moles of solute per kilogram of solvent. (Ch. 11)
molar absorptivity (ε) A measure of how well a compound or ion absorbs light. (Ch. 4)
molar enthalpy of fusion (ΔH_{fus}) The energy required to convert 1 mole of a solid substance at its melting point into the liquid state. (Ch. 6)
molar enthalpy of vaporization (ΔH_{vap}) The energy required to convert 1 mole of a liquid substance at its boiling point to the vapor state. (Ch. 6)
molar heat capacity (c_P) The quantity of energy required to raise the temperature of 1 mole of a substance by 1°C at constant pressure. (Ch. 6)
molar mass ($\mathcal{M}$) The mass of 1 mole of a substance; the molar mass of an element in grams per mole is numerically equal to that element's average atomic mass in unified atomic mass units. (Ch. 3)
molar volume The volume occupied by 1 mole of an ideal gas at STP; 22.4 L. (Ch. 5)
molarity (M) The concentration of a solution expressed in moles of solute per liter of solution ($M = n/V$). (Ch. 4)
mole (mol) The SI base unit for expressing quantities of substances. One mole contains $6.02214076 \times 10^{23}$ particles: atoms, molecules, or formula units. (Ch. 3)
mole fraction (X_x) The ratio of the number of moles of a component in a mixture to the total number of moles in the mixture. (Ch. 5)
molecular compound A compound composed of molecules that contain the atoms of two or more elements covalently bonded together. (Ch. 2)
molecular equation An equation describing a reaction in solution in which the reactants and products are written as neutral compounds. (Ch. 4)
molecular formula A notation showing the number and type of atoms present in one molecule of a molecular compound. (Ch. 2)
molecular geometry The three-dimensional arrangement of the atoms in a molecule. (Ch. 9)
molecular ion (M^+) The peak of highest mass in a mass spectrum; it has the same mass as the molecule from which it came. (Ch. 3)
molecular mass The mass of one molecule of a molecular compound. (Ch. 3)
molecular orbital A region of characteristic shape and energy where electrons in a molecule are delocalized over two or more atoms in a molecule. (Ch. 9)
molecular orbital diagram An energy-level diagram showing the relative energies and electron occupancy of the molecular orbitals for a molecule. (Ch. 9)
molecular orbital (MO) theory A bonding theory based on the mixing of atomic orbitals of similar shapes and energies to form molecular orbitals that extend across two or more atoms. (Ch. 9)
molecular recognition The process by which molecules interact with other molecules in living tissues to produce a biological effect. (Ch. 9)
molecular solid A solid formed by intermolecular attractive forces among neutral, covalently bonded molecules. (Ch. 12)
molecularity The number of ions, atoms, or molecules involved in an elementary step in a reaction. (Ch. 13)
molecule A collection of atoms chemically bonded together in characteristic proportions. (Ch. 1)
monodentate ligand A species that forms only a single coordinate bond to a metal ion in a complex. (Ch. 22)
monomer A small molecule that bonds with others like it to form polymers. (Ch. 12)
monoprotic acid An acid that has one ionizable hydrogen atom per molecule. (Ch. 15)
monosaccharide A single-sugar unit and the simplest carbohydrate. (Ch. 20)

N

nanoparticle An approximately spherical sample of matter with dimensions smaller than 100 nm (1×10^{-7} m). (Ch. 12)
natural abundance The proportion of an isotope, usually expressed in percent, relative to all the isotopes of that element in a natural sample. (Ch. 2)
Nernst equation An equation relating the potential of a cell (or half-cell) reaction to its standard potential ($E°$) and to the concentrations of its reactants and products. (Ch. 18)
net ionic equation An equation describing a reaction in solution that contains only the species changed by the reaction; it is obtained by eliminating the spectator ions from the overall ionic equation. (Ch. 4)
neutralization reaction A reaction that takes place when an acid reacts with a base and produces a solution of a salt in water. (Ch. 4)
neutron An electrically neutral (uncharged) subatomic particle found in the nucleus of an atom. (Ch. 2)

neutron capture The absorption of a neutron by a nucleus. (Ch. 2)
noble gases The elements in group 18 of the periodic table. (Ch. 2)
node A location in a standing wave that experiences no displacement; in the context of orbitals, nodes are locations at which electron density goes to zero. (Ch. 7)
nonelectrolyte A substance that does *not* dissociate into ions and therefore does *not* result in conductivity when dissolved in water. (Ch. 4)
nonessential element An element present in humans that has no known function. (Ch. 21)
nonmetals The elements with properties opposite those of metals, including poor conductivity of heat and electricity. (Ch. 2)
nonspontaneous process A process that occurs only as long as energy is continually added to the system. (Ch. 17)
normal boiling point The temperature at which the vapor pressure of a liquid equals 1 atm (760 torr). (Ch. 10)
n-type semiconductor A semiconductor containing electron-rich dopant atoms that contribute excess electrons. (Ch. 9)
nuclear chemistry A subdiscipline of chemistry that explores the properties and reactivity of atomic nuclei. (Ch. 19)
nuclear fission A nuclear reaction in which the nucleus of an element splits into two lighter nuclei; the process is usually accompanied by the release of one or more neutrons and energy. (Ch. 19)
nuclear fusion A nuclear reaction in which subatomic particles or atomic nuclei collide with each other at very high speeds and fuse, forming more massive nuclei and releasing energy. (Ch. 19)
nuclear reaction A process that changes the number of neutrons or protons in the nucleus of an atom. (Ch. 19)
nucleic acid One of a family of large molecules; nucleic acids include deoxyribonucleic acid (DNA) and ribonucleic acid (RNA), molecules that store the genetic blueprint of an organism and control the production of proteins. (Ch. 20)
nucleon Either a proton or a neutron in a nucleus. (Ch. 2)
nucleosynthesis The natural formation of nuclei because of fusion and other nuclear processes. (Ch. 2)
nucleotide A monomer unit from which nucleic acids are made. (Ch. 20)
nucleus The positively charged center of an atom that contains nearly all the atom's mass. (Ch. 2)
nuclide An atom with particular numbers of neutrons and protons in its nucleus. (Ch. 2)

O

octahedral The molecular geometry about a central atom with a steric number of 6 and no lone pairs of electrons, in which all six sites are equivalent. (Ch. 9)
octet rule The tendency of atoms of main group elements to make bonds by gaining, losing, or sharing electrons to achieve a valence shell containing eight electrons (i.e., four electron pairs). (Ch. 8)
oil A liquid triglyceride containing primarily unsaturated fatty acids. (Ch. 20)
open system A system that exchanges both energy and matter with the surroundings. (Ch. 6)
optical activity A property of a compound that refers to the rotation of a beam of plane-polarized light when it passes through a solution of the compound. (Ch. 20)
optical isomers Molecules that are not superimposable on their mirror images. (Ch. 20)
orbital A region around the nucleus of an atom where the probability of finding an electron is high; each orbital is defined by the square of the wave function (ψ^2) and is identified by a unique combination of three quantum numbers. (Ch. 7)
orbital diagram Depiction of the arrangement of electrons in an atom or ion, using boxes to represent orbitals. (Ch. 7)
organic chemistry The study of organic compounds. (Ch. 2)
organic compound A compound composed of molecules containing carbon combined with hydrogen and other elements, such as nitrogen, oxygen, phosphorous, and sulfur. (Ch. 2)
organometallic compound A molecule containing direct carbon–metal covalent bonds. (Ch. 22)
osmosis The flow of a fluid through a semipermeable membrane to balance the concentration of solutes in solutions on the two sides of the membrane. The solvent particles' flow proceeds from the more dilute solution into the more concentrated one. (Ch. 11)
osmotic pressure (π) The pressure applied across a semipermeable membrane to stop the flow of solvent from the compartment containing pure solvent or a less concentrated solution to the compartment containing a more concentrated solution. The osmotic pressure of a solution increases with solute concentration, M, and with solution temperature, T. (Ch. 11)
outlier A data point that is distant from the other observations. (Ch. 1)
overall ionic equation An equation describing a reaction in solution that shows all the species, both ionic and molecular, present in the reaction mixture. (Ch. 4)
overall reaction order The sum of the exponents of the concentration terms in the rate law. (Ch. 13)
overlap The degree to which atomic orbitals on adjoining atoms occupy the same region of space. (Ch. 9)
oxidation A chemical change in which a species loses electrons; the oxidation number of the species increases. (Ch. 4)
oxidation number (O.N.) (also called *oxidation state*) A positive or negative number based on the number of electrons an atom gains or loses when it forms an ion, or that it shares when it forms a covalent bond with another element; pure elements have an oxidation number of zero. (Ch. 4)
oxidation state See *oxidation number (O.N.)*.
oxidizing agent A substance in a redox reaction that contains the element being reduced. (Ch. 4)
oxoanion A polyatomic ion that contains oxygen in combination with one or more other elements. (Ch. 2)

P

packing efficiency The percentage of the total volume of a unit cell occupied by the atoms, ions, or molecules. (Ch. 12)
paramagnetic A substance with one or more unpaired electrons that is attracted to a magnetic field. (Ch. 9)
partial pressure The contribution to the total pressure made by one gas in a mixture of gases. (Ch. 5)
Pauli exclusion principle The principle that no two electrons in an atom can have the same set of four quantum numbers. (Ch. 7)
peptide A compound of two or more amino acids joined by peptide bonds; small peptides containing up to 20 amino acids are oligopeptides, whereas polypeptides are chains longer than 20 amino acids but shorter than proteins. (Ch. 20)

peptide bond The result of a condensation reaction between the carboxylic acid group of one amino acid and the amine group of another. (Ch. 20)

percent composition The composition of a compound expressed in terms of the percentage by mass of each element in the compound. (Ch. 3)

percent ionization See *degree of ionization.*

percent yield The ratio, expressed as a percentage, of the actual yield of a chemical reaction to the theoretical yield. (Ch. 3)

period A horizontal row in the periodic table. (Ch. 2)

periodic table of the elements A chart of the elements arranged in order of their atomic numbers and in a pattern based on their physical and chemical properties. (Ch. 2)

pH The negative logarithm of the hydronium ion concentration in an aqueous solution. (Ch. 15)

pH buffer A solution that resists changes in pH when acids or bases are added to it; typically a solution of a weak acid and its conjugate base or a weak base and its conjugate acid. (Ch. 16)

pH indicator A water-soluble weak organic acid that changes color as it ionizes. (Ch. 16)

phase diagram A graphical representation of how the stabilities of the physical states of a substance depend on temperature and pressure. (Ch. 10)

phospholipid A molecule of glycerol with two fatty acid chains and one polar group containing a phosphate; phospholipids are major constituents of cell membranes. (Ch. 20)

phosphorylation A reaction resulting in the addition of a phosphate group to an organic molecule. (Ch. 17)

photochemical smog A mixture of gases formed in the lower atmosphere when sunlight interacts with compounds produced in internal combustion engines and other pollution sources. (Ch. 13)

photoelectric effect The phenomenon of light striking a metal surface and producing an electric current (a flow of electrons). (Ch. 7)

photon A quantum of electromagnetic radiation. (Ch. 7)

physical process A transformation of a sample of matter, such as a change in its physical state, that does not alter the chemical identity of any substance in the sample. (Ch. 1)

physical property A property of a substance that can be observed without changing it into another substance. (Ch. 1)

p*I* (also called *isoelectric point*) The pH at which molecules of the amino acid have, on average, zero charge. (Ch. 20)

pi (π) bond A covalent bond in which electron density is greatest around—not along—the bonding axis; it forms from side-on overlap of two parallel *p* orbitals. (Ch. 9)

pi (π) molecular orbitals Molecular orbitals formed by the mixing of atomic orbitals oriented above and below, or in front of and behind, the bonding axis. (Ch. 9)

Planck constant (*h*) The proportionality constant between the energy and frequency of electromagnetic radiation: $E = h\nu$; $h = 6.626 \times 10^{-34}\ \text{J} \cdot \text{s}$. (Ch. 7)

pOH The negative logarithm of the hydroxide ion concentration in an aqueous solution. (Ch. 15)

polar covalent bond A bond characterized by unequal sharing of bonding pairs of electrons between atoms. (Ch. 8)

polarizability The relative ease with which the electron clouds surrounding the nuclei of individual atoms or molecules can be distorted, thereby inducing a temporary dipole. (Ch. 10)

polyatomic ion A charged group of two or more atoms joined by covalent bonds. (Ch. 2)

polydentate ligand A species that can form more than one coordinate bond per molecule. (Ch. 22)

polymer (or *macromolecule*) A very large molecule with high molar mass formed by bonding many small molecules of low molecular mass. (Ch. 12)

polyprotic acid An acid that has two or more ionizable hydrogen atoms per molecule. (Ch. 15)

polysaccharide A polymer of monosaccharides. (Ch. 20)

porphyrin A type of tetradentate macrocyclic ligand. (Ch. 22)

positron ($^{0}_{1}\beta$) A particle with the mass of an electron but with a positive charge. (Ch. 19)

positron emission The spontaneous emission of a positron from a neutron-poor nucleus. (Ch. 19)

potential energy (PE) The energy stored in an object because of its position or composition. (Ch. 6)

precipitate A solid product formed from a reaction in solution. (Ch. 4)

precipitation reaction A reaction that produces a slightly soluble product upon mixing two solutions. (Ch. 4)

precision Agreement between the results of multiple measurements that were carried out in the same way. (Ch. 1)

pressure (*P*) The ratio of a force to the surface area over which the force is applied. (Ch. 5)

pressure–volume (*P–V*) work The work associated with the expansion or compression of a gas. (Ch. 6)

primary (1°) structure The sequence in which amino acid monomers occur in a protein chain. (Ch. 20)

principal quantum number (*n*) A positive integer describing the relative size and energy of an atomic orbital or group of orbitals in an atom. (Ch. 7)

product A substance formed during a chemical reaction. (Ch. 3)

protein A biological polymer made of amino acids. (Ch. 20)

proton A positively charged subatomic particle present in the nucleus of an atom. (Ch. 2)

p-type semiconductor A semiconductor containing electron-poor dopant atoms that cause a reduction in the number of electrons, which is equivalent to the presence of positively charged holes. (Ch. 9)

pure substance Matter that has a constant composition and cannot be broken down into simpler matter by any physical process. (Ch. 1)

Q

quantized Having values restricted to whole-number multiples of a specific base value. (Ch. 7)

quantum (plural *quanta*) The smallest discrete quantity of a form of energy. (Ch. 7)

quantum mechanics See *wave mechanics.*

quantum number A number that specifies the energy, the probable location or orientation of an orbital, or the spin of an electron within an orbital. (Ch. 7)

quantum theory A model based on the idea that energy is absorbed and emitted in discrete quantities of energy called quanta. (Ch. 7)

quarks Elementary particles that combine to form neutrons and protons. (Ch. 2)

quaternary (4°) structure The larger structure functioning as a single unit that results when two or more proteins associate. (Ch. 20)

R

racemic mixture A sample containing equal amounts of both enantiomers of a compound. (Ch. 20)

radioactive decay The spontaneous disintegration of unstable particles accompanied by the release of radiation. (Ch. 19)

radioactivity The spontaneous emission of high-energy radiation and particles by materials. (Ch. 2)

radiocarbon dating A method for establishing the age of a carbon-containing object by measuring the amount of radioactive carbon-14 remaining in the object. (Ch. 19)

radiometric dating A method for determining the age of an object on the basis of the quantity of a radioactive nuclide, the products of its decay that the object contains, or both. (Ch. 19)

radionuclide An unstable nuclide that undergoes radioactive decay. (Ch. 19)

Raoult's law The principle that the vapor pressure of the solvent in a solution is equal to the vapor pressure of the pure solvent multiplied by the mole fraction of the solvent in the solution. (Ch. 11)

rate constant The proportionality constant that relates the rate of a reaction to the concentrations of reactants at a particular temperature. (Ch. 13)

rate law An equation that defines the experimentally determined relation between the concentrations of reactants in a chemical reaction and the rate of that reaction. (Ch. 13)

rate-determining step The slowest step in a multistep reaction mechanism. (Ch. 13)

reactant A substance consumed during a chemical reaction. (Ch. 3)

reaction mechanism A set of steps describing how a reaction occurs at the molecular level; the mechanism must be consistent with the experimentally determined rate law for the reaction. (Ch. 13)

reaction order An experimentally determined number defining the dependence of the reaction rate on the concentration of a reactant. (Ch. 13)

reaction quotient (Q) The numerical value of the mass action expression based on the concentrations or partial pressures of the reactants and products present at any time during a reaction; at equilibrium, $Q = K$. (Ch. 14)

redox flow battery An electrochemical cell consisting of two solutions separated by an ion-exchange membrane. (Ch. 18)

reducing agent A substance in a redox reaction that contains the element being oxidized. (Ch. 4)

reduction A chemical change in which a species gains electrons; the oxidation number of the species decreases. (Ch. 4)

relative biological effectiveness (RBE) A factor that accounts for the differences in physical damage caused by different types of radiation. (Ch. 19)

replication The process by which one double-stranded DNA forms two new DNA molecules, each one containing one strand from the original molecule and one new strand. (Ch. 20)

representative elements See *main group elements*.

resonance A characteristic of electron distributions when two or more equivalent Lewis structures can be drawn for one compound. (Ch. 8)

resonance stabilization The stability of a molecular structure due to delocalization of its electrons. (Ch. 9)

resonance structure One of two or more Lewis structures with the same arrangement of atoms but different arrangements of bonding pairs of electrons. (Ch. 8)

reverse osmosis A process in which solvent is forced through semipermeable membranes, leaving a more concentrated solution behind. (Ch. 11)

reversible process A process that can be run in the reverse direction in such a way that, once the system has been restored to its original state, no net energy has flowed either to the system or to its surroundings. (Ch. 17)

root-mean-square speed (u_{rms}) The square root of the average of the squared speeds of all the molecules in a population of gas molecules; a molecule possessing the average kinetic energy moves at this speed. (Ch. 5)

S

salt The product of a neutralization reaction; it is made up of the cation of the base in the reaction plus the anion of the acid. (Ch. 4)

saturated hydrocarbon An alkane; compounds containing the maximum ratio of hydrogen atoms to carbon atoms. (Ch. 6)

saturated solution A solution that contains the maximum concentration of a solute possible at a given temperature. (Ch. 4)

Schrödinger wave equation A description of how the electron matter wave varies with location and time around the nucleus of a hydrogen atom. (Ch. 7)

scientific method An approach to acquiring knowledge based on observation of phenomena, development of a testable hypothesis, and additional experiments that test the validity of the hypothesis. (Ch. 1)

scientific theory (model) A general explanation of a widely observed phenomenon that has been extensively tested and validated. (Ch. 1)

scintillation counter An instrument that determines the level of radioactivity in samples by measuring the intensity of light emitted by phosphors in contact with the samples. (Ch. 19)

second law of thermodynamics The principle that the total entropy of the universe increases in any spontaneous process. (Ch. 17)

secondary (2°) structure The pattern of arrangement of segments of a protein chain. (Ch. 20)

seesaw The molecular geometry about a central atom with a steric number of 5 and one lone pair of electrons in an equatorial position. (Ch. 9)

semiconductor A metalloid with electrical conductivity between that of metals and insulators that can be chemically altered to increase its electrical conductivity. (Ch. 9)

semimetals See *metalloids*.

sievert (Sv) The SI unit used to express the amount of biological damage caused by ionizing radiation. (Ch. 19)

sigma (σ) bond A covalent bond in which the highest electron density lies between the two atoms along the bond axis. (Ch. 9)

sigma (σ) molecular orbital A molecular orbital in which the greatest electron density is concentrated along an imaginary line drawn through the bonded atom centers. (Ch. 9)

significant figures All the certain digits in a measured value plus one estimated digit; the greater the number of significant figures, the greater the certainty with which the value is known. (Ch. 1)

simple cubic (sc) unit cell An array of atoms or molecules with one particle at the eight corners of a cube. (Ch. 12)

single bond A chemical bond that results when two atoms share one pair of electrons. (Ch. 8)

solid A form of matter that has a definite volume and shape. (Ch. 1)

solubility The maximum amount of a substance that dissolves in a given volume of solution at a given temperature. (Ch. 4)

solubility product (K_{sp}) (also called *solubility-product constant*) An equilibrium constant that describes the formation of a saturated solution of a slightly soluble salt. (Ch. 16)

solubility-product constant See *solubility product* (K_{sp}).

solute Any component in a solution other than the solvent. A solution may contain one or more solutes. (Ch. 4)

solution Another name for homogeneous mixture; solutions are often liquids, but they may also be solids or gases. (Ch. 1)

solvent The component of a solution that is present in the largest number of moles. (Ch. 4)

***sp* hybrid orbitals** Two hybrid orbitals oriented 180° from one another, formed by mixing one *s* and one *p* orbital. (Ch. 9)

sp^2 hybrid orbitals Three hybrid orbitals in a trigonal planar orientation, formed by mixing one *s* and two *p* atomic orbitals. (Ch. 9)

sp^3 hybrid orbitals A set of four hybrid orbitals with a tetrahedral orientation, formed by mixing one *s* and three *p* atomic orbitals. (Ch. 9)

specific heat (c_s) The quantity of energy required to raise the temperature of 1 g of a substance by 1°C at constant pressure. (Ch. 6)

spectator ion An ion present in a reaction mixture in solution that is unchanged by the reaction; spectator ions appear in overall ionic equations but not in net ionic equations. (Ch. 4)

spectrochemical series A list of ligands rank-ordered by their ability to split the energies of the *d* orbitals of transition metal ions. (Ch. 22)

sphere of hydration The cluster of water molecules surrounding an ion in aqueous solution; the general term applied to such a cluster forming in any solvent is *sphere of solvation*. (Ch. 10)

spin magnetic quantum number (m_s) Either $+\frac{1}{2}$ or $-\frac{1}{2}$, indicating that the spin orientation of an electron is either up or down, respectively. (Ch. 7)

spontaneous process A process that occurs without outside intervention. (Ch. 17)

square planar The molecular geometry about a central atom with a steric number of 6 and two lone pairs of electrons that occupy axial sites; the atoms occupy four equatorial positions. (Ch. 9)

square pyramidal The molecular geometry about a central atom with a steric number of 6 and one lone pair of electrons; as typically drawn, the atoms occupy four equatorial and one axial site. (Ch. 9)

standard cell potential (E°_{cell}) A measure of how forcefully an electrochemical cell in which all reactants and products in their standard states can pump electrons through an external circuit. (Ch. 18)

standard conditions In thermodynamics: a pressure of 1 bar (~1 atm) and some specified temperature, assumed to be 25°C unless otherwise stated; for solutions, a concentration of 1 *M* is specified. (Ch. 6)

standard deviation (*s*) A measure of the amount of variation, or dispersion, in a set of related values. The lowercase Greek letter sigma (σ) is also used as the symbol for standard deviation. (Ch. 1)

standard enthalpy of formation (ΔH°_f) The enthalpy change of a formation reaction; also called *standard heat of formation* or *heat of formation*. (Ch. 6)

standard enthalpy of reaction (ΔH°_{rxn}) The energy associated with a reaction that takes place under standard conditions; also called *standard heat of reaction*. (Ch. 6)

standard free energy of formation (ΔG°_f) The change in free energy associated with the formation of 1 mole of a compound in its standard state from its component elements. (Ch. 17)

standard hydrogen electrode (SHE) A reference electrode based on the half-reaction $2\ H^+(aq) + 2\ e^- \rightarrow H_2(g)$ that produces a standard electrode potential of 0.000 V. (Ch. 18)

standard molar entropy (S°) The absolute entropy of 1 mole of a substance in its standard state. (Ch. 17)

standard reduction potential (E°) The potential of a reduction half-reaction in which all reactants and products are in their standard states at 25°C. (Ch. 18)

standard solution A solution of known concentration used in titrations. (Ch. 4)

standard state The most stable form of a substance under 1 bar pressure and some specified temperature (25°C unless otherwise stated). (Ch. 6)

standard temperature and pressure (STP) 0°C and 1 bar as defined by IUPAC; 0°C and 1 atm are commonly used in the United States and in this textbook. (Ch. 5)

standing wave A wave confined to a given space, with a wavelength (λ) related to the length *L* of the space by $L = n(\lambda/2)$, where *n* is a whole number. (Ch. 7)

state function A property of an entity based solely on its chemical or physical state or both, but not on how it achieved that state. (Ch. 6)

stereoisomers Isomers created by differences in the orientations of the bonds in molecules. (Ch. 20)

steric number (SN) The sum of both the number of atoms bonded to a central atom and the number of lone pairs of electrons on the central atom. (Ch. 9)

stimulatory effect Increased activity, growth, or other biological response to the presence of a nonessential element. (Ch. 21)

stock solution A concentrated solution of a substance used to prepare solutions of lower concentration. (Ch. 4)

stoichiometry The quantitative relationship between reactants and products in a chemical reaction. (Ch. 3)

straight-chain hydrocarbon A hydrocarbon in which the carbon atoms are bonded together in one continuous carbon chain. (Ch. 6)

strong acid An acid that completely dissociates into ions in aqueous solution. (Ch. 4)

strong base A base that completely dissociates into ions in aqueous solution. (Ch. 4)

strong electrolyte A substance that dissociates completely into ions when it dissolves in water. (Ch. 4)

strong nuclear force The fundamental force of nature that keeps nucleons together in atomic nuclei. (Ch. 19)

sublimation Transformation of a solid directly into a vapor (gas). (Ch. 1)

substitutional alloy An alloy in which atoms of the nonhost metal replace host atoms in the crystal lattice. (Ch. 12)

substrate The reactant that binds to the active site in an enzyme-catalyzed reaction. (Ch. 20)

supercritical fluid A substance at conditions above its critical temperature and pressure, where the liquid and vapor phases are indistinguishable and have some characteristics of both a liquid and a gas. (Ch. 10)

supersaturated solution A solution that contains more than the maximum quantity of solute predicted to be soluble in a given volume of solution at a given temperature. (Ch. 4)

surface tension The energy needed to separate the molecules at the surface of a liquid. (Ch. 10)

surroundings Everything that is not part of the system. (Ch. 6)

system The part of the universe that is the focus of a thermochemical study. (Ch. 6)

T

termolecular step A step in a reaction mechanism involving a collision among three molecules. (Ch. 13)

tertiary (3°) structure The three-dimensional, biologically active structure of a protein that arises because of interactions between the R groups on its amino acids. (Ch. 20)

tetrahedral The molecular geometry about a central atom with a steric number of 4 and no lone pairs of electrons; the bond angles are all 109.5°. (Ch. 9)

theoretical yield The maximum amount of product possible in a chemical reaction for given quantities of reactants; also called *stoichiometric yield*. (Ch. 3)

thermal energy The kinetic energy of atoms, ions, and molecules. (Ch. 6)

thermal equilibrium A condition in which temperature is uniform throughout a material and no energy flows from one point to another. (Ch. 6)

thermochemical equation The chemical equation of a reaction that includes the change in enthalpy that accompanies that reaction. (Ch. 6)

thermochemistry The study of energy changes that occur during chemical reactions. (Ch. 6)

thermodynamics The study of energy and its transformations. (Ch. 6)

third law of thermodynamics The entropy of a perfect crystal is zero at absolute zero. (Ch. 17)

threshold frequency (ν_0) The minimum frequency of light required to produce the photoelectric effect. (Ch. 7)

titrant The standard solution added to the sample in a titration. (Ch. 4)

titration An analytical method for determining the concentration of a solute in a sample by reacting the solute with a standard solution of known concentration. (Ch. 4)

torr A unit of pressure. There are exactly 760 torr in one atm of pressure. (Ch. 5)

trace essential element An essential element present in the body in average concentrations between 1 and 1000 μg of element per gram of body mass. (Ch. 21)

trans isomer (also called *E isomer*) A molecule with two like groups (such as two R groups or two hydrogen atoms) on opposite sides of the molecule. (Ch. 20)

transcription The process of copying the information in DNA to RNA. (Ch. 20)

transfer RNA (tRNA) The form of the nucleic acid RNA that delivers amino acids, one at a time, to polypeptide chains being assembled by the ribosome–mRNA complex. (Ch. 20)

transition metals The elements in groups 3 through 12 of the periodic table. (Ch. 2)

transition state A high-energy state between reactants and products in a chemical reaction. (Ch. 13)

translation The process of assembling proteins from the information encoded in RNA. (Ch. 20)

tricarboxylic acid (TCA) cycle A series of reactions that continue the oxidation of pyruvate formed in glycolysis. (Ch. 20)

trigonal bipyramidal The molecular geometry about a central atom with a steric number of 5 and no lone pairs of electrons; three atoms occupy equatorial sites (bond angle 120°) and two other atoms occupy axial sites (bond angle 180°) above and below the equatorial plane; the bond angle between the axial and equatorial bonds is 90°. (Ch. 9)

trigonal planar The molecular geometry about a central atom with a steric number of 3 and no lone pairs of electrons; the bond angles are all 120°. (Ch. 9)

trigonal pyramidal The molecular geometry about a central atom with a steric number of 4 and one lone pair of electrons. (Ch. 9)

triple bond A chemical bond in which two atoms share three pairs of electrons. (Ch. 8)

triple point The temperature and pressure at which all three phases of a substance coexist. (Ch. 10)

T-shaped The molecular geometry about a central atom with a steric number of 5 and two lone pairs of electrons that occupy equatorial positions; the outer atoms occupy two axial sites and one equatorial site. (Ch. 9)

U

ultratrace essential element An essential element present in the body in average concentrations less than 1 μg of element per gram of body mass. (Ch. 21)

unified atomic mass unit (u) The unit used to express the relative masses of atoms and subatomic particles; it is exactly 1/12 the mass of one atom of carbon with six protons and six neutrons in its nucleus. (Ch. 2)

unimolecular step A step in a reaction mechanism involving only one molecule on the reactant side. (Ch. 13)

unit cell The repeating unit of the arrangement of atoms, ions, or molecules in a crystal lattice. (Ch. 12)

universal gas constant The constant R in the ideal gas equation; its value and units depend on the units used for the variables in the equation. (Ch. 5)

V

valence band A band of orbitals that are filled or partially filled by valence electrons. (Ch. 9)

valence bond theory A quantum mechanics–based theory of bonding incorporates the assumption that covalent bonds form when half-filled orbitals on different atoms overlap or occupy the same region in space. (Ch. 9)

valence electrons Electrons in the outermost occupied shell of an atom; the electrons that are transferred or shared in chemical reactions. (Ch. 7)

valence shell The outermost occupied shell of an atom. (Ch. 7)

valence-shell electron-pair repulsion (VSEPR) theory A model predicting that the arrangement of valence electron pairs around a central atom minimizes their mutual repulsion to produce the lowest-energy orientations. (Ch. 9)

van der Waals equation An equation that includes experimentally determined factors a and b that quantify the contributions of non-negligible molecular volume and non-negligible intermolecular interactions to the behavior of real gases with respect to changes in P, V, and T. (Ch. 5)

van 't Hoff factor (also called *i factor*) The ratio of the experimentally measured value of a colligative property to the theoretical value expected for that property if the solute were a nonelectrolyte. (Ch. 11)

vapor pressure The pressure exerted by a gas at a given temperature in equilibrium with its liquid phase. (Ch. 10)

vinyl group The subgroup $CH_2{=}CH{-}$. (Ch. 12)

viscosity The resistance to flow of a liquid. (Ch. 10)

voltaic cell An electrochemical cell in which chemical energy is transformed into electrical work by a spontaneous cell reaction. (Ch. 18)

W

wave function (ψ) A solution to the Schrödinger wave equation. (Ch. 7)

wave mechanics (also called *quantum mechanics*) A mathematical description of the wavelike behavior of particles on the atomic level. (Ch. 7)

wavelength (λ) The distance from crest to crest or trough to trough on a wave. (Ch. 7)

wave–particle duality The behavior of an object that exhibits the properties of both a wave and a particle. (Ch. 7)

weak acid An acid that is a weak electrolyte and so has a limited capacity to donate protons to the medium. (Ch. 4)

weak base A base that is a weak electrolyte and so has a limited capacity to accept protons. (Ch. 4)

weak electrolyte A substance that dissolves in water with most molecules staying intact, but with a small fraction dissociating into ions. (Ch. 4)

work A form of energy—specifically, the energy required to move an object through a given distance. (Ch. 6)

work function (ϕ) The amount of energy needed to remove an electron from the surface of a metal. (Ch. 7)

Z

***Z* isomer** See *cis isomer.*

zeolites Natural crystalline minerals or synthetic materials consisting of three-dimensional networks of channels that contain sodium or other 1+ cations. (Ch. 11)

zwitterion A molecule that has both negatively and positively charged groups in its structure. (Ch. 20)

Answers
to Particulate Review, Concept Tests, and Practice Exercises

Chapter 1

Particulate Review

8 hydrogen atoms
4 hydrogen molecules
5 helium atoms
Molecules are composed of atoms.

Concept Tests

p. 6 (b) and (c) show a physical process (a solid substance that melts does not change its identity), whereas (a) and (d) show a chemical reaction (burning fuel produces different substances).
p. 8 1:1
p. 12 (b)
p. 15 Energy is absorbed.
p. 18 (b) Decreasing
p. 23 (1) 3 significant figures; (2) 3 significant figures; (4) 4 significant figures
p. 26 The volume value (it has 4 significant figures) is the weak link because the mass value has 6.
p. 31 2.805 g. No penny would have a mass equal to the average mass.
p. 36 Hypothesis, because his explanation had not been thoroughly tested yet.

Practice Exercises

1.1. Properties (a) and (c) are physical properties; property (d) is a chemical property; property (b) is both physical and chemical.
1.2. a. The diatomic gas on the left undergoes deposition as it becomes a solid on the right.
b. Sublimation
1.3. 0.324 km; 3.24×10^4 cm
1.4. 9.45×10^{12} km (using 365 days/yr)
1.5. 1.14
1.6. Statistics (a) and (e) are exact numbers; statistics (b)–(d) have inherent uncertainty.
1.7. Mean = 36.38 mg/dL
STDEV = 0.38 mg/dL
95% CL = 36.38 ± 0.47 mg/dL or 36.4 ± 0.5 mg/dL
1.8. The mean and standard deviation are 194 ± 18.6 mg/dL. Testing 215 mg/dL as an outlier gives us a calculated Z value of 1.1 mg/dL, which is less than the reference Z values for $n = 3$ from Table 1.8 at both the 95% and 99% confidence levels. The high value of 215 should *not* be considered an outlier.
1.9. −233°C = −387°F = 40 K
123°C = 253°F = 396 K

Chapter 2

Particulate Review

Helium, He, element
Nitrogen, N_2, element
Gold, Au, element
Ethanol, CH_3CH_2OH, compound
Sodium chloride, NaCl, compound

Concept Tests

p. 51 Matter is neutral, so some positive particle must be present to provide electrical neutrality.
p. 55 ${}^{87}_{38}X$ is an isotope of strontium because its atomic number is 38 (each of its nuclei contains 38 protons). ${}^{90}_{40}X$ is an isotope of zirconium with a mass number of 90 (40 protons, 50 neutrons), and ${}^{234}_{90}X$ is an isotope of thorium because it has an atomic number of 90.
p. 58 Because sometimes the next heaviest element that was known in the early 1870s did not have chemical properties similar to the other elements in the next column to be filled. For example, the next heaviest element after calcium (a group II element in Mendeleev's table) was titanium, whose properties fit those of the elements in group IV, not group III. So, Mendeleev skipped the cell in group III and placed titanium in group IV. The missing group III element, scandium, was discovered (by a Scandinavian chemist) in 1879.
p. 64 (c) CsN
p. 73 (a) aldehyde; (b) ether; (c) amine

Practice Exercises

2.1. a. ${}^{56}Fe$
b. ${}^{15}N$
c. ${}^{37}Cl$
d. ${}^{39}K$
2.2. (a) 27 p, 33 n; (b) 53 p, 78 n; (c) 77 p, 115 n
2.3. ${}^{107}Ag$ = 51.85%; ${}^{109}Ag$ = 48.15%
2.4. a. As, arsenic
b. K, potassium
c. Cd, cadmium
d. Se, selenium
2.5. 40.0 g
2.6. (a)–(d) are molecular; (e) is ionic.
2.7. a. Tetraphosphorus decoxide
b. Nitrogen monoxide
c. Nitrogen trichloride
d. SF_6
e. ICl
f. Br_2O

2.8. a. $SrCl_2$
b. MgO
c. NaF
d. $CaBr_2$

2.9. $MnCl_2$ and MnO_2

2.10. a. $Sr(NO_3)_2$
b. K_2SO_4
c. Sodium hypochlorite
d. Potassium permanganate

2.11. a. Hypobromous acid
b. Iodous acid
c. Selenous acid

Chapter 3

Particulate Review

CO_2, carbon dioxide
H_2O, water
N_2, nitrogen
CO, carbon monoxide
H_2S, hydrogen sulfide

Concept Tests

p. 93 Both the mole and a gross represent a specific number of particles independent of the size, mass, or identity of an individual particle.

p. 95 One gram of Ag has more atoms because Ag has a smaller molar mass: fewer grams per mole mean more moles (or atoms) per gram.

p. 104 Changing the formula of a reactant or product changes its identity. In this reaction oxygen gas consists of molecules of O_2, not individual atoms of O.

p. 122 (a) C_2H_4 and (c) $C_{20}H_{40}$ have the same empirical formula (CH_2) and, therefore, the same percent composition (85.6% C, 14.4% H); (b) C_2H_2 and (d) C_6H_6 have the same empirical formula (CH) and are 92.3% C and 7.7% H.

p. 125 The empirical formula of (b) C_3H_8O is identical to its molecular formula; (a) $C_2H_6O_2$ and (c) $C_6H_{12}O_6$ have empirical formulas that differ from the given molecular formulas. The empirical formula of $C_2H_6O_2$ is CH_3O, and the empirical formula of $C_6H_{12}O_6$ is CH_2O.

p. 127 No, because they have different molecular formulas.

p. 128 Molecular formula

Practice Exercises

3.1. 1.5×10^{10} atoms

3.2. 6.80×10^{24} electrons

3.3. 5.41×10^{-2} mol C

3.4. 49.2 g Au

3.5. $CO_2 = 44.01$ g/mol; $O_2 = 32.00$ g/mol; $C_6H_{12}O_6 = 180.16$ g/mol

3.6. 5.00×10^{-3} mol; 3.01×10^{21} formula units

3.7. 72.2 g Cr, 89.3 g Cr

3.8. + →

3.9. (a) $P_4(s) + 5\ O_2(g) \rightarrow P_4O_{10}(s)$; (b) $P_4O_{10}(s) + 6\ H_2O(\ell) \rightarrow 4\ H_3PO_4(aq)$

3.10. $2\ CO(g) + O_2(g) \rightarrow 2\ CO_2(g)$

3.11. $C_3H_8(g) + 5\ O_2(g) \rightarrow 3\ CO_2(g) + 4\ H_2O(g)$

3.12. $2\ C_4H_{10}(g) + 13\ O_2(g) \rightarrow 8\ CO_2(g) + 10\ H_2O(g)$; 3.03 g CO_2 produced

3.13. 321 g Cu_2S; 129 g SO_2

3.14. This combustion mixture is fuel lean.

3.15. 3.74 g $C_2H_4N_2$

3.16. 70.4%

3.17. 1.10×10^3 g Al_2O_3; 194 g C

3.18. 21.32% C; 78.68% F

3.19. NO

3.20. Li_2CO_3

3.21. $C_{20}H_{40}$

3.22. C_5H_8 and $C_{20}H_{32}$

3.23. $C_8H_8O_3$

Chapter 4

Particulate Review

Ethylene glycol is a covalent compound with covalent bonds, whereas sodium sulfate is an ionic compound with ionic bonds. The space-filling representations of the molecules of ethylene glycol remain "intact" and do not separate into smaller particles, whereas the particles shown for sodium sulfate are separate sodium cations and sulfate anions. (Note that the space-filling sulfate anions contain covalent bonds between the sulfur and oxygen atoms.)

Concept Tests

p. 151 (c) Clear cough syrup; (d) Filtered dry air

p. 154 (d) 56,000 nM NaCl is the least concentrated solution.

p. 162 Solvent is added to dilute the stock solution, which decreases the concentration.

p. 165 Equivalent molar concentrations mean that the same number of moles of solute are dissolved in a given solution volume. Differences in conductivity result from the different numbers of ions the solutes make when they dissolve.

p. 165 Drawing (a) represents a weak electrolyte and (b) represents a strong electrolyte.

p. 169 (a) H_2E^{2-}; (b) H_4E

p. 170 (a) weak acid; (b) neither; (c) weak base; (d) strong acid

p. 180 Mix solutions containing equal numbers of moles of $Pb(NO_3)_2(aq)$ and $Na_2CrO_4(aq)$. Let the mixture stand for 10 min; filter off the yellow $PbCrO_4$ precipitate; wash the precipitate with water; let dry overnight and remove from the filter.

p. 184 In the reaction between Fe_3O_4 between O_2 that forms Fe_2O_3, iron is oxidized because Fe_2O_3 contains more O atoms per Fe atom (1.50) than Fe_3O_4 does (1.33). As Fe is oxidized, O is reduced.

p. 187 Reactions (a) and (b) are redox reactions. In reaction (a), the O.N. of Br goes from 0 to −1 as the O.N. of Sn goes from +2 to +4. In reaction (b), the O.N. of F goes from 0 to −1 as the O.N. of O goes from −2 to 0.

Practice Exercises

4.1. The arsenic concentration in the well water is 120 times the WHO standard.

4.2. 1.88 M $MgCl_2$

4.3. 0.109 M KCl

4.4. 4.48 g $NaC_3H_5O_3$

4.5. $V_{initial} = 12.5$ mL

4.6. 1.16×10^{-4} *M*

4.7. a. $CH_3COOH(aq) + NaOH(aq) \rightarrow CH_3COONa(aq) + H_2O(\ell)$

b. $CH_3COOH(aq) + Na^+(aq) + OH^-(aq) \rightarrow Na^+(aq) + CH_3COO^-(aq) + H_2O(\ell)$

c. $CH_3COOH(aq) + OH^-(aq) \rightarrow CH_3COO^-(aq) + H_2O(\ell)$

4.8. (a) strong electrolyte, neither acidic nor basic; (b) strong electrolyte, strong acid; (c) weak electrolyte; weak base, (d) nonelectrolyte, neither acidic nor basic

4.9. The lemon juice is 0.238 *M* $C_6H_8O_7$; 100.0 mL of juice contains 4.57 g $C_6H_8O_7$.

4.10. 0.0987 *M*

4.11. (a) no; (b) yes, Hg_2Cl_2 precipitates (c) $Hg_2^{2+}(aq) + 2\,Cl^-(aq) \rightarrow Hg_2Cl_2(s)$

4.12. $3\,Ba^{2+}(aq) + 6\,OH^-(aq) + 2\,H_3PO_4(aq) \rightarrow Ba_3(PO_4)_2(s) + 6\,H_2O(\ell)$

4.13. 0.174 g HgS

4.14. 1.31×10^{-4} *M* SO_4^{2-}

4.15. Yes

4.16. a. +3

b. +6

c. +6

4.17. Oxygen is reduced and is the oxidizing agent; S is oxidized and SO_2 is the reducing agent.

4.18. a. Yes, because the oxidation numbers for iron and palladium change from reactants to products.

b. $2\,Fe(s) + 3\,Pd^{2+}(aq) \rightarrow 2\,Fe^{3+}(aq) + 3\,Pd(s)$

4.19. $Au^{3+}(aq)$ is most easily reduced. Li(*s*) is most easily oxidized.

4.20. The products are $Al(NO_3)_3(aq)$ and Ag(*s*). $Al(s) + 3\,Ag^+(aq) \rightarrow Al^{3+}(aq) + 3\,Ag(s)$

4.21. $3\,HO_2^-(aq) + 2\,MnO_4^-(aq) + H_2O(\ell) \rightarrow 3\,O_2(g) + 2\,MnO_2(s) + 5\,OH^-(aq)$

Chapter 5

Particulate Review

The balanced chemical equation is $2\,H_2O(\ell) \rightarrow 2\,H_2(g) + O_2(g)$. The product is a mixture of two elements.

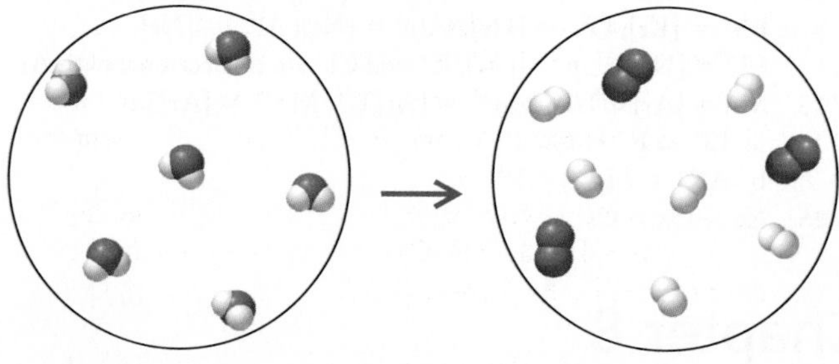

Concept Tests

p. 222 (c)

p. 224 The two graphs have different slopes because volume is inversely proportional to pressure. The slope of graph 1 is twice the slope of graph 2.

p. 226 (c)

p. 230 a and c

p. 235 Kr

p. 236 (a) i; (b) ii

p. 239 Any quantity of a gas has only one pressure that can be measured directly. For a mixture, this is P_{total}.

p. 241 Yes

p. 245 Lower

p. 247 $UF_6 < SF_6 < Kr < CO_2 < Ar < H_2$

p. 248 The change in temperature produces the same proportionate change in the u_{rms} values of the two gases.

p. 252 CO_2, CH_4

Practice Exercises

5.1. $P = 410$ Pa $= 0.41$ kPa

5.2. $\Delta h = 72$ mm. The mercury level in the right-hand (atmosphere) side of the manometer would still be higher.

5.3. $V_2 = 10.5$ L

5.4. $V_1/V_2 = 0.622$

5.5. $P = 2.5$ atm

5.6. $V_2 = 2.10 \times 10^3$ L

5.7. $V = 1.8 \times 10^3$ L

5.8. 1.10×10^2 g

5.9. 259 mL of N_2O

5.10. The balloon will sink to the floor.

5.11. $\mathcal{M} = 44.0$ g/mol;
$HCl(aq) + NaHCO_3(aq) \rightarrow NaCl(aq) + CO_2(g) + H_2O(\ell)$

5.12. $O_2 = 0.1201$, He $= 0.8799$

5.13. $X_{O_2} = 0.042$

5.14. 2.2 mg of H_2

5.15. $u_{rms,He} = 1.367 \times 10^3$ m/s, or 2.645 times faster than N_2

5.16. Ar

5.17. $P_{He} = 25.2$ atm, which is slightly higher than that of an ideal gas, whereas P_{N_2} was slightly lower. The lower value for N_2 results from the stronger interactions between its molecules, compared to the much smaller atoms of He, which offset the increase in *PV/RT* that both gases experience because of the incompressibility of their atoms and molecules.

Chapter 6

Particulate Review

The molecule is H_2O (water). The ions are H_3O^+ (hydronium), Cl^- (chloride), Na^+ (sodium), and OH^- (hydroxide). The buret contains H_2O molecules, OH^- ions, and Na^+ ions. The solution in the flask initially contains H_2O molecules, Cl^- ions, and H_3O^+ ions. After the acid in the flask has been neutralized, the flask contains H_2O molecules, Cl^- ions, and Na^+ ions. The net ionic equation is $H_3O^+(aq) + OH^-(aq) \rightarrow 2\,H_2O(\ell)$.

Concept Tests

p. 275 No; rider 1 has more potential energy:
$m_1gh = (PE)_{rider\,1} > (PE)_{rider\,2} = m_2gh$

p. 275 The potential energy of rider 1 is greater than the potential energy of rider 2; rider 1 has greater kinetic energy:

$$\tfrac{1}{2}m_1u^2 = (KE)_{rider\,1} > (KE)_{rider\,2} = \tfrac{1}{2}m_2u^2$$

p. 277 There is far more water in the pool than in the cup. The thermal energy in the cup is “less than,” even if the pool temperature is, say, 20°C.

p. 280 (a) open; (b, c) closed, because the bottle and the sandwich wrap can conduct heat; (d) open

p. 295 The mass of the aluminum is much less than the mass of the water, and the molar heat capacity of aluminum is much less than that of water.

p. 314 Each alkane has two terminal $-CH_3$ groups, which have a mass of 15 g/mol each. Therefore, the mass of the $-CH_2-$ groups is 114 g/mol − 2(15 g/mol) = 84 g/mol. Each $-CH_2-$ group has a mass of 14 g/mol. So, 84 g/mol would be the mass of the $-CH_2-$ units, so there are (84 g/mol)/(14 g/mol) = 6 $-CH_2-$ units. Therefore, $n = 2 + 6 = 8$.

p. 316 (a) 1 mole CH_4; (b) 1 g H_2

Practice Exercises

6.1. About 13% slower

6.2. a. The match is the system, $q < 0$, and the process is exothermic.
b. The wax is the system, $q < 0$, and the process is exothermic.
c. The liquid is the system, $q > 0$, and the process is endothermic.

6.3. $\Delta E = 32$ J

6.4. $w = 1.58 \times 10^6$ kJ

6.5. 1.13×10^3 g or 1.13 kg

6.6. ΔH_{sys} is negative; q_{surr} is positive.

6.7. −321 kJ

6.8. 16.2°C

6.9. 118 J

6.10. $C_{calorimeter} = 4.390$ kJ/°C. Energy released = 16.4 kJ.

6.11. −4.6 kJ/mol

6.12. Reverse B and then add the three equations.

6.13.

$$2\,CH_4(g) + 3\,O_2(g) \rightarrow \cancel{2\,CO(g)} + 4\,H_2O(g) \qquad \Delta H_A = -1038 \text{ kJ}$$

$$\cancel{2\,CO(g)} + O_2(g) \rightarrow 2\,CO_2(g) \qquad \Delta H_C = -566 \text{ kJ}$$

$$2\,CH_4(g) + 4\,O_2(g) \rightarrow 2\,CO_2(g) + 4\,H_2O(g) \qquad 2\,\Delta H_B = -1604 \text{ kJ}$$

For 1 mole CH_4, $\Delta H_{comb} = -802$ kJ

6.14. a. $Ca(s) + C(s, \text{graphite}) + \frac{3}{2}\,O_2(g) \rightarrow CaCO_3(s)$
b. $2\,C(s, \text{graphite}) + 2\,H_2(g) + O_2(g) \rightarrow CH_3COOH(\ell)$
c. $K(s) + Mn(s) + 2\,O_2(g) \rightarrow KMnO_4(s)$

6.15. $\Delta H^\circ_{rxn} = -41.2$ kJ

6.16. Fuel value of kerosene = 43.0 kJ/g; fuel density of kerosene = 35.3 kJ/mL

6.17. $C_{calorimeter} = 11.2$ kJ/°C

Chapter 7

Particulate Review

The ions from left to right are K^+, S^{2-}, Al^{3+}, and F^-. Potassium has 19 electrons in a neutral atom and 18 electrons in its cation, K^+. Sulfur has 16 electrons in a neutral atom and 18 electrons in its anion, S^{2-}. Aluminum has 13 electrons in a neutral atom and 10 electrons in its cation, Al^{3+}. Fluorine has 9 electrons in a neutral atom and 10 electrons in its anion, F^-.

Concept Tests

p. 338 (c)

p. 340 Higher frequency

p. 342 (a) continuous; (b) quantized; (c) continuous; (d) quantized

p. 345 Yes; the excess energy increases the kinetic energy of the electron.

p. 349 (c) > (a) > (b) > (d)

p. 350 No, because He atoms have two electrons.

p. 361 Five

p. 372 Ground-state potassium atom: $[Ar]4s^1$; excited-state potassium atom: $[Ar]3p^1$ (others would be possible as well); potassium ion, K^+: [Ar]. The potassium atom has one more electron than the potassium ion.

p. 373 A half-filled set of *f* orbitals is more stable because adding another *f* electron requires pairing energy.

p. 376 $1s > 2s > 2p > 3s > 3p > 4s > 4p$

p. 379 Because the valence electrons are farther from the nucleus and shielded from it by more inner-shell electrons

p. 380 The magnitude of IE and EA both increase with increasing Z across a row (except for group 18). EA values do not display clear trends within groups, whereas IE values decrease with increasing Z.

Practice Exercises

7.1. $\lambda = 3.30$ m

7.2. $E_{453} = 4.39 \times 10^{-19}$ J; $E_{482} = 4.12 \times 10^{-19}$ J

7.3. $\lambda = 2.62 \times 10^{-7}$ m, or 262 nm

7.4. 486 nm

7.5. Prediction: Less energy is required to remove an electron from the hydrogen atom in the $n = 3$ state than for a hydrogen atom in the $n = 1$ state. Calculated value: 2.420×10^{-19} J

7.6. $\lambda = 3.3 \times 10^{-10}$ m

7.7. $\Delta x \geq 7 \times 10^{-11}$ m

7.8. 25

7.9.

n	ℓ	m_ℓ	m_s
3	1	−1	$+\frac{1}{2}$
3	1	0	$+\frac{1}{2}$
3	1	1	$+\frac{1}{2}$

7.10. Ga: $[Ar]3d^{10}4s^24p^1$ As: $[Ar]3d^{10}4s^24p^3$

7.11. Co = $[Ar]3d^74s^2$

7.12. K^+ = [Ar]; I^- = $[Kr]4d^{10}5s^25p^6$ = [Xe]; Ba^{2+} = [Xe]; Rb^+ = [Kr]; O^{2-} = $[He]2s^22p^6$ = [Ne]; Al^{3+} = [Ne]; Cl^- = $[Ne]3s^23p^6$ = [Ar]. K^+ and Cl^- are isoelectronic with Ar.

7.13. Mn = $[Ar]3d^54s^2$; Mn^{3+} = $[Ar]3d^4$; Mn^{4+} = $[Ar]3d^3$

7.14. a. $Li^+ < F^- < Cl^-$
b. $Al^{3+} < Mg^{2+} < P^{3-}$

7.15. Ne > Ca > Cs

Chapter 8

Particulate Review

$$CH_4(g) + 2\,O_2(g) \rightarrow CO_2(g) + 2\,H_2O(g)$$

For each mole of CH_4 that is combusted:
4 moles of C—H bonds and 2 moles of O=O bonds are broken;
2 moles of C=O bonds and 4 moles of O—H bonds are formed.

Concept Tests

p. 407 $\leftarrow\!\!+ \quad +\!\!\rightarrow$

$:\ddot{O}=C=\ddot{O}:$

p. 409 No, stretching in N≡N or O═O does *not* result in infrared absorptions because the bonds are nonpolar, so no change in polarity occurs when they stretch.

p. 417 −1

p. 426 $N_2O > NO_2 > NO$

p. 430 The O═O bond is not as strong as the N≡N bond, so the O═O bond is more easily broken.

Practice Exercises

8.1. (i) gold, metallic bonds attract each gold atom to the gold atoms directly surrounding it.
(ii) calcium carbonate, ionic bonds attract calcium cations (Ca^{2+}) and carbonate anions (CO_3^{2-}) to each other. The C and O atoms in each carbonate ion are covalently bonded together.
(iii) phosphorus, covalent bonds attract each of the four P atoms to the other three in a P_4 molecule.

8.2. H—C—H (with H above and H below C, each joined by a single bond)

8.3. $:\ddot{Cl}-\ddot{P}-\ddot{Cl}:$ with $:\ddot{Cl}:$ bonded below P

8.4. $:\ddot{O}=C=\ddot{O}:$

8.5. $[:\ddot{F}:]^- \; Mg^{2+} \; [:\ddot{F}:]^-$

8.6. Be—Cl; the bond is a polar covalent bond. There would be no ionic bonds formed among any of these pairs.

8.7. $:\ddot{O}=\ddot{S}-\ddot{O}: \leftrightarrow :\ddot{O}-\ddot{S}=\ddot{O}:$

8.8. Resonance forms for N_3^-:

$[:\ddot{N}=N=\ddot{N}:]^- \leftrightarrow [:N\equiv N-\ddot{N}:]^- \leftrightarrow [:\ddot{N}-N\equiv N:]^-$

Resonance forms for NO_2^+:

$[:\ddot{O}=N=\ddot{O}:]^+ \leftrightarrow [:O\equiv N-\ddot{O}:]^+ \leftrightarrow [:\ddot{O}-N\equiv O:]^+$

8.9. $[:\ddot{O}=N=\ddot{O}:]^+$

8.10. Six resonance structures of NO_3^--type arrangement (N bonded to three O atoms, with the double bond and formal charges distributed differently in each), joined by $\leftrightarrow$.

The structure with O═N (formal charges: O 0, N +1, O 0, O −1) (with 6 resonance structures) is preferred because it has lower formal charges than the structure with single bonds only (formal charges: O 0, N +2, O −1, O −1) (with 3 resonance structures)

8.11. $\left[:\ddot{O}=\ddot{Se}-\ddot{O}:\right]^{2-}$ (with $:\ddot{O}:$ bonded below Se) $\leftrightarrow \left[:\ddot{O}-\ddot{Se}-\ddot{O}:\right]^{2-}$ (with $\ddot{O}$ double-bonded below Se) $\leftrightarrow \left[:\ddot{O}-\ddot{Se}=\ddot{O}:\right]^{2-}$ (with $:\ddot{O}:$ bonded below Se)

8.12. Sodium atoms do not have an octet.

Na—$\ddot{Cl}$:
| |
$:\ddot{Cl}$—Na

8.13. $\Delta H_{rxn} = -79$ kJ

Chapter 9

Particulate Review

Malic acid is $C_4H_6O_5$ and contains two carboxylic acid functional groups (one at each end of the molecule) and one alcohol functional group in the middle. Putrescine is $C_4H_{12}N_2$ and contains two amine functional groups.

Concept Tests

p. 447 Molecular geometry is derived from electron-pair geometry. They are the same if there are no lone pairs around the central atom but different if there are lone pairs around the central atom.

p. 451 (a) > (c) > (b)

p. 456 The bond angles will be slightly smaller because of greater repulsion between the lone pair of electrons and the bonding pairs in the seesaw shape.

p. 460 The lone pair on the N atom in NCl_3 gives it a steric number of 4, which means it has a trigonal pyramidal molecular shape. The asymmetry of this shape means the N—Cl bond dipoles do not completely offset each other so that NCl_3 has a permanent dipole. The Al atom in $AlCl_3$ has no lone pairs of valence electrons, so its SN = 3 giving $AlCl_3$ a symmetrical trigonal planar molecular shape. All its bond dipoles offset one another, and it has no permanent dipole.

p. 461 The electronegativity difference between H and S is less than the difference between H and O.

p. 469 It needs unhybridized *p* orbitals to form π bonds.

p. 473 (a) and (c), because there are alternating double and single bonds at sp^2 hybridized carbons

p. 476 (b) and (c) are chiral

p. 478 They are similar in that they are the products of mixing atomic orbitals, but different in that MOs are delocalized over the entire molecule.

p. 483 Yes, because O_2 is paramagnetic and N_2 is not.

p. 486 No, the bond orders in the ground states are higher because each has one more electron in a bonding orbital and one fewer in an antibonding orbital.

p. 491 Overlapping conduction and valence bands, because the filled valence band formed by magnesium atoms' filled 3*s* orbitals overlaps the conduction band formed by their empty *p* orbitals.

Practice Exercises

9.1. Tetrahedral:

Cl
H—C···Cl
Cl

9.2. The bond angles in SO_3 are 120°; the bond angle in SO_2 should be slightly less than 120°.

9.3. (a) Tetrahedral; bond angles ~109.5° (b) T-shaped; bond angles slightly less than 90° and 180° (c) Square planar; bond angles = 90° and 180°

9.4. No, because the molecular geometry of CS_2 is linear. Yes, because the C—O—C bond angle is about 109.5°, which means the two C—O bond dipoles do not completely offset one another, giving the molecule a permanent dipole.

9.5. 5*p* on I and 4*p* on Br

9.6. (a) CCl_4 and (d) PH_3

9.7. Hexane: (zigzag chain); heptane: $CH_3(CH_2)_5CH_3$

9.8. H—N̈=N̈—H (Diazene); H—N̈—N̈—H with H on each N (Hydrazine)

Each N in diazene is trigonal planar. With one lone pair on each N atom, the molecular geometry around the N atoms is bent with H—N—N bond angles of less than 120°. The N atoms are sp^2 hybridized and the molecule is flat. Each N in hydrazine is tetrahedral. With one lone pair on each N atom, the molecular geometry around the N atoms is trigonal pyramidal with H—N—N bond angles of less than 109.5°. The N atoms are sp^3 hybridized and the molecule is nonplanar.

9.9. H_2^+ can exist; its bond order is 0.5.

9.10. The bond order increases with the addition of an electron to B_2 and C_2.

9.11.

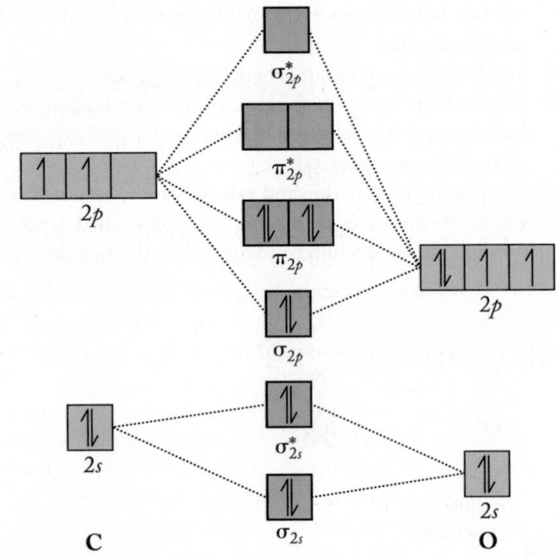

The C—O bond order is 3.

9.12. Selenium

Chapter 10

Particulate Review

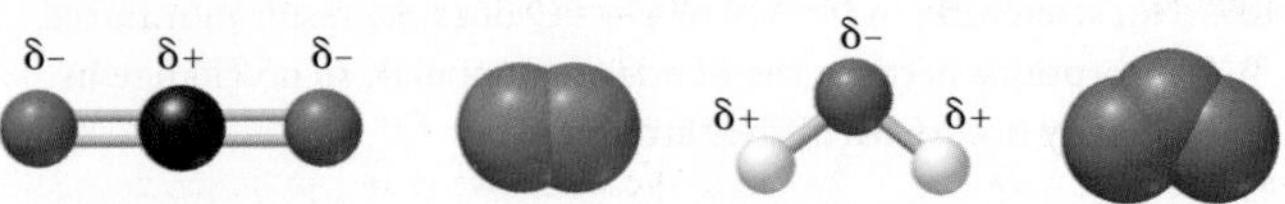

Carbon dioxide is not polar despite containing polar bonds. Water is polar and contains polar bonds. Ozone is polar even though it contains nonpolar bonds.

Concept Tests

p. 510 H < Ne < Ar < Kr

p. 510 Molecules of CCl_4 are larger and experience larger dispersion forces.

p. 513 Acetone has the higher boiling point. Its larger dipole moment results in stronger dipole–dipole interactions, meaning that a higher temperature (more energy) is required to overcome them to convert liquid acetone to acetone vapor.

p. 515 (a) CH_3Br; (b) CH_3CH_2OH; (c) $(CH_3)_2NH$

p. 516 Yes. Ion–induced dipole forces should be slightly stronger than dipole–induced dipole forces, but weaker than dipole–dipole interactions.

p. 521 Gasoline

p. 524 Yes, it's possible that increased pressure will cause the ice to melt, according to the slope of the blue line in Figure 10.23 and the footprint of the tires.

p. 526 No, the slope of the line separating solid and liquid CO_2 indicates that the solid expands when it melts, which means it is denser than liquid CO_2.

p. 528 No, the adhesive forces are weak.

Practice Exercises

10.1. As pure substances, all five compounds experience dispersion forces. Compounds (b)–(e) are polar and experience dipole–dipole forces. Only (e) can participate in hydrogen bonding.

10.2. To vaporize, molecules of liquid ethylene glycol each need to break as many as six hydrogen bonds compared to a maximum of three hydrogen bonds per molecule of 2-propanol. Therefore, ethylene glycol has a higher boiling point.

10.3. 28.4 kJ/mol

10.4. As its temperature is raised from −100°C to 50°C at $P = 10$ atm, solid CO_2 melts at about −55°C and liquid CO_2 vaporizes at about −35°C.

10.5. Helium, being smaller and with fewer electrons than nitrogen, is less soluble in blood because it is less polarizable and experiences weaker dipole–induced dipole interactions.

10.6. Methanol. Both methanol and hexanol have an –OH group that can form hydrogen bonds with the oxygen on acetone. Hexanol has a much longer hydrocarbon chain than methanol, and its molecules would interact via dispersion forces, which might cause hexanol to interact more strongly with itself than with acetone and hence cause it to be less soluble in acetone.

10.7. 9.5×10^{-5} mol/L

Chapter 11

Particulate Review

Potassium and sodium are both metals; sulfur and fluorine are nonmetals. We learned in Chapter 7 that atomic radii increase down a group (so potassium is larger than sodium) and across a period (so metals are larger than nonmetals). We also know that cations are smaller than their neutral parent atoms, and anions are larger than their neutral parent atoms. The image is drawn with neutral atoms above their ions, so the atoms and ions are

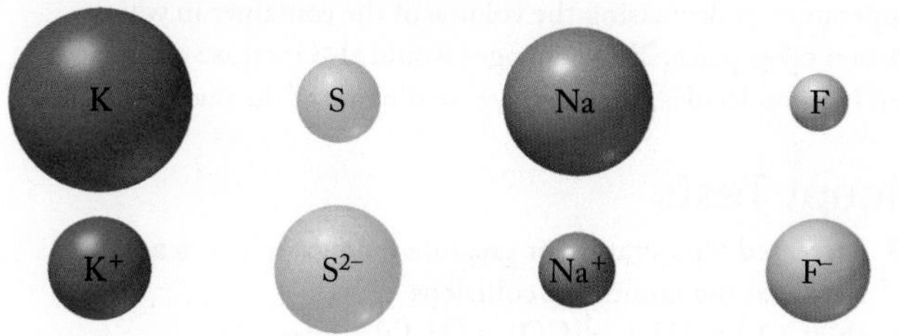

Concept Tests

p. 553 (a) $LiF < NaF < KF$; (b) $CaF_2 < CaCl_2 < CaBr_2$

p. 562 The vapor pressure of a pure solvent is an intensive property. The vapor pressure of a solution is an extensive property.

p. 566 Dimethyl ether will distill first because its smaller dipole moment means its intermolecular attractions are weaker than those of acetone.

p. 569 (c)

p. 571 The solution is mostly water, which has a density of 1 kg/L, so $\text{volume}_{H_2O} \approx \text{mass}_{H_2O}$.

p. 576 (b)

p. 578 (c)

p. 581 Toward the KCl

p. 581 Doubling the concentration

p. 585 Although more opposing pressure is needed at 50°C than at 20°C, the greater molecular motion would cause osmosis to occur faster.

Practice Exercises

11.1. $BaO > CaCl_2 > KCl$

11.2. TiO_2

11.3. $U = -3792$ kJ/mol

11.4. $U = -670$ kJ/mol

11.5. $P_{solution} = 0.848$ atm or 644 torr

11.6.

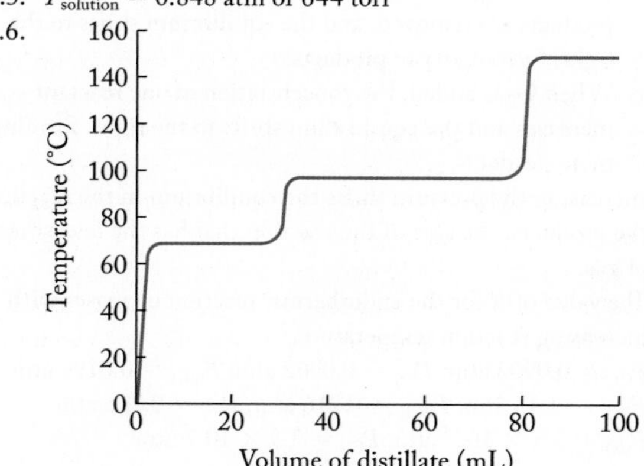

11.7. 1.4

11.8. 0.84 *m*

11.9. 4.4 *m*

11.10. −15.2°C

11.11. $i = 3$; 102.7°C

11.12. −0.12°C

11.13. 28.1 atm

11.14. 3.7 atm

11.15. 27.5 atm

11.16. 6.40×10^4 g/mol

Chapter 12

Particulate Review

Platinum is a metal with delocalized electrons. Reacting elemental potassium with elemental iodine results in the transfer of electrons from potassium atoms to iodine atoms. The K^+ cations and I^- anions that form potassium iodide are held together by ion–ion electrostatic attractions. Platinum atoms are attracted to one another through London forces resulting from electrostatic attractions between nuclei in neighboring platinum atoms and the valence electrons in those same atoms.

Concept Tests

p. 607 A crystal lattice is a three-dimensional repeating array of particles in a crystalline solid; a unit cell is the smallest repeating pattern within a crystal lattice.

p. 613 Yes, because it is homogeneous; the tin atoms are uniformly distributed among the copper atoms.

p. 614 Both alloys could be either.

p. 621 A group 15 element such as N or P

p. 623 Different molecular structures and properties mean LDPE and HDPE must be recycled separately.

p. 628 The longer the hydrocarbon portion of the chain, the lower the water solubility.

Practice Exercises

12.1. The atomic radii of both silver and gold are 144 pm (see Appendix 3). For silver, $d = 10.57$ g/mL, close to the 10.50 g/mL value for the density of silver from Appendix 3; for gold, $d = 19.41$ g/mL, close to the 19.3 g/mL value for the density of gold from Appendix 3.

12.2. The sizes of the titanium and molybdenum atoms relative to niobium both fall within the 15% radii guideline. However, niobium and molybdenum both form body-centered cubic structures, whereas titanium is hexagonal closest-packed. Therefore, molybdenum will form an alloy, but titanium will not.

12.3. TlBr will crystallize in the CsCl structure rather than the rock salt structure because the anion and cation have nearly the same ionic radii. The ratio of Tl^{3+} to Br^+ in $TlBr_3$ does not match the stoichiometry of the unit cell in the CsCl structure.

12.4. 101 pm; 2.16 g/cm^3

12.5. The carbon skeleton of the monomer is

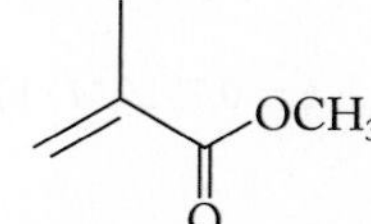

The condensed structure of the monomer is $H_2C{=}C(CH_3)C(O)OCH_3$

12.6. London dispersion and dipole–induced dipole interactions

12.7.

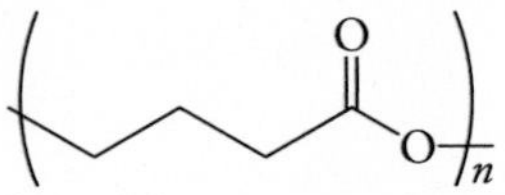

12.8. The polar fibers of cotton and polyester repel very nonpolar greases and oils but attract water molecules, so perspiration wicks out of the gloves to cool the skin.

12.9. The carbon-skeleton structures of the monomers are

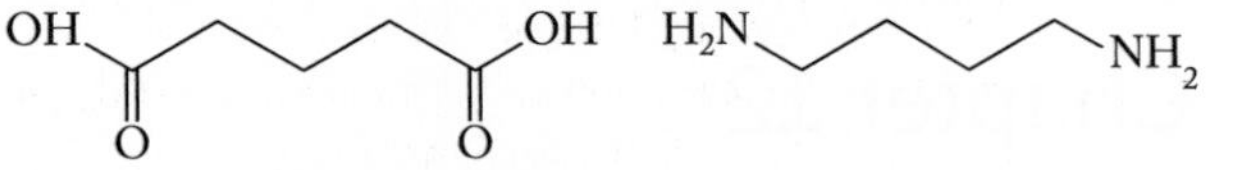

The repeating unit in the polymer is

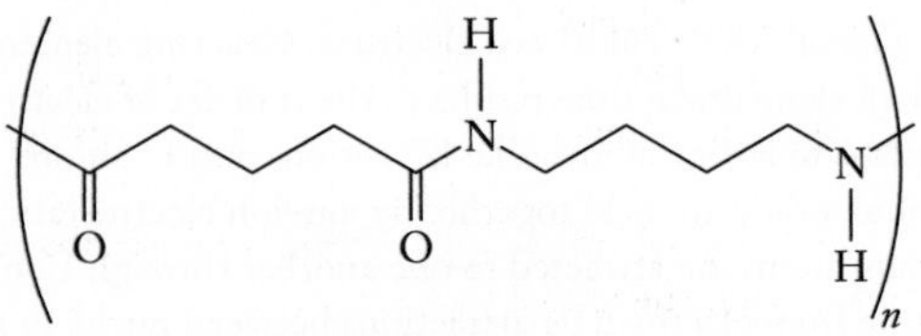

Chapter 13

Particulate Review

Cylinder 1 contains the more concentrated solution (0.01 *M*). The concentration of cylinder 2 is 0.005 *M*. The two cylinders contain equivalent amounts of dissolved solid.

Concept Tests

p. 658 (d)
p. 662 (a) 1; (b) 0
p. 663 Four: rate = $k[A][B]^2$; rate = $k[A]^2[B]$; rate = $k[A]^3$; rate = $k[B]^3$
p. 669 Fast
p. 674 Molecules are moving faster, so the likelihood of collisions is greater.
p. 680 (c)
p. 682 n
p. 684 No
p. 687 Similar rate laws indicate similar reaction mechanisms.
p. 689 (a)
p. 689 Intermediate

Practice Exercises

13.1. The rate of formation of CO_2 (a product) is twice the rate of consumption of O_2.

13.2. $\frac{\Delta[N_2]}{\Delta t} = 10.8\ M/s$ and $\frac{\Delta[H_2O]}{\Delta t} = 21.5\ M/s$

13.3. $1.2 \times 10^{-6}\ M/s$

13.4. Rate = $k[NO][NO_3]$; $k = 1.57 \times 10^{10}/(M \cdot s)$

13.5. The decomposition of H_2O_2 is first order; $k = 8.30 \times 10^{-4}\ s^{-1}$

13.6. 0.430 *M*

13.7. $k = 2.5 \times 10^{-2}$/day

13.8. This reaction is second order in $[NO_2]$; $k = 0.751/(M \cdot h)$

13.9. 6.9 kJ/mol

13.10. 306 K

13.11. Rate = $k_{overall}[A]^2[B]^2$

13.12. Because none of the rate laws possible with this mechanism match the experimental rate law, this proposed mechanism cannot be valid.

13.13. Yes, NO_2 acts as a catalyst in this reaction.

Chapter 14

Particulate Review

NO and O_2 combine to produce NO_2; NO and O_2 are produced when NO_2 molecules collide. The chemical equation for the two simultaneous reactions is

$$2\,NO(g) + O_2(g) \rightleftharpoons 2\,NO_2(g)$$

The rate would be increased by any change that increases the number of collisions between NO molecules and O_2 molecules, such as increasing temperature or decreasing the volume of the container in which the reaction takes place. These changes would also increase collisions between NO_2 molecules, that is, increase the rate of the reverse reaction.

Concept Tests

p. 710 Increased temperature or pressure for a gas-phase reaction increases the number of collisions.
p. 714 (b) $[CO_2] = [H_2] > [CO] = [H_2O]$
p. 726 Zone (a)
p. 733 The number of moles of gaseous reactants and products is the same.

Practice Exercises

14.1. $K_c = \frac{[CO][H_2]^3}{[CH_4][H_2O]}$ $K_p = \frac{(P_{CO})(P_{H_2})^3}{(P_{CH_4})(P_{H_2O})}$

14.2. $K_c = \frac{[CH_3OH]}{[CO][H_2]^2} = 2.9 \times 10^2$

14.3. $K_p = 2.7 \times 10^4$

14.4. $K_c = 3.6 \times 10^8$

14.5. $K_{p,reverse} = 2.3 \times 10^2$

14.6. $K_c = 0.13$

14.7. $K_{c,overall} = 1.7 \times 10^2$

14.8. $Q < K$, so this reaction is not at equilibrium and it proceeds to the right.

14.9. a. $K_p = \frac{(P_{CO})^2}{(P_{CO_2})}$

b. $K_p = \frac{(P_{CO})}{(P_{CO_2})(P_{H_2})}$

14.10. a. When the reaction is cooled and water vapor condenses, one product is removed from the reaction mixture and the equilibrium shifts to the right, forming more SO_2.

b. When SO_2 gas dissolves in liquid water as it condenses, products are removed, and the equilibrium shifts to the right, forming more products.

c. When O_2 is added, the concentration of one reactant increases and the equilibrium shifts to the right, forming more products.

14.11. Increasing the pressure shifts the equilibrium in the reaction to the products, the side of the reaction that has the fewest moles of gas.

14.12. The value of K for the endothermic reaction increases with increasing reaction temperature.

14.13. $P_{Cl_2} = 0.0902$ atm; $P_{Br_2} = 0.0902$ atm; $P_{BrCl} = 0.0196$ atm

14.14. $P_{PCl_5} = 1.13$ atm; $P_{PCl_3} = 0.216$ atm; $P_{Cl_2} = 0.216$ atm

14.15. $P_{CO} = 3.5 \times 10^{-5}$ atm; $P_{Cl_2} = 3.5 \times 10^{-5}$ atm

Chapter 15

Particulate Review

Only HCl and CH_3COOH produce H^+ (H_3O^+) ions when dissolved in water. HCl molecules dissociate completely, whereas only a small fraction of the CH_3COOH molecules dissociate. The acidic hydrogen in CH_3COOH is the hydrogen atom bonded to oxygen; the hydrogen atoms in the $-CH_3$ group are not acidic.

Concept Tests

p. 757 As CO_2 is removed more rapidly than it can be produced, both equilibria shift to the left.
p. 769 pH 0.22 = strongly acidic; 4.37 = weakly acidic; 7.00 = neutral; 10.03 = weakly basic; 13.77 = strongly basic
p. 771 Higher
p. 773 The larger its K_a, the greater the percent ionization for a given concentration of HA.
p. 775 Most: C; least: A
p. 783 C and D
p. 784 $H_3PO_4 > H_2PO_4^- > HPO_4^{2-} > PO_4^{3-}$
p. 787 Phosphoric acid: no. Citric acid: yes, the second ionization step.

Practice Exercises

15.1. $HClO_2 > CH_2BrCOOH >$ Lactic $> C_6H_5COOH > HN_3$
15.2. Strongest: H_2SO_4; weakest: H_2SeO_3
15.3. $\underset{\text{acid}}{CH_3COOH(aq)} + \underset{\text{base}}{H_2O(\ell)} \rightleftharpoons \underset{\text{conjugate base}}{CH_3COO^-(aq)} + \underset{\text{conjugate acid}}{H_3O^+(aq)}$
15.4. $S^{2-} > ClO^- > CH_3COO^- > HSO_3^- > Br^-$
15.5. Greatest increase: b; greatest decrease: d
15.6. The acidity ($[H_3O^+]$) of the ocean now is $10^{-8.07} = 8.5 \times 10^{-9}\ M$. It was $6.6 \times 10^{-9}\ M$, which means it has increased by $\frac{(8.5 - 6.6) \times 10^{-9}\ M}{6.6 \times 10^{-9}\ M} = \frac{1.9}{6.6} = 27\%$
15.7. $[H_3O^+] = 2.0 \times 10^{-12}\ M$; $[OH^-] = 5.0 \times 10^{-3}\ M$
15.8. pH = 2.23; percent ionization = 12%; $K_a = 7.9 \times 10^{-4}$
15.9. 7.2% ionized, $K_b = 5.7 \times 10^{-4}$
15.10. 12.65
15.11. pH = 3.10
15.12. pH = 11.97
15.13. pH = 7.52
15.14. pH = 0.68. Note that doubling the concentration decreased the pH by only 0.28 units. Had $[H_3O^+]$ doubled, pH would have decreased by $-\log 2 = 0.30$ units.
15.15. pH = 5.60
15.16. $SO_4^{2-}(aq) + H_2O(\ell) \rightleftharpoons HSO_4^-(aq) + OH^-(aq)$
15.17. pH = 11.66
15.18. pH = 5.62

Chapter 16

Particulate Review

The compounds are (a) PbS, (b) Li_2S, (c) CaF_2, and (d) KF; PbS and CaF_2 are much less soluble than Li_2S and KF.

Concept Tests

p. 816 (b) because there is a 1:1 ratio of base to acid.
p. 820 (c) is the equivalence point. The titration starts at (b) and proceeds through (d), (c), and then (a).
p. 821 (b) depicts $HClO(aq)$ before the titration, (a) depicts the midpoint, and (c) depicts the equivalence point.
p. 822 Titrating with a weak base would produce too small a change in pH at the equivalence point to be detected precisely.
p. 823 10.64
p. 824 (c) depicts $NH_3(aq)$ before the titration, (b) the buffer region, and (a) the equivalence point.
p. 827 Because neutralization of the CO_3^{2-} produces more HCO_3^-, which adds to the HCO_3^- present initially in the sample, which makes the second pH plateau wider than the first
p. 833 $Fe^{2+}(aq)$ should be a weaker Lewis acid than $Fe^{3+}(aq)$ because of its smaller charge and larger size. Therefore the K_a for $Fe^{2+}(aq)$ would be below that of (smaller than) $Fe^{3+}(aq)$.
p. 835 $K_{sp} = 1.7 \times 10^{-6}$
p. 838 Decrease: NaF; increase: HNO_3; no effect: neither
p. 840 The value $[OH^-]$ is defined by the initial $[Ca^{2+}]$ and the K_{sp} of $Ca(OH)_2$.

Practice Exercises

16.1. pH = 3.25
16.2. pH = 10.46
16.3. H_2SO_3 ($pK_a = 1.77$)
16.4. $(CH_3CH_2)_2NH$ ($pK_b = 3.07$)
16.5. 930 g of sodium ascorbate + 51 g of ascorbic acid
16.6. pH = 4.97; pH = 4.98
16.7. (a) change in pH of +0.11; (b) change in pH of +0.22
16.8. Both buffer A and buffer B decrease in pH by 0.15.
16.9. pH = 11.39
16.10. (a) pH = 3.45; (b) pH = 3.75; (c) pH = 4.05
16.11. No, not always, because HCO_3^- ions initially present in the sample cause the midpoint pH to be lower than pK_{a_2}.
16.12. 10.3 mL followed by an additional 13.7 mL.
16.13. CaO acts as a Lewis base, whereas CO_2 acts as a Lewis acid.
16.14. $[Ag^+(aq)] = 1.6 \times 10^{-8}\ M$
16.15. $3.1 \times 10^{-6}\ M$
16.16. $8.2 \times 10^{-4}\ M$
16.17. $1.1 \times 10^{-5}\ M$
16.18. Yes
16.19. Yes, Ba^{2+} and Ca^{2+} ions in solution can be completely separated by selective precipitation with F^-.

Chapter 17

Particulate Review

Combustion is an exothermic process resulting from the endothermic breaking of bonds in propane (C—H single bonds and C—C single bonds) and oxygen (O=O double bonds) along with the exothermic formation of bonds in carbon dioxide (C=O double bonds) and water (H—O single bonds).

Concept Tests

p. 855 (a), (c), and (d)
p. 858 $\Delta S_{univ} > 0$
p. 865 (a) Yes, it is spontaneous because $\Delta G < 0$; (b) thermodynamics says nothing about the rate of reaction.
p. 867 The reaction is very slow, and the activation energy is large.
p. 868 No. The sum of $\Delta G^\circ_{f,prod}$ are all the same, but the reactants have different ΔG°_f values, so each reaction will have a different ΔG° value.
p. 870

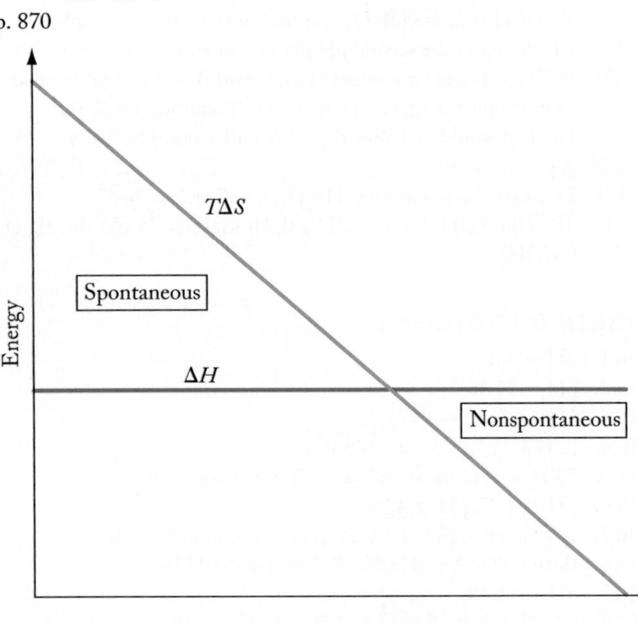

p. 876 (e)
p. 886 (e)

Practice Exercises

17.1. (a) increase in entropy; (b) decrease in entropy
17.2. (a) > (b) > (d) > (c)
17.3. Prediction: S_{sys} decreases; $\Delta S^\circ_{rxn} = -243$ J/K
17.4. a. ΔS_{rxn} is expected to be negative.
b. $\Delta S^\circ_{rxn} = -326$ J/K
c. $\Delta H^\circ_{rxn} = -571.6$ kJ/mol
d. $\Delta G^\circ_{rxn} = -474.3$ kJ, so the reaction is spontaneous under standard conditions.
17.5. 142.2 kJ/mol
17.6. $T < 5939$ K
17.7. The reaction is spontaneous only at high temperatures.
17.8. 6.09×10^5
17.9. 2.95×10^{-37} at 298 K; 9.20×10^{-13} at 2000 K
17.10. $\Delta G^\circ = -196$ kJ

Chapter 18

Particulate Review

In the reaction $4\ Fe(s) + 3\ O_2(g) \rightarrow 2\ Fe_2O_3(s)$, iron is oxidized and oxygen is reduced as electrons transfer from iron to oxygen.

Concept Tests

p. 903 No, because the atomic masses of Cu and Zn are different.
p. 904 The cell consists of a zinc anode, which is oxidized to Zn^{2+} ions as Cu^{2+} ions are reduced to Cu metal at the cathode.
p. 908 The most reactive metals listed at the top of the activity series form ions that are the least likely to be reduced, which is reflected in their very negative standard reduction potentials that are listed at the bottom of Table A6.1.
p. 908 (a)
p. 912 $E^\circ_{cell} < 0$, negative; $\Delta G^\circ_{cell} > 0$, positive
p. 916 SHE: 0.257 V; Ag/AgCl: 0.479 V
p. 917 Less than 0.800 V
p. 924 C is oxidized; Co is reduced.
p. 930 H_2
p. 934 Redox flow batteries are designed to meet very large electrical energy needs, so they need very large volumes of dissolved electrolytes to produce that energy.

Practice Exercises

18.1. $O_2(g) + 2\ NO_2^-(aq) \rightarrow 2\ NO_3^-(aq)$
18.2. The balanced redox reaction is

$$3\ Cu^{2+}(aq) + 2\ Al(s) \rightarrow 3\ Cu(s) + 2\ Al^{3+}(aq)$$

The cell diagram is

$$Al(s)\ |\ Al^{3+}(aq)\ ||\ Cu^{2+}(aq)\ |\ Cu(s)$$

18.3. The net ionic equation is

$$Cd(s) + 2\ NiO(OH)(s) + 2\ H_2O(\ell) \rightarrow Cd(OH)_2(s) + 2\ Ni(OH)_2(s) \quad E^\circ_{cell} = 1.33\ V$$

18.4. $\Delta G_{cell} = -2.90 \times 10^2$ kJ
18.5. $E_{cell} = 1.63$ V
18.6. $K = 2 \times 10^{62}$
18.7. 1.30×10^2 g
18.8. 1.5 g

Chapter 19

Particulate Review

From left to right: Nuclide 1 has a mass number of 11 (5 protons + 6 neutrons; 5 electrons), as does nuclide 2 (6 protons + 5 neutrons; 6 electrons). Nuclide 3 has a mass number of 13 (6 protons + 7 neutrons; 6 electrons). Nuclides 2 and 3 are isotopes.

Concept Tests

p. 950 The loss in mass becomes an energy that must be added to separate the nucleons that are bound together.

p. 955 The neutron-to-proton ratio for stable isotopes increases with atomic number. For heavy isotopes it is about 1.5 to 1. Losing α particles increases this ratio, creating neutron-rich nuclides that undergo β decay.

p. 961 Increased CO_2 (^{14}C depleted) in the air from burned fossil fuels will reduce the $^{14}C/^{12}C$ ratio in the air and in plant tissues.

p. 966 More: 200 Bq/m^3 = 5.4 pCi/L.

p. 974 To fuse nuclei, their coulombic repulsion must be overcome, which requires they collide at very high velocities, requiring very high temperatures. Human-made fusion has not been carried out except in a hydrogen bomb. We cannot make a fusion reactor until the process itself can be safely carried out.

Practice Exercises

19.1. β decay; $^{28}_{14}Si$

19.2. $A = 6.91 \times 10^9$ Bq; $A = 187$ mCi

19.3. 98.6%

19.4. 4900 years old

19.5. 1×10^{11} X-rays

19.6. 1.70×10^9 kJ/mol

Chapter 20

Particulate Review

The functional groups from left to right in the top row are carboxylic acid, alcohol, and amine. In the second row from left to right, the functional groups are an ether, amide, and aldehyde. In the bottom row from left to right they are a ketone and ester. The carboxylic acid and the amine function as a Brønsted–Lowry acid and base, respectively. When they react, they form the amide.

Concept Tests

p. 995 (e)

p. 1003 (a) alkane; (b) aromatic; (c) alcohol; (d) amide; (e) carboxylic acid; (f) amine

p. 1003 Glycine is not chiral because the α-carbon is not bonded to four different groups (there are two hydrogen atoms). Isoleucine and threonine each have two chiral centers.

p. 1010 Those with nonpolar R groups, such as alanine, leucine, and isoleucine.

p. 1012 No, a change in the 2° structure is involved.

p. 1019

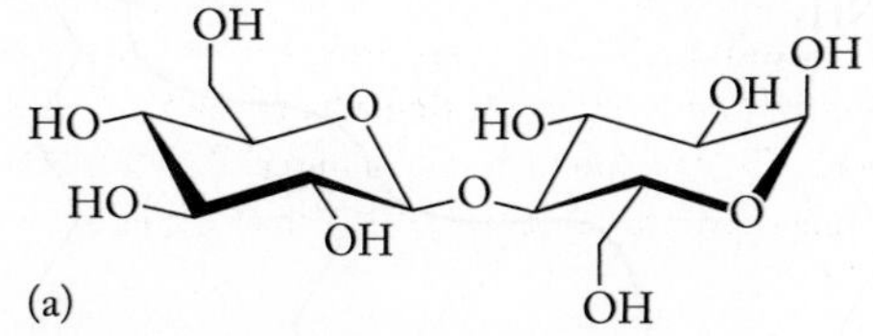

(a)

p. 1023 At low temperatures the unsaturated fatty acids remain liquid.

p. 1023 Olestra is a much larger molecule with many more C—C and C—H bonds, so on a per-mole basis it would give off much more energy.

p. 1029 No, the direction in which the code is read matters; UUG codes for leucine.

Practice Exercises

20.1. (a) Constitutional isomers; (b) two different compounds; (c) constitutional isomers

20.2.

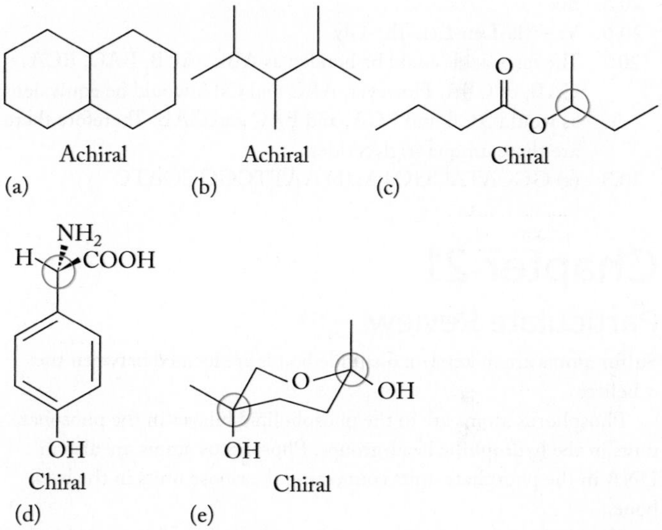

(a) Achiral (b) Achiral (c) Chiral

(d) Chiral (e) Chiral

20.3.

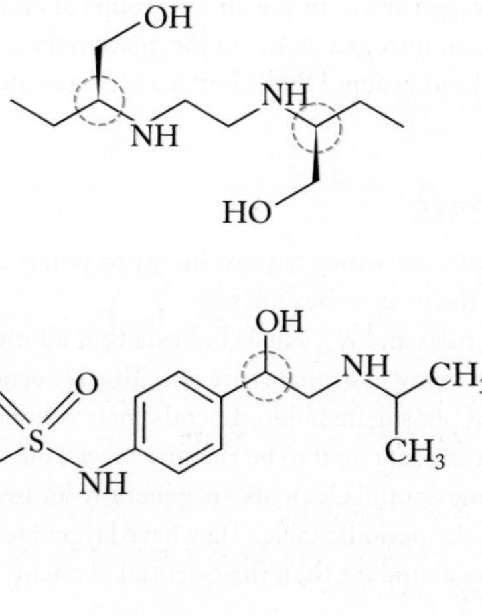

20.4.

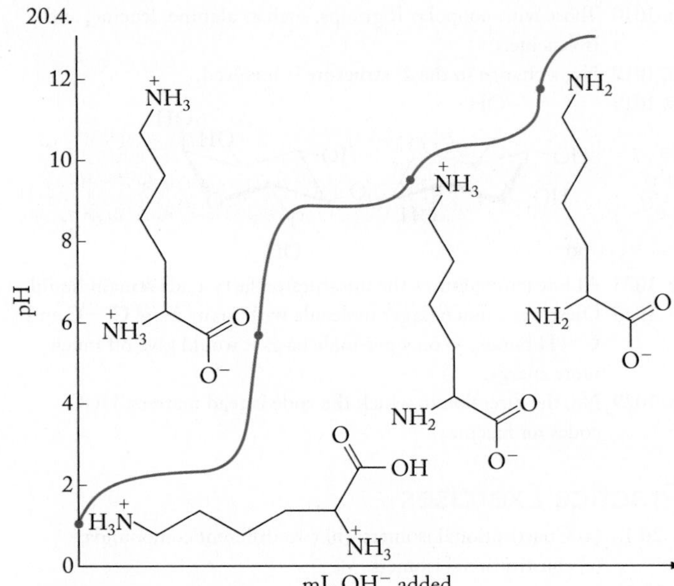

20.5. Six

20.6. Val-Ala-Leu-Leu-Thr-Gly

20.7. The fatty acids could be bonded as ABC, ACB, BAC, BCA, CAB, or CBA. However, ABC and CBA would be equivalent, as would ACB and BCA, and BAC and CAB. Therefore there are three unique triglycerides.

20.8. (a) GCCATAGGCTA; (b) AATTCGGCGATC

Chapter 21

Particulate Review

Sulfur atoms are in keratin: disulfide bonds are located between the α helices.

Phosphorus atoms are in the phospholipid bilayer in the phosphate units in the hydrophilic head-groups. Phosphorus atoms are also in DNA in the phosphate units connecting the ribose units in the backbone.

Keratin has nitrogen atoms in the amide groups of amino acids; phospholipids contain nitrogen atoms in the quaternary amine groups in the hydrophilic head group; DNA contains nitrogen atoms in the nucleotide bases.

Concept Tests

p. 1054 Nonspontaneous—they require energy to pump ions.

p. 1057 $K > 1$ for the drug to be effective.

p. 1060 Solubility rules and K_{sp} values indicate that all nitrates are soluble but many phosphates are not. This is borne out by the presence of phosphate in biominerals such as bone and teeth.

p. 1068 Essential elements tend to be the first or second elements in a group. Nonessential elements are generally located toward the bottom of the periodic table. They have larger atomic numbers and are less abundant than the essential elements in the same group.

p. 1069 To avoid tissue damage, the nuclide should not emit high-energy α or β particles, which have greater RBE.

Practice Exercises

21.1. $E_{K^+} = -179$ mV; $\Delta G_{transport} = +17$ kJ/mol; because $\Delta G > 0$, the transport is nonspontaneous.

21.2. pH = 1.194; the volume of $Mg(OH)_2$ solution required to neutralize the acid solution is 0.584 mL.

21.3. $N_2(g) + 10\ H^+(aq) + 8\ e^- \rightarrow 2\ NH_4^+(aq) + H_2(g)$

21.4.

O=N–O ↔ O–N=O

$[O=N–O]^-$ ↔ $[O–N=O]^-$

Both species have bent molecular geometry about the nitrogen atoms; the nitrogen atoms are sp^2 hybridized, and lone pairs are in sp^2 hybrid orbitals.

21.5. Bismuth-212: 2.0×10^{-6} mg; bismuth-213: 9.5×10^{-9} mg

Chapter 22

Particulate Review

The Lewis structures for ammonia (NH_3), borane (BH_3), and water (H_2O) are

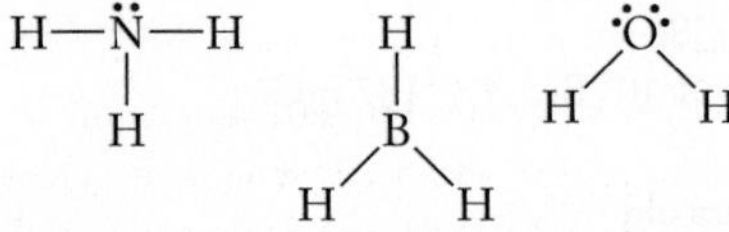

Ammonia has one lone pair on its central atom, and water has two lone pairs on its central atom; therefore both can function as Lewis bases. Borane has no lone pairs on the central boron atom and an incomplete octet; it can function as a Lewis acid.

Concept Tests

p. 1086 Solutions of the orange compound will have greater electrical conductivity and a lower freezing point than solutions of the reddish purple compound because they form four and three ions, respectively, in aqueous solution.

p. 1087 CN^- ions occupy the inner coordination of Fe^{3+}; Na^+ is the counterion. $Na_3[Fe(CN)_6]$ would have the same conductivity as $[Co(NH_3)_6]Cl_3$.

p. 1092 Tetrachloroplatinate(II)

p. 1102 Green

p. 1104 (a) CN^- is a stronger field ligand than pyridine. (b) Ru^{2+} ions are larger than Fe^{2+} and have a larger Δ_o; their 4*d* electrons interact more with ligand lone pairs than do the 3*d* electrons of Fe^{2+}.

p. 1106 No for square planar; yes for tetrahedral

p. 1109 Yes. Manganese(IV) has three unpaired *d* electrons with only one possible distribution; there is no other option. Manganese(III) has four *d* electrons; depending on the strength of the ligands, it can be either high spin (4 unpaired electrons) or low spin (2 unpaired electrons).

p. 1111 H_2

p. 1116 A longer half-life means slower decay; collecting a sufficient signal to get an image will take longer.

Practice Exercises

22.1. $[Ru(NH_3)_4Cl_2]^+$ is the complex ion; Ru is present as Ru^{3+}; chloride is the counterion.

22.2.

		LIGAND								
Compound	**Counterion**	**Formula**	**Name**	**Number**	**Prefix**	**Formula**	**Name**	**Number**	**Prefix**	M^{n+}
$[Zn(NH_3)_4]Cl_2$	Cl^- (chloride)	NH_3	ammine	4	tetra-					2^+
$[Co(NH_3)_4(H_2O)_2](NO_2)_2$	NO_2^- (nitrite)	NH_3	ammine	4	tetra-	H_2O	aqua	2	di-	2^+

a. $[Zn(NH_3)_4]Cl_2$ = tetraamminezinc(II) chloride
b. $[Co(NH_3)_4(H_2O)_2](NO_2)_2$ = tetraamminediaquacobalt(II) nitrite

22.3. Six, if we count the six oxygens in the $-CO_2^-$ groups or three if we consider each $-CO_2^-$ as one group.

22.4. Ni^{3+} can have either a high-spin or a low-spin configuration; none of the ions is diamagnetic.

22.5.

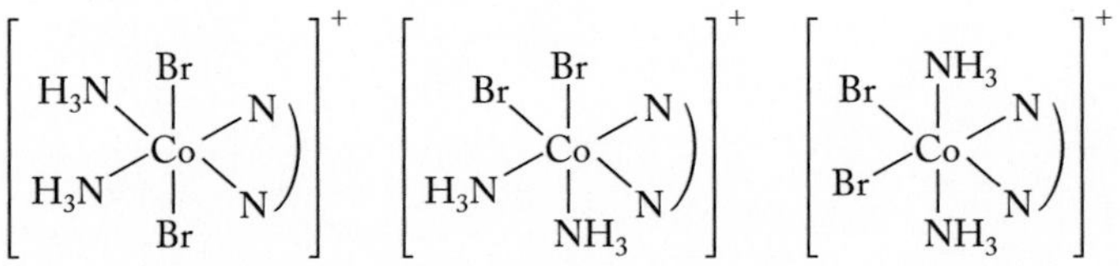

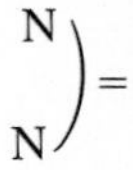

$N{\frown}N = H_2NCH_2CH_2NH_2$

cis-Diammine-*trans*-dibromo(ethylenediamine)cobalt(III)
cis-Diammine-*cis*-dibromo(ethylenediamine)cobalt(III)
trans-Diammine-*cis*-dibromo(ethylenediamine)cobalt(III)

22.6. 2.5×10^6

Answers
to Selected End-of-Chapter Questions and Problems

Chapter 1

1.1. (a) A pure compound in the gas phase; (b) a heterogeneous mixture of blue-element atoms and red-element atoms: blue atoms are in the gas phase and red atoms are in the liquid phase.

1.3. (b)

1.5. CO_2 would be a solid at 180 K, and a solid is characterized by having a rigid three-dimensional array in which the molecules are in close proximity to one another.

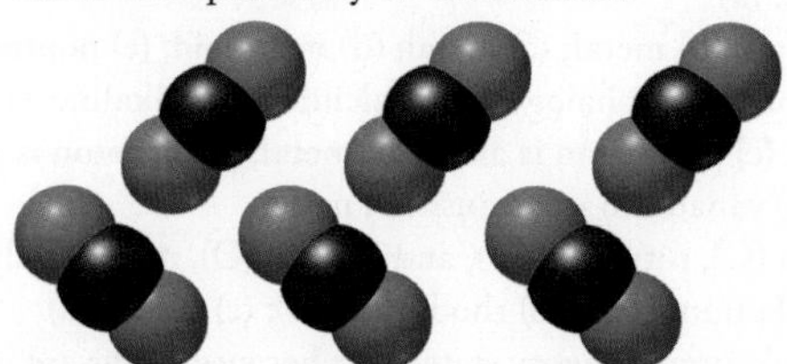

1.7. (a) $C_4H_{10}O$; (b) $C_2H_6O_2$; (c) CCl_4

1.9. Sample A is both accurate and precise. Sample B is precise but not accurate.

1.11. (a); (b); (e)

1.13. (a)

1.15. (a)

1.17. Nitrogen is N_2 and oxygen is O_2.

1.19. (a) N_2O; (b) NO_2; (c) NO

1.21. $H_2O(g) + SO_3(g) \rightarrow H_2SO_4(\ell)$

1.23. Density, melting point, thermal and electrical conductivity, and softness (a–d) are all physical properties, whereas both tarnishing and reaction with water (e and f) are chemical properties.

1.25. Parts (a), (b), (c), (e), and (f) are physical properties; part (d) is a chemical property.

1.27. (c) and (f)

1.29. (b)

1.31. Particles are most free to move about in the gas phase and least free in the solid phase.

1.33. The snow sublimed to form water vapor.

1.35. Extensive properties will change with the size of the sample and therefore cannot be used to identify a substance.

1.37. To form a hypothesis we need at least one observation, experiment, or idea (from examining nature).

1.39. yes

1.41. *Theory* in normal conversation refers to someone's idea or opinion or speculation that can be changed. Scientifically, *theory* refers to an idea that has been validated by experimentation.

1.43. SI units can be easily converted into a larger or smaller unit by multiplying or dividing by multiples of 10. USC units are based on other number multiples and thus are more cumbersome to manipulate.

1.45. 23 g

1.47. (a) $\frac{1\ \text{ps}}{1 \times 10^3\ \text{fs}}$; (b) $\frac{1\ \text{kg}}{1 \times 10^6\ \text{mg}}$; (c) $\frac{1\ \text{kg Ti}}{2.20 \times 10^{-4}\ \text{m}^3}$ or $\frac{4540\ \text{kg Ti}}{1\ \text{m}^3}$

1.49. 1.99 m

1.51. 3.27 m/s

1.53. 941 Calories

1.55. 58.0 cm^3

1.57. 73.8 mL

1.59. (a) 5.8×10^3 g; (b) 2.1×10^2 oz.

1.61. Only one data point may be eliminated.

1.63. The standard deviation (s) is larger than the range described by the 95% confidence interval. The formula for calculating confidence interval in part relies on the standard deviation multiplied by the ratio of the t value divided by the square root of n (the number of data points). In this case the t value is less than the square root of the number of data points so the 95% confidence interval will be less than the standard deviation.

1.65. (b) and (c), depending on the significance of the zero in (c).

1.67. (b) and (c), depending on the significance of the zero in (c).

1.69. (b), (c), and (d), depending on the significance of the zero in (d).

1.71. (a) 17.4; (b) 7×10^{-14}; (c) 5.70×10^{-23}; (d) 3.58×10^{-3}

1.73. (a) 7.0×10^5 g/mL; (b) 7.88 g; (c) 4.29×10^{-21} J; (d) 461 m/s

1.75. (a) Manufacturer 1 has $\bar{x} = 0.511$ and $s = 0.005$; manufacturer 2 has $\bar{x} = 0.513$ and $s = 0.001$; manufacturer 3 has $\bar{x} = 0.501$ and $s = 0.001$; (b) manufacturer 3; (c) manufacturer 3 is precise and accurate; manufacturer 2 is precise but not accurate.

1.77. Yes, it is an outlier.

1.79. A scale where all possible values of temperature are above zero. The coldest possible temperature would correspond to a value of 0.

1.81. 1235 K and 1337 K

1.83. Converting all temperatures to Kelvin for comparison: The first is already in Kelvin, 93.0 K; 250.0°C converts to 23.2 K; and −231.1°F converts to 127.0 K. The highest critical temperature is 127.0 K for the sample originally measured in °F.

1.85. −269.0°C

1.87. 39.2°C

1.89. −89.2°C; 183.9 K

1.91. Both mixtures (a) and (b) react so that there is neither sodium nor chlorine left over.

1.93. a. Peanuts $\bar{x}$ = 63%, s = 3.9; raisins $\bar{x}$ = 37%, s = 3.9.
b. Yes, peanuts μ = 63 ± 4.6%; raisins μ = 37 ± 4.6%.

1.95. The procedure that scientists use to measure values such as the age of a dinosaur bone doesn't provide a precision that allows the measurement to be reliable to within a single year. A range of values that includes the uncertainty of the measurement in question (this bone is between 67 million and 69 million years old) might be more relatable.

1.97. 5.2×10^2 mg of NaF

1.99. 10 tablets

Chapter 2

2.1. The positively charged alpha particle will be attracted to the negative (left) plate, whereas the negatively charged beta particle will be attracted to the positive (right) plate.

2.3. (a) $^{11}_{5}B$; (b) $^{11}_{6}C$; (c) $^{13}_{6}C$

2.5. (c)

2.7. (a) chlorine (Cl_2) (yellow); (b) neon (Ne) (red); (c) sodium (Na) (dark blue)

2.9. (a) Mg (green) will form MgO; (b) K (red) will form K_2O; (c) Ti (yellow) will form TiO_2; (d) Al (dark blue) will form Al_2O_3.

2.11. The element shaded in red (Zn) is too heavy to have been produced by stellar fusion.

2.13. Because some of the alpha particles were bounced back toward their source and some were scattered at large angles of deflection, Rutherford concluded that the positive charge in the atom could not be spread out (as described in the plum-pudding model) in the atom, but must be concentrated in the center of the atom (the nucleus). Because most of the particles passed directly through the gold foil, he reasoned that the nucleus must be small compared to the entire atom. The negatively charged electrons do not deflect the particles, and Rutherford reasoned that the electrons took up the remainder of the space of the atom outside of the nucleus.

2.15. The fact that cathode rays were deflected by a magnetic field indicated that the rays were streams of charged particles.

2.17. A *weighted average* takes into account the proportion of each value in the group of values to be averaged.

2.19. The element symbol (X) and the atomic number (*Z*) provide the same information. Any change to the number of protons would change the identity of the element.

2.21. (a) ^{11}B; (b) ^{7}Li; (c) ^{14}N; (d) ^{20}Ne

2.23.

	Atom	Mass Number	Atomic Number = Number of Protons	Number of Neutrons = Mass Number − Atomic Number	Number of Electrons = Number of Protons
(a)	^{14}C	14	6	8	6
(b)	^{59}Fe	59	26	33	26
(c)	^{90}Sr	90	38	52	38
(d)	^{210}Pb	210	82	128	82

2.25.

Symbol	^{16}O	^{56}Fe	^{118}Sn	^{197}Au
Number of Protons	8	26	50	79
Number of Neutrons	8	30	68	118
Number of Electrons	8	26	50	79
Mass Number	16	56	118	197

2.27. Average atomic mass for the Australian volcanic crater lake sample is 10.82 u. This is slightly higher than the textbook value of 10.811 u.

2.29. The average atomic mass of magnesium is slightly higher on Mars, 24.32 u.

2.31. 47.95 u

2.33. Mendeleev knew only the masses of the elements at the time he arranged the elements in his periodic table.

2.35. group 2, RO; group 3, R_2O_3; group 4, RO_2

2.37. (d)

2.39. (a), (b), (d)

2.41. (a) metal; (b) metal; (c) metal; (d) metalloid; (e) nonmetal

2.43. (a) Bromine is a halogen; (b) calcium is an alkaline earth metal; (c) potassium is an alkali metal; (d) krypton is a noble gas; (e) vanadium is a transition metal.

2.45. carbon (C), nitrogen (N), and oxygen (O), respectively

2.47. (a) palladium (Pd); (b) rhodium (Rh); (c) gold (Au)

2.49. Dalton's atomic theory states that because atoms are indivisible, the ratio of the atoms in a compound must be a ratio of whole numbers. When water decomposes the ratio of the volume of hydrogen to oxygen is 2:1, a whole-number ratio, because the atoms in water are present in that same 2:1 ratio.

2.51. Molecular compounds are composed of nonmetals; ionic compounds are formed between a metal (or metals) and a nonmetal (or nonmetals).

2.53. 1:1.5 (or 2:3) for $CoS:Co_2S_3$

2.55. 7.5 g O

2.57.

Symbol	$^{37}S^{2-}$	$^{19}F^{-}$	$^{90}Sr^{2+}$	$^{120}Sn^{4+}$
Number of Protons	16	9	38	50
Number of Neutrons	18	10	52	70
Number of Electrons	18	10	36	46
Mass Number	34	19	90	120

2.59. Compounds (a) and (d) consist of molecules; compounds (b) and (c) consist of ions.

2.61. (a) ionic bonds; (b) covalent bonds; (c) covalent bonds; (d) ionic bonds

2.63. (a) 9 atoms; (b) 5 atoms; (c) 14 atoms; (d) 13 atoms

2.65. XO_2^{2-}

2.67. Roman numerals indicate the charge on the transition metal cation in the compound.

2.69. (a) phosphorus monoxide; (b) diphosphorus trioxide; (c) diphosphorus tetroxide; (d) diphosphorus pentoxide

2.71. (a) Na_2S, sodium sulfide; (b) $SrCl_2$, strontium chloride; (c) Al_2O_3, aluminum oxide; (d) LiH, lithium hydride

2.73. magnesium hydroxide

2.75. (a) sodium oxide; (b) sodium sulfide; (c) sodium sulfate; (d) sodium nitrate; (e) sodium nitrite

2.77. (a) K_2S; (b) K_2Se; (c) Rb_2SO_4; (d) $RbNO_2$; (e) $MgSO_4$

2.79. (a) $NaBrO$; (b) K_2SO_4; (c) $LiIO_3$; (d) $Mg(NO_2)_2$
2.81. (b)
2.83. (b), lithium sulfate should be Li_2SO_4
2.85.

	Na^+ Cation	K^+ Cation	Mg^{2+} Cation	Ca^{2+} Cation
Cl^- Anion	$NaCl$	KCl	$MgCl_2$	$CaCl_2$
$H_2PO_4^-$ Anion	NaH_2PO_4	KH_2PO_4	$Mg(H_2PO_4)_2$	$Ca(H_2PO_4)_2$

2.87. (a) chromium(III) telluride; (b) vanadium(III) sulfate; (c) iron(I) chromate; (d) manganese(II) oxide
2.89. (a) $ZnCr_2O_7$; (b) $Fe(CH_3CO_2)_3$; (c) Hg_2O_2; (d) $Sc(SCN)_3$
2.91. (a) HI; (b) HBrO; (c) HIO_3; (d) HBr
2.93. (a) hydrofluoric acid; (b) bromic acid; (c) hydrobromic acid; (d) periodic acid
2.95. alkanes, alkenes, and alkynes
2.97. Amines and amides contain nitrogen atoms. Amides contain both nitrogen and oxygen atoms.
2.99. (a) alkanes; (b) alkynes
2.101. RCO_2R', the ester functional group
2.103. Electrons (d) formed first; deuterons (a) formed last.
2.105. The density of the universe is decreasing with time.
2.107. $HClO_3$, chloric acid
2.109. ${}^{21}_{10}Ne + {}^{4}_{2}\alpha \rightarrow {}^{1}_{0}n + {}^{24}_{12}Mg$
2.111. a. two
b. The beta particle would be deflected toward the top—away from the negative plate (toward the positive plate). The alpha particle would be deflected down—away from the positive plate (toward the negative plate).
2.113. (a) 12 H atoms for every 1 He atom; (b) the ratio is smaller so there is more helium present now; (c) stars consume hydrogen and produce helium; (d) evaluate the elemental composition of older galaxies and compare to our galaxy.
2.115. 14 Cu:1 Sn
2.117. (a) Scandium (Sc) with an average mass of 44.956 u, gallium (Ga) with an average mass of 69.723 u, and germanium (Ge) with an average mass of 72.61 u; (b) ekaaluminum is gallium, ekaboron is scandium, and ekasilicon is germanium; (c) scandium was discovered in 1879 by Lars Fredrik Nilson in Sweden, gallium was discovered in 1875 by Paul-Émile Lecoq de Boisbaudran in France, and germanium was discovered in 1886 by Clemens Winkler in Germany.
2.119. 60.11%
2.121. (a) ${}^{79}Br$–${}^{79}Br$ = 157.8366 u, ${}^{79}Br$–${}^{81}Br$ = 159.8346 u, ${}^{81}Br$–${}^{81}Br$ = 161.8326 u; (b) ${}^{79}Br$–${}^{79}Br$, 25.41% abundant, ${}^{79}Br$–${}^{81}Br$, 50.00% abundant, ${}^{81}Br$–${}^{81}Br$, 24.59% abundant
2.123. (a) CO_2, carbon dioxide; (b) Li_3N, lithium nitride
2.125. Radium will adopt a 2+ charge to form Ra^{2+} ions. It will likely be malleable, be relatively dense, conduct heat and electric current, and melt at a fairly high temperature in its metallic state.

	Melting Point (°C)		Melting Point (°C)
$CaCl_2$	772	CaO	2572
$SrCl_2$	874	SrO	2531
$BaCl_2$	962	BaO	1923
$RaCl_2$	(950 to 1050)	RaO	(1700 to 2000)

2.127. Despite being heavier (on average), argon contains 18 protons, whereas potassium contains 19 protons. Because the modern periodic table is organized by increasing atomic number, argon is placed before potassium.

Chapter 3

3.1. (a) $X(g) + Y(g) \rightarrow XY(g)$; (b) $X(g) + Y(g) \rightarrow XY(s)$; (c) $2\,X(g) + 2\,Y(g) \rightarrow XY_2(g)$; (d) $X_2(g) + Y_2(g) \rightarrow 2\,XY(g)$
3.3. (b)
3.5. (a) and (c) have the same empirical formula (NO_2), and (b) and (d) have the same empirical formula (N_2O).
3.7. (a) water; (b) carbon dioxide
3.9. 75%
3.11. All would be combination reactions (any two or more substances combine to form one product).
3.13. (c) $Fe(\ell)$
3.15. Dozen is too small of a unit to conveniently express the very large number of atoms, ions, or molecules present in laboratory quantities.
3.17. No, the molar mass of a substance does not directly correlate to the number of atoms in a molecular compound. The statement would be true only if the two compounds were composed of the same element.
3.19. (a) 7.3×10^{-10} mol Ne; (b) 7.0×10^{-11} mol CH_4; (c) 4.2×10^{-12} mol O_3; (d) 8.1×10^{-15} mol NO_2
3.21. (a) 7.53×10^{22} Ti atoms; (b) 7.53×10^{22} Ti atoms; (c) 1.51×10^{23} Ti atoms; (d) 2.26×10^{23} Ti atoms
3.23. (a) 4 moles of Fe_2S_3; (b) 4 moles of CaS; (c) each quantity contains the same number of moles of S
3.25. (a) 3.00 mol; (b) 3.00 mol; (c) 4.50 mol
3.27. 41.63 mol
3.29. 7×10^{19} Ir atoms
3.31. (a) 64.06 g/mol; (b) 48.00 g/mol; (c) 44.01 g/mol; (d) 108.01 g/mol
3.33. (a) 152.16 g/mol; (b) 164.22 g/mol; (c) 148.22 g/mol; (d) 132.17 g/mol
3.35. (a) 55.85 g Fe; (b) 111.70 g Fe; (c) 55.85 g Fe; (d) 167.55 g Fe
3.37. 0.752 mol SiO_2
3.39. H_2O
3.41. 10.3 g
3.43. diamond
3.45. no
3.47. no
3.49.

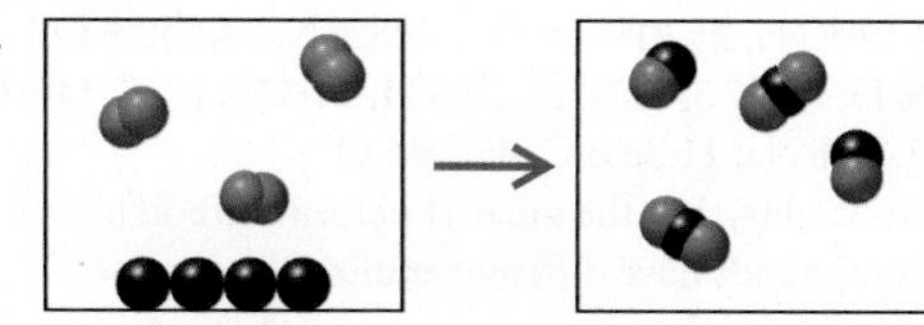

3.51. 2
3.53. a. $3\,FeSiO_3(s) + 4\,H_2O(\ell) \rightarrow Fe_3Si_2O_5(OH)_4(s) + H_4SiO_4(aq)$
b. $Fe_2SiO_4(s) + 2\,CO_2(g) + 2\,H_2O(\ell) \rightarrow 2\,FeCO_3(s) + H_4SiO_4(aq)$
c. $Fe_3Si_2O_5(OH)_4(s) + 3\,CO_2(g) + 2\,H_2O(\ell) \rightarrow 3\,FeCO_3(s) + 2\,H_4SiO_4(aq)$
3.55. a. $N_2(g) + O_2(g) \rightarrow 2\,NO(g)$
b. $2\,NO(g) + O_2(g) \rightarrow 2\,NO_2(g)$
c. $NO(g) + NO_3(g) \rightarrow 2\,NO_2(g)$
d. $2\,N_2(g) + O_2(g) \rightarrow 2\,N_2O(g)$

3.57. a. $N_2O_5(g) + Na(s) \rightarrow NaNO_3(s) + NO_2(g)$
b. $H_2O(\ell) + N_2O_4(g) \rightarrow HNO_3(aq) + HNO_2(aq)$
c. $3\ NO(g) \rightarrow N_2O(g) + NO_2(g)$
d. $2\ C_2H_2(g) + 5\ O_2(g) \rightarrow 4\ CO_2(g) + 2\ H_2O(g)$

3.59. Yes; if the masses were unequal, then the equation would be missing either some reactants or products.

3.61. 1.17 kg

3.63. 31.5 g

3.65. 60. g of SO_2

3.67. $2\ C_8H_{18} + 25\ O_2 \rightarrow 16\ CO_2 + 18\ H_2O$; $C_2H_6O + 3\ O_2 \rightarrow 2\ CO_2 + 3\ H_2O$; octane produces more CO_2 per gram

3.69. 346 g

3.71. (c) less than the sum of the masses of Fe and S at the start

3.73. Reactions do not always go to completion because the reaction may be slow or may have, for a portion of the reaction, yielded different products than expected.

3.75. The lowered mass of reactant will lead to a lowered mass of decomposition products and therefore a lower actual yield. Because percent yield compares actual yield to theoretical yield (which would be based on the mass of $NaHCO_3$ before the spill) it will also be lower.

3.77. 3 cups

3.79. (a) 5.0 g Li_3N; (b) 34.5 g H_3PO_4; (c) 8.0 g SO_3

3.81. $NH_3(g) + HCl(g) \rightarrow NH_4Cl(aq)$; 0.7 g NH_3

3.83. (a) $2\ PbO(s) + 2\ NaCl(aq) + H_2O(\ell) + CO_2(g) \rightarrow Pb_2Cl_2CO_3(s) + 2\ NaOH(aq)$; (b) 12.2 g; (c) 22.3%

3.85. 56%

3.87. (a) $C_6H_{12}O_6(aq) \rightarrow 2\ C_2H_5OH(\ell) + 2\ CO_2(g)$; (b) 77.1%

3.89. An empirical formula shows the lowest whole-number ratio of atoms in a substance. A molecular formula shows the actual numbers of each kind of atom that compose one molecule of the substance.

3.91. Yes, an empirical formula gives the simplest whole-number ratio of the elements. The molecular formula has the same ratio of elements but may be a whole-number multiple of the empirical formula.

3.93. Yes. All of these have the same empirical and molecular formulas. These three molecules are isomers (molecules with the same molecular formula but different connections between atoms).

3.95.

Molecular Formula	C_6H_{14}	C_7H_{16}	C_8H_{18}	C_9H_{20}
Empirical Formula	C_3H_7	C_7H_{16}	C_4H_9	C_9H_{20}

3.97. (a) 74.19% Na, 25.81% O; (b) 57.48% Na, 40.00% O, 2.52% H; (c) 27.37% Na, 1.20% H, 14.30% C, 57.14% O; (d) 43.38% Na, 11.33% C, 45.28% O

3.99. Pyrene, $C_{16}H_{10}$, has the greatest percent carbon by mass. All four compounds have different empirical formulas.

3.101. CH_4

3.103. Ti_6Al_4V

3.105. citric acid

3.107. $Mg_3Si_2H_4O_9$

3.109. The empirical formula is CHN; the molecular formula is $C_5H_5N_5$.

3.111. The excess of oxygen is required in combustion analysis to ensure the complete reaction of the hydrogen and carbon to form water and carbon dioxide.

3.113. the molar mass of the compound

3.115. NO_2

3.117. The empirical formula is C_2H_3; the molecular formula is $C_{20}H_{30}$.

3.119. $C_{10}H_{18}O$ (empirical and molecular)

3.121. $C_9H_8O_4$ (empirical and molecular)

3.123. (a) 36.08%; (b) 36.08%

3.125. (a) 870 g; (b) 1020 g; (c) 1400 g

3.127. (a) $a = 1$, $b = 3$, charge on U is 6+; (b) $c = 3$, $d = 8$, charge on U is 5.33+; (c) $x = 2$, $y = 2$, $z = 6$

3.129. (a) 5.838×10^{20} molecules of $C_{13}H_{18}O_2$; (b) 3.008×10^{21} formula units of $CaCO_3$; (c) 9×10^{18} molecules of $C_{16}H_{19}ClN_2$

3.131. (a) Mn_2O_3 is manganese(III) oxide, and MnO_2 is manganese(IV) oxide; (b) % Mn in MnO_2 = 69.60%, % Mn in Mn_2O_3 = 63.19%; (c) these compounds contain the same elements but in different atom ratios

3.133. (a) 0.966 g; (b) $C_{10}H_{20}O_{10}$

3.135. 1×10^{-8} mol

3.137. 55 mol of ethanol

3.139. Re

3.141. 82.4%

3.143. (a) 6.0 metric tons; (b) $2\ SO_2(g) + 2\ H_2O(g) + O_2(g) \rightarrow 2\ H_2SO_4(\ell)$; (c) 9.2 metric tons

3.145. 3.06 g H_2SO_4

3.147. (a) KCl; (b) NaCl; (c) Li_2CO_3

3.149. (a) 3.05×10^3 g KO_2; (b) more Na_2O_2; (c) KO_2

Chapter 4

4.1. yellow

4.3. (a) Cl (purple); (b) S (orange); (c) N (green); (d) P (blue)

4.5. strong electrolyte: (a) and (c); weak electrolyte: (b); strong acid: (c); weak acid: (b); nonelectrolyte: (d)

4.7. Hydronium and nitrate ions will remain in solution.

4.9. The solvent is usually the liquid component of the solution. If both the solvent and solute are liquids or solids, the solvent is the component present in the greatest amount.

4.11. 1.00 *M*

4.13. (a) 5.6 *M* $BaCl_2$; (b) 1.00 *M* Na_2CO_3; (c) 1.30 *M* $C_6H_{12}O_6$; (d) 5.92 *M* KNO_3

4.15. (a) 0.14 *M* Na^+; (b) 0.11 *M* Cl^-; (c) 0.096 *M* SO_4^{2-}; (d) 0.20 *M* Ca^{2+}

4.17. (a) 11.7 g NaCl; (b) 4.99 g $CuSO_4$; (c) 6.41 g CH_3OH

4.19. 0.590 g of dissolved ions in 2.75 L

4.21. (a) 0.0096 mol; (b) 7.80×10^{-4} mol; (c) 8.8×10^{-2} mol; (d) 4.22 mol

4.23. orchard sample: 3.4×10^{-4} mmol/L, 0.12 ppm; residential area sample: 5.6×10^{-6} mmol/L, 0.020 ppm; after storm sample: 0.032 mmol/L, 11 ppm

4.25. 4.54×10^{-7} mg NF_3/kg air

4.27. 4×10^{-5} *M*

4.29. (a) 0.0210 *M* K^+; (b) 1.06 m*M* LiCl; (c) 0.0434 m*M* Zn^{2+}

4.31. 58.6 mL

4.33. 0.40 *M*

4.35. 12.3 mL

4.37. $A = 0.75$

4.39. Table salt produces Na^+ and Cl^- ions in solution when it dissolves. Sugar does not dissociate into ions because it is a molecular compound with only covalent bonds. Ions are required to conduct electricity.

4.41. The lack of ions in methanol means that the liquid is nonconductive. Molten NaOH, however, has freely moving Na^+ and OH^- ions, which can conduct electricity.

4.43. The conductivity of the more concentrated solution will be higher than that of the dilute solution.
4.45. In order of decreasing conductivity: 1.0 M Na_2SO_4 > 1.2 M KCl > 1.0 M NaCl > 0.75 M LiCl
4.47. (a) 0.025 M; (b) 0.050 M; (c) 0.075 M
4.49. (b)
4.51. (a) < (c) < (b)
4.53. acid
4.55. (a) strong acid; (b) weak acid; (c) weak acid; (d) strong acid
4.57. base
4.59. (a) strong base; (b) weak base; (c) weak base; (d) strong base
4.61. (a) Ionic and net ionic equations are the same: $H^+(aq) + SO_4^-(aq) + Ca^{2+}(aq) + OH^-(aq) \rightarrow CaSO_4(s) + H_2O(\ell)$, the acid is H_2SO_4 and the base is $Ca(OH)_2$; (b) ionic and net ionic equations are the same: $PbCO_3(s) + H^+(aq) + HSO_4^-(aq) \rightarrow PbSO_4(s) + CO_2(g) + H_2O(\ell)$, the acid is H_2SO_4 and the base is $PbCO_3$; (c) ionic equation: $Ca^{2+}(aq) + 2\ OH^-(aq) + 2\ CH_3COOH(aq) \rightarrow Ca^{2+}(aq) + 2\ CH_3COO^-(aq) + 2\ H_2O(\ell)$, net ionic equation ($Ca^{2+}$ is a spectator ion): $OH^-(aq) + CH_3COOH(aq) \rightarrow CH_3COO^-(aq) + H_2O(\ell)$, CH_3COOH is the acid and $Ca(OH)_2$ is the base
4.63. (a) molecular equation: $Mg(OH)_2(s) + H_2SO_4(aq) \rightarrow MgSO_4(aq) + 2\ H_2O(\ell)$, net ionic equation: $Mg(OH)_2(s) + HSO_4^-(aq) + H^+(aq) \rightarrow Mg^{2+}(aq) + SO_4^{2-}(aq) + 2\ H_2O(\ell)$; (b) molecular equation: $MgCO_3(s) + 2\ HCl(aq) \rightarrow MgCl_2(aq) + H_2O(\ell) + CO_2(g)$, net ionic equation: $MgCO_3(s) + 2\ H^+(aq) \rightarrow Mg^{2+}(aq) + H_2O(\ell) + CO_2(g)$; (c) molecular equation and net ionic equation are the same: $NH_3(g) + HCl(g) \rightarrow NH_4Cl(s)$; (d) molecular equation: $SO_3(g) + 2\ NaOH(aq) \rightarrow Na_2SO_4(aq) + H_2O(\ell)$, net ionic equation: $SO_3(g) + 2\ OH^-(aq) \rightarrow SO_4^{2-}(aq) + H_2O(\ell)$
4.65. $PbCO_3(s) + 2\ H^+(aq) \rightarrow Pb^{2+}(s) + CO_2(g) + H_2O(\ell)$; $Pb(OH)_2(s) + 2\ H^+(aq) \rightarrow Pb^{2+}(s) + 2\ H_2O(\ell)$
4.67. (a) 18.0 mL, (b) 31.6 mL, (c) 114 mL
4.69. 0.72 g HI
4.71. 0.499 L
4.73. 280 mL
4.75. A saturated solution contains the maximum concentration of a solute. A supersaturated solution *temporarily* contains *more* than the maximum concentration of a solute at a given temperature.
4.77. $CaCO_3$
4.79. A saturated solution may not be a concentrated solution if the solute is only sparingly or slightly soluble in the solvent. In that case, the solution is saturated but dilute.
4.81. (a) Barium sulfate is insoluble, (e) lead(II) hydroxide is insoluble, and (f) calcium phosphate is insoluble
4.83. (a) Balanced equation: $Pb(NO_3)_2(aq) + Na_2SO_4(aq) \rightarrow PbSO_4(s) + 2\ NaNO_3(aq)$, overall ionic equation: $Pb^{2+}(aq) + 2\ NO_3^-(aq) + 2\ Na^+(aq) + SO_4^{2-}(aq) \rightarrow PbSO_4(s) + 2\ Na^+(aq) + 2\ NO_3^-(aq)$, net ionic equation: $Pb^{2+}(aq) + SO_4^{2-}(aq) \rightarrow PbSO_4(s)$; (b) no precipitation reaction occurs; (c) balanced equation: $FeCl_2(aq) + Na_2S(aq) \rightarrow FeS(s) + 2\ NaCl(aq)$, overall ionic equation: $Fe^{2+}(aq) + 2\ Cl^-(aq) + 2\ Na^+(aq) + S^{2-}(aq) \rightarrow FeS(s) + 2\ Na^+(aq) + 2\ Cl^-(aq)$, net ionic equation: $Fe^{2+}(aq) + S^{2-}(aq) \rightarrow FeS(s)$; (d) balanced equation: $MgSO_4(aq) + BaCl_2(aq) \rightarrow MgCl_2(aq) + BaSO_4(s)$, overall ionic equation: $Mg^{2+}(aq) + SO_4^{2-}(aq) + Ba^{2+}(aq) + 2\ Cl^-(aq) \rightarrow Mg^{2+}(aq) + 2\ Cl^-(aq) + BaSO_4(s)$, net ionic equation: $SO_4^{2-}(aq) + Ba^{2+}(aq) \rightarrow BaSO_4(s)$
4.85. 0.0211 g
4.87. 0.054 g
4.89. 1.3×10^5 g
4.91. (a) $[Na^+]$ and $[NO_3^-]$ remain the same, and $[Ag^+]$ and $[Cl^-]$ decrease; (b) $[Na^+]$ and $[Cl^-]$ remain the same, and $[H^+]$ and $[OH^-]$ decrease; (c) all ionic concentrations remain the same.
4.93. The number of electrons gained or lost is directly related to the change in oxidation number of a species.
4.95. (a) −1, (b) +1, (c) −2, (d) −3
4.97. Silver
4.99. $Na^+(\ell) + e^- \rightarrow Na(s)$, $2\ Cl^-(\ell) \rightarrow Cl_2(g) + 2\ e^-$
4.101. (a) +3, (b) +3, (c) +3
4.103. (a) $2\ e^- + Br_2(\ell) \rightarrow 2\ Br^-(aq)$, reduction; (b) $Pb(s) + 2\ Cl^-(aq) \rightarrow PbCl_2(s) + 2\ e^-$, oxidation; (c) $2\ e^- + O_3(g) + 2\ H^+(aq) \rightarrow O_2(g) + H_2O(\ell)$, reduction; (d) $2\ H_2SO_3(aq) + H^+(aq) + 2\ e^- \rightarrow HS_2O_4^-(aq) + 2\ H_2O(\ell)$, reduction
4.105. (a) $2\ H^+(aq) + MnO_2(s) + 2\ HCl(aq) \rightarrow Mn^{2+}(aq) + Cl_2(g) + 2\ H_2O(\ell)$, HCl is the reducing agent and MnO_2 is the oxidizing agent, manganese is reduced and chlorine is oxidized. (b) $I_2(s) + 2\ S_2O_3^{2-}(aq) \rightarrow S_4O_6^{2-}(aq) + 2\ I^-(aq)$, $S_2O_3^{2-}$ is the reducing agent and I_2 is the oxidizing agent, sulfur is oxidized and iodine is reduced; (c) $8\ H^+(aq) + MnO_4^-(aq) + Fe^{2+}(aq) \rightarrow Mn^{2+}(aq) + Fe^{3+}(aq) + 4\ H_2O(\ell)$, Fe^{2+} is the reducing agent and MnO_4^- is the oxidizing agent, iron is oxidized and manganese is reduced
4.107. a.

Reactants	Products
SiO_2: Si = +4, O = −2	Fe_2SiO_4: Fe = +2, Si = +4, O = −2
Fe_3O_4: Fe = 2 Fe^{3+}, 1 Fe^{2+}, O = −2	O_2: O = 0
Oxygen is oxidized (O^{2-} to O_2) and iron is reduced (Fe^{3+} to Fe^{2+})	

b.

Reactants	Products
SiO_2: Si = +4, O = −2	Fe_2SiO_4: Fe = +2, Si = +4, O = −2
Fe: Fe = 0	
O_2: O = 0	
Iron is oxidized (Fe^0 to Fe^{2+}) and oxygen is reduced (O_2 to O^{2-})	

c.

Reactants	Products
FeO: Fe = +2, O = −2	$Fe(OH)_3$: Fe = +3, O = −2, H = +1
O_2: O = 0	
H_2O: H = +1, O = −2	
Iron is oxidized (Fe^{2+} to Fe^{3+}) and oxygen is reduced (O_2 to O^{2-})	

4.109. (a) $O_2(aq) + 4\ FeCO_3(s) \rightarrow 2\ Fe_2O_3(s) + 4\ CO_2(g)$; (b) $O_2(aq) + 6\ FeCO_3\ (s) \rightarrow 2\ Fe_3O_4(s) + 6\ CO_2(g)$; (c) $O_2(aq) + 4\ Fe_3O_4(s) \rightarrow 6\ Fe_2O_3(s)$
4.111. $NH_4^+(aq) + 2\ O_2(g) \rightarrow NO_3^-(aq) + 2\ H^+(aq) + H_2O\ (\ell)$

4.113. a.

Reactants	Products
$HCrO_4^-$: H = +1, Cr = +6, O = −2	Cr_2O_3: Cr = +3, O = −2
H_2S: H = +1, S = −2	SO_4^{2-}: S = +6, O = −2

b. $3\ H_2S(aq) + 8\ HCrO_4^-(aq) + 2\ H^+ \rightarrow 3\ SO_4^{2-}(aq) + 4\ Cr_2O_3(s) + 8\ H_2O(\ell)$

c. 3

4.115. $2\ Fe(OH)_2^+(aq) + Mn^{2+}(aq) \rightarrow 2\ Fe^{2+}(aq) + 2\ H_2O(\ell) + MnO_2(s)$

4.117. zinc and aluminum

4.119. Vanadium is placed below aluminum and scandium is placed above aluminum on the activity series. We could use magnesium to test scandium's position. If magnesium is oxidized by Sc^{3+} ions, scandium must lie between aluminum and magnesium.

4.121. (a) $Cr_2O_7^{2-}(aq) + 14\ H^+(aq) + 6\ Fe^{2+}(aq) \rightarrow 2\ Cr^{3+}(aq) + 7\ H_2O(\ell) + 6\ Fe^{3+}(aq)$; (b) 0.123 *M*

4.123. 1.95 *M*

4.125. (a) 11.7 *M*; (b) 42.7 mL; (c) 1.72 kg

4.127. a.

Reactants	Products
NaCl: Na = +1, Cl = −1	Na_2SO_4: Na = +1, S = +6, O = −2
H_2SO_4: H = +1, S = +6, O = −2	$MnCl_2$: Mn = +2, Cl = −1
MnO_2: Mn = +4, O = −2	H_2O: H = +1, O = −2
	Cl_2: Cl = 0
$4\ NaCl(aq) + 2\ H_2SO_4(aq) + MnO_2(s) \rightarrow 2\ Na_2SO_4(aq) + MnCl_2(aq) + 2\ H_2O(\ell) + Cl_2(g)$	

b. $2\ Cl^-(aq) + 4\ H^+(aq) + MnO_2(s) \rightarrow Mn^{2+}(aq) + 2\ H_2O(\ell) + Cl_2(g)$

c. $Cl_2(g) + H_2O(\ell) \rightarrow HOCl(aq) + HCl(aq)$

4.129. a.

Reactants	Products
FeO_4^{2-}: Fe = +6, O = −2	FeO(OH)(s): Fe = +3, O = −2, H = +1
H_2O: H = +1, O = −2	O_2: O = 0
	OH^-: O = −2, H = +1
$4\ FeO_4^{2-}(aq) + 6\ H_2O(\ell) \rightarrow 4\ FeO(OH)(s) + 3\ O_2(g) + 8\ OH^-(aq)$	

b.

Reactants	Products
FeO_4^{2-}: Fe = +6, O = −2	Fe_2O_3: Fe = +3, O = −2
H_2O: H = +1, O = −2	O_2: O = 0
	OH^-: O = 0, H = +1
$4\ FeO_4^{2-}(aq) + 4\ H_2O(\ell) \rightarrow 2\ Fe_2O_3(s) + 3\ O_2(g) + 8\ OH^-(aq)$	

4.131. (a) H_3PO_4, phosphoric acid; (b) H_2SeO_3, selenous acid; (c) H_3BO_3, boric acid

4.133. (a) molecular equation: $NaOH(aq) + HF(aq) \rightarrow NaF(aq) + H_2O(\ell)$, net ionic equation ($Na^+$ is a spectator ion): $OH^-(aq) + HF(aq) \rightarrow F^-(aq) + H_2O(\ell)$; (b) molecular equation: $K_2SO_4(aq) + Sr(NO_3)_2(aq) \rightarrow SrSO_4(s) + 2\ KNO_3(aq)$, net ionic equation ($K^+$ and NO_3^- are spectator ions): $SO_4^{2-}(aq) + Sr^{2+}(aq) \rightarrow SrSO_4(s)$; (c) molecular and net ionic equation: $2\ KClO_3(s) \rightarrow 2\ KCl(s) + 3\ O_2(g)$

4.135. (a) $Ca_{10}(PO_4)_6(OH)_2(s) + 2\ F^-(aq) \rightarrow Ca_{10}(PO_4)_6F_2(s) + 2\ OH^-(aq)$; (b) $2 \times 10^{-4}\ M\ F^-$; (c) 0.6 mg F^-

4.137. (a) $3\ CH_2O \rightarrow CO_2 + C_2H_5OH$; (b) $C_2H_5OH + O_2 \rightarrow CH_3COOH + H_2O$; (c) 0 in CH_2O and CH_3COOH; +4 in CO_2; −2 in C_2H_5OH; (d) 66.7 g acetic acid

4.139. 4.95%

4.141. (a) graph b; (b) graph b; (c) graph c

4.143. Silver carbonate is insoluble in water and the bulb will not glow. Upon addition of one equivalent of HCl the following reaction occurs: $Ag_2CO_3(s) + 2\ H^+(aq) + 2\ Cl^-(aq) \rightarrow 2\ AgCl(s) + 2\ H_2O(\ell) + CO_2(g)$. Because there are no soluble ions after the reaction the lightbulb will not glow.

4.145. a.

Reactants	Products
REACTION 1	
N_2: N = 0	NH_3: N = −3
REACTION 2	
NH_4^+: N = −3	NO_2^-: N = +3

b. For the first reaction, H^+ and N_2 are being reduced. For the second reaction the metal (M^{3+}) is being reduced.

c. reaction 1: $N_2(g) + 8\ H^+(aq) + 8\ M^{2+}(aq) \rightarrow 2\ NH_3(aq) + H_2(g) + 8\ M^{3+}(aq)$; reaction 2: $2\ H_2O(\ell) + NH_4^+(aq) + 6\ M^{3+}(aq) \rightarrow NO_2^-(aq) + 8\ H^+(aq) + 6\ M^{2+}(aq)$

4.147. balanced equation: $BrO_3^-(aq) + 5\ Br^-(aq) + 6\ H^+(aq) \rightarrow 3\ Br_2(aq) + 3\ H_2O(\ell)$; (a) yes; (b) Br_2; (c) yes; (d) Br_2

4.149. 87.3 g CaF_2

Chapter 5

5.1. The height of mercury in a barometer depends on the atmospheric pressure: the higher the pressure the higher the column of mercury. Because Denver, CO, has lower atmospheric pressure (as a result of being at a higher elevation), the barometer labeled (a) represents the pressure in Denver, CO.

5.3. Situation 1 (atmospheric pressure is increased at constant temperature) could be represented by either (a) or (c). Situation 2 (temperature is increased at constant pressure) would be represented by (b).

5.5. (c)

5.7. line 1; no

5.9. line 2

5.11. The right-hand tube contains N_2.

5.13. (b)

5.15. Curve 1 represents SO_2. Curve 2 could represent CO_2, or C_3H_8.

5.17. Br_2 (orange)

5.19. (c)

5.21. (a)

5.23. Force is the product of the mass of an object and the acceleration due to gravity. Pressure uses force in its definition: it is the force exerted over a given area.
5.25. The weight of the column of mercury lowers the overall height of the mercury in the column, creating a space at the closed end of the column that is under vacuum (no gas molecules). As the pressure of the atmosphere increases over the open pool of mercury, the column of mercury rises in the tube to indicate rising pressure.
5.27. 10 atm
5.29. 1 millibar = 100 pascals
5.31. A sharpened skate blade has a smaller area over which the force is distributed than that of a dull skate blade.
5.33. As we go up in altitude, the overlying mass of the atmosphere above us decreases, which causes the pressure to decrease.
5.35. The column of air exerts a pressure of 3780 Pa or 0.0385 kg/cm^2, which is far less than the pressure exerted by the tower itself.
5.37. (a) 0.020 atm; (b) 0.739 atm
5.39. (a) 814.6 mmHg; (b) 1.072 atm; (c) 1086 mbar
5.41. (a) 91 atm; (b) CO_2 is a heavier molecule than N_2 so the atmosphere of Venus has a greater mass than the atmosphere of Earth.
5.43. The volume increases by a factor of 5.
5.45. decrease
5.47. increase
5.49. (a) 2.00 atm; (b) 0.664 atm; (c) 1.67 atm
5.51. 2.30 atm; 13.0 m
5.53.

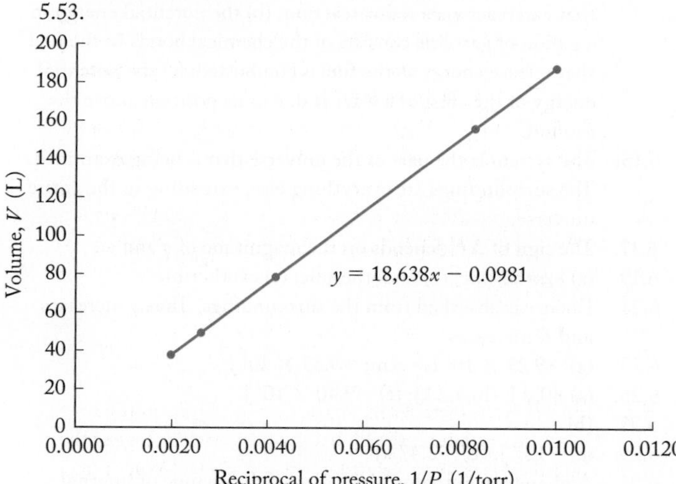

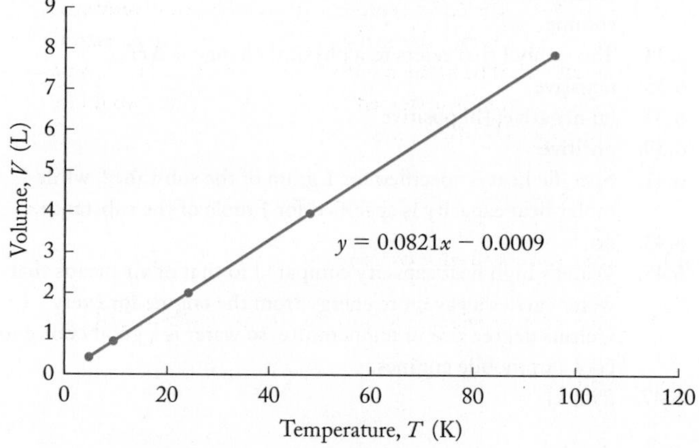

a. There would be no difference in the graphs if 1 mole of argon gas was used.
b. The general relationship remains the same (volume is inversely related to pressure) but the slope of the line would be smaller (slope would equal nRT).
c. The general relationship remains the same (volume is directly proportional to temperature) but the slope of the line will decrease (slope would equal nR/P).

5.55. 596 K, or 323°C
5.57. (a) 4.27 L; (b) 2.02 L; (c) 2.41 L
5.59. (b)
5.61. (a) no change; (b) decrease to 1/4 the original volume; (c) increase of 17%
5.63. 1.39×10^5 L
5.65. 6.6 atm
5.67. Standard temperature and pressure (STP) is defined as 1 atm and 0°C (273 K); $V = 22.4$ L
5.69. the product of the number of moles of gas in the sample, the temperature, and the gas constant
5.71. 0.67 mol
5.73. 1.50 atm
5.75. (a) 1.8×10^3 L; (b) 800% change; (c) 1.8×10^5 L; (d) the balloon will break under both sets of conditions
5.77. (a) 0.0419 mol/h; (b) 10.7 g
5.79. 1.5×10^3 g $(NH_4)_2Cr_2O_7$ and 1.2×10^2 g $KClO_3$
5.81. 715 g
5.83. The densities of different gases are not necessarily the same for a particular temperature and pressure. Gases with different molar masses will exhibit different densities.
5.85. Density (a) increases with increasing pressure and (b) increases with decreasing temperature.
5.87. (a) 9.08 g/L; (b) in the basement
5.89. SO_2
5.91. 78.1 g/mol
5.93. the pressure that a particular gas individually contributes to the total pressure of a mixture
5.95. (c)
5.97. oxygen
5.99. $P_{total} = 5.56$ atm and $P_{N_2} = 1.12$ atm
5.101. (a) $4\,NH_3(g) + 7\,O_2(g) \rightarrow 4\,NO_2(g) + 6\,H_2O(g)$; (b) Partial pressures are directly proportional to the number of moles of gas present so the ratio of partial pressures needed would match the stoichiometry of the reaction: 4 NH_3:7 O_2.
5.103. (a) 0.0190 mol; (b) no, there will be fewer moles of O_2 if collected over ethanol
5.105. (a) greater than; (b) less than; (c) greater than; (d) greater than
5.107. 1.7 times more
5.109. 680 torr
5.111. 25%
5.113. The speed of a molecule in a gas that has the average kinetic energy of all the molecules in the sample.
5.115. (a) As the molar mass increases the u_{rms} decreases; (b) as temperature increases the u_{rms} increases.
5.117. To determine the molar mass of an unknown gas, measure the rate of its effusion (r_x) relative to the rate of effusion of a known gas (r_y). Because we know the molar mass of the known gas, $\mathcal{M}_y$, we can use the equation for Graham's law to solve for the unknown $\mathcal{M}_x$.
5.119. Group 18 elements are all gases, but this is not true for the elements of group 15—only nitrogen, N_2, is a gas in its elemental form.
5.121. The rank order in terms of increasing root-mean-square speed is $SO_2 < NO_2 < CO_2$.

5.123. (a) Gas C is O_2; (b) the other gases are heavier than O_2; (c) gas A matches the molar mass of F_2.

5.125. No, the graphed points curve away from the line (the points are below the line at low and high temperatures and above the line at intermediate temperatures) as would be expected due to the right side of the root-mean-square speed equation being proportional to the square root of T ($u_{rms} = \sqrt{3RT/\mathcal{M}}$). A second plot below of u_{rms} versus $\sqrt{T}$ using the same data is shown to be a perfectly straight line.

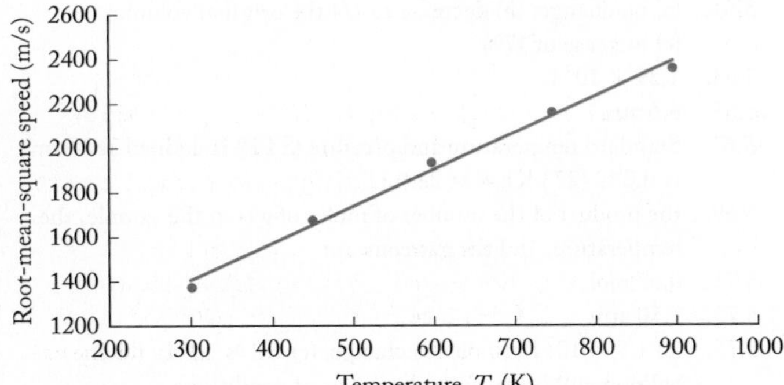

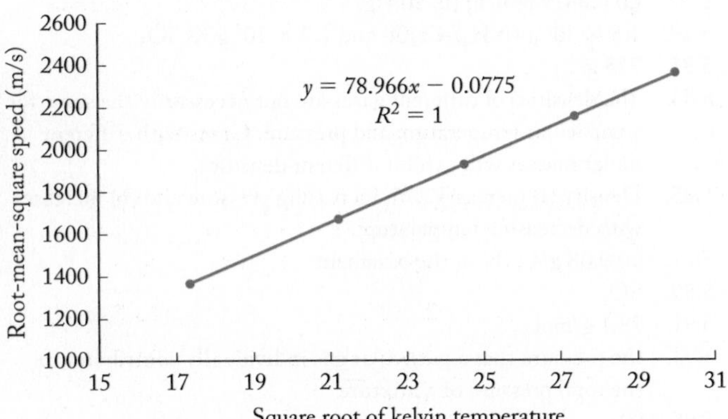

5.127. 0.707

5.129. No, the molar masses are nearly identical.

5.131. 128.0 g/mol

5.133. (a) $r(^{12}CO_2)/r(^{13}CO_2) = 1.01$; (b) $^{12}CO_2$

5.135. the smaller balloon

5.137. $P = \dfrac{nRT}{(V - nb)} - \dfrac{n^2a}{V^2}$

Real gases exhibit lower pressures than we would expect for an ideal gas because of the slight attractive forces between particles. Other particles exert a force that cancels out a small amount of the force exerted by the gas particle on the container wall, resulting in a lower pressure.

5.139. krypton (Kr)

5.141. H_2

5.143. (a) $P = 910$ atm; (b) $P = 476$ atm; (c) at higher pressures the gas molecules are closer together and attractive forces (variable a in the van der Waals equation) can play a larger role.

5.145. 126 L

5.147. 2835 psi

5.149. 3.31 L

5.151. 120 mmHg = (a) 120 torr = (b) 0.158 atm = (c) 0.160 bar = (d) 16.0 kPa; 80 mmHg = (a) 80 torr = (b) 0.11 atm = (c) 0.11 bar = (d) 11 kPa

5.153. 21 L

5.155. $V = 2.59$ L; $d = 7.33$ g/L

5.157. (a) CH_3CO_2H and $(CH_3)_3N$; (b) HCl and $(CH_3)_3N$; (c) no

5.159. 305 atm, or 4490 psi (well below the unsafe pressure range)

5.161. 79.3 g/mol

5.163. 137 g/mol

5.165. $u_{rms} = \sqrt{\dfrac{3P}{d}}$

5.167. 2.7 atm

5.169. 38.4 L of N_2; 192 L of CO_2; 1.74 g/L

5.171. 382 L

5.173. 54.9 L

5.175. 0.00781 mol

Chapter 6

6.1. At 35 ft above street level, KE = 150 J; just before hitting the street, KE = 500 J.

6.3. (a) The piston is higher in the cylinder; (b) yes; (c) the system did work on the surroundings.

6.5. (a) No; (b) the internal energy of the system increases.

6.7. (a) $2\ SO_2(g) + O_2(g) \rightarrow 2\ SO_3(g)$; (b) positive; (c) −98.9 kJ/mol

6.9. Energy makes work possible.

6.11. The value of a state function is independent of the path—only the initial and final values are important.

6.13. (a) The potential energy in a battery consists of the chemicals that can react via a redox reaction; (b) the potential energy in a gallon of gasoline consists of the chemical bonds in the fuel that release energy as the fuel is combusted; (c) the potential energy of the crest of a wave is due to its position above the ground.

6.15. The system is the part of the universe that is being examined. The surroundings are everything else, extending to the entire universe.

6.17. The sign of ΔE depends on the magnitude of q and ω.

6.19. (a) exothermic; (b) endothermic; (c) exothermic

6.21. Energy is absorbed from the surroundings. Thus q increases and E increases.

6.23. (a) -9.25×10^5 L · atm; -9.37×10^7 J

6.25. (a) 80.0 J; (b) 9.2 kJ; (c) -9.40×10^2 J

6.27. (b)

6.29. (a) −127 J; (b) −2.47 kJ

6.31. A change in enthalpy is the sum of the change of internal energy and the product of the system's pressure and change in volume.

6.33. The symbol that refers to a physical change is ΔH_{fus}.

6.35. negative

6.37. (a) negative; (b) positive

6.39. positive

6.41. Specific heat is specified for 1 gram of the substance, while molar heat capacity is specified for 1 mole of the substance.

6.43. no

6.45. Water's high heat capacity compared to that of air means that water carries away more energy from the engine for every Celsius degree rise in temperature, so water is a good choice to cool automobile engines.

6.47. 29.3 kJ

6.49.

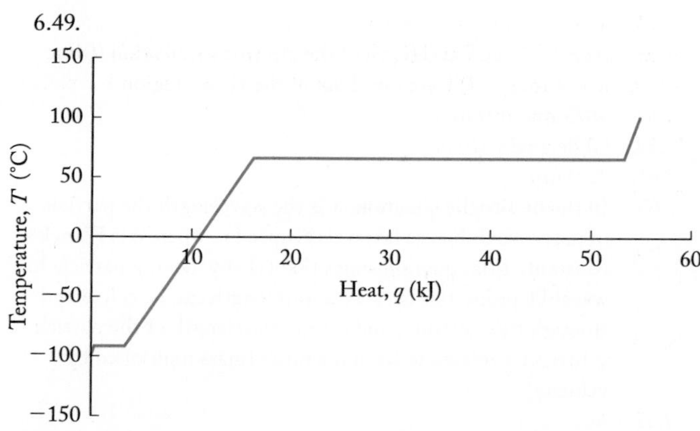

6.51. 886.2 g
6.53. gold
6.55. $\Delta T = -47.5$°C
6.57. It is necessary to know the heat capacity of a calorimeter to know how much energy (generated or absorbed by the system) is required to change the temperature of the surroundings (the calorimeter) in order to calculate the heat capacity or final temperature of the system in an experiment.
6.59. The heat capacity of the new liquid is different from that of water. The liquid is part of the calorimeter and therefore part of the surroundings. The calorimeter constant depends on the heat capacity of the surroundings, so the heat capacity must be determined.
6.61. NH_4NO_3 provides the greatest temperature drop.
6.63. 6.12 kJ/°C
6.65. −5127 kJ/mol
6.67. 23.29°C
6.69. (a) $Mg(s) + 2\,H^+(aq) \rightarrow H_2(g) + Mg^{2+}(aq)$; (b) −44 kJ
6.71. 0.145 g
6.73. When we apply Hess's law all the heat is accounted for in the reaction—energy is neither created nor destroyed.
6.75. Reverse the direction of the second reaction and then add it to the first reaction to obtain the third reaction (after canceling species common to both sides).
6.77. 54.08 kJ
6.79. If we write out the chemical equations for the ΔH°_f for the reactants and products, these formation reactions will add up to the overall reaction.
6.81. No, allotropes do not have the same standard enthalpy of formation. Because each allotrope has a different bonding arrangement, there will be different amounts of energy contained in the bonding of each allotrope. The most stable is defined as having a standard enthalpy of formation of 0 kJ/mol; any less stable allotrope will have a higher value for its standard enthalpy of formation.
6.83. (a) and (d)
6.85. 52 kJ
6.87. $CO_2(g) + 4\,H_2(g) \rightarrow CH_4(g) + 2\,H_2O(\ell)$; $\Delta H^\circ_{rxn} = -252.9$ kJ
6.89. −7198.5 kJ
6.91. Reverse the direction of the combustion reaction (change the sign of ΔH_{comb}) and add the reaction describing the formation of $CO_2(g)$ from pure $C(s)$ and $O_2(g)$ to obtain the reaction for the formation of $CO(g)$: $\Delta H^\circ_{f,CO} = \Delta H^\circ_{f,CO_2} - \Delta H_{comb,CO}$
6.93. −18 kJ/mol
6.95. The energy per gram a fuel releases on burning
6.97. The fuel value (kJ/g) is obtained by dividing the molar enthalpy of combustion (kJ/mol) by the molar mass (g/mol).
6.99. (a) gasoline; (b) gasoline
6.101. (a) 36.10 kJ/g; (b) 36,100 kJ; (c) 6.95 g
6.103. Diethyl ether has a fuel value of 36.781 kJ/g and a fuel density of 2.624×10^4 kJ/L. Both of these values are lower than those of diesel oil.
6.105. 13 kJ
6.107. 3.54×10^4 mol
6.109. (a) −92.2 kJ; (b) 46.1 kJ
6.111. 58.1 g
6.113. 33.3°C
6.115. (a) $6\,FeO(s) + O_2(g) \rightarrow 2\,Fe_3O_4(s)$; (b) $\Delta H_{rxn} = -636$ kJ
6.117. (a) The reaction is exothermic; (b) bonds must be broken, which requires more energy than for breaking intermolecular forces; (c) 793 kJ
6.119. The balanced chemical equation for the reaction of iron(II) oxide with oxygen to form iron (III) oxide is $4\,FeO(s) + O_2(g) \rightarrow 2\,Fe_2O_3(s)$; $\Delta H^\circ_{rxn} = -560.8$ kJ
6.121. (a) −1167.1 kJ; (b) no, less energy would be released if hydrazine was in the liquid state.
6.123. (a) exothermic; (b) −509.8 kJ; (c) −955 kJ; (d) the reaction would be less exothermic
6.125. −74.9 kJ/mol
6.127. −1388 kJ
6.129. −1234.7 kJ/mol; ethanol has a greater fuel value
6.131. 39°C
6.133.

Element	$\mathcal{M}$ (g/mol)	c_s[J/(g · °C)]	$\mathcal{M} \times c_s$
Bismuth	*210.8*	0.120	*25.3*
Lead	207.2	0.123	*25.5*
Gold	197.0	0.125	*24.6*
Platinum	195.1	*0.130*	*25.3*
Tin	118.7	0.215	*25.5*
Silver	*108.6*	0.233	*25.3*
Zinc	65.4	0.388	*25.4*
Copper	63.5	0.397	*25.2*
Cobalt/nickel	*58.4*	0.433	*25.3*
Iron	55.8	0.460	*25.7*
Sulfur	32.1	*0.788*	*25.3*
Average value in column 4			*25.3*

(a) The units are J/(mol · °C); (c) although silver is the likely element with a mass of 108.6 g/mol, there is more uncertainty about the element with a mass of 58.4 g/mol. This value lies between the accepted molar masses for cobalt and nickel.
6.135. exothermic, $\Delta H° = -1.9$ kJ
6.137. The reaction is endothermic, $\Delta H_{comb} = 68$ kJ.
6.139. a. For $H_2(g)$ the fuel density is 10.78 kJ/L, and for $H_2(\ell)$ the fuel density is 8492 kJ/L.
b. $H_3NBH_3(g) \rightarrow NH_3(g) + BH_3(g)$ $\Delta H^\circ_{rxn} = 102.2$ kJ
$H_3NBH_3(g) \rightarrow H_2(g) + H_2NBH_2(g)$ $\Delta H^\circ_{rxn} = -28.40$ kJ
$H_2NBH_2(g) \rightarrow H_2(g) + HNBH(g)$ $\Delta H^\circ_{rxn} = 123.4$ kJ
c. 76.6 kg
6.141. yes
6.143. exothermic

Chapter 7

7.1. (a) purple (Na), red (Cr), and orange (Au); (b) blue (Ne); (c) orange (Au); (d) red (Cr); (e) blue (Ne) and green (Cl)

7.3. (a) One electron would be lost from the element shaded blue (Rb) to become the [Kr] core electron configuration; (b) two electrons would be lost from the element shaded green (Sr) to become the [Kr] core electron configuration; (c) three electrons would be lost from the element shaded orange (Y) to become the [Kr] core electron configuration; (d) the element shaded gray (I) would gain one electron to become the [Xe] core electron configuration; (e) the element shaded red (Te) would gain two electrons to become the [Xe] core electron configuration.

7.5. (a) gray (I) < red (Te) < orange (Y) < green (Sr) < blue (Rb); (b) orange (Y^{3+}) < green (Sr^{2+}) < blue (Rb^+) < gray (I^-) < red (Te^{2-})

7.7. (a) Na; (b) K; (c) Na^+

7.9. (a) wave (b); (b) wave (a); (c) wave (b)

7.11. The hydrogen absorption spectrum consists of dark lines at wavelengths specific to hydrogen. The emission spectrum has bright lines on a dark background with the lines appearing at the exact same wavelengths as the dark lines in the absorption spectrum.

7.13. Because each element shows distinctive and unique absorption and emission lines, the bright emission lines observed for the pure elements could be matched to the many dark absorption lines in the spectrum of sunlight. This approach can be used to deduce the Sun's elemental composition.

7.15. All of these forms of light have perpendicular, oscillating electric and magnetic fields that travel together through space.

7.17. The lead shield protects the parts of the body that might be exposed to the highly energetic X-rays but are not being imaged.

7.19. X-rays and ultraviolet radiation

7.21. (a) Ultraviolet radiation; (b) gamma rays; (c) microwave radiation

7.23. The energy of the UV photon is twice that of the red light photon.

7.25. $6.06 \times 10^{14}\ s^{-1}$ (or Hz)

7.27. (a) 0.13 m; (b) 0.354 m

7.29. (a)

7.31. 1.28 s

7.33. A quantum is the smallest indivisible amount of radiant energy that an atom can absorb or emit. A photon is the smallest individual packet or particle of light energy.

7.35. As the intensity (amplitude) of the light increases, more photons strike the surface as a function of time resulting in more electrons being emitted.

7.37. (b) is quantized, as the elevator can stop only at specified floors. Both the height of the escalator treads (a) and the speed of an automobile (c) are continuously variable and so are not considered quantized.

7.39. 6.52×10^{-19} J; this photon could break the C—C bond.

7.41. potassium; 8.04×10^5 m/s

7.43. no

7.45. 3.17×10^{18} photons/s

7.47. The single electron interacts only with the proton in the nucleus; there are no other electrons to repel it.

7.49. The difference between n levels determines emission energy.

7.51. (a)

7.53. no

7.55. At $n = 7$, the wavelength of the electron's transition (from $n = 7$ to $n = 2$) has moved out of the visible region.

7.57. 1875 nm; infrared

7.59. (a) decrease; (b) no

7.61. 72.9 nm

7.63. In the de Broglie equation, λ is the wavelength the particle of mass m exhibits as it travels at speed u, where h is Planck's constant. This equation states that (1) any moving particle has wavelike properties because a wavelength can be calculated through the equation, and (2) the wavelength of the particle is inversely related to its momentum (mass multiplied by velocity).

7.65. no

7.67. (a) 10.8 nm; (b) 0.180 nm; (c) 1.3×10^{-27} nm; (d) 3.7×10^{-54} nm

7.69. (c)

7.71. (a) 1.77×10^3 m/s; (b) 0.0565 nm

7.73. $\Delta x \geq 1.3 \times 10^{-13}$ m

7.75. The Bohr model orbit showed the quantized nature of the electron in the atom as a particle moving around the nucleus in concentric orbits. In quantum theory, an orbital is a region of space where the probability of finding the electron is high. The electron is not viewed as a particle, but as a wave, and it is not confined to a clearly defined orbit; rather, we refer to the probability of the electron being at various locations around the nucleus.

7.77. A larger n means the electron is more likely to be found further from the nucleus so the orbital must also be larger.

7.79. (a) 1; (b) 4; (c) 9; (d) 16; (e) 25

7.81. 8 electrons

7.83. (a) 2s, two electrons; (b) 3p, six electrons; (c) 4d, ten electrons; (d) 1s, two electrons

7.85. (b)

7.87. Degenerate orbitals have the same energy and are indistinguishable from each other.

7.89. As we start from an argon core of electrons we move to potassium and calcium, which are located in the s block of the periodic table. It is not until Sc, Ti, V, and so on that we begin to fill electrons into the 3d shell.

7.91. Electrons may be promoted to many different orbitals with energy higher than the ground state; each of these corresponds to an excited state with a different energy.

7.93. (c) 3s < (a) 3d < (d) 4p < (b) 5p

7.95. Li: [He]$2s^1$; Li^+: $1s^2$ or [He]; Ca: [Ar]$4s^2$; F^-: [He]$2s^22p^6$ or [Ne]; Na^+: [He]$2s^22p^6$ or [Ne]; Mg^{2+}: [He]$2s^22p^6$ or [Ne]; Al^{3+}: [He]$2s^22p^6$ or [Ne]

7.97. The highest-energy electron in each atom is the first electron to occupy the s subshell—each atom's electron configuration ends with ns^1.

7.99. (a) K: [Ar]$4s^1$, one unpaired electron; (b) K^+: [Ar], zero unpaired electrons; (c) S^{2-}: [Ar], zero unpaired electrons; (d) N: [He]$2s^22p^3$, 3 unpaired electrons; (e) Ba: [Xe]$6s^2$, zero unpaired electrons; (f) Ti^{4+}: [Ar], zero unpaired electrons; (g) Al: [Ne]$3s^23p^1$, one unpaired electron

7.101. (a) Ti, 2 unpaired electrons; (b) Cr, 6 unpaired electrons; (c) Cu, 1 unpaired electron; (d) Pd, zero unpaired electrons.

7.103. carbon: $1s^22s^22p^2$

7.105. no

7.107. Al^{3+}, N^{3-}, Mg^{2+}, Cs^+

7.109. (a) and (d)

7.111. (a) B: $1s^22s^22p^1$ or [He]$2s^22p^1$; O: $1s^22s^22p^4$ or [He]$2s^22p^4$; (b) H (+1), B (+3), O (−2); H^+: $1s^0$; B^{3+}: $1s^22s^0$ or [He]; O^{2-}: $1s^22s^22p^6$ or [Ne]

7.113. (a) $2\ Mg(s) + O_2(g) \rightarrow 2\ MgO(s)$; (b) Mg: $1s^22s^22p^63s^2$ or [Ne]$3s^2$; Mg^{2+}: $1s^22s^22p^6$ or [Ne]; O^{2-}: $1s^22s^22p^6$ or [Ne]; (c) Mg is oxidized and O is reduced

7.115. (c)

7.117. (a) Ionization energy would correspond to removing a 2*s* electron not a 1*s* electron; (b) no

7.119. If electrons do not repel each other as much in Na^+ as they do in Na, they will have a lower energy and be, on average, closer to the nucleus resulting in a smaller size. When electrons are added to an atom (Cl), the electron–electron repulsion increases so the electrons have a higher energy and they will be, on average, farther from the nucleus thereby creating a larger species (Cl^-).

7.121. Rb; the size of the atoms increases down a group because electrons have been added to higher *n* levels.

7.123. (a) Al > P > Cl > Ar; (b) Sn > Ge > Si > C; (c) K > Na > Li > Li^+; (d) Cl^- > Cl > F > Ne

7.125. (a) As the atomic number increases down a group, electrons are added to higher *n* levels leading to a decrease in ionization energy; (b) as the atomic number increases across a period, the effective nuclear charge increases leading to an increase in ionization energy across a period of elements.

7.127. Fluorine, due to its higher nuclear charge, exerts a higher Z_{eff} on the 2*p* electrons than does boron resulting in higher ionization energy.

7.129. Sr

7.131. (a) I < Br < Cl < F; (b) Na < Mg < Li < Be; (c) N < O < F < Ne (O is actually slightly lower than N but this is an anomaly in the general trend.)

7.133. No, the statement about all atoms with negative EA values existing as anions is false. For example, fluorine atoms are less stable than the fluoride anion and exhibit a negative value for EA, but fluorine exists in nature as F_2, a homonuclear diatomic molecule.

7.135. As we move down a group, electron–electron repulsions decrease as a result of the valence shells being farther from the nucleus. Minimizing the energy penalty of adding an electron results in a greater electron affinity.

7.137. (a) $\lambda = 0.242$ m, $E = 8.22 \times 10^{-25}$ J; (b) no

7.139. 2.3 GHz corresponds to a higher energy, shorter wavelength, and higher frequency than 162 MHz.

7.141. (a) Moving across a period in the periodic table, the energy separation of the 2*s* and 2*p* orbitals would increase. Because energy is inversely proportional to wavelength, the wavelength of this for the $2p \rightarrow 2s$ transition would decrease; (b) moving down a column in the periodic table, the energy separation between the 2*s* and 2*p* orbitals would again increase so the wavelength of this for the $2p \rightarrow 2s$ transition would decrease.

7.143. Both configurations have one unpaired electron so the two configurations would be expected to exhibit the same effect when exposed to a magnetic field.

7.145. (a) true; (b) false; (c) true; (d) true

7.147. (a) No, the second ionization energy for Ga would involve removing an electron from a lower-energy subshell (the 3*d*) than the first ionization energy (the 4*p*); for the other elements, both the first and second ionization energies would be for removing electrons from the 4*p* subshell. This difference results in the second ionization energy for Ga being significantly higher than those of Ge and As, which would not match the trend for first ionization energies. (b) Rb; its higher nuclear charge exerts a greater Z_{eff} on the electron being lost for a second ionization than that of Kr.

7.149. 319 nm

7.151. (a) La: [Xe] $6s^25d^1$; Ce: [Xe]$6s^24f^15d^1$; Pr: [Xe]$6s^24f^25d^1$; Nd: [Xe]$6s^24f^35d^1$; Pm: [Xe]$6s^24f^45d^1$; Sm: [Xe]$6s^24f^55d^1$; Eu: [Xe]$6s^24f^65d^1$; Gd: [Xe]$6s^24f^75d^1$; Tb: [Xe]$6s^24f^85d^1$; Dy: [Xe]$6s^24f^95d^1$; Ho: [Xe]$6s^24f^{10}5d^1$; Er: [Xe]$6s^24f^{11}5d^1$; Tm: [Xe]$6s^24f^{12}5d^1$; Yb: [Xe]$6s^24f^{13}5d^1$; Lu: [Xe]$6s^24f^{14}5d^1$; (b) La: same; Ce: same; Pr: [Xe]$6s^24f^3$; Nd: [Xe]$6s^24f^4$; Pm: [Xe]$6s^24f^5$; Sm: [Xe]$6s^24f^6$; Eu: [Xe]$6s^24f^7$; Gd: same; Tb: [Xe]$6s^24f^9$; Dy: [Xe]$6s^24f^{10}$; Ho: [Xe]$6s^24f^{11}$; Er: [Xe]$6s^24f^{12}$; Tm: [Xe]$6s^24f^{13}$; Yb: [Xe]$6s^24f^{14}$; Lu: same; (c) the 5*d* orbital could be higher in energy than the 4*f* orbital; (d) we can for Tb, Dy, Ho, Er, Tm, and Yb.

7.153. (a) Ne, 5.76; Ar, 6.76; (b) the outermost electron in argon is a 3*p* electron, which is mostly shielded by the electrons in the $n = 2$ level (10 electrons) and the $n = 1$ level (2 electrons), whereas the outermost electron in neon is a 2*p* electron, which is shielded only by the electrons in the $n = 1$ level (2 electrons).

7.155. When we think of the electron as a wave we can envision the node between the two lobes as a wave of zero amplitude and the *p* orbital as a standing wave.

7.157. No, it will be in the infrared portion of the electromagnetic spectrum.

Chapter 8

8.1. (a) Be (group 2); (b) C (group 14); (c) O (group 16); (d) Ne (group 18)

8.3. Carbon (blue) has the capacity to make four bonds.

8.5. fluorine (yellow)

8.7. Mg^{2+}

8.9. (b)

8.11. (a); SO_2 exhibits a bent molecular geometry. The central sulfur atom has a lower electronegativity than that of the flanking oxygen atoms so the electron density should be lower at the sulfur and higher at the oxygens.

8.13. The red line represents KF; the blue line represents KCl.

8.15. The arrangement of the atoms in two of the structures is S—O—S, and in the other two structures it is S—S—O. Because the arrangement of atoms differs they are not resonance structures. Also, for each arrangement, the structures do not show a different positioning of electrons on the atoms, just a change in the angle between the bonds. The "bent form" and the "linear form" are not resonance forms of each other if the numbers of lone pairs and bonding pairs of electrons on each atom are the same.

8.17. The structures do not have the correct number of valence electrons. H_2NO_3S: should have 31 valence electrons but only 25 are shown. NO_2Cl: should have 24 valence electrons but only 22 are shown. S_2F_2: should have 26 valence electrons but only 24 are shown. Br_3^-: should have 22 valence electrons but only 18 are shown.

8.19. Bond order and bond length are inversely proportional (blue line). Bond order and bond energy are directly proportional, and this relationship would look like the red line in the diagram.

8.21. The positively charged protons in each atom are attracted to the negatively charged electron in the other atom.

8.23. The sodium cations and the chloride anions are attracted to each other. Particles with the same charge repel each other; sodium cations will repel other sodium cations and chloride anions will repel other chloride anions.

8.25. (a) Ni(*s*): metallic bonding; $S_8(s)$: covalent bonding. Both types of bonding result from shared electrons. In metallic bonding those electrons are delocalized to form a "sea" of mobile electrons; in covalent bonding the electrons are shared between the atoms. (b) Ni(*s*): metallic bonding; $Na_2S(s)$: ionic bonding. Both types of bonding involve movement of electrons. In metallic bonding electrons are delocalized to form a "sea" of mobile electrons, in ionic bonding electrons are transferred from the cation to the anion. (c) $SO_2(g)$: covalent bonding; $Na_2S(s)$: ionic bonding. Both types of bonding rely on atoms reaching a more stable electron arrangement. In covalent bonding electrons are shared to reach the more stable arrangement; in ionic bonding electrons are transferred from cation to anion to reach the more stable arrangement. (d) Ni(*s*): metallic bonding, $Na_2SO_4(s)$: ionic bonding between the sodium cation and the sulfate anion, covalent bonding within the sulfate ion. (e) $S_8(s)$: covalent bonding; $Na_2SO_4(s)$: ionic bonding between the sodium cation and the sulfate anion; covalent bonding within the sulfate ion. (f) $Na_2S(s)$: ionic bonding; $Na_2SO_4(s)$: ionic bonding between the sodium cation and the sulfate anion; covalent bonding within the sulfate ion.

8.27. yes, for H and He

8.29. yes

8.31. groups 1, 13, 15, and 17

8.33. No, sharing of electrons will result in fewer than 16 electrons in diatomic molecules.

8.35. Li• •Mg• •Ȧl•

8.37. The corrected Lewis symbols are $[K]^+$; $[\cdot Pb\cdot]^{2+}$; $[:\ddot{S}:]^{2-}$; $[Al]^{3+}$

8.39. $[\cdot In\cdot]^+$ $[:\ddot{I}:]^-$ $[Ca]^{2+}$ $[\cdot Sn\cdot]^{2+}$

I^- and Ca^{2+} have complete valence-shell octet.

8.41. (a) Ne; (b) the element identity is different but the complement of electrons is identical.

8.43. (a) $[X\cdot]^+$; (b) $[X]^{3+}$

8.45. (a) 8; (b) 8; (c) 8; (d) 10

8.47. (a) :C≡O:; (b) Ö=Ö; (c) $[:\ddot{Cl}-\ddot{O}:]^-$; (d) $[:C≡N:]^-$

8.49. (a) :F̈—C(—:C̈l:)(—:F̈:)—C̈l: (b) :F̈—C(:C̈l:)(:C̈l:)—C(:F̈:)(:C̈l:)—F̈: (c) (:C̈l)(:F̈)C=C(:F̈)(:F̈)

8.51. H—C(H)(H)—C(H)(H)—C(H)(H)—C(H)(H)—S̈—H H—S̈—H

8.53. :C̈l—C̈l—Ö: $[:\ddot{O}-\ddot{Cl}(-:\ddot{O}:)-\ddot{O}:]^-$

8.55. If there is an electronegativity difference of 2.0 or greater, the bond between the atoms is ionic; below 2.0 the bond is covalent.

8.57. The size of the atom is the result of the nucleus pulling on the electrons. The higher the nuclear charge, the stronger the pull on the electrons within a given valence shell. This is why the size of atoms generally decreases across a period. A small atom will form a shorter bond with another atom, and the electrons in the bond will feel a stronger pull from the nucleus of a smaller atom because the bonding electrons will be "closer" to the nucleus. This stronger pull results in a higher electronegativity for smaller atoms.

8.59. A polar covalent bond is one in which the electrons are shared, but not equally, by the atoms.

8.61. Like the panes of glass in a greenhouse, the greenhouse gases in the atmosphere are transparent to visible light. Once the visible light warms the surface of Earth and is reemitted as infrared (lower energy) light, the greenhouse gases absorb the infrared light; this prevents some of the warmth from escaping the atmosphere in the same way that the panes of glass do not allow the energy from inside the greenhouse to escape.

8.63. An infrared-active stretch must change the electric field of the molecule. Stretching the nonpolar N–N bond will not alter the dipole moment whereas stretching the N–O bond will. Only the N–O bond stretch contributes to the absorption of infrared radiation by N_2O.

8.65. The polar bonds and the atoms with the greater electronegativity (underlined) are: <u>C</u>—Se, C—<u>O</u>, <u>N</u>—H, and <u>C</u>—H.

8.67. CsI

8.69. the I—Cl bond

8.71. the O—H bond

8.73. PF_3

8.75. Resonance occurs when two or more valid Lewis structures may be drawn for a molecular species. The true structure of the species is a hybrid of the structures drawn.

8.77. A molecule or ion shows resonance when there is more than one correct Lewis structure; that is, when the electrons in the correct Lewis structure may be distributed in more than one way. Often, when the central atom has both a single and a double bond, resonance is possible.

8.79. Either N–O bond in the NO_2 structure could be double-bonded, and the formal charges for each structure are identical, so there is more than one correct Lewis structure, and NO_2 will exhibit resonance.

$:\ddot{O}^{-1}-\dot{N}^{+1}=\ddot{O}^{+0} \leftrightarrow \ddot{O}^{+0}=\dot{N}^{+1}-\ddot{O}:^{-1}$

The resonance forms of CO_2 show that one is dominant (the one in which all formal charges are zero), and so the other forms contribute little to the true structure of CO_2.

$\ddot{O}^{0}=C^{0}=\ddot{O}^{0} \leftrightarrow :O^{+1}≡C^{0}-\ddot{O}:^{-1} \leftrightarrow :\ddot{O}^{-1}-C^{0}≡O:^{+1}$

8.81.

H—C≡N—Ö: ⟷ H—C̈—N≡O: ⟷ H—C̈=N=Ö

8.83. N_2O_2:

Ö=N̈—N̈=Ö ⟷ :O≡N—N̈—Ö: ⟷ Ö=N=N̈—Ö: ⟷

:Ö—N≡N—Ö: ⟷ :Ö—N̈—N≡O: ⟷ :Ö—N̈=N=Ö

N_2O_3:

Ö=N̈—N(=Ö)(—:Ö:) ⟷ Ö=N̈—N(—:Ö:)(=Ö) ⟷

Ö=N=N(—:Ö:)(—:Ö:) ⟷ :O≡N—N̈(—:Ö:)(—:Ö:)

8.85. a. :F̤—Ö—Ö—F̤:; H—Ö—Ö—H
b. The presence of an O=O double bond shortens the bond length.

:F̤—Ö—Ö—F̤: ⟶ :F̤—Ö—Ö⁺ :F̤:⁻ ⟶ :F̤—Ö=Ö :F̤:⁻

Incomplete octet on oxygen

8.87. :C̤l—Se—N=S—Ö: ⟷ :C̤l—Se=N—S—Ö:

:C̤l—Se—N—S=Ö: ⟷ :Cl=Se—N—S—Ö:

:C̤l—Se—N=S=Ö:

8.89. The best possible structure for a molecule judging by formal charges is the structure in which the formal charges are minimized and the negative formal charges are on the most electronegative atoms in the structure.

8.91. No. The electronegativity of oxygen (3.5) is higher than that of sulfur (2.5) so the negative formal charge must be on the O atom in the structure that contributes to the bonding.

8.93. H—N≡C: (0, +1, −1) H—C≡N: (0, 0, 0)

The formal charges are zero for all the atoms in HCN, whereas in HNC the carbon atom, with a lower electronegativity than N, has a −1 formal charge.

8.95. H_2N=C=N̤: ⟷ H_2N̈—C≡N:

The preferred structure is the one with the C triple-bonded to N.

8.97. (a) $H_2C=O$ and $H_2O^{2+}=C^{2-}$

The Lewis structure with carbon as the central atom is more likely as it minimizes formal charges; (b) no

8.99. :N̤—O≡N: (−2, +2, 0) ⟷ :N=O=N: (−1, +2, −1) ⟷ :N≡O—N̤: (0, +2, −2)

Because oxygen is more electronegative than nitrogen, none of these structures is likely to be stable because the formal charge on O is positive.

8.101. yes

8.103. In order for the atom to accommodate more than 8 electrons in covalently bonded molecules, it would require the use of orbitals beyond *s* and *p*. The *d* orbitals are not available to the small elements in the second period.

8.105. (a), (b), and (c)

8.107. (a) 6; (b) 6; (c) 12; (d) 10

8.109. NOF_3: O (−1), N (+1), F (0) single bonds; POF_3: O (0), P (0), F (0), P=O double bond

In POF_3 there is a double bond and no formal charges; in NOF_3 there are only single bonds and formal charges are present on N and O.

8.111. SeF_4 and $[SeF_5]^-$

In both structures Se has more than eight valence electrons.

8.113. O_2Cl—Cl̤: (all formal charges 0)

The central chlorine atom has 12 electrons.

8.115. (c), (d), and (e)

8.117. (d)

8.119. The formation of a dimer completes the octet on Al.

2 H_3C—Al(CH_3)—C̤l: ⟶ dimer with bridging Cl atoms (Cl⁺, Al⁻)

8.121. BF_3 + BCl_3 ⟶ bridged F/Cl dimer ⟶ :F̤—B(F) + :F̤:—B(Cl)—C̤l:

8.123. Each bond in the nitrate ion is 1.33 bonds because of resonance.

$[NO_3]^-$ ⟷ $[NO_3]^-$ ⟷ $[NO_3]^-$

Each bond in the nitrite ion is 1.5 bonds because of resonance.

[:Ö=N̈—Ö:]⁻ ⟷ [:Ö—N̈=Ö:]⁻

No, the nitrogen–oxygen bond lengths in NO_3^- and NO_2^- are not the same, they are different.

8.125. The nitrogen–oxygen bond in N_2O_4 has a bond order of 1.5 due to four equivalent resonance forms:

O₂N—NO₂ ⟷ O₂N—NO₂ ⟷ O₂N—NO₂ ⟷ O₂N—NO₂

The nitrogen–oxygen bond in N_2O has a bond order of 1.5 due to resonance between three resonance forms (where the last resonance structure shown does not significantly contribute to the structure of the molecule because of the buildup of too much formal charge):

$$\overset{-1}{:\ddot{N}}=\overset{+1}{N}=\overset{0}{\ddot{O}}: \longleftrightarrow \overset{0}{:N}\equiv\overset{+1}{N}-\overset{-1}{:\ddot{O}}: \longleftrightarrow \overset{-2}{:\ddot{N}}-\overset{+1}{N}\equiv\overset{+1}{O}:$$

Therefore N_2O_4 and N_2O are expected to have nearly equal bond lengths.

8.127. (a) $NO^+ < NO_2^- < NO_3^-$; (b) $NO_3^- < NO_2^- < NO^+$

8.129. no

8.131. We must account for all the bonds that break and all the bonds that form in the reaction. In order to do so, we must have a balanced chemical reaction.

8.133. If the compounds are in the solid or liquid phase, interactions between molecules may slightly change the bond energy for a given bond.

8.135. (a) 862 kJ; (b) 98 kJ; (c) 93 kJ

8.137. 1068 kJ/mol

8.139. 278 kJ less

8.141. −667 kJ

8.143. 552 kJ/mol

8.145. a. $:\ddot{O}=C=C=C=\ddot{O}:$

b. The two carbon oxygen bonds in this molecule would have the same length.

8.147. A potassium ion has a charge of +1. The potassium ion would have lost the single valence electron that would be present in the neutral potassium atom.

8.149. $\overset{0}{:\ddot{S}}=\overset{0}{C}=\overset{0}{\ddot{S}:}$ $\overset{0}{:\ddot{S}}=\overset{+2}{S}=\overset{-2}{\ddot{C}:}$

The preferred structure would be the one with the carbon atom in the center.

8.151. a. [Lewis structure: $COCl_2$ with C=O double bond and two C–Cl bonds]

b. $H-\ddot{O}-H + COCl_2 \rightarrow :\ddot{O}=C=\ddot{O}: + 2H-\ddot{Cl}:$

8.153. (a) Each arrangement has a structure where all atoms have a zero formal charge:

[Lewis structure: HN=S(=O)=O] $H-\ddot{O}-\ddot{N}=\ddot{S}=\ddot{O}:$

(b) In both structures the S–N bond is a double bond. (c) Formal charges cannot be used to predict the preferred arrangement because both arrangements have all zero formal charges. (d) 143 pm for the sulfur–oxygen bond in each structure.

8.155. For Cl_2O_6 with a Cl—Cl bond:

[Lewis structure: $O_3Cl-ClO_3$, all formal charges 0]

For Cl_2O_6 with Cl—O—Cl bonds:

[Lewis structure: $O_3Cl-O-ClO_2$... all formal charges 0]

For ClO_2:

$$\overset{0}{:\ddot{O}}=\overset{0}{\dot{\ddot{Cl}}}=\overset{0}{\ddot{O}:}$$

8.157. (a) $\cdot C\equiv N:$ The more likely structure for cyanogen is the one with the C—C bond. (b) It would be expected that oxalic acid would retain the C—C bond from the cyanogen from which it is formed in the reaction of cyanogen with water. This is consistent with the structure for cyanogen predicted from formal charges of the two possible structures.

8.159. [Lewis structure: $N\equiv C-SF_3$, all formal charges 0]

8.161. [Lewis structure: $[TeOF_5]^{2-}$, O formal charge −1, Te formal charge −1]

8.163. (a) 2; (b) 4; (c) 1; (d) 5; (e) 8

8.165. (a) Assuming A and X have different electronegativities, we would be able to distinguish X—A—A from A—X—A. Bond lengths would only be useful if we were able to look for a characteristic A—A bond that would be found in X—A—A but not in A—X—A. (b) Electrostatic potential mapping cannot distinguish resonance forms. If the resonance forms for A—X—A shown are all equally weighted (none is more preferred than another), we would expect the average X—A bond to be a double bond as in A=X=A. If one resonance form contributes more to the bonding in a molecule, we would see that resonance form reflected in the bond distances being shortened or lengthened.

8.167. $[NH_4]^+$ $[:\ddot{S}-H]^-$

There cannot be a nitrogen–sulfur covalent bond because the nitrogen atom in NH_4^+ has a complete octet through its bonding with hydrogen and because it cannot expand its octet as it is a second period element.

8.169. a. [Five resonance structures of the pentazolide ring N_5^-, connected by ↔]

b. Bond order is 1.4 (or 7/5). There are seven bonds between any two nitrogen atoms spread over five structures. The bond distance is ~140 pm.

8.171. $\text{F}^{0}\text{—Al}^{0}(\text{—Cl}^{0})\text{—Cl}^{0}$

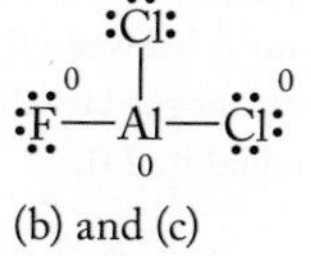

8.173. (b) and (c)

8.175. (a) and (b)

$[\text{N}^{-2}\text{—N}^{0}\text{=N}^{0}\text{—N}^{+1}\equiv\text{N}^{0}]^- \leftrightarrow [\text{N}^{-2}\text{—N}^{0}\text{=N}^{+1}\text{=N}^{+1}\text{=N}^{-1}]^- \leftrightarrow$

$[\text{N}^{-2}\text{—N}^{+1}\equiv\text{N}^{+1}\text{—N}^{0}\text{=N}^{-1}]^- \leftrightarrow [\text{N}^{-1}\text{=N}^{0}\text{—N}^{+1}\equiv\text{N}^{+1}\text{—N}^{-2}]^- \leftrightarrow$

$[\text{N}^{0}\equiv\text{N}^{+1}\text{—N}^{0}\text{=N}^{0}\text{—N}^{-2}]^- \leftrightarrow [\text{N}^{-1}\text{=N}^{0}\text{—N}^{0}\text{=N}^{+1}\text{=N}^{-1}]^- \leftrightarrow$

$[\text{N}^{-1}\text{=N}^{+1}\text{=N}^{0}\text{—N}^{0}\text{=N}^{-1}]^- \leftrightarrow [\text{N}^{0}\equiv\text{N}^{+1}\text{—N}^{-1}\text{—N}^{0}\text{=N}^{-1}]^- \leftrightarrow$

$[\text{N}^{-1}\text{=N}^{0}\text{—N}^{-1}\text{—N}^{+1}\equiv\text{N}^{0}]^-$

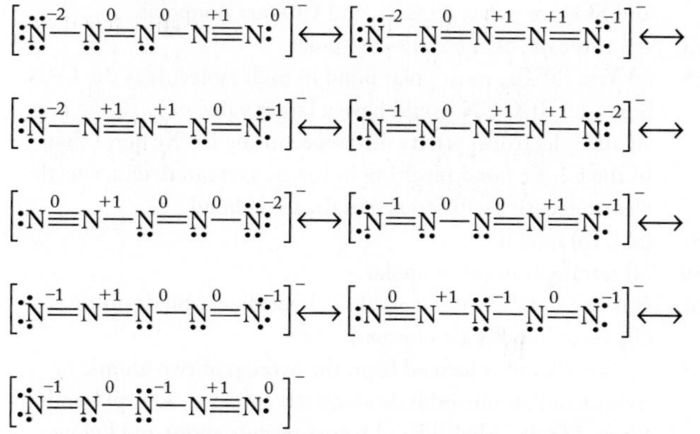

The structures that contribute the most are the ones with the lowest formal charges (the last four of the structures shown).

(c) N_3^- has the Lewis structures:

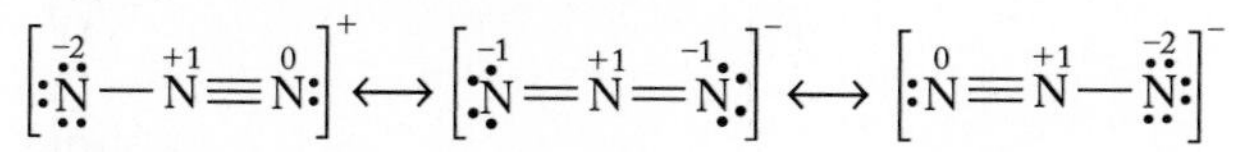

$[\text{N}^{-2}\text{—N}^{+1}\equiv\text{N}^{0}]^- \leftrightarrow [\text{N}^{-1}\text{=N}^{+1}\text{=N}^{-1}]^- \leftrightarrow [\text{N}^{0}\equiv\text{N}^{+1}\text{—N}^{-2}]^-$

From these resonance structures we see that each bond is predicted to be of double-bond character in N_3^-. Therefore in N_5^- there are two longer N–N bonds than in N_3^-. N_3^- has the higher average bond order.

8.177.

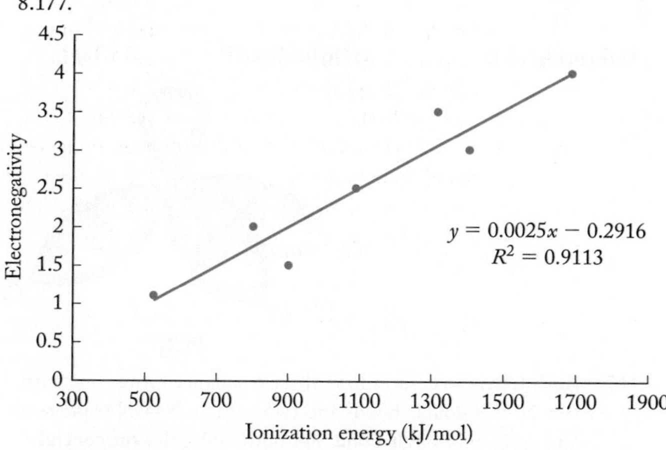

Using the equation for the best-fit line where x = the ionization energy of neon gives a value of y (electronegativity) of neon: $y = 4.9$

8.179. (a) Isoelectronic means that the two species have the same number of electrons.

(b) 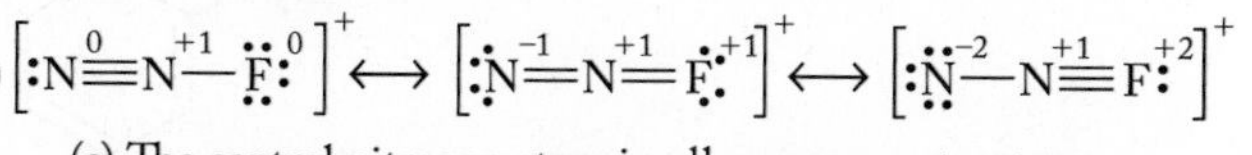 $[\text{N}^{0}\equiv\text{N}^{+1}\text{—F}^{0}]^+ \leftrightarrow [\text{N}^{-1}\text{=N}^{+1}\text{=F}^{+1}]^+ \leftrightarrow [\text{N}^{-2}\text{—N}^{+1}\equiv\text{F}^{+2}]^+$

(c) The central nitrogen atom in all resonance structures always carries a +1 formal charge. (d) The second and third resonance forms shown are unacceptable because they have greater than the minimal formal charges on the atoms.

(e) Structures can be drawn with the fluorine as the central atom in the molecule, but this would place a +3 formal charge on the fluorine atom (the most electronegative element) so these structures are very unlikely:

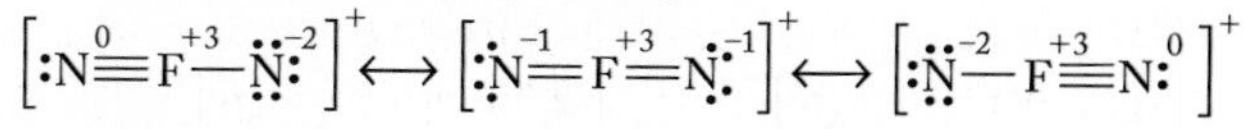

$[\text{N}^{0}\equiv\text{F}^{+3}\text{—N}^{-2}]^+ \leftrightarrow [\text{N}^{-1}\text{=F}^{+3}\text{=N}^{-1}]^+ \leftrightarrow [\text{N}^{-2}\text{—F}^{+3}\equiv\text{N}^{0}]^+$

8.181. $\text{F}_2\text{C=CF}_2$

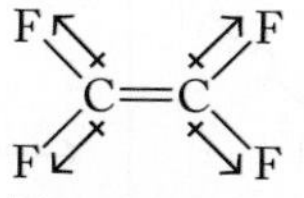

This molecule is nonpolar overall because the individual bond dipoles are equal in magnitude and, as vectors, they cancel each other out.

8.183. a.

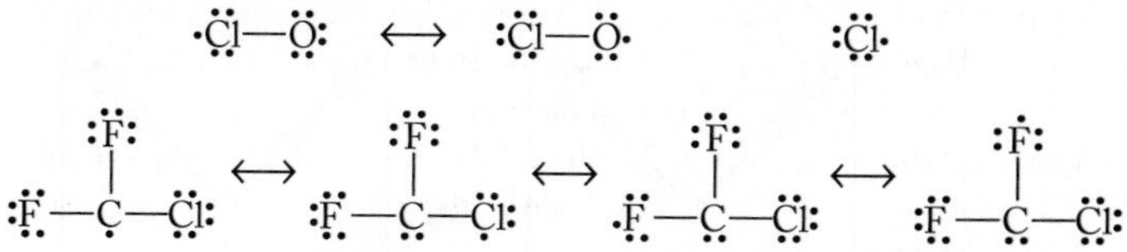

$\cdot\text{Cl—O:} \leftrightarrow \text{:Cl—O}\cdot \qquad \text{:Cl}\cdot$

$\text{F—C(F)—Cl:} \leftrightarrow \text{F—C(F)—Cl:} \leftrightarrow \cdot\text{F—C(F)—Cl:} \leftrightarrow \text{F—C(F)—Cl:}$

In order of increasing reactivity: $CF_2Cl\cdot < ClO\cdot < Cl\cdot$.

b. The products are

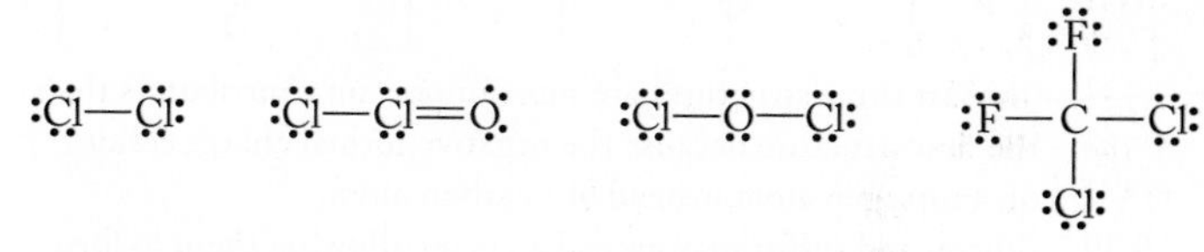

$\text{Cl—Cl} \qquad \text{Cl—Cl=O} \qquad \text{Cl—O—Cl} \qquad \text{F—C(F)(Cl)—Cl}$

Chapter 9

9.1. (b)

9.3. (a) trigonal planar, trigonal planar; (b) tetrahedral, trigonal pyramidal; (c) trigonal bipyramidal, T-shaped

9.5. All three molecules have polar bonds. BF_3 has no dipole moment.

9.7. (a) trigonal planar and tetrahedral; (b) trigonal planar: sp^2; tetrahedral: sp^3; (c) One hybridized orbital is required for each electron domain around a central atom. A trigonal planar structure has three electron domains and therefore requires three hybridized orbitals. A tetrahedral structure has four electron domains and therefore requires four hybridized orbitals.

9.9. N_2F_2 and NCCN are planar; there are no delocalized π electrons in any of these molecules.

9.11. more

9.13. 72° and 90°

9.15. Because the electrons take up most of the space in the atom and because the nucleus is located in the center of the electron cloud, the electron clouds repel each other before the nuclei get close to each other.

9.17. Because the lone pair is attracted by only one nucleus, it is less confined than bonding pairs and occupies more space around the nucleus of the central N atom.

9.19. The seesaw geometry has only two lone pair–bond pair interactions at 90° (compared to a trigonal bipyramidal–derived trigonal pyramidal, which would have three), so it has lower energy.

9.21. (b) < (c) < (a)

9.23. trigonal bipyramidal and seesaw

9.25. (a)

9.27. Pentagonal pyramid and distorted octahedral

9.29. (a) tetrahedral; (b) trigonal pyramidal; (c) bent; (d) tetrahedral

9.31. (a) 109.5°; (b) <109.5°; (c) ~120°; (d) 90°, 180°

9.33. (a) tetrahedral; (b) tetrahedral; (c) trigonal planar; (d) linear

9.35. O_3, SO_2, and S_2O are bent (120°), whereas N_2O and CO_2 are linear.

9.37.

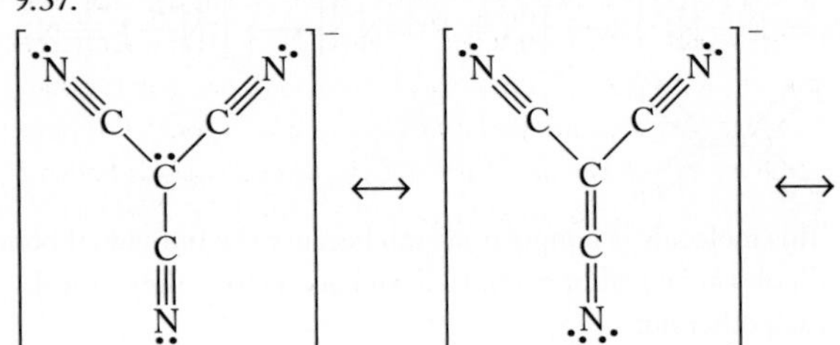

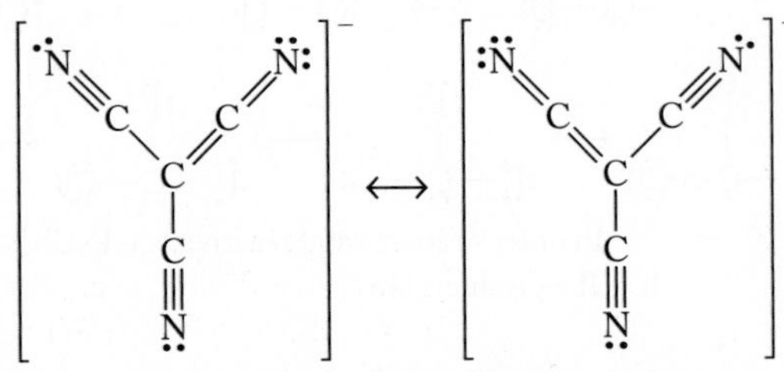

The last three structures are more important contributors than the first structure because the negative formal charge resides on a nitrogen atom instead of a carbon atom.

9.39. Silicon and sulfur may exceed an octet allowing them to form a double bond with the central nitrogen making the nitrogen trigonal planar in $N(SiH_3)_3$ and $N(SCF_3)_3$. Carbon cannot exceed an octet so the carbon atoms can only form single bonds in $N(CH_3)_3$ resulting in a trigonal pyramidal geometry.

9.41.

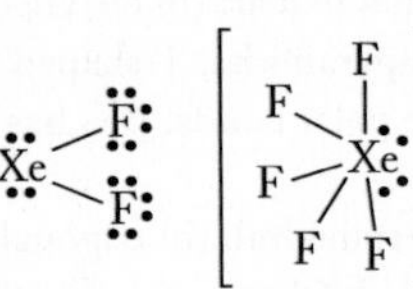

Square planar

Pentagonal bipyramidal

9.43.

ClO_2^+

SN = 3
Electron-pair geometry = trigonal planar
One lone pair
Molecular geometry = bent (122°)
Bond order = 2

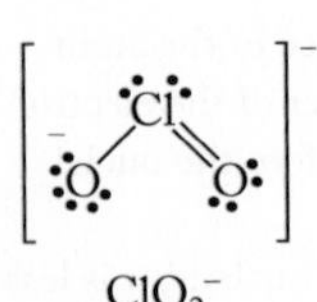

ClO_2^-

SN = 4
Electron-pair geometry = tetrahedral
Two lone pair
Molecular geometry = bent (110°)
Bond order = 1.5

ClO_2

SN = 4
Electron-pair geometry = tetrahedral
One lone pair + one unpaired electron
Molecular geometry = bent (118°)
Bond order = 1.5

9.45. Moving down the periodic table from S to Se to Te, the lone-pair electrons reside in sequentially larger orbitals. With those electrons spread over a larger volume, the remaining bonds in the seesaw structure would be pushed closer together.

9.47. Bond types are defined on the basis of the electronegativity differences of the atoms involved. Bonds are classified as nonpolar covalent if the atoms involved have the same electronegativity. Bonds are classified as polar covalent if the electronegativity difference of the atoms is between 0 and 2.0. Bonds are classified as ionic if the electronegativity difference between the atoms is greater than or equal to 2.0.

9.49. yes

9.51. (a) All five molecules contain polar bonds; (b) $CHCl_3$, H_2S, and SO_2 are polar; (c) CCl_4 and CO_2 are nonpolar

9.53. All three of the molecules are polar.

9.55. (a) Yes. (b) The most polar bond in each molecule is the C–N bond. (c) BrXeCN would have a larger value of μ. There is a smaller electronegativity difference in the Br–Xe bond than in the Cl–Xe bond resulting in less of a counterbalance to the electronegativity difference in the C–N bond.

9.57. (a) i; (b) i; (c) ii

9.59. (a) tetrahedral; (b) nonpolar

9.61. It is energetically unfavorable to hybridize orbitals unless a chemical bond is also formed.

9.63. A sigma bond is formed from the overlap of two atomic or hybridized atomic orbitals along a bonding axis. A pi bond is formed from unhybridized *p* or *d* orbitals above and below a bonding axis.

9.65.

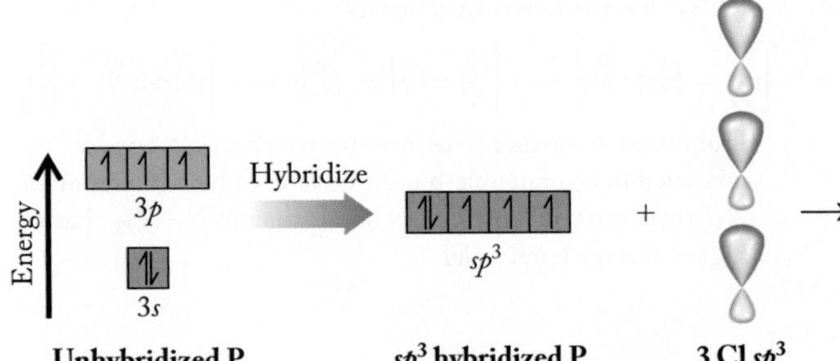

P
Cl
Cl
Cl

PCl_3

9.67. Both Lewis structures have three resonance forms: one with two N=N double bonds and two with an N≡N triple bond and an N—N single bond. For both molecules the central nitrogen atoms are *sp* hybridized.

9.69. SF_2 (sp^3), SF_4 (sp^4), and SF_6 (sp).

9.71. The N—O bond is a single bond. There is also an N—C single bond and an N=C double bond.

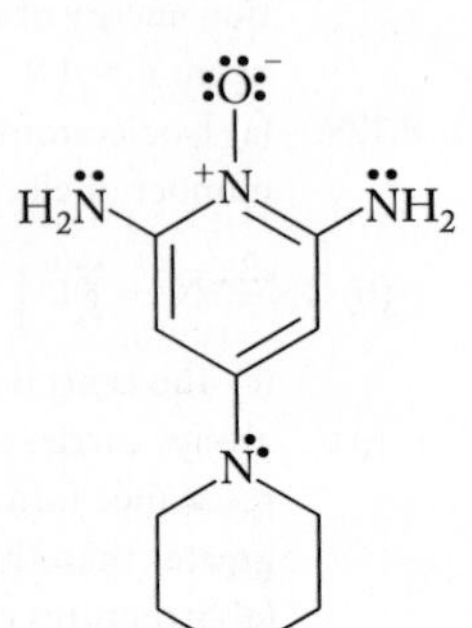

9.73. The molecular geometry around the phosphorus atom is bent and its hybridization is sp^2.

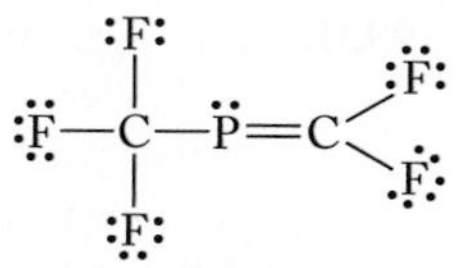

9.75. All types of structure convey the correct atom stoichiometry within the molecule, though the condensed structure does not show connectivity. Lewis, Kekulé, and carbon-skeleton structures are all capable of displaying resonance structures, and functional groups may be identified in each. A certain degree of chemical knowledge is required to correctly interpret a carbon-skeleton structure because the identity or presence of C and H atoms is to be assumed.

9.77. Yes; in resonance structures the electron distribution is blurred across all of the resonance forms. Resonance forms are Lewis structures that show alternative (yet still valid) electron distributions in a molecule.

9.79. (b) and (c)

9.81. $CH_3(CH_2)_2CH{=}CH{-}CH{=}CH(CH_2)_9OH$

9.83. (a) Progesterone has 20 carbon atoms, testosterone has 19 carbon atoms; (b) progesterone has four sp^2 carbon atoms, testosterone has three sp^2 carbon atoms; (c) both molecules contain alkane, alkene, and ketone functional groups.

9.85. Progesterone has six chiral carbons, and testosterone has six chiral carbons.

9.87. $CH_3(CH_2)_7CH{=}CH(CH_2)_7C(O)OH$

9.89.

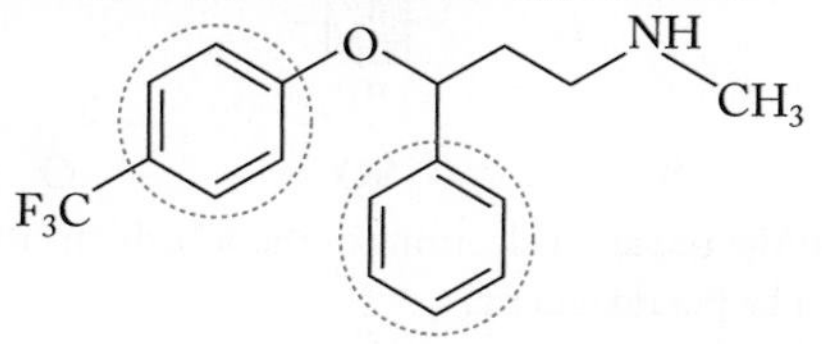

9.91.

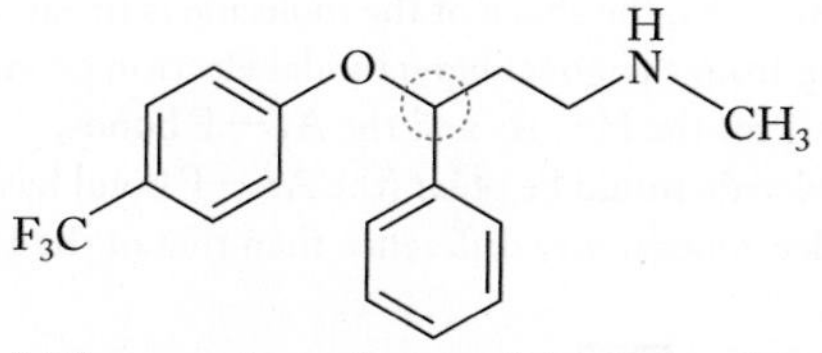

9.93. (a) Starting from the top of the ring and progressing clockwise: S is tetrahedral, N is bent, all three carbons in the six-membered ring are trigonal planar, and O is bent. (b) C–O bond: sp^2 C orbital with sp^3 O orbital; C–N bond: sp^2 C orbital with sp^3 N orbital; (c) the extra electron is in an sp^3 hybrid orbital; (d) no delocalized electrons.

9.95. (a) and (c) are chiral; none of the molecules have delocalized electrons.

9.97. A bonding molecular orbital has its highest region of electron density between the atom centers. An antibonding molecular orbital has its highest region of electron density outside of the space between the atom centers.

9.99. No; the overlap of 1*s* and 2*s* orbitals is not as efficient as 1*s*–1*s* or 2*s*–2*s* overlaps—there is better overlap for orbitals of the same size and energy.

9.101. If a molecule or ion has an even number of valence electrons distributed over an odd number of orbitals or if the highest occupied molecular orbital is a π orbital, paramagnetic species could result.

9.103. In the sea-of-electrons model, application of an electrical potential across a metal causes its mobile valence electrons to move toward the positive potential. In band theory there is no energy gap between the valence and the conduction band so electrons are free to move into the empty band.

9.105. Doping a metalloid generates a semiconductor with a smaller bandgap than that of the starting material. The n-type semiconductors have "extra" electrons higher in energy than the valence band of the undoped material. The p-type semiconductors have "extra" holes lower in energy than the conduction band of the undoped material. In both types of doped semiconductor, the effect is similar; the band gap is lowered and the electrons flow with a smaller energy barrier.

9.107.

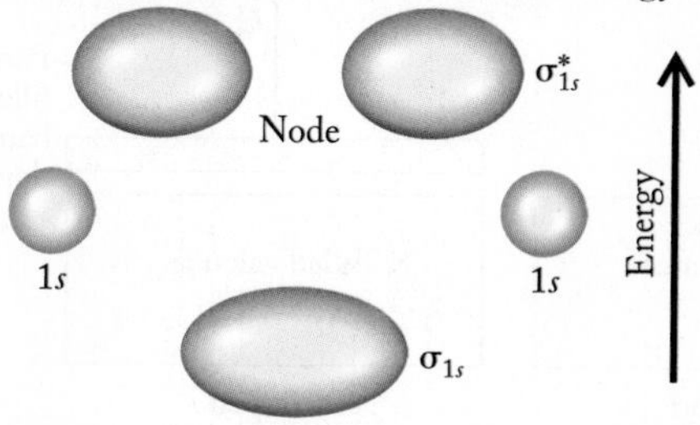

9.109. (a) N_2^+: $(\sigma_{2s})^2(\sigma^*_{2s})^2(\pi_{2p})^4(\sigma_{2p})^1$; O_2^+: $(\sigma_{2s})^2(\sigma^*_{2s})^2(\sigma_{2p})^2(\pi_{2p})^4(\pi^*_{2p})^1$; C_2^+: $(\sigma_{2s})^2(\sigma^*_{2s})^2(\pi_{2p})^3$; Br_2^{2-}: $(\sigma_{4s})^2(\sigma^*_{4s})^2(\sigma_{4p})^2(\pi_{4p})^4(\pi^*_{4p})^4(\sigma^*_{4p})^2$; (b) N_2^+: BO = 2.5; O_2^+: BO = 2.5; C_2^+: BO = 1.5; Br_2^{2-}: BO = 0; (c) all species with nonzero bond order (N_2^+, O_2^+, C_2^+) are expected to exist.

9.111. (a), (b), (c), (e), (g), and (h)

9.113.

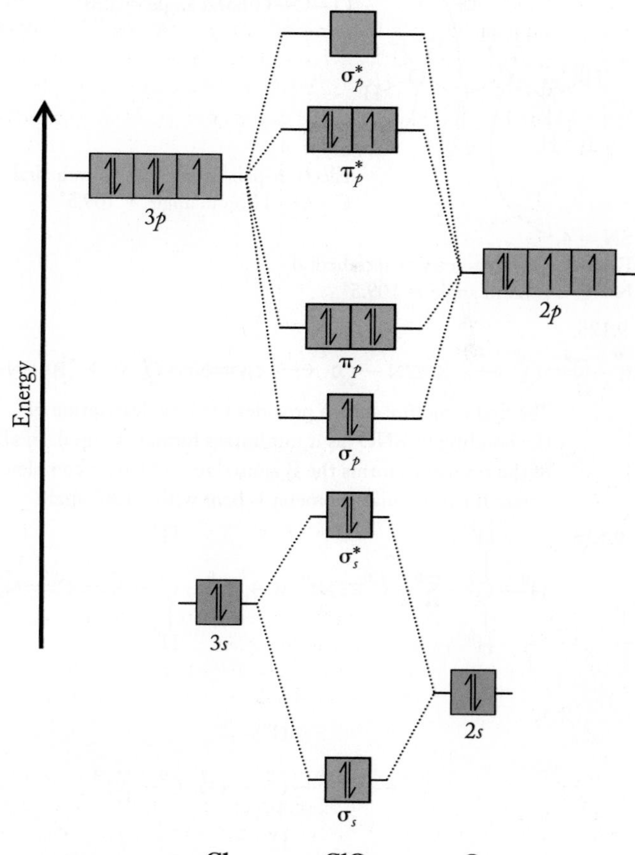

The odd electron is located in an antibonding orbital.

9.115. (a) and (b)

9.117. no

9.119. a. n-type

b.

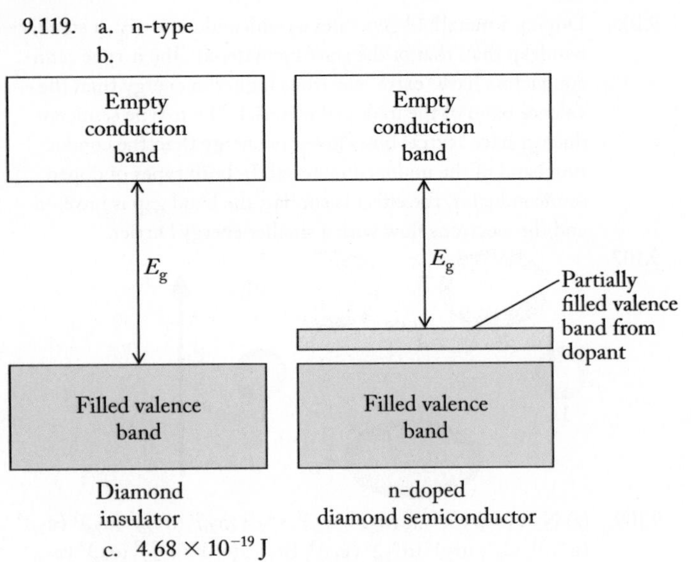

c. 4.68×10^{-19} J

9.121. (a) $[:\ddot{Cl}{=}\ddot{O}:]^{+}$; (b) bond order = 2

9.123.

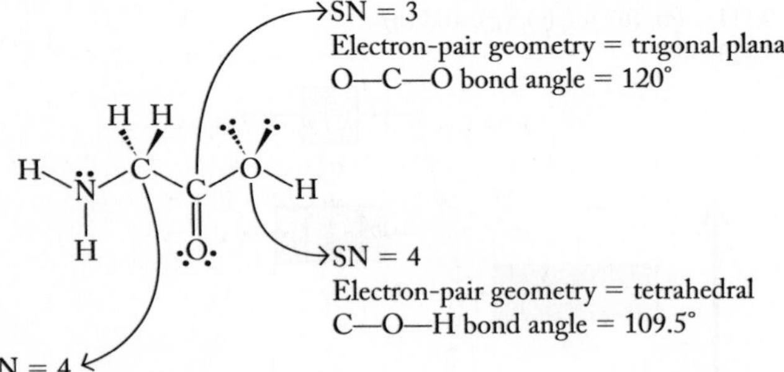

9.125.

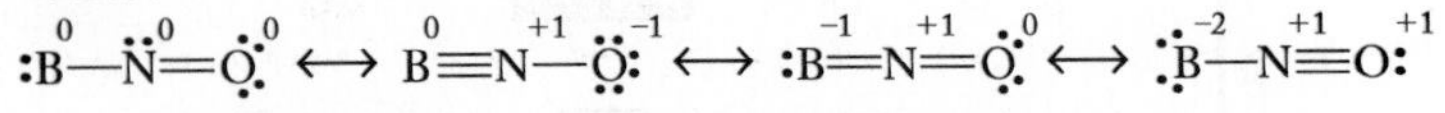

$\overset{0}{:B}—\overset{0}{\ddot{N}}=\overset{0}{\ddot{O}}: \longleftrightarrow \overset{0}{B}\equiv\overset{+1}{N}—\overset{-1}{\ddot{O}}: \longleftrightarrow \overset{-1}{:B}=\overset{+1}{N}=\overset{0}{\ddot{O}}: \longleftrightarrow \overset{-2}{:\ddot{B}}—\overset{+1}{N}\equiv\overset{+1}{O}:$

The first structure shown provides the best description of the bonding in BNO (as it minimizes formal charges). In all of the resonance forms the B atom does not have a complete octet. The molecular geometry is bent with a 120° angle.

9.127.

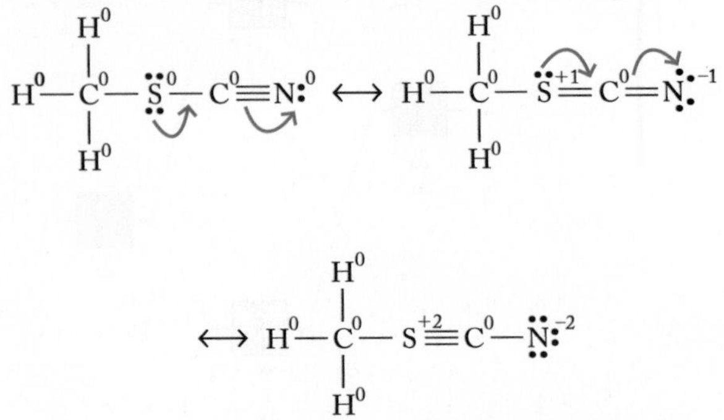

$H^0_3C^0—\ddot{S}^0—C^0\equiv N:^0 \longleftrightarrow H^0_3C^0—\overset{+1}{S}=C^0=\ddot{N}:^{-1}$

$\longleftrightarrow H^0_3C^0—\overset{+2}{S}\equiv C^0—\ddot{N}:^{-2}$

The first structure would be the most stable as it minimizes formal charges. The carbon attached to the three hydrogens and the sulfur is tetrahedral and the carbon bonded to the sulfur and the nitrogen is linear.

9.129.

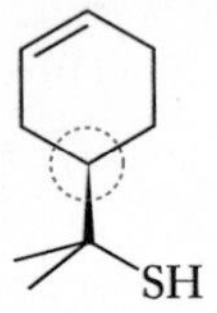

SH

9.131.

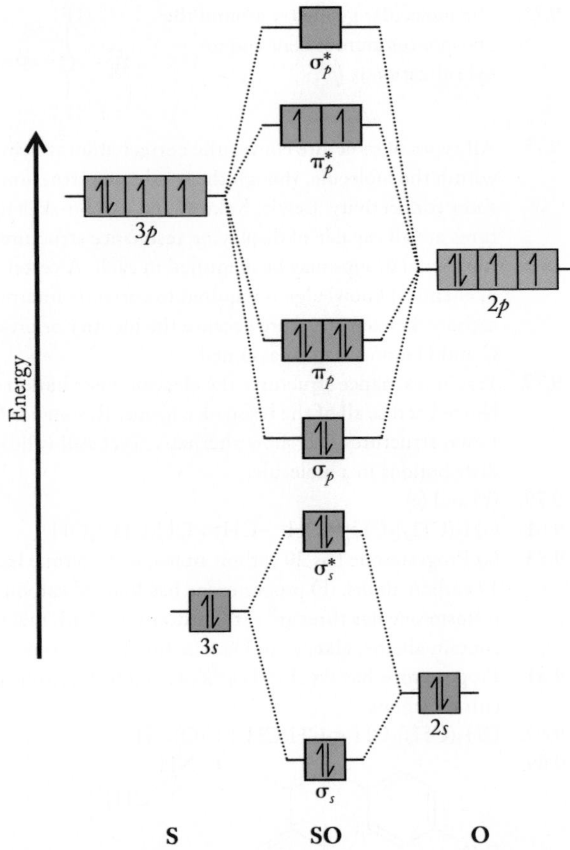

Because of the unpaired electrons in the π^* orbitals the molecule would be paramagnetic.

9.133. (a) $H—\ddot{Ar}—\ddot{F}:$, the shape of the molecule is linear (stemming from a trigonal bipyramidal electron geometry); (b) BO = 1 for the H—Ar and the Ar—F bonds; (c) the molecule would be polar (the Ar—F bond has a larger electronegativity difference than that of the H—Ar bond).

9.135. diamagnetic

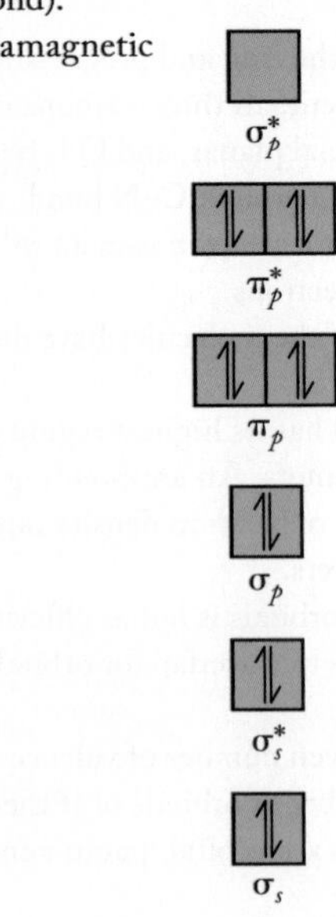

9.137.

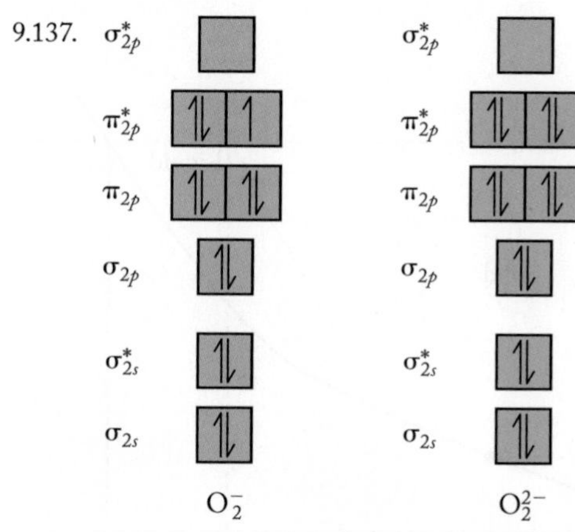

Superoxide, O_2^-: $(\sigma_{2s})^2\,(\sigma^*_{2s})^2\,(\sigma_{2p})^2\,(\pi_{2p})^4\,(\pi^*_{2p})^3$, BO = 1.5; peroxide, O_2^{2-}: $(\sigma_{2s})^2\,(\sigma^*_{2s})^2\,(\sigma_{2p})^2\,(\pi_{2p})^4\,(\pi^*_{2p})^4$, BO = 1. The Lewis structures predict a BO = 1 for both compounds:

$$[\cdot\ddot{\underset{\cdot\cdot}{O}}-\ddot{\underset{\cdot\cdot}{O}}:]^- \qquad [:\ddot{\underset{\cdot\cdot}{O}}-\ddot{\underset{\cdot\cdot}{O}}:]^{2-}$$

The predicted bond order for peroxide is the same in the molecular orbital diagram and the Lewis structure, but the bond order for superoxide is different: 1.5 from the molecular orbital diagram and 1 from the Lewis structure.

9.139. Valence bond theory without hybrid orbitals would provide a better description of the bonding in H_2S, H_2Se, and H_2Te because unhybridized *p* orbitals are 90° from each other. VSEPR and valence bond theory with hybrid orbitals would predict an angle of 109.5° for these molecules.

Chapter 10

10.1. The figure represents Xe in the solid state. The attractions between Xe atoms are the result of dispersion forces. As temperature increases, the solid will melt into a liquid (the particles would still be close together but will not have a rigid, ordered arrangement) and eventually vaporize into a gas (the particles would be far apart).

10.3. All three would be vanishingly soluble in water but (b) would be the least soluble. As the hydrocarbon chain gets longer, the dispersion interactions between the molecules within its pure liquid form will get stronger; also, the longer the chain, the more its dissolution would disrupt the hydrogen bonding in water.

10.5. The slope of the trend line in each graph is proportional to the size of ΔH_{vap}, so the graph with the larger slope has a larger enthalpy of vaporization. A large value of ΔH_{vap} corresponds to strong intermolecular forces, as seen in the graph for the liquid on the left.

10.7. (a) The freezing point increases. (b) No; as pressure is increased, the liquid phase is converted into solid at a given temperature. This means that the solid is more dense than the liquid and so the solid will not float in its liquid phase.

10.9. dispersion forces

10.11. Because real gases have nonzero volume, and when cooled they have less kinetic energy to overcome their intermolecular forces.

10.13. (a) CCl_4; (b) C_3H_8

10.15. PCl_5 is a solid at room temperature, whereas PCl_3 is a liquid. Despite being nonpolar, PCl_5 has a higher molar mass and a larger surface area, so the dispersion forces for PCl_5 outweigh the weaker dispersion forces and dipole–dipole interactions for PCl_3.

10.17. Because of the full positive or negative charge on the ion, the ion–dipole interaction is stronger than the dipole–dipole interaction.

10.19. The charge buildup on H (partially positive) and the electronegative element (partially negative) means that the X–H bond is polar. It is still a dipole–dipole interaction except that its strength is noticeably higher than in other dipole–dipole interactions.

10.21. The greater dispersion forces of CH_2Cl_2 add to the dipole–dipole interactions to give stronger intermolecular forces between the CH_2Cl_2 molecules than those between the CH_2F_2 molecules.

10.23. Dimethylsulfide has stronger intermolecular forces. Sulfur is larger and more polarizable than oxygen, which leads to stronger dispersion forces.

10.25. (a) Ethanol has a higher boiling point because it is capable of hydrogen bonding, whereas ethanethiol is not; (b) 1.968 times faster

10.27. (b)

10.29. (a) and (d)

10.31. When the average kinetic energy of the liquid molecules increases, more of the molecules can escape the liquid phase and enter the gas phase. More molecules in the gas phase increase the vapor pressure.

10.33. intensive

10.35. (a) $<$ (b) $<$ (c)

10.37. (a) 41.0 kJ/mol; (b) 4.32 torr

10.39. 27.3 kJ/mol

10.41. (a) solid, liquid, and gas; (b) liquid and gas

10.43. (a) solid phase; (b) gas phase

10.45. Yes. From the phase diagram for water we see that above the triple point, the solid phase must change to the liquid phase to enter the gas phase. Below the triple point, changing the temperature at a given pressure will sublime solid water into the gas phase.

10.47. approximately 55°C to 60°C

10.49. Ethylene glycol has stronger intermolecular forces. We can tell this from the higher boiling point of ethylene glycol and explain it based on a second O–H bond that can hydrogen-bond to nearby molecules in the liquid phase.

10.51. Reduce the temperature from 25°C to 0.01°C then reduce the pressure from 1 atm to 0.006 atm.

10.53. Water would be in the liquid phase initially and would vaporize to the gas phase when the pressure was lowered.

10.55. (a) Yes, as pressure increases the melting point of ice decreases. (b) It would take approximately 100 atm of pressure to reduce the melting point by 0.5°C.

10.57. A needle floats on water but not on methanol because of the high surface tension of water. This is because water can hydrogen-bond through two O–H bonds with other water molecules whereas methanol has only one O–H bond through which to form hydrogen bonds.

10.59. The expansion of the water in the pipes on freezing may create sufficient pressure on the walls of the pipes to cause them to burst.

10.61. The cohesive forces in mercury are stronger than the adhesive forces of the mercury to the glass. In water the adhesive forces of the water to the glass are strong enough that the water can climb upward on the glass.

10.63. Molecules in the bulk liquid are "pulled" by all the other liquid molecules surrounding them and they are therefore "suspended" in the bulk liquid. Molecules on the surface of a liquid, however, are pulled only by the molecules under and beside them creating a tight film of molecules on the surface.

10.65. As the temperature increases, the surface tension decreases because the intermolecular forces acting on the molecular "film" on the surface of the liquid are weaker. Likewise, the viscosity decreases as the temperature increases because molecules have more energy to readily break the intermolecular forces to enable them to slide past each other more freely.

10.67. Water is able to form more hydrogen bonds to the glass surface (composed of Si—O—H bonds) than ethanol so water rises higher in the capillary tube.

10.69. Liquid B is expected to have higher surface tension and higher viscosity because its higher boiling point is indicative of stronger intermolecular forces.

10.71. Miscible solutes and solvents dissolve completely in each other; an insoluble solute does not dissolve at all.

10.73. 1,1-Dichloroethane is polar and therefore soluble in water. 1,2-Dichloroethane is nonpolar and insoluble in water.

10.75. Hydrophilic substances dissolve in water. Hydrophobic substances do not dissolve, or are immiscible, in water.

10.77. (a) $CHCl_3$; (b) CH_3OH; (c) NaF; (d) BaF_2

10.79. (a), (b), (c), and (d)

10.81. (b)

10.83. Both of these compounds can dissociate when added to water. Dissociation will result in ion–dipole interactions with water, which are even stronger than the hydrogen-bonding interactions that would be present in the nonionic form of the drugs.

10.85. (b) and (c) are ethers, whereas (a) and (d) are alcohols. The expected ordering of boiling points is (b) < (c) < (a) < (d).

10.87. At lower temperature, the solubility of a gas increases because fewer gas molecules dissolved in a solvent have sufficient (kinetic) energy to overcome the intermolecular forces between the solvent molecules and escape the surface of the liquid.

10.89. No, the Henry's law constant for air is not simply the sum of k_H for N_2 and O_2.

$$\begin{aligned} C_{air} &= k_{H,air}P_{air} = C_{O_2} + C_{N_2} \\ k_{H,air}P_{air} &= k_{H,O_2}P_{O_2} + k_{H,N_2}P_{N_2} \\ &= k_{H,O_2}(0.22P_{air}) + k_{H,N_2}(0.78P_{air}) \\ &= 0.22k_{H,O_2}P_{air} + 0.78k_{H,N_2}P_{air} \\ &= (0.22k_{H,O_2} + 0.78k_{H,N_2})P_{air} \\ k_{H,air} &= 0.22k_{H,O_2} + 0.78k_{H,N_2} \end{aligned}$$

10.91. SO_3 and CO_2 (because they are nonpolar).

10.93. 3.7×10^{-2} mol/(L · atm)

10.95. (a) 2.7×10^{-3} *M*; (b) 2.3×10^{-2} *M*

10.97. 2.18 g

10.99. Alcohols exhibit stronger intermolecular forces (hydrogen bonds) than do ethers (dipole–dipole).

10.101. Increase; the equilibrium line between the solid and vapor phases of water increases upward and to the right of the phase diagram. As the pressure increases, the temperature at which sublimation occurs will also increase until the ice reaches the triple point (0.0060 atm and 0.010°C).

10.103. yes

10.105.

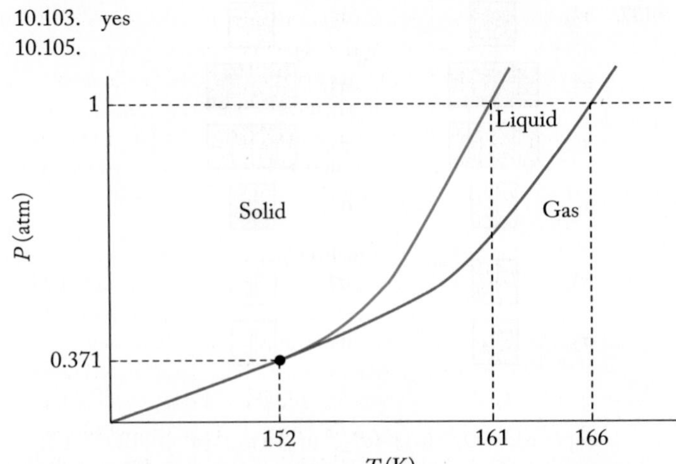

10.107. Inside a narrow capillary tube, the mass of water that rises to a particular height in the column equals the mass of water that can rise in a wider-diameter tube, but the level in the test tube is lower because the volume (mass) is spread out across a wider area. It is the same mass that feels the force of gravity being balanced by the adhesive and cohesive forces.

10.109. As the temperature of the pond water increases, the oxygen becomes less soluble and the fish do not have enough oxygen to survive.

10.111. (a)

Chapter 11

11.1. (a)

11.3. (a) (I) = NaCl, (II) = $MgCl_2$, (III) = $C_6H_{12}O_6$, (IV) = K_3PO_4; (b) (i) (III), (ii) (IV), (iii) (III), (iv) (IV)

11.5. (a) If the substances were electrolytes we would expect to observe deviations from linearity at higher concentrations due to ion-pair formation. Because both substances exhibit a linear relationship between temperature change in the freezing point of the solution and molality, they are nonelectrolytes. (b) The freezing-point-depression constant (K_f) would correspond to the slope of the line: because the slope is the same for both solutions, we can conclude that the K_f value is the same for both solutions.

11.7. (a) In diagram (i) water will pass through the membrane from the pure water side to the protein solution side, and in diagram (ii) water will pass from the 0.2 *M* KCl solution side to the protein solution side. (b) In diagram (i) Na^+ ions will flow from the 0.5 *M* NaCl (protein solution) side to the pure water side of the membrane, and in diagram (ii) Na^+ ions will flow from the 0.5 *M* NaCl (protein solution) side to the aqueous 0.2 *M* KCl side of the membrane. (c) In diagram (i) K^+ ions will flow from the 0.1 *M* KCl (protein solution) side to the pure water side of the membrane, and in diagram (ii) K^+ ions will flow from the 0.2 *M* KCl (aqueous solution) side to the 0.1 *M* KCl (protein solution) side of the membrane.

11.9. (a) The sample is a mixture. (b) (i) The mixture contains three substances; (ii) the first (lowest boiling) material has a volume of 20 mL, the second (middle boiling) material has a volume of 50 mL, and the third (highest boiling) material has a volume of 30 mL; (iii) the first material boils at 20°C, the second at 70°C, and the third at 130°C.

11.11. (c)

11.13. The ion–ion bond in $CaSO_4$ is stronger than the ion–ion bond in NaCl because of the higher charges on the cation and anion. For $CaSO_4$ this is greater than the ion–dipole interactions that would occur when Ca^{2+} and SO_4^{2-} dissolve, so $CaSO_4$ is not very soluble in water. NaCl has a lower ion–ion bond strength and its ion–dipole interactions with water are strong, so it dissolves in water.

11.15. $CsBr < KBr < SrBr_2$

11.17. MgO will have a higher melting point than LiF.

11.19. The enthalpy of solution ($\Delta H_{solution}$) is the overall change in enthalpy when an ionic solute is dissolved in a polar solvent. This is the sum of the ion–ion interactions between ions, dipole–dipole interaction between solvent molecules, and ion–dipole interactions between the solute and the solvent. The enthalpy of hydration ($\Delta H_{hydration}$) is the energy associated only with the formation of solvated ions. This is the sum of the dipole–dipole interaction between solvent molecules and the ion–dipole interactions between the solute and the solvent. The lattice energy (U) is the change in energy when free ions in the gas phase combine to form 1 mole of a solid ionic compound.

11.21. A large negative lattice enthalpy (U) means that $\Delta H_{solution}$ is more likely to be positive (because $\Delta H_{solution} = \Delta H_{hydration} - U$). Therefore, it will be less favorable to separate the ions.

11.23. Melting point decreases as the atomic number of X increases.

11.25. (a)

11.27. −723 kJ/mol

11.29. -6.90×10^2 kJ/mol

11.31. A nonvolatile solute is a compound that dissolves into a solvent and does not, under conditions to maintain the solution, enter appreciably into the gas phase.

11.33. When the average kinetic energy of the liquid molecules increases, more of the molecules can escape the liquid phase and enter the gas phase. More molecules in the gas phase increases the vapor pressure.

11.35. The vapor pressure of pure ethanol is greater that the vapor pressure of the sugar solution of ethanol. The rate of evaporation will be higher for the pure ethanol. On the other hand the rate of condensation in the beaker containing the sugar solution is higher than the rate of evaporation, so eventually the beaker containing the sugar solution will overflow.

11.37. Mole fraction of water in solution: $X = 0.70$; vapor pressure of solution at 25°C: $P_{solution} = 17$ torr

11.39.

$$P_{soln} = X_{solvent} \cdot P^{\circ}_{solvent}$$
$$X_{solvent} + X_{solute} = 1$$
$$X_{solvent} = 1 - X_{solute}$$
$$P_{soln} = (1 - X_{solute}) \cdot P^{\circ}_{solvent}$$
$$P_{soln} = P^{\circ}_{solvent} - P^{\circ}_{solvent} \cdot (X_{solute})$$
$$P^{\circ}_{solvent} \cdot (X_{solute}) = P^{\circ}_{solvent} - P_{soln}$$
$$X_{solute} = \frac{(P^{\circ}_{solvent} - P_{soln})}{P^{\circ}_{solvent}}$$

11.41. C_5H_{12}

11.43. Both components of the solution are of similar size and both are nonpolar. It is reasonable that the magnitude of the intermolecular forces would be similar and that the solution would behave ideally.

11.45. 18 torr

11.47. (a) The vapor phase would be 54% styrene and 46% ethylbenzene for an ethylbenzene-to-styrene ratio of 27:23; (b) (ii)

11.49. The van't Hoff factor is an experimentally determined value and cannot be used to determine the molality (and thus the formula mass) for an ionic solute.

11.51. As the solute of the sap evaporates, the remaining solution becomes more concentrated. This will lower the vapor pressure and raise the boiling point.

11.53. Whereas a molecular compound will not dissociate, an ionic compound may dissociate in the solvent. This dissociation yields two or more particles in solution from one dissolved solute particle. This results in greater changes in the melting and boiling points than for those of a molecular solute that does not dissociate.

11.55. The theoretical value of i for CH_3OH is 1 because methanol is molecular and does not dissociate in a solvent such as water. NaBr has a theoretical value of $i = 2$ because it dissociates into two particles on dissolution (Na^+ and Br^-). K_2SO_4 has a theoretical value of $i = 3$ because it dissociates into three particles on dissolution (2 K^+ and SO_4^{2-}).

11.57. A semipermeable membrane is a boundary between two solutions through which some molecules may pass but others cannot. Usually small molecules may pass through but large molecules are excluded.

11.59. Solvent flows across a semipermeable membrane from the more dilute solution side to the more concentrated solution side to balance the concentration of solutes on both sides of the membrane.

11.61. Reverse osmosis transfers solvent across a semipermeable membrane from a region of higher solute concentration to a region of lower solute concentration. Because reverse osmosis goes against the natural flow of solvent across the membrane, the key component needed is a pump to apply pressure to the more concentrated side of the membrane. Other components needed include a containment system, piping to introduce and remove the solutions, and a tough semipermeable membrane that can withstand the high pressures needed.

11.63. Because NaCl(*aq*) exists in solution as $Na^+(aq)$ and $Cl^-(aq)$, we need twice as many moles of dextrose to achieve the same molar concentration.

11.65. (a) 0.21 *m*; (b) 0.572 *m*; (c) 0.440 *m*

11.67. (a) 749 g; (b) 31.7 g; (c) 104.6 g

11.69. 6.5×10^{-5} *m* NH_3; 8.7×10^{-6} *m* NO_2^-; 2.19×10^{-2} *m* NO_3^-

11.71. 3.80°C

11.73. 2.52×10^{-2} *m*

11.75. −1.89°C

11.77. (a) < (b) < (c)

11.79. (a) from side A to side B; (b) from side B to side A; (c) from side A to side B

11.81. (a) 57.5 atm; (b) 0.682 atm; (c) 52.9 atm; (d) 46 atm

11.83. (a) 2.76×10^{-2} *M*; (b) 1.11×10^{-3} *M*; (c) 1.00×10^{-2} *M*

11.85. 108 mL

11.87. 1.41 atm

11.89. A cation exchanger can replace dissolved cations with H^+ and an anion exchanger can replace dissolved anions with OH^-.

11.91. (a) 1. $Ag^+(aq) + Cl^-(aq) \rightarrow AgCl(s)$; 2. NR; 3. $Mg^{2+}(aq) + 2\,OH^-(aq) \rightarrow Mg(OH)_2(s)$; (b) reactions 1 and 3

11.93. (a) osmotic pressure increases; (b) freezing point decreases; (c) boiling point increases

11.95. concentrations between 0.014 and 0.022 *M* (or 850–1300 g of hemoglobin/L)

11.97. Molar mass = 164 g/mol; the molecular formula of eugenol is $C_{10}H_{12}O_2$

11.99. Step 1: Sodium metal, Na(*s*), absorbs energy and is sublimated, Na(*g*). Step 2: Energy is added to $Cl_2(g)$ causing the Cl–Cl bond to be broken producing Cl(*g*). Step 3: Energy is added to Na(*g*) causing the loss of an electron and the formation of gas-phase sodium cations, $Na^+(g)$. Step 4: Energy is released as the gas-phase chlorine atom, Cl(*g*), gains an electron to form the gas-phase ion, $Cl^-(g)$. Step 5: Energy is released as the gas-phase ions, Na^+(g) and Cl^-(g), react to form the ionic solid NaCl(*s*).

11.101. yes

11.103. For 0.0935 *m* NH_4Cl, $i = 1.85$; for 0.0378 *m* $(NH_4)_2SO_4$, $i = 2.46$

11.105. 2.3 atm

11.107. 4.26×10^2 g/mol

11.109. Cations: $[Na^+]_{saline} = [Na^+]_{Ringer's} + [K^+]_{Ringer's} + [Ca^{2+}]_{Ringer's}$; anions: $[Cl^-]_{saline} = [Cl^-]_{Ringer's} + [C_3H_5O_3^-]_{Ringer's}$

Chapter 12

12.1. (b) and (d) are crystalline, (a) and (c) are amorphous

12.3.

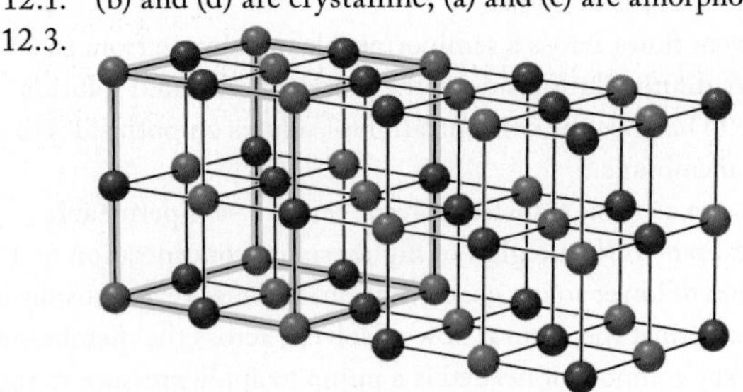

The chemical formula is A_4B_4 or AB.

12.5. A_4B_2C

12.7. Li_2S

12.9. empirical formula: NiTi; density: 5.78 g/cm^3

12.11. MgB_2

12.13. (c)

12.15. (a) Hydrogen bonds. (b) Each strand contains amide functional groups. Amides can be formed by a condensation reaction between an amine and a carboxylic acid.

12.17. Cubic closest-packed structures have an *abcabc* . . . pattern, and hexagonal closest-packed structures have an *abab* . . . pattern.

12.19. body-centered cubic

12.21. These structural forms are not allotropes because iron is not molecular.

12.23. 104.2 pm

12.25. 513 pm

12.27. (c)

12.29. 1.05 g/cm^3

12.31. 4.50 g/cm^3

12.33. In a substitutional alloy some of the atoms of one lattice are replaced with atoms of another element. One example of a substitutional alloy is bronze, in which up to 30% of the Cu atoms in a copper lattice have been replaced with tin atoms. The tin atoms in this alloy are randomly distributed throughout the lattice occupying any lattice position normally occupied by a copper atom. An interstitial alloy contains solutes in the spaces (or "holes") between atoms when the solvent lattice forms. One example of an interstitial alloy is austenite, an alloy of carbon in the octahedral holes of the iron fcc lattice.

12.35. The edge length will be smaller.

12.37. The significant increase in mass when hafnium is substituted for magnesium will result in an alloy of greater density than that of pure magnesium.

12.39. Yes, both unit cells contain the same ratio of Ni to Ti so both are valid.

12.41. (a) 90 pm; (b) substitutional alloy

12.43. a. substitutional alloy

b.

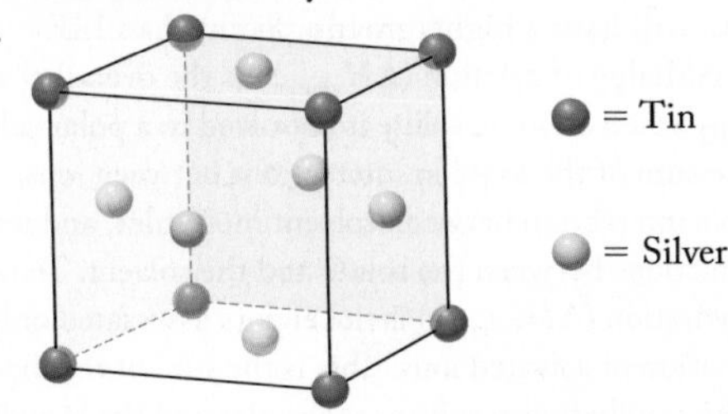

12.45. V_2C

12.47. (a) MoW; (b) yes, the heavier tungsten atom is occupying the body center; (c) 14.1 g/cm^3

12.49. (a) a substitutional alloy with a bcc lattice; (b) $CuZn_3$

12.51. Yes, the alloy would be more dense. Tin has a greater molar mass than copper so substituting tin into the unit cell would result in a greater mass.

12.53. K^+ is large and so does not fit well into the octahedral holes of the fcc lattice of the Cl^- ions.

12.55. The radius of Cl^- is 181 pm and the radius of Cs^+ is 170 pm so their radii are very similar. Figure P12.55 shows the Cs^+ occupying the center of the cubic cell, so CsCl could be viewed as a body-centered cubic structure when taking into account the ions' slight difference in size. The CsCl structure could also be viewed as a simple cubic lattice of Cl^- ions with a Cs^+ in the cubic holes.

12.57. The rock salt structure would not provide the correct ratio of ions for $CaCl_2$.

12.59. less than

12.61. $MgFe_2O_4$

12.63. (a) octahedral; (b) half

12.65. (a) The rock salt structure is an fcc lattice of anions with cations in the octahedral holes. The cesium chloride structure is a simple cubic lattice of anions with cations in the cubic holes. (b) Both structures have the same ratio of Mg^{2+} to Se^{2-} ions (1:1).

12.67. a.

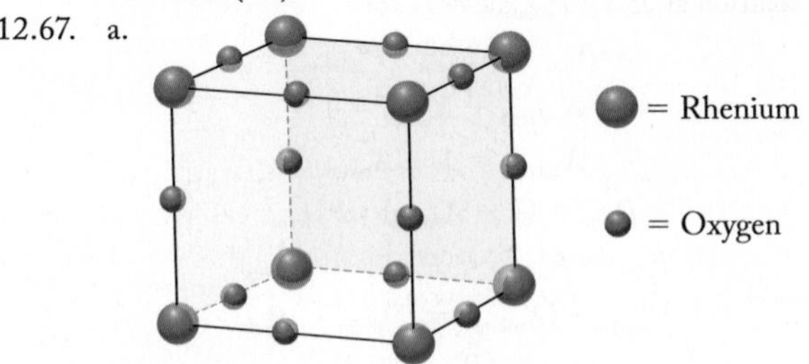

b. 5.25 g/cm^3

12.69. 421 pm

12.71. The hybridization of the phosphorus atom in black phosphorus is *sp*^{3} with bond angles of 102° to give a puckered ring. In graphene, the carbon atoms are *sp*^{2} hybridized with bond angles of 120° which gives graphene a flat geometry.

12.73. (a) 1.71 g/cm^3; (b) 498.6 pm

12.75. (a) 3.81 g/cm^3; (b) because of the increased distance between P atoms (238 pm compared to 200 pm if the P atoms are touching along the cell edge), either B or Si could fit into a cubic hole; (c) as both B and Si have fewer valence electrons than P does, these structures would be equivalent to p-type semiconductors.

12.77. Monomers must have appropriate functional groups to react with one another in a polymerization. Provided they have appropriate functional groups, different monomers can react to form a polymer.

12.79. Polyethylene and $C_{24}H_{50}$ are both composed primarily of long chains of $-CH_2-$ groups (with $-CH_3$ groups at the ends of the chains). The difference between the structures is that polyethylene would contain chains of varying lengths and molecular weights, whereas all of the chains in $C_{24}H_{50}$ have the same length and molecular weight.

12.81. 3571 monomer units

12.83.

12.85. HO(C=O)(CH$_2$)$_{12}$CH$_3$ HO–CH(CH$_2$OH)CH$_2$OH

12.87. a. Polymer I monomers:

$H_2N(CH_2)_8NH_2$ and $HOC(O)(CH_2)_8C(O)OH$

Polymer II monomers:

$H_2N(CH_2)_6NH_2$ and $HOC(O)(CH_2)_{10}C(O)OH$

b. Because of the longer chain length between C=O bonds in polymer II, we might expect that this polymer would be more flexible along that portion of the polymer chain.

12.89. (a) HCl; (b) the CO_3 or $-OCOO-$ linkage found in the Lexan structure is what qualifies it as a polycarbonate.

12.91. XYZ_3

12.93. (a) bcc; (b) two phase changes, hcp $\rightarrow$ bcc $\rightarrow$ hex (3).

12.95. (a) 2.3×10^5 unit cells/particle; (b) 9.1×10^5 Ag atoms; (c) 8.4×10^{12} Ag particles

12.97. (a) 7.53 g/cm^3; (b) 3.42 g/cm^3; (c) 3.36 g/cm^3

12.99. (a) The sizes of the neutral atoms are Li, 152 pm; and Cl, 99 pm (compared to Li^+, 76 pm; and Cl^-, 181 pm). The cubic unit cell for the neutral atoms would therefore have a shorter edge length of 502 pm (514 pm for the ionic version). Because the masses within the unit cell would be the same, this version of the compound would have a higher density. (b) 2.23 g/cm^3

12.101. The maximum size of the atomic radius to fit into the tetrahedral hole in P_4 would be 45.1 pm. Hydrogen (37 pm) or helium (32 pm) would be able to fit.

12.103. A bcc Cu lattice with four corner atoms replaced by Al is consistent with the formula.

12.105. a.

Repeating unit depicted in Problem 12.105(a)

b.

Repeating unit depicted in Problem 12.105(b)

12.107. This polymer has one monomeric unit with seven carbon atoms; nylon-6 has a single monomeric unit as well, but it consists of six carbon atoms.

12.109. Cross-linking increases the strength, hardness, melting (softening) point, and chemical resistance of a polymer.

12.111.

Chapter 13

13.1. $[N_2O]$: green line; $[O_2]$: red line

13.3. (b)

13.5. (d)

13.7. (c)

13.9. (a) 3*; (b) arrow a (red); (c) arrow b (black); (d) arrow c (blue)

13.11. Reaction 2 has the larger slope and therefore the larger activation energy. The rate of reaction 2 will be more sensitive to changes in temperature.

13.13. As the concentration of NO increases in the atmosphere, more NO_2 is generated. This process is not immediate and it results in a decrease in the amount of NO in the atmosphere as it proceeds. The maximum concentration of NO_2 will occur several hours after the maximum concentration of NO is reached as a result of this delay.

13.15. O atoms are more reactive as a result of the incomplete octet and unpaired electrons on each atom.

13.17. nitrogen, N (light blue)

13.19. The rate of a reaction is the overall change in the concentration of reactants and products, whereas the rate constant for a reaction is the proportionality between the rate of the reaction and the concentration of each reactant raised to the order of each reactant.

13.21. Tracking the partial pressure of CH_3CHO, CH_4, or CO. The collected rate data would be the same for all alternatives. A fourth possibility, as there are different numbers of moles of gas on the reactant and product sides of the equation, would be to track changes in the total pressure as the reaction proceeds.

13.23. In solution, ionic compounds can dissociate and mix with one another. As solids, ions from the two reactants are still bound to one another so it is very unlikely that a reaction will occur on any significant scale.

13.25. The rate of formation of B and consumption of A are equal but are opposite in sign.

13.27. decrease

13.29. (a) $\frac{\Delta[N_2]}{\Delta t} = \frac{\Delta[O_2]}{\Delta t}$; (b) $\frac{\Delta[N_2]}{\Delta t} = -\frac{1}{2}\frac{\Delta[NO]}{\Delta t}$

13.31. 5.8×10^{-6} *M*/s

13.33. (a) rate $= \frac{\Delta[CO_2]}{\Delta t} = -\frac{2}{3}\frac{\Delta[CO]}{\Delta t}$;
(b) rate $= \frac{\Delta[COS]}{\Delta t} = -\frac{\Delta[SO_2]}{\Delta t}$;
(c) rate $= \frac{\Delta[CO]}{\Delta t} = 3\frac{\Delta[SO_2]}{\Delta t}$

13.35. (a) 1.2×10^7 *M*/s; (b) 2.9×10^4 *M*/s

13.37. between 0 and 100 μs: 1.4×10^{-5} *M*/μs; between 200 and 300 μs: 5.5×10^{-6} *M*/μs

13.39.

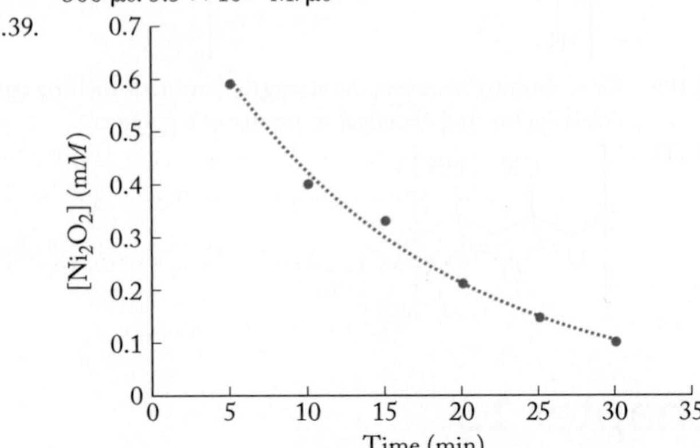

Instantaneous rate changes: at 10 minutes = 3.8 m*M*/min; at 15 minutes = 2.7 m*M*/min; at 25 minutes = 1.3 m*M*/min.

13.41. The concentration of reactants decreases over time so the likelihood of collisions between reactants decreases and the rate of reaction will decrease.

13.43. zero order

13.45. If the $[B]_0 = 50[A]_0$ then even if all of A reacts, the amount and concentration of B will appear unchanged due to the large excess of B. Because [B] is essentially constant in the reaction, it will appear to be part of the rate constant, *k*, and the rate law will appear to be rate = *k*[A].

13.47. (a) first order in [A], first order in [B], second order overall; (b) second order in [A], first order in [B], third order overall; (c) first order in [A], second order in [B], third order overall; (d) first order in [A], −1 order in [B], zero order overall

13.49. (a) rate $= k[O][NO_2]$; *k* units $= M^{-1}\,s^{-1}$; (b) rate $= k[NO]^2[Cl_2]$; *k* units $= M^{-2}\,s^{-1}$; (c) rate $= k[CHCl_3][Cl_2]^{1/2}$; *k* units $= M^{-1/2}\,s^{-1}$; (d) rate $= k[O_3]^2[O]^{-1}$; *k* units $= s^{-1}$

13.51. (a) rate = *k*[BrO]; (b) rate $= k[BrO]^2$; (c) rate = *k*[BrO]; (d) rate $= k[BrO]^0 = k$

13.53. No; the same behavior would be observed if the reaction were second order in one reactant and zero order in the other.

13.55. (a) rate $= k[NO_2][O_3]$; (b) 4.9×10^{-11} *M*/s; (c) 4.9×10^{-11} *M*/s; (d) the rate doubles

13.57. (c)

13.59. rate $= 3.4 \times 10^{-5}\,s^{-1}\,[N_2O_5]$

13.61. rate $= k[ClO_2][OH^-]$; $k = 14\,M^{-1}\,s^{-1}$

13.63. rate $= k[NO]^2[H_2]$; $k = 6.34\,M^{-2}\,s^{-1}$

13.65. (a) rate $= k[Ni_2O_2]$ (a plot of $\ln[Ni_2O_2]$ vs. time is a straight line indicating a first-order reaction); (b) $k = 0.0701\ \text{min}^{-1}$; (c) $t_{1/2} = 9.89$ min

13.67. (a) $0.0740\ \text{min}^{-1}$; (b) 0.11 (or 11%)

13.69. (a) $k = 4.99 \times 10^{-3}\ \text{min}^{-1}$ (or $8.31 \times 10^{-5}\,s^{-1}$); (b) 0.616 *M*

13.71. (a) rate $= k[N_2O]$; (b) 4

13.73. (a) rate $= k[^{32}P]$ and $0.0485\ \text{day}^{-1}$; (b) 14.3 days

13.75. $k = 5.40 \times 10^{-12}\ \text{cm}^3\ \text{molecules}^{-1}\,s^{-1}$; $t_{1/2} = 0.712$ s

13.77. reverse

13.79. An increase in temperature increases the frequency and the kinetic energy at which the reactants collide. This speeds up the reaction. The order of the reaction is unaffected.

13.81. the reaction with the larger activation energy (150 kJ/mol)

13.83. Lowering the temperature slows the undesired reaction that leads to loss of the aromas.

13.85. $E_a = 14.5$ kJ/mol; $A = 1.74 \times 10^9$

13.87. (a) $E_a = 314$ kJ/mol; (b) $A = 5.03 \times 10^{10}$; (c) $k = 1.06 \times 10^{-44}\ M^{-1/2}\,s^{-1}$

13.89. $E_a = 39.1$ kJ/mol; $A = 1.27 \times 10^{12}$

13.91. Yes, though it doesn't have to be one step. If a catalyst or intermediate is involved, the reaction could proceed in two or more steps.

13.93. No. If a reactant is zero order, its involvement in the mechanism (collision with other reactants or intermediates) must occur after the rate-determining step.

13.95.

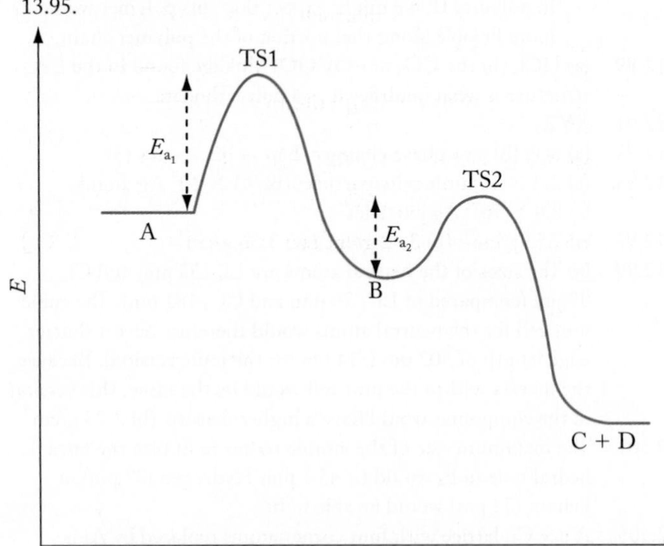

13.97. (a) rate $= k[SO_2Cl_2]$; unimolecular; (b) rate $= k[NO_2][CO]$; bimolecular; (c) rate $= k[NO_2]^2$; bimolecular

13.99. $N_2O_5(g) + O(g) \rightarrow 2\,NO_2(g) + O_2(g)$

13.101. the second step

13.103. (a) The second step is rate determining; (b)

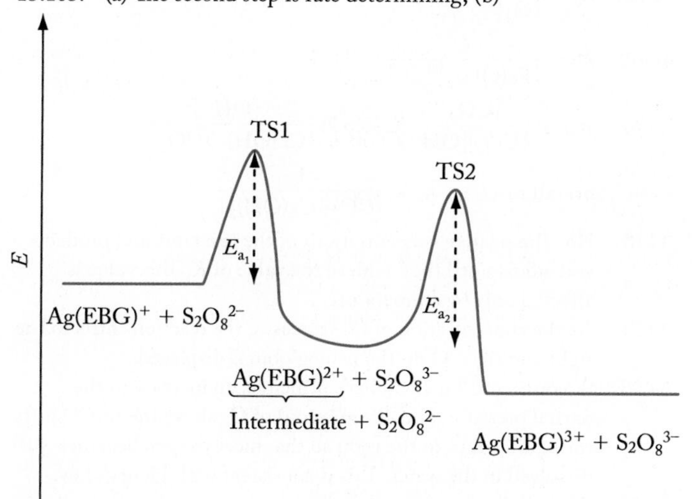

13.105. (b) or (c)

13.107. Yes. Because the rate of the reaction is faster (affecting the rate) and the activation energy is lowered (affecting the value of k), a catalyst affects both the rate of the reaction and the value of the rate constant.

13.109. yes

13.111. no

13.113. NO is the catalyst.

13.115. the reaction of O_3 with Cl

13.117. When the concentration of a reactant (O_2 for the combustion reaction) increases, the rate of the reaction increases.

13.119. The bodily reactions that use O_2 are slower at colder temperatures.

13.121. Yes, we could use other times, not just $t = 0$, as long as the rate of the reverse reaction is still much slower than that of the forward reaction; reactant concentrations would still be used.

13.123. An elementary step with a molecularity of zero would have no molecules colliding.

13.125. 5.2 M/s

13.127. (a) rate $= k[NH_3][HNO_2]^2$; the units of k are $M^{-2}\,s^{-1}$.
(b) step 1: $NH_3(g) + HNO_2(aq) \rightarrow NH_4^+(aq) + NO_2^-(aq)$; for this step $k_1[NH_3][HNO_2] = k_{-1}[NH_4^+][NO_2^-]$; rearranging: $[NH_3][HNO_2] = \frac{k_{-1}}{k_1}[NH_4^+][NO_2^-]$; plugging into the rate law from part (a): rate $= k[NH_4^+][NO_2^-]\,[HNO_2]$.
(c) −482.4 kJ.

(d)

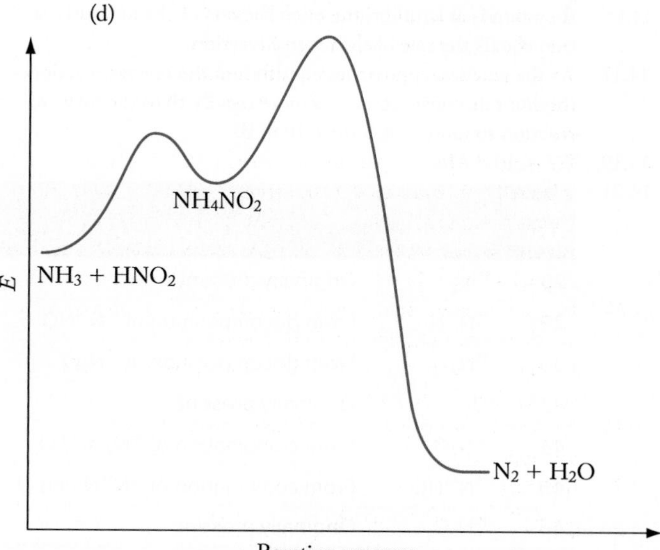

13.129. a. rate $= k[NO][ONOO^-]$; $k = 1.30 \times 10^{-3}\,M^{-1}\,s^{-1}$
b.

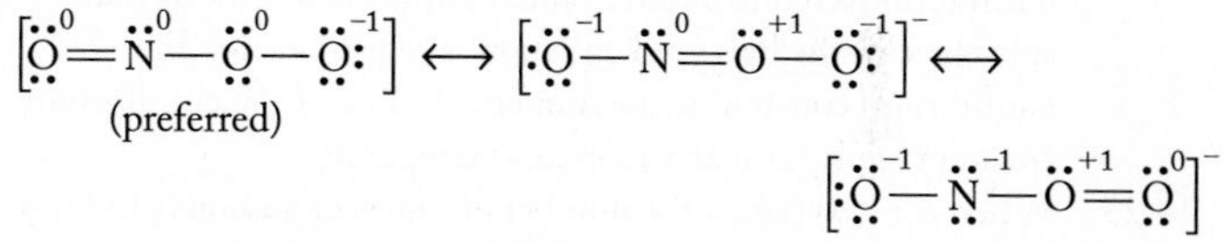

$$\left[\overset{0}{\ddot{O}}=\overset{0}{\ddot{N}}-\overset{0}{\ddot{O}}-\overset{-1}{\ddot{O}}:\right]^- \longleftrightarrow \left[:\overset{-1}{\ddot{O}}-\overset{0}{\ddot{N}}=\overset{+1}{\ddot{O}}-\overset{-1}{\ddot{O}}:\right]^- \longleftrightarrow$$

(preferred)

$$\left[:\overset{-1}{\ddot{O}}-\overset{-1}{\ddot{N}}-\overset{+1}{\ddot{O}}=\overset{0}{\ddot{O}}\right]^-$$

c. −55 kJ

Chapter 14

14.1. (a) $K_c = \frac{[A_3B]^2}{[A_2]^3[B_2]}$; (b) $K_c = 0.87$

14.3. This reaction is not at equilibrium because Q (0.038) $\neq K$. Because $Q < K$, the reaction would shift to the right (product side).

14.5. No, at 20 μs the concentrations of A and B are still changing.

14.7. (a)

14.9.

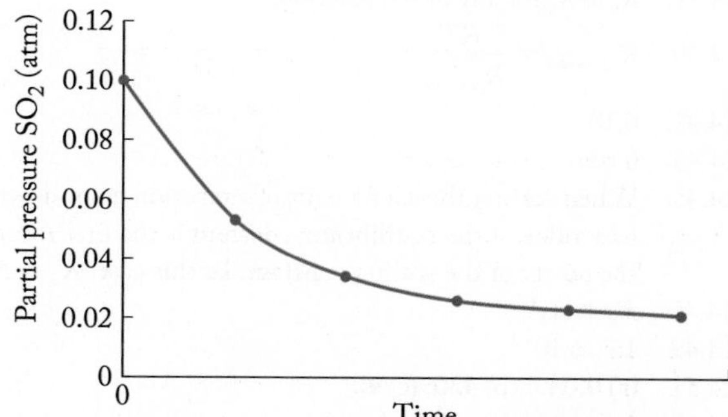

Because $K_c = \frac{1}{P_{SO_2}}$, the equilibrium pressure of SO_2 will be 0.02 atm.

14.11. A graph describing the progress of a reaction to equilibrium will reach a point where the concentrations of the reactants and products do not change over time. Kinetics evaluates the change in concentration of a reactant over time.

14.13. No. The equilibrium may favor either the products or the reactants depending on the thermodynamic energy difference. The rate at which the reversible reaction occurs does not change the energy of the products or the reactants.

14.15. A system is at equilibrium when the rate of the forward reaction equals the rate of the reverse reaction.

14.17. As the reaction approaches equilibrium the reverse reaction (higher rate constant) occurs more rapidly than the forward reaction so more A is formed than B.

14.19. 2.7×10^{-3} *M*/s

14.21.

Molar Mass	Compound	How Present
28	$^{14}N_2$	Originally present
29	$^{15}N^{14}N$	From decomposition of $^{15}N^{14}NO$
30	$^{15}N_2$	From decomposition of $^{15}N_2O$
32	O_2	Originally present
44	$^{14}N_2O$	From combination of $^{14}N_2$ and O_2
45	$^{15}N^{14}NO$	From combination of $^{15}N^{14}N$ and O_2
46	$^{15}N_2O$	Originally present

14.23. An equilibrium constant expression is defined as the ratio of the concentrations or partial pressures of products to reactants (each term raised to a power equal to the coefficient of that substance in the balanced equation) at equilibrium. The equilibrium constant is the numerical value of the equilibrium constant expression at a particular temperature.

14.25. When $\Delta n = 0$ (when the number of moles of gaseous products is equal to the number of moles of gaseous reactants).

14.27. (a) $K_c = \dfrac{[H_2O]^2[Cl_2]^2}{[HCl]^4[O_2]}$, $K_p = \dfrac{P_{H_2O}{}^2 P_{Cl_2}{}^2}{P_{HCl}{}^4 P_{O_2}}$;

(b) $K_c = \dfrac{[CH_4][H_2S]^2}{[CS_2][H_2]^4}$; $K_p = \dfrac{P_{CH_4} P_{H_2S}{}^2}{P_{CS_2} P_{H_2}{}^4}$

14.29. $K_p = \dfrac{P_{SO_2}}{P_S P_{O_2}}$, $K_c = \dfrac{[SO_2]}{[S][O_2]}$; $K_p = K_c(RT)^{\Delta n}$ with $\Delta n = 1 - 2 = -1$ for this reaction.

14.31. 0.068

14.33. 0.50

14.35. 27

14.37. $K_p \neq K_c$ for any of the reactions.

14.39. $K_{c,\text{forward}} = \dfrac{RT}{K_{p,\text{reverse}}}$

14.41. 0.10

14.43. 0.65

14.45. When scaling the coefficients of a reaction up or down, the new value of the equilibrium constant is the first *K* raised to the power of the scaling constant. In this case: $K_2 = K_1{}^{1/2}$.

14.47. $K_2 = K_1{}^{1/2}$

14.49. 1.7×10^{-2}

14.51. (a) 0.049; (b) 420; (c) 20

14.53. 1.1×10^{-18}

14.55. The reaction quotient, *Q*, is for a reaction mixture not necessarily at equilibrium. Only if the system is at equilibrium will *Q* be equal to *K*.

14.57. No, $Q < K$ so the reaction proceeds to the right (toward products) to reach equilibrium.

14.59. $Q > K$ so the reaction will proceed to the left (toward reactants).

14.61. The reaction is not at equilibrium. It will proceed to the left (toward reactants).

14.63. $K_c = \dfrac{1}{[O_2]}$

14.65. $K_c = \dfrac{1}{[Fe(OH)_2]^4[O_2]}$

14.67. step 1: $\dfrac{[CO_3{}^{2-}]}{[CO_2][OH^-]^2}$; step 2: $\dfrac{[OH^-]^2}{[Ca(OH)_2][CO_3{}^{2-}]}$;

overall reaction: $K_c = \dfrac{1}{[CO_2][Ca(OH)_2]}$

14.69. No. The relative concentrations of the reactants and products will adjust until they achieve the value of *K*. This value is affected only by temperature.

14.71. As the concentration of O_2 increases, the reaction shifts to the right and the CO on the hemoglobin is displaced.

14.73. According to Le Châtelier's principle, an increase in the partial pressure (or concentration) of O_2 above the water shifts the equilibrium to the right so that more oxygen becomes dissolved in the water. This is consistent with Henry's law.

14.75. (b) and (d)

14.77. (a) Increasing the concentration of the reactant O_3 shifts the equilibrium to the right, increasing the concentration of the product O_2; (b) increasing the concentration of the product O_2 shifts the equilibrium to the left, increasing the concentration of the reactant O_3; (c) decreasing the volume of the reaction to 1/10 its original volume shifts the equilibrium to the left, increasing the concentration of the reactant O_3.

14.79. The equilibrium shifts to the left.

14.81. (b)

14.83. When *K* is small, the amount of reactants that are transformed into products may be so small that at equilibrium the concentrations of the reactants are approximately equal to the initial concentrations. This means we can make an approximation in the *K* expression to make our calculations easier.

14.85. (a) $P_{PCl_5} = 0.024$ atm, $P_{PCl_3} = 1.036$ atm, $P_{PCl_2} = 0.536$ atm; (b) the partial pressure of PCl_3 decreases and the partial pressure of PCl_5 increases.

14.87. $[H_2O] = [Cl_2O] = 3.76 \times 10^{-3}$ *M*, $[HOCl] = 1.13 \times 10^{-3}$ *M*

14.89. 6.9×10^5:1

14.91. $P_{CO} = 2.4$ atm; $P_{CO_2} = 3.8$ atm

14.93. (a) $P_{NO} = 0.272$ atm, $P_{NO_2} = 7.98 \times 10^{-3}$ atm; (b) $P_{\text{total}} = 0.416$ atm

14.95. $P_{O_2} = 0.17$ atm, $P_{N_2} = 0.75$ atm, $P_{NO} = 0.080$ atm

14.97. 5.75 *M*

14.99. $P_{CO} = P_{Cl_2} = 0.258$ atm, $P_{COCl_2} = 0.00680$ atm

14.101. $[CO] = [H_2O] = 0.031$ *M*, $[CO_2] = [H_2] = 0.069$ *M*

14.103. The reaction is endothermic; at 1500 K, $K_p = 5.5 \times 10^{-11}$; at 2500 K, $K_p = 4.0 \times 10^{-3}$; and at 3000 K, $K_p = 0.40$. This reaction does not favor products even at very high temperatures so it is not a viable source of CO and is not a remedy to decrease CO_2 as a contributor to global warming. Also, the process produces poisonous CO gas.

14.105. 9×10^{-22} *M*

14.107. (a) $P_{SO_2} = 9.2 \times 10^{-74}$ atm; (b) 2.63×10^{-47} molecules, so essentially all the SO_2 has been removed from the reaction mixture.

14.109. (a) $K_p = P_{CO_2}$ for both reactions; (b) $5\ CaCO_3(s) + 2\ MgCO_3(s) \rightarrow 5\ CaO(s) + 2\ MgO(s) + 7\ CO_2(g)$, $K_p = [CO]^7$; (c) In an open container the gaseous CO_2 can leave the reaction vessel and the reaction can continue to form more products until the reactants run out. In a closed container the CO_2 builds up and an equilibrium could be established limiting the production of the products; (d) No, the value of K_p depends only on the temperature.

Chapter 15

15.1. red line

15.3. the right (smallest) bar

15.5. (a) basic; (b) the amine group (R_2NH)

15.7. Piperidine is the stronger base. The electronegative O atom in morpholine pulls electron density away from the N atom rendering this the less basic molecule. By comparison, piperidine with a $-CH_2-$ group instead of an O atom nearby is a stronger base.

15.9. The electron-withdrawing Cl atoms pull electron density away from the O–H bond leading to greater ionization.

15.11. Water acts as the Brønsted–Lowry acid and LiOH acts as the Brønsted–Lowry base.

15.13. Water acts as the Brønsted–Lowry acid and methylamine acts as the Brønsted–Lowry base.

15.15. (a) HCl is the acid and NaOH is the base; (b) HCl is the acid and $MgCO_3$ is the base; (c) H_2SO_4 is the acid and NH_3 is the base.

15.17. Sulfur is more electronegative than selenium. The higher electronegativity on the sulfur atom stabilizes the anion HSO_4^- more than the anion $HSeO_4^-$.

15.19. The $-CH_3$ group on CH_3NH_2 is electron-donating compared to –H and results in a greater electron density on the N in CH_3NH_2.

15.21. 0.65 *M*

15.23. 0.0410 *M*

15.25. HF conjugate base: F^-; $HBrO_2$ conjugate base: BrO_2^-; H_2SO_4 conjugate base: HSO_4^-; $HClO_4$ conjugate base: ClO_4^-

15.27. Conjugate acid: H_3PO_4; conjugate base: HPO_4^{2-}

15.29. Because the pH function is a −log function, as $[H_3O^+]$ increases the value of $-\log[H_3O^+]$ decreases.

15.31. $[H_3O^+]$ must be greater than 1 *M*. One example is 2.5 *M* HCl, which has a pH of −0.40.

15.33.

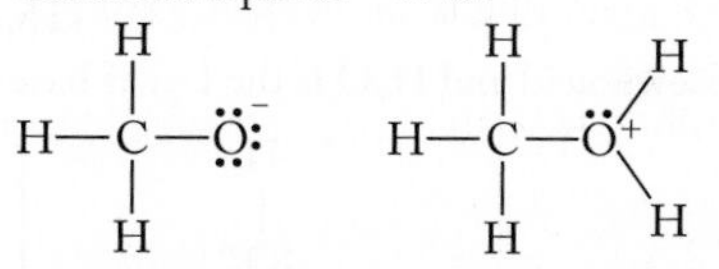

```
     H                  H      H
     |                  |     /
  H—C—Ö:⁻           H—C—Ö⁺
     |                  |     \
     H                  H      H
```

15.35. (a) pH = 3.62, pOH = 10.38, acidic; (b) pH = 7.26, pOH = 6.74, basic; (c) pH = 12.41, pOH = 1.59, basic; (d) pH = 0, pOH =14, basic

15.37. (a) 7.7×10^{-4} *M*; (b) 0.012 *M*; (c) 1.9×10^{-2} *M*; (d) 4.4×10^{-4} *M*

15.39. (a) pH = 0.810, pOH = 13.190; (b) pH = 2.301, pOH = 11.699; (c) pH = 1.903, pOH = 12.097; (d) pH = 1.204, pOH = 12.796

15.41. 7.003

15.43. (a) HCl > HNO_2 > CH_3COOH > HClO; (b) HClO < CH_3COOH < HNO_2 < HCl

15.45. As a strong acid, HCl completely dissociates in aqueous solution; because HF is a weak acid, it only partially dissociates. The 1.0 *M* HCl solution contains significantly more ions to carry electrical current than does the HF solution.

15.47.

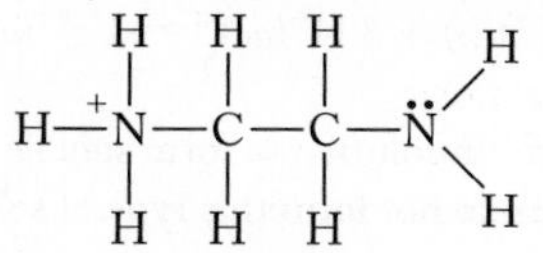

```
     H  H  H        H
     |  |  |       /
  H—N⁺—C—C—N̈
     |  |  |       \
     H  H  H        H
```

15.49. (a) water; (b) water

15.51. 8.9×10^{-4}

15.53. 1.34%

15.55. 2.3 times

15.57. (a) weaker; (b) 10.86; (c) 1.0×10^{-4}

15.59. (a) 9.73; (b) 9.30

15.61. 9.50

15.63. With each successive ionization it becomes more difficult to remove H^+ from a species that is more negatively charged.

15.65. Ge is less electronegative than C, so even less electron density is pulled away from the acidic proton in H_2GeO_3 than in H_2CO_3 resulting in a lower percent ionization and K_a for H_2GeO_3.

15.67. −0.304

15.69. 2.80

15.71. 10.27

15.73. ammonium nitrate

15.75. The citric acid in the lemon juice neutralizes the volatile trimethylamine to make a nonvolatile dissolved salt.

15.77. 3.77

15.79. 7.35

15.81. (a) $HClO_4$ and CH_3COOH; (b) $Ca(OH)_2$; (c) $HClO_4$ and CH_3COOH; (d) $Ca(OH)_2$ and CH_3NH_2

15.83. Yes, all Arrhenius bases are Brønsted–Lowry bases. No, not all Brønsted–Lowry bases are Arrhenius bases. For example, NH_3 is a Brønsted–Lowry base but not an Arrhenius base.

15.85. In burning S-containing fuels:

$$S_8(s) + 8\,O_2(g) \rightarrow 8\,SO_2(g) \xrightarrow{+4\,O_2(g)} 8\,SO_3(g)$$

In the presence of water these produce a weak acid, H_2SO_3, and a strong acid, H_2SO_4:

$$SO_2(g) + H_2O(\ell) \rightarrow H_2SO_3(aq)$$

$$SO_3(g) + H_2O(\ell) \rightarrow H_2SO_4(aq)$$

These acids dissolve $CaCO_3$:

$$2\,H^+(aq) + CaCO_3(s) \rightarrow H_2CO_3(aq) + Ca^{2+}$$

15.87. 1.4×10^9 L

15.89. (a) In water, amines show basic character. They pick up a proton from water to form ammonium cations and OH^-. Therefore dissolving Prozac in water gives a slightly basic solution. (b) The secondary amine (N atom) on Prozac is more likely to react with HCl than is the O atom. (c) The HCl salt of Prozac is more soluble because it is charged, and water molecules form ion–dipole forces around the molecule that are stronger than the dipole–induced dipole forces between the neutral molecule and water.

15.91. (a) $[C_5F_5]^-$ (resonance structures: ring with lone pair on carbon ⟷ ring with circle)

F, F, F, F, F ⟷ F, F, F, F, F

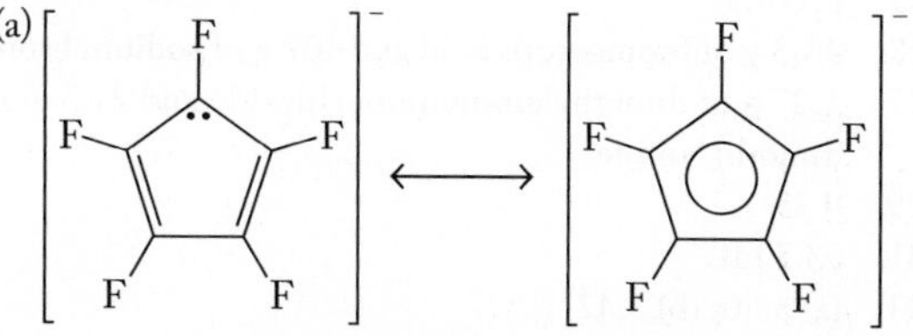

$4n + 2 = 6\ \pi$ electrons so this ring is aromatic. (b) C_5F_5H is very acidic because the presence of five very electronegative F atoms on the carbon ring stabilizes the anion formed when the proton is lost.

15.93. (a)

$$\overset{+1+5-2}{HNO_3} + 2\,\overset{+1+6-2}{H_2SO_4} \rightarrow \overset{+5-2}{NO_2^+} + \overset{+1-2}{H_3O^+} + 2\,\overset{+1+6-2}{HSO_4^-}$$

No oxidation numbers change over the course of this reaction so it is not a redox reaction.

(b)

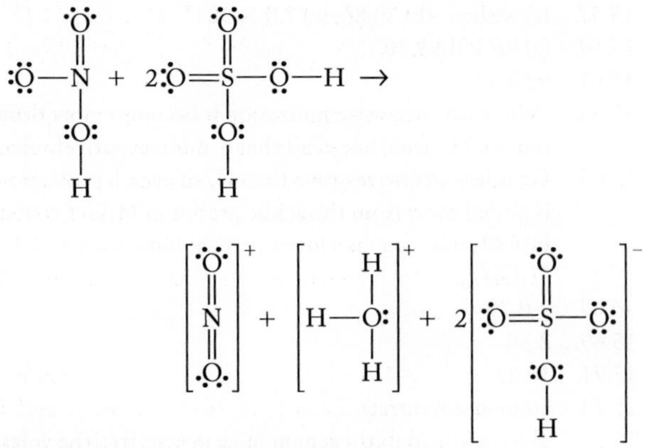

$:\ddot{O}{-}N(=O)(-OH) + 2\,O{=}S(=O)(-OH){-}\ddot{O}{-}H \rightarrow$

$[O{=}N{=}O]^+ + [H{-}\ddot{O}(H){-}H]^+ + 2\,[O{=}S(=O)(-OH){-}\ddot{O}:]^-$

acid = H_2SO_4, conjugate base = HSO_4^-, base = HNO_3 (the –OH group), conjugate acid = H_3O^+

15.95. increase

Chapter 16

16.1. The blue titration curve represents the titration of a 1 M solution of strong acid; the red curve represents the titration of a 1 M solution of weak acid.

16.3. the indicator with a pK_a of 9.0

16.5. The red titration curve represents the titration of Na_2CO_3; the blue titration curve represents the titration of $NaHCO_3$.

16.7. NH_4Cl is dissolved in the yellow solution, $NaC_2H_3O_2$ is dissolved in the blue solution, and $NH_4C_2H_3O_2$ is dissolved in the light green solution.

16.9. (a) buffer; (b) some of the weak acid, HCOOH, will be converted to conjugate base form, $HCOO^-$.

16.11. A buffer is formed that can resist changes in pH as a result of adding acid or base to the solution. A weak acid can only moderate pH changes when base is added.

16.13. iodoacetic acid–sodium iodoacetate

16.15. The quantity of an external acid or base that can be added to a buffer while maintaining a pH that is ±1 unit from the pK_a of the acid component

16.17. The pH will not change.

16.19. 0.54

16.21. 2.46

16.23. 0.178:1

16.25. 46.3 g of bromoacetic acid and 107 g of sodium bromoacetate

16.27. 2.27 g of dimethylammonium chloride and 21.3 g of dimethylamine

16.29. 9.25

16.31. 53.7 mL

16.33. (a) 3.50; (b) 3.42

16.35. Yes; strong acid–strong base neutralizations generate water and a neutral salt, which would have a pH of 7.

16.37. The end point is a highly visible, colored to colorless transition that occurs at pK_a of 8.8, near the first equivalence point of pH = 8.5.

16.39. 4.44

16.41. After 10.0 mL of OH^- has been added, the pH = 4.75; after 15.0 mL of OH^- has been added, the pH = 5.23; after 20.0 mL of OH^- has been added, the pH = 8.78; after 25.0 mL of OH^- has been added, the pH = 12.10; after 35.0 mL of OH^- has been added, the pH = 12.50.

16.43. (a) 2.945 M; (b) no effect

16.45.

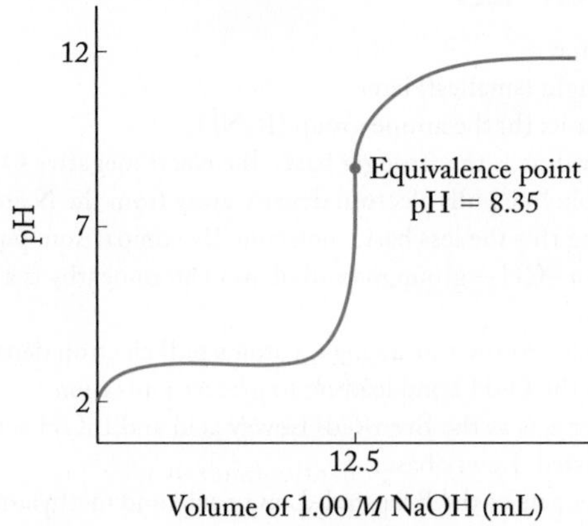

pH: 12, 7, 2
Equivalence point pH = 8.35
12.5
Volume of 1.00 M NaOH (mL)

16.47. (a) less than 7; (b) greater than 7; (c) equal to 7

16.49. two equivalence points; phenolphthalein and alizarin yellow R

16.51. No; some Lewis bases donate an electron pair but do not accept a proton while doing so.

16.53. Yes; Brønsted–Lowry acids donate a proton, meaning that they can also accept an electron pair.

16.55.

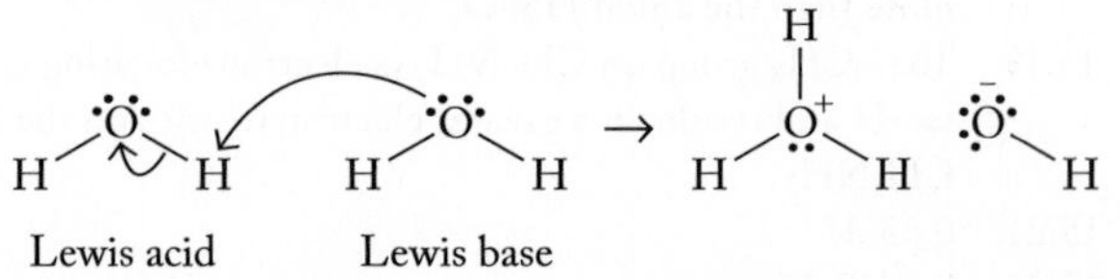

$H_2O + H_2O \rightarrow H_3O^+ + OH^-$

Lewis acid Lewis base

16.57. The sulfur atom in SO_2 and the hydrogen atom in H_2O are acting as Lewis acids while the oxygen atoms in SO_2 and H_2O are acting as Lewis bases.

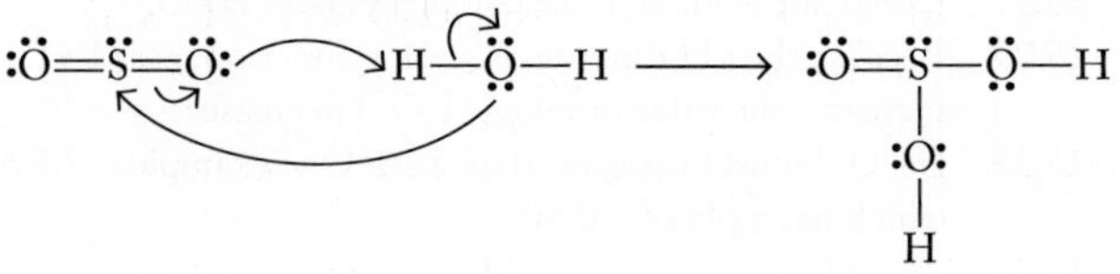

$:\ddot{O}{=}\ddot{S}{=}\ddot{O}: + H{-}\ddot{O}{-}H \longrightarrow :\ddot{O}{=}\ddot{S}(-\ddot{O}H){-}\ddot{O}{-}H$

16.59. $B(OH)_3$ is the Lewis acid and H_2O is the Lewis base.

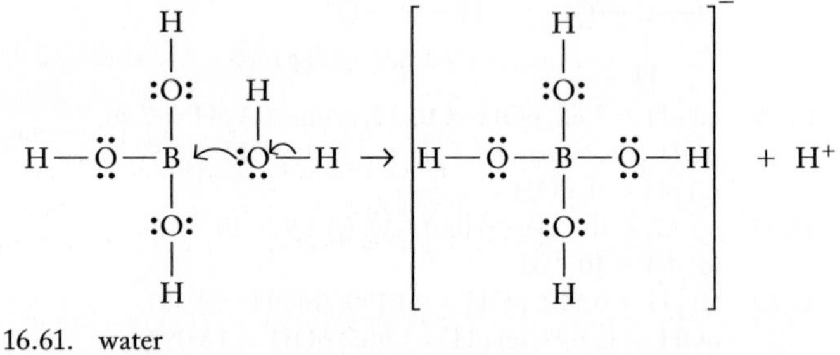

$B(OH)_3 + H_2O \rightarrow [B(OH)_4]^- + H^+$

16.61. water

16.63. Ag^+ forms a soluble complex with NH_3 removing Ag^+ from solution and shifting the equilibrium for the dissolution of AgCl to the right.

16.65. 1.25×10^{-10} M

16.67. 2.48×10^{-13} M

16.69. (b) and (d)

16.71. The solution will become more acidic.

16.73. In basic solution: $Cr(OH)_3(s) + OH^-(aq) \rightleftharpoons Cr(OH)_4^-(aq)$; in acidic solution: $Cr(OH)_3(s) + 3\,H^+(aq) \rightleftharpoons Cr^{3+}(aq) + 3\,H_2O(\ell)$

16.75. $Al(OH)_3$ reacts with OH^- in solution to form soluble $Al(OH)_4^-$. The other ions do not form this type of soluble complex ion.

16.77. 2.80

16.79. 1.80

16.81.

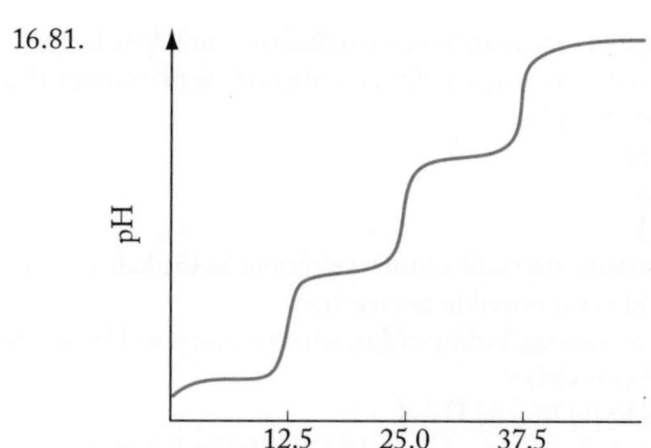

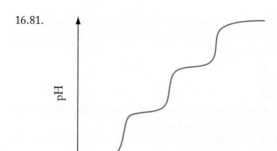

16.83. Molar solubility is the quantity (moles) of substance that dissolves in a liter of solution. The solubility product is the equilibrium constant for the dissociation of a substance.
16.85. Mg^{2+}
16.87. endothermic
16.89. Acidic substances react with the OH^- released on dissolution of hydroxyapatite. The equilibrium is shifted to the right dissolving more hydroxyapatite.
16.91. 1.08×10^{-10}
16.93. $[Cu^+] = [Cl^-] = 1.01 \times 10^{-3}\ M$
16.95. 10.151
16.97. (d)
16.99. no
16.101. yes
16.103. (a) SO_4^{2-}; (b) $1.34 \times 10^{-4}\ M$
16.105. (a) Because HF is weak, the $[F^-]$ is lower than that of water and so HF reacts with H_2O to form F^- as the major anionic species; (b) $K_{overall} = 1.8 \times 10^{-4}$; (c) pH = 2.01; $[HF_2^-]_{eq} = 3.64 \times 10^{-4}\ M$
16.107. 6.6×10^{-16}

Chapter 17

17.1. The gas in the balloon on the right is under greater pressure and has a greater entropy.
17.3. No. This is a low probability because each gas would then be confined to a smaller volume and would have more order, so this change would involve a decrease in entropy. The tank is an isolated system as long as no heat is transferred through the cylinder wall.
17.5. Entropy would decrease as the system goes from 3 moles of gaseous reactants to 2 moles of gaseous products.
17.7. Entropy would increase as the system goes from 3 moles of gaseous reactants to 7 moles of gaseous products.
17.9. at low temperature
17.11. (a) and (c) are spontaneous because both have a negative value for ΔG_{rxn}. The balanced equations are (a) $2\ C_6H_6(\ell) + 15\ O_2(g) \rightarrow 12\ CO_2(g) + 6\ H_2O(\ell)$ and (c) $2\ H_2(g) + O_2(g) \rightarrow 2\ H_2O(\ell)$.
17.13. There are more microstates available in (b) than in (a).
17.15. They are equal in magnitude but the sign is reversed.
17.17. ΔS_{sys} is positive and ΔS_{surr} is negative.
17.19. (a) and (b)
17.21. ΔS_{sys} is negative and ΔS_{univ} is negative.
17.23. ΔS_{surr} must be less than +66.0 J/K.
17.25. (a) decreases; (b) increases; (c) approximately the same
17.27. (a) $S_8(g)$; (b) $S_2(g)$; (c) $O_3(g)$; $O_2(g)$
17.29. fullerenes
17.31. (a) $CH_4(g) < CF_4(g) < CCl_4(g)$; (b) $CH_2O(g) < CH_3CHO\ (g) < CH_3CH_2CHO(g)$; (c) $HF(g) < H_2O(g) < NH_3(g)$
17.33. positive
17.35. Solids have a lower entropy (are a more ordered phase) than that of particles dissolved in aqueous solution. As a result of the precipitation reaction, fewer particles are present in any phase of the reaction. Both of these factors lead to the ΔS_{rxn} values being less than zero.
17.37. (a) 24.9 J/K; (b) −146.4 J/K; (c) −73.2 J/K; (d) −175.8 J/K
17.39. 534 J/(K · mol)
17.41. (c)
17.43. In general, exothermic reactions have a corresponding increase in entropy. Both an increase in entropy and decrease in enthalpy lead to a spontaneous reaction, though now we realize that it is possible for an exothermic reaction to be spontaneous at low temperatures if disfavored by entropy.
17.45. ΔS is positive, ΔH is positive, and ΔG is negative.
17.47. (b) and (d)
17.49. For NaBr, −18 kJ/mol; for NaI, −30. kJ/mol
17.51. +91.4 kJ
17.53. +56.5 kJ
17.55. −90.3 kJ
17.57. No; if ΔS_{rxn} were positive then the exothermic reaction would be spontaneous at all temperatures.
17.59. high temperature
17.61. 981.3 K
17.63. ΔH° = 51.5 kJ; ΔS° = 123.1 J/K or 0.1231 kJ/K; 418 K or 145°C
17.65. (a) low temperatures; (b) low temperatures; (c) all temperatures
17.67. positive
17.69. Yes. In a circumstance where the product concentrations are zero, the value of the reaction quotient, Q, will be zero. The equilibrium constant must be larger than zero, so in this circumstance $Q < K$ and this reaction will shift toward the products until it reaches the point where $Q = K$.
17.71. (c)
17.73. 27.1 kJ/mol
17.75. exothermic
17.77. exothermic
17.79. 1.3×10^{-31}
17.81. −115 kJ/mol
17.83. Some of the products of one reaction must be reactants of the second reaction and the overall value of ΔG when the reactions are summed must be negative.
17.85. The bond arrangements are only slightly different between the two structures.
17.87. (a) +50.8 kJ and −394.4 kJ; (b) $CH_4(g) + O_2(g) \rightarrow 2\ H_2(g) + CO_2(g)$, $\Delta G^\circ_{rxn} = -343.6$ kJ; = spontaneous
17.89. 3.81×10^{-3}
17.91. −73 kJ/mol
17.93. 24
17.95. 4.47×10^3
17.97. 83.5 J/(K · mol)
17.99. (a) −90.7 J/(K · mol); (b) a large negative value of ΔH will cause ΔG to be negative (despite the expected negative ΔS) and thus the reaction is spontaneous.
17.101. (c) and (d)
17.103. −1308.4 kJ
17.105. below 386.0 K
17.107. above 896 K

17.109. (a) Cu_2O: copper(I) oxide; CuO: copper(II) oxide; (b) never spontaneous; (c) Cu_2O is more complex than CuO because it has more bonds.

17.111. (a) Trouton's rule predicts that ΔS°_{vap} for liquids is relatively constant because the change in entropy from liquid to gas should be similar for similar molecules; (b) helium exhibits a lower ΔS°_{vap} because its intermolecular forces are weak causing more randomness than usual in the liquid phase. Methanol and water deviate because these liquids form hydrogen bonds causing more order in the liquid state.

17.113. (a) ΔH is negative and ΔS is positive. (b) No; the reverse reaction with a positive ΔH and a negative ΔS would never be spontaneous.

Chapter 18

18.1. With careful layering, the electrode in each half-cell is immersed in a solution of its cations. Ions can migrate across the interface between the two solutions, which allows the cell reaction to proceed without a porous bridge.

18.3. Ag is the cathode; Pt in the standard hydrogen electrode (SHE) is the anode; electrons flow from the SHE to Ag.

18.5. blue line

18.7. (a) $2\,H^+(aq) + 2\,e^- \rightarrow H_2(g)$, $E^\circ_{cathode} = -0.000$ V; $2\,H_2O(\ell) \rightarrow O_2(g) + 4\,H^+(aq) + 4\,e^-$, $E^\circ_{anode} = 1.229$ V; (b) sulfuric acid increases the conductivity of the solution.

18.9. (a) cathode; (b) chromium forms a protective surface layer of Cr_2O_3, which prevents the metal beneath it from being oxidized

18.11. (b)

18.13. (a) The separator allows the transport of ionic charge carriers that are needed to complete the circuit during the passage of current through an electrochemical cell; (b) the wire would allow electrons to pass, but not ions, and because electrons cannot travel through the solution, the wire would not complete the circuit.

18.15. $Cr_2O_7^{2-}(aq) + 14\,H^+(aq) + 6\,Fe^{2+}(aq) \rightarrow 2\,Cr^{3+}(aq) + 7\,H_2O(\ell) + 6\,Fe^{3+}(aq)$

18.17. (a) cathode: $Ni^{2+}(aq) + 2\,e^- \rightarrow Ni(s)$; anode: $Cd(s) \rightarrow Cd^{2+}(aq) + 2\,e^-$; (b) $Ni^{2+}(aq) + Cd(s) \rightarrow Ni(s) + Cd^{2+}(aq)$; (c) $Cd(s) \mid Cd^{2+}(aq) \parallel Ni^{2+}(aq) \mid Ni(s)$

18.19. (a) cathode: $O_2(g) + 4\,H^+ + 4\,e^- \rightarrow 2\,H_2O(\ell)$; anode: $4\,Li(s) \rightarrow 4\,Li^+(aq) + 4\,e^-$; $H^+(aq)/O_2(g) \mid H_2O(\ell) \parallel Li^+(aq) \mid Li(s)$; (b) cathode: $O_2(g) + 2\,H_2O(\ell) + 4\,e^- \rightarrow 4\,OH^-(aq)$; anode: $4\,Li(s) \rightarrow 4\,Li^+(aq) + 4\,e^-$; $H_2O(\ell)/O_2(g) \mid OH^-(aq) \parallel Li^+(aq) \mid Li(s)$

18.21. (a) For M = Ag or Au: $4\,M(s) + 2\,H_2O(\ell) + O_2(g) \rightarrow 4\,M^+(aq) + 4\,OH^-(aq)$; (b) 4 moles of electrons are transferred

18.23. Cl_2 ($\Delta E^\circ_{red} = 1.3583$ V) is a better oxidizing agent than O_2 ($\Delta E^\circ_{red} = 1.229$ V). The higher the reduction potential, the more powerful the oxidizing agent.

18.25. no

18.27. $O_2(g) + 2\,H_2O(\ell) + 2\,Zn(s) + 4\,OH^-(aq) \rightarrow 2\,Zn(OH)_4^{2-}(aq)$

18.29. (a) less than 1.10 V; (b) Zn

18.31. (a) $NiO(OH)(s) + TiZr_2H(s) \rightarrow TiZr_2(s) + Ni(OH)_2(s)$; (b) 1.32 V

18.33. The negative sign is an indication that the work done by the voltaic cell on its surroundings corresponds to free energy that is lost by the cell.

18.35. −48.3 kJ

18.37. −290 kJ

18.39. −116 kJ

18.41. The platinum electrode transfers electrons to the half-cell; it is inert and not involved in the reaction.

18.43. it requires passing hydrogen gas, which is combustible, across the elctrode surface

18.45. (a) 2.37 V; (b) Mg; (c) $H_2(g)$

18.47. Voltage of a battery is governed by the Nernst equation:

$$E_{cell} = E^\circ_{cell} - \frac{RT}{nF}\ln Q$$

The logarithmic relationship between E_{cell} and Q means that large changes in the concentrations of products and reactants as the battery discharges produce only small changes E_{cell} (battery voltage).

18.49. (a) 0.130 V; (b) 2.47×10^4

18.51. $E_{rxn} = 1.55$ V; decrease

18.53. At constant pH the reaction will have a positive E_{cell} (and be spontaneous) until the $[Pb^{2+}]$ reaches 1.5×10^{-6} M.

18.55. (c) and (f)

18.57. Al–O_2

18.59. Li–MnO_2

18.61. Teflon is a polymeric material composed of poly(tetrafluoroethylene). Because no metals or ionizable atoms are present in Teflon (the C–F bond is extremely stable), no redox activity is expected. Teflon is also water repellent, which keeps the metal structure drier than it would be with asbestos mats.

18.63. The material must have a more negative standard reduction potential than aluminum. Magnesium or an aluminum–magnesium alloy could serve as a sacrificial anode for aluminum.

18.65. In a voltaic cell the electrons are produced at the anode so a negative (−) charge builds up there; in an electrolytic cell electrons are forced onto the cathode so that it builds up negative (−) charge.

18.67. Br_2

18.69. The student's best choices are Na_2SO_4, H_2SO_4, and Na_2CO_3. Electrolysis of NaBr and NaI solutions may form Br_2 and I_2 at the anodes.

18.71. 27.0 minutes

18.73. (a) 5.78×10^{-3} L; (b) no, because some Cl_2 and Br_2 would be produced

18.75. −0.270 V

18.77. Hybrid vehicles are less expensive, have greater driving ranges between refueling stops, and burn gasoline, which is cheaper and more widely available than hydrogen. However, cars powered by fuel cells and electric motors emit fewer pollutants and convert the chemical energy in their hydrogen fuel to mechanical work more efficiently.

18.79. Electric engines are more efficient, converting more of the energy into motion instead of losing it as heat.

18.81. (a) $\overset{-4+1}{CH_4}(g) + \overset{+1}{H_2}O(g) \rightarrow \overset{+2}{C}O(g) + 3\,\overset{0}{H_2}(g)$; $\overset{+2}{C}O(g) + \overset{+1}{H_2}O(g) \rightarrow \overset{0}{H_2}(g) + \overset{+4}{C}O_2(g)$. (b) For the reaction of CH_4 with H_2O, $\Delta G^\circ_{rxn} = 142.2$ kJ/mol; for the reaction of CO with H_2O, $\Delta G^\circ_{rxn} = -28.6$ kJ/mol; and for the overall reaction, $\Delta G^\circ_{overall} = \Delta G^\circ_{rxn_1} + \Delta G^\circ_{rxn_2} = 113.6$ kJ.
18.83. oxidation of Cr^{3+}
18.85. 0.617 V
18.87. 8.56×10^{19}
18.89. (a) cathode; (b) no, because Mg^{2+} has a lower (less negative) reduction potential than Na^+; (c) no; (d) H_2, Cl_2 and O_2
18.91. (a) −0.87 V; (b) Mo_3S_4: Mo = +2.67; $MgMoS_4$: Mo = +2; (c) Mg^{2+} is added to the electrolyte to better carry the charge in the cell. This cation is produced at the anode and consumed at the cathode.
18.93. 0.067 V
18.95. (a) $Fe(s) + 3\,OH^-(aq) \rightarrow Fe(OH)_3(s) + 3\,e^-$; $O_2(g) + 2\,H_2O(\ell) + 4\,e^- \rightarrow 4\,OH^-(aq)$; (b) 1.176 V; (c) −1484.4 kJ; 1.28 V
18.97. (a) 0.21 V; (b) 1.2×10^7; (c) $[OCl^-] = 1.0 \times 10^{-7}$ *M* or 0.0051 mg/L

Chapter 19

19.1. (a) purple (lithium); (b) red (radium); (c) red (radium)
19.3. (c)
19.5. (d)
19.7. (c), (d), and (e)
19.9. positive
19.11. The *mass defect* is the difference between the mass of the nucleus of an isotope and the sum of the masses of the individual nuclear particles that make up that isotope. The *binding energy* is the energy released when individual nucleons combine to form the nucleus of an isotope.
19.13. For all nuclides up to iron, the binding energy per nucleon for the reactants is less than that of the products, therefore the products are more stable and energy is released.
19.15. 3.47×10^{-13} J
19.17. 1.09×10^{-12} J
19.19. greater than 1
19.21. Alpha decay increases the neutron-to-proton ratio, which produces less stable isotopes that become more stable through β decay, which decreases the neutron-to-proton ratio.
19.23. Both of these processes are β decay.
19.25. Increasing neutron:proton ratio: $^{25}Al < {}^{24}Na < {}^{11}B < {}^{14}C$; ^{24}Na has the most neutrons.
19.27. (a) ^{64}Ni for ^{64}Cu, ^{89}Y for ^{89}Zr; (b) $^{64}_{28}Ni + {}^1_1p \rightarrow {}^{64}_{29}Cu + {}^1_0n$; $^{89}_{39}Y + {}^1_1p \rightarrow {}^{89}_{40}Zr + {}^1_0n$; (c) not necessarily, as long as the decay occurs slowly enough to ensure the desired isotope is present
19.29. (a) electron capture or positron emission; (b) electron capture or positron emission; (c) β decay
19.31. ^{56}Co has 27 protons and 29 neutrons and is neutron-poor; it may undergo electron capture or positron emission. ^{44}Ti has 22 protons and 22 neutrons and is neutron-poor; it may undergo electron capture or positron emission.
19.33. (a) ^{32}Cl, ^{33}Cl, and ^{34}Cl will emit positrons; (b) ^{38}Cl and ^{39}Cl will emit β particles; (c) ^{36}Cl will emit either positrons or β particles.
19.35. The rate of a radioactive decay is dependent only on the amount of the reactant.
19.37. the ability of radioactive particles to ionize atoms
19.39. 0.181 day^{-1}
19.41. 131 years
19.43. After 8.762 half-lives (50,000 years), the ratio of ^{14}C present to that originally in an artifact is $N_t/N_0 = 0.50^{8.726} = 0.00236$, or 0.236%. This is too low of a level to detect.
19.45. After 0.00023 half-lives (300,000 years), the ratio of ^{40}K present to that in the original sample is $N_t/N_0 = 0.50^{0.00023} = 0.9998$, or 99.98%. This value represents a decrease of 0.02%, which is the smallest change in ^{40}K activity that can be determined accurately.
19.47. 35%
19.49. The oldest ring would have a 15% lower $^{14}C/^{12}C$ ratio.
19.51. 36,600 years
19.53. The level of radioactivity is the amount of radioactive particles present in a given instant of time. The dose is the accumulation of exposure over a length of time.
19.55. When radon-222 decays to polonium-218 while in the lungs, it emits an α particle; the ^{218}Po, a reactive solid that is chemically similar to oxygen, lodges in the lung tissue where it continues to emit α radiation. Alpha radiation is one of the most damaging kinds of radiation when in contact with biological tissues. The result of exposure to high levels of radon is an increased risk for lung cancer.
19.57. 5 μSv = 5 μGy; 250 μJ
19.59. (a) $^{90}_{38}Sr \rightarrow {}^{\ 0}_{-1}\beta + {}^{90}_{39}Y$; (b) 3.28×10^8 ^{90}Sr atoms; (c) strontium-90 is more concentrated in milk than in many other foods because it is chemically similar to calcium and milk is rich in calcium.
19.61. (a) 0.15 decays/s; (b) 7.0×10^4 atoms
19.63. (a) The half-life should be long enough to effect treatment of the cancerous cells but not so long as to cause damage to healthy tissues; (b) because α radiation does not penetrate far beyond a tumor, the α decay mode is best; (c) products should be nonradioactive, if possible, or have short half-lives and be able to be flushed from the body by normal cellular and biological processes.
19.65. No. Imaging should use isotopes with short half-lives so as to produce a measurable signal and not cause too much damage to cells. Imaging isotopes should produce radiation that can be detected outside of the body (β particles or γ rays). Radionuclides used for cancer treatments can have longer half-lives and produce radiation (such as α particles) that severely damage nearby cells.
19.67. (a) positron emission or electron capture; (b) positron emission or electron capture; (c) positron emission or electron capture
19.69. 74.3 days
19.71. 51%
19.73. 136 min
19.75. (a) $^{10}_5B + {}^1_0n \rightarrow {}^7_3Li + {}^4_2\alpha$; (b) 4.42×10^{-13} J; (c) α particles have high relative biological effectiveness and they do not penetrate into healthy tissue if the radionuclide is placed inside a tumor.
19.77. Control rods made of boron or cadmium are used to absorb excess neutrons to control the rate of energy release.
19.79. The neutron-to-proton ratio for heavy nuclei is high, and when the nuclide undergoes fission to form smaller nuclides it must emit neutrons because the fission products require a lower neutron-to-proton ratio for stability.
19.81. (a) $^{138}_{52}Te$; (b) $^{133}_{51}Sb$; (c) $^{143}_{55}Cs$
19.83. (a) no; (b) none of the reactions produce neutrons; (c) 1.50×10^{-10} J

19.85. Today's Sun has fewer neutrons than the primordial Sun. The primordial Sun formed α particles via the fusion of protons and neutrons:

$$2\,{}^{1}_{1}\text{p} + 2\,{}^{1}_{0}\text{n} \rightarrow 2\,{}^{2}_{1}\text{H} \rightarrow {}^{4}_{2}\text{He}$$

Today's Sun forms α particles via the fusion of two protons:

$$2\,{}^{1}_{1}\text{p} \rightarrow {}^{2}_{1}\text{H} + {}^{0}_{+1}\beta$$

$$2\,{}^{2}_{1}\text{H} + 2\,{}^{1}_{1}\text{p} \rightarrow 2\,{}^{3}_{2}\text{He} \rightarrow {}^{4}_{2}\text{He} + 2\,{}^{1}_{1}\text{p}$$

19.87. (a) 4.37×10^{-12} J; (b) 6.80×10^{-12} J; (c) 2.69×10^{-12} J; (d) 1.60×10^{-12} J

19.89. (a) 7.67×10^{-13} J; (b) -3.96×10^{-13} J

19.91. (a) The annihilation when antihydrogen reacts with hydrogen would release a large amount of energy to power the starship *Enterprise*; hydrogen is an abundant fuel in the universe and therefore could easily be procured to react with any antihydrogen produced in the warp engine. (b) If any of the antimatter fuel came into contact with conventional matter (such as the ship, the crew, or the warp engine), the two would annihilate each other releasing a large quantity of energy in the form of an explosion.

19.93. The energy released in the fusion reaction is $\Delta E = 9.91 \times 10^{-13}$ J/nucleon. The energy released in the fission reaction is $\Delta E = 1.4 \times 10^{-13}$ J/nucleon. On a per-nucleon basis, the fusion reaction generates more energy.

19.95. (a) Geiger counter; (b) 2880 years; (c) ^{241}Am is an α emitter, and α particles do not travel more than a few inches in air and cannot penetrate the first layer of skin.

19.97. (a) ${}^{249}_{98}\text{Cf} + {}^{48}_{20}\text{Ca} \rightarrow {}^{294}_{118}\text{Og} + 3\,{}^{1}_{0}\text{n}$; (b) ${}^{290}_{116}\text{Lv}$; (c) ${}^{286}_{114}\text{Fl}$; (d) ${}^{282}_{112}\text{Cn}$; (e) because ^{294}Og is a member of the noble gas family, it should have chemical and physical properties similar to those of radon.

19.99. ^{210}Pb

19.101. (a) ${}^{64}_{28}\text{Ni} + {}^{124}_{50}\text{Sn} \rightarrow {}^{188}_{78}\text{Pt}$; (b) ${}^{196}_{78}\text{Pt}$

19.103. 3.35:1

19.105. (a) ${}^{40}_{19}\text{K} \rightarrow {}^{40}_{18}\text{Ar} + {}^{0}_{+1}\beta$; (b) because the half-life of ^{40}K is so much longer than that of ^{14}C.

Chapter 20

20.1.

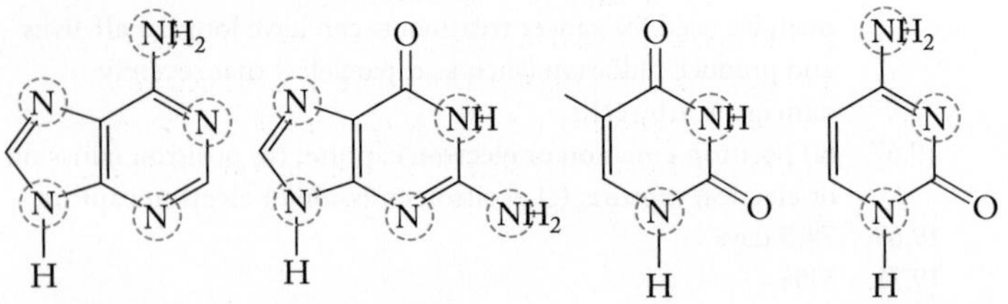

Adenine Guanine Thymine Cytosine

20.3. (a) palmitic acid; (b) stearic acid

20.5. Tyrosine, glycine, glycine, phenylalanine, and methionine

20.7. Trans fats exhibit a type of stereoisomerism around the C=C bond where similar groups on the two carbon atoms are situated on opposite sides of the double bond. Structures (a) and (c) contain trans fats.

20.9. Sucrose; the difference in the structures is that in sucralose three –OH groups on sucrose have been replaced by Cl atoms. Being derived from sucrose implies that the sugar is natural, but the presence of Cl atoms on sugars is not natural.

20.11. No. The molecular formulas are different (C_6H_{14} for hexane and C_6H_{12} for cyclohexane).

20.13. No; the terms *enantiomer* and *optically active* can be used to describe the same chiral molecule, but that molecule would not be *achiral.*

20.15. Constitutional isomers have the same molecular formula but different bonds to the atoms, possibly even different functional groups. Stereoisomers have the same bonds to each atom but different arrangements of those bonds.

20.17. Glycine has no chiral carbon centers.

20.19. *sp*^{3}

20.21.

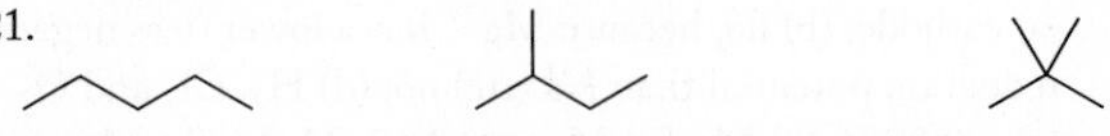

Pentane 2-methylbutane 2,2-dimethylpropane

20.23. (b), (c), (d), and (e)

20.25. With the C=C bond at the end of the carbon chain, a potential rotation about the C=C bond doesn't produce a different stereoisomer. (Another way to look at it: if one of the carbon atoms that are part of the double bond has the same two groups attached—like 2 H atoms in these examples—there will not be cis and trans isomers.)

20.27. (a) trans, *E*; (b) cis, *Z*

20.29. (a)

20.31.

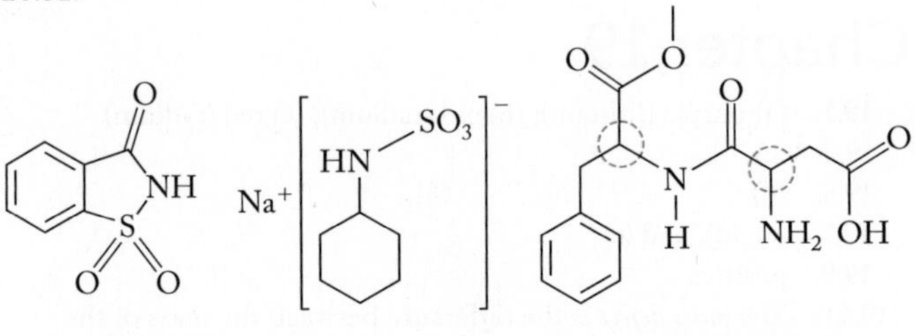

Saccharin Sodium cyclamate Aspartame

Neither saccharin nor sodium cyclamate contains a chiral center.

20.33. (a) Yes, there are 2 chiral centers (circled below); (b) yes, in addition to the 2 chiral centers there is a trans double bond (highlighted below).

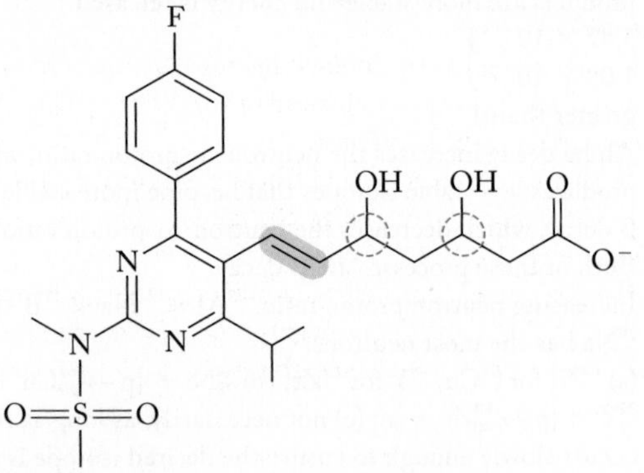

20.35. decreases

20.37. Both amino acids have acidic, polar side chains, with a C=O bond approximately the same distance from the amine and acid functional groups.

20.39. D- and L- refer to how the four groups on a chiral carbon are oriented.

20.41. (a) and (c)

20.43. Most amino acids are zwitterionic at pH ≈ 7.4 because the amino group will be protonated and the carboxylic acid group will be deprotonated, giving $H_2\overset{+}{N}$–CH(R)–COO$^-$.

20.45. hydrogen bonding

20.47. Lysine contains two amino groups, one of which is on a long carbon tail. This can react with the carboxylic acid on the carbon tail of glutamic acid to form a salt bridge.

20.49. a.

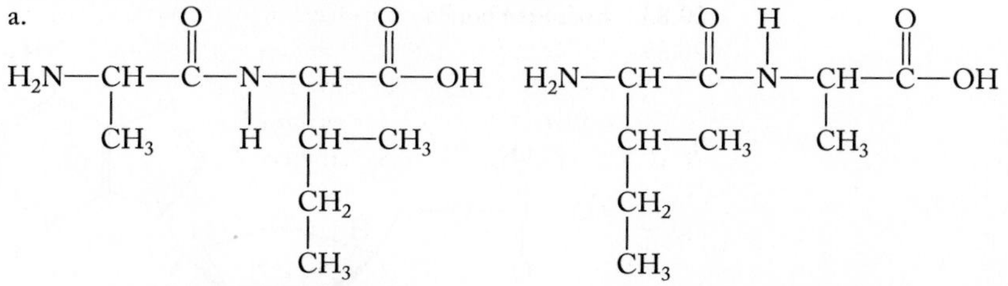

Alanylisoleucine (AlaIle) Isoleucylalanine (IleAla)

b.

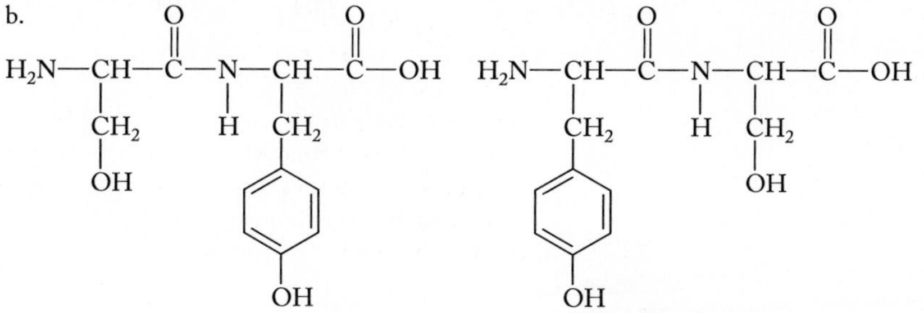

Seryltyrosine (SerTyr) Tyrosylserine (TyrSer)

c.

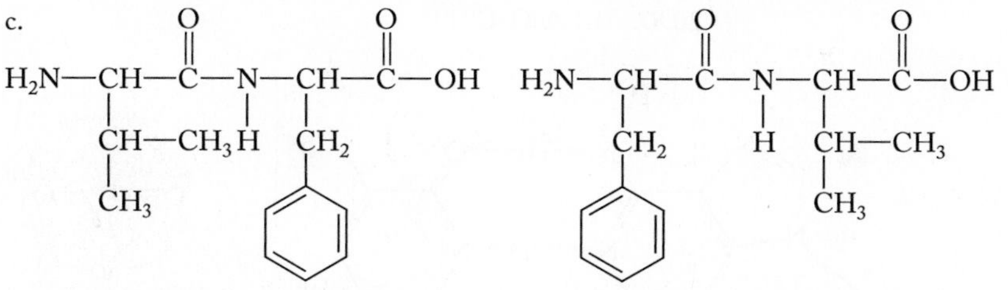

Valylphenylalanine (ValPhe) Phenylalanylvaline (PheVal)

20.51. (a) alanine + glycine; (b) leucine + leucine; (c) tyrosine + phenylalanine

20.53. NH_3

20.55. Starch has α-glycosidic bonds, but cellulose has β-glycosidic bonds. Starch coils into granules, but cellulose forms linear molecules.

20.57. No, they are slightly different because the linear structure has one C═O double bond instead of two C—O single bonds in each of the cyclic structures.

20.59. The bonding in glucose and fructose is nearly the same.

20.61. To calculate the free-energy change for a two-step process, we sum the individual ΔG values for each reaction.

20.63. The position of the hydroxyl group on carbon 1 differs between the α and β forms of galactose. The relative positions of the hydroxyl groups on carbons 2, 3, and 4 are the same on both isomers.

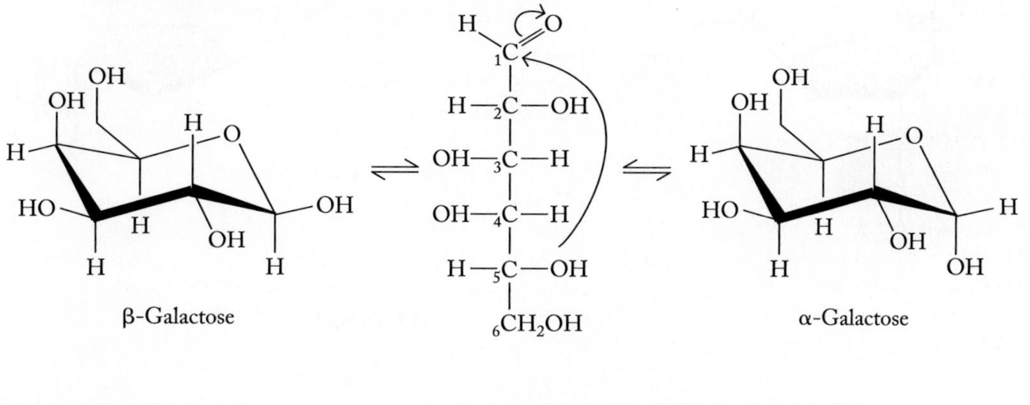

β-Galactose ⇌ (open-chain form, C-1 to C-6) ⇌ α-Galactose

20.65. (c)

20.67. (b)

20.69. −16.4 kJ

20.71. Saturated fatty acids have all carbon–carbon single bonds in their structures but unsaturated fatty acids have carbon–carbon double bonds.

20.73. Fatty acids have a high fuel value, and eating sticks of butter affords Arctic explorers more energy per gram of food compared to that of carbohydrates or protein.

20.75. If the two fatty acids linked to the glycerol at C-1 and C-3 are different then, yes, the triglyceride has a chiral center.

20.77. 0.50 kg of α-linolenic acid would consume more hydrogen than would 1.0 kg of oleic acid; (b) no; the hydrogenation product of both species is stearic acid.

20.79.

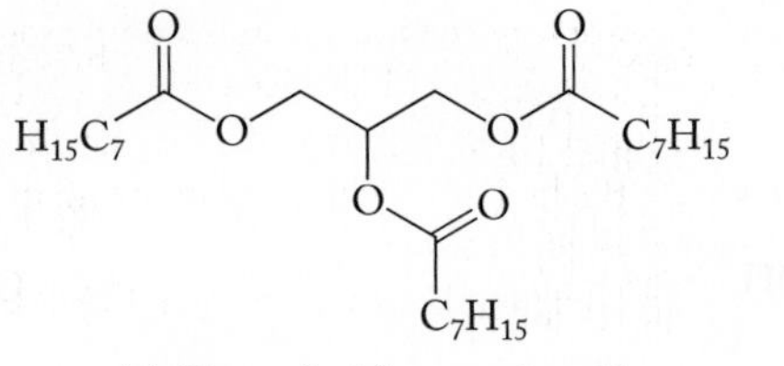

(a) Glycerol with octanoic acid

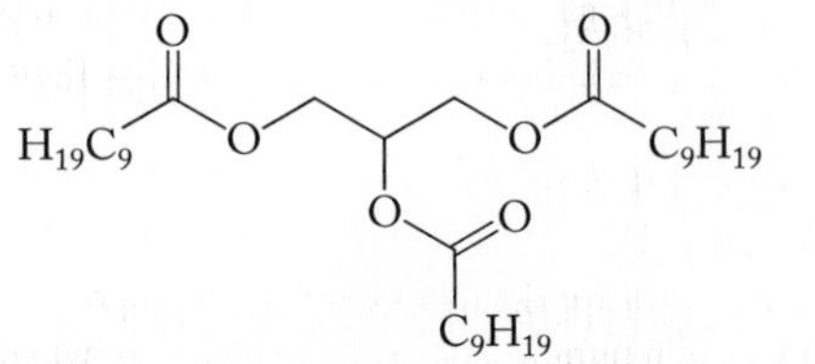

(b) Glycerol with decanoic acid

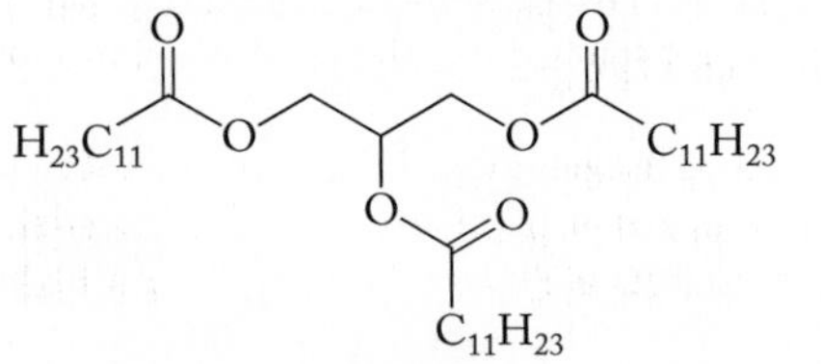

(c) Glycerol with dodecanoic acid

20.81. A phosphate group, a five-carbon sugar, and a nitrogen base; alternating sugar residues and phosphate groups form the backbone.

20.83. hydrogen bonds

20.85.

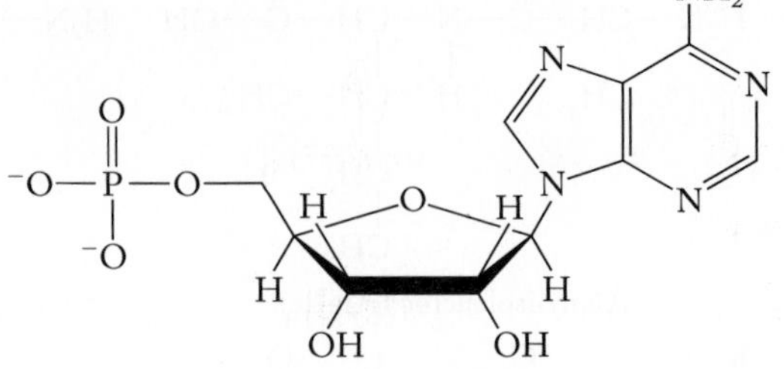

20.87. a. A-G-C

b.

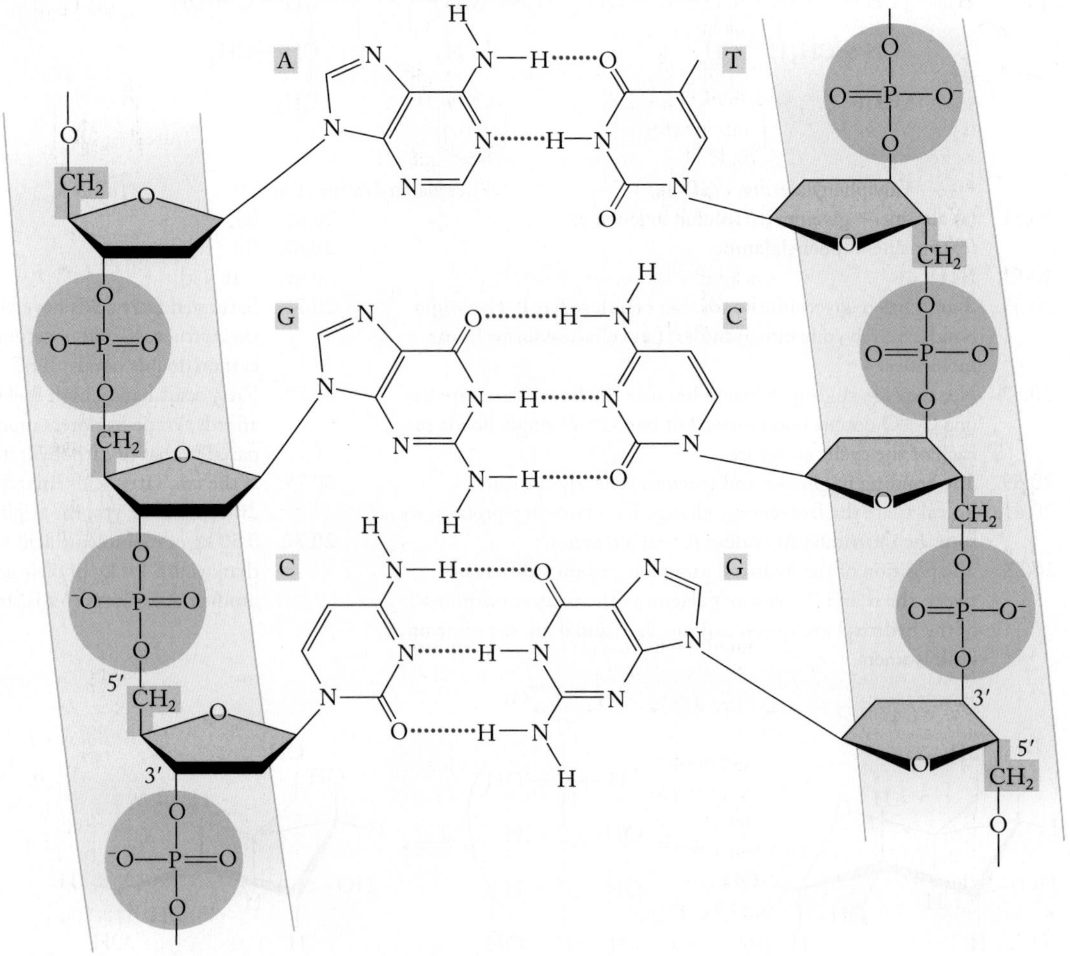

20.89. (a) 12; (b) alkene, aldehyde; (c) cis/trans isomerism
20.91. (a) sucrose; (b) esters; (c) $C_{15}H_{31}COOH$
20.93. (a) There is an extra $-CH_2-$ group in homocysteine's sulfur-containing side chain; (b) yes
20.95. yes
20.97.

$H_2N-CH(-CH(CH_3)-CH_3)-C(=O)-OH$ (Valine) $H_2N-CH(-CH_2-CH(CH_3)-CH_3)-C(=O)-OH$ (Leucine) $H_2N-CH(-CH(CH_3)-CH_2-CH_3)-C(=O)-OH$ (Isoleucine)

Valine   Leucine   Isoleucine

20.99. glutamic acid, cysteine, and glycine
20.101. Yes; because there is no difference in the number of C—C, C—H, C═O, C—O, or N—H bonds between the two compounds, we expect on the basis of average bond energies that the fuel values of leucine and isoleucine should be identical. Isoleucine might have a lower fuel value because the $-CH_3$ group is close to the $-COOH$ and $-NH_2$ groups, and this difference in shape contributes to the slightly different fuel values.
20.103. 16

COOH

20.105. Alanine, aspartic acid, cysteine, glutamic acid, glutamine, glycine, isoleucine, leucine, methionine, phenylalanine, proline, serine, threonine, tryptophan, tyrosine, and valine all have a p*I* value below 7.4 and would be primarily in the zwitterionic form at pH 7.4. Arginine, asparagine, histidine, and lysine have p*I* values above 7.4 and will not reach the zwitterionic form until higher pH values.
20.107. Both molecules have a N atom with a bent shape that has a lone pair of electrons oriented away from the ring it is bound in and an additional N atom located on a flexible chain of carbon atoms.

Chapter 21

21.1. (d)
21.3. group 16 (lavender)
21.5. −6.7 kJ
21.7. trigonal planar
21.9. phosphate
21.11. Without an essential element, biological processes that rely on that element would shut down or deteriorate. If a nonessential element is missing, there would not be severe deleterious effects.
21.13. The main criterion that distinguishes major, trace, and ultratrace essential elements from one another is their concentration in the body. Ultratrace elements are present at $<1\ \mu g/g$, trace elements are between 1 and 1000 μg/g, and major elements are present at >1 mg/g.
21.15. (a) 1.6 ppm; (b) 525 ppm; (c) 0.43 ppm
21.17. (a) oxygen; (b) oxygen; (c) carbon
21.19. The valence electrons for the neutral atoms from groups 1 and 2 reside in *s* sublevels (thus *s*-block elements); the valence electrons for neutral atoms in groups 13–18 reside in *s* and *p* sublevels with the higher-energy electrons being in the *p* sublevel (thus *p*-block elements).
21.21. Li_2O is an ionic compound whereas the bonding in CO is covalent.
21.23. electrical conductivity
21.25. Ca^{2+} is larger than Mg^{2+} and may not fit into the active site of a biomolecule as easily as Be^{2+}, which is smaller than Mg^{2+}.
21.27. K^+
21.29. $Al^{3+} < Mg^{2+} < Li^+ < Cl^-$
21.31. $K < Mg < S < F$
21.33. The two processes are the reverse of one another.
21.35. ion channels, ion pumps, and through diffusion when part of a complex ion
21.37. The hydrophobic interior of the cell membrane makes it difficult to transport charged ions through the cell membrane.
21.39. K^+
21.41. $CaCO_3$ is more insoluble (less soluble) than $CaSO_4$, which makes calcium carbonate a better structural material. Also, the partial pressure of CO_2 in the atmosphere is higher than that of SO_3.
21.43. 0.56 atm
21.45. 0.072 V
21.47. (a) 0.059 V; (b) 0.34 mol
21.49. a.

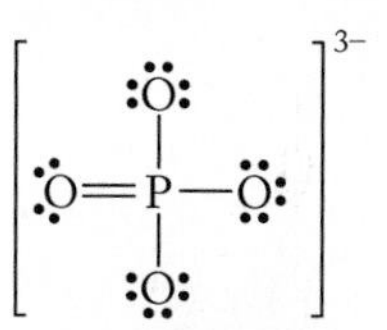

Phosphate
oxidation state on P = +5

O=P—O, H (2−)

Phosphite
oxidation state on P = +3

b. Fewer weeds would grow when phosphite fertilizer is used so less herbicide would be needed.

21.51. $^{137}Cs^+$ may substitute for K^+ in cells; as a β emitter with a relatively long half-life, it would lead to an increase in cancer risk.
21.53. Δ*S* is probably positive (more disorder when solid is dissolved) and Δ*H* is probably negative (would require breaking bonds).
21.55. The superoxide anion contains an unpaired electron and an incomplete octet on one of the oxygen atoms. This radical will react to complete the octet and pair the electron, which would cause it to be reduced in the process.
21.57. a.

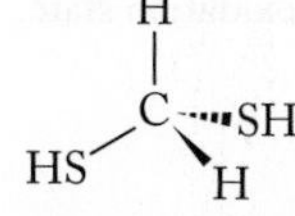

Methanedithiol would have the highest vapor pressure.

b. Vapor pressure would decrease. The presence of oxygen means the resulting molecules would be capable of hydrogen-bonding interactions, which are significantly stronger than the dipole–dipole interactions involving sulfur atoms. The stronger interactions would lead to higher boiling points and lower vapor pressure.

21.59. 3.06
21.61. $K = [OH^-]/[F^-] = 8.48$; the equilibrium lies slightly to the right under standard conditions.
21.63. (a) More soluble; (b) 8.5×10^{-8} mol/L; (c) 1.5×10^{-7} mol/L

21.65. (a) 2.7×10^{-7} *M*; (b) no, the solubility corresponds to 56 ppb; (c) $PbCO_3$ will be more soluble in acidic solutions as the CO_3^{2-} ion can be protonated to HCO_3^-, which would shift the solubility equilibrium forward.

21.67. H_2O is the most polar and H_2Te is the least polar.

21.69. We must consider the decay mode to ensure that α emitters are not used inside the body. The γ radiation produced by imaging radionuclides has relatively low energy and can easily escape the body minimizing damage to tissues and organs.

21.71. Because β particles have low charge and relatively small mass, they penetrate tissues better than α particles do. When β particles exit the body, they are imaged on a detector screen to create the image of the organ.

21.73. ^{111}In is proton-rich (or neutron-poor), so electron capture, which changes a proton into a neutron, helps the nucleus become more stable. ^{213}Bi is proton-poor (or neutron-rich), so β decay, which changes a neutron to a proton, helps the nucleus become more stable.

21.75. dispersion forces and dipole–induced dipole interactions

21.77. 41 h

21.79. (a) $Bi^{3+}(aq) + 7\,H_2O(\ell) \rightleftharpoons Bi(H_2O)_5(OH)^{2+}(aq) + H_3O^+(aq)$; (b) pH = 2.5; (c) more acidic than vinegar

21.81. $Al(OH)_3(s) + 3\,H^+(aq) \rightarrow 3\,H_2O(\ell) + Al^{3+}(aq)$

21.83. (a) $Al(OH)_3$; (b) yes; (c) 2.5 g of $Mg(OH)_2$, 2.2 g of $Al(OH)_3$

21.85. a. group 15

b. :C̈l—S̈b—C̈l: with :C̈l: bonded below Sb

c. Lewis base

Chapter 22

22.1. (a) chromium (green) and cobalt (yellow); (b) vanadium (black) and zinc (blue); (c) zinc (blue)

22.3. (d)

22.5. All of the compounds have some ligands that are in cis arrangements and some in trans arrangements.

22.7. flask (a) (orange): $[Co(NH_3)_6]^{3+}$; flask (b) (blue): $[Co(H_2O)_6]^{3+}$; flask (c) (slight yellow/colorless): $[Co(CN)_6]^{3-}$

22.9. Max absorption is in the yellow to green portion of the visible light spectrum so the solution should be purple.

22.11. (a) blue: CoI_4^{2-}; (c) black: $CoCl_4^{2-}$

22.13. water molecules

22.15. Iron in the anion is in the Fe^{2+} oxidation state (for an overall charge of −4 for each complex anion). The iron atoms not in the complex anion are in the Fe^{3+} oxidation state.

22.17.

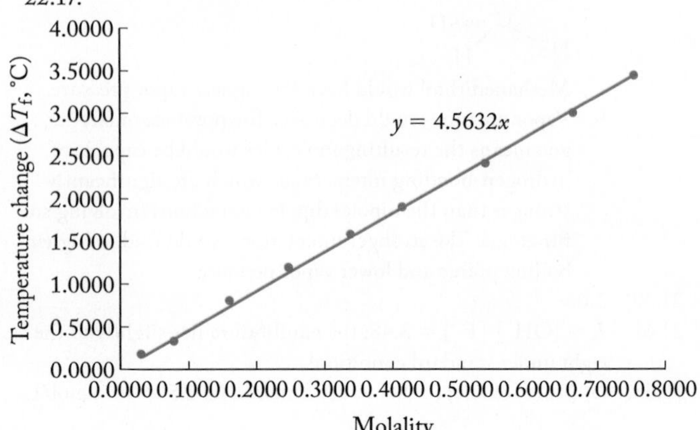

In the equation of the line, the slope is equal to the van't Hoff factor times the freezing-point-depression constant. For an aqueous solution, the freezing-point-depression constant for water is 1.86°C/molal, which would mean the value of the van't Hoff factor would be 2.45 for $K_3[Fe(CN)_6]$ indicating there is significant ion-pair formation occurring (an ideal solution of $K_3[Fe(CN)_6]$ would have a van't Hoff factor approaching 4).

22.19. $[Pt(NH_3)_6]Cl_4$; $[Pt(NH_3)_5Cl]Cl_3$; $[Pt(NH_3)_4Cl_2]Cl_2$; $[Pt(NH_3)_3Cl_3]Cl$; $[Pt(NH_3)_2Cl_4]$

22.21. $[Pt(NH_3)_6]Cl_4$: hexaammineplatinum(IV) chloride; $[Pt(NH_3)_5Cl]Cl_3$: pentaamminechloroplatinum(IV) chloride; $[Pt(NH_3)_4Cl_2]Cl_2$: tetraamminedichloroplatinum(IV) chloride; $[Pt(NH_3)_3Cl_3]Cl$: triamminetrichloroplatinum(IV) chloride; $[Pt(NH_3)_2Cl_4]$: diamminetetrachloroplatinum(IV)

22.23. (a) hexaamminechromium(III); (b) hexaaquacobalt(III); (c) pentacyanonickelate(II)

22.25. (a) potassium tetrabromocobaltate(II); (b) ammonium aquatrihydroxozincate(II); (c) pentaamminechloroiron(I) dimer

22.27. (a) bis(ethylenediamine)zinc(II) sulfate; (b) pentaammineaquanickel(II) chloride; (c) potassium hexacyanoferrate(II)

22.29. A sequestering agent is a multidentate ligand that separates metal ions from other substances so that they can no longer react. Properties that make a sequestering agent effective include strong bonds formed between the metal and the ligand and large formation constants.

22.31. As pH increases the chelating ability increases because OH^- removes the H atoms on the carboxylic acid groups, providing an additional site for binding to the metal cation.

22.33. The energy difference between *d* orbitals lies within the visible range of the spectrum. For most transition metals, transitions occur between *d* orbitals, and thus these complexes appear colored.

22.35. The repulsions due to the ligands in a square planar crystal field are higher for the d_{xy} orbital than for the d_{xz} and d_{yz} orbitals. The d_{xy} orbital's energy is higher because this orbital lies in the plane of the ligands.

22.37. The yellow solution contains (b) $Cr(NH_3)_6^{3+}$ and the violet solution contains (a) $Cr(H_2O)_6^{3+}$.

22.39. colorless

22.41. $NiCl_4^{2-}$

22.43. Ligand field strength determines whether a complex is high or low spin. Stronger ligand fields induce low-spin arrangements.

22.45. Fe^{2+} has four unpaired electrons; Cu^{2+} has one unpaired electron; Co^{2+} has three unpaired electrons; Mn^{3+} has four unpaired electrons

22.47. For the Fe^{2+} (d^6) to have no unpaired electron in an octahedral crystal field, it would have to be low spin. The change from 298 to 80 K must have caused the compound to transition from high spin to low spin.

22.49. (a) Mn^{4+} in MnO_2, two Mn^{3+} and one Mn^{2+} in Mn_3O_4; (b) both low-spin and high-spin configurations are possible in Mn_3O_4 (d^4 and d^5) but not in MnO_2 (d^3).

22.51. paramagnetic

22.53. For an octahedral geometry, *cis-* means that two ligands are side by side and have a 90° bond angle between them. Ligands that are *trans-* to each other have 180° bond angles between them.

22.55. at least two different ligands

22.57. yes

22.59. $\left[\begin{array}{c}\text{Br} \quad \text{Cl} \\ \text{Cu} \\ \text{Br} \quad \text{Cl}\end{array}\right]^{2-}$ $\left[\begin{array}{c}\text{Cl} \quad \text{Br} \\ \text{Cu} \\ \text{Br} \quad \text{Cl}\end{array}\right]^{2-}$

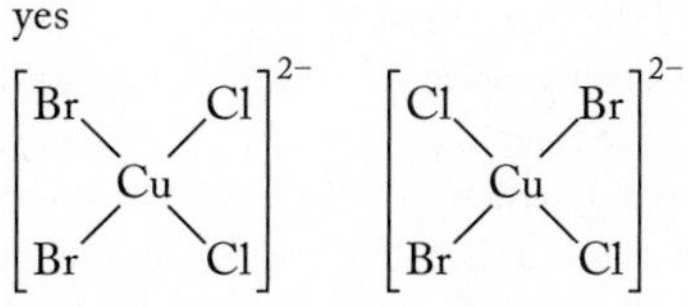

Cis Trans

Neither of these isomers is chiral.

22.61. (a) Enzymes catalyze biochemical reactions; (b) no

22.63. Enzymes lower the activation energy.

22.65. much greater than 1

22.67. positive

22.69. 1.59×10^8 times faster

22.71. zero order

22.73. Toxic metals that are sensitive imaging agents may be used at extremely low concentrations such that we could essentially ignore any toxicity. The complex ion could also exhibit different toxicity from that of the metal ion alone. The metal ion might also be taken up preferentially by, for example, cancer cells, thus targeting any damage to unhealthy cells.

22.75. γ radiation

22.77. by binding to the nitrogen atoms in DNA to stop the division of cells

22.79. paramagnetic

22.81. $K = 130$

22.83. $[\text{penicillamine}]_{eq} = 4.43 \times 10^{-4}$ *M* and $[\text{cysteine}]_{eq} = 5.6 \times 10^{-4}$ *M*

22.85. (a) The ammonia and azide ligands are each cis to one another in one molecule and trans to one another in the other molecule; (b) *cis*-diamminediazidodihydroxyplatinum(IV) and *trans*-diamminediazidodihydroxyplatinum(IV); (c) yes;
(d) 

22.87. (a) yes; (b) diamagnetic

22.89. a. The yellow complex has larger splitting energy. We can conclude that it contains Co^{3+} due to its having no unpaired electrons.

b. [↑][↑] / [↑↓][↑↓][↑] — Red/purple Co^{2+} solution

[ ][ ] / [↑↓][↑↓][↑↓] — Yellow Co^{3+} solution

22.91. Ag^{2+} would be d^9—with an odd number of electrons it will always have one unpaired electron and be paramagnetic. Ag^+ would most likely be $5s^0 4d^{10}$ (instead of $5s^1 4d^9$). With a completely filled *d* shell there would be no unpaired electrons so the ion would be diamagnetic. Ag^{3+} would be d^8. The silver ions in AgO are square planar, which results in the d_{xz}, d_{yz}, d_{z2}, and d_{xy} orbitals being filled and the much higher in energy d_{x2-y2} orbital being empty resulting in all electrons being paired and the ion being diamagnetic.

22.93. a.

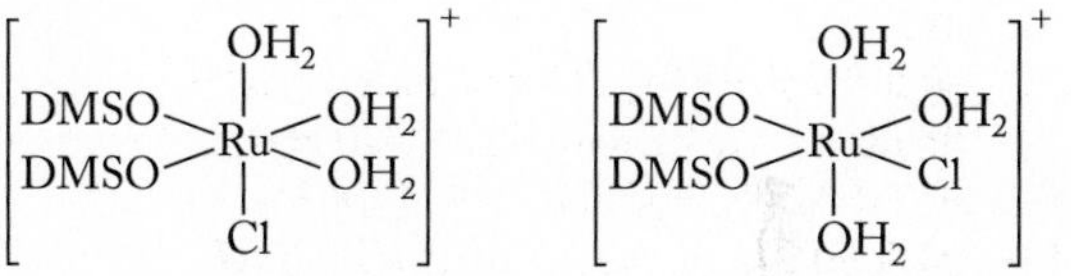

$[Ru(DMSO)_2(OH_2)_3Cl]^+$: fac isomer — DMSO, DMSO, Cl, OH₂ (top), OH₂, OH₂ around Ru

$[Ru(DMSO)_2(OH_2)_3Cl]^+$: mer isomer — DMSO, DMSO, OH₂ (top), OH₂, Cl, OH₂ (bottom) around Ru

fac isomer mer isomer

When three identical ligands occupy one face of an octahedron, the isomer is described as facial, or *fac*. If the three identical ligands and the metal ion are in one plane, the isomer is described as meridional, or *mer*.

b. Linkage isomers are possible because DMSO can attach to a transition metal through the S or the O:

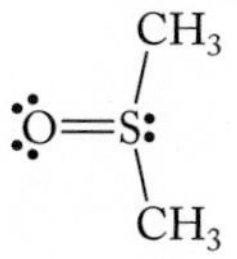

$:\ddot{O}=\ddot{S}(CH_3)_2$

22.95. five unpaired electrons

Credits

Front Matter

p. 5 (top): Russ Heinl/Shutterstock; (bottom): Kian Khoon Tan/Alamy Stock Photo; p. 6 (top): AP Photo/Noah Berger; (bottom): Ron Sanford/Science Source; p. 7 (top): NASA's Goddard Space Flight Center/Conceptual Image Lab/Michael Lentz; (bottom): Tom Walker/Alamy Stock Photo; p. 8 (top): Thegoodly/Getty Images RF; (center): W. Perry Conway/Getty Images; (bottom): Vadim Petrakov/Shutterstock; p. 9 (top): AP Photo/Charlie Riedel; (bottom): MicroDiscovery/Corbis/Getty Images; p. 10 (top): Stan Sholik/Alamy Stock Photo; (center): Kyodo/Newscom; (bottom): Design Pics Inc/Alamy Stock Photo; p. 11 (top): iStock/Getty Images; (bottom): WaterFrame/Alamy Stock Photo; p. 12 (top): Garry Webber/Shutterstock; (center): Courtesy, Jim Motavalli; (bottom): Dennis Flaherty/Science Source; p.13 (top): saied shahin kiya/Shutterstock; (bottom): Michael Neelon(misc)/Alamy Stock Photo; p. 14: addkm/Shutterstock; p. 19 (from top, down): Chrissy Gilbert; Paige Scannell; Youngstown State University; Kathryn Dyer; p. 20: JGI/Jamie Grill/ageFotostock.

Chapter 1

pp. 2–3: NASA, ESA, P. Oesch (Yale University), G. Brammer (STScI), P. van Dokkum (Yale University), and G. Illingworth (University of California, Santa Cruz); p. 5 (gold): Photographer's Choice/Punchstock; (ice): Shutterstock; (vinegar) Getty Images; (dressing): Bloomberg/Contributor/Getty Images; p. 6 (top): Sami Sarkis/Getty Images; (bottom): Charles D. Winters/Science Source; p. 7 (a): Andrew Dunn/Alamy Stock Photo; (c): Panther Media GmbH/Alamy Stock Photo; p. 9: Lester V. Bergman/Getty Images; p. 10: Amélie Benoist/Science Source; p. 11 (a): Nicholas Rjabow/Shutterstock; (b): John Vlahidis/Alamy Stock Photo; (c): Huntstock/Getty Images; (bottom): Andrea Pistolesi/ZUMA Press/Newscom; p. 14: Russ Heinl/Shutterstock; p. 16: W. Langdon Kihin/National Geographic Creative; p. 17: STScI and NASA; p. 21: World History Archive/Ann Ronan Collection/Agefototstock; p. 22: Cappi Thompson/Getty Images; p. 23 (both): Courtesy of Ohaus; p. 26: Turtle Rock Scientific/Science Source; p. 34: (top): Bettmann/Getty Images; (bottom): Science Photo Library/Alamy; p. 35 (top): AP Images; (a): NASA/WMAP Science Team; (b): NASA/WMAP Science Team; p. 36: NASA/JPL-Caltech MSSS; p. 37 (d): Amélie Benoist/Science Source; (e): Russ Heinl/Shutterstock; p. 40: (c): Shutterstock; (e): Charles D. Winters / Science Source; (g): Charles D Winters/Science Source.

Chapter 2

pp. 46–47: Kian Khoon Tan/Alamy Stock Photo; p. 48 (both): Missouri Historical Society, St Louis; p. 49: The Royal Institution, London/Bridgeman Images; p. 51 (top): Astrid & Hanns-Frieder Michler/Science Source; (bottom): Archive PL/Alamy Stock Photo; p. 53: Grasseto/iStockphoto/Getty Images; p. 59 (foil): Andrey Kuzmin/Alamy Stock Photo; (copper): Igor Stevanovic/Shutterstock; (mercury): Shutterstock; (sulfur): Corbin17/Alamy Stock Photo; (bromine): Martyn F. Chillmaid/Science Source; (iodine): Dorling Kindersley ltd/Alamy Stock Photo; (silicon): Dorling Kindersley ltd/Alamy Stock Photo; (germanium): Alfred Pasieka/Science Source; (antimony): Dirk Wiersma/Science Source; p. 63: Dirk Wiersma/Science Photo Library/Science Source; p. 66: Maksym Yemelyanov/ageFotostock; p. 67: hein teh/Alamy Stock Photo; p. 76 (top): NASA/CXC/SAO; (bottom): NASA/HST/J. Morse/K. Davidson; p. 77: ISM/Phototake/ISM/Medical Images/Diomedia; p. 78: NASA/HST/J. Morse/K. Davidson.

Chapter 3

pp. 86–87: AP Photo/Noah Berger; p. 88: Roger Ressmeyer/Corbis/VCG/Getty; p. 90: Gennady Teplitskiy/Shutterstock; p. 92 (top): Science Source; (bottom): icefront/iStockphoto; p. 93 (top): Zach Holmes/Alamy; (center): KarSol/Shutterstock; (bottom): RuslanK/Shutterstock; p. 94: Francois Lenoir/Reuters; p. 95: Kei Shooting/Shutterstock; p. 97 (top): Charles D. Winters/Science Source; (bottom): Private Collection/Photo © Zev Radovan/Bridgeman Images; p. 100: Roger Ressmeyer/Corbis/VCG/Getty; p. 105: vovan13/iStockphoto; p. 111: Davids1993 Dreamtime.com; p. 119: Gilbert S Grant/Photo Researchers/Getty Images; p. 120: vupulepe/Shutterstock; p. 123 (top): AB Forces News Collection/Alamy Stock Photo; (a): Biophoto Associates/Science Source; (b): Krzysztof Stanek/Dreamstime.com; (c): FocusTechnology/Alamy; p. 126 (a): Rojo Images/Shutterstock; (b): Thinkstock; p. 130: pikselstock/Shutterstock; p. 131 (top): Pawe Borówka/Shutterstock; (a): Marina Onokhina/Shutterstock; (b): Bochkarev Photography/Shutterstock; p. 132: Leithan Partnership t/a The Picture Pantry/Alamy Stock Photo; p. 133: Robin Bush/Oxford Scientific/Getty Images; p. 134: Science Source; p. 138 (P3.8): From *Interpretation of Mass Spectra, 3rd Edition* by Fred W. McLafferty and Frantisek Turecek, © 1980. Reproduced with permission from University Science Books, all rights reserved; p. 138: (c): Shutterstock; (d): Cherkas/Shutterstock; (e): iStockphoto; (g): Shane Maritch/Shutterstock; p. 139: Photo Researchers/Science Source; p. 143: aneese/Thinkstock; p. 146: Smithsonian Institution, Washington, D.C./Bridgeman Images.

Chapter 4

pp. 148–149 (top): Ron Sanford/Science Source; p. 149 (bottom): Turtle Rock Scientific/Science Source; p. 151 (top a): Time Life Pictures/Getty Images; (top b): NASA; (top c): Steve Schmeissner/Science Source; (bottom a): NASA/JPL-Caltech/MSSS and PSI; (bottom b): NASA/JPL-Caltech/MSSS and PSI; p. 152 (left): Photo Researchers/Science Source; (right): Photo Researchers/Science Source; p. 153: Colin Anderson/Getty Images/Brand X; p. 157: Lipskiy/Shutterstock; p. 159 (a-d): Turtle Rock Scientific/Science Source; (bottom): Ton Koene/Agefotostock/Design Pics; p. 162: Used by permission of Thermo Fisher Scientific, the copyright owner; p. 163: Turtle Rock Scientific/Science Source; p. 164 (all): Turtle Rock Scientific/Science Source; p. 168 (left): Richard Thom/Visuals Unlimited; (bottom left) TopFoto/The Image Works; (bottom right) TopFoto/The Image Works; p. 173 (all): Turtle Rock/Science Source; p. 174: Dirk Ercken/Shutterstock; p. 176: Turtle Rock Scientific/Science Source; p. 177 (both): Turtle Rock/Science Source; p. 182: Turtle Rock Scientific/Science Source; p. 183 (all): Turtle Rock Scientific/Science Source; p. 184: NASA; p. 187 (both): Charles D. Winters/Science Source; p. 190 (all): Turtle Rock Scientific/Science

Source; p. 193 (top): Bill Ross/Getty Images; (bottom): USDA Natural Resources Conservation Service; p. 194 (all): Turtle Rock Scientific/Science Source; p. 198 (left and right): Turtle Rock/Scientific/Science Source; (bottom right, both): Charles D. Winters/Science Source; p. 201 (a): Charles D. Winters/Science Source; (e left): GIPhotostock/Science Source; (e right): GIPhotostock/Science Source; (I) Andrew Lambert Photography/Science Source; p. 204: David R. Frazer/Photo Library, Inc./Alamy; p. 205: B. Murton/Southampton Oceanography Centre/Science Source; p. 209: Richard Sugarek/Environmental Protection Agency.

Chapter 5

pp. 212–213: NASA's Goddard Space Flight Center/Conceptual Image Lab/Michael Lentz; p. 213 (bottom): JGI/Jamie Grill/ageFotostock; p. 214: SPL/Science Source; p. 215: Science Source; p. 216 (a): Sergey Borisov/Alamy Stock Photo; (b): Bill Ross/Getty Images (c): Christian Kohler/Shutterstock; (d): Ignacio Salaverria/iStockphoto; p. 218: NASA/JPL-Caltech/Space Science Institute; p. 219: Sam Ogden/Science Source; p. 220: David R. Frazier Photography, Inc./Alamy; p. 229: Spencer Sutton/Science Source; p. 233: Thinkstock/Getty Images; p. 234 (top): Romilly Lockyer/Getty Images; (bottom): Jim Zuckerman/Alamy Stock Photo; p. 235: David Samuel Robbins/Getty Images; p. 236: BHammond/Alamy Stock Photo; p. 237 (all): Martyn F. Chillmaid/Science Source; p. 242: Turtle Rock Scientific/Science Source; p. 250 (both): Martyn F. Chillmaid/Science Source; p. 254: Cultura Creative (RF)/Alamy Stock Photo; p. 255: Turtle Rock Scientific/Science Source; p. 256: JGI/Jamie Grill/ageFotostock; p. 258: Lester V. Bergman/Getty Images; p. 259: Lester V. Bergman/Getty Images; p. 261 (top): Arnold Media/Getty Images (bottom): Marc Bruxelle/Alamy Stock Photo; p. 262: Jonathan Blair/Getty Images; p. 263: Turtle Rock Scientific/Science Source; p. 264: F. Jack Jackson/Alamy; p. 265: Bary Bishop/National Geographic Creative; p. 269: Stocktrek/Getty Images RF.

Chapter 6

pp. 270–271: Tom Walker/Alamy Stock Photo; p. 270 (bottom): Turtle Rock Scientific/Science Source; p. 275 (left): Yutaka/AFLO/Newscom; (right): AP Photo/Matthias Schrader; p. 276: Turtle Rock Scientific/Science Source; p. 279 (left) Marmaduke St. John/Alamy; (right): NASA; p. 286: POOL/REUTERS/Newscom; p. 287: Sprokop/Dreamstime; p. 289 (top): Science Source; (bottom): Johner Images/Getty Images; p. 294: Rtimages/Dreamstime; p. 315 (a): Dr. Keith Wheeler/Science Source; (b): AP Photo; p. 318 (a): SKrow/iStock-Photo; (b): nito/Shutterstock; (c): jfmdesign/iStockPhoto; p. 324: (b) Charles D. Winters/Science Source; (f) Samo Trebizan/Shutterstock; (g): hraska/Shutterstock; (i): Thinkstock/Getty Images; p. 326: (a-b): Photo Researchers/Science Source.

Chapter 7

pp. 334–335: Thegoodly/Getty Images; p. 336: North Wind Picture Archives/Alamy Stock Photo; p. 337 (left a): Richard Treptow/Science Source; (right a): Ted Kinsman/Science Source; (left b): Richard Treptow/Science Source; (right b): Ted Kinsman/Science Source; (left c): Richard Treptow/Science Source; (right c): Ted Kinsman/Science Source; p. 338 (x ray hand): Sergey Tarasov/Alamy Stock Photo; (blue hand): Andrew Ennis/Stockimo/Alamy Stock Photo; (house): Avalon/Construction Photography/Alamy Stock Photo; (microwave): vaeenma/Istockphoto; (tower): Alex Genovese/Alamy Stock Photo; p. 341 (b-d): Ted Kinsman/Science Source; (bottom): New York Public Library/Science Source; p. 342: Kampol Taepanich/Shutterstock; p. 352: Westend61 GmbH/Alamy Stock Photo; p. 357: Ted Kinsman/Science Source; p. 358: Margrethe Bohr Collection/American Institute of Physics/Science Photo Library/Science Source; p. 359: Christian Ohde/ageFotostock; p. 366 (a): David Taylor/Science Source; (b): Dorling Kindersley/Getty Images; p. 369: Doug Martin/Science Source; p. 380: suthiwat/Shutterstock; p. 384 (left): Marmaduke St. John/Alamy Stock Photo; (right): medeni/Alamy Stock Photo; p. 385: Arterra Picture Library/Alamy Stock Photo; p. 388 (top): SPL/Science Source; (bottom): Turtle Rock Scientific/Science Source; p. 390: SPL/Science Source.

Chapter 8

pp. 392–393: NicePrsopects-Bali/Alamy Stock Photo; p. 394: image-BROKER/Alamy Stock Photo; p. 396 (bottle): Charles D. Winter/Science Source; (salt): Charles D. Winter/Science Source; (wire): Charles D. Winter/Science Source; p. 410: W. Perry Conway/Getty Images; p. 418: age fotostock/Alamy Stock Photo; p. 420: Bob Rowan/Getty Images; p. 437 (left): GIPhotoStock/Science Source; (middle): Ed Endicott/Alamy Stock Photo; (right): GIPhotoStock/Science Source; p. 438: Education Images/UIG via Getty Images; p. 442 (left) B.A.E. Inc./Alamy Stock Photo; (right) NASA/Damian Peach.

Chapter 9

pp. 444–445: Steven Morris/Alamy Stock Photo; p. 449 (all): Editorial Image, LLC/Alamy Stock Photo; p. 461: Thomas Gilbert; p. 471: Jan Hanus/Alamy Stock Photo; p. 474: Thomas Gilbert; p. 477: Vadim Petrakov/Shutterstock; p. 483: Yoav Levy/Medical Images/Diomedia; p. 494 (a): John t. Fowler/Alamy Stock Photo; (b): Jack Thomas/Alamy Stock Photo; p. 495: Editorial Image, LLC/Alamy Stock Photo; p. 504 (left): Smithsonian Institution, Washington, D.C./Bridgeman Images; (right): Johnny Habell/Shutterstock.

Chapter 10

pp. 506–507: AP Photo/Charlie Riedel; p. 520 (a): W. W. Norton & Company, Inc.; (b): sciencephotos/Alamy Stock Photo; p. 526 (a): sciencephotos/Alamy Stock Photo; (b): Martin Shields/Science Source; p. 527(top): Sinclair Stammers/Science Source; (a): Sinclair Stammers/Science Source; (b): Peter Gardner/Dorling Kindersley/Science Source; p. 528: Paul Rapson/Science Source; p. 529: iStockphoto; p. 531: Photo Researchers/Science Source; p. 532: fStop Images GmbH/Alamy Stock Photo; p. 538: Martin Shields/Science Source; p. 542 (left): Shutterstock; (right): Steve Hawkins Photography/Alamy Stock Photo.

Chapter 11

pp. 548–549: MicroDiscovery/Corbis/Getty Images; p. 556 (left): Andrew Lambert Photography/Science Source; (middle): Charles D. Winters/Science Source; (right): Paul Whitehall/Science Photo Library/Science Source; p. 579 (all): Dr. David M. Phillips/Visuals Unlimited; p. 583: Richard T. Nowitz/Getty Images; p. 585 (a): Eddie Gerald/Alamy Stock Photo; (b): Rainman Illawong Bay 3; (c): John Kasawa/Dreamstime; p. 588 (a): Joel Arem/Science Source; (b): W. W. Norton & Company, Inc.; p. 589 (all): Thomas Gilbert; p. 590: Dr. David M. Phillips/Visuals Unlimited.

Chapter 12

pp. 600–601: Joanna Nassar; p. 602 (a): © Goruppa/Dreamstime; (b): © Raja Rc/Dreamstime; (c): Photo Researchers/Science Source; p. 605 (a): Dorling Kindersley/Getty Images; (b): image-broker/Alamy; p. 606: Stan Sholik/Alamy Stock Photo; p. 607: Tom Gilbert; p. 612 (a): Andrew Taylor/Alamy Stock Photo; (b): Alonso Aguilar/Getty Images/iStockphoto; (bottom): Kris Mercer/Alamy Stock Photo; p. 613 (a): Philippe Plailly/Science Source; (b): Philippe Plailly/Science Source; p. 616: Photo Researchers/Science Source; p. 618: The Natural History Museum/Alamy Stock Photo; p. 619: Çagla Köhserli/Alamy Stock Photo; p. 620 (a): Jon Stokes/Science Source; (b): Anneka/Shutterstock; (c): Andre Geim & Kostya Novoselov/Science Source; p. 622: Andrew Payne/Alamy Stock Photo; p. 623: BSIP SA/Alamy Stock Photo; p. 624 (top): Jeffrey Hamilton/Getty Images; (bottom): Alexandra Grablewski/Photolibrary/Getty Images; p. 625 (bag): Emily Spence/Lexington Herald-Leader/MCT Newscom; (carpet): Fotosearch; (rope): studiomode/Alamy Stock Photo; (jacket): ZUMA Press Inc/Alamy Stock Photo; p. 626 (left): W. W. Norton & Company, Inc; (right): Photo Researchers/Science Source; p. 629: Southern Illinois University/Science Source; p. 631 Creatas/Jupiterimages; p. 633: ©Photo12/The Image Works; p. 634: Pablo Alcala/KRT/Newscom; p. 639: Don Johnston_IH/Alamy Stock Photo; p. 642: The Picture Art Collection/Alamy Stock Photo; p. 646 (P12.94): Republished with permission of University of California Press, from *Phase Diagrams of the Elements* by David A. Young, 1991; permission conveyed through Copyright Clearance Center, Inc.; p. 646: Greenshoots Communications/Alamy Stock Photo.

Chapter 13

pp. 648–649: Kyodo/Newscom; p. 648 (bottom): Turtle Rock Scientific/Science Source; p. 650: Mirrorpix/Newscom; p. 652 (a): Kateleigh/Dreamstime; (b): SPL/Science Source; (c): Charles D. Winters/Science Source; p. 688: Image courtesy the TOMS science team & and the Scientific Visualization Studio, NASA GSFC; p. 691: Dorling Kindersley ltd/Alamy Stock Photo; p. 694: Image courtesy the TOMS science team & and the Scientific Visualization Studio, NASA GSFC; p. 697 (c, f, i): Image courtesy the TOMS science team & and the Scientific Visualization Studio, NASA GSFC; p. 701: Lewis Houghton/Science Source; p. 706: Andrew Lambert Photography/Science Photo Library/ Science Source.

Chapter 14

pp. 708–709: Design Pics Inc/Alamy Stock Photo; p. 717: FOR ALAN/Alamy Stock Photo; p. 727 (a): Stan Pritchard/Alamy Stock Photo; (b): Robert Jones/Alamy Stock Photo; p. 729: Turtle Rock Scientific/Science Source; p. 732 (all): Turtle Rock Scientific/Science Source; p. 735: Charles D. Winters/Science Source; p. 744 (all): Turtle Rock Scientific/Science Source; p. 746: Pat Bagley/Salt Lake Tribune; p. 747 (b, d, f): Turtle Rock Scientific/Science Source; (h): Photo Researchers/Science Source.

Chapter 15

pp. 754–755: iStock/Getty Images; p. 778: Jeffrey B. Banke/Shutterstock; p. 779: Henrik Larsson/Shutterstock; p. 789: dcwcreations/Shutterstock.

Chapter 16

pp. 802–803: WaterFrame/Alamy Stock Photo; p. 817 (all): Tom Gilbert; p. 818: Turtle Rock Scientific/Science Source; p. 826: Alexey Kamenskly/Shutterstock; p. 831: Turtle Rock Scientific/Science Source; p. 844: GIPhotoStock/Science Source; p. 845 (a): SPL/Science Source; (d, f): Tom Gilbert; (i): SPL/Science Source.

Chapter 17

pp. 850–851: Gary Webber/Shutterstock; p. 853: Turtle Rock Scientific/Science Source; p. 854: Photo Researchers/Science Source; p. 858: B. & C. Alexander/Science Source; p. 881 (a): Reinhard H./picture alliance/Arco Images G/Newscom; (b): Scimat/Science Source; p; 887: © Kenneth Garrett; p. 892 (b): Reika/Shutterstock; (d): StockFood Ltd./Alamy Stock Photo; (f): novzzvon/iStockphoto/Getty Images; (h): Fullempty/Dreamstime.

Chapter 18

pp. 898–899: Jim Motavalli; p. 898 (inset): Jeff Greenberg/UIG via Getty Images; p. 901: GIPhotoStock/Science Source; p. 903 (top) Mary Evans Picture Library/Alamy; (bottom) Pictorial Press Ltd/Alamy Stock Photo; p. 921: Paul Mogford/Alamy Stock Photo; p. 922: MAILLAC/REA/Redux; p. 926: Istockphoto; p. 928: Richard Goldberg/Dreamstime; p. 930: John Gress/REUTERS/Newscom; p. 932: Stephen Barnes/Energy/Alamy Stock Photo; p. 934: Uli Deck/picture-alliance/dpa/AP Images; p. 936: GIPhotoStock/Science Source; p. 938: (top): Gaspar R Avila/Alamy Stock Photo; (middle): Science Photos/Alamy; (bottom): David Huntley Creative/Shutterstock; p. 939 (a): Getty Images/Thinkstock; (c): Konstantin Kirillov/Alamy Stock Photo; (e): Charles D. Winters/Science Source; (g): Science Source/Science Source; (i): Science Source/Science Source; p. 943: age fotostock/Alamy Stock Photo; p. 945 (top): Glenda Powers/Dreamstime; (middle): Mike Egerton/ZUMA Press/Newscom; (bottom): Emory Kristof, National Geographic Image Collection.

Chapter 19

pp. 946–947: www.BibleLandPictures.com/Alamy Stock Photo; p. 956: Joseph White/Alamy Stock Photo; p. 961: Dennis Flaherty/Science Source; p. 964 (center): Igor Kostin/Sygma via Getty Images; (bottom): LLNL/Science Source; p. 965: Xinhua/eyevine/Redux; p. 968: Dr. Robert Friedland/Science Source; p. 970 (a); Astrid & Hanns-Frieder Michler/Science Source; (b): Tom Tracy Photography/Alamy Stock Photo; (c): MEIGNEUX/SIPA/Newscom; (bottom): Argonne National Laboratory, US DOE Office of Environmental Management; p. 973: US Dept. of Energy/Science Source; p. 975: Jeff J Daly/Alamy Stock Photo; p. 976: The American Weekly, 1926; p. 977: Dennis Flaherty/Science Source; p. 982 (top): © musée du quai Branly - Jacques Chirac, Dist. RMN-Grand Palais/Art Resource, NY; (center): blickwinkel/Alamy Stock Photo; (bottom): Pavlov et al. 2001. Human presence in the European Arctic nearly 40,000 years ago. Nature Sept 2001. Copyright © 2001 Springer Nature.

Chapter 20

pp. 986–987: saied shahin kiya/Shutterstock; p. 1003: Valueline/Punchstock; p. 1010 (a): Dennis Kunkel Microscopy/Science Source; (b): Omikron/Science Source; p. 1012: Kokodrill/Dreamstime; p. 1013 (top): N.I.S.T; (bottom): Chemical Design/Science Source; p. 1025: Biophoto Associates/Science Source; p. 1030 (top): NASA; (bottom): W.R. Normak, USGS.

Chapter 21

pp. 1044–1045; Michael Neelon (misc)/Alamy Stock Photo; p. 1048 (both): Photo Researchers/Science Source; p. 1051: Don W. Fawcett/Science Source; p. 1064 (left): Tom Brakefield/Stockbyte/Getty Images; (right): Dreamstime; p. 1068: Photo Researchers/Science Source; p. 1069 (left): Simon Fraser/Medical Physics, RVI, Newcastle upon Tyne/Science Source; (right): Southern Illinois University/Science Source; p. 1073 (all): Photo Researchers/Science Source; p. 1074: Simon Fraser/Medical Physics, RVI, Newcastle upon Tyne/Science Source; p. 1080: Laura McDermott/The New York Times/Redux.

Chapter 22

pp. 1082–1083: addkm/Shutterstock; p. 1085: Charles D. Winters/Science Source; p. 1086 (a-b): Courtesy of Paige Scannell; p. 1087: (both): Courtesy of Paige Scannell; p. 1088: Image copyright © The Metropolitan Museum of Art. Image source: Art Resource, NY; p. 1094 (a-b): Turtle Rock Scientific/Science Source; p. 1097: Turtle Rock Scientific/Science Source; p. 1099: GIPhotoStock/Science Source; p. 1101: Turtle Rock Scientific/Science Source; p. 1102: Vaughan Fleming/Science Source; p. 1105 (a-b): Courtesy of Paige Scannell; p. 1111: Phantatomix/Science Source; p. 1115: Zephry/Science Photo Library/Science Source; p. 1116: "Small molecular gadolinium (III) complexes as MRI contrast agents for diagnostic imaging," Chan Kannie Wai-Yan and Wong wing-Tak, Coordination Chemistry Reviews, Sept. 2007. Copyright © 2007 Elsevier B.V. All rights reserved; p. 1120: Mike Egerton/ZUMA Press/Newscom; p. 1124 (all): Courtesy of Paige Scannell; p. 1125 (e both): SPL/Science Source.

Index

Note: Material in figures or tables is indicated by *italic* page numbers. Footnotes are indicated by n after the page number.

A

absolute entropy, 859–61, 862–64, 886
absolute temperature
  average kinetic energy of gas molecules and, 244, 245, 247–48
  Kelvin scale of, 32, 33–34
  pressure of gas and, 226–28
  volume of gas and, 223–25
absolute zero (0 K), 33
  determined from Charles's law, 225
  zero entropy of crystal at, 859–60, 886
absorbance (A), 162–63
absorbed dose, 963
acceptor band, 492
accuracy, 27, 28–30
acetaldehyde, in photochemical smog, 651
acetamide, *632*
acetate, at hydrothermal vents, 1030
acetic acid
  condensation reaction with ammonia, *632*
  converted to methane by bacteria, 304
  dimer of, *514*
  endothermic reaction with sodium bicarbonate, 853, 854
  titration with sodium hydroxide, 818–19
  vinegar as solution of, 173–74, 589, 853
  as weak acid, 169, 759–60, 771
  as weak electrolyte, 164–65
acetone
  boiling point, 513
  dipole moment, 513
  hydrogen bonding with water, 516
acetyl-coenzyme A, 1020, 1025, 1112
acetylene (ethyne), 72
  hybrid orbitals in, 467
  Lewis structure of, 403
  mass spectrum of, 128
  torch using, 288
acid rain
  chemical weathering caused by, 167–68, 183
  on early Earth, 91
  nonmetal oxides producing acid in, 168, *169*
  sulfur trioxide and, 91, 103, *168*, 783, 1061–62
  sulfur-containing fuels and, 103, 783
acid–base equilibria
  autoionization of water and, 767–68, 781–82, 786
  in blood, 756–57, 792–93
  of complex ions, 830–34
  conjugate acid–base pairs and, 764–67, 789–90, 806–9
  ocean acidification and, 804–6, 841–42
  See also K_a; K_b; pH
acid–base reactions, 152, 165–71
  according to Lewis model, 829
  antacid tablets and, 153, 174–75, 1056–57
  precipitate formed in, 176, 179
  See also acid–base titrations; neutralization reactions
acid–base titrations, 171–75, 818–27
  of amino acids, 1004–6
  indicators for, 173, 816–17, 818
  midpoint of, 821, 823
  with multiple equivalence points, 824–27
  principal steps of, 818
  of strong acid with strong base, 171–72, 818–20
  of weak acid with strong base, 173–74, 818–19, 820–22
  of weak or strong base with strong acid, 174–75, 823
acidic salts, 788–90, 791–92
acids
  Arrhenius concept of, 756
  Brønsted–Lowry concept of, 165, 756–57, 764, 765, 789, 827–29, 832–34
  carboxylic, *73*, 169, 628–34
  formulas of, 69–70, *70*
  gastric, 153, 770, 771–72, 1056–57
  hydrated metal ions as, 830, 832–34, 838, 1085
  ionization reactions of, 758–60
  Lewis concept of, 827–30, 1085
  naming of, 69–70, *70*
  oxoacids, 69–70, *70*, 760–62, 783–87, 832
  polyprotic, 169, 783–87
  See also amino acids; strong acids; weak acids
acrolein, 445, 472
actinides, 60, 368, 373
activated complex, 676, 681
activation energy (E_a), 674–80
  calculation of, 676–78
  of catalyzed reaction, 736
  of enzyme-catalyzed reaction, 692
  for ozone decomposition, 688–89
active site
  of biomolecular structure, 446, 473
  of enzyme, 692, 1014–15
activity series, 191–93
actual yield, 116–19
addition polymers, 622, *634*
addition reactions, 622
adenine, 1025–29, 1111
adenosine diphosphate (ADP), 882–84, 1054, 1061
adenosine triphosphate (ATP), 882–84
  calcium ions and, 1054
  conversion between ADP and, 1061
  cytochromes and, 1111
  in formation of peptide bond, 1006
  gastric acid and, 1056
  in glucose metabolism, 1014
  magnesium ions and, 1054
  Na^+–K^+ pumps and, 1052
  phosphate ions and, 1061
adenosine-5′-phosphosulfate, 1062
adhesive forces, 527–28
adipic acid, *631*, 632
adsorption, by heterogeneous catalyst, 689
aerosol cans, 227–28
air, 214, 215
  composition of, *214*
  density of, 215
  See also atmosphere of Earth
air bags, automobile, 232, 234
airplanes
  oxygen generators on, 233–34
  oxygen partial pressure outside of, 241
alanine
  as chiral compound, 997
  titration curve for, 1004
  as zwitterion at physiological pH, 1003–4
alcohols, 72, *73*
  in ester formation, 628
  hydrogen bonding in, 519, 520–21, 528
  in polyester formation, 628–30
  polymers of, 626–28
  viscosity of, 528
  water solubility of, 529–30
  See also ethanol; methanol
aldehydes, *73*
  Lewis structures of, 402
  See also acetaldehyde; formaldehyde
alkali metals (group 1), 60, 1049, 1050
  bcc unit cells of, *606*, 607
  colors in Bunsen burner flames, *366*
  hydroxides of, 763, 778–79
alkaline batteries, 909–10

alkaline earth metals (group 2), 60, 1049, 1050
hydroxides of, 763, 778, 779
alkalinity titration, 824–27
alkanes, 71, 313–15
boiling points, *314*, 510–11
isomers of, 991–93
melting points, *314*
nomenclature of, 313
prefixes for naming, *313*
standard molar entropies of, 862
valence bond theory of, 467
alkenes, 71–72
isomers of, 993–95
polymers of, 622–25
valence bond theory of, 467
alkyl hydroperoxides, 1066
alkynes, 72
valence bond theory of, 467
allotropes, 410
of carbon, 620–21
alloys, 604, 612–16
definition of, 612
electrochemical properties, 935–36
interstitial, 614–16
substitutional, 613–14, 615–16
alpha (α) decay, 954
in uranium-238 decay series, 954, 955
alpha (α) particles
biological effects of, 963, *964*, 965
in BNCT for brain tumors, 1072
from breathing radon, 965
early experiments with, 51–53, 959
in Geiger counter, 956
in nuclear fusion, 972–75
from radium, 975
symbols and mass of, *949*
α helix, *1009*, 1011, 1012, 1013
altitude
atmospheric pressure and, 217, 535, 536
partial pressure of oxygen and, 241, 242, 535, 536
alumina, 319–20
aluminum
in $AlGaAs_2$ semiconductors, 492–93
alloys with magnesium, 927
electron configuration of, 366
in human body, 1072
production from ore, 319–20
recycling of, 319–20
in sacrificial anode, 927, *928*
surface layer of oxide on, 927
toxicity of, 1067, 1072
aluminum carbonate, 1072
aluminum chloride
bonding in, 424, 426
as Lewis acid, 829–30
aluminum hydride, 424–25
aluminum hydroxide, 756, 833, 838, 1072
aluminum ions in soils, 756
aluminum nitrate, 838
Alzheimer's disease, 968, 1012, 1072, 1116
amides, *73*, 632
peptide bond and, 1006
polyamides, 631–34
amine (N-) terminus, 1006
amines, *73*
amides from carboxylic acids and, 632
polyamides from acids and, 631–34
as weak bases, 170
amino acid residues, 1006
amino acids, 1001–3
acid–base properties of, 1003–6
α-carbon of, 997, 1001
biosynthesis of, 1057–58
chiral center of, 997
D- and L- enantiomers of, 1003, 1030
essential, 1003
genetic code for, 1027–28, *1028*
as ligands in medical applications, 1113
in meteorites, 88, 1030
origin of life and, 88, 100, 1029
in peptides, 1006–9
primary structure of proteins and, *1009*, 1010–11
in protein synthesis, 1027–29
selenocysteine, 1066, 1067
in space, 1029–30
structures and abbreviations of, *1002*
sulfur-containing, 1012, 1013, 1062, 1112
titration curves of, 1004–6
transition metal cations bonding to, 1109
aminocarboxylic acids, 1096–97
ammonia
in amino acid biosynthesis, 1057–58
from animal waste, 588
condensation reaction with carboxylic acid, *632*
conjugate acid of, 765
in fertilizer production, 710, 1058
hydrogen bonds in, 513–14, *514*, 531
K_b value of, 776
as Lewis base, 827–29, 1094
Lewis structure of, 401
as ligand in complex ions, 830–32, 1086, *1087*, 1094, 1099–1100, 1104–6, 1107
in nitric acid synthesis, 743–44
in nitrogen cycle, 1058, 1111
pH of solution of, 780–81
as polar molecule, 459
solubility in water, 531
sp^3 hybrid orbitals in, 464
titration with strong acid, 823
trigonal pyramidal structure of, 453
from urea in plants, 1060, 1112
as weak base, 169–70, 763–64, 776, 780–81, 808–9
as weak electrolyte, 170
ammonia synthesis, 710
catalyst in, 736
equilibrium constant for, 879–80
free energy change in, 866–67
relative rates in, 654–55
shifting the equilibrium of, 731, *733*, 734–35
temperature change and, 734–35
See also Haber–Bosch process
ammonium acetate, as neutral salt, 789
ammonium chloride, weakly acidic solutions of, 788, 789, 791–92
ammonium cyanate, 89, 991
ammonium ions
in amino acid biosynthesis, 1057
in fertilizers, 1058
hydrolysis of, 1060
in nitrogen cycle, *1058*, 1060
ammonium nitrate, as fertilizer, 710
ammonium nitrate, dissolution in water
in cold pack for injuries, 288, 854
as endothermic process, 854, 864
enthalpy change in, 303–4
entropy change in, 863–64
Amontons, Guillaume, 227
Amontons's law, 226–28, 244, 245
amorphous solids, 602, *603*
ampere (A), 920–21
amphiprotic substance, 170, 767
amplitude of wave, 339
amyloid β, 1012
anabolism, 1013
analyte, 818
angioplasty, 612–13
angles. *See* bond angles
angular molecular geometry. *See* bent (angular) molecular geometry
angular momentum quantum number (l), 355–56, *357*
aniline, 764
anions
definition of, 9
enthalpy of hydration of, *560*
ionic radii of, 375–76
oxoanions, 68–69, 760–61, *762*, 785–87
of p block elements, 370–71
separating in solution, 840–41
anode
of cathode-ray tube, 49
of cell diagram, 904
of electrochemical cell, 902
sacrificial, 927, *928*
of voltaic cell vs. electrolytic cell, 930
antacid tablets, 153, 174–75
aluminum in, 1072
bismuth subsalicylate in, *1073*
calcium carbonate in, 94, 1056–57
antibacterial properties
of copper, 612
of gold–platinum nanoparticles, 615
of silver compounds, 1119
antibonding orbitals, 477–78
anti-fluorite structure, 618
antifreeze solutions, 521, 563–64, 569, 574–75
antimatter, 952

antimony
  in human body, 1068
  in leishmaniasis treatment, 1047, 1072
  stimulatory effect of, 1047, 1068
  toxicity of, 1068, 1072
antioxidant, selenocysteine as, 1066
aquamarine, 1102
aquatic life
  calcium carbonate in marine organisms, 805, 842, 1048, 1054
  marine mammals, 148–49
  phosphate fertilizers and, 1061
  solubility of oxygen and, 534
  species metabolizing bromide, 1068
  water and, 528–29, 533
aqueous solutions, 151
  acid–base reactions in, 152, 153, 165–71, 174–75, 176, 179
  acid–base titrations in, 171–75, 816–27
  activity series of metals in, 191–93
  boiling point elevation of, 569–70, 572–73, 575–77
  dipole–dipole interactions in, 512–13
  on Earth's surface, 152
  electrolytes and nonelectrolytes, 163–65, 575–78
  equilibrium constant expressions for reactions in, 728
  freezing point depression of, 569–70, 573–78
  ion–dipole interactions in, 511–12, 529, 554, 560
  limiting reactants in, 183
  in living things, 152
  phase diagram for, *569*, 569–70
  phase symbol for, 10
  precipitation reactions in, 152, 176–82, 839–41
  of salts, acid–base properties of, 788–92
  solubilities of biopolymers in, 550
  See also concentration; hydrated ions; solutions
aragonite, 842
argon
  banned for athletes, 1071
  electron configuration of, 366
arithmetic mean, 28
aromatic compounds, 414, 473
  polymers of, 625–26, 633–34
  as weak bases, 764
Arrhenius, Svante, 675, 756
Arrhenius acids, 756
Arrhenius bases, 756
Arrhenius equation, 675–79
arrow
  bond polarity and, 405
  in chemical equation, 8
  of equilibrium, 165
arthritis, transition metal compounds for, 602, 1118, 1119
asparagus, and urine odor, 1063–64
aspartame, 1007
aspartic acid, 1001, *1002*, 1003, 1004
  titration curve of, 1005–6
asphalt, *314*
astatine, *102*, 1049
asthma, inhalers for, 120
Aston, Francis W., 54, 55, 56
atmosphere (1 atm), 216–17
  different pressure units and, *217*
atmosphere of Earth
  auroras in, 477, 486
  composition of, *214*
  earliest, 88, 90–91, 107–8, 150
  as gaseous solution, 151
  origin of life and, 88
  thickness of, 215
  See also carbon dioxide; greenhouse gases; oxygen
atmospheric pollutants. *See* acid rain; nitrogen dioxide; nitrogen monoxide; smog; sulfur dioxide; sulfur trioxide
atmospheric pressure (P_{atm}), 215–18
  altitude and, 217, 535, 536
  calculation of, 218
  definition of, 215
  normal boiling point and, 520
  oxygen in blood and, 536
  as sum of partial pressures, 238
  on Titan, 218
atom, definition of, 7
atomic absorption spectra, 337–38
atomic emission spectra, 337, 338
atomic mass, average, 56–58
atomic mass units, unified (u), 53
atomic number (Z), 54–56
  changed in radioactive decay, *954*
atomic orbitals. *See* orbitals, atomic
atomic radii
  alloy formation and, 613, 614, *615*, 615–16
  periodic trends in, 373–76, 1049
  unit cells and, 608–11
atomic solids, 602, *603*
atomic theory
  Dalton's theory, 53–54, 61
  Rutherford's model, 52–53, 347
  Thomson's plum-pudding model, 51–52
ATP. *See* adenosine triphosphate (ATP)
ATP sulfurylase, 1062
aufbau principle, 362, 366–67
  molecular orbitals and, 481
auranofin, 1119
auroras, 477, 486
austenite, 614–15
autoionization of water, 767–68
  in rainwater, 786
  in very dilute strong acid, 781–82
automobiles
  air bags of, 232, 234
  air filters of, 10
  all-electric, 123–24, 900, 921, 933
  catalytic converters of, 652, 653, 690–91
  ethylene glycol in cooling system of, 521, 563–64, 574–75
  fuel cells for, 900, 931–33
  hybrid, 123–24, 900, 921–23
  hydrogen-powered, 279, 931–33
  lead–acid batteries of, 917–18, 928, *929*
  nitrogen monoxide formed in, 418, 650, 651, 653–54, 656, 690–91, 722, 1058
  useful work done by engines of, 869
average
  calculation of, 27–28
  weighted, 56–57
average atomic mass, 56–58
average speed of molecules, 246
Avogadro, Amedeo, 91, 226, 243–44
Avogadro constant (N_A), 91–93
  molar mass and, 94, 96–97
Avogadro's law, 225–26, 229
  kinetic molecular theory and, 244–45
  reaction of gases and, 234
axial atoms and bonds, 448, 454, 456, 457
azurin, 1112

B

Baby Tooth Survey, 48–49
background electrolyte
  in lithium-ion battery, 923–24
  in zinc/copper voltaic cell, 903
Bacon, Francis, 17
bad breath, 1063
baking soda. *See* sodium bicarbonate
balanced chemical equations, 91, 100–105
  for combustion reactions, 105–7, 109–10
  for electrochemical cell, 906–7
  equilibrium constant and, 713, 721
  for formation reactions, 308–9
  multiplied or divided by a number, 721–22
  for oxidation–reduction reactions, 188–91, 194–97, 906–7, 1059–60
  reaction mechanism and, 681, 682
  relative reaction rates and, 654–56
  in RICE table, 737
ball-and-stick molecular models, 9
balloon angioplasty, 612–13
balloons
  Avogadro's law and, 225–26
  Boyle's law and, *221*, 222–23
  Charles's law and, 223–25
  density of gas and, 236
  effusion through holes in, 249–50
  helium-filled, 222–23, 225, *236*, 249, 286
  hot-air, 224–25, 236, 284–85
  mimicking electron-pair geometry, 449
  weather balloons, 220, 228–29, 230–31
Balmer, Johann, 345–46, 347, 348, 357
band gap (E_g), 491–92
band theory, 490–93
barium
  electron configuration of, 367
  in human body, 1067

barium sulfate
in exoskeletons, 552, 1067
ion–ion interactions in, 552–53
for medical procedures, 179
solubility of, 835–36
barograph, *219*
barometer, 215–17, 219
bases
Arrhenius concept of, 756
Brønsted–Lowry concept of, 165, 169, 764, 765, 776, 827–29
Lewis concept of, 827–30, 1051, 1085, 1093–94, 1111
monobasic, 818
See also acid–base equilibria; acid–base reactions; acid–base titrations; strong bases; weak bases
basic salts, 788–89, 790–91
batteries
alkaline, 909–10
button batteries, 912–13
capacity of, 920–25
decrease in potential with usage, 917
invented by Volta, 903
lead–acid, 917–18, 928, *929*
lithium-ion, 123–24, 923–24, 928, 974
need for better vehicle batteries, 900
nickel–metal hydride, 921–23, 928, 929–30
rechargeable, 903, 921–23, 928–30
redox flow batteries, 934
work done by, 920–21
zinc–air batteries, 910–11
zinc/copper voltaic cell, 901–4, 908–9, 911–12, 918
BCNU (1,3-bis(2-chloroethyl)-1-nitrosourea), 537–38
becquerel (Bq), 956
Becquerel, Henri, 51, 956
Beer's law, 162–63
bell curve, 30
belt of stability, 951–52, 954
bent (angular) molecular geometry
hybrid orbitals and, *468*
with one lone pair, 452, *455*
of ozone, 460
of sulfur dioxide, 452
with two lone pairs, 453, *455*
of water, 454, 459
benzene
boiling-point-elevation constant, *572*
freezing-point-depression constant, *572*
mass spectrum of, 128
pi (π) bonds in, 472, 477
resonance in, 414, 472
solubility properties of, 534
benzoic acid
calorimeter constant determined with, 301, 302
slight water solubility, 810
beryllium, electron configuration of, 364
beryllium chloride, 424
beryllium toxicity, 1048
Berzelius, Jöns Jacob, 89
beta (β) decay, 951–52
atomic number and mass number in, *954*
of carbon isotopes, *953*
of free neutrons, 959
of massive nuclides, 954, 955
of molybdenum-99, 77
of phosphorus-32, 953–54
in stellar nucleosynthesis, 76
beta (β) particles, 951–52
biological effects of, 963, *964*
early experiments with, 51
in Geiger counter, 956
in medical imaging, 968
in radiation therapy, *967*
symbols and mass of, *949*
β-oxidation of fatty acids, 1024
β-pleated sheet, 1011–12
bicarbonate
in alkalinity titration, 824–27
in blood, 152, 756–57, 792–93, 1056
carbonic anhydrase and, 792–93, 1109–10
in seawater, 805, 842
See also sodium bicarbonate
Big Bang, 4, 17–18, 32, 34–36
primordial nucleosynthesis and, 74–75
bimolecular elementary step, 681
binary acids, 69, *70*
binary ionic compounds
crystal structures, 616–20
formulas, 62–64, 66–67
Lewis structures, 404
naming, 66–67
See also ionic compounds
binary molecular compounds
names and formulas, 65–66
prefixes for naming, *65*
binding energy (BE), 949–50
biocatalysis, 692, 1014
biochemistry, 988
See also carbohydrates; human body; life; lipids; nucleic acids; proteins
biofuel, 1019
biomass, 1019
biomolecules, 988
biopolymers, 989, 1006
bismuth compounds, for diarrhea, 1046–47, 1072
bismuth subsalicylate, 1072
blackbody radiation, 341
blood
acid–base equilibria in, 756–57, 792–93
drug transport in, 531, 537
as heterogeneous mixture, 6–7
osmosis and, *579*, 581, 582–83
oxygen in, 536
pH of, 792–93, 806
separation of, 9–10
solutes in, 152, 153, 154, *155*, 550
zeolites absorbing water from, 588
See also red blood cells
blood type, 1016
blue cheese, 471
bluing, 927
body-centered cubic (bcc) unit cell, *605*, 607, *608*, 609–10
of ferrite, 615
molten iron crystallizing in, 614
of tantalum, 610–11
Bohr, Niels, 347, 354, *358*
Bohr model, 347–50
de Broglie wavelength and, 351–52
uncertainty principle and, 354
boiling
entropy change in, *861*
spontaneous, 853, 854
boiling point
of binary hydrides, 513–14
of ethers, 530–31
of halogens, 510
hydrogen bonding and, 513–14, 519, 520–21
intermolecular forces and, 510–11, 513–14, 518–19, 520–21
molecular shape and, 510–11
of noble gases, 509, 510
normal, 520
periodic trends in, 1049, *1050*
on phase diagram, 524
of polar substances, 513
of straight-chain hydrocarbons, *314*, 510
temperature scales and, 33
boiling point elevation, 569–70, 572–73
of automobile coolant, 563–64
constants for selected solvents, *572*
van't Hoff factor and, 575–77
Boltwood, Bertram, 959
Boltzmann, Ludwig, 885, 886
Boltzmann constant, 886
Boltzmann's definition of entropy, 886
bomb calorimeter, 300–302
food value and, 317–19
bond. *See* chemical bonds
bond angles
in carbon dioxide, 447
definition of, 447
double bond and, 451, 452
hybrid orbitals and, *468*
lone pairs and, 452–53, *455*
in methane, 447, 453
VSEPR theory and, 448–57
bond dipole, 405, 457–61
See also polar covalent bonds
bond energies, 427–30
in hydrogen molecule, 396
of selected covalent bonds, *427*, 427–28
bond length, 396, 426, *427*
bond order, 426
bond energy and, 429
fractional, 486–87
in molecular orbital theory, 479–80, 482, 483–84, 485–87

bond polarity, 405–9
  See also polar covalent bonds
bond strength. *See* bond energies
bonding capacity, 398–99
  formal charge and, 418
  resonance and, 411, 418
bonding orbitals, 477–78
bonding pairs
  equal and unequal sharing of, 405
  in Lewis structures, 399
  in VSEPR theory, 447
bones, mammalian
  calcium phosphate in, 552–53
  as composite material, 1054
  strontium and barium in, 1067
  See also hydroxyapatite
Born, Max, 355
Born–Haber cycle
  enthalpy of hydration and, 559–61
  lattice energy and, 556–58
boron
  electron configuration of, 364
  in human body, 1072
  paramagnetic diatomic molecule of, 483
boron neutron-capture therapy (BNCT), 1072
boron trifluoride
  electronegativity difference in, 423, 426
  as Lewis acid, 829
  resonance structures, 423–24, 425–26
  trigonal planar geometry, 449
Bosch, Carl, 736
boundary–surface representation, of *s* orbital, 360
Boyle, Robert, 220–21
Boyle's law, 220–23, 229, 244
  Le Châtelier's principle and, 731
branched-chain hydrocarbons, 314, *315*
brass, 604, 612
  specific heat of, 298–99
breath-hold diver, 254, 255
breathing, and gas laws, *230*
breeder reactor, 970–71
bristlecone pines, 961
bromcresol green, 826
bromine
  in human body, 1068
  as liquid element, 1049
  vapor pressure of, 520
bromochlorofluoromethane, 475
Brønsted–Lowry acids, 165, 756–57
  of conjugate pair, 764, 765, 789
  hydrated metal ions as, 832–34
  Lewis acids and, 827–29
Brønsted–Lowry bases, 165, 756–57
  of conjugate pair, 764, 765, 789
  hydroxide ions as, 169
  Lewis bases and, 827–29
  weak, 776
bronze, 612, 613
buckyballs, 621
budotitane, 1117
buffer capacity, 812–16
buffers. *See* pH buffers
Bunsen, Robert Wilhelm, 337
bupropion, 1000–1001
buret, 172, 818
butadiene, *473*
butane, 313, 314, *315*
  standard molar entropy of, *862*
butanethiol, 1064
butanone, 470
button batteries, 912–13
butyric acid, 628

C

cadmium ions, in drinking water, 155–56
calcium
  electron configuration of, 366–67
  in living organisms, 1047, 1054
  strontium-90 substituting for, 46–47, 1067
calcium carbonate
  in antacid tablet, 94, 1056–57
  decomposition of, 219–20, 727
  dissolved by carbonic acid, 167–68
  dissolved by sulfuric acid in rain, 168, 183
  reaction with aqueous acetic acid, 589–90
  in skeletons of marine organisms, 805, 842, 1048, 1054
calcium fluoride
  crystal structure, 618
  lattice energy, 557–58
  pH effect on solubility of, 837–38
calcium hydroxide
  neutralization reaction with nitric acid, 167
  pH of solution of, 778, 779
  preparing solution of, 158–59
calcium ions
  in living organisms, 1047, 1054
  separating from magnesium ions, 840
  water softening and, 587–88
calcium oxide
  from decomposition of calcium carbonate, 219, 727
  Lewis structure of, 404
  reaction with sulfur dioxide, 829
calcium phosphate, 552–53
  See also hydroxyapatite
calcium sulfate
  acid rain promoting production of, 168
  lattice energy of, 560–61
  solubility in water, 550–51, 552
calibration, 27
calibration curve, in spectrophotometry, 162, 163
calorie (cal), 285
Calorie (Cal), 285, 318
calorimeter, 297
  bomb, 300–302, 317–19
  coffee-cup, 302–4
calorimeter constant, 301–2
calorimetry, 297–304
  coffee-cup, 302–4
  constant-volume, 301–2
  enthalpy of reaction measured with, 300–303
  food value and, 317–19
camping lanterns, 634–35
cancer
  cell membrane potentials in, 1053
  chromate ions contributing to, 1113
  cisplatin treatment for, 1104–5, 1118
  colonoscopy to screen for, 1073–74
  gallium and indium compounds in imaging of, 1070
  gallium compounds with activity against, 1072
  ionizing radiation contributing to, 963, 964, 966, 975–76, 1067
  lung cancer caused by radon, 966
  polycyclic aromatic hydrocarbons and, 473
  radiation therapy for, 967, 1072
  in Radium Girls, 975–76
  solubilities of drug candidates, 537–38
  transition metals in imaging of, 1113–16
  transition metals in treatment of, 1104–5, 1114, 1115, 1117–19
candle wax, *314*, 315
capillary action, 527–28
caraway, 446, 475, 476
carbohydrates, 1016–20
  See also glucose
carbon
  allotropes of, 620–21
  as charcoal, 233, 305
  chiral compounds of, 475–76, 494
  diatomic molecule of, 483
  electron configuration of, 364–65
  entropy of forms of, 862
  isotopes of, *953*
  pi (π) bonds formed by, 466, 469
  radioactive isotopes of, *953*
  with *sp* hybrid orbitals, 467–68, 469
  with sp^2 hybrid orbitals, 465–66
  with sp^3 hybrid orbitals, 463–64
  in steel, 614–15, *615*
  stellar nucleosynthesis of, 76
  variety in compounds of, 989
  See also organic compounds
carbon-11
  electron capture in, 953
  positron emission from, 952, 968, 1069
carbon-14
  beta (β) decay of, 951–52, 960
  radiocarbon dating with, 946–47, 960–61, *962*
carbon cycle, 108
carbon dioxide
  in amino acid biosynthesis, 1057
  from antacid tablets, 153, 175
  atmospheric increase in, 109, 235, 394, 804–6
  in blood, 152, 756–57, 792–93
  bond angle of, 447
  bond energy in, 428

carbon dioxide *(cont.)*
bond length of, 426
bonding in, 417
in carbonated beverages, 535
from combustion of charcoal, 305
from combustion of fossil fuels, 108–11, 394, 804–5
from combustion of hydrocarbons, 105–7
from decomposition of calcium carbonate, 219–20, 727
density of, 235, 236
dissolved in rainwater, 167–68, 785–86
dissolved in river water, 807–8
as dry ice, 525
of early atmosphere, 90
elimination from human body, 1056
enzyme-catalyzed hydrolysis of, 692, 1109
for fighting fires, 235
formal charge calculations for, 417
in glucose metabolism, 1020
as greenhouse gas, 394, 409
at hydrothermal vents, 1030
infrared absorption by, 408–9
linear geometry of, 447, 449, 469
natural sources of, 105
as nonpolar molecule, 458
ocean acidification and, 804–6, 842
phase diagram of, 524, 525
polar bonds in, 458
polymer impenetrable to, 626
from production of hydrogen fuel, 933
radiocarbon dating and, 960
from reaction of baking soda and vinegar, *853*, 854
stability at 25°C, 717
supercritical, 525
valence bond representation of, 469
from volcanoes, 90, 235
See also water–gas shift reaction
carbon monoxide
atmospheric, 105
bond length in, 426
catalytic converters and, 690–91
from combustion of charcoal, 305
from combustion of hydrocarbons, 307, 690–91
of early atmosphere, 90
ethanol additive to gasoline and, 1019
reaction with oxygen, 105
See also water–gas shift reaction
carbon monoxide dehydrogenase, 1112
carbon nanotubes, 621
carbon tetrachloride
boiling-point-elevation constant, *572*
freezing-point-depression constant, *572*
limited water solubility, 531
tetrahedral geometry, 450
carbon tetrafluoride, 458
carbonate
alkalinity titration and, 824–27
ocean acidification and, 805–6
carbonic acid
in alkalinity titration, 824–27
in blood, 756–57, 792–93
dissolving calcium carbonate, 167–68
ocean acidification and, 804–5, 841–42
pH of solution of, 785–86
in rainwater, 167–68, 785–86
as weak acid, 169
carbonic anhydrase, 692, 792–93
structure of, 1012, *1013*
zinc ion in, 1109–10
carbonyl group, 402
polarization of, 407–8
sp^2 hybrid orbitals for, 465–66
carboplatin, *1118*
carboxylic acid (C-) terminus, 1007
carboxylic acids, *73*, 169
esterification of, 628
polyamides from amines and, 631–34
polyesters from alcohols and, 628–30
carotene, 1054, *1055*
cars. *See* automobiles
carvone, *475*
catabolism, 1013
catalases, *652*, 1110, 1112
catalysts, 653, 687–93
bacterial, 1030
definition of, 687
equilibrium and, 736
in fuel cells, 931–32
in Haber–Bosch process, 736
heterogeneous, 689, 692
homogeneous, 689
in hydrogen gas production, 710, 711
in Ostwald process, 743
ozone layer and, 687–90
rate constant and, 662
RNA acting as, 1030
See also enzymes
catalytic converters, 652, 653, 690–91
cathode
of cathode-ray tube, 49
of cell diagram, 904
of electrochemical cell, 902
of voltaic cell vs. electrolytic cell, 930
cathode rays, 49–50, 54
cathode-ray tubes (CRTs), 49–50, 51n, 54
cathodic protection, 927
cations
definition of, 9
ionic radii, 375–76
metals that form more than one, 67–68
of *s* block elements, 370, 371
separating in solution, 840
of transition metals, 372–73
See also complex ions; hydrated ions
cell, electrochemical. *See* electrochemical cells
cell cytoplasm, pH of, 811
cell diagrams, 903–4, 906–7
with standard hydrogen electrode, 914
cell membrane
attacked by alkyl hydroperoxides, 1066
electrochemical potentials of, 1052–53
hydrophobic substances passing through, 531, 537
ion channels in, 1051, 1053, 1056
origin of life and, 1030
osmosis and, 579, *579*, 581, 582–83
structure of, 1020, *1024*, 1024–25, *1051*
cell potential (E_{cell})
definition of, 916
effect of concentration on, 916–18
See also standard cell potential (E°_{cell})
cellobiose, 1019
cellulases, 1019
cellulose, 1016, 1018, 1019
Celsius temperature scale, 32–34
cembrene A, 131
central dogma of molecular biology, 1028
centrifugation, 10
cesium, in human body, 1067
cesium chloride, crystal structure of, 617–18
ceviche, 1006
Chadwick, James, 53
chain reaction, nuclear, 969, 970
chalcocite, 887
changes of state
at constant temperature and pressure, 290, 291
energy transfers in, *281*, 282
enthalpy changes in, 287
entropy changes in, 858, 861, 861n
on heating curve, 289–90
intermolecular forces and, 292, 524, 525
spontaneous, 853, 854, 869–70
See also phase diagrams
charcoal, combustion of, 233, 305
Charles, Jacques, 223
Charles's law, 223–25, 229, 244, 245
chelate effect, 1096
chelating agents, 1095
chelating ligands, 1113
chelation, 1095–97
medical applications of, *1113*, 1116–17
chemical bonds
coordinate bonds, 830, 1085–88
definition of, 7
energies of, 396, *427*, 427–30
energy of reactions and, 279
limitations of models of, 425–26, 493
in molecular models, 9
stability of compounds and, 395
in structural formulas, 9
types of, 395–97
valence bond theory, 461–69, 474
See also bond angles; bond order; covalent bonds; ionic bonds; Lewis structures
chemical equations, 8
phase symbols in, 8, 10
reactants and products in, 90–91
See also balanced chemical equations

chemical equilibrium
basic concept of, 710–12
catalysts and, 736
in complex ion formation, 830–32
dynamic nature of, 712
far to right or left, 717, 741–43
free energy and, 865, 872–77, 916
heterogeneous, 727–28
homogeneous, 726–27
Le Châtelier's principle and, 728–36
summary of response to stresses, *736*
in voltaic cell, 917, 919
See also equilibrium constant (*K*)
chemical formulas
of common acids, 69–70, *70*
of complex ions, 1086, 1088–93
definition of, 8
of ionic compounds, 62–64, 66–68
of molecular compounds, 8, 62, 65–66
of polyatomic ions, *68*, 68–69
of transition metal compounds, 67–68, 1086, 1088–93
chemical kinetics
definition of, 652
half-life in, 668–69, 672, 673, *673*
spontaneity and, 853
See also reaction mechanisms; reaction order; reaction rates
chemical nomenclature, 64
See also names of complex ions; names of compounds; names of polyatomic ions
chemical properties, 13
chemical reactions
bond energies and, 429–30
definition of, 6
energy changes in, 272, 273, 279
energy profile for, 676, 681–82, 692–93, 736
entropy changes in, 863–64, 866–67
equations for, 8, 90–91
gases in, 232–35, 242–43
in living systems, 880–84
spontaneity and, 870–72
See also acid–base reactions; addition reactions; combustion reactions; condensation reactions; formation reactions; oxidation–reduction reactions; precipitation reactions; stoichiometry
chemical weathering, 167–68, 183
chemistry, 5
Chernobyl nuclear disaster, 964–65, 1067
Chevrolet Bolt, 898–99, 923
chiral centers, 997, 998–99
chirality, 995–1001
of amino acids, 1003, 1030
aroma of grapefruit and, 1064
of carbon in japonilure, 494
in living systems, 1000–1001
molecular recognition and, 475–76, 997, 1000
of monosaccharides, 1016, 1017
in octahedral complex ions, 1095, 1106
recognizing, 998–99
See also enantiomers
chloric acid, *762*
chloride ions
determining concentration of, 180, 181–82
in human body, 1055–57
size of, 376
chlorine
as catalyst for ozone destruction, 688–89
oxoacids of, 761, *762*
chlorine monoxide
decomposition of, 671–72, 677–78
photodecomposition of ozone and, 688
chlorins, 1108
chlorofluorocarbons (CFCs), 687–88
chloroform, Lewis structure of, 400
chlorophyll *b*, 342–43, *1055*
chlorophylls, 1054, 1108
chlorous acid, *762*
cholesterol, 153, 1025
chrome yellow, 180
chromium
complex ions of, 833, 1097–99, 1102–3
diabetes and, 1113, 1118
electron configuration of, 367
as essential dietary element, 1112–13
in stainless steel, 614, 927
toxic and carcinogenic oxidation state of, 1113
chromium hydroxide, 833
chromodulin, 1112–13
cilantro, 444, 473
cisplatin, 1104–5, 1118
cis–trans isomers, 994
in coordination compounds, 1104–6, 1107, 1120
in lipids, 1020, 1023
citrate cycle, 1020, 1112
citrate ion, medical applications of, *1113*
citric acid, 174, 786, *787*, 1097
Clausius, Rudolf, 884–85
Clausius–Clapeyron equation, 521–23
climate change
fuel-cell–powered cars and, 933
See also global warming
close packing, 605–7, *608*
holes in, 614
See also cubic closest-packed (ccp) structure; hexagonal closest-packed (hcp) structure
closed system, 280
clusters, 621
coal, sulfur impurities in, 650, 727, 783, 829
coal mines, acid drainage after abandonment, 171–72
COAST problem-solving framework, 11–12
cobalt(II) chloride, 735, 1085–86, 1087
cobalt(II) complex ions, 1085, 1087
cobalt(III)
in coenzyme B_{12}, 1112
in complex ions, 1086, *1087*, 1105–7
cobalt(IV) oxide, in lithium-ion battery, 923
codons, 1028–29
coenzyme A, 1020
coenzyme B_{12}, 1112
coenzymes, 1112
coffee-cup calorimetry, 302–4
cohesive forces, 527–28
coinage metals, 1118–19
cold pack, chemical, 288, 854
colligative properties, 569–86
boiling point elevation, 563–64, 569–70, *572*, 572–73, 575–77
definition of, 569
freezing point depression, 569–70, *572*, 573–78
molality units used for, 570–71
molar mass of solute and, 585–86
osmotic pressure, 579–83, 585–86, 589–90
reverse osmotic pressure, 583–85
van't Hoff factor and, 575–78, 580–82, 583
vapor pressure as, 562–64, 569
colonoscopy, 1073–74
color wheel, *1099*
colors
of auroras, 477, 486
complementary, 1099
of plant pigments, 756
spectrophotometry and, 162–63
of transition metal solutions, 1097–1102, 1120, 1121
combination reaction, 90
combined gas law, 230–31
combustion analysis, 128–32
combustion reactions, 105–7
of alkane fuels, 314, *315*
carbon monoxide generated in, 305, 307, 690–91
of charcoal, 233, 305
definition of, 105
of ethanol, 868
free energy change in, 868–69
of gasoline, 109–10, 869
of hydrogen, 279, 717, 933
as oxidation–reduction reactions, 184
as spontaneous reactions, 853
See also enthalpy of combustion (ΔH_{comb}); hydrocarbon fuels; methane combustion
comet, amino acid found in, 1029–30
common-ion effect, 806–9
pH buffers and, 809
solubility and, 835–36
complementary colors, 1099
complete proteins, 1003
complex ions
basic concepts of, 1085–88
crystal field theory and, 1097–1102
formation of, 830–32
formulas of, 1086, 1088–93
hydrated, 830, 832–34, 838, 1085, 1093–94, 1098–99, 1104
membrane transport by means of, 1051
names of, 1088–92

complex ions *(cont.)*
  with negative charge, 1090–92
  with positive charge, 1088, 1090
  See also coordination compounds; ligands
complexes, 1085
  See also complex ions
composite material, 1054
compounds
  definition of, 6
  See also chemical formulas; coordination compounds; ionic compounds; molecular compounds; names of compounds; organic compounds
concentrated solution, 153
concentration, 153–59
  buffer capacity and, 813–15
  cell potential and, 916–18
  definition of, 153
  determining by titration, 171–75, 818–27
  determining with precipitation reaction, 180, 181–82
  determining with spectrophotometry, 162–63
  dilution and, 159–62
  in equilibrium constant expressions, 713–16, 727–28
  Le Châtelier's principle and, 729–31
  of major constituents of seawater and serum, 154–55
  mole fraction as expression of, 239
  preparing solution with desired molarity, 158–59
  of pure liquid, 727–28
  of pure solid, 727
  reaction rates and, 652, 653–54, 659–74
  units of, 153–55, 570–71
condensation, 15, *281*
  enthalpy of, 287
  in equilibrium with vaporization, 519–20
  as exothermic process, 281
  free energy and, 870
  phase diagram and, 524
condensation polymers, 628–34, *634*
condensation reactions, 628, 632, 1017
condensed electron configuration, 364
condensed structures, 470, 471, 472
conduction band, 491, 492
conductivity. *See* electrical conductivity; thermal conductivity
confidence intervals, 29–32
conformers, *998*
conjugate acid–base pair, 764–67
  pH of solution containing, 806–9
  product of K_a and K_b for, 789–90, 808
  See also pH buffers
conservation of energy, 275, 283
conservation of mass, 91, 99
constant composition, law of, 8
constant pressure
  enthalpy change at, 286–87, 864
  of phase changes, 290
constant temperature, of phase changes, 290, 291
constant-volume calorimetry, 301–2
constitutional isomers, 991–94
  classification of, *998*
  linkage isomers, 1106–7
  mass spectra of, 1001
  of triglycerides, 1021–22, 1031
control samples, 27–29
conversion factors, 20–22
coordinate bonds, 830, 1085–88
coordination compounds, 1086–88
  in biochemistry, 1108–13
  isomerism in, 1104–7
  in medicine, 1113–19
  names of, 1088, 1092–93
  oxidation states in, 1086, 1088, 1090, 1091
  See also complex ions
coordination number, 1086
  shapes corresponding to, *1087*
copolymer, 626
copper
  alloys of, *612*, 613
  antibacterial properties of, 612
  bonding in, 490–91
  as coinage metal, 1118
  corrosion of iron in contact with, 925–27, 935–36
  electron configuration of, 367
  malleability of, 612
  in photosynthesis, 1109
  proteins containing, 1112
  reaction with air, 612
  redox reaction between zinc and, 900–903
  refining of, 111–12, 887
  See also zinc/copper voltaic cell
copper ions (Cu^{2+})
  in complex ions, 1099–1100, 1103
  electron configuration of, 372
  spectrophotometric measurement of, 162, 163
  spin state of, 1103
copper(II) chloride complexes, 1120–21
copper(II) sulfate
  complex formation with ammonia, 830–32
  to kill algae in pool, 1120
  zinc immersed in solution of, 901–3
core electrons, 364
corrosion, 925–27
  protection from, 927
  as rust, 193, 850, 853, 925
  in seawater, 925, 926, 927, *928*, 935–36
cosmic microwave background radiation, 34–35
cotransport, of sodium and chloride ions, 1056
cotton, 630, *631*, 1016
coulomb (C), 51, 911
Coulomb, Charles Augustin de, 277
coulombic interaction. *See* electrostatic potential energy (E_{el})
Coulomb's law, 277
counterions, 1086, 1088, 1092
coupled reactions
  in copper refining, 887
  in living systems, 880–84
covalent bonds, 62, 395–97
  coordinate bonds as, 830, 1085–88
  electrostatic potential energy and, 395–96
  lengths of, 396, 426, *427*
  polar, 405–9, 426, 457–61, 758
  in solids, 602, *603*, 620–21
  strengths of, 396, *427*, 427–30
  See also Lewis structures; valence bond theory
covalent network solids, 602, *603*
  allotropes of carbon, 620–21
covalent radius, *373*
critical mass, 969
critical point, 525
crude oil, 314, 316
  distillation of, 564
  formation water accompanying, 573
  nonpolar solvents derived from, 532
  Raoult's law and, 567, 568
crystal field splitting energy (Δ), 1098–99
  factors affecting size of, 1103–4
  magnetic properties and, 1102–4
  spectrochemical series and, 1101, 1121
crystal field theory, 1097–1102
crystal lattice
  holes in, 614, *615*, 617–18, 635
  lattice energy, 554–61
  stacking patterns, 604–7, *608*
  See also unit cells
crystalline solids, 602, *603*
  zero entropy at 0 K, 859–60, 886
C-terminus, 1007
cubic closest-packed (ccp) structure, *605*, 607, 611
  of anions in anti-fluorite structure, 618
  of anions in sodium chloride, 617
  bronze and, 613
  of gold and platinum, 615–16
  holes in, *615*
  malleability of metals and, 612
  packing efficiency of, 607, 611
  summary of, *608*
  See also face-centered cubic (fcc) unit cell
cubic holes, *615*, 618
cubic packing. *See* simple cubic (sc) unit cell
cuprite, 111–12
cuproine, 163
curie (Ci), 956
Curie, Marie, 51, 948, 956, 962, 975, 977
Curie, Pierre, 51, 956, 975, 977
cycles per second, 339
cymene, 126–27
cysteine, 1012, 1013, 1062
cystic fibrosis, 1056
cytochromes, 1110–11, 1120, 1121
cytosine, 1025–29

d block elements, 367, 369
  See also transition metals

d orbitals, 355–56, 362
colors of transition metal complexes and, 1097, 1099, 1120–21
crystal field theory and, 1097–1104
filling of, 367–70
hybridization using, 488
magnetic properties and, 1102–4
D5W, 583
da Vinci, Leonardo, 184
Dacron, 630, *631*
dalton (Da), 53
Dalton, John, 53–54, 61, 62
Dalton's law of partial pressures, 238, 242–43, 244
dating methods, radiometric, 959–61
dative covalent bonds, 830
de Broglie, Louis, 350, 351, 352
de Broglie wavelength, 350–52
Dead Sea Scrolls, 946–47, 960
debye (D), 457
deferoxamine (DFO), 1116–17
degenerate orbitals, 364–65
degree of ionization
of weak acids, 772–75
of weak bases, 776–78
delocalized electrons
in anions of oxoacids, 761, *762*, 832
in aromatic compounds, 473
in benzene, 472
in compounds with double bonds, *473*
in graphite, 621
in metallic solid, 396
in molecular orbitals, 477, 487
delta (Δ)
as crystal field splitting energy, 1098
meaning change, 283
delta (δ), as partial electrical charge, 405
density (*d*)
calculating molarity from, 157–58
calculation of, 23–26
of common materials, *20*
definition of, 13
of gases, 235–38
of gases vs. liquids or solids, 215
of metals, 607, 610–11
unit cell dimensions and, 607, 610–11, 619–20, 634–35
units of, 19–20
dental fillings, 93
deoxyribose, 1025
deposition, 15, *281*
enthalpy of, 287
on phase diagram, 524
desalination of water, 10, 583–85
detergents, 532, 588
detritus, 108
deuterium, 973–74
deuterons, 74, *949*, 971–72, 973
dextrorotatory enantiomer, 997, 1003
dextrose, 583, 1016
See also glucose
diabetes, 1113, 1118
diagnostic radiology. *See* medical imaging
diamagnetic substances, 483
diamond, 620–21
standard molar entropy of, 862
diamond crystal lattice, *606*, 620, 621
diastereomers, *998*
diatomic molecules
elements existing as, 7–8, 101–2, 373, 1049
heteronuclear, 484–86
homonuclear, 480–84
magnetic properties of, 482–83
molecular orbitals of, 480–86
quantized energy levels of, 885–86
diazene, hybrid orbitals in, *466*
dichloromethane, 460, 461
Dicke, Robert, 34–35
Dieng Plateau disaster, 235, 236
diesel oil, 314–15, *315*, 316–17
dietary reference intake (DRI), 1048, 1065, 1108
diethyl ether, 520, 530–31
diethylenetriamine ligand, 1095
diethylenetriaminepentaacetate ($DTPA^{5-}$), *1113*
diffusion, 250
difunctional molecules, 628, 629–30, 631
digested food, pH of, 770, 771–72
dihydrogen phosphate/hydrogen phosphate buffer, 811–12
dilute solution, 153
dilutions, 159–62
dimensional analysis, 20–22
dimethyl ether, 518–19, 1063
dimethyl sulfide, 1063, 1064–65
dimethyl sulfoxide, 1064–65
dimethylamine, 170, 764
dinitrogen monoxide
as greenhouse gas, 668, 1058
half-life for decomposition of, 668
production of, 235, 415
structure of, 414–17, 418
dinitrogen pentoxide
decomposition of, 666–68, 669, 685–86
hydrolysis of, 168, 1060
dipeptides, 1006, 1007–8
1,3-diphosphoglycerate^{4-} (1,3-DPG^{4-}), 883–84
dipole moment (μ), 457, 458, *459*
See also polar molecules
dipole–dipole interactions, 512–13, *515*
boiling points and, 1065
ionization of strong acids and, 758
of polar solutes in polar solvents, 529–30
in solids, *603*
viscosity and, 528
between water and glass, 527
See also hydrogen bonds
dipole–induced dipole interactions, 513, *515*
solubility and, 529, 530, 531
between water and gases, 534
dipoles
bond dipole, 405, 457–61
induced, 509–11
interaction between induced and permanent, 513, *515*, 529, 530, 534
permanent, 457–58, 459–61
See also ion–dipole interactions
diprotic acids, 168–69, 783–86
disaccharides, 1017, 1019
dispersion forces (London forces), 509–11, *515*
boiling points and, 510–11, 1049, 1065
in graphite, 621
of long hydrocarbon chains, *533*, 534
molecular shape and, 510
in protein structures, 1012, 1013
in solids, *603*
in solutions of volatile liquids, 568
in uranium hexafluoride, 969
viscosity and, 528
disphenoidal geometry, 454
dissociation
compared to dissolving, 165
of strong electrolyte, 163
of weak electrolyte, 164
dissolution of ionic compounds
enthalpy change in polar solvent, 553–55, 559–60
entropy change in water, 863–64
ion–dipole interactions and, 511–12, 529, 554, 560
See also solubility
distillation, 10, 281–82, 564–67
disulfide linkage, 1012, 1013
DNA (deoxyribonucleic acid), 1025–28
antitumor gallium compounds and, 1072
antitumor transition metal compounds and, 1104–5, 1119
beryllium toxicity and, 1048
chromate toxicity and, 1113
hydrogen bonds in, 516, 1026, *1027*
polycyclic aromatic hydrocarbons and, 473
radiation damage to, 964
dogs, panting of, 522–23
dopant, 492
doping, 492
double bonds
bond angles and, 451, 452
bond energy of, 429
bond length of, 426
definition of, 401
hybridization and, 465, 467, *468*
Lewis structures with, 401–3
restricted rotation around, 994–95
double helix, 1026
doublets, 357
drain cleaner, 778–79
drawing
of larger molecules, 470–72
of Lewis structures, 399–401, 411
of three-dimensional structures, 450

drinking water
 contaminated by rock salt, 180
 fluoride added to, 1066
 produced by desalination, 583–85
 safe levels for contaminants in, 155–57
 treated with zeolites, 588
drugs
 anticancer, 537–38, 1072
 chiral, 997, 1000–1001, 1014
 fullerenes used for transport of, 621
 lithium-containing, 96–97, 1072
 partition coefficients of, 537
 PEGs attached to, 627
 produced by biocatalysis, 692
 producing nitric oxide, 418
 protease inhibitors, 1015
 proton pump inhibitor, 1057
 shelf-stability of, 197–98
 transition metals in, 1114–15, 1116–19
dry gas, 242
dry ice, 525
ductility of metals, 59
dynamic equilibrium
 of glucose structures in solution, 1017
 in weak acid solution, 169
 in weak electrolyte solution, 165
dystrophin, 989

E

E isomers, 994
 See also cis–trans isomers
Earth
 elemental composition of, 89, *90*
 See also atmosphere of Earth
Earth's crust
 age of oldest rocks, 960
 chemical weathering of, 167
 elemental composition of, 1047–48
 insoluble compounds in, 175
 ionic solids in, 616
 redox processes in rocks and soils, 193–94
Edison, Thomas, 853
effective dose of radiation
 biological effects and, 963, 964
 from radon gas, 966
 units of, 963
 from X-rays, 967
effective nuclear charge (Z_{eff}), 363, 364
 ionization energies and, 377–79
 periodic trends and, 1049
 size of atoms and, 375
effusion, 249–50
egg, osmotic pressure experiment with, 589–90
Einstein, Albert
 de Broglie theory and, 352
 Galileo and, 4
 mass-energy equivalence, 18, 350, 949, 969, 974–75
 photoelectric effect and, 343, 344
 quantum indeterminacy and, 359
electric field, dipoles aligned in, 457
electrical charge, 911–12
 See also electrostatic potential energy (E_{el}); partial electrical charge
electrical conductivity
 of diamond, 620
 of graphite, 621
 of metals, 491
 of semiconductors, 491–92
electrical power
 carbon dioxide from generation of, 110–11
 nuclear fission used for, 970–71
 nuclear fusion and, 971–75
electrical work, 911–12, 920–21
electrochemical cells, 901–3
 diagrams of, 903–4, 906–7, 914
 electrolytic, 903, 930
 free-energy change in, 911–13
 See also standard cell potential (E°_{cell}); voltaic cells
electrochemistry
 basic concepts of, 900–903
 See also corrosion; electrochemical cells
electrode potentials, 914–16
electrodes
 of electrochemical cell, 902, 904, 908–9
 in electrolyte, 163
 hydrogen electrode, 768, 914–16
 reference electrodes, 914–16
 of voltaic cell vs. electrolytic cell, 930
electrolysis, 903, 930
 of water, 905–6
electrolytes, 163–65
 background electrolyte, 903, 923–24
 colligative properties and, 575–78
 corrosion promoted by, 925, 926
 definition of, 163
 See also strong electrolytes; weak electrolytes
electrolytic cell, 903, 930
electromagnetic radiation, 339–40
 emitted and absorbed by elements, 337–38
 emitted by hot objects, 341
 emitted by semiconductors, 492–93
 Planck's quantum theory of, 341–43
electromagnetic spectrum, *338*, 339
electron affinity (EA), 379–80
 Born–Haber cycle for calculation of, 557
 periodic trends in, 1049
electron capture, 953, *954*
 in gallium and indium nuclide decay, 1070
electron configurations of atoms, 362–70
 condensed, 364
 definition of, 363
electron configurations of ions, 370–73
 condensed, 373–73
 of isoelectronic species, 371
 of main group elements, 370–72
 of transition metals, 372–73
electron density, 359–61
electron spin, 357–58
electron transitions, 348–49
 between molecular orbitals, 477, 486
 in sodium atom, 366
electron transport in cells, 1110–11
electron waves, 350–53
electron-deficient compounds, 398
electronegativity, 405–7
 formal charge and, 416
 heteronuclear diatomic molecules and, 485
 hypervalency and, 421
 Lewis structures and, 407, 411–12
 periodic trends in, 1049
electronegativity difference (ΔEN), 407–8
 in boron trifluoride, 423, 426
 limitations of bonding models and, 426
 between oxygen and sulfur, 1063
 in PDB (*para*-dichlorobenzene), 431
 in polar molecules, 459, 460
electron-pair geometry, 447, *448*
 summary of, *455*
electrons
 Big Bang and, 74
 as cathode rays, 50
 charge on one mole of, 911
 mass-to-charge ratio of, 49–51
 Millikan's determination of mass, 50–51
 properties of, *53*, *949*
 symbols for, *949*
 See also beta (β) decay; beta (β) particles
electrophoresis, 10
electroplating, 930
electrostatic attractions in solids, *603*
electrostatic potential energy (E_{el}), 277
 bond formation and, 395–96
 energy levels of diatomic molecule and, 885
 of ion–ion interactions, 277, *278*, 551–53
 of sodium chloride structure, 616
elementary steps, 681–83
 in enzyme kinetics, 692
elements
 allotropes of, 410, 620–21
 atoms of, 7, 53–54
 definition of, 6
 emission and absorption spectra, 337–38
 existing as diatomic molecules, 7–8, 101–2, 373, 1049
 isotopes of, 54–58
 origin of, 74–76
 standard states, 308
 transmutation of, 52
 See also periodic table
emission spectrum
 atomic, 337, 338
 of heated metal filament, 341
 of hydrogen, 345–50, 357
 of sodium, 366
empirical formulas, 63
 compared to molecular formulas, 122, 124–27
 from percent composition data, 121–24
 See also ionic compounds

enantiomers, 997–99
of amino acids, 1003, 1030
of complex ions, 1106
of lipids, 1020, 1021–22, 1031
in living systems, 1000–1001
of pharmaceuticals, 997, 1000–1001, 1014
racemic mixture of, 1000, 1014
See also chirality
end point of titration, 173
endothermic processes, 281–82
bond breaking as, 427
enthalpy changes in, 287
equilibrium constant for, 878
free energy changes in, 871
spontaneous, 853–54, 864, 871
energy
of Big Bang, 18
of chemical reactions, 8, 272, 279
conservation of, 275, 283
definition of, 8
from dietary glucose, 1020, 1110–11
from dietary lipids, 1022, 1024
dispersion of, 855, 856, 860, 861, 869, 884–85, 886
Einstein's equivalence to mass, 18, 350, 949, 969, 974–75
flowing from warm to cooler object, 273, 857–58
from hydrolysis of ATP, 1061
of photon, 342–43, 348
thermodynamics as study of, 273
See also activation energy (E_a); bond energies; Gibbs free energy (G); internal energy (E); kinetic energy (KE); potential energy (PE); work
energy microstates, 886
energy profile for reaction, 676
with catalyst, 692–93, 736
with two maxima, 681–82
energy transfer, 273, 281
See also heat (q)
energy-level diagram
of electron transitions in hydrogen, 348
of subshells in multielectron atoms, *367*
enthalpy (H)
definition of, 286
as a state function, 287, 306
enthalpy change (ΔH), 286–88
bond energy as, 427–30
in dissolution of ionic compounds, 553–55, 559–60
free energy change and, 864–72
heat at constant pressure and, 286–87, 864
sign of, 287
units of, 287
useful work as part of, 869
enthalpy of combustion (ΔH_{comb}), 299–300, 301–2
Hess's law and, 305, 311–12
methane bond energy and, 429
standard, 311–12, *315*, 316
enthalpy of formation. *See* standard enthalpy of formation (ΔH°_f)
enthalpy of fusion, molar (ΔH_{fus}), 291, 292, 293, 870
enthalpy of hydration ($\Delta H_{hydration}$), 554–55, 559–61
for selected cations and anions, *560*
enthalpy of reaction (ΔH_{rxn}), 299–300
from bond energies, 429–30
from calorimetry, 300–303
from Hess's law, 304–8, 312
See also standard enthalpy of reaction (ΔH°_{rxn})
enthalpy of solution ($\Delta H_{solution}$), 303–4, 554–55, 559–61
enthalpy of solvation ($\Delta H_{solvation}$), 554
enthalpy of vaporization (ΔH_{vap}), 287, 522
molar, 291, 292
entropy (S)
absolute, 859–61, 862–64, 886
Boltzmann's definition of, 886
calculating changes in, 863–64
chelate effect and, 1096
of different phases, 860–61
factors affecting changes in, 861
living systems and, 881
macroscopic thermodynamic view of, 884, 886
microstate view of, 885–86
of reaction, 863–64, 866–67, 878–79
spontaneous processes and, 855–56, 858–59, 865–67, 869–72, 1096
standard molar, *860*, 860–61, 862–64
as state function, 863
temperature and, 856–58, 859–60, 861
units of, 857
enzymes, 691–93, 1013–15
acid–base characteristics of, 1006
biocatalysis in pharmaceutical industry, 692
carbonic anhydrase, 692, 792–93, 1012, *1013*, 1109–10
catalases, *652*, 1110, 1112
inhibitors of, 1014–15
in nitrogen cycle, 1058, 1059–60
in photosynthesis, 1110, 1112
transition metals in, 1109–13
enzyme–substrate complex, 692–93, 1014
equations. *See* chemical equations; net ionic equations; nuclear equations
equatorial atoms and bonds, 448, 454, 456, 457
equilibrium. *See* acid–base equilibria; chemical equilibrium; phase equilibria; solubility equilibria
equilibrium arrow, 165
equilibrium constant (K)
calculations based on, 736–43
catalyst and, 736
definition of, 713
determining whether reaction is at equilibrium and, 726, 736
equilibrium concentrations or pressures and, 736–43
for ionization reactions of acids, 759–60
for ionization reactions of bases, 764
for redox reactions, 919–20
at specific temperature, 879–80
standard free energy of reaction and, 874–77, 878, 879–80
summary for manipulation of, 724
temperature dependence of, 713, 734–35, 877–80
as unitless expression, 714
very large, 717, 920
very small, 717, 741–43, 920
See also chemical equilibrium; K_a; K_b; K_c; K_f; K_p; K_{sp}; K_w
equilibrium constant expressions, 713–15
compared to reaction quotient, 724–26
definition of, 713
for heterogeneous equilibria, 726–28
multiplying or dividing an equation and, 721–22
pure liquids in, 727–28
pure solids in, 727
relationships between K_c and K_p, 717–19
for reverse reactions, 719–21
for sum of two or more reactions, 722–24
units and, 714
equilibrium lines, 523
equilibrium potential (E_{ion}), 1052–53
equimolar amounts, 112
equivalence point of titration, 172–73, 818–21, 823
multiple, 824–27
essential amino acids, 1003, 1058
essential elements, 1046
major, 1046, 1048, 1050–65
trace, 1046, 1048, 1065, 1066–67, *1084*, 1108, 1109–11
transition metals among, *1084*, 1084–85, 1108, 1109–13
ultratrace, 1046, 1048, 1065–66, *1084*, 1108, 1111–13
esters, *73*, 628
polyesters, 628–30
ethambutol, 1000
ethane, 71, 313, 314
standard molar entropy of, *862*
valence bond theory and, 474–75
ethanethiol, 1063
ethanol, 72
absolute entropy of, 862
arrangement of atoms in, 9
as biofuel, 1019
boiling point, 518–19, 520–21
boiling-point-elevation constant, *572*
in esterification of acid, 628
from fermentation of glucose, 1019, 1020
freezing-point-depression constant, *572*
hydrogen bonding in, 519, 520–21
as nonelectrolyte, 164
standard free energy of combustion, 868
vapor pressure of solution in water, 567–68
viscosity of, 528

ethene. *See* ethylene (ethene)
ethers, *73*, 530–31
diethyl ether, 520, 530–31
dimethyl ether, 518–19, 1063
polymers of, 626–27
ethylene (ethene), 71–72
copolymer with vinyl acetate, 626
polymers of, 622–23, *624*
ripening of fruits and vegetables and, 473, 475
valence bond theory and, 473–75
ethylene carbonate, 924
ethylene glycol
in antifreeze, 521, 563–64, 569, 574–75
boiling point of, 521
freezing point of solution of, 574–75
hydrogen bonding in, 521
miscibility in water, 534
polymer of, 626–27
vapor pressure of, 520
vapor pressure of solution of, 563
ethylene oxide, polymer of, 626–27
ethylenediamine ligand, 1094–96, 1100, 1106
ethylenediaminetetraacetic acid (EDTA), 1096, *1113*
eugenol, 131–32
evaporation, 281, 521
See also vaporization
evaporative cooling, 522–23
exact values, 26–27
excited states
of hydrogen atom, 348
indeterminate lifetimes of, 359
of sodium atom, 366
exoskeletons, ionic compounds in, 552–53, 1067
exothermic processes, 281–82
bond formation as, 427
enthalpy changes in, 287
entropy of surroundings and, 856
equilibrium constant for, 878
free energy changes in, 871
spontaneous processes and, 853, 865, 871
extensive properties, 12

F

f block elements, 368, 369, 373
f orbitals, 355, *356*, 362
filling of, 368, 373
face-centered cubic (fcc) unit cell, *605*, 607, 608–9, 611
in anti-fluorite structure, 618
of austenite, 614, 615
in fluorite structure, 618, *619*, 635
of gold and platinum, 615–16
of lithium chloride, 619–20
of sodium chloride, 616–17
of sphalerite, 618
See also cubic closest-packed (ccp) structure
Fahrenheit temperature scale, 32–34
families of elements, 59
Faraday, Michael, 911
Faraday constant (F), 911–13
fats, 1022
fatty acids, 1020–24
β-oxidation of, 1024
names and formulas of, *1021*
oxidizing agents from metabolism of, 1066
of phospholipids, 1024
plants' degradation of, 1110
femtochemistry, 680, 680n
ferrite, 615
fertilizers
nitrogen-based, 708, 710, 1058
phosphates in, 1061
field strength, in spectrochemical series, 1101, 1103–4
filled shell, 363
filtration of mixtures, 10
fireworks
magnesium oxide in, 558
red, 380
first harmonic, 352
first ionization energy (IE_1), 377–79
electronegativity and, 406
periodic trends in, 1049
first law of thermodynamics, 284, 285, 852
first-order reactions
distinguishing from second-order, 671–72
half-life of, 668–69
integrated rate law for, 664–68
radioactive decay as, 956–59
summary of how to distinguish, 673–74
fluids, 14
fluorapatite, 1066
fluoride
electron configuration of, 370–71
in water, 155, 1066
fluorine
as strong oxidizing agent, 907
as trace essential element, 1065, 1066
fluorite, 618, *619*
fluorite structure, 618, *619*, 634–35
5-fluorouracil (5-FU), 537–38
food value, 317–19
force
work and, 273
See also intermolecular forces
formal charge (FC), 414–18
in exceptions to octet rule, 419–24
as model with limitations, 425–26
formaldehyde
bond energy in, 428
bond lengths in, 426
dipole moment, 460
formula, 125
Lewis structure, 402
polar covalent bond in, 408
sp^2 hybrid orbitals for, 465–66
trigonal planar geometry of, 451, 465–66
formation constant (K_f), 831–32
chelation therapy and, 1116–17
formation reactions, 308–9
formation water, 573
formic acid, 774–75, 779–80
formula mass, 95–96
formula unit, 63
formulas. *See* chemical formulas; empirical formulas; molecular formulas; structural formulas
fossil fuels
vs. alternative sources of electricity, 933–34
carbon dioxide from combustion of, 108–11, 394, 804–5
coal with sulfur content, 650, 727, 783, 829
definition of, 108
greenhouse gases from combustion of, 394
origin of, 108
production of hydrogen from, 933
vehicle technologies for replacement of, 123–24, 900, 921–23, 931–33
See also gasoline; hydrocarbon fuels; natural gas
fractional distillation, 564–67
francium, 1049
Fraunhofer, Joseph von, 337
Fraunhofer lines, *336*, 337, 338
free energy. *See* Gibbs free energy (G)
free energy of formation. *See* standard free energy of formation (ΔG°_f)
free energy of reaction. *See* standard free energy of reaction (ΔG°_{rxn})
free radicals, 419
photochemical smog and, 651
produced by ionizing radiation, 962
freezing, 15, *281*
enthalpy change in, 287
entropy change in, 858
free energy and, 870
See also melting
freezing point. *See* melting point
freezing point depression, 569–70, 573–75
constants for selected solvents, *572*
van't Hoff factor and, 575–78
frequency (ν), 339–40
energy of a quantum and, 341
frequency factor, in rate constant, 675
freshwater
pH of, 788, 825
See also alkalinity titration; drinking water
fructose, 1016, 1017
5-FU (5-fluorouracil), 537–38
fuel cells, 310, 900, 931–33
fuel density, 316–17
fuel value, *315*, 316–17
fuels. *See* fossil fuels; hydrocarbon fuels
Fukushima nuclear disaster, 965
Fuller, R. Buckminster, 272, 621
fullerenes, 621
functional groups, 72–73, *73*, 989, *990*
multiple, 472–73
fundamental, of standing wave, 352

G

gadolinium contrast agents, 1069, 1114, 1115–16
Galileo Galilei, 4
gallium
  cancer therapy with compounds of, 1072
  medical imaging with radioisotopes of, 1070–71, 1114–15
  silicon doped with, 492
gallium arsenide (GaAs), 492–93
Galvani, Luigi, 903
galvanic cells, 903
  See also voltaic cells
gamma (γ) rays, 952–53
  biological effects of, 962, 962n, 963, *964*
  in electromagnetic spectrum, *338*
  in Geiger counter, 956
  in medical imaging, 968, 1069, 1070, 1114
  in radiation therapy, 967
gas constant (*R*), 229
gas discharge tubes, 337
gas laws, 220–32
  Amontons's law, 226–28, 244, 245
  Avogadro's law, 225–26, 229, 234, 244–45
  Boyle's law, 220–23, 229, 244
  Charles's law, 223–25, 229, 244, 245
  ideal gas law, 228–36, *230*
  kinetic molecular theory and, 244–45
gaseous solution, 151
gases
  in chemical reactions, 232–35, 242–43
  collected over water, 242–43
  compared to liquids and solids, 14–16, 214–15, 508–9
  definition of, 14
  density of, 235–38
  intermolecular forces and, 508, 509
  kinetic molecular theory of, 243–50
  mixtures of, 238–43
  molecular speeds in, 245–50
  phase symbol for, 8
  real, 250–53
  standard states of, *860*
  water solubility of, 534–37
  wet vs. dry, 242
  See also greenhouse gases; noble gases (group 18); pressure
gasoline
  alkanes in, 314–15
  carbon dioxide from combustion of, 109–10
  from distillation of crude oil, 564
  energy available from combustion of, 869
  ethanol added to, 1019
  ethanol as alternative to, 1019
  fuel value and fuel density of, *315*, 316–17
  hydrogen fuel compared to, 933
  pollution from MTBE added to, 669
gastric acid, 153, 770, 771–72, 1056–57
gauge pressure, 219
Gaussian distribution, 30
Geiger, Hans, 51–52
Geiger counters, 956
general gas equation, 230–31
genetic code, 1027–28, *1028*
Gerlach, Walther, 357
germanium, 493, 621, 1068
Gibbs, J. Willard, 865
Gibbs free energy (*G*), 864–69
  chemical equilibrium and, 865, 872–77, 916
  in coupled reactions, 882–84
  reactions in living systems and, 880–84
  temperature effect on spontaneity and, 869–72
Gibbs free energy change (ΔG)
  additive, 883–84
  as electrical work, 911–13
  electrochemical equilibrium constant and, 920
  for ion transport across cell membrane, 1052, 1053
  as maximum available work, 865, 869, 886
  standard cell potential and, 920
  See also standard free energy of formation (ΔG°_f); standard free energy of reaction (ΔG°_{rxn})
global warming, 394, 430
  See also climate change; greenhouse gases
glucose
  α and β isomers of, 1016–17
  anaerobic bacterial metabolism of, 304
  chromium in regulation of, 1112–13, 1118
  cytochromes in metabolism of, 1110–11
  in D5W solution, 583
  diabetes and, 1113, 1118
  empirical formula of, 125
  enthalpy of combustion, 300
  ethanol from fermentation of, 1019, 1020
  food value of, 318
  glycosidic bonds involving, 1017–18, 1019
  magnesium ions in metabolism of, 1054
  metabolic pathways of, 881–84, 1014, 1019–20, 1024, 1054
  in osmium-based anti-inflammatory agents, 1119
  osmotic pressure and, 581
  polysaccharides formed from, 1016, 1017–19
  in positron emission tomography, 968
  redox reaction in breakdown of, 152
  structure of, 1016–17
  in sucrose structure, 1017
  See also photosynthesis
glucose-6-phosphate, 881–82, 1014, 1061
glutamate, 1112
glutamic acid, 1001, *1002*, 1003, 1004
glycerides, 1020, 1021–22
glycerol, 1020, 1024
glycine, 100, 115, 1030, 1057
glycolaldehyde, 124–25
glycolic acid, copolymer with lactic acid, 629
glycolysis, 881–83, 1014, 1019–20
glycoproteins, 1016
glycosidic bonds, 1017–18, 1019
gold, 602, 606
  alloyed with platinum, 615–16
  antitumor complexes of, 1118–19
  arthritis drugs containing, 602, 1119
  printed circuits using, 600
gold-foil experiments, 51–52
Goudsmit, Samuel, 357
Graham, Thomas, 249
Graham's law of diffusion, 250
Graham's law of effusion, 249–50
graphene, *620*, 621
graphite, 620, 621, 862
gravitational potential energy, 274
gray (Gy), 963
Great Pacific Garbage Patch, 622
greenhouse effect, 394, 408–9
greenhouse gases, 394
  carbon dioxide as, 394, 409
  climate stability and, 430
  dinitrogen monoxide as, 668, 1058
  infrared absorption by, 394, 408–9, 430
  methane as, 392
  nitrous oxide as, 415, 668
  sulfur hexafluoride as, 420
ground state, of hydrogen atom, 348
groups of elements, 59, 60–61
Grubbs' test, 31–32
guanine, 1025–29, 1111
Guldberg, Cato, 713

H

Haber, Fritz, 736
Haber–Bosch process, 736
  equilibrium constant of, 879–80
  fertilizer produced by, 1058
  hydrogen production for, 739–40
  nitric acid production and, 743
  See also ammonia synthesis
half-life ($t_{1/2}$) of radionuclide, 957–59
  radiometric dating and, 959–61
half-life ($t_{1/2}$) of reaction
  first-order, 668–69
  second-order, 672
  summary of, for single reactant, *673*
  zero-order, 673
half-reactions, 188–91
  balancing redox equations with, 188–91, 194–97, 906–7, 1059–60
  combined to give net ionic equation, 901, 905–6
  in electrochemistry, 900–901
  pH conditions of, 194–96, 905–6, 932–33
  See also electrochemical cells
halide ions
  attempted separation of, 840–41
  as Lewis bases, 829–30
halite structure, 122
Hall–Héroult process, 319

halogens (group 17)
  in algae, lichens, and fungi, 1111
  atomic radii, 374
  boiling points, 510
  common acids containing, 69, *70*
  as diatomic molecules, 1049
  electron affinities, 379
  in periodic table, 60
  polar bonds with hydrogen, 407
  properties of, 1049, 1050
haloperoxidases, 1111
hard water, softening of, 587–88
heat (q)
  from car engine, 869
  at constant pressure, 286–87, 864
  definition of, 273
  as energy transfer, 281
  enthalpy change and, 864
  entropy change and, 856–57, 864
  infrared radiation and, 341
  internal energy and, 283–85
  sign of, 281, 284–85
  as wasted energy, 869
heat capacity (C_P)
  of calorimeter, 300–301
  See also molar heat capacity (c_P)
heat of reaction. *See* enthalpy of reaction (ΔH_{rxn})
heat sinks, 293
heat transfer, 273
heating curves
  of solution of volatile liquids, 564–66
  of water, 289–90
Heisenberg, Werner, 353, 354
Heisenberg uncertainty principle, 352–54
Heliox, 241
helium
  balloons inflated with, 222–23, 225, *236*, 249, 286
  binding energy of nucleus, 949
  electron configuration of, 363
  molecular orbital theory and, 479
  in nuclear fusion reactors, 973–74
  primordial nucleosynthesis of, 74–75, 971–72
  in scuba diving mixture, 215, 240–41, 254–55
  in solar fusion, 972–73
  stellar nucleosynthesis of, 75, 948–49
helium-3
  enthalpy of fusion of, 869n
  in magnetic resonance imaging, 1071
  in nuclear fusion, 972–73
hematite, 91, 193
  See also iron(III) oxide
heme groups, in cytochromes, 1110
heme iron, dietary, 1084
hemoglobin
  iron deficiency or excess and, 1116
  oxygen binding sites on, 536
  porphyrins in, 1108
  quaternary structure of, 1013
  sickle-cell, 1010–11
Henderson–Hasselbalch equation, 807–9
  buffers and, 809, 811, 812–13, 814, 815
  ocean acidification and, 842
  titration of weak acid with strong base and, 821
  titration of weak base with strong acid and, 823
Henry, William, 535
Henry's law, 535–37
heptan-2-one, *471*
heptane
  combustion of, 109–10
  distillation with octane, 564–66
hertz (Hz), 339
Hess's law, 304–8
heteroatoms, 71, 72, 989
  in amino acids, 1001
heterogeneous alloys, 604, 612
heterogeneous catalysts, 689
  in biocatalysis, 692
heterogeneous equilibria, 727–28
heterogeneous mixtures, 7
heteronuclear diatomic molecules, molecular orbitals of, 484–86
heteropolymer, 626
hexagonal closest-packed (hcp) structure, 605–6, 607, *608*
  holes in, *615*
hexagonal unit cell, *605*, 606
hexamethylenediamine, *631*
high-density polyethylene (HDPE), 622–23, *624*
high-spin states, 1102–4
Hindenburg, *717*
HIV protease, 621
  inhibitors of, 1015
holes in crystal lattice, 614, *615*, 617–18, 635
homocysteine, 1112
homogeneous alloys, 604, 612, 613
homogeneous catalysts, 689
homogeneous equilibria, 726–27
homogeneous mixtures, 7, 151
homonuclear diatomic molecules
  elements existing as, 7–8, 101–2, 373
  magnetic properties of, 482–83
  molecular orbitals of, 480–84
homopolymer, 622
Honda Clarity FCV, *932*
hot-air balloon, 224–25, 236, 284–85
Hoyle, Fred, 18
Hubble, Edwin, 17
human body
  classes of chemical reactions in, 152
  coupled reactions in, 880–84
  elemental composition of, 1047–48
  gastric acid in, 153, 770, 771–72, 1056–57
  nitric oxide in, 418
  See also blood; cancer; essential elements; medical imaging; medical therapy
Hund, Friedrich, 364
Hund's rule, 364–65
  crystal field splitting and, 1098, 1102, 1103, 1104
  molecular orbitals and, 481, 482
hybrid atomic orbitals, 463
  in larger molecules, 475
  sp, 467–69
  sp^2, 465–66
  sp^3, 463–65
  summary of, *468*
hybrid vehicles, 123–24, 900, 921–23
hybridization, 463
hydrated complex ions, 830, 832–34, 838, 1085, 1093–94, 1098–99
  spin states of, 1104
  in swimming pool water, 1120–21
hydrated ions, 512
  enthalpy of hydration and, 554–55, 559–61, *560*
  enthalpy of solution and, 554–55, 559, 560
hydration, sphere of, 512
  of complex ion, *833*
hydride ion, 1050
hydrocarbon fuels
  alkanes as, 314, *315*
  balanced equations for combustion of, 105–7
  carbon monoxide from combustion of, 307–8, 690–91
  enthalpies of combustion of, 299–300, 307–8
  enthalpies of formation of, 309–10
  ethanol added to, 1019
  fuel values of, *315*, 316–17
  oxidized by catalytic converters, 690–91
  photochemical smog and, 651
  polycyclic aromatic hydrocarbons from burning of, 473
  See also crude oil; fossil fuels; natural gas
hydrocarbons, 71–72
  atmospheric, 119
  branched-chain, 314, *315*
  combustion analysis of, 128–31
  as nonpolar solutes, 530
  standard molar entropies of, 862, 863
  straight-chain, 314, *314*, *315*, 510
  See also alkanes; hydrocarbon fuels
hydrochloric acid, 165–67
  activity series of metals and, 192–93
  gastric, 153, 770, 771–72, 1056–57
  ionization reaction leading to, 758
  neutralization reaction with sodium hydroxide, 166–67
  pH of concentrated solution, 778
  pH of very dilute solution, 782
  titration with sodium hydroxide, 818–20
  See also hydrogen chloride
hydrofluoric acid, burns caused by, 180–81

hydrogen
  1*s* orbital of, 359–60
  in activity series, 192
  bacterial production of, 304
  Bohr model of, 347–50
  combustion of, 279, 717, 933
  covalent H–H bond formed by, 395–96
  deuterium isotope of, 973–74
  electron configuration of, 363, 364
  elemental properties of, 1050
  enzyme-catalyzed oxidation of, 1112
  in fertilizer production, 710
  as fuel, 272, 279, 310, 931–33
  fuel value of, 316
  Hindenburg disaster and, *717*
  ionization energy of, 348, 349
  molecular anion of, 479–80
  molecular orbitals of, 478–79
  primordial nucleosynthesis of, 74–75
  produced for ammonia synthesis, 710–12
  solar fusion of, 971–73
  valence bond theory of, 462
  See also steam–methane reforming reaction; water–gas shift reaction
hydrogen bonds, 513–19, *515*
  of acetone in water, 516
  in alcohols, 519, 520–21, 528
  in amides, 632
  in ammonia, 513–14, *514*, 531
  boiling points and, 513–14, 519, 520–21
  of glucose in water, 1016
  in Kevlar, 633–34
  in nucleic acids, 516, 1026, *1027*
  in protein structures, 516, 1006, 1011–12
  in solids, *603*
  viscosity and, 528
  in water, 513–15, 521, 526–27, 560
hydrogen chloride
  absolute entropy of, 862
  in aqueous solution, 165, 758
  enthalpy of reaction from component gases, 429–30
  as gaseous molecular compound, 165
  overlapping orbitals in, 462
  polar covalent bond in, 405
  See also hydrochloric acid
hydrogen electrode, 768, 914–16
hydrogen emission spectrum, 345–46
  Bohr model and, 347–50
  pair of red lines in, 357
hydrogen fluoride
  bond polarity in, 407, 457
  hydrogen bonds in, 513–14
  solubility in water, 531
hydrogen halides, polarity of bonds in, 407, 457, 459
hydrogen iodide, equilibrium in formation of, 737–38
hydrogen ions, meaning hydronium ions, 165–66
hydrogen peroxide
  Lewis structure of, 403
  in plants' degradation of fatty acids, 1110
  rate of decomposition, *652*
  superoxide dismutases and, 1112
  xanthine oxidase and, 1111
hydrogen sulfate anion, 169
hydrogen sulfide, 1062–63, *1063*
  from cytochrome activity in pool water, 1120, 1121
  at hydrothermal vents, 1030
hydrogenases, 1112
hydrogenation of oils, 1022–23, 1031
hydrolysis
  of nitrogen compounds, 1060
  of nonmetal oxides, 168
hydronium ions, 152, 165–66
  acid strength and, 760
  Arrhenius acids and, 756
  in blood, 756–57
  as conjugate acid of water, 766
  ionization of acids and, 758, 759–60
  pathways for conversions involving, *778*
  pH as measurement of, 768–70
  sodium ion channels and, 1051
  in very dilute strong acid solution, 781–82
hydrophilic interactions of proteins, 1013
hydrophilic properties, 532, 537, 538
hydrophobic interactions in proteins, 1013
hydrophobic properties, 532, 537
  of lipids, 1020
hydrothermal vents, deep-ocean, 150, 1030
hydroxide ions
  Arrhenius bases and, 756
  Brønsted–Lowry bases and, 169, 764
  as conjugate base of water, 766
  from dissociation of strong bases, 763
  as Lewis bases, 829
  pathways for conversions involving, *778*
  redox reactions involving, 193–94, 195–96
hydroxyapatite, 152, 1054, 1060
  fluoride and, 1066
hydroxyl radicals
  photochemical smog and, 651
  produced by ionizing radiation, 962
hypertonic solution, *579*, 583
hypervalency, 420–21
hypobromous acid, 761, *763*
hypochlorous acid, 761, *762*, *763*
hypohalous acids, 761
  relative strengths of, *763*, 832
hypoiodous acid, 761, *763*
hypothesis, 17
hypotonic solution, *579*, 583
hypoxia, 535

I

i factor, 575–78
  osmotic pressure and, 580–82, 583
ICA table
  for buffer capacity, 813, 814
  for titration of weak acid with strong base, 822
  See also RICE table
ice
  aquatic life and, 528–29
  chilling beverages with, 294–97
  energy in melting of, 281, 282, 287, 289–91, 292, 293, 294–97
  entropy change in freezing and, 858
  entropy change in melting of, 857–58, 861
  hydrogen bonds in, 524, 528
  less dense than water, 524, 528–29
  molecular motion in, 854, *855*
  phase diagram of water and, 523–24, 525–26
  temperature effect on melting/freezing of, 869–70
ice-resurfacing machine, 287–88
ideal gas
  definition of, 228
  kinetic molecular theory of, 243–50
ideal gas law, 228–32
  breathing and, *230*
  chemical reactions and, 232–35
  density of a gas and, 235–36
  deviations from, 250–53
ideal solutions
  Raoult's law and, 562–64, 568
  theoretical values of K_a in, 841
  van't Hoff factor and, 577
inches of mercury, 217
indicators, for acid–base titrations, 173, 816–17, 818
indium-111, 1070–71
induced dipoles, 509–11
  interacting with permanent dipoles, 513, *515*, 529, 530, 534
induced-fit model of enzyme action, 1014–15
inertness, chemical, 365
infrared active vibrations, 409
infrared inactive vibrations, 409
infrared radiation
  in electromagnetic spectrum, *338*
  from gallium arsenide, 492
  greenhouse effect and, 394, 408–9, 430
  from hot objects, 341
  in hydrogen emission spectrum, 346, 348
  photoelectric applications of, 344
  vibration of bonds and, 408–9, 430
inhibitors of enzymes, 1014–15
initial rate of a reaction, 659
inner coordination sphere, *833*, 1085
inorganic compounds
  names and formulas of, 64–70
  occurrence on Earth, 64–65
insect pheromones, 127, 494
insignificant digits, 24
integrated rate law
  definition of, 665–66
  for first-order reaction, 664–68
  for second-order reaction, 670–72

integrated rate law *(cont.)*
  summary of, for single reactant, 673–74
  for zero-order reaction, 673
intensive properties, 12
  reduction potential as, 911
intercalation in DNA, 473
intermediate, 681, 684
intermetallic compound, 612–13
intermolecular forces, 508–9
  boiling points and, 510–11, 513–14, 518–19, 520–21
  capillary action and, 527–28
  changes of state and, 292, 524, 525
  in enzyme action, 1014
  identifying, 517–18
  of polar molecules, 511–19
  in protein structures, 1011–12, 1013
  in real gases, 251–52
  relative strengths of, *515*
  in solids, 602, *603*, 604
  surface tension and, 526–27
  vaporization and, 519
  viscosity and, 528
  in water, 511–12, 526–28
  See also dipole–dipole interactions; dipole–induced dipole interactions; dispersion forces; hydrogen bonds; ion–dipole interactions
internal energy (E)
  changes in, 283–85
  definition of, 283
  enthalpy and, 286–87
  useful work and, 869
International Union of Pure and Applied Chemistry (IUPAC), 71–72
interstitial alloys, 614–16
intramolecular forces, 508
intravenous solutions, 550, 583
iodine
  in human health, 1065, 1066–67
  as volatile solid, 1049
iodine-131
  for thyroid cancer, 967
  thyroid cancer caused by, 964
  in thyroid imaging, 1069
ion channels, 1051, 1053, 1056
ion exchange, 10, 587–88
ion pairs, 576–78
  apparent values of K_a and, 841
  in concentrated solutions, 841
  in physiological saline, 583
ion product, 839
ion pumps, 1052, 1054, 1056
ion–dipole interactions, 511–12, *515*
  coordinate bonds and, 830, 1085
  dissolution of ionic compounds and, 511–12, 529, 554, 560
  in ion channels, 1051
  membrane transport and, 1051
  protein solubility and, 550
  protein structures and, 1012
ionic bonds, 395, *396*, 397
  electronegativity difference and, 407
  equilibrium distance of, *278*
  in solids, *603*
  strength of, 551–53
  of transition metal ions, 1086
ionic character, 406
ionic compounds, 9
  binary, 62–64, 66–67, 404, 616–20
  enthalpy of solution, 303–4, 554–55, 559–61
  formulas of, 62–64, 66–67, 122
  inter-ion distance in, 555
  lattice energy of, 554–61
  Lewis structures of, 404
  naming of, 66–67
  solubility rules in water, *176*, *177*
  of transition metals, 67–68
  See also dissolution of ionic compounds; salts
ionic equations, 166–67
  See also net ionic equations
ionic radii, 373–76
  in crystal structures, 616–20, 634–35
  lattice energies and, 555
  of main group elements, 375–76, *551*
  melting points and, 555
  of polyatomic ions, *552*
  strength of ion–ion interactions and, 551–53
ionic solids, 602, *603*, 604, 616–20
ion–ion interactions
  coulombic, 277, *278*
  in protein structures, 1012
  strength of, 551–53, 554
  See also lattice energy (U)
ionization energy (IE), 376–79
  electronegativity and, 406
  of hydrogen atom, 348, 349–50
  metallic bonding and, 490
  periodic trends in, 1049
ionization reactions of acids, 758–61
ionizing radiation
  biological effects of, 962–67, 975–76, 1067
  dosage of, 963–65
  radon gas and, 965–66
  sources of U.S. exposure to, *965*
  therapeutic uses of, 967
  units of, 963, *963*
ions, 9
  charges of transition metal ions, *1091*
  electrostatic potential energy between, 277, 551–53
  mass of, 94
  monatomic, 63, 67–68, 375–76
  See also anions; cations; complex ions; electron configurations of ions; molecular ions; polyatomic ions
iron
  available to the body, 1084
  in catalyst for Haber–Bosch process, 736
  chelation therapy for excess of, 1116–17
  complexed with heme, 1084, 1110, *1111*
  corroding in contact with copper, 925–27, 935–36
  deficiency of, 1116
  enzymes containing, 1110–11, 1112
  redox reactions of compounds of, 193–96
  steel as alloy of, 614–15, 927
  stellar nucleosynthesis and, 76, 948–49, 950
  See also corrosion
iron carbide, 615
iron ions as Fe^{2+}
  in hemoglobin, 1013
  spectrophotometric measurement of, 162–63
iron ions as Fe^{3+}
  electron configuration of, 372–73
  spin states of, 1102
iron(III) hydroxide, 194, 833
iron(III) oxide, 193, 850, 853
iron(III) sulfate, 91
isobars, 217
isoelectric point (pI), 1004–6
isoelectronic species, 371
isolated system, 279–80
isomers
  classification of, *998*
  of coordination compounds, 1104–7
  definition of, 89, 991
  See also chirality; constitutional isomers; stereoisomers
isotonic solution, *579*, 583
isotopes, 54–58
  See also radionuclides

J

Japanese beetle, 494
japonilure, 494
joints, artificial
  polyethylene in knee replacement, 623
  tantalum metal in, 607
Joliot-Curie, Irène, 962
joule (J), 285

K

K. *See* equilibrium constant (K)
K_a, 758–60
  apparent values of, 841–42
  for diprotic acid, 783–86
  of hydrated metal ions, *833*
  K_b of conjugate base and, 789–90
  theoretical values of, 841
  for triprotic acid, 786–87
  for weak acid, 773–75, 779–80
K_b, 764
  K_a of conjugate acid and, 789–90
  for weak base, 776–77, 780–81
K_c, 714–16, 717
  compared to Q_c, 725
  free energy change and, 877
  K_p and, 717–19

K_f (formation constant), 831–32
  chelation therapy and, 1116–17
K_p, 714–15, 716–17
  free energy change and, 877, 878–80
  K_c and, 717–19
  Le Châtelier's principle and, 732
K_{sp} (solubility product), 834–41
  reaction quotient and, 839–40
K_w, 768, 771, *778*
  for conjugate acid–base pairs, 790, 808
  very dilute strong acid and, 781–82
Kalkwasser
  pH of, 779
  preparing solution of, 158–59
Kekulé, August, 470
Kekulé structures, 470, 471
kelvin (K), 33
Kelvin temperature scale, 32–34
  Charles's law and, 223
keratins, 1011, 1013
kerosene, 314–15, *315*, 316
ketones, *73*, 470
Kevlar, 633–34
kilowatt-hour, 921
kinetic energy (KE)
  average, 244, 245, 246, 247–48, 276–77, 653, 674
  definition of, 275
  dispersion of, 855, 856, 859, 860, 861
  entropy and, 859–60, 861
  of gas molecules, 244, 245–48
  of gas-phase ions, 278
  at molecular level, 276–78
  as part of internal energy, 283
  of photoelectron, 343
  potential energy converted to, 274–75
  reaction rates and, 653, 674
  vaporization and, 519
kinetic molecular theory of gases, 243–50
  reaction rates and, 659, 674
kinetics, chemical
  definition of, 652
  half-life in, 668–69, 672, 673, *673*
  spontaneity and, 853
  See also reaction mechanisms; reaction order; reaction rates
Kirchhoff, Gustav Robert, 337
krypton-83, in magnetic resonance imaging, 1071
Kwolek, Stephanie, 633

L

lactase, 1013–14
lactic acid, *883*, 884
  copolymer with glycolic acid, 629
lactose intolerance, 1013–14
landfill, gases emitted by, 238
lanterns, for camping, 634–35
lanthanides, 60, 368, 373
  in scintigraphic imaging, 1114–15
lasers, 359
lattice energy (U), 554–55
  calculating, 556–58
  of common binary ionic compounds, *554*
  enthalpy of hydration and, 559–61
laughing gas. *See* dinitrogen monoxide
law of conservation of energy, 275, 283
law of conservation of mass, 99
law of constant composition, 8
law of mass action, 713
law of multiple proportions, 61–62
Le Châtelier, Henri Louis, 729
Le Châtelier's principle, 728–36
  acid–base equilibria and, 757, 806, 820
  adding or removing reactants or products, 729–31
  pressure and volume changes, 731–34
  solubility equilibria and, 838
  summary of, *736*
  temperature changes and, 734–36, 878
lead
  electron configuration of, 369
  in solder, 612
  toxicity of, 1048, 1116
  in uranium-238 decay series, 955, 959, 965
lead–acid battery, 917–18
  recharging of, 928, *929*
lead(II) chloride, precipitation of, 839, 840–41
lead(II) chromate, 180
lead(II) iodide, precipitation of, 177–78
lean mixture, 115
Lemaître, Georges-Henri, 18
length, units of, 18–19, *20*
Leonardo da Vinci, 184
leveling effect, 766–67
levorotatory enantiomer, 997, 1003
Lewis, Gilbert N., 398, 405, 426, 828
Lewis acids, 827–30, 1085
Lewis bases, 827–30
  as ligands, 830, 1085, 1093–94, 1111
  in membrane transport, 1051
  water as, 1094
Lewis dot symbols, 398–99
Lewis structures, 398–404
  of binary ionic compounds, 404
  compared to other representations, 470
  definition of, 399
  with double and triple bonds, 401–3
  drawing by five-step approach, 399–401, 411
  electronegativity and, 407, 411–12
  exceptions to octet rule and, 418–26
  formal charge and, 414–18, 419–24, 425–26
  with less than an octet, 423–25
  of molecules in sulfur cycle, 1064–65
  with more than an octet, 420–23
  octet rule and, 398, 399–400
  with odd number of electrons, 418–20
  overview of, 398
  resonance and, 410–14
  steric number determined from, 448, 449
  three dimensions not shown by, 446–47
Lewis symbols, 398–99
Libby, Willard, 960
life
  built from 40 or 50 small molecules, 988
  classes of chemical reactions in, 152
  coordination compounds in, 1108–13
  origin of, 88–89, 100, 150, 1029–30
  thermodynamics of, 880–84
  water on Mars and, 150
  See also aquatic life; human body; photosynthesis
ligand displacement reaction, 1094
ligand exchange reaction, 1116–17
ligands, 830, 1085
  of biologically active coordination complexes, 1113
  as Lewis bases, 830, 1085, 1093–94, 1111
  macrocyclic, 1108
  names and structures of, *1089*
  polydentate, 1095–97, 1106, 1108–13
  prefixes for number of, *1089*, 1106
  spectrochemical series of, 1101, 1103–4, 1121
  spin states of ions and, 1103–4
light
  emitted by hot objects, 341
  spectrum of, 336–39, *338*
  speed of, 339
  wave–particle duality of, 344
  See also colors; photosynthesis; sunlight
lightning, ozone production by, 410
like dissolves like, 532, 534
lime. *See* calcium oxide
limestone
  decomposed into lime and CO_2, 727
  dissolved by carbonic acid, 167–68
  See also calcium carbonate
limiting reactants, 112–16
  in aqueous phase reactions, 183
  in gas phase reactions, 234
limonene, 130–31, 997
linear complex ions, 1087
linear geometry
  hybrid orbitals and, 467–69, *468*
  Lewis structure and, 447
  with no lone pairs, 448, 449
  with three lone pairs, *455*, 456
linkage isomerism, 1106–7
lipid bilayer, *1024*, 1025, *1051*
  attacked by alkyl hydroperoxides, 1066
lipids, 1020–25
liquids
  compared to solids and gases, 14–16, 214–15, 508–9
  definition of, 14
  equilibrium constant expressions and, 727–28
  intermolecular forces in, 508, 509
  phase symbol for, 8
  standard states of, *860*
  vapor pressure of, 519–23, 562–69
  volatilities of, 10, 521
liter (L), 19, *20*

lithium
  electron configuration of, 363–64
  in nuclear fusion reactors, 974
  as strong reducing agent, 908
  therapeutic applications of, 1072
lithium aluminum hydride, 424–25
lithium carbonate
  for bipolar disorder, 96–97, 1072
  of red fireworks, 380
lithium chloride, crystal structure of, 619–20
lithium perchlorate, freezing point depression by, 576
lithium-ion batteries, 123–24, 923–24, 928, 974
local maxima, of electron density, 360–61
lock-and-key model of enzyme action, 1014
London, Fritz, 509, 510
London forces. *See* dispersion forces
lone pairs
  Lewis acids and bases and, 827–29, 830, 1085
  in Lewis structures, 399
  VSEPR theory and, 447, 452–57
low-density polyethylene (LDPE), 622–23, *624*
low-spin states, 1102–4
lutetium-177, 1115
lycopene, 126–27
Lyman, Theodore, 346, 348

M

macrocyclic ligands, 1108
macromolecules, 622
  See also polymers
macroscopic phenomena, 4–5
magnesium
  alloys with aluminum, 927
  electron configuration of, 366
  ionization energies of, 377
  in living organisms, 1054
magnesium citrate, 1073–74
magnesium hydroxide, 834–35, 840
magnesium ions
  replaced by toxic beryllium, 1048
  separating from calcium ions, 840
  water softening and, 587–88
magnesium oxide, lattice energy of, 558
magnetic properties
  molecular orbital theory and, 482–83
  of transition metal ions, 1102–4
magnetic quantum number (m_l), 355–56, *357*
magnetic resonance imaging (MRI), 1069, 1071, 1113–14, 1115–16
magnetite, bonded to steel, 927
main chain of alkane, 991
main group elements, 59
  for diagnosis and therapy, 1068–73
  electron configurations of ions of, 370–72
  human health and, 1046–49
  ionic radii of, 375–76, *551*
  ionization energies of, 377–79
  Lewis symbols of, 398
  periodic trends in, 1049–50, *1050*
  properties of, 1049–50
  See also major essential elements
major essential elements, 1046, 1048, 1050–65
malic acid, 444
malleability, 59, 604, 612, 613
maltose, as human energy source, 306
manganese
  MRI contrast agents and, 1116
  photosynthesis and, 1109
manganese dioxide
  in alkaline batteries, 909–10
  hydrogen peroxide decomposition and, *652*
manganese ions, spin states of, 1103
manometers, 218–20
marble, chemical weathering of, 167–68, 183
margarine, 1022
Mars, water on, 150
Marsden, Ernest, 51–52
marsh gas, 314
mass
  atomic, 53, 56–58
  conservation of, 91, 99
  definition of, 5
  density and, 13
  Einstein's equivalence to energy, 18, 350, 949, 969, 974–75
  molecular, 95, 127–28
  units of, 19, *20*
mass action, law of, 713
mass action expression, 713
  as reaction quotient, 724–26
  See also reaction quotient (*Q*)
mass defect (Δm), 949
mass number (*A*), 54–56
  changed in radioactive decay, *954*
mass of product, from reactant mass, 110–11
mass of reactant, from product mass, 111–12
mass spectra, 127–28
  of constitutional isomers, 1001
  in radiometric dating, 960
mass spectrometers, 56, 127
mass-to-mass ratios, 155–56
mass-to-volume ratios, 153
matter
  classes of, 5–7
  definition of, 5
  properties of, 12–13
  states of, 14–16
matter waves, 350–53
Maxwell, James Clerk, 339, 341
mean, 28
measurements
  accuracy of, 27, 28–30
  precision of, 27–32
  significant figures in, 22–27
  units of, 18–22
medical imaging
  gadolinium contrast agents in, 1069, 1114, 1115–16
  of heart, 1070, 1114, *1115*
  main group elements used in, 1068–71
  with MRI, 1069, 1071, 1113–14, 1115–16
  with PET, 968, 1069, 1070, 1071, 1113
  radionuclides used in, 968, *968*
  with technetium, 77, 1114
  of thyroid, 77, 1069
  transition metals in, 1113–16
medical therapy
  arterial stents, 612–13
  artificial joints, 607, 623
  for burn patients, 629
  diamond films on prosthetic implants, 621
  dissolving sutures, 629
  PETE in, 629
  with radiation, 967
  Teflon tubing in vascular surgery, 624
  transition metals in, 1113, 1116–19
  See also drugs
melting, 15, *281*
  as endothermic process, 281
  enthalpy change in, 287, 291
  entropy change in, 857–58, 861
  spontaneous, 853, 854
  See also freezing
melting point
  calculated from enthalpy and entropy of fusion, 870
  lattice energy and, 555
  periodic trends in, 1049, *1050*
  on phase diagram for water, 524
  of straight-chain alkanes, *314*
  temperature scales and, 32–33
  types of solids and, *603*, 604
  See also freezing point depression
membrane. *See* cell membrane; osmotic pressure (Π)
membrane potential ($E_{membrane}$), 1052–53
Mendeleev, Dmitri, 58, 60, 61, 77
meniscus, 527, 537
mercury
  determining ionic concentrations of, 919–20
  emission and absorption spectra of, 338
  meniscus of, 527
  therapy for poisoning from, 1116, 1117
messenger RNA (mRNA), 1028
metabolic pathways
  enzymes of, 692, 1013, 1014–15
  of fatty acids, 1024
  of glucose, 881–84, 1014, 1019–20, 1024, 1054
metal hydrides, 1050
metallic bonds, 396, 1049
metallic radius, 373
metallic solids, 602, *603*
metalloenzymes, 1109–13
metalloids (semimetals)
  in periodic table, 59, 1049
  as semiconductors, 491–93
  unit cells in, *606*

metals
 activity series for, 191–93
 conductivity of, 491
 densities of, 607, 610–11
 major metals in human body, 1050–54
 in periodic table, 59, 60, 61, 1049
 photoelectric effect of, 343–45
 properties of, 59, 602
 structures of, 604–11
 unit cells of, 606–7
 See also alkali metals (group 1); alkaline earth metals (group 2); transition metals
metastable nucleus, 1114
meteorites
 age of, 960
 amino acids in, 88, 1030
 heating of early atmosphere by, 90
 Martian, 150
 organic compounds in, 125
meter (m), 18–19, 20
methane
 as alkane, 71, 313, 314
 atmospheric, sources of, 105, 119
 bacterial generation of, 304, 1112
 bond angles in, 447
 bond length in, 426
 from decomposing solid waste, 238
 from degradation of organic matter, 314, *315*
 diamond films grown from, 621
 as greenhouse gas, 392
 in hydrogen gas production, 310–11, 710–11, 718–19, 933, 1031
 sp^3 hybrid orbitals in, 463–64
 standard molar entropy of, *862*
 See also natural gas; steam–methane reforming reaction
methane combustion, 105–6
 atmospheric carbon dioxide and, 110–11
 bond energies in, 427, *428*, 428–29
 entropy change in, 856
 incomplete, 307–8
 as oxidation, 184
methanethiol, 1063, 1064, *1064*, 1065
methanol, 72
 absolute entropy of, 862
 gas-phase reaction with water, 871–72
 miscibility with water, 529–30
methionine, 1062, 1112
methyl group, 313
 enzymatic transfer of, 1112
methyl mercaptan, 445
methylamine, 170, 764, 771
methylaspartate, 1112
methylene group, 313
2-methylpropane, boiling point of, 513
metric system. *See* SI units
micromolarity, 154
microstates, 886
microwave background radiation, 34–35
milk of magnesia, 834
Miller, Stanley, 88, 100, 101, 1029
millibars, 217
Millikan, Robert, 50–51, 54
millimeters of mercury (mmHg), 216–17
millimolarity, 154
miscibility
 definition of, 215
 of gases, 215
 of methanol and water, 529–30
 predicting, 534
mixtures, 6–7
 of gases, 238–43
 separation of, 9–10
 of volatile solutes, 564–69
models
 molecular, 9
 scientific, 17
molality (*m*), 570–71
molar absorptivity (ε), 162–63
molar enthalpy of fusion (ΔH_{fus}), 291, 292, 293, 870
molar enthalpy of vaporization (ΔH_{vap}), 291, 292
molar heat capacity (c_P), 290–91
 absolute entropy determined from, 860
 measurement of, 297–98
 of selected substances, *290*
 of water, 290, 291, 293
molar mass ($\mathcal{M}$), 93–98
 from colligative properties, 585–86
 definition of, 94
 gas density and, 235–38
 Graham's law of effusion and, 249–50
 of ionic compound, 95–96
 molecular formula and, 125–26
 from osmotic pressure, 585–86
 root-mean-square speed and, 246–50
 van der Waals constants and, 252
molar solubility, 834–38
 of glucose in water, 1016
molar volume, 229
molarity (*M*), 154, 154n
 appropriate for reactions in solution, 570
 calculated from density, 157–58
 in calculating osmotic pressure, 580
mole (mol), 91–93
mole fraction (X_x)
 definition of, 239
 of gases, 239–42
 of solvent, 562
mole ratio, 113–15
molecular compounds, 8, 62
 binary, 65–66
 prefixes for naming, *65*
molecular equations, 166, 167
molecular formulas, 8, 62
 compared to empirical formulas, 122, 124–27
 names of compounds and, 65–66
molecular geometry, 447
 of complex ions, 1086–87
 drawing to represent three dimensions, 450
 hybrid orbitals and, *468*
 summary of, *455*
molecular ions, M^+ in mass spectrometry, 127–28
molecular mass, 95
 from mass spectrum, 127–28
molecular models, 9
molecular motion, types of, *283*, *855*
molecular orbital, definition of, 477
molecular orbital diagram, 478
molecular orbital (MO) theory, 477–93
 basic concepts of, 477–78
 colors of auroras and, 477, 486
 electron transitions in, 477, 486
 fractional bond orders and, 486–87, 488, 490
 guidelines for, 480–81
 of heteronuclear diatomic molecules, 484–86
 of homonuclear diatomic molecules, 480–84
 of hydrogen and helium, 478–80
 metallic bonding and, 490–91
 semiconductors and, 491–93
molecular recognition
 carbohydrates in, 1016
 chirality and, 475–76, 997, 1000
 definition of, 473
 insect pheromones and, 494
 by sense of smell, 446, 475, 476, 997, 1063
molecular solids, 602, *603*, 604
molecularity of elementary step, 681
molecules, 7–9
Molina, Mario, 687
molybdenum
 in enzymes, 1111
 in technetium generator, 77, 1114
monatomic ions, 63
 ionic radii of, 375–76
 metals that form more than one, 67–68
monobasic bases, 818
monodentate ligands, 1095
monomers, 622
monoprotic acids, 168, 783
monosaccharides, 1016–17
 phosphate groups bonded to, 1061
 See also glucose
most probable speed of molecules, 246
moth balls, 431
motor vehicles
 fuel cells for, 900, 931–33
 See also automobiles
multiple proportions, law of, 61–62
muriatic acid, 778
muscle contraction, 1054
muscular dystrophy (MD), 989
mutases, 1112
myochrysine, 1119

N

Na^+–K^+ ion pumps, 1052, 1054
names of complex ions, 1088–93
 with polydentate ligands, 1106

names of compounds
common acids, 69–70, *70*
coordination compounds, 1088, 1092–93
ionic compounds, 66–68
molecular compounds, 65–66
organic compounds, 71–72, *73*
peptides, 1007–8
Roman numerals in, 67–68, 1088
transition metal compounds, 67–68, 1088, 1092–93
names of polyatomic ions, *68*, 68–69
nanomolarity, 154
nanoparticles
of gold, 602
of gold–platinum alloy, 615–16
nanotubes, carbon, 621
naproxen, 997
natural abundances of isotopes, 56–58
natural gas
combustion of, 105–7
compounds in, 71, 314
electric power plants and, 110–11
entropies of alkanes in, 862
incomplete combustion of, 307–8
odorant added to, 1063
See also methane; methane combustion
natural products, 71
neon
electron configuration of, 365
isotopes of, 54, 55, 56–57
neon-19 in PET imaging, 1071
Nernst, Walther, 916
Nernst equation, 916–19
cell membrane potential and, 1052
nerve impulses, 1052, 1053, 1054
net ionic equations, 166–67
activity series of metals as list of, 193
for alkaline batteries, 909–10
for button batteries, 912–13
by combining half-reactions, 901, 905–6
for precipitation reactions, 178–79
for redox reactions, 190, 195
without ions, 913
neutral salts, 788–89, *789*
neutralization reactions, 166–68
measurement of enthalpy of reaction, 302–3
precipitate formed in, 176
titrations using, 171–72
See also acid–base reactions
neutron capture, 76
neutrons
beta decay of free neutrons, 959
Big Bang and, 74
discovery of, 53
in nuclear fission, 968–70
in nuclear fusion reactors, 973–74
in nucleosynthesis, 74, 76, 949
properties of, *53*
in Rutherford's atomic model, 53
symbol for, 74
See also radioactive decay
Newton, Isaac, 336
nickel
as catalyst, 710
enzymes containing, 1112
in nitinol, 612–13
nickel ions (Ni^{2+}), electron configuration of, 372–73
nickel(II) complex ions, 1093–95, 1099, 1100–1101, 1106, 1107
nickel–metal hydride (NiMH) batteries, 921–23, 928, 929–30
niobium, in complex with antitumor activity, 1117
NiTi (nitinol), 612–13
nitrate ions
in fertilizer, 1058
in nitrogen cycle, 1057–60, 1111
reduction to nitrite, 1058, 1059, 1111
resonance structures of, 413–14, 761
in soil and water, 1057
nitrate reductases, 1058, 1059, 1111
nitric acid
from hydrolysis of dinitrogen pentoxide, 168, 1060
industrial production of, 743–44
neutralization reaction with calcium hydroxide, 167
as strong acid, 168, 760–61
nitric oxide. *See* nitrogen monoxide
nitrilotriacetic acid (NTA), 1097
nitrite ions
enzyme-catalyzed reduction of, 1058, 1110
from enzyme-catalyzed reduction of nitrate, 1058, 1059, 1111
hydrolysis of, 1060
as ligands, 1107
resonance structures of, 761
strength of nitrous acid and, 760–61
nitrogen
auroras and, 486
in automobile air bags, 232, 234
bonding in diatomic molecule of, 429, 468
deviation from ideality, 252–53
electron configuration of, 365
in living systems, 1057–60
as main component of air, 215, 238–39
molecular orbitals for, 481–82
molecular speed of, 247–48
oxidation numbers in compounds of, 1058
in scuba diving mixture, 215, 240–41, 242, 254–55
with *sp* hybrid orbitals, 468
with sp^2 hybrid orbitals, 466
with sp^3 hybrid orbitals, 464
nitrogen cycle, 1058, 1111
nitrogen dioxide
catalytic converters and, 690
dimerization of, 715–16, 717–18, 719, 728–29, 731–32, 734, 873–74, 876
equilibrium constant for decomposition of, 720–21
equilibrium constant for formation of, 720–21, 877
formed in engines, 1058
hydrolysis of, 1060
as odd-electron molecule, 418, 419–20
in Ostwald process, 743–44
in photochemical smog, 418, 650, 651, 693–94, 714–15, 722
rate of formation of, 655, 656–61, 662, 683–84
thermal decomposition of, 670, 672, 680–83
zero-order reaction with carbon monoxide, 672–73, 686
nitrogen monoxide (nitric oxide)
bond length, 485
as catalyst in ozone decomposition, 689–90
catalytic converters and, 690–91
equilibrium in production of, 726, 741–42
equilibrium in reaction with oxygen, 714–15
formed in engines, 418, 650, 651, 653–54, 656, 690–91, 722, 1058
important roles in biology and medicine, 418
mechanism of reaction with hydrogen, 684–85
molecular orbital theory of, 484–86
as odd-electron molecule, 418–19, 484, 485
in Ostwald process, 743–44
photochemical smog and, 650, 651–52, 693–94, 722
rate of production of, 663–64, 670
rate of reaction with oxygen, 655, 656–60, 662, 683–84
rate of reaction with ozone, 660–61, 675–77
reaction with transition metals, 485
nitrogen narcosis, 240, 242, 254
nitrogenases, 1058, 1060
nitrous acid
conjugate base of, 765
percent ionization of, 774
as weak acid, 760–61, 772–73
nitrous oxide. *See* dinitrogen monoxide
Nobel Prize
Arrhenius, Svante, 756
de Broglie, Louis, 352
Einstein, Albert, 344
Planck, Max, *341*
Urey, Harold, 88
Werner, Alfred, 1085
Zewail, Ahmed H., 680n
noble gases (group 18), 60, 1049
as atomic solids, 602
banned for athletes, 1071
boiling points of, 509, 510
electron configurations of, 365
in medical imaging, 1071
radon exposure, 965–66, 1048
nodes
in *p* orbital, 361–62
in *s* orbital, 360–61
of standing wave, 352, 362

nomenclature, 64
See also names of complex ions; names of compounds; names of polyatomic ions
nonactin, 1051
nonbonding molecular orbital, 487, 488
nonelectrolytes, 164
colligative properties and, 575, 576
nonessential elements in human body, 1046–47, 1067–68
nonmetal oxides, forming acids in water, 168, *169*
nonmetals
in human body, 1050
in periodic table, 59, 1049
nonpolar covalent bond, 405
electronegativity difference (ΔEN) and, 407
nonpolar solvents, 532–33
nonspontaneous processes
definition of, 852
in electrolytic cells, 930
entropy changes in, 856, 858–59
free energy changes in, 865, 873
in living systems, 880–82, 884
recharging of batteries, 928
normal boiling point, 520
normal distribution, 30
N-terminus, 1006
n-type semiconductor, 492, 493
nuclear charge. *See* atomic number (Z); effective nuclear charge (Z_{eff})
nuclear chemistry, 948
nuclear equations, 74
nuclear fission, 968–71
nuclear fusion
efforts to produce energy with, 973–74
in primordial nucleosynthesis, 74, 971–72
solar, 971–73
stellar, 75–76, 948–49, 950
nuclear power plants, 970–71
nuclear reactions, 951–55
nuclear submarine, sinking of, 935–36
nuclear weapons, 969, 970–71
nuclear weapons testing, 48–49, 1067
nucleic acids, 1025–29
enzyme-catalyzed degradation of, 1111
phosphate incorporated into, 1061
See also DNA (deoxyribonucleic acid); RNA (ribonucleic acid)
nucleons, 54
See also neutrons; protons
nucleosynthesis
primordial, 74–75, 971–72
stellar, 75–76, 948–49, 950
nucleotides, 1025–27
enzyme-catalyzed degradation of, 1111
self-assembly of, 1030
nucleus
metastable, 1114
in Rutherford atomic model, 52–53
nuclides, 55–56
See also radionuclides
nylon, *631*, 632–33

O

obsidian, absolute entropy of, 862
ocean acidification, 804–6, 841–42
oceans
of early Earth, 88, 150
as heat sinks, 293
See also seawater
octahedral complex ions, 1086–87, *1087*
chirality in, 1095, 1106
cis–trans isomers in, 1105–6
crystal field splitting and, 1098, 1102–4
with macrocyclic ligands, 1108–9
octahedral coordination compounds, 1107, 1120–21
octahedral electron-pair geometry, 448–49, *455*, 456
octahedral holes, 614, *615*
in ferrite, 615
in lithium chloride, 619–20
in sodium chloride, 616–17
octahedral molecular geometry, 448–49, *455*
octane
distillation with heptane, 564–66
heating curve of, 564
limited solubility in water, 530
miscibility with octanol, 534
standard free energy of combustion, 868
octanoic acid, 810
octanol
miscibility with octane, 534
partition coefficient and, 537
solubilities of alcohols in, 533
solubility of drug candidates in, 537–38
viscosity of, 528
octet rule, 398
exceptions to, 418–26
limitations of models using, 425–26
odd-electron molecules, 418–20
oils, 1022
hydrogenation of, 1022–23, 1031
olestra, 1023
oligomers, 622
oligopeptides, 1006
omega-3 fatty acids, 1020–21, 1022
open system, 280
optical activity, 995–97
optical isomers, 475–76, 995–97, 1000–1001, 1003
See also chirality
orbital diagrams, 363
orbitals, atomic
degenerate, 364–65
diagram of filling sequence, *368*
filling of, 362–70
hybrid, 463–69, *468*, 475
overlap of, 462
periodic table and, 362–70
quantum numbers and, 355–58
sizes and shapes of, 359–62
wave function and, 355
orbitals, molecular. *See* molecular orbital (MO) theory
organic chemistry, 71, 988
organic compounds, 65, 70–73
functional groups of, 72–73, *73*, 472–73, 989, *990*
large number of, 1050
names of, 71–72, *73*
as polar solvents, 924
Wöhler's early synthesis of, 89
See also hydrocarbons
organometallic compounds, 1118
osmarins, 1119
osmium compounds, anti-inflammatory, 1119
osmosis, 579–83
reverse, 583–85
osmotic pressure (Π), 579–83
in egg experiment, 589–90
molar mass determined from, 585–86
reverse, 584–85
in small intestine, 1073–74
Ostwald, Wilhelm, 743
Ostwald process, 743–44
Ötzi the Iceman, 887
outlier, 30–32
overall ionic equation, 166, 167
overall reaction order, 662–64
overlap of atomic orbitals, 462
oxalate ion, *473*
oxidation
definition of, 187
original meaning of, 184
oxidation half-reaction, 900–901
oxidation number (O.N.), 184–86
of main group elements in nature, *1050*
of metal in complex ion, 1086, 1088, 1090, 1091, 1104
in nitrogen cycle, 1058
oxidation state. *See* oxidation number (O.N.)
oxidation–reduction reactions, 184–98
in acidic or neutral solutions, 192–93, 194–95, 905
activity series and, 191–93
annotation of equations for, 186, 190
balancing equations for, 188–91, 194–97, 906–7, 1059–60
in basic solutions, 193–94, 195–96, 905–6, 1059
changes in oxidation number in, 186–87
coenzyme B_{12} in, 1112
cytochromes as catalysts for, 1110–11
electrochemistry and, 900–903
electron transfer in, 187–88
equilibrium constants for, 919–20
glucose metabolism and, 152, 1110–11
in nature, 193–97
net ionic equations for, 190, 195, 901, 905–6
in nitric acid synthesis, 743–44
in nitrogen cycle, 1058, 1059–60
original definition of, 184
photosynthesis as, 1109
spontaneous, 907, 909, 920, 925, 928, 930
See also electrochemical cells; half-reactions

oxide ion, 766, 829
oxidizing agents, 187–88
  free radicals, 419
  permanganate ion, 194
  superoxide, 1112
oxoacids, 69–70, *70*
  acid strength of, 760–62, 783–87, 832
  of chlorine, 761, *762*
oxoanions, *68*, 68–70, 760–61, *762*, 785–87
oxygen
  as component of air, 215, 238–39
  in glucose metabolism, 1020
  hemoglobin binding of, 536
  at high altitude, 241–42, 535, 536
  in human body, 1047, 1050
  molecular energy states of, 885–86
  molecular orbitals for, 481–82
  as paramagnetic substance, 483
  in scuba diving mixture, 215, 240–41, 242, 254–55
  with sp^2 hybrid orbitals, 466
  toxic levels of, 242, 254
  water solubility of, 531, 534–35, 536–37
oxygen generators on airplanes, 232, 233–34
oxygen transport in mollusks, 1112
ozone, 409–11
  bond lengths in, 410, 426
  molecular orbital theory of, 486–87
  permanent dipole of, *459*, 460
  photochemical decomposition of, 665–66, 687–89
  in photochemical smog, 651, 693–94
  reaction with nitrogen monoxide, 660–61, 675–77, 689–90
ozone depletion in stratosphere
  catalysts and, 687–90
  chlorine monoxide and, 671
  Rowland and Molina's prediction of, 687

P

p block elements, 367, 369
  monatomic anions of, 370–71
p orbitals, 355–56, 361–62
packing efficiency, 607, *608*, 611
paclitaxel (Taxol), 132–33
paraffins, 313, *314*
paramagnetic substances
  molecular orbital theory and, 483
  transition metal ions and complexes, 1102–4, 1120
partial electrical charge
  on induced dipole, 509
  of polar covalent bond, 405, 457
partial ionic character. *See* polar covalent bonds
partial pressure
  calculating equilibrium value of, 739–43
  calculation of, 239
  Dalton's law of, 238, 242–43, 244
  definition of, 238
  in equilibrium constant expressions, 714–15, 716–17, 718–19
  of gas collected over water, 242–43
  Le Châtelier's principle and, 731–34
  solubility of gas in liquid and, 535–37
  of water vapor at selected temperatures, *242*
  of water vapor in equilibrium with water, 519–20
particulate representations, 4–5
partition coefficient, 537
parts per billion (ppb), 155–56
parts per million (ppm), 155–56
pascal (Pa), 217
Pascal, Blaise, 217
Paschen, Friedrich, 346, 348
Patterson, Claire, 960
Pauli, Wolfgang, 358
Pauli exclusion principle, 358
Pauling, Linus, 405, 407, 462
PDB (*para*-dichlorobenzene), 431
pentane, 470, 510, *511*
2,4-pentanedione, *473*
Penzias, Arno A., 35
peptide bond, 1006, 1017
peptides, 1006–9
percent composition, 119–24
percent ionization
  of weak acids, 772–75
  of weak bases, 776–78
percent yield, 117–19
perchloric acid, *762*
periodic table, 54, 58–61
  electron configurations and, 362–70
  electronegativity and, 406–7
  orbital filling sequence in, *368*
  symbols in, 54–55
  unit cells of metals and metalloids in, *606*
periodic trends, 1049–50
  in atomic and ionic radii, 373–76, 1049
  in boiling point, 1049, *1050*
  in electron affinity, 1049
  in electronegativity, 1049
  in first ionization energy, 1049
  in melting point, 1049, *1050*
periods, 59–60
permanent dipoles, 457–58, 459–61
  interacting with induced dipoles, 513, *515*, 529, 530, 534
  See also polar molecules
permanganate ion, as oxidizing agent, 194
permeability of cell membrane, 1052
peroxidases, 1110
peroxyacetyl nitrate (PAN), 651, 693–94
petroleum, 108, 109
  cleaning up spills of, 532
  plant-based substitutes for, 623, 627
petroleum distillate, 531
pH, 768–72
  of aqueous salt solutions, 788–92
  calculation for acidic and basic solutions, 778–83
  charge on amino acid and, 1004–6
  of common aqueous solutions, *769*
  of digested food, 770, 771–72
  of human blood, 792–93, 806, 1004
  ionization of weak acid and, 774–75
  ionization of weak base and, 776–77
  pathways for conversions involving, *778*
  of seawater, 788, 805
  significant figures in, 770
  solubility and, 836–38
  of solution of conjugate acid–base pair, 806–9
  of strong acid solution, 778
  of strong diprotic acid solution, 783–84
  of swimming pool water, 816–17
  of very dilute strong acid, 781–82
  of weak acid solution, 779–80
  of weak base solution, 780–81
  of weak diprotic acid solution, 785–86, 792–93
pH buffers, 809–16
  buffer capacity of, 812–16
  preparing with desired pH, 811–12
  selecting components for, 809–10
  water solubility required for, 810
pH electrode, 818
pH indicators, 816–17, *817*
  for alkalinity titration, 826
pH meter, 818
phase changes. *See* changes of state
phase diagrams, 523–26
  for aqueous solution of nonvolatile solute, 569–70
  for carbon dioxide, 524, 525
  for water, 523–26, 569–70
phase equilibria, 869–70
phase symbols, 8
  for aqueous solution, 10
phenanthroline, 162–63
phenol red, 816–17, 826
phenolphthalein, 173, 826
pheromones, of insects, 127, 494
phosgene, equilibrium in decomposition of, 742–43
phosphate esters, 1060, *1061*
phosphate ion, 422–23, 1060–61
phosphine, 1060
3-phosphoglycerate^{3-} (3-PG^{3-}), 883–84
phospholipids, *1024*, 1024–25
phosphoric acid, 786–87, 1060
phosphors, 956
phosphorus, 1060–61
phosphorus cycle, 1060
phosphorus pentachloride, 420–21, 487, 488–89, 740–41
phosphorus pentafluoride, trigonal bipyramidal geometry of, 450
phosphorus trichloride, 460, 740–41
phosphorus-32, radioactive decay of, 953–54
phosphorus-doped silicon, 492
phosphorylation of glucose, 882–83, 1014

photochemical smog, 418, 650–52, 693–94, 714–15, 722
photoelectric effect, 343–45, 490
photon, 342, 350
photosynthesis, 107–8, 112
  annual turnover of water column and, 529
  carbon dioxide consumed in, 394
  copper-containing enzymes in, 1112
  cytochromes in, 1110
  light-absorbing molecules in, 1054, *1055*
  manganese and, 1109
  as nonspontaneous process, 881
  radiocarbon dating and, 960
phototube, *344*
phycoerythrobilin, *1055*
physical process, definition of, 5
physical properties, 13
physiological pH
  charge on amino acid at, 1004
  charge on peptide at, 1007
physiological saline, 161, 583, *583*
pI (isoelectric point), 1004–6
pi (π) bonds
  in benzene, 472
  between carbon atoms, 467, 621
  definition of, 466
  delocalization of electrons in, 472–73
  in ethylene, 473–74
  in formaldehyde, *465*, 466
  formed by nitrogen, 466
  in stereoisomers, 994–95
pi (π) molecular orbitals, 481
picomolarity, 154
pinene, 119–20
piston and cylinder, 284, 285
pitchblende, 51, 969
pK_a, 771
  Henderson-Hasselbalch equation and, 807–9
  pH indicators and, 816–17
  titration curves of amino acids and, 1004–6
pK_b, 771
  calculating from pK_a, 808
  Henderson-Hasselbalch equation and, 808–9
pK_w, 771, *778*
Planck, Max, 341, 345
Planck constant, 341–42, 348, 350, 353
plane-polarized light, 475, 995–97, 1000–1001, 1003
plaque
  of cholesterol deposits, *1025*
  of protein deposits, 1012
plaster of Paris, 560–61
plastocyanin, 1112
platinum
  in cisplatin anticancer agent, 1104–5, 1118
  gold alloyed with, 615–16
  microscope image of atoms, *7*
platinum group metals, 1118, 1119
platinum–nickel catalyst, in fuel cell, 931–32
platinum–rhodium catalyst, in Ostwald process, 743
β-pleated sheet, 1011–12
plum-pudding model, 51–52
plutonium-239, 970–71
pOH, 771–72
  ionization of weak base and, 776–77
  pathways for conversions involving, *778*
polar amino acids, 1001, *1002*, 1003
polar covalent bonds, 405–9, 457–61
  electronegativity difference and, 407–8, 426
  in strong acids, 758
polar molecules
  bond dipoles in, 457–61
  intermolecular forces in, 511–19
  solubility and, 529–34
polar solvents
  dissolution of ionic compounds in, 553–55
  organic, 924
polarizability, of neutral atoms and molecules, 509–10, 513, 1049
polarized light, 475, 995–97, 1000–1001, 1003
polonium, 607, 948, 955, 965
polyamides, 631–34
polyamines, 1095
polyatomic ions, 68–69
  common examples of, *68*
  formal charge of, 416
  in human body, 1050, 1061
  i factor for, 576
  Lewis structures of, 399, 403
  oxoanions, *68*, 68–70, 760–61, *762*, 785–87
  in phosphorus cycle, 1060–61
  radii of, *552*
  resonance structures of, 413–14, 416
polycyclic aromatic hydrocarbons (PAHs), 473
polydentate ligands, 1095–97
  biological, 1108–13
  macrocyclic, 1108
  prefixes for number of, 1106
polyesters, 628–30
polyethylene (PE), 622–23, *624*
poly(ethylene glycol) (PEG), 626–27
poly(ethylene oxide) (PEO), 626–27
poly(ethylene terephthalate) (PETE), 626, 629
poly(ethylene-co-vinyl alcohol) (EVAL), 626
polymers, 604, 622–34
  of alcohols and ethers, 626–28
  of alkenes, 622–25
  with aromatic rings, 625–26, 633–34
  biopolymers, 989, 1006
  compared to small molecules, 622
  copolymers, 626
  gold circuit printed on, 600
  heteropolymers, 626
  medical applications of, 623, 624, 629
  molar masses of, 622
  polyamides, 631–34
  polyesters, 628–30
  polyvinyl chloride, 156–57
  summary of common polymers, *634*
polypeptides, 1006
polypropylene, 624–25
polyprotic acids, 169, 783–87
polysaccharides, 1016, 1017–19
polystyrene (PS), 625–26
polytetrafluoroethylene (Teflon), 623–24
polyunsaturated fatty acids, 1020–21
poly(vinyl alcohol) (PVAL), 626
poly(vinyl chloride), pipes made of, 156–57
pool test kits, 816, *817*
porphyrins, 1108
  of cytochromes, 1110–11
positron, *949*, 952
positron emission, 952
  atomic number and mass number in, *954*
  by carbon isotopes, *953*
  in gallium nuclide decay, 1070
  in nuclear fusion, 972, 973, 974–75
positron emission tomography (PET), 968, 1069, 1070, 1071, 1113
potassium
  electron configuration of, 366–67
  in living organisms, 1051–52, 1053
potassium ion channel, 1051
potential energy (PE), 273–75
  gravitational, 274
  at molecular level, 276–79
  as part of internal energy, 283
  See also electrostatic potential energy (E_{el})
pounds per square inch (psi), 217
power, units of, 921
precipitate, 176
precipitation
  selective, 840–41
  solubility and, 182
precipitation reactions, 152, 176–82
  in analytical chemistry, 180–82
  definition of, 176
  net ionic equations for, 178–79
  solubility and, 176–78, 839–41
  for synthesizing slightly soluble salts, 179
  tooth enamel formed by, 152
precision, 27–32
pressure
  boiling point and, 524
  definition of, 215
  kinetic molecular theory and, 244–45
  measurement of, 215–20
  phase diagrams and, 523–26
  units of, 216–17
  volume of gas and, 220–23
  water solubility of gases and, 534, 535
  See also atmospheric pressure (P_{atm}); partial pressure; vapor pressure
pressure–volume (P–V) work, 284–86
  by car engine, 869
  enthalpy and, 286
  sign of, 284
  units of, 285–86
primary (1°) structure of proteins, *1009*, 1010–11
primordial nucleosynthesis, 74–75, 971–72
principal quantum number (n), 355–56, *357*
problem-solving framework, 11–12

product, 8, 90–91
  mass of reactant and, 110–12
  removing or adding, to shift reaction, 729–31
propane
  as alkane, 313, 314
  boiling point, 518–19
  enthalpy of combustion, 299–300, 311–12
  entropy of combustion, 861
  in natural gas, 314
  standard molar entropy of, *862*
propene. *See* propylene (propene)
propylene (propene), 71–72
  polymer of, 624–25
  simulated smog and, 693–94
propylene carbonate, 924
propyne, 72
protease inhibitors, 1015
proteins, 1001–15
  abundance of, 1001
  acid–base characteristics of, 1106
  in bone, 1054
  copper-containing, 1112
  functional classes of, *1010*
  glycoproteins, 1016
  hydrogen bonds in, 516, 1006, 1011–12
  inherited diseases and, 989
  number of amino acid residues in, 1006
  primary (1°) structure of, *1009*, 1010–11
  quaternary (4°) structure of, *1009*, 1013
  salting out from aqueous solutions, 550
  secondary (2°) structure of, *1009*, 1011–12
  separation by electrophoresis, 10
  structure of, 1009–13
  synthesis of, 1027–29
  tertiary (3°) structure of, *1009*, 1012–13
  See also amino acids; enzymes
proton acceptor (base), 165
proton donor (acid), 165
proton pump inhibitor, 1057
proton-exchange membrane (PEM), 931–32
protons
  Big Bang and, 74
  meaning hydronium ions, 165–66
  in nuclear fusion, 971–75
  properties of, *53*
  in Rutherford's atomic model, 52–53
  symbol for, 74, *949*
Prussian blue, 1088
p-type semiconductor, 492, 493
pure substances, 5–6
putrescine, 444
P–V work. *See* pressure–volume (*P–V*) work
pyridine, 764
pyrites, 171
pyruvate ions, *881*, 1019–20, 1025, 1054

Q

quantized property, 342, 348
quantum, definition of, 341
quantum mechanics, 354
  See also quantum theory
quantum numbers, 354–58
  periodic table and, 368–69
quantum theory
  electron waves and, 350–53
  entropy and, 885–86
  Heisenberg uncertainty principle, 352–54
  hydrogen spectrum and, 345–50
  indeterminacy in, 359
  lasers and, 359
  photoelectric effect and, 343–45
  of Planck, 341–43
  quantum numbers in, 354–58, 368–69
  summary of development of, *359*
  valence bond theory and, 462
quarks, 74
quartz, absolute entropy of, 862
quaternary (4°) structure of proteins, *1009*, 1013

R

racemic mixture, 1000, 1014
rad (radiation absorbed dose), 963
radial distribution profiles
  of 1*s* orbitals, 360, *361*
  of 2*s* and 2*p* orbitals, 363–64
  relative sizes of *s* orbitals, 360, *361*
radiation. *See* electromagnetic radiation; gamma (γ) rays; infrared radiation; ionizing radiation; ultraviolet (UV) radiation
radioactive decay, 950–55, *954*
  See also alpha (α) decay; beta (β) decay; electron capture; gamma (γ) rays; positron emission
radioactive decay series, 954–55
radioactivity
  of actinides, 60
  biological effects of, 948, 962–67
  definition of, 48
  early experiments on, 51–52
  measurement of, 956–58
  medical procedures using, 77, 948
  from nuclear weapons testing, 48–49, 1067
  sources of exposure to, *965*
  units of, 956
radiocarbon dating, 946–47, 960–61
  calibration curve for, 961, *962*
radiology, 967–68, *968*
  See also medical imaging
radiometric dating, 959–61
radionuclides, 951–55
  medical applications of, 77, 967–68, 1114–15
  medical imaging with, 1068–71, 1113–15
radium, 948, 955, 957–58, 975–76
Radium Girls, 975–76
radon gas, 965–66, 1048
rainbows, 336, 340
rainwater
  carbonic acid in, 167–68, 785–86
  pH changes produced by soils, 788
  See also acid rain
randomly distributed data, 30
Raoult, François Marie, 562
Raoult's law, 562–64
  deviations from, 568–69
  for mixtures of volatile compounds, 567–68
rate constant, 661, 662
  Arrhenius equation for, 675–79
  calculating activation energy from, 677–78
  for first-order reaction, 666–67
  half-life and, 668–69
  from initial reaction rate data, 662–64
  for radioactive decay, 956–58
  for second-order reaction, 670–72
  temperature and, 662, 677, 678–79
  units of, 662–63
  for zero-order reaction, 673
rate law
  definition of, 661
  for first-order reaction, 664–68
  from initial reaction rate data, 662–64
  reaction mechanisms and, 682–86
  for second-order reaction, 670–72
  summary of, for single reactant, 673–74
  for zero-order reaction, 672–73
  See also reaction order; reaction rates
rate-determining step, 682–86
  zero-order reactions and, 686
reactants, 8, 90–91
  limiting reactant, 112–16
  mass of product and, 110–12
  removing or adding, to shift reaction, 729–31
reaction mechanisms, 652, 680–87
  difficulty of proving, 684
  elementary steps in, 681–83
  rate laws and, 682–86
  rate-determining step in, 682–86
  zero-order reactions and, 686
reaction order, 659–64
  definition of, 659
  determination of, 661–62
  distinguishing between first and second order, 671–72
  first-order, 664–69, 671–72
  overall, 662–64
  second-order, 670–72
  summary of, for single reactant, 673–74
  zero-order, 672–73, 686
reaction quotient (*Q*), 724–26
  cell potential and, 916–17, 918–19
  free energy change and, 872–74
  for ionization of weak acid, 774
  Le Châtelier's principle and, 730, 732, 734
  solubility equilibria and, 839–40
reaction rates, 652–54
  average, 656
  concentration and, 652, 653–54, 659–74
  at equilibrium, 712
  experimentally determined, 654–56
  factors affecting, 652–53
  instantaneous, 656–59
  rate-determining step and, 682–86
  relative, 654–56

reaction rates *(cont.)*
sign of, 653–54
temperature and, 653, 662, 674–79, 690
units of, 654
See also rate constant; rate law
reactions. *See* chemical reactions; nuclear reactions
real gases, 250–53
receptors, biological, 473, 475
chirality and, 1000
nasal, 475, 997, 1063
recommended dietary allowance (RDA), 1048, 1108
red beans and rice, 1003
red blood cells
carbon dioxide in, 1056
iron excess and, 1116
osmosis and, *579*, 581, 582–83
sickle-cell anemia and, 1010–11
red fireworks, 380
redox flow batteries, 934
redox reactions, 152, 184
See also oxidation–reduction reactions
reducing agent, 188
reductases, 1058, 1059, 1111
reduction
definition of, 187
original meaning of, 184
See also oxidation–reduction reactions
reduction half-reaction, 900–901
reduction potentials ($E°$), 907–11
corrosion and, 925–26
relative biological effectiveness (RBE), 963
rem (roentgen equivalent man), 963
remote-control device, 344
replication fork, 1026, *1028*
replication of DNA, 1026–27, *1028*
representative elements. *See* main group elements
residues, amino acid, 1006
resonance, 409–14
bond lengths and, 426
exceptions to octet rule and, 421–26
formal charge and, 414–18, 419–20, 421–24, 425–26
molecular orbital theory and, 486–87
in PDB (*para*-dichlorobenzene), 431
resonance stabilization, 472
respiration, 108, 114–15, 306
respiratory acidosis, 757
reversal potential (E_{ion}), 1052–53
reverse osmosis, 583–85
reverse reactions, K values for, 719–21
reversible process, 857, 858, 869
reversible reactions, 711–12
See also chemical equilibrium
rhenium, 1114
rhodium
antitumor complexes of, *1118*, 1119
in catalyst for Ostwald process, 743
ribose, 1025
ribosome, 1028
RICE table
common-ion effect, 806, 836
equilibrium concentrations or pressures, 737–43
ionization of weak acid, 775
ionization of weak base, 777
molar solubility, 835, 836
pH of salt solutions, 791, 792
pH of strong diprotic acid solution, 784
pH of very dilute strong acid, 782
pH of weak acid, 779–80
pH of weak base, 780–81
pH of weak diprotic acid solution, 786, 793
uncomplexed metal ion concentration, 832
See also ICA table
rich mixture, 115
Ringer's lactate solution, 159
RNA (ribonucleic acid), 1025–26, 1028–29
beryllium toxicity and, 1048
RNA world hypothesis, 1030
rock candy, 182
rock salt, used to melt snow, 180
rock salt structure, 634–35
Roman numerals, 67–68, 1088
Röntgen, Wilhelm Conrad, 51, 51n
root-mean-square speed (u_{rms}), 246–49
rotational motion
energy levels of, 885–86
internal energy and, *283*
temperature and, 859
of water molecules, 854, *855*
rounding off, 24
See also significant figures
Rowland, Sherwood, 687
rubidium, in human body, 1047, 1067
rust, 193, 850, 853, 925
See also corrosion
ruthenium, antitumor complexes of, *1118*, 1119
Rutherford, Ernest, 51–53, 54, 347, 959
Rydberg, Johannes Robert, 346, 347

S

s block elements, 367, 369
monatomic cations of, 370
s orbitals, 355–56, 359–61
sacrificial anodes, 927, *928*
saline solution, 161, 163, 583, *583*
salt bridge, 902–3
salting out effect, 550
salts
acid–base properties of, 788–92
definition of, 166
solubilities of, 176–77, 550–51
See also ionic compounds
sanochrysin, 1119
Sapphire Pool in Yellowstone National Park, 826–27
saturated fatty acids, 1020, *1021*, 1022
saturated hydrocarbons, 313
saturated solution, 182–83
scale, produced by hard water, 587
scandium, 367, 372, 1114
scanning tunneling microscope (STM), *7*
Schrödinger, Erwin, 354, 355
Schrödinger wave equation, 354–55, 357, 358
scientific method, 16–18
scientific theory (or model), 17
scintigraphic imaging, 1114–15
scintillation counters, 956, 1114
scuba diving, 215, 240–41, 242, 254–55
seawater
acidification of, 804–6, 841–42
apparent K_a values in, 841–42
bromide ions in, 1068
common-ion effect on solubility in, 835–36
concentrations of major constituents, 154, *155*
corrosion in, 925, 926, 927, *928*, 935–36
desalination of, 10, 583–84
deuterium in, 974
diluted in estuaries, 161
elemental composition of, 1047–48
ion pairs in, 841
marine mammals in, 148–49
osmosis and, 580, 581–82, 584
pH of, 788, 805
selective separation of ions from, 840
vapor pressure of, 561–62
zeolites formed by, 588
second ionization energy (IE_2), 377, 378
second law of thermodynamics, 852, 855–56, 858
secondary (2°) structure of proteins, *1009*, 1011–12
second-order reactions
distinguishing from first-order, 671–72
half-life of, 672
integrated rate law for, 670–72
summary of how to distinguish, 673–74
seed crystal, 182
seesaw molecular geometry, 454, *455*, 457
selective precipitation, 840–41
selenium, 1065–66
selenocysteine, 1066, 1067
semiconductors, 491–93
carbon allotropes with properties of, 621
semimetals (metalloids)
in periodic table, 59, 1049
as semiconductors, 491–93
unit cells in, *606*
semipermeable membrane, 579
of chicken egg, 589
walls of small intestine, 1073, 1074
See also osmosis; reverse osmosis
sequestering agent, 1096
serum, human, major constituents of, 154–55
shape memory alloys, 613
shapes
of larger molecules, 470–71
London dispersion forces and, 510–11
See also molecular geometry; molecular recognition
shells, 355–56, 368–69
filled, 363
valence shell, 364, 365, 366

SI units, 18–20
base units, *19*
conversions involving, 20–22
derived units, 19–20
of ionizing radiation, *963*
prefixes for, *19*
U.S. Customary units and, 19, *20*
sickle-cell anemia, 1010–11
side chain of alkane, 991
side-chain groups, of amino acids, 1001
sievert (Sv), 963
sigma (σ) bonds
between carbon atoms, 467, 620, 621
definition of, 462
between hydrogen atoms, 462
pi (π) bonds and, 466
sigma (σ) molecular orbital, 478
significant figures, 22–27
of pH values, 770
silica, amorphous, 1067
silicic acid, 1067
silicon
in biological systems, 1065, 1067
diamond crystal lattice of, 621
as semiconductor, 491–92
silicon dioxide
absolute entropy of, 862
adhesive forces of glass and, 527
biological roles of, 1067
silk, 1012
silver
antitumor complexes of, 1118–19
drugs containing, 1119
electron configuration of, 370
electron configuration of cation of, 372
oxidation in seawater, 935–36
silver sulfadiazine, 1119
silver–silver chloride electrode, 915–16
simple cubic (sc) unit cell, *605*, 607, 608
of cesium chloride, 617–18
holes in, *615*, 617–18
single bond, 399
hybridization and, 467, *468*
skeletal structure, 399, 470–72
skunk odor, 1064
slightly soluble ionic compounds
guidelines for, *176*
synthesis of, 179
smell
molecular recognition by, 446, 475, 476, 997, 1063
of sulfur compounds, 1062–64
smog
pea-soup, 650
photochemical, 418, 650–52, 693–94, 714–15, 722
simulations of, 693–94
sodium
electron configuration of, 365–66
emission spectrum of, 337, 340, 366
in living organisms, 1051–54
sodium-24, in medical imaging, 968
sodium azide, in air bags, 232, 234
sodium bicarbonate
in antacid tablets, 153, 174–75
endothermic reaction with acetic acid, 853, 854
sodium carbonate, added to swimming pool, 816
sodium chlorate, in oxygen generators, 232, 233–34
sodium chloride
boiling point elevation due to, 575
Born–Haber cycle for formation of, 556–57
crystal structure, 616–17
dissolving in water, 511–12
electrolysis of, 930
empirical formula of, 62–63
enthalpy of hydration, 559–60
freezing point depression due to, 575–76, 577, 578
lattice energy, 556–57
Lewis structure of, 404
molality vs. molarity of solution, 570
neutral solution of, 788
osmotic pressure and, 580, *580*, 581–82, 583
solubility in water, 550–51, 552
as strong electrolyte, 163, 575, 577
van't Hoff factor for, 575–76, 577–78
See also saline solution; seawater
sodium fluoride
lattice energy of, 557–58
weakly basic solution of, 788
sodium hydroxide
neutralization reaction with hydrochloric acid, 166–67
pH of solution of, 778–79
titration with strong acid, 823
sodium hypochlorite, acid–base properties of, 789, 790–91
sodium ion channel, 1051
sodium ions
electron configuration of, 370, 371
in living organisms, 1051–54
size of, 375–76
solar fusion, 971–73
solar power, 933–34
solder, 612
solid solutions, 151, 604, 612
solids
classification of, 602, *603*
compared to liquids and gases, 14–16, 214–15, 508–9
definition of, 14
equilibrium constant expressions and, 727
intermolecular forces in, 508–9
phase symbol for, 8
standard states of, *860*
solubility
of alcohols in water, 529–30
biological availability of metals and, 1084
of common ionic compounds, 176–77
common-ion effect and, 835–36
definition of, 182, 834
of gases in water, 534–37
of glucose in water, 1016
molar, 834–38
pH and, 836–38
polarity and, 529–34
of polymer PEG, 627
rules for determining, *176*, *177*
saturated solutions and, 182–83
separating ions in solution and, 841–42
See also dissolution of ionic compounds
solubility equilibria, 834–41
reaction quotient and, 839
solubility product (K_{sp}), 834–41
reaction quotient and, 839–40
solute, definition of, 151
solutions
components of, 151
concentrations of, 153–59
definition of, 7, 151
dilution of, 159–62
enthalpy of solution, 303–4, 554–55, 559–61
ideal, 562–64, 568, 577, 841
preparing with known concentration, 158–59
saturated, 182
solid, 151, 604, 612
standard states of, *860*
vapor pressure of, 561–64
See also aqueous solutions; colligative properties; concentration; dissolution of ionic compounds; solubility
solvated ions, 512, 554
See also hydrated ions
solvent
definition of, 151
mole fraction of, 562
polar vs. nonpolar, 532–33
Sørensen, Søren, 768
sotalol, 1000
sp hybrid orbitals, 467–69
sp^2 hybrid orbitals, 465–66
sp^3 hybrid orbitals, 463–65
space-filling molecular models, 9
spearmint, 446, 475, 476
specific heat (c_s), 290
absolute entropy determined from, 860
measurement of, 297–99
spectator ions, 166, 167
in electrochemical cell, 901
in redox reactions, 190
spectrochemical series, 1101, 1103–4, 1121
spectrophotometry, 162–63
spectrum
electromagnetic, *338*
of emission and absorption by elements, 337–38
of heated metal rod, 341
speed of light, 339
sphalerite, 618
sphalerite structure, 618

sphere of hydration, 512
  of complex ion, *833*
sphere of solvation, 512
spider silk, 1012
spin magnetic quantum number (m_s), 357–58, 482
spin states of transition metal ions, 1102–4
spin-paired electrons, 363
spodumene, 123–24
spontaneous processes, 852–54
  chelate effect and, 1096
  definition of, 852
  endothermic, 853–54, 864, 871–72
  entropy changes in, 855–56, 858–59, 864–67, 869–72, 1096
  equilibrium constant and, 875–76
  free energy changes in, 865–67, 869–72, 873
  in living systems, 880–84, 1053–54
  temperature and, 869–72
spontaneous redox reactions
  corrosion as, 925
  standard cell potential and, 909, 920
  standard reduction potential and, 907
  in voltaic cells, 928, 930
square packing, *605*, 607
square planar complex ions, 1087, *1087*, 1099–1100
  cis–trans isomers in, 1104–5
  with macrocyclic ligands, 1108
square planar molecular geometry, *455*, 456
square pyramidal molecular geometry, *455*, 456
stacking patterns, 604–7, *608*
stainless steel, 614, 927
stalactites and stalagmites, 168, 182
standard cell potential (E°_{cell}), 908–10
  equilibrium constant and, 919–20
  of fuel cells, 932–33
  hydrogen electrode and, 914–16
  nonstandard conditions and, 916–19
  standard free energy change and, 912–13
  Statue of Liberty and, 926
  See also cell potential (E_{cell})
standard conditions, 308
standard deviation (s), 28–32
standard enthalpy of combustion (ΔH°_{comb})
  for common fuels, *315*
  fuel value and, 316
  for propane, 311–12
  See also enthalpy of combustion (ΔH_{comb})
standard enthalpy of formation (ΔH°_f), 308–13
  lattice energy calculated from, 556
  for selected substances, *308*
standard enthalpy of reaction (ΔH°_{rxn}), 308
  equilibrium constant as function of temperature and, 878–79
  free energy change and, 866–67
  for reverse reaction, 311
  from standard enthalpies of formation, 308–13
  See also enthalpy of reaction (ΔH_{rxn})
standard entropy of reaction (ΔS°_{rxn}), 863–64
  equilibrium constant as function of temperature and, 878–79
  free energy change and, 866–67
standard free energy of formation (ΔG°_f), 867–68
  equilibrium constant and, 873, 877
standard free energy of reaction (ΔG°_{rxn}), 866–68
  equilibrium constant and, 874–77, 878, 879–80
  glucose metabolism and, 882–84
  reaction quotient and, 873–74
  temperature effect on spontaneity and, 871–72
standard heat of formation. *See* standard enthalpy of formation (ΔH°_f)
standard heat of reaction. *See* standard enthalpy of reaction (ΔH°_{rxn})
standard hydrogen electrode (SHE), 914–16
standard molar entropy (S°), 860–61
  chemical reaction and, 863–64
  molecular structure and, 862–63
  selected values of, *860*
standard reduction potentials (E°), 907–11
  corrosion and, 925–26
  as intensive property, 911
standard solution, definition of, 172
standard states, 308
  free energy changes and, 877
  of pure substances and solutions, *860*
standard temperature and pressure (STP), 229
standing wave, 352
starch, 1016, 1018, 1019
state function
  definition of, 274
  enthalpy as, 287, 306
  entropy as, 863
  internal energy as, 283
  potential energy as, 274
states of matter, 14–16
  reaction rates and, 652
  standard entropy values of, 860–61
  See also changes of state
Statue of Liberty, 925–27
steam. *See* water vapor
steam–methane reforming reaction, 310–11, 710–11, 718–19, 1031
steel, 614–15
  inhibiting corrosion of, 927, *928*
  properties of, *615*
  stainless, 614, 927
stellar nucleosynthesis, 75–76, 948–49, 950
stents, arterial, 612–13
stereoisomers, 994–1001
  classification of, *998*
  of coordination compounds, 1104–6, 1107
  of lipids, 1020, 1021–22
  See also chirality; cis–trans isomers; enantiomers
steric number (SN), 448, 449
  of complex ions, *1087*
  greater than 4, 487–90
  hybridization and, 464, 465, 467, *468*
Stern, Otto, 357
stimulatory effect, 1047, 1068
stock solutions, dilution of, 159–62
stoichiometric yield, 114
stoichiometry
  balancing chemical equations in, 100–107
  definition of, 99
  gases in, 233–35
  limiting reactants in, 112–16
  masses of products and reactants in, 110–12
  mole ratio in, 113–15
  moles in, 91–93
  percent yield in, 117–19
  See also balanced chemical equations
straight-chain hydrocarbons, 314, *314*, *315*, 510
strong acids, 168, *169*
  activity series of metals and, 192–93
  common examples of, 758, *758*
  conjugate bases of, 766
  definition of, 168
  diprotic, 783–84
  ionization reactions in water, 758
  leveling effect of, 766
  oxoacid strength and, 760–62
  pH of nearly all solutions, 778
  pH of very dilute solutions, 781–82
strong bases, 169, 763
  calculating pH of, 778–79
  conjugate acids of, 766
  leveling effect of, 766–67
strong electrolytes, 163
  strong acids as, 168
  strong bases as, 169
  van't Hoff factor for, 575–78
strong nuclear force, 949
strontium
  electron configuration of, 369
  in human body, 1067
  isotopic abundances of, 57–58
strontium sulfate, 552–53, 1067
strontium-90, 48–49, 1067
structural formulas, 9
structural isomers, 991–94
styrene, 625
subatomic particles, 49–54, *53*, 74, *949*
  See also electrons; neutrons; protons
sublimation, 15, *281*
  enthalpy of, 287, 556
  on phase diagram, 524
sublimation points, 524
subshells, 355–56, 368–69
substitutional alloys, 613–14, 615–16
substrate, 692–93, 1014–15
sucrose, 1017
  in olestra, 1023

sulfate ions
  ion channels for, 1113
  Lewis structure of, 421
  in sulfur cycle, 1062, 1111
sulfite ions
  enzyme-catalyzed oxidation of, 1111
  enzyme-catalyzed reduction of, 1110
  produced by Lewis base, 829
sulfite oxidase, 1111
sulfur, 1060, 1061–65
  odiferous compounds of, 1062–64
sulfur cycle, 1061–62, 1064–65, 1111
sulfur dioxide
  from coal-burning power plants, 727, 783, 829
  geometry of, 452
  as Lewis acid, 829
  reaction with calcium oxide, 727, 728
  in sulfur cycle, 1061–62
  sulfur trioxide produced from, 103–4, 721–22, 783
  from volcanic activity, 90
sulfur hexafluoride, 420–21
  bonding theory for, 487, 489–90
  as greenhouse gas, *420*
  octahedral geometry of, 450, 489
sulfur tetrafluoride, 456–57
sulfur trioxide
  acid rain and, 91, 103, *168*, 783, 1061–62
  reaction with water vapor, 90, 98–99, 168, 783
  resonance structures of, 412
  in sulfur cycle, 1061–62
  from sulfur dioxide and oxygen, 103–4, 721–22, 783
sulfuric acid
  from abandoned coal mines, 171–72
  in acid rain, 168, 183, 783
  determining concentration by titration, 171–73
  as diprotic acid, 168–69, 783
  in early atmosphere, 90–91
  Lewis structure of, 421–22
  pH of solution of, 783–84
  as strong acid, 168
  structural reasons for strength of, 761
  from sulfur trioxide and water, 90, 91, 98–99, 168, 783, 1061–62
sunlight
  chemical energy from, 272
  energy in food derived from, 881
  photochemical decomposition of ozone and, 665–66, 687–89
  spectrum of, 336–37, 338
  See also photochemical smog; photosynthesis
supercritical fluid, 525
supercritical region, 523
supernovas, 76, 950
superoxide dismutases, 1112
superoxide ion, 1112
supersaturated solution, 182
surface area, and evaporation, 519
surface tension, 526–27
surfactants, 532
surgical steel, 614
surroundings, 279
  entropy and, 855–59, 864
  first law of thermodynamics and, 284, 285
swamp gas, 314, *315*
swimming pool water
  at 2016 Olympics, 1120–21
  pH of, 816, *817*
system
  definition of, 279
  entropy and, 855–59, 864–65
  first law of thermodynamics and, 284, 285
  types of, 279–80

T

tantalum, 607, 610–11
Taxol (paclitaxel), 132–33
t-distribution, 29
technetium, 77, 1114, *1115*
teeth
  compounds used to fill caries, 93
  enamel of, 152, 1066
  fluoride and, 155, 1066
  hydroxyapatite in, 152, 1054, 1066
  strontium-90 in, 48–49
Teflon (polytetrafluoroethylene), 623–24
temperature
  density of a gas and, 235–38
  of Earth, 394
  energy of particles and, 276–77
  entropy change and, 856–58, 859–60, 861
  equilibria shifting due to changes in, 734–36
  equilibrium constant and, 713, 734–35, 877–80
  heat and, 273, 281
  of interstellar space, 32, 34
  kinetic molecular theory and, 245–47
  molar heat capacity and, 290
  phase diagrams and, 523–26
  radiation from hot objects and, 341
  rate constant and, 662, 677, 678–79
  reaction rates and, 653, 662, 674–79, 690
  solubility and, 182
  spontaneous processes and, 869–72
  vapor pressure and, 519, 520–23
  volume of a gas and, 223
  water solubility of gases and, 534–35
temperature scales, 32–34
termolecular elementary step, 681
tertiary (3°) structure of proteins, *1009*, 1012–13
tetrahedral complex ions, 1087, *1087*, 1100, 1104
tetrahedral geometry, 448, 450, 453, 454, *455*, 461
  in β-pleated sheet, 1011
  nonpolar substances with, 458
  sp^3 hybrid orbitals and, 463–65, *468*
tetrahedral holes, 614, *615*
  in binary ionic solids, 618, 635
tetrahydrofuran, 924
thalassemia, 1116–17
thalidomide, 1014
thallium-201, 1070
theoretical yield, 114, 115–16
  vs. actual yield, 116–19
theory, scientific, 17
therapeutic radiology, 967
thermal conductivity, 621
thermal energy, 277
thermal equilibrium, 273
thermochemical equations, 300
thermochemistry, 273
thermocline, 529
thermodynamics
  definition of, 273
  first law of, 284, 285, 852
  of living systems, 880–84
  second law of, 852, 855–56, 858
  third law of, 860, 886
thermos bottle, 279–80
thiocyanate ligand, 1107
third law of thermodynamics, 860, 886
Thomson, Joseph John (J. J.), 49–50, 51, 54
thorite, 634–35
thorium
  oxide of, 634–35
  in radioactive decay series, 954–55
Thresher, USS, 935–36
threshold frequency (ν_0), 343, 344–45
thymine, 1025–28
thyroid hormones, 1066–67
tin, 612, 613, 621
Titan, atmospheric pressure on, 218
titanium
  cation of, 372
  in complexes with antitumor activity, 1117–18
  electron configuration of, 367
  hcp crystal structure of, 605–6
  in nitinol, 612–13
titrant, 172, 818
titrations. *See* acid–base titrations
tokamak, 973–74
torr, 216–17
Torricelli, Evangelista, 217
Toyota Prius, 921–22
trace essential elements, 1046, 1048, 1065, 1066–67
  transition metals among, *1084*, 1108, 1109–11
tracer, radioactive, 968
trans fats, 1023
trans isomers, 994
  See also cis–trans isomers
transcription, 1028
transfer RNA (tRNA), 1028–29
transferrin, 1072, 1116–17, 1118

transition metals
  bcc unit cells in, *606*, 607
  common ionic charges of, *1091*
  coordinate bonds and, 1085–87
  densities of, 607
  electron configurations of, 367–68, 370
  electron configurations of cations of, 372–73
  in enzymes, 1109–13
  essential for human health, *1084*, 1084–85, 1108, 1109–13
  in medicine, 1113–19
  naming compounds of, 67–68, 1088, 1092–93
  nitrogen monoxide reactions with, 485–86
  in periodic table, 59
  writing formulas containing, 67–68, 1086, 1088–93
  See also complex ions; coordination compounds
transition state, 676
  of enzyme-catalyzed reaction, 692–93, 1014
  two-step reaction and, 681
translation, 1028–29
translational motion
  energy levels of, 885, 886
  internal energy and, *283*
  temperature and, 859–60
  of water molecules, 854, *855*
transmutation of elements, 52
transport rate constant, 669
tree rings, carbon-14 content of, 961
tricarboxylic acid (TCA) cycle, 1020, 1112
triglycerides, 1020, 1021–22, 1024
  hydrogenation of, 1031
trigonal bipyramidal geometry
  with lone pairs, 454–56, 457
  with no lone pairs, 448
  summary of, *455*
trigonal planar geometry, 448, 449, *455*
  of ethylene, 473, *474*
  of formaldehyde, 451, 460
  of ozone electron pairs, 460
  sp^2 hybrid orbitals and, 465–66, *468*
trigonal pyramidal molecular geometry, 453, *455*
  permanent dipoles and, 459–60
  sp^3 hybrid orbitals and, *468*
trimethylamine, 170, 776–77
Trimix, 240–41, 254
tripeptides, 1006
triple bonds
  bond energy of, 429
  bond length of, 426
  definition of, 401
  hybridization and, 467, *468*
  Lewis structures with, 401–3
triple point, 524–25
triprotic acids, 169, 786–87
tritium
  helium-3 from decay of, 1071
  in nuclear fusion reactors, 973–74
T-shaped molecular geometry, *455*, 456
turnover number of enzyme, 692

U

Uhlenbeck, George, 357
ultrahigh molecular weight polyethylene (UHMWPE), 623
ultratrace essential elements, 1046, 1048, 1065–66
  transition metals among, *1084*, 1108, 1111–13
ultraviolet (UV) radiation, *338*
  in hydrogen emission spectrum, 346, 348
  molecular orbitals and, 477
  See also sunlight
uncertainty principle, 352–54
unified atomic mass units (u), 53
unimolecular elementary step, 681
unit cells, 605–7
  of bronze, *613*
  contributions of atoms to, *609*, *610*
  definition of, 606
  diamond type of, *606*, 620
  dimensions of, 607–11
  periodic table and, *606*
  summary of, *608*
  See also body-centered cubic (bcc) unit cell; face-centered cubic (fcc) unit cell; simple cubic (sc) unit cell
unit factor method, 20–22
units
  conversions of, 20–22
  of energy changes, 285
  of ionizing radiation, *963*
  of rate constant, 662–63
  of reaction rates, 654
  SI units, 18–20
  U.S. Customary units, 19, *20*
universal gas constant (*R*), 229
universe
  elemental composition of, 1047–48
  entropy of, 855–56, 857–58, 864–65
  origin of elements in, 74–76
  timeline of energy and matter in, *75*
  See also Big Bang
unsaturated fatty acids, 1020–21, *1021*, 1022–23
uracil, 1025, 1029
uraninite, 51
uranium
  dating geological samples containing, 959–60
  early research with, 51, 948, 956, 959–60
  stellar nucleosynthesis and, 76
uranium hexafluoride, 969
uranium-235, fission reactions of, 968–70
uranium-238
  in breeder reactor, 970
  decay series of, 954–55, *955*
  in geological samples, 959–60
  radon from decay of, 965
urea
  as fertilizer, 710
  in living systems, 1060, 1112
  Wöhler's synthesis of, 89, 991
ureases, 1060, 1112
Urey, Harold, 88, 100, 101, 1029
uric acid, 1111
U.S. Customary units, 19, *20*

V

valence band, 491–93
valence bond theory, 461–69
  comparing ethane and ethylene, 474
valence electrons, 364
  formal charge and, 415–16
  in Lewis structures, 398, 399
  octet rule and, 398
  repulsion between, 375
  See also Lewis structures
valence shell, 364, 365, 366
valence-shell electron-pair repulsion (VSEPR) theory, 447–57
  bond dipoles and, 458–61
  central atoms with lone pairs, 452–57
  central atoms with no lone pairs, 448–51
  overview of, 447
van der Waals constants, *252*
van der Waals equation, 252–53
van't Hoff, Jacobus, 575, 675, 879
van't Hoff equation, 879–80
van't Hoff factor (*i* factor), 575–78
  osmotic pressure and, 580–82, 583
vanadium
  in complex with antitumor activity, 1117–18
  diabetes and, 1118
  electron configuration of, 367
  in haloperoxidases, 1111
  in redox flow battery, 934
vanillin, 132
vapor, 14
vapor pressure
  as colligative property, 562–64, 569
  definition of, 520
  of mixtures of volatile solutes, 564–69
  of pure liquids, 519–23
  of solutions of nonvolatile solutes, 561–64
  temperature and, 519, 520–23
vaporization, 15, *281*
  enthalpy of, 287, 291, 292, 522
  in equilibrium with condensation, 519–20
  free energy and, 870
  molar enthalpy of, 291
  rate of, 519
  See also evaporation
vermilion, 181
vibrational motion
  energy levels of, 885–86
  infrared frequencies and, 408–9
  internal energy and, *283*
  standard molar entropy and, 862
  temperature and, 859
  of water molecules, 854, *855*

vinegar
chicken egg submerged in, 589
concentration of acetic acid in, 173–74
reaction with sodium bicarbonate, 853
See also acetic acid
vinyl acetate, copolymer with ethylene, 626
vinyl alcohol, polymer of, 626
vinyl chloride, in drinking water, 156–57
vinyl group, 625
viscosity, 528
vitamin B_{12}, 1112
vitamin C
iron absorption and, 1084
shielded by EDTA, 1096
volatile organic compounds (VOCs), in smog, 651
volatile sulfur compounds (VSCs), 1063
volatility, 10, 521
volcanic activity, 88, 90, 235
Volta, Alessandro, 903
voltaic cells, 903, *904*
chemical equilibrium in, 917, 919
work done by, 911–12, 920–21
See also batteries; electrochemical cells; fuel cells; zinc/copper voltaic cell
voltmeter, 909
volume
Boyle's law and, 220–23, 229
Charles's law and, 223–25, 229
density and, 13, 23–26
entropy changes and, 861
gas-phase equilibria and, 731–34
units of, 19, *20*

W

Waage, Peter, 713
water
from abandoned coal mines, 171–72
acid–base behavior of, 765
alkalinity of sample of, 824–27
as amphiprotic substance, 170, 767
aquatic life and, 528–29, 533
autoionization of, 767–68, 781–82, 786
bent (angular) geometry of, 454, 459
boiling point of, 520, 521
boiling-point-elevation constant, *572*, 572–73
corrosion promoted by, 925–27
density changes with temperature, 524, 528–29
distillation of, 281–82
electrolysis of, 905–6
entropy changes in changes of state, 858, 860–61, 861n
freezing-point-depression constant, *572*, 574–75
group 16 dihydrides compared to, 1062–63, *1063*
hard, 587–88
heat capacity of, 290, 291, 293, 294–95, 296
heating curve of, 289–90
heating of, 289–93
hydrated complex ions and, 830, 832–34, 838, 1085, 1093–94, 1098–99, 1104, 1120–21
hydrated ions and, 512, 554–55, 559–61
hydrogen bonding in, 513–15, 521, 526–27, 560
hydrogen bonds with acetone, 516
in hydrolysis reactions, 168, 1060
intermolecular interactions of, 511–12, 526–28
ionizing radiation and, 962, 962n
as Lewis base, 1094
on Mars, 150
melting point of, 524
melting/freezing behavior, 524, 525–26
molecular motion in phases of, 854, *855*
normal boiling point of, 520
origin of life and, 150
partition coefficient between octanol and, 537
phase diagrams for, 523–26, 569–70
phase equilibria of, 869–70
as polar molecule, 459, 511–12
purification of, 10, 281–82, 588
from reaction between hydrogen and oxygen, 8, 272–73, 279
remarkable properties of, 526–29
solubilities in, 531–34
solubility of gases in, 534–37
sp^3 hybrid orbitals in, 464–65
spontaneous phase changes of, 854, 869–70
states of, 14–15
supercritical, 525
swimming pool pH and, 816–17
temporary dipole induced by, *513*
viscosity of, 528
See also aqueous solutions; drinking water; freshwater; ice; seawater
water pollution
cell membranes and, 533
with MTBE, 669
water softening, 587–88
water vapor
from combustion of hydrocarbons, 105–7
of early atmosphere, 88, 90, 150
entropy of, 860–61
gaseous product collected over water and, 242–43
partial pressure of, *242*
reaction with sulfur trioxide, 90, 98–99, 168, 783
See also steam–methane reforming reaction
water–gas shift reaction
achievement of equilibrium in, 711–12
equilibrium constant for, 712–13
equilibrium partial pressure calculation, 739–40
K_p equals K_c for, 718
reaction quotient for, 724–26
removing carbon dioxide to shift equilibrium, 729–30
watt (W), 921
wave functions (ψ), 354–55
wave mechanics, 354
wave number ($1/\lambda$), 346
wavelength (λ), 339–40
de Broglie, 350–52
energy of a quantum and, 342
wave–particle duality, 344–45, 352–53, 354
waves
characteristics of, 339
electromagnetic, 339–40
wax, *314*, 315
weak acids, 169
amides functioning as, 632
buffers based on, 809–10
calculating pH of, 779–80
common examples of, 758–59, *759*
conjugate bases of, 766
definition of, 169
degree of ionization of, 772–75
diprotic, 785–86
ionization reactions in water, 758–59
K_a values of, 773–75, 779–80
oxoacid strength and, 785–86
pH of solution of, 779–80
pH of solution of diprotic acid, 785–86
pH of solution with conjugate base and, 807–8
salt solutions with property of, 791–92
weak bases, 169, 763–64
amines as, 170
ammonia as, 169–70, 763–64, 776, 780–81, 808–9
buffers based on, 810
common examples of, *763*
conjugate acids of, 766
degree of ionization of, 776–78
pH of solution of, 780–81
pH of solution of conjugate acid and, 808–9
salt solutions with property of, 788, 789–91
weak electrolytes, 164–65, 170
weak-link principle, 24–26
weather balloons, 220, 228–29, 230–31
weighted average, 56–57
weighting factor, 56
welding, nitric oxide produced in, 418
Werner, Alfred, 1085–86
wet gas, 242
wetlands, soil color in, 193–94, 195–96
Wilson, Robert W., 35
wind power, 933–34
Wöhler, Friedrich, 65, 89, 991
Wollaston, William Hyde, 337
work
definition of, 273
electrical, 911–12, 920–21
energy and, 8, 273, 852
internal energy and, 283–85
maximum energy available for, 865, 869, 886
pressure–volume (P–V) work, 284–86, 869
work function (ϕ), 343, 344–45

X

xanthine oxidase, 1111, 1112
xenon, banned for athletes, 1071
xenon difluoride, 488
xenon-129, in magnetic resonance imaging, 1071
X-rays
  biological effects of, 962, 964
  discovery of, 51, 51n

Y

yellowcake, 969
yield
  actual, 116–19
  percent, 117–19
  theoretical, 114, 115–16
yttrium, 1114

Z

Z isomers, 994
  See also cis–trans isomers
zeolites, 588
zero-order reactions, 672–73
  half-life of, 673
  reaction mechanisms and, 686
  summary of how to distinguish, 673–74
Zevalin, 1070
Zewail, Ahmed H., 680n
zinc
  band theory and, 491
  in brass, 612
  in carbonic anhydrase, 1109–10
  conductivity of, 491
  electron configuration of cation of, 372
  redox reaction between copper and, 900–903
  in sacrificial anodes, 927
  in superoxide dismutase, 1112
zinc hydroxide, 834
zinc sulfide
  crystal structure of, 618
  in fluorescent paint, 975, 976
zinc–air battery, 910–11
zinc/copper voltaic cell, 901–3
  cell diagram of, 903–4
  ion concentrations and, 918
  standard cell potential of, 908–9, 911, 912
  standard free-energy change of, 912
  work done by, 911–12
zircon, in dating geological samples, 960
zwitterions, 1003–6

ATOMIC COLOR PALETTE

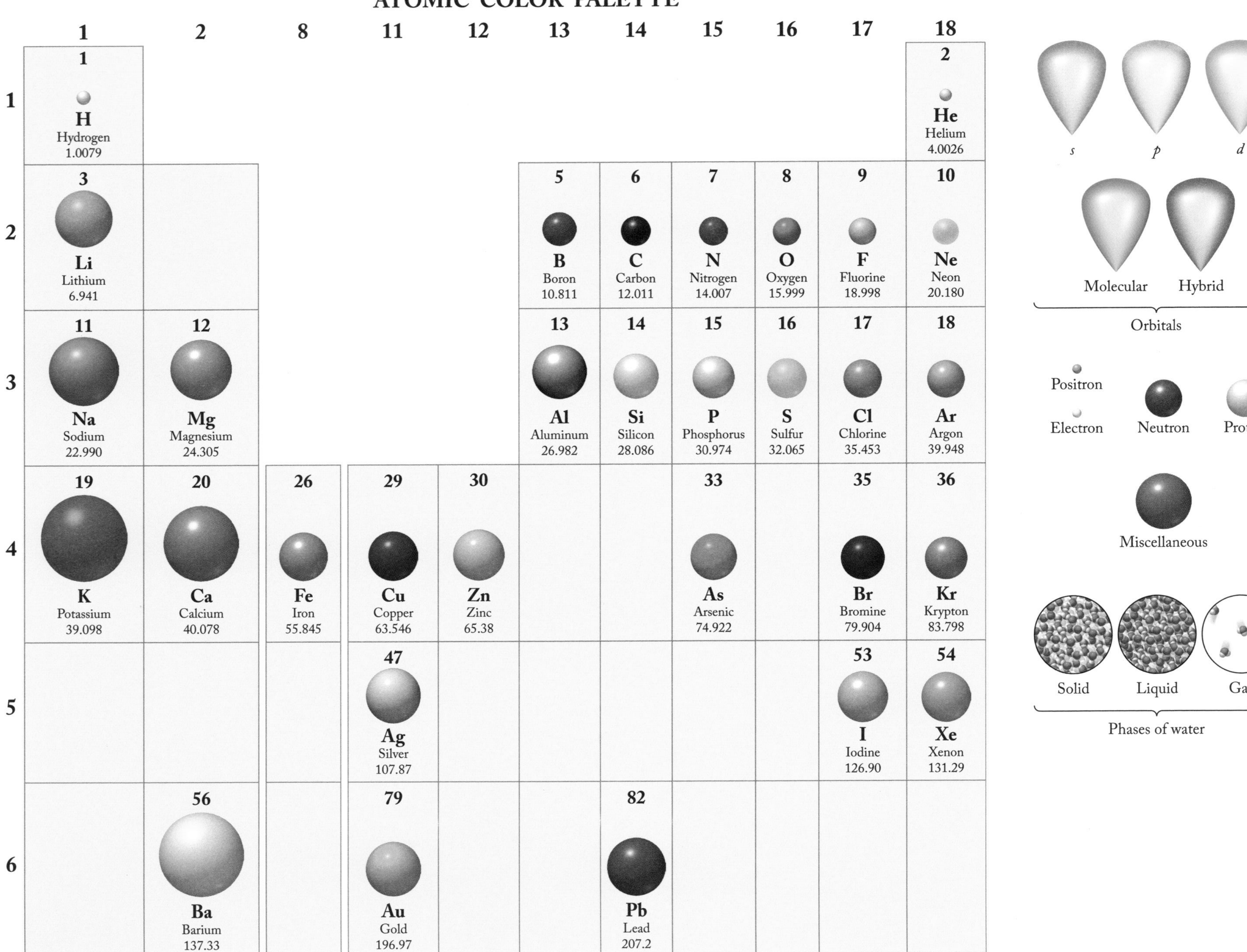

Special Units and Conversion Factors

Quantity	Unit	Symbol	Conversion
energy[a]	electron-volt	eV	1 eV = 1.6022×10^{-19} J
energy[a]	kilowatt-hour	kWh	1 kWh = 3600 kJ
energy	calorie	cal	1 cal = 4.184 J
mass	pound	lb	1 lb = 453.59 g
mass[a]	unified atomic mass unit	u	1 u = 1.66054×10^{-27} kg
length	angstrom	Å	1 Å = 10^{-8} cm = 10^{-10} m
length	inch	in	1 in = 2.54 cm
length	mile	mi	1 mi = 5280 ft = 1.6093 km
pressure	atmosphere	atm	1 atm = 1.01325×10^{5} Pa
pressure	torr	torr	1 torr = 1/760 atm
temperature	Celsius scale	°C	$T(°C) = T(K) - 273.15$
temperature	Fahrenheit scale	°F	$T(°F) = \frac{9}{5}T(°C) + 32$
volume	liter	L	1 L = 1 dm^3 = 10^{-3} m^3
volume	cubic centimeter	cm^3, cc	1 cm^3 = 1 mL = 10^{-3} L
volume	cubic foot	ft^3	1 ft^3 = 7.4805 gal
volume	gallon (U.S.)	gal	1 gal = 3.785 L

[a]From http://physics.nist.gov/cuu/constants/.

Physical Constants[a]

Quantity	Symbol	Value
Avogadro constant	N_A	6.0221×10^{23} mol^{-1}
Bohr radius	a_0	5.29×10^{-11} m
Boltzmann constant	k_B	1.3806×10^{-23} J/K
electron charge-to-mass ratio	$-e/m_e$	1.7588×10^{11} C/kg
elementary charge	e	1.602×10^{-19} C
Faraday constant	F	9.65×10^{4} C/mol
mass of an electron	m_e	9.10938×10^{-31} kg
mass of a neutron	m_n	1.67493×10^{-27} kg
mass of a proton	m_p	1.67262×10^{-27} kg
molar volume of ideal gas at 0°C and 1 atm	V_m	22.4 L/mol
Planck constant	h	6.626×10^{-34} J · s
speed of light in vacuum	c	2.998×10^{8} m/s
universal gas constant	R	8.314 J/(mol · K) 0.08206 L · atm/(mol · K)

[a]From http://physics.nist.gov/cuu/constants/.